Statistical
and Adaptive
Signal Processing

McGraw-Hill Series in Electrical and Computer Engineering

Senior Consulting Editor
Stephen W. Director, University of Michigan, Ann Arbor

Circuits and Systems
Communications and Signal Processing
Computer Engineering
Control Theory and Robotics
Electromagnetics
Electronics and VLSI Circuits
Introductory
Power
Antennas, Microwaves, and Radar

Previous Consulting Editors

Ronald N. Bracewell, Colin Cherry, James F. Gibbons, Willis W. Harman, Hubert Heffner, Edward W. Herold, John G. Linvill, Simon Ramo, Ronald A. Rohrer, Anthony E. Siegman, Charles Susskind, Frederick E. Terman, John G. Truxal, Ernst Weber, and John R. Whinnery

COMMUNICATIONS AND SIGNAL PROCESSING

Senior Consulting Editor
Stephen W. Director, University of Michigan, Ann Arbor

Auñón/Chandrasekar: *Introduction to Probability and Random Processes*
Antoniou: *Digital Filters: Analysis and Design*
Bose: *Neural Network Fundamentals with Graphs, Algorithms, and Applications*
Carlson: *Communication Systems: An Introduction to Signals and Noise in Electrical Communication*
Cassandras: *Discrete Event Systems Modeling and Performance Analysis*
Cherin: *An Introduction to Optical Fibers*
Childers: *Probability and Random Processes Using MATLAB*
Collin: *Antennas and Radiowave Propagation*
Collin: *Foundations for Microwave Engineering*
Cooper and McGillem: *Modern Communications and Spread Spectrum*
Davenport: *Probability and Random Processes: An Introduction for Applied Scientists and Engineers*
Drake: *Fundamentals of Applied Probability Theory*
Gardner: *Introduction to Random Processes*
Jong: *Method of Discrete Signal and System Analysis*
Keiser: *Local Area Networks*
Keiser: *Optical Fiber Communications*
Kershenbaum: *Telecommunication Network Design and Algorithms*
Kraus: *Antennas*
Kuc: *Introduction to Digital Signal Processing*
Lee: *Mobile Communications Engineering*
Lindner: *Introduction to Signals and Systems*
Mitra: *Digital Signal Processing: A Computer-Based Approach*
Papoulis: *Probability, Random Variables, and Stochastic Processes*
Papoulis: *Signal Analysis*
Papoulis: *The Fourier Integral and Its Applications*
Parsons: *Voice and Speech Processing*
Peebles: *Probability, Random Variables, and Random Signal Principles*
Powers: *An Introduction to Fiber Optic Systems*
Proakis: *Digital Communications*
Russell: *Telecommunications Protocols*
Schwartz: *Information Transmission, Modulation, and Noise*
Siebert: *Circuits, Signals, and Systems*
Smith: *Modern Communication Circuits*
Taub and Schilling: *Principles of Communication Systems*
Taylor: *Principles of Signals and Systems*
Taylor: *Hands-On Digital Signal Processing*
Viterbi and Omura: *Principles of Digital Communications and Coding*
Viniotis: *Probability and Random Processes*
Walrand: *Communications Networks*
Waters: *Active Filter Design*

Statistical and Adaptive Signal Processing

Spectral Estimation, Signal Modeling, Adaptive
Filtering and Array Processing

Dimitris G. Manolakis

Massachusetts Institute of Technology
Lincoln Laboratory

Vinay K. Ingle

Northeastern University

Stephen M. Kogon

Massachusetts Institute of Technology
Lincoln Laboratory

Boston   Burr Ridge, IL   Dubuque, IA   Madison, WI   New York   San Francisco
St. Louis   Bangkok   Bogotá   Caracas   Lisbon   London   Madrid
Mexico City   Milan   New Delhi   Seoul   Singapore   Sidney   Taipei   Toronto

McGraw-Hill Higher Education ⚛
*A Division of The **McGraw-Hill** Companies*

STATISTICAL AND ADAPTIVE SIGNAL PROCESSING

Spectral Estimation, Signal Modeling,
Adaptive Filtering and Array Processing

1 2 3 4 5 6 7 8 9 0 DOC/DOC 9 0 9 8 7 6 5 4 3 2 1 0 9

ISBN 0-07-040051-2

Vice president/Editor-in-Chief: *Kevin T. Kane*
Publisher: *Thomas Casson*
Sponsoring editor: *Catherine Fields*
Editorial assistant: *Michelle L. Flomenhoft*
Senior marketing manager: *John T. Wannemacher*
Project manager: *Amy Hill*
Senior production supervisor: *Rose Hepburn*
Freelance design coordinator: *Pam Verros*
Supplement coordinator: *Betty Hadala*
Compositor: *Techsetters, Inc.*
Typeface: *10/12 Times Roman*
Printer: *R. R. Donnelley & Sons Company*

Library of Congress Cataloging-in-Publication Data
Manolakis, Dimitris G.
    Statistical and adaptive signal processing : spectral estimation,
signal modeling, adaptive filtering and array processing / D.
Manolakis, V. Ingle, S. Kogon.
        p.   cm. — (McGraw-Hill series in electrical and computer
engineering. Computer engineering.)
    ISBN 0-07-040051-2
    1. Signal processing—Statistical methods.   2. Adaptive signal
processing. I. Ingle, Vinay K. II. Kogon, Stephen M.
III. Title. IV. Series
TK5102.9.M36 2000
621.382'2—dc21   99-30380

http://www.mhhe.com

ABOUT THE AUTHORS

DIMITRIS G. MANOLAKIS, a native of Greece, received his education (B.S. in physics and Ph.D. in electrical engineering) from the University of Athens, Greece. He is currently a member of the technical staff at MIT Lincoln Laboratory, in Lexington, Massachusetts. Previously, he was a Principal Member, Research Staff, at Riverside Research Institute. Dr. Manolakis has taught at the University of Athens, Northeastern University, Boston College, and Worcester Polytechnic Institute; and he is coauthor of the textbook *Digital Signal Processing: Principles, Algorithms, and Applications* (Prentice-Hall, 1996, 3d ed.). His research experience and interests include the areas of digital signal processing, adaptive filtering, array processing, pattern recognition, and radar systems.

VINAY K. INGLE is Associate Professor of Electrical and Computer Engineering at Northeastern University. He received his Ph.D. in electrical and computer engineering from Rensselaer Polytechnic Institute in 1981. He has broad research experience and has taught courses on topics including signal and image processing, stochastic processes, and estimation theory. Professor Ingle is coauthor of the textbooks *DSP Laboratory Using the ADSP-2101 Microprocessor* (Prentice-Hall, 1991) and *DSP Using Matlab* (PWS Publishing Co., Boston, 1996).

STEPHEN M. KOGON received the Ph.D. degree in electrical engineering from Georgia Institute of Technology. He is currently a member of the technical staff at MIT Lincoln Laboratory in Lexington, Massachusetts. Previously, he has been associated with Raytheon Co., Boston College, and Georgia Tech Research Institute. His research interests are in the areas of adaptive processing, array signal processing, radar, and statistical signal modeling.

To my beloved wife, Anna, and to the loving memory of my father, Gregory.
DGM

To my beloved wife, Usha, and adoring daughters, Natasha and Trupti.
VKI

To my wife and best friend, Lorna, and my children, Gabrielle and Matthias.
SMK

BRIEF CONTENTS

CONTENTS

PREFACE

The principal goal of this book is to provide a unified introduction to the theory, implementation, and applications of statistical and adaptive signal processing methods. We have focused on the key topics of spectral estimation, signal modeling, adaptive filtering, and array processing, whose selection was based on the grounds of theoretical value and practical importance. The book has been primarily written with students and instructors in mind. The principal objectives are to provide an introduction to basic concepts and methodologies that can provide the foundation for further study, research, and application to new problems. To achieve these goals, we have focused on topics that we consider fundamental and have either multiple or important applications.

APPROACH AND PREREQUISITES

The adopted approach is intended to help both students and practicing engineers understand the fundamental mathematical principles underlying the operation of a method, appreciate its inherent limitations, and provide sufficient details for its practical implementation. The academic flavor of this book has been influenced by our teaching whereas its practical character has been shaped by our research and development activities in both academia and industry. The mathematical treatment throughout this book has been kept at a level that is within the grasp of upper-level undergraduate students, graduate students, and practicing electrical engineers with a background in digital signal processing, probability theory, and linear algebra.

ORGANIZATION OF THE BOOK

Chapter 1 introduces the basic concepts and applications of statistical and adaptive signal processing and provides an overview of the book. Chapters 2 and 3 review the fundamentals of discrete-time signal processing, study random vectors and sequences in the time and frequency domains, and introduce some basic concepts of estimation theory. Chapter 4 provides a treatment of parametric linear signal models (both deterministic and stochastic) in the time and frequency domains. Chapter 5 presents the most practical methods for the estimation of correlation and spectral densities. Chapter 6 provides a detailed study of the theoretical properties of optimum filters, assuming that the relevant signals can be modeled as stochastic processes with known statistical properties; and Chapter 7 contains algorithms and structures for optimum filtering, signal modeling, and prediction. Chapter

8 introduces the principle of least-squares estimation and its application to the design of practical filters and predictors. Chapters 9, 10, and 11 use the theoretical work in Chapters 4, 6, and 7 and the practical methods in Chapter 8, to develop, evaluate, and apply practical techniques for signal modeling, adaptive filtering, and array processing. Finally, Chapter 12 introduces some advanced topics: definition and properties of higher-order moments, blind deconvolution and equalization, and stochastic fractional and fractal signal models with long memory. Appendix A contains a review of the matrix inversion lemma, Appendix B reviews optimization in complex space, Appendix C contains a list of the MATLAB functions used throughout the book, Appendix D provides a review of useful results from matrix algebra, and Appendix E includes a proof for the minimum-phase condition for polynomials.

USE OF THE BOOK

This book can be used in a number of different ways to teach either one-semester or two-semester graduate courses in a typical electrical engineering curriculum. Most topics in the book can be covered in a two-semester course on statistical and adaptive signal processing. Typical one-term courses are outlined in the following table:

Courses	Chapters
Discrete-time random signals and statistical signal processing	Review of Chapters 1, 2, and 3; Sections 4.1–4.4, 6.1–6.6
Spectrum estimation and signal modeling	Review of Chapters 1, 2, and 3; Chapters 4 and 5, Sections 6.1–6.5, 7.4–7.5, 8.1–8.9, Chapter 9, and possibly Sections 12.1, 12.5–12.6
Adaptive filtering	Review of Chapters 1, 2, and 3; Chapters 6, 7, 8, 10, and Sections 12.1–12.4
Introduction to array processing	Review of Chapter 3, Sections 6.1–6.5, review of Section 2.2, Sections 5.1, 11.1–11.4, Chapter 8, Sections 9.5, 9.6, 11.5, 11.6, 11.7

THEORY AND PRACTICE

It is our belief that sound theoretical understanding goes hand-in-hand with practical implementation and application to real-world problems. Therefore, the book includes a large number of computer experiments that illustrate important concepts and help the reader to easily implement the various methods. Every chapter includes examples, problems, and computer experiments that facilitate the comprehension of the material. To help the reader understand the theoretical basis and limitations of the various methods and apply them to real-world problems, we provide MATLAB functions for all major algorithms and examples illustrating their use. The MATLAB files and additional material about the book can be found at http://www.mhhe.com/catalogs/0070400512.mhtml. A Solutions Manual with detailed solutions to all the problems is available to the instructors adopting the book for classroom use.

FEEDBACK

Although we are fully aware that there always exists room for improvement, we believe that this book is a big step forward for an introductory textbook in statistical and adaptive signal processing. However, as engineers, we know that every search for the optimum requires the will to change and quest for additional improvement. Thus, we would appreciate feedback from teachers, students, and engineers using this book for self-study at vingle@lynx.neu.edu.

sl# ACKNOWLEDGMENTS

We are indebted to a number of individuals who have contributed in different, but important, ways to the shaping of our knowledge in general and the preparation of this book in particular. In this respect we wish to extend our appreciation to C. Caroubalos, G. Carayannis, J. Makhoul (DGM), M. Schetzen (VKI), and J. Holder, S. Krich, and D. Williams (SMK). We are grateful to E. Baranoski, G. Borsari, J. Schodorf, S. Smith, A. Steinhardt, and J. Ward for helping us to improve the presentation in various parts of the book.

We express our sincere gratitude to Kevin Donohue, University of Kentucky; Amro El-Jaroudi, University of Pittsburgh; Edward A. Lee, University of California-Berkeley; Ray Liu, University of Maryland; Randy Moses, The Ohio State University; and Kristina Ropella, Marquette University, for their constructive and helpful reviews.

Lynn Cox persuaded us to choose McGraw-Hill, Inc., as our publisher, and we have not regretted that decision. We are grateful to Lynn for her enthusiasm and her influence in shaping the scope and the objectives of our book. The fine team at McGraw-Hill, including Michelle Flomenhoft, Catherine Fields, Betsy Jones, and Nina Kreiden, has made the publication of this book an exciting and pleasant experience. We also thank N. Bulock and Mathworks, Inc., for promptly providing various versions of MATLAB and A. Turcotte for helping with some of the drawings in the book.

Last, but not least, we would like to express our sincere appreciation to our families for their full-fledged support and understanding over the past several years. We fully realize that the completion of such a project would not be possible without their continual sustenance and encouragement.

Dimitris G. Manolakis
Vinay K. Ingle
Stephen M. Kogon

CHAPTER 1

Introduction

This book is an introduction to the theory and algorithms used for the analysis and processing of random signals and their applications to real-world problems. The fundamental characteristic of random signals is captured in the following statement: Although random signals are evolving in time in an unpredictable manner, their average statistical properties exhibit considerable regularity. This provides the ground for the description of random signals using statistical averages instead of explicit equations. When we deal with random signals, the main objectives are the statistical description, modeling, and exploitation of the dependence between the values of one or more discrete-time signals and their application to theoretical and practical problems.

Random signals are described mathematically by using the theory of probability, random variables, and stochastic processes. However, in practice we deal with random signals by using statistical techniques. Within this framework we can develop, at least in principle, theoretically *optimum signal processing* methods that can inspire the development and can serve to evaluate the performance of practical *statistical signal processing* techniques. The area of *adaptive signal processing* involves the use of optimum and statistical signal processing techniques to design signal processing systems that can modify their characteristics, during normal operation (usually in real time), to achieve a clearly predefined application-dependent objective.

The purpose of this chapter is twofold: to illustrate the nature of random signals with some typical examples and to introduce the four major application areas treated in this book: *spectral estimation*, *signal modeling*, *adaptive filtering*, and *array processing*. Throughout the book, the emphasis is on the application of techniques to actual problems in which the theoretical framework provides a foundation to motivate the selection of a specific method.

1.1 RANDOM SIGNALS

A *discrete-time signal* or *time series* is a set of observations taken sequentially in time, space, or some other independent variable. Examples occur in various areas, including engineering, natural sciences, economics, social sciences, and medicine.

A discrete-time signal $x(n)$ is basically a sequence of real or complex numbers called samples. Although the integer index n may represent any physical variable (e.g., time, distance), we shall generally refer to it as *time*. Furthermore, in this book we consider only time series with observations occurring at equally spaced intervals of time.

Discrete-time signals can arise in several ways. Very often, a discrete-time signal is obtained by periodically sampling a continuous-time signal, that is, $x(n) = x_c(nT)$, where $T = 1/F_s$ (seconds) is the sampling period and F_s (samples per second or hertz) is the sampling frequency. At other times, the samples of a discrete-time signal are obtained

1

by accumulating some quantity (which does not have an instantaneous value) over equal intervals of time, for example, the number of cars per day traveling on a certain road. Finally, some signals are inherently discrete-time, for example, daily stock market prices. *Throughout the book, except if otherwise stated, the terms* signal, time series, *or* sequence *will be used to refer to a discrete-time signal.*

The key characteristics of a time series are that the observations are *ordered* in time and that adjacent observations are *dependent* (related). To see graphically the relation between the samples of a signal that are l sampling intervals away, we plot the points $\{x(n), x(n+l)\}$ for $0 \leq n \leq N - 1 - l$, where N is the length of the data record. The resulting graph is known as the *l lag scatter plot*. This is illustrated in Figure 1.1, which shows a speech signal and two scatter plots that demonstrate the correlation between successive samples. We note that for adjacent samples the data points fall close to a straight line with a positive slope. This implies high correlation because every sample is followed by a sample with about the same amplitude. In contrast, samples that are 20 sampling intervals apart are much less correlated because the points in the scatter plot are randomly spread.

When successive observations of the series are dependent, we may use past observations to predict future values. If the prediction is exact, the series is said to be *deterministic*. However, in most practical situations we cannot predict a time series exactly. Such time

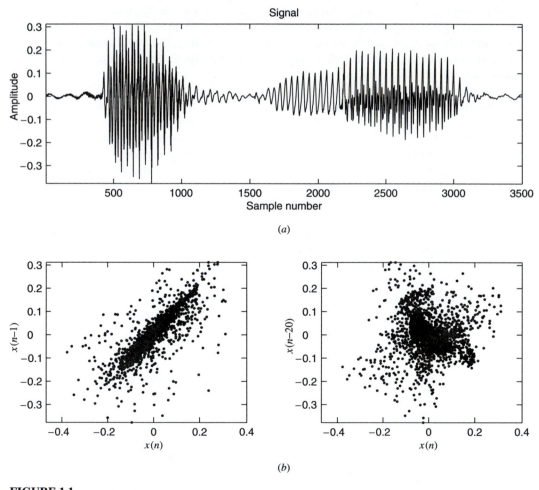

FIGURE 1.1
(*a*) The waveform for the speech signal "signal"; (*b*) two scatter plots for successive samples and samples separated by 20 sampling intervals.

series are called *random* or *stochastic*, and the degree of their predictability is determined by the dependence between consecutive observations. The ultimate case of randomness occurs when every sample of a random signal is independent of all other samples. Such a signal, which is completely unpredictable, is known as *white noise* and is used as a building block to simulate random signals with different types of dependence. To summarize, the fundamental characteristic of a random signal is the inability to precisely specify its values. In other words, a random signal is not predictable, it never repeats itself, and we cannot find a mathematical formula that provides its values as a function of time. As a result, random signals can only be mathematically described by using the theory of stochastic processes (see Chapter 3).

This book provides an introduction to the fundamental theory and a broad selection of algorithms widely used for the processing of discrete-time random signals. Signal processing techniques, dependent on their main objective, can be classified as follows (see Figure 1.2):

- **Signal analysis.** The primary goal is to extract useful information that can be used to understand the signal generation process or extract features that can be used for signal classification purposes. Most of the methods in this area are treated under the disciplines of *spectral estimation* and *signal modeling*. Typical applications include detection and classification of radar and sonar targets, speech and speaker recognition, detection and classification of natural and artificial seismic events, event detection and classification in biological and financial signals, efficient signal representation for data compression, etc.
- **Signal filtering.** The main objective of signal filtering is to improve the quality of a signal according to an acceptable criterion of performance. Signal filtering can be subdivided into the areas of frequency selective filtering, adaptive filtering, and array processing. Typical applications include noise and interference cancelation, echo cancelation, channel equalization, seismic deconvolution, active noise control, etc.

We conclude this section with some examples of signals occurring in practical applications. Although the desciption of these signals is far from complete, we provide sufficient information to illustrate their random nature and significance in signal processing applications.

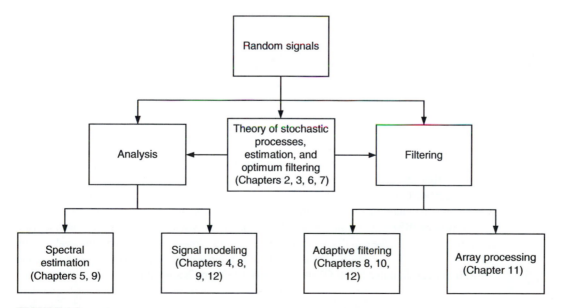

FIGURE 1.2
Classification of methods for the analysis and processing of random signals.

Speech signals. Figure 1.3 shows the spectrogram and speech waveform corresponding to the utterance "signal." The spectrogram is a visual representation of the distribution of the signal energy as a function of time and frequency. We note that the speech signal has significant changes in both amplitude level and spectral content across time. The waveform contains segments of voiced (quasi-periodic) sounds, such as "e," and unvoiced or fricative (noiselike) sounds, such as "g."

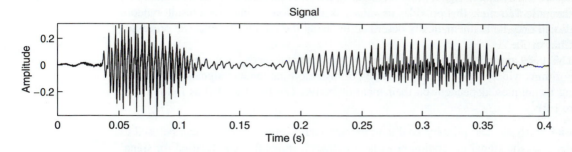

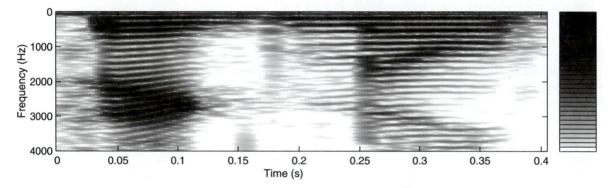

FIGURE 1.3
Spectrogram and acoustic waveform for the utterance "signal." The horizontal dark bands show the resonances of the vocal tract, which change as a function of time depending on the sound or phoneme being produced.

Speech production involves three processes: generation of the sound excitation, articulation by the vocal tract, and radiation from the lips and/or nostrils. If the excitation is a quasi-periodic train of air pressure pulses, produced by the vibration of the vocal cords, the result is a voiced sound. Unvoiced sounds are produced by first creating a constriction in the vocal tract, usually toward the mouth end. Then we generate turbulence by forcing air through the constriction at a sufficiently high velocity. The resulting excitation is a broadband noiselike waveform.

The spectrum of the excitation is shaped by the vocal tract tube, which has a frequency response that resembles the resonances of organ pipes or wind instruments. The resonant frequencies of the vocal tract tube are known as *formant frequencies*, or simply *formants*. Changing the shape of the vocal tract changes its frequency response and results in the generation of different sounds. Since the shape of the vocal tract changes slowly during continuous speech, we usually assume that it remains almost constant over intervals on the order of 10 ms. More details about speech signal generation and processing can be found in Rabiner and Schafer 1978; O'Shaughnessy 1987; and Rabiner and Juang 1993.

Electrophysiological signals. Electrophysiology was established in the late eighteenth century when Galvani demonstrated the presence of electricity in animal tissues. Today, electrophysiological signals play a prominent role in every branch of physiology, medicine, and biology. Figure 1.4 shows a set of typical signals recorded in a sleep laboratory (Rechtschaffen and Kales 1968). The most prominent among them is the electroencephalogram (EEG), whose spectral content changes to reflect the state of alertness and the mental activity of the subject. The EEG signal exhibits some distinctive waves, known as rhythms, whose dominant spectral content occupies certain bands as follows: delta (δ), 0.5 to 4 Hz; theta (θ), 4 to 8 Hz; alpha (α), 8 to 13 Hz; beta (β), 13 to 22 Hz; and gamma (γ), 22 to 30 Hz. During sleep, if the subject is dreaming, the EEG signal shows rapid low-amplitude fluctuations similar to those obtained in alert subjects, and this is known as rapid eye movement (REM) sleep. Some other interesting features occurring during nondreaming sleep periods resemble alphalike activity and are known as sleep spindles. More details can be found in Duffy et al. 1989 and Niedermeyer and Lopes Da Silva 1998.

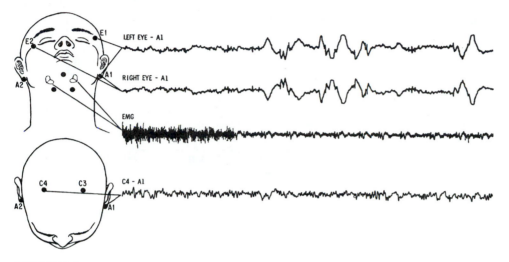

FIGURE 1.4
Typical sleep laboratory recordings. The two top signals show eye movements, the next one illustrates EMG (electromyogram) or muscle tonus, and the last one illustrates brain waves (EEG) during the onset of a REM sleep period (*from Rechtschaffen and Kales 1968*).

The beat-to-beat fluctuations in heart rate and other cardiovascular variables, such as arterial blood pressure and stroke volume, are mediated by the joint activity of the sympathetic and parasympathetic systems. Figure 1.5 shows time series for the heart rate and systolic arterial blood pressure. We note that both heart rate and blood pressure fluctuate in a complex manner that depends on the mental or physiological state of the subject. The individual or joint analysis of such time series can help to understand the operation of the cardiovascular system, predict cardiovascular diseases, and help in the development of drugs and devices for cardiac-related problems (Grossman et al. 1996; Malik and Camm 1995; Saul 1990).

Geophysical signals. Remote sensing systems use a variety of electro-optical sensors that span the infrared, visible, and ultraviolet regions of the spectrum and find many civilian and defense applications. Figure 1.6 shows two segments of infrared scans obtained by a space-based radiometer looking down at earth (Manolakis et al. 1994). The shape of the profiles depends on the transmission properties of the atmosphere and the objects in the radiometer's field-of-view (terrain or sky background). The statistical characterization and modeling of infrared backgrounds are critical for the design of systems to detect missiles against such backgrounds as earth's limb, auroras, and deep-space star fields (Sabins 1987; Colwell 1983). Other geophysical signals of interest are recordings of natural and man-made seismic events and seismic signals used in geophysical prospecting (Bolt 1993; Dobrin 1988; Sheriff 1994).

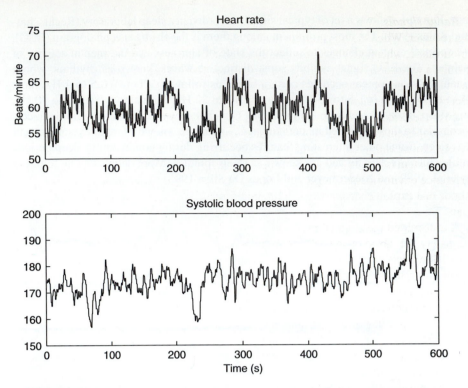

FIGURE 1.5
Simultaneous recordings of the heart rate and systolic blood pressure signals for a subject at rest.

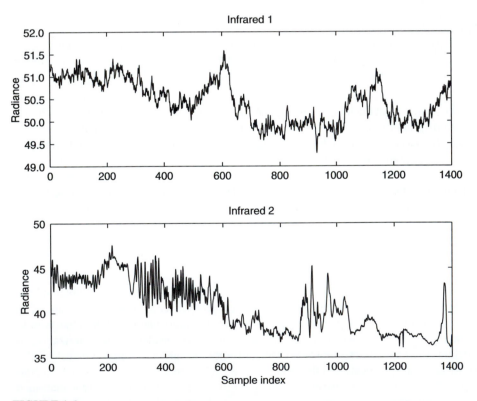

FIGURE 1.6
Time series of infrared radiation measurements obtained by a scanning radiometer.

Radar signals. We conveniently define a radar system to consist of both a transmitter and a receiver. When the transmitter and receiver are colocated, the radar system is said to be monostatic, whereas if they are spatially separated, the system is bistatic. The radar first transmits a waveform, which propagates through space as electromagnetic energy, and then measures the energy returned to the radar via reflections. When the returns are due to an object of interest, the signal is known as a *target*, while undesired reflections from the earth's surface are referred to as *clutter*. In addition, the radar may encounter energy transmitted by a hostile opponent attempting to *jam* the radar and prevent detection of certain targets. Collectively, clutter and jamming signals are referred to as *interference*. The challenge facing the radar system is how to extract the targets of interest in the presence of sometimes severe interference environments. Target detection is accomplished by using adaptive processing methods that exploit characteristics of the interference in order to suppress these undesired signals.

A transmitted radar signal propagates through space as electromagnetic energy at approximately the speed of light $c = 3 \times 10^8$ m/s. The signal travels until it encounters an object that reflects the signal's energy. A portion of the reflected energy returns to the radar receiver along the same path. The round-trip delay of the reflected signal determines the distance or *range* of the object from the radar. The radar has a certain receive aperture, either a continuous aperture or one made up of a series of sensors. The relative delay of a signal as it propagates across the radar aperture determines its angle of arrival, or bearing. The extent of the aperture determines the accuracy to which the radar can determine the direction of a target. Typically, the radar transmits a series of pulses at a rate known as the *pulse repetition frequency*. Any target motion produces a phase shift in the returns from successive pulses caused by the *Doppler effect*. This phase shift across the series of pulses is known as the Doppler frequency of the target, which in turn determines the target radial velocity. The collection of these various parameters (range, angle, and velocity) allows the radar to locate and track a target.

An example of a radar signal as a function of range in kilometers (km) is shown in Figure 1.7. The signal is made up of a target, clutter, and thermal noise. All the signals have been normalized with respect to the thermal noise floor. Therefore, the normalized noise has unit variance (0 dB). The target signal is at a range of 100 km with a signal-to-noise ratio (SNR) of 15 dB. The clutter, on the other hand, is present at all ranges and is highly nonstationary. Its power levels vary from approximately 40 dB at near ranges down to the thermal noise floor (0 dB) at far ranges. Part of the nonstationarity in the clutter is due to the range falloff of the clutter as its power is attenuated as a function of range. However, the rises and dips present between 100 and 200 km are due to terrain-specific artifacts. Clearly, the target is not visible, and the clutter interference must be removed or canceled in order to

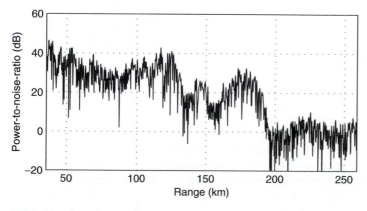

FIGURE 1.7
Example of a radar return signal, plotted as relative power with respect to noise versus range.

detect the target. The challenge here is how to cancel such a nonstationary signal in order to extract the target signal and motivate the use of adaptive techniques that can adapt to the rapidly changing interference environment. More details about radar and radar signal processing can be found in Skolnik 1980; Skolnik 1990; and Nathanson 1991.

1.2 SPECTRAL ESTIMATION

The central objective of signal analysis is the development of quantitative techniques to study the properties of a signal and the differences and similarities between two or more signals from the same or different sources. The major areas of random signal analysis are (1) statistical analysis of signal amplitude (i.e., the sample values); (2) analysis and modeling of the correlation among the samples of an individual signal; and (3) joint signal analysis (i.e., simultaneous analysis of two signals in order to investigate their interaction or interrelationships). These techniques are summarized in Figure 1.8. The prominent tool in signal analysis is spectral estimation, which is a generic term for a multitude of techniques used to estimate the distribution of energy or power of a signal from a set of observations. Spectral estimation is a very complicated process that requires a deep understanding of the underlying theory and a great deal of practical experience. Spectral analysis finds many applications in areas such as medical diagnosis, speech analysis, seismology and geophysics, radar and sonar, nondestructive fault detection, testing of physical theories, and evaluating the predictability of time series.

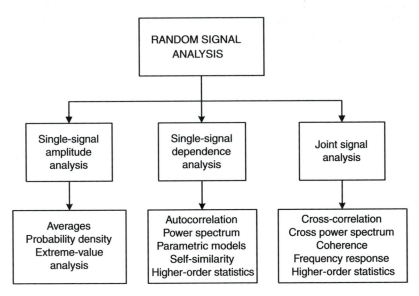

FIGURE 1.8
Summary of random signal analysis techniques.

Amplitude distribution. The range of values taken by the samples of a signal and how often the signal assumes these values together determine the signal variability. The signal variability can be seen by plotting the time series and is quantified by the histogram of the signal samples, which shows the percentage of the signal amplitude values within a certain range. The numerical description of signal variability, which depends only on the value of the signal samples and not on their ordering, involves quantities such as mean value, median, variance, and dynamic range.

Figure 1.9 shows the one-step increments, that is, the first difference $x_d(n) = x(n) - x(n-1)$, or approximate derivative of the infrared signals shown in Figure 1.6, whereas Figure 1.10 shows their histograms. Careful examination of the shape of the histogram curves indicates that the second signal jumps quite frequently between consecutive samples with large steps. In other words, the probability of large increments is significant, as exemplified

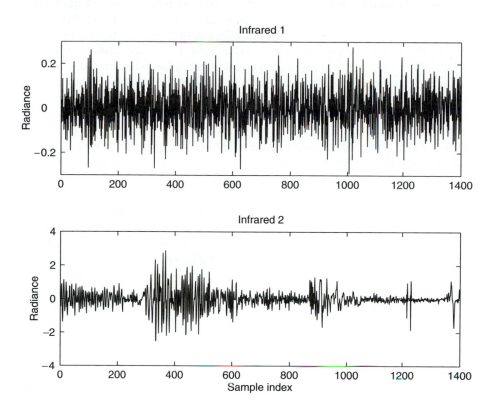

FIGURE 1.9
One-step-increment time series for the infrared data shown in Figure 1.6.

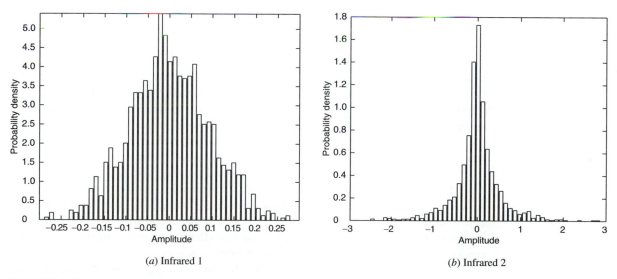

(*a*) Infrared 1   (*b*) Infrared 2

FIGURE 1.10
Histograms for the infrared increment signals.

by the fat tails of the histogram in Figure 1.10(*b*). The knowledge of the probability of extreme values is essential in the design of detection systems for digital communications, military surveillance using infrared and radar sensors, and intensive care monitoring. In general, the shape of the histogram, or more precisely the probability density, is very important in applications such as signal coding and event detection. Although many practical signals follow a Gaussian distribution, many other signals of practical interest have distributions that are non-Gaussian. For example, speech signals have a probability density that can be reasonably approximated by a gamma distribution (Rabiner and Schafer 1978).

The significance of the Gaussian distribution in signal processing stems from the following facts. First, many physical signals can be described by Gaussian processes. Second, the central limit theorem (see Chapter 3) states that any process that is the result of the combination of many elementary processes will tend, under quite general conditions, to be Gaussian. Finally, linear systems preserve the Gaussianity of their input signals. To understand the last two statements, consider N independent random quantities $x_1, x_2, \ldots, x_N$ with the same probability density $p(x)$ and pose the following question: When does the probability distribution $p_N(x)$ of their sum $x = x_1 + x_2 + \cdots + x_N$ have the same shape (within a scale factor) as the distribution $p(x)$ of the individual quantities? The standard answer is that $p(x)$ should be Gaussian, because the sum of N Gaussian random variables is again a Gaussian, but with variance equal to N times that of the individual signals. However, if we allow for distributions with infinite variance, additional solutions are possible. The resulting probability distributions, known as *stable* or *Levy distributions*, have infinite variance and are characterized by a thin main lobe and fat tails, resembling the shape of the histogram in Figure 1.10(*b*). Interestingly enough, the Gaussian distribution is a stable distribution with finite variance (actually the only one). Because Gaussian and stable non-Gaussian distributions are invariant under linear signal processing operations, they are very important in signal processing.

Correlation and spectral analysis. Although scatter plots (see Figure 1.1) illustrate nicely the existence of correlation, to obtain quantitative information about the correlation structure of a time series $x(n)$ with zero mean value, we use the *empirical normalized autocorrelation sequence*

$$\hat{\rho}(l) = \frac{\displaystyle\sum_{n=l}^{N-1} x(n)x^*(n-l)}{\displaystyle\sum_{n=0}^{N-1} |x(n)|^2} \tag{1.2.1}$$

which is an estimate of the theoretical normalized autocorrelation sequence. For lag $l = 0$, the sequence is perfectly correlated with itself and we get the maximum value of 1. If the sequence does not change significantly from sample to sample, the correlation of the sequence with its shifted copies, though diminished, is still close to 1. Usually, the correlation decreases as the lag increases because distant samples become less and less dependent. Note that reordering the samples of a time series changes its autocorrelation but not its histogram.

We say that signals whose empirical autocorrelation decays fast, such as an exponential, have short-memory or short-range dependence. If the empirical autocorrelation decays very slowly, as a hyperbolic function does, we say that the signal has long-memory or long-range dependence. These concepts will be formulated in a theoretical framework in Chapter 3. Furthermore, we shall see in the next section that effective modeling of time series with short or long memory requires different types of models.

The spectral density function shows the distribution of signal power or energy as a function of frequency (see Figure 1.11). The autocorrelation and the spectral density of a signal form a Fourier transform pair and hence contain the same information. However, they present this information in different forms, and one can reveal information that cannot

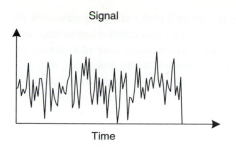
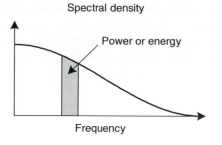

FIGURE 1.11
Illustration of the concept of power or energy spectral density function of a random signal.

be easily extracted from the other. It is fair to say that the spectral density is more widely used than the autocorrelation.

Although the correlation and spectral density functions are the most widely used tools for signal analysis, there are applications that require the use of correlations among three or more samples and the corresponding spectral densities. These quantities, which are useful when we deal with non-Gaussian processes and nonlinear systems, belong to the area of higher-order statistics and are described in Chapter 12.

Joint signal analysis. In many applications, we are interested in the relationship between two different random signals. There are two cases of interest. In the first case, the two signals are of the same or similar nature, and we want to ascertain and describe the similarity or interaction between them. For example, we may want to investigate if there is any similarity in the fluctuation of infrared radiation in the two profiles of Figure 1.6.

In the second case, we may have reason to believe that there is a *causal relationship* between the two signals. For example, one signal may be the input to a system and the other signal the output. The task in this case is to find an accurate description of the system, that is, a description that allows accurate estimation of future values of the output from the input. This process is known as *system modeling* or *identification* and has many practical applications, including understanding the operation of a system in order to improve the design of new systems or to achieve better control of existing systems.

In this book, we will study joint signal analysis techniques that can be used to understand the dynamic behavior between two or more signals. An interesting example involves using signals, like the ones in Figure 1.5, to see if there is any coupling between blood pressure and heart rate. Some interesting results regarding the effect of respiration and blood pressure on heart rate are discussed in Chapter 5.

1.3 SIGNAL MODELING

In many theoretical and practical applications, we are interested in generating random signals with certain properties or obtaining an efficient representation of real-world random signals that captures a desired set of their characteristics (e.g., correlation or spectral features) in the best possible way. We use the term *model* to refer to a mathematical description that provides an efficient representation of the "essential" properties of a signal.

For example, a finite segment $\{x(n)\}_{n=0}^{N-1}$ of any signal can be approximated by a linear combination of constant ($\lambda_k = 1$) or exponentially fading ($0 < \lambda_k < 1$) sinusoids

$$x(n) \simeq \sum_{k=1}^{M} a_k \lambda_k^n \cos\left(\omega_k n + \phi_k\right) \tag{1.3.1}$$

where $\{a_k, \lambda_k, \omega_k, \phi_k\}_{k=1}^{M}$ are the model parameters. A good model should provide an accurate description of the signal with $4M \ll N$ parameters. From a practical viewpoint, we are most interested in *parametric models*, which assume a given functional form completely specified by a finite number of parameters. In contrast, *nonparametric models* do not put any restriction on the functional form or the number of model parameters.

If any of the model parameters in (1.3.1) is random, the result is a random signal. The most widely used model is given by

$$x(n) = \sum_{k=1}^{M} a_k \cos\left(\omega_k n + \phi_k\right)$$

where the amplitudes $\{a_k\}_1^N$ and the frequencies $\{\omega_k\}_1^N$ are constants and the phases $\{\phi_k\}_1^N$ are random. This model is known as the *harmonic process model* and has many theoretical and practical applications (see Chapters 3 and 9).

Suppose next that we are given a sequence $w(n)$ of independent and identically distributed observations. We can create a time series $x(n)$ with dependent observations, by linearly combining the values of $w(n)$ as

$$x(n) = \sum_{k=-\infty}^{\infty} h(k)w(n-k) \tag{1.3.2}$$

which results in the widely used *linear random signal model*. The model specified by the convolution summation (1.3.2) is clearly nonparametric because, in general, it depends on an infinite number of parameters. Furthermore, the model is a linear, time-invariant system with impulse response $h(k)$ that determines the *memory* of the model and, therefore, the dependence properties of the output $x(n)$. By properly choosing the weights $h(k)$, we can generate a time series with almost any type of dependence among its samples.

In practical applications, we are interested in linear parametric models. As we will see, parametric models exhibit a dependence imposed by their structure. However, if the number of parameters approaches the range of the dependence (in number of samples), the model can mimic any form of dependence. The list of desired features for a good model includes these: (1) the number of model parameters should be as small as possible (*parsimony*), (2) estimation of the model parameters from the data should be easy, and (3) the model parameters should have a physically meaningful interpretation.

If we can develop a successful parametric model for the behavior of a signal, then we can use the model for various applications:

1. To achieve a better understanding of the physical mechanism generating the signal (e.g., earth structure in the case of seismograms).
2. To track changes in the source of the signal and help identify their cause (e.g., EEG).
3. To synthesize artificial signals similar to the natural ones (e.g., speech, infrared backgrounds, natural scenes, data network traffic).
4. To extract parameters for pattern recognition applications (e.g., speech and character recognition).
5. To get an efficient representation of signals for data compression (e.g., speech, audio, and video coding).
6. To forecast future signal behavior (e.g., stock market indexes) (Pindyck and Rubinfeld 1998).

In practice, signal modeling involves the following steps: (1) selection of an appropriate model, (2) selection of the "right" number of parameters, (3) fitting of the model to the actual data, and (4) model testing to see if the model satisfies the user requirements for the particular application. As we shall see in Chapter 9, this process is very complicated and depends heavily on the understanding of the theoretical model properties (see Chapter 4), the amount of familiarity with the particular application, and the experience of the user.

1.3.1 Rational or Pole-Zero Models

Suppose that a given sample $x(n)$, at time n, can be approximated by the previous sample weighted by a coefficient a, that is, $x(n) \approx ax(n-1)$, where a is assumed constant over the signal segment to be modeled. To make the above relationship exact, we add an excitation term $w(n)$, resulting in

$$x(n) = ax(n-1) + w(n) \tag{1.3.3}$$

where $w(n)$ is an excitation sequence. Taking the z-transform of both sides (discussed in Chapter 2), we have

$$X(z) = az^{-1}X(z) + W(z) \tag{1.3.4}$$

which results in the following system function:

$$H(z) = \frac{X(z)}{W(z)} = \frac{1}{1 - az^{-1}} \tag{1.3.5}$$

By using the identity

$$H(z) = \frac{1}{1 - az^{-1}} = 1 + az^{-1} + a^2z^{-2} + \cdots \qquad -1 < a < 1 \tag{1.3.6}$$

the single-parameter model in (1.3.3) can be expressed in the following nonparametric form

$$x(n) = w(n) + aw(n-1) + a^2w(n-2) + \cdots \tag{1.3.7}$$

which clearly indicates that the model generates a time series with *exponentially* decaying dependence.

A more general model can be obtained by including a linear combination of the P previous values of the signal and of the Q previous values of the excitation in (1.3.3), that is,

$$x(n) = \sum_{k=1}^{P} (-a_k) x(n-k) + \sum_{k=0}^{Q} d_k w(n-k) \tag{1.3.8}$$

The resulting system function

$$H(z) = \frac{X(z)}{W(z)} = \frac{\sum_{k=0}^{Q} d_k z^{-k}}{1 + \sum_{k=1}^{P} a_k z^{-k}} \tag{1.3.9}$$

is *rational*, that is, a ratio of two polynomials in the variable z^{-1}, hence the term *rational models*. We will show in Chapter 4 that *any* rational model has a dependence structure or memory that decays exponentially with time. Because the roots of the numerator polynomial are known as *zeros* and the roots of the denominator polynomial as *poles*, these models are also known as *pole-zero models*. In the time-series analysis literature, these models are known as *autoregressive moving-average (ARMA)* models.

Modeling the vocal tract. An example of the application of the pole-zero model is for the characterization of the speech production system. Most generally, speech sounds are classified as either voiced or unvoiced. For both of these types of speech, the production is modeled by exciting a linear system, the vocal tract, with an excitation having a flat, that is, constant, spectrum. The vocal tract, in turn, is modeled by using a pole-zero system, with the poles modeling the vocal tract resonances and the zeros serving the purpose of dampening the spectral response between pole frequencies. In the case of voiced speech, the input to the vocal tract model is a quasi-periodic pulse waveform, whereas for unvoiced speech the source is modeled as random noise. The system model of the speech production process is shown in Figure 1.12. The parameters of this model are the voiced/unvoiced

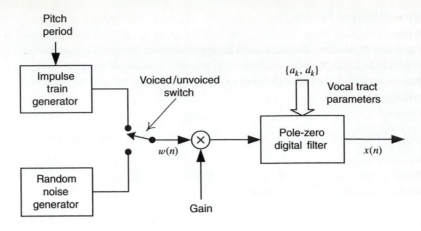

FIGURE 1.12
Speech synthesis system based on pole-zero modeling.

classification, the pitch period for voiced sounds, the gain parameter, and the coefficients $\{d_k\}$ and $\{a_k\}$ of the vocal tract filter (1.3.9). This model is widely used for low-bit-rate (less than 2.4 kbits/s) speech coding, synthetic speech generation, and extraction of features for speech and speaker recognition (Rabiner and Schafer 1978; Rabiner and Juang 1993; Furui 1989).

1.3.2 Fractional Pole-Zero Models and Fractal Models

Although the dependence in (1.3.7) becomes stronger as the pole $a \to 1$, it cannot effectively model time series whose autocorrelation decays asymptotically as a power law. For $a = 1$, that is, for a pole on the unit circle (unit pole), we obtain an everlasting constant dependence, but the output of the model increases without limit and the model is said to be unstable. However, we can obtain a stable model with long memory by creating a *fractional unit pole*, that is, by raising (1.3.6) by a fractional power. Indeed, using the identity

$$H(z) = \frac{1}{(1 - z^{-1})^d} = 1 + dz^{-1} + \frac{d(d+1)}{2!}z^{-2} + \cdots \qquad -\frac{1}{2} < d < \frac{1}{2} \qquad (1.3.10)$$

we have
$$x(n) = w(n) + dw(n-1) + \frac{d(d+1)}{2!}w(n-2) + \cdots \qquad (1.3.11)$$

The weights $h_d(n)$ in (1.3.11) decay according to n^{d-1} as $n \to \infty$; that is, the dependence decays asymptotically as a power law or hyperbolically. Even if the model (1.3.11) is specified by one parameter, its implementation involves an infinite-order convolution summation. Therefore, its practical realization requires an approximation by a rational model that can be easily implemented by using a difference equation. If $w(n)$ is a sequence of independent Gaussian random variables, the process generated by (1.3.11) is known as *fractionally differenced Gaussian noise*. Rational models including one or more fractional poles are known in time-series analysis as *fractional autoregressive integrated moving-average* models and are studied in Chapter 12. The short-term dependence of these models is exponential, whereas their long-term dependence is hyperbolic.

In continuous time, we can create long dependence by using a fractional pole. This is illustrated by the following Laplace transform pair

$$\mathcal{L}\{t^{\beta-1}\} \propto \frac{1}{s^\beta} \qquad \beta > 0 \qquad (1.3.12)$$

which corresponds to an integrator for $\beta = 1$ and a *fractional integrator* for $0 < \beta < 1$. Clearly, the memory of a continuous-time system with impulse response $h_\beta(t) = t^{\beta-1}$ for

$t \geq 0$ and $h_\beta(t) = 0$ for $t < 0$ decays hyperbolically. The response of such a system to white Gaussian noise results in a nonstationary process called *fractional Brownian motion*. Sampling the fractional Brownian motion process at equal intervals and computing the one-step increments result in a stationary discrete-time process known as *fractional Gaussian noise*. Both processes exhibit long memory and are of great theoretical and practical interest and their properties and applications are discussed in Chapter 12.

Exciting a rational model with fractional Gaussian noise leads to a very flexible class of models that exhibit exponential short-range dependence and hyperbolic long-range dependence. The excitation of fractional models (either discrete-time or continuous-time) with statistically independent inputs whose amplitude changes are distributed according to a stable probability law leads to random signal models with long dependence and high amplitude variability. Such models have many practical applications and are also discussed in Chapter 12.

If we can reproduce an object by magnifying some portion of it, we say that the object is *scale-invariant* or *self-similar*. Thus, self-similarity is invariance with respect to scaling. Self-similar geometric objects are known as *fractals*. More specifically, a signal $x(t)$ is self-similar if $x(ct) = c^H x(t)$ for some $c > 0$. The constant H is known as the *self-similarity index*. It can easily be seen that a signal described by a power law, say, $x(t) = \alpha t^\beta$, is self-similar. However, such signals are of limited interest. A more interesting and useful type of signal is one that exhibits a weaker statistical version of self-similarity. A random signal is called *(statistically) self-similar* if its statistical properties are scale-invariant, that is, its statistics do not change under magnification or minification. Self-similar random signals are also known as *random fractals*. Figure 1.13 provides a visual illustration of the self-similar

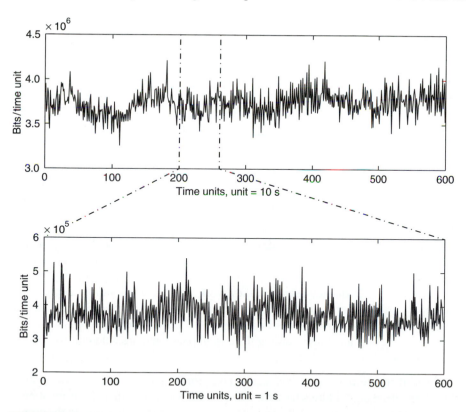

FIGURE 1.13
Pictorial illustration of self-similarity for the variable bit rate video traffic time series. The bottom series is obtained from the top series by expanding the segment between the two vertical lines. Although the two series have lengths of 600 and 60 s, they are remarkably similar visually and statistically (*Courtesy of M. Garrett and M. Vetterli*).

behavior of the variable bit rate video traffic time series. The analysis and modeling of such time series find extensive applications in Internet traffic applications (Michiel and Laevens 1997; Garrett and Willinger 1994).

A classification of the various signal models described previously is given in Figure 1.14, which also provides information about the chapters of the book where these signals are discussed.

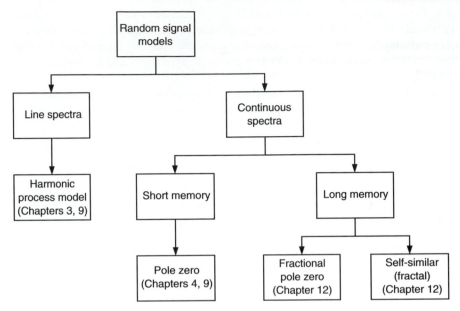

FIGURE 1.14
Classification of random signal models.

1.4 ADAPTIVE FILTERING

Conventional frequency-selective digital filters with *fixed* coefficients are designed to have a given frequency response chosen to alter the spectrum of the input signal in a desired manner. Their key features are as follows:

1. The filters are linear and time-invariant.
2. The design procedure uses the desired passband, transition bands, passband ripple, and stopband attenuation. We do *not* need to know the sample values of the signals to be processed.
3. Since the filters are frequency-selective, they work best when the various components of the input signal occupy nonoverlapping frequency bands. For example, it is easy to separate a signal and additive noise when their spectra do not overlap.
4. The filter coefficients are chosen during the design phase and are held constant during the normal operation of the filter.

However, there are many practical application problems that cannot be successfully solved by using fixed digital filters because either we do not have sufficient information to design a digital filter with fixed coefficients or the design criteria change during the normal operation of the filter. Most of these applications can be successfully solved by using special "smart" filters known collectively as *adaptive filters*. The distinguishing feature of adaptive filters is that they can modify their response to improve performance during operation without any intervention from the user.

1.4.1 Applications of Adaptive Filters

The best way to introduce the concept of adaptive filtering is by describing some typical application problems that can be effectively solved by using an adaptive filter. The applications of adaptive filters can be sorted for convenience into four classes: (1) system identification, (2) system inversion, (3) signal prediction, and (4) multisensor interference cancelation (see Figure 1.15 and Table 1.1). We next describe each class of applications and provide a typical example for each case.

TABLE 1.1
Classification of adaptive filtering applications.

Application class	Examples
System identification	Echo cancelation
	Adaptive control
	Channel modeling
System inversion	Adaptive equalization
	Blind deconvolution
Signal prediction	Adaptive predictive coding
	Change detection
	Radio frequency interference cancelation
Multisensor interference cancelation	Acoustic noise control
	Adaptive beamforming

System Identification

This class of applications, known also as system modeling, is illustrated in Figure 1.15(*a*). The system to be modeled can be either real, as in control system applications, or some hypothetical signal transmission path (e.g., the echo path). The distinguishing characteristic of the system identification application is that the input of the adaptive filter is noise-free and the desired response is corrupted by additive noise that is uncorrelated with the input signal. Applications in this class include echo cancelation, channel modeling, and identification of systems for control applications (Gitlin et al. 1992; Ljung 1987; Åström and Wittenmark 1990). In control applications, the purpose of the adaptive filter is to estimate the parameters or the state of the system and then to use this information to design a controller. In signal processing applications, the goal is to obtain a good estimate of the desired response according to the adopted criterion of performance.

Acoustic echo cancelation. Figure 1.16 shows a typical audio teleconferencing system that helps two groups of people, located at two different places, to communicate effectively. However, the performance of this system is degraded by the following effects: (1) The *reverberations* of the room result from the fact that the microphone picks up not only the speech coming from the talker but also reflections from the walls and furniture in the room. (2) *Echoes* are created by the acoustic coupling between the microphone and the loudspeaker located in the same room. Speech from room B not only is heard by the listener in room A but also is picked up by the microphone in room A, and unless it is prevented, will return as an echo to the speaker in room B.

Several methods to deal with acoustic echoes have been developed. However, the most effective technique to prevent or control echoes is adaptive echo cancelation. The basic idea is very simple: To cancel the echo, we generate a replica or pseudo-echo and then subtract it from the real echo. To synthesize the echo replica, we pass the signal at the loudspeaker through a device designed to duplicate the reverberation and echo properties of the room (*echo path*), as is illustrated in Figure 1.17.

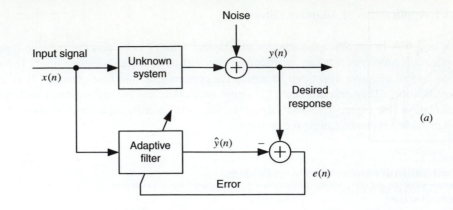

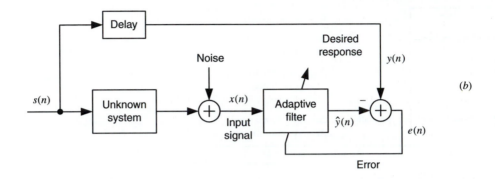

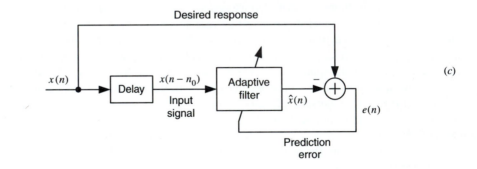

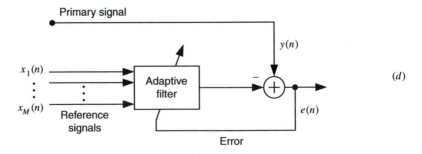

FIGURE 1.15
The four basic classes of adaptive filtering applications: (*a*) system identification, (*b*) system inversion, (*c*) signal prediction, and (*d*) multisensor interference cancelation.

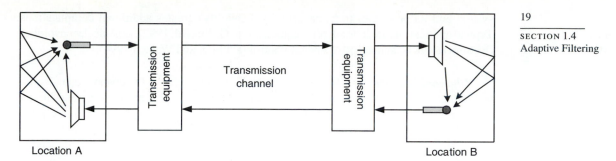

FIGURE 1.16
Typical teleconferencing system without echo control.

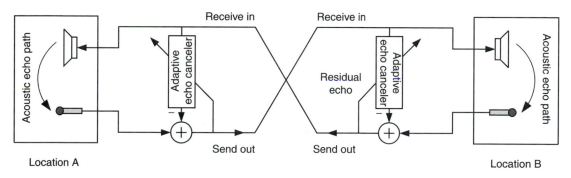

FIGURE 1.17
Principle of acoustic echo cancelation using an adaptive echo canceler.

In practice, there are two obstacles to this approach. (1) The echo path is usually unknown before actual transmission begins and is quite complex to model. (2) The echo path is changing with time, since even the move of a talker alters the acoustic properties of the room. Therefore, we *cannot* design and use a *fixed* echo canceler with satisfactory performance for all possible connections. There are two possible ways around this problem:

1. Design a compromise *fixed* echo canceler based on some "average" echo path, assuming that we have sufficient information about the connections to be seen by the canceler.
2. Design an *adaptive echo canceler* that can "learn" the echo path when it is first turned on and afterward "tracks" its variations without any intervention from the designer. Since an adaptive canceler matches the echo patch for any given connection, it performs better than a fixed compromise canceler.

We stress that the main task of the canceler is to estimate the echo signal with sufficient accuracy; the estimation of the echo path is simply the means for achieving this goal. The performance of the canceler is measured by the attenuation of the echo. The adaptive echo canceler achieves this goal, by modifying its response, using the residual echo signal in an as-yet-unspecified way. More details about acoustic echo cancelation can be found in Gilloire et al. (1996).

System inversion

This class of applications, which is illustrated in Figure 1.15(*b*), is also known as inverse system modeling. The goal of the adaptive filter is to estimate and apply the inverse of the system. Dependent on the application, the input of the adaptive filter may be corrupted by additive noise, and the desired response may not be available. The existence of the inverse system and its properties (e.g., causality and stability) creates additional complications. Typical applications include adaptive equalization (Gitlin et al. 1992), seismic deconvolution (Robinson 1984), and adaptive inverse control (Widrow and Walach 1994).

Channel equalization. To understand the basic principles of the channel equalization techniques, we consider a binary data communication system that transmits a band-limited analog pulse with amplitudes A (symbol 1) or $-A$ (symbol 0) every T_b s (see Figure 1.18). Here T_b is known as the *symbol interval* and $\mathcal{R}_b = 1/T_b$ as the *baud rate*. As the signal propagates through the channel, it is delayed and attenuated in a frequency-dependent manner. Furthermore, it is corrupted by additive noise and other natural or man-made interferences. The goal of the receiver is to measure the amplitude of each arriving pulse and to determine which one of the two possible pulses has been sent. The received signal is sampled once per symbol interval after filtering, automatic gain control, and carrier removal. The sampling time is adjusted to coincide with the "center" of the received pulse. The shape of the pulse is chosen to attain the maximum rate at which the receiver can still distinguish the different pulses. To achieve this goal, we usually choose a band-limited pulse that has periodic zero crossings every T_b s.

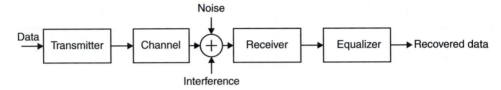

FIGURE 1.18
Simple model of a digital communications system.

If the periodic zero crossings of the pulse are preserved after transmission and reception, we can measure its amplitude without interference from overlapping adjacent pulses. However, channels that deviate from the ideal response (constant magnitude and linear phase) destroy the periodic zero-crossing property and the shape of the peak of the pulse. As a result, the tails of adjacent pulses interfere with the measurement of the current pulse and can lead to an incorrect decision. This type of degradation, which is known as *intersymbol interference (ISI)*, is illustrated in Figure 1.19.

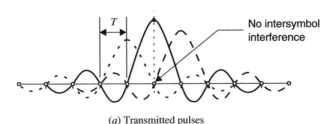

(*a*) Transmitted pulses

FIGURE 1.19
Pulse trains (*a*) without intersymbol interference and (*b*) with intersymbol interference.

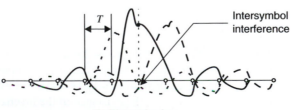

(*b*) Distorted pulses

We can compensate for the ISI distortion by using a linear filter called an *equalizer*. The goal of the equalizer is to restore the received pulse, as closely as possible, to its original shape. The equalizer transforms the channel to a near-ideal one if its response resembles the *inverse* of the channel. Since the channel is unknown and possibly time-varying, there are two ways to approach the problem: (1) Design a fixed compromise equalizer to obtain satisfactory performance over a broad range of channels, or (2) design an equalizer that can "learn" the inverse of the particular channel and then "track" its variation in real time.

The characteristics of the equalizer are adjusted by some algorithm that attempts to attain the best possible performance. The most appropriate criterion of performance for data transmission systems is the probability of symbol error. However it cannot be used for two reasons: (1) The "correct" symbol is unknown to the receiver (otherwise there would be no reason to communicate), and (2) the number of decisions (observations) needed to estimate the low probabilities of error is extremely large. Thus, practical equalizers assess their performance by using some function of the difference between the "correct" symbol and the output. The operation of practical equalizers involves two modes[†] of operation, dependent on how we substitute for the unavailable correct symbol sequence. (1) A known *training sequence* is transmitted, and the equalizer attempts to improve its performance by comparing its output to a synchronized replica of the training sequence stored at the receiver. Usually this mode is used when the equalizer starts a transmission session. (2) At the end of the training session, when the equalizer starts making reliable decisions, we can replace the training sequence with the equalizer's own decisions.

Adaptive equalization is a mature technology that has had the greatest impact on digital communications systems, including voiceband, microwave and troposcatter radio, and cable TV modems (Qureshi 1985; Lee and Messerschmitt 1994; Gitlin et al. 1992; Bingham 1988; Treichler et al. 1996).

Signal prediction

In the next class of applications, the goal is to estimate the value $x(n_0)$ of a random signal by using a set of consecutive signal samples $\{x(n), n_1 \leq n \leq n_2\}$. There are three cases of interest: (1) forward prediction, when $n_0 > n_2$; (2) backward "prediction," when $n_0 < n_1$; and (3) smoothing or interpolation, when $n_1 < n_0 < n_2$. Clearly, in the last case the value at $n = n_0$ is not used in the computation of the estimate. The most widely used type is forward linear prediction or simply *linear prediction*[‡] [see Figure 1.15(c)], where the estimate is formed by using a linear combination of past samples (Makhoul 1975).

Linear predictive coding (LPC). The efficient storage and transmission of analog signals using digital systems requires the minimization of the number of bits necessary to represent the signal while maintaining the quality to an acceptable level according to a certain criterion of performance. The conversion of an analog (continuous-time, continuous-amplitude) signal to a digital (discrete-time, discrete-amplitude) signal involves two processes: sampling and quantization. Sampling converts a continuous-time signal to a discrete-time signal by measuring its amplitude at equidistant intervals of time. Quantization involves the representation of the measured continuous amplitude using a finite number of symbols and always creates some amount of distortion (quantization noise).

For a fixed number of bits, decreasing the dynamic range of the signal (and therefore the range of the quantizer) decreases the required quantization step and therefore the average quantization error power. Therefore, we can decrease the quantization noise by reducing the dynamic range or equivalently the variance of the signal. If the signal samples are

---

[†] Another mode of operation, where the equalizer can operate without the benefit of a training sequence (blind or self-recovering mode), is discussed in Chapter 12.

[‡] As we shall see in Chapters 4 and 6, linear prediction is closely related, but not identical, to all-pole signal modeling.

significantly correlated, the variance of the difference between adjacent samples is smaller than the variance of the original signal. Thus, we can improve quality by quantizing this difference instead of the original signal. This idea is exploited by the linear prediction system shown in Figure 1.20. This system uses a linear predictor to form an estimate (prediction) $\hat{x}(n)$ of the present sample $x(n)$ as a linear combination of the M past samples, that is,

$$\hat{x}(n) = \sum_{k=1}^{M} a_k x(n - k) \tag{1.4.1}$$

The coefficients $\{a_k\}_1^M$ of the linear predictor are determined by exploiting the correlation between adjacent samples of the input signal with the objective of making the prediction error

$$e(n) = x(n) - \hat{x}(n) \tag{1.4.2}$$

as small as possible. If the prediction is good, the dynamic range of $e(n)$ should be smaller than the dynamic range of $x(n)$, resulting in a smaller quantization noise for the same number of bits or the same quantization noise with a smaller number of bits. The performance of the LPC system depends on the accuracy of the predictor. Since the statistical properties of the signal $x(n)$ are unknown and change with time, we *cannot* design an optimum fixed predictor. The established practical solution is to use an *adaptive* linear predictor that automatically adjusts its coefficients to compute a "good" prediction at each time instant. A detailed discussion of adaptive linear prediction and its application to audio, speech, and video signal coding is provided in Jayant and Noll (1984).

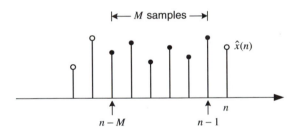

FIGURE 1.20
Illustration of the linear prediction of a signal $x(n)$ using a finite number of past samples.

Multisensor interference cancelation

The key feature of this class of applications is the use of multiple sensors to remove undesired interference and noise. Typically, a *primary* signal contains both the signal of interest and the interference. Other signals, known as *reference* signals, are available for the purposes of canceling the undesired interference [see Figure 1.15(d)]. These reference signals are collected using other sensors in which the signal of interest is not present or is so weak that it can be ignored. The amount of correlation between the primary and reference signals is measured and used to form an estimate of the interference in the primary signal, which is subsequently removed. Had the signal of interest been present in the reference signal(s), then this process would have resulted in the removal of the desired signal as well. Typical applications in which interference cancelation is employed include array processing for radar and communications, biomedical sensing systems, and active noise control (Widrow et al. 1975; Kuo and Morgan 1996).

Active noise control (ANC). The basic idea behind an ANC system is the cancelation of acoustic noise using destructive wave interference. To create destructive interference that cancels an acoustic noise wave (primary) at a point P, we can use a loudspeaker that creates, at the same point P, another wave (secondary) with the same frequency, the same amplitude, and 180° phase difference. Therefore, with appropriate control of the peaks and troughs of the secondary wave, we can produce zones of destructive interference (quietness). ANC systems using digital signal processing technology find applications in air-conditioning ducts, aircraft, cars, and magnetic resonance imaging (MRI) systems (Elliott and Nelson 1993; Kuo and Morgan 1996).

Figure 1.21 shows the key components of an adaptive ANC system described in Crawford et al. 1997. The task of the loudspeaker is to generate an acoustic wave that is an 180° phase-inverted version of the signal $y(t)$ when it arrives at the error microphone. In this case the error signal $e(t) = y(t) + \hat{y}(t) = 0$, and we create a "quiet zone" around the microphone. If the acoustic paths (1) from the noise source to the reference microphone (G_x), (2) from the noise source to the error microphone (G_y), (3) from the secondary loudspeaker to the reference microphone (H_x), and (4) from the secondary loudspeaker to the error microphone ($H_{\hat{y}}$) are linear, time-invariant, and known, we can design a linear filter H such that $e(n) = 0$. For example, if the effects of H_x and $H_{\hat{y}}$ are negligible, the filter H should invert G_x to obtain $v(t)$ and then replicate G_y to synthesize $\hat{y}(t) \simeq y(t)$. The quality of cancellation depends on the accuracy of these two modeling processes.

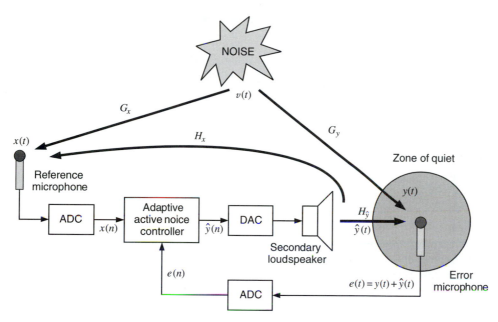

FIGURE 1.21
Block diagram of the basic components of an active noise control system.

In practice, the acoustic environment is unknown and time-varying. Therefore, we cannot design a fixed ANC filter with satisfactory performance. The only feasible solution is to use an adaptive filter with the capacity to identify and track the variation of the various acoustic paths and the spectral characteristics of the noise source in real time. The adaptive ANC filter adjusts its characteristics by trying to minimize the energy of the error signal $e(n)$. Adaptive ANC using digital signal processing technology is an active area of research, and despite several successes many problems remain to be solved before such systems find their way to more practical applications (Crawford et al. 1997).

1.4.2 Features of Adaptive Filters

Careful inspection of the applications discussed in the previous section indicates that every adaptive filter consists of the following three modules (see Figure 1.22).

1. **Filtering structure.** This module forms the output of the filter using measurements of the input signal or signals. The filtering structure is *linear* if the output is obtained as a linear combination of the input measurements; otherwise, it is said to be *nonlinear*. The structure is fixed by the designer, and its parameters are adjusted by the adaptive algorithm.

2. **Criterion of performance (COP).** The output of the adaptive filter and the desired response (when available) are processed by the COP module to assess its quality with respect to the requirements of the particular application.

3. **Adaptive algorithm.** The adaptive algorithm uses the value of the criterion of performance, or some function of it, and the measurements of the input and desired response (when available) to decide how to modify the parameters of the filter to improve its performance.

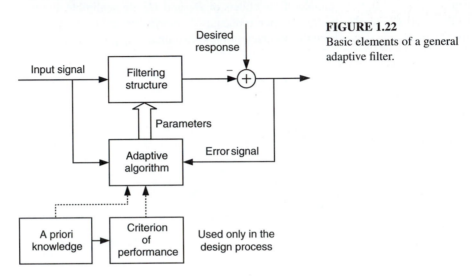

FIGURE 1.22
Basic elements of a general adaptive filter.

Every adaptive filtering application involves one or more input signals and a desired response signal that may or may not be accessible to the adaptive filter. We collectively refer to these relevant signals as the *signal operating environment* (*SOE*) of the adaptive filter. The design of any adaptive filter requires a great deal of *a priori information* about the SOE and a deep understanding of the particular application (Claasen and Mecklenbrauker 1985). This information is needed by the designer to choose the filtering structure and the criterion of performance and to design the adaptive algorithm. To be more specific, adaptive filters are designed for a specific type of input signal (speech, binary data, etc.), for specific types of interferences (additive white noise, sinusoidal signals, echoes of the input signals, etc.), and for specific types of signal transmission paths (e.g., linear time-invariant or time-varying). After the proper design decisions have been made, the only unknowns, when the adaptive filter starts its operation, are a set of parameters that are to be determined by the adaptive algorithm using signal measurements. Clearly, unreliable a priori information and/or incorrect assumptions about the SOE can lead to serious performance degradations or even unsuccessful adaptive filter applications.

If the characteristics of the relevant signals are constant, the goal of the adaptive filter is to find the parameters that give the best performance and then to stop the adjustment. However, when the characteristics of the relevant signals change with time, the adaptive filter should first find and then continuously readjust its parameters to track these changes.

A very influential factor in the design of adaptive algorithms is the availability of a desired response signal. We have seen that for certain applications, the desired response may not be available for use by the adaptive filter. In this book we focus on supervised adaptive filters that require the use of a desired response signal and we simply call them adaptive filters (Chapter 10). Unsupervised adaptive filters are discussed in Chapter 12.

Suppose now that the relevant signals can be modeled by stochastic processes with known statistical properties. If we adopt the minimum mean square error as a criterion of performance, we can design, at least in principle, an optimum filter that provides the ultimate solution. From a theoretical point of view, the goal of the adaptive filter is to replicate the performance of the optimum filter without the benefit of knowing and using the *exact* statistical properties of the relevant signals. In this sense, the theory of optimum filters (see Chapters 6 and 7) is a prerequisite for the understanding, design, performance evaluation, and successful application of adaptive filters.

1.5 ARRAY PROCESSING

Array processing deals with techniques for the analysis and processing of signals collected by a group of sensors. The collection of sensors makes up the array, and the manner in which the signals from the sensors are combined and handled constitutes the processing. The type of processing is dictated by the needs of the particular application. Array processing has found widespread application in a large number of areas, including radar, sonar, communications, seismology, geophysical prospecting for oil and natural gas, diagnostic ultrasound, and multichannel audio systems.

1.5.1 Spatial Filtering or Beamforming

Generally, an array receives spatially propagating signals and processes them to emphasize signals arriving from a certain direction; that is, it acts as a spatially discriminating filter. This spatial filtering operation is known as *beamforming*, because essentially it emulates the function of a mechanically steered antenna. An array processor steers a beam to a particular direction by computing a properly weighted sum of the individual sensor signals. An example of the spatial response of the beamformer, known as the *beampattern*, is shown in Figure 1.23. The beamformer emphasizes signals in the direction to which it is steered while attenuating signals from other directions.

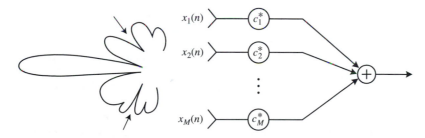

FIGURE 1.23
Example of the spatial response of an array, known as a beampattern, that emphasizes signals from a direction of interest, known as the look direction.

In the case of an array with sensors equally spaced on a line, known as a *uniform linear array (ULA)*, there is a direct analogy between beamforming and the frequency-selective filtering of a discrete-time signal using a *finite impulse response (FIR)* filter. This analogy between a beamformer and an FIR filter is illustrated in Figure 1.24. The array of sensors spatially samples the impinging waves so that in the case of a ULA, the sampling

FIR filter   Uniform linear array

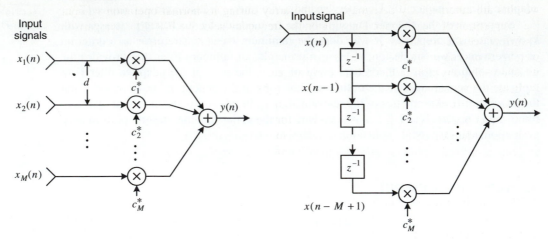

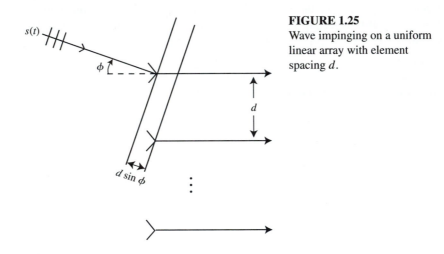

FIGURE 1.24
Analogy between beamforming and frequency-selective FIR filtering.

FIGURE 1.25
Wave impinging on a uniform
linear array with element
spacing d.

is performed at equal spatial increments. By contrast, an FIR filter uses a uniformly time-sampled signal as its input. Consider a plane wave impinging on an array as in Figure 1.25. The spatial signal arrives at each sensor with a delay determined by the angle of arrival ϕ. In the case of a narrowband signal, this delay corresponds to an equal phase shift from sensor to sensor that results in a spatial frequency across the ULA of

$$u = \frac{d}{\lambda}\sin\phi \qquad (1.5.1)$$

where λ is the wavelength of the signal and d is the uniform spacing of the sensors. This spatial frequency is analogous to the temporal frequency encountered in discrete-time signals. In the beamforming operation, the sensor signals are combined with weights on each of the sensor signals just as an FIR filter produces an output that is the weighted sum of time samples. As a frequency-selective FIR filter extracts signals at a frequency of interest, a beamformer seeks to emphasize signals with a certain spatial frequency (i.e., signals arriving from a particular angle). Thus, it is often beneficial to view a beamformer as a spatial frequency-selective filter.

Many times an array must contend with undesired signals arriving from other directions, which may prevent it from successfully extracting the signal of interest for which it was designed. In this case, the array must adjust its response to the data it receives to reject signals from these other directions. The resulting array is an *adaptive array* as the beamforming weights are automatically determined by the array during its normal operation without the intervention of the designer. Drawing on the frequency-selective FIR filter comparison again, we see that an adaptive array is analogous to an adaptive FIR filter that adjusts its weights to pass signals at the desired frequency or signals with certain statistical properties while rejecting any signals that do not satisfy these requirements. Again, if we can model the SOE, using stationary processes with known statistical properties, we can design an *optimum* beamformer that minimizes or maximizes a certain criterion of performance. The optimum beamformer can be used to provide guidelines for the design of adaptive beamformers and used as a yardstick for their performance evaluation. The analysis, design, and performance evaluation of fixed, optimum, and adaptive beamformers are discussed in Chapter 11.

1.5.2 Adaptive Interference Mitigation in Radar Systems

The goal of an airborne surveillance radar system is to determine the presence of target signals. These targets can be either airborne or found on the ground below. Typical targets of interest are other aircraft, ground moving vehicles, or hostile missiles. The desired information from these targets is their relative distance from our airborne platform, known as the *range*, their angle with respect to the platform, and their relative speed. The processing of the radar consists of the following sequence:

- Filter out undesired signals through adaptive processing.
- Determine the presence of targets, a process known as *detection*.
- Estimate the parameters of all detected targets.

To sense these targets, the radar system transmits energy in the direction it is searching for targets. The transmitted energy propagates from the airborne radar to the target that reflects the radar signal. The reflection then propagates from the target back to the radar. Since the radar signal travels at the speed of light (3×10^8 m/s), the round-trip delay between transmission and reception of this signal determines the range of the target. The received signal is known as the *return*. The angle of the target is determined through the use of beamforming or spatial filtering using an array of sensor elements. To this end, the radar forms a bank of spatial filters evenly spaced in angle and determines which filter contains the target. For example, we might be interested in the angular sector between $-1° \leq \phi \leq 1°$. Then we might set up a bank of beamformers in this angular region with a spacing of $0.5°$. If these spatial filters perform this operation nonadaptively, it is often referred to as *conventional beamforming*.

The detection of target signals is inhibited by the presence of other undesired signals known as *interference*. Two common types of interference are the reflections of the radar signal from the ground, known as *clutter*, and other transmitted energy at the same operating frequency as the radar, referred to as *jamming*. Jamming can be the hostile transmission of energy to prevent us from detecting certain signals, or it may be incidental, for example, from another radar. Such an interference scenario for an airborne surveillance radar is depicted in Figure 1.26. The interference signals are typically much larger than the target return. Thus, when a nonadaptive beamformer is used, interference leaks in through the sidelobes of the beamformer and prevents us from detecting the target. However, we can adjust the beamformer weights such that signals from the directions of the interference are rejected while other directions are searched for targets. If the weights are adapted to the received data in this way, then the array is known as an *adaptive array* and the operation is called *adaptive beamforming*. The use of an adaptive beamformer is also illustrated in Figure 1.26. We show the spatial response or *beampattern* of the adaptive array. Note that the peak gain of the beamformer is in the direction of the target. On the other hand, the clutter and jamming are rejected by placing *nulls* in the beampattern.

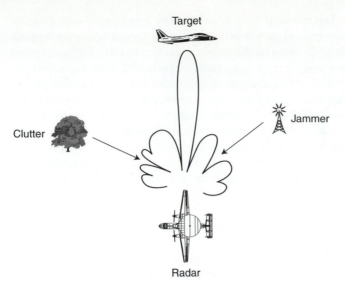

FIGURE 1.26
Example of adaptive beamformer used with an airborne
surveillance radar for interference mitigation.

In practice, we do not know the directions of the interferers. Therefore, we need an adaptive beamformer that can determine its weights by estimating the statistics of the interference environment. If we can model the SOE using stochastic processes with known statistical properties, we can design an optimum beamformer that provides the ultimate performance. The discussion about adaptive filters in Section 1.4.2 applies to adaptive beamformers as well.

Once we have determined the presence of the target signal, we want to get a better idea of the exact angle it was received from. Recall that the beamformers have angles associated with them, so the angle of the beamformer in which the target was detected can serve as a rough estimate of the angle of the target. The coarseness of our initial estimate is governed by the spacing in angle of the filter bank of beamformers, for example, 1°. This resolution in angle of the beamformer is often called a *beamwidth*. To get a better estimate, we can use a variety of angle estimation methods. If the angle estimate can refine the accuracy down to one-tenth of a beamwidth, for example, 0.1°, then the angle estimator is said to achieve 10 : 1 beamsplitting. Achieving an angle accuracy better than the array beamwidth is often called *superresolution*.

1.5.3 Adaptive Sidelobe Canceler

Consider the scenario in Figure 1.26 from the adaptive beamforming example for interference mitigation in a radar system. However, instead of an array of sensors, consider a fixed (i.e., nonadaptive) channel that has high gain in the direction of the target. This response may have been the result of a highly directive dish antenna or a nonadaptive beamformer. Sometimes it is necessary to perform beamforming nonadaptively to limit the number of channels. One such case arises for very large arrays for which it is impractical to form channels by digitally sampling every element. The array is partitioned into subarrays that all form nonadaptive beams in the same direction. Then the subarray outputs form the spatial channels that are sampled. Each channel is highly directive, though with a lower resolution than the entire array. In the case of interference, it is then present in all these subarray channels and must be removed in some way. To restore its performance to the interference-free case, the radar system must employ a spatially adaptive method that removes the interference in the main channel. The *sidelobe canceler* is one such method and is illustrated in Figure 1.27.

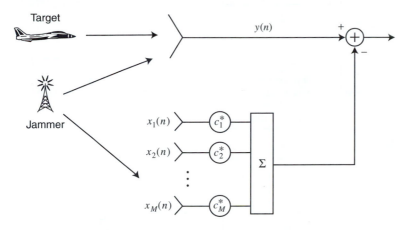

FIGURE 1.27
Sidelobe canceler with a highly directive main channel and auxiliary channels.

Note that the signal of interest is received from a particular direction in which we assume the main channel has a large gain. On the other hand, the jamming signal is received from another direction, and since it has much higher power than the attenuation of the antenna sidelobes, the jamming interference obscures the signals of interest. This high-gain channel is known as the main channel that contains both the signal of interest and the jamming interference. The sidelobe canceler uses one or more auxiliary channels in order to cancel the main-channel interference. These auxiliary channels typically have much lower gain in the direction in which the main channel is directed so that they contain only the interference. The signal of interest is weak enough that it is below the thermal noise floor in these auxiliary channels. Examples of these auxiliary channels would be omnidirectional sensors or even directive sensors pointed in the direction of the interference. Note that for very strong signals, the signal of interest may be present in the auxiliary channel, in which case signal cancelation can occur. Clearly, this application belongs to the class of multisensor interference cancelation shown in Figure 1.15.

The sidelobe canceler uses the auxiliary channels to form an estimate of the interference in the main channel. The estimate is computed by weighting the auxiliary channel in an adaptive manner dependent on the cross-correlation between the auxiliary channels and the main channel. The estimate of the main-channel interference is subtracted from the main channel. The result is an overall antenna response with a spatial null directed at the interference source while maintaining high gain in the direction of interest. Clearly, if we had sufficient a priori information, the problem could be solved by designing a fixed canceler. However, the lack of a priori information and the changing properties of the environment make an adaptive canceler the only viable solution.

1.6 ORGANIZATION OF THE BOOK

In this section we provide an overview of the main topics covered in the book so as to help the reader navigate through the material and understand the interdependence among the various chapters (see Figure 1.28).

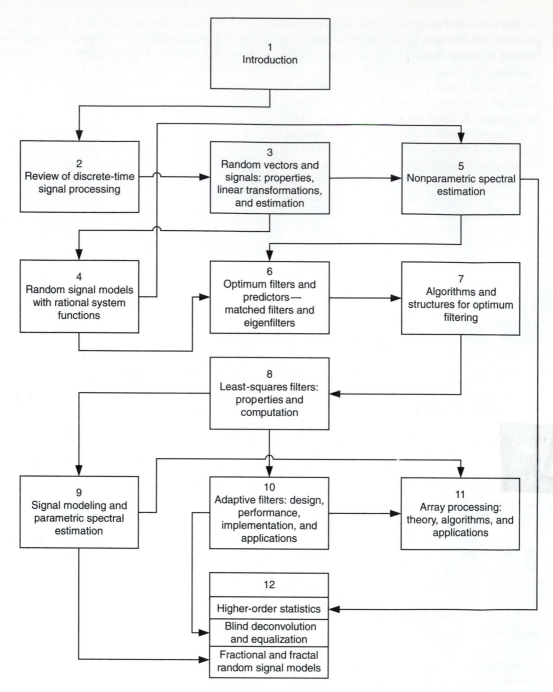

FIGURE 1.28
Flowchart organization of the book's chapters.

In Chapter 2, we review the fundamental topics in discrete-time signal processing that can be used for both deterministic and random signals. Chapter 3 provides a concise review of the theory of random variables and random sequences and elaborates on certain topics that are crucial to developments in subsequent chapters. Reading these chapters is essential to familiarize the reader with notation and properties that are repeatedly used throughout the rest of the book. Chapter 5 presents the most practical methods for nonparametric estimation of correlation and spectral densities. The use of these techniques for exploratory investigation of the relevant signal characteristics before performing any modeling or adaptive filtering is invaluable.

Chapters 4 and 6 provide a detailed study of the theoretical properties of signal models and optimum filters, assuming that the relevant signals can be modeled by stochastic processes with known statistical properties. In Chapter 7, we develop algorithms and structures for optimum filtering and signal modeling and prediction.

Chapter 8 introduces the general method of least squares and shows how to use it for the design of filters and predictors from actual signal observations. The statistical properties and the numerical computation of least-squares estimates are also discussed in detail.

Chapters 9, 10, and 11 use the theoretical work in Chapters 4, 6, and 7 and the practical methods in Chapter 8 to develop, evaluate, and apply practical techniques for signal modeling, adaptive filtering, and array processing. Finally, Chapter 12 illustrates the use of higher-order statistics, presents the basic ideas of blind deconvolution and equalization, and concludes with a concise introduction to fractional and random fractal signal models.

CHAPTER 2

Fundamentals of Discrete-Time Signal Processing

In many disciplines, signal processing applications nowadays are almost always implemented using digital hardware operating on digital signals. The basic foundation of this modern approach is based on discrete-time system theory. This book also deals with statistical analysis and processing of discrete-time signals, and modeling of discrete-time systems. Therefore, the purpose of this chapter is to focus attention on some important issues of discrete-time signal processing that are of fundamental importance to signal processing, in general, and to this book, in particular. The intent of this chapter is not to teach topics in elementary digital signal processing but to review material that will be used throughout this book and to establish a consistent notation for it. There are several textbooks on these topics, and it is assumed that the reader is familiar with the theory of digital signal processing as found in Oppenheim and Schafer (1989); Proakis and Manolakis (1996).

We begin this chapter with a description and classification of signals in Section 2.1. Representation of deterministic signals from the frequency-domain viewpoint is presented in Section 2.2. In Section 2.3, discrete-time systems are defined, but the treatment is focused on *linear*, *time-invariant* (*LTI*) systems, which are easier to deal with mathematically and hence are widely used in practice. Section 2.4 on minimum-phase systems and system invertibility is an important section in this chapter that should be reviewed prior to studying the rest of the book. The last section, Section 2.5, is devoted to lattice and lattice/ladder structures for discrete-time systems (or filters). A brief summary of the topics discussed in this chapter is provided in Section 2.6.

2.1 DISCRETE-TIME SIGNALS

The physical world is replete with *signals*, that is, physical quantities that change as a function of time, space, or some other independent variable. Although the physical nature of signals arising in various applications may be quite different, there are signals that have some basic features in common. These attributes make it possible to classify signals into families to facilitate their analysis. On the other hand, the mathematical description and analysis of signals require *mathematical signal models* that allow us to choose the appropriate mathematical approach for analysis. Signal characteristics and the classification of signals based upon either such characteristics or the associated mathematical models are the subject of this section.

2.1.1 Continuous-Time, Discrete-Time, and Digital Signals

If we assume that to every set of assigned values of independent variables there corresponds a unique value of the physical quantity (dependent variable), then every signal can be viewed as a function. The dependent variable may be real, in which case we have a *real-valued signal*; or it may be complex, and then we talk about a *complex-valued signal*. The independent variables are always real.

Any signal whose samples are a single-valued function of one independent variable is referred to as a *scalar one-dimensional* signal. We will refer to it simply as a signal. These signals involve one dependent variable and one independent variable and are the signals that we mainly deal with in this book. The speech signal shown in Figure 1.1 provides a typical example of a scalar signal.

Let us now look at both the dependent and independent variables of a signal from a different perspective. Every signal variable may take on values from either a continuous set of values (*continuous* variable) or a discrete set of values (*discrete* variable). Signals whose dependent and independent variables are continuous are usually referred to as *continuous-time* signals, and we will denote these signals by the subscript c, such as $x_c(t)$. In contrast, signals where both the dependent and the independent variables are discrete are called *digital* signals. If only the independent variables are specified to be discrete, then we have a *discrete signal*. We note that a discrete signal is defined only at discrete values of the independent variables, but it may take on any value. Clearly, digital signals are a subset of the set of discrete signals.

In this book, we mainly deal with scalar discrete signals in which the independent variable is time. We refer to them as *discrete-time signals*. Such signals usually arise in practice when we *sample* continuous-time signals, that is, when we select values at discrete-time instances. In all practical applications, the values of a discrete-time signal can only be described by binary numbers with a finite number of bits. Hence, only a discrete set of values is possible; strictly speaking, this means that, in practice, we deal with only digital signals. Clearly, digital signals are the only signals amenable to direct digital computation. Any other signal has to be first converted to digital form before numerical processing is possible.

Because the discrete nature of the dependent variable complicates the analysis, the usual practice is to deal with discrete-time signals and then to consider the effects of the discrete amplitude as a separate issue. Obviously, these effects can be reduced to any desirable level by accordingly increasing the number of bits (or word length) in the involved numerical processing operations. Hence, in the remainder of the book, we limit our attention to discrete-time signals.

2.1.2 Mathematical Description of Signals

The mathematical analysis of a signal requires the availability of a mathematical description for the signal itself. The type of description, usually referred to as a *signal model*, determines the most appropriate mathematical approach for the analysis of the signal. We use the term *signal* to refer to either the signal itself or its mathematical description, that is, the signal model. The exact meaning will be apparent from the context. Clearly, this distinction is necessary if a signal can be described by more than one model. We start with the most important classification of signal models as either deterministic or random.

Deterministic signals

Any signal that can be described by an explicit mathematical relationship is called *deterministic*. In the case of continuous-time signals, this relationship is a given function of time, for example, $x_c(t) = A \cos(2\pi F_0 t + \theta)$, $-\infty < t < \infty$. For discrete-time signals that, mathematically speaking, are sequences of numbers, this relationship may be either a functional expression, for example, $x(n) = a^n$, $-\infty < n < \infty$, or a table of values.

In general, we use the notation $x(n)$ to denote the sequence of numbers that represent a discrete-time signal. Furthermore, we use the term *nth sample* to refer to the value of this sequence for a specific value of n. Strictly speaking, the terminology is correct only if the discrete-time signal has been obtained by sampling a continuous-time signal $x_c(t)$. In the case of periodic sampling with sampling period T, we have $x(n) = x_c(nT)$, $-\infty < n < \infty$; that is, $x(n)$ is the nth sample of $x_c(t)$. Sometimes, just for convenience, we may plot $x_c(t)$ even if we deal with the signal $x(n)$. Finally, we note that sometimes it is convenient to form and manipulate complex-valued signals using a pair of real-valued signals as the real and imaginary components.

Basic signals. There are some basic discrete-time signals that we will repeatedly use throughout this book:

- The *unit sample* or *unit impulse* sequence $\delta(n)$, defined as

$$\delta(n) = \begin{cases} 1 & n = 0 \\ 0 & n \neq 0 \end{cases} \tag{2.1.1}$$

- The *unit step* sequence $u(n)$, defined as

$$u(n) = \begin{cases} 1 & n \geq 0 \\ 0 & n < 0 \end{cases} \tag{2.1.2}$$

- The exponential sequence of the form

$$x(n) = a^n \qquad -\infty < n < \infty \tag{2.1.3}$$

If a is a complex number, that is, $a = re^{j\omega_0}$, $r > 0$, $\omega \neq 0, \pi$, then $x(n)$ is complex-valued, that is,

$$x(n) = r^n e^{j\omega_0 n} = x_R(n) + jx_I(n) \tag{2.1.4}$$

where $\qquad x_R(n) = r^n \cos \omega_0 n \qquad$ and $\qquad x_I(n) = r^n \sin \omega_0 n \tag{2.1.5}$

are the real and imaginary parts of $x(n)$, respectively. The complex exponential signal $x(n)$ and the real sinusoidal signals $x_R(n)$ and $x_I(n)$, which have a decaying (growing) envelope if $r < 1 (r > 1)$, are very useful in the analysis of discrete-time signals and systems.

Signal classification. Deterministic signals can be classified as energy or power, periodic or aperiodic, of finite or infinite duration, causal or noncausal, and even or odd signals. Although we next discuss these concepts for discrete-time signals, a similar discussion applies to continuous-time signals as well.

- The total energy or simply the *energy* of a signal $x(n)$ is given by

$$E_x = \sum_{n=-\infty}^{\infty} |x(n)|^2 \geq 0 \tag{2.1.6}$$

The energy is zero if and only if $x(n) = 0$ for all n. The average power or simply the *power* of a signal $x(n)$ is defined as

$$P_x = \lim_{N \to \infty} \frac{1}{2N+1} \sum_{n=-N}^{N} |x(n)|^2 \geq 0 \tag{2.1.7}$$

A signal with finite energy, that is, $0 < E_x < \infty$, is called an *energy signal*. Signals with finite power, that is, $0 < P_x < \infty$, are referred to as *power signals*. Clearly, energy signals have zero power, and power signals have infinite energy.

- A discrete-time signal $x(n)$ is called *periodic* with fundamental period N if $x(n+N) = x(n)$ for all n. Otherwise it is called *aperiodic*. It can be seen that the complex exponential in (2.1.4) is periodic if and only if $\omega_0/(2\pi) = k/N$, that is, if $\omega_0/(2\pi)$ is a rational number. Clearly, a periodic signal is a power signal with power P given by

$$P_x = \frac{1}{N}\sum_{n=0}^{N-1}|x(n)|^2 \tag{2.1.8}$$

- We say that a signal $x(n)$ has *finite duration* if $x(n) = 0$ for $n < N_1$ and $n > N_2$, where N_1 and N_2 are finite integer numbers with $N_1 \leq N_2$. If $N_1 = -\infty$ and/or $N_2 = \infty$, the signal $x(n)$ has *infinite duration*.
- A signal $x(n)$ is said to be *causal* if $x(n) = 0$ for $n < 0$. Otherwise, it is called *noncausal*.
- Finally, a real-valued signal $x(n)$ is called *even* if $x(-n) = x(n)$ and *odd* if $x(-n) = -x(n)$.

Other classifications for deterministic signals will be introduced in subsequent sections.

Random signals

In contrast to the deterministic signals discussed so far, there are many other signals in practice that cannot be described to any reasonable accuracy by explicit mathematical relationships. The lack of such an explicit relationship implies that the signal evolves in time in an unpredictable manner from the point of view of the observer. Such signals are called *random*. The output of a noise generator, the height of waves in a stormy sea, and the acoustic pressures generated by air rushing through the human vocal tract are examples of random signals. At this point one could say that complete knowledge of the physics of the signal could provide an explicit mathematical relationship, at least within the limits of the uncertainty principle. However, such relationships are typically too complex to be of any practical use.

In general, although random signals are evolving in time in an unpredictable manner, their average properties can often be assumed to be deterministic; that is, they can be specified by explicit mathematical formulas. This concept is key to the modeling of a random signal as a stochastic process.

Thus, random signals are mathematically described by *stochastic processes* and can be analyzed by using *statistical* methods instead of explicit equations. The theory of probability, random variables, and stochastic processes provides the mathematical framework for the theoretical study of random signals.

2.1.3 Real-World Signals

The classification of various physical data as being either deterministic or random might be debated in many cases. For example, it might be argued that no physical data in practice can be truly deterministic since there is always a possibility that some unforeseen event in the future might influence the phenomenon producing the data in a manner that was not originally considered. On the other hand, it might be argued that no physical data are truly random since exact mathematical descriptions might be possible if sufficient knowledge of the basic mechanisms of the phenomenon producing the data were known. In practical terms, the decision as to whether physical data are deterministic or random is usually based upon the ability to reproduce the data by controlled experiments. If an experiment producing specific data of interest can be repeated many times with identical results (within the limits of experimental error), then the data can generally be considered deterministic. If an experiment cannot be designed that will produce identical results when the experiment is repeated, then the data must usually be considered random in nature.

In the deterministic signal model, signals are assumed to be explicitly known for all time from $-\infty$ to $+\infty$. In this sense, no uncertainty exists regarding their past, present, or future amplitude values. The simplest description of any signal is an amplitude-versus-time plot. This "time history" of the signal is very useful for visual analysis because it helps in the identification of specific patterns, which can subsequently be used to extract useful information from the signal. However, quite often, information present in a signal becomes more evident by transformation of the signal into another domain. In this section, we review some transforms for the representation and analysis of discrete-time signals.

2.2.1 Fourier Transforms and Fourier Series

Frequency analysis is, roughly speaking, the process of decomposing a signal into frequency components, that is, complex exponential signals or sinusoidal signals. Although the physical meaning of frequency analysis is almost the same for any signal, the appropriate mathematical tools depend upon the type of signal under consideration. The two characteristics that specify the frequency analysis tools for deterministic signals are

- The nature of time: continuous-time or discrete-time signals.
- The existence of harmony: periodic or aperiodic signals.

Thus, we have the following four types of frequency analysis tools.

Fourier series for continuous-time periodic signals

If a continuous-time signal $x_c(t)$ is periodic with fundamental period T_p, it can be expressed as a linear combination of harmonically related complex exponentials

$$x_c(t) = \sum_{k=-\infty}^{\infty} \check{X}_c(k) e^{j2\pi k F_0 t} \tag{2.2.1}$$

where $F_0 = 1/T_p$ is the *fundamental frequency*, and

$$\check{X}_c(k) = \frac{1}{T_p} \int_0^{T_p} x_c(t) e^{-j2\pi k F_0 t} \, dt \tag{2.2.2}$$

which are termed the Fourier coefficients,[†] or the *spectrum* of $x_c(t)$.

It can be shown that the power of the signal $x_c(t)$ is given by Parseval's relation

$$P_x = \frac{1}{T_p} \int_0^{T_p} |x_c(t)|^2 \, dt = \sum_{k=-\infty}^{\infty} \left| \check{X}_c(k) \right|^2 \tag{2.2.3}$$

Since $|\check{X}_c(k)|^2$ represents the power in the kth frequency component, the sequence $|\check{X}_c(k)|^2$, $-\infty < k < \infty$, is called the *power spectrum* of $x_c(t)$ and shows the distribution of power within various frequency components. Since the power of $x_c(t)$ is confined to the discrete frequencies $0, \pm F_0, \pm 2F_0, \ldots$, we say that $x_c(t)$ has a *line* or *discrete spectrum*.

Fourier transform for continuous-time aperiodic signals

The frequency analysis of a continuous-time, aperiodic signal can be done by using the Fourier transform

$$X_c(F) = \int_{-\infty}^{\infty} x_c(t) e^{-j2\pi F t} \, dt \tag{2.2.4}$$

[†] We use the notation $\check{X}_c(k)$ instead of $X_c(k)$ to distinguish it from the Fourier transform $X_c(F)$ introduced in (2.2.4).

which exists if $x_c(t)$ satisfies the *Dirichlet conditions*, which require that $x_c(t)$: (1) have a finite number of maxima or minima within any finite interval, (2) have a finite number of discontinuities within any finite interval, and (3) be absolutely integrable, that is,

$$\int_{-\infty}^{\infty} |x_c(t)|\, dt < \infty \tag{2.2.5}$$

The signal $x_c(t)$ can be synthesized from its *spectrum* $X_c(F)$ by using the following inverse Fourier transform formula

$$x_c(t) = \int_{-\infty}^{\infty} X_c(F)e^{j2\pi Ft}\, dF \tag{2.2.6}$$

The energy of $x_c(t)$ can be computed in either the time or frequency domain using Parseval's relation

$$E_x = \int_{-\infty}^{\infty} |x_c(t)|^2\, dt = \int_{-\infty}^{\infty} |X_c(F)|^2\, dF \tag{2.2.7}$$

The function $|X_c(F)|^2 \geq 0$ shows the distribution of energy of $x_c(t)$ as a function of frequency. Hence, it is called the *energy spectrum* of $x_c(t)$. We note that continuous-time, aperiodic signals have *continuous* spectra.

Fourier series for discrete-time periodic signals

Any discrete-time periodic signal $x(n)$ with fundamental period N can be expressed by the following Fourier series

$$x(n) = \sum_{k=0}^{N-1} X_k e^{j(2\pi/N)kn} \tag{2.2.8}$$

where

$$X_k = \frac{1}{N} \sum_{n=0}^{N-1} x(n)e^{-j(2\pi/N)kn} \tag{2.2.9}$$

are the corresponding Fourier coefficients. The basis sequences $s_k(n) \triangleq e^{j(2\pi/N)kn}$ are periodic with fundamental period N in both time and frequency, that is, $s_k(n+N) = s_k(n)$ and $s_{k+N}(n) = s_k(n)$.

The sequence $X_k, k = 0, \pm1, \pm2, \ldots$, is called the *spectrum* of the periodic signal $x(n)$. We note that $X_{k+N} = X_k$; that is, the spectrum of a discrete-time periodic signal is discrete and periodic with the same period.

The power of the periodic signal $x(n)$ can be determined by Parseval's relation

$$P_x = \frac{1}{N} \sum_{n=0}^{N-1} |x(n)|^2 = \sum_{k=0}^{N-1} |X_k|^2 \tag{2.2.10}$$

The sequence $|X_k|^2$ is known as the *power spectrum* of the periodic sequence $x(n)$.

Fourier transform for discrete-time aperiodic signals

Any discrete-time signal that is absolutely summable, that is,

$$\sum_{n=-\infty}^{\infty} |x(n)| < \infty \tag{2.2.11}$$

can be described by the discrete-time Fourier transform (DTFT)

$$X(e^{j\omega}) \triangleq \mathcal{F}[x(n)] = \sum_{n=-\infty}^{\infty} x(n)e^{-j\omega n} \tag{2.2.12}$$

where $\omega = 2\pi f$ is the frequency variable in *radians per sampling interval* or simply in *radians per sample* and f is the frequency variable in *cycles per sampling interval* or simply in *cycles per sample*. The signal $x(n)$ can be synthesized from its *spectrum* $X(e^{j\omega})$ by the inverse Fourier transform

$$x(n) = \frac{1}{2\pi} \int_{-\pi}^{\pi} X(e^{j\omega})e^{j\omega n}\, d\omega \qquad (2.2.13)$$

We will say that $x(n)$ and $X(e^{j\omega})$ form a *Fourier transform pair* denoted by

$$x(n) \overset{\mathcal{F}}{\longleftrightarrow} X(e^{j\omega}) \qquad (2.2.14)$$

The function $X(e^{j\omega})$ is periodic with fundamental period 2π. If $x(n)$ is real-valued, then $|X(e^{j\omega})| = |X(e^{-j\omega})|$ (even function) and $\angle X(e^{-j\omega}) = -\angle X(e^{j\omega})$ (odd function).

The energy of the signal can be computed in either the time or frequency domain using Parseval's relation

$$E_x = \sum_{n=-\infty}^{\infty} |x(n)|^2 = \frac{1}{2\pi} \int_{-\pi}^{\pi} |X(e^{j\omega})|^2\, d\omega \qquad (2.2.15)$$

$$= \int_{-\pi}^{\pi} \frac{|X(e^{j\omega})|^2}{2\pi}\, d\omega \qquad (2.2.16)$$

The function $|X(e^{j\omega})|^2/(2\pi) \geq 0$ and describes the distribution of the energy of the signal at various frequencies. Therefore, it is called the *energy spectrum* of $x(n)$.

Spectral classification of deterministic signals

So far we have discussed frequency analysis methods for periodic power signals and aperiodic energy signals. However, there are deterministic aperiodic signals with finite power. One such class of signals is the complex exponential $Ae^{j(\omega_0 n+\theta_0)}$ sequence [or equivalently, the sinusoidal sequence $A\cos(\omega_0 n + \theta_0)$], in which $\omega_0/(2\pi)$ is not a rational number. This sequence is not periodic, as discussed in Section 2.1.2; however it has a line spectrum at $\omega = \omega_0 + 2\pi k$, for any integer k, since

$$x(n) = Ae^{j(\omega_0 n+\theta_0)} = Ae^{j[(\omega_0+2\pi k)n+\theta_0]} \qquad k = 0, \pm 1, \pm 2, \ldots$$

(or at $\omega = \pm\omega_0 + 2\pi k$ for the sinusoidal sequence). Hence such sequences are termed as *almost periodic* and can be treated in the frequency domain in almost the same fashion.

Another interesting class of aperiodic power signals is those consisting of a linear combination of complex exponentials with nonharmonically related frequencies $\{\omega_l\}_{l=1}^{L}$, for example,

$$x(n) = \sum_{l=1}^{L} X_l e^{j\omega_l n} \qquad (2.2.17)$$

Clearly, these signals have discrete (or line) spectra, but the lines are not uniformly distributed on the frequency axis. Furthermore, the distances between the various lines are not harmonically related. We will say that these signals have *discrete nonharmonic spectra*. Note that periodic signals have discrete *harmonic* spectra.

There is yet another class of power signals, for example, the unit-step signal $u(n)$ defined in (2.1.2). The Fourier transform of such signals exists only in the context of the theory of generalized functions, which allows the use of impulse functions in the frequency domain (Papoulis 1977); for example, the Fourier transform of the unit step $u(n)$ is given by

$$\mathcal{F}[u(n)] = \frac{1}{1 - e^{-j\omega}} + \sum_{k=-\infty}^{\infty} \pi\delta(\omega - 2\pi k) \qquad (2.2.18)$$

Such signals have *mixed spectra*. The use of impulses also implies that the line spectrum can be represented by an impulse train in the frequency domain as a continuous spectrum. Figure 2.1 provides a classification of deterministic signals (with finite power or energy) in the frequency domain.

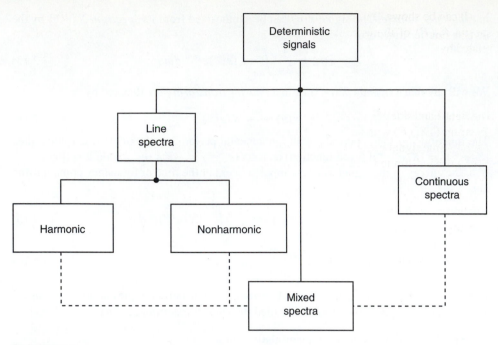

FIGURE 2.1
Spectral classification of deterministic (finite power or energy) signals.

2.2.2 Sampling of Continuous-Time Signals

In most practical applications, discrete-time signals are obtained by sampling continuous-time signals periodically in time. If $x_c(t)$ is a continuous-time signal, the discrete-time signal $x(n)$ obtained by periodic sampling is given by

$$x(n) = x_c(nT) \qquad -\infty < n < \infty \tag{2.2.19}$$

where T is the *sampling period*. The quantity $F_s = 1/T$, the number of samples taken per unit of time, is called the *sampling rate* or *sampling frequency*.

Since (2.2.19) established a relationship between the signals $x_c(t)$ and $x(n)$, there should be a corresponding relation between the spectra

$$X_c(F) = \int_{-\infty}^{\infty} x_c(t) e^{-j2\pi F t}\, dt \tag{2.2.20}$$

and

$$X(e^{j\omega}) = \sum_{n=-\infty}^{\infty} x(n) e^{-j\omega n} \tag{2.2.21}$$

of these signals.

To establish a relationship between $X_c(F)$ and $X(e^{j\omega})$, first we need to find a relation between the frequency variables F and ω. To this end, we note that periodic sampling imposes a relationship between t and n, namely, $t = nT = n/F_s$. Substituting $t = n/F_s$ into (2.2.20) and comparing with the exponentials in (2.2.20) and (2.2.21), we see that

$$2\pi \frac{F}{F_s} = \omega = 2\pi f \quad \text{or} \quad f = \frac{F}{F_s} \tag{2.2.22}$$

Since f appears to be a ratio frequency, it is also called a *relative* frequency. The term *normalized frequency* is also sometimes used for the discrete-time frequency variable f.

It can be shown (Proakis and Manolakis 1996; Oppenheim and Schafer 1989) that the spectra $X_c(F)$ of the continuous-time signal and $X(e^{j\omega})$ of the discrete-time signal are related by

$$X(e^{j2\pi F/F_s}) = F_s \sum_{k=-\infty}^{\infty} X_c(F - kF_s) \qquad (2.2.23)$$

The right-hand side of (2.2.23) consists of a periodic repetition of the scaled continuous-time spectrum $F_s X_c(F)$ with period F_s. This periodicity is necessary because the spectrum of any discrete-time signal has to be periodic. To see the implications of (2.2.23), let us assume that $X_c(F)$ is band-limited, that is, $X_c(F) = 0$ for $|F| > B$, as shown in Figure 2.2. According to (2.2.23), the spectrum $X(F)$ is the superposition of an infinite number of replications of $X_c(F)$ at integer multiples of the sampling frequency F_s. Figure 2.2(b) illustrates the situation when $F_s \geq 2B$, whereas Figure 2.2(c) shows what happens if $F_s < 2B$. In the latter case, high-frequency components take on the identity of lower frequencies, a phenomenon known as *aliasing*. Obviously, aliasing can be avoided only if the sampled continuous-time signal is band-limited and the sampling frequency F_s is equal to at least twice the

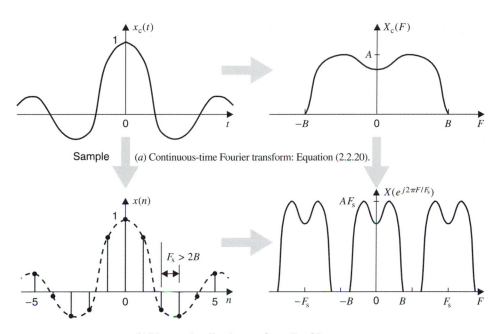

(a) Continuous-time Fourier transform: Equation (2.2.20).

(b) Discrete-time Fourier transform: $F_s > 2B$.

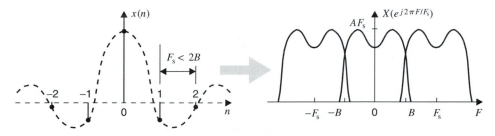

(c) Discrete-time Fourier transform: $F_s < 2B$.

FIGURE 2.2
Sampling operation.

bandwidth ($F_s \geq 2B$). This leads to the well-known sampling theorem, which can be stated as follows:

SAMPLING THEOREM. A band-imited, real-valued, continuous-time signal with bandwidth B can be uniquely recovered from its samples, provided that the sampling rate F_s is at least equal to twice the bandwidth, that is, provided that $F_s \geq 2B$.

If the conditions of the sampling theorem are fulfilled, that is, if $X_c(F) = 0$ for $|F| > B$ and $F_s \geq 2B$, then the signal $x_c(t)$ can be recovered from its samples $x(n) = x_c(nT)$ by using the following interpolation formula

$$x_c(t) = \sum_{n=-\infty}^{\infty} x_c(nT) \frac{\sin\left[(\pi/T)(t-nT)\right]}{(\pi/T)(t-nT)} \tag{2.2.24}$$

The minimum sampling rate of $F_s = 2B$ is called the *Nyquist rate*. In practice, the infinite summation in (2.2.24) has to be substituted by a finite one. Hence, only approximate reconstruction is possible.

2.2.3 The Discrete Fourier Transform

The N-point *discrete Fourier transform* (*DFT*) of an N-point sequence $\{x(n), n = 0, 1, \ldots, N-1\}$ is defined by[†]

$$\tilde{X}(k) = \sum_{n=0}^{N-1} x(n)e^{-j(2\pi/N)kn} \qquad k = 0, 1, \ldots, N-1 \tag{2.2.25}$$

The N-point sequence $\{x(n), n = 0, 1, \ldots, N-1\}$ can be recovered from its DFT coefficients $\{\tilde{X}(k), k = 0, 1, \ldots, N-1\}$ by the following *inverse* DFT formula:

$$x(n) = \frac{1}{N} \sum_{k=0}^{N-1} \tilde{X}(k)e^{j(2\pi/N)kn} \qquad n = 0, 1, \ldots, N-1 \tag{2.2.26}$$

We note that by its definition, the N-point DFT requires or provides information only for N samples of a discrete-time signal. Hence, it does not provide a frequency decomposition of the signal because any discrete-time signal must be specified for all discrete-time instances, $-\infty < n < \infty$. The use of DFT for frequency analysis depends on the signal values outside the interval $0 \leq n \leq N-1$. Depending on these values, we can obtain various interpretations of the DFT. The value of the DFT lies exactly in these interpretations.

DFT of finite-duration signals. Let $x(n)$ be a finite-duration signal with nonzero values over the range $0 \leq n \leq N-1$ and zero values elsewhere. If we evaluate $X(e^{j\omega})$ at N equidistant frequencies, say, $\omega_k = (2\pi/N)k, 0 \leq k \leq N-1$, we obtain

$$X(e^{j\omega_k}) = X(e^{j2\pi k/N}) = \sum_{n=0}^{N-1} x(n)e^{-j(2\pi/N)kn} = \tilde{X}(k) \tag{2.2.27}$$

which follows by comparing the last equation with (2.2.25). This implies that *the N-point DFT of a finite-duration signal with length N is equal to the Fourier transform of the signal at frequencies $\omega_k = (2\pi/N)k, 0 \leq k \leq N-1$.* Hence, in this case, the N-point DFT corresponds to the uniform sampling of the Fourier transform of a discrete-time signal at N equidistant points, that is, sampling in the frequency domain.

[†]In many traditional textbooks, the DFT is denoted by $X(k)$. We will use the notation $\tilde{X}(k)$ to distinguish the DFT from the DTFT $X(e^{j\omega})$ function or its samples.

DFT of periodic signals. Suppose now that $x(n)$ is a periodic sequence with fundamental period N. This sequence can be decomposed into frequency components by using the Fourier series in (2.2.8) and (2.2.9). Comparison of (2.2.26) with (2.2.8) shows that

$$\tilde{X}(k) = N X_k \quad k = 0, 1, \ldots, N - 1 \tag{2.2.28}$$

that is, *the DFT of one period of a periodic signal is given by the Fourier series coefficients of the signal scaled by the fundamental period.* Obviously, computing the DFT of a fraction of a period will lead to DFT coefficients that are not related to the Fourier series coefficients of the periodic signal.

The DFT can be efficiently computed by using a family of fast algorithms, referred to as *fast Fourier transform (FFT)* algorithms, with complexity proportional to $N \log_2 N$. Due to the efficiency offered by these algorithms, the DFT is widely used for the computation of spectra, correlations, and convolutions and for the implementation of digital filters.

2.2.4 The z-Transform

The z-transform of a sequence is a very powerful tool for the analysis of linear and time-invariant systems. It is defined by the following pair of equations:

$$X(z) \triangleq \mathcal{Z}[x(n)] = \sum_{n=-\infty}^{\infty} x(n) z^{-n} \tag{2.2.29}$$

$$x(n) = \frac{1}{2\pi j} \oint_C X(z) z^{n-1} \, dz \tag{2.2.30}$$

Equation (2.2.29) is known as the *direct transform*, whereas equation (2.2.30) is referred to as the *inverse transform*. The set of values of z for which the power series in (2.2.29) converges is called the *region of convergence (ROC)* of $X(z)$. A sufficient condition for convergence is

$$\sum_{n=-\infty}^{\infty} |x(n)||z^{-n}| < \infty \tag{2.2.31}$$

In general, the ROC is a ring in the complex plane; that is, $R_1 < |z| < R_2$. The values of R_1 and R_2 depend on the nature of the signal $x(n)$. For finite-duration signals, $X(z)$ is a polynomial in z^{-1}, and the ROC is the entire z-plane with a possible exclusion of the points $z = 0$ and/or $z = \pm\infty$. For causal signals with infinite duration, the ROC is, in general, $R_1 < |z| < \infty$, that is, the exterior of a circle. For anticausal signals $[x(n) = 0, n > 0]$, the ROC is the interior of a circle, that is, $0 < |z| < R_2$. For two-sided infinite-duration signals, the ROC is, in general, a ring $R_1 < |z| < R_2$. The contour of integration in the inverse transform in (2.2.30) can be any counterclockwise closed path that encloses the origin and is inside the ROC.

If we compute the z-transform on the unit circle of the z-plane, that is, if we set $z = e^{j\omega}$ in (2.2.29) and (2.2.30), we obtain

$$X(z)|_{z=e^{j\omega}} = X(e^{j\omega}) = \sum_{n=-\infty}^{\infty} x(n) e^{-j\omega n} \tag{2.2.32}$$

$$x(n) = \frac{1}{2\pi} \int_{-\pi}^{\pi} X(e^{j\omega}) e^{j\omega n} \, d\omega \tag{2.2.33}$$

which are the Fourier transform and inverse Fourier transform relating the signals $x(n)$ and $X(e^{j\omega})$. This relation holds only if the unit circle is inside the ROC.

The z-transform has many properties that are useful for the study of discrete-time signals and systems. Some of these properties are given in Table 2.1. Assuming that the involved Fourier transform exists, setting $z = e^{j\omega}$ in each of the properties of Table 2.1 gives a corresponding table of properties for the Fourier transform.

An important family of z-transforms is those for which $X(z)$ is a rational function, that is, a ratio of two polynomials in z or z^{-1}. The roots of the numerator polynomial, that is, the values of z for which $X(z) = 0$, are referred to as the *zeros* of $X(z)$. The roots of the denominator polynomial, that is, the values of z for which $X(z) = \pm\infty$, are referred to as the *poles* of $X(z)$. Although zeros and poles may occur at $z = 0$ or $z = \pm\infty$, we usually do not count them. As will be seen throughout this book, the locations of poles and zeros play an important role in the analysis of signals and systems. To display poles and zeros in the z-plane, we use the symbols $\times$ and $\circ$, respectively.

The inverse z-transform—that is, determining the signal $x(n)$ given its z-transform $X(z)$—involves the computation of the contour integral in (2.2.30). However, most practical applications involve rational z-transforms that can be easily inverted using partial fraction expansion techniques. Finally, we note that a working familiarity with the z-transform technique is necessary for the complete understanding of the material in subsequent chapters.

2.2.5 Representations of Narrowband Signals

A signal is known as a *narrowband* signal if it is band-limited to a band whose width is small compared to the band center frequency. Such a narrowband signal transform $X_c(F)$ is shown in Figure 2.3(a), and the corresponding signal waveform $x_c(t)$ that it may represent is shown in Figure 2.3(b). The center frequency of $x_c(t)$ is F_0, and its bandwidth is B, which is much less than F_0. It is informative to note that the signal $x_c(t)$ appears to be a sinusoidal waveform whose amplitude and phase are both *varying slowly* with respect to the variations of the cosine wave. Therefore, such a signal can be represented by

$$x_c(t) = a(t) \cos [2\pi F_0 t + \theta(t)] \tag{2.2.34}$$

where $a(t)$ describes the amplitude variation (or envelope modulation) and $\theta(t)$ describes the phase modulation of a carrier wave of frequency F_0 Hz. Although (2.2.34) can be used to describe any arbitrary signal, the concepts of envelope and phase modulation are

TABLE 2.1
Properties of z-Transform.

Property	Time domain	z-Domain	ROC
Notation	$x(n)$	$X(z)$	ROC : $R_l < \|z\| < R_u$
	$x_1(n)$	$X_1(z)$	ROC$_1$: $R_{1l} < \|z\| < R_{1u}$
	$x_2(n)$	$X_2(z)$	ROC$_2$: $R_{2l} < \|z\| < R_{2u}$
Linearity	$a_1 x_1(n) + a_2 x_2(n)$	$a_1 X_1(z) + a_2 X_2(z)$	ROC$_1 \cap$ ROC$_2$
Time shifting	$x(n-k)$	$z^{-k} X(z)$	$R_l < \|z\| < R_u$, except $z = 0$ if $k > 0$
Scaling in the z-domain	$a^n x(n)$	$X(a^{-1}z)$	$\|a\|R_l < \|z\| < \|a\|R_u$
Time reversal	$x(-n)$	$X(z^{-1})$	$\dfrac{1}{R_l} < \|z\| < \dfrac{1}{R_u}$
Conjugation	$x^*(n)$	$X^*(z^*)$	ROC
Differentiation	$nx(n)$	$-z\dfrac{dX(z)}{dz}$	ROC
Convolution	$x_1(n) * x_2(n)$	$X_1(z)X_2(z)$	ROC$_1 \cap$ ROC$_2$
Multiplication	$x_1(n)x_2(n)$	$\dfrac{1}{2\pi j}\oint_C X_1(v)X_2\left(\dfrac{z}{v}\right)v^{-1}dv$	$R_{1l}R_{2l} < \|z\| < R_{1u}R_{2u}$
Parseval's relation	$\displaystyle\sum_{n=-\infty}^{\infty} x_1(n)x_2^*(n) = \dfrac{1}{2\pi j}\oint_C X_1(v)X_2^*\left(\dfrac{1}{v^*}\right)v^{-1}dv$		

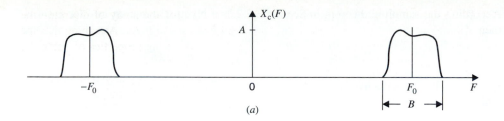

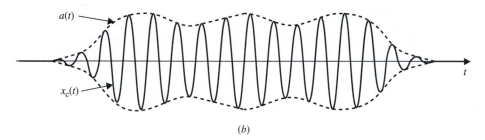

FIGURE 2.3
Narrowband signal: (*a*) Fourier transform and (*b*) waveform.

meaningless unless $a(t)$ and $\theta(t)$ vary slowly in comparison to $\cos 2\pi F_0 t$, or equivalently, unless $B \ll F_0$.

In literature, two approaches are commonly used to describe a narrowband signal. In the first approach, the signal is represented by using a *complex envelope*, while in the second approach the *quadrature component* representation is used. By using Euler's identity, it is easy to verify that (2.2.34) can be put in the form

$$x_c(t) = \text{Re}[a(t)e^{j[2\pi F_0 t + \theta(t)]}] = \text{Re}[a(t)e^{j\theta(t)}e^{j2\pi F_0 t}] \tag{2.2.35}$$

Let
$$\tilde{x}_c(t) \triangleq a(t)e^{j\theta(t)} \tag{2.2.36}$$

Then from (2.2.35) we obtain

$$x_c(t) = \text{Re}[\tilde{x}_c(t)e^{j2\pi F_0 t}] \tag{2.2.37}$$

The complex-valued signal $\tilde{x}_c(t)$ contains both the amplitude and phase variations of $x_c(t)$, and hence it is referred to as the *complex envelope* of the narrowband signal $x_c(t)$. Similarly, again starting with (2.2.34) and this time using the trigonometric identity, we can write

$$x_c(t) = a(t)\cos 2\pi F_0 t \cos \theta(t) - a(t)\sin 2\pi F_0 t \sin \theta(t) \tag{2.2.38}$$

Let
$$x_{cI}(t) \triangleq a(t)\cos \theta(t) \tag{2.2.39}$$

$$x_{cQ}(t) \triangleq a(t)\sin \theta(t) \tag{2.2.40}$$

which are termed the *in-phase* and the *quadrature* components of narrowband signal $x_c(t)$, respectively. Then (2.2.38) can be written as

$$x_c(t) = x_{cI}(t)\cos 2\pi F_0 t - x_{cQ}(t)\sin 2\pi F_0 t \tag{2.2.41}$$

Clearly, the above two representations are related. If we expand (2.2.36), then we obtain

$$\tilde{x}_c(t) = x_{cI}(t) + jx_{cQ}(t) \tag{2.2.42}$$

which implies that the in-phase and quadrature components are, respectively, the real and imaginary parts of the complex envelope $\tilde{x}_c(t)$. These representations will be used extensively in Chapter 11.

Bandpass sampling theorem. One application of the complex-envelope representation lies in the optimum sampling of narrowband signals. In a general sense, the narrowband signal $x_c(t)$ is also a bandpass signal that is approximately band-limited to $(F_0 + B/2)$ Hz.

According the sampling theorem in Section 2.2.2, the Nyquist sampling rate for $x_c(t)$ is then

$$F_s = 2\left(F_0 + \frac{B}{2}\right) \approx 2F_0 \quad \text{for } B \ll F_0$$

However, since the effective bandwidth of $x_c(t)$ is $B/2$ Hz, the optimum rate should be B, which is much smaller than $2F_0$. To obtain this optimum rate, consider (2.2.34), which we can write as

$$
\begin{aligned}
x_c(t) &= a(t)\cos\left[2\pi F_0 t + \theta(t)\right] = a(t)\frac{e^{j[2\pi F_0 t + \theta(t)]} + e^{-j[2\pi F_0 t + \theta(t)]}}{2} \\
&= \frac{a(t)e^{j\theta(t)}}{2}e^{j2\pi F_0 t} + \frac{a(t)e^{-j\theta(t)}}{2}e^{-j2\pi F_0 t} \\
&= \frac{1}{2}\tilde{x}_c(t)e^{j2\pi F_0 t} + \frac{1}{2}\tilde{x}_c^*(t)e^{-j2\pi F_0 t}
\end{aligned}
\tag{2.2.43}
$$

Using the transform properties from Table 2.1, we see that the Fourier transform of $x_c(t)$ is given by

$$X_c(F) = \tfrac{1}{2}[\tilde{X}_c(F - F_0) + \tilde{X}_c^*(-F - F_0)] \tag{2.2.44}$$

The first term in (2.2.44) is the Fourier transform of $\tilde{x}_c(t)$ shifted by F_0, and hence it must be the positive band-limited portion of $X_c(F)$. Similarly, the second term in (2.2.44) is the Fourier transform of $\tilde{x}_c^*(t)$ shifted by $-F_0$ (or shifted left by F_0). Now the Fourier transform of $\tilde{x}_c^*(t)$ is $X_c^*(-F)$, and hence the second term must be the negative band-limited portion of $X_c(F)$.

We thus conclude that $\tilde{x}_c(t)$ is a *baseband* complex-valued signal limited to the band of width B, as shown in Figure 2.4. Furthermore, note that the sampling theorem of Section 2.2.2 is applicable to real- as well as complex-valued signals. Therefore, we can sample the complex envelope $\tilde{x}_c(t)$ at the Nyquist rate of B sampling intervals per second; and, by extension, we can sample the narrowband signal $x_c(t)$ at the same rate without aliasing. From (2.2.24), the sampling representation of $\tilde{x}_c(t)$ is given by

$$\tilde{x}_c(t) = \sum_{n=-\infty}^{\infty} \tilde{x}_c\left(\frac{n}{B}\right)\frac{\sin\left[\pi B(t - n/B)\right]}{\pi B(t - n/B)} \tag{2.2.45}$$

Substituting (2.2.45) and (2.2.36) in (2.2.37), we obtain

$$
\begin{aligned}
x_c(t) &= \mathrm{Re}\left\{\sum_{n=-\infty}^{\infty} \tilde{x}_c\left(\frac{n}{B}\right)\frac{\sin\left[\pi B(t - n/B)\right]}{\pi B(t - n/B)}e^{j2\pi F_0 t}\right\} \\
&= \mathrm{Re}\left\{\sum_{n=-\infty}^{\infty} a\left(\frac{n}{B}\right)e^{j\theta(n/B)}e^{j2\pi F_0 t}\frac{\sin\left[\pi B(t - n/B)\right]}{\pi B(t - n/B)}\right\} \\
&= \sum_{n=-\infty}^{\infty} a\left(\frac{n}{B}\right)\cos\left[2\pi F_0 t + \theta\left(\frac{n}{B}\right)\right]\frac{\sin\left[\pi B(t - n/B)\right]}{\pi B(t - n/B)}
\end{aligned}
\tag{2.2.46}
$$

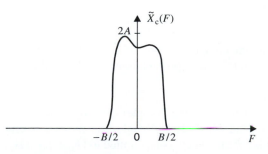

FIGURE 2.4
Fourier transform of a complex envelope $\tilde{x}_c(t)$.

which is the amplitude-phase form of the bandpass sampling theorem. Using trigonometric identity, the quadrature-component form of the theorem is given by

$$x_c(t) = \sum_{n=-\infty}^{\infty} \left[x_{cI}\left(\frac{n}{B}\right) \cos 2\pi F_0 t - x_{cQ}\left(\frac{n}{B}\right) \sin 2\pi F_0 t \right] \frac{\sin\left[\pi B(t - n/B)\right]}{\pi B(t - n/B)} \quad (2.2.47)$$

Applications of this theorem are considered in Chapter 11.

2.3 DISCRETE-TIME SYSTEMS

In this section, we review the basics of linear, time-invariant systems by emphasizing those aspects of particular importance to this book. For our purposes, a *system* is defined to be any physical device or algorithm that transforms a signal, called the *input* or *excitation*, into another signal, called the *output* or *response*. When the system is simply an algorithm, it may be realized in either hardware or software. Although a system can be specified from its parts and their functions, it will often turn out to be more convenient to characterize a system in terms of its response to specific signals. The mathematical relationships between the input and output signals of a system will be referred to as a (system) *model*. In the case of a *discrete-time system*, the model is simply a transformation that uniquely maps the input signal $x(n)$ to an output signal $y(n)$. This is denoted by

$$y(n) = H[x(n)] \qquad -\infty < n < \infty \quad (2.3.1)$$

and is graphically depicted as in Figure 2.5.

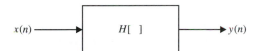

FIGURE 2.5
Block diagram representation of a discrete-time system.

2.3.1 Analysis of Linear, Time-Invariant Systems

The systems we shall deal with in this book are linear and time-invariant and are always assumed to be initially at rest. No initial conditions or other information will affect the output signal.

Time-domain analysis. The output of a linear, time-invariant system can always be expressed as the *convolution* summation between the input sequence $x(n)$ and the *impulse response* or *unit sample response* sequence $h(n) \triangleq H[\delta(n)]$ of the system, that is,

$$y(n) = x(n) * h(n) \triangleq \sum_{k=-\infty}^{\infty} x(k)h(n - k) \quad (2.3.2)$$

where $*$ denotes the convolution operation. It can easily be shown that an equivalent expression is

$$y(n) = \sum_{k=-\infty}^{\infty} h(k)x(n - k) = h(n) * x(n) \quad (2.3.3)$$

Thus, given the input $x(n)$ to a linear, time-invariant system, the output $y(n)$ can be computed by using the impulse response $h(n)$ of the system and either formula (2.3.2) or (2.3.3).

If $x(n)$ and $h(n)$ are arbitrary sequences of finite duration, then the above convolution can also be computed by using a *matrix-vector multiplication* operation. Let $x(n), 0 \leq n \leq N - 1$, and $h(n), 0 \leq n \leq M - 1$, be two finite-duration sequences of lengths N

and $M(< N)$ respectively.[†] Then from (2.3.3), the sequence $y(n)$ is also a finite-duration sequence over $0 \leq n \leq L - 1$ with $L \triangleq N + M - 1$ samples. If the samples of $y(n)$ and $h(n)$ are arranged in the column vectors $\mathbf{y}$ and $\mathbf{h}$, respectively, then from (2.3.3) we obtain

$$
\begin{bmatrix} y(0) \\ \vdots \\ \vdots \\ y(M-1) \\ \vdots \\ y(N-1) \\ \vdots \\ \vdots \\ y(L-1) \end{bmatrix} = \begin{bmatrix} x(0) & 0 & \cdots & 0 \\ \vdots & \ddots & \ddots & \vdots \\ \vdots & & \ddots & 0 \\ x(M-1) & \cdots & \cdots & x(0) \\ \vdots & \ddots & \ddots & \vdots \\ x(N-1) & \cdots & \cdots & x(N-M) \\ 0 & & \ddots & \vdots \\ \vdots & & \ddots & \vdots \\ 0 & \cdots & 0 & x(N-1) \end{bmatrix} \begin{bmatrix} h(0) \\ h(1) \\ \vdots \\ h(M-1) \end{bmatrix}
\tag{2.3.4}
$$

or
$$
\mathbf{y} = \mathbf{X}\mathbf{h} \tag{2.3.5}
$$

where the $L \times M$ matrix $\mathbf{X}$ contains linear shifts in $x(n-k)$ for $n = 0, \ldots, N-1$, which are arranged as rows. The matrix $\mathbf{X}$ is termed an *input data* matrix. It has an interesting property that all the elements along any diagonal are equal. Such a matrix is called a *Toeplitz* matrix, and thus $\mathbf{X}$ has a *Toeplitz* structure. Note that the first and the last $M - 1$ rows of $\mathbf{X}$ contain zero (or boundary) values. Therefore, the first and the last $M - 1$ samples of $y(n)$ contain transient boundary effects. In passing, we note that the vector $\mathbf{y}$ can also be obtained as

$$
\mathbf{y} = \mathbf{H}\mathbf{x} \tag{2.3.6}
$$

in which $\mathbf{H}$ is a Toeplitz matrix obtained from (2.3.2). However, we will emphasize the approach given in (2.3.5) in subsequent chapters.

MATLAB provides a built-in function called `conv` that computes the convolution of two finite-duration sequences and is invoked by `y = conv(h,x)`. Alternatively, the convolution can also be implemented using (2.3.4) in which the Toeplitz data matrix $\mathbf{X}$ is obtained using the function `toeplitz` (see Problem 2.4).

A system is called *causal* if the present value of the output signal depends only on the present and/or past values of the input signal. Although causality is necessary for the real-time implementation of discrete-time systems, it is not really a problem in off-line applications where the input signal has already been recorded. A necessary and sufficient condition for a linear, time-invariant system to be causal is that the impulse response $h(n) = 0$ for $n < 0$.

Stability is another important system property. There are various types of stability criteria. A system is called *bounded-input bounded-output (BIBO) stable* or simply *stable* if and only if every bounded input, namely, $|x(n)| \leq M_x < \infty$ for all n, produces a bounded output, that is, $|y(n)| \leq M_y < \infty$ for all n. Clearly, unstable systems generate unbounded output signals and, hence, are not useful in practical applications because they will result in an overflow in the output. It can be shown that an LTI system is BIBO stable if and only if

$$
\sum_{n=-\infty}^{\infty} |h(n)| < \infty \tag{2.3.7}
$$

Transform-domain analysis. In addition to the time-domain convolution approach, the output of a linear, time-invariant system can be determined by using transform techniques. Indeed, by using the convolution property of the z-transform (see Table 2.1), (2.3.2) yields

$$
Y(z) = H(z)X(z) \tag{2.3.8}
$$

[†]For the purpose of this illustration, we assume that the sequences begin at $n = 0$, but they may have any arbitrary finite duration.

where $X(z)$, $Y(z)$, and $H(z)$ are the z-transforms of the input, output, and impulse response sequences, respectively. The z-transform $H(z) = \mathcal{Z}[h(n)]$ of the impulse response is called the *system function* and plays a very important role in the analysis and characterization of linear, time-invariant systems. If the unit circle is inside the ROC of $H(z)$, the system is stable and $H(e^{j\omega})$ provides its frequency response.

Evaluating (2.3.8) on the unit circle gives

$$Y(e^{j\omega}) = H(e^{j\omega})X(e^{j\omega}) \tag{2.3.9}$$

where $H(e^{j\omega})$ is the *frequency response* function of the system. Since, in general, $H(e^{j\omega})$ is complex-valued, we have

$$H(e^{j\omega}) = |H(e^{j\omega})|e^{j\angle H(e^{j\omega})} \tag{2.3.10}$$

and $|H(e^{j\omega})|$ is the *magnitude response*, and $\angle H(e^{j\omega})$ is the *phase response* of the system. For a system with a real impulse response, $|H(e^{j\omega})|$ has even symmetry and $\angle H(e^{j\omega})$ has odd symmetry. The *group delay* response of a system with frequency response $H(e^{j\omega})$ is defined as

$$\tau(e^{j\omega}) = -\frac{d}{d\omega}\angle H(e^{j\omega}) \tag{2.3.11}$$

and provides a measure of the average delay of the system as a function of frequency.

Systems described by linear, constant-coefficient difference equations. A discrete-time system is called *practically realizable* if it satisfies the following conditions: (1) It requires a finite amount of memory, and (2) the amount of arithmetic operations required for the computation of each output sample is finite. Clearly, any system that does not satisfy either of these conditions cannot be implemented in practice.

If, in addition to being linear and time-invariant, we require a system to be causal and practically realizable, then the most general input/output description of such a system takes the form of a constant-coefficient, linear difference equation

$$y(n) = -\sum_{k=1}^{P} a_k y(n-k) + \sum_{k=0}^{Q} d_k x(n-k) \tag{2.3.12}$$

In case the system parameters $\{a_k, d_k\}$ depend on time, the system is linear and *time-varying*. If, however, the system parameters depend on either the input or output signals, then the system becomes *nonlinear*.

By limiting our attention to constant parameters and evaluating the z-transform of both sides of (2.3.12), we obtain

$$H(z) = \frac{Y(z)}{X(z)} = \frac{\sum_{k=0}^{Q} d_k z^{-k}}{1 + \sum_{k=1}^{P} a_k z^{-k}} \triangleq \frac{D(z)}{A(z)} \tag{2.3.13}$$

Clearly, a system with a rational system function can be described, within a gain factor, by the locations of its poles and zeros in the complex z-plane

$$H(z) = \frac{D(z)}{A(z)} = G \frac{\prod_{k=1}^{Q}(1 - z_k z^{-1})}{\prod_{k=1}^{P}(1 - p_k z^{-1})} \tag{2.3.14}$$

The system described by (2.3.12) or equivalently by (2.3.13) or (2.3.14) is stable if its poles, that is, the roots of the denominator polynomial $A(z)$, are all inside the unit circle.

The difference equation in (2.3.12) is implemented in MATLAB using the `filter` function. In its simplest form, this function is invoked by `y = filter(d,a,x)` where `d = [d0,d1,...,dQ]` and `a = [1,a1,...,aP]` are the numerator and denominator coefficient arrays in (2.3.13), respectively.

If the coefficients a_k in (2.3.12) are zero, we have

$$y(n) = \sum_{k=0}^{Q} d_k x(n-k) \tag{2.3.15}$$

which compared to (2.3.3) yields

$$h(n) = \begin{cases} d_n & 0 \le n \le Q \\ 0 & \text{elsewhere} \end{cases} \tag{2.3.16}$$

that is, the system in (2.3.15) has an impulse response with finite duration and is called a *finite impulse response* (*FIR*) system. From (2.3.13), it follows that the system function of an FIR system is a polynomial in z^{-1}, and thus $H(z)$ has Q trivial poles at $z = 0$ and Q zeros. For this reason, FIR systems are also referred to as *all-zero* (*AZ*) systems. Figure 2.6 shows a straightforward block diagram realization of the FIR system (2.3.15) in terms of unit delays, adders, and multipliers.

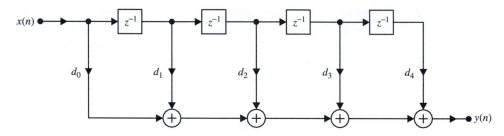

FIGURE 2.6
FIR filter realization (direct form).

In MATLAB, FIR filters are represented either by the values of the impulse response $h(n)$ or by the difference equation coefficients d_n. Therefore, for computational purposes, we can use either the `y = conv(h,x)` function or the `y = filter(d,[1],x)` function. There is a difference in the outputs of these two implementations that should be noted. The `conv` function produces all values of $y(n)$ in (2.3.4), while the output sequence from the filter function provides $y(0), \ldots, y(N-1)$. This can be seen by referring to matrix $\mathbf{X}$ in (2.3.4). The input data matrix $\mathbf{X}$ contains only the first N rows; that is, the output of the `filter` function contains transient effects from the boundary at $n = 0$. For signal processing applications, the use of the `filter` function is strongly encouraged.

When a system has both poles and zeros, $H(z)$ can be expressed using partial fraction expansion form as follows

$$H(z) = \sum_{k=1}^{P} \frac{A_k}{1 - p_k z^{-k}} \tag{2.3.17}$$

if the poles are distinct and $Q < P$. The corresponding impulse response is then given by

$$h(n) = \sum_{k=1}^{P} A_k (p_k)^n u(n) \tag{2.3.18}$$

that is, each pole contributes an exponential mode of infinite duration to the impulse response. We conclude that the presence of any nontrivial pole in a system implies an infinite-duration impulse response. We refer to such systems as *infinite impulse response* (*IIR*) systems. If $d_k = 0$ for $k = 1, 2, \ldots, Q$, the system has only poles, with zeros at $z = 0$, and is called an *all-pole* (*AP*) system. It should be stressed that although all-pole and pole-zero systems are IIR, not all IIR systems are *pole-zero* (*PZ*) systems. Indeed, there are many useful systems, for example, an ideal low-pass filter, that cannot be described by rational system functions of finite order. Figures 2.7 and 2.8 show direct-form realizations of an all-pole and a pole-zero system.

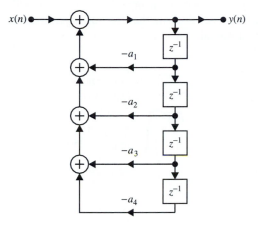

FIGURE 2.7
All-pole system realization (direct form).

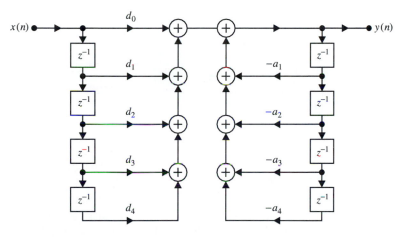

FIGURE 2.8
Pole-zero system realization (direct form).

2.3.2 Response to Periodic Inputs

Although the convolution summation formula can be used to compute the response of a stable system to any input signal, (2.3.8) cannot be used with periodic inputs because periodic signals do not possess a z-transform. However, a frequency domain formula similar to (2.3.9) can be developed for periodic inputs.

Let $x(n)$ be a periodic signal with fundamental period N. This signal can be expanded in a Fourier series as

$$x(n) = \sum_{k=0}^{N-1} X_k e^{j2\pi kn/N} \qquad n = 0, 1, \ldots, N-1 \qquad (2.3.19)$$

where X_k are the Fourier series coefficients. Substituting (2.3.19) into (2.3.3) gives

$$y(n) = \sum_{k=0}^{N-1} X_k H(e^{j2\pi k/N}) e^{j2\pi kn/N} \qquad (2.3.20)$$

where $H(e^{j2\pi k/N})$ are samples of $H(e^{j\omega})$. But (2.3.20) is just the Fourier series expansion of $y(n)$, hence

$$Y_k = H(e^{j2\pi k/N}) X_k \qquad k = 0, 1, \ldots, N-1 \qquad (2.3.21)$$

Thus, the response of a linear, time-invariant system to a periodic input is also periodic with the same period. Figure 2.9 illustrates, in the frequency domain, the effect of an LTI system on the spectrum of aperiodic and periodic input signals.

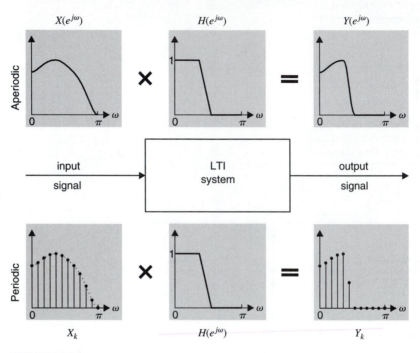

FIGURE 2.9
LTI system operation in the frequency domain.

EXAMPLE 2.3.1. Consider the system

$$y(n) = ay(n-1) + x(n) \qquad 0 < a < 1$$

If we restrict the inputs of the system to be only periodic signals with fundamental period N, determine the impulse response of an equivalent FIR system that will provide an identical output to the system described above.

Solution. The system output can be described by (2.3.21), where

$$H(z) = \frac{Y(z)}{X(z)} = \frac{1}{1 - az^{-1}} = \mathcal{Z}\{a^n u(n)\}$$

From Figure 2.9, it is clearly seen that every system whose frequency response is identical to $H(e^{j\omega})$ at the sampling points $\omega_k = (2\pi/N)k, 0 \le k \le N-1$, provides the same output when excited by a periodic signal having fundamental period N. An FIR system having this property can be obtained by taking the inverse N-point DFT of $\tilde{H}(k), 0 \le k \le N-1$. The resulting impulse response $\tilde{h}(n)$ is simply the N-point periodic extension of $h(n) = a^n u(n)$, that is,

$$\tilde{h}(n) = \sum_{l=-\infty}^{\infty} h(n + lN) = \sum_{l=0}^{\infty} a^{n+lN} = \frac{a^n}{1 - a^N} \qquad 0 \le n \le N-1 \qquad (2.3.22)$$

since $h(n + lN)$ for $l < 0$ does not contribute to the sum for $0 \le n \le N-1$.

The example above looked simple enough. Unfortunately, for somewhat more complicated all-pole filters, it becomes very difficult to evaluate the infinite summation in (2.3.22) in closed form, even if $h(n)$ is available, which is often not the case.

2.3.3 Correlation Analysis and Spectral Density

The investigation of system responses to specific input signals requires either the explicit computation of the output signal or measurements to relate characteristic properties of the output signal to corresponding characterisitics of the system and the input signal. A fundamental tool needed for such analysis is the *correlation* between two signals that provides a quantitative measure of similarity between two signals. The *correlation sequence* between two discrete-time signals $x(n)$ and $y(n)$ is defined by

$$r_{xy}(l) = \begin{cases} \displaystyle\sum_{n=-\infty}^{\infty} x(n)y^*(n-l) & : \text{energy signals} \\ \displaystyle\lim_{N\to\infty} \frac{1}{2N+1} \sum_{n=-N}^{N} x(n)y^*(n-l) & : \text{power signals} \end{cases} \tag{2.3.23}$$

where l is termed the *lag* (or shift) variable. The *autocorrelation* sequence of a signal is obtained by assuming that $y(n) = x(n)$, that is, if we correlate a signal with itself. Thus

$$r_{xx}(l) = \begin{cases} \displaystyle\sum_{n=-\infty}^{\infty} x(n)x^*(n-l) & : \text{energy signal} \\ \displaystyle\lim_{N\to\infty} \frac{1}{2N+1} \sum_{n=-N}^{N} x(n)x^*(n-l) & : \text{power signal} \end{cases} \tag{2.3.24}$$

In this case, we use the simplified notation $r_x(l)$ or even $r(l)$ if there is no possibility of confusion.

The autocorrelation sequence $r_x(l)$ and the energy spectrum of a signal $x(n)$ form a Fourier transform pair

$$r_x(l) \xleftrightarrow{\mathcal{F}} R_x(e^{j\omega}) \tag{2.3.25}$$

Since, $R_x(e^{j\omega}) = |X(e^{j\omega})|^2$, the Wiener-Khintchine theorem (2.3.25) is usually used to define the energy spectral density function, $R_x(e^{j\omega})$. Clearly, $r_x(l)$ and $R_x(e^{j\omega})$ do not contain any phase information.

In many instances, we need to evaluate the cross-correlation between the input and output signals and the autocorrelation of the output signals. It can be easily shown that

$$r_{yx}(l) = h(l) * r_x(l) \tag{2.3.26}$$

$$r_y(l) = h^*(-l) * r_{xy}^*(l) = r_h(l) * r_x(l) \tag{2.3.27}$$

where
$$r_h(l) = \sum_{n=-\infty}^{\infty} h(n)h^*(n-l) = h(l) * h^*(-l) \tag{2.3.28}$$

is the autocorrelation of the impulse response. Taking the z-transform of both sides in the above equations, we obtain

$$R_{yx}(z) = H(z)R_x(z) \tag{2.3.29}$$

$$R_y(z) = H^*\left(\frac{1}{z^*}\right) R_{yx}(z) = R_h(z)R_x(z) \tag{2.3.30}$$

and
$$R_h(z) \triangleq H(z)H^*\left(\frac{1}{z^*}\right) \tag{2.3.31}$$

where $R_x(z)$, $R_y(z)$, and $R_h(z)$ are known as *complex spectral density* functions. Evaluating (2.3.30) on the unit circle $z = e^{j\omega}$ gives

$$R_y(e^{j\omega}) = R_h(e^{j\omega})R_x(e^{j\omega}) = |H(e^{j\omega})|^2 R_x(e^{j\omega}) \tag{2.3.32}$$

The output correlations $r_{xy}(l)$ and $r_y(l)$ for a periodic input with fundamental period N are computed via their spectral densities using the Fourier series. For example, it can be easily shown that

$$R_k^{(y)} = |H(e^{j2\pi k/N})|^2 R_k^{(x)} \qquad 0 \le k \le N - 1 \tag{2.3.33}$$

where $R_k^{(x)}$, $R_k^{(y)}$ are the power spectral densities of $x(n)$ and $y(n)$, respectively.

In exploring the properties of the various system models, we shall need to excite them by some input. Of particular interest are deterministic inputs that have constant power spectrum values (such as the unit sample sequence) or inputs that have constant power spectrum envelopes (such as all-pass signals). Since we have already discussed the unit sample response, we next focus on all-pass signals.

All-pass signals have a flat-spectrum, that is,

$$R_x(e^{j\omega}) = |X(e^{j\omega})|^2 = G^2 \qquad -\pi < \omega \le \pi \tag{2.3.34}$$

and, therefore, $r_x(l) = G^2\delta(l)$. The simplest example is $x(n) = \delta(n - k)$. A more interesting case is that of all-pass signals with nonlinear phase characteristic (see Section 2.4.2). The autocorrelation and the spectral density of the output $y(n)$ of LTI systems to all-pass excitations can be computed by the formulas used for unit impulse excitations, that is,

$$r_y(l) = G^2 r_h(l) = G^2 \sum_{n=-\infty}^{\infty} h(n)h^*(n - 1) \tag{2.3.35}$$

and

$$R_y(z) = G^2 H(z)H^*\left(\frac{1}{z^*}\right) \tag{2.3.36}$$

By properly choosing G, we can always assume that $h(0) = 1$.

2.4 MINIMUM PHASE AND SYSTEM INVERTIBILITY

In this section, we introduce the concept of minimum phase and show how it is related to the invertibility of linear, time-invariant systems. Several properties of all-pass and minimum-phase systems are also discussed.

2.4.1 System Invertibility and Minimum-Phase Systems

A system $H[\cdot]$ with input $x(n)$, $-\infty < n < \infty$, and output $y(n)$, $-\infty < n < \infty$, is called *invertible* if we can uniquely determine its input signal from the output signal. This is possible if the correspondence between the input and output signals is one-to-one. The system that produces $x(n)$, when excited by $y(n)$, is denoted by H_{inv} and is called the *inverse* of system H. Obviously, the cascade of H and H_{inv} is the identity system. Obtaining the inverse of an arbitrary system is a very difficult problem. However, if a system is linear and time-invariant, then if its inverse exists, the inverse is also linear and time-invariant. Hence, if $h(n)$ is the impulse response of a linear, time-invariant system and $h_{\text{inv}}(n)$ that of its inverse, we have

$$[x(n) * h(n)] * h_{\text{inv}}(n) = x(n)$$

or

$$h(n) * h_{\text{inv}}(n) = \delta(n) \tag{2.4.1}$$

Thus, given $h(n)$, $-\infty < n < \infty$, we can obtain $h_{\text{inv}}(n)$, $-\infty < n < \infty$, by solving the convolution equation (2.4.1), which is not an easy task in general. However, (2.4.1) can be converted to a simpler algebraic equation using the z-transform. Indeed, using the convolution theorem, we obtain

$$H_{\text{inv}}(z) = \frac{1}{H(z)} \qquad (2.4.2)$$

where $H_{\text{inv}}(z)$ is the system function of the inverse system. If $H(z)$ is a pole-zero system, that is,

$$H(z) = \frac{D(z)}{A(z)} \qquad (2.4.3)$$

then

$$H_{\text{inv}}(z) = \frac{A(z)}{D(z)} \qquad (2.4.4)$$

Thus, the zeros of the system become the poles of its inverse, and vice versa. Furthermore, the inverse of an all-pole system is all-zero, and vice versa.

EXAMPLE 2.4.1. Consider a system with impulse response

$$h(n) = \delta(n) - \tfrac{1}{4}\delta(n-1)$$

Determine impulse response of the inverse system.

Solution. The system function of its inverse is

$$H_{\text{inv}}(z) = \frac{1}{1 - \tfrac{1}{4}z^{-1}}$$

which has a pole at $z = \tfrac{1}{4}$. If we choose the ROC as $|z| > \tfrac{1}{4}$, the inverse system is causal and stable, and

$$h_{\text{inv}}(n) = (\tfrac{1}{4})^n u(n)$$

However, if we choose the ROC as $|z| < \tfrac{1}{4}$, the inverse system is noncausal and unstable

$$h_{\text{inv}}(n) = -(\tfrac{1}{4})^n u(-n-1)$$

This simple example illustrates that the knowledge of the impulse response of a linear, time-invariant system does not uniquely specify its inverse. Additional information such as causality and stability would be helpful in many cases. This leads us to the concept of minimum-phase systems.

A discrete-time, linear, time-invariant system with impulse response $h(n)$ is called *minimum-phase* if both the system and its inverse system $h_{\text{inv}}(n)$ are causal and stable, that is,

$$h(n) * h_{\text{inv}}(n) = \delta(n) \qquad (2.4.5)$$

$$h(n) = 0 \quad n < 0 \quad \text{and} \quad h_{\text{inv}}(n) = 0 \quad n < 0 \qquad (2.4.6)$$

$$\sum_{n=0}^{\infty} |h(n)| < \infty \quad \text{and} \quad \sum_{n=0}^{\infty} |h_{\text{inv}}(n)| < \infty \qquad (2.4.7)$$

We note that if a system is minimum-phase, its inverse is also minimum-phase. This is very important in deconvolution problems, where the inverse system has to be causal and stable for implementation purposes.

Sometimes, especially in geophysical applications, the stability requirements (2.4.7) are replaced by the less restrictive[†] finite energy conditions

$$\sum_{n=0}^{\infty} |h(n)|^2 < \infty \quad \text{and} \quad \sum_{n=0}^{\infty} |h_{\text{inv}}(n)|^2 < \infty \qquad (2.4.8)$$

which are implied by (2.4.7). However, note that (2.4.8) does not necessarily imply (2.4.7).

[†]This definition of minimum phase allows singularities (poles or zeros) on the unit circle.

Clearly, a PZ system is minimum-phase if all its poles and zeros are inside the unit circle. Indeed, if all roots of $A(z)$ and $D(z)$ are inside the unit circle, the system $H(z)$ in (2.4.3) and its inverse $H_{inv}(z)$ in (2.4.4) are both causal and stable.

In an analogous manner, we can define a *maximum-phase system* as one in which both the system and its inverse are noncausal and stable. A PZ system then is maximum-phase if all its poles and zeros are outside the unit circle. Clearly, if $H(z)$ is minimum-phase, then $H(z^{-1})$ is maximum-phase. A system that is neither minimum-phase nor maximum-phase is called a *mixed-phase* system.

2.4.2 All-Pass Systems

We shall say that a linear, time-invariant system is *all-pass*, denoted by $H_{ap}(e^{j\omega})$, if

$$|H_{ap}(e^{j\omega})| = 1 \qquad -\pi < \omega \le \pi \tag{2.4.9}$$

The simplest all-pass system is characterized by

$$H_{ap}(z) = z^k$$

which simply time-shifts (delay $k < 0$, advance $k > 0$) the input signal.

A more interesting, nontrivial family of all-pass systems is characterized by the system function (dispersive all-pass systems)

$$H_{ap}(z) = \frac{a_P^* + a_{P-1}^* z^{-1} + \cdots + z^{-P}}{1 + a_1 z^{-1} + \cdots + a_P z^{-P}} = \frac{z^{-P} A^*(1/z^*)}{A(z)} \tag{2.4.10}$$

Indeed, it can be easily seen that

$$|H_{ap}(e^{j\omega})|^2 = H_{ap}(z) H_{ap}^* \left(\frac{1}{z^*}\right)\bigg|_{z=e^{j\omega}} = 1 \tag{2.4.11}$$

In the case of real-valued coefficients, (2.4.10) takes the form

$$H_{ap}(z) = \frac{a_P + a_{P-1} z^{-1} + \cdots + z^{-P}}{1 + a_1 z^{-1} + \cdots + a_P z^{-P}} = \frac{z^{-P} A(z^{-1})}{A(z)} \tag{2.4.12}$$

The poles and zeros of an all-pass system are conjugate reciprocals of one another; that is, they are conjugate symmetric with respect to the unit circle. Indeed, if p_0 is a root of $A(z)$, then $1/p_0^*$ is a root of $A^*(1/z^*)$. Thus, if $p_0 \triangleq re^{j\theta}$ is a pole of $H_{ap}(z)$, then $1/p_0^* = (1/r)e^{j\theta}$ is a zero of the system. This typical pattern is illustrated in Figure 2.10

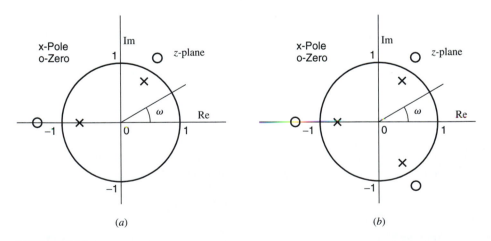

(a)   (b)

FIGURE 2.10
Typical pole-zero patterns of a PZ, all-pass system: (a) complex-valued coefficients and (b) real-valued coefficients.

for system functions with both complex and real coefficients. Therefore, the system function of any pole-zero all-pass system can be expressed as

$$H_{ap}(z) = \prod_{k=1}^{P} \frac{p_k^* - z^{-1}}{1 - p_k z^{-1}} \qquad (2.4.13)$$

The similar expressions $(z^{-1} - p_k^*)/(1 - p_k z^{-1})$ and $(1 - p_k z^{-1})/(z^{-1} - p_k^*)$ [the negative and inverse of (2.4.13), respectively] are often used in the literature. For systems with real parameters, singularities should appear in complex conjugate pairs.

Properties of all-pass systems. All-pass systems have some interesting properties. We list these properties without proofs. Some of these proofs are trivial, and others are explored in problems.

1. The output energy of a stable all-pass system is equal to the input energy; that is,

$$E_y = \sum_{n=-\infty}^{\infty} |y(n)|^2 = \frac{1}{2\pi} \int_{-\pi}^{\pi} \left| H_{ap}(e^{j\omega}) X(e^{j\omega}) \right|^2 d\omega = E_x \qquad (2.4.14)$$

due to (2.4.9). This leads to a very interesting property for the cumulative energy of a causal all-pass system (see Problem 2.6).

2. A causal, stable, PZ, all-pass system with P poles has a phase response $\angle H_{ap}(e^{j\omega})$ that decreases monotonically from $\angle H_{ap}(e^{j0})$ to $\angle H_{ap}(e^{j0}) - 2\pi P$ as ω increases from 0 to 2π (see Problem 2.7).

3. All-pass systems have nonnegative group delay, which is defined as the negative of the first derivative of the phase response, that is,

$$\tau_{ap}(\omega) \triangleq -\frac{d}{d\omega} \angle H_{ap}(e^{j\omega}) \geq 0 \qquad (2.4.15)$$

This property is a direct result of the second property.

4. The all-pass system function $H_{ap}(z)$

$$H_{ap}(z) = \frac{1 - \alpha z^{-1}}{z^{-1} - \alpha^*} \qquad |\alpha| < 1 \qquad (2.4.16)$$

satisfies
$$|H_{ap}(z)| \begin{cases} < 1 & \text{if} \quad |z| < 1 \\ = 1 & \text{if} \quad |z| = 1 \\ > 1 & \text{if} \quad |z| > 1 \end{cases} \qquad (2.4.17)$$

For proof see Problem 2.10.

2.4.3 Minimum-Phase and All-Pass Decomposition

We next show that any causal, PZ system that has no poles or zeros on the unit circle can be expressed as

$$H(z) = H_{min}(z) H_{ap}(z) \qquad (2.4.18)$$

where $H_{min}(z)$ is minimum-phase and $H_{ap}(z)$ is all-pass, as shown in Figure 2.11. Indeed, let $H(z)$ be a non-minimum-phase system with one zero $z = 1/a$, $|a| < 1$, outside the unit circle and all other poles and zeros inside the unit circle. Then $H(z)$ can be factored as

$$H(z) = H_1(z)(a - z^{-1}) \qquad (2.4.19)$$

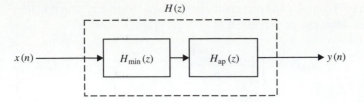

FIGURE 2.11
Minimum phase and all-pass decomposition.

where $H_1(z)$ is minimum-phase. Equivalently, (2.4.19) can be expressed as

$$H(z) = H_1(z)(a - z^{-1})\frac{1 - a^* z^{-1}}{1 - a^* z^{-1}}$$

$$= [H_1(z)(1 - a^* z^{-1})]\frac{a - z^{-1}}{1 - a^* z^{-1}} \qquad (2.4.20)$$

$$= H_{\min}(z)\frac{a - z^{-1}}{1 - a^* z^{-1}}$$

where $H_{\min}(z)$ is minimum-phase and the factor $(a - z^{-1})/(1 - a^* z^{-1})$ is all-pass, because $|a| < 1$. Note that the minimum-phase system was obtained from $H(z)$ by reflecting the zero $z = 1/a$, which was outside the unit circle, to the zero $z = a^*$ inside the unit circle. This approach can clearly be generalized for any PZ system. Thus, given a non-minimum-phase PZ system, we can create a minimum-phase one with the same magnitude response (or equivalently the same impulse response autocorrelation) by reflecting all poles and zeros that are outside the unit circle inside the unit circle. From the previous discussion it follows that there are 2^Q Qth-order AZ systems with the same magnitude response. This is illustrated in the following example.

EXAMPLE 2.4.2. For $Q = 2$, determine all four second-order AZ systems with the same magnitude response.

Solution. For a second-order all-zero system $(0 < a < 1, 0 < b < 1)$ we obtain the following systems

$$\begin{aligned} H_{\min}(z) &= (1 - az^{-1})(1 - bz^{-1}) & H_{\max}(z) &= (1 - az)(1 - bz) \\ H_{\mathrm{mix}1}(z) &= (1 - az)(1 - bz^{-1}) & H_{\mathrm{mix}2}(z) &= (1 - az^{-1})(1 - bz) \end{aligned} \qquad (2.4.21)$$

that have the same spectrum

$$R(z) = H(z)H(z^{-1}) = (1 - az^{-1})(1 - bz^{-1})(1 - az)(1 - bz) \qquad (2.4.22)$$

and the same autocorrelation

$$r(l) = \begin{cases} 1 + a^2 b^2 + (a + b)^2 & l = 0 \\ -(a + b)(1 + ab) & l = 1, -1 \\ ab & l = 2, -2 \\ 0 & \text{otherwise} \end{cases} \qquad (2.4.23)$$

but different impulse and phase responses, as shown in Figure 2.12.

EXAMPLE 2.4.3. Consider the following all-zero minimum-phase system:

$$\begin{aligned} H_{\min}(z) &= (1 - 0.8e^{j0.6\pi} z^{-1})(1 - 0.8e^{-j0.6\pi} z^{-1}) \\ &\quad \times (1 - 0.8e^{j0.9\pi} z^{-1})(1 - 0.8e^{-j0.9\pi} z^{-1}) \end{aligned} \qquad (2.4.24)$$

Determine the maximum- and mixed-phase systems with the same magnitude response.

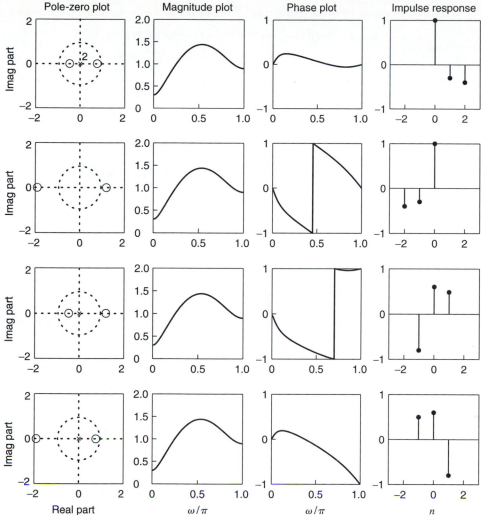

FIGURE 2.12

Pole-zero, frequency response, and impulse response plots for minimum-phase (row 1), maximum-phase (row 2), mixed-phase 1 (row 3), and mixed-phase 2 (row 4) systems in Example 2.4.2. Note that the abscissa in Phase plots are labeled in units of π radians.

Solution. To obtain a maximum-phase system with the same magnitude response, we reflect the zeros of $H_{\min}(z)$ from inside the unit circle to their conjugate reciprocal locations that are outside the unit circle by using the transformation $z_0 \to 1/z_0^*$. This leads to the following transformation for each first-order factor:

$$(1 - re^{j\theta}z^{-1}) \to r(1 - \frac{1}{r}e^{j\theta}z^{-1}) \tag{2.4.25}$$

The scaling factor r in the right-hand side is included to guarantee that the transformation does not scale the magnitude response. The resulting maximum-phase system is

$$H_{\max}(z) = (0.8)^4(1 - 1.25e^{j0.6\pi}z^{-1})(1 - 1.25e^{-j0.6\pi}z^{-1})$$
$$\times (1 - 1.25e^{j0.9\pi}z^{-1})(1 - 1.25e^{-j0.9\pi}z^{-1}) \tag{2.4.26}$$

If we reflect only the zero at $0.8e^{\pm j0.6\pi}$, we obtain the mixed-phase system

$$H_1(z) = (0.8)^2(1 - 1.25e^{j0.6\pi}z^{-1})(1 - 1.25e^{-j0.6\pi}z^{-1})$$
$$\times (1 - 0.8e^{j0.9\pi}z^{-1})(1 - 0.8e^{-j0.9\pi}z^{-1}) \tag{2.4.27}$$

Similarly, if we reflect only the zero at $0.8e^{\pm j0.9\pi}$, we obtain the second mixed-phase system

$$H_2(z) = (0.8)^2(1 - 0.8e^{j0.6\pi}z^{-1})(1 - 0.8e^{-j0.6\pi}z^{-1})$$
$$\times (1 - 1.25e^{j0.9\pi}z^{-1})(1 - 1.25e^{-j0.9\pi}z^{-1}) \tag{2.4.28}$$

Figure 2.13 shows the pole-zero, magnitude response, phase response, and group delay plots for all four systems. Clearly, the minimum-phase system has the smallest group delay, the maximum-phase system has the largest group delay, while the mixed-phase systems have in-between amounts of group delay across all frequencies. Finally, it can be easily shown that the system $H_{\max}(z)/H_{\min}(z)$ is an all-pass system.

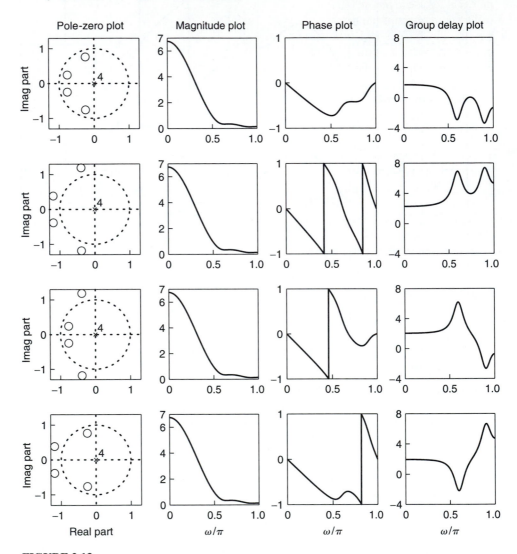

FIGURE 2.13
Pole-zero and frequency response plots for minimum-phase (row 1), maximum-phase (row 2), mixed-phase 1 (row 3), and mixed-phase 2 (row 4) systems in Example 2.4.3. Note that the abscissa in Phase plots are labeled in units of π radians while those in Group delay plots are labeled in sampling intervals.

The minimum- (maximum-) phase AZ system has all its zeros inside (outside) the unit circle. From (2.4.12), it follows that an all-pass system can be expressed as

$$H_{ap}(z) = \frac{H_{\max}(z)}{H_{\min}(z)} \tag{2.4.29}$$

where $H_{\min}(z)$ and $H_{\max}(z)$ are the Pth-order minimum-phase and maximum-phase systems, respectively, with the same magnitude response. Indeed, it can be easily seen that

$$H_{\max}(z) = z^{-P} H^*_{\min}\left(\frac{1}{z^*}\right) \tag{2.4.30}$$

or $h_{\max}(n) = h^*_{\min}(P - n)$.

In practice, it is very important to find out if a given system is minimum-phase. Clearly, the definition cannot be used in practice because either the system $h(n)$ or its inverse is going to be IIR. Furthermore, most of the above properties using either $h(n)$ or $H(e^{j\omega})$ are not practical for use in real-world systems. However, if we deal with PZ systems, we can check if they are minimum-phase by computing the poles and zeros and check if they are inside the unit circle. This is, however, a computationally expensive procedure, especially for high-order systems. Fortunately, there are several tests that allow us to find out if the zeros of a polynomial are inside the unit circle without computing them. See Theorem 2.3.

Properties of minimum-phase systems. Minimum-phase systems have some very interesting properties. Next we list some of these properties without proofs. More details can be found in Oppenheim and Schafer (1989) and Proakis and Manolakis (1996).

1. For causal, stable systems with the same magnitude response, the minimum-phase system has *algebraically* the smallest group delay response at every frequency, that is, $\tau_{\min}(e^{j\omega}) \leq \tau(e^{j\omega})$, for all ω. Thus, strictly speaking, minimum-phase systems are *minimum group delay systems*. However, the term *minimum-phase* has been established in the engineering literature.

2. Of all causal and stable systems with the same magnitude response, the minimum-phase system minimizes the "energy delay"

$$\sum_{n=k}^{\infty} |h(n)|^2 \quad \text{for all } k = 0, 1, \ldots, \infty \tag{2.4.31}$$

where $h(n)$ is the system impulse response.

3. The system $H(z)$ is minimum-phase if $\log|H(e^{j\omega})|$ and $\angle H(e^{j\omega})$ form a Hilbert transform pair.

> **EXAMPLE 2.4.4.** In this example we illustrate the energy delay property of minimum-phase systems. Consider the all-zero minimum-phase system (2.4.24) given in Example 2.4.3 and repeated here:
>
> $$H_{\min}(z) = (1 - 0.8e^{j0.6\pi}z^{-1})(1 - 0.8e^{-j0.6\pi}z^{-1})$$
> $$\times (1 - 0.8e^{j0.9\pi}z^{-1})(1 - 0.8e^{-j0.9\pi}z^{-1})$$
>
> In the top row of four plots in Figure 2.14, we depict the impulse responses of the minimum-, maximum-, and mixed-phase systems. The bottom plot contains the graph of the energy delay $\sum_{n=k}^{\infty} |h(n)|^2$ for $k = 0, 1, \ldots, 4$, for each of the systems. As expected, the minimum-phase system has the least amount of energy delay while the maximum-phase system has the greatest amount of energy delay at each n. The graphs of the energy delays for mixed-phase systems are somewhere in between the above two graphs.

Additional properties of minimum-phase systems are explored in the problems.

2.4.4 Spectral Factorization

One interesting and practically useful question is the following: Can we completely determine the system $H(z)$ by computing the autocorrelation $r_y(l)$ or, equivalently, the spectral density $R_y(e^{j\omega})$? The answer is not a unique one since all we know either from $r_y(l)$ or from $R_y(e^{j\omega})$ is the magnitude response $|H(e^{j\omega})|$, but not the phase response $\angle H(e^{j\omega})$. To obtain a unique system from (2.3.35) or (2.3.36), we have to impose additional

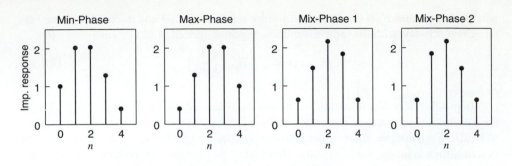

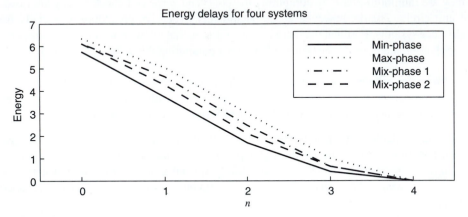

FIGURE 2.14
Impulse response plots of the four systems in the top row and the energy delay plots in the bottom row in Example 2.4.4.

conditions on $H(z)$. One such condition is that of a minimum-phase system. The process of obtaining the minimum-phase system that produces the signal $y(n)$ with autocorrelation $r_y(l)$ or spectral density $R_y(z)$ is called *spectral factorization*. Equivalently, the spectral factorization problem can be stated as the determination of a minimum-phase system from its magnitude response or from the autocorrelation of its impulse response.

Solving the spectral factorization problem by finding roots of $R_y(z)$ is known as the *root method*, and besides its practical utility, it illustrates some basic principles.

1. Every rational power spectral density has, within a scale factor, a unique minimum-phase factorization.
2. There are 2^{P+Q} rational systems with the same power spectral density, where Q and P are numerator and denominator polynomial degrees, respectively.
3. Not all possible rational functions are valid power spectral densities since for a valid $R_y(z)$ the roots should appear in pairs, z_k and $1/z_k^*$.

These principles can be generalized to any power spectral density by extending $P + Q \to \infty$. The spectral factorization procedure is guaranteed by the following theorem.

THEOREM 2.1. If $\ln R_y(z)$ is analytic in an open ring $\alpha < |z| < 1/\alpha$ in the z-plane and the ring includes the unit circle, then $R_y(z)$ can be factored as

$$R_y(z) = G^2 H_{\min}(z) H_{\min}^* \left(\frac{1}{z^*} \right) \qquad (2.4.32)$$

where $H_{\min}(z)$ is a minimum-phase system.

Proof. Using the analyticity of $\ln R_y(z)$, we can expand $\ln R_y(z)$ in a Laurent series (Churchill and Brown 1984) as

$$\ln R_y(z) = \sum_{-\infty}^{\infty} g(l) z^{-l} \qquad (2.4.33)$$

where the sequence $g(l)$ is known as the *cepstrum* of the sequence $r_y(l)$ (Oppenheim and Schafer 1989). Evaluating (2.4.33) on the unit circle, we obtain

$$\ln R_y(e^{j\omega}) = \sum_{-\infty}^{\infty} g(l)e^{-j\omega l} \tag{2.4.34}$$

or

$$g(l) = \frac{1}{2\pi} \int_{-\pi}^{\pi} \ln R_y(e^{j\omega})e^{j\omega l}\, d\omega \tag{2.4.35}$$

Since $R_y(e^{j\omega}) = |Y(e^{j\omega})|^2$ is a real, nonnegative function, the sequence $g(l)$ is a conjugate symmetric sequence, that is,

$$g(l) = g^*(-l) \tag{2.4.36}$$

and

$$G^2 \triangleq \exp g(0) = \exp\left[\frac{1}{2\pi} \int_{-\pi}^{\pi} \ln R_y(e^{j\omega})\, d\omega\right] \geq 0 \tag{2.4.37}$$

From (2.4.33), we can express $R_y(z)$ in a factored form as

$$R_y(z) = \exp\left[\sum_{-\infty}^{\infty} g(l)z^{-l}\right] = \exp\left[\sum_{-\infty}^{-1} g(l)z^{-l} + g(0) + \sum_{1}^{\infty} g(l)z^{-l}\right]$$

$$= \exp g(0) \exp\left[\sum_{1}^{\infty} g(l)z^{-l}\right] \exp\left[\sum_{-\infty}^{-1} g(l)z^{-l}\right] \tag{2.4.38}$$

$$= G^2 \exp\left[\sum_{1}^{\infty} g(l)z^{-l}\right] \exp\left[\sum_{1}^{\infty} g^*(l)z^{l}\right]$$

where we used (2.4.36). After defining

$$H(z) \triangleq \exp\left[\sum_{1}^{\infty} g(l)z^{-l}\right] \qquad |z| > \alpha \tag{2.4.39}$$

so that

$$H^*\left(\frac{1}{z^*}\right) = \exp\left[\sum_{1}^{\infty} g^*(l)z^{l}]\right] \qquad |z| < \frac{1}{\alpha} \tag{2.4.40}$$

we obtain the spectral factorization (2.3.36). Furthermore, from (2.4.37) we note that the constant G^2 is equal to the geometric mean of $R_y(e^{j\omega})$. From (2.4.39), note that $H(z)$ is the z-transform of a causal and stable sequence, hence it can be expanded as

$$H(z) = 1 + h(1)z^{-1} + h(2)z^{-2} + \cdots \tag{2.4.41}$$

where $h(0) = \lim_{z\to\infty} H(z) = 1$. Also from (2.4.39) $H(z)$ corresponds to a minimum-phase system so that from (2.4.40) $H^*(1/z^*)$ is a stable, anticausal, and maximum-phase system.

The analyticity of $\ln R_y(z)$ is guaranteed by the Paley-Wiener theorem given below without proof (see Papoulis 1991).

THEOREM 2.2 (PALEY-WIENER THEOREM). The spectral factorization in (2.4.32) is possible if $R_y(z)$ satisfies the Paley-Wiener condition

$$\int_{-\infty}^{\infty} |\ln R_y(e^{j\omega})|\, d\omega < \infty$$

If $H(z)$ is known to be minimum-phase, the spectral factorization is unique.

In general, the solution of the spectral factorization problem is difficult. However, it is quite simple in the case of signals with rational spectral densities. Suppose that $R_y(z)$ is a rational complex spectral density function. Since $r_y(l) = r_y^*(-l)$ implies that $R_y(z) = R_y^*(1/z^*)$, if z_i is a root, then $1/z_i^*$ is also a root. If z_i is inside the unit circle, then $1/z_i^*$ is outside. To obtain the minimum-phase system $H(z)$ corresponding to $R_y(z)$, we determine the poles and zeros of $R_y(z)$ and form $H(z)$ by choosing all poles and zeros that are inside the unit circle, that is,

$$H(z) = G \frac{\displaystyle\prod_{k=1}^{Q}(1 - z_k z^{-1})}{\displaystyle\prod_{k=1}^{P}(1 - p_k z^{-1})} \qquad (2.4.42)$$

where $|z_k| < 1, k = 1, 2, \ldots, Q$ and $|p_k| < 1, k = 1, 2, \ldots, P$.

Before we illustrate this by an example, it should be emphasized that for real-valued coefficients $R_y(e^{j\omega})$ is a rational function of $\cos\omega$. Indeed, we have from (2.3.36) and (2.3.13)

$$R_y(z) = G^2 H(z) H^*\left(\frac{1}{z^*}\right) = G^2 \frac{D(z)D^*(1/z^*)}{A(z)A^*(1/z^*)} \qquad (2.4.43)$$

where

$$D(z) = \sum_{k=0}^{Q} d_k z^{-k} \quad \text{and} \quad A(z) = 1 + \sum_{k=1}^{P} a_k z^{-k} \qquad (2.4.44)$$

Clearly, (2.4.43) can be written as

$$R_y(e^{j\omega}) = G^2 \frac{R_d(e^{j\omega})}{R_a(e^{j\omega})} = G^2 \frac{r_d(0) + 2 \displaystyle\sum_{l=1}^{Q} r_d(l) \cos l\omega}{r_a(0) + 2 \displaystyle\sum_{l=1}^{P} r_a(l) \cos l\omega} \qquad (2.4.45)$$

where $r_d(l) = r_d^*(-l)$ and $r_a(l) = r_a^*(-l)$ are the autocorrelations of the coefficient sequences $\{d_0, d_1, \ldots, d_Q\}$ and $\{1, a_1, \ldots, a_P\}$, respectively. Since $\cos l\omega$ can be expressed as a polynomial

$$\cos l\omega = \sum_{i=0}^{l} \alpha_i (\cos\omega)^i$$

it follows that $R_y(e^{j\omega})$ is a rational function of $\cos\omega$.

EXAMPLE 2.4.5. Let

$$R_y(e^{j\omega}) = \frac{1.04 + 0.4\cos\omega}{1.25 + \cos\omega}$$

Determine the minimum-phase system corresponding to $R_y(e^{j\omega})$.

Solution. Replacing $\cos\omega$ by $(e^{j\omega} + e^{-j\omega})/2$ or directly by $(z + z^{-1})/2$ gives

$$R_y(z) = \frac{1.04 + 0.2z + 0.2z^{-1}}{1.25 + 0.5z + 0.5z^{-1}} = 0.4\frac{(z + 5)(z + 0.2)}{(z + 2)(z + 0.5)}$$

The required minimum-phase system $H(z)$ is

$$H(z) = \frac{z + 0.2}{z + 0.5} = \frac{1 + 0.2z^{-1}}{1 + 0.5z^{-1}} \qquad (2.4.46)$$

2.5 LATTICE FILTER REALIZATIONS

In Section 2.3, we described simple FIR and IIR filter realizations using block diagram elements. These realizations are called filter structures for which there are many different types available for implementation (Proakis and Manolakis 1996). In this section, we discuss the lattice and lattice-ladder filters. The lattice filter is an implementation of a digital filter with rational system functions. This structure is used extensively in digital speech processing and in the implementation of adaptive filters, which are discussed in Chapter 10.

2.5.1 All-Zero Lattice Structures

In Section 2.3, we discussed a direct-form realization of an AZ filter (see Figure 2.6). In this section, we present *lattice* structures for the realization of AZ filters. These structures will be used extensively throughout this book.

The basic AZ lattice is shown in Figure 2.15. Because the AZ lattice is often used to implement the inverse of an AP filter, we begin our introduction to the lattice by a realization of the AZ filter

$$A(z) = 1 + \sum_{l=1}^{P} a_l z^{-l} \tag{2.5.1}$$

The lattice in Figure 2.15(*a*) is the two-multiplier, or Itakura-Saito, lattice. The lattice has P parameters $\{k_m, 1 \le m \le P\}$ that map to the a_l direct-form parameters via a recursive relation that is derived below.

At the mth stage of the lattice, shown in Figure 2.15(*a*), we have the relations

$$f_m(n) = f_{m-1}(n) + k_m g_{m-1}(n-1) \qquad 1 \le m \le P \tag{2.5.2}$$
$$g_m(n) = k_m^* f_{m-1}(n) + g_{m-1}(n-1) \qquad 1 \le m \le P \tag{2.5.3}$$

and from Figure 2.15(*b*), we have

$$f_0(n) = g_0(n) = x(n) \tag{2.5.4}$$
$$y(n) = f_P(n) \tag{2.5.5}$$

Taking the z-transform of $f_m(n)$ and $g_m(n)$, we have

$$F_m(z) = F_{m-1}(z) + k_m z^{-1} G_{m-1}(z) \tag{2.5.6}$$
$$G_m(z) = k_m^* F_{m-1}(z) + z^{-1} G_{m-1}(z) \tag{2.5.7}$$

Dividing both equations by $X(z)$ and denoting the transfer functions from the input $x(n)$ to the outputs of the mth stage by $A_m(z)$ and $B_m(z)$, where

$$A_m(z) \triangleq \frac{F_m(z)}{F_0(z)} \qquad B_m(z) \triangleq \frac{G_m(z)}{G_0(z)} \tag{2.5.8}$$

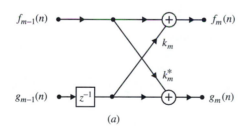

(*a*)

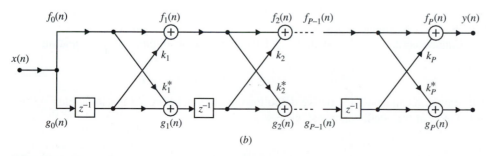

(*b*)

FIGURE 2.15
All-zero lattice structure.

we have
$$A_m(z) = A_{m-1}(z) + k_m z^{-1} B_{m-1}(z) \tag{2.5.9}$$

$$B_m(z) = k_m^* A_{m-1}(z) + z^{-1} B_{m-1}(z) \tag{2.5.10}$$

with
$$A_0(z) = B_0(z) = 1 \tag{2.5.11}$$

and
$$A(z) = A_P(z) \tag{2.5.12}$$

Thus, the desired $A(z)$ is obtained as the transfer function $A_P(z)$ at the Pth stage of the lattice. Now (2.5.9) and (2.5.10) can be written in matrix form as

$$\begin{bmatrix} A_m(z) \\ B_m(z) \end{bmatrix} = \begin{bmatrix} 1 & k_m z^{-1} \\ k_m^* & z^{-1} \end{bmatrix} \begin{bmatrix} A_{m-1}(z) \\ B_{m-1}(z) \end{bmatrix} \tag{2.5.13}$$

$$= \mathbf{Q}_m(z) \begin{bmatrix} A_{m-1}(z) \\ B_{m-1}(z) \end{bmatrix} \tag{2.5.14}$$

where
$$\mathbf{Q}_m(z) \triangleq \begin{bmatrix} 1 & k_m z^{-1} \\ k_m^* & z^{-1} \end{bmatrix} \tag{2.5.15}$$

Then, using the recursive relation (2.5.13), we obtain

$$\begin{bmatrix} A_P(z) \\ B_P(z) \end{bmatrix} = \prod_{m=1}^{P} \mathbf{Q}_m(z) \begin{bmatrix} 1 \\ 1 \end{bmatrix} \tag{2.5.16}$$

If we write $A_m(z)$ as

$$A_m(z) = \sum_{l=0}^{m} a_k^{(m)} z^{-l} \tag{2.5.17}$$

then we can show that

$$a_0^{(m)} = 1 \qquad \text{for all } m \tag{2.5.18}$$

and that
$$B_m(z) = \sum_{l=0}^{m} b_k^{(m)} z^{-l} = z^{-m} A_m^* \left(\frac{1}{z^*} \right) \tag{2.5.19}$$

that is, for $m = 1, 2, \ldots, P$

$$b_l^{(m)} = \begin{cases} a_{m-l}^{(m)*} & l = 1, 2, \ldots, m-1 \\ 1 & l = m \end{cases} \tag{2.5.20}$$

The polynomial $B_m(z)$ is known as the conjugate *reverse* polynomial of $A_m(z)$ because its coefficients are the conjugates of those of $A_m(z)$ except that they are in reverse order. So since

$$A_m(z) = 1 + a_1^{(m)} z^{-1} + a_2^{(m)} z^{-2} + \cdots + a_m^{(m)} z^{-m} \tag{2.5.21}$$

then
$$B_m(z) = a_m^{(m)*} + a_{m-1}^{(m)*} z^{-1} + \cdots + a_1^{(m)*} z^{-1} + z^{-m} \tag{2.5.22}$$

If z_0 is a zero of $A_m(z)$, then z_0^{-1} is a zero of $B_m(z)$. Therefore, if $A_m(z)$ is minimum-phase, then $B_m(z)$ is maximum-phase.

Equations (2.5.19), (2.5.9), and (2.5.10) can be combined into a single equation

$$A_m(z) = A_{m-1}(z) + k_m z^{-m} A_{m-1}^* \left(\frac{1}{z^*} \right) \tag{2.5.23}$$

This equation can be used to derive the following relation between the coefficients at stage m in terms of the coefficients at stage $m - 1$:

$$a_l^{(m)} = \begin{cases} 1 & l = 0 \\ a_l^{(m-1)} + k_m a_{m-l}^{(m-1)*} & l = 1, 2, \ldots, m-1 \\ k_m & l = m \end{cases} \tag{2.5.24}$$

To solve for the coefficients of the transfer function of the complete P-stage lattice, compute (2.5.24) recursively, starting with $m = 1$ until $m = P$. The final coefficients a_l of the desired filter $A(z)$ are then given by

$$a_l = a_l^{(P)} \qquad 0 \le l \le P \tag{2.5.25}$$

By substituting $m - l$ for l in (2.5.24), we have

$$a_{m-l}^{(m)} = a_{m-l}^{(m-1)} + k_m a_l^{(m-1)*} \tag{2.5.26}$$

Therefore, $a_l^{(m)}$ and $a_{m-l}^{(m)}$ can be computed simultaneously using $a_l^{(m-1)}$, $a_{m-l}^{(m-1)}$, and k_m.

The lattice parameters k_m can be recovered from the coefficients a_l by a backward recursion. Eliminating $z^{-1} B_{m-1}(z)$ from (2.5.9) and (2.5.10) and using (2.5.19), we obtain

$$A_{m-1}(z) = \frac{A_m(z) - k_m z^{-m} A_m^*(1/z^*)}{1 - |k_m|^2} \tag{2.5.27}$$

The recursion can be started by setting $a_l^{(P)} = a_l, 0 \le l \le P$. Then, with $m = P, P - 1, \ldots, 1$, we compute from (2.5.27)

$$k_m = a_m^{(m)}$$

$$a_l^{(m-1)} = \begin{cases} 1 & l = 0 \\ \dfrac{a_l^{(m)} - k_m a_{m-l}^{(m)*}}{1 - |k_m|^2} & 1 \le l \le m - 1 \end{cases} \tag{2.5.28}$$

This is the backward recursion to compute k_m from a_l. The computation in (2.5.28) is always possible except when some $|k_m| = 1$. Except for this indeterminate case, the mapping between the lattice parameters k_m and the coefficients a_l of the corresponding all-zero filter is unique.

The MATLAB function [k] = df2latcf(a) computes lattice coefficients k_m from polynomial coefficients a_k using (2.5.28). Similarly, the function [a] = latcf2df(k) computes the direct-form coefficients from the lattice form.

Although the AZ lattice filters are highly modular, their software implementation is more complex than the direct-form structures. To understand this implementation, we will consider the steps involved in determining one output sample in a P-stage AZ lattice. Assume that $x(n)$ is available over $1 \le n \le N$.

Input stage: The describing equation is

$$f_0(n) = g_0(n) = x(n) \qquad 1 \le n \le N$$

Thus in the implementation, $f_0(n)$ and $g_0(n)$ can be replaced by the input sample $x(n)$, which is assumed to be available in array x.

Stage 1: The describing equations are

$$f_1(n) = f_0(n) + k_1 g_0(n-1) = x(n) + k_1 x(n-1)$$
$$g_1(n) = k_1^* f_0(n) + g_0(n-1) = k_1^* x(n) + x(n-1)$$

Assuming that we have two arrays f and g of length P available to store $f_m(n)$ and $g_m(n)$ at each n, respectively, and two arrays k and ck of length P to store k_m and k_m^*, respectively, then the MATLAB fragment is

```
f(1) = x(n) + k(1)*x(n-1);
g(1) = ck(1)*x(n) + x(n-1);
```

At $n = 1$, we need $x(0)$ in the above equations. This is an initial condition and is assumed to be zero. Hence in the implementation, we need to augment the x array by prepending it with a zero. This should be done in the initialization part. Similarly, arrays f and g should be initialized to zero.

Stages 2 through P: The describing equations are

$$f_m(n) = f_{m-1}(n) + k_m g_{m-1}(n-1)$$
$$g_m(n) = k_m^* f_{m-1}(n) + g_{m-1}(n-1)$$

Note that we need old (i.e., at $n-1$) values of array g in $g_{m-1}(n-1)$. Although it is possible to avoid an additional array, for programming simplicity, we will assume that $g_m(n-1)$ is available in an array g_old of length P. This array should also be initialized to zero. The MATLAB fragment is

```
f(m)  =  f(m-1)  +  k(m)*g_old(m-1);
g(m)  =  ck*f(m-1)  +  g_old(m-1);
```

Output stage: The describing equation is

$$y(n) = f_P(n)$$

Also we need to store the current $g_m(n)$ values in the g_old array for use in the calculations of the next output value. Thus the MATLAB fragment is

```
g_old = g;
y = f(P);
```

Now we can go back to stage 1 with new input value and recursively compute the remaining output values.

The complete procedure is implemented in the function y = latcfilt(k,x).

2.5.2 All-Pole Lattice Structures

The AZ lattice in Figure 2.15 can be restructured quite simply to yield a corresponding all-pole (AP) lattice structure. Let an AP system function be given by

$$H(z) = \frac{1}{1 + \displaystyle\sum_{l=1}^{P} a_l z^{-l}} = \frac{1}{A(z)} \qquad (2.5.29)$$

which clearly is the *inverse* system of the AZ lattice of Figure 2.15. The difference equation corresponding to (2.5.29) is

$$y(n) + \sum_{l=1}^{P} a_l y(n-l) = x(n) \qquad (2.5.30)$$

If we interchange $x(n)$ with $y(n)$ in (2.5.30), we will obtain the AZ system of (2.5.1). Therefore, the lattice structure of the AP system can be obtained from Figure 2.15(b) by interchanging $x(n)$ with $y(n)$. This lattice structure with P stages is shown in Figure 2.16(b). To determine the mth stage of the AP lattice, we consider (2.5.4) and (2.5.5) and interchange $x(n)$ with $y(n)$. Thus the lattice structure shown in Figure 2.16(b) has

$$f_P(n) = x(n) \qquad (2.5.31)$$

as the input and

$$f_0(n) = g_0(n) = y(n) \qquad (2.5.32)$$

as the output. The signal quantities $\{f_m(n)\}_{m=0}^{P}$ then must be computed in descending order, which can be obtained by rearranging (2.5.2) but not (2.5.3). Thus we obtain

$$f_{m-1}(n) = f_m(n) - k_m g_{m-1}(n-1) \qquad (2.5.33)$$

and

$$g_m(n) = k_m^* f_{m-1}(n) + g_{m-1}(n-1) \qquad (2.5.34)$$

These two equations represent the mth stage of the all-pole lattice, shown in Figure 2.16(a), where $f_m(n)$ and $g_{m-1}(n)$ are now the inputs to the mth stage and $f_{m-1}(n)$ and $g_m(n)$ are

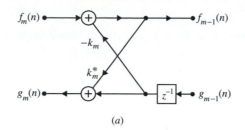

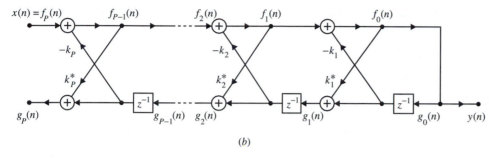

FIGURE 2.16
All-pole lattice structure.

the outputs. The transfer function from the input to the output is the same as that from $f_P(n)$ to $f_0(n)$. This transfer function is the inverse of the transfer function from $f_0(n)$ to $f_P(n)$. From (2.5.8), we conclude that the transfer function from $x(n)$ to $y(n)$ in Figure 2.16 is equal to

$$H(z) = \frac{Y(z)}{X(z)} = \frac{F_0(z)}{F_P(z)} = \frac{1}{A_P(z)} \tag{2.5.35}$$

where $A_P(z) = A(z)$ in (2.5.29). To multiply (2.5.35) by the gain G, we simply multiply either $x(n)$ or $y(n)$ by G in Figure 2.16(b).

Stability of all-pole systems. A causal LTI system is stable if all its poles are inside the unit circle. For all-pole systems described by the denominator polynomial $A_P(z)$, this implies that all its p roots are inside the unit circle, or alternatively, stability implies that $A_P(z)$ is a minimum-phase polynomial. Numerical implementation of polynomial root-finding operation is time-consuming. However, the following theorem shows how the lattice coefficients $\{k_m\}_{m=1}^P$ can be used for stability purposes.

THEOREM 2.3. The polynomial

$$A_P(z) = 1 + a_1^{(P)} z^{-1} + \cdots + a_P^{(P)} z^{-P} \tag{2.5.36}$$

is minimum-phase, that is, has all its zeros inside the unit circle if and only if

$$|k_m| < 1 \qquad 1 \leq m \leq P \tag{2.5.37}$$

Proof. See Appendix E.

Therefore, if the lattice parameters k_m in Figure 2.16 are less than unity in magnitude, then the all-pole filter $H(z)$ in (2.5.35) is minimum-phase and stable since $A(z)$ is guaranteed to have all its zeros inside the unit circle.

Since the AP lattice coefficients are derived from the same procedure used for the AZ lattice filter, we can use the `k = df2latcf(a)` function in MATLAB. Care must be taken to ignore the k_0 coefficient in the k array. Similarly, the `a = latcf2df(k)` function can be used to convert the lattice k_m coefficients to the direct-form coefficients a_k provided that $k_0 = 1$ is used as the first element of the k array.

All-pass lattice

The transfer function from $f_P(n)$ to $g_P(n)$ in Figure 2.16(b) can be written as

$$\frac{G_P(z)}{F_P(z)} = \frac{G_P(z)}{G_0(z)}\frac{F_0(z)}{F_P(z)} \tag{2.5.38}$$

where we used the fact that $F_0(z) = G_0(z)$. From (2.5.8) and (2.5.19), we conclude that

$$\frac{G_P(z)}{F_P(z)} = \frac{B_P(z)}{A_P(z)} = \frac{z^{-P}A^*(1/z^*)}{A(z)} = \frac{a_P^* + a_{P-1}^* z^{-1} + \cdots + z^{-P}}{1 + a_1 z^{-1} + \cdots + a_P z^{-P}} \tag{2.5.39}$$

which is the transfer function of an all-pass filter, since its magnitude on the unit circle is unity at all frequencies.

2.6 SUMMARY

In this chapter we have reviewed the fundamental concepts of discrete-time signal processing in both the time and frequency domains. We introduced usual definitions and descriptions of signals, and we provided the analytical tools for linear system operations. Significant attention was also given to those topics that will be used extensively in the rest of the book. These topics include minimum-phase systems, inverse systems, and spectral factorization. Finally, filters, which will be used in the chapter on adaptive filters, were discussed in greater detail. It is important to grasp the material discussed in this chapter since it is fundamental to understanding concepts presented in the remaining chapters. Therefore, the reader should also consult any one of the widely used references on this subject (Proakis and Manolakis 1996; Oppenheim and Schafer 1989).

PROBLEMS

2.1 A continuous-time signal $x_c(t)$ is sampled by an A/D converter to obtain the sequence $x(n)$. It is processed by a digital filter $h(n) = 0.8^n u(n)$ to obtain the sequence $y(n)$, which is further reconstructed using an ideal D/A converter to obtain the continuous-time output $y_c(t)$. The sampling frequency of A/D and D/A converters is 100 sampling intervals per second.

(a) If $x_c(t) = 2\cos(40\pi t + \pi/3)$, what is the digital frequency ω_0 in $x(n)$?
(b) If $x_c(t)$ is as given above, determine the steady-state response $y_{c,ss}(t)$.
(c) Determine two different $x_c(t)$ signals that would give the same steady-state response $y_{c,ss}(t)$ above.

2.2 Let $x(n)$ be a sinusoidal sequence of frequency ω_0 and of finite length N, that is,

$$x(n) = \begin{cases} A\cos\omega_0 n & 0 \le n \le N-1 \\ 0 & \text{otherwise} \end{cases}$$

Thus $x(n)$ can be thought of as an infinite-length sinusoidal sequence multiplied by a rectangular window of length N.

(a) If the DTFT of $x(n)$ is expressed in terms of the real and imaginary parts as

$$X(e^{j\omega}) \triangleq X_R(\omega) + jX_I(\omega)$$

determine analytical expressions for $X_R(\omega)$ and $X_I(\omega)$. Express $\cos\omega$ in terms of complex exponentials and use the modulation property of the DTFT to arrive at the result.
(b) Choose $N = 32$ and $\omega_0 = \pi/4$, and plot $X_R(\omega)$ and $X_I(\omega)$ for $\omega \in [-\pi, \pi]$.
(c) Compute the 32-point DFT of $x(n)$, and plot its real and imaginary samples. Superimpose the above DTFT plots on the DFT plots. Comment on the results.
(d) Repeat the above two parts for $N = 32$ and $\omega_0 = 1.1\pi/4$. Why are the plots so markedly different?

2.3 Let $x(n) = \cos(\pi n/4)$, and assume that we have only 16 samples available for processing.

(a) Compute the 16-point DFT of these 16 samples, and plot their magnitudes. (Make sure that this is a stem plot.)

(b) Now compute the 32-point DFT of the sequence formed by appending the above 16 samples with 16 zero-valued samples. This is called *zero padding*. Now plot the magnitudes of the DFT samples.

(c) Repeat part (b) for the 64-point sequence by padding 48 zero-valued samples.

(d) Explain the effect and hence the purpose of the zero padding operation on the DTFT spectrum.

2.4 Let $x(n) = \{1, 2, 3, 4, 3, 2, 1\}$ and $h(n) = \{-1, 0, 1\}$.

(a) Determine the convolution $y(n) = x(n) * h(n)$ using the matrix-vector multiplication approach given in (2.3.5).

(b) Develop a MATLAB function to implement the convolution using the Toeplitz matrix in (2.3.4). The form of the function should be y = convtoep(x,h).

(c) Verify your function, using the sequences given in part (a) above.

2.5 Let $x(n) = (0.9)^n u(n)$.

(a) Determine $x(n) * x(n)$ analytically, and plot its first 101 samples.

(b) Truncate $x(n)$ to the first 51 samples. Compute and plot the convolution $x(n) * x(n)$, using the conv function.

(c) Assume that $x(n)$ is the impulse response of an LTI system. Determine the filter function coefficient vectors a and b. Using the filter function, compute and plot the first 101 samples of the convolution $x(n) * x(n)$.

(d) Comment on your plots. Which MATLAB approach is best suited for infinite-length sequences and why?

2.6 Let $H_{ap}(z)$ be a causal and stable all-pass system excited by a causal input $x(n)$ producing the response $y(n)$. Show that for any time n_0,

$$\sum_{n=0}^{n_0} |y(n)|^2 \le \sum_{n=0}^{n_0} |x(n)|^2 \qquad (P.1)$$

2.7 This problem examines monotone phase-response property of a causal and stable PZ all-pass system.

(a) Consider the pole-zero diagram of a real first-order all-pass system

$$H(z) = \frac{p - z^{-1}}{1 - pz^{-1}}$$

Show that its phase response decreases monotonically from π (at $\omega = 0$) to $-\pi$ (at $\omega = 2\pi$).

(b) Consider the pole-zero diagram of a real second-order all-pass system

$$H(z) = \left[\frac{(r\angle\theta) - z^{-1}}{1 - (r\angle\theta)^* z^{-1}} \right] \left[\frac{(r\angle\theta)^* - z^{-1}}{1 - (r\angle\theta)z^{-1}} \right]$$

Show that its phase response decreases monotonically as ω increases from 0 to π.

(c) Generalize the results of parts (a) and (b) to show that the phase response of a causal and stable PZ all-pass system decreases monotonically from $\angle[H(e^{j0})]$ to $\angle[H(e^{j0})] - 2\pi p$ as ω increases from 0 to π.

2.8 This problem explores the minimum group delay property of the minimum-phase systems.

(a) Consider the following causal and stable minimum-, maximum-, and mixed-phase systems

$$H_{min}(z) = (1 - 0.25z^{-1})(1 + 0.5z^{-1})$$
$$H_{max}(z) = (0.25 - z^{-1})(0.5 + z^{-1})$$
$$H_{mix}(z) = (1 - 0.25z^{-1})(0.5 + z^{-1})$$

which have the same magnitude response. Compute and plot group delay responses. Observe that the minimum-phase system has the minimum group delay.

(b) Using (2.4.18) and Problem 2.7, prove the minimum group delay property of the minimum-phase systems.

2.9 Given the following spectral density functions, express them in minimum- and maximum-phase components.

(a) $R_y(z) = \dfrac{1 - 2.5z^{-1} + z^{-2}}{1 - 2.05z^{-1} + z^{-2}}$

(b) $R_y(z) = \dfrac{3z^2 - 10 + 3z^{-2}}{3z^2 + 10 + 3z^{-2}}$

2.10 Consider the all-pass system function $H_{\text{ap}}(z)$ given by

$$H_{\text{ap}}(z) = \frac{1 - \alpha z^{-1}}{z^{-1} - \alpha^*} \qquad |\alpha| < 1 \qquad\qquad (\text{P.2})$$

(a) Determine $|H_{\text{ap}}(z)|^2$ as a ratio of polynomials in z.
(b) Show that

$$D^2_{|H|}(z) - A^2_{|H|}(z) = (|z|^2 - 1)(1 - |\alpha|^2)$$

where

$$|H_{\text{ap}}(z)|^2 = \frac{D^2_{|H|}(z)}{A^2_{|H|}(z)}$$

(c) Using $|\alpha| < 1$ and the above result, show that

$$|H_{\text{ap}}(z)| \begin{cases} < 1 & \text{if} \quad |z| < 1 \\ = 1 & \text{if} \quad |z| = 1 \\ > 1 & \text{if} \quad |z| > 1 \end{cases}$$

2.11 Consider the system function of a stable system of the form

$$H(z) = \frac{a + bz^{-1} + cz^{-2}}{c + bz^{-1} + az^{-2}}$$

(a) Show that the magnitude of the frequency response function $|H(e^{j\omega})|$ is equal to 1 for all frequencies, that is, it is an all-pass system.
(b) Let

$$H(z) = \frac{3 - 2z^{-1} + z^{-2}}{1 - 2z^{-1} + 3z^{-2}}$$

Determine both the magnitude and the phase of the frequency response $H(e^{j\omega})$, and plot these functions over $[0, \pi]$.

2.12 Consider the system function of a third-order FIR system

$$H(z) = 12 + 28z^{-1} - 29z^{-2} - 60z^{-3}$$

(a) Determine the system functions of all other FIR systems whose magnitude responses are identical to that of $H(z)$.
(b) Which of these systems is a minimum-phase system and which one is a maximum-phase system?
(c) Let $h_k(n)$ denote the impulse response of the kth FIR system determined in part (a) and define the energy delay of the kth system by

$$\mathcal{E}_k(n) \triangleq \sum_{m=n}^{\infty} |h_k(m)|^2 \qquad 0 \le n \le 3$$

for all values of k. Show that

$$\mathcal{E}_{\min}(n) \le \mathcal{E}_k(n) \le \mathcal{E}_{\max}(n) \qquad 0 \le n \le 3$$

and

$$\mathcal{E}_{\min}(\infty) = \mathcal{E}_k(\infty) = \mathcal{E}_{\max}(\infty) = 0$$

where $\mathcal{E}_{\min}(n)$ and $\mathcal{E}_{\max}(n)$ are energy delays of the minimum-phase and maximum-phase systems, respectively.

2.13 Consider the system function

$$H(z) = \frac{1 + z^{-1} - 6z^{-2}}{1 + \frac{1}{4}z^{-1} - \frac{1}{8}z^{-2}}$$

(a) Show that the system $H(z)$ is not minimum-phase.

(b) Construct a minimum-phase system $H_{\min}(z)$ such that $|H_{\min}(e^{j\omega})| = |H(e^{j\omega})|$.

(c) Is $H(z)$ a maximum-phase system? If yes, explain why. If not, then construct a minimum-phase system $H_{\max}(z)$ such that $|H_{\max}(e^{j\omega})| = |H(e^{j\omega})|$.

2.14 Implement the following system as a parallel connection of two all-pass systems:

$$H(z) = \frac{3 + 9z^{-1} + 9z^{-2} + 3z^{-3}}{12 + 10z^{-1} + 2z^{-2}}$$

2.15 Determine the impulse response of an all-pole system with lattice parameters

$$k_1 = 0.2 \quad k_2 = 0.3 \quad k_3 = 0.5 \quad k_4 = 0.7$$

Draw the direct- and lattice form structures of the above system.

CHAPTER 3

Random Variables, Vectors, and Sequences

So far we have dealt with deterministic signals, that is, signals whose amplitude is uniquely specified by a mathematical formula or rule. However, there are many important examples of signals whose precise description (i.e., as deterministic signals) is extremely difficult, if not impossible. As mentioned in Section 2.1, such signals are called *random signals*. Although random signals are evolving in time in an unpredictable manner, their average properties can be often assumed to be deterministic; that is, they can be specified by explicit mathematical formulas. This is the key for the modeling of a random signal as a stochastic process.

Our aim in the subsequent discussions is to present some basic results from the theory of random variables, random vectors, and discrete-time stochastic processes that will be useful in the chapters that follow. We assume that most readers have some basic knowledge of these topics, and so parts of this chapter may be treated as a review exercise. However, some specific topics are developed in greater depth with a viewpoint that will serve as a foundation for the rest of the book. A more complete treatment can be found in Papoulis (1991), Helstrom (1992), and Stark and Woods (1994).

3.1 RANDOM VARIABLES

The concept of random variables begins with the definition of probability. Consider an experiment with a finite or infinite number of unpredictable outcomes from a *universal set*, denoted by $\mathcal{S} = \{\zeta_1, \zeta_2, \ldots\}$. A collection of subsets of $\mathcal{S}$ containing $\mathcal{S}$ itself and that is closed under countable set operations is called a σ *field* and denoted by $\mathcal{F}$. Elements of $\mathcal{F}$ are called *events*. The unpredictability of these events is measured by a nonnegative set function $\Pr\{\zeta_k\}$, $k = 1, 2, \ldots$, called the *probability* of event ζ_k. This set function satisfies three well-known and intuitive axioms (Papoulis 1991) such that the probability of any event produced by set-theoretic operations on the events of $\mathcal{S}$ can be uniquely determined. Thus, any situation of random nature, abstract or otherwise, can be studied using the axiomatic definition of probability by defining an appropriate probability space $(\mathcal{S}, \mathcal{F}, \Pr)$.

In practice it is often difficult, if not impossible, to work with this probability space for two reasons. First, the basic space contains abstract events and outcomes that are difficult to manipulate. In engineering applications, we want random outcomes that can be measured and manipulated in a meaningful way by using numerical operations. Second, the probability function $\Pr\{\cdot\}$ is a set function that again is difficult, if not impossible, to manipulate by using calculus. These two problems are addressed through the concept of the random variable.

DEFINITION 3.1 (RANDOM VARIABLE). A random variable $x(\zeta)$ is a mapping that assigns a real number x to every outcome ζ from an abstract probability space. This mapping should satisfy the following two conditions: (1) the interval $\{x(\zeta) \le x\}$ is an event in the abstract probability space for every x; (2) $\Pr\{x(\zeta) = \infty\} = 0$ and $\Pr\{x(\zeta) = -\infty\} = 0$.

A complex-valued random variable is defined by $x(\zeta) = x_R(\zeta) + jx_I(\zeta)$ where $x_R(\zeta)$ and $x_I(\zeta)$ are real-valued random variables. We will discuss complex-valued random variables in Section 3.2. Strictly speaking, a random variable is neither random nor a variable but is a function or a mapping. As shown in Figure 3.1, the domain of a random variable is the universal set $\mathcal{S}$, and its range is the real line $\mathbb{R}$. Thus, events and outcomes are now numbers, which can be added, subtracted, or manipulated otherwise.

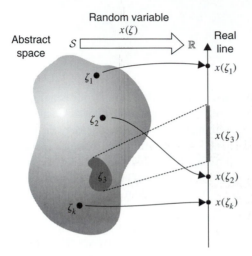

FIGURE 3.1

Graphical illustration of random variable mapping.

An important comment on notation. We will use $x(\zeta)$, $y(\zeta)$, ..., to denote random variables and the corresponding lowercase alphabet without parentheses to denote their values; for example, $x(\zeta) = x$ means that the random variable $x(\zeta)$ takes value equal to x. We believe that this notation will not cause any confusion because the meaning of the lowercase variable will be clear from the context.[†] A specific value of the random variable realization will be denoted by $x(\zeta^0) = x^0$ (corresponding to a particular event ζ^0 in the original space).

A random variable is called *discrete-valued* if x takes a discrete set of values $\{x_k\}$; otherwise, it is termed a *continuous-valued* random variable. A *mixed* random variable takes both discrete and continuous values.

3.1.1 Distribution and Density Functions

The probability set function $\Pr\{x(\zeta) \le x\}$ is a function of the set $\{x(\zeta) \le x\}$, but it is also a number that varies with x. Hence it is also a function of a point x on the real line $\mathbb{R}$. This point function is the well-known *cumulative distribution function* (cdf) $F_x(x)$ of a random variable $x(\zeta)$ and is defined by

$$F_x(x) \triangleq \Pr\{x(\zeta) < x\} \tag{3.1.1}$$

The second important probability function is the *probability density function* (pdf) $f_x(x)$,

---

[†]Traditionally, the uppercase alphabet is used to denote random variables. We have reserved the use of uppercase alphabet for transform-domain quantities.

which is defined as a formal derivative

$$f_x(x) \triangleq \frac{dF_x(x)}{dx} \tag{3.1.2}$$

Note that the pdf $f_x(x)$ is not the probability, but must be multiplied by a certain interval Δx to obtain a probability, that is,

$$f_x(x)\Delta x \approx \Delta F_x(x) \triangleq F_x(x + \Delta x) - F_x(x) = \Pr\{x < x(\zeta) \le x + \Delta x\} \tag{3.1.3}$$

Integrating both sides of (3.1.2), we obtain

$$F_x(x) = \int_{-\infty}^{x} f_x(v)\,dv \tag{3.1.4}$$

For discrete-valued random variables, we use the *probability mass function* (pmf) p_k, defined as the probability that random variable $x(\zeta)$ takes a value equal to x_k, or

$$p_k \triangleq \Pr\{x(\zeta) = x_k\} \tag{3.1.5}$$

These probability functions satisfy several important properties (Papoulis 1991), such as

$$0 \le F_x(x) \le 1 \qquad F_x(-\infty) = 0 \qquad F_x(\infty) = 1 \tag{3.1.6}$$

$$f_x(x) \ge 0 \qquad \int_{-\infty}^{\infty} f_x(x)\,dx = 1 \tag{3.1.7}$$

Using these functions and their properties, we can compute the probabilities of any event (or interval) on $\mathbb{R}$. For example,

$$\Pr\{x_1 < x(\zeta) \le x_2\} = F_x(x_2) - F_x(x_1) = \int_{x_1}^{x_2} f_x(x)\,dx \tag{3.1.8}$$

3.1.2 Statistical Averages

To completely characterize a random variable, we have to know its probability density function. In practice, it is desirable to summarize some of the key aspects of a density function by using a few numbers rather than to specify the entire density function. These numbers, which are called *statistical averages* or *moments*, are evaluated by using the mathematical expectation operation. Although density functions are needed to theoretically compute moments, in practice, moments are easily estimated without the explicit knowledge of density functions.

Mathematical expectation

This is one of the most important operations in the theory of random variables. It is generally used to describe various statistical averages, and it is also needed in estimation theory. The *expected* or *mean value* of a random variable $x(\zeta)$ is given by

$$E\{x(\zeta)\} \triangleq \mu_x = \begin{cases} \displaystyle\sum_k x_k p_k & x(\zeta) \text{ discrete} \\[2ex] \displaystyle\int_{-\infty}^{\infty} x f_x(x)\,dx & x(\zeta) \text{ continuous} \end{cases} \tag{3.1.9}$$

Although, strictly speaking, to compute $E\{x(\zeta)\}$ we need the definitions for both the discrete and continuous random variables, we will follow the engineering practice of using the expression for the continuous random variable (which can also describe a discrete random variable if we allow impulse functions in its pdf). The expectation operation computes a statistical average by using the density $f_x(x)$ as a weighting function. Hence, the mean μ_x can be regarded as the "location" (or the "center of gravity") of the density $f_x(x)$, as shown in Figure 3.2(a). If $f_x(x)$ is symmetric about $x = a$, then $\mu_x = a$ and, in particular, if $f_x(x)$

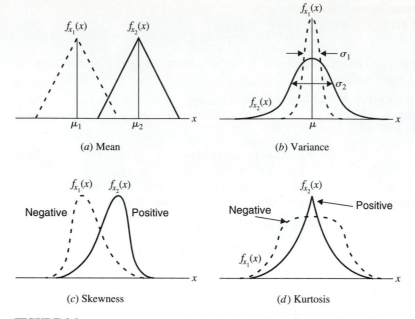

(a) Mean

(b) Variance

(c) Skewness

(d) Kurtosis

FIGURE 3.2
Illustration of mean, standard deviation, skewness, and kurtosis.

is an even function, then $\mu_x = 0$. One important property of expectation is that it is a linear operation, that is,

$$E\{\alpha x(\zeta) + \beta\} = \alpha \mu_x + \beta \tag{3.1.10}$$

Let $y(\zeta) = g[x(\zeta)]$ be a random variable obtained by transforming $x(\zeta)$ through a suitable function.† Then the expectation of $y(\zeta)$ is given by

$$E\{y(\zeta)\} \triangleq E\{g[x(\zeta)]\} = \int_{-\infty}^{\infty} g(x) f_x(x)\, dx \tag{3.1.11}$$

Moments

Using the expectation operations (3.1.9) and (3.1.11), we can define various moments of the random variable $x(\zeta)$ that describe certain useful aspects of the density function. Let $g[x(\zeta)] = x^m(\zeta)$. Then

$$r_x^{(m)} \triangleq E\{x^m(\zeta)\} = \int_{-\infty}^{\infty} x^m f_x(x)\, dx \tag{3.1.12}$$

is called the mth-order *moment* of $x(\zeta)$. In particular, $r_x^{(0)} = 1$, and the first-order moment $r_x^{(1)} = \mu_x$. The second-order moment $r_x^{(2)} = E\{x^2(\zeta)\}$ is called the *mean-squared value*, and it plays an important role in estimation theory. Note that

$$E\{x^2(\zeta)\} \neq E^2\{x(\zeta)\} \tag{3.1.13}$$

Corresponding to these moments we also have central moments. Let $g[x(\zeta)] = [x(\zeta) - \mu_x]^m$, then

$$\gamma_x^{(m)} \triangleq E\{[x(\zeta) - \mu_x]^m\} = \int_{-\infty}^{\infty} (x - \mu_x)^m f_x(x)\, dx \tag{3.1.14}$$

is called the mth-order *central moment* of $x(\zeta)$. In particular, $\gamma_m^{(0)} = 1$ and $\gamma_x^{(1)} = 0$, which is obvious. Clearly, a random variable's moments and central moments are identical if its

---

†Such a function $g(\cdot)$ is called a *Baire function* (Papoulis 1991).

mean value is zero. The second central moment is of considerable importance and is called the *variance* of $x(\zeta)$, denoted by σ_x^2. Thus

$$\text{var}[x(\zeta)] \triangleq \sigma_x^2 \triangleq \gamma_x^{(2)} = E\{[x(\zeta) - \mu_x]^2\} \tag{3.1.15}$$

The quantity $\sigma_x = \sqrt{\gamma_x^{(2)}}$ is called the *standard deviation* of $x(\zeta)$ and is a measure of the spread (or dispersion) of the observed values of $x(\zeta)$ around its mean μ_x [see Figure 3.2(b)]. The relation between a random variable's moments and central moments is given by (see Problem 3.3)

$$\gamma_x^{(m)} = \sum_{k=0}^{m} \binom{m}{k} (-1)^k \mu_x^k r_x^{(n-k)} \tag{3.1.16}$$

In particular, and also from (3.1.15), we have

$$\sigma_x^2 = r_x^{(2)} - \mu_x^2 = E\{x^2(\zeta)\} - E^2\{x(\zeta)\} \tag{3.1.17}$$

The quantity *skewness* is related to the third-order central moment and characterizes the degree of asymmetry of a distribution around its mean, as shown in Figure 3.2(c). It is defined as a *normalized* third-order central moment, that is,

$$\text{Skew} \triangleq \tilde{\kappa}_x^{(3)} \triangleq E\left\{ \left[\frac{x(\zeta) - \mu_x}{\sigma_x} \right]^3 \right\} = \frac{1}{\sigma_x^3} \gamma_x^{(3)} \tag{3.1.18}$$

and is a *dimensionless* quantity. It is a pure number that attempts to describe leaning of the shape of the distribution. The skewness is zero if the density function is symmetric about its mean value, is positive if the shape leans towards the right, or is negative if it leans towards the left.

The quantity related to the fourth-order central moment is called *kurtosis*, which is also a dimensionless quantity. It measures the relative flatness or peakedness of a distribution about its mean as shown in Figure 3.2(d). This relative measure is with respect to a normal distribution, which will be introduced in the next section. The kurtosis is defined as

$$\text{Kurtosis} \triangleq \tilde{\kappa}_x^{(4)} \triangleq E\left\{ \left[\frac{x(\zeta) - \mu_x}{\sigma_x} \right]^4 \right\} - 3 = \frac{1}{\sigma_x^4} \gamma_x^{(4)} - 3 \tag{3.1.19}$$

where the term -3 makes the kurtosis $\tilde{\kappa}_x^{(4)} = 0$ for the normal distribution [see (3.1.40) for explanation].

Chebyshev's inequality. A useful result in the interpretation and use of the mean μ and the variance σ^2 of a random variable is given by Chebyshev's inequality. Given a random variable $x(\zeta)$ with its mean μ_x and variance σ_x^2, we have the inequality

$$\text{Pr}\{|x(\zeta) - \mu_x| \geq k\sigma_x\} \leq \frac{1}{k^2} \qquad k > 0 \tag{3.1.20}$$

The interpretation of the above inequality is that regardless of the shape of $f_x(x)$, the random variable $x(\zeta)$ deviates from its mean by k times its standard deviation with probability less than or equal to $1/k^2$.

Characteristic functions

The Fourier and Laplace transforms find many uses in probability theory through the concepts of characteristic and moment generating functions. The *characteristic function* of a random variable $x(\zeta)$ is defined by the integral

$$\Phi_x(\xi) \triangleq E\{e^{j\xi x(\zeta)}\} = \int_{-\infty}^{\infty} f_x(x) e^{j\xi x} \, dx \tag{3.1.21}$$

which can be interpreted as the Fourier transform of $f_x(x)$ with sign reversal in the complex exponential. To avoid confusion with the cdf, we do not use $F_x(\xi)$ to denote this Fourier transform. Furthermore, the variable ξ in $\Phi_x(\xi)$ is not and should not be interpreted as frequency. When $j\xi$ in (3.1.21) is replaced by a complex variable s, we obtain the *moment generating function* defined by

$$\bar{\Phi}_x(s) \triangleq E\{e^{sx(\zeta)}\} = \int_{-\infty}^{\infty} f_x(x) e^{sx}\, dx \tag{3.1.22}$$

which again can be interpreted as the Laplace transform of $f_x(x)$ with sign reversal. Expanding e^{sx} in (3.1.22) in a Taylor series at $s = 0$, we obtain

$$\bar{\Phi}_x(s) = E\{e^{sx(\zeta)}\} = E\left\{1 + sx(\zeta) + \frac{[sx(\zeta)]^2}{2!} + \cdots + \frac{[sx(\zeta)]^m}{m!} + \cdots\right\} \tag{3.1.23}$$

$$= 1 + s\mu_x + \frac{s^2}{2!}r_x^{(2)} + \cdots + \frac{s^m}{m!}r_x^{(m)} + \cdots$$

provided every moment $r_x^{(m)}$ exists. Thus from (3.1.23) we infer that if all moments of $x(\zeta)$ are known (and exist), then we can assemble $\bar{\Phi}_x(s)$ and upon inverse Laplace transformation, we can determine the density function $f_x(x)$. If we differentiate $\bar{\Phi}_x(s)$ with respect to s, we obtain

$$r_x^{(m)} = \frac{d^m[\bar{\Phi}_x(s)]}{ds^m}\bigg|_{s=0} = (-j)^m \frac{d^m[\Phi_x(\xi)]}{d\xi^m}\bigg|_{\xi=0} \qquad m = 1, 2, \ldots \tag{3.1.24}$$

which provides the mth-order moment of the random variable $x(\zeta)$.

The functions $\Phi_x(\xi)$ and $\bar{\Phi}_x(s)$ possess all the properties associated with the Fourier and Laplace transforms, respectively. Thus, since $f_x(x)$ is always a real-valued function, $\Phi_x(\xi)$ is conjugate symmetric; and if $f_x(x)$ is also an even function, then $\Phi_x(\xi)$ is a real-valued even function. In addition, they possess several properties due to the basic nature of the pdf. Therefore, the characteristic function $\Phi_x(\xi)$ always exists[†] since

$$\int |f_x(x)|\, dx = \int f_x(x)\, dx = 1$$

and $\Phi_x(\xi)$ is maximum at the origin, that is,

$$|\Phi_x(\xi)| \le \Phi_x(0) = 1 \tag{3.1.25}$$

since $f_x(x) \ge 0$.

Cumulants

These statistical descriptors are similar to the moments, but provide better information for higher-order moment analysis, which we will consider in detail in Chapter 12. The cumulants are derived by considering the moment generating function's natural logarithm. This logarithm is commonly referred to as the *cumulant generating function* and is given by

$$\bar{\Psi}_x(s) \triangleq \ln \bar{\Phi}_x(s) = \ln E\{e^{sx(\zeta)}\} \tag{3.1.26}$$

When s is replaced by $j\xi$ in (3.1.26), the resulting function is known as the *second characteristic function* and is denoted by $\Psi_x(\xi)$.

The *cumulants* $\kappa_x^{(m)}$ of a random variable $x(\zeta)$ are defined as the derivatives of the cumulant generating function, that is,

$$\kappa_x^{(m)} \triangleq \frac{d^m[\bar{\Psi}_x(s)]}{ds^m}\bigg|_{s=0} = (-j)^m \frac{d^m[\Psi_x(\xi)]}{d\xi^m}\bigg|_{\xi=0} \qquad m = 1, 2, \ldots \tag{3.1.27}$$

[†] We will generally choose the characteristic function over the moment generating function.

Clearly, $\kappa_x^{(0)} = 0$. It can be shown that (see Problem 3.4) for a zero-mean random variable, the first five cumulants as functions of the central moments are given by

$$\kappa_x^{(1)} = r_1^{(x)} = \mu_x = 0 \tag{3.1.28}$$
$$\kappa_x^{(2)} = \gamma_x^{(2)} = \sigma_x^2 \tag{3.1.29}$$
$$\kappa_x^{(3)} = \gamma_x^{(3)} \tag{3.1.30}$$
$$\kappa_x^{(4)} = \gamma_x^{(4)} - 3\sigma_x^4 \tag{3.1.31}$$
$$\kappa_x^{(5)} = \gamma_x^{(5)} - 10\gamma_x^{(3)}\sigma_x^2 \tag{3.1.32}$$

which show that the first two cumulants are identical to the first two central moments. Clearly due to the logarithmic function in (3.1.26), cumulants are useful for dealing with products of characteristic functions (see Section 3.2.4).

3.1.3 Some Useful Random Variables

Random variable models are needed to describe (or approximate) complex physical phenomena using simple parameters. For example, the random phase of a sinusoidal carrier can be described by a uniformly distributed random variable so that we can study its statistical properties. This approximation allows us to investigate random signals in a sound mathematical way. We will describe three continuous random variable models although there are several other known continuous as well as discrete models available in the literature.

Uniformly distributed random variable. This is an appropriate model in situations in which random outcomes are "equally likely." Here $x(\zeta)$ assumes values on $\mathbb{R}$ according to the pdf

$$f_x(x) = \begin{cases} \dfrac{1}{b-a} & a \leq x \leq b \\ 0 & \text{elsewhere} \end{cases} \tag{3.1.33}$$

where $a < b$ are specified parameters. This pdf is shown in Figure 3.3. The corresponding

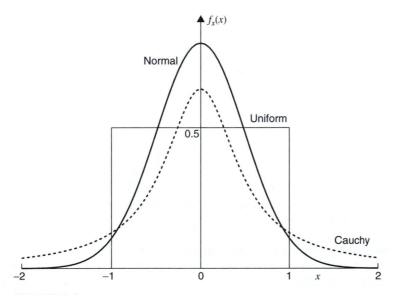

FIGURE 3.3
Probability density functions of useful random variables.

cdf is given by

$$F_x(x) = \int_{-\infty}^{x} f_x(v)\,dv = \begin{cases} 0 & x < a \\ \dfrac{x-a}{b-a} & a \le x \le b \\ 1 & x > a \end{cases} \qquad (3.1.34)$$

and the characteristic function is given by

$$\Phi_x(\xi) = \frac{e^{j\xi b} - e^{j\xi a}}{j\xi(b-a)} \qquad (3.1.35)$$

The mean and the variance of this random variable are given by, respectively,

$$\mu_x = \frac{a+b}{2} \quad \text{and} \quad \sigma_x^2 = \frac{(b-a)^2}{12} \qquad (3.1.36)$$

Normal random variable. This is the most useful and convenient model in many applications, as we shall see later. It is also known as a *Gaussian* random variable, and we will use both terms interchangeably. The pdf of a normally distributed random variable $x(\zeta)$ with mean μ_x and standard deviation σ_x is given by

$$f_x(x) = \frac{1}{\sqrt{2\pi\sigma_x^2}} \exp\left[-\frac{1}{2} \left(\frac{x - \mu_x}{\sigma_x} \right)^2 \right] \qquad (3.1.37)$$

where $-\infty < \mu < \infty$ and $\sigma \ge 0$ (see Figure 3.3). The characteristic function of the normal random variable is given by

$$\Phi_x(\xi) = \exp(j\mu_x\xi - \tfrac{1}{2}\sigma_x^2\xi^2) \qquad (3.1.38)$$

Clearly, the pdf of a normal random variable is completely described by its mean μ_x and standard deviation σ_x and is denoted by $\mathcal{N}(\mu_x, \sigma_x^2)$. We note that *all higher-order moments of a normal random variable can be determined in terms of the first two moments*, that is,

$$\gamma_x^{(m)} = E\{[x(\zeta) - \mu_x]^m\} = \begin{cases} 1 \cdot 3 \cdot 5 \cdots (m-1)\sigma_x^m & \text{if } m \text{ even} \\ 0 & \text{if } m \text{ odd} \end{cases} \qquad (3.1.39)$$

In particular, we obtain the fourth moment as

$$\gamma_x^{(4)} = 3\sigma_x^4 \qquad (3.1.40)$$

or from (3.1.19), kurtosis = 0, which explains the term -3 in (3.1.19).

From (3.1.37), we observe that the Gaussian random variable is completely determined by its first two moments (mean μ_x and variance σ_x^2), which means that the higher moments do not provide any additional information about the Gaussian density function. In fact, all higher-order moments can be obtained in terms of the first two moments [see Equation (3.1.39)]. Thus for a non-Gaussian random variable, we would like to know how different that random variable is from a Gaussian random variable (this is also known as a departure from the Gaussian-ness). This measurement of the deviation from being Gaussian is given by the cumulants that were defined in (3.1.27). Roughly speaking, the cumulants are like central moments (which measure deviations from the mean) of non-Gaussian random variables for Gaussian departure. Also from (3.1.30) and (3.1.31), we see that all higher-order (that is, $m > 2$) cumulants of a Gaussian random variable are zero. This fact is used in the analysis and estimation of non-Gaussian random variables (and later for non-Gaussian random processes).

Cauchy random variable. This is an appropriate model in which a random variable takes large values with significant probability (heavy-tailed distribution). The Cauchy pdf with parameters μ and β is given by

$$f_x(x) = \frac{\beta}{\pi} \frac{1}{(x-\mu)^2 + \beta^2} \qquad (3.1.41)$$

and is shown in Figure 3.3. The corresponding cdf is given by

$$F_x(x) = 0.5 + \frac{1}{\pi} \arctan \frac{x - \mu}{\beta} \tag{3.1.42}$$

and the characteristic function is given by

$$\Phi_x(\xi) = \exp(j\mu\xi - \beta|\xi|) \tag{3.1.43}$$

The Cauchy random variable has mean $\mu_x = \mu$. However, its variance does not exist because $E\{x^2\}$ fails to exist in any sense, and hence the moment generating function does not exist. It has the property that the sum of M independent Cauchy random variables is also Cauchy (see Example 3.2.3). Thus a Cauchy random variable is an example of an infinite-variance random variable.

Random number generators. Random numbers, by definition, are truly unpredictable, and hence it is not possible to generate them by using a well-defined algorithm on a computer. However, in many simulation studies, we need to use sequences of numbers that appear to be random and that possess required properties, for example, Gaussian random numbers in a Monte Carlo analysis. These numbers are called *pseudo random numbers*, and many excellent algorithms are available to generate them on a computer (Park and Miller 1988). In MATLAB, the function `rand` generates numbers that are uniformly distributed over $(0, 1)$ while the function `randn` generates $\mathcal{N}(0, 1)$ pseudo random numbers.

3.2 RANDOM VECTORS

In many applications, a group of signal observations can be modeled as a collection of random variables that can be grouped to form a *random vector*. This is an extension of the concept of random variable and generalizes many scalar quantities to vectors and matrices. One example of a random vector is the case of a complex-valued random variable $x(\zeta) = x_R(\zeta) + jx_I(\zeta)$, which can be considered as a group of $x_R(\zeta)$ and $x_I(\zeta)$. In this section, we provide a review of the basic properties of random vectors and related results from linear algebra. We first begin with real-valued random vectors and then extend their concepts to complex-valued random vectors.

3.2.1 Definitions and Second-Order Moments

A real-valued vector containing M random variables

$$\mathbf{x}(\zeta) = [x_1(\zeta), x_2(\zeta), \dots, x_M(\zeta)]^T \tag{3.2.1}$$

is called a *random M vector* or a random vector when dimensionality is unimportant. As usual, superscript T denotes the transpose of the vector. We can think of a real-valued random vector as a mapping from an abstract probability space to a vector-valued, real space $\mathbb{R}^M$. Thus the range of this mapping is an M-dimensional space.

Distribution and density functions

A random vector is completely characterized by its *joint* cumulative distribution function, which is defined by

$$F_{\mathbf{x}}(x_1, \dots, x_M) \triangleq \Pr\{x_1(\zeta) \le x_1, \dots, x_M(\zeta) \le x_M\} \tag{3.2.2}$$

and is often written as

$$F_{\mathbf{x}}(\mathbf{x}) = \Pr\{\mathbf{x}(\zeta) \le \mathbf{x}\} \tag{3.2.3}$$

for convenience. A random vector can be also characterized by its *joint* probability density function, which is defined by

$$f_{\mathbf{x}}(\mathbf{x}) = \lim_{\substack{\Delta x_1 \to 0 \\ \vdots \\ \Delta x_M \to 0}} \frac{\Pr\{x_1 < x_1(\zeta) \le x_1 + \Delta x_1, \ldots, x_M < x_M(\zeta) \le x_M + \Delta x_M\}}{\Delta x_1 \cdots \Delta x_M}$$

$$\triangleq \frac{\partial}{\partial x_1} \cdots \frac{\partial}{\partial x_M} F_{\mathbf{x}}(\mathbf{x})$$

$$(3.2.4)$$

The function

$$f_{x_j}(x_j) = \int \cdots \int_{(M-1)} f_{\mathbf{x}}(\mathbf{x}) \, dx_1 \cdots dx_{j-1} \, dx_{j+1} \cdots dx_M \qquad (3.2.5)$$

is known as a *marginal* density function and describes individual random variables. Thus the probability functions defined for a random variable in the previous section are more appropriately called marginal functions. The joint pdf $f_{\mathbf{x}}(\mathbf{x})$ must be multiplied by a certain M-dimensional region $\Delta\mathbf{x}$ to obtain a probability. From (3.2.4) we obtain

$$F_{\mathbf{x}}(\mathbf{x}) = \int_{-\infty}^{x_1} \cdots \int_{-\infty}^{x_M} f_{\mathbf{x}}(\boldsymbol{v}) \, dv_1 \cdots dv_M = \int_{-\infty}^{\mathbf{x}} f_{\mathbf{x}}(\boldsymbol{v}) \, d\boldsymbol{v} \qquad (3.2.6)$$

These joint probability functions also satisfy several important properties that are similar to (3.1.6) through (3.1.8) for random variables. In particular, note that both $f_{\mathbf{x}}(\mathbf{x})$ and $F_{\mathbf{x}}(\mathbf{x})$ are positive multidimensional functions.

The joint [and conditional probability (see Papoulis 1991)] functions can also be used to define the concept of independent random variables. Two random variables $x_1(\zeta)$ and $x_2(\zeta)$ are *independent* if the events $\{x_1(\zeta) \le x_1\}$ and $\{x_2(\zeta) \le x_2\}$ are jointly independent, that is, if

$$\Pr\{x_1(\zeta) \le x_1, x_2(\zeta) \le x_2\} = \Pr\{x_1(\zeta) \le x_1\} \Pr\{x_2(\zeta) \le x_2\}$$

which implies that

$$F_{x_1, x_2}(x_1, x_2) = F_{x_1}(x_1) F_{x_2}(x_2) \qquad \text{and} \qquad f_{x_1, x_2}(x_1, x_2) = f_{x_1}(x_1) f_{x_2}(x_2) \qquad (3.2.7)$$

Complex-valued random variables and vectors

As we shall see in later chapters, in applications such as channel equalization, array processing, etc., we encounter complex signal and noise models. To formulate these models, we need to describe complex random variables and vectors, and then extend our standard definitions and results to the complex case. A complex random variable is defined as[†] $x(\zeta) = x_R(\zeta) + jx_I(\zeta)$, where $x_R(\zeta)$ and $x_I(\zeta)$ are real-valued random variables. Thus we can think of $x(\zeta)$ as a mapping from an abstract probability space S to a complex space $\mathbb{C}$. Alternatively, $x(\zeta)$ can be thought of as a real-valued random vector $[x_R(\zeta), x_I(\zeta)]^T$ with a joint cdf $F_{x_R, x_I}(x_R, x_I)$ or a joint pdf $f_{x_1, x_2}(x_1, x_2)$ that will allow us to define its statistical averages. The mean of $x(\zeta)$ is defined as

$$E\{x(\zeta)\} = \mu_x = E\{x_R(\zeta) + jx_I(\zeta)\} = \mu_{x_R} + j\mu_{x_I} \qquad (3.2.8)$$

and the variance is defined as

$$\sigma_x^2 = E\{|x(\zeta) - \mu_x|^2\} \qquad (3.2.9)$$

which can be shown to be equal to

$$\sigma_x^2 = E\{|x(\zeta)|^2\} - |\mu_x|^2 \qquad (3.2.10)$$

[†]We will not make any distinction in *notation* between a real-valued and a complex-valued random variable. The actual type should be evident from the context.

A complex-valued random vector is given by

$$\mathbf{x}(\zeta) = \mathbf{x}_R(\zeta) + j\mathbf{x}_I(\zeta) = \begin{bmatrix} x_{R1}(\zeta) \\ \vdots \\ x_{RM}(\zeta) \end{bmatrix} + j \begin{bmatrix} x_{I1}(\zeta) \\ \vdots \\ x_{IM}(\zeta) \end{bmatrix} \qquad (3.2.11)$$

and we can think of a complex-valued random vector as a mapping from an abstract probability space to a vector-valued, complex space $\mathbb{C}^M$. The cdf for the complex-valued random
vector $\mathbf{x}(\zeta)$ is then defined as

$$F_{\mathbf{x}}(\mathbf{x}) \triangleq \Pr\{\mathbf{x}(\zeta) \le \mathbf{x}\} \triangleq \Pr\{\mathbf{x}_R(\zeta) \le \mathbf{x}_R, \mathbf{x}_I(\zeta) \le \mathbf{x}_I\} \qquad (3.2.12)$$

while its marginal pdf is defined as

$$\begin{aligned} f_{\mathbf{x}}(\mathbf{x}) = & \lim_{\substack{\Delta x_{R1} \to 0 \\ \vdots \\ \Delta x_{IM} \to 0}} \frac{\Pr\{\mathbf{x}_R(\zeta) \le \mathbf{x}_R, \mathbf{x}_I(\zeta) \le \mathbf{x}_I\}}{\Delta x_{R1} \Delta x_{I1} \cdots \Delta x_{RM} \Delta x_{IM}} \\[1em] & \triangleq \frac{\partial}{\partial x_{R1}} \frac{\partial}{\partial x_{I1}} \cdots \frac{\partial}{\partial x_{RM}} \frac{\partial}{\partial x_{IM}} F_{\mathbf{x}}(\mathbf{x}) \end{aligned} \qquad (3.2.13)$$

From (3.2.13), the cdf is obtained by integrating the pdf over all real and imaginary parts,
that is

$$F_{\mathbf{x}}(\mathbf{x}) = \int_{-\infty}^{x_{R1}} \cdots \int_{-\infty}^{x_{IM}} f_{\mathbf{x}}(\mathbf{v}) \, dv_{R1} \cdots dv_{IM} = \int_{-\infty}^{\mathbf{x}} f_{\mathbf{x}}(\mathbf{v}) \, d\mathbf{v} \qquad (3.2.14)$$

where the single integral in the last expression is used as a compact notation for multidimensional integrals and should not be confused with a complex-contour integral. These
probability functions for a complex-valued random vector possess properties similar to
those of the real-valued random vectors. In particular,

$$\int_{-\infty}^{\infty} f_{\mathbf{x}}(\mathbf{x}) \, d\mathbf{x} = 1 \qquad (3.2.15)$$

Statistical description

Clearly the above probability functions require an enormous amount of information
that is not easy to obtain or is too complex mathematically for practical use. In practical
applications, random vectors are described by less complete but more manageable statistical
averages.

Mean vector. As we have seen before, the most important statistical operation is the
expectation operation. The marginal expectation of a random vector $\mathbf{x}(\zeta)$ is called the *mean
vector* and is defined by

$$\boldsymbol{\mu}_{\mathbf{x}} = E\{\mathbf{x}(\zeta)\} = \begin{bmatrix} E\{x_1(\zeta)\} \\ \vdots \\ E\{x_M(\zeta)\} \end{bmatrix} = \begin{bmatrix} \mu_1 \\ \vdots \\ \mu_M \end{bmatrix} \qquad (3.2.16)$$

where the integral is taken over the entire $\mathbb{C}^M$ space. The components of $\boldsymbol{\mu}$ are the means
of the individual random variables.

Correlation and covariance matrices. The second-order moments of a random vector
$\mathbf{x}(\zeta)$ are given as matrices and describe the spread of its distribution. The *autocorrelation
matrix* is defined by

$$\mathbf{R}_{\mathbf{x}} \triangleq E\{\mathbf{x}(\zeta)\mathbf{x}^H(\zeta)\} = \begin{bmatrix} r_{11} & \cdots & r_{1M} \\ \vdots & \ddots & \vdots \\ r_{M1} & \cdots & r_{MM} \end{bmatrix} \qquad (3.2.17)$$

where superscript H denotes the conjugate transpose operation, the diagonal terms

$$r_{ii} \triangleq E\{|x_i(\zeta)|^2\} \quad i = 1, \ldots, M \tag{3.2.18}$$

are the second-order moments, denoted earlier as $r_{x_i}^{(2)}$, of random variables $x_i(\zeta)$, and the off-diagonal terms

$$r_{ij} \triangleq E\{x_i(\zeta)x_j^*(\zeta)\} = r_{ji}^* \quad i \neq j \tag{3.2.19}$$

measure the *correlation*, that is, the statistical similarity between the random variables $x_i(\zeta)$ and $x_j(\zeta)$. From (3.2.19) we note that the correlation matrix $\mathbf{R_x}$ is conjugate symmetric or *Hermitian*, that is, $\mathbf{R_x} = \mathbf{R_x}^H$.

The *autocovariance matrix* is defined by

$$\mathbf{\Gamma_x} \triangleq E\{[\mathbf{x}(\zeta) - \boldsymbol{\mu_x}][\mathbf{x}(\zeta) - \boldsymbol{\mu_x}]^H\} \triangleq \begin{bmatrix} \gamma_{11} & \cdots & \gamma_{1M} \\ \vdots & \ddots & \vdots \\ \gamma_{M1} & \cdots & \gamma_{MM} \end{bmatrix} \tag{3.2.20}$$

where the diagonal terms

$$\gamma_{ii} = E\{|x_i(\zeta) - \mu_i|^2\} \quad i = 1, \ldots, M \tag{3.2.21}$$

are the (self-)variances of $x_i(\zeta)$ denoted earlier as $\sigma_{x_i}^2$ while the off-diagonal terms

$$\gamma_{ij} = E\{[x_i(\zeta) - \mu_i][x_j(\zeta) - \mu_j]^*\} = E\{x_i(\zeta)x_j^*(\zeta)\} - \mu_i\mu_j^* = \gamma_{ji}^* \quad i \neq j \tag{3.2.22}$$

are the values of the *covariance* between $x_i(\zeta)$ and $x_j(\zeta)$. The covariance matrix $\mathbf{\Gamma_x}$ is also a Hermitian matrix. The covariance γ_{ij} can also be expressed in terms of standard deviations of $x_i(\zeta)$ and $x_j(\zeta)$ as $\gamma_{ij} = \rho_{ij}\sigma_i\sigma_j$, where

$$\rho_{ij} \triangleq \frac{\gamma_{ij}}{\sigma_i\sigma_j} = \rho_{ji} \tag{3.2.23}$$

is called the *correlation coefficient* between $x_i(\zeta)$ and $x_j(\zeta)$. Note that

$$|\rho_{ij}| \leq 1 \quad i \neq j \qquad \rho_{ii} = 1 \tag{3.2.24}$$

The correlation coefficient measures the degree of statistical similarity between two random variables. If $|\rho_{ij}| = 1$, then random variables are said to be *perfectly correlated*; but if $\rho_{ij} = 0$ (that is, when the covariance $\gamma_{ij} = 0$), then $x_i(\zeta)$ and $x_j(\zeta)$ are said to *uncorrelated*.

The autocorrelation and autocovariance matrices are related. Indeed, we can easily see that

$$\mathbf{\Gamma_x} \triangleq E\{[\mathbf{x}(\zeta) - \boldsymbol{\mu_x}][\mathbf{x}(\zeta) - \boldsymbol{\mu_x}]^H\} = \mathbf{R_x} - \boldsymbol{\mu_x}\boldsymbol{\mu_x}^H \tag{3.2.25}$$

which shows that these two moments have essentially the same amount of information. In fact, if $\boldsymbol{\mu_x} = \mathbf{0}$, then $\mathbf{\Gamma_x} = \mathbf{R_x}$. The autocovariance measures a *weaker* form of interaction between random variables called *correlatedness* that should be contrasted with the *stronger* form of independence that we described in (3.2.7). If random variables $x_i(\zeta)$ and $x_j(\zeta)$ are independent, then they are also uncorrelated since (3.2.7) implies that

$$E\{x_i(\zeta)x_j^*(\zeta)\} = E\{x_i(\zeta)\}E\{x_j^*(\zeta)\} \quad \text{or} \quad \gamma_{ij} = 0 \tag{3.2.26}$$

but uncorrelatedness does not imply independence unless random variables are jointly Gaussian (see Problem 3.15). The autocorrelation also measures another weaker form of interaction called *orthogonality*. Random variables $x_i(\zeta)$ and $x_j(\zeta)$ are *orthogonal* if their correlation

$$r_{ij} = E\{x_i(\zeta)x_j^*(\zeta)\} = 0 \quad i \neq j \tag{3.2.27}$$

Clearly, from (3.2.26) if one or both random variables have zero means, then uncorrelatedness also implies orthogonality.

We can also define correlation and covariance functions between two random vectors.

Let $\mathbf{x}(\zeta)$ and $\mathbf{y}(\zeta)$ be random M- and L-vectors, respectively. Then the $M \times L$ matrix

$$\mathbf{R_{xy}} \triangleq E\{\mathbf{xy}^H\} = \begin{bmatrix} E\{x_1(\zeta)y_1^*(\zeta)\} & \cdots & E\{x_1(\zeta)y_L^*(\zeta)\} \\ \vdots & \ddots & \vdots \\ E\{x_M(\zeta)y_1^*(\zeta)\} & \cdots & E\{x_M(\zeta)y_L^*(\zeta)\} \end{bmatrix} \quad (3.2.28)$$

is called a *cross-correlation* matrix whose elements r_{ij} are the correlations between random variables $x_i(\zeta)$ and $y_j(\zeta)$. Similarly the $M \times L$ matrix

$$\mathbf{\Gamma_{xy}} \triangleq E\{[\mathbf{x}(\zeta)-\boldsymbol{\mu_x}][\mathbf{y}(\zeta)-\boldsymbol{\mu_y}]^H\} = \mathbf{R_x} - \boldsymbol{\mu_x}\boldsymbol{\mu_y}^H \quad (3.2.29)$$

is called a *cross-covariance* matrix whose elements c_{ij} are the covariances between $x_i(\zeta)$ and $y_j(\zeta)$. In general the cross-matrices are not square matrices, and even if $M = L$, they are not necessarily symmetric. Two random vectors $\mathbf{x}(\zeta)$ and $\mathbf{y}(\zeta)$ are said to be

- Uncorrelated if

$$\mathbf{\Gamma_{xy}} = \mathbf{0} \Rightarrow \mathbf{R_{xy}} = \boldsymbol{\mu_x}\boldsymbol{\mu_y}^H \quad (3.2.30)$$

- Orthogonal if

$$\mathbf{R_{xy}} = \mathbf{0} \quad (3.2.31)$$

Again, if $\boldsymbol{\mu_x}$ or $\boldsymbol{\mu_y}$ or both are zero vectors, then (3.2.30) implies (3.2.31).

3.2.2 Linear Transformations of Random Vectors

Many signal processing applications involve linear operations on random vectors. Linear transformations are relatively simple mappings and are given by the matrix operation

$$\mathbf{y}(\zeta) = g[\mathbf{x}(\zeta)] = \mathbf{Ax}(\zeta) \quad (3.2.32)$$

where $\mathbf{A}$ is an $L \times M$ (not necessarily square) matrix. The random vector $\mathbf{y}(\zeta)$ is completely described by the density function $f_{\mathbf{y}}(\mathbf{y})$. If $L > M$, then only M $y_i(\zeta)$ random variables can be independently determined from $\mathbf{x}(\zeta)$. The remaining $(L - M)$ $y_i(\zeta)$ random variables can be obtained from the first $y_i(\zeta)$ random variables. Thus we need to determine $f_{\mathbf{y}}(\mathbf{y})$ for M random variables from which we can determine $f_{\mathbf{y}}(\mathbf{y})$ for all L random variables. If $M > L$, then we can augment $\mathbf{y}$ into an M-vector by introducing auxiliary random variables

$$y_{L+1}(\zeta) = x_{L+1}(\zeta), \ldots, y_M(\zeta) = x_M(\zeta) \quad (3.2.33)$$

to determine $f_{\mathbf{y}}(\mathbf{y})$ for M random variables from which we can determine $f_{\mathbf{y}}(\mathbf{y})$ for the original L random variables. Therefore, for the determination of the pdf $f_{\mathbf{y}}(\mathbf{y})$, we will assume that $L = M$ and that $\mathbf{A}$ is nonsingular.

Furthermore, we will first consider the case in which both $\mathbf{x}(\zeta)$ and $\mathbf{y}(\zeta)$ are real-valued random vectors, which also implies that $\mathbf{A}$ is a real-valued matrix. This approach is necessary because the complex case leads to a slightly different result. Then the pdf $f_{\mathbf{y}}(\mathbf{y})$ is given by

$$f_{\mathbf{y}}(\mathbf{y}) = \frac{f_{\mathbf{x}}(g^{-1}(\mathbf{y}))}{|\mathbf{J}|} \quad (3.2.34)$$

where $\mathbf{J}$ is called the *Jacobian* of the transformation (3.2.32), given by

$$\mathbf{J} = \det \begin{bmatrix} \dfrac{\partial y_1}{\partial x_1} & \cdots & \dfrac{\partial y_M}{\partial x_1} \\ \vdots & \ddots & \vdots \\ \dfrac{\partial y_1}{\partial x_M} & \cdots & \dfrac{\partial y_M}{\partial y_M} \end{bmatrix} = \det \mathbf{A} \quad (3.2.35)$$

From (3.2.34) and (3.2.35), the pdf of $\mathbf{y}(\zeta)$ is given by

$$f_{\mathbf{y}}(\mathbf{y}) = \frac{f_{\mathbf{x}}(\mathbf{A}^{-1}\mathbf{y})}{|\det \mathbf{A}|} \quad \text{real-valued random vector} \quad (3.2.36)$$

from which moment computations of any order of $\mathbf{y}(\zeta)$ can be performed. Now we consider the case of the complex-valued random vectors. Then by applying the above approach to both real and imaginary parts, the result (3.2.36) becomes

$$f_{\mathbf{y}}(\mathbf{y}) = \frac{f_{\mathbf{x}}(\mathbf{A}^{-1}\mathbf{y})}{|\det \mathbf{A}|^2} \quad \text{complex-valued random vector} \quad (3.2.37)$$

This shows that sometimes we can get different results depending upon whether we assume real- or complex-valued random vectors in our analysis.

Determining $f_{\mathbf{y}}(\mathbf{y})$ is, in general, tedious except in the case of Gaussian random vectors, as we shall see later. In practice, the knowledge of $\boldsymbol{\mu}_{\mathbf{y}}$, $\boldsymbol{\Gamma}_{\mathbf{y}}$, $\boldsymbol{\Gamma}_{\mathbf{xy}}$, or $\boldsymbol{\Gamma}_{\mathbf{yx}}$ is sufficient in many applications. If we take the expectation of both sides of (3.2.32), we find that the mean vector is given by

$$\boldsymbol{\mu}_{\mathbf{y}} = E\{\mathbf{y}(\zeta)\} = E\{\mathbf{A}\mathbf{x}(\zeta)\} = \mathbf{A}E\{\mathbf{x}(\zeta)\} = \mathbf{A}\boldsymbol{\mu}_{\mathbf{x}} \quad (3.2.38)$$

The autocorrelation matrix of $\mathbf{y}(\zeta)$ is given by

$$\mathbf{R}_{\mathbf{y}} = E\{\mathbf{y}\mathbf{y}^H\} = E\{\mathbf{A}\mathbf{x}\mathbf{x}^H\mathbf{A}^H\} = \mathbf{A}E\{\mathbf{x}\mathbf{x}^H\}\mathbf{A}^H = \mathbf{A}\mathbf{R}_{\mathbf{x}}\mathbf{A}^H \quad (3.2.39)$$

Similarly, the autocovariance matrix of $\mathbf{y}(\zeta)$ is given by

$$\boldsymbol{\Gamma}_{\mathbf{y}} = \mathbf{A}\boldsymbol{\Gamma}_{\mathbf{x}}\mathbf{A}^H \quad (3.2.40)$$

Consider the cross-correlation matrix

$$\mathbf{R}_{\mathbf{xy}} = E\{\mathbf{x}(\zeta)\mathbf{y}^H(\zeta)\} = E\{\mathbf{x}(\zeta)\mathbf{x}^H(\zeta)\mathbf{A}^H\} \quad (3.2.41)$$

$$= E\{\mathbf{x}(\zeta)\mathbf{x}^H(\zeta)\}\mathbf{A}^H = \mathbf{R}_{\mathbf{x}}\mathbf{A}^H \quad (3.2.42)$$

and hence $\mathbf{R}_{\mathbf{yx}} = \mathbf{A}\mathbf{R}_{\mathbf{x}}$. Similarly, the cross-covariance matrices are

$$\boldsymbol{\Gamma}_{\mathbf{xy}} = \boldsymbol{\Gamma}_{\mathbf{x}}\mathbf{A}^H \quad \text{and} \quad \boldsymbol{\Gamma}_{\mathbf{yx}} = \mathbf{A}\boldsymbol{\Gamma}_{\mathbf{x}} \quad (3.2.43)$$

3.2.3 Normal Random Vectors

If the components of the random vector $\mathbf{x}(\zeta)$ are jointly normal, then $\mathbf{x}(\zeta)$ is a normal random M-vector. Again, the pdf expressions for the real- and complex-valued cases are slightly different, and hence we consider these cases separately. The real-valued normal random vector has the pdf

$$f_{\mathbf{x}}(\mathbf{x}) = \frac{1}{(2\pi)^{M/2}|\boldsymbol{\Gamma}_{\mathbf{x}}|^{1/2}} \exp\left[-\frac{1}{2}(\mathbf{x} - \boldsymbol{\mu}_{\mathbf{x}})^T \boldsymbol{\Gamma}_{\mathbf{x}}^{-1}(\mathbf{x} - \boldsymbol{\mu}_{\mathbf{x}})\right] \quad \text{real} \quad (3.2.44)$$

with mean $\boldsymbol{\mu}_{\mathbf{x}}$ and covariance $\boldsymbol{\Gamma}_{\mathbf{x}}$. It will be denoted by $\mathcal{N}(\boldsymbol{\mu}_{\mathbf{x}}, \boldsymbol{\Gamma}_{\mathbf{x}})$. The term in the exponent $(\mathbf{x} - \boldsymbol{\mu}_{\mathbf{x}})^T \boldsymbol{\Gamma}_{\mathbf{x}}^{-1}(\mathbf{x} - \boldsymbol{\mu}_{\mathbf{x}})$ is a positive definite quadratic function of x_i and is also given by

$$(\mathbf{x} - \boldsymbol{\mu}_{\mathbf{x}})^T \boldsymbol{\Gamma}_{\mathbf{x}}^{-1}(\mathbf{x} - \boldsymbol{\mu}_{\mathbf{x}}) = \sum_{i=1}^{M}\sum_{j=1}^{M} \langle\boldsymbol{\Gamma}_{\mathbf{x}}^{-1}\rangle_{ij}(x_i - \mu_i)(x_j - \mu_j) \quad (3.2.45)$$

where $\langle\boldsymbol{\Gamma}_{\mathbf{x}}^{-1}\rangle_{ij}$ denotes the (i, j)th element of $\boldsymbol{\Gamma}_{\mathbf{x}}^{-1}$. The characteristic function of the normal random vector is given by

$$\Phi_{\mathbf{x}}(\boldsymbol{\xi}) = \exp(j\boldsymbol{\xi}^T \boldsymbol{\mu}_x - \tfrac{1}{2}\boldsymbol{\xi}^T \boldsymbol{\Gamma}_{\mathbf{x}}\boldsymbol{\xi}) \quad (3.2.46)$$

where $\boldsymbol{\xi}^T = [\xi_1, \ldots, \xi_M]$.

The complex-valued normal random vector has the pdf

$$f_{\mathbf{x}}(\mathbf{x}) = \frac{1}{\pi^M |\mathbf{\Gamma_x}|} \exp[-(\mathbf{x} - \boldsymbol{\mu_x})^H \mathbf{\Gamma_x}^{-1}(\mathbf{x} - \boldsymbol{\mu_x})] \qquad \text{complex} \qquad (3.2.47)$$

with mean $\boldsymbol{\mu_x}$ and covariance $\mathbf{\Gamma_x}$. This pdf will be denoted by $\mathcal{CN}(\boldsymbol{\mu_x}, \mathbf{\Gamma_x})$. If $\mathbf{x}(\zeta)$ is a scalar complex-valued random variable $x(\zeta)$ with mean μ_x and variance σ_x^2, then (3.2.47) reduces to

$$f_x(x) = \frac{1}{\pi \sigma_x^2} \exp\left(-\frac{|x - \mu|^2}{\sigma_x^2}\right) \qquad (3.2.48)$$

which should be compared with the pdf given in (3.1.37). Note that the pdf in (3.1.37) is not obtained by setting the imaginary part of $x(\zeta)$ in (3.2.48) equal to zero. For a more detailed discussion on this aspect, see Therrien (1992) or Kay (1993). The term $(\mathbf{x} - \boldsymbol{\mu_x})^H \mathbf{\Gamma_x}^{-1}(\mathbf{x} - \boldsymbol{\mu_x})$ in the exponent of (3.2.47) is also a positive definite quadratic function and is given by

$$(\mathbf{x} - \boldsymbol{\mu_x})^H \mathbf{\Gamma_x}^{-1}(\mathbf{x} - \boldsymbol{\mu_x}) = \sum_{i=1}^{M} \sum_{j=1}^{M} \langle \mathbf{\Gamma_x}^{-1} \rangle_{ij} (x_i - \mu_i)^* (x_j - \mu_j) \qquad (3.2.49)$$

The characteristic function for the complex-valued normal random vector is given by

$$\Phi_{\mathbf{x}}(\boldsymbol{\xi}) = \exp[j \operatorname{Re}(\boldsymbol{\xi}^H \boldsymbol{\mu_x}) - \tfrac{1}{4} \boldsymbol{\xi}^H \mathbf{\Gamma_x} \boldsymbol{\xi}] \qquad (3.2.50)$$

The normal distribution is a useful model of a random vector because of its many important properties:

1. The pdf is completely specified by the mean vector and the covariance matrix, which are relatively easy to estimate in practice. All other higher-order moments can be obtained from these parameters.
2. If the components of $\mathbf{x}(\zeta)$ are mutually uncorrelated, then they are also independent. (See Problem 3.15.) This is useful in many derivations.
3. A linear transformation of a normal random vector is also normal. This can be easily seen by using (3.2.38), (3.2.40), and (3.2.44) in (3.2.36); that is, for the real-valued case we obtain

$$f_{\mathbf{y}}(\mathbf{y}) = \frac{1}{(2\pi)^{M/2} |\mathbf{\Gamma_y}|^{1/2}} \exp\left[-\frac{1}{2}(\mathbf{y} - \boldsymbol{\mu_y})^T \mathbf{\Gamma_y}^{-1}(\mathbf{y} - \boldsymbol{\mu_y})\right] \qquad \text{real} \qquad (3.2.51)$$

This result can also be proved by using the moment generating function in (3.2.46) (see Problem 3.6). Similarly for the complex-valued case, from (3.2.37) and (3.2.47) we obtain

$$f_{\mathbf{y}}(\mathbf{y}) = \frac{1}{\pi^M |\mathbf{\Gamma_y}|} \exp[-(\mathbf{y} - \boldsymbol{\mu_y})^H (\mathbf{A}^{-1})^H \mathbf{\Gamma_x}^{-1} \mathbf{A}^{-1}(\mathbf{y} - \boldsymbol{\mu_y})] \qquad \text{complex} \qquad (3.2.52)$$

4. The fourth-order moment of a normal random vector

$$\mathbf{x}(\zeta) = [x_1(\zeta) \; x_2(\zeta) \; x_3(\zeta) \; x_4(\zeta)]^T$$

can be expressed in terms of its second-order moments. For the real case, that is, when $\mathbf{x}(\zeta) \sim \mathcal{N}(\mathbf{0}, \mathbf{\Gamma_x})$, we have

$$\begin{aligned} E\{x_1(\zeta)x_2(\zeta)x_3(\zeta)x_4(\zeta)\} = \; & E\{x_1(\zeta)x_2(\zeta)\}E\{x_3(\zeta)x_4(\zeta)\} \\ & + E\{x_1(\zeta)x_3(\zeta)\}E\{x_2(\zeta)x_4(\zeta)\} \\ & + E\{x_1(\zeta)x_4(\zeta)\}E\{x_2(\zeta)x_3(\zeta)\} \end{aligned} \qquad (3.2.53)$$

For the complex case, that is, when $\mathbf{x}(\zeta) \sim \mathcal{CN}(\mathbf{0}, \mathbf{\Gamma_x})$, we have

$$
\begin{aligned}
E\{x_1^*(\zeta)x_2(\zeta)x_3^*(\zeta)x_4(\zeta)\} &= E\{x_1^*(\zeta)x_2(\zeta)\}E\{x_3^*(\zeta)x_4(\zeta)\} \\
&\quad + E\{x_1^*(\zeta)x_4(\zeta)\}E\{x_2(\zeta)x_3^*(\zeta)\}
\end{aligned}
\tag{3.2.54}
$$

The proof of (3.2.53) is tedious but straightforward. However, the proof of (3.2.54) is complicated and is discussed in Kay (1993).

3.2.4 Sums of Independent Random Variables

In many applications, a random variable $y(\zeta)$ can be expressed as a linear combination of M statistically independent random variables $\{x_k(\zeta)\}_1^M$, that is,

$$
y(\zeta) = c_1 x_1(\zeta) + c_2 x_2(\zeta) + \cdots + c_M x_M(\zeta) = \sum_{k=1}^{M} c_k x_k(\zeta)
\tag{3.2.55}
$$

where $\{c_k\}_1^M$ is a set of fixed coefficients. In these situations, we would like to compute the first two moments and the pdf of $y(\zeta)$. The moment computation is straightforward, but the pdf computation requires the use of characteristic functions. When these results are extended to the sum of an infinite number of statistically independent random variables, we obtain a powerful theorem called the *central limit theorem* (*CLT*). Another interesting concept develops when the sum of IID random variables preserves their distribution, which results in stable distributions.

Mean. Using the linearity of the expectation operator and taking the expectation of both sides of (3.2.55), we obtain

$$
\mu_y = \sum_{k=1}^{M} c_k \mu_{x_k}
\tag{3.2.56}
$$

Variance. Again by using independence, the variance of $y(\zeta)$ is given by

$$
\sigma_y^2 = E\left\{ \left| \sum_{k=1}^{M} c_k [x_k(\zeta) - \mu_{x_k}] \right|^2 \right\} = \sum_{k=1}^{M} |c_k|^2 \sigma_{x_k}^2
\tag{3.2.57}
$$

where we have used the statistical independence between random variables.

Probability density function. Before we derive the pdf of $y(\zeta)$ in (3.2.55), we consider two special cases. First, let

$$
y(\zeta) = x_1(\zeta) + x_2(\zeta)
\tag{3.2.58}
$$

where $x_1(\zeta)$ and $x_2(\zeta)$ are statistically independent. Then its characteristic function is given by

$$
\Phi_y(\xi) = E\{e^{j\xi y(\zeta)}\} = E\{e^{j\xi[x_1(\zeta)+x_2(\zeta)]}\} = E\{e^{j\xi x_1(\zeta)}\}E\{e^{j\xi x_2(\zeta)}\}
\tag{3.2.59}
$$

where the last equality follows from the independence. Hence

$$
\Phi_y(\xi) = \Phi_{x_1}(\xi)\Phi_{x_2}(\xi)
\tag{3.2.60}
$$

or from the convolution property of the Fourier transform

$$
f_y(y) = f_{x_1}(y) * f_{x_2}(y)
\tag{3.2.61}
$$

From (3.2.60) the second characteristic function of $y(\zeta)$ is given by

$$
\Psi_y(\xi) = \Psi_{x_1}(\xi) + \Psi_{x_2}(\xi)
\tag{3.2.62}
$$

or the mth-order cumulant of $y(\zeta)$ is given by

$$\kappa_m^{(y)} = \kappa_m^{(x_1)} + \kappa_m^{(x_2)} \tag{3.2.63}$$

These results can be easily generalized to the sum of M independent random variables.

EXAMPLE 3.2.1. Let $\{x_k(\zeta)\}_{k=1}^4$ be four IID random variables uniformly distributed over $[-0.5, 0.5]$. Compute and plot the pdfs of $y_M(\zeta) \triangleq \sum_{k=1}^M x_k$ for $M = 2, 3$, and 4. Compare these pdfs with that of a zero-mean Gaussian random variable.

Solution. Let $f(x)$ be the pdf of a uniform random variable over $[-0.5, 0.5]$, that is,

$$f(x) = \begin{cases} 1 & -0.5 \le x \le 0.5 \\ 0 & \text{otherwise} \end{cases} \tag{3.2.64}$$

Then from (3.2.61)

$$f_{y_2}(y) = f(y) * f(y) = \begin{cases} 1+y & -1 \le y \le 0 \\ 1-y & 0 \le y \le 1 \\ 0 & \text{otherwise} \end{cases} \tag{3.2.65}$$

Similarly, we have

$$f_{y_3}(y) = f_{y_2}(y) * f(y) = \begin{cases} \frac{1}{2}(y+\frac{3}{2})^2 & -\frac{3}{2} \le y \le -\frac{1}{2} \\ \frac{3}{4} - y^2 & -\frac{1}{2} \le y \le \frac{1}{2} \\ \frac{1}{2}(y-\frac{3}{2})^2 & \frac{1}{2} \le y \le \frac{3}{2} \\ 0 & \text{otherwise} \end{cases} \tag{3.2.66}$$

and

$$f_{y_4}(y) = f_{y_3}(y) * f(y) = \begin{cases} \frac{1}{6}(y+2)^3 & -2 \le y \le -1 \\ -\frac{1}{2}y^3 - y^2 + \frac{2}{3} & -1 \le y \le 0 \\ \frac{2}{3} + \frac{1}{2}y^3 - y^2 & 0 \le y \le 1 \\ -\frac{1}{6}(-2+y)^3 & 1 \le y \le 2 \\ 0 & \text{otherwise} \end{cases} \tag{3.2.67}$$

The plots of $f_{y_2}(y)$, $f_{y_3}(y)$, and $f_{y_4}(y)$ are shown in Figure 3.4 along with the zero-mean Gaussian pdf. The variance of the Gaussian random variable is chosen so that 99.92 percent of the pdf area is over $[-2, 2]$. We observe that as M increases, the pdf plots appear to get closer to the shape of the Gaussian pdf. This observation will be explored in detail in the CLT.

Next, let $y(\zeta) = ax(\zeta) + b$; then the characteristic function of $y(\zeta)$ is

$$\Phi_y(\xi) = E\{e^{j[ax(\zeta)+b]\xi}\} = E\{e^{ja\xi x(\zeta)} e^{jb\xi}\} = \Phi_x(a\xi)e^{jb\xi} \tag{3.2.68}$$

and by using the properties of the Fourier transform, the pdf of $y(\zeta)$ is given by

$$f_y(y) = \frac{1}{|a|} f_x\left(\frac{y-b}{a}\right) \tag{3.2.69}$$

From (3.2.68), the second characteristic function is given by

$$\Psi_y(\xi) = \Psi_x(a\xi) + jb\xi \tag{3.2.70}$$

and the cumulants are given by

$$\begin{aligned} \kappa_y^{(m)} &= (-j)^m \left.\frac{d^m \Psi_y(\xi)}{d\xi^m}\right|_{\xi=0} = a^m(-j)^m \left.\frac{d^m \Psi_x(a\xi)}{d\xi^m}\right|_{\xi=0} \\ &= a^m \kappa_x^{(m)} \quad m > 1 \end{aligned} \tag{3.2.71}$$

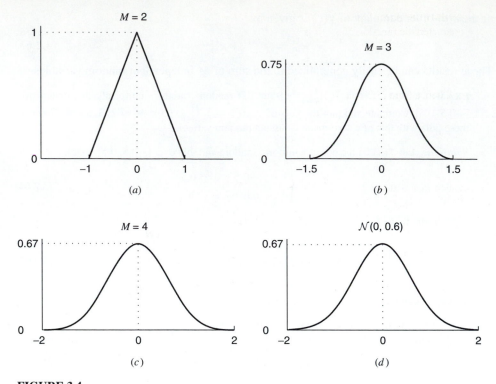

FIGURE 3.4
The pdf plots of (a) sum of two, (b) sum of three, (c) sum of four, and (d) Gaussian random variables in Example 3.2.1.

Finally, consider $y(\zeta)$ in (3.2.55). Using the results in (3.2.60) and (3.2.68), we have

$$\Phi_y(\xi) = \prod_{k=1}^{M} \Phi_{x_k}(c_k \xi) \tag{3.2.72}$$

from which the pdf of $y(\zeta)$ is given by

$$f_y(y) = \frac{1}{|c_1|} f_{x_1}\left(\frac{y}{c_1}\right) * \frac{1}{|c_2|} f_{x_2}\left(\frac{y}{c_2}\right) * \cdots * \frac{1}{|c_M|} f_{x_M}\left(\frac{y}{c_M}\right) \tag{3.2.73}$$

From (3.2.62) and (3.2.70), the second characteristic function is given by

$$\Psi_y(\xi) = \sum_{k=1}^{M} \Psi_{x_k}(c_k \xi) \tag{3.2.74}$$

and hence from (3.2.63) and (3.2.71), the cumulants of $y(\zeta)$ are

$$\kappa_y^{(m)} = \sum_{k=1}^{M} c_k^m \kappa_{x_k}^{(m)} \tag{3.2.75}$$

where c_k^m is the mth power of c_k.

In the following two examples, we consider two interesting cases in which the sum of IID random variables retains their original distribution. The first case concerns Gaussian random variables that have finite variances while the second case involves Cauchy random variables that possess infinite variance.

EXAMPLE 3.2.2. Let $x_k(\zeta) \sim \mathcal{N}(\mu_k, \sigma_k^2)$, $k = 1, \ldots, M$ and let $y(k) = \sum_1^M x_k(\zeta)$. The characteristic function of $x_k(\zeta)$ is

$$\Phi_{x_k}(\xi) = \exp\left(j\mu_k\xi - \frac{\xi^2\sigma_k^2}{2}\right)$$

and hence from (3.2.72), we have

$$\Phi_y(\xi) = \exp\left(j\xi \sum_{k=1}^M \mu_k - \frac{\xi^2 \sum_{k=1}^M \sigma_k^2}{2}\right)$$

which means that $y(\zeta)$ is also a Gaussian random variable with mean $\sum_{k=1}^M \mu_k$ and variance $\sum_{k=1}^M \sigma_k^2$, that is, $y(\zeta) \sim \mathcal{N}(\sum_{k=1}^M \mu_k, \sum_{k=1}^M \sigma_k^2)$. In particular, if the $x_k(\zeta)$ are IID with a pdf $\mathcal{N}(\mu, \sigma^2)$, then

$$\Phi_y(\xi) = \exp\left(jM\mu\xi - \frac{\xi^2 M\sigma^2}{2}\right) = \exp\left[M\left(j\xi\mu - \frac{\xi^2\sigma^2}{2}\right)\right] \qquad (3.2.76)$$

This behavior of $y(\zeta)$ is in contrast with that of the sum of the IID random variables in Example 3.2.1 in which the uniform pdf changed its form after M-fold convolutions.

EXAMPLE 3.2.3. As a second case, consider M IID random variables $\{x_k(\zeta)\}_{k=1}^M$ with Cauchy distribution

$$f_{x_k}(x) = \frac{\beta}{\pi} \frac{1}{(x-\alpha)^2 + \beta^2}$$

and let $y(k) = \sum_1^M x_k(\zeta)$. Then from (3.1.43), we have

$$\Phi_x(\xi) = \exp(j\alpha\xi - \beta|\xi|)$$

and hence

$$\Phi_y(\xi) = \exp(jM\alpha\xi - M\beta|\xi|) = \exp[M(j\alpha\xi - \beta|\xi|)] \qquad (3.2.77)$$

This once again shows that the sum random variable has the same distribution (up to a scale factor) as that of the individual random variables, which in this case is the Cauchy distribution.

From these examples, we note that the Gaussian and the Cauchy random variables are *invariant*, or that they have a "self-reproducing" property under linear transformations. These two examples also raise some interesting questions. Are there any other random variables that possess this *invariance* property? If such random variables exist, what is the form of their pdfs or, alternatively, of their characteristic functions, and what can we say about their means and variances? From (3.2.76) and (3.2.77), observe that if the characteristic function has a general form

$$\Phi_{x_k}(\xi) = a^{\theta(\xi)} \qquad (3.2.78)$$

where a is some constant and $\theta(\xi)$ is some function of ξ, then we have

$$\Phi_y(s) = a^{M\theta(s)} \qquad (3.2.79)$$

that is, the characteristic function of the sum has the same *functional form* except for a change in scale. Are Gaussian and Cauchy both special cases of some general situation? These questions are answered by the concept of *stable* (more appropriately, linearly invariant or self-reproducing) *distributions*.

Stable distributions. These distributions satisfy the "stability" property, which in simple terms means that the distributions are preserved (or that they self-reproduce) under convolution. The only stable distribution that has finite variance is the Gaussian distribution, which has been well understood and is used extensively in the literature and in practice.

The remaining stable distributions have infinite variances (and in some cases, infinite means) which means that the corresponding random variables exhibit large fluctuations. These distributions can then be used to model signals with large variability and hence are finding increasing use in many diverse applications such as the gravitational fields of stars, temperature distributions in a nuclear reaction, or stock market fluctuations (Lamperti 1996; Samorodnitsky and Taqqu 1994; Feller 1966).

Before we formally define stable distributions, we introduce the following notation for convenience

$$y(\zeta) \overset{\mathrm{d}}{=} x(\zeta) \tag{3.2.80}$$

to indicate that the random variables $x(\zeta)$ and $y(\zeta)$ have the same distribution. For example, if $y(\zeta) = ax(\zeta) + b$, we have

$$F_y(y) = F_x\left(\frac{y - b}{a}\right) \tag{3.2.81}$$

and therefore $x(\zeta) \overset{\mathrm{d}}{=} ax(\zeta) + b$.

DEFINITION 3.2. Let $x_1(\zeta), x_2(\zeta), \ldots, x_M(\zeta)$ be IID random variables with a common distribution $F_x(x)$ and let $s_M(\zeta) = x_1(\zeta) + \cdots + x_M(\zeta)$ be their sum. The distribution $F_x(x)$ is said to be *stable* if for each M there exist constants $a_M > 0$ and b_M such that

$$s_M(\zeta) \overset{\mathrm{d}}{=} a_M x(\zeta) + b_M \tag{3.2.82}$$

and that $F_x(x)$ is not concentrated at one point.

If (3.2.82) holds for $b_M = 0$, we say that $F_x(x)$ is stable in the *strict sense*. The condition that $F_x(x)$ is not concentrated at one point is necessary because such a distribution is always stable. Thus it is a degenerate case that is of no practical interest. A stable distribution is called *symmetric stable* if the distribution is symmetric, which also implies that it is strictly stable.

It can be shown that for any stable random variable $x(\zeta)$ there is a number $\alpha, 0 < \alpha \le 2$, such that the constant a_M in (3.2.82) is $a_M = M^{1/\alpha}$. The number α is known as the *index of stability* or *characteristic exponent*. A stable random variable $x(\zeta)$ with index α is called α *stable*.

Since there is no closed-form expression for the probability density function of stable random variables, except in special cases, they are specified by their characteristic function $\Phi(\xi)$. This characteristic function is given by

$$\Phi(\xi) = \begin{cases} \exp\{j\mu\xi - |\sigma\xi|^\alpha \cdot [1 - j\beta \, \mathrm{sign}(\xi) \tan\left(\dfrac{\pi\alpha}{2}\right)]\} & \alpha \ne 1 \\[2mm] \exp\{j\mu\xi - |\sigma\xi|^\alpha \cdot [1 - j\beta\left(\dfrac{2}{\pi}\right) \mathrm{sign}(\xi) \ln|\xi|]\} & \alpha = 1 \end{cases} \tag{3.2.83}$$

where $\mathrm{sign}(\xi) = \xi/|\xi|$ if $\xi \ne 0$ and zero otherwise. We shall use the notation $S_\alpha(\sigma, \beta, \mu)$ to denote the stable random variable defined by (3.2.83). The parameters in (3.2.83) have the following meaning:

1. The characteristic exponent $\alpha, 0 < \alpha \le 2$, determines the shape of the distribution and hence the flatness of the tails.
2. The skewness (or alternatively, symmetry) parameter $\beta, -1 < \beta < 1$, determines the symmetry of the distribution: $\beta = 0$ specifies a symmetric distribution, $\beta < 0$ a left-skewed distribution, and $\beta > 0$ a right-skewed distribution.
3. The scale parameter $\sigma, 0 \le \sigma < \infty$, determines the range or dispersion of the stable distribution.
4. The location parameter $\mu, -\infty < \mu < \infty$, determines the center of the distribution.

We next list some useful properties of stable random variables.

1. For $0 < \alpha < 2$, the tails of a stable distribution decay as a power law, that is,

$$\Pr[|x(\zeta) - \mu| \geq x] \simeq \frac{C}{x^\alpha} \qquad \text{as } x \to \infty \qquad (3.2.84)$$

where C is a constant that depends on the scale parameter σ. As a result of this behavior, α-stable random variables have infinite second-order moments. In particular,

$$\begin{aligned} E\{|x(\zeta)|^p\} < \infty \qquad &\text{for any } 0 < p \leq \alpha \\ E\{|x(\zeta)|^p\} = \infty \qquad &\text{for any } p > \alpha \end{aligned} \qquad (3.2.85)$$

Also $\mathrm{var}[x(\zeta)] = \infty$ for $0 < \alpha < 2$, and $E\{|x(\zeta)|\} = \infty$ if $0 < \alpha < 1$.

2. A stable distribution is symmetric about μ iff $\beta = 0$. A symmetric α-stable distribution is denoted as $S\alpha S$, and its characteristic function is given by

$$\Phi(\xi) = \exp(j\mu\xi - |\sigma\xi|^\alpha) \qquad (3.2.86)$$

3. If $x(\zeta)$ is $S\alpha S$ with $\alpha = 2$ in (3.2.83), we have a Gaussian distribution with variance equal to $2\sigma^2$, that is, $\mathcal{N}(\mu, 2\sigma^2)$, whose tails decay exponentially and not as a power law. Thus, the Gaussian is the only stable distribution with finite variance.

4. If $x(\zeta)$ is $S\alpha S$ with $\alpha = 1$, we have a Cauchy distribution with density

$$f_x(x) = \frac{\sigma/\pi}{(x - \mu)^2 + \sigma^2} \qquad (3.2.87)$$

A standard ($\mu = 0$, $\sigma = 1$) Cauchy random variable $x(\zeta)$ can be generated from a $[0, 1]$ uniform random variable $u(\zeta)$, by using the transformation $x = \tan[\pi(u - \frac{1}{2})]$.

5. If $x(\zeta)$ is $S\alpha S$ with $\alpha = \frac{1}{2}$, we have a *Levy distribution*, which has both infinite variance and infinite mean. The pdf of this distribution does not have a functional form and hence must be computed numerically.

In Figure 3.5, we display characteristic and density functions of Gaussian, Cauchy, and Levy random variables. The density plots were computed numerically using the MATLAB function `stablepdf`.

Infinitely divisible distributions. A distribution $F_x(x)$ is infinitely divisible if and only if for each M there exists a distribution $F_M(x)$ such that

$$f_x(x) = f_M(x) * f_M(x) * \cdots * f_M(x) \qquad (3.2.88)$$

or by using the convolution theorem,

$$\Phi_x(\xi) = \Phi_M(\xi)\, \Phi_M(\xi) \cdots \Phi_M(\xi) = \Phi_M^M(\xi) \qquad (3.2.89)$$

that is, for each M the random variable $x(\zeta)$ can be represented as the sum $x(\zeta) = x_1(\zeta) + \cdots + x_M(\zeta)$ of M IID random variables with a *common* distribution $F_M(x)$. Clearly, all stable distributions are infinitely divisible, and $F_M(x)$ differs from $F_x(x)$ only by the location parameter. An example of infinitely divisible pdf is shown in Figure 3.6 for $M = 4$, $\alpha = 1.5$, and $\beta = 1$.

Central limit theorem. Consider the random variable $y(\zeta)$ defined in (3.2.55). We would like to know about the convergence of its distribution as $M \to \infty$. If $y(\zeta)$ is a sum of IID random variables with a stable distribution, the distribution of $y(\zeta)$ also converges to a stable distribution. What result should we expect if the individual distributions are not stable and, in particular, are of finite variance? As we observed in Example 3.2.1, the sum of uniformly distributed independent random variables appears to converge to a Gaussian distribution. Is this result valid for any other distribution? The following version of the CLT answers these questions.

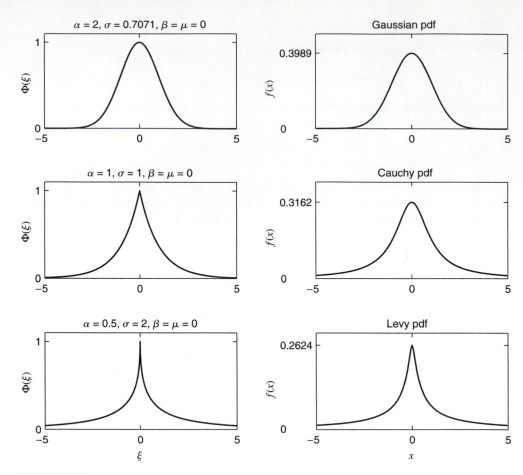

FIGURE 3.5
The characteristic and density function plots of Gaussian, Cauchy, and Levy random variables.

THEOREM 3.1 (CENTRAL LIMIT THEOREM). Let $\{x_k(\zeta)\}_{k=1}^{M}$ be a collection of random variables such that $x_1(\zeta), x_2(\zeta), \ldots, x_M(\zeta)$ (a) are mutually independent and (b) have the same distribution, and (c) the mean and variance of each random variable exist and are finite, that is, $\mu_{x_k} < \infty$ and $\sigma_{x_k}^2 < \infty$ for all $k = 1, 2, \ldots, M$. Then, the distribution of the normalized sum

$$y_M(\zeta) = \frac{\displaystyle\sum_{k=1}^{M} x_k(\zeta) - \mu_{y_M}}{\sigma_{y_M}}$$

approaches that of a normal random variable with zero mean and unit standard deviation as $M \to \infty$.

Proof. See Borkar (1995).

Comments. The following important comments are in order regarding the CLT.

1. Since we are assuming IID components in the normalized sum, the above theorem is known as the *equal-component case of the CLT*.
2. It should be emphasized that the convergence in the above theorem is in *distribution* (cdf) and not necessarily in density (pdf). Suppose we have M discrete and IID random variables. Then their normalized sum will always remain discrete no matter how large M is, but the distribution of the sum will converge to the the integral of the Gaussian pdf.

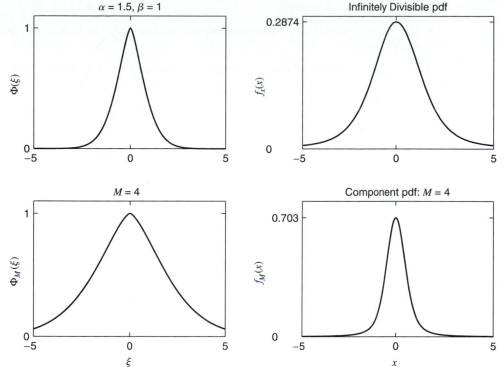

FIGURE 3.6
The characteristic and density function plots of an infinitely divisible distribution.

3. The word *central* in the CLT is a reminder that the distribution converges to the Gaussian distribution *around the center*, that is, around the mean. Note that while the limit distribution is found to be Gaussian, frequently the Gaussian limit gives a poor approximation for the tails of the actual distribution function of the sum when M is finite, even though the actual value under consideration might seem to be quite large.

4. As a final point, we note that in the above theorem the assumption of finite variance is critical to obtain a Gaussian limit. This implies that *all* distributions with finite variance will converge to the Gaussian when independent copies of their random variables are added. What happens if the variance is infinite? Then in this case the sum converges to one of the stable distributions. For example, as shown in Example 3.2.3, the sum of Cauchy random variables converges to a Cauchy distribution.

3.3 DISCRETE-TIME STOCHASTIC PROCESSES

After studying random variables and vectors, we can now extend these concepts to discrete-time signals (or sequences). Many natural sequences can be characterized as random signals because we cannot determine their values precisely, that is, they are unpredictable. A natural mathematical framework for the description of these discrete-time random signals is provided by discrete-time stochastic processes.

To obtain a formal definition, consider an experiment with a finite or infinite number of unpredictable outcomes from a sample space $S = \{\zeta_1, \zeta_2, \ldots\}$, each occurring with a probability $\Pr\{\zeta_k\}, k = 1, 2, \ldots$. By some rule we assign to each element ζ_k of S a deterministic sequence $x(n, \zeta_k), -\infty < n < \infty$. The sample space S, the probabilities $\Pr\{\zeta_k\}$, and the sequences $x(n, \zeta_k), -\infty < n < \infty$, constitute a *discrete-time stochastic process* or random sequence. Formally,

$x(n, \zeta)$, $-\infty < n < \infty$, is a random sequence if for a fixed value n_0 of n, $x(n_0, \zeta)$ is a random variable.

The set of all possible sequences $\{x(n, \zeta)\}$ is called an *ensemble*, and each individual sequence $x(n, \zeta_k)$, corresponding to a specific value of $\zeta = \zeta_k$, is called a *realization* or a *sample sequence* of the ensemble.

There are four possible interpretations of $x(n, \zeta)$, depending on the character of n and ζ, as illustrated in Figure 3.7:

- $x(n, \zeta)$ is a random variable if n is *fixed* and ζ is a variable.
- $x(n, \zeta)$ is a sample sequence if ζ is *fixed* and n is a variable.
- $x(n, \zeta)$ is a number if both n and ζ are *fixed*.
- $x(n, \zeta)$ is a stochastic process if both n and ζ are variables.

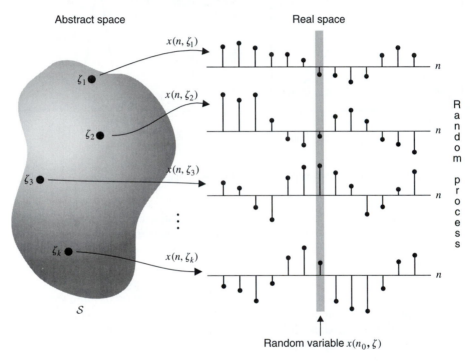

FIGURE 3.7
Graphical description of random sequences.

A random sequence is also called a *time series* in the statistics literature. It is a sequence of random variables, or it can be thought of as an *infinite-dimensional* random vector. As with any collection of infinite objects, one has to be careful with the asymptotic (or convergence) properties of a random sequence. If n is a continuous variable taking values in $\mathbb{R}$, then $x(n, \zeta)$ is an *uncountable* collection of random variables or an ensemble of waveforms. This ensemble is called a continuous-time stochastic process or a *random process*. Although these processes can be handled similarly to sequences, they are more difficult to deal with in a rigorous mathematical manner than sequences are. Furthermore, practical signal processing requires discrete-time signals. Hence in this book we consider random sequences rather than random waveforms.

Finally, in passing we note that the word *stochastic* is derived from the Greek word *stochasticos*, which means skillful in aiming or guessing. Hence, the terms *random process* and *stochastic process* will be used interchangeably throughout this book.

As mentioned before, a deterministic signal is by definition exactly predictable. This assumes that there exists a certain functional relationship that completely describes the signal, even if this relationship is not available. The unpredictability of a random process is, in general, the combined result of two things. First, the selection of a single realization is based on the outcome of a random experiment. Second, no functional description is available for all realizations of the ensemble. However, in some special cases, such a functional relationship is available. This means that after the occurrence of a specific realization, its future values can be predicted exactly from its past ones. If the future samples of any realization of a stochastic process can be predicted from the past ones, the process is called *predictable* or *deterministic*; otherwise, it is said to be a *regular process*. For example, the process $x(n, \zeta) = c$, where c is a random variable, is a predictable stochastic process because every realization is a discrete-time signal with constant amplitude. In practice, we most often deal with regular stochastic processes.

The simplest description of any random signal is provided by an amplitude-versus-time plot. Inspection of this plot provides qualitative information about some significant features of the signal that are useful in many applications. These features include, among others, the following:

1. The frequency of occurrence of various signal amplitudes, described by the probability distribution of samples.
2. The degree of dependence between two signal samples, described by the correlation between them.
3. The existence of "cycles" or quasi-periodic patterns, obtained from the signal power spectrum (which will be described in Section 3.3.6).
4. Indications of variability in the mean, variance, probability density, or spectral content.

The first feature above, the amplitude distribution, is obtained by plotting the histogram, which is an estimate of the first-order probability density of the underlying stochastic process. The probability density indicates waveform features such as "spikiness" and boundedness. Its form is crucial in the design of reliable estimators, quantizers, and event detectors.

The dependence between two signal samples (which are random variables) is given theoretically by the autocorrelation sequence and is quantified in practice by the empirical correlation (see Chapter 1), which is an estimate of the autocorrelation sequence of the underlying process. It affects the rate of amplitude change from sample to sample.

Cycles in the data are related to sharp peaks in the power spectrum or periodicity in the autocorrelation. Although the power spectrum and the autocorrelation contain the same information, they present it in different fashions.

Variability in a given quantity (e.g., variance) can be studied by evaluating this quantity for segments that can be assumed locally stationary and then analyzing the segment-to-segment variation. Such short-term descriptions should be distinguished from long-term ones, where the whole signal is analyzed as a single segment.

All the above features, to a lesser or greater extent, are interrelated. Therefore, it is impossible to point out exactly the effect of each one upon the visual appearance of the signal. However, a lot of insight can be gained by introducing the concepts of signal variability and signal memory, which are discussed in Sections 3.3.5 and 3.4.3 respectively.

3.3.1 Description Using Probability Functions

From Figure 3.7, it is clear that at $n = n_0$, $x(n_0, \zeta)$ is a random variable that requires a first-order probability function, say cdf $F_x(x; n_0)$, for its description. Similarly, $x(n_1, \zeta)$ and $x(n_2, \zeta)$ are joint random variables at instances n_1 and n_2, respectively, requiring a joint cdf $F_x(x_1, x_2; n_1, n_2)$. Stochastic processes contain infinitely many such random variables. Hence they are completely described, in a statistical sense, if their kth-order distribution function

$$F_x(x_1, \ldots, x_k; n_1, \ldots, n_k) = \Pr\{x(n_1) \le x_1, \ldots, x(n_k) \le x_k\} \quad (3.3.1)$$

is known for every value of $k \ge 1$ and for all instances $n_1, n_2, \ldots, n_k$. The kth-order pdf is given by

$$f_x(x_1, \ldots, x_k; n_1, \ldots, n_k) \triangleq \frac{\partial^{2k} F_x(x_1, \ldots, x_K; n_1, \ldots, n_k)}{\partial x_{R1} \cdots \partial x_{Ik}} \quad k \ge 1 \quad (3.3.2)$$

Clearly, the probabilistic description requires a lot of information that is difficult to obtain in practice except for simple stochastic processes. However, many (but not all) properties of a stochastic process can be described in terms of *averages* associated with its first- and second-order densities.

For simplicity, in the rest of the book, we will use a compact notation $x(n)$ to represent either a random process $x(n, \zeta)$ or a single realization $x(n)$, which is a member of the ensemble. Thus we will drop the variable ζ from all notations involving random variables, vectors, or processes. We believe that this will not cause any confusion and that the exact meaning will be clear from the context. Also the random process $x(n)$ is assumed to be complex-valued unless explicitly specified as real-valued.

3.3.2 Second-Order Statistical Description

The second-order statistic of $x(n)$ at time n is specified by its *mean value* $\mu_x(n)$ and its *variance* $\sigma_x^2(n)$, defined by

$$\mu_x(n) = E\{x(n)\} = E\{x_R(n) + jx_I(n)\} \quad (3.3.3)$$

and
$$\sigma_x^2(n) = E\{|x(n) - \mu_x(n)|^2\} = E\{|x(n)|^2\} - |\mu_x(n)|^2 \quad (3.3.4)$$

respectively. Note that both $\mu_x(n)$ and $\sigma_x(n)$ are, in general, deterministic sequences.

The second-order statistics of $x(n)$ at two different times n_1 and n_2 are given by the two-dimensional autocorrelation (or autocovariance) sequences. The *autocorrelation sequence* of a discrete-time random process is defined as the joint moment of the random variables $x(n_1)$ and $x(n_2)$, that is,

$$r_{xx}(n_1, n_2) = E\{x(n_1)x^*(n_2)\} \quad (3.3.5)$$

It provides a measure of the dependence between values of the process at two different times. In this sense, it also provides information about the time variation of the process. The *autocovariance* sequence of $x(n)$ is defined by

$$\gamma_{xx}(n_1, n_2) = E\{[x(n_1) - \mu_x(n_1)][x(n_2) - \mu_x(n_2)]^*\}$$
$$= r_{xx}(n_1, n_2) - \mu_x(n_1)\mu_x^*(n_2) \quad (3.3.6)$$

We will use notations such as $\gamma_x(n_1, n_2)$, $r_x(n_1, n_2)$, $\gamma(n_1, n_2)$, or $r(n_1, n_2)$ when there is no confusion as to which signal we are referring. Note that, in general, the second-order statistics are defined on a two-dimensional grid of integers.

The statistical relation between two stochastic processes $x(n)$ and $y(n)$ that are jointly distributed (i.e., they are defined on the same sample space $\mathcal{S}$) can be described by their *cross-correlation* and *cross-covariance* functions, defined by

$$r_{xy}(n_1, n_2) = E\{x(n_1)y^*(n_2)\} \quad (3.3.7)$$

and
$$\gamma_{xy}(n_1, n_2) = E\{[x(n_1) - \mu_x(n_1)][y(n_2) - \mu_y(n_2)]^*\}$$
$$= r_{xy}(n_1, n_2) - \mu_x(n_1)\mu_y^*(n_2) \quad (3.3.8)$$

The *normalized cross-correlation* of two random processes $x(n)$ and $y(n)$ is defined by

$$\rho_{xy}(n_1, n_2) = \frac{\gamma_{xy}(n_1, n_2)}{\sigma_x(n_1)\sigma_y(n_2)} \quad (3.3.9)$$

Some definitions

We now describe some useful types of stochastic processes based on their statistical properties. A random process is said to be

- An *independent* process if

$$f_x(x_1, \ldots, x_k; n_1, \ldots, n_k) = f_1(x_1; n_1) \cdots f_k(x_k; n_k) \qquad \forall k, n_i, i = 1, \ldots, k \quad (3.3.10)$$

that is, $x(n)$ is a sequence of independent random variables. If all random variables have the same pdf $f(x)$ for all k, then $x(n)$ is called an IID (independent and identically distributed) random sequence.

- An *uncorrelated* process if $x(n)$ is a sequence of uncorrelated random variables, that is,

$$\gamma_x(n_1, n_2) = \begin{cases} \sigma_x^2(n_1) & n_1 = n_2 \\ 0 & n_1 \neq n_2 \end{cases} = \sigma_x^2(n_1)\delta(n_1 - n_2) \qquad (3.3.11)$$

Alternatively, we have

$$r_x(n_1, n_2) = \begin{cases} \sigma_x^2(n_1) + |\mu_x(n_1)|^2 & n_1 = n_2 \\ \mu_x(n_1)\mu_x^*(n_2) & n_1 \neq n_2 \end{cases} \qquad (3.3.12)$$

- An *orthogonal* process if it is a sequence of orthogonal random variables, that is,

$$r_x(n_1, n_2) = \begin{cases} \sigma_x^2(n_1) + |\mu_x(n_1)|^2 & n_1 = n_2 \\ 0 & n_1 \neq n_2 \end{cases} = E\{|x(n_1)|^2\}\delta(n_1 - n_2) \quad (3.3.13)$$

- An *independent increment* process if $\forall k > 1$ and $\forall n_1 < n_2 < \cdots < n_k$, the *increments*

$$\{x(n_1)\}, \{x(n_2) - x(n_1)\}, \ldots, \{x(n_k) - x(n_{k-1})\}$$

are jointly independent. For such sequences, the kth-order probability function can be constructed as products of the probability functions of its increments.

- A *wide-sense periodic* (WSP) process with period N if

$$\mu_x(n) = \mu_x(n + N) \qquad \forall n \qquad (3.3.14)$$

and $\quad r_x(n_1, n_2) = r_x(n_1 + N, n_2) = r_x(n_1, n_2 + N) = r_x(n_1 + N, n_2 + N) \quad (3.3.15)$

Note that in the above definition, $\mu_x(n)$ is periodic in one dimension while $r_x(n_1, n_2)$ is periodic in two dimensions.

- A *wise-sense cyclostationary* process if there exists an integer N such that

$$\mu_x(n) = \mu_x(n + N) \qquad \forall n \qquad (3.3.16)$$

and $\qquad\qquad\qquad r_x(n_1, n_2) = r_x(n_1 + N, n_2 + N) \qquad\qquad\qquad (3.3.17)$

Note that in the above definition, $r_x(n_1, n_2)$ is *not* periodic in a two-dimensional sense. The correlation sequence is invariant to shift by N in *both* of its arguments.

- If all kth-order distributions of a stochastic process are jointly Gaussian, then it is called a Gaussian random sequence.

We can also extend some of these definitions to the case of two joint stochastic processes. The random processes $x(n)$ and $y(n)$ are said to be

- *Statistically independent* if for all values of n_1 and n_2

$$f_{xy}(x, y; n_1, n_2) = f_x(x; n_1)f_y(y; n_2) \qquad (3.3.18)$$

- *Uncorrelated* if for every n_1 and n_2 ($n_1 \neq n_2$)

$$\gamma_{xy}(n_1, n_2) = 0 \quad \text{or} \quad r_{xy}(n_1, n_2) = \mu_x(n_1)\mu_y^*(n_2) \qquad (3.3.19)$$

- *Orthogonal* if for every n_1 and n_2 ($n_1 \neq n_2$)

$$r_{xy}(n_1, n_2) = 0 \qquad (3.3.20)$$

3.3.3 Stationarity

A random process $x(n)$ is called *stationary* if statistics determined for $x(n)$ are equal to those for $x(n + k)$, for every k. More specifically, we have the following definition.

> **DEFINITION 3.3 (STATIONARY OF ORDER N).** A stochastic process $x(n)$ is called *stationary of order N* if
>
> $$f_x(x_1, \ldots, x_N; n_1, \ldots, n_N) = f_x(x_1, \ldots, x_N; n_{1+k}, \ldots, n_{N+k}) \qquad (3.3.21)$$
>
> for any value of k. If $x(n)$ is stationary for all orders $N = 1, 2, \ldots$, it is said to be *strict-sense stationary* (SSS).

An IID sequence is SSS. However, SSS is more restrictive than necessary for most practical applications. A more relaxed form of stationarity, which is sufficient for practical problems, occurs when a random process is *stationary up to order* 2, and it is also known as wide-sense stationarity.

> **DEFINITION 3.4 (WIDE-SENSE STATIONARITY).** A random signal $x(n)$ is called *wide-sense stationary* (WSS) if
>
> 1. Its mean is a constant independent of n, that is,
>
> $$E\{x(n)\} = \mu_x \qquad (3.3.22)$$
>
> 2. Its variance is also a constant independent of n, that is,
>
> $$\text{var}[x(n)] = \sigma_x^2 \qquad (3.3.23)$$
>
> and
> 3. Its autocorrelation depends only on the distance $l = n_1 - n_2$, called *lag*, that is,
>
> $$r_x(n_1, n_2) = r_x(n_1 - n_2) = r_x(l) = E\{x(n + l)x^*(n)\} = E\{x(n)x^*(n - l)\} \qquad (3.3.24)$$

From (3.3.22), (3.3.24), and (3.3.5) it follows that the autocovariance of a WSS signal also depends only on $l = n_1 - n_2$, that is,

$$\gamma_x(l) = r_x(l) - |\mu_x|^2 \qquad (3.3.25)$$

> **EXAMPLE 3.3.1.** Let $w(n)$ be a zero-mean, uncorrelated Gaussian random sequence with variance $\sigma^2(n) = 1$.
>
> *a.* Characterize the random sequence $w(n)$.
> *b.* Define $x(n) = w(n) + w(n - 1)$, $-\infty < n < \infty$. Determine the mean and autocorrelation of $x(n)$. Also characterize $x(n)$.
>
> ***Solution.*** Note that the variance of $w(n)$ is a constant.
>
> *a.* Since uncorrelatedness implies independence for Gaussian random variables, $w(n)$ is an independent random sequence. Since its mean and variance are constants, it is at least stationary in the first order. Furthermore, from (3.3.12) or (3.3.13) we have
>
> $$r_w(n_1, n_2) = \sigma^2 \delta(n_1 - n_2) = \delta(n_1 - n_2)$$
>
> Hence $w(n)$ is also a WSS random process.
> *b.* The mean of $x(n)$ is zero for all n since $w(n)$ is a zero-mean process. Consider
>
> $$\begin{aligned}
> r_x(n_1, n_2) &= E\{x(n_1)x(n_2)\} \\
> &= E\{[w(n_1) + w(n_1 - 1)][w(n_2) + w(n_2 - 1)]\} \\
> &= r_w(n_1, n_2) + r_w(n_1, n_2 - 1) + r_w(n_1 - 1, n_2) \\
> &\quad + r_w(n_1 - 1, n_2 - 1) \\
> &= \sigma^2 \delta(n_1 - n_2) + \sigma^2 \delta(n_1 - n_2 + 1) \\
> &\quad + \sigma^2 \delta(n_1 - 1 - n_2) + \sigma^2 \delta(n_1 - 1 - n_2 + 1) \\
> &= 2\delta(n_1 - n_2) + \delta(n_1 - n_2 + 1) + \delta(n_1 - n_2 - 1)
> \end{aligned}$$

Clearly, $r_x(n_1, n_2)$ is a function of $n_1 - n_2$. Hence

$$r_x(l) = 2\delta(l) + \delta(l+1) + \delta(l-1)$$

Therefore, $x(n)$ is a WSS sequence. However, it is not an independent random sequence since both $x(n)$ and $x(n+1)$ depend on $w(n)$.

EXAMPLE 3.3.2 (WIENER PROCESS). Toss a fair coin at each n, $-\infty < n < \infty$. Let

$$w(n) = \begin{cases} +S & \text{if heads is outcome} & \Pr(\text{H}) = 1 \\ -S & \text{if tails is outcome} & \Pr(\text{T}) = 1 \end{cases}$$

where S is a step size. Clearly, $w(n)$ is an independent random process with

$$E\{w(n)\} = 0$$

and

$$E\{w^2(n)\} = \sigma_w^2 = S^2 \left(\tfrac{1}{2}\right) + S^2 \left(\tfrac{1}{2}\right) = S^2$$

Define a new random process $x(n)$, $n \geq 1$, as

$$x(1) = w(1)$$
$$x(2) = x(1) + w(2) = w(1) + w(2)$$
$$\vdots$$
$$x(n) = x(n-1) + w(n) = \sum_{i=1}^{n} w(i)$$

Note that $x(n)$ is a running sum of independent steps or increments; thus it is an independent increment process. Such a sequence is called a *discrete Wiener process* or *random walk*. We can easily see that

$$E\{x(n)\} = E\left\{\sum_{i=1}^{n} w(i)\right\} = 0$$

and

$$E\{x^2(n)\} = E\left\{\sum_{i=1}^{n} w(i) \sum_{k=1}^{n} w(k)\right\} = E\left\{\sum_{i=1}^{n} \sum_{k=1}^{n} w(i)w(k)\right\}$$

$$= \sum_{i=1}^{n} \sum_{k=1}^{n} E\{w(i)w(k)\} = \sum_{i=1}^{n} E\{w^2(i)\} = nS^2$$

Therefore, random walk is a nonstationary (or *evolutionary*) process with zero mean and variance that grows with n, the number of steps taken.

It should be stressed at this point that although any strict-sense stationary signal is wide-sense stationary, the inverse is not always true, except if the signal is Gaussian. However in practice, it is very rare to encounter a signal that is stationary in the wide sense but not stationary in the strict sense (Papoulis 1991).

Two random signals $x(n)$ and $y(n)$ are called *jointly wide-sense stationary* if each is wide-sense stationary and their cross-correlation depends only on $l = n_1 - n_2$

$$r_{xy}(l) = E\{x(n)y^*(n-l)\} = r_{xy}(l) - \mu_x \mu_y^* \tag{3.3.26}$$

Note that as a consequence of wide-sense stationarity the two-dimensional correlation and covariance sequences become one-dimensional sequences. This is a very important result that ultimately allows for a nice spectral description of stationary random processes.

Properties of autocorrelation sequences

The autocorrelation sequence of a stationary process has many important properties (which also apply to autocovariance sequences, but we will discuss mostly correlation sequences). Vector versions of these properties are discussed extensively in Section 3.4.4, and their proofs are explored in the problems.

PROPERTY 3.3.1. The average power of a WSS process $x(n)$ satisfies

$$r_x(0) = \sigma_x^2 + |\mu_x|^2 \geq 0 \tag{3.3.27}$$

and

$$r_x(0) \geq |r_x(l)| \quad \text{for all } l \tag{3.3.28}$$

Proof. See Problem 3.21 and Property 3.3.6.

This property implies that the correlation attains its maximum value at zero lag and this value is nonnegative. The quantity $|\mu_x|^2$ is referred to as the *average dc power*, and the quantity $\sigma_x^2 = \gamma_x(0)$ is referred to as the *average ac power* of the random sequence. The quantity $r_x(0)$ then is the *total* average power of $x(n)$.

PROPERTY 3.3.2. The autocorrelation sequence $r_x(l)$ is a conjugate symmetric function of lag l, that is,

$$r_x^*(-l) = r_x(l) \tag{3.3.29}$$

Proof. It follows from Definition 3.4 and from (3.3.24).

PROPERTY 3.3.3. The autocorrelation sequence $r_x(l)$ is nonnegative definite; that is, for any $M > 0$ and any α_k, α_m

$$\sum_{k=1}^{M} \sum_{m=1}^{M} \alpha_k r_x(k - m)\alpha_m^* \geq 0 \tag{3.3.30}$$

This is a necessary and sufficient condition for a sequence $r_x(l)$ to be the autocorrelation sequence of a random sequence.

Proof. See Problem 3.22.

Since in this book we exclusively deal with wide-sense stationary processes, we will use the term *stationary* to mean wide-sense stationary. The properties of autocorrelation and cross-correlation sequences of jointly stationary processes, $x(n)$ and $y(n)$, are summarized in Table 3.1.

Although SSS and WSS forms are widely used in practice, there are processes with different forms of stationarity. Consider the following example.

EXAMPLE 3.3.3. Let $x(n)$ be a real-valued random process generated by the system

$$x(n) = \alpha x(n - 1) + w(n) \quad n \geq 0 \quad x(-1) = 0 \tag{3.3.31}$$

where $w(n)$ is a stationary random process with mean μ_w and $r_w(l) = \sigma_w^2 \delta(l)$. The process $x(n)$ generated using (3.3.31) is known as a *first-order autoregressive*, or AR(1), process,[†] and the process $w(n)$ is known as a *white noise* process (defined in Section 3.3.6). Determine the mean $\mu_x(n)$ of $x(n)$ and comment on its stationarity.

Solution. To compute the mean of $x(n)$, we express it as a function of $\{w(n), w(n-1), \ldots, w(0)\}$ as follows

$$x(0) = \alpha x(-1) + w(0) = w(0)$$
$$x(1) = \alpha x(0) + w(1) = \alpha w(0) + w(1)$$
$$\vdots$$
$$x(n) = \alpha^n w(0) + \alpha^{n-1} w(1) + \cdots + w(n) = \sum_{k=0}^{n} \alpha^k w(n - k)$$

[†] Note that from (3.3.31), $x(n-1)$ completely determines the distribution for $x(n)$, and $x(n)$ completely determines the distribution for $x(n + 1)$, and so on. If

$$f_{x(n)|x(n-1)\ldots}(x_n|x_{n-1}\ldots) = f_{x(n)|x(n-1)}(x_n|x_{n-1})$$

then the process is termed a *Markov process*.

Hence the mean of $x(n)$ is given by

$$\mu_x(n) = E\left\{\sum_{k=0}^{n}\alpha^k w(n-k)\right\} = \mu_w\left(\sum_{k=0}^{n}\alpha^k\right) = \begin{cases} \dfrac{1-\alpha^{n+1}}{1-\alpha}\mu_w & \alpha \neq 1 \\ (n+1)\mu_w & \alpha = 1 \end{cases}$$

Clearly, the mean of $x(n)$ depends on n, and hence it is nonstationary. However, if we assume that $|\alpha| < 1$ (which implies that the system is BIBO stable), then as $n \to \infty$, we obtain

$$\mu_x(n) = \mu_w \frac{1-\alpha^{n+1}}{1-\alpha} \xrightarrow[n\to\infty]{} \frac{\mu_x}{1-\alpha}$$

Thus $x(n)$ approaches first-order stationarity for large n. Similar analysis for the autocorrelation of $x(n)$ shows that $x(n)$ approaches wide-sense stationarity for large n (see Problem 3.23).

The above example illustrates a form of stationarity called *asymptotic* stationarity. A stochastic process $x(n)$ is *asymptotically stationary* if the statistics of random variables $x(n)$ and $x(n+k)$ become stationary as $k \to \infty$. When LTI systems are driven by zero-mean uncorrelated-component random processes, the output process becomes asymptotically stationary in the *steady state*. Another useful form of stationarity is given by stationary increments. If the increments $\{x(n) - x(n-k)\}$ of a process $x(n)$ form a stationary process for every k, we say that $x(n)$ is a process with *stationary increments*. Such processes can be used to model data in various practical applications (see Chapter 12).

The simplest way, to examine in practice if a real-world signal is stationary, is to investigate the physical mechanism that produces the signal. If this mechanism is time-invariant, then the signal is stationary. In case it is impossible to draw a conclusion based on physical considerations, we should rely on statistical methods (Bendat and Piersol 1986; Priestley 1981). Note that stationarity in practice means that a random signal has statistical properties that do not change over the time interval we observe the signal. For evolutionary signals the statistical properties change continuously with time. An example of a highly nonstationary random signal is the signals associated with the vibrations induced in space vehicles during launch and reentry. However, there is a kind of random signal whose statistical properties change slowly with time. Such signals, which are stationary over short periods, are called *locally stationary* signals. Many signals of great practical interest, such as speech, EEG, and ECG, belong to this family of signals.

Finally, we note that general techniques for the analysis of nonstationary signals do not exist. Thus only special methods that apply to specific types of nonstationary signals can be developed. Many such methods remove the nonstationary component of the signal, leaving behind another component that can be analyzed as stationary (Bendat and Piersol 1986; Priestley 1981).

3.3.4 Ergodicity

A stochastic process consists of the ensemble and a probability law. If this information is available, the statistical properties of the process can be determined in a quite straightforward manner. However, in the real world, we have access to only a limited number (usually one) of realizations of the process. The question that arises then is, Can we infer the statistical characteristics of the process from a single realization?

This is possible for the class of random processes that are called *ergodic* processes. Roughly speaking, ergodicity implies that all the statistical information can be obtained from any single representative member of the ensemble.

Time averages

All the statistical averages that we have defined up to this point are known as *ensemble averages* because they are obtained by "freezing" the time variable and averaging over the ensemble (see Fig. 3.7). Averages of this type are formally defined by using the expectation operator $E\{\ \}$. Ensemble averaging is not used frequently in practice, because it is impractical to obtain the number of realizations needed for an accurate estimate. Thus the need for a different kind of average, based on only one realization, naturally arises. Obviously such an average can be obtained only by time averaging.

The *time average* of a quantity, related to a discrete-time random signal, is defined as

$$\langle(\cdot)\rangle \triangleq \lim_{N\to\infty} \frac{1}{2N+1} \sum_{n=-N}^{N} (\cdot) \tag{3.3.32}$$

Note that, owing to its dependence on a single realization, any time average is itself a random variable. The time average is taken over all time because all realizations of a random process exist for all time; that is, they are power signals.

For every ensemble average we can define a corresponding time average. The following time averages are of special interest:

$$\text{Mean value} = \langle x(n)\rangle$$
$$\text{Mean square} = \langle |x(n)|^2\rangle$$
$$\text{Variance} = \langle |x(n) - \langle x(n)\rangle|^2\rangle$$
$$\text{Autocorrelation} = \langle x(n)x^*(n-l)\rangle \tag{3.3.33}$$
$$\text{Autocovariance} = \langle [x(n) - \langle x(n)\rangle][x(n-l) - \langle x(n)\rangle]^*\rangle$$
$$\text{Cross-correlation} = \langle x(n)y^*(n-l)\rangle$$
$$\text{Cross-covariance} = \langle [x(n) - \langle x(n)\rangle][y(n-l) - \langle y(n)\rangle]^*\rangle$$

It is necessary to mention at this point the remarkable similarity between time averages and the correlation sequences for deterministic power signals. Although this is just a formal similarity, due to the fact that random signals are power signals, both quantities have the same properties. However, we should always keep in mind that although time averages are random variables (because they are functions of ζ), the corresponding quantities for deterministic power signals are fixed numbers or deterministic sequences.

Ergodic random processes

As we have already mentioned, in many practical applications only one realization of a random signal is available instead of the entire ensemble. In general, a single member of the ensemble does not provide information about the statistics of the process. However, if the process is stationary and ergodic, then all statistical information can be derived from only one typical realization of the process.

A random signal $x(n)$ is called *ergodic*[†] if its ensemble averages equal appropriate time averages. There are several degrees of ergodicity (Papoulis 1991). We will discuss two of them: ergodicity in the mean and ergodicity in correlation.

DEFINITION 3.5 (ERGODIC IN THE MEAN). A random process $x(n)$ is ergodic *in the mean* if

$$\langle x(n)\rangle = E\{x(n)\} \tag{3.3.34}$$

DEFINITION 3.6 (ERGODIC IN CORRELATION). A random process $x(n)$ is *ergodic in correlation* if

$$\langle x(n)x^*(n-l)\rangle = E\{x(n)x^*(n-l)\} \tag{3.3.35}$$

Note that since $\langle x(n)\rangle$ is constant and $\langle x(n)x^*(n-l)\rangle$ is a function of l, if $x(n)$ is ergodic in both the mean and correlation, then it is also WSS. Thus only stationary signals can be ergodic. On the other hand, WSS does not imply ergodicity of any kind. Fortunately,

---

[†] Strictly speaking, the form of ergodicity that we will use is called *mean-square ergodicity* since the underlying convergence of random variables is in the mean-square sense (Stark and Woods 1994). Therefore, equalities in the definitions are in the mean-square sense.

in practice almost all stationary processes are also ergodic, which is very useful for the estimation of their statistical properties. From now on we will use the term *ergodic* to mean both ergodicity in the mean and ergodicity in correlation.

DEFINITION 3.7 (JOINT ERGODICITY). Two random signals are called *jointly ergodic* if they are individually ergodic and in addition

$$\langle x(n)y^*(n-l) \rangle = E\{x(n)y^*(n-l)\} \tag{3.3.36}$$

A physical interpretation of ergodicity is that one realization of the random signal $x(n)$, as time n tends to infinity, takes on values with the same statistics as the value $x(n_1)$, corresponding to all samples of the ensemble members at a given time $n = n_1$.

In practice, it is of course impossible to use the time-average formulas introduced above, because only finite records of data are available. In this case, it is common practice to replace the operator (3.3.32) by the operator

$$\langle (\cdot) \rangle_N = \frac{1}{2N+1} \sum_{n=-N}^{N} (\cdot) \tag{3.3.37}$$

to obtain *estimates* of the true quantities. Our desire in such problems is to find estimates that become increasingly accurate (in a sense to be defined in Section 3.6) as the length $2N + 1$ of the record of used data becomes larger.

Finally, to summarize, we note that whereas stationarity ensures the time invariance of the statistics of a random signal, ergodicity implies that any statistics can be calculated either by averaging over all members of the ensemble at a fixed time or by time-averaging over any single representative member of the ensemble.

3.3.5 Random Signal Variability

If we consider a stationary random sequence $w(n)$ that is IID with zero mean, its key characteristics depend on its first-order density. Figure 3.8 shows the probability density functions and sample realizations for IID processes with uniform, Gaussian, and Cauchy probability distributions. In the case of the uniform distribution, the amplitude of the random variable is limited to a range, with values occurring outside this interval with zero probability. On the other hand, the Gaussian distribution does not have a finite interval of support, allowing for the possibility of any value. The same is true of the Cauchy distribution, but its characteristics are dramatically different from those of the Gaussian distribution. The center lobe of the density is much narrower while the tails that extend out to infinity are significantly higher. As a result, the realization of the Cauchy random process contains numerous spikes or extreme values while the remainder of the process is more compact about the mean. Although the Gaussian random process allows for the possibility of large values, the probability of their occurrence is so small that they are not found in realizations of the process.

The major difference between the Gaussian and Cauchy distributions lies in the area found under the tails of the density as it extends out to infinity. This characteristic is related to the *variability* of the process. The heavy tails, as found in the Cauchy distribution, result in an abundance of spikes in the process, a characteristic referred to as *high variability*. On the other hand, a distribution such as the Gaussian does not allow for extreme values and indicates *low variability*. The extent of the variability of a given distribution is determined by the heaviness of the tails. Distributions with heavy tails are called *long-tailed* distributions and have been used extensively as models of impulsive random processes.

DEFINITION 3.8. A distribution is called *long-tailed* if its tails decay hyperbolically or algebraically as

$$Pr\{|x(n)| \geq x\} \sim Cx^{-\alpha} \qquad \text{as } x \to \infty \tag{3.3.38}$$

where C is a constant and the variable α determines the rate of decay of the distribution.

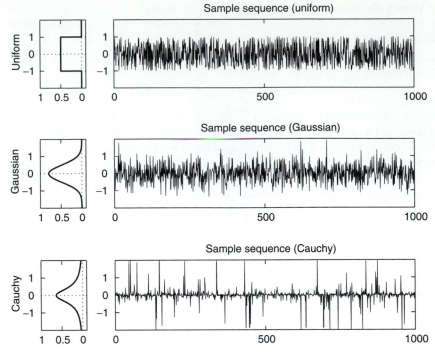

FIGURE 3.8
Probability density functions and sample realizations of an IID process with uniform,
Gaussian, and Cauchy distributions.

By means of comparison, the Gaussian distribution has an exponential rate of decay.
The implication of the algebraically decaying tail is that the process has infinite variance,
that is,

$$\sigma_x^2 = E\{|x(n)|^2\} = \infty$$

and therefore lacks second-order moments. The lack of second-order moments means that, in
addition to the variance, the correlation functions of these processes do not exist. Since most
signal processing algorithms are based on second-order moment theory, infinite variance
has some extreme implications for the way in which such processes are treated.

In this book, we shall model high variability, and hence infinite variance, using the
family of symmetric stable distributions. The reason is twofold: First, a linear combination of
stable random variables is stable. Second, stable distributions appear as limits in central limit
theorems (see stable distributions in Section 3.2.4). Stable distributions are characterized
by a parameter $\alpha, 0 < \alpha \le 2$. They are Cauchy when $\alpha = 1$ and Gaussian when $\alpha = 2$.
However, they have finite variance only when $\alpha = 2$.

In practice, the type of data under consideration governs the variability of the modeling
distribution. Random signals restricted to a certain interval, such as the phase of complex
random signals, are well suited for uniform distributions. On the other hand, signals allowing
for any possible value but generally confined to a region are better suited for Gaussian
models. However, if a process contains spikes and therefore has high variability, it is best
characterized by a long-tailed distribution such as the Cauchy distribution. Impulsive signals
have been found in a variety of applications, such as communication channels, radar signals,
and electronic circuit noise. In all cases, the variability of the process dictates the appropriate
model.

3.3.6 Frequency-Domain Description of Stationary Processes

Discrete-time stationary random processes have correlation sequences that are functions of a single index. This leads to nice and powerful representations in both the frequency and the z-transform domains.

Power spectral density

The *power spectral density* (PSD, or more appropriately autoPSD) of a stationary stochastic process $x(n)$ is a Fourier transformation of its autocorrelation sequence $r_x(l)$. If $r_x(l)$ is periodic (which corresponds to a wide-sense periodic stochastic process) in l, then the DTFS discussed in Section 2.2.1 can be used to obtain the PSD, which has the form of a *line spectrum*. If $r_x(l)$ is nonperiodic, the DTFT discussed in Section 2.2.1 can be used provided that $r_x(l)$ is absolutely summable. This means that the process $x(n)$ must be a zero-mean process. In general, a stochastic process can be a mixture of periodic and nonperiodic components.[†]

If we allow impulse functions in the DTFT to represent periodic (or almost periodic) sequences and non-zero-mean processes (see Section 2.2.1), then we can define the PSD as

$$R_x(e^{j\omega}) = \sum_{l=-\infty}^{\infty} r_x(l)e^{-j\omega l} \tag{3.3.39}$$

where ω is the frequency in radians per sample. If the process $x(n)$ is a zero-mean nonperiodic process, then (3.3.39) is enough to determine the PSD. If $x(n)$ is periodic (including nonzero mean) or almost periodic, then the PSD is given by

$$R_x(e^{j\omega}) = \sum_{i} 2\pi A_i \delta(\omega - \omega_i) \tag{3.3.40}$$

where the A_i are amplitudes of $r_x(l)$ at frequencies ω_i. For discussion purposes we will assume that $x(n)$ is a zero-mean nonperiodic process. The autocorrelation $r_x(l)$ can be recovered from the PSD by using the inverse DTFT as

$$r_x(l) = \frac{1}{2\pi} \int_{-\pi}^{\pi} R_x(e^{j\omega})e^{j\omega l}\, d\omega \tag{3.3.41}$$

EXAMPLE 3.3.4. Determine the PSD of a zero-mean WSS process $x(n)$ with $r_x(l) = a^{|l|}$, $-1 < a < 1$.

Solution. From (3.3.39) we have

$$R_x(e^{j\omega}) = \sum_{l=-\infty}^{\infty} a^{|l|}e^{-j\omega l} \qquad -1 < a < 1$$

$$= \frac{1}{1 - ae^{j\omega}} + \frac{1}{1 - ae^{-j\omega}} - 1 \tag{3.3.42}$$

$$= \frac{1 - a^2}{1 + a^2 - 2a\cos\omega} \qquad -1 < a < 1$$

which is a real-valued, even, and nonnegative function of ω.

Properties of the autoPSD. The power spectral density $R_x(e^{j\omega})$ has three key properties that follow from corresponding properties of the autocorrelation sequence and the DTFT.

[†]Periodic components are predictable processes as discussed before. However, some nonperiodic components can also be predictable. Hence nonperiodic components are not always regular processes.

PROPERTY 3.3.4. The autoPSD $R_x(e^{j\omega})$ is a real-valued periodic function of frequency with period 2π for any (real- or complex-valued) process $x(n)$. If $x(n)$ is real-valued, then $R_x(e^{j\omega})$ is also an even function of ω, that is,

$$R_x(e^{j\omega}) = R_x(e^{-j\omega}) \tag{3.3.43}$$

Proof. It follows from autocorrelation and DTFT properties.

PROPERTY 3.3.5. The autoPSD is nonnegative definite, that is,

$$R_x(e^{j\omega}) \geq 0 \tag{3.3.44}$$

Proof. This follows from the nonnegative definiteness of the autocorrelation sequence [see also discussions leading to (3.4.27)].

PROPERTY 3.3.6. The area under $R_x(e^{j\omega})$ is *nonnegative* and it equals the average power of $x(n)$. Indeed, from (3.3.41) it follows with $l = 0$ that

$$\frac{1}{2\pi} \int_{-\pi}^{\pi} R_x(e^{j\omega}) \, d\omega = r_x(0) = E\{|x(n)|^2\} \geq 0 \tag{3.3.45}$$

Proof. It follows from Property 3.3.5.

White noise. A random sequence $w(n)$ is called a (second-order) *white noise process* with mean μ_w and variance σ_w^2, denoted by

$$w(n) \sim \text{WN}(\mu_w, \sigma_w^2) \tag{3.3.46}$$

if and only if $E\{w(n)\} = \mu_x$ and

$$r_w(l) = E\{w(n)w^*(n-l)\} = \sigma_w^2 \delta(l) \tag{3.3.47}$$

which implies that $\qquad R_w(e^{j\omega}) = \sigma_w^2 \qquad -\pi \leq \omega \leq \pi \tag{3.3.48}$

The term *white noise* is used to emphasize that all frequencies contribute the same amount of power, as in the case of white light, which is obtained by mixing all possible colors by the same amount. If, in addition, the pdf of $x(n)$ is Gaussian, then the process is called a (second-order) *white Gaussian noise* process, and it will be denoted by $\text{WGN}(\mu_w, \sigma_w^2)$.

If the random variables $w(n)$ are independently and identically distributed with mean μ_w and variance σ_w^2, then we shall write

$$w(n) \sim \text{IID}(\mu_w, \sigma_w^2) \tag{3.3.49}$$

This is sometimes referred to as a *strict* white noise.

We emphasize that the conditions of uncorrelatedness or independence do not put any restriction on the form of the probability density function of $w(n)$. Thus we can have an IID process with any type of probability distribution. Clearly, white noise is the simplest random process because it does not have any structure. However, we will see that it can be used as the basic building block for the construction of processes with more complicated dependence or correlation structures.

Harmonic processes. A *harmonic process* is defined by

$$x(n) = \sum_{k=1}^{M} A_k \cos(\omega_k n + \phi_k) \qquad \omega_k \neq 0 \tag{3.3.50}$$

where M, $\{A_k\}_1^M$, and $\{\omega_k\}_1^M$ are constants and $\{\phi_k\}_1^M$ are pairwise independent random variables uniformly distributed in the interval $[0, 2\pi]$. It can be shown (see Problem 3.9) that $x(n)$ is a stationary process with mean

$$E\{x(n)\} = 0 \qquad \text{for all } n \tag{3.3.51}$$

$$r_x(l) = \frac{1}{2} \sum_{k=1}^{N} A_k^2 \cos \omega_k l \qquad -\infty < l < \infty \tag{3.3.52}$$

We note that $r_x(l)$ consists of a sum of "in-phase" cosines with the same frequencies as in $x(n)$.

If $\omega_k/(2\pi)$ are rational numbers, $r_x(l)$ is periodic and can be expanded as a Fourier series. These series coefficients provide the power spectrum $R_x(k)$ of $x(n)$. However, because $r_x(l)$ is a linear superposition of cosines, it always has a line spectrum with $2M$ lines of strength $A_k^2/4$ at frequencies $\pm\omega_k$. If $r_x(l)$ is periodic, then the lines are equidistant (i.e., harmonically related), hence the name *harmonic process*. If $\omega/(2\pi)$ is irrational, then $r_x(l)$ is almost periodic and can be treated in the frequency domain in almost the same fashion. Hence the power spectrum of a harmonic process is given by

$$R_x(e^{j\omega}) = \sum_{k=-M}^{M} 2\pi \left(\frac{A_k^2}{4} \right) \delta(\omega - \omega_k) = \sum_{k=-M}^{M} \frac{\pi}{2} A_k^2 \delta(\omega - \omega_k) \tag{3.3.53}$$

EXAMPLE 3.3.5. Consider the following harmonic process

$$x(n) = \cos(0.1\pi n + \phi_1) + 2\sin(1.5n + \phi_2)$$

where ϕ_1 and ϕ_2 are IID random variables uniformly distributed in the interval $[0, 2\pi]$. The first component of $x(n)$ is periodic with $\omega_1 = 0.1\pi$ and period equal to 20 while the second component is almost periodic with $\omega_2 = 1.5$. Thus the sequence $x(n)$ is almost periodic. A sample function realization of $x(n)$ is shown in Figure 3.9(a). The mean of $x(n)$ is

$$\mu_x(n) = E\{x(n)\} = E\{\cos(0.1\pi n + \phi_1) + 2\sin(1.5n + \phi_2)\} = 0$$

and the autocorrelation sequence (using mutual independence between ϕ_1 and ϕ_2) is

$$\begin{aligned} r_x(n_1, n_2) &= E\{x(n_1)x_2^*(n_2)\} \\ &= E\{\cos(0.1\pi n_1 + \phi_1)\cos(0.1\pi n_2 + \phi_1)\} \\ &\quad + E\{2\sin(1.5n_1 + \phi_2)2\sin(1.5n_2 + \phi_2)\} \\ &= \tfrac{1}{2}\cos[0.1\pi(n_1 - n_2)] + 2\cos[1.5(n_1 - n_2)] \end{aligned}$$

or $\qquad r_x(l) = \tfrac{1}{2}\cos 0.1\pi l + 2\cos 1.5l \qquad l = n_1 - n_2$

Thus the line spectrum $R_{\omega_k}^{(x)}$ is given by

$$R_{\omega_k}^{(x)} = \begin{cases} 1 & \omega_1 = -1.5 \\ \frac{1}{4} & \omega_2 = -0.1\pi \\ \frac{1}{4} & \omega_3 = 0.1\pi \\ 1 & \omega_4 = 1.5 \end{cases}$$

and the power spectrum $R_x(e^{j\omega})$ is given by

$$R_x(e^{j\omega}) = 2\pi\delta(\omega + 1.5) + \frac{\pi}{2}\delta(\omega + 0.1\pi) + \frac{\pi}{2}\delta(\omega - 0.1\pi) + 2\pi\delta(\omega - 1.5)$$

The line spectrum of $x(n)$ is shown in Figure 3.9(b) and the corresponding power spectrum in Figure 3.9(c).

The harmonic process is predictable because any given realization is a sinusoidal sequence with fixed amplitude, frequency, and phase. We stress that the independence of the phases is required to guarantee the stationarity of $x(n)$ in (3.3.50). The uniform distribution of the phases is necessary to make $x(n)$ a stationary process (see Problem 3.9). The harmonic process (3.3.50), in general, is non-Gaussian; however, it becomes Gaussian if the amplitudes A_k are random variables with a Rayleigh distribution (Porat 1994).

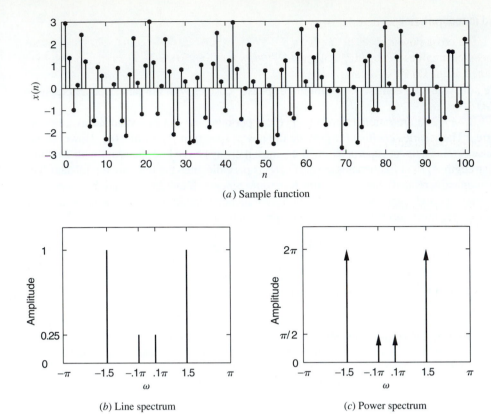

(a) Sample function

(b) Line spectrum

(c) Power spectrum

FIGURE 3.9
The time and frequency-domain description of the harmonic process in Example 3.3.5.

EXAMPLE 3.3.6. Consider a complex-valued process given by

$$x(n) = Ae^{j\omega_0 n} = |A|e^{j(\omega_0 n + \phi)}$$

where A is a complex-valued random variable and ω_0 is constant. The mean of $x(n)$

$$E\{x(n)\} = E\{A\}e^{j\omega_0 n}$$

can be constant only if $E\{A\} = 0$. If $|A|$ is constant and ϕ is uniformly distributed on $[0, 2\pi]$, then we have $E\{A\} = |A|E\{e^{j\phi}\} = 0$. In this case the autocorrelation is

$$r_x(n_1, n_2) = E\{Ae^{j(\omega_0 n_1 + \phi)}A^* e^{-j(\omega_0 n_2 + \phi)}\} = |A|^2 e^{j(n_1 - n_2)\omega_0}$$

Since the mean is constant and the autocorrelation depends on the difference $l \triangleq n_1 - n_2$, the process is wide-sense stationary.

The above example can be generalized to harmonic processes of the form

$$x(n) = \sum_{k=1}^{M} A_k e^{j(\omega_k n + \phi_k)} \tag{3.3.54}$$

where M, $\{A_k\}_1^M$, and $\{\omega_k\}_1^M$ are constants and $\{\phi_k\}_1^M$ are pairwise independent random variables uniformly distributed in the interval $[0, 2\pi]$. The autocorrelation sequence is

$$r_x(l) = \sum_{k=1}^{M} |A_k|^2 e^{j\omega_k l} \tag{3.3.55}$$

and the power spectrum consists of M impulses with amplitudes $2\pi|A_k|^2$ at frequencies ω_k. If the amplitudes $\{A_k\}_{k=1}^M$ are random variables, mutually independent of the random phases, the quantity $|A_k|^2$ is replaced by $E\{|A_k|^2\}$.

Cross-power spectral density

The cross-power spectral density of two zero-mean and jointly stationary stochastic processes provides a description of their statistical relations in the frequency domain and is defined as the DTFT of their cross-correlation, that is,

$$R_{xy}(e^{j\omega}) = \sum_{l=-\infty}^{\infty} r_{xy}(l)e^{-j\omega l} \tag{3.3.56}$$

The cross-correlation $r_{xy}(l)$ can be recovered by the inverse DTFT

$$r_{xy}(l) = \frac{1}{2\pi} \int_{-\pi}^{\pi} R_{xy}(e^{j\omega})e^{j\omega l} \, d\omega \tag{3.3.57}$$

The cross-spectrum $R_{xy}(e^{j\omega})$ is, in general, a complex function of ω. From $r_{xy}(l) = r_{yx}^*(-l)$ it follows that

$$R_{xy}(e^{j\omega}) = R_{yx}^*(e^{j\omega}) \tag{3.3.58}$$

This implies that $R_{xy}(e^{j\omega})$ and $R_{yx}(e^{j\omega})$ have the same magnitude but opposite phase.

The normalized cross-spectrum

$$\mathcal{G}_{xy}(e^{j\omega}) \triangleq \frac{R_{xy}(e^{j\omega})}{\sqrt{R_x(e^{j\omega})}\sqrt{R_y(e^{j\omega})}} \tag{3.3.59}$$

is called the *coherence function*. Its squared magnitude

$$|\mathcal{G}_{xy}(e^{j\omega})|^2 = \frac{|R_{xy}(e^{j\omega})|^2}{R_x(\omega)R_y(e^{j\omega})} \tag{3.3.60}$$

is known as the *magnitude square coherence* (*MSC*) and can be thought of as a sort of correlation coefficient in the frequency domain. If $x(n) = y(n)$, then $\mathcal{G}_{xy}(e^{j\omega}) = 1$ (maximum correlation) whereas if $x(n)$ and $y(n)$ are uncorrelated, then $R_{xy}(l) = 0$ and hence $\mathcal{G}_{xy}(e^{j\omega}) = 0$. In other words, $0 \leq |\mathcal{G}_{xy}(e^{j\omega})| \leq 1$.

Complex spectral density functions

If the sequences $r_x(l)$ and $r_{xy}(l)$ are absolutely summable within a certain ring of the complex z plane, we can obtain their z-transforms

$$R_x(z) = \sum_{l=-\infty}^{\infty} r_x(l)z^{-l} \tag{3.3.61}$$

$$R_{xy}(z) = \sum_{l=-\infty}^{\infty} r_{xy}(l)z^{-l} \tag{3.3.62}$$

which are known as the *complex spectral density* and *complex cross-spectral density* functions, respectively. If the unit circle, defined by $z = e^{j\omega}$, is within the region of convergence of the above summations, then

$$R_x(e^{j\omega}) = R_x(z)|_{z=e^{j\omega}} \tag{3.3.63}$$

$$R_{xy}(e^{j\omega}) = R_{xy}(z)|_{z=e^{j\omega}} \tag{3.3.64}$$

The correlation and power spectral density properties of random sequences are summarized in Table 3.1.

EXAMPLE 3.3.7. Consider the random sequence given in Example 3.3.4 with autoPSD in (3.3.42)

$$R_x(e^{j\omega}) = \frac{1 - a^2}{1 + a^2 - 2a \cos \omega} \qquad |a| < 1$$

Determine the complex autoPSD $R_x(z)$.

Solution. The complex autoPSD is given by $R_x(z) = R_x(e^{j\omega})|_{e^{j\omega}=z}$. Since

$$\cos\omega = \frac{e^{j\omega} + e^{-j\omega}}{2} = \left.\frac{z + z^{-1}}{2}\right|_{z=e^{j\omega}}$$

we obtain

$$R_x(z) = \frac{1 - a^2}{1 + a^2 - 2a\left(\dfrac{z + z^{-1}}{2}\right)} = \frac{(a - a^{-1})z^{-1}}{1 - (a + a^{-1})z^{-1} + z^{-2}} \qquad |a| < |z| < \frac{1}{|a|}$$

Now the inverse z-transform of $R_x(z)$ determines the autocorrelation sequence $r_x(l)$, that is,

$$R_x(z) = \frac{(a - a^{-1})z^{-1}}{1 - (a + a^{-1})z^{-1} + z^{-2}} = \frac{(a - a^{-1})z^{-1}}{(1 - az^{-1})(1 - a^{-1}z^{-1})}$$

$$= \frac{1}{(1 - az^{-1})} - \frac{1}{(1 - a^{-1}z^{-1})} \qquad |a| < |z| < |a|^{-1}$$

or

$$r_x(l) = a^l u(l) + (a^{-1})^l u(-l - 1) = a^{|l|} \qquad (3.3.65)$$

This approach can be used to determine autocorrelation sequences from autoPSD functions.

Table 3.1 provides a summary of correlation and spectral properties of stationary random sequences.

TABLE 3.1
Summary of correlation and spectral properties of stationary random sequences.

Definitions	
Mean value	$\mu_x = E\{x(n)\}$
Autocorrelation	$r_x(l) = E\{[x(n)x^*(n - l)\}$
Autocovariance	$\gamma_x(l) = E\{[x(n) - \mu_x][x(n - l) - \mu_x]^*\}$
Cross-correlation	$r_{xy}(l) = E\{x(n)y^*(n - l)\}$
Cross-covariance	$\gamma_{xy}(l) = E\{[x(n) - \mu_x][y(n - l) - \mu_y]^*\}$
Power spectral density	$R_x(e^{j\omega}) = \sum_{l=-\infty}^{\infty} r_x(l)e^{-j\omega l}$
Cross-power spectral density	$R_{xy}(e^{j\omega}) = \sum_{l=-\infty}^{\infty} r_{xy}(l)e^{-j\omega l}$
Magnitude square coherence	$|\mathcal{G}_{xy}(e^{j\omega})|^2 = |R_{xy}(e^{j\omega})|^2/[R_x(e^{j\omega})R_y(e^{j\omega})]$

Interrelations
$\gamma_x(l) = r_x(l) - |\mu_x|^2$
$\gamma_{xy}(l) = r_{xy}(l) - \mu_x\mu_y^*$

Properties	
Autocorrelation	**Auto-PSD**
$r_x(l)$ is nonnegative definite	$R_x(e^{j\omega}) \geq 0$ and real
$r_x(l) = r_x^*(-l)$	$R_x(e^{j\omega}) = R_x(e^{-j\omega})$ [real $x(n)$]
$|r_x(l)| \leq r_x(0)$	$R_x(z) = R_x^*(1/z^*)$
$|\rho_x(l)| \leq 1$	$R_x(z) = R_x(z^{-1})$ [real $x(n)$]
Cross-correlation	**Cross-PSD**
$r_{xy}(l) = r_{yx}^*(-l)$	
$|r_{xy}(l)| \leq [r_x(0)r_y(0)]^{1/2} \leq$	$R_{xy}(z) = R_{yx}^*(1/z^*)$
$\quad \frac{1}{2}[r_x(0) + r_y(0)]$	$0 \leq |\mathcal{G}_{xy}(e^{j\omega})| \leq 1$
$|\rho_{xy}(l)| \leq 1$	

This section deals with the processing of stationary random sequences using linear, time-invariant (LTI) systems. We focus on expressing the second-order statistical properties of the output in terms of the corresponding properties of the input and the characteristics of the system.

3.4.1 Time-Domain Analysis

The first question to ask when we apply a random signal to a system is, Just what is the meaning of such an operation? We ask this because a random process is not just a single sequence but an ensemble of sequences (see Section 3.3). However, since each realization of the stochastic process is a deterministic signal, it is an acceptable input producing an output that is clearly a single realization of the output stochastic process. For an LTI system, each pair of input-output realizations is described by the convolution summation

$$y(n, \zeta) = \sum_{k=-\infty}^{\infty} h(k)x(n - k, \zeta) \qquad (3.4.1)$$

If the sum in the right side of (3.4.1) exists for all ζ such that $\Pr\{\zeta\} = 1$, then we say that we have almost-everywhere convergence or convergence with probability 1 (Papoulis 1991). The existence of such convergence is ruled by the following theorem (Brockwell and Davis 1991).

THEOREM 3.2. If the process $x(n, \zeta)$ is stationary with $E\{|x(n, \zeta)|\} < \infty$ and if the system is BIBO-stable, that is, $\sum_{-\infty}^{\infty} |h(k)| < \infty$, then the output $y(n, \zeta)$ of the system in (3.4.1) converges absolutely with probability 1, or

$$y(n, \zeta) = \sum_{k=-\infty}^{\infty} h(k)x(n - k, \zeta) \qquad \text{for all } \zeta \in \mathcal{A}, \Pr\{\mathcal{A}\} = 1 \qquad (3.4.2)$$

and is stationary. Furthermore, if $E\{|x(n, \zeta)|^2\} < \infty$, then $E\{|y(n, \zeta)|^2\} < \infty$ and $y(n, \zeta)$ converges in the mean square to the same limit and is stationary.

A less restrictive condition of finite power on the system impulse response $h(n)$ also guarantees the mean square existence of the output process, as stated in the following theorem.

THEOREM 3.3. If the process $x(n, \zeta)$ is zero-mean and stationary with $\sum_{l=-\infty}^{\infty} |r_x(l)| < \infty$, and if the system (3.4.1) satisfies the condition

$$\sum_{n=-\infty}^{\infty} |h(n)|^2 = \frac{1}{2\pi} \int_{-\pi}^{\pi} |H(e^{j\omega})|^2 \, d\omega < \infty \qquad (3.4.3)$$

then the output $y(n, \zeta)$ converges in the mean square sense and is stationary.

The above two theorems are applicable when input processes have finite variances. However, IID sequences with α-stable distributions have infinite variances. If the impulse response of the system in (3.4.1) decays fast enough, then the following theorem (Brockwell and Davis 1991) guarantees the absolute convergence of $y(n, \zeta)$ with probability 1. These issues are of particular importance for inputs with high variability and are discussed in Section 3.3.5.

THEOREM 3.4. Let $x(n, \zeta)$ be an IID sequence of random variables with α-stable distribution, $0 < \alpha < 2$. If the impulse response $h(n)$ satisfies

$$\sum_{n=-\infty}^{\infty} |h(n)|^{\delta} < \infty \qquad \text{for some } \delta \in (0, \alpha)$$

then the output $y(n, \zeta)$ in (3.4.1) converges absolutely with probability 1.

Clearly, a complete description of the output stochastic process $y(n)$ requires the computation of an infinite number of convolutions. Thus, a better alternative would be to determine the statistical properties of $y(n)$ in terms of the statistical properties of the input and the characteristics of the system. For Gaussian signals, which are used very often in practice, first- and second-order statistics are sufficient.

Output mean value. If $x(n)$ is stationary, its first-order statistic is determined by its mean value μ_x. To determine the mean value of the output, we take the expected value of both sides of (3.4.1):

$$\mu_y = \sum_{k=-\infty}^{\infty} h(k)E\{x(n-k)\} = \mu_x \sum_{k=-\infty}^{\infty} h(k) = \mu_x H(e^{j0}) \qquad (3.4.4)$$

Since μ_x and $H(e^{j0})$ are constant, μ_y is also constant. Note that $H(e^{j0})$ is the dc gain of the spectrum.

Input-output cross-correlation. If we take complex conjugate of (3.4.1), premultiply it by $x(n+l)$, and take the expectation of both sides, we have

$$E\{x(n+l)y^*(l)\} = \sum_{k=-\infty}^{\infty} h^*(k)E\{x(n+l)x^*(n-k)\}$$

or

$$r_{xy}(l) = \sum_{k=-\infty}^{\infty} h^*(k)r_{xx}(l+k) = \sum_{m=-\infty}^{\infty} h^*(-m)r_{xx}(l-m)$$

Hence,

$$r_{xy}(l) = h^*(-l) * r_{xx}(l) \qquad (3.4.5)$$

Similarly,

$$r_{yx}(l) = h(l) * r_{xx}(l) \qquad (3.4.6)$$

Output autocorrelation. Postmultiplying both sides of (3.4.1) by $y^*(n-l)$ and taking the expectation, we obtain

$$E\{y(n)y^*(n-l)\} = \sum_{k=-\infty}^{\infty} h(k)E\{x(n-k)y^*(n-l)\} \qquad (3.4.7)$$

or

$$r_{yy}(l) = \sum_{k=-\infty}^{\infty} h(k)r_{xy}(l-k) = h(l) * r_{xy}(l) \qquad (3.4.8)$$

From (3.4.5) and (3.4.8) we get

$$r_y(l) = h(l) * h^*(-l) * r_x(l) \qquad (3.4.9)$$

or

$$r_y(l) = r_h(l) * r_x(l) \qquad (3.4.10)$$

where

$$r_h(l) \triangleq h(l) * h^*(-l) = \sum_{n=-\infty}^{\infty} h(n)h^*(n-l) \qquad (3.4.11)$$

is the autocorrelation of the impulse response and is called the system correlation sequence.

Since μ_y is constant and $r_y(l)$ depends only on the lag l, the response of a stable system to a stationary input is also a stationary process. A careful examination of (3.4.10) shows that *when a signal $x(n)$ is filtered by an LTI system with impulse response $h(n)$ its autocorrelation is "filtered" by a system with impulse response equal to the autocorrelation of its impulse response*, as shown in Figure 3.10.

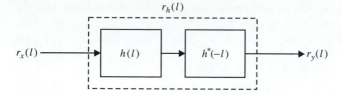

FIGURE 3.10
An equivalent LTI system for autocorrelation filtration.

Output power. The power $E\{|y(n)|^2\}$ of the output process $y(n)$ is equal to $r_y(0)$, which from (3.4.9) and (3.4.10) and the symmetry property of $r_x(l)$ is

$$P_y = r_y(0) = r_h(l) * r_x(l)|_{l=0}$$

$$= \sum_{k=-\infty}^{\infty} r_h(k)r_x(-k) = \sum_{k=-\infty}^{\infty} [h(k) * h^*(-k)]r_x(k)$$

$$= \sum_{k=-\infty}^{\infty} \sum_{m=-\infty}^{\infty} h(m)h^*(m-k)r_x(k) \tag{3.4.12}$$

$$= \sum_{k=-\infty}^{\infty} r_h(k)r_x(k) \tag{3.4.13}$$

or for FIR filters with $\mathbf{h} = [h(0)\ h(1)\ \cdots\ h(M-1)]^T$, (3.4.12) can be written as

$$P_y = \mathbf{h}^H \mathbf{R}_x \mathbf{h} \tag{3.4.14}$$

Finally, we note that when $\mu_x = 0$, we have $\mu_y = 0$ and $\sigma_y^2 = P_y$.

Output probability density function. Finding the probability density of the output of an LTI system is very difficult, except in some special cases. Thus, if $x(n)$ is a Gaussian process, then the output is also a Gaussian process with mean and autocorrelation given by (3.4.4) and (3.4.10). Also if $x(n)$ is IID, the probability density of the output is obtained by noting that $y(n)$ is a weighted sum of independent random variables. Indeed, the probability density of the sum of independent random variables is the convolution of their probability densities or the products of their characteristic functions. Thus if the input process is an IID stable process then the output process is also stable whose probability density can be computed by using characteristic functions.

3.4.2 Frequency-Domain Analysis

To obtain the output autoPSD and complex autoPSD, we recall that if $H(z) = \mathcal{Z}\{h(n)\}$, then, for real $h(n)$,

$$\mathcal{Z}\{h^*(-n)\} = H^*\left(\frac{1}{z^*}\right) \tag{3.4.15}$$

From (3.4.5), (3.4.6), and (3.4.7) we obtain

$$R_{xy}(z) = H^*\left(\frac{1}{z^*}\right) R_x(z) \tag{3.4.16}$$

$$R_{yx}(z) = H(z)R_x(z) \tag{3.4.17}$$

and

$$R_y(z) = H(z)H^*\left(\frac{1}{z^*}\right) R_x(z) \tag{3.4.18}$$

For a stable system, the unit circle $z = e^{j\omega}$ lies within the ROCs of $H(z)$ and $H(z^{-1})$. Thus,

$$R_{xy}(e^{j\omega}) = H^*(e^{j\omega})R_x(e^{j\omega}) \tag{3.4.19}$$

$$R_{yx}(e^{j\omega}) = H(e^{j\omega})R_x(e^{j\omega}) \tag{3.4.20}$$

and

$$R_y(e^{j\omega}) = H(e^{j\omega})H^*(e^{j\omega})R_x(e^{j\omega}) \tag{3.4.21}$$

or

$$R_y(e^{j\omega}) = |H(e^{j\omega})|^2 R_x(e^{j\omega}) \tag{3.4.22}$$

Thus, if we know the input and output autocorrelations or autospectral densities, we can determine the magnitude response of a system, but not its phase response. Only cross-correlation or cross-spectral densities can provide phase information [see (3.4.19) and (3.4.20)].

It can easily be shown that the power of the output is

$$E\{|y(n)|^2\} = r_{yy}(0) = \frac{1}{2\pi} \int_{-\pi}^{\pi} |H(e^{j\omega})|^2 R_x(e^{j\omega})\, d\omega \tag{3.4.23}$$

$$= \sum_{l=-\infty}^{l} r_x(l)r_h(l) \tag{3.4.24}$$

which is equivalent to (3.4.13).

Consider now a narrowband filter with frequency response

$$H(e^{j\omega}) = \begin{cases} 1 & \omega_c - \dfrac{\Delta\omega}{2} \le \omega \le \omega_c + \dfrac{\Delta\omega}{2} \\ 0 & \text{elsewhere} \end{cases} \tag{3.4.25}$$

The power of the filter output is

$$E\{|y(n)|^2\} = \frac{1}{2\pi} \int_{\omega_c-\Delta\omega/2}^{\omega_c+\Delta\omega/2} R_x(e^{j\omega})\, d\omega \simeq R_x(e^{j\omega_c}) \tag{3.4.26}$$

assuming that $\Delta\omega$ is sufficiently small and that $R_x(e^{j\omega})$ is continuous at $\omega = \omega_c$. Since $E\{|y(n)|^2\} \ge 0$, $R_x(e^{j\omega_c})$ is also nonnegative for all ω_c and $\Delta\omega$, hence

$$R_x(e^{j\omega}) \ge 0 \qquad -\pi \le \omega \le \pi \tag{3.4.27}$$

Hence, the PSD $R_x(e^{j\omega})$ is nonnegative definite for any random sequence $x(n)$ real (or complex). Furthermore, $R_x(e^{j\omega})\, d\omega/(2\pi)$, has the interpretation of power, or $R_x(e^{j\omega})$ is a power density as a function of frequency (in cycles per second). Table 3.2 shows various input-output relationships in both the time and frequency domains.

TABLE 3.2

Second-order moments of stationary random sequences processed by linear, time-invariant systems.

Time domain	Frequency domain	z Domain
$y(n) = h(n) * x(n)$	Not available	Not available
$r_{yx}(l) = h(l) * r_x(l)$	$R_{yx}(e^{j\omega}) = H(e^{j\omega})R_x(e^{j\omega})$	$R_{yx}(z) = H(z)R_x(z)$
$r_{xy}(l) = h^*(-l) * r_x(l)$	$R_{xy}(e^{j\omega}) = H^*(e^{j\omega})R_x(e^{j\omega})$	$R_{xy}(z) = H^*(1/z^*)R_x(z)$
$r_y(l) = h(l) * r_{xy}(l)$	$R_y(e^{j\omega}) = H(e^{j\omega})R_{xy}(e^{j\omega})$	$R_y(z) = H(z)R_{xy}(z)$
$r_y(l) = h(l) * h^*(-l) * r_x(l)$	$R_x(e^{j\omega}) = |H(e^{j\omega})|^2 R_x(e^{j\omega})$	$R_y(z) = H(z)H^*(1/z^*)R_x(z)$

3.4.3 Random Signal Memory

Given the "zero-memory" process $w(n) \sim \text{IID}(0, \sigma_w^2)$, we can introduce dependence by passing it though an LTI system. The extent and degree of the imposed dependence are dictated by the shape of the system's impulse response. The probability density of $w(n)$ is not explicitly involved. Suppose now that we are given the resulting linear process $x(n)$, and we want to quantify its memory. For processes with finite variance we can use the *correlation length*

$$L_c = \frac{1}{r_x(0)} \sum_{l=0}^{\infty} r_x(l) = \sum_{l=0}^{\infty} \rho_x(l)$$

which equals the area under the normalized autocorrelation sequence curve and shows the maximum distance at which two samples are significantly correlated.

An IID process has no memory and is completely described by its first-order density. A linear process has memory introduced by the impulse response of the generating system. If $w(n)$ has finite variance, the memory of the process is determined by the autocorrelation of the impulse response because $r_x(l) = \sigma_w^2 r_h(l)$. Also, the higher-order densities of the process are nonzero. Thus, the variability of the output—that is, what amplitudes takes the signal, how often, and how fast the amplitude changes from sample to sample—is the combined effect of the input probability density and the system memory.

> **DEFINITION 3.9.** A stationary process $x(n)$ with finite variance is said to have *long memory* if there exist constants α, $0 < \alpha < 1$, and $C_r > 0$ such that

$$\lim_{l \to \infty} \frac{1}{C_r \sigma_x^2} r_x(l) l^{\alpha} = 1$$

This implies that the autocorrelation has fat or heavy tails, that is, asymptotically decays as a power law

$$\rho_x(l) \simeq C_r |l|^{-\alpha} \qquad \text{as } l \to \infty$$

and slowly enough that

$$\sum_{l=-\infty}^{\infty} \rho_x(l) = \infty$$

that is, a long-memory process has infinite correlation length. If

$$\sum_{l=-\infty}^{\infty} \rho_x(l) < \infty$$

we say that that the process has *short memory*. This is the case for autocorrelations that decay exponentially, for example, $\rho_x(l) = a^{|l|}$, $-1 < a < 1$.

An equivalent definition of long memory can be formulated in terms of the power spectrum (Beran 1994; Samorodnitsky and Taqqu 1994).

> **DEFINITION 3.10.** A stationary process $x(n)$ with finite variance is said to have *long memory* if there exist constants β, $0 < \beta < 1$, and $C_R > 0$ such that

$$\lim_{\omega \to 0} \frac{1}{C_R \sigma_x^2} R_x(e^{j\omega}) |\omega|^{\beta} = 1$$

This asymptotic definition implies that

$$R_x(e^{j\omega}) \simeq \frac{C_R \sigma_x^2}{|\omega|^{\beta}} \qquad \text{as } \omega \to 0$$

and

$$R_x(0) = \sum_{l=-\infty}^{\infty} r_x(l) = \infty$$

The first-order density determines the mean value and the variance of a process, whereas the second-order density determines the autocorrelation and power spectrum. There is a coupling between the probability density and the autocorrelation or power spectrum of a process. However, this coupling is not extremely strong because there are processes that have different densities and the same autocorrelation. Thus, we can have random signal models with short or long memory and low or high variability. Random signal models are discussed in Chapters 4 and 12.

3.4.4 General Correlation Matrices

We first begin with the properties of general correlation matrices. Similar properties apply to covariance matrices.

PROPERTY 3.4.1. The correlation matrix of a random vector $\mathbf{x}$ is conjugate symmetric or Hermitian, that is,

$$\mathbf{R}_\mathbf{x} = \mathbf{R}_\mathbf{x}^H \tag{3.4.28}$$

Proof. This follows easily from (3.2.19).

PROPERTY 3.4.2. The correlation matrix of a random vector $\mathbf{x}$ is nonnegative definite (n.n.d.); or for every nonzero complex vector $\mathbf{w} = [w_1 \ w_2 \ \cdots \ w_M]^T$, the quadratic form $\mathbf{w}^H \mathbf{R}_\mathbf{x} \mathbf{w}$ is nonnegative, that is,

$$\mathbf{w}^H \mathbf{R}_\mathbf{x} \mathbf{w} \geq 0 \tag{3.4.29}$$

Proof. To prove (3.4.29), we define the dot product

$$\alpha = \mathbf{w}^H \mathbf{x} = \mathbf{x}^T \mathbf{w}^* = \sum_{k=1}^{M} w_k^* x_k \tag{3.4.30}$$

The mean square value of the random variable α is

$$E\{|\alpha|^2\} = E\{\mathbf{w}^H \mathbf{x}\mathbf{x}^H \mathbf{w}\} = \mathbf{w}^H E\{\mathbf{x}\mathbf{x}^H\}\mathbf{w} = \mathbf{w}^H \mathbf{R}_\mathbf{x} \mathbf{w} \tag{3.4.31}$$

Since $E\{|\alpha|^2\} \geq 0$, if follows that $\mathbf{w}^H \mathbf{R}_\mathbf{x} \mathbf{w} \geq 0$. We also note that a matrix is called *positive definite* (p.d.) if $\mathbf{w}^H \mathbf{R}_\mathbf{x} \mathbf{w} > 0$.

Eigenvalues and eigenvectors of R

For a Hermitian matrix $\mathbf{R}$ we wish to find an $M \times 1$ vector $\mathbf{q}$ that satisfies the condition

$$\mathbf{R}\mathbf{q} = \lambda \mathbf{q} \tag{3.4.32}$$

where λ is a constant. This condition implies that the linear transformation performed by matrix $\mathbf{R}$ does not change the direction of vector $\mathbf{q}$. Thus $\mathbf{R}\mathbf{q}$ is a *direction-invariant* mapping. To determine the vector $\mathbf{q}$, we write (3.4.32) as

$$(\mathbf{R} - \lambda \mathbf{I})\mathbf{q} = \mathbf{0} \tag{3.4.33}$$

where $\mathbf{I}$ is the $M \times M$ identity matrix and $\mathbf{0}$ is an $M \times 1$ vector of zeros. Since $\mathbf{q}$ is arbitrary, the only way (3.4.33) is satisfied is if the determinant of $\mathbf{R} - \lambda \mathbf{I}$ equals zero, that is,

$$\det(\mathbf{R} - \lambda \mathbf{I}) = 0 \tag{3.4.34}$$

This equation is an Mth-order polynomial in λ and is called the *characteristic equation* of $\mathbf{R}$. It has M roots $\{\lambda_i\}_{i=1}^{M}$, called *eigenvalues*, which, in general, are distinct. If (3.4.34) has repeated roots, then $\mathbf{R}$ is said to have *degenerate* eigenvalues. For each eigenvalue λ_i we can satisfy (3.4.32)

$$\mathbf{R}\mathbf{q}_i = \lambda_i \mathbf{q}_i \qquad i = 1, \dots, M \tag{3.4.35}$$

where the $\mathbf{q}_i$ are called *eigenvectors* of $\mathbf{R}$. Therefore, the $M \times M$ matrix $\mathbf{R}$ has M eigenvectors. To uniquely determine $\mathbf{q}_i$, we use (3.4.35) along with the normality condition that $\|\mathbf{q}_i\| = 1$. A MATLAB function [Lambda,Q] = eig(R) is available to compute eigenvalues and eigenvectors of $\mathbf{R}$.

There are further properties of the autocorrelation matrix $\mathbf{R}$ based on its eigenanalysis, which we describe below. Consider a matrix $\mathbf{R}$ that is Hermitian and nonnegative definite ($\mathbf{w}^H \mathbf{R}\mathbf{w} \geq 0$) with eigenvalues $\{\lambda_i\}_{i=1}^M$ and eigenvectors $\{\mathbf{q}_i\}_{i=1}^M$.

PROPERTY 3.4.3. The matrix $\mathbf{R}^k$ ($k = 1, 2, \ldots$) has eigenvalues $\lambda_1^k, \lambda_2^k, \ldots, \lambda_M^k$.

Proof. See Problem 3.16.

PROPERTY 3.4.4. If the eigenvalues $\lambda_1, \lambda_2, \ldots, \lambda_M$ are distinct, the corresponding eigenvectors $\{\mathbf{q}_i\}_{i=1}^M$ are *linearly independent*.

Proof. This property can be proved by using Property 3.4.3. Given M not-all-zero scalars $\{\alpha_i\}_{i=1}^M$, if

$$\sum_{i=1}^M \alpha_i \mathbf{q}_i = \mathbf{0} \qquad (3.4.36)$$

then the eigenvectors $\{\mathbf{q}_i\}_{i=1}^M$ are said to be *linearly dependent*. Assume that (3.4.36) is true for some not-all-zero scalars $\{\alpha_i\}_{i=1}^M$ and that the eigenvalues $\{\lambda_i\}_{i=1}^M$ are distinct. Now multiply (3.4.36) repeatedly by $\mathbf{R}^k$, $k = 0, \ldots, M - 1$ and use Property 3.4.3 to obtain

$$\sum_{i=1}^M \alpha_i \mathbf{R}^k \mathbf{q}_i = \sum_{i=1}^M \alpha_i \lambda_i^k \mathbf{q}_i = \mathbf{0} \qquad k = 0, \ldots, M - 1 \qquad (3.4.37)$$

which can be arranged in a matrix format for $i = 1, \ldots, M$ as

$$\begin{bmatrix} \alpha_1 \mathbf{q}_1 & \alpha_2 \mathbf{q}_2 & \alpha_3 \mathbf{q}_3 & \cdots & \alpha_M \mathbf{q}_M \end{bmatrix} \begin{bmatrix} 1 & \lambda_1 & \lambda_1^2 & \cdots & \lambda_1^{M-1} \\ 1 & \lambda_2 & \lambda_2^2 & \cdots & \lambda_2^{M-1} \\ \vdots & \vdots & \vdots & \ddots & \vdots \\ 1 & \lambda_M & \lambda_M^2 & \cdots & \lambda_M^{M-1} \end{bmatrix} = \mathbf{0} \qquad (3.4.38)$$

Since all the λ_i are distinct, the matrix containing the λ_i in (3.4.38) above is nonsingular. This matrix is called a *Vandermonde* matrix. Therefore, premultiplying both sides of (3.4.38) by the inverse of the Vandermonde matrix, we obtain

$$[\alpha_1 \mathbf{q}_1 \; \alpha_2 \mathbf{q}_2 \; \alpha_3 \mathbf{q}_3 \; \cdots \; \alpha_M \mathbf{q}_M] = \mathbf{0} \qquad (3.4.39)$$

Since eigenvectors $\{\mathbf{q}_i\}_{i=1}^M$ are not zero vectors, the only way (3.4.39) can be satisfied is if all $\{\alpha_i\}_{i=1}^M$ are zero. This implies that (3.4.36) cannot be satisfied for any set of not-all-zero scalars $\{\alpha_i\}_{i=1}^M$, which further implies that $\{\mathbf{q}_i\}_{i=1}^M$ are linearly independent.

PROPERTY 3.4.5. The eigenvalues $\{\lambda_i\}_{i=1}^M$ are real and *nonnegative*.

Proof. From (3.4.35), we have

$$\mathbf{q}_i^H \mathbf{R} \mathbf{q}_i = \lambda_i \mathbf{q}_i^H \mathbf{q}_i \qquad i = 1, 2, \ldots, M \qquad (3.4.40)$$

Since R is positive semidefinite, the quadratic form $\mathbf{q}_i^H \mathbf{R} \mathbf{q}_i \geq 0$. Also since $\mathbf{q}_i^H \mathbf{q}_i$ is an inner product, $\mathbf{q}_i^H \mathbf{q}_i > 0$. Hence

$$\lambda_i = \frac{\mathbf{q}_i^H \mathbf{R} \mathbf{q}_i}{\mathbf{q}_i^H \mathbf{q}_i} \geq 0 \qquad i = 1, 2, \ldots, M \qquad (3.4.41)$$

Furthermore, if $\mathbf{R}$ is positive definite, then $\lambda_i > 0$ for all $1 \leq i \leq M$. The quotient in (3.4.41) is a useful quantity and is known as the *Raleigh quotient* of vector $\mathbf{q}_i$.

PROPERTY 3.4.6. If the eigenvalues $\{\lambda_i\}_{i=1}^M$ are distinct, then the corresponding eigenvectors are orthogonal to one another, that is,

$$\lambda_i \neq \lambda_j \Rightarrow \mathbf{q}_i^H \mathbf{q}_j = 0 \qquad \text{for } i \neq j \qquad (3.4.42)$$

Proof. Consider (3.4.35). We have

$$\mathbf{R}\mathbf{q}_i = \lambda_i \mathbf{q}_i \tag{3.4.43}$$

and

$$\mathbf{R}\mathbf{q}_j = \lambda_j \mathbf{q}_j \tag{3.4.44}$$

for some $i \neq j$. Premultiplying both sides of (3.4.43) by $\mathbf{q}_j^H$, we obtain

$$\mathbf{q}_j^H \mathbf{R}\mathbf{q}_i = \mathbf{q}_j^H \lambda_i \mathbf{q}_i = \lambda_i \mathbf{q}_j^H \mathbf{q}_i \tag{3.4.45}$$

Taking the conjugate transpose of (3.4.44), using the Hermitian property (3.4.35) of $\mathbf{R}$, and using the realness Property 3.4.5 of eigenvalues, we get

$$\mathbf{q}_j^H \mathbf{R} = \lambda_j \mathbf{q}_j^H \tag{3.4.46}$$

Now postmultiplying (3.4.46) by q_i and comparing with (3.4.45), we conclude that

$$\lambda_i \mathbf{q}_j^H \mathbf{q}_i = \lambda_j \mathbf{q}_j^H \mathbf{q}_i \quad \text{or} \quad (\lambda_i - \lambda_j)\mathbf{q}_j^H \mathbf{q}_i = 0 \tag{3.4.47}$$

Since the eigenvalues are assumed to be distinct, the only way (3.4.47) can be satisfied is if $\mathbf{q}_j^H \mathbf{q}_i = 0$ for $i \neq j$, which further proves that the corresponding eigenvectors are orthogonal to one another.

PROPERTY 3.4.7. Let $\{\mathbf{q}_i\}_{i=1}^{M}$ be an orthonormal set of eigenvectors corresponding to the distinct eigenvalues $\{\lambda_i\}_{i=1}^{M}$ of an $M \times M$ correlation matrix $\mathbf{R}$. Then $\mathbf{R}$ can be diagonalized as follows:

$$\mathbf{\Lambda} = \mathbf{Q}^H \mathbf{R} \mathbf{Q} \tag{3.4.48}$$

where the orthonormal matrix $\mathbf{Q} \triangleq [\mathbf{q}_1 \ \cdots \ \mathbf{q}_M]$ is known as an *eigenmatrix* and $\mathbf{\Lambda}$ is an $M \times M$ diagonal eigenvalue matrix, that is,

$$\mathbf{\Lambda} \triangleq \text{diag}(\lambda_1, \dots, \lambda_M) \tag{3.4.49}$$

Proof. Arranging the vectors in (3.4.35) in a matrix format, we obtain

$$[\mathbf{R}\mathbf{q}_1 \ \mathbf{R}\mathbf{q}_2 \ \cdots \ \mathbf{R}\mathbf{q}_M] = [\lambda_1 \mathbf{q}_1 \ \lambda_2 \mathbf{q}_2 \ \cdots \ \lambda_M \mathbf{q}_M]$$

which, by using the definitions of $\mathbf{Q}$ and $\mathbf{\Lambda}$, can be further expressed as

$$\mathbf{R}\mathbf{Q} = \mathbf{Q}\mathbf{\Lambda} \tag{3.4.50}$$

Since $\mathbf{q}_i, i = 1, \dots, M$, is an orthonormal set of vectors, the eigenmatrix $\mathbf{Q}$ is unitary, that is, $\mathbf{Q}^{-1} = \mathbf{Q}^H$. Now premultiplying both sides of (3.4.50) by $\mathbf{Q}^H$, we obtain the desired result.

This diagonalization of the autocorrelation matrix plays an important role in filtering and estimation theory, as we shall see later. From (3.4.48) the correlation matrix $\mathbf{R}$ can also be written as

$$\mathbf{R} = \mathbf{Q}\mathbf{\Lambda}\mathbf{Q}^H = \lambda_1 \mathbf{q}_1 \mathbf{q}_1^H + \cdots + \lambda_M \mathbf{q}_M \mathbf{q}_M^H = \sum_{m=1}^{M} \lambda_m \mathbf{q}_m \mathbf{q}_m^H \tag{3.4.51}$$

which is known as the *spectral theorem*, or *Mercer's theorem*. If $\mathbf{R}$ is positive definite (and hence invertible), its inverse is given by

$$\mathbf{R}^{-1} = (\mathbf{Q}\mathbf{\Lambda}\mathbf{Q}^H)^{-1} = \mathbf{Q}\mathbf{\Lambda}^{-1}\mathbf{Q}^H = \sum_{m=1}^{M} \frac{1}{\lambda_m} \mathbf{q}_m \mathbf{q}_m^H \tag{3.4.52}$$

because $\mathbf{\Lambda}$ is a diagonal matrix.

PROPERTY 3.4.8. The trace of $\mathbf{R}$ is the summation of all eigenvalues, that is,

$$\text{tr}(\mathbf{R}) = \sum_{i=1}^{M} \lambda_i \tag{3.4.53}$$

Proof. See Problem 3.17.

PROPERTY 3.4.9. The determinant of $\mathbf{R}$ is equal to the product of all eigenvalues, that is,

$$\det \mathbf{R} = |\mathbf{R}| = \prod_{i=1}^{M} \lambda_i = |\mathbf{\Lambda}| \tag{3.4.54}$$

Proof. See Problem 3.18.

PROPERTY 3.4.10. Determinants of $\mathbf{R}$ and $\mathbf{\Gamma}$ are related by

$$|\mathbf{R}| = |\mathbf{\Gamma}|(1 + \boldsymbol{\mu}_x^H \mathbf{\Gamma}_\mathbf{x} \boldsymbol{\mu}_x) \tag{3.4.55}$$

Proof. See Problem 3.19.

3.4.5 Correlation Matrices from Random Processes

A stochastic process can also be represented as a random vector, and its second-order statistics given by the mean vector and the correlation matrix. Obviously, these quantities are functions of the index n. Let an $M \times 1$ random vector $\mathbf{x}(n)$ be derived from the random process $x(n)$ as follows:

$$\mathbf{x}(n) \triangleq [x(n)\, x(n-1)\, \cdots\, x(n-M+1)]^T \tag{3.4.56}$$

Then its mean is given by an $M \times 1$ vector

$$\boldsymbol{\mu}_x(n) = [\mu_x(n)\, \mu_x(n-1)\, \cdots\, \mu_x(n-M+1)]^T \tag{3.4.57}$$

and the correlation by an $M \times M$ matrix

$$\mathbf{R}_x(n) = \begin{bmatrix} r_x(n, n) & \cdots & r_x(n, n-M+1) \\ \vdots & \ddots & \vdots \\ r_x(n-M+1, n) & \cdots & r_x(n-M+1, n-M+1) \end{bmatrix} \tag{3.4.58}$$

Clearly, $\mathbf{R}_x(n)$ is Hermitian since $r_x(n-i, n-j) = r_x^*(n-j, n-i), 0 \le i, j \le M-1$. This vector representation will be useful when we discuss optimum filters.

Correlation matrices of stationary processes

The correlation matrix $\mathbf{R}_x(n)$ of a general stochastic process $x(n)$ is a Hermitian $M \times M$ matrix defined in (3.4.58) with elements $r_x(n-i, n-j) = E\{x(n-i)x^*(n-j)\}$. For stationary processes this matrix has an interesting additional structure. First, $\mathbf{R}_x(n)$ is a constant matrix $\mathbf{R}_x$; then using (3.3.24), we have

$$r_x(n-i, n-j) = r_x(j-i) = r_x(l \triangleq j-i) \tag{3.4.59}$$

Finally, by using conjugate symmetry $r_x(l) = r_x^*(-l)$, the matrix $\mathbf{R}_x$ is given by

$$\mathbf{R}_x = \begin{bmatrix} r_x(0) & r_x(1) & r_x(2) & \cdots & r_x(M-1) \\ r_x^*(1) & r_x(0) & r_x(1) & \cdots & r_x(M-2) \\ r_x^*(2) & r_x^*(1) & r_x(0) & \cdots & r_x(M-3) \\ \vdots & \vdots & \vdots & \ddots & \vdots \\ r_x^*(M-1) & r_x^*(M-2) & r_x^*(M-3) & \cdots & r_x(0) \end{bmatrix} \tag{3.4.60}$$

It can be easily seen that $\mathbf{R}_x$ is Hermitian and Toeplitz.[†] Thus, the autocorrelation matrix of a stationary process is Hermitian, nonnegative definite, and Toeplitz. Note that $\mathbf{R}_x$ is not persymmetric because elements along the main antidiagonal are not equal, in general.

[†] A matrix is called *Toeplitz* if the elements along each diagonal, parallel to the main diagonal, are equal.

Eigenvalue spread and spectral dynamic range

The ill conditioning of a matrix $\mathbf{R}_x$ increases with its condition number $\mathcal{X}(\mathbf{R}_x) = \lambda_{\max}/\lambda_{\min}$. When $\mathbf{R}_x$ is a correlation matrix of a stationary process, then $\mathcal{X}(R_x)$ is bounded from above by the dynamic range of the PSD $R_x(e^{j\omega})$ of the process $x(n)$. The larger the spread in eigenvalues, the wider (or less flat) the variation of the PSD function. This is also related to the dynamic range or to the data spread in $x(n)$ and is a useful measure in practice. This result is given by the following theorem, in which we have dropped the subscript of $R_x(e^{j\omega})$ for clarity.

THEOREM 3.5. Consider a zero-mean stationary random process with autoPSD

$$R(e^{j\omega}) = \sum_{l=-\infty}^{\infty} r(l)e^{-j\omega l}$$

then
$$\min_{\omega} R(e^{j\omega}) \le \lambda_i \le \max_{\omega} R(e^{j\omega}) \quad \text{for all } i = 1, 2, \ldots, M \quad (3.4.61)$$

Proof. From (3.4.41) we have

$$\lambda_i = \frac{\mathbf{q}_i^H \mathbf{R} \mathbf{q}_i}{\mathbf{q}_i^T \mathbf{q}_i} \quad (3.4.62)$$

Consider the quadratic form

$$\mathbf{q}_i^H \mathbf{R} \mathbf{q}_i = \sum_{k=1}^{M} \sum_{l=1}^{M} q_i(k)r(l-k)q_i(l)$$

where $\mathbf{q}_i = [q_i(1) \ q_i(2) \ \cdots \ q_i(M)]^T$. Using (3.3.41) and the stationarity of the process, we obtain

$$\mathbf{q}_i^H \mathbf{R} \mathbf{q}_i = \frac{1}{2\pi} \sum_k \sum_l q_i^*(k)q_i(l) \int_{-\pi}^{\pi} R(e^{j\omega})e^{j\omega(l-k)} \, d\omega$$

$$= \frac{1}{2\pi} \int_{-\pi}^{\pi} R(e^{j\omega}) \left[\sum_{k=1}^{M} q_i^*(k)e^{-j\omega k}\right] \left[\sum_{l=1}^{M} q_i(l)e^{j\omega l}\right] d\omega \quad (3.4.63)$$

or
$$\mathbf{q}_i^H \mathbf{R} \mathbf{q}_i = \frac{1}{2\pi} \int_{-\pi}^{\pi} R(e^{j\omega})|Q(e^{j\omega})|^2 d\omega \quad (3.4.64)$$

Similarly, we have

$$\mathbf{q}_i^T \mathbf{q}_i = \frac{1}{2\pi} \int_{-\pi}^{\pi} |Q(e^{j\omega})|^2 \, d\omega \quad (3.4.65)$$

Substituting (3.4.64) and (3.4.65) in (3.4.62), we obtain

$$\lambda_i = \frac{\int_{-\pi}^{\pi} |Q(e^{j\omega})|^2 R(e^{j\omega}) \, d\omega}{\int_{-\pi}^{\pi} |Q(e^{j\omega})|^2 \, d\omega} \quad (3.4.66)$$

However, since $R(e^{j\omega}) \ge 0$, we have the following inequality:

$$\min_{\omega} R(e^{j\omega}) \int_{-\pi}^{\pi} |Q(e^{j\omega})|^2 d\omega \le \int_{-\pi}^{\pi} |Q(e^{j\omega})|^2 R(e^{j\omega}) \, d\omega$$

$$\le \max_{\omega} R(e^{j\omega}) \int_{-\pi}^{\pi} |Q(e^{j\omega})|^2 d\omega$$

from which we easily obtain the desired result. The above result also implies that

$$\mathcal{X}(\mathbf{R}) \triangleq \frac{\lambda_{\max}}{\lambda_{\min}} \le \frac{\max_{\omega} R(e^{j\omega})}{\min_{\omega} R(e^{j\omega})} \quad (3.4.67)$$

which becomes equality as $M \to \infty$.

In many practical and theoretical applications, it is desirable to represent a random vector (or sequence) with a linearly equivalent vector (or sequence) consisting of uncorrelated components. If $\mathbf{x}$ is a correlated random vector and if $\mathbf{A}$ is a nonsingular matrix, then the linear transformation

$$\mathbf{w} = \mathbf{A}\mathbf{x} \tag{3.5.1}$$

results in a random vector $\mathbf{w}$ that contains the same "information" as $\mathbf{x}$, and hence random vectors $\mathbf{x}$ and $\mathbf{w}$ are said to be linearly equivalent. Furthermore, if $\mathbf{w}$ is an uncorrelated random vector, then each component w_i of $\mathbf{w}$ can be thought of as *adding* "new" information (or *innovation*) to $\mathbf{w}$ that is not present in the remaining components. Such a representation is called an *innovations representation* and provides additional insight into the understanding of random vectors and sequences. Additionally, it can simplify many theoretical derivations and can result in computationally efficient implementations.

Since $\mathbf{\Gamma_w}$ must be a diagonal matrix, we need to diagonalize the Hermitian, positive definite matrix $\mathbf{\Gamma_x}$ through the transformation matrix $\mathbf{A}$. There are two approaches to this diagonalization. One approach is to use the eigenanalysis presented in Section 3.4.4, which results in the well-known Karhunen-Loève (KL) transform. The other approach is to use triangularization methods from linear algebra, which leads to the LDU (UDL) and LU (UL) decompositions. These vector techniques can be further extended to random sequences that give us the KL expansion and the spectral factorizations, respectively.

3.5.1 Transformations Using Eigendecomposition

Let $\mathbf{x}$ be a random vector with mean vector $\boldsymbol{\mu}_\mathbf{x}$ and covariance matrix $\mathbf{\Gamma_x}$. The linear transformation

$$\mathbf{x}_0 = \mathbf{x} - \boldsymbol{\mu}_\mathbf{x} \tag{3.5.2}$$

results in a zero-mean vector $\mathbf{x}_0$ with correlation (and covariance) matrix equal to $\mathbf{\Gamma_x}$. This transformation shifts the origin of the M-dimensional coordinate system to the mean vector. We will now consider the zero-mean random vector $\mathbf{x}_0$ for further transformations.

Orthonormal transformation

Let $\mathbf{Q_x}$ be the eigenmatrix of $\mathbf{\Gamma_x}$, and let us choose $\mathbf{Q_x}^H$ as our linear transformation matrix $\mathbf{A}$ in (3.2.32). Consider

$$\mathbf{w} = \mathbf{Q_x}^H \mathbf{x}_0 = \mathbf{Q_x}^H (\mathbf{x} - \boldsymbol{\mu}_\mathbf{x}) \tag{3.5.3}$$

Then

$$\boldsymbol{\mu}_\mathbf{w} = \mathbf{Q_x}^H (E\{\mathbf{x}_0\}) = \mathbf{0} \tag{3.5.4}$$

and from (3.2.39) and (3.4.48)

$$\mathbf{\Gamma_w} = \mathbf{R_w} = E\{\mathbf{Q_x}^H \mathbf{x}_0 \mathbf{x}_0^H \mathbf{Q_x}\} = \mathbf{Q_x}^H \mathbf{\Gamma_x} \mathbf{Q_x} = \mathbf{\Lambda_x} \tag{3.5.5}$$

Since $\mathbf{\Lambda_x}$ is diagonal, $\mathbf{\Gamma_w}$ is also diagonal, and hence this transformation has some interesting properties:

1. The random vector $\mathbf{w}$ has zero mean, and its components are mutually uncorrelated (and hence orthogonal). Furthermore, if $\mathbf{x}$ is $\mathcal{N}(\boldsymbol{\mu}_\mathbf{x}, \mathbf{\Gamma_x})$, then $\mathbf{w}$ is $\mathcal{N}(\mathbf{0}, \mathbf{\Lambda_x})$ with independent components.
2. The variances of random variables w_i, $i = 1, \ldots, M$, are equal to the eigenvalues of $\mathbf{\Gamma_x}$.
3. Since the transformation matrix $\mathbf{A} = \mathbf{Q_x}^H$ is orthonormal, the transformation is called an *orthonormal transformation* and the distance measure

$$d^2(\mathbf{x}_0) \triangleq \mathbf{x}_0^H \mathbf{\Gamma_x}^{-1} \mathbf{x}_0 \tag{3.5.6}$$

is preserved under the transformation. This distance measure is also known as the *Mahalanobis distance*; and in the case of normal random vectors, it is related to the log-likelihood function.

4. Since $\mathbf{w} = \mathbf{Q}_{\mathbf{x}}^H (\mathbf{x} - \boldsymbol{\mu}_{\mathbf{x}})$, we have

$$w_i = \mathbf{q}_i^H (\mathbf{x} - \boldsymbol{\mu}_{\mathbf{x}}) = \|\mathbf{x} - \boldsymbol{\mu}_{\mathbf{x}}\| \cos[\measuredangle(\mathbf{x} - \boldsymbol{\mu}_{\mathbf{x}}, \mathbf{q}_i)] \qquad i = 1, \dots, M \tag{3.5.7}$$

which is the projection of $\mathbf{x} - \boldsymbol{\mu}_{\mathbf{x}}$ onto the unit vector $\mathbf{q}_i$. Thus $\mathbf{w}$ represents $\mathbf{x}$ in a new coordinate system that is shifted to $\boldsymbol{\mu}_{\mathbf{x}}$ and spanned by $\mathbf{q}_i, i = 1, \dots, M$. A geometric interpretation of this transformation for a two-dimensional case is shown in Figure 3.11, which shows a contour of $d^2(\mathbf{x}_0) = \mathbf{x}^H \boldsymbol{\Gamma}_{\mathbf{x}}^{-1} \mathbf{x} = \mathbf{w}^H \boldsymbol{\Lambda}_{\mathbf{x}}^{-1} \mathbf{w}$ in the $\mathbf{x}$ and $\mathbf{w}$ coordinate systems ($\mathbf{w} = Q_{\mathbf{x}}^H \mathbf{x}$).

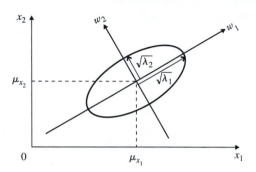

FIGURE 3.11
Orthogonal transformation in two dimensions.

Isotropic transformation

In the above orthonormal transformation, the autocorrelation matrix $\mathbf{R}_{\mathbf{w}}$ is diagonal but not an identity matrix $\mathbf{I}$. This can be achieved by an additional linear mapping of $\boldsymbol{\Lambda}_{\mathbf{x}}^{-1/2}$. Let

$$\mathbf{y} = \boldsymbol{\Lambda}_{\mathbf{x}}^{-1/2} \mathbf{w} = \boldsymbol{\Lambda}_{\mathbf{x}}^{-1/2} \mathbf{Q}_{\mathbf{x}}^H \mathbf{x}_0 = \boldsymbol{\Lambda}_{\mathbf{x}}^{-1/2} \mathbf{Q}_{\mathbf{x}}^H (\mathbf{x} - \boldsymbol{\mu}_{\mathbf{x}}) \tag{3.5.8}$$

Then

$$\mathbf{R}_{\mathbf{y}} = \boldsymbol{\Lambda}_{\mathbf{x}}^{-1/2} \mathbf{Q}_{\mathbf{x}}^H \boldsymbol{\Gamma}_{\mathbf{x}} \mathbf{Q}_{\mathbf{x}} \boldsymbol{\Lambda}_{\mathbf{x}}^{-1/2} = \boldsymbol{\Lambda}_{\mathbf{x}}^{-1/2} \boldsymbol{\Lambda}_{\mathbf{x}} \boldsymbol{\Lambda}_{\mathbf{x}}^{-1/2} = \mathbf{I} \tag{3.5.9}$$

This is called an *isotropic transformation* because *all* components of $\mathbf{y}$ are zero-mean, uncorrelated random variables with unit variance.[†] The geometric interpretation of this transformation for a two-dimensional case is shown in Figure 3.12. It clearly shows that there is not only a shift and rotation but also a scaling of the coordinate axis so that the distribution is equal in all directions, that is, it is direction-invariant. Because the transformation $\mathbf{A} = \boldsymbol{\Lambda}_{\mathbf{x}}^{-1/2} \mathbf{Q}_{\mathbf{x}}^H$ is orthogonal but not orthonormal, the distance measure $d^2(\mathbf{x}_0)$ is not preserved under this mapping. Since the correlation matrix after this transformation is an identity matrix $\mathbf{I}$, it is invariant under any orthonormal mapping, that is,

$$\mathbf{Q}^H \mathbf{I} \mathbf{Q} = \mathbf{Q}^H \mathbf{Q} = \mathbf{I} \tag{3.5.10}$$

This fact can be used for simultaneous diagonalization of two Hermitian matrices.

> **EXAMPLE 3.5.1.** Consider a stationary sequence with correlation matrix
>
> $$\mathbf{R}_x = \begin{bmatrix} 1 & a \\ a & 1 \end{bmatrix}$$
>
> where $-1 < a < 1$. The eigenvalues
>
> $$\lambda_1 = 1 + a \qquad \lambda_2 = 1 - a$$

[†] In the literature, an isotropic transformation is also known as a *whitening* transformation. We believe that this terminology is not accurate because both vectors $\mathbf{Q}_{\mathbf{x}}^H \mathbf{x}_0$ and $\boldsymbol{\Lambda}_{\mathbf{x}}^{-1/2} \mathbf{Q}_{\mathbf{x}}^H \mathbf{x}_0$ have uncorrelated coefficients.

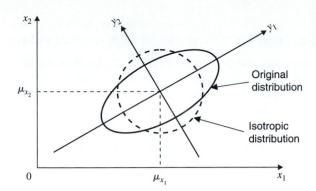

FIGURE 3.12
Isotropic transformation in two
dimensions.

127

SECTION 3.5
Innovations Representation
of Random Vectors

are obtained from the characteristic equation

$$\det(\mathbf{R}_x - \lambda\mathbf{I}) = \det\begin{bmatrix} 1-\lambda & a \\ a & 1-\lambda \end{bmatrix} = (1-\lambda)^2 - a^2 = 0$$

To find the eigenvector $\mathbf{q}_1$, we solve the linear system

$$\begin{bmatrix} 1 & a \\ a & 1 \end{bmatrix}\begin{bmatrix} q_1^{(1)} \\ q_2^{(1)} \end{bmatrix} = (1+a)\begin{bmatrix} q_1^{(1)} \\ q_2^{(1)} \end{bmatrix}$$

which gives $q_1^{(1)} = q_2^{(1)}$. Similarly, we find that $q_1^{(2)} = -q_2^{(2)}$. If we normalize both vectors to unit length, we obtain the eigenvectors

$$\mathbf{q}_1 = \frac{1}{\sqrt{2}}\begin{bmatrix} 1 \\ 1 \end{bmatrix} \qquad \mathbf{q}_2 = \frac{1}{\sqrt{2}}\begin{bmatrix} 1 \\ -1 \end{bmatrix}$$

From the above results we see that $\det\mathbf{R}_x = 1 - a^2 = \lambda_1\lambda_2$ and $\mathbf{Q}^H\mathbf{Q} = \mathbf{I}$, where $\mathbf{Q} = [\mathbf{q}_1\ \mathbf{q}_2]$.

3.5.2 Transformations Using Triangular Decomposition

The linear transformations discussed above were based on diagonalization of hermitian matrices through eigenvalue-eigenvector decomposition. These are useful in many detection and estimation problems. Triangular matrix decomposition leads to transformations that result in *causal* or *anticausal* linear filtering of associated sequences. Hence these mappings play an important role in linear filtering. There are two such decompositions: the *lower-diagonal-upper* (*LDU*) one leads to causal filtering while the *upper-diagonal-lower* (*UDL*) one results in anticausal filtering.

Lower-diagonal-upper decomposition

Any Hermitian, positive definite matrix $\mathbf{R}$ can be factored as (Goulob and Van Loan 1989)

$$\mathbf{R} = \mathbf{L}\mathbf{D}_L\mathbf{L}^H \tag{3.5.11}$$

or equivalently

$$\mathbf{L}^{-1}\mathbf{R}\mathbf{L}^{-H} = \mathbf{D}_L \tag{3.5.12}$$

where $\mathbf{L}$ is a *unit lower triangular* matrix, $\mathbf{D}_L$ is a diagonal matrix with positive elements, and $\mathbf{L}^H$ is a *unit upper triangular* matrix. The MATLAB function [L,D]=ldlt(R), given in Section 5.2, computes the LDU decomposition.

Since $\mathbf{L}$ is unit lower triangular, we have $\det\mathbf{R} = \prod_{i=1}^{M}\xi_i^l$, where $\xi_1^l, \ldots, \xi_M^l$ are the diagonal elements of $\mathbf{D}_L$. If we define the linear transformation

$$\mathbf{w} = \mathbf{L}^{-1}\mathbf{x} \triangleq \mathbf{B}\mathbf{x} \tag{3.5.13}$$

we find that

$$\mathbf{R_w} = E\{\mathbf{w}\mathbf{w}^H\} = \mathbf{L}^{-1}E\{\mathbf{x}\mathbf{x}^H\}\mathbf{L}^{-H} = \mathbf{L}^{-1}\mathbf{R}\mathbf{L}^{-H} = \mathbf{D}_L \tag{3.5.14}$$

Clearly, the components of $\mathbf{w}$ are orthogonal, and the elements $\xi_1^l, \ldots, \xi_M^l$ are their second moments. Therefore, this transformation appears to be similar to the orthogonal one. However, the vector $\mathbf{w}$ is not obtained as a simple rotation of $\mathbf{x}$. To understand this mapping, we first note that $\mathbf{B} = \mathbf{L}^{-1}$ is also a unit lower triangular matrix (Goulob and Van Loan 1989). Then we can write (3.5.13) as

$$\begin{bmatrix} w_1 \\ \vdots \\ w_i \\ \vdots \\ w_M \end{bmatrix} = \begin{bmatrix} 1 & \cdots & 0 & \cdots & 0 \\ \vdots & \ddots & & & \vdots \\ b_{i1} & \cdots & 1 & \cdots & 0 \\ \vdots & & & \ddots & \vdots \\ b_{M1} & \cdots & b_{Mi} & \cdots & 1 \end{bmatrix} \begin{bmatrix} x_1 \\ \vdots \\ x_i \\ \vdots \\ x_M \end{bmatrix} \tag{3.5.15}$$

where b_{ik} are elements of $\mathbf{B}$. From (3.5.15) we conclude that w_i is a linear combination of $x_k, k \le i$, that is,

$$w_i = \sum_{k=1}^{i} b_{ik} x_k \qquad 1 \le i \le M \tag{3.5.16}$$

If the signal vector $\mathbf{x}$ consists of consecutive samples of a discrete-time stochastic process $x(n)$, that is,

$$\mathbf{x} = [x(n)\, x(n-1)\, \cdots\, x(n-M+1)]^T \tag{3.5.17}$$

then (3.5.16) can be interpreted as a causal linear filtering of the random sequence (see Chapter 2). This transformation will be used extensively in optimum linear filtering and prediction problems.

A similar LDU decomposition of autocovariance matrices can be performed by following the identical steps above. In this case, the components of the transformed vector $\mathbf{w}$ are uncorrelated, and the elements ξ_i^l, $1 \le i \le M$, of $\mathbf{D}_L$ are variances.

Upper-diagonal-lower decomposition

This diagonalization is almost identical to the previous one and involves factorization of a Hermitian, positive definite matrix into an upper-diagonal-lower form. It is given by

$$\mathbf{R} = \mathbf{U}\mathbf{D}_U\mathbf{U}^H \tag{3.5.18}$$

or equivalently

$$\mathbf{U}^{-1}\mathbf{R}\mathbf{U}^{-H} = \mathbf{D}_U = \text{diag}(\xi_1^u, \ldots, \xi_M^u) \tag{3.5.19}$$

in which the matrix $\mathbf{U}$ is unit upper triangular, the matrix $\mathbf{U}^H$ is unit lower triangular, and the matrix $\mathbf{D}_U$ is diagonal with positive elements. Note that $\mathbf{U}^H \ne \mathbf{L}$ and $\mathbf{D}_U \ne \mathbf{D}_L$. Following the same analysis as above, we have $\det \mathbf{R} = \det \mathbf{D}_U = \prod_{i=1}^{M} \xi_i^u$. Since $\mathbf{A} = \mathbf{U}^{-1}$ is unit upper triangular in the transformation $\mathbf{w} = \mathbf{U}^{-1}\mathbf{x}$, the components of $\mathbf{w}$ are orthogonal and are obtained by linear combinations of $x_k, k \ge i$, that is,

$$w_i = \sum_{k=i}^{M} l_{ik} x_k \qquad 1 \le i \le M \tag{3.5.20}$$

This represents an anticausal filtering of a random sequence if $\mathbf{x}$ is a signal vector. Table 3.3 compares and contrasts orthogonal and triangular decompositions. We note that the LDU decomposition does not have the nice geometric interpretation (rotation of the coordinate system) of the eigendecomposition transformation.

Generation of real-valued random vectors with given second-order moments. Suppose that we want to generate M samples, say, $x_1, x_2, \ldots, x_M$, of a real-valued random vector $\mathbf{x}$ with mean $\mathbf{0}$ and a given symmetric and positive definite autocorrelation matrix

TABLE 3.3

129

SECTION 3.5
Innovations Representation
of Random Vectors

Comparison of orthogonal and triangular decompositions for zero-mean random vectors.

Orthogonal decomposition	Triangular decomposition
$\mathbf{R} = E\{\mathbf{xx}^H\}$	$\mathbf{R} = E\{\mathbf{xx}^H\}$
$\mathbf{Rq}_i = \lambda_i \mathbf{q}_i$	
$\mathbf{Q} = [\mathbf{q}_1, \mathbf{q}_2, \ldots, \mathbf{q}_M]$	$\mathbf{L}$ = unit lower triangular
$\mathbf{\Lambda} = \mathrm{diag}\{\lambda_1, \lambda_2, \ldots, \lambda_M\}$	$\mathbf{D} = \mathrm{diag}\{\xi_1, \xi_2, \ldots, \xi_M\}$
$\mathbf{R} = \mathbf{Q\Lambda Q}^H = \sum_{i=1}^{M} \lambda_i \mathbf{q}_i \mathbf{q}_i^H$	$\mathbf{R} = \mathbf{LDL}^H$
$\mathbf{\Lambda} = \mathbf{Q}^H \mathbf{RQ}$	$\mathbf{D} = \mathbf{L}^{-1}\mathbf{RL}^{-H}$
$\mathbf{R}^{-1} = \mathbf{Q\Lambda}^{-1}\mathbf{Q}^H = \sum_{i=1}^{M} \frac{1}{\lambda_i} \mathbf{q}_i \mathbf{q}_i^H$	$\mathbf{R}^{-1} = \mathbf{L}^{-H}\mathbf{D}^{-1}\mathbf{L}^{-1}$
$\mathbf{\Lambda}^{-1} = \mathbf{Q}^H \mathbf{R}^{-1}\mathbf{Q}$	$\mathbf{D}^{-1} = \mathbf{L}^{-H}\mathbf{R}^{-1}\mathbf{L}^{-1}$
$\det \mathbf{R} = \det \mathbf{\Lambda} = \prod_{i=1}^{M} \lambda_i$	$\det \mathbf{R} = \det \mathbf{D} = \prod_{i=1}^{M} \xi_i$
$\mathrm{tr}\,\mathbf{R} = \mathrm{tr}\,\mathbf{\Lambda} = \sum_{i=1}^{M} \lambda_i$	
Whitening (noncausal)	**Whitening (causal)**
$\mathbf{w} = \mathbf{Q}^H \mathbf{x}$	$\mathbf{w} = \mathbf{L}^{-1}\mathbf{x}$
$E\{\mathbf{ww}^H\} = \mathbf{\Lambda}$	$E\{\mathbf{ww}^H\} = \mathbf{D}$

$\mathbf{R_x}$. The innovations representation given in this section suggests three approaches to generate samples of such a random vector. The general approach is to factor $\mathbf{R_x}$, using either the orthonormal or the triangularization transformation, to obtain the diagonal matrix ($\mathbf{\Lambda_x}$ or $\mathbf{D}_L^{(x)}$ or $\mathbf{D}_U^{(x)}$), generate M samples of an IID sequence with the obtained diagonal variances, and then transform these samples by using the inverse transformation matrix ($\mathbf{Q_x}$ or $\mathbf{L_x}$ or $\mathbf{U_x}$). We hasten to add that, in general, the original distribution of the IID samples will not be preserved unless the samples are jointly normal. Therefore, in the following discussion, we assume that a normal pseudorandom number generator is used to generate M independent samples of $\mathbf{w}$. The three methods are as follows.

Eigendecomposition approach. First factor $\mathbf{R_x}$ as $\mathbf{R_x} = \mathbf{Q_x \Lambda_x Q_x}^H$. Then generate $\mathbf{w}$, using the distribution $\mathcal{N}(\mathbf{0}, \mathbf{\Lambda_x})$. Finally, compute the desired vector $\mathbf{x}$, using $\mathbf{x} = \mathbf{Q_x w}$.

LDU triangularization approach. First factor $\mathbf{R_x}$ as $\mathbf{R_x} = \mathbf{L_x D}_L^{(x)}\mathbf{L_x}^H$. Then generate $\mathbf{w}$, using the distribution $\mathcal{N}(\mathbf{0}, \mathbf{D}_L^{(x)})$. Finally, compute the desired vector $\mathbf{x}$, using $\mathbf{x} = \mathbf{L_x w}$.[†]

UDL triangularization approach. First factor $\mathbf{R_x}$ as $\mathbf{R_x} = \mathbf{U_x D}_U^{(x)}\mathbf{U_x}^H$. Then generate $\mathbf{w}$, using the distribution $\mathcal{N}(\mathbf{0}, \mathbf{D}_U^{(x)})$. Finally, compute the desired vector $\mathbf{x}$, using $\mathbf{x} = \mathbf{U_x w}$.

Additional discussion and more complete treatment on the generation of random vectors are given in Johnson (1994).

3.5.3 The Discrete Karhunen-Loève Transform

In many signal processing applications, it is convenient to represent the samples of a random signal in another set of numbers (or coefficients) so that this new representation possesses some useful properties. For example, for coding purposes we want to transform a signal

[†]If we use the Cholesky decomposition $\mathbf{R_x} = \tilde{\mathbf{L}}_x \tilde{\mathbf{L}}_x^H$, where $\tilde{\mathbf{L}}_x = \{\mathbf{D}_L^{(x)}\}^{1/2}\mathbf{L_x}$, then $\mathbf{w} = \mathcal{N}(\mathbf{0}, \mathbf{I})$ will generate $\mathbf{x}$ with the given correlation $\mathbf{R}_x$, using $\mathbf{x} = \tilde{\mathbf{L}}_x \mathbf{w}$.

so that its energy is concentrated in only a few coefficients (which are then transmitted); or for optimal filtering purposes we may want uncorrelated samples so that the filtering complexity is reduced or the signal-to-noise ratio is enhanced. A general approach is to expand a signal as a linear combination of orthogonal basis functions so that components of the signal with respect to basis functions do not interfere with one another. There are several such basis functions; the most widely known is the set of complex exponentials used in DTFT (or DFT) that are used in linear filtering, as we discussed in Section 3.4. Other examples are functions used in discrete cosine transform, discrete sine transform, Haar transform, etc., which are useful in coding applications (Jain 1989).

As discussed in this section, a set of orthogonal basis functions for which the signal components are statistically uncorrelated to one another is based on the second-order properties of the random process and, in particular, on the diagonalization of its covariance matrix. It is also an optimal representation of the signal in the sense that it provides a representation with the *smallest mean square error* among all other orthogonal transforms. This has applications in the analysis of random signals as well as in coding. This transform was first suggested by Karhunen and Loève for continuous random processes. It was extended to discrete random signals by Hotelling and is also known as the Hotelling transform. In keeping with the current nomenclature, we will call it the *discrete Karhunen-Loève transform (DKLT)* (Fukunaga 1990).

Development of the DKLT

Let $\mathbf{x} = [x_1 \ x_2 \ \cdots \ x_M]^T$ be a zero-mean[†] random vector with autocorrelation matrix $\mathbf{R_x}$. We want to represent $\mathbf{x}$ using the linear transformation

$$\mathbf{w} = \mathbf{A}^H \mathbf{x} \qquad \mathbf{A}^{-1} = \mathbf{A}^H \qquad (3.5.21)$$

where $\mathbf{A}$ is a unitary matrix. Then

$$\mathbf{x} = \mathbf{A}\mathbf{w} = \sum_{i=1}^{M} w_i \mathbf{a}_i \qquad \mathbf{a}_i^H \mathbf{a}_j = 0 \qquad i \neq j \qquad (3.5.22)$$

Let us represent $\mathbf{x}$ using the first m, $1 \leq m \leq M$, components of $\mathbf{w}$, that is,

$$\hat{\mathbf{x}} \triangleq \sum_{i=1}^{m} w_i \mathbf{a}_i \qquad 1 \leq m \leq M \qquad (3.5.23)$$

Then from (3.5.22) and (3.5.23), the error between $\mathbf{x}$ and $\hat{\mathbf{x}}$ is given by

$$\mathbf{e}_m \triangleq \mathbf{x} - \hat{\mathbf{x}} = \sum_{i=1}^{M} w_i \mathbf{a}_i - \sum_{i=1}^{m} w_i \mathbf{a}_i = \sum_{i=m+1}^{M} w_i \mathbf{a}_i \qquad (3.5.24)$$

and hence the mean-squared error (MSE) is

$$E_m \triangleq E\{\mathbf{e}_m^H \mathbf{e}_m\} = \sum_{i=m+1}^{M} \mathbf{a}_i^H E\{|w_i|^2\}\mathbf{a}_i = \sum_{i=m+1}^{M} E\{|w_i|^2\}\mathbf{a}_i^H \mathbf{a}_i \qquad (3.5.25)$$

Since from (3.5.21) $w_i = \mathbf{a}_i^H \mathbf{x}$, we have $E\{|w_i|^2\} = \mathbf{a}_i^H \mathbf{R_x} \mathbf{a}_i$. Now we want to determine the matrix $\mathbf{A}$ that will minimize the MSE E_m subject to $\mathbf{a}_i^H \mathbf{a}_i = 1, i = m+1, \ldots, M$ so that from (3.5.25)

$$E_m = \sum_{i=m+1}^{M} E\{|w_i|^2\} = \sum_{i=m+1}^{M} \mathbf{a}_i^H \mathbf{R_x} \mathbf{a}_i \qquad \mathbf{a}_i^H \mathbf{a}_i = 1 \qquad i = m+1, \ldots, M \quad (3.5.26)$$

[†]If the mean is not zero, then we perform the transformation on the mean-subtracted vector, using the covariance matrix.

This optimization can be done by using the Lagrange multiplier approach (Appendix B); that is, we minimize

$$\sum_{i=m+1}^{M} \mathbf{a}_i^H \mathbf{R}_x \mathbf{a}_i + \sum_{i=m+1}^{M} \lambda_i (1 - \mathbf{a}_i^H \mathbf{a}_i) \quad i = m+1, \ldots, M$$

Hence after setting the gradient equal to zero,

$$\nabla_{\mathbf{a}_i} \left[\sum_{i=m+1}^{M} \mathbf{a}_i^H \mathbf{R}_x \mathbf{a}_i + \sum_{i=m+1}^{M} \lambda_i (1 - \mathbf{a}_i^H \mathbf{a}_i) \right] = (\mathbf{R}_x \mathbf{a}_i)^* - (\lambda_i \mathbf{a}_i)^* = 0 \quad (3.5.27)$$

we obtain
$$\mathbf{R}_x \mathbf{a}_i = \lambda_i \mathbf{a}_i \quad i = m+1, \ldots, M$$

which is equivalent to (3.4.35) in the eigenanalysis of Section 3.4.4. Hence λ_i is the eigenvalue, and the corresponding $\mathbf{a}_i$ is the eigenvector of $\mathbf{R}_x$. Clearly, since $1 \le m \le M$, the transformation matrix $\mathbf{A}$ should be chosen as the eigenmatrix $\mathbf{Q}$. Hence

$$\begin{bmatrix} \uparrow \\ \mathbf{w} \\ \downarrow \end{bmatrix} = \begin{bmatrix} \leftarrow & \mathbf{q}_1^H & \rightarrow \\ \leftarrow & \mathbf{q}_2^H & \rightarrow \\ \vdots & \vdots & \vdots \\ \leftarrow & \mathbf{q}_M^H & \rightarrow \end{bmatrix} \begin{bmatrix} \uparrow \\ \mathbf{x} \\ \downarrow \end{bmatrix}$$

or more concisely
$$\mathbf{w} = \mathbf{Q}^H \mathbf{x} \quad (3.5.28)$$

provides an orthonormal transformation so that the transformed vector $\mathbf{w}$ is a zero-mean, uncorrelated random vector with autocorrelation $\mathbf{\Lambda}$. This transformation is called the DKLT, and its inverse relationship (or synthesis) is given by

$$\begin{bmatrix} \uparrow \\ \mathbf{x} \\ \downarrow \end{bmatrix} = \begin{bmatrix} \uparrow & \uparrow & \cdots & \uparrow \\ \mathbf{q}_1 & \mathbf{q}_2 & \cdots & \mathbf{q}_M \\ \downarrow & \downarrow & \cdots & \downarrow \end{bmatrix} \begin{bmatrix} \uparrow \\ \mathbf{w} \\ \downarrow \end{bmatrix} \quad (3.5.29)$$

or
$$\mathbf{x} = \mathbf{Q}\mathbf{w} = \mathbf{q}_1 w_1 + \mathbf{q}_2 w_2 + \cdots + \mathbf{q}_M w_M \quad (3.5.30)$$

From Section 3.5.1, the geometric interpretation of this transformation is that $\{w_k\}_1^M$ are projections of the vector $\mathbf{x}$ with respect to the rotated coordinate system of $\{\mathbf{q}_k\}_1^M$. The eigenvalues λ_i also have an interesting interpretation, as we shall see in the following representation.

Optimal reduced-basis representation

Generally we would expect any transformation to provide only few meaningful components so that we can use only those basis vectors resulting in a smaller representation error. To determine this *reduced-basis representation* property of the DKLT, let us use first $K < M$ eigenvectors (instead of all $\mathbf{q}_i$). Then from (3.5.26), we have

$$E_K = \sum_{i=K+1}^{M} \lambda_i \quad (3.5.31)$$

In other words, the MSE in the reduced-basis representation, when the first K basis vectors are used, is the sum of the remaining eigenvalues (which are never negative). Therefore, to obtain a minimum MSE (that is, an optimum) representation, the procedure is to choose K eigenvectors corresponding to the K *largest* eigenvalues.

Application in data compression. The DKLT is a transformation on a random vector that produces a zero-mean, uncorrelated vector and that can minimize the mean square representation error. One of its popular applications is data compression in communications and, in particular, in speech and image coding. Suppose we want to send a sample function of a speech process $x_c(t)$. If we sample this waveform and obtain M samples $\{x(n)\}_0^{M-1}$, then we need to send M data values. Instead, if we analyze the correlation of $\{x(n)\}_0^{M-1}$ and determine that M values can be approximated by a smaller K numbers of w_i and the corresponding $\mathbf{q}_i$, then we can compute these K data values $\{w_i\}_1^K$ at the transmitter and send them to the receiver through the communication channel. At the receiver, we can reconstruct $\{x(n)\}_0^{M-1}$ by using (3.5.23), as shown in Figure 3.13. Obviously, both the transmitter and receiver must have the information about the eigenvectors $\{\mathbf{q}_i\}_1^M$. A considerable amount of compression is achieved if K is much smaller than M.

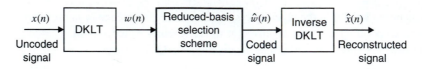

FIGURE 3.13
Signal coding scheme using the DKLT.

Periodic random sequences

As we noted in the previous section, the correlation matrix of a stationary process is Toeplitz. If the autocorrelation sequence of a random process is periodic with fundamental period M, its correlation matrix becomes *circulant*. All rows (columns) of a circulant matrix are obtained by circular rotation of its first row (column). Using (3.4.60) and the periodicity relation $r_x(l) = r_x(l - M)$, we obtain

$$\mathbf{R}_x = \begin{bmatrix} r_x(0) & r_x(1) & r_x(2) & \cdots & r_x(M-1) \\ r_x(M-1) & r_x(0) & r_x(1) & \cdots & r_x(M-2) \\ r_x(M-2) & r_x(M-1) & r_x(0) & \cdots & r_x(M-3) \\ \vdots & \vdots & \vdots & \ddots & \vdots \\ r_x(1) & r_x(2) & r_x(3) & \cdots & r_x(0) \end{bmatrix} \tag{3.5.32}$$

which is a circulant matrix. We note that a circulant matrix is Toeplitz but not vice versa.

If we define the M-point DFT of the periodic sequence $r_x(l)$

$$\tilde{R}_x(k) = \sum_{l=0}^{M-1} r_x(l) W_M^{kl} \tag{3.5.33}$$

where $W_M \triangleq e^{-j2\pi/M}$, and the vector

$$\mathbf{w}_k \triangleq \frac{1}{\sqrt{M}} [1 \ W_M^k \ W_M^{2k} \ \cdots \ W_M^{(M-1)k}]^T \qquad 0 \leq k \leq M - 1 \tag{3.5.34}$$

we can easily see that multiplying the first row of $\mathbf{R}_x$ by the vector $\mathbf{w}_k$ results in $\tilde{R}_x(k)/\sqrt{M}$. Using $W_M^{-k} = W_M^{(M-1)k}$, we find that the product of the second row by $\mathbf{w}_k$ is equal to $\tilde{R}_x(k) W_M^k/\sqrt{M}$. In general, the ith row by $\mathbf{w}_k$ gives $\tilde{R}_x(k) W_M^{(i-1)k}/\sqrt{M}$. Therefore, we have

$$\mathbf{R}_x \mathbf{w}_k = \tilde{R}_x(k) \mathbf{w}_k \qquad 0 \leq k \leq M - 1 \tag{3.5.35}$$

which shows that the normalized DFT vectors $\mathbf{w}_k$ are the eigenvectors of the circulant matrix $\mathbf{R}_x$ with as corresponding eigenvalues the DFT coefficients $\tilde{R}_x(k)$. Therefore, *the DFT provides the DKLT of periodic random sequences*. We recall that $\tilde{R}_x(k)$ are samples of the DTFT $R_x(e^{j2\pi k/M})$ of the finite-length sequence $r_x(l), 0 \leq l \leq M - 1$.

If we define the $M \times M$ matrix

$$\mathbf{W} \triangleq [\mathbf{w}_0 \ \mathbf{w}_1 \ \cdots \ \mathbf{w}_{M-1}] \tag{3.5.36}$$

we can show that

$$\mathbf{W}^H \mathbf{W} = \mathbf{W} \mathbf{W}^H = \mathbf{I} \tag{3.5.37}$$

that is, the matrix $\mathbf{W}$ is unitary. The set of equations (3.5.35) can be written as

$$\mathbf{W}^H \mathbf{R}_x \mathbf{W} = \text{diag}\{\tilde{R}_x(0), \tilde{R}_x(1), \ldots, \tilde{R}_x(M-1)\} \tag{3.5.38}$$

which shows that the DFT performs the diagonalization of circulant matrices. Although there is no fast algorithm for the diagonalization of general Toeplitz matrices, in many cases we can use the DFT to approximate the DKLT of stationary random sequences. The approximation is adequate if the correlation becomes negligible for $|l| > M$, which is the case for many stationary processes. This explains the fact that the eigenvectors of a Toeplitz matrix resemble complex exponentials for large values of M. The DKLT also can be extended to handle the representation of random sequences. These issues are further explored in Therrien (1992), Gray (1972), and Fukunaga (1990).

3.6 PRINCIPLES OF ESTIMATION THEORY

The key assumption underlying our discussion up to this point was that the probability distributions associated with the problem under consideration were known. As a result, all required probabilities, autocorrelation sequences, and PSD functions either could be derived from a set of assumptions about the involved random processes or were given a priori. However, in most practical applications, this is the exception rather than the rule. Therefore, the properties and parameters of random variables and random processes should be obtained by collecting and analyzing finite sets of measurements. In this section, we introduce some basic concepts of estimation theory that will be used repeatedly in the rest of the book. Complete treatments of estimation theory can be found in Kay (1993), Helstrom (1995), Van Trees (1968), and Papoulis (1991).

3.6.1 Properties of Estimators

Suppose that we collect N observations $\{x(n)\}_0^{N-1}$ from a stationary stochastic process and use them to estimate a parameter θ (which we assume to be real-valued) of the process using some function $\hat{\theta}[\{x(n)\}_0^{N-1}]$. The same results can be used for a set of measurements $\{x_k(n)\}_1^N$ obtained from N sensors sampling stochastic processes with the same distributions. The function $\hat{\theta}[\{x(n)\}_0^{N-1}]$ is known as an *estimator* whereas the value taken by the estimator, using a particular set of observations, is called a *point estimate* or simply an *estimate*. The intention of the estimator design is that the estimate should be as close to the true value of the parameter as possible. However, if we use another set of observations or a different number of observations from the same set, it is highly unlikely that we will obtain the same estimate. As an example of an estimator, consider estimating the mean μ_x of a stationary process $x(n)$ from its N observations $\{x(n)\}_0^{N-1}$. Then the natural estimator is a simple arithmetic average of these observations, given by

$$\hat{\mu}_x = \hat{\theta}[\{x(n)\}_0^{N-1}] = \frac{1}{N} \sum_{n=0}^{N-1} x(n) \tag{3.6.1}$$

Similarly, a natural estimator of the variance σ_x^2 of the process $x(n)$ would be

$$\sigma_x^2 = \hat{\theta}[\{x(n)\}_0^{N-1}] = \frac{1}{N} \sum_{n=0}^{N-1} [x(n) - \hat{\mu}_x]^2 \tag{3.6.2}$$

If we repeat this procedure a large number of times, we will obtain a large number of estimates, which can be used to generate a histogram showing the distribution of the estimates. Before the collection of observations, we would like to describe all sets of data that can be obtained by using the random variables $\{x(n, \zeta)\}_0^{N-1}$. The obtained set of N observations $\{x(n)\}_0^{N-1}$ can thus be regarded as one realization of the random variables $\{x(n, \zeta)\}_0^{N-1}$ defined on an N-dimensional sample space. In this sense, the estimator $\hat{\theta}[\{x(n, \zeta)\}_0^{N-1}]$ becomes a random variable whose distribution can be obtained from the joint distribution of the random variables $\{x(n, \zeta)\}_0^{N-1}$. This distribution is called the *sampling distribution* of the estimator and is a fundamental concept in estimation theory because it provides all the information we need to evaluate the quality of an estimator.

The sampling distribution of a "good" estimator should be concentrated as closely as possible about the parameter that it estimates. To determine how "good" an estimator is and how different estimators of the same parameter compare with one another, we need to determine their sampling distributions. Since it is *not* always possible to derive the exact sampling distributions, we have to resort to properties that use the lower-order moments (mean, variance, mean square error) of the estimator.

Bias of estimator. The *bias* of an estimator $\hat{\theta}$ of a parameter θ is defined as

$$B(\hat{\theta}) \triangleq E[\hat{\theta}] - \theta \tag{3.6.3}$$

while the *normalized* bias is defined as

$$\varepsilon_b \triangleq \frac{B(\hat{\theta})}{\theta} \qquad \theta \neq 0 \tag{3.6.4}$$

When $B(\hat{\theta}) = 0$, the estimator is said to be *unbiased* and the pdf of the estimator is centered exactly at the true value θ. Generally, one should select estimators that are unbiased such as the mean estimator in (3.6.1) or very nearly unbiased such as the variance estimator in (3.6.2). However, it is not always wise to select an unbiased estimator, as we will see below and in Section 5.2 on the estimation of autocorrelation sequences.

Variance of estimator. The *variance* of the estimator $\hat{\theta}$ is defined by

$$\text{var}(\hat{\theta}) = \sigma_{\hat{\theta}}^2 \triangleq E\{|\hat{\theta} - E\{\hat{\theta}\}|^2\} \tag{3.6.5}$$

which measures the spread of the pdf of $\hat{\theta}$ around its average value. Therefore, one would select an estimator with the smallest variance. However, this selection is not always compatible with the small bias requirement. As we will see below, reducing variance may result in an increase in bias. Therefore, a balance between these two conflicting requirements is required, which is provided by the mean square error property. The *normalized standard deviation* (also called the coefficient of variation) is defined by

$$\varepsilon_r \triangleq \frac{\sigma_{\hat{\theta}}}{\theta} \qquad \theta \neq 0 \tag{3.6.6}$$

Mean square error. The *mean square error* (MSE) of the estimator is given by

$$\text{MSE}(\theta) = E\{|\hat{\theta} - \theta|^2\} = \sigma_{\hat{\theta}}^2 + |B|^2 \tag{3.6.7}$$

Indeed, we have

$$\text{MSE}(\theta) = E\{|\theta - E\{\hat{\theta}\} - (\hat{\theta} - E\{\hat{\theta}\})|^2\}$$

$$= E\{|\theta - E\{\hat{\theta}\}|^2\} + E\{|\hat{\theta} - E\{\hat{\theta}\}|^2\} \tag{3.6.8}$$

$$- (\theta - E\{\hat{\theta}\})E\{(\hat{\theta} - E\{\hat{\theta}\})^*\} - (\theta - E\{\hat{\theta}\})^* E\{\hat{\theta} - E\{\hat{\theta}\}\}$$

$$= |\theta - E\{\hat{\theta}\}|^2 + E\{|\hat{\theta} - E\{\hat{\theta}\}|^2\} \tag{3.6.9}$$

which leads to (3.6.7) by using (3.6.3) and (3.6.5). Ideally, we would like to minimize the MSE, but this minimum is not always zero. Hence minimizing variance can increase the bias. The *normalized MSE* is defined as

$$\varepsilon \triangleq \frac{\text{MSE}(\theta)}{\theta} \qquad \theta \neq 0 \tag{3.6.10}$$

Cramér-Rao lower bound. If it is possible to minimize the MSE when the bias is zero, then clearly the variance is also minimized. Such estimators are called *minimum variance unbiased* estimators, and they attain an important minimum bound on the variance of the estimator, called the *Cramér-Rao lower bound* (CRLB), or *minimum variance bound*. If $\hat{\theta}$ is unbiased, then it follows that $E\{\hat{\theta} - \theta\} = 0$, which may be expressed as

$$\int_{-\infty}^{\infty} \cdots \int (\hat{\theta} - \theta) f_{\mathbf{x};\theta}(\mathbf{x}; \theta) \, d\mathbf{x} = 0 \tag{3.6.11}$$

where $\mathbf{x}(\zeta) = [x_1(\zeta), x_2(\zeta), \ldots, x_N(\zeta)]^T$ and $f_{\mathbf{x};\theta}(\mathbf{x}; \theta)$ is the joint density of $\mathbf{x}(\zeta)$, which depends on a fixed but unknown parameter θ. If we differentiate (3.6.11) with respect to θ, assuming real-valued $\hat{\theta}$, we obtain

$$0 = \int_{-\infty}^{\infty} \cdots \int \frac{\partial}{\partial \theta}[(\hat{\theta} - \theta) f_{\mathbf{x};\theta}(\mathbf{x}; \theta)] \, d\mathbf{x} = \int_{-\infty}^{\infty} \cdots \int (\hat{\theta} - \theta) \frac{\partial f_{\mathbf{x};\theta}(\mathbf{x}; \theta)}{\partial \theta} \, d\mathbf{x} - 1 \tag{3.6.12}$$

Using the fact

$$\frac{\partial \ln[f_{\mathbf{x};\theta}(\mathbf{x}; \theta)]}{\partial \theta} = \frac{1}{f_{\mathbf{x};\theta}(\mathbf{x}; \theta)} \frac{\partial f_{\mathbf{x};\theta}(\mathbf{x}; \theta)}{\partial \theta}$$

or

$$\frac{\partial f_{\mathbf{x};\theta}(\mathbf{x}; \theta)}{\partial \theta} = \frac{\partial \ln[f_{\mathbf{x};\theta}(\mathbf{x}; \theta)]}{\partial \theta} f_{\mathbf{x};\theta}(\mathbf{x}; \theta) \tag{3.6.13}$$

and substituting (3.6.13) in (3.6.12), we get

$$\int_{-\infty}^{\infty} \cdots \int \left\{ (\hat{\theta} - \theta) \frac{\partial \ln[f_{\mathbf{x};\theta}(\mathbf{x}; \theta)]}{\partial \theta} \right\} f_{\mathbf{x};\theta}(\mathbf{x}; \theta) \, d\mathbf{x} = 1 \tag{3.6.14}$$

Clearly, the left side of (3.6.14) is simply the expectation of the expression inside the brackets, that is,

$$E \left\{ (\hat{\theta} - \theta) \frac{\partial \ln[f_{\mathbf{x};\theta}(\mathbf{x}; \theta)]}{\partial \theta} \right\} = 1 \tag{3.6.15}$$

Using the *Cauchy-Schwarz inequality* (Papoulis 1991; Stark and Woods 1994) $|E\{x(\zeta)y(\zeta)\}|^2 \leq E\{|x(\zeta)|^2\} E\{|y(\zeta)|^2\}$, we obtain

$$E\{(\hat{\theta} - \theta)^2\} E \left\{ \left(\frac{\partial \ln[f_{\mathbf{x};\theta}(\mathbf{x}; \theta)]}{\partial \theta} \right)^2 \right\} \geq E^2 \left\{ (\hat{\theta} - \theta) \frac{\partial \ln[f_{\mathbf{x};\theta}(\mathbf{x}; \theta)]}{\partial \theta} \right\} = 1 \tag{3.6.16}$$

The first term on the left-hand side is the variance of the estimator $\hat{\theta}$ since it is unbiased. Hence

$$\text{var}(\hat{\theta}) \geq \frac{1}{E\{[\partial \ln f_{\mathbf{x};\theta}(\mathbf{x}; \theta)/\partial \theta]^2\}} \tag{3.6.17}$$

which is one form of the CRLB and can also be expressed as

$$\text{var}(\hat{\theta}) \geq -\frac{1}{E\{\partial^2 \ln f_{\mathbf{x};\theta}(\mathbf{x}; \theta)/\partial \theta^2\}} \tag{3.6.18}$$

The function $\ln f_{\mathbf{x};\theta}(\mathbf{x}; \theta)$ is called the *log likelihood function* of θ. The CRLB expresses the minimum error variance of any estimator $\hat{\theta}$ of θ in terms of the joint density $f_{\mathbf{x};\theta}(\mathbf{x}; \theta)$

of observations. Hence every unbiased estimator must have a variance greater than a certain number. An unbiased estimate that satisfies the CRLB (3.6.18) with equality is called an *efficient* estimate. If such an estimate exists, then it can be obtained as a unique solution to the likelihood equation

$$\frac{\partial \ln f_{\mathbf{x};\theta}(\mathbf{x};\theta)}{\partial \theta} = 0 \tag{3.6.19}$$

The solution of (3.6.19) is called the *maximum likelihood* (ML) estimate. Note that if the efficient estimate does not exist, then the ML estimate will not achieve the lower bound and hence it is difficult to ascertain how closely the variance of any estimate will approach the bound. The CRLB can be generalized to handle the estimation of vector parameters (Therrien 1992).

Consistency of estimator. If the MSE of the estimator can be made to approach zero as the sample size N becomes large, then from (3.6.7) both the bias and the variance will tend to zero. Then the sampling distribution will tend to concentrate about θ, and eventually as $N \to \infty$, the sampling distribution will become an impulse at θ. This is an important and desirable property, and the estimator that possesses it is called a *consistent* estimator.

Confidence interval. If we know the sampling distribution of an estimator, we can use the observations to compute an interval that has a specified probability of covering the unknown true parameter value. This interval is called a *confidence interval*, and the coverage probability is called the *confidence level*. When we interpret the meaning of confidence intervals, it is important to remember that it is the interval that is the random variable, and not the parameter. This concept will be explained in the sequel by means of specific examples.

3.6.2 Estimation of Mean

The natural estimator of the mean μ_x of a stationary sequence $x(n)$ from the observations $\{x(n)\}_0^{N-1}$ is the *sample mean*, given by

$$\hat{\mu}_x = \frac{1}{N}\sum_{n=0}^{N-1} x(n) \tag{3.6.20}$$

The estimate $\hat{\mu}_x$ is a random variable that depends on the number and values of the observations. Changing N or the set of observations will lead to another value for $\hat{\mu}_x$. Since the mean of the estimator is given by

$$E\{\hat{\mu}_x\} = \mu_x \tag{3.6.21}$$

the estimator $\hat{\mu}_x$ is unbiased. If $x(n) \sim \text{WN}(\mu_x, \sigma_x^2)$, we have

$$\text{var}(\hat{\mu}_x) = \frac{\sigma_x^2}{N} \tag{3.6.22}$$

because the samples of the process are uncorrelated random variables. This variance, which is a measure of the estimator's quality, increases if $x(n)$ is nonwhite.

Indeed, for a correlated random sequence, the variance of $\hat{\mu}_x$ is given by (see Problem 3.30)

$$\text{var}(\hat{\mu}_x) = N^{-1}\sum_{l=-N}^{N}\left(1-\frac{|l|}{N}\right)\gamma_x(l) \le N^{-1}\sum_{l=-N}^{N}|\gamma_x(l)| \tag{3.6.23}$$

where $\gamma_x(l)$ is the covariance sequence of $x(n)$. If $\gamma_x(l) \to 0$ as $l \to \infty$, then $\text{var}(\hat{\mu}_x) \to 0$ as $N \to \infty$ and hence $\hat{\mu}_x$ is a consistent estimator of μ_x. If $\sum_{l=-\infty}^{\infty}|\gamma_x(l)| < \infty$, then

$$\lim_{N\to\infty} N\,\mathrm{var}(\hat{\mu}_x) = \lim_{N\to\infty} \sum_{l=-N}^{N}\left(1-\frac{|l|}{N}\right)\gamma_x(l) = \sum_{l=-\infty}^{\infty}\gamma_x(l) \qquad (3.6.24)$$

The expression for $\mathrm{var}(\hat{\mu}_x)$ in (3.6.23) can also be put in the form (see Problem 3.30)

$$\mathrm{var}(\hat{\mu}_x) = \frac{\sigma_x^2}{N}[1 + \Delta_N(\rho_x)] \qquad (3.6.25)$$

where
$$\Delta_N(\rho_x) = 2\sum_{l=0}^{N}\left(1-\frac{l}{N}\right)\rho_x(l) \qquad \rho_x(l) = \frac{\gamma_x(l)}{\sigma_x^2} \qquad (3.6.26)$$

Since $\Delta_N(\rho_x) \geq 0$, the variance of the estimator increases as the amount of correlation among the samples of $x(n)$ increases. This implies that as the correlation increases, we need more samples to retain the quality of the estimate because each additional sample carries "less information." For this reason the estimation of long-memory processes and processes with infinite variance is extremely difficult.

Sampling distribution. If we know the joint pdf of the random variables $\{x(n)\}_0^{N-1}$, we can determine, at least in principle, the pdf of $\hat{\mu}_x$. For example, if it is assumed that the observations are IID as $\mathcal{N}(\mu_x, \sigma_x^2)$ then from (3.6.21) and (3.6.23), it can be seen that $\hat{\mu}_x$ is normal with mean μ_x and variance σ_x^2/N, that is,

$$f_{\hat{\mu}_x}(\hat{\mu}_x) = \frac{1}{\sqrt{2\pi}(\sigma_x/\sqrt{N})}\exp\left[-\frac{1}{2}\left(\frac{\hat{\mu}_x - \mu_x}{\sigma_x/\sqrt{N}}\right)^2\right] \qquad (3.6.27)$$

which is the sampling distribution of the mean. If N is large, then from the central limit theorem, the sampling distribution of the sample mean (3.6.27) is usually very close to the normal distribution, even if the individual distributions are not normal.

If we know the standard deviation σ_x, we can compute the probability

$$\mathrm{Pr}\left\{\mu_x - k\frac{\sigma_x}{\sqrt{N}} < \hat{\mu}_x < \mu_x + k\frac{\sigma_x}{\sqrt{N}}\right\} \qquad (3.6.28)$$

that the random variable $\hat{\mu}_x$ is within a certain interval specified by two fixed quantities. A simple rearrangement of the above inequality leads to

$$\mathrm{Pr}\left\{\hat{\mu}_x - k\frac{\sigma_x}{\sqrt{N}} < \mu_x < \hat{\mu}_x + k\frac{\sigma_x}{\sqrt{N}}\right\} \qquad (3.6.29)$$

which gives the probability that the fixed quantity μ_x lies between the two random variables $\hat{\mu}_x - k\sigma_x/\sqrt{N}$ and $\hat{\mu}_x + k\sigma_x/\sqrt{N}$. Hence (3.6.29) provides the probability that an interval with fixed length $2k\sigma_x/\sqrt{N}$ and randomly centered at the estimated mean includes the true mean. If we choose k so that the probability defined by (3.6.29) is equal to 0.95, the interval is known as the 95 percent confidence interval. To understand the meaning of this reasoning, we stress that for each set of measurements we compute a confidence interval that either contains or does not contain the true mean. However, if we repeat this process for a large number of observation sets, about 95 percent of the obtained confidence intervals will include the true mean. We stress that by no means does this imply that a confidence interval includes the true mean with probability 0.95.

If the variance σ_x^2 is unknown, then it has to be determined from the observations. This results in two modifications of (3.6.29). First, σ_x is replaced by

$$\hat{\sigma}_x^2 = \frac{1}{N-1}\sum_{n=0}^{N-1}[x(n) - \hat{\mu}_x]^2 \qquad (3.6.30)$$

which implies that the center and the length of the confidence interval are different for each set of observations. Second, the random variable $(\hat{\mu}_x - \mu_x)/(\hat{\sigma}_x/\sqrt{N})$ is distributed according to *Student's t distribution with $v = N - 1$ degrees of freedom* (Parzen 1960), which tends to a Gaussian for large values of N. In these cases, the factor k in (3.6.29) is replaced by the appropriate value t of Student's distribution, using $N - 1$ degrees of freedom, for the desired level of confidence.

If the observations are normal but not IID, then from (3.6.25), the mean estimator $\hat{\mu}_x$ is normal with mean μ and variance $(\sigma_x^2/N)[1 + \Delta_N(\rho_x)]$. It is now easy to construct exact confidence intervals for $\hat{\mu}_x$ if $\rho_x(l)$ is known, and approximate confidence intervals if $\rho_x(l)$ is to be estimated from the observations. For large N, the variance $\text{var}(\hat{\mu}_x)$ can be approximated by

$$\text{var}(\hat{\mu}_x) = \frac{\sigma_x^2}{N}[1 + \Delta_N(\rho_x)]$$

$$\simeq \frac{\sigma_x^2}{N}\left[1 + 2\sum_1^N \rho_x(l)\right] \tag{3.6.31}$$

$$\triangleq \frac{v}{N} \qquad v = \sigma_x^2\left\{1 + 2\sum_1^N \rho_x(l)\right\}$$

and hence an approximate 95 percent confidence interval for $\hat{\mu}_x$ is given by

$$\left(\hat{\mu}_x - 1.96\sqrt{\frac{v}{N}},\, \hat{\mu}_x + 1.96\sqrt{v/N}\right) \tag{3.6.32}$$

This means that, on average, the above interval will enclose the true value μ_x on 95 percent of occasions. For many practical random processes (especially those modeled as ARMA processes), the result in (3.6.32) is a good approximation.

EXAMPLE 3.6.1. Consider the AR(1) process

$$x(n) = ax(n-1) + w(n) \qquad -1 < a < 1$$

where $w(n) \sim \text{WN}(0, \sigma_w^2)$. We wish to compute the variance of the mean estimator $\hat{\mu}_x$ of the process $x(n)$. Using straightforward calculations, we obtain

$$\mu_x = 0 \qquad \sigma_x^2 = \frac{\sigma_w^2}{1 - a^2} \qquad \text{and} \qquad \rho_x(l) = a^{|l|}$$

From (3.6.26) we evaluate the term

$$\Delta_N(\rho) = \frac{2a}{1-a}\left[1 - \frac{1}{N(1-a)} + \frac{a^N}{N(1-a)}\right] \simeq \frac{2a}{1-a} \qquad \text{for } N \gg 1$$

When $a \to 1$, that is, when the dependence between the signal samples increases, then the factor $\Delta_N(\rho)$ takes large values and the quality of estimator decreases drastically. Similar conclusions can be drawn using the approximation (3.6.31)

$$v = \left(1 + 2\sum_1^\infty a^l\right)\frac{\sigma_w^2}{1 - a^2} = \frac{\sigma_w^2}{(1-a)^2}$$

We will next verify these results using two Monte Carlo simulations: one for $a = 0.9$, which represents high correlations among samples, and the other for $a = 0.1$. Using a Gaussian pseudorandom number generator with mean 0 and variance $\sigma_w^2 = 1$, we generated $N = 100$ samples of the AR(1) process $x(n)$. Using v in (3.6.31) and (3.6.32), we next computed the confidence intervals. For $a = 0.9$, we obtain

$$v = 100 \qquad \text{and} \qquad \text{confidence interval: } (\hat{\mu}_x - 0.98, \hat{\mu}_x + 0.98)$$

and for $a = 0.1$, we obtain

$$v = 1.2345 \qquad \text{and} \qquad \text{confidence interval: } (\hat{\mu}_x - 0.2178, \hat{\mu}_x + 0.2178)$$

Clearly, when the dependence between signal samples increases, the quality of the estimator decreases drastically and hence the confidence interval is wider. To have the same confidence interval, we should increase the number of samples N.

We next estimate the mean, using (3.6.20), and we repeat the experiment 10,000 times. Figure 3.14 shows histograms of the computed means for $a = 0.9$ and $a = 0.1$. The confidence intervals are also shown as dotted lines around the true mean. The histograms are approximately Gaussian in shape. The histogram for the high-correlation case is wider than that for the low-correlation case, which is to be expected. The 95 percent confidence intervals also indicate that very few estimates are outside the interval.

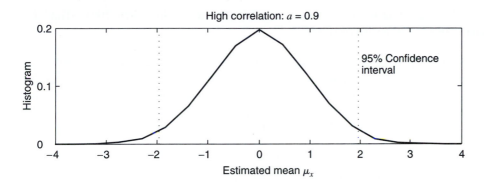

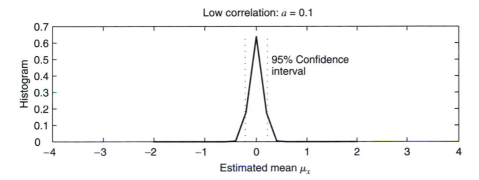

FIGURE 3.14
Histograms of mean estimates in Example 3.6.1.

3.6.3 Estimation of Variance

The natural estimator of the variance σ_x of a stationary sequence $x(n)$ from the observations $\{x(n)\}_0^{N-1}$ is the *sample variance*, given by

$$\hat{\sigma}_x^2 \triangleq \frac{1}{N} \sum_{n=0}^{N-1} \{x(n) - \hat{\mu}_x\}^2 \tag{3.6.33}$$

By using the mean estimate $\hat{\mu}_x$ from (3.6.20), the mean of the variance estimator can be shown to equal (see Problem 3.31)

$$E\{\hat{\sigma}_x^2\} = \sigma_x^2 - \text{var}(\hat{\mu}_x) = \sigma_x^2 - \frac{1}{N} \sum_{l=-N}^{N} \left(1 - \frac{|l|}{N}\right) \gamma_x(l) \tag{3.6.34}$$

If the sequence $x(n)$ is uncorrelated, then

$$E\{\hat{\sigma}_x^2\} = \sigma_x^2 - \frac{\sigma_x^2}{N} = \left(\frac{N-1}{N}\right) \sigma_x^2 \tag{3.6.35}$$

From (3.6.34) or (3.6.35), it is obvious that the estimator in (3.6.33) is biased. If $\gamma_x(l) \to 0$ as $l \to \infty$, then $\text{var}(\hat{\mu}_x) \to 0$ as $N \to \infty$ and hence $\hat{\sigma}_x^2$ is an asymptotically unbiased estimator of σ_x^2. In practical applications, the variance estimate is nearly unbiased for large N. Note that if we use the actual mean μ_x in (3.6.33), then the resulting estimator is unbiased.

The general expression for the variance of the variance estimator is fairly complicated and requires higher-order moments. It can be shown that for either estimators

$$\text{var}(\hat{\sigma}_x^2) \approx \frac{\gamma_x^{(4)}}{N} \quad \text{for large } N \tag{3.6.36}$$

where $\gamma_x^{(4)}$ is the fourth central moment of $x(n)$ (Brockwell and Davis 1991). Thus the estimator in (3.6.33) is also consistent.

Sampling distribution. In the case of the mean estimator, the sampling distribution involved the distribution of sums of random variables. The variance estimator involves the sum of the squares of random variables, for which the sampling distribution computation is complicated. For example, if there are N independent measurements from an $\mathcal{N}(0, 1)$ distribution, then the sampling distribution of the random variable

$$\chi_N^2 = x_1^2 + x_2^2 + \cdots + x_N^2 \tag{3.6.37}$$

is given by the *chi-squared distribution with N degrees of freedom*. The general form of χ_N^2 with ν degrees of freedom is

$$f_{\chi_\nu^2}(x) = \frac{1}{2^{\nu/2}\Gamma(\nu/2)} x^{\nu/2-1} \exp\left(-\frac{x}{2}\right) \quad 0 \leq x \leq \infty \tag{3.6.38}$$

where $\Gamma(\nu/2) = \int_0^\infty e^{-t} t^{\nu/2-1}\, dt$ is the gamma function with argument $\nu/2$.

For the variance estimator in (3.6.33), it can be shown (Parzen 1960) that $N\hat{\sigma}_x^2$ is distributed as chi squared with $\nu = N - 1$ *degrees of freedom*. This means that, for any set of N observations, there will only be $N - 1$ independent deviations $\{x(n) - \hat{\mu}_x\}$, since their sum is zero from the definition of the mean. Assuming that the observations are $\mathcal{N}(\mu, \sigma^2)$, the random variables $x(n)/\sigma$ will be $\mathcal{N}(\mu/\sigma, 1)$ and hence the random variable

$$\frac{N\hat{\sigma}_x^2}{\sigma^2} = \frac{1}{\sigma^2} \sum_{n=0}^{N-1} [x(n) - \hat{\mu}_x]^2 \tag{3.6.39}$$

will be chi squared distributed with $\nu = N - 1$. Therefore, using values of the chi-squared distribution, confidence intervals for the variance estimator can be computed. In particular, since $N\hat{\sigma}_x^2/\sigma^2$ is distributed as χ_ν^2, the 95 percent limits of the form

$$\Pr\left\{\chi_\nu\left(\frac{0.05}{2}\right) < N\hat{\sigma}_x^2/\sigma^2 \leq \chi_\nu\left(1 - \frac{0.05}{2}\right)\right\} = 0.95 \tag{3.6.40}$$

can be obtained from chi-squared tables (Fisher and Yates 1938). By rearranging (3.6.40), the random variable $\sigma^2/\hat{\sigma}_x^2$ satisfies

$$\Pr\left\{\frac{N}{\chi_\nu(0.975)} < \frac{\sigma^2}{\hat{\sigma}_x^2} \leq \frac{N}{\chi_\nu(0.025)}\right\} = 0.95 \tag{3.6.41}$$

Using $l_1 = N/\chi_\nu(0.975)$ and $l_2 = N/\chi_\nu(0.025)$, we see that (3.6.41) implies that

$$\Pr\{l_2\hat{\sigma}_x^2 \geq \sigma^2 \quad \text{and} \quad l_1\hat{\sigma}_x^2 < \sigma^2\} = 0.95 \tag{3.6.42}$$

Thus the 95 percent confidence interval based on the estimate $\hat{\sigma}_x^2$ is $(l_1\hat{\sigma}_x^2, l_2\hat{\sigma}_x^2)$. Note that this interval is sensitive to the validity of the normal assumption of random variables leading to (3.6.39). This is not the case for the confidence intervals for the mean estimates because, thanks to the central limit theorem, the computation of the interval can be based on the normal assumption.

EXAMPLE 3.6.2. Consider again the AR(1) process given in Example 3.6.1:

$$x(n) = ax(n-1) + w(n) \qquad -1 < a < 1 \qquad w(n) \sim \text{WN}(0, 1)$$

with
$$\mu_x = 0 \qquad \sigma_x^2 = \frac{\sigma_w^2}{1 - a^2} \qquad \text{and} \qquad \rho_x(l) = a^{|l|} \tag{3.6.43}$$

We wish to compute the mean of the variance estimator $\hat{\sigma}_x^2$ of the process $x(n)$. From (3.6.34), we obtain

$$E[\hat{\sigma}_x^2] = \sigma_x^2 \left[1 - \frac{1}{N} \sum_{l=-N}^{N} \left(1 - \frac{|l|}{N} \right) a^{|l|} \right] \tag{3.6.44}$$

When $a \to 1$, that is, when the dependence between the signal samples increases, the mean of the estimate deviates significantly from the true value σ_x^2 and the quality of the estimator decreases drastically. For small dependence, the mean is very close to σ_x^2. These conclusions can be verified using two Monte Carlo simulations as before: one for $a = 0.9$, which represents high correlations among samples, and the other for $a = 0.1$. Using a Gaussian pseudorandom number generator with mean 0 and unit variance, we generated $N = 100$ samples of the AR(1) process $x(n)$. The computed parameters according to (3.6.43) and (3.6.44) are

$$a = 0.9: \qquad \sigma_x^2 = 5.2632 \qquad E\{\hat{\sigma}_x^2\} = 4.3579$$

$$a = 0.1: \qquad \sigma_x^2 = 1.0101 \qquad E\{\hat{\sigma}_x^2\} = 0.9978$$

We next estimate the variance by using (3.6.33) and repeat the experiment 10,000 times. Figure 3.15 shows histograms of computed variances for $a = 0.9$ and for $a = 0.1$. The computed

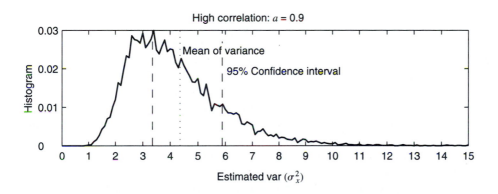

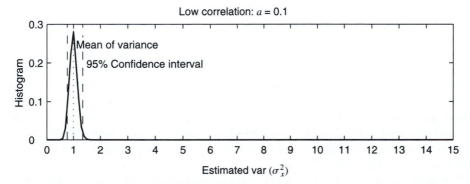

FIGURE 3.15
Histograms of variance estimates in Example 3.6.2.

means of the variance estimates are also shown as dotted lines. Clearly, the histogram is much wider for the high-correlation case and much narrower (almost symmetric and Gaussian) for the low-correlation case.

The 95 percent confidence intervals are given by $(l_1\hat{\sigma}_x^2, l_2\hat{\sigma}_x^2)$, where $l_1 = N/\chi_\nu(0.975)$ and $l_2 = N/\chi_\nu(0.025)$. The values of l_1 and l_2 are obtained from the chi-squared distribution curves (Jenkins and Watts 1968). For $N = 100$, $l_1 = 0.77$ and $l_2 = 1.35$; hence the 95 percent confidence intervals for σ_x^2 are

$$(0.77\hat{\sigma}_x^2, 1.35\hat{\sigma}_x^2)$$

also shown as dashed lines around the mean value $E\{\hat{\sigma}_x^2\}$. The confidence interval for the high-correlation case, $a = 0.9$, does not appear to be a good interval, which implies that the approximation leading to (3.6.42) is not a good one for this case. Such is not the case for $a = 0.1$.

3.7 SUMMARY

In this chapter we provided an overview of the basic theory of discrete-time stochastic processes. We began with the notion of a random variable as a mapping from the abstract probability space to the real space, extended it to random vectors as a collection of random variables, and introduced discrete-time stochastic processes as an indexed family (or time series) of random variables. A complete probabilistic description of these random objects requires the knowledge of joint distribution or density functions, which is difficult to acquire except in simple cases. Therefore, the emphasis was placed on description using joint moments of distributions, and, in particular, the emphasis was placed on the second-order moments, which are relatively easy to estimate or compute in practice.

We defined the mean and the variance to describe random variables, and we provided three useful models of random variables. For random vector description, we defined the mean vector and the autocorrelation matrix. Linear transformations of random vectors were discussed, using densities and correlation matrices. The normal random vector was then introduced as a useful model of a random vector. A particularly simple linear transformation, namely, the sum of independent random variables, was used to introduce random variables with stable and infinitely divisible distributions. To describe stochastic processes, we proceeded to define mean and autocorrelation sequences. In many applications, the concept of stationary of random processes is a useful one that reduces the computational complexity. Assuming time invariance on the first two moments, we defined a wide-sense stationary (WSS) process in which the mean is a constant and correlation between random variables at two distinct times is a function of time difference or lag. The rest of the chapter was devoted to the analysis of WSS processes.

A stochastic process is generally observed in practice as a single sample function (a speech signal or a radar signal) from which it is necessary to estimate the first- and the second-order moments. This requires the notion of ergodicity, which provides a framework for the computation of statistical averages using time averages over a single realization. Although this framework requires theoretical results using mean square convergence, we provided a simple approach of using appropriate time averages. An important random signal characteristic called variability was introduced. The WSS processes were then described in the frequency domain using the power spectral density function, which is a physical quantity that can be measured in practice. Some random processes exhibiting flat spectral envelopes were analyzed including one of white noise. Since random processes are generally processed using linear systems, we described linear system operations with random inputs in both the time and frequency domains.

The properties of correlation matrices and sequences play an important role in filtering and estimation theory and were discussed in detail, including eigenanalysis. Another important random signal characteristic called memory was also introduced. Stationary random signals were modeled using autocorrelation matrices, and the relationship between spectral flatness and eigenvalue spread was explored. These properties were used in an alternate representation of random vectors as well as processes using uncorrelated components which were based on diagonalization and triangularization of correlation matrices. This resulted in the discrete KL transform and KL expansion. These concepts will also be useful in later chapters on optimal filtering and adaptive filtering.

Finally, we concluded this chapter with the introduction of elementary estimation theory. After discussion of properties of estimators, two important estimators of mean and variance were treated in detail along with their sampling distributions. These topics will be useful in many subsequent chapters.

PROBLEMS

3.1 The exponential density function is given by

$$f_x(x) = \frac{1}{a} e^{-x/a} u(x) \tag{P.1}$$

where a is a parameter and $u(x)$ is a unit step function.

(a) Plot the density function for $a = 1$.

(b) Determine the mean, variance, skewness, and kurtosis of the Rayleigh random variable with $a = 1$. Comment on the significance of these moments in terms of the shape of the density function.

(c) Determine the characteristic function of the exponential pdf.

3.2 The Rayleigh density function is given by

$$f_x(x) = \frac{x}{\sigma^2} e^{-x^2/(2\sigma^2)} u(x) \tag{P.2}$$

where σ is a parameter and $u(x)$ is a unit step function. Repeat Problem 3.1 for $\sigma = 1$.

3.3 Using the binomial expansion of $\{x(\zeta) - \mu_x\}^m$, show that the mth central moment is given by

$$M_m^{(x)} = \sum_{k=0}^{m} \binom{m}{k} (-1)^k \mu_x^k \xi_{m-k}^{(x)}$$

Similarly, show that

$$\xi_m^{(x)} = \sum_{k=0}^{m} \binom{m}{k} \mu_x^k M_{m-k}^{(x)}$$

3.4 Consider a zero-mean random variable $x(\zeta)$. Using (3.1.26), show that the first four cumulants of $x(\zeta)$ are given by (3.1.28) through (3.1.31).

3.5 A random vector $\mathbf{x}(\zeta) = [x_1(\zeta) \ x_2(\zeta)]^T$ has mean vector $\mu_{\mathbf{x}} = [1 \ 2]^T$ and covariance matrix

$$\Gamma_{\mathbf{x}} = \begin{bmatrix} 4 & 0.8 \\ 0.8 & 1 \end{bmatrix}$$

This vector is transformed to another random vector $\mathbf{y}(\zeta)$ by the following linear transformation:

$$\begin{bmatrix} y_1(\zeta) \\ y_2(\zeta) \\ y_3(\zeta) \end{bmatrix} = \begin{bmatrix} 1 & 3 \\ -1 & 2 \\ 2 & 3 \end{bmatrix} \begin{bmatrix} x_1(\zeta) \\ x_2(\zeta) \end{bmatrix}$$

Determine (a) the mean vector $\mu_{\mathbf{y}}$, (b) the autocovariance matrix $\Gamma_{\mathbf{y}}$, and (c) the cross-correlation matrix $\mathbf{R_{xy}}$.

3.6 Using the moment generating function, show that the linear transformation of a Gaussian random vector is also Gaussian.

3.7 Let $\{x_k(\zeta)\}_{k=1}^4$ be four IID random variables with exponential distribution (P.1) with $a = 1$. Let

$$y_k(\zeta) = \sum_{l=1}^{k} x_l(\zeta) \qquad 1 \le k \le 4$$

(a) Determine and plot the pdf of $y_2(\zeta)$.
(b) Determine and plot the pdf of $y_3(\zeta)$.
(c) Determine and plot the pdf of $y_4(\zeta)$.
(d) Compare the pdf of $y_4(\zeta)$ with that of the Gaussian density.

3.8 For each of the following, determine whether the random process is (1) WSS or (2) m.s. ergodic in the mean.

(a) $X(t) = A$, where A is a random variable uniformly distributed between 0 and 1.
(b) $X_n = A \cos \omega_0 n$, where A is a Gaussian random variable with mean 0 and variance 1.
(c) A Bernoulli process with $\Pr[X_n = 1] = p$ and $\Pr[X_n = -1] = 1 - p$.

3.9 Consider the harmonic process $x(n)$ defined in (3.3.50).

(a) Determine the mean of $x(n)$.
(b) Show that the autocorrelation sequence is given by

$$r_x(l) = \frac{1}{2} \sum_{k=1}^{N} |c_k|^2 \cos \omega_k l \qquad -\infty < l < \infty$$

3.10 Suppose that the random variables ϕ_k in the real-valued harmonic process model are distributed with a pdf $f_{\phi_k}(\phi_k) = (1 + \cos \phi_k)/(2\pi)$, $-\pi \le \phi_k \le \pi$. Is the resulting stochastic process stationary?

3.11 A stationary random sequence $x(n)$ with mean $\mu_x = 4$ and autocovariance

$$\gamma_x(n) = \begin{cases} 4 - |n| & |n| \le 3 \\ 0 & \text{otherwise} \end{cases}$$

is applied as an input to a linear shift-invariant (LSI) system whose impulse response $h(n)$ is

$$h(n) = u(n) - u(n - 4)$$

where $u(n)$ is a unit step sequence. The output of this system is another random sequence $y(n)$. Determine (a) the mean sequence $\mu_y(n)$, (b) the cross-covariance $\gamma_{xy}(n_1, n_2)$ between $x(n_1)$ and $y(n_2)$, and (c) the autocovariance $\gamma_y(n_1, n_2)$ of the output process $y(n)$.

3.12 A causal LTI system, which is described by the difference equation

$$y(n) = \tfrac{1}{2}y(n - 1) + x(n) + \tfrac{1}{3}x(n - 1)$$

is driven by a zero-mean WSS process with autocorrelation $r_x(l) = 0.5^{|l|}$.

(a) Determine the PSD and the autocorrelation of the output sequence $y(n)$.
(b) Determine the cross-correlation $r_{xy}(l)$ and cross-PSD $R_{xy}(e^{j\omega})$ between the input and output signals.

3.13 A WSS process with PSD $R_x(e^{j\omega}) = 1/(1.64 + 1.6 \cos \omega)$ is applied to a causal system described by the following difference equation

$$y(n) = 0.6y(n - 1) + x(n) + 1.25x(n - 1)$$

Compute (a) the PSD of the output and (b) the cross-PSD $R_{xy}(e^{j\omega})$ between input and output.

3.14 Determine whether the following matrices are valid correlation matrices:

$$\text{(a)} \quad \mathbf{R}_1 = \begin{bmatrix} 1 & 1 \\ 1 & 1 \end{bmatrix} \qquad\qquad \text{(b)} \quad \mathbf{R}_2 = \begin{bmatrix} 1 & \frac{1}{2} & \frac{1}{4} \\ \frac{1}{2} & 1 & \frac{1}{2} \\ \frac{1}{4} & \frac{1}{2} & 1 \end{bmatrix}$$

$$\text{(c)} \quad \mathbf{R}_3 = \begin{bmatrix} 1 & 1-j \\ 1+j & 1 \end{bmatrix} \qquad \text{(d)} \quad \mathbf{R}_4 = \begin{bmatrix} 1 & \frac{1}{2} & 1 \\ \frac{1}{2} & 2 & \frac{1}{2} \\ 1 & 1 & 1 \end{bmatrix}$$

3.15 Consider a normal random vector $\mathbf{x}(\zeta)$ with components that are mutually uncorrelated, that is, $\rho_{ij} = 0$. Show that (a) the covariance matrix $\mathbf{\Gamma_x}$ is diagonal and (b) the components of $\mathbf{x}(\zeta)$ are mutually independent.

3.16 Show that if a real, symmetric, and nonnegative definite matrix $\mathbf{R}$ has eigenvalues $\lambda_1, \lambda_2, \ldots, \lambda_M$, then the matrix $\mathbf{R}^k$ has eigenvalues $\lambda_1^k, \lambda_2^k, \ldots, \lambda_M^k$.

3.17 Prove that the trace of $\mathbf{R}$ is given by

$$\operatorname{tr} \mathbf{R} = \sum \lambda_i$$

3.18 Prove that the determinant of $\mathbf{R}$ is given by

$$\det \mathbf{R} = |\mathbf{R}| = \prod \lambda_i = |\mathbf{\Lambda}|$$

3.19 Show that the determinants of $\mathbf{R}$ and $\mathbf{\Gamma}$ are related by

$$\det \mathbf{R} = \det \mathbf{\Gamma}(1 + \boldsymbol{\mu}^H \mathbf{\Gamma} \boldsymbol{\mu})$$

3.20 Let $\mathbf{R_x}$ be the correlation matrix of the vector $\mathbf{x} = [x(0)\, x(2)\, x(3)]^T$, where $x(n)$ is a zero-mean WSS process.

(a) Check whether the matrix $\mathbf{R_x}$ is Hermitian, Toeplitz, and nonnegative definite.
(b) If we know the matrix $\mathbf{R_x}$, can we determine the correlation matrix of the vector $\bar{\mathbf{x}} = [x(0)\, x(1)\, x(2)\, x(3)]^T$?

3.21 Using the nonnegativeness of $E\{[x(n+l) \pm x(n)]^2\}$, show that $r_x(0) \geq |r_x(l)|$ for all l.

3.22 Show that $r_x(l)$ is nonnegative definite, that is,

$$\sum_{l=1}^{M} \sum_{k=1}^{M} a_l r_x(l-k) a_k^* \geq 0 \qquad \forall M, \ \forall a_1, \ldots, a_M$$

3.23 Let $x(n)$ be a random process generated by the AP(1) system

$$x(n) = \alpha x(n-1) + w(n)] \qquad n \geq 0 \qquad x(-1) = 0$$

where $w(n)$ is an IID$(0, \sigma_w^2)$ process.

(a) Determine the autocorrelation $r_x(n_1, n_2)$ function.
(b) Show that $r_x(n_1, n_2)$ asymptotically approaches $r_x(n_1 - n_2)$, that is, it becomes shift-invariant.

3.24 Let $\mathbf{x}$ be a random vector with mean $\boldsymbol{\mu_x}$ and autocorrelation $\mathbf{R_x}$.

(a) Show that $\mathbf{y} = \mathbf{Q}^T \mathbf{x}$ transforms $\mathbf{x}$ to an uncorrelated component vector $\mathbf{y}$ if $\mathbf{Q}$ is the eigenmatrix of $\mathbf{R_x}$.
(b) Comment on the geometric interpretation of this transformation.

3.25 The mean and the covariance of a Gaussian random vector $\mathbf{x}$ are given by, respectively,

$$\boldsymbol{\mu}_{\mathbf{x}} = \begin{bmatrix} 1 \\ 2 \end{bmatrix} \quad \text{and} \quad \boldsymbol{\Gamma}_{\mathbf{x}} = \begin{bmatrix} 1 & \frac{1}{2} \\ \frac{1}{2} & 1 \end{bmatrix}$$

Plot the 1σ, 2σ, and 3σ concentration ellipses representing the contours of the density function in the (x_1, x_2) plane. *Hints:* The radius of an ellipse with major axis a (along x_1) and minor axis $b < a$ (along x_2) is given by

$$r^2 = \frac{a^2 b^2}{a^2 \sin^2 \theta + b^2 \cos^2 \theta}$$

where $0 \le \theta \le 2\pi$. Compute the 1σ ellipse specified by $a = \sqrt{\lambda_1}$ and $b = \sqrt{\lambda_2}$ and then rotate and translate each point $\mathbf{x}^{(i)} = [x_1^{(i)} \; x_2^{(i)}]^T$ using the transformation $\mathbf{w}^{(i)} = \mathbf{Q}_x \mathbf{x}^{(i)} + \boldsymbol{\mu}_x$.

3.26 Consider the process $x(\mu) = ax(\mu - 1) + w(\mu)$, where $w(\mu) \sim \text{WN}(0, \sigma_w^2)$.

(a) Show that the $M \times M$ correlation matrix of the process is symmetric Toeplitz and is given by

$$\mathbf{R}_{\mathbf{x}} = \frac{\sigma_w^2}{1 - a^2} \begin{bmatrix} 1 & a & \cdots & a^{m-1} \\ a & 1 & \cdots & a^{m-2} \\ \vdots & \vdots & \ddots & \vdots \\ a^{m-1} & a^{m-2} & \cdots & 1 \end{bmatrix}$$

(b) Verify that

$$\mathbf{R}_{\mathbf{x}}^{-1} = \frac{1}{\sigma_w^2} \begin{bmatrix} 1 & -a & 0 & \cdots & 0 \\ -a & 1+a^2 & -a & \cdots & 0 \\ 0 & -a & \ddots & \vdots & \vdots \\ \vdots & \vdots & & 1+a^2 & -a \\ 0 & 0 & \cdots & -a & 1 \end{bmatrix}$$

(c) Show that if

$$\mathbf{L}_{\mathbf{x}} = \begin{bmatrix} 1 & 0 & \cdots & 0 \\ -a & 1 & \cdots & 0 \\ \vdots & \vdots & \ddots & 0 \\ 0 & 0 & -a & 1 \end{bmatrix}$$

then $\mathbf{L}_x^T \mathbf{R}_x \mathbf{L}_x = (1 - a^2)\mathbf{I}$.

(d) For $\sigma_w^2 = 1$, $a = 0.95$, and $M = 8$ compute the DKLT and the DFT.

(e) Plot the eigenvalues of each transform in the same graph of the PSD of the process. Explain your findings.

(f) Plot the eigenvectors of each transform and compare the results.

(g) Repeat parts (e) and (f) for $M = 16$ and $M = 32$. Explain the obtained results.

(h) Repeat parts (e) to (g) for $a = 0.5$ and compare with the results obtained for $a = 0.95$.

3.27 Determine three different innovations representations of a zero-mean random vector $\mathbf{x}$ with correlation matrix

$$\mathbf{R}_x = \begin{bmatrix} 1 & \frac{1}{4} \\ \frac{1}{4} & 1 \end{bmatrix}$$

3.28 Verify that the eigenvalues and eigenvectors of the $M \times M$ correlation matrix of the process $x(\mu) = w(\mu) + bw(n-1)$, where $w(n) \sim \text{WN}(0, \sigma_w^2)$ are given by $\lambda_k = R_x(e^{j\omega_k})$, $q_n^{(k)} = \sin \omega_k n$, $\omega_k = \pi k/(M+1)$, where $k = 1, 2, \ldots, M$, (a) analytically and (b) numerically for $\sigma_w^2 = 1$ and $M = 8$. *Hint:* Plot the eigenvalues on the same graph with the PSD.

3.29 Consider the process $x(n) = w(n) + bw(n-1)$.

(a) Compute the DKLT for $M = 3$.

(b) Show that the variances of the DKLT coefficients are $\sigma_x^2(1 + \sqrt{2}b)$, σ_x^2, and $\sigma_x^2(1 - \sqrt{2}b)$.

3.30 Let $x(n)$ be a stationary random process with mean μ_x and covariance $\gamma_x(l)$. Let $\hat{\mu}_x = 1/N \sum_{n=0}^{N-1} x(n)$ be the sample mean from the observations $\{x(n)\}_{n=0}^{N-1}$.

(a) Show that the variance of $\hat{\mu}_x$ is given by

$$\text{var}(\hat{\mu}_x) = N^{-1} \sum_{l=-N}^{N} \left(1 - \frac{|l|}{N}\right) \gamma_x(l) \leq N^{-1} \sum_{l=-N}^{N} |\gamma_x(l)| \qquad \text{(P.3)}$$

(b) Show that the above result (P.3) can be expressed as

$$\text{var}(\hat{\mu}_x) = \frac{\sigma_x^2}{N}[1 + \Delta_N(\rho_x)] \qquad \text{(P.4)}$$

where
$$\Delta_N(\rho_x) = 2 \sum_{l=0}^{N} \left(1 - \frac{l}{N}\right) \rho_x(l) \qquad \rho_x(l) = \frac{\gamma_x(l)}{\sigma_x^2}$$

(c) Show that (P.3) reduces to $\text{var}(\hat{\mu}_x) = \sigma_x^2/N$ for a $\text{WN}(\mu_x, \sigma_x^2)$ process.

3.31 Let $x(n)$ be a stationary random process with mean μ_x, variance σ_x^2, and covariance $\gamma_x(l)$. Let

$$\hat{\sigma}_x^2 \triangleq \frac{1}{N} \sum_{n=0}^{N-1} [x(n) - \hat{\mu}_x]^2$$

be the sample variance from the observations $\{x(n)\}_{n=0}^{N-1}$.

(a) Show that the mean of $\hat{\sigma}_x^2$ is given by

$$E\{\hat{\sigma}_x^2\} = \sigma_x^2 - \text{var}(\hat{\mu}_x) = \sigma_x^2 - \frac{1}{N} \sum_{l=-N}^{N} \left(1 - \frac{|l|}{N}\right) \gamma_x(l)$$

(b) Show that the above result reduces to $\text{var}(\hat{\mu}_x) = (N-1)\sigma_x^2/N$ for a $\text{WN}(\mu_x, \sigma_x^2)$ process.

3.32 The Cauchy distribution with mean μ is given by

$$f_x(x) = \frac{1}{\pi} \frac{1}{1 + (x - \mu)^2} \qquad -\infty < x < \infty$$

Let $\{x_k(\zeta)\}_{i=k}^{N}$ be N IID random variables with the above distribution. Consider the mean estimator based on $\{x_k(\zeta)\}_{i=k}^{N}$

$$\hat{\mu}(\zeta) = \frac{1}{N} \sum_{k=1}^{N} x_k(\zeta)$$

Determine whether $\hat{\mu}(\zeta)$ is a consistent estimator of μ.

Linear Signal Models

In this chapter we introduce and analyze the properties of a special class of stationary random sequences that are obtained by driving a linear, time-invariant system with white noise. We focus on filters having a system function that is rational, that is, the ratio of two polynomials. The power spectral density of the resulting process is also rational, and its shape is completely determined by the filter coefficients. We will use the term *pole-zero* models when we want to emphasize the system viewpoint and the term *autoregressive moving-average* models to refer to the resulting random sequences. The latter term is not appropriate when the input is a harmonic process or a deterministic signal with a flat spectral envelope. We discuss the impulse response, autocorrelation, power spectrum, partial autocorrelation, and cepstrum of all-pole, all-zero, and pole-zero models. We express all these quantities in terms of the model coefficients and develop procedures to convert from one parameter set to another. Low-order models are studied in detail, because they are easy to analyze analytically and provide insight into the behavior and properties of higher-order models. An understanding of the correlation and spectral properties of a signal model is very important for the selection of the appropriate model in practical applications. Finally, we investigate a special case of pole-zero models with one or more unit poles. Pole-zero models are widely used for the modeling of stationary signals with short memory whereas models with unit poles are useful for the modeling of certain nonstationarity processes with trends.

4.1 INTRODUCTION

In Chapter 3 we defined and studied random processes as a mathematical tool to analyze random signals. In practice, we also need to generate random signals that possess certain known, second-order characteristics, or we need to describe observed signals in terms of the parameters of known random processes.

The simplest random signal model is the wide sense stationary white noise sequence $w(n) \sim \mathrm{WN}(0, \sigma_w^2)$ that has uncorrelated samples and a flat PSD. It is also easy to generate in practice by using simple algorithms. If we filter white noise with a stable LTI filter, we can obtain random signals with almost any arbitrary aperiodic correlation structure or continuous PSD. If we wish to generate a random signal with a line PSD using the previous approach, we need an LTI filter with "line" frequency response; that is, we need an oscillator. Unfortunately, such a system is not *stable*, and its output cannot be stationary. Fortunately, random signals with line PSDs can be easily generated by using the *harmonic process* model (linear combination of sinusoidal sequences with statistically independent random phases) discussed in Section 3.3.6. Figure 4.1 illustrates the filtering of white noise and "white " (flat spectral envelope) harmonic process by an LTI filter. Signal models with mixed PSDs can be obtained by combining the above two models, a process justified by a powerful result known as the *Wold decomposition*.

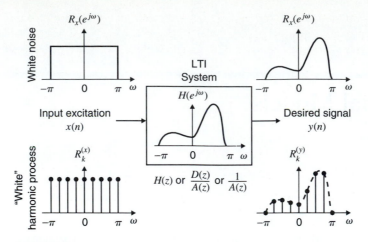

FIGURE 4.1
Signal models with continuous and discrete (line) power spectrum densities.

When the LTI filter is specified by its impulse response, we have a *nonparametric* signal model because there is no restriction regarding the form of the model and the number of parameters is infinite. However, if we specify the filter by a finite-order rational system function, we have a *parametric* signal model described by a finite number of parameters. We focus on parametric models because they are simpler to deal with in practical applications. The two major topics we address in this chapter are (1) the derivation of the second-order moments of AP, AZ, and PZ models, given the coefficients of their system function, and (2) the design of an AP, AZ, or PZ system that produces a random signal with a given autocorrelation sequence or PSD function. The second problem is known as *signal modeling* and theoretically is equivalent to the spectral factorization procedure developed in Section 2.4.4. The modeling of harmonic processes is theoretically straightforward and does not require the use of a linear filter to change the amplitude of the spectral lines. The challenging problem in this case is the identification of the filter by observing its response to a harmonic process with a flat PSD. The modeling problem for continuous PSDs has a solution, at least in principle, for every regular random sequence.

In practical applications, the second-order moments of the signal to be modeled are not known a priori and have to be estimated from a set of signal observations. This element introduces a new dimension and additional complications to the signal modeling problem, which are discussed in Chapter 9. In this chapter we primarily focus on parametric models that replicate the second-order properties (autocorrelation or PSD) of stationary random sequences. If the sequence is Gaussian, the model provides a complete statistical characterization. The characterization of non-Gaussian processes, which requires the use of higher-order moments, is discussed in Chapter 12.

4.1.1 Linear Nonparametric Signal Models

Consider a stable LTI system with impulse response $h(n)$ and input $w(n)$. The output $x(n)$ is given by the convolution summation

$$x(n) = \sum_{k=-\infty}^{\infty} h(k)w(n-k) \tag{4.1.1}$$

which is known as a *nonrecursive* system representation because the output is computed by linearly weighting samples of the input signal.

Linear random signal model. If the input $w(n)$ is a zero-mean white noise process with variance σ_w^2, autocorrelation $r_w(l) = \sigma_w^2 \delta(l)$, and PSD $R_w(e^{j\omega}) = \sigma_w^2$, $-\pi < \omega \leq \pi$, then from Table 3.2 the autocorrelation, complex PSD, and PSD of the output $x(n)$ are given by, respectively,

$$r_x(l) = \sigma_w^2 \sum_{k=-\infty}^{\infty} h(k)h^*(k-l) = \sigma_w^2 r_h(l) \tag{4.1.2}$$

$$R_x(z) = \sigma_w^2 H(z)H^*\left(\frac{1}{z^*}\right) \tag{4.1.3}$$

$$R_x(e^{j\omega}) = \sigma_w^2 |H(e^{j\omega})|^2 = \sigma_w^2 R_h(e^{j\omega}) \tag{4.1.4}$$

We notice that when the input is a white noise process, the shape of the autocorrelation and the power spectrum (*second-order moments*) of the output signal are completely characterized by the system. We use the term *system-based signal model* to refer to the signal generated by a system with a white noise input. If the system is linear, we use the term *linear random signal model*. In the statistical literature, the resulting model is known as the *general linear process model*. However, we should mention that in some applications it is more appropriate to use a deterministic input with flat spectral envelope or a "white" harmonic process input.

Recursive representation. Suppose now that the inverse system $H_I(n) = 1/H(z)$ is *causal and stable*. If we assume, without any loss of generality, that $h(0) = 1$, then $h_I(n) = \mathcal{Z}^{-1}\{H_I(n)\}$ has $h_I(0) = 1$. Therefore the input $w(n)$ can be obtained by

$$w(n) = x(n) + \sum_{k=1}^{\infty} h_I(k)x(n-k) \tag{4.1.5}$$

Solving for $x(n)$, we obtain the following recursive representation for the output signal

$$x(n) = -\sum_{k=1}^{\infty} h_I(k)x(n-k) + w(n) \tag{4.1.6}$$

We use the term *recursive* representation to emphasize that the present value of the output is obtained by a linear combination of all past output values, plus the present value of the input. By construction the nonrecursive and recursive representations of system $h(n)$ are equivalent; that is, they produce the same output when they are excited by the same input signal.

Innovations representation. If the system $H(z)$ is minimum-phase, then both $h(n)$ and $h_I(n)$ are causal and stable. Hence, the output signal can be expressed nonrecursively by

$$x(n) = \sum_{k=0}^{\infty} h(k)w(n-k) = \sum_{k=-\infty}^{n} h(n-k)w(k) \tag{4.1.7}$$

or recursively by (4.1.6).

From (4.1.7) we obtain

$$x(n+1) = \sum_{k=-\infty}^{n} h(n+1-k)w(k) + w(n+1)$$

or by using (4.1.5)

$$x(n+1) = \underbrace{\sum_{k=-\infty}^{n} h(n+1-k)\sum_{j=-\infty}^{n} h_I(k-j)x(j)}_{\text{past information: linear combination of } x(n),\ x(n-1),\ldots} + \underbrace{w(n+1)}_{\text{new information}} \tag{4.1.8}$$

Careful inspection of (4.1.8) indicates that if the system generating $x(n)$ is minimum-phase, the sample $w(n+1)$ brings all the new information (*innovation*) to be carried by the sample $x(n+1)$. All other information can be predicted from the past samples $x(n), x(n-1), \dots$ of the signal (see Section 6.6). *We stress that this interpretation holds only if $H(z)$ is minimum-phase.*

The system $H(z)$ generates the signal $x(n)$ by introducing dependence in the white noise input $w(n)$ and is known as the *synthesis* or *coloring* filter. In contrast, the inverse system $H_I(z)$ can be used to recover the input $w(n)$ and is known as the *analysis* or *whitening* filter. In this sense the innovations sequence and the output process are completely equivalent. The synthesis and analysis filters are shown in Figure 4.2.

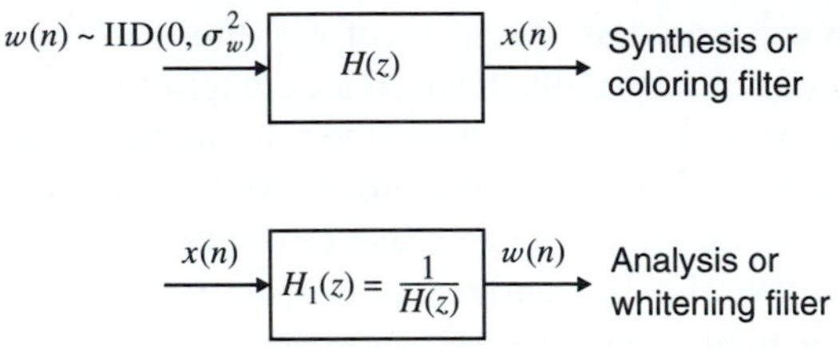

FIGURE 4.2
Synthesis and analysis filters used in innovations representation.

Spectral factorization

Most random processes with a continuous PSD $R_x(e^{j\omega})$ can be generated by exciting a minimum-phase system $H_{\min}(z)$ with white noise. The PSD of the resulting process is given by

$$R_x(e^{j\omega}) = \sigma_w^2 \, |H_{\min}(e^{j\omega})|^2 \tag{4.1.9}$$

The process of obtaining $H_{\min}(z)$ from $R_x(e^{j\omega})$ or $r_x(l)$ is known as *spectral factorization*.

If the PSD $R_x(e^{j\omega})$ satisfies the Paley-Wiener condition

$$\int_{-\pi}^{\pi} |\ln R_x(e^{j\omega})| d\omega < \infty \tag{4.1.10}$$

then the process $x(n)$ is called *regular* and its complex PSD can be factored as follows (see Section 2.4.4)

$$R_x(z) = \sigma_w^2 H_{\min}(z) H_{\min}^* \left(\frac{1}{z^*} \right) \tag{4.1.11}$$

where

$$\sigma_w^2 = \exp \left\{ \frac{1}{2\pi} \int_{-\pi}^{\pi} \ln[R_x(e^{j\omega})] \, d\omega \right\} \tag{4.1.12}$$

is the variance of the white noise input and can be interpreted as the *geometric mean* of $R_x(e^{j\omega})$. Consider the inverse Fourier transform of $\ln R_x(e^{j\omega})$:

$$c(k) \triangleq \frac{1}{2\pi} \int_{-\pi}^{\pi} \ln[R_x(e^{j\omega})] \, e^{jk\omega} \, d\omega \tag{4.1.13}$$

which is a sequence known as the *cepstrum* of $r_x(l)$. Note that $c(0) = \sigma_w^2$. Thus in the cepstral domain, the multiplicative factors $H_{\min}(z)$ and $H_{\min}^*(1/z^*)$ are now *additively separable* due to the natural logarithm of $R_x(e^{j\omega})$. Define

$$c_+(k) \triangleq \frac{c(0)}{2} + c(k)u(k-1) \tag{4.1.14}$$

and

$$c_-(k) \triangleq \frac{c(0)}{2} + c(k)u(-k-1) \tag{4.1.15}$$

as the positive- and negative-axis projections of $c(k)$, respectively, with $c(0)$ distributed equally between them. Then we obtain

$$h_{\min}(n) = \mathcal{F}^{-1}\{\exp \mathcal{F}[c_+(k)]\} \tag{4.1.16}$$

as the impulse response of the minimum-phase system $H_{\min}(z)$. Similarly,

$$h_{\max}(n) = \mathcal{F}^{-1}\{\exp \mathcal{F}[c_-(k)]\} \tag{4.1.17}$$

is the corresponding maximum-phase system. This completes the spectral factorization procedure for an arbitrary PSD $R_x(e^{j\omega})$, which, in general, is a complicated task. However, it is straightforward if $R_x(z)$ is a rational function, as we discussed in Section 2.4.2.

Spectral flatness measure

The spectral flatness measure (SFM) of a zero-mean process with PSD $R_x(e^{j\omega})$ is defined by (Makhoul 1975)

$$\mathrm{SFM}_x \triangleq \frac{\exp\left\{\dfrac{1}{2\pi}\displaystyle\int_{-\pi}^{\pi} \ln[R_x(e^{j\omega})]\,d\omega\right\}}{\dfrac{1}{2\pi}\displaystyle\int_{-\pi}^{\pi} R_x(e^{j\omega})\,d\omega} = \frac{\sigma_w^2}{\sigma_x^2} \tag{4.1.18}$$

where the second equality follows from (4.1.12). It describes the shape (or more appropriately, flatness) of the PSD by a single number. If $x(n)$ is a white noise process, then $R_x(e^{j\omega}) = \sigma_x^2$ and $\mathrm{SFM}_x = 1$. More specifically, we can show that

$$0 \le \mathrm{SFM}_x \le 1 \tag{4.1.19}$$

Observe that the numerator of (4.1.18) is the geometric mean while the denominator is the arithmetic mean of a real-valued, nonnegative continuous waveform $R_x(e^{j\omega})$. Since $x(n)$ is a regular process satisfying (4.1.10), these means are always positive. Furthermore, their ratio, by definition, is never greater than unity and is equal to unity if the waveform is constant. This, then, proves (4.1.19). A detailed proof is given in Jayant and Noll (1984).

When $x(n)$ is obtained by filtering the zero-mean white noise process $w(n)$ through the filter $H(z)$, then the coloring of $R_x(e^{j\omega})$ is due to $H(z)$. In this case, $R_x(e^{j\omega}) = \sigma_w^2|H(e^{j\omega})|^2$ from (4.1.9), and we obtain

$$\mathrm{SFM}_x = \frac{\sigma_w^2}{\sigma_x^2} = \frac{\sigma_w^2}{\dfrac{1}{2\pi}\displaystyle\int_{-\pi}^{\pi}\sigma_w^2|H(e^{j\omega})|^2\,d\omega} = \frac{1}{\dfrac{1}{2\pi}\displaystyle\int_{-\pi}^{\pi}|H(e^{j\omega})|^2\,d\omega} \tag{4.1.20}$$

Thus SFM_x is the inverse of the filter power (or *power transfer factor*) if $h(0)$ is normalized to unity.

4.1.2 Parametric Pole-Zero Signal Models

Parametric models describe a system with a finite number of parameters. The major subject of this chapter is the treatment of parametric models that have rational system functions. To this end, consider a system described by the following linear constant-coefficient difference equation

$$x(n) + \sum_{k=1}^{P} a_k\, x(n-k) = \sum_{k=0}^{Q} d_k\, w(n-k) \tag{4.1.21}$$

where $x(n)$ and $w(n)$ are the input and output signals, respectively. Taking the z-transform of both sides, we find that the system function is

$$H(z) = \frac{X(z)}{W(z)} = \frac{\displaystyle\sum_{k=0}^{Q} d_k z^{-k}}{1 + \displaystyle\sum_{k=1}^{P} a_k z^{-k}} \triangleq \frac{D(z)}{A(z)} \tag{4.1.22}$$

We can express $H(z)$ in terms of the poles and zeros of the system as follows:

$$H(z) = d_0 \frac{\displaystyle\prod_{k=1}^{Q}(1 - z_k z^{-1})}{\displaystyle\prod_{k=1}^{P}(1 - p_k z^{-1})} \tag{4.1.23}$$

The system has Q zeros $\{z_k\}$ and P poles $\{p_k\}$ (zeros and poles at $z = 0$ are not considered here). The term d_0 is the system gain. For the rest of the book, we assume that the polynomials $D(z)$ and $A(z)$ do not have any common roots, that is, common poles and zeros have already been canceled.

Types of pole-zero models

There are three cases of interest:

- For $P > 0$ and $Q > 0$, we have a *pole-zero* model, denoted by PZ(P, Q). If the model is assumed to be causal, its output is given by

$$x(n) = -\sum_{k=1}^{P} a_k x(n - k) + \sum_{k=0}^{Q} d_k w(n - k) \tag{4.1.24}$$

- For $P = 0$, we have an *all-zero* model, denoted by AZ(Q). The input-output difference equation is

$$x(n) = \sum_{k=0}^{Q} d_k w(n - k) \tag{4.1.25}$$

- For $Q = 0$, we have an *all-pole* model, denoted by AP(P). The input-output difference equation is

$$x(n) = -\sum_{k=1}^{P} a_k x(n - k) + d_0 w(n) \tag{4.1.26}$$

If we excite a parametric model with white noise, we obtain a signal whose second-order moments are determined by the parameters of the model. Indeed, from Section 3.4.2, we recall that if $w(n) \sim \text{IID}\{0, \sigma_w^2\}$ with finite variance,[†] then

$$r_x(l) = \sigma_w^2 r_h(l) = \sigma_w^2 h(l) * h^*(-l) \tag{4.1.27}$$

$$R_x(z) = \sigma_w^2 R_h(z) = \sigma_w^2 H(z) H^*\left(\frac{1}{z^*}\right) \tag{4.1.28}$$

$$R_x(e^{j\omega}) = \sigma_w^2 R_h(e^{j\omega}) = \sigma_w^2 |H(e^{j\omega})|^2 \tag{4.1.29}$$

Such signal models are of great practical interest and have special names in the statistical literature:

- The AZ(Q) is known as the *moving-average* model, denoted by MA(Q).
- The AP(P) is known as the *autoregressive* model, denoted by AR(P).
- The PZ(P, Q) is known as the *autoregressive moving-average* model, denoted by ARMA (P, Q).

We specify a parametric signal model by normalizing $d_0 = 1$ and setting the variance of the input to σ_w^2. The defining set of model parameters is given by $\{a_1, a_2, \ldots, a_P, d_1, \ldots, d_Q, \sigma_w^2\}$ (see Figure 4.3). An alternative is to set $\sigma_w^2 = 1$ and leave d_0 arbitrary. We stress that these models assume the resulting processes are stationary, which is ensured if the corresponding systems are BIBO stable.

[†]The case of infinite variance is discussed in Chapter 12.

FIGURE 4.3

155

SECTION 4.1
Introduction

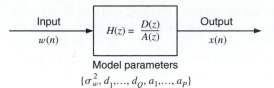

Block diagram representation of a parametric, rational signal model.

Short-memory behavior

To find the memory behavior of pole-zero models, we investigate the nature of their impulse response. To this end, we recall that for $Q \geq P$, (4.1.23) can be expanded as

$$H(z) = \sum_{j=0}^{Q-P} B_j z^{-j} + \sum_{k=1}^{P} \frac{A_k}{1 - p_k z^{-1}} \tag{4.1.30}$$

where for simplicity we assume that the model has P distinct poles. The first term in (4.1.30) disappears if $P > Q$. The coefficients B_j can be obtained by long division:

$$A_k = (1 - p_k z^{-1})H(z)|_{z=p_k} \tag{4.1.31}$$

If the model is causal, taking the inverse z-transform results in an impulse response that is a linear combination of impulses, real exponentials, and damped sinusoids (produced by the combination of complex exponentials)

$$h(n) = \sum_{j=0}^{Q-P} B_j \delta(n-j) + \sum_{k=1}^{P_1} A_k (p_k)^n u(n) + \sum_{i=1}^{P_2} C_i r_i^n \cos(\omega_i n + \phi_i) u(n) \tag{4.1.32}$$

where $p_i = r_i e^{\pm j \omega_i}$ and $P = P_1 + 2P_2$. Recall that $u(n)$ and $\delta(n)$ are the unit step and unit impulse functions, respectively. We note that the memory of any all-pole model decays exponentially with time and that the rate of decay is controlled by the pole closest to the unit circle. The contribution of multiple poles at the same location is treated in Problem 4.1.

Careful inspection of (4.1.32) leads to the following conclusions:

1. For AZ(Q) models, the impulse response has finite duration and, therefore, can have any shape.
2. The impulse response of causal AP(P) and PZ(P, Q) models with single poles consists of a linear combination of damped real exponentials (produced by the real poles) and exponentially damped sinusoids (produced by complex conjugate poles). The rate of decay decreases as the poles move closer to the unit circle and is determined by the pole closest to the unit circle.
3. The model is stable if and only if $h(n)$ is absolutely summable, which, due to (4.1.32), is equivalent to $|p_k| < 1$ for all k. In other words, a causal pole-zero model is BIBO stable if and only if all the poles are inside the unit circle.[†]

We conclude that causal, stable PZ(P, Q) models with $P > 0$ have an exponentially fading memory because their impulse response decays exponentially with time. Therefore, the autocorrelation $r_h(l) = h(l) * h^*(-l)$ also decays exponentially (see Example 4.2.2), and pole-zero models have short memory according to the definition given in Section 3.4.3.

Generation of random signals with rational power spectra

Sample realizations of random sequences with rational power spectra can be easily generated by using the difference equation (4.1.24) and a random number generator. In most applications, we use a Gaussian excitation because the generated sequence will also be Gaussian. For non-Gaussian inputs, it is difficult to predict the type of distribution of the output signal. If, on one hand, we specify the frequency response of the model,

---

[†]Poles on the unit circle are discussed in Section 4.5.

the coefficients of the difference equation can be obtained by using a digital filter design package. If, on the other hand, the power spectrum or the autocorrelation is given, the coefficients of the model are determined via spectral factorization. If we wish to avoid the transient effects that make some of the initial output samples nonstationary, we should consider the response of the model only after the initial transients have died out.

4.1.3 Mixed Processes and Wold Decomposition

An arbitrary stationary random process can be constructed to possess a continuous PSD $R_x(e^{j\omega})$ and a discrete power spectrum $R_x(k)$. Such processes are called *mixed* processes because the continuous PSD is due to regular processes while the discrete spectrum is due to harmonic (or almost periodic) processes. A further interpretation of mixed processes is that the first part is an *unpredictable* process while the second part is a *predictable* process (in the sense that past samples can be used to exactly determine future samples). This interpretation is due to the Wold decomposition theorem.

> **THEOREM 4.1 (WOLD DECOMPOSITION).** A general stationary random process can be written as a sum
>
> $$x(n) = x_r(n) + x_p(n) \tag{4.1.33}$$
>
> where $x_r(n)$ is a regular process possessing a continuous spectrum and $x_p(n)$ is a predictable process possessing a discrete spectrum. Furthermore, $x_r(n)$ is orthogonal to $x_p(n)$; that is,
>
> $$E\{x_r(n_1)x_p^*(n_2)\} = 0 \quad \text{for all } n_1, n_2 \tag{4.1.34}$$

The proof of this theorem is very involved, but a good approach to it is given in Therrien (1992). Using (4.1.34), the correlation sequence of $x(n)$ in (4.1.33) is given by

$$r_x(l) = r_{x_r}(l) + r_{x_p}(l)$$

from which we obtain the continuous and discrete spectra. As discussed above, the regular process has an innovations representation $w(n)$ that is uncorrelated but *not* independent. For example, $w(n)$ can be the output of an all-pass filter driven by an IID sequence. To determine if this is the case, we need to use higher-order moments (see Section 12.1).

4.2 ALL-POLE MODELS

We start our discussion of linear signal models with all-pole models because they are the easiest to analyze and the most often used in practical applications. We assume an all-pole model of the form

$$H(z) = \frac{d_0}{A(z)} = \frac{d_0}{1 + \sum_{k=1}^{P} a_k z^{-k}} = \frac{d_0}{\prod_{k=1}^{P}(1 - p_k z^{-1})} \tag{4.2.1}$$

where d_0 is the system gain and P is the order of the model. The all-pole model can be implemented using either a direct or a lattice structure. The conversion between the two sets of parameters can be done by using the step-up and step-down recursions described in Section 2.5.

4.2.1 Model Properties

In this section, we derive analytic expressions for various properties of the all-pole model, namely, the impulse response, the autocorrelation, and the spectrum. We determine the system-related properties $r_h(l)$ and $R_h(e^{j\omega})$ because the results can be readily applied to obtain the signal model properties for inputs with both continuous and discrete spectra.

The impulse response $h(n)$ can be specified by first rewriting (4.2.1) as

$$H(z) + \sum_{k=1}^{P} a_k H(z) z^{-k} = d_0$$

and then taking the inverse z-transform to obtain

$$h(n) + \sum_{k=1}^{P} a_k h(n-k) = d_0 \delta(n) \tag{4.2.2}$$

If the system is causal, then

$$h(n) = -\sum_{k=1}^{P} a_k h(n-k) + d_0 \delta(n) \tag{4.2.3}$$

If $H(z)$ has all its poles inside the unit circle, then $h(n)$ is a causal, stable sequence and the system is minimum-phase. From (4.2.3) we have

$$h(0) = d_0 \tag{4.2.4}$$

$$h(n) = -\sum_{k=1}^{P} a_k h(n-k) \qquad n > 0 \tag{4.2.5}$$

and owing to causality we have

$$h(n) = 0 \qquad n < 0 \tag{4.2.6}$$

Thus, except for the value at $n = 0$, $h(n)$ can be obtained recursively as a linearly weighted summation of its previous values $h(n-1), \ldots, h(n-P)$. One can say that $h(n)$ can be *predicted* (with zero error for $n \neq 0$) from the past P values. Thus, the coefficients $\{a_k\}$ are often referred to as *predictor coefficients*. Note that there is a close relationship between all-pole models and linear prediction that will be discussed in Section 4.2.2.

From (4.2.4) and (4.2.5), we can also write the inverse relation

$$a_n = -\frac{h(n)}{h(0)} - \sum_{k=1}^{n-1} a_k \frac{h(n-k)}{h(0)} \qquad n > 0 \tag{4.2.7}$$

with $a_0 = 1$. From (4.2.7) and (4.2.4), we conclude that if we are given the first $P + 1$ values of the impulse response $h(n)$, $0 \leq n \leq P$, then the parameters of the all-pole filter are completely specified.

Finally, we note that a causal $H(z)$ can be written as a one-sided, infinite polynomial $H(z) = \sum_{n=0}^{\infty} h(n)z^{-n}$. This representation of $H(z)$ implies that any finite-order, all-pole model can be represented equivalently by an infinite number of zeros. In general, a single pole can be represented by an infinite number of zeros, and conversely a single zero can be represented by an infinite number of poles. If the poles are inside the unit circle, so are the corresponding zeros, and vice versa.

EXAMPLE 4.2.1. A single pole at $z = a$ can be represented by

$$H(z) = \frac{1}{1 - az^{-1}} = \sum_{n=0}^{\infty} a^n z^{-n} \qquad |a| < 1 \tag{4.2.8}$$

The question is, where are the infinite number of zeros located? To find the answer, let us consider the finite polynomial

$$H_N(z) = \sum_{n=0}^{N} a^n z^{-n} \tag{4.2.9}$$

where we have truncated $H(z)$ at $n = N$. Thus $H_N(z)$ is a geometric series that can be written in closed form as

$$H_N(z) = \frac{1 - a^{N+1}z^{-(N+1)}}{1 - az^{-1}} \tag{4.2.10}$$

And $H_N(z)$ has a single pole at $z = a$ and $N + 1$ zeros at

$$z_i = ae^{j2\pi i/(N+1)} \quad i = 0, 1, \ldots, N \tag{4.2.11}$$

The $N + 1$ zeros are equally distributed on the circle $|z| = a$ with one of the zeros (for $i = 0$) located at $z = a$. But the zero at $z = a$ cancels the pole at the same location. Therefore, $H_N(z)$ has the remaining N zeros:

$$z_i = ae^{j2\pi i(N+1)} \quad i = 1, 2, \ldots, N \tag{4.2.12}$$

The transfer function $H(z)$ of the single-pole model is obtained from $H_N(z)$ by letting N go to infinity. In the limit, $H_\infty(z)$ has an infinite number of zeros equally distributed on the circle $|z| = a$; the zeros are everywhere on that circle except at the point $z = a$. Similarly, the denominator from (4.2.8), a polynomial with a single zero at $z = a$, can be written as

$$A(z) = 1 - az^{-1} = \frac{1}{H(z)} = \frac{1}{1 + \displaystyle\sum_{n=0}^{\infty} a^n z^{-n}} \quad |a| < 1 \tag{4.2.13}$$

that is, a single zero can also be represented by an infinite number of poles. In this case, the poles are equally distributed on a circle that passes through the location of the zero; the poles are everywhere on the circle except at the actual location of the zero.

Autocorrelation. The impulse response $h(n)$ of an all-pole model has infinite duration so that its autocorrelation involves an infinite summation, which is not practical to write in closed form except for low-order models. However, the autocorrelation function obeys a recursive relation that relates the autocorrelation values to the model parameters. Multiplying (4.2.2) by $h^*(n - l)$ and summing over all n, we have

$$\sum_{n=-\infty}^{\infty} \sum_{k=0}^{P} a_k h(n - k)h^*(n - l) = d_0 \sum_{n=-\infty}^{\infty} h^*(n - l)\delta(n) \tag{4.2.14}$$

where $a_0 = 1$. Interchanging the order of summations in the left-hand side, we obtain

$$\sum_{k=0}^{P} a_k r_h(l - k) = d_0 h^*(-l) \quad -\infty < l < \infty \tag{4.2.15}$$

where $r_h(l)$ is the autocorrelation of $h(n)$. Equation (4.2.15) is true for all l, but because $h(l) = 0$ for $l < 0$, $h(-l) = 0$ for $l > 0$, and we have

$$\sum_{k=0}^{P} a_k r_h(l - k) = 0 \quad l > 0 \tag{4.2.16}$$

From (4.2.4) and (4.2.15), we also have for $l = 0$,

$$\sum_{k=0}^{P} a_k r_h(k) = |d_0|^2 \tag{4.2.17}$$

where we used the fact that $r_h^*(-l) = r_h(l)$. Equation (4.2.16) can be rewritten as

$$r_h(l) = -\sum_{k=1}^{P} a_k r_h(l - k) \quad l > 0 \tag{4.2.18}$$

which is a recursive relation for $r_h(l)$ in terms of past values of the autocorrelation and $\{a_k\}$. Relation (4.2.18) for $r_h(l)$ is similar to relation (4.2.5) for $h(n)$, but with one important difference: (4.2.5) for $h(n)$ is true for all $n \neq 0$ while (4.2.18) for $r_h(l)$ is true only if $l > 0$; for $l < 0$, $r_h(l)$ obeys (4.2.15).

If we define the *normalized* autocorrelation coefficients as

$$\rho_h(l) = \frac{r_h(l)}{r_h(0)} \qquad (4.2.19)$$

then we can divide (4.2.17) by $r_h(0)$ and deduce the following relation for $r_h(0)$

$$r_h(0) = \frac{|d_0|^2}{1 + \displaystyle\sum_{k=1}^{P} a_k \rho_h(k)} \qquad (4.2.20)$$

which is the energy of the output of the all-pole filter when excited by a single impulse.

Autocorrelation in terms of poles. The complex spectrum of the AP(P) model is

$$R_h(z) = H(z)H\left(\frac{1}{z^*}\right) = |d_0|^2 \prod_{k=1}^{P} \frac{1}{(1 - p_k z^{-1})(1 - p_k z^*)} \qquad (4.2.21)$$

Therefore, the autocorrelation sequence can be expressed in terms of the poles by taking the inverse z-transform of $R_h(z)$, that is, $r_h(l) = \mathcal{Z}^{-1}\{R_h(z)\}$. The poles p_k of the minimum-phase model $H(z)$ contribute causal terms in the partial fraction expansion, whereas the poles $1/p_k$ of the nonminimum-phase model $H(1/z^*)$ contribute noncausal terms. This is best illustrated with the following example.

EXAMPLE 4.2.2. Consider the following minimum-phase AP(1) model

$$H(z) = \frac{1}{1 + az^{-1}} \qquad -1 < a < 1 \qquad (4.2.22)$$

Owing to causality, the ROC of $H(z)$ is $|z| > |a|$. The z-transform

$$H(z^{-1}) = \frac{1}{1 + az} \qquad -1 < a < 1 \qquad (4.2.23)$$

corresponds to the noncausal sequence $h(-n) = (-a)^{-n}u(-n)$, and its ROC is $|z| < 1/|a|$. Hence,

$$R_h(z) = H(z)H(z^{-1}) = \frac{1}{(1 + az^{-1})(1 + az)} \qquad (4.2.24)$$

which corresponds to a two-sided sequence because its ROC, $|a| < |z| < 1/|a|$, is a ring in the z-plane. Using partial fraction expansion, we obtain

$$R_h(z) = \frac{-a}{1 - a^2} \frac{z^{-1}}{1 + az^{-1}} + \frac{1}{1 - a^2} \frac{1}{1 + az} \qquad (4.2.25)$$

The pole $p = -a$ corresponds to the causal sequence $[1/(1 - a^2)](-a)^l u(l - 1)$, and the pole $p = -1/a$ to the noncausal sequence $[1/(1 - a^2)](-a)^{-l}u(-l)$. Combining the two terms, we obtain

$$r_h(l) = \frac{1}{1 - a^2}(-a)^{|l|} \qquad -\infty < l < \infty \qquad (4.2.26)$$

or

$$\rho_h(l) = (-a)^{|l|} \qquad -\infty < l < \infty \qquad (4.2.27)$$

Note that complex conjugate poles will contribute two-sided damped sinusoidal terms like the ones described in Section 4.1.2 for the AP(2) model.

Impulse train excitations. The response of an AP(P) model to a periodic impulse train with period L is periodic with the same period and is given by

$$\tilde{h}(n) + \sum_{k=1}^{P} a_k \tilde{h}(n-k) = d_0 \sum_{m=-\infty}^{\infty} \delta(n+Lm)$$

$$= \begin{cases} d_0 & n+Lm=0 \\ 0 & n+Lm \neq 0 \end{cases} \tag{4.2.28}$$

which shows that the prediction error is zero for samples inside the period and d_0 at the beginning of each period. If we multiply both sides of (4.2.28) by $\tilde{h}(n-l)$ and sum over a period $0 \leq n \leq L-1$, we obtain

$$\tilde{r}_h(l) + \sum_{k=1}^{P} a_k \tilde{r}_h(l-k) = \frac{d_0}{L} \tilde{h}^*(-l) \qquad \text{all } l \tag{4.2.29}$$

where $\tilde{r}_h(l)$ is the periodic autocorrelation of $\tilde{h}(n)$. Since, in contrast to $h(n)$ in (4.2.15), $\tilde{h}(n)$ is not necessarily zero for $n < 0$, the periodic autocorrelation $\tilde{r}_h(l)$ will not in general obey the linear prediction equation anywhere. Similar results can be obtained for harmonic process excitations.

Model parameters in terms of autocorrelation. Equations (4.2.15) for $l = 0, 1, \ldots, P$ comprise $P+1$ equations that relate the $P+1$ parameters of $H(z)$, namely, d_0 and $\{a_k, \ 1 \leq k \leq P\}$, to the first $P+1$ autocorrelation coefficients $r_h(0), \ r_h(1), \ldots, r_h(P)$. These $P+1$ equations can be written in matrix form as

$$\begin{bmatrix} r_h(0) & r_h(1) & \cdots & r_h(P) \\ r_h^*(1) & r_h(0) & \cdots & r_h(P-1) \\ \vdots & \vdots & \ddots & \vdots \\ r_h^*(P) & r_h^*(P-1) & \cdots & r_h(0) \end{bmatrix} \begin{bmatrix} 1 \\ a_1 \\ \vdots \\ a_P \end{bmatrix} = \begin{bmatrix} |d_0|^2 \\ 0 \\ \vdots \\ 0 \end{bmatrix} \tag{4.2.30}$$

If we are given the first $P+1$ autocorrelations, (4.2.30) comprises a system of $P+1$ linear equations, with a Hermitian Toeplitz matrix that can be solved for d_0 and $\{a_k\}$.

Because of the special structure in (4.2.30), the model parameters are found from the autocorrelations by using the last set of P equations in (4.2.30), followed by the computation of d_0 from the first equation, which is the same as (4.2.17). From (4.2.30), we can write in matrix notation

$$\mathbf{R}_h \mathbf{a} = -\mathbf{r}_h \tag{4.2.31}$$

where $\mathbf{R}_h$ is the autocorrelation matrix, $\mathbf{a}$ is the vector of the model parameters, and $\mathbf{r}_h$ is the vector of autocorrelations. Since $r_x(l) = \sigma_w^2 r_h(l)$, we can also express the model parameters in terms of the autocorrelation $r_x(l)$ of the output process $x(n)$ as follows:

$$\mathbf{R}_x \mathbf{a} = -\mathbf{r}_x \tag{4.2.32}$$

These equations are known as the *Yule-Walker equations* in the statistics literature. In the sequel, we drop the subscript from the autocorrelation sequence or matrix whenever the analysis holds for both the impulse response and the model output.

Because of the Toeplitz structure and the nature of the right-hand side, the linear systems (4.2.31) and (4.2.32) can be solved recursively by using the algorithm of Levinson-Durbin (see Section 7.4). After $\mathbf{a}$ is solved for, the system gain d_0 can be computed from (4.2.17).

Therefore, given $r(0), r(1), \ldots, r(P)$, we can completely specify the parameters of the all-pole model by solving a set of linear equations. Below, we will see that the converse is also true: Given the model parameters, we can find the first $P+1$ autocorrelations by solving a set of linear equations. *This elegant solution of the spectral factorization problem is unique to all-pole models.* In the case in which the model contains zeros ($Q \neq 0$), the spectral factorization problem requires the solution of a nonlinear system of equations.

Autocorrelation in terms of model parameters. If we normalize the autocorrelations in (4.2.31) by dividing throughout by $r(0)$, we obtain the following system of equations

$$\mathbf{Pa} = -\boldsymbol{\rho} \tag{4.2.33}$$

where $\mathbf{P}$ is the normalized autocorrelation matrix and

$$\boldsymbol{\rho} = [\rho(1) \ \rho(2) \ \cdots \ \rho(P)]^H \tag{4.2.34}$$

is the vector of normalized autocorrelations. This set of P equations relates the P model coefficients with the first P (normalized) autocorrelation values. If the poles of the all-pole filter are strictly inside the unit circle, the mapping between the P-dimensional vectors $\mathbf{a}$ and $\boldsymbol{\rho}$ is *unique*. If, in fact, we are given the vector $\mathbf{a}$, then the normalized autocorrelation vector $\boldsymbol{\rho}$ can be computed from $\mathbf{a}$ by using the set of equations that can be deduced from (4.2.33)

$$\mathbf{A}\boldsymbol{\rho} = -\mathbf{a} \tag{4.2.35}$$

where $\langle\mathbf{A}\rangle_{ij} = a_{i-j} + a_{i+j}$, assuming $a_m = 0$ for $m < 0$ and $m > P$ (see Problem 4.6).

Given the set of coefficients in $\mathbf{a}$, $\boldsymbol{\rho}$ can be obtained by solving (4.2.35). We will see that, under the assumption of a stable $H(z)$, a solution always exists. Furthermore, there exists a simple, recursive solution that is efficient (see Section 7.5). If, in addition to $\mathbf{a}$, we are given d_0, we can evaluate $r(0)$ with (4.2.20) from $\boldsymbol{\rho}$ computed by (4.2.35). Autocorrelation values $r(l)$ for lags $l > P$ are found by using the recursion in (4.2.18) with $r(0), r(1), \ldots, r(P)$.

EXAMPLE 4.2.3. For the AP(3) model with real coefficients we have

$$\begin{bmatrix} r(0) & r(1) & r(2) \\ r(1) & r(0) & r(1) \\ r(2) & r(1) & r(0) \end{bmatrix} \begin{bmatrix} a_1 \\ a_2 \\ a_3 \end{bmatrix} = - \begin{bmatrix} r(1) \\ r(2) \\ r(3) \end{bmatrix} \tag{4.2.36}$$

$$d_0^2 = r(0) + a_1 r(1) + a_2 r(2) + a_3 r(3) \tag{4.2.37}$$

Therefore, given $r(0), r(1), r(2), r(3)$, we can find the parameters of the all-pole model by solving (4.2.36) and then substituting into (4.2.37).

Suppose now that instead we are given the model parameters d_0, a_1, a_2, a_3. If we divide both sides of (4.2.36) by $r(0)$ and solve for the normalized autocorrelations $\rho(1)$, $\rho(2)$, and $\rho(3)$, we obtain

$$\begin{bmatrix} 1+a_2 & a_3 & 0 \\ a_1+a_3 & 1 & 0 \\ a_2 & a_1 & 1 \end{bmatrix} \begin{bmatrix} \rho(1) \\ \rho(2) \\ \rho(3) \end{bmatrix} = - \begin{bmatrix} a_1 \\ a_2 \\ a_3 \end{bmatrix} \tag{4.2.38}$$

The value of $r(0)$ is obtained from

$$r(0) = \frac{d_0^2}{1 + a_1\rho(1) + a_2\rho(2) + a_3\rho(3)} \tag{4.2.39}$$

If $r(0) = 2$, $r(1) = 1.6$, $r(2) = 1.2$, and $r(3) = 1$, the Toeplitz matrix in (4.2.36) is positive definite because it has positive eigenvalues. Solving the linear system gives $a_1 = -0.9063$, $a_2 = 0.2500$, and $a_3 = -0.1563$. Substituting these values in (4.2.37), we obtain $d_0 = 0.8329$. Using the last two relations, we can recover the autocorrelation from the model parameters.

Correlation matching. All-pole models have the unique distinction that the model parameters are completely specified by the first $P + 1$ autocorrelation coefficients via a set of linear equations. We can write

$$\begin{bmatrix} d_0 \\ \mathbf{a} \end{bmatrix} \leftrightarrow \begin{bmatrix} r(0) \\ \boldsymbol{\rho} \end{bmatrix} \tag{4.2.40}$$

that is, the mapping of the model parameters $\{d_0, a_1, a_2, \ldots, a_P\}$ to the autocorrelation coefficients specified by the vector $\{r(0), \rho(1), \ldots, \rho(P)\}$ is reversible and unique. This statement implies that given any set of autocorrelation values $r(0), r(1), \ldots, r(P)$, we can always find an all-pole model whose first $P + 1$ autocorrelation coefficients are equal to the given autocorrelations. This *correlation matching* of all-pole models is quite remarkable. This property is not shared by all-zero models and is true for pole-zero models only under certain conditions, as we will see in Section 4.4.

Spectrum. The z-transform of the autocorrelation $r(l)$ of $H(z)$ is given by

$$R(z) = H(z)H\left(\frac{1}{z^*}\right) \tag{4.2.41}$$

The spectrum is then equal to

$$R(e^{j\omega}) = |H(e^{j\omega})|^2 = \frac{|d_0|^2}{|A(e^{j\omega})|^2} \tag{4.2.42}$$

The right-hand side of (4.2.42) suggests a method for computing the spectrum: First compute $A(e^{j\omega})$ by taking the Fourier transform of the sequence $\{1, a_1, \ldots, a_P\}$, then take the squared of the magnitude and divide $|d_0|^2$ by the result. The fast Fourier transform (FFT) can be used to this end by appending the sequence $\{1, a_1, \ldots, a_P\}$ with as many zeros as needed to compute the desired number of frequency points.

Partial autocorrelation and lattice structures. We have seen that an AP(P) model is completely described by the first $P + 1$ values of its autocorrelation. However, we cannot determine the order of the model by using the autocorrelation sequence because it has infinite duration. Suppose that we start fitting models of increasing order m, using the autocorrelation sequence of an AP(P) model and the Yule-Walker equations

$$\begin{bmatrix} 1 & \rho(1) & \cdots & \rho(m-1) \\ \rho^*(1) & 1 & \cdots & \vdots \\ \vdots & \vdots & \ddots & \rho(1) \\ \rho^*(m-1) & \cdots & \rho^*(1) & 1 \end{bmatrix} \begin{bmatrix} a_1^{(m)} \\ a_2^{(m)} \\ \vdots \\ a_m^{(m)} \end{bmatrix} = - \begin{bmatrix} \rho^*(1) \\ \rho^*(2) \\ \vdots \\ \rho^*(m) \end{bmatrix} \tag{4.2.43}$$

Since $a_m^{(m)} = 0$ for $m > P$, we can use the sequence $a_m^{(m)}$, $m = 1, 2, \ldots$, which is known as the *partial autocorrelation sequence (PACS)*, to determine the order of the all-pole model. Recall from Section 2.5 that

$$a_m^{(m)} = k_m \tag{4.2.44}$$

that is, the PACS is identical to the lattice parameters. A statistical definition and interpretation of the PACS are also given in Section 7.2. The PACS can be defined for any valid (i.e., positive definite) autocorrelation sequence and can be efficiently computed by using the algorithms of Levinson-Durbin and Schur (see Chapter 7).

Furthermore, it has been shown (Burg 1975) that

$$r(0) \prod_{m=1}^{P} \frac{1 - |k_m|}{1 + |k_m|} \leq R(e^{j\omega}) \leq r(0) \prod_{m=1}^{P} \frac{1 + |k_m|}{1 - |k_m|} \tag{4.2.45}$$

which indicates that the spectral dynamic range increases if some lattice parameter moves close to 1 or equivalently some pole moves close to the unit circle.

Equivalent model representations. From the previous discussions (see also Chapter 7) we conclude that a minimum-phase AP(P) model can be uniquely described by any one of the following representations:

1. Direct structure: $\{d_0, a_1, a_2, \ldots, a_P\}$
2. Lattice structure: $\{d_0, k_1, k_2, \ldots, k_P\}$
3. Autocorrelation: $\{r(0), r(1), \ldots, r(P)\}$

where we assume, without loss of generality, that $d_0 > 0$. Note that the minimum-phase property requires that all poles be inside the unit circle or all $|k_m| < 1$ or that $\mathbf{R}_{P+1}$ be positive definite. The transformation from any of the above representations to any other can be done by using the algorithms developed in Section 7.5.

Minimum-phase conditions. As we will show in Section 7.5, if the Toeplitz matrix $\mathbf{R}_h$ (or equivalently $\mathbf{R}_x$) is positive definite, then $|k_m| < 1$ for all $m = 1, 2, \ldots, P$. Therefore, the AP(P) model obtained by solving the Yule-Walker equations is minimum-phase. Therefore, the Yule-Walker equations provide a simple and elegant solution to the spectral factorization problem for all-pole models.

> **EXAMPLE 4.2.4.** The poles of the model obtained in Example 4.2.3 are 0.8316, $0.0373+0.4319i$, and $0.0373 - 0.4319i$. We see that the poles are inside the unit circle and that the autocorrelation sequence is positive definite. If we set $r_h(2) = -1.2$, the autocorrelation becomes negative definite and the obtained model $\mathbf{a} = [1 \ -1.222 \ 1.1575]^T$, $d_0 = 2.2271$, is nonminimum-phase.

Pole locations. The poles of $H(z)$ are the zeros $\{p_k\}$ of the polynomial $A(z)$. If the coefficients of $A(z)$ are assumed to be real, the poles are either real or come in complex conjugate pairs. In order for $H(z)$ to be minimum-phase, all poles must be inside the unit circle, that is, $|p_k| < 1$. The model parameters a_k can be written as sums of products of the poles p_k. In particular, it is easy to see that

$$a_1 = -\sum_{k=1}^{P} p_k \tag{4.2.46}$$

$$a_P = \prod_{k=1}^{P} (-p_k) \tag{4.2.47}$$

Thus, the first coefficient a_1 is the negative of the sum of the poles, and the last coefficient a_P is the product of the negative of the individual poles. Since $|p_k| < 1$, we must have $|a_P| < 1$ for a minimum-phase polynomial for which $a_0 = 1$. However, note that the reverse is not necessarily true: $|a_P| < 1$ does not guarantee minimum phase. The roots p_k can be computed by using any number of standard root-finding routines.

4.2.2 All-Pole Modeling and Linear Prediction

Consider the AP(P) model

$$x(n) = -\sum_{k=1}^{P} a_k x(n-k) + w(n) \tag{4.2.48}$$

Now recall from Chapter 1 that the Mth-order linear predictor of $x(n)$ and the corresponding prediction error $e(n)$ are

$$\hat{x}(n) = -\sum_{k=1}^{M} a_k^0 x(n-k) \tag{4.2.49}$$

$$e(n) = x(n) - \hat{x}(n) = x(n) + \sum_{k=1}^{M} a_k^0 x(n-k) \tag{4.2.50}$$

or

$$x(n) = -\sum_{k=1}^{M} a_k^0 x(n-k) + e(n) \tag{4.2.51}$$

Notice that if the order of the linear predictor equals the order of the all-pole model ($M = P$) and if $a_k^0 = a_k$, then the prediction error is equal to the excitation of the all-pole model, that is, $e(n) = w(n)$. Since all-pole modeling and FIR linear prediction are closely related, many properties and algorithms developed for one of them can be applied to the other. Linear prediction is extensively studied in Chapters 6 and 7.

4.2.3 Autoregressive Models

Causal all-pole models excited by white noise play a major role in practical applications and are known as *autoregressive (AR) models*. An AR(P) model is defined by the difference equation

$$x(n) = -\sum_{k=1}^{P} a_k x(n-k) + w(n) \tag{4.2.52}$$

where $\{w(n)\} \sim \mathrm{WN}(0, \sigma_w^2)$. An AR($P$) model is valid only if the corresponding AP(P) system is stable. In this case, the output $x(n)$ is a stationary sequence with a mean value of zero. Postmultiplying (4.2.52) by $x^*(n-l)$ and taking the expectation, we obtain the following recursive relation for the autocorrelation:

$$r_x(l) = -\sum_{k=1}^{P} a_k r_x(l-k) + E\{w(n)x^*(n-l)\} \tag{4.2.53}$$

Similarly, using (4.1.1), we can show that $E\{w(n)x^*(n-l)\} = \sigma_w^2 h^*(-l)$. Thus, we have

$$r_x(l) = -\sum_{k=1}^{P} a_k r_x(l-k) + \sigma_w^2 h^*(-l) \qquad \text{for all } l \tag{4.2.54}$$

The variance of the output signal is

$$\sigma_x^2 = r_x(0) = -\sum_{k=1}^{P} a_k r_x(k) + \sigma_w^2$$

or

$$\sigma_x^2 = \frac{\sigma_w^2}{1 + \sum_{k=1}^{P} a_k \rho_x(k)} \tag{4.2.55}$$

If we substitute $l = 0, 1, \ldots, P$ in (4.2.55) and recall that $h(n) = 0$ for $n < 0$, we obtain the following set of Yule-Walker equations:

$$\begin{bmatrix} r_x(0) & r_x(1) & \cdots & r_x(P) \\ r_x^*(1) & r_x(0) & \cdots & r_x(P-1) \\ \vdots & \vdots & \ddots & \vdots \\ r_x^*(P) & r_x^*(P-1) & \cdots & r_x(0) \end{bmatrix} \begin{bmatrix} 1 \\ a_1 \\ \vdots \\ a_P \end{bmatrix} = \begin{bmatrix} \sigma_w^2 \\ 0 \\ \vdots \\ 0 \end{bmatrix} \tag{4.2.56}$$

Careful inspection of the above equations reveals their similarity to the corresponding relationships developed previously for the AP(P) model. This should be no surprise since the power spectrum of the white noise is flat. However, there is one important difference we should clarify: AP(P) models were specified with a gain d_0 and the parameters $\{a_1, a_2, \ldots, a_P\}$, but for AR($P$) models we set the gain $d_0 = 1$ and define the model by the variance of the white excitation σ_w^2 and the parameters $\{a_1, a_2, \ldots, a_P\}$. In other words, we incorporate the gain of the model into the power of the input signal. Thus, the power spectrum of the output is $R_x(e^{j\omega}) = \sigma_w^2 |H(e^{j\omega})|^2$. Similar arguments apply to all parametric models driven by white noise. We just rederived some of the relationships to clarify these issues and to provide additional insight into the subject.

4.2.4 Lower-Order Models

In this section, we derive the properties of lower-order all-pole models, namely, first- and second-order models, with real coefficients.

First-order all-pole model: AP(1)

An AP(1) model has a transfer function

$$H(z) = \frac{d_0}{1 + az^{-1}} \tag{4.2.57}$$

with a single pole at $z = -a$ on the real axis. It is clear that $H(z)$ is minimum-phase if

$$-1 < a < 1 \tag{4.2.58}$$

From (4.2.18) with $P = 1$ and $l = 1$, we have

$$a_1 = -\frac{r(1)}{r(0)} = -\rho(1) \tag{4.2.59}$$

Similarly, from (4.2.44) with $m = 1$,

$$a_1^{(1)} = a = -\rho(1) = k_1 \tag{4.2.60}$$

Since from (4.2.4), $h(0) = d_0$, and from (4.2.5) $h(n) = -a_1 h(n-1)$ for $n > 0$, the impulse response of a single-pole filter is given by

$$h(n) = d_0(-a)^n u(n) \tag{4.2.61}$$

The same result can, of course, be obtained by taking the inverse z-transform of $H(z)$.

The autocorrelation is found in a similar fashion. From (4.2.18) and by using the fact that the autocorrelation is an even function,

$$r(l) = r(0)(-a)^{|l|} \quad \text{for all } l \tag{4.2.62}$$

and from (4.2.20)

$$r(0) = \frac{d_0^2}{1 - a^2} = \frac{d_0^2}{1 - k_1^2} \tag{4.2.63}$$

Therefore, if the energy $r(0)$ in the impulse response is set to unity, then the gain must be set to

$$d_0 = \sqrt{1 - k_1^2} \quad r(0) = 1 \tag{4.2.64}$$

The z-transform of the autocorrelation is then

$$R(z) = \frac{d_0^2}{(1 + az^{-1})(1 + az)} = r(0) \sum_{l=-\infty}^{\infty} (-a)^{|l|} z^{-l} \tag{4.2.65}$$

and the spectrum is

$$R(e^{j\omega}) = |H(e^{j\omega})|^2 = \frac{d_0^2}{|1 + ae^{-j\omega}|^2} = \frac{d_0^2}{1 + 2a\cos\omega + a^2} \tag{4.2.66}$$

Figures 4.4 and 4.5 show a typical realization of the output, the impulse response, autocorrelation, and spectrum of two AP(1) models. The sample process realizations were obtained by driving the model with white Gaussian noise of zero mean and unit variance. When the positive pole ($p = -a = 0.8$) is close to the unit circle, successive samples

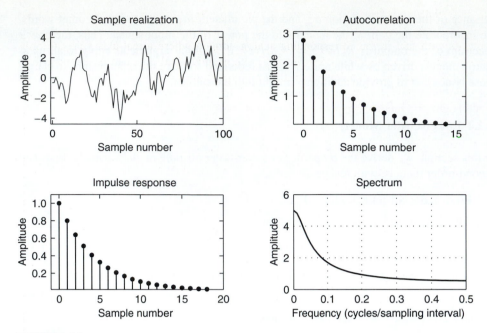

FIGURE 4.4
Sample realization of the output process, impulse response, autocorrelation, and spectrum of
an AP(1) model with $a = -0.8$.

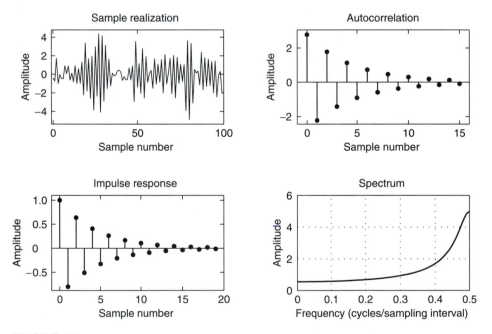

FIGURE 4.5
Sample realization of the output process, impulse response, autocorrelation, and spectrum of an
AP(1) model with $a = 0.8$.

of the output process are similar, as dictated by the slowly decaying autocorrelation and
the corresponding low-pass spectrum. In contrast, a negative pole close to the unit circle
results in a rapidly oscillating sequence. This is clearly reflected in the alternating sign of
the autocorrelation sequence and the associated high-pass spectrum.

Note that a positive real pole is a type of low-pass filter, while a negative real pole has the spectral characteristics of a high-pass filter. (This situation in the digital domain contrasts with that in the corresponding analog domain where a real-axis pole can only have low-pass characteristics.) The discrete-time negative real pole can be thought of as one-half of two conjugate poles at half the sampling frequency. Notice that both spectra are even and have zero slope at $\omega = 0$ and $\omega = \pi$. These propositions are true of the spectra of all parametric models (i.e., pole-zero models) with real coefficients (see Problem 4.13).

Consider now the real-valued AR(1) process $x(n)$ generated by

$$x(n) = -ax(n-1) + w(n) \tag{4.2.67}$$

where $\{w(n)\} \sim \mathrm{WN}(0, \sigma_w^2)$. Using the formula $R_x(z) = \sigma_w^2 H(z) H^*(1/z^*)$ and previous results, we can see that the autocorrelation and the PSD of $x(n)$ are given by

$$r_x(l) = \frac{\sigma_w^2}{1 - a^2}(-a)^{|l|}$$

and

$$R_x(e^{j\omega}) = \sigma_w^2 \frac{1 - a^2}{1 + a^2 + 2a \cos \omega}$$

respectively. Since $\sigma_x^2 = r_x(0) = \sigma_w^2/(1 - a^2)$, the SFM of $x(n)$ is [see (Section 4.1.18)]

$$\mathrm{SFM}_x = \frac{\sigma_w^2}{\sigma_x^2} = 1 - a^2 \tag{4.2.68}$$

Clearly, if $a = 0$, then from (4.2.67), $x(n)$ is a white noise process and from (4.2.68), $\mathrm{SFM}_x = 1$. If $a \to 1$, then $\mathrm{SFM}_x \to 0$; and in the limit when $a = 1$, the process becomes a random walk process, which is a nonstationary process with linearly increasing variance $E\{x^2(n)\} = n\sigma_w^2$. The correlation matrix is Toeplitz, and it is a rare exception in which eigenvalues and eigenvectors can be described by analytical expressions (Jayant and Noll 1984).

Second-order all-pole model: AP(2)

The system function of an AP(2) model is given by

$$H(z) = \frac{d_0}{1 + a_1 z^{-1} + a_2 z^{-2}} = \frac{d_0}{(1 - p_1 z^{-1})(1 - p_2 z^{-1})} \tag{4.2.69}$$

From (4.2.46) and (4.2.47), we have

$$a_1 = -(p_1 + p_2)$$
$$a_2 = p_1 p_2 \tag{4.2.70}$$

Recall that $H(z)$ is minimum-phase if the two poles p_1 and p_2 are inside the unit circle. Under these conditions, a_1 and a_2 lie in a triangular region defined by

$$-1 < a_2 < 1$$
$$a_2 - a_1 > -1 \tag{4.2.71}$$
$$a_2 + a_1 > -1$$

and shown in Figure 4.6. The first condition follows from (4.2.70) since $|p_1| < 1$ and $|p_2| < 1$. The last two conditions can be derived by assuming real roots and setting the larger root to less than 1 and the smaller root to greater than -1. By adding the last two conditions, we obtain the redundant condition $a_2 > -1$.

Complex roots occur in the region

$$\frac{a_1^2}{4} < a_2 \le 1 \qquad \text{complex poles} \tag{4.2.72}$$

with $a_2 = 1$ resulting in both roots being on the unit circle. Note that, in order to have complex poles, a_2 cannot be negative. If the complex poles are written in polar form

$$p_i = re^{\pm j\theta} \qquad 0 \le r \le 1 \tag{4.2.73}$$

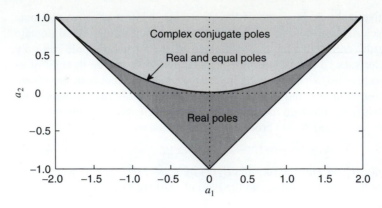

FIGURE 4.6
Minimum-phase region (triangle) for the AP(2) model in the (a_1, a_2)
parameter space.

then
$$a_1 = -2r \cos \theta \qquad a_2 = r^2 \qquad \text{(4.2.74)}$$

and
$$H(z) = \frac{d_0}{1 - (2r \cos \theta)z^{-1} + r^2 z^{-2}} \qquad \text{complex poles} \qquad \text{(4.2.75)}$$

Here, r is the radius (magnitude) of the poles, and θ is the angle or normalized frequency
of the poles.

Impulse response. The impulse response of an AP(2) model can be written in terms
of its two poles by evaluating the inverse z-transform of (4.2.69). The result is

$$h(n) = \frac{d_0}{p_1 - p_2} (p_1^{n+1} - p_2^{n+1})u(n) \qquad \text{(4.2.76)}$$

for $p_1 \neq p_2$. Otherwise, for $p_1 = p_2 = p$,

$$h(n) = d_0(n + 1)p^n u(n) \qquad \text{(4.2.77)}$$

In the special case of a complex conjugate pair of poles $p_1 = re^{j\theta}$ and $p_2 = re^{-j\theta}$,
Equation (4.2.76) reduces to

$$h(n) = d_0 \, r^n \frac{\sin[(n + 1)\theta]}{\sin \theta} u(n) \qquad \text{complex poles} \qquad \text{(4.2.78)}$$

Since $0 < r < 1$, $h(n)$ is a damped sinusoid of frequency θ.

Autocorrelation. The autocorrelation can also be written in terms of the two poles as

$$r(l) = \frac{d_0^2}{(p_1 - p_2)(1 - p_1 p_2)} \left(\frac{p_1^{l+1}}{1 - p_1^2} - \frac{p_2^{l+1}}{1 - p_2^2} \right) \qquad l \geq 0 \qquad \text{(4.2.79)}$$

from which we can deduce the energy

$$r(0) = \frac{d_0^2(1 + p_1 p_2)}{(1 - p_1 p_2)(1 - p_1^2)(1 - p_2^2)} \qquad \text{(4.2.80)}$$

For the special case of a complex conjugate pole pair, (4.2.79) can be rewritten as

$$r(l) = \frac{d_0^2 r^l \{\sin[(l + 1)\theta] - r^2 \sin[(l - 1)\theta]\}}{[(1 - r^2) \sin \theta](1 - 2r^2 \cos 2\theta + r^4)} \qquad l \geq 0 \qquad \text{(4.2.81)}$$

Then from (4.2.80) we can write an expression for the energy in terms of the polar coordinates of the complex conjugate pole pair

$$r(0) = \frac{d_0^2(1 + r^2)}{(1 - r^2)(1 - 2r^2 \cos 2\theta + r^4)} \qquad \text{(4.2.82)}$$

The normalized autocorrelation is given by

$$\rho(l) = \frac{r^l\{\sin[(l+1)\theta] - r^2 \sin[(l-1)\theta]\}}{(1+r^2)\,\sin\theta} \qquad l \geq 0 \tag{4.2.83}$$

which can be rewritten as

$$\rho(l) = \frac{1}{\cos\beta}\, r^l \cos(l\theta - \beta) \qquad l \geq 0 \tag{4.2.84}$$

where

$$\tan\beta = \frac{(1-r^2)\cos\theta}{(1+r^2)\sin\theta} \tag{4.2.85}$$

Therefore, $\rho(l)$ is a damped cosine wave with its maximum amplitude at the origin.

Spectrum. By setting the two poles equal to

$$p_1 = r_1 e^{j\theta_1} \qquad p_2 = r_2 e^{j\theta_2} \tag{4.2.86}$$

the spectrum of an AP(2) model can be written as

$$R(e^{j\omega}) = \frac{d_0^2}{[1 - 2r_1\cos(\omega - \theta_1) + r_1^2][1 - 2r_2\cos(\omega - \theta_2) + r_2^2]} \tag{4.2.87}$$

There are four cases of interest

Pole locations	Peak locations	Type of $R(e^{j\omega})$
$p_1 > 0,\ p_2 > 0$	$\omega = 0$	Low-pass
$p_1 < 0,\ p_2 < 0$	$\omega = \pi$	High-pass
$p_1 > 0,\ p_2 < 0$	$\omega = 0,\ \omega = \pi$	Stopband
$p_{1,2} = re^{\pm j\theta}$	$0 < \omega < \pi$	Bandpass

and they depend on the location of the poles on the complex plane.

We concentrate on the fourth case of complex conjugate poles, which is of greatest interest. The other three cases are explored in Problem 4.15. The spectrum is given by

$$R(e^{j\omega}) = \frac{d_0^2}{[1 - 2r\cos(\omega - \theta) + r^2][1 - 2r\cos(\omega + \theta) + r^2]} \tag{4.2.88}$$

The peak of this spectrum can be shown to be located at a frequency ω_c, given by

$$\cos\omega_c = \frac{1 + r^2}{2r}\cos\theta \tag{4.2.89}$$

Since $1 + r^2 > 2r$ for $r < 1$, and we have

$$\cos\omega_c > \cos\theta \tag{4.2.90}$$

the spectral peak is lower than the pole frequency for $0 < \theta < \pi/2$ and higher than the pole frequency for $\pi/2 < \theta < \pi$.

This behavior is illustrated in Figure 4.7 for an AP(2) model with $a_1 = -0.4944$, $a_2 = 0.64$, and $d_0 = 1$. The model has two complex conjugate poles with $r = 0.8$ and $\theta = \pm 2\pi/5$. The spectrum has a single peak and displays a passband type of behavior. The impulse response is a damped sine wave while the autocorrelation is a damped cosine. The typical realization of the output shows clearly a pseudoperiodic behavior that is explained by the shape of the autocorrelation and the spectrum of the model. We also notice that if the poles are complex conjugates, the autocorrelation has pseudoperiodic behavior.

Equivalent model descriptions. We now write explicit formulas for a_1 and a_2 in terms of the lattice parameters k_1 and k_2 and the autocorrelation coefficients. From the step-up

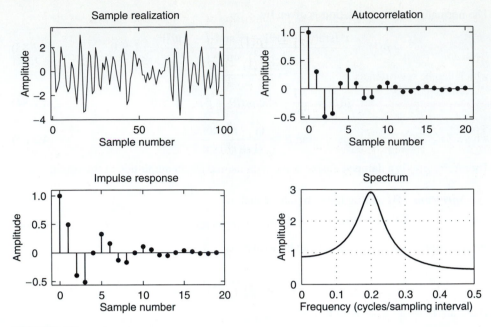

FIGURE 4.7
Sample realization of the output process, impulse response, autocorrelation, and spectrum of an
AP(2) model with complex conjugate poles.

and step-down recursions in Section 2.5, we have

$$a_1 = k_1(1 + k_2)$$
$$a_2 = k_2$$

(4.2.91)

and the inverse relations

$$k_1 = \frac{a_1}{1 + a_2}$$
$$k_2 = a_2$$

(4.2.92)

From the Yule-Walker equations (4.2.18), we can write the two equations

$$a_1 r(0) + a_2 r(1) = -r(1)$$
$$a_1 r(1) + a_2 r(0) = -r(2)$$

(4.2.93)

which can be solved for a_1 and a_2 in terms of $\rho(1)$ and $\rho(2)$

$$a_1 = -\rho(1)\frac{1 - \rho(2)}{1 - \rho^2(1)}$$

$$a_2 = \frac{\rho^2(1) - \rho(2)}{1 - \rho^2(1)}$$

(4.2.94)

or for $\rho(1)$ and $\rho(2)$ in terms of a_1 and a_2

$$\rho(1) = -\frac{a_1}{1 + a_2}$$

$$\rho(2) = -a_1\rho(1) - a_2 = \frac{a_1^2}{1 + a_2} - a_2$$

(4.2.95)

From the equations above, we can also write the relation and inverse relation between the coefficients k_1 and k_2 and the normalized autocorrelations $\rho(1)$ and $\rho(2)$ as

$$k_1 = -\rho(1)$$

$$k_2 = \frac{\rho^2(1) - \rho(2)}{1 - \rho^2(1)} \tag{4.2.96}$$

and

$$\rho(1) = -k_1$$

$$\rho(2) = k_1(1 + k_2) - k_2 \tag{4.2.97}$$

The gain d_0 can also be written in terms of the other coefficients. From (4.2.20), we have

$$d_0^2 = r(0)[1 + a_1\rho(1) + a_2\rho(2)] \tag{4.2.98}$$

which can be shown to be equal to

$$d_0^2 = r(0)(1 - k_1)(1 - k_2) \tag{4.2.99}$$

Minimum-phase conditions. In (4.2.71), we have a set of conditions on a_1 and a_2 so that the AP(2) model is minimum-phase, and Figure 4.6 shows the corresponding *admissible* region for minimum-phase models. Similar relations and regions can be derived for the other types of parameters, as we will show below. In terms of k_1 and k_2, the AP(2) model is minimum-phase if

$$|k_1| < 1 \qquad |k_2| < 1 \tag{4.2.100}$$

This region is depicted in Figure 4.8(a). Shown also is the region that results in complex roots, which is specified by

$$0 < k_2 < 1 \tag{4.2.101}$$

$$k_1 < \frac{4k_2}{(1 + k_2)^2} \tag{4.2.102}$$

Because of the correlation matching property of all-pole models, we can find a minimum-phase all-pole model for every positive definite sequence of autocorrelation values. Therefore, the admissible region of autocorrelation values coincides with the positive definite region. The positive definite condition is equivalent to having all the principal minors of the autocorrelation matrix in (4.2.30) be positive definite; that is, the corresponding determinants are positive. For $P = 2$, there are two conditions:

$$\det \begin{bmatrix} 1 & \rho(1) \\ \rho(1) & 1 \end{bmatrix} < 1 \qquad \det \begin{bmatrix} 1 & \rho(1) & \rho(2) \\ \rho(1) & 1 & \rho(1) \\ \rho(2) & \rho(1) & 1 \end{bmatrix} < 1 \tag{4.2.103}$$

These two conditions reduce to

$$|\rho(1)| < 1 \tag{4.2.104}$$

$$2\rho^2(1) - 1 < \rho(2) < 1 \tag{4.2.105}$$

which determine the admissible region shown in Figure 4.8(b). Conditions (4.2.105) can also be derived from (4.2.71) and (4.2.95). The first condition in (4.2.105) is equivalent to

$$\left| \frac{a_1}{1 + a_2} \right| < 1 \tag{4.2.106}$$

which can be shown to be equivalent to the last two conditions in (4.2.71).

It is important to note that the region in Figure 4.8(b) is the admissible region for *any* positive definite autocorrelation, including the autocorrelation of mixed-phase signals. This is reasonable since the autocorrelation does not contain phase information and allows the

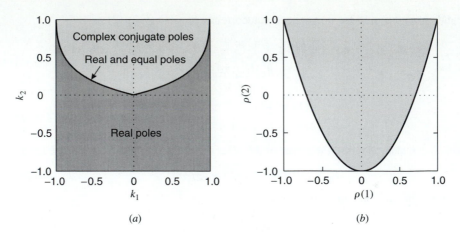

FIGURE 4.8
Minimum-phase and positive definiteness regions for the AP(2) model in the (*a*) (k_1, k_2) space and (*b*) ($\rho(1)$, $\rho(2)$) space.

signal to have minimum- and maximum-phase components. What we are claiming here, however, is that for every autocorrelation sequence in the positive definite region, we can find a minimum-phase all-pole model with the same autocorrelation values. Therefore, for this problem, the positive definite region is identical to the admissible minimum-phase region.

4.3 ALL-ZERO MODELS

In this section, we investigate the properties of the all-zero model. The output of the all-zero model is the weighted average of delayed versions of the input signal

$$x(n) = \sum_{k=0}^{Q} d_k w(n - k) \tag{4.3.1}$$

where Q is the order of the model. The system function is

$$H(z) = D(z) = \sum_{k=0}^{Q} d_k z^{-k} \tag{4.3.2}$$

The all-zero model can be implemented by using either a direct or a lattice structure. The conversion between the two sets of parameters can be done by using the step-up and step-down recursions described in Chapter 7 and setting $A(z) = D(z)$. Notice that the same set of parameters can be used to implement either an all-zero or an all-pole model by using a different structure.

4.3.1 Model Properties

We next provide a brief discussion of the properties of the all-zero model.

Impulse response. It can be easily seen that the AZ(Q) model is an FIR system with an impulse response

$$h(n) = \begin{cases} d_n & 0 \le n \le Q \\ 0 & \text{elsewhere} \end{cases} \tag{4.3.3}$$

Autocorrelation. The autocorrelation of the impulse response is given by

$$r_h(l) = \sum_{n=-\infty}^{\infty} h(n)h^*(n-l) = \begin{cases} \sum_{k=0}^{Q-l} d_k d_{k+l}^* & 0 \le l \le Q \\ 0 & l > Q \end{cases} \tag{4.3.4}$$

and
$$r_h^*(-l) = r_h(l) \quad \text{all } l \tag{4.3.5}$$

We usually set $d_0 = 1$, which implies that

$$r_h(l) = d_l^* + d_1 d_{l+1}^* + \cdots + d_{Q-l} d_Q^* \quad l = 0, 1, \ldots, Q \tag{4.3.6}$$

hence, the normalized autocorrelation is

$$\rho_h(l) = \begin{cases} \dfrac{d_l^* + d_1 d_{l+1}^* + \cdots + d_{Q-l} d_Q^*}{1 + |d_1|^2 + \cdots + |d_Q|^2} & l = 1, 2, \ldots, Q \\ 0 & l > Q \end{cases} \tag{4.3.7}$$

We see that the autocorrelation of an AZ(Q) model is zero for lags $|l|$ exceeding the order Q of the model. If $\rho_h(1), \rho_h(2), \ldots, \rho_h(Q)$ are known, then the Q equations (4.3.7) can be solved for model parameters $d_1, d_2, \ldots, d_q$. However, unlike the Yule-Walker equations for the AP(P) model, which are linear, Equations (4.3.7) are nonlinear and their solution is quite complicated (see Section 9.3).

Spectrum. The spectrum of the AZ(Q) model is given by

$$R_h(e^{j\omega}) = D(z)D(z^{-1})|_{z=e^{j\omega}} = |D(e^{j\omega})|^2 = \sum_{l=-Q}^{Q} r_h(l)e^{-j\omega l} \tag{4.3.8}$$

which is basically a trigonometric polynomial.

Impulse train excitations. The response $\tilde{h}(n)$ of the AZ(Q) model to a periodic impulse train with period L is periodic with the same period, and its spectrum is a sampled version of (4.3.8) at multiples of $2\pi/L$ (see Section 2.3.2). Therefore, to recover the autocorrelation $r_h(l)$ and the spectrum $R_h(e^{j\omega})$ from the autocorrelation or spectrum of $\tilde{h}(n)$, we should have $L \ge 2Q + 1$ in order to avoid aliasing in the autocorrelation lag domain. Also, if $L > Q$, the impulse response $h(n), 0 \le n \le Q$, can be recovered from the response $\tilde{h}(n)$ (no time-domain aliasing) (see Problem 4.24).

Partial autocorrelation and lattice-ladder structures. The PACS of an AZ(Q) model is computed by fitting a series of AP(P) models for $P = 1, 2, \ldots$, to the autocorrelation sequence (4.3.7) of the AZ(Q) model. Since the AZ(Q) model is equivalent to an AP(∞) model, the PACS of an all-zero model has infinite extent and behaves as the autocorrelation sequence of an all-pole model. This is illustrated later for the low-order AZ(1) and AZ(2) models.

4.3.2 Moving-Average Models

A moving-average model is an AZ(Q) model with $d_0 = 1$ driven by white noise, that is,

$$x(n) = w(n) + \sum_{k=1}^{Q} d_k w(n-k) \tag{4.3.9}$$

where $\{w(n)\} \sim \text{WN}(0, \sigma_w^2)$. The output $x(n)$ has zero mean and variance of

$$\sigma_x^2 = \sigma_w^2 \sum_{k=0}^{Q} |d_k|^2 \tag{4.3.10}$$

The autocorrelation and power spectrum are given by $r_x(l) = \sigma_w^2 \, r_h(l)$ and $R_x(e^{j\omega}) = \sigma_w^2 |D(e^{j\omega})|^2$, respectively. Clearly, observations that are more than Q samples apart are uncorrelated because the autocorrelation is zero after lag Q.

4.3.3 Lower-Order Models

To familiarize ourselves with all-zero models, we next investigate in detail the properties of the AZ(1) and AZ(2) models with real coefficients.

The first-order all-zero model: AZ(1).   For generality, we consider an AZ(1) model whose system function is

$$H(z) = G(1 + d_1 z^{-1}) \tag{4.3.11}$$

The model is stable for any value of d_1 and minimum-phase for $-1 < d_1 < 1$. The autocorrelation is the inverse z-transform of

$$R_h(z) = H(z)H(z^{-1}) = G^2[d_1 z + (1 + d_1^2) + d_1 z^{-1}] \tag{4.3.12}$$

Hence, $r_h(0) = G^2(1 + d_1^2)$, $r_h(1) = r_h(-1) = G^2 d_1$, and $r_h(l) = 0$ elsewhere. Therefore, the normalized autocorrelation is

$$\rho_h(l) = \begin{cases} 1 & l = 0 \\ \dfrac{d_1}{1 + d_1^2} & l = \pm 1 \\ 0 & |l| \geq 2 \end{cases} \tag{4.3.13}$$

The condition $-1 < d_1 < 1$ implies that $|\rho_h(1)| \leq \frac{1}{2}$ for a minimum-phase model. From $\rho_h(1) = d_1/(1 + d_1^2)$, we obtain the quadratic equation

$$\rho_h(1)d_1^2 - d_1 + \rho_h(1) = 0 \tag{4.3.14}$$

which has the following two roots:

$$d_1 = \frac{1 \pm \sqrt{1 - 4\rho_h^2(1)}}{2\rho_h(1)} \tag{4.3.15}$$

Since the product of the roots is 1, if d_1 is a root, then $1/d_1$ must also be a root. Hence, only one of these two roots can satisfy the minimum-phase condition $-1 < d_1 < 1$.

The spectrum is obtained by setting $z = e^{j\omega}$ in (4.3.12), or from (4.3.8)

$$R_h(e^{j\omega}) = G^2(1 + d_1^2 + 2d_1 \cos \omega) \tag{4.3.16}$$

The autocorrelation is positive definite if $R_h(e^{j\omega}) > 0$, which holds for all values of d_1. Note that if $d_1 > 0$, then $\rho_h(1) > 0$ and the spectrum has low-pass behavior (see Figure 4.9), whereas a high-pass spectrum is obtained when $d_1 < 0$ (see Figure 4.10).

The first lattice parameter of the AZ(1) model is $k_1 = d_1$. The PACS can be obtained from the Yule-Walker equations by using the autocorrelation sequence (4.3.13). Indeed, after some algebra we obtain

$$k_m = \frac{(-d_1)^m (1 - d_1^2)}{1 - d_1^{2(m+1)}} \qquad m = 1, 2, \ldots, \infty \tag{4.3.17}$$

(see Problem 4.25). Notice the duality between the ACS and PACS of AP(1) and AZ(1) models.

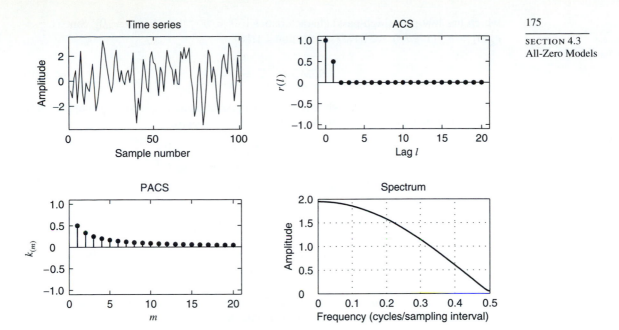

FIGURE 4.9
Sample realization of the output process, ACS, PACS, and spectrum of an AZ(1) model with $d_1 = 0.95$.

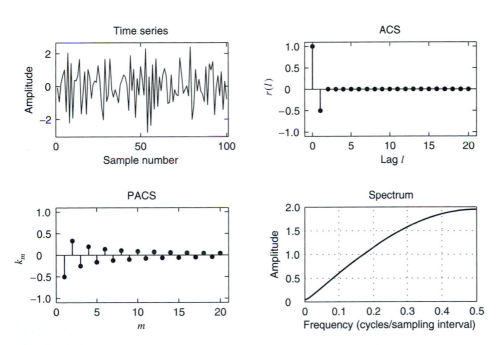

FIGURE 4.10
Sample realization of the output process, ACS, PACS, and spectrum of an AZ(1) model with $d_1 = -0.95$.

Consider now the MA(1) real-valued process $x(n)$ generated by

$$x(n) = w(n) + bw(n-1)$$

where $\{w(n)\} \sim \text{WN}(0, \sigma_w^2)$. Using $R_x(z) = \sigma_w^2 H(z) H^*(1/z^*)$, we obtain the PSD function

$$R_x(e^{j\omega}) = \sigma_w^2 (1 + b^2 + 2b \cos \omega)$$

which has low-pass (high-pass) characteristics if $0 < b \leq 1$ ($-1 \leq b < 0$). Since $\sigma_x^2 = r_x(0) = \sigma_w^2(1 + b^2)$, we have (see Section 4.1.18)

$$\text{SFM}_x = \frac{\sigma_w^2}{\sigma_x^2} = \frac{1}{1 + b^2} \tag{4.3.18}$$

which is maximum for $b = 0$ (white noise). The correlation matrix is banded Toeplitz (only a number of diagonals close to the main diagonal are nonzero)

$$\mathbf{R}_x = \sigma_w^2(1 + b^2) \begin{bmatrix} 1 & b & 0 & \cdots & 0 \\ b & 1 & b & \cdots & 0 \\ 0 & b & 1 & \cdots & 0 \\ \vdots & \vdots & \vdots & \ddots & \vdots \\ 0 & 0 & 0 & \cdots & 1 \end{bmatrix} \tag{4.3.19}$$

and its eigenvalues and eigenvectors are given by $\lambda_k = R_x(e^{j\omega_k})$, $q_n^{(k)} = \sin \omega_k n$, $\omega_k = \pi k/(M+1)$, where $k = 1, 2, \ldots, M$ (see Problem 4.30).

The second-order all-zero model: AZ(2). Now let us consider the second-order all-zero model. The system function of the AZ(2) model is

$$H(z) = G(1 + d_1 z^{-1} + d_2 z^{-2}) \tag{4.3.20}$$

The system is stable for all values of d_1 and d_2, and minimum-phase [see the discussion for the AP(2) model] if

$$\begin{aligned} -1 &< d_2 < 1 \\ d_2 - d_1 &> -1 \\ d_2 + d_1 &> -1 \end{aligned} \tag{4.3.21}$$

which is a triangular region identical to that shown in Figure 4.6. The normalized autocorrelation and the spectrum are

$$\rho_h(l) = \begin{cases} 1 & l = 0 \\ \dfrac{d_1(1 + d_2)}{1 + d_1^2 + d_2^2} & l = \pm 1 \\ \dfrac{d_2}{1 + d_1^2 + d_2^2} & l = \pm 2 \\ 0 & |l| \geq 3 \end{cases} \tag{4.3.22}$$

and $\quad R_h(e^{j\omega}) = G^2[(1 + d_1^2 + d_2^2) + 2d_1(1 + d_2)\cos \omega + 2d_2 \cos 2\omega] \tag{4.3.23}$

respectively.

The minimum-phase region in the autocorrelation domain is shown in Figure 4.11 and is described by the equations

$$\begin{aligned} \rho(2) + \rho(1) &= -0.5 \\ \rho(2) - \rho(1) &= -0.5 \\ \rho^2(1) &= 4\rho(2)[1 - 2\rho(2)] \end{aligned} \tag{4.3.24}$$

derived in Problem 4.26. The formula for the PACS is quite involved. The important thing is the duality between the ACS and the PACS of AZ(2) and AP(2) models (see Problem 4.27).

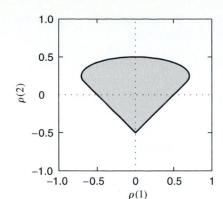

FIGURE 4.11
Minimum-phase region in the autocorrelation domain for the AZ(2) model.

4.4 POLE-ZERO MODELS

We will focus on causal pole-zero models with a recursive input-output relationship given by

$$x(n) = -\sum_{k=1}^{P} a_k x(n-k) + \sum_{k=0}^{Q} d_k w(n-k) \tag{4.4.1}$$

where we assume that $P > 0$ and $Q \geq 1$. The models can be implemented using either direct-form or lattice-ladder structures (Proakis and Manolakis 1996).

4.4.1 Model Properties

In this section, we present some of the basic properties of pole-zero models.

Impulse response. The impulse response of a causal pole-zero model can be written in recursive form from (4.4.1) as

$$h(n) = -\sum_{k=1}^{P} a_k h(n-k) + d_n \qquad n \geq 0 \tag{4.4.2}$$

where

$$d_n = 0 \qquad n > Q$$

and $h(n) = 0$ for $n < 0$. Clearly, this formula is useful if the model is stable. From (4.4.2), it is clear that

$$h(n) = -\sum_{k=1}^{P} a_k h(n-k) \qquad n > Q \tag{4.4.3}$$

so that the impulse response obeys the linear prediction equation for $n > Q$. Thus if we are given $h(n)$, $0 \leq n \leq P + Q$, we can compute $\{a_k\}$ from (4.4.3) by using the P equations specified by $Q + 1 \leq n \leq Q + P$. Then we can compute $\{d_k\}$ from (4.4.2), using $0 \leq n \leq Q$. Therefore, the first $P + Q + 1$ values of the impulse response completely specify the pole-zero model.

If the model is minimum-phase, the impulse response of the inverse model $h_I(n) = \mathcal{Z}^{-1}\{A(z)/D(z)\}$, $d_0 = 1$ can be computed in a similar manner.

Autocorrelation. The complex spectrum of $H(z)$ is given by

$$R_h(z) = H(z)H\left(\frac{1}{z^*}\right) = \frac{D(z)D(1/z^*)}{A(z)A(1/z^*)} \triangleq \frac{R_d(z)}{R_a(z)} \tag{4.4.4}$$

where $R_d(z)$ and $R_a(z)$ are both finite two-sided polynomials. In a manner similar to the all-pole case, we can write a recursive relation between the autocorrelation, impulse response, and parameters of the model. Indeed, from (4.4.4) we obtain

$$A(z)R_h(z) = D(z)H\left(\frac{1}{z^*}\right) \tag{4.4.5}$$

Taking the inverse z-transform of (4.4.5) and noting that the inverse z-transform of $H(1/z^*)$ is $h^*(-n)$, we have

$$\sum_{k=0}^{P} a_k r_h(l-k) = \sum_{k=0}^{Q} d_k h^*(k-l) \qquad \text{for all } l \tag{4.4.6}$$

Since $h(n)$ is causal, we see that the right-hand side of (4.4.6) is zero for $l > Q$:

$$\sum_{k=0}^{P} a_k r_h(l-k) = 0 \qquad l > Q \tag{4.4.7}$$

Therefore, the autocorrelation of a pole-zero model obeys the linear prediction equation for $l > Q$.

Because the impulse response $h(n)$ is a function of a_k and d_k, the set of equations in (4.4.6) is nonlinear in terms of parameters a_k and d_k. However, (4.4.7) is linear in a_k; therefore, we can compute $\{a_k\}$ from (4.4.7), using the set of equations for $l = Q + 1, \ldots, Q + P$, which can be written in matrix form as

$$\begin{bmatrix} r_h(Q) & r_h(Q+1) & \cdots & r_h(Q+P-1) \\ r_h(Q-1) & r_h(Q) & \cdots & r_h(Q+P-2) \\ \vdots & \vdots & \ddots & \vdots \\ r_h(Q-P+1) & r_h(Q-P+2) & \cdots & r_h(Q) \end{bmatrix} \begin{bmatrix} a_1 \\ a_2 \\ \vdots \\ a_P \end{bmatrix} = - \begin{bmatrix} r_h(Q-1) \\ r_h(Q-2) \\ \vdots \\ r_h(Q-P) \end{bmatrix} \tag{4.4.8}$$

or

$$\bar{\mathbf{R}}_h \mathbf{a} = -\bar{\mathbf{r}}_h \tag{4.4.9}$$

Here, $\bar{\mathbf{R}}_h$ is a non-Hermitian Toeplitz matrix, and the linear system (4.4.8) can be solved by using the algorithm of Trench (Trench 1964; Carayannis et al. 1981).

Even after we solve for $\mathbf{a}$, (4.4.6) continues to be nonlinear in d_k. To compute d_k, we use (4.4.4) to find $R_d(z)$

$$R_d(z) = R_a(z)R_h(z) \tag{4.4.10}$$

where the coefficients of $R_a(z)$ are given by

$$r_a(l) = \sum_{k=0}^{P-|l|} a_k a_{k+|l|}^* \qquad -P \le l \le P \tag{4.4.11}$$

From (4.4.10), $r_d(l)$ is the convolution of $r_a(l)$ with $r(l)$, given by

$$r_d(l) = \sum_{k=-P}^{P} r_a(k) r_h(l-k) \tag{4.4.12}$$

If $r(l)$ was originally the autocorrelation of a PZ(P, Q) model, then $r_d(l)$ in (4.4.12) will be zero for $|l| > Q$. Since $R_d(z)$ is specified, it can be factored into the product of two polynomials $D(z)$ and $D(1/z^*)$, where $D(z)$ is minimum-phase, as shown in Section 2.4.

Therefore, we have seen that, given the values of the autocorrelation $r_h(l)$ of a PZ(P, Q) model in the range $0 \le l \le P + Q$, we can compute the values of the parameters $\{a_k\}$ and $\{d_k\}$ such that $H(z)$ is minimum-phase. Now, given the parameters of a pole-zero model, we can compute its autocorrelation as follows. Equation (4.4.4) can be written as

$$R_h(z) = R_a^{-1}(z)R_d(z) \tag{4.4.13}$$

where $R_a^{-1}(z)$ is the spectrum of the all-pole model $1/A(z)$, that is, $1/R_a(z)$. The coefficients of $R_a^{-1}(z)$ can be computed from $\{a_k\}$ by using (4.2.20) and (4.2.18). The coefficients of $R_d(z)$ are computed from (4.3.8). Then $R_h(z)$ is the convolution of the two autocorrelations thus computed, which is equivalent to multiplying the two polynomials in (4.4.13) and equating equal powers of z on both sides of the equation. Since $R_d(z)$ is finite, the summations used to obtain the coefficients of $R_h(z)$ are also finite.

EXAMPLE 4.4.1. Consider a signal that has autocorrelation values of $r_h(0) = 19$, $r_h(1) = 9$, $r_h(2) = -5$, and $r_h(3) = -7$. The parameters of the PZ(2, 1) model are found in the following manner. First form the equation from (4.4.8)

$$\begin{bmatrix} 9 & 19 \\ -5 & 9 \end{bmatrix} \begin{bmatrix} a_1 \\ a_2 \end{bmatrix} = \begin{bmatrix} 5 \\ 7 \end{bmatrix}$$

which yields $a_1 = -\frac{1}{2}$, $a_2 = \frac{1}{2}$. Then we compute the coefficients from (4.4.11), $r_a(0) = \frac{3}{2}$, $r_a(\pm 1) = -\frac{3}{4}$, and $r_a(\pm 2) = \frac{1}{2}$. Computing the convolution in (4.4.12) for $l \leq Q = 1$, we obtain the following polynomial:

$$R_d(z) = 4z + 10 + 4z^{-1} = 4\left(1 + \frac{1}{2z^{-1}}\right)(z + 2)$$

Therefore, $D(z)$ is obtained by taking the causal part, that is, $D(z) = 2[1 + 1/(2z^{-1})]$, and $d_1 = \frac{1}{2}$.

Spectrum. The spectrum of $H(z)$ is given by

$$R_h(e^{j\omega}) = |H(e^{j\omega})|^2 = \frac{|D(e^{j\omega})|^2}{|A(e^{j\omega})|^2} \tag{4.4.14}$$

Therefore, $R_h(e^{j\omega})$ can be obtained by dividing the spectrum of $D(z)$ by the spectrum of $A(z)$. Again, the FFT can be used to advantage in computing the numerator and denominator of (4.4.14). If the spectrum $R_h(e^{j\omega})$ of a PZ(P, Q) model is given, then the parameters of the (minimum-phase) model can be recovered by first computing the autocorrelation $r_h(l)$ as the inverse Fourier transform of $R_h(e^{j\omega})$ and then using the procedure outlined in the previous section to compute the sets of coefficients $\{a_k\}$ and $\{d_k\}$.

Partial autocorrelation and lattice-ladder structures. Since a PZ(P, Q) model is equivalent to an AP(∞) model, its PACS has infinite extent and behaves, after a certain lag, as the PACS of an all-zero model.

4.4.2 Autoregressive Moving-Average Models

The autoregressive moving-average model is a PZ(P, Q) model driven by white noise and is denoted by ARMA(P, Q). Again, we set $d_0 = 1$ and incorporate the gain into the variance (power) of the white noise excitation. Hence, a causal ARMA(P, Q) model is defined by

$$x(n) = -\sum_{k=1}^{P} a_k x(n-k) + w(n) + \sum_{k=1}^{Q} d_k w(n-k) \tag{4.4.15}$$

where $\{w(n)\} \sim WN(0, \sigma_w^2)$. The ARMA(P, Q) model parameters are $\{\sigma_w^2, a_1, \ldots, a_P, d_1, \ldots, d_Q\}$. The output has zero mean and variance of

$$\sigma_x^2 = -\sum_{k=1}^{P} a_k r_x(k) + \sigma_w^2 [1 + \sum_{k=1}^{Q} d_k h(k)] \tag{4.4.16}$$

where $h(n)$ is the impulse response of the model. The presence of $h(n)$ in (4.4.16) makes the dependence of σ_x^2 on the model parameters highly nonlinear. The autocorrelation of

$x(n)$ is given by

$$\sum_{k=0}^{P} a_k r_x(l-k) = \sigma_w^2 \left[1 + \sum_{k=1}^{Q} d_k h(k-l) \right] \quad \text{for all } l \qquad (4.4.17)$$

and the power spectrum by

$$R_x(e^{j\omega}) = \sigma_w^2 \frac{|D(e^{j\omega})|^2}{|A(e^{j\omega})|^2} \qquad (4.4.18)$$

The significance of ARMA(P, Q) models is that they can provide more accurate representations than AR or MA models with the same number of parameters. *The ARMA model is able to combine the spectral peak matching of the AR model with the ability of the MA model to place nulls in the spectrum.*

4.4.3 The First-Order Pole-Zero Model: PZ(1, 1)

Consider the PZ(1, 1) model with the following system function

$$H(z) = G \frac{1 + d_1 z^{-1}}{1 + a_1 z^{-1}} \qquad (4.4.19)$$

where d_1 and a_1 are real coefficients. The model is minimum-phase if

$$-1 < d_1 < 1$$
$$-1 < a_1 < 1 \qquad (4.4.20)$$

which correspond to the rectangular region shown in Figure 4.12(a).

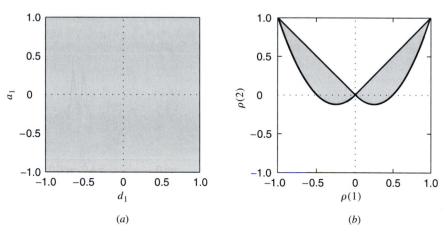

(a) (b)

FIGURE 4.12
Minimum-phase and positive definiteness regions for the PZ(1, 1) model in the
(a) (d_1, a_1) space and (b) ($\rho(1)$, $\rho(2)$) space.

For the minimum-phase case, the impulse responses of the direct and the inverse models are

$$h(n) = \mathcal{Z}^{-1}\{H(z)\} = \begin{cases} 0 & n < 0 \\ G & n = 0 \\ G(-a_1)^{n-1}(d_1 - a_1) & n > 0 \end{cases} \qquad (4.4.21)$$

and
$$h_I(n) = \mathcal{Z}^{-1}\left\{\frac{1}{H(z)}\right\} = \begin{cases} 0 & n < 0 \\ G & n = 0 \\ G(-d_1)^{n-1}(a_1 - d_1) & n > 0 \end{cases} \qquad (4.4.22)$$

respectively. We note that as the pole $p = -a_1$ gets closer to the unit circle, the impulse response decays more slowly and the model has "longer memory." The zero $z = -d_1$ controls the impulse response of the inverse model in a similar way. The PZ(1, 1) model is equivalent to the AZ(∞) model

$$x(n) = Gw(n) + G\sum_{k=1}^{\infty} h(k)w(n-k) \qquad (4.4.23)$$

or the AP(∞) model

$$x(n) = -\sum_{k=1}^{\infty} h_I(k)x(n-k) + Gw(n) \qquad (4.4.24)$$

If we wish to approximate the PZ(1, 1) model with a finite-order AZ(Q) model, the order Q required to achieve a certain accuracy increases as the pole moves closer to the unit circle. Likewise, in the case of an AP(P) approximation, better fits to the PZ(P, Q) model require an increased order P as the zero moves closer to the unit circle.

To determine the autocorrelation, we recall from (4.4.6) that for a causal model

$$r_h(l) = -a_1 r_h(l-1) + Gh(-l) + Gd_1 h(1-l) \qquad \text{all } l \qquad (4.4.25)$$

or
$$r_h(0) = -a_1 r_h(1) + G + Gd_1(d_1 - a_1)$$

$$r_h(1) = -a_1 r_h(0) + Gd_1 \qquad (4.4.26)$$

$$r_h(l) = -a_1 r_h(l-1) \qquad l \geq 2$$

Solving the first two equations for $r_h(0)$ and $r_h(1)$, we obtain

$$r_h(0) = G\frac{1 + d_1^2 - 2a_1 d_1}{1 - a_1^2} \qquad (4.4.27)$$

and
$$r_h(1) = G\frac{(d_1 - a_1)(1 - a_1 d_1)}{1 - a_1^2} \qquad (4.4.28)$$

The normalized autocorrelation is given by

$$\rho_h(1) = \frac{(d_1 - a_1)(1 - a_1 d_1)}{1 + d_1^2 - 2a_1 d_1} \qquad (4.4.29)$$

and
$$\rho_h(l) = (-a_1)^{l-1}\rho_h(l-1) \qquad l \geq 2 \qquad (4.4.30)$$

Note that given $\rho_h(1)$ and $\rho_h(2)$, we have a nonlinear system of equations that must be solved to obtain a_1 and d_1. By using Equations (4.4.20), (4.4.29), and (4.4.30), it can be shown (see Problem 4.28) that the PZ(1, 1) is minimum-phase if the ACS satisfies the conditions

$$|\rho(2)| < |\rho(1)|$$
$$\rho(2) > \rho(1)[2\rho(1) + 1] \qquad \rho(1) < 0 \qquad (4.4.31)$$
$$\rho(2) > \rho(1)[2\rho(1) - 1] \qquad \rho(1) > 0$$

which correspond to the admissible region shown in Figure 4.12(b).

4.4.4 Summary and Dualities

Table 4.1 summarizes the key properties of all-zero, all-pole, and pole-zero models. These properties help to identify models for empirical discrete-time signals. Furthermore, the table shows the duality between AZ and AP models. More specifically, we see that

1. An invertible AZ(Q) model is equivalent to an AP(∞) model. Thus, it has a finite-extent autocorrelation and an infinite-extent partial autocorrelation.
2. A stable AP(P) model is equivalent to an AZ(∞) model. Thus, it has an infinite-extent autocorrelation and a finite-extent partial autocorrelation.
3. The autocorrelation of an AZ(Q) model behaves as the partial autocorrelation of an AP(P) model, and vice versa.
4. The spectra of an AP(P) model and an AZ(Q) model are related through an inverse relationship.

TABLE 4.1
Summary of all-pole, all-zero, and pole-zero model properties

Model	AP(P)	AZ(Q)	PZ(P, Q)
Input-output description	$x(n) + \sum_{k=1}^{P} a_k x(n-k) = w(n)$	$x(n) = d_0 w(n) + \sum_{k=1}^{Q} d_k w(n-k)$	$x(n) + \sum_{k=1}^{P} a_k x(n-k)$
			$= d_0 w(n) + \sum_{k=1}^{Q} d_k w(n-k)$
System function	$H(z) = \dfrac{1}{A(z)} = \dfrac{d_0}{1 + \sum_{k=1}^{P} a_k z^{-k}}$	$H(z) = D(z) = d_0 + \sum_{k=1}^{Q} d_k z^{-k}$	$H(z) = \dfrac{D(z)}{A(z)}$
Recursive representation	Finite summation	Infinite summation	Infinite summation
Nonrecursive representation	Infinite summation	Finite summation	Infinite summation
Stablity conditions	Poles inside unit circle	Always	Poles inside unit circle
Invertiblity conditions	Always	Zeros inside unit circle	Zeros inside unit circle
Autocorrelation sequence	Infinite duration (damped exponentials and/or sine waves)	Finite duration	Infinite duration (damped exponentials and/or sine waves after $Q - P$ lags)
	Tails off	Cuts off	Tails off
Partial autocorrelation	Finite duration	Infinite duration (damped exponentials and/or sine waves)	Infinite duration (dominated by damped exponentials and/or sine waves after $Q - P$ lags)
	Cuts off	Tails off	Tails off
Spectrum	Good peak matching	Good "notch" matching	Good peak and valley matching

These dualities and properties have been shown and illustrated for low-order models in the previous sections.

4.5 MODELS WITH POLES ON THE UNIT CIRCLE

In this section, we show that by restricting some poles to being on the unit circle, we obtain models that are useful for modeling certain types of nonstationary behavior.

Pole-zero models with poles on the unit circle are unstable. Hence, if we drive them with stationary white noise, the generated process is nonstationary. However, as we will see in the sequel, placing a small number of real poles at $z = 1$ or complex conjugate poles at $z_k = e^{\pm j\theta_k}$ provides a class of models useful for modeling certain types of nonstationary behavior. The system function of a pole-zero model with d poles at $z = 1$, denoted as PZ(P, d, Q), is

$$H(z) = \frac{D(z)}{A(z)} \frac{1}{(1 - z^{-1})^d} \tag{4.5.1}$$

and can be viewed as PZ(P, Q) model, $D(z)/A(z)$, followed by a dth-order accumulator. The accumulator $y(n) = y(n-1) + x(n)$ has the system function $1/(1 - z^{-1})$ and can be thought of as a discrete-time integrator. The presence of the unit poles makes the PZ(P, d, Q) model non-minimum-phase. Since the model is unstable, we cannot use the convolution summation to represent it because, in practice, only finite-order approximations are possible. This can be easily seen if we recall that the impulse response of the model PZ($0, d, 0$) equals $u(n)$ for $d = 1$ and $(n + 1)u(n)$ for $d = 2$. However, if $D(z)/A(z)$ is minimum-phase, the inverse model $H_I(z) = 1/H(z)$ is stable, and we can use the recursive form (see Section 4.1) to represent the model. Indeed, we always use this representation when we apply this model in practice.

The spectrum of the PZ($0, d, 0$) model is

$$R_d(e^{j\omega}) = \frac{1}{[2\sin(\omega/2)]^{2d}} \tag{4.5.2}$$

and since $R_d(0) = \sum_{l=-\infty}^{\infty} r_d(l) = \infty$, the autocorrelation does not exist.

In the case of complex conjugate poles, the term $(1 - z^{-1})^d$ in (4.5.1) is replaced by $(1 - 2\cos\theta_k z^{-1} + z^{-2})^d$, that is,

$$H(z) = \frac{D(z)}{A(z)} \frac{1}{(1 - 2\cos\theta_k z^{-1} + z^{-2})^d} \tag{4.5.3}$$

The second term is basically a cascade of AP(2) models with complex conjugate poles on the unit circle. This model exhibits strong periodicity in its impulse response, and its "resonance-like" spectrum diverges at $\omega = \theta_k$.

With regard to the partial autocorrelation, we recall that the presence of poles on the unit circle results in some lattice parameters taking on the values ± 1.

EXAMPLE 4.5.1. Consider the following causal PZ(1, 1, 1) model

$$H(z) = \frac{1 + d_1 z^{-1}}{1 + a_1 z^{-1}} \frac{1}{1 - z^{-1}} = \frac{1 + d_1 z^{-1}}{1 - (1 - a_1)z^{-1} - a_1 z^{-2}} \tag{4.5.4}$$

with $-1 < a_1 < 1$ and $-1 < d_1 < 1$.

The difference equation representation of the model uses previous values of the output and the present and previous values of the input. It is given by

$$y(n) = (1 - a_1)y(n-1) + a_1 y(n-2) + x(n) + d_1 x(n-1) \tag{4.5.5}$$

To express the output in terms of the present and previous values of the input (nonrecursive representation), we find the impulse response of the model

$$h(n) = \mathcal{Z}^{-1}\{H(z)\} = A_1 u(n) + A_2(-a_1)^n u(n) \tag{4.5.6}$$

where $A_1 = (1 + d_1)/(1 + a_1)$ and $A_2 = (a_1 - d_1)/(1 + a_1)$. Note that the model is unstable, and it cannot be approximated by an FIR system because $h(n) \to A_1 u(n)$ as $n \to \infty$.

Finally, we can express the output as a weighted sum of previous outputs and the present input, using the impulse response of the inverse model $G(z) = 1/H(z)$

$$h_I(n) = \mathcal{Z}^{-1}\{H_I(z)\} = B_1 \delta(n) + B_2 \delta(n-1) + B_3(-d_1)^n u(n) \tag{4.5.7}$$

where $B_1 = (a_1 - d_1 + a_1 d_1)/d_1^2$, $B_2 = -a_1/d_1$, and $B_3 = (-a_1 + d_1 - a_1 d_1 + d_1^2)/d_1^2$. Since $-1 < d_1 < 1$, the sequence $h_I(n)$ decays at a rate governed by the value of d_1. If $h_I(n) \simeq 0$ for $n \geq p_d$, the recursive formula

$$y(n) = -\sum_{k=1}^{p_d} h_I(k)y(n-k) + x(n) \tag{4.5.8}$$

provides a good representation of the PZ(1, 1, 1) model. For example, if $a_1 = 0.3$ and $d_1 = 0.5$, we find that $|h_I(n)| \leq 0.0001$ for $n \geq 12$, which means that the current value of the model output can be computed with sufficient accuracy from the 12 most recent values of signal $y(n)$. This is illustrated in Figure 4.13, which also shows a realization of the output process if the model is driven by white Gaussian noise.

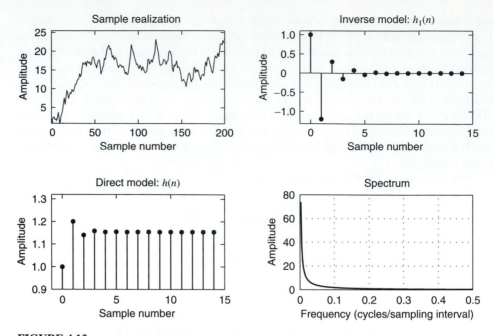

FIGURE 4.13
Sample realization of the output process, impulse response, impulse response of the inverse model, and spectrum of a PZ(1, 1, 1) model with $a_1 = 0.3$, $d_1 = 0.5$, and $d = 1$. The value $R(e^{j0}) = \infty$ is not plotted.

Autoregressive integrated moving-average models. In Section 3.3.2 we discussed discrete-time random signals with stationary increments. Clearly, driving a PZ(P, d, Q) model with white noise generates a random signal whose dth difference is a stationary ARMA(P, Q) process. Such time series are known in the statistical literature as *autoregressive integrated moving-average models*, denoted ARIMA (P, d, Q). They are useful in modeling signals with certain *stochastic trends* (e.g., random changes in the level and slope of the signal). Indeed, many empirical signals (e.g., infrared background measurements and stock prices) exhibit this type of behavior (see Figure 1.6). Notice that the ARIMA(0, 1, 0) process, that is, $x(n) = x(n-1) + w(n)$, where $\{w(n)\} \sim \text{WN}(0, \sigma_w^2)$, is the discrete-time equivalent of the *random walk* or *Brownian motion* process (Papoulis 1991).

When the unit poles are complex conjugate, the model is known as a *harmonic PZ model*. This model produces random sequences that exhibit "random periodic behavior" and are known as *seasonal time series* in the statistical literature. Such signals repeat themselves cycle by cycle, but there is some randomness in both the length and the pattern of each cycle. The identification and estimation of ARIMA and seasonal models and their applications can be found in Box, Jenkins, and Reinsel (1994); Brockwell and Davis (1991); and Hamilton (1994).

4.6 CEPSTRUM OF POLE-ZERO MODELS

In this section we determine the cepstrum of pole-zero models and its properties, and we develop algorithms to convert between direct structure model parameters and cepstral coefficients. The cepstrum has been proved a valuable tool in speech coding and recognition applications and has been extensively studied in the corresponding literature (Rabiner and Schafer 1978; Rabiner and Juang 1993; Furui 1989). For simplicity, we consider models with real coefficients.

4.6.1 Pole-Zero Models

The cepstrum of the impulse response $h(n)$ of a pole-zero model is the inverse z-transform of

$$\log H(z) = \log D(z) - \log A(z) \tag{4.6.1}$$

$$= \log d_0 + \sum_{i=1}^{Q} \log (1 - z_i z^{-1}) - \sum_{i=1}^{P} \log (1 - p_i z^{-1}) \tag{4.6.2}$$

where $\{z_i\}$ and $\{p_i\}$ are the zeros and poles of $H(z)$, respectively. If we assume that $H(z)$ is minimum-phase and use the power series expansion

$$\log (1 - \alpha z^{-1}) = -\sum_{n=1}^{\infty} \frac{\alpha^n}{n} z^{-n} \qquad |z| > |\alpha|$$

we find that the cepstrum $c(n)$ is given by

$$c(n) = \begin{cases} 0 & n < 0 \\ \log d_0 & n = 0 \\ \dfrac{1}{n} \left(\sum_{i=1}^{P} p_i^n - \sum_{i=1}^{Q} z_i^n \right) & n > 0 \end{cases} \tag{4.6.3}$$

Since the poles and zeros are assumed to be inside the unit circle, (4.6.3) implies that $c(n)$ is bounded by

$$-\frac{P+Q}{n} \le c(n) \le \frac{P+Q}{n} \tag{4.6.4}$$

with equality if and only if all the roots are appropriately at $z = 1$ or $z = -1$.

If $H(z)$ is minimum-phase, then there exists a unique mapping between the cepstrum and the impulse response, given by the recursive relations (Oppenheim and Schafer 1989)

$$c(0) = \log h(0) = \log d_0$$

$$c(n) = \frac{h(n)}{h(0)} - \frac{1}{n} \sum_{m=0}^{n-1} m c(m) \frac{h(n-m)}{h(0)} \qquad n > 0 \tag{4.6.5}$$

and

$$h(0) = e^{c(0)}$$

$$h(n) = h(0)c(n) + \frac{1}{n} \sum_{m=0}^{n-1} m c(m) h(n-m) \qquad n > 0 \tag{4.6.6}$$

where we have assumed $d_0 > 0$ without loss of generality. Therefore, given the cepstrum $c(n)$ in the range $0 \le n \le P+Q$, we can completely recover the parameters of the pole-zero model as follows. From (4.6.6) we can compute $h(n)$, $0 \le n \le P+Q$, and from (4.4.2) and (4.4.3) we can recover $\{a_k\}$ and $\{d_k\}$.

4.6.2 All-Pole Models

The cepstrum of a minimum-phase all-pole model is given by (4.6.2) and (4.6.3) with $Q = 0$. Since $H(z)$ is minimum-phase, the cepstrum $c(n)$ of $1/A(z)$ is simply the negative of the cepstrum of $A(z)$, which can be written in terms of a_k (see also Problem 4.34). As a result, the cepstrum can be obtained from the direct-form coefficients by using the following recursion

$$c(n) = \begin{cases} -a_n - \dfrac{1}{n}\displaystyle\sum_{k=1}^{n-1}(n-k)\,a_k\,c(n-k) & 1 \le n \le P \\[4mm] -\dfrac{1}{n}\displaystyle\sum_{k=1}^{P}(n-k)\,a_k\,c(n-k) & n > P \end{cases} \tag{4.6.7}$$

The inverse relation is

$$a_n = -c(n) - \frac{1}{n}\sum_{k=1}^{n-1}(n-k)\,a_k\,c(n-k) \qquad n > 0 \tag{4.6.8}$$

which shows that the first P cepstral coefficients completely determine the model parameters (Furui 1981).

From (4.6.7) it is evident that the cepstrum generally decays as $1/n$. Therefore, it may be desirable sometimes to consider

$$c'(n) = nc(n) \tag{4.6.9}$$

which is known as the *ramp cepstrum* since it is obtained by multiplying the cepstrum by a ramp function. From (4.6.9) and (4.6.4), we note that the ramp cepstrum of an AP(P) model is bounded by

$$|c'(n)| \le P \qquad n > 0 \tag{4.6.10}$$

with equality if and only if all the poles are at $z = 1$ or $z = -1$. Also $c'(n)$ is equal to the negative of the inverse z-transform of the derivative of $\log H(z)$. From the preceding equations, we can write

$$c'(n) = -na_n - \sum_{k=1}^{n-1} a_k c'(n-k) \qquad 1 \le n \le P \tag{4.6.11}$$

$$c'(n) = -\sum_{k=1}^{P} a_k c'(n-k) \qquad n > P \tag{4.6.12}$$

and

$$a_n = \frac{1}{n}\left[c'(n) + \sum_{k=1}^{n-1} a_k c'(n-k) \right] \qquad n > 0 \tag{4.6.13}$$

It is evident that the first P values of $c'(n)$, $1 \le n \le P$, completely specify the model coefficients. However, since $c'(0) = 0$, the information about the gain d_0 is lost in the ramp cepstrum. Equation (4.6.12) for $n > P$ is reminiscent of similar equations for the impulse response in (4.2.5) and the autocorrelation in (4.2.18), with the major difference that for the ramp cepstrum the relation is only true for $n > P$, while for the impulse response and the autocorrelation, the relations are true for $n > 0$ and $k > 0$, respectively.

Since $R(z) = H(z)H(z^{-1})$, we have

$$\log R(z) = \log H(z) + \log H(z^{-1}) \tag{4.6.14}$$

and if $c_r(n)$ is the real cepstrum of $R(e^{j\omega})$, we conclude that

$$c_r(n) = c(n) + c(-n) \tag{4.6.15}$$

For minimum-phase $H(z)$, $c(n) = 0$ for $n < 0$. Therefore,

$$c_r(n) = \begin{cases} c(-n) & n < 0 \\ 2c(0) & n = 0 \\ c(n) & n > 0 \end{cases} \tag{4.6.16}$$

and
$$c(n) = \begin{cases} 0 & n < 0 \\ \dfrac{c_r(0)}{2} & n = 0 \\ c_r(n) & n > 0 \end{cases} \qquad (4.6.17)$$

In other words, the cepstrum $c(n)$ can be obtained simply by taking the inverse Fourier transform of $\log R(e^{j\omega})$ to obtain $c_r(n)$ and then applying (4.6.17).

EXAMPLE 4.6.1. From (4.6.7) we find that the cepstrum of the AP(1) model is given by

$$c(n) = \begin{cases} 0 & n < 0 \\ \log d_0 & n = 0 \\ \dfrac{1}{n}(-a)^n & n > 0 \end{cases} \qquad (4.6.18)$$

From (4.2.18) with $P = 1$ and $k = 1$, we have $a_1^{(1)} = -r(1)/r(0) = k_1$; and from (4.6.7) we have $a_1 = -c(1)$. These results are summarized below:

$$a_1^{(1)} = a = -\rho(1) = k_1 = -c(1) \qquad (4.6.19)$$

The fact that $\rho(1) = c(1)$ here is peculiar to a single-pole spectrum and is not true in general for arbitrary spectra. And $\rho(1)$ is the integral of a cosine-weighted spectrum while $c(1)$ is the integral of a cosine-weighted log spectrum.

EXAMPLE 4.6.2. From (4.6.7), the cepstrum for an AP(2) model is equal to

$$c(n) = \begin{cases} 0 & n < 0 \\ \log d_0 & n = 0 \\ \dfrac{1}{n}(p_1^n + p_2^n) & n > 0 \end{cases} \qquad (4.6.20)$$

For a complex conjugate pole pair, we have

$$c(n) = \frac{2}{n} r^n \cos n\theta \qquad n > 0 \qquad (4.6.21)$$

where $p_{1,2} = r \exp(\pm j\theta)$. Therefore, the cepstrum of a damped sine wave is a damped cosine wave. The cepstrum and autocorrelation are similar in that they are both damped cosines, but the cepstrum has an additional $1/n$ weighting. From (4.6.7) and (4.6.8) we can relate the model parameters and the cepstral coefficients:

$$\begin{aligned} a_1 &= -c(1) \\ a_2 &= -c(2) + \tfrac{1}{2}c^2(1) \end{aligned} \qquad (4.6.22)$$

and

$$\begin{aligned} c(1) &= -a_1 \\ c(2) &= -a_2 + \tfrac{1}{2}a_1^2 \end{aligned} \qquad (4.6.23)$$

Using (4.2.71) and the relations for the cepstrum, we can derive the conditions on the cepstrum for $H(z)$ to be minimum-phase:

$$\begin{aligned} c(2) &> \frac{c^2(1)}{2} - 1 \\ c(2) &< \frac{c^2(1)}{2} - c(1) + 1 \\ c(2) &< \frac{c^2(1)}{2} + c(1) + 1 \end{aligned} \qquad (4.6.24)$$

The corresponding admissible region is shown in Figure 4.14. The region corresponding to complex roots is given by

$$\frac{1}{2}c^2(1) - 1 < c(2) < \frac{c^2(1)}{4} \qquad (4.6.25)$$

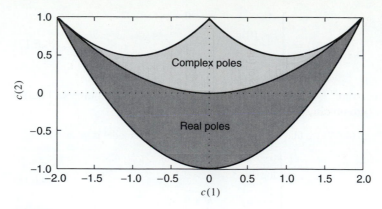

FIGURE 4.14
Minimum-phase region of the AP(2) model in the cepstral domain.

In comparing Figures 4.6, 4.8, and 4.14, we note that the admissible regions for the PACS and ACS are convex while that for the cepstral coefficients is not. (A region is convex if a straight line drawn between any two points in the region lies completely in the region.) In general, the PACS and the ACS span regions or spaces that are convex. The admissible region in Figure 4.14 for the model coefficients is also convex. However, for $P > 2$ the admissible regions for the model coefficients are not convex, in general.

Cepstral distance. A measure of the difference between two signals, which has many applications in speech coding and recognition, is the distance between their log spectra (Rabiner and Juang 1993). It is known as the *cepstral distance* and is defined as

$$\text{CD} \triangleq \frac{1}{2\pi} \int_{-\pi}^{\pi} |\log R_1(e^{j\omega}) - \log R_2(e^{j\omega})|^2 \, d\omega \qquad (4.6.26)$$

$$= \sum_{n=-\infty}^{\infty} [c_1(n) - c_2(n)]^2 \qquad (4.6.27)$$

where $c_1(n)$ and $c_2(n)$ are the cepstral coefficients of $R_1(e^{j\omega})$ and $R_2(e^{j\omega})$, respectively (see Problem 4.36). Since for minimum-phase sequences the cepstrum decays fast, the summation (4.6.27) can be computed with sufficient accuracy using a small number of terms, usually 20 to 30. For minimum-phase all-pole models, which are mostly used in speech processing, the cepstral coefficients are efficiently computed using the recursion (4.6.7).

4.6.3 All-Zero Models

The cepstrum of a minimum-phase all-zero model is given by (4.6.2) and (4.6.3) with $P = 0$. The cepstrum corresponding to a minimum-phase AZ(Q) model is related to its real cepstrum by

$$c(n) = \begin{cases} 0 & n < 0 \\ \dfrac{c_r(n)}{2} & n = 0 \\ c_r(n) & n > 0 \end{cases} \qquad (4.6.28)$$

Since we found $c(n)$, the coefficients of a minimum-phase AZ(Q) model $D(z)$ can be evaluated recursively from

$$d_k = \begin{cases} e^c d_0 & k = 0 \\ c(k)d_0 + \dfrac{1}{k}\displaystyle\sum_{m=0}^{k-1} mc(m)d_{k-m} & 1 \le k \le Q \end{cases} \qquad (4.6.29)$$

This procedure for finding a minimum-phase polynomial $D(z)$ from the autocorrelation consists in first computing the cepstrum from the log spectrum, then applying (4.6.28) and the recursion (4.6.29) to compute the coefficients d_k. This approach to the spectral factorization of AZ(Q) models is preferable because finding the roots of $R(z)$ for large Q may be cumbersome.

Mixed pole-zero model representations. In the previous sections we saw that the $P + Q + 1$ parameters of the minimum-phase PZ(P, Q) model can be represented equivalently and uniquely by $P + Q + 1$ values of the impulse response, the autocorrelation, or the cepstrum. A question arises as to whether PZ(P, Q) can be represented uniquely by a mixture of representations, as long as the total number of representative values is $P+Q+1$. For example, could we have a unique representation that consists of, say, Q autocorrelation values and $P + 1$ impulse response values, or some other mixture? The answer to this question has not been explored in general; the relevant equations are sufficiently nonlinear that a totally different approach would appear to be needed to solve the general problem.

4.7 SUMMARY

In this chapter we introduced the class of pole-zero signal models and discussed their properties. Each model consists of two components: an excitation source and a system. In our treatment, we emphasized that the properties of a signal model are shaped by the properties of both components; and we tried, whenever possible, to attribute each property to its originator. Thus, for uncorrelated random inputs, which by definition are the excitations for ARMA models, the second-order moments of the signal model and its minimum-phase characteristics are completely determined by the system. For excitations with line spectra, properties such as minimum phase are meaningful only when they are attributed to the underlying system. If the goal is to model a signal with a line PSD, the most appropriate approach is to use a harmonic process.

We provided a detailed description of the autocorrelation, power spectrum density, partial correlation, and cepstral properties of all AZ, AP, and PZ models for the general case and for first- and second-order models. An understanding of these properties is very important for model selection in practical applications.

PROBLEMS

4.1 Show that a second-order pole p_i contributes the term $np_i^n u(n)$ and a third-order pole the terms $np_i^n u(n) + n^2 p_i^n u(n)$ to the impulse response of a causal PZ model. The general case is discussed in Oppenheim et al. (1997).

4.2 Consider a zero-mean random sequence $x(n)$ with PSD

$$R_x(e^{j\omega}) = \frac{5 + 3\cos\omega}{17 + 8\cos\omega}$$

(a) Determine the innovations representation of the process $x(n)$.
(b) Find the autocorrelation sequence $r_x(l)$.

4.3 We want to generate samples of a Gaussian process with autocorrelation $r_x(l) = (\frac{1}{2})^{|l|} + (-\frac{1}{2})^{|l|}$ for all l.

(a) Find the difference equation that generates the process $x(n)$ when excited by $w(n) \sim$ WGN(0, 1).

(b) Generate $N = 1000$ samples of the process and estimate the pdf, using the histogram and the normalized autocorrelation $\rho_x(l)$ using $\hat{\rho}_x(l)$ [see Section (1.2.1)].

(c) Check the validity of the model by plotting on the same graph (i) the true and estimated pdf of $y(n)$ and (ii) the true and estimated autocorrelation.

4.4 Compute and compare the autocorrelations of the following processes:

(a) $x_1(n) = w(n) + 0.3w(n-1) - 0.4w(n-1)$ and

(b) $x_1(n) = w(n) - 1.2w(n-1) - 1.6w(n-1)$ where $w(n) \sim$ WGN(0, 1).

Explain your findings.

4.5 Compute and plot the impulse response and the magnitude response of the systems $H(z)$ and $H_N(z)$ in Example 4.2.1 for $a = 0.7, 0.95$ and $N = 8, 16, 64$. Investigate how well the all-zero systems approximate the single-pole system.

4.6 Prove Equation (4.2.35) by writing explicitly Equation (4.2.33) and rearranging terms. Then show that the coefficient matrix $\mathbf{A}$ can be written as the sum of a triangular Toeplitz matrix and a triangular Hankel matrix (recall that a matrix $\mathbf{H}$ is Hankel if the matrix $\mathbf{JHJ}^H$ is Toeplitz).

4.7 Use the Yule-Walker equations to determine the autocorrelation and partial autocorrelation coefficients of the following AR models, assuming that $w(n) \sim$ WN(0, 1).

(a) $x(n) = 0.5x(n-1) + w(n)$.

(b) $x(n) = 1.5x(n-1) - 0.6x(n-2) + w(n)$.

What is the variance σ_x^2 of the resulting process?

4.8 Given the AR process $x(n) = x(n-1) - 0.5x(n-2) + w(n)$, complete the following tasks.

(a) Determine $\rho_x(1)$.

(b) Using $\rho_x(0)$ and $\rho_x(1)$, compute $\{\rho_x(l)\}_2^{15}$ by the corresponding difference equation.

(c) Plot $\rho_x(l)$ and use the resulting graph to estimate its period.

(d) Compare the period obtained in part (c) with the value obtained using the PSD of the model. (*Hint:* Use the frequency of the PSD peak.)

4.9 Given the parameters d_0, a_1, a_2, and a_3 of an AP(3) model, compute its ACS analytically and verify your results, using the values in Example 4.2.3. (*Hint:* Use Cramer's rule.)

4.10 Consider the following AP(3) model: $x(n) = 0.98x(n-3) + w(n)$, where $w(n) \sim$ WGN(0, 1).

(a) Plot the PSD of $x(n)$ and check if the obtained process is going to exhibit a pseudoperiodic behavior.

(b) Generate and plot 100 samples of the process. Does the graph support the conclusion of part (a)? If yes, what is the period?

(c) Compute and plot the PSD of the process $y(n) = \frac{1}{3}[x(n-1) + x(n) + x(n+1)]$.

(d) Repeat part (b) and explain the difference between the behavior of processes $x(n)$ and $y(n)$.

4.11 Consider the following AR(2) models: (i) $x(n) = 0.6x(n-1) + 0.3x(n-2) + w(n)$ and (ii) $x(n) = 0.8x(n-1) - 0.5x(n-2) + w(n)$, where $w(n) \sim$ WGN(0, 1).

(a) Find the general expression for the normalized autocorrelation sequence $\rho(l)$, and determine σ_x^2.

(b) Plot $\{\rho(l)\}_0^{15}$ and check if the models exhibit pseudoperiodic behavior.

(c) Justify your answer in part (b) by plotting the PSD of the two models.

4.12 (a) Derive the formulas that express the PACS of an AP(3) model in terms of its ACS, using the Yule-Walker equations and Cramer's rule.

(b) Use the obtained formulas to compute the PACS of the AP(3) model in Example 4.2.3.

(c) Check the results in part (b) by recomputing the PACS, using the algorithm of Levinson-Durbin.

4.13 Show that the spectrum of any PZ model with real coefficients has zero slope at $\omega = 0$ and $\omega = \pi$.

4.14 Derive Equations (4.2.71) describing the minimum-phase region of the AP(2) model, starting from the conditions

(a) $|p_1| < 1, |p_2| < 1$ and

(b) $|k_1| < 1, |k_2| < 1$.

4.15 (a) Show that the spectrum of an AP(2) model with real poles can be obtained by the cascade connection of two AP(1) models with real coefficients.

(b) Compute and plot the impulse response, ACS, PACS, and spectrum of the AP models with $p_1 = 0.6$, $p_2 = -0.9$, and $p_1 = p_2 = 0.9$.

4.16 Prove Equation (4.2.89) and demonstrate its validity by plotting the spectrum (4.2.88) for various values of r and θ.

4.17 Prove that if the AP(P) model $A(z)$ is minimum-phase, then

$$\frac{1}{2\pi} \int_{-\pi}^{\pi} \log \frac{1}{|A(e^{j\omega})|^2} \, d\omega = 0$$

4.18 (a) Prove Equations (4.2.101) and (4.2.102) and recreate the plot in Figure 4.8(a).

(b) Determine and plot the regions corresponding to complex and real poles in the autocorrelation domain by recreating Figure 4.8(b).

4.19 Consider an AR(2) process $x(n)$ with $d_0 = 1$, $a_1 = -1.6454$ $a_2 = 0.9025$, and $w(n) \sim$ WGN(0, 1).

(a) Generate 100 samples of the process and use them to estimate the ACS $\hat{\rho}_x(l)$, using Equation (1.2.1).

(b) Plot and compare the estimated and theoretical ACS values for $0 \leq l \leq 10$.

(c) Use the estimated values of $\hat{\rho}_x(l)$ and the Yule-Walker equations to estimate the parameters of the model. Compare the estimated with the true values, and comment on the accuracy of the approach.

(d) Use the estimated parameters to compute the PSD of the process. Plot and compare the estimated and true PSDs of the process.

(e) Compute and compare the estimated with the true PACS.

4.20 Find a minimum-phase model with autocorrelation $\rho(0) = 1$, $\rho(\pm 1) = 0.25$, and $\rho(l) = 0$ for $|l| \geq 2$.

4.21 Consider the MA(2) model $x(n) = w(n) - 0.1w(n-1) + 0.2w(n-2)$.

(a) Is the process $x(n)$ stationary? Why?

(b) Is the model minimum-phase? Why?

(c) Determine the autocorrelation and partial autocorrelation of the process.

4.22 Consider the following ARMA models: (i) $x(n) = 0.6x(n-1) + w(n) - 0.9w(n-1)$ and (ii) $x(n) = 1.4x(n-1) - 0.6x(n-2) + w(n) - 0.8w(n-1)$.

(a) Find a general expression for the autocorrelation $\rho(l)$.

(b) Compute the partial autocorrelation k_m for $m = 1, 2, 3$.

(c) Generate 100 samples from each process, and use them to estimate $\{\hat{\rho}(l)\}_0^{20}$ using Equation (1.2.1).

(d) Use $\hat{\rho}(l)$ to estimate $\{\hat{k}_m\}_1^{20}$.

(e) Plot and compare the estimates with the theoretically obtained values.

4.23 Determine the coefficients of a PZ(2, 1) model with autocorrelation values $r_h(0) = 19, r_h(1) = 9, r_h(2) = -5$, and $r_h(3) = -7$.

4.24 (a) Show that the impulse response of an AZ(Q) model can be recovered from its response $\tilde{h}(n)$ to a periodic train with period L if $L > Q$.

(b) Show that the ACS of an AZ(Q) model can be recovered from the ACS or spectrum of $\tilde{h}(n)$ if $L \geq 2Q + 1$.

4.25 Prove Equation (4.3.17) and illustrate its validity by computing the PACS of the model $H(z) = 1 - 0.8z^{-1}$.

4.26 Prove Equations (4.3.24) that describe the minimum-phase region of the AZ(2) model.

4.27 Consider an AZ(2) model with $d_0 = 2$ and zeros $z_{1,2} = 0.95e^{\pm j\pi/3}$.

(a) Compute and plot $N = 100$ output samples by exciting the model with the process $w(n) \sim$ WGN(0, 1).

(b) Compute and plot the ACS, PACS, and spectrum of the model.

(c) Repeat parts (a) and (b) by assuming that we have an AP(2) model with poles at $p_{1,2} = 0.95e^{\pm j\pi/3}$.

(d) Investigate the duality between the ACS and PACS of the two models.

4.28 Prove Equations (4.4.31) and use them to reproduce the plot shown in Figure 4.12(b). Indicate which equation corresponds to each curve.

4.29 Determine the spectral flatness measure of the following processes:

(a) $x(n) = a_1 x(n-1) + a_2 x(n-2) + w(n)$ and

(b) $x(n) = w(n) + b_1 w(n-1) + b_2 w(n-2)$, where $w(n)$ is a white noise sequence.

4.30 Consider a zero-mean wide-sense stationary (WSS) process $x(n)$ with PSD $R_x(e^{j\omega})$ and an $M \times M$ correlation matrix with eigenvalues $\{\lambda_k\}_1^M$. Szegö's theorem (Grenander and Szegö 1958) states that if $g(\cdot)$ is a continuous function, then

$$\lim_{M \to \infty} \frac{g(\lambda_1) + g(\lambda_2) + \cdots + g(\lambda_M)}{M} = \frac{1}{2\pi} \int_{-\pi}^{\pi} g[R_x(e^{j\omega})] \, d\omega$$

Using this theorem, show that

$$\lim_{M \to \infty} (\det \mathbf{R}_x)^{1/M} = \exp\left\{ \frac{1}{2\pi} \int_{-\pi}^{\pi} \ln[R_x(e^{j\omega})] \, d\omega \right\}$$

4.31 Consider two linear random processes with system functions

$$\text{(i)} \ \ H(z) = \frac{1 - 0.81z^{-1} - 0.4z^{-2}}{(1 - z^{-1})^2} \quad \text{and} \quad \text{(ii)} \ \ H(z) = \frac{1 - 0.5z^{-1}}{1 - z^{-1}}$$

(a) Find a difference equation that leads to a numerically stable simulation of each process.

(b) Generate and plot 100 samples from each process, and look for indications of nonstationarity in the obtained records.

(c) Compute and plot the second difference of (i) and the first difference of (ii). Comment about the stationarity of the obtained records.

4.32 Generate and plot 100 samples for each of the linear processes with system functions

(a) $H(z) = \dfrac{1}{(1 - z^{-1})(1 - 0.9z^{-1})}$

(b) $H(z) = \dfrac{1 - 0.5z^{-1}}{(1 - z^{-1})(1 - 0.9z^{-1})}$

and then estimate and examine the values of the ACS $\{\hat{\rho}(l)\}_0^{20}$ and the PACS $\{\hat{k}_m\}_1^{20}$.

4.33 Consider the process $y(n) = d_0 + d_1 n + d_2 n^2 + x(n)$, where $x(n)$ is a stationary process with known autocorrelation $r_x(l)$.

    (a) Show that the process $y^{(2)}(n)$ obtained by passing $y(n)$ through the filter $H(z) = (1-z^{-1})^2$ is stationary.

    (b) Express the autocorrelation $r_y^{(2)}(l)$ of $y^{(2)}(n)$ in terms of $r_x(l)$. *Note:* This process is used in practice to remove quadratic trends from data before further analysis.

4.34 Prove Equation (4.6.7), which computes the cepstrum of an AP model from its coefficients.

4.35 Consider a minimum-phase AZ(Q) model $D(z) = \sum_{k=0}^{Q} d_k z^{-k}$ with complex cepstrum $c(k)$. We create another AZ model with coefficients $\tilde{d}_k = \alpha^k d_k$ and complex cepstrum $\tilde{c}(k)$.

    (a) If $0 < \alpha < 1$, find the relation between $\tilde{c}(k)$ and $c(k)$.
    (b) Choose α so that the new model has no minimum phase.
    (c) Choose α so that the new model has a maximum phase.

4.36 Prove Equation (4.6.27), which determines the cepstral distance in the frequency and time domains.

CHAPTER 5

Nonparametric Power Spectrum Estimation

The essence of frequency analysis is the representation of a signal as a superposition of sinusoidal components. In theory, the exact form of this decomposition (spectrum) depends on the assumed signal model. In Chapters 2 and 3 we discussed the mathematical tools required to define and compute the spectrum of signals described by deterministic and stochastic models, respectively. In practical applications, where only a finite segment of a signal is available, we cannot obtain a complete description of the adopted signal model. Therefore, we can only compute an approximation (estimate) of the spectrum of the adopted signal model ("true" or theoretical spectrum). The quality of the estimated spectrum depends on

- How well the assumed signal model represents the data.
- What values we assign to the unavailable signal samples.
- Which spectrum estimation method we use.

Clearly, meaningful application of spectrum estimation in practical problems requires sufficient a priori information, understanding of the signal generation process, knowledge of theoretical concepts, and experience.

In this chapter we discuss the most widely used correlation and spectrum estimation methods, as well as their properties, implementation, and application to practical problems. We discuss only *nonparametric* techniques that do *not* assume a particular functional form, but allow the form of the estimator to be determined *entirely* by the data. These methods are based on the discrete Fourier transform of either the signal segment or its autocorrelation sequence. In contrast, parametric methods assume that the available signal segment has been generated by a specific parametric model (e.g., a pole-zero or harmonic model). Since the choice of an inappropriate signal model will lead to erroneous results, the successful application of parametric techniques, without sufficient a priori information, is very difficult in practice. These methods are discussed in Chapter 9.

We begin this chapter with an introductory discussion on the purpose of, and the DSP approach to, spectrum estimation. We explore various errors involved in the estimation of finite-length data records (i.e., based on partial information). We also outline conventional techniques for deterministic signals, using concepts developed in Chapter 2. Also in Section 3.6, we presented important concepts and results from the estimation theory that are used extensively in this chapter. Section 5.3 is the main section of this chapter in which we discuss various nonparametric approaches to the power spectrum estimation of stationary random signals. This analysis is extended to joint stationary (bivariate) random signals for the computation of the cross-spectrum in Section 5.4. The computation of auto and cross-spectra using Thomson's multiple windows (or multitapers) is discussed in Section

5.5. Finally, in Section 5.6 we summarize important topics and concepts from this chapter. A classification of the various spectral estimation methods that are discussed in this book is provided in Figure 5.1.

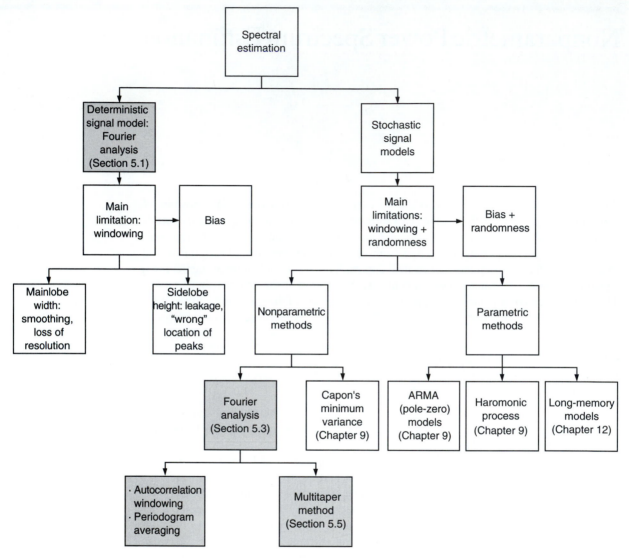

FIGURE 5.1
Classification of various spectrum estimation methods.

5.1 SPECTRAL ANALYSIS OF DETERMINISTIC SIGNALS

If we adopt a deterministic signal model, the mathematical tools for spectral analysis are the Fourier series and the Fourier transforms summarized in Section 2.2.1. It should be stressed at this point that applying any of these tools requires that the signal values in the entire time interval from $-\infty$ to $+\infty$ be available. If it is known a priori that a signal is periodic, then only one period is needed. The rationale for defining and studying various spectra for deterministic signals is threefold. First, we note that every realization (or sample function) of a stochastic process is a deterministic function. Thus we can use the Fourier series and transforms to compute a spectrum for stationary processes. Second, deterministic functions and sequences are used in many aspects of the study of stationary processes, for example, the autocorrelation sequence, which is a deterministic sequence. Third, the various spectra that can be defined for deterministic signals can be used to summarize important features of stationary processes.

Most practical applications of spectrum estimation involve continuous-time signals. For example, in speech analysis we use spectrum estimation to determine the pitch of the glottal excitation and the formants of the vocal tract (Rabiner and Schafer 1978). In electroencephalography, we use spectrum estimation to study sleep disorders and the effect of medication on the functioning of the brain (Duffy, Iyer, and Surwillo 1989). Another application is in Doppler radar, where the frequency shift between the transmitted and the received waveform is used to determine the radial velocity of the target (Levanon 1988).

The numerical computation of the spectrum of a continuous-time signal involves three steps:

1. Sampling the continuous-time signal to obtain a sequence of samples.
2. Collecting a finite number of contiguous samples (data segment or block) to use for the computation of the spectrum. This operation, which usually includes weighting of the signal samples, is known as *windowing,* or *tapering.*
3. Computing the values of the spectrum at the desired set of frequencies. This step is usually implemented using some efficient implementation of the DFT.

The above processing steps, which are necessary for DFT-based spectrum estimation, are shown in Figure 5.2. The continuous-time signal is first processed through a low-pass (antialiasing) filter and then sampled to obtain a discrete-time signal. Data samples of frame length N with *frame overlap* N_0 are selected and then conditioned using a window. Finally, a suitable-length DFT of the windowed data is taken as an estimate of its spectrum, which is then analyzed. In this section, we discuss in detail the effects of each of these operations on the accuracy of the computed spectrum. The understanding of the implications of these effects is very important in all practical applications of spectrum estimation.

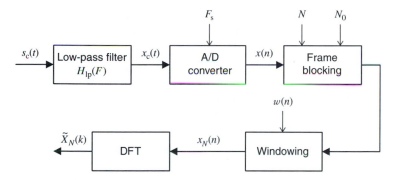

FIGURE 5.2
DFT-based Fourier analysis system for continuous-time signals.

5.1.1 Effect of Signal Sampling

The continuous-time signal $s_c(t)$, whose spectrum we seek to estimate, is first passed through a low-pass filter, also known as an *antialiasing* filter $H_{lp}(F)$, in order to minimize the aliasing error after sampling. The antialiased signal $x_c(t)$ is then sampled through an analog-to-digital converter[†] (ADC) to produce the discrete-time sequence $x(n)$, that is,

$$x(n) = x_c(t)|_{t=n/F_s} \qquad (5.1.1)$$

[†]We will ignore the quantization of discrete-time signals as discussed in Chapter 2.

From the sampling theorem in Section 2.2.2, we have

$$X(e^{j2\pi F/F_s}) = F_s \sum_{l=-\infty}^{\infty} X_c(F - lF_s) \tag{5.1.2}$$

where $X_c(F) = H_{lp}(F)S_c(F)$. We note that the spectrum of the discrete-time signal $x(n)$ is a periodic replication of $X_c(F)$. Overlapping of the replicas $X_c(F - lF_s)$ results in aliasing. Since any practical antialiasing filter does not have infinite attenuation in the stopband, some nonzero overlap of frequencies higher than $F_s/2$ should be expected within the band of frequencies of interest in $x(n)$. These aliased frequencies give rise to the aliasing error, which, in any practical signal, is unavoidable. It can be made negligible by a properly designed antialiasing filter $H_{lp}(F)$.

5.1.2 Windowing, Periodic Extension, and Extrapolation

In practice, we compute the spectrum of a signal by using a finite-duration segment. The reason is threefold:

1. The spectral composition of the signal changes with time.   or
2. We have only a finite set of data at our disposal.   or
3. We wish to keep the computational complexity to an acceptable level.

Therefore, it is necessary to partition $x(n)$ into blocks (or *frames*) of data prior to processing. This operation is called *frame blocking*, and it is characterized by two parameters: the *length* of frame N and the *overlap* between frames N_0 (see Figure 5.2). Therefore, the central problem in practical frequency analysis can be stated as follows:

> Determine the spectrum of a signal $x(n)$, $-\infty < n < \infty$, from its values in a finite interval $0 \leq n \leq N - 1$, that is, from a finite-duration segment.

Since $x(n)$ is unknown for $n < 0$ and $n \geq N$, we cannot say, without having sufficient a priori information, whether the signal is periodic or aperiodic. If we can reasonably assume that the signal is periodic with fundamental period N, we can easily determine its spectrum by computing its Fourier series, using the DFT (see Section 2.2.1).

However, in most practical applications, we cannot make this assumption because the available block of data could be either part of the period of a periodic signal or a segment from an aperiodic signal. In such cases, the spectrum of the signal *cannot* be determined without *assigning* values to the signal samples outside the available interval. There are three ways to deal with this issue:

1. *Periodic extension.* We assume that $x(n)$ is periodic with period N, that is, $x(n) = x(n + N)$ for all n, and we compute its Fourier series, using the DFT.
2. *Windowing.* We assume that the signal is zero outside the interval of observation, that is, $x(n) = 0$ for $n < 0$ and $n \geq N$. This is equivalent to multiplying the signal with the rectangular window

$$w_R(n) \triangleq \begin{cases} 1 & 0 \leq n \leq N - 1 \\ 0 & \text{elsewhere} \end{cases} \tag{5.1.3}$$

   The resulting sequence is aperiodic, and its spectrum is obtained by the discrete-time Fourier transform (DTFT).
3. *Extrapolation.* We use a priori information about the signal to extrapolate (i.e., determine its values for $n < 0$ and $n \geq N$) outside the available interval and then determine its spectrum by using the DTFT.

Periodic extension and windowing can be considered the simplest forms of extrapolation. It should be obvious that a successful extrapolation results in better spectrum estimates than periodic extension or windowing. Periodic extension is a straightforward application of the DFT, whereas extrapolation requires some form of a sophisticated signal model. As we shall see, most of the signal modeling techniques discussed in this book result in some kind of extrapolation. We first discuss, in the next section, the effect of spectrum sampling as imposed by the application of DFT (and its side effect—the periodic extension) before we provide a detailed analysis of the effect of windowing.

5.1.3 Effect of Spectrum Sampling

In many real-time spectrum analyzers, as illustrated in Figure 5.2, the spectrum is computed (after signal conditioning) by using the DFT. From Section 2.2.3, we note that this computation samples the continuous spectrum at equispaced frequencies. Theoretically, if the number of DFT samples is greater than or equal to the frame length N, then the exact continuous spectrum (based on the given frame) can be obtained by using the frequency-domain reconstruction (Oppenheim and Schafer 1989; Proakis and Manolakis 1996). This reconstruction, which requires a periodic sinc function [defined in (5.1.9)], is not a practical function to implement, especially in real-time applications. Hence a simple linear interpolation is used for plotting or display purposes. This linear interpolation can lead to misleading results even though the computed DFT sample values are correct. It is possible that there may not be a DFT sample precisely at a frequency where a peak of the DTFT is located. In other words, the DFT spectrum misses this peak, and the resulting linearly interpolated spectrum provides the wrong location and height of the DTFT spectrum peak. This error can be made smaller by sampling the DTFT spectrum at a finer grid, that is, by increasing the size of the DFT. The denser spectrum sampling is implemented by an operation called *zero padding* and is discussed later in this section.

Another effect of the application of DFT for spectrum calculations is the periodic extension of the sequence in the time domain. From our discussion in Section 2.2.3, it follows that the N-point DFT

$$\tilde{X}(k) = \sum_{n=0}^{N-1} x(n)e^{-j(2\pi/N)kn} \tag{5.1.4}$$

is periodic with period N. This should be expected given the relationship of the DFT to the Fourier transform or the Fourier series of discrete-time signals, which are periodic in ω with period 2π. A careful look at the inverse DFT

$$x(n) = \frac{1}{N} \sum_{k=0}^{N-1} \tilde{X}(k)e^{j(2\pi/N)kn} \tag{5.1.5}$$

reveals that $x(n)$ is also periodic with period N. This is a somewhat surprising result since no assumption about the signal $x(n)$ outside the interval $0 \le n \le N - 1$ has been made. However, this periodicity in the time domain can be easily justified by recalling that sampling in the time domain results in a periodicity in the frequency domain, and vice versa.

To understand these effects of spectrum sampling, consider the following example in which a continuous-time sinusoidal signal is sampled and then is truncated by a rectangular window before its DFT is performed.

EXAMPLE 5.1.1. A continuous-time signal $x_c(t) = 2 \cos 2\pi t$ is sampled with a sampling frequency of $F_s = 1/T = 10$ samples per second, to obtain the sequence $x(n)$. It is windowed by an N-point rectangular window $w_R(n)$ to obtain the sequence $x_N(n)$. Determine and plot $|\tilde{X}_N(k)|$, the magnitude of the DFT of $x_N(n)$, for (a) $N = 10$ and (b) $N = 15$. Comment on the shapes of these plots.

Solution. The discrete-time signal $x(n)$ is a sampled version of $x_c(t)$ and is given by

$$x(n) = x_c(t = nT) = 2\cos\frac{2\pi n}{F_s} = 2\cos 0.2\pi n \qquad T = 0.1 \text{ s}$$

Then, $x(n)$ is a periodic sequence with fundamental period $N = 10$.

a. For $N = 10$, we obtain $x_N(n) = 2\cos 0.4\pi n, 0 \leq n \leq 9$, which contains one period of $x(n)$. The periodic extension of $x_N(n)$ and the magnitude plot of its DFT are shown in the top row of Figure 5.3. For comparison, the DTFT $X_N(e^{j\omega})$ of $x_N(n)$ is also superimposed on the DFT samples. We observe that the DFT has only two nonzero samples, which together constitute the correct frequency of the analog signal $x_c(t)$. The DTFT has a mainlobe and several sidelobes due to the windowing effect. However, the DFT samples the sidelobes at their zero values, as illustrated in the DFT plot. Another explanation for this behavior is that since the samples in $x_N(n)$ for $N = 10$ constitute one full period of $\cos 0.4\pi n$, the 10-point periodic extension of $x_N(n)$, shown in the top left graph of Figure 5.3, results in the original sinusoidal sequences $x(n)$. Thus what the DFT "sees" is the exact sampled signal $x_c(t)$. In this case, the choice of N is a desirable one.

b. For $N = 15$, we obtain $x_N(n) = 2\cos 0.4\pi n, 0 \leq n \leq 14$, which contains $1\frac{1}{2}$ periods of $x(n)$. The periodic extension of $x_N(n)$ and the magnitude plot of its DFT are shown in the bottom row of Figure 5.3. Once again for comparison, the DTFT $X_N(e^{j\omega})$ of $x_N(n)$ is superimposed on the DFT samples. In this case, the DFT plot looks markedly different

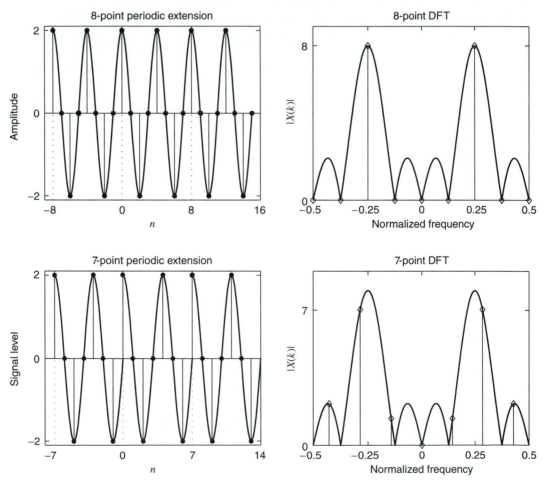

FIGURE 5.3
Effect of window length L on the DFT spectrum shape.

from that for $N = 10$ although the DTFT plot appears to be similar. In this case, the DFT does not sample two peaks at the exact frequencies; hence if the resulting DFT samples are joined by the linear interpolation, then we will get a misleading result. Since the sequence $x_N(n)$ does not contain full periods of $\cos 0.4\pi n$, the periodic extension of $x_N(n)$ contains discontinuities at $n = lN, l = 0, \pm1, \pm2, \ldots$, as shown in the bottom left graph of Figure 5.3. This discontinuity results in higher-order harmonics in the DFT values. The DTFT plot also has mainlobes and sidelobes, but the DFT samples these sidelobes at nonzero values. Therefore, the length of the window is an important consideration in spectrum estimation. The sidelobes are the source of the problem of leakage that gives rise to bias in the spectral values, as we will see in the following section. The suppression of the sidelobes is controlled by the window shape, which is another important consideration in spectrum estimation.

A quantitative description of the above interpretations and arguments related to the capacities and limitations of the DFT is offered by the following result (see Proakis and Manolakis 1996).

THEOREM 5.1 (DFT SAMPLING THEOREM). Let $x_c(t), -\infty < t < \infty$, be a continuous-time signal with Fourier transform $X_c(F), -\infty < F < \infty$. Then, the N-point sequences $\{Tx_p(n), 0 \le n \le N-1\}$ and $\{\tilde{X}_p(k), 0 \le k \le N-1\}$ form an N-point DFT pair, that is,

$$x_p(n) \triangleq \sum_{m=-\infty}^{\infty} x_c(nT - mNT) \xrightarrow[N]{\text{DFT}} \tilde{X}_p(k) \triangleq F_s \sum_{l=-\infty}^{\infty} X_c\left(k\frac{F_s}{N} - lF_s\right) \qquad (5.1.6)$$

where $F_s = 1/T$ is the sampling frequency.

Proof. The proof is explored in Problem 5.1.

Thus, given a continuous-time signal $x_c(t)$ and its spectrum $X_c(F)$, we can create a DFT pair by sampling and aliasing in the time and frequency domains. Obviously, this DFT pair provides a "faithful" description of $x_c(t)$ and $X_c(F)$ if both the time-domain aliasing and the frequency-domain aliasing are insignificant. The meaning of relation (5.1.6) is graphically illustrated in Figure 5.4. In this figure, we show the time-domain signals in the left column and their Fourier transforms in the right column. The top row contains continuous-time signals, which are shown as nonperiodic and of infinite extent in both domains, since many real-world signals exhibit this behavior. The middle row contains the sampled version of the continuous-time signal and its periodic Fourier transform (the nonperiodic transform is shown as a dashed curve). Clearly, aliasing in the frequency domain is evident. Finally, the bottom row shows the sampled (periodic) Fourier transform and its corresponding time-domain periodic sequence. Again, aliasing in the time domain should be expected. Thus we have sampled and periodic signals in both domains with the certainty of aliasing one domain and the possibility in both domains. This figure should be recalled any time we use the DFT for the analysis of sampled signals.

Zero padding

The N-point DFT values of an N-point sequence $x(n)$ are samples of the DTFT $X(e^{j\omega})$, as discussed in Chapter 2. These samples can be used to reconstruct the DTFT $X(e^{j\omega})$ by using the periodic sinc interpolating function. Alternatively, one can obtain more (i.e., dense) samples of the DTFT by computing a larger N_{FFT}-point DFT of $x(n)$, where $N_{FFT} \gg N$. Since the number of samples of $x(n)$ is fixed, the only way we can treat $x(n)$ as an N_{FFT}-point sequence is by appending $N_{FFT} - N$ zeros to it. This procedure is called the *zero padding* operation, and it is used for many purposes including the augmentation of the sequence length so that a power-of-2 FFT algorithm can be used. In spectrum estimation, zero padding is primarily used to provide a *better-looking* plot of the spectrum of a finite-length sequence. This is shown in Figure 5.5 where the magnitude of an N_{FFT}-point DFT of the eight-point sequence $x(n) = \cos(2\pi n/4)$ is plotted for $N_{FFT} = 8, 16, 32$, and 64. The DTFT magnitude $|X(e^{j\omega})|$ is also shown for comparison. It can be seen that as more zeros are appended (by increasing N_{FFT}), the resulting larger-point DFT provides more closely spaced samples of the DTFT, thus giving a better-looking plot. Note, however, that the zero padding *does not increase* the resolution of the spectrum; that is, there are no new peaks and valleys in the display, just a better display of the available information. This type of plot is called a *high-density spectrum*. For a *high-resolution spectrum*, we have to collect more information by increasing N. The DTFT plots shown in Figures 5.3 and 5.5 were obtained by using a very large amount of zero padding.

5.1.4 Effects of Windowing: Leakage and Loss of Resolution

To see the effect of the window on the spectrum of an arbitrary deterministic signal $x(n)$, defined over the entire range $-\infty < n < \infty$, we notice that the available data record can be expressed as

$$x_N(n) = x(n)w_{\text{R}}(n) \tag{5.1.7}$$

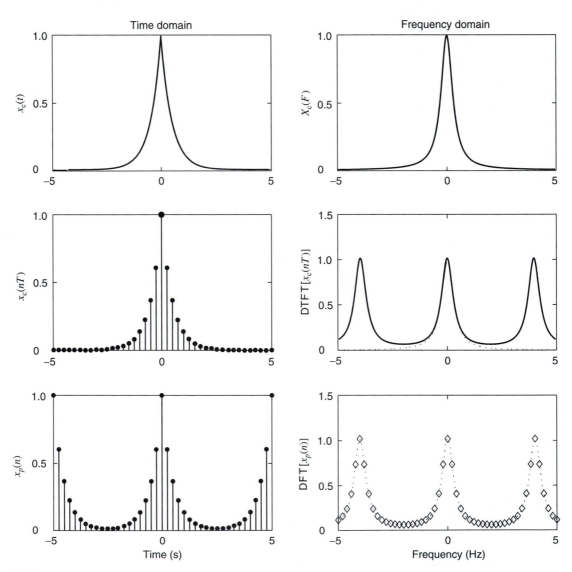

FIGURE 5.4
Graphical illustration of the DFT sampling theorem.

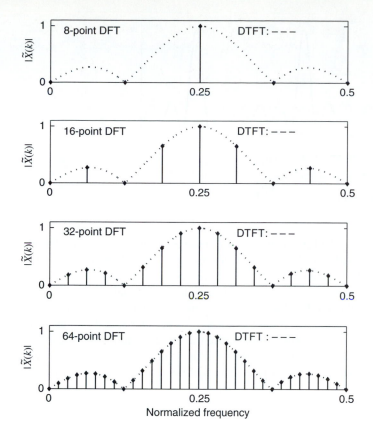

FIGURE 5.5
Effect of zero padding.

where $w_R(n)$ is the rectangular window defined in (5.1.3). Thus, a finite segment of the signal can be thought of as a product of the actual signal $x(n)$ and a *data window* $w(n)$. In (5.1.7), $w(n) = w_R(n)$, but $w(n)$ can be any arbitrary finite-duration sequence. The Fourier transform of $x_N(n)$ is

$$X_N(e^{j\omega}) = X(e^{j\omega}) \otimes W(e^{j\omega}) \triangleq \frac{1}{2\pi} \int_{-\pi}^{\pi} X(e^{j\theta}) W(e^{j(\omega-\theta)}) \, d\theta \qquad (5.1.8)$$

that is, $X_N(e^{j\omega})$ equals the periodic convolution of the actual Fourier transform with the Fourier transform $W(e^{j\omega})$ of the data window. For the rectangular window, $W(e^{j\omega}) = W_R(e^{j\omega})$, where

$$W_R(e^{j\omega}) = \left[\frac{\sin(\omega N/2)}{\sin(\omega/2)} \right] e^{-j\omega(N-1)/2} \triangleq A(\omega) e^{-j\omega(N-1)/2} \qquad (5.1.9)$$

The function $A(\omega)$ is a periodic function in ω with fundamental period equal to 2π and is called a *periodic sinc* function. Figure 5.6 shows three periods of $A(\omega)$ for $N = 11$. We note that $W_R(e^{j\omega})$ consists of a mainlobe (ML).

$$W_{ML}(e^{j\omega}) = \begin{cases} W_R(e^{j\omega}) & |\omega| < \dfrac{2\pi}{N} \\ 0 & \dfrac{2\pi}{N} < |\omega| \le \pi \end{cases} \qquad (5.1.10)$$

and the sidelobes $W_{SL}(e^{j\omega}) = W_R(e^{j\omega}) - W_{ML}(e^{j\omega})$. Thus, (5.1.8) can be written as

$$X_N(e^{j\omega}) = X(e^{j\omega}) \otimes W_{ML}(e^{j\omega}) + X(e^{j\omega}) \otimes W_{SL}(e^{j\omega}) \qquad (5.1.11)$$

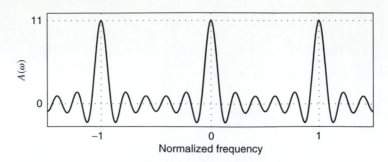

FIGURE 5.6
Plot of $A(\omega) = \sin(\omega N/2)/\sin(\omega/2)$ for $N = 11$.

The first convolution in (5.1.11) smoothes rapid variations and suppresses narrow peaks in $X(e^{j\omega})$, whereas the second convolution introduces ripples in smooth regions of $X(e^{j\omega})$ and can create "false" peaks. Therefore, the spectrum we observe is the convolution of the actual spectrum with the Fourier transform of the data window. The only way to improve the estimate is to increase the window length N or to choose another window shape. For the rectangular window, increasing N results in a narrower mainlobe, and the distortion is reduced. As $N \to \infty$, $W_R(e^{j\omega})$ tends to an impulse train with period 2π and $X_N(e^{j\omega})$ tends to $X(e^{j\omega})$, as expected. Since in practice the value of N is always finite, the only way to improve the estimate $X_N(e^{j\omega})$ is by properly choosing the shape of the window $w(n)$. The only restriction on $w(n)$ is that it be of finite duration.

It is known that any time-limited sequence $w(n)$ has a Fourier transform $W(e^{j\omega})$ that is nonzero except at a finite number of frequencies. Thus, from (5.1.8) we see that the estimated value $X_N(e^{j\omega_0})$ is computed by using all values of $X(e^{j\omega})$ weighted by $W(e^{j(\omega_0-\theta)})$. The contribution of the sinusoidal components with frequencies $\omega \neq \omega_0$ to the value $X_N(e^{j\omega_0})$ introduces an error known as *leakage*. As the name suggests, energy from one frequency range "leaks" into another, giving the wrong impression of stronger or weaker frequency components.

To illustrate the effect of the window shape and duration on the estimated spectrum, consider the signal

$$x(n) = \cos 0.35\pi n + \cos 0.4\pi n + 0.25 \cos 0.8\pi n \tag{5.1.12}$$

which has a line spectrum with lines at frequencies $\omega_1 = 0.35\pi$, $\omega_2 = 0.4\pi$, and $\omega_3 = 0.8\pi$. This line spectrum (normalized so that the magnitude is between 0 and 1) is shown in the top graph of Figure 5.7 over $0 \leq \omega \leq \pi$. The spectrum $X_N(e^{j\omega})$ of $x_N(n)$ using the rectangular window is given by

$$X_N(e^{j\omega}) = \tfrac{1}{2}[W(e^{j(\omega+\omega_1)}) + W(e^{j(\omega-\omega_1)}) + W(e^{j(\omega+\omega_2)}) + W(e^{j(\omega-\omega_2)}) \\ + 0.25W(e^{j(\omega+\omega_3)}) + 0.25W(e^{j(\omega-\omega_3)})] \tag{5.1.13}$$

The second and the third plots in Figure 5.7 show 2048-point DFTs of $x_N(n)$ for a rectangular data window with $N = 21$ and $N = 81$. We note that the ability to pick out peaks (*resolvability*) depends on the duration $N-1$ of the data window.[†] To resolve two spectral lines at $\omega = \omega_1$ and $\omega = \omega_2$ using a rectangular window, we should have the difference $|\omega_1 - \omega_2|$ greater than the mainlobe width $\Delta\omega$, which is approximately equal to $2\pi/(N-1)$, in radians per sampling interval, from the plot of $A(\omega)$ in Figure 5.6, that is,

$$|\omega_1 - \omega_2| > \Delta\omega \approx \frac{2\pi}{N-1} \quad \text{or} \quad N > \frac{2\pi}{|\omega_1 - \omega_2|} + 1$$

[†] Since there are N samples in a data window, the number of intervals or durations is $N-1$.

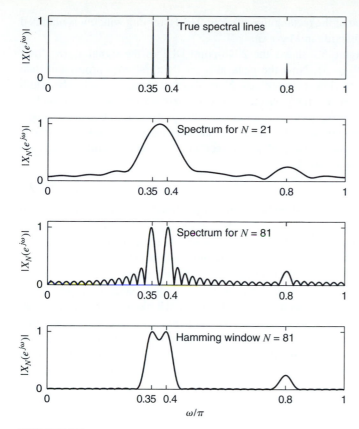

FIGURE 5.7
Spectrum of three sinusoids using rectangular and Hamming windows.

For a rectangular window of length N, the exact value of $\Delta\omega$ is equal to $1.81\pi/(N-1)$. If N is too small, the two peaks at $\omega = 0.35\pi$ and $\omega = 0.4\pi$ are fused into one, as shown in the $N = 21$ plot. When $N = 81$, the corresponding plot shows a resolvable separation; however, the peaks have shifted somewhat from their true locations. This is called *bias*, and it is a direct result of the leakage from sidelobes. In both cases, the peak at $\omega = 0.8\pi$ can be distinguished easily (but also has a bias).

Another important observation is that the sidelobes of the data window introduce false peaks. For a rectangular window, the peak sidelobe level is 13 dB below zero, which is not a good attenuation. Thus these false peaks have values that are comparable to that of the true peak at $\omega = 0.8\pi$, as shown in Figure 5.7. These peaks can be minimized by reducing the amplitudes of the sidelobes. The rectangular window cannot help in this regard because of Gibb's well-known phenomenon associated with it. We need a different window shape. However, any window other than the rectangular window has a wider mainlobe; hence this reduction can be achieved only at the expense of the resolution. To illustrate this, consider the Hamming (Hm) data window, given by

$$w_{\text{Hm}}(n) = \begin{cases} 0.54 - 0.46\cos\dfrac{2\pi n}{N-1} & 0 \le n \le N-1 \\ 0 & \text{otherwise} \end{cases} \qquad (5.1.14)$$

with the approximate width of the mainlobe equal to $8\pi/(N-1)$ and the exact mainlobe width equal to $6.27\pi/(N-1)$. The peak sidelobe level is 43 dB below zero, which is considerably better than that of the rectangular window. The Hamming window is obtained by using the `hamming(N)` function in MATLAB.

The bottom plot in Figure 5.7 shows the 2048-point DFT of the signal $x_N(n)$ for a Hamming window with $N = 81$. Now the peak at $\omega = 0.8\pi$ is more prominent than before, and the sidelobes are almost suppressed. Note also that since the mainlobe width of the Hamming window is wider, the peaks have a wider base—so much so that the first two frequencies are barely recognized. We can correct this problem by choosing a larger window length. This interplay between the shape and the duration of a window function is one of the important issues and, as we will see in Section 5.3, produces similar effects in the spectral analysis of random signals.

Some useful windows

The design of windows for spectral analysis applications has drawn a lot of attention and is examined in detail in Harris (1978). We have already discussed two windows, namely, the rectangular and the Hamming window. Another useful window in spectrum analysis is due to Hann and is mistakenly known as the Hanning window. There are several such windows with varying degrees of tradeoff between resolution (mainlobe width) and leakage (peak sidelobe level). These windows are known as *fixed* windows since each provides a fixed amount of leakage that is independent of the length N. Unlike fixed windows, there are windows that contain a design parameter that can be used to trade between resolution and leakage. Two such windows are the Kaiser window and the Dolph-Chebyshev window, which are widely used in spectrum estimation. Figure 5.8 shows the time-domain window functions and their corresponding frequency-domain log-magnitude plots in decibels for these five windows. The important properties such as peak sidelobe level and mainlobe width of these windows are compared in Table 5.1.

TABLE 5.1

Comparison of properties of commonly used windows. Each window is assumed to be of length N.

Window type	Peak sidelobe level (dB)	Approximate mainlobe width	Exact mainlode width
Rectangular	-13	$\dfrac{4\pi}{N-1}$	$\dfrac{1.81\pi}{N-1}$
Hanning	-32	$\dfrac{8\pi}{N-1}$	$\dfrac{5.01\pi}{N-1}$
Hamming	-43	$\dfrac{8\pi}{N-1}$	$\dfrac{6.27\pi}{N-1}$
Kaiser	$-A$	—	$\dfrac{A-8}{2.285N-1}$
Dolph-Chebyshev	$-A$	—	$\cos^{-1}\left[\left(\cosh\dfrac{\cosh^{-1}10^{A/20}}{N-1}\right)^{-1}\right]$

Hanning window. This window is given by the function

$$w_{\text{Hn}}(n) = \begin{cases} 0.5 - 0.5\cos\dfrac{2\pi n}{N-1} & 0 \le n \le N-1 \\ 0 & \text{otherwise} \end{cases} \tag{5.1.15}$$

which is a raised cosine function. The peak sidelobe level is 32 dB below zero, and the approximate mainlobe width is $8\pi/(N-1)$ while the exact mainlobe width is $5.01\pi/(N-1)$. In MATLAB this window function is obtained through the function `hanning(N)`.

Kaiser window. This window function is due to J. F. Kaiser and is given by

$$
w_K(n) = \begin{cases} \dfrac{I_0\left\{\beta\sqrt{1 - [1 - 2n/(N-1)]^2}\right\}}{I_0(\beta)} & 0 \le n \le N-1 \\ 0 & \text{otherwise} \end{cases} \qquad (5.1.16)
$$

where $I_0(\cdot)$ is the modified zero-order Bessel function of the first kind and β is a window shape parameter that can be chosen to obtain various peak sidelobe levels and the

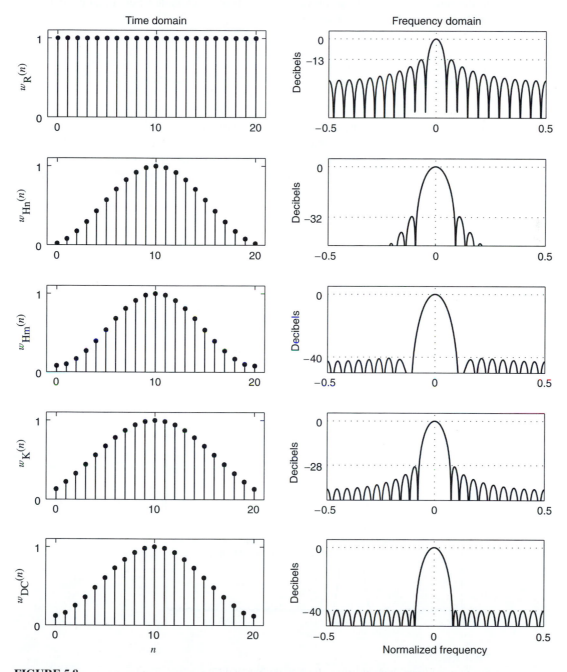

FIGURE 5.8
Time-domain window functions and their frequency-domain characteristics for rectangular, Hanning, Hamming, Kaiser, and Dolph-Chebyshev windows.

corresponding mainlobe widths. Clearly, $\beta = 0$ results in the rectangular window while $\beta > 0$ results in lower sidelobe leakage at the expense of a wider mainlobe. Kaiser has developed approximate design equations for β. Given a peak sidelobe level of A dB below the peak value, the approximate value of β is given by

$$\beta \simeq \begin{cases} 0 & A \le 21 \\ 0.5842(A-21)^{0.4} + 0.07886(A-21) & 21 < A \le 50 \\ 0.1102(A-8.7) & A > 50 \end{cases} \tag{5.1.17}$$

Furthermore, to achieve the given values of the peak sidelobe level of A and the mainlobe width $\Delta\omega$, the length N must satisfy

$$\Delta\omega = \frac{A-8}{2.285(N-1)} \tag{5.1.18}$$

In MATLAB this window is given by the function kaiser(N,beta).

Dolph-Chebyshev window. This window is characterized by the property that the peak sidelobe levels are constant; that is, it has an "equiripple" behavior. The window $w_{\mathrm{DC}}(n)$ is obtained as the inverse DFT of the Chebyshev polynomial evaluated at N equally spaced frequencies around the unit circle. The details of this window function computation are available in Harris (1978). The parameters of the Dolph-Chebyshev window are the constant sidelobe level A in decibels, the window length N, and the mainlobe width $\Delta\omega$. However, only two of the three parameters can be independently specified. In spectrum estimation, parameters N and A are generally specified. Then $\Delta\omega$ is given by

$$\Delta\omega = \cos^{-1}\left[\left(\cosh\frac{\cosh^{-1} 10^{A/20}}{N-1}\right)^{-1}\right] \tag{5.1.19}$$

In MATLAB this window is obtained through the function chebwin(N,A).

To illustrate the usefulness of these windows, consider the same signal containing three frequencies given in (5.1.12). Figure 5.9 shows the spectrum of $x_N(n)$ using the

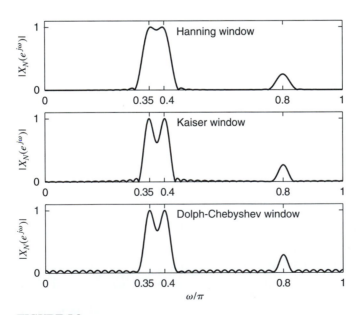

FIGURE 5.9
Spectrum of three sinusoids using Hanning, Kaiser, and Chebyshev windows.

Hanning, Kaiser, and Chebyshev windows for length $N = 81$. The Kaiser and Chebyshev window parameters are adjusted so that the peak sidelobe level is 40 dB or below. Clearly, these windows have suppressed sidelobes considerably compared to that of the rectangular window but the main peaks are wider with negligible bias. The two peaks in the Hanning window spectrum are barely resolved because the mainlobe width of this window is much wider than that of the rectangular window. The Chebyshev window spectrum has uniform sidelobes while the Kaiser window spectrum shows decreasing sidelobes away from the mainlobes.

5.1.5 Summary

In conclusion, the frequency analysis of deterministic signals requires a careful study of three important steps. First, the continuous-time signal $x_c(t)$ is sampled to obtain samples $x(n)$ that are collected into blocks or frames. The frames are "conditioned" to minimize certain errors by multiplying by a window sequence $w(n)$ of length N. Finally the windowed frames $x_N(n)$ are transformed to the frequency domain using the DFT. The resulting DFT spectrum $\tilde{X}_N(k)$ is a *faithful* replica of the actual spectrum $X_c(F)$ if the following errors are sufficiently small.

Aliasing error. This is an error due to the sampling operation. If the sampling rate is sufficiently high and if the antialiasing filter is properly designed so that most of the frequencies of interest are represented in $x(n)$, then this error can be made smaller. However, a certain amount of aliasing should be expected. The sampling principle and aliasing are discussed in Section 2.2.2.

Errors due to finite-length window. There are several errors such as resolution loss, bias, and leakage that are attributed to the windowing operation. Therefore, a careful design of the window function and its length is necessary to minimize these errors. These topics were discussed in Section 5.1.4. In Table 5.1 we summarize key properties of five windows discussed in this section that are useful for spectrum estimation.

Spectrum reconstruction error. The DFT spectrum $\tilde{X}_N(k)$ is a number sequence that must be reconstructed into a continuous function for the purpose of plotting. A practical choice for this reconstruction is the first-order polynomial interpolation. This reconstruction error can be made smaller (and in fact comparable to the screen resolution) by choosing a large number of frequency samples, which can be achieved by the *zero padding* operation in the DFT. It was discussed in Section 5.1.3.

With the understanding of frequency analysis concepts developed in this section, we are now ready to tackle the problem of spectral analysis of stationary random signals. From Chapter 3, we recognize that the true spectral values can only be obtained as estimates. This requires some understanding of key concepts from estimation theory, which is developed in Section 3.6.

5.2 ESTIMATION OF THE AUTOCORRELATION OF STATIONARY RANDOM SIGNALS

The second-order moments of a stationary random sequence—that is, the mean value μ_x, the autocorrelation sequence $r_x(l)$, and the PSD $R_x(e^{j\omega})$—play a crucial role in signal analysis and signal modeling. In this section, we discuss the estimation of the autocorrelation sequence $r_x(l)$ using a finite data record $\{x(n)\}_0^{N-1}$ of the process.

For a stationary process $x(n)$, the most widely used estimator of $r_x(l)$ is given by the *sample autocorrelation sequence*

$$\hat{r}_x(l) \triangleq \begin{cases} \dfrac{1}{N} \displaystyle\sum_{n=0}^{N-l-1} x(n+l)x^*(n) & 0 \le l \le N-1 \\ \hat{r}_x^*(-l) & -(N-1) \le l < 0 \\ 0 & \text{elsewhere} \end{cases} \tag{5.2.1}$$

or, equivalently,

$$\hat{r}_x(l) \triangleq \begin{cases} \dfrac{1}{N} \displaystyle\sum_{n=l}^{N-1} x(n)x^*(n-l) & 0 \le l \le N-1 \\ \hat{r}_x^*(-l) & -(N-1) \le l < 0 \\ 0 & \text{elsewhere} \end{cases} \tag{5.2.2}$$

which is a random sequence. Note that without further information beyond the observed data $\{x(n)\}_0^{N-1}$, it is not possible to provide reasonable estimates of $r_x(l)$ for $|l| \ge N$. Even for lag values $|l|$ close to N, the correlation estimates are unreliable since very few $x(n+|l|)x(n)$ pairs are used. A good rule of thumb provided by Box and Jenkins (1976) is that N should be at least 50 and that $|l| \le N/4$. The sample autocorrelation $\hat{r}_x(l)$ given in (5.2.1) has a desirable property that for each $l \ge 1$, the sample autocorrelation matrix

$$\hat{\mathbf{R}}_x = \begin{bmatrix} \hat{r}_x(0) & \hat{r}_x^*(1) & \cdots & \hat{r}_x^*(N-1) \\ \hat{r}_x(1) & \hat{r}_x(0) & \cdots & \hat{r}_x^*(N-2) \\ \vdots & \vdots & \ddots & \vdots \\ \hat{r}_x(N-1) & \hat{r}_x(N-2) & \cdots & \hat{r}_x(0) \end{bmatrix} \tag{5.2.3}$$

is nonnegative definite (see Section 3.5.1). This property is explored in Problem 5.5. MATLAB provides functions to compute the correlation matrix $\hat{\mathbf{R}}_x$ (for example, corr), given the data $\{x(n)\}_{n=0}^{N-1}$; however, the book toolbox function rx = autoc(x,L); computes $\hat{r}_x(l)$ according to (5.2.1) very efficiently.

The estimate of covariance $\gamma_x(l)$ from the data record $\{x(n)\}_0^{N-1}$ is given by the sample autocovariance sequence

$$\hat{\gamma}_x(l) = \begin{cases} \dfrac{1}{N} \displaystyle\sum_{n=0}^{N-l-1} [x(n+l) - \hat{\mu}_x][x^*(n) - \hat{\mu}_x^*] & 0 \le l \le N-1 \\ \hat{\gamma}_x^*(-l) & -(N-1) \le l < 0 \\ 0 & \text{elsewhere} \end{cases} \tag{5.2.4}$$

so that the corresponding autocovariance matrix $\hat{\mathbf{\Gamma}}_x$ is nonnegative definite. Similarly, the sample autocorrelation coefficient sequence $\hat{\rho}_x(l)$ is given by

$$\hat{\rho}_x(l) = \frac{\hat{\gamma}_x(l)}{\hat{\sigma}_x^2} \tag{5.2.5}$$

In the rest of this section, we assume that $x(n)$ is a zero-mean process and hence $\hat{r}_x(l) = \hat{\gamma}_x(l)$, so that we can discuss the autocorrelation estimate in detail.

To determine the statistical quality of this estimator, we now consider its mean and variance.

Mean of $\hat{r}_x(l)$. We first note that (5.2.1) can be written as

$$\hat{r}_x(l) = \frac{1}{N} \sum_{n=-\infty}^{\infty} x(n+l)w(n+l)x^*(n)w(n) \qquad |l| \ge 0 \tag{5.2.6}$$

where
$$w(n) = w_R(n) = \begin{cases} 1 & 0 \le n \le N-1 \\ 0 & \text{elsewhere} \end{cases} \tag{5.2.7}$$

is the rectangular window. The expected value of $\hat{r}_x(l)$ is

$$E\{\hat{r}_x(l)\} = \frac{1}{N} \sum_{n=-\infty}^{\infty} E\{x(n+l)x^*(n)\}w(n+l)w(n) \qquad l \ge 0$$

and
$$E\{\hat{r}_x(-l)\} = E\{\hat{r}_x^*(l)\} \qquad -l \le 0$$

Therefore
$$E\{\hat{r}_x(l)\} = \frac{1}{N} r_x(l) r_w(l) \tag{5.2.8}$$

where
$$r_w(l) = w(l) * w(-l) = \sum_{n=-\infty}^{\infty} w(n)w(n+l) \tag{5.2.9}$$

is the autocorrelation of the window sequence. For the rectangular window

$$r_w(l) = w_B(n) \triangleq \begin{cases} N - |l| & |l| \le N-1 \\ 0 & \text{elsewhere} \end{cases} \tag{5.2.10}$$

which is the unnormalized triangular or Bartlett window. Thus

$$E\{\hat{r}_x(l)\} = \frac{1}{N} r_x(l) w_B(n) = r_x(l) \left(1 - \frac{|l|}{N}\right) w_R(n) \tag{5.2.11}$$

Therefore, we conclude that the relation (5.2.1) provides a *biased* estimate of $r_x(l)$ because the expected value of $\hat{r}_x(l)$ from (5.2.11) is not equal to the true autocorrelation $r_x(l)$. However, $\hat{r}_x(l)$ is an *asymptotically unbiased* estimator since if $N \to \infty$, $E\{\hat{r}_x(l)\} \to r_x(l)$. Clearly, the bias is small if $\hat{r}_x(l)$ is evaluated for $|l| \le L$, where L is the maximum desired lag and $L \ll N$.

Variance of $\hat{r}_x(l)$. An approximate expression for the covariance of $\hat{r}_x(l)$ is given by Jenkins and Watts (1968)

$$\text{cov}\{\hat{r}_x(l_1), \hat{r}_x(l_2)\} \simeq \frac{1}{N} \sum_{l=-\infty}^{\infty} [r_x(l)r_x(l+l_2-l_1) + r_x(l+l_2)r_x(l-l_1)] \tag{5.2.12}$$

This indicates that successive values of $\hat{r}_x(l)$ may be highly correlated and that $\hat{r}_x(l)$ may fail to die out even if it is expected to. This makes the interpretation of autocorrelation graphs quite challenging because we do not know whether the variation is real or statistical.

The variance of $\hat{r}_x(l)$, which can be obtained by setting $l_1 = l_2$ in (5.2.12), tends to zero as $N \to \infty$. Thus, $\hat{r}_x(l)$ provides a good estimate of $r_x(l)$ if the lag $|l|$ is much smaller than N. However, as $|l|$ approaches N, fewer and fewer samples of $x(n)$ are used to evaluate $\hat{r}_x(l)$. As a result, the estimate $\hat{r}_x(l)$ becomes worse and its variance increases.

Nonnegative definiteness of $\hat{r}_x(l)$. An alternative estimator for the autocorrelation sequence is given by

$$\check{r}_x(l) = \begin{cases} \dfrac{1}{N-l} \displaystyle\sum_{n=0}^{N-l-1} x(n+l)x^*(n) & 0 \le l \le L < N \\ \check{r}_x^*(-l) & -N < -L \le l < 0 \\ 0 & \text{elsewhere} \end{cases} \tag{5.2.13}$$

Although this estimator is unbiased, it is not used in spectral estimation because of its negative definiteness. In contrast, the estimator $\hat{r}_x(l)$ from (3.6.33) is nonnegative definite, and any spectral estimates based on it do not have any negative values. Furthermore, the estimator $\hat{r}_x(l)$ has smaller variance and mean square error than the estimator $\check{r}_x(l)$ (Jenkins and Watts 1968). Thus, in this book we use the estimator $\hat{r}_x(l)$ defined in (5.2.1).

5.3 ESTIMATION OF THE POWER SPECTRUM OF STATIONARY RANDOM SIGNALS

From a practical point of view, most stationary random processes have continuous spectra. However, harmonic processes (i.e., processes with line spectra) appear in several applications either alone or in mixed spectra (a mixture of continuous and line spectra). We first discuss the estimation of continuous spectra in detail. The estimation of line spectra is considered in Chapter 9.

The power spectral density of a zero-mean stationary stochastic process was defined in (3.3.44) as

$$R_x(e^{j\omega}) \triangleq \sum_{l=-\infty}^{\infty} r_x(l)e^{-j\omega l} \tag{5.3.1}$$

assuming that the autocorrelation sequence $r_x(l)$ is absolutely summable. We will deal with the problem of estimating the power spectrum $R_x(e^{j\omega})$ of a stationary process $x(n)$ from a finite record of observations $\{x(n)\}_0^{N-1}$ of a single realization. The ideal goal is to devise an estimate that will faithfully characterize the power-versus-frequency distribution of the stochastic process (i.e., all the sequences of the ensemble) using only a segment of a single realization. For this to be possible, the estimate should typically involve some kind of averaging among several realizations or along a single realization.

In some practical applications (e.g., interferometry), it is possible to directly measure the autocorrelation $r_x(l)$, $|l| \leq L < N$ with great accuracy. In this case, the spectrum estimation problem can be treated as a deterministic one, as described in Section 5.1. We will focus on the "stochastic" version of the problem, where $R_x(e^{j\omega})$ is estimated from the available data $\{x(n)\}_0^{N-1}$. A natural estimate of $R_x(e^{j\omega})$, suggested by (5.3.1), is to estimate $r_x(l)$ from the available data and then transform it by using (5.3.1).

5.3.1 Power Spectrum Estimation Using the Periodogram

The periodogram is an estimator of the power spectrum, introduced by Schuster (1898) in his efforts to search for hidden periodicities in solar sunspot data. The *periodogram* of the data segment $\{x(n)\}_0^{N-1}$ is defined by

$$\hat{R}_x(e^{j\omega}) \triangleq \frac{1}{N} \left| \sum_{n=0}^{N-1} v(n)e^{-j\omega n} \right|^2 = \frac{1}{N} |V(e^{j\omega})|^2 \tag{5.3.2}$$

where $V(e^{j\omega})$ is the DTFT of the windowed sequence

$$v(n) = x(n)w(n) \qquad 0 \leq n \leq N-1 \tag{5.3.3}$$

The above definition of the periodogram stems from Parseval's relation (2.2.10) on the power of a signal. The window $w(n)$, which has length N, is known as the *data window*. Usually, the term *periodogram* is used when $w(n)$ is a rectangular window. In contrast, the term *modified periodogram* is used to stress the use of nonrectangular windows. The values of the periodogram at the discrete set of frequencies $\{\omega_k = 2\pi k/N\}_0^{N-1}$ can be calculated by

$$\tilde{\hat{R}}_x(k) \triangleq \hat{R}_x(e^{j2\pi k/N}) = \frac{1}{N} |\tilde{V}(k)|^2 \qquad k = 0, 1, \ldots, N-1 \tag{5.3.4}$$

where $\tilde{V}(k)$ is the N-point DFT of the windowed segment $v(n)$. In MATLAB, the modified periodogram computation is implemented by using the function

```
Rx = psd(x,Nfft,Fs,window(N),'none');
```

where `window` is the name of any MATLAB-provided window function (e.g., hamming); `Nfft` is the size of the DFT, which is chosen to be larger than N to obtain a high-density spectrum (see zero padding in Section 5.1.1); and `Fs` is the sampling frequency, which is used for plotting purposes. If the window `boxcar` is used, then we obtain the periodogram estimate.

The periodogram can be expressed in terms of the autocorrelation estimate $\hat{r}_v(l)$ of the windowed sequence $v(n)$ as (see Problem 5.9)

$$\hat{R}_x(e^{j\omega}) = \sum_{l=-(N-1)}^{N-1} \hat{r}_v(l)\, e^{-j\omega l} \tag{5.3.5}$$

which shows that $\hat{R}_x(e^{j\omega})$ is a "natural" estimate of the power spectrum. From (5.3.2) it follows that $\hat{R}_x(e^{j\omega})$ is nonnegative for all frequencies ω. This results from the fact that the autocorrelation sequence $\hat{r}(l)$, $0 \le |l| \le N-1$, is nonnegative definite. If we use the estimate $\check{r}_x(l)$ from (5.2.13) in (5.3.5) instead of $\hat{r}_x(l)$, the obtained periodogram may assume negative values, which implies that $\check{r}_x(l)$ is not guaranteed to be nonnegative definite.

The inverse Fourier transform of $\hat{R}_x(e^{j\omega})$ provides the estimated autocorrelation $\hat{r}_v(l)$, that is,

$$\hat{r}_v(l) = \frac{1}{2\pi} \int_{-\pi}^{\pi} \hat{R}_x(e^{j\omega}) e^{j\omega l} \, d\omega \tag{5.3.6}$$

because $\hat{r}_v(l)$ and $\hat{R}_x(e^{j\omega})$ form a DTFT pair. Using (5.3.6) and (5.2.1) for $l=0$, we have

$$\hat{r}_v(0) = \frac{1}{N} \sum_{n=0}^{N-1} |v(n)|^2 = \frac{1}{2\pi} \int_{-\pi}^{\pi} \hat{R}_x(e^{j\omega}) \, d\omega \tag{5.3.7}$$

Thus, the periodogram $\hat{R}_x(e^{j\omega})$ shows how the power of the segment $\{v(n)\}_0^{N-1}$, which provides an estimate of the variance of the process $x(n)$, is distributed as a function of frequency.

Filter bank interpretation. The above assertion that the periodogram describes a distribution of power as a function of frequency can be interpreted in a different way, in which the power estimate over a narrow frequency band is attributed to the output power of a narrow-bandpass filter. This leads to the well-known *filter bank interpretation* of the periodogram. To develop this interpretation, consider the basic (unwindowed) periodogram estimator $\hat{R}_x(e^{j\omega})$ in (5.3.2), evaluated at a frequency $\omega_k \triangleq k\Delta\omega \triangleq 2\pi k/N$, which can be expressed as

$$\hat{R}_x(e^{j\omega_k}) = \frac{1}{N} \left| \sum_{n=0}^{N-1} x(n)e^{-j\omega_k n} \right|^2 = \frac{1}{N} \left| \sum_{n=0}^{N-1} x(n)e^{j2\pi k - j\omega_k n} \right|^2$$

$$= \frac{1}{N} \left| \sum_{n=0}^{N-1} x(n)e^{j\omega_k(N-n)} \right|^2 \qquad \text{since } \omega_k N = 2\pi k \tag{5.3.8}$$

$$= \frac{1}{N} \left| \sum_{m=0}^{N-1} x(N-1-m)e^{j\omega_k m} \right|^2$$

Clearly, the term inside the absolute value sign in (5.3.8) can be interpreted as a convolution of $x(n)$ and $e^{j\omega_k n}$, evaluated at $n = N - 1$. Define

$$h_k(n) \triangleq \begin{cases} \dfrac{1}{N} e^{j\omega_k n} & 0 \le n \le N-1 \\[2mm] 0 & \text{otherwise} \end{cases} \tag{5.3.9}$$

as the impulse response of a linear system whose frequency response is given by

$$H_k(e^{j\omega}) = \mathcal{F}[h_k(n)] = \frac{1}{N} \sum_{n=0}^{N-1} e^{j\omega_k n} e^{-j\omega n}$$

$$= \frac{1}{N} \sum_{n=0}^{N-1} e^{-j(\omega-\omega_k)n} = \frac{1}{N} \frac{e^{-jN(\omega-\omega_k)} - 1}{e^{-j(\omega-\omega_k)} - 1} \tag{5.3.10}$$

$$= \frac{1}{N} \frac{\sin[N(\omega-\omega_k)/2]}{\sin[(\omega-\omega_k)/2]} e^{-j(N-1)(\omega-\omega_k)/2}$$

which is a linear-phase, narrow-bandpass filter centered at $\omega = \omega_k$. The 3-dB bandwidth of this filter is proportional to $2\pi/N$ rad per sampling interval (or $1/N$ cycles per sampling interval). A plot of the magnitude response $|H_k(e^{j\omega})|$, for $\omega_k = \pi/2$ and $N = 50$, is shown in Figure 5.10, which evidently shows the narrowband nature of the filter.

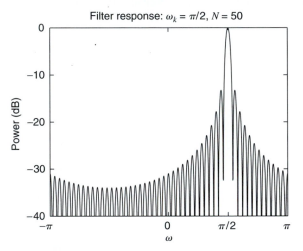

Filter response: $\omega_k = \pi/2$, $N = 50$

FIGURE 5.10
The magnitude of the frequency response of the narrow-bandpass filter for $\omega_k = \pi/2$ and $N = 50$.

Continuing, we also define the output of the filter $h_k(n)$ by $y_k(n)$, that is,

$$y_k(n) \triangleq h_k(n) * x(n) = \frac{1}{N} \sum_{m=0}^{N-1} x(n-m) e^{j\omega_k m} \tag{5.3.11}$$

Then (5.3.8) can be written as

$$\hat{R}_x(e^{j\omega_k}) = N|y_k(N-1)|^2 \tag{5.3.12}$$

Now consider the average power in $y_k(n)$, which can be evaluated using the spectral density as

$$E\{|y_k(n)|^2\} = \frac{1}{2\pi} \int_{-\pi}^{\pi} R_x(e^{j\omega})|H_k(e^{j\omega})|^2 \, d\omega$$

$$\approx \frac{\Delta\omega}{2\pi} R_x(e^{j\omega_k}) = \frac{1}{N} R_x(e^{j\omega_k}) \tag{5.3.13}$$

since $H(e^{j\omega})$ is a narrowband filter. If we estimate the average power $E\{|y_k(n)|^2\}$ using one sample $y_k(N-1)$, then from (5.3.13) the estimated spectral density is the periodogram given by (5.3.12), which says that the kth DFT sample of the periodogram [see (5.3.4)] is given by the average power of a *single* $(N-1)$st output sample of the ω_k-centered narrow-bandpass filter. Now imagine one such filter for each ω_k, $k = 0, \ldots, N-1$, frequencies. Thus we have a bank of filters, each tuned to the discrete frequency (based on the data record length), providing the periodogram estimates every N samples. This filter bank is inherently built into the periodogram and hence need not be explicitly implemented. The block diagram of this filter bank approach to the periodogram computation is shown in Figure 5.11.

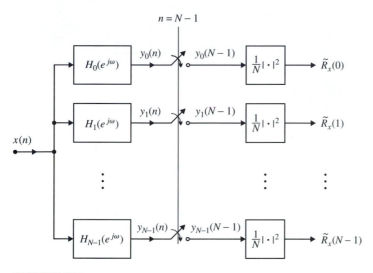

FIGURE 5.11
The filter bank approach to the periodogram computation.

In Section 5.1, we observed that the periodogram of a deterministic signal approaches the true energy spectrum as the number of observations $N \to \infty$. To see how the power spectrum of random signals is related to the number observations, we consider the following example.

EXAMPLE 5.3.1 (PERIODOGRAM OF A SIMULATED WHITE NOISE SEQUENCE). Let $x(n)$ be a stationary white Gaussian noise with zero-mean and unit variance. The theoretical spectrum of $x(n)$ is

$$R_x(e^{j\omega}) = \sigma_x^2 = 1 \qquad -\pi < \omega \le \pi$$

To study the periodogram estimate, 50 different N-point records of $x(n)$ were generated using a pseudorandom number generator. The periodogram $\hat{R}_x(e^{j\omega})$ of each record was computed for $\omega = \omega_k = 2\pi k/1024$, $k = 0, 1, \ldots, 512$, that is, with $N_{\text{FFT}} = 1024$, from the available data using (5.3.4) for $N = 32$, 128, and 256. These results in the form of periodogram overlays (a Monte Carlo simulation) and their averages are shown in Figure 5.12. We notice that $\hat{R}_x(e^{j\omega})$ fluctuates so erratically that it is impossible to conclude from its observation that the signal has a flat spectrum. Furthermore, the size of the fluctuations (as seen from the ensemble average) is not reduced by increasing the segment length N. In this sense, we should not expect the periodogram $\hat{R}_x(e^{j\omega})$ to converge to the true spectrum $R_x(e^{j\omega})$ in some statistical sense as $N \to \infty$. Since $R_x(e^{j\omega})$ is constant over frequency, the fluctuations of $\hat{R}_x(e^{j\omega})$ can be characterized by their mean, variance, and mean square error over frequency for each N and are given in Table 5.2. It can be seen that although the mean value tends to 1 (true value), the standard deviation is not reduced as N increases. In fact, it is close to 1; that is, it is of the order of the size of the quantity to be estimated. This illustrates that the periodogram is not a good estimate of the power spectrum.

Since for each value of ω, $\hat{R}_x(e^{j\omega})$ is a random variable, the erratic behavior of the periodogram estimator, which is illustrated in Figure 5.12, can be explained by considering its mean, covariance, and variance.

TABLE 5.2

Performance of periodogram for white Gaussian noise signal in Example 5.3.1.

N	32	128	256
$E[R_x(e^{j\omega_k})]$	0.7829	0.8954	0.9963
$\mathrm{var}[R_x(e^{j\omega_k})]$	0.7232	1.0635	1.1762
MSE	0.7689	1.07244	1.1739

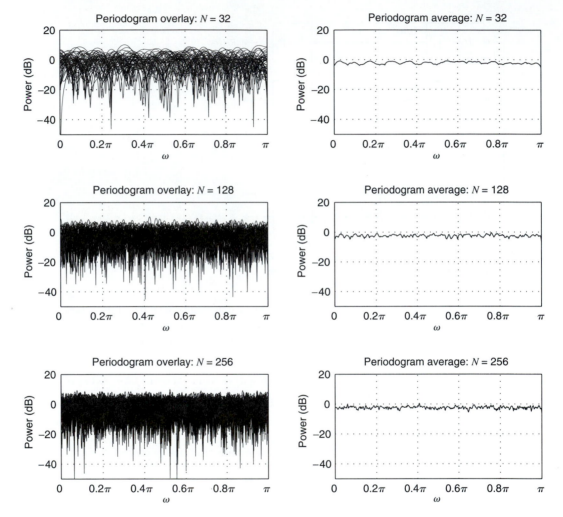

FIGURE 5.12
Periodograms of white Gaussian noise in Example 5.3.1.

Mean of $\hat{R}_x(e^{j\omega})$. Taking the mathematical expectation of (5.3.5) and using (5.2.8), we obtain

$$E\{\hat{R}_x(e^{j\omega})\} = \sum_{l=-(N-1)}^{N-1} E\{\hat{r}_x(l)\}e^{-j\omega l} = \frac{1}{N}\sum_{l=-(N-1)}^{N-1} r_x(l)r_w(l)e^{-j\omega l} \qquad (5.3.14)$$

Since $E\{\hat{R}_x(e^{j\omega})\} \neq R_x(e^{j\omega})$, the periodogram is a biased estimate of the true power spectrum $R_x(e^{j\omega})$.

Equation (5.3.14) can be interpreted in the frequency domain as a periodic convolution. Indeed, using the frequency domain convolution theorem, we have

$$E\{\hat{R}_x(e^{j\omega})\} = \frac{1}{2\pi N} \int_{-\pi}^{\pi} R_x(e^{j\theta}) R_w(e^{j(\omega-\theta)}) \, d\theta \qquad (5.3.15)$$

where
$$R_w(e^{j\omega}) = |W(e^{j\omega})|^2 \qquad (5.3.16)$$

is the spectrum of the window. Thus, the expected value of the periodogram is obtained by convolving the true spectrum $R_x(e^{j\omega})$ with the spectrum $R_w(e^{j\omega})$ of the window. This is equivalent to windowing the true autocorrelation $r_x(l)$ with the *correlation* or *lag* window $r_w(l) = w(l) * w(-l)$, where $w(n)$ is the data window.

To understand the implications of (5.3.15), consider the rectangular data window (5.2.7). Using (5.2.11), we see that (5.3.14) becomes

$$E\{\hat{R}_x(e^{j\omega})\} = \sum_{l=-(N-1)}^{N-1} \left(1 - \frac{|l|}{N}\right) r_x(l) e^{-j\omega l} \qquad (5.3.17)$$

For nonperiodic autocorrelations, the value of $r_x(l)$ becomes negligible for large values of $|l|$. Hence, as the record length N increases, the term $(1 - |l|/N) \to 1$ for all l, which implies that

$$\lim_{N \to \infty} E\{\hat{R}_x(e^{j\omega})\} = R_x(e^{j\omega}) \qquad (5.3.18)$$

that is, the periodogram is an *asymptotically unbiased* estimator of $R_x(e^{j\omega})$. In the frequency domain, we obtain

$$R_w(e^{j\omega}) = \mathcal{F}\{w_R(l) * w_R(-l)\} = |W_R(e^{j\omega})|^2 = \left[\frac{\sin(\omega N/2)}{\sin(\omega/2)}\right]^2 \qquad (5.3.19)$$

where
$$W_R(e^{j\omega}) = e^{-j\omega(N-1)/2} \frac{\sin(\omega N/2)}{\sin(\omega/2)} \qquad (5.3.20)$$

is the Fourier transform of the rectangular window. The spectrum $R_w(e^{j\omega})$, in (5.3.19), of the correlation window $r_w(l)$ approaches a periodic impulse train as the window length increases.[†] As a result, $E\{\hat{R}_x(e^{j\omega})\}$ approaches the true power spectrum $R_x(e^{j\omega})$ as N approaches ∞.

The result (5.3.18) holds for any window that satisfies the following two conditions:

1. The window is normalized such that

$$\sum_{n=0}^{N-1} |w(n)|^2 = N \qquad (5.3.21)$$

This condition is obtained by noting that, for asymptotic unbiasedness, we want $R_w(e^{j\omega})/N$ in (5.3.15) to be an approximation of an impulse in the frequency domain. Since the area under the impulse function is unity, using (5.3.16) and Parseval's theorem, we have

$$\frac{1}{2\pi N} \int_{-\pi}^{\pi} |W(e^{j\omega})|^2 \, d\omega = \frac{1}{N} \sum_{n=0}^{N-1} |w(n)|^2 = 1 \qquad (5.3.22)$$

2. The width of the mainlobe of the spectrum $R_w(e^{j\omega})$ of the correlation window decreases as $1/N$. This condition guarantees that the area under $R_w(e^{j\omega})$ is concentrated at the origin as N becomes large.

[†]This spectrum is sometimes referred to as the *Fejer kernel*.

The bias is introduced by the sidelobes of the correlation window through leakage, as illustrated in Section 5.1. Therefore, we can reduce the bias by using the modified periodogram and a "better" window. Bias can be avoided if either $N = \infty$, in which case the spectrum of the window is a periodic train of impulses, or $R_x(e^{j\omega}) = \sigma_x^2$, that is, $x(n)$ has a flat power spectrum. Thus, for white noise, $\hat{R}_x(e^{j\omega})$ is unbiased for all N. This fact was apparent in Example 5.3.1 and is very important for practical applications. In the following example, we illustrate that the bias becomes worse as the dynamic range of the spectrum increases.

EXAMPLE 5.3.2 (BIAS AND LEAKAGE PROPERTIES OF THE PERIODOGRAM). Consider an AR(2) process with

$$\mathbf{a}_2 = [1 \ -0.75 \ 0.5]^T \qquad d_0 = 1 \tag{5.3.23}$$

and an AR(4) process with

$$\mathbf{a}_4 = [1 \ -2.7607 \ 3.8106 \ -2.6535 \ 0.9238]^T \qquad d_0 = 1 \tag{5.3.24}$$

where $w(n) \sim \text{WN}(0, 1)$. Both processes have been used extensively in the literature for power spectrum estimation studies (Percival and Walden 1993). Their power spectrum is given by (see Chapter 4)

$$R_x(e^{j\omega}) = \frac{\sigma_w^2 d_0}{|A(e^{j\omega})|^2} = \frac{\sigma_w^2}{\left| \sum\limits_{k=0}^{p} a_k e^{j\omega k} \right|^2} \tag{5.3.25}$$

For simulation purposes, $N = 1024$ samples of each process were generated. The sample realizations and the shapes of the two power spectra in (5.3.25) are shown in Figure 5.13. The dynamic range of the two spectra, that is, $\max\limits_{\omega} R_x(e^{j\omega}) / \min\limits_{\omega} R_x(e^{j\omega})$, is about 15 and 65 dB, respectively.

From the sample realizations, periodograms and modified periodograms, based on the Hanning window, were computed by using (5.3.4) at $N_{\text{FFT}} = 1024$ frequencies. These are shown in Figure 5.14. The periodograms for the AR(2) and AR(4) processes, respectively, are shown in the

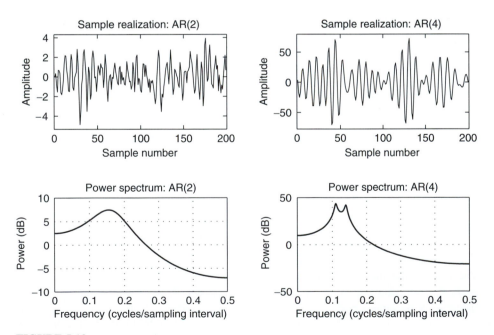

FIGURE 5.13
Sample realizations and power spectra of the AR(2) and AR(4) processes used in Example 5.3.2.

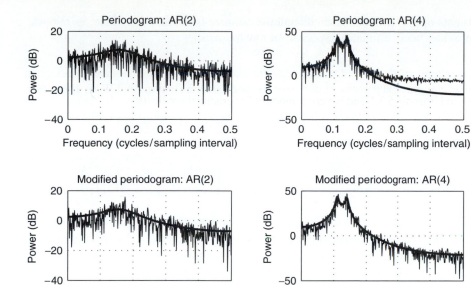

FIGURE 5.14
Illustration of properties of periodogram as a power spectrum estimator.

top row while the modified periodograms for the same processes are shown in the bottom row. These plots illustrate that the periodogram is a biased estimator of the power spectrum. In the case of the AR(2) process, since the spectrum has a small dynamic range (15 dB), the bias in the periodogram estimate is not obvious; furthermore, the windowing in the modified periodogram did not show much improvement. On the other hand, the AR(4) spectrum has a large dynamic range, and hence the bias is clearly visible at high frequencies. This bias is clearly reduced by windowing of the data in the modified periodogram. In both cases, the random fluctuations are not reduced by the data windowing operation.

EXAMPLE 5.3.3 (FREQUENCY RESOLUTION PROPERTY OF THE PERIODOGRAM). Consider two unit-amplitude sinusoids observed in unit variance white noise. Let

$$x(n) = \cos(0.35\pi n + \phi_1) + \cos(0.4\pi n + \phi_2) + \nu(n)$$

where ϕ_1 and ϕ_2 are jointly independent random variables uniformly distributed over $[-\pi, \pi]$ and $\nu(n)$ is a unit-variance white noise. Since two frequencies, 0.35π and 0.4π, are close, we will need (see Table 5.1)

$$N - 1 > \frac{1.81\pi}{0.4\pi - 0.35\pi} \quad \text{or} \quad N > 37$$

To obtain a periodogram ensemble, 50 realizations of $x(n)$ for $N = 32$ and $N = 64$ were generated, and their periodograms were computed. The plots of these periodogram overlays and the corresponding ensemble average for $N = 32$ and $N = 64$ are shown in Figure 5.15. For $N = 32$, frequencies in the periodogram cannot be resolved, as expected; but for $N = 64$ it is possible to separate the two sinusoids with ease. Note that the modified periodogram (i.e., data windowing) will not help since windowing increases smoothing and smearing of peaks.

The case of nonzero mean. In the periodogram method of spectrum analysis in this section, we assumed that the random signal has zero mean. If a random signal has nonzero mean, it should be estimated using (3.6.20) and then removed from the signal prior to computing its periodogram. This is because the power spectrum of a nonzero mean signal has an impulse at the zero frequency. If this mean is relatively large, then because of the leakage inherent in the periodogram, this mean will obscure low-amplitude, low-frequency components of the spectrum. Even though the estimate is not an exact value, its removal often provides better estimates, especially at low frequencies.

Covariance of $\hat{R}_x(e^{j\omega})$. Obtaining an expression for the covariance of the periodogram is a rather complicated process. However, it has been shown (Jenkins and Watts 1968) that

$$
\text{cov}\{\hat{R}_x(e^{j\omega_1}), \hat{R}_x(e^{j\omega_2})\} \simeq R_x(e^{j\omega_1}) R_x(e^{j\omega_2}) \left(\left\{ \frac{\sin\left[(\omega_1 + \omega_2)N/2\right]}{N \sin\left[(\omega_1 + \omega_2)/2\right]} \right\}^2 \right.
$$
$$
\left. + \left\{ \frac{\sin\left[(\omega_1 - \omega_2)N/2\right]}{N \sin\left[(\omega_1 - \omega_2)/2\right]} \right\}^2 \right) \tag{5.3.26}
$$

This expression applies to stationary random signals with zero mean and Gaussian probability density. The approximation becomes exact if the signal has a flat spectrum (white noise). Although this approximation deteriorates for non-Gaussian probability densities, the qualitative results that one can draw from this approximation appear to hold for a rather broad range of densities.

From (5.3.26), for $\omega_1 = (2\pi/N)k_1$ and $\omega_2 = (2\pi/N)k_2$ with k_1, k_2 integers, we have

$$
\text{cov}\{\hat{R}_x(e^{j\omega_1}) \hat{R}_x(e^{j\omega_2})\} \simeq 0 \quad \text{for } k_1 \neq k_2 \tag{5.3.27}
$$

Thus, values of the periodogram spaced in frequency by integer multiples of $2\pi/N$ are approximately uncorrelated. As the record length N increases, these uncorrelated periodogram samples come closer together, and hence the rate of fluctuations in the periodogram increases. This explains the results in Figure 5.12.

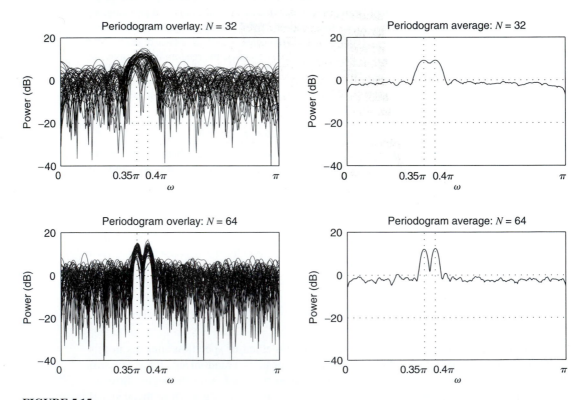

FIGURE 5.15
Illustration of the frequency resolution property of the periodogram in Example 5.3.3.

Variance of $\hat{R}_x(e^{j\omega})$. The variance of the periodogram at a particular frequency $\omega = \omega_1 = \omega_2$ can be obtained from (5.3.26)

$$\text{var}\{\hat{R}_x(e^{j\omega})\} \simeq R_x^2(e^{j\omega})\left[1 + \left(\frac{\sin \omega N}{N \sin \omega}\right)^2\right] \qquad (5.3.28)$$

For large values of N, the variance of $\hat{R}_x(e^{j\omega})$ can be approximated by

$$\text{var}\{\hat{R}_x(e^{j\omega})\} \simeq \begin{cases} R_x^2(e^{j\omega}) & 0 < \omega < \pi \\ 2R_x^2(e^{j\omega}) & \omega = 0, \pi \end{cases} \qquad (5.3.29)$$

This result is crucial, because it shows that the variance of the periodogram (estimate) remains at the level of $R_x^2(e^{j\omega})$ (quantity to be estimated), independent of the record length N used. Furthermore, since the variance does not tend to zero as $N \rightarrow \infty$, the periodogram is not a consistent estimator; that is, its distribution does not tend to cluster more closely around the true spectrum as N increases.[†]

This behavior was illustrated in Example 5.3.1. The variance of $\hat{R}_x(e^{j\omega_k})$ fails to decrease as N increases because the number of periodogram values $\hat{R}_x(e^{j\omega_k})$, $k = 0, 1, \ldots, N - 1$, is always equal to the length N of the data record.

EXAMPLE 5.3.4 (COMPARISON OF PERIODOGRAM AND MODIFIED PERIODOGRAM).
Consider the case of three sinusoids discussed in Section 5.1.4. In particular, we assume that these sinusoids are observed in white noise with

$$x(n) = \cos(0.35\pi n + \phi_1) + \cos(0.4\pi n + \phi_1) + 0.25\cos(0.8\pi n + \phi_1) + v(n)$$

where ϕ_1, ϕ_2, and ϕ_3 are jointly independent random variables uniformly distributed over $[-\pi, \pi]$ and $v(n)$ is a unit-variance white noise. An ensemble of 50 realizations of $x(n)$ was generated using $N = 128$. The periodograms and the Hamming window–based modified periodograms of these realizations were computed, and the results are shown in Figure 5.16. The top row of the figure contains periodogram overlays and the corresponding ensemble average for the unwindowed periodogram, and the bottom row shows the same for the modified periodogram. Spurious peaks (especially near the two close frequencies) in the periodogram have been suppressed by the data windowing operation in the modified periodogram; hence the peak corresponding to 0.8π is sufficiently enhanced. This enhancement is clearly at the expense of the frequency resolution (or smearing of the true peaks), which is to be expected. The overall variance of the noise floor is still not reduced.

Failure of the periodogram

To conclude, we note that the periodogram in its "basic form" is a very poor estimator of the power spectrum function. The failure of the periodogram when applied to random signals is uniquely pointed out in Jenkins and Watts (1968, p. 213):

> The basic reason why Fourier analysis breaks down when applied to time series is that it is based on the assumption of *fixed* amplitudes, frequencies and phases. Time series, on the other hand, are characterized by *random* changes of frequencies, amplitudes and phases. Therefore it is not surprising that Fourier methods need to be adapted to account for the random nature of a time series.

The attempt at improving the periodogram by windowing the available data, that is, by using the modified periodogram in Example 5.3.4, showed that the presence and the length of the window had no effect on the variance. The major problems with the periodogram lie in its variance, which is on the order of $R_x^2(e^{j\omega})$, as well as in its erratic behavior. Thus, to obtain a better estimator, we should reduce its variance; that is, we should "smooth" the periodogram.

[†]The definition of the PSD by $R_x(e^{j\omega}) = \lim_{N\to\infty} \hat{R}_x(e^{j\omega})$ is not valid because even if $\lim_{N\to\infty} E\{\hat{R}_x(e^{j\omega})\} = R_x(e^{j\omega})$, the variance of $\hat{R}_x(e^{j\omega})$ does not tend to zero as $N \rightarrow \infty$ (Papoulis 1991).

From the previous discussion, it follows that the sequence $\widetilde{\hat{R}}_x(k)$, $k = 0, 1, \ldots, N - 1$, of the harmonic periodogram components can be reasonably assumed to be a sequence of uncorrelated random variables. Furthermore, it is well known that the variance of the sum of K uncorrelated random variables with the same variance is $1/K$ times the variance of one of these individual random variables. This suggests two ways of reducing the variance, which also lead to smoother spectral estimators:

- Average contiguous values of the periodogram.
- Average periodograms obtained from multiple data segments.

It should be apparent that owing to stationarity, the two approaches should provide comparable results under similar circumstances.

5.3.2 Power Spectrum Estimation by Smoothing a Single Periodogram— The Blackman-Tukey Method

The idea of reducing the variance of the periodogram through smoothing using a moving-average filter was first proposed by Daniel (1946). The estimator proposed by Daniel is a zero-phase moving-average filter, given by

$$\hat{R}_x^{(PS)}(e^{j\omega_k}) \triangleq \frac{1}{2M+1} \sum_{j=-M}^{M} \hat{R}_x(e^{j\omega_{k-j}}) \triangleq \sum_{j=-M}^{M} W(e^{j\omega_j}) \hat{R}_x(e^{j\omega_{k-j}}) \tag{5.3.30}$$

where $\omega_k = (2\pi/N)k$, $k = 0, 1, \ldots, N - 1$, $W(e^{j\omega_j}) \triangleq 1/(2M + 1)$, and the superscript (PS) denotes periodogram smoothing. Since the samples of the periodogram are approxi-

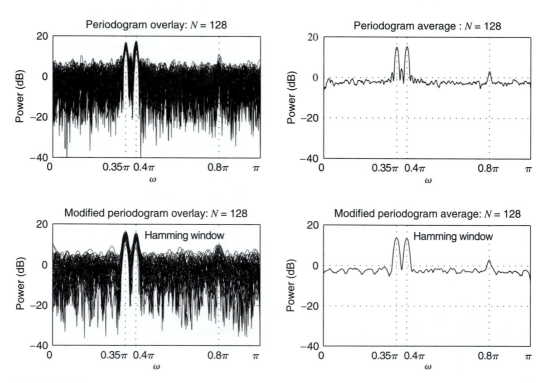

FIGURE 5.16
Comparison of periodogram and modified periodogram in Example 5.3.4.

mately uncorrelated,

$$\text{var}\{\hat{R}_x^{(PS)}(e^{j\omega_k})\} \simeq \frac{1}{2M+1}\,\text{var}\{\hat{R}_x(e^{j\omega_k})\} \tag{5.3.31}$$

that is, averaging $2M + 1$ consecutive spectral lines reduces the variance by a factor of $2M + 1$. The quantity $\Delta\omega \approx (2\pi/N)(2M+1)$ determines the frequency resolution, since any peaks within the $\Delta\omega$ range are smoothed over the entire interval $\Delta\omega$ into a single peak and cannot be resolved. Thus, increasing M reduces the variance (resulting in a smoother spectrum estimate), at the expense of spectral resolution. This is the fundamental tradeoff in practical spectral analysis.

Blackman-Tukey approach

The discrete moving average in (5.3.30) is computed in the frequency domain. We now introduce a better and simpler way to smooth the periodogram by operating on the estimated autocorrelation sequence. To this end, we note that the continuous frequency equivalent of the discrete convolution formula (5.3.30) is the periodic convolution

$$\hat{R}_x^{(PS)}(e^{j\omega}) = \frac{1}{2\pi}\int_{-\pi}^{\pi}\hat{R}_x(e^{j(\omega-\theta)})W_a(e^{j\theta})\,d\theta = \hat{R}_x(e^{j\omega}) \otimes W_a(e^{j\omega}) \tag{5.3.32}$$

where $W_a(e^{j\omega})$ is a periodic function of ω with period 2π, given by

$$W_a(e^{j\omega}) = \begin{cases} \dfrac{1}{\Delta\omega} & |\omega| < \dfrac{\Delta\omega}{2} \\ 0 & \dfrac{\Delta\omega}{2} \le \omega \le \pi \end{cases} \tag{5.3.33}$$

By using the convolution theorem, (5.3.32) can be written as

$$\hat{R}_x^{(PS)}(e^{j\omega}) = \sum_{l=-(L-1)}^{L-1}\hat{r}_x(l)w_a(l)e^{-j\omega l} \tag{5.3.34}$$

where $w_a(l)$ is the inverse Fourier transform of $W_a(e^{j\omega})$ and $L < N$. As we have already mentioned, the window $w_a(l)$ is known as the correlation or lag window.[†] The correlation window corresponding to (5.3.33) is

$$w_a(l) = \frac{\sin(l\Delta\omega/2)}{\pi l} \qquad -\infty < l < \infty \tag{5.3.35}$$

Since $w_a(l)$ has infinite duration, its truncation at $|l| = L \le N$ creates ripples in $W_a(e^{j\omega})$ (Gibbs effect). To avoid this problem, we use correlation windows with finite duration, that is, $w_a(l) = 0$ for $|l| > L \le N$. For real sequences, where $\hat{r}_x(l)$ is real and even, $w_a(l)$ [and hence $W_a(e^{j\omega})$] should be real and even. Given that $\hat{R}_x(e^{j\omega})$ is nonnegative, a sufficient (but not necessary) condition that $\hat{R}_x^{(PS)}(e^{j\omega})$ be nonnegative is that $W_a(e^{j\omega}) \ge 0$ for all ω. This condition holds for the Bartlett (triangular) and Parzen (see Problem 5.11) windows, but it does not hold for the Hamming, Hanning, or Kaiser window.

Thus, we note that *smoothing the periodogram* $\hat{R}_x(e^{j\omega})$ *by convolving it with the spectrum* $W_a(e^{j\omega}) = \mathcal{F}\{w_a(l)\}$ *is equivalent to windowing the autocorrelation estimate* $\hat{r}_x(l)$ *with the correlation window* $w_a(l)$. This approach to power spectrum estimation, which was introduced by Blackman and Tukey (1959), involves the following steps:

[†]The term *spectral window* is quite often used for $W_a(e^{j\omega}) = \mathcal{F}\{w_a(l)\}$, the Fourier transform of the correlation window. However, this term is misleading because $W_a(e^{j\omega})$ is essentially a frequency-domain impulse response. We use the term *correlation window* for $w_a(l)$ and the term *Fourier transform of the correlation window* for $W_a(e^{j\omega})$.

1. Estimate the autocorrelation sequence from the unwindowed data.
2. Window the obtained autocorrelation samples.
3. Compute the DTFT of the windowed autocorrelation as given in (5.3.34).

A pictorial comparison between the theoretical [i.e., using (5.3.32)] and the above practical computation of power spectrum using the single-periodogram smoothing is shown in Figure 5.17.

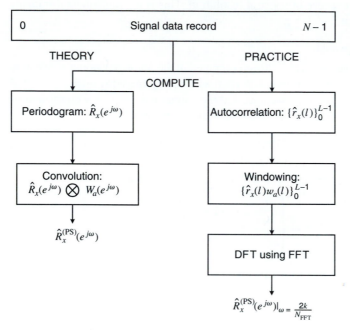

FIGURE 5.17
Comparison of the theory and practice of the Blackman-Tukey method.

The resolution of the Blackman-Tukey power spectrum estimator is determined by the duration $2L - 1$ of the correlation window. For most correlation windows, the resolution is measured by the 3-dB bandwidth of the mainlobe, which is on the order of $2\pi/L$ rad per sampling interval.

The statistical quality of the Blackman-Tukey estimate $\hat{R}_x^{(PS)}(e^{j\omega})$ can be evaluated by examining its mean, covariance, and variance.

Mean of $\hat{R}_x^{(PS)}(e^{j\omega})$. The expected value of the smoothed periodogram $\hat{R}_x^{(PS)}(e^{j\omega})$ can be obtained by using (5.3.34) and (5.2.11). Indeed, we have

$$
\begin{aligned}
E\{\hat{R}_x^{(PS)}(e^{j\omega})\} &= \sum_{l=-(L-1)}^{L-1} E\{\hat{r}_x(l)\} w_a(l) e^{-j\omega l} \\
&= \sum_{l=-(L-1)}^{L-1} r_x(l) \left(1 - \frac{|l|}{N}\right) w_a(l) e^{-j\omega l}
\end{aligned}
\tag{5.3.36}
$$

or, using the frequency convolution theorem, we have

$$
E\{\hat{R}_x^{(PS)}(e^{j\omega})\} = R_x(e^{j\omega}) \otimes W_B(e^{j\omega}) \otimes W_a(e^{j\omega})
\tag{5.3.37}
$$

where
$$W_B(e^{j\omega}) = \mathcal{F}\left\{\left(1 - \frac{|l|}{N}\right) w_R(n)\right\} = \frac{1}{N}\left[\frac{\sin(\omega N/2)}{\sin(\omega/2)}\right]^2 \qquad (5.3.38)$$

is the Fourier transform of the Bartlett window. Since $E\{\hat{R}_x^{(PS)}(e^{j\omega})\} \neq R_x(e^{j\omega})$, $\hat{R}_x^{(PS)}(e^{j\omega})$ is a *biased* estimate of $R_x(e^{j\omega})$.

For $L \ll N$, $(1 - |l|/N) \simeq 1$ and hence we obtain

$$E\{\hat{R}_x^{(PS)}(e^{j\omega})\} = \sum_{l=-(L-1)}^{L-1} r_x(l)\left(1 - \frac{|l|}{N}\right) w_a(l)e^{-j\omega l}$$

$$\simeq R_x(e^{j\omega}) \otimes W_a(e^{j\omega}) \qquad (5.3.39)$$

$$= \frac{1}{2\pi}\int_{-\pi}^{\pi} R_x(e^{j\theta}) W_a(e^{j(\omega-\theta)})\, d\theta$$

If L is sufficiently large, the correlation window $w_a(l)$ consists of a narrow mainlobe. If $R_x(e^{j\omega})$ can be assumed to be constant within the mainlobe, we have

$$E\{\hat{R}_x^{(PS)}(e^{j\omega})\} \simeq R_x(e^{j\omega})\frac{1}{2\pi}\int_{-\pi}^{\pi} W_a(e^{j(\omega-\theta)})\, d\theta$$

which implies that $\hat{R}_x^{(PS)}(e^{j\omega})$ is *asymptotically unbiased* if

$$\frac{1}{2\pi}\int_{-\pi}^{\pi} W_a(e^{j\omega})\, d\omega = w_a(0) = 1 \qquad (5.3.40)$$

that is, if the spectrum of the correlation window has unit area. Under this condition, if both L and N tend to infinity, then $W_a(e^{j\omega})$ and $W_B(e^{j\omega})$ become periodic impulse trains and the convolution (5.3.37) reproduces $R_x(e^{j\omega})$.

Covariance of $\hat{R}_x^{(PS)}(e^{j\omega})$. The following approximation

$$\text{cov}\{\hat{R}_x^{(PS)}(e^{j\omega_1}), \hat{R}_x^{(PS)}(e^{j\omega_2})\} \simeq \frac{1}{2\pi N}\int_{-\pi}^{\pi} R_x^2(e^{j\theta}) W_a(e^{j(\omega_1-\theta)}) W_a(e^{j(\omega_2-\theta)})\, d\theta \qquad (5.3.41)$$

derived in Jenkins and Watts (1968), holds under the assumptions that (1) N is sufficiently large that $W_B(e^{j\omega})$ behaves as a periodic impulse train and (2) L is sufficiently large that $W_a(e^{j\omega})$ is sufficiently narrow that the product $W_a(e^{j(\omega_1+\theta)})W_a(e^{j(\omega_2-\theta)})$ is negligible. Hence, the covariance increases proportionally to the width of $W_a(e^{j\omega})$, and the amount of overlap between the windows $W_a(e^{j(\omega_1-\theta)})$ (centered at ω_1) and $W_a(e^{j(\omega_2-\theta)})$ (centered at ω_2) increases.

Variance of $\hat{R}_x^{(PS)}(e^{j\omega})$. When $\omega = \omega_1 = \omega_2$, (5.3.41) gives

$$\text{var}\{\hat{R}_x^{(PS)}(e^{j\omega})\} \simeq \frac{1}{2\pi N}\int_{-\pi}^{\pi} R_x^2(e^{j\omega}) W_a^2(e^{j(\omega-\theta)})\, d\theta \qquad (5.3.42)$$

If $R_x(e^{j\omega})$ is smooth within the width of $W_a(e^{j\omega})$, then

$$\text{var}\{\hat{R}_x^{(PS)}(e^{j\omega})\} \simeq R_x^2(e^{j\omega})\frac{1}{2\pi N}\int_{-\pi}^{\pi} W_a^2(e^{j\omega})\, d\omega \qquad (5.3.43)$$

or
$$\text{var}\{\hat{R}_x^{(PS)}(e^{j\omega})\} \simeq \frac{E_w}{N} R_x^2(e^{j\omega}) \qquad 0 < \omega < \pi \qquad (5.3.44)$$

where
$$E_w = \frac{1}{2\pi}\int_{-\pi}^{\pi} W_a^2(e^{j\omega})\, d\omega = \sum_{l=-(L-1)}^{L-1} w_a^2(l) \qquad (5.3.45)$$

is the energy of the correlation window. From (5.3.29) and (5.3.44) we have

$$\frac{\text{var}\{\hat{R}_x^{(\text{PS})}(e^{j\omega})\}}{\text{var}\{\hat{R}_x(e^{j\omega})\}} \simeq \frac{E_w}{N} \qquad 0 < \omega < \pi \tag{5.3.46}$$

which is known as the *variance reduction factor* or *variance ratio* and provides the reduction in variance attained by smoothing the periodogram.

In the beginning of this section, we explained the variance reduction in terms of frequency-domain averaging. An alternative explanation can be provided by considering the windowing of the estimated autocorrelation. As discussed in Section 5.2, the variance of the autocorrelation estimate increases as $|l|$ approaches N because fewer and fewer samples are used to compute the estimate. Since every value of $\hat{r}_x(l)$ affects the value of $\hat{R}_x(\omega)$ at all frequencies, the less reliable values affect the quality of the periodogram everywhere. Thus, we can reduce the variance of the periodogram by minimizing the contribution of autocorrelation terms with large variance, that is, with lags close to N, by proper windowing.

As we have already stressed, there is a tradeoff between resolution and variance. For the variance to be small, we must choose a window that contains a small amount of energy E_w. Since $|w_a(l)| \le 1$, we have $E_w \le 2L$. Thus, to reduce the variance, we must have $L \ll N$. The bias of $\hat{R}_x^{(\text{PS})}(e^{j\omega})$ is directly related to the resolution, which is determined by the mainlobe width of the window, which in turn is proportional to $1/L$. Hence, to reduce the bias, $W_a(e^{j\omega})$ should have a narrow mainlobe that demands a large L. The requirements for high resolution (small bias) and low variance can be simultaneously satisfied only if N is sufficiently large. The variance reduction for some commonly used windows is examined in Problem 5.12. Empirical evidence suggests that use of the Parzen window is a reasonable choice.

Confidence intervals. In the interpretation of spectral estimates, it is important to know whether the spectral details are real or are due to statistical fluctuations. Such information is provided by the confidence intervals (Chapter 3). When the spectrum is plotted on a logarithmic scale, the $(1 - \alpha) \times 100$ percent confidence interval is constant at every frequency, and it is given by (Koopmans 1974)

$$\left(10 \log \hat{R}_x^{(\text{PS})}(e^{j\omega}) - 10 \log \frac{\chi_\nu^2(1 - \alpha/2)}{\nu}, \ 10 \log \hat{R}_x^{(\text{PS})}(e^{j\omega}) + 10 \log \frac{\nu}{\chi_\nu^2(\alpha/2)} \right) \tag{5.3.47}$$

where

$$\nu = \frac{2N}{\displaystyle\sum_{k=-(L-1)}^{L} w_a^2(l)} \tag{5.3.48}$$

is the degrees of freedom of a χ_ν^2 distribution.

Computation of $\hat{R}_x^{(\text{PS})}(e^{j\omega})$ using the DFT. In practice, the Blackman-Tukey power spectrum estimator is computed by using an N-point DFT as follows:

1. Estimate the autocorrelation $r_x(l)$, using the formula

$$\hat{r}_x(l) = \hat{r}_x^*(-l) = \frac{1}{N} \sum_{n=0}^{N+l-1} x(n+l)x^*(n) \qquad l = 0, 1, \ldots, L-1 \tag{5.3.49}$$

For $L > 100$, indirect computation of $\hat{r}_x(l)$ by using DFT techniques is usually more efficient (see Problem 5.13).

2. Form the sequence

$$f(l) = \begin{cases} \hat{r}_x(l)w_a(l) & 0 \le l \le L-1 \\ 0 & L \le l \le N-L \\ \hat{r}_x^*(N-l)w_a(N-l) & N-L+1 \le l \le N-1 \end{cases} \quad (5.3.50)$$

3. Compute the power spectrum estimate

$$\hat{R}_x^{(PS)}(e^{j\omega})|_{\omega=(2\pi/N)k} = F(k) = \text{DFT}\{f(l)\} \quad 0 \le k \le N-1 \quad (5.3.51)$$

as the N-point DFT of the sequence $f(l)$.

MATLAB does not provide a direct function to implement the Blackman-Tukey method. However, such a function can be easily constructed by using built-in MATLAB functions and the above approach. The book toolbox function

$$\text{Rx = bt\_psd(x,Nfft,window,L);}$$

implements the above algorithm in which `window` is any available MATLAB window and `Nfft` is chosen to be larger than N to obtain a high-density spectrum.

> **EXAMPLE 5.3.5 (BLACKMAN-TUKEY METHOD).** Consider the spectrum estimation of three sinusoids in white noise given in Example 5.3.4, that is,
>
> $$x(n) = \cos(0.35\pi n + \phi_1) + \cos(0.4\pi n + \phi_2) + 0.25\cos(0.8\pi n + \phi_3) + v(n) \quad (5.3.52)$$
>
> where ϕ_1, ϕ_2, and ϕ_3 are jointly independent random variables uniformly distributed over $[-\pi, \pi]$ and $v(n)$ is a unit-variance white noise. An ensemble of 50 realizations of $x(n)$ was generated using $N = 512$. The autocorrelations of these realizations were estimated up to lag $L = 64$, 128, and 256. These autocorrelations were windowed using the Bartlett window, and then their 1024-point DFT was computed as the spectrum estimate. The results are shown in Figure 5.18. The top row of the figure contains estimate overlays and the corresponding ensemble average for $L = 64$, the middle row for $L = 128$, and the bottom row for $L = 256$. Several observations can be made from these plots. First, the variance in the estimate has considerably reduced over the periodogram estimate. Second, the lower the lag distance L, the lower the variance and the resolution (i.e., the higher the smoothing of the peaks). This observation is consistent with our discussion above about the effect of L on the quality of estimates. Finally, all the frequencies including the one at 0.8π are clearly distinguishable, something that the basic periodogram could not achieve.

5.3.3 Power Spectrum Estimation by Averaging Multiple Periodograms— The Welch-Bartlett Method

As mentioned in Section 5.3.1, in general, the variance of the sum of K IID random variables is $1/K$ times the variance of each of the random variables. Thus, to reduce the variance of the periodogram, we could average the periodograms from K different realizations of a stationary random signal. However, in most practical applications, only a single realization is available. In this case, we can subdivide the existing record $\{x(n), 0 \le n \le N-1\}$ into K (possibly overlapping) smaller segments as follows:

$$x_i(n) = x(iD + n)w(n) \quad 0 \le n \le L-1, 0 \le i \le K-1 \quad (5.3.53)$$

where $w(n)$ is a window of duration L and D is an *offset* distance. If $D < L$, the segments overlap; and for $D = L$, the segments are contiguous. The periodogram of the ith segment is

$$\hat{R}_{x,i}(e^{j\omega}) \triangleq \frac{1}{L}|X_i(e^{j\omega})|^2 = \frac{1}{L}\left|\sum_{n=0}^{L-1} x_i(n)e^{-j\omega n}\right|^2 \quad (5.3.54)$$

We remind the reader that the window $w(n)$ in (5.3.53) is called a data window because it is applied directly to the data, in contrast to a correlation window that is applied to the autocorrelation sequence [see (5.3.34)]. Notice that there is no need for the data window to have an even shape or for its Fourier transform to be nonnegative. The purpose of using the data window is to control spectral leakage.

The spectrum estimate $\hat{R}_x^{(PA)}(e^{j\omega})$ is obtained by averaging K periodograms as follows:

$$\hat{R}_x^{(PA)}(e^{j\omega}) \triangleq \frac{1}{K} \sum_{i=0}^{K-1} \hat{R}_{x,i}(e^{j\omega}) = \frac{1}{KL} \sum_{i=0}^{K-1} |X_i(e^{j\omega})|^2 \qquad (5.3.55)$$

where the superscript (PA) denotes periodogram averaging. To determine the bias and variance of $\hat{R}_x^{(PA)}(e^{j\omega})$, we let $D = L$ so that the segments do not overlap. The so-computed estimate $\hat{R}_x^{(PA)}(e^{j\omega})$ is known as the *Bartlett* estimate. We also assume that $r_x(l)$ is very small for $|l| > L$. This implies that the signal segments can be assumed to be approximately uncorrelated. To show that the simple periodogram averaging in Bartlett's method reduces the periodogram variance, we consider the following example.

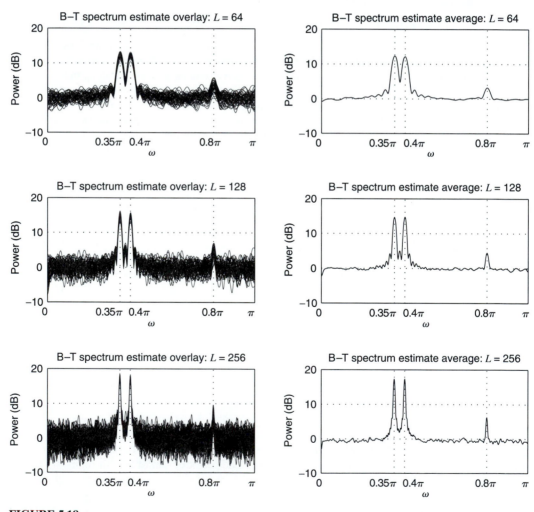

FIGURE 5.18
Spectrum estimation of three sinusoids in white noise using the Blackman-Tukey method in Example 5.3.5.

EXAMPLE 5.3.6 (PERIODOGRAM AVERAGING). Let $x(n)$ be a stationary white Gaussian noise with zero mean and unit variance. The theoretical spectrum of $x(n)$ is

$$R_x(e^{j\omega}) = \sigma_x^2 = 1 \qquad -\pi < \omega \le \pi$$

An ensemble of 50 different 512-point records of $x(n)$ was generated using a pseudorandom number generator. The Bartlett estimate of each record was computed for $K = 1$ (i.e., the basic periodogram), $K = 4$ (or $L = 128$), and $K = 8$ (or $L = 64$). The results in the form of estimate overlays and averages are shown in Figure 5.19. The effect of periodogram averaging is clearly evident.

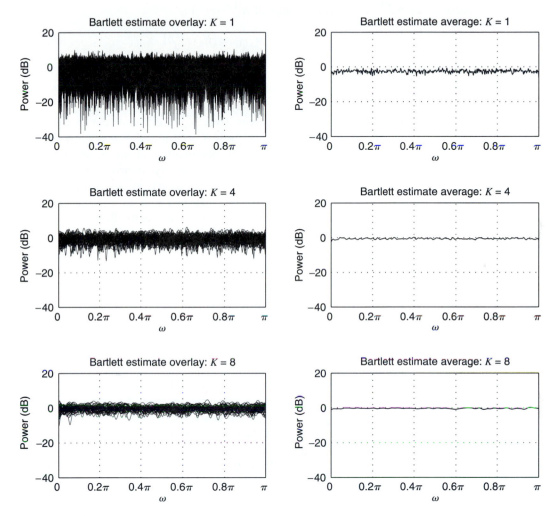

FIGURE 5.19
Spectral estimation of white noise using Bartlett's method in Example 5.3.6.

Mean of $\hat{R}_x^{(PA)}(e^{j\omega})$. The mean value of $\hat{R}_x^{(PA)}(e^{j\omega})$ is

$$E\{\hat{R}_x^{(PA)}(e^{j\omega})\} = \frac{1}{K}\sum_{i=0}^{K-1} E\{\hat{R}_{x,i}(e^{j\omega})\} = E\{\hat{R}_x(e^{j\omega})\} \qquad (5.3.56)$$

where we have assumed that $E\{\hat{R}_{x,i}(e^{j\omega})\} = E\{\hat{R}_x(e^{j\omega})\}$ because of the stationarity assumption. From (5.3.56) and (5.3.15), we have

$$E\{\hat{R}_x^{(PA)}(e^{j\omega})\} = E\{\hat{R}_x(e^{j\omega})\} = \frac{1}{2\pi L}\int_{-\pi}^{\pi} R_x(e^{j\theta})R_w(e^{j(\omega-\theta)})\, d\theta \qquad (5.3.57)$$

where $R_w(e^{j\omega})$ is the spectrum of the data window $w(n)$. Hence, $\hat{R}_x^{(PA)}(e^{j\omega})$ is a biased estimate of $R_x(e^{j\omega})$. However, if the data window is normalized such that

$$\sum_{n=0}^{L-1} w^2(n) = L \tag{5.3.58}$$

the estimate $\hat{R}_x^{(PA)}(e^{j\omega})$ becomes asymptotically unbiased [see the discussion following equation (5.3.15)].

Variance of $\hat{R}_x^{(PA)}(e^{j\omega})$. The variance of $\hat{R}_x^{(PA)}(e^{j\omega})$ is

$$\text{var}\{\hat{R}_x^{(PA)}(e^{j\omega})\} = \frac{1}{K}\,\text{var}\{\hat{R}_x(e^{j\omega})\} \tag{5.3.59}$$

or using (5.3.29) gives

$$\text{var}\{\hat{R}_x^{(PA)}(e^{j\omega})\} \simeq \frac{1}{K} R_x^2(e^{j\omega}) \tag{5.3.60}$$

Clearly, as K increases, the variance tends to zero. Thus, $\hat{R}_x^{(PA)}(e^{j\omega})$ provides an asymptotically unbiased and consistent estimate of $R_x(e^{j\omega})$. If N is fixed and $N = KL$, we see that increasing K to reduce the variance (or equivalently obtain a smoother estimate) results in a decrease in L, that is, a reduction in resolution (or equivalently an increase in bias).

When $w(n)$ in (5.3.53) is the rectangular window of duration L, the square of its Fourier transform is equal to the Fourier transform of the triangular sequence $w_T(n) \triangleq L - |l|$, $|l| < L$, which when combined with the $1/L$ factor in (5.3.57), results in the Bartlett window

$$w_B(l) = \begin{cases} 1 - \dfrac{|l|}{L} & |l| < L \\ 0 & \text{elsewhere} \end{cases} \tag{5.3.61}$$

with

$$W_B(e^{j\omega}) = \frac{1}{L}\left[\frac{\sin(\omega L/2)}{\sin(\omega/2)}\right]^2 \tag{5.3.62}$$

This special case of averaging multiple nonoverlapping periodograms was introduced by Bartlett (1953).

The method has been extended to modified overlapping periodograms by Welch (1970), who has shown that the shape of the window does not affect the variance formula (5.3.59). Welch showed that overlapping the segments by 50 percent reduces the variance by about a factor of 2, owing to doubling the number of segments. More overlap does not result in additional reduction of variance because the data segments become less and less independent. Clearly, the nonoverlapping segments can be uncorrelated only for white noise signals. However, the data segments can be considered approximately uncorrelated if they do not have sharp spectral peaks or if their autocorrelations decay fast.

Thus, the variance reduction factor for the spectral estimator $\hat{R}_x^{(PA)}(e^{j\omega})$ is

$$\frac{\text{var}\{\hat{R}_x^{(PA)}(e^{j\omega})\}}{\text{var}\{\hat{R}_x(e^{j\omega})\}} \simeq \frac{1}{K} \qquad 0 < \omega < \pi \tag{5.3.63}$$

and is reduced by a factor of 2 for 50 percent overlap.

Confidence intervals. The $(1-\alpha)\times 100$ percent confidence interval on a logarithmic scale may be shown to be (Jenkins and Watts 1968)

$$\left(10\log\hat{R}_x^{(PA)}(e^{j\omega}) - 10\log\frac{\chi_{2K}^2(1-\alpha/2)}{2K},\ 10\log\hat{R}_x^{(PA)}(e^{j\omega}) + 10\log\frac{2K}{\chi_{2K}^2(\alpha/2)}\right)$$

$$\tag{5.3.64}$$

where χ_{2K}^2 is a chi-squared distribution with $2K$ degrees of freedom.

Computation of $\hat{R}_x^{(PA)}(e^{j\omega})$ using the DFT. In practice, to compute $\hat{R}_x^{(PA)}(e^{j\omega})$ at L equally spaced frequencies $\omega_k = 2\pi k/L, 0 \le k \le L - 1$, the method of periodogram averaging can be easily and efficiently implemented by using the DFT as follows (we have assumed that L is even):

1. Segment data $\{x(n)\}_0^{N-1}$ into K segments of length L, each offset by D duration using

$$\bar{x}_i(n) = x(iD + n) \qquad 0 \le i \le K - 1, 0 \le n \le L - 1 \qquad (5.3.65)$$

If $D = L$, there is no overlap; and if $D = L/2$, the overlap is 50 percent.

2. Window each segment, using data window $w(n)$

$$x_i(n) = \bar{x}_i(n)w(n) = x(iD + n)w(n) \qquad 0 \le i \le K - 1, 0 \le n \le L - 1 \qquad (5.3.66)$$

3. Compute the N-point DFTs $X_i(k)$ of the segments $x_i(n), 0 \le i \le K - 1$,

$$\tilde{X}_i(k) = \sum_{n=0}^{L-1} x_i(n)e^{-j(2\pi/L)kn} \qquad 0 \le k \le L - 1, 0 \le i \le K - 1 \qquad (5.3.67)$$

4. Accumulate the squares $|\tilde{X}_i(k)|^2$

$$\tilde{S}_i(k) \triangleq \sum_{i=0}^{K-1} |\tilde{X}_i(k)|^2 \qquad 0 \le k \le L/2 \qquad (5.3.68)$$

5. Finally, normalize by KL to obtain the estimate $\hat{R}_x^{(PA)}(k)$:

$$\hat{R}_x^{(PA)}(k) = \frac{1}{KL} \sum_{i=0}^{K-1} \tilde{S}_i(k) \qquad 0 \le k \le N/2 \qquad (5.3.69)$$

At this point we emphasize that the spectrum estimate $\hat{R}_x^{(PA)}(k)$ is always nonnegative. A pictorial description of this computational algorithm is shown in Figure 5.20. A more efficient way to compute $\hat{R}_x^{(PA)}(k)$ is examined in Problem 5.14.

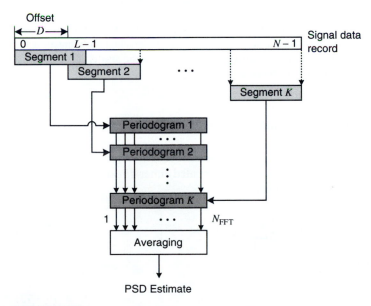

FIGURE 5.20
Pictorial description of the Welch-Bartlett method.

In MATLAB the Welch-Bartlett method is implemented by using the function

$$Rx = psd(x,Nfft,Fs,window(L),Noverlap,'none');$$

where $window$ is the name of any MATLAB-provided window function (e.g., hamming); Nfft is the size of the DFT, which is chosen to be larger than N to obtain a high-density spectrum; Fs is the sampling frequency, which is used for plotting purposes; and Noverlap specifies the number of overlapping samples. If the boxcar window is used along with Noverlap=0, then we obtain Bartlett's method of periodogram averaging. (Note that Noverlap is different from the offset parameter D given above.) If Noverlap=L/2 is used, then we obtain Welch's averaged periodogram method with 50 percent overlap.

A biased estimate $\hat{r}_x(l)$, $|l| < L$, of the autocorrelation sequence of $x(n)$ can be obtained by taking the inverse N-point DFT of $\hat{R}_x^{(PA)}(k)$ if $N \geq 2L - 1$. Since only samples of the continuous spectrum $\hat{R}_x^{(PA)}(e^{j\omega})$ are available, the obtained autocorrelation sequence $\hat{r}_x^{(PA)}(l)$ is an aliased version of the true autocorrelation $r_x(l)$ of the signal $x(n)$ (see Problem 5.15).

> **EXAMPLE 5.3.7 (BARTLETT'S METHOD).** Consider again the spectrum estimation of three sinusoids in white noise given in Example 5.3.4, that is,
>
> $$x(n) = \cos(0.35\pi n + \phi_1) + \cos(0.4\pi n + \phi_2) + 0.25\cos(0.8\pi n + \phi_3) + v(n) \qquad (5.3.70)$$
>
> where ϕ_1, ϕ_2, and ϕ_3 are jointly independent random variables uniformly distributed over $[-\pi, \pi]$ and $v(n)$ is a unit-variance white noise. An ensemble of 50 realizations of $x(n)$ was generated using $N = 512$. The Bartlett estimate of each ensemble was computed for $K = 1$ (i.e., the basic periodogram), $K = 4$ (or $L = 128$), and $K = 8$ (or $L = 64$). The results in the form of estimate overlays and averages are shown in Figure 5.21. Observe that the variance in the estimate has consistently reduced over the periodogram estimate as the number of averaging segments has increased. However, this reduction has come at the price of broadening of the spectral peaks. Since no window is used, the sidelobes are very prominent even for the $L = 8$ segment. Thus confidence in the $\omega = 0.8\pi$ spectral line is not very high for the $L = 8$ case.

> **EXAMPLE 5.3.8 (WELCH'S METHOD).** Consider Welch's method for the random process in the above example for $N = 512$, 50 percent overlap, and a Hamming window. Three different values for L were considered; $L = 256$ (3 segments), $L = 128$ (7 segments), and $L = 64$ (15 segments). The estimate overlays and averages are shown in Figure 5.22. In comparing these results with those in Figure 5.21, note that the windowing has considerably reduced the spurious peaks in the spectra but has also further smoothed the peaks. Thus the peak at 0.8π is recognizable with high confidence, but the separation of two close peaks is not so clear for $L = 64$. However, the $L = 128$ case provides the best balance between separation and detection. On comparing the Blackman-Tukey (Figure 5.18) and Welch estimates, we observe that the results are comparable in terms of variance reduction and smoothing aspects.

5.3.4 Some Practical Considerations and Examples

The periodogram and its modified version, which is the basic tool involved in the estimation of the power spectrum of stationary signals, can be computed either *directly* from the signal samples $\{x(n)\}_0^{N-1}$ using the DTFT formula

$$\hat{R}_x(e^{j\omega}) = \frac{1}{N}\left|\sum_{n=0}^{N-1} w(n)x(n)e^{-j\omega n}\right|^2 \qquad (5.3.71)$$

or *indirectly* using the autocorrelation sequence

$$\hat{R}_x(e^{j\omega}) = \sum_{l=-(N-1)}^{N-1} \hat{r}_x(l)e^{-j\omega l} \qquad (5.3.72)$$

where $\hat{r}_x(l)$ is the estimated autocorrelation of the windowed segment $\{w(n)x(n)\}_0^{N-1}$. The periodogram $\hat{R}_x(e^{j\omega})$ provides an unacceptable estimate of the power spectrum because

1. it has a bias that depends on the length N and the shape of the data window $w(n)$ and
2. its variance is equal to the true spectrum $R_x(e^{j\omega})$.

Given a data segment of fixed duration N, there is no way to reduce the bias, or equivalently to increase the resolution, because it depends on the length and the shape of the window. However, we can reduce the variance either by averaging the single periodogram of the data (method of Blackman-Tukey) or by averaging multiple periodograms obtained by partitioning the available record into smaller overlapping segments (method of Bartlett-Welch).

The method of Blackman-Tukey is based on the following modification of the indirect periodogram formula

$$\hat{R}_x^{(PS)}(e^{j\omega}) = \sum_{l=-(L-1)}^{L-1} \hat{r}_x(l)w_a(l)e^{-j\omega l} \tag{5.3.73}$$

which basically involves windowing of the estimated autocorrelation sequence with a proper

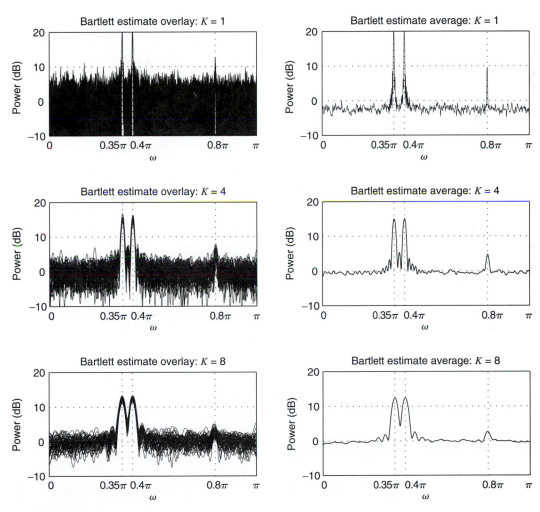

FIGURE 5.21
Estimation of three sinusoids in white noise using Bartlett's method in Example 5.3.7.

correlation window. Using only the first $L \ll N$ more-reliable values of the autocorrelation sequence reduces the variance of the spectrum estimate by a factor of approximately L/N. However, at the same time, this reduces the resolution from about $1/N$ to about $1/L$. The recommended range for L is between $0.1N$ and $0.2N$.

The method of Bartlett-Welch is based on partitioning the available data record into windowed overlapping segments of length L, computing their periodograms by using the direct formula (5.3.71), and then averaging the resulting periodograms to compute the estimate

$$\hat{R}_x^{(\text{PA})}(e^{j\omega}) = \frac{1}{KL} \left| \sum_{n=0}^{L-1} x_i(n) e^{-j\omega n} \right|^2 \tag{5.3.74}$$

whose resolution is reduced to approximately $1/L$ and whose variance is reduced by a factor of about $1/K$, where K is the number of segments.

The reduction in resolution and variance of the Blackman-Tukey estimate is achieved by "averaging" the values of the spectrum at consecutive frequency bins by windowing the estimated autocorrelation sequence. In the Bartlett-Welch method, the same effect is achieved by averaging the values of multiple shorter periodograms at the same frequency

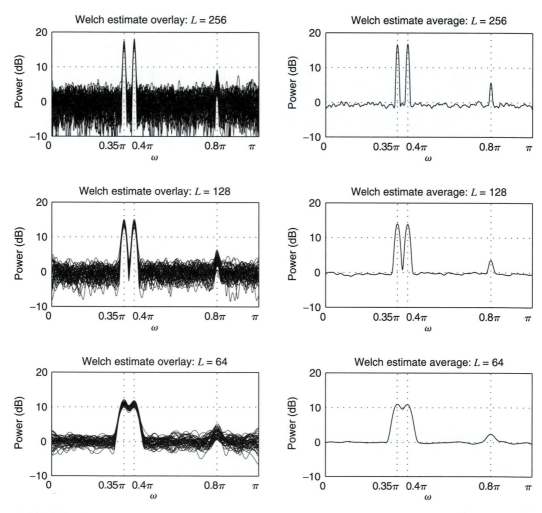

FIGURE 5.22
Estimation of three sinusoids in white noise using Welch's method in Example 5.3.8.

bin. The PSD estimation methods and their properties are summarized in Table 5.3. The multitaper spectrum estimation method given in the last column of Table 5.3 is discussed in Section 5.5.

235

SECTION 5.3
Estimation of the Power
Spectrum of Stationary
Random Signals

TABLE 5.3
Comparison of PSD estimation methods.

	Periodogram $\hat{R}_x(e^{j\omega})$	Single-periodogram smoothing (Blackman-Tukey): $\hat{R}_x^{(PS)}(e^{j\omega})$	Multiple-periodogram averaging (Bartlett-Welch): $\hat{R}_x^{(PA)}(e^{j\omega})$	Multitaper (Thomson): $\hat{R}_x^{(MT)}(e^{j\omega})$
Description of the method	Compute DFT of data record	Compute DFT of windowed autocorrelation estimate (see Figure 5.17)	Split record into K segments and average their modified periodograms (see Figure 5.20)	Window data record using K orthonormal tapers and average their periodograms (see Figure 5.30)
Basic idea	Natural estimator of $R_x(e^{j\omega})$; the error $\|r_x(l) - \hat{r}_x(l)\|$ is large for large $\|l\|$	Local smoothing of $\hat{R}_x(e^{j\omega})$ by weighting $\hat{r}_x(l)$ with a lag window $w_a(l)$	Overlap data records to create more segments; window segments to reduce bias; average periodograms to reduce variance	For properly designed orthogonal tapers, periodograms are independent at each frequency. Hence averaging reduces variance
Bias	Severe for small N; negligible for large N	Asymptotically unbiased	Asymptotically unbiased	Negligible for properly designed tapers
Resolution	$\propto \dfrac{1}{N}$	$\propto \dfrac{1}{L}$, L is maximum lag	$\propto \dfrac{1}{L}$ L is segment length	$\propto \dfrac{1}{N}$
Variance	Unacceptable: about $R_x^2(e^{j\omega})$ for all N	$R_x^2(e^{j\omega}) \times \dfrac{E_w}{N}$	$\dfrac{R_x^2(e^{j\omega})}{K}$ K is number of segments	$\dfrac{R_x^2(e^{j\omega})}{K}$ K is number of tapers

EXAMPLE 5.3.9 (COMPARISON OF BLACKMAN-TUKEY AND WELCH-BARTLETT METHODS).
Figure 5.23 illustrates the properties of the power spectrum estimators based on autocorrelation windowing and periodogram averaging using the AR(4) model (5.3.24). The top plots show the power spectrum of the process. The left column plots show the power spectrum obtained by windowing the data with a Hanning window and the autocorrelation with a Parzen window of length $L = 64$, 128, and 256. We notice that as the length of the window increases, the resolution decreases and the variance increases. We see a similar behavior with the method of averaged periodograms as the segment length L increases from 64 to 256. Clearly, both methods give comparable results if their parameters are chosen properly.

Example of ocean wave data. To apply spectrum estimation techniques discussed in this chapter to real data, we will use two real-valued time series that are obtained by recording the height of ocean waves as a function of time, as measured by two wave gages of different designs. These two series are shown in Figure 5.24. The top graph shows the wire wave gage data while the bottom graph shows the infrared wave gage data. The frequency responses of these gages are such that—mainly because of its inertia—frequencies higher than 1 Hz cannot be reliably measured. The frequency range between 0.2 and 1 Hz is also important because the rate at which the spectrum decreases has a physical model associated with it. Both series were collected at a rate of 30 samples per second. There are 4096 samples in each series.[†] We will also use these data to study joint signal analysis in the next section.

[†]These data were collected by A. Jessup, Applied Physics Laboratory, University of Washington. It was obtained from StatLib, a statistical archive maintained by Carnegie Mellon University.

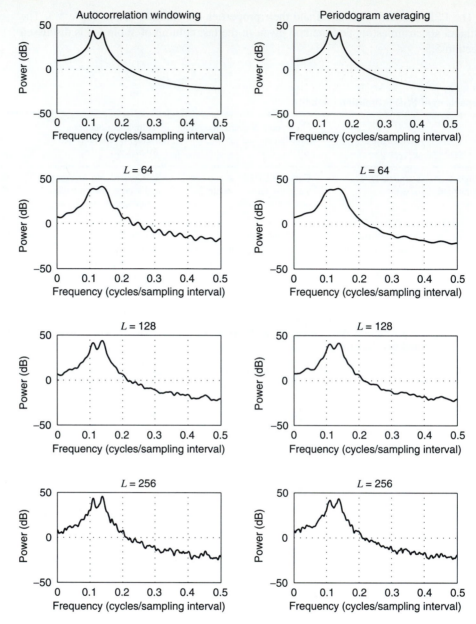

FIGURE 5.23
Illustration of the properties of the power spectrum estimators using autocorrelation
windowing (left column) and periodogram averaging (right column) in Example 5.3.9.

EXAMPLE 5.3.10 (ANALYSIS OF THE OCEAN WAVE DATA). Figure 5.25 depicts the periodogram averaging and smoothing estimates of the wire wave gage data. The top row of plots shows the Welch estimate using a Hamming window, $L = 256$, and 50 percent overlap between segments. The bottom row shows the Blackman-Tukey estimate using a Bartlett window and a lag length of $L = 256$. In both cases, a zoomed view of the plots between 0 and 1 Hz is shown in the right column to obtain a better view of the spectra. Both spectral estimates provide a similar spectral behavior, especially over the frequency range of 0 to 1 Hz. Furthermore, both show a broad, low-frequency peak at 0.13 Hz, corresponding to a period of about 8 s. The dominant features of the time series thus can be attributed to this peak and other features in the 0- to 0.2-Hz range. The shape of the spectrum between 0.2 and 1 Hz is a decaying exponential and is consistent with the physical model. Similar results were obtained for the infrared wave gauge data.

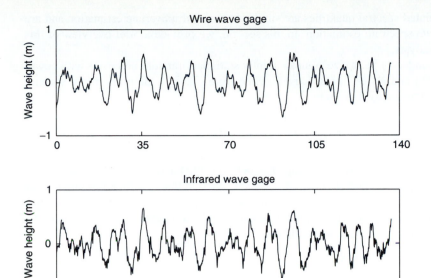

FIGURE 5.24
Display of ocean wave data.

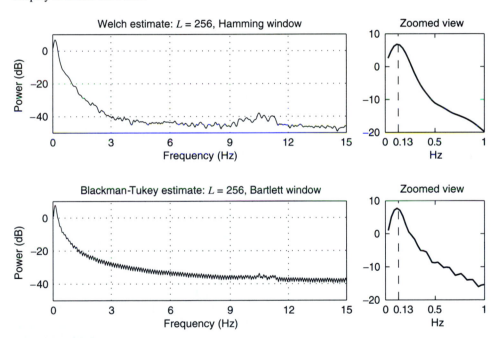

FIGURE 5.25
Spectrum estimation of the ocean wave data using the Welch and Blackman-Tukey methods.

5.4 JOINT SIGNAL ANALYSIS

Until now, we discussed estimation techniques for the computation of the power spectrum of one random process $x(n)$, which is also known as *univariate* spectral estimation. In many practical applications, we have two jointly stationary random processes and we wish to study the correlation between them. The analysis and computation of this correlation and the associated spectral quantities are similar to those of univariate estimation and are called *bivariate* spectrum estimation. In this section, we provide a brief overview of this joint signal analysis.

Let $x(n)$ and $y(n)$ be two zero-mean, jointly stationary random processes with power spectra $R_x(e^{j\omega})$ and $R_y(e^{j\omega})$, respectively. Then from (3.3.61), the cross-power spectral density of $x(n)$ and $y(n)$ is given by

$$R_{xy}(e^{j\omega}) = \sum_{l=-\infty}^{\infty} r_{xy}(l)e^{-j\omega l} \tag{5.4.1}$$

where $r_{xy}(l)$ is the cross-correlation sequence between $x(n)$ and $y(n)$. The cross-spectral density $R_{xy}(e^{j\omega})$ is, in general, a complex-valued function that is difficult to interpret or plot in its complex form. Therefore, we need to express it by using real-valued functions that are easier to deal with. It is customary to express the conjugate of $R_{xy}(e^{j\omega})$ in terms of its real and imaginary components, that is,

$$R_{xy}(e^{j\omega}) = C_{xy}(\omega) - jQ_{xy}(\omega) \tag{5.4.2}$$

where
$$C_{xy}(\omega) \triangleq \text{Re}\,[R_{xy}(e^{j\omega})] \tag{5.4.3}$$

is called the *cospectrum* and

$$Q_{xy}(\omega) \triangleq \text{Im}\,[R_{xy}^*(e^{j\omega})] = -\text{Im}[R_{xy}(e^{j\omega})] \tag{5.4.4}$$

is called the *quadrature spectrum*. Alternately, the most popular approach is to express $R_{xy}(e^{j\omega})$ in terms of its magnitude and angle components, that is,

$$R_{xy}(e^{j\omega}) = A_{xy}(\omega)\exp[j\Phi_{xy}(\omega)] \tag{5.4.5}$$

where
$$A_{xy}(\omega) = |R_{xy}(e^{j\omega})| = \sqrt{C_{xy}^2(\omega) + Q_{xy}^2(\omega)} \tag{5.4.6}$$

and
$$\Phi_{xy}(\omega) = \angle R_{xy}(e^{j\omega}) = \tan^{-1}\{-Q_{xy}(\omega)/C_{xy}(\omega)\} \tag{5.4.7}$$

The magnitude $A_{xy}(\omega)$ is called the *cross-amplitude spectrum*, and the angle $\Phi_{xy}(\omega)$ is called the *phase spectrum*. All these derived functions are real-valued and hence can be examined graphically. However, the phase spectrum has the 2π ambiguity in its computation, which makes its interpretation somewhat problematic.

From (3.3.64) the normalized cross-spectrum, called the *complex coherence function*, is given by

$$\mathcal{G}_{xy}(\omega) = \frac{R_{xy}(e^{j\omega})}{\sqrt{R_x(e^{j\omega})R_y(e^{j\omega})}} \tag{5.4.8}$$

which is a complex-valued frequency-domain correlation coefficient that measures the correlation between the random amplitudes of the complex exponentials with frequency ω in the spectral representations of $x(n)$ and $y(n)$. Hence to interpret this coefficient, its magnitude $|\mathcal{G}_{xy}(\omega)|$ is computed, which is referred to as the *coherency spectrum*. Recall that in Chapter 3, we called $|\mathcal{G}_{xy}(\omega)|^2$ the *magnitude-squared coherence* (MSC). Clearly, $0 \leq |\mathcal{G}_{xy}(\omega)| \leq 1$. Since the coherency spectrum captures the amplitude spectrum but completely ignores the phase spectrum, in practice, the coherency and the phase spectrum are useful real-valued summaries of the cross-spectrum.

5.4.1 Estimation of Cross Power Spectrum

Now we apply the techniques developed in Section 5.3 to the problem of estimating the cross-spectrum and its associated real-valued functions. Let $\{x(n), y(n)\}_0^{N-1}$ be the data record available for estimation. By using the periodogram (5.3.5) as a guide, the estimator for $R_{xy}(e^{j\omega})$ is the *cross-periodogram* given by

$$\hat{R}_{xy}(e^{j\omega}) \triangleq \sum_{l=-(N-1)}^{N-1} \hat{r}_{xy}(l)e^{-j\omega l} \tag{5.4.9}$$

$$\text{where} \quad \hat{r}_{xy}(l) = \begin{cases} \dfrac{1}{N}\displaystyle\sum_{n=0}^{N-l-1} x(n+l)y^*(n) & 0 \le l \le N-1 \\[4mm] \dfrac{1}{N}\displaystyle\sum_{n=0}^{N+l-1} x(n)y^*(n-l) & -(N-1) \le l \le -1 \\[4mm] 0 & l \le -N \text{ or } l \ge N \end{cases} \tag{5.4.10}$$

In analogy to (5.3.2), the cross-periodogram can also be written as

$$\hat{R}_{xy}(e^{j\omega}) = \frac{1}{N} \left[\sum_{n=0}^{N-1} x(n)e^{-j\omega n} \right] \left[\sum_{n=0}^{N-1} y(n)e^{-j\omega n} \right]^* \tag{5.4.11}$$

Once again, it can be shown that the bias and variance properties of the cross-periodogram are as poor as those of the periodogram. Another disturbing result of these periodograms is that from (5.4.11) and (5.3.2), we obtain

$$|\hat{R}_{xy}(e^{j\omega})|^2 = \left(\frac{1}{N}\right)^2 \left| \sum_{n=0}^{N-1} x(n)e^{-j\omega n} \right|^2 \left| \sum_{n=0}^{N-1} y(n)e^{-j\omega n} \right|^2 = \hat{R}_x(e^{j\omega})\hat{R}_y(e^{j\omega})$$

which implies that if we estimate the MSC from the "raw" autoperiodograms as well as cross-periodograms, then the result is always unity for all frequencies. This seemingly unreasonable result is due to the fact that the frequency-domain correlation coefficient at each frequency ω is estimated by using only one single pair of observations from the two signals. Therefore, a reasonable amount of smoothing in the periodogram is necessary to reduce the inherent variability of the cross-spectrum and to improve the accuracy of the estimated coherency. This variance reduction can be achieved by straightforward extensions of various techniques discussed in Section 5.3 for power spectra. These methods include periodogram smoothing across frequencies and the various modified periodogram averaging techniques. In practice, Welch's approach to modified periodogram averaging, based on overlapped segments, is preferred owing to its superior performance. For illustration purposes, we describe Welch's approach in a brief fashion. More details can be found in Carter (1987).

In this approach, we subdivide the existing data records $\{x(n), y(n); 0 \le n \le N-1\}$ into K overlapping smaller segments of length L as follows:

$$\begin{aligned} x_i(n) &= x(iD+n)w(n) \\ y_i(n) &= y(iD+n)w(n) \end{aligned} \qquad 0 \le n \le L-1, 0 \le i \le K-1 \tag{5.4.12}$$

where $w(n)$ is a data window of length L and $D = L/2$ for 50 percent overlap. The cross-periodogram of the ith segment is given by

$$\hat{R}_i(e^{j\omega}) = \frac{1}{L} X_i(e^{j\omega})Y_i^*(e^{j\omega}) = \frac{1}{L} \left[\sum_{n=0}^{L-1} x_i(n)e^{-j\omega n} \right] \left[\sum_{n=0}^{L-1} y_i(n)e^{-j\omega n} \right]^* \tag{5.4.13}$$

Finally, the smoothed cross-spectrum $\hat{R}_{xy}^{(PA)}(e^{j\omega})$ is obtained by averaging K cross-periodograms as follows:

$$\hat{R}_{xy}^{(PA)}(e^{j\omega}) = \frac{1}{K} \sum_{i=0}^{K-1} \hat{R}_i(e^{j\omega}) = \frac{1}{KL} \sum_{i=0}^{K-1} X_i(e^{j\omega})Y_i^*(e^{j\omega}) \tag{5.4.14}$$

Similar to (5.3.51), the DFT computation of $\hat{R}_{xy}^{(PA)}(e^{j\omega})$ is given by

$$\tilde{\hat{R}}_{xy}^{(PA)}(k) = \frac{1}{KL} \sum_{i=0}^{K-1} \left[\sum_{n=0}^{L-1} x_i(n)e^{-j2\pi kn/N} \right] \left[\sum_{n=0}^{L-1} y_i(n)e^{-j2\pi kn/N} \right]^* \qquad (5.4.15)$$

where $0 \leq k \leq N-1$, $N > L$.

Estimation of cospectra and quadrature spectra. Once the cross-spectrum $R_{xy}(e^{j\omega})$ has been estimated, we can compute the estimates of all the associated real-valued spectra by replacing $R_{xy}(e^{j\omega})$ with its estimate $\hat{R}_{xy}^{(PA)}(e^{j\omega})$ in the definitions of these functions. To estimate the cospectrum, we use

$$\hat{C}_{xy}^{(PA)}(\omega) = \text{Re}[\hat{R}_{xy}^{(PA)}(e^{j\omega})] = \text{Re}\left[\frac{1}{KL} \sum_{i=0}^{K-1} X_i(e^{j\omega})Y_i^*(e^{j\omega}) \right] \qquad (5.4.16)$$

and to estimate the quadrature spectrum, we use

$$\hat{Q}_{xy}^{(PA)}(\omega) = -\text{Im}[\hat{R}_{xy}^{(PA)}(e^{j\omega})] = -\text{Im}\left[\frac{1}{KL} \sum_{i=0}^{K-1} X_i(e^{j\omega})Y_i^*(e^{j\omega}) \right] \qquad (5.4.17)$$

The analyses of bias, variance, and covariance of these estimates are similar in complexity to those of the autocorrelation spectral estimates, and the details can be found in Goodman (1957) and Jenkins and Watts (1968).

Estimation of cross-amplitude and phase spectra. Following the definitions in (5.4.6) and (5.4.7), we may estimate the cross-amplitude spectrum $A_{xy}(\omega)$ and the phase spectrum $\Phi_{xy}(\omega)$ between the random processes $x(n)$ and $y(n)$ by

$$\hat{A}_{xy}^{(PA)}(\omega) = \sqrt{[\hat{C}_{xy}^{(PA)}(\omega)]^2 + [\hat{Q}_{xy}^{(PA)}(\omega)]^2} \qquad (5.4.18)$$

and

$$\hat{\Phi}_{xy}^{(PA)}(\omega) = \tan^{-1}\{-\hat{Q}_{xy}^{(PA)}(\omega)/\hat{C}_{xy}^{(PA)}(\omega)\} \qquad (5.4.19)$$

where the estimates $\hat{C}_{xy}^{(PA)}(e^{j\omega})$ and $\hat{Q}_{xy}^{(PA)}(e^{j\omega})$ are given by (5.4.16) and (5.4.17), respectively. Since the cross-amplitude and phase spectral estimates are nonlinear functions of the cospectral and quadrature spectral estimates, their analysis in terms of bias, variance, and covariance is much more complicated. Once again, the details are available in Jenkins and Watts (1968).

Estimation of coherency spectrum. The coherency spectrum is given by the magnitude of the complex coherence $\mathcal{G}_{xy}(\omega)$. Replacing $R_{xy}(e^{j\omega})$, $R_x(e^{j\omega})$, and $R_y(e^{j\omega})$ by their estimates in (5.4.8), we see the estimate for the coherency spectrum is given by

$$|\hat{\mathcal{G}}_{xy}^{(PA)}(\omega)| = \frac{|\hat{R}_{xy}^{(PA)}(e^{j\omega})|}{\sqrt{\hat{R}_x^{(PA)}(e^{j\omega})\hat{R}_y^{(PA)}(e^{j\omega})}} = \left\{ \frac{[\hat{C}_{xy}^{(PA)}(\omega)]^2 + [\hat{Q}_{xy}^{(PA)}(\omega)]^2}{\hat{R}_x^{(PA)}(e^{j\omega})\hat{R}_y^{(PA)}(e^{j\omega})} \right\}^{1/2} \qquad (5.4.20)$$

with bias and variance properties similar to those of the cross-amplitude spectrum.

In MATLAB the function

```
Rxy=csd(x,y,Nfft,Fs,window(L),Noverlap);
```

is available, which is similar to the psd function described in Section 5.3.3. It estimates the cross-spectral density of signal vectors x and y by using Welch's method. The *window* parameter specifies a window function, Fs is the sampling frequency for plotting purposes,

`Nfft` is the size of the FFT used, and `Noverlap` specifies the number of overlapping samples. The function

$$\text{cohere}(x,y,\text{Nfft},\text{Fs},window(\text{L}),\text{Noverlap});$$

estimates the coherency spectrum between two vectors `x` and `y`. Its values are between 0 and 1.

5.4.2 Estimation of Frequency Response Functions

When random processes $x(n)$ and $y(n)$ are the input and output of some physical system, the bivariate spectral estimation techniques discussed in this section can be used to estimate the system characteristics, namely, its frequency response. Problems of this kind arise in many applications including communications, industrial control, and biomedical signal processing. In communications applications, we need to characterize a channel over which signals are transmitted. In this situation, a known training signal is transmitted, and the channel response is recorded. By using the statistics of these two signals, it is possible to estimate channel characteristics within a reasonable accuracy. In the industrial applications such as a gas furnace, the classical methods using step (or sinusoidal) inputs may be inappropriate because of large disturbances generated within the system. Hence, it is necessary to use statistical methods that take into account noise generated in the system.

From Chapter 3, we know that if $x(n)$ and $y(n)$ are input and output signals of an LTI system characterized by the impulse response $h(n)$, then

$$y(n) = h(n) * x(n) \tag{5.4.21}$$

The impulse response $h(n)$, in principle, can be computed through the deconvolution operation. However, deconvolution is not always computationally feasible. If the input and output processes are jointly stationary, then from Chapter 3 we know that the cross-correlation between these two processes is given by

$$r_{yx}(l) = h(l) * r_x(l) \tag{5.4.22}$$

and the cross-spectrum is given by

$$R_{yx}(e^{j\omega}) = H(e^{j\omega})R_x(e^{j\omega}) \tag{5.4.23}$$

or

$$H(e^{j\omega}) = \frac{R_{yx}(e^{j\omega})}{R_x(e^{j\omega})} \tag{5.4.24}$$

Hence, if we can estimate the auto power spectrum and cross power spectrum with reasonable accuracy, then we can determine the frequency response of the system.

Consider next an LTI system with additive output noise,[†] as shown in Figure 5.26. This model situation applies to many practical problems where the input measurements $x(n)$ are essentially without noise while the output measurements $y(n)$ can be modeled by the sum of the ideal response $y_o(n)$ due to $x(n)$ and an additive noise $v(n)$, which is statistically independent of $x(n)$. If we observe the input $x(n)$ and the ideal output $y_o(n)$, the frequency response can be obtained by

$$H(e^{j\omega}) = \frac{R_{y_o x}(e^{j\omega})}{R_x(e^{j\omega})} \tag{5.4.25}$$

where all signals are assumed stationary with zero mean (see Section 5.3.1). Since $x(n)$ and $v(n)$ are independent, we can easily show that

$$R_{y_o x}(e^{j\omega}) = R_{yx}(e^{j\omega}) \tag{5.4.26}$$

[†]More general situations involving both additive input noise and additive output noise are discussed in Bendat and Piersol (1980).

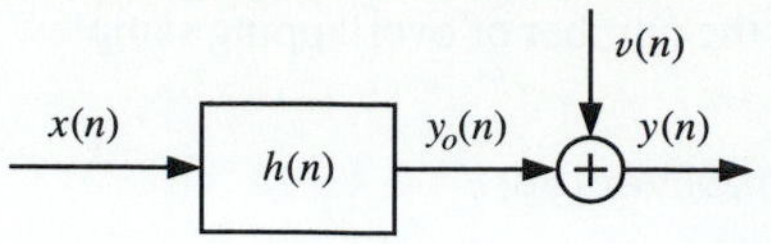

$v(n)$

$x(n)$ → | $h(n)$ | → $y_o(n)$ → (+) → $y(n)$

FIGURE 5.26
Input-output LTI system model with output noise.

and

$$R_y(e^{j\omega}) = R_{y_o}(e^{j\omega}) + R_v(e^{j\omega})$$ (5.4.27)

where

$$R_{y_o}(e^{j\omega}) = |H(e^{j\omega})|^2 R_x(e^{j\omega})$$ (5.4.28)

is the ideal output PSD produced by the input. From (5.4.25) and (5.4.26), we have

$$H(e^{j\omega}) = \frac{R_{yx}(e^{j\omega})}{R_x(e^{j\omega})}$$ (5.4.29)

which shows that we can determine the frequency response by using the cross power spectral density between the noisy output and the input signals. Given a finite record of input-output data $\{x(n), y(n)\}_0^{N-1}$, we estimate $\hat{R}_{yx}^{(\text{PA})}(e^{j\omega_k})$ and $\hat{R}_x^{(\text{PA})}(e^{j\omega_k})$ by using one of the previously discussed methods and then estimate $H(e^{j\omega})$ at a set of equidistant frequencies $\{\omega_k = 2\pi k/K\}_0^{K-1}$, that is,

$$\hat{H}(e^{j\omega_k}) = \frac{\hat{R}_{yx}^{(\text{PA})}(e^{j\omega_k})}{\hat{R}_x^{(\text{PA})}(e^{j\omega_k})}$$ (5.4.30)

The coherence function, which measures the linear correlation between two signals $x(n)$ and $y(n)$ in the frequency domain, is given by

$$\mathcal{G}_{xy}^2(\omega) = \frac{|R_{xy}(e^{j\omega})|^2}{R_x(e^{j\omega})R_y(e^{j\omega})}$$ (5.4.31)

and satisfies the inequality $0 \leq \mathcal{G}_{xy}^2(\omega) \leq 1$ (see Section 3.3.6). If $R_{xy}(e^{j\omega}) = 0$ for all ω, then $\mathcal{G}_{xy}^2(\omega) = 0$. On the other hand, if $y(n) = h(n) * x(n)$, then $\mathcal{G}_{xy}^2(\omega) = 1$ because $R_y(e^{j\omega}) = |H(e^{j\omega})|^2 R_x(e^{j\omega})$ and $R_{xy}(e^{j\omega}) = H^*(e^{j\omega})R_x(e^{j\omega})$. Furthermore, we can show that the coherence function is invariant under linear transformations. Indeed, if $x_1(n) = h_1(n)*x(n)$ and $y_1(n) = h_2(n)*y(n)$, then $\mathcal{G}_{xy}^2(\omega) = \mathcal{G}_{x_1 y_1}^2(\omega)$ (see Problem 5.16). To avoid delta function behavior at $\omega = 0$, we should remove the mean value from the data before we compute $\mathcal{G}_{xy}^2(\omega)$. Also $\mathcal{G}_{xy}^2(\omega)$ should be greater than 0 to avoid division by 0.

In practice, the coherence function is usually greater than 0 and less than 1. This may result from one or more of the following reasons (Bendat and Piersol 1980):

1. Excessive measurement noise.
2. Significant resolution bias in the spectral estimates.
3. The system relating $y(n)$ to $x(n)$ is nonlinear.
4. The output $y(n)$ is *not* produced exclusively by the input $x(n)$.

Using (5.4.28), (5.4.25), $R_{xy_o}(e^{j\omega}) = H^*(e^{j\omega})R_x(e^{j\omega})$, and (5.4.31), we obtain

$$R_{y_o}(e^{j\omega}) = \mathcal{G}_{xy}^2(\omega)R_y(e^{j\omega})$$ (5.4.32)

which is known as the *coherent output PSD*. Combining the last equation with (5.4.27), we have

$$R_v(e^{j\omega}) = [1 - \mathcal{G}_{xy}^2(\omega)]R_y(e^{j\omega})$$ (5.4.33)

which can be interpreted as the part of the output PSD that cannot be produced from the input by using linear operations.

Substitution of (5.4.27) into (5.4.32) results in

$$G_{xy}^2(\omega) = 1 - \frac{R_v(e^{j\omega})}{R_y(e^{j\omega})} \tag{5.4.34}$$

which shows that $G_{xy}^2(\omega) \to 1$ as $R_v(e^{j\omega})/R_y(e^{j\omega}) \to 0$ and $G_{xy}^2(\omega) \to 0$ as $R_v(e^{j\omega})/R_y(e^{j\omega}) \to 1$. Typically, the coherence function between input and output measurements reveals the presence of errors and helps to identify their origin and magnitude. Therefore, the coherence function provides a useful tool for evaluating the accuracy of frequency response estimates.

In MATLAB the function

```
H = tfe(x,y,Nfft,Fs,window(L),Noverlap)
```

is available that estimates the transfer function of the system with input signal x and output y using Welch's method. The `window` parameter specifies a window function, Fs is the sampling frequency for plotting purposes, Nfft is the size of the FFT used, and Noverlap specifies the number of overlapping samples.

We next provide two examples that illustrate some of the problems that may arise when we estimate frequency response functions by using input and output measurements.

EXAMPLE 5.4.1. Consider the AP(4) system

$$H(z) = \frac{1}{1 - 2.7607z^{-1} + 3.8106z^{-2} - 2.6535z^{-3} + 0.9238z^{-4}}$$

discussed in Example 5.3.2. The input is white Gaussian noise, and the output of this system is corrupted by additive white Gaussian noise, as shown in Figure 5.27. We wish to estimate the frequency response of the system from a set of measurements $\{x(n), y(n)\}_0^{N-1}$. Since the input is white, when the output signal-to-noise ratio (SNR) is very high, we can estimate the magnitude response of the system by computing the PSD of the output signal. However, to compute the phase response or a more accurate estimate of the magnitude response, we should use the joint measurements of the input and output signals, as explained above.

Figure 5.27 shows estimates of the MSC function, magnitude response functions (in linear and log scales), and phase response functions for two different levels of output SNR: 32 and 0 dB. When SNR = 32 dB, we note that $|G_{xy}(\omega)|$ is near unity at almost all frequencies, as we theoretically expect for ideal LTI input-output relations. The estimated magnitude and phase responses are almost identical to the theoretical ones with the exception at the two sharp peaks of $|H(e^{j\omega})|$. Since the SNR is high, the two notches in $|G_{xy}(\omega)|$ at the same frequencies suggest a bias error due to the lack of sufficient frequency resolution. When SNR = 0 dB, we see that $|G_{xy}(\omega)|$ falls very sharply for frequencies above 0.2 cycle per sampling interval. We notice that the presence of noise increases the random errors in the estimates of magnitude and phase response in this frequency region, and the bias error in the peaks of the magnitude response. Finally, we note that the uncertainty fluctuations in $|G_{xy}(\omega)|$ increase as $|G_{xy}(\omega)| \to 0$, as predicted by the formula

$$\frac{\text{std}[|\hat{G}_{xy}^{(PA)}(\omega)|]}{|G_{xy}(\omega)|} = \sqrt{2}\frac{1 - |\hat{G}_{xy}^{(PA)}(\omega)|^2}{|G_{xy}^{(PA)}(\omega)|\sqrt{K}} \tag{5.4.35}$$

where $\text{std}(\cdot)$ means standard deviation and K is the number of averaged segments (Bendat and Piersol 1980).

EXAMPLE 5.4.2. In this example we illustrate the use of frequency response estimation to study the effect of respiration and blood pressure on heart rate. Figure 5.28 shows the systolic blood pressure (mmHg), heart rate (beats per minute), and the respiration (mL) signals with their corresponding PSD functions (Grossman 1998). The sampling frequency is $F_s = 5$ Hz, and the PSDs were estimated using the method of averaged periodograms with 50 percent overlap. Note the corresponding quasiperiodic oscillations of blood pressure and heart rate occurring approximately every 12 s (0.08 Hz). Close inspection of the heart rate time series will also reveal

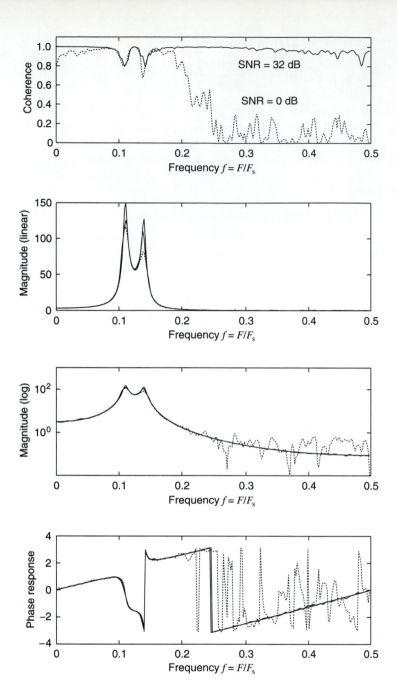

FIGURE 5.27
Estimated coherence, magnitude response, and phase response for the
AP(4) system. The solid lines show the ideal magnitude and phase
responses.

another rhythm corresponding to the respiratory period (about 4.3 s, or 0.23 Hz). These rhythms reflect nervous system mechanisms that control the activity of the heart and the circulation under most circumstances.

The left column of Figure 5.29 shows the coherence, magnitude response, and phase response between respiration as input and heart rate as output. Heart rate fluctuates clearly at the respiratory frequency (here at 0.23 Hz); this is indicated by the large amount of heart rate power and the high degree of coherence at the respiratory frequency. Heart function is largely controlled

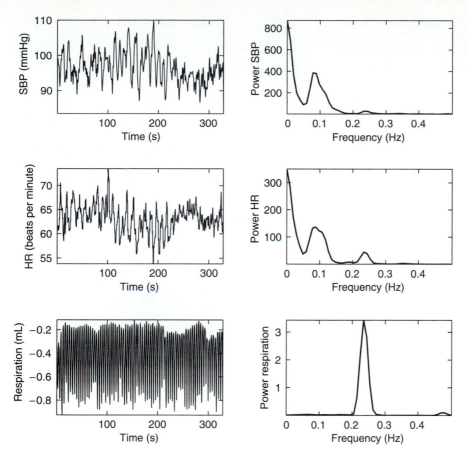

FIGURE 5.28
Continuous systolic blood pressure (SBP), heart rate (HR), and respiration of a young man during quiet upright tilt and their estimated PSDs.

by two branches of the autonomic nervous system, the parasympathetic and sympathetic. Frequency analysis of cardiovascular signals may improve our understanding of the manner in which these two branches interact under varied circumstances. Heart rate fluctuations at the respiratory frequency (termed *respiratory sinus arrhythmia*) are primarily mediated by the parasympathetic branch of the autonomic nervous system. Increases in respiratory sinus arrhythmia indicate enhanced parasympathetic influence upon the heart. Sympathetic oscillations of heart rate occur only at slower frequencies (below 0.10 Hz) owing to the more sluggish frequency response characteristics of the sympathetic branch of the autonomic nervous system.

The right column of Figure 5.29 shows the coherence, magnitude response, and phase response between systolic blood pressure as input and heart rate as output. Coherent oscillations among cardiac and blood pressure signals can often be discerned in a frequency band with a typical center frequency of 0.10 Hz (usual range, 0.07 to 0.12 Hz). This phenomenon has been tied to the cardiovascular baroreflex system, which involves baroreceptors, that is, bodies of cells in the carotid arteries and aorta that are sensitive to stretch. When blood pressure is increased, these baroreceptors fire proportionally to stretch and pressure changes, sending commands via the brain to the heart and circulatory system. This baroreflex system is the only known physiological system acting to buffer rapid and extreme surges or falls in blood pressure. Increased baroreceptor stretch, for example, slows the heart rate by means of increased parasympathetic activity; decreased baroreceptor stretch will elicit cardiovascular sympathetic activation that will speed the heart and constrict arterial vessels. Thus pressure drops due to a decrease in flow. The 0.10-Hz blood pressure oscillations (see PSD in Figure 5.28) are sympathetic in origin and are produced by periodic sympathetic constriction of arterial blood vessels.

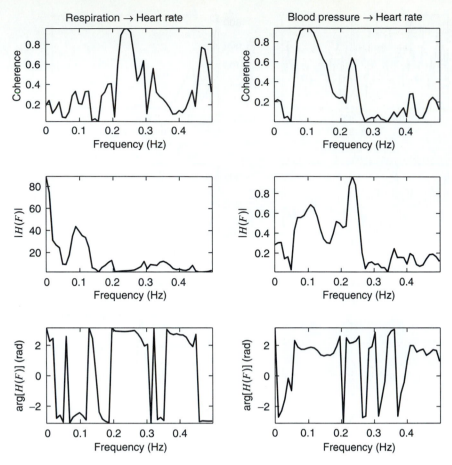

FIGURE 5.29
Coherence, magnitude response, and phase response between respiration as input and
heart rate as output, and systolic blood pressure as input and heart rate as output.

5.5 MULTITAPER POWER SPECTRUM ESTIMATION

Tapering is another name for the data windowing operation in the time domain. The periodogram estimate of the power spectrum, discussed in Section 5.3, is an operation on a data
record $\{x(n)\}_{n=0}^{N-1}$. One interpretation of this finite-duration data record is that it is obtained
by truncating an infinite-duration process $x(n)$ with a rectangular window (or taper). Since
bias and variance properties of the periodogram estimate are unacceptable, methods for
bias and variance reduction were developed either by smoothing estimates in the frequency
domain (using lag windows) or by averaging periodograms computed over several short
segments (data windows). Since these window functions (other than the rectangular one)
typically taper the response toward both ends of the data record, windows are also referred
to as *tapers*.

   In 1982, Thomson suggested an alternate approach for producing a "direct" (or "raw"
periodogram-based) spectral estimator. In this method, rather than use a single rectangular
data taper as in the periodogram estimate, several data tapers are used on the same data
record to compute several modified periodograms. These modified periodograms are then
averaged (with or without weighting) to produce the multitaper spectral estimate. The central
premise of this multitaper approach is that if the data tapers are properly designed orthogonal
functions, then, under mild conditions, the spectral estimates would be independent of each
other at every frequency. Thus, averaging would reduce the variance while proper design of

full-length windows would reduce bias and loss of resolution. Thomson suggested windows based on discrete prolate spheroidal sequences (DPSSs) that form an orthonormal set, although any other orthogonal set with desirable properties can also be used. This DPSS set is also known as the set of *Slepian* tapers. The multitaper method is different in spirit from the other methods in that it does not seek to produce highly smoothed spectra. Detailed discussions of the multitaper approach are given in Thomson (1982) and in Percival and Walden (1993). In this section, we provide a brief sketch of the algorithm.

5.5.1 Estimation of Auto Power Spectrum

Given a data record $\{x(n)\}_{n=0}^{N-1}$ of length N, consider a set of K data tapers $\{w_k(n); 0 \le n \le N-1, 0 \le k \le K-1\}$. These tapers are assumed to be orthonormal, that is,

$$\sum_{n=0}^{N-1} w_k(n)w_l(n) = \begin{cases} 1 & k = l \\ 0 & k \ne l \end{cases} \tag{5.5.1}$$

Let $\hat{R}_{k,x}(e^{j\omega})$ be the periodogram estimator based on kth taper. Then, similar to (5.3.2), we obtain

$$\hat{R}_{k,x}(e^{j\omega}) = \frac{1}{N} \left| \sum_{n=0}^{N-1} w_k(n)x(n)e^{-j\omega n} \right|^2 \tag{5.5.2}$$

The simple averaged multitaper (MT) estimator is then defined by

$$\hat{R}_x^{(\mathrm{MT})}(e^{j\omega}) = \frac{1}{K} \sum_{k=0}^{K-1} \hat{R}_{k,x}(e^{j\omega}) \tag{5.5.3}$$

A pictorial description of this multitaper algorithm is shown in Figure 5.30. Another approach, suggested by Thomson, is to apply adaptive weights (both frequency- and data-dependent) prior to averaging to protect against the biasing degradations of different tapers.

In either case, the multitaper estimator is an average of direct spectral estimators (called eigenspectra by Thomson) employing an orthonormal set of tapers. Thomson (1982) showed that under mild conditions, the orthonormality of the tapers results in an approximate independence of each individual $\hat{R}_{k,x}(e^{j\omega})$ at every frequency ω. This approximate independence further implies that the equivalent degrees of freedom for $\hat{R}_x^{(\mathrm{MT})}(e^{j\omega})$ are equal to twice the number of data tapers. This increase in degrees of freedom is enough to shrink the width of the 95 percent confidence interval for $\hat{R}_x^{(\mathrm{MT})}(e^{j\omega})$ and to reduce the variability to the point at which the overall shape of the spectrum is easily recognizable even though the spectrum is not highly smoothed.

Clearly, the success of this approach lies in the selection of K orthonormal tapers. To understand the rationale behind the selection of these tapers, consider the bias or mean of $\hat{R}_{k,x}(e^{j\omega})$. Following (5.3.15), we obtain

$$E\{\hat{R}_{k,x}(e^{j\omega})\} = \frac{1}{2\pi N} \int_{-\pi}^{\pi} R_x(e^{j\theta}) R_{k,w}(e^{j(\omega-\theta)}) \, d\theta \tag{5.5.4}$$

where

$$R_{k,w}(e^{j\omega}) = \mathcal{F}\{w_k(n) * w_k(-n)\} = |W_k(e^{j\omega})|^2 \tag{5.5.5}$$

It follows, then, from (5.5.3) that

$$E\{\hat{R}_x^{(\mathrm{MT})}(e^{j\omega})\} = \frac{1}{2\pi N} \int_{-\pi}^{\pi} R_x(e^{j\theta}) \bar{R}_w(e^{j(\omega-\theta)}) \, d\theta \tag{5.5.6}$$

where

$$\bar{R}_w(e^{j\omega}) \triangleq \frac{1}{K} \sum_{k=0}^{K-1} |W_k(e^{j\omega})|^2 \tag{5.5.7}$$

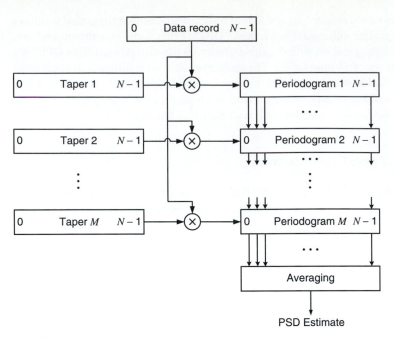

FIGURE 5.30
A pictorial description of the multitaper approach to power spectrum
estimation.

The function $\bar{R}_w(e^{j\omega})$ is the spectral window of the averaged multitaper estimator, which is obtained by averaging spectra of the individual tapers. Hence, for $\bar{R}_w(e^{j\omega})$ to produce a good leakage-free estimate $\hat{R}_x^{(MT)}(e^{j\omega})$, all K spectral windows must provide good protection against leakage. Therefore, each taper must have low sidelobe levels. Furthermore, the averaging of K individual periodograms also reduces the overall variance of $\hat{R}_x^{(MT)}(e^{j\omega})$. The reduction in variance is possible if the $\hat{R}_{k,x}(e^{j\omega})$ are pairwise uncorrelated with common variance, in which case the variance reduces by a factor of $1/K$.

Thus, we need K orthonormal data tapers such that each one provides a good protection against leakage and such that the resulting individual spectral estimates are nearly uncorrelated. One such set is obtained by using DPSS with parameter W and of orders $k = 0, \ldots, K-1$, where K is chosen to be less than or equal to the number $2W$ (called the *Shannon number*, which is also a fixed-resolution bandwidth). The design of these sequences is discussed in detail in Thomson (1982) and in Percival and Walden (1993). In MATLAB these tapers are generated by using the [w]=dpss(L,W) function, where L is the length of $2W$ tapers computed in matrix w.

The first four 21-point DPSS tapers with $W = 4$ and their Fourier transforms are shown in Figure 5.31 while the next four DPSS tapers are shown in Figure 5.32. It can be seen that higher-order tapers assume both positive and negative values. The zeroth-order taper (like other windows) heavily attenuates data values near $n = 0$ and $n = L$. The higher-order tapers successively give greater weights to these values to the point that tapers for $k \geq K$ have very poor bias properties and hence are not used. This behavior is quite evident in the frequency domain where as the taper order increases, mainlobe width and sidelobe attenuation decrease. The multitapering approach can be interpreted as a technique in which higher-order tapers capture information that is "lost" when only the first taper is used.

In MATLAB the function

$$[\text{Pxx}, \text{Pxxc}, \text{F}] = \text{PMTM}(x, W, \text{Nfft}, \text{Fs})$$

estimates the power spectrum of the data vector x in the array Pxx, using the multitaper approach. The function uses DPSS tapers with parameter W and adaptive weighted averaging

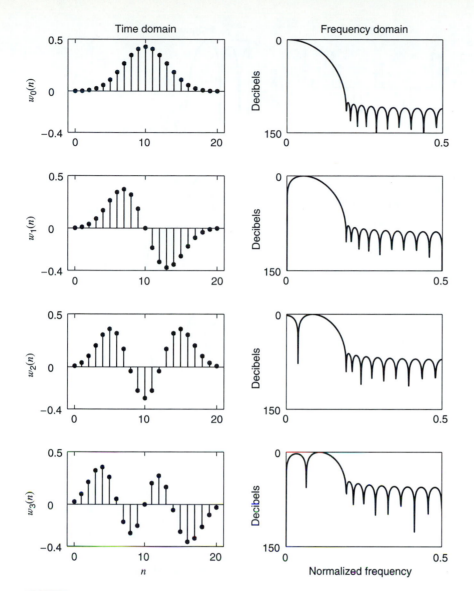

FIGURE 5.31
DPSS data tapers for $k = 0, 1, 2, 3$ in the time and frequency domains.

as the default method. The 95 percent confidence interval is available in Pxxc. The size of
the DFT used is Nfft, the sampling frequency is Fs, and the frequency values are returned
in the vector F.

Another much simpler set of orthonormal tapers was suggested by Reidel and Siderenko
(1995). This particular set contains harmonically related sinusoidal tapers. One important
aspect of multitapering is to reduce the periodogram variance without reducing resolution
caused by smoothing across frequencies. If the spectrum is changing slowly across the band
so that sidelobe bias is not severe (recall the argument given for the unbiasedness of the
periodogram for the white noise process), then sine tapers can reduce the variance. The kth
taper in this set of $k = 0, 1, \ldots, N - 1$ tapers is given by

$$w_k(n) = \sqrt{\frac{2}{N+1}} \sin \frac{\pi (k+1)(n+1)}{N+1} \qquad n = 0, 1, \ldots, N - 1 \qquad (5.5.8)$$

where the amplitude term on the right is a normalization factor that ensures orthonormality of
the tapers. These sine tapers have much narrower mainlobe but also much higher sidelobes

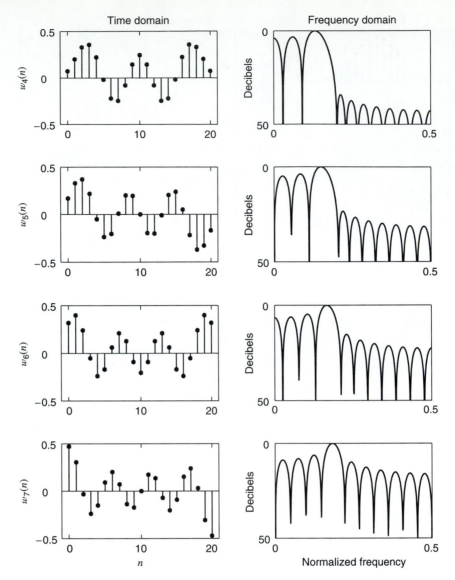

FIGURE 5.32
DPSS data tapers for $k = 4, 5, 6, 7$ in the time and frequency domains.

(recall the rectangular window) than the DPSS tapers. Thus they achieve a smaller bias due to smoothing by the mainlobe than the DPSS tapers, but at the expense of sidelobe suppression. Clearly this performance is acceptable if the spectrum is varying slowly. Owing to their simple nature, these tapers can be analyzed analytically, and it can be shown that (Reidel and Siderenko 1995) the kth sinusoidal taper has its spectral energy concentrated in the frequency bands

$$\frac{\pi k}{N+1} \le |\omega| \le \frac{\pi(k+2)}{N+1} \qquad k = 0, 1, \ldots, N-1 \tag{5.5.9}$$

If the first $K < N$ tapers are used, then the multitaper estimator has the spectral window concentrated in the band

$$\left[-\frac{K+1}{N+1}, \frac{K+1}{N+1} \right] \tag{5.5.10}$$

A summary of the multitaper algorithm performance and its comparison with other PSD estimation methods are given in Table 5.3.

EXAMPLE 5.5.1 (THREE SINUSOIDS IN WHITE NOISE). Consider the random process $x(n)$ containing three sinusoids in white noise discussed earlier, that is,

$$x(n) = \cos(0.35\pi n + \phi_1) + \cos(0.4\pi n + \phi_2) + 0.25\cos(0.8\pi n + \phi_3) + v(n)$$

Fifty realizations of $x(n)$, $0 \le n \le N - 1$, were processed using the PMTM function to obtain multitaper spectrum estimates for $K = 3$, 5, and 7 Slepian tapers. The results are shown in Figure 5.33 in the form of overlays and averages. Several interesting observations and comparisons with the previous methods can be made. The number of tapers used in the estimation determines the variance and the smearing of the spectrum. When fewer tapers are used, the peaks are sharper and narrower but the noise variance is larger. After increasing the number of tapers, the variance is decreased but the peaks become wider. When these estimates are compared with those from Welch's method, an interesting feature can be noticed. The broadening of the peaks is not just at the base but is present along the entire length of the peak. Therefore, even with seven tapers, peaks are distinguishable. This feature is due to the bandwidth of the average spectral window due to K tapers.

EXAMPLE 5.5.2 (OCEAN WAVE DATA). Consider the wire gage wave data of Figure 5.34. The multitaper estimate $\hat{R}_x^{(\text{MT})}(e^{j\omega})$ of these 4096-point data is obtained using the PMTM function in which the parameter W is set to 4. The plots are shown in Figure 5.34. The upper graph shows the spectrum over 0 to 2 Hz while the lower graph shows a zoomed plot over 0 to 0.5 Hz

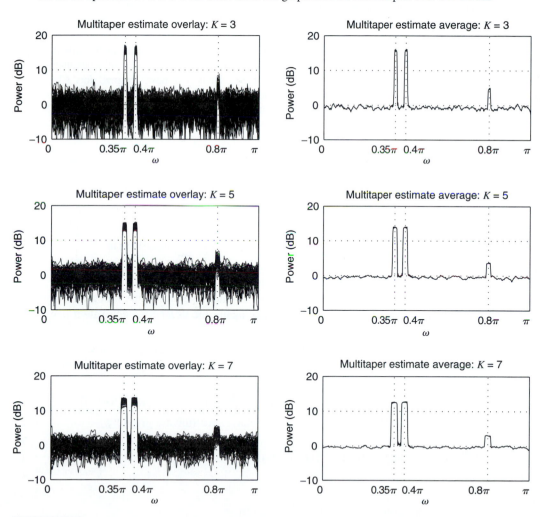

FIGURE 5.33
Spectrum estimation of three sinusoids in white noise using the multitaper method in Example 5.5.1.

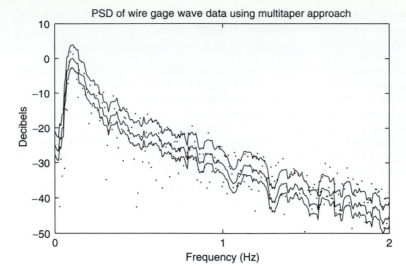

FIGURE 5.34
Spectrum estimation of the wire gage wave data using the multitaper method
in Example 5.5.2.

for clarity. In each graph, the middle solid is the spectral estimate in decibels while the upper
and lower solid curves are the upper and lower limits of the 95 percent confidence interval. For
comparison purposes, the "raw" periodogram estimate is also shown as small dots. Clearly, the
periodogram has a large variability that is reduced in the multitaper estimate. At the same time,
the multitaper estimate is not smooth, but its variability is small enough to follow the shape of
the overall structure.

5.5.2 Estimation of Cross Power Spectrum

The multitapering approach can also be extended to the estimation of the cross power spectrum. Following (5.4.11), the multitaper estimator of the cross power spectrum is given by

$$\hat{R}_{xy}^{(\mathrm{MT})}(e^{j\omega}) = \frac{1}{KL_{xy}} \sum_{k=0}^{K-1} \left[\sum_{n=0}^{L_{xy}-1} w_k(n)x(n)e^{-j\omega n} \right] \left[\sum_{n=0}^{L_{xy}-1} w_k(n)y(n)e^{-j\omega n} \right]^* \quad (5.5.11)$$

where $w_k(n)$ is the kth-order data taper of length L_{xy} and a fixed-resolution bandwidth of
$2W$. As with the auto power spectrum, the use of multitaper averaging reduces the variability of the cross-periodogram $R_{xy}(e^{j\omega})$. Once again, the number of equivalent degrees of
freedom for $\hat{R}_{xy}(e^{j\omega})$ is equal to $2K$.

The real-valued functions associated with the cross power spectrum can also be estimated by using the multitaper approach in a similar fashion. The cospectrum and the
quadrature spectrum are given by

$$\hat{C}_{xy}^{(\mathrm{MT})}(\omega) = \mathrm{Re}[\hat{R}_{xy}^{(\mathrm{MT})}(e^{j\omega})] \quad \text{and} \quad \hat{Q}_{xy}^{(\mathrm{MT})}(\omega) = -\mathrm{Im}[\hat{R}_{xy}^{(\mathrm{MT})}(e^{j\omega})] \quad (5.5.12)$$

while the cross-amplitude spectrum and the phase spectrum are given by

$$\hat{A}_{xy}^{(\mathrm{MT})}(\omega) = \sqrt{[\hat{C}_{xy}^{(\mathrm{MT})}(\omega)]^2 + [\hat{Q}_{xy}^{(\mathrm{MT})}(\omega)]^2} \quad \text{and} \quad \hat{\Phi}_{xy}^{(\mathrm{MT})}(\omega) = \tan^{-1}\left[-\frac{\hat{Q}_{xy}^{(\mathrm{MT})}(\omega)}{\hat{C}_{xy}^{(\mathrm{MT})}(\omega)} \right]$$

$$(5.5.13)$$

Finally, the coherency spectrum is given by

$$|\hat{\mathcal{G}}_{xy}^{(\mathrm{MT})}(\omega)| = \left\{ \frac{[\hat{C}_{xy}^{(\mathrm{MT})}(\omega)]^2 + [\hat{Q}_{xy}^{(\mathrm{MT})}(\omega)]^2}{\hat{R}_x^{(\mathrm{MT})}(e^{j\omega})\hat{R}_y^{(\mathrm{MT})}(e^{j\omega})} \right\}^{1/2} \quad (5.5.14)$$

MATLAB does not provide a function for cross power spectrum estimation using the multitaper approach. However, by using the DPSS function, it is relatively straightforward to implement the simple averaging method of (5.5.11).

EXAMPLE 5.5.3. Again consider the wire gage and the infrared gage wave data of Figure 5.24. The multitaper estimate of the cross power spectrum $\hat{R}_{xy}^{(MT)}(e^{j\omega})$ of these two 4096-point sequences is obtained by using (5.5.11) in which the parameter W is set to 4. Figure 5.35 shows plots of the estimates of the auto power spectra of the two data sets in solid lines. The cross power spectrum of the two signals is shown with a dotted line. It is interesting to note that the two auto power spectra agree almost perfectly over the band up to 0.3 Hz and then reasonably well up to 0.9 Hz, beyond which point the spectrum due to the infrared gage is consistently higher due to high-frequency noise inherent in the measurements. The cross power spectrum agrees with the two auto power spectra at low frequencies up to 0.2 Hz. Figure 5.36 contains two graphs; the

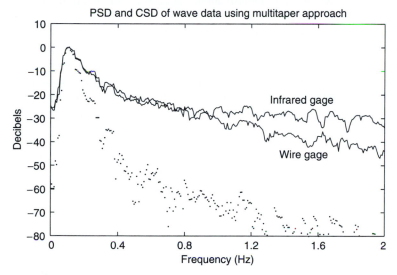

FIGURE 5.35
Cross power spectrum estimation of the wave data using the multitaper approach.

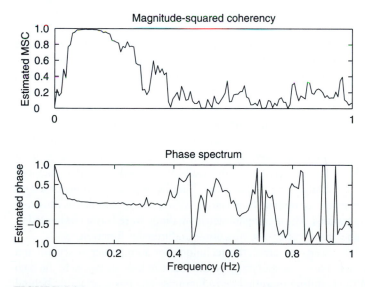

FIGURE 5.36
Coherency and phase spectrum of the wave data using the multitaper approach.

upper graph is for the MSC while the lower one is for the phase spectrum. Consistent with our observation of the cross power spectrum in Figure 5.36, the MSC is almost one over these lower frequencies. The phase spectrum is almost a linear function over the range over which the two auto power spectra agree. Thus, the multitaper approach provides estimates that agree with the conventional techniques.

5.6 SUMMARY

In this chapter, we presented many different nonparametric methods for estimating the power spectrum of a wide-sense stationary random process. Nonparametric methods do not depend on any particular model of the process but use estimators that are determined entirely by the data. Therefore, one has to be very careful about the data and the interpretation of results based on them.

We began by revisiting the topic of frequency analysis of deterministic signals. Since the spectrum estimation of random processes is based on the Fourier transformation of data, the purpose of this discussion was to identify and study errors associated with the practical implementation. In this regard, three problems—the sampling of the continuous signal, windowing of the sampled data, and the sampling of the spectrum—were isolated and discussed in detail. Some useful data windows and their characteristics were also given. This background was necessary to understand more complex spectrum estimation methods and their results.

An important topic of autocorrelation estimation was considered next. Although this discussion was not directly related to spectrum estimation, its inclusion was appropriate since one important method (i.e., that of Blackman and Tukey) was based on this estimation. The statistical properties of the estimator and its implementation completed this topic.

The major part of this chapter was devoted to the section on the auto power spectrum estimation. The classical approach was to develop an estimator from the Fourier transform of the given values of the process. This was called the periodogram method, and it resulted in a natural PSD estimator as a Fourier transform of an autocorrelation estimate. Unfortunately, the statistical analysis of the periodogram showed that it was not an unbiased estimator or a consistent estimator; that is, its variability did not decrease with increasing data record length. The modification of the periodogram using the data window lessened the spectral leakage and improved the unbiasedness but did not decrease the variance. Several examples were given to verify these aspects.

To improve the statistical performance of the simple periodogram, we then looked at several possible improvements to the basic technique. Two main directions emerged for reducing the variance: periodogram smoothing and periodogram averaging. These approaches produced consistent and asymptotically unbiased estimates. The periodogram smoothing was obtained by applying the lag window to the autocorrelation estimate and then Fourier-transforming it. This method was due to Blackman and Tukey, and results of its mean and variance were given. The periodogram averaging was done by segmenting the data to obtain several records, followed by windowing to reduce spectral leakage, and finally by averaging their periodograms to reduce variance. This was the well-known Welch-Bartlett method, and the results of its statistical analysis were also given. Finally, implementations based on the DFT and MATLAB were given for both methods along with several examples to illustrate the performance of their estimates. These nonparametric methods were further extended to estimate the cross power spectrum, coherence functions, and transfer function.

Finally, we presented a newer nonparametric technique for auto power spectrum and cross power spectrum that was based on applying several data windows or tapers to the data followed by averaging of the resulting modified periodograms. The basic principle behind this method was that if the tapers are orthonormal and properly designed (to reduce leakage), then the resulting periodograms can be considered to be independent at each frequency and hence their average would reduce the variance. Two orthogonal sets of data taper, namely, the Slepian and sinusoidal, were provided. The implementation using MATLAB was given, and examples were given to complete the chapter.

PROBLEMS

5.1 Let $x_c(t)$, $-\infty < t < \infty$, be a continuous-time signal with Fourier transform $X_c(F)$, $-\infty < F < \infty$, and let $x(n)$ be obtained by sampling $x_c(t)$ every T per sampling interval with its DTFT $X(e^{j\omega})$.

(a) Show that the DTFT $X(e^{j\omega})$ is given by

$$X(e^{j\omega}) = F_s \sum_{l=-\infty}^{\infty} X_c(f F_s - l F_s) \qquad \omega = 2\pi f \cdot \quad F_s = \frac{1}{T}$$

(b) Let $\tilde{X}_p(k)$ be obtained by sampling $X(e^{j\omega})$ every $2\pi/N$ rad per sampling interval, that is,

$$\tilde{X}_p(k) = X(e^{j2\pi k/N}) = F_s \sum_{l=-\infty}^{\infty} X_c\left(\frac{kF_s}{N} - lF_s\right)$$

Then show that inverse DFT($\tilde{X}_p$) is given by

$$x_p(n) \triangleq \text{IDFT}(\tilde{X}_p) = x_p(n) \triangleq \sum_{m=-\infty}^{\infty} x_c(nT - mNT)$$

5.2 MATLAB provides two functions to generate triangular windows, namely, bartlett and triang. These two functions actually generate two slightly different coefficients.

(a) Use bartlett to generate $N = 11$, 31, and 51 length windows $w_B(n)$, and plot their samples, using the stem function.
(b) Compute the DTFTs $W_B(e^{j\omega})$, and plot their magnitudes over $[-\pi, \pi]$. Determine experimentally the width of the mainlobe as a function of N. Repeat part (a) using the triang function. How are the lengths and the mainlobe widths different in this case? Which window function is an appropriate one in terms of nonzero samples?
(c) Determine the length of the bartlett window that has the same mainlobe width as that of a 51-point rectangular window.

5.3 Sidelobes of the window transform contribute to the spectral leakage due to the frequency-domain convolution. One measure of this leakage is the maximum sidelobe height, which generally occurs at the first sidelobe for all windows except the Dolph-Chebyshev window.

(a) For simple windows such as the rectangular, Hanning, or Hamming window, the maximum sidelobe height is independent of window length N. Choose $N = 11$, 31, and 51, and determine the maximum sidelobe height in decibels for the above windows.
(b) For the Kaiser window, the maximum sidelobe height is controlled by the shape parameter β and is proportional to $\beta/\sinh\beta$. Using several values of β and N, verify the relationship between β and the maximum sidelobe height.
(c) Determine the value of β that gives the maximum sidelobe height nearly the same as that of the Hamming window of the same length. Compare the mainlobe widths and the window coefficients of these two windows.
(d) For the Dolph-Chebyshev window, all sidelobes have the same height A in decibels. For $A = 40$, 50, and 60 dB, determine the 3-dB mainlobe widths for $N = 31$ length window.

5.4 Let $x(n)$ be given by

$$y(n) = \cos\omega_1 n + \cos(\omega_2 n + \phi) \quad \text{and} \quad x(n) = y(n)w(n)$$

where $w(n)$ is a length-N data window. The $|X(e^{j\omega})|^2$ is computed using MATLAB and is plotted over $[0, \pi]$.

(a) Let $w(n)$ be a rectangular window. For $\omega_1 = 0.25\pi$ and $\omega_2 = 0.3\pi$, determine the minimum length N so that the two frequencies in the $|X(e^{j\omega})|^2$ plot are barely separable for any arbitrary $\phi \in [-\pi, \pi]$. (You may want to consider the worst possible value of ϕ or experiment, using several values of ϕ.)

(b) Repeat part (a) for a Hamming window.

(c) Repeat part (a) for a Blackman window.

5.5 In this problem we will prove that the autocorrelation matrix $\hat{\mathbf{R}}_x$ given in (5.2.3), in which the sample correlations are defined by (5.2.1), is a nonnegative definite matrix, that is,

$$\mathbf{x}^H \hat{\mathbf{R}}_x \mathbf{x} \geq 0 \quad \text{for every } \mathbf{x} \geq \mathbf{0}$$

(a) Show that $\hat{\mathbf{R}}_x$ can be decomposed into the product $\mathbf{X}^H \mathbf{X}$, where $\mathbf{X}$ is called a data matrix. Determine the form of $\mathbf{X}$.

(b) Using the above decomposition, now prove that $\mathbf{x}^H \hat{\mathbf{R}}_x \mathbf{x} \geq 0$, for every $\mathbf{x} \geq \mathbf{0}$.

5.6 An alternative autocorrelation estimate $\check{r}_x(l)$ is given in (5.2.13) and is repeated below.

$$\check{r}_x(l) = \begin{cases} \dfrac{1}{N-l} \displaystyle\sum_{n=0}^{N-l-1} x(n+l)x^*(n) & 0 \leq l \leq L < N \\ \check{r}_x^*(-l) & -N < -L \leq l < 0 \\ 0 & \text{elsewhere} \end{cases}$$

(a) Show that the mean of $\check{r}_x(l)$ is equal to $r_x(l)$ and an approximate expression for the variance of $\check{r}_x(l)$.

(b) Show that the mean of the corresponding periodogram [that is, $\check{R}_x(e^{j\omega}) \triangleq \mathcal{F}[\check{r}_x(l)]$] is given by

$$E\{\check{R}_x(e^{j\omega})\} = \frac{1}{2\pi} \int_{-\pi}^{\pi} R_x(e^{j\varphi}) W_R(e^{j\omega-\varphi}) d\varphi$$

where $W_R(e^{\omega})$ is the DTFT of the rectangular window and is sometimes called the *Dirichlet kernel*.

5.7 Consider the above unbiased autocorrelation estimator $\check{r}_x(l)$ of a zero-mean white Gaussian process with variance σ_x^2.

(a) Determine the variance of $\check{r}_x(l)$. Compute its limiting value as $l \to \infty$.

(b) Repeat part (a) for the biased estimator $\hat{r}_x(l)$. Comment on any differences in the results.

5.8 Show that the autocorrelation matrix $\check{\mathbf{R}}_x$ formed by using $\check{r}_x(l)$ is not nonnegative definite, that is,

$$\mathbf{x}^H \hat{\mathbf{R}}_x \mathbf{x} < 0 \quad \text{for some } \mathbf{x} \geq \mathbf{0}$$

5.9 In this problem, we will show that the periodogram $\hat{R}_x(e^{j\omega})$ can also be expressed as a DTFT of the autocorrelation estimate $\hat{r}_x(l)$ given in (5.2.1).

(a) Let $v(n) = x(n)w_R(n)$, where $w_R(n)$ is a rectangular window of length N. Show that

$$\hat{r}_x(l) = \frac{1}{N} v(l) * v^*(-l) \tag{P.1}$$

(b) Take the DTFT of (P.1) to show that

$$\hat{R}_x(e^{j\omega}) = \sum_{l=-N+1}^{N-1} \hat{r}_x(l) e^{-j\omega l}$$

5.10 Consider the following simple windows over $0 \leq n \leq N - 1$: rectangular, Bartlett, Hanning, and Hamming.

(a) Determine analytically the DTFT of each of the above windows.

(b) Sketch the magnitude of these Fourier transforms for $N = 31$.

(c) Verify your sketches by performing a numerical computation of the DTFT using MATLAB.

5.11 The Parzen window is given by

$$
w_P(l) \triangleq
\begin{cases}
1 - 6\left(\dfrac{l}{L}\right)^2 + 6\left(\dfrac{l}{L}\right)^3 & 0 \le |l| \le \dfrac{L}{2} \\[2ex]
2\left(1 - \dfrac{l}{L}\right)^3 & \dfrac{L}{2} < |l| < L \\[2ex]
0 & \text{elsewhere}
\end{cases}
\tag{P.2}
$$

(a) Show that its DTFT is given by

$$
W_P(e^{j\omega}) \simeq \left[\frac{\sin(\omega L/4)}{\sin(\omega/4)}\right]^4 \ge 0
\tag{P.3}
$$

Hence using the Parzen window as a correlation window always produces nonnegative spectrum estimates.

(b) Using MATLAB, compute and plot the time-domain window $w_P(l)$ and its frequency-domain response $W_P(e^{j\omega})$ for $L = 5$, 10, and 20.

(c) From the frequency-domain plots in part (b) experimentally determine the 3-dB mainlobe width $\Delta\omega$ as a function of L.

5.12 The variance reduction ratio of a correlation window $w_a(l)$ is defined as

$$
\frac{\mathrm{var}\{\hat{R}_x^{(PS)}(e^{j\omega})\}}{\mathrm{var}\{\hat{R}_x(e^{j\omega})\}} \simeq \frac{E_w}{N} \qquad 0 < \omega < \pi
$$

where

$$
E_w = \frac{1}{2\pi}\int_{-\pi}^{\pi} W_a^2(e^{j\omega})\, d\omega = \sum_{l=-(L-1)}^{L-1} w_a^2(l)
$$

(a) Using MATLAB, compute and plot E_w as a function of L for the following windows: rectangular, Bartlett, Hanning, Hamming, and Parzen.

(b) Using your computations above, show that for $L \gg 1$, the variance reduction ratio for each window is given by the formula in the following table.

Window name	Variance reduction factor
Rectangular	$2L/N$
Bartlett	$0.667L/N$
Hanning	$0.75L/N$
Hamming	$0.7948L/N$
Parzen	$0.539L/N$

5.13 For $L > 100$, the direct computation of $\hat{r}_x(l)$ using (5.3.49) is time-consuming; hence an indirect computation using the DFT can be more efficient. This computation is implemented by the following steps:

- Given the sequence $\{x(n)\}_{n=0}^{N-1}$, pad enough zeros to make it a $(2N-1)$-point sequence.
- Compute the N_{FFT}-point FFT of $x(n)$ to obtain $\tilde{X}(k)$, where N_{FFT} is equal to the next power-of-2 number that is greater than or equal to $2N - 1$.
- Compute $1/N|\tilde{X}(k)|^2$ to obtain $\tilde{\hat{R}}(k)$.
- Compute the N_{FFT}-point IFFT of $\tilde{\hat{R}}(k)$ to obtain $\hat{r}_x(l)$.

Develop a MATLAB function rx = autocfft(x,L) which computes $\hat{r}_x(l)$, over $-L \le l \le L$. Compare this function with the autoc function discussed in the chapter in terms of the execution time for $L \ge 100$.

5.14 The Welch-Bartlett estimate $\hat{R}_x^{(PA)}(k)$ is given by

$$\hat{R}_x^{(PA)}(k) = \frac{1}{KL} \sum_{i=0}^{K-1} |X_i(e^{j\omega})|^2$$

If $x(n)$ is real-valued, then the sum in the above expression can be evaluated more efficiently. Let K be an even number. Then we will combine two real-valued sequences into one complex-valued sequence and compute one FFT, which will reduce the overall computations. Specifically, let

$$g_r(n) \triangleq x_{2r}(n) + jx_{2r+1}(n) \quad n = 0, 1, \ldots, L-1, r = 0, 1, \ldots, \frac{K}{2} - 1$$

Then the L-point DFT of $g_r(n)$ is given by

$$\tilde{G}_r(k) = \tilde{X}_{2r}(k) + j\tilde{X}_{2r+1}(k) \quad k = 0, 1, \ldots, L-1, r = 0, 1, \ldots, \frac{K}{2} - 1$$

(a) Show that

$$|\tilde{G}_r(k)|^2 + |\tilde{G}_r(L-k)|^2 = 2[|\tilde{X}_{2r}(k)|^2 + |\tilde{X}_{2r+1}(k)|^2] \quad k, r = 0, \ldots, \frac{K}{2} - 1$$

(b) Determine the resulting expression for $\hat{R}_x^{(PA)}(k)$ in terms of $\tilde{G}(k)$.

(c) What changes are necessary if K is an odd number? Provide detailed steps for this case.

5.15 Since $\hat{R}_x^{(PA)}(e^{j\omega})$ is a PSD estimate, one can determine autocorrelation estimate $\hat{r}_x^{(PA)}(l)$ from Welch's method as

$$\hat{r}_x^{(PA)}(l) = \frac{1}{2\pi} \int_{-\pi}^{\pi} \hat{R}_x^{(PA)}(e^{j\omega}) e^{j\omega l} \, d\omega \tag{P.4}$$

Let $\tilde{\hat{R}}_x^{(PA)}(k)$ be the samples of $\hat{R}_x^{(PA)}(e^{j\omega})$ according to

$$\tilde{\hat{R}}_x^{(PA)}(k) \triangleq \hat{R}_x^{(PA)}(e^{j2\pi k/N_{FFT}}) \quad 0 \le k \le N_{FFT} - 1$$

(a) Show that the IDFT $\tilde{\hat{r}}_x^{(PA)}(l)$ of $\tilde{\hat{R}}_x^{(PA)}(k)$ is an aliased version of the autocorrelation estimate $\hat{r}_x^{(PA)}(l)$.

(b) If the length of the overlapping data segment in Welch's method is L, how should N_{FFT} be chosen to avoid aliasing in $\tilde{\hat{r}}_x^{(PA)}(l)$?

5.16 Show that the coherence function $\mathcal{G}_{xy}^2(\omega)$ is invariant under linear transformation, that is, if $x_1(n) = h_1(n) * x(n)$ and $y_1(n) = h_2(n) * y(n)$, then

$$\mathcal{G}_{xy}^2(\omega) = \mathcal{G}_{x_1 y_2}^2(\omega)$$

5.17 Bartlett's method is a special case of Welch's method in which nonoverlapping sections of length L are used without windowing in the periodogram averaging operation.

(a) Show that the ith periodogram in this method can be expressed as

$$\hat{R}_{x,i}(e^{j\omega}) = \sum_{l=-L}^{L} \hat{r}_{x,i}(l) w_B(l) e^{-j\omega l} \tag{P.5}$$

where $w_B(l)$ is a $(2L-1)$-length Bartlett window.

(b) Let $\mathbf{u}(e^{j\omega}) \triangleq [1 \quad e^{j\omega} \quad \cdots \quad e^{j(M-1)\omega}]^T$. Show that $\hat{R}_{x,i}(e^{j\omega})$ in (P.5) can be expressed as a quadratic product

$$\hat{R}_{x,i}(e^{j\omega}) = \mathbf{u}^H(e^{j\omega}) \hat{\mathbf{R}}_{x,i} \mathbf{u}(e^{\omega}) \tag{P.6}$$

where $\hat{\mathbf{R}}_{x,i}$ is the autocorrelation matrix of $\hat{r}_{x,i}(l)$ values.

(c) Finally, show that the Bartlett estimate is given by

$$\hat{R}_x^{(B)}(e^{j\omega}) = \frac{1}{K} \sum_{i=1}^{K} \mathbf{u}^H(e^{j\omega}) \hat{\mathbf{R}}_{x,i} \mathbf{u}(e^{\omega}) \tag{P.7}$$

5.18 In this problem, we will explore a spectral estimation technique that uses combined data and correlation weighting (Carter and Nuttall 1980). In this technique, the following steps are performed:

- Given $\{x(n)\}_{n=0}^{N-1}$, compute the Welch-Bartlett estimate $\hat{R}_x^{(PA)}(e^{j\omega})$ by choosing the appropriate values of L and D.
- Compute the autocorrelation estimate $\hat{r}_x^{(PA)}(l)$, $-L \leq l \leq L$, using the approach described in Problem 5.15.
- Window $\hat{r}_x^{(PA)}(l)$, using a lag window $w_a(n)$ to obtain $\hat{r}_x^{(CN)}(l) \triangleq \hat{r}_x^{(PA)}(l) w_a(n)$.
- Finally, compute the DTFT of $\hat{r}_x^{(CN)}(l)$ to obtain the new spectrum estimate $\hat{R}_x^{(CN)}(e^{j\omega})$.

(a) Determine the bias of $\hat{R}_x^{(CN)}(e^{j\omega})$.
(b) Comment on the effect of additional windowing on the variance and resolution of the estimate.
(c) Implement this technique in MATLAB, and compute spectral estimates of the process containing three sinusoids in white noise, which was discussed in the chapter. Experiment with various values of L and with different windows. Compare your results to those given for the Welch-Bartlett and Blackman-Tukey methods.

5.19 Explain why we use the scaling factor

$$\sum_{n=0}^{L-1} w^2(n)$$

which is the energy of the data window in the Welch-Bartlett method.

5.20 Consider the basic periodogram estimator $\hat{R}_x(e^{j\omega})$ at the zero frequency, that is, at $\omega = 0$.

(a) Show that

$$\hat{R}_x(e^{j0}) = \frac{1}{N} \left| \sum_{n=0}^{N-1} x(n) e^{j0} \right|^2 = \frac{1}{N} \left| \sum_{n=0}^{N-1} x(n) \right|^2$$

(b) If $x(n)$ is a real-valued white Gaussian process with variance σ_x^2, determine the mean and variance of $\hat{R}_x(e^{j0})$.
(c) Determine if $\hat{R}_x(e^{j0})$ is a consistent estimator by evaluating the variance as $N \to \infty$.

5.21 Consider Bartlett's method for estimating $R_x(e^{j0})$ using $L = 1$; that is, we use nonoverlapping segments of single samples. The periodogram of one sample $x(n)$ is simply $|x(n)|^2$. Thus we have

$$\hat{R}_x^{(B)}(e^{j0}) = \frac{1}{N} \sum_{n=0}^{N-1} \hat{R}_{x,n}(e^{j0}) = \frac{1}{N} \sum_{n=0}^{N-1} |x(n)|^2$$

Again assume that $x(n)$ is a real-valued white Gaussian process with variance σ_x^2.

(a) Determine the mean and variance of $\hat{R}_x^{(B)}(e^{j0})$.
(b) Compare the above result with those in Problem 5.20. Comment on any differences.

5.22 One desirable property of lag or correlation windows is that their Fourier transforms are nonnegative.

(a) Formulate a procedure to generate a symmetric lag window of length $2L+1$ with nonnegative Fourier transform.
(b) Using the Hanning window as a prototype in the above procedure, determine and plot a 31-length lag window. Also plot its Fourier transform.

5.23 Consider the following random process

$$x(n) = \sum_{k=1}^{4} A_k \sin(\omega_k n + \phi_k) + v(n)$$

where
$$A_1 = 1 \qquad A_2 = 0.5 \qquad A_3 = 0.5 \qquad A_4 = 0.25$$
$$\omega_1 = 0.1\pi \qquad \omega_2 = 0.6\pi \qquad \omega_3 = 0.65\pi \qquad \omega_4 = 0.8\pi$$

and the phases $\{\phi_i\}_{i=1}^4$ are IID random variables uniformly distributed over $[-\pi, \pi]$. Generate 50 realizations of $x(n)$ for $0 \le n \le 256$.

(a) Compute the Blackman-Tukey estimates for $L = 32$, 64, and 128, using the Bartlett lag window. Plot your results, using overlay and averaged estimates. Comment on your plots.
(b) Repeat part (a), using the Parzen window.
(c) Provide a qualitative comparison between the above two sets of plots.

5.24 Consider the random process given in Problem 5.23.

(a) Compute the Bartlett estimate, using $L = 16$, 32, and 64. Plot your results, using overlay and averaged estimates. Comment on your plots.
(b) Compute the Welch estimate, using 50 percent overlap, Hamming window, and $L = 16$, 32, and 64. Plot your results, using overlay and averaged estimates. Comment on your plots.
(c) Provide a qualitative comparison between the above two sets of plots.

5.25 Consider the random process given in Problem 5.23.

(a) Compute the multitaper spectrum estimate, using $K = 3$, 5, and 7 Slepian tapers. Plot your results, using overlay and averaged estimates. Comment on your plots.
(b) Make a qualitative comparison between the above plots and those obtained in Problems 5.23 and 5.24.

5.26 Generate 1000 samples of an AR(1) process using $a = -0.9$. Determine its theoretical PSD.

(a) Determine and plot the periodogram of the process along with the true spectrum. Comment on the plots.
(b) Compute the Blackman-Tukey estimates for $L = 10$, 20, 50, and 100. Plot these estimates along with the true spectrum. Comment on your results.
(c) Compute the Welch estimates for 50 percent overlap, Hamming window, and $L = 10$, 20, 50, and 100. Plot these estimates along with the true spectrum. Comment on your results.

5.27 Generate 1000 samples of an AR(1) process using $a = 0.9$. Determine its theoretical PSD.

(a) Determine and plot the periodogram of the process along with the true spectrum. Comment on the plots.
(b) Compute the Blackman-Tukey estimates for $L = 10$, 20, 50, and 100. Plot these estimates along with the true spectrum. Comment on your results.
(c) Compute the Welch estimates for 50 percent overlap, Hamming window, and $L = 10$, 20, 50, and 100. Plot these estimates along with the true spectrum. Comment on your results.

5.28 Multitaper estimation technique requires a properly designed orthonormal set of tapers for the desired performance. One set discussed in the chapter was that of harmonically related sinusoids given in (5.5.8).

(a) Design a MATLAB function [tapers] = sine_tapers(N,K) that generates $K < N$ sinusoidal tapers of length N.
(b) Using the above function, compute and plot the Fourier transform magnitudes of the first 5 tapers of length 51.

5.29 Design a MATLAB function Pxx = psd_sinetaper(x,K) that determines the multitaper estimates using the sine tapers.

(a) Apply the function psd_sinetaper to the AR(1) process given in Problem 5.26, and compare its performance.
(b) Apply the function psd_sinetaper to the AR(1) process given in Problem 5.27, and compare its performance.

CHAPTER 6

Optimum Linear Filters

In this chapter, we present the theory and application of optimum linear filters and predictors. We concentrate on linear filters that are optimum in the sense of minimizing the mean square error (MSE). The minimum MSE (MMSE) criterion leads to a theory of linear filtering that is elegant and simple, involves only second-order statistics, and is useful in many practical applications. The optimum filter designed for a given set of second-order moments can be used for any realizations of stochastic processes with the same moments.

We start with the general theory of linear MMSE estimators and their computation, using the triangular decomposition of Hermitian positive definite matrices. Then we apply the general theory to the design of optimum FIR filters and linear predictors for both nonstationary and stationary processes (Wiener filters). We continue with the design of nonparametric (impulse response) and parametric (pole-zero) optimum IIR filters and predictors for stationary processes. Then we present the design of optimum filters for inverse system modeling, blind deconvolution, and their application to equalization of data communication channels. We conclude with a concise introduction to optimum matched filters and eigenfilters that maximize the output SNR. These signal processing methods find extensive applications in digital communication, radar, and sonar systems.

6.1 OPTIMUM SIGNAL ESTIMATION

As we discussed in Chapter 1, the solution of many problems of practical interest depends on the ability to accurately estimate the value $y(n)$ of a signal (*desired response*) by using a set of values (*observations* or *data*) from another related signal or signals. Successful estimation is possible if there is significant statistical dependence or correlation between the signals involved in the particular application. For example, in the linear prediction problem we use the M past samples $x(n-1), x(n-2), \ldots, x(n-M)$ of a signal to estimate the current sample $x(n)$. The echo canceler in Figure 1.17 uses the transmitted signal to form a replica of the received echo. The radar signal processor in Figure 1.27 uses the signals $x_k(n)$ for $1 \leq k \leq M$ received by the linear antenna array to estimate the value of the signal $y(n)$ received from the direction of interest. Although the signals in these and other similar applications have different physical origins, the mathematical formulations of the underlying signal processing problems are very similar.

In array signal processing, the data are obtained by using M different sensors. The situation is simpler for filtering applications, because the data are obtained by delaying a single discrete-time signal; that is, we have $x_k(n) = x(n+1-k), 1 \leq k \leq M$ (see Figure 6.1). Further simplifications are possible in linear prediction, where both the desired response and the data are time samples of the same signal, for example, $y(n) = x(n)$ and

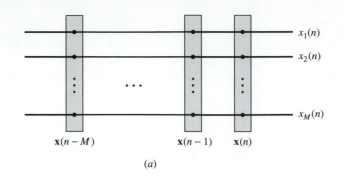

(a)

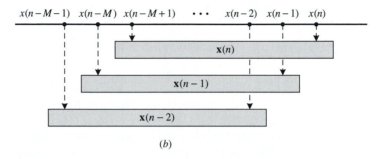

(b)

FIGURE 6.1
Illustration of the data vectors for (a) array processing (multiple sensors)
and (b) FIR filtering or prediction (single sensor) applications.

$x_k(n) = x(n - k)$, $1 \leq k \leq M$. As a result, the design and implementation of optimum filters and predictors are simpler than those for an optimum array processor.

Since array processing problems are the most general ones, we will formulate and solve the following estimation problem: Given a set of data $x_k(n)$ for $1 \leq k \leq M$, determine an estimate $\hat{y}(n)$, of the desired response $y(n)$, using the rule (*estimator*)

$$\hat{y}(n) \triangleq H\{x_k(n), 1 \leq k \leq M\} \tag{6.1.1}$$

which, in general, is a nonlinear function of the data. When $x_k(n) = x(n + 1 - k)$, the estimator takes on the form of a discrete-time filter that can be linear or nonlinear, time-invariant or time-varying, and with a finite- or infinite-duration impulse response. Linear filters can be implemented using any direct, parallel, cascade, or lattice-ladder structure (see Section 2.5 and Proakis and Manolakis 1996).

The difference between the estimated response $\hat{y}(n)$ and the desired response $y(n)$, that is,

$$e(n) \triangleq y(n) - \hat{y}(n) \tag{6.1.2}$$

is known as the *error signal*. We want to find an estimator whose output approximates the desired response as closely as possible according to a certain performance criterion. We use the term *optimum estimator* or *optimum signal processor* to refer to such an estimator. We stress that *optimum* is not used as a synonym for *best*; it simply means the best under the given set of assumptions and conditions. If either the criterion of performance or the assumptions about the statistics of the processed signals change, the corresponding optimum filter will change as well. Therefore, an optimum estimator designed for a certain performance metric and set of assumptions may perform poorly according to some other criterion or if the actual statistics of the processed signals differ from the ones used in the design. For this reason, the sensitivity of the performance to deviations from the assumed statistics is very important in practical applications of optimum estimators.

Therefore, the design of an optimum estimator involves the following steps:

1. Selection of a computational structure with well-defined parameters for the implementation of the estimator.
2. Selection of a criterion of performance or cost function that measures the performance of the estimator under some assumptions about the statistical properties of the signals to be processed.
3. Optimization of the performance criterion to determine the parameters of the optimum estimator.
4. Evaluation of the optimum value of the performance criterion to determine whether the optimum estimator satisfies the design specifications.

Many practical applications (e.g., speech, audio, and image coding) require subjective criteria that are difficult to express mathematically. Thus, we focus on criteria of performance that (1) only depend on the estimation error $e(n)$, (2) provide a sufficient measure of the user satisfaction, and (3) lead to a mathematically tractable problem. We generally select a criterion of performance by compromising between these objectives.

Since, in most applications, negative and positive errors are equally harmful, we should choose a criterion that weights both negative and positive errors equally. Choices that satisfy this requirement include the absolute value of the error $|e(n)|$, or the squared error $|e(n)|^2$, or some other power of $|e(n)|$ (see Figure 6.2). The emphasis put on different values of the error is a key factor when we choose a criterion of performance. For example, the squared-error criterion emphasizes the effect of large errors much more than the absolute error criterion. Thus, the squared-error criterion is more sensitive to outliers (occasional large values) than the absolute error criterion is.

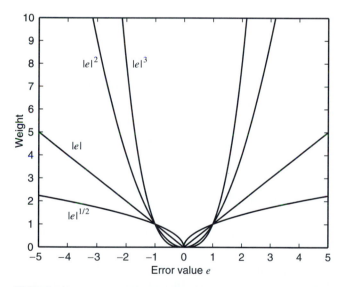

FIGURE 6.2
Graphical illustration of various error-weighting functions.

To develop a mathematical theory that will help to design and analyze the performance of optimum estimators, we assume that the desired response and the data are realizations of stochastic processes. Furthermore, although in practice the estimator operates on specific realizations of the input and desired response signals, we wish to design an estimator with good performance across all members of the ensemble, that is, an estimator that "works well on average." Since, at any fixed time n, the quantities $y(n)$, $x_k(n)$ for $1 \leq k \leq M$, and $e(n)$ are random variables, we should choose a criterion that involves the ensemble or time averaging of some function of $|e(n)|$. Here is a short list of potential criteria of performance:

1. The mean square error criterion

$$P(n) \triangleq E\{|e(n)|^2\} \tag{6.1.3}$$

   which leads, in general, to a nonlinear optimum estimator.

2. The mean αth-order error criterion $E\{|e(n)|^\alpha\}$, $\alpha \neq 2$. Using a lower- or higher-order moment of the absolute error is more appropriate for certain types of non-Gaussian statistics than the MSE (Stuck 1978).

3. The sum of squared errors (SSE)

$$E(n_i, n_f) \triangleq \sum_{n=n_i}^{n_f} |e(n)|^2 \tag{6.1.4}$$

   which, if it is divided by $n_f - n_i + 1$, provides an estimate of the MSE.

The MSE criterion (6.1.3) and the SSE criterion (6.1.4) are the most widely used because they (1) are mathematically tractable, (2) lead to the design of useful systems for practical applications, and (3) can serve as a yardstick for evaluating estimators designed with other criteria (e.g., signal-to-noise ratio, maximum likelihood). In most practical applications, we use linear estimators, which further simplifies their design and evaluation.

   Mean square estimation is a rather vast field that was originally developed by Gauss in the nineteenth century. The current theories of estimation and optimum filtering started with the pioneering work of Wiener and Kolmogorov that was later extended by Kalman, Bucy, and others. Some interesting historical reviews are given in Kailath (1974) and Sorenson (1970).

6.2 LINEAR MEAN SQUARE ERROR ESTIMATION

In this section, we develop the theory of linear MSE estimation. We concentrate on linear estimators for various reasons, including mathematical simplicity and ease of implementation. The problem can be stated as follows:

> Design an estimator that provides an estimate $\hat{y}(n)$ of the desired response $y(n)$ using a *linear* combination of the data $x_k(n)$ for $1 \leq k \leq M$, such that the MSE $E\{|y(n) - \hat{y}(n)|^2\}$ is minimized.

More specifically, the linear estimator is defined by

$$\hat{y}(n) \triangleq \sum_{k=1}^{M} c_k^*(n) x_k(n) \tag{6.2.1}$$

and the goal is to determine the coefficients $c_k(n)$ for $1 \leq k \leq M$ such that the MSE (6.1.3) is minimized. In general, a new set of optimum coefficients should be computed for each time instant n. Since we assume that the desired response and the data are realizations of stochastic processes, the quantities $y(n)$, $x_1(n)$, ..., $x_M(n)$ are random variables at any fixed time n. For convenience, we formulate and solve the estimation problem at a fixed time instant n. Thus, we drop the time index n and restate the problem as follows:

> Estimate a random variable y (the desired response) from a set of related random variables $x_1, x_2, \ldots, x_M$ (data) using the linear estimator

$$\hat{y} \triangleq \sum_{k=1}^{M} c_k^* x_k = \mathbf{c}^H \mathbf{x} \tag{6.2.2}$$

where
$$\mathbf{x} = [x_1 \ x_2 \ \cdots \ x_M]^T \tag{6.2.3}$$

is the *input data vector* and

$$\mathbf{c} = [c_1 \ c_2 \ \cdots \ c_M]^T \tag{6.2.4}$$

is the *parameter* or *coefficient vector* of the estimator.

Unless otherwise stated, *all* random variables are assumed to have *zero*-mean values. The number M of data components used is called the *order* of the estimator. The linear estimator (6.2.2) is represented graphically as shown in Figure 6.3 and involves a computational structure known as the *linear combiner*. The MSE

$$P \triangleq E\{|e|^2\} \tag{6.2.5}$$

where
$$e \triangleq y - \hat{y} \tag{6.2.6}$$

is a function of the parameters c_k. Minimization of (6.2.5) with respect to parameters c_k leads to a linear estimator $\mathbf{c}_o$ that is optimum in the MSE sense. The parameter vector $\mathbf{c}_o$ is known as the *linear MMSE (LMMSE) estimator* and $\hat{y}_o$ as the LMMSE estimate.

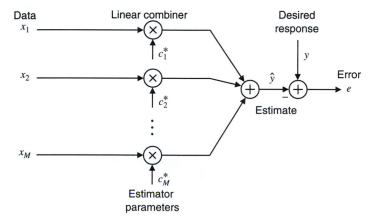

FIGURE 6.3
Block diagram representation of the linear estimator.

6.2.1 Error Performance Surface

To determine the linear MMSE estimator, we seek the value of the parameter vector $\mathbf{c}$ that minimizes the function (6.2.5). To this end, we want to express the MSE as a function of the parameter vector $\mathbf{c}$ and to understand the nature of this dependence.

By using (6.2.5), (6.2.6), (6.2.2), and the linearity property of the expectation operator, the MSE is given by

$$P(\mathbf{c}) = E\{|e|^2\} = E\{(y - \mathbf{c}^H \mathbf{x})(y^* - \mathbf{x}^H \mathbf{c})\}$$

$$= E\{|y|^2\} - \mathbf{c}^H E\{\mathbf{x}y^*\} - E\{y\mathbf{x}^H\}\mathbf{c} + \mathbf{c}^H E\{\mathbf{x}\mathbf{x}^H\}\mathbf{c}$$

or more compactly,

$$P(\mathbf{c}) = P_y - \mathbf{c}^H \mathbf{d} - \mathbf{d}^H \mathbf{c} + \mathbf{c}^H \mathbf{R} \mathbf{c} \tag{6.2.7}$$

where
$$P_y \triangleq E\{|y|^2\} \tag{6.2.8}$$

is the power of the desired response,

$$\mathbf{d} \triangleq E\{\mathbf{x}y^*\} \tag{6.2.9}$$

is the cross-correlation vector between the data vector $\mathbf{x}$ and the desired response y, and

$$\mathbf{R} \triangleq E\{\mathbf{x}\mathbf{x}^H\} \tag{6.2.10}$$

is the correlation matrix of the data vector $\mathbf{x}$. The matrix $\mathbf{R}$ is guaranteed to be Hermitian and nonnegative definite (see Section 3.4.4).

The function $P(\mathbf{c})$ is known as the *error performance surface* of the estimator. Equation (6.2.7) shows that the MSE $P(\mathbf{c})$ (1) depends *only* on the second-order moments of the desired response and the data and (2) is a quadratic function of the estimator coefficients and represents an $(M+1)$-dimensional surface with M degrees of freedom. We will see that if $\mathbf{R}$ is positive definite, then the quadratic function $P(\mathbf{c})$ is bowl-shaped and has a unique minimum that corresponds to the optimum parameters. The next example illustrates this fact for the second-order case.

EXAMPLE 6.2.1. If $M=2$ and the random variables y, x_1, and x_2 are real-valued, the MSE is

$$P(c_1, c_2) = P_y - 2d_1 c_1 - 2d_2 c_2 + r_{11}c_1^2 + 2r_{12}c_1 c_2 + r_{22}c_2^2$$

because $r_{12} = r_{21}$. And $P(c_1, c_2)$ is a second-order function of coefficients c_1 and c_2, and Figure 6.4 shows two plots of the function $P(c_1, c_2)$ that are quite different in appearance. The surface in Figure 6.4(a) looks like a bowl and has a unique extremum that is a minimum. The values for the error surface parameters are $P_y = 0.5$, $r_{11} = r_{22} = 4.5$, $r_{12} = r_{21} = -0.1545$, $d_1 = -0.5$, and $d_2 = -0.1545$. On the other hand, in Figure 6.4(b), we have a saddle point that is neither a

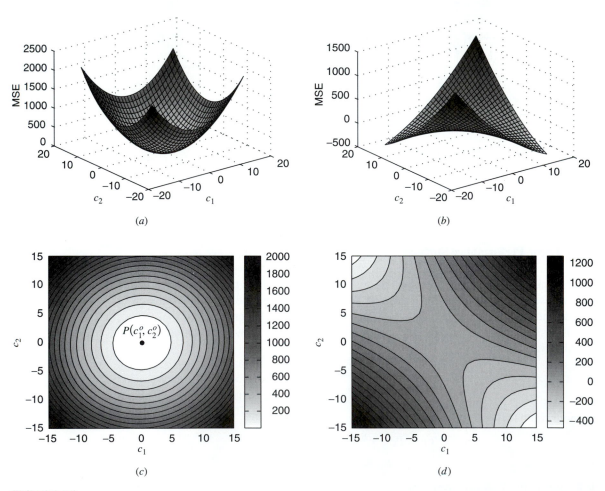

(a)

(b)

(c)

(d)

FIGURE 6.4
Representative surface and contour plots for positive definite and negative definite quadratic error performance surfaces.

minimum nor a maximum (here only the matrix elements have changed to $r_{11} = r_{22} = 1, r_{12} = r_{21} = 2$). If we cut the surfaces with planes parallel to the (c_1, c_2) plane, we obtain *contours* of constant MSE that are shown in Figure 6.4(*c*) and (*d*). In conclusion, the error performance surface is bowl-shaped and has a unique minimum only if the matrix **R** is positive definite (the determinants of the two matrices are 20.23 and -3, respectively). Only in this case can we obtain an estimator that minimizes the MSE, and the contours are concentric ellipses whose center corresponds to the optimum estimator. The bottom of the bowl is determined by setting the partial derivatives with respect to the unknown parameters to zero, that is,

$$\frac{\partial P(c_1, c_2)}{\partial c_1} = 0 \quad \text{which results in} \quad r_{11}c_1^o + r_{12}c_2^o = d_1$$

$$\frac{\partial P(c_1, c_2)}{\partial c_2} = 0 \quad \text{which results in} \quad r_{12}c_1^o + r_{22}c_2^o = d_2$$

This is a linear system of two equations with two unknowns whose solution provides the coefficients c_1^o and c_2^o that minimize the MSE function $P(c_1, c_2)$.

When the optimum filter is specified by a rational system function, the error performance surface may be nonquadratic. This is illustrated in the following example.

EXAMPLE 6.2.2. Suppose that we wish to estimate the real-valued output $y(n)$ of the "unknown" system (see Figure 6.5)

$$G(z) = \frac{0.05 - 0.4z^{-1}}{1 - 1.1314z^{-1} + 0.25z^{-2}}$$

using the pole-zero filter

$$H(z) = \frac{b}{1 - az^{-1}}$$

by minimizing the MSE $E\{e^2(n)\}$ (Johnson and Larimore 1977). The input signal $x(n)$ is white noise with zero mean and variance σ_x^2. The MSE is given by

$$E\{e^2(n)\} = E\{[y(n) - \hat{y}(n)]^2\} = E\{y^2(n)\} - 2E\{y(n)\hat{y}(n)\} + E\{\hat{y}^2(n)\}$$

and is a function of parameters b and a. Since the impulse response $h(n) = ba^n u(n)$ of the optimum filter has infinite duration, we cannot use (6.2.7) to compute $E\{e^2(n)\}$ and to plot the error surface. The three components of $E\{e^2(n)\}$ can be evaluated as follows, using Parseval's theorem: The power of the desired response

$$E\{y^2(n)\} = \sigma_x^2 \sum_{n=0}^{\infty} g^2(n) = \frac{\sigma_x^2}{2\pi j} \oint G(z)G(z^{-1})z^{-1} \, dz \triangleq \sigma_x^2 \sigma_g^2$$

is constant and can be computed either numerically by using the first M "nonzero" samples of $g(n)$ or analytically by evaluating the integral using the residue theorem. The power of the optimum filter output is

$$E\{\hat{y}^2(n)\} = E\{x^2(n)\} \sum_{n=0}^{\infty} h^2(n) = \frac{\sigma_x^2}{2\pi j} \oint H(z)H(z^{-1})z^{-1} \, dz = \sigma_x^2 \frac{b^2}{1 - a^2}$$

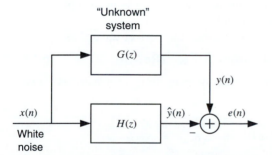

"Unknown"
system

$G(z)$

$y(n)$

$x(n)$

White
noise

$H(z)$

$\hat{y}(n)$

$e(n)$

$+$

$-$

FIGURE 6.5
Identification of an "unknown" system using an optimum filter.

which is a function of parameters b and a. The middle term is

$$E\{y(n)\hat{y}(n)\} = E\left\{\sum_{k=0}^{\infty} g(k)x(n-k)\sum_{m=0}^{\infty} h(m)x(n-m)\right\}$$

$$= \sigma_x^2 \sum_{k=0}^{\infty} g(k)h(k) = \frac{\sigma_x^2}{2\pi j}\oint G(z)H(z^{-1})z^{-1}\,dz = bG(z)|_{z^{-1}=a}$$

because $E\{x(n-k)x(n-m)\} = \sigma_x^2\delta(m-k)$. For convenience we compute the normalized MSE

$$P(b,a) \triangleq \frac{E\{e^2(n)\}}{\sigma_g^2} = \sigma_x^2 - \frac{2b}{\sigma_g^2}G(z)\bigg|_{z^{-1}=a} + \frac{\sigma_x^2}{\sigma_g^2}\frac{b^2}{1-a^2}$$

whose surface and contour plots are shown in Figure 6.6. We note that the resulting error performance surface is bimodal with a global minimum $P = 0.277$ at $(b,a) = (-0.311, 0.906)$ and a local minimum $P = 0.976$ at $(b,a) = (0.114, -0.519)$. As a result, the determination of the optimum filter requires the use of nonlinear optimization techniques with all associated drawbacks.

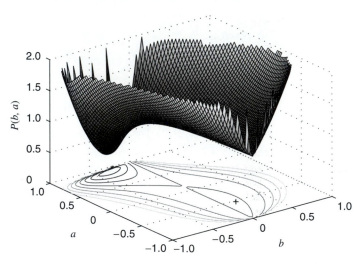

FIGURE 6.6
Illustration of the nonquadratic form of the error performance surface of a pole-zero optimum filter specified by the coefficients of its difference equation.

6.2.2 Derivation of the Linear MMSE Estimator

The approach in Example 6.2.1 can be generalized to obtain the necessary and sufficient conditions that determine the linear MMSE estimator.[†] Here, we present a simpler matrix-based approach that is sufficient for the scope of this chapter.

We first notice that we can put (6.2.7) into the form of a "perfect square" as

$$P(\mathbf{c}) = P_y - \mathbf{d}^H\mathbf{R}^{-1}\mathbf{d} + (\mathbf{Rc} - \mathbf{d})^H\mathbf{R}^{-1}(\mathbf{Rc} - \mathbf{d}) \tag{6.2.11}$$

where only the third term depends on $\mathbf{c}$. If $\mathbf{R}$ is positive definite, the inverse matrix $\mathbf{R}^{-1}$ exists

---

[†]For complex-valued random variables, there are some complications that should be taken into account because $|e|^2$ is not an *analytic* function. This topic is discussed in Appendix B.

and is positive definite; that is, $\mathbf{z}^H \mathbf{R}^{-1} \mathbf{z} > 0$ for all $\mathbf{z} \neq \mathbf{0}$. Therefore, if $\mathbf{R}$ is positive definite, the term $\mathbf{d}^H \mathbf{R}^{-1} \mathbf{d} > 0$ decreases the cost function by an amount determined exclusively by the second-order moments. In contrast, the term $(\mathbf{Rc} - \mathbf{d})^H \mathbf{R}^{-1}(\mathbf{Rc} - \mathbf{d}) > 0$ increases the cost function depending on the choice of the estimator parameters. Thus, the best estimator is obtained by setting $\mathbf{Rc} - \mathbf{d} = \mathbf{0}$.

Therefore, the *necessary and sufficient* conditions that determine the linear MMSE estimator $\mathbf{c}_o$ are

$$\mathbf{Rc}_o = \mathbf{d} \tag{6.2.12}$$

and

$$\mathbf{R} \text{ is positive definite} \tag{6.2.13}$$

In greater detail, (6.2.12) can be written as

$$\begin{bmatrix} r_{11} & r_{12} & \cdots & r_{1M} \\ r_{21} & r_{22} & \cdots & r_{2M} \\ \vdots & \vdots & \ddots & \vdots \\ r_{M1} & r_{M2} & \cdots & r_{MM} \end{bmatrix} \begin{bmatrix} c_1 \\ c_2 \\ \vdots \\ c_M \end{bmatrix} = \begin{bmatrix} d_1 \\ d_2 \\ \vdots \\ d_M \end{bmatrix} \tag{6.2.14}$$

where

$$r_{ij} \triangleq E\{x_i x_j^*\} = r_{ji}^* \tag{6.2.15}$$

and

$$d_i \triangleq E\{x_i y^*\} \tag{6.2.16}$$

and are known as the set of *normal equations*. The invertibility of the correlation matrix $\mathbf{R}$—and hence the existence of the optimum estimator—is guaranteed if $\mathbf{R}$ is positive definite. In theory, $\mathbf{R}$ is guaranteed to be nonnegative definite, but in physical applications it will almost always be positive definite. The normal equations can be solved by using any general-purpose routine for a set of linear equations.

Using (6.2.11) and (6.2.12), we find that the MMSE P_o is

$$P_o = P_y - \mathbf{d}^H \mathbf{R}^{-1} \mathbf{d} = P_y - \mathbf{d}^H \mathbf{c}_o \tag{6.2.17}$$

where we can easily show that the term $\mathbf{d}^H \mathbf{c}_o$ is equal to $E\{|\hat{y}_o|^2\}$, the power of the optimum estimate. If $\mathbf{x}$ and y are uncorrelated ($\mathbf{d} = \mathbf{0}$), we have the worst situation ($P_o = P_y$) because there is no linear estimator that can reduce the MSE. If $\mathbf{d} \neq \mathbf{0}$, there is always going to be some reduction in the MSE owing to the correlation between the data vector $\mathbf{x}$ and the desired response y, assuming that $\mathbf{R}$ is positive definite. The best situation corresponds to $\hat{y} = y$, which gives $P_o = 0$. Thus, for comparison purposes, we use the *normalized MSE*

$$\mathcal{E} \triangleq \frac{P_o}{P_y} = 1 - \frac{P_{\hat{y}_o}}{P_y} \tag{6.2.18}$$

because it is bounded between 0 and 1, that is,

$$0 \leq \mathcal{E} \leq 1 \tag{6.2.19}$$

If $\tilde{\mathbf{c}}$ is the deviation from the optimum vector $\mathbf{c}_o$, that is, if $\mathbf{c} = \mathbf{c}_o + \tilde{\mathbf{c}}$, then substituting into (6.2.11) and using (6.2.17), we obtain

$$P(\mathbf{c}_o + \tilde{\mathbf{c}}) = P(\mathbf{c}_o) + \tilde{\mathbf{c}}^H \mathbf{R} \tilde{\mathbf{c}} \tag{6.2.20}$$

Equation (6.2.20) shows that if $\mathbf{R}$ is positive definite, any deviation $\tilde{\mathbf{c}}$ from the optimum vector $\mathbf{c}_o$ increases the MSE by an amount $\tilde{\mathbf{c}}^H \mathbf{R} \tilde{\mathbf{c}} > 0$, which is known as the *excess MSE*, that is,

$$\text{Excess MSE} \triangleq P(\mathbf{c}_o + \tilde{\mathbf{c}}) - P(\mathbf{c}_o) = \tilde{\mathbf{c}}^H \mathbf{R} \tilde{\mathbf{c}} \tag{6.2.21}$$

We emphasize that the excess MSE depends only on the input correlation matrix and not on the desired response. This fact has important implications because any deviation from the optimum can be detected by monitoring the MSE.

For nonzero-mean random variables, we use the estimator $\hat{y} \triangleq c_0 + \mathbf{c}^H \mathbf{x}$. The elements of $\mathbf{R}$ and $\mathbf{d}$ are replaced by the corresponding covariances and $c_0 = E\{y\} - \mathbf{c}^H E\{\mathbf{x}\}$ (see Problem 6.1). In the sequel, unless otherwise explicitly stated, we assume that all random variables have zero mean or have been reduced to zero mean by replacing y by $y - E\{y\}$ and $\mathbf{x}$ by $\mathbf{x} - E\{\mathbf{x}\}$.

6.2.3 Principal-Component Analysis of the Optimum Linear Estimator

The properties of optimum linear estimators and their error performance surfaces depend on the correlation matrix $\mathbf{R}$. We can learn a lot about the nature of the optimum estimator if we express $\mathbf{R}$ in terms of its eigenvalues and eigenvectors. Indeed, from Section 3.4.4, we have

$$\mathbf{R} = \mathbf{Q}\mathbf{\Lambda}\mathbf{Q}^H = \sum_{i=1}^{M} \lambda_i \mathbf{q}_i \mathbf{q}_i^H \quad \text{and} \quad \mathbf{\Lambda} = \mathbf{Q}^H \mathbf{R} \mathbf{Q} \tag{6.2.22}$$

where

$$\mathbf{\Lambda} = \text{diag}\{\lambda_1, \lambda_2, \ldots, \lambda_M\} \tag{6.2.23}$$

are the eigenvalues of $\mathbf{R}$, assumed to be distinct, and

$$\mathbf{Q} = [\mathbf{q}_1 \ \mathbf{q}_2 \ \cdots \ \mathbf{q}_M] \tag{6.2.24}$$

are the eigenvectors of $\mathbf{R}$. The *modal* matrix $\mathbf{Q}$ is unitary, that is,

$$\mathbf{Q}^H \mathbf{Q} = \mathbf{I} \tag{6.2.25}$$

which implies that $\mathbf{Q}^{-1} = \mathbf{Q}^H$. The relationship (6.2.22) between $\mathbf{R}$ and $\mathbf{\Lambda}$ is known as a *similarity transformation*.

In general, the multiplication of a vector by a matrix changes both the length and the direction of the vector. We define a coordinate transformation of the optimum parameter vector by

$$\mathbf{c}_o' \triangleq \mathbf{Q}^H \mathbf{c}_o \quad \text{or} \quad \mathbf{c}_o \triangleq \mathbf{Q}\mathbf{c}_o' \tag{6.2.26}$$

Since

$$\|\mathbf{c}_o\| = (\mathbf{Q}\mathbf{c}_o')^H \mathbf{Q}\mathbf{c}_o' = \mathbf{c}_o'^H \mathbf{Q}^H \mathbf{Q} \mathbf{c}_o' = \|\mathbf{c}_o'\| \tag{6.2.27}$$

the transformation (6.2.26) changes the direction of the transformed vector but not its length.

If we substitute (6.2.22) into the normal equations (6.2.12), we obtain

$$\mathbf{Q}\mathbf{\Lambda}\mathbf{Q}^H \mathbf{c}_o = \mathbf{d} \quad \text{or} \quad \mathbf{\Lambda}\mathbf{Q}^H \mathbf{c}_o = \mathbf{Q}^H \mathbf{d}$$

which results in

$$\mathbf{\Lambda}\mathbf{c}_o' = \mathbf{d}' \tag{6.2.28}$$

where

$$\mathbf{d}' \triangleq \mathbf{Q}^H \mathbf{d} \quad \text{or} \quad \mathbf{d} \triangleq \mathbf{Q}\mathbf{d}' \tag{6.2.29}$$

is the transformed "decoupled" cross-correlation vector.

Because $\mathbf{\Lambda}$ is diagonal, the set of M equations (6.2.28) can be written as

$$\lambda_i c_{o,i}' = d_i' \quad 1 \le i \le M \tag{6.2.30}$$

where $c_{o,i}'$ and d_i' are the components of $\mathbf{c}_o'$ and $\mathbf{d}'$, respectively. This is an uncoupled set of M first-order equations. If $\lambda_i \ne 0$, then

$$c_{o,i}' = \frac{d_i'}{\lambda_i} \quad 1 \le i \le M \tag{6.2.31}$$

and if $\lambda_i = 0$, the value of $c_{o,i}'$ is indeterminate.

The MMSE becomes

$$P_o = P_y - \mathbf{d}^H \mathbf{c}_o$$
$$= P_y - (\mathbf{Q}\mathbf{d}')^H \mathbf{Q}\mathbf{c}'_o = P_y - \mathbf{d}'^H \mathbf{c}'_o$$
$$= P_y - \sum_{i=1}^{M} d_i'^* c'_{o,i} = P_y - \sum_{i=1}^{M} \frac{|d_i'|^2}{\lambda_i}$$

(6.2.32)

which shows how the eigenvalues and the decoupled cross-correlations affect the performance of the optimum filter. The advantage of (6.2.31) and (6.2.32) is that we can study the behavior of each parameter of the optimum estimator independently of all the remaining ones.

To appreciate the significance of the principal-component transformation, we will discuss the error surface of a second-order estimator. However, all the results can be easily generalized to estimators of order M, whose error performance surface exists in a space of $M + 1$ dimensions. Figure 6.7 shows the contours of constant MSE for a positive definite, second-order error surface. The contours are concentric ellipses centered at the tip of the optimum vector $\mathbf{c}_o$. We define a new coordinate system with origin at $\mathbf{c}_o$ and axes determined by the major axis $\tilde{\mathbf{v}}_1$ and the minor axis $\tilde{\mathbf{v}}_2$ of the ellipses. The two axes are orthogonal, and the resulting system is known as the *principal coordinate system*. The transformation from the "old" system to the "new" system is done in two steps:

$$\text{Translation:} \quad \tilde{\mathbf{c}} = \mathbf{c} - \mathbf{c}_o$$
$$\text{Rotation:} \quad \tilde{\mathbf{v}} = \mathbf{Q}^H \tilde{\mathbf{c}}$$

(6.2.33)

where the rotation changes the axes of the space to match the axes of the ellipsoid. The excess MSE (6.2.21) becomes

$$\Delta P(\tilde{\mathbf{v}}) = \tilde{\mathbf{c}}^H \mathbf{R} \tilde{\mathbf{c}} = \tilde{\mathbf{c}}^H \mathbf{Q} \mathbf{\Lambda} \mathbf{Q}^H \tilde{\mathbf{c}} = \tilde{\mathbf{v}}^H \mathbf{\Lambda} \tilde{\mathbf{v}} = \sum_{i=1}^{M} \lambda_i |\tilde{v}_i|^2$$

(6.2.34)

which shows that the penalty paid for the deviation of a parameter from its optimum value is proportional to the corresponding eigenvalue. Clearly, changes in uncoupled parameters (which correspond to $\lambda_i = 0$) do not affect the excess MSE.

Using (6.2.22), we have

$$\mathbf{c}_o = \mathbf{R}^{-1} \mathbf{d} = \mathbf{Q} \mathbf{\Lambda}^{-1} \mathbf{Q}^H \mathbf{d} = \sum_{i=1}^{M} \frac{\mathbf{q}_i^H \mathbf{d}}{\lambda_i} \mathbf{q}_i = \sum_{i=1}^{M} \frac{d_i'}{\lambda_i} \mathbf{q}_i$$

(6.2.35)

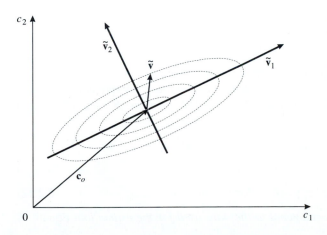

FIGURE 6.7
Contours of constant MSE and principal-component axes for a second-order quadratic error surface.

and the optimum estimate can be written as

$$\hat{y}_o = \mathbf{c}_o^H \mathbf{x} = \sum_{i=1}^{M} \frac{d_i'}{\lambda_i} (\mathbf{q}_i^H \mathbf{x}) \tag{6.2.36}$$

which leads to the representation of the optimum estimator shown in Figure 6.8. The eigen-filters $\mathbf{q}_i$ decorrelate the data vector $\mathbf{x}$ into its principal components, which are weighted and added to produce the optimum estimate.

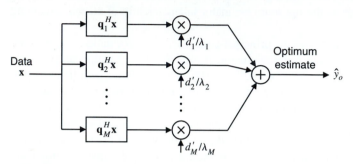

FIGURE 6.8
Principal-components representation of the optimum linear estimator.

6.2.4 Geometric Interpretations and the Principle of Orthogonality

It is convenient and pedagogic to think of random variables with zero mean value and finite variance as vectors in an abstract vector space with an inner product (i.e., a Hilbert space) defined by their correlation

$$\langle x, y \rangle \triangleq E\{xy^*\} \tag{6.2.37}$$

and the length of a vector by

$$\|x\|^2 \triangleq \langle x, x \rangle = E\{|x|^2\} < \infty \tag{6.2.38}$$

From the definition of the correlation coefficient in Section 3.2.1 and the above definitions, we obtain

$$|\langle x, y \rangle|^2 \leq \|x\| \|y\| \tag{6.2.39}$$

which is known as the Cauchy-Schwartz inequality. Two random variables are *orthogonal*, denoted by $x \perp y$, if

$$\langle x, y \rangle = E\{xy^*\} = 0 \tag{6.2.40}$$

which implies they are uncorrelated since they have zero mean.

This geometric viewpoint offers an illuminating and intuitive interpretation for many aspects of MSE estimation that we will find very useful. Indeed, using (6.2.9), (6.2.10), and (6.2.12), we have

$$E\{\mathbf{x}e_o^*\} = E\{\mathbf{x}(y^* - \mathbf{x}^H \mathbf{c}_o)\} = E\{\mathbf{x}y^*\} - E\{\mathbf{x}\mathbf{x}^H\}\mathbf{c}_o = \mathbf{d} - \mathbf{R}\mathbf{c}_o = \mathbf{0}$$

Therefore
$$E\{\mathbf{x}e_o^*\} = \mathbf{0} \tag{6.2.41}$$

or
$$E\{x_m e_o^*\} = 0 \quad \text{for } 1 \leq m \leq M \tag{6.2.42}$$

that is, the *estimation error is orthogonal to the data used for the estimation.* Equations

(6.2.41), or equivalently (6.2.42), are known as the *orthogonality principle* and are widely used in linear MMSE estimation.

To illustrate the use of the orthogonality principle, we note that any linear combination $c_1^* x_1 + \cdots + c_M^* x_M$ lies in the subspace defined by the vectors[†] $x_1, \ldots, x_M$. Therefore, the estimate $\hat{y}$ that minimizes the squared length of the error vector e, that is, the MSE, is determined by the foot of the perpendicular from the tip of the vector y to the "plane" defined by vectors $x_1, \ldots, x_M$. This is illustrated in Figure 6.9 for $M = 2$. Since e_o is perpendicular to every vector in the plane, we have $x_m \perp e_o$, $1 \leq m \leq M$, which leads to the orthogonality principle (6.2.42). Conversely, we can start with the orthogonality principle (6.2.41) and derive the normal equations. This interpretation has led to the name normal equations for (6.2.12). We will see several times that the concept of orthogonality has many important theoretical and practical implications. As an illustration, we apply the Pythagorean theorem to the orthogonal triangle formed by vectors $\hat{y}_o$, e_o, and y, in Figure 6.9, to obtain

$$\|y\|^2 = \|\hat{y}_o\|^2 + \|e_o\|^2$$

or
$$E\{|y|^2\} = E\{|\hat{y}_o|^2\} + E\{|e_o|^2\} \tag{6.2.43}$$

which decomposes the power of the desired response into two components, one that is correlated to the data and one that is uncorrelated to the data.

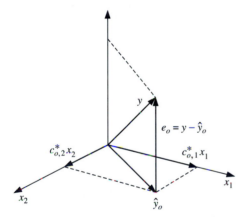

FIGURE 6.9
Pictorial illustration of the orthogonality principle.

6.2.5 Summary and Further Properties

We next summarize, for emphasis and future reference, some important properties of optimum, in the MMSE sense, linear estimators.

1. Equations (6.2.12) and (6.2.17) show that the optimum estimator and the MMSE depend *only* on the second-order moments of the desired response and the data. The dependence on the second-order moments is a consequence of both the linearity of the estimator and the use of the MSE criterion.
2. The error performance surface of the optimum estimator is a quadratic function of its coefficients. If the data correlation matrix is positive definite, this function has a unique minimum that determines the optimum set of coefficients. The surface can be visualized as a bowl, and the optimum estimator corresponds to the bottom of the bowl.

[†]We should be careful to avoid confusing vector random variables, that is, vectors whose components are random variables, and random variables interpreted as vectors in the abstract vector space defined by Equations (6.2.37) to (6.2.39).

3. If the data correlation matrix $\mathbf{R}$ is positive definite, any deviation from the optimum increases the MMSE according to (6.2.21). The resulting excess MSE depends on $\mathbf{R}$ only. This property is very useful in the design of adaptive filters.

4. When the estimator operates with the optimum set of coefficients, the error e_o is uncorrelated (orthogonal) to both the data $x_1, x_2, \ldots, x_M$ and the optimum estimate $\hat{y}_o$. This property is very useful if we want to monitor the performance of an optimum estimator in practice and is used also to design adaptive filters.

5. The MMSE, the optimum estimator, and the optimum estimate can be expressed in terms of the eigenvalues and eigenvectors of the data correlation matrix. See (6.2.32), (6.2.35), and (6.2.36).

6. The general (unconstrained) estimator

$$\hat{y} \triangleq h(\mathbf{x}) = h(x_1, x_2, \ldots, x_M)$$

that minimizes the MSE

$$P = E\{|y - h(\mathbf{x})|^2\}$$

with respect to $h(\mathbf{x})$ is given by the *mean of the conditional density*, that is,

$$\hat{y}_o \triangleq h_o(\mathbf{x}) = E\{y|\mathbf{x}\} = \int_{-\infty}^{\infty} y p_y(y|\mathbf{x})\, dy$$

and clearly is a nonlinear function of $x_1, \ldots, x_M$. If the desired response and the data are jointly Gaussian, the linear MMSE estimator is the best in the MMSE sense; that is, we *cannot* find a nonlinear estimator that produces an estimate with smaller MMSE (Papoulis 1991).

6.3 SOLUTION OF THE NORMAL EQUATIONS

In this section, we present a numerical method for the solution of the normal equations and the computation of the minimum error, using a slight modification of the Cholesky decomposition of Hermitian positive definite matrices known as the *lower-diagonal-upper decomposition*, or *LDL^H decomposition* for short.

Hermitian positive definite matrices can be uniquely decomposed into the product of a lower triangular and a diagonal and an upper triangular matrix as

$$\mathbf{R} = \mathbf{L}\mathbf{D}\mathbf{L}^H \tag{6.3.1}$$

where $\mathbf{L}$ is a unit lower-triangular matrix

$$\mathbf{L} \triangleq \begin{bmatrix} 1 & 0 & \cdots & 0 \\ l_{10} & 1 & \cdots & 0 \\ \vdots & \vdots & \ddots & \vdots \\ l_{M-1,0} & l_{M-1,1} & \cdots & 1 \end{bmatrix} \tag{6.3.2}$$

and

$$\mathbf{D} = \operatorname{diag}\{\xi_1, \xi_2, \ldots, \xi_M\} \tag{6.3.3}$$

is a diagonal matrix with strictly real, positive elements. When the decomposition (6.3.1) is known, we can solve the normal equations

$$\mathbf{R}\mathbf{c}_o = \mathbf{L}\mathbf{D}(\mathbf{L}^H\mathbf{c}_o) = \mathbf{d} \tag{6.3.4}$$

by solving the lower triangular system

$$\mathbf{L}\mathbf{D}\mathbf{k} \triangleq \mathbf{d} \tag{6.3.5}$$

for the intermediate vector $\mathbf{k}$ and the upper triangular system

$$\mathbf{L}^H\mathbf{c}_o = \mathbf{k} \tag{6.3.6}$$

for the optimum estimator $\mathbf{c}_o$. The advantage is that the solution of triangular systems of equations is trivial.

We next provide a constructive proof of the LDL^H decomposition by example and illustrate its application to the solution of the normal equations for $M = 4$. The generalization to an arbitrary order is straightforward and is given in Section 7.1.4.

EXAMPLE 6.3.1. Writing the decomposition (6.3.1) explicitly for $M = 4$, we have

$$
\begin{bmatrix} r_{11} & r_{12} & r_{13} & r_{14} \\ r_{21} & r_{22} & r_{23} & r_{24} \\ r_{31} & r_{32} & r_{33} & r_{34} \\ r_{41} & r_{42} & r_{43} & r_{44} \end{bmatrix} = \begin{bmatrix} 1 & 0 & 0 & 0 \\ l_{10} & 1 & 0 & 0 \\ l_{20} & l_{21} & 1 & 0 \\ l_{30} & l_{31} & l_{32} & 1 \end{bmatrix} \begin{bmatrix} \xi_1 & 0 & 0 & 0 \\ 0 & \xi_2 & 0 & 0 \\ 0 & 0 & \xi_3 & 0 \\ 0 & 0 & 0 & \xi_4 \end{bmatrix} \begin{bmatrix} 1 & l_{10}^* & l_{20}^* & l_{30}^* \\ 0 & 1 & l_{21}^* & l_{31}^* \\ 0 & 0 & 1 & l_{32}^* \\ 0 & 0 & 0 & 1 \end{bmatrix}
$$

(6.3.7)

where $r_{ij} = r_{ji}$ and $\xi_i > 0$, by assumption. If we perform the matrix multiplications on the right-hand side of (6.3.7) and equate the matrix elements on the left and right sides, we obtain

$$r_{11} = \xi_1 \qquad\qquad \Rightarrow \quad \xi_1 = r_{11}$$

$$r_{21} = \xi_1 l_{10} \qquad\qquad \Rightarrow \quad l_{10} = \frac{r_{21}}{\xi_1}$$

$$r_{22} = \xi_1 |l_{10}|^2 + \xi_2 \qquad\qquad \Rightarrow \quad \xi_2 = r_{22} - \xi_1 |l_{10}|^2$$

$$r_{31} = \xi_1 l_{20} \qquad\qquad \Rightarrow \quad l_{20} = \frac{r_{31}}{\xi_1}$$

$$r_{32} = \xi_1 l_{20} l_{10}^* + \xi_2 l_{21} \qquad\qquad \Rightarrow \quad l_{21} = \frac{r_{32} - \xi_1 l_{20} l_{10}^*}{\xi_2}$$

$$r_{33} = \xi_1 |l_{20}|^2 + \xi_2 |l_{21}|^2 + \xi_3 \qquad\qquad \Rightarrow \quad \xi_3 = r_{33} - \xi_1 |l_{20}|^2 - \xi_2 |l_{21}|^2$$

$$r_{41} = \xi_1 l_{30} \qquad\qquad \Rightarrow \quad l_{30} = \frac{r_{41}}{\xi_1}$$

$$r_{42} = \xi_1 l_{30} l_{10}^* + \xi_2 l_{31} \qquad\qquad \Rightarrow \quad l_{31} = \frac{r_{42} - \xi_1 l_{30} l_{10}^*}{\xi_2}$$

$$r_{43} = \xi_1 l_{30} l_{20}^* + \xi_2 l_{31} l_{21}^* + \xi_3 l_{32} \qquad\qquad \Rightarrow \quad l_{32} = \frac{r_{43} - \xi_1 l_{30} l_{20}^* - \xi_2 l_{31} l_{21}^*}{\xi_3}$$

$$r_{44} = \xi_1 |l_{30}|^2 + \xi_2 |l_{31}|^2 + \xi_3 |l_{32}|^2 + \xi_4 \quad \Rightarrow \quad \xi_4 = r_{44} - \xi_1 |l_{30}|^2 - \xi_2 |l_{31}|^2 - \xi_3 |l_{32}|^2$$

(6.3.8)

which provides a row-by-row computation of the elements of the LDL^H decomposition. We note that the computation of the next row does not change the already computed rows.

The lower unit triangular system in (6.3.5) becomes

$$
\begin{bmatrix} 1 & 0 & 0 & 0 \\ l_{10} & 1 & 0 & 0 \\ l_{20} & l_{21} & 1 & 0 \\ l_{30} & l_{31} & l_{32} & 1 \end{bmatrix} \begin{bmatrix} k_1 \\ k_2 \\ k_3 \\ k_4 \end{bmatrix} = \begin{bmatrix} d_1/\xi_1 \\ d_2/\xi_2 \\ d_3/\xi_3 \\ d_4/\xi_4 \end{bmatrix}
$$

(6.3.9)

and can be solved by forward substitution, starting with the first equation. Indeed, we obtain

$$k_1 = \frac{d_1}{\xi_1} \qquad\qquad \Rightarrow \quad k_1 = \frac{d_1}{\xi_1}$$

$$l_{10} k_1 + k_2 = \frac{d_2}{\xi_2} \qquad\qquad \Rightarrow \quad k_2 = \frac{d_2}{\xi_2} - l_{10} k_1$$

$$l_{20} k_1 + l_{21} k_2 + k_3 = \frac{d_3}{\xi_3} \qquad\qquad \Rightarrow \quad k_3 = \frac{d_3}{\xi_3} - l_{20} k_1 - l_{21} k_2$$

(6.3.10)

$$l_{30} k_1 + l_{31} k_2 + l_{32} k_3 + k_4 = \frac{d_4}{\xi_4} \qquad \Rightarrow \quad k_4 = \frac{d_4}{\xi_4} - l_{30} k_1 - l_{31} k_2 - l_{32} k_3$$

which compute the coefficients k_i in "forward" order. Then, the optimum estimator is obtained by solving the upper unit triangular system in (6.3.6) by backward substitution, starting from the last equation. Indeed, we have

$$\begin{bmatrix} 1 & l_{10}^* & l_{20}^* & l_{30}^* \\ 0 & 1 & l_{21}^* & l_{31}^* \\ 0 & 0 & 1 & l_{32}^* \\ 0 & 0 & 0 & 1 \end{bmatrix} \begin{bmatrix} c_1^{(4)} \\ c_2^{(4)} \\ c_3^{(4)} \\ c_4^{(4)} \end{bmatrix} = \begin{bmatrix} k_1 \\ k_2 \\ k_3 \\ k_4 \end{bmatrix} \Rightarrow \begin{aligned} c_4^{(4)} &= k_4 \\ c_3^{(4)} &= k_3 - l_{32}^* c_4 \\ c_2^{(4)} &= k_2 - l_{21}^* c_3 - l_{31}^* c_4 \\ c_1^{(4)} &= k_1 - l_{10}^* c_2 - l_{20}^* c_3 - l_{30}^* c_4 \end{aligned} \tag{6.3.11}$$

that is, the coefficients of the optimum estimator are computed in "backward" order. As a result of this backward substitution, computing one more coefficient for the optimum estimator changes *all* the previously computed coefficients. Indeed, the coefficients of the third-order estimator are

$$\begin{bmatrix} 1 & l_{10}^* & l_{20}^* \\ 0 & 1 & l_{21}^* \\ 0 & 0 & 1 \end{bmatrix} \begin{bmatrix} c_1^{(3)} \\ c_2^{(3)} \\ c_3^{(3)} \end{bmatrix} = \begin{bmatrix} k_1 \\ k_2 \\ k_3 \end{bmatrix} \Rightarrow \begin{aligned} c_3^{(3)} &= k_3 \\ c_2^{(3)} &= k_2 - l_{21}^* c_3^{(3)} \\ c_1^{(3)} &= k_1 - l_{10}^* c_2^{(3)} - l_{20}^* c_3^{(3)} \end{aligned} \tag{6.3.12}$$

which are different from the first three coefficients of the fourth-order estimator.

Careful inspection of the formulas for r_{11}, r_{22}, r_{33}, and r_{44} shows that the diagonal elements of $\mathbf{R}$ provide an upper bound for the elements of $\mathbf{L}$ and $\mathbf{D}$, which is the reason for the good numerical properties of the LDL^H decomposition algorithm. The general formulas for the row-by-row computation of the triangular decomposition, forward substitution, and backward substitution are given in Table 6.1 and can be easily derived by generalizing the results of the previous example. The triangular decomposition requires $M^3/6$ operations, and the solution of each triangular system requires $M(M+1)/2 \approx M^2/2$ operations.

TABLE 6.1
Solution of normal equations using triangular decomposition.

For $i = 1, 2, \ldots, M$ and for $j = 0, 1, \ldots, i - 1$,

$$l_{ij} = \frac{1}{\xi_i} \left(r_{i+1, j+1} - \sum_{m=0}^{j-1} \xi_{m+1} l_{im} l_{jm}^* \right) \qquad \text{(not executed when } i = M)$$

$$\xi_i = r_{ii} - \sum_{m=1}^{i-1} \xi_m |l_{i-1, m-1}|^2$$

For $i = 1, 2, \ldots, M$,

$$k_i = \frac{d_i}{\xi_i} - \sum_{m=0}^{i-2} l_{i-1, m} k_{m+1}$$

For $i = M, M-1, \ldots, 1$,

$$c_i = k_i - \sum_{m=i+1}^{M} l_{m-1, i-1}^* c_m$$

The decomposition (6.3.1) leads to an interesting and practical formula for the computation of the MMSE *without* using the optimum estimator coefficients. Indeed, using (6.2.17), (6.3.6), and (6.3.1), we obtain

$$P_o = P_y - \mathbf{c}_o^H \mathbf{R} \mathbf{c}_o = P_y - \mathbf{k}^H \mathbf{L}^{-1} \mathbf{R} (\mathbf{L}^{-1})^H \mathbf{k} = P_y - \mathbf{k}^H \mathbf{D} \mathbf{k} \tag{6.3.13}$$

or in scalar form

$$P_o = P_y - \sum_{i=1}^{M} \xi_i |k_i|^2 \tag{6.3.14}$$

since $\mathbf{D}$ is diagonal. Equation (6.3.14) shows that because $\xi_i > 0$, *increasing the order of the filter can only reduce the minimum error and hence leads to a better estimate.* Another important application of (6.3.14) is in the computation of arbitrary positive definite quadratic forms. Such problems arise in various statistical applications, such as detection and hypothesis testing, involving the correlation matrix of Gaussian processes (McDonough and Whalen 1995).

Since the determinant of a unit lower triangular matrix equals 1, from (6.3.1) we obtain

$$\det \mathbf{R} = \prod_{i=1}^{M} \xi_i \qquad (6.3.15)$$

which shows that if $\mathbf{R}$ is positive definite, $\xi_i > 0$ for all i, and vice versa.

The triangular decomposition of symmetric, positive definite matrices is numerically stable. The function [L,D]=ldlt(R) implements the first part of the algorithm in Table 6.1, and it fails only if matrix $\mathbf{R}$ is *not* positive definite. Therefore, it can be used as an efficient test to find out whether a symmetric matrix is positive definite. The function [co,Po]=lduneqs(L,D,d) computes the MMSE estimator using the last formula in Table 6.1 and the corresponding MMSE using (6.3.14).

To summarize, linear MMSE estimation involves the following computational steps

$$
\begin{array}{lll}
1.\ \mathbf{R} = E\{\mathbf{x}\mathbf{x}^H\}, \mathbf{d} = E\{\mathbf{x}y^*\} & \text{Normal equations } \mathbf{R}\mathbf{c}_o = \mathbf{d} & \\
2.\ \mathbf{R} = \mathbf{LDL}^H & \text{Triangular decomposition} & \\
3.\ \mathbf{LDk} = \mathbf{d} & \text{Forward substitution } \rightarrow \mathbf{k} & \\
4.\ \mathbf{L}^H \mathbf{c}_o = \mathbf{k} & \text{Backward substitution } \rightarrow \mathbf{c}_o & (6.3.16)\\
5.\ P_o = P_y - \mathbf{k}^H \mathbf{D}\mathbf{k} & \text{MMSE computation} & \\
6.\ e = y - \mathbf{c}_o^H \mathbf{x} & \text{Computation of residuals} &
\end{array}
$$

The vector $\mathbf{k}$ can also be obtained using the LDL^H decomposition of an augmented correlation matrix as follows. To this end, consider now the augmented vector

$$\bar{\mathbf{x}} = \begin{bmatrix} \mathbf{x} \\ y \end{bmatrix} \qquad (6.3.17)$$

and its correlation matrix

$$\bar{\mathbf{R}} = E\{\bar{\mathbf{x}}\bar{\mathbf{x}}^H\} = \begin{bmatrix} E\{\mathbf{x}\mathbf{x}^H\} & E\{\mathbf{x}y^*\} \\ E\{y\mathbf{x}^H\} & E\{|y|^2\} \end{bmatrix} = \begin{bmatrix} \mathbf{R} & \mathbf{d} \\ \mathbf{d}^H & P_y \end{bmatrix} \qquad (6.3.18)$$

We can easily show that the LDL^H decomposition of $\bar{\mathbf{R}}$ is

$$\bar{\mathbf{R}} = \begin{bmatrix} \mathbf{L} & \mathbf{0} \\ \mathbf{k}^H & 1 \end{bmatrix} \begin{bmatrix} \mathbf{D} & \mathbf{0} \\ \mathbf{0}^H & P_o \end{bmatrix} \begin{bmatrix} \mathbf{L}^H & \mathbf{k}^H \\ \mathbf{0}^H & 1 \end{bmatrix} \qquad (6.3.19)$$

which provides the MMSE P_o and the quantities $\mathbf{L}$ and $\mathbf{k}$ required to obtain the optimum estimator $\mathbf{c}_o$ by solving $\mathbf{L}^H \mathbf{c}_o = \mathbf{k}$.

EXAMPLE 6.3.2. Compute, using the LDL^H method, the optimum estimator and the MMSE specified by the following second-order moments:

$$
\mathbf{R} = \begin{bmatrix} 1 & 3 & 2 & 4 \\ 3 & 12 & 18 & 21 \\ 2 & 18 & 54 & 48 \\ 4 & 21 & 48 & 55 \end{bmatrix} \qquad \mathbf{d} = \begin{bmatrix} 1 \\ 2 \\ 1.5 \\ 4 \end{bmatrix} \qquad \text{and} \qquad P_y = 100
$$

Solution. We first compute the triangular factors

$$
\mathbf{L} = \begin{bmatrix} 1 & 0 & 0 & 0 \\ 3 & 1 & 0 & 0 \\ 2 & 4 & 1 & 0 \\ 4 & 3 & 2 & 1 \end{bmatrix} \qquad \mathbf{D} = \begin{bmatrix} 1 & 0 & 0 & 0 \\ 0 & 3 & 0 & 0 \\ 0 & 0 & 2 & 0 \\ 0 & 0 & 0 & 4 \end{bmatrix}
$$

using (6.3.8), and the vector $\mathbf{k}$

$$
\mathbf{k} = [1 \ -\tfrac{1}{3} \ 1.75 \ -1]^T
$$

using (6.3.9). Then we determine the optimum estimator

$$
\mathbf{c} = [34.5 \ -12\tfrac{1}{3} \ 3.75 \ -1]^T
$$

by solving the triangular system (6.3.11). The corresponding MMSE

$$
P_o = 88.5
$$

can be evaluated by using either (6.2.17) or (6.3.14). The reader can easily verify that the LDL^H decomposition of $\bar{\mathbf{R}}$ provides the elements of $\mathbf{L}$, $\mathbf{k}$, and P_o.

Since the diagonal elements ξ_k are positive, the matrix

$$
\mathcal{L} \triangleq \mathbf{LD}^{1/2} \tag{6.3.20}
$$

is lower triangular with positive diagonal elements. Then (6.3.1) can be written as

$$
\mathbf{R} = \mathcal{L}\mathcal{L}^H \tag{6.3.21}
$$

which is known as the *Cholesky decomposition* of $\mathbf{R}$ (Golub and Van Loan 1996). The computation of $\mathcal{L}$ requires $M^3/6$ multiplications and additions and M square roots and can be done by using the function `L=chol(R)'`. The function `[L,D]=ldltchol(R)` computes the LDL^H decomposition using the function `chol`.

6.4 OPTIMUM FINITE IMPULSE RESPONSE FILTERS

In the previous section, we presented the theory of general linear MMSE estimators [see Figure 6.1(*a*)]. In this section, we apply these results to the design of optimum linear filters, that is, filters whose performance is the best possible when measured according to the MMSE criterion [see Figure 6.1(*b*)]. The general formulation of the optimum filtering problem is shown in Figure 6.10. The optimum filter forms an estimate $\hat{y}(n)$ of the desired response $y(n)$ by using samples from a related input signal $x(n)$. The theory of optimum filters was developed by Wiener (1942) in continuous time and Kolmogorov (1939) in discrete time. Levinson (1947) reformulated the theory for FIR filters and stationary processes and developed an efficient algorithm for the solution of the normal equations that exploits the Toeplitz structure of the autocorrelation matrix $\mathbf{R}$ (see Section 7.4). For this reason, linear MMSE filters are often referred to as Wiener filters.

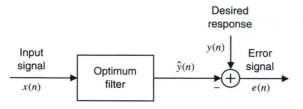

FIGURE 6.10

Block diagram representation of the optimum filtering problem.

We consider a linear FIR filter specified by its impulse response $h(n, k)$. The output of the filter is determined by the superposition summation

$$\hat{y}(n) = \sum_{k=0}^{M-1} h(n, k)x(n - k) \tag{6.4.1}$$

$$\triangleq \sum_{k=1}^{M} c_k^*(n)x(n - k + 1) \triangleq \mathbf{c}^H(n)\mathbf{x}(n) \tag{6.4.2}$$

where
$$\mathbf{c}(n) \triangleq [c_1(n) \ c_2(n) \ \cdots \ c_M(n)]^T \tag{6.4.3}$$

and
$$\mathbf{x}(n) \triangleq [x(n) \ x(n - 1) \ \cdots \ x(n - M + 1)]^T \tag{6.4.4}$$

are the filter *coefficient vector*[†] and the *input data vector*, respectively. Equation (6.4.1) becomes a convolution if $h(n, k)$ does not depend on n, that is, when the filter is time-invariant. The objective is to find the coefficient vector that minimizes the MSE $E\{|e(n)|^2\}$.

We prefer FIR over IIR filters because (1) any stable IIR filter can be approximated to any desirable degree by an FIR filter and (2) optimum FIR filters are easily obtained by solving a linear system of equations.

6.4.1 Design and Properties

To determine the optimum FIR filter $\mathbf{c}_o(n)$, we note that at every time instant n, the optimum filter is the linear MMSE estimator of the desired response $y(n)$ based on the data $\mathbf{x}(n)$. Since for any fixed n the quantities $y(n), x(n), \ldots, x(n - M + 1)$ are random variables, we can determine the optimum filter either from (6.2.12) by replacing $\mathbf{x}$ by $\mathbf{x}(n)$, y by $y(n)$, and $\mathbf{c}_o$ by $\mathbf{c}_o(n)$; or by applying the orthogonality principle (6.2.41). Indeed, using (6.2.41), (6.1.2), and (6.4.2), we have

$$E\{\mathbf{x}(n)[y^*(n) - \mathbf{x}^H(n)\mathbf{c}_o(n)]\} = \mathbf{0} \tag{6.4.5}$$

which leads to the following set of normal equations

$$\mathbf{R}(n)\mathbf{c}_o(n) = \mathbf{d}(n) \tag{6.4.6}$$

where
$$\mathbf{R}(n) \triangleq E\{\mathbf{x}(n)\mathbf{x}^H(n)\} \tag{6.4.7}$$

is the correlation matrix of the input data vector and

$$\mathbf{d}(n) \triangleq E\{\mathbf{x}(n)y^*(n)\} \tag{6.4.8}$$

is the cross-correlation vector between the desired response and the input data vector, that is, the input values stored currently in the filter memory and used by the filter to estimate the desired response. We see that, at every time n, the coefficients of the optimum filter are obtained as the solution of a linear system of equations. The filter $\mathbf{c}_o(n)$ is optimum if and only if the Hermitian matrix $\mathbf{R}(n)$ is positive definite.

To find the MMSE, we can use either (6.2.17) or the orthogonality principle (6.2.41). Using the orthogonality principle, we have

$$\begin{aligned} P_o(n) &= E\{e_o(n)[y^*(n) - \mathbf{x}^H(n)\mathbf{c}_o(n)]\} \\ &= E\{e_o(n)y^*(n)\} \quad \text{due to orthogonality} \\ &= E\{[y(n) - \mathbf{x}^H(n)\mathbf{c}_o(n)]y^*(n)\} \end{aligned}$$

[†] We define $c_k(n) \triangleq h^*(n, k)$ in order to comply with the definition $\mathbf{R}(n) \triangleq E\{\mathbf{x}(n)\mathbf{x}^H(n)\}$ of the correlation matrix.

which can be written as

$$P_o(n) = P_y(n) - \mathbf{d}^H(n)\mathbf{c}_o(n) \tag{6.4.9}$$

The first term

$$P_y(n) \triangleq E\{|y(n)|^2\} \tag{6.4.10}$$

is the power of the desired response signal and represents the MSE in the absence of filtering. The second term $\mathbf{d}^H(n)\mathbf{c}_o(n)$ is the reduction in the MSE that is obtained by using the optimum filter.

In many practical applications, we need to know the performance of the optimum filter in terms of MSE reduction prior to computing the coefficients of the filter. Then we can decide if it is preferable to (1) use an optimum filter (assuming we can design one), (2) use a simpler suboptimum filter with adequate performance, or (3) not use a filter at all. Hence, the performance of the optimum filter can serve as a yardstick for other competing methods.

The optimum filter consists of (1) a linear system solver that determines the optimum set of coefficients from the normal equations formed, using the known second-order moments, and (2) a discrete-time filter that computes the estimate $\hat{y}(n)$ (see Figure 6.11). The solution of (6.4.6) can be obtained by using standard linear system solution techniques. In MATLAB, we solve (6.4.6) by copt=R\d and compute the MMSE by Popt=Py-dot(conj(d),copt). The optimum filter is implemented by yest=filter(copt,1,x). We emphasize that the optimum filter only needs the input signal for its operation, that is, to form the estimate of $y(n)$; the desired response, if it is available, may be used for other purposes.

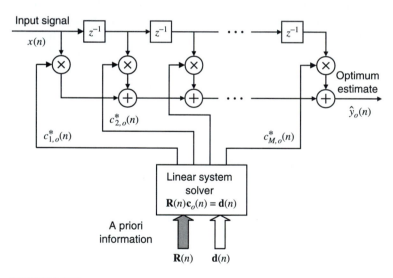

FIGURE 6.11
Design and implementation of a time-varying optimum FIR filter.

Conventional frequency-selective filters are designed to shape the spectrum of the input signal within a specific frequency band in which it operates. In this sense, these filters are effective only if the components of interest in the input signal have their energy concentrated within *nonoverlapping* bands. To design the filters, we need to know the limits of these bands, not the values of the sequences to be filtered. Note that such filters do not depend on the values of the data (values of the samples) to be filtered; that is, they are *not* data-adaptive. In contrast, optimum filters are designed using the second-order moments of the processed signals and have the same effect on all classes of signals with the same second-order moments. Optimum filters are effective even if the signals of interest have overlapping spectra. Although the actual data values also do not affect optimum filters, that is, they are also not data-adaptive, these filters are optimized to the statistics of the data and thus provide superior performance when judged by the statistical criterion.

The dependence of the optimum filter *only* on the second-order moments is a consequence of the linearity of the filter and the use of the MSE criterion. Phase information about the input signal or non-second-order moments of the input and desired response processes is not needed; even if the moments are known, they are not used by the filter. Such information is useful only if we employ a nonlinear filter or use another criterion of performance.

The error performance surface of the optimum direct-form FIR filter is a quadratic function of its impulse response. If the input correlation matrix is positive definite, this function has a unique minimum that determines the optimum set of coefficients. The surface can be visualized as a bowl, and the optimum filter corresponds to the bottom of the bowl. The bottom is moving if the processes are nonstationary and fixed if they are stationary. In general, the shape of the error performance surface depends on the criterion of performance and the structure of the filter. Note that the use of another criterion of performance or another filter structure may lead to error performance surfaces with multiple local minima or saddle points.

6.4.2 Optimum FIR Filters for Stationary Processes

Further simplifications and additional insight into the operation of optimum linear filters are possible when the input and desired response stochastic processes are *jointly wide-sense stationary*. In this case, the correlation matrix of the input data and the cross-correlation vector do *not* depend on the time index n. Therefore, the optimum filter and the MMSE are *time-invariant* (i.e., they are independent of the time index n) and are determined by

$$\mathbf{R}\mathbf{c}_o = \mathbf{d} \tag{6.4.11}$$

and

$$P_o = P_y - \mathbf{d}^H \mathbf{c}_o \tag{6.4.12}$$

Owing to stationarity, the autocorrelation matrix is

$$\mathbf{R} \triangleq \begin{bmatrix} r_x(0) & r_x(1) & \cdots & r_x(M-1) \\ r_x^*(1) & r_x(0) & \cdots & r_x(M-2) \\ \vdots & \vdots & \ddots & \vdots \\ r_x^*(M-1) & r_x^*(M-2) & \cdots & r_x(0) \end{bmatrix} \tag{6.4.13}$$

determined by the autocorrelation $r_x(l) = E\{x(n)x^*(n-l)\}$ of the input signal. The cross-correlation vector between the desired response and the input data vector is

$$\mathbf{d} \triangleq [d_1 \; d_2 \; \cdots \; d_M]^T \triangleq [r_{yx}(0) \; r_{yx}(1) \; \cdots \; r_{yx}(M-1)]^T \tag{6.4.14}$$

and P_y is the power of the desired response. For stationary processes, the matrix $\mathbf{R}$ is Toeplitz and positive definite unless the components of the data vector are linearly dependent.

Since the optimum filter is time-invariant, it is implemented by using convolution

$$\hat{y}_o(n) = \sum_{k=0}^{M-1} h_o(k) \, x(n-k) \tag{6.4.15}$$

where $h_o(n) = c_{o,n}^*$ is the impulse response of the optimum filter.

Using (6.4.13), (6.4.14), $h_o(n) = c_{o,n}^*$, and $r(l) = r^*(-l)$, we can write the normal equations (6.4.11) more explicitly as

$$\sum_{k=0}^{M-1} h_o(k) r(m-k) = r_{yx}(m) \qquad 0 \le m \le M-1 \tag{6.4.16}$$

which is the discrete-time counterpart of the *Wiener-Hopf* integral equation, and its solution determines the impulse response of the optimum filter. We notice that the cross-correlation between the input signal and the desired response (right-hand side) is equal to the convolution between the autocorrelation of the input signal and the optimum filter (left-hand side). Thus, to obtain the optimum filter, we need to solve a convolution equation.

The MMSE is given by

$$P_o = P_y - \sum_{k=0}^{M-1} h_o(k) r_{yx}^*(k) \tag{6.4.17}$$

which is obtained by substituting (6.4.14) into (6.4.12). Table 6.2 summarizes the information required for the design of an optimum (in the MMSE sense) linear time-invariant filter, the Wiener-Hopf equations that define the filter, and the resulting MMSE.

TABLE 6.2

Specification of optimum linear filters for stationary signals. The limits 0 and $M-1$ on the summations can be replaced by any values M_1 and M_2.

Filter and Error Definitions	$e(n) \triangleq y(n) - \sum_{k=0}^{M-1} h(k)x(n-k)$
Criterion of Performance	$P \triangleq E\{|e(n)|^2\} \to$ minimum
Wiener-Hopf Equations	$\sum_{k=0}^{M-1} h_o(k) r_x(m-k) = r_{yx}(m), 0 \le m \le M-1$
Minimum MSE	$P_o = P_y - \sum_{k=0}^{M-1} h_o(k) r_{yx}^*(k)$
Second-Order Statistics	$r_x(l) = E\{x(n)x^*(n-l)\}, P_y = \{|y(n)|^2\}$
	$r_{yx}(l) = E\{y(n)x^*(n-l)\}$

To summarize, for nonstationary processes $\mathbf{R}(n)$ is Hermitian and nonnegative definite, and the optimum filter $\mathbf{h}_o(n)$ is time-varying. For stationary processes, $\mathbf{R}$ is Hermitian, nonnegative definite, and Toeplitz, and the optimum filter is time-invariant. A Toeplitz autocorrelation matrix is positive definite if the power spectrum of the input satisfies $R_x(e^{j\omega}) > 0$ for all frequencies ω. In both cases, the filter is used for all realizations of the processes. If $M = \infty$, we have a causal IIR optimum filter determined by an infinite-order linear system of equations that can only be solved in the stationary case by using analytical techniques (see Section 6.6).

EXAMPLE 6.4.1. Consider a harmonic random process

$$y(n) = A \cos(\omega_0 n + \phi)$$

with fixed, but unknown, amplitude and frequency, and random phase ϕ, uniformly distributed on the interval from 0 to 2π. This process is corrupted by additive white Gaussian noise $v(n) \sim N(0, \sigma_v^2)$ that is uncorrelated with $y(n)$. The resulting signal $x(n) = y(n) + v(n)$ is available to the user for processing. Design an optimum FIR filter to remove the corrupting noise $v(n)$ from the observed signal $x(n)$.

Solution. The input of the optimum filter is $x(n)$, and the desired response is $y(n)$. The signal $y(n)$ is obviously unavailable, but to design the filter, we only need the second-order moments $r_x(l)$ and $r_{yx}(l)$. We first note that since $y(n)$ and $v(n)$ are uncorrelated, the autocorrelation of the input signal is

$$r_x(l) = r_y(l) + r_v(l) = \tfrac{1}{2}A^2 \cos \omega_0 l + \sigma_v^2 \delta(l)$$

where $r_y(l) = \tfrac{1}{2}A^2 \cos \omega_0 l$ is the autocorrelation of $y(n)$. The cross-correlation between the desired response $y(n)$ and the input signal $x(n)$ is

$$r_{yx}(l) = E\{y(n)[y(n-l) + v(n-l)]\} = r_y(l)$$

Therefore, the autocorrelation matrix $\mathbf{R}$ is symmetric Toeplitz and is determined by the elements $r(0), r(1), \ldots, r(M-1)$ of its first row. The right-hand side of the Wiener-Hopf equations is $\mathbf{d} = [r_y(0)\ r_y(1) \cdots r_y(M-1)]^T$. If we know $r_y(l)$ and σ_v^2, we can numerically determine the optimum filter and the MMSE from (6.4.11) and (6.4.12). For example, suppose that $A = 0.5$, $f_0 = \omega_0/(2\pi) = 0.05$, and $\sigma_v^2 = 0.5$. The input signal-to-noise ratio (SNR) is

$$\text{SNR}_I = 10 \log \frac{A^2/2}{\sigma_v^2} = -6.02 \text{ dB}$$

The *processing gain* (PG), defined as the ratio of signal-to-noise ratios at the output and input of a signal processing system

$$\text{PG} \triangleq \frac{\text{SNR}_O}{\text{SNR}_I}$$

provides another useful measure of performance.

The first problem we encounter is how to choose the order M of the filter. In the absence of any a priori information, we compute $\mathbf{h}_o$ and P_o^h for $1 \le M \le M_{\max} = 50$ and PG and plot both results in Figure 6.12. We see that an $M = 20$ order filter provides satisfactory performance. Figure 6.13 shows a realization of the corrupted and filtered signals. Another useful approach to evaluate how well the optimum filter enhances a harmonic signal is to compute the spectra of the input and output signals and the frequency response of the optimum filter. These are shown in Figure 6.14, where we see that the optimum filter has a sharp bandpass about frequency f_0, as expected (for details see Problem 6.5).

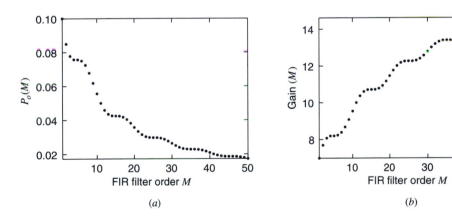

(a)   (b)

FIGURE 6.12
Plots of (a) the MMSE and (b) the processing gain as a function of the filter order M.

To illustrate the meaning of the estimator's optimality, we will use a Monte Carlo simulation. Thus, we generate $K = 100$ realizations of the sequence $x(\zeta_i, n)$, $0 \le n \le N - 1 (N = 1000)$; we compute the output sequence $\hat{y}(\zeta_i, n)$, using (6.4.15); and then the error sequence $e(\zeta_i, n) = y(\zeta_i, n) - \hat{y}(\zeta_i, n)$ and its variance $\hat{P}(\zeta_i)$. Figure 6.15 shows a plot of $\hat{P}(\zeta_i)$, $1 \le \zeta_i \le K$. We

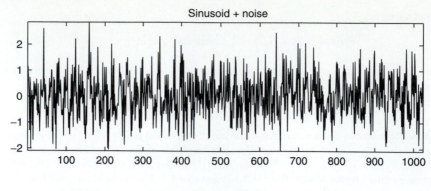

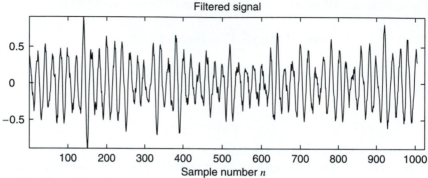

FIGURE 6.13
Example of the noise-corrupted and filtered sinusoidal signals.

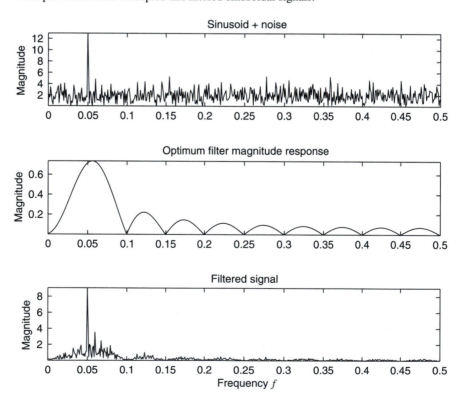

FIGURE 6.14
PSD of the input signal, magnitude response of the optimum filter, and PSD of the output signal.

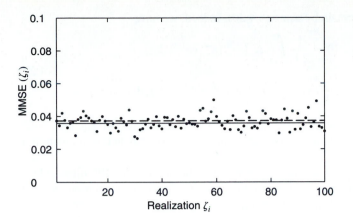

FIGURE 6.15
Results of Monte Carlo simulation of the optimum filter.

notice that although the filter performs better or worse than the optimum in particular cases, on average its performance is close to the theoretically predicted one. This is exactly the meaning of the MMSE criterion: *optimum performance on the average (in the MMSE sense)*.

For a certain realization, the optimum filter may not perform as well as some other linear filters; however, on average, it performs better than any other linear filter of the same order when all possible realizations of $x(n)$ and $y(n)$ are considered.

6.4.3 Frequency-Domain Interpretations

We will now investigate the performance of the optimum filter, for stationary processes, in the frequency domain. Using (6.2.7), (6.4.13), and (6.4.14), we can easily show that the MSE of an FIR filter $h(n)$ is given by

$$P = E\{|e(n)|^2\} = r_y(0) - \sum_{k=0}^{M-1} h(k)r_{yx}^*(k) - \sum_{k=0}^{M-1} h^*(k)r_{yx}(k) + \sum_{k=0}^{M-1}\sum_{l=0}^{M-1} h(k)r(l-k)h^*(l)$$
(6.4.18)

The frequency response function of the FIR filter is

$$H(e^{j\omega}) \triangleq \sum_{k=0}^{M-1} h(k)e^{-j\omega k}$$
(6.4.19)

Using Parseval's theorem,

$$\sum_{n=-\infty}^{\infty} x_1(n)x_2^*(n) = \frac{1}{2\pi}\int_{-\pi}^{\pi} X_1(e^{j\omega})X_2^*(e^{j\omega})\,d\omega$$
(6.4.20)

we can show that the MSE (6.4.18) can be expressed in the frequency domain as

$$P = r_y(0) - \frac{1}{2\pi}\int_{-\pi}^{\pi}[H(e^{j\omega})R_{yx}^*(e^{j\omega}) + H^*(e^{j\omega})R_{yx}(e^{j\omega}) - H(e^{j\omega})H^*(e^{j\omega})R_x(e^{j\omega})]\,d\omega$$
(6.4.21)

where $R_x(e^{j\omega})$ is the PSD of $x(n)$ and $R_{yx}(e^{j\omega})$ is the cross-PSD of $y(n)$ and $x(n)$ (see Problem 6.10). This formula holds for both FIR and IIR filters.

If we minimize (6.4.21) with respect to $H(e^{j\omega})$, we obtain the system function of the optimum filter and the MMSE. However, we leave this for Problem 6.11 and instead express

(6.4.17) in the frequency domain by using (6.4.20). Indeed, we have

$$P_o = r_y(0) - \frac{1}{2\pi} \int_{-\pi}^{\pi} H_o(e^{j\omega}) R_{yx}^*(e^{j\omega}) \, d\omega$$

$$= \frac{1}{2\pi} \int_{-\pi}^{\pi} [R_y(e^{j\omega}) - H_o(e^{j\omega}) R_{yx}^*(e^{j\omega})] \, d\omega \tag{6.4.22}$$

where $H_o(e^{j\omega})$ is the frequency response of the optimum filter. The above equation holds for any filter, FIR or IIR, as long as we use the proper limits to compute the summation in (6.4.19).

We will now obtain a formula for the MMSE that holds only for IIR filters whose impulse response extends from $-\infty$ to ∞. In this case, (6.4.16) is a convolution equation that holds for $-\infty < m < \infty$. Using the convolution theorem of the Fourier transform, we obtain

$$H_o(e^{j\omega}) = \frac{R_{yx}(e^{j\omega})}{R_x(e^{j\omega})} \tag{6.4.23}$$

which, we again stress, holds for noncausal IIR filters *only*. Substituting into (6.4.22), we obtain

$$P_o = \frac{1}{2\pi} \int_{-\pi}^{\pi} [1 - \frac{|R_{yx}(e^{j\omega})|^2}{R_y(e^{j\omega}) R_x(e^{j\omega})}] R_y(e^{j\omega}) \, d\omega$$

or

$$P_o = \frac{1}{2\pi} \int_{-\pi}^{\pi} [1 - \mathcal{G}_{yx}(e^{j\omega})] R_y(e^{j\omega}) \, d\omega \tag{6.4.24}$$

where $\mathcal{G}_{yx}(e^{j\omega})$ is the coherence function between $x(n)$ and $y(n)$.

This important equation indicates that the performance of the optimum filter depends on the coherence between the input and desired response processes. As we recall from Section 5.4, the coherence is a measure of both the noise disturbing the observations and the relative linearity between $x(n)$ and $y(n)$. The optimum filter can reduce the MMSE at a certain band only if there is significant coherence, that is, $\mathcal{G}_{yx}(e^{j\omega}) \simeq 1$. Thus, the optimum filter $H_o(z)$ constitutes the best, in the MMSE sense, linear relationship between the stochastic processes $x(n)$ and $y(n)$. These interpretations apply to causal IIR and FIR optimum filters, even if (6.4.23) and (6.4.24) only hold approximately in these cases (see Section 6.6).

6.5 LINEAR PREDICTION

Linear prediction plays a prominent role in many theoretical, computational, and practical areas of signal processing and deals with the problem of estimating or predicting the value $x(n)$ of a signal at the time instant $n = n_0$, by using a set of other samples from the *same* signal. Although linear prediction is a subject useful in itself, its importance in signal processing is also due, as we will see later, to its use in the development of fast algorithms for optimum filtering and its relation to all-pole signal modeling.

6.5.1 Linear Signal Estimation

Suppose that we are given a set of values $x(n), x(n-1), \ldots, x(n-M)$ of a stochastic process and we wish to estimate the value of $x(n-i)$, using a linear combination of the remaining samples. The resulting estimate and the corresponding estimation error are given by

$$\hat{x}(n-i) \triangleq -\sum_{\substack{k=0 \\ k\neq i}}^{M} c_k^*(n)x(n-k) \tag{6.5.1}$$

and
$$e^{(i)}(n) \triangleq x(n-i) - \hat{x}(n-i)$$

$$= \sum_{k=0}^{M} c_k^*(n)x(n-k) \quad \text{with } c_i(n) \triangleq 1 \tag{6.5.2}$$

where $c_k(n)$ are the coefficients of the estimator as a function of discrete-time index n. The process is illustrated in Figure 6.16.

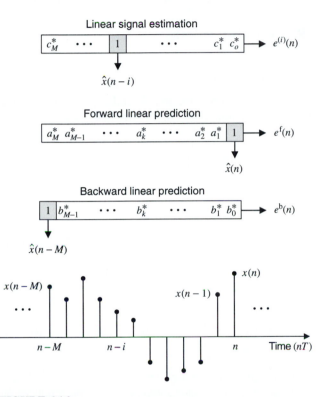

FIGURE 6.16
Illustration showing the samples, estimates, and errors used in
linear signal estimation, forward linear prediction, and backward
linear prediction.

To determine the MMSE signal estimator, we partition (6.5.2) as

$$e^{(i)}(n) = \sum_{k=0}^{i-1} c_k^*(n)x(n-k) + x(n-i) + \sum_{k=i+1}^{M} c_k^*(n)x(n-k)$$

$$\triangleq \mathbf{c}_1^H(n)\mathbf{x}_1(n) + x(n-i) + \mathbf{c}_2^H(n)\mathbf{x}_2(n) \tag{6.5.3}$$

$$\triangleq [\bar{\mathbf{c}}^{(i)}(n)]^H \bar{\mathbf{x}}(n)$$

where the partitions of the coefficient and data vectors, around the ith component, are easily defined from the context. To obtain the normal equations and the MMSE for the optimum *linear signal estimator*, we note that

$$\text{Desired response} = x(n-i) \qquad \text{data vector} = \begin{bmatrix} \mathbf{x}_1(n) \\ \mathbf{x}_2(n) \end{bmatrix}$$

Using (6.4.6) and (6.4.9) or the orthogonality principle, we have

$$\begin{bmatrix} \mathbf{R}_{11}(n) & \mathbf{R}_{12}(n) \\ \mathbf{R}_{12}^T(n) & \mathbf{R}_{22}(n) \end{bmatrix} \begin{bmatrix} \mathbf{c}_1(n) \\ \mathbf{c}_2(n) \end{bmatrix} = - \begin{bmatrix} \mathbf{r}_1(n) \\ \mathbf{r}_2(n) \end{bmatrix} \tag{6.5.4}$$

or more compactly[†]

$$\mathbf{R}^{(i)}(n)\mathbf{c}_o^{(i)}(n) = -\mathbf{d}^{(i)}(n) \tag{6.5.5}$$

and

$$P_o^{(i)}(n) = P_x(n-i) + \mathbf{r}_1^H(n)\mathbf{c}_1(n) + \mathbf{r}_2^H(n)\mathbf{c}_2(n) \tag{6.5.6}$$

where for $i, j = 1, 2$

$$\mathbf{R}_{ij}(n) \triangleq E\{\mathbf{x}_i(n)\mathbf{x}_j^H(n)\} \tag{6.5.7}$$

$$\mathbf{r}_j(n) \triangleq E\{\mathbf{x}_j(n)x^*(n-i)\} \tag{6.5.8}$$

$$P_x(n) = E\{|x(n)|^2\} \tag{6.5.9}$$

For various reasons, to be seen later, we will combine (6.5.4) and (6.5.6) into a single equation. To this end, we note that the correlation matrix of the extended vector

$$\bar{\mathbf{x}}(n) = \begin{bmatrix} \mathbf{x}_1(n) \\ x(n-i) \\ \mathbf{x}_2(n) \end{bmatrix} \tag{6.5.10}$$

can be partitioned as

$$\bar{\mathbf{R}}(n) = E\{\bar{\mathbf{x}}(n)\bar{\mathbf{x}}^H(n)\} = \begin{bmatrix} \mathbf{R}_{11}(n) & \mathbf{r}_1(n) & \mathbf{R}_{12}(n) \\ \mathbf{r}_1^H(n) & P_x(n-i) & \mathbf{r}_2^H(n) \\ \mathbf{R}_{12}^H(n) & \mathbf{r}_2(n) & \mathbf{R}_{22}(n) \end{bmatrix} \tag{6.5.11}$$

with respect to its ith row and ith column. Using (6.5.4), (6.5.6), and (6.5.11), we obtain

$$\bar{\mathbf{R}}(n)\bar{\mathbf{c}}_o^{(i)}(n) = \begin{bmatrix} \mathbf{0} \\ P_o^{(i)}(n) \\ \mathbf{0} \end{bmatrix} \leftarrow i\text{th row} \tag{6.5.12}$$

which completely determines the linear signal estimator $\mathbf{c}^{(i)}(n)$ and the MMSE $P_o^{(i)}(n)$.

If $M = 2L$ and $i = L$, we have a *symmetric linear smoother* $\bar{\mathbf{c}}(n)$ that produces an estimate of the middle sample by using the L past and the L future samples. The above formulation suggests an easy procedure for the computation of the linear signal estimator for any value of i, which is outlined in Table 6.3 and implemented by the function `[ci,Pi]=lsigest(R,i)`. We next discuss two types of linear signal estimation that are of special interest and have their own dedicated notation.

6.5.2 Forward Linear Prediction

One-step *forward linear prediction* (FLP) involves the estimation or prediction of the value $x(n)$ of a stochastic process by using a linear combination of the past samples $x(n-1), \ldots, x(n-M)$ (see Figure 6.16). We should stress that in signal processing applications

---

[†]The minus sign on the right-hand side of the normal equations is the result of arbitrarily setting the coefficient $c_i(n) \triangleq 1$.

TABLE 6.3
Steps for the computation of optimum signal estimators.

1. Determine the matrix $\bar{\mathbf{R}}(n)$ of the extended data vector $\bar{\mathbf{x}}(n)$.
2. Create the $M \times M$ submatrix $\mathbf{R}^{(i)}(n)$ of $\bar{\mathbf{R}}(n)$ by removing its ith row and its ith column.
3. Create the $M \times 1$ vector $\mathbf{d}^{(i)}(n)$ by extracting the ith column $\bar{\mathbf{d}}^{(i)}(n)$ of $\bar{\mathbf{R}}(n)$ and removing its ith element.
4. Solve the linear system $\mathbf{R}^{(i)}(n)\mathbf{c}_o^{(i)}(n) = -\mathbf{d}^{(i)}(n)$ to obtain $\mathbf{c}_o^{(i)}(n)$.
5. Compute the MMSE $P_o^{(i)}(n) = [\bar{\mathbf{d}}^{(i)}(n)]^H \bar{\mathbf{c}}_o^{(i)}(n)$.

of linear prediction, what is important is the ability to obtain a good estimate of a sample, pretending that it is unknown, instead of forecasting the future. Thus, the term *prediction* is used more with the signal estimation than forecasting in mind. The forward predictor is a linear signal estimator with $i = 0$ and is denoted by

$$e^{\mathrm{f}}(n) \triangleq x(n) + \sum_{k=1}^{M} a_k^*(n)x(n-k) \tag{6.5.13}$$

$$= x(n) + \mathbf{a}^H(n)\mathbf{x}(n-1)$$

where
$$\mathbf{a}(n) \triangleq [a_1(n)\, a_2(n)\, \cdots\, a_M(n)]^T \tag{6.5.14}$$

is known as the *forward linear predictor* and $a_k(n)$ with $a_0(n) \triangleq 1$ as the FLP error filter. To obtain the normal equations and the MMSE for the optimum FLP, we note that for $i = 0$, (6.5.11) can be written as

$$\bar{\mathbf{R}}(n) = \begin{bmatrix} P_x(n) & \mathbf{r}^{\mathrm{f}H}(n) \\ \mathbf{r}^{\mathrm{f}}(n) & \mathbf{R}(n-1) \end{bmatrix} \tag{6.5.15}$$

where
$$\mathbf{R}(n) = E\{\mathbf{x}(n)\mathbf{x}^H(n)\} \tag{6.5.16}$$

and
$$\mathbf{r}^{\mathrm{f}}(n) = E\{\mathbf{x}(n-1)x^*(n)\} \tag{6.5.17}$$

Therefore, (6.5.5) and (6.5.6) give

$$\mathbf{R}(n-1)\mathbf{a}_o(n) = -\mathbf{r}^{\mathrm{f}}(n) \tag{6.5.18}$$

and
$$P_o^{\mathrm{f}}(n) = P_x(n) + \mathbf{r}^{\mathrm{f}H}(n)\mathbf{a}_o(n) \tag{6.5.19}$$

or
$$\bar{\mathbf{R}}(n) \begin{bmatrix} 1 \\ \mathbf{a}_o(n) \end{bmatrix} = \begin{bmatrix} P_o^{\mathrm{f}}(n) \\ \mathbf{0} \end{bmatrix} \tag{6.5.20}$$

which completely specifies the FLP parameters.

6.5.3 Backward Linear Prediction

In this case, we want to estimate the sample $x(n - M)$ in terms of the future samples $x(n), x(n-1), \ldots, x(n-M+1)$ (see Figure 6.16). The term *backward linear prediction* (*BLP*) is not accurate but is used since it is an established convention. A more appropriate name might be *postdiction* or *hindsight*. The BLP is basically a linear signal estimator with $i = M$ and is denoted by

$$e^{\mathrm{b}}(n) \triangleq \sum_{k=0}^{M-1} b_k^*(n)x(n-k) + x(n-M) \tag{6.5.21}$$

$$= \mathbf{b}^H(n)\mathbf{x}(n) + x(n-M)$$

where
$$\mathbf{b}(n) \triangleq [b_0(n)\, b_1(n)\, \cdots\, b_{M-1}(n)]^T \tag{6.5.22}$$

is the BLP and $b_k(n)$ with $b_M(n) \triangleq 1$ is the *backward prediction error filter* (*BPEF*). For $i = M$, (6.5.11) gives

$$\bar{\mathbf{R}}(n) = \begin{bmatrix} \mathbf{R}(n) & \mathbf{r}^{\mathrm{b}}(n) \\ \mathbf{r}^{\mathrm{b}H}(n) & P_x(n-M) \end{bmatrix} \tag{6.5.23}$$

where

$$\mathbf{r}^{\mathrm{b}}(n) \triangleq E\{\mathbf{x}(n)x^*(n-M)\} \tag{6.5.24}$$

The optimum backward linear predictor is specified by

$$\mathbf{R}(n)\mathbf{b}_o(n) = -\mathbf{r}^{\mathrm{b}}(n) \tag{6.5.25}$$

and the MMSE is

$$P_o^{\mathrm{b}}(n) = P_x(n-M) + \mathbf{r}^{\mathrm{b}H}(n)\mathbf{b}_o(n) \tag{6.5.26}$$

and can be put in a single equation as

$$\bar{\mathbf{R}}(n) \begin{bmatrix} \mathbf{b}_o(n) \\ 1 \end{bmatrix} = \begin{bmatrix} \mathbf{0} \\ P_o^{\mathrm{b}}(n) \end{bmatrix} \tag{6.5.27}$$

In Table 6.4, we summarize the definitions and design equations for optimum FIR filtering and prediction. Using the entries in this table, we can easily obtain the normal equations and the MMSE for the FLP and BLP from those of the optimum filter.

TABLE 6.4
Summary of the design equations for optimum FIR filtering and prediction.

	Optimum filter	FLP	BLP
Input data vector	$\mathbf{x}(n)$	$\mathbf{x}(n-1)$	$\mathbf{x}(n)$
Desired response	$y(n)$	$x(n)$	$x(n-M)$
Coefficient vector	$\mathbf{h}(n)$	$\mathbf{a}(n)$	$\mathbf{b}(n)$
Estimation error	$e(n) = y(n) - \mathbf{c}^H(n)\mathbf{x}(n)$	$e^{\mathrm{f}}(n) = x(n) - \mathbf{a}^H(n)\mathbf{x}(n-1)$	$e^{\mathrm{b}}(n) = x(n-M) - \mathbf{b}^H(n)\mathbf{x}(n)$
Normal equations	$\mathbf{R}(n)\mathbf{h}_o(n) = \mathbf{d}(n)$	$\mathbf{R}(n-1)\mathbf{a}_o(n) = -\mathbf{r}^{\mathrm{f}}(n)$	$\mathbf{R}(n)\mathbf{b}_o(n) = -\mathbf{r}^{\mathrm{b}}(n)$
MMSE	$P_o^c(n) = P_y(n) - \mathbf{c}_o^H(n)\mathbf{d}(n)$	$P_o^{\mathrm{f}}(n) = P_x(n) + \mathbf{a}^H(n)\mathbf{r}_o^{\mathrm{f}}(n)$	$P_o^{\mathrm{b}}(n) = P_x(n-M) + \mathbf{b}^H(n)\mathbf{r}_o^{\mathrm{b}}(n)$
Required moments	$\mathbf{R}(n) = E\{\mathbf{x}(n)\mathbf{x}^H(n)\}$	$\mathbf{r}^{\mathrm{f}}(n) = E\{\mathbf{x}(n-1)x^*(n)\}$	$\mathbf{r}^{\mathrm{b}}(n) = E\{\mathbf{x}(n)x^*(n-M)\}$
	$\mathbf{d}(n) = E\{\mathbf{x}(n)y^*(n)\}$		
Stationary processes	$\mathbf{R}\mathbf{c}_o = \mathbf{d}$, $\mathbf{R}$ is Toeplitz	$\mathbf{R}\mathbf{a}_o = -\mathbf{r}^*$	$\mathbf{R}\mathbf{b}_o = -\mathbf{J}\mathbf{r} \Rightarrow \mathbf{b}_o = \mathbf{J}\mathbf{a}_o^*$

6.5.4 Stationary Processes

If the process $x(n)$ is stationary, then the correlation matrix $\bar{\mathbf{R}}(n)$ does *not* depend on the time n and it is Toeplitz

$$\bar{\mathbf{R}} = \begin{bmatrix} r(0) & r(1) & \cdots & r(M) \\ r^*(1) & r(0) & \cdots & r(M-1) \\ \vdots & \vdots & \ddots & \vdots \\ r^*(M) & r^*(M-1) & \cdots & r(0) \end{bmatrix} \tag{6.5.28}$$

Therefore, all the resulting linear MMSE signal estimators are *time-invariant*. If we define the correlation vector

$$\mathbf{r} \triangleq [r(1)\, r(2)\, \cdots\, r(M)]^T \qquad (6.5.29)$$

where $r(l) = E\{x(n)x^*(n-l)\}$, we can easily see that the cross-correlation vectors for the FLP and the BLP are

$$\mathbf{r}^f = E\{\mathbf{x}(n-1)x^*(n)\} = \mathbf{r}^* \qquad (6.5.30)$$

and

$$\mathbf{r}^b = E\{\mathbf{x}(n)x^*(n-M)\} = \mathbf{J}\mathbf{r} \qquad (6.5.31)$$

where

$$\mathbf{J} = \begin{bmatrix} 0 & 0 & \cdots & 1 \\ \vdots & \vdots & \ddots & \vdots \\ 0 & 1 & \cdots & 0 \\ 1 & 0 & \cdots & 0 \end{bmatrix} \qquad (6.5.32)$$

is the exchange matrix that simply reverses the order of the vector elements. Therefore,

$$\mathbf{R}\mathbf{a}_o = -\mathbf{r}^* \qquad (6.5.33)$$

$$P_o^f = r(0) + \mathbf{r}^H \mathbf{a}_o \qquad (6.5.34)$$

$$\mathbf{R}\mathbf{b}_o = -\mathbf{J}\mathbf{r} \qquad (6.5.35)$$

$$P_o^b = r(0) + \mathbf{r}^H \mathbf{J}\mathbf{b}_o \qquad (6.5.36)$$

where the Toeplitz matrix $\mathbf{R}$ is obtained from $\bar{\mathbf{R}}$ by deleting the last column and row. Using the centrosymmetry property of symmetric Toeplitz matrices

$$\mathbf{R}\mathbf{J} = \mathbf{J}\mathbf{R}^* \qquad (6.5.37)$$

and (6.5.33), we have

$$\mathbf{J}\mathbf{R}^*\mathbf{a}_o^* = -\mathbf{J}\mathbf{r} \quad \text{or} \quad \mathbf{R}\mathbf{J}\mathbf{a}_o^* = -\mathbf{J}\mathbf{r} \qquad (6.5.38)$$

Comparing the last equation with (6.5.35), we have

$$\mathbf{b}_o = \mathbf{J}\mathbf{a}_o^* \qquad (6.5.39)$$

that is, the BLP coefficient vector is the reverse of the conjugated FLP coefficient vector. Furthermore, from (6.5.34), (6.5.36), and (6.5.39), we have

$$P_o \triangleq P_o^f = P_o^b \qquad (6.5.40)$$

that is, the forward and backward prediction error powers are equal.

This remarkable symmetry between the MMSE forward and backward linear predictors holds for stationary processes but disappears for nonstationary processes. Also, we do not have such a symmetry if a criterion other than the MMSE is used and the process to be predicted is non-Gaussian (Weiss 1975; Lawrence 1991).

EXAMPLE 6.5.1. To illustrate the basic ideas in FLP, BLP, and linear smoothing, we consider the second-order estimators for stationary processes.

The augmented equations for the first-order FLP are

$$\begin{bmatrix} r(0) & r(1) \\ r^*(1) & r(0) \end{bmatrix} \begin{bmatrix} a_0^{(1)} \\ a_1^{(1)} \end{bmatrix} = \begin{bmatrix} P_1^f \\ 0 \end{bmatrix}$$

and they can be solved by using Cramer's rule. Indeed, we obtain

$$a_0^{(1)} = \frac{\det \begin{bmatrix} P_1^f & r(1) \\ 0 & r(0) \end{bmatrix}}{\det \mathbf{R}_2} = \frac{r(0)P_1^f}{\det \mathbf{R}_2} = 1 \Rightarrow P_1^f = \frac{\det \mathbf{R}_2}{\det \mathbf{R}_1} = \frac{r^2(0) - |r(1)|^2}{r(0)}$$

and
$$a_1^{(1)} = \frac{\det \begin{bmatrix} r(0) & P_1^f \\ r^*(1) & 0 \end{bmatrix}}{\det \mathbf{R}_2} = \frac{-P_1^f r^*(1)}{\det \mathbf{R}_2} = -\frac{r^*(1)}{r(0)}$$

for the MMSE and the FLP. For the second-order case we have

$$\begin{bmatrix} r(0) & r(1) & r(2) \\ r^*(1) & r(0) & r(1) \\ r^*(2) & r^*(1) & r(0) \end{bmatrix} \begin{bmatrix} a_0^{(2)} \\ a_1^{(2)} \\ a_2^{(2)} \end{bmatrix} = \begin{bmatrix} P_2^f \\ 0 \\ 0 \end{bmatrix}$$

whose solution is

$$a_0^{(2)} = \frac{P_2^f \det \mathbf{R}_2}{\det \mathbf{R}_3} = 1 \Rightarrow P_2^f = \frac{\det \mathbf{R}_3}{\det \mathbf{R}_2}$$

$$a_1^{(2)} = \frac{-P_2^f \det \begin{bmatrix} r^*(1) & r(1) \\ r^*(2) & r(0) \end{bmatrix}}{\det \mathbf{R}_3} = \frac{-\det \begin{bmatrix} r^*(1) & r(1) \\ r^*(2) & r(0) \end{bmatrix}}{\det \mathbf{R}_2} = \frac{r(1)r^*(2) - r(0)r^*(1)}{r^2(0) - |r(1)|^2}$$

and

$$a_2^{(2)} = \frac{P_2^f \det \begin{bmatrix} r^*(1) & r(0) \\ r^*(2) & r^*(1) \end{bmatrix}}{\det \mathbf{R}_3} = \frac{\det \begin{bmatrix} r^*(1) & r(0) \\ r^*(2) & r^*(1) \end{bmatrix}}{\det \mathbf{R}_2} = \frac{[r^*(1)]^2 - r(0)r^*(2)}{r^2(0) - |r(1)|^2}$$

Similarly, for the BLP

$$\begin{bmatrix} r(0) & r(1) \\ r^*(1) & r(0) \end{bmatrix} \begin{bmatrix} b_0^{(1)} \\ b_1^{(1)} \end{bmatrix} = \begin{bmatrix} 0 \\ P_1^b \end{bmatrix}$$

where $b_1^{(1)} = 1$, we obtain

$$P_1^b = \frac{\det \mathbf{R}_2}{\det \mathbf{R}_1} \quad \text{and} \quad b_0^{(1)} = -\frac{r(1)}{r(0)}$$

$$\begin{bmatrix} r(0) & r(1) & r(2) \\ r^*(1) & r(0) & r(1) \\ r^*(2) & r^*(1) & r(0) \end{bmatrix} \begin{bmatrix} b_0^{(2)} \\ b_1^{(2)} \\ b_2^{(2)} \end{bmatrix} = \begin{bmatrix} 0 \\ 0 \\ P_2^b \end{bmatrix}$$

$$P_2^b = \frac{\det \mathbf{R}_3}{\det \mathbf{R}_2} \quad b_1^{(2)} = \frac{r^*(1)r(2) - r(0)r(1)}{r^2(0) - |r(1)|^2} \quad b_0^{(2)} = \frac{r^2(1) - r(0)r(2)}{r^2(0) - |r(1)|^2}$$

We note that
$$P_1^f = P_1^b \qquad a_1^{(1)} = b_1^{(1)*}$$

and
$$P_2^f = P_2^b \qquad a_1^{(2)} = b_1^{(2)*} \qquad a_2^{(2)} = b_0^{(2)*}$$

which is a result of the stationarity of $x(n)$ or equivalently of the Toeplitz structure of $\mathbf{R}_m$.

For the linear signal estimator, we have

$$\begin{bmatrix} r(0) & r(1) & r(2) \\ r^*(1) & r(0) & r(1) \\ r^*(2) & r^*(1) & r(0) \end{bmatrix} \begin{bmatrix} c_0^{(2)} \\ c_1^{(2)} \\ c_2^{(2)} \end{bmatrix} = \begin{bmatrix} 0 \\ P_2 \\ 0 \end{bmatrix}$$

with $c_1^{(2)} = 1$. Using Cramer's rule, we obtain

$$P_2 = \frac{\det \mathbf{R}_3}{\det \mathbf{R}_3^{(2)}}$$

$$c_0^{(2)} = \frac{-P_2 \det \begin{bmatrix} r(1) & r(2) \\ r^*(1) & r(0) \end{bmatrix}}{\det \mathbf{R}_3} = -\frac{\det \begin{bmatrix} r(1) & r(2) \\ r^*(1) & r(0) \end{bmatrix}}{\det \mathbf{R}_3^{(2)}} = \frac{r^*(1)r(2) - r(0)r(1)}{r^2(0) - |r(1)|^2}$$

$$c_2^{(2)} = \frac{-P_2 \det \begin{bmatrix} r(0) & r(1) \\ r^*(2) & r^*(1) \end{bmatrix}}{\det \mathbf{R}_3} = -\frac{\det \begin{bmatrix} r(0) & r(1) \\ r^*(2) & r^*(1) \end{bmatrix}}{\det \mathbf{R}_3^{(2)}} = \frac{r(1)r^*(2) - r(0)r^*(1)}{r^2(0) - |r(1)|^2}$$

from which we see that $c_0^{(2)} = c_2^{(2)*}$; that is, we have a linear phase estimator.

6.5.5 Properties

Linear signal estimators and predictors have some interesting properties that we discuss next.

PROPERTY 6.5.1. If the process $x(n)$ is stationary, then the symmetric, linear smoother has linear phase.

Proof. Using the centrosymmetry property $\bar{\mathbf{R}}\mathbf{J} = \mathbf{J}\bar{\mathbf{R}}^*$ and (6.5.12) for $M = 2L$, $i = L$, we obtain

$$\bar{\mathbf{c}} = \mathbf{J}\bar{\mathbf{c}}^* \tag{6.5.41}$$

that is, the symmetric, linear smoother has even symmetry and, therefore, has linear phase (see Problem 6.12).

PROPERTY 6.5.2. If the process $x(n)$ is stationary, the forward prediction error filter (PEF) $1, a_1, a_2, \ldots, a_M$ is minimum-phase and the backward PEF $b_0, b_1, \ldots, b_{M-1}, 1$ is maximum-phase.

Proof. The system function of the Mth-order forward PEF can be factored as

$$A(z) = 1 + \sum_{k=1}^{M} a_k^* z^{-k} = G(z)(1 - qz^{-1})$$

where q is a zero of $A(z)$ and

$$G(z) = 1 + \sum_{k=1}^{M-1} g_k z^{-k}$$

is an $(M-1)$st-order filter. The filter $A(z)$ can be implemented as the cascade connection of the filters $G(z)$ and $1 - qz^{-1}$ (see Figure 6.17). The output $s(n)$ of $G(z)$ is

$$s(n) = x(n) + g_1 x(n-1) + \cdots + g_{M-1} x(n-M+1)$$

and it is easy to see that

$$E\{s(n-1)e^{f*}(n)\} = 0 \tag{6.5.42}$$

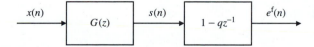

FIGURE 6.17
The prediction error filter with one zero factored out.

because $E\{x(n-k)e^{f*}(n)\} = 0$ for $1 \leq k \leq M$. Since the output of the second filter can be expressed as

$$e^f(n) = s(n) - qs(n-1)$$

we have

$$E\{s(n-1)e^{f*}(n)\} = E\{s(n-1)s^*(n)\} - q^* E\{s(n-1)s^*(n-1)\} = 0$$

which implies that

$$q = \frac{r_s(-1)}{r_s(0)} \Rightarrow |q| \leq 1$$

because q is equal to the normalized autocorrelation of $s(n)$. If the process $x(n)$ is not predictable, that is, $E\{|e^f(n)|^2\} \neq 0$, we have

$$\begin{aligned} E\{|e^f(n)|^2\} &= E\{e^f(n)[s^*(n) - q^*s^*(n-1)]\} \\ &= E\{e^f(n)s^*(n)\} \quad \text{due to (6.5.42)} \\ &= E\{[s(n) - qs(n-1)]s^*(n)\} \\ &= r_s(0)(1 - |q|^2) \neq 0 \end{aligned}$$

which implies that

$$|q| < 1$$

that is, the zero q of the forward PEF filter is strictly inside the unit circle. Repeating this process, we can show that all zeros of $A(z)$ are inside the unit circle; that is, $A(z)$ is minimum-phase. This proof was presented in Vaidyanathan et al. (1996). The property $\mathbf{b} = \mathbf{J}\mathbf{a}^*$ is equivalent to

$$B(z) = z^{-M} A^* \left(\frac{1}{z^*} \right)$$

which implies that $B(z)$ is a maximum-phase filter (see Section 2.4).

PROPERTY 6.5.3. The forward and backward prediction error filters can be expressed in terms of the eigenvalues $\bar{\lambda}_i$ and the eigenvectors $\bar{\mathbf{q}}_i$ of the correlation matrix $\bar{\mathbf{R}}(n)$ as follows

$$\begin{bmatrix} 1 \\ \mathbf{a}_o(n) \end{bmatrix} = P_o^f(n) \sum_{i=1}^{M+1} \frac{1}{\bar{\lambda}_i} \bar{\mathbf{q}}_i \bar{q}_{i,1}^* \qquad (6.5.43)$$

and

$$\begin{bmatrix} \mathbf{b}_o(n) \\ 1 \end{bmatrix} = P_o^b(n) \sum_{i=1}^{M+1} \frac{1}{\bar{\lambda}_i} \bar{\mathbf{q}}_i \bar{q}_{i,M+1}^* \qquad (6.5.44)$$

where $\bar{q}_{i,1}$ and $\bar{q}_{i,M+1}$ are the first and last components of $\bar{\mathbf{q}}_i$. The first equation of (6.5.43) and the last equation in (6.5.44) can be solved to provide the MMSEs $P_o^f(n)$ and $P_o^b(n)$, respectively.

Proof. See Problem 6.13.

PROPERTY 6.5.4. Let $\bar{\mathbf{R}}^{-1}(n)$ be the inverse of the correlation matrix $\bar{\mathbf{R}}(n)$. Then, the inverse of the ith element of the ith column of $\bar{\mathbf{R}}^{-1}(n)$ is equal to the MMSE $P^{(i)}(n)$, and the ith column normalized by the ith element is equal to $\mathbf{c}^{(i)}(n)$.

Proof. See Problem 6.14.

PROPERTY 6.5.5. The MMSE prediction errors can be expressed as

$$P_o^f(n) = \frac{\det \bar{\mathbf{R}}(n)}{\det \mathbf{R}(n-1)} \qquad P_o^b(n) = \frac{\det \bar{\mathbf{R}}(n)}{\det \mathbf{R}(n)} \qquad (6.5.45)$$

Proof. Problem 6.17.

The previous concepts are illustrated in the following example.

EXAMPLE 6.5.2. A random sequence $x(n)$ is generated by passing the white Gaussian noise process $w(n) \sim WN(0, 1)$ through the filter

$$x(n) = w(n) + \tfrac{1}{2}w(n-1)$$

Determine the second-order FLP, BLP, and symmetric linear signal smoother.

Solution. The complex power spectrum is

$$R(z) = H(z)H(z^{-1}) = (1 + \tfrac{1}{2}z^{-1})(1 + \tfrac{1}{2}z) = \tfrac{1}{2}z + \tfrac{5}{4} + \tfrac{1}{2}z^{-1}$$

Therefore, the autocorrelation sequence is equal to $r(0) = \tfrac{5}{4}, r(\pm 1) = \tfrac{1}{2}, r(l) = 0$ for $|l| \geq 2$. Since the power spectrum $R(e^{j\omega}) = \tfrac{5}{4} + \cos\omega > 0$ for all ω, the autocorrelation matrix is positive definite. The same is true of any principal submatrix. To determine the second-order linear signal estimators, we start with the matrix

$$\bar{\mathbf{R}} = \begin{bmatrix} \tfrac{5}{4} & \tfrac{1}{2} & 0 \\ \tfrac{1}{2} & \tfrac{5}{4} & \tfrac{1}{2} \\ 0 & \tfrac{1}{2} & \tfrac{5}{4} \end{bmatrix}$$

and follow the procedure outlined in Section 6.5.1 or use the formulas in Table 6.3. The results are

Forward linear prediction $(i = 0)$: $\{a_k\} \rightarrow \{1, -0.476, 0.190\}$ $P_o^f = 1.0119$
Symmetric linear smoothing $(i = 1)$: $\{c_k\} \rightarrow \{-0.4, 1, -0.4\}$ $P_o^s = 0.8500$
Backward linear prediction $(i = 2)$: $\{b_k\} \rightarrow \{0.190, -0.476, 1\}$ $P_o^b = 1.0119$

The inverse of the correlation matrix $\bar{\mathbf{R}}$ is

$$\bar{\mathbf{R}}^{-1} = \begin{bmatrix} 0.9882 & -0.4706 & 0.1882 \\ -0.4706 & 1.1765 & -0.4706 \\ 0.1882 & -0.4706 & 0.9882 \end{bmatrix}$$

and we see that dividing the first, second, and third columns by 0.9882, 1.1765, and 0.9882 provides the forward PEF, the symmetric linear smoothing filter, and the backward PEF, respectively. The inverses of the diagonal elements provide the MMSEs P_o^f, P_o^s, and P_o^b. The reader can easily see, by computing the zeros of the corresponding system functions, that the FLP is minimum-phase, the BLP is maximum-phase, and the symmetric linear smoother is mixed-phase. It is interesting to note that the smoother performs better than either of the predictors.

6.6 OPTIMUM INFINITE IMPULSE RESPONSE FILTERS

So far we have dealt with optimum FIR filters and predictors for nonstationary and stationary processes. In this section, we consider the design of optimum IIR filters for stationary stochastic processes. For nonstationary processes, the theory becomes very complicated. The Wiener-Hopf equations for optimum IIR filters are the same for FIR filters; only the limits in the convolution summation and the range of values for which the normal equations hold are different. Both are determined by the limits of summation in the filter convolution equation. We can easily see from (6.4.16) and (6.4.17), or by applying the orthogonality principle (6.2.41), that the optimum IIR filter

$$\hat{y}(n) = \sum_k h_o(k)x(n-k) \tag{6.6.1}$$

is specified by the Wiener-Hopf equations

$$\sum_k h_o(k)r_x(m-k) = r_{yx}(m) \tag{6.6.2}$$

and the MMSE is given by

$$P_o = r_y(0) - \sum_k h_o(k) r_{yx}^*(k) \tag{6.6.3}$$

where $r_x(l)$ is the autocorrelation of the input stochastic process $x(n)$ and $r_{yx}(l)$ is the cross-correlation between $x(n)$ and desired response process $y(n)$. We assume that the processes $x(n)$ and $y(n)$ are jointly wide-sense stationary with zero mean values.

The range of summation in the above equations includes all the nonzero coefficients of the impulse response of the filter. The range of k in (6.6.1) determines the number of unknowns and the number of equations, that is, the range of m. For IIR filters, we have an infinite number of equations and unknowns, and thus only analytical solutions for (6.6.2) are possible. The key to analytical solutions is that the left-hand side of (6.6.2) can be expressed as the convolution of $h_o(m)$ with $r_x(m)$, that is,

$$h_o(m) * r_x(m) = r_{yx}(m) \tag{6.6.4}$$

which is a *convolutional equation* that can be solved by using the z-transform. The complexity of the solution depends on the range of m.

The formula for the MMSE is the same for any filter, either FIR or IIR. Indeed, using Parseval's theorem and (6.6.3), we obtain

$$P_o = r_y(0) - \frac{1}{2\pi j} \oint_C H_o(z) R_{yx}^*\left(\frac{1}{z^*}\right) z^{-1}\, dz \tag{6.6.5}$$

where $H_o(z)$ is the system function of the optimum filter and $R_{yx}(z) = \mathcal{Z}\{r_{yx}(l)\}$. The power P_y can be computed by

$$P_y = r_y(0) = \frac{1}{2\pi j} \oint_C R_y(z) z^{-1}\, dz \tag{6.6.6}$$

where $R_y(z) = \mathcal{Z}\{r_y(l)\}$. Combining (6.6.5) with (6.6.6), we obtain

$$P_o = \frac{1}{2\pi j} \oint_C [R_y(z) - H_o(z) R_{yx}^*\left(\frac{1}{z^*}\right)] z^{-1}\, dz \tag{6.6.7}$$

which expresses the MMSE in terms of z-transforms. To obtain the MMSE in the frequency domain, we replace z by $e^{j\omega}$. For example, (6.6.5) becomes

$$P_o = r_y(0) - \frac{1}{2\pi} \int_{-\pi}^{\pi} H_o(e^{j\omega}) R_{yx}^*(e^{j\omega})\, d\omega$$

where $H_o(e^{j\omega})$ is the frequency response of the optimum filter.

6.6.1 Noncausal IIR Filters

For the noncausal IIR filter

$$\hat{y}(n) = \sum_{k=-\infty}^{\infty} h_{\text{nc}}(k) x(n-k) \tag{6.6.8}$$

the range of the Wiener-Hopf equations (6.6.2) is $-\infty < m < \infty$ and can be easily solved by using the convolution property of the z-transform. This gives

$$H_{\text{nc}}(z) R_x(z) = R_{yx}(z)$$

or

$$H_{\text{nc}}(z) = \frac{R_{yx}(z)}{R_x(z)} \tag{6.6.9}$$

where $H_{\text{nc}}(z)$ is the system function of the optimum filter, $R_x(z)$ is the complex PSD of $x(n)$, and $R_{yx}(z)$ is the complex cross-PSD between $y(n)$ and $x(n)$.

6.6.2 Causal IIR Filters

For the causal IIR filter

$$\hat{y}(n) = \sum_{k=0}^{\infty} h_c(k)x(n-k) \tag{6.6.10}$$

the Wiener-Hopf equations (6.6.2) hold only for m in the range $0 \le m < \infty$. Since the sequence $r_y(m)$ can be expressed as the convolution of $h_o(m)$ and $r_x(m)$ *only* for $m \ge 0$, we cannot solve (6.6.2) using the z-transform. However, a simple solution is possible using the spectral factorization theorem.[†] This approach was introduced for continuous-time processes in Bode and Shannon (1950) and Zadeh and Ragazzini (1950). It is based on the following two observations:

1. The solution of the Wiener-Hopf equations is trivial if the input is white.
2. Any regular process can be transformed to an equivalent white process.

White input processes. We first note that if the process $x(n)$ is white noise, the solution of the Wiener-Hopf equations is trivial. Indeed, if

$$r_x(l) = \sigma_x^2 \delta(l)$$

Then Equation (6.6.4) gives

$$h_c(m) * \delta(m) = \frac{r_{yx}(m)}{\sigma_x^2} \qquad 0 \le m < \infty$$

which implies that

$$h_c(m) = \begin{cases} \dfrac{1}{\sigma_x^2} r_{yx}(m) & 0 \le m < \infty \\ 0 & m < 0 \end{cases} \tag{6.6.11}$$

because the filter is causal. The system function of the optimum filter is given by

$$H_c(z) = \frac{1}{\sigma_x^2} [R_{yx}(z)]_+ \tag{6.6.12}$$

where

$$[R_{yx}(z)]_+ \triangleq \sum_{l=0}^{\infty} r_{yx}(l)z^{-l} \tag{6.6.13}$$

is the one-sided z-transform of the two-sided sequence $r_{yx}(l)$. The MMSE is given by

$$P_c = r_y(0) - \frac{1}{\sigma_x^2} \sum_{k=0}^{\infty} |r_{yx}(k)|^2 \tag{6.6.14}$$

which follows from (6.6.3) and (6.6.11).

Regular input processes. The PSD of a regular process can be factored as

$$R_x(z) = \sigma_x^2 H_x(z)H_x^*\left(\frac{1}{z^*}\right) \tag{6.6.15}$$

where $H_x(z)$ is the innovations filter (see Section 4.1). The innovations process

$$w(n) = x(n) - \sum_{k=1}^{\infty} h_x(k)w(n-k) \tag{6.6.16}$$

[†] An analogous matrix-based approach is extensively used in Chapter 7 for the design and implementation of optimum FIR filters.

is white and *linearly equivalent* to the input process $x(n)$. Therefore, linear estimation of $y(n)$ based on $x(n)$ is equivalent to linear estimation of $y(n)$ based on $w(n)$. The optimum filter that estimates $y(n)$ from $x(n)$ is obtained by cascading the whitening filter $1/H_x(z)$ with the optimum filter that estimates $y(n)$ from $w(n)$ (see Figure 6.18). Since $w(n)$ is white, the optimum filter for estimating $y(n)$ from $w(n)$ is

$$H_c'(z) = \frac{1}{\sigma_x^2}[R_{yw}(z)]_+ \tag{6.6.17}$$

where $[R_{yw}(z)]_+$ is the one-sided z-transform of $r_{yw}(l)$. To express $H_c'(z)$ in terms of $R_{yx}(z)$, we need the relationship between $R_{yw}(z)$ and $R_{yx}(z)$. From

$$x(n) = \sum_{k=0}^{\infty} h_x(k)w(n-k)$$

we obtain

$$E\{y(n)x^*(n-l)\} = \sum_{k=0}^{\infty} h_x^*(k)E\{y(n)w^*(n-l-k)\}$$

or

$$r_{yx}(l) = \sum_{k=0}^{\infty} h_x^*(k)r_{yw}(l+k) \tag{6.6.18}$$

Taking the z-transform of the above equation leads to

$$R_{yw}(z) = \frac{R_{yx}(z)}{H_x^*(1/z^*)} \tag{6.6.19}$$

which, combined with (6.6.17), gives

$$H_c'(z) = \frac{1}{\sigma_x^2}\left[\frac{R_{yx}(z)}{H_x^*(1/z^*)}\right]_+ \tag{6.6.20}$$

which is the causal optimum filter for the estimation of $y(n)$ from $w(n)$. The optimum filter for estimating $y(n)$ from $x(n)$ is

$$H_c(z) = \frac{1}{\sigma_x^2 H_x(z)}\left[\frac{R_{yx}(z)}{H_x^*(1/z^*)}\right]_+ \tag{6.6.21}$$

which is causal since it is the cascade connection of two causal filters [see Figure 6.19(a)].

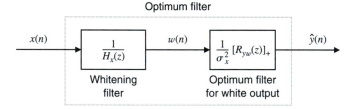

Optimum filter

$x(n)$ → Whitening filter $\frac{1}{H_x(z)}$ → $w(n)$ → Optimum filter for white output $\frac{1}{\sigma_x^2}[R_{yw}(z)]_+$ → $\hat{y}(n)$

FIGURE 6.18
Optimum causal IIR filter design by the spectral factorization method.

The MMSE from (6.6.3) can also be expressed as

$$P_c = r_y(0) - \frac{1}{\sigma_x^2}\sum_{k=0}^{\infty}|r_{yw}(k)|^2 \tag{6.6.22}$$

which shows that the MMSE decreases as we increase the order of the filter. Table 6.5 summarizes the equations required for the design of optimum FIR and IIR filters.

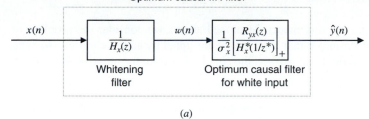

Optimum causal IIR filter

(a)

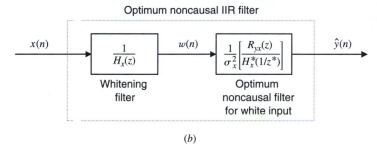

Optimum noncausal IIR filter

(b)

FIGURE 6.19
Comparison of causal and noncausal IIR optimum filters.

TABLE 6.5
Design of FIR and IIR optimum filters for stationary processes.

Filter type	Solution	Required quantities
FIR	$e(n) = y(n) - \mathbf{c}_o^H \mathbf{x}(n)$	$\mathbf{R} = [r_x(m-k)], \mathbf{d} = [r_{yx}(m)]$
	$\mathbf{c}_o = \mathbf{R}^{-1} \mathbf{d}$	$0 \le k, m \le M-1, M =$ finite
	$P_o = r_y(0) - \mathbf{d}^H \mathbf{c}_o$	
Noncausal IIR	$H_{\mathrm{nc}}(z) = \dfrac{R_{yx}(z)}{R_x(z)}$	$R_x(z) = \mathcal{Z}\{r_x(l)\}$
		$R_{yx}(z) = \mathcal{Z}\{r_{xy}^*(l)\}$
	$P_{\mathrm{nc}} = r_y(0) - \displaystyle\sum_{k=-\infty}^{\infty} h_{\mathrm{nc}}(k) r_{yx}^*(k)$	
Causal IIR	$H_{\mathrm{c}}(z) = \dfrac{1}{\sigma_x^2 H_x(z)} \left[\dfrac{R_{yx}(z)}{H_x^*(1/z^*)} \right]_+$	$R_x(z) = \sigma_x^2 H_x(z) H_x^*(1/z^*)$
	$P_{\mathrm{c}} = r_y(0) - \displaystyle\sum_{k=0}^{\infty} h_{\mathrm{nc}}(k) r_{yx}^*(k)$	$R_{yx}(z) = \mathcal{Z}\{r_{xy}(l)\}$

Finally, since the equation for the noncausal IIR filter can be written as

$$H_{\mathrm{nc}}(z) = \frac{1}{\sigma_x^2 H_x(z)} \frac{R_{yx}(z)}{H_x^*(1/z^*)} \qquad (6.6.23)$$

we see that the only difference from the causal filter is that the noncausal filter includes both the causal and noncausal parts of $R_{yx}(z)/H_x(z^{-1})$ [see Figure 6.19(b)]. By using the innovations process $w(n)$, the MMSE can be expressed as

$$P_{\mathrm{nc}} = r_y(0) - \frac{1}{\sigma_x^2} \sum_{k=-\infty}^{\infty} |r_{yw}(k)|^2 \qquad (6.6.24)$$

and is known as the *irreducible MMSE* because it is the best performance that can be achieved by a linear filter. Indeed, since $|r_{yw}(k)| \ge 0$, every coefficient we add to the optimum filter can help to reduce the MMSE.

6.6.3 Filtering of Additive Noise

To illustrate the optimum filtering theory developed above, we consider the problem of estimating a "useful" or desired signal $y(n)$ that is corrupted by additive noise $v(n)$. The goal is to find an optimum filter that extracts the signal $y(n)$ from the noisy observations

$$x(n) = y(n) + v(n) \tag{6.6.25}$$

given that $y(n)$ and $v(n)$ are uncorrelated processes with known autocorrelation sequences $r_y(l)$ and $r_v(l)$.

To design the optimum filter, we need the autocorrelation $r_x(l)$ of the input signal $x(n)$ and the cross-correlation $r_{yx}(l)$ between the desired response $y(n)$ and the input signal $x(n)$. Using (6.6.25), we find

$$r_x(l) = E\{x(n)x^*(n-l)\} = r_y(l) + r_v(l) \tag{6.6.26}$$

and

$$r_{yx}(l) = E\{y(n)x^*(n-l)\} = r_y(l) \tag{6.6.27}$$

because $y(n)$ and $v(n)$ are uncorrelated.

The design of optimum IIR filters requires the functions $R_x(z)$ and $R_{yx}(z)$. Taking the z-transform of (6.6.26) and (6.6.27), we obtain

$$R_x(z) = R_y(z) + R_v(z) \tag{6.6.28}$$

and

$$R_{yx}(z) = R_y(z) \tag{6.6.29}$$

The noncausal optimum filter is given by

$$H_{nc}(z) = \frac{R_{yx}(z)}{R_x(z)} = \frac{R_y(z)}{R_y(z) + R_v(z)} \tag{6.6.30}$$

which for $z = e^{j\omega}$ shows that, for those values of ω for which $|R_y(e^{j\omega})| \gg |R_v(e^{j\omega})|$, that is, for high SNR, we have $|H_{nc}(e^{j\omega})| \approx 1$. In contrast, if $|R_y(e^{j\omega})| \ll |R_v(e^{j\omega})|$, that is, for low SNR, we have $|H_{nc}(e^{j\omega})| \approx 0$. Thus, the optimum filter "passes" its input in bands with high SNR and attenuates it in bands with low SNR, as we would expect intuitively.

Substituting (6.6.30) into (6.6.7), we obtain

$$P_{nc} = \frac{1}{2\pi j} \oint_C \frac{R_y(z)R_v(z)}{R_y(z) + R_v(z)} z^{-1}\, dz \tag{6.6.31}$$

which provides an expression for the MMSE that does not require knowledge of the optimum filter.

We next illustrate the design of optimum filters for the reduction of additive noise with a detailed numerical example.

EXAMPLE 6.6.1. In this example we illustrate the design of an optimum IIR filter to extract a random signal with known autocorrelation sequence

$$r_y(l) = \alpha^{|l|} \qquad -1 < \alpha < 1 \tag{6.6.32}$$

which is corrupted by additive white noise with autocorrelation

$$r_v(l) = \sigma_v^2 \delta(l) \tag{6.6.33}$$

The processes $y(n)$ and $v(n)$ are uncorrelated.

Required statistical moments. The input to the filter is the signal $x(n) = y(n) + v(n)$ and the desired response, the signal $y(n)$. The first step in the design is to determine the required second-order moments, that is, the autocorrelation of the input process and the cross-correlation between input and desired response. Substituting into (6.6.26) and (6.6.27), we have

$$r_x(l) = \alpha^{|l|} + \sigma_v^2 \delta(l) \tag{6.6.34}$$

and

$$r_{yx}(l) = \alpha^{|l|} \tag{6.6.35}$$

To simplify the derivations and deal with "nice, round" numbers, we choose $\alpha = 0.8$ and $\sigma_v^2 = 1$.
Then the complex power spectral densities of $y(n)$, $v(n)$, and $x(n)$ are

$$R_y(z) = \frac{(\frac{3}{5})^2}{(1 - \frac{4}{5}z^{-1})(1 - \frac{4}{5}z)} \qquad \frac{4}{5} < |z| < \frac{5}{4} \tag{6.6.36}$$

$$R_v(z) = \sigma_v^2 = 1 \tag{6.6.37}$$

and
$$R_x(z) = \frac{8}{5}\frac{(1 - \frac{1}{2}z^{-1})(1 - \frac{1}{2}z)}{(1 - \frac{4}{5}z^{-1})(1 - \frac{4}{5}z)} \tag{6.6.38}$$

respectively.

Noncausal filter. Using (6.6.9), (6.6.29), (6.6.36), and (6.6.38), we obtain

$$H_{nc}(z) = \frac{R_{yx}(z)}{R_x(z)} = \frac{9}{40}\frac{1}{(1 - \frac{1}{2}z^{-1})(1 - \frac{1}{2}z)} \qquad \frac{1}{2} < |z| < 2$$

Evaluating the inverse the z-transform we have

$$h_{nc}(n) = \frac{3}{10}(\frac{1}{2})^{|n|} \qquad -\infty < n < \infty$$

which clearly corresponds to a noncausal filter. From (6.6.3), the MMSE is

$$P_{nc} = 1 - \frac{3}{10}\sum_{k=-\infty}^{\infty}(\frac{1}{2})^{|k|}(\frac{4}{5})^{|k|} = \frac{3}{10} \tag{6.6.39}$$

and provides the irreducible MMSE.

Causal filter. To find the optimum causal filter, we need to perform the spectral factorization

$$R_x(z) = \sigma_x^2 H_x(z)H_x(z^{-1})$$

which is provided by (6.6.38) with

$$\sigma_x^2 = \frac{8}{5} \tag{6.6.40}$$

and
$$H_x(z) = \frac{1 - \frac{1}{2}z^{-1}}{1 - \frac{4}{5}z^{-1}} \tag{6.6.41}$$

Thus, $\quad R_{yw}(z) = \frac{R_{yx}(z)}{H_x(z^{-1})} = \frac{0.36}{(1 - \frac{4}{5}z^{-1})(1 - \frac{1}{2}z)} = \frac{0.6}{1 - \frac{4}{5}z^{-1}} + \frac{0.3z}{1 - \frac{1}{2}z} \tag{6.6.42}$

where the first term (causal) converges for $|z| > \frac{4}{5}$ and the second term (noncausal) converges for $|z| < 2$. Hence, taking the causal part

$$\left[\frac{R_{yx}(z)}{H_x(z^{-1})}\right]_+ = \frac{\frac{3}{5}}{1 - \frac{4}{5}z^{-1}}$$

and substituting into (6.6.21), we obtain the causal optimum filter

$$H_c(z) = \frac{5}{8}\left(\frac{1 - \frac{4}{5}z^{-1}}{1 - \frac{1}{2}z^{-1}}\frac{\frac{3}{5}}{1 - \frac{4}{5}z^{-1}}\right) = \frac{3}{8}\left(\frac{1}{1 - \frac{1}{2}z^{-1}}\right) \qquad |z| < \frac{1}{2} \tag{6.6.43}$$

The impulse response is

$$h_c(n) = \frac{3}{8}(\frac{1}{2})^n u(n)$$

which corresponds to a causal and stable IIR filter. The MMSE is

$$P_c = r_y(0) - \sum_{k=0}^{\infty} h_c(k)r_{yx}(k) = 1 - \frac{3}{8}\sum_{k=0}^{\infty}(\frac{1}{2})^k(\frac{4}{5})^k = \frac{3}{8} \tag{6.6.44}$$

which is, as expected, larger than P_{nc}.

From (6.6.43), we see that the optimum causal filter is a first-order recursive filter that can be implemented by the difference equation

$$\hat{y}(n) = \tfrac{1}{2}\hat{y}(n-1) + \tfrac{3}{8}x(n)$$

In general, this is possible only when $H_C(z)$ is a rational function.

Computation of MMSE using the innovation. We next illustrate how to find the MMSE by using the cross-correlation sequence $r_{yw}(l)$. From (6.6.42), we obtain

$$r_{yw}(l) = \begin{cases} \tfrac{3}{5}(\tfrac{4}{5})^l & l \geq 0 \\ \tfrac{3}{5}2^l & l < 0 \end{cases} \tag{6.6.45}$$

which, in conjunction with (6.6.22) and (6.6.24), gives

$$P_c = r_y(0) - \frac{1}{\sigma_x^2}\sum_{k=0}^{\infty} r_{yw}^2(k) = 1 - \tfrac{5}{8}(\tfrac{3}{5})^2\sum_{k=0}^{\infty}(\tfrac{4}{5})^{2k} = \tfrac{3}{8}$$

and

$$P_{nc} = r_y(0) - \frac{1}{\sigma_x^2}\left[\sum_{k=0}^{\infty} r_{yw}^2(k) - \sum_{k=-\infty}^{-1} r_{yw}^2(k)\right] = \frac{3}{10}$$

which agree with (6.6.44) and (6.6.39).

Noncausal smoothing filter. Suppose now that we want to estimate the value $y(n+D)$ of the desired response from the data $x(n)$, $-\infty < n < \infty$. Since

$$E\{y(n+D)x(n-l)\} = r_{yx}(n+D) \tag{6.6.46}$$

and

$$\mathcal{Z}\{r_{yx}(n+D)\} = z^D R_{yx}(z) \tag{6.6.47}$$

the noncausal Wiener smoothing filter is

$$H_{nc}^D(z) = \frac{z^D R_{yx}(z)}{R_x(z)} = \frac{z^D R_y(z)}{R_x(z)} = z^D H_{nc}(z) \tag{6.6.48}$$

$$h_{nc}^D(n) = h_{nc}(n+D) \tag{6.6.49}$$

The MMSE is

$$P_{nc}^D = r_y(0) - \sum_{k=-\infty}^{\infty} h_{nc}(k+D)r_{yx}(k+D) = P_{nc} \tag{6.6.50}$$

which is independent of the time shift D.

Causal prediction filter. We estimate the value $y(n+D)$ $(D>0)$ of the desired response using the data $x(k)$, $-\infty < k \leq n$. The whitening part of the causal prediction filter does not depend on $y(n)$ and is still given by (6.6.41). The coloring part depends on $y(n+D)$ and is given by $R'_{yw}(z) = z^D R_{yw}(z)$ or $r'_{yw}(l) = r_{yw}(l+D)$. Taking into consideration that $D>0$, we can show (see Problem 6.31) that the system function and the impulse response of the causal Wiener predictor are

$$H_c^{[D]}(z) = \frac{5}{8}\left(\frac{1-\tfrac{4}{5}z^{-1}}{1-\tfrac{1}{2}z^{-1}}\right)\left[\frac{\tfrac{3}{5}(\tfrac{4}{5})^D}{1-\tfrac{4}{5}z^{-1}}\right] = \frac{\tfrac{3}{8}(\tfrac{4}{5})^D}{1-\tfrac{1}{2}z^{-1}} \tag{6.6.51}$$

and

$$h_c^{[D]}(n) = \tfrac{3}{8}(\tfrac{4}{5})^D(\tfrac{1}{2})^n u(n) \tag{6.6.52}$$

respectively. This shows that as $D \to \infty$, the impulse response $h_c^{[D]}(n) \to 0$, which is consistent with our intuition that the prediction is less and less reliable. The MMSE is

$$P_c^{[D]} = 1 - \tfrac{3}{8}(\tfrac{4}{5})^{2D}\sum_{k=0}^{\infty}(\tfrac{2}{5})^k = 1 - \tfrac{5}{8}(\tfrac{4}{5})^{2D} \tag{6.6.53}$$

and $P_c^{[D]} \to r_y(0) = 1$ as $D \to \infty$, which agrees with our earlier observation. For $D = 2$, the MMSE is $P_c^{[2]} = 93/125 = 0.7440 > P_c$, as expected.

Causal smoothing filter. To estimate the value $y(n + D)$ $(D < 0)$ of the desired response using the data $x(n)$, $-\infty < k \le n$, we need a smoothing Wiener filter. The derivation, which is straightforward but somewhat involved, is left for Problem 6.32. The system function of the optimum smoothing filter is

$$H_c^{[D]}(z) = \frac{3}{8} \left(\frac{z^D}{1 - \frac{1}{2}z^{-1}} + \frac{2^D \sum_{l=0}^{-D-1} 2^l z^{-l}}{1 - \frac{1}{2}z^{-1}} - \frac{4}{5} \frac{2^D \sum_{l=0}^{-D-1} 2^l z^{-l-1}}{1 - \frac{1}{2}z^{-1}} \right) \qquad (6.6.54)$$

where $D < 0$. To find the impulse response for $D = -2$, we invert (6.6.54). This gives

$$h_c^{[-2]}(k) = \frac{3}{32}\delta(k) + \frac{51}{320}\delta(k-1) + \frac{39}{128}\left(\frac{1}{2}\right)^{k-2}u(k-2) \qquad (6.6.55)$$

and if we express $r_{yx}(k-2)$ in a similar form, we can compute the MMSE

$$P_c^{[-2]} = 1 - \frac{3}{50} - \frac{51}{400} - \left(\frac{39}{128}\right)\frac{5}{3} = \frac{39}{128} = 0.3047 \qquad (6.6.56)$$

which is less than $P_c = 0.375$. This should be expected since the smoothing Wiener filter uses more information than the Wiener filter (i.e., when $D = 0$). In fact it can be shown that

$$\lim_{D \to -\infty} P_c^{[D]} = P_{nc} \quad \text{and} \quad \lim_{D \to -\infty} h_c^{[D]}(n) = h_{nc}(n) \qquad (6.6.57)$$

which is illustrated in Figure 6.20. Figure 6.21 shows the impulse responses of the various optimum IIR filters designed in this example. Interestingly, all are obtained by shifting and truncating the impulse response of the optimum noncausal IIR filter.

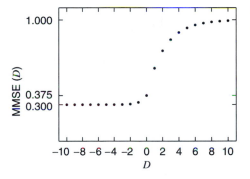

FIGURE 6.20
MMSE as a function of the time shift D.

FIR filter. The Mth-order FIR filter is obtained by solving the linear system

$$\mathbf{Rh} = \mathbf{d}$$

where

$$\mathbf{R} = \text{Toeplitz}(1 + \sigma_v^2, \alpha, \ldots, \alpha^{M-1})$$

and

$$\mathbf{d} = [1\, \alpha\, \cdots\, \alpha^{M-1}]^T$$

The MMSE is

$$P_o = r_y(0) - \sum_{k=0}^{M-1} h_o(k)r_{yx}(k)$$

and is shown in Figure 6.22 as a function of the order M together with P_c and P_{nc}. We notice that an optimum FIR filter of order $M = 4$ provides satisfactory performance. This can be explained by noting that the impulse response of the causal optimum IIR filter is negligible for $n > 4$.

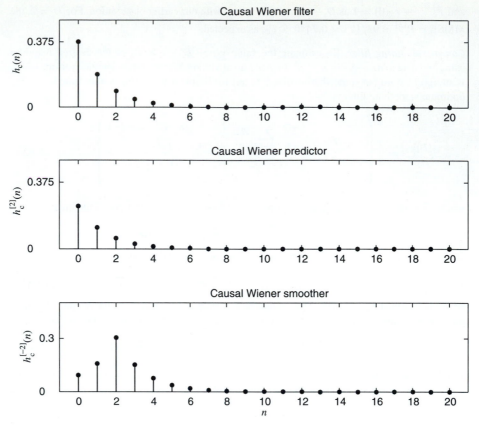

FIGURE 6.21
Impulse response of optimum filters for pure filtering, prediction, and smoothing.

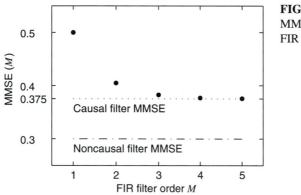

FIGURE 6.22
MMSE as a function of the optimum
FIR filter order M.

6.6.4 Linear Prediction Using the Infinite Past—Whitening

The one-step forward IIR linear predictor is a causal IIR optimum filter with desired response $y(n) \triangleq x(n+1)$. The prediction error is

$$e^{\mathrm{f}}(n+1) = x(n+1) - \sum_{k=0}^{\infty} h_{\mathrm{lp}}(k)x(n-k) \qquad (6.6.58)$$

where

$$H_{\mathrm{lp}}(z) = \sum_{k=0}^{\infty} h_{\mathrm{lp}}(k)z^{-k} \qquad (6.6.59)$$

is the system function of the optimum predictor. Since $y(n) = x(n+1)$, we have $r_{yx}(l) = r_x(l+1)$ and $R_{yx}(z) = z R_x(z)$. Hence, the optimum predictor is

$$H_{lp}(z) = \frac{1}{\sigma_x^2 H_x(z)} \left[\frac{z \sigma_x^2 H_x(z) H_x(z^{-1})}{H_x(z^{-1})} \right]_+ = \frac{[z H_x(z)]_+}{H_x(z)} = \frac{z H_x(z) - z}{H_x(z)}$$

and the prediction error filter (PEF) is

$$H_{PEF}(z) = \frac{E^f(z)}{X(z)} = 1 - z^{-1} H_{lp}(z) = \frac{1}{H_x(z)} \tag{6.6.60}$$

that is, *the one-step IIR linear predictor of a regular process is identical to the whitening filter of the process.* Therefore, the prediction error process is white, and the prediction error filter is minimum-phase. We will see that the efficient solution of optimum filtering problems includes as a prerequisite the solution of a linear prediction problem. Furthermore, algorithms for linear prediction provide a convenient way to perform spectral factorization in practice.

The MMSE is

$$P_o^f = \frac{1}{2\pi j} \oint_C \left\{ R_x(z) - z \left[1 - \frac{1}{H_x(z)} \right] z^{-1} R_x^* \left(\frac{1}{z^*} \right) \right\} z^{-1} dz$$

$$= \frac{1}{2\pi j} \oint_C R_x(z) \frac{1}{H_x(z)} z^{-1} dz \tag{6.6.61}$$

$$= \sigma_x^2 \frac{1}{2\pi j} \oint_C H_x^* \left(\frac{1}{z^*} \right) z^{-1} dz = \sigma_x^2$$

because

$$\frac{1}{2\pi j} \oint_C H_x^* \left(\frac{1}{z^*} \right) z^{-1} dz = h_x(0) = 1$$

From Section 2.4.4 and (6.6.61) we have

$$P_o^f = \sigma_x^2 = \exp \left[\frac{1}{2\pi} \int_{-\pi}^{\pi} \ln R_x(e^{j\omega}) \, d\omega \right] \tag{6.6.62}$$

which is known as the *Kolmogorov-Szegö* formula.

We can easily see that the D-step predictor $(D > 0)$ is given by

$$H_D(z) = \frac{[z^D H_x(z)]_+}{H_x(z)} = \frac{1}{H_x(z)} \sum_{k=D}^{\infty} h_x(k) z^{-k+D} \tag{6.6.63}$$

but is not guaranteed to be minimum-phase for $D \neq 1$.

EXAMPLE 6.6.2. Consider a minimum-phase AR(2) process

$$x(n) = a_1 x(n-1) + a_2 x(n-2) + w(n)$$

where $w(n) \sim \text{WN}(0, \sigma_w^2)$. The complex PSD of the process is

$$R_x(z) = \frac{\sigma_x^2}{A(z) A(z^{-1})} \triangleq \sigma_x^2 H_x(z) H_x(z^{-1})$$

where $A(z) \triangleq 1 - a_1 z^{-1} - a_2 z^{-2}$ and $\sigma_x^2 = \sigma_w^2$. The one-step forward predictor is given by

$$H_{lp}(z) = z - \frac{z}{H_x(z)} = z - z A(z) = a_1 + a_2 z^{-1}$$

or

$$\hat{x}(n+1) = a_1 x(n) + a_2 x(n-1)$$

as should be expected because the present value of the process depends only on the past two values. Since the excitation $w(n)$ is white and cannot be predicted from the present or previous values of the signal $x(n)$, it is equal to the prediction error $e^f(n)$. Therefore, $\sigma_{ef}^2 = \sigma_w^2$, as expected from (6.6.62). This shows that the MMSE of the one-step linear predictor depends on the SFM of the process $x(n)$. It is maximum for a white noise process, which is clearly unpredictable.

Predictable processes. A random process $x(n)$ is said to be (exactly) *predictable* if $P_e = E\{|e^{\mathrm{f}}(n)|^2\} = 0$. We next show that a process $x(n)$ is predictable if and only if its PSD consists of impulses, that is,

$$R_x(e^{j\omega}) = \sum_k A_k \delta(\omega - \omega_k) \tag{6.6.64}$$

or in other words, $x(n)$ is a harmonic process. For this reason harmonic processes are also known as *deterministic* processes. From (6.6.60) we have

$$P_e = E\{|e^{\mathrm{f}}(n)|^2\} = \int_{-\pi}^{\pi} |H_{\mathrm{PEF}}(e^{j\omega})|^2 R_x(e^{j\omega})\,d\omega \tag{6.6.65}$$

where $H_{\mathrm{PEF}}(e^{j\omega})$ is the frequency response of the prediction error filter. Since $R_x(e^{j\omega}) \geq 0$, the integral in (6.6.65) is zero if and only if $|H_{\mathrm{PEF}}(e^{j\omega})|^2 R_x(e^{j\omega}) = 0$. This is possible only if $R_x(e^{j\omega})$ is a linear combination of impulses, as in (6.6.64), and $e^{j\omega_k}$ are the zeros of $H_{\mathrm{PEF}}(z)$ on the unit circle (Papoulis 1985).

From the Wold decomposition theorem (see Section 4.1.3) we know that every random process can be decomposed into two components that are mutually orthogonal: (1) a regular component with continuous PSD that can be modeled as the response of a minimum-phase system to white noise and (2) a predictable process that can be exactly predicted from a linear combination of past values. This component has a line PSD and is essentially a harmonic process. A complete discussion of this subject can be found in Papoulis (1985, 1991) and Therrien (1992).

6.7 INVERSE FILTERING AND DECONVOLUTION

In many practical applications, a signal of interest passes through a distorting system whose output may be corrupted by additive noise. When the distorting system is linear and time-invariant, the observed signal is the convolution of the desired input with the impulse response of the system. Since in most cases we deal with linear and time-invariant systems, the terms *filtering* and *convolution* are often used interchangeably.

Deconvolution is the process of retrieving the *unknown* input of a *known* system by using its observed output. If the system is also unknown, which is more common in practical applications, we have a problem of *blind deconvolution*. The term *blind deconvolution* was introduced in Stockham et al. (1975) for a method used to restore old records. Other applications include estimation of the vocal tract in speech processing, equalization of communication channels, deconvolution of seismic data for the elimination of multiple reflections, and image restoration.

The basic problem is illustrated in Figure 6.23. The output of the unknown LTI system $G(z)$, which is assumed BIBO stable, is given by

$$x(n) = \sum_{k=-\infty}^{\infty} g(k)w(n - k) \tag{6.7.1}$$

where $w(n) \sim \mathrm{IID}(0, \sigma_w^2)$ is a white noise sequence. Suppose that we observe the output $x(n)$ and that we wish to recover the input signal $w(n)$, and possibly the system $G(z)$, using the output signal and some statistical information about the input.

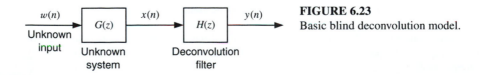

FIGURE 6.23
Basic blind deconvolution model.

If we know the system $G(z)$, the inverse system $H(z)$ is obtained by noticing that perfect retrieval of the input is possible if

$$h(n) * g(n) * w(n) = b_0 w(n - n_0) \quad (6.7.2)$$

where b_0 and n_0 are constants. From (6.7.2), we have $h(n) * g(n) = b_0\delta(n - n_0)$, or equivalently

$$H(z) = b_0\frac{z^{-n_0}}{G(z)} \quad (6.7.3)$$

which provides the system function of the inverse system. The input can be recovered by convolving the output with the inverse system $H(z)$. Therefore, the terms *inverse filtering* and *deconvolution* are equivalent for LTI systems.

There are three approaches for blind deconvolution:

- Identify the system $G(z)$, design its inverse system $H(z)$, and then compute the input $w(n)$.
- Identify directly the inverse $H(z) = 1/G(z)$ of the system, and then determine the input $w(n)$.
- Estimate directly the input $w(n)$ from the output $x(n)$.

Any of the above approaches requires either directly or indirectly the estimation of both the magnitude response $|G(e^{j\omega})|$ and the phase response $\angle G(e^{j\omega})$ of the unknown system. In practice, the problem becomes more complicated because the output $x(n)$ is usually corrupted by additive noise. If this noise is uncorrelated with the input signal and the required second-order moments are available, we show how to design an optimum inverse filter that provides an optimum estimate of the input in the presence of noise. In Section 6.8 we apply these results to the design of optimum equalizers for data transmission systems. The main blind identification and deconvolution problem, in which only statistical information about the output is known, is discussed in Chapter 12.

We now discuss the design of optimum inverse filters for linearly distorted signals observed in the presence of additive output noise. The typical configuration is shown in Figure 6.24. Ideally, we would like the optimum filter to restore the distorted signal $x(n)$ to its original value $y(n)$. However, the ability of the optimum filter to attain ideal performance is limited by three factors. First, there is additive noise $v(n)$ at the output of the system. Second, if the physical system $G(z)$ is causal, its output $s(n)$ is delayed with respect to the input, and we may need some delay z^{-D} to improve the performance of the system. When $G(z)$ is a non-minimum-phase system, the inverse system is either noncausal or unstable and should be approximated by a causal and stable filter. Third, the inverse system may be IIR and should be approximated by an FIR filter.

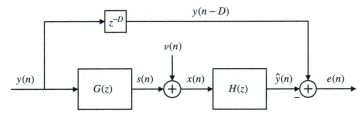

FIGURE 6.24
Typical configuration for optimum inverse system modeling.

The optimum inverse filter is the noncausal Wiener filter

$$H_{nc}(z) = \frac{z^{-D}R_{yx}(z)}{R_x(z)} \quad (6.7.4)$$

where the term z^{-D} appears because the desired response is $y_D(n) \triangleq y(n - D)$. Since $y(n)$

and $v(n)$ are uncorrelated, we have

$$R_{yx}(z) = R_{ys}(z) \tag{6.7.5}$$

and

$$R_x(z) = G(z)G^*\left(\frac{1}{z^*}\right)R_y(z) + R_v(z) \tag{6.7.6}$$

The cross-correlation between $y(n)$ and $s(n)$

$$R_{ys}(z) = G^*\left(\frac{1}{z^*}\right)R_y(z) \tag{6.7.7}$$

is obtained by using Equation (6.6.18). Therefore, the optimum inverse filter is

$$H_{\text{nc}}(z) = \frac{z^{-D}G^*(1/z^*)R_y(z)}{G(z)G^*(1/z^*)R_y(z) + R_v(z)} \tag{6.7.8}$$

which, in the absence of noise, becomes

$$H_{\text{nc}}(z) = \frac{z^{-D}}{G(z)} \tag{6.7.9}$$

as expected. The behavior of the optimum inverse system is illustrated in the following example.

EXAMPLE 6.7.1. Let the system $G(z)$ be an all-zero non-minimum-phase system given by

$$G(z) = \tfrac{1}{5}(-3z + 7 - 2z^{-1}) = -\tfrac{3}{5}(1 - \tfrac{1}{3}z^{-1})(z - 2)$$

Then the inverse system is given by

$$H(z) = G^{-1}(z) = \frac{5}{-3z + 7 - 2z^{-1}} = \frac{1}{1 - \tfrac{1}{3}z^{-1}} - \frac{1}{1 - 2z^{-1}}$$

which is stable if the ROC is $-\tfrac{1}{3} < |z| < 2$. Therefore, the impulse response of the inverse system is

$$h(n) = \begin{cases} (\tfrac{1}{3})^n & n \geq 0 \\ 2^n & n < 0 \end{cases}$$

which is noncausal and stable.

Following the discussion given in this section, we want to design an optimum inverse system given that $G(z)$ is driven by a white noise sequence $y(n)$ and that the additive noise $v(n)$ is white, that is, $R_y(z) = \sigma_y^2$ and $R_v(z) = \sigma_v^2$. From (6.7.8), the optimum noncausal inverse filter is given by

$$H_{\text{nc}}(z) = \frac{z^{-D}}{G(z) + [1/G(z^{-1})](\sigma_v^2/\sigma_y^2)}$$

which can be computed by assuming suitable values for variances σ_y^2 and σ_v^2. Note that if $\sigma_v^2 \ll \sigma_y^2$, that is, for very large SNR, we obtain (6.7.9).

A more interesting case occurs when the optimum inverse filter is FIR, which can be easily implemented. To design this FIR filter, we will need the autocorrelation $r_x(l)$ and the cross-correlation $r_{y_D x}(l)$, where $y_D(n) = y(n - D)$ is the delayed system input sequence. Since

$$R_x(z) = \sigma_y^2 G(z)G(z^{-1}) + \sigma_v^2$$

and

$$R_{y_D x}(l) = \sigma_y^2 z^{-D} G(z^{-1})$$

we have (see Section 3.4.1)

$$r_x(l) = g(l) * g(-l) * r_y(l) + r_v(l) = \sigma_y^2[g(l) * g(-l)] + \sigma_v^2\delta(l)$$

and

$$r_{y_D x}(l) = g(-l) * r_y(l - D) = \sigma_y^2 g(-l + D)$$

respectively. Now we can determine the optimum FIR filter $\mathbf{h}_D$ of length M by constructing an $M \times M$ Toeplitz matrix $\mathbf{R}$ from $r_x(l)$ and an $M \times 1$ vector $\mathbf{d}$ from $r_{y_D}(l)$ and then solving

$$\mathbf{R}\mathbf{h}_D = \mathbf{d}$$

for various values of D. We can then plot the MMSE as a function of D to determine the best value of D (and the corresponding FIR filter) which will give the smallest MMSE. For example, if $\sigma_y^2 = 1$, $\sigma_v^2 = 0.1$, and $M = 10$, the correlation functions are

$$r_x(l) = \left[\underset{\substack{\uparrow \\ l=0}}{\frac{6}{25}, -\frac{7}{5}, \frac{129}{50}, -\frac{7}{5}, \frac{6}{25}} \right] \quad \text{and} \quad r_{y_Dx}(l) = \left[\underset{\substack{\uparrow \\ l=D}}{-\frac{2}{5}, \frac{7}{5}, -\frac{3}{5}} \right]$$

The resulting MMSE as a function of D is shown in Figure 6.25, which indicates that the best value of D is approximately $M/2$. Finally, plots of impulse responses of the inverse system are shown in Figure 6.26. The first plot shows the noncausal $h(n)$, the second plot shows the causal

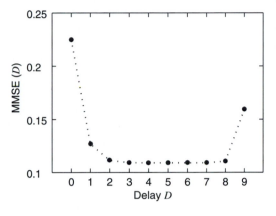

FIGURE 6.25
The inverse filtering MMSE as a function of delay D.

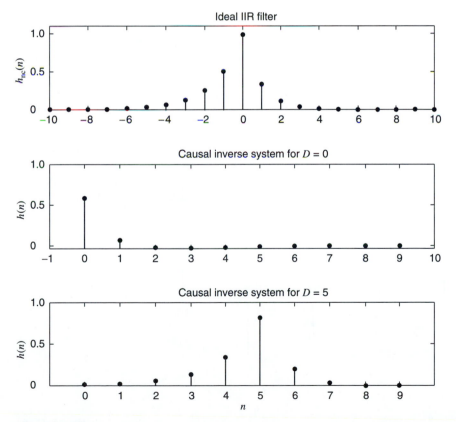

FIGURE 6.26
Impulse responses of optimum inverse filters.

FIR system $h_0(n)$ for $D = 0$, and the third plot shows the causal FIR system $h_D(n)$ for $D = 5$. It is clear that the optimum delayed FIR inverse filter for $D \simeq M/2$ closely matches the impulse response of the inverse filter $h(n)$.

6.8 CHANNEL EQUALIZATION IN DATA TRANSMISSION SYSTEMS

The performance of data transmission systems through channels that can be approximated by linear systems is limited by factors such as finite bandwidth, intersymbol interference, and thermal noise (see Section 1.4). Typical examples include telephone lines, microwave line-of-sight radio links, satellite channels, and underwater acoustic channels. When the channel frequency response deviates form the ideal of flat magnitude and linear phase, both (left and right) tails of a transmitted pulse will interfere with neighboring pulses. Hence, the value of a sample taken at the center of a pulse will contain components from the tails of the other pulses. The distortion caused by the overlapping tails is known as *intersymbol interference (ISI)*, and it can lead to erroneous decisions that increase the probability of error. For band-limited channels with low background noise (e.g., voice band telephone channel), ISI is the main performance limitation for high-speed data transmission. In radio and undersea channels, ISI is the result of multipath propagation (Siller 1984).

Intersymbol interference occurs in all pulse modulation systems, including frequency-shift keying (FSK), phase-shift keying (PSK), and quadrature amplitude modulation (QAM). However, to simplify the presentation, we consider a baseband pulse amplitude modulation (PAM) system. This does not result in any loss of generality because we can obtain an equivalent baseband model for any linear modulation scheme (Proakis 1996). We consider the K-ary ($K = 2^L$) PAM communication system shown in Figure 6.27(a). The binary

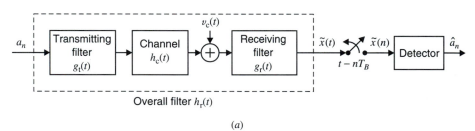

(a)

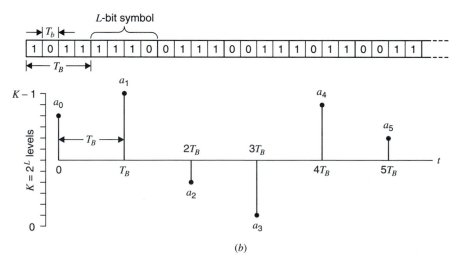

(b)

FIGURE 6.27
(a) Baseband pulse amplitude modulation data transmission system model and (b) input symbol sequence a_n.

input sequence is subdivided into L-bit blocks, or *symbols*, and each symbol is mapped to one of the K amplitude levels, as shown in Figure 6.27(b). The interval T_B is called the symbol or *baud interval* while the interval T_b is called the *bit interval*. The quantity $\mathcal{R}_B = 1/T_B$ is known as the *baud rate*, and the quantity $\mathcal{R}_b = L\mathcal{R}_B$ is the *bit rate*.

The resulting symbol sequence $\{a_n\}$ modulates the transmitted pulse $g_t(t)$. For analysis purposes, the symbol sequence $\{a_n\}$ can be represented by an equivalent continuous-time signal using an impulse train, that is,

$$\{a_n\}_{-\infty}^{\infty} \Leftrightarrow \sum_{n=-\infty}^{\infty} a_n \delta(t - nT_B) \tag{6.8.1}$$

The modulated pulses are transmitted over the channel represented by the impulse response $h_c(t)$ and the additive noise $v_c(t)$. The received signal is filtered by the receiving filter $g_r(t)$ to obtain $\tilde{x}(t)$. Using (6.8.1), the signal $\tilde{x}(t)$ at the output of the receiving filter is given by

$$\tilde{x}(t) = \sum_{k=-\infty}^{\infty} a_k \{\delta(t - kT_B) * g_t(t) * h_c(t) * g_r(t)\} + v_c(t) * g_r(t)$$

$$\triangleq \sum_{k=-\infty}^{\infty} a_k \tilde{h}_r(t - kT_B)\tilde{v}(t) \tag{6.8.2}$$

where

$$\tilde{h}_r(t) \triangleq g_t(t) * h_c(t) * g_r(t) \tag{6.8.3}$$

is the impulse response of the combined system of transmitting filter, channel, and receiving filter, and

$$\tilde{v}(t) \triangleq g_r(t) * v_c(t) \tag{6.8.4}$$

is the additive noise at the output of the receiving filter.

6.8.1 Nyquist's Criterion for Zero ISI

If we sample the received signal $x(t)$ at the time instant $t_0 + nT_B$, we obtain

$$\tilde{x}(t_0 + nT_B) = \sum_{k=-\infty}^{\infty} a_k \tilde{h}_r(t_0 + nT_B - kT_B) + \tilde{v}(t_0 + nT_B)$$

$$= a_n \tilde{h}_r(t_0) + \sum_{\substack{k=-\infty \\ k \neq n}}^{\infty} a_k \tilde{h}_r(t_0 + nT_B - kT_B) + \tilde{v}(t_0 + nT_B) \tag{6.8.5}$$

where t_0 accounts for the channel delay and the sampler phase. The first term in (6.8.5) is the desired signal term while the third term is the noise term. The middle term in (6.8.5) represents the ISI, and it will be zero if and only if

$$\tilde{h}_r(t_0 + nT_B - kT_B) = 0 \quad n \neq k \tag{6.8.6}$$

As was first shown by Nyquist (Gitlin, Hayes, and Weinstein 1992), a time-domain pulse $\tilde{h}_r(t)$ will have zero crossings once every T_B s, that is,

$$\tilde{h}_r(nT_B) = \begin{cases} 1 & n = 0 \\ 0 & n \neq 0 \end{cases} \tag{6.8.7}$$

if its Fourier transform satisfies the condition

$$\sum_{l=-\infty}^{\infty} \tilde{H}_r\left(F + \frac{l}{T_B}\right) = T_B \tag{6.8.8}$$

This condition is known as the *Nyquist criterion for zero ISI* and its basic meaning is illustrated in Figure 6.28.

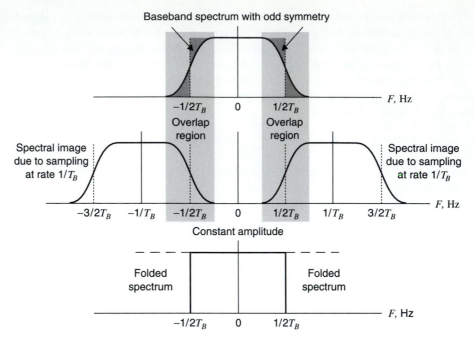

FIGURE 6.28
Frequency-domain Nyquist criterion for zero ISI.

A pulse shape that satisfies (6.8.8) and that is widely used in practice is of the *raised cosine family*

$$\tilde{h}_{\rm rc}(t) = \frac{\sin(\pi t/T_B)}{\pi t/T_B} \frac{\cos(\pi \alpha t/T_B)}{1 - 4\alpha^2 t^2/T_B^2} \tag{6.8.9}$$

where $0 \le \alpha \le 1$ is known as the *rolloff factor*. This pulse and its Fourier transform for $\alpha = 0, 0.5$, and 1 are shown in Figure 6.29. The choice of $\alpha = 0$ reduces $\tilde{h}_{\rm rc}(t)$ to the unrealizable sinc pulse and $\mathcal{R}_B = 1/T_B$, whereas for $\alpha = 1$ the symbol rate is $\mathcal{R}_B = 1/(2T_B)$. In practice, we can see the effect of ISI and the noise if we display the received signal on the vertical axis of an oscilloscope and set the horizontal sweep rate at $1/T_B$. The resulting display is known as *eye pattern* because it resembles the human eye. The closing of the eye increases with the increase in ISI.

6.8.2 Equivalent Discrete-Time Channel Model

Referring to Figure 6.27(a), we note that the input to the data transmission system is a discrete-time sequence $\{a_n\}$ at the symbol rate $1/T_B$ symbols per second, and the input to the detector is also a discrete-time sequence $\tilde{x}(nT_B)$ at the symbol rate. Thus the overall system between the input symbols and the equalizer can be modeled as a discrete-time channel model for further analysis. From (6.8.2), after sampling at the symbol rate, we obtain

$$\tilde{x}(nT_B) = \sum_{k=-\infty}^{\infty} a_k \tilde{h}_{\rm r}(nT_B - kT_B) + \tilde{v}(nT_B) \tag{6.8.10}$$

where $\tilde{h}_{\rm r}(t)$ is given in (6.8.3) and $\tilde{v}(t)$ is given in (6.8.4). The first term in (6.8.10) can be interpreted as a discrete-time IIR filter with impulse response[†] $\tilde{h}_{\rm r}(n) \triangleq h_{\rm r}(nT_B)$ with input

---

[†] Here we have abused the notation to avoid a new symbol.

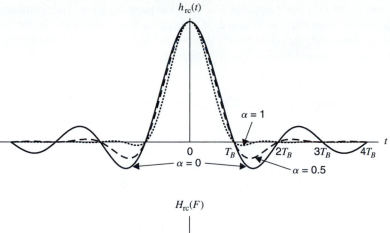

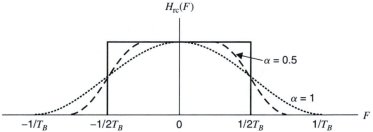

FIGURE 6.29
Pulses with a raised cosine spectrum.

a_k. In a practical data transmission system, it is not unreasonable to assume that $\tilde{h}_r(n) = 0$ for $|n| \geq L$, where L is some arbitrary positive integer. Then we obtain

$$\tilde{x}(n) = \sum_{k=-L}^{L} a_k \tilde{h}_r(n-k) + \tilde{v}(n) \tag{6.8.11}$$

$$\tilde{x}(n) \triangleq \tilde{x}(nT_B) \qquad \tilde{v}(n) \triangleq \tilde{v}(nT_B)$$

which is an FIR filter of length $2L + 1$, shown in Figure 6.30.

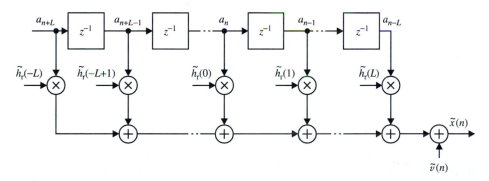

FIGURE 6.30
Equivalent discrete-time model of data transmission system with ISI.

There is one difficulty with this model. If we assume that the additive channel noise $v_c(t)$ is zero-mean white, then the equivalent noise sequence $\tilde{v}(n)$ is not white. This can be seen from the definition of $\tilde{v}(t)$ in (6.8.4). Thus the autocorrelation of $\tilde{v}(n)$ is given by

$$r_{\tilde{v}}(l) = \sigma_v^2 r_{g_r}(l) \tag{6.8.12}$$

where σ_v^2 is the variance of the samples of $v_c(t)$ and $r_{g_r}(l)$ is the sampled autocorrelation of

$g_r(t)$. This nonwhiteness of $\tilde{v}(t)$ poses a problem in the subsequent design and performance evaluation of equalizers. Therefore, in practice, it is necessary to whiten this noise by designing a whitening filter and placing it after the sampler in Figure 6.27(a). The whitening filter is designed by using spectral factorization of $\mathcal{Z}[r_{g_r}(l)]$. Let

$$R_{g_r}(z) = \mathcal{Z}[r_{g_r}(l)] = R_{g_r}^+(z)R_{g_r}^-(z) \tag{6.8.13}$$

where $R_{g_r}^+(z)$ is the minimum-phase factor and $R_{g_r}^-(z)$ is the maximum-phase factor. Choosing

$$W(z) \triangleq \frac{1}{R_{g_r}^+(z)} \tag{6.8.14}$$

as a causal, stable, and recursive filter and applying the sampled sequence $\tilde{x}(n)$ to this filter, we obtain

$$x(n) \triangleq w(n) * \tilde{x}(n) = \sum_{k=0}^{\infty} a_k h_r(n-k) + v(n) \tag{6.8.15}$$

where
$$h_r(n) \triangleq \tilde{h}_r(n) * w(n) \tag{6.8.16}$$
and
$$v(n) \triangleq w(n) * \tilde{v}(n) \tag{6.8.17}$$

The spectral density of $v(n)$, from (6.8.12), (6.8.13), and (6.8.14), is given by

$$R_v(z) = R_w(z)R_{\tilde{v}}(z) = \frac{1}{R_{g_r}^+(z)R_{g_r}^-(z)}\sigma_v^2 R_{g_r}^+(z)R_{g_r}^-(z) = \sigma_v^2 \tag{6.8.18}$$

which means that $v(n)$ is a white sequence. Once again, assuming that $h_r(n) = 0, n > L$, where L is an arbitrary positive integer, we obtain an equivalent discrete-time channel model with white noise

$$x(n) = \sum_{k=0}^{L} a_k h_r(n-k) + v(n) \tag{6.8.19}$$

This equivalent model is shown in Figure 6.31. An example to illustrate the use of this model in the design and analysis of an equalizer is given in the next section.

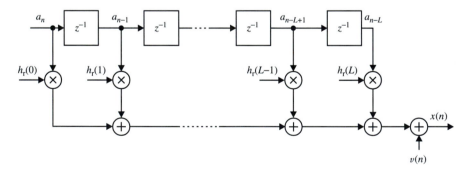

FIGURE 6.31
Equivalent discrete-time model of data transmission system with ISI and WGN.

6.8.3 Linear Equalizers

If we know the characteristics of the channel, that is, the magnitude response $|H_c(F)|$ and the phase response $\angle H_c(F)$, we can design optimum transmitting and receiving filters that will maximize the SNR and will result in zero ISI at the sampling instant. However, in practice we have to deal with channels whose characteristics are either unknown (dial-up telephone channels) or time-varying (ionospheric radio channels). In this case, we usually use a receiver that consists of a fixed filter $g_r(t)$ and an adjustable *linear equalizer*, as shown in Figure 6.32. The response of the fixed filter either is matched to the transmitted pulse or is designed as a compromise equalizer for an "average" channel typical of the given application. In principle, to eliminate the ISI, we should design the equalizer so that the overall pulse shape satisfies Nyquist's criterion (6.8.6) or (6.8.8).

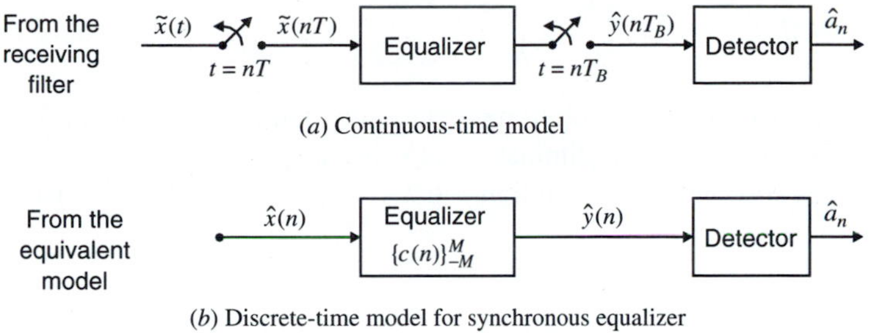

(a) Continuous-time model

(b) Discrete-time model for synchronous equalizer

FIGURE 6.32
Equalizer-based receiver model.

The most widely used equalizers are implemented using digital FIR filters. To this end, as shown in Figure 6.32(a), we sample the received signal $\tilde{x}(t)$ periodically at times $t = t_0 + nT$, where t_0 is the sampling phase and T is the sampling period. The sampling period should be less or equal to the symbol interval T_B because the output of the equalizer should be sampled once every symbol interval (the case $T > T_B$ creates aliasing). For digital implementation T should be chosen as a rational fraction of the symbol interval, that is, $T = L_1 T_B / L_2$, with $L_1 \leq L_2$ (typical choices are $T = T_B$, $T = T_B/2$, or $T = 2T_B/3$). If the sampling interval $T = T_B$, we have a *synchronous* or *symbol equalizer* (SE)[†] and if $T < T_B$ a *fractionally spaced equalizer* (FSE).[‡] The output of the equalizer is quantized to obtain the decision $\hat{a}_n$.

The goal of the equalizer is to determine the coefficients $\{c_k\}_{-M}^{M}$ so as to minimize the ISI according to some criterion of performance. The most meaningful criterion for data transmission is the average probability of error. However, this criterion is a nonlinear function of the equalizer coefficients, and its minimization is extremely difficult.

We next discuss two criteria that are used in practical applications. For this discussion we assume a synchronous equalizer, that is, $T = T_B$. The FSE is discussed in Chapter 12. For the synchronous equalizer, the equivalent discrete-time model given in Figure 6.31 is applicable in which the input is $x(n)$, given by

$$x(n) = \sum_{l=0}^{L} a_l h_r(n - l) + v(n) \tag{6.8.20}$$

The output of the equalizer is given by

$$\hat{y}(n) = \sum_{k=-M}^{M} c^*(k)x(n-k) \triangleq \mathbf{c}^H \mathbf{x}(n) \tag{6.8.21}$$

where

$$\mathbf{c} = [c(-M) \cdots c(0) \cdots c(M)]^T \tag{6.8.22}$$

$$\mathbf{x}(n) = [x(n+M) \cdots x(n) \cdots x(n-M)]^T \tag{6.8.23}$$

This equalizer model is shown in Figure 6.32(b).

[†] Also known as a baud-spaced equalizer (BSE).

[‡] The most significant difference between SE and FSE is that by properly choosing T we can completely avoid aliasing at the input of the FSE. Thus, the FSE can provide better compensation for timing phase and asymmetries in the channel response without noise enhancement (Qureshi 1985).

6.8.4 Zero-Forcing Equalizers

Zero-forcing (zf) equalization (Lucky, Saltz, and Weldon 1968) requires that the response of the equalizer to the combined pulse $\tilde{h}_r(t)$ satisfy the Nyquist criterion (6.8.7). For the FIR equalizer in (6.8.21), in the absence of noise we have

$$\sum_{k=-M}^{M} c_{zf}(k) h_r(n-k) = \begin{cases} 1 & n = 0 \\ 0 & n = \pm 1, \pm 2, \ldots, \pm M \end{cases} \tag{6.8.24}$$

which is a linear system of equations whose solution provides the required coefficients. The zero-forcing equalizer does not completely eliminate the ISI because it has finite duration. If $M = \infty$, Equation (6.8.24) becomes a convolution equation that can be solved by using the z-transform. The solution is

$$C_{zf}(z) = \frac{1}{H_r(z)} \tag{6.8.25}$$

where $H_r(z)$ is the z-transform of $h_r(n)$. Thus, the zero-forcing equalizer is an inverse filter that inverts the frequency-folded (aliased) response of the overall channel. When M is finite, then it is generally impossible to eliminate the ISI at the output of the equalizer because there are only $2M + 1$ adjustable parameters to force zero ISI outside of $[-M, M]$. Then the equalizer design problem reverts to minimizing the *peak distortion*

$$D \triangleq \sum_{n \neq 0} \left| \sum_{k=-M}^{M} c_{zf}(k) h_r(n-k) \right| \tag{6.8.26}$$

This distortion function can be shown to be a convex function (Lucky 1965), and its minimization, in general, is difficult to obtain except when the input ISI is less than 100 percent (i.e., the eye pattern is open). This minimization and the determination of $\{c_{zf}\}$ can be obtained by using the steepest descent algorithm, which is discussed in Chapter 9.

Zero-forcing equalizers have two drawbacks: (1) They ignore the presence of noise and therefore amplify the noise appearing near the spectral nulls of $H_r(e^{j\omega})$, and (2) they minimize the peak distortion or *worst-case ISI* only when the eye is open. For these reasons they are not currently used for bad channels or high-speed modems (Qureshi 1985). The above two drawbacks are eliminated if the equalizers are designed using the MSE criterion.

6.8.5 Minimum MSE Equalizers

It has been shown (Saltzberg 1968) that the error rate $\Pr\{\hat{a}_n \neq a_n\}$ decreases monotonically with the MSE defined by

$$\text{MSE} = E\{|e(n)|^2\} \tag{6.8.27}$$

where
$$e(n) = y(n) - \hat{y}(n) = a_n - \hat{y}(n) \tag{6.8.28}$$

is the difference between the desired response $y(n) \triangleq a_n$ and the actual response $\hat{y}(n)$ given in (6.8.21). Therefore, if we minimize the MSE in (6.8.27), we take into consideration both the ISI and the noise at the output of the equalizer. For $M = \infty$, following the arguments similar to those leading to (6.8.25), the minimum MSE equalizer is specified by

$$C_{\text{MSE}}(z) = \frac{H_r^*(1/z^*)}{H_r(z) H_r^*(1/z^*) + \sigma_v^2} \tag{6.8.29}$$

where σ_v^2 is the variance of the sampled channel noise $v_c(kT_B)$. Clearly, (6.8.29) reduces to the zero-forcing equalizer if $\sigma_v^2 = 0$. Also (6.8.29) is the classical Wiener filter. For finite M, the minimum MSE equalizer is specified by

$$\mathbf{R}\mathbf{c}_o = \mathbf{d} \tag{6.8.30}$$

$$P_o = P_a - \mathbf{c}_o^H \mathbf{d} \qquad (6.8.31)$$

where $\mathbf{R} = E\{\mathbf{x}(n)\mathbf{x}^H(n)\}$ and $\mathbf{d} = E\{a_n^*\mathbf{x}(n)\}$. The data sequence $y(n) = a_n$ is assumed to be white with zero mean and power $P_a = E\{|a_n|^2\}$, and uncorrelated with the additive channel noise. Under these assumptions, the elements of the correlation matrix $\mathbf{R}$ and the cross-correlation vector $\mathbf{d}$ are given by

$$r_{ij} \triangleq E\{x(n-i)x^*(n-j)\}$$
$$= P_a \sum_m h_{\mathrm{r}}(m-i)h_{\mathrm{r}}^*(m-j) + \sigma_v^2 \delta_{ij} \qquad -M \le i, j \le M \qquad (6.8.32)$$

and
$$d_i \triangleq E\{x(n-i)y^*(n)\} = P_a h_{\mathrm{r}}(-i) \qquad -M \le i, j \le M \qquad (6.8.33)$$

that is, in terms of the overall (equivalent) channel response $h_{\mathrm{r}}(n)$ and the noise power σ_v^2. We hasten to stress that matrix $\mathbf{R}$ is Toeplitz if $T = T_B$; otherwise, for $T \ne T_B$, matrix $\mathbf{R}$ is Hermitian but not Toeplitz.

Since MSE equalizers, in contrast to zero-forcing equalizers, take into account both the statistical properties of the noise and the ISI, they are more robust to both noise and large amounts of ISI.

EXAMPLE 6.8.1. Consider the model of the data communication system shown in Figure 6.33. The input symbol sequence $\{a(n)\}$ is a Bernoulli sequence $\{\pm 1\}$, with $\Pr\{1\} = \Pr\{-1\} = 0.5$. The channel (including the receiving and whitening filter) is modeled as

$$h(n) = \begin{cases} 0.5\left[1 + \cos\dfrac{2\pi(n-2)}{W}\right] & n = 1, 2, 3 \\ 0 & \text{otherwise} \end{cases} \qquad (6.8.34)$$

where W controls the amount of amplitude distortion introduced by the channel. The channel impulse response values are

$$h(n) = \left\{0, \underset{\uparrow}{0.5}\left(1 + \cos\frac{2\pi}{W}\right), 1, 0.5\left(1 + \cos\frac{2\pi}{W}\right), 0\right\} \qquad (6.8.35)$$

which is a symmetric channel, and its frequency response is

$$H(e^{j\omega}) = e^{-j2\omega}\left[1 + \left(1 + \cos\frac{2\pi}{W}\right)\cos\omega\right]$$

The channel noise $v(n)$ is modeled as white Gaussian noise (WGN) with zero mean and variance σ_v^2. The equalizer is an 11-tap FIR filter whose optimum tap weights $\{c(n)\}$ are obtained using either optimum filter theory (nonadaptive approach) or adaptive algorithms that will be described in Chapter 10. The input to the equalizer is

$$x(n) = s(n) + v(n) = h(n) * a(n) + v(n) \qquad (6.8.36)$$

where $s(n)$ represents the distorted pulse sequence. The output of the equalizer is $\hat{y}(n)$, which is an estimate of $a(n)$. In practical modem implementations, the equalizer is initially designed using a training sequence that is known to the receiver. It is shown in Figure 6.33 as the sequence $y(n)$. It is reasonable to introduce a delay D in the training sequence to account for delays introduced in the channel and in the equalizer; that is, $y(n) = a(n - D)$ during the training phase. The error sequence $e(n)$ is further used to design the equalizer $c(n)$. The aim of this example is to study the effect of the delay D and to determine its optimum value for proper operation.

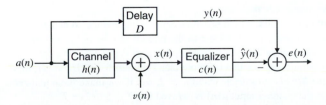

FIGURE 6.33

Data communication model used in Example 6.8.1.

To obtain an optimum equalizer $c(n)$, we will need the autocorrelation matrix $\mathbf{R}_x$ of the input sequence $x(n)$ and the cross-correlation vector $\mathbf{d}$ between $x(n)$ and $y(n)$. Consider the autocorrelation $r_x(l)$ of $x(n)$. From (6.8.36), assuming real-valued quantities, we obtain

$$r_x(l) = E\{[s(n) + v(n)][s(n-l) + v(n-l)]\} = r_s(l) + \sigma_v^2 \delta(l) \tag{6.8.37}$$

where we have assumed that $s(n)$ and $v(n)$ are uncorrelated. Since $s(n)$ is a convolution between $\{a_n\}$ and $h(n)$, the autocorrelation $r_s(l)$ is given by

$$r_s(l) = r_a(l) * r_h(l) = r_h(l) \tag{6.8.38}$$

where $r_a(l) = \delta(l)$ since $\{a(n)\}$ is a Bernoulli sequence, and $r_h(l)$ is the autocorrelation of the channel response $h(n)$ and is given by

$$r_h(l) = h(l) * h(-l)$$

Using the symmetric channel response values in (6.8.35), we find that the autocorrelation $r_x(l)$ in (6.8.37) is given by

$$r_x(0) = h^2(1) + h^2(2) + h^2(3) + \sigma_v^2 = 1 + 0.5 \left(1 + \cos\frac{2\pi}{W}\right)^2 + \sigma_v^2$$

$$r_x(\pm 1) = h(1)h(2) + h(2)h(3) = 1 + \cos\frac{2\pi}{W} \tag{6.8.39}$$

$$r_x(\pm 2) = h(1)h(3) = 0.25 \left(1 + \cos\frac{2\pi}{W}\right)^2$$

$$r_x(l) = 0 \quad |l| \geq 3$$

Since the equalizer is an 11-tap FIR filter, the autocorrelation matrix $\mathbf{R}_x$ is an 11×11 matrix. However, owing to few nonzero values of $r_x(l)$ in (6.8.39), it is also a *quintdiagonal* matrix with the main diagonal containing $r_x(0)$ and two upper and lower non-zero diagonals. The cross-correlation between $x(n)$ and $y(n) = a(n - D)$ is given by

$$\begin{aligned} d(l) &= E\{a(n-D)x(n-l)\} = E\{a(n-D)[s(n-l) + v(n-l)]\} \\ &= E\{a(n-D)s(n-l)\} + E\{a(n-D)v(n-l)\} \\ &= E\{a(n-D)[h(n-l) * a(n-l)]\} \\ &= h(D-l) * r_a(D-l) = h(D-l) \end{aligned} \tag{6.8.40}$$

where we have used (6.8.36). The last step follows from the fact that $r_a(l) = \delta(l)$. Using the channel impulse response values in (6.8.35), we obtain

$$\begin{aligned} D &= 0 & d(l) &= h(-l) = 0 & l &\geq 0 \\ D &= 1 & d(l) &= h(1-l) \Rightarrow d(0) = h(1) & d(l) &= 0 \quad l > 0 \\ D &= 2 & d(l) &= h(2-l) \Rightarrow d(0) = h(2) & d(1) &= h(1) \quad d(l) = 0 \quad l > 1 \\ &\;\vdots & &\quad\vdots \end{aligned} \tag{6.8.41}$$

$$\begin{aligned} D &= 7 & d(l) &= h(7-l) \Rightarrow d(4) = h(3) \quad d(5) = h(2) \quad d(6) = h(1) \\ & & d(l) &= 0 \quad \text{elsewhere} \end{aligned}$$

Remarks. There are some interesting observations that we can make from (6.8.41) in which the delay D turns the estimation problem into a filtering, prediction, or smoothing problem.

1. When $D = 0$, we have a filtering case. The cross-correlation vector $\mathbf{d} = \mathbf{0}$, hence the equalizer taps are all zeros. This means that if we do not provide any delay in the system, the cross-correlation is zero and equalization is not possible because $\mathbf{c}_o = \mathbf{0}$.
2. When $D = 1$, we have a one-step prediction case.
3. When $D \geq 2$, we have a smoothing filter, which provides better performance. When $D = 7$, we note that the vector $\mathbf{d}$ is symmetric [with respect to the middle sample $d(5)$] and hence we should expect the best performance because the channel is also symmetric. We can also show that $D = 7$ is the optimum delay for this example (see Problem 6.40). However, this should not be a surprise since $h(n)$ is symmetric about $n = 2$, and if we make the equalizer symmetric about $n = 5$, then the channel input $a(n)$ is delayed by $D = 5 + 2 = 7$.

Figure 6.34 shows the channel impulse response $h(n)$ and the equalizer $c(n)$ for $D = 7$, $\sigma_v^2 = 0.001$, and $W = 2.9$ and $W = 3.1$.

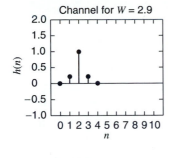

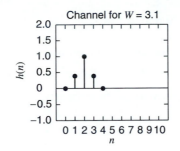

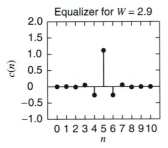

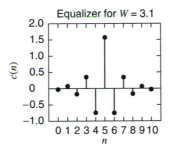

FIGURE 6.34
Channel impulse response $h(n)$ and the equalizer $c(n)$ for $D = 7$,
$\sigma_v^2 = 0.001$, and $W = 2.9$ and $W = 3.1$.

6.9 MATCHED FILTERS AND EIGENFILTERS

In this section we discuss the design of optimum filters that maximize the output signal-to-noise power ratio. Such filters are used to detect signals in additive noise in many applications, including digital communications and radar. First we discuss the case of a known deterministic signal in noise, and then we extend the results to the case of a random signal in noise.

Suppose that the observations obtained by sampling the output of a single sensor at M instances, or M sensors at the same instant, are arranged in a vector $\mathbf{x}(n)$. Furthermore, we assume that the available signal $\mathbf{x}(n)$ consists of a desired signal $\mathbf{s}(n)$ plus an additive noise plus interference signal $\mathbf{v}(n)$, that is,

$$\mathbf{x}(n) = \mathbf{s}(n) + \mathbf{v}(n) \tag{6.9.1}$$

where $\mathbf{s}(n)$ can be one of two things. It can be a *deterministic* signal of the form $\mathbf{s}(n) = \alpha \mathbf{s}_0$, where $\mathbf{s}_0$ is the completely known shape of $\mathbf{s}(n)$ and α is a complex random variable with power $P_\alpha = E\{|\alpha|^2\}$. The argument $\angle\alpha$ provides the unknown initial phase, and the modulus $|\alpha|$, the amplitude of the signal, respectively. It can also be a *random* signal with known correlation matrix $\mathbf{R}_s(n)$. The signals $\mathbf{s}(n)$ and $\mathbf{v}(n)$ are assumed to be uncorrelated with zero means.

The output of a linear processor (combiner or FIR filter) with coefficients $\{c_k^*\}_1^M$ is

$$y(n) = \mathbf{c}^H \mathbf{x}(n) = \mathbf{c}^H \mathbf{s}(n) + \mathbf{c}^H \mathbf{v}(n) \tag{6.9.2}$$

and its power
$$P_y(n) = E\{|y(n)|^2\} = E\{\mathbf{c}^H \mathbf{x}(n)\mathbf{x}^H(n)\mathbf{c}\} = \mathbf{c}^H \mathbf{R}_x(n)\mathbf{c} \tag{6.9.3}$$

is a quadratic function of the filter coefficients.

The output noise power is
$$P_v(n) = E\{|\mathbf{c}^H \mathbf{v}(n)|^2\} = E\{\mathbf{c}^H \mathbf{v}(n)\mathbf{v}^H(n)\mathbf{c}\} = \mathbf{c}^H \mathbf{R}_v(n)\mathbf{c} \tag{6.9.4}$$

where $\mathbf{R}_v(n)$ is the noise correlation matrix. The determination of the output SNR, and hence the subsequent optimization, depends on the nature of the signal $\mathbf{s}(n)$.

6.9.1 Deterministic Signal in Noise

In the deterministic signal case, the power of the signal is

$$P_s(n) = E\{|\alpha \mathbf{c}^H \mathbf{s}_0|^2\} = P_\alpha |\mathbf{c}^H \mathbf{s}_0|^2 \tag{6.9.5}$$

and therefore the output SNR can be written as

$$\text{SNR}(\mathbf{c}) = P_\alpha \frac{|\mathbf{c}^H \mathbf{s}_0|^2}{\mathbf{c}^H \mathbf{R}_v(n) \mathbf{c}} \tag{6.9.6}$$

White noise case. If the correlation matrix of the additive noise is given by $\mathbf{R}_v(n) = P_v \mathbf{I}$, the SNR becomes

$$\text{SNR}(\mathbf{c}) = \frac{P_\alpha}{P_v} \frac{|\mathbf{c}^H \mathbf{s}_0|^2}{\mathbf{c}^H \mathbf{c}} \tag{6.9.7}$$

which simplifies the maximization process. Indeed, from the Cauchy-Schwartz inequality

$$\mathbf{c}^H \mathbf{s}_0 \leq (\mathbf{c}^H \mathbf{c})^{1/2} (\mathbf{s}_0^H \mathbf{s}_0)^{1/2} \tag{6.9.8}$$

we conclude that the SNR in (6.9.7) attains its maximum value

$$\text{SNR}_{\max} = \frac{P_\alpha}{P_v} \mathbf{s}_0^H \mathbf{s}_0 \tag{6.9.9}$$

if the optimum filter $\mathbf{c}_o$ is chosen as

$$\mathbf{c}_o = \kappa \mathbf{s}_0 \tag{6.9.10}$$

that is, when the filter is a scaled *replica* of the known signal shape. This property resulted in the term *matched filter*, which is widely used in communications and radar applications.[†] We note that if a vector $\mathbf{c}_o$ maximizes the SNR (6.9.7), then any constant κ times $\mathbf{c}_o$ maximizes the SNR as well. Therefore, we can choose this constant in any way we want. In this section, we choose κ_o so that $\mathbf{c}_0^H \mathbf{s}_0 = 1$.

Colored noise case. Using the Cholesky decomposition $\mathbf{R}_v = \mathbf{L}_v \mathbf{L}_v^H$ of the noise correlation matrix, we can write the SNR in (6.9.6) as

$$\text{SNR}(\mathbf{c}) = P_\alpha \frac{|(\mathbf{L}_v^H \mathbf{c})^H (\mathbf{L}_v^{-1} \mathbf{s}_0)|^2}{(\mathbf{L}_v^H \mathbf{c})^H (\mathbf{L}_v^H \mathbf{c})} \tag{6.9.11}$$

which, according to the Cauchy-Schwartz inequality, attains its maximum

$$\text{SNR}_{\max} = P_\alpha \|\mathbf{L}_v^{-1} \mathbf{s}_0\|^2 = P_\alpha \mathbf{s}_0^H \mathbf{R}_v^{-1} \mathbf{s}_0 \tag{6.9.12}$$

when the optimum filter satisfies $\mathbf{L}_v^H \mathbf{c}_o = \kappa \mathbf{L}_v^{-1} \mathbf{s}_0$, or equivalently

$$\mathbf{c}_o = \kappa \mathbf{R}_v^{-1} \mathbf{s}_0 \tag{6.9.13}$$

which provides the optimum matched filter for color additive noise. Again, the optimum filter can be scaled in any desirable way. We choose $\mathbf{c}_o^H \mathbf{s}_0 = 1$ which implies $\kappa = (\mathbf{s}_0^H \mathbf{R}_v^{-1} \mathbf{s}_0)^{-1}$.

If we pass the observed signal through the preprocessor $\mathbf{L}_v^{-1}$, we obtain a signal $\mathbf{L}_v^{-1} \mathbf{s}$ in additive white noise $\tilde{\mathbf{v}} = \mathbf{L}_v^{-1} \mathbf{v}$ because $E\{\tilde{\mathbf{v}} \tilde{\mathbf{v}}^H\} = E\{\mathbf{L}_v^{-1} \mathbf{v} \mathbf{v}^H \mathbf{L}_v^{-H}\} = \mathbf{I}$. Therefore, the optimum matched filter in additive color noise is the cascade of a whitening filter followed by a matched filter for white noise (compare with a similar decomposition for the optimum

---

[†] We note that the matched filter $\mathbf{c}_o$ in (6.9.10) is not a complex conjugate reversed version of the signal $\mathbf{s}$. This happens when we define the matched filter as a convolution that involves a reversal of the impulse response (Therrien 1992).

Wiener filter in Figure 6.19). The application of the optimum matched filter is discussed in Section 11.3, which provides a more detailed treatment.

EXAMPLE 6.9.1. Consider a finite-duration deterministic signal $s(n) = a^n$, $0 \leq n \leq M - 1$, corrupted by additive noise $v(n)$ with autocorrelation sequence $r_v(l) = \sigma_0^2 \rho^{|l|}/(1 - \rho^2)$. We determine and plot the impulse response of an Mth-order matched filter for $a = 0.6$, $M = 8$, $\sigma_0^2 = 0.25$, and (a) $\rho = 0.1$ and (b) $\rho = -0.8$. We first note that the signal vector is $\mathbf{s} = [1 \ a \ a^2 \ \cdots \ a^7]^T$ and that the noise correlation matrix $\mathbf{R}_v$ is Toeplitz with first row $[r_v(0) \ r_v(1) \ \cdots \ r_v(7)]$. The optimum matched filters are determined by $\mathbf{c} = \mathbf{R}_v^{-1}\mathbf{s}_0$ and are shown in Figure 6.35. We notice that for $\rho = 0.1$ the matched filter looks like the signal because the correlation between the samples of the interference is very small; that is, the additive noise is close to white. For $\rho = -0.8$ the correlation increases, and the shape of the optimum filter differs more from the shape of the signal. However, as a result of the increased noise correlation, the optimum SNR increases.

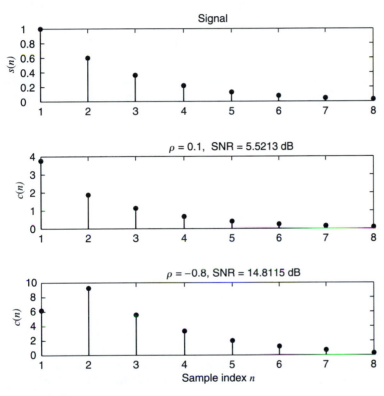

FIGURE 6.35
Signal and impulse responses of the optimum matched filter that maximizes the SNR in the presence of additive color noise.

6.9.2 Random Signal in Noise

In the case of a random signal with known correlation matrix $\mathbf{R}_s$, the SNR is

$$\text{SNR}(\mathbf{c}) = \frac{\mathbf{c}^H \mathbf{R}_s \mathbf{c}}{\mathbf{c}^H \mathbf{R}_v \mathbf{c}} \tag{6.9.14}$$

that is, the ratio of two quadratic forms. We again distinguish two cases.

White noise case. If the correlation matrix of the noise is given by $\mathbf{R}_v = P_v \mathbf{I}$, we have

$$\text{SNR}(\mathbf{c}) = \frac{1}{P_v} \frac{\mathbf{c}^H \mathbf{R}_s \mathbf{c}}{\mathbf{c}^H \mathbf{c}} \tag{6.9.15}$$

which has the form of Rayleigh's quotient (Strang 1980; Leon 1998). By using the innovations transformation $\tilde{\mathbf{c}} = \mathbf{Q}^H \mathbf{c}$, where the unitary matrix $\mathbf{Q}$ is obtained from the eigendecomposition $\mathbf{R}_s = \mathbf{Q}\boldsymbol{\Lambda}\mathbf{Q}^H$, the SNR can be expressed as

$$\text{SNR}(\mathbf{c}) = \frac{1}{P_v}\frac{\tilde{\mathbf{c}}^H \boldsymbol{\Lambda} \tilde{\mathbf{c}}}{\tilde{\mathbf{c}}^H \tilde{\mathbf{c}}} = \frac{1}{P_v}\frac{\lambda_1|\tilde{c}_1|^2 + \cdots + \lambda_M|\tilde{c}_M|^2}{|\tilde{c}_1|^2 + \cdots + |\tilde{c}_M|^2} \tag{6.9.16}$$

where $0 \le \lambda_1 \le \cdots \le \lambda_M$ are the eigenvalues of the signal correlation matrix. The SNR is maximized if we choose $\tilde{c}_M = 1$ and $\tilde{c}_1 = \cdots = \tilde{c}_{M-1} = 0$ and is minimized if we choose $\tilde{c}_1 = 1$ and $\tilde{c}_2 = \cdots = \tilde{c}_M = 0$. Therefore, for any positive definite matrix $\mathbf{R}_s$, we have

$$\lambda_{\min} \le \frac{\mathbf{c}^H \mathbf{R}_s \mathbf{c}}{\mathbf{c}^H \mathbf{c}} \le \lambda_{\max} \tag{6.9.17}$$

which is known as *Rayleigh's quotient* (Strang 1980). This implies that the optimum filter $\mathbf{c} = \mathbf{Q}\tilde{\mathbf{c}}$ is the eigenvector corresponding to the maximum eigenvalue of $\mathbf{R}_s$, that is,

$$\mathbf{c} = \mathbf{q}_{\max} \tag{6.9.18}$$

and provides a maximum SNR

$$\text{SNR}_{\max} = \frac{\lambda_{\max}}{P_v} \tag{6.9.19}$$

where $\lambda_{\max} = \lambda_M$. The obtained optimum filter is sometimes known as an *eigenfilter* (Makhoul 1981). The following example provides a geometric interpretation of these results for a second-order filter.

EXAMPLE 6.9.2. Suppose that the signal correlation matrix $\mathbf{R}_s$ is given by (see Example 3.5.1)

$$\mathbf{R} \triangleq \begin{bmatrix} 1 & \rho \\ \rho & 1 \end{bmatrix} = \frac{1}{\sqrt{2}}\begin{bmatrix} 1 & 1 \\ -1 & 1 \end{bmatrix}\begin{bmatrix} 1-\rho & 0 \\ 0 & 1+\rho \end{bmatrix}\frac{1}{\sqrt{2}}\begin{bmatrix} 1 & 1 \\ -1 & 1 \end{bmatrix}^H = \mathbf{Q}\boldsymbol{\Lambda}\mathbf{Q}^H$$

where $\rho = 0.81$. To obtain a geometric interpretation, we fix $\mathbf{c}^H \mathbf{c} = 1$ and try to maximize the numerator $\mathbf{c}^H \mathbf{R} \mathbf{c} > 0$ (we assume that $\mathbf{R}$ is positive definite). The relation $c_1^2 + c_2^2 = 1$ represents a circle in the (c_1, c_2) plane. The plot can be easily obtained by using the parametric description $c_1 = \cos\phi$ and $c_2 = \sin\phi$. To obtain the plot of $\mathbf{c}^H \mathbf{R} \mathbf{c} = 1$, we note that

$$\mathbf{c}^H \mathbf{R} \mathbf{c} = \mathbf{c}^H \mathbf{Q}\boldsymbol{\Lambda}\mathbf{Q}^H \mathbf{c} = \tilde{\mathbf{c}}^H \boldsymbol{\Lambda} \tilde{\mathbf{c}} = \lambda_1^2 \tilde{c}_1^2 + \lambda_2^2 \tilde{c}_2^2 = 1$$

where $\tilde{\mathbf{c}} \triangleq \mathbf{Q}^H \mathbf{c}$. To plot $\lambda_1^2 \tilde{c}_1^2 + \lambda_2^2 \tilde{c}_2^2 = 1$, we use the parametric description $\tilde{c}_1 = \cos\phi/\sqrt{\lambda_1}$ and $\tilde{c}_2 = \cos\phi/\sqrt{\lambda_2}$. The result is an ellipse in the $(\tilde{c}_1, \tilde{c}_2)$ plane. For $\tilde{c}_2 = 0$ we have $\tilde{c}_1 = 1/\sqrt{\lambda_1}$, and for $\tilde{c}_1 = 0$ we have $\tilde{c}_2 = 1/\sqrt{\lambda_2}$. Since $\lambda_1 < \lambda_2$, $2/\sqrt{\lambda_1}$ provides the length of the major axis determined by the eigenvector $\mathbf{q}_1 = [1 \ -1]^T/\sqrt{2}$. Similarly, $2/\sqrt{\lambda_2}$ provides the length of the minor axis determined by the eigenvector $\mathbf{q}_2 = [1 \ 1]^T/\sqrt{2}$. The coordinates of the ellipse in the (c_1, c_2) plane are obtained by the rotation transformation $\mathbf{c} = \mathbf{Q}\tilde{\mathbf{c}}$. The resulting circle and ellipse are shown in Figure 6.36. The maximum value of $\mathbf{c}^H \mathbf{R} \mathbf{c} = \lambda_1^2 \tilde{c}_1^2 + \lambda_2^2 \tilde{c}_2^2$ on the circle $\tilde{c}_1^2 + \tilde{c}_2^2 = 1$ is obtained for $\tilde{c}_1 = 0$ and $\tilde{c}_2 = 1$, that is, at the endpoint of eigenvector $\mathbf{q}_2$, and is equal to the largest eigenvalue λ_2. Similarly, the minimum is λ_1 and is obtained at the tip of eigenvector $\mathbf{q}_1$ (see Figure 6.36). Therefore, the optimum filter is $\mathbf{c} = \mathbf{q}_2$ and the maximum SNR is λ_2/P_v.

Colored noise case. Using the Cholesky decomposition $\mathbf{R}_v = \mathbf{L}_v \mathbf{L}_v^H$ of the noise correlation matrix, we process the observed signal with the transformation $\mathbf{L}_v^{-1}$, that is, we obtain

$$\begin{aligned} \mathbf{x}_v(n) &\triangleq \mathbf{L}_v^{-1}\mathbf{x}(n) = \mathbf{L}_v^{-1}\mathbf{s}(n) + \mathbf{L}_v^{-1}\mathbf{v}(n) \\ &= \mathbf{s}_v(n) + \tilde{\mathbf{v}}(n) \end{aligned} \tag{6.9.20}$$

where $\tilde{\mathbf{v}}(n)$ is white noise with $E\{\tilde{\mathbf{v}}(n)\tilde{\mathbf{v}}^H(n)\} = \mathbf{I}$ and $E\{\mathbf{s}_v(n)\mathbf{s}_v^H(n)\} = \mathbf{L}_v^{-1}\mathbf{R}_s\mathbf{L}_v^{-H}$. Therefore, the optimum matched filter is determined by the eigenvector corresponding to the maximum eigenvalue of matrix $\mathbf{L}_v^{-1}\mathbf{R}_s\mathbf{L}_v^{-H}$, that is, the correlation matrix of the transformed signal $\mathbf{s}_v(n)$.

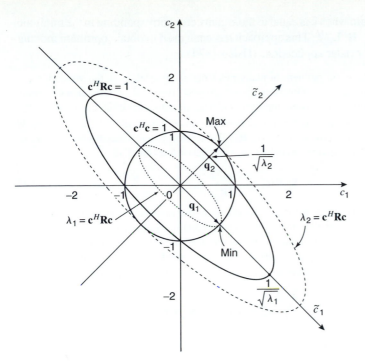

FIGURE 6.36
Geometric interpretation of the optimization process for the
derivation of the optimum eigenfilter using isopower contours for
$\lambda_1 < \lambda_2$.

The problem can also be solved by using the simultaneous diagonalization of the signal
and noise correlation matrices $\mathbf{R}_s$ and $\mathbf{R}_v$, respectively. Starting with the decomposition
$\mathbf{R}_v = \mathbf{Q}_v \mathbf{\Lambda}_v \mathbf{Q}_v^H$, we compute the isotropic transformation

$$\mathbf{x}_v(n) \triangleq \mathbf{\Lambda}_v^{-1/2} \mathbf{Q}_v^H \mathbf{x}(n)$$
$$= \mathbf{\Lambda}_v^{-1/2} \mathbf{Q}_v^H \mathbf{s}(n) + \mathbf{\Lambda}_v^{-1/2} \mathbf{Q}_v^H \mathbf{v}(n) \triangleq \tilde{\mathbf{s}}(n) + \tilde{\mathbf{v}}(n) \tag{6.9.21}$$

where $E\{\tilde{\mathbf{v}}(n)\tilde{\mathbf{v}}^H(n)\} = \mathbf{I}$ and $E\{\tilde{\mathbf{s}}(n)\tilde{\mathbf{s}}^H(n)\} = \mathbf{\Lambda}_v^{-1/2} \mathbf{Q}_v^H \mathbf{R}_s \mathbf{Q}_v \mathbf{\Lambda}_v^{-1/2} \triangleq \mathbf{R}_{\tilde{s}}$. Since the
noise vector is white, the optimum matched filter is determined by the eigenvector corre-
sponding to the maximum eigenvalue of matrix $\mathbf{R}_{\tilde{s}}$.

Finally, if $\mathbf{R}_{\tilde{s}} = \mathbf{Q}_{\tilde{s}} \mathbf{\Lambda}_{\tilde{s}} \mathbf{Q}_{\tilde{s}}^H$, the transformation

$$\mathbf{x}_{vs}(n) \triangleq \mathbf{Q}_{\tilde{s}}^H \mathbf{x}_v(n) = \mathbf{Q}_{\tilde{s}}^H \tilde{\mathbf{s}}(n) + \mathbf{Q}_{\tilde{s}}^H \tilde{\mathbf{v}}(n) \triangleq \bar{\mathbf{s}}(n) + \bar{\mathbf{v}}(n) \tag{6.9.22}$$

results in new signal and noise vectors with correlation matrices

$$E\{\bar{\mathbf{s}}(n)\bar{\mathbf{s}}^H(n)\} = \mathbf{Q}_{\tilde{s}}^H \mathbf{R}_{\tilde{s}} \mathbf{Q}_{\tilde{s}} = \mathbf{\Lambda}_{\tilde{s}} \tag{6.9.23}$$

$$E\{\bar{\mathbf{v}}(n)\bar{\mathbf{v}}^H(n)\} = \mathbf{Q}_{\tilde{s}}^H \mathbf{I} \mathbf{Q}_{\tilde{s}} = \mathbf{I} \tag{6.9.24}$$

Therefore, the transformation matrix

$$\mathbf{Q} \triangleq \mathbf{Q}_{\tilde{s}}^H \mathbf{\Lambda}_v^{-1/2} \mathbf{Q}_v^H \tag{6.9.25}$$

diagonalizes matrices $\mathbf{R}_s$ and $\mathbf{R}_v$ simultaneously (Fukunaga 1990).

The maximization of (6.9.14) can also be obtained by whitening the signal, that is, by
using the Cholesky decomposition $\mathbf{R}_s = \mathbf{L}_s \mathbf{L}_s^H$ of the signal correlation matrix. Indeed,
using the transformation $\tilde{\mathbf{c}} \triangleq \mathbf{L}_s^H \mathbf{c}$, we have

$$\text{SNR}(\mathbf{c}) = \frac{\mathbf{c}^H \mathbf{R}_s \mathbf{c}}{\mathbf{c}^H \mathbf{R}_v \mathbf{c}} = \frac{\mathbf{c}^H \mathbf{L}_s \mathbf{L}_s^H \mathbf{c}}{\mathbf{c}^H \mathbf{L}_s \mathbf{L}_s^{-1} \mathbf{R}_v \mathbf{L}_s^{-H} \mathbf{L}_s^H \mathbf{c}} = \frac{\tilde{\mathbf{c}}^H \tilde{\mathbf{c}}}{\tilde{\mathbf{c}}^H \mathbf{L}_s^{-1} \mathbf{R}_v \mathbf{L}_s^{-H} \tilde{\mathbf{c}}} \tag{6.9.26}$$

which attains its maximum when $\tilde{\mathbf{c}}$ is equal to the eigenvector corresponding to the minimum eigenvalue of matrix $\mathbf{L}_s^{-1}\mathbf{R}_v\mathbf{L}_s^{-H}$. This approach has been used to obtain optimum moving-target indicator filters for radar applications (Hsiao 1974).

EXAMPLE 6.9.3. The basic problem in many radar detection systems is the separation of a useful signal from colored noise or interference background. In several cases the signal is a point target (i.e., it can be modeled as a unit impulse) or is random with a flat PSD, that is, $\mathbf{R}_s = P_a \mathbf{I}$. Suppose that the background is colored with correlation $r_v(i, j) = \rho^{(i-j)^2}, 1 \le i, j \le M$, which leads to a Toeplitz correlation matrix $\mathbf{R}_v$. We determine and compare three filters for interference rejection. The first is a matched filter that maximizes the SNR

$$\text{SNR}(\mathbf{c}) = \frac{P_a \mathbf{c}^H \mathbf{c}}{\mathbf{c}^H \mathbf{R}_v \mathbf{c}} \tag{6.9.27}$$

by setting $\mathbf{c}$ equal to the eigenvector corresponding to the minimum eigenvalue of $\mathbf{R}_v$. The second approach is based on the method of linear prediction. Indeed, if we assume that the interference $v_k(n)$ is much stronger than the useful signal $s_k(n)$, we can obtain an estimate $\hat{v}_1(n)$ of $v_1(n)$ using the observed samples $\{x_k(n)\}_2^M$ and then subtract $\hat{v}_1(n)$ from $x_1(n)$ to cancel the interference. The Wiener filter with desired response $y(n) = v_1(n)$ and input data $\{x_k(n)\}_2^M$ is

$$\hat{y}(n) = -\sum_{k=1}^{M-1} a_k^* x_{k+1}(n) \triangleq -\mathbf{a}^H \tilde{\mathbf{x}}(n)$$

and is specified by the normal equations

$$\tilde{\mathbf{R}}_x \mathbf{a} = -\tilde{\mathbf{d}}$$

and the MMSE

$$P^f = E\{|v_1|^2\} + \tilde{\mathbf{d}}^H \mathbf{a}$$

where

$$\langle \tilde{\mathbf{R}}_x \rangle_{ij} = E\{x_{i+1}(n)x_{i+1}^*(n)\} \simeq E\{v_{i+1}(n)v_{i+1}^*(n)\}$$

and

$$\tilde{d}_i = E\{v_1 x_{i+1}^*(n)\} \simeq E\{v_1(n)v_{i+1}^*(n)\}$$

because the interference is assumed much stronger than the signal. Using the last four equations, we obtain

$$\mathbf{R}_v \begin{bmatrix} 1 \\ \mathbf{a} \end{bmatrix} = \begin{bmatrix} E^f \\ \mathbf{0} \end{bmatrix} \tag{6.9.28}$$

which corresponds to the forward linear prediction error (LPE) filter discussed in Section 6.5.2. Finally, for the sake of comparison, we consider the binomial filters $H_M(z) = (1 - z^{-1})^M$ that are widely used in radar systems for the elimination of stationary (i.e., nonmoving) clutter. Figure 6.37 shows the magnitude response of the three filters for $\rho = 0.9$ and $M = 4$. We emphasize that

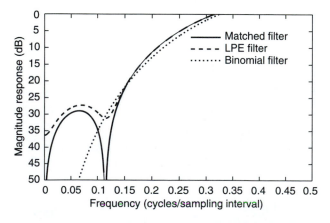

FIGURE 6.37
Comparison of frequency responses of matched filter, prediction error filter, and binomial interference rejection filter.

the FLP method is suboptimum compared to matched filtering. However, because the frequency response of the FLP filter does not have the deep zero notch, we use it if we do not want to lose useful signals in that band (Chiuppesi et al. 1980).

325

PROBLEMS

6.10 SUMMARY

In this chapter, we discussed the theory and application of optimum linear filters designed by minimizing the MSE criterion of performance. Our goal was to explain the characteristics of each criterion, emphasize when its use made sense, and illustrate its meaning in the context of practical applications.

We started with linear processors that formed an estimate of the desired response by combining a set of different signals (data) and showed that the parameters of the optimum processor can be obtained by solving a linear system of equations (normal equations). The matrix and the right-hand side vector of the normal equations are completely specified by the second-order moments of the input data and the desired response. Next, we used the developed theory to design optimum FIR filters, linear signal estimators, and linear predictors.

We emphasized the case of stationary stochastic processes and showed that the resulting optimum estimators are time-invariant. Therefore, we need to design only one optimum filter that can be used to process all realizations of the underlying stochastic processes. Although another filter may perform better for some realizations, that is, the estimated MSE is smaller than the MMSE, on average (i.e., when we consider all possible realizations), the optimum filter is the best.

We showed that the performance of optimum linear filters improves as we increase the number of filter coefficients. Therefore, the noncausal IIR filter provides the best possible performance and can be used as a yardstick to assess other filters. Because IIR filters involve an infinite number of parameters, their design involves linear equations with an infinite number of unknowns. For stationary processes, these equations take the form of a convolution equation that can be solved using z-transform techniques. If we use a pole-zero structure, the normal equations become nonlinear and the design of the optimum filter is complicated by the presence of multiple local minima.

Then we discussed the design of optimum filters for inverse system modeling and blind deconvolution, and we provided a detailed discussion of their use in the important practical application of channel equalization for data transmission systems.

Finally, we provided a concise introduction to the design of optimum matched filters and eigenfilters that maximize the output SNR and find applications for the detection of signals in digital communication and radar systems.

PROBLEMS

6.1 Let $\mathbf{x}$ be a random vector with mean $E\{\mathbf{x}\}$. Show that the linear MMSE estimate $\hat{y}$ of a random variable y using the data vector $\mathbf{x}$ is given by $\hat{y} = y_o + \mathbf{c}^H\mathbf{x}$, where $y_o = E\{y\} - \mathbf{c}^H E\{\mathbf{x}\}$, $\mathbf{c} = \mathbf{R}^{-1}\mathbf{d}$, $\mathbf{R} = E\{\mathbf{x}\mathbf{x}^H\}$, and $\mathbf{d} = E\{xy^*\}$.

6.2 Consider an optimum FIR filter specified by the input correlation matrix $\mathbf{R} = \text{Toeplitz}\{1, \frac{1}{4}\}$ and cross-correlation vector $\mathbf{d} = [1 \; \frac{1}{2}]^T$.

(a) Determine the optimum impulse response $\mathbf{c}_o$ and the MMSE P_o.

(b) Express $\mathbf{c}_o$ and P_o in terms of the eigenvalues and eigenvectors of $\mathbf{R}$.

6.3 Repeat Problem 6.2 for a third-order optimum FIR filter.

6.4 A process $y(n)$ with the autocorrelation $r_y(l) = a^{|l|}$, $-1 < a < 1$, is corrupted by additive, uncorrelated white noise $v(n)$ with variance σ_v^2. To reduce the noise in the observed process $x(n) = y(n) + v(n)$, we use a first-order Wiener filter.

    (a) Express the coefficients $c_{o,1}$ and $c_{o,2}$ and the MMSE P_o in terms of parameters a and σ_v^2.

    (b) Compute and plot the PSD of $x(n)$ and the magnitude response $|C_o(e^{j\omega})|$ of the filter when $\sigma_v^2 = 2$, for both $a = 0.8$ and $a = -0.8$, and compare the results.

    (c) Compute and plot the processing gain of the filter for $a = -0.9, -0.8, -0.7, \ldots, 0.9$ as a function of a and comment on the results.

6.5 Consider the harmonic process $y(n)$ and its noise observation $x(n)$ given in Example 6.4.1.

    (a) Show that $r_y(l) = \frac{1}{2}A^2 \cos \omega_0 l$.

    (b) Write a Matlab function h = opt_fir(A,f0,var_v,M) to design an Mth-order optimum FIR filter impulse response $h(n)$. Use the toeplitz function from MATLAB to generate correlation matrix **R**.

    (c) Determine the impulse response of a 20th-order optimum FIR filter for $A = 0.5$, $f_0 = 0.05$, and $\sigma_v^2 = 0.5$.

    (d) Using MATLAB, determine and plot the magnitude response of the above-designed filter, and verify your results with those given in Example 6.4.1.

6.6 Consider a "desired" signal $s(n)$ generated by the process $s(n) = -0.8w(n-1) + w(n)$, where $w(n) \sim \text{WN}(0, \sigma_w^2)$. This signal is passed through the causal system $H(z) = 1 - 0.9z^{-1}$ whose output $y(n)$ is corrupted by additive white noise $v(n) \sim \text{WN}(0, \sigma_v^2)$. The processes $w(n)$ and $v(n)$ are uncorrelated with $\sigma_w^2 = 0.3$ and $\sigma_v^2 = 0.1$.

    (a) Design a second-order optimum FIR filter that estimates $s(n)$ from the signal $x(n) = y(n) + v(n)$ and determine $\mathbf{c}_o$ and P_o.

    (b) Plot the error performance surface, and verify that it is quadratic and that the optimum filter points to its minimum.

    (c) Repeat part (a) for a third-order filter, and see whether there is any improvement.

6.7 Repeat Problem 6.6, assuming that the desired signal is generated by $s(n) = -0.8s(n-1) + w(n)$.

6.8 Repeat Problem 6.6, assuming that $H(z) = 1$.

6.9 A stationary process $x(n)$ is generated by the difference equation $x(n) = \rho x(n-1) + w(n)$, where $w(n) \sim \text{WN}(0, \sigma_w^2)$.

    (a) Show that the correlation matrix of $x(n)$ is given by

$$\mathbf{R}_x = \frac{\sigma_w^2}{1 - \rho^2} \text{Toeplitz}\{1, \rho, \rho^2, \ldots, \rho^{M-1}\}$$

    (b) Show that the Mth-order FLP is given by $a_1^{(M)} = -\rho$, $a_k^{(M)} = 0$ for $k > 1$ and the MMSE is $P_M^f = \sigma_w^2$.

6.10 Using Parseval's theorem, show that (6.4.18) can be written as (6.4.21) in the frequency domain.

6.11 By differentiating (6.4.21) with respect to $H(e^{j\omega})$, derive the frequency response function $H_o(e^{j\omega})$ of the optimum filter in terms of $R_{yx}(e^{j\omega})$ and $R_x(e^{j\omega})$.

6.12 A conjugate symmetric linear smoother is obtained from (6.5.12) when $M = 2L$ and $i = L$. If the process $x(n)$ is stationary, then, using $\bar{\mathbf{R}}\mathbf{J} = \mathbf{J}\bar{\mathbf{R}}^*$, show that $\bar{\mathbf{c}} = \mathbf{J}\bar{\mathbf{c}}^*$.

6.13 Let $\bar{\mathbf{Q}}$ and $\bar{\mathbf{\Lambda}}$ be the matrices from the eigendecomposition of $\bar{\mathbf{R}}$, that is, $\bar{\mathbf{R}} = \bar{\mathbf{Q}}\bar{\mathbf{\Lambda}}\bar{\mathbf{Q}}^H$.

    (a) Substitute **R** into (6.5.20) and (6.5.27) to prove (6.5.43) and (6.5.44).

(b) Generalize the above result for a jth-order linear signal estimator $\mathbf{c}^{(j)}(n)$; that is, prove that

$$\mathbf{c}^{(j)}(n) = P_o^{(j)}(n) \sum_{i=1}^{M+1} \frac{1}{\lambda_i} \tilde{\mathbf{q}}_i \tilde{q}_{i,j}$$

6.14 Let $\tilde{\mathbf{R}}(n)$ be the inverse of the correlation matrix $\bar{\mathbf{R}}(n)$ given in (6.5.11).

(a) Using (6.5.12), show that the diagonal elements of $\tilde{\mathbf{R}}(n)$ are given by

$$\langle \tilde{\mathbf{R}}(n) \rangle_{i,i} = \frac{1}{P^{(i)}(n)} \qquad 1 \leq i \leq M+1$$

(b) Furthermore, show that

$$\mathbf{c}^{(i)}(n) = \frac{\tilde{\mathbf{r}}_i(n)}{\langle \tilde{\mathbf{R}}(n) \rangle_{i,i}} \qquad 1 \leq i \leq M+1$$

where $\tilde{\mathbf{R}}_i(n)$ is the i-th column of $\tilde{\mathbf{R}}(n)$.

6.15 The first five samples of the autocorrelation sequence of a signal $x(n)$ are $r(0) = 1, r(1) = 0.8, r(2) = 0.6, r(3) = 0.4$, and $r(4) = 0.3$. Compute the FLP, the BLP, the optimum symmetric smoother, and the corresponding MMSE (a) by using the normal equations method and (b) by using the inverse of the normal equations matrix.

6.16 For the symmetric, Toeplitz autocorrelation matrix $\mathbf{R} = \text{Toeplitz}\{r(0), r(1), r(2)\} = r(0) \times \text{Toeplitz}\{1, \rho_1, \rho_2\}$ with $\mathbf{R} = \mathbf{LDL}^H$ and $\mathbf{D} = \text{diag}\{\xi_1, \xi_2, \xi_3\}$, the following conditions are equivalent:

- $\mathbf{R}$ is positive definite.
- $\xi_i > 0$ for $1 \leq i \leq 3$.
- $|k_i| < 1$ for $1 \leq i \leq 3$.

Determine the values of ρ_1 and ρ_2 for which $\mathbf{R}$ is positive definite, and plot the corresponding area in the (ρ_1, ρ_2) plane.

6.17 Prove the first equation in (6.5.45) by rearranging the FLP normal equations in terms of the unknowns $P_o^f(n), a_1(n), \ldots, a_M(n)$ and then solve for $P_o^f(n)$, using Cramer's rule. Repeat the procedure for the second equation.

6.18 Consider the signal $x(n) = y(n) + v(n)$, where $y(n)$ is a useful random signal corrupted by noise $v(n)$. The processes $y(n)$ and $v(n)$ are uncorrelated with PSDs

$$R_y(e^{j\omega}) = \begin{cases} 1 & 0 \leq |\omega| \leq \dfrac{\pi}{2} \\ 0 & \dfrac{\pi}{2} < |\omega| \leq \pi \end{cases}$$

and

$$R_v(e^{j\omega}) = \begin{cases} 1 & \dfrac{\pi}{4} \leq |\omega| \leq \dfrac{\pi}{2} \\ 0 & 0 \leq |\omega| < \dfrac{\pi}{4} \text{ and } \dfrac{\pi}{2} < |\omega| \leq \pi \end{cases}$$

respectively. (a) Determine the optimum IIR filter and find the MMSE. (b) Determine a third-order optimum FIR filter and the corresponding MMSE. (c) Determine the noncausal optimum FIR filter defined by

$$\hat{y}(n) = h(-1)x(n+1) + h(0)x(n) + h(1)x(n-1)$$

and the corresponding MMSE.

6.19 Consider the ARMA(1, 1) process $x(n) = 0.8x(n-1) + w(n) + 0.5w(n-1)$, where $w(n) \sim$ WGN(0, 1). (a) Determine the coefficients and the MMSE of (1) the one-step ahead FLP $\hat{x}(n) = a_1 x(n-1) + a_2 x(n-2)$ and (2) the two-step ahead FLP $\hat{x}(n+1) = a_1 x(n-1) + a_2 x(n-2)$. (b) Check if the obtained prediction error filters are minimum-phase, and explain your findings.

6.20 Consider a random signal $x(n) = s(n) + v(n)$, where $v(n) \sim \text{WGN}(0, 1)$ and $s(n)$ is the AR(1) process $s(n) = 0.9s(n - 1) + w(n)$, where $w(n) \sim \text{WGN}(0, 0.64)$. The signals $s(n)$ and $v(n)$ are uncorrelated. (a) Determine and plot the autocorrelation $r_s(l)$ and the PSD $R_s(e^{j\omega})$ of $s(n)$. (b) Design a second-order optimum FIR filter to estimate $s(n)$ from $x(n)$. What is the MMSE? (c) Design an optimum IIR filter to estimate $s(n)$ from $x(n)$. What is the MMSE?

6.21 A useful signal $s(n)$ with PSD $R_s(z) = [(1 - 0.9z^{-1})(1 - 0.9z)]^{-1}$ is corrupted by additive uncorrelated noise $v(n) \sim \text{WN}(0, \sigma^2)$. (a) The resulting signal $x(n) = s(n) + v(n)$ is passed through a causal filter with system function $H(z) = (1 - 0.8z^{-1})^{-1}$. Determine (1) the SNR at the input, (2) the SNR at the output, and (3) the processing gain, that is, the improvement in SNR. (b) Determine the causal optimum filter and compare its performance with that of the filter in (a).

6.22 A useful signal $s(n)$ with PSD $R_s(z) = 0.36[(1 - 0.8z^{-1})(1 - 0.8z)]^{-1}$ is corrupted by additive uncorrelated noise $v(n) \sim \text{WN}(0, 1)$. Determine the optimum noncausal and causal IIR filters, and compare their performance by examining the MMSE and their magnitude response. *Hint:* Plot the magnitude responses on the same graph with the PSDs of signal and noise.

6.23 Consider a process with PSD $R_x(z) = \sigma^2 H_x(z)H_x(z^{-1})$. Determine the D-step ahead linear predictor, and show that the MMSE is given by $P^{(D)} = \sigma^2 \sum_{n=0}^{D-1} h_x^2(n)$. Check your results by using the PSD $R_x(z) = (1 - a^2)[(1 - az^{-1})(1 - az)]^{-1}$.

6.24 Let $x(n) = s(n) + v(n)$ with $R_v(z) = 1$, $R_{sv}(z) = 0$, and

$$R_s(z) = \frac{0.75}{(1 - 0.5z^{-1})(1 - 0.5z)}$$

Determine the optimum filters for the estimation of $s(n)$ and $s(n - 2)$ from $\{x(k)\}_{-\infty}^n$ and the corresponding MMSEs.

6.25 For the random signal with PSD

$$R_x(z) = \frac{(1 - 0.2z^{-1})(1 - 0.2z)}{(1 - 0.9z^{-1})(1 - 0.9z)}$$

determine the optimum two-step ahead linear predictor and the corresponding MMSE.

6.26 Repeat Problem 6.25 for

$$R_x(z) = \frac{1}{(1 - 0.2z^{-1})(1 - 0.2z)(1 - 0.9z^{-1})(1 - 0.9z)}$$

6.27 Let $x(n) = s(n) + v(n)$ with $v(n) \sim \text{WN}(0, 1)$ and $s(n) = 0.6s(n - 1) + w(n)$, where $w(n) \sim \text{WN}(0, 0.82)$. The processes $s(n)$ and $v(n)$ are uncorrelated. Determine the optimum filters for the estimation of $s(n)$, $s(n + 2)$, and $s(n - 2)$ from $\{x(k)\}_{-\infty}^n$ and the corresponding MMSEs.

6.28 Repeat Problem 6.27 for $R_s(z) = [(1 - 0.5z^{-1})(1 - 0.5z)]^{-1}$, $R_v(z) = 5$, and $R_{sv}(z) = 0$.

6.29 Consider the random sequence $x(n)$ generated in Example 6.5.2

$$x(n) = w(n) + \frac{1}{2}w(n - 1)$$

where $w(n)$ is WN(0, 1). Generate K sample functions $\{w_k(n)\}_{n=0}^N$, $k = 1, \ldots, K$ of $w(n)$, in order to generate K sample functions $\{x_k(n)\}_{n=0}^N$, $k = 1, \ldots, K$ of $x(n)$.

(a) Use the second-order FLP a_k to obtain predictions $\{\hat{x}_k^f(n)\}_{n=2}^N$ of $x_k(n)$, for $k = 1, \ldots, K$. Then determine the average error

$$\hat{P}^f = \frac{1}{N - 1} \sum_{n=2}^N |x_k(n) - \hat{x}_k^f(n)|^2 \qquad k = 1, \ldots, K$$

and plot it as a function of k. Compare it with P_o^f.

(b) Use the second-order BLP b_k to obtain predictions $\{\hat{x}_k^b(n)\}_{n=0}^{N-2}$, $k = 1, \ldots, K$ of $x_k(n)$.
Then determine the average error

$$\hat{P}^b = \frac{1}{N-1} \sum_{n=0}^{N-2} |x_k(n) - \hat{x}_k^b(n)|^2 \qquad k = 1, \ldots, K$$

and plot it as a function of k. Compare it with P_o^b.

(c) Use the second-order symmetric linear smoother c_k to obtain smooth estimates $\{\hat{x}_k^c(n)\}_{n=0}^{N-2}$ of $x_k(n)$ for $k = 1, \ldots, K$. Determine the average error

$$\hat{P}^s = \frac{1}{N-1} \sum_{n=1}^{N-1} |x_k(n) - \hat{x}_k^c(n)|^2 \qquad k = 1, \ldots, K$$

and plot it as a function of k. Compare it with P_o^s.

6.30 Let $x(n) = y(n) + v(n)$ be a wide-sense stationary process. The linear, symmetric smoothing filter estimator of $y(n)$ is given by

$$\hat{y}(n) = \sum_{k=-L}^{L} h(k)x(n-k)$$

(a) Determine the normal equations for the optimum MMSE filter.
(b) Show that the smoothing filter $\mathbf{c}_o^s$ has linear phase.
(c) Use the Lagrange multiplier method to determine the MMSE Mth-order estimator $\hat{y}(n) = \mathbf{c}^H \mathbf{x}(n)$, where $M = 2L + 1$, when the filter vector $\mathbf{c}$ is constrained to be conjugate symmetric, that is, $\mathbf{c} = \mathbf{J}\mathbf{c}^*$. Compare the results with those obtained in part (a).

6.31 Consider the causal prediction filter discussed in Example 6.6.1. To determine $H_c^{[D]}(z)$, first compute the causal part of the z-transform $[R'_{yw}(z)]_+$. Next compute $H_c^{[D]}(z)$ by using (6.6.21).

(a) Determine $h_c^{[D]}(n)$.
(b) Using the above $h_c^{[D]}(n)$, show that

$$P_c^{[D]} = 1 - \frac{5}{8}(\frac{4}{5})^{2D}$$

6.32 Consider the causal smoothing filter discussed in Example 6.6.1.

(a) Using $[r'_{yw}(l)]_+ = r_{yw}(l+D)u(l)$, $D < 0$, show that $[r'_{yw}(l)]_+$ can be put in the form

$$[r'_{yw}(l)]_+ = \frac{3}{5}(\frac{4}{5})^{l+D}u(l+D) + \frac{3}{5}(2^{l+D})[u(l) - u(l+D)] \qquad D < 0$$

(b) Hence, show that $[R'_{yw}(z)]_+$ is given by

$$[R'_{yw}(z)]_+ = \frac{3}{5}\frac{z^D}{1 - \frac{4}{5}z^{-1}} + \frac{3}{5}(2^D)\sum_{l=0}^{-D-1} 2^l z^{-l}$$

(c) Finally using (6.6.21), prove (6.6.54).

6.33 In this problem, we will prove (6.6.57)

(a) Starting with (6.6.42), show that $[R'_{yw}(z)]_+$ can also be put in the form

$$[R'_{yw}(z)]_+ = \frac{3}{5}\left(\frac{z^D}{1 - \frac{4}{5}z^{-1}} + \frac{2^D - z^D}{1 - 2z^{-1}}\right)$$

(b) Now, using (6.6.21), show that

$$H_c^{[D]}(z) = \frac{3}{8}\left[\frac{2^D(1 - \frac{4}{5}z^{-1}) + \frac{3}{5}z^{D-1}}{(1 - \frac{4}{5}z^{-1})(1 - 2z^{-1})}\right]$$

hence, show that

$$\lim_{D \to -\infty} H_c^{[D]}(z) = \frac{9}{40} \left[\frac{z^D}{(1 - \frac{4}{5}z^{-1})(1 - 2z^{-1})} \right] = z^D H_{nc}(z)$$

(c) Finally, show that $\lim_{D \to \infty} P_c^{[D]} = P_{nc}$.

6.34 Consider the block diagram of a simple communication system shown in Figure 6.38. The information resides in the signal $s(n)$ produced by exciting the system $H_1(z) = 1/(1+0.95z^{-1})$ with the process $w(n) \sim \text{WGN}(0, 0.3)$. The signal $s(n)$ propagates through the channel $H_2(z) = 1/(1 - 0.85z^{-1})$, and is corrupted by the additive noise process $v(n) \sim \text{WGN}(0, 0.1)$, which is uncorrelated with $w(n)$. (a) Determine a second-order optimum FIR filter ($M = 2$) that estimates the signal $s(n)$ from the received signal $x(n) = z(n) + v(n)$. What is the corresponding MMSE P_o? (b) Plot the error performance surface and verify that the optimum filter corresponds to the bottom of the "bowl." (c) Use a Monte Carlo simulation (100 realizations with a 1000-sample length each) to verify the theoretically obtained MMSE in part (a). (d) Repeat part (a) for $M = 3$ and check if there is any improvement. *Hint:* To compute the autocorrelation of $z(n)$, notice that the output of $H_1(z)H_2(z)$ is an AR(2) process.

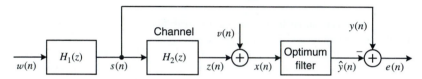

FIGURE 6.38
Block diagram of simple communication system used in Problem 6.34.

6.35 Write a program to reproduce the results shown in Figure 6.35 of Example 6.9.1. (a) Produce plots for $\rho = 0.1, -0.8, 0.8$. (b) Repeat part (a) for $M = 16$. Compare the plots obtained in (a) and (b) and justify any similarities or differences.

6.36 Write a program to reproduce the plot shown in Figure 6.36 of Example 6.9.2. Repeat for $\rho = -0.81$ and explain the similarities and differences between the two plots.

6.37 In this problem we study in greater detail the interference rejection filters discussed in Example 6.9.3. (a) Shows that SNRs for the matched filter and FLP filter are given by

M	Matched filter	FLP filter
2	$\dfrac{1}{1-\rho}$	$\dfrac{1-\rho^2}{1+\rho^2}$
3	$\dfrac{2}{2+\rho^4(1-\sqrt{1+8\rho^{-6}})}$	$\dfrac{1+\rho^2+3\rho^4+\rho^6}{(\rho^2-1)(\rho^4-1)}$

and check the results numerically. (b) Compute and plot the SNRs and compare the performance of both filters for $M = 2, 3, 4$ and $\rho = 0.6, 0.8, 0.9, 0.95, 0.99$, and 0.995. For what values of ρ and M do the two methods give similar results? Explain your conclusions. (c) Plot the magnitude response of the matched, FLP, and binomial filters for $M = 3$ and $\rho = 0.9$. Why does the optimum matched filter always have some nulls in its frequency response?

6.38 Determine the matched filter for the deterministic pulse $s(n) = \cos \omega_0 n$ for $0 \le n \le M-1$ and zero elsewhere when the noise is (a) white with variance σ_v^2 and (b) colored with autocorrelation $r_v(l) = \sigma_v^2 \rho^{|l|}/(1-\rho^2), -1 < \rho < 1$. Plot the frequency response of the filter and superimpose it on the noise PSD, for $\omega_0 = \pi/6$, $M = 12$, $\sigma_v^2 = 1$, and $\rho = 0.9$. Explain the shape of the obtained response. (c) Study the effect of the SNR in part (a) by varying the value of σ_v^2. (d) Study the effect of the noise correlation in part (c) by varying the value of ρ.

6.39 Consider the equalization experiment in Example 6.8.1 with $M = 11$ and $D = 7$. (a) Compute and plot the magnitude response $|H(e^{j\omega})|$ of the channel and $|C_o(e^{j\omega})|$ of the optimum equalizer for $W = 2.9, 3.1, 3.3$, and 3.5 and comment upon the results. (b) For the same values of W, compute the spectral dynamic range $|H(e^{j\omega})|_{max}/|H(e^{j\omega})|_{min}$ of the channel and the eigenvalue spread $\lambda_{max}/\lambda_{min}$ of the $M \times M$ input correlation matrix. Explain how the variation in one affects the other.

6.40 In this problem we clarify some of the properties of the MSE equalizer discussed in Example 6.8.1. (a) Compute and plot the MMSE P_o as a function of M, and recommend how to choose a "reasonable" value. (b) Compute and plot P_o as a function of the delay D for $0 \le D \le 11$. What is the best value of D? (c) Study the effect of input SNR upon P_o for $M = 11$ and $D = 7$ by fixing $\sigma_y^2 = 1$ and varying σ_v^2.

6.41 In this problem we formulate the design of optimum linear signal estimators (LSE) using a constrained optimization framework. To this end we consider the estimator $e(n) = c_0^* x(n) + \cdots + c_M^* x(n - M) \triangleq \mathbf{c}^H \mathbf{x}(n)$ and we wish to minimize the output power $E\{|e(n)|^2\} = \mathbf{c}^H \mathbf{R} \mathbf{c}$. To prevent the trivial solution $\mathbf{c} = \mathbf{0}$ we need to impose some constraint on the filter coefficients and use Lagrange multipliers to determine the minimum. Let $\mathbf{u}_i$ be an $M \times 1$ vector with one at the ith position and zeros elsewhere. (a) Show that minimizing $\mathbf{c}^H \mathbf{R} \mathbf{c}$ under the linear constraint $\mathbf{u}_i^T \mathbf{c} = 1$ provides the following estimators: FLP if $i = 0$, BLP if $i = M$, and linear smoother if $i \ne 0, M$. (b) Determine the appropriate set of constraints for the L-steps ahead linear predictor, defined by $c_0 = 1$ and $\{c_k = 0\}_1^{L-1}$, and solve the corresponding constrained optimization problem. Verify your answer by obtaining the normal equations using the orthogonality principle. (c) Determine the optimum linear estimator by minimizing $\mathbf{c}^H \mathbf{R} \mathbf{c}$ under the quadratic constraints $\mathbf{c}^H \mathbf{c} = 1$ and $\mathbf{c}^H \mathbf{W} \mathbf{c} = 1$ (**W** is a positive definite matrix) which impose a constraint on the length of the filter vector.

CHAPTER 7

Algorithms and Structures
for Optimum Linear Filters

The design and application of optimum filters involves (1) the solution of the normal equations to determine the optimum set of coefficients, (2) the evaluation of the cost function to determine whether the obtained parameters satisfy the design requirements, and (3) the implementation of the optimum filter, that is, the computation of its output that provides the estimate of the desired response.

The normal equations can be solved by using any general-purpose routine for linear simultaneous equations. However, there are several important reasons to study the normal equations in greater detail in order to develop efficient, special-purpose algorithms for their solution. First, the throughput of several real-time applications can only be served with serial or parallel algorithms that are obtained by exploiting the special structure (e.g., Toeplitz) of the correlation matrix. Second, sometimes we can develop order-recursive algorithms that help us to choose the correct filter order or to stop the algorithm before the manifestation of numerical problems. Third, some algorithms lead to intermediate sets of parameters that have physical meaning, provide easy tests for important properties (e.g., minimum phase), or are useful in special applications (e.g., data compression). Finally, sometimes there is a link between the algorithm for the solution of the normal equations and the structure for the implementation of the optimum filter.

In this chapter, we present different algorithms for the solution of the normal equations, the computation of the minimum mean square error (MMSE), and the implementation of the optimum filter. We start in Section 7.1 with a discussion of some results from matrix algebra that are useful for the development of order-recursive algorithms and introduce an algorithm for the order-recursive computation of the LDL^H decomposition, the MMSE, and the optimum estimate in the general case. In Section 7.2, we present some interesting interpretations for the various introduced algorithmic quantities and procedures that provide additional insight into the optimum filtering problem.

The only assumption we have made so far is that we know the required second-order statistics; hence, the results apply to any linear estimation problem: array processing, filtering, and prediction of nonstationary or stationary processes. In the sequel, we impose additional constraints on the input data vector and show how to exploit them in order to simplify the general algorithms and structures or specify new ones. In Section 7.3, we explore the shift invariance of the input data vector to develop a time-varying lattice-ladder structure for the optimum filter. However, to derive an order-recursive algorithm for the computation of either the direct or lattice-ladder structure parameters of the optimum time-varying filter, we need an analytical description of the changing second-order statistics of the nonstationary input process. Recall that in the simplest case of stationary processes, the correlation matrix is constant and Toeplitz. As a result, the optimum FIR filters and predictors are time-invariant, and their direct or lattice-ladder structure parameters can be computed (only once) using efficient, order-recursive algorithms due to Levinson and Durbin (Section 7.4) or Schür (Section 7.6). Section 7.5 provides a derivation of the lattice-ladder structures for optimum filtering and prediction, their structural and statistical properties, and algorithms for transformations between the various sets of parameters. Section 7.7 deals with efficient, order-recursive algorithms for the triangularization and inversion of Toeplitz matrices.

The chapter concludes with Section 7.8 which provides a concise introduction to the Kalman filtering algorithm. The Kalman filter provides a recursive solution to the minimum MSE filtering problem when the input stochastic process is described by a known state space model. This is possible because the state space model leads to a recursive formula for the updating of the required second-order moments.

7.1 FUNDAMENTALS OF ORDER-RECURSIVE ALGORITHMS

In Section 6.3, we introduced a method to solve the normal equations and compute the MMSE using the LDL^H decomposition. The optimum estimate is computed as a sum of products using a linear combiner supplied with the optimum coefficients and the input data. The key characteristic of this approach is that the order of the estimator should be fixed initially, and in case we choose a different order, we have to repeat *all* the computations. Such computational methods are known as *fixed-order algorithms*.

When the order of the estimator becomes a design variable, we need to modify our notation to take this into account. For example, the mth-order estimator $\mathbf{c}_m(n)$ is obtained by minimizing $E\{|e_m(n)|^2\}$, where

$$e_m(n) \triangleq y(n) - \hat{y}_m(n) \tag{7.1.1}$$

$$\hat{y}_m(n) \triangleq \mathbf{c}_m^H(n)\mathbf{x}_m(n) \tag{7.1.2}$$

$$\mathbf{c}_m(n) \triangleq [c_1^{(m)}(n) \ c_2^{(m)}(n) \ \cdots \ c_m^{(m)}(n)]^T \tag{7.1.3}$$

$$\mathbf{x}_m(n) \triangleq [x_1(n) \ x_2(n) \ \cdots \ x_m(n)]^T \tag{7.1.4}$$

In general, we use the subscript m to denote the order of a matrix or vector and the superscript m to emphasize that a scalar is a component of an $m \times 1$ vector. We note that these quantities are functions of time n, but sometimes we do not explicitly show this dependence for the sake of simplicity.

If the mth-order estimator $\mathbf{c}_m(n)$ has been computed by solving the normal equations, it seems to be a waste of computational power to start from scratch to compute the $(m+1)$st-order estimator $\mathbf{c}_{m+1}(n)$. Thus, we would like to arrange the computations so that the results for order m, that is, $\mathbf{c}_m(n)$ or $\hat{y}_m(n)$, can be used to compute the estimates for order $m+1$, that is, $\mathbf{c}_{m+1}(n)$ or $\hat{y}_{m+1}(n)$. The resulting procedures are called *order-recursive algorithms* or *order-updating relations*. Similarly, procedures that compute $\mathbf{c}_m(n+1)$ from $\mathbf{c}_m(n)$ or $\hat{y}_m(n+1)$ from $\hat{y}_m(n)$ are called *time-recursive algorithms* or *time-updating relations*. Combined order and time updates are also possible. All these updates play a central role in the design and implementation of many optimum and adaptive filters.

In this section, we derive order-recursive algorithms for the computation of the LDL^H decomposition, the MMSE, and the MMSE optimal estimate. We also show that there is no order-recursive algorithm for the computation of the estimator parameters.

7.1.1 Matrix Partitioning and Optimum Nesting

We start by introducing some notation that is useful for the discussion of order-recursive algorithms.[†] Notice that if the order of the estimator increases from m to $m+1$, then the input data vector is augmented with one additional observation x_{m+1}. We use the notation

---

[†] All quantities in Sections 7.1 and 7.2 are functions of the time index n. However, for notational simplicity we do not explicitly show this dependence.

$\mathbf{x}_{m+1}^{\lceil m \rceil}$ to denote the vector that consists of the first m components and $\mathbf{x}_{m+1}^{\lfloor m \rfloor}$ for the last m components of vector $\mathbf{x}_{m+1}$. The same notation can be generalized to matrices. The $m \times m$ matrix $\mathbf{R}_{m+1}^{\lceil m \rceil}$, obtained by the intersection of the first m rows and columns of $\mathbf{R}_{m+1}$, is known as the mth-order *leading principal submatrix* of $\mathbf{R}_{m+1}$. In other words, if r_{ij} are the elements of $\mathbf{R}_{m+1}$, then the elements of $\mathbf{R}_{m+1}^{\lceil m \rceil}$ are r_{ij}, $1 \le i, j \le m$. Similarly, $\mathbf{R}_{m+1}^{\lfloor m \rfloor}$ denotes the matrix obtained by the intersection of the last m rows and columns of $\mathbf{R}_{m+1}$. For example, if $m = 3$ we obtain

$$\mathbf{R}_4 = \begin{bmatrix} \begin{array}{cccc} r_{11} & r_{12} & r_{13} & r_{14} \\ r_{21} & r_{22} & r_{23} & r_{24} \\ r_{31} & r_{32} & r_{33} & r_{34} \\ r_{41} & r_{42} & r_{43} & r_{44} \end{array} \end{bmatrix} \qquad \mathbf{R}_4^{[3]} \ \ (\text{top}) \qquad \mathbf{R}_4^{\lfloor 3 \rfloor} \ (\text{bottom}) \tag{7.1.5}$$

which illustrates the *upper left corner* and *lower right corner* partitionings of matrix $\mathbf{R}_4$.

Since $\mathbf{x}_{m+1}^{\lceil m \rceil} = \mathbf{x}_m$, we can easily see that the correlation matrix can be partitioned as

$$\mathbf{R}_{m+1} = E\left\{ \begin{bmatrix} \mathbf{x}_m \\ x_{m+1} \end{bmatrix} \begin{bmatrix} \mathbf{x}_m^H & x_{m+1}^* \end{bmatrix} \right\} = \begin{bmatrix} \mathbf{R}_m & \mathbf{r}_m^b \\ \mathbf{r}_m^{bH} & \rho_m^b \end{bmatrix} \tag{7.1.6}$$

where

$$\mathbf{r}_m^b \triangleq E\{\mathbf{x}_m x_{m+1}^*\} \tag{7.1.7}$$

and

$$\rho_m^b \triangleq E\{|x_{m+1}|^2\} \tag{7.1.8}$$

The result

$$\mathbf{x}_{m+1}^{\lceil m \rceil} = \mathbf{x}_m \Rightarrow \mathbf{R}_m = \mathbf{R}_{m+1}^{\lceil m \rceil} \tag{7.1.9}$$

is known as the *optimum nesting property* and is instrumental in the development of order-recursive algorithms. Similarly, we can show that $\mathbf{x}_{m+1}^{\lceil m \rceil} = \mathbf{x}_m$ implies

$$\mathbf{d}_{m+1} = E\{\mathbf{x}_{m+1} y^*\} = E\left\{ \begin{bmatrix} \mathbf{x}_m \\ x_{m+1} \end{bmatrix} y^* \right\} = \begin{bmatrix} \mathbf{d}_m \\ d_{m+1} \end{bmatrix} \tag{7.1.10}$$

or

$$\mathbf{x}_{m+1}^{\lceil m \rceil} = \mathbf{x}_m \Rightarrow \mathbf{d}_m = \mathbf{d}_{m+1}^{\lceil m \rceil} \tag{7.1.11}$$

that is, the right-hand side of the normal equations also has the optimum nesting property.

Since (7.1.9) and (7.1.11) hold for all $1 \le m \le M$, the correlation matrix $\mathbf{R}_M$ and the cross-correlation vector $\mathbf{d}_M$ contain the information for the computation of all the optimum estimators $\mathbf{c}_m$ for $1 \le m \le M$.

7.1.2 Inversion of Partitioned Hermitian Matrices

Suppose now that we know the inverse $\mathbf{R}_m^{-1}$ of the leading principal submatrix $\mathbf{R}_{m+1}^{\lceil m \rceil} = \mathbf{R}_m$ of matrix $\mathbf{R}_{m+1}$ and we wish to use it to compute $\mathbf{R}_{m+1}^{-1}$ without having to repeat all the work. Since the inverse $\mathbf{Q}_{m+1}$ of the Hermitian matrix $\mathbf{R}_{m+1}$ is also Hermitian, it can be partitioned as

$$\mathbf{Q}_{m+1} = \begin{bmatrix} \mathbf{Q}_m & \mathbf{q}_m \\ \mathbf{q}_m^H & q_m \end{bmatrix} \tag{7.1.12}$$

Using (7.1.6), we obtain

$$\mathbf{R}_{m+1}\mathbf{Q}_{m+1} = \begin{bmatrix} \mathbf{R}_m & \mathbf{r}_m^b \\ \mathbf{r}_m^{bH} & \rho_m^b \end{bmatrix} \begin{bmatrix} \mathbf{Q}_m & \mathbf{q}_m \\ \mathbf{q}_m^H & q_m \end{bmatrix} = \begin{bmatrix} \mathbf{I}_m & \mathbf{0}_m \\ \mathbf{0}_m^H & 1 \end{bmatrix} \tag{7.1.13}$$

After performing the matrix multiplication, we get

$$\mathbf{R}_m \mathbf{Q}_m + \mathbf{r}_m^b \mathbf{q}_m^H = \mathbf{I}_m \tag{7.1.14}$$

$$\mathbf{r}_m^{bH} \mathbf{Q}_m + \rho_m^b \mathbf{q}_m^H = \mathbf{0}_m^H \tag{7.1.15}$$

$$\mathbf{R}_m \mathbf{q}_m + \mathbf{r}_m^b q_m = \mathbf{0}_m \tag{7.1.16}$$

$$\mathbf{r}_m^{bH} \mathbf{q}_m + \rho_m^b q_m = 1 \tag{7.1.17}$$

where $\mathbf{0}_m$ is the $m \times 1$ zero vector. If matrix $\mathbf{R}_m$ is invertible, we can solve (7.1.16) for $\mathbf{q}_m$

$$\mathbf{q}_m = -\mathbf{R}_m^{-1} \mathbf{r}_m^b q_m \tag{7.1.18}$$

and then substitute into (7.1.17) to obtain q_m as

$$q_m = \frac{1}{\rho_m^b - \mathbf{r}_m^{bH} \mathbf{R}_m^{-1} \mathbf{r}_m^b} \tag{7.1.19}$$

assuming that the scalar quantity $\rho_m^b - \mathbf{r}_m^{bH} \mathbf{R}_m^{-1} \mathbf{r}_m^b \neq 0$. Substituting (7.1.19) into (7.1.18), we obtain

$$\mathbf{q}_m = \frac{-\mathbf{R}_m^{-1} \mathbf{r}_m^b}{\rho_m^b - \mathbf{r}_m^{bH} \mathbf{R}_m^{-1} \mathbf{r}_m^b} \tag{7.1.20}$$

which, in conjunction with (7.1.14), yields

$$\mathbf{Q}_m = \mathbf{R}_m^{-1} - \mathbf{R}_m^{-1} \mathbf{r}_m^b \mathbf{q}_m^H = \mathbf{R}_m^{-1} + \frac{\mathbf{R}_m^{-1} \mathbf{r}_m^b (\mathbf{R}_m^{-1} \mathbf{r}_m^b)^H}{\rho_m^b - \mathbf{r}_m^{bH} \mathbf{R}_m^{-1} \mathbf{r}_m^b} \tag{7.1.21}$$

We note that (7.1.19) through (7.1.21) express the parts of the inverse matrix $\mathbf{Q}_{m+1}$ in terms of known quantities. For our purposes, we express the above equations in a more convenient form, using the quantities

$$\mathbf{b}_m \triangleq [b_0^{(m)} \, b_1^{(m)} \, \cdots \, b_{m-1}^{(m)}]^T \triangleq -\mathbf{R}_m^{-1} \mathbf{r}_m^b \tag{7.1.22}$$

and

$$\alpha_m^b \triangleq \rho_m^b - \mathbf{r}_m^{bH} \mathbf{R}_m^{-1} \mathbf{r}_m^b = \rho_m^b + \mathbf{r}_m^{bH} \mathbf{b}_m \tag{7.1.23}$$

Thus, if matrix $\mathbf{R}_m$ is invertible and $\alpha_m^b \neq 0$, combining (7.1.13) with (7.1.19) through (7.1.23), we obtain

$$\mathbf{R}_{m+1}^{-1} = \begin{bmatrix} \mathbf{R}_m & \mathbf{r}_m^b \\ \mathbf{r}_m^{bH} & \rho_m^b \end{bmatrix}^{-1} = \begin{bmatrix} \mathbf{R}_m^{-1} & \mathbf{0}_m \\ \mathbf{0}_m^H & 0 \end{bmatrix} + \frac{1}{\alpha_m^b} \begin{bmatrix} \mathbf{b}_m \\ 1 \end{bmatrix} \begin{bmatrix} \mathbf{b}_m^H & 1 \end{bmatrix} \tag{7.1.24}$$

which determines $\mathbf{R}_{m+1}^{-1}$ from $\mathbf{R}_m^{-1}$ by using a simple rank-one modification known as the *matrix inversion by partitioning lemma* (Noble and Daniel 1988).

Another useful expression for α_m^b is

$$\alpha_m^b = \frac{\det \mathbf{R}_{m+1}}{\det \mathbf{R}_m} \tag{7.1.25}$$

which reinforces the importance of the quantity α_m^b for the invertibility of matrix $\mathbf{R}_{m+1}$ (see Problem 7.1).

EXAMPLE 7.1.1. Given the matrix

$$\mathbf{R}_3 = \begin{bmatrix} 1 & \frac{1}{2} & \frac{1}{3} \\ \frac{1}{2} & 1 & \frac{1}{2} \\ \frac{1}{3} & \frac{1}{2} & 1 \end{bmatrix} = \begin{bmatrix} \mathbf{R}_2 & \mathbf{r}_2^b \\ \mathbf{r}_2^{bH} & \rho_2^b \end{bmatrix}$$

and the inverse matrix

$$\mathbf{R}_2^{-1} = \begin{bmatrix} 1 & \frac{1}{2} \\ \frac{1}{2} & 1 \end{bmatrix}^{-1} = \frac{1}{3} \begin{bmatrix} 4 & -2 \\ -2 & 4 \end{bmatrix}$$

compute matrix $\mathbf{R}_3^{-1}$, using the matrix inversion by partitioning lemma.

Solution. To determine $\mathbf{R}_3^{-1}$ from the order-updating formula (7.1.24), we first compute

$$\mathbf{b}_2 = -\mathbf{R}_2^{-1}\mathbf{r}_2^{b} = -\frac{1}{3}\begin{bmatrix} 4 & -2 \\ -2 & 4 \end{bmatrix}\begin{bmatrix} \frac{1}{3} \\ \frac{1}{2} \end{bmatrix} = -\frac{1}{9}\begin{bmatrix} 1 \\ 4 \end{bmatrix}$$

and

$$\alpha_2^{b} = \rho_2^{b} + \mathbf{r}_2^{bH}\mathbf{b}_2 = 1 - \frac{1}{9}\begin{bmatrix} 1 & 1 \\ 3 & 2 \end{bmatrix}\begin{bmatrix} 1 \\ 4 \end{bmatrix} = \frac{20}{27}$$

using (7.1.22) and (7.1.23). Then we compute

$$\mathbf{R}_3^{-1} = \frac{1}{3}\left[\begin{array}{cc|c} 4 & -2 & 0 \\ -2 & 4 & 0 \\ \hline 0 & 0 & 0 \end{array}\right] + \frac{27}{20}\begin{bmatrix} -\frac{1}{9} \\ -\frac{4}{9} \\ 1 \end{bmatrix}\left[\begin{array}{cc|c} -\frac{1}{9} & -\frac{4}{9} & 1 \end{array}\right] = \frac{1}{20}\begin{bmatrix} 27 & -12 & -3 \\ -12 & 32 & -12 \\ -3 & -12 & 27 \end{bmatrix}$$

using (7.1.24). The reader can easily verify the above calculations using MATLAB.

Following a similar approach, we can show (see Problem 7.2) that the inverse of the lower right corner partitioned matrix $\mathbf{R}_{m+1}$ can be expressed as

$$\mathbf{R}_{m+1}^{-1} \triangleq \begin{bmatrix} \rho_m^{f} & \mathbf{r}_m^{fH} \\ \mathbf{r}_m^{f} & \mathbf{R}_m^{f} \end{bmatrix}^{-1} = \begin{bmatrix} 0 & \mathbf{0}_m^{H} \\ \mathbf{0}_m & (\mathbf{R}_m^{f})^{-1} \end{bmatrix} + \frac{1}{\alpha_m^{f}}\begin{bmatrix} 1 \\ \mathbf{a}_m \end{bmatrix}\begin{bmatrix} 1 & \mathbf{a}_m^{H} \end{bmatrix} \tag{7.1.26}$$

where

$$\mathbf{a}_m \triangleq [a_1^{(m)} \; a_2^{(m)} \; \cdots \; a_m^{(m)}]^{T} \triangleq -(\mathbf{R}_m^{f})^{-1}\mathbf{r}_m^{f} \tag{7.1.27}$$

$$\alpha_m^{f} \triangleq \rho_m^{f} - \mathbf{r}_m^{fH}(\mathbf{R}_m^{f})^{-1}\mathbf{r}_m^{f} = \rho_m^{f} + \mathbf{r}_m^{fH}\mathbf{a}_m = \frac{\det \mathbf{R}_{m+1}}{\det \mathbf{R}_m^{f}} \tag{7.1.28}$$

and the relationship (7.1.26) exists if matrix $\mathbf{R}_m^{f}$ is invertible and $\alpha_m^{f} \neq 0$. A similar set of formulas can be obtained for arbitrary matrices (see Problem 7.3).

Interpretations. The vector $\mathbf{b}_m$, defined by (7.1.22), is the MMSE estimator of observation x_{m+1} from data vector $\mathbf{x}_m$. Indeed, if

$$e_m^{b} = x_{m+1} - \hat{x}_{m+1} = x_{m+1} + \mathbf{b}_m^{H}\mathbf{x}_m \tag{7.1.29}$$

we can show, using the orthogonality principle $E\{\mathbf{x}_m e_m^{b*}\} = \mathbf{0}$, that $\mathbf{b}_m$ results in the MMSE given by

$$P_m^{b} = \rho_m^{b} + \mathbf{b}_m^{H}\mathbf{r}_m^{b} = \alpha_m^{b} \tag{7.1.30}$$

Similarly, we can show that $\mathbf{a}_m$, defined by (7.1.27), is the optimum estimator of x_1 based on $\tilde{\mathbf{x}}_m \triangleq [x_2 \; x_3 \; \cdots \; x_{m+1}]^{T}$. By using the orthogonality principle, $E\{\mathbf{x}_m e_m^{f*}\} = \mathbf{0}$, the MMSE is

$$P_m^{f} = \rho_m^{f} + \mathbf{r}_m^{fH}\mathbf{a}_m = \alpha_m^{f} \tag{7.1.31}$$

If $\mathbf{x}_{m+1} = [x(n) \; x(n-1) \; \cdots \; x(n-m)]^{T}$, then $\mathbf{b}_m$ provides the *backward linear predictor* (BLP) and $\mathbf{a}_m$ the *forward linear predictor* (FLP) of the process $x(n)$ from Section 6.5. For convenience, we always use this terminology even if, strictly speaking, the linear prediction interpretation is not applicable.

7.1.3 Levinson Recursion for the Optimum Estimator

We now illustrate how to use (7.1.24) to express the optimum estimator $\mathbf{c}_{m+1}$ in terms of the estimator $\mathbf{c}_m$. Indeed, using (7.1.24), (7.1.10), and the normal equations $\mathbf{R}_m\mathbf{c}_m = \mathbf{d}_m$,

we have

$$\mathbf{c}_{m+1} = \mathbf{R}_{m+1}^{-1} \mathbf{d}_{m+1}$$

$$= \begin{bmatrix} \mathbf{R}_m^{-1} & \mathbf{0}_m \\ \mathbf{0}_m^T & 0 \end{bmatrix} \begin{bmatrix} \mathbf{d}_m \\ d_{m+1} \end{bmatrix} + \frac{1}{\alpha_m^b} \begin{bmatrix} \mathbf{b}_m \\ 1 \end{bmatrix} \begin{bmatrix} \mathbf{b}_m^H & 1 \end{bmatrix} \begin{bmatrix} \mathbf{d}_m \\ d_{m+1} \end{bmatrix}$$

$$= \begin{bmatrix} \mathbf{R}_m^{-1} \mathbf{d}_m \\ 0 \end{bmatrix} + \begin{bmatrix} \mathbf{b}_m \\ 1 \end{bmatrix} \frac{\mathbf{b}_m^H \mathbf{d}_m + d_{m+1}}{\alpha_m^b}$$

or more concisely

$$\mathbf{c}_{m+1} = \begin{bmatrix} \mathbf{c}_m \\ 0 \end{bmatrix} + \begin{bmatrix} \mathbf{b}_m \\ 1 \end{bmatrix} k_m^c \tag{7.1.32}$$

where the quantities

$$k_m^c \triangleq \frac{\beta_m^c}{\alpha_m^b} \tag{7.1.33}$$

and

$$\beta_m^c \triangleq \mathbf{b}_m^H \mathbf{d}_m + d_{m+1} \tag{7.1.34}$$

contain the "new information" d_{m+1} (the new component of $\mathbf{d}_{m+1}$). By using (7.1.22) and $\mathbf{R}_m \mathbf{c}_m = \mathbf{d}_m$, alternatively β_m^c can be written as

$$\beta_m^c = -\mathbf{r}_m^{bH} \mathbf{c}_m + d_{m+1} \tag{7.1.35}$$

We will use the term *Levinson recursion* for the order-updating relation (7.1.32) because a similar recursion was introduced as part of the celebrated algorithm due to Levinson (see Section 7.3). However, we stress that even though (7.1.32) is order-recursive, the parameter vector $\mathbf{c}_{m+1}$ *does not* have the optimum nesting property, that is, $\mathbf{c}_{m+1}^{[m]} \neq \mathbf{c}_m$.

Clearly, if we know the vector $\mathbf{b}_m$, we can determine $\mathbf{c}_{m+1}$, using (7.1.32); however, its practical utility depends on how easily we can obtain the vector $\mathbf{b}_m$. In general, $\mathbf{b}_m$ requires the solution of an $m \times m$ linear system of equations, and the computational savings compared to direct solution of the $(m + 1)$st-order normal equations is insignificant. For the Levinson recursion to be useful, we need an order recursion for vector $\mathbf{b}_m$. Since matrix $\mathbf{R}_{m+1}$ has the optimum nesting property, we need to check whether the same is true for the right-hand side vector in $\mathbf{R}_{m+1} \mathbf{b}_{m+1} = -\mathbf{r}_{m+1}^b$. From the definition $\mathbf{r}_m^b \triangleq E\{\mathbf{x}_m x_{m+1}^*\}$, we can easily see that $\mathbf{r}_{m+1}^{b[m]} \neq \mathbf{r}_m^b$ and $\mathbf{r}_{m+1}^{b\lfloor m \rfloor} \neq \mathbf{r}_m^b$. Hence, in general, we cannot find a Levinson recursion for vector $\mathbf{b}_m$. This is possible *only* in optimum filtering problems in which the input data vector $\mathbf{x}_m(n)$ has a shift-invariance structure (see Section 7.3).

EXAMPLE 7.1.2. Use the Levinson recursion to determine the optimum linear estimator $\mathbf{c}_3$ specified by the matrix

$$\mathbf{R}_3 = \begin{bmatrix} 1 & \frac{1}{2} & \frac{1}{3} \\ \frac{1}{2} & 1 & \frac{1}{2} \\ \frac{1}{3} & \frac{1}{2} & 1 \end{bmatrix}$$

in Example 7.1.1 and the cross-correlation vector

$$\mathbf{d}_3 = [1 \ 2 \ 4]^T$$

Solution. For $m = 1$ we have $r_{11} c_1^{(1)} = d_1$, which gives $c_1^{(1)} = 1$. Also, from (7.1.32) and (7.1.34) we obtain $k_0^c = c_1^{(1)} = 1$ and $\beta_0^c = d_1 = 1$. Finally, from $k_0^c = \beta_0^c / \alpha_0^b$, we get $\alpha_0^b = 1$.

To obtain $\mathbf{c}_2$, we need $b_1^{(1)}$, k_1^c, β_1^c, and α_1. We have

$$r_{11}b_1^{(1)} = -r_1^b \Rightarrow b_1^{(1)} = -\frac{\frac{1}{2}}{1} = -\frac{1}{2}$$

$$\beta_1^c = b_1^{(1)}d_1 + d_2 = -\frac{1}{2}(1) + 2 = \frac{3}{2}$$

$$\alpha_1^b = \rho_1^b + r_1^b b_1^{(1)} = 1 + \frac{1}{2}\left(-\frac{1}{2}\right) = \frac{3}{4}$$

$$k_1^c = \frac{\beta_1^c}{\alpha_1^b} = 2$$

and therefore

$$\mathbf{c}_2 = \begin{bmatrix} \mathbf{c}_1 \\ 0 \end{bmatrix} + \begin{bmatrix} \mathbf{b}_1 \\ 1 \end{bmatrix} k_1^c = \begin{bmatrix} 1 \\ 0 \end{bmatrix} + \begin{bmatrix} -\frac{1}{2} \\ 1 \end{bmatrix} 2 = \begin{bmatrix} 0 \\ 2 \end{bmatrix}$$

To determine $\mathbf{c}_3$, we need $\mathbf{b}_2$, β_2^c, and α_2^b. To obtain $\mathbf{b}_2$, we solve the linear system

$$\mathbf{R}_2\mathbf{b}_2 = -\mathbf{r}_2^b \quad \text{or} \quad \begin{bmatrix} 1 & \frac{1}{2} \\ \frac{1}{2} & 1 \end{bmatrix} \begin{bmatrix} b_1^{(2)} \\ b_2^{(2)} \end{bmatrix} = -\begin{bmatrix} \frac{1}{3} \\ \frac{1}{2} \end{bmatrix} \Rightarrow \mathbf{b}_2 = -\frac{1}{9}\begin{bmatrix} 1 \\ 4 \end{bmatrix}$$

and then compute

$$\beta_2^c = \mathbf{b}_2^T\mathbf{d}_2 + d_3 = -\frac{1}{9}\begin{bmatrix} 1 & 4 \end{bmatrix}\begin{bmatrix} 1 \\ 2 \end{bmatrix} + 4 = 3$$

$$\alpha_2^b = \rho_2^b + \mathbf{r}_2^{bT}\mathbf{b}_2 = 1 + \begin{bmatrix} \frac{1}{3} & \frac{1}{2} \end{bmatrix}\begin{bmatrix} 1 \\ 4 \end{bmatrix}\left(-\frac{1}{9}\right) = \frac{20}{27}$$

$$k_2^c = \frac{\beta_2^c}{\alpha_2^b} = \frac{81}{20}$$

The desired solution $\mathbf{c}_3$ is obtained by using the Levinson recursion

$$\mathbf{c}_3 = \begin{bmatrix} \mathbf{c}_2 \\ 0 \end{bmatrix} + \begin{bmatrix} \mathbf{b}_2 \\ 1 \end{bmatrix} k_2^c \Rightarrow \mathbf{c}_3 = \begin{bmatrix} 0 \\ 2 \\ 0 \end{bmatrix} + \begin{bmatrix} -\frac{1}{9} \\ -\frac{4}{9} \\ 1 \end{bmatrix} \frac{81}{20} = \frac{1}{20}\begin{bmatrix} -9 \\ 4 \\ 81 \end{bmatrix}$$

which agrees with the solution obtained by solving $\mathbf{R}_3\mathbf{c}_3 = \mathbf{d}_3$ using the function c3=R3\d3. We can also solve this linear system by developing an algorithm using the lower partitioning (7.1.26) as discussed in Problem 7.4.

Matrix inversion and the linear system solution for $m = 1$ are trivial (scalar division only). If $\mathbf{R}_M$ is strictly positive definite, that is, $\mathbf{R}_m = \mathbf{R}_M^{[m]}$ is positive definite for all $1 \leq m \leq M$, the inverse matrices $\mathbf{R}_m^{-1}$ and the solutions of $\mathbf{R}_m\mathbf{c}_m = \mathbf{d}_m$, $2 \leq m \leq M$, can be determined using (7.1.22) and the Levinson recursion (7.1.32) for $m = 1, 2, \ldots, M-1$. However, in practice using the LDL^H provides a better method for performing these computations.

7.1.4 Order-Recursive Computation of the LDL^H Decomposition

We start by showing that the LDL^H decomposition can be computed in an order-recursive manner. The procedure is developed as part of a formal proof of the LDL^H decomposition using induction.

For $M = 1$, the matrix $\mathbf{R}_1$ is a positive number r_{11} and can be written uniquely in the form $r_{11} = 1 \cdot \xi_1 \cdot 1 > 0$. As we increment the order m, the $(m+1)$st-order principal submatrix of $\mathbf{R}_m$ can be partitioned as in (7.1.6). By the induction hypothesis, there are unique matrices $\mathbf{L}_m$ and $\mathbf{D}_m$ such that

$$\mathbf{R}_m = \mathbf{L}_m \mathbf{D}_m \mathbf{L}_m^H \tag{7.1.36}$$

We next form the matrices

$$\mathbf{L}_{m+1} = \begin{bmatrix} \mathbf{L}_m & \mathbf{0} \\ \mathbf{l}_m^H & 1 \end{bmatrix} \qquad \mathbf{D}_{m+1} = \begin{bmatrix} \mathbf{D}_m & \mathbf{0} \\ \mathbf{0}^H & \xi_{m+1} \end{bmatrix} \tag{7.1.37}$$

and try to determine the vector $\mathbf{l}_m$ and the positive number ξ_{m+1} so that

$$\mathbf{R}_{m+1} = \mathbf{L}_{m+1} \mathbf{D}_{m+1} \mathbf{L}_{m+1}^H \tag{7.1.38}$$

Using (7.1.6) and (7.1.36) through (7.1.38), we see that

$$(\mathbf{L}_m \mathbf{D}_m) \mathbf{l}_m = \mathbf{r}_m^b \tag{7.1.39}$$

$$\rho_m^b = \mathbf{l}_m^H \mathbf{D}_m \mathbf{l}_m + \xi_{m+1}, \qquad \xi_{m+1} > 0 \tag{7.1.40}$$

Since
$$\det \mathbf{R}_m = \det \mathbf{L}_m \det \mathbf{D}_m \det \mathbf{L}_m^H = \xi_1 \xi_2 \cdots \xi_m > 0 \tag{7.1.41}$$

then $\det \mathbf{L}_m \mathbf{D}_m \neq 0$ and (7.1.39) has a unique solution $\mathbf{l}_m$. Finally, from (7.1.41) we obtain $\xi_{m+1} = \det \mathbf{R}_{m+1}/\det \mathbf{R}_m$, and therefore $\xi_{m+1} > 0$ because $\mathbf{R}_{m+1}$ is positive definite. Hence, ξ_{m+1} is uniquely computed from (7.1.41), which completes the proof.

Because the triangular matrix $\mathbf{L}_m$ is generated row by row using (7.1.39) and because the diagonal elements of matrix $\mathbf{D}_m$ are computed sequentially using (7.1.40), both matrices have the optimum nesting property, that is, $\mathbf{L}_m = \mathbf{L}^{\lceil m \rceil}$, $\mathbf{D}_m = \mathbf{D}^{\lceil m \rceil}$. The optimum filter $\mathbf{c}_m$ is then computed by solving

$$\mathbf{L}_m \mathbf{D}_m \mathbf{k}_m \triangleq \mathbf{d}_m \tag{7.1.42}$$

$$\mathbf{L}_m^H \mathbf{c}_m = \mathbf{k}_m \tag{7.1.43}$$

Using (7.1.42), we can easily see that $\mathbf{k}_m$ has the optimum nesting property, that is, $\mathbf{k}_m = \mathbf{k}^{\lceil m \rceil}$ for $1 \leq m \leq M$. This is a consequence of the lower triangular form of $\mathbf{L}_m$. The computation of $\mathbf{L}_m$, $\mathbf{D}_m$, and $\mathbf{k}_m$ can be done in a simple, order-recursive manner, which is all that is needed to compute $\mathbf{c}_m$ for $1 \leq m \leq M$. However, the optimum estimator does not have the optimum nesting property, that is, $\mathbf{c}_{m+1}^{[m]} \neq \mathbf{c}_m$, because of the backward substitution involved in the solution of the upper triangular system (7.1.43) (see Example 6.3.1).

Using (7.1.42) and (7.1.43), we can write the MMSE for the mth-order linear estimator as

$$P_m = P_y - \mathbf{c}_m^H \mathbf{d}_m = P_y - \mathbf{k}_m^H \mathbf{D}_m \mathbf{k}_m \tag{7.1.44}$$

which, owing to the optimum nesting property of $\mathbf{D}_m$ and $\mathbf{k}_m$, leads to

$$P_m = P_{m-1} - \xi_m |k_m|^2 \tag{7.1.45}$$

which is initialized with $P_0 = P_y$. Equation (7.1.45) provides an *order-recursive algorithm* for the computation of the MMSE.

7.1.5 Order-Recursive Computation of the Optimum Estimate

The computation of the optimum linear estimate $\hat{y}_m = \mathbf{c}_m^H \mathbf{x}_m$, using a linear combiner, requires m multiplications and $m-1$ additions. Therefore, if we want to compute $\hat{y}_m$, for $1 \leq m \leq M$, we need M linear combiners and hence $M(M+1)/2$ operations.

We next provide an alternative, more efficient order-recursive implementation that exploits the triangular decomposition of $\mathbf{R}_{m+1}$. We first notice that using (7.1.43), we obtain

$$\hat{y}_m = \mathbf{c}_m^H \mathbf{x}_m = (\mathbf{k}_m^H \mathbf{L}_m^{-1}) \mathbf{x}_m = \mathbf{k}_m^H (\mathbf{L}_m^{-1} \mathbf{x}_m) \tag{7.1.46}$$

Next, we define vector $\mathbf{w}_m$ as

$$\mathbf{L}_m \mathbf{w}_m \triangleq \mathbf{x}_m \qquad (7.1.47)$$

which can be found by using forward substitution in order to solve the triangular system. Therefore, we obtain

$$\hat{y}_m = \mathbf{k}_m^H \mathbf{w}_m = \sum_{i=1}^{m} k_i^* w_i \qquad (7.1.48)$$

which provides the estimate $\hat{y}_m$ in terms of $\mathbf{k}_m$ and $\mathbf{w}_m$, that is, without using the estimator vector $\mathbf{c}_m$. Hence, *if the ultimate goal is the computation of $\hat{y}_m$ we do not need to compute the estimator* $\mathbf{c}_m$.

For an order-recursive algorithm to be possible, the vector $\mathbf{w}_m$ must have the optimum nesting property, that is, $\mathbf{w}_m = \mathbf{w}_{m+1}^{[m]}$. Indeed, using (7.1.37) and the matrix inversion by partitioning lemma for nonsymmetric matrices (see Problem 7.3), we obtain

$$\mathbf{L}_{m+1}^{-1} = \begin{bmatrix} \mathbf{L}_m & \mathbf{0} \\ \mathbf{l}_m^H & 1 \end{bmatrix}^{-1} = \begin{bmatrix} \mathbf{L}_m^{-1} & \mathbf{0} \\ \mathbf{v}_m^H & 1 \end{bmatrix}$$

where

$$\mathbf{v}_m = -\mathbf{L}_m^{-H} \mathbf{l}_m = -(\mathbf{L}_m^H)^{-1} \mathbf{D}_m^{-1} \mathbf{L}_m^{-1} \mathbf{r}_m^b = -\mathbf{R}_m^{-1} \mathbf{r}_m^b = \mathbf{b}_m$$

due to (7.1.22). Therefore,

$$\mathbf{w}_{m+1} = \mathbf{L}_{m+1}^{-1} \mathbf{x}_{m+1} = \begin{bmatrix} \mathbf{L}_m^{-1} & \mathbf{0} \\ \mathbf{b}_m^H & 1 \end{bmatrix} \begin{bmatrix} \mathbf{x}_m \\ x_{m+1} \end{bmatrix} = \begin{bmatrix} \mathbf{w}_m \\ w_{m+1} \end{bmatrix} \qquad (7.1.49)$$

where

$$w_{m+1} = \mathbf{b}_m^H \mathbf{x}_m + x_{m+1} = e_m^b \qquad (7.1.50)$$

from (7.1.29). In this case, we can derive order-recursive algorithms for the computation of $\hat{y}_m$ and e_m, for all $1 \leq m \leq M$. Indeed, using (7.1.48) and (7.1.49), we obtain

$$\hat{y}_m = \hat{y}_{m-1} + k_m^* w_m \qquad (7.1.51)$$

with $\hat{y}_0 = 0$. From (7.1.51) and $e_m = y - \hat{y}_m$, we have

$$e_m = e_{m-1} - k_m^* w_m \qquad (7.1.52)$$

for $m = 1, 2, \ldots, M$ with $e_0 = y$. The quantity w_m can be computed in an order-recursive manner by solving (7.1.47) using forward substitution. Indeed, from the mth row of (7.1.47) we obtain

$$w_m = x_m - \sum_{i=1}^{m-1} l_{i-1}^{(m-1)} w_i \qquad (7.1.53)$$

which provides a *recursive* computation of w_m for $m = 1, 2, \ldots, M$. To comply with the order-oriented notation, we use $l_{i-1}^{(m-1)}$ instead of $l_{m-1,i-1}$. Depending on the application, we use either (7.1.51) or (7.1.52).

For MMSE estimation, all the quantities are functions of the time index n, and therefore, the triangular decomposition of $\mathbf{R}_m$ and the recursions (7.1.51) through (7.1.53) should be repeated for every new set of observations $y(n)$ and $\mathbf{x}(n)$.

EXAMPLE 7.1.3. A linear estimator is specified by the correlation matrix $\mathbf{R}_4$ and the cross-correlation vector $\mathbf{d}_4$ in Example 6.3.2. Compute the estimates $\hat{y}_m$, $1 \leq m \leq 4$, if the input data vector is given by $\mathbf{x}_4 = [1\ 2\ 1\ -1]^T$.

Solution. Using the triangular factor $\mathbf{L}_4$ and the vector $\mathbf{k}_4$ found in Example 6.3.2 and (7.1.53), we find

$$\mathbf{w}_4 = [1\ -1\ 3\ -8]^T$$

and

$$\hat{y}_1 = 1 \qquad \hat{y}_2 = \tfrac{4}{3} \qquad \hat{y}_3 = 6.6 \qquad \hat{y}_4 = 14.6$$

which the reader can verify by computing $\mathbf{c}_m$ and $\hat{y}_m = \mathbf{c}_m^T \mathbf{x}_m$, $1 \leq m \leq 4$.

If we compute the matrix

$$\mathbf{B}_{m+1} \overset{\triangle}{=} \mathbf{L}_{m+1}^{-1} = \begin{bmatrix} 1 & 0 & \cdots & 0 \\ b_0^{(1)} & 1 & \cdots & 0 \\ \vdots & \vdots & \ddots & \vdots \\ b_0^{(m)} & b_1^{(m)} & \cdots & 1 \end{bmatrix} \tag{7.1.54}$$

then (7.1.49) can be written as

$$\mathbf{w}_{m+1} = \mathbf{e}_{m+1}^{b} = \mathbf{B}_{m+1}\mathbf{x}_{m+1} \tag{7.1.55}$$

where

$$\mathbf{e}_{m+1}^{b} \overset{\triangle}{=} [e_0^b \ e_1^b \ \cdots \ e_m^b]^T \tag{7.1.56}$$

is the BLP error vector. From (7.1.22), we can easily see that the rows of $\mathbf{B}_{m+1}$ are formed by the optimum estimators $\mathbf{b}_m$ of x_{m+1} from $\mathbf{x}_m$. Note that the elements of matrix $\mathbf{B}_{m+1}$ are denoted by using the order-oriented notation $b_i^{(m)}$ introduced in Section 7.1 rather than the conventional b_{mi} matrix notation. Equation (7.1.55) provides an alternative computation of $\mathbf{w}_{m+1}$ as a matrix-vector multiplication. Each component of $\mathbf{w}_{m+1}$ can be computed independently, and hence in parallel, by the formula

$$w_j = x_j + \sum_{i=1}^{j-1} b_{i-1}^{(j-1)*} x_i \qquad 1 \le j \le m \tag{7.1.57}$$

which, in contrast to (7.1.53), is nonrecursive. Using (7.1.57) and (7.1.51), we can derive the order-recursive MMSE estimator implementation shown in Figure 7.1.

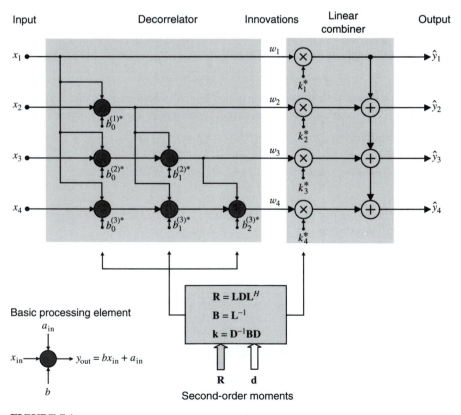

FIGURE 7.1
Orthogonal order-recursive structure for linear MMSE estimation.

Finally, we notice that matrix $\mathbf{B}_m$ provides the UDU^H decomposition of the inverse correlation matrix $\mathbf{R}_m$. Indeed, from (7.1.36) we obtain

$$\mathbf{R}_m^{-1} = (\mathbf{L}_m^H)^{-1}\mathbf{D}_m^{-1}\mathbf{L}_m^{-1} = \mathbf{B}_m^H\mathbf{D}_m^{-1}\mathbf{B}_m \tag{7.1.58}$$

because inversion and transposition are interchangeable and the UDU^H decomposition is unique. This formula provides a practical method to compute the inverse of the correlation matrix by using the LDL^H decomposition because computing the inverse of a triangular matrix is simple (see Problem 7.5).

7.2 INTERPRETATIONS OF ALGORITHMIC QUANTITIES

We next show that various intermediate quantities that appear in the linear MMSE estimation algorithms have physical and statistical interpretations that, besides their intellectual value, facilitate better understanding of the operation, performance, and numerical properties of the algorithms.

7.2.1 Innovations and Backward Prediction

The correlation matrix of $\mathbf{w}_m$ is

$$E\{\mathbf{w}_m\mathbf{w}_m^H\} = \mathbf{L}_m^{-1}E\{\mathbf{x}_m\mathbf{x}_m^H\}\mathbf{L}_m^{-H} = \mathbf{D}_m \tag{7.2.1}$$

where we have used (7.1.47) and the triangular decomposition (7.1.36). Therefore, the components of $\mathbf{w}_m$ are uncorrelated, random variables with variances

$$\xi_i = E\{|w_i|^2\} \tag{7.2.2}$$

since $\xi_i \geq 0$. Furthermore, the two sets of random variables $\{w_1, w_2, \ldots, w_M\}$ and $\{x_1, x_2, \ldots, x_M\}$ are *linearly equivalent* because they can be obtained from each other through the linear transformation (7.1.47). This transformation removes all the redundant correlation among the components of $\mathbf{x}$ and is known as a *decorrelation* or *whitening* operation (see Section 3.5.2). Because the random variables w_i are uncorrelated, each of them adds "new information" or innovation. In this sense, $\{w_1, w_2, \ldots, w_m\}$ is the *innovations representation* of the random variables $\{x_1, x_2, \ldots, x_m\}$. Because $\mathbf{x}_m = \mathbf{L}_m\mathbf{w}_m$, the random vector $\mathbf{w}_m = \mathbf{e}_m^b$ is the *innovations representation*, and $\mathbf{x}_m$ and $\mathbf{w}_m$ are *linearly equivalent* as well, (see Section 3.5).

The cross-correlation matrix between $\mathbf{x}_m$ and $\mathbf{w}_m$ is

$$E\{\mathbf{x}_m\mathbf{w}_m^H\} = E\{\mathbf{L}_m\mathbf{w}_m\mathbf{w}_m^H\} = \mathbf{L}_m\mathbf{D}_m \tag{7.2.3}$$

which shows that, owing to the lower triangular form of $\mathbf{L}_m$, $E\{x_i w_j^*\} = 0$ for $j > i$. We will see in Section 7.6 that these factors are related to the gapped functions and the algorithm of Schür.

Furthermore, since $e_m^b = w_{m+1}$, from (7.1.50) we have

$$P_m^b = \xi_{m+1} = E\{|w_{m+1}|^2\}$$

which also can be shown algebraically by using (7.1.41), (7.1.40), and (7.1.30). Indeed, we have

$$\xi_{m+1} = \frac{\det \mathbf{R}_{m+1}}{\det \mathbf{R}_m} = \rho_m^b - \mathbf{l}_m^H\mathbf{D}_m\mathbf{l}_m = \rho_m^b - \mathbf{r}_m^{bH}\mathbf{R}_m^{-1}\mathbf{r}_m^b = P_m^b \tag{7.2.4}$$

and, therefore,

$$\mathbf{D}_m = \mathrm{diag}\{P_0^b, P_1^b, \ldots, P_{m-1}^b\} \tag{7.2.5}$$

7.2.2 Partial Correlation

In general, the random variables $y, x_1, \ldots, x_m, x_{m+1}$ are correlated. The correlation between y and x_{m+1}, after the influence from the components of the vector $\mathbf{x}_m$ has been removed, is known as *partial correlation*. To remove the correlation due to $\mathbf{x}_m$, we extract from y and x_{m+1} the components that can be predicted from $\mathbf{x}_m$. The remaining correlation is from the estimation errors e_m and e_m^{b}, which are both uncorrelated with $\mathbf{x}_m$ because of the orthogonality principle. Therefore, the partial correlation of y and x_{m+1} is

$$
\begin{aligned}
\text{PARCOR}(y; \; x_{m+1}) \triangleq E\{e_m e_m^{\mathrm{b}*}\} &= E\{(y - \mathbf{c}_m^H \mathbf{x}_m) e_m^{\mathrm{b}*}\} \\
&= E\{y e_m^{\mathrm{b}*}\} = E\{y(x_{m+1}^* + \mathbf{x}_m^H \mathbf{b}_m)\} \\
&= E\{y x_{m+1}^*\} + E\{y \mathbf{x}_m^H\}\mathbf{b}_m \\
&= d_{m+1}^* + \mathbf{d}_m^H \mathbf{b}_m \triangleq \beta_m^{c*}
\end{aligned}
\tag{7.2.6}
$$

where we have used the orthogonality principle $E\{\mathbf{x}_m e_m^{\mathrm{b}*}\} = \mathbf{0}$ and (7.1.10), (7.1.50), and (7.1.34).

The partial correlation $\text{PARCOR}(y; \; x_{m+1})$ is also related to the parameters k_m obtained from the LDL^H decomposition. Indeed, from (7.1.42) and (7.1.54), we obtain the relation

$$
\mathbf{k}_{m+1} = \mathbf{D}_{m+1}^{-1} \mathbf{B}_{m+1} \mathbf{d}_{m+1}
\tag{7.2.7}
$$

whose last row is

$$
k_{m+1} = \frac{\mathbf{b}_m^H \mathbf{d}_m + d_{m+1}}{\xi_{m+1}} = \frac{\beta_m^c}{P_m^{\mathrm{b}}} = k_m^c
\tag{7.2.8}
$$

owing to (7.2.4) and (7.2.6).

EXAMPLE 7.2.1. The LDL^H decomposition of matrix $\mathbf{R}_3$ in Example 7.1.2 is given by

$$
\mathbf{L} = \begin{bmatrix} 1 & 0 & 0 \\ \frac{1}{2} & 1 & 0 \\ \frac{1}{3} & \frac{4}{9} & 1 \end{bmatrix} \qquad \mathbf{D} = \begin{bmatrix} 1 & 0 & 0 \\ 0 & \frac{3}{4} & 0 \\ 0 & 0 & \frac{20}{27} \end{bmatrix}
$$

and can be found by using the function `[L,D]=ldlt(R)`. Comparison with the results obtained in Example 7.1.2 shows that the rows of the matrix

$$
\mathbf{L}^{-1} = \begin{bmatrix} 1 & 0 & 0 \\ -\frac{1}{2} & 1 & 0 \\ -\frac{1}{9} & -\frac{4}{9} & 1 \end{bmatrix}
$$

provide the elements of the backward predictors, whereas the diagonal elements of $\mathbf{D}$ are equal to the scalars α_m. Using (7.2.7), we obtain $\mathbf{k} = [1 \; 2 \; \frac{81}{20}]^T$ whose elements are the quantities k_0^c, k_1^c, and k_2^c computed in Example 7.1.2 using the Levinson recursion.

7.2.3 Order Decomposition of the Optimum Estimate

The equation $\hat{y}_{m+1} = \hat{y}_m + k_{m+1}^* w_{m+1}$, with $k_{m+1} = \beta_m^c / P_m^{\mathrm{b}} = k_m^c$, shows that the improvement in the estimate when we include one more observation x_{m+1}, that is, when we increase the order by 1, is proportional to the innovation w_{m+1} contained in x_{m+1}. The innovation is the part of x_{m+1} that cannot be linearly estimated from the already used data $\mathbf{x}_m$. The term w_{m+1} is scaled by the ratio of the partial correlation between y and the "new" observation x_{m+1} and the power of the innovation P_m^{b}.

Thus, *the computation of the $(m+1)$st-order estimate of y based on $\mathbf{x}_{m+1} = [\mathbf{x}_m^T \; x_{m+1}]$* *can be reduced to two mth-order estimation problems: the estimation of y based on $\mathbf{x}_m$ and the estimation of the new observation x_{m+1} based on $\mathbf{x}_m$.* This decomposition of linear estimation problems into smaller ones has very important applications to the development of efficient algorithms and structures for MMSE estimation.

We use the term *direct* for the implementation of the MMSE linear combiner as a sum of products, involving the optimum parameters $c_i^{(m)}$, $1 \leq i \leq m$, to emphasize the direct use of these coefficients. Because the random variables w_i used in the implementation of Figure 7.1 are orthogonal, that is, $\langle w_i, w_j \rangle = 0$ for $i \neq j$, we refer to this implementation as the *orthogonal implementation* or the *orthogonal structure*. These two structures appear in every type of linear MMSE estimation problem, and their particular form depends on the specifics of the problem and the associated second-order moments. In this sense, they play a prominent role in linear MMSE estimation in general, and in this book in particular.

We conclude our discussion with the following important observations:

1. The direct implementation combines correlated, that is, redundant information, and it is *not* order-recursive because increasing the order of the estimator *destroys* the optimality of the existing coefficients. Again, the reason is that the direct-form optimum filter coefficients do not possess the optimal nesting property.
2. The orthogonal implementation consists of a decorrelator and a linear combiner. The estimator combines the innovations of the data (nonredundant information) and is order-recursive because it does not use the optimum coefficient vector. Hence, increasing the order of the estimator preserves the optimality of the existing lower-order part. The resulting structure is modular such that each additional term improves the estimate by an amount proportional to the included innovation w_m.
3. Using the vector interpretation of random variables, the transformation $\tilde{\mathbf{x}}_m = \mathbf{F}_m \mathbf{x}_m$ is just a change of basis. The choice $\mathbf{F}_m = \mathbf{L}_m^{-1}$ converts from the *oblique* set $\{x_1, x_2, \ldots, x_m\}$ to the *orthogonal* basis $\{w_1, w_2, \ldots, w_m\}$. The advantage of working with orthogonal bases is that adding new components does not affect the optimality of previous ones.
4. The LDL^H decomposition for random vectors is the matrix equivalent of the spectral factorization theorem for discrete-time, stationary, stochastic processes. Both approaches facilitate the design and implementation of optimum FIR and IIR filters (see Sections 6.3 and 6.6).

7.2.4 Gram-Schmidt Orthogonalization

We next combine the geometric interpretation of the random variables with the Gram-Schmidt procedure used in linear algebra. The Gram-Schmidt procedure produces the innovations $\{w_1, w_2, \ldots, w_m\}$ by orthogonalizing the original set $\{x_1, x_2, \ldots, x_m\}$.

We start by choosing w_1 to be in the direction of x_1, that is,

$$w_1 = x_1$$

The next "vector" w_2 should be orthogonal to w_1. To determine w_2, we subtract from x_2 its component along w_1 [see Figure 7.2(a)], that is,

$$w_2 = x_2 - l_0^{(1)} w_1$$

where $l_0^{(1)}$ is obtained from the condition $w_2 \perp w_1$ as follows:

$$\langle w_2, w_1 \rangle = \langle x_2, w_1 \rangle - l_0^{(1)} \langle w_1, w_1 \rangle = 0$$

or

$$l_0^{(1)} = \frac{\langle x_2, w_1 \rangle}{\langle w_1, w_1 \rangle}$$

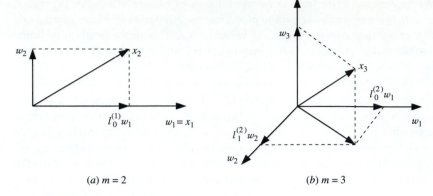

(a) m = 2   (b) m = 3

FIGURE 7.2
Illustration of the Gram-Schmidt orthogonalization process.

Similarly, to determine x_3, we subtract from x_3 its components along w_1 and w_2, that is,

$$w_3 = x_3 - l_0^{(2)} w_1 - l_1^{(2)} w_2$$

as illustrated in Figure 7.2(b). Using the conditions $w_3 \perp w_1$ and $w_3 \perp w_2$, we can easily see that

$$l_0^{(2)} = \frac{\langle x_3, w_1 \rangle}{\langle w_1, w_1 \rangle} \qquad l_1^{(2)} = \frac{\langle x_3, w_2 \rangle}{\langle w_2, w_2 \rangle}$$

This approach leads to the following *classical Gram-Schmidt algorithm*:

- Define $w_1 = x_1$.
- For $2 \le m \le M$, compute

$$w_m = x_m - l_0^{(m-1)} w_1 \cdots - l_{m-2}^{(m-1)} w_{m-1} \tag{7.2.9}$$

where
$$l_i^{(m-1)} = \frac{\langle x_{m-1}, w_i \rangle}{\langle w_i, w_i \rangle} \tag{7.2.10}$$

assuming that $\langle w_i, w_i \rangle \ne 0$.

From the derivation of the algorithm it should be clear that the sets $\{x_1, \ldots, x_m\}$ and $\{w_1, \ldots, w_m\}$ are linearly equivalent for $m = 1, 2, \ldots, M$. Using (7.2.11), we obtain

$$\mathbf{x}_m = \mathbf{L}_m \mathbf{w}_m \tag{7.2.11}$$

where
$$\mathbf{L}_m \triangleq \begin{bmatrix} 1 & 0 & \cdots & 0 \\ l_0^{(1)} & 1 & \cdots & 0 \\ \vdots & \vdots & \ddots & \vdots \\ l_0^{(m-1)} & l_1^{(m-1)} & \cdots & 1 \end{bmatrix} \tag{7.2.12}$$

is a unit lower triangular matrix. Since, by construction, the components of $\mathbf{w}_m$ are uncorrelated, its correlation matrix $\mathbf{D}_m$ is diagonal with elements $\xi_i = E\{|w_i|^2\}$. Using (7.2.11), we obtain

$$\mathbf{R}_m = E\{\mathbf{x}_m \mathbf{x}_m^H\} = \mathbf{L}_m E\{\mathbf{w}_m \mathbf{w}_m^H\} \mathbf{L}_m^H = \mathbf{L}_m \mathbf{D}_m \mathbf{L}_m^H \tag{7.2.13}$$

which is precisely the unique LDL^H decomposition of the correlation matrix $\mathbf{R}_m$. There-fore, *the Gram-Schmidt orthogonalization of the data vector $\mathbf{x}_m$ provides an alternative approach to obtain the LDL^H decomposition of its correlation matrix $\mathbf{R}_m = E\{\mathbf{x}_m \mathbf{x}_m^H\}$.*

7.3 ORDER-RECURSIVE ALGORITHMS FOR OPTIMUM FIR FILTERS

347

SECTION 7.3
Order-Recursive
Algorithms for Optimum
FIR Filters

The key difference between a linear combiner and an FIR filter is the nature of the input data vector. The input data vector for FIR filters consists of *consecutive samples* from the *same* discrete-time stochastic process, that is,

$$\mathbf{x}_m(n) = [x(n)\, x(n-1)\, \cdots\, x(n-m+1)]^T \tag{7.3.1}$$

instead of samples from m different processes $x_i(n)$. This *shift invariance* of the input data vector allows for the development of simpler, order-recursive algorithms and structures for optimum FIR filtering and prediction compared to those for general linear estimation. Furthermore, the quest for order-recursive algorithms leads to a natural, elegant, and unavoidable interconnection between optimum filtering and the BLP and FLP problems.

We start with the following upper and lower partitioning of the input data vector

$$\mathbf{x}_{m+1}(n) = \begin{bmatrix} x(n) \\ x(n-1) \\ \vdots \\ x(n-m+1) \\ x(n-m) \end{bmatrix} = \begin{bmatrix} \mathbf{x}_m(n) \\ x(n-m) \end{bmatrix} = \begin{bmatrix} x(n) \\ \mathbf{x}_m(n-1) \end{bmatrix} \tag{7.3.2}$$

which shows that $\mathbf{x}_{m+1}^{[m]}(n)$ and $\mathbf{x}_{m+1}^{\lfloor m \rfloor}(n)$ are simply shifted versions (by one sample delay) of the same vector $\mathbf{x}_m(n)$. The shift invariance of $\mathbf{x}_{m+1}(n)$ results in an analogous shift invariance for the correlation matrix $\mathbf{R}_{m+1}(n) = E\{\mathbf{x}_{m+1}(n)\mathbf{x}_{m+1}^H(n)\}$. Indeed, we can easily show that the upper-lower partitioning of the correlation matrix is

$$\mathbf{R}_{m+1}(n) = \begin{bmatrix} \mathbf{R}_m(n) & \mathbf{r}_m^b(n) \\ \mathbf{r}_m^{bH}(n) & P_x(n-m) \end{bmatrix} \tag{7.3.3}$$

and the lower-upper partitioning is

$$\mathbf{R}_{m+1}(n) = \begin{bmatrix} P_x(n) & \mathbf{r}_m^{fH}(n) \\ \mathbf{r}_m^f(n) & \mathbf{R}_m(n-1) \end{bmatrix} \tag{7.3.4}$$

where

$$\mathbf{r}_m^b(n) = E\{\mathbf{x}_m(n)x^*(n-m)t\} \tag{7.3.5}$$

$$\mathbf{r}_m^f(n) = E\{\mathbf{x}_m(n-1)x^*(n)\} \tag{7.3.6}$$

$$P_x(n) = E\{|x(n)|^2\} \tag{7.3.7}$$

We note that, in contrast to the general case (7.1.5) where the matrix $\mathbf{R}_m^f(n) = \mathbf{R}_{m+1}^{\lfloor m \rfloor}(n)$ is *unrelated* to $\mathbf{R}_m(n)$, here the matrix $\mathbf{R}_{m+1}^{\lfloor m \rfloor}(n) = \mathbf{R}_m(n-1)$. This is a by-product of the shift-invariance property of the input data vector and takes the development of order-recursive algorithms one step further. We begin our pursuit of an order-recursive algorithm with the development of a Levinson order recursion for the optimum FIR filter coefficients.

7.3.1 Order-Recursive Computation of the Optimum Filter

Suppose that at time n we have already computed the optimum FIR filter $\mathbf{c}_m(n)$ specified by

$$\mathbf{c}_m(n) = \mathbf{R}_m^{-1}(n)\mathbf{d}_m(n) \tag{7.3.8}$$

and the MMSE is

$$P_m^c(n) = P_y(n) - \mathbf{d}_m^H(n)\mathbf{c}_m(n) \tag{7.3.9}$$

where

$$\mathbf{d}_m(n) = E\{\mathbf{x}_m(n)y^*(n)\} \tag{7.3.10}$$

We wish to compute the optimum filter

$$\mathbf{c}_{m+1}(n) = \mathbf{R}_{m+1}^{-1}(n)\mathbf{c}_{m+1}(n)$$

by modifying $\mathbf{c}_m(n)$ using an order-recursive algorithm. From (7.3.3), we see that matrix $\mathbf{R}_{m+1}(n)$ has the optimum nesting property. Using the upper partitioning in (7.3.2), we obtain

$$\mathbf{d}_{m+1}(n) = E\left\{\begin{bmatrix}\mathbf{x}_m(n)\\ x(n-m)\end{bmatrix}y^*(n)\right\} = \begin{bmatrix}\mathbf{d}_m(n)\\ d_{m+1}(n)\end{bmatrix} \quad (7.3.11)$$

which shows that $\mathbf{d}_{m+1}(n)$ also has the optimum nesting property. Therefore, we can develop a Levinson order recursion using the upper left matrix inversion by partitioning lemma

$$\mathbf{R}_{m+1}^{-1}(n) = \begin{bmatrix}\mathbf{R}_m^{-1}(n) & \mathbf{0}\\ \mathbf{0}^T & 0\end{bmatrix} + \frac{1}{P_m^b(n)}\begin{bmatrix}\mathbf{b}_m(n)\\ 1\end{bmatrix}\begin{bmatrix}\mathbf{b}_m^H(n) & 1\end{bmatrix} \quad (7.3.12)$$

where

$$\mathbf{b}_m(n) = -\mathbf{R}_m^{-1}(n)\mathbf{r}_m^b(n) \quad (7.3.13)$$

is the optimum BLP, and

$$P_m^b(n) = \frac{\det \mathbf{R}_{m+1}(n)}{\det \mathbf{R}_m(n)} = P_x(n-m) + \mathbf{r}_m^{bH}(n)\mathbf{b}_m(n) \quad (7.3.14)$$

is the corresponding MMSE. Equations (7.3.12) through (7.3.14) follow easily from (7.1.22), (7.1.23), and (7.1.24). It is interesting to note that $\mathbf{b}_m(n)$ is the optimum estimator for the additional observation $x(n-m)$ used by the optimum filter $\mathbf{c}_{m+1}(n)$. Substituting (7.3.11) and (7.3.12) into (7.3.8), we obtain

$$\mathbf{c}_{m+1}(n) = \begin{bmatrix}\mathbf{c}_m(n)\\ 0\end{bmatrix} + \begin{bmatrix}\mathbf{b}_m(n)\\ 1\end{bmatrix}k_m^c(n) \quad (7.3.15)$$

where

$$k_m^c(n) \triangleq \frac{\beta_m^c(n)}{P_m^b(n)} \quad (7.3.16)$$

and

$$\beta_m^c(n) \triangleq \mathbf{b}_m^H(n)\mathbf{d}_m(n) + d_{m+1}(n) \quad (7.3.17)$$

Thus, if we know the BLP $\mathbf{b}_m(n)$, we can determine $\mathbf{c}_{m+1}(n)$ by using the Levinson recursion in (7.3.15).

Levinson recursion for the backward predictor. For the order recursion in (7.3.15) to be useful, we need an order recursion for the BLP $\mathbf{b}_m(n)$. This is possible if the linear systems

$$\mathbf{R}_m(n)\mathbf{b}_m(n) = -\mathbf{r}_m^b(n)$$
$$\mathbf{R}_{m+1}(n)\mathbf{b}_{m+1}(n) = -\mathbf{r}_{m+1}^b(n) \quad (7.3.18)$$

are nested. Since the matrices are nested [see (7.3.3)], we check whether the right-hand side vectors are nested. We can easily see that no optimum nesting is possible if we use the upper partitioning in (7.3.2). However, if we use the lower-upper partitioning, we obtain

$$\mathbf{r}_{m+1}^b(n) = E\left\{\begin{bmatrix}x(n)\\ \mathbf{x}_m(n-1)\end{bmatrix}x^*(n-m-1)\right\} \triangleq \begin{bmatrix}r_{m+1}^b(n)\\ \mathbf{r}_m^b(n-1)\end{bmatrix} \quad (7.3.19)$$

which provides a partitioning that includes the wanted vector $\mathbf{r}_m^b(n)$ delayed by one sample as a result of the shift invariance of $\mathbf{x}_m(n)$. To explore this partitioning, we use the lower-upper corner matrix inversion by partitioning lemma

$$\mathbf{R}_{m+1}^{-1}(n) = \begin{bmatrix}0 & \mathbf{0}^H\\ \mathbf{0} & \mathbf{R}_m^{-1}(n-1)\end{bmatrix} + \frac{1}{P^f(n)}\begin{bmatrix}1\\ \mathbf{a}_m(n)\end{bmatrix}\begin{bmatrix}1 & \mathbf{a}_m^H(n)\end{bmatrix} \quad (7.3.20)$$

where
$$\mathbf{a}_m(n) \triangleq -\mathbf{R}_m^{-1}(n-1)\mathbf{r}_m^f(n) \qquad (7.3.21)$$

is the optimum FLP and

$$P_m^f(n) = \frac{\det \mathbf{R}_{m+1}(n)}{\det \mathbf{R}_m(n-1)} = P_x(n) + \mathbf{r}_m^{fH}(n)\mathbf{a}_m(n) \qquad (7.3.22)$$

is the forward linear prediction MMSE. Equations (7.3.20) through (7.3.22) follow easily from (7.1.26) through (7.1.28). Substituting (7.3.20) and (7.3.19) into

$$\mathbf{b}_{m+1}(n) = -\mathbf{R}_{m+1}^{-1}(n)\mathbf{r}_{m+1}^b(n)$$

we obtain the recursion

$$\mathbf{b}_{m+1}(n) = \begin{bmatrix} 0 \\ \mathbf{b}_m(n-1) \end{bmatrix} + \begin{bmatrix} 1 \\ \mathbf{a}_m(n) \end{bmatrix} k_m^b(n) \qquad (7.3.23)$$

where
$$k_m^b(n) \triangleq -\frac{\beta_m^b(n)}{P_m^f(n)} \qquad (7.3.24)$$

and
$$\beta_m^b(n) \triangleq r_{m+1}^b(n) + \mathbf{a}_m^H(n)\mathbf{r}_m^b(n-1) \qquad (7.3.25)$$

To proceed with the development of the order-recursive algorithm, we clearly need an order recursion for the optimum FLP $\mathbf{a}_m(n)$.

Levinson recursion for the forward predictor. Following a similar procedure for the Levinson recursion of the BLP, we can derive the Levinson recursion for the FLP. If we use the upper-lower partitioning in (7.3.2), we obtain

$$\mathbf{r}_{m+1}^f(n) = E\{\mathbf{x}_{m+1}(n-1)x^*(n)\} = \begin{bmatrix} \mathbf{r}_m^f(n) \\ r_{m+1}^f(n) \end{bmatrix} \qquad (7.3.26)$$

which in conjunction with (7.3.12) and (7.3.21) leads to the following order recursion

$$\mathbf{a}_{m+1}(n) = \begin{bmatrix} \mathbf{a}_m(n) \\ 0 \end{bmatrix} + \begin{bmatrix} \mathbf{b}_m(n-1) \\ 1 \end{bmatrix} k_m^f(n) \qquad (7.3.27)$$

where
$$k_m^f(n) \triangleq -\frac{\beta_m^f(n)}{P_m^b(n-1)} \qquad (7.3.28)$$

and
$$\beta_m^f(n) \triangleq \mathbf{b}_m^H(n-1)\mathbf{r}_m^f(n) + r_{m+1}^f(n) \qquad (7.3.29)$$

Is an order-recursive algorithm feasible? For $m = 1$, we have a scalar equation $r_{11}(n)c_1^{(1)}(n) = d_1(n)$ whose solution is $c_1^{(1)}(n) = d_1(n)/r_{11}(n)$. Using the Levinson order recursions for $m = 1, 2, \ldots, M-1$, we can find $\mathbf{c}_M(n)$ if the quantities $\mathbf{b}_m(n-1)$ and $P_m^b(n-1)$, $1 \le m < M$, required by (7.3.27) and (7.3.28) are known. The lack of this information prevents the development of a complete order-recursive algorithm for the solution of the normal equations for optimum FIR filtering or prediction. The need for time updates arises because each order update requires both the upper left corner and the lower right corner partitionings

$$\mathbf{R}_{m+1}(n) = \begin{bmatrix} \mathbf{R}_m(n) & \times \\ \times & \times \end{bmatrix} = \begin{bmatrix} \times & \times \\ \times & \mathbf{R}_m(n-1) \end{bmatrix}$$

of matrix $\mathbf{R}_{m+1}$. The presence of $\mathbf{R}_m(n-1)$, which is a result of the nonstationarity of the input signal, creates the need for a time updating of $\mathbf{b}_m(n)$. This is possible only for certain types of nonstationarity that can be described by simple relations between $\mathbf{R}_m(n)$ and $\mathbf{R}_m(n-1)$. The simplest case occurs for stationary processes where $\mathbf{R}_m(n) = \mathbf{R}_m(n-1) = \mathbf{R}_m$. Another very useful case occurs for nonstationary processes generated by linear state-space models, which results in the Kalman filtering algorithm (see Section 7.8).

Partial correlation interpretation. The partial correlation between $y(n)$ and $x(n-m)$, after the influence of the intermediate samples $x(n), x(n-1), \ldots, x(n-m+1)$ has been removed, is

$$E\{e_m^b(n)e_m^*(n)\} = \mathbf{b}_m^H(n)\mathbf{d}_m(n) + d_{m+1}(n) = \beta_m^c(n) \tag{7.3.30}$$

which is obtained by working as in the derivation of (7.2.6). It can be shown, following a procedure similar to that leading to (7.2.8), that the $k_m(n)$ parameters in the Levinson recursions can be obtained from

$$\begin{aligned}
\mathbf{R}_m(n) &= \mathbf{L}_m(n)\mathbf{D}_m(n)\mathbf{L}_m^H(n) \\
\mathbf{L}_m(n)\mathbf{D}_m(n)\mathbf{k}_m^c(n) &= \mathbf{d}_m(n) \\
\mathbf{L}_m(n)\mathbf{D}_m(n)\mathbf{k}_m^f(n) &= \mathbf{r}_m^b(n) \\
\mathbf{L}_m(n-1)\mathbf{D}_m(n-1)\mathbf{k}_m^b(n) &= \mathbf{r}_m^f(n)
\end{aligned} \tag{7.3.31}$$

that is, as a by-product of the LDL^H decomposition.

Similarly, if we consider the sequence $x(n), x(n-1), \ldots, x(n-m), x(n-m-1)$, we can show that the partial correlation between $x(n)$ and $x(n-m-1)$ is given by (see Problem 7.6)

$$E\{e_m^b(n-1)e_m^{f*}(n)\} = r_{m+1}^f(n) + \mathbf{b}_m^H(n-1)\mathbf{r}_m^f(n) = \beta_m^f(n) \tag{7.3.32}$$

Because $r_{m+1}^f(n) = r_{m+1}^{b*}(n)$, we have the following simplification

$$\begin{aligned}
\beta_m^f(n) &= \mathbf{b}_m^H(n-1)\mathbf{R}_m(n-1)\mathbf{R}_m^{-1}(n-1)\mathbf{r}_m^f(n) + r_{m+1}^f(n) \\
&= \mathbf{r}_m^{bH}(n-1)\mathbf{a}_m(n) + r_{m+1}^{b*}(n) = \beta_m^{b*}(n)
\end{aligned}$$

which is known as *Burg's lemma* (Burg 1975). In order to simplify the notation, we define

$$\beta_m(n) \triangleq \beta_m^f(n) = \beta_m^{b*}(n) \tag{7.3.33}$$

Using (7.3.24), (7.3.28), and (7.3.30), we obtain

$$k_m^b(n)k_m^f(n) = \frac{|\beta_m(n)|^2}{P_m^f(n)P_m^b(n-1)} = \frac{|E\{e_m^b(n-1)e_m^{f*}(n)\}|^2}{E\{|e_m^f(n)|^2\}E\{|e_m^b(n-1)|^2\}} \tag{7.3.34}$$

which implies that

$$0 \leq k_m^f(n)k_m^b(n) \leq 1 \tag{7.3.35}$$

because the last term in (7.3.34) is the squared magnitude of the correlation coefficient of the random variables $e_m^f(n)$ and $e_m^b(n-1)$.

Order recursions for the MMSEs. Using the Levinson order recursions, we can obtain order-recursive formulas for the computation of $P_m^f(n)$, $P_m^b(n)$, and $P_m^c(n)$. Indeed, using (7.3.26), (7.3.27), and (7.3.29), we have

$$\begin{aligned}
P_{m+1}^f(n) &= P_x(n) + \mathbf{r}_{m+1}^{fH}(n)\mathbf{a}_{m+1}(n) \\
&= P_x(n) + [\mathbf{r}_m^{fH}(n)r_{m+1}^{f*}(n)]\left\{\begin{bmatrix}\mathbf{a}_m(n) \\ 0\end{bmatrix} + \begin{bmatrix}\mathbf{b}_m(n-1) \\ 1\end{bmatrix}k_m^f(n)\right\} \\
&= P_x(n) + \mathbf{r}_m^{fH}(n)\mathbf{a}_m(n) + [\mathbf{r}_m^{fH}(n)\mathbf{b}_m(n-1) + r_{m+1}^{f*}(n)]k_m^f(n)
\end{aligned}$$

or
$$P_{m+1}^f(n) = P_m^f(n) + \beta_m^*(n)k_m^f(n) = P_m^f(n) - \frac{|\beta_m(n)|^2}{P_m^b(n-1)} \tag{7.3.36}$$

If we work in a similar manner, we obtain

$$P_{m+1}^b(n) = P_m^b(n-1) + \beta_m(n)k_m^b(n) = P_m^b(n-1) - \frac{|\beta_m(n)|^2}{P_m^f(n)} \tag{7.3.37}$$

and
$$P^c_{m+1}(n) = P^c_m(n) - \beta^{c*}_m(n)k^c_m(n) = P^c_m(n) - \frac{|\beta^c_m(n)|^2}{P^b_m(n)} \qquad (7.3.38)$$

If the subtrahends in the previous recursions are nonzero, increasing the order of the filter always improves the estimates, that is, $P^c_{m+1}(n) \le P^c_m(n)$. Also, the conditions $P^f_m(n) \ne 0$ and $P^b_m(n) \ne 0$ are critical for the invertibility of $\mathbf{R}_m(n)$ and the computation of the optimum filters. The above relations are special cases of (7.1.45) and can be derived from the LDL^H decomposition (see Problem 7.7). The presence of vectors with mixed optimum nesting (upper-lower and lower-upper) in the definitions of $\beta_m(n)$ and $\beta^c_m(n)$ does not lead to similar order recursions for these quantities. However, for stationary processes we can break the dot products in (7.3.17) and (7.3.25) into scalar recursions, using an algorithm first introduced by Schür (see Section 7.6).

7.3.2 Lattice-Ladder Structure

We saw that the shift invariance of the input data vector made it possible to develop the Levinson recursions for the BLP and the FLP. We next show that these recursions can be used to simplify the triangular order-recursive estimation structure of Figure 7.1 by reducing it to a more efficient (linear instead of triangular), lattice-ladder filter structure that simultaneously provides the FLP, BLP, and FIR filtering estimates.

The computation of the estimation errors using direct-form structures is based on the following equations:

$$\begin{aligned}
e^f_m(n) &= x(n) + \mathbf{a}^H_m(n)\mathbf{x}_m(n-1) \\
e^b_m(n) &= x(n-m) + \mathbf{b}^H_m(n)\mathbf{x}_m(n) \\
e_m(n) &= y(n) - \mathbf{c}^H_m(n)\mathbf{x}_m(n)
\end{aligned} \qquad (7.3.39)$$

Using (7.3.2), (7.3.27), and (7.3.39), we obtain

$$e^f_{m+1}(n) = x(n) + \left\{ \begin{bmatrix} \mathbf{a}_m(n) \\ 0 \end{bmatrix} + \begin{bmatrix} \mathbf{b}_m(n-1) \\ 1 \end{bmatrix} k^f_m(n) \right\}^H \begin{bmatrix} \mathbf{x}_m(n-1) \\ x(n-1-m) \end{bmatrix}$$

$$= x(n) + \mathbf{a}^H_m(n)\mathbf{x}_m(n-1) + [\mathbf{b}^H_m(n-1)\mathbf{x}_m(n-1) + x(n-1-m)]k^{f*}_m(n)$$

or
$$e^f_{m+1}(n) = e^f_m(n) + k^{f*}_m(n)e^b_m(n-1) \qquad (7.3.40)$$

In a similar manner, we obtain

$$e^b_{m+1}(n) = e^b_m(n-1) + k^{b*}_m(n)e^f_m(n) \qquad (7.3.41)$$

using (7.3.2), (7.3.23), and (7.3.39). Relations (7.3.40) and (7.3.41) are executed for $m = 0, 1, \ldots, M-2$, with $e^f_0(n) = e^b_0(n) = x(n)$, and constitute a lattice filter that implements the FLP and the BLP.

Using (7.3.2), (7.3.15), and (7.3.39), we can show that the optimum filtering error can be computed by

$$e_{m+1}(n) = e_m(n) - k^{c*}_m(n)e^b_m(n) \qquad (7.3.42)$$

which is executed for $m = 0, 1, \ldots, M-1$, with $e_0(n) = y(n)$. The last equation provides the ladder part, which is coupled with the lattice predictor to implement the optimum filter. The result is the time-varying lattice-ladder structure shown in Figure 7.3. Notice that a new set of lattice-ladder coefficients has to be computed for every n, using $\mathbf{R}_m(n)$ and $\mathbf{d}_m(n)$. The parameters of the lattice-ladder structure can be obtained by LDL^H decomposition using (7.3.31). Suppose now that we know $P^f_0(n) = P^b_0(n) = P_x(n)$, $P^b_0(n-1)$, $P^c_0(n) = P_y(n)$, $\{\beta_m(n)\}^{M-1}_0$, and $\{\beta^c_m(n)\}^M_0$. Then we can determine $P^f_m(n)$, $P^b_m(n)$, and $P^c_m(n)$ for all m, using (7.3.36) through (7.3.38), and all filter coefficients, using (7.3.16), (7.3.24), and

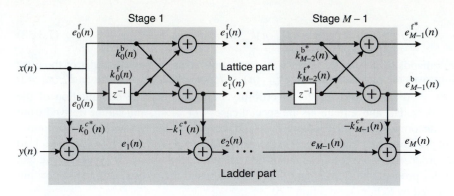

FIGURE 7.3
Lattice-ladder structure for FIR optimum filtering and prediction.

(7.3.28). However, to obtain a completely time-recursive updating algorithm, we need time updatings for $\beta_m(n)$ and $\beta_m^c(n)$. As we will see later, this is possible if $\mathbf{R}(n)$ and $\mathbf{d}(n)$ are fixed or are defined by known time-updating formulas.

We recall that the BLP error vector $e_{m+1}^b(n)$ is the innovations vector of the data $\mathbf{x}_{m+1}(n)$. Notice that as a result of the shift invariance of the input data vector, the triangular decorrelator of the general linear estimator (see Figure 7.1) is replaced by a simpler, "linear" lattice structure. For stationary processes, the lattice-ladder filter is time-invariant, and we need to compute only one set of coefficients that can be used for all signals with the same $\mathbf{R}$ and $\mathbf{d}$ (see Section 7.5).

7.3.3 Simplifications for Stationary Stochastic Processes

When $x(n)$ and $y(n)$ are jointly wide-sense stationary (WSS), the optimum estimators are time-invariant and we have the following simplifications:

- All quantities are independent of n; thus we do not need time recursions for the BLP parameters.
- $\mathbf{b}_m = \mathbf{J}\mathbf{a}_m^*$ (see Section 6.5.4), and thus we do not need the Levinson recursion for the BLP $\mathbf{b}_m$.

Both simplifications are a consequence of the Toeplitz structure of the correlation matrix $\mathbf{R}_m$. Indeed, comparing the partitionings

$$\mathbf{R}_{m+1}(n) = \begin{bmatrix} \mathbf{R}_m & \mathbf{J}\mathbf{r}_m \\ \mathbf{r}_m^H \mathbf{J} & r(0) \end{bmatrix} = \begin{bmatrix} r(0) & \mathbf{r}_m^T \\ \mathbf{r}_m^* & \mathbf{R}_m \end{bmatrix} \tag{7.3.43}$$

where
$$\mathbf{r}_m \triangleq [r(1) \, r(2) \, \cdots \, r(m)]^T \tag{7.3.44}$$

with (7.3.3) and (7.3.4), we have

$$\mathbf{R}_m(n) = \mathbf{R}_m(n-1) = \mathbf{R}_m$$
$$\mathbf{r}_m^f(n) = \mathbf{r}_m^* \tag{7.3.45}$$
$$\mathbf{r}_m^b(n) = \mathbf{J}\mathbf{r}_m$$

which can be used to simplify the order recursions derived for nonstationary processes. Indeed, we can easily show that

$$\mathbf{a}_{m+1} = \begin{bmatrix} \mathbf{a}_m \\ 0 \end{bmatrix} + \begin{bmatrix} \mathbf{b}_m \\ 1 \end{bmatrix} k_m \tag{7.3.46}$$

where
$$\mathbf{b}_m = \mathbf{J}\mathbf{a}_m^* \tag{7.3.47}$$

$$k_m \triangleq k_m^{\text{f}} = k_m^{\text{b}*} = -\frac{\beta_m}{P_m} \tag{7.3.48}$$

$$\beta_m \triangleq \beta_m^{\text{f}} = \beta_m^{\text{b}*} = \mathbf{b}_m^H \mathbf{r}_m^* + r^*(m+1) \tag{7.3.49}$$

$$P_m \triangleq P_m^{\text{b}} = P_m^{\text{f}} = P_{m-1} + \beta_{m-1}^* k_{m-1} = P_{m-1} + \beta_{m-1} k_{m-1}^* \tag{7.3.50}$$

This recursion provides a complete order-recursive algorithm for the computation of the FLP $\mathbf{a}_m$ for $1 \le m \le M$ from the autocorrelation sequence $r(l)$ for $0 \le l \le M$.

The optimum filters $\mathbf{c}_m$ for $1 \le m \le M$ can be obtained from the quantities $\mathbf{a}_m$ and P_m for $1 \le m \le M-1$ and $\mathbf{d}_M$, using the following Levinson recursion

$$\mathbf{c}_{m+1} = \begin{bmatrix} \mathbf{c}_m \\ 0 \end{bmatrix} + \begin{bmatrix} \mathbf{J}\mathbf{a}_m \\ 1 \end{bmatrix} k_m^c \tag{7.3.51}$$

where
$$k_m^c \triangleq \frac{\beta_m^c}{P_m} \tag{7.3.52}$$

and
$$\beta_m^c = \mathbf{b}_m^H \mathbf{d}_m + d_{m+1} \tag{7.3.53}$$

The MMSE P_m^c is then given by

$$P_m^c = P_{m-1}^c - \beta_m^c k_m^c \tag{7.3.54}$$

and although it is not required by the algorithm, P_m^c is useful for selecting the order of the optimum filter. Both algorithms are discussed in greater detail in Section 7.4.

7.3.4 Algorithms Based on the UDU^H Decomposition

Hermitian positive definite matrices can also be factorized as

$$\mathbf{R} = \mathbf{U}\bar{\mathbf{D}}\mathbf{U}^H \tag{7.3.55}$$

where $\mathbf{U}$ is a unit upper triangular matrix and $\bar{\mathbf{D}}$ is a diagonal matrix with positive elements $\bar{\xi}_i$, using the function [U,D]=udut(R) (see Problem 7.8). Using the decomposition (7.3.55), we can obtain the solution of the normal equations by solving the triangular systems, first for $\bar{\mathbf{k}}$

$$(\mathbf{U}\bar{\mathbf{D}})\bar{\mathbf{k}} \triangleq \mathbf{d} \tag{7.3.56}$$

and then for $\mathbf{c}$
$$\mathbf{U}^H\mathbf{c} = \bar{\mathbf{k}} \tag{7.3.57}$$

by backward and forward substitution, respectively. The MMSE estimate can be computed by

$$\hat{y} = \mathbf{c}^H \mathbf{x} = \bar{\mathbf{k}}^H \bar{\mathbf{w}} \tag{7.3.58}$$

where
$$\bar{\mathbf{w}} \triangleq \mathbf{U}^{-1}\mathbf{x} \tag{7.3.59}$$

is an *innovations* vector for the data vector $\mathbf{x}$. It can be shown that the rows of $\mathbf{A} \triangleq \mathbf{U}^{-1}$ are the linear MMSE estimator of x_m based on $[x_{m+1} \, x_{m+2} \, \cdots \, x_M]^T$. Furthermore, the UDU^H factorization (7.3.55) can be obtained by the Gram-Schmidt algorithm, starting with x_M and proceeding "backward" to x_1 (see Problem 7.9). The various triangular decompositions of the correlation matrix $\mathbf{R}$ are summarized in Table 7.1.

If we define the reversed vectors $\tilde{\mathbf{x}} \triangleq \mathbf{J}\mathbf{x}$ and $\tilde{\mathbf{w}} \triangleq \mathbf{J}\mathbf{w}$, we obtain

$$\tilde{\mathbf{x}} = \mathbf{J}\mathbf{x} = \mathbf{J}\mathbf{L}\mathbf{J}\mathbf{J}\mathbf{w} = \mathbf{J}\mathbf{L}\mathbf{J}\tilde{\mathbf{w}} \triangleq \tilde{\mathbf{U}}\tilde{\mathbf{w}} \tag{7.3.60}$$

TABLE 7.1

Summary of the triangular decompositions of the correlation matrix.

Decomposition / Matrix			
$\mathbf{R}$	$\mathbf{LDL}^H$	$\mathbf{U\bar{D}U}^H$	$\mathbf{A} = \mathbf{U}^{-1}$
$\mathbf{R}^{-1}$	$\mathbf{A}^H \bar{\mathbf{D}} \mathbf{A}$	$\mathbf{B}^H \mathbf{D}^{-1} \mathbf{B}$	$\mathbf{B} = \mathbf{L}^{-1}$

because $\mathbf{J}^2 = \mathbf{I}$ and $\tilde{\mathbf{U}} = \mathbf{JLJ}$ is upper triangular. The correlation matrix of $\tilde{\mathbf{x}}$ is

$$\tilde{\mathbf{R}} = E\{\tilde{\mathbf{x}}\tilde{\mathbf{x}}^H\} = \tilde{\mathbf{U}}\tilde{\mathbf{D}}\tilde{\mathbf{U}}^H \tag{7.3.61}$$

where $\tilde{\mathbf{D}} \triangleq E\{\tilde{\mathbf{w}}\tilde{\mathbf{w}}^H\}$ is the diagonal correlation matrix of $\tilde{\mathbf{w}}$. Equation (7.3.61) provides the UDU^H decomposition of $\tilde{\mathbf{R}}$.

A natural question arising at this point is whether we can develop order-recursive algorithms and structures for optimum filtering using the UDU^H instead of the LDL^H decomposition of the correlation matrix. The UDU^H decomposition is coupled to a partitioning of $\mathbf{R}_{m+1}(n)$ starting at the lower right corner and moving to the upper left corner that provides the following sequence of submatrices

$$\mathbf{R}_1(n - m) \rightarrow \mathbf{R}_2(n - m + 1) \rightarrow \cdots \rightarrow \mathbf{R}_m(n) \tag{7.3.62}$$

which, in turn, are related to the FLPs

$$\mathbf{a}_1(n - m) \rightarrow \mathbf{a}_2(n - m + 1) \rightarrow \cdots \rightarrow \mathbf{a}_m(n) \tag{7.3.63}$$

and the FLP errors

$$e_1^{\mathrm{f}}(n - m + 1) \rightarrow e_2^{\mathrm{f}}(n - m + 2) \rightarrow \cdots \rightarrow e_m^{\mathrm{f}}(n) \tag{7.3.64}$$

If we define the FLP error vector

$$\mathbf{e}_{m+1}^{\mathrm{f}}(n) = [e_m^{\mathrm{f}}(n)\ e_{m-1}^{\mathrm{f}}(n - 1)\ \cdots\ e_0^{\mathrm{f}}(n - m)]^T \tag{7.3.65}$$

we see that

$$\mathbf{e}_{m+1}^{\mathrm{f}}(n) = \mathbf{A}_{m+1}(n)\mathbf{x}_{m+1}(n) \tag{7.3.66}$$

where

$$\mathbf{A}_{m+1}(n) \triangleq \begin{bmatrix} 1 & a_1^{(m)}(n) & a_2^{(m)}(n) & \cdots & a_m^{(m)}(n) \\ 0 & 1 & a_1^{(m-1)}(n - 1) & \cdots & a_{m-1}^{(m-1)}(n - 1) \\ \vdots & \vdots & \vdots & \ddots & \vdots \\ 0 & 0 & 0 & 1 & a_1^{(1)}(n - m + 1) \\ 0 & 0 & 0 & 0 & 1 \end{bmatrix} \tag{7.3.67}$$

The elements of the vector $\mathbf{e}_{m+1}^{\mathrm{f}}(n)$ are uncorrelated, and the LDL^H decomposition of the inverse correlation matrix (see Problem 7.10) is given by

$$\mathbf{R}_{m+1}^{-1}(n) = \mathbf{A}_{m+1}^H(n)\bar{\mathbf{D}}_{m+1}^{-1}(n)\mathbf{A}_{m+1}(n) \tag{7.3.68}$$

where $\bar{\mathbf{D}}_{m+1}(n)$ is the correlation matrix of $\mathbf{e}_{m+1}^{\mathrm{f}}(n)$. Using $\mathbf{e}_{m+1}^{\mathrm{f}}(n)$ as an orthogonal basis instead of $\mathbf{e}_{m+1}^{\mathrm{b}}(n)$ results in a complicated lattice structure because of the additional delay elements required for the forward prediction errors. Thus, the LDL^H decomposition is the method of choice in practical applications for linear MMSE estimation.

Since the correlation matrix of a stationary, stochastic process is Toeplitz, we can explore its special structure to develop efficient, order-recursive algorithms for the linear system solution, matrix triangularization, and matrix inversion. Although we develop such algorithms in the context of optimum FIR filtering and prediction, the results apply to other applications involving Toeplitz matrices (Golub and van Loan 1996).

Suppose that we know the optimum filter $\mathbf{c}_m$ is given by

$$\mathbf{c}_m = \mathbf{R}_m^{-1}\mathbf{d}_m \tag{7.4.1}$$

and we wish to use it to compute the optimum filter $\mathbf{c}_{m+1}$

$$\mathbf{c}_{m+1} = \mathbf{R}_{m+1}^{-1}\mathbf{d}_{m+1} \tag{7.4.2}$$

We first notice that the matrix $\mathbf{R}_{m+1}$ and the vector $\mathbf{d}_{m+1}$ can be partitioned as follows

$$\mathbf{R}_{m+1} = \begin{bmatrix} r(0) & \cdots & r(m-1) & r(m) \\ \vdots & \ddots & \vdots & \vdots \\ r^*(m-1) & \cdots & r(0) & r(1) \\ \hline r^*(m) & \cdots & r^*(1) & r(0) \end{bmatrix} = \begin{bmatrix} \mathbf{R}_m & \mathbf{Jr}_m \\ \mathbf{r}_m^H\mathbf{J} & r(0) \end{bmatrix} \tag{7.4.3}$$

$$\mathbf{d}_{m+1} = \begin{bmatrix} \mathbf{d}_m \\ d_{m+1} \end{bmatrix} \tag{7.4.4}$$

which shows that both quantities have the optimum nesting property, that is, $\mathbf{R}_{m+1}^{[m]} = \mathbf{R}_m$ and $\mathbf{d}_{m+1}^{[m]} = \mathbf{d}_m$.

Using the matrix inversion by partitioning lemma (7.1.24), we obtain

$$\mathbf{R}_{m+1}^{-1} = \begin{bmatrix} \mathbf{R}_m^{-1} & \mathbf{0} \\ \mathbf{0}^H & 0 \end{bmatrix} + \frac{1}{P_m^b}\begin{bmatrix} \mathbf{b}_m \\ 1 \end{bmatrix}\begin{bmatrix} \mathbf{b}_m^H & 1 \end{bmatrix} \tag{7.4.5}$$

where

$$\mathbf{b}_m = -\mathbf{R}_m^{-1}\mathbf{Jr}_m \tag{7.4.6}$$

and

$$P_m^b = r(0) + \mathbf{r}_m^H\mathbf{Jb}_m \tag{7.4.7}$$

Substitution of (7.4.4) and (7.4.5) into (7.4.2) gives

$$\mathbf{c}_{m+1} = \begin{bmatrix} \mathbf{c}_m \\ 0 \end{bmatrix} + \begin{bmatrix} \mathbf{b}_m \\ 1 \end{bmatrix}k_m^c \tag{7.4.8}$$

where

$$k_m^c \triangleq \frac{\beta_m^c}{P_m^b} \tag{7.4.9}$$

and

$$\beta_m^c \triangleq \mathbf{b}_m^H\mathbf{d}_m + d_{m+1} = -\mathbf{c}_m^H\mathbf{Jr}_m + d_{m+1} \tag{7.4.10}$$

Equations (7.4.8) through (7.4.10) constitute a Levinson recursion for the optimum filter and have been obtained without making use of the Toeplitz structure of $\mathbf{R}_{m+1}$.

The development of a complete order-recursive algorithm is made possible by exploiting the Toeplitz structure. Indeed, when the correlation matrix $\mathbf{R}_m$ is Toeplitz, we have

$$\mathbf{b}_m = \mathbf{Ja}_m^* \tag{7.4.11}$$

and

$$P_m \triangleq P_m^b = P_m^f \tag{7.4.12}$$

as we recall from Section 6.5. Since we can determine $\mathbf{b}_m$ from $\mathbf{a}_m$, we need to perform only *one* Levinson recursion, either for $\mathbf{b}_m$ or for $\mathbf{a}_m$.

To avoid the use of the lower right corner partitioning, we develop an order recursion for the FLP $\mathbf{a}_m$. Indeed, to compute $\mathbf{a}_{m+1}$ from $\mathbf{a}_m$, recall that

$$\mathbf{a}_{m+1} = -\mathbf{R}_{m+1}^{-1}\mathbf{r}_{m+1}^* \tag{7.4.13}$$

which, when combined with (7.4.5) and

$$\mathbf{r}_{m+1} = \begin{bmatrix} \mathbf{r}_m \\ r(m+1) \end{bmatrix} \tag{7.4.14}$$

leads to the Levinson recursion

$$\mathbf{a}_{m+1} = \begin{bmatrix} \mathbf{a}_m \\ 0 \end{bmatrix} + \begin{bmatrix} \mathbf{b}_m \\ 1 \end{bmatrix} k_m \tag{7.4.15}$$

where

$$k_m \triangleq -\frac{\beta_m}{P_m} \tag{7.4.16}$$

$$\beta_m \triangleq \mathbf{b}_m^H \mathbf{r}_m^* + r^*(m+1) = \mathbf{a}_m^T \mathbf{J} \mathbf{r}_m^* + r^*(m+1) \tag{7.4.17}$$

and

$$P_m = r(0) + \mathbf{r}_m^H \mathbf{a}_m^* = r(0) + \mathbf{a}_m^T \mathbf{r}_m \tag{7.4.18}$$

Also, using (7.1.46) and (7.2.6), we can show that

$$\det \mathbf{R}_m = \prod_{m=0}^{M-1} P_m \quad \text{with } P_0 = r(0) \tag{7.4.19}$$

which emphasizes the importance of P_m for the invertibility of the autocorrelation matrix. The MMSE P_m for either the forward or the backward predictor of order m can be computed recursively as follows:

$$P_{m+1} = r(0) + [\mathbf{r}_m^H \ r^*(m+1)] \left\{ \begin{bmatrix} \mathbf{a}_m \\ 0 \end{bmatrix} + \begin{bmatrix} \mathbf{b}_m \\ 1 \end{bmatrix} k_m \right\}^* \tag{7.4.20}$$

$$= r(0) + \mathbf{r}_m^H \mathbf{a}_m^* + [\mathbf{r}_m^H \mathbf{b}_m^* + r^*(m+1)]k_m^*$$

or

$$P_{m+1} = P_m + \beta_m k_m^* = P_m + \beta_m^* k_m \tag{7.4.21}$$

The following recursive formula for the computation of the MMSE

$$P_{m+1}^c = P_m^c - \beta_m^c k_m^{c*} = P_m^c - \beta_m^{c*} k_m^c \tag{7.4.22}$$

can be found by using (7.4.8).

Therefore, the algorithm of Levinson consists of two parts: a set of recursions that compute the optimum FLP or BLP and a set of recursions that use this information to compute the optimum filter. The part that computes the linear predictors is known as the Levinson-Durbin algorithm and was pointed out by Durbin (1960). From a linear system solution point of view, the algorithm of Levinson solves a Hermitian Toeplitz system with *arbitrary* right-hand side vector $\mathbf{d}$; the Levinson-Durbin algorithm deals with the special case $\mathbf{d} = \mathbf{r}^*$ or $\mathbf{Jr}$. Additional interpretations are discussed in Section 7.7.

Algorithm of Levinson-Durbin

The algorithm of Levinson-Durbin, which takes as input the autocorrelation sequence $r(0), r(1), \ldots, r(M)$ and computes the quantities $\mathbf{a}_m$, P_m, and k_{m-1} for $m = 1, 2, \ldots, M$, is illustrated in the following examples.

EXAMPLE 7.4.1. Determine the FLP $\mathbf{a}_2 = [a_1^{(2)} \ a_2^{(2)}]^T$ and the MMSE P_2 from the autocorrelation values $r(0)$, $r(1)$, and $r(2)$.

Solution. To initialize the algorithm, we determine the first-order predictor by solving the normal equations $r(0)a_1^{(1)} = -r^*(1)$. Indeed, we have

$$a_1^{(1)} = -\frac{r^*(1)}{r(0)} = k_0 = -\frac{\beta_0}{P_0}$$

which implies that $\quad \beta_0 = r^*(1) \quad P_0 = r(0)$

To update to order 2, we need k_1 and hence β_1 and P_1, which can be obtained by

$$\beta_1 = a_1^{(1)}r^*(1) + r^*(2) = \frac{r(0)r^*(2) - [r^*(1)]^2}{r(0)}$$

$$P_1 = P_0 + \beta_0 k_0^* = \frac{r^2(0) - |r(1)|^2}{r(0)}$$

as $\quad k_1 = \frac{[r^*(1)]^2 - r(0)r^*(2)}{r^2(0) - |r(1)|^2}$

Therefore, using Levinson's recursion, we obtain

$$a_1^{(2)} = a_1^{(1)} + a_1^{(1)*}k_1 = \frac{r(1)r^*(2) - r(0)r^*(1)}{r^2(0) - |r(1)|^2}$$

and $\quad a_2^{(2)} = k_1$

which agree with the results obtained in Example 6.5.1. The resulting MMSE can be found by using $P_2 = P_1 + \beta_1 k_1^*$.

EXAMPLE 7.4.2. Use the Levinson-Durbin algorithm to compute the third-order forward predictor for a signal $x(n)$ with autocorrelation sequence $r(0) = 3$, $r(1) = 2$, $r(2) = 1$, and $r(3) = \frac{1}{2}$.

Solution. To initialize the algorithm, we notice that the first-order predictor is given by $r(0)a_1^{(1)} = -r(1)$ and that for $m = 0$, (7.4.15) gives $a_1^{(1)} = k_0$. Hence, we have

$$a_1^{(1)} = -\frac{r(1)}{r(0)} = -\frac{2}{3} = k_1 = \frac{\beta_0}{P_0}$$

which implies $\quad P_0 = r(0) = 3 \quad \beta_0 = r(1) = 2$

To compute $\mathbf{a}_2$ by (7.4.15), we need $a_1^{(1)}$, $b_1^{(1)} = a_1^{(1)}$, and $k_1 = -\beta_1/P_1$. From (7.4.21), we have

$$P_1 = P_0 + \beta_0 k_0 = 3 + 2(-\tfrac{2}{3}) = \tfrac{5}{3}$$

and from (7.4.17)

$$\beta_1 = \mathbf{r}_1^T \mathbf{J}\mathbf{a}_1 + r(2) = 2(-\tfrac{2}{3}) + 1 = -\tfrac{1}{3}$$

Hence, $\quad k_1 = -\frac{\beta_1}{P_1} = -\frac{-\tfrac{1}{3}}{\tfrac{5}{3}} = \frac{1}{5}$

and $\quad \mathbf{a}_2 = \begin{bmatrix} -\tfrac{2}{3} \\ 0 \end{bmatrix} + \frac{1}{5}\begin{bmatrix} -\tfrac{2}{3} \\ 1 \end{bmatrix} = \begin{bmatrix} -\tfrac{4}{5} \\ \tfrac{1}{5} \end{bmatrix}$

Continuing in the same manner, we obtain

$$P_2 = P_1 + \beta_1 k_1 = \tfrac{5}{3} + (-\tfrac{1}{3})(\tfrac{1}{5}) = \tfrac{8}{5}$$

$$\beta_2 = \mathbf{r}_2^T \mathbf{J}\mathbf{a}_2 + r(3) = \begin{bmatrix} 2 & 1 \end{bmatrix}\begin{bmatrix} \tfrac{1}{5} \\ -\tfrac{4}{5} \end{bmatrix} + \frac{1}{2} = \frac{1}{10}$$

$$k_2 = -\frac{\beta_2}{P_2} = -\frac{\frac{1}{10}}{\frac{8}{5}} = -\frac{1}{16}$$

$$\mathbf{a}_3 = \begin{bmatrix} \mathbf{a}_2 \\ 0 \end{bmatrix} + \begin{bmatrix} \mathbf{Ja}_2 \\ 1 \end{bmatrix} k_3 = \begin{bmatrix} -\frac{4}{5} \\ \frac{1}{5} \\ 0 \end{bmatrix} - \begin{bmatrix} \frac{1}{5} \\ -\frac{4}{5} \\ 1 \end{bmatrix} \frac{1}{16} = \begin{bmatrix} -\frac{13}{16} \\ \frac{1}{4} \\ -\frac{1}{16} \end{bmatrix}$$

$$P_3 = P_2 + \beta_2 k_2 = \frac{8}{5} + \frac{1}{10}\left(-\frac{1}{16}\right) = \frac{51}{32}$$

The algorithm of Levinson-Durbin, summarized in Table 7.2, requires M^2 operations and is implemented by the function [a,k,Po]=durbin(r,M).

TABLE 7.2
Summary of the Levinson-Durbin algorithm.

1. **Input:** $r(0), r(1), r(2), \ldots, r(M)$

2. **Initialization**
   (a) $P_0 = r(0), \beta_0 = r^*(1)$
   (b) $k_0 = -r^*(1)/r(0), a_1^{(1)} = k_0$

3. **For** $m = 1, 2, \ldots, M-1$
   (a) $P_m = P_{m-1} + \beta_{m-1}k_{m-1}^*$
   (b) $\mathbf{r}_m = [r(1)\, r(2)\, \cdots\, r(m)]^T$
   (c) $\beta_m = \mathbf{a}_m^T \mathbf{Jr}_m^* + r^*(m+1)$
   (d) $k_m = -\dfrac{\beta_m}{P_m}$
   (e) $\mathbf{a}_{m+1} = \begin{bmatrix} \mathbf{a}_m \\ 0 \end{bmatrix} + \begin{bmatrix} \mathbf{Ja}_m^* \\ 1 \end{bmatrix} k_m$

4. $P_M = P_{M-1} + \beta_M k_M^*$

5. **Output:** $\mathbf{a}_M, \{k_m\}_0^{M-1}, \{P_m\}_1^M$

Algorithm of Levinson

The next example illustrates the algorithm of Levinson that can be used to solve a system of linear equations with a Hermitian Toeplitz matrix and arbitrary right-hand side vector.

EXAMPLE 7.4.3. Consider an optimum filter with input $x(n)$ and desired response $y(n)$. The autocorrelation of the input signal is $r(0) = 3, r(1) = 2$, and $r(2) = 1$. The cross-correlation between the desired response and input is $d_1 = 1, d_2 = 2$, and $d_3 = \frac{5}{2}$; and the power of $y(n)$ is $P_y = 3$. Design a third-order optimum FIR filter, using the algorithm of Levinson.

Solution. We start initializing the algorithm by noticing that for $m = 0$ we have $r(0)a_1^{(1)} = -r(1)$, which gives

$$a_1^{(1)} = k_0 = -\frac{r(1)}{r(0)} = -\frac{2}{3}$$

$$P_0 = r(0) = 3 \qquad \beta_0 = r(1) = 2$$

and

$$P_1 = P_0 + \beta_0 k_0 = 3 + 2(-\tfrac{2}{3}) = \tfrac{5}{3}$$

Next, we compute the Levinson recursion for the first-order optimum filter

$$P_0^c = 5 \qquad \beta_0^c = d_1 = 1$$

$$k_0^c = c_1^{(1)} = \frac{d_1}{r(0)} = \frac{1}{3}$$

$$P_1^c = P_0^c - \beta_0^c k_0^c = 3 - 1(\tfrac{1}{3}) = \tfrac{8}{3}$$

Then we carry the Levinson recursion for $m = 1$ to obtain

$$\beta_1 = \mathbf{r}_1^T \mathbf{J}\mathbf{a}_1 + r(2) = 2(-\tfrac{2}{3}) + 1 = -\tfrac{1}{3}$$

$$k_1 = -\frac{\beta_1}{P_1} = -\frac{-\tfrac{1}{3}}{\tfrac{5}{3}} = \frac{1}{5}$$

$$\mathbf{a}_2 = \begin{bmatrix} -\tfrac{2}{3} \\ 0 \end{bmatrix} + \frac{1}{5}\begin{bmatrix} -\tfrac{2}{3} \\ 1 \end{bmatrix} = \begin{bmatrix} -\tfrac{4}{5} \\ \tfrac{1}{5} \end{bmatrix}$$

$$P_2 = P_1 + \beta_1 k_1 = \tfrac{5}{3} + (-\tfrac{1}{3})(\tfrac{1}{5}) = \tfrac{8}{5}$$

for the optimum predictor, and

$$\beta_1^c = \mathbf{a}_1^T \mathbf{J}\mathbf{d}_1 + d_2 = -\tfrac{2}{3}(1) + 2 = \tfrac{4}{3}$$

$$k_1^c = \frac{\beta_1^c}{P_1} = \frac{\tfrac{4}{3}}{\tfrac{5}{3}} = \frac{4}{5}$$

$$\mathbf{c}_2 = \begin{bmatrix} \tfrac{1}{3} \\ 0 \end{bmatrix} + \frac{4}{5}\begin{bmatrix} -\tfrac{2}{3} \\ 1 \end{bmatrix} = \begin{bmatrix} -\tfrac{1}{5} \\ \tfrac{4}{5} \end{bmatrix}$$

$$P_2^c = P_1^c - \beta_1^c k_1^c = \tfrac{8}{3} - \tfrac{4}{3}(\tfrac{4}{5}) = \tfrac{8}{5}$$

for the optimum filter. The last recursion ($m = 2$) is carried out only for the optimum filter and gives

$$\beta_2^c = \mathbf{a}_2^T \mathbf{J}\mathbf{d}_2 + d_3 = \begin{bmatrix} \tfrac{1}{5} & -\tfrac{4}{5} \end{bmatrix}\begin{bmatrix} 1 \\ 2 \end{bmatrix} + \frac{5}{2} = \frac{11}{10}$$

$$k_2^c = \frac{\beta_2^c}{P_2} = \frac{\tfrac{11}{10}}{\tfrac{8}{5}} = \frac{11}{16}$$

$$\mathbf{c}_3 = \begin{bmatrix} \mathbf{c}_2 \\ 0 \end{bmatrix} + \begin{bmatrix} \mathbf{J}\mathbf{a}_2 \\ 1 \end{bmatrix} k_2^c = \begin{bmatrix} -\tfrac{1}{5} \\ \tfrac{4}{5} \\ 0 \end{bmatrix} + \frac{11}{16}\begin{bmatrix} \tfrac{1}{5} \\ -\tfrac{4}{5} \\ 1 \end{bmatrix} = \begin{bmatrix} -\tfrac{1}{16} \\ \tfrac{1}{4} \\ \tfrac{11}{16} \end{bmatrix}$$

$$P_3^c = P_2^c - \beta_2^c k_2^c = \tfrac{8}{5} - \tfrac{11}{10}(\tfrac{11}{16}) = \tfrac{27}{32}$$

The algorithm of Levinson, summarized in Table 7.3, is implemented by the MATLAB function [c,k,kc,Pc]=levins(R,d,Py,M) and requires $2M^2$ operations because it involves two dot products and two scalar-vector multiplications. A parallel processing implementation of the algorithm is not possible because the dot products involve additions that cannot be executed simultaneously. Notice that adding $M = 2^q$ numbers using $M/2$ adders requires $q = \log_2 M$ steps. This bottleneck can be avoided by using the algorithm of Schür (see Section 7.6).

Minimum phase and autocorrelation extension

Using (7.4.16), we can also express the recursion (7.4.21) as

$$P_{m+1} = P_m(1 - |k_m|^2) = P_m - \frac{|\beta_m|^2}{P_m} \qquad (7.4.23)$$

TABLE 7.3
Summary of the algorithm of Levinson.

1. **Input:** $\{r(l)\}_0^M, \{d_m\}_1^M, P_y$

2. **Initialization**
   (a) $P_0 = r(0), \beta_0 = r^*(1), P_0^c = P_y$
   (b) $k_0 = -\beta_0/P_0, a_1^{(1)} = k_0$
   (c) $\beta_0^c = d_1$
   (d) $k_0^c = -\beta_0^c/P_0, c_1^{(1)} = k_0^c$
   (e) $P_1^c = P_0^c + \beta_0^c k_0^{c*}$

3. **For** $m = 1, 2, \ldots, M - 1$
   (a) $\mathbf{r}_m = [r(1) \; r(2) \; \cdots \; r(m)]^T$
   (b) $\beta_m = \mathbf{a}_m^T \mathbf{J} \mathbf{r}_m^* + r^*(m+1)$
   (c) $P_m = P_{m-1} + \beta_{m-1} k_{m-1}^*$
   (d) $k_m = -\dfrac{\beta_m}{P_m}$
   (e) $\mathbf{a}_{m+1} = \begin{bmatrix} \mathbf{a}_m \\ 0 \end{bmatrix} + \begin{bmatrix} \mathbf{J}\mathbf{a}_m^* \\ 1 \end{bmatrix} k_m$
   (f) $\beta_m^c = -\mathbf{c}_m^H \mathbf{J} \mathbf{r}_m + d_{m+1}$
   (g) $k_m^c = \dfrac{\beta_m^c}{P_m}$
   (h) $\mathbf{c}_{m+1} = \begin{bmatrix} \mathbf{c}_m \\ 0 \end{bmatrix} + \begin{bmatrix} \mathbf{J}\mathbf{a}_m^* \\ 1 \end{bmatrix} k_m^c$
   (i) $P_{m+1}^c = P_m^c + \beta_m^c k_m^{c*}$

4. **Output:** $\mathbf{a}_M, \mathbf{c}_M, \{k_m, k_m^c\}_0^{M-1}, \{P_m, P_m^c\}_0^M$

which, since $P_m \geq 0$, implies that

$$P_{m+1} \leq P_m \tag{7.4.24}$$

and since the matrix $\mathbf{R}_m$ is positive definite, then $P_m > 0$ and (7.4.23) implies that

$$|k_m| \leq 1 \tag{7.4.25}$$

for all $1 \leq m < M$. If

$$P_0 > \cdots > P_{M-1} > P_M = 0 \tag{7.4.26}$$

then the process $x(n)$ is predictable and (7.4.23) implies that

$$k_M = \pm 1 \quad \text{and} \quad |k_m| < 1 \quad 1 \leq k < M \tag{7.4.27}$$

(see Section 6.6.4). Also if

$$P_{M-1} > P_M = \cdots = P_\infty = P > 0 \tag{7.4.28}$$

from (7.4.23) we have

$$k_m = 0 \quad \text{for } m > M \tag{7.4.29}$$

which implies that the process $x(n)$ is AR(M) and $e_M^f(n) \sim \text{WN}(0, P_M)$ (see Section 4.2.3). Finally, we note that since the sequence $P_0, P_1, P_2, \ldots$ is nonincreasing, its limit as $m \to \infty$ exists and is nonnegative. A regular process must satisfy $|k_m| < 1$ for all m, because $|k_m| = 1$ implies that $P_m = 0$, which contradicts the regularity assumption.

For $m = 0$, (7.4.19) gives $P_0 = r(0)$. Carrying out (7.4.23) from $m = 0$ to $m = M$, we obtain

$$P_M = r(0) \prod_{m=1}^M (1 - |k_{m-1}|^2) \tag{7.4.30}$$

which converges, as $M \to \infty$, if $|k_m| < 1$.

7.5 LATTICE STRUCTURES FOR OPTIMUM FIR FILTERS AND PREDICTORS

To compute the forward prediction error of an FLP of order m, we use the formula

$$e_m^f(n) = x(n) + \mathbf{a}_m^H \mathbf{x}_m(n-1) = x(n) + \sum_{k=1}^{m} a_k^{(m)*} x(n-k) \qquad (7.5.1)$$

Similarly, for the BLP we have

$$e_m^b(n) = x(n-m) + \mathbf{b}_m^H \mathbf{x}_m(n) = x(n-m) + \sum_{k=0}^{m-1} b_k^{(m)*} x(n+1-k) \qquad (7.5.2)$$

Both filters can be implemented using the direct-form filter structure shown in Figure 7.4. Since $\mathbf{a}_m$ and $\mathbf{b}_m$ do not have the optimum nesting property, we cannot obtain order-recursive direct-form structures for the computation of the prediction errors. However, next we show that we can derive an order-recursive lattice-ladder structure for the implementation of optimum predictors and filters using the algorithm of Levinson.

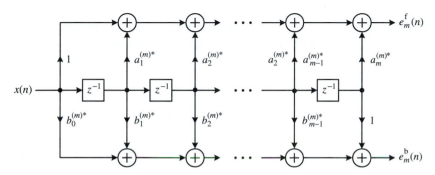

FIGURE 7.4
Direct-form structure for the computation of the mth-order forward and backward prediction errors.

7.5.1 Lattice-Ladder Structures

We note that the data vector for the $(m + 1)$st-order predictor can be partitioned in the following ways:

$$
\begin{aligned}
\mathbf{x}_{m+1}(n) &= [x(n)\ x(n-1)\ \cdots\ x(n-m+1)\ x(n-m)]^T \\
&= [\mathbf{x}_m^T(n)\ x(n-m)]^T \qquad (7.5.3) \\
&= [x(n)\ \mathbf{x}_m^T(n-1)]^T \qquad (7.5.4)
\end{aligned}
$$

Using (7.5.1), (7.5.3), (7.4.15), and (7.5.2), we obtain

$$e_{m+1}^f(n) = x(n) + \left\{ \begin{bmatrix} \mathbf{a}_m \\ 0 \end{bmatrix} + \begin{bmatrix} \mathbf{b}_m \\ 1 \end{bmatrix} k_m \right\}^H \begin{bmatrix} \mathbf{x}_m(n-1) \\ x(n-m-1) \end{bmatrix}$$

$$= x(n) + \mathbf{a}_m^H \mathbf{x}_m(n-1) + k_m^* [\mathbf{b}_m^H \mathbf{x}_m(n-1) + x(n-1-m)]$$

or

$$e_{m+1}^f(n) = e_m^f(n) + k_m^* e_m^b(n-1) \qquad (7.5.5)$$

Using (7.4.11) and (7.4.15), we obtain the following Levinson-type recursion for the backward predictor:

$$\mathbf{b}_{m+1} = \begin{bmatrix} 0 \\ \mathbf{b}_m \end{bmatrix} + \begin{bmatrix} 1 \\ \mathbf{a}_m \end{bmatrix} k_m^*$$

The backward prediction error is

$$e_{m+1}^b(n) = x(n-m-1) + \left\{ \begin{bmatrix} 0 \\ \mathbf{b}_m \end{bmatrix} + \begin{bmatrix} 1 \\ \mathbf{a}_m \end{bmatrix} k_m^* \right\}^H \begin{bmatrix} x(n) \\ \mathbf{x}_m(n-1) \end{bmatrix}$$

$$= x(n-m-1) + \mathbf{b}_m^H \mathbf{x}_m(n-1) + k_m[x(n) + \mathbf{a}_m^H \mathbf{x}_m(n-1)]$$

or
$$e_{m+1}^b(n) = e_m^b(n-1) + k_m e_m^f(n) \tag{7.5.6}$$

Recursions (7.5.5) and (7.5.6) can be computed for $m = 0, 1, \ldots, M-1$. The initial conditions $e_0^f(n)$ and $e_0^b(n)$ are easily obtained from (7.5.1) and (7.5.2). The recursions also lead to the following all-zero lattice algorithm

$$
\begin{aligned}
e_0^f(n) &= e_0^b(n) = x(n) \\
e_m^f(n) &= e_{m-1}^f(n) + k_{m-1}^* e_{m-1}^b(n-1) \qquad m = 1, 2, \ldots, M \\
e_m^b(n) &= k_{m-1} e_{m-1}^f(n) + e_{m-1}^b(n-1) \qquad m = 1, 2, \ldots, M \\
e(n) &= e_M^f(n)
\end{aligned}
\tag{7.5.7}
$$

that is implemented using the structure shown in Figure 7.5. The *lattice parameters* k_m are known as *reflection coefficients* in the speech processing and geophysics areas.

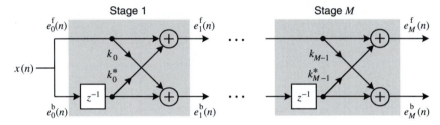

FIGURE 7.5
All-zero lattice structure for the implementation of the forward and backward prediction error filters.

The Levinson recursion for the optimum filter, (7.4.8) through (7.4.10), adds a *ladder* part to the lattice structure for the forward and backward predictors. Using (7.4.8), (7.5.7), and the partitioning in (7.5.3), we can express the filtering error of order $m+1$ in terms of $e_m(n)$ and $e_m^b(n)$ as follows

$$e_{m+1}(n) = y(n) - \mathbf{c}_{m+1}^H \mathbf{x}_{m+1}(n) = e_m(n) - k_m^{c*} e_m^b(n) \tag{7.5.8}$$

for $m = 0, 1, \ldots, M-1$. The resulting lattice-ladder structure is similar to the one shown in Figure 7.3. However, owing to stationarity all coefficients are constant, and $k_m^f(n) = k_m^b(n) = k_m$. We note that the efficient solution of the Mth-order optimum filtering problem is derived from the solution of the $(M-1)$st-order forward and backward prediction problems of the input process. In fact, the lattice part serves to decorrelate the samples $x(n), x(n-1), \ldots, x(n-M)$, producing the uncorrelated samples $e_0^b(n), e_1^b(n), \ldots, e_M^b(n)$ (innovations), which are then linearly combined ("recorrelated") to obtain the optimum estimate of the desired response.

System functions. We next express the various lattice relations in terms of z-transforms. Taking the z-transform of (7.5.1) and (7.5.2), we obtain

$$E_m^f(z) = \left(1 + \sum_{k=1}^{M} a_k^{(m)*} z^{-k}\right) X(z) \triangleq A_m(z) X(z) \tag{7.5.9}$$

$$E_m^b(z) = \left(z^{-m} + \sum_{k=1}^{M} b_k^{(m)*} z^{-k+1}\right) X(z) \triangleq B_m(z) X(z) \tag{7.5.10}$$

where $A_m(z)$ and $B_m(z)$ are the system functions of the paths from the input to the outputs of the mth stage of the lattice. Using the symmetry relation $\mathbf{a}_m = \mathbf{J}\mathbf{b}_m^*$, $1 \leq m \leq M$, we obtain

$$B_m(z) = z^{-m} A_m^* \left(\frac{1}{z^*}\right) \tag{7.5.11}$$

Note that if z_0 is a zero of $A_m(z)$, then z_0^{-1} is a zero of $B_m(z)$. Therefore, if $A_m(z)$ is minimum-phase, then $B_m(z)$ is maximum-phase.

Taking the z-transform of the lattice equations (7.5.7), we have for the mth stage

$$E_m^f(z) = E_{m-1}^f(z) + k_{m-1}^* z^{-1} E_{m-1}^b(z) \tag{7.5.12}$$

$$E_m^b(z) = k_{m-1} E_{m-1}^f(z) + z^{-1} E_{m-1}^b(z) \tag{7.5.13}$$

Dividing both equations by $X(z)$ and using (7.5.9) and (7.5.10), we have

$$A_m(z) = A_{m-1}(z) + k_{m-1}^* z^{-1} B_{m-1}(z) \tag{7.5.14}$$

$$B_m(z) = k_{m-1} A_{m-1}(z) + z^{-1} B_{m-1}(z) \tag{7.5.15}$$

which, when initialized with

$$A_0(z) = B_0(z) = 1 \tag{7.5.16}$$

describe the lattice filter in the z domain.

The z-transform of the ladder-part (7.5.8) is given by

$$E_{m+1}(z) = E_m(z) - k_m^{c*} E_m^b(z) \tag{7.5.17}$$

where $E_m(z)$ is the z-transform of the error sequence $e_m(n)$.

All-pole or "inverse" lattice structure. If we wish to recover the input $x(n)$ from the prediction error $e(n) = e_M^f(n)$, we can use the following all-pole lattice filter algorithm

$$\begin{aligned} e_M^f(n) &= e(n) \\ e_{m-1}^f(n) &= e_m^f(n) - k_{m-1}^* e_{m-1}^b(n-1) & m &= M, M-1, \ldots, 1 \\ e_m^b(n) &= e_{m-1}^b(n-1) + k_{m-1} e_{m-1}^f(n) & m &= M, M-1, \ldots, 1 \\ x(n) &= e_0^f(n) = e_0^b(n) \end{aligned} \tag{7.5.18}$$

which is derived as explained in Section 2.5 and is implemented by using the structure in Figure 7.6. Although the system functions of the all-zero lattice in (7.5.7) and the all-pole lattice in (7.5.18) are $H_{AZ}(z) = A(z)$ and $H_{AP}(z) = 1/A(z)$, the two lattice structures are described by the same set of lattice coefficients. The difference is the signal flow (see feedback loops in the all-pole structure). This structure is used in speech processing applications (Rabiner and Schafer 1978).

7.5.2 Some Properties and Interpretations

Lattice filters have some important properties and interesting interpretations that make them a useful tool in optimum filtering and signal modeling.

Optimal nesting. The all-zero lattice filter has an *optimal nesting* property when it is used for the implementation of an FLP. Indeed, if we use the lattice parameters obtained via the algorithm of Levinson-Durbin, the all-zero lattice filter driven by the signal $x(n)$

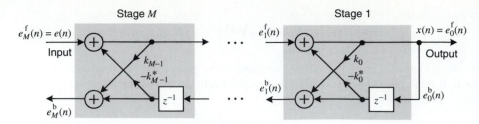

FIGURE 7.6
All-pole lattice structure for recovering the input signal from the forward prediction error.

produces prediction errors $e_m^f(n)$ and $e_m^b(n)$ at the output of the mth stage for all $1 \leq m \leq M$. This implies that we can increase the order of the filter by attaching additional stages without destroying the optimality of the previous stages. In contrast, the direct-form filter structure implementation requires the computation of the entire predictor for each stage. However, the nesting property does not hold for the all-pole lattice filter because of the feedback path.

Orthogonality. The backward prediction errors $e_m^b(n)$ for $0 \leq m \leq M$ are uncorrelated (see Section 7.2), that is,

$$E\{e_m^b(n)e_k^{b*}(n)\} = \begin{cases} P_m & k = m \\ 0 & k \neq m \end{cases} \qquad (7.5.19)$$

and constitute the innovations representation of the input samples $x(n), x(n-1), \ldots,$ $x(n-m)$. We see that at a given time instant n, the backward prediction errors for orders $m = 0, 1, 2, \ldots, M$ are uncorrelated and are part of a nonstationary sequence because the variance $E|e_m^b(n)|^2\} = P_m$ depends on n. This should be expected because, for a given n, each $e_m^b(n)$ is computed using a different set of predictor coefficients. In contrast, for a given m, the sequence $e_m^b(n)$ is stationary for $-\infty < n < \infty$.

Reflection coefficients. The all-pole lattice structure is very useful in the modeling of layered media, where each stage of the lattice models one layer or section of the medium. Traveling waves in geophysical layers, in acoustic tubes of varying cross-sections, and in multisectional transmission lines have been modeled in this fashion. The modeling is performed such that the wave travel time through each section is the same, but the sections may have different impedances. The mth section is modeled with the signals $e_m^f(n)$ and $e_m^b(n)$ representing the forward and backward traveling waves, respectively.

If Z_m and Z_{m-1} are the characteristic impedances at sections m and $m-1$, respectively, then k_m represents the reflection coefficients between the two sections, given by

$$k_m = \frac{Z_m - Z_{m-1}}{Z_m + Z_{m-1}} \qquad (7.5.20)$$

For this reason, the lattice parameters k_m are often known as *reflection coefficients*. As reflection coefficients, it makes good sense that their magnitudes not exceed unity. The termination of the lattice assumes a perfect reflection, and so the reflected wave $e_0^b(n)$ is equal to the transmitted wave $e_0^f(n)$. The result of this specific termination is an overall all-pole model (Rabiner and Schafer 1978).

Partial correlation coefficients. The *partial correlation coefficient* (*PCC*) between $x(n)$ and $x(n-m-1)$ (see also Section 7.2.2) is defined as the correlation coefficient between $e_m^f(n)$ and $e_m^b(n-1)$, that is,

$$\mathrm{PCC}\{x(n-m-1); x(n)\} \triangleq \frac{\mathrm{PARCOR}\{x(n-m-1); x(n)\}}{\sqrt{E\{|e_m^b(n-1)|^2\}E\{|e_m^f(n)|^2\}}} \qquad (7.5.21)$$

and, therefore, it takes values in the range $[-1, 1]$ (Kendall and Stuart 1979).

Working as in Section 7.2, we can show that

$$E\{e_m^b(n-1)e_m^{f*}(n)\} = \mathbf{b}_m^H \mathbf{r}_m + r(m+1) = \beta_m \qquad (7.5.22)$$

which in conjunction with

$$E\{|e_m^b(n-1)|^2\} = E\{|e_m^f(n)|^2\} = P_m \qquad (7.5.23)$$

and (7.4.16), results in

$$k_m = -\frac{\beta_m}{P_m} = -\mathrm{PCC}\{x(n-m-1); x(n)\} \qquad (7.5.24)$$

That is, for stationary processes the lattice parameters are the negative of the partial auto-correlation sequence and satisfy the relation

$$|k_m| \leq 1 \qquad \text{for all } 0 \leq m \leq M-1 \qquad (7.5.25)$$

derived also for (7.4.25) using an alternate approach.

Minimum phase. According to Theorem 2.3 (Section 2.5), the roots of the polynomial $A(z)$ are inside the unit circle if and only if

$$|k_m| < 1 \qquad \text{for all } 0 \leq m \leq M-1 \qquad (7.5.26)$$

which implies that the filters with system functions $A(z)$ and $1/A(z)$ are minimum-phase. The strict inequalities (7.5.26) are satisfied if the stationary process $x(n)$ is nonpredictable, which is the case when the Toeplitz autocorrelation matrix $\mathbf{R}$ is positive definite.

Lattice-ladder optimization. As we saw in Section 2.5, the output of an FIR lattice filter is a nonlinear function of the lattice parameters. Hence, if we try to design an optimum lattice filter by minimizing the MSE with respect to the lattice parameters, we end up with a nonlinear optimization problem (see Problem 7.11). In contrast, the Levinson algorithm leads to a lattice-ladder realization of the optimum filter through the order-recursive solution of a linear optimization problem. This subject is of interest to signal modeling and adaptive filtering (see Chapters 9 and 10).

7.5.3 Parameter Conversions

We have shown that the Mth-order forward linear predictor of a stationary process $x(n)$ is uniquely specified by a set of linear equations in terms of the autocorrelation sequence and the prediction error filter is minimum-phase. Furthermore, it can be implemented using either a direct-form structure with coefficients $a_1^{(M)}, a_2^{(M)}, \ldots, a_M^{(M)}$ or a lattice structure with parameters $k_1, k_2, \ldots, k_M$. Next we show how to convert between the following equivalent representations of a linear predictor:

1. Direct-form filter structure: $\{P_M, a_1, a_2, \ldots, a_M\}$.
2. Lattice filter structure: $\{P_M, k_0, k_1, \ldots, k_{M-1}\}$.
3. Autocorrelation sequence: $\{r(0), r(1), \ldots, r(M)\}$.

The transformation between the above representations is performed using the algorithms shown in Figure 7.7.

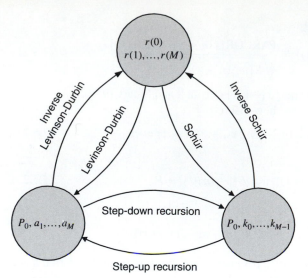

FIGURE 7.7
Equivalent representations for
minimum-phase linear prediction
error filters.

Lattice-to-direct (step-up) recursion. Given the lattice parameters $k_1, k_2, \ldots, k_M$ and the MMSE error P_M, we can compute the forward predictor $\mathbf{a}_M$ by using the following recursions

$$\mathbf{a}_m = \begin{bmatrix} \mathbf{a}_{m-1} \\ 0 \end{bmatrix} + \begin{bmatrix} \mathbf{J}\mathbf{a}_{m-1}^* \\ 1 \end{bmatrix} k_{m-1} \tag{7.5.27}$$

$$P_m = P_{m-1}(1 - |k_{m-1}|^2) \tag{7.5.28}$$

for $m = 1, 2, \ldots, M$. This conversion is implemented by the function `a=stepup(k,PM)`.

Direct-to-lattice (step-down) recursion. Using the partitioning

$$\begin{aligned} \bar{\mathbf{a}}_m &= [a_1^{(m)} \ a_2^{(m)} \ \cdots \ a_{m-1}^{(m)}]^T \\ k_m &= a_m^{(m)} \end{aligned} \tag{7.5.29}$$

we can write recursion (7.5.27) as

$$\bar{\mathbf{a}}_m = \mathbf{a}_{m-1} + \mathbf{J}\mathbf{a}_{m-1}^* k_{m-1}$$

or by taking the complex conjugate and multiplying both sides by $\mathbf{J}$

$$\mathbf{J}\bar{\mathbf{a}}_m^* = \mathbf{J}\mathbf{a}_{m-1}^* + \mathbf{a}_{m-1}k_{m-1}^*$$

Eliminating $\mathbf{J}\mathbf{a}_{m-1}^*$ from the last two equations and solving for $\mathbf{a}_{m-1}$, we obtain

$$\mathbf{a}_{m-1} = \frac{\bar{\mathbf{a}}_m - \mathbf{J}\bar{\mathbf{a}}_m^* k_{m-1}}{1 - |k_{m-1}|^2} \tag{7.5.30}$$

From (7.5.28), we have

$$P_{m-1} = \frac{P_m}{1 - |k_{m-1}|^2} \tag{7.5.31}$$

Given $\mathbf{a}_M$ and P_M, we can obtain k_m and P_m for $1 \le m \le M$ by computing the last two recursions for $m = M, M-1, \ldots, 2$. We should stress that both recursions break down if $|k_m| = \pm 1$. The step-down algorithm is implemented by the function `[k, P0]=stepdown(a,PM)`.

EXAMPLE 7.5.1. Given the third-order FLP coefficients $a_1^{(3)}, a_2^{(3)}, a_3^{(3)}$, compute the lattice parameters k_0, k_1, k_2.

Solution. With the help of (7.5.29) the vector relation (7.5.30) can be written in scalar form as

$$k_{m-1} = a_m^{(m)} \tag{7.5.32}$$

and
$$a_i^{(m-1)} = \frac{a_i^{(m)} - a_{m-i}^{(m)*} k_{m-1}}{1 - |k_{m-1}|^2} \tag{7.5.33}$$

which can be used to implement the step-down algorithm for $m = M, M-1, \ldots, 2$ and $i = 1, 2, \ldots, m$. Starting with $m = 3$ and $i = 1, 2$, we have

$$k_2 = a_3^{(3)} \qquad a_1^{(2)} = \frac{a_1^{(3)} - a_2^{(3)*} k_2}{1 - |k_2|^2} \qquad a_2^{(2)} = \frac{a_2^{(3)} - a_1^{(3)*} k_2}{1 - |k_2|^2}$$

Similarly, for $m = 2$ and $i = 1$, we obtain

$$k_1 = a_2^{(2)} \qquad a_1^{(1)} = \frac{a_1^{(2)} - a_1^{(2)*} k_1}{1 - |k_1|^2} = k_0$$

which completes the solution.

The step-up and step-down recursions also can be expressed in polynomial form as

$$A_m(z) = A_{m-1}(z) + k_{m-1}^* z^{-m} A_{m-1}^* \left(\frac{1}{z^*}\right) \tag{7.5.34}$$

and
$$A_{m-1}(z) = \frac{A_m(z) - k_{m-1}^* z^{-m} A_m^*(1/z^*)}{1 - |k_{m-1}|^2} \tag{7.5.35}$$

respectively.

Lattice parameters to autocorrelation. If we know the lattice parameters $k_1, k_2, \ldots,$ k_M and P_M, we can compute the values $r(0), r(1), \ldots, r(M)$ of the autocorrelation sequence using the formula

$$r(m+1) = -k_m^* P_m - \mathbf{a}_m^H \mathbf{J} \mathbf{r}_m \tag{7.5.36}$$

which follows from (7.4.16) and (7.4.17), in conjunction with (7.5.27) and (7.4.21) for $m = 1, 2, \ldots, M$. Equation (7.5.36) is obtained by eliminating β_m from (7.4.9) and (7.4.10). This algorithm is used by the function r=k2r(k,PM). Another algorithm that computes the autocorrelation sequence from the lattice coefficients and does not require the intermediate computation of $\mathbf{a}_m$ is provided in Section 7.6.

EXAMPLE 7.5.2. Given P_0, k_0, k_1, and k_2, compute the autocorrelation values $r(0), r(1), r(2),$ and $r(3)$.

Solution. Using $r(0) = P_0$ and

$$r(m+1) = -k_m^* P_m - \mathbf{a}_m^H \mathbf{J} \mathbf{r}_m$$

for $m = 0$, we have

$$r(1) = -k_0^* P_0$$

For $m = 1$

$$r(2) = -k_1^* P_1 - a_1^{(1)*} r(1)$$

where
$$P_1 = P_0(1 - |k_0|^2)$$

Finally, for $m = 2$ we obtain

$$r(3) = -k_2^* P_2 - [a_1^{(2)*} r(2) + k_1^* r(1)]$$

where
$$P_2 = P_1(1 - |k_1|^2)$$

and
$$a_1^{(2)} = a_1^{(1)} + a_1^{(1)*} k_1 = k_0 + k_0^* k_1$$

from the Levinson recursion.

Direct parameters to autocorrelation. Given $\mathbf{a}_M$ and P_M, we can compute the autocorrelation sequence $r(0), r(1), \ldots, r(M)$ by using (7.5.29) through (7.5.36). This method is known as the *inverse Levinson algorithm* and is implemented by the function r=a2r(a,PM).

7.6 ALGORITHM OF SCHÜR

The algorithm of Schür is an order-recursive procedure for the computation of the lattice parameters $k_1, k_2, \ldots, k_M$ of the optimum forward predictor from the autocorrelation sequence $r(0), r(1), \ldots, r(M)$ without computing the direct-form coefficients $\mathbf{a}_m, m = 1, 2, \ldots, M$. The reverse process is known as the *inverse Schür algorithm*. The algorithm also can be extended to compute the ladder parameters of the optimum filter and the LDL^H decomposition of a Toeplitz matrix. The algorithm has its roots in the original work of Schür (Schür 1917), who developed a procedure to test whether a polynomial is analytic and bounded in the unit disk.

7.6.1 Direct Schür Algorithm

We start by defining the cross-correlation sequences between $e_m^f(n)$, $e_m^b(n)$, and $x(n)$

$$\xi_m^f(l) \triangleq E\{x(n - l)e_m^{f*}(n)\} \quad \text{with } \xi_m^f(l) = 0, \text{ for } 1 \le l \le m \tag{7.6.1}$$

$$\xi_m^b(l) \triangleq E\{x(n - l)e_m^{b*}(n)\} \quad \text{with } \xi_m^b(l) = 0, \text{ for } 0 \le l < m \tag{7.6.2}$$

which are also known as *gapped functions* because of the regions of zeros created by the orthogonality principle (Robinson and Treitel 1980).

Multiplying the direct-form equations (7.5.1) and (7.5.2) by $x^*(n - l)$ and taking the mathematical expectation of both sides, we obtain

$$\xi_m^f(l) = r(l) + \mathbf{a}_m^H \tilde{\mathbf{r}}_m(l - 1) \tag{7.6.3}$$

and

$$\xi_m^b(l) = r(l - m) + \mathbf{b}_m^H \tilde{\mathbf{r}}_m(l) \tag{7.6.4}$$

where

$$\tilde{\mathbf{r}}_m(l) \triangleq [r(l) \, r(l - 1) \, \cdots \, r(l - m + 1)]^T \tag{7.6.5}$$

We notice that $\xi_m^f(l)$ and $\xi_m^b(l)$ can be interpreted as forward and backward autocorrelation prediction errors, because they occur when we feed the sequence $r(0), r(1), \ldots, r(m + 1)$ through the optimum predictors $\mathbf{a}_m$ and $\mathbf{b}_m$ of the process $x(n)$. Using the property $\mathbf{b}_m = \mathbf{J}\mathbf{a}_m^*$, we can show that (see Problem 7.30)

$$\xi_m^b(l) = \xi_m^{f*}(m - l) \tag{7.6.6}$$

If we set $l = m + 1$ in (7.6.3) and $l = m$ in (7.6.4), and notice that $\tilde{\mathbf{r}}_m(m) = \mathbf{J}\mathbf{r}_m^*$, then we have

$$\xi_m^f(m + 1) = r(m + 1) + \mathbf{a}_m^H \mathbf{J}\mathbf{r}_m^* = \beta_m^* \tag{7.6.7}$$

and

$$\xi_m^b(m) = r(0) + \mathbf{r}_m^H \mathbf{J}\mathbf{b}_m = P_m \tag{7.6.8}$$

respectively. Therefore, we have

$$k_m = -\frac{\beta_m}{P_m} = -\frac{\xi_m^f(m + 1)}{\xi_m^b(m)} \tag{7.6.9}$$

that is, we can compute k_{m+1} in terms of $\xi_m^f(l)$ and $\xi_m^b(l)$.

Multiplying the lattice recursions (7.5.7) by $x^*(n - l)$ and taking the mathematical expectation of both sides, we obtain

$$\xi_0^f(l) = \xi_0^b(l) = r(l)$$

$$\xi_m^f(l) = \xi_{m-1}^f(l) + k_{m-1}^* \xi_{m-1}^b(l - 1) \quad m = 1, 2, \ldots, M \tag{7.6.10}$$

$$\xi_m^b(l) = k_{m-1}\xi_{m-1}^f(l) + \xi_{m-1}^b(l - 1) \quad m = 1, 2, \ldots, M$$

which provides a lattice structure for the computation of the cross-correlations $\xi_m^f(l)$ and $\xi_m^b(l)$. In contrast, (7.6.7) and (7.6.8) provide a computation using a direct-form structure.

In the next example we illustrate how to use the lattice structure (7.6.10) to compute the lattice parameters $k_1, k_2, \ldots, k_M$ from the autocorrelation sequence $r(0), r(1), \ldots, r(M)$ without the intermediate explicit computation of the predictor coefficients $\mathbf{a}_m$.

EXAMPLE 7.6.1. Use the algorithm of Schür to compute the lattice parameters $\{k_0, k_1, k_2\}$ and the MMSE P_3 from the autocorrelation sequence coefficients

$$r(0) = 3 \qquad r(1) = 2 \qquad r(2) = 1 \qquad r(3) = \tfrac{1}{2}$$

Solution. Starting with (7.6.9) for $m = 0$, we have

$$k_0 = -\frac{\xi_0^f(1)}{\xi_0^b(0)} = -\frac{r(1)}{r(0)} = -\frac{2}{3}$$

because $\xi_0^f(l) = \xi_0^b(l) = r(l)$. To compute k_1, we need $\xi_1^f(2)$ and $\xi_1^b(1)$, which are obtained from (7.6.10) by setting $l = 2$. Indeed, we have

$$\xi_1^f(2) = \xi_0^f(2) + k_0 \xi_0^b(1) = 1 + (-\tfrac{2}{3})2 = -\tfrac{1}{3}$$

$$\xi_1^b(1) = \xi_0^b(0) + k_0 \xi_0^f(1) = 3 + (-\tfrac{2}{3})2 = \tfrac{5}{3} = P_1$$

and

$$k_1 = -\frac{\xi_1^f(2)}{\xi_1^b(1)} = -\frac{-\tfrac{1}{3}}{\tfrac{5}{3}} = \tfrac{1}{5}$$

The computation of k_2 requires $\xi_2^f(3)$ and $\xi_2^b(2)$, which in turn need $\xi_1^f(3)$ and $\xi_1^b(2)$. These quantities are computed by

$$\xi_1^f(3) = \xi_0^f(3) + k_0 \xi_0^b(2) = \tfrac{1}{2} + (-\tfrac{2}{3})1 = -\tfrac{1}{6}$$

$$\xi_1^b(2) = \xi_0^b(1) + k_0 \xi_0^f(2) = 2 + (-\tfrac{2}{3})1 = \tfrac{4}{3}$$

$$\xi_2^f(3) = \xi_1^f(3) + k_1 \xi_1^b(2) = -\tfrac{1}{6} + \tfrac{1}{5} \cdot \tfrac{4}{3} = \tfrac{1}{10}$$

$$\xi_2^b(2) = \xi_1^b(1) + k_1 \xi_1^f(2) = \tfrac{4}{3} + \tfrac{1}{5}(-\tfrac{1}{6}) = \tfrac{8}{5} = P_2$$

and the lattice coefficient is

$$k_2 = -\frac{\xi_2^f(3)}{\xi_2^b(2)} = -\frac{\tfrac{1}{10}}{\tfrac{8}{5}} = -\frac{1}{16}$$

The final MMSE is computed by

$$P_3 = P_2(1 - |k_2|^2) = \tfrac{8}{5}(1 - \tfrac{1}{256}) = \tfrac{51}{32}$$

although we could use the formula $\xi_m^b(m) = P_m$ as well. Therefore the lattice coefficients and the MMSE are found to be

$$k_0 = -\tfrac{2}{3} \qquad k_1 = \tfrac{1}{5} \qquad k_2 = -\tfrac{1}{16} \qquad P_3 = \tfrac{51}{32}$$

It is worthwhile to notice that the k_m parameters can be obtained by "feeding" the sequence $r(0), r(1), \ldots, r(M)$ through the lattice filter as a signal and switching on the stages one by one after computing the required lattice coefficient. The value of k_m is computed at time $n = m$ from the inputs to stage m (see Problem 7.30).

The procedure outlined in the above example is known as the algorithm of Schür and has good numerical properties because the quantities used in the lattice structure (7.6.10) are bounded. Indeed, from (7.6.1) and (7.6.2) we have

$$|\xi_m^f(l)|^2 \leq |E\{|e_m^f(n)|^2\}||E\{|x(n-l)|^2\}| \leq P_m r(0) \leq r^2(0) \tag{7.6.11}$$

$$|\xi_m^b(l)|^2 \leq |E\{|e_m^b(n)|^2\}||E\{|x(n-l)|^2\}| \leq P_m r(0) \leq r^2(0) \tag{7.6.12}$$

because $P_m \leq P_0 = r(0)$. As a result of this fixed dynamic range, the algorithm of Schür can be easily implemented with fixed-point arithmetic. The numeric stability of the Schür algorithm provided the motivation for its use in speech processing applications (LeRoux and Gueguen 1977).

7.6.2 Implementation Considerations

Figure 7.8 clarifies the computational steps in Example 7.4.2, using three decomposition trees that indicate the quantities needed to compute k_0, k_1, and k_2 when we use the lattice recursions (7.6.10) for real-valued signals. We can easily see that the computations for k_0 are part of those for k_1, which in turn are part of the computations for k_2. Thus, the tree for k_2 includes also the quantities needed to compute k_0 and k_1. The computations required to compute k_0, k_1, k_2, and k_3 are

$$1. \quad k_0 = -\frac{\xi_0^f(1)}{\xi_0^b(0)}$$

$$9. \quad \xi_2^f(4) = \xi_1^f(4) + k_1 \xi_1^b(3)$$

$$2. \quad \xi_1^f(4) = \xi_0^f(4) + k_0 \xi_0^b(3)$$

$$10. \quad \xi_2^b(3) = \xi_1^b(2) + k_1 \xi_1^f(3)$$

$$3. \quad \xi_1^b(3) = \xi_0^b(2) + k_0 \xi_0^f(3)$$

$$11. \quad \xi_2^f(3) = \xi_1^f(3) + k_1 \xi_1^b(2)$$

$$4. \quad \xi_1^f(3) = \xi_0^f(3) + k_0 \xi_0^b(2)$$

$$12. \quad \xi_2^b(2) = \xi_1^b(1) + k_1 \xi_1^f(2)$$

$$5. \quad \xi_1^b(2) = \xi_0^b(1) + k_0 \xi_0^f(2)$$

$$13. \quad k_2 = -\frac{\xi_2^f(3)}{\xi_2^b(2)}$$

$$6. \quad \xi_1^f(2) = \xi_0^f(2) + k_0 \xi_0^b(1)$$

$$14. \quad \xi_3^f(4) = \xi_2^f(4) + k_2 \xi_2^b(3)$$

$$7. \quad \xi_1^b(1) = \xi_0^b(0) + k_0 \xi_0^f(1)$$

$$15. \quad \xi_3^b(3) = \xi_2^b(2) + k_2 \xi_2^f(3)$$

$$8. \quad k_1 = -\frac{\xi_1^f(2)}{\xi_1^b(1)}$$

$$16. \quad k_3 = -\frac{\xi_3^f(4)}{\xi_3^b(3)}$$

With the help of the corresponding tree decomposition diagram, this can be arranged as shown in Figure 7.9. The obtained computational structure was named the *superlattice* because it consists of a triangular array of latticelike stages (Carayannis et al. 1985). Note that the superlattice has no redundancy and is characterized by local interconnections; that is, the quantities needed at any given node are available from the immediate neighbors.

The two-dimensional layout of the superlattice suggests various algorithms to perform the computations.

1. *Parallel algorithm.* We first note that all equations involving the coefficient k_m constitute one stage of the superlattice and can be computed in parallel after the computation of k_m because all inputs to the current stage are available from the previous one. This algorithm can be implemented by $2(M-1)$ processors in $M-1$ "parallel" steps (Kung and Hu 1983). Since each step involves one division to compute k_m and then $2(M-m)$ multiplications and additions for the parallel computations, the number of utilized processors decreases from $2(M-1)$ to 1. The algorithm is not order-recursive because the order M must be known before the superlattice structure is set up.

2. *Sequential algorithm.* A sequential implementation of the parallel algorithm is essentially equivalent to the version introduced for speech processing applications (LeRoux and Gueguen 1977). This algorithm, which is implemented by the function k=schurlg(r,M) and summarized in Table 7.4, starts with Equation (1) and computes sequentially Equations (2), (3), etc.

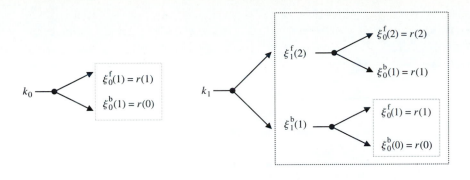

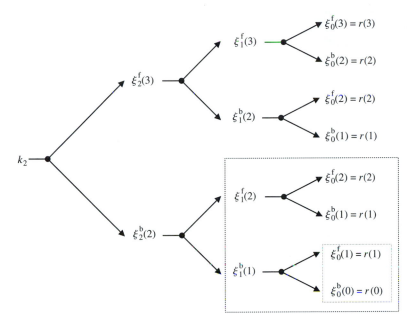

FIGURE 7.8
Tree decomposition for the computations required by the algorithm of Schür.

3. *Sequential order-recursive algorithm.* The parallel algorithm starts at the left of the superlattice and performs the computations within the vertical strips in parallel. Clearly,
the order M should be fixed before we start, and the algorithm is not order-recursive.
Careful inspection of the superlattice reveals that we can obtain an order-recursive algorithm by organizing the computations in terms of the *slanted* shadowed strips shown
in Figure 7.9. Indeed, we start with k_0 and then perform the computations in the first
slanted strip to determine the quantities $\xi_1^f(2)$ and $\xi_1^b(1)$ needed to compute k_1. We
proceed with the next slanted strip, compute k_2, and conclude with the computation
of the last strip and k_4. The computations within each slanted strip are performed
sequentially.

4. *Partitioned-parallel algorithm.* Suppose that we have P processors with $P < M$. This algorithm partitions the superlattice into groups of P consecutive slanted strips (partitions)
and performs the computations of each partition, in parallel, using the P processors. It
turns out that by storing some intermediate quantities, we have everything needed by the
superlattice to compute all the partitions, one at a time (Koukoutsis et al. 1991). This
algorithm provides a very convenient scheme for the implementation of the superlattice
using multiprocessing (see Problem 7.31).

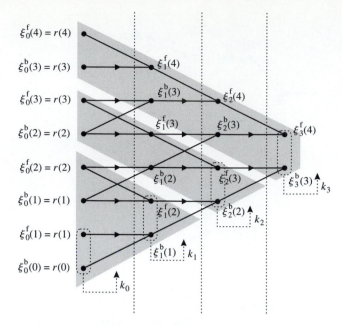

FIGURE 7.9
Superlattice structure organization of the algorithm of Schür. The
input is the autocorrelation sequence and the output the lattice
parameters.

TABLE 7.4
Summary of the algorithm of Schür.

1. **Input:** $\{r(l)\}_0^M$

2. **Initialization**
   (a) For $n = 0, 1, \ldots, M$
   $$\xi_0^f(l) = \xi_0^b(l) = r(l)$$
   (b) $k_0 = -\dfrac{\xi_0^f(1)}{\xi_0^b(0)}$
   (c) $P_1 = r(0)(1 - |k_1|^2)$

3. **For** $m = 1, 2, \ldots, M - 1$
   (a) **For** $l = m, m + 1, \ldots, M$
   $$\xi_m^f(l) = \xi_{m-1}^f(l) + k_{m-1}^* \xi_{m-1}^b(l - 1)$$
   $$\xi_m^b(l) = k_{m-1}\xi_{m-1}^f(l) + \xi_{m-1}^b(l - 1)$$
   (b) $k_m = -\dfrac{\xi_m^f(m)}{\xi_m^b(m)}$
   (c) $P_{m+1} = P_m(1 - |k_m|^2)$

4. **Output:** $\{k_m\}_0^{M-1}, \{P_m\}_1^M$

Extended Schür algorithm. To extend the Schür algorithm for the computation of the
ladder parameters k_m^c, we define the cross-correlation sequence

$$\xi_m^c(l) \triangleq E\{x(n - l)e_m^*(l)\} \qquad \text{with } \xi_m^c(l) = 0, \text{ for } 0 \leq l < m \tag{7.6.13}$$

due to the orthogonality principle. Multiplying (7.5.8) by $x^*(n - l)$ and taking the mathematical expectation, we obtain a direct form

$$\xi_m^c(l) = d_{l+1} - \mathbf{c}_m^T \tilde{\mathbf{r}}_m(l) \tag{7.6.14}$$

and a ladder-form equation

$$\xi_{m+1}^c(l) = \xi_m^c(l) - k_m^{c*}\xi_m^b(l) \tag{7.6.15}$$

For $l = m$, we have

$$\xi_m^c(m) = d_{l+1} - \mathbf{c}_m^H \mathbf{J}\mathbf{r}_m = \beta_m^c \tag{7.6.16}$$

and

$$k_m^c = \frac{\beta_m^c}{P_m} = \frac{\xi_m^c(m)}{\xi_m^b(m)} \tag{7.6.17}$$

that is, we can compute the sequence k_m^c using a lattice-ladder structure.

The computations can be arranged in the form of a *superladder* structure, shown in Figure 7.10 (Koukoutsis et al. 1991). See also Problem 7.32. In turn, (7.6.17) can be used in conjunction with the superlattice to determine the lattice-ladder parameters of the optimum FIR filter. The superladder structure is illustrated in the following example.

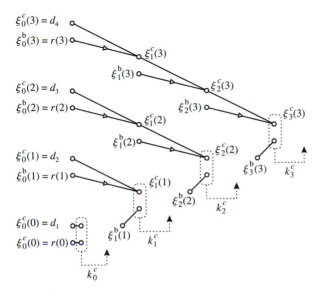

FIGURE 7.10
Graphical illustration of the superladder structure.

EXAMPLE 7.6.2. Determine the lattice-ladder parameters of an optimum FIR filter with input autocorrelation sequence given in Example 7.6.1 and cross-correlation sequence $d_1 = 1$, $d_2 = 2$, and $d_3 = \frac{5}{2}$, using the extended Schür algorithm.

Solution. Since the lattice parameters were obtained in Example 7.6.1, we only need to find the ladder parameters. Hence, using (7.6.15), (7.6.17), and the values of $\xi_m^b(l)$ computed in Example 7.6.1, we have

$$k_0^c = -\frac{\xi_0^c(0)}{\xi_0^b(0)} = -\frac{d_1}{r(0)} = -\frac{1}{3}$$

$$\xi_1^c(1) = \xi_0^c(1) + k_0^c\xi_0^b(1) = 2 - \frac{1}{3}(2) = \frac{4}{3}$$

$$\xi_1^c(2) = \xi_0^c(2) + k_0^c\xi_0^b(2) = \frac{5}{2} - \frac{1}{3}(1) = \frac{13}{6}$$

$$k_1^c = -\frac{\xi_1^c(1)}{\xi_1^b(1)} = -\frac{\frac{4}{3}}{\frac{5}{3}} = -\frac{4}{5}$$

$$\xi_2^c(2) = \xi_1^c(2) + k_1^c\xi_1^b(2) = \frac{13}{6} - \frac{4}{5}(\frac{4}{3}) = \frac{11}{10}$$

$$k_2^c = -\frac{\xi_2^c(2)}{\xi_2^b(2)} = -\frac{\frac{11}{10}}{\frac{8}{5}} = -\frac{11}{16}$$

which provide the values of the ladder parameters. These values are identical to those obtained in Example 7.4.3.

7.6.3 Inverse Schür Algorithm

The inverse Schür algorithm computes the autocorrelation sequence coefficients $r(0)$, $r(1), \ldots, r(m)$ from the lattice parameters $k_0, k_1, \ldots, k_M$ and the MMSE P_M of the linear predictor. The organization of computations is best illustrated by the following example.

EXAMPLE 7.6.3. Given the lattice filter coefficients

$$k_0 = -\frac{2}{3} \quad k_1 = \frac{1}{5} \quad k_2 = -\frac{1}{16}$$

and the MMSE $P_3 = 51/32$, compute the autocorrelation samples $r(0), r(1), r(2)$, and $r(3)$, using the inverse Schür algorithm.

Solution. We base our approach on the part of the superlattice structure shown in Figure 7.9 that is enclosed by the nodes $\xi_0^b(0), \xi_0^f(3), \xi_2^f(3)$, and $\xi_2^b(2)$. To start at the lower left corner, we compute $r(0)$, using (7.4.30):

$$r(0) = \frac{P_3}{\prod_{m=0}^{2}(1-k_m^2)} = \frac{\frac{51}{32}}{(1-\frac{4}{9})(1-\frac{1}{25})(1-\frac{1}{256})} = 3$$

This also follows from (7.5.31). Then, continuing the computations from the line defined by $r(0)$ and $\xi_2^b(2)$ to the node defined by $\xi_0^f(3) = r(3)$, we have

$$r(1) = -k_0 r(0) = -(-\tfrac{2}{3})3 = 2$$
$$\xi_1^b(1) = \xi_0^b(0) + k_0\xi_0^f(1) = 3 + (-\tfrac{2}{3})2 = \tfrac{5}{3}$$
$$\xi_1^f(2) = -k_1\xi_1^b(1) = -\tfrac{1}{5}(\tfrac{5}{3}) = -\tfrac{1}{3}$$
$$r(2) = \xi_0^f(2) = \xi_1^f(2) - k_0\xi_0^b(1) = -\tfrac{1}{3} - (-\tfrac{2}{3})2 = 1$$
$$\xi_1^b(2) = \xi_0^b(1) + k_0\xi_0^f(2) = 2 + (-\tfrac{2}{3})1 = \tfrac{4}{3}$$
$$\xi_2^b(2) = \xi_1^b(1) + k_1\xi_1^f(2) = \tfrac{5}{3} + \tfrac{1}{5}(-\tfrac{1}{3}) = \tfrac{8}{5}$$
$$\xi_2^f(3) = -k_2\xi_2^b(2) = -(-\tfrac{1}{16})(\tfrac{8}{5}) = \tfrac{1}{10}$$
$$\xi_1^f(3) = \xi_2^f(3) - k_1\xi_1^b(2) = \tfrac{1}{10} - \tfrac{1}{5}(\tfrac{4}{3}) = -\tfrac{1}{6}$$
$$r(3) = \xi_0^f(3) = \xi_1^f(3) - k_0\xi_0^b(2) = -\tfrac{1}{6} - (-\tfrac{2}{3})1 = \tfrac{1}{2}$$

as can be easily verified by the reader. Thus, the autocorrelation sequence is

$$r(0) = 3 \quad r(1) = 2 \quad r(2) = 1 \quad r(3) = \tfrac{1}{2}$$

which agree with the autocorrelation sequence coefficients used in Example 7.6.1 with the direct Schür algorithm.

The inverse Schür algorithm is implemented by the function `r=invschur(k,PM)`, which follows the same procedure as the previous example.

7.7 TRIANGULARIZATION AND INVERSION OF TOEPLITZ MATRICES

In this section, we develop LDL^H decompositions for both Toeplitz matrices and the inverse of Toeplitz matrices, followed by a recursion for the computation of the inverse of a Toeplitz matrix.

7.7.1 LDL^H Decomposition of Inverse of a Toeplitz Matrix

Since $\mathbf{R}_m$ is a Hermitian Toeplitz matrix that also happens to be persymmetric, that is, $\mathbf{JR}_m\mathbf{J} = \mathbf{R}_m^*$, taking its inverse, we obtain

$$\mathbf{JR}_m^{-1}\mathbf{J} = (\mathbf{R}_m^*)^{-1} \tag{7.7.1}$$

The last equation shows that the inverse of a Toeplitz matrix, although not Toeplitz, is persymmetric. From (7.1.58), we recall that the BLP coefficients and the MMSE P_m^b provide the quantities for the $\mathbf{UDU}^T$ decomposition of $\mathbf{R}_{m+1}^{-1}$, that is,

$$\mathbf{R}_{m+1}^{-1} = \mathbf{B}_{m+1}^H\mathbf{D}_{m+1}^{-1}\mathbf{B}_{m+1} \tag{7.7.2}$$

where
$$\mathbf{B}_{m+1} = \begin{bmatrix} 1 & 0 & \cdots & 0 & 0 \\ b_0^{(1)} & 1 & \cdots & 0 & 0 \\ \vdots & \vdots & \ddots & \vdots & \vdots \\ b_0^{(m-1)} & b_1^{(m-1)} & \cdots & 1 & 0 \\ b_0^{(m)} & b_1^{(m)} & \cdots & b_{m-1}^{(m)} & 1 \end{bmatrix} \tag{7.7.3}$$

and
$$\mathbf{D}_{m+1} = \operatorname{diag}\{P_0^b, P_1^b, \ldots, P_m^b\} \tag{7.7.4}$$

For a Toeplitz matrix $\mathbf{R}_{m+1}$, we can obtain the LDL^H decomposition of its inverse by using (7.7.2) and the property $\mathbf{J} = \mathbf{J}^{-1}$ of the exchange matrix. Starting with (7.7.1), we obtain

$$(\mathbf{R}_{m+1}^*)^{-1} = \mathbf{JR}_{m+1}^{-1}\mathbf{J} = (\mathbf{JB}_{m+1}^H\mathbf{J})(\mathbf{JD}_{m+1}^{-1}\mathbf{J})(\mathbf{JB}_{m+1}\mathbf{J}) \tag{7.7.5}$$

If we define

$$\mathbf{A}_{m+1} \triangleq \mathbf{JB}_{m+1}^*\mathbf{J} \tag{7.7.6}$$

and
$$\bar{\mathbf{D}}_{m+1} \triangleq \mathbf{JD}_{m+1}^{-1}\mathbf{J} = \operatorname{diag}\{P_m, P_{m-1}, \ldots, P_0\} \tag{7.7.7}$$

then (7.7.2) gives
$$\mathbf{R}_{m+1}^{-1} = \mathbf{A}_{m+1}^H\bar{\mathbf{D}}_{m+1}^{-1}\mathbf{A}_{m+1} \tag{7.7.8}$$

which provides the unique LDL^H decomposition of the matrix $\mathbf{R}_{m+1}^{-1}$. Indeed, using the property $\mathbf{a}_j = \mathbf{Jb}_j^*$ for $1 \le j \le m$, or equivalently $a_i^{(j)} = b_{j-i}^{(j)*}$, we can write matrix $\mathbf{A}_{m+1} = \mathbf{JB}_{m+1}^*\mathbf{J}$ as

$$\mathbf{A}_{m+1} = \begin{bmatrix} 1 & a_1^{(m)*} & a_2^{(m)*} & \cdots & a_m^{(m)*} \\ 0 & 1 & a_1^{(m-1)*} & \cdots & a_{m-1}^{(m-1)*} \\ \vdots & \vdots & \vdots & \ddots & \vdots \\ 0 & 0 & \cdots & \ddots & a_1^{(1)*} \\ 0 & 0 & 0 & \cdots & 1 \end{bmatrix} \tag{7.7.9}$$

which is an upper unit triangular matrix. We stress that the property $\mathbf{JB}_{m+1}^*\mathbf{J} = \mathbf{A}_{m+1}$ and the above derivation of (7.7.8) hold for Toeplitz matrices only. However, the decomposition in (7.7.2) holds for any Hermitian, positive definite matrix (see Section 7.1.4).

As we saw in Section 6.3, the solution of the normal equations $\mathbf{Rc} = \mathbf{d}$ can be obtained in three steps as

$$\mathbf{R} = \mathbf{LDL}^H \Rightarrow \mathbf{LDk}^c = \mathbf{d} \Rightarrow \mathbf{L}^H\mathbf{c} = \mathbf{k}^c \tag{7.7.10}$$

where the LDL^H decomposition requires about $M^3/6$ flops and the solution of each triangular system $M^2/2$ flops. Since $\mathbf{R}^{-1} = \mathbf{B}^H\mathbf{D}^{-1}\mathbf{B}$, the Levinson-Durbin algorithm performs the UDU^H decomposition of $\mathbf{R}^{-1}$ when $\mathbf{R}$ is Toeplitz, at a cost of M^2 flops; that is, it reduces the computational complexity by an order of magnitude. The Levinson recursion for the optimum filter is equivalent to the solution of the two triangular systems and requires M^2 operations.

EXAMPLE 7.7.1. Compute the lattice-ladder parameters of an MMSE finite impulse response filter specified by the normal equations

$$\begin{bmatrix} 3 & 2 & 1 \\ 2 & 3 & 2 \\ 1 & 2 & 3 \end{bmatrix} \begin{bmatrix} h(0) \\ h(1) \\ h(2) \end{bmatrix} = \begin{bmatrix} 1 \\ 2 \\ \frac{5}{2} \end{bmatrix}$$

using two different approaches: the LDL^H decomposition and the algorithm of Levinson.

Solution. The LDL^H decomposition of $\mathbf{R}$ is

$$\mathbf{L} = \begin{bmatrix} 1 & 0 & 0 \\ \frac{2}{3} & 1 & 0 \\ \frac{1}{3} & \frac{4}{5} & 1 \end{bmatrix} \quad \mathbf{D} = \begin{bmatrix} 3 & 0 & 0 \\ 0 & \frac{5}{3} & 0 \\ 0 & 0 & \frac{8}{5} \end{bmatrix} \quad \mathbf{L}^{-1} = \begin{bmatrix} 1 & 0 & 0 \\ -\frac{2}{3} & 1 & 0 \\ \frac{1}{5} & -\frac{4}{5} & 1 \end{bmatrix}$$

and using (7.3.31), we have

$$\mathbf{k}_3^c = \mathbf{D}^{-1}\mathbf{L}^{-1}\mathbf{d} = \begin{bmatrix} \frac{1}{3} & \frac{4}{5} & \frac{11}{16} \end{bmatrix}^T$$

which gives the three ladder parameters. The two lattice parameters are obtained by solving the system

$$\mathbf{L}^{[2]}\mathbf{D}^{[2]}\mathbf{k}_2 = \mathbf{r}_2^b \quad \text{with } \mathbf{r}_2^b = [1 \ 2]^T$$

which gives $k_0 = \frac{1}{3}$ and $k_1 = \frac{4}{5}$. The results agree with those obtained in Example 7.4.3 using the algorithm of Levinson. We also note that the rows of $\mathbf{L}^{-1}$ provide the first- and second-order forward and backward linear predictors. This is the case because the matrix is Toeplitz. For symmetric matrices the LDL^H decomposition provides the backward predictors only.

7.7.2 LDL^H Decomposition of a Toeplitz Matrix

The computation of the LDL^H decomposition of a symmetric, positive definite matrix requires on the order of M^3 computations. In Section 7.1, we saw that the cross-correlation between $x(n)$ and $e_m^b(n)$ is related to the LDL^H decomposition of the correlation matrix $\mathbf{R}_m$. We next show that we can extend the Schür algorithm to compute the LDL^H decomposition of a Toeplitz matrix with $O(M^2)$ computations using the cross-correlations $\xi_m^b(l)$.

To illustrate the basic process, we note that evaluating the product on the left with the help of (7.6.4), we obtain

$$\begin{bmatrix} r(0) & r(1) & r(2) & r(3) \\ r(1) & r(0) & r(1) & r(2) \\ r(2) & r(1) & r(0) & r(1) \\ r(3) & r(2) & r(1) & r(0) \end{bmatrix} \begin{bmatrix} 1 & b_0^{(1)*} & b_0^{(2)*} & b_0^{(3)*} \\ 0 & 1 & b_1^{(2)*} & b_1^{(3)*} \\ 0 & 0 & 1 & b_2^{(3)*} \\ 0 & 0 & 0 & 1 \end{bmatrix} = \begin{bmatrix} \xi_0^{b(0)} & 0 & 0 & 0 \\ \xi_0^b(1) & \xi_1^b(1) & 0 & 0 \\ \xi_0^b(2) & \xi_1^b(2) & \xi_2^b(2) & 0 \\ \xi_0^b(3) & \xi_1^b(3) & \xi_2^b(3) & \xi_3^b(3) \end{bmatrix}$$

that is, a lower triangular matrix $\tilde{\mathbf{L}}$, which can be written as

$$\tilde{\mathbf{L}} = \begin{bmatrix} 1 & 0 & 0 & 0 \\ \dfrac{\xi_0^b(1)}{P_0} & 1 & 0 & 0 \\ \dfrac{\xi_0^b(2)}{P_0} & \dfrac{\xi_1^b(2)}{P_1} & 1 & 0 \\ \dfrac{\xi_0^b(3)}{P_0} & \dfrac{\xi_1^b(3)}{P_1} & \dfrac{\xi_2^b(3)}{P_2} & 1 \end{bmatrix} \begin{bmatrix} P_0 & 0 & 0 & 0 \\ 0 & P_1 & 0 & 0 \\ 0 & 0 & P_2 & 0 \\ 0 & 0 & 0 & P_3 \end{bmatrix} = \mathbf{LD}$$

because $P_m = \xi^b_m(m) \geq 0$. Therefore, $\mathbf{RB}^H = \mathbf{LD}$ and since $\mathbf{R}$ is Hermitian, we have $\mathbf{R} = \mathbf{LDB}^{-H} = \mathbf{B}^{-1}\mathbf{DL}^H$, which implies that $\mathbf{B}^{-1} = \mathbf{L}$. This results in the following LDL^H factorization of the $(M+1) \times (M+1)$ symmetric Toeplitz matrix $\mathbf{R}$

$$\mathbf{R} = \mathbf{LDL}^H \tag{7.7.11}$$

where

$$\mathbf{L} = \mathbf{B}^{-1} = \begin{bmatrix} 1 & 0 & \cdots & 0 & 0 \\ \bar{\xi}^b_0(1) & 1 & \cdots & 0 & 0 \\ \bar{\xi}^b_0(2) & \bar{\xi}^b_1(2) & \cdots & 0 & 0 \\ \vdots & \vdots & \ddots & \vdots & \vdots \\ \bar{\xi}^b_0(M) & \bar{\xi}^b_1(M) & \cdots & \bar{\xi}^b_{M-1}(M) & 1 \end{bmatrix} \tag{7.7.12}$$

$$\bar{\xi}^b_m(l) = \frac{\xi^b_m(l)}{\xi^b_m(m)} = \frac{\xi^b_m(l)}{P_m} \tag{7.7.13}$$

and

$$\mathbf{D} = \mathrm{diag}\{P_0, P_1, \ldots, P_M\} \tag{7.7.14}$$

The basic recursion (7.6.10) in the algorithm of Schür can be extended to compute the elements of $\tilde{\mathbf{L}}$ and hence the LDL^H factorization of the Toeplitz matrix $\mathbf{R}$ (see Problem 7.33).

Since a Toeplitz matrix is persymmetric, that is, $\mathbf{JRJ} = \mathbf{R}^*$, we have

$$\mathbf{R} = \mathbf{JR}^*\mathbf{J} = (\mathbf{JL}^*\mathbf{J})(\mathbf{JDJ})(\mathbf{JL}^H\mathbf{J}) \triangleq \mathbf{U}\bar{\mathbf{D}}\mathbf{U}^H \tag{7.7.15}$$

which provides the UDU^H decomposition of $\mathbf{R}$. Notice that the relation $\mathbf{U} = \mathbf{JL}^*\mathbf{J}$ also can be obtained from $\mathbf{A} = \mathbf{JB}^*\mathbf{J}$ [see (7.4.11)], which in turn is a consequence of the symmetry between forward and backward prediction for stationary processes.

The validity of (7.6.10) also can be shown by computing the product

$$\mathbf{RA}^H = \begin{bmatrix} r(0) & r(1) & r(2) & r(3) \\ r(1) & r(0) & r(1) & r(2) \\ r(2) & r(1) & r(0) & r(1) \\ r(3) & r(2) & r(1) & r(0) \end{bmatrix} \begin{bmatrix} 1 & 0 & 0 & 0 \\ a^{(3)*}_1 & 1 & 0 & 0 \\ a^{(3)*}_2 & a^{(2)*}_1 & 1 & 0 \\ a^{(3)*}_3 & a^{(2)*}_2 & a^{(1)*}_1 & 1 \end{bmatrix} \tag{7.7.16}$$

$$= \begin{bmatrix} \xi^f_3(0) & \xi^f_2(-1) & \xi^f_1(-2) & \xi^f_0(-3) \\ 0 & \xi^f_2(0) & \xi^f_1(-1) & \xi^f_0(-2) \\ 0 & 0 & \xi^f_1(0) & \xi^f_0(-1) \\ 0 & 0 & 0 & \xi^f_0(0) \end{bmatrix} \tag{7.7.17}$$

with the help of (7.6.3) and $r(-l) = r^*(l)$. The formula $\mathbf{U} = \mathbf{JL}^*\mathbf{J}$ relates $\xi^f_m(l)$ and $\xi^b_m(l)$, as expected by (7.6.10).

7.7.3 Inversion of Real Toeplitz Matrices

From the discussion in Section 7.1, it follows from (7.1.12) that the inverse $\mathbf{Q}_M$ of a symmetric, positive definite matrix $\mathbf{R}_M$ is given by

$$\mathbf{Q}_M \triangleq \begin{bmatrix} \mathbf{Q} & \mathbf{q} \\ \mathbf{q}^T & q \end{bmatrix} \tag{7.7.18}$$

with

$$\mathbf{q} = \frac{\mathbf{b}}{P} \tag{7.7.19}$$

$$q = \frac{1}{P} \tag{7.7.20}$$

and
$$\mathbf{Q} = \mathbf{R}^{-1} + \frac{1}{P}\mathbf{b}\mathbf{b}^T \tag{7.7.21}$$

as given by (7.1.18), (7.1.19), and (7.1.21). The matrix $\mathbf{Q}$ is an $(M-1) \times (M-1)$ matrix, and $\mathbf{b}$ is the $(M-1)$st-order BLP. Next we show that for Toeplitz matrices we can compute $\mathbf{Q}_M$ with $O(M^2)$ computations.

First, we note that the last column and the last row of $\mathbf{Q}_M$ can be obtained by solving the Toeplitz system $\mathbf{Rb} = -\mathbf{Jr}$ using the Levinson-Durbin algorithm. Then we show that we can compute the elements of $\mathbf{Q}$ by exploiting the persymmetry property of Toeplitz matrices, moving from the known edges to the interior. Indeed, since $\mathbf{R}$ is persymmetric, that is, $\mathbf{R} = \mathbf{JRJ}$, we have $\mathbf{R}^{-1} = \mathbf{JR}^{-1}\mathbf{J}$, that is, $\mathbf{R}^{-1}$ is also persymmetric. From (7.7.21), we have

$$\langle\mathbf{Q}\rangle_{ij} = \langle\mathbf{R}^{-1}\rangle_{ij} + \frac{q_i q_j}{P} = \langle\mathbf{R}^{-1}\rangle_{M-j,M-i} + \frac{q_i q_j}{P} \tag{7.7.22}$$

because $\mathbf{R}^{-1}$ is persymmetric, and

$$\langle\mathbf{R}^{-1}\rangle_{M-j,M-i} = \langle\mathbf{Q}\rangle_{M-j,M-i} - \frac{q_{M-j} q_{M-i}}{P} \tag{7.7.23}$$

Combining (7.7.22) and (7.7.23), we obtain

$$\langle\mathbf{Q}\rangle_{ij} = \langle\mathbf{Q}\rangle_{M-j,M-i} - \frac{1}{P}(q_i q_j - q_{M-j} q_{M-i}) \tag{7.7.24}$$

which in conjunction with persymmetry makes possible the computation of the elements of $\mathbf{Q}$ from $\mathbf{q}$ and q. The process is illustrated for $M = 6$ in the following diagram

$$\mathbf{Q}_6 = \begin{bmatrix} p_1 & p_1 & p_1 & p_1 & p_1 & k \\ p_1 & p_2 & p_2 & p_2 & u_1 & k \\ p_1 & p_2 & p_3 & u_2 & u_1 & k \\ p_1 & p_2 & u_2 & u_2 & u_1 & k \\ p_1 & u_1 & u_1 & u_1 & u_1 & k \\ k & k & k & k & k & k \end{bmatrix}$$

where we start with the known elements k and then compute the u elements by using the updating property (7.7.22) and the elements p by using the persymmetry property (7.7.24) in the following order: $k \rightarrow p_1 \rightarrow u_1 \rightarrow p_2 \rightarrow u_2 \rightarrow p_3$. Clearly, because the matrix $\mathbf{Q}_M = \mathbf{R}_M^{-1}$ is both symmetric and persymmetric, we need to compute only the elements in the following wedge:

$$
\begin{array}{cccccc}
p_1 & p_1 & p_1 & p_1 & p_1 & k \\
 & p_2 & p_2 & p_2 & u_1 & \\
 & & p_3 & u_2 & &
\end{array}
$$

which can be easily extended to the general case. This algorithm, which was introduced by Trench (1964), requires $O(M^2)$ operations and is implemented by the function

```
Q=invtoepl(r,M)
```

The algorithm is generalized for complex Toeplitz matrices in Problem 7.40.

7.8 KALMAN FILTER ALGORITHM

The various optimum linear filter algorithms and structures that we discussed so far in this chapter provide us with the determination of filter coefficients or optimal estimates using some form of recursive update. Some algorithms and structures are order-recursive while others are time-recursive. In effect, they tell us how the past values should be updated to determine the present values. Unfortunately, these techniques do not lend themselves very well to the more complicated nonstationary problems. Readers will note carefully that the only case in which we obtained efficient order-recursive algorithms and structures was in the stationary environment, using the approaches of Levinson and Schür.

In 1960, R. E. Kalman provided an alternative approach to formulating the MMSE linear filtering problem using dynamic models. This "Kalman filter" technique was quickly hailed as a practical solution to a number of problems that were intractable using the more established Wiener methods. As we see in this section, the Kalman filter algorithm is actually a special case of the optimal linear filter algorithms that we have studied. However, it is used in a number of fields such as aerospace and navigation, where a signal trajectory can be well defined. Its use in statistical signal processing is somewhat limited (adaptive filters discussed in Chapter 10 are more appropriate). The two main features of the Kalman filter formulation and solution are the dynamic (or state-space) modeling of the random processes under consideration and the time-recursive processing of the input data.

In this section, we discuss only the discrete-time Kalman filter. The continuous-time version is covered in several texts including Gelb (1977) and Brown and Hwang (1997). As a motivation to this approach, we begin with the following estimation problem.

7.8.1 Preliminary Development

Suppose that we want to obtain a linear MMSE estimate of a random variable y using the related random variables (observations) $\{x_1, x_2, \ldots, x_m\}$, that is,

$$\hat{y}_m \triangleq E\{y|x_1, x_2, \ldots, x_m\} \tag{7.8.1}$$

as described in Section 7.1.5. Furthermore, we want to obtain this estimate in an order-recursive fashion, that is, determine $\hat{y}_m$ in terms of $\hat{y}_{m-1}$. We considered and solved this problem in Section 7.1. Our approach, which is somewhat different from that in Section 7.1, is as follows: Assume that we have computed the corresponding estimate $\hat{y}_{m-1}$, we have the observations $\{x_1, x_2, \ldots, x_m\}$, and we wish to determine the estimate $\hat{y}_m$. Then we carry out the following steps:

1. We first determine the optimal *one-step prediction* of x_m, that is,

$$
\begin{aligned}
\hat{x}_{m|m-1} &\triangleq \{x_m|x_1, x_2, \ldots, x_{m-1}\} \\
&= [\mathbf{R}_{m-1}^{-1} \mathbf{r}_{m-1}^b]^H \mathbf{x}_{m-1} = -\mathbf{b}_{m-1}^H \mathbf{x}_{m-1} \\
&= -\sum_{k=1}^{m-1} [b_k^{(m-1)}]^* x_k
\end{aligned}
\tag{7.8.2}
$$

where the vector and matrix quantities are as defined in Section 7.1.
2. When the new data value x_m is received, we determine the optimal *prediction error*

$$e_m^b \triangleq x_m - \hat{x}_{m|m-1} = w_m \tag{7.8.3}$$

which is the new information or innovations contained in the new data.
3. Determine a linear MMSE estimate of y, given the new information w_m:

$$E\{y|w_m\} = E\{y_m w_m^*\}(E\{w_m w_m^*\})^{-1} w_m \tag{7.8.4}$$

4. Finally, form a linear estimate $\hat{y}_m$ of the form

$$\hat{y}_m = \hat{y}_{m-1} + E\{y|w_m\} = \hat{y}_{m-1} + E\{y_m w_m^*\}(E\{w_m w_m^*\})^{-1} w_m \tag{7.8.5}$$

The algorithm is initialized with $\hat{y}_0 = 0$. Note that the quantity $E\{y_m w_m^*\}(E\{w_m w_m^*\})^{-1}$ is equal to the coefficient k_m^* and that we have rederived (7.1.51). For the implementation of (7.8.5), see Figure 7.1.

EXAMPLE 7.8.1. Let the observed random data be obtained from a stationary random process; that is, the data are of the form

$$\{x(1), x(2), \ldots, x(n), \ldots\} \qquad r(n, l) = r(n - l)$$

Also instead of estimating a single random variable, we want to estimate the sample $y(n)$ of a random process $\{y(n)\}$ that is jointly stationary with $x(n)$. Then, following the analysis leading to (7.8.5), we obtain

$$\hat{y}(n) = \hat{y}(n - 1) + k_n^* w(n) = \hat{y}(n - 1) + k_n^*[x(n) + \sum_{k=0}^{n-1} [b_k^{(n-1)}]^* x(k)] \qquad (7.8.6)$$

It is interesting to note that, because of stationarity, we have a time-recursive algorithm in (7.8.6). The coefficients $\{k_n^*\}$ can be obtained recursively by using the algorithms of Levinson or Schür. However, the data prediction term does require a growing memory. Indeed, if we define the vector

$$\mathbf{x}(n) = [x(1)\, x(2)\, \cdots\, x(n)]^T$$

whose order is equal to time index n, we have

$$\hat{y}(n) = \sum_{k=1}^{n} [c_k^{(n)}]^* x(k) \triangleq \mathbf{c}_n^H \mathbf{x}(n)$$

The optimum estimator is given by

$$\mathbf{R}_n \mathbf{c}_n = \mathbf{d}_n$$

where
$$\mathbf{R}_n \triangleq E\{\mathbf{x}(n)\mathbf{x}^H(n)\} \qquad \mathbf{d}_n \triangleq E\{\mathbf{x}(n)y^*(n)\}$$

Since, owing to stationarity, the matrix $\mathbf{R}_n$ is Toeplitz, we can derive a lattice-ladder structure $\{k_n, k_n^c\}$ that solves this problem recursively (see Section 7.4). When each new observation $\{y(n + 1)\}$ is received, we use the moments $r(n + 1)$ and $d(n + 1)$ to compute new lattice-ladder parameters $\{k_{n+1}, k_{n+1}^c\}$ and we add a new stage to the "growing-order" (and, therefore, growing-memory) filter.

The above example underscores two problems with our estimation technique if we were to obtain a true time-recursive algorithm with finite memory. The first problem concerns the time-recursive update for the k_m^* term or, in particular, for $E\{y_m w_m^*\}$ and $(E\{w_m w_m^*\})^{-1}$. We alluded to this problem in Section 7.1. In the example, we solved this problem by assuming a stationary signal environment. The second problem deals with the infinite memory in (7.8.2). This problem can be solved if we are able to compute the data prediction term also in a time-recursive fashion. In the stationary case, this problem can be solved by using the Levinson-Durbin or Schür algorithm. For nonstationary situations, the above two problems are solved by the Kalman filter by assuming appropriate dynamic models for the process to be estimated and for the observation data.

Consider the optimal one-step prediction term in (7.8.2), defined as

$$\hat{x}(n|n - 1) \triangleq E\{x(n)|x(0), \ldots, x(n - 1)\} \qquad (7.8.7)$$

which requires growing memory. If we assume the following linear data relation model

$$x(n) = H(n)y(n) + v(n) \qquad (7.8.8)$$

with
$$E\{v(n)y^*(l)\} = 0 \qquad \text{for all } n, l \qquad (7.8.9)$$

$$E\{v(n)v^*(l)\} = r_v(n)\delta_{n,l} \qquad \text{for all } n, l \qquad (7.8.10)$$

then (7.8.7) becomes

$$\hat{x}(n|n - 1) = E\{[H(n)y(n) + v(n)]|x(0), \ldots, x(n - 1)\}$$
$$= H(n)\hat{y}(n|n - 1) \qquad (7.8.11)$$

where we have used the notation

$$\hat{y}(n|n - 1) \triangleq E\{y(n)|x(0), \ldots, x(n - 1)\} \qquad (7.8.12)$$

Thus, we will be successful in obtaining a finite-memory computation for $\hat{x}(n|n-1)$ if we can obtain a recursion for $\hat{y}(n|n-1)$ in terms of $\hat{y}(n-1|n-1)$. This is possible if we assume the following linear signal model

$$y(n) = a(n-1)y(n-1) + \eta(n) \tag{7.8.13}$$

with appropriate statistical assumptions on the random process $\eta(n)$. Thus it is now possible to complete the development of the Kalman filter. The *signal model* (7.8.13) provides the dynamics of the time evolution of the signal to be estimated while (7.8.8) is known as the *observation model*, since it relates the signal $y(n)$ with the observation $x(n)$. These models are formally defined in the next section.

7.8.2 Development of Kalman Filter

Since the Kalman filter is also well suited for vector processes, we begin by assuming that the random process to be estimated can be modeled in the form

$$\mathbf{y}(n) = \mathbf{A}(n-1)\mathbf{y}(n-1) + \mathbf{B}(n)\boldsymbol{\eta}(n) \tag{7.8.14}$$

which is known as the *signal* (or *state vector*) *model* where

    $\mathbf{y}(n) = k \times 1$ signal *state vector* at time n

  $\mathbf{A}(n-1) = k \times k$ matrix that relates $\mathbf{y}(n-1)$ to $\mathbf{y}(n)$ in absence of a forcing function

    $\boldsymbol{\eta}(n) = k \times 1$ zero-mean white noise sequence with covariance matrix $\mathbf{R}_\eta(n)$

    $\mathbf{B}(n) = k \times k$ *input* matrix

$$\tag{7.8.15}$$

The matrix $\mathbf{A}(n-1)$ is known as the *state-transition matrix* while $\boldsymbol{\eta}(n)$ is also known as the *modeling error vector*.

    The *observation* (or *measurement*) *model* is described using the linear relationship

$$\mathbf{x}(n) = \mathbf{H}(n)\mathbf{y}(n) + \mathbf{v}(n) \tag{7.8.16}$$

where

    $\mathbf{x}(n) = m \times 1$ signal state vector at time n

    $\mathbf{H}(n) = m \times k$ matrix that gives ideal linear relationship between $\mathbf{y}(n)$ and $\mathbf{x}(n)$

    $\mathbf{v}(n) = k \times 1$ zero-mean white noise sequence with covariance matrix $\mathbf{R}_v(n)$

$$\tag{7.8.17}$$

The matrix $\mathbf{H}(n)$ is known as the *output* matrix, and the sequence $\mathbf{v}(n)$ is known as the *observation error*.

    We further assume the following statistical properties:

$$E\{\mathbf{y}(n)\mathbf{v}^H(l)\} = \mathbf{0} \quad \text{for all } n, l \tag{7.8.18}$$

$$E\{\boldsymbol{\eta}(n)\mathbf{v}^H(l)\} = \mathbf{0} \quad \text{for all } n, l \tag{7.8.19}$$

$$E\{\boldsymbol{\eta}(n)\mathbf{y}^H(-1)\} = \mathbf{0} \quad \text{for all } n \tag{7.8.20}$$

$$E\{\mathbf{y}(-1)\} = \mathbf{0} \tag{7.8.21}$$

$$E\{\mathbf{y}(-1)\mathbf{y}^H(-1)\} = \mathbf{R}_y(-1) \tag{7.8.22}$$

The first three relations, (7.8.18) to (7.8.20), imply orthogonality between respective random variables while the last two, (7.8.21) and (7.8.22), establish the mean and covariance of the initial-condition vector $\mathbf{y}(-1)$.

    From (7.8.14) and (7.8.21) the mean of $\mathbf{y}(n) = \mathbf{0}$ for all n, and the evolution of its correlation matrix is given by

$$\mathbf{R}_y(n) = \mathbf{A}(n-1)\mathbf{R}_y(n-1)\mathbf{A}^H(n-1) + \mathbf{B}(n)\mathbf{R}_\eta(n)\mathbf{B}^H(n) \tag{7.8.23}$$

From (7.8.16), the mean of $\mathbf{x}(n) = \mathbf{0}$ for all n, and from (7.8.23) the evolution of its correlation matrix is given by

$$\begin{aligned}
\mathbf{R}_x(n) &= \mathbf{H}(n)[\mathbf{A}(n-1)\mathbf{R}_y(n-1)\mathbf{A}^H(n-1) \\
&\quad + \mathbf{B}(n)\mathbf{R}_\eta(n)\mathbf{B}^H(n)]\mathbf{H}^H(n) + \mathbf{R}_v(n)
\end{aligned} \tag{7.8.24}$$

Evolution of optimal estimates

We now assume that we have available the MMSE estimate $\hat{\mathbf{y}}(n-1|n-1)$ of $\mathbf{y}(n-1)$ based on the observations up to and including time $n-1$. Using (7.8.14) and (7.8.20), the one-step prediction of $\mathbf{y}(n)$ is given by

$$\hat{\mathbf{y}}(n|n-1) = \mathbf{A}(n-1)\hat{\mathbf{y}}(n-1|n-1) \tag{7.8.25}$$

with initial condition $\hat{\mathbf{y}}(-1|-1) = \mathbf{y}(-1)$. From (7.8.16), the one-step prediction of $\mathbf{x}(n)$ is given by

$$\hat{\mathbf{x}}(n|n-1) = \mathbf{H}(n)\hat{\mathbf{y}}(n|n-1) = \mathbf{H}(n)\mathbf{A}(n-1)\hat{\mathbf{y}}(n-1|n-1) \tag{7.8.26}$$

Thus we have a recursive formula to compute the predicted observation. The prediction error (7.8.3) from (7.8.16) is now given by

$$\begin{aligned}
\mathbf{w}(n) &= \mathbf{x}(n) - \hat{\mathbf{x}}(n|n-1) \\
&= \mathbf{H}(n)\mathbf{y}(n) + \mathbf{v}(n) - \mathbf{H}(n)\hat{\mathbf{y}}(n|n-1) \\
&= \mathbf{H}(n)\tilde{\mathbf{y}}(n|n-1) + \mathbf{v}(n)
\end{aligned} \tag{7.8.27}$$

where we have defined the signal prediction error

$$\tilde{\mathbf{y}}(n|n-1) \triangleq \mathbf{y}(n) - \hat{\mathbf{y}}(n|n-1) \tag{7.8.28}$$

Now the quantity corresponding to $E\{w_m w_m^*\}$ in (7.8.5) is given by

$$\mathbf{R}_w(n) = E\{\mathbf{w}(n)\mathbf{w}^H(n)\} = \mathbf{H}(n)\mathbf{R}_{\tilde{y}}(n|n-1)\mathbf{H}^H(n) + \mathbf{R}_v(n) \tag{7.8.29}$$

where

$$\mathbf{R}_{\tilde{y}}(n|n-1) \triangleq E\{\tilde{\mathbf{y}}(n|n-1)\tilde{\mathbf{y}}^H(n|n-1)\} \tag{7.8.30}$$

is called the prediction (*a priori*) error covariance matrix. Similarly, from (7.8.27) the quantity corresponding to $E\{y_m w_m^*\}$ in (7.8.5) is given by

$$\begin{aligned}
E\{\mathbf{y}(n)\mathbf{w}^H(n)\} &= E\{\mathbf{y}(n)[\tilde{\mathbf{y}}^H(n|n-1)\mathbf{H}^H(n) + \mathbf{v}^H(n)]\} \\
&= E\{[\tilde{\mathbf{y}}(n|n-1) + \hat{\mathbf{y}}(n|n-1)] \\
&\quad \times [\tilde{\mathbf{y}}^H(n|n-1)\mathbf{H}^H(n) + \mathbf{v}^H(n)]\} \\
&= E\{\tilde{\mathbf{y}}(n|n-1)\tilde{\mathbf{y}}^H(n|n-1)\}\mathbf{H}^H(n) \\
&= \mathbf{R}_{\tilde{y}}(n|n-1)\mathbf{H}^H(n)
\end{aligned} \tag{7.8.31}$$

since the optimal prediction error $\tilde{\mathbf{y}}(n|n-1)$ is orthogonal to the optimal prediction $\hat{\mathbf{y}}(n|n-1)$. Now the updated MMSE estimate (which is also known as the *filtered* estimate) corresponding to (7.8.5) is

$$\begin{aligned}
\hat{\mathbf{y}}(n|n) &= \hat{\mathbf{y}}(n|n-1) + \mathbf{R}_{\tilde{y}}(n|n-1)\mathbf{H}^H(n)\mathbf{R}_w^{-1}(n)\{\mathbf{x}(n) - \hat{\mathbf{x}}(n|n-1)\} \\
&= \hat{\mathbf{y}}(n|n-1) + \mathbf{K}(n)\{\mathbf{x}(n) - \mathbf{H}(n)\hat{\mathbf{y}}(n|n-1)\}
\end{aligned} \tag{7.8.32}$$

where we have defined a new quantity

$$\mathbf{K}(n) \triangleq \mathbf{R}_{\tilde{y}}(n|n-1)\mathbf{H}^H(n)\mathbf{R}_w^{-1}(n) \tag{7.8.33}$$

which is known as the *Kalman gain matrix* and where $\hat{\mathbf{y}}(n|n-1)$ is given in terms of $\hat{\mathbf{y}}(n-1|n-1)$ using (7.8.25). Thus we have

$$\begin{aligned}
\text{Prediction:} \quad & \hat{\mathbf{y}}(n|n-1) = \mathbf{A}(n-1)\hat{\mathbf{y}}(n-1|n-1) \\
\text{Filter:} \quad & \hat{\mathbf{y}}(n|n) = \hat{\mathbf{y}}(n|n-1) + \mathbf{K}(n)\{\mathbf{x}(n) - \mathbf{H}(n)\hat{\mathbf{y}}(n|n-1)\}
\end{aligned} \tag{7.8.34}$$

and we have succeeded in obtaining a time-updating algorithm for recursively computing the MMSE estimates. All that remains is a time evolution of the gain matrix $\mathbf{K}(n)$. Since $\mathbf{R}_w(n)$ from (7.8.29) also depends on $\mathbf{R}_{\tilde{y}}(n|n-1)$, what we need is an update equation for the error covariance matrix.

Evolution of error covariance matrices

First we define the filtered error as

$$
\begin{aligned}
\tilde{\mathbf{y}}(n|n) &\triangleq \mathbf{y}(n) - \hat{\mathbf{y}}(n|n) \\
&= \mathbf{y}(n) - \hat{\mathbf{y}}(n|n-1) - \mathbf{K}(n)\{\mathbf{x}(n) - \mathbf{H}(n)\hat{\mathbf{y}}(n|n-1)\} \\
&= \tilde{\mathbf{y}}(n|n-1) - \mathbf{K}(n)\mathbf{w}(n)
\end{aligned}
\tag{7.8.35}
$$

where we have used (7.8.27) and (7.8.34). Then the filtered error covariance is given by

$$
\begin{aligned}
\mathbf{R}_{\tilde{y}}(n|n) &\triangleq E\{\tilde{\mathbf{y}}(n|n)\tilde{\mathbf{y}}^H(n|n)\} \\
&= \mathbf{R}_{\tilde{y}}(n|n-1) - \mathbf{K}(n)\mathbf{R}_w(n)\mathbf{K}^H(n) \\
&= \mathbf{R}_{\tilde{y}}(n|n-1) - \mathbf{K}(n)\mathbf{R}_w(n)\mathbf{R}_w^{-1}(n)\mathbf{H}(n)\mathbf{R}_{\tilde{y}}(n|n-1) \\
&= [\mathbf{I} - \mathbf{K}(n)\mathbf{H}(n)]\mathbf{R}_{\tilde{y}}(n|n-1)
\end{aligned}
\tag{7.8.36}
$$

where in the second-to-last step we substituted (7.8.33) for $\mathbf{K}^H(n)$. The error covariance $\mathbf{R}_{\tilde{y}}(n|n)$ is also known as the *a posteriori* error covariance. Finally, we need to determine the a priori prediction error covariance at time n from $\mathbf{R}_{\tilde{y}}(n-1|n-1)$ to complete the recursive calculations. From the prediction equation in (7.8.34), we obtain the prediction error at time n as

$$
\begin{aligned}
\mathbf{y}(n) - \hat{\mathbf{y}}(n|n-1) &= \mathbf{A}(n-1)\mathbf{y}(n-1) + \mathbf{B}(n)\boldsymbol{\eta}(n) - \mathbf{A}(n-1)\hat{\mathbf{y}}(n-1|n-1) \\
\tilde{\mathbf{y}}(n|n-1) &= \mathbf{A}(n-1)\tilde{\mathbf{y}}(n-1|n-1) + \mathbf{B}(n)\boldsymbol{\eta}(n)
\end{aligned}
\tag{7.8.37}
$$

or $\quad \mathbf{R}_{\tilde{y}}(n|n-1) = \mathbf{A}(n-1)\mathbf{R}_{\tilde{y}}(n-1|n-1)\mathbf{A}^H(n-1) + \mathbf{B}(n)\mathbf{R}_\eta(n)\mathbf{B}^H(n) \quad (7.8.38)$

with initial condition $\mathbf{R}_{\tilde{y}}(-1|-1) = \mathbf{R}_y(-1)$. Thus we have

A priori error covariance: $\quad \mathbf{R}_{\tilde{y}}(n|n-1) = \mathbf{A}(n-1)\mathbf{R}_{\tilde{y}}(n-1|n-1)\mathbf{A}^H(n-1)$
$$+ \mathbf{B}(n)\mathbf{R}_\eta(n)\mathbf{B}^H(n)$$

Kalman gain: $\quad \mathbf{K}(n) = \mathbf{R}_{\tilde{y}}(n|n-1)\mathbf{H}^H(n)\mathbf{R}_w^{-1}(n)$

A posteriori error covariance: $\quad \mathbf{R}_{\tilde{y}}(n|n) = [\mathbf{I} - \mathbf{K}(n)\mathbf{H}(n)]\mathbf{R}_{\tilde{y}}(n|n-1)$

$$\tag{7.8.39}$$

The complete Kalman Filter algorithm is given in Table 7.5, and the block diagram description is provided in Figure 7.11.

EXAMPLE 7.8.2. Let $y(n)$ be an AR(2) process described by

$$
y(n) = 1.8y(n-1) - 0.81y(n-2) + 0.1\eta(n) \qquad n \geq 0
\tag{7.8.40}
$$

where $\eta(n) \sim \text{WGN}(0, 1)$ and $y(-1) = y(-2) = 0$. We want to determine the linear MMSE estimate of $y(n)$, $n \geq 0$, by observing

$$
x(n) = y(n) + \sqrt{10}v(n) \qquad n \geq 0
\tag{7.8.41}
$$

where $v(n) \sim \text{WGN}(0, 10)$ and orthogonal to $\eta(n)$.

Solution. From (7.8.40) and (7.8.41), we first formulate the state vector and observation equations:

$$
\mathbf{y}(n) \triangleq \begin{bmatrix} y(n) \\ y(n-1) \end{bmatrix} = \begin{bmatrix} 1.8 & -0.81 \\ 1 & 0 \end{bmatrix} \begin{bmatrix} y(n-1) \\ y(n-2) \end{bmatrix} + \begin{bmatrix} 0.1 \\ 0 \end{bmatrix} \eta(n)
\tag{7.8.42}
$$

and

$$
x(n) = \begin{bmatrix} 1 & 0 \end{bmatrix} \begin{bmatrix} y(n) \\ y(n-1) \end{bmatrix} + \sqrt{10}v(n)
\tag{7.8.43}
$$

TABLE 7.5

Summary of the Kalman filter algorithm.

1. **Input:**

    (a) Signal model parameters: $\mathbf{A}(n-1), \mathbf{B}(n), \mathbf{R}_\eta(n); n = 0, 1, 2, \ldots$

    (b) Observation model parameters: $\mathbf{H}(n), \mathbf{R}_v(n); n = 0, 1, 2, \ldots$

    (c) Observation data: $\mathbf{y}(n); n = 0, 1, 2, \ldots$

2. **Initialization:** $\hat{\mathbf{y}}(0|-1) = \mathbf{y}(-1) = \mathbf{0}; \mathbf{R}_{\tilde{y}}(-1|-1) = \mathbf{R}_y(-1)$

3. **Time recursion:**   For $n = 0, 1, 2, \ldots$

    (a) Signal prediction: $\hat{\mathbf{y}}(n|n-1) = \mathbf{A}(n-1)\hat{\mathbf{y}}(n-1|n-1)$

    (b) Data prediction: $\hat{\mathbf{x}}(n|n-1) = \mathbf{H}(n)\hat{\mathbf{y}}(n|n-1)$

    (c) A priori error covariance:
    $$\mathbf{R}_{\tilde{y}}(n|n-1) = \mathbf{A}(n-1)\mathbf{R}_{\tilde{y}}(n-1|n-1)\mathbf{A}^H(n-1) + \mathbf{B}(n)\mathbf{R}_\eta(n)\mathbf{B}^H(n)$$

    (d) Kalman gain:
    $$\mathbf{K}(n) = \mathbf{R}_{\tilde{y}}(n|n-1)\mathbf{H}^H(n)\mathbf{R}_w^{-1}(n)$$
    $$\mathbf{R}_w(n) = \mathbf{H}(n)\mathbf{R}_{\tilde{y}}(n|n-1)\mathbf{H}^H(n + \mathbf{R}_v(n)$$

    (e) Signal update: $\hat{\mathbf{y}}(n|n) = \hat{\mathbf{y}}(n|n-1) + \mathbf{K}(n)[\mathbf{x}(n) - \hat{\mathbf{x}}(n|n-1)]$

    (f) A posteriori error covariance:
    $$\mathbf{R}_{\tilde{y}}(n|n) = [\mathbf{I} - \mathbf{K}(n)\mathbf{H}(n)]\mathbf{R}_{\tilde{y}}(n|n-1)$$

4. **Output:**   Filtered estimate $\hat{\mathbf{y}}(n|n), n = 0, 1, 2, \ldots$

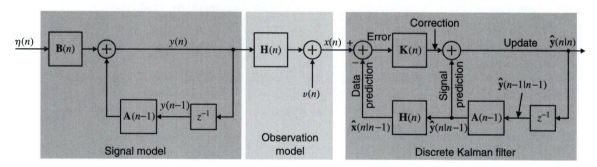

FIGURE 7.11
The block diagram of the Kalman filter model and algorithm.

Hence the relevant matrix quantities are

$$\mathbf{A}(n) = \begin{bmatrix} 1.8 & -0.81 \\ 1 & 0 \end{bmatrix} \qquad \mathbf{B}(n) = \begin{bmatrix} 0.1 \\ 0 \end{bmatrix} \qquad \mathbf{R}_\eta(n) = 1$$

and
$$\mathbf{H}(n) = [1 \quad 0] \qquad \mathbf{R}_v(n) = 10 \tag{7.8.44}$$

Now the Kalman filter equation from Table 7.5 can be implemented with zero initial conditions. Note that since the system matrices are constant, the processes $x(n)$ and $y(n)$ are asymptotically stationary.

Using (7.8.40) and (7.8.41), we generated 100 samples of $y(n)$ and $x(n)$. The observation $x(n)$ was processed using the Kalman filter equations to obtain $\hat{y}_f(n) = \hat{y}(n|n)$, and the results are shown in Figure 7.12. Owing to a large observation noise variance, the $x(n)$ values are very noisy around the signal $y(n)$ values. However, the Kalman filter was able to track $x(n)$ closely and reduce the noise $v(n)$ degradation. In Figure 7.13 we show the evolution of Kalman filter gain values $K_1(n)$ and $K_2(n)$ along with the estimation error variance. The filter reaches its steady state in about 20 samples and becomes a stationary filter as expected. In such situations, the gain and error covariance equations can be implemented off-line (since these equations are data-independent) to obtain a constant-gain matrix. The data then can be filtered using this constant gain to reduce on-line computational complexity.

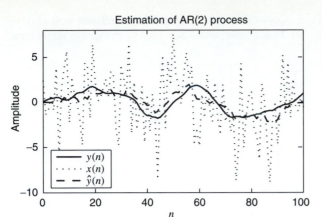

Estimation of AR(2) process

FIGURE 7.12
Estimation of AR(2) process
using Kalman filter in
Example 7.8.2.

385

SECTION 7.8
Kalman Filter Algorithm

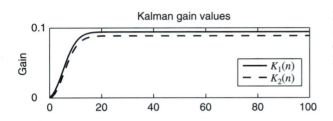

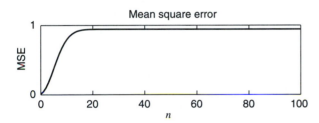

FIGURE 7.13
Kalman filter gains and
estimation error covariance in
Example 7.8.2.

In the next example, we consider the case of the estimation of position of an object in a linear motion subjected to random acceleration.

EXAMPLE 7.8.3. Consider an object traveling in a straight-line motion that is perturbed by random acceleration. Let $y_p(n) = y_c(nT)$ be the true position of the object at the nth sampling instant, where T is the sampling interval in seconds and $y_c(t)$ is the instantaneous position. This position is measured by a sensor that records noisy observations. Let $x(n)$ be the measured position at the nth sampling instant. Then we can model the observation as

$$x(n) = y_p(n) + v(n) \qquad n \geq 0 \tag{7.8.45}$$

where $v(n) \sim \text{WGN}(0, \sigma_v^2)$. To derive the state dynamic equation, we assume that the object is in a steady-state motion (except for the random acceleration). Let $y_v(n) = \dot{y}_c(nT)$ be the true velocity at the nth sampling instant, where $\dot{y}_c(t)$ is the instantaneous velocity. Then we have the following equations of motion

$$y_v(n) = y_v(n-1) + y_a(n-1)T \tag{7.8.46}$$

$$y_p(n) = y_p(n-1) + y_v(n-1)T + \frac{1}{2}y_a(n-1)T^2 \tag{7.8.47}$$

where we have assumed that the acceleration $\ddot{y}_c(t)$ is constant over the sampling interval and that $y_a(n-1)$ is the acceleration over $(n-1)T \leq t < nT$. We now define the state vector as

$$\mathbf{y}(n) \triangleq \begin{bmatrix} y_p(n) \\ y_v(n) \end{bmatrix} \tag{7.8.48}$$

and the modeling error as $\eta(n) \triangleq y_a(n - 1)$, which is assumed to be random with $\eta(n) \sim$ WGN$(0, \sigma_\eta^2)$ and orthogonal to $v(n)$. Thus (7.8.46) and (7.8.47) can be arranged in vector form as

$$\mathbf{y}(n) = \begin{bmatrix} 1 & T \\ 0 & 1 \end{bmatrix} \mathbf{y}(n - 1) + \begin{bmatrix} \dfrac{T^2}{2} \\ T \end{bmatrix} \eta(n) \qquad n \geq 0 \qquad (7.8.49)$$

Thus we have

$$\mathbf{A} = \begin{bmatrix} 1 & T \\ 0 & 1 \end{bmatrix} \qquad \text{and} \qquad \mathbf{B} = \begin{bmatrix} \dfrac{T^2}{2} \\ T \end{bmatrix}$$

Similarly, the observation (7.8.45) is given by

$$x(n) = [1 \; 0] \mathbf{y}(n) + v(n) \qquad n \geq 0 \qquad (7.8.50)$$

and hence $\mathbf{H} = [1 \; 0]$. Let the initial conditions be $y_p(-1)$ and $y_v(-1)$. Now given the noisy observations $\{x(n)\}$ and all the necessary information $[T, \sigma_v^2, \sigma_\eta^2, y_p(-1),$ and $y_v(-1)]$, we can recursively estimate the position and velocity of the object at each sampling instance. An approach similar to this is used in aircraft navigation systems.

Using the following values

$$T = 0.1 \qquad \sigma_v^2 = \sigma_\eta^2 = 0.25 \qquad y_p(-1) = 0 \qquad y_v(-1) = 1$$

we simulated the trajectory of the object over [0, 10] second interval. From Table 7.5 Kalman filter equations were obtained, and the true positions as well as velocities were estimated using the noisy positions. Figure 7.14 shows the estimation results. The top graph shows the true, noisy, and estimated positions. The bottom graph shows the true and estimated velocities. Due to random acceleration values (which are moderate), the true velocity has small deviations from the constant value of 1 while the true position trajectory is approximately linear. The estimates of the position follow the true values very closely. However, the velocity estimates have more errors around the true velocities. This is because no direct measurements of velocities are available; therefore, the velocity of the object can be inferred only from position measurements.

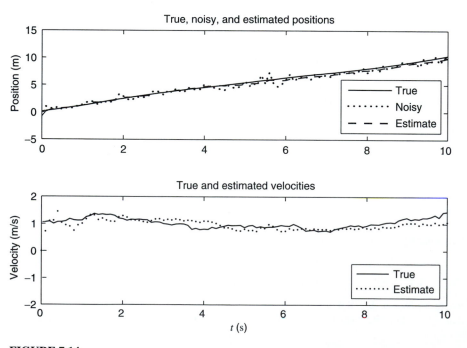

FIGURE 7.14
Estimation of positions and velocities using Kalman filter in Example 7.8.3.

In Figure 7.15, we show the trajectories of Kalman gain values and trace of the error covariance matrices. The top graph contains the gain values corresponding to position (K_p) and velocity (K_v). The steady state of the filter is reached in about 3 s. The bottom left graph contains the a priori and a posteriori error covariances, which also reach the steady-state values in 3 s and which appear to be very close to each other. Therefore, in the bottom right graph we show an exploded view of the steady-state region over a 1-s interval. It is interesting to note that the steady-state error covariances before and after processing an observation are not the same. As a result of making an observation, the a posteriori errors are reduced from the a priori ones. However, owing to random acceleration, the errors increase during the intervals between observations. This is shown as dotted lines in Figure 7.15. The steady state is reached when the decrease in errors achieved by each observation is canceled by the increase between observations.

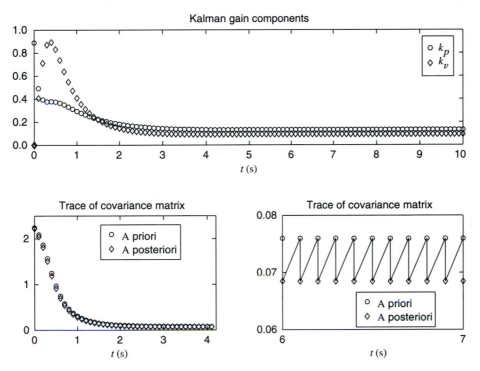

FIGURE 7.15
Kalman filter gains and estimation error variances in Example 7.8.3.

It should be clear from the above two examples that the Kalman filter can recursively estimate signal values because of the assumption of dynamic models (7.8.14) and (7.8.16). Therefore, in this sense, the Kalman filter approach is a special case of the more general Wiener filter problem that we considered earlier. In many signal processing applications (e.g., data communication systems), assumption of such models is difficult to justify, which limits the use of Kalman filters.

7.9 SUMMARY

The application of optimum FIR filters and linear combiners involves the following two steps.

• *Design*. In this step, we determine the optimum values of the estimator parameters by solving the normal equations formed by using the *known* second-order moments. For stationary processes the design step is done only once. For nonstationary processes, we repeat the design when the statistics change.

- *Implementation*. In this step, we use the optimum parameters and the input data to compute the optimum estimate.

The type and complexity of the algorithms and structures available for the design and implementation of linear MMSE estimators depend on two factors:

- The shift invariance of the input data vector.
- The stationarity of the signals that determine the second-order moments in the normal equations.

As we introduce more structure (shift invariance or stationarity), the algorithms and structures become simpler. From a mathematical point of view, this is reflected in the structure of the correlation matrix, which starting from general Hermitian at one end becomes Toeplitz at the other.

Linear combiners

The input vector is not shift-invariant because the optimum estimate is computed by using samples from M different signals. The correlation matrix $\mathbf{R}$ is Hermitian and usually positive definite. The normal equations are solved by using the LDL^H decomposition, and the optimum estimate is computed by using the obtained parameters. However, in many applications where we need the optimum estimate and not the coefficients of the optimum combiner, we can implement the MMSE linear combiner, using the *orthogonal order-recursive structure* shown in Figure 7.1. This structure consists of two parts: (1) a triangular decorrelator (orthogonalizer) that decorrelates the input data vector and produces its innovations vector and (2) a linear combiner that combines the uncorrelated innovations to compute the optimum estimates for *all orders* $1 \leq m \leq M$.

FIR filters and predictors

In this case the input data vector is shift-invariant, which leads to simplifications, whose extent depends on the stationarity of the involved signals.

Nonstationary case. In general, the correlation matrix is Hermitian and positive definite with no additional structure, and the LDL^H decomposition is the recommended method to solve the normal equations. However, the input shift invariance leads to a remarkable coupling between FLP, BLP, and FIR filtering, resulting in a simplified orthogonal order-recursive structure, which now takes the form of a lattice ladder filter (see Figure 7.3). The backward prediction errors of all orders $1 \leq m \leq M$ provide the innovations of the input data vector. The parameters of lattice structure (decorrelator) are specified by the components of the LDL^H decomposition of the input correlation matrix. The coefficients of the ladder part (correlator) depend on both the input correlation matrix and the cross-correlation between the desired response and the input data vector.

Stationary case. In this case, the addition of stationarity to the shift invariance makes the correlation matrix Toeplitz. The presence of the Toeplitz structure has the following consequences:

1. The development of efficient order-recursive algorithms, with computational complexity proportional to M^2, for the solution of the normal equations and the triangularization of the correlation matrix.

   a. Levinson algorithm solves $\mathbf{Rc} = \mathbf{d}$ for arbitrary right-hand side vector $\mathbf{d}$ ($2M^2$ operations).
   b. Levinson-Durbin algorithm solves $\mathbf{Ra} = -\mathbf{r}^*$ when the right-hand side has special structure (M^2 operations).
   c. Schür algorithm computes directly the lattice-ladder parameters from the autocorrelation and cross-correlation sequences.

The MMSE FLP, BLP, and FIR filters are time-invariant; that is, their coefficients (direct-form or lattice-ladder structures) are constant and should be computed only once.

The algorithms for MMSE filtering and prediction of stationary processes are the simplest ones. However, we can also develop efficient algorithms for nonstationary processes that have special structure. There are two cases of interest:

- The Kalman filtering algorithm that can be used for processes generated by a state-space model with *known* parameters.
- Algorithms for α-stationary processes, that is, processes whose correlation matrix is near to Toeplitz, as measured by a special distance known as the *displacement rank* (Morf et al. 1977).

PROBLEMS

7.1 By first computing the matrix product

$$
\begin{bmatrix} \mathbf{R}_m & \mathbf{r}_m^b \\ \mathbf{r}_m^{bH} & \rho_m^b \end{bmatrix} \begin{bmatrix} \mathbf{I}_m & \mathbf{0}_m \\ -\dfrac{\mathbf{r}_m^{bH}}{\rho_m^b} & \dfrac{1}{\rho_m^b} \end{bmatrix}
$$

and then the determinants of both sides, prove Equation (7.1.25). Another proof, obtained using the LDL^H decomposition, is given by Equation (7.2.4).

7.2 Prove the matrix inversion lemma for lower right corner partitioned matrices, which is described by Equations (7.1.26) and (7.1.28).

7.3 This problem generalizes the matrix inversion lemmas to nonsymmetric matrices.

    (*a*) Show that if $\mathbf{R}^{-1}$ exists, the inverse of an upper left corner partitioned matrix is given by

$$
\begin{bmatrix} \mathbf{R} & \mathbf{r} \\ \tilde{\mathbf{r}}^T & \sigma \end{bmatrix}^{-1} = \frac{1}{\alpha} \begin{bmatrix} \alpha \mathbf{R}^{-1} + \mathbf{w}\mathbf{v}^T & \mathbf{w} \\ \mathbf{v}^T & 1 \end{bmatrix}
$$

where

$$\mathbf{R}\mathbf{w} \triangleq -\mathbf{r}$$
$$\mathbf{R}^T\mathbf{v} \triangleq -\tilde{\mathbf{r}}$$
$$\alpha \triangleq \sigma - \tilde{\mathbf{r}}^T \mathbf{R}^{-1}\mathbf{r} = \sigma + \mathbf{v}^T\mathbf{r} = \sigma + \tilde{\mathbf{r}}^T\mathbf{w}$$

    (*b*) Show that if $\mathbf{R}^{-1}$ exists, the inverse of a lower right corner partitioned matrix is given by

$$
\begin{bmatrix} \sigma & \tilde{\mathbf{r}}^T \\ \mathbf{r} & \mathbf{R} \end{bmatrix}^{-1} = \frac{1}{\alpha} \begin{bmatrix} 1 & \mathbf{v}^T \\ \mathbf{w} & \alpha \mathbf{R}^{-1} + \mathbf{w}\mathbf{v}^T \end{bmatrix}
$$

where

$$\mathbf{R}\mathbf{w} \triangleq -\mathbf{r}$$
$$\mathbf{R}^T\mathbf{v} \triangleq -\tilde{\mathbf{r}}$$
$$\alpha \triangleq \sigma - \tilde{\mathbf{r}}^T \mathbf{R}^{-1}\mathbf{r} = \sigma + \mathbf{v}^T\mathbf{r} = \sigma + \tilde{\mathbf{r}}^T\mathbf{w}$$

    (*c*) Check the validity of the lemmas in parts (*a*) and (*b*), using MATLAB.

7.4 Develop an order-recursive algorithm to solve the linear system in Example 7.1.2, using the lower right corner partitioning lemma (7.1.26).

7.5 In this problem we consider two different approaches for inversion of symmetric and positive definite matrices by constructing an arbitrary fourth-order positive definite correlation matrix $\mathbf{R}$ and comparing their computational complexities.

    (*a*) Given that the inverse of a lower (upper) triangular matrix is itself lower (upper) triangular, develop an algorithm for triangular matrix inversion.

    (*b*) Compute the inverse of $\mathbf{R}$, using the algorithm in part (*a*) and Equation (7.1.58).

(c) Build up the inverse of **R**, using the recursion (7.1.24).

(d) Estimate the number of operations for each method as a function of order M, and check their validity for $M = 4$, using MATLAB.

7.6 Using the appropriate orthogonality principles and definitions, prove Equation (7.3.32).

7.7 Prove Equations (7.3.36) to (7.3.38), using Equation (7.1.45).

7.8 Working as in Example 6.3.1, develop an algorithm for the upper-lower decomposition of a symmetric positive definite matrix. Then use it to factorize the matrix in Example 6.3.1, and verify your results, using the function [U,D]=udut(R).

7.9 In this problem we explore the meaning of the various quantities in the decomposition $\mathbf{R} = \mathbf{U}\bar{\mathbf{D}}\mathbf{U}^H$ of the correlation matrix.

(a) Show that the rows of $\mathbf{A} = \mathbf{U}^{-1}$ are the MMSE estimator of x_m from $x_{m+1}, x_{m+2}, \ldots, x_M$.

(b) Show that the decomposition $\mathbf{R} = \mathbf{U}\bar{\mathbf{D}}\mathbf{U}^H$ can be obtained by the Gram-Schmidt orthogonalization process, starting with the random variable x_M and ending with x_1, that is, proceeding backward.

7.10 In this problem we clarify the various quantities and the form of the partitionings involved in the UDU^H decomposition, using an $m = 4$ correlation matrix.

(a) Prove that the components of the forward prediction error vector (7.3.65) are uncorrelated.

(b) Writing explicitly the matrix **R**, identify and express the quantities in Equations (7.3.62) through (7.3.67).

(c) Using the matrix **R** in Example 6.3.1, compute the predictors in (7.3.67) by using the corresponding normal equations, verify your results, comparing them with the rows of matrix **A** computed directly from the LDL^H decomposition of $\mathbf{R}^{-1}$ or the UDU^H decomposition of **R** (see Table 7.1).

7.11 Given an all-zero lattice filter with coefficients k_0 and k_1, determine the MSE $P(k_0, k_1)$ as a function of the required second-order moments, assumed jointly stationary, and plot the error performance surface. Use the statistics in Example 6.2.1.

7.12 Given the autocorrelation $r(0) = 1, r(1) = r(2) = \frac{1}{2}$, and $r(3) = \frac{1}{4}$, determine all possible representations for the third-order prediction error filter (see Figure 7.7).

7.13 Repeat Problem 7.12 for $k_1 = k_2 = k_3 = \frac{1}{3}$ and $P_3 = (\frac{2}{3})^3$.

7.14 Use Levinson's algorithm to solve the normal equations $\mathbf{R}\mathbf{c} = \mathbf{d}$ where $\mathbf{R} = $ Toeplitz$\{3, 2, 1\}$ and $\mathbf{d} = [6\ 6\ 2]^T$.

7.15 Consider a random sequence with autocorrelation $\{r(l)\}_0^3 = \{1, 0.8, 0.6, 0.4\}$. (a) Determine the FLP $\mathbf{a}_m$ and the corresponding error P_m^f for $m = 1, 2, 3$. (b) Determine and draw the flow diagram of the third-order lattice prediction error filter.

7.16 Using the Levinson-Durbin algorithm, determine the third-order linear predictor $\mathbf{a}_3$ and the MMSE P_3 for a signal with autocorrelation $r(0) = 1, r(1) = r(2) = \frac{1}{2}$, and $r(3) = \frac{1}{4}$.

7.17 Given the autocorrelation sequence $r(0) = 1, r(1) = r(2) = \frac{1}{2}$, and $r(3) = \frac{1}{4}$, compute the lattice and direct-form coefficients of the prediction error filter, using the algorithm of Schür.

7.18 Determine ρ_1 and ρ_2 so that the matrix $\mathbf{R} = $ Toeplitz$\{1, \rho_1, \rho_2\}$ is positive definite.

7.19 Suppose that we want to fit an AR(2) model to a sinusoidal signal with random phase in additive noise. The autocorrelation sequence is given by

$$r(l) = P_0 \cos \omega_0 l + \sigma_v^2 \delta(l)$$

(a) Determine the model parameters $a_1^{(2)}$, $a_2^{(2)}$, and σ_w^2 in terms of P_0, ω_0, and σ_v^2. (b) Determine the lattice parameters of the model. (c) What are the limiting values of the direct and lattice parameters of the model when $\sigma_v^2 \to 0$?

7.20 Given the parameters $r(0) = 1$, $k_1 = k_2 = \frac{1}{2}$, and $k_3 = \frac{1}{4}$, determine all other equivalent representations of the prediction error filter (see Figure 7.7).

7.21 Let $\{r(l)\}_0^P$ be samples of the autocorrelation sequence of a stationary random signal $x(n)$. (a) Is it possible to extend $r(l)$ for $|l| > P$ so that the PSD

$$R(e^{j\omega}) = \sum_{l=-\infty}^{\infty} r(l)e^{-j\omega l}$$

is valid, that is, $R(e^{j\omega}) \geq 0$? (b) Using the algorithm of Levinson-Durbin, develop a procedure to check if a given autocorrelation extension is valid. (c) Use the algorithm in part (b) to find the necessary and sufficient conditions so that $r(0) = 1$, $r(1) = \rho_1$, and $r(2) = \rho_2$ are a valid autocorrelation sequence. Is the resulting extension unique?

7.22 Justify the following statements. (a) The whitening filter for a stationary process $x(n)$ is time-varying. (b) The filter in part (a) can be implemented by using a lattice structure and switching its stages on one by one with the arrival of each new sample. (c) If $x(n)$ is AR(P), the whitening filter becomes time-invariant $P + 1$ sampling intervals after the first sample is applied. *Note:* We assume that the input is applied to the filter at $n = 0$. If the input is applied at $n = -\infty$, the whitening filter of a stationary process is always time-invariant.

7.23 Given the parameters $r(0) = 1$, $k_1 = \frac{1}{2}$, $k_2 = \frac{1}{3}$, and $k_3 = \frac{1}{4}$, compute the determinant of the matrix $\mathbf{R}_4 = \text{Toeplitz}\{r(0), r(1), r(2), r(3)\}$.

7.24 (a) Determine the lattice second-order prediction error filter (PEF) for a sequence $x(n)$ with autocorrelation $r(l) = (\frac{1}{2})^{|l|}$. (b) Repeat part (a) for the sequence $y(n) = x(n) + v(n)$, where $v(n) \sim \text{WN}(0, 0.2)$ is uncorrelated to $x(n)$. (c) Explain the change in the lattice parameters using frequency domain reasoning (think of the PEF as a whitening filter).

7.25 Consider a prediction error filter specified by $P_3 = (\frac{15}{16})^2$, $k_1 = \frac{1}{4}$, $k_2 = \frac{1}{2}$, and $k_3 = \frac{1}{4}$. (a) Determine the direct-form filter coefficients. (b) Determine the autocorrelation values $r(1)$, $r(2)$, and $r(3)$. (c) Determine the value $r(4)$ so that the MMSE P_4 for the corresponding fourth-order filter is the minimum possible.

7.26 Consider a prediction error filter $A_M(z) = 1 + a_1^{(M)} z^{-1} + \cdots + a_M^{(M)} z^{-M}$ with lattice parameters $k_1, k_2, \ldots, k_M$. (a) Show that if we set $\hat{k}_m = (-1)^m k_m$, then $\hat{a}_m^{(M)} = (-1)^m a_m^{(M)}$. (b) What are the new filter coefficients if we set $\hat{k}_m = \rho^m k_m$, where ρ is a complex number with $|\rho| = 1$? What happens if $|\rho| < 1$?

7.27 Suppose that we are given the values $\{r(l)\}_{-m+1}^{m-1}$ of an autocorrelation sequence such that the Toeplitz matrix $\mathbf{R}_m$ is positive definite. (a) Show that the values of $r(m)$ such that $\mathbf{R}_{m+1}$ is positive definite determine a disk in the complex plane. Find the center α_m and the radius ζ_m of this disk. (b) By induction show that there are infinitely many extensions of $\{r(l)\}_{-m+1}^{m-1}$ that make $\{r(l)\}_{-\infty}^{\infty}$ a valid autocorrelation sequence.

7.28 Consider the MA(1) sequence $x(n) = w(n) + d_1 w(n-1)$, $w(n) \sim \text{WN}(0, \sigma_w^2)$. (a) Show that

$$\det \mathbf{R}_m = r(0) \det \mathbf{R}_{m-1} - |r(1)|^2 \mathbf{R}_{m-2} \qquad m \geq 2$$

(b) Show that $k_m = -r^m(1)/\det \mathbf{R}_m$ and that

$$\frac{1}{k_m} = -\frac{r(0)}{r(1)} \frac{1}{k_{m-1}} - \frac{r^*(1)}{r(1)} \frac{1}{k_{m-2}}$$

(c) Determine the initial conditions and solve the recursion in (b) to show that

$$k_m = \frac{(1 - |d_1|^2)(-d_1)^m}{1 - |d_1|^{2m+2}}$$

which tends to zero as $m \to \infty$.

7.29 Prove Equation (7.6.6) by exploiting the symmetry property $\mathbf{b}_m = \mathbf{J}\mathbf{a}_m^*$.

7.30 In this problem we show that the lattice parameters can be obtained by "feeding" the autocorrelation sequence through the lattice filter as a signal and switching on the stages one by one after the required lattice coefficient is computed. The value of k_m is computed at time $n = m$ from the inputs to stage m. (a) Using (7.6.10), draw the flow diagram of a third-order lattice filter that implements this algorithm. (b) Using the autocorrelation sequence in Example 7.6.1, "feed" the sequence $\{r(n)\}_0^3 = \{3, 2, 1, \frac{1}{2}\}$ through the filter one sample at a time, and compute the lattice parameters. *Hint:* Use Example 7.6.1 for guidance.

7.31 Draw the supperlattice structure for $M = 8$, and show how it can be partitioned to distribute the computations to three processors for parallel execution.

7.32 Derive the superladder structure shown in Figure 7.10.

7.33 Extend the algorithm of Schür to compute the LDL^H decomposition of a Hermitian Toeplitz matrix, and write a MATLAB function for its implementation.

7.34 Given the matrix $\mathbf{R}_3 = \text{Toeplitz}\{1, \frac{1}{2}, \frac{1}{2}\}$, use the appropriate order-recursive algorithms to compute the following: (a) The LDL^H and UDU^H decompositions of $\mathbf{R}$, (b) the LDL^H and UDU^H decompositions of $\mathbf{R}^{-1}$, and (c) the inverse matrix $\mathbf{R}^{-1}$.

7.35 Consider the AR(1) process $x(n) = \rho x(n - 1) + w(n)$, where $w(n) \sim \text{WN}(0, \sigma_w^2)$ and $-1 < \rho < 1$. (a) Determine the correlation matrix $\mathbf{R}_{M+1}$ of the process. (b) Determine the Mth-order FLP, using the algorithm of Levinson-Durbin. (c) Determine the inverse matrix $\mathbf{R}_{M+1}^{-1}$, using the triangular decomposition discussed in Section 7.7.

7.36 If $r(l) = \cos \omega_0 l$, determine the second-order prediction error filter and check whether it is minimum-phase.

7.37 Show that the MMSE linear predictor of $x(n + D)$ in terms of $x(n - 1), x(n - 2), \ldots, x(n - M)$ for $k \ge 1$ is given by

$$\mathbf{R}\mathbf{a}^{(D)} = -\mathbf{r}^{(D)}$$

where $\mathbf{r}^{(D)} = [r(D) \; r(D+1) \; \cdots \; r(D+M-1)]^T$. Develop a recursion that computes $\mathbf{a}^{(D+1)}$ from $\mathbf{a}^{(D)}$ by exploring the shift invariance of the vector $\mathbf{r}^{(D)}$. See Manolakis et al. (1983).

7.38 The normal equations for the optimum symmetric signal smoother (see Section 6.5.1) can be written as

$$\mathbf{R}_{2m+1}\mathbf{c}_{2m+1} = \begin{bmatrix} \mathbf{0} \\ P_{2m+1} \\ \mathbf{0} \end{bmatrix}$$

where P_{2m+1} is the MMSE, $\mathbf{c}_{2m+1} = \mathbf{J}\mathbf{c}_{2m+1}^*$, and $c_m^{(2m+1)} = 1$. (a) Using a "central" partitioning of $\mathbf{R}_{2m+3}$ and the persymmetry property of Toeplitz matrices, develop a recursion to determine $\mathbf{c}_{2m+3}$ from $\mathbf{c}_{2m+1}$. (b) Develop a complete order-recursive algorithm for the computation of $\{\mathbf{c}_{2m+1}, P_{2m+1}\}_0^M$ (see Kok et al. 1993).

7.39 Using the triangular decomposition of a Toeplitz correlation matrix, show that (a) the forward prediction errors of various orders and at the same time instant, that is,

$$\mathbf{e}^{\text{f}}(n) = [e_0^{\text{f}}(n) \; e_1^{\text{f}}(n) \; \cdots \; e_m^{\text{f}}(n)]^T$$

are correlated and (b) the forward prediction errors

$$\bar{\mathbf{e}}^{\mathrm{f}}(n) = [e_M^{\mathrm{f}}(n)\ e_{M-1}^{\mathrm{f}}(n-1)\ \cdots\ e_0^{\mathrm{f}}(n-M)]^T$$

are uncorrelated.

7.40 Generalize the inversion algorithm described in Section 7.7.3 to handle Hermitian Toeplitz matrices.

7.41 Consider the estimation of a constant α from its noisy observations. The signal and observation models are given by

$$y(n+1) = y(n) \qquad n > 0 \qquad y(0) = \alpha$$
$$x(n) = y(n) + v(n) \qquad v(n) \sim \mathrm{WGN}(0, \sigma_v^2)$$

(a) Develop scalar Kalman filter equations, assuming the initial condition on the a posteriori error variance $R_{\tilde{y}}(0|0)$ equal to r_0.

(b) Show that the a posteriori error variance $R_{\tilde{y}}(n|n)$ is given by

$$R_{\tilde{y}}(n|n) = \frac{r_0}{1 + (r_0/\sigma_v^2)n} \tag{P.1}$$

(c) Show that the optimal filter for the estimation of the constant α is given by

$$\hat{y}(n) = \hat{y}(n-1) + \frac{r_0/\sigma_v^2}{1 + (r_0/\sigma_v^2)n}[x(n) - \hat{y}(n-1)]$$

7.42 Consider a random process with PDF given by

$$R_s(e^{j\omega}) = \frac{4}{2.4661 - 1.629\cos\omega + 0.81\cos 2\omega}$$

(a) Using MATLAB, plot the PSD $R_s(e^{j\omega})$ and determine the resonant frequency ω_0.

(b) Using spectral factorization, develop a signal model for the process of the form

$$\mathbf{y}(n) = \mathbf{A}\mathbf{y}(n-1) + \mathbf{B}\eta(n)$$
$$s(n) = [1\ \ 0]\mathbf{y}(n)$$

where $\mathbf{y}(n)$ is a 2×1 vector, $\eta(n) \sim \mathrm{WGN}(0, 1)$, and $\mathbf{A}$ and $\mathbf{B}$ are matrices with appropriate dimensions.

(c) Let $x(n)$ be the observed values of $s(n)$ given by

$$x(n) = s(n) + v(n) \qquad v(n) \sim \mathrm{WGN}(0, 1)$$

Assuming reasonable initial conditions, develop Kalman filter equations and implement them, using MATLAB. Study the performance of the filter by simulating a few sample functions of the signal process $s(n)$ and its observation $x(n)$.

7.43 *Alternative form of the Kalman filter.* A number of different identities and expressions can be obtained for the quantities defining the Kalman filter.

(a) By manipulating the last two equations in (7.8.39) show that

$$\mathbf{R}_{\tilde{y}}(n|n) = \mathbf{R}_{\tilde{y}}(n|n-1) - \mathbf{R}_{\tilde{y}}(n|n-1)\mathbf{H}^H(n)$$
$$\times [\mathbf{H}(n)\mathbf{R}_{\tilde{y}}(n|n-1)\mathbf{H}^H(n) + \mathbf{R}_v(n)]^{-1}\mathbf{H}\mathbf{R}_{\tilde{y}}(n|n-1) \tag{P.2}$$

(b) If the inverses of $\mathbf{R}_{\tilde{y}}(n|n)$, $\mathbf{R}_{\tilde{y}}(n|n-1)$, and $\mathbf{R}_v$ exist, then show that

$$\mathbf{R}_{\tilde{y}}^{-1}(n|n) = \mathbf{R}_{\tilde{y}}^{-1}(n|n-1) + \mathbf{H}^H(n)\mathbf{R}_v^{-1}(n)\mathbf{H}(n) \tag{P.3}$$

This shows that the update of the error covariance matrix does not require the Kalman gain matrix (but does require matrix inverses).

(c) Finally show that the gain matrix is given by

$$\mathbf{K}(n) = \mathbf{R}_{\tilde{y}}(n|n)\mathbf{H}^H(n)\mathbf{R}_v^{-1}(n) \tag{P.4}$$

which is computed by using the a posteriori error covariance matrix.

7.44 In Example 7.8.3 we assumed that only the position measurements were available for estimation. In this problem we will assume that we also have a noisy sensor to measure velocity measurements. Hence the observation model is

$$\mathbf{x}(n) \triangleq \begin{bmatrix} x_p(n) \\ x_v(n) \end{bmatrix} = \begin{bmatrix} y_p(n) + v_1(n) \\ y_v(n) + v_2(n) \end{bmatrix} \tag{P.5}$$

where $v_1(n)$ and $v_2(n)$ are two independent zero-mean white Gaussian noise sources with variances $\sigma_{v_1}^2$ and $\sigma_{v_2}^2$, respectively.

(a) Using the state vector model given in Example 7.8.3 and the observation model in (P.5), develop Kalman filter equations to estimate position and velocity of the object at each n.

(b) Using the parameter values

$$T = 0.1 \qquad \sigma_{v_1}^2 = \sigma_{v_2}^2 = \sigma_\eta^2 = 0.25 \qquad y_p(-1) = 0 \qquad y_v(-1) = 1$$

simulate the true and observed positions and velocities of the object. Using your Kalman filter equations, generate plots similar to the ones given in Figures 7.14 and 7.15.

(c) Discuss the effects of velocity measurements on the estimates.

7.45 In this problem, we will assume that the acceleration $y_a(n)$ is an AR(1) process rather than a white noise process. Let $y_a(n)$ be given by

$$y_a(n) = \alpha y_a(n-1) + \eta(n) \qquad \eta(n) \sim \text{WGN}(0, \sigma_\eta^2) \qquad y_a(-1) = 0 \tag{P.6}$$

(a) Augment the state vector $\mathbf{y}(n)$ in (7.8.48), using variable $y_a(n)$, and develop the state vector as well as the observation model, assuming that only the position is measured.

(b) Using the above model and the parameter values

$$T = 0.1 \qquad \alpha = 0.9 \qquad \sigma_v^2 = \sigma_\eta^2 = 0.25$$
$$y_p(-1) = 0 \qquad y_v(-1) = 1$$

simulate the linear motion of the object. Using Kalman filter equations, estimate the position, velocity, and acceleration values of the object at each n. Generate performance plots similar to the ones given in Figures 7.14 and 7.15.

(c) Now assume that noisy measurements of $y_v(n)$ and $y_a(n)$ are also available, that is, the observation model is

$$\mathbf{x}(n) \triangleq \begin{bmatrix} x_p(n) \\ x_v(n) \\ x_a(n) \end{bmatrix} = \begin{bmatrix} y_p(n) + v_1(n) \\ y_v(n) + v_2(n) \\ y_a(n) + v_3(n) \end{bmatrix} \tag{P.7}$$

where $v_1(n)$, $v_2(n)$, and $v_3(n)$ are IID zero-mean white Gaussian noise sources with variance σ_v^2. Repeat parts (a) and (b) above.

CHAPTER 8

Least-Squares Filtering and Prediction

In this chapter, we deal with the design and properties of linear combiners, finite impulse response (FIR) filters, and linear predictors that are optimum in the least-squares error (LSE) sense. The principle of least squares is widely used in practice because second-order moments are rarely known. In the first part of this chapter (Sections 8.1 through 8.4), we concentrate on the design, properties, and applications of least-squares (LS[†]) estimators. Section 8.1 discusses the principle of LS estimation. The unique aspects of the different implementation structures, starting with the general linear combiner followed by the FIR filter and predictor, are treated in Sections 8.2 to 8.4. In the second part (Sections 8.5 to 8.7), we discuss various numerical algorithms for the solution of the LSE normal equations and the computation of LSE estimates including QR decomposition techniques (House-holder reflections, Givens rotations, and modified Gram-Schmidt orthogonalization) and the singular value decomposition (SVD).

8.1 THE PRINCIPLE OF LEAST SQUARES

The *principle of least squares* was introduced by the German mathematician Carl Friedrich Gauss, who used it to determine the orbit of the asteroid Ceres in 1821 by formulating the estimation problem as an optimization problem.

The design of optimum filters in the minimum mean square error (MMSE) sense, discussed in Chapter 6, requires the a priori knowledge of second-order moments. However, such statistical information is simply *not* available in most practical applications, for which we can only obtain measurements of the input and desired response signals. To avoid this problem, we can (1) estimate the required second-order moments from the available data (see Chapter 5), if possible, to obtain an estimate of the optimum MMSE filter, or (2) design an optimum filter by minimizing a criterion of performance that is a function of the available data.

In this chapter, we use the minimization of the sum of the squares of the estimation error as the criterion of performance for the design of optimum filters. This method, known as *least-squares error (LSE) estimation*, requires the measurement of *both* the input signal and the desired response signal. A natural question arising at this point is, What is the purpose of estimating the values of a *known,* desired response signal? There are several answers:

[†] A note about abbreviations used throughout the chapter: The two acronyms *LSE* and *LS* will be used almost interchangably. Although LSE is probably the more accurate term, LS has become a standard reference to LSE estimators.

1. In system modeling applications, the goal is to obtain a mathematical model describing the input-output behavior of an actual system. A quality estimator provides a good model for the system. The desired result is the estimator or system model, not the actual estimate.
2. In linear predictive coding, the useful result is the prediction error or the respective predictor coefficients.
3. In many applications, the desired response is not available (e.g., digital communications). Therefore, we do not always have a complete set of data from which to design the LSE estimator. However, if the data do not change significantly over a number of sets, then one special complete set, the training set, is used to design the estimator. The resulting estimator is then applied to the processing of the remaining incomplete sets.

The use of measured signal values to determine the coefficients of the estimator leads to some fundamental differences between MMSE and LSE estimation that are discussed where appropriate.

To summarize, depending on the available information, there are two ways to design an optimum estimator: (1) If we know the second-order moments, we use the MMSE criterion and design a filter that is optimum for all possible sets of data with the same statistics. (2) If we only have a block of data, we use the LSE criterion to design an estimator that is optimum for the given block of data. Optimum MMSE estimators are obtained by using ensemble averages, whereas LSE estimators are obtained by using finite-length time averages. For example, an MMSE estimator, designed using ensemble averages, is optimum for all realizations. In contrast, an LSE estimator, designed using a block of data from a particular realization, depends on the numerical values of samples used in the design. If the processes are ergodic, the LSE estimator approaches the MMSE estimator as the block length of the data increases toward infinity.

8.2 LINEAR LEAST-SQUARES ERROR ESTIMATION

We start with the derivation of general linear LS filters that are implemented using the linear combiner structure described in Section 6.2. A set of measurements of the desired response $y(n)$ and the input signals $x_k(n)$ for $1 \leq k \leq M$ has been taken for $0 \leq n \leq N - 1$. As in optimum MMSE estimation, the problem is to estimate the desired response $y(n)$ using the linear combination

$$\hat{y}(n) = \sum_{k=1}^{M} c_k^*(n) x_k(n) = \mathbf{c}^H(n) \mathbf{x}(n) \tag{8.2.1}$$

We define the estimation error as

$$e(n) = y(n) - \hat{y}(n) = y(n) - \mathbf{c}^H(n) \mathbf{x}(n) \tag{8.2.2}$$

and the coefficients of the combiner are determined by minimizing the sum of the squared errors

$$E \triangleq \sum_{n=0}^{N-1} |e(n)|^2 \tag{8.2.3}$$

that is, the *energy* of the error signal. *For this minimization to be possible, the coefficient vector $\mathbf{c}(n)$ should be held constant over the measurement time interval* $0 \leq n \leq N - 1$. The constant vector $\mathbf{c}_{ls}$ resulting from this optimization depends on the measurement set and is known as the *linear LSE estimator*. In the statistical literature, LSE estimation is known as linear regression, where (8.2.2) is called a *regression function*, $e(n)$ are known as *residuals* (leftovers), and $\mathbf{c}(n)$ is the *regression vector* (Montgomery and Peck 1982).

The system of equations in (8.2.2), or equivalently $e^*(n) = y^*(n) - \mathbf{x}^H(n)\,\mathbf{c}$, can be written in matrix form as

$$
\begin{bmatrix} e^*(0) \\ e^*(1) \\ \vdots \\ e^*(N-1) \end{bmatrix} = \begin{bmatrix} y^*(0) \\ y^*(1) \\ \vdots \\ y^*(N-1) \end{bmatrix}
$$

$$
- \begin{bmatrix} x_1^*(0) & x_2^*(0) & \cdots & x_M^*(0) \\ x_1^*(1) & x_2^*(1) & \cdots & x_M^*(1) \\ \vdots & \vdots & \ddots & \vdots \\ x_1^*(N-1) & x_2^*(N-1) & \cdots & x_M^*(N-1) \end{bmatrix} \begin{bmatrix} c_1 \\ c_2 \\ \vdots \\ c_M \end{bmatrix}
$$

(8.2.4)

or more compactly as

$$
\mathbf{e} = \mathbf{y} - \mathbf{X}\mathbf{c} \tag{8.2.5}
$$

where

$$
\begin{aligned}
\mathbf{e} &\triangleq [e(0)\ e(1)\ \cdots\ e(N-1)]^H && \text{error data vector } (N \times 1) \\
\mathbf{y} &\triangleq [y(0)\ y(1)\ \cdots\ y(N-1)]^H && \text{desired response vector } (N \times 1) \\
\mathbf{X} &\triangleq [\mathbf{x}(0)\ \mathbf{x}(1)\ \cdots\ \mathbf{x}(N-1)]^H && \text{input data matrix } (N \times M) \\
\mathbf{c} &\triangleq [c_1\ c_2\ \cdots\ c_M]^T && \text{combiner parameter vector } (M \times 1)
\end{aligned}
$$

(8.2.6)

are defined by comparing (8.2.4) to (8.2.5). The input data matrix $\mathbf{X}$ can be partitioned either columnwise or rowwise as follows:

$$
\mathbf{X} \triangleq [\tilde{\mathbf{x}}_1, \tilde{\mathbf{x}}_2, \ldots, \tilde{\mathbf{x}}_M] = \begin{bmatrix} \mathbf{x}^H(0) \\ \mathbf{x}^H(1) \\ \vdots \\ \mathbf{x}^H(N-1) \end{bmatrix} \tag{8.2.7}
$$

where the columns $\tilde{\mathbf{x}}_k$ of $\mathbf{X}$

$$
\tilde{\mathbf{x}}_k \triangleq [x_k(0)\ x_k(1)\ \cdots\ x_k(N-1)]^H
$$

will be called *data records* and the rows

$$
\mathbf{x}(n) \triangleq [x_1(n)\ x_2(n)\ \cdots\ x_M(n)]^T
$$

will be called *snapshots*. Both of these partitionings of the data matrix, which are illustrated in Figure 8.1, are useful in the derivation, interpretation, and computation of LSE estimators.

The LSE estimator operates in a block processing mode; that is, it processes a frame of N snapshots using the steps shown in Figure 8.2. The input signals are blocked into frames of N snapshots with successive frames overlapping by N_0 samples. The values of N and N_0 depend on the application. The required estimate or residual signals are unblocked at the final stage of the processor.

If we set $\mathbf{e} = \mathbf{0}$, we have a set of N equations with M unknowns. If $N = M$, then (8.2.4) usually has a unique solution. For $N > M$, we have an overdetermined system of linear equations that typically has no solution. Conversely, if $N < M$, we have an underdetermined system that has an infinite number of solutions. However, even if $M > N$ or $N > M$, the system (8.2.4) has a natural, unique, least-squares solution. We next focus our attention on overdetermined systems since they play a very important role in practical applications. The underdetermined least-squares problem is examined in Section 8.7.2.

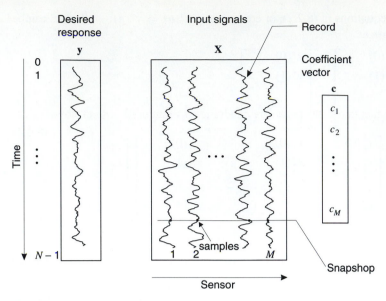

FIGURE 8.1
The columns of the data matrix are the records of data collected at each
input (sensor), whereas each row contains the samples from all inputs at the
same instant.

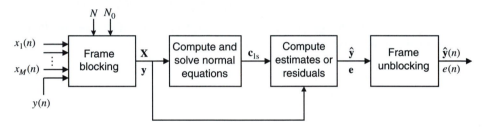

FIGURE 8.2
Block processing implementation of a general linear LSE estimator.

8.2.1 Derivation of the Normal Equations

We provide an algebraic and a geometric solution to the LSE estimation problem; a calculus-based derivation is given in Problem 8.1.

Algebraic derivation. The energy of the error can be written as

$$E = \mathbf{e}^H \mathbf{e} = (\mathbf{y}^H - \mathbf{c}^H \mathbf{X}^H)(\mathbf{y} - \mathbf{X}\mathbf{c})$$

$$= \mathbf{y}^H \mathbf{y} - \mathbf{c}^H \mathbf{X}^H \mathbf{y} - \mathbf{y}^H \mathbf{X}\mathbf{c} + \mathbf{c}^H \mathbf{X}^H \mathbf{X}\mathbf{c} \qquad (8.2.8)$$

$$= E_y - \mathbf{c}^H \hat{\mathbf{d}} - \hat{\mathbf{d}}^H \mathbf{c} + \mathbf{c}^H \hat{\mathbf{R}} \mathbf{c}$$

where

$$E_y \triangleq \mathbf{y}^H \mathbf{y} = \sum_{n=0}^{N-1} |y(n)|^2 \qquad (8.2.9)$$

$$\hat{\mathbf{R}} \triangleq \mathbf{X}^H \mathbf{X} = \sum_{n=0}^{N-1} \mathbf{x}(n)\mathbf{x}^H(n) \qquad (8.2.10)$$

$$\hat{\mathbf{d}} \triangleq \mathbf{X}^H \mathbf{y} = \sum_{n=0}^{N-1} \mathbf{x}(n)y^*(n) \qquad (8.2.11)$$

Note that these quantities can be viewed as time-average estimates of the desired response power, correlation matrix of the input data vector, and the cross-correlation vector between the desired response and the data vector, when these quantities are divided by the number of data samples N.

We emphasize that all formulas derived for the MMSE criterion hold for the LSE criterion if we replace the expectation operator $E\{(\cdot)\}$ with the time-average operator $\sum_{n=0}^{N-1}(\cdot)$. This results from the fact that both criteria are quadratic cost functions. Therefore, working as in Section 6.2.2, we conclude that if the time-average correlation matrix $\hat{\mathbf{R}}$ is positive definite, the LSE estimator $\mathbf{c}_{ls}$ is provided by the solution of the normal equations

$$\hat{\mathbf{R}}\mathbf{c}_{ls} = \hat{\mathbf{d}} \tag{8.2.12}$$

and the minimum sum of squared errors is given by

$$E_{ls} = E_y - \hat{\mathbf{d}}^H\hat{\mathbf{R}}^{-1}\hat{\mathbf{d}} = E_y - \hat{\mathbf{d}}^H\mathbf{c}_{ls} \tag{8.2.13}$$

Since $\hat{\mathbf{R}}$ is Hermitian, we only need to compute the elements

$$\hat{r}_{ij} = \tilde{\mathbf{x}}_i^H\tilde{\mathbf{x}}_j \tag{8.2.14}$$

in the upper triangular part, which requires $M(M+1)/2$ dot products. The right-hand side requires M dot products

$$\hat{d}_i = \tilde{\mathbf{x}}_i^H\mathbf{y} \tag{8.2.15}$$

Note that each dot product involves N arithmetic operations, each consisting of one multiplication and one addition. Thus, to form the normal equations requires a total of

$$\tfrac{1}{2}M(M+1)N + MN = \tfrac{1}{2}M^2N + \tfrac{3}{2}MN \tag{8.2.16}$$

arithmetic operations. When $\hat{\mathbf{R}}$ is nonsingular, which is the case when $\hat{\mathbf{R}}$ is positive definite, we can solve the normal equations using either the LDL^H or the Cholesky decomposition (see Section 6.3). However, it should be stressed at this point that most of the computational work lies in forming the normal equations rather than their solution. The formulation of the overdetermined LS equations and the normal equations is illustrated graphically in Figure 8.3. The solution of LS problems has been extensively studied in various application areas and in numerical analysis. The basic methods for the solution of the LS problem, which are discussed in this book, are shown in Figure 8.4. We just stress here that for overdetermined LS problems, well-behaved data, and sufficient numerical precision, all these methods provide comparable results.

Geometric derivation. We may think of the desired response record $\mathbf{y}$ and the data records $\tilde{\mathbf{x}}_k$, $1 \le k \le M$, as vectors in an N-dimensional vector space, with the dot product and length defined by

$$\langle \tilde{\mathbf{x}}_i, \tilde{\mathbf{x}}_j \rangle \triangleq \tilde{\mathbf{x}}_i^H\tilde{\mathbf{x}}_j = \sum_{n=0}^{N-1} x_i(n)\, x_j^*(n) \tag{8.2.17}$$

and

$$\|\tilde{\mathbf{x}}\|^2 \triangleq \langle \tilde{\mathbf{x}}, \tilde{\mathbf{x}} \rangle = \sum_{n=0}^{N-1} |x(n)|^2 = E_x \tag{8.2.18}$$

respectively. The estimate of the desired response record can be expressed as

$$\hat{\mathbf{y}} = \mathbf{X}\mathbf{c} = \sum_{k=1}^{M} c_k\tilde{\mathbf{x}}_k \tag{8.2.19}$$

that is, as a linear combination of the data records.

The M vectors $\tilde{\mathbf{x}}_k$ form an M-dimensional subspace, called the *estimation space*, which is the column space of data matrix $\mathbf{X}$. Clearly, any estimate $\hat{\mathbf{y}}$ must lie in the estimation space. The desired response record $\mathbf{y}$, in general, lies outside the estimation space. The estimation

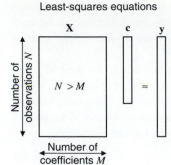

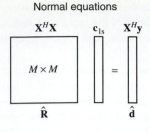

Least-squares equations   Normal equations

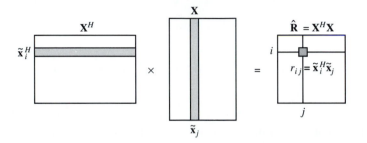

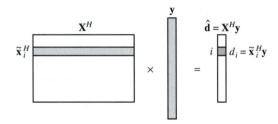

FIGURE 8.3
The LS problem and computation of the normal equations.

space for $M = 2$ and $N = 3$ is illustrated in Figure 8.5. The error vector $\mathbf{e}$ points from the tip of $\hat{\mathbf{y}}$ to the tip of $\mathbf{y}$. The squared length of $\mathbf{e}$ is minimum when $\mathbf{e}$ is perpendicular to the estimation space, that is, $\mathbf{e} \perp \tilde{\mathbf{x}}_k$ for $1 \leq k \leq M$.

Therefore, we have the orthogonality principle

$$\langle \tilde{\mathbf{x}}_k, \mathbf{e} \rangle = \tilde{\mathbf{x}}_k^H \mathbf{e} = 0 \qquad 1 \leq k \leq M \tag{8.2.20}$$

or more compactly

$$\mathbf{X}^H \mathbf{e} = \mathbf{X}^H (\mathbf{y} - \mathbf{X}\mathbf{c}_{ls}) = \mathbf{0}$$

or

$$(\mathbf{X}^H \mathbf{X})\mathbf{c}_{ls} = \mathbf{X}^H \mathbf{y} \tag{8.2.21}$$

which we recognize as the LSE normal equations from (8.2.12).

The LS solution splits the desired response $\mathbf{y}$ into two orthogonal components, namely, $\hat{\mathbf{y}}_{ls}$ and $\mathbf{e}_{ls}$. Therefore,

$$\|\mathbf{y}\|^2 = \|\hat{\mathbf{y}}_{ls}\|^2 + \|\mathbf{e}_{ls}\|^2 \tag{8.2.22}$$

and, using (8.2.18) and (8.2.19), we have

$$E_{ls} = E_y - \mathbf{c}_{ls}^H \mathbf{X}^H \mathbf{X}\mathbf{c}_{ls} = E_y - \mathbf{c}_{ls}^H \mathbf{X}^H \mathbf{y} \tag{8.2.23}$$

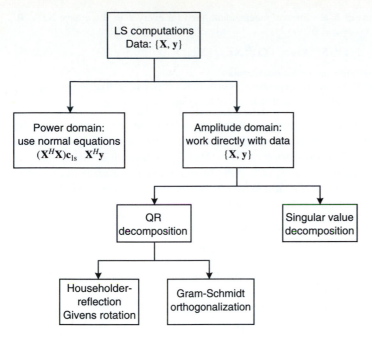

FIGURE 8.4
Classification of different computational algorithms for the solution of
the LS problem.

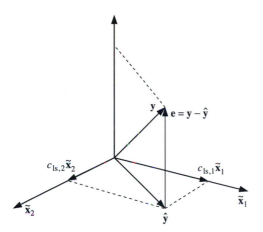

FIGURE 8.5
Vector space interpretation of LSE estimation
for $N = 3$ (dimension of data space) and
$M = 2$ (dimension of estimation subspace).

which is identical to (8.2.13). The normalized total squared error is

$$\mathcal{E} \triangleq \frac{E_{\text{ls}}}{E_y} = 1 - \frac{E_{\hat{y}}}{E_y} \tag{8.2.24}$$

which is in the range $0 \leq \mathcal{E} \leq 1$, with limits of 0 and 1, which correspond to the worst and
best cases, respectively.

Uniqueness. The solution of the LSE normal equations exists and is unique if the
time-average correlation matrix $\hat{\mathbf{R}}$ is invertible. We shall prove the following:

THEOREM 8.1. The time-average correlation matrix $\hat{\mathbf{R}} = \mathbf{X}^H \mathbf{X}$ is invertible if and only if the
columns $\tilde{\mathbf{x}}_k$ of $\mathbf{X}$ are linearly independent, or equivalently if and only if $\hat{\mathbf{R}}$ is positive definite.

Proof. If the columns of $\mathbf{X}$ are linearly independent, then for every $\mathbf{z} \neq \mathbf{0}$ we have $\mathbf{Xz} \neq \mathbf{0}$. This implies that for every $\mathbf{z} \neq \mathbf{0}$

$$\mathbf{z}^H(\mathbf{X}^H\mathbf{X})\mathbf{z} = (\mathbf{Xz})^H\mathbf{Xz} = \|\mathbf{Xz}\|^2 > 0 \tag{8.2.25}$$

that is, $\hat{\mathbf{R}}$ is positive definite and hence nonsingular.

If the columns of $\mathbf{X}$ are linearly dependent, then there is a vector $\mathbf{z}_0 \neq \mathbf{0}$ such that $\mathbf{Xz}_0 = \mathbf{0}$. Therefore, $\mathbf{X}^H\mathbf{Xz}_0 = \mathbf{0}$, which implies that $\hat{\mathbf{R}} = \mathbf{X}^H\mathbf{X}$ is singular.

For a matrix to have linearly independent columns, the number of rows should be equal to or larger than the number of columns; that is, we must have more equations than unknowns. To summarize, *the overdetermined ($N > M$) LS problem has a unique solution provided by the normal equations in (8.2.12) if the time-average correlation matrix $\hat{\mathbf{R}}$ is positive definite, or equivalently if the data matrix $\mathbf{X}$ has linearly independent columns.*

In this case, the LS solution can be expressed as

$$\mathbf{c}_{ls} = \mathbf{X}^+\mathbf{y} \tag{8.2.26}$$

where

$$\mathbf{X}^+ \triangleq (\mathbf{X}^H\mathbf{X})^{-1}\mathbf{X}^H \tag{8.2.27}$$

is an $M \times N$ matrix known as the *pseudo-inverse* or the *Moore-Penrose generalized inverse* of matrix $\mathbf{X}$ (Golub and Van Loan 1996; Strang 1980).

The LS estimate $\hat{\mathbf{y}}_{ls}$ of $\mathbf{y}$ can be expressed as

$$\hat{\mathbf{y}}_{ls} = \mathbf{Py} \tag{8.2.28}$$

where

$$\mathbf{P} \triangleq \mathbf{X}(\mathbf{X}^H\mathbf{X})^{-1}\mathbf{X}^H \tag{8.2.29}$$

is known as the *projection matrix* because it projects the data vector $\mathbf{y}$ onto the column space of $\mathbf{X}$ to provide the LS estimate $\hat{\mathbf{y}}_{ls}$ of $\mathbf{y}$. Similarly, the LS error vector $\mathbf{e}_{ls}$ can be expressed as

$$\mathbf{e}_{ls} = (\mathbf{I} - \mathbf{P})\mathbf{y} \tag{8.2.30}$$

where $\mathbf{I}$ is the $N \times N$ identity matrix. The projection matrix $\mathbf{P}$ is Hermitian and *idempotent*, that is,

$$\mathbf{P} = \mathbf{P}^H \tag{8.2.31}$$

and

$$\mathbf{P}^2 = \mathbf{P}^H\mathbf{P} = \mathbf{P} \tag{8.2.32}$$

respectively.

When the columns of $\mathbf{X}$ are linearly dependent, the LS problem has many solutions. Since all these solutions satisfy the normal equations and the orthogonal projection of $\mathbf{y}$ onto the column space of $\mathbf{X}$ is unique, all these solutions produce an error vector $\mathbf{e}$ of equal length, that is, the same LSE. This subject is discussed in Section 8.6.2 (minimum-norm solution).

EXAMPLE 8.2.1. Suppose that we wish to estimate the sequence $\mathbf{y} = [1\ 2\ 3\ 2]^T$ from the observation vectors $\tilde{\mathbf{x}}_1 = [1\ 2\ 1\ 1]^T$ and $\tilde{\mathbf{x}}_2 = [2\ 1\ 2\ 3]^T$. Determine the optimum filter, the error vector $\mathbf{e}_{ls}$, and the LSE E_{ls}.

Solution. We first compute the quantities

$$\hat{\mathbf{R}} = \mathbf{X}^T\mathbf{X} = \begin{bmatrix} 1 & 2 \\ 2 & 1 \\ 1 & 2 \\ 1 & 3 \end{bmatrix}^T \begin{bmatrix} 1 & 2 \\ 2 & 1 \\ 1 & 2 \\ 1 & 3 \end{bmatrix} = \begin{bmatrix} 7 & 9 \\ 9 & 18 \end{bmatrix} \qquad \hat{\mathbf{d}} = \mathbf{X}^T\mathbf{y} = \begin{bmatrix} 1 & 2 \\ 2 & 1 \\ 1 & 2 \\ 1 & 3 \end{bmatrix}^T \begin{bmatrix} 1 \\ 2 \\ 3 \\ 2 \end{bmatrix} = \begin{bmatrix} 10 \\ 16 \end{bmatrix}$$

and we then solve the normal equations $\hat{\mathbf{R}}\mathbf{c}_{ls} = \hat{\mathbf{d}}$ to obtain the LS estimator

$$\mathbf{c}_{ls} = \hat{\mathbf{R}}^{-1}\hat{\mathbf{d}} = \begin{bmatrix} \frac{2}{5} & -\frac{1}{5} \\ -\frac{1}{5} & \frac{7}{45} \end{bmatrix} \begin{bmatrix} 10 \\ 16 \end{bmatrix} = \begin{bmatrix} \frac{4}{5} \\ \frac{22}{45} \end{bmatrix}$$

and the LSE

$$E_{ls} = E_y - \hat{\mathbf{d}}^T \mathbf{c}_{ls} = 18 - \begin{bmatrix} 10 \\ 16 \end{bmatrix}^T \begin{bmatrix} \frac{4}{5} \\ \frac{22}{45} \end{bmatrix} = \frac{98}{45}$$

The projection matrix is

$$\mathbf{P} = \mathbf{X}(\mathbf{X}^T\mathbf{X})^{-1}\mathbf{X}^T = \begin{bmatrix} \frac{2}{9} & \frac{1}{9} & \frac{2}{9} & \frac{1}{3} \\ \frac{1}{9} & \frac{43}{45} & \frac{1}{9} & -\frac{2}{15} \\ \frac{2}{9} & \frac{1}{9} & \frac{2}{9} & \frac{1}{3} \\ \frac{1}{3} & -\frac{2}{15} & \frac{1}{3} & \frac{3}{5} \end{bmatrix}$$

which can be used to determine the error vector

$$\mathbf{e}_{ls} = \mathbf{y} - \mathbf{P}\mathbf{y} = [-\frac{7}{9} \quad -\frac{4}{45} \quad \frac{11}{9} \quad -\frac{4}{15}]^T$$

whose squared norm is equal to $\|\mathbf{e}_{ls}\|^2 = \frac{98}{45} = E_{ls}$, as expected. We can also easily verify the orthogonality principle $\mathbf{e}_{ls}^T\tilde{\mathbf{x}}_1 = \mathbf{e}_{ls}^T\tilde{\mathbf{x}}_2 = 0$.

Weighted least-squares estimation. The previous results were derived by using an LS criterion that treats every error $e(n)$ equally. However, based on a priori information, we may wish to place greater importance on different errors, using the weighted LS criterion

$$E_w = \sum_{n=0}^{N-1} w(n)|e(n)|^2 = \mathbf{e}^H\mathbf{W}\mathbf{e} \tag{8.2.33}$$

where

$$\mathbf{W} \triangleq \text{diag}\{w(0), w(1), \ldots, w(N-1)\} \tag{8.2.34}$$

is a diagonal weighting matrix with positive elements. Usually, we choose small weights where the errors are expected to be large, and vice versa. Minimization of E_w with respect to $\mathbf{c}$ yields the *weighted LS (WLS) estimator*

$$\mathbf{c}_{wls} = (\mathbf{X}^H\mathbf{W}\mathbf{X})^{-1}\mathbf{X}^H\mathbf{W}\mathbf{y} \tag{8.2.35}$$

assuming that the inverse of the matrix $\mathbf{X}^H\mathbf{W}\mathbf{X}$ exists. We can easily see that when $\mathbf{W} = \mathbf{I}$, then $\mathbf{c}_{wls} = \mathbf{c}_{ls}$. The criterion in (8.2.33) can be generalized by choosing $\mathbf{W}$ to be any Hermitian, positive definite matrix (see Problem 8.2).

8.2.2 Statistical Properties of Least-Squares Estimators

A useful approach for evaluating the quality of an LS estimator is to study its statistical properties. Toward this end, we assume that the obtained measurements $\mathbf{y}$ actually have been generated by

$$\mathbf{y} = \mathbf{X}\mathbf{c}_o + \mathbf{e}_o \tag{8.2.36}$$

where $\mathbf{e}_o$ is the random measurement error vector. We may think of $\mathbf{c}_o$ as the "true" parameter vector. Using (8.2.36), we see that (8.2.21) gives

$$\mathbf{c}_{ls} = \mathbf{c}_o + (\mathbf{X}^H\mathbf{X})^{-1}\mathbf{X}^H\mathbf{e}_o \tag{8.2.37}$$

We make the following assumptions about the random measurement error vector $\mathbf{e}_o$:

1. The error vector $\mathbf{e}_o$ has zero mean

$$E\{\mathbf{e}_o\} = \mathbf{0} \tag{8.2.38}$$

2. The error vector $\mathbf{e}_o$ has uncorrelated components with constant variance $\sigma_{e_o}^2$; that is, the correlation matrix is given by

$$\mathbf{R}_{e_o} = E\{\mathbf{e}_o\mathbf{e}_o^H\} = \sigma_{e_o}^2 \mathbf{I} \qquad (8.2.39)$$

3. There is no information about $\mathbf{e}_o$ contained in data matrix $\mathbf{X}$; that is,

$$E\{\mathbf{e}_o|\mathbf{X}\} = E\{\mathbf{e}_o\} = \mathbf{0} \qquad (8.2.40)$$

4. If $\mathbf{X}$ is a deterministic $N \times M$ matrix, then it has rank M. This means that $\mathbf{X}$ is a full-column rank and that $\mathbf{X}^H\mathbf{X}$ is invertible. If $\mathbf{X}$ is a stochastic $N \times M$ matrix, then $E\{(\mathbf{X}^H\mathbf{X})^{-1}\}$ exists.

In the following analysis, we consider two possibilities: $\mathbf{X}$ is deterministic and stochastic. Under these conditions, the LS estimator $\mathbf{c}_{ls}$ has several desirable properties.

Deterministic data matrix

In this case, we assume that the LS estimators are obtained from the deterministic data values; that is, the matrix $\mathbf{X}$ is treated as a matrix of constants. Then the properties of the LS estimators can be derived from the statistical properties of the random measurement error vector $\mathbf{e}_o$.

PROPERTY 8.2.1. The LS estimator $\mathbf{c}_{ls}$ is an unbiased estimator of $\mathbf{c}_o$, that is,

$$E\{\mathbf{c}_{ls}\} = \mathbf{c}_o \qquad (8.2.41)$$

Proof. Taking the expectation of both sides of (8.2.37), we have

$$E\{\mathbf{c}_{ls}\} = E\{\mathbf{c}_o\} + (\mathbf{X}^H\mathbf{X})^{-1}\mathbf{X}^H E\{\mathbf{e}_o\} = \mathbf{c}_o$$

because $\mathbf{X}$ is deterministic and $E\{\mathbf{e}_o\} = \mathbf{0}$.

PROPERTY 8.2.2. The covariance matrix of $\mathbf{c}_{ls}$ corresponding to the error $\mathbf{c}_{ls} - \mathbf{c}_o$ is

$$\mathbf{\Gamma}_{ls} \triangleq E\{(\mathbf{c}_{ls} - \mathbf{c}_o)(\mathbf{c}_{ls} - \mathbf{c}_o)^H\} = \sigma_{e_o}^2(\mathbf{X}^H\mathbf{X})^{-1} = \sigma_{e_o}^2\hat{\mathbf{R}}^{-1} \qquad (8.2.42)$$

Proof. Using (8.2.37), (8.2.39), and the definition (8.2.42), we easily obtain

$$\mathbf{\Gamma}_{ls} = (\mathbf{X}^H\mathbf{X})^{-1}\mathbf{X}^H E\{\mathbf{e}_o\mathbf{e}_o^H\}\mathbf{X}(\mathbf{X}^H\mathbf{X})^{-1} = \sigma_{e_o}^2(\mathbf{X}^H\mathbf{X})^{-1}$$

Note that the diagonal elements of matrix $\sigma_e^2\hat{\mathbf{R}}^{-1}$ are also equal to the variance of the LS combiner vector $\mathbf{c}_{ls}$.

PROPERTY 8.2.3. An unbiased estimate of the error variance $\sigma_{e_o}^2$ is given by

$$\hat{\sigma}_{e_o}^2 = \frac{E_{ls}}{N - M} \qquad (8.2.43)$$

where N is the number of observations, M is the number of parameters, and E_{ls} is the LS error.

Proof. Using (8.2.30) and (8.2.36), we obtain

$$\mathbf{e}_{ls} = (\mathbf{I} - \mathbf{P})\mathbf{y} = (\mathbf{I} - \mathbf{P})\mathbf{e}_o$$

which results in

$$E_{ls} = \mathbf{e}_{ls}^H\mathbf{e}_{ls} = \mathbf{e}_o^H(\mathbf{I} - \mathbf{P})^H(\mathbf{I} - \mathbf{P})\mathbf{e}_o = \mathbf{e}_o^H(\mathbf{I} - \mathbf{P})\mathbf{e}_o$$

because of (8.2.32). Since E_{ls} depends on $\mathbf{e}_o$, it is a random variable whose expected value is

$$E\{E_{ls}\} = E\{\mathbf{e}_o^H(\mathbf{I} - \mathbf{P})\mathbf{e}_o\} = E\{\text{tr}[(\mathbf{I} - \mathbf{P})\mathbf{e}_o\mathbf{e}_o^H]\}$$

$$= \text{tr}[(\mathbf{I} - \mathbf{P})E\{\mathbf{e}_o\mathbf{e}_o^H\}] = \sigma_e^2 \text{ tr}(\mathbf{I} - \mathbf{P})$$

since $\text{tr}(\mathbf{AB}) = \text{tr}(\mathbf{BA})$, where tr is the trace function. However,

$$\text{tr}(\mathbf{I} - \mathbf{P}) = \text{tr}[\mathbf{I} - \mathbf{X}(\mathbf{X}^H\mathbf{X})^{-1}\mathbf{X}^H]$$

$$= \text{tr}[\mathbf{I}_{N \times N} - (\mathbf{X}^H\mathbf{X})^{-1}\mathbf{X}^H\mathbf{X}]$$

$$= \text{tr}(\mathbf{I}_{N \times N}) - \text{tr}[(\mathbf{X}^H\mathbf{X})^{-1}\mathbf{X}^H\mathbf{X}]$$

$$= \text{tr}(\mathbf{I}_{N \times N}) - \text{tr}(\mathbf{I}_{M \times M}) = N - M$$

therefore

$$\sigma_{e_o}^2 = \frac{E\{E_{\text{ls}}\}}{N - M} \tag{8.2.44}$$

which proves that $\hat{\sigma}_e^2$ is an unbiased estimate of $\sigma_{e_o}^2$.

Similar to (8.2.41), the mean value of $\mathbf{c}_{\text{wls}}$ is

$$E\{\mathbf{c}_{\text{wls}}\} = E\{\mathbf{c}_o\} + (\mathbf{X}^H \mathbf{W} \mathbf{X})^{-1} \mathbf{X}^H \mathbf{W} E\{\mathbf{e}_o\} = E\{\mathbf{c}_o\} \tag{8.2.45}$$

that is, the WLS estimator is an unbiased estimate of $\mathbf{c}_o$. The covariance matrix of $\mathbf{c}_{\text{wls}}$ is

$$\mathbf{\Gamma}_{\text{wls}} = (\mathbf{X}^H \mathbf{W} \mathbf{X})^{-1} \mathbf{X}^H \mathbf{W} \mathbf{R}_{e_o} \mathbf{W} \mathbf{X} (\mathbf{X}^H \mathbf{W} \mathbf{X})^{-1} \tag{8.2.46}$$

where $\mathbf{R}_{e_o}$ is the correlation matrix of $\mathbf{e}_o$. It is easy to see that when $\mathbf{R}_{e_o} = \sigma_{e_o}^2 \mathbf{I}$ and $\mathbf{W} = \mathbf{I}$, we obtain (8.2.42).

PROPERTY 8.2.4. The trace of $\mathbf{\Gamma}_{\text{wls}}$ attains its minimum when $\mathbf{W} = \mathbf{R}_{e_o}^{-1}$. The resulting estimator

$$\mathbf{c}_{\text{mv}} = (\mathbf{X}^H \mathbf{R}_{e_o}^{-1} \mathbf{X})^{-1} \mathbf{X}^H \mathbf{R}_{e_o}^{-1} \mathbf{y} \tag{8.2.47}$$

is known as the *minimum variance* or *Markov estimator* and is the *best linear unbiased estimator* (*BLUE*).

Proof. The proof is somewhat involved. Interested readers can see Goodwin and Payne (1977) and Scharf (1991).

PROPERTY 8.2.5. If $\mathbf{R}_{e_o} = \sigma_{e_o}^2 \mathbf{I}$, the LS estimator $\mathbf{c}_{\text{ls}}$ is also the best linear unbiased estimator.

Proof. It follows from (8.2.47) with the substitution $\mathbf{R}_{e_o} = \sigma_{e_o}^2 \mathbf{I}$.

PROPERTY 8.2.6. When the random observation vector $\mathbf{e}_o$ has a normal distribution with mean zero and correlation matrix $\mathbf{R}_{e_o} = \sigma_{e_o}^2 \mathbf{I}$, that is, when its components are uncorrelated, the LS estimator $\mathbf{c}_{\text{ls}}$ is also the maximum likelihood estimator.

Proof. Since the components of vector $\mathbf{e}_o$ are uncorrelated and normally distributed with zero mean and variance σ_e^2, the likelihood function for real-valued $\mathbf{e}_o$ is given by

$$L(\mathbf{c}) = \prod_{n=0}^{N-1} \frac{1}{\sqrt{2\pi} \sigma_{e_o}} \exp \left[-\frac{|e_o(n)|^2}{2\sigma_{e_o}^2} \right] \tag{8.2.48}$$

and its logarithm by

$$\ln L(\mathbf{c}) = -\frac{1}{2\sigma_{e_o}^2} \mathbf{e}_o^H \mathbf{e}_o - \frac{N}{2} \ln(2\pi \sigma_{e_o}^2) = -\frac{1}{2\sigma_{e_o}^2} (\mathbf{y} - \mathbf{X}\mathbf{c})^H (\mathbf{y} - \mathbf{X}\mathbf{c}) + \text{const} \tag{8.2.49}$$

For complex-valued $\mathbf{e}_o$, the terms $\sqrt{2\pi}\sigma_{e_o}$ and $2\sigma_{e_o}^2$ in (8.2.48) are replaced by $\pi\sigma_{e_o}^2$ and $\sigma_{e_o}^2$, respectively. Since the logarithm is a monotonic function, maximization of $L(\mathbf{c})$ is equivalent to minimization of $\ln L(\mathbf{c})$. It is easy to see, by comparison with (8.2.8), that the LS solution maximizes this likelihood function.

Stochastic data matrix

We now extend the statistical properties of $\mathbf{c}_{\text{ls}}$ from the preceding section to the situation in which the data values in $\mathbf{X}$ are obtained from a random source with a known probability distribution. This situation is best handled by first obtaining the desired results conditioned on $\mathbf{X}$, which is equivalent to the deterministic case. We then determine the unconditional results by (statistical) averaging over the conditional distributions using the following properties of the conditional averages.

The *conditional mean* and the *conditional covariance* of a random vector $\mathbf{x}(\zeta)$, given another random vector $\mathbf{y}(\zeta)$, are defined by

$$\boldsymbol{\mu}_{x|y} \triangleq E\{\mathbf{x}(\zeta)|\mathbf{y}(\zeta)\}$$

and

$$\mathbf{\Gamma}_{x|y} \triangleq E\{[\mathbf{x}(\zeta) - \boldsymbol{\mu}_{x|y}][\mathbf{x}(\zeta) - \boldsymbol{\mu}_{x|y}]^H \mid \mathbf{y}(\zeta)\}$$

respectively. Since both quantities are random objects, it can be shown that

$$\mu_x = E\{\mathbf{x}(\zeta)\} = E_{\mathbf{y}}\{E\{\mathbf{x}(\zeta)|\mathbf{y}(\zeta)\}\}$$

which is known as the *law of iterated expectations* and that

$$\Gamma_x = \Gamma_{\mu_{x|y}}^{(y)} + \mu_{\Gamma_{x|y}}^{(y)}$$

which is called the *decomposition of the covariance rule*. This rule states that the covariance of a random vector $\mathbf{x}(\zeta)$ decomposes into the covariance of the conditional mean plus the mean of the conditional covariance. The covariance of the conditional mean, $\mu_{x|y}$, is given by

$$\Gamma_{\mu_{x|y}}^{(y)} \triangleq E_{\mathbf{y}}\{[\mu_{x|y} - \mu_x][\mu_{x|y} - \mu_x]^H\}$$

where the notation $\Gamma_{[\cdot]}^{(y)}$ indicates the covariance over the distribution of $\mathbf{y}(\zeta)$. More details can be found in Greene (1993).

PROPERTY 8.2.7. The LS estimator $\mathbf{c}_{ls}$ is an unbiased estimator of $\mathbf{c}_o$.

Proof. Taking the conditional expectation with respect to $\mathbf{X}$ of both sides of (8.2.37), we obtain

$$E\{\mathbf{c}_{ls}|\mathbf{X}\} = E\{\mathbf{c}_o|\mathbf{X}\} + (\mathbf{X}^H\mathbf{X})^{-1}\mathbf{X}^H E\{\mathbf{e}_o|\mathbf{X}\} \qquad (8.2.50)$$

Now using the law of iterated expectations, we get

$$E\{\mathbf{c}_{ls}\} = E_{\mathbf{X}}\{E\{\mathbf{c}_{ls}|\mathbf{X}\}\} = \mathbf{c}_o + E\{(\mathbf{X}^H\mathbf{X})^{-1}\mathbf{X}^H E\{\mathbf{e}_o|\mathbf{X}\}\}$$

Since $E\{\mathbf{e}_o|\mathbf{X}\} = \mathbf{0}$, from assumption 3, we have $E\{\mathbf{c}_{ls}\} = \mathbf{c}_o$. Thus $\mathbf{c}_{ls}$ is also unconditionally unbiased.

PROPERTY 8.2.8. The covariance matrix of $\mathbf{c}_{ls}$ corresponding to the error $\mathbf{c}_{ls} - \mathbf{c}_o$ is

$$\Gamma_{ls} \triangleq E\{(\mathbf{c}_{ls} - \mathbf{c}_o)(\mathbf{c}_{ls} - \mathbf{c}_o)^H\} = \sigma_{e_o}^2 E\{(\mathbf{X}^H\mathbf{X})^{-1}\} \qquad (8.2.51)$$

Proof. From (8.2.42), the conditional covariance matrix of $\mathbf{c}_{ls}$, conditional on $\mathbf{X}$, is

$$E\{(\mathbf{c}_{ls} - \mathbf{c}_o)(\mathbf{c}_{ls} - \mathbf{c}_o)^H|\mathbf{X}\} = \sigma_{e_o}^2(\mathbf{X}^H\mathbf{X})^{-1} \qquad (8.2.52)$$

For the unconditional covariance, we use the decomposition of covariance rule to obtain

$$E\{(\mathbf{c}_{ls} - \mathbf{c}_o)(\mathbf{c}_{ls} - \mathbf{c}_o)^H\} = E_{\mathbf{X}}\{E\{(\mathbf{c}_{ls} - \mathbf{c}_o)(\mathbf{c}_{ls} - \mathbf{c}_o)^H|\mathbf{X}\}\}$$
$$+ E_{\mathbf{X}}\{(E\{\mathbf{c}_{ls}|\mathbf{X}\} - \mathbf{c}_o)(E\{\mathbf{c}_{ls}|\mathbf{X}\} - \mathbf{c}_o)^H\}$$

The second term on the right-hand side above is equal to zero since $E\{\mathbf{c}_{ls}|\mathbf{X}\} = \mathbf{c}_o$ and hence

$$E\{(\mathbf{c}_{ls} - \mathbf{c}_o)(\mathbf{c}_{ls} - \mathbf{c}_o)^H\} = E_{\mathbf{X}}\{E\{(\mathbf{c}_{ls} - \mathbf{c}_o)(\mathbf{c}_{ls} - \mathbf{c}_o)^H|\mathbf{X}\}\}$$
$$= E_{\mathbf{X}}\{\sigma_{e_o}^2(\mathbf{X}^H\mathbf{X})^{-1}\} = \sigma_{e_o}^2 E\{(\mathbf{X}^H\mathbf{X})^{-1}\}$$

Thus the earlier result in (8.2.42) is modified by the expected value (or averaging) of $(\mathbf{X}^H\mathbf{X})^{-1}$.

One important conclusion about the statistical properties of the LS estimator is that the results obtained for the deterministic data matrix $\mathbf{X}$ are also valid for the stochastic case. This conclusion also applies for the Markov estimators and maximum likelihood estimators (Greene 1993).

8.3 LEAST-SQUARES FIR FILTERS

We will now apply the theory of linear LS error estimation to the design of FIR filters. The treatment closely follows the notation and approach in Section 6.4. Recall that the filtering error is

$$e(n) = y(n) - \sum_{k=0}^{M-1} h(k)\, x(n-k) \triangleq y(n) - \mathbf{c}^H\mathbf{x}(n) \qquad (8.3.1)$$

where $y(n)$ is the desired response,

$$\mathbf{x}(n) = [x(n)\, x(n-1)\, \cdots\, x(n-M+1)]^T \tag{8.3.2}$$

is the input data vector, and

$$\mathbf{c} = [c_0\, c_1\, \cdots\, c_{M-1}]^T \tag{8.3.3}$$

is the filter coefficient vector related to impulse response by $c_k = h^*(k)$. Suppose that we take measurements of the desired response $y(n)$ and the input signal $x(n)$ over the time interval $0 \le n \le N-1$. We hold the coefficients $\{c_k\}_0^{M-1}$ of the filter constant within this period and set any other required data samples equal to zero. For example, at time $n = 0$, that is, when we take the first measurement $x(0)$, the filter needs the samples $x(0), x(-1), \ldots, x(-M+1)$ to compute the output sample $\hat{y}(0)$. Since the samples $x(-1), \ldots, x(-M+1)$ are not available, to operate the filter, we should replace them with *arbitrary* values or start the filtering operation at time $n = M-1$. Indeed, for $M-1 \le n \le N-1$, all the input samples of $x(n)$ required by the filter to compute the output $\{\hat{y}(n)\}_{M-1}^{N-1}$ are available. If we want to compute the output while the last sample $x(N-1)$ is still in the filter memory, we must continue the filtering operation until $n = N+M-2$. Again, we need to assign arbitrary values to the unavailable samples $x(N), \ldots, x(N+M-2)$. Most often, we set the unavailable samples equal to zero, which can be thought of as windowing the sequences $x(n)$ and $y(n)$ with a rectangular window. To simplify the illustration, suppose that $N = 7$ and $M = 3$. Writing (8.3.1) for $n = 0, 1, \ldots, N+M-1$ and arranging in matrix form, we obtain

$$
\begin{array}{c}
0 \rightarrow \\[4pt]
\\
M-1 \rightarrow \\[4pt]
\\
\\
\\
N-1 \rightarrow \\[4pt]
\\
N+M-2 \rightarrow
\end{array}
\begin{bmatrix}
e^*(0) \\
e^*(1) \\
\hline
e^*(2) \\
e^*(3) \\
e^*(4) \\
e^*(5) \\
e^*(6) \\
\hline
e^*(7) \\
e^*(8)
\end{bmatrix}
=
\begin{bmatrix}
y^*(0) \\
y^*(1) \\
y^*(2) \\
y^*(3) \\
y^*(4) \\
y^*(5) \\
y^*(6) \\
0 \\
0
\end{bmatrix}
-
\begin{bmatrix}
x^*(0) & 0 & 0 \\
x^*(1) & x^*(0) & 0 \\
x^*(2) & x^*(1) & x^*(0) \\
x^*(3) & x^*(2) & x^*(1) \\
x^*(4) & x^*(3) & x^*(2) \\
x^*(5) & x^*(4) & x^*(3) \\
x^*(6) & x^*(5) & x^*(4) \\
0 & x^*(6) & x^*(5) \\
0 & 0 & x^*(6)
\end{bmatrix}
\begin{bmatrix}
c_0 \\
c_1 \\
c_2
\end{bmatrix}
\tag{8.3.4}
$$

or, in general,

$$\mathbf{e} = \mathbf{y} - \mathbf{X}\mathbf{c} \tag{8.3.5}$$

where the exact form of $\mathbf{e}$, $\mathbf{y}$, and $\mathbf{X}$ depends on the range $N_i \le n \le N_f$ of measurements to be used, which in turn determines the range of summation

$$E = \sum_{n=N_i}^{N_f} |e(n)|^2 = \mathbf{e}^H\mathbf{e} \tag{8.3.6}$$

in the LS criterion. The LS FIR filter is found by solving the LS normal equations

$$(\mathbf{X}^H\mathbf{X})\mathbf{c}_{ls} = \mathbf{X}^H\mathbf{y} \tag{8.3.7}$$

or

$$\hat{\mathbf{R}}\mathbf{c}_{ls} = \hat{\mathbf{d}} \tag{8.3.8}$$

with an LS error of

$$E_{ls} = E_y - \hat{\mathbf{d}}^H\mathbf{c}_{ls} \tag{8.3.9}$$

where E_y is the energy of the desired response signal. The elements of the time-average correlation matrix $\hat{\mathbf{R}}$ are given by

$$\hat{r}_{ij} = \tilde{\mathbf{x}}_i^H\tilde{\mathbf{x}}_j = \sum_{n=N_i}^{N_f} x(n+1-i)x^*(n+1-j) \qquad 1 \le i,\, j \le M \tag{8.3.10}$$

where $\tilde{x}_i$ are the columns of data matrix $\mathbf{X}$. A simple manipulation of (8.3.10) leads to

$$\hat{r}_{i+1,j+1} = \hat{r}_{ij} + x(N_i-i)x^*(N_i-j) - x(N_f+1-i)x^*(N_f+1-j) \qquad 1 \le i,j < M$$
(8.3.11)

which relates the elements of matrix $\hat{\mathbf{R}}$ that are located on the same diagonal. This property holds because the columns of $\mathbf{X}$ are obtained by shifting the first column. The recursion in (8.3.11) suggests the following way of efficiently computing $\hat{\mathbf{R}}$:

1. Compute the first row of $\hat{\mathbf{R}}$ by using (8.3.10). This requires M dot products and a total of about $M(N_f - N_i)$ operations.
2. Compute the remaining elements in the upper triangular part of $\hat{\mathbf{R}}$, using (8.3.11). This required number of operations is proportional to M^2.
3. Compute the lower triangular part of $\hat{\mathbf{R}}$, using the Hermitian symmetry relation $\hat{r}_{ji} = \hat{r}_{ij}^*$.

Notice that direct computation of the upper triangular part of $\hat{\mathbf{R}}$ using (8.3.10), that is, without the recursion, requires approximately $M^2N/2$ operations, which increases significantly for moderate or large values of M.

There are four ways to select the summation range $N_i \le n \le N_f$ that are used in LS filtering and prediction:

No windowing. If we set $N_i = M-1$ and $N_f = N-1$, we only use the available data and there are no distortions caused by forcing the data at the borders to artificial values.

Prewindowing. This corresponds to $N_i = 0$ and $N_f = N-1$ and is equivalent to setting the samples $x(0), x(-1), \ldots, x(-M+1)$ equal to zero. As a result, the term $x(M-i)x(M-j)$ does not appear in (8.3.11). This method is widely used in LS adaptive filtering.

Postwindowing. This corresponds to $N_i = M-1$ and $N_f = N+M-2$ and is equivalent to setting the samples $x(N), \ldots, x(N+M-2)$ equal to zero. As a result, the term $x(M-i)x(M-j)$ does not appear in (8.3.11). This method is not used very often for practical applications without prewindowing.

Full windowing. In this method, we impose both prewindowing and postwindowing (full windowing) to the input data and postwindowing to the desired response. The range of summation is from $N_i = 0$ to $N_f = N+M-2$, and as a result of full windowing, Eq. (8.3.11) becomes $\hat{r}_{i+1,j+1} = \hat{r}_{ij}$. Therefore, the elements $\hat{r}_{ij}$, depend on $i-j$, and matrix $\hat{\mathbf{R}}$ is Toeplitz. In this case, the normal equations (8.2.12) can be obtained from the Wiener-Hopf equations (6.4.11) by replacing the theoretical autocorrelations with their estimated values (see Section 5.2).

Clearly, as $N \gg M$ the performance difference between the various methods becomes insignificant. The no-windowing and full-windowing methods are known in the signal processing literature as the *autocorrelation* and *covariance methods*, respectively (Makhoul 1975b). We avoid these terms because they can lead to misleading statistical interpretations. We notice that in the LS filtering problem, the data matrix $\mathbf{X}$ is Toeplitz and the normal equations matrix $\hat{\mathbf{R}} = \mathbf{X}^H\mathbf{X}$ is the product of two Toeplitz matrices. However, $\hat{\mathbf{R}}$ is Toeplitz only in the full-windowing case when $\mathbf{X}$ is banded Toeplitz. In all other cases $\hat{\mathbf{R}}$ is *near to Toeplitz* or $\hat{\mathbf{R}}$ is *close to Toeplitz* in a sense made precise in Morf, et al. (1977).

The matrix $\hat{\mathbf{R}}$ and vector $\hat{\mathbf{d}}$, for the various windowing methods, are computed by using the MATLAB function [R,d]=lsmatvec(x,M,method,y), which is based on (8.3.10) and (8.3.11). Then the LS filter is computed by cls=R\d. Figure 8.6 shows an FIR LSE filter operating in block processing mode.

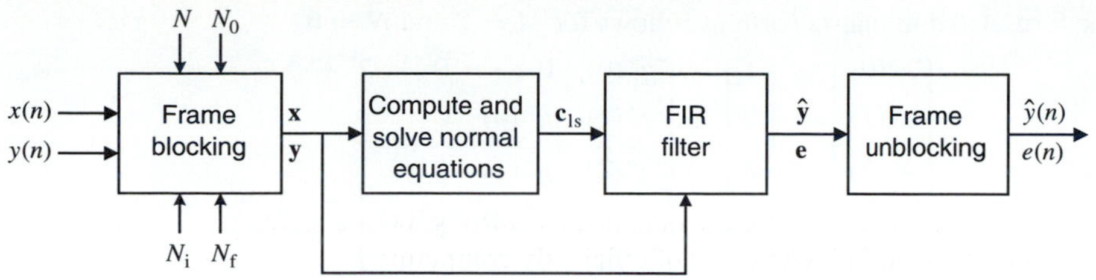

FIGURE 8.6
Block processing implementation of an FIR LSE filter.

EXAMPLE 8.3.1. To illustrate the design of least-squares FIR filters, suppose that we have a set of measurements of $x(n)$ and $y(n)$ for $0 \le n \le N - 1$ with $N = 100$ that have been generated by the difference equation

$$y(n) = 0.5x(n) + 0.5x(n-1) + v(n)$$

The input $x(n)$ and the additive noise $v(n)$ are uncorrelated processes from a normal (Gaussian) distribution with mean $E\{x(n)\} = E\{v(n)\} = 0$ and variance $\sigma_x^2 = \sigma_v^2 = 1$. Fitting the model

$$\hat{y}(n) = h(0)x(n) + h(1)x(n-1)$$

to the measurements with the no-windowing LS criterion, we obtain

$$\mathbf{c}_{ls} = \begin{bmatrix} 0.5361 \\ 0.5570 \end{bmatrix} \qquad \hat{\sigma}_e^2 = 1.0419 \qquad \hat{\sigma}_e^2 \hat{\mathbf{R}}^{-1} = \begin{bmatrix} 0.0073 & -0.0005 \\ -0.0005 & 0.0071 \end{bmatrix}$$

using (8.3.7), (8.3.9), (8.2.44), and (8.2.42). If the mean of the additive noise is nonzero, for example, if $E\{v(n)\} = 1$, we get

$$\mathbf{c}_{ls} = \begin{bmatrix} 0.4889 \\ 0.5258 \end{bmatrix} \qquad \hat{\sigma}_e^2 = 1.8655 \qquad \hat{\sigma}_e^2 \hat{\mathbf{R}}^{-1} = \begin{bmatrix} 0.0131 & -0.0009 \\ -0.0009 & 0.0127 \end{bmatrix}$$

which shows that the variance of the estimates, that is, the diagonal elements of $\hat{\sigma}_e^2 \hat{\mathbf{R}}^{-1}$, increases significantly. Suppose now that the recording device introduces an outlier in the input data at $x(30) = 20$. The estimated LS model and its associated statistics are given by

$$\mathbf{c}_{ls} = \begin{bmatrix} 0.1796 \\ 0.1814 \end{bmatrix} \qquad \hat{\sigma}_e^2 = 1.6270 \qquad \hat{\sigma}_e^2 \hat{\mathbf{R}}^{-1} = \begin{bmatrix} 0.0030 & 0.0000 \\ 0.0000 & 0.0030 \end{bmatrix}$$

Similarly, when an outlier is present in the output data, for example, at $y(30) = 20$, then the LS model and its statistics are

$$\mathbf{c}_{ls} = \begin{bmatrix} 0.6303 \\ 0.4653 \end{bmatrix} \qquad \hat{\sigma}_e^2 = 5.0979 \qquad \hat{\sigma}_e^2 \hat{\mathbf{R}}^{-1} = \begin{bmatrix} 0.0357 & -0.0025 \\ -0.0025 & 0.0347 \end{bmatrix}$$

In general, LS estimates are very sensitive to colored additive noise and outliers (Ljung 1987). Note that all the LS solutions in this example were produced with one sample realization $x(n)$ and that the results will vary for any other realizations.

LS inverse filters. Given a causal filter with impulse response $g(n)$, its inverse filter $h(n)$ is specified by $g(n) * h(n) = \delta(n - n_0)$, $n_0 \ge 0$. We focus on causal inverse filters, which are often infinite impulse response (IIR), and we wish to approximate them by some FIR filter $c_{ls}(n) = h^*(n)$ that is optimum according to the LS criterion. In this case, the actual impulse response $g(n) * c_{ls}^*(n)$ of the combined system deviates from the desired response $\delta(n - n_0)$, resulting in an error $e(n)$. The convolution equation

$$e(n) = \delta(n - n_0) - \sum_{k=0}^{M} c_{ls}^*(k) \, g(n - k) \qquad (8.3.12)$$

can be formulated in matrix form as follows for $M = 2$ and $N = 6$

$$
\begin{bmatrix} e^*(0) \\ e^*(1) \\ \hline e^*(2) \\ e^*(3) \\ e^*(4) \\ e^*(5) \\ e^*(6) \\ \hline e^*(7) \\ e^*(8) \end{bmatrix} = \begin{bmatrix} 1 \\ 0 \\ \hline 0 \\ 0 \\ 0 \\ 0 \\ 0 \\ \hline 0 \\ 0 \end{bmatrix} - \begin{bmatrix} g^*(0) & 0 & 0 \\ g^*(1) & g^*(0) & 0 \\ \hline g^*(2) & g^*(1) & g^*(0) \\ g^*(3) & g^*(2) & g^*(1) \\ g^*(4) & g^*(3) & g^*(2) \\ g^*(5) & g^*(4) & g^*(3) \\ g^*(6) & g^*(5) & g^*(4) \\ \hline 0 & g^*(6) & g^*(5) \\ 0 & 0 & g^*(6) \end{bmatrix} \begin{bmatrix} c_{ls}(0) \\ c_{ls}(1) \\ c_{ls}(2) \end{bmatrix}
$$

assuming that $n_0 = 0$. In general,

$$ \mathbf{e} = \boldsymbol{\delta}_i - \mathbf{G}\mathbf{c}_{ls}^{(i)} \tag{8.3.13} $$

where $\boldsymbol{\delta}_i$ is a vector whose ith element is 1 and whose remaining elements are all zero. The LS inverse filter and the corresponding error are given by

$$ (\mathbf{G}^H\mathbf{G})\mathbf{c}_{ls}^{(i)} = \mathbf{G}^H\boldsymbol{\delta}_i \tag{8.3.14} $$

and

$$ E_{ls}^{(i)} = 1 - \boldsymbol{\delta}_i^H\mathbf{G}\mathbf{c}_{ls}^{(i)} = 1 - g^*(i)c_{ls}^{(i)}(i) \qquad 0 \le i \le M + N \tag{8.3.15} $$

respectively.

Using the projection operators (8.2.29) and (8.2.30), we can express the LS error as

$$ E_{ls}^{(i)} = \boldsymbol{\delta}_i^H(\mathbf{P} - \mathbf{I})^H(\mathbf{P} - \mathbf{I})\boldsymbol{\delta}_i \tag{8.3.16} $$

where

$$ \mathbf{P} = \mathbf{G}(\mathbf{G}^H\mathbf{G})^{-1}\mathbf{G}^H \tag{8.3.17} $$

The total error for all possible delays $0 \le i \le N + M$ can be written as

$$ E_{\text{total}} = \sum_{i=0}^{N+M} E_{ls}^{(i)} = \text{tr}[\mathbf{D}^H(\mathbf{P} - \mathbf{I})^H(\mathbf{P} - \mathbf{I})\mathbf{D}] \tag{8.3.18} $$

where

$$ \mathbf{D} \triangleq [\boldsymbol{\delta}_0 \ \boldsymbol{\delta}_1 \ \boldsymbol{\delta}_2 \ \cdots \ \boldsymbol{\delta}_{N+M}] = \mathbf{I} $$

is the $(N + M + 1) \times (N + M + 1)$ identity matrix. Since $\mathbf{D} = \mathbf{I}$, $\mathbf{P} = \mathbf{P}^H$, and $\mathbf{P}^2 = \mathbf{P}$, we obtain

$$ E_{\text{total}} = \text{tr}[\mathbf{D}^H(\mathbf{P} - \mathbf{I})^H(\mathbf{P} - \mathbf{I})\mathbf{D}] = \text{tr}(\mathbf{I} - \mathbf{P}) = \text{tr}(\mathbf{I}) - \text{tr}(\mathbf{P}) $$

or

$$ E_{\text{total}} = N \tag{8.3.19} $$

because $\text{tr}(\mathbf{I}) = N + M + 1$ and

$$ \text{tr}(\mathbf{P}) = \text{tr}[\mathbf{G}(\mathbf{G}^H\mathbf{G})^{-1}\mathbf{G}^H] = \text{tr}[\mathbf{G}^H\mathbf{G}(\mathbf{G}^H\mathbf{G})^{-1}] = M + 1 \tag{8.3.20} $$

Hence, E_{total} depends on the length $N + 1$ of the filter $g(n)$ and is independent of the length $M + 1$ of the inverse filter $c_{ls}(n)$. If the minimum $E_{ls}^{(i)}$, for a given N, occurs at delay $i = i_0$, we have

$$ E_{ls}^{(i_0)} \le \frac{N}{N + M + 1} \tag{8.3.21} $$

which shows that $E_{ls}^{(i_0)} \to 0$ as $M \to \infty$ (Claerbout and Robinson 1963).

EXAMPLE 8.3.2. Suppose that $g(n) = \delta(n) - \alpha\delta(n - 1)$, where α is a real constant. The exact inverse filter is

$$ H(z) = \frac{1}{1 - \alpha z^{-1}} - h(n) = \alpha^n u(n) $$

and is minimum-phase only if $-1 < \alpha < 1$. The inverse LS filter for $M = 1$ and $N \geq 2$ is obtained by applying (8.3.14) with

$$
\mathbf{G} = \begin{bmatrix} 1 & 0 \\ -\alpha & 1 \\ 0 & -\alpha \end{bmatrix} \quad \text{and} \quad \boldsymbol{\delta} = \begin{bmatrix} 1 \\ 0 \\ 0 \end{bmatrix}
$$

The normal equations are

$$
\begin{bmatrix} 1+\alpha^2 & -\alpha \\ -\alpha & 1+\alpha^2 \end{bmatrix} \begin{bmatrix} c_{\text{ls}}(0) \\ c_{\text{ls}}(1) \end{bmatrix} = \begin{bmatrix} 1 \\ 0 \end{bmatrix}
\tag{8.3.22}
$$

leading to the LS inverse filter

$$
c_{\text{ls}}(0) = \frac{1+\alpha^2}{1+\alpha^2+\alpha^4} \qquad c_{\text{ls}}(1) = \frac{\alpha}{1+\alpha^2+\alpha^4}
$$

with LS error

$$
E_{\text{ls}} = 1 - c_{\text{ls}}(0) = \frac{\alpha^4}{1+\alpha^2+\alpha^4}
$$

The system function of the LS inverse filter is

$$
H_{\text{ls}}(z) = \frac{1+\alpha^2}{1+\alpha^2+\alpha^4} \left(1 + \frac{\alpha}{1+\alpha^2} z^{-1} \right)
$$

and has a zero at $z_1 = -\alpha/(1+\alpha^2) = -1/(\alpha + \alpha^{-1})$. Since $|z_1| < 1$ for any value of α, the LS inverse filter is minimum-phase even if $g(n)$ is not. This stems from the fact that the normal equations (8.3.22) specify a one-step forward linear predictor with a correlation matrix that is Toeplitz and positive definite for any value of α (see Section 7.4).

8.4 LINEAR LEAST-SQUARES SIGNAL ESTIMATION

We now discuss the application of the LS method to general signal estimation, FLP, BLP, and combined forward and backward linear prediction. The reader is advised to review Section 6.5, which provides a detailed discussion of the same problems for the MMSE criterion. The presentation in this section closely follows the viewpoint and notation in Section 6.5.

8.4.1 Signal Estimation and Linear Prediction

Suppose that we wish to compute the linear LS signal estimator $c_k^{(i)}$ defined by

$$
e^{(i)}(n) = \sum_{k=0}^{M} c_k^{(i)*} x(n-k) = \mathbf{c}^{(i)H} \bar{\mathbf{x}}(n) \qquad \text{with } c_i^{(i)} \triangleq 1
\tag{8.4.1}
$$

from the data $x(n)$, $0 \leq n \leq N-1$. Using (8.4.1) and following the process that led to (8.3.4), we obtain

$$
\mathbf{e}^{(i)} = \bar{\mathbf{X}} \mathbf{c}^{(i)}
\tag{8.4.2}
$$

where

$$
\bar{\mathbf{X}} = \begin{bmatrix}
x^*(0) & 0 & \cdots & 0 \\
x^*(1) & x^*(0) & \cdots & 0 \\
\vdots & \vdots & \ddots & \vdots \\
x^*(M) & x^*(M-1) & \cdots & x^*(0) \\
\vdots & \vdots & & \vdots \\
x^*(N-1) & x^*(N-2) & \cdots & x^*(N-M-1) \\
0 & x^*(N-1) & \cdots & x^*(N-M) \\
\vdots & \vdots & \ddots & \vdots \\
0 & 0 & \cdots & x^*(N-1)
\end{bmatrix}
\tag{8.4.3}
$$

is the combined data and desired response matrix with all the unavailable samples set equal to zero (full windowing). Matrix $\bar{\mathbf{X}}$ can be partitioned columnwise as

$$\bar{\mathbf{X}} = [\mathbf{X}_1 \ \mathbf{y} \ \mathbf{X}_2] \tag{8.4.4}$$

where $\mathbf{y}$, the desired response, is the ith column of $\bar{\mathbf{X}}$. Using (8.4.4), we can easily show that the LS signal estimator $\mathbf{c}_{ls}^{(i)}$ and the associated LS error $E_{ls}^{(i)}$ are determined by

$$(\bar{\mathbf{X}}^H\bar{\mathbf{X}})\mathbf{c}_{ls}^{(i)} = \begin{bmatrix} \mathbf{0} \\ E_{ls}^{(i)} \\ \mathbf{0} \end{bmatrix} \tag{8.4.5}$$

where $E_{ls}^{(i)}$ is the ith element of the right-hand side vector (see Problem 8.3). If we define the time-average correlation matrix

$$\bar{\mathbf{R}} \triangleq \bar{\mathbf{X}}^H\bar{\mathbf{X}} \tag{8.4.6}$$

and use the augmented normal equations in (8.4.5), we obtain a set of equations that have the same form as (6.5.12), the equations for the MMSE signal estimator. Therefore, after we have computed $\bar{\mathbf{R}}$, using the command `Rbar=lsmatvec(x,M+1,method)`, we can use the steps in Table 6.3 to compute the LS forward linear predictor (FLP), the backward linear predictor (BLP), the symmetric smoother, or any other signal estimator with delay i. Again, we use the standard notation $E_{ls}^{(0)} = E^f$ and $\mathbf{c}_{ls}^{(0)} = \mathbf{a}$ for the FLP and $E_{ls}^{(M)} = E^b$ and $\mathbf{c}_{ls}^{(M)} = \mathbf{b}$ for the BLP.

All formulas given in Section 6.5 hold for LS signal estimators if the matrix $\mathbf{R}(n)$ is replaced by $\bar{\mathbf{R}}$. However, we stress that although the optimum MMSE signal estimator $\mathbf{c}_o^{(i)}(n)$ is a deterministic vector, the LS signal estimator $\mathbf{c}_{ls}^{(i)}$ is a random vector that is a function of the random measurements $x(n)$, $0 \le n \le N-1$. In the full-windowing case, matrix $\bar{\mathbf{R}}$ is Toeplitz; if it is also positive definite, then the FLP is minimum-phase. Although the use of full windowing leads to these nice properties, it also creates some "edge effects" and bias in the estimates because we try to estimate some signal values using values that are not part of the signal by forcing the samples leading and lagging the available data measurements to zero.

EXAMPLE 8.4.1. Suppose that we are given the signal segment $x(n) = \alpha^n$, $0 \le n \le N$, where α is an arbitrary complex-valued constant. Determine the first-order one-step forward linear predictor, using the full-windowing and no-windowing methods.

Solution. We start by forming the combined desired response and data matrix

$$\bar{\mathbf{X}}^H = \begin{bmatrix} x(0) & x(1) & \cdots & x(N) & 0 \\ 0 & x(0) & \cdots & x(N-1) & x(N) \end{bmatrix}$$

For the full-windowing method, the matrix

$$\bar{\mathbf{R}} = \bar{\mathbf{X}}^H\bar{\mathbf{X}} = \begin{bmatrix} \hat{r}_x(0) & \hat{r}_x(1) \\ \hat{r}_x^*(1) & \hat{r}_x(0) \end{bmatrix}$$

is Toeplitz with elements

$$\hat{r}_x(0) = \sum_{n=0}^{N} |x(n)|^2 = \sum_{n=0}^{N} |\alpha|^{2n} = \frac{1 - |\alpha|^{2(N+1)}}{1 - |\alpha|^2}$$

and

$$\hat{r}_x(1) = \sum_{n=1}^{N} x(n)\,x^*(n-1) = \sum_{n=1}^{N} \alpha^n(\alpha^*)^{n-1} = \alpha^* \frac{1 - |\alpha|^{2N}}{1 - |\alpha|^2}$$

Therefore, we have

$$\begin{bmatrix} \hat{r}_x(0) & \hat{r}_x(1) \\ \hat{r}_x^*(1) & \hat{r}_x(0) \end{bmatrix} \begin{bmatrix} 1 \\ a_1^{(1)} \end{bmatrix} = \begin{bmatrix} E_1^f \\ 0 \end{bmatrix}$$

whose solution gives

$$a_1^{(1)} = -\frac{\hat{r}_x^*(1)}{\hat{r}_x(0)} = -\alpha\frac{1 - |\alpha|^{2N}}{1 - |\alpha|^{2(N+1)}}$$

and

$$E_1^f = \hat{r}_x(0) + \hat{r}_x(1)a_1^{(1)} = \frac{1 - |\alpha|^{2(2N+1)}}{1 - |\alpha|^{2(N+1)}}$$

Since for every sequence $|\hat{r}_x(l)| \le |\hat{r}_x(0)|$, we have $|a_1^{(1)}| \le 1$; that is, the obtained prediction error filter always is minimum-phase. Furthermore, if $|\alpha| < 1$, then $\lim_{N \to \infty} a_1^{(1)} = -\alpha$ and $\lim_{N \to \infty} E_1^f = 1 = x(0)$. In the no-windowing case, the matrix

$$\bar{\mathbf{R}} = \bar{\mathbf{X}}^H\bar{\mathbf{X}} = \begin{bmatrix} \hat{r}_{11} & \hat{r}_{12} \\ \hat{r}_{12}^* & \hat{r}_{22} \end{bmatrix}$$

is Hermitian but not Toeplitz with elements

$$\hat{r}_{11} = \sum_{n=1}^{N} |x(n)|^2 = |\alpha|^2\frac{1 - |\alpha|^{2N}}{1 - |\alpha|^2} \qquad \hat{r}_{22} = \sum_{n=0}^{N-1} |x(n)|^2 = \frac{1 - |\alpha|^{2N}}{1 - |\alpha|^2}$$

$$\hat{r}_{12} = \sum_{n=1}^{N} x(n)\,x^*(n-1) = \alpha^*\frac{1 - |\alpha|^{2N}}{1 - |\alpha|^2}$$

Solving the linear system

$$\begin{bmatrix} \hat{r}_{11} & \hat{r}_{12} \\ \hat{r}_{12}^* & \hat{r}_{22} \end{bmatrix}\begin{bmatrix} 1 \\ \bar{a}_1^{(1)} \end{bmatrix} = \begin{bmatrix} \bar{E}_1^f \\ 0 \end{bmatrix}$$

we obtain

$$\bar{a}_1^{(1)} = -\frac{\hat{r}_{12}^*}{\hat{r}_{22}} = -\alpha$$

and

$$\bar{E}_1^f = \hat{r}_{11} + \hat{r}_{12}\,\bar{a}_1^{(1)} = 0$$

We see that the no-windowing method provides a perfect linear predictor because there is no distortion due to windowing. However, the obtained prediction error filter is minimum-phase only when $|\alpha| < 1$.

EXAMPLE 8.4.2. To illustrate the statistical properties of least-squares FLP, we generate $K = 500$ realizations of the MA(1) process $x(n) = w(n) + \frac{1}{2}w(n-1)$, where $w(n) \sim \text{WN}(0, 1)$ (see Example 6.5.2). Each realization $x(\zeta_i, n)$ has duration $N = 100$ samples. We use these data to design an $M = 2$ order FLP, using the no-windowing LS method. The estimated mean and variance of the obtained K FLP vectors are

$$\text{Mean}\{\mathbf{a}(\zeta_i)\} = \begin{bmatrix} -0.4695 \\ 0.1889 \end{bmatrix} \quad \text{and} \quad \text{var}\{\mathbf{a}(\zeta_i)\} = \begin{bmatrix} 0.0086 \\ 0.0092 \end{bmatrix}$$

whereas the average of the variances $\hat{\sigma}_e^2$ is 0.9848. We notice that both means are close to the theoretical values obtained in Example 6.5.2. The covariance matrix of a given LS estimate $\mathbf{a}_{ls}$ was found to be

$$\hat{\sigma}_e^2\hat{\mathbf{R}}^{-1} = \begin{bmatrix} 0.0099 & -0.0043 \\ -0.0043 & 0.0099 \end{bmatrix}$$

whose diagonal elements are close to the components of var$\{\mathbf{a}\}$, as expected. The bias in the estimate $\mathbf{a}_{ls}$ results from the fact that the residuals in the LS equations are correlated with each other (see Problem 8.4).

8.4.2 Combined Forward and Backward Linear Prediction (FBLP)

For stationary stochastic processes, the optimum MMSE forward and backward linear predictors have even conjugate symmetry, that is,

$$\mathbf{a}_o = \mathbf{J}\mathbf{b}_o^* \tag{8.4.7}$$

because both directions of time have the same second-order statistics. Formally, this property stems from the Toeplitz structure of the autocorrelation matrix (see Section 6.5). However, we could possibly improve performance by minimizing the total forward and backward squared error

$$E^{\mathrm{fb}} = \sum_{n=N_{\mathrm{i}}}^{N_{\mathrm{f}}} \{|e^{\mathrm{f}}(n)|^2 + |e^{\mathrm{b}}(n)|^2\} = (\mathbf{e}^{\mathrm{f}})^H \mathbf{e}^{\mathrm{f}} + (\mathbf{e}^{\mathrm{b}})^H \mathbf{e}^{\mathrm{b}} \tag{8.4.8}$$

under the constraint

$$\mathbf{a}^{\mathrm{fb}} \triangleq \mathbf{a} = \mathbf{J}\mathbf{b}^* \tag{8.4.9}$$

The FLP and BLP overdetermined sets of equations are

$$\mathbf{e}^{\mathrm{f}} = \bar{\mathbf{X}} \begin{bmatrix} 1 \\ \mathbf{a} \end{bmatrix} \quad \text{and} \quad \mathbf{e}^{\mathrm{b}} = \bar{\mathbf{X}} \begin{bmatrix} \mathbf{b} \\ 1 \end{bmatrix} \tag{8.4.10}$$

or

$$\mathbf{e}^{\mathrm{f}} = \bar{\mathbf{X}} \begin{bmatrix} 1 \\ \mathbf{a}^{\mathrm{fb}} \end{bmatrix} \quad \text{and} \quad \mathbf{e}^{\mathrm{b}*} = \bar{\mathbf{X}}^* \begin{bmatrix} \mathbf{b}^* \\ 1 \end{bmatrix} = \bar{\mathbf{X}}^* \mathbf{J} \begin{bmatrix} 1 \\ \mathbf{a}^{\mathrm{fb}} \end{bmatrix} \tag{8.4.11}$$

where we have used (8.4.9) and the property $\mathbf{J}\mathbf{J} = \mathbf{I}$ of the exchange matrix. If we combine the above two equations as

$$\begin{bmatrix} \mathbf{e}^{\mathrm{f}} \\ \mathbf{e}^{\mathrm{b}*} \end{bmatrix} = \begin{bmatrix} \bar{\mathbf{X}} \\ \bar{\mathbf{X}}^* \mathbf{J} \end{bmatrix} \begin{bmatrix} 1 \\ \mathbf{a}^{\mathrm{fb}} \end{bmatrix} \tag{8.4.12}$$

then the forward-backward linear predictor that minimizes E^{fb} is given by (see Problem 8.5)

$$\begin{bmatrix} \bar{\mathbf{X}} \\ \bar{\mathbf{X}}^* \mathbf{J} \end{bmatrix}^H \begin{bmatrix} \bar{\mathbf{X}} \\ \bar{\mathbf{X}}^* \mathbf{J} \end{bmatrix} \begin{bmatrix} 1 \\ \mathbf{a}_{\mathrm{ls}}^{\mathrm{fb}} \end{bmatrix} = \begin{bmatrix} E_{\mathrm{ls}}^{\mathrm{fb}} \\ \mathbf{0} \end{bmatrix}$$

or

$$(\bar{\mathbf{X}}^H \bar{\mathbf{X}} + \mathbf{J}\bar{\mathbf{X}}^T \bar{\mathbf{X}}^* \mathbf{J}) \begin{bmatrix} 1 \\ \mathbf{a}_{\mathrm{ls}}^{\mathrm{fb}} \end{bmatrix} = \begin{bmatrix} E_{\mathrm{ls}}^{\mathrm{fb}} \\ \mathbf{0} \end{bmatrix} \tag{8.4.13}$$

which can be solved by using the steps described in Table 6.3. The time-average forward-backward correlation matrix

$$\hat{\mathbf{R}}_{\mathrm{fb}} \triangleq \bar{\mathbf{X}}^H \bar{\mathbf{X}} + \mathbf{J}\bar{\mathbf{X}}^T \bar{\mathbf{X}}^* \mathbf{J} \tag{8.4.14}$$

with elements

$$\hat{r}_{ij}^{\mathrm{fb}} = \hat{r}_{ij} + \hat{r}_{M-i,M-j}^* \qquad 0 \le i,\ j \le M \tag{8.4.15}$$

is persymmetric; that is, $\mathbf{J}\hat{\mathbf{R}}_{\mathrm{fb}}\mathbf{J} = \hat{\mathbf{R}}_{\mathrm{fb}}^*$ and its elements are conjugate symmetric about both main diagonals. In MATLAB we compute $\hat{\mathbf{R}}_{\mathrm{fb}}$ by these commands:

```
Rbar=lsmatvec(x,M+1,method)
Rfb=Rbar+flipud(fliplr(conj(Rbar)))
```

The FBLP method is used with *no windowing* and was originally introduced independently by Ulrych and Clayton (1976) and Nuttall (1976) as a spectral estimation technique under the name *modified covariance method* (see Section 9.2). If we use full windowing, then $\mathbf{a}^{\mathrm{fb}} = (\mathbf{a} + \mathbf{J}\mathbf{b}^*)/2$ (see Problem 8.6).

8.4.3 Narrowband Interference Cancelation

Several practical applications require the removal of *narrowband interference* (*NBI*) from a wideband desired signal corrupted by additive white noise. For example, ground and foliage-penetrating radars operate from 0.01 to 1 GHz and use either an impulse or a chirp waveform. To achieve high resolution, these waveforms are extremely wideband, occupying at least 100 MHz within the range of 0.01 to 1 GHz. However, these frequency ranges are extensively used by TV and FM stations, cellular phones, and other relatively narrowband (less than 1 MHz) radio-frequency (RF) sources. Clearly, these sources spoil the radar returns with narrowband RF interference (Miller et al. 1997). Since the additive noise is often due to the sensor circuitry, it will be referred to as *sensor thermal noise*. Next we provide a practical solution to this problem, using an LS linear predictor. Suppose that the corrupted signal $x(n)$ is given by

$$x(n) = s(n) + y(n) + v(n) \tag{8.4.16}$$

where
$$s(n) = \text{signal of interest} \tag{8.4.17}$$

$$y(n) = \text{narrowband interference}$$

$$v(n) = \text{thermal (white) noise}$$

are the individual components, assumed to be stationary stochastic processes.

We wish to design an NBI canceler that estimates and rejects the interference signal $y(n)$ from the signal $x(n)$, while preserving the signal of interest $s(n)$. Since signals $y(n)$ and $x(n)$ are correlated, we can form an estimate of the NBI using the optimum linear estimator

$$\hat{y}(n) = \mathbf{c}_o^H \mathbf{x}(n - D) \tag{8.4.18}$$

where
$$\mathbf{R}\mathbf{c}_o = \mathbf{d} \tag{8.4.19}$$

$$\mathbf{R} = E\{\mathbf{x}(n - D)\mathbf{x}^H(n - D)\} \tag{8.4.20}$$

$$\mathbf{d} = E\{\mathbf{x}(n - D)\, y^*(n)\} \tag{8.4.21}$$

and D is an integer delay whose use will be justified shortly. Note that if $D = 1$, then (8.4.18) is the LS forward linear predictor. If $\hat{y}(n) = y(n)$, the output of the canceler is $x(n) - \hat{y}(n) = s(n) + v(n)$; that is, the NBI is completely excised, and the desired signal is corrupted by white noise only and is said to be thermal noise–limited.

Since, in practice, the required second-order moments are not available, we need to use an LS estimator instead. However, the quantity $\mathbf{X}^H\mathbf{y}$ in (8.2.21) requires the NBI signal $y(n)$, which is also not available. To overcome this obstacle, consider the optimum MMSE D-step forward linear predictor

$$e^{\mathrm{f}}(n) = x(n) + \mathbf{a}^H \mathbf{x}(n - D) \tag{8.4.22}$$

$$\mathbf{R}\mathbf{a} = -\mathbf{r}^{\mathrm{f}} \tag{8.4.23}$$

where $\mathbf{R}$ is given by (8.4.20) and

$$\mathbf{r}^{\mathrm{f}} = E\{\mathbf{x}(n - D)x^*(n)\} \tag{8.4.24}$$

In many NBI cancelation applications, the components of the observed signal have the following properties:

1. The desired signal $s(n)$, the NBI $y(n)$, and the thermal noise $v(n)$ are mutually uncorrelated.
2. The thermal noise $v(n)$ is white; that is, $r_v(l) = \sigma_v^2 \delta(l)$.
3. The desired signal $s(n)$ is wideband and therefore has a short correlation length; that is, $r_v(l) = 0$ for $|l| \geq D$.
4. The NBI has a long correlation length; that is, its autocorrelation takes significant values over the range $0 \leq |l| \leq M$ for $M > D$.

In practice, the second and third properties mean that the desired signal and the thermal noise are approximately uncorrelated after a certain small lag. These are precisely the properties exploited by the canceler to separate the NBI from the desired signal and the background noise.

As a result of the first assumption, we have

$$E\{x(n-k)y^*(n)\} = E\{y(n-k)y^*(n)\} = r_y(k) \qquad \text{for all } k \tag{8.4.25}$$

and

$$r_x(l) = r_s(l) + r_y(l) + r_v(l) \tag{8.4.26}$$

Making use of the second and third assumptions, we have

$$r_x(l) = r_y(l) \qquad \text{for } l \neq 0, 1, \dots, D-1 \tag{8.4.27}$$

The exclusion of the lags for $l \neq 0, 1, \dots, D-1$ in $\mathbf{r}$ and $\mathbf{d}$ is critical, and we have arranged for that by forcing the filter and the predictor to form their estimates using the *delayed* data vector $\mathbf{x}(n-D)$. From (8.4.21), (8.4.24), and (8.4.27), we conclude that $\mathbf{d} = \mathbf{r}^f$ and therefore $\mathbf{c}_o = \mathbf{a}_o$. Thus, the optimum NBI estimator $\mathbf{c}_o$ is equal to the D-step linear predictor $\mathbf{a}_o$, which can be determined exclusively from the input signal $x(n)$. The cleaned signal is

$$x(n) - \hat{y}(n) = x(n) + \mathbf{a}_o^H \mathbf{x}(n-D) = e^f(n) \tag{8.4.28}$$

which is identical to the D-step forward prediction error. This leads to the linear prediction NBI canceler shown in Figure 8.7.

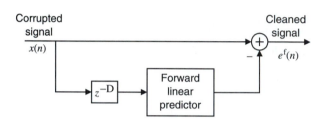

Corrupted signal $x(n)$ — Cleaned signal $e^f(n)$

FIGURE 8.7
Block diagram of linear prediction NBI canceler.

To illustrate the performance of the linear prediction NBI canceler, we consider an impulse radar operating in a location with commercial radio and TV stations. The desired signal is a short-duration impulse corrupted by additive thermal noise and NBI (see Figure 8.8). The spectrum of the NBI is shown in Figure 8.9. We use a block of data ($N = 4096$) to design an FBLP with $D = 1$ and $M = 100$ coefficients, using the LS criterion with no windowing. Then we compute the cleaned signal, using (8.4.28). The cleaned signal, its spectrum, and the magnitude response of the NBI canceler are shown in Figures 8.8 and 8.9. We see that the canceler acts as a notch filter that optimally puts notches at the peaks of the NBI. A detailed description of the design of optimum least-squares NBI cancelers is given in Problem 8.27.

8.5 LS COMPUTATIONS USING THE NORMAL EQUATIONS

The solution of the normal equations for both MMSE and LSE estimation problems is computed by using the same algorithms. The key difference is that in MMSE estimation $\mathbf{R}$ and $\mathbf{d}$ are known, whereas in LSE estimation they need to be computed from the observed input and desired response signal samples. Therefore, it is natural to want to take advantage of the same algorithms developed for MMSE estimation in Chapter 7, whenever possible. However, keep in mind that despite algorithmic similarities, there are fundamental differences between the two classes of estimators that are dictated by the different nature of the the criteria of performance (see Section 8.1). In this section, we show how the computational algorithms and structures developed for linear MMSE estimation can be applied to linear LSE estimation, relying heavily on the material presented in Chapter 7.

8.5.1 Linear LSE Estimation

The computation of a general linear LSE estimator requires the solution of a linear system

$$\hat{\mathbf{R}}\mathbf{c}_{ls} = \hat{\mathbf{d}} \qquad (8.5.1)$$

where the time-average correlation matrix $\hat{\mathbf{R}}$ is Hermitian and positive definite [see (8.2.25)]. We can solve (8.5.1) by using the LDL^H or the Cholesky decomposition introduced in Section 6.3. The computation of linear LSE estimators involves the steps summarized in Table 8.1. We again stress that the major computational effort is involved in the computation of $\hat{\mathbf{R}}$ and $\hat{\mathbf{d}}$.

Steps 2 and 3 in (6.3.16) can be facilitated by a single extended LDL^H decomposition. To this end, we form the augmented data matrix

$$\bar{\mathbf{X}} = [\mathbf{X} \; \mathbf{y}] \qquad (8.5.2)$$

and compute its time-average correlation matrix

$$\bar{\mathbf{R}} = \bar{\mathbf{X}}^H \bar{\mathbf{X}} = \begin{bmatrix} \mathbf{X}^H \mathbf{X} & \mathbf{X}^H \mathbf{y} \\ \mathbf{y}^H \mathbf{X} & \mathbf{y}^H \mathbf{y} \end{bmatrix} = \begin{bmatrix} \hat{\mathbf{R}} & \hat{\mathbf{d}} \\ \hat{\mathbf{d}}^H & E_y \end{bmatrix} \qquad (8.5.3)$$

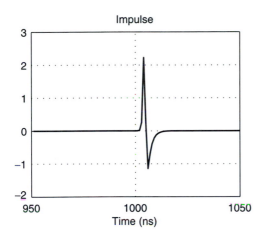

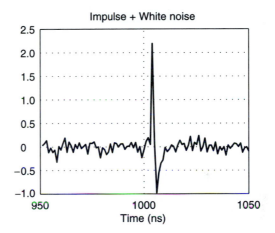

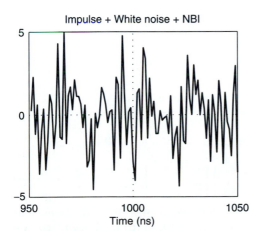

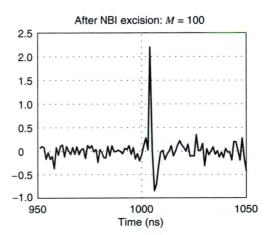

FIGURE 8.8
NBI cancelation: time-domain results.

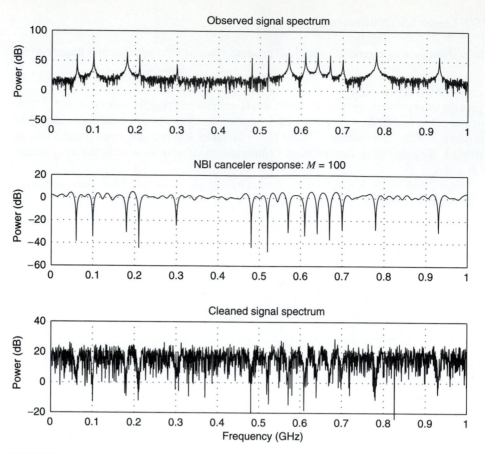

FIGURE 8.9
NBI cancelation: frequency-domain results.

TABLE 8.1
**Comparison between the LDL^H and Cholesky decomposition methods for the
solution of normal equations.**

Step	LDL^H decomposition	Cholesky decomposition	Description
1	$\hat{\mathbf{R}} = \mathbf{X}^H\mathbf{X}, \hat{\mathbf{d}} = \mathbf{X}^H\mathbf{y}$		Normal equations $\hat{\mathbf{R}}\mathbf{c}_{ls} = \hat{\mathbf{d}}$
2	$\hat{\mathbf{R}} = \mathbf{L}\mathbf{D}\mathbf{L}^H$	$\hat{\mathbf{R}} = \mathcal{L}\mathcal{L}^H$	Triangular decomposition
3	$\mathbf{L}\mathbf{D}\mathbf{k} = \hat{\mathbf{d}}$	$\mathcal{L}\tilde{\mathbf{k}} = \hat{\mathbf{d}}$	Forward substitution $\rightarrow \mathbf{k}$ or $\tilde{\mathbf{k}}$
4	$\mathbf{L}^H\mathbf{c}_{ls} = \mathbf{k}$	$\mathcal{L}^H\mathbf{c}_{ls} = \tilde{\mathbf{k}}$	Backward substitution $\rightarrow \mathbf{c}_{ls}$
5	$E_{ls} = E_y - \mathbf{k}^H\mathbf{D}\mathbf{k}$	$E_{ls} = E_y - \tilde{\mathbf{k}}^H\tilde{\mathbf{k}}$	LSE computation
6	$\mathbf{e}_{ls} = \mathbf{y} - \mathbf{X}\mathbf{c}_{ls}$	$\mathbf{e}_{ls} = \mathbf{y} - \mathbf{X}\mathbf{c}_{ls}$	Computation of residuals

We then can show (see Problem 8.9) that the LDL^H decomposition of $\bar{\mathbf{R}}$ is given by

$$\bar{\mathbf{R}} = \begin{bmatrix} \mathbf{L} & \mathbf{0} \\ \mathbf{k}^H & 1 \end{bmatrix} \begin{bmatrix} \mathbf{D} & \mathbf{0} \\ \mathbf{0}^H & E_{ls} \end{bmatrix} \begin{bmatrix} \mathbf{L}^H & \mathbf{k}^H \\ \mathbf{0}^H & 1 \end{bmatrix} \quad (8.5.4)$$

and thus provides the vector $\mathbf{k}$ and the LSE E_{ls}. Therefore, we can solve the normal equations
(8.5.1), using the LDL^H decomposition of $\bar{\mathbf{R}}$ to compute $\mathbf{L}$ and $\mathbf{k}$ and then solving $\mathbf{L}^H\mathbf{c}_{ls} = \mathbf{k}$
to compute $\mathbf{c}_{ls}$.

A careful inpection of the design equations for the general, mth-order, MMSE and
LSE estimators, derived in Chapter 6 and summarized in Table 8.2, shows that the LSE

TABLE 8.2

419

SECTION 8.5
LS Computations Using the
Normal Equations

Comparison between the MMSE and LSE normal equations for general linear estimation.

	MMSE	LSE
Available information	$\mathbf{R}_m(n)$, $\mathbf{d}_m(n)$	$\{\mathbf{x}_m(n),\ y(n), n_i \le n \le n_f\}$
Normal equations	$\mathbf{R}_m(n)\mathbf{c}_m(n) = \mathbf{d}_m(n)$	$\hat{\mathbf{R}}_m\mathbf{c}_m = \hat{\mathbf{d}}_m$
Minimum error	$P_m(n) = P_y(n) - \mathbf{d}_m^H(n)\mathbf{c}_m(n)$	$E_m = E_y - \hat{\mathbf{d}}_m^H\mathbf{c}_m$
Correlation matrix	$\mathbf{R}_m(n) \triangleq E\{\mathbf{x}_m(n)\mathbf{x}_m^H(n)\}$	$\hat{\mathbf{R}}_m = \mathbf{X}_m^H\mathbf{X}_m = \displaystyle\sum_{n=0}^{N-1} \mathbf{x}_m(n)\mathbf{x}_m^H(n)$
Cross-correlation vector	$\mathbf{d}_m(n) \triangleq E\{\mathbf{x}_m(n)y^*(n)\}$	$\hat{\mathbf{d}}_m = \mathbf{X}_m^H\mathbf{y} = \displaystyle\sum_{n=0}^{N-1} \mathbf{x}_m(n)y^*(n)$
Power	$P_y(n) = E\{|y(n)|^2\}$	$E_y = \mathbf{y}^H\mathbf{y} = \displaystyle\sum_{n=0}^{N-1} |y(n)|^2$

equations can be obtained from the MMSE equations by replacing the linear operator $E\{\cdot\}$ by the linear operator $\sum_n(\cdot)$. As a result, all algorithms developed in Sections 7.1 and 7.2 can be used for linear LSE estimation problems.

For example, we can easily see that $\hat{\mathbf{R}}_M$, $\hat{\mathbf{d}}_M$, $\mathbf{L}_M$, $\mathbf{D}_M$, and $\mathbf{k}_M$ have the optimum nesting property described in Section 7.1.1, that is, $\hat{\mathbf{R}}_m = \hat{\mathbf{R}}_M^{[m]}$ and so on. As a result, the factors of the LDL^H decomposition have the optimum nesting property, and we can obtain an order-recursive structure for the computation of the LSE estimate $\hat{y}_m(n)$. Indeed, if we define

$$\mathbf{w}_m(n) = \mathbf{L}_m^{-1}\mathbf{x}_m(n) \qquad 0 \le n \le N - 1 \tag{8.5.5}$$

then $\quad \hat{\mathbf{R}}_m = \displaystyle\sum_{n=0}^{N-1} \mathbf{x}_m(n)\mathbf{x}_m^H(n) = \mathbf{L}_m\left[\displaystyle\sum_{n=0}^{N-1} \mathbf{w}_m(n)\mathbf{w}_m^H(n)\right]\mathbf{L}_m \triangleq \mathbf{L}_m\mathbf{D}_m\mathbf{L}_m^H \tag{8.5.6}$

where the matrix $\mathbf{D}_m$ is diagonal because the LDL^H decomposition is unique. If we define the record vectors

$$\tilde{\mathbf{w}}_j \triangleq [w_j(0)\ w_j(1)\ \cdots\ w_j(N-1)]^H \tag{8.5.7}$$

and the data matrix

$$\mathbf{W}_m \triangleq [\tilde{\mathbf{w}}_1\ \tilde{\mathbf{w}}_2\ \cdots\ \tilde{\mathbf{w}}_m] \tag{8.5.8}$$

then

$$\mathbf{D}_m = \mathbf{W}_m^H\mathbf{W}_m = \text{diag}\{\xi_1, \xi_2, \ldots, \xi_m\} \tag{8.5.9}$$

where

$$\xi_i = \sum_{n=0}^{N-1} |w_i(n)|^2 = \tilde{\mathbf{w}}_i^H\tilde{\mathbf{w}}_i \tag{8.5.10}$$

From (8.5.9), we have

$$\tilde{\mathbf{w}}_i^H\tilde{\mathbf{w}}_j = 0 \qquad \text{for } i \ne j \tag{8.5.11}$$

that is, the columns of $\mathbf{W}_m$ are orthogonal and, in this sense, are the innovation vectors of the columns of data matrix $\mathbf{X}_m$, according to the LS interpretation of orthogonality introduced in Section 8.2.

Following the approach in Section 7.1.5, we can show that the following order-recursive algorithm

$$w_m(n) = x_m(n) - \sum_{i=1}^{m-1} l_{i-1}^{(m-1)*}w_i(n) \tag{8.5.12}$$

$$\hat{y}_m(n) = \hat{y}_{m-1}(n) + k_m^*w_m(n)$$

or

$$e_m(n) = e_{m-1}(n) - k_m^* w_m(n)$$

computed for $n = 0, 1, \ldots, N-1$ and $m = 1, 2, \ldots, M$, provides the LSE estimates for orders $1 \leq m \leq M$.

The statistical interpretations of innovation and partial correlation for $w_m(n)$ and k_{m+1} hold now in a deterministic LSE sense. For example, the *partial correlation* between $\tilde{\mathbf{y}}$ and $\tilde{\mathbf{x}}_{m+1}$ is defined by using the residual records $\tilde{\mathbf{e}}_m = \tilde{\mathbf{y}} - \mathbf{X}_m \mathbf{c}_m$ and $\tilde{\mathbf{e}}_m^b = \tilde{\mathbf{x}}_{m+1} + \mathbf{X}_m \mathbf{b}_m$, where $\mathbf{b}_m$ is the least-squares error BLP. Indeed, if $\beta_{m+1} \triangleq \tilde{\mathbf{e}}_m^H \tilde{\mathbf{e}}_m^b$, we can show that $k_{m+1} = \beta_{m+1}/\xi_{m+1}$ (see Problem 8.11).

EXAMPLE 8.5.1. Solve the LS problem with the following data matrix and desired response signal:

$$\mathbf{X} = \begin{bmatrix} 1 & 1 & 1 \\ 2 & 2 & 1 \\ 3 & 1 & 3 \\ 1 & 0 & 1 \end{bmatrix} \qquad \mathbf{y} = \begin{bmatrix} 1 \\ 2 \\ 4 \\ 3 \end{bmatrix}$$

Solution. We start by computing the time-average correlation matrix and cross-correlation vector

$$\hat{\mathbf{R}} = \begin{bmatrix} 15 & 8 & 13 \\ 8 & 6 & 6 \\ 13 & 6 & 12 \end{bmatrix} \qquad \hat{\mathbf{d}} = \begin{bmatrix} 20 \\ 9 \\ 18 \end{bmatrix}$$

followed by the LDL^H decomposition of $\hat{\mathbf{R}}$ using the MATLAB function [L,D]=ldlt(X). This gives

$$\mathbf{L} = \begin{bmatrix} 1 & 0 & 0 \\ 0.5333 & 1 & 0 \\ 0.8667 & -0.5385 & 1 \end{bmatrix} \qquad \mathbf{D} = \begin{bmatrix} 15 & 0 & 0 \\ 0 & 1.7333 & 0 \\ 0 & 0 & 0.2308 \end{bmatrix}$$

and working through the steps in Table 8.1, we find the LS solution and LSE to be

$$\mathbf{c}_{ls} = [3.0 \quad -1.5 \quad -1.0]^T \qquad E_{ls} = 1.5$$

using the following sequence of MATLAB commands

```
k=L\ dhat;
cls=L'\ k;
Els=sum((y-X'*cls).^2);
```

These results can be verified by using the command cls=Rhat\dhat.

8.5.2 LSE FIR Filtering and Prediction

As we stressed in Section 7.3, the fundamental difference between general linear estimation and FIR filtering and prediction, which is the key to the development of efficient order-recursive algorithms, is the shift invariance of the input data vector

$$\mathbf{x}_{m+1}(n) = [x(n)\, x(n-1) \, \cdots \, x(n-m+1)\, x(n-m)]^T \tag{8.5.13}$$

The input data vector can be partitioned as

$$\mathbf{x}_{m+1}(n) = \begin{bmatrix} \mathbf{x}_m(n) \\ x(n-m) \end{bmatrix} = \begin{bmatrix} x(n) \\ \mathbf{x}_m(n-1) \end{bmatrix} \tag{8.5.14}$$

which shows that samples from different times are incorporated as the order is increased. This creates a coupling between order and time updatings that has significant implications in the development of efficient algorithms. Indeed, we can easily see that the matrix

$$\hat{\mathbf{R}}_{m+1} = \sum_{n=N_i}^{N_f} \mathbf{x}_{m+1}(n)\mathbf{x}_{m+1}^H(n) \tag{8.5.15}$$

can be partitioned as

$$\hat{\mathbf{R}}_{m+1} = \begin{bmatrix} \hat{\mathbf{R}}_m & \hat{\mathbf{r}}_m^b \\ \hat{\mathbf{r}}_m^{bH} & E_m^b \end{bmatrix} = \begin{bmatrix} E_m^f & \hat{\mathbf{r}}_m^{fH} \\ \hat{\mathbf{r}}_m^f & \hat{\mathbf{R}}_m^f \end{bmatrix} \tag{8.5.16}$$

where

$$\hat{\mathbf{R}}_m^f = \hat{\mathbf{R}}_m + \mathbf{x}_m(N_i - 1)\mathbf{x}_m^H(N_i - 1) - \mathbf{x}_m(N_f)\mathbf{x}_m^H(N_f) \tag{8.5.17}$$

is the matrix equivalent of (8.2.28). We notice that the relationship between $\hat{\mathbf{R}}_m^f$ and $\hat{\mathbf{R}}_m$, which allows for the development of a complete set of order-recursive algorithms for FIR filtering and prediction, depends on the choice of N_i and N_f, that is, the windowing method selected.

As we discussed in Section 8.3, there are four cases of interest. In the *full-windowing* case ($N_i = 0$, $N_f = N + M - 2$), we have $\hat{\mathbf{R}}_m^f = \hat{\mathbf{R}}_m$ and $\hat{\mathbf{R}}_m$ is Toeplitz. Therefore, all the algorithms and structures developed in Chapter 7 for Toeplitz matrices can be utilized.

In the *prewindowing* case ($N_i = 0$, $N_f = N - 1$), Equation (8.5.17) becomes

$$\hat{\mathbf{R}}_m^f = \hat{\mathbf{R}}_m - \mathbf{x}_m(N - 1)\mathbf{x}_m^H(N - 1) \tag{8.5.18}$$

Since $\mathbf{x}_m(n) = \mathbf{0}$ for $n \leq 0$ (prewindowing), $\hat{\mathbf{R}}_m$ is a function of N. If we use the definition

$$\hat{\mathbf{R}}_m(N) \triangleq \sum_{n=0}^{N-1} \mathbf{x}_m(n)\mathbf{x}_m^H(n) \tag{8.5.19}$$

then the time-updating (8.5.18) can be written as

$$\hat{\mathbf{R}}_m^f = \hat{\mathbf{R}}_m(N - 1) = \hat{\mathbf{R}}_m(N) - \mathbf{x}_m(N - 1)\mathbf{x}_m^H(N - 1) \tag{8.5.20}$$

and the order-updating (8.5.16) as

$$\hat{\mathbf{R}}_{m+1}(N) = \begin{bmatrix} \hat{\mathbf{R}}_m(N) & \hat{\mathbf{r}}_m^b(N) \\ \hat{\mathbf{r}}_m^{bH}(N) & E_m^b(N) \end{bmatrix} = \begin{bmatrix} E_m^f(N) & \hat{\mathbf{r}}_m^{fH}(N) \\ \hat{\mathbf{r}}_m^f(N) & \hat{\mathbf{R}}_m(N - 1) \end{bmatrix} \tag{8.5.21}$$

which has the same form as (7.3.3). Therefore, all order recursions developed in Section 7.3 can be applied in the prewindowing case. However, to get a complete algorithm, we need recursions for the time updatings of the BLP $\mathbf{b}_m(N - 1) \rightarrow \mathbf{b}_m(N)$ and $E_m^b(N - 1) \rightarrow E_m^b(N)$, which can be developed by using the time-recursive algorithms developed in Chapter 10 for LS adaptive filters. The *postwindowing* case can be developed in a similar fashion, but it is of no particular practical interest.

In the *no-windowing* case ($N_i = M - 1$, $N_f = N - 1$), matrices $\hat{\mathbf{R}}_m$ and $\hat{\mathbf{R}}_m^f$ depend on both M and N. Thus, although the development of order recursions can be done as in the prewindowing case, the time updatings are more complicated due to (8.5.17) (Morf et al. 1977). Setting the lower limit to $N_i = M - 1$ means that all filters $\mathbf{c}_m$, $1 \leq m \leq M$, are optimized over the interval $M - 1 \leq n \leq N - 1$, which makes the optimum nesting property possible. If we set $N_i = m - 1$, each filter $\mathbf{c}_m$ is optimized over the interval $m - 1 \leq n \leq N - 1$; that is, it utilizes all the available data. However, in this case, the optimum nesting property $\hat{\mathbf{R}}_m = \hat{\mathbf{R}}_M^{\lceil m \rceil}$ does not hold, and the resulting order-recursive algorithms are slightly more complicated (Kalouptsidis et al. 1984).

The development of order-recursive algorithms for FBLP least-squares filters and predictors with linear phase constraints, for example, $\mathbf{c}_m = \pm \mathbf{J} \mathbf{c}_m^*$, is more complicated, in general. A review of existing algorithms and more references can be found in Theodoridis and Kalouptsidis (1993).

In conclusion, we notice that order-recursive algorithms are more efficient than the LDL^H decomposition–based solutions only if N is much larger than M. Furthermore, their numerical properties are inferior to those of the LDL^H decomposition methods; therefore, a bit of extra caution needs to be exercised when order-recursive algorithms are employed.

8.6 LS COMPUTATIONS USING ORTHOGONALIZATION TECHNIQUES

When we use the LDL^H or Cholesky decomposition for the computation of LSE filters, we first must compute the time-average correlation matrix $\hat{\mathbf{R}} = \mathbf{X}^H\mathbf{X}$ and the time-average cross-correlation vector $\hat{\mathbf{d}} = \mathbf{X}^H\mathbf{y}$ from the data $\mathbf{X}$ and $\mathbf{y}$. Although this approach is widely used in practice, there are certain applications that require methods with better numerical properties. When numerical considerations are a major concern, the orthogonalization techniques, discussed in this section, and the singular value decomposition, discussed in Section 8.7, are the methods of choice for the solution of LS problems.

Orthogonal transformations are linear changes of variables that preserve length. In matrix notation

$$\mathbf{y} = \mathbf{Q}^H\mathbf{x} \tag{8.6.1}$$

where $\mathbf{Q}$ is an orthogonal matrix, that is,

$$\mathbf{Q}^{-1} = \mathbf{Q}^H \quad \Rightarrow \quad \mathbf{Q}\mathbf{Q}^H = \mathbf{I} \tag{8.6.2}$$

From this property, we can easily see that

$$\|\mathbf{y}\|^2 = \mathbf{y}^H\mathbf{y} = \mathbf{x}^H\mathbf{Q}\mathbf{Q}^H\mathbf{x} = \mathbf{x}^H\mathbf{x} = \|\mathbf{x}\|^2 \tag{8.6.3}$$

that is, multiplying a vector by an orthogonal matrix does not change the length of the vector.[†] As a result, algorithms that use orthogonal transformations do not amplify roundoff errors, resulting in more accurate numerical algorithms. There are two ways to look at the solution of LS problems using orthogonalization techniques:

- Use orthogonal matrices to transform the data matrix $\mathbf{X}$ to a form that simplifies the solution of the normal equations without affecting the time-average correlation matrix $\hat{\mathbf{R}} = \mathbf{X}^H\mathbf{X}$. For any orthogonal matrix $\mathbf{Q}$, we have

$$\hat{\mathbf{R}} = \mathbf{X}^H\mathbf{X} = \mathbf{X}^H\mathbf{Q}\mathbf{Q}^H\mathbf{X} = (\mathbf{Q}^H\mathbf{X})^H\mathbf{Q}^H\mathbf{X} \tag{8.6.4}$$

Clearly, we can repeat this process as many times as we wish until the matrix $\mathbf{X}^H \mathbf{Q}_1 \mathbf{Q}_2 \cdots$ is in a form that simplifies the solution of the LS problem.

- Since orthogonal transformations preserve the length of a vector, multiplying the residual $\mathbf{e} = \mathbf{y} - \mathbf{X}\mathbf{c}$ by an orthogonal matrix does not change the total squared error. Hence, multiplying the residuals by $\mathbf{Q}^H$ gives

$$\min_{\mathbf{c}} \|\mathbf{e}\| = \min_{\mathbf{c}} \|\mathbf{y} - \mathbf{X}\mathbf{c}\| = \min_{\mathbf{c}} \|\mathbf{Q}^H(\mathbf{y} - \mathbf{X}\mathbf{c})\| \tag{8.6.5}$$

Thus, the goal is to find a matrix $\mathbf{Q}$ that simplifies the solution of the LS problem.

Suppose that we have already found an $N \times N$ orthogonal matrix $\mathbf{Q}$ such that

$$\mathbf{X} = \mathbf{Q}\begin{bmatrix}\mathcal{R}\\\mathbf{O}\end{bmatrix} \tag{8.6.6}$$

where, in practice, $\mathbf{Q}$ is constructed to make the $M \times M$ matrix $\mathcal{R}$ upper triangular.[‡] Using (8.6.5), we have

$$\|\mathbf{e}\| = \|\mathbf{Q}^H\mathbf{e}\| = \|\mathbf{Q}^H\mathbf{y} - \mathbf{Q}^H\mathbf{X}\mathbf{c}\| \tag{8.6.7}$$

Using the partitioning

$$\mathbf{Q} \triangleq [\mathbf{Q}_1\ \mathbf{Q}_2] \tag{8.6.8}$$

where $\mathbf{Q}_1$ has M columns, we obtain

$$\mathbf{X} = \mathbf{Q}_1\mathcal{R} \tag{8.6.9}$$

[†] Matrix $\mathbf{Q}$ is an arbitrary unitary matrix and should not be confused with the eigenvector matrix of $\mathbf{R}$.

[‡] The symbol $\mathcal{U}$ would be more appropriate for the upper triangular matrix $\mathcal{R}$ which can also be mistaken for the correlation matrix $\mathbf{R}$. However, we chose $\mathcal{R}$ because, otherwise, it would be difficult to use the well-established term *QR factorization*.

which is known as the *"thin"* QR decomposition. Similarly,

$$\mathbf{z} \triangleq \mathbf{Q}^H \mathbf{y} = \begin{bmatrix} \mathbf{Q}_1^H \mathbf{y} \\ \mathbf{Q}_2^H \mathbf{y} \end{bmatrix} \triangleq \begin{bmatrix} \mathbf{z}_1 \\ \mathbf{z}_2 \end{bmatrix} \tag{8.6.10}$$

where $\mathbf{z}_1$ has M components and $\mathbf{z}_2$ has $N - M$ components. Substitution of (8.6.9) and (8.6.10) into (8.6.7) gives

$$\|\mathbf{e}\| = \left\| \begin{bmatrix} \mathcal{R}\mathbf{c} \\ \mathbf{0} \end{bmatrix} - \begin{bmatrix} \mathbf{Q}_1^H \mathbf{y} \\ \mathbf{Q}_2^H \mathbf{y} \end{bmatrix} \right\| = \left\| \begin{bmatrix} \mathcal{R}\mathbf{c} - \mathbf{z}_1 \\ -\mathbf{z}_2 \end{bmatrix} \right\| \tag{8.6.11}$$

Since the term $\mathbf{z}_2 = \mathbf{Q}_2^H \mathbf{y}$ does not depend on the parameter vector $\mathbf{c}$, the length of $\|\mathbf{e}\|$ becomes minimum if we set $\mathbf{c} = \mathbf{c}_{ls}$, that is,

$$\mathcal{R}\mathbf{c}_{ls} = \mathbf{z}_1 \tag{8.6.12}$$

and

$$E_{ls} = \|\mathbf{Q}_2^H \mathbf{y}\|^2 = \|\mathbf{z}_2\| \tag{8.6.13}$$

where the upper triangular system in (8.6.12) can be solved for $\mathbf{c}_{ls}$ by back substitution.

The steps for the solution of the LS problem using the QR decomposition are summarized in Table 8.3.

TABLE 8.3
Solution of the LS problem using the QR decomposition method.

Step	Computations	Description
1	$\mathbf{X} = \mathbf{Q} \begin{bmatrix} \mathcal{R} \\ \mathbf{0} \end{bmatrix}$	QR decomposition
2	$\mathbf{z} = \mathbf{Q}^H \mathbf{y} = \begin{bmatrix} \mathbf{z}_1 \\ \mathbf{z}_2 \end{bmatrix}$	Transformation and partitioning of $\mathbf{y}$
3	$\mathcal{R}\mathbf{c}_{ls} = \mathbf{z}_1$	Backward substitution $\rightarrow \mathbf{c}_{ls}$
4	$E_{ls} = \|\mathbf{z}_2\|^2$	Computation of LS error
5	$\mathbf{e}_{ls} = \mathbf{Q} \begin{bmatrix} \mathbf{0} \\ \mathbf{z}_2 \end{bmatrix}$	Back transformation of residuals

Using the QR decomposition (8.6.6), we have

$$\hat{\mathbf{R}} = \mathbf{X}^H \mathbf{X} = \mathcal{R}^H \mathcal{R} \tag{8.6.14}$$

which, in conjunction with the unique Cholesky decomposition $\hat{\mathbf{R}} = \mathcal{L}\mathcal{L}^H$, gives

$$\mathcal{R} = \mathcal{L}^H \tag{8.6.15}$$

that is, the QR factorization computes the Cholesky factor $\mathcal{R}$ directly from data matrix $\mathbf{X}$. Also, since $\mathcal{L}^H \mathbf{c}_{ls} = \tilde{\mathbf{k}}$, we have

$$\tilde{\mathbf{k}} = \mathbf{z}_1 \tag{8.6.16}$$

which, owing to the Cholesky decomposition, leads to

$$E_{ls} = E_y - \tilde{\mathbf{k}}^H \tilde{\mathbf{k}} = \|\mathbf{z}_2\|^2 \tag{8.6.17}$$

because $\|\mathbf{y}\|^2 = \|\mathbf{Q}^H \mathbf{y}\|^2 = \|\mathbf{z}_1\|^2 + \|\mathbf{z}_2\|^2$.

If we form the augmented matrix

$$\bar{\mathbf{X}} = [\mathbf{X} \; \mathbf{y}] \tag{8.6.18}$$

the QR decomposition of $\bar{\mathbf{X}}$ provides the triangular factor

$$\begin{bmatrix} \mathcal{R} & \tilde{\mathbf{k}} \\ \mathbf{0}^H & \tilde{\xi} \end{bmatrix} \tag{8.6.19}$$

which is identical to the one obtained from the Cholesky decomposition of $\bar{\mathbf{R}} = \bar{X}^H \bar{X}$ with $\mathcal{R} = \mathcal{L}^H$ and $\tilde{\xi}^2 = E_{ls}$ (see Problem 8.14).

EXAMPLE 8.6.1. Solve the LS problem in Example 8.5.1

$$\mathbf{X} = \begin{bmatrix} 1 & 1 & 1 \\ 2 & 2 & 1 \\ 3 & 1 & 3 \\ 1 & 0 & 1 \end{bmatrix} \qquad \mathbf{y} = \begin{bmatrix} 1 \\ 2 \\ 4 \\ 3 \end{bmatrix}$$

using the QR decomposition approach.

Solution. Using the MATLAB function [Q,R]=qr(X), we obtain

$$\mathbf{Q} = \begin{bmatrix} -0.2582 & -0.3545 & 0.8006 & 0.4082 \\ -0.5164 & -0.7089 & -0.4804 & 0.0000 \\ -0.7746 & 0.4557 & 0.1601 & -0.4082 \\ -0.2582 & 0.4051 & -0.3203 & 0.8165 \end{bmatrix}$$

$$\mathcal{R} = \begin{bmatrix} -3.8730 & -2.0656 & -3.5666 \\ 0 & -1.3166 & 0.7089 \\ 0 & 0 & 0.4804 \\ 0 & 0 & 0 \end{bmatrix}$$

and following the steps in Table 8.3, we find the LS solution and the LSE to be

$$\mathbf{c}_{ls} = [3.0 \ -1.5 \ -1.0]^T \qquad E_{ls} = 1.5$$

using the sequence of MATLAB commands

```
z=Q'*y;
cls=R(1:3,1:3)'\z(1:3);
Els=sum(z(4).2);
```

In applications that require only the error (or residual) vector $\mathbf{e}_{ls}$, we do not need to solve the triangular system $\mathcal{R}\mathbf{c}_{ls} = \mathbf{z}_1$. Instead, we can compute directly the error by $\mathbf{e}_{ls} = \mathbf{Q}[\begin{smallmatrix} 0 \\ \mathbf{z}_2 \end{smallmatrix}]$ or the MATLAB command e=Q*[zeros(1,M) z2']'. This approach is known as *direct error (or residual) extraction* and plays an important role in LS adaptive filtering algorithms and architectures (see Chapter 10).

It is generally agreed in numerical analysis that orthogonal decomposition methods applied directly to data matrix $\mathbf{X}$ are *preferable* to the computation and solution of the normal equations whenever numerical stability is important (Hager 1988; Golub and Van Loan 1996). The sensitivity of the solution $\mathbf{c}_{ls}$ to perturbations in the data $\mathbf{X}$ and $\mathbf{y}$ depends on the ratio of the largest to the smallest eigenvalues of $\hat{\mathbf{R}}$ and does not depend on the algorithm used to compute the solution. Furthermore, the numerical accuracy required to compute $\mathcal{L}$ directly from $\mathbf{X}$ is one-half of that required to compute $\mathcal{L}$ from $\hat{\mathbf{R}}$. The "squaring" $\hat{\mathbf{R}} = \mathbf{X}^H \mathbf{X}$ of the data to form the time-average correlation matrix results in a loss of information and should be avoided if the numerical precision is not deemed sufficient. Algorithms that compute $\mathcal{L}$ directly from $\mathbf{X}$ are known as *square root methods*. However, by paraphrasing Rader (1996), we use the terms *amplitude-domain techniques* for methods that compute $\mathcal{L}$ directly from $\mathbf{X}$ and *power-domain techniques* for methods that compute $\mathcal{L}$ indirectly from $\hat{\mathbf{R}} = \mathbf{X}^H \mathbf{X}$. These ideas are illustrated in the following example.

EXAMPLE 8.6.2. Let

$$\mathbf{X} = \begin{bmatrix} 1 & 1 \\ \epsilon & 0 \\ 0 & \epsilon \end{bmatrix} \qquad \hat{\mathbf{R}} = \mathbf{X}^T \mathbf{X} = \begin{bmatrix} 1+\epsilon^2 & 1 \\ 1 & 1+\epsilon^2 \end{bmatrix}$$

where $\mathbf{X}^T \mathbf{X}$ is clearly positive definite and nonsingular. Let the desired signal be $\mathbf{y} = [2 \ \epsilon \ \epsilon]^T$ so that $\hat{\mathbf{d}} = [2+\epsilon^2 \ 2+\epsilon^2]^T$. If ϵ is such that $1 + \epsilon^2 = 1$, due to limited numerical precision, the matrix $\mathbf{X}^T\mathbf{X}$ becomes singular. If we set $\epsilon = 10^{-8}$, solving the LS equations for $\mathbf{c}_{ls}$ using the MATLAB command $\texttt{cls=Rhat\backslash dhat}$ is not possible since $\hat{\mathbf{R}}$ is singular to the working precision of MATLAB. However, if the problem is solved using the QR decomposition as shown in Example 8.6.1, we find $\mathbf{c}_{ls} = [1\ 1]^T$. Note that even for slightly larger values of ϵ the MATLAB command $\texttt{cls=Rhat\backslash dhat}$ is able to find a solution that differs from the true LS solution since $\hat{\mathbf{R}}$ is ill conditioned.

There are two classes of orthogonal decomposition algorithms:

1. Methods that compute the orthogonal matrix $\mathbf{Q}$: Householder reflections and Givens rotations
2. Methods that compute $\mathbf{Q}_1$: classical and modified Gram-Schmidt orthogonalizations

These decompositions are illustrated in Figure 8.10. The cost of the QR decomposition using the Givens rotations is twice the cost of using Householder reflections or the Gram-Schmidt orthogonalization. The standard method for the computation of the QR decomposition and the solution of LS problems employs the Householder transformation. The Givens rotations are preferred for the implementation of adaptive LS filters (see Chapter 10).

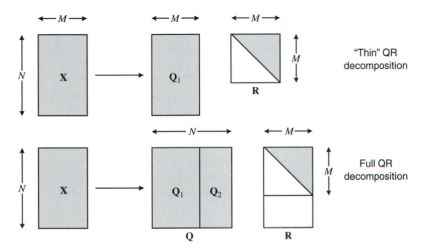

FIGURE 8.10
Pictorial illustration of the differences between thin and full QR decompositions.

8.6.1 Householder Reflections

Consider a vector $\mathbf{x}$ and a fixed line l in the plane (see Figure 8.11). If we reflect $\mathbf{x}$ about the line l, we obtain a vector $\mathbf{y}$ that is the mirror image of $\mathbf{x}$. Clearly, the vector $\mathbf{x}$ and its *reflection* $\mathbf{y}$ have the same length. We define a unit vector $\mathbf{w}$ in the direction of $\mathbf{x} - \mathbf{y}$ as

$$\mathbf{w} \triangleq \frac{1}{\|\mathbf{x} - \mathbf{y}\|}(\mathbf{x} - \mathbf{y}) \tag{8.6.20}$$

assuming that $\mathbf{x}$ and $\mathbf{y}$ are nonzero vectors.

Since the projection of $\mathbf{x}$ on $\mathbf{w}$ is $(\mathbf{w}^H\mathbf{x})\mathbf{w}$, simple inspection of Figure 8.11 gives

$$\mathbf{y} = \mathbf{x} - 2(\mathbf{w}^H\mathbf{x})\mathbf{w} = \mathbf{x} - 2(\mathbf{w}\mathbf{w}^H)\mathbf{x} = (\mathbf{I} - 2\mathbf{w}\mathbf{w}^H)\mathbf{x} \triangleq \mathbf{H}\mathbf{x}$$

where
$$\mathbf{H} \triangleq \mathbf{I} - 2\mathbf{w}\mathbf{w}^H \tag{8.6.21}$$

In general, any matrix $\mathbf{H}$ of the form (8.6.21) with $\|\mathbf{w}\| = 1$ is known as a *Householder reflection* or *Householder transformation* (Householder 1958) and has the following properties

$$\mathbf{H}^H = \mathbf{H} \quad \mathbf{H}^H\mathbf{H} = \mathbf{I} \quad \mathbf{H}^{-1} = \mathbf{H}^H \tag{8.6.22}$$

that is, the matrix $\mathbf{H}$ is unitary.

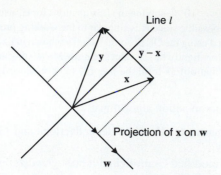

FIGURE 8.11
The Householder reflection vector.

We can build a Householder matrix $\mathbf{H}_k$ that leaves intact the first $k - 1$ components of a given vector $\mathbf{x}$, changes the kth component, and annihilates (zeros out) the remaining components, that is,

$$y_i = (\mathbf{Hx})_i = \begin{cases} x_i & i = 1, 2, \ldots, k - 1 \\ y_k & i = k \\ 0 & i = k + 1, \ldots, N \end{cases} \tag{8.6.23}$$

where y_k is to be determined. If we set

$$y_k = \pm \left(\sum_{i=k}^{N} |x_i|^2 \right)^{1/2} e^{j\theta_k} \tag{8.6.24}$$

where θ_k is the angle part of x_k (if complex-valued), then both $\mathbf{x}$ and $\mathbf{y}$ have the same length. There are two choices for the sign of y_k. Since the computation of $\mathbf{w}$ by (8.6.20) involves subtraction (which can lead to severe numerical problems when two numbers are nearly equal), we choose the negative sign so that y_k and x_k have opposite signs. Hence, $y_k - x_k$ is never the difference between nearly equal numbers. Therefore, using (8.6.20), we find that $\mathbf{w}$ is given by

$$\mathbf{w} = \frac{1}{\sqrt{2 s_k (s_k + |x_k|)}} \begin{bmatrix} 0 \\ \vdots \\ 0 \\ (|x_k| + s_k)e^{j\theta_k} \\ x_{k+1} \\ \vdots \\ x_N \end{bmatrix} \tag{8.6.25}$$

where

$$s_k \triangleq \left(\sum_{i=k}^{N} |x_i|^2 \right)^{1/2} \tag{8.6.26}$$

In general, an $N \times M$ matrix $\mathbf{X}$ with $N > M$ can be diagonalized with a sequence of M Householder transformations

$$\mathbf{H}_M \cdots \mathbf{H}_2 \mathbf{H}_1 \mathbf{X} = \mathcal{R} \tag{8.6.27}$$

or

$$\mathbf{X} = \mathbf{Q}\mathcal{R} \tag{8.6.28}$$

where

$$\mathbf{Q} \triangleq \mathbf{H}_1 \mathbf{H}_2 \cdots \mathbf{H}_M \tag{8.6.29}$$

Note that for $M = N$ we need only $M - 1$ reflections.

We next illustrate by an example how to compute the QR decomposition of a rectangular matrix by using a sequence of Householder transformations.

EXAMPLE 8.6.3. Find the QR decomposition of the data matrix

$$\mathbf{X} = \begin{bmatrix} 1 & 2 \\ 2 & 3 \\ 6 & 7 \end{bmatrix}$$

using Householder reflections.

Solution. Using (8.6.25), we compute the vector $\mathbf{w}_1 = [0.7603\ 0.2054\ 0.6162]^T$ and the Householder reflection matrix $\mathbf{H}_1$ for the first column of $\mathbf{X}$. The modified data matrix is

$$\mathbf{H}_1\mathbf{X} = \begin{bmatrix} -6.4031 & -7.8087 \\ 0 & 0.3501 \\ 0 & -0.9496 \end{bmatrix}$$

Similarly, we compute the vector $\mathbf{w}_2 = [0\ 0.8203\ -0.5719]^T$ and matrix $\mathbf{H}_2$ for the second column of $\mathbf{H}_1\mathbf{X}$, which results in the desired QR decomposition

$$\mathbf{H}_2\mathbf{H}_1\mathbf{X} = \mathcal{R} = \begin{bmatrix} -6.4031 & -7.8087 \\ 0 & -1.0121 \\ 0 & 0 \end{bmatrix}$$

$$\mathbf{Q} = \mathbf{H}_1\mathbf{H}_2 = \begin{bmatrix} -0.1562 & -0.7711 & -0.6172 \\ -0.3123 & -0.5543 & -0.7715 \\ -0.9370 & 0.3133 & -0.1543 \end{bmatrix}$$

This result can be verified by using the MATLAB function $[\mathtt{Q},\mathtt{R}]\mathtt{=qr(X)}$, which implements the Householder transformation.

8.6.2 The Givens Rotations

The second elementary transformation that does not change the length of a vector is a rotation about an axis (see Figure 8.12). To describe the method of Givens, we assume for simplicity that the vectors are real-valued. The components of the rotated vector $\mathbf{y}$ in terms of the components of the original vector $\mathbf{x}$ are

$$y_1 = r\ \cos(\phi + \theta) = x_1\ \cos\theta - x_2\ \sin\theta$$
$$y_2 = r\ \sin(\phi + \theta) = x_1\ \sin\theta + x_2\ \cos\theta$$

or in matrix form

$$\begin{bmatrix} y_1 \\ y_2 \end{bmatrix} = \begin{bmatrix} \cos\theta & -\sin\theta \\ \sin\theta & \cos\theta \end{bmatrix}\begin{bmatrix} x_1 \\ x_2 \end{bmatrix} \triangleq \mathbf{G}(\theta)\begin{bmatrix} x_1 \\ x_2 \end{bmatrix} \qquad (8.6.30)$$

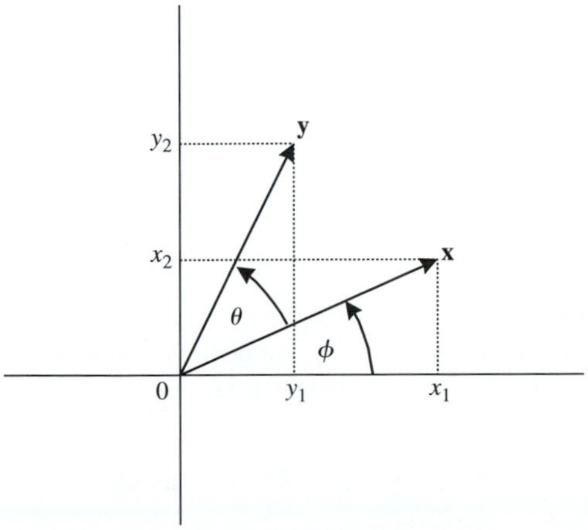

FIGURE 8.12
The Givens rotation.

where θ is the angle of rotation. We can easily show that the rotation matrix $\mathbf{G}(\theta)$ in (8.6.30) is orthogonal and has a determinant $\det \mathbf{G}(\theta) = 1$.

Any matrix of the form

$$\mathbf{G}_{ij}(\theta) \triangleq \begin{bmatrix} 1 & \cdots & 0 & \cdots & 0 & \cdots & 0 \\ \vdots & \ddots & \vdots & & \vdots & & \vdots \\ 0 & \cdots & c & \cdots & -s & \cdots & 0 \\ \vdots & & \vdots & \ddots & \vdots & & \vdots \\ 0 & \cdots & s & \cdots & c & \cdots & 0 \\ \vdots & & \vdots & & \vdots & \ddots & \vdots \\ 0 & \cdots & 0 & \cdots & 0 & \cdots & 1 \end{bmatrix} \begin{matrix} \\ \\ \leftarrow i \\ \\ \leftarrow j \\ \\ \\ \end{matrix} \tag{8.6.31}$$

$$\underset{i}{\uparrow} \qquad \underset{j}{\uparrow}$$

with

$$c^2 + s^2 = 1 \tag{8.6.32}$$

is known as a *Givens rotation*. When this matrix is applied to a vector $\mathbf{x}$, it rotates the components x_i and x_j through an angle $\theta = \arctan(s/c)$ while leaving all other components intact (Givens 1958). Because of (8.6.30), we can write $c = \cos\theta$ and $s = \sin\theta$ for some angle θ. It can easily be shown that the matrix $\mathbf{G}_{ij}(\theta)$ is orthogonal.

The Givens rotations have two attractive features. First, performing the rotation $\mathbf{y} = \mathbf{G}_{ij}(\theta)\mathbf{x}$ as

$$\begin{aligned} y_i &= c x_i - s x_j \\ y_j &= s x_i + c x_j \\ y_k &= x_k \qquad k \neq i, j \end{aligned} \tag{8.6.33}$$

requires only four multiplications and two additions. Second, we can choose c and s to annihilate the jth component of a vector. Indeed, if we set

$$c = \frac{x_i}{\sqrt{x_i^2 + x_j^2}} \qquad s = -\frac{x_j}{\sqrt{x_i^2 + x_j^2}} \tag{8.6.34}$$

in (8.6.31), then

$$y_i = \sqrt{x_i^2 + x_j^2} \qquad \text{and} \qquad y_j = 0 \tag{8.6.35}$$

Using a sequence of Givens rotations, we can annihilate (zero out) all elements of a matrix $\mathbf{X}$ below the main diagonal to obtain the upper triangular matrix of the QR decomposition. The product of all the Givens rotation matrices provides matrix $\mathbf{Q}$. We stress that the order of rotations cannot be arbitrary because later rotations can destroy zeros introduced earlier. A version of the Givens algorithm without square roots, which is known as the *fast Givens QR*, is discussed in Golub and Van Loan (1996).

We illustrate this procedure with the next example.

EXAMPLE 8.6.4. The QR decomposition can be found in order to find the LS solution using the Givens rotations. Given the same data matrix $\mathbf{X}$ as in Example 8.6.3

$$\mathbf{X} = \begin{bmatrix} 1 & 2 \\ 2 & 3 \\ 6 & 7 \end{bmatrix}$$

we first zero the last element of the first column, that is, element (3, 1), using the Givens rotation matrix $\mathbf{G}_{31}$ with $c = -0.1664$ and $s = 0.9864$. Indeed, using (8.6.34), we have

$$\mathbf{G}_{31}\mathbf{X} = \begin{bmatrix} -6.0828 & -7.2336 \\ 2 & 3 \\ 0 & 0.8220 \end{bmatrix}$$

Then the element (2, 1) is eliminated by using the Givens rotation matrix $\mathbf{G}_{21}$ with $c = 0.9550$ and $s = 0.3123$, resulting in

$$\mathbf{G}_{21}\mathbf{G}_{31}\mathbf{X} = \begin{bmatrix} -6.4031 & -7.8087 \\ 0 & 0.5905 \\ 0 & 0.8220 \end{bmatrix}$$

Finally, the QR factorization is found after applying the Givens rotation matrix $\mathbf{G}_{32}$ with $c = -0.5834$ and $s = 0.8122$:

$$\mathcal{R} = \mathbf{G}_{32}\mathbf{G}_{21}\mathbf{G}_{31}\mathbf{X} = \begin{bmatrix} -6.4031 & -7.8087 \\ 0 & -1.0121 \\ 0 & 0 \end{bmatrix}$$

$$\mathbf{Q} = \mathbf{G}_{31}^T \mathbf{G}_{21}^T \mathbf{G}_{32}^T = \begin{bmatrix} -0.1562 & -0.7711 & -0.6172 \\ -0.3123 & -0.5543 & -0.7715 \\ -0.9370 & 0.3133 & -0.1543 \end{bmatrix}$$

which, as expected, agrees with the QR decomposition found in Example 8.6.3.

In the case of complex-valued vectors, the components of rotated vector $\mathbf{y}$ in (8.6.30) are given by

$$\begin{bmatrix} y_1 \\ y_2 \end{bmatrix} = \begin{bmatrix} \cos\theta & -e^{-j\psi}\sin\theta \\ e^{j\psi}\sin\theta & \cos\theta \end{bmatrix} \begin{bmatrix} x_1 \\ x_2 \end{bmatrix} \tag{8.6.36}$$

where $c \triangleq \cos\theta$ and $s \triangleq e^{j\psi}\sin\theta$. The element $-s$ of the rotation matrix $\mathbf{G}_{ij}(\theta)$ is replaced by $-s^*$, where $c^2 + |s|^2 = 1$ instead of (8.6.32).

8.6.3 Gram-Schmidt Orthogonalization

If we are given a set of M linearly independent vectors $\mathbf{x}_1, \mathbf{x}_2, \ldots, \mathbf{x}_M$, we can create an orthonormal basis $\mathbf{q}_1, \mathbf{q}_2, \ldots, \mathbf{q}_M$ that spans the same space by using a systematic procedure known as the *classical Gram-Schmidt (GS) othogonalization method* (see also Section 7.2.4). The GS method starts by choosing

$$\mathbf{q}_1 = \frac{\mathbf{x}_1}{\|\mathbf{x}_1\|} \tag{8.6.37}$$

as the first basis vector. To obtain $\mathbf{q}_2$, we express $\mathbf{x}_2$ as the sum of two components: its projection $(\mathbf{q}_1^H\mathbf{x}_2)\mathbf{q}_1$ onto $\mathbf{q}_1$ and a vector $\mathbf{p}_2$ that is perpendicular to $\mathbf{q}_1$. Hence,

$$\mathbf{p}_2 = \mathbf{x}_2 - (\mathbf{q}_1^H\mathbf{x}_2)\mathbf{q}_1 \tag{8.6.38}$$

and $\mathbf{q}_2$ is obtained by normalizing $\mathbf{p}_2$, that is,

$$\mathbf{q}_2 = \frac{\mathbf{p}_2}{\|\mathbf{p}_2\|} \tag{8.6.39}$$

The vectors $\mathbf{q}_1$ and $\mathbf{q}_2$ have unit length, are orthonormal, and span the same space as $\mathbf{x}_1$ and $\mathbf{x}_2$. In general, the orthogonal basis vector $\mathbf{q}_j$ is obtained by removing from $\mathbf{x}_j$ its projections onto the already computed vectors $\mathbf{q}_1$ to $\mathbf{q}_{j-1}$. Therefore, we have

$$\mathbf{p}_j = \mathbf{x}_j - \sum_{i=1}^{j-1}(\mathbf{q}_i^H\mathbf{x}_j)\mathbf{q}_i \quad \text{and} \quad \mathbf{q}_j = \frac{\mathbf{p}_j}{\|\mathbf{p}_j\|} \tag{8.6.40}$$

for all $1 \leq j \leq M$.

The GS algorithm can be used to obtain the "thin" $\mathbf{Q}_1\mathcal{R}$ factorization. Indeed, if we define

$$r_{ij} \triangleq \mathbf{q}_i^H\mathbf{x}_j \quad r_{jj} \triangleq \|\mathbf{p}_j\| \tag{8.6.41}$$

we have
$$\mathbf{p}_j = r_{jj}\mathbf{q}_j = \mathbf{x}_j - \sum_{i=1}^{j-1} r_{ij}\mathbf{q}_i \tag{8.6.42}$$

or by solving for $\mathbf{x}_j$

$$\mathbf{x}_j = \sum_{i=1}^{j} \mathbf{q}_i r_{ij} \quad j = 1, 2, \ldots, M \tag{8.6.43}$$

Using matrix notation, we can express this relation as $\mathbf{X} = \mathbf{Q}_1 \mathcal{R}$, which is exactly the thin $\mathbf{Q}_1 \mathcal{R}$ factorization in (8.6.9).

Major drawbacks of the GS procedure are that it does not produce accurate results and that the resulting basis may not be orthogonal when implemented using finite-precision arithmetic. However, we can achieve better numerical behavior if we reorganize the computations in a form known as the *modified Gram-Schmidt (MGS) algorithm* (Björck 1967). We start the first step by defining $\mathbf{q}_1$ as before

$$\mathbf{q}_1 = \frac{\mathbf{x}_1}{\|\mathbf{x}_1\|} \tag{8.6.44}$$

However, all the remaining vectors $\mathbf{x}_2, \ldots, \mathbf{x}_M$ are modified to be orthogonal to $\mathbf{q}_1$ by subtracting from each vector its projection onto $\mathbf{q}_1$, that is,

$$\mathbf{x}_i^{(1)} = \mathbf{x}_i - (\mathbf{q}_1^H \mathbf{x}_i)\mathbf{q}_1 \quad i = 2, \ldots, M \tag{8.6.45}$$

At the second step, we define the vector

$$\mathbf{q}_2 = \frac{\mathbf{x}_2^{(1)}}{\|\mathbf{x}_2^{(1)}\|} \tag{8.6.46}$$

which is already orthogonal to $\mathbf{q}_1$. Then we modify the remaining vectors to make them orthogonal to $\mathbf{q}_2$

$$\mathbf{x}_i^{(2)} = \mathbf{x}_i^{(1)} - (\mathbf{q}_2^H \mathbf{x}_i^{(1)})\mathbf{q}_2 \quad i = 3, \ldots, M \tag{8.6.47}$$

Continuing in a similar manner, we compute $\mathbf{q}_m$ and the updated vectors $\mathbf{x}_i^{(m)}$ by

$$\mathbf{q}_m = \frac{\mathbf{x}_m^{(m-1)}}{\|\mathbf{x}_m^{(m-1)}\|} \tag{8.6.48}$$

and
$$\mathbf{x}_i^{(m)} = \mathbf{x}_i^{(m-1)} - (\mathbf{q}_m^H \mathbf{x}_i^{(m-1)})\mathbf{q}_m \quad i = m+1, \ldots, M \tag{8.6.49}$$

The MGS algorithm involves the following steps, outlined in Table 8.4 and is implemented by the function `Q=mgs(X)`. The superior numerical properties of the modified algorithm

TABLE 8.4

Orthogonalization of a set of vectors using the modified Gram-Schmidt algorithm.

Modified GS Algorithm
For $m = 1$ to M
$\quad r_{mm} = \|\mathbf{x}_m\|^2$
$\quad \mathbf{q}_m = \mathbf{x}_m / r_{mm}$
$\quad$ For $i = m + 1$ to M
$\quad\quad r_{mi} = \mathbf{q}_m^H \mathbf{x}_i$
$\quad\quad \mathbf{x}_i \leftarrow \mathbf{x}_i - r_{mi}\mathbf{q}_m$
$\quad$ next i
next m

stem from the fact that successive $\mathbf{x}_i^{(m)}$ generated by (8.6.49) decrease in size and that the dot product $\mathbf{q}_m^H \mathbf{x}_i^{(m-1)}$ can be computed more accurately than the dot product $\mathbf{q}_m^H \mathbf{x}_i$.

EXAMPLE 8.6.5. Consider an LS problem (Dahlquist and Björck 1974) with

$$\mathbf{X} = \begin{bmatrix} 1 & 1 & 1 \\ \epsilon & 0 & 0 \\ 0 & \epsilon & 0 \\ 0 & 0 & \epsilon \end{bmatrix} \qquad \mathbf{y} = \begin{bmatrix} 1 \\ 0 \\ 0 \\ 0 \end{bmatrix}$$

where $\epsilon^2 \ll 1$, that is, ϵ^2 can be neglected compared to 1. We first compute $\mathbf{X}^T \mathbf{X}$ and $\mathbf{X}^T \mathbf{y}$ to determine the normal equations

$$\begin{bmatrix} 1+\epsilon^2 & 1 & 1 \\ 1 & 1+\epsilon^2 & 1 \\ 1 & 1 & 1+\epsilon^2 \end{bmatrix} \mathbf{c}_{\text{ls}} = \begin{bmatrix} 1 \\ 1 \\ 1 \end{bmatrix}$$

which provide the exact solution $\mathbf{c}_{\text{ls}} = [1\ 1\ 1]^T/(3+\epsilon^2)$. Numerically, the matrix $\mathbf{X}^T \mathbf{X}$ is singular on any computer with accuracy such that $1+\epsilon^2$ is rounded to 1. Applying the MGS algorithm to the column vectors of the augmented matrix $[\mathbf{X}\ \ \mathbf{y}]$, and taking into consideration that $1+\epsilon^2$ is rounded to 1, we obtain

$$\mathbf{Q} = \begin{bmatrix} 1 & 0 & 0 \\ \epsilon & -\epsilon & -\dfrac{\epsilon}{2} \\ 0 & \epsilon & -\dfrac{\epsilon}{2} \\ 0 & 0 & \epsilon \end{bmatrix} \qquad \mathbf{e} = -\dfrac{\epsilon}{3}\begin{bmatrix} 0 \\ 1 \\ 1 \\ 1 \end{bmatrix}$$

$$\mathcal{R} = \begin{bmatrix} 1 & 1 & 1 \\ 0 & 1 & \frac{1}{2} \\ 0 & 0 & 1 \end{bmatrix} \qquad \mathbf{z} = \begin{bmatrix} 1 \\ \frac{1}{2} \\ \frac{1}{3} \end{bmatrix}$$

which corresponds to the thin QR decomposition. Solving $\mathcal{R}\mathbf{c}_{\text{ls}} = \mathbf{z}$, we obtain $\mathbf{c}_{\text{ls}} = [1\ 1\ 1]^T/3$, which agrees with the exact solution under the assumption that $1+\epsilon^2$ is rounded to 1.

8.7 LS COMPUTATIONS USING THE SINGULAR VALUE DECOMPOSITION

The *singular value decomposition* (*SVD*) plays a prominent role in the theoretical analysis and practical solution of LS problems because (1) it provides a unified framework for the solution of overdetermined and underdetermined LS problems with full rank or that are rank-deficient and (2) it is the best numerical method to solve LS problems in practice. In this section, we discuss the existence and fundamental properties of the SVD, show how to use it for solving the LS problem, and apply it to determine the numerical rank of a matrix. More details are given in Golub and Van Loan (1996), Leon (1990), Stewart (1973), Watkins (1991), and Klema and Laub (1980).

8.7.1 Singular Value Decomposition

The eigenvalue decomposition reduces a *Hermitian* matrix to a diagonal matrix by premultiplying and postmultiplying it by a single unitary matrix. The singular value decomposition, introduced in the next theorem, reduces a *general* matrix to a diagonal one by premultiplying and postmultiplying it by two different orthogonal matrices.

THEOREM 8.2. Any real $N \times M$ matrix $\mathbf{X}$ with rank r (recall that r is defined as the number of linearly independent columns of a matrix) can be written as

$$\mathbf{X} = \mathbf{U}\boldsymbol{\Sigma}\mathbf{V}^H \tag{8.7.1}$$

where $\mathbf{U}$ is an $N \times N$ unitary matrix, $\mathbf{V}$ is an $M \times M$ unitary matrix, and $\boldsymbol{\Sigma}$ is an $N \times M$ matrix with $\langle \boldsymbol{\Sigma} \rangle_{ij} = 0$, $i \neq j$, and $\langle \boldsymbol{\Sigma} \rangle_{ii} = \sigma_i > 0$, $i = 1, 2, \ldots, r$. The numbers σ_i are known as the *singular values* of $\mathbf{X}$ and are usually arranged in decreasing order as $\sigma_1 \geq \sigma_2 \geq \cdots \geq \sigma_r > 0$.

Proof. We follow the derivation given in Stewart (1973). Since the matrix $\mathbf{X}^H\mathbf{X}$ is positive semidefinite, it has nonnegative eigenvalues $\sigma_1^2, \sigma_2^2, \ldots, \sigma_M^2$ such that $\sigma_1 \geq \sigma_2 \geq \cdots \geq \sigma_r > 0 = \sigma_{r+1} = \cdots = \sigma_M$ for $0 \leq r \leq M$. Let $\mathbf{v}_1, \mathbf{v}_2, \ldots, \mathbf{v}_M$ be the eigenvectors corresponding to the eigenvalues $\sigma_1^2, \sigma_2^2, \ldots, \sigma_M^2$. Consider the partitioning $\mathbf{V} = [\mathbf{V}_1 \ \mathbf{V}_2]$, where $\mathbf{V}_1$ consists of the first r columns of $\mathbf{V}$. If $\boldsymbol{\Sigma}_r = \text{diag}\{\sigma_1, \sigma_2, \ldots, \sigma_r\}$, then we obtain $\mathbf{V}_1^H\mathbf{X}^H\mathbf{X}\mathbf{V}_1 = \boldsymbol{\Sigma}_r^2$ and

$$\boldsymbol{\Sigma}_r^{-1}\mathbf{V}_1^H\mathbf{X}^H\mathbf{X}\mathbf{V}_1\boldsymbol{\Sigma}_r^{-1} = \mathbf{I} \tag{8.7.2}$$

Since $\mathbf{V}_2^H\mathbf{X}^H\mathbf{X}\mathbf{V}_2 = \mathbf{0}$, we have

$$\mathbf{X}\mathbf{V}_2 = \mathbf{0} \tag{8.7.3}$$

If we define

$$\mathbf{U}_1 \triangleq \mathbf{X}\mathbf{V}_1\boldsymbol{\Sigma}_r^{-1} \tag{8.7.4}$$

then (8.7.2) gives $\mathbf{U}_1^H\mathbf{U}_1 = \mathbf{I}$; that is, the columns of $\mathbf{U}_1$ are unitary. A unitary matrix $\mathbf{U} \triangleq [\mathbf{U}_1 \ \mathbf{U}_2]$ is found by properly choosing the components of $\mathbf{U}_2$, that is, $\mathbf{U}_2^H\mathbf{U}_1 = \mathbf{0}$ and $\mathbf{U}_2^H\mathbf{U}_2 = \mathbf{I}$. Then

$$\mathbf{U}^H\mathbf{X}\mathbf{V} = \begin{bmatrix} \mathbf{U}_1^H \\ \mathbf{U}_2^H \end{bmatrix} \mathbf{X} [\mathbf{V}_1 \ \mathbf{V}_2] = \begin{bmatrix} \mathbf{U}_1^H\mathbf{X}\mathbf{V}_1 & \mathbf{U}_1^H(\mathbf{X}\mathbf{V}_2) \\ \mathbf{U}_2^H\mathbf{X}\mathbf{V}_1 & \mathbf{U}_2^H(\mathbf{X}\mathbf{V}_2) \end{bmatrix} = \begin{bmatrix} \boldsymbol{\Sigma}_r & \mathbf{0} \\ \mathbf{0} & \mathbf{0} \end{bmatrix} \tag{8.7.5}$$

because of (8.7.2), (8.7.3), and $\mathbf{U}_2^H\mathbf{X}\mathbf{V}_1 = (\mathbf{U}_2^H\mathbf{U}_1)\boldsymbol{\Sigma}_r = \mathbf{0}$.

The SVD of a matrix, which is illustrated in Figure 8.13, provides a wealth of information about the structure of the matrix. Figure 8.14 provides a geometric interpretation of the SVD of a 2×2 matrix $\mathbf{X}$ (see Problem 8.23 for details).

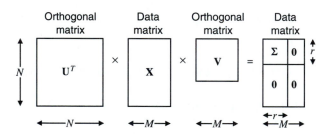

FIGURE 8.13
Pictorial representation of the singular value decomposition of a matrix.

Properties and interpretations. We next provide a summary of interpretations and properties whose proofs are given in the references and the problems.

1. Postmultiplying (8.7.1) by $\mathbf{V}$ and equating columns, we obtain

$$\mathbf{X}\mathbf{v}_i = \begin{cases} \sigma_i\mathbf{u}_i & i = 1, 2, \ldots, r \\ 0 & i = r+1, \ldots, M \end{cases} \tag{8.7.6}$$

that is, $\mathbf{v}_i$ (columns of $\mathbf{V}$) are the *right singular vectors* of $\mathbf{X}$.

2. Premultiplying (8.7.1) by $\mathbf{U}^H$ and equating rows, we obtain

$$\mathbf{u}_i^H\mathbf{X} = \begin{cases} \sigma_i\mathbf{v}_i^H & i = 1, 2, \ldots, r \\ 0 & i = r+1, \ldots, N \end{cases} \tag{8.7.7}$$

that is, $\mathbf{u}_i$ (columns of $\mathbf{U}$) are the *left singular vectors* of $\mathbf{X}$.

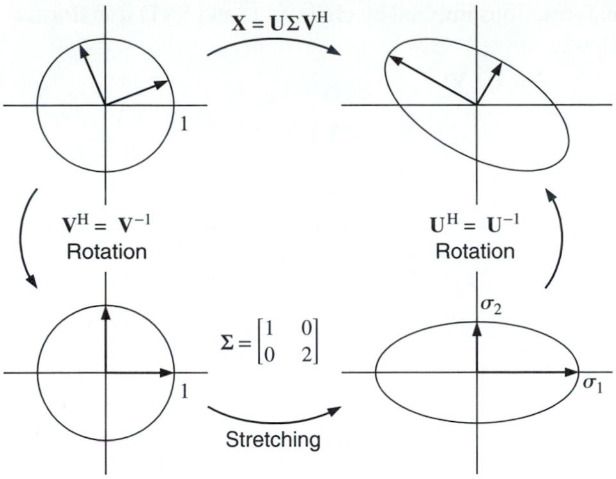

FIGURE 8.14
The SVD of a 2×2 matrix maps the unit circle into an ellipse whose semimajor and semiminor axes are equal to the singular values of the matrix.

433

SECTION 8.7
LS Computations Using the Singular Value Decomposition

3. Let $\lambda_i(\cdot)$ and $\sigma_i^2(\cdot)$ denote the ith largest eigenvalue and singular value of a given matrix, respectively. The vectors $\mathbf{v}_1, \ldots, \mathbf{v}_M$ are eigenvectors of $\mathbf{X}^H\mathbf{X}$; $\mathbf{u}_1, \ldots, \mathbf{u}_N$ are eigenvectors of $\mathbf{X}\mathbf{X}^H$, for which the squares of the singular values $\sigma_1^2, \ldots, \sigma_r^2$ of $\mathbf{X}$ are the first r nonzero eigenvalues of $\mathbf{X}^H\mathbf{X}$ and $\mathbf{X}\mathbf{X}^H$, that is,

$$\lambda_i(\mathbf{X}^H\mathbf{X}) = \lambda_i(\mathbf{X}\mathbf{X}^H) = \sigma_i^2(\mathbf{X}) \tag{8.7.8}$$

4. In the product $\mathbf{X} = \mathbf{U}\mathbf{\Sigma}\mathbf{V}^H$, the last $N - r$ columns of $\mathbf{U}$ and $M - r$ columns of $\mathbf{V}$ are superfluous because they interact only with blocks of zeros in $\mathbf{\Sigma}$. This leads to the following *thin SVD* representation of $\mathbf{X}$

$$\mathbf{X} = \mathbf{U}_r \mathbf{\Sigma}_r \mathbf{V}_r^H \tag{8.7.9}$$

where $\mathbf{U}_r$ and $\mathbf{V}_r$ consist of the first r columns of $\mathbf{U}$ and $\mathbf{V}$, respectively, and $\mathbf{\Sigma}_r = \text{diag}\{\sigma_1, \sigma_2, \ldots, \sigma_r\}$.

5. The SVD can be expressed as

$$\mathbf{X} = \sum_{i=1}^{r} \sigma_i \mathbf{u}_i \mathbf{v}_i^H \tag{8.7.10}$$

that is, as a sum of cross products weighted by the singular values.

6. If the matrix $\mathbf{X}$ has rank r, then:

 a. The first r columns of $\mathbf{U}$ form an orthonormal basis for the space spanned by the columns of $\mathbf{X}$ (*range space* or *column space* of $\mathbf{X}$).
 b. The first r columns of $\mathbf{V}$ form an orthonormal basis for the space spanned by the rows of $\mathbf{X}$ (*range space* of $\mathbf{X}^H$ or *row space* of $\mathbf{X}$).
 c. The last $M - r$ columns of $\mathbf{V}$ form an orthonormal basis for the space of vectors orthogonal to the rows of $\mathbf{X}$ (*null space* of $\mathbf{X}$).
 d. The last $N - r$ columns of $\mathbf{U}$ form an orthonormal basis for the *null space* of $\mathbf{X}^H$.

7. The Euclidean norm of $\mathbf{X}$ is

$$\|\mathbf{X}\| = \sigma_1 \tag{8.7.11}$$

8. The Frobenius norm of $\mathbf{X}$, that is, the square root of the sum of the squares of its elements, is

$$\|\mathbf{X}\|_F \triangleq \sqrt{\sum_{i=1}^{N} \sum_{j=1}^{M} |x_{ij}|^2} = \sqrt{\sigma_1^2 + \sigma_2^2 + \cdots + \sigma_r^2} \tag{8.7.12}$$

9. The difference between the transformations implied by eigenvalue and SVD transformations can be summarized as follows:

Eigenvalue decomposition $\qquad\qquad$ SVD

$\mathbf{R} = \mathbf{Q}\mathbf{\Lambda}\mathbf{Q}^H \qquad\qquad\qquad \mathbf{X} = \mathbf{U}\mathbf{\Sigma}\mathbf{V}^H$

$$
\begin{array}{ccccccc}
 & & & \mathbf{X} & & \mathbf{X}^H & \\
\mathbf{q}_1 & \xrightarrow{\lambda_1} & \mathbf{q}_1 & \mathbf{v}_1 \xrightarrow{\sigma_1} & \mathbf{u}_1 & \xrightarrow{\sigma_1} & \mathbf{v}_1 \\
\mathbf{q}_2 & \xrightarrow{\lambda_2} & \mathbf{q}_2 & \mathbf{v}_2 \xrightarrow{\sigma_2} & \mathbf{u}_2 & \xrightarrow{\sigma_2} & \mathbf{v}_2 \\
\vdots & \vdots & \vdots & \vdots \quad \vdots & \vdots & \vdots & \vdots \\
\mathbf{q}_M & \xrightarrow{\lambda_M} & \mathbf{q}_M & \mathbf{v}_r \xrightarrow{\sigma_r} & \mathbf{u}_r & \xrightarrow{\sigma_r} & \mathbf{v}_r \\
\end{array}
$$

$$
\left.\begin{array}{c}\mathbf{v}_{r+1} \\ \vdots \\ \mathbf{v}_M\end{array}\right\} \to 0 \qquad \left.\begin{array}{c}\mathbf{u}_{r+1} \\ \vdots \\ \mathbf{u}_M\end{array}\right\} \to 0
$$

This illustrates the need for left and right singular values and vectors.

We can compute the SVD of a matrix $\mathbf{X}$ by forming the matrices $\mathbf{X}^H\mathbf{X}$ and $\mathbf{X}\mathbf{X}^H$ and computing their eigenvalues and eigenvectors (see Problem 8.21). However, we should avoid this approach because the "squaring" of $\mathbf{X}$ to form these correlation matrices results in a *loss of information* and should be avoided (see Example 8.6.2).

In practice, the SVD is computed by using the algorithm of Golub and Reinsch (1970) or the R-SVD algorithm described in Chan (1982), which for $N \gg M$ is twice as fast. The state of the art in SVD research is provided in Golub and Van Loan (1996), whereas reliable numerical algorithms and code are given in LA-PACK, LINPACK, and Numerical Recipes in C (Press et al. 1992).

8.7.2 Solution of the LS Problem

So far, we have discussed the solution of the overdetermined ($N > M$) LS problem with full-rank ($r = M$) data matrices using the normal equations and the QR decomposition techniques. We next show how the SVD can be used to solve the LS problem *without* making any assumptions about the dimensions N and M or the rank r of data matrix $\mathbf{X}$.

Suppose that we know the exact SVD of data matrix $\mathbf{X} = \mathbf{U}\mathbf{\Sigma}\mathbf{V}^H$. Since $\mathbf{U}$ is orthogonal,

$$\|\mathbf{y} - \mathbf{X}\mathbf{c}\| = \|\mathbf{y} - \mathbf{U}\mathbf{\Sigma}\mathbf{V}^H\mathbf{c}\| = \|\mathbf{U}^H\mathbf{y} - \mathbf{\Sigma}\mathbf{V}^H\mathbf{c}\| \tag{8.7.13}$$

If we define $\qquad\qquad \mathbf{y}' \triangleq \mathbf{U}^H\mathbf{y} \qquad \mathbf{c}' \triangleq \mathbf{V}^H\mathbf{c}$

we obtain the LSE

$$\|\mathbf{y} - \mathbf{X}\mathbf{c}\|^2 = \|\mathbf{y}' - \mathbf{\Sigma}\mathbf{c}'\| = \sum_{i=1}^{r} |y_i' - \sigma_i c_i'|^2 + \sum_{i=r+1}^{N} |y_i'|^2 \tag{8.7.14}$$

which is minimized if and only if $c_i' = y_i'/\sigma_i$ for $i = 1, 2, \ldots, r$. We notice that when $r < M$, the terms $c_{r+1}', \ldots, c_M'$ do not appear in (8.7.14). Therefore, they have no effect on the residual and can be chosen arbitrarily. To illustrate this point, consider the geometric interpretation in Figure 8.5. There is only one linear combination of the linearly independent vectors $\tilde{\mathbf{x}}_1$ and $\tilde{\mathbf{x}}_2$ that determines the optimum LS estimate. If the data matrix has one more column $\tilde{\mathbf{x}}_3$ that lies in the same plane, then there are an infinite number of linear combinations $c_1\tilde{\mathbf{x}}_1 + c_2\tilde{\mathbf{x}}_2 + c_3\tilde{\mathbf{x}}_3$ that satisfy the LSE criterion. To obtain a *unique* LS solution from all solutions $\mathbf{c}$ that minimize $\|\mathbf{y} - \mathbf{X}\mathbf{c}\|$, we choose the one with the minimum length $\|\mathbf{c}\|$. Since the matrix $\mathbf{V}$ is orthogonal, we have $\|\mathbf{c}'\| = \|\mathbf{V}^H\mathbf{c}\| = \|\mathbf{c}\|$, and the norm $\|\mathbf{c}\|$ is minimized when the norm $\|\mathbf{c}'\|$ is minimized. Hence, choosing $c'_{r+1} = \cdots = c'_M = 0$ provides the *minimum-norm solution* to the LS problem. In summary, the unique, minimum-norm solution to the LS problem is

$$\mathbf{c}_{\text{ls}} = \sum_{i=1}^{r} \frac{\mathbf{u}_i^H \mathbf{y}}{\sigma_i} \mathbf{v}_i \qquad (8.7.15)$$

where

$$c'_i = \begin{cases} \dfrac{y'_i}{\sigma_i} = \dfrac{\mathbf{u}_i^H \mathbf{y}}{\sigma_i} & i = 1, \ldots, r \\ 0 & i = r+1, \ldots, M \end{cases} \qquad (8.7.16)$$

and

$$E_{\text{ls}} = \|\mathbf{y} - \mathbf{X}\mathbf{c}_{\text{ls}}\|^2 = \sum_{i=r+1}^{N} |y'_i|^2 = \sum_{i=r+1}^{N} |\mathbf{u}_i^H \mathbf{y}|^2 \qquad (8.7.17)$$

is the corresponding LS error.

We next express the unique minimum-norm solution to the LS problem in terms of the pseudoinverse of data matrix $\mathbf{X}$ using the SVD. To this end, we note that (8.7.16) can be written in matrix form

$$\mathbf{c}' = \boldsymbol{\Sigma}^+ \mathbf{y}' \qquad (8.7.18)$$

where

$$\boldsymbol{\Sigma}^+ \triangleq \begin{bmatrix} \boldsymbol{\Sigma}_r^{-1} & \mathbf{0} \\ \mathbf{0} & \mathbf{0} \end{bmatrix} \qquad (8.7.19)$$

is an $N \times N$ matrix with $\boldsymbol{\Sigma}_r^{-1} = \text{diag}\{1/\sigma_1, \ldots, 1/\sigma_r\}$. Therefore, using (8.7.15) and (8.7.19), we obtain

$$\mathbf{c}_{\text{ls}} = \mathbf{V}\boldsymbol{\Sigma}^+\mathbf{U}^H\mathbf{y} = \mathbf{X}^+\mathbf{y} \qquad (8.7.20)$$

where

$$\mathbf{X}^+ \triangleq \mathbf{V}\boldsymbol{\Sigma}^+\mathbf{U}^H = \sum_{i=1}^{r} \frac{1}{\sigma_i} \mathbf{v}_i \mathbf{u}_i^H \qquad (8.7.21)$$

is the pseudoinverse of matrix $\mathbf{X}$. For full-rank matrices, the pseudoinverse is defined as $\mathbf{X}^+ = (\mathbf{X}^H\mathbf{X})^{-1}\mathbf{X}^H$ (Golub and Van Loan 1996), so that using (8.7.21) leads to the LS solution in (8.2.21). If $N = M = \text{rank}(\mathbf{X})$, then $\mathbf{X}^+ = \mathbf{X}^{-1}$. Therefore, (8.7.21) holds for any rectangular or square matrix that is either full rank or rank-deficient. Formally, $\mathbf{X}^+$ can be defined independently of the LS problem as the unique $M \times N$ matrix $\mathbf{A}$ that satisfies the four Moore-Penrose conditions

$$\begin{array}{ll} \mathbf{XAX} = \mathbf{X} & (\mathbf{XA})^H = \mathbf{XA} \\ \mathbf{AXA} = \mathbf{A} & (\mathbf{AX})^H = \mathbf{AX} \end{array} \qquad (8.7.22)$$

which implies that $\mathbf{XX}^+$ and $\mathbf{X}^+\mathbf{X}$ are orthogonal projections onto the range space of $\mathbf{X}$ and $\mathbf{X}^H$ (see Problem 8.25). However, we stress that the pseudoinverse is, for the most part, a theoretical tool, and there is seldom any reason for its use in practice.

In summary, the computation of the LS estimator using the SVD involves the steps shown in Table 8.5. The vector $\mathbf{c}_{\text{ls}}$ is unique and satisfies two requirements: (1) It minimizes the sum of the errors, and (2) it has the smallest Euclidean norm.

The following example illustrates the use of the SVD for the computation of the LS estimator.

EXAMPLE 8.7.1. Solve the LS problem with the following data matrix and desired response signal:

$$\mathbf{X} = \begin{bmatrix} 1 & 1 & 1 \\ 2 & 2 & 1 \\ 3 & 1 & 3 \\ 1 & 0 & 1 \end{bmatrix} \qquad \mathbf{y} = \begin{bmatrix} 1 \\ 2 \\ 4 \\ 3 \end{bmatrix}$$

TABLE 8.5

Solution of the LS problem using the SVD method.

Step	Description
1	Compute the SVD $\mathbf{X} = \mathbf{U}\boldsymbol{\Sigma}\mathbf{V}^H$
2	Determine the rank r of $\mathbf{X}$
3	Compute $y'_i = \mathbf{u}_i^H \mathbf{y}, i = 1, \ldots, N$
4	Compute $\mathbf{c}_{ls} = \sum_{i=1}^{r} \dfrac{y'_i}{\sigma_i} \mathbf{v}_i$
5	Compute $E_{ls} = \sum_{i=r+1}^{N} |y'_i|^2$

Solution. We start by computing the SVD of $\mathbf{X} = \mathbf{U}\boldsymbol{\Sigma}\mathbf{V}^T$ by using the MATLAB function [U,S,V]=svd(X). This gives

$$\mathbf{U} = \begin{bmatrix} 0.3041 & 0.2170 & 0.8329 & 0.4082 \\ 0.4983 & 0.7771 & -0.3844 & 0.0000 \\ 0.7768 & -0.4778 & 0.0409 & -0.4082 \\ 0.2363 & -0.3474 & -0.3960 & 0.8165 \end{bmatrix}$$

$$\boldsymbol{\Sigma} = \begin{bmatrix} 5.5338 & 0 & 0 \\ 0 & 1.5139 & 0 \\ 0 & 0 & 0.2924 \\ 0 & 0 & 0 \end{bmatrix} \quad \mathbf{V} = \begin{bmatrix} 0.6989 & 0.3754 & -0.60882 \\ -0.0063 & 0.8544 & -0.5196 \\ -0.7152 & 0.3593 & 0.5994 \end{bmatrix}^T$$

which implies that the data matrix has rank $r = 3$. Next we compute

$$\mathbf{y}' = \mathbf{U}^T\mathbf{y} = \begin{bmatrix} 5.1167 \\ -1.1821 \\ -0.9602 \\ 1.2247 \end{bmatrix} \quad \mathbf{c}_{ls} = \begin{bmatrix} 3.0 \\ -1.5 \\ -1.0 \end{bmatrix} \quad E_{ls} = 1.5$$

by the MATLAB commands

```
yp=U'*y;
cls=V*(yp(1:r)./diag(S));
Els=sum(yp(r+1:N).^2);
```

which implement steps 3, 4, and 5 in Table 8.5. The LS solution also can be obtained from cls=X\y. If we set $\langle \mathbf{X} \rangle_{23} = 2$, the first and last columns of $\mathbf{X}$ become linearly dependent, the SVD has only two nonzero singular values, and the svd function warns that $\mathbf{X}$ is rank-deficient.

Table 8.6 shows the numerical operations required by the various LS solution methods (Golub and Van Loan 1996). For full-rank (nonsingular) data matrices, all other methods are simpler than the SVD. However, these methods are inaccurate when $\mathbf{X}$ is rank-deficient (nearly singular). In such cases, the SVD reveals the near singularity of the data matrix and is the method of choice because it provides a reliable computation of the numerical rank (see the next section).

Normal equations versus QR decomposition. The squaring of $\mathbf{X}$ to form the time-average correlation matrix $\hat{\mathbf{R}} = \mathbf{X}^H\mathbf{X}$ results in a loss of information and should be avoided. Since $\|\mathbf{X}^{-1}\| = 1/\sigma_{\min}$, the condition number of $\mathbf{X}$ is

$$\kappa(\mathbf{X}) = \|\mathbf{X}\|\|\mathbf{X}^{-1}\| = \frac{\sigma_{\max}}{\sigma_{\min}} \tag{8.7.23}$$

TABLE 8.6
Computational complexity of LS computation algorithms.

437

SECTION 8.7
LS Computations Using the
Singular Value
Decomposition

LS Algorithm	FLOPS (floating point operations)
Normal equations	$NM^2 + M^3/3$
Householder orthogonalization	$2NM^2 - 2M^3/3$
Givens orthogonalization	$3NM^2 - M^3$
Modified Gram-Schmidt	$2NM^2$
Golub-Reinsch SVD	$4NM^2 + 8M^3$
R-SVD	$2NM^2 + 11M^3$

which is analogous to the eigenvalue ratio for square Hermitian matrices. Hence,

$$\kappa(\mathbf{X}^H \mathbf{X}) = \frac{\lambda_{\max}}{\lambda_{\min}} = \frac{\sigma_{\max}^2}{\sigma_{\min}^2} = \kappa^2(\mathbf{X}) \tag{8.7.24}$$

which shows that squaring a matrix can only worsen its condition.

The study of the sensitivity of the LS problem is complicated. However, the following conclusions (Golub and Van Loan 1996; Van Loan 1997) can be drawn:

1. The sensitivity of the LS solution is roughly proportional to the quantity $\kappa(\mathbf{X}) + \sqrt{E_{\mathrm{ls}}}\kappa^2(\mathbf{X})$. Hence, any method produces inaccurate results when applied to ill-conditioned problems with large E_{ls}.
2. The method of normal equations produces a solution $\mathbf{c}_{\mathrm{ls}}$ whose relative error is approximately eps $\cdot \kappa^2(\mathbf{X})$, where eps is the machine precision.
3. The QR method (Householder, Givens, MGS) produces a solution $\mathbf{c}_{\mathrm{ls}}$ whose relative error is approximately eps $\cdot [\kappa(\mathbf{X}) + \sqrt{E_{\mathrm{ls}}}\kappa^2(\mathbf{X})]$.

In general, QR methods are more accurate than and can be used for a wider class of data matrices than the normal equations approach, even if the latter is about twice as fast.

In many practical applications, we need to update the Cholesky or QR decomposition after the original data matrix has been modified by the addition or deletion of a row or column (rank 1 modifications). Techniques for the efficient computation of these decompositions by updating the existing ones can be found in Golub and Van Loan (1996) and Gill et al. (1974).

8.7.3 Rank-Deficient LS Problems

In theory, it is relatively easy to determine the rank of a matrix or that a matrix is rank-deficient. However, both tasks become complicated in practice when the elements of the matrix are specified with inadequate accuracy or the matrix is near singular. The SVD provides the means of determining how close a matrix is to being rank-deficient, which in turn leads to the concept of numerical rank. To this end, suppose that the elements of matrix $\mathbf{X}$ are known with an accuracy of order ϵ, and its computed singular values $\sigma_1 \geq \sigma_2 \geq \cdots \geq \sigma_M$ are such that

$$\sigma_{r+1}^2 + \sigma_{r+2}^2 + \cdots + \sigma_M^2 < \epsilon^2 \tag{8.7.25}$$

Then if we set $\mathbf{\Sigma}_r \triangleq \mathrm{diag}\{\sigma_1, \ldots, \sigma_r, 0, \ldots, 0\}$ and

$$\mathbf{X}_r \triangleq \mathbf{U} \begin{bmatrix} \mathbf{\Sigma}_r \\ \mathbf{0} \end{bmatrix} \mathbf{V}^H \tag{8.7.26}$$

we have
$$\|\mathbf{X} - \mathbf{X}_r\|_F = \sqrt{\sigma_{r+1}^2 + \sigma_{r+2}^2 + \cdots + \sigma_M^2} < \epsilon \tag{8.7.27}$$

and matrix $\mathbf{X}$ is said to be near a matrix of rank r or $\mathbf{X}$ has *numerical rank* r. It can be shown that $\mathbf{X}_r$ is the matrix of rank r that is nearest to $\mathbf{X}$ in the Frobenius norm sense (Leon

1990; Stewart 1973). This result has important applications in signal modeling and data compression.

Computing the LS solution for rank-deficient data matrices requires extra care. When a singular value is equal to a very small number, its reciprocal, which is a singular value of the pseudoinverse $\mathbf{X}^+$, is a very large number. As a result, the LS solution deviates substantially from the "true" solution.

One way to handle this problem is to replace each singular value below a certain cutoff value (*thresholding*) with zero. A typical threshold is a fraction of σ_1 determined by either the machine precision available or the accuracy of the elements in the data matrix (measurement accuracy). For example, if the data matrix is accurate to six decimal places, we set the threshold at $10^{-6}\sigma_1$ (Golub and Van Loan 1996).

Another way is to replace the LS criterion (8.7.14) by

$$E\{\mathbf{c}, \ \psi\} = \|\mathbf{y} - \mathbf{Xc}\|^2 + \psi \|\mathbf{c}\|^2 \tag{8.7.28}$$

where the constant $\psi > 0$ reflects the importance of the norm of the solution vector. The term $\|\mathbf{c}\|$ acts a stabilizer, that is, prevents the solution $\mathbf{c}_\psi$ from becoming too large (*regularization*). Indeed, using the method of Lagrange multipliers, we can show that

$$\mathbf{c}_\psi = \sum_{i=1}^{r} \frac{\sigma_i}{\sigma_i^2 + \psi}(\mathbf{u}_i^H \mathbf{y})\mathbf{v}_i \tag{8.7.29}$$

which is known as the *regularized solution*. We note that $\mathbf{c}_\psi = \mathbf{c}_{ls}$ when $\psi = 0$. However, when $\psi > 0$, as $\sigma_i \to 0$ the term $\sigma_i/(\sigma_i^2 + \psi)$ in (8.7.29) tends to zero while the term $1/\sigma_i \to \infty$ in (8.7.15) tends to infinity. Furthermore, it can be shown that $\|\mathbf{c}_{ls}\| \le \|\mathbf{y}\|/\sigma_r$ and $\|\mathbf{c}_\psi\| \le \|\mathbf{y}\|/\sqrt{\psi}$ (Hager 1988).

Since the minimum-norm LS solution requires only the first r columns of $\mathbf{U}$, where r is the numerical rank of $\mathbf{X}$, we can use the thin SVD. If $N \gg M$, the computation of either $\mathbf{U}_r$ or $\mathbf{U}$ is expensive. However, in practical SVD algorithms, $\mathbf{U}$ is computed as the product of many reflections and rotations. Hence, we can compute $\mathbf{y}' = \mathbf{U}^H \mathbf{y}$ by updating $\mathbf{y}$ at each step i with each orthogonal transformation, that is, $\mathbf{U}_i^H \mathbf{y} \to \mathbf{y}$.

8.8 SUMMARY

In this chapter we discussed the theory, implementation, and application of linear estimators (combiners, filters, and predictors) that are optimum according to the LSE criterion of performance. The fundamental differences between linear MMSE and LSE estimators are as follows:

- MMSE estimators are designed using ensemble average second-order moments $\mathbf{R}$ and $\mathbf{d}$; they can be designed prior to operation, and during their normal operation they need only the input signals.
- LSE estimators are designed using time-average estimates $\hat{\mathbf{R}}$ and $\hat{\mathbf{d}}$ of the second-order moments or data matrix $\mathbf{X}$ and the desired response vector $\mathbf{y}$. For this reason LSE estimators are sometimes said to be *data-adaptive*. The design and operation of LSE estimators are *coupled* and are usually accomplished by using either of the following approaches:

    – Collect a block of training data $\mathbf{X}_{tr}$ and $\mathbf{y}_{tr}$ and use them to design an LSE estimator; use it to process subsequent blocks. Clearly, this approach is meaningful if all blocks have statistically similar characteristics.
    – For each collected block of data $\mathbf{X}$ and $\mathbf{y}$, compute the LSE filter $\mathbf{c}_{ls}$ or the LSE estimate $\hat{\mathbf{y}}$ (whatever is needed).

There are various numerical algorithms designed to compute LSE estimators and estimates. For well-behaved data and sufficient numerical precision, all these methods produce the same results and therefore provide the same LSE performance, that is, the same total squared error.

However, when ill-conditioned data, finite precision, or computational complexity is a concern, the choice of the LS computational algorithm is very important.

We saw that there are two major families of numerical algorithms for dealing with LS problems:

Power-domain techniques solve LS estimation problems using the time-average moments $\hat{\mathbf{R}} = \mathbf{X}^H\mathbf{X}$ and $\hat{\mathbf{d}} = \mathbf{X}^H\mathbf{y}$. The most widely used methods are the LDL^H and Cholesky decompositions.

Amplitude-domain techniques operate directly on data matrix $\mathbf{X}$ and the desired response vector. In general, they require more computations and have better numerical properties than power-domain methods. This group includes the QR orthogonalization methods (Householder, Givens, and modified Gram-Schmidt) and the SVD method.

The QR decomposition methods apply a unitary transformation to the data matrix to reduce it to an upper triangular one, whereas the GS methods apply an upper triangular matrix transformation to orthogonalize the columns of the data matrix.

In conclusion, we emphasize that there are various ways to compute the coefficients of an optimum estimator and the value of the optimum estimate. We stress that the performance of any optimum estimator, as measured by the MMSE or LSE, does *not* depend on the particular implementation as long as we have sufficient numerical precision. Therefore, if we want to investigate how well an optimum estimator performs in a certain application, we can use any implementation, as long as computational complexity is not a consideration.

PROBLEMS

8.1 By differentiating (8.2.8) with respect to the vector $\mathbf{c}$, show that the LSE estimator $\mathbf{c}_{ls}$ is given by the solution of the normal equations (8.2.12).

8.2 Let the weighted LSE be given by $E_w = \mathbf{e}^H\mathbf{We}$, where $\mathbf{W}$ is a Hermitian positive definite matrix.

(a) By minimizing E_w with respect to the vector $\mathbf{c}$, show that the wieghted LSE estimator is given by (8.2.35).

(b) Using the LDL^H decomposition $\mathbf{W} = \mathbf{LDL}^H$, show that the weighted LS criterion corresponds to prefiltering the error or the data.

8.3 Using direct substitution of (8.4.4) into (8.4.5), show that the LS estimator $\mathbf{c}_{ls}^{(i)}$ and the associated LS error $E_{ls}^{(i)}$ are determined by (8.4.5).

8.4 Consider a linear system described by the difference equation $y(n) = 0.9y(n-1) + 0.1x(n-1) + v(n)$, where $x(n)$ is the input signal, $y(n)$ is the output signal, and $v(n)$ is an output disturbance. Suppose that we have collected $N = 1000$ samples of input-output data and that we wish to estimate the system coefficients, using the LS criterion with no windowing. Determine the coefficients of the model $y(n) = ay(n-1) + dx(n-1)$ and their estimated covariance matrix $\hat{\sigma}_e^2\hat{\mathbf{R}}^{-1}$ when

(a) $x(n) \sim \text{WGN}(0, 1)$ and $v(n) \sim \text{WGN}(0, 1)$ and

(b) $x(n) \sim \text{WGN}(0, 1)$ and $v(n) = 0.8v(n-1) + w(n)$ is an AR(1) process with $w(n) \sim \text{WGN}(0, 1)$. Comment upon the quality of the obtained estimates by comparing the matrices $\hat{\sigma}_e^2\hat{\mathbf{R}}^{-1}$ obtained in each case.

8.5 Use Lagrange multipliers to show that Equation (8.4.13) provides the minimum of (8.4.8) under the constraint (8.4.9).

8.6 If full windowing is used in LS, then the autocorrelation matrix is Toeplitz. Using this fact, show that in the combined FBLP the predictor is given by

$$\mathbf{a}^{fb} = \frac{1}{2}(\mathbf{a} + \mathbf{Jb}^*)$$

8.7 Consider the noncausal "middle" sample linear signal estimator specified by (8.4.1) with $M = 2L$ and $i = L$.

(a) Show that if we apply full windowing to the data matrix, the resulting signal estimator is conjugate symmetric, that is, $\mathbf{c}^{(L)} = \mathbf{Jc}^{(L)*}$. This property does not hold for any other windowing method.

(b) Derive the normal equations for the signal estimator that minimizes the total squared error $E^{(L)} = \|\mathbf{e}^{(L)}\|^2$ under the constraint $\mathbf{c}^{(L)} = \mathbf{Jc}^{(L)*}$.

(c) Show that if we enforce the normal equation matrix to be centro-Hermitian, that is, we use the normal equations

$$(\bar{\mathbf{X}}^H \bar{\mathbf{X}} + \mathbf{J}\bar{\mathbf{X}}^T \bar{\mathbf{X}}^* \mathbf{J})\mathbf{c}^{(L)} = \begin{bmatrix} \mathbf{0} \\ E^{(L)} \\ \mathbf{0} \end{bmatrix}$$

then the resulting signal smoother is conjugate symmetric.

(d) Illustrate parts (a) to (c), using the data matrix

$$\mathbf{X} = \begin{bmatrix} 1 & 1 & 1 \\ 2 & 2 & 1 \\ 3 & 1 & 3 \\ 1 & 0 & 1 \\ 1 & 2 & 1 \end{bmatrix}$$

and check which smoother provides the smallest total squared error. Try to justify the obtained answer.

8.8 A useful impulse response for some geophysical signal processing applications is the Mexican hat wavelet

$$g(t) = \frac{2}{\sqrt{3}}\pi^{-1/4}(1 - t^2)e^{-t^2/2}$$

which is the second derivative of a Gaussian pulse.

(a) Plot the wavelet $g(t)$ and the magnitude and phase of its Fourier transform.

(b) By examining the spectrum of the wavelet, determine a reasonable sampling frequency F_s.

(c) Design an optimum LS inverse FIR filter for the discrete-time wavelet $g(nT)$, where $T = 1/F_s$. Determine a reasonable value for M by plotting the LSE E_M as a function of order M. Investigate whether we can improve the inverse filter by introducing some delay n_0. Determine the best value of n_0 and plot the impulse response of the resulting filter and the combined impulse response $g(n) * h(n - n_0)$, which should resemble an impulse.

(d) Repeat part (c) by increasing the sampling frequency by a factor of 2 and comparing with the results obtained in part (c).

8.9 (a) Prove Equation (8.5.4) regarding the LDL^H decomposition of the augmented matrix $\bar{\mathbf{R}}$.

(b) Solve the LS estimation problem in Example 8.5.1, using the LDL^H decomposition of $\bar{\mathbf{R}}$ and the partitionings in (8.5.4).

8.10 Prove the order-recursive algorithm described by the relations given in (8.5.12). Demonstrate the validity of this approach, using the data in Example 8.5.1.

8.11 In this problem, we wish to show that the statistical interpretations of innovation and partial correlation for $w_m(n)$ and k_{m+1} in (8.5.12) hold in a deterministic LSE sense. To this end, suppose that the "partial correlation" between $\tilde{\mathbf{y}}$ and $\tilde{\mathbf{x}}_{m+1}$ is defined using the residual records $\tilde{\mathbf{e}}_m = \tilde{\mathbf{y}} - \mathbf{X}_m \mathbf{c}_m$ and $\tilde{\mathbf{e}}_m^b = \tilde{\mathbf{x}}_{m+1} + \mathbf{X}_m \mathbf{b}_m$, where $\mathbf{b}_m$ is the LSE BLP. Show that $k_{k+1} = \beta_{m+1}/\xi_{m+1}$, where $\beta_{m+1} \triangleq \tilde{\mathbf{e}}_m^H \tilde{\mathbf{e}}_m^b$ and $\xi_{m+1} = \tilde{\mathbf{e}}_m^{bH} \tilde{\mathbf{e}}_m^b$. Demonstrate the validity of these formulas using the data in Example 8.5.1.

8.12 Show that the Cholesky decomposition of a Hermitian positive definite matrix $\mathbf{R}$ can be computed by using the following algorithm

$$\text{for } j = 1 \text{ to } M$$
$$l_{ij} = \left(r_{ij} - \sum_{k=1}^{j-1} |l_{jk}|^2\right)^{1/2}$$
$$\text{for } i = j+1 \text{ to } M$$
$$l_{ij} = \left(r_{ij} - \sum_{k=1}^{j-1} l_{ik}^* l_{jk}\right)/l_{jj}$$
$$\text{end } i$$
$$\text{end } j$$

and write a MATLAB function for its implementation. Test your code using the built-in MATLAB function chol.

8.13 Compute the LDL^T and Cholesky decompositions of the following matrices:

$$\mathbf{X}_1 = \begin{bmatrix} 9 & 3 & -6 \\ 3 & 4 & 1 \\ -6 & 1 & 9 \end{bmatrix} \quad \text{and} \quad \mathbf{X}_2 = \begin{bmatrix} 6 & 4 & -2 \\ 4 & 5 & 3 \\ -2 & 3 & 6 \end{bmatrix}$$

8.14 Solve the LS problem in Example 8.6.1,

(a) using the QR decomposition of the augmented data matrix $\bar{\mathbf{X}} = [\mathbf{X} \ \mathbf{y}]$ and
(b) using the Cholesky decomposition of the matrix $\bar{\mathbf{R}} = \bar{\mathbf{X}}^H \bar{\mathbf{X}}$.
   Note: Use MATLAB built-in functions for the QR and Cholesky decompositions.

8.15 (a) Show that a unit vector $\mathbf{w}$ is an eigenvector of the matrix $\mathbf{H} = \mathbf{I} - 2\mathbf{w}\mathbf{w}^H$. What is the corresponding eigenvalue?
(b) If a vector $\mathbf{z}$ is orthogonal to $\mathbf{w}$, show that $\mathbf{z}$ is an eigenvector of $\mathbf{H}$. What is the corresponding eigenvalue?

8.16 Solve the LS problem

$$\mathbf{X} = \begin{bmatrix} 1 & 2 \\ 1 & 3 \\ 1 & 2 \\ 1 & -1 \end{bmatrix} \quad \mathbf{y} = \begin{bmatrix} -3 \\ 10 \\ 3 \\ 6 \end{bmatrix}$$

using the Householder transformation.

8.17 Solve Problem 8.16 by using the Givens transformation.

8.18 Compute the QR decomposition of the data matrix

$$\mathbf{X} = \begin{bmatrix} 4 & 2 & 1 \\ 2 & 0 & 1 \\ 2 & 0 & -1 \\ 1 & 2 & 1 \end{bmatrix}$$

using the GS and MGS methods, and compare the obtained results.

8.19 Solve the following LS problem

$$\mathbf{X} = \begin{bmatrix} 1 & -2 & -1 \\ 2 & 0 & 1 \\ 2 & -4 & 2 \\ 4 & 0 & 0 \end{bmatrix} \quad \mathbf{y} = \begin{bmatrix} -1 \\ 1 \\ 1 \\ -2 \end{bmatrix}$$

by computing the QR decomposition using the GS algorithm.

8.20 Show that the computational organization of the MGS algorithm shown in Table 8.4 can be used to compute the GS algorithm if we replace the step $r_{im} = \mathbf{q}_i^H \mathbf{x}_m$ by $r_{im} = \mathbf{q}_i^H \mathbf{q}_m$.

8.21 Compute the SVD of $\mathbf{X} = \begin{bmatrix} 1 & 1 \\ 1 & 1 \\ 0 & 0 \end{bmatrix}$ by computing the eigenvalues and eigenvectors of $\mathbf{X}^H \mathbf{X}$ and $\mathbf{X}\mathbf{X}^H$. Check with the results obtained using the svd function.

8.22 Repeat Problem 8.21 for

(a) $\mathbf{X} = \begin{bmatrix} 6 & 2 \\ -7 & 6 \end{bmatrix}$ and

(b) $\mathbf{X} = \begin{bmatrix} 0 & 1 & 1 \\ 1 & 1 & 0 \end{bmatrix}$.

8.23 Write a MATLAB program to produce the plots in Figure 8.14, using the matrix $\mathbf{X} = \begin{bmatrix} 6 & 2 \\ -7 & 6 \end{bmatrix}$. *Hint:* Use a parametric description of the circle in polar coordinates.

8.24 For the matrix $\mathbf{X} = \begin{bmatrix} 0 & 1 & 1 \\ 1 & 1 & 0 \end{bmatrix}^T$ determine $\mathbf{X}^+$ and verify that $\mathbf{X}$ and $\mathbf{X}^+$ satisfy the four Moore-Penrose conditions (8.7.22).

8.25 Prove the four Moore-Penrose conditions in (8.7.22) and explain why $\mathbf{X}\mathbf{X}^+$ and $\mathbf{X}^+\mathbf{X}$ are orthogonal projections onto the range space of $\mathbf{X}$ and $\mathbf{X}^H$.

8.26 In this problem we examine in greater detail the radio-frequency interference cancelation experiment discussed in Section 8.4.3. We first explain the generation of the various signals and then proceed with the design and evaluation of the LS interference canceler.

(a) The useful signal is a pointlike target defined by

$$ s(t) = \frac{d}{dt} \left(\frac{1}{e^{-\alpha t/t_r} + e^{\alpha t/t_f}} \right) \triangleq \frac{dg(t)}{dt} $$

where $\alpha = 2.3$, $t_r = 0.4$, and $t_f = 2$. Given that $F_s = 2$ GHz, determine $s(n)$ by computing the samples $g(n) = g(nT)$ in the interval $-2 \le nT \le 6$ ns and then computing the first difference $s(n) = g(n) - g(n-1)$. Plot the signal $s(n)$ and its Fourier transform (magnitude and phase), and check whether the pointlike and wideband assumptions are justified.

(b) Generate $N = 4096$ samples of the narrowband interference using the formula

$$ z(n) = \sum_{i=1}^{L} A_i \sin(\omega_i n + \phi_i) $$

and the following information:

```
Fs=2; % All frequencies are measured in GHz.
F=0.1*[0.6 1 1.8 2.1 3 4.8 5.2 5.7 6.1 6.4 6.7 7 7.8 9.3]';
L=length(F);
om=2*pi*F/Fs;
A=[0.5 1 1 0.5 0.1 0.3 0.5 1 1 0.5 0.3 1.5 0.5]';
rand('seed',1954);
phi=2*pi*rand(L,1);
```

(c) Compute and plot the the periodogram of $z(n)$ to check the correctness of your code.

(d) Generate N samples of white Gaussian noise $v(n) \sim \text{WGN}(0, 0.1)$ and create the observed signal $x(n) = 5s(n - n_0) + z(n) + v(n)$, where $n_0 = 1000$. Compute and plot the periodogram of $x(n)$.

(e) Design a one-step ahead ($D = 1$) linear predictor with $M = 100$ coefficients using the FBLP method with no windowing. Then use the obtained FBLP to clean the corrupted signal $x(n)$ as shown in Figure 8.7. To evaluate the performance of the canceler, generate the plots shown in Figures 8.8 and 8.9.

8.27 Careful inspection of Figure 8.9 indicates that the the D-step prediction error filter, that is, the system with input $x(n)$ and output $e^f(n)$, acts as a whitening filter. In this problem, we try to solve Problem 8.26 by designing a practical whitening filter using a power spectral density (PSD) estimate of the corrupted signal $x(n)$.

(a) Estimate the PSD $\hat{R}_x^{(PA)}(e^{j\omega_k})$, $\omega_k = 2\pi k/N_{FFT}$, of the signal $x(n)$, using the method of averaged periodograms. Use a segment length of $L = 256$ samples, 50 percent overlap, and $N_{FFT} = 512$.

(b) Since the PSD does not provide any phase information, we shall design a whitening FIR filter with linear phase by

$$\tilde{H}(k) = \frac{1}{\sqrt{\hat{R}_x^{(PA)}(e^{j\omega_k})}} e^{-j\frac{2\pi}{N_{FFT}}\frac{N_{FFT}-1}{2}k}$$

where $\tilde{H}(k)$ is the DFT of the impulse response of the filter, that is,

$$\tilde{H}(k) = \sum_{n=0}^{N_{FFT}-1} h(n) e^{-j\frac{2\pi}{N_{FFT}}nk}$$

with $0 \le k \le N_{FFT} - 1$.

(c) Use the obtained whitening filter to clean the corrupted signal $x(n)$, and compare its performance with the FBLP canceler by generating plots similar to those shown in Figures 8.8 and 8.9.

(d) Repeat part (c) with $L = 128$, $N_{FFT} = 512$ and $L = 512$, $N_{FFT} = 1024$ and check whether spectral resolution has any effect upon the performance. *Note:* Information about the design and implementation of FIR filters using the DFT can be found in Proakis and Manolakis (1996).

8.28 Repeat Problem 8.27, using the multitaper method of PSD estimation.

8.29 In this problem we develop an RFI canceler using a symmetric linear smoother with guard samples defined by

$$e(n) = x(n) - \hat{x}(n) \triangleq x(n) + \sum_{k=D}^{M} c_k[x(n-k) + x(n-k)]$$

where $1 \le D < M$ prevents the use of the D adjacent samples to the estimation of $x(n)$.

(a) Following the approach used in Section 8.4.3, demonstrate whether such a canceler can be used to mitigate RFI and under what conditions.

(b) If there is theoretical justification for such a canceler, estimate its coefficients, using the method of LS with no windowing for $M = 50$ and $D = 1$ for the situation described in Problem 8.26.

(c) Use the obtained filter to clean the corrupted signal $x(n)$, and compare its performance with the FBLP canceler by generating plots similar to those shown in Figures 8.8 and 8.9.

(d) Repeat part (c) for $D = 2$.

8.30 In Example 6.7.1 we studied the design and performance of an optimum FIR inverse system. In this problem, we design and analyze the performance of a similar FIR LS inverse filter, using training input-output data.

(a) First, we generate $N = 100$ observations of the input signal $y(n)$ and the noisy output signal $x(n)$. We assume that $x(n) \sim WGN(0, 1)$ and $v(n) \sim WGN(0, 0.1)$. To avoid transient effects, we generate 200 samples and retain the last 100 samples to generate the required data records.

(b) Design an LS inverse filter with $M = 10$ for $0 \le D < 10$, using no windowing, and choose the best value of delay D.

(c) Repeat part (b) using full windowing.

(d) Compare the LS filters obtained in parts (b) and (c) with the optimum filter designed in Example 6.7.1. What are your conclusions?

8.31 In this problem we estimate the equalizer discussed in Example 6.8.1, using input-output training data, and we evaluate its performance using Monte Carlo simulation.

(a) Generate $N = 1000$ samples of input-desired response data $\{x(n), \ a(n)\}_0^{N-1}$ and use them to estimate the correlation matrix $\hat{\mathbf{R}}_x$ and the cross-correlation vector $\hat{\mathbf{d}}$ between $\mathbf{x}(n)$ and $y(n - D)$. Use $D = 7$, $M = 11$, and $W = 2.9$. Solve the normal equations to determine the LS FIR equalizer and the corresponding LSE.

(b) Repeat part (a) 500 times; by changing the seed of the random number generators, compute the average (over the realizations) coefficient vector and average LSE, and compare with the optimum MSE equalizer obtained in Example 6.8.1. What are your conclusions?

(c) Repeat parts (a) and (b) by setting $W = 3.1$.

CHAPTER 9

Signal Modeling and Parametric Spectral Estimation

This chapter is a transition from theory to practice. It focuses on the selection of an appropriate model for a given set of data, the estimation of the model parameters, and how well the model actually "fits the data." Although the development of parameter estimation techniques requires a strong theoretical background, the selection of a good model and its subsequent evaluation require the user to have sufficient practical experience and a familiarity with the intended application. We provide complete, detailed algorithms for fitting pole-zero models to data using least-squares techniques. The estimation of all-pole model parameters involves the solution of a linear system of equations, whereas pole-zero modeling requires nonlinear least-squares optimization. The chapter is roughly organized into two separate but related parts.

In the first part, we begin in Section 9.1 by explaining the steps that are required in the model-building process. Then, in Section 9.2, we introduce various least-squares algorithms for the estimation of parameters of direct and lattice all-pole models, provide different interpretations, and discuss some order selection criteria. For pole-zero models we provide, in Section 9.3, a nonlinear optimization algorithm that estimates the parameters of the model by minimizing the least-squares criterion. We conclude this part with Section 9.4 in which we discuss the applications of pole-zero models to spectral estimation and speech processing.

In the second part, we begin with the method of minimum-variance spectral estimation (Capon's method). Then we describe frequency estimation methods based on the harmonic model: the Pisarenko harmonic decomposition and the MUSIC, minimum-norm, and ESPRIT algorithms. These methods are suitable for applications in which the signals of interest can be represented by *complex exponential* or *harmonic models*. Signals consisting of complex exponentials are found in a variety of applications including as formant frequencies in speech processing, moving targets in radar, and spatially propagating signals in array processing.

9.1 THE MODELING PROCESS: THEORY AND PRACTICE

In this section, we discuss the modeling of *real-world* signals using parametric pole-zero (PZ) signal models, whose theoretical properties were discussed in Chapter 4. We focus on PZ (P, Q) models with white input sequences, which are also known as ARMA (P, Q) random signal models. These models are defined by the linear constant-coefficient difference equation

$$x(n) = -\sum_{k=1}^{P} a_k x(n-k) + w(n) + \sum_{k=1}^{Q} d_k w(n-k) \qquad (9.1.1)$$

where $w(n) \sim \text{WN}(0, \sigma_w^2)$ with $\sigma_w^2 < \infty$. The power spectral density (PSD) of the output signal is

$$R(e^{j\omega}) = \sigma_w^2 \left| \frac{1 + \sum_{k=1}^{Q} d_k e^{-j\omega k}}{1 + \sum_{k=1}^{P} a_k e^{-j\omega k}} \right|^2 = \sigma_w^2 \frac{|D(e^{-j\omega})|^2}{|A(e^{-j\omega})|^2} \qquad (9.1.2)$$

which is a rational function completely specified by the parameters, $\{a_1, a_2, \ldots, a_P\}$, $\{d_1, \ldots, d_Q\}$, and σ_w^2. We stress that since these models are linear, time-invariant (LTI), the resulting process $x(n)$ is stationary, which is ensured if the corresponding systems are BIBO stable.

The essence of signal modeling and of the resulting parametric spectrum estimation is the following: Given finite-length data $\{x(n)\}_{n=0}^{N-1}$, which can be regarded as a sample sequence of the signal under consideration, we want to estimate signal model parameters $\{\hat{a}_k\}_1^P$, $\{\hat{b}_k\}_1^Q$, and $\hat{\sigma}_w^2$, to satisfy a prescribed criterion. Furthermore, if the parameter estimates are sufficiently accurate, then the following formula

$$\hat{R}(e^{j\omega}) = \hat{\sigma}_w^2 \left| \frac{1 + \sum_{k=1}^{Q} \hat{d}_k e^{-j\omega k}}{1 + \sum_{k=1}^{P} \hat{a}_k e^{-j\omega k}} \right|^2 = \hat{\sigma}_w^2 \frac{|\hat{D}(e^{-j\omega})|^2}{|\hat{A}(e^{-j\omega})|^2} \qquad (9.1.3)$$

should provide a reasonable estimate of the signal PSD. A similar argument applies to harmonic signal models and harmonic spectrum estimation in which the model parameters are the amplitudes and frequencies of complex exponentials (see Section 3.3.6).

The development of such models involves the steps shown in Figure 9.1. In this chapter, we assume that we have removed trends, seasonal variations, and other nonstationarities from the data. We further assume that unit poles have been removed from the data by using the differencing approach discussed in Box et al. (1994).

Model selection

In this step, we basically select the structure of the model (direct or lattice), and we make a preliminary decision on the orders P and Q of the model. The most important aid to model selection is the insight and understanding of the signal and the physical mechanism that generates it. Hence, in some applications (e.g., speech processing) physical considerations point to the type and order of the model; when we lack a priori information or we have insufficient knowledge of the mechanism generating the signal, we resort to data analysis methods.

In general, *to select a candidate model, we estimate the autocorrelation, partial autocorrelation, and power spectrum from the available data, and we compare them to the corresponding quantities obtained from the theoretical models* (see Table 4.1). This preliminary data analysis provides sufficient information to choose a PZ model and some initial estimate for P and Q to start a model building process. Several *order selection criteria* have been developed that penalize both model misfit and a large number of parameters. Although theoretically interesting and appealing, these criteria are of limited value when we deal with actual signals.

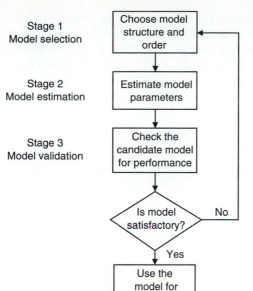

Stage 1
Model selection — Choose model structure and order

Stage 2
Model estimation — Estimate model parameters

Stage 3
Model validation — Check the candidate model for performance

Is model satisfactory? — No

Yes

Use the model for your application

FIGURE 9.1
Steps in the signal model building process.

The model structure influences (1) the complexity of the algorithm that estimates the model parameters and (2) the shape of the criterion function (quadratic or nonquadratic). Therefore, the structure (direct or lattice) is not critical to the performance of the model, and its choice is not as crucial as the choice of the order of the model.

Model estimation

In this step, also known as *model fitting*, we use the available data $\{x(n)\}_0^{N-1}$ to estimate the parameters of the selected model, using optimization of some criterion. Although there are several criteria (e.g., maximum likelihood, spectral matching) that can be used to measure the performance or quality of a PZ model, we concentrate on the least-squares (LS) error criterion. As we shall see, the estimation of all-pole (AP) models leads to linear optimization problems whereas the estimation of all-zero (AZ) and PZ models requires the solution of nonlinear optimization problems. Parameter estimation for PZ models using other criteria can be found in Kay (1988), Box et al. (1994), Porat (1994), and Ljung (1987).

Model validation

Here we investigate how well the obtained model captures the key features of the data. We then take corrective actions, if necessary, by modifying the order of the model, and repeat the process until we get an acceptable model. The goal of the model validation process is to find out whether the model

- Agrees sufficiently with the observed data
- Describes the "true" signal generation system
- Solves the problem that initiated the design process

Of course, the ultimate test is whether the model satisfies the requirements of the intended application, that is, the objective and subjective criteria that specify the performance of the model, computational complexity, cost, etc. In this discussion, we concentrate on how well the model fits the observed data in an LS error statistical sense.

The existence of any structure in the residual or prediction error signal indicates a misfit between the model and the data. Hence, a key validation technique is to check whether the residual process, which is generated by the inverse of the fitted model, is a realization of white noise. This can be checked by using, among others, the following statistical techniques (Brockwell and Davis 1991; Bendat and Piersol 1986):

Autocorrelation test. It can be shown (Kendall and Stuart 1983) that when N is suffi-ciently large, the distribution of the estimated autocorrelation coefficients $\hat{\rho}(l) = \hat{r}(l)/\hat{r}(0)$ is approximately Gaussian with zero mean and variance of $1/N$. The approximate 95 per-cent confidence limits are $\pm 1.96/\sqrt{N}$. Any estimated values of $\hat{\rho}(l)$ that fall outside these limits are "significantly" different from zero with 95 percent confidence. Values well beyond these limits indicate nonwhiteness of the residual signal.

Power spectrum density test. Given a set of data $\{x(n)\}_{n=0}^{N-1}$, the *standardized cumu-lative periodogram* is defined by

$$\tilde{I}(k) \triangleq \begin{cases} 0 & k < 1 \\ \dfrac{\displaystyle\sum_{i=1}^{k} \hat{R}(e^{j2\pi i/N})}{\displaystyle\sum_{i=1}^{K} \hat{R}(e^{j2\pi i/N})} & 1 \le k \le K \\ 1 & k > K \end{cases} \tag{9.1.4}$$

where K is the integer part of $N/2$. If the process $x(n)$ is white Gaussian noise (WGN), then the random variables $\tilde{I}(k), k = 1, 2, \ldots, K$, are independently and uniformly distributed in the interval $(0, 1)$, and the plot of $\tilde{I}(k)$ should be approximately linear with respect to k (Jenkins and Watts 1968). The hypothesis is rejected at level 0.05 if $\tilde{I}(k)$ exits the boundaries specified by

$$\tilde{I}^{(b)}(k) = \frac{k-1}{K-1} \pm 1.36(K-1)^{-1/2} \qquad 1 \le k \le K \tag{9.1.5}$$

Partial autocorrelation test. This test is similar to the autocorrelation test. Given the residual process $x(n)$, it can be shown (Kendall and Stuart 1983) that when N is sufficiently large, the partial autocorrelation sequence (PACS) values $\{k_l\}$ for lag l [defined in (4.2.44)] are approximately independent with distribution WN $(0, 1/N)$. This means that roughly 95 percent of the PACS values fall within the bounds $\pm 1.96/\sqrt{N}$. If we observe values consistently well beyond this range for N sufficiently large, it may indicate nonwhiteness of the signal.

> **EXAMPLE 9.1.1.** To apply the above tests and interpret their results, we consider a WGN sequence $x(n)$. By using the randn function, 100 samples of $x(n)$ with zero mean and unit variance were generated. These samples are shown in Figure 9.2. From these samples, the autocorrelation estimates up to lag 40, denoted by $\{\hat{r}(l)\}_{l=0}^{40}$, were computed using the autoc function, from which the the correlation coefficients $\hat{\rho}(l)$ were obtained. The first 10 coefficients are shown in Figure 9.2 along with the appropriate confidence limits. As expected, the first coefficient at lag 0 is unity while the remaining coefficients are within the limits.
>
> Next, using the psd function, a periodogram based on 100 samples was computed, from which the cumulative periodogram $\tilde{I}(k)$ was obtained and plotted as a function of the normal-ized frequency, as shown in Figure 9.2. The confidence limits are also shown. The computed cumulative periodogram is a monotonic increasing function lying within the limits.
>
> Finally, using the durbin function, PACS sequence $\{k_l\}_{l=1}^{40}$ was computed from the esti-mated correlations and plotted in Figure 9.2. Again all the values for lags $l \ge 1$ are within the confidence limits. Thus all three tests suggest that the 100-point data are almost surely from a white noise sequence.

Although the whiteness of the residuals is a good test for model fitting, it does not provide a definite answer to the problem. Some additional procedures include checking whether

- The criterion of performance decreases (fast enough) as we increase the order of the model.

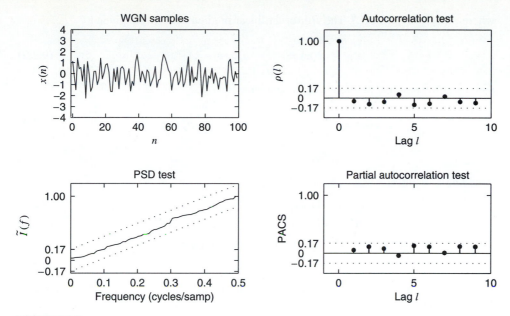

FIGURE 9.2
Validation tests on white Gaussian noise in Example 9.1.1.

- The estimate of the variance of the residual decreases as the number N of observations increases.
- Some estimated parameters that have physical meaning (e.g., reflection coefficients) assume values that make sense.
- The estimated parameters have sufficient accuracy for the intended application.

Finally, to demonstrate that the model is sufficiently accurate for the purpose for which it was designed, we can use a method known as *cross-validation*. Basically, in cross-validation we use one set of data to fit the model and another, statistically independent set of data to test it. Cross-validation is of paramount importance when we build models for control, forecasting, and pattern recognition (Ljung 1987). However, in signal processing applications, such as spectral estimation and signal compression, where the goal is to provide a good fit of the model to the analyzed data, cross-validation is not as useful.

9.2 ESTIMATION OF ALL-POLE MODELS

We next use the principle of least squares to estimate parameters of all-pole signal models assuming both white and periodic excitations. We also discuss criteria for model order selection, techniques for estimation of all-pole lattice parameters, and the relationship between all-pole estimation methods using the methods of least squares and maximum entropy. The relationship between all-pole model estimation and minimum-variance spectral estimation is explored in Section 9.5.

9.2.1 Direct Structures

Consider the $\text{AR}(P_0)$ model

$$x(n) = -\sum_{k=1}^{P_0} a_k^* x(n-k) + w(n) \tag{9.2.1}$$

where $w(n) \sim \text{WN}(0, \sigma_w^2)$. The Pth-order linear predictor of $x(n)$ is given by

$$\hat{x}(n) = -\sum_{k=1}^{P} \hat{a}_k^* x(n-k) \tag{9.2.2}$$

and the corresponding prediction error sequence is

$$e(n) = x(n) - \hat{x}(n) = x(n) + \sum_{k=1}^{P} \hat{a}_k^* x(n-k) \tag{9.2.3}$$

$$= \hat{\mathbf{a}}^H \mathbf{x}(n) \tag{9.2.4}$$

where $\hat{a}_0 = 1$ and

$$\hat{\mathbf{a}} = [1 \; \hat{a}_1 \; \cdots \; \hat{a}_P]^T \tag{9.2.5}$$

$$\mathbf{x}(n) = [x(n) \; x(n-1) \; \cdots \; x(n-P)]^T \tag{9.2.6}$$

Thus the error over the range $N_i \leq n \leq N_f$ can be expressed as a vector

$$\mathbf{e} = \bar{\mathbf{X}} \hat{\mathbf{a}} \tag{9.2.7}$$

where $\bar{\mathbf{X}}$ is the data matrix defined in (8.4.3). For the full-windowing case, the data matrix $\bar{\mathbf{X}}$ is given by

$$\bar{\mathbf{X}}^H = \begin{bmatrix} x(0) & x(1) & \cdots & x(P) & \cdots & 0 & \cdots & 0 \\ 0 & x(0) & \cdots & x(P-1) & \cdots & x(N-1) & \cdots & 0 \\ \vdots & \vdots & \ddots & \vdots & & \vdots & \vdots & \ddots & \vdots \\ 0 & 0 & \cdots & x(0) & & \cdots & x(N-P) & \cdots & x(N-1) \end{bmatrix} \tag{9.2.8}$$

while for the no-windowing case the data matrix $\bar{\mathbf{X}}$ is

$$\bar{\mathbf{X}}^H = \begin{bmatrix} x(P) & x(P+1) & \cdots & x(N-2) & x(N-1) \\ x(P-1) & x(P) & \cdots & x(N-3) & x(N-2) \\ \vdots & \vdots & \ddots & \vdots & \vdots \\ x(0) & x(1) & \cdots & x(N-P-2) & x(N-P-1) \end{bmatrix} \tag{9.2.9}$$

Notice that if $P = P_0$ and $\hat{a}_k = a_k$, the prediction error $e(n)$ is identical to the white noise excitation $w(n)$. Furthermore, if $\text{AR}(P_0)$ is minimum-phase, then $w(n)$ is the innovation process of $x(n)$ and $\hat{x}(n)$ is the MMSE prediction of $x(n)$. Thus, we can obtain a good estimate of the model parameters by minimizing some function of the prediction error.

In theory, we minimize the MSE $E\{|e(n)|^2\}$. In practice, since this is not possible, we estimate $\{a_k\}_1^P$ for a given P by minimizing the total squared error

$$\mathcal{E}_P = \sum_{n=N_i}^{N_f} |e(n)|^2 = \sum_{n=N_i}^{N_f} \left| x(n) + \sum_{k=1}^{P} \hat{a}_k^* x(n-k) \right|^2 \tag{9.2.10}$$

$$= \sum_{n=N_i}^{N_f} |\hat{\mathbf{a}}^H \mathbf{x}(n)|^2 = \hat{\mathbf{a}}^H \bar{\mathbf{X}}^H \bar{\mathbf{X}} \hat{\mathbf{a}} \tag{9.2.11}$$

over the range $N_i \leq n \leq N_f$. Hence, we can use the methods discussed in Section 8.4 for the computation of LS linear predictors. In particular, the forward linear predictor coefficient $\{\hat{a}_k\}_{k=1}^P$ and the associated LS error $\hat{\mathcal{E}}_P$ are obtained by solving the normal equations

$$(\bar{\mathbf{X}}^H \bar{\mathbf{X}}) \hat{\mathbf{a}} = \begin{bmatrix} \hat{\mathcal{E}}_P \\ \mathbf{0} \end{bmatrix} \tag{9.2.12}$$

The solution of (9.2.12) is discussed extensively in Chapter 8.

The least-squares AP(P) parameter estimates have properties similar to those of linear prediction. For example, if the process $w(n)$ is Gaussian, the least-squares no-windowing estimates are also maximum-likelihood estimates (Jenkins and Watts 1968). The variance of the excitation process can be obtained from the LS error $\hat{\mathcal{E}}_P$ by

$$\hat{\sigma}_w^2 = \frac{1}{N+P}\hat{\mathcal{E}}_P = \frac{1}{N+P}\sum_{n=0}^{N+P-1}|e(n)|^2 \quad \text{full windowing} \quad (9.2.13)$$

or

$$\hat{\sigma}_w^2 = \frac{1}{N-P}\hat{\mathcal{E}}_P = \frac{1}{N-P}\sum_{n=P}^{N-1}|e(n)|^2 \quad \text{no windowing} \quad (9.2.14)$$

for the full-windowing or no-windowing methods, respectively. Furthermore, in the full-windowing case, if the Toeplitz correlation matrix is positive definite, the obtained model is guaranteed to be minimum-phase (see Section 7.4). MATLAB functions

[ahat,e,V] = arwin(x,P) and [ahat,e,V] = arls(x,P)

are provided that compute the model parameters, the error sequence, and the modeling error using the full-windowing and no-windowing methods, respectively.

We present three examples below to illustrate the all-pole model determination and its use in PSD estimation. The first example uses real data consisting of water-level measurements of Lake Huron from 1875 to 1972. The second example also uses real data containing sunspot numbers for 1770 through 1869. These sunspot numbers have an approximate cycle of period around 10 to 12 years. The Lake Huron and sunspot data are shown in Figure 9.3. The third example generates simulated AR(4) data to estimate model parameters and through them the PSD values. In each case, the mean was computed and removed from the data prior to processing.

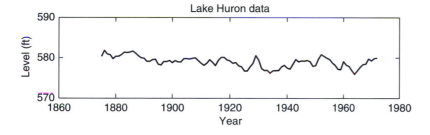

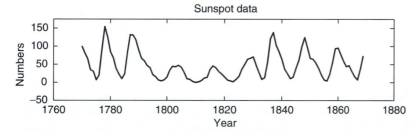

FIGURE 9.3
The Lake Huron and sunspot data used in Examples 9.2.1 and 9.2.2.

EXAMPLE 9.2.1. A careful examination of Lake Huron water-level measurement data indicates that a low-order all-pole model might be a suitable representation of the data. To test this hypothesis, first- and second-order models were considered. Using the full-windowing method, model parameters were computed:

$$\text{First-order} \qquad \hat{a}_1 = -0.791, \qquad \hat{\sigma}_w^2 = 0.5024$$

$$\text{Second-order} \qquad \hat{a}_1 = -1.002, \qquad \hat{a}_2 = 0.2832, \qquad \hat{\sigma}_w^2 = 0.4460$$

Using these model parameters, the data were filtered and the residuals were computed. Three tests for checking the whiteness of the residuals as described in Section 9.1 were performed to ascertain the validity of models. In Figure 9.4, we show the residuals, the autocorrelation test, the PSD test, and the partial correlation test for the first-order model. The partial correlation test indicates that the PACS coefficient at lag 1 is outside the confidence limits and thus the first-order model is a poor fit. In Figure 9.5 we show the same plots for the second-order model. Clearly, these tests show that the residuals are approximately white. Therefore, the AR(2) model appears to be a good match to the data.

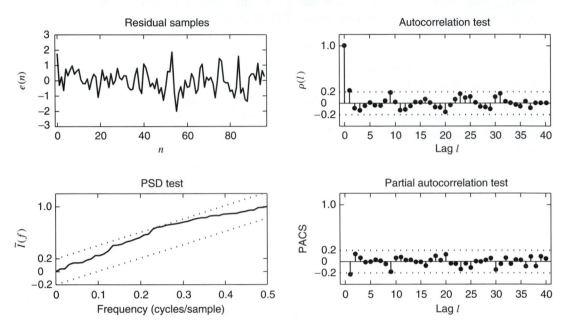

FIGURE 9.4
Validation tests on the first-order model fit to the Lake Huron water-level measurement data in Example 9.2.1.

EXAMPLE 9.2.2. Figure 9.6 shows the PACS coefficients of the sunspot numbers along with the 95 percent confidence limits. Since all PACS values beyond lag 2 fall well inside the limits, a second-order model is a possible candidate for the data. Therefore, the second-order model parameters were estimated from the data to obtain the model

$$x(n) = 1.318x(n-1) - 0.634x(n-2) + w(n) \qquad \hat{\sigma}_w^2 = 289.2$$

In Figure 9.7 we show the residuals obtained by filtering the data along with three tests for its whiteness. The plots show that the estimated model is a reasonable fit to the data. Finally, in Figure 9.8 we show the PSD estimated from the AR(2) model as well as from the periodogram. The periodogram is very noisy and is devoid of any structure. The AR(2) spectrum is smoother and distinctly shows a peak at 0.1 cycle per sampling interval. Since the sampling rate is 1 sampling interval per year, the peak corresponds to 10 years per cycle, which agrees with the observations. Thus the parametric approach to PSD estimation was appropriate.

EXAMPLE 9.2.3. We illustrate the least-squares algorithms described above, using the AR(4) process $x(n)$ introduced in Example 5.3.2. The system function of the model is given by

$$H(z) = \frac{1}{1 - 2.7607z^{-1} + 3.8106z^{-2} - 2.6535z^{-3} + 0.9238z^{-4}}$$

and the excitation is a zero-mean Gaussian white noise with unit variance. Suppose that we are given the $N = 250$ samples of $x(n)$ shown in Figure 9.9 and we wish to model the underlying process by using an all-pole model. To identify a candidate model, we compute the autocorrelation, partial autocorrelation, and periodogram, using the available data. Careful inspection of Figure 9.9 and the signal model characteristics given in Table 4.1 suggests an AR model. Since the PACS plot cuts off around $P = 5$, we choose $P = 4$ and fit an AR(4) model to the data, using both the full-windowing and no-windowing methods. Figure 9.10 shows the actual spectrum of the process, the spectra of the estimated models, and the periodogram. Clearly, the no-windowing estimate provides a better fit because it does not impose any windowing on the data. Figure 9.11 shows the residual, autocorrelation, partial autocorrelation, and periodogram for the no-windowing-based model. We see that the residuals can be assumed uncorrelated with reasonable confidence, which implies that the model captures the second-order statistics of the data.

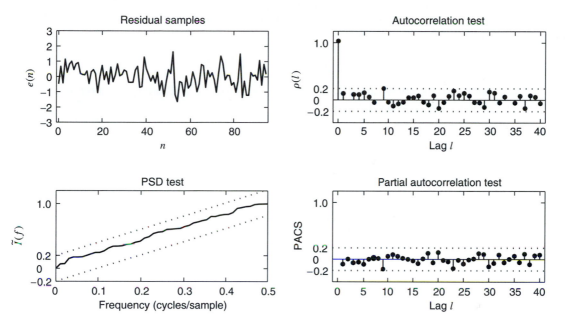

FIGURE 9.5
Validation tests on the second-order model fit to the Lake Huron water-level measurement data in Example 9.2.1.

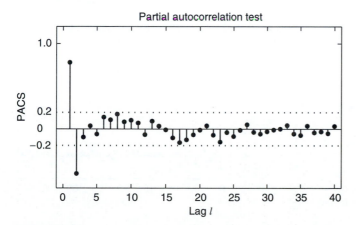

FIGURE 9.6
The PACS values of the sunspot numbers in Example 9.2.2.

Modified covariance method. The LS method described above to estimate model parameters uses the forward linear predictor and prediction error. There is also another approach that is based on the backward linear predictor. Recall that the backward linear predictor derived from the known correlations is the complex conjugate of the forward predictor (and likewise, the corresponding errors are identical). However, the LS estimators and errors based on the actual data are different because the data read in each direction are *different* from a statistical viewpoint. Hence, it is much more reasonable to consider both forward and backward predictors and to minimize the combined error

$$\mathcal{E}_P^{\text{fb}} \triangleq \sum_{n=N_{\text{i}}}^{N_{\text{f}}} [|e^f(n)|^2 + |e^b(n)|^2]$$

$$= \sum_{n=N_{\text{i}}}^{N_{\text{f}}} [|\hat{\mathbf{a}}^H \mathbf{x}(n)|^2 + |\hat{\mathbf{a}}^T \mathbf{x}^*(n)|^2] \tag{9.2.15}$$

$$= \hat{\mathbf{a}}^H \bar{\mathbf{X}}^H \bar{\mathbf{X}} \hat{\mathbf{a}} + \hat{\mathbf{a}}^H \bar{\mathbf{X}}^T \bar{\mathbf{X}}^* \hat{\mathbf{a}}$$

subject to the constraint that the first component of $\hat{\mathbf{a}}$ is 1. The minimization of $\mathcal{E}_p^{\text{fb}}$ leads to the set of normal equations

$$(\bar{\mathbf{X}}^H \bar{\mathbf{X}} + \bar{\mathbf{X}}^T \bar{\mathbf{X}}^*)\hat{\mathbf{a}} = \begin{bmatrix} \hat{\mathcal{E}}_P^{\text{fb}} \\ \mathbf{0} \end{bmatrix} \tag{9.2.16}$$

which can be solved efficiently to obtain the model parameters (see Section 8.4.2). This method of using the forward-backward predictors is called the *modified covariance method*. Not only does it have the advantage of minimizing the combined global error, but also since it uses more data in (9.2.16), it gives better estimates and lower error. A similar minimization approach, but implemented at each local stage, is used in Burg's method, which is discussed in Section 9.2.2.

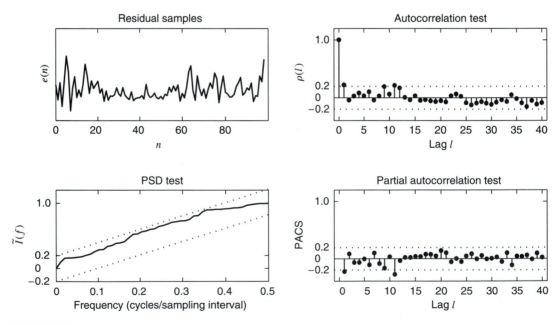

FIGURE 9.7
Validation tests on the second-order model fit to the sunspot numbers in Example 9.2.2.

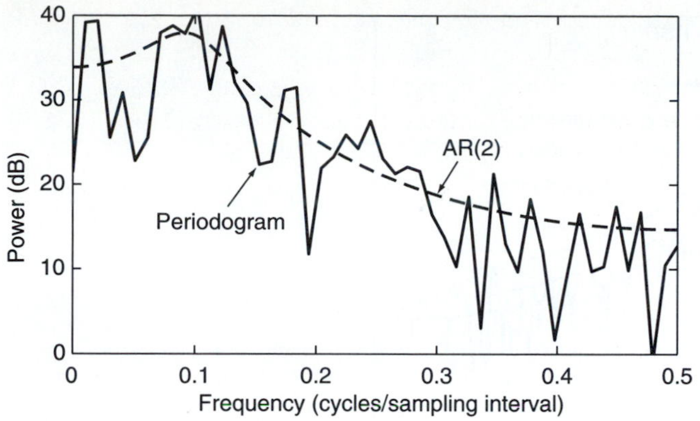

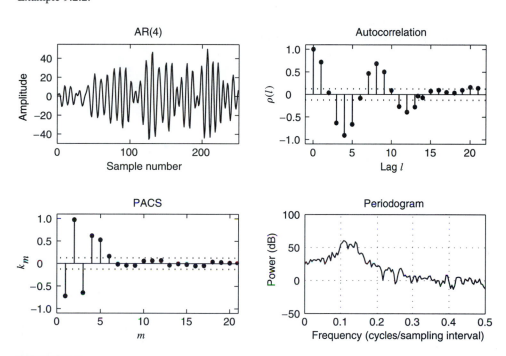

FIGURE 9.8
Comparison of the periodogram and the AR(2) spectrum in
Example 9.2.2.

FIGURE 9.9
Data segment from an AR(4) process, and the corresponding autocorrelation, partial
autocorrelation, and periodogram.

Frequency-domain interpretation. In the case of full windowing, by using Parseval's
theorem, the error energy can be written as

$$\mathcal{E} = \sum_{n=-\infty}^{\infty} |e(n)|^2 = \frac{1}{2\pi} \int_{-\pi}^{\pi} \frac{|X(e^{j\omega})|^2}{|\hat{H}(e^{j\omega})|^2} \, d\omega \tag{9.2.17}$$

where $|X(e^{j\omega})|^2$ is the spectrum of the modeled windowed signal segment and $\hat{H}(e^{j\omega})$ is
the frequency response of the estimated all-pole model [or estimated spectrum of $x(n)$].
This expression is a good approximation for the other windowing methods if $N \gg P$. Since
the integrand in (9.2.17) is positive, minimizing the error $\mathcal{E}$ is equivalent to minimizing the

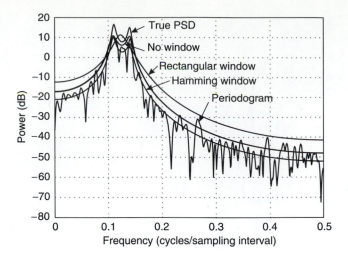

FIGURE 9.10
Periodogram, theoretical AR(4) spectrum, and AR(4) model
spectra using full windowing, Hamming windowing, and no
windowing.

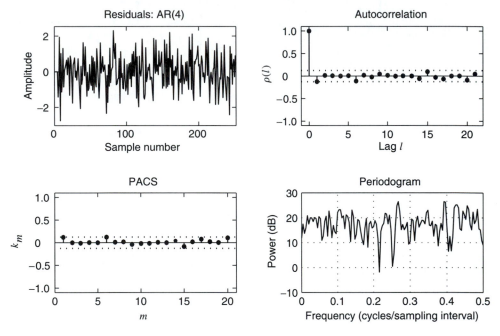

FIGURE 9.11
Residual sequence for the AR(4) data, and the corresponding autocorrelation, partial
autocorrelation, and periodogram.

integrated ratio of the energy spectrum of the modeled signal segment to its all-pole-based
spectrum.

The presence of this ratio in (9.2.17) has three additional consequences. (1) The quality
of the spectral matching is *uniform* over the whole frequency range, irrespective of the
shape of the spectrum. (2) Since regions where $|X(e^{j\omega})| > |\hat{H}(e^{j\omega})|$ contribute more to
the total error than regions where $|X(e^{j\omega})| < |\hat{H}(e^{j\omega})|$ do, the match is better near spectral
peaks than near spectral valleys. (3) The all-pole model provides a good estimate of the
envelope of the signal spectrum $|X(e^{j\omega})|^2$. These properties are apparent in Figure 9.12,

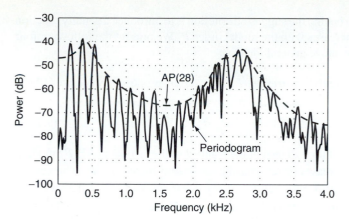

FIGURE 9.12
Illustration of the spectral envelope matching property of all-pole
models.

which shows a comparison between $20 \log |X(e^{j\omega})|$ (obtained using the periodogram) and
$20 \log |\hat{H}(e^{j\omega})|$ [obtained by an AP(28) model fitted using full windowing] for a 20-ms,
Hamming windowed, speech signal sampled at 20 kHz. Note that the slope of $|\hat{H}(e^{j\omega})|$ is
always zero at frequencies $\omega = 0$ and $\omega = \pi$, as expected. More details on these issues can
be found in Makhoul (1975b).

The error energy (9.2.17) is also related to the Itakura-Saito (IS) distortion measure,
which is given by

$$d_{IS}(R_1, R_2) \triangleq \frac{1}{2\pi} \int_{-\pi}^{\pi} [\exp V(e^{j\omega}) - V(e^{j\omega}) - 1] \, d\omega \qquad (9.2.18)$$

where $R_1(e^{j\omega})$ and $R_2(e^{j\omega})$ are two spectra, and

$$V(e^{j\omega}) \triangleq \log R_1(e^{j\omega}) - \log R_2(e^{j\omega}) \qquad (9.2.19)$$

Indeed, we can show that

$$d_{IS}(R_1, R_2) = \frac{1}{2\pi} \int_{-\pi}^{\pi} \frac{R_1(e^{j\omega})}{R_2(e^{j\omega})} \, d\omega - \log \frac{\sigma_1^2}{\sigma_2^2} - 1 \qquad (9.2.20)$$

where σ_1^2 and σ_2^2 are the variances of the innovation sequences corresponding to the factorization of spectra $R_1(e^{j\omega})$ and $R_2(e^{j\omega})$, respectively. More details can be found in Rabiner
and Juang (1993).

Order selection criteria. The order of an all-pole signal model plays an important role
in the modeling problem. It determines the number of parameters to be estimated and hence
the computational complexity of the algorithm. But more importantly, it affects the quality
of the spectrum estimates. If a much lower order is selected, then the resulting spectrum
will be smooth and will display poor resolution. If a much larger order is used, then the
spectrum may contain spurious peaks at best and a phenomenon called *spectrum splitting*
at worst, in which a single peak is split into two separate and distinct peaks (Hayes 1996).

Several criteria have been proposed over the years for model order selection; however,
in practice nothing surpasses the graphical approach outlined in Examples 9.2.1 and 9.2.2
combined with the experience of the user. Therefore, we only provide a brief summary of
some well-known criteria and refer the interested reader to Kay (1988), Porat (1994), and
Ljung (1987) for more details. The simplest approach would be to monitor the modeling
error and then select the order at which this error enters a steady state. However, for all-pole
models, the modeling error is monotonically decreasing, which makes this approach all but

impossible. The general idea behind the suggested criterion is to introduce a penalty function in the modeling error that increases with the model order P. We present the following four criteria that are based on the above general idea.

FPE criterion. The *final prediction error (FPE)* criterion, proposed by Akaike (1970), is based on the function

$$\text{FPE}(P) = \frac{N + P}{N - P} \, \hat{\sigma}_P^2 \qquad (9.2.21)$$

where $\hat{\sigma}_P^2$ is the modeling error [or variance of the residual of the estimated AP(P) model]. We note that the term $\hat{\sigma}_P^2$ decreases or remains the same with increasing P, whereas the term $(N + P)/(N - P)$ accounts for the increase in $\hat{\sigma}_P^2$ due to inaccuracies in the estimated parameters and increases with P. Clearly, FPE(P) is an inflated version of $\hat{\sigma}_P^2$. The FPE order selection criterion is to choose P that will minimize the function in (9.2.21).

AIC. The *Akaike information criterion (AIC)*, also introduced by Akaike (1974), is based on the function

$$\text{AIC}(P) = N \, \log \, \hat{\sigma}_P^2 + 2P \qquad (9.2.22)$$

It is a very general criterion that provides an estimate of the Kullback-Leibler distance (Kullback 1959) between an assumed and the true probability density function of the data. The performances of the FPE criterion and the AIC are quite similar.

MDL criterion. The *minimum description length (MDL)* criterion was proposed by Risannen (1978) and uses the function

$$\text{MDL}(P) = N \log \hat{\sigma}_P^2 + P \log N \qquad (9.2.23)$$

The first term in (9.2.23) decreases with P, but the second penalty term increases. It has been shown (Risannen 1978) that this criterion provides a consistent order estimate in that as the probability that the estimated order is equal to the true order approaches 1, the data length N tends to infinity.

CAT. This criterion is based on Parzen's *criterion autoregressive transfer (CAT)* function (Parzen 1977), which is given by

$$\text{CAT}(P) = \frac{1}{N} \sum_{k=1}^{P} \frac{N - k}{N \hat{\sigma}_k^2} - \frac{N - P}{N \hat{\sigma}_P^2} \qquad (9.2.24)$$

This criterion is asymptotically equivalent to the AIC and the MDL criteria.

Basically, all order selection criteria add to the variance of the residuals a term that grows with the order of the model and estimate the order of the model by minimizing the resulting criterion. However, when $P \ll N$, which is the case in many practical applications, the criterion does not exhibit a clear minimum that makes the order selection process difficult (see Problem 9.1).

9.2.2 Lattice Structures

We noted in Section 7.5 that a prediction error filter, and hence the AP model, can also be implemented by using a lattice structure. The Pth-order forward prediction error $e(n) = e_P^f(n)$ and the total squared error

$$\mathcal{E}_P = \sum_{n=N_i}^{N_f} |e(n)|^2 \qquad (9.2.25)$$

are nonlinear functions of the lattice parameters k_m, $0 \leq m \leq P - 1$. For example, if $P = 2$, we have

$$e_2^f(n) = x(n) + (k_0^* + k_0 k_1^*)x(n-1) + k_1^* x(n-2)$$

which shows that $e_2^f(n)$ depends on the product $k_0 k_1^*$. Thus, fitting an all-pole lattice model by minimizing $\mathcal{E}_P$ with respect to k_m, $0 \leq m \leq P - 1$, leads to a difficult nonlinear optimization problem.

We can avoid this problem by replacing the above "global" optimization with P "local" optimizations from $m = 1$ to P, one for each stage of the lattice. From the lattice equations

$$e_m^f(n) = e_{m-1}^f(n) + k_{m-1}^* e_{m-1}^b(n-1) \tag{9.2.26}$$

$$e_m^b(n) = e_{m-1}^b(n-1) + k_{m-1} e_{m-1}^f(n) \tag{9.2.27}$$

we see that the mth-order prediction errors depend on the coefficient k_{m-1} only. Furthermore, the values of $e_{m-1}^f(n)$ and $e_{m-1}^b(n)$ have been computed by using k_{m-2}, which has been determined from the optimization step at the previous stage.

Hence, to minimize the forward prediction error

$$\mathcal{E}_m^f = \sum_{n=N_i}^{N_f} |e_m^f(n)|^2 \tag{9.2.28}$$

we substitute (9.2.26) into (9.2.28) and differentiate[†] with respect to k_{m-1}^*. This leads to the following optimum value of k_{m-1}

$$k_{m-1}^{FP} = -\frac{\beta_{m-1}^{fb}}{\mathcal{E}_{m-1}^b} \tag{9.2.29}$$

where

$$\beta_m^{fb} = \sum_{n=N_i}^{N_f} [e_m^f(n)]^* e_m^b(n-1) \tag{9.2.30}$$

and

$$\mathcal{E}_m^b = \sum_{n=N_i}^{N_f} |e_m^b(n-1)|^2 \tag{9.2.31}$$

Similarly, minimization of the backward prediction error (9.2.31) gives

$$k_{m-1}^{BP} = -\frac{\beta_{m-1}^{fb}}{\mathcal{E}_{m-1}^f} \tag{9.2.32}$$

Burg (1967) suggested the estimation of k_{m-1} by minimizing

$$\mathcal{E}_m^{fb} = \sum_{n=N_i}^{N_f} \{|e_m^f(n)|^2 + |e_m^b(n)|^2\} \tag{9.2.33}$$

at each stage of the lattice.[‡] Indeed, substituting (9.2.26) and (9.2.27) in the last equation, we obtain the relationship

$$\mathcal{E}_m^{fb} = (1 + |k_{m-1}|^2)\mathcal{E}_{m-1}^f + 4\operatorname{Re}(k_{m-1}^* \beta_{m-1}^{fb}) + (1 + |k_{m-1}|^2)\mathcal{E}_{m-1}^b \tag{9.2.34}$$

[†] See Appendix B for a discussion of how to find an optimum of a real-valued function of a complex variable and its conjugate.

[‡] This approach should not be confused with the maximum entropy method introduced also by Burg and discussed later.

If we set $\partial \mathcal{E}_m^{\text{fb}} / \partial k_{m-1}^* = 0$, we obtain the following estimate of k_{m-1}:

$$k_{m-1}^{\text{B}} = -\frac{\beta_{m-1}^{\text{fb}}}{\frac{1}{2}(\mathcal{E}_{m-1}^{\text{f}} + \mathcal{E}_{m-1}^{\text{b}})} = \frac{2k_{m-1}^{\text{FM}} k_{m-1}^{\text{BM}}}{k_{m-1}^{\text{FM}} + k_{m-1}^{\text{BM}}} \tag{9.2.35}$$

We note that k_{m-1}^{B} is the harmonic mean of k_{m-1}^{FP} and k_{m-1}^{BP}. We also stress that the obtained model is *different* from the one resulting from the forward-backward least-squares (FBLS) method through global optimization [see (9.2.11)].

Itakura and Saito (1971) proposed an estimate of k_{m-1} based on replacing the theoretical ensemble averages in (7.5.24) by time averages. Their estimate is given by

$$k_{m-1}^{\text{IS}} = -\frac{\beta_{m-1}^{\text{fb}}}{\sqrt{\mathcal{E}_{m-1}^{\text{f}} \mathcal{E}_{m-1}^{\text{b}}}} = \text{sign}(k_{m-1}^{\text{FP}} \text{ or } k_{m-1}^{\text{BP}})\sqrt{k_{m-1}^{\text{FP}} k_{m-1}^{\text{BP}}} \tag{9.2.36}$$

and is also known as the geometric mean method. Since it can be shown that

$$|k_{m-1}^{\text{B}}| \leq |k_{m-1}^{\text{IS}}| \leq 1 \tag{9.2.37}$$

both estimates result in minimum-phase models (see Problem 9.2). From (9.2.36) and (9.2.37) we conclude that if $|k_{m-1}^{\text{FP}}| < 1$, then $|k_{m-1}^{\text{BP}}| > 1$ and vice versa; that is, if the FLP is minimum-phase, then the BLP is maximum-phase and vice versa. Several other estimates are discussed in Makhoul (1977) and Viswanathan and Makhoul (1975).

In all previous methods, we use *no windowing*; that is, we set $N_{\text{i}} = m$ and $N_{\text{f}} = N - 1$. If we use data windowing, all the above estimates are identical to the data windowing estimates obtained using the algorithm of Levinson-Durbin (see Problem 9.3).

The variance of the residuals can be estimated by

$$\bar{\sigma}_m^2 = \frac{1}{2} \frac{E_m^{\text{fb}}}{N - m} \tag{9.2.38}$$

which for large values of N (see Problem 9.12) can be approximated by

$$\hat{\sigma}_m^2 = \hat{\sigma}_{m-1}^2 (1 - |k_{m-1}|^2) \tag{9.2.39}$$

where

$$\hat{\sigma}_0^2 = \frac{1}{N} \sum_{n=0}^{N-1} |x(n)|^2 \tag{9.2.40}$$

The computations for the lattice estimation methods are summarized in Table 9.1, and the algorithms are implemented by the function `[k,var] = aplatest(x,P)`.

9.2.3 Maximum Entropy Method

We next show how LS all-pole modeling is related to Burg's method of maximum entropy. To this end, suppose that $x(n)$ is a normal, stationary process with zero mean. The M-dimensional complex-valued vector $\mathbf{x} \triangleq \mathbf{x}_M(n)$ obeys a normal distribution

$$p(\mathbf{x}) = \frac{1}{\pi^M \det \mathbf{R}} \exp(-\mathbf{x}^H \mathbf{R}^{-1} \mathbf{x}) \tag{9.2.41}$$

where $\mathbf{R}$ is a Toeplitz correlation matrix. By definition, its entropy is given by

$$\mathcal{H}(\mathbf{x}) \triangleq -E\{\log p(\mathbf{x})\} = M \log \pi + \log(\det \mathbf{R}) + M \tag{9.2.42}$$

because $E\{\mathbf{x}^H \mathbf{R}^{-1} \mathbf{x}\} = M$. If the process $x(n)$ is regular, that is, $|k_m| < 1$ for all m, we have

$$\det \mathbf{R} = \prod_{m=0}^{M-1} P_m \quad \text{and} \quad P_m = r(0) \prod_{j=1}^{m} (1 - |k_j|^2) \tag{9.2.43}$$

TABLE 9.1
Algorithm for estimation of AP lattice parameters.

1. **Input:** $x(n)$ for $N_i \leq n \leq N_f$

2. **Initialization**
   a. $e_0^f(n) = e_0^b(n) = x(n)$.
   b. Compute β_0^{fb}, E_0^f, and E_0^b from $x(n)$.
   c. Compute k_1^{FP} and k_1^{BP}.
   d. Compute either k_1^{IS} or k_1^B from k_1^{FP} and k_1^{BP}.
   e. Apply the first stage of the lattice to $x(n)$ using either k_1^{IS} or k_1^B to obtain $e_1^f(n)$ and $e_1^b(n)$.

3. **For** $m = 2, 3, \ldots, M$
   a. Compute β_{m-1}^{fb}, E_{m-1}^f, and E_{m-1}^b from $e_{m-1}^f(n)$ and $e_{m-1}^b(n)$.
   b. Compute k_m^{FP} and k_m^{BP}.
   c. Compute either k_m^{IS} or k_m^B from k_m^{FP} and k_m^{BP}.
   d. Apply the mth stage of the lattice to $e_{m-1}^f(n)$ and $e_{m-1}^b(n)$ using either k_m^{IS} or k_m^B to obtain $e_m^f(n)$ and $e_m^b(n)$.

4. **Output:** Either k_m^{IS} or k_m^B for $m = 1, 2, \ldots, M$ and $e_m^f(n)$ and $e_m^b(n)$.

where $P_m = P_m^f = P_m^b$ (see Section 7.4). If we substitute (9.2.43) into (9.2.42), we obtain

$$\mathcal{H}(\mathbf{x}) = M \log \pi + M + M \log r(0) + \sum_{m=1}^{M-1} (M-m) \log(1 - |k_m|^2) \qquad (9.2.44)$$

which expresses the entropy in terms of $r(0)$ and the PACS k_m, $1 \leq m \leq M \leq \infty$ [recall that any parametric model can be specified by $r(0)$ and the PACS]. Suppose now that we are given the first $P+1$ values $r(0), r(1), \ldots, r(P)$ of the autocorrelation sequence and we wish to find a model, by choosing the remaining values $r(l), l > P$, so that the entropy is maximized. From (9.2.44), we see that the entropy is maximized if we choose $k_m = 0$ for $m > P$, that is, by modeling the process $x(n)$ by an AR(P) model. In conclusion, among all regular Gaussian processes with the same first $P+1$ autocorrelation values, the AR(P) process has the maximum entropy. Any other choices for k_m, $m > P$, that satisfy the condition $|k_m| < 1$ lead to a valid extension of the autocorrelation sequence. The "extended" values $r(l), l > P$, can be obtained by using the inverse Levinson-Durbin or the inverse Schür algorithm (see Chapter 7). The relation between autoregressive modeling and the principle of maximum entropy, known as the *maximum entropy method*, was introduced by Burg (1967, 1975). We note that the above proof, given in Porat (1994), is different from the original proof provided by Burg (Burg 1975; Therrien 1992). An interesting discussion of various arguments in favor of and against the maximum entropy method can be found in Makhoul (1986).

9.2.4 Excitations with Line Spectra

When the excitation of a parametric model has a spectrum with lines at L frequencies ω_m, the spectrum of the output signal provides information about the frequency response of the model at these frequencies only. For simplicity, assume equidistant samples at frequencies $\omega_m = 2\pi m/L, 0 \leq m \leq L-1$. Given a set of values $R_x(e^{j\omega_m}) = |X(e^{j\omega_m})|$, we wish to find an AP($P$) model whose spectrum $\hat{R}_h(e^{j\omega})$ matches $R_x(\omega_m)$ at the given frequencies, by minimizing the criterion

$$\tilde{\mathcal{E}} = \frac{d_0}{L} \sum_{m=1}^{L} \frac{R_x(e^{j\omega_m})}{\hat{R}_h(e^{j\omega_m})} \qquad (9.2.45)$$

which is the discrete version of (9.2.17) and d_0 is the gain of the model (see Section 4.2). The minimization of (9.2.45) with respect to the model parameters $\{a_k\}$ results in the Yule-Walker equations

$$\sum_{k=0}^{P} a_k^* \tilde{r}(i-k) = \begin{cases} \tilde{\mathcal{E}} & i = 0 \\ 0 & 1 \le i \le P \end{cases} \tag{9.2.46}$$

where

$$\tilde{r}(l) = \frac{1}{L} \sum_{m=1}^{L} R_x(e^{j\omega_m}) e^{j\omega_m} \tag{9.2.47}$$

For continuous spectra, linear prediction uses the autocorrelation

$$r(l) = \frac{1}{2\pi} \int_{-\pi}^{\pi} R_x(e^{j\omega}) e^{j\omega} d\omega \tag{9.2.48}$$

which is related to $\tilde{r}(l)$ by

$$\tilde{r}(l) = \sum_{m=-\infty}^{\infty} r(l - Lm) \tag{9.2.49}$$

that is, $\tilde{r}(l)$ is an aliased version of $r(l)$. We have seen that linear prediction equates the autocorrelation of the AP(P) model to the autocorrelation of the modeled signal for the first $P+1$ lags. Hence, when we use linear prediction for a signal with line spectra, the autocorrelation of the all-pole model will be matched to $\tilde{r}(l) \ne r(l)$ and will always result in a model different from the original. Clearly, the correlation matching condition cannot compensate for the autocorrelation aliasing, which becomes more pronounced as L decreases. This phenomenon, which is severe for voiced sounds with high pitch, is illustrated in Problem 9.13. A method that provides better estimates, by minimizing a discrete version of the Itakura-Saito error measure, has been developed for both AP and PZ models by El-Jaroudi and Makhoul (1991, 1989).

9.3 ESTIMATION OF POLE-ZERO MODELS

The estimation of PZ(P, Q) model parameters for $Q \ne 0$ leads to a nonlinear LS optimization problem. As a result, a vast number of suboptimum methods, with reduced computational complexity, have been developed to avoid this problem. For example, some techniques estimate the AP(P) and AZ(Q) parameters separately. However, today the availability of high-speed computers has made exact least-squares the method of choice. Since the nonlinear LS optimization with respect to complex vectors and its conjugate is inherently difficult, and since this optimization does not provide any additional insight into the solution technique, we assume, in this section, that the quantities are real-valued. Furthermore, most of the real-world applications of pole-zero models almost always involve real-valued signals and systems. The extension to the complex-valued case is straightforward.

Consider the PZ(P, Q) model

$$x(n) = -\sum_{k=1}^{P} a_k x(n-k) + w(n) + \sum_{k=1}^{Q} d_k w(n-k) \tag{9.3.1}$$

where $w(n) \sim \text{WN}(0, \sigma_w^2)$. Using vector notation, we can express (9.3.1) as

$$x(n) = \mathbf{z}^T(n-1)\mathbf{c}_{\text{pz}} + w(n) \tag{9.3.2}$$

where

$$\mathbf{z}(n) \triangleq [-x(n) \cdots -x(n-P+1)\ w(n) \cdots w(n-Q+1)]^T \tag{9.3.3}$$

and

$$\mathbf{c}_{\text{pz}} = [\mathbf{a}^T\ \mathbf{d}^T] = [a_1 \cdots a_P\ d_1 \cdots d_Q]^T \tag{9.3.4}$$

9.3.1 Known Excitation

Assume for a moment that the excitation $w(n)$ is known. Then we can predict $x(n)$ from past values, using the following linear predictor

$$\hat{x}(n) = \mathbf{z}^T(n-1)\mathbf{c} \tag{9.3.5}$$

where

$$\mathbf{c} = [\hat{a}_1 \cdots \hat{a}_P \, \hat{d}_1 \cdots \hat{d}_Q]^T \tag{9.3.6}$$

are the predictor parameters. The prediction error

$$e(n) = x(n) - \hat{x}(n) = x(n) - \mathbf{z}^T(n-1)\mathbf{c} \tag{9.3.7}$$

equals $w(n)$ if $\mathbf{c} = \mathbf{c}_{\text{pz}}$. Minimization of the total squared error

$$\mathcal{E}(\mathbf{c}) \triangleq \sum_{n=N_i}^{N_f} e^2(n) \tag{9.3.8}$$

leads to the following linear system of equations

$$\hat{\mathbf{R}}_z \mathbf{c} = \hat{\mathbf{r}}_z \tag{9.3.9}$$

where

$$\hat{\mathbf{R}}_z = \sum_{n=N_i}^{N_f} \mathbf{z}(n-1)\mathbf{z}^T(n-1) \tag{9.3.10}$$

and

$$\hat{\mathbf{r}}_z = \sum_{n=N_i}^{N_f} \mathbf{z}(n-1)x(n) \tag{9.3.11}$$

Usually, we use residual windowing, which implies that $N_i = \max(P, Q)$ and $N_f = N-1$. Since the matrix $\hat{\mathbf{R}}_z$ is symmetric and positive semidefinite, we can solve (9.3.9) using LDL^H decomposition. Thus, if we know the excitation $w(n)$, the least-squares estimation of the $\text{PZ}(P, Q)$ model parameters reduces to the solution of a linear system of equations. An estimate of the input variance is given by

$$\hat{\sigma}_w^2 = \frac{1}{N - \max(P, Q)} \sum_{n=\max(P,Q)}^{N-1} e^2(n) \tag{9.3.12}$$

This method, which is implemented by the function pzls.m, is known as the *equation-error method* and can be used to identify a system from input-output data (Ljung 1987) (see Problem 9.14).

9.3.2 Unknown Excitation

In most applications, the excitation $w(n)$ is never known. However, we can obtain a good estimate of $x(n)$ by replacing $w(n)$ by $e(n)$ in (9.3.5). This makes a natural choice if the model used to obtain $e(n)$ is reasonably accurate. The prediction error is then given by

$$e(n) = x(n) - \hat{x}(n) = x(n) - \hat{\mathbf{z}}^T(n-1)\mathbf{c} \tag{9.3.13}$$

where

$$\hat{\mathbf{z}}(n) \triangleq [-x(n) \cdots -x(n-P+1) \, e(n) \cdots e(n-Q+1)]^T \tag{9.3.14}$$

If we write (9.3.13) explicitly

$$e(n) = -\sum_{k=1}^{Q} \hat{d}_k e(n-k) + x(n) + \sum_{k=1}^{P} \hat{a}_k x(n-k) \tag{9.3.15}$$

we see that the prediction error is obtained by exciting the inverse model with the signal $x(n)$. Hence, the inverse model has to be stable. To satisfy this condition, we require the estimated model to be minimum-phase.

The recursive computation of $e(n)$ by (9.3.15) makes the prediction error a nonlinear function of the model parameters. To illustrate this, consider the prediction error for a first-order model, that is, for $P = Q = 1$

$$e(n) = x(n) + \hat{a}_1 x(n-1) - \hat{d}_1 e(n-1)$$

Assuming $e(0) = 0$, we have for $n = 1, 2, 3$

$$e(1) = x(1) + \hat{a}_1 x(0)$$

$$e(2) = x(2) + \hat{a}_1 x(1) - \hat{d}_1 e(1)$$

$$= x(2) + (\hat{a}_1 - \hat{d}_1) x(1) - \hat{a}_1 \hat{d}_1 x(0)$$

$$e(3) = x(3) + \hat{a}_1 x(2) - \hat{d}_1 e(2)$$

$$= x(3) + (\hat{a}_1 - \hat{d}_1) x(2) - (\hat{a}_1 - \hat{d}_1)\hat{d}_1 x(1) + \hat{a}_1 \hat{d}_1^2 x(0)$$

which shows that $e(n)$ is a nonlinear function of the model parameters if $Q \neq 0$. Thus, the total squared error

$$\mathcal{E}(\mathbf{c}) = \sum_{n=N_i}^{N_f} e^2(n) \tag{9.3.16}$$

expressed in terms of the signal values $x(0), x(1), \ldots, x(N-1)$, is a nonquadratic function of the model parameters. Sometimes, $\mathcal{E}(\mathbf{c})$ has several local minima. The model parameters can be obtained by minimizing the total square error using nonlinear optimization techniques.

9.3.3 Nonlinear Least-Squares Optimization

We next outline such a technique that is based on the method of Gauss-Newton. More details can be found in Scales (1985); Luenberger (1984); and Gill, Murray, and Wright (1981). To this end, we expand the function $\mathcal{E}(\mathbf{c})$ as a Taylor series

$$\mathcal{E}(\mathbf{c}_0 + \Delta \mathbf{c}) = \mathcal{E}(\mathbf{c}_0) + (\Delta \mathbf{c})^T \nabla \mathcal{E}(\mathbf{c}_0) + \tfrac{1}{2}(\Delta \mathbf{c})^T [\nabla^2 \mathcal{E}(\mathbf{c}_0)](\Delta \mathbf{c}) + \cdots \tag{9.3.17}$$

where

$$\nabla \mathcal{E}(\mathbf{c}) \triangleq \begin{bmatrix} \dfrac{\partial \mathcal{E}}{\partial c_1} & \dfrac{\partial \mathcal{E}}{\partial c_2} & \cdots & \dfrac{\partial \mathcal{E}}{\partial c_{p+q}} \end{bmatrix}^T \tag{9.3.18}$$

is the vector of the first partial derivatives or *gradient vector* and $\nabla^2 \mathcal{E}(\mathbf{c})$, whose (i, j)th element is $\partial^2 \mathcal{E}/(\partial c_i \partial c_j)$, is the (symmetric) matrix of second partial derivatives (*Hessian matrix*).

The Taylor expansion of a quadratic function has only the first three terms. Indeed, for the known excitation case we have

$$\nabla \mathcal{E}(\mathbf{c}) = 2 \sum_{n=N_i}^{N_f} \mathbf{z}(n-1)e(n) = 2(\mathbf{r}_z - \hat{\mathbf{R}}_z \mathbf{c}) \tag{9.3.19}$$

and

$$\nabla^2 \mathcal{E}(\mathbf{c}) = 2 \sum_{n=N_i}^{N_f} \mathbf{z}(n-1)\mathbf{z}^T(n-1) = 2\hat{\mathbf{R}}_z \tag{9.3.20}$$

Higher-order terms are zero, and if $\mathbf{c}_0$ is the minimum, then $\nabla \mathcal{E}(\mathbf{c}_0) = \mathbf{0}$. In this case, (9.3.17) becomes

$$\mathcal{E}(\mathbf{c}_0 + \Delta \mathbf{c}) = \mathcal{E}(\mathbf{c}_0) + (\Delta \mathbf{c})^T \hat{\mathbf{R}}_z (\Delta \mathbf{c})$$

which shows that if $\hat{\mathbf{R}}_z$ is positive definite, that is, $(\Delta\mathbf{c})^T\hat{\mathbf{R}}_z(\Delta\mathbf{c}) \geq 0$, then any deviation from the minimum results in an increase in the total squared error.

This relationship holds approximately for nonquadratic functions, as long as $\mathbf{c}_0$ is close to a minimum. Thus, if we are at a point $\mathbf{c}_i$ with total squared error $\mathcal{E}(\mathbf{c}_i)$, we can move to a point $\mathbf{c}_{i+1}$ with total squared error $\mathcal{E}(\mathbf{c}_{i+1}) \leq \mathcal{E}(\mathbf{c}_i)$ by moving in the direction of $-\nabla\mathcal{E}(\mathbf{c}_i)$. This suggests the following iterative procedure

$$\mathbf{c}_{i+1} = \mathbf{c}_i - \mu_i\mathbf{G}_i\nabla\mathcal{E}(\mathbf{c}_i) \tag{9.3.21}$$

where the positive scalar μ_i controls the length of the descent and matrix $\mathbf{G}_i$ modifies the direction of the descent, as is specified by the gradient vector. Various choices for these quantities lead to various optimization algorithms. For quadratic functions, choosing $\mathbf{c}_0 = \mathbf{0}$, $\mu_0 = 1$, and $\mathbf{G}_0 = (2\hat{\mathbf{R}}_z)^{-1}$ (inverse of the Hessian matrix) gives $\mathbf{c}_1 = \hat{\mathbf{R}}_z^{-1}\hat{\mathbf{r}}_z$; that is, we find the unique minimum in one step. This provides the motivation for modifying the direction of the gradient using the inverse of the Hessian matrix, even for nonquadratic functions. This choice is justified as long as we are close to a minimum.

Using (9.3.13), we compute the Hessian as follows

$$\nabla^2\mathcal{E}(\mathbf{c}) = \nabla[\nabla\mathcal{E}(\mathbf{c})]^T = 2\sum_{n=N_i}^{N_f}\boldsymbol{\psi}(n)\boldsymbol{\psi}^T(n) + 2\sum_{n=N_i}^{N_f}[\nabla\boldsymbol{\psi}^T(n)]e(n) \tag{9.3.22}$$

where

$$\boldsymbol{\psi}(n) \triangleq \nabla e(n) = \left[\frac{\partial e(n)}{\partial\hat{a}_1} \cdots \frac{\partial e(n)}{\partial\hat{a}_P} \frac{\partial e(n)}{\partial\hat{d}_1} \cdots \frac{\partial e(n)}{\partial\hat{d}_Q}\right]^T \tag{9.3.23}$$

We usually approximate the Hessian with the first summation in (9.3.22), that is,

$$\mathbf{H} = 2\sum_{n=N_i}^{N_f}\boldsymbol{\psi}(n)\boldsymbol{\psi}^T(n) \tag{9.3.24}$$

Similarly, the gradient is given by

$$\nabla\mathcal{E}(\mathbf{c}) \triangleq \mathbf{v} = 2\sum_{n=N_i}^{N_f}\boldsymbol{\psi}(n)\,e(n) \tag{9.3.25}$$

If we set $\mathbf{G} = \mathbf{H}^{-1}$, the direction vector $\mathbf{g} = \mathbf{G}\mathbf{v} = \mathbf{H}^{-1}\mathbf{v}$ can be obtained by solving the following linear system of equations:

$$\mathbf{H}\mathbf{g} = \mathbf{v} \tag{9.3.26}$$

Clearly, the factor 2 in the definitions of $\mathbf{H}$ and $\mathbf{v}$ does not affect the solution $\mathbf{g}$, and can be dropped. Although the matrix $\mathbf{H}$ is guaranteed by (9.3.24) to be positive semidefinite, in practice it may be singular or close to singular. To avoid such problems in solving (9.3.26), we *regularize* the matrix by adding a small positive constant δ to its diagonal; that is, we approximate the Hessian by $\mathbf{H} + \delta\mathbf{I}$, where $\mathbf{I}$ is the identity matrix. This approach is known as the *Levenberg-Marquard regularization* (Dennis and Schnabel 1983; Ljung 1987).

We next compute the gradient $\boldsymbol{\psi}(n) = \nabla e(n)$, using (9.3.23) and (9.3.15). Indeed, we have

$$\frac{\partial e(n)}{\partial\hat{a}_j} = x(n-j) - \sum_{k=1}^{Q}\hat{d}_k\frac{\partial e(n-k)}{\partial\hat{a}_j} \qquad j = 1, 2, \ldots, P \tag{9.3.27}$$

and

$$\frac{\partial e(n)}{\partial\hat{d}_j} = -e(n-j) - \sum_{k=1}^{Q}\hat{d}_k\frac{\partial e(n-k)}{\partial\hat{d}_j} \qquad j = 1, 2, \ldots, Q \tag{9.3.28}$$

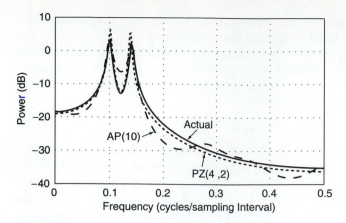

FIGURE 9.13
Illustration of the capability of a PZ(4, 2) and AP(10) model to
estimate the PSD of an ARMA(4, 2) process from a 300-sample
segment.

Thus, the components of the gradient vector are obtained by driving the all-pole filter

$$H_\psi(z) = \frac{1}{D(z)} = \frac{1}{1 + \sum_{k=1}^{Q} \hat{d}_k z^{-k}} \tag{9.3.29}$$

with the signals $x(n)$ and $-e(n)$, respectively. This filter is stable if the estimated model is
minimum-phase.

The above development leads to the following iterative algorithm, implemented in the
MATLAB function `armals.m`, which computes the parameters of a PZ(P, Q) model from
the data $x(0), x(1), \ldots, x(N-1)$ by minimizing the LS error. The LS pole-zero modeling
algorithm consists of the following steps:

1. Fit an AP($P + Q$) model to the data, using the no-windowing LS method, and compute
   the prediction error $e(n)$ (see Section 9.2).
2. Fit a PZ(P, Q) model to the data $\{x(n), e(n), 0 \le n \le N - 1\}$, using the known
   excitation method. Convert the model to minimum-phase, if necessary. Use Equations
   (9.3.9) to (9.3.11).
3. Start the iterative minimization procedure, which involves the following steps:

   a. Compute the gradient $\psi(n)$, using (9.3.27) and (9.3.28).
   b. Compute the Hessian **H** and the gradient **v**, using (9.3.24) and (9.3.25).
   c. Solve (9.3.26) to compute the search vector **g**. If necessary, use the Levenberg-
   Marquard regularization technique.
   d. For $\mu = 1, \frac{1}{2}, \ldots, \frac{1}{10}$, compute $\mathbf{c} \leftarrow \mathbf{c} + \mu\mathbf{g}$, convert the model to minimum-phase,
   if necessary, and compute the corresponding value of $\mathcal{E}(\mathbf{c})$. Choose the value of **c** that
   gives the smaller total squared error.[†]
   e. Stop if $\mathcal{E}(\mathbf{c})$ does not change significantly or if a certain number of iterations have
   been exceeded.

4. Compute the estimate of the input variance, using (9.3.15) and (9.3.12).

The application of the LS PZ(P, Q) model estimation algorithm is illustrated in Figure 9.13,
which shows the actual PSD of a PZ(4, 2) model and the estimated PSDs, using an LS PZ(4,

---

[†]This approach was suggested in Ljung (1987), problem 10S.1.

2) and an AP(10) model fitted to a 300-sample segment of the output process. We notice that, in contrast to the PZ model, the AP model does not provide a good match at the spectral zero. More details are provided in Problem 9.15.

9.4 APPLICATIONS

Pole-zero modeling has many applications in such fields as spectral estimation, speech processing, geophysics, biomedical signal processing, and general time series analysis and forecasting (Marple 1987; Kay 1988; Robinson and Treitel 1980; Box, Jenkins, and Reinsel 1994). In this section, we discuss the application of pole-zero models to spectral estimation and speech processing.

9.4.1 Spectral Estimation

After we have estimated the parameters of a PZ model, we can compute the PSD of the analyzed process by

$$
\hat{R}(e^{j\omega}) = \hat{\sigma}_w^2 \frac{\left| 1 + \sum_{k=1}^{Q} \hat{d}_k e^{j\omega k} \right|^2}{\left| 1 + \sum_{k=1}^{P} \hat{a}_k e^{j\omega k} \right|^2}
\tag{9.4.1}
$$

In practice, we mainly use AP models because (1) the all-zero PSD estimator is essentially identical to the Blackman-Tukey one (see Problem 9.16) and (2) the application of pole-zero PSD estimators is limited by computational and other practical difficulties. Also, any continuous PSD can be approximated arbitrarily well by the PSD of an AP(P) model if P is chosen large enough (Anderson 1971). However, in practice, the value of P is limited by the amount of available data (usually $P < N/3$). The statistical properties of all-pole PSD estimators are difficult to obtain; however, it has been shown that the estimator is consistent only if the analyzed process is AR(P_0) with $P_0 \leq P$. Furthermore, the quality of the estimator degrades if the process is contaminated by noise. More details about pole-zero PSD estimation can be found in Kay (1988), Porat (1994), and Percival and Walden (1993).

The performance of all-pole PSD estimators depends on the method used to estimate the model parameters, the order of the model, and the presence of noise. The effect of model mismatch is shown in Figure 9.13 and is further investigated in Problem 9.17. Order selection in all-pole PSD estimation is absolutely critical: If P is too large, the obtained PSD exhibits spurious peaks; if P is too small, the structure of the PSD is smoothed over. The increased resolution of the parametric techniques, compared to the nonparametric PSD estimation methods, is basically the result of imposing structure on the data (i.e., a model). The model makes possible the extrapolation of the ACS, which in turns leads to better resolution. However, if the adopted model is inaccurate, that is, if it does not match the data, then the "gained" resolution reflects the model and not the data! As a result, despite their popularity and their "success" with simulated signals, the application of parametric PSD estimation techniques to actual experimental data is rather limited.

Figure 9.14 shows the results of a Monte Carlo simulation of various all-pole PSD estimation techniques. We see that, except for the windowing approach that results in a significant loss of resolution, all other techniques have similar performance. However, we should mention that the forward/backward LS all-pole modeling method is considered to provide the best results (Marple 1987).

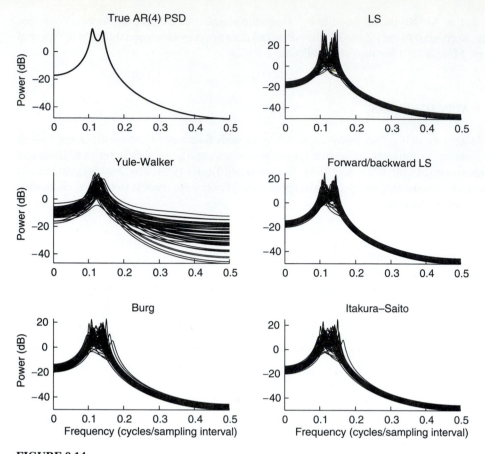

FIGURE 9.14

Monte Carlo simulation for the comparison of all-pole PSD estimation techniques, using 50
realizations of a 50-sample segment from an AR(4) process using fourth-order AP models.

In practice, it is our experience that the best way to estimate the PSD of an actual signal
is to combine parametric *prewhitening* with nonparametric PSD estimation methods. The
process is illustrated in Figure 9.15 and involves the following steps:

1. Fit an AP(P) model to the data using the forward LS, forward/backward LS, or Burg's
   method with no windowing.
2. Compute the residual (prediction error)

$$e(n) = x(n) + \sum_{k=1}^{P} a_k^* x(n-k) \qquad P \le n \le N-1 \tag{9.4.2}$$

and then compute and plot its ACS, PACS, and cumulative periodogram (see Figure 9.2)
to see if it is reasonably white. The goal is not to completely whiten the residual but to
reduce its spectral dynamic range, that is, to increase its spectral flatness to *avoid* spectral
leakage.

3. Compute the PSD $\hat{R}_e(e^{j\omega_k})$, using one of the nonparametric techniques discussed in
   Chapter 5.
4. Compute the PSD of $x(n)$ by

$$\hat{R}_x(e^{j\omega_k}) = \frac{\hat{R}_e(e^{j\omega_k})}{|A(e^{j\omega_k})|^2} \tag{9.4.3}$$

that is, by applying postcoloring to "undo" the prewhitening.

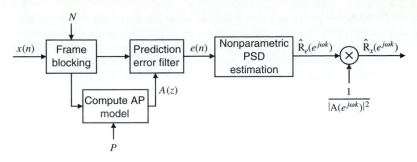

FIGURE 9.15
Block diagram of nonparametric PSD estimation using linear prediction prewhitening.

The main goal of AP modeling here is to reduce the spectral dynamic range to avoid leakage. In other words, we need a good linear predictor regardless of whether the process is true AR(P). Therefore, very accurate order selection and model fit are not critical, because all spectral structure not captured by the model is still in the residuals. Needless to say, if the periodogram of $x(n)$ has a small dynamic range, we do not need prewhitening. Another interesting application of prewhitening is for the detection of outliers in practical data (Martin and Thomson 1982).

> **EXAMPLE 9.4.1.** To illustrate the effectiveness of the above prewhitening and postcoloring method, consider the AR(4) process $x(n)$ used in Example 9.2.3. This process has a large dynamic range, and hence the nonparametric methods such as Welch's periodogram averaging method will suffer from leakage problems. Using the system function of the model
>
> $$H(z) = \frac{1}{A(z)} = \frac{1}{1 - 2.7607z^{-1} + 3.8106z^{-2} - 2.6535z^{-3} + 0.9238z^{-4}}$$
>
> and WGN $(0, 1)$ input sequence, we generated 256 samples of $x(n)$. These samples were then used to obtain the all-pole LS predictor coefficients using the arwin function. The spectrum $|A(e^{j\omega})|^{-2}$ corresponding to this estimated model is shown in Figure 9.16 as a dashed curve. The signal samples were prewhitened using the model to obtain the residuals $e(n)$. The nonparametric PSD estimate $\hat{R}_e(e^{j\omega})$ of $e(n)$ was computed by using Welch's method with $L = 64$ and 50 percent overlap. Finally, $\hat{R}_e(e^{j\omega})$ was postcolored using the spectrum $|A(e^{j\omega})|^{-2}$ to obtain $\hat{R}_x(e^{j\omega})$, which is shown in Figure 9.16 as a solid line. For comparison purposes, the Welch

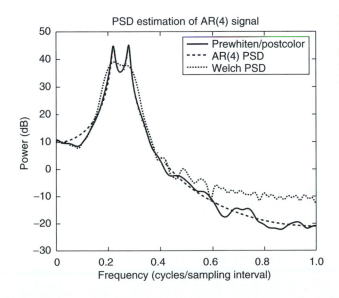

FIGURE 9.16
Spectral estimation of AR(4) process using prewhitening and postcoloring method in Example 9.4.1.

PSD estimate of $x(n)$ is also shown as a dotted line. As expected, the nonparametric estimate does not resolve the two peaks in the true spectrum and suffers from leakage at high frequencies. However, the combined nonparametric and parametric estimate resolves two peaks with ease and also follows the true spectrum quite well. Therefore, the use of the parametric method as a preprocessor is highly recommended especially in large-dynamic-range situations.

9.4.2 Speech Modeling

All-pole modeling using LS linear prediction is widely employed in speech processing applications because (1) it provides a good approximation to the vocal tract for voiced sounds and adequate approximation for unvoiced and transient sounds, (2) it results in a good separation between source (fine spectral structure) and vocal tract (spectral envelop), and (3) it is analytically tractable and leads to efficient software and hardware implementations.

Figure 9.17 shows a typical AP modeling system, also known as the *linear predictive coding (LPC)* processor, that is used in speech synthesis, coding, and recognition applications. The processor operates in a block processing mode; that is, it processes a frame of N samples and computes a vector of model parameters using the following basic steps:

1. *Preemphasis.* The digitized speech signal is filtered by the high-pass filter

$$H_1(z) = 1 - \alpha z^{-1} \qquad 0.9 \le \alpha \le 1 \tag{9.4.4}$$

to reduce the dynamic range of the spectrum, that is, to flatten the spectral envelope, and make subsequent processing less sensitive to numerical problems (Makhoul 1975a). Usually $\alpha = 0.95$, which results in about a 32 dB boost in the spectrum at $\omega = \pi$ over that at $\omega = 0$. The preemphasizer can be made adaptive by setting $\alpha = \rho(1)$, where $\rho(l)$ is the normalized autocorrelation of the frame, which corresponds to a first-order optimum prediction error filter.

2. *Frame blocking.* Here the preemphasized signal is blocked into frames of N samples with successive frames overlapping by $N_0 \simeq N/3$ samples. In speech recognition $N = 300$ with a sampling rate $F_s = 6.67$ Hz, which corresponds to 45-ms frames overlapping by 15 ms.

3. *Windowing.* Each frame is multiplied by an N-sample window (usually Hamming) to smooth the discontinuities at the beginning and the end of the frame.

4. *Autocorrelation computation.* Here the LPC processor computes the first $P + 1$ values of the autocorrelation sequence. Usually, $P = 8$ in speech recognition and $P = 12$ in speech coding applications. The value of $r(0)$ provides the energy of the frame, which is useful for speech detection.

5. *LPC analysis.* In this step the processor uses the $P + 1$ autocorrelations to compute an LPC parameter set for each speech frame. Depending on the required parameters, we

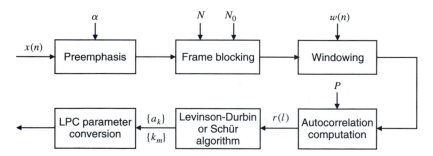

FIGURE 9.17
Block diagram of an AP modeling processor for speech coding and recognition.

can use the algorithm of Levinson-Durbin or the algorithm of Schür. The most widely used parameters are

$$a_m = a_m^{(P)} \qquad \text{LPC coefficients}$$

$$k_m \qquad \text{PACS}$$

$$g_m = \frac{1}{2} \log \frac{1 - k_m}{1 + k_m} = \tanh^{-1} k_m \qquad \text{log area ratio coefficients}$$

$$c(m) \qquad \text{cepstral coefficients}$$

$$\omega_m \qquad \text{line spectrum pairs}$$

where $1 \le m \le P$, except for the cepstrum, which is computed up to about $3P/2$. The line spectrum pair parameters, which are pole angles of the singular filters, were discussed in Section 2.5.8, and their application to speech processing is considered in Furui (1989).

The log area ratio and the line spectrum pair coefficients have good quantization properties and are used for speech coding (Rabiner and Schafer 1978; Furui 1989); the cepstral coefficients provide an excellent discriminant for speech and speaker recognition applications (Rabiner and Juang 1993; Mammone et al. 1996). AP models are extensively used for the modeling of speech sounds. However, the AP model does not provide an accurate description of the speech spectral envelope when the speech production process resembles a PZ system (Atal and Schroeder 1978). This can happen when (1) the nasal tract is coupled to the main vocal tract through the velar opening, for example, during the generation of nasals and nasalized sounds, (2) the source of excitation is not at the glottis but is in the interior of the vocal tract (Flanagan 1972), and (3) the transmission or recording channel has zeros in its response. Although a zero can be approximated with arbitrary precision by a number of poles, this approximation is usually inefficient and leads to spectral distortion and other problems. These problems can be avoided by using pole-zero modeling, as illustrated in the following example. More details about pole-zero speech modeling can be found in Atal and Schroeder (1978).

Figure 9.18(a) shows a Hamming window segment from an artificial nasal speech signal sampled at $F_s = 10$ kHz. According to acoustic theory, such sounds require both poles and zeros in the vocal tract system function. Before the fitting of the model, the data are passed though a preemphasis filter with $\alpha = 0.95$. Figure 9.18(b) shows the periodogram of the speech segment, the spectrum of an AP(16) model using data windowing, and the spectrum of a PZ(12, 6) model using the least-squares algorithm described in Section 9.3.3 (see Problem 9.18 for details). We see that the pole-zero model matches zeros ("valleys") in the periodogram of the data better than other models do.

9.5 MINIMUM-VARIANCE SPECTRUM ESTIMATION

Spectral estimation methods were discussed in Chapter 5 that are based on the discrete Fourier transform (DFT) and are data-independent; that is, the processing does not depend on the actual values of the samples to be analyzed. Window functions can be employed to cut down on sidelobe leakage, at the expense of resolution. These methods have, as a rule of thumb, an approximate resolution of $\Delta f \approx 1/N$ cycles per sampling interval. Thus, for all these methods, resolution performance is limited by the number of available data samples N. This problem is only accentuated when the data must be subdivided into segments to reduce the variance of the spectrum estimate by averaging periodograms. The effective resolution is then on the order of $1/M$, where M is the window length of the segments. For many applications the amount of data available for spectrum estimation may be limited

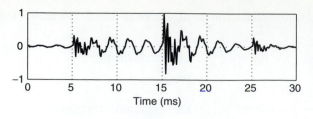

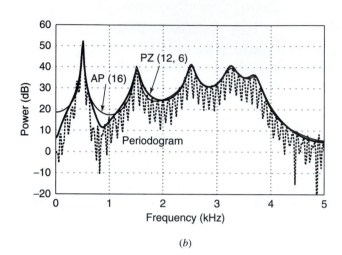

FIGURE 9.18
(*a*) Speech segment and
(*b*) periodogram, spectrum of a
data windowing-based AP(16)
model, and spectrum of a
residual windowing-based
PZ(12, 6) model.

either because the signal may only be considered stationary over limited intervals of time or may only be collected over a short finite interval.

Many times, it may be necessary to resolve spectral peaks that are spaced closer than the $1/M$ limit imposed by the amount of data available. All the DFT-based methods use a predetermined, fixed processing that is independent of the values of the data. However, there are methods, termed *data-adaptive spectrum estimation* (Lacoss 1971), that can exploit actual characteristics of the data to offer significant improvements over the data-independent, DFT-based methods, particularly in the case of limited data samples. *Minimum-variance spectral estimation* is one such technique (Capon 1969). Like the methods from Chapter 5, the minimum-variance spectral estimator is nonparametric; that is, it does not assume an underlying model for the data. However, the spectral estimator adapts itself to the characteristics of the data in order to reject as much out-of-band energy, that is, leakage, as possible. In addition, minimum-variance spectral estimation provides improved resolution—better than the $\Delta f \approx 1/N$ associated with the DFT-based methods. As a result, the minimum-variance method is commonly referred to as a *high-resolution spectral estimator*. Note that model-based data-adaptive methods, such as the LS all-pole method, also have high resolving capabilities when the model adequately represents the data.

Theory

We derive the minimum-variance spectral estimator by using a filter bank structure in which each of the filters adapts its response to the data. Recall that the goal of a power spectrum estimator is to determine the power content of a signal at a certain frequency. To this end, we would like to measure $R(e^{j2\pi f})$ at the frequency of interest only and not have our estimate influenced by energy present at other frequencies. Thus, we might interpret spectral estimation as a methodology in determining the ideal, frequency-selective filter for each frequency. Recall the filter bank interpretation of a power spectral estimator from Chapter 5. This ideal filter for f_k should pass energy within its bandwidth Δf but reject all other energy, that is,

$$|H_k(e^{j2\pi f})|^2 = \begin{cases} \Delta f & |f - f_k| \leq \dfrac{\Delta f}{2} \\ 0 & \text{otherwise} \end{cases} \qquad (9.5.1)$$

where the factor $\Delta f \sim 1/M$ accounts for the filter bandwidth.[†] Therefore, the filter does not impart a gain across the bandwidth of the filter, and the output of the filter is a measure of power in the frequency band around f_k. However, since such an ideal filter does not exist in practice, we need to design one that passes energy at the center frequency while rejecting as much out-of-band energy as possible.

A filter bank–based spectral estimator should have filters at all frequencies of interest. The filters have equal spacing in frequency, spanning the fundamental frequency range $-\frac{1}{2} \leq f < \frac{1}{2}$. Let us denote the total number of frequencies by K and the center frequency of the kth filter as

$$f_k = \frac{k-1}{K} - \frac{1}{2} \qquad (9.5.2)$$

for $k = 1, 2, \ldots, K$. The output of the kth filter is the convolution of the signal $x(n)$ with the impulse response of the filter $h_k(n)$, which can also be expressed in vector form as

$$y_k(n) = h_k(n) * x(n) = \sum_{m=0}^{M-1} h_k(m)x(n-m) = \mathbf{c}_k^H \mathbf{x}(n) \qquad (9.5.3)$$

where

$$\mathbf{c}_k = [h_k^*(0)\, h_k^*(1)\, \cdots\, h_k^*(M-1)]^T \qquad (9.5.4)$$

is the impulse response of the kth filter, and

$$\mathbf{x}(n) = [x(n)\, x(n-1)\, \cdots\, x(n-M+1)]^T \qquad (9.5.5)$$

is the input data vector. In addition, we define the *frequency vector* $\mathbf{v}(f)$ as a vector of complex exponentials at frequency f within the time-window vector from (9.5.5)

$$\mathbf{v}(f) = [1\, e^{-j2\pi f}\, \cdots\, e^{-j2\pi f(M-1)}]^T \qquad (9.5.6)$$

When the frequency vector $\mathbf{v}(f)$ is chosen as the filter weight vector in (9.5.4), then the filter will pass signals at frequency f. Note that if we have $\mathbf{c}_k = \mathbf{v}(f_k)$, then the resulting filter bank performs a DFT since $\mathbf{v}(f)$ is a column vector in the DFT matrix. Thus, all the DFT-based methods, when interpreted using a filter bank structure, use a form of $\mathbf{v}(f)$, possibly with a window, as filter weights. See Chapter 5 for the filter bank interpretation of the DFT.

The output $y_k(n)$ of the kth filter should ideally give an estimate of the power spectrum at f_k. The output power of the kth filter is

$$E\{|y_k(n)|^2\} = \mathbf{c}_k^H \mathbf{R}_x \mathbf{c}_k \qquad (9.5.7)$$

where $\mathbf{R}_x = E\{\mathbf{x}(n)\mathbf{x}^H(n)\}$ is the correlation matrix of the input data vector from (9.5.5). Since the ideal filter response from (9.5.1) cannot be realized, we instead constrain our filter $\mathbf{c}_k$ to have a response at the center frequency f_k of

$$H_k(f_k) = |\mathbf{c}_k^H \mathbf{v}(f_k)|^2 = \frac{1}{M} \qquad (9.5.8)$$

This constraint ensures that the center frequency of our bandpass filter is at the frequency f_k. To eliminate as much out-of-band energy as possible, the filter is formulated as the filter that minimizes its output power subject to the center frequency constraint in (9.5.8), that is,

$$\min \mathbf{c}_k^H \mathbf{R}_x \mathbf{c}_k \qquad \text{subject to} \qquad \mathbf{c}_k^H \mathbf{v}(f_k) = \frac{1}{\sqrt{M}} \qquad (9.5.9)$$

[†] A similar normalization was performed for all the DFT-based methods. Note that the same is not true of a sinusoidal signal that has zero bandwidth. See Example 9.5.2.

This constraint requires the filter to have a response of $1/\sqrt{M}$ to a frequency vector from (9.5.6) at the frequency of interest while rejecting (minimizing) energy from all other frequencies. The solution to this constrained optimization problem can be found via Lagrange multipliers (see Appendix B and Problem 9.22) to be

$$\mathbf{c}_k = \frac{\sqrt{M}\mathbf{R}_x^{-1}\mathbf{v}(f_k)}{\mathbf{v}^H(f_k)\mathbf{R}_x^{-1}\mathbf{v}(f_k)} \tag{9.5.10}$$

By substituting (9.5.10) into (9.5.3), we obtain the output of the kth filter. The power of this signal, from (9.5.7), is the minimum-variance spectral estimate

$$\hat{R}_M^{(mv)}(e^{j2\pi f_k}) = E\{|y_k(n)|^2\} = \frac{M}{\mathbf{v}^H(f_k)\mathbf{R}_x^{-1}\mathbf{v}(f_k)} \tag{9.5.11}$$

where the subscript M denotes the length of the data vector used to compute the spectral estimate. Note that in order to compute the minimum-variance spectral estimate, we need to find the inverse of the correlation matrix, which is a Toeplitz matrix since $x(n)$ is stationary. Efficient techniques for computing the inverse of a Toeplitz matrix were discussed in Chapter 7.

Implementation

A spectral estimator attempts to determine the power of a random process as a function of frequency based on a finite set of observations. Since the minimum-variance estimate of the spectrum involves the correlation matrix of the input data vector, which is unknown in practice, the correlation matrix must be estimated from the data. An estimate of the $M \times M$ correlation matrix, known as the *sample correlation matrix*, is given by[†]

$$\hat{\mathbf{R}}_x = \frac{1}{N - M + 1}\mathbf{X}^H\mathbf{X} \tag{9.5.12}$$

where
$$\mathbf{X}^H = [\mathbf{x}(M) \quad \mathbf{x}(M+1) \quad \cdots \quad \mathbf{x}(N)]$$

$$= \begin{bmatrix} x(M) & x(M+1) & \cdots & x(N) \\ x(M-1) & x(M) & \cdots & x(N-1) \\ \vdots & \vdots & \ddots & \vdots \\ x(1) & x(2) & \cdots & x(N-M+1) \end{bmatrix} \tag{9.5.13}$$

is the data matrix formed from $x(n)$ for $0 \le n \le N - 1$. Any of the other methods of forming a data matrix discussed in Chapter 8 can also be employed. Note that the data matrix in (9.5.13) does not produce a Toeplitz matrix $\hat{\mathbf{R}}_x$ in (9.5.12), though other methods from Chapter 8 will produce a Toeplitz sample correlation matrix.

An estimate of the spectrum based on the sample correlation matrix is found by substituting $\hat{\mathbf{R}}_x$ for the true correlation matrix $\mathbf{R}_x$ in (9.5.11). Note that, in practice, the sample correlation matrix is not actually computed. The form of the sample correlation matrix resembles the product of the data matrices in the least-squares (LS) problem that is addressed in Chapter 8. Therefore, we might compute the upper triangular factor of the data matrix $\mathcal{R}_x$ by using one of the techniques discussed in Chapter 8, such as a QR factorization. Indeed, if we compute the QR factorization made up of the orthonormal matrix $\mathbf{Q}_x$ and the upper triangular factor $\mathcal{R}_x$

$$\mathbf{X} = \mathbf{Q}_x\mathcal{R}_x \tag{9.5.14}$$

then the minimum-variance spectrum estimator based on the sample correlation matrix is

$$\hat{R}_M^{(scmv)}(e^{j2\pi f_k}) = \frac{M}{|\mathbf{v}^H(f_k)\mathcal{R}_x^{-H}|^2} \tag{9.5.15}$$

[†]We have normalized by $N - M + 1$ the number of realizations of the time-window vector $\mathbf{x}(n)$ in the data matrix **X**. This normalization is necessary so that the output of the filter bank corresponds to an estimate of power.

Note that the conjugation of the upper triangular matrix comes about through the formulation of the data matrix in (9.5.13).

We have not addressed the issue of choosing the filter length M. Ideally, M is chosen to be as large as possible in order to maximize the rejection of out-of-band energy. However, from a practical point of view, we must place a limit on the filter length. As the filter length increases, the size of the data matrix grows, which increases the amount of computation necessary. In addition, since we are inherently estimating the correlation matrix, reducing the variance of this estimator requires averaging over a set of realizations of the input data vector $\mathbf{x}(n)$. Thus, for a fixed data record size of N, we must balance the length of the time window M against the number of realizations of the input data vector $N - M + 1$.

As we will demonstrate in the following example, the minimum-variance spectrum estimator provides a means of achieving high resolution, certainly better than the $\Delta f \sim 1/M$ limit of the DFT-based methods. High resolving capability essentially means that the minimum-variance spectrum estimator can better distinguish complex exponential signals closely spaced in frequency. This topic is explored further in Section 9.6. However, high resolution does not come without a cost. In practice, the spectrum cannot be estimated over a continuous frequency interval and must be computed at a finite set of discrete frequency points. Since the minimum-variance estimator is based on $\mathbf{R}_x^{-1}$, it is very sensitive to the exact frequency points at which the spectrum is estimated. Therefore, the minimum-variance spectrum needs to be computed at a very fine frequency spacing in order to accurately measure the power of such a complex exponential. In some applications where computational cost is a concern, the DFT-based methods are probably preferred, as long as they provide the necessary resolution and sidelobe leakage is properly controlled.

EXAMPLE 9.5.1. In this example, we explore the resolving capability of the minimum-variance spectrum estimator and compare its performance to that of a DFT-based method (Bartlett) and the all-pole method. Two closely spaced complex exponentials, both with an amplitude of $\sqrt{10}$, at discrete-time frequencies of $f = 0.1$ and $f = 0.12$ are contained in noise with unit power $\sigma_w^2 = 1$. We apply the spectrum estimators with time-window lengths (or order) $M = 16, 32, 64$, and 128 to signals consisting of 500 time samples. The estimated spectra were then averaged over 100 realizations. The resulting average spectrum estimates are shown in Figure 9.19. Note that the frequency spacing of the two complex exponentials is $\Delta f = 0.02$, suggesting a time-window length of at least $M = 50$ to resolve them with a DFT-based method. The minimum-variance spectrum estimator, however, is able to resolve them at the $M = 32$ window length, for which they are clearly not distinguishable using the DFT-based method. On the other hand, the all-pole spectrum estimate is able to resolve the two complex exponentials even for as low an order as $M = 16$, for which the minimum-variance spectrum was not successful. In general, the superior resolving capability of the all-pole model over the minimum-variance spectrum estimator is due to an averaging effect that comes about through the nonparametric nature of the minimum-variance method. This subject is explored following the next example. Note that the estimated noise level is most accurately measured by the minimum-variance method in all cases. Recall that the signal amplitude was $\sqrt{10}$, yet the estimated power at the frequencies of the complex exponentials increases as the window length M increases. In the filter bank interpretation of the minimum-variance spectrum estimator, the normalization assumed a constant signal power level across the bandwidth of the frequency-selective filter. However, the complex exponential is actually an impulse in frequency and has zero bandwidth. Therefore, the estimated power will grow with the length of the time window used for the spectrum estimator as a result of this bandwidth normalization. The gain imparted on a complex exponential signal is explored in Example 9.5.2.

EXAMPLE 9.5.2. Consider the complex exponential signal with frequency f_1 contained in noise

$$x(n) = \alpha_1 e^{j2\pi f_1 n} + w(n)$$

where $\alpha_1 = |\alpha_1| e^{j\psi_1}$ is a complex number with constant amplitude $|\alpha_1|$ and random phase ψ_1 with uniform distribution over $[0, 2\pi]$. The correlation matrix of $x(n)$ is

$$\mathbf{R}_x = |\alpha_1|^2 \mathbf{v}(f_1)\mathbf{v}^H(f_1) + \sigma_w^2 \mathbf{I}$$

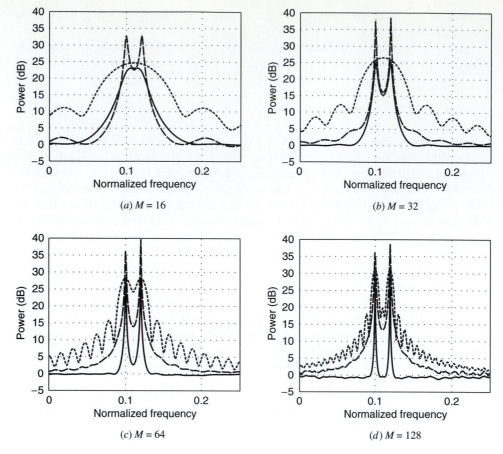

(a) M = 16

(b) M = 32

(c) M = 64

(d) M = 128

FIGURE 9.19

Comparison of the minimum-variance (solid line), all-pole (large dashed line), and Fourier-based (small dashed line) spectrum estimators for different time window lengths M.

Using the matrix inversion lemma from Appendix B, we can write the inverse of the correlation matrix as

$$\mathbf{R}_x^{-1} = \frac{1}{\sigma_w^2} \mathbf{I} - \frac{|\alpha_1|^2 \mathbf{v}(f_1)\mathbf{v}^H(f_1)}{\sigma_w^2[\sigma_w^2 + |\alpha_1|^2 \mathbf{v}(f_1)\mathbf{v}^H(f_1)]} = \frac{1}{\sigma_w^2}\left[\mathbf{I} - \frac{|\alpha_1|^2}{\sigma_w^2 + M|\alpha_1|^2}\mathbf{v}(f_1)\mathbf{v}^H(f_1)\right]$$

Substituting this expression for the inverse of the correlation matrix into (9.5.11) for the minimum-variance spectrum estimate, we have

$$\hat{R}_M^{(\text{mv})}(e^{j2\pi f_1}) = \frac{M}{\mathbf{v}^H(f_1)\,\mathbf{R}_x^{-1}\,\mathbf{v}(f_1)} = \frac{\sigma_w^2}{1 - \dfrac{|\alpha_1|^2/M}{\sigma_w^2 + M|\alpha_1|^2}|\mathbf{v}^H(f_1)\,\mathbf{v}(f_1)|^2}$$

Recall that the norm of the frequency vector $\mathbf{v}(f)$ from (9.5.6) is $\mathbf{v}^H(f_1)\,\mathbf{v}(f_1) = M$. Therefore, the minimum-variance power spectrum estimate at $f = f_1$ is

$$\hat{R}_M^{(\text{mv})}(e^{j2\pi f_1}) = \sigma_w^2 + M|\alpha_1|^2$$

that is, the sum of the noise power and the signal power times the time-window length. This gain of M on the signal power comes about through the normalization we imposed on our filter in (9.5.8). This normalization assumed the signal had equal amplitude across the passband of the filter. A complex exponential, on the other hand, has no bandwidth and thus this normalization imparts a gain of M on the signal. Therefore, if an estimate of the amplitude of a complex exponential is desired, this gain must be accounted for. Last, let us examine the behavior of the minimum variance spectrum estimator at the other frequencies that contain only noise. In the case of $M \gg 1$, then $\mathbf{v}^H(f)\,\mathbf{v}(f_1) \approx 0$ and

$$\hat{R}_M^{(\text{mv})}(e^{j2\pi f}) \approx \sigma_w^2$$

Relationship between the minimum-variance and all-pole spectrum estimation methods

The minimum-variance spectrum estimator has an interesting relation to the all-pole spectrum estimator discussed in Section 9.4. Recall from (9.5.11) that the minimum-variance spectrum estimate is a function of $\mathbf{R}_x^{-1}$. The inverse of a Toeplitz correlation matrix was studied in Chapter 7 and from (7.7.8) can be written as an LDL^H decomposition

$$\mathbf{R}_x^{-1} = \mathbf{A}^H \bar{\mathbf{D}}^{-1} \mathbf{A} \tag{9.5.16}$$

where the upper triangular matrix $\mathbf{A}$ from (7.7.9) is given by

$$\mathbf{A} = \begin{bmatrix} 1 & a_1^{(M-1)*} & a_2^{(M-1)*} & \cdots & a_{M-1}^{(M-1)*} \\ 0 & 1 & a_1^{(M-2)*} & \cdots & a_{M-2}^{(M-2)*} \\ \vdots & \vdots & \vdots & \ddots & \vdots \\ 0 & 0 & 0 & \cdots & a_1^{(1)*} \\ 0 & 0 & 0 & \cdots & 1 \end{bmatrix} \tag{9.5.17}$$

and the diagonal matrix $\bar{\mathbf{D}}$ is

$$\bar{\mathbf{D}} = \text{diag}\{P_M, P_{M-1}, \ldots, P_1\} \tag{9.5.18}$$

Recall from Chapter 7 that the columns of the lower triangular factor $\mathbf{L} = \mathbf{A}^H$ are the coefficients of the forward linear predictors of orders $m = 1, 2, \ldots, M-1$ for the signal $x(n)$ with correlation matrix $\mathbf{R}_x$. P_m is the residual output power resulting from the application of this mth-order forward linear predictor to the signal $x(n)$. In turn, the forward linear predictor coefficients form the mth-order all-pole model. The model orders are found in descending order as the column index increases. Let us denote the column vector of coefficients for the mth-order all-pole model as

$$\mathbf{a}_m = [1 \; a_m^{(1)} \; a_m^{(2)} \; \cdots \; a_m^{(m)}]^T \tag{9.5.19}$$

We can write the estimate of the spectrum derived from an mth-order all-pole model in vector notation as

$$\hat{R}_m^{(\text{ap})}(e^{j2\pi f}) = \frac{P_m}{|\mathbf{v}_m^H(f)\mathbf{a}_m|^2} \tag{9.5.20}$$

where $\mathbf{v}_m(f)$ is the frequency vector from (9.5.6) of order $M = m$. Then we can substitute (9.5.16) into the minimum-variance spectrum estimator from (9.5.11) to obtain

$$\hat{R}_M^{(\text{mv})}(e^{j2\pi f}) = \frac{M}{\mathbf{v}_M^H(f)\mathbf{R}_x^{-1}\mathbf{v}_M(f)} = \frac{M}{\mathbf{v}_M^H(f)\mathbf{A}^H\bar{\mathbf{D}}^{-1}\mathbf{A}\mathbf{v}_M(f)} \tag{9.5.21}$$

Therefore, we can write the following relationship between the reciprocals of the minimum-variance and all-pole model spectrum estimators

$$\frac{1}{\hat{R}_M^{(\text{mv})}(e^{j2\pi f})} = \sum_{m=1}^{M} \frac{|\mathbf{v}_m^H(f)\mathbf{a}_m|^2}{M P_m} = \frac{1}{M}\sum_{m=1}^{M} \frac{1}{\hat{R}_m^{(\text{ap})}(e^{j2\pi f})} \tag{9.5.22}$$

where the subscripts denote the order of the respective spectrum estimators. Thus, the minimum-variance spectrum estimator for a filter of length M is formed by averaging spectrum estimates from all-pole models of orders 1 through M. Note that the resolving capabilities of the all-pole model improve with increasing model order. As a result, the resolution of the minimum-variance spectrum estimator must be worse than that of the

Mth-order all-pole model as we observed in Example 9.5.1. However, on the other hand, this averaging of all-pole model spectra indicates a lower variance for the minimum-variance spectrum estimator.

9.6 HARMONIC MODELS AND FREQUENCY ESTIMATION TECHNIQUES

The pole-zero models we have discussed so far assume a linear time-invariant system that is excited by white noise. However, in many applications, the signals of interest are complex exponentials contained in white noise for which a *sinusoidal* or *harmonic model* is more appropriate. Signals consisting of complex exponentials are found as formant frequencies in speech processing, moving targets in radar, and spatially propagating signals in array processing.[†] For real signals, complex exponentials make up a complex conjugate pair (sinusoids), whereas for complex signals, they may occur at a single frequency.

For complex exponentials found in noise, the parameters of interest are the frequencies of the signals. Therefore, our goal is to estimate these frequencies from the data. One might consider estimating the power spectrum by using the nonparametric methods discussed in Chapter 5 or the minimum-variance spectral estimate from Section 9.5. The frequency estimates of the complex exponentials are then the frequencies at which peaks occur in the spectrum. Certainly, the use of these nonparametric methods seems appropriate for complex exponential signals since they make no assumptions about the underlying process. We might also consider making use of an all-pole model for the purposes of spectrum estimation as discussed in Section 9.4.1, also known as the maximum entropy method (MEM) spectral estimation technique. Even though some of these methods can achieve very fine resolution, none of these methods accounts for the underlying model of complex exponentials in noise. As in all modeling problems, the use of the appropriate model is desirable from an intuitive point of view and advantageous in terms of performance. We begin by describing the harmonic signal model, deriving the model in a vector notation, and looking at the eigendecomposition of the correlation matrix of complex exponentials in noise. Then we describe frequency estimation methods based on the harmonic model: the Pisarenko harmonic decomposition, and the MUSIC, minimum-norm, and ESPRIT algorithms.

These methods have the ability to resolve complex exponentials closely spaced in frequency and has led to the name *superresolution* commonly being associated with them. However, a word of caution on the use of these harmonic models. The high level of performance in terms of resolution is achieved by assuming an underlying model of the data. As with all other parametric methods, the performance of these techniques depends upon how closely this mathematical model matches the actual physical process that produced the signals. Deviations from this assumption result in model mismatch and will produce frequency estimates for a signal that may not have been produced by complex exponentials. In this case, the frequency estimates have little meaning.

9.6.1 Harmonic Model

Consider the signal model that consists of P complex exponentials in noise

$$x(n) = \sum_{p=1}^{P} \alpha_p e^{j2\pi n f_p} + w(n) \tag{9.6.1}$$

[†] In array processing, a spatially propagating wave produces a complex exponential signal as measured across uniformly spaced sensors in an array. The frequency of the complex exponential is determined by the angle of arrival of the impinging, spatially propagating signal. Thus, in array processing the frequency estimation problem is known as *angle-of-arrival (AOA)* or *direction-of-arrival (DOA)* estimation. This topic is discussed in Section 11.7.

The normalized, discrete-time frequency of the pth component is

$$f_p = \frac{\omega_p}{2\pi} = \frac{F_p}{F_s} \tag{9.6.2}$$

where ω_p is the discrete-time frequency in radians, F_p is the actual frequency of the pth complex exponential, and F_s is the sampling frequency. The complex exponentials may occur either individually or in complex conjugate pairs, as in the case of real signals. In general, we want to estimate the frequencies and possibly also the amplitudes of these signals. Note that the phase of each complex exponential is contained in the amplitude, that is,

$$\alpha_p = |\alpha_p| e^{j\psi_p} \tag{9.6.3}$$

where the phases ψ_p are uncorrelated random variables uniformly distributed over $[0, 2\pi]$. The magnitude $|\alpha_p|$ and the frequency f_p are deterministic quantities. If we consider the spectrum of a harmonic process, we note that it consists of a set of impulses with a constant background level at the power of the white noise $\sigma_w^2 = E\{|w(n)|^2\}$. As a result, the power spectrum of complex exponentials is commonly referred to as a *line spectrum*, as illustrated in Figure 9.20.

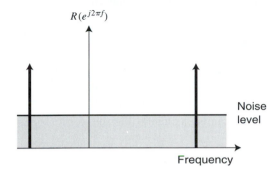

$R(e^{j2\pi f})$

Noise level

Frequency

FIGURE 9.20
The spectrum of complex exponentials in noise.

Since we will make use of matrix methods based on a certain time window of length M, it is useful to characterize the signal model in the form of a vector over this time window consisting of the sample delays of the signal. Consider the signal $x(n)$ from (9.6.1) at its current and future $M - 1$ values. This time window can be written as

$$\mathbf{x}(n) = [x(n) \; x(n+1) \; \cdots \; x(n+M-1)]^T \tag{9.6.4}$$

We can then write the signal model consisting of complex exponentials in noise from (9.6.1) for a length-M time-window vector as

$$\mathbf{x}(n) = \sum_{p=1}^{P} \alpha_p \mathbf{v}(f_p) e^{j2\pi n f_p} + \mathbf{w}(n) = \mathbf{s}(n) + \mathbf{w}(n) \tag{9.6.5}$$

where $\mathbf{w}(n) = [w(n) \; w(n+1) \; \cdots \; w(n+M-1)]^T$ is the time-window vector of white noise and

$$\mathbf{v}(f) = [1 \; e^{j2\pi f} \; \cdots \; e^{j2\pi(M-1)f}]^T \tag{9.6.6}$$

is the time-window frequency vector. Note that $\mathbf{v}(f)$ is simply a length-M DFT vector at frequency f. We differentiate here between the signal $\mathbf{s}(n)$, consisting of the sum of complex exponentials, and the noise component $\mathbf{w}(n)$, respectively.

Consider the time-window vector model consisting of a sum of complex exponentials in noise from (9.6.5). The autocorrelation matrix of this model can be written as the sum of signal and noise autocorrelation matrices

$$\mathbf{R}_x = E\{\mathbf{x}(n)\mathbf{x}^H(n)\} = \mathbf{R}_s + \mathbf{R}_w$$
$$= \sum_{p=1}^{P} |\alpha_p|^2 \mathbf{v}(f_p)\mathbf{v}^H(f_p) + \sigma_w^2 \mathbf{I} = \mathbf{V}\mathbf{A}\mathbf{V}^H + \sigma_w^2 \mathbf{I} \tag{9.6.7}$$

where
$$\mathbf{V} = [\mathbf{v}(f_1) \, \mathbf{v}(f_2) \, \cdots \, \mathbf{v}(f_P)] \tag{9.6.8}$$

is an $M \times P$ matrix whose columns are the time-window frequency vectors from (9.6.6) at frequencies f_p of the complex exponentials and

$$\mathbf{A} = \begin{bmatrix} |\alpha_1|^2 & 0 & \cdots & 0 \\ 0 & |\alpha_2|^2 & \ddots & \vdots \\ \vdots & \ddots & \ddots & 0 \\ 0 & \cdots & 0 & |\alpha_P|^2 \end{bmatrix} \tag{9.6.9}$$

is a diagonal matrix of the powers of each of the respective complex exponentials. The autocorrelation matrix of the white noise is

$$\mathbf{R}_w = \sigma_w^2 \mathbf{I} \tag{9.6.10}$$

which is full rank, as opposed to $\mathbf{R}_s$ which is rank-deficient for $P < M$. In general, we will always choose the length of our time window M to be greater than the number of complex exponentials P.

The autocorrelation matrix can also be written in terms of its eigendecomposition

$$\mathbf{R}_x = \sum_{m=1}^{M} \lambda_m \mathbf{q}_m \mathbf{q}_m^H = \mathbf{Q}\mathbf{\Lambda}\mathbf{Q}^H \tag{9.6.11}$$

where λ_m are the eigenvalues in descending order, that is, $\lambda_1 \geq \lambda_2 \geq \cdots \geq \lambda_M$, and $\mathbf{q}_m$ are their corresponding eigenvectors. Here $\mathbf{\Lambda}$ is a diagonal matrix made up of the eigenvalues found in descending order on the diagonal, while the columns of $\mathbf{Q}$ are the corresponding eigenvectors. The eigenvalues due to the signals can be written as the sum of the signal power in the time window and the noise:

$$\lambda_m = M|\alpha_m|^2 + \sigma_w^2 \quad \text{for} \quad m \leq P \tag{9.6.12}$$

The remaining eigenvalues are due to the noise only, that is,

$$\lambda_m = \sigma_w^2 \quad \text{for} \quad m > P \tag{9.6.13}$$

Therefore, the P largest eigenvalues correspond to the signal made up of complex exponentials and the remaining eigenvalues have equal value and correspond to the noise. Thus, we can partition the correlation matrix into portions due to the signal and noise eigenvectors

$$\mathbf{R}_x = \sum_{m=1}^{P} (M|\alpha_m|^2 + \sigma_w^2)\mathbf{q}_m \mathbf{q}_m^H + \sum_{m=P+1}^{M} \sigma_w^2 \mathbf{q}_m \mathbf{q}_m^H \tag{9.6.14}$$

$$= \mathbf{Q}_s \mathbf{\Lambda}_s \mathbf{Q}_s^H + \sigma_w^2 \mathbf{Q}_w \mathbf{Q}_w^H$$

where
$$\mathbf{Q}_s = [\mathbf{q}_1 \, \mathbf{q}_2 \, \cdots \, \mathbf{q}_P] \quad \mathbf{Q}_w = [\mathbf{q}_{P+1} \, \cdots \, \mathbf{q}_M] \tag{9.6.15}$$

are matrices whose columns consist of the signal and noise eigenvectors, respectively. The matrix $\mathbf{\Lambda}_s$ is a $P \times P$ diagonal matrix containing the signal eigenvalues from (9.6.12). Thus, the M-dimensional subspace that contains the observations of the time-window signal vector from (9.6.5) can be split into two subspaces spanned by the signal and noise eigenvectors, respectively. These two subspaces, known as the *signal subspace* and the *noise subspace*, are orthogonal to each other since the correlation matrix is Hermitian symmetric.[†] All the subspace methods discussed later in this section rely on the partitioning of the vector space into signal and noise subspaces. Recall from Chapter 8 in (8.2.29) that the projection matrix

---

[†]The eigenvectors of a Hermitian symmetric matrix are orthogonal.

from an M-dimensional space onto an L-dimensional subspace ($L < M$) spanned by a set of vectors $\mathbf{Z} = [\mathbf{z}_1 \ \mathbf{z}_2 \ \cdots \ \mathbf{z}_L]$ is

$$\mathbf{P} = \mathbf{Z}(\mathbf{Z}^H\mathbf{Z})^{-1}\mathbf{Z}^H \tag{9.6.16}$$

Therefore, we can write the matrices that project an arbitrary vector onto the signal and noise subspaces as

$$\mathbf{P}_s = \mathbf{Q}_s\mathbf{Q}_s^H \qquad \mathbf{P}_w = \mathbf{Q}_w\mathbf{Q}_w^H \tag{9.6.17}$$

since the eigenvectors of the correlation matrix are orthonormal ($\mathbf{Q}_s^H\mathbf{Q}_s = \mathbf{I}$ and $\mathbf{Q}_w^H\mathbf{Q}_w = \mathbf{I}$). Since the two subspaces are orthogonal

$$\mathbf{P}_w\mathbf{Q}_s = \mathbf{0} \qquad \mathbf{P}_s\mathbf{Q}_w = \mathbf{0} \tag{9.6.18}$$

then all the time-window frequency vectors from (9.6.5) must lie completely in the signal subspace, that is,

$$\mathbf{P}_s\mathbf{v}(f_p) = \mathbf{v}(f_p) \qquad \mathbf{P}_w\mathbf{v}(f_p) = \mathbf{0} \tag{9.6.19}$$

These concepts are central to the subspace-based frequency estimation methods discussed in Sections 9.6.2 through 9.6.5.

Note that in our analysis, we are considering the theoretical or true correlation matrix $\mathbf{R}_x$. In practice, the correlation matrix is not known and must be estimated from the measured data samples. If we have a time-window signal vector from (9.6.4), then we can form the data matrix by stacking the rows with measurements of the time-window data vector at a time n

$$\mathbf{X} = \begin{bmatrix} \mathbf{x}^T(0) \\ \mathbf{x}^T(1) \\ \vdots \\ \mathbf{x}^T(n) \\ \vdots \\ \mathbf{x}^T(N-2) \\ \mathbf{x}^T(N-1) \end{bmatrix} = \begin{bmatrix} x(0) & x(1) & \cdots & x(M-1) \\ x(1) & x(2) & \cdots & x(M) \\ \vdots & \vdots & \vdots & \vdots \\ x(n) & x(n+1) & \cdots & x(n+M-1) \\ \vdots & \vdots & \vdots & \vdots \\ x(N-2) & x(N-1) & \cdots & x(N+M-3) \\ x(N-1) & x(N) & \cdots & x(N+M-2) \end{bmatrix} \tag{9.6.20}$$

which has dimensions of $N \times M$, where N is the number of data records or snapshots and M is the time-window length. From this matrix, we can form an estimate of the correlation matrix, referred to as the sample correlation matrix

$$\hat{\mathbf{R}}_x = \frac{1}{N}\mathbf{X}^H\mathbf{X} \tag{9.6.21}$$

In the case of an estimated sample correlation matrix, the noise eigenvalues are no longer equal because of the finite number of samples used to compute $\hat{\mathbf{R}}$. Therefore, the nice, clean threshold between signal and noise eigenvalues, as described in (9.6.12) and (9.6.13), no longer exists. The model order estimation techniques discussed in Section 9.2 can be employed to attempt to determine the number of complex exponentials P present. In practice, these methods are best used as rough estimates, as their performance is not very accurate, especially for short data records.

For several of the frequency estimation techniques described in this section, the analysis considers the use of eigenvalues and eigenvectors of the correlation matrix for the purposes of defining signal and noise subspaces.[†] In practice, we estimate the signal and noise subspaces by using the eigenvectors and eigenvalues of the sample correlation matrix. Note that for notational expedience we will not differentiate between eigenvectors and eigenvalues of the

---

[†] The ESPRIT method uses a singular value decomposition of data matrix $\mathbf{X}$.

true and sample correlation matrices. However, the reader should always keep in mind that the sample correlation matrix eigendecomposition is what must be used for implementation. We note that use of an estimate rather than the true correlation matrix will result in a degradation in performance, the analysis of which is beyond the scope of this book.

9.6.2 Pisarenko Harmonic Decomposition

The *Pisarenko harmonic decomposition* (PHD) was the first frequency estimation method proposed that was based on the eigendecomposition of the correlation matrix and its partitioning into signal and noise subspaces (Pisarenko 1973). This method uses the eigenvector associated with the smallest eigenvalue to estimate the frequencies of the complex exponentials. Although this method has limited practical use owing to its sensitivity to noise, it is of great theoretical interest because it was the first method based on signal and noise subspace principles and it helped to fuel the development of many well-known subspace methods, such as MUSIC and ESPRIT.

Consider the model of complex exponentials contained in noise in (9.6.5) and the eigendecomposition of its correlation matrix in (9.6.14). The eigenvector corresponding to the minimum eigenvalue must be orthogonal to all the eigenvectors in the signal subspace. Thus, we choose the time window to be of length

$$M = P + 1 \tag{9.6.22}$$

that is, 1 greater than the number of complex exponentials. Therefore, the noise subspace consists of a single eigenvector

$$\mathbf{Q}_w = \mathbf{q}_M \tag{9.6.23}$$

corresponding to the minimum eigenvalue λ_M. By virtue of the orthogonality between the signal and noise subspaces, each of the P complex exponentials in the time-window signal vector model in (9.6.5) is orthogonal to this eigenvector

$$\mathbf{v}^H(f_p)\mathbf{q}_M = \sum_{k=1}^{M} q_M(k)e^{-j2\pi f_p(k-1)} = 0 \quad \text{for} \quad m \leq P \tag{9.6.24}$$

Making use of this property, we can compute

$$\bar{R}_{\text{phd}}(e^{j2\pi f}) = \frac{1}{|\mathbf{v}^H(f)\mathbf{q}_M|^2} = \frac{1}{|Q_M(e^{j2\pi f})|^2} \tag{9.6.25}$$

which is commonly referred to as a *pseudospectrum*. The frequencies are then estimated by observing the P peaks in $\bar{R}_{\text{phd}}(e^{j2\pi f})$. Note that since (9.6.25) requires a search of all frequencies $-0.5 \leq f \leq 0.5$, in practice a dense sampling of the frequencies is generally necessary. The quantity

$$Q_M(e^{j2\pi f}) = \mathbf{v}^H(f)\mathbf{q}_M = \sum_{k=1}^{M} q_M(k)e^{-j2\pi f(k-1)} \tag{9.6.26}$$

is simply the Fourier transform of the Mth eigenvector corresponding to the minimum eigenvalue. Thus, the pseudospectrum for the Pisarenko harmonic decomposition $\bar{R}_{\text{phd}}(e^{j2\pi f})$ can be efficiently implemented by computing the FFT of $\mathbf{q}_M$ with sufficient zero padding to provide the necessary frequency resolution. Then $\bar{R}_{\text{phd}}(e^{j2\pi f})$ is simply the reciprocal of the spectrum of the noise eigenvector, that is, the squared magnitude of its Fourier transform. Note that $\bar{R}_{\text{phd}}(e^{j2\pi f})$ is not an estimate of the true power spectrum since it contains no information about the powers of the complex exponentials $|\alpha_p|^2$ or the background noise level σ_w^2. However, these amplitudes can be found by using the estimated frequencies and the corresponding time-window frequency vectors along with the relationship of eigenvalues and eigenvectors. See Problem 9.24 for details.

Alternately, the frequencies of the complex exponentials can be found by computing the zeros of the Fourier transform of the Mth eigenvector in (9.6.23). The z-transform of this eigenvector is

$$Q_M(z) = \sum_{k=1}^{M} q_M(k) z^{-k} = \prod_{k=1}^{M-1} (1 - e^{j2\pi f_k} z^{-1}) \tag{9.6.27}$$

where the phases of the $P = M - 1$ roots of this polynomial are the frequencies f_k of the $P = M - 1$ complex exponentials.

As we stated up front, the significance of the Pisarenko harmonic decomposition is seen mostly from a theoretical perspective. The limitations of its practical use stem from the fact that it uses a single noise eigenvector and, as a result, lacks the necessary robustness needed for most applications. Since the correlation matrix is not known and must be estimated from data, the resulting noise eigenvector of the estimated correlation matrix is only an estimate of the actual noise eigenvector. Because we only use one noise eigenvector, this method is very sensitive to any errors in the estimation of the noise eigenvector.

EXAMPLE 9.6.1. We demonstrate the use of the Pisarenko harmonic decomposition with a sinusoid in noise. The amplitude and frequency of the sinusoid are $\alpha = 1$ and $f = 0.2$, respectively. The additive noise has unit power ($\sigma_w^2 = 1$). Using MATLAB, this signal is generated:

```
x = sin(2*pi*f*[0:(N-1)]') + (randn(N,1)+j*randn(N,1))/sqrt(2);
```

Since the number of complex exponentials is equal to $P = 2$ (a complex conjugate pair for a sinusoid), the time-window length is chosen to be $M = 3$. After forming the $N \times M$ data matrix $\mathbf{X}$ and computing the sample correlation matrix $\hat{\mathbf{R}}_x$, we can compute the pseudospectrum as follows:

```
[Q0,D] = eig(R); % eigendecomposition
[lambda,index] = sort(abs(diag(D))); % order by eigenvalue magnitude
lambda = lambda(M:-1:1); Q=Q0(:,index(M:-1:1));
Rbar = 1./abs(fftshift(fft(Q(:,M),Nfft))).^2;
```

Figure 9.21 shows the pseudospectrum of the Pisarenko harmonic decomposition for a single realization with an FFT size of 1024. Note the two peaks near $f = \pm 0.2$. Recall that

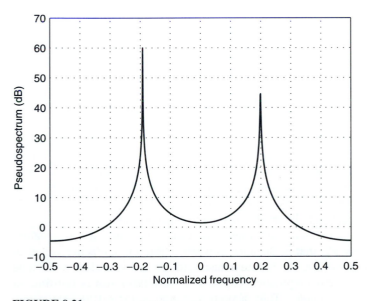

FIGURE 9.21

Pseudospectrum for the Pisarenko harmonic decomposition of a sinusoid in noise with frequency $f = 0.2$.

this is a pseudospectrum, so that the actual values do not correspond to an estimate of power. A MATLAB routine for estimating frequencies using the Pisarenko harmonic decomposition is provided in phd.m.

9.6.3 MUSIC Algorithm

The *multiple signal classification (MUSIC)* frequency estimation method was proposed as an improvement on the Pisarenko harmonic decomposition (Bienvenu and Kopp 1983; Schmidt 1986). Like the Pisarenko harmonic decomposition, the M-dimensional space is split into signal and noise components using the eigenvectors of the correlation matrix from (9.6.15). However, rather than limit the length of the time window to $M = P + 1$, that is, 1 greater than the number of complex exponentials, allow the size of the time window to be $M > P + 1$. Therefore, the noise subspace has a dimension greater than 1. Using this larger dimension allows for averaging over the noise subspace, providing an improved, more robust frequency estimation method than Pisarenko harmonic decomposition.

Because of the orthogonality between the noise and signal subspaces, all the time-window frequency vectors of the complex exponentials are orthogonal to the noise subspace from (9.6.19). Thus, for each eigenvector ($P < m \leq M$)

$$\mathbf{v}^H(f_p)\mathbf{q}_m = \sum_{k=1}^{M} q_m(k)e^{-j2\pi f_p(k-1)} = 0 \qquad (9.6.28)$$

for all the P frequencies f_p of the complex exponentials. Therefore, if we compute a pseudospectrum for each noise eigenvector as

$$\bar{R}_m(e^{j2\pi f}) = \frac{1}{|\mathbf{v}^H(f)\mathbf{q}_m|^2} = \frac{1}{|Q_m(e^{j2\pi f})|^2} \qquad (9.6.29)$$

the polynomial $Q_m(e^{j2\pi f})$ has $M-1$ roots, P of which correspond to the frequencies of the complex exponentials. These roots produce P peaks in the pseudospectrum from (9.6.29). Note that the pseudospectra of all $M - P$ noise eigenvectors share these roots that are due to the signal subspace. The remaining roots of the noise eigenvectors, however, occur at different frequencies. There are no constraints on the location of these roots, so that some may be close to the unit circle and produce extra peaks in the pseudospectrum. A means of reducing the levels of these spurious peaks in the pseudospectrum is to average the $M - P$ pseudospectra of the individual noise eigenvectors

$$\bar{R}_{\text{music}}(e^{j2\pi f}) = \frac{1}{\displaystyle\sum_{m=P+1}^{M} |\mathbf{v}^H(f)\mathbf{q}_m|^2} = \frac{1}{\displaystyle\sum_{m=P+1}^{M} |Q_m(e^{j2\pi f})|^2} \qquad (9.6.30)$$

which is known as the MUSIC pseudospectrum. The frequency estimates of the P complex exponentials are then taken as the P peaks in this pseudospectrum. Again, the term *pseudospectrum* is used because the quantity in (9.6.30) does not contain information about the powers of the complex exponentials or the background noise level. Note that for $M = P+1$, the MUSIC method is equivalent to Pisarenko harmonic decomposition.

The implicit assumption in the MUSIC pseudospectrum is that the noise eigenvalues all have equal power $\lambda_m = \sigma_w^2$, that is, the noise is white. However, in practice, when an estimate is used in place of the actual correlation matrix, the noise eigenvalues will not be equal. The differences become more pronounced when the correlation matrix is estimated from a small number of data samples. Thus, a slight variation on the MUSIC algorithm, known as the *eigenvector (ev) method*, was proposed to account for the potentially different noise eigenvalues (Johnson and DeGraaf 1982). For this method, the pseudospectrum is

$$\bar{R}_{\text{ev}}(e^{j\omega}) = \frac{1}{\displaystyle\sum_{m=P+1}^{M} \frac{1}{\lambda_m} |\mathbf{v}^H(f)\mathbf{q}_m|^2} = \frac{1}{\displaystyle\sum_{k=P+1}^{M} \frac{1}{\lambda_m} |Q_m(e^{j2\pi f})|^2} \qquad (9.6.31)$$

where λ_m is the eigenvalue corresponding to the eigenvector $\mathbf{q}_m$. The pseudospectrum of each eigenvector is normalized by its corresponding eigenvalue. In the case of equal noise eigenvalues ($\lambda_m = \sigma_w^2$) for $P + 1 \leq m \leq M$, the eigenvector and MUSIC methods are identical.

The peaks in the MUSIC pseudospectrum correspond to the frequencies at which the denominator in (9.6.30) $\sum_{m=P+1}^{M} |Q_m(e^{j2\pi f})|^2$ approaches zero. Therefore, we might want to consider the z-transform of this denominator

$$\bar{P}_{\text{music}}(z) = \sum_{m=P+1}^{M} Q_m(z) Q_m^* \left(\frac{1}{z^*}\right) \qquad (9.6.32)$$

which is the sum of the z-transforms of the pseudospectrum due to each noise eigenvector. This $(2M - 1)$th-order polynomial has $M - 1$ pairs of roots with one inside and one outside the unit circle. Since we assume that the complex exponentials are not damped, their corresponding roots must lie on the unit circle. Thus, if we have found the $M - 1$ roots of (9.6.32), the P closest roots to the unit circle will correspond to the complex exponentials. The phases of these roots are then the frequency estimates. This method of rooting the polynomial corresponding to the MUSIC pseudospectrum is known as *root-MUSIC* (Barabell 1983). Note that in many cases, a rooting method is more efficient than computing a pseudospectrum at a very fine frequency resolution that may require a very large FFT. Statistical performance analyses of the MUSIC algorithm can be found in Kaveh and Barabell (1986) and Stoica and Nehorai (1989). For the performance of the root-MUSIC method see Rao and Hari (1989). A routine for the MUSIC algorithm is provided in `music.m` and a routine for the root-MUSIC algorithm is provided in `rootmusic.m`.

> **EXAMPLE 9.6.2.** In this example, we demonstrate the use of the MUSIC algorithm and examine its performance in terms of resolution with respect to that of the minimum-variance spectral estimator. Consider the following scenario: Two complex exponentials in unit power noise ($\sigma_w^2 = 1$) with normalized frequencies $f = 0.1, 0.2$ both with amplitudes of $\alpha = 1$. We generate $N = 128$ samples of the signal and use a frequency vector of length $M = 8$. Proceeding as we did in Example 9.6.1, we compute the eigendecomposition and partition it into signal and noise subspaces. The MUSIC pseudospectrum is computed as
>
> ```
> Qbar = zeros(Nfft,1);
> for n = 1:(M-P)
>    Qbar = Qbar + abs(fftshift(fft(Q(:,M-(n-1)),Nfft))).^2;
> end
> Rbar = 1./Qbar;
> ```
>
> The minimum-variance spectral estimate and the MUSIC pseudospectrum are computed and averaged over 1000 realizations using an FFT size of 1024. The result is shown in Figure 9.22. The two exponentials have been clearly resolved using the MUSIC algorithm, whereas they are not very clear using the minimum-variance spectral estimate. Since the minimum-variance spectral estimator is nonparametric and makes no assumptions about the underlying model, it cannot achieve the resolution of the MUSIC algorithm.

9.6.4 Minimum-Norm Method

The minimum-norm method (Kumaresan and Tufts 1983), like the MUSIC algorithm, uses a time-window vector of length $M > P + 1$ for the purposes of frequency estimation. For MUSIC, a larger time window is used than for Pisarenko harmonic decomposition, resulting

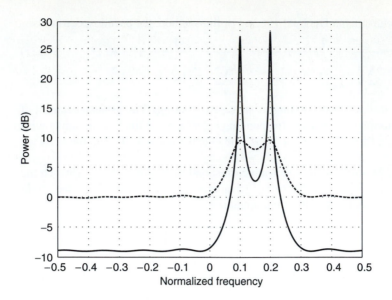

FIGURE 9.22

Comparison of the minimum-variance spectral estimate (dashed line)
and the MUSIC pseudospectrum (solid line) for two complex
exponentials in noise.

in a larger noise subspace. The use of a larger subspace provides the necessary robustness
for frequency estimation when an estimated correlation matrix is used. The same principle is
applied in the minimum-norm frequency estimation method. However, rather than average
the pseudospectra of all the noise subspace eigenvectors to reduce spurious peaks, as in the
case of the MUSIC algorithm, a different approach is taken.

Consider a single vector $\mathbf{u}$ contained in the noise subspace. The pseudospectrum of
this vector is given by

$$\bar{R}(e^{j2\pi f}) = \frac{1}{|\mathbf{v}^H(f)\mathbf{u}|^2} \tag{9.6.33}$$

Since the vector $\mathbf{u}$ lies in the noise subspace, its pseudospectrum in (9.6.33) has P peaks
corresponding to the complex exponentials in the signal subspace. However, $\mathbf{u}$ is length M so
that its pseudospectrum may exhibit an additional $M - P - 1$ peaks that do not correspond
to the frequencies of the complex exponentials. These spurious peaks lead to frequency
estimation errors. In the case of Pisarenko harmonic decomposition, spurious peaks were
not a concern since $M = P + 1$ and therefore its pseudospectrum in (9.6.25) only had
P peaks. On the other hand, the MUSIC algorithm diluted the strength of these spurious
peaks since its pseudospectrum in (9.6.30) is produced by averaging the pseudospectra of
the $M - P$ noise eigenvectors.

Recall the projection onto the noise subspace from (9.6.17) is

$$\mathbf{P}_w = \mathbf{Q}_w\mathbf{Q}_w^H \tag{9.6.34}$$

where $\mathbf{Q}_w$ is the matrix of noise eigenvectors. Therefore, for any vector $\mathbf{u}$ that lies in the
noise subspace

$$\mathbf{P}_w\mathbf{u} = \mathbf{u} \qquad \mathbf{P}_s\mathbf{u} = \mathbf{0} \tag{9.6.35}$$

where $\mathbf{P}_s$ is the signal subspace projection matrix and $\mathbf{0}$ is the length-P zero vector. Now
let us consider the z-transform of the coefficients of $\mathbf{u} = [u(1)\, u(2)\, \cdots\, u(M)]^T$

$$U(z) = \sum_{k=0}^{M-1} u(k)z^{-k} = \prod_{k=1}^{P}(1 - e^{j2\pi f_k}) \prod_{k=P+1}^{M-1}(1 - z_kz^{-1}) \tag{9.6.36}$$

This polynomial is the product of the P roots corresponding to complex exponentials that lie on the unit circle and the $M - P - 1$ roots that in general do not lie directly on the unit circle but can potentially produce spurious peaks in the pseudospectrum of $\mathbf{u}$. Therefore, we want to choose $\mathbf{u}$ so that it minimizes the spurious peaks due to these other roots of its associated polynomial $U(z)$.

The minimum-norm method, as its name implies, seeks to minimize the norm of $\mathbf{u}$ in order to avoid spurious peaks in its pseudospectrum. Using (9.6.35), the norm of a vector $\mathbf{u}$ contained in the noise subspace is

$$\|\mathbf{u}\|^2 = \mathbf{u}^H \mathbf{u} = \mathbf{u}^H \mathbf{P}_w \mathbf{u} \tag{9.6.37}$$

However, an unconstrained minimization of this norm will produce the zero vector. Therefore, we place the constraint that the first element of $\mathbf{u}$ must equal 1.[†] This constraint can be expressed as

$$\boldsymbol{\delta}_1^H \mathbf{u} = 1 \tag{9.6.38}$$

where $\boldsymbol{\delta}_1 = [1 \ 0 \ \cdots \ 0]^T$. Then the determination of the minimum-norm vector comes down to solving the following constrained minimization problem:

$$\min \|\mathbf{u}\|^2 = \mathbf{u}^H \mathbf{P}_w \mathbf{u} \quad \text{subject to} \quad \boldsymbol{\delta}_1^H \mathbf{u} = 1 \tag{9.6.39}$$

The solution can be found by using Lagrange multipliers (see Appendix B) and is given by

$$\mathbf{u}_{mn} = \frac{\mathbf{P}_w \boldsymbol{\delta}_1}{\boldsymbol{\delta}_1^H \mathbf{P}_w \boldsymbol{\delta}_1} \tag{9.6.40}$$

The frequency estimates are then obtained from the peaks in the pseudospectrum of the minimum-norm (mn) vector, $\mathbf{u}_{mn}$

$$\bar{R}_{mn}(e^{j2\pi f}) = \frac{1}{|\mathbf{v}^H(f)\mathbf{u}_{mn}|^2} \tag{9.6.41}$$

The performance of the minimum-norm frequency estimation method is similar to that of MUSIC. For a performance comparison see Kaveh and Barabell (1986). Note that it is also possible to implement the minimum-norm method by rooting a polynomial rather than computing a psuedospectrum (see Problem 9.25).

EXAMPLE 9.6.3. In this example, we illustrate the use of the minimum-norm method and compare its performance to that of the other three frequency estimation methods discussed in this chapter: Pisarenko harmonic decomposition, the MUSIC algorithm, and the eigenvector method. The pseudospectrum of the minimum-norm method is found by first computing the minimum-norm vector $\mathbf{u}_{mn}$ and then finding its pseudospectrum, that is,

```
delta1 = zeros(M,1); delta1(1) = 1;
Pn=Q(:,(P+1):M)*Q(:,(P+1):M)'; % noise subspace projection matrix
u = (Pn*e1)/(e1'*Pn*e1); % minimum-norm vector
Rbar = 1./abs(fftshift(fft(u,Nfft))).^2; % pseudospectrum
```

Consider the case of $P = 4$ complex exponentials in noise with frequencies $f = 0.1, 0.25,$ 0.4, and -0.1, all with an amplitude of $\alpha = 1$. The power of the noise is set to $\alpha_w^2 = 1$ with 100 realizations. The time-window length used was $M = 8$ for all the methods except Pisarenko harmonic decomposition, which is constrained to use $M = P + 1 = 5$. The pseudospectra are shown in Figure 9.23 with an FFT size of 1024, where we have not averaged in order to demonstrate the variance of the various methods. Here we see the large variance in the frequency estimates that is produced by Pisarenko harmonic decomposition compared to the other methods, which is a direct result of using a one-dimensional noise subspace. The other methods all perform comparably in terms of estimating the frequencies of the complex exponentials. Note the fluctuations in the pseudospectrum of the eigenvector method that result from the normalization

---

[†]The choice of a value of 1 is somewhat arbitrary, since any nonzero constant will result in a similar solution.

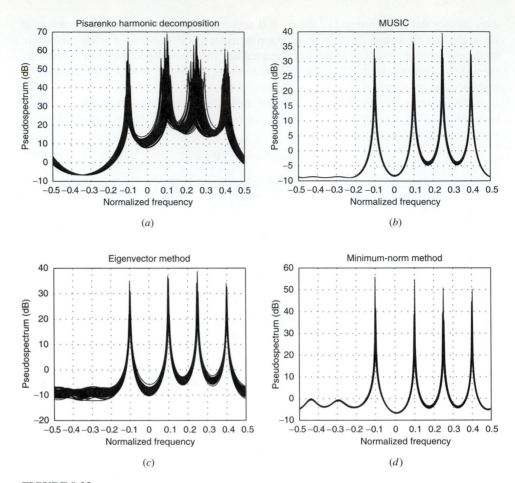

FIGURE 9.23

Comparison of the eigendecomposition-based frequency estimation methods: (*a*) Pisarenko
harmonic decomposition, (*b*) MUSIC, (*c*) eigenvector method, and (*d*) minimum-norm method.

by the eigenvalues. Since these eigenvalues vary over realizations, the pseudospectra will also
reflect a similar variation. Routines for the eigenvector method and the minimum-norm method
are provided in `ev_method.m` and `minnorm.m`, respectively.

9.6.5 ESPRIT Algorithm

A frequency estimation technique that is built upon the same principles as other subspace
methods but further exploits a deterministic relationship between subspaces is the *estimation
of signal parameters via rotational invariance techniques (ESPRIT)* algorithm. This method
differs from the other subspace methods discussed so far in this chapter in that the signal
subspace is estimated from the data matrix $\mathbf{X}$ rather than the estimated correlation matrix
$\hat{\mathbf{R}}_x$. The essence of ESPRIT lies in the rotational property between staggered subspaces
that is invoked to produce the frequency estimates. In the case of a discrete-time signal or
time series, this property relies on observations of the signal over two identical intervals
staggered in time. This condition arises naturally for discrete-time signals, provided that the
sampling is performed uniformly in time.[†] Extensions of the ESPRIT method to a spatial

---

[†]This condition is violated in the case of a nonuniformly sampled time series.

array of sensors, the application for which it was originally proposed, will be discussed in Chapter 11 in Section 11.7. We first describe the original, least-squares version of the algorithm (Roy et al. 1986) and then extend the derivation to total least-squares ESPRIT (Roy and Kailath 1989), which is the preferred method for use. Since the derivation of the algorithm requires an extensive amount of formulation and matrix manipulations, we have included a block diagram in Figure 9.24 to be used as a guide through this process.

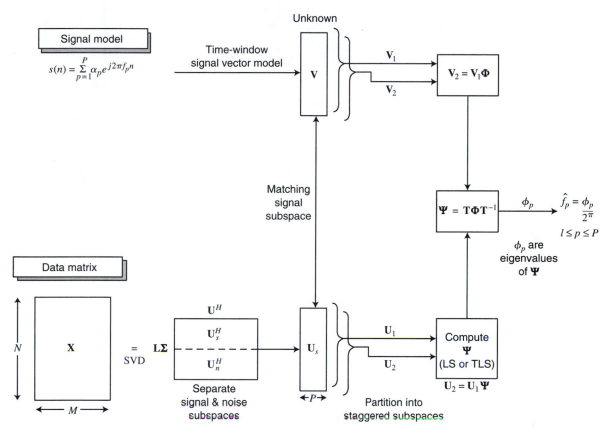

FIGURE 9.24
Block diagram demonstrating the flow of the ESPRIT algorithm starting from the data matrix through the frequency estimates.

Consider a single complex exponential $s_0(n) = e^{j2\pi fn}$ with complex amplitude α and frequency f. This signal has the following property

$$s_0(n+1) = \alpha e^{j2\pi f(n+1)} = s_0(n)e^{j2\pi f} \tag{9.6.42}$$

that is, the next sample value is a phase-shifted version of the current value. This phase shift can be represented as a rotation on the unit circle $e^{j2\pi f}$. Recall the time-window vector model from (9.6.4) consisting of a signal $\mathbf{s}(n)$, made up of complex exponentials, and the noise component $\mathbf{w}(n)$

$$\mathbf{x}(n) = \sum_{p=1}^{P} \alpha_p \mathbf{v}(f_p)e^{j2\pi nf_p} + \mathbf{w}(n) = \mathbf{V}\mathbf{\Phi}^n\boldsymbol{\alpha} + \mathbf{w}(n) = \mathbf{s}(n) + \mathbf{w}(n) \tag{9.6.43}$$

where the P columns of matrix $\mathbf{V}$ are length-M time-window frequency vectors of the complex exponentials

$$\mathbf{V} = [\mathbf{v}(f_1)\,\mathbf{v}(f_2)\,\cdots\,\mathbf{v}(f_P)] \qquad (9.6.44)$$

The vector $\boldsymbol{\alpha}$ consists of the amplitudes of the complex exponentials α_p. On the other hand, matrix $\boldsymbol{\Phi}$ is the diagonal matrix of phase shifts between neighboring time samples of the individual, complex exponential components of $\mathbf{s}(n)$

$$\boldsymbol{\Phi} = \mathrm{diag}\,\{\phi_1, \phi_2, \ldots, \phi_P\} = \begin{bmatrix} e^{j2\pi f_1} & 0 & \cdots & 0 \\ 0 & e^{j2\pi f_2} & \cdots & 0 \\ \vdots & \vdots & \ddots & \vdots \\ 0 & \cdots & 0 & e^{j2\pi f_P} \end{bmatrix} \qquad (9.6.45)$$

where $\phi_p = e^{j2\pi f_p}$ for $p = 1, 2, \ldots, P$. Since the frequencies of the complex exponentials f_p completely describe this rotation matrix, frequency estimates can be obtained by finding $\boldsymbol{\Phi}$. Let us consider two overlapping subwindows of length $M - 1$ within the length M time-window vector. This subwindowing operation is illustrated in Figure 9.25. Consider the signal consisting of the sum of complex exponentials

$$\mathbf{s}(n) = \begin{bmatrix} \mathbf{s}_{M-1}(n) \\ s(n + M - 1) \end{bmatrix} = \begin{bmatrix} s(n) \\ \mathbf{s}_{M-1}(n + 1) \end{bmatrix} \qquad (9.6.46)$$

where $\mathbf{s}_{M-1}(n)$ is the length-$(M - 1)$ subwindow of $\mathbf{s}(n)$, that is,

$$\mathbf{s}_{M-1}(n) = \mathbf{V}_{M-1}\boldsymbol{\Phi}^n\boldsymbol{\alpha} \qquad (9.6.47)$$

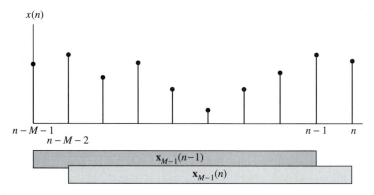

FIGURE 9.25
Time-staggered, overlapping windows used by the ESPRIT algorithm.

Matrix $\mathbf{V}_{M-1}$ is constructed in the same manner as $\mathbf{V}$ except its time-window frequency vectors are of length $M - 1$, denoted as $\mathbf{v}_{M-1}(f)$,

$$\mathbf{V}_{M-1} = [\mathbf{v}_{M-1}(f_1)\,\mathbf{v}_{M-1}(f_2)\,\cdots\,\mathbf{v}_{M-1}(f_P)] \qquad (9.6.48)$$

Recall that $s(n)$ is the scalar signal made up of the sum of complex exponentials at time n. Using the relation in (9.6.47), we can define the matrices

$$\mathbf{V}_1 = \mathbf{V}_{M-1}\boldsymbol{\Phi}^n \quad \mathbf{V}_2 = \mathbf{V}_{M-1}\boldsymbol{\Phi}^{n+1} \qquad (9.6.49)$$

where $\mathbf{V}_1$ and $\mathbf{V}_2$ correspond to the unstaggered and staggered windows, that is,

$$\mathbf{V} = \begin{bmatrix} \mathbf{V}_1 \\ *\ *\ \cdots\ * \end{bmatrix} = \begin{bmatrix} *\ *\ \cdots\ * \\ \mathbf{V}_2 \end{bmatrix} \qquad (9.6.50)$$

Clearly, by examining (9.6.49), these two matrices of time-window frequency vectors are related as

$$\mathbf{V}_2 = \mathbf{V}_1 \mathbf{\Phi} \tag{9.6.51}$$

Note that each of these two matrices spans a different, though related, $(M-1)$-dimensional subspace.

Now suppose that we have a data matrix $\mathbf{X}$ from (9.6.20) with N data records of the length-M time-window vector signal $\mathbf{x}(n)$. Using the singular value decomposition (SVD) discussed in Chapter 8, we can write the data matrix as[†]

$$\mathbf{X} = \mathbf{L}\mathbf{\Sigma}\mathbf{U}^H \tag{9.6.52}$$

where $\mathbf{L}$ is an $N \times N$ matrix of left singular vectors and $\mathbf{U}$ is an $M \times M$ matrix of right singular vectors. Both of these matrices are unitary; that is, $\mathbf{V}^H\mathbf{V} = \mathbf{I}$ and $\mathbf{U}^H\mathbf{U} = \mathbf{I}$. The matrix $\mathbf{\Sigma}$ has dimensions $N \times M$ consisting of singular values on the main diagonal ordered in descending magnitude. The squared magnitudes of the singular values are equal to the eigenvalues of $\hat{\mathbf{R}}$ scaled by a factor of N from (9.6.21), and the columns of $\mathbf{U}$ are their corresponding eigenvectors. Thus, $\mathbf{U}$ forms an orthonormal basis for the underlying M-dimensional vector space. This subspace can be partitioned into signal and noise subspaces as

$$\mathbf{U} = [\mathbf{U}_s | \mathbf{U}_n] \tag{9.6.53}$$

where $\mathbf{U}_s$ is the matrix of right-hand singular vectors corresponding to the singular values with the P largest magnitudes. Note that since the signal portion consists of the sum of complex exponentials modeled as time-window frequency vectors $\mathbf{v}(f)$, all these frequency vectors, for $f = f_1, f_2, \ldots, f_P$, must also lie in the signal subspace. As a result, the matrices $\mathbf{V}$ and $\mathbf{U}_s$ span the same subspace. Therefore, there exists an invertible transformation $\mathbf{T}$ that maps $\mathbf{U}_s$ into $\mathbf{V}$, that is,

$$\mathbf{V} = \mathbf{U}_s\mathbf{T} \tag{9.6.54}$$

The transformation $\mathbf{T}$ is never solved for in this derivation, but instead is only formulated as a mapping between these two matrices within the signal subspace.

Proceeding as we did with the matrix $\mathbf{V}$ in (9.6.50), we can partition the signal subspace into two smaller $(M-1)$-dimensional subspaces as

$$\mathbf{U}_s = \begin{bmatrix} \mathbf{U}_1 \\ * \ * \ \cdots \ * \end{bmatrix} = \begin{bmatrix} * \ * \ \cdots \ * \\ \mathbf{U}_2 \end{bmatrix} \tag{9.6.55}$$

where $\mathbf{U}_1$ and $\mathbf{U}_2$ correspond to the unstaggered and staggered subspaces, respectively. Since $\mathbf{V}_1$ and $\mathbf{V}_2$ correspond to the same subspaces, the relation from (9.6.54) must also hold for these subspaces

$$\mathbf{V}_1 = \mathbf{U}_1\mathbf{T} \qquad \mathbf{V}_2 = \mathbf{U}_2\mathbf{T} \tag{9.6.56}$$

The staggered and unstaggered components of the matrix $\mathbf{V}$ in (9.6.50) are related through the subspace rotation $\mathbf{\Phi}$ in (9.6.51). Since the matrices $\mathbf{U}_1$ and $\mathbf{U}_2$ also span these respective, related subspaces, a similar, though different, rotation must exist that relates (rotates) $\mathbf{U}_1$ to $\mathbf{U}_2$

$$\mathbf{U}_2 = \mathbf{U}_1\mathbf{\Psi} \tag{9.6.57}$$

where $\mathbf{\Psi}$ is this rotation matrix.

Recall that frequency estimation comes down to solving for the subspace rotation matrix $\mathbf{\Phi}$. We can estimate $\mathbf{\Phi}$ by making use of the relations in (9.6.56) together with the

---

[†] Our notation differs slightly from that introduced in Chapter 8 in order to avoid confusion with the matrix of time-window frequency vectors $\mathbf{V}$.

rotations between the staggered signal subspaces in (9.6.51) and (9.6.57). In this process, the matrices $\mathbf{U}_1$ and $\mathbf{U}_2$ are known from the SVD on data matrix $\mathbf{X}$. First, we solve for $\boldsymbol{\Psi}$ from the relation in (9.6.57), using the method of least-squares (LS) from Chapter 8

$$\boldsymbol{\Psi} = (\mathbf{U}_1^H \mathbf{U}_1)^{-1} \mathbf{U}_1^H \mathbf{U}_2 \tag{9.6.58}$$

Substituting (9.6.57) into (9.6.56), we have

$$\mathbf{V}_2 = \mathbf{U}_2 \mathbf{T} = \mathbf{U}_1 \boldsymbol{\Psi} \mathbf{T} \tag{9.6.59}$$

Similarly, we can also solve for $\mathbf{V}_2$, using the relation in (9.6.51) and substituting (9.6.56) for $\mathbf{V}_1$

$$\mathbf{V}_2 = \mathbf{V}_1 \boldsymbol{\Phi} = \mathbf{U}_1 \mathbf{T} \boldsymbol{\Phi} \tag{9.6.60}$$

Thus, equating the two right-hand sides of (9.6.59) and (9.6.60), we have the following relation between the two subspace rotations

$$\boldsymbol{\Psi} \mathbf{T} = \mathbf{T} \boldsymbol{\Phi} \tag{9.6.61}$$

or equivalently

$$\boldsymbol{\Psi} = \mathbf{T} \boldsymbol{\Phi} \mathbf{T}^{-1} \tag{9.6.62}$$

Equations (9.6.61) and (9.6.62) should be recognized as the relationship between eigenvectors and eigenvalues of the matrix $\boldsymbol{\Psi}$ (Golub and Van Loan 1996). Therefore, the diagonal elements of $\boldsymbol{\Phi}$, ϕ_p for $p = 1, 2, \ldots, P$, are simply the eigenvalues of $\boldsymbol{\Psi}$. As a result, the estimates of the frequencies are

$$\hat{f}_p = \frac{\angle \phi_p}{2\pi} \tag{9.6.63}$$

where $\angle \phi_p$ is the phase of ϕ_p. Although the principle behind the ESPRIT algorithm, namely, the use of subspace rotations, is quite simple, one can easily get lost in the details of the derivation of the algorithm. Note that we have only used simple matrix relationships. An illustrative example of the implementation of ESPRIT in MATLAB is given in Example 9.6.4 to help clarify the details of the algorithm. However, first we give a total least-squares version of the algorithm, which is the preferred method for use.

Note that the subspaces $\mathbf{U}_1$ and $\mathbf{U}_2$ are *both* only estimates of the true subspaces that correspond to $\mathbf{V}_1$ and $\mathbf{V}_2$, respectively, obtained from the data matrix $\mathbf{X}$. The estimate of the subspace rotation was obtained by solving (9.6.57) using the LS criterion

$$\boldsymbol{\Psi}_{ls} = (\mathbf{U}_1^H \mathbf{U}_1)^{-1} \mathbf{U}_1^H \mathbf{U}_2 \tag{9.6.64}$$

This LS solution is obtained by minimizing the errors in an LS sense from the following formulation

$$\mathbf{U}_2 + \mathbf{E}_2 = \mathbf{U}_1 \boldsymbol{\Psi} \tag{9.6.65}$$

where $\mathbf{E}_2$ is a matrix consisting of errors between $\mathbf{U}_2$ and the true subspace corresponding to $\mathbf{V}_2$. Note that this LS formulation assumes errors only on the estimation of $\mathbf{U}_2$ and no errors between $\mathbf{U}_1$ and the true subspace that it is attempting to estimate corresponding to $\mathbf{V}_1$. Therefore, since $\mathbf{U}_1$ is also an estimated subspace, a more appropriate formulation is

$$\mathbf{U}_2 + \mathbf{E}_2 = (\mathbf{U}_1 + \mathbf{E}_1) \boldsymbol{\Psi} \tag{9.6.66}$$

where $\mathbf{E}_1$ is the matrix representing the errors between $\mathbf{U}_1$ and the true subspace corresponding to $\mathbf{V}_1$. A solution to this problem, known as *total least squares (TLS)*, is obtained by minimizing the Frobenius norm of the two error matrices

$$\| \mathbf{E}_1 \quad \mathbf{E}_2 \|_F \tag{9.6.67}$$

Since the principles of TLS are beyond the scope of this book, we simply give the procedure to obtain the TLS solution of $\boldsymbol{\Psi}$ and refer the interested reader to Golub and Van Loan (1996).

First, form a matrix made up of the staggered signal subspace matrices $\mathbf{U}_1$ and $\mathbf{U}_2$ placed side by side,[†] and perform an SVD

$$[\mathbf{U}_1\ \mathbf{U}_2] = \tilde{\mathbf{L}}\tilde{\boldsymbol{\Sigma}}\tilde{\mathbf{U}}^H \qquad (9.6.68)$$

We then operate on the $2P \times 2P$ matrix $\tilde{\mathbf{U}}$ of right singular vectors. This matrix is partitioned into $P \times P$ quadrants

$$\tilde{\mathbf{U}} = \begin{bmatrix} \tilde{\mathbf{U}}_{11} & \tilde{\mathbf{U}}_{12} \\ \tilde{\mathbf{U}}_{21} & \tilde{\mathbf{U}}_{22} \end{bmatrix} \qquad (9.6.69)$$

The TLS solution for the subspace rotation matrix $\boldsymbol{\Psi}$ is then

$$\boldsymbol{\Psi}_{\text{tls}} = -\tilde{\mathbf{U}}_{12}\tilde{\mathbf{U}}_{22}^{-1} \qquad (9.6.70)$$

The frequency estimates are then obtained from (9.6.62) and (9.6.63) by using $\boldsymbol{\Psi}_{\text{tls}}$ from (9.6.70). Although the TLS version of ESPRIT involves slightly more computations, it is generally preferred over the LS version based on formulation in (9.6.66). A statistical analysis of the performance of the ESPRIT algorithms is given in Ottersten et al. (1991).

EXAMPLE 9.6.4. In this illustrative example, we demonstrate the use of both the LS and TLS versions of the ESPRIT algorithm on a set of complex exponentials in white noise using MATLAB. First, generate a signal $s(n)$ of length $N = 128$ consisting of complex exponential signals at normalized frequencies $f = 0.1, 0.15, 0.4,$ and -0.15, all with amplitude $\alpha = 1$. Each of the complex exponentials is generated by `exp(j*2*pi*f*[0:(N-1)]');`. The overall signal in white noise with unit power ($\sigma_w^2 = 1$) is then

```
x = s + (randn(N,1)+j*randn(N,1))/sqrt(2);
```

We form the data matrix corresponding to (9.6.20) for a time window of length $M = 8$. The least-squares ESPRIT algorithm is then performed as follows:

```
[L,S,U] = svd(X);
Us = U(:,1:P); % signal subspace
U1 = Us(1:(M-1),:); U2 = Us(2:M,:); % signal subspaces
Psi = U1\U2; % LS solution for Psi
```

If we are using the TLS version of ESPRIT, then solve for

```
[LL,SS,UU] = svd([U1 U2]); UU12 = UU(1:P,(P+1):(2*P));
UU22 = UU((P+1):(2*P),(P+1):(2*P));
Psi = -UU12*inv(UU22); % TLS solution for Psi
```

The frequencies are found by computing the phases of the eigenvalues of $\boldsymbol{\Psi}$, that is,

```
phi = eig(Psi); % eigenvalues of Psi
fhat = angle(phi)/(2*pi); % frequency estimates
```

In both cases, we average over 1000 realizations and obtain average estimated frequencies very close to the true values $f = 0.1, 0.15, 0.4,$ and -0.15 used to generate the signals. Routines for both the LS and TLS versions of ESPRIT are provided in `esprit_ls.m` and `esprit_tls.m`.

9.7 SUMMARY

In this chapter, we have examined the modeling process for both pole-zero and harmonic signal models. As for all signal modeling problems, the procedure begins with the selection of the appropriate model for the signal under consideration. Then the signal model is applied by estimating the model parameters from a collection of data samples. However, as we

---

[†] Note that this matrix $[\mathbf{U}_1\ \mathbf{U}_2] \neq \mathbf{U}_s = [\mathbf{U}_1^T\ \mathbf{U}_2^T]^T$ from (9.6.55).

have stressed throughout this chapter, nothing is more valuable in the modeling process than specific knowledge of the signal and its underlying process in order to assess the validity of the model for a particular signal. For this reason, we began the chapter with a discussion of a model building procedure, starting with the choice of the appropriate model and the estimation of its parameters, and concluding with the validation of the model. Clearly, if the model is not well-suited for the signal, the application of the model becomes meaningless.

In the first part of the chapter, we considered the application of the parametric signal models that were discussed in Chapter 4. The estimation of all-pole models was presented for both direct and lattice structures. Within this context, we used various model order selection criteria to determine the order of the all-pole model. However, these criteria are not necessarily limited to all-pole models. In addition, the relationship was given between the all-pole model and Burg's method of maximum entropy. Next, we considered the pole-zero modeling. Using a nonlinear least-squares technique, a method was presented for estimating the parameters of the pole-zero model. The use of pole-zero models for the purposes of spectral estimation along with their application to speech modeling was also considered.

The latter part of the chapter focused on harmonic signal models, that is, modeling signals using the sum of complex exponentials. The harmonic modeling problem becomes one of estimating the frequency of the complex exponentials. As a bridge between these pole-zero and harmonic models, we discussed the topic of minimum-variance spectral estimation. As will be explored in the problems that follow, there are several interesting relations between the minimum-variance spectrum and the harmonic models. In addition, a relationship between the minimum-variance spectral estimator and the all-pole model was established. Then, we discuss some of the more popular harmonic modeling methods. Starting with the Pisarenko harmonic decomposition, the first such model, we discuss the MUSIC, eigenvector, root-MUSIC, and minimum-norm methods for frequency estimation. All of these methods are based on computing a pseudospectrum or a rooting polynomial from an estimated correlation matrix. Finally, we give a brief derivation of the ESPRIT algorithm, both in its original LS form and the more commonly used TLS form.

PROBLEMS

9.1 Consider the random process $x(n)$ described in Example 9.2.3 that is simulated by exciting the system function

$$H(z) = \frac{1}{1 - 2.7607z^{-1} + 3.8108z^{-2} - 2.6535z^{-3} + 0.9238z^{-4}}$$

using a WGN(0, 1) process. Generate $N = 250$ samples of the process $x(n)$.

(a) Write a MATLAB function that implements the modified covariance method to obtain AR(P) model coefficients and the modeling error variance $\hat{\sigma}_P^2$ as a function of P, using N samples of $x(n)$.

(b) Compute and plot the variance $\hat{\sigma}_P^2$, FPE(P), AIC(P), MDL(P), and CAT(P) for $P = 1, 2, \ldots, 15$.

(c) Comment on your results and the usefulness of model selection criteria for the process $x(n)$.

9.2 Consider the Burg approach of minimizing forward-backward LS error $\mathcal{E}_m^{\text{fb}}$ in (9.2.33).

(a) Show that by using (9.2.26) and (9.2.27), $\mathcal{E}_m^{\text{fb}}$ can be put in the form of (9.2.34).

(b) By minimizing $\mathcal{E}_m^{\text{fb}}$ with respect to k_{m-1}, show that the expression for the optimum k_{m-1}^{B} is given by (9.2.35).

(c) Show that $|k_{m-1}^{\text{B}}| < 1$.

(d) Show that $|k_{m-1}^{\text{B}}| < |k_{m-1}^{\text{IS}}| \leq 1$ where k_{m-1}^{IS} is defined in (9.2.36).

9.3 Generate an AR(2) process using the system function

$$H(z) = \frac{1}{1 - 0.9z^{-1} + 0.81z^{-1}}$$

excited by a WGN(0, 1) process. Illustrate numerically that if we use the full-windowing method, that is, the matrix $\bar{\mathbf{X}}$ in (9.2.8), then the PACS estimates $\{k_m^{FM}\}_{m=0}^1$, $\{k_m^{BM}\}_{m=0}^1$, and $\{k_m^B\}_{m=0}^1$ of Section 9.2 are identical and hence can be obtained by using the Levinson-Durbin algorithm.

9.4 Generate sample sequences of an AR(2) process

$$x(n) = w(n) - 1.5857x(n-1) - 0.9604x(n-2)$$

where $w(n) \sim$ WGN(0, 1). Choose $N = 256$ samples for each realization.

(a) Design a first-order optimum linear predictor, and compute the prediction error $e_1(n)$. Test the whiteness of the error sequence $e_1(n)$ using the autocorrelation, PSD, and partial correlation methods, discussed in Section 9.1. Show your results as an overlay plot using 20 realizations.

(b) Repeat the above part, using second- and third-order linear predictors.

(c) Comment on your plots.

9.5 Generate sample functions of the process

$$x(n) = 0.5w(n) + 0.5w(n-1)$$

where $w(n) \sim$ WGN(0, 1). Choose $N = 256$ samples for each realization.

(a) Test the whiteness of $x(n)$ and show your results, using overlay plots based on 10 realizations.

(b) Process $x(n)$ through the AR(1) filter

$$H(z) = \frac{1}{1 + 0.95z^{-1}}$$

to obtain $y(n)$. Test the whiteness of $y(n)$ and show your results, using overlay plots based on 10 realizations.

9.6 The process $x(n)$ contains a complex exponential in white noise, that is,

$$x(n) = Ae^{j(\omega_0 n + \theta)} + w(n)$$

where A is a real positive constant, θ is a random variable uniformly distributed over $[0, 2\pi]$, ω_0 is a constant between 0 and π, and $w(n) \sim$ WGN(0, σ_w^2). The purpose of this problem is to analytically obtain a maximum entropy method (MEM) estimate by fitting an AR(P) model and then evaluating $\{a_k\}_0^P$ model coefficients.

(a) Show that the $(P+1) \times (P+1)$ autocorrelation matrix of $x(n)$ is given by

$$\mathbf{R}_x = A^2 \mathbf{e}\mathbf{e}^H + \sigma_w^2 \mathbf{I}$$

where $\mathbf{e} = [1\ e^{-j\omega_0}\ \cdots\ e^{-jP\omega_0}]^T$.

(b) By solving autocorrelation normal equations, show that

$$\mathbf{a}_P \triangleq [1\ a_1\ \cdots\ a_P]^T$$

$$= \left(1 + \frac{A^2}{\sigma_w^2 + A^2 P}\right)\left[\mathbf{e} - \frac{A^2}{\sigma_w^2 + (P+1)A^2}[1\ 0\ \cdots\ 0]^T\right]$$

(c) Show that the MEM estimate based on the above coefficients is given by

$$\hat{R}_x(e^{j\omega}) = \frac{\sigma_w^2 \left[1 - \dfrac{A^2}{\sigma_w^2 + (P+1)A^2}\right]}{\left|1 - \dfrac{A^2}{\sigma_w^2 + (P+1)A^2} W_R(e^{j(\omega-\omega_0)})\right|^2}$$

where $W_R(e^{j\omega})$ is the DTFT of the $(P+1)$ length rectangular window.

9.7 An AR(2) process $y(n)$ is observed in noise $v(n)$ to obtain $x(n)$, that is,

$$x(n) = y(n) + v(n) \qquad v(n) \sim \text{WGN}(0, \sigma_v^2)$$

where $v(n)$ is uncorrelated with $y(n)$ and

$$y(n) = 1.27y(n-1) - 0.81y(n-2) + w(n) \qquad w(n) \sim \text{WGN}(0, 1)$$

(a) Determine and plot the true power spectrum $R_x(e^{j\omega})$.

(b) Generate 10 realizations of $x(n)$, each with $N = 256$ samples. Using the LS approach with forward-backward linear predictor, estimate the power spectrum for $P = 2$ and $\sigma_v^2 = 1$. Obtain an overlay plot of this estimate, and compare it with the true spectrum.

(c) Repeat part (b), using $\sigma_v^2 = 10$. Comment on the effect of increasing noise variance on spectrum estimates.

(d) Since the noise variance σ_v^2 affects only $r_x(0)$, investigate the effect of subtracting a small amount from $r_x(0)$ on the spectrum estimates in part (c).

9.8 Let $x(n)$ be a random process whose correlation is estimated. The values for the first five lags are $r_x(0) = 1, r_x(1) = 0.7, r_x(2) = 0.5, r_x(3) = 0.3$, and $r_x(4) = 0$.

(a) Determine and plot the Blackman-Tukey power spectrum estimate.

(b) Assume that $x(n)$ is modeled by an AP(2) model. Determine and plot its spectrum estimate.

(c) Now assume that PZ(1, 1) is an appropriate model for $x(n)$. Determine and plot the spectrum estimate.

9.9 The narrowband process $x(n)$ is generated using the AP(4) model

$$H(z) = \frac{1}{1 + 0.98z^{-1} + 1.92z^{-2} + 0.94z^{-3} + 0.92z^{-4}}$$

driven by WGN(0, 0.001). Generate 10 realizations, each with $N = 256$ samples, of this process.

(a) Determine and plot the true power spectrum $R_x(e^{j\omega})$.

(b) Using the LS approach with forward linear predictor, estimate the power spectrum for $P = 4$. Obtain an overlay plot of this estimate, and compare it with the true spectrum.

(c) Repeat part (b) with $P = 8$ and 12. Provide a qualitative description of your results with respect to model order size.

(d) Using the LS approach with forward-backward linear predictor, estimate the power spectrum for $P = 4$. Obtain an overlay plot of this estimate. Compare it with the plot in part (b).

9.10 Consider the following PZ(4, 2) model

$$H(z) = \frac{1 - z^{-2}}{1 + 0.41z^{-4}}$$

driven by WGN(0, 1) to obtain a broadband ARMA process $x(n)$. Generate 10 realizations, each with $N = 256$ samples, of this process.

(a) Determine and plot the true power spectrum $R_x(e^{j\omega})$.

(b) Using the LS approach with forward-backward linear predictor, estimate the power spectrum for $P = 12$. Obtain an overlay plot of this estimate, and compare it with the true spectrum.

(c) Using the nonlinear LS pole-zero modeling algorithm of Section 9.3.3, estimate the power spectrum for $P = 4$ and $Q = 2$. Obtain an overlay plot of this estimate, and compare it with the plot in part (b).

9.11 A random process $x(n)$ is given by

$$x(n) = \cos\left(\frac{\pi n}{3} + \theta_1\right) + w(n) - w(n-2) + \cos\left(\frac{2\pi n}{3} + \theta_2\right)$$

where $w(n) \sim \text{WGN}(0, 1)$ and θ_1 and θ_2 are IID random variables uniformly distributed between 0 and 2π. Generate a sample sequence with $N = 256$ samples.

(a) Determine and plot the true spectrum $R_x(e^{j\omega})$.

(b) Using the LS approach with forward-backward linear predictor, estimate the power spectrum for $P = 10, 20$, and 40 from the generated sample sequence. Compare it with the true spectrum.

(c) Using the nonlinear LS pole-zero modeling algorithm of Section 9.3.3, estimate the power spectrum for $P = 4$ and $Q = 2$. Compare it with the true spectrum and with the plot in part (b).

9.12 Show that, for large values of N, the modeling error variance estimate given by Equation (9.2.38) can be approximated by the estimate given by Equation (9.2.39).

9.13 This problem investigates the effect of correlation aliasing observed in LS estimation of model parameters when the AP model is excited by discrete spectra. Consider an AP(1) model with pole at $z = \alpha$ excited by a periodic sequence of period N. Let $x(n)$ be the output sequence.

(a) Show that the correlation at lag 1 satisfies

$$r_x(1) = \frac{\alpha^{N-1} + \alpha}{1 + \alpha^N} r_x(0) \tag{P.1}$$

(b) Using the LS approach, determine the estimate $\hat{\alpha}$ as a function of α and N. Compute $\hat{\alpha}$ for $\alpha = 0.9$ and $N = 10$.

(c) Generate $x(n)$, using $\alpha = 0.95$ and the periodic impulse train with $N = 10$. Compute and plot the correlation sequence $r_x(l)$, $0 \le l \le N - 1$, of $x(n)$. Compare your plot with the AP(1) model correlation for $\alpha = 0.95$. Comment on your observations and discuss why they explain the discrepancy between α and $\hat{\alpha}$.

(d) Repeat part (c) for $N = 100$ and 1000. Show analytically and numerically that $\hat{\alpha} \to \alpha$ as $N \to \infty$.

9.14 In this problem, we investigate the equation error method of Section 9.3.1. Consider the PZ(2, 2) model

$$x(n) = 0.3x(n-1) + 0.4x(n-2) + w(n) + 0.25w(n-2)$$

Generate $N = 200$ samples of $x(n)$, using $w(n) \sim \text{WGN}(0, \sqrt{10})$. Record values of both $x(n)$ and $w(n)$.

(a) Using the residual windowing method, that is, $N_i = \max(P, Q)$ and $N_f = N - 1$, compute the estimates of the above model parameters.

(b) Compute the input variance estimate $\hat{\sigma}_w^2$ from your estimated values in part (a). Compare it with the actual value σ_w^2 and with (9.3.12).

9.15 Consider the following PZ(4, 2) model

$$x(n) = 1.8766x(n-1) - 2.6192x(n-2) + 1.6936x(n-3) - 0.8145x(n-4)$$
$$+ w(n) + 0.05w(n-1) - 0.855w(n-2)$$

excited by $w(n) \sim \text{WGN}(0, \sqrt{10})$. Generate 300 samples of $x(n)$.

(a) Using the nonlinear LS pole-zero modeling algorithm of Section 9.3.3, estimate the parameters of the above model from the $x(n)$ data segment.

(b) Assuming the AP(10) model for the data segment, estimate its parameters by using the LS approach described in Section 9.2.

(c) Generate a plot similar to Figure 9.13 by computing spectra corresponding to the true PZ(4, 2), estimated PZ(4, 2), and estimated AP(10) models. Compare and comment on your results.

9.16 Using matrix notation, show that AZ power spectrum estimation is equivalent to the Blackman-Tukey method discussed in Chapter 5.

9.17 Consider the PZ(4, 2) model given in Problem 9.15. Generate 300 samples of $x(n)$.

(a) Fit an AP(5) model to the data and plot the resulting spectrum.

(b) Fit an AP(10) model to the data and plot the resulting spectrum.

(c) Fit an AP(50) model to the data and plot the resulting spectrum.

(d) Compare your plots with the true spectrum, and discuss the effect of model mismatch on the quality of the spectrum.

9.18 Use the supplied (about 50-ms) segment of a speech signal sampled at 8192 samples per second.

    (*a*) Compute a periodogram of the speech signal (see Chapter 5).

    (*b*) Using data windowing, fit an AP(16) model to the speech data and compute the spectrum.

    (*c*) Using the residual windowing, fit a PZ(12, 6) model to the speech data and compute the spectrum.

    (*d*) Plot the above three spectra on one graph, and comment on the performance of each method.

9.19 One practical approach to spectrum estimation discussed in Section 9.4 is the prewhitening and postcoloring method.

    (*a*) Develop a MATLAB function to implement this method. Use the forward/backward LS method to determine AP(P) parameters and the Welch method for nonparametric spectrum estimation.

    (*b*) Verify your function on the short segment of the speech segment from Problem 9.18.

    (*c*) Compare your results with those obtained in Problem 9.18.

9.20 Consider a white noise process with variance σ_w^2. Find its minimum-variance power spectral estimate.

9.21 Find the minimum-variance spectrum of a first-order all pole model, that is,

$$x(n) = -a_1 x(n-1) + w(n)$$

9.22 The filter coefficient vector for the minimum-variance spectrum estimator is given in (9.5.10). Using Lagrange multipliers, discussed in Appendix B, solve this constrained optimization to find this weight vector.

9.23 Using the relationship between the minimum-variance and the all-pole model spectrum estimators in (9.5.22), generate a recursive relationship for the minimum-variance spectrum estimators of increasing window length. In other words, write $\hat{R}_{M+1}^{(mv)}(e^{j2\pi f})$ in terms of $\hat{R}_M^{(mv)}(e^{j2\pi f})$ and the all-pole model spectrum estimator $\hat{R}_M^{(ap)}(e^{j2\pi f})$ in (9.5.20).

9.24 In Pisarenko harmonic decomposition, discussed in Section 9.6.2, we determine the frequencies of the complex exponentials in white noise through the use of the pseudospectrum. The word *pseudospectrum* was used because its value does not correspond to an estimated power. Find a set of linear equations that can be solved to find the powers of the complex exponentials. *Hint:* Use the relationship of eigenvalues and eigenvectors $\mathbf{R}_x \mathbf{q}_m = \lambda_m \mathbf{q}_m$ for $m = 1, 2, \ldots, M$.

9.25 For the MUSIC algorithm, we showed a means of using the MUSIC pseudospectrum to derive a polynomial that could be rooted to obtain frequency estimates, which is known as root-MUSIC. Find a similar rooting method for the minimum-norm frequency estimation procedure.

9.26 The Pisarenko harmonic decomposition, MUSIC, and minimum-norm algorithms yield frequency estimates by computing a pseudospectrum using the Fourier transforms of the eigenvectors. However, these pseudospectra do not actually estimate a power. Derive the minimum-variance spectral estimator in terms of the Fourier transforms of the eigenvectors and the associated eigenvalues. Relate this result to the MUSIC and eigenvector method pseudospectra.

9.27 Show that the pseudospectrum for the MUSIC algorithm is equivalent to the minimum-variance spectrum in the case of an infinite signal-to-noise ratio.

9.28 Find a relationship between the minimum-norm pseudospectrum and the MUSIC pseudospectrum. What are the implications of this relationship?

9.29 In (9.5.22), we derived a relationship between the minimum-variance spectral estimator and spectrum estimators derived from all-pole models of orders 1 to M. Find a similar relationship between the pseudospectra of the MUSIC and minimum-norm algorithms that shows that the MUSIC pseudospectrum is a weighted average of minimum-norm pseudospectra.

CHAPTER 10

Adaptive Filters

In Chapter 1, we discussed different practical applications that demonstrated the need for adaptive filters, pointed out the key aspects of the underlying *signal operating environment* (*SOE*), and illustrated the key features and types of adaptive filters. The defining characteristic of an adaptive filter is its ability to operate satisfactorily, according to a criterion of performance acceptable to the user, in an unknown and possibly time-varying environment without the intervention of the designer. In Chapter 6, we developed the theory of optimum filters under the assumption that the filter designer has complete knowledge of the statistical properties (usually second-order moments) of the SOE. However, in real-world applications such information is seldom available, and the most practical solution is to use an adaptive filter. Adaptive filters can improve their performance, during normal operation, by learning the statistical characteristics through processing current signal observations.

In this chapter, we develop a mathematical framework for the design and performance evaluation of adaptive filters, both theoretically and by simulation. The goal of an adaptive filter is to "find and track" the optimum filter corresponding to the same signal operating environment with complete knowledge of the required statistics. In this context, optimum filters provide both guidance for the development of adaptive algorithms and a yardstick for evaluating the theoretical performance of adaptive filters. We start in Section 10.1 with discussion of a few typical application problems that can be effectively solved by using an adaptive filter. The performance of adaptive filters is evaluated using the concepts of stability, speed of adaptation, quality of adaptation, and tracking capabilities. These issues and the key features of an adaptive filter are discussed in Section 10.2. Since most adaptive algorithms originate from deterministic optimization methods, in Section 10.3 we introduce the family of steepest-descent algorithms and study their properties. Sections 10.4 and 10.5 provide a detailed discussion of the derivation, properties, and applications of the two most important adaptive filtering algorithms: the least mean square (LMS) and the recursive least-squares (RLS) algorithms. The conventional RLS algorithm, introduced in Section 10.5, can be used for either array processing (multiple-sensor or general input data vector) applications or FIR filtering (single-sensor or shift-invariant input data vector) applications. Section 10.6 deals with different implementations of the RLS algorithm for array processing applications, whereas Section 10.7 provides fast implementations of the RLS algorithm for the FIR filtering case. The development of the later algorithms is a result of the shift invariance of the data stored in the memory of the FIR filter. Finally, in Section 10.8 we provide a concise introduction to the tracking properties of the LMS and the RLS algorithms.

10.1 TYPICAL APPLICATIONS OF ADAPTIVE FILTERS

As we have already seen in Chapter 1, many practical applications cannot be successfully solved by using fixed digital filters because either we do not have sufficient information to design a digital filter with fixed coefficients or the design criteria change during the normal operation of the filter. Most of these applications can be successfully solved by using a special type of "smart" filters known collectively as *adaptive filters*. The distinguishing feature of adaptive filters is that they can modify their response to improve their performance during operation without any intervention from the user.

The best way to introduce adaptive filters is with some applications for which they are well suited. These and other applications are discussed in greater detail in the sequel as we develop the necessary background and tools.

10.1.1 Echo Cancelation in Communications

An echo is the delayed and distorted version of an original signal that returns to its source. In some applications (radar, sonar, or ultrasound), the echo is the wanted signal; however, in communication applications, the echo is an unwanted signal that must be eliminated. There are two types of echoes in communication systems: (1) *electrical* or *line echoes*, which are generated electrically due to impedance mismatches at points along the transmission medium, and (2) *acoustic echoes*, which result from the reflection of sound waves and acoustic coupling between a microphone and a loudspeaker.

Here we focus on electrical echoes in voice communications; electrical echoes in data communications are discussed in Section 10.4.4, and acoustic echoes in teleconferencing and hands-free telephony were discussed in Section 1.4.1.

Electrical echoes are observed on long-distance telephone circuits. A simplified form of such a circuit, which is sufficient for the present discussion, is shown in Figure 10.1. The local links from the customer to the telephone office consist of bidirectional two-wire connections, whereas the connection between the telephone offices is a four-wire carrier facility that may include a satellite link. The conversion between two-wire and four-wire links is done by special devices known as *hybrids*. An ideal hybrid should pass (1) the incoming signal to the two-wire output without any leakage into its output port and (2) the signal from the two-wire circuit to its output port without reflecting any energy back to the two-wire line (Sondhi and Berkley 1980). In practice, due to impedance mismatches, the hybrids do not operate perfectly. As a result, some energy on the incoming branch of the four-wire circuit leaks into the outgoing branch and returns to the source as an echo (see Figure 10.1). This echo, which is usually 11 dB down from the original signal, makes it difficult to carry on a conversation if the round-trip delay is larger than 40 ms. Satellite links, as a consequence of high altitude, involve round-trip delays of 500 to 600 ms.

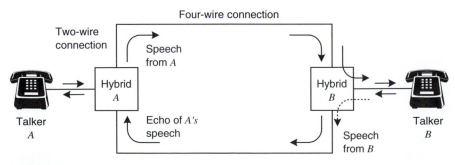

FIGURE 10.1
Echo generation in a long-distance telephone network.

The first devices used by telephone companies to control voice echoes were echo suppressors. Basically, an *echo suppressor* is a voice-activated switch that attempts to impose an open circuit on the return path from listener to talker when the listener is silent (see Figure 10.2). The main problems with these devices are speech clipping during double-talking and the inability to effectively deal with round-trip delays longer than 100 ms (Weinstein 1977).

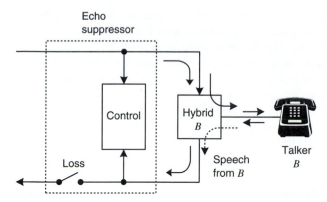

FIGURE 10.2
Principle of echo suppression.

The problems associated with echo suppressors could be largely avoided if we could estimate the *transmission path* from point C to point D (see Figure 10.3), which is known as the *echo path*. If we knew the echo path, we could design a filter that produced a copy or replica of the echo signal when driven by the signal at point C. Subtraction of the echo replica from the signal at point D will eliminate the echo without distorting the speech of the second talker that may be present at point D. The resulting device, shown in Figure 10.3, is known as an *echo canceler*.

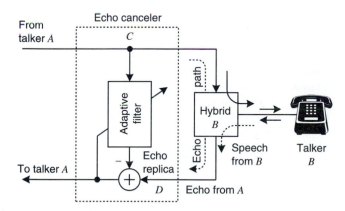

FIGURE 10.3
Principle of echo cancelation.

In practice, the channel characteristics are generally *not* known. For dial-up telephone lines, the channel differs from call to call, and the characteristics of radio and microwave channels (phase perturbations, fading, etc.) change significantly with time. Therefore, we *cannot* design and use a *fixed* echo canceler with satisfactory performance for all possible connections. There are two possible ways around this problem:

1. Design a compromise *fixed* echo canceler based on some "average" echo path, assuming that we have sufficient information about the connections to be seen by the canceler.
2. Design an *adaptive echo canceler* that can "learn" the echo path when it is first turned on and afterward "tracks" its variations without any intervention from the designer. Since an adaptive canceler matches the echo path for any given connection, it performs better than a compromise fixed canceler.

We stress that the main task of the canceler is to estimate the echo signal with sufficient accuracy; the estimation of the echo path is simply the means of achieving this goal. The performance of the canceler is measured by the attenuation, in decibels, of the echo, which is known as *echo return loss enhancement*. The adaptive echo canceler achieves this goal by modifying its response, using the residual echo signal in an as yet unspecified way.

Adaptive echo cancelers are widely used in voice telecommunications, and the international standards organization CCITT has issued a set of recommendations (CCITT G. 165) that outlines the basic requirements for echo cancelers. More details can be found in Weinstein (1977) and Murano et al. (1990).

10.1.2 Equalization of Data Communication Channels

Channel equalization, which is probably the most widely employed technique in practical data transmission systems, was first introduced in Section 1.4.1. In Section 6.8 we discussed the design of symbol rate zero-forcing and optimum MSE equalizers. As we recall, every pulse propagating through the channel suffers a certain amount of *time dispersion* because the frequency response of the channel deviates from the ideal one of constant magnitude and linear phase. Some typical sources of dispersion for practical communication channels are summarized in Table 10.1. As a result, the tails of adjacent pulses interfere with the measurement of the current pulse (intersymbol interference) and can lead to an incorrect decision.

TABLE 10.1
Summary of causes of dispersion in various communications systems.

Transmission system	Causes of dispersion
Cable TV	Transmitter filtering; coaxial-cable dispersion; cable amplifiers; reflections from impedance mismatches; bandpass filters
Microwave radio	Transmitter filtering; reflections from impedance mismatches; multipath propagation; scattering; input bandpass filtering
Voiceband modems	Digital-to-analog image suppression; channel filtering; twisted-pair transmission line; multiplexing and demultiplexing filters; hybrids; antialias lowpass filters
Troposcatter radio	Transmitter filtering; atmospheric dispersion; scattering at interface between troposphere and stratosphere; receiver bandpass filtering; input amplifiers

Source: From Treichler et al. 1996.

Since the channel can be modeled as a linear system, assuming that the receiver and the transmitter do not include any nonlinear operations, we can compensate for its distortion by using a linear equalizer. The goal of the equalizer is to restore the received pulse, as closely as possible, to its original shape. The equalizer transforms the channel to a near-ideal one if its response resembles the *inverse* of the channel. Since the channel is unknown and possibly time-varying, there are two ways to approach the problem: (1) Design a compromise fixed equalizer to obtain satisfactory performance over a broad range of channels, or (2) design an equalizer that can learn the inverse of the particular channel and then track its variation in real time.

The characteristics of the equalizer are adjusted by some algorithm that attempts to attain the best possible performance. The most appropriate criterion of performance for data transmission systems is the probability of error. However, it cannot be used for two reasons: (1) the "correct" symbol is unknown to the receiver (otherwise there would be no reason to communicate), and (2) the number of decisions needed to estimate the low probabilities of error is extremely large. Thus, practical equalizers assess their performance by using some function of the difference between the correct symbol and their output. The operation of practical equalizers involves three modes of operation, dependent on how we substitute for the unavailable correct symbol sequence.

Training mode: A known *training sequence* is transmitted, and the equalizer attempts to improve its performance by comparing its output to a synchronized replica of the training sequence stored at the receiver. Usually this mode is used when the equalizer starts a transmission session.

Decision-directed mode: At the end of the training session, when the equalizer starts making reliable decisions, we can replace the training sequence with the equalizer's own decisions.

"Blind" or self-recovering mode: There are several applications in which the use of a training sequence is not desired or feasible. This may occur in multipoint networks for computer communications or in wideband digital systems over coaxial facilities during rerouting (Godard 1980; Sato 1975). Also when the decision-directed mode of a microwave channel equalizer fails, after deep fades, we do not have a reverse channel to call for retraining (Foschini 1985). In such cases, where the equalizer should be able to learn or recover the characteristics of the channel without the benefit of a training sequence, we say that the equalizer operates in blind or self-recovering mode.

Adaptive equalization is a mature technology that has had the greatest impact on digital communications systems, including voiceband, microwave and troposcatter radio, and cable TV modems (Qureshi 1985; Lee and Messerschmitt 1994; Gitlin et al. 1992; Bingham 1988; Treichler et al. 1996, 1998).

10.1.3 Linear Predictive Coding

The efficient storage and transmission of analog signals using digital systems requires the minimization of the number of bits necessary to represent the signal while maintaining the quality to an acceptable level according to a certain criterion of performance. The conversion of an analog (continuous-time, continuous-amplitude) signal to a digital (discrete-time, discrete-amplitude) signal involves two processes: sampling and quantization. Sampling converts a continuous-time signal to a discrete-time signal by measuring its amplitude at equidistant intervals of time. Quantization involves the representation of the measured continuous amplitude by using a finite number of symbols. Therefore, a small range of amplitudes will use the same symbol (see Figure 10.4). A code word is assigned to each symbol by the coder. When the digital representation is used for digital signal processing, the quantization levels and the corresponding code words are uniformly distributed. However, for coding applications, levels may be nonuniformly distributed to match the distribution of the signal amplitudes.

For all practical purposes, the range of a quantizer is equal to $R_Q = \Delta \cdot 2^B$, where Δ is the quantization step size and B is the number of bits, and should cover the dynamic range of the signal. The difference between the unquantized sample $x(n)$ and the quantized sample $\hat{x}(n)$, that is,

$$e(n) \triangleq \hat{x}(n) - x(n) \tag{10.1.1}$$

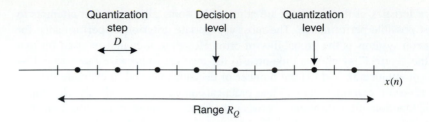

FIGURE 10.4
Partitioning of the range of a 3-bit (eight-level) uniform quantizer.

is known as the *quantization error* and is always in the range $-\Delta/2 \leq e(n) \leq \Delta/2$. If we define the signal-to-noise ratio by

$$\text{SNR} \triangleq \frac{E\{x^2(n)\}}{E\{e^2(n)\}} \tag{10.1.2}$$

it can be shown (Rabiner and Schafer 1978; Jayant and Noll 1984) that

$$\text{SNR(dB)} \simeq 6B \tag{10.1.3}$$

which states that each added binary digit increases the SNR by 6 dB.

For a fixed number of bits, decreasing the dynamic range of the signal (and therefore the range of the quantizer) decreases the required quantization step and therefore the average quantization error power. Therefore, we can increase the SNR by reducing the dynamic range, or equivalently the variance of the signal. If the signal samples are significantly correlated, the variance of the difference between adjacent samples is smaller than the variance of the original signal. Thus, we can improve the SNR by quantizing this difference instead of the original signal.

The differential quantization concept is exploited by the *linear predictive coding (LPC)* system illustrated in Figure 10.5. The quantized signal is the difference

$$d(n) = x(n) - \tilde{x}(n) \tag{10.1.4}$$

where $\tilde{x}(n)$ is an estimate or prediction of the signal $x(n)$ obtained by the predictor using a quantized version

$$\hat{x}(n) = \tilde{x}(n) + \hat{d}(n) \tag{10.1.5}$$

of the original signal (see Figure 10.5). If the quantization error of the difference signal is

$$e_d(n) = \hat{d}(n) - d(n) \tag{10.1.6}$$

we obtain

$$\hat{x}(n) = x(n) + e_d(n) \tag{10.1.7}$$

using (10.1.4) and (10.1.5). The significance of (10.1.7) is that the quantization error of the original signal is equal to the quantization error of the difference signal, independently of the properties of the predictor. Note that if $c'(n) = c(n)$, that is, there are no transmission or storage errors, then the signal reconstructed by the decoder is $\hat{x}'(n) = \hat{x}(n)$. If the prediction is good, the dynamic range of $d(n)$ should be smaller than the dynamic range of $x(n)$, resulting in a smaller quantization noise for the same number of bits or the same quantization noise with a smaller number of bits. The performance of the LPC system depends on the accuracy of the predictor. In most practical applications, we use a linear predictor that forms an estimate (prediction) $\tilde{x}(n)$ of the present sample $x(n)$ as a linear combination of the M past samples, that is,

$$\tilde{x}(n) = \sum_{k=1}^{M} a_k \hat{x}(n-k) \tag{10.1.8}$$

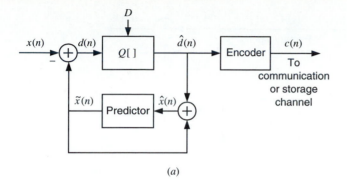

(a)

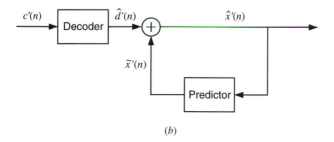

(b)

FIGURE 10.5
Block diagram of a linear predictive coding system: (a) coder
and (b) decoder.

The coefficients $\{a_k\}_1^M$ of the linear predictor are determined by exploiting the correlation
between adjacent samples of the input signal with the objective to make the prediction error
as small as possible. Since the statistical properties of the signal $x(n)$ are unknown and
change with time, we *cannot* design an optimum fixed predictor. The established practical solution uses an *adaptive* linear predictor that automatically adjusts its coefficients to
compute a "good" prediction at each time instant. A detailed discussion of adaptive linear
prediction and its application to audio, speech, and video signal coding is provided in Jayant
and Noll (1984).

10.1.4 Noise Cancelation

In Section 1.4.1 we discussed the concept of active noise control using adaptive filters.
We now provide a theoretical explanation for the general problem of noise canceling using
multiple sensors. The principle of general noise cancelation is illustrated in Figure 10.6. The
signal of interest $s(n)$ is corrupted by uncorrelated additive noise $v_1(n)$, and the combined
signal $s(n) + v_1(n)$ provides what is known as *primary input*. A second sensor, located
at a different point, acquires a noise $v_2(n)$ (*reference input*) that is uncorrelated with the
signal $s(n)$ but correlated with the noise $v_1(n)$. If we can design a filter that provides a good
estimate $\hat{y}(n)$ of the noise $v_1(n)$, by exploiting the correlation between $v_1(n)$ and $v_2(n)$,
then we could recover the desired signal by subtracting $\hat{y}(n) \approx v_1(n)$ from the primary
input.

Let us assume that the signals $s(n)$, $v_1(n)$, and $v_2(n)$ are jointly wide-sense stationary
with zero mean values. The "clean" signal is given by the error

$$e(n) = s(n) + [v_1(n) - \hat{y}(n)]$$

where $\hat{y}(n)$ depends on the filter structure and parameters. The MSE is given by

$$E\{|e(n)|^2\} = E\{|s(n)|^2\} + E\{|v_1(n) - \hat{y}(n)|^2\}$$

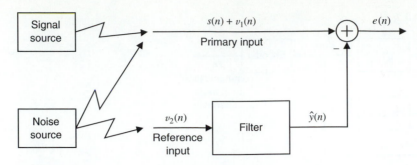

FIGURE 10.6
Principle of adaptive noise cancelation using a reference input.

because the signals $s(n)$ and $v_1(n) - \hat{y}(n)$ are uncorrelated. Since the signal power is not influenced by the filter, if we design a filter that minimizes the total output power $E\{|e(n)|^2\}$, then that filter will minimize the output noise power $E\{|v_1(n) - \hat{y}(n)|^2\}$. Therefore, $\hat{y}(n)$ will be the MMSE estimate of the noise $v_1(n)$, and the canceler maximizes the output signal-to-noise ratio. If we know the second-order moments of the primary and reference inputs, we can design an optimum linear canceler using the techniques discussed in Chapter 6. However, in practice, the design of an optimum canceler is not feasible because the required statistical moments are either unknown or time-varying. Once again, a successful solution can be obtained by using an adaptive filter that automatically adjusts its parameters to obtain the best possible estimate of the interfering noise (Widrow et al. 1975).

10.2 PRINCIPLES OF ADAPTIVE FILTERS

In this section, we discuss a mathematical framework for the analysis and performance evaluation of adaptive algorithms. The goal is to develop design guidelines for the application of adaptive algorithms to practical problems. The need for adaptive filters and representative applications that can benefit from their use have been discussed in Sections 1.4.1 and 10.1.

10.2.1 Features of Adaptive Filters

The applications we have discussed are only a sample from a multitude of practical problems that can be successfully solved by using adaptive filters, that is, filters that automatically change their characteristics to attain the right response at the right time. Every adaptive filtering application involves one or more input signals and a desired response signal that may or may not be accessible to the adaptive filter. We collectively refer to these signals as the *signal operating environment* (*SOE*) of the adaptive filter. Every adaptive filter consists of three modules (see Figure 10.7):

> *Filtering structure.* This module forms the output of the filter using measurements of the input signal or signals. The filtering structure is *linear* if the output is obtained as a linear combination of the input measurements; otherwise it is said to be *nonlinear*. For example, the filtering module can be an adjustable finite impulse response (FIR) digital filter implemented with a direct or lattice structure or a recursive filter implemented using a cascade structure. The structure is fixed by the designer, and its parameters are adjusted by the adaptive algorithm.
>
> *Criterion of performance (COP).* The output of the adaptive filter and the desired response (when available) are processed by the COP module to assess its quality with respect to the requirements of the particular application. The choice of the

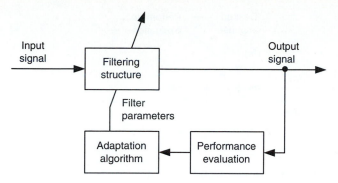

FIGURE 10.7
Basic elements of a general adaptive filter.

criterion is a balanced compromise between what is acceptable to the user of the application and what is mathematically tractable; that is, it can be manipulated to derive an adaptive algorithm. Most adaptive filters use some average form of the square error because it is mathematically tractable and leads to the design of useful practical systems.

Adaptation algorithm. The adaptive algorithm uses the value of the criterion of performance, or some function of it, and the measurements of the input and desired response (when available) to decide how to modify the parameters of the filter to improve its performance. The complexity and the characteristics of the adaptive algorithm are functions of the filtering structure and the criterion of performance.

The design of any adaptive filter requires some generic *a priori information* about the SOE and a deep understanding of the particular application. This information is needed by the designer to choose the criterion of performance and the filtering structure. Clearly, unreliable a priori information and/or incorrect assumptions about the SOE can lead to serious performance degradations or even unsuccessful adaptive filter applications. The conversion of the performance assessment to a successful parameter adjustment strategy, that is, the design of an adaptive algorithm, is the most difficult step in the design and application of adaptive filters.

If the characteristics of the SOE are constant, the goal of the adaptive filter is to find the parameters that give the best performance and then stop the adjustment. The initial period, from the time the filter starts its operation until the time it gets reasonably close to its best performance, is known as the *acquisition* or *convergence mode*. However, when the characteristics of the SOE change with time, the adaptive filter should first find and then continuously readjust its parameters to track these changes. In this case, the filter starts with an acquisition phase that is followed by a *tracking mode*.

A very influential factor in the design of adaptive algorithms is the availability of a desired response signal. We have seen that for certain applications, the desired response may not be available for use by the adaptive filter. Therefore, the adaptation must be performed in one of two ways:

Supervised adaptation. At each time instant, the adaptive filter knows in advance the desired response, computes the error (i.e., the difference between the desired and actual response), evaluates the criterion of performance, and uses it to adjust its coefficients. In this case, the structure in Figure 10.7 is simplified to that of Figure 10.8.

Unsupervised adaptation. When the desired response is unavailable, the adaptive filter cannot explicitly form and use the error to improve its behavior. In some applications, the input signal has some measurable property (i.e., constant envelope) that is lost by the time it reaches the adaptive filter. The adaptive filter adjusts its parameters in such a way as to restore the lost property of the input signal. The *property restoral*

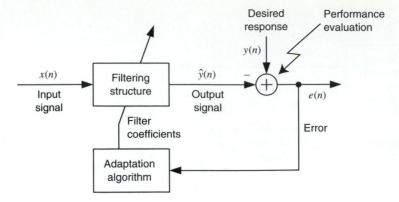

FIGURE 10.8
Basic elements of a supervised adaptive filter.

approach to adaptive filtering was introduced in Treichler et al. (1987). In some other applications (e.g., digital communications) the basic task of the adaptive filter is to classify each received pulse to one of a finite set of symbols. In this case we basically have a problem of unsupervised classification (Fukunaga 1990).

In this chapter we focus our discussion on supervised adaptive filters, that is, filters that have access to a desired response signal; unsupervised adaptive filters, which operate without the benefit of a desired response, are discussed in Section 12.3, in the context of blind equalization.

10.2.2 Optimum versus Adaptive Filters

We have mentioned several times that the theory of stochastic processes provides the mathematical framework for the design and analysis of optimum filters. In Chapter 6, we introduced filters that are optimum according to the MSE criterion of performance; and in Chapter 7, we developed algorithms and structures for their efficient design and implementation. However, optimum filters are a theoretical tool and cannot be used in practical applications because we do not know the statistical quantities (e.g., second-order moments) that are required for their design. Adaptive filters can be thought as the practical counterpart of optimum filters: They try to reach the performance of optimum filters by processing measurements of the SOE in real time, which makes up for the lack of a priori statistics.

For this analysis, we consider the general case of a linear combiner that includes filtering and prediction as special cases. However, for convenience we use the terms *filters* and *filtering*. We remind the reader that, from a mathematical point of view, the key difference between a linear combiner and an FIR filter or predictor is the shift invariance (temporal ordering) of the input data vector. This difference, which is illustrated in Figure 10.9, also has important implications in the implementation of adaptive filters. To this end, suppose that the SOE is comprised of M input signals $x_k(n, \zeta)$ and a desired response signal $y(n, \zeta)$, which are sample realizations of random sequences.[†]

Then the estimate of $y(n, \zeta)$ is computed by using the linear combiner

$$\hat{y}(n, \zeta) = \sum_{k=1}^{M} c_k^*(n) x_k(n, \zeta) \triangleq \mathbf{c}^H(n)\mathbf{x}(n, \zeta) \qquad (10.2.1)$$

where
$$\mathbf{c}(n) = [c_1(n)\ c_2(n)\ \cdots\ c_M(n)]^T \qquad (10.2.2)$$

[†] For clarity, in this section only, we include the dependence on ζ to denote random variables.

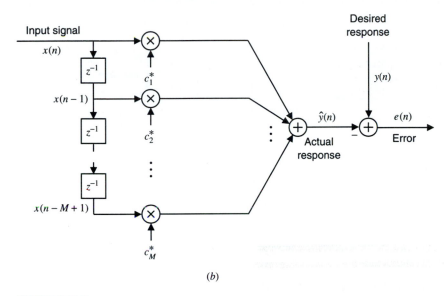

FIGURE 10.9
Illustration of the difference of the input signal between (a) a multiple-input linear
combiner and (b) a single-input FIR filter.

is the coefficient vector and

$$\mathbf{x}(n, \zeta) = [x_1(n, \zeta) \ x_2(n, \zeta) \ \cdots \ x_M(n, \zeta)]^T \tag{10.2.3}$$

is the input data vector. For single-sensor applications, the input data vector is shift-invariant

$$\mathbf{x}(n) = [x(n, \zeta) \ x(n - 1, \zeta) \ \cdots \ x(n - M + 1, \zeta)]^T \tag{10.2.4}$$

and the linear combiner takes the form of the FIR filter

$$\hat{y}(n, \zeta) = \sum_{k=0}^{M-1} h(n, k)x(n - k, \zeta) \triangleq \mathbf{c}^H(n)\mathbf{x}(n, \zeta) \tag{10.2.5}$$

where $c_k(n) = h^*(n, k)$ are the samples of the impulse response at time n.

Optimum filters. If we know the second-order moments of the SOE, we can design an
optimum filter $\mathbf{c}_o(n)$ by solving the normal equations

$$\mathbf{R}(n)\mathbf{c}_o(n) = \mathbf{d}(n) \tag{10.2.6}$$

where

$$\mathbf{R}(n) = E\{\mathbf{x}(n, \zeta)\mathbf{x}^H(n, \zeta)\} \tag{10.2.7}$$

and
$$\mathbf{d}(n) = E\{\mathbf{x}(n, \zeta)y^*(n, \zeta)\} \tag{10.2.8}$$

are the correlation matrix of the input data vector and the cross-correlation between the input data vector and the desired response, respectively. During its normal operation, the optimum filter works with specific realizations of the SOE, that is,

$$\hat{y}_o(n, \zeta) = \mathbf{c}_o^H(n)\mathbf{x}(n, \zeta) \tag{10.2.9}$$

$$\varepsilon_o(n, \zeta) = y(n, \zeta) - \hat{y}_o(n, \zeta) \tag{10.2.10}$$

where $\hat{y}_o(n, \zeta)$ is the optimum estimate and $\varepsilon_o(n, \zeta)$ is the optimum instantaneous error [see Figure 10.10(a)]. However, the filter is optimized with respect to its average performance across all possible realizations of the SOE, and the MMSE

$$P_o(n) = E\{|\varepsilon_o(n, \zeta)|^2\} = P_y(n) - \mathbf{d}^H(n)\mathbf{c}_o(n) \tag{10.2.11}$$

shows how well the filter performs on average. Also, we emphasize that the optimum coefficient vector is a nonrandom quantity and that the desired response is not essential for the operation of the optimum filter [see Equation (10.2.9)].

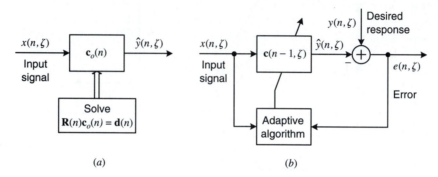

(a)   (b)

FIGURE 10.10
Illustration of the difference in operation between (a) optimum filters and (b) adaptive filters.

If the SOE is stationary, the optimum filter is computed once and is used with *all* realizations $\{\mathbf{x}(n, \zeta), y(n, \zeta)\}$. For nonstationary environments, the optimum filter design is repeated at every time instant n because the optimum filter is time-varying.

Adaptive filters. In most practical applications, where the second-order moments $\mathbf{R}(n)$ and $\mathbf{d}(n)$ are *unknown*, the use of an adaptive filter is the best solution. If the SOE is ergodic, we have

$$\mathbf{R} = \lim_{N \to \infty} \frac{1}{2N+1} \sum_{n=-N}^{N} \mathbf{x}(n, \zeta)\mathbf{x}^H(n, \zeta) \tag{10.2.12}$$

$$\mathbf{d} = \lim_{N \to \infty} \frac{1}{2N+1} \sum_{n=-N}^{N} \mathbf{x}(n, \zeta)y^*(n, \zeta) \tag{10.2.13}$$

because ensemble averages are equal to time averages (see Section 3.3). If we collect a sufficient amount of data $\{\mathbf{x}(n, \zeta), y(n, \zeta)\}_0^{N-1}$, we can obtain an acceptable estimate of the optimum filter by computing the estimates

$$\hat{\mathbf{R}}_N(\zeta) = \frac{1}{N} \sum_{n=0}^{N-1} \mathbf{x}(n, \zeta)\mathbf{x}^H(n, \zeta) \tag{10.2.14}$$

$$\hat{\mathbf{d}}_N(\zeta) = \frac{1}{N} \sum_{n=0}^{N-1} \mathbf{x}(n, \zeta) y^*(n, \zeta) \tag{10.2.15}$$

by time-averaging and then solving the linear system

$$\hat{\mathbf{R}}_N(\zeta) \mathbf{c}_N(\zeta) = \hat{\mathbf{d}}_N(\zeta) \tag{10.2.16}$$

The obtained coefficients can be used to filter the data in the interval $0 \le n \le N - 1$ or to start filtering the data for $n \ge N$, on a sample-by-sample basis, in real time. This procedure, which we called *block adaptive filtering* in Chapter 8, should be repeated each time the properties of the SOE change significantly. Clearly, block adaptive filters cannot track statistical variations within the operating block and cannot be used in all applications.

Indeed, there are applications, for example, adaptive equalization, in which each input sample should be processed immediately after its observation and before the arrival of the next sample. In such cases, we should use a sample-by-sample adaptive filter that starts filtering immediately after the observation of the pair $\{\mathbf{x}(0), y(0)\}$ using a "guess" $\mathbf{c}(-1)$ for the adaptive filter coefficients. Usually, the initial guess $\mathbf{c}(-1)$ is a very poor estimate of the optimum filter $\mathbf{c}_o$. However, this estimate is improved with time as the filter processes additional pairs of observations.

As we discussed in Section 10.2.1, an adaptive filter consists of three key modules: an adjustable filtering structure that uses input samples to compute the output, the criterion of performance that monitors the performance of the filter, and the adaptive algorithm that updates the filter coefficients. The key component of any adaptive filter is the *adaptive algorithm*, which is a rule to determine the filter coefficients from the available data $\mathbf{x}(n, \zeta)$ and $y(n, \zeta)$ [see Figure 10.10(b)]. The dependence of $\mathbf{c}(n, \zeta)$ on the input signal makes the adaptive filter a nonlinear and time-varying stochastic system.

The data available to the adaptive filter at time n are the input data vector $\mathbf{x}(n, \zeta)$, the desired response $y(n, \zeta)$, and the most recent update $\mathbf{c}(n - 1, \zeta)$ of the coefficient vector. The adaptive filter, at each time n, performs the following computations:

1. *Filtering:*

$$\hat{y}(n, \zeta) = \mathbf{c}^H(n - 1, \zeta) \mathbf{x}(n, \zeta) \tag{10.2.17}$$

2. *Error formation:*

$$e(n, \zeta) = y(n, \zeta) - \hat{y}(n, \zeta) \tag{10.2.18}$$

3. *Adaptive algorithm:*

$$\mathbf{c}(n, \zeta) = \mathbf{c}(n - 1, \zeta) + \Delta\mathbf{c}\{\mathbf{x}(n, \zeta), e(n, \zeta)\} \tag{10.2.19}$$

where the increment or correction term $\Delta\mathbf{c}(n, \zeta)$ is chosen to bring $\mathbf{c}(n, \zeta)$ close to $\mathbf{c}_o$, with the passage of time. If we can successively determine the corrections $\Delta\mathbf{c}(n, \zeta)$ so that $\mathbf{c}(n, \zeta) \simeq \mathbf{c}_o$, that is, $\|\mathbf{c}(n, \zeta) - \mathbf{c}_o\| < \delta$, for some $n > N_\delta$, we obtain a good approximation for $\mathbf{c}_o$ by avoiding the explicit averagings (10.2.14), (10.2.15), and the solution of the normal equations (10.2.16). A key requirement is that $\Delta\mathbf{c}(n, \zeta)$ must vanish if the error $e(n, \zeta)$ vanishes. Hence, $e(n, \zeta)$ plays a major role in determining the increment $\Delta\mathbf{c}(n, \zeta)$.

We notice that the estimate $\hat{y}(n, \zeta)$ of the desired response $y(n, \zeta)$ is evaluated using the *current* input vector $\mathbf{x}(n, \zeta)$ and the *past* filter coefficients $\mathbf{c}(n - 1, \zeta)$. The estimate $\hat{y}(n, \zeta)$ and the corresponding error $e(n, \zeta)$ can be considered as *predicted* estimates compared to the *actual* estimates that would be evaluated using the *current* coefficient vector $\mathbf{c}(n, \zeta)$. Coefficient updating methods that use the predicted error $e(n, \zeta)$ are known as *a priori type adaptive algorithms*.

If we use the *actual* estimates, obtained using the current estimate $\mathbf{c}(n, \zeta)$ of the adaptive filter coefficients, we have

1. *Filtering:*

$$\hat{y}_a(n, \zeta) = \mathbf{c}^H(n, \zeta) \mathbf{x}(n, \zeta) \tag{10.2.20}$$

2. *Error formation:*

$$\varepsilon(n, \zeta) = y(n, \zeta) - \hat{y}_a(n, \zeta) \tag{10.2.21}$$

3. *Adaptive algorithm:*

$$\mathbf{c}(n, \zeta) = \mathbf{c}(n - 1, \zeta) + \Delta\mathbf{c}\{\mathbf{x}(n, \zeta), \varepsilon(n, \zeta)\} \tag{10.2.22}$$

which are known as *a posteriori type adaptive algorithms*. The terms *a priori* and *a posteriori* were introduced in Carayannis et al. (1983) to emphasize the use of estimates evaluated before or after the updating of the filter coefficients. The difference between a priori and a posteriori errors and their meanings will be further clarified when we discuss adaptive least-squares filters in Section 10.5. The timing diagram for the above two algorithms is shown in Figure 10.11.

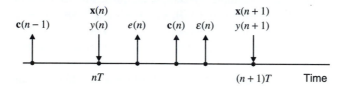

FIGURE 10.11
Timing diagrams for a priori and a posteriori adaptive algorithms.

In conclusion, the objective of an adaptive filter is to use the available data at time n, namely, $\{\mathbf{x}(n, \zeta), y(n, \zeta), \mathbf{c}(n - 1, \zeta)\}$, to update the "old" coefficient vector $\mathbf{c}(n - 1, \zeta)$ to a "new" estimate $\mathbf{c}(n, \zeta)$ so that $\mathbf{c}(n, \zeta)$ is closer to the optimum filter vector $\mathbf{c}_o(n)$ and the output $\hat{y}(n)$ is a better estimate of the desired response $y(n)$. Most adaptive algorithms have the following form:

$$\begin{bmatrix} \text{New} \\ \text{coefficient} \\ \text{vector} \end{bmatrix} = \begin{bmatrix} \text{old} \\ \text{coefficient} \\ \text{vector} \end{bmatrix} + \begin{bmatrix} \text{adaptation} \\ \text{gain} \\ \text{vector} \end{bmatrix} \cdot \begin{pmatrix} \text{error} \\ \text{signal} \end{pmatrix} \tag{10.2.23}$$

where the error signal is the difference between the desired response and the predicted or actual outputs of the adaptive filter. One of the fundamental differences among the various algorithms is the optimality of the used adaptation gain vector and the amount of computation required for its evaluation.

10.2.3 Stability and Steady-State Performance of Adaptive Filters

We now address the issues of stability and performance of adaptive filters. Since the goal of an adaptive filter $\mathbf{c}(n, \zeta)$ is *first to find and then track* the optimum filter $\mathbf{c}_o(n)$ as *quickly and accurately* as possible, we can evaluate its performance by measuring some function of its deviation

$$\tilde{\mathbf{c}}(n, \zeta) \triangleq \mathbf{c}(n, \zeta) - \mathbf{c}_o(n) \tag{10.2.24}$$

from the corresponding optimum filter. Clearly, an acceptable adaptive filter should be stable in the bounded-input bounded-output (BIBO) sense, and its performance should be close to that of the associated optimum filter. The analysis of BIBO stability is extremely difficult because adaptive filters are nonlinear, time-varying systems working in a random SOE. The performance of adaptive filters is primarily measured by investigating the value of the MSE as a function of time. To discuss these problems, first we consider an adaptive filter working in a stationary SOE, and then we extend our discussion to a nonstationary SOE.

The adaptive filter starts its operation at time, say, $n = 0$, and by processing the observations $\{x(n, \zeta), y(n, \zeta)\}_0^\infty$ generates a sequence of vectors $\{\mathbf{c}(n, \zeta)\}_0^\infty$ using the adaptive algorithm. Since the FIR filtering structure is always stable, the output or the error of the adaptive filter will be bounded if its coefficients are always kept close to the coefficients of the associated optimum filter. However, the presence of the feedback loop in every adaptive filter (see Figure 10.10) raises the issue of stability. In a stationary SOE, where the optimum filter $\mathbf{c}_o$ is constant, convergence of $\mathbf{c}(n, \zeta)$ to $\mathbf{c}_o$ as $n \to \infty$ will guarantee the BIBO stability of the adaptive filter. For a specific realization ζ, the kth component $c_k(n, \zeta)$ or the norm $\|\mathbf{c}(n, \zeta)\|$ of the vector $\mathbf{c}(n, \zeta)$ is a sequence of numbers that might or might not converge.[†] Since the coefficients $c_k(n, \zeta)$ are random, we must use the concept of *stochastic convergence* (Papoulis 1991).

We say that a random sequence converges *everywhere* if the sequence $c_k(n, \zeta)$ converges for every ζ, that is,

$$\lim_{n \to \infty} c_k(n, \zeta) = c_{o,k}(\zeta) \tag{10.2.25}$$

where the limit $c_{o,k}(\zeta)$ depends, in general, on ζ. Requiring the adaptive filter to converge to $\mathbf{c}_o$ for every possible realization of the SOE is both hard to guarantee and not necessary, because some realizations may have very small or zero probability of occurrence.

If we wish to ensure that the adaptive filter converges for the realizations of the SOE that may actually occur, we can use the concept of convergence almost everywhere. We say that the random sequence $c_k(n, \zeta)$ *converges almost everywhere* or *with probability 1* if

$$P\{\lim_{n \to \infty} |c_k(n, \zeta) - c_{o,k}(\zeta)| = 0\} = 1 \tag{10.2.26}$$

which implies that there can be some sample sequences that do not converge, which must occur with probability zero.

Another type of stochastic convergence that is used in adaptive filtering is defined by

$$\lim_{n \to \infty} E\{|c_k(n, \zeta) - c_{o,k}|^2\} = \lim_{n \to \infty} E\{|\tilde{c}_k(n, \zeta)|^2\} = 0 \tag{10.2.27}$$

and is known as *convergence in the MS sense*. The primary reason for the use of mean square (MS) convergence is that unlike the almost-everywhere convergence, it uses only one sequence of numbers that takes into account the averaging effect of all sample sequences. Furthermore, it uses second-order moments for verification and has an interpretation in terms of power. Convergence in MS does not imply—nor is implied by—convergence with probability 1. Since

$$\frac{E\{|\tilde{c}_k(n, \zeta)|^2\}}{\delta} = \frac{|E\{\tilde{c}_k(n, \zeta)\}|^2}{\delta} + \frac{\text{var}\{\tilde{c}_k(n, \zeta)\}}{\delta^2} \tag{10.2.28}$$

if we can show that $E\{\tilde{c}_k(n)\} \to 0$ as $n \to \infty$ and $\text{var}\{\tilde{c}_k(n, \zeta)\}$ is bounded for all n, we can ensure convergence in MS. In this case, we can say that an adaptive filter that operates in a stationary SOE is an asymptotically stable filter.

Performance measures

In theoretical investigations, any quantity that measures the deviation of an adaptive filter from the corresponding optimum filter can be used to evaluate its performance.

The *mean square deviation (MSD)*

$$\mathcal{D}(n) \triangleq E\{\|\mathbf{c}(n, \zeta) - \mathbf{c}_o(n)\|^2\} = E\{\|\tilde{\mathbf{c}}(n, \zeta)\|^2\} \tag{10.2.29}$$

[†] We recall that a sequence of real nonrandom numbers $a_0, a_1, a_2, \ldots$ converges to a number a if and only if for every positive number δ there exists a positive integer N_δ such that for all $n > N_\delta$, we have $|a_n - a| < \delta$. This is abbreviated by $\lim_{n \to \infty} a_n = a$.

measures the average distance between the coefficient vectors of the adaptive and optimum filters. Although the MSD is not measurable in practice, it is useful in analytical studies. Adaptive algorithms that minimize $\mathcal{D}(n)$ for each value of n are known as algorithms with *optimum learning*.

In Section 6.2.2 we showed that if the input correlation matrix is positive definite, any deviation, say, $\tilde{\mathbf{c}}(n)$, of the optimum filter coefficients from their optimum setting increases the mean square error (MSE) by an amount equal to $\tilde{\mathbf{c}}^H(n)\mathbf{R}\tilde{\mathbf{c}}(n)$, known as *excess MSE* (*EMSE*). In adaptive filters, the random deviation $\tilde{\mathbf{c}}(n, \zeta)$ from the optimum results in an EMSE, which is measured by the ensemble average of $\tilde{\mathbf{c}}^H(n, \zeta)\mathbf{R}\tilde{\mathbf{c}}(n, \zeta)$. For a posteriori adaptive filters, the MSE can be decomposed as

$$P'(n) \triangleq E\{|\varepsilon(n, \zeta)|^2\} \triangleq P'_o(n) + P'_{\text{ex}}(n) \tag{10.2.30}$$

where $P'_{\text{ex}}(n)$ is the EMSE and $P'_o(n)$ is the MMSE given by

$$P'_o(n) \triangleq E\{|\varepsilon_o(n, \zeta)|^2\} \tag{10.2.31}$$

with

$$\varepsilon_o(n, \zeta) \triangleq y(n, \zeta) - \mathbf{c}_o^H(n)\mathbf{x}(n, \zeta) \tag{10.2.32}$$

as the a posteriori optimum filtering error. Clearly, the a posteriori EMSE $P'_{\text{ex}}(n)$ is given by

$$P'_{\text{ex}}(n) \triangleq P'(n) - P'_o(n) \tag{10.2.33}$$

For a priori adaptive algorithms, where we use the "old" coefficient vector $\mathbf{c}(n-1, \zeta)$, it is more appropriate to use the a priori EMSE given by

$$P_{\text{ex}}(n) \triangleq P(n) - P_o(n) \tag{10.2.34}$$

where

$$P(n) \triangleq E\{|e(n, \zeta)|^2\} \tag{10.2.35}$$

and

$$P_o(n) \triangleq E\{|e_o(n, \zeta)|^2\} \tag{10.2.36}$$

with

$$e_o(n, \zeta) \triangleq y(n, \zeta) - \mathbf{c}_o^H(n-1)\mathbf{x}(n, \zeta) \tag{10.2.37}$$

as the a priori optimum filtering error. If the SOE is stationary, we have $\varepsilon_o(n, \zeta) = e_o(n, \zeta)$; that is, the optimum a priori and a posteriori errors are identical.

The dimensionless ratio

$$\mathcal{M}(n) \triangleq \frac{P_{\text{ex}}(n)}{P_o(n)} \quad \text{or} \quad \mathcal{M}'(n) \triangleq \frac{P'_{\text{ex}}(n)}{P'_o(n)} \tag{10.2.38}$$

known as *misadjustment*, is a useful measure of the quality of adaptation. Since the EMSE is always positive, *there is no adaptive filter that can perform (on the average) better than the corresponding optimum filter*. In this sense, we can say that the excess MSE or the misadjustment measures the cost of adaptation.

Acquisition and tracking

Plots of the MSD, MSE, or $\mathcal{M}(n)$ as a function of n, which are known as *learning curves*, characterize the performance of an adaptive filter and are widely used in theoretical and experimental studies. When the adaptive filter starts its operation, its coefficients provide a poor estimate of the optimum filter and the MSD or the MSE is very large. As the number of observations processed by the adaptive filter increases with time, we expect the quality of the estimate $\mathbf{c}(n, \zeta)$ to improve, and therefore the MSD and the MSE to decrease. The property of an adaptive filter to bring the coefficient vector $\mathbf{c}(n, \zeta)$ close to the optimum filter $\mathbf{c}_o$, independently of the initial condition $\mathbf{c}(-1)$ and the statistical properties of the SOE, is called *acquisition*. During the acquisition phase, we say that the adaptive filter is in a transient mode of operation.

A natural requirement for any adaptive algorithm is that adaptation stops after the algorithm has found the optimum filter $\mathbf{c}_o$. However, owing to the randomness of the SOE and the finite amount of data used by the adaptive filter, its coefficients continuously fluctuate about their optimum settings, that is, about the coefficients of the optimum filter, in a random manner. As a result, the adaptive filter reaches a steady-state mode of operation, after a certain time, and its performance stops improving.

The transient and steady-state modes of operation in a stationary SOE are illustrated in Figure 10.12(a). The duration of the acquisition phase characterizes the *speed of adaptation* or *rate of convergence* of the adaptive filter, whereas the steady-state EMSE or misadjustment characterizes the *quality of adaptation*. These properties depend on the SOE, the filtering structure, and the adaptive algorithm.

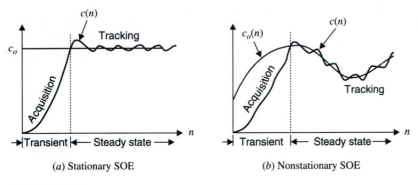

(a) Stationary SOE   (b) Nonstationary SOE

FIGURE 10.12
Modes of operation in a stationary and nonstationary SOE.

At each time n, any adaptive filter computes an estimate of the optimum filter using a finite amount of data. The error resulting from the finite amount of data is known as *estimation error*. An additional error, known as the *lag error*, results when the adaptive filter attempts to track a time-varying optimum filter $\mathbf{c}_o(n)$ in a nonstationary SOE. The modes of operation of an adaptive filter in a nonstationary SOE are illustrated in Figure 10.12(b). The SOE of the adaptive filter becomes nonstationary if $\mathbf{x}(n, \zeta)$ or $y(n, \zeta)$ or both are nonstationary. The nonstationarity of the input is more severe than that of the desired response because it may affect the invertibility of $\mathbf{R}(n)$. Since the adaptive filter has to first acquire and then track the optimum filter, tracking is a steady-state property. Therefore, in general, the speed of adaptation (a transient-phase property) and the tracking capability (a steady-state property) are two different characteristics of the adaptive filter. Clearly, tracking is feasible only if the statistics of the SOE change "slowly" compared to the speed of tracking of the adaptive filter. These concepts will become more precise in Section 10.8, where we discuss the tracking properties of adaptive filters.

10.2.4 Some Practical Considerations

The complexity of the hardware or software implementation of an adaptive filter is basically determined by the following factors: (1) the number of instructions per time update or computing time required to complete one time updating; (2) the number of memory locations required to store the data and the program instructions; (3) the structure of information flow in the algorithm, which is very important for implementations using parallel processing, systolic arrays, or VLSI chips; and (4) the investment in hardware design tools and software development. We focus on implementations for general-purpose computers or special-purpose digital signal processors that basically involve programming in a high level or assembly language. More details about DSP software development can be found in Embree and Kimble (1991) and in Lapsley et al. (1997).

The digital implementation of adaptive filters implies the use of finite-word-length arithmetic. As a result, the performance of the practical (finite-precision) adaptive filters deviates from the performance of ideal (infinite-precision) adaptive filters. Finite-precision implementation affects the performance of adaptive filters in several complicated ways. The major factors are (1) the quantization of the input signal(s) and the desired response, (2) the quantization of filter coefficients, and (3) the roundoff error in the arithmetic operations used to implement the adaptive filter. The nonlinear nature of adaptive filters coupled with the nonlinearities introduced by the finite-word-length arithmetic makes the performance evaluation of practical adaptive filters extremely difficult. Although theoretical analysis provides insight and helps to clarify the behavior of adaptive filters, the most effective way is to simulate the filter and measure its performance.

Finite precision affects two important properties of adaptive filters, which, although related, are not equivalent. Let us denote by $\mathbf{c}_{ip}(n)$ and $\mathbf{c}_{fp}(n)$ the coefficient vectors of the filter implemented using infinite- and finite-precision arithmetic, respectively. An adaptive filter is said to be *numerically stable* if the difference vector $\mathbf{c}_{ip}(n) - \mathbf{c}_{fp}(n)$ remains always bounded, that is, the roundoff error propagation system is stable. Numerical stability is an inherent property of the adaptive algorithm and cannot be altered by increasing the numerical precision. Indeed, increasing the word length or reorganizing the computations will simply delay the divergence of an adaptive filter; only actual change of the algorithm can stabilize an adaptive filter by improving the properties of the roundoff error propagation system (Ljung and Ljung 1985; Cioffi 1987).

The *numerical accuracy* of an adaptive filter measures the deviation, at steady state, of any obtained estimates from theoretically expected values, due to roundoff errors. Numerical accuracy results in an increase of the output error without catastrophic problems and can be reduced by increasing the word length. In contrast, lack of numerical stability leads to catastrophic overflow (divergence or blowup of the algorithm) as a result of roundoff error accumulation. Numerically unstable algorithms converging before "explosion" may provide good numerical accuracy. Therefore, although the two properties are related, one does not imply the other.

Two other important issues are the sensitivity of an algorithm to bad or abnormal input data (e.g., poorly exciting input) and its sensitivity to initialization. All these issues are very important for the application of adaptive algorithms to real-world problems and are further discussed in the context of specific algorithms.

10.3 METHOD OF STEEPEST DESCENT

Most adaptive filtering algorithms are obtained by simple modifications of iterative methods for solving deterministic optimization problems. Studying these techniques helps one to understand several aspects of the operation of adaptive filters. In this section we discuss gradient-based optimization methods because they provide the ground for the development of the most widely used adaptive filtering algorithms.

As we discussed in Section 6.2.1, the error performance surface of an optimum filter, in a stationary SOE, is given by

$$P(\mathbf{c}) = P_y - \mathbf{c}^H \mathbf{d} - \mathbf{d}^H \mathbf{c} + \mathbf{c}^H \mathbf{R} \mathbf{c} \tag{10.3.1}$$

where $P_y = E\{|y(n)|^2\}$. Equation (10.3.1) is a quadratic function of the coefficients and represents a bowl-shaped surface (when $\mathbf{R}$ is positive definite) and has a unique minimum at $\mathbf{c}_o$ (optimum filter). There are two distinct ways to find the minimum of (10.3.1):

1. Solve the normal equations $\mathbf{R}\mathbf{c} = \mathbf{d}$, using a direct linear system solution method.
2. Find the minimum of $P(\mathbf{c})$, using an iterative minimization algorithm.

Although direct methods provide the solution in a finite number of steps, sometimes we prefer iterative methods because they require less numerical precision, are computationally less expensive, work when $\mathbf{R}$ is not invertible, and are the only choice for nonquadratic performance functions.

In all iterative methods, we start with an approximate solution (a guess), which we keep changing until we reach the minimum. Thus, to find the optimum $\mathbf{c}_o$, we start at some arbitrary point $\mathbf{c}_0$, usually the null vector $\mathbf{c}_0 = \mathbf{0}$, and then start a search for the "bottom of the bowl." The key is to choose the steps in a systematic way so that each step takes us to a lower point until finally we reach the bottom. What differentiates various optimization algorithms is how we choose the direction and the size of each step.

Steepest-descent algorithm (SDA)

If the function $P(\mathbf{c})$ has continuous derivatives, it is possible to approximate its value at an arbitrary neighboring point $\mathbf{c} + \Delta\mathbf{c}$ by using the Taylor expansion

$$P(\mathbf{c} + \Delta\mathbf{c}) = P(\mathbf{c}) + \sum_{i=1}^{M} \frac{\partial P(\mathbf{c})}{\partial c_i}\Delta c_i + \frac{1}{2}\sum_{i=1}^{M}\sum_{j=1}^{M}\Delta c_i \frac{\partial^2 P(\mathbf{c})}{\partial c_i \partial c_j}\Delta c_j + \cdots \quad (10.3.2)$$

or more compactly

$$P(\mathbf{c} + \Delta\mathbf{c}) = P(\mathbf{c}) + (\Delta\mathbf{c})^T \nabla P(\mathbf{c}) + \tfrac{1}{2}(\Delta\mathbf{c})^T[\nabla^2 P(\mathbf{c})](\Delta\mathbf{c}) + \cdots \quad (10.3.3)$$

where $\nabla P(\mathbf{c})$ is the gradient vector, with elements $\partial P(\mathbf{c})/\partial c_i$, and $\nabla^2 P(\mathbf{c})$ is the Hessian matrix, with elements $\partial^2 P(\mathbf{c})/(\partial c_i \partial c_j)$. For simplicity we consider filters with real coefficients, but the conclusions apply when the coefficients are complex. For the quadratic function (10.3.1), we have

$$\nabla P(\mathbf{c}) = 2(\mathbf{Rc} - \mathbf{d}) \quad (10.3.4)$$

$$\nabla^2 P(\mathbf{c}) = 2\mathbf{R} \quad (10.3.5)$$

and the higher-order terms are zero. For nonquadratic functions, higher-order terms are nonzero, but if $\|\Delta\mathbf{c}\|$ is small, we can use a quadratic approximation. We note that if $\nabla P(\mathbf{c}_o) = \mathbf{0}$ and $\mathbf{R}$ is positive definite, then $\mathbf{c}_o$ is the minimum because $(\Delta\mathbf{c})^T[\nabla^2 P(\mathbf{c}_o)]\cdot(\Delta\mathbf{c}) > 0$ for any nonzero $\Delta\mathbf{c}$. Hence, if we choose the step Δc so that $(\Delta\mathbf{c})^T \nabla P(\mathbf{c}) < 0$, we will have $P(\mathbf{c} + \Delta\mathbf{c}) < P(\mathbf{c})$, that is, we make a step to a point closer to the minimum. Since $(\Delta\mathbf{c})^T \nabla P(\mathbf{c}) = \|\Delta\mathbf{c}\|\|\nabla P(\mathbf{c})\|\cos\theta$, the reduction in MSE is maximum when $\Delta\mathbf{c} = -\nabla P(\mathbf{c})$. For this reason, the direction of the negative gradient is known as the direction of *steepest descent*. This leads to the following iterative minimization algorithm

$$\mathbf{c}_k = \mathbf{c}_{k-1} + \mu[-\nabla P(\mathbf{c}_{k-1})] \quad k \geq 0 \quad (10.3.6)$$

which is known as the *method of steepest descent* (Scales 1985). The positive constant μ, known as the *step-size parameter*, controls the size of the descent in the direction of the negative gradient. The algorithm is usually initialized with $\mathbf{c}_0 = \mathbf{0}$. The steepest-descent algorithm (SDA) is illustrated in Figure 10.13 for a single-parameter case.

For the cost function in (10.3.1), the SDA becomes

$$\mathbf{c}_k = \mathbf{c}_{k-1} + 2\mu(\mathbf{d} - \mathbf{Rc}_{k-1}) = (\mathbf{I} - 2\mu\mathbf{R})\mathbf{c}_{k-1} + 2\mu\mathbf{d} \quad (10.3.7)$$

which is a recursive difference equation. Note that k denotes an iteration in the SDA and has nothing to do with time. However, this iterative optimization can be combined with filtering to obtain a type of "asymptotically" optimum filter defined by

$$e(n, \zeta) = y(n, \zeta) - \mathbf{c}_{n-1}^H \mathbf{x}(n, \zeta) \quad (10.3.8)$$

$$\mathbf{c}_n = \mathbf{c}_{n-1} + 2\mu(\mathbf{d} - \mathbf{Rc}_{n-1}) \quad (10.3.9)$$

and is further discussed in Problem 10.2.

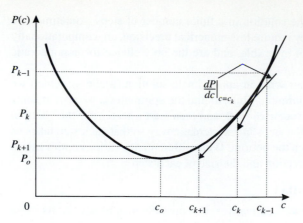

FIGURE 10.13
Illustration of gradient search of the MSE surface for the minimum error point.

There are two key performance factors in the design of iterative optimization algorithms: stability and rate of convergence.

Stability

An algorithm is said to be *stable* if it converges to the minimum regardless of the starting point. To investigate the stability of SDA, we rewrite (10.3.7) in terms of the coefficient error vector

$$\tilde{\mathbf{c}}_k \triangleq \mathbf{c}_k - \mathbf{c}_o \quad k \geq 0 \tag{10.3.10}$$

as

$$\tilde{\mathbf{c}}_k = (\mathbf{I} - 2\mu\mathbf{R})\tilde{\mathbf{c}}_{k-1} \quad k \geq 0 \tag{10.3.11}$$

which is a homogeneous difference equation. Using the principal-components transformation $\mathbf{R} = \mathbf{Q}\boldsymbol{\Lambda}\mathbf{Q}^H$ (see Section 3.5), we can write (10.3.11) as

$$\tilde{\mathbf{c}}'_k = (\mathbf{I} - 2\mu\boldsymbol{\Lambda})\tilde{\mathbf{c}}'_{k-1} \quad k \geq 0 \tag{10.3.12}$$

where

$$\tilde{\mathbf{c}}'_k = \mathbf{Q}^H \tilde{\mathbf{c}}_k \quad k \geq 0 \tag{10.3.13}$$

is the transformed coefficient error vector. Since $\boldsymbol{\Lambda}$ is diagonal, (10.3.12) consists of a set of M decoupled first-order difference equations

$$\tilde{c}'_{k,i} = (1 - 2\mu\lambda_i)\tilde{c}'_{k-1,i} \quad i = 1, 2, \ldots, M, k \geq 0 \tag{10.3.14}$$

with each describing a *natural mode* of the SDA. The solutions of (10.3.12) are given by

$$\tilde{c}'_{k,i} = (1 - 2\mu\lambda_i)^k \tilde{c}'_{0,i} \quad k \geq 0 \tag{10.3.15}$$

If for all $1 \leq i \leq M$

$$-1 < 1 - 2\mu\lambda_i < 1 \tag{10.3.16}$$

or equivalently

$$0 < \mu < \frac{1}{\lambda_i} \tag{10.3.17}$$

then $\tilde{c}'_{k,i}$, $1 \leq i \leq M$, tends to zero as $k \to \infty$. This implies that $\mathbf{c}_k$ converges *exponentially* to $\mathbf{c}_o$ as $k \to \infty$ because $\|\tilde{\mathbf{c}}'_k\| = \|\mathbf{Q}^T \tilde{\mathbf{c}}_k\| = \|\tilde{\mathbf{c}}_k\|$. If $\mathbf{R}$ is positive definite, its eigenvalues are positive and

$$0 < \mu < \frac{1}{\lambda_{\max}} \tag{10.3.18}$$

provides a necessary and sufficient condition for the convergence of SDA.

To investigate the transient behavior of the SDA as a function of k, we note that using (10.3.10), (10.3.11), and (10.3.14), we have

$$c_{k,i} = c_{o,i} + \sum_{i=1}^{M} q_{ik}\tilde{c}'_{0,i}(1 - 2\mu\lambda_i)^k \tag{10.3.19}$$

where $c_{o,i}$ are the optimum coefficients and q_{ik} the elements of the eigenvector matrix $\mathbf{Q}$. The MSE at step k is

$$P_k = P_o + \sum_{i=1}^{M} \lambda_i (1 - 2\mu\lambda_i)^{2k} |\tilde{c}'_{0,i}|^2 \tag{10.3.20}$$

and can be obtained by substituting (10.3.19) in (10.3.1). If μ satisfies (10.3.18), we have $\lim_{k\to\infty} P_k = P_o$ and the MSE converges exponentially to the optimum value. The curve obtained by plotting the MSE P_k as a function of the number of iterations k is known as the *learning curve*.

Rate of convergence

The rate (or speed) of convergence depends upon the algorithm and the nature of the performance surface. The most influential effect is inflicted by the condition number of the Hessian matrix that determines the shape of the contours of $P(\mathbf{c})$. When $P(\mathbf{c})$ is quadratic, it can be shown (Luenberger 1984) that

$$P(\mathbf{c}_k) \leq \left[\frac{\mathcal{X}(\mathbf{R}) - 1}{\mathcal{X}(\mathbf{R}) + 1} \right]^2 P(\mathbf{c}_{k-1}) \tag{10.3.21}$$

where $\mathcal{X}(\mathbf{R}) = \lambda_{\max}/\lambda_{\min}$ is the condition number of $\mathbf{R}$. If we recall that the eigenvectors corresponding to $\lambda_{\min}$ and $\lambda_{\max}$ point to the directions of minimum and maximum curvature, respectively, we see that the convergence slows down as the contours become more eccentric (flattened). For circular contours, that is, when $\mathcal{X}(\mathbf{R}) = 1$, the algorithm converges in one step. We stress that even if the $M - 1$ eigenvalues of $\mathbf{R}$ are equal and the remaining one is far away, still the convergence of the SDA is very slow.

The rate of convergence can be characterized by using the *time constant* τ_i defined by

$$1 - 2\mu\lambda_i = \exp\left(-\frac{1}{\tau_i}\right) \simeq 1 - \frac{1}{\tau_i} \tag{10.3.22}$$

which provides the time (or number of iterations) it takes for the ith mode $c_{k,i}$ of (10.3.19) to decay to $1/e$ of its initial value $c_{0,i}$. When $\mu \ll 1$, we obtain

$$\tau_i \simeq \frac{1}{2\mu\lambda_i} \tag{10.3.23}$$

In a similar fashion, the time constant $\tau_{i,\mathrm{mse}}$ for the MSE P_k can be shown to be

$$\tau_{i,\mathrm{mse}} \simeq \frac{1}{4\mu\lambda_i} \tag{10.3.24}$$

by using (10.3.20) and (10.3.22).

Thus, for all practical purposes, the time constant (for coefficient $\mathbf{c}_k$ or for MSE P_k) of the SDA is $\tau \simeq 1/(\mu\lambda_{\min})$, which in conjunction with $\mu < 1/\lambda_{\max}$ results in $\tau > \lambda_{\max}/\lambda_{\min}$. Hence, *the larger the eigenvalue spread of the input correlation matrix $\mathbf{R}$, the longer it takes for the SDA to converge.*

In the following example, we illustrate above-discussed properties of the SDA by using it to compute the parameters of a second-order forward linear predictor.

EXAMPLE 10.3.1. Consider a signal generated by the second-order autoregressive AR(2) process

$$x(n) + a_1 x(n - 1) + a_2 x(n - 2) = w(n) \tag{10.3.25}$$

where $w(n) \sim \mathrm{WGN}(0, \sigma_w^2)$. Parameters a_1 and a_2 are chosen so that the system (10.3.25) is minimum-phase. We want to design an adaptive filter that uses the samples $x(n-1)$ and $x(n-2)$ to predict the value $x(n)$ (desired response).

If we multiply (10.3.25) by $x(n-k)$, for $k = 0, 1, 2$, and take the mathematical expectation of both sides, we obtain a set of linear equations

$$r(0) + a_1 r(1) + a_2 r(2) = \sigma_w^2 \tag{10.3.26}$$

$$r(1) + a_1 r(0) + a_2 r(1) = 0 \tag{10.3.27}$$

$$r(2) + a_1 r(1) + a_2 r(0) = 0 \tag{10.3.28}$$

which can be used to express the autocorrelation of $x(n)$ in terms of model parameters a_1, a_2, and σ_w^2. Indeed, solving (10.3.26) through (10.3.28), we obtain

$$r(0) = \sigma_x^2 = \frac{1 + a_2}{1 - a_2} \frac{\sigma_w^2}{(1 + a_2)^2 - a_1^2}$$

$$r(1) = \frac{-a_1}{1 + a_2} r(0) \tag{10.3.29}$$

$$r(2) = \left(-a_2 + \frac{a_1^2}{1 + a_2} \right) r(0)$$

We choose $\sigma_x^2 = 1$, so that

$$\sigma_w^2 = \frac{(1 - a_2)[(1 + a_2)^2 - a_1^2]}{1 + a_2} \sigma_x^2 \tag{10.3.30}$$

The coefficients of the optimum predictor

$$\hat{y}(n) = \hat{x}(n) = c_{o,1} x(n-1) + c_{o,2} x(n-2) \tag{10.3.31}$$

are given by (see Section 6.5)

$$r(0) c_{o,1} + r(1) c_{o,2} = r(1) \tag{10.3.32}$$

$$r(1) c_{o,1} + r(0) c_{o,2} = r(2) \tag{10.3.33}$$

with

$$P_o^f = r(0) + r(1) c_{o,1} + r(0) c_{o,2} \tag{10.3.34}$$

whose comparison with (10.3.26) through (10.3.28) shows that $c_{o,1} = -a_1$, $c_{o,2} = -a_2$, and $P_o^f = \sigma_w^2$, as expected.

The eigenvalues of the input correlation matrix

$$\mathbf{R} = \begin{bmatrix} r(0) & r(1) \\ r(1) & r(0) \end{bmatrix} \tag{10.3.35}$$

are

$$\lambda_{1,2} = \left(1 \mp \frac{a_1}{1 + a_2} \right) \sigma_x^2 \tag{10.3.36}$$

from which the eigenvalue spread is

$$\mathcal{X}(\mathbf{R}) = \frac{\lambda_1}{\lambda_2} = \frac{1 - a_1 + a_2}{1 + a_1 + a_2} \tag{10.3.37}$$

which, if $a_2 > 0$ and $a_1 < 0$, is larger than 1.

Now we perform MATLAB experiments with varying eigenvalue spread $\mathcal{X}(\mathbf{R})$ and step-size parameter μ. In these experiments, we choose σ_w^2 so that $\sigma_x^2 = 1$. The SDA is given by

$$\mathbf{c}_k \triangleq [c_{k,1} \; c_{k,2}]^T = \mathbf{c}_{k-1} + 2\mu(\mathbf{d} - \mathbf{R}\mathbf{c}_{k-1})$$

where

$$\mathbf{d} = [r(1) \; r(2)]^T \quad \text{and} \quad \mathbf{c}_0 = [0 \; 0]^T$$

We choose two different sets of values for a_1 and a_2, one for a small and the other for a large eigenvalue spread. These values are shown in Table 10.2 along with the corresponding eigenvalue spread $\mathcal{X}(\mathbf{R})$ and the MMSE σ_w^2.

TABLE 10.2

Parameter values used in the SDA for the second-order forward prediction problem.

Eigenvalue spread	a_1	a_2	λ_1	λ_2	$\mathcal{X}(R)$	σ_w^2
Small	−0.1950	0.95	1.1	0.9	1.22	0.0965
Large	−1.5955	0.95	1.818	0.182	9.99	0.0322

Using each set of parameter values, the SDA is implemented starting with the null coefficient vector $\mathbf{c}_0$ with two values of step-size parameters. To describe the transient behavior of the algorithm, it is informative to plot the trajectory of $c_{k,1}$ versus $c_{k,2}$ as a function of the iteration index k along with the contours of the error surface $P(\mathbf{c}_k)$. The trajectory of $\mathbf{c}_k$ begins at the origin $\mathbf{c}_0 = \mathbf{0}$ and ends at the optimum value $\mathbf{c}_o = -[a_1 \ a_2]^T$. This illustration of the transient behavior can also be obtained in the domain of the transformed error coefficients $\tilde{\mathbf{c}}'_k$. Using (10.3.15), we see these coefficients are given by

$$\tilde{\mathbf{c}}'_k = \begin{bmatrix} \tilde{c}'_{k,1} \\ \tilde{c}'_{k,2} \end{bmatrix} = \begin{bmatrix} (1 - 2\mu\lambda_1)^k \tilde{c}'_{0,1} \\ (1 - 2\mu\lambda_2)^k \tilde{c}'_{0,2} \end{bmatrix} \tag{10.3.38}$$

where $\tilde{\mathbf{c}}'_0$ from (10.3.10) and (10.3.13) is given by

$$\tilde{\mathbf{c}}'_0 = \begin{bmatrix} \tilde{c}'_{0,1} \\ \tilde{c}'_{0,1} \end{bmatrix} = \mathbf{Q}^T \tilde{\mathbf{c}}_0 = \mathbf{Q}^T (\mathbf{c}_0 - \mathbf{c}_o) = -\mathbf{Q}^T \mathbf{c}_o = \mathbf{Q}^T \begin{bmatrix} a_1 \\ a_2 \end{bmatrix} \tag{10.3.39}$$

Thus the trajectory of $\tilde{\mathbf{c}}'_k$ begins at $\tilde{\mathbf{c}}'_0$ and ends at the origin $\tilde{\mathbf{c}}'_k = \mathbf{0}$. The contours of the MSE function in the transformed domain are given by $P_k - P_o$. From (10.3.20), these contours are given by

$$P_k - P_o^f = \sum_{i=1}^{2} \lambda_i (\tilde{c}'_k)^2 = \lambda_1 (\tilde{c}'_{k,1})^2 + \lambda_2 (\tilde{c}'_{k,2})^2 \tag{10.3.40}$$

Small eigenvalue spread and overdamped response. For this experiment, the parameter values were selected to obtain the eigenvalue spread approximately equal to 1 $[\mathcal{X}(\mathbf{R}) = 1.22]$. The step size selected was $\mu = 0.15$, which is less than $1/\lambda_{max} = 1/1.1 = 0.9$ for convergence. For this value of μ, the transient response is overdamped. Figure 10.14 shows four graphs indicating the

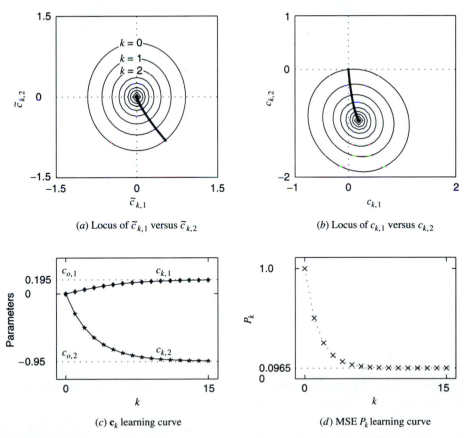

(a) Locus of $\tilde{c}_{k,1}$ versus $\tilde{c}_{k,2}$

(b) Locus of $c_{k,1}$ versus $c_{k,2}$

(c) $\mathbf{c}_k$ learning curve

(d) MSE P_k learning curve

FIGURE 10.14
Performance curves for the steepest-descent algorithm used in the linear prediction problem with step-size parameter $\mu = 0.15$ and eigenvalue spread $\mathcal{X}(\mathbf{R}) = 1.22$.

behavior of the algorithm. In the graph (a), the trajectory of $\tilde{\mathbf{c}}'_k$ is shown for $0 \le k \le 15$ along with the corresponding loci $\tilde{\mathbf{c}}'_k$ for a fixed value of $P_k - P_o$. The first two loci for $k = 0$ and 1 are numbered to show the direction of the trajectory. Graph (b) shows the corresponding trajectory and the contours for $\mathbf{c}_k$. Graph (c) shows plots of $c_{k,1}$ and $c_{k,2}$ as a function of iteration step k, while graph (d) shows a similar learning curve for the MSE P_k. Several observations can be made about these plots. The contours of constant $\tilde{\mathbf{c}}'_k$ are almost circular since the spread is approximately 1, while those of $\mathbf{c}_k$ are somewhat elliptical, which is to be expected. The trajectories of $\tilde{\mathbf{c}}'_k$ and $\mathbf{c}_k$ as a function of k are normal to the contours. The coefficients converge to their optimum values in a monotonic fashion, which confirms the overdamped nature of the response. Also this convergence is rapid, in about 15 steps, which is to be expected for a small eigenvalue spread.

Large eigenvalue spread and overdamped response. For this experiment, the parameter values were selected so that the eigenvalue spread was approximately equal to 10 $[\mathcal{X}(\mathbf{R}) = 9.99]$. The step size was again selected as $\mu = 0.15$. Figure 10.15 shows the performance plots for this experiment, which are similar to those of Figure 10.14. The observations are also similar except for those due to the larger spread. First, the contours, even in the transformed domain, are elliptical; second, the convergence is slow, requiring about 60 steps in the algorithm. The transient response is once again overdamped.

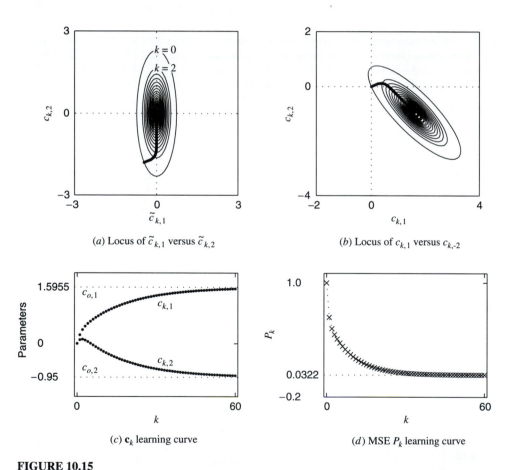

(a) Locus of $\tilde{c}_{k,1}$ versus $\tilde{c}_{k,2}$

(b) Locus of $c_{k,1}$ versus $c_{k,-2}$

(c) $\mathbf{c}_k$ learning curve

(d) MSE P_k learning curve

FIGURE 10.15
Performance curves for the steepest-descent algorithm used in the linear prediction problem with step-size parameter $\mu = 0.15$ and eigenvalue spread $\mathcal{X}(\mathbf{R}) = 10$.

Large eigenvalue spread and underdamped response. Finally, in the third experiment, we consider the model parameters of the above case and increase the step size to $\mu = 0.5$ ($< 1/\lambda_{\max} = 0.55$) so that the transient response is underdamped. Figure 10.16 shows the corresponding plots.

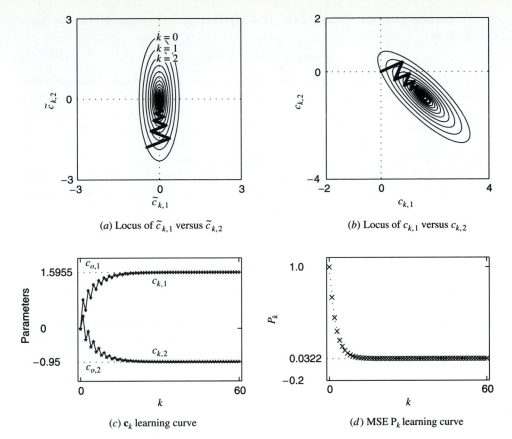

(a) Locus of $\tilde{c}_{k,1}$ versus $\tilde{c}_{k,2}$

(b) Locus of $c_{k,1}$ versus $c_{k,2}$

(c) $\mathbf{c}_k$ learning curve

(d) MSE P_k learning curve

FIGURE 10.16
Performance curves for the steepest-descent algorithm used in the linear prediction problem with eigenvalue spread $\mathcal{X}(\mathbf{R}) = 10$ and varying step-size parameters $\mu = 0.15$ and $\mu = 0.5$.

Note that the coefficients converge in an oscillatory fashion; however, the convergence is fairly rapid compared to that of the overdamped case. Thus the selection of the step size is an important design issue.

Newton's type of algorithms

Another family of algorithms with a faster rate of convergence includes Newton's method and its modifications. The basic idea of Newton's method is to achieve convergence in one step when $P(\mathbf{c})$ is quadratic. Thus, if $\mathbf{c}_k$ is to be the minimum of $P(\mathbf{c})$, the gradient $\nabla P(\mathbf{c}_k)$ of $P(\mathbf{c})$ evaluated at $\mathbf{c}_k$ (10.2.19) should be zero. From (10.2.19), we can write

$$\nabla P(\mathbf{c}_k) = \nabla P(\mathbf{c}_{k-1}) + \nabla^2 P(\mathbf{c}_{k-1}) \Delta \mathbf{c}_k = \mathbf{0} \qquad (10.3.41)$$

Thus $\nabla P(\mathbf{c}_k) = 0$ leads to the step increment

$$\Delta \mathbf{c}_k = -[\nabla^2 P(\mathbf{c}_{k-1})]^{-1} \nabla P(\mathbf{c}_{k-1}) \qquad (10.3.42)$$

and hence the adaptive algorithm is given by

$$\mathbf{c}_k = \mathbf{c}_{k-1} - \mu [\nabla^2 P(\mathbf{c}_{k-1})]^{-1} \nabla P(\mathbf{c}_{k-1}) \qquad (10.3.43)$$

where $\mu > 0$ is the step size. For quadratic error surfaces, from (10.3.4) and (10.3.5), we obtain with $\mu = 1$

$$\mathbf{c}_k = \mathbf{c}_{k-1} - [\nabla^2 P(\mathbf{c}_{k-1})]^{-1} \nabla P(\mathbf{c}_{k-1}) = \mathbf{c}_{k-1} - (\mathbf{c}_{k-1} - \mathbf{R}^{-1}\mathbf{d}) = \mathbf{c}_o \qquad (10.3.44)$$

which shows that indeed the algorithm converges in one step.

For the quadratic case, since $\nabla^2 P(\mathbf{c}_{k-1}) = 2\mathbf{R}$ from (10.3.1), we can express Newton's algorithm as

$$\mathbf{c}_k = \mathbf{c}_{k-1} - \mu \mathbf{R}^{-1} \nabla P(\mathbf{c}_{k-1}) \tag{10.3.45}$$

where μ is the step size that regulates the convergence rate. Other modified Newton methods replace the Hessian matrix $\nabla^2 P(\mathbf{c}_{k-1})$ with another matrix, which is guaranteed to be positive definite and, in some way, close to the Hessian. These Newton-type algorithms generally provide faster convergence. However, in practice, the inversion of $\mathbf{R}$ is numerically intensive and can lead to a numerically unstable solution if special care is not taken. Therefore, the SDA is more popular in adaptive filtering applications.

When the function $P(\mathbf{c})$ is nonquadratic, it is approximated locally by a quadratic function that is minimized exactly. However, the step obtained in (10.3.42) does not lead to the minimum of $P(\mathbf{c})$, and the iteration should be repeated several times. A more detailed treatment of linear and nonlinear optimization techniques can be found in Scales (1985) and in Luenberger (1984).

10.4 LEAST-MEAN-SQUARE ADAPTIVE FILTERS

In this section, we derive, analyze the performance, and present some practical applications of the *least-mean-square* (*LMS*) adaptive algorithm. The LMS algorithm, introduced by Widrow and Hoff (1960), is widely used in practice due to its simplicity, computational efficiency, and good performance under a variety of operating conditions.

10.4.1 Derivation

We first present two approaches to the derivation of the LMS algorithm that will help the reader to understand its operation. The first approach uses approximation to the gradient function while the second approach uses geometric arguments.

Optimization approach. The SDA uses the second-order moments $\mathbf{R}$ and $\mathbf{d}$ to iteratively compute the optimum filter $\mathbf{c}_o = \mathbf{R}^{-1}\mathbf{d}$, starting with an initial guess, usually $\mathbf{c}_0 = \mathbf{0}$, and then obtaining better approximations by taking steps in the direction of the negative gradient, that is,

$$\mathbf{c}_k = \mathbf{c}_{k-1} + \mu[-\nabla P(\mathbf{c}_{k-1})] \tag{10.4.1}$$

where

$$\nabla P(\mathbf{c}_{k-1}) = 2(\mathbf{R}\mathbf{c}_{k-1} - \mathbf{d}) \tag{10.4.2}$$

is the gradient of the performance function (10.3.1). In practice, where only the input $\{\mathbf{x}(j)\}_0^n$ and the desired response $\{y(j)\}_0^n$ are known, we can only compute an estimate of the "true" or exact gradient (10.4.2) using the available data. To develop an adaptive algorithm from (10.4.1), we take the following steps: (1) replace the iteration subscript k by the time index n; and (2) replace $\mathbf{R}$ and $\mathbf{d}$ by their instantaneous estimates $\mathbf{x}(n)\mathbf{x}^H(n)$ and $\mathbf{x}(n)y^*(n)$, respectively. The instantaneous estimate of the gradient (10.4.2) becomes

$$\nabla P(\mathbf{c}_{k-1}) = 2\mathbf{R}\mathbf{c}_{k-1} - 2\mathbf{d} \simeq 2\mathbf{x}(n)\mathbf{x}^H(n)\mathbf{c}(n-1) - 2\mathbf{x}(n)y^*(n) = -2\mathbf{x}(n)e^*(n) \tag{10.4.3}$$

where

$$e(n) = y(n) - \mathbf{c}^H(n-1)\mathbf{x}(n) \tag{10.4.4}$$

is the a priori filtering error. The estimate (10.4.3) also can be obtained by starting with the approximation $P(\mathbf{c}) \simeq |e(n)|^2$ and taking its gradient. The coefficient adaptation algorithm is

$$\mathbf{c}(n) = \mathbf{c}(n-1) + 2\mu\mathbf{x}(n)e^*(n) \tag{10.4.5}$$

which is obtained by substituting (10.4.3) and (10.4.4) in (10.4.1). The step-size parameter 2μ is also known as the *adaptation gain*.

The LMS algorithm, specified by (10.4.5) and (10.4.4), has both important similarities to and important differences from the SDA (10.3.7). The SDA contains deterministic quantities while the LMS operates on random quantities. The SDA is *not* an adaptive algorithm because it only depends on the second-order moments $\mathbf{R}$ and $\mathbf{d}$ and not on the SOE $\{\mathbf{x}(n, \zeta), y(n, \zeta)\}$. Also, the iteration index k has nothing to do with time. Simply stated, the SDA provides an iterative solution to the linear system $\mathbf{Rc} = \mathbf{d}$.

Geometric approach. Suppose that an adaptive filter operates in a stationary signal environment seeking the optimum filter $\mathbf{c}_o$. At time n the filter has access to input vector $\mathbf{x}(n)$, the desired response $y(n)$, and the previous or old coefficient estimate $\mathbf{c}(n-1)$. Its goal is to use this information to determine a new estimate $\mathbf{c}(n)$ that is closer to the optimum vector $\mathbf{c}_o$ or equivalently to choose $\mathbf{c}(n)$ so that $\|\tilde{\mathbf{c}}(n)\| < \|\tilde{\mathbf{c}}(n-1)\|$, where $\tilde{\mathbf{c}}(n) = \mathbf{c}(n) - \mathbf{c}_o$ is the coefficient error vector given by (10.2.24). Eventually, we want $\|\tilde{\mathbf{c}}(n)\|$ to become negligible as $n \to \infty$.

The vector $\tilde{\mathbf{c}}(n-1)$ can be decomposed into two orthogonal components

$$\tilde{\mathbf{c}}(n-1) = \tilde{\mathbf{c}}_x(n-1) + \tilde{\mathbf{c}}_x^{\perp}(n-1) \tag{10.4.6}$$

one parallel and one orthogonal to the input vector $\mathbf{x}(n)$, as shown in Figure 10.17(a). The response of the *error filter* $\tilde{\mathbf{c}}(n-1)$ to the input $\mathbf{x}(n)$ is

$$\tilde{y}(n) = \tilde{\mathbf{c}}^H(n-1)\mathbf{x}(n) = \tilde{\mathbf{c}}_x^H(n-1)\mathbf{x}(n) \tag{10.4.7}$$

which implies that

$$\tilde{\mathbf{c}}_x(n-1) = \frac{\tilde{y}^*(n)}{\|\mathbf{x}(n)\|^2}\mathbf{x}(n) \tag{10.4.8}$$

which can be verified by direct substitution in (10.4.7). Note that $\mathbf{x}(n)/\|\mathbf{x}(n)\|$ is a unit vector along the direction of $\mathbf{x}(n)$.

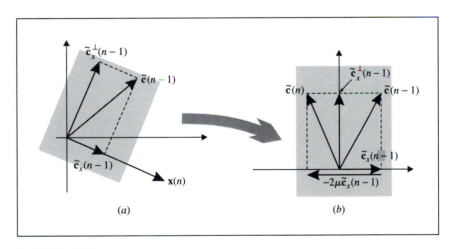

(a)   (b)

FIGURE 10.17
The geometric approach for the derivation of the LMS algorithm.

If we only know $\mathbf{x}(n)$ and $\tilde{y}(n)$, the best strategy to decrease $\tilde{\mathbf{c}}(n)$ is to choose $\tilde{\mathbf{c}}(n) = \tilde{\mathbf{c}}_x^{\perp}(n-1)$, or equivalently subtract $\tilde{\mathbf{c}}_x(n-1)$ from $\tilde{\mathbf{c}}(n-1)$. From Figure 10.17(a) note that as long as $\tilde{\mathbf{c}}_x(n-1) \neq \mathbf{0}$, $\|\tilde{\mathbf{c}}(n)\| = \|\tilde{\mathbf{c}}_x^{\perp}(n-1)\| < \|\tilde{\mathbf{c}}_x(n-1)\|$. This suggests the following adaptation algorithm

$$\tilde{\mathbf{c}}(n) = \tilde{\mathbf{c}}(n-1) - \tilde{\mu}\frac{\tilde{y}^*(n)}{\|\mathbf{x}(n)\|^2}\mathbf{x}(n) \tag{10.4.9}$$

which guarantees that $\|\tilde{\mathbf{c}}(n)\| < \|\tilde{\mathbf{c}}(n-1)\|$ as long as $0 < \tilde{\mu} < 2$ and $\tilde{y}(n) \neq 0$, as shown in Figure 10.17(b). The best choice clearly is $\tilde{\mu} = 1$.

Unfortunately, the signal $\tilde{y}(n)$ is not available, and we have to replace it with some reasonable approximation. From (10.2.18) and (10.2.10) we obtain

$$\tilde{e}(n) \triangleq e(n) - e_o(n) = y(n) - \hat{y}(n) - y(n) + \hat{y}_o(n) = \hat{y}_o(n) - \hat{y}(n)$$

$$= [\mathbf{c}_o^H - \mathbf{c}^H(n-1)]\mathbf{x}(n) = -\tilde{\mathbf{c}}^H(n-1)\mathbf{x}(n) = -\tilde{y}(n) \qquad (10.4.10)$$

where we have used (10.4.7). Using the approximation

$$\tilde{e}(n) = e(n) - e_o(n) \simeq e(n)$$

we combine it with (10.4.10) to get

$$\mathbf{c}(n) = \mathbf{c}(n-1) + \tilde{\mu}\frac{e^*(n)}{\|\mathbf{x}(n)\|^2}\mathbf{x}(n) \qquad (10.4.11)$$

which is known as the *normalized LMS algorithm*. Note that the effective step size $\tilde{\mu}/\|\mathbf{x}(n)\|^2$ is time-varying. The LMS algorithm in (10.4.5) follows if we set $\|\mathbf{x}(n)\| = 1$ and choose $\tilde{\mu} = 2\mu$.

LMS algorithm. The LMS algorithm can be summarized as

$$\hat{y}(n) = \mathbf{c}^H(n-1)\mathbf{x}(n) \qquad \qquad \text{filtering}$$

$$e(n) = y(n) - \hat{y}(n) \qquad \qquad \text{error formation} \qquad (10.4.12)$$

$$\mathbf{c}(n) = \mathbf{c}(n-1) + 2\mu\mathbf{x}(n)e^*(n) \qquad \text{coefficient updating}$$

where μ is adaptation step size. The algorithm requires $2M + 1$ complex multiplications and $2M$ complex additions. Figure 10.18 shows an implementation of an FIR adaptive filter using the LMS algorithm, which is implemented in MATLAB using the function [yhat,c]=firlms(x,y,M,mu). The a posteriori form of the LMS algorithm is developed in Problem 10.9.

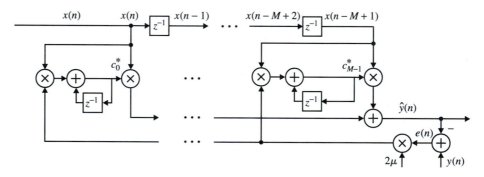

FIGURE 10.18
An FIR adaptive filter realization using the LMS algorithm.

10.4.2 Adaptation in a Stationary SOE

In the sequel, we study the stability and steady-state performance of the LMS algorithm in a stationary SOE; that is, we assume that the input and the desired response processes are jointly stationary. In theory, the goal of the LMS adaptive filter is to identify the optimum filter $\mathbf{c}_o = \mathbf{R}^{-1}\mathbf{d}$ from observations of the input $\mathbf{x}(n)$ and the desired response

$$y(n) = \mathbf{c}_o^H\mathbf{x}(n) + e_o(n) \qquad (10.4.13)$$

The optimum error $e_o(n)$ is orthogonal to the vector $\mathbf{x}(n)$; that is, $E\{\mathbf{x}(n)e^*(n)\} = \mathbf{0}$ and acts as measurement or output noise, as shown in Figure 10.19.

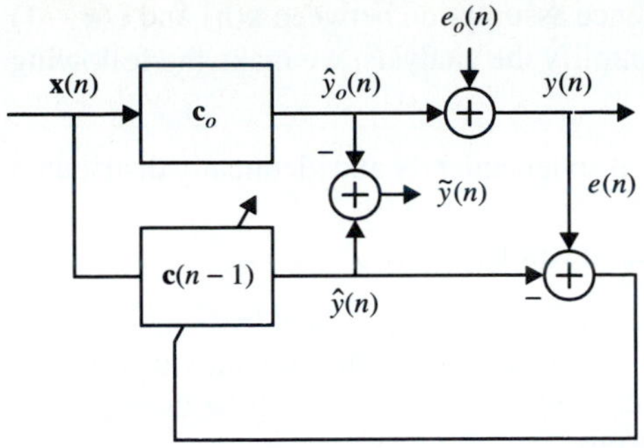

FIGURE 10.19
LMS algorithm in a stationary SOE.

527

SECTION 10.4
Least-Mean-Square
Adaptive Filters

The first step in the statistical analysis of the LMS algorithm is to determine a difference equation for the coefficient error vector $\tilde{\mathbf{c}}(n)$. To this end, we subtract $\mathbf{c}_o$ from both sides of (10.4.5), to obtain

$$\tilde{\mathbf{c}}(n) = \tilde{\mathbf{c}}(n-1) + 2\mu\mathbf{x}(n)e^*(n) \tag{10.4.14}$$

which expresses the LMS algorithm in terms of the coefficient error vector. We next use (10.4.12) and (10.4.13) in (10.4.14) to eliminate $e(n)$ by expressing it in terms of $\tilde{\mathbf{c}}(n-1)$ and $e_o(n)$. The result is

$$\tilde{\mathbf{c}}(n) = [\mathbf{I} - 2\mu\mathbf{x}(n)\mathbf{x}^H(n)]\tilde{\mathbf{c}}(n-1) + 2\mu\mathbf{x}(n)e_o^*(n) \tag{10.4.15}$$

which is a time-varying forced or nonhomogeneous stochastic difference equation. The irreducible error $e_o(n)$ accounts for measurement noise, modeling errors, unmodeled dynamics, quantization effects, and other disturbances. The presence of $e_o(n)$ prevents convergence because it forces $\tilde{\mathbf{c}}(n)$ to fluctuate around zero. Therefore, the important issue is the BIBO stability of the system (10.4.15). From (10.2.28), we see that $\|\tilde{\mathbf{c}}(n)\|$ is bounded in mean square if we can show that $E\{\tilde{\mathbf{c}}(n)\} \to 0$ as $n \to \infty$ and $\mathrm{var}\{\tilde{c}_k(n)\}$ is bounded for all n. To this end, we develop difference equations for the mean value $E\{\tilde{\mathbf{c}}(n)\}$ and the correlation matrix

$$\boldsymbol{\Phi}(n) \triangleq E\{\tilde{\mathbf{c}}(n)\tilde{\mathbf{c}}^H(n)\} \tag{10.4.16}$$

of the coefficient error vector $\tilde{\mathbf{c}}(n)$. As we shall see, the MSD and the EMSE can be expressed in terms of matrices $\boldsymbol{\Phi}(n)$ and $\mathbf{R}$. The time evolution of these quantities provides sufficient information to evaluate the stability and steady-state performance of the LMS algorithm.

Convergence of the mean coefficient vector

If we take the expectation of (10.4.15), we have

$$E\{\tilde{\mathbf{c}}(n)\} = E\{\tilde{\mathbf{c}}(n-1)\} - 2\mu E\{\mathbf{x}(n)\mathbf{x}^H(n)\tilde{\mathbf{c}}(n-1)\} \tag{10.4.17}$$

because $E\{\mathbf{x}(n)e_o^*(n)\} = \mathbf{0}$ owing to the orthogonality principle. The computation of the second term in (10.4.17) requires the correlation between the input signal and the coefficient error vector.

If we assume that $\mathbf{x}(n)$ and $\tilde{\mathbf{c}}(n-1)$ are statistically independent, (10.4.17) simplifies to

$$E\{\tilde{\mathbf{c}}(n)\} = (\mathbf{I} - 2\mu\mathbf{R})E\{\tilde{\mathbf{c}}(n-1)\} \tag{10.4.18}$$

which has the same form as (10.3.11) for the SDA. Therefore, $\tilde{\mathbf{c}}(n)$ converges in the MS sense, that is, $\lim_{n\to\infty} E\{\tilde{\mathbf{c}}(n)\} = \mathbf{0}$, if the eigenvalues of the system matrix $(\mathbf{I} - 2\mu\mathbf{R})$ are less than 1. Hence, if $\mathbf{R}$ is positive definite and $\lambda_{\max}$ is its maximum eigenvalue, the condition

$$0 < 2\mu < \frac{1}{\lambda_{\max}} \tag{10.4.19}$$

ensures that the LMS algorithm converges in the MS sense [see the discussion following (10.2.27)].

Independence assumption. The independence assumption between $\mathbf{x}(n)$ and $\tilde{\mathbf{c}}(n-1)$ was critical to the derivation of (10.4.18). To simplify the analysis, we make the following *independence assumptions* (Gardner 1984):

A1 The sequence of input data vectors $\mathbf{x}(n)$ is independently and identically distributed with zero mean and correlation matrix $\mathbf{R}$.

A2 The sequences $\mathbf{x}(n)$ and $e_o(n)$ are independent for all n.

From (10.4.15), we see that $\tilde{\mathbf{c}}(n-1)$ depends on $\tilde{\mathbf{c}}(0)$, $\{\mathbf{x}(k)\}_0^{n-1}$, and $\{e_o(k)\}_0^{n-1}$. Since the sequence $\mathbf{x}(n)$ is IID and the quantities $\mathbf{x}(n)$ and $e_o(n)$ are independent, we conclude that $\mathbf{x}(n)$, $e_0(n)$, and $\tilde{\mathbf{c}}(n-1)$ are mutually independent. This result will be used several times to simplify the analysis of the LMS algorithm.

The independence assumption A1, first introduced in Widrow et al. (1976) and in Mazo (1979), ignores the statistical dependence among successive input data vectors; however, it preserves sufficient statistical information about the adaptation process to lead to useful design guidelines. Clearly, for FIR filtering applications, the independence assumption is violated because two successive input data vectors $\mathbf{x}(n)$ and $\mathbf{x}(n+1)$ have $M-1$ common elements (shift-invariance property).

Evolution of the coefficient error correlation matrix

The MSD can be expressed in terms of the trace of the correlation matrix[†] $\boldsymbol{\Phi}(n)$, that is,

$$\mathcal{D}(n) = \mathrm{tr}[\boldsymbol{\Phi}(n)] \tag{10.4.20}$$

which can be easily seen by using (10.2.29) and the definition of trace. If we postmultiply both sides of (10.4.15) by their respective Hermitian transposes and take the mathematical expectation, we obtain

$$
\begin{aligned}
\boldsymbol{\Phi}(n) &= E\{\tilde{\mathbf{c}}(n)\tilde{\mathbf{c}}^H(n)\} \\
&= E\{[\mathbf{I} - 2\mu\mathbf{x}(n)\mathbf{x}^H(n)]\tilde{\mathbf{c}}(n-1)\tilde{\mathbf{c}}^H(n-1)[\mathbf{I} - 2\mu\mathbf{x}(n)\mathbf{x}^H(n)]^H\} \\
&\quad + 2\mu E\{[\mathbf{I} - 2\mu\mathbf{x}(n)\mathbf{x}^H(n)]\tilde{\mathbf{c}}(n-1)e_o(n)\mathbf{x}^H(n)\} \\
&\quad + 2\mu E\{\mathbf{x}(n)e_o^*(n)\tilde{\mathbf{c}}^H(n-1)[\mathbf{I} - 2\mu\mathbf{x}(n)\mathbf{x}^H(n)]^H\} \\
&\quad + 4\mu^2 E\{\mathbf{x}(n)e_o^*(n)e_o(n)\mathbf{x}^H(n)\}
\end{aligned}
\tag{10.4.21}
$$

From the independence assumptions, $e_o(n)$ is independent with $\tilde{\mathbf{c}}(n-1)$ and $\mathbf{x}(n)$. Therefore, the second and third terms in (10.4.21) vanish, and the fourth term is equal to $4\mu^2 P_o\mathbf{R}$. If we expand the first term, we obtain

$$\boldsymbol{\Phi}(n) = \boldsymbol{\Phi}(n-1) - 2\mu[\mathbf{R}\boldsymbol{\Phi}(n-1) + \boldsymbol{\Phi}(n-1)\mathbf{R}] + 4\mu^2\mathbf{A} + 4\mu^2 P_o\mathbf{R} \tag{10.4.22}$$

where

$$\mathbf{A} \triangleq E\{\mathbf{x}(n)\mathbf{x}^H(n)\tilde{\mathbf{c}}(n-1)\tilde{\mathbf{c}}^H(n-1)\mathbf{x}(n)\mathbf{x}^T(n)\} \tag{10.4.23}$$

and the terms $\mathbf{R}\boldsymbol{\Phi}(n-1)$ and $\boldsymbol{\Phi}(n-1)\mathbf{R}$ have been computed by using the mutual independence of $\mathbf{x}(n)$, $\tilde{\mathbf{c}}(n-1)$, and $e_o(n)$.

The computation of matrix $\mathbf{A}$ can be simplified if we make additional assumptions about the statistical properties of $\mathbf{x}(n)$. As shown in Gardner (1984), development of a recursive relation for the elements of $\boldsymbol{\Phi}(n)$ using only the independent assumptions requires the products with and the inversion of a $M^2 \times M^2$ matrix, where M is the size of $\mathbf{x}(n)$.

The evaluation of this term when $\mathbf{x}(n) \sim$ IID, an assumption that is more appropriate for data transmission applications, is discussed in Gardner (1984). The computation for $\mathbf{x}(n)$ being a *spherically invariant random process* (*SIRP*) is discussed in Rupp (1993). SIRP models, which include the Gaussian distribution as a special case, provide a good

---

[†]Note that when (10.4.19) holds, $\lim_{n\to\infty} E\{\tilde{\mathbf{c}}(n)\} = \mathbf{0}$, and therefore $\boldsymbol{\Phi}(n)$ provides asymptotically the covariance of $\tilde{\mathbf{c}}(n)$.

characterization of speech signals. However, independently of the assumption used, the basic conclusions remain the same.

Assuming that $\mathbf{x}(n)$ is normally distributed, that is, $\mathbf{x}(n) \sim \mathcal{N}(\mathbf{0}, \mathbf{R})$, a significant amount of simplification can be obtained. Indeed, in this case we can use the moment factorization property for normal random variables to express fourth-order moments in terms of second-order moments (Papoulis 1991). As we showed in Section 3.2.3, if z_1, z_2, z_3, and z_4 are complex-valued, zero-mean, and jointly distributed normal random variables, then

$$E\{z_1 z_2^* z_3 z_4^*\} = E\{z_1 z_2^*\} E\{z_3 z_4^*\} + E\{z_1 z_4^*\} E\{z_2^* z_3\} \tag{10.4.24}$$

or if they are real-valued, then

$$E\{z_1 z_2 z_3 z_4\} = E\{z_1 z_2\} E\{z_3 z_4\} + E\{z_1 z_3\} E\{z_2 z_4\} + E\{z_1 z_4\} E\{z_2 z_3\} \tag{10.4.25}$$

Using direct substitution of (10.4.24) or (10.4.25) in (10.4.23), we can show that

$$\mathbf{A} = \begin{cases} \mathbf{R}\boldsymbol{\Phi}(n-1)\mathbf{R} + \mathbf{R}\,\mathrm{tr}[\mathbf{R}\boldsymbol{\Phi}(n-1)] & \text{complex case} \\ 2\mathbf{R}\boldsymbol{\Phi}(n-1)\mathbf{R} + \mathbf{R}\,\mathrm{tr}[\mathbf{R}\boldsymbol{\Phi}(n-1)] & \text{real case} \end{cases} \tag{10.4.26}$$

Finally, substituting (10.4.26) in (10.4.22), we obtain a difference equation for $\boldsymbol{\Phi}(n)$. This is summarized in the following property:

PROPERTY 10.4.1. Using the independence assumptions A1 and A2, and the normal distribution assumption of $\mathbf{x}(n)$, the correlation matrix of the coefficient error vector $\tilde{\mathbf{c}}(n)$ satisfies the difference equation

$$\begin{aligned} \boldsymbol{\Phi}(n) = \boldsymbol{\Phi}(n-1) &- 2\mu[\mathbf{R}\boldsymbol{\Phi}(n-1) + \boldsymbol{\Phi}(n-1)\mathbf{R}] \\ &+ 4\mu^2 \mathbf{R}\boldsymbol{\Phi}(n-1)\mathbf{R} + 4\mu^2 \mathbf{R}\,\mathrm{tr}[\mathbf{R}\boldsymbol{\Phi}(n-1)] + 4\mu^2 P_o \mathbf{R} \end{aligned} \tag{10.4.27}$$

in the complex case and

$$\begin{aligned} \boldsymbol{\Phi}(n) = \boldsymbol{\Phi}(n-1) &- 2\mu[\mathbf{R}\boldsymbol{\Phi}(n-1) + \boldsymbol{\Phi}(n-1)\mathbf{R}] \\ &+ 8\mu^2 \mathbf{R}\boldsymbol{\Phi}(n-1)\mathbf{R} + 4\mu^2 \mathbf{R}\,\mathrm{tr}[\mathbf{R}\boldsymbol{\Phi}(n-1)] + 4\mu^2 P_o \mathbf{R} \end{aligned} \tag{10.4.28}$$

in the real case. Both relations are matrix difference equations driven by the constant term $4\mu^2 P_o \mathbf{R}$.

The presence of the term $4\mu^2 P_o \mathbf{R}$ in (10.4.27) or (10.4.28) implies that $\boldsymbol{\Phi}(n)$ will never become zero, and as a result the coefficients of the LMS adaptive filter will always fluctuate about their optimum settings, which prevents convergence. It has been shown (Bucklew et al. 1993) that asymptotically $\tilde{\mathbf{c}}(n)$ follows a zero-mean normal distribution. The amount of fluctuation is measured by matrix $\boldsymbol{\Phi}(n)$. In contrast, the absence of a driving term in (10.4.18) allows the convergence of $E\{\mathbf{c}(n)\}$ to the optimum vector $\mathbf{c}_o$.

Since there are two distinct forms for the difference equation of $\boldsymbol{\Phi}(n)$, we will consider the *real* case (10.4.28) for further discussion. Similar analysis can be done for the complex case (10.4.27), which is undertaken in Problem 10.11. To further simplify the analysis, we transform $\boldsymbol{\Phi}(n)$ to the principal coordinate space of $\mathbf{R}$ using the spectral decomposition

$$\mathbf{Q}^T \mathbf{R} \mathbf{Q} = \boldsymbol{\Lambda}$$

by defining the matrix

$$\boldsymbol{\Theta}(n) \triangleq \mathbf{Q}^T \boldsymbol{\Phi}(n) \mathbf{Q} \tag{10.4.29}$$

which is symmetric and positive definite [when $\boldsymbol{\Phi}(n)$ is positive definite].

If we pre- and postmultiply (10.4.28) by $\mathbf{Q}^T$ and $\mathbf{Q}$ and use $\mathbf{Q}^T \mathbf{Q} = \mathbf{Q}\mathbf{Q}^T = \mathbf{I}$, we obtain

$$\begin{aligned} \boldsymbol{\Theta}(n) = \boldsymbol{\Theta}(n-1) &- 2\mu[\boldsymbol{\Lambda}\boldsymbol{\Theta}(n-1) + \boldsymbol{\Theta}(n-1)\boldsymbol{\Lambda}] \\ &+ 8\mu^2 \boldsymbol{\Lambda}\boldsymbol{\Theta}(n-1)\boldsymbol{\Lambda} + 4\mu^2 \boldsymbol{\Lambda}\,\mathrm{tr}[\boldsymbol{\Lambda}\boldsymbol{\Theta}(n-1)] + 4\mu^2 P_o \boldsymbol{\Lambda} \end{aligned} \tag{10.4.30}$$

which is easier to work with because of the diagonal nature of $\boldsymbol{\Lambda}$. For any symmetric and positive definite matrix $\boldsymbol{\Theta}$, we have $|\theta_{ij}(n)|^2 \leq \theta_{ii}\theta_{jj}$. Hence, the convergence of the diagonal elements ensures the convergence of the off-diagonal elements. This observation and (10.4.30) suggest that to analyze the LMS algorithm, we should extract from (10.4.30) the equations for the diagonal elements

$$\boldsymbol{\theta}(n) \triangleq [\theta_1(n)\, \theta_2(n)\, \cdots\, \theta_M(n)]^T \tag{10.4.31}$$

of $\boldsymbol{\Theta}(n)$ and form a difference equation for the vector $\boldsymbol{\theta}(n)$. Indeed, we can easily show that

$$\boldsymbol{\theta}(n) = \mathbf{B}\boldsymbol{\theta}(n-1) + 4\mu^2 P_o \boldsymbol{\lambda} \tag{10.4.32}$$

where

$$\mathbf{B} \triangleq \boldsymbol{\Lambda}(\rho) + 4\mu^2 \boldsymbol{\lambda}\boldsymbol{\lambda}^T \tag{10.4.33}$$

$$\boldsymbol{\lambda} \triangleq [\lambda_1\, \lambda_2\, \cdots\, \lambda_M]^T \tag{10.4.34}$$

$$\boldsymbol{\Lambda}(\rho) \triangleq \text{diag}\{\rho_1, \rho_2, \ldots, \rho_M\} \tag{10.4.35}$$

$$\rho_k = 1 - 4\mu\lambda_k + 8\mu^2\lambda_k^2 = (1 - 2\mu\lambda_k)^2 + 4\mu^2\lambda_k^2 > 0 \qquad 1 \le k \le M \tag{10.4.36}$$

and λ_k are the eigenvalues of $\mathbf{R}$. The solution of the vector difference equation (10.4.32) is

$$\boldsymbol{\theta}(n) = \mathbf{B}^n \boldsymbol{\theta}(0) + 4\mu^2 P_o \sum_{j=0}^{n-1} \mathbf{B}^j \boldsymbol{\lambda} \tag{10.4.37}$$

and can be easily found by recursion.

The stability of the linear system (10.4.32) is determined by the eigenvalues of the symmetric matrix $\mathbf{B}$. Using (10.4.33) and (10.4.35), for an arbitrary vector $\mathbf{z}$, we obtain

$$\mathbf{z}^T \mathbf{B} \mathbf{z} = \mathbf{z}^T \boldsymbol{\Lambda}(\rho)\mathbf{z} + 4\mu^2(\boldsymbol{\lambda}^T \mathbf{z})^2 = \sum_{k=1}^{M} \rho_k z_k^2 + 4\mu^2(\boldsymbol{\lambda}^T \mathbf{z})^2 \tag{10.4.38}$$

where we have used (10.4.36). Hence (10.4.38), for $\mathbf{z} \ne \mathbf{0}$, implies that $\mathbf{z}^T \mathbf{B} \mathbf{z} > 0$, that is, the matrix $\mathbf{B}$ is positive definite. Since matrix $\mathbf{B}$ is symmetric and positive definite, its eigenvalues $\lambda_k(\mathbf{B})$ are real and positive. The system (10.4.37) will be BIBO stable if and only if

$$0 < \lambda_k(\mathbf{B}) < 1 \qquad 1 \le k \le M \tag{10.4.39}$$

To find the range of μ that ensures (10.4.39), we use the Gerschgorin circles theorem (Noble and Daniel 1988), which states that *each eigenvalue of an $M \times M$ matrix $\mathbf{B}$ lies in at least one of the disks with center at the diagonal element b_{kk} and radius equal to the sum of absolute values $|b_{kj}|$, $j \ne k$, of the remaining elements of the row*. Since the elements of $\mathbf{B}$ are positive, we can easily see that

$$\lambda_k(\mathbf{B}) - b_{kk} < \sum_{\substack{j=1 \\ j \ne k}}^{M} b_{ki} \qquad \text{or} \qquad \lambda_k(\mathbf{B}) < \rho_k + 4\mu^2\lambda_k \sum_{j=1}^{M} \lambda_i$$

using (10.4.33). Hence using (10.4.36), we see the eigenvalues of $\mathbf{B}$ satisfy (10.4.39) if

$$1 - 4\mu\lambda_k + 8\mu^2\lambda_k^2 + 4\mu^2\lambda_k \,\text{tr}\mathbf{R} < 1$$

or

$$-\mu\lambda_k + 2\mu^2\lambda_k^2 + \mu^2\lambda_k \,\text{tr}\mathbf{R} < 0$$

which implies that $\mu > 0$ and

$$2\mu < \frac{1}{\lambda_k + \text{tr}\mathbf{R}} < \frac{1}{\text{tr}\mathbf{R}}$$

because $\lambda_k > 0$ for all k. In conclusion, if the adaptation step μ satisfies the condition

$$0 < 2\mu < \frac{1}{\text{tr}\mathbf{R}} \tag{10.4.40}$$

then the system (10.4.37) is stable and therefore the sequence $\boldsymbol{\theta}(n)$ converges.

PROPERTY 10.4.2. When the stability condition (10.4.40) holds, the solution (10.4.37) of the difference equation (10.4.32) can be written as

$$\boldsymbol{\theta}(n) = \mathbf{B}^n[\boldsymbol{\theta}(0) - \boldsymbol{\theta}(\infty)] + \boldsymbol{\theta}(\infty) \qquad (10.4.41)$$

where $\boldsymbol{\theta}(0)$ is the initial value and $\boldsymbol{\theta}(\infty)$ is the steady-state value of $\boldsymbol{\theta}(n)$.

Proof. Using the identity

$$\sum_{j=0}^{n-1} \mathbf{B}^j = (\mathbf{I} - \mathbf{B}^n)(\mathbf{I} - \mathbf{B})^{-1} = (\mathbf{I} - \mathbf{B})^{-1} - \mathbf{B}^n(\mathbf{I} - \mathbf{B})^{-1}$$

the solution (10.4.37) can be written as

$$\boldsymbol{\theta}(n) = \mathbf{B}^n[\boldsymbol{\theta}(0) - 4\mu^2 P_o(\mathbf{I} - \mathbf{B})^{-1}\boldsymbol{\lambda}] + 4\mu^2 P_o(\mathbf{I} - \mathbf{B})^{-1}\boldsymbol{\lambda} \qquad (10.4.42)$$

When the eigenvalues of $\mathbf{B}$ are inside the unit circle, we have

$$\lim_{n \to \infty} \boldsymbol{\theta}(n) \triangleq \boldsymbol{\theta}(\infty) = 4\mu^2 P_o(\mathbf{I} - \mathbf{B})^{-1}\boldsymbol{\lambda} \qquad (10.4.43)$$

because the first term converges to zero. Substituting (10.4.43) in (10.4.42), we obtain (10.4.41).

Evolution of the mean square error

We next express the MSE as a function of $\boldsymbol{\lambda}$ and $\boldsymbol{\theta}$. Using (10.2.10) and (10.2.18), we have

$$e(n) = y(n) - \mathbf{c}^H(n-1)\mathbf{x}(n) = e_o(n) - \tilde{\mathbf{c}}^H(n-1)\mathbf{x}(n) \qquad (10.4.44)$$

where $e_o(n)$ is the optimum filtering error and $\tilde{\mathbf{c}}(n)$ is the coefficient error vector. The (a priori) MSE of the adaptive filter at time n is

$$P(n) \triangleq E\{|e(n)|^2\}$$
$$= E\{|e_o(n)|^2\} - E\{\tilde{\mathbf{c}}^H(n-1)\mathbf{x}(n)e_o^*(n)\} - E\{e_o(n)\mathbf{x}^H(n)\tilde{\mathbf{c}}(n-1)\} \qquad (10.4.45)$$
$$+ E\{\tilde{\mathbf{c}}^H(n-1)\mathbf{x}(n)\mathbf{x}^H(n)\tilde{\mathbf{c}}(n-1)\}$$

Since $\tilde{\mathbf{c}}(n)$ is a random vector, the evaluation of the MSE (10.4.45) requires the correlation between $\mathbf{x}(n)$ and $\tilde{\mathbf{c}}(n-1)$. Using the independence assumptions A1 and A2, we see that the second and third terms in (10.4.45) become zero, as explained before, and the excess MSE is given by the last term

$$P_{\text{ex}}(n) = E\{\tilde{\mathbf{c}}^H(n-1)\mathbf{x}(n)\mathbf{x}^H(n)\tilde{\mathbf{c}}(n-1)\} \qquad (10.4.46)$$

If we define the quantities

$$\mathbf{A} \triangleq \tilde{\mathbf{c}}^H(n-1) \quad \text{and} \quad \mathbf{B} \triangleq \mathbf{x}(n)\mathbf{x}^H(n)\tilde{\mathbf{c}}(n-1) \qquad (10.4.47)$$

and notice that $\mathbf{AB} = \text{tr}(\mathbf{AB})$ (because $\mathbf{AB}$ is a scalar) and $\text{tr}(\mathbf{AB}) = \text{tr}(\mathbf{BA})$, we obtain

$$P_{\text{ex}}(n) = E\{\text{tr}(\mathbf{AB})\} = E\{\text{tr}(\mathbf{BA})\} = \text{tr}(E\{\mathbf{BA}\})$$
$$= \text{tr}(E\{\mathbf{x}(n)\mathbf{x}^H(n)\}E\{\tilde{\mathbf{c}}(n-1)\tilde{\mathbf{c}}^H(n-1)\})$$

because expectation is a linear operation and $\mathbf{x}(n)$ and $\tilde{\mathbf{c}}(n-1)$ have been assumed statistically independent. Therefore, the excess MSE can be expressed as

$$P_{\text{ex}}(n) = \text{tr}[\mathbf{R}\boldsymbol{\Phi}(n-1)] \qquad (10.4.48)$$

where $\boldsymbol{\Phi}(n) = E\{\tilde{\mathbf{c}}(n)\tilde{\mathbf{c}}^H(n)\}$ is the correlation matrix of the coefficient error vector. This expression simplifies to

$$P_{\text{ex}}(n) = M\sigma_x^2\sigma_c^2 \qquad (10.4.49)$$

if $\mathbf{R} = \sigma_x^2\mathbf{I}$ and $\boldsymbol{\Phi}(n) = \sigma_c^2\mathbf{I}$.

If $\mathbf{R}$ and $\mathbf{\Phi}(n)$ are both positive definite, relation (10.4.48) shows that $P_{\text{ex}}(n) > 0$, that is, the MSE attained by the adaptive filter is larger than the optimum MSE P_o of the optimum filter (cost of adaptation).

Next we develop a difference equation for $P_{\text{ex}}(n)$, using, for convenience, the principal coordinate system of the input correlation matrix $\mathbf{R}$. Since the trace of a matrix remains invariant under an orthogonal transformation, we have

$$P_{\text{ex}}(n) = \text{tr}[\mathbf{R}\mathbf{\Phi}(n)] = \text{tr}[\mathbf{\Lambda}\mathbf{\Theta}(n)] = \lambda^T \boldsymbol{\theta}(n) \tag{10.4.50}$$

where the elements of λ are the eigenvalues of $\mathbf{R}$ and the elements of $\boldsymbol{\theta}(n)$ are the diagonal elements of $\mathbf{\Theta}(n)$.

Since the most often observable and important quantity for the operation of an adaptive filter is the MSE, we use our previous results to determine the value of MSE as a function of n, that is, the learning curve of the LMS adaptive filter. To this end, we use the orthogonal decomposition $\mathbf{B} = \mathbf{Q}(\mathbf{B})\mathbf{\Lambda}(\mathbf{B})\mathbf{Q}^H(\mathbf{B})$ to express $\mathbf{B}^n$ as

$$\mathbf{B}^n = \mathbf{Q}(\mathbf{B})\mathbf{\Lambda}^n(\mathbf{B})\mathbf{Q}^H(\mathbf{B}) = \sum_{k=1}^{M} \lambda_k^n(\mathbf{B})\mathbf{q}_k(\mathbf{B})\mathbf{q}_k^H(\mathbf{B}) \tag{10.4.51}$$

where $\lambda_k(\mathbf{B})$ are the eigenvalues and $\mathbf{q}_k(\mathbf{B})$ are the eigenvectors of matrix $\mathbf{B}$. Substituting (10.4.41) and (10.4.51) into (10.4.50) and recalling that $P(n) = P_o + P_{\text{ex}}(n)$, we obtain

$$P(n) = P_o + P_{\text{tr}}(n) + P_{\text{ex}}(\infty) \tag{10.4.52}$$

where $P_{\text{ex}}(\infty)$ is termed the *steady-state excess MSE* and

$$P_{\text{tr}}(n) \triangleq \sum_{k=1}^{M} \gamma_k(\mathbf{R}, \mathbf{B}) \lambda_k^n(\mathbf{B}) \tag{10.4.53}$$

is termed the *transient MSE* because it dies out exponentially when $0 < \lambda_k(\mathbf{B}) < 1, 1 \leq k \leq M$. The constants

$$\gamma_k(\mathbf{R}, \mathbf{B}) \triangleq \lambda^T(\mathbf{R})\mathbf{q}_k(\mathbf{B})\mathbf{q}_k^H(\mathbf{B})[\boldsymbol{\theta}(0) - \boldsymbol{\theta}(\infty)] \tag{10.4.54}$$

are determined by the eigenvalues $\lambda_k(\mathbf{R})$ of matrix $\mathbf{R}$ and the eigenvectors $\mathbf{q}_k(\mathbf{B})$ of matrix $\mathbf{B}$. Since the minimum MSE P_o is available, we need to determine the steady-state excess MSE $P_{\text{ex}}(\infty)$.

PROPERTY 10.4.3. When the LMS adaptive algorithm converges, the steady-state excess MSE is given by

$$P_{\text{ex}}(\infty) = P_o \frac{C(\mu)}{1 - C(\mu)} \tag{10.4.55}$$

where

$$C(\mu) \triangleq \sum_{k=1}^{M} \frac{\mu \lambda_k}{1 - 2\mu \lambda_k} \tag{10.4.56}$$

and λ_k are the eigenvalues of the input correlation matrix.

Proof. Using (10.4.32) and (10.4.35), we obtain the difference equation

$$\theta_k(n) = \rho_k \theta_k(n-1) + 4\mu^2 \lambda_k P_{\text{ex}}(n-1) + 4\mu^2 P_o \lambda_k \tag{10.4.57}$$

When (10.4.40) holds, (10.4.57) attains the following steady-state form

$$\theta_k(\infty) = \rho_k \theta_k(\infty) + 4\mu^2 \lambda_k P_{\text{ex}}(\infty) + 4\mu^2 P_o \lambda_k$$

whose solution, in conjunction with (10.4.36), gives

$$\theta_k(\infty) = \mu \frac{P_o + P_{\text{ex}}(\infty)}{1 - 2\mu \lambda_k}$$

and
$$P_{\text{ex}}(\infty) = \sum_{k=1}^{M} \lambda_k \theta_k(\infty) = [P_o + P_{\text{ex}}(\infty)] \sum_{k=1}^{M} \frac{\mu \lambda_k}{1 - 2\mu \lambda_k}$$

Solving the last equation for $P_{\text{ex}}(\infty)$, we obtain (10.4.55) and (10.4.56).

Solving (10.4.55) for $C(\mu)$ gives

$$C(\mu) = \frac{P_{\text{ex}}(\infty)}{P_o + P_{\text{ex}}(\infty)} \tag{10.4.58}$$

which implies that
$$0 < C(\mu) < 1 \tag{10.4.59}$$

because P_o and $P_{\text{ex}}(\infty)$ are positive quantities. It has been shown that (10.4.59) leads to the tighter bound $0 < 2\mu < 2/(3\,\text{tr}\,\mathbf{R})$ for the adaptation step μ (Horowitz and Senne 1981; Feuer and Weinstein 1985). Therefore, convergence in the MSE imposes a stronger constraint on the step size μ than does (10.4.40), which ensures convergence in the mean.

10.4.3 Summary and Design Guidelines

There are many theoretical and simulation analyses of the LMS adaptive algorithm under a variety of assumptions. In this book, we have focused on results that help us to understand its operation and performance and to develop design guidelines for its practical application. The operation and performance of the LMS adaptive filter are determined by its stability and the properties of its learning curve, which shows the evolution of the MSE as a function of time. The MSE produced by the LMS adaptive algorithm consists of three components [see (10.4.52)]

$$P(n) = P_o + P_{\text{tr}}(n) + P_{\text{ex}}(\infty)$$

where P_o is the optimum MSE, $P_{\text{tr}}(n)$ is the transient MSE, and $P_{\text{ex}}(\infty)$ is the steady-state excess MSE. This equation provides the basis for understanding and evaluating the operation of the LMS adaptive algorithm in a stationary SOE. For convenience, the LMS adaptive filtering algorithm is summarized in Table 10.3.

TABLE 10.3
Summary of the LMS algorithm.

Design parameters
$\mathbf{x}(n) = $ input data vector at time n
$y(n) = $ desired response at time n
$\mathbf{c}(n) = $ filter coefficient vector at time n
$M = $ number of coefficients
$\mu = $ step-size parameter
$0 < \mu \ll \dfrac{1}{\sum\limits_{k=1}^{M} E\{|x_k(n)|^2\}}$

Initialization
$\mathbf{c}(-1) = \mathbf{x}(-1) = \mathbf{0}$

Computation
For $n = 0, 1, 2, \ldots,$ compute
$\hat{y}(n) = \mathbf{c}^H(n-1)\mathbf{x}(n)$
$e(n) = y(n) - \hat{y}(n)$
$\mathbf{c}(n) = \mathbf{c}(n-1) + 2\mu \mathbf{x}(n) e^*(n)$

Stability. The LMS adaptive filter converges in the mean-square sense, that is, the transient MSE dies out, if the adaptation step μ satisfies the condition

$$0 < 2\mu < \frac{K}{\text{tr}\mathbf{R}} \tag{10.4.60}$$

where $\text{tr}\mathbf{R}$ is the trace of the input correlation matrix and K is a constant that depends weakly on the statistics of the input data vector. For example, when $\mathbf{x}(n) \sim \mathcal{N}(\mathbf{0}, \mathbf{R})$, we proved that $K = 1$ or $\frac{2}{3}$. In addition, this condition ensures that on average the LMS adaptive filter converges to the optimum filter. We stress that in most practical applications, where the independence assumption does not hold, the *step size μ should be much smaller than* $K/\text{tr}\mathbf{R}$. Therefore, the exact value of K is not important in practice.

Rate of convergence. The transient MSE dies out exponentially without exhibiting any oscillations. This follows from (10.4.53) because when μ satisfies (10.4.40), the eigenvalues of matrix $\mathbf{B}$ are positive and less than 1. The *settling time*, that is, the time taken for the transients to die out, is proportional to the average time constant

$$\tau_{\text{lms,av}} = \frac{1}{\mu\lambda_{\text{av}}} \tag{10.4.61}$$

where $\lambda_{\text{av}} = (\sum_{k=1}^{M} \lambda_k)/M$ is the average eigenvalue of $\mathbf{R}$ (Widrow et al. 1976). The quantity $P_{\text{tr}}^{\text{total}} = \sum_{n=0}^{\infty} P_{\text{tr}}(n)$, which provides the total transient MSE, can be used as a measure of the speed of adaptation. When $\mu\lambda_k \ll 1$ (see Problem 10.12), we have

$$P_{\text{tr}}^{\text{total}} \triangleq \sum_{n=0}^{\infty} P_{\text{tr}}(n) \simeq \frac{1}{4\mu} \sum_{k=1}^{M} \Delta\theta_k(0) \tag{10.4.62}$$

where $\Delta\theta_k(0)$ is the initial distance of a coefficient from its optimum setting measured in principal coordinates. As is intuitively expected, *the smaller the step size and the farther the initial coefficients are from their optimum settings, the more iterations it takes for the LMS algorithm to converge.* Furthermore, from the discussion in Section 10.3, it follows that the LMS algorithm will converge faster if the contours of the error surface are circles, that is, when the input correlation matrix is $\mathbf{R} = \sigma_x^2 \mathbf{I}$.

Steady-state excess MSE. The excess MSE after the adaptation has been completed (i.e., the steady-state value) is given by (10.4.55). When $\mu\lambda_k \ll 1$, we may approximate (10.4.55) as follows

$$P_{\text{ex}}(\infty) \simeq P_o \frac{\mu\,\text{tr}\mathbf{R}}{1 - \mu\,\text{tr}\mathbf{R}}$$

which allows a much easier interpretation. Solving for $\mu\,\text{tr}\mathbf{R}$, we obtain $\mu\,\text{tr}\mathbf{R} \simeq P_{\text{ex}}(\infty)/[P_{\text{ex}}(\infty) + P_o]$ which implies that $0 < \mu\,\text{tr}\mathbf{R} < 1$. Since $\mu\,\text{tr}\mathbf{R} \ll 1$, we often use the approximation

$$P_{\text{ex}}(\infty) \simeq \mu P_o\,\text{tr}\mathbf{R} \tag{10.4.63}$$

which implies that $P_{\text{ex}}(\infty) \ll P_o$, that is, for small values of the step size the excess MSE is much smaller than the optimum MSE. Note that the presence of the irreducible error $e_o(n)$ prevents perfect adaptation as $n \to \infty$ because $P_o > 0$.

Speed versus quality of adaptation. From the previous discussion we see that there is a tradeoff between rate of convergence (speed of adaptation) and steady-state excess MSE (quality of adaptation, or accuracy of the adaptive filter). The first requirement for an adaptive filter is stability, which is ensured by choosing μ to satisfy (10.4.60). Within this range, decreasing μ to reduce the desired level of misadjustment, according to (10.4.63), decreases the speed of convergence; see (10.4.62). Conversely, if μ is increased to increase the speed of convergence; this results in an increase in misadjustment. This tradeoff between speed of convergence and misadjustment is a fundamental feature of the LMS algorithm.

FIR filters. In this case, the input is a stationary process $x(n)$ with a Toeplitz correlation matrix $\mathbf{R}$. Therefore, we have

$$\text{tr}\mathbf{R} = Mr(0) = ME\{|x(n)|^2\} = MP_x \tag{10.4.64}$$

where MP_x is called the *tap input power*. Substituting (10.4.40) into (10.4.64), we obtain

$$0 < 2\mu < \frac{1}{MP_x} = \frac{1}{\text{tap input power}} \tag{10.4.65}$$

which shows that the selection of the step size depends on the input power. Using (10.4.63) and (10.4.64), we see that misadjustment $\mathcal{M}$ is given by

$$\mathcal{M} = \frac{P_{\text{ex}}(\infty)}{P_o} \simeq \mu MP_x \tag{10.4.66}$$

which shows that for given M and P_x the value of misadjustment is proportional to μ. We emphasize that the misadjustment provides a measure of how close an LMS adaptive filter is to the corresponding optimum filter.

The statistical properties of the SOE, that is, the correlation of the input signal and the cross-correlation between input and desired response signals, play a key role in the performance of the LMS adaptive filter.

- First, we should make sure that the relation between $x(n)$ and $y(n)$ can be accurately modeled by a linear FIR filter with M coefficients. Inadequacy of the FIR structure, output observation noise, or lack of correlation between $x(n)$ and $y(n)$ increases the magnitude of the irreducible error. If M is very large, we may want to use a pole-zero IIR filter (Shynk 1989; Treichler et al. 1987). If the relationship between $x(n)$ and $y(n)$ is nonlinear, we certainly need a nonlinear filtering structure (Mathews 1991).
- The LMS algorithm uses a "noisy" instantaneous estimate of the gradient vector. However, when the correlation between input and desired response is weak, the algorithm should make more cautious steps ("wait and average"). Such algorithms update their coefficients every L samples, using all samples between successive updatings to determine the gradient (gradient averaging).
- The eigenvalue structure of $\mathbf{R}$ as measured by its eigenvalue spread ($\lambda_{\max}/\lambda_{\min}$) or equivalently by the *spectral flatness measure* (*SFM*) (see Section 4.1) has a strong effect on the rate of convergence of the LMS algorithm. In general, the rate of convergence decreases as the eigenvalue spread increases, that is, as the contours of the cost function become more elliptical, or equivalently the input spectrum becomes more nonwhite.

Normalized LMS algorithm. According to (10.4.60), the selection of μ in practical applications is complicated because the power of the input signal either is unknown or varies with time. This problem can be addressed by using the *normalized LMS (NLMS)* algorithm [see (10.4.11)]

$$\mathbf{c}(n) = \mathbf{c}(n-1) + \frac{\tilde{\mu}}{E_M(n)}\mathbf{x}(n)e^*(n) \tag{10.4.67}$$

where $E_M(n) = \|\mathbf{x}(n)\|^2$ and $0 < \tilde{\mu} < 1$. It can be shown that the NLMS algorithm converges in the mean square if $0 < \tilde{\mu} < 1$ (Rupp 1993; Slock 1993), which makes the selection of the step size $\tilde{\mu}$ much easier than the selection of μ in the LMS algorithm.

For FIR filters, the quantity $E_M(n)$ provides an estimate of $ME\{|x(n)|^2\}$ and can be computed recursively by using the sliding-window formula

$$E_M(n) = E_M(n-1) + |x(n)|^2 - |x(n-M)|^2 \tag{10.4.68}$$

where $E_M(-1) = 0$ or a first-order recursive filter estimator. In practice, to avoid division by zero, if $\mathbf{x}(n) = \mathbf{0}$, we set $E_M(n) = \delta + \|\mathbf{x}(n)\|^2$, where δ is a small positive constant.

Other approaches and analyses. The analysis of the LMS algorithm presented in this section is simple, clarifies its performance, and provides useful design guidelines. However, there are many other approaches, which are beyond the scope of this book, that differ in terms of complexity, accuracy, and objectives. Major efforts to remove the independence assumption and replace it with the more realistic statistically dependent input assumption are documented in Macchi (1995), Solo (1997), and Butterweck (1995) and the references therein. Convergence analysis of the LMS algorithm using the *stochastic approximation approach* and a deterministic approach using the *method of ordinary differential equations* are discussed in Solo and Kong (1995), Sethares (1993), and Benveniste et al. (1987). Other types of analyses deal with the determination of the probability densities and the probability of large excursions of the adaptive filter coefficients for various types of input signals (Rupp 1995). The analysis of the convergence properties of the LMS algorithm and its variations is still an active area of research, and new results appear continuously.

10.4.4 Applications of the LMS Algorithm

We now discuss three practical applications in which the LMS algorithm has made a significant impact. In the first case, we consider the previously discussed linear prediction problem and compare the performance of the LMS algorithm with that of the SDA. Table 10.4 provides a summary of the key differences between the SDA and the LMS algorithms. In the second case, we study echo cancelation in full-duplex data transmission, which employs the LMS algorithm in its implementation. In the third case, we discuss the application of adaptive equalization, which is used to minimize intersymbol interference (ISI) in a dispersive channel environment.

TABLE 10.4
Comparison between the SDA and LMS algorithms.

SDA	LMS
Deterministic algorithm:	Stochastic algorithm:
$\lim_{n \to \infty} \mathbf{c}(n) = \mathbf{c}_o$	$\lim_{n \to \infty} E\{\mathbf{c}(n)\} = \mathbf{c}_o$
If converges, it terminates to $\mathbf{c}_o$	If converges, it fluctuates about $\mathbf{c}_o$
	The size of fluctuations is proportional to μ
Noiseless gradient estimate	Noisy gradient estimate
Deterministic steps	Random steps
We can only compare the ensemble average behavior of LMS with the SDA.	

Linear prediction

In Example 10.3.1, the AR(2) model given in (10.3.25) was considered, and the SDA was used to determine the corresponding linear predictor coefficients. We also analyzed the performance of the SDA. In the following example, we perform a similar acquisition of predictor coefficients using the LMS algorithm, and we study the effects of the eigenvalue spread of the input correlation matrix on the convergence of the LMS adaptive algorithm when it is used to update the coefficients.

EXAMPLE 10.4.1. The second-order system in (10.3.25) is repeated here, which generates the signal $x(n)$:

$$x(n) + a_1 x(n-1) + a_2 x(n-2) = w(n)$$

where $w(n) \sim \text{WGN}(0, \sigma_w^2)$ and where the coefficients are selected from Table 10.2 for two different eigenvalue spreads. A Gaussian pseudorandom number generator was used to obtain 1000 realizations of $x(n)$ using each set of parameter values given in Table 10.2. These sample realizations were used for statistical analysis.

The second-order LMS adaptive predictor with coefficients $\mathbf{c}(n) = [c_1(n)\ c_2(n)]^T$ is given by [see (10.4.12)]

$$e(n) = x(n) - c_1(n-1)x(n-1) - c_2(n-2)x(n-2) \qquad n \geq 0$$

$$c_1(n) = c_1(n-1) + 2\mu e(n)x(n-1)$$

$$c_2(n) = c_2(n-1) + 2\mu e(n)x(n-2)$$

where μ is the step-size parameter. The adaptive predictor was initialized by setting $x(-1) = x(-2) = 0$ and $c_1(-1) = c_2(-1) = 0$. The above adaptive predictor was implemented with $\mu = 0.04$, and the predictor coefficients as well as the MSE were recorded for each realization. These quantities were averaged to study the behavior of the LMS algorithm. These calculations were repeated for $\mu = 0.01$.

In Figure 10.20 we show several plots obtained for $\mathcal{X}(\mathbf{R}) = 1.22$. In plot ($a$) we show the ensemble averaged trajectory $\{\mathbf{c}(n)\}_{n=0}^{150}$ superimposed on the MSE contours. A trajectory of a simple realization is also shown to illustrate its randomness. In plot (b) the $\mathbf{c}(n)$ learning curve for the averaged value as well as for one single realization is shown. In plot (c) the corresponding learning curves for the MSE are depicted. Finally, in plot (d) we show the effect of step size μ on the MSE learning curve. Similar plots are shown in Figure 10.21 for $\mathcal{X}(\mathbf{R}) = 10$.

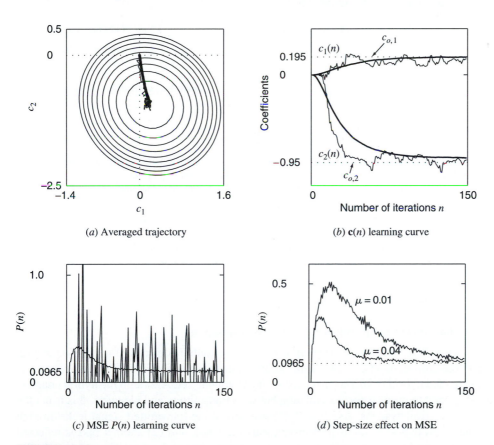

(a) Averaged trajectory

(b) $\mathbf{c}(n)$ learning curve

(c) MSE $P(n)$ learning curve

(d) Step-size effect on MSE

FIGURE 10.20
Performance curves for the LMS used in the linear prediction problem with step-size parameter $\mu = 0.04$ and eigenvalue spread $\mathcal{X}(\mathbf{R}) = 1.22$.

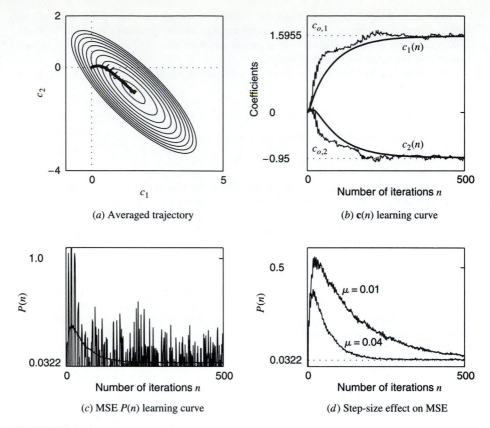

(a) Averaged trajectory

(b) $\mathbf{c}(n)$ learning curve

(c) MSE $P(n)$ learning curve

(d) Step-size effect on MSE

FIGURE 10.21
Performance curves for the LMS used in the linear prediction problem with step-size parameter $\mu = 0.04$ and eigenvalue spread $\mathcal{X}(\mathbf{R}) = 10$.

Several observations can be made from these plots:

- The trajectories and the learning curves for a simple realization are clearly random or "noisy," while the averaging over the ensemble clearly has a smoothing effect.
- The averaged quantities (coefficients and the MSE) converge to the true values, and this convergence rate is in accordance with theory.
- The rate of convergence of the LMS algorithm depends on the step size μ. The smaller the step size, the slower the rate.
- The rate of convergence also depends on the eigenvalue spread $\mathcal{X}(\mathbf{R})$. The larger the spread, the slower the rate. For $\mathcal{X}(\mathbf{R}) = 1.22$, the algorithm converges in about 150 steps while for $\mathcal{X}(\mathbf{R}) = 10$ it requires about 500 steps.

Clearly these observations compare well with the theory.

Echo cancelation in full-duplex data transmission

Figure 10.22 illustrates a system that achieves simultaneous data transmission in both directions (full-duplex) over two-wire circuits using the special two-wire to four-wire interfaces (called *hybrid couplers*) that exist in any telephone set. Although the hybrid couplers are designed to provide perfect isolation between transmitters and receivers, this is not the case in practical systems. As a result, (1) one part of the transmitted signal leaks through the near-end hybrid to its own receiver (near-end echo), and (2) another part is reflected by the far-end hybrid and ends up at its own receiver (far-end echo). The combined echo signal, which can be 30 dB stronger than the signal received from the other end, increases the number of errors. We note that in contrast with acoustic echo cancelation, the delay of echoes in data transmission is immaterial.

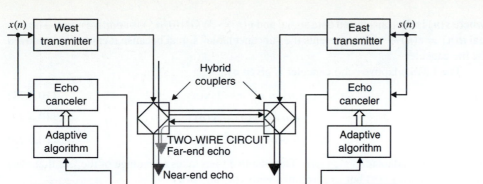

FIGURE 10.22
Model of a full-duplex data transmission system that uses an echo canceler
in the modems.

The best way to address this problem is to form a replica of the echo and then subtract it from the incoming signal. We can model the echoes as the result of an "echo" path between the transmitter and the receiver. For baseband data transmission this echo path is basically linear and varies very slowly with time. Therefore, we can obtain a replica of the echo signal using an FIR LMS adaptive filter (*echo canceler*), as shown in Figure 10.22. The inclusion of the transmitter in the echo path, as long as it involves linear operations, simplifies the implementation and improves the speed of adaptation because the input is an IID binary data sequence of values $+1$ and -1 with equal probability (Verhoeckx et al. 1979).

Referring to Figure 10.23, if we assume that the echo path has an FIR impulse response, the echo signal is given by

$$y(n) = \mathbf{c}_o^T \mathbf{x}(n) \tag{10.4.69}$$

where $$\mathbf{c}_o = [c_o(0) \, c_o(1) \, \cdots \, c_o(M-1)]^T$$

If $g(n)$ is the impulse response of the transmission path from the far-end transmitter to the near-end receiver, the received signal is given by

$$s_r(n) = y(n) + z(n) + v(n) \triangleq y(n) + u(n) \tag{10.4.70}$$

$$z(n) = \sum_{k=0}^{\infty} g(k)s(n-k)$$

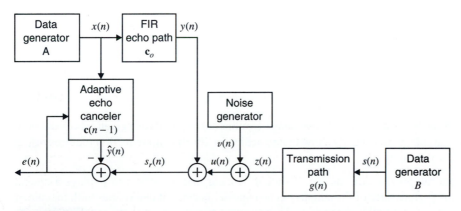

FIGURE 10.23
Block diagram of a system for investigating the performance of adaptive echo canceler.

where $s(n)$ is the transmitted data signal and $v(n) \sim \text{WGN}(0, \sigma_v^2)$ is additive noise. The signal $u(n) = z(n) + v(n)$ represents the "uncancelable" signal because it cannot be removed by the canceler.

The LMS adaptive echo canceler is given by

$$\hat{y}(n) = \mathbf{c}^T(n-1)\mathbf{x}(n) \tag{10.4.71}$$

$$e(n) = y(n) - \hat{y}(n) \tag{10.4.72}$$

$$\mathbf{c}(n) = \mathbf{c}(n-1) + 2\mu e(n)\mathbf{x}(n) \tag{10.4.73}$$

where μ is the adaptation step size. The adaptive filter takes advantage of the fact that $x(n)$ is correlated with $y(n)$ but uncorrelated with $s(n)$ and $v(n)$.

The residual (uncanceled) echo is

$$e_r(n) \triangleq y(n) - \hat{y}(n) = [\mathbf{c}_o - \mathbf{c}(n-1)]^T \mathbf{x}(n) \triangleq -\tilde{\mathbf{c}}^T(n-1)\mathbf{x}(n) \tag{10.4.74}$$

and if we assume that $\tilde{\mathbf{c}}(n-1)$ and $\mathbf{x}(n)$ are independent, then

$$P_r(n) = E\{e_r^2(n)\} = E\{\tilde{\mathbf{c}}^T(n-1)\tilde{\mathbf{c}}(n-1)\}$$

because $\mathbf{R} = E\{\mathbf{x}(n)\mathbf{x}^T(n)\} = \mathbf{I}$. Using (10.4.69), (10.4.71), and (10.4.72), we can easily show that

$$\tilde{\mathbf{c}}(n) = \tilde{\mathbf{c}}(n-1) - 2\mu\mathbf{x}(n)\mathbf{x}^T(n)\tilde{\mathbf{c}}(n-1) + 2\mu\mathbf{x}(n)u(n) \tag{10.4.75}$$

If we premultiply (10.4.75) by its transpose and take the mathematical expectation, we obtain

$$P_r(n+1) = (1 - 4\mu + 4\mu^2 M)P_r(n) + 4\mu M \sigma_u^2 \tag{10.4.76}$$

using the independence assumption and the relation $\mathbf{x}^T(n)\mathbf{x}(n) = M$. The solution of (10.4.76), in terms of the residual echo ratio $P_r(n)/\sigma_u^2$, is

$$\frac{P_r(n)}{\sigma_u^2} = (1 - 4\mu + 4\mu^2 M)^n \left[\frac{P_r(0)}{\sigma_u^2} - \frac{\mu M}{1 - \mu M} \right] + \frac{\mu M}{1 - \mu M} \tag{10.4.77}$$

and describes completely the operation of the LMS adaptive echo canceler. Indeed, we draw the following conclusions:

1. The algorithm converges if

$$|1 - 4\mu + 4\mu^2 M| < 1 \quad \text{or} \quad 0 < \mu < \frac{1}{M} \tag{10.4.78}$$

   which agrees with (10.4.40) because $\text{tr}\,\mathbf{R} = M$.
2. After convergence we have

$$P_r(\infty) = \frac{\mu M}{1 - \mu M}\sigma_u^2 \simeq \mu M \sigma_u^2 \tag{10.4.79}$$

   which again is in agreement with (10.4.63).
3. If $P_r(n)/\sigma_u^2 \gg \mu M/(1 - \mu M)$, we have

$$\frac{P_r(n)}{P_r(0)} \simeq (1 - 4\mu + 4\mu^2 M)^n \tag{10.4.80}$$

   which can be used to find out how many iterations are required for a given echo reduction. For example, we can easily show that to achieve a 20-dB echo reduction requires $n_{20} \simeq 1.15/\mu$ iterations.

From the previous discussion, it should be clear that the step size μ plays a crucial role in the performance of the adaptive echo canceler because it determines both the rate of convergence and the minimum residual echo cancelation that can be attained. Furthermore, we clearly see the tradeoff between fast adaptation and residual echo power.

EXAMPLE 10.4.2. Consider the system shown in Figure 10.23 for investigating the performance of the LMS algorithm in adaptive echo cancelation and to verify the above conclusions. The data generators A (in modem A) and B (in modem B) output symbols $+1$ or -1 with equal probability (i.e., Bernoulli sequence). The FIR filter following data generator A models the echo path, which is assumed to be

$$c_o(n) = -\frac{5}{3}(\frac{1}{2})^n + \frac{8}{3}(\frac{4}{5})^n \qquad 0 \le n \le M - 1$$

where $M = 20$ is the total length of the echo path. The filter following data generator B models the transmission path between the far-end transmitter and the near-end receiver, which we will assume to be

$$g(n) = \frac{4}{5}(\frac{3}{5})^n \quad n \ge 0$$

The noise generator is a white Gaussian source with $\sigma_v^2 = 1$ and models the transmission noise. Using the equations $\sigma_y^2 = \sum_{k=0}^{N-1} c_o^2(k)$ and $\sigma_u^2 = \sum_{k=0}^{\infty} g^2(k) + \sigma_v^2$, we scale $u(n)$ so that $10 \log(\sigma_y^2/\sigma_u^2) = 30$ dB. The adaptive echo canceler employs the LMS algorithm with $\mathbf{c}(0) = \mathbf{0}$. We perform Monte Carlo simulations on this system. Figure 10.24 shows the residual echo ratio $P_r(n)/\sigma_u^2$ evaluated by ensemble averaging over 200 independent trials of the experiment, for two different step sizes in the LMS algorithm [which satisfy (10.4.78)], superimposed on the corresponding theoretical curves computed by using (10.4.79) and (10.4.80). Clearly, the simulations support the theoretical results quite accurately. More detailed discussions of adaptive echo cancelation techniques for both baseband and passband data transmission systems can be found in Gitlin et al. (1992) and in Ling (1993a).

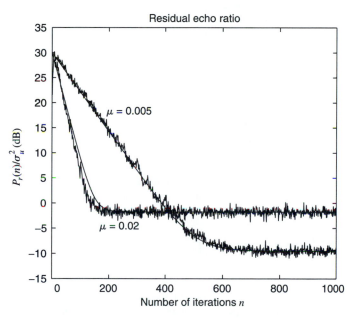

FIGURE 10.24
Performance analysis of the LMS algorithm in the adaptive echo cancelation that clearly shows the tradeoff between rate of convergence and residual echo power.

Adaptive equalization

In Section 6.8, we discussed the theory and implementation of channel equalization in data transmission systems. When data are transmitted below 2400 bits/s, the ISI is relatively small and does not pose a problem in the operation of a modem. However, for high-speed communication over 2400 bits/s, an equalizer is needed in the modem to compensate for the channel distortion. Since channel characteristics are generally unknown and time-varying, an adaptive algorithm is required that leads to adaptive equalization. Figure 10.25 describes an application of adaptive filtering to adaptive channel equalization. Initially, coefficients of the equalizer are adjusted, by means of the LMS algorithm, by transmitting a known training sequence of short duration. After this short training period, the actual data sequence $\{y(n)\}$ is transmitted. The slow variation in channel characteristics is then continuously tracked by adjusting coefficients of the equalizer, using the decisions in place of the known training sequence. This approach works well when decision errors are infrequent.

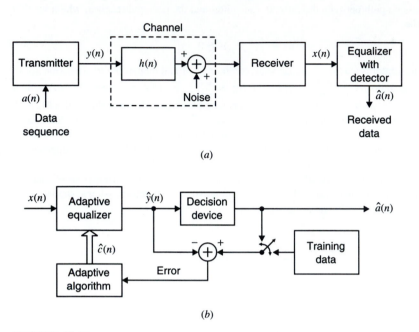

FIGURE 10.25
Model of an adaptive equalizer in a data transmission system.

EXAMPLE 10.4.3. Figure 10.26 shows the block diagram of the system used in the experimental investigation of the performance of the LMS algorithm used in the adaptive equalizer. The data source generates Bernoulli sequence $\{y(n)\}$ with symbols $+1$ and -1 having zero mean and unit variance. The channel following the source is modeled by the raised cosine impulse response

$$h(n) = \begin{cases} 0.5\left\{1 + \cos\left[\dfrac{2\pi}{W}(n-2)\right]\right\} & n = 1, 2, 3 \\ 0 & \text{otherwise} \end{cases} \tag{10.4.81}$$

where parameter W is used to control the amount of channel distortion. The amount of channel distortion increases with W. The random noise generator outputs white Gaussian sequence $v(n)$ which models the noise in the channel. The equalizer input is

$$x(n) = \sum_{k=1}^{3} h(k)y(n-k) + v(n) \tag{10.4.82}$$

Since $y(n)$ is an independent sequence and since $v(n)$ is uncorrelated with $y(n)$, the maximum lag that produces nonzero correlation is 2. Thus the correlation of $x(n)$ is given by

$$r_x(0) = h^2(1) + h^2(2) + h^2(3) + \sigma_v^2$$

$$r_x(1) = h(1)h(2) + h(2)h(3)$$

$$r_x(2) = h(1)h(3)$$

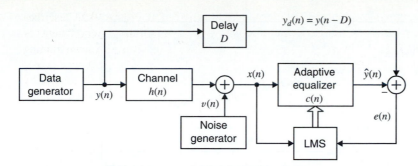

FIGURE 10.26
Block diagram of a system for investigating the performance of an adaptive equalizer.

from which an $M \times M$ autocorrelation matrix $\mathbf{R}$ can be constructed for an equalizer of length M. Clearly, parameter W also controls the eigenvalues of $\mathbf{R}$ and hence the ratio $\mathcal{X}(\mathbf{R})$. The design of an MSE equalizer has been discussed in Example 6.8.1. Here we study the performance of the corresponding LMS adaptive equalizer.

The training signal $y(n)$ is delayed by an amount equal to the combined delay introduced by the channel and the equalizer for the desired signal. The impulse response $h(n)$ in (10.4.81) is symmetric with respect to $n = 2$, and assuming that the equalizer is a linear-phase FIR filter, the total delay is equal to $\Delta = (M - 1)/2 + 2$. The error signal $e(n) = y(n - \Delta) - \hat{y}(n)$ is used along with $x(n)$ to implement the LMS algorithm in the adaptive equalizer with $\mathbf{c}(0) = \mathbf{0}$. We performed Monte Carlo simulations using 100 realizations of random sequences with $M = 11$; $\Delta = 7$; $\sigma_v^2 = 0.001$; $W = 2.9$ and $W = 3.5$; and $\mu = 0.01, 0.04$, and 0.08. The results are shown in Figures 10.27 and 10.28.

Effect of eigenvalue spread. Performance plots of the LMS algorithm for $W = 2.9$ and $W = 3.5$ are shown in Figure 10.27. In plot (*a*) we depict MSE learning curves from which we observe that the convergence rate of the MSE decreases with W [or equivalently with increase in $\mathcal{X}(\mathbf{R})$], which is to be expected. The steady-state error, on the other hand, increases with W. In plots (*b*) and (*c*) we show the ensemble averaged equalizer coefficients. Clearly, the responses are symmetric with respect to $n = 5$ as assumed. Also equalizer coefficients converge to different inverses due to changes in the channel characteristics.

Effect of step size μ. In Figure 10.28 we show the MSE learning curves obtained for $W = 2.9$ and with three different step-size parameter values of 0.01, 0.04, and 0.08. It indicates that μ affects the rate of convergence as well as the steady-state value. For $\mu = 0.08$, the algorithm converges in about 100 iterations but has higher steady-state value than the case for $\mu = 0.04$, which requires about 275 iterations for convergence. For $\mu = 0.01$ more than 500 iterations are needed. Finally, Figure 10.29 shows sample realizations of the transmitted, received, and equalized sequences using the discussed LMS equalizer.

10.4.5 Some Practical Considerations

The LMS is the most widely known and used adaptive algorithm because of its simplicity and robustness to disturbances and model errors. We next discuss some issues related to its robustness, finite-word-length effects, and implementation.

Robustness

If we assume the model in Figure 10.19, an adaptive filter is said to be *robust* if the effect of the disturbances $\{\mathbf{c}(-1), e_o(n)\}$ on the resulting estimation errors $\{\tilde{\mathbf{c}}(n), e(n)\}$ (or $\{\tilde{\mathbf{c}}(n), \varepsilon(n)\}$), as measured by their energy, is small (Sayed and Rupp 1998). Basically a robust adaptive filter should be insensitive to the initial conditions $\mathbf{c}(-1)$ and the optimum

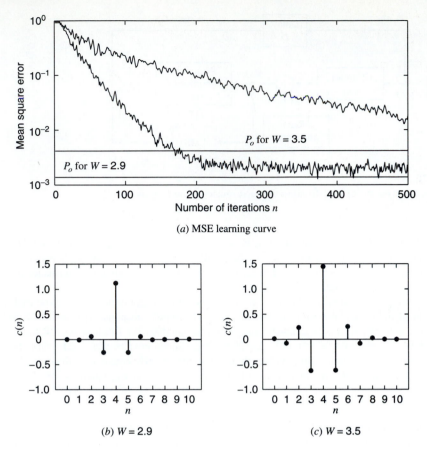

(a) MSE learning curve

(b) W = 2.9

(c) W = 3.5

FIGURE 10.27
Performance analysis curves of the LMS algorithm in the adaptive equalizer:
$\mu = 0.04$.

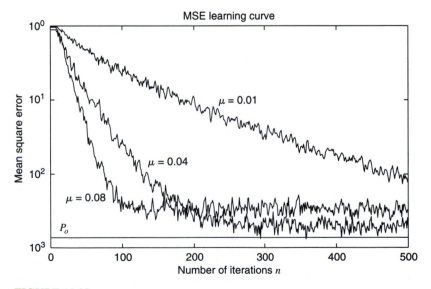

FIGURE 10.28
MSE learning curves of the LMS algorithm in the adaptive equalizer: $W = 2.9$.

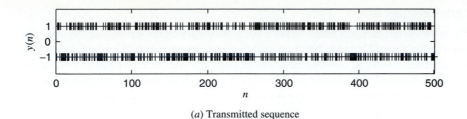

(a) Transmitted sequence

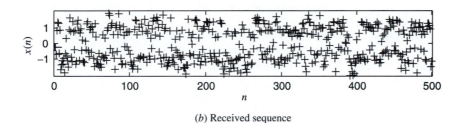

(b) Received sequence

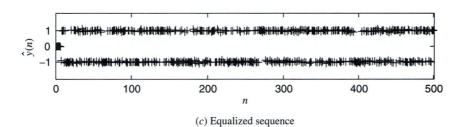

(c) Equalized sequence

FIGURE 10.29
Sample realizations of the transmitted, received, and equalized sequences using an FIR
LMS equalizer.

residual error $e_o(n)$, which acts as measurement noise. These inputs are collectively called
disturbances. In practice, $e_o(n)$ accounts not only for measurement noise but also for model
mismatching, quantization errors, and other inaccuracies.

If we define the energies of the disturbances and the estimation errors by

$$E_{\text{dist}}(n) = \frac{1}{2\mu}\|\tilde{\mathbf{c}}(-1)\|^2 + \sum_{j=0}^{n}|e_o(n)|^2 \tag{10.4.83}$$

and

$$E_{\text{error}}(n) = \frac{1}{2\mu}\|\tilde{\mathbf{c}}(n)\|^2 + \sum_{j=0}^{n}|\tilde{y}(n)|^2 \tag{10.4.84}$$

it can be shown that the coefficient vectors determined by the LMS algorithm satisfy the
condition

$$E_{\text{error}}(n) \le E_{\text{dist}}(n) \tag{10.4.85}$$

assuming that $0 < 2\mu \le 1/\|\mathbf{x}(n)\|^2$ (Sayed and Kailath 1994; Sayed and Rupp 1996).
Equation (10.4.85) shows that the energy of the residuals is always upper-bounded by the
energy of the disturbances, which explains the robust behavior of the LMS algorithm.

Furthermore, it can be shown that the LMS algorithm minimizes the maximum possible difference between these two energies, over all disturbances with finite energy, and is
optimum according to the H^{∞} (or minimax) criterion (Sayed and Rupp 1998; Hassibi et
al. 1996).

Finite-precision effects

When we design an LMS adaptive filter for a stationary SOE, we choose the step size μ to provide the desired balance between speed of convergence and misadjustment. If we are not concerned about fast convergence, we can reduce μ so much as to obtain practically insignificant misadjustment. However, in a digital implementation, the adaptation of the LMS algorithm stops (stalls) when the correction term becomes smaller in magnitude than one-half of the least significant bit (LSB), that is,

$$|2\mu e^*(n)x(n-k)| \leq \frac{\text{LSB}}{2} \tag{10.4.86}$$

Therefore, a decrease in μ may result in a performance degradation, unless we increase the number of bits (i.e., the precision) of the filter coefficients. If X_{rms} is the root mean square (rms) amplitude of the input signal, to a good approximation we have

$$|e(n)| \leq \frac{\text{LSB}}{4\mu X_{\text{rms}}} \triangleq \text{DRE} \tag{10.4.87}$$

where DRE is known as the *digital residual error* (Gitlin et al. 1973). We note that for a given number of bits the DRE increases as we decrease the step size μ.

The roundoff numerical errors contribute to the steady-state EMSE a term that is inversely proportional to μ, whereas the quantization of the input data and the filter output contributes a second term that is independent of the step size (Caraiscos and Liu 1984). Hence, in practice the step size of the LMS algorithm cannot be decreased below the level where the degradation effects of quantization and finite-precision arithmetic become significant. Also, the finite-precision effects become more pronounced as the ill conditioning of the input increases (Alexander 1987).

When one or more eigenvalues of the input correlation matrix are zero, the corresponding adaptation modes either do not converge or may result in overflow due to nonlinear quantization effects (Gitlin et al. 1982). These effects can be prevented by using a technique known as *leakage*. The *leaky* LMS algorithm is given by

$$\mathbf{c}(n) = (1 - \gamma\mu)\mathbf{c}(n-1) + \mu e^*(n)\mathbf{x}(n) \tag{10.4.88}$$

where γ is the leakage coefficient. Since μ and γ are very small positive constants, $1 - \gamma\mu$ is slightly less than 1. The updating (10.4.88) is obtained by minimizing the cost function

$$P(n) = |e(n)|^2 + \gamma\|\mathbf{c}(n)\|^2 \tag{10.4.89}$$

which includes a penalty term proportional to the size of the coefficient vector. The price of leakage is an increase in computational complexity and some bias in the obtained estimates (see Problem 10.17). More details and practical applications of the leaky LMS algorithm to adaptive equalization are discussed in Gitlin et al. (1992, 1982).

We can simplify the hardware implementation of LMS adaptive filters by using nonlinearities to avoid the multiplications involved in the updating of the filter coefficients. These simplified LMS algorithms update the filter coefficients by using quantized correction terms such as $\mu\,\text{sign}\{e(n)\}x(n-k)$, $\mu e(n)\text{sign}\{x(n-k)\}$, or $\mu\,\text{sign}\{e(n)x(n-k)\}$; and their performance is degraded by the lower precision. Various signum-based LMS adaptive algorithms are discussed in Claasen and Mecklenbrauker (1981), Duttweiler (1982), and Treichler et al. (1987).

Transform-domain and block LMS algorithms

The LMS algorithm attains its best rate of convergence when the input correlation matrix is diagonal with equal eigenvalues. In the case of FIR filters, this implies that the input signal is white noise. When the components of the input data vector are correlated, we can improve the convergence by using an isotropic decorrelating transformation, as shown in Figure 10.30. The transformation matrix can be obtained by using either the triangular or the

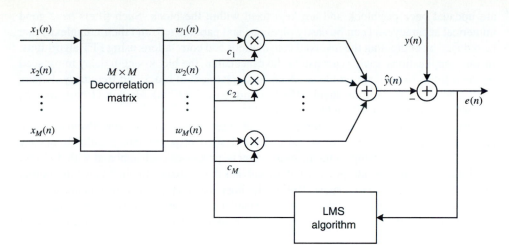

FIGURE 10.30
Transform domain LMS adaptive filter structure.

orthogonal decomposition of the input correlation matrix as explained in Section 3.5. Since the innovations vector used by the LMS algorithm has uncorrelated components with unit variance, the error performance surface is a hypersphere, and the *transform-domain LMS algorithm* attains its best rate of convergence. In practice, when the input correlation matrix is unknown and possibly time-varying, we can only use suboptimum transforms such as the DFT, the *discrete cosine transform (DCT)*, the *discrete wavelet transform (DWT)*, or some other orthogonal transform. The performance of the obtained adaptive filter depends on the decorrelation properties of the transform, which in turn depends on the properties of the input correlation matrix. Another approach to overcome the problem of slow convergence for highly correlated inputs is found in the family of *affine projection algorithms* discussed in Ozeki and Umeda (1984), Rupp (1995), and Morgan and Kratzer (1996) and the references therein.

In applications that require adaptive filters with a very large number of coefficients, real-time implementation of the LMS algorithm becomes quite involved. For example, acoustic echo cancelers with 8000 coefficients (500 ms sampled at 16 kHz) are typical for teleconference applications (Gilloire et al. 1996). The complexity of such applications can be reduced by using block adaptive filters (see Figure 10.31) that process one block of data at a time in either the time or the frequency domain. The adaptive filter coefficients

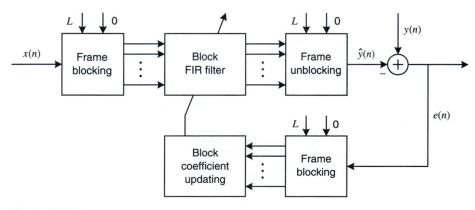

FIGURE 10.31
Block adaptive filter structure.

are updated once per block and are kept fixed within the block. Such filters have good numerical accuracy, and can be easily pipelined and parallelized, and their complexity can be reduced by computing the involved convolutions and correlations using FFT algorithms. In some applications, such as acoustic echo cancelation, the block-length delay introduced by these filters may create problems. A detailed treatment of block and frequency-domain LMS algorithms is given in Shynk (1992), Gilloire et al. (1996), Haykin (1996), Jenkins and Marshall (1998), and Treichler et al. (1987).

Another approach to reduce complexity and improve convergence is *subband adaptive filtering*, which splits the input signal and the desired response into smaller frequency bands (subbands), subsamples the resulting signals, processes each subband with different LMS filters, and finally interpolates and recombines the subbands to obtain the filter output (Shynk 1992; Gilloire and Vetterli 1992). The improved convergence results because the spectral dynamic range of each subband is smaller than that of the full band. However, the performance of subband adaptive filters is degraded by the cross-talk between adjacent subbands.

10.5 RECURSIVE LEAST-SQUARES ADAPTIVE FILTERS

In this section we use the method of LS to develop adaptive filters, we determine their rate of convergence and misadjustment, and we introduce the *conventional recursive least-squares (CRLS) algorithm* for their implementation. The CRLS algorithm does *not* impose any restrictions on the input data vector; therefore, it can be used for both array processing and FIR filtering applications.

10.5.1 LS Adaptive Filters

LS adaptive filters are designed so that the updating of their coefficients always attains the minimization of the total squared error from the time the filter initiated operation up to the current time. Therefore, the filter coefficients at time index n are chosen to minimize the cost function

$$E(n) = \sum_{j=0}^{n} \lambda^{n-j} |e(j)|^2 = \sum_{j=0}^{n} \lambda^{n-j} |y(j) - \mathbf{c}^H \mathbf{x}(j)|^2 \tag{10.5.1}$$

where $e(j)$ is the instantaneous error and the constant λ, $0 < \lambda \leq 1$, is the *forgetting factor*. Note that since the filter coefficients are held constant during the observation interval $0 \leq j \leq n$, the a priori and a posteriori errors are identical. The coefficient vector obtained by minimizing (10.5.1) is denoted by $\mathbf{c}(n)$ and provides the optimum LSE filter at time n. When $\lambda = 1$, we say that the algorithm has *growing memory* because the values of the filter coefficients are a function of all the past input values. The forgetting factor (see Figure 10.32) is used to ensure that data in the distant past are paid less attention ("forgotten") in order to provide the filter with tracking capability when it operates in a varying SOE (see Section 10.8).

The filter coefficients that minimize the total squared error (10.5.1) are specified by the normal equations

$$\hat{\mathbf{R}}(n)\mathbf{c}(n) = \hat{\mathbf{d}}(n) \tag{10.5.2}$$

where

$$\hat{\mathbf{R}}(n) \triangleq \sum_{j=0}^{n} \lambda^{n-j} \mathbf{x}(j) \mathbf{x}^H(j) \tag{10.5.3}$$

and

$$\hat{\mathbf{d}}(n) \triangleq \sum_{j=0}^{n} \lambda^{n-j} \mathbf{x}(j) y^*(j) \tag{10.5.4}$$

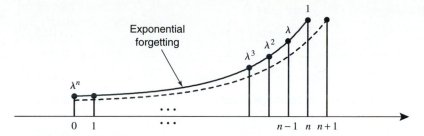

FIGURE 10.32
Exponential weighting of observations at times n and $n + 1$. Older data are more heavily discounted by the algorithm.

provide exponentially weighted estimates of the input correlation matrix and the cross-correlation vector between input and desired response due to the presence of λ^{n-j} in the cost function (10.5.1). The minimum total squared error is

$$E_{\min}(n) = E_y(n) - \hat{\mathbf{d}}^H(n)\mathbf{c}(n) \tag{10.5.5}$$

where

$$E_y(n) \triangleq \sum_{j=0}^{n} \lambda^{n-j}|y(j)|^2 \tag{10.5.6}$$

is the energy of the weighted desired response signal. These formulas have been derived in Section 8.2.1.

Suppose now that we wait for some $n > M$, where $\hat{\mathbf{R}}(n)$ is usually nonsingular, we compute $\hat{\mathbf{R}}(n)$ and $\hat{\mathbf{d}}(n)$, and then we solve the normal equations (10.5.2) to determine the filter coefficients $\mathbf{c}(n)$. This approach, which is time-consuming, should be repeated with the arrival of new pairs of observations $\{\mathbf{x}(n), y(n)\}$, that is, at times $n + 1, n + 2$, etc.

A first reduction in computational complexity can be obtained by noticing that (10.5.3) can be expressed as

$$\hat{\mathbf{R}}(n) = \lambda\hat{\mathbf{R}}(n - 1) + \mathbf{x}(n)\mathbf{x}^H(n) \tag{10.5.7}$$

which shows that the "new" correlation matrix $\hat{\mathbf{R}}(n)$ can be updated by weighting the "old" correlation matrix $\hat{\mathbf{R}}(n - 1)$ with the forgetting factor λ and then incorporating the "new information" $\mathbf{x}(n)\mathbf{x}^H(n)$. Since the outer product $\mathbf{x}(n)\mathbf{x}^H(n)$ is a matrix of rank 1, (10.5.7) provides a rank 1 modification of the correlation matrix. Similarly, using (10.5.4), we can show that

$$\hat{\mathbf{d}}(n) = \lambda\hat{\mathbf{d}}(n - 1) + \mathbf{x}(n)y^*(n) \tag{10.5.8}$$

which provides a time update of the cross-correlation vector.

We next show that using these two updatings, we can determine the new coefficient vector $\mathbf{c}(n)$ from the old coefficient vector $\mathbf{c}(n-1)$ and the new observation pair $\{\mathbf{x}(n), y(n)\}$ without solving the normal equations (10.5.2) from scratch.

A priori adaptive LS algorithm. If we solve (10.5.7) for $\hat{\mathbf{R}}(n - 1)$ and (10.5.8) for $\hat{\mathbf{d}}(n - 1)$ and use the normal equations (10.5.2), we have

$$[\hat{\mathbf{R}}(n) - \mathbf{x}(n)\mathbf{x}^H(n)]\mathbf{c}(n - 1) = \hat{\mathbf{d}}(n) - \mathbf{x}(n)y^*(n)$$

or after some simple manipulations

$$\hat{\mathbf{R}}(n)\mathbf{c}(n - 1) + \mathbf{x}(n)e^*(n) = \hat{\mathbf{d}}(n) \tag{10.5.9}$$

where

$$e(n) = y(n) - \mathbf{c}^H(n - 1)\mathbf{x}(n) \tag{10.5.10}$$

is the a priori estimation error. If the matrix $\hat{\mathbf{R}}(n)$ is invertible, by multiplying both sides of (10.5.9) by $\hat{\mathbf{R}}^{-1}(n)$ and using (10.5.2), we obtain

$$\mathbf{c}(n - 1) + \hat{\mathbf{R}}^{-1}(n)\mathbf{x}(n)e^*(n) = \hat{\mathbf{R}}^{-1}(n)\hat{\mathbf{d}}(n) = \mathbf{c}(n) \tag{10.5.11}$$

If we define the *adaptation gain vector* $\mathbf{g}(n)$ by

$$\hat{\mathbf{R}}(n)\mathbf{g}(n) \triangleq \mathbf{x}(n) \tag{10.5.12}$$

Equation (10.5.11) can be written as

$$\mathbf{c}(n) = \mathbf{c}(n-1) + \mathbf{g}(n)e^*(n) \tag{10.5.13}$$

which shows how to update the old coefficient vector $\mathbf{c}(n-1)$ to obtain the current vector $\mathbf{c}(n)$.

> **EXAMPLE 10.5.1.** It is instructive at this point to derive the LS adaptive filter with a single coefficient. Indeed, since for $M = 1$ the correlation matrix $\hat{\mathbf{R}}(n)$ becomes the scalar $E_x(n)$, we obtain
>
> $$E_x(n) = \lambda E_x(n-1) + |x(n)|^2$$
> $$e(n) = y(n) - c^*(n-1)x(n)$$
> $$c(n) = c(n-1) + \frac{1}{E_x(n)}x(n)e^*(n)$$
>
> which is like an LMS algorithm with time-varying gain $\mu(n) = 1/E_x(n)$. However, the present algorithm is optimum in the LS sense.

A posteriori adaptive LS algorithm. If we substitute (10.5.7) and (10.5.8) into the normal equations (10.5.2), after some simple manipulations, we obtain

$$\lambda\hat{\mathbf{R}}(n-1)\mathbf{c}(n) - \mathbf{x}(n)\varepsilon^*(n) = \lambda\hat{\mathbf{d}}(n-1) \tag{10.5.14}$$

where

$$\varepsilon(n) = y(n) - \mathbf{c}^H(n)\mathbf{x}(n) \tag{10.5.15}$$

is the a posteriori estimation error. If the matrix $\hat{\mathbf{R}}(n-1)$ is invertible, (10.5.14) gives

$$\mathbf{c}(n) - \lambda^{-1}\hat{\mathbf{R}}^{-1}(n-1)\mathbf{x}(n)\varepsilon^*(n) = \hat{\mathbf{R}}^{-1}(n-1)\hat{\mathbf{d}}(n-1) = \mathbf{c}(n-1)$$

or

$$\mathbf{c}(n) = \mathbf{c}(n-1) + \bar{\mathbf{g}}(n)\varepsilon^*(n) \tag{10.5.16}$$

where

$$\lambda\hat{\mathbf{R}}(n-1)\bar{\mathbf{g}}(n) \triangleq \mathbf{x}(n) \tag{10.5.17}$$

determines the *alternative adaptation gain vector* $\bar{\mathbf{g}}(n)$.

Since recursions (10.5.15) and (10.5.16) are coupled, the a posteriori algorithm is not applicable. However, if we substitute (10.5.16) into (10.5.15), we obtain

$$\varepsilon(n) = y(n) - [\mathbf{c}^H(n-1) + \varepsilon(n)\bar{\mathbf{g}}^H(n)]\mathbf{x}(n)$$
$$= e(n) - \varepsilon(n)\bar{\mathbf{g}}^H(n)\mathbf{x}(n)$$

or

$$\varepsilon(n) = \frac{e(n)}{\bar{\alpha}(n)} \tag{10.5.18}$$

where

$$\bar{\alpha}(n) \triangleq 1 + \bar{\mathbf{g}}^H(n)\mathbf{x}(n) = 1 + \lambda^{-1}\mathbf{x}^H(n)\hat{\mathbf{R}}^{-1}(n-1)\mathbf{x}(n) \tag{10.5.19}$$

is known as the *conversion factor*. Hence, we can use (10.5.19) and (10.5.18) to compute the a posteriori error $\varepsilon(n)$ before we update the filter coefficient vector. This trick makes possible the realization and use of the a posteriori LS adaptive filter algorithm. If $\hat{\mathbf{R}}(n-1)$ is positive definite, we have $\bar{\alpha}(n) > 1$ and $|\varepsilon(n)| < |e(n)|$ for all n. Therefore,

$$\sum_n |\varepsilon(n)|^2 < \sum_n |e(n)|^2 \tag{10.5.20}$$

which should be expected[†] because the adaptive filter is designed by minimizing, at each time n, the total squared a posteriori error $\varepsilon(n)$.

[†] The computation of the quantity $\sum_{j=0}^n \lambda^{n-j}|y(j) - \mathbf{c}^H\mathbf{x}(j)|^2$ for $\mathbf{c} = \mathbf{c}(n)$, $\mathbf{c}(j)$, or $\mathbf{c}(j-1)$ gives the block, a posteriori, or a priori total squared error. Clearly, only the block filter performs optimum LS filtering for all data in the interval $0 \le j \le n$ (see Problem 10.22).

Also, from (10.5.13), (10.5.16), and (10.5.18) we obtain

$$\mathbf{g}(n) = \frac{\bar{\mathbf{g}}(n)}{\bar{\alpha}(n)} \tag{10.5.21}$$

which shows that the two adaptation gains have the same direction but different lengths. However, from (10.5.13) and (10.5.16) we see that the corrections $\mathbf{g}(n)e^*(n)$ and $\bar{\mathbf{g}}(n)\varepsilon^*(n)$ are equal.

Another conversion factor, defined in terms of the gain vector $\mathbf{g}(n)$, is

$$\alpha(n) \triangleq 1 - \mathbf{x}^H(n)\hat{\mathbf{R}}^{-1}(n)\mathbf{x}(n) = 1 - \mathbf{x}^H(n)\mathbf{g}(n) \tag{10.5.22}$$

and has some interesting interpretations. Using (10.5.21), we have

$$\alpha(n) = 1 - \frac{\mathbf{x}^H(n)\bar{\mathbf{g}}(n)}{\bar{\alpha}(n)}$$

$$\alpha(n)\bar{\alpha}(n) = \bar{\alpha}(n) + 1 - [1 + \mathbf{x}^H(n)\bar{\mathbf{g}}(n)] = 1$$

or

$$\alpha(n) = \frac{1}{\bar{\alpha}(n)} \tag{10.5.23}$$

which shows that the two conversion factors are inverses of each other. Since the input correlation matrix is nonnegative definite, that is, $\mathbf{x}^H(n)\hat{\mathbf{R}}^{-1}(n)\mathbf{x}(n) \geq 0$, (10.5.22) implies

$$0 < \alpha(n) \leq 1 \tag{10.5.24}$$

that is, the conversion factor $\alpha(n)$ is bounded by 0 and 1. This bound allows the interpretation of $\alpha(n)$ as an angle variable (Lee et al. 1981), and its monitoring can provide information about the proper operation of RLS algorithms. Also the quantity $1 - \alpha(n)$ can be interpreted as a *likelihood variable* (Lee et al. 1981). It can be shown (see Problem 10.23) that

$$\alpha(n) = \lambda^M \frac{\det \hat{\mathbf{R}}(n-1)}{\det \hat{\mathbf{R}}(n)} \tag{10.5.25}$$

which shows the importance of $\alpha(n)$ or $\bar{\alpha}(n)$ for the invertibility for the estimated correlation matrix.

The computational organization of the a priori and a posteriori LS adaptive algorithms is summarized in Table 10.5.

TABLE 10.5
Summary of a priori and a posteriori LS adaptive filter approaches.

	A priori LS adaptive filter	A posteriori LS adaptive filter
Correlation matrix	$\hat{\mathbf{R}}(n) = \lambda\hat{\mathbf{R}}(n-1) + \mathbf{x}(n)\mathbf{x}^H(n)$	$\hat{\mathbf{R}}(n) = \lambda\hat{\mathbf{R}}(n-1) + \mathbf{x}(n)\mathbf{x}^H(n)$
Adaptation gain	$\hat{\mathbf{R}}(n)\mathbf{g}(n) = \mathbf{x}(n)$	$\lambda\hat{\mathbf{R}}(n-1)\bar{\mathbf{g}}(n) = \mathbf{x}(n)$
A priori error	$e(n) = y(n) - \mathbf{c}^H(n-1)\mathbf{x}(n)$	$e(n) = y(n) - \mathbf{c}^H(n-1)\mathbf{x}(n)$
Conversion factor	$\alpha(n) = 1 - \mathbf{g}^H(n)\mathbf{x}(n)$	$\bar{\alpha}(n) = 1 + \bar{\mathbf{g}}^H(n)\mathbf{x}(n)$
A posteriori error	$\varepsilon(n) = \alpha(n)e(n)$	$\varepsilon(n) = \dfrac{e(n)}{\bar{\alpha}(n)}$
Coefficient updating	$\mathbf{c}(n) = \mathbf{c}(n-1) + \mathbf{g}(n)e^*(n)$	$\mathbf{c}(n) = \mathbf{c}(n-1) + \bar{\mathbf{g}}(n)\varepsilon^*(n)$

Figure 10.33 shows a block diagram representation of the a priori LS adaptive filter. There are two important points to be made:

- The adaptation gain is strictly a function of the input signal. The desired response only affects the magnitude and sign of the coefficient correction term through the error.

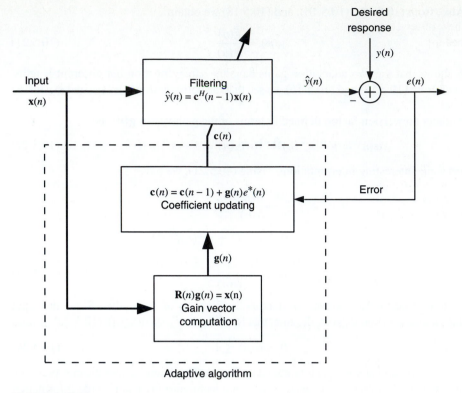

FIGURE 10.33
Basic elements of the a priori LS adaptive filter. Note that the filtering process has no effect on the computation of the gain vector.

- The most demanding computational task in RLS filtering is the computation of the adaptation gain. This involves the solution of a linear system of equations, which requires $O(M^3)$ operations per time update.

10.5.2 Conventional Recursive Least-Squares Algorithm

The major computational load in LS adaptive filters, that is, the computation of the gain vectors

$$\mathbf{g}(n) = \hat{\mathbf{R}}^{-1}(n)\mathbf{x}(n) \tag{10.5.26}$$

or

$$\bar{\mathbf{g}}(n) = \lambda^{-1}\hat{\mathbf{R}}^{-1}(n-1)\mathbf{x}(n) \tag{10.5.27}$$

can be reduced if we can find a recursive formula to update the inverse

$$\mathbf{P}(n) \triangleq \hat{\mathbf{R}}^{-1}(n) \tag{10.5.28}$$

of the correlation matrix. We can develop such an updating by using the rank 1 updating (10.5.7) and the matrix inversion lemma

$$(\lambda\mathbf{R} + \mathbf{x}\mathbf{x}^H)^{-1} = \lambda^{-1}\mathbf{R}^{-1} - \frac{(\lambda^{-1}\mathbf{R}^{-1}\mathbf{x})(\lambda^{-1}\mathbf{R}^{-1}\mathbf{x})^H}{1 + \lambda^{-1}\mathbf{x}^H\mathbf{R}^{-1}\mathbf{x}} \tag{10.5.29}$$

discussed in Appendix A.

Indeed, using (10.5.29), (10.5.7), (10.5.26), and (10.5.19), we can easily show that

$$\mathbf{P}(n) = \lambda^{-1}\mathbf{P}(n-1) - \mathbf{g}(n)\bar{\mathbf{g}}^H(n) \tag{10.5.30}$$

which provides the desired updating formula. Indeed, given the old matrix $\mathbf{P}(n-1)$ and the new observations $\{\mathbf{x}(n), y(n)\}$ we compute the new matrix $\mathbf{P}(n)$, using the following procedure

$$\bar{\mathbf{g}}(n) = \lambda^{-1}\mathbf{P}(n-1)\mathbf{x}(n)$$

$$\alpha(n) = 1 + \bar{\mathbf{g}}^H(n)\mathbf{x}(n)$$

$$\mathbf{g}(n) = \frac{\bar{\mathbf{g}}(n)}{\bar{\alpha}(n)} \qquad (10.5.31)$$

$$\mathbf{P}(n) = \lambda^{-1}\mathbf{P}(n-1) - \mathbf{g}(n)\bar{\mathbf{g}}^H(n)$$

which is known as the *conventional recursive LS (CRLS) algorithm*. We again stress that *the CRLS algorithm is valid for both linear combiners and FIR filters because it does not make any assumptions about the nature of the input data vector*. However, for FIR filters we usually assume prewindowing, that is, $\mathbf{x}(-1) = \mathbf{0}$, or equivalently $x(n) = 0$ for $-M \leq n \leq -1$.

Updating of the minimum total squared error. We next derive an update recursion for the minimum total squared error (10.5.5). Using (10.5.6), we can easily see that

$$E_y(n) = \lambda E_y(n-1) + y(n)y^*(n) \qquad (10.5.32)$$

which provides a recursive updating for the energy of the desired response. Substituting (10.5.32) and (10.5.13) into (10.5.5), we obtain

$$E_{\min}(n) = \lambda E_y(n-1) + y(n)y^*(n) - \hat{\mathbf{d}}^H(n)\mathbf{c}(n-1) - \hat{\mathbf{d}}^H(n)\mathbf{g}(n)e^*(n)$$

or by using (10.5.8)

$$E_{\min}(n) = \lambda E_y(n-1) + y(n)y^*(n) - \hat{\mathbf{d}}^H(n)\mathbf{g}(n)e^*(n)$$
$$- y(n)\mathbf{x}^H(n)\mathbf{c}(n-1) - \lambda\hat{\mathbf{d}}^H(n-1)\mathbf{c}(n-1)$$

Rearranging the terms of the last equation and using (10.5.5), we have

$$E_{\min}(n) = \lambda[E_y(n-1) - \hat{\mathbf{d}}^H(n-1)\mathbf{c}(n-1)] + [y(n) - \hat{\mathbf{d}}^H(n)\mathbf{g}(n)]e^*(n)$$
$$= \lambda E_{\min}(n-1) + \{y(n) - [\hat{\mathbf{d}}^H(n)\hat{\mathbf{R}}^{-1}(n)][\hat{\mathbf{R}}(n)\mathbf{g}(n)]\}e^*(n)$$
$$= \lambda E_{\min}(n-1) + [y(n) - \mathbf{c}^H(n)\mathbf{x}(n)]e^*(n)$$

where the last equation is obtained because the matrix $\hat{\mathbf{R}}(n)$ and its inverse are Hermitian. The last equation leads to

$$E_{\min}(n) = \lambda E_{\min}(n-1) + \varepsilon(n)e^*(n) \qquad (10.5.33)$$

$$= \lambda E_{\min}(n-1) + \bar{\alpha}(n)|\varepsilon(n)|^2 \qquad (10.5.34)$$

$$= \lambda E_{\min}(n-1) + \frac{|e(n)|^2}{\alpha(n)} \qquad (10.5.35)$$

which provide the desired updating formulas. Since the product $\varepsilon(n)e^*(n)$ is by necessity real, we have $\varepsilon(n)e^*(n) = \varepsilon^*(n)e(n)$. The value of $E_{\min}(n)$ increases with time and reaches a finite limit value only if $\lambda < 1$.

10.5.3 Some Practical Considerations

In the practical implementation of CRLS adaptive filters, we have to deal with the issues of computational complexity, initialization, and finite-word-length effects.

Computational complexity. The complete CRLS algorithm is summarized in Table 10.6. A measure of the computational complexity of the CRLS algorithm is provided by

TABLE 10.6

Practical implementation of the RLS algorithm. To update P(n), we only compute its upper (low) triangular part and determine the other part using Hermitian symmetry.

Initialization

$$\mathbf{c}(-1) = \mathbf{0} \qquad \mathbf{P}(-1) = \delta^{-1}\mathbf{I}$$
δ = small positive constant

For each $n = 0, 1, 2, \ldots$ compute:

Adaptation gain computation

$$\bar{\mathbf{g}}_\lambda(n) = \mathbf{P}(n-1)\mathbf{x}(n)$$

$$\alpha_\lambda(n) = \lambda + \bar{\mathbf{g}}_\lambda^H(n)\mathbf{x}(n)$$

$$\mathbf{g}(n) = \frac{\bar{\mathbf{g}}_\lambda(n)}{\alpha_\lambda(n)}$$

$$\mathbf{P}(n) = \lambda^{-1}[\mathbf{P}(n-1) - \mathbf{g}(n)\bar{\mathbf{g}}_\lambda^H(n)]$$

Filtering

$$e(n) = y(n) - \mathbf{c}^H(n-1)\mathbf{x}(n)$$

Coefficient updating

$$\mathbf{c}(n) = \mathbf{c}(n-1) + \mathbf{g}(n)e^*(n)$$

the number of operations (one operation consists of one multiplication and one addition) required to perform one updating. Since $\mathbf{P}(n)$ is Hermitian, it is possible to implement the algorithm so that it will require $2M^2 + 4M$ operations per time updating. The computation of $\bar{\mathbf{g}}_\lambda(n)$ and the updating of $\mathbf{P}(n)$ require $O(M^2)$ operations. In contrast, all remaining formulas, which involve dot products and vector-by-scalar multiplications, require $O(M)$ operations. The inversion of the correlation matrix $\hat{\mathbf{R}}(n)$ is essentially replaced by the scalar division used to compute $\mathbf{g}(n)$.

Initialization. There are two ways to obtain the values $\mathbf{P}(-1)$ and $\mathbf{c}(-1)$ required to initialize the CRLS algorithm. The most obvious way is to collect an initial block of data $\{\mathbf{x}(n), y(n)\}_{-n_0}^{-1}$, $n_0 > M$, and then compute the exact inverse matrix $\mathbf{P}(-1)$ and the exact LS solution $\mathbf{c}(-1)$.

The approach used in practice is to set $\mathbf{P}(-1) = \delta^{-1}\mathbf{I}$, where δ is a very small positive number (on the order of $0.01\sigma_x^2$) and $\mathbf{c}(-1) = \mathbf{0}$. For FIR filters this corresponds to setting $x(-M + 1) = \sqrt{\delta}$ and $x(n) = 0$ for $-M + 2 \leq n \leq -1$. For any $n > M$, the normal equations matrix is $\delta\mathbf{I} + \hat{\mathbf{R}}(n)$ and results in a biased estimate of $\mathbf{c}(n)$. However, for large n the choice of δ is unimportant because the algorithm has exponentially forgetting memory for $\lambda < 1$.

It can be shown (see Problem 10.24) that this approach provides a set of coefficients that minimizes the modified cost function

$$E(n) = \delta\lambda^{n+1}\|\mathbf{c}\|^2 + \sum_{j=0}^{n} \lambda^{n-j}|y(j) - \mathbf{c}^H\mathbf{x}(j)|^2 \qquad (10.5.36)$$

instead of (10.5.1). This approach amounts to regularization of the LS solution (see Section 8.7.3) and is further discussed in Hubing and Alexander (1991). Note that if we turn off the input, that is, we set $\mathbf{x}(n) = \mathbf{0}$, then (10.5.30) becomes $\mathbf{P}(n) = \lambda^{-1}\mathbf{P}(n-1)$, which is an unstable recursion when $\lambda < 1$.

Finite-word-length effects. There are different RLS algorithms that are algebraically equivalent; that is, they solve the same set of normal equations. Therefore, they have the same rate of convergence and the same insensitivity to variations in the eigenvalue spread of the input correlation matrix with the CRLS algorithm. All RLS algorithms are obtained by exploiting exact mathematical relations between various algorithmic quantities to obtain better computational or numerical properties. Many of these algorithmic quantities have certain physical meanings or theoretical properties. For example, in the CRLS algorithm, the matrix $\mathbf{P}(n)$ is Hermitian and positive definite, the angle variable satisfies $0 < \alpha(n) \leq 1$, and energy $E(n)$ should be always positive. However, when we use finite precision, some of these exact relations, properties, or acceptable ranges for certain algorithmic variables may be violated.

The numerical instability of RLS algorithms can be traced to such forms of *numerical inconsistencies* (Verhaegen 1989; Yang and Böhme 1992; Haykin 1996). The crucial part of the CRLS algorithm is the updating of the inverse correlation matrix $\mathbf{P}(n)$ via (10.5.30). The CRLS algorithm becomes numerically unstable when the matrix $\mathbf{P}(n) = \hat{\mathbf{R}}^{-1}(n)$ loses its Hermitian symmetry or its positive definiteness (Verhaegen 1989). In practice, we can preserve the Hermitian symmetry of $\mathbf{P}(n)$ by computing only its lower (or upper) triangular part, using (10.5.30), and then filling the other part, using the relation $p_{ij}(n) = p_{ji}^*(n)$. Another approach is to replace $\mathbf{P}(n)$ by $[\mathbf{P}(n) + \mathbf{P}^H(n)]/2$ after updating from $\mathbf{P}(n-1)$ to $\mathbf{P}(n)$.

It has been shown that the CRLS algorithm is numerically stable for $\lambda < 1$ and diverges for $\lambda = 1$ (Ljung and Ljung 1985).

10.5.4 Convergence and Performance Analysis

The purpose of any LS adaptive filter, in a stationary SOE, is to identify the optimum filter $\mathbf{c}_o = \mathbf{R}^{-1}\mathbf{d}$ from observations of the input vector $\mathbf{x}(n)$ and the desired response

$$y(n) = \mathbf{c}_o^H \mathbf{x}(n) + e_o(n) \tag{10.5.37}$$

To simplify the analysis we adopt the independence assumptions discussed in Section 10.4.2. The results of the subsequent analysis hold for any LS adaptive filter implemented using the CRLS method or any other algebraically equivalent algorithm. We derive separate results for the growing memory and the fading memory (exponential forgetting) algorithms.

Growing memory ($\lambda = 1$)

In this case all the values of the error signal, from the time the filter starts its operation to the present, have the same influence on the cost function. As a result, the filter loses its tracking ability, which is not important if the filter is used in a stationary SOE.

Convergence in the mean. For $n > M$ the coefficient vector $\mathbf{c}(n)$ is identical to the block LS solution discussed in Section 8.2.2. Therefore

$$E\{\mathbf{c}(n)\} = \mathbf{c}_o \quad \text{for } n > M \tag{10.5.38}$$

that is, the RLS algorithm converges in the mean for $n > M$, where M is the number of coefficients.

Mean square deviation. For $n > M$ we have

$$\mathbf{\Phi}(n) = \sigma_o^2 E\{\hat{\mathbf{R}}^{-1}(n)\} \tag{10.5.39}$$

because $\mathbf{c}(n)$ is an exact LS estimate (see Section 8.2.2). The correlation matrix $\hat{\mathbf{R}}(n)$ is described by a complex Wishart distribution, and the expectation of its inverse is

$$E\{\hat{\mathbf{R}}^{-1}(n)\} = \frac{1}{n-M}\mathbf{R}^{-1} \quad n > M \tag{10.5.40}$$

as shown in Muirhead (1982) and Haykin (1996). Hence

$$\mathbf{\Phi}(n) = \frac{\sigma_o^2}{n - M}\mathbf{R}^{-1} \qquad n > M \tag{10.5.41}$$

and the MSD is

$$\mathcal{D}(n) = \mathrm{tr}[\mathbf{\Phi}(n)] = \frac{\sigma_o^2}{n - M}\sum_{i=1}^{M}\frac{1}{\lambda_i} \qquad n > M \tag{10.5.42}$$

where λ_i, the eigenvalues of $\mathbf{R}$, should not be confused with the forgetting factor λ. From (10.5.42) we conclude that (1) the MSD is magnified by the smallest eigenvalue of $\mathbf{R}$ and (2) the MSD decays almost linearly with time.

A priori excess MSE. We now focus on the a priori LS algorithm because it is widely used in practice and to facilitate a fairer comparison with the (a priori) LMS algorithm. To this end, we note that the a priori excess MSE formula (10.4.48)

$$P_{\mathrm{ex}}(n) = \mathrm{tr}[\mathbf{R}\mathbf{\Phi}(n-1)] \tag{10.5.43}$$

derived in Section 10.4.2, under the independence assumption, holds for any a priori adaptive algorithm. Hence, substituting (10.5.41) into (10.5.43), we obtain

$$P_{\mathrm{ex}}(n) = \frac{M}{n - M - 1}\sigma_o^2 \qquad n > M \tag{10.5.44}$$

which shows that $P_{\mathrm{ex}}(n)$ tends to zero as $n \to \infty$.

Exponentially decaying memory ($0 < \lambda < 1$)

In this case the most recent values of the observations have greater influence on the formation of the LS estimate of the filter coefficients. The memory of the filter, that is, the *effective* number of samples used to form the various estimates, is about $1/(1 - \lambda)$ for $0.95 < \lambda < 1$ (see Section 10.8).

Convergence in the mean. We start by multipying both sides of (10.5.11) by $\hat{\mathbf{R}}(n)$, and then we use (10.5.7) and (10.5.10) to obtain

$$\hat{\mathbf{R}}(n)\mathbf{c}(n) = \lambda\hat{\mathbf{R}}(n-1)\mathbf{c}(n-1) + \mathbf{x}(n)y^*(n) \tag{10.5.45}$$

If we multiply (10.5.7) by $\mathbf{c}_o$ and subtract the resulting equation from (10.5.45), we get

$$\hat{\mathbf{R}}(n)\tilde{\mathbf{c}}(n) = \lambda\hat{\mathbf{R}}(n-1)\tilde{\mathbf{c}}(n-1) + \mathbf{x}(n)e_o^*(n) \tag{10.5.46}$$

where $\tilde{\mathbf{c}}(n) = \mathbf{c}(n) - \mathbf{c}_o$ is the coefficient error vector. Solving (10.5.46) by recursion, we obtain

$$\tilde{\mathbf{c}}(n) = \lambda^n\hat{\mathbf{R}}^{-1}(n)\hat{\mathbf{R}}(0)\tilde{\mathbf{c}}(0) + \hat{\mathbf{R}}^{-1}(n)\sum_{j=0}^{n}\lambda^{n-j}\mathbf{x}(j)e_o^*(j) \tag{10.5.47}$$

which depends on the initial conditions and the optimum error $e_o(n)$. If we assume that $\hat{\mathbf{R}}(n)$, $\mathbf{x}(j)$, and $e_o(j)$ are independent and we take the expectation of (10.5.47), we obtain

$$E\{\tilde{\mathbf{c}}(n)\} = \delta\lambda^n E\{\hat{\mathbf{R}}^{-1}(n)\}\tilde{\mathbf{c}}(0) \tag{10.5.48}$$

where, as usual, we have set $\hat{\mathbf{R}}(0) = \delta\mathbf{I}$, $\delta > 0$. If the matrix $\hat{\mathbf{R}}(n)$ is positive definite and $0 < \lambda < 1$, then the mean vector $E\{\tilde{\mathbf{c}}(n)\} \to \mathbf{0}$ as $n \to \infty$. Hence, the RLS algorithm with exponential forgetting converges asymptotically in the mean to the optimum filter.

Mean square deviation. Using (10.5.46), we obtain the following difference equation for the coefficient error vector

$$\tilde{\mathbf{c}}(n) = \lambda\hat{\mathbf{R}}^{-1}(n)\hat{\mathbf{R}}(n-1)\tilde{\mathbf{c}}(n-1) + \hat{\mathbf{R}}^{-1}(n)\mathbf{x}(n)e_o^*(n)$$

or
$$\tilde{\mathbf{c}}(n) \simeq \lambda \tilde{\mathbf{c}}(n-1) + \hat{\mathbf{R}}^{-1}(n)\mathbf{x}(n)e_o^*(n)$$

because $\hat{\mathbf{R}}^{-1}(n)\hat{\mathbf{R}}(n-1) \simeq \mathbf{I}$ for large n. If we neglect the dependence among $\tilde{\mathbf{c}}(n-1)$, $\hat{\mathbf{R}}(n)$, $\mathbf{x}(n)$, and $e_o(n)$, we have

$$\mathbf{\Phi}(n) \simeq \lambda^2 \mathbf{\Phi}(n-1) + \sigma_o^2 E\{\hat{\mathbf{R}}^{-1}(n)\mathbf{x}(n)\mathbf{x}^H(n)\hat{\mathbf{R}}^{-1}(n)\} \qquad (10.5.49)$$

where $\sigma_o^2 = E\{|e_o(n)|^2\}$.

To make the analysis mathematically tractable, we need an approximation for the inverse matrix $\hat{\mathbf{R}}^{-1}(n)$. To this end, using (10.5.3), we have

$$E\{\hat{\mathbf{R}}(n)\} = \sum_{j=0}^{n} \lambda^{n-j} E\{\mathbf{x}(n)\mathbf{x}^H(n)\} = \frac{1-\lambda^{n+1}}{1-\lambda}\mathbf{R} \simeq \frac{1}{1-\lambda}\mathbf{R} \qquad (10.5.50)$$

where the last approximation holds for $n \gg 1$. If we use the approximation $E\{\hat{\mathbf{R}}(n)\} \simeq \hat{\mathbf{R}}(n)$, we obtain

$$\hat{\mathbf{R}}^{-1}(n) \simeq (1-\lambda)\mathbf{R}^{-1} \qquad (10.5.51)$$

which is more rigorously justified in Eleftheriou and Falconer (1986). Using the last approximation, (10.5.50) becomes

$$\mathbf{\Phi}(n) \simeq \lambda^2 \mathbf{\Phi}(n-1) + (1-\lambda)^2\sigma_o^2\mathbf{R}^{-1} \qquad (10.5.52)$$

which converges because $\lambda^2 < 1$. At steady state we have

$$(1-\lambda^2)\mathbf{\Phi}(\infty) \simeq (1-\lambda)^2\sigma_o^2\mathbf{R}^{-1}$$

because $\mathbf{\Phi}(n) \simeq \mathbf{\Phi}(n-1)$ for $n \gg 1$. Hence

$$\mathbf{\Phi}(\infty) \simeq \frac{1-\lambda}{1+\lambda}\sigma_o^2\mathbf{R}^{-1} \qquad (10.5.53)$$

and therefore

$$\mathcal{D}_\lambda(\infty) = \text{tr}[\mathbf{\Phi}(\infty)] = \frac{1-\lambda}{1+\lambda}\sigma_o^2\sum_{i=1}^{M}\frac{1}{\lambda_i} \qquad (10.5.54)$$

which in contrast to (10.5.42) does not converge to zero as $n \to \infty$. This is explained by noticing that when $\lambda < 1$, the RLS algorithm has finite memory and does not use effectively all the data to form its estimate.

Steady-state a priori excess MSE. From (10.5.43) and (10.5.53) we obtain

$$P_{\text{ex}}(\infty) = \text{tr}[\mathbf{R}\mathbf{\Phi}(\infty)] \simeq \frac{1-\lambda}{1+\lambda}M\sigma_o^2 \qquad (10.5.55)$$

which shows that as a result of finite memory, there is a steady-state excess MSE that decreases as λ approaches 1, that is, as the effective memory of the algorithm increases.

Summary

The results of the above analysis are summarized in Table 10.7 for easy reference. We stress at this point that all RLS algorithms, independent of their implementation, have the same performance, assuming that we use sufficient numerical precision (e.g., double-precision floating-point arithmetic). Sometimes, RLS algorithms are said to have optimum learning because at every time instant they minimize the weighted error energy from the start of the operation (Tsypkin 1973). These properties are illustrated in the following example.

EXAMPLE 10.5.2. Consider the adaptive equalizer of Example 10.4.3 shown in block diagram form in Figure 10.26. In this example, we replace the LMS block in Figure 10.26 by the RLS block, and we study the performance of the RLS algorithm and compare it with that of the LMS

TABLE 10.7

Summary of RLS and LMS performance in a stationary SOE.

Property	Growing memory RLS algorithm	Exponential memory RLS algorithm	LMS algorithm
Convergence in the mean	For all $n > M$	Asymptotically for $n \to \infty$	Asymptotically for $n \to \infty$
Convergence in MS	Independent of the eigenvalue spread	Independent of the eigenvalue spread	Depends on the eigenvalue spread
Excess MSE	$P_{ex}(n) = \dfrac{M\sigma_o^2}{n - M - 1} \to 0$	$P_{ex}(\infty) = \dfrac{1-\lambda}{1+\lambda} M\sigma_o^2$	$P_{ex}(\infty) \simeq \mu\sigma_o^2 \, \text{tr} \, \mathbf{R}$

algorithm. The input data source is a Bernoulli sequence $\{y(n)\}$ with symbols $+1$ and -1 having zero mean and unit variance. The channel impulse response is a raised cosine

$$
h(n) = \begin{cases} 0.5 \left\{ 1 + \cos\left[\dfrac{2\pi}{W}(n-2) \right] \right\} & n = 1, 2, 3 \\ 0 & \text{otherwise} \end{cases} \tag{10.5.56}
$$

where the parameter W controls the amount of channel distortion [or the eigenvalue spread $\mathcal{X}(\mathbf{R})$ produced by the channel]. The channel noise sequence $v(n)$ is white Gaussian with $\sigma_v^2 = 0.001$. The adaptive equalizer has $M = 11$ coefficients, and the input signal $y(n)$ is delayed by $\Delta = 7$ samples. The error signal $e(n) = y(n - \Delta) - \hat{y}(n)$ is used along with $x(n)$ to implement the RLS algorithm given in Table 10.6 with $\mathbf{c}(0) = \mathbf{0}$ and $\delta = 0.001$. We performed Monte Carlo simulations on 100 realizations of random sequences with $W = 2.9$ and $W = 3.5$, and $\lambda = 1$ and 0.8. The results are shown in Figures 10.34 and 10.35.

Effect of eigenvalue spread. Performance plots of the RLS algorithm for $W = 2.9$ and $W = 3.5$ are shown in Figure 10.34. In plot (a) we depict MSE learning curves along with the steady-state (or minimum) error. We observe that the MSE convergence rate of the RLS, unlike that for the LMS, does not change with W [or equivalently with change in $\mathcal{X}(\mathbf{R})$]. The steady-state error, on the other hand, increases with W. The important difference between the two algorithms is that the convergence rate is faster for the RLS (compare Figures 10.34 and 10.27). Clearly, this faster convergence of the RLS algorithm is achieved by an increase in computational complexity. In plots (b) and (c) we show the ensemble averaged equalizer coefficients. Clearly, the responses are symmetric with respect to $n = 5$, as assumed. Also equalizer coefficients converge to different inverses due to changes in the channel characteristics.

Effect of forgetting factor λ. In Figure 10.35 we show the MSE learning curves obtained for $W = 2.9$ and with two different factors of 1 and 0.8. For $\lambda = 1$, as explained before, the algorithm has infinite memory and hence the steady-state excess MSE is zero. This fact can be verified in the plot for $\lambda = 1$ in which the MSE converges to the minimum error. For $\lambda = 0.8$, the effective memory is $1/(1 - \lambda) = 5$, which clearly is inadequate for the accurate estimation of the required statistics, resulting in increased excess MSE. Therefore, the algorithm should produce a nonzero excess MSE. This fact can be observed from the plot for $\lambda = 0.8$.

There are two practical issues regarding the RLS algorithm that need an explanation. The first issue relates to the practical value of λ. Although λ can take any value in the interval $0 \le \lambda \le 1$, since it influences the effective memory size, the value of λ should be closer to 1. This value is determined by the number of parameters to be estimated and the desired size of the effective memory. Typical values used are between 0.99 and 1 (not 0.8, as we used in this example for demonstration). The second issue deals with the actual computation of matrix $\mathbf{P}(n)$. This matrix must be conjugate symmetric and positive definite. However, an implementation of the CRLS algorithm of Table 10.6 on a finite-precision processor will eventually disturb this symmetry and positive definiteness and would result in an unstable performance. Therefore, it is necessary to force this symmetry either by computing only its lower (or upper) triangular values or by using $\mathbf{P}(n) \leftarrow [\mathbf{P}(n) + \mathbf{P}^H(n)]/2$. Failure to do so generally affects the algorithm performance for $\lambda < 1$.

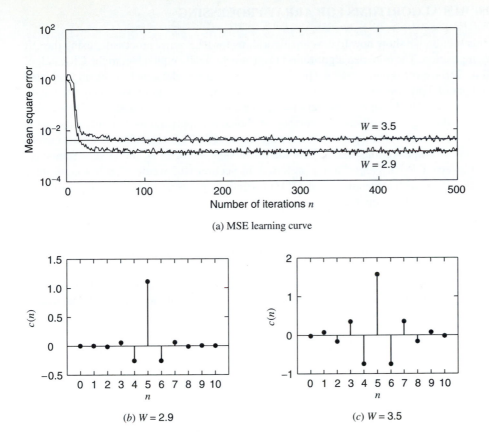

(a) MSE learning curve

(b) $W = 2.9$

(c) $W = 3.5$

FIGURE 10.34
Performance analysis curves of the RLS algorithm in the adaptive equalizer: $\lambda = 1$.

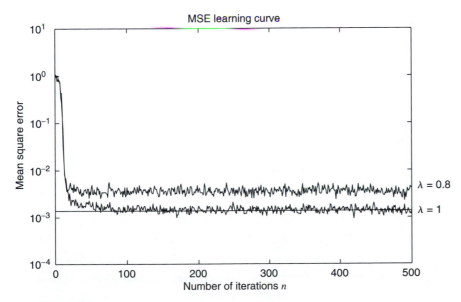

FIGURE 10.35
MSE learning curves of the RLS algorithm in the adaptive equalizer: $W = 2.9$.

10.6 RLS ALGORITHMS FOR ARRAY PROCESSING

In this section we show how to develop algorithms for RLS array processing using the QR decomposition. The obtained algorithms (1) are algebraically equivalent to the CRLS algorithm, (2) have very good numerical properties, and (3) are modular and can be implemented using parallel processing. Since there are no restrictions on the input data vector, the algorithms require $O(M^2)$ operations per time update and can be used for both array processing and FIR filtering applications. The method of choice for applications that only require the a priori error $e(n)$ or the a posteriori error $\varepsilon(n)$ is the QR-RLS algorithm using the Givens rotations. For applications that require the coefficient vector $\mathbf{c}(n)$, the Givens rotations–based inverse QR-RLS algorithm is preferred. In Section 10.7 we develop fast algorithms for FIR filters, with a complexity of $O(M)$ operations per time update, by exploiting the shift invariance of the input data vector.

10.6.1 LS Computations Using the Cholesky and QR Decompositions

We start by reformulating the exponentially weighted LS filtering problem in terms of data matrices, as discussed in Section 8.2. If $\mathbf{c}(n)$ is the LS filter coefficient vector at time instant n, we have

$$\varepsilon(j) = y(j) - \mathbf{c}^H(n)\mathbf{x}(j) \qquad 0 \leq j \leq n \tag{10.6.1}$$

where
$$\mathbf{x}(j) = [x_1(j)\, x_2(j)\, \cdots\, x_M(j)]^T \tag{10.6.2}$$

for array processing and

$$\mathbf{x}(j) = [x(j)\, x(j-1)\, \cdots\, x(j-M+1)]^T \tag{10.6.3}$$

for FIR filtering. We stress that $\mathbf{c}(n)$ should be held constant during the optimization interval $0 \leq j \leq n$. Using the $(n+1) \times M$ data matrix

$$\mathbf{X}^H(n) \triangleq [\mathbf{x}(0)\, \mathbf{x}(1)\, \cdots\, \mathbf{x}(n)]$$

$$= \begin{bmatrix} x_1(0) & x_1(1) & \cdots & x_1(n) \\ x_2(0) & x_2(1) & \cdots & x_2(n) \\ \vdots & \vdots & \ddots & \vdots \\ x_M(0) & x_M(1) & \cdots & x_M(n) \end{bmatrix} \tag{10.6.4}$$

the $(n+1) \times 1$ desired response vector

$$\mathbf{y}(n) \triangleq [y(0)\, y(1)\, \cdots\, y(n)]^H \tag{10.6.5}$$

and the $(n+1) \times 1$ a posteriori error vector

$$\boldsymbol{\varepsilon}(n) \triangleq [\varepsilon(0)\, \varepsilon(1)\, \cdots\, \varepsilon(n)]^H \tag{10.6.6}$$

we can combine the $n+1$ equations (10.6.1) in a single equation as

$$\boldsymbol{\varepsilon}(n) = \mathbf{y}(n) - \mathbf{X}(n)\mathbf{c}(n) \tag{10.6.7}$$

If we define the $(n+1) \times (n+1)$ exponential weighting matrix

$$\boldsymbol{\Lambda}^2(n) \triangleq \text{diag}\,\{\lambda^n, \lambda^{n-1}, \ldots, 1\} \tag{10.6.8}$$

we can express the total squared error (10.5.1) and the normal equations (10.5.2) in the form required to apply orthogonal decomposition techniques (see Section 8.6). Indeed, we can easily see that the total squared error can be written as

$$E(n) = \sum_{j=0}^{n} \lambda^{n-j} |\varepsilon(j)|^2 = \|\boldsymbol{\Lambda}(n)\boldsymbol{\varepsilon}(n)\|^2 \tag{10.6.9}$$

and the LS filter coefficients are determined by the normal equations

$$\hat{\mathbf{R}}(n)\mathbf{c}(n) = \hat{\mathbf{d}}(n) \tag{10.6.10}$$

where
$$\hat{\mathbf{R}}(n) = \sum_{j=0}^{n} \lambda^{n-j}\mathbf{x}(j)\mathbf{x}^H(j) = [\mathbf{\Lambda}(n)\mathbf{X}(n)]^H[\mathbf{\Lambda}(n)\mathbf{X}(n)] \tag{10.6.11}$$

and
$$\hat{\mathbf{d}}(n) = \sum_{j=0}^{n} \lambda^{n-j}\mathbf{x}(j)y^*(j) = [\mathbf{\Lambda}(n)\mathbf{X}(n)]^H[\mathbf{\Lambda}(n)\mathbf{y}(n)] \tag{10.6.12}$$

are expressed as a function of the weighted data matrix $[\mathbf{\Lambda}(n)\mathbf{X}(n)]$ and the weighted desired response vector $[\mathbf{\Lambda}(n)\mathbf{y}(n)]$.

In Chapter 6 we discussed how to solve the normal equations (10.6.10) by using either the Cholesky decomposition

$$\hat{\mathbf{R}}(n) = \tilde{\mathbf{L}}(n)\tilde{\mathbf{L}}^H(n) \tag{10.6.13}$$

or the LDU decomposition

$$\hat{\mathbf{R}}(n) = \mathbf{L}(n)\mathbf{D}(n)\mathbf{L}^H(n) \tag{10.6.14}$$

where $\tilde{\mathbf{L}}(n) = \mathbf{D}^{1/2}(n)\mathbf{L}(n)$.

The Cholesky factor $\tilde{\mathbf{L}}(n)$ can be computed either from matrix $\hat{\mathbf{R}}(n)$ using the Cholesky decomposition algorithm (see Section 6.3) or from data matrix $[\mathbf{\Lambda}(n)\mathbf{X}(n)]$ using one of the QR decomposition methods (Givens, Householder, or MGS) discussed in Chapter 8.

Suppose now that the QR decomposition[†] is

$$\mathbf{Q}(n)[\mathbf{\Lambda}(n)\mathbf{X}(n)] = \begin{bmatrix} \tilde{\mathbf{R}}(n) \\ \mathbf{0} \end{bmatrix} \tag{10.6.15}$$

where $\tilde{\mathbf{R}}(n)$ is a unique upper triangular matrix with positive diagonal elements and $\mathbf{Q}(n)$ is a unitary matrix. From (10.6.11) and (10.6.15) we have

$$\hat{\mathbf{R}}(n) = \tilde{\mathbf{R}}^H(n)\tilde{\mathbf{R}}(n) \tag{10.6.16}$$

which implies, owing to the uniqueness of the Cholesky decomposition, that $\tilde{\mathbf{L}}(n) = \tilde{\mathbf{R}}^H(n)$. Although the two approaches are *algebraically* equivalent, the QR decomposition (QRD) methods have superior numerical properties because they avoid the squaring operation (10.6.11) (see Section 8.6).

Given the Cholesky factor $\tilde{\mathbf{R}}(n)$, we first solve the lower triangular system

$$\tilde{\mathbf{R}}^H(n)\tilde{\mathbf{k}}(n) \triangleq \hat{\mathbf{d}}(n) \tag{10.6.17}$$

to obtain the partial correlation vector $\tilde{\mathbf{k}}(n)$, using forward elimination. In the case of QRD the vector $\tilde{\mathbf{k}}(n)$ is obtained by transforming $\mathbf{\Lambda}(n)\mathbf{y}(n)$ and retaining its first M components, that is,

$$\mathbf{Q}(n)[\mathbf{\Lambda}(n)\mathbf{y}(n)] = \mathbf{z}(n) \triangleq \begin{bmatrix} \tilde{\mathbf{k}}(n) \\ \mathbf{z}_2(n) \end{bmatrix} \tag{10.6.18}$$

where $\tilde{\mathbf{k}}(n) = \mathbf{z}^{\lceil M \rceil}(n)$ (see Section 8.6). The minimum LSE is given by

$$E(n) = E_y(n) - \hat{\mathbf{d}}^H(n)\mathbf{c}(n) = \|\mathbf{y}(n)\|^2 - \|\tilde{\mathbf{k}}(n)\|^2 \tag{10.6.19}$$

which was also proved in Section 8.6.

To compute the filter parameters, we can solve the upper triangular system

$$\tilde{\mathbf{R}}(n)\mathbf{c}(n) = \tilde{\mathbf{k}}(n) \tag{10.6.20}$$

by backward elimination. As we discussed in Section 6.3, the solution of (10.6.20) is *not* order-recursive.

[†]To comply with adaptive filtering literature, we express the QR decomposition as $\mathbf{QX} = \tilde{\mathbf{R}}$ instead of $\mathbf{Q}^H\mathbf{X} = \mathcal{R}$, which we used in Chapter 8 and is widely used in numerical analysis.

In applications that only require the a posteriori or a priori errors, we can avoid the solution of (10.6.20). Indeed, if we define the LS innovations vector $\tilde{\mathbf{w}}(n)$ by

$$\tilde{\mathbf{R}}^H(n)\tilde{\mathbf{w}}(n) \triangleq \mathbf{x}(n) \tag{10.6.21}$$

we obtain

$$\varepsilon(n) = y(n) - \mathbf{c}^H(n)\mathbf{x}(n) = y(n) - \tilde{\mathbf{k}}^H(n)\tilde{\mathbf{w}}(n) \tag{10.6.22}$$

and

$$e(n) = y(n) - \mathbf{c}^H(n-1)\mathbf{x}(n) = y(n) - \tilde{\mathbf{k}}^H(n-1)\tilde{\mathbf{w}}(n) \tag{10.6.23}$$

which can be used to compute the errors without knowledge of the parameter vector $\mathbf{c}(n)$. Furthermore, since the lower triangular systems (10.6.17) and (10.6.21) satisfy the optimum nesting property, we can compute both errors in an order-recursive manner.

If we know the factors $\mathbf{L}(n)$ and $\mathbf{D}(n)$ of $\hat{\mathbf{R}}(n)$ at each time instant n, we can use the orthogonal triangular structure shown in Figure 7.1 (see Sections 7.1.5 and 8.5) to compute all $e_m(n)$ and $\varepsilon_m(n)$ for all $1 \le m \le M$. A similar structure can be obtained by using the Cholesky factor $\tilde{\mathbf{L}}(n)$ (see Problem 10.26).

From the discussion in Section 10.5.1 we saw that the key part of the CRLS algorithm is the computation of the gain vector

$$\hat{\mathbf{R}}(n)\mathbf{g}(n) = \mathbf{x}(n) \tag{10.6.24}$$

or the alternative gain vector $\lambda\hat{\mathbf{R}}(n-1)\bar{\mathbf{g}}(n) = \mathbf{x}(n)$. Using (10.6.16), (10.6.21), and (10.6.24), we obtain

$$\tilde{\mathbf{R}}(n)\mathbf{g}(n) = \tilde{\mathbf{w}}(n) \tag{10.6.25}$$

which expresses the gain vector in terms of the Cholesky factor $\tilde{\mathbf{R}}(n)$ and the innovations vector $\tilde{\mathbf{w}}(n)$. Similarly with (10.6.20), (10.6.25) lacks the optimum nesting property that is required to obtain an order-recursive algorithm.

To summarize, if we can update the Cholesky factors of either $\hat{\mathbf{R}}(n)$ or $\hat{\mathbf{R}}^{-1}(n)$, we can develop exact RLS algorithms that provide both the filtering errors and the coefficient vector or the filtering error only. The relevant relations are shown in Table 10.8. We stress that the Cholesky decomposition method determines the factors $\tilde{\mathbf{R}}(n)$ and $\tilde{\mathbf{k}}(n)$ by factoring the matrix

$$\hat{\mathbf{R}}(n) = [\mathbf{X}(n)\,\mathbf{y}(n)]^H \mathbf{\Lambda}^2(n)[\mathbf{X}(n)\,\mathbf{y}(n)]$$

whereas the QRD methods factor the data matrix $\mathbf{\Lambda}(n)[\mathbf{X}(n)\,\mathbf{y}(n)]$. Since all these algorithms propagate the square roots $\tilde{\mathbf{R}}(n)$ or $\tilde{\mathbf{R}}^{-1}(n)$, the matrices determined by $\hat{\mathbf{R}}(n) = \tilde{\mathbf{R}}^H(n)\tilde{\mathbf{R}}(n)$ and $\hat{\mathbf{R}}^{-1}(n) = \tilde{\mathbf{R}}^{-1}(n)\tilde{\mathbf{R}}^{-H}(n)$ are guaranteed to be Hermitian and are more likely to preserve their positive definiteness. Hence, such algorithms have better numerical properties than the CRLS method.

TABLE 10.8

Triangular decomposition RLS algorithms using coefficient updating and direct error extraction.

Error and coefficients updating	Error-only updating
$\tilde{\mathbf{R}}^H(n)\tilde{\mathbf{w}}(n) = \mathbf{x}(n)$	$\tilde{\mathbf{R}}^H(n)\tilde{\mathbf{k}}(n) \triangleq \hat{\mathbf{d}}(n)$
$\tilde{\mathbf{R}}(n)\mathbf{g}(n) = \tilde{\mathbf{w}}(n)$	$\tilde{\mathbf{R}}^H(n)\tilde{\mathbf{w}}(n) = \mathbf{x}(n)$
$e(n) = y(n) - \mathbf{c}^H(n-1)\mathbf{x}(n)$	$e(n) = y(n) - \tilde{\mathbf{k}}^H(n-1)\tilde{\mathbf{w}}(n)$
$\mathbf{c}(n) = \mathbf{c}(n-1) + \mathbf{g}(n)e^*(n)$	

10.6.2 Two Useful Lemmas

We next prove two lemmas that are very useful in the development of RLS algorithms using QRD methods. We start with the first lemma, which stems from the algebraic equivalence between the Cholesky and QR decompositions.

LEMMA 10.1. Computing the QRD of the $(n + 1) \times M$ data matrix $\boldsymbol{\Lambda}(n)\mathbf{X}(n)$ is equivalent to evaluating the QRD of the $(M + 1) \times M$ matrix

$$\begin{bmatrix} \sqrt{\lambda}\tilde{\mathbf{R}}(n-1) \\ \mathbf{x}^H(n) \end{bmatrix}$$

Proof. Indeed, if we express $\boldsymbol{\Lambda}(n)\mathbf{X}(n)$ as

$$\boldsymbol{\Lambda}(n)\mathbf{X}(n) = \begin{bmatrix} \sqrt{\lambda}\boldsymbol{\Lambda}(n-1)\mathbf{X}(n-1) \\ \mathbf{x}^H(n) \end{bmatrix} \tag{10.6.26}$$

and define a matrix

$$\bar{\mathbf{Q}}(n-1) \triangleq \begin{bmatrix} \mathbf{Q}(n-1) & \mathbf{0} \\ \mathbf{0}^H & 1 \end{bmatrix} \tag{10.6.27}$$

we obtain

$$\bar{\mathbf{Q}}(n-1)\boldsymbol{\Lambda}(n)\mathbf{X}(n) = \begin{bmatrix} \sqrt{\lambda}\tilde{\mathbf{R}}(n-1) \\ \mathbf{0} \\ \mathbf{x}^H(n) \end{bmatrix} \tag{10.6.28}$$

by using (10.6.15). If we can construct a matrix $\hat{\mathbf{Q}}(n)$ that performs the QRD of the right-hand side of (10.6.28), then the unitary matrix $\mathbf{Q}(n) \triangleq \hat{\mathbf{Q}}(n)\bar{\mathbf{Q}}(n-1)$ performs the QRD of $\boldsymbol{\Lambda}(n)\mathbf{X}(n)$. Since the block of zeros in (10.6.28) has no effect on the construction of matrix $\hat{\mathbf{Q}}(n)$, the construction of $\hat{\mathbf{Q}}(n)$ is equivalent to finding a unitary matrix that performs the QRD of

$$\begin{bmatrix} \sqrt{\lambda}\tilde{\mathbf{R}}(n-1) \\ \mathbf{x}^H(n) \end{bmatrix}$$

The second lemma, known as the *matrix factorization lemma* (Golub and Van Loan 1996; Sayed and Kailath 1994), provides an elegant tool for the derivation of QRD-based RLS algorithms.

LEMMA 10.2. If $\mathbf{A}$ and $\mathbf{B}$ are any two $N \times M (N \leq M)$ matrices, then

$$\mathbf{A}^H\mathbf{A} = \mathbf{B}^H\mathbf{B} \tag{10.6.29}$$

if and only if there exists an $N \times N$ unitary matrix $\mathbf{Q}$ $(\mathbf{Q}^H\mathbf{Q} = \mathbf{I})$ such that

$$\mathbf{Q}\mathbf{A} = \mathbf{B} \tag{10.6.30}$$

Proof. From (10.6.30) we have $\mathbf{B}^H\mathbf{B} = \mathbf{A}^H\mathbf{Q}^H\mathbf{Q}\mathbf{A} = \mathbf{A}^H\mathbf{A}$, which proves (10.6.29). To prove the converse, we use the singular value decomposition (SVD) of matrices $\mathbf{A}$ and $\mathbf{B}$

$$\mathbf{A} = \mathbf{U}_A\boldsymbol{\Sigma}_A\mathbf{V}_A^H \tag{10.6.31}$$

$$\mathbf{B} = \mathbf{U}_B\boldsymbol{\Sigma}_B\mathbf{V}_B^H \tag{10.6.32}$$

where $\mathbf{U}_A$ and $\mathbf{U}_B$ are $N \times N$ unitary matrices, $\mathbf{V}_A$ and $\mathbf{V}_B$ are $M \times M$ unitary matrices, and $\boldsymbol{\Sigma}_A$ and $\boldsymbol{\Sigma}_B$ are $N \times M$ matrices consisting of the nonnegative singular values of $\mathbf{A}$ and $\mathbf{B}$. Using (10.6.29) in conjunction with (10.6.31) and (10.6.32), we obtain

$$\mathbf{V}_A = \mathbf{V}_B \tag{10.6.33}$$

and

$$\boldsymbol{\Sigma}_A = \boldsymbol{\Sigma}_B \tag{10.6.34}$$

If we now define the matrix

$$\mathbf{Q} \triangleq \mathbf{U}_B\mathbf{U}_A^H$$

and use (10.6.33) and (10.6.34), we have

$$\mathbf{Q}\mathbf{A} = \mathbf{U}_B\mathbf{U}_A^H\mathbf{U}_A\boldsymbol{\Sigma}_A\mathbf{V}_A^H = \mathbf{U}_B\boldsymbol{\Sigma}_B\mathbf{V}_B^H = \mathbf{B}$$

which proves the converse of the lemma.

10.6.3 The QR-RLS Algorithm

We next show how to update the factors $\tilde{\mathbf{R}}(n)$ and $\tilde{\mathbf{k}}(n)$ of the extended data matrix $\mathbf{\Lambda}(n)[\mathbf{X}(n)\,\mathbf{y}(n)]$ and then compute the a priori error $e(n)$ or the a posteriori error $\varepsilon(n)$. The findings hold independently of the method we use to construct the orthogonalizing matrix $\mathbf{Q}(n)$.

Suppose now that at time n we know the old Cholesky factors $\tilde{\mathbf{R}}(n-1)$ and $\tilde{\mathbf{k}}(n-1)$, we receive the new data $\{\mathbf{x}(n), y(n)\}$, and we wish to determine the new factors $\tilde{\mathbf{R}}(n)$ and $\tilde{\mathbf{k}}(n)$ without repeating all the work. To this end, we show that if there exists a unitary matrix $\mathbf{Q}(n)$ that annihilates the vector $\mathbf{x}^H(n)$ from the last row of the left-hand side matrix in the relation

$$\mathbf{Q}(n)\begin{bmatrix} \sqrt{\lambda}\tilde{\mathbf{R}}(n-1) & \sqrt{\lambda}\tilde{\mathbf{k}}(n-1) & \mathbf{0} \\ \mathbf{x}^H(n) & y^*(n) & 1 \end{bmatrix} = \begin{bmatrix} \tilde{\mathbf{R}}(n) & \tilde{\mathbf{k}}(n) & \tilde{\mathbf{w}}(n) \\ \mathbf{0}^H & \tilde{e}^*(n) & \tilde{\alpha}(n) \end{bmatrix} \quad (10.6.35)$$

then the right-hand side matrix provides the required updates and errors. The scalar $\tilde{\alpha}(n)$ is real-valued because it is equal to the last diagonal element of $\mathbf{Q}(n)$. The meaning and use of $\tilde{\alpha}(n)$ and $\tilde{\mathbf{w}}(n)$, which comprise the last column of $\mathbf{Q}(n)$, will be explained in the sequel.

If we apply Lemma 10.2 with

$$\mathbf{A} = \begin{bmatrix} \sqrt{\lambda}\tilde{\mathbf{R}}(n-1) & \sqrt{\lambda}\tilde{\mathbf{k}}(n-1) & \mathbf{0} \\ \mathbf{x}^H(n) & y^*(n) & 1 \end{bmatrix} \quad \text{and} \quad \mathbf{B} = \begin{bmatrix} \tilde{\mathbf{R}}(n) & \tilde{\mathbf{k}}(n) & \tilde{\mathbf{w}}(n) \\ \mathbf{0}^H & \tilde{e}^*(n) & \tilde{\alpha}(n) \end{bmatrix}$$

we obtain[†]

$$\langle\mathbf{B}^H\mathbf{B}\rangle_{11} = \tilde{\mathbf{R}}^H(n)\tilde{\mathbf{k}}(n) = \lambda\tilde{\mathbf{R}}^H(n-1)\lambda\tilde{\mathbf{R}}(n-1) + \mathbf{x}(n)\mathbf{x}^H(n) = \langle\mathbf{A}^H\mathbf{A}\rangle_{11} \quad (10.6.36)$$

$$\langle\mathbf{B}^H\mathbf{B}\rangle_{12} = \tilde{\mathbf{R}}^H(n)\tilde{\mathbf{k}}(n) = \lambda\tilde{\mathbf{R}}^H(n-1)\tilde{\mathbf{k}}(n-1) + \mathbf{x}(n)y^*(n) = \langle\mathbf{A}^H\mathbf{A}\rangle_{12} \quad (10.6.37)$$

$$\langle\mathbf{B}^H\mathbf{B}\rangle_{13} = \tilde{\mathbf{R}}^H(n)\tilde{\mathbf{w}}(n) = \mathbf{x}(n) = \langle\mathbf{A}^H\mathbf{A}\rangle_{13} \quad (10.6.38)$$

$$\langle\mathbf{B}^H\mathbf{B}\rangle_{23} = \tilde{\mathbf{k}}^H(n)\tilde{\mathbf{w}}(n) + \tilde{e}(n)\tilde{\alpha}(n) = y(n) = \langle\mathbf{A}^H\mathbf{A}\rangle_{23} \quad (10.6.39)$$

$$\langle\mathbf{B}^H\mathbf{B}\rangle_{33} = \tilde{\mathbf{w}}^H(n)\tilde{\mathbf{w}}(n) + \tilde{\alpha}^2(n) = 1 = \langle\mathbf{A}^H\mathbf{A}\rangle_{33} \quad (10.6.40)$$

We first note that (10.6.36) is identical to the time updating (10.5.7) of the correlation matrix. Hence, $\tilde{\mathbf{R}}(n)$ is the Cholesky factor of $\hat{\mathbf{R}}(n)$. Also (10.6.37) is identical, due to (10.6.17), to the time updating (10.5.8) of the cross-correlation vector $\hat{\mathbf{d}}(n)$, and (10.6.38) is the definition (10.6.21) of the innovations vector. To uncover the physical meaning of $\tilde{e}(n)$ and $\tilde{\alpha}(n)$, we note that comparing (10.6.39) to (10.6.22) gives

$$\varepsilon(n) = \tilde{e}(n)\tilde{\alpha}(n) \quad (10.6.41)$$

which shows that $\tilde{e}(n)$ is a scaled version of the a posteriori error. Starting with (10.6.40) and using (10.6.20), (10.6.16), and (10.5.22), we obtain

$$\tilde{\alpha}^2(n) = 1 - \tilde{\mathbf{w}}^H(n)\tilde{\mathbf{w}}(n) = 1 - \mathbf{x}^H(n)\mathbf{R}^{-1}(n)\mathbf{x}(n) = \alpha(n) \quad (10.6.42)$$

or
$$\tilde{\alpha}(n) = \sqrt{\alpha(n)} \quad (10.6.43)$$

which shows that $\tilde{\alpha}(n)$ is a normalized conversion factor. Since

$$\varepsilon(n) = \alpha(n)e(n) = \tilde{\alpha}^2(n)e(n) \quad (10.6.44)$$

using (10.6.41), we obtain

$$\tilde{e}(n) = \sqrt{e(n)\varepsilon(n)} \quad (10.6.45)$$

[†] $\langle\ \rangle_{ij}$ denotes the ijth element of a block matrix.

that is, $\tilde{e}(n)$ is the geometric mean of the a priori and a posteriori LS errors. Furthermore, (10.6.41) and (10.6.44) give

$$e(n) = \frac{\tilde{e}(n)}{\tilde{\alpha}(n)} \qquad (10.6.46)$$

which also can be proved from (10.6.35) directly (see Problem 10.45).

In summary, to determine the updates of $\tilde{\mathbf{R}}(n)$ and $\tilde{\mathbf{k}}(n)$ of the Cholesky factors and the a priori error $e(n)$ we simply need to determine a unitary matrix $\mathbf{Q}(n)$ that annihilates the vector $\mathbf{x}^H(n)$ in (10.6.35). The construction of the matrix $\mathbf{Q}(n)$ is discussed later in Section 10.6.6.

10.6.4 Extended QR-RLS Algorithm

In applications that require the coefficient vector, we need to solve the upper triangular system $\tilde{\mathbf{R}}(n)\mathbf{c}(n) = \tilde{\mathbf{k}}(n)$ by back substitution. This method is not order-recursive and cannot be implemented in parallel. An alternative approach can be chosen by appending one more column to the matrices of the QR algorithm (10.6.35). To simplify the derivation, we combine the first column of (10.6.35) and the new column to construct the formula

$$\mathbf{Q}(n) \begin{bmatrix} \sqrt{\lambda}\tilde{\mathbf{R}}(n-1) & \tilde{\mathbf{R}}^{-H}(n-1)/\sqrt{\lambda} \\ \mathbf{x}^H(n) & \mathbf{0}^H \end{bmatrix} = \begin{bmatrix} \tilde{\mathbf{R}}(n) & \mathbf{D}(n) \\ \mathbf{0}^H & \tilde{\mathbf{g}}^H(n) \end{bmatrix} \qquad (10.6.47)$$

where $\mathbf{D}(n)$ and $\tilde{\mathbf{g}}(n)$ are yet to be determined. Using Lemma 10.2, we obtain

$$\langle \mathbf{B}^H \mathbf{B} \rangle_{12} = \tilde{\mathbf{R}}^H(n)\mathbf{D}(n) = \mathbf{I} = \langle \mathbf{A}^H \mathbf{A} \rangle_{12} \qquad (10.6.48)$$

which implies that $\mathbf{D}(n) = \tilde{\mathbf{R}}^{-H}(n)$ is the Cholesky factor of $\mathbf{R}^{-1}(n)$ and can be updated by using the same orthogonal transformation $\mathbf{Q}(n)$. Furthermore, we have

$$\langle \mathbf{B}^H \mathbf{B} \rangle_{22} = \tilde{\mathbf{R}}^{-1}(n)\tilde{\mathbf{R}}^{-H}(n) + \tilde{\mathbf{g}}(n)\tilde{\mathbf{g}}^H(n) = \frac{1}{\lambda}\tilde{\mathbf{R}}(n-1)\tilde{\mathbf{R}}^{-H}(n-1) = \langle \mathbf{A}^H \mathbf{A} \rangle_{22}$$

which, using (10.6.16), gives

$$\hat{\mathbf{R}}^{-1}(n) = \mathbf{P}(n) = \frac{1}{\lambda}\mathbf{P}(n-1) - \tilde{\mathbf{g}}(n)\tilde{\mathbf{g}}^H(n)$$

Comparing the last equation to (10.5.30) gives

$$\tilde{\mathbf{g}}(n) = \frac{\mathbf{g}(n)}{\sqrt{\alpha(n)}} = \frac{\mathbf{g}(n)}{\tilde{\alpha}(n)} \qquad (10.6.49)$$

that is, $\tilde{\mathbf{g}}(n)$ is a scaled version of the RLS gain vector. Using (10.5.13) gives

$$\mathbf{c}(n) = \mathbf{c}(n-1) + \tilde{\mathbf{g}}(n)\tilde{e}^*(n) \qquad (10.6.50)$$

which provides a time-updating formula for the filter coefficient vector. This method of updating the coefficient vector $\mathbf{c}(n)$ is known as the *extended QR-RLS algorithm* (Yang and Böhme 1992; Sayed and Kailath 1994). This algorithm is not widely used because the propagation of both $\tilde{\mathbf{R}}(n)$ and $\tilde{\mathbf{R}}^{-H}(n)$ may lead to numerical problems, especially in finite-precision implementations. This problem may be avoided by using the inverse QR-RLS algorithm, discussed next. Other methods of extracting the coefficient vector are discussed in Shepherd and McWhirter (1993).

10.6.5 Inverse QR-RLS Algorithm

From the CRLS algorithm we obtain

$$1 + \frac{1}{\lambda}\mathbf{x}^H(n)\mathbf{P}(n-1)\mathbf{x}(n) = \frac{1}{\alpha(n)} \tag{10.6.51}$$

$$\frac{1}{\lambda}\mathbf{P}(n-1)\mathbf{x}(n) = \frac{\mathbf{g}(n)}{\alpha(n)} \tag{10.6.52}$$

$$\frac{1}{\lambda}\mathbf{P}(n-1) = \mathbf{P}(n) + \frac{\mathbf{g}(n)}{\sqrt{\alpha(n)}}\frac{\mathbf{g}^H(n)}{\sqrt{\alpha(n)}} \tag{10.6.53}$$

which combined with the Cholesky decomposition

$$\mathbf{P}(n) = \hat{\mathbf{R}}^{-1}(n) = \tilde{\mathbf{R}}^{-1}(n)\tilde{\mathbf{R}}^{-H}(n) \tag{10.6.54}$$

leads to the following identity

$$\begin{bmatrix} \frac{1}{\sqrt{\lambda}}\mathbf{x}^H(n)\tilde{\mathbf{R}}^{-1}(n-1) & 1 \\ \frac{1}{\sqrt{\lambda}}\tilde{\mathbf{R}}^{-1}(n-1) & \mathbf{0} \end{bmatrix} \begin{bmatrix} \frac{1}{\sqrt{\lambda}}\tilde{\mathbf{R}}^{-H}(n-1)\mathbf{x}(n) & \frac{1}{\sqrt{\lambda}}\tilde{\mathbf{R}}^{-H}(n-1) \\ 1 & \mathbf{0}^H \end{bmatrix}$$

$$= \begin{bmatrix} \mathbf{0}^H & \frac{1}{\sqrt{\alpha(n)}} \\ \tilde{\mathbf{R}}^{-1}(n) & \frac{\mathbf{g}(n)}{\sqrt{\alpha(n)}} \end{bmatrix} \begin{bmatrix} \mathbf{0} & \tilde{\mathbf{R}}^{-H}(n) \\ \frac{1}{\sqrt{\alpha(n)}} & \frac{\mathbf{g}^H(n)}{\sqrt{\alpha(n)}} \end{bmatrix}$$

$$\tag{10.6.55}$$

where $\tilde{\mathbf{R}}^{-1}(n)$ is an upper triangular matrix. From (10.6.55) and Lemma 10.2 there is a unitary matrix $\mathbf{Q}(n)$ such that

$$\mathbf{Q}(n)\begin{bmatrix} \frac{1}{\sqrt{\lambda}}\tilde{\mathbf{R}}^{-H}(n-1)\mathbf{x}(n) & \frac{1}{\sqrt{\lambda}}\tilde{\mathbf{R}}^{-H}(n-1) \\ 1 & \mathbf{0}^H \end{bmatrix} = \begin{bmatrix} \mathbf{0} & \tilde{\mathbf{R}}^{-H}(n) \\ \frac{1}{\sqrt{\alpha(n)}} & \frac{\mathbf{g}^H(n)}{\sqrt{\alpha(n)}} \end{bmatrix} \tag{10.6.56}$$

This shows that annihilating the vector $\tilde{\mathbf{R}}^{-H}(n-1)\mathbf{x}(n)/\sqrt{\lambda} = \tilde{\mathbf{w}}(n)/\sqrt{\lambda}$ updates the Cholesky factor $\tilde{\mathbf{R}}^{-H}(n)$, the normalized gain vector $\tilde{\mathbf{g}}(n)$, and the conversion factor $\tilde{\alpha}(n)$. Again, the only requirement of matrix $\mathbf{Q}(n)$ is to annihilate the row vector $\tilde{\mathbf{w}}^H(n)/\sqrt{\lambda}$. This algorithm, like the CRLS method, is initialized by setting $\hat{\mathbf{P}}^H(-1) = \tilde{\mathbf{R}}^{-H}(-1) = \delta^{-1}\mathbf{I}$, where δ is a very small positive number.

10.6.6 Implementation of QR-RLS Algorithm Using the Givens Rotations

To develop a complete QRD-based RLS algorithm, we need to construct the matrix $\mathbf{Q}(n)$ that annihilates the vector $\mathbf{x}^H(n)$ on the left-hand side of (10.6.35). Since we do not need the vector $\tilde{\mathbf{w}}(n)$ and we can compute $\tilde{\alpha}(n)$ from matrix $\mathbf{Q}(n)$, as we shall see later, we work with the following part

$$\mathbf{Q}(n)\underbrace{\begin{bmatrix} \sqrt{\lambda}\tilde{\mathbf{R}}(n-1) & \sqrt{\lambda}\tilde{\mathbf{k}}(n-1) \\ \mathbf{x}^H(n) & y^*(n) \end{bmatrix}}_{\bar{\mathbf{R}}(n)} = \begin{bmatrix} \tilde{\mathbf{R}}(n) & \tilde{\mathbf{k}}(n) \\ \mathbf{0}^H & \tilde{e}^*(n) \end{bmatrix} \tag{10.6.57}$$

and show how to annihilate the elements of $\mathbf{x}^H(n)$, one by one, using a sequence of M Givens rotations. We remind the reader that the matrix $\tilde{\mathbf{R}}(n-1)$ is upper triangular. We start by constructing a Givens rotation matrix $\mathbf{G}^{(1)}(n)$ that operates on the first and last rows of $\bar{\mathbf{R}}(n)$ to annihilate the first element of $\mathbf{x}^H(n)$. More specifically, we wish to find a Givens rotation such that

$$\begin{bmatrix} c_1 & \mathbf{0}^H & s_1^* \\ \mathbf{0} & \mathbf{I} & \mathbf{0} \\ -s_1 & \mathbf{0}^H & c_1 \end{bmatrix} \begin{bmatrix} \sqrt{\lambda}\tilde{r}_{11}(n-1) & \sqrt{\lambda}\tilde{r}_{12}(n-1) & \cdots & \sqrt{\lambda}\tilde{r}_{1M}(n-1) & \sqrt{\lambda}\tilde{k}_1(n-1) \\ \mathbf{0} & \vdots & \ddots & \vdots & \vdots \\ x_1^*(n) & x_2^*(n) & \cdots & x_M^*(n) & y^*(n) \end{bmatrix}$$

$$= \begin{bmatrix} \tilde{r}_{11}(n) & \tilde{r}_{12}(n) & \cdots & \tilde{r}_{1M}(n) & \tilde{k}_1(n) \\ \mathbf{0} & \vdots & \ddots & \vdots & \vdots \\ 0 & x_2^{(2)}(n) & \cdots & x_M^{(2)}(n) & y^{(2)}(n) \end{bmatrix}$$

To this end, we use the first element of the first row and the first element of the last row to determine the rotation parameters c_1 and s_1, and then we apply the rotation to the remaining M pairs of the two rows. Note that for consistency of notation we define $x_k^{(1)}(n) \triangleq x_k(n)$ and $y^{(1)}(n) \triangleq y(n)$. Then using $\sqrt{\lambda}\tilde{r}_{22}(n)$ and $x_2^{(2)}(n)$, we determine $\mathbf{G}^{(2)}(n)$ and annihilate the second element of the last row by operating on the $M-1$ pairs of the second row and the last row of the matrix $\mathbf{G}^{(2)}(n)\bar{\mathbf{R}}(n)$.

In general, we use the elements $\sqrt{\lambda}\tilde{r}_{ii}(n)$ and $x_i^{(i)}(n)$ to determine the Givens rotation matrix $\mathbf{G}^{(i)}(n)$ that operates on the ith row and the last row of the rotated matrix $\mathbf{G}^{(i-1)}(n)\cdots\mathbf{G}^{(1)}(n)\bar{\mathbf{R}}(n)$ to annihilate the element $x_i^{(i)}(n)$. Therefore,

$$\begin{bmatrix} c_i & s_i^* \\ -s_i & c_i \end{bmatrix} \begin{bmatrix} 0 & \cdots & 0 & \sqrt{\lambda}\tilde{r}_{ii}(n-1) & \cdots & \sqrt{\lambda}\tilde{r}_{iM}(n-1) & \sqrt{\lambda}\tilde{k}_i(n-1) \\ 0 & \cdots & 0 & x_i^{(i)}(n) & \cdots & x_M^*(n) & y^{(i)}(n) \end{bmatrix}$$

$$= \begin{bmatrix} 0 & \cdots & 0 & \tilde{r}_{ii}(n) & \tilde{r}_{i,i+1}(n) & \cdots & \tilde{r}_{iM}(n) & \tilde{k}_i(n) \\ 0 & \cdots & 0 & 0 & x_{i+1}^{(i+1)}(n) & \cdots & x_M^{(i+1)}(n) & y^{(i+1)}(n) \end{bmatrix}$$

$$(10.6.58)$$

where
$$c_i = \frac{\sqrt{\lambda}\tilde{r}_{ii}(n-1)}{\tilde{r}_{ii}(n)} \qquad s_i = \frac{x_i^{(i)}(n)}{\tilde{r}_{ii}(n)} \tag{10.6.59}$$

and
$$\tilde{r}_{ii}(n) = [\lambda\tilde{r}_{ii}^2(n-1) + |x_i^{(i)}(n)|^2]^{1/2} \tag{10.6.60}$$

Thus, if we perform (10.6.58) for $i = 1, 2, \ldots, M$, we annihilate the first M elements in the last row of $\bar{\mathbf{R}}(n)$ and convert $\bar{\mathbf{R}}(n)$ to the triangular matrix shown in (10.6.57). This process requires a total of $M(M+1)/2$ Givens rotations. The orthogonalization matrix is

$$\mathbf{Q}(n) = \mathbf{G}^{(M)}(n)\cdots\mathbf{G}^{(2)}(n)\mathbf{G}^{(1)}(n) \tag{10.6.61}$$

where
$$\mathbf{G}^{(i)}(n) = \begin{bmatrix} 1 & & & & & & & \\ & \ddots & & & & & & \\ & & 1 & & & & & \\ & & & c_i(n) & \cdots & & s_i^*(n) & \\ & & & & 1 & & & \\ & & & \vdots & & \ddots & \vdots & \\ & & & & & & 1 & \\ & & & -s_i(n) & \cdots & & c_i(n) \end{bmatrix} \tag{10.6.62}$$

are $(M+1) \times (M+1)$ rotation matrices. Note that all off-diagonal elements, except those in the $(i, M+1)$ and $(M+1, i)$ locations, are zero.

From (10.6.35) we can easily see that $\tilde{\alpha}(n)$ equals the last diagonal element of $\mathbf{Q}(n)$. Furthermore, taking into consideration the special structure of $\mathbf{G}^{(i)}(n)$ and (10.6.61), we obtain

$$\tilde{\alpha}(n) = \prod_{i=1}^{M} c_i(n) \qquad (10.6.63)$$

that is, $\tilde{\alpha}(n)$ is the product of the cosine terms in the M Givens rotations. This justifies the interpretation of $\tilde{\alpha}(n)$ and $\alpha(n) = \tilde{\alpha}^2(n)$ as angle variables.

Although the LS solution is not defined if $n < M$, the RLS Givens algorithm may be initialized by setting $\tilde{\mathbf{R}}(0) = \mathbf{0}$ and $\tilde{\mathbf{k}}(n) = \mathbf{0}$. The Givens rotation–based RLS algorithm is summarized in Table 10.9. The algorithm requires about $2M^2$ multiplications, $2M$ divisions, and M square roots per time update.

TABLE 10.9
The Givens rotation–based RLS algorithm.

Initialization
Set all elements $\tilde{r}_{ij}(-1) = 0, \tilde{k}_i(-1) = 0$

Time Recursion: $n = 0, 1, \ldots$

$\tilde{e}(n) = y(n) \qquad \tilde{\alpha}(n) = 1$

For $i = 1$ to M do

$\qquad \tilde{r}_{ii}(n) = \{\lambda \tilde{r}_{ii}^2(n-1) + |x_i(n)|^2\}^{1/2}$

$\qquad c = \dfrac{\sqrt{\lambda}\tilde{r}_{ii}(n-1)}{\tilde{r}_{ii}(n)} \qquad s = \dfrac{x_i(n)}{\tilde{r}_{ii}(n)}$

$\qquad$ [If $\tilde{r}_{ii}(n) = 0$, set $c = 1$ and $s = 0$]

$\qquad$ For $j = i + 1$ to M do

$\qquad\qquad \bar{x} = cx_j(n) - s\tilde{r}_{ij}(n-1)$

$\qquad\qquad \tilde{r}_{ij}(n) = c\tilde{r}_{ij}(n-1) + s^*x_j(n)$

$\qquad\qquad x_j(n) = \bar{x}$

$\qquad$ End

$\qquad \bar{e} = c\tilde{e}(n) - s\tilde{k}_i(n-1)$

$\qquad \tilde{k}_i(n) = c\tilde{k}_i(n-1) + s^*\tilde{e}(n)$

$\qquad \tilde{e}(n) = \bar{e}$

$\qquad \tilde{\alpha}(n) = c\tilde{\alpha}(n)$

End

$\varepsilon(n) = \tilde{e}(n)\tilde{\alpha}(n)$ or $e(n) = \dfrac{\tilde{e}(n)}{\tilde{\alpha}(n)}$

The algorithm in Table 10.9 may be implemented in parallel using a triangular array of processors, as illustrated in Figure 10.36 for $M = 3$. At time $n - 1$, the elements of $\tilde{\mathbf{R}}(n-1)$ and $\tilde{\mathbf{k}}(n-1)$ are stored in the array elements. The arriving new input data $[\mathbf{x}^H(n)\ y^*(n)]$ are fed from the top and propagate downward. The Givens rotation parameters are calculated in the boundary cells and propagate from left to right. The internal cells receive the rotation parameters from the left, perform the rotation on the data from the top, and pass results to the cells at right and below. The angle variable $\tilde{\alpha}(n)$ is computed along the boundary cells and the a priori or a posteriori error at the last cell. This updating procedure is repeated at each time step upon the arrival of the new data. This structure was derived in McWhirter (1983) by eliminating the linear part used to determine the coefficient vector, by back substitution, from the systolic array introduced in Gentleman and Kung (1981) for the solution of general LS problems. Clearly, the array in Figure 10.36 performs two distinct functions: It propagates the matrix $\tilde{\mathbf{R}}(n)$ and the vector $\tilde{\mathbf{k}}(n)$ that define the LS array processor, and it performs,

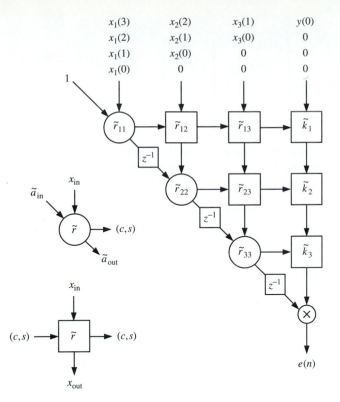

FIGURE 10.36
Systolic array implementation of the QR-RLS algorithm and
functional description of its processing elements.

although in a not-so-obvious way, the filtering operation by providing at the output the error
$\varepsilon(n)$ or $e(n)$. Figure 10.36 provides a functional description of the processing elements
only. In practice, there are different hardware and software implementations using systolic
arrays, wavefront arrays, and CORDIC processors. More detailed descriptions can be found
in McWhirter and Proudler (1993), Shepherd and McWhirter (1993), and Haykin (1996).

10.6.7 Implementation of Inverse QR-RLS Algorithm Using the Givens Rotations

If we define the vector

$$\bar{\mathbf{w}}(n) \triangleq \frac{1}{\sqrt{\lambda}} \tilde{\mathbf{R}}^{-H}(n-1)\mathbf{x}(n) \tag{10.6.64}$$

and the scalar

$$\hat{\alpha}^2(n) \triangleq \frac{1}{\alpha(n)} = 1 + \bar{\mathbf{w}}^H(n)\bar{\mathbf{w}}(n) \tag{10.6.65}$$

we can express (10.6.56) as

$$\mathbf{Q}(n) \begin{bmatrix} \bar{\mathbf{w}}(n) & \frac{1}{\sqrt{\lambda}}\tilde{\mathbf{R}}^{-H}(n-1) \\ 1 & \mathbf{0}^H \end{bmatrix} = \begin{bmatrix} \mathbf{0} & \tilde{\mathbf{R}}^{-H}(n) \\ \hat{\alpha}(n) & \tilde{\mathbf{g}}^H(n) \end{bmatrix} \tag{10.6.66}$$

where $\tilde{\mathbf{g}}(n)$ is the normalized gain vector (10.6.49). The matrix $\mathbf{Q}(n)$ will be chosen as a
sequence of Givens rotation matrices $\mathbf{G}^{(i)}(n)$ defined in (10.6.62).

We first show that we can determine the angle parameters $c_i(n)$ and $s_i(n)$ of $\mathbf{G}^{(i)}(n)$
using only the elements of $\bar{\mathbf{w}}(n)$. To this end, we choose the angle parameters of the rotation

matrix $\mathbf{G}^{(1)}(n)$ in

$$\mathbf{G}^{(1)}(n)\begin{bmatrix} \bar{w}_1(n) \\ \bar{w}_2(n) \\ \vdots \\ \bar{w}_M(n) \\ 1 \end{bmatrix} = \begin{bmatrix} 0 \\ \bar{w}_2(n) \\ \vdots \\ \bar{w}_M(n) \\ \hat{\alpha}_1(n) \end{bmatrix} \qquad (10.6.67)$$

to annihilate the first element $\bar{w}_1(n)$. Note that owing to the structure of $\mathbf{G}^{(1)}(n)$ the remaining elements of $\bar{\mathbf{w}}(n)$ are left unaffected. Since unitary transformations preserve the Euclidean norm of a vector, we can easily see that

$$\hat{\alpha}_1^2(n) = 1 + |\bar{w}_1(n)|^2$$

which expresses $\hat{\alpha}_1(n)$ in terms of $\bar{\mathbf{w}}(n)$. From the first and last equations in (10.6.67), we have the system

$$c_1(n)\bar{w}_1(n) + s_1^*(n) = 0$$
$$-s_1(n)\bar{w}_1(n) + c_1(n) = \hat{\alpha}_1(n)$$

whose solution $\qquad c_1(n) = \dfrac{1}{\hat{\alpha}_1(n)} \qquad s_1(n) = -\dfrac{\bar{w}_1^*(n)}{\hat{\alpha}_1(n)}$

provides the required parameters. Similarly, we can determine the rotation $\mathbf{G}^{(2)}(n)$ to annihilate the element $\bar{w}_2(n)$ of the vector on the right-hand side of (10.6.67). The required rotation parameters are

$$c_2(n) = \frac{\hat{\alpha}_1(n)}{\hat{\alpha}_2(n)} \qquad s_2(n) = -\frac{\bar{w}_2^*(n)}{\hat{\alpha}_2(n)}$$

where $\qquad \hat{\alpha}_2^2(n) = 1 + |\bar{w}_1(n)|^2 + |\bar{w}_2(n)|^2 = \hat{\alpha}_1^2(n) + |\bar{w}_2(n)|^2$

provides a recursive formula for the computation of $\hat{\alpha}_i(n)$. The remaining elements of $\bar{\mathbf{w}}(n)$ can be annihilated by continuing in a similar way. In general, for $i = 1, 2, \ldots, M$, we have

$$\hat{\alpha}_i(n) = [\hat{\alpha}_{i-1}^2(n) + |\bar{w}_i(n)|^2]^{1/2} \qquad \hat{\alpha}_0(n) = 1 \qquad (10.6.68)$$

$$c_i(n) = \frac{\hat{\alpha}_{i-1}(n)}{\hat{\alpha}_i(n)} \qquad s_i(n) = -\frac{\bar{w}_i^*(n)}{\hat{\alpha}_i(n)} \qquad (10.6.69)$$

and $\hat{\alpha}(n) = \hat{\alpha}_M(n)$.

Let us denote by $\tilde{p}_{ij}(n)$ the elements of matrix $\tilde{\mathbf{R}}^{-H}(n)$ and by $g_j^{(i)}(n)$ the elements of vector $\tilde{\mathbf{g}}^H(n)$ after the ith rotation. The first rotation updates the first element of the matrix $\tilde{\mathbf{R}}^{-H}(n-1)/\sqrt{\lambda}$ and modifies the first element of $\tilde{\mathbf{g}}^H(n)$. Indeed, from

$$\mathbf{G}^{(1)}(n)\begin{bmatrix} \dfrac{1}{\sqrt{\lambda}}\tilde{\mathbf{R}}^{-H}(n-1) \\ \mathbf{0}^H \end{bmatrix} = \begin{bmatrix} \tilde{p}_{11}(n) & 0 & \cdots & 0 \\ \vdots & \vdots & \ddots & \vdots \\ g_1^{(1)}(n) & 0 & \cdots & 0 \end{bmatrix} \qquad (10.6.70)$$

we obtain $\qquad \tilde{p}_{11}(n) = \dfrac{1}{\sqrt{\lambda}}c_1(n)\tilde{p}_{11}(n-1)$

$$g_1^{(1)}(n) = -\frac{1}{\sqrt{\lambda}}s_1(n)\tilde{p}_{11}(n-1)$$

Multiplication of (10.6.70) by $\mathbf{G}^{(2)}(n)$ updates the second row of $\tilde{\mathbf{R}}^{-H}(n-1)/\sqrt{\lambda}$ and modifies the first two elements of $\tilde{\mathbf{g}}^H(n)$. In general, the ith rotation updates the ith row of

$\tilde{\mathbf{R}}^{-H}(n-1)/\sqrt{\lambda}$ and modifies the i first elements of $\tilde{\mathbf{g}}^H(n)$ using the formulas

$$\tilde{p}_{ij}(n) = \frac{1}{\sqrt{\lambda}}c_i(n)\tilde{p}_{ij}(n-1) + s_i^*(n)g_j^{(i-1)}(n) \qquad (10.6.71)$$

$$g_j^{(i)}(n) = c_i(n)g_j^{(i-1)}(n) - \frac{1}{\sqrt{\lambda}}s_i(n)\tilde{p}_{ij}(n-1) \qquad (10.6.72)$$

for $1 \le i \le M$ and $1 \le j \le i$. These recursions are initialized with $g_j^{(i)}(n) = 0, 1 \le j \le M$, $i \le j$, and provide the required quantities after M rotations. The complete inverse QR-RLS algorithm is summarized in Table 10.10, whereas a systolic array implementation is discussed in Alexander and Ghirnikar (1993).

TABLE 10.10
Summary of the inverse QR-RLS Givens algorithm.[†]

Initialization
$\mathbf{c}(-1) = \mathbf{x}(-1) = \mathbf{0}$ $\tilde{p}_{ij}(-1) = \delta \gg 1$

Time Recursion: $n = 0, 1, \dots$

$e(n) = y(n) - \mathbf{c}^H(n-1)\mathbf{x}(n)$

$g_j^{(i)}(n) = 0 \qquad 1 \le j \le M, i \le j$

$\hat{\alpha}_0(n) = 1$

For $i = 1$ to M do

  $\bar{w}_i(n) = \dfrac{1}{\sqrt{\lambda}}\sum_{j=1}^{i}\tilde{p}_{ij}(n-1)x_i(n)$

  $\hat{\alpha}_i(n) = [\hat{\alpha}_{i-1}^2(n) + |\bar{w}_i(n)|^2]^{1/2}$

  $c_i(n) = \dfrac{\hat{\alpha}_{i-1}(n)}{\hat{\alpha}_i(n)} \qquad s_i(n) = -\dfrac{\bar{w}_i^*(n)}{\hat{\alpha}_i(n)}$

  For $j = 1$ to i do

    $\tilde{p}_{ij}(n) = \dfrac{1}{\sqrt{\lambda}}c_i(n)\tilde{p}_{ij}(n-1) + s_i^*(n)g_j^{(i-1)}(n)$

    $g_j^{(i)}(n) = c_i(n)g_j^{(i-1)}(n) - \dfrac{1}{\sqrt{\lambda}}s_i(n)\tilde{p}_{ij}(n-1)$

  End

End

$\tilde{e}(n) = \dfrac{e(n)}{\hat{\alpha}_M(n)}$

For $m = 1$ to M do

  $c_m(n) = c_m(n-1) + g_j^{(i)*}(n)\tilde{e}^*(n)$

End

[†]The computations can be done "in-place" using temporary variables as shown in Table 10.9.

10.6.8 Classification of RLS Algorithms for Array Processing

Whereas the CRLS algorithm provides the basis for the introduction and performance evaluation of exact LS adaptive filters for array processing, the Givens rotation–based QR-RLS algorithms provide the most desirable implementation in terms of numerical behavior and ease of hardware implementation. However, there are many more algorithms that have interesting theoretical interpretations or may better serve the needs of particular applications. In general, we have the following types of RLS algorithms.

1. The CRLS algorithm, which is a fixed-order algorithm, updates the inverse $\mathbf{P}(n) = \hat{\mathbf{R}}^{-1}(n)$ of the correlation matrix and then computes the gain vector through a matrix-by-vector multiplication (see Section 10.5).

2. Power-domain square root algorithms propagate either $\tilde{\mathbf{R}}(n)$ or its inverse $\tilde{\mathbf{P}}(n) \triangleq \tilde{\mathbf{R}}^{-1}(n)$, using formulas derived from the Cholesky decomposition of $\hat{\mathbf{R}}(n)$ or $\mathbf{P}(n) = \hat{\mathbf{R}}^{-1}(n)$, respectively. They include two types:

   a. Algorithms that propagate $\{\tilde{\mathbf{R}}(n), \tilde{\mathbf{k}}(n)\}$ (information filter approach) or $\{\tilde{\mathbf{R}}^{-1}(n), \tilde{\mathbf{k}}(n)\}$ (covariance filter approach[†]) and provide the a priori or a posteriori errors only.

   b. Algorithms that propagate $\tilde{\mathbf{R}}(n)$ and compute $\mathbf{g}(n)$ by solving (10.6.25) or propagate $\tilde{\mathbf{R}}^{-1}(n)$ and compute $\mathbf{g}(n)$ in a matrix-by-vector multiplication. Both algorithms compute the parameter vector $\mathbf{c}(n)$ and the error $e(n)$ or $\varepsilon(n)$.

3. Amplitude-domain square root algorithms that propagate either $\tilde{\mathbf{R}}(n)$ (*QRD-based RLS*) or its inverse $\tilde{\mathbf{P}}(n) \triangleq \tilde{\mathbf{R}}^{-1}(n)$ (*inverse QRD–based RLS*) working *directly* with the data matrix $\mathbf{\Lambda}(n)[\mathbf{X}(n)\,\mathbf{y}(n)]$. In both cases, we can develop algorithms providing only the error $e(n)$ or $\varepsilon(n)$ or both the errors and the parameter vector $\mathbf{c}(n)$.

Algorithms that propagate the Cholesky factor $\tilde{\mathbf{R}}(n)$ avoid the loss-of-symmetry problem and have better numerical properties because the condition number of $\tilde{\mathbf{R}}(n)$ equals the square root of the condition number of $\hat{\mathbf{R}}(n)$. Because QRD-based algorithms have superior numerical properties to their Cholesky counterparts, we have focused on RLS algorithms based on the QRD of the data set $\mathbf{\Lambda}(n)[\mathbf{X}(n)\,\mathbf{y}(n)]$. More specifically we discussed QRD-based RLS algorithms using the Givens rotations. Other QRD-based RLS algorithms using the MGS (Ling et al. 1986) and Householder transformations (Liu et al. 1992; Steinhardt 1988; Rader and Steinhardt 1986) also have been developed but are not as widely used.

It is generally accepted that QR decomposition leads to the best methods for solving the LS problem (Golub and Van Loan 1996). It has been shown by simulation that the Givens rotation–based QR-RLS algorithm is numerically stable for $\lambda < 1$ and diverges for $\lambda = 1$ (Yang and Böhme 1992; Haykin 1996). This is the algorithm of choice for applications that require only the a priori or a posteriori errors. Since the extended QR-RLS algorithm propagates both $\tilde{\mathbf{R}}(n)$ and $\tilde{\mathbf{R}}^{-H}(n)$ independently from each other, in finite-precision implementations, the computed values of $\tilde{\mathbf{R}}(n)$ and $\tilde{\mathbf{R}}^{-H}(n)$ deviate from each other's Hermitian inverse. As a result of this numerical inconsistency, the algorithm becomes numerically unstable (Haykin 1996). To avoid this problem, we can use either the QR-RLS algorithm with back substitution "on the fly" or the inverse QR-RLS algorithm (Alexander and Ghirnikar 1993; Pan and Plemmons 1989). The updating of $\mathbf{c}(n)$ with this last algorithm can be implemented in systolic array form without interrupting the adaptation process.

If we factor out the diagonal elements of matrix $\tilde{\mathbf{R}}(n)$, obtained by QRD, we can express $\tilde{\mathbf{R}}(n)$ as

$$\tilde{\mathbf{R}}(n) = \mathbf{D}^{1/2}(n)\tilde{\mathbf{R}}_1(n) \qquad (10.6.73)$$

where $\tilde{\mathbf{R}}_1(n)$ is an upper triangular matrix with unit diagonal elements, and

$$\mathbf{D}(n) \triangleq \text{diag}\{\tilde{r}_{11}^2(n), \tilde{r}_{22}^2(n), \ldots, \tilde{r}_{MM}^2(n)\} \qquad (10.6.74)$$

is a diagonal matrix with positive elements. We can easily see that $\tilde{\mathbf{R}}_1^H(n)$ and $\mathbf{D}(n)$ provide the factors of the LDU decomposition (10.6.14). It turns out that (10.6.73) provides the basis for various QRD-based RLS algorithms that do not require square root operations. In similar manner, the LDU decomposition makes possible the square root–free triangularization of $\hat{\mathbf{R}}(n)$ (see Section 6.3). All algorithms that use the Cholesky factor $\tilde{\mathbf{R}}(n)$ or its inverse $\tilde{\mathbf{R}}^{-1}(n)$ require square root operations, which we can avoid if we use the LDU decomposition factors $\tilde{\mathbf{R}}_1(n)$ and $\mathbf{D}(n)$. Because such algorithms have inferior numerical properties to their square

---

[†] The terms *information* and *covariance* filtering-type algorithms are used in the context of Kalman filter theory (Bierman 1977; Kailath 1981).

root counterparts and are more prone to overflow and underflow problems, and because square root operations are within the reach of current digital hardware, we concentrate on RLS algorithms that propagate the Cholesky factor or its inverse (Stewart and Chapman 1990). However, square root–free algorithms are very useful for VLSI implementations. The interested reader can find information about such algorithms in Bierman and Thornton (1977), Ljung and Soderstrom (1983), Bierman and Thornton (1977), and Hsieh et al. (1993).

A unified derivation of the various RLS algorithms using a state-space formulation and their correspondence with related Kalman filtering algorithms is given in Sayed and Kailath (1994, 1998) and in Haykin (1996).

All algorithms mentioned above hold for *arbitrary* input data vectors and require $O(M^2)$ arithmetic operations per time update. However, if the input data vector has a shift-invariant structure, all algorithms lead to simplified versions that require $O(M)$ arithmetic operations per time update. These algorithms, which can be used for LS FIR filtering and prediction applications, are discussed in the following section.

10.7 FAST RLS ALGORITHMS FOR FIR FILTERING

In Section 7.3 we exploited the shift invariance of the input data vector

$$\mathbf{x}_{m+1}(n) = \begin{bmatrix} \mathbf{x}_m(n) \\ x(n-m) \end{bmatrix} = \begin{bmatrix} x(n) \\ \mathbf{x}_m(n-1) \end{bmatrix} \tag{10.7.1}$$

to develop a lattice-ladder structure for optimum FIR filters and predictors. The determination of the optimum parameters (see Figure 7.3) required the LDL^H decomposition of the correlation matrix $\mathbf{R}(n)$ and the solution of three triangular systems at each time n. However, for stationary signals the optimum filter is time-invariant, and the coefficients of its direct or lattice-ladder implementation structure are evaluated only once, using the algorithm of Levinson.

The key for the development of order-recursive algorithms was the following order partitioning of the correlation matrix

$$\mathbf{R}_{m+1}(n) = \begin{bmatrix} \mathbf{R}_m(n) & \mathbf{r}_m^b(n) \\ \mathbf{r}_m^{bH}(n) & P_x(n-m) \end{bmatrix} = \begin{bmatrix} P_x(n) & \mathbf{r}_m^{fH}(n) \\ \mathbf{r}_m^f(n) & \mathbf{R}_m(n-1) \end{bmatrix} \tag{10.7.2}$$

which is a result of the shift-invariance property (10.7.1). The same partitioning can be obtained for the LS correlation matrix $\hat{\mathbf{R}}_m(n)$

$$\hat{\mathbf{R}}_{m+1}(n) = \sum_{j=0}^{n} \lambda^{n-j} \mathbf{x}_{m+1}(n) \mathbf{x}_{m+1}^H(n)$$

$$= \begin{bmatrix} \hat{\mathbf{R}}_m(n) & \hat{\mathbf{r}}_m^b(n) \\ \hat{\mathbf{r}}_m^{bH}(n) & E_x(n-m) \end{bmatrix} = \begin{bmatrix} E_x(n) & \hat{\mathbf{r}}_m^{fH}(n) \\ \hat{\mathbf{r}}_m^f(n) & \hat{\mathbf{R}}_m(n-1) \end{bmatrix} \tag{10.7.3}$$

if we assume that $\mathbf{x}_m(-1) = \mathbf{0}$, a condition known as *prewindowing* (see Section 8.3). This condition is neccesary to ensure the presence of the term $\hat{\mathbf{R}}_m(n-1)$ in the lower right corner partitioning of $\hat{\mathbf{R}}_{m+1}(n)$.

The identical forms of (10.7.2) and (10.7.3) imply that the order-recursive relations and the lattice-ladder structure developed in Section 7.3 for optimum FIR filters can be used for *prewindowed* LS FIR filters. Simply, the expectation operator $E\{(\cdot)\}$ should be replaced by the time-averaging operator $\sum_{j=0}^{n} \lambda^{n-j}(\cdot)$, and the term *power* should be replaced by the term *energy*, when we go from the optimum MSE to the LSE formulation.

In this section we exploit the shift invariance (10.7.1) and the time updating

$$\hat{\mathbf{R}}_m(n) = \lambda \hat{\mathbf{R}}_m(n-1) + \mathbf{x}_m(n) \mathbf{x}_m^H(n) \tag{10.7.4}$$

to develop the following types of fast algorithms with $O(M)$ complexity:

1. Fast *fixed-order* algorithms for RLS direct-form FIR filters by explicitly updating the gain vectors $\mathbf{g}(n)$ and $\bar{\mathbf{g}}(n)$.
2. Fast *order-recursive* algorithms for RLS FIR lattice-ladder filters by indirect or direct updating of their coefficients.
3. QR decomposition–based RLS lattice-ladder algorithms using the Givens rotation.

All relationships in Section 7.3 are valid for the prewindowed LS problem, but we replace P by E to emphasize the energy interpretation of the cost function. The quantities appearing in the partitionings given by (10.7.3) specify a prewindowed LS forward linear predictor $-\mathbf{a}_m$ and an LS backward linear predictor $-\mathbf{b}_m$. Table 10.11 shows the correspondences between general FIR filtering, FLP, and BLP. Using these correspondences and the normal equations for LS filtering, we can easily obtain the normal equations and the total LSE for the FLP and the BLP, which are also summarized in Table 10.11 (see Problem 10.28). We stress that the predictor parameters $\mathbf{a}_m(n)$ and $\mathbf{b}_m(n)$ are held fixed over the optimization interval $0 \le j \le n$.

TABLE 10.11

Summary and correspondences between LS FIR filtering, forward linear prediction, and backward linear prediction.

	FIR filter	FLP	BLP
Input data vector	$\mathbf{x}_m(n)$	$\mathbf{x}_m(n-1)$	$\mathbf{x}_m(n)$
Desired response	$y(n)$	$x(n)$	$x(n-m)$
Coefficient vector	$\mathbf{c}_m(n)$	$-\mathbf{a}_m(n)$	$-\mathbf{b}_m(n)$
Error	$\varepsilon_m(n) = y(n) - \mathbf{c}_m^H(n)\mathbf{x}_m(n)$	$\varepsilon_m^{\mathrm{f}}(n) = x(n) + \mathbf{a}_m^H(n)\mathbf{x}_m(n-1)$	$\varepsilon_m^{\mathrm{b}}(n) = x(n-m) + \mathbf{b}_m^H(n)\mathbf{x}_m(n)$
Cost function	$E_m(n) = \sum_{j=0}^{n} \lambda^{n-j}\|\varepsilon_m(j)\|^2$	$E_m^{\mathrm{f}}(n) = \sum_{j=0}^{n} \lambda^{n-j}\|\varepsilon_m^{\mathrm{f}}(j)\|^2$	$E_m^{\mathrm{b}}(n) = \sum_{j=0}^{n} \lambda^{n-j}\|\varepsilon_m^{\mathrm{b}}(j)\|^2$
Normal equations	$\hat{\mathbf{R}}_m(n)\mathbf{c}_m(n) = \hat{\mathbf{d}}_m(n)$	$\hat{\mathbf{R}}_m(n-1)\mathbf{a}_m(n) = -\hat{\mathbf{r}}_m^{\mathrm{f}}(n)$	$\hat{\mathbf{R}}_m(n)\mathbf{b}_m(n) = -\hat{\mathbf{r}}_m^{\mathrm{b}}(n)$
LSE	$E_m(n) = E_y(n) - \mathbf{c}_m^H(n)\hat{\mathbf{d}}_m(n)$	$E_m^{\mathrm{f}}(n) = E_x(n) + \mathbf{a}_m^H(n)\hat{\mathbf{r}}_m^{\mathrm{f}}(n)$	$E_m^{\mathrm{b}}(n) = E_x(n-m) + \mathbf{b}_m^H(n)\hat{\mathbf{r}}_m^{\mathrm{b}}(n)$
Correlation matrix	$\hat{\mathbf{R}}_m(n) = \sum_{j=0}^{n} \lambda^{n-j}\mathbf{x}_m(j)\mathbf{x}_m^H(j)$	$\hat{\mathbf{R}}_m(n-1)$	$\hat{\mathbf{R}}_m(n)$
Cross-correlation vectors	$\hat{\mathbf{d}}_m(n) = \sum_{j=0}^{n} \lambda^{n-j}\mathbf{x}_m(j)y^*(j)$	$\hat{\mathbf{r}}_m^{\mathrm{f}}(n) = \sum_{j=0}^{n} \lambda^{n-j}\mathbf{x}_m(j-1)x^*(j)$	$\hat{\mathbf{r}}_m^{\mathrm{b}}(n) = \sum_{j=0}^{n} \lambda^{n-j}\mathbf{x}_m(j)x^*(j-m)$

Table 10.12 summarizes the a priori and a posteriori time updates for the LS FIR filter derived in Section 10.5. If we use the correspondences between general FIR filtering and linear prediction, we can easily deduce similar time-updating recursions for the FLP and the BLP. These updates, which are also discussed in Problem 10.29, are summarized in Table 10.12.

10.7.1 Fast Fixed-Order RLS FIR Filters

The major computational task in RLS filters is the computation of the gain vector $\mathbf{g}(n)$ or $\bar{\mathbf{g}}(n)$. The CRLS algorithm updates the inverse matrix $\hat{\mathbf{R}}^{-1}(n)$ and then determines the gain vector via a matrix-by-vector multiplication that results in $O(M^2)$ complexity. The only way to reduce the complexity from $O(M^2)$ to $O(M)$ is by directly updating the gain vectors. We next show how to develop such algorithms by exploiting the shift-invariant structure of the input data vector shown in (10.7.1).

TABLE 10.12
Summary of LS time-updating relations using a priori and a posteriori errors.

	Equation	A priori time updating	A posteriori time updating
Gain	(a)	$\hat{\mathbf{R}}_m(n)\mathbf{g}_m(n) = \mathbf{x}_m(n)$	$\lambda\hat{\mathbf{R}}_m(n-1)\bar{\mathbf{g}}_m(n) = \mathbf{x}_m(n)$
Filter	(b)	$e_m(n) = y(n) - \mathbf{c}_m^H(n-1)\mathbf{x}_m(n)$	$\varepsilon_m(n) = y(n) - \mathbf{c}_m^H(n)\mathbf{x}_m(n)$
	(c)	$\mathbf{c}_m(n) = \mathbf{c}_m(n-1) + \mathbf{g}_m(n)e_m^*(n)$	$\mathbf{c}_m(n) = \mathbf{c}_m(n-1) + \bar{\mathbf{g}}_m(n)\varepsilon_m^*(n)$
	(d)	$E_m(n) = \lambda E_m(n-1) + \alpha_m(n)|e_m(n)|^2$	$E_m(n) = \lambda E_m(n-1) + \dfrac{|\varepsilon_m(n)|^2}{\alpha_m(n)}$
FLP	(e)	$e_m^f(n) = x(n) + \mathbf{a}_m^H(n-1)\mathbf{x}_m(n-1)$	$\varepsilon_m^f(n) = x(n) + \mathbf{a}_m^H(n)\mathbf{x}_m(n-1)$
	(f)	$\mathbf{a}_m(n) = \mathbf{a}_m(n-1) - \mathbf{g}_m(n-1)e_m^{f*}(n)$	$\mathbf{a}_m(n) = \mathbf{a}_m(n-1) - \bar{\mathbf{g}}_m(n-1)\varepsilon_m^{f*}(n)$
	(g)	$E_m^f(n) = \lambda E_m^f(n-1) + \alpha_m(n-1)|e_m^f(n)|^2$	$E_m^f(n) = \lambda E_m^f(n-1) + \dfrac{|\varepsilon_m^f(n)|^2}{\alpha_m(n-1)}$
BLP	(h)	$e_m^b(n) = x(n-m) + \mathbf{b}_m^H(n-1)\mathbf{x}_m(n)$	$\varepsilon_m^b(n) = x(n-m) + \mathbf{b}_m^H(n)\mathbf{x}_m(n)$
	(i)	$\mathbf{b}_m(n) = \mathbf{b}_m(n-1) - \mathbf{g}_m(n)e_m^{b*}(n)$	$\mathbf{b}_m(n) = \mathbf{b}_m(n-1) - \bar{\mathbf{g}}_m(n)\varepsilon_m^{b*}(n)$
	(j)	$E_m^b(n) = \lambda E_m^b(n-1) + \alpha_m(n)|e_m^b(n)|^2$	$E_m^b(n) = \lambda E_m^b(n-1) + \dfrac{|\varepsilon_m^b(n)|^2}{\alpha_m(n)}$

Fast Kalman algorithm: Updating the gain g(n)

Suppose that we know the gain

$$\mathbf{g}_m(n-1) = \hat{\mathbf{R}}_m^{-1}(n-1)\mathbf{x}_m(n-1) \tag{10.7.5}$$

and we wish to compute the gain

$$\mathbf{g}_m(n) = \hat{\mathbf{R}}_m^{-1}(n)\mathbf{x}_m(n) \tag{10.7.6}$$

at the next time instant by "adjusting" $\mathbf{g}_m(n-1)$, using the new data $\{\mathbf{x}_m(n), y(n)\}$.

If we use the matrix inversion by partitioning formulas (7.1.24) and (7.1.26) for matrix $\hat{\mathbf{R}}_{m+1}(n)$, we have

$$\hat{\mathbf{R}}_{m+1}(n) = \begin{bmatrix} \hat{\mathbf{R}}_m^{-1}(n) & \mathbf{0}_m \\ \mathbf{0}_m^H & 0 \end{bmatrix} + \frac{1}{E_m^b(n)}\begin{bmatrix} \mathbf{b}_m(n) \\ 1 \end{bmatrix}\begin{bmatrix} \mathbf{b}_m^H(n) & 1 \end{bmatrix} \tag{10.7.7}$$

and

$$\hat{\mathbf{R}}_{m+1}(n) = \begin{bmatrix} 0 & \mathbf{0}_m^H \\ \mathbf{0}_m & \hat{\mathbf{R}}_m^{-1}(n) \end{bmatrix} + \frac{1}{E_m^f(n)}\begin{bmatrix} 1 \\ \mathbf{a}_m(n) \end{bmatrix}\begin{bmatrix} 1 & \mathbf{a}_m^H(n) \end{bmatrix} \tag{10.7.8}$$

as was shown in Section 7.1.

Using (10.7.7), the first partitioning in (10.7.1), and the definition of $\varepsilon_m^b(n)$ from Table 10.12, we obtain

$$\mathbf{g}_{m+1}(n) = \begin{bmatrix} \mathbf{g}_m(n) \\ 0 \end{bmatrix} + \frac{\varepsilon_m^b(n)}{E_m^b(n)}\begin{bmatrix} \mathbf{b}_m(n) \\ 1 \end{bmatrix} \tag{10.7.9}$$

which provides a *pure order update* of the gain vector $\mathbf{g}_m(n)$. Similarly, using (10.7.8), the second partitioning in (10.7.1), and the definition of $\varepsilon_m^f(n)$ from Table 10.12, we have

$$\mathbf{g}_{m+1}(n) = \begin{bmatrix} 0 \\ \mathbf{g}_m(n-1) \end{bmatrix} + \frac{\varepsilon_m^f(n)}{E_m^f(n)}\begin{bmatrix} 1 \\ \mathbf{a}_m(n) \end{bmatrix} \tag{10.7.10}$$

which provides a *combined order and time update* of the gain vector $\mathbf{g}_m(n)$. This is the key to the development of fast algorithms for updating the gain vector.

Given the gain $\mathbf{g}_m(n-1)$, first we compute $\mathbf{g}_{m+1}(n)$, using (10.7.10). Then we compute $\mathbf{g}_m(n)$ from the first m equations of (10.7.9) as

$$\mathbf{g}_m(n) = \mathbf{g}_{m+1}^{[m]}(n) - g_{m+1}^{(m+1)}(n)\mathbf{b}_m(n) \tag{10.7.11}$$

because
$$g_{m+1}^{(m+1)}(n) = \frac{\varepsilon_m^b(n)}{E_m^b(n)} \qquad (10.7.12)$$

from the last equation in (10.7.9). The updatings (10.7.9) and (10.7.10) require time updatings for the predictors $\mathbf{a}_m(n)$ and $\mathbf{b}_m(n)$ and the minimum error energies $E_m^f(n)$ and $E_m^b(n)$, which are given in Table 10.12. The only remaining problem is the coupling between $\mathbf{g}_m(n)$ in (10.7.11) and $\mathbf{b}_m(n)$ in

$$\mathbf{b}_m(n) = \mathbf{b}_m(n-1) - \mathbf{g}_m(n)e_m^{b*}(n) \qquad (10.7.13)$$

which can be avoided by eliminating $\mathbf{b}_m(n)$. Carrying out the elimination, we obtain

$$\mathbf{g}_m(n) = \frac{\mathbf{g}_{m+1}^{[m]}(n) - g_{m+1}^{(m+1)}(n)\mathbf{b}_m(n-1)}{1 - g_{m+1}^{(m+1)}(n)e_m^{b*}(n)} \qquad (10.7.14)$$

which provides the last step required to complete the updating. This approach, which is known as the *fast Kalman algorithm*, was developed in Falconer and Ljung (1978) using the ideas introduced by Morf (1974). To emphasize the fixed-order nature of the algorithm, we set $m = M$ and drop the order subscript for all quantities of order M. The computational organization of the algorithm, which requires $9M$ operations per time updating, is summarized in Table 10.13.

TABLE 10.13

Fast Kalman algorithm for time updating of LS FIR filters.

Equation	Computation
	Old estimates: $\mathbf{a}(n-1), \mathbf{b}(n-1), \mathbf{g}(n-1), \mathbf{c}(n-1), E^f(n-1)$ New data: $\{\mathbf{x}(n),\ y(n)\}$
	Gain and predictor update
(a)	$e^f(n) = x(n) + \mathbf{a}^H(n-1)\mathbf{x}(n-1)$
(b)	$\mathbf{a}(n) = \mathbf{a}(n-1) - \mathbf{g}(n-1)e^{f*}(n)$
(c)	$\varepsilon^f(n) = x(n) + \mathbf{a}^H(n)\mathbf{x}(n-1)$
(d)	$E^f(n) = \lambda E^f(n-1) + \varepsilon^f(n)e^{f*}(n)$
(e)	$\mathbf{g}_{M+1}(n) = \begin{bmatrix} 0 \\ \mathbf{g}(n-1) \end{bmatrix} + \frac{\varepsilon^f(n)}{E^f(n)}\begin{bmatrix} 1 \\ \mathbf{a}(n) \end{bmatrix}$
(f)	$e^b(n) = x(n-M) + \mathbf{b}^H(n-1)\mathbf{x}(n)$
(g)	$\mathbf{g}(n) = \dfrac{\mathbf{g}_{M+1}^{[M]}(n) - g_{M+1}^{(M+1)}(n)\mathbf{b}(n-1)}{1 - g_{M+1}^{(M+1)}(n)e^{b*}(n)}$
(h)	$\mathbf{b}(n) = \mathbf{b}(n-1) - \mathbf{g}(n)e^{b*}(n)$
	Filter update
(i)	$e(n) = y(n) - \mathbf{c}^H(n-1)\mathbf{x}(n)$
(j)	$\mathbf{c}(n) = \mathbf{c}(n-1) + \mathbf{g}(n)e^*(n)$

The FAEST algorithm: Updating the gain $\bar{\mathbf{g}}(n)$

In a similar way we can update the gain vector

$$\bar{\mathbf{g}}_m(n) = \frac{1}{\lambda}\hat{\mathbf{R}}_m^{-1}(n-1)\mathbf{x}_m(n) \qquad (10.7.15)$$

by using (10.7.9) and (10.7.10). Indeed, using (10.7.10) with the lower partitioning (10.7.1) and (10.7.9) with the upper partitioning (10.7.1), we obtain

$$\bar{\mathbf{g}}_{m+1}(n) = \begin{bmatrix} 0 \\ \bar{\mathbf{g}}_m(n-1) \end{bmatrix} + \frac{e_m^f(n)}{\lambda E_m^f(n-1)}\begin{bmatrix} 1 \\ \mathbf{a}_m(n-1) \end{bmatrix} \qquad (10.7.16)$$

and
$$\bar{\mathbf{g}}_{m+1}(n) = \begin{bmatrix} \bar{\mathbf{g}}_m(n) \\ 0 \end{bmatrix} + \frac{e_m^b(n)}{\lambda E_m^b(n-1)} \begin{bmatrix} \mathbf{b}_m(n-1) \\ 1 \end{bmatrix} \qquad (10.7.17)$$

which provide a link between $\bar{\mathbf{g}}_m(n-1)$ and $\bar{\mathbf{g}}_m(n)$. From (10.7.17) we obtain

$$\bar{\mathbf{g}}_m(n) = \bar{\mathbf{g}}_{m+1}^{[m]}(n) - \bar{g}_{m+1}^{(m+1)}(n)\mathbf{b}_m(n-1) \qquad (10.7.18)$$

because
$$\bar{g}_{m+1}^{(m+1)}(n) = \frac{e_m^b(n)}{\lambda E_m^b(n-1)} \qquad (10.7.19)$$

from the last row of (10.7.17). The fundamental difference between (10.7.9) and (10.7.17) is that the presence of $\mathbf{b}_m(n-1)$ in the latter breaks the coupling between gain vector and backward predictor. Furthermore, (10.7.19) can be used to compute $e_m^b(n)$ by

$$e_m^b(n) = \lambda E_m^b(n-1)\bar{g}_{m+1}^{(m+1)}(n) \qquad (10.7.20)$$

with only two multiplications.

The time updatings of the predictors using the gain $\bar{\mathbf{g}}_m(n)$, which are given in Table 10.12, require the a posteriori errors that can be computed from the a priori errors by using the conversion factor

$$\bar{\alpha}_m(n) = 1 + \bar{\mathbf{g}}_m^H(n)\mathbf{x}_m(n) \qquad (10.7.21)$$

which should be updated in time as well. This can be achieved by a two-step procedure as follows. First, using (10.7.16) and the lower partitioning (10.7.1), we obtain

$$\bar{\alpha}_{m+1}(n) = \bar{\alpha}_m(n-1) + \frac{|e_m^f(n)|^2}{\lambda E_m^f(n-1)} \qquad (10.7.22)$$

which is a combined time and order updating. Then we use (10.7.17) and the upper partitioning (10.7.1) to obtain

$$\bar{\alpha}_m(n) = \bar{\alpha}_{m+1}(n) - \bar{g}_{m+1}^{(m+1)}(n)e_m^{b*}(n) \qquad (10.7.23)$$

or
$$\bar{\alpha}_m(n) = \bar{\alpha}_{m+1}(n) - \frac{|e_m^b(n)|^2}{\lambda E_m^b(n-1)} \qquad (10.7.24)$$

which in conjunction with (10.7.22) provides the required time update $\bar{\alpha}_m(n-1) \to \bar{\alpha}_{m+1}(n) \to \bar{\alpha}_m(n)$.

This leads to the *fast a posteriori error sequential technique (FAEST)* algorithm presented in Table 10.14, which was introduced in Carayannis et al. (1983). The FAEST algorithm requires only $7M$ operations per time update and is the most efficient known algorithm for prewindowed RLS FIR filters.

Fast transversal filter (FTF) algorithm. This is an a posteriori type of algorithm obtained from the FAEST by using the conversion factor

$$\alpha_m(n) = 1 - \mathbf{g}_m^H(n)\mathbf{x}_m(n) \qquad (10.7.25)$$

instead of the conversion factor $\bar{\alpha}_m(n) = 1/\alpha_m(n)$. Using the Levinson recursions (10.7.9) and (10.7.10) in conjunction with the upper and lower partitionings in (10.7.1), we obtain

$$\alpha_{m+1}(n) = \alpha_m(n) - \frac{|\varepsilon_m^b(n)|^2}{E_m^b(n)} \qquad (10.7.26)$$

and
$$\alpha_{m+1}(n) = \alpha_m(n-1) - \frac{|\varepsilon_m^f(n)|^2}{E_m^f(n)} \qquad (10.7.27)$$

respectively. To obtain the FTF algorithm, we replace $\bar{\alpha}_m(n)$ in Table 10.14 by $1/\alpha_m(n)$ and Equation (h) by (10.7.27). To obtain $\alpha_m(n)$ from $\alpha_{m+1}(n)$, we cannot use (10.7.26) because

TABLE 10.14

FAEST algorithm for time updating of LS FIR filters.

Equation	Computation
	Old estimates: $\mathbf{a}(n-1)$, $\mathbf{b}(n-1)$, $\mathbf{c}(n-1)$, $\bar{\mathbf{g}}(n-1)$, $E^f(n-1)$, $E^b(n-1)$, $\bar{\alpha}(n-1)$
	New data: $\{\mathbf{x}(n),\ y(n)\}$

Gain and predictor update

(a)	$e^f(n) = x(n) + \mathbf{a}^H(n-1)\mathbf{x}(n-1)$
(b)	$\varepsilon^f(n) = \dfrac{e^f(n)}{\bar{a}(n-1)}$
(c)	$\mathbf{a}(n) = \mathbf{a}(n-1) - \bar{\mathbf{g}}(n-1)\varepsilon^{f*}(n)$
(d)	$E^f(n) = \lambda E^f(n-1) + \varepsilon^f(n)e^{f*}(n)$
(e)	$\bar{\mathbf{g}}_{M+1}(n) = \begin{bmatrix} 0 \\ \bar{\mathbf{g}}(n-1) \end{bmatrix} + \dfrac{e^f(n)}{\lambda E^f(n-1)}\begin{bmatrix} 1 \\ \mathbf{a}(n-1) \end{bmatrix}$
(f)	$e^b(n) = \lambda E^b(n-1)\bar{g}_{M+1}^{(M+1)}(n)$
(g)	$\bar{\mathbf{g}}(n) = \bar{\mathbf{g}}_{M+1}^{[M]}(n) - \bar{g}_{M+1}^{(M+1)}(n)\mathbf{b}(n-1)$
(h)	$\bar{\alpha}_{M+1}(n) = \bar{\alpha}(n-1) + \dfrac{|e^f(n)|^2}{\lambda E^f(n-1)}$
(i)	$\bar{\alpha}(n) = \bar{\alpha}_{M+1}(n) - \bar{g}_{M+1}^{(M+1)*}(n)e^b(n)$
(j)	$\mathbf{b}(n) = \mathbf{b}(n-1) - \bar{\mathbf{g}}(n)\varepsilon^{b*}(n)$
(k)	$\varepsilon^b(n) = \dfrac{e^b(n)}{\bar{\alpha}(n)}$
(l)	$E^b(n) = \lambda E^b(n-1) + \varepsilon^b(n)e^{b*}(n)$

Filter update

(m)	$e(n) = y(n) - \mathbf{c}^H(n-1)\mathbf{x}(n)$
(n)	$\varepsilon(n) = \dfrac{e(n)}{\bar{\alpha}(n)}$
(o)	$\mathbf{c}(n) = \mathbf{c}(n-1) + \bar{\mathbf{g}}(n)\varepsilon^*(n)$

it requires quantities dependent on $\alpha_m(n)$. To avoid this problem, we replace Equation (i) by the following relation

$$\alpha_m(n) = \frac{\alpha_{m+1}(n)}{1 - \alpha_{m+1}(n)\bar{g}_{m+1}^{(m+1)}(n)e_m^{b*}(n)} \tag{10.7.28}$$

obtained by combining (10.7.24), (10.7.19), and $\bar{\alpha}_m(n) = 1/\alpha_m(n)$. This algorithm, which has the same complexity as FAEST, was introduced in Cioffi and Kailath (1984) using a geometric derivation, and is known as the *fast transversal filter (FTF) algorithm*.

An alternative updating to (10.7.27) can be obtained by noticing that

$$\alpha_{m+1}(n) = \alpha_m(n-1) - \alpha_m^2(n-1)\frac{|e_m^f(n)|^2}{E_m^f(n)}$$

$$= \frac{\alpha_m(n-1)}{E_m^f(n)}[E_m^f(n) - \alpha_m(n-1)|e_m^f(n)|^2]$$

or equivalently

$$\alpha_{m+1}(n) = \alpha_m(n-1)\frac{\lambda E_m^f(n-1)}{E_m^f(n)} \tag{10.7.29}$$

which can be used instead of (10.7.27) in the FTF algorithm. In a similar way, we can show that

$$\alpha_{m+1}(n) = \alpha_m(n) \frac{\lambda E_m^{\mathrm{b}}(n-1)}{E_m^{\mathrm{b}}(n)} \qquad (10.7.30)$$

which will be used later.

Some practical considerations

Figure 10.37 shows the realization of an adaptive RLS filter using the direct-form structure. The coefficient updating can be done using any of the introduced fast RLS algorithms. Some issues related to the implementation of these filters using multiprocessing are discussed in Problem 10.48.

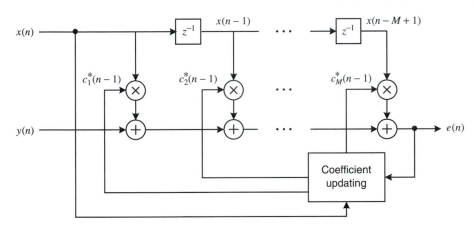

FIGURE 10.37
Implementation of an adaptive FIR filter using a direct-form structure.

In practice, the fast direct-form RLS algorithms are initialized at $n = 0$ by setting

$$E^{\mathrm{f}}(-1) = E^{\mathrm{b}}(-1) = \delta > 0$$
$$\alpha(-1) = 1 \quad \text{or} \quad \bar{\alpha}(-1) = 1 \qquad (10.7.31)$$

and all other quantities equal to zero. The constant δ is chosen as a small positive number on the order of $0.01\sigma_x^2$ (Hubing and Alexander 1991). For $\lambda < 1$, the effects of the initial conditions are quickly "forgotten." An exact initialization method is discussed in Problem 10.31.

Although the fast direct-form RLS algorithms have the lowest computational complexity, they suffer from numerical instability when $\lambda < 1$ (Ljung and Ljung 1985). When these algorithms are implemented with finite precision, the exact algebraic relations used for their derivation breakdown and lead to numerical problems.

There are two ways to deal with stabilization of the fast direct-form RLS algorithms. In the first approach, we try to identify precursors of ill behavior (warnings) and then use appropriate rescue operations to restore the normal operation of the algorithm (Lin 1984; Cioffi and Kailath 1984). One widely used rescue variable is

$$\eta_m(n) \triangleq \frac{\alpha_{m+1}(n)}{\alpha_m(n)} = \frac{\lambda E_m^{\mathrm{b}}(n-1)}{E_m^{\mathrm{b}}(n)} \qquad (10.7.32)$$

which satisfies $0 \le \eta_m(n) \le 1$ for infinite-precision arithmetic (see Problem 10.33 for more details).

In the second approach, we exploit the fact that certain algorithmic quantities can be computed in two different ways. Therefore, we could use their difference, which provides a measure of the numerical errors, to change the dynamics of the error propagation system and stabilize the algorithm. For example, both $e_m^b(n)$ and $\alpha_m(n)$ can be computed either using their definition or simpler order-recursions. This approach has been used to obtain stabilized algorithms with complexities $9M$ and $8M$; however, their performance is highly dependent on proper initialization (Slock and Kailath 1991, 1993).

10.7.2 RLS Lattice-Ladder Filters

The lattice-ladder structure[†] derived in Section 7.3 using the MSE criterion, due to the similarity of (10.7.2) and (10.7.3), holds for the prewindowed LSE criterion as well. This structure, which is depicted in Figure 10.38 for the a posteriori error case, is described by the following equations

$$\varepsilon_0^f(n) = \varepsilon_0^b(n) = x(n)$$

$$\varepsilon_{m+1}^f(n) = \varepsilon_m^f(n) + k_m^{f*}(n)\varepsilon_m^b(n-1) \qquad 0 \le m < M-1 \qquad (10.7.33)$$

$$\varepsilon_{m+1}^b(n) = \varepsilon_m^b(n-1) + k_m^{b*}(n)\varepsilon_m^f(n) \qquad 0 \le m < M-1 \qquad (10.7.34)$$

for the lattice part and

$$\varepsilon_0(n) = y(n)$$

$$\varepsilon_{m+1}(n) = \varepsilon_m(n) - k_m^{c*}(n)\varepsilon_m^b(n) \qquad 0 \le m \le M-1 \qquad (10.7.35)$$

for the ladder part. The lattice parameters are given by

$$k_m^f(n) = -\frac{\beta_m(n)}{E_m^b(n-1)} \qquad (10.7.36)$$

and

$$k_m^b(n) = -\frac{\beta_m^*(n)}{E_m^f(n)} \qquad (10.7.37)$$

and the ladder parameters by

$$k_m^c(n) = \frac{\beta_m^c(n)}{E_m^b(n)} \qquad (10.7.38)$$

where

$$\beta_m(n) = \mathbf{b}_m^H(n-1)\mathbf{r}_m^f(n) + r_{m+1}^f(n) \qquad (10.7.39)$$

and

$$\beta_m^c(n) = \mathbf{b}_m^H(n)\mathbf{d}_m(n) + d_{m+1}(n) \qquad (10.7.40)$$

are the partial correlation parameters.

However, as we recall, the time updating of the minimum LSE energies and the partial correlations is possible only if there is a time update for the correlation matrix $\mathbf{R}_m(n)$ and the cross-correlation vector $\mathbf{d}_m(n)$.

The minimum LSE energies can be updated in time using

$$E_m^f(n) = \lambda E_m^f(n-1) + e_m^f(n)\varepsilon_m^{f*}(n) \qquad (10.7.41)$$

$$E_m^b(n) = \lambda E_m^b(n-1) + e_m^b(n)\varepsilon_m^{b*}(n) \qquad (10.7.42)$$

or their variations, given in Table 10.12.

To update the partial correlation $\beta_m(n)$, we start with the definition (10.7.39) and then use the time-updating formulas for all involved quantities, rearranging and recombining

---

[†]In Chapter 7 we used the symbol $e(n)$ because we had no need to distinguish between a priori and a posteriori errors. However, since the error $e(n)$ in Section 7.3 is an a posteriori error, we now use the symbol $\varepsilon(n)$.

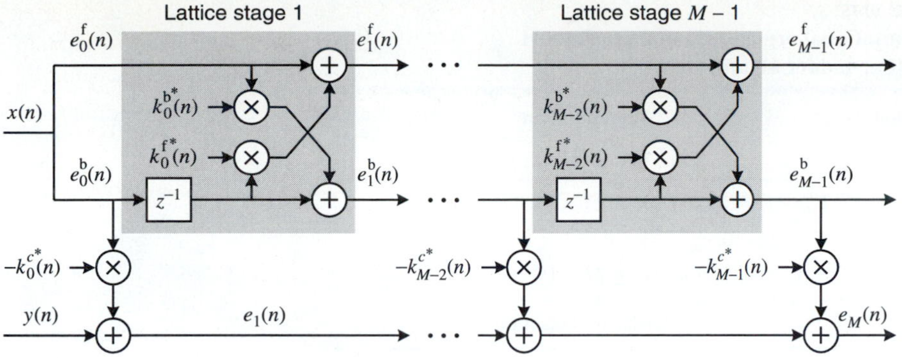

FIGURE 10.38
A posteriori error RLS lattice-ladder filter.

terms as follows:

$$\beta_m(n+1) = \mathbf{b}_m^H(n)\mathbf{r}_m^f(n+1) + r_{m+1}^f(n+1)$$

$$= \mathbf{b}_m^H(n)[\lambda \mathbf{r}_m^f(n) + \mathbf{x}_m(n)x^*(n+1)]$$

$$\quad + [\lambda r_{m+1}^f(n) + x(n-m)x^*(n+1)]$$

$$= \lambda \mathbf{b}_m^H(n)\mathbf{r}_m^f(n) + \varepsilon_m^b(n)x^*(n+1) + \lambda r_{m+1}^f(n)$$

$$= \lambda[\mathbf{b}_m^H(n-1) - \varepsilon_m^b(n)\bar{\mathbf{g}}_m(n)]\mathbf{r}_m^f(n)$$

$$\quad + \lambda r_{m+1}^f(n) + \varepsilon_m^b(n)x^*(n+1)$$

$$= \lambda \beta_m(n) + \varepsilon_m^b(n)[x^*(n+1) - \lambda \bar{\mathbf{g}}_m(n)\mathbf{r}_m^f(n)]$$

$$= \lambda \beta_m(n) + \varepsilon_m^b(n)[x^*(n+1) - \mathbf{x}_m^H(n)\mathbf{R}_m^{-1}(n-1)\mathbf{r}_m^f(n)]$$

$$= \lambda \beta_m(n) + \varepsilon_m^b(n)[x^*(n+1) + \mathbf{x}_m^H(n)\mathbf{a}(n)]$$

$$= \lambda \beta_m(n) + \varepsilon_m^b(n)e_m^{f*}(n+1)$$

which provides the desired update formula. The updating

$$\beta_m(n) = \lambda \beta_m(n-1) + \varepsilon_m^b(n-1)e_m^{f*}(n) \tag{10.7.43}$$

$$= \lambda \beta_m(n-1) + \frac{1}{\alpha_m(n-1)}\varepsilon_m^b(n-1)\varepsilon_m^{f*}(n) \tag{10.7.44}$$

is feasible because the right-hand side involves already-known quantities.

In a similar way (see Problem 10.36), we can show that

$$\beta_m^c(n) = \lambda \beta_m^c(n-1) + \varepsilon_m^b(n)e_m^*(n) \tag{10.7.45}$$

$$= \lambda \beta_m^c(n-1) + \frac{1}{\alpha_m(n)}\varepsilon_m^b(n)\varepsilon_m^*(n) \tag{10.7.46}$$

which facilitates the updating of the ladder parameters.

To obtain an a posteriori algorithm, we need the conversion factor $\alpha_m(n)$, which can be obtained using the order-recursive formula (10.7.26). A detailed organization of the a posteriori LS lattice-ladder algorithm, which requires about $20M$ operations per time update, is given in Table 10.15. The initialization of the algorithm is easily obtained from the definitions of the corresponding quantities. The condition $\alpha_0(n-1) = 1$ follows from (10.7.25), and the positive constant δ is chosen to ensure the inveribility of the LS correlation matrix $\mathbf{R}(n)$ (see Section 10.5). The time-updating recursions (c) and (d) can be replaced by order recursions, as explained in Problem 10.37.

TABLE 10.15

Computational organization of a posteriori LS lattice-ladder algorithm.

Equation	Computation
	Time initialization ($n = 0$)
	$E_m^f(-1) = E_m^b(-1) = \delta > 0 \qquad 0 \le m < M - 1$
	$\beta_m(-1) = 0, \varepsilon_m^b(-1) = 0 \qquad 0 \le m < M - 1$
	$\beta_m^c(-1) = 0 \qquad 0 \le m \le M - 1$
	Order initialization
(a)	$\varepsilon_0^f(n) = \varepsilon_0^b(n) = x(n) \qquad \varepsilon_0(n) = y(n) \qquad \alpha_0(n-1) = 1$
	Lattice part: $m = 0, 1, \ldots, M - 1$
(b)	$\beta_m(n) = \lambda \beta_m(n-1) + \dfrac{\varepsilon_m^b(n-1)\varepsilon_m^{f*}(n)}{\alpha_m(n-1)}$
(c)	$E_m^f(n) = \lambda E_m^f(n-1) + \dfrac{|\varepsilon_m^f(n)|^2}{\alpha_m(n-1)}$
(d)	$E_m^b(n) = \lambda E_m^b(n-1) + \dfrac{|\varepsilon_m^b(n)|^2}{\alpha_m(n)}$
(e)	$k_m^f(n) = \dfrac{-\beta_m(n)}{E_m^b(n-1)}$
(f)	$k_m^b(n) = \dfrac{-\beta_m^*(n)}{E_m^f(n)}$
(g)	$\varepsilon_{m+1}^f(n) = \varepsilon_m^f(n) + k_m^{f*}(n)\varepsilon_m^b(n-1)$
(h)	$\varepsilon_{m+1}^b(n) = \varepsilon_m^b(n-1) + k_m^{b*}(n)\varepsilon_m^f(n)$
(i)	$\alpha_{m+1}(n) = \alpha_m(n) - \dfrac{|\alpha_m(n)e_m^b(n)|^2}{E_m^b(n)}$
	Ladder part: $m = 1, 2, \ldots, M$
(j)	$\beta_m^c(n) = \lambda \beta_m^c(n-1) + \varepsilon_m^b(n)\varepsilon_m^*(n)/\alpha_m(n)$
(k)	$k_m^c(n) = \dfrac{\beta_m^c(n)}{E_m^b(n)}$
(l)	$\varepsilon_{m+1}(n) = \varepsilon_m(n) - k_m^{c*}(n)\varepsilon_m^b(n)$

If instead of the a posteriori errors we use the a priori ones, we obtain the following recursions

$$e_0^f(n) = e_0^b(n) = x(n)$$

$$e_{m+1}^f(n) = e_m^f(n) + k_m^{f*}(n-1)e_m^b(n-1) \qquad 0 \le m < M - 1 \qquad (10.7.47)$$

$$e_{m+1}^b(n) = e_m^b(n-1) + k_m^{b*}(n-1)e_m^f(n) \qquad 0 \le m < M - 1 \qquad (10.7.48)$$

for the lattice part and

$$e_0(n) = y(n)$$
$$e_{m+1}(n) = e_m(n) - k_m^{c*}(n-1)e_m^b(n) \qquad 1 \le m \le M \qquad (10.7.49)$$

for the ladder part (see Problem 10.38). As expected, the a priori structure uses the old LS estimates of the lattice-ladder parameters. Based on these recursions, we can develop the a priori error RLS lattice-ladder algorithm shown in Table 10.16, which requires about $20M$ operations per time update.

TABLE 10.16

Computational organization of a priori LS lattice-ladder algorithm.

Equation	Computation
	Time initialization
	$E_m^f(-1) = E_m^b(-1) = \delta > 0$
	$\beta_m(-1) = 0 \quad e_m^b(-1) = 0 \quad 0 \le m < M - 1$
	$\beta_m^c(-1) = 0 \quad 0 \le m \le M - 1$
	Order initialization
(a)	$e_0^f(n) = e_0^b(n) = x(n) \quad e_0(n) = y(n) \quad \alpha_0(n-1) = 1$
	Lattice Part: $m = 0, 1, \ldots, M-2$
(b)	$e_{m+1}^f(n) = e_m^f(n) + k_m^{f*}(n-1)e_m^b(n-1)$
(c)	$e_{m+1}^b(n) = e_m^b(n-1) + k_m^{b*}(n-1)e_m^f(n)$
(d)	$\beta_m(n) = \lambda\beta_m(n-1) + \alpha_m(n-1)\,e_m^b(n-1)\,e_m^{f*}(n)$
(e)	$E_m^f(n) = \lambda E_m^f(n-1) + \alpha_m(n-1)|e_m^f(n)|^2$
(f)	$E_m^b(n) = \lambda E_m^b(n-1) + \alpha_m(n)|e_m^b(n)|^2$
(g)	$k_m^f(n) = \dfrac{-\beta_m(n)}{E_m^b(n-1)}$
(h)	$k_m^b(n) = \dfrac{-\beta_m^*(n)}{E_m^f(n)}$
(i)	$\alpha_m(n) = \alpha_{m-1}(n) - \dfrac{|e_{m-1}^b(n)|^2}{E_{m-1}^b(n)}$
	Ladder part: $m = 1, 2, \ldots, M$
(j)	$\beta_m^c(n) = \lambda\beta_m^c(n-1) + \alpha_m(n)e_m^b(n)e_m^*(n)$
(k)	$k_m^c(n) = \dfrac{\beta_m^c(n)}{E_m^b(n)}$
(l)	$e_{m+1}(n) = e_m(n) - k_m^{c*}(n-1)e_m^b(n)$

10.7.3 RLS Lattice-Ladder Filters Using Error Feedback Updatings

The LS lattice-ladder algorithms introduced in the previous section update the partial correlations $\beta_m(n)$ and $\beta_m^c(n)$ and the minimum error energies $E_m^f(n)$ and $E_m^b(n)$, and then compute the coefficients of the LS lattice-ladder filter by division. We next develop two algebraically equivalent algorithms, that is, algorithms that solve the same LS problem, which update the lattice-ladder coefficients *directly*. These algorithms, introduced in Ling et al. (1986), have good numerical properties when implemented with finite-word-length arithmetic.

Starting with (10.7.38) and (10.7.45) we have

$$k_m^c(n) = \frac{\beta_m^c(n)}{E_m^b(n)} = \lambda\frac{\beta_m^c(n-1)}{E_m^b(n-1)}\frac{E_m^b(n-1)}{E_m^b(n)} + \frac{\alpha_m(n)e_m^b(n)e_m^*(n)}{E_m^b(n)}$$

$$= \frac{1}{E_m^b(n)}[k_m^c(n-1)\lambda E_m^b(n-1) + \alpha_m(n)e_m^b(n)e_m^*(n)] \qquad (10.7.50)$$

or using

$$\lambda E_m^b(n-1) = E_m^b(n) - \alpha_m(n)e_m^b(n)e_m^{b*}(n)$$

we obtain

$$k_m^c(n) = k_m^c(n-1) + \frac{\alpha_m(n)e_m^b(n)}{E_m^b(n)}[e_m^*(n) - k_m^c(n-1)e_m^{b*}(n)]$$

or

$$k_m^c(n) = k_m^c(n-1) + \frac{\alpha_m(n)e_m^b(n)e_{m+1}^*(n)}{E_m^b(n)} \tag{10.7.51}$$

using (10.7.49). Equation (10.7.51) provides a *direct* updating of the ladder parameters. Similar direct updating formulas can be obtained for the lattice coefficients (see Problem 10.39). Using these updatings, we obtain the a priori RLS lattice-ladder algorithm with error feedback shown in Table 10.17.

TABLE 10.17
Computational organization of a priori RLS lattice-ladder algorithm with direct updating of its coefficients using error feedback formula.

Equation	Computation
	Time initialization
	$E_m^f(-1) = E_m^b(-1) = \delta > 0$
	$k_m^f(-1) = k_m^b(-1) = 0$
	$e_m^b(-1) = 0 \qquad k_m^c(-1) = 0$
	Order initialization
(a)	$e_0^f(n) = e_0^b(n) = x(n) \qquad e_0(n) = y(n) \qquad \alpha_0(n) = 1$
	Lattice part: $m = 0, 1, \ldots, M-2$
(b)	$e_{m+1}^f(n) = e_m^f(n) + k_m^{f*}(n-1)e_m^b(n-1)$
(c)	$e_{m+1}^b(n) = e_m^b(n-1) + k_m^{b*}(n-1)e_m^f(n)$
(d)	$E_m^f(n) = \lambda E_m^f(n-1) + \alpha_m(n-1)|e_m^f(n)|^2$
(e)	$E_m^b(n) = \lambda E_m^b(n-1) + \alpha_m(n)|e_m^b(n)|^2$
(f)	$k_m^f(n) = k_m^f(n-1) - \dfrac{\alpha_m(n-1)e_m^b(n-1)e_{m+1}^{f*}(n)}{E_m^b(n-1)}$
(g)	$k_m^b(n) = k_m^b(n-1) - \dfrac{\alpha_m(n-1)e_m^f(n)e_{m+1}^{b*}(n)}{E_m^f(n)}$
(h)	$\alpha_{m+1}(n) = \alpha_m(n) - \dfrac{|\alpha_m(n)e_m^b(n)|^2}{E_m^b(n)}$
	Ladder part: $m = 0, 1, \ldots, M-1$
(i)	$e_{m+1}(n) = e_m(n) - k_m^{c*}(n-1)e_m^b(n)$
(j)	$k_m^c(n) = k_m^c(n-1) + \dfrac{\alpha_m(n)e_m^b(n)e_{m+1}^*(n)}{E_m^b(n)}$

We note that we first use the coefficient $k_m^c(n-1)$ to compute the higher-order error $e_{m+1}(n)$ by (10.7.49) and then use that error to update the coefficient using (10.7.51). This updating has a feedback-like structure that is sometimes referred to as *error feedback form*. An a posteriori form of the RLS lattice-ladder algorithm with error feedback can be easily obtained as shown in Problem 10.40. Simulation studies (Ling et al. 1986) have shown that when we use finite-precision arithmetic, the algorithms with direct updating of the lattice coefficients have better numerical properties than the algorithms with indirect updating.

10.7.4 Givens Rotation–Based LS Lattice-Ladder Algorithms

We next show how to implement the LS lattice-ladder computations by using the Givens rotation (see Section 8.6) with and without square roots. The resulting algorithms explore the shift invariance of the input data to reduce the computational complexity from $O(M^2)$ to $O(M)$ operations (Ling 1991; Proudler et al. 1989).

We start by introducing the *angle normalized errors*

$$\tilde{e}_m(n) \triangleq \sqrt{e_m(n)\varepsilon_m(n)} = e_m(n)\sqrt{\alpha_m(n)} \tag{10.7.52}$$

$$\tilde{e}_m^f(n) \triangleq \sqrt{e_m^f(n)\varepsilon_m^f(n)} = e_m^f(n)\sqrt{\alpha_m(n-1)} \tag{10.7.53}$$

$$\tilde{e}_m^b(n) \triangleq \sqrt{e_m^b(n)\varepsilon_m^b(n)} = e_m^b(n)\sqrt{\alpha_m(n)} \tag{10.7.54}$$

which are basically the geometric mean of the corresponding a priori and a posteriori errors [see the discussion following (10.5.24) for the interpretation of $\alpha_m(n)$ as an angle variable]. If we formulate the LS problem in terms of these errors, we do not need to distinguish between a priori and a posteriori error algorithms.

Using the a priori lattice equation (10.7.47) for the forward predictor and the definitions of the angle normalized errors, we obtain

$$\tilde{e}_{m+1}^f(n) = \sqrt{\frac{\alpha_{m+1}(n-1)}{\alpha_m(n-1)}}\tilde{e}_m^f(n) - \frac{\beta_m^*(n-1)}{\sqrt{E_m^b(n-2)}}\sqrt{\frac{\alpha_{m+1}(n-1)}{\alpha_m(n-1)}}\frac{\tilde{e}_m^b(n-1)}{\sqrt{E_m^b(n-2)}}$$

or by using (10.7.30)

$$\tilde{e}_{m+1}^f(n) = \sqrt{\frac{\lambda E_m^b(n-2)}{E_m^b(n-1)}}\tilde{e}_m^f(n) - \frac{\beta_m^*(n-1)}{\sqrt{E_m^b(n-2)}}\sqrt{\lambda}\frac{\tilde{e}_m^b(n-1)}{\sqrt{E_m^b(n-1)}} \tag{10.7.55}$$

If we define the quantities

$$\tilde{c}_m^b(n) \triangleq \sqrt{\frac{\lambda E_m^b(n-1)}{E_m^b(n)}} \tag{10.7.56}$$

$$\tilde{s}_m^b(n) \triangleq \frac{\tilde{e}_m^b(n)}{\sqrt{E_m^b(n)}} \tag{10.7.57}$$

and $$\tilde{k}_m^f(n) \triangleq -\frac{\beta_m^*(n)}{\sqrt{E_m^b(n-1)}} = k_m^f(n)\sqrt{E_m^b(n-1)} \tag{10.7.58}$$

we obtain

$$\tilde{e}_{m+1}^f(n) = \tilde{c}_m^b(n-1)\tilde{e}_m^f(n) + \sqrt{\lambda}\tilde{s}_m^b(n-1)\tilde{k}_m^f(n-1) \tag{10.7.59}$$

which provides the order updating of the angle normalized forward prediction error.

To obtain the update equation for the normalized coefficient $\tilde{k}_m^f(n)$, we start with

$$\beta_m(n) = \lambda\beta_m(n-1) + \alpha_m(n-1)e_m^b(n-1)e_m^{f*}(n) \tag{10.7.60}$$

and using (10.7.58), (10.7.53), and (10.7.54), we obtain

$$\tilde{k}_m^f(n) = \sqrt{\lambda}\tilde{k}_m^f(n-1)\sqrt{\frac{\lambda E_m^b(n-2)}{E_m^b(n-1)}} - \frac{\tilde{e}_m^{b*}(n-1)}{\sqrt{E_m^b(n-1)}}\tilde{e}_m^f(n)$$

or finally

$$\tilde{k}_m^{\mathrm{f}}(n) = \sqrt{\lambda}\,\tilde{c}_m^{\mathrm{b}}(n-1)\tilde{k}_m^{\mathrm{f}}(n-1) - \tilde{s}_m^{\mathrm{b}*}(n-1)\tilde{e}_m^{\mathrm{f}}(n) \tag{10.7.61}$$

with the help of (10.7.56) and (10.7.57).

Using the a priori lattice equation (10.7.48) for the backward predictor and the definitions of the angle normalized errors, we obtain

$$\tilde{e}_{m+1}^{\mathrm{b}}(n) = \sqrt{\frac{\alpha_{m+1}(n)}{\alpha_m(n-1)}}\,\tilde{e}_m^{\mathrm{b}}(n-1) - \frac{\beta_m(n-1)}{\sqrt{E_m^{\mathrm{f}}(n-1)}}\sqrt{\frac{\alpha_{m+1}(n)}{\alpha_m(n-1)}}\,\frac{\tilde{e}_m^{\mathrm{f}}(n)}{\sqrt{E_m^{\mathrm{f}}(n-1)}}$$

or

$$\tilde{e}_{m+1}^{\mathrm{b}}(n) = \sqrt{\frac{\lambda E_m^{\mathrm{f}}(n-1)}{E_m^{\mathrm{f}}(n)}}\,\tilde{e}_m^{\mathrm{b}}(n-1) - \frac{\beta_m(n-1)}{\sqrt{E_m^{\mathrm{f}}(n-1)}}\sqrt{\lambda}\,\frac{\tilde{e}_m^{\mathrm{f}}(n)}{\sqrt{E_m^{\mathrm{f}}(n)}} \tag{10.7.62}$$

by using (10.7.29). If we define the quantities

$$c_m^{\mathrm{f}}(n) \triangleq \frac{\lambda E_m^{\mathrm{f}}(n-1)}{E_m^{\mathrm{f}}(n)} \tag{10.7.63}$$

$$\tilde{s}_m^{\mathrm{f}}(n) \triangleq \frac{\tilde{e}_m^{\mathrm{f}}(n)}{\sqrt{E_m^{\mathrm{f}}(n)}} \tag{10.7.64}$$

and

$$\tilde{k}_m^{\mathrm{b}}(n) \triangleq -\frac{\beta_m^{\mathrm{b}}(n)}{\sqrt{E_m^{\mathrm{b}}(n)}} = k_m^{\mathrm{b}}(n)\sqrt{E_m^{\mathrm{f}}(n)} \tag{10.7.65}$$

we obtain

$$\tilde{e}_{m+1}^{\mathrm{b}}(n) = \tilde{c}_m^{\mathrm{f}}(n)\tilde{e}_m^{\mathrm{b}}(n-1) + \sqrt{\lambda}\,\tilde{s}_m^{\mathrm{f}}(n)\tilde{k}_m^{\mathrm{b}*}(n-1) \tag{10.7.66}$$

which provides the update equation for the angle normalized backward prediction error. The updating of $\tilde{k}_m^{\mathrm{b}}(n)$ is given by

$$\tilde{k}_m^{\mathrm{b}*}(n) = \sqrt{\lambda}\,\tilde{c}_m^{\mathrm{f}}(n)\tilde{k}_m^{\mathrm{b}*}(n-1) - \tilde{s}_m^{\mathrm{f}*}(n)\tilde{e}_m^{\mathrm{b}}(n-1) \tag{10.7.67}$$

and can be easily obtained, like (10.7.61), by combining (10.7.60) with (10.7.63) through (10.7.65).

Similar updatings can be easily derived for the ladder part of the filter. Indeed, using (10.7.49), the definitions of the angle normalized errors, and (10.7.30), we have

$$\tilde{e}_{m+1}(n) = \sqrt{\frac{\lambda E_m^{\mathrm{b}}(n-1)}{E_m^{\mathrm{b}}(n)}}\,\tilde{e}_m(n) - \frac{\beta_m^{c*}(n-1)}{\sqrt{E_{m-1}^{\mathrm{b}}(n-1)}}\sqrt{\lambda}\,\frac{\tilde{e}_m^{\mathrm{b}}(n)}{\sqrt{E_m^{\mathrm{b}}(n)}}$$

or

$$\tilde{e}_{m+1}(n) = \tilde{c}_m^{\mathrm{b}}(n)\tilde{e}_m(n) - \sqrt{\lambda}\,\tilde{s}_m^{\mathrm{b}}(n)\tilde{k}_m^{c*}(n-1) \tag{10.7.68}$$

where

$$\tilde{k}_m^c(n) \triangleq \frac{\beta_m^c(n)}{\sqrt{E_m^{\mathrm{b}}(n)}} = k_m^c(n)\sqrt{E_m^{\mathrm{b}}(n)} \tag{10.7.69}$$

is a normalized ladder coefficient. This coefficient can be updated by using the recursion

$$\tilde{k}_m^c(n) = \sqrt{\lambda}\,\tilde{c}_m^{\mathrm{b}}(n)\tilde{k}_m^c(n-1) + \tilde{s}_m^{\mathrm{b}}(n)\tilde{e}_m^*(n) \tag{10.7.70}$$

which can be obtained, like (10.7.61) and (10.7.67), by using (10.7.45) and related definitions.

If we define the normalized energies

$$\tilde{E}_m^{\mathrm{f}}(n) \triangleq \sqrt{E_m^{\mathrm{f}}(n)} \tag{10.7.71}$$

and

$$\tilde{E}_m^{\mathrm{b}}(n) \triangleq \sqrt{E_m^{\mathrm{b}}(n)} \tag{10.7.72}$$

we can easily show, using (10.7.41) and (10.7.42), that

$$\tilde{E}_m^{\mathrm{f}}(n) = \sqrt{\lambda}\tilde{c}_m^{\mathrm{f}}(n)\tilde{E}_m^{\mathrm{f}}(n-1) + \tilde{s}_m^{\mathrm{f}}(n)\tilde{e}_m^{\mathrm{f}*}(n) \tag{10.7.73}$$

and

$$\tilde{E}_m^{\mathrm{b}}(n) = \sqrt{\lambda}\tilde{c}_m^{\mathrm{b}}(n)\tilde{E}_m^{\mathrm{b}}(n-1) + \tilde{s}_m^{\mathrm{b}}(n)\tilde{e}_m^{\mathrm{b}*}(n) \tag{10.7.74}$$

which provide time updates for the normalized minimum energies. However, the following recursions

$$\tilde{E}_m^{\mathrm{f}}(n) = \{\lambda[\tilde{E}_m^{\mathrm{f}}(n-1)]^2 + |\tilde{e}_m^{\mathrm{f}}(n)|^2\}^{1/2} \tag{10.7.75}$$

$$\tilde{E}_m^{\mathrm{b}}(n) = \{\lambda[\tilde{E}_m^{\mathrm{b}}(n-1)]^2 + |\tilde{e}_m^{\mathrm{b}}(n)|^2\}^{1/2} \tag{10.7.76}$$

obtained from (10.7.41) and (10.7.42), provide more convenient updatings.

We now have a complete formulation of the LS lattice-ladder recursions using angle normalized errors. To see the meaning and significance of these recursions, we express them in matrix form as

$$\begin{bmatrix} \tilde{e}_{m+1}^{\mathrm{f}}(n) \\ \tilde{k}_m^{\mathrm{f}}(n) \end{bmatrix} = \begin{bmatrix} \tilde{c}_m^{\mathrm{b}}(n-1) & \tilde{s}_m^{\mathrm{b}}(n-1) \\ -\tilde{s}_m^{\mathrm{b}*}(n-1) & \tilde{c}_m^{\mathrm{b}}(n-1) \end{bmatrix} \begin{bmatrix} \tilde{e}_m^{\mathrm{f}}(n) \\ \sqrt{\lambda}\tilde{k}_m^{\mathrm{f}}(n-1) \end{bmatrix} \tag{10.7.77}$$

$$\begin{bmatrix} \tilde{e}_{m+1}^{\mathrm{b}}(n) \\ \tilde{k}_m^{\mathrm{b}*}(n) \end{bmatrix} = \begin{bmatrix} \tilde{c}_m^{\mathrm{f}}(n) & \tilde{s}_m^{\mathrm{f}}(n) \\ -\tilde{s}_m^{\mathrm{f}*}(n) & \tilde{c}_m^{\mathrm{f}}(n) \end{bmatrix} \begin{bmatrix} \tilde{e}_m^{\mathrm{b}}(n-1) \\ \sqrt{\lambda}\tilde{k}_m^{\mathrm{b}*}(n-1) \end{bmatrix} \tag{10.7.78}$$

$$\begin{bmatrix} \tilde{e}_{m+1}(n) \\ \tilde{k}_m^{c*}(n) \end{bmatrix} = \begin{bmatrix} \tilde{c}_m^{\mathrm{b}}(n) & -\tilde{s}_m^{\mathrm{b}}(n) \\ \tilde{s}_m^{\mathrm{b}*}(n) & \tilde{c}_m^{\mathrm{b}}(n) \end{bmatrix} \begin{bmatrix} \tilde{e}_m(n) \\ \sqrt{\lambda}\tilde{k}_m^{c*}(n-1) \end{bmatrix} \tag{10.7.79}$$

where we see that the updating of the forward predictor parameters and the ladder parameters involves the same matrix delayed by one sample. The different position of the minus sign, due to the different sign used in the definitions of $\tilde{k}_m^{\mathrm{f}}(n)$ and $\tilde{k}_m^c(n)$, is immaterial. Furthermore, it is straightforward to show that

$$|\tilde{c}_m^{\mathrm{f}}(n)|^2 + |\tilde{s}_m^{\mathrm{f}}(n)|^2 = 1 \tag{10.7.80}$$

and

$$|\tilde{c}_m^{\mathrm{b}}(n)|^2 + |\tilde{s}_m^{\mathrm{b}}(n)|^2 = 1 \tag{10.7.81}$$

which imply that the matrices in (10.7.77) through (10.7.79) are the Givens rotation matrices. Therefore, we have obtained a formulation of the LS lattice-ladder algorithm that updates the angle normalized errors and a set of normalized lattice-ladder coefficients using the Givens rotations. Using (10.7.76) and definitions of $\tilde{c}_m^{\mathrm{b}}(n)$ and $\tilde{s}_m^{\mathrm{b}}(n)$, we can show that

$$\begin{bmatrix} \tilde{E}_m^{\mathrm{b}}(n) \\ 0 \end{bmatrix} = \begin{bmatrix} \tilde{c}_m^{\mathrm{b}}(n) & \tilde{s}_m^{\mathrm{b}*}(n) \\ -\tilde{s}_m^{\mathrm{b}}(n) & \tilde{c}_m^{\mathrm{b}}(n) \end{bmatrix} \begin{bmatrix} \sqrt{\lambda}\tilde{E}_m^{\mathrm{b}}(n-1) \\ \tilde{e}_m^{\mathrm{b}}(n) \end{bmatrix} \tag{10.7.82}$$

which shows that we can use the BLP Givens rotation to update the normalized energy $\tilde{E}_m^{\mathrm{b}}(n)$. A similar transformation can be obtained for $\tilde{E}_m^{\mathrm{f}}(n)$. However, the energy updatings are usually performed using (10.7.75) and (10.7.76).

The square root–free version of the Givens LS lattice-ladder filter is basically a simple modification of the error feedback form of the a priori LS lattice-ladder algorithm. Indeed, using (10.7.50), we have

$$k_m^c(n) = \frac{\lambda E_m^{\mathrm{b}}(n-1)}{E_m^{\mathrm{b}}(n)}k_m^c(n-1) + \frac{\alpha_m(n)e_m^{\mathrm{b}}(n)}{E_m^{\mathrm{b}}(n)}e_m^*(n)$$

or if we define the quantities

$$c_m^{\mathrm{b}}(n) \triangleq \frac{\lambda E_m^{\mathrm{b}}(n-1)}{E_m^{\mathrm{b}}(n)} = \left|\tilde{c}_m^{\mathrm{b}}(n)\right|^2 \tag{10.7.83}$$

and

$$s_m^{\mathrm{b}}(n) \triangleq \frac{\alpha_m(n)e_m^{\mathrm{b}}(n)}{E_m^{\mathrm{b}}(n)} \tag{10.7.84}$$

we obtain

$$k_m^c(n) = c_m^b(n)k_m^c(n-1) + s_m^b(n)e_m^*(n) \tag{10.7.85}$$

which provides the required updating for the ladder parameters.

Similarly, using the error feedback a priori updatings for the lattice parameters, we obtain the recursions

$$k_m^f(n) = c_m^b(n-1)k_m^f(n-1) - s_m^b(n-1)e_m^{f*}(n) \tag{10.7.86}$$

and

$$k_m^b(n) = c_m^f(n)k_m^b(n-1) - s_m^f(n)e_m^{b*}(n-1) \tag{10.7.87}$$

where

$$c_m^f(n) \triangleq \frac{\lambda E_m^f(n-1)}{E_m^f(n)} = |\tilde{c}_m^f(n)|^2 \tag{10.7.88}$$

and

$$s_m^f(n) \triangleq \frac{\alpha_m(n-1)e_m^f(n)}{E_m^f(n)} \tag{10.7.89}$$

are the forward rotation parameters. These recursions constitute the basis for the square root–free Givens LS lattice-ladder algorithm.

Table 10.18 provides the complete computational organizations of the Givens LS lattice-ladder algorithms with and without square roots. The square root algorithm is initialized as

TABLE 10.18
Summary of the Givens LS lattice-ladder adaptive filter algorithms.

Equation	Square root form	Square root–free form
	Forward rotation parameters	
(a)	$\tilde{E}_m^f(n) = \{\lambda[\tilde{E}_m^f(n-1)]^2 + |\tilde{e}_m^f(n)|^2\}^{1/2}$	$E_m^f(n) = \lambda E_m^f(n-1) + \alpha_m(n-1)|e_m^f(n)|^2$
(b)	$\tilde{c}_m^f(n) = \dfrac{\sqrt{\lambda}\tilde{E}_m^f(n-1)}{\tilde{E}_m^f(n)}$	$c_m^f(n) = \dfrac{\lambda E_m^f(n-1)}{E_m^f(n)}$
(c)	$\tilde{s}_m^f(n) = \dfrac{\tilde{e}_m^f(n)}{\tilde{E}_m^f(n)}$	$s_m^f(n) = \dfrac{\alpha_m(n-1)e_m^f(n)}{E_m^f(n)}$
	Backward Rotation Parameters	
(d)	$\tilde{E}_m^b(n) = \{\lambda[\tilde{E}_m^b(n-1)]^2 + |\tilde{e}_m^b(n)|^2\}^{1/2}$	$E_m^b(n) = \lambda E_m^b(n-1) + \alpha_m(n)|e_m^b(n)|^2$
(e)	$\tilde{c}_m^b(n) = \dfrac{\sqrt{\lambda}\tilde{E}_m^b(n-1)}{\tilde{E}_m^b(n)}$	$c_m^b(n) = \dfrac{\lambda E_m^b(n-1)}{E_m^b(n)}$
(f)	$\tilde{s}_m^b(n) = \dfrac{\tilde{e}_m^b(n)}{\tilde{E}_m^b(n)}$	$s_m^b(n) = \dfrac{\alpha_m(n)e_m^b(n)}{E_m^b(n)}$
	Forward predictor rotator	
(g)	$\tilde{e}_{m+1}^f(n) = \tilde{c}_m^b(n-1)\tilde{e}_m^f(n) + \sqrt{\lambda}\tilde{s}_m^b(n-1)\tilde{k}_m^f(n-1)$	$e_{m+1}^f(n) = e_m^f(n) + k_m^{f*}(n-1)e_m^b(n-1)$
(h)	$\tilde{k}_m^f(n) = \sqrt{\lambda}\tilde{c}_m^b(n-1)\tilde{k}_m^f(n-1) - \tilde{s}_m^{b*}(n-1)\tilde{e}_m^f(n)$	$k_m^f(n) = c_m^b(n-1)k_m^f(n-1) - s_m^b(n-1)e_m^{f*}(n)$
	Backward predictor rotator	
(i)	$\tilde{e}_{m+1}^b(n) = \tilde{c}_m^f(n)\tilde{e}_m^b(n-1) + \sqrt{\lambda}\tilde{s}_m^f(n)\tilde{k}_m^{b*}(n-1)$	$e_{m+1}^b(n) = e_m^b(n-1) + k_m^{b*}(n-1)e_m^f(n)$
(j)	$\tilde{k}_m^{b*}(n) = \sqrt{\lambda}\tilde{c}_m^f(n)\tilde{k}_m^{b*}(n-1) - \tilde{s}_m^{f*}(n)\tilde{e}_m^b(n-1)$	$k_m^b(n) = c_m^f(n)k_m^b(n-1) - s_m^f(n)e_m^{b*}(n-1)$
	Filter rotator	
(k)	$\tilde{e}_{m+1}(n) = \tilde{c}_m^b(n)\tilde{e}_m(n) - \sqrt{\lambda}\tilde{s}_m^b(n)\tilde{k}_m^{c*}(n-1)$	$e_{m+1}(n) = e_m(n) - k_m^{c*}(n-1)e_m^b(n)$
(l)	$\tilde{k}_m^c(n) = \sqrt{\lambda}\tilde{c}_m^b(n)\tilde{k}_m^c(n-1) + \tilde{s}_m^b(n)\tilde{e}_m^*(n)$	$k_m^c(n) = c_m^b(n)k_m^c(n-1) + s_m^b(n)e_m^*(n)$

usual with $E_m^f(-1) = E_m^b(-1) = \delta > 0$, $\tilde{e}_0^f(n) = \tilde{e}_0^b(n) = x(n)$, $\tilde{e}_0(n) = y(n)$, $\alpha_0(n) = 1$, and all other variables set to zero. The square root–free algorithm is initialized as the a priori algorithm with error feedback. Figure 10.39 shows a single stage of the LS lattice-ladder filter based on Givens rotations with square roots.

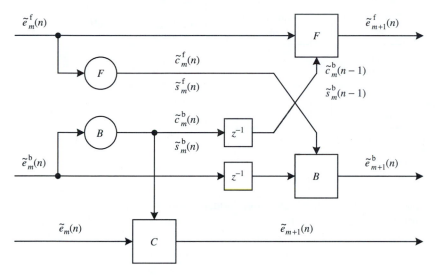

FIGURE 10.39
Block diagram representation of the Givens RLS lattice-ladder stage. Circles denote computing elements that calculate the rotation parameters and squares denote computing elements that perform the rotations.

10.7.5 Classification of RLS Algorithms for FIR Filtering

Every exact RLS algorithm discussed in this section consists of two parts: a part that computes the LS forward and backward predictors of the input signal and a part that uses information from the linear prediction part to compute the LS filter. In all cases, information flows from the prediction part to the filtering part, but not vice versa. Therefore, all critical numerical operations take place in the linear prediction section.

For direct-form structures, the prediction problems facilitate the fast computation of the RLS gain vectors.

In the case of lattice-ladder structures, the lattice part (which again solves the linear prediction problem) decorrelates (or orthogonalizes in the LS sense) the input signal vector and creates an orthogonal base consisting of the backward prediction errors $\{e_m^b(n)\}_0^{M-1}$. This orthogonal basis is used by the ladder part to form the LS filtering error. Essentially, the LS lattice part facilitates the triangular UDL decomposition of the inverse correlation matrix $\hat{\mathbf{R}}^{-1}(n)$ or the Gram-Schmidt orthogonalization of the columns of data matrix $\mathbf{X}(n)$. This property makes the RLS lattice-ladder algorithm order-recursive, like its minimum MSE counterpart (see Section 7.3).

The QRD-RLS lattice-ladder algorithms also consist of a lattice part that solves the linear prediction problem and a ladder part that uses information from the lattice to form the LS filtering estimate. The LS lattice produces the triangularization of the inverse correlation matrix $\hat{\mathbf{R}}^{-1}(n)$ whereas the QRD LS lattice produces the upper triangular Cholesky factor of $\hat{\mathbf{R}}(n)$ by applying an orthogonal transformation to data matrix $\mathbf{X}(n)$.

The correspondence of these algorithms to their counterparts for RLS array processing, discussed in Section 10.6, is summarized in Figure 10.40.

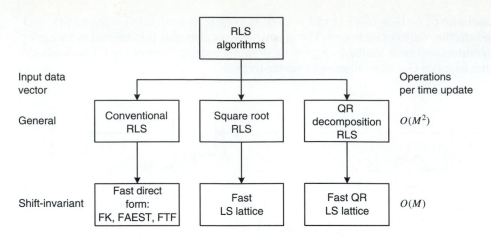

FIGURE 10.40
Classification of RLS algorithms for array processing and FIR filtering.

It is interesting to note that the RLS lattice-ladder algorithms with error feedback are identical in form to the square root–free Givens rotation–based QRD-RLS lattice-ladder algorithms. This similarity explains the excellent numerical properties of both structures.

The RLS lattice-ladder algorithms (both UDL^H-decomposition based and QR-decomposition based) share the following highly desirable characteristics:

- *Good numerical properties* that originate from the square root decomposition (Cholesky or QR) part of the algorithms.
- *Good convergence properties*, which are inherited from the exact LS minimization performed by all algorithms.
- *Modularity and regularity* that make possible their VLSI and multiprocessing implementation.

It has been shown (Ljung and Ljung 1985) that all RLS lattice-ladder algorithms are numerically stable for $\lambda < 1$. However, they differ in terms of numerical accuracy. It turns out that the lattice-ladder algorithms with error feedback (which are basically equivalent to the square root–free QRD lattice ladder) and the QRD lattice-ladder algorithms have the best numerical accuracy.

10.8 TRACKING PERFORMANCE OF ADAPTIVE ALGORITHMS

Tracking of a time-varying system is an important problem in many areas of application. Consider, for example, a digital communications system in which the channel characteristics may change with time for various reasons. If we want to incorporate an echo canceler in such a system, then clearly the echo canceler must monitor the changing impulse response of the echo path so that it can generate an accurate replica of the echo. This will require the adaptive algorithm of an echo canceler to possess an acceptable tracking capability. Similar situations arise in adaptive equalization, adaptive prediction, adaptive noise canceling, and so on. In all these applications, adaptive filters are forced to operate in a *nonstationary* SOE. In this section, we examine the ability and performance of the LMS and RLS algorithms to track the ever-changing minimum point of the error surface.

As discussed earlier, the tracking mode is a steady-state operation of the adaptive algorithm, and it follows the acquisition mode, which is a transient phenomenon. Therefore, the algorithm must acquire the system parameters before tracking can commence. This has two implications. First, the rate of convergence is generally not related to the tracking behavior, and as such, we analyze the tracking behavior when the number of iterations (or steps) is relatively large. Second, the time variation of the parameter change should be small enough compared to the rate of convergence that the algorithm can perform adequate tracking; otherwise, it is constantly acquiring the parameters.

10.8.1 Approaches for Nonstationary SOE

To effectively track a nonstationary SOE, adaptive algorithms should use only *local* statistics. There are three practical ways in which this can be achieved.

Exponentially growing window

In this approach, the current data are artificially emphasized by exponentially weighting past data values, as shown in Figure 10.41(a). The error function that is minimized is given by

$$E(n) = \sum_{j=0}^{n} \lambda^{n-j} |y(j) - \mathbf{c}^H \mathbf{x}(j)|^2 = \lambda E(n-1) + |y(n) - \mathbf{c}^H \mathbf{x}(n)|^2 \qquad (10.8.1)$$

where $0 < \lambda < 1$. Clearly, this is the cost function we used in the development of the RLS algorithm, given in Table 10.6, in which λ is termed the forgetting factor. The effective window length is given by

$$L_{\text{eff}} \triangleq \frac{\displaystyle\sum_{n=0}^{\infty} \lambda^n}{\lambda^0} = \frac{1}{1 - \lambda} \qquad (10.8.2)$$

Hence for good tracking performance λ should be in the range $0.9 \leq \lambda < 1$. Note that $\lambda = 1$ results in a *rectangularly growing* window that uses global statistics and hence will not be able to track parameter changes. Thus the RLS algorithm with exponential forgetting is capable of using the local information needed to adapt in a nonstationary SOE.

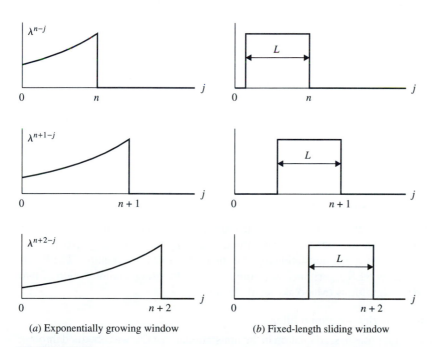

(a) Exponentially growing window   (b) Fixed-length sliding window

FIGURE 10.41
Illustration of exponentially growing and fixed-length sliding windows.

Fixed-length sliding window

The basic feature of this approach is that the parameter estimates are based only on a finite number of past data values, as shown in Figure 10.41(b). Let us consider a rectangular window of fixed length $L > M$. Then the cost function that is minimized is given by

$$E(n, L) \triangleq \sum_{j=n-L+1}^{n} |y(j) - \mathbf{c}^H \mathbf{x}(j)|^2 \tag{10.8.3}$$

When a new data value at $n+1$ is added to the sum in (10.8.3), the old data value is discarded, that is, all old data values beyond $n - L + 1$ are discarded. Thus the active number of data values is always a constant equal to L, which makes this as a constant-memory adaptive algorithm. By following the steps given for the RLS adaptive filter in Section 10.5, it is possible to derive a recursive algorithm to determine the filter $\mathbf{c}(n)$ that minimizes the error function in (10.8.3).

Let $\mathbf{c}_{\{n-L\}}(n - 1)$ denote the estimate of $\mathbf{c}(n - 1)$ based on L data values between $n - L$ and $n - 1$. After the new data value at n is observed, the RLS algorithm in Table 10.6 is applicable with $\lambda = 1$ and with obvious extension of notation. Hence we obtain the algorithm

$$\mathbf{c}_{\{n-L\}}(n) = \mathbf{c}_{\{n-L\}}(n - 1) + \mathbf{g}_{\{n-L\}}(n)e^*(n) \tag{10.8.4}$$

$$e(n) = y(n) - \mathbf{c}_{\{n-L\}}^H(n - 1)\mathbf{x}(n) \tag{10.8.5}$$

$$\mathbf{g}_{\{n-L\}}(n) = \frac{\bar{\mathbf{g}}_{\{n-L\}}(n)}{\alpha_{\{n-L\}}(n)} \tag{10.8.6}$$

$$\bar{\mathbf{g}}_{\{n-L\}}(n) = \mathbf{P}_{\{n-L\}}(n - 1)\mathbf{x}(n) \tag{10.8.7}$$

$$\alpha_{\{n-L\}}(n) = 1 + \bar{\mathbf{g}}_{\{n-L\}}^H(n)\mathbf{x}(n) \tag{10.8.8}$$

$$\mathbf{P}_{\{n-L\}}(n) = \mathbf{P}_{\{n-L\}}(n - 1) - \mathbf{g}_{\{n-L\}}(n)\bar{\mathbf{g}}_{\{n-L\}}^H(n) \tag{10.8.9}$$

The above algorithm is based on $L + 1$ data values. To maintain the data window at fixed length L, we have to discard the observation at $n - L$. By using the matrix inversion lemma given in Appendix A, it can be shown that (see Problem 10.51)

$$\mathbf{c}_{\{n-L+1\}}(n) = \mathbf{c}_{\{n-L\}}(n) - \mathbf{g}_{\{n-L+1\}}(n)e^*(n - L) \tag{10.8.10}$$

$$e(n - L) = y(n - L) - \mathbf{c}_{\{n-L\}}^H(n)\mathbf{x}(n - L) \tag{10.8.11}$$

$$\mathbf{g}_{\{n-L+1\}}(n) = \frac{\bar{\mathbf{g}}_{\{n-L+1\}}(n)}{\alpha_{\{n-L+1\}}(n)} \tag{10.8.12}$$

$$\bar{\mathbf{g}}_{\{n-L+1\}}(n) = \mathbf{P}_{\{n-L\}}(n)\mathbf{x}(n - L) \tag{10.8.13}$$

$$\alpha_{\{n-L+1\}}(n) = 1 - \bar{\mathbf{g}}_{\{n-L+1\}}^H(n)\mathbf{x}(n - L) \tag{10.8.14}$$

$$\mathbf{P}_{\{n-L+1\}}(n) = \mathbf{P}_{\{n-L\}}(n) + \mathbf{g}_{\{n-L+1\}}(n)\bar{\mathbf{g}}_{\{n-L+1\}}^H(n) \tag{10.8.15}$$

The overall algorithm for the fixed-memory rectangular window adaptive algorithm is given by (10.8.4) through (10.8.15), which recursively update $\mathbf{c}_{\{n-L\}}(n - 1)$ to $\mathbf{c}_{\{n-L+1\}}(n)$. Thus, this algorithm can adapt to the nonstationary SOE using the local information. The fixed-length sliding-window RLS algorithm can be implemented by using a combination of two prewindowed RLS algorithms (Manolakis et al. 1987).

Evolutionary model—Kalman filter

In the first two approaches, adaptation in the nonstationarity SOE was obtained through the local information, either by discarding old data or by deemphasizing it. In the third approach, we assume that we have a statistical model that describes the nonstationarity of the SOE. This model is in the form of a stochastic difference equation together with appropriate statistical properties. This leads to the well-known Kalman filter formulation in which we assume that the parameter variations are modeled by

$$\mathbf{c}(n) = \boldsymbol{\Xi}(n)\mathbf{c}(n-1) + \boldsymbol{v}(n) \tag{10.8.16}$$

where $\boldsymbol{v}(n)$ is a random vector with zero mean and correlation matrix $\boldsymbol{\Sigma}(n)$, and $\boldsymbol{\Xi}(n)$ is the state-transition matrix known for all n. The desired signal $y(n)$ is modeled as

$$y(n) = \mathbf{c}^H(n)\mathbf{x}(n) + \varepsilon(n) \tag{10.8.17}$$

where $\varepsilon(n)$ is the a posteriori estimation error assumed to be zero-mean with variance σ_ε^2. Thus in this formulation, the parameter vector $\mathbf{c}(n)$ acts as the state of a system while the input data vector $\mathbf{x}(n)$ acts as the time-varying output vector. Now the best linear unbiased estimate $\hat{\mathbf{c}}(n)$ of $\mathbf{c}(n)$ based on past observations $\{y(i)\}_{i=0}^n$ can be obtained by using the Kalman filter equations (Section 7.8). These recursive equations are given by

$$\hat{\mathbf{c}}(n) = \boldsymbol{\Xi}(n)\hat{\mathbf{c}}(n-1) + \mathbf{g}(n)[y(n) - \hat{\mathbf{c}}^H(n-1)\boldsymbol{\Xi}^H(n)\mathbf{x}(n)] \tag{10.8.18}$$

$$\mathbf{g}(n) = \frac{\boldsymbol{\Xi}(n)\mathbf{P}(n-1)\mathbf{x}(n)}{\sigma_\varepsilon^2 + \mathbf{x}^H(n)\mathbf{P}(n-1)\mathbf{x}(n)} \tag{10.8.19}$$

$$\mathbf{P}(n) = \boldsymbol{\Xi}(n)\mathbf{P}(n-1)\boldsymbol{\Xi}^H(n) + \boldsymbol{\Sigma}(n)$$

$$-\boldsymbol{\Xi}(n)\mathbf{P}(n-1)\frac{\mathbf{x}(n)\mathbf{x}^H(n)}{\sigma_\varepsilon^2 + \mathbf{x}^H(n)\mathbf{P}(n-1)\mathbf{x}(n)}\mathbf{P}(n-1)\boldsymbol{\Xi}^H(n) \tag{10.8.20}$$

where $\mathbf{g}(n)$ is the Kalman gain matrix and $\mathbf{P}(n)$ is the error covariance matrix. This approach implies that if the time-varying parameters are modeled as state equations, then the Kalman filter rather than the adaptive filter is a proper solution.

Furthermore, it can be shown that the Kalman filter has a close similarity to the RLS adpative filters if we make the following appropriate substitutions:

Exponential memory: If we substitute

$$\boldsymbol{\Xi}(n) = \mathbf{I} \qquad \sigma_\varepsilon^2 = \lambda \qquad \boldsymbol{\Sigma}(n) = \frac{1-\lambda}{\lambda}[\mathbf{I} - \mathbf{g}(n)\mathbf{x}^H(n)]\mathbf{P}(n-1) \tag{10.8.21}$$

then we obtain the exponential memory RLS algorithm given in Table 10.6.
Rectangularly growing memory: If we substitute

$$\boldsymbol{\Xi}(n) = \mathbf{I} \qquad \sigma_\varepsilon^2 = 1 \qquad \boldsymbol{\Sigma}(n) = 0 \tag{10.8.22}$$

then we obtain the rectangularly growing memory RLS algorithm.

10.8.2 Preliminaries in Performance Analysis

In Sections 10.4 and 10.5.4, we developed and analyzed the LMS and RLS algorithms in stationary environments, respectively. However, these algorithms are generally used in applications (e.g., modems) that are intended to operate continuously in SOE whose characteristics change with time. Therefore, we need to discuss the performance of these two widely used algorithms in such situations. Although we provided various adaptive filtering approaches for time-varying environments above, we now discuss, in the remainder of this section, the ability of these two algorithms to track time-varying parameters. We provide both analytical results, assuming a model of parameter variation, and experimental results, using simulations.

A popular approach for this analytical assessment is to assume a first-order AR model with finite variance [that is we set $\boldsymbol{\Xi}(n) = \rho\mathbf{I}$ in (10.8.16)]. Although higher-order models are also possible, only a few results on the tracking performance using these models are currently available. It is ironic that most analytical results on the tracking performance have been obtained for the random-walk model (a special case of the first-order AR model), which is unrealistic because of the infinite variance. A tutorial review of the latest results for the general case and additional references are available in Macchi (1996).

In our analysis of tracking characteristics of the LMS and RLS algorithms, we use the first-order AR model and discuss its effect on the tracking performance. The closed-form results will be given using the random-walk model and confirmed using simulated experiments.

Analysis setup

In the tracking analysis, it is desirable to use the *a priori* adaptive filter. Hence we assume that the desired response is generated by the following filter model[†]

$$y(n) = \mathbf{c}_o^H(n-1)\mathbf{x}(n) + v(n) \tag{10.8.23}$$

where $v(n)$ is assumed to be WGN$(0, \sigma_v^2)$ with $\sigma_v^2 < \infty$. The random processes $\mathbf{x}(n)$ and $v(n)$ are assumed to be independent and stationary. The variation of $\mathbf{c}_o(n)$ is modeled by the first-order AR (or Markov) process

$$\mathbf{c}_o(n) = \rho\mathbf{c}_o(n-1) + \boldsymbol{\psi}(n) \tag{10.8.24}$$

with $0 < \rho < 1$ and creates the nonstationarity of the SOE. The quantity $\boldsymbol{\psi}(n)$ is the uncertainty in the model and assumed to be independent of $\mathbf{x}(n)$ and $v(n)$, with mean $E\{\boldsymbol{\psi}(n)\} = \mathbf{0}$ and correlation $E\{\boldsymbol{\psi}(n)\boldsymbol{\psi}^H(n)\} = \mathbf{R}_\psi$. Tracking is generally achievable if ρ is close to 1. The random-walk model is obtained by using $\rho = 1$ in (10.8.24).

Conjugate transposing and premultiplying both sides of (10.8.23) by $\mathbf{x}(n)$, taking the expectation, and using independence between $\mathbf{x}(n)$ and $v(n)$, we obtain

$$\mathbf{R}\mathbf{c}_o(n-1) = \mathbf{d}(n) \tag{10.8.25}$$

Hence, $\mathbf{c}_o(n-1)$ is the optimum a priori filter and

$$e_o(n) = y(n) - \mathbf{c}_o^H(n-1)\mathbf{x}(n) = v(n) \tag{10.8.26}$$

is the *optimum a priori error*. If $\mathbf{R}_\psi = \mathbf{0}$ and $\rho = 1$, we have $\mathbf{c}_o(n) = \mathbf{c}_o$ for all n, and therefore $y(n)$ is wide-sense stationary (WSS). In this case, we have a stationary environment, and the goal of the adaptive filter is to find the optimum filter $\mathbf{c}_o$. For $\mathbf{R}_\psi \neq \mathbf{0}$, the adaptive filter should find and track the optimum a priori filter $\mathbf{c}_o(n)$. This setup, which is widely used to analyze the properties of adaptive algorithms, is illustrated in Figure 10.42.

Assumptions

To analyze the tracking performance of adaptive algorithms, we use the assumptions discussed elsewhere and repeated below for convenience.

A1 The sequence of input data vectors $\mathbf{x}(n)$ is WGN$(\mathbf{0}, \mathbf{R})$.

A2 The desired response $y(n)$ can be modeled as

$$y(n) = \mathbf{c}_o^H(n-1)\mathbf{x}(n) + e_o(n) \tag{10.8.27}$$

where $e_o(n)$ is WGN$(0, \sigma_o^2)$.

A3 The time variation of $\mathbf{c}_o(n)$ is described by

$$\mathbf{c}_o(n) = \rho\mathbf{c}_o(n-1) + \boldsymbol{\psi}(n) \tag{10.8.28}$$

where $0 \leq \rho \leq 1$ and $\boldsymbol{\psi}(n)$ is WGN$(\mathbf{0}, \mathbf{R}_\Psi)$.

A4 The random sequences $\mathbf{x}(n)$, $e_o(n)$, and $\boldsymbol{\psi}(n)$ are mutually independent.

Through these assumptions, we want to stress that the nonstationarity of the SOE is created solely by $\mathbf{c}_o(n)$ and not by $\mathbf{x}(n)$, which is WSS.

[†]We use this model to make a fair comparison between the adaptive and the optimum filter.

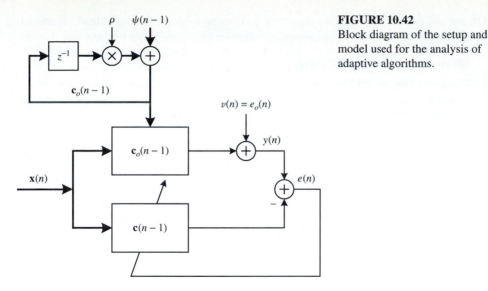

FIGURE 10.42
Block diagram of the setup and
model used for the analysis of
adaptive algorithms.

595

SECTION 10.8
Tracking Performance of
Adaptive Algorithms

Although we provide analysis for (10.8.27), many results are given for the random walk model ($\rho = 1$). The case $0 < \rho < 1$, which is straightforward but complicated, is discussed in Solo and Kong (1995). Before we delve into this analysis, we discuss criteria that are used for evaluating the tracking performance.

Degree of nonstationarity

To determine whether an adaptive algorithm can adequately track the changing SOE, one needs to define the speed of variation of the statistics of the adaptive filter environment. This speed is quantified in terms of the *degree of nonstationarity* (DNS), introduced in Macchi (1995, 1996), and is defined by

$$\eta(n) \triangleq \sqrt{\frac{E\{|y_{o,\text{incr}}(n)|^2\}}{P_o(n)}} \tag{10.8.29}$$

where

$$y_{o,\text{incr}}(n) = [\mathbf{c}_o(n) - \mathbf{c}_o(n-1)]^H \mathbf{x}(n) \tag{10.8.30}$$

is the output of the *incremental* filter. The numerator is the power introduced by the variation of the optimum filter, and the denominator is the MMSE, which in the context of (10.8.26) is equal to the power of the output noise. Assuming $\rho = 1$ in (10.8.28), we see that (10.8.30) is given by

$$y_{o,\text{incr}}(n) = \mathbf{\Psi}^H \mathbf{x}(n)$$

and hence the numerator in (10.8.29) is given by

$$\begin{aligned} E\{|y_{o,\text{incr}}(n)|^2\} &= E\{\mathbf{\Psi}^H \mathbf{x}(n)\mathbf{x}^H(n)\mathbf{\Psi}\} = \text{tr}[E\{\mathbf{\Psi}^H \mathbf{x}(n)\mathbf{x}^H(n)\mathbf{\Psi}\}] \\ &= \text{tr}[E\{\mathbf{\Psi}\mathbf{\Psi}^H \mathbf{x}(n)\mathbf{x}^H t\}] = \text{tr}[E\{\mathbf{\Psi}\mathbf{\Psi}^H\}E\{\mathbf{x}(n)\mathbf{x}^H\}] \\ &= \text{tr}[\mathbf{R}_\Psi \mathbf{R}] = \text{tr}[\mathbf{R}\mathbf{R}_\Psi] \end{aligned} \tag{10.8.31}$$

where we have used the independence assumption A4. Substituting (10.8.31) in (10.8.29), we obtain

$$\eta(n) \triangleq \sqrt{\frac{\text{tr}[\mathbf{R}\mathbf{R}_\Psi]}{P_o(n)}} \tag{10.8.32}$$

Smaller values of η ($\ll 1$) imply that the adaptive algorithm can track time variations of the nonstationary SOE. On the contrary, if $\eta > 1$, then the statistical variations of the

SOE are too fast for the adaptive algorithm to keep up with the SOE and lead to massive misadjustment errors. In such situations, an adaptive filter should not be used.

Mean square deviation (MSD)

We defined the MSD $\mathcal{D}(n)$ in (10.2.29) as a performance measure for adaptive filters in the steady-state environment. It is also used for measuring the tracking performance. Consider the coefficient error vector $\tilde{\mathbf{c}}(n)$, which can be written as

$$\tilde{\mathbf{c}}(n) = \mathbf{c}(n) - \mathbf{c}_o(n)$$
$$= [\mathbf{c}(n) - E\{\mathbf{c}(n)\}] + [E\{\mathbf{c}(n)\} - \mathbf{c}_o(n)] \tag{10.8.33}$$
$$\triangleq \tilde{\mathbf{c}}_1(n) + \tilde{\mathbf{c}}_2(n) \tag{10.8.34}$$

where $\tilde{\mathbf{c}}_1(n)$ is the fluctuation of the adaptive filter parameter vector about its mean (estimation error) and $\tilde{\mathbf{c}}_2(n)$ is the bias of $\mathbf{c}(n)$ with respect to the true vector $\mathbf{c}_o(n)$ (systematic or lag error). Using the independence assumption of the previous section that $\mathbf{x}(n)$ and $\mathbf{c}(n-1)$ are statistically independent, we can show that (Macchi 1996)

$$E\{\tilde{\mathbf{c}}_1^H(n)\tilde{\mathbf{c}}_2(n)\} = 0 \tag{10.8.35}$$

which by using (10.2.29) and (10.8.34) leads to

$$\mathcal{D}(n) = \mathcal{D}_1(n) + \mathcal{D}_2(n) \tag{10.8.36}$$

The first MSD term is due to the parameter estimation error and is called the *estimation variance*. The second MSD term is due to the parameter lag error and is termed *lag variance*, and its presence indicates the nonstationary environment.

Misadjustment and lowest excess MSE

The second performance measure, defined in (10.2.38), is the (a priori) misadjustment $\mathcal{M}(n)$, which is the ratio of the excess MSE $P_{\text{ex}}(n)$ to the MMSE $P_o(n)$. The a priori excess MSE is given by

$$P_{\text{ex}}(n) = E\{|\tilde{\mathbf{c}}^H(n-1)\mathbf{x}(n)|^2\} = E\{|\tilde{\mathbf{c}}_1^H(n-1)\mathbf{x}(n) + \tilde{\mathbf{c}}_2^H(n-1)\mathbf{x}(n)|^2\} \tag{10.8.37}$$

which under the independence assumption and (10.8.35) can be written as

$$P_{\text{ex}}(n) = P_{\text{ex},1}(n) + P_{\text{ex},2}(n) \tag{10.8.38}$$

where the first term, $P_{\text{ex},1}(n)$, is excess MSE due to estimation error and is termed the *estimation noise* while the second term, $P_{\text{ex},2}(n)$, is the excess MSE due to lag error and is called the *lag noise*. Therefore, we can also write the misadjustment $\mathcal{M}(n)$ as

$$\mathcal{M}(n) = \mathcal{M}_1(n) + \mathcal{M}_2(n) \tag{10.8.39}$$

where $\mathcal{M}_1(n)$ is the *estimation misadjustment* and $\mathcal{M}_2(n)$ is the *lag misadjustment*.

In the context of the first-order Markov model, the best performance obtained by any a priori adaptive filter occurs if $\mathbf{c}(n) = \rho\mathbf{c}_o(n-1)$. This observation makes possible the computation of a lower bound for the excess MSE of any a priori adaptive algorithm. From (10.8.34) and (10.8.24), we have

$$\tilde{\mathbf{c}}(n) = \mathbf{c}(n) - \mathbf{c}_o(n) = [\mathbf{c}(n) - \rho\mathbf{c}_o(n-1)] - \boldsymbol{\psi}(n)$$
$$\triangleq \hat{\mathbf{c}}(n) - \boldsymbol{\psi}(n) \tag{10.8.40}$$

and hence

$$P_{\text{ex}}(n) = E\{|\tilde{\mathbf{c}}^H(n-1)\mathbf{x}(n)|^2\}$$
$$= E\{|\hat{\mathbf{c}}^H(n-1)\mathbf{x}(n) - \boldsymbol{\psi}^H(n-1)\mathbf{x}(n)|^2\} \tag{10.8.41}$$
$$= E\{|\hat{\mathbf{c}}^H(n-1)\mathbf{x}(n)|^2\} + E\{|\boldsymbol{\psi}^H(n-1)\mathbf{x}(n)|^2\}$$
$$+ 2E\{\hat{\mathbf{c}}^H(n-1)\mathbf{x}(n)\mathbf{x}^H(n)\boldsymbol{\psi}(n-1)\} \tag{10.8.42}$$

Since the term $\hat{\mathbf{c}}(n)$ does not depend on $\boldsymbol{\psi}(n)$ and since the random sequences $\mathbf{x}(n)$ and $\boldsymbol{\psi}(n-1)$ are assumed independent, the last term in (10.8.42) is zero. Hence,

$$P_{\text{ex}}(n) \geq E\{|\boldsymbol{\psi}^H(n-1)\mathbf{x}(n)|^2\} \tag{10.8.43}$$

which provides a lower bound for the excess MSE of any a priori adaptation algorithm. Because $\boldsymbol{\psi}(n)$ and $\mathbf{x}(n)$ are assumed independent, we obtain

$$E\{|\boldsymbol{\psi}^H(n-1)\mathbf{x}(n)|^2\} = \text{tr}(\mathbf{R}\mathbf{R}_\psi) \tag{10.8.44}$$

Similarly, neglecting the dependence between $\mathbf{x}(n)$ and $\tilde{\mathbf{c}}(n-1)$, we have

$$E\{|\tilde{\mathbf{c}}^H(n-1)\mathbf{x}(n)|^2\} = \text{tr}[\mathbf{R}\boldsymbol{\Phi}(n-1)] \tag{10.8.45}$$

which provides the a priori excess MSE. Furthermore, it can be shown that the DNS places a lower limit on the misadjustment, that is,

$$\mathcal{M}(n) = \frac{P_{\text{ex}}(n)}{P_o(n)} \geq \frac{E\{|\boldsymbol{\psi}^H(n-1)\mathbf{x}(n)|^2\}}{P_o(n)} = \frac{\text{tr}(\mathbf{R}\mathbf{R}_\psi)}{\sigma_v^2} = \eta^2(n) \tag{10.8.46}$$

10.8.3 LMS Algorithm

Using the LMS algorithm (10.4.12), the error vector in (10.8.34), and the Markov model in (10.8.28) with $\rho = 1$, we can easily obtain

$$\tilde{\mathbf{c}}(n) = [\mathbf{I} - 2\mu\mathbf{x}(n)\mathbf{x}^H(n)]\tilde{\mathbf{c}}(n-1) + 2\mu\mathbf{x}(n)e_o^*(n) - \boldsymbol{\psi}(n) \tag{10.8.47}$$

which, compared to (10.4.15), has one extra input. Since $\mathbf{x}(n)$, $e_o(n)$, and $\boldsymbol{\psi}(n)$ are mutually independent, $\boldsymbol{\psi}(n)$ adds only an extra term $\sigma_\psi^2 \mathbf{I}$ to the correlation of $\tilde{\mathbf{c}}(n)$.

Misadjustment. To determine the misadjustment, we perform orthogonal transformation of the correlation matrix of $\tilde{\mathbf{c}}(n)$. When we transform (10.4.28) to (10.4.30), using the orthogonal transformation (10.4.29), the presence of the diagonal matrix $\sigma_\psi^2 \mathbf{I}$ changes only the diagonal components with the addition of the term σ_ψ^2. Indeed, we can easily show that

$$\theta_k(n) = \rho_k\theta_k(n-1) + 4\mu^2\lambda_k P_{\text{ex}}(n-1) + 4\mu^2 P_o\lambda_k + \sigma_\psi^2 \tag{10.8.48}$$

where $P_o(n) = P_o = \sigma_v^2$ for large n. Clearly, (10.8.48) converges under the same conditions as (10.4.40). At steady state we have

$$\theta_k(\infty) = \rho_k\theta_k(\infty) + 4\mu^2\lambda_k P_{\text{ex}}(\infty) + 4\mu^2 P_o\lambda_k + \sigma_\psi^2 \tag{10.8.49}$$

or using (10.4.36), we have

$$\theta_k(\infty) = \mu\frac{P_o + P_{\text{ex}}(\infty)}{1 - 2\mu\lambda_k} + \frac{1}{4\mu\lambda_k}\frac{\sigma_\psi^2}{1 - 2\mu\lambda_k} \tag{10.8.50}$$

which in conjunction with (10.4.55) and (10.4.56) gives

$$P_{\text{ex}}(\infty) = \frac{C(\mu)}{1 - C(\mu)}\sigma_v^2 + \frac{1}{4\mu}\frac{D(\mu)}{1 - C(\mu)}\sigma_\psi^2 \tag{10.8.51}$$

where

$$D(\mu) \triangleq \sum_{k=1}^M \frac{1}{1 - 2\mu\lambda_k} \tag{10.8.52}$$

If $\mu\lambda_k \ll 1$, we have $C(\mu) \simeq \mu\,\text{tr}(\mathbf{R})$ and $D(\mu) \simeq M$, which lead to

$$P_{\text{ex}}(\infty) \simeq \mu\sigma_v^2\,\text{tr}(\mathbf{R}) + \frac{1}{4\mu}M\sigma_\psi^2 \tag{10.8.53}$$

or

$$\mathcal{M}(\infty) \simeq \mu\,\text{tr}(\mathbf{R}) + \frac{1}{4\mu}M\frac{\sigma_\psi^2}{\sigma_v^2} \tag{10.8.54}$$

Hence in the steady state, the misadjustment can be approximated by two terms. The first term is estimation misadjustment, which increases with μ, while the second term is the lag misadjustment, which decreases with μ. Therefore, an optimum value of μ exists that minimizes $\mathcal{M}(\infty)$, given by

$$\mu_{\mathrm{opt}} \simeq \frac{\sigma_\psi}{2\sigma_\nu} \sqrt{\frac{M}{\mathrm{tr}(\mathbf{R})}} \tag{10.8.55}$$

or

$$\mathcal{M}_{\min}(\infty) \simeq \frac{\sigma_\psi}{\sigma_\nu} \sqrt{M \, \mathrm{tr}(\mathbf{R})} \tag{10.8.56}$$

MSD. To determine the MSD, consider (10.8.47). For small step size μ, the system matrix $[\mathbf{I} - 2\mu\mathbf{x}(n)\mathbf{x}^H(n)]$ is very close to the identity matrix. Hence using the direct averaging method due to Kushner (1984), we can obtain a close solution of $\tilde{\mathbf{c}}(n)$ by solving (10.8.47) in which the system matrix is replaced by its average $[\mathbf{I} - 2\mu\mathbf{R}]$, that is,

$$\tilde{\mathbf{c}}(n) = [\mathbf{I} - 2\mu\mathbf{R}]\tilde{\mathbf{c}}(n-1) + 2\mu\mathbf{x}(n)e_o^*(n) - \boldsymbol{\psi}(n) \tag{10.8.57}$$

where we have kept the same notation. Taking the covariance of both sides of (10.8.57), we obtain

$$\boldsymbol{\Phi}(n) = [\mathbf{I} - 2\mu\mathbf{R}]\boldsymbol{\Phi}(n-1)[\mathbf{I} - 2\mu\mathbf{R}] + 4\mu^2\sigma_\nu^2\mathbf{R} + \mathbf{R}_\psi \tag{10.8.58}$$

The approximate steady-state solution of (10.8.58) is given by

$$\mathbf{R}\boldsymbol{\Phi} + \boldsymbol{\Phi}\mathbf{R} \simeq 2\mu\sigma_\nu^2\mathbf{R} + \frac{\mathbf{R}_\psi}{2\mu} \tag{10.8.59}$$

where the second-order term $4\mu^2\mathbf{R}\boldsymbol{\Phi}\mathbf{R}$ is ignored for small values of μ. After premultiplying (10.8.59) by $\mathbf{R}^{-1}$, we obtain

$$\boldsymbol{\Phi} + \mathbf{R}^{-1}\boldsymbol{\Phi}\mathbf{R} \simeq 2\mu\sigma_\nu^2 + \frac{\mathbf{R}^{-1}\mathbf{R}_\psi}{2\mu} \tag{10.8.60}$$

Taking the trace of (10.8.60) and using $\mathrm{tr}(\mathbf{R}^{-1}\boldsymbol{\Phi}\mathbf{R}) = \mathrm{tr}(\boldsymbol{\Phi})$, we obtain

$$\mathrm{tr}(\boldsymbol{\Phi}) \simeq \mu M \sigma_\nu^2 + \frac{\mathrm{tr}(\mathbf{R}^{-1}\mathbf{R}_\psi)}{4\mu} \tag{10.8.61}$$

By following the development in (10.8.28), it can be shown that (Problem 10.52) $\mathcal{D}(\infty) = \mathrm{tr}(\boldsymbol{\Phi})$. Hence

$$\mathcal{D}(\infty) \simeq \mu M \sigma_\nu^2 + \frac{\mathrm{tr}(\mathbf{R}^{-1}\mathbf{R}_\psi)}{4\mu} \tag{10.8.62}$$

As expected, the MSD has two terms: The estimation deviation is linearly proportional to μ while the lag deviation is inversely proportional to μ. The optimum value of the step size μ is obtained when both deviations are equal and is given by

$$\mu_{\mathrm{opt}} \simeq \frac{1}{2} \sqrt{\frac{\mathrm{tr}(\mathbf{R}^{-1}\mathbf{R}_\psi)}{M\sigma_\nu^2}} \tag{10.8.63}$$

or

$$\mathcal{D}_{\min}(\infty) = \sqrt{M\sigma_\nu^2 \, \mathrm{tr}(\mathbf{R}^{-1}\mathbf{R}_\psi)} \tag{10.8.64}$$

EXAMPLE 10.8.1. To study the tracking performance of the LMS algorithm, we will simulate a slowly time-varying SOE whose parameters follow an almost random-walk behavior. The simulation setup is shown in Figure 10.42 and given by (10.8.27) and (10.8.28). The simulation parameters are as follows:

$\mathbf{c}_o(n)$ model parameters: $\quad \mathbf{c}_o(0) = \begin{bmatrix} -0.8 \\ 0.95 \end{bmatrix} \quad M = 2 \quad \rho = 0.999$

$$\boldsymbol{\psi}(n) \sim \mathrm{WGN}(\mathbf{0}, \mathbf{R}_\psi) \quad \mathbf{R}_\psi = (0.01)^2\mathbf{I}$$

Signal $\mathbf{x}(n)$ parameters: $\mathbf{x}(n) \sim \mathrm{WGN}(\mathbf{0}, \mathbf{R})$ $\mathbf{R} = \mathbf{I}$

Noise $v(n)$ parameters: $v(n) \sim \mathrm{WGN}(0, \sigma_v^2)$ $\sigma_v = 0.1$

For these values, the degree of nonstationarity from (10.8.32) is given by

$$\eta(n) = \frac{\sqrt{\mathrm{tr}[\mathbf{R}\mathbf{R}_\psi]}}{\sigma_v} = 0.1414 < 1$$

which means that the LMS can track the time variations of the SOE.

Three different adaptations (slow, matched, and fast) of the LMS algorithm were designed. Their adaptation results are shown in Figures 10.43 through 10.48. From (10.8.55) and (10.8.63), the optimum performance is obtained when

$$\mu_{\mathrm{opt}} = 0.05$$

for which $\mathcal{M}_{\min}(\infty) = 0.2$ and $\mathcal{D}_{\min}(\infty) = 0.002$. Hence, the following values for μ were selected for simulation:

Slow: $\mu = 0.01$

Matched: $\mu = 0.1$

Fast: $\mu = 0.3$

Figure 10.43 shows the matched adaptation of parameter coefficients while Figure 10.44 shows the resulting $\mathcal{D}(n)$ and $\mathcal{M}(n)$. Clearly, the LMS tracks the varying coefficients nicely with expected small misregistration and deviation errors. Figure 10.45 shows the slow adaptation of parameter coefficients while Figure 10.46 shows the resulting $\mathcal{D}(n)$ and $\mathcal{M}(n)$. In this case, although the LMS algorithm tracks with bounded error variance, the tracking is not very good and the resulting misregistration errors are large. Finally, Figure 10.47 shows the fast adaptation of parameter coefficients while Figure 10.48 shows the resulting $\mathcal{D}(n)$ and $\mathcal{M}(n)$. In this case, although the algorithm is able to keep track of the slowly varying coefficients, the resulting variance is large and hence the estimation errors are large. Once again, the total errors are large compared to those for the matched case.

10.8.4 RLS Algorithm with Exponential Forgetting

Consider again the model given in Figure 10.42 and described in the analysis setup.

Misadjustment. To determine the misadjustment in tracking, we first evaluate the excess MSE caused by lag, that is, by the deviation between $E\{\mathbf{c}(n)\}$ and the optimum a priori filter $\mathbf{c}_o(n)$. Combining

$$\mathbf{c}(n) = \mathbf{c}(n - 1) + \hat{\mathbf{R}}^{-1}(n)\mathbf{x}(n)e^*(n) \tag{10.8.65}$$

with $$e^*(n) = e_o^*(n) - \mathbf{x}^H(n)[\mathbf{c}(n-1) - \mathbf{c}_o(n-1)] \tag{10.8.66}$$

and taking the expectation result in

$$E\{\mathbf{c}(n)\} = E\{\mathbf{c}(n-1)\} + E\{\hat{\mathbf{R}}^{-1}(n)\mathbf{x}(n)\mathbf{x}^H(n)\}[E\{\mathbf{c}(n-1)\} - \mathbf{c}_o(n-1)] \tag{10.8.67}$$

because the expectation of $\hat{\mathbf{R}}^{-1}(n)\mathbf{x}(n)e_o^*(n)$ vanishes. Using the approximation $E\{\hat{\mathbf{R}}^{-1}(n) \cdot \mathbf{x}(n)\mathbf{x}^H(n)\} \simeq (1 - \lambda)\mathbf{I}$, we have

$$\tilde{\mathbf{c}}_{\mathrm{lag}}(n) \simeq \lambda\tilde{\mathbf{c}}_{\mathrm{lag}}(n) + \mathbf{c}_o(n-1) - \mathbf{c}_o(n) \tag{10.8.68}$$

or $$\tilde{\mathbf{c}}_{\mathrm{lag}}(n) \simeq \lambda\tilde{\mathbf{c}}_{\mathrm{lag}}(n-1) - \boldsymbol{\psi}(n) \tag{10.8.69}$$

for the random-walk ($\rho = 1$) model. The covariance matrix is

$$\boldsymbol{\Phi}_{\mathrm{lag}}(n) \simeq \lambda^2 \boldsymbol{\Phi}_{\mathrm{lag}}(n-1) + \mathbf{R}_\psi \tag{10.8.70}$$

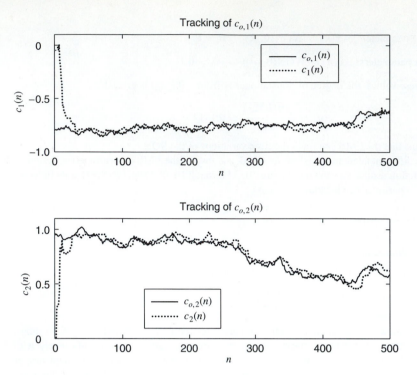

FIGURE 10.43
Matched adaptation of slowly time-varying parameters: LMS algorithm with $\mu = 0.1$.

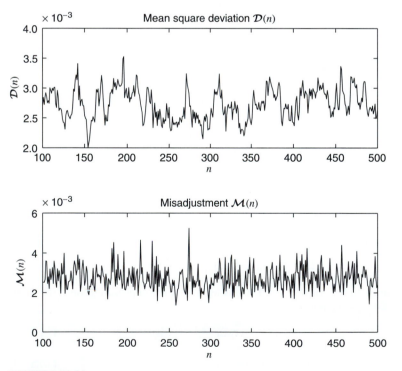

FIGURE 10.44
Learning curves of LMS algorithm with matched adaptation.

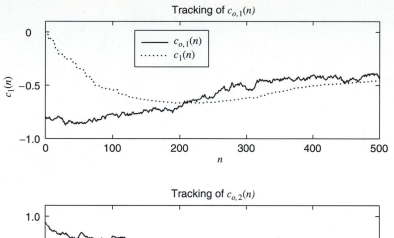

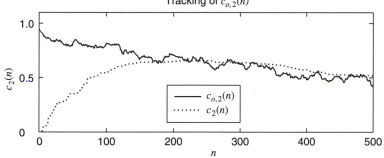

FIGURE 10.45
Slow adaptation of slowly time-varying parameters: LMS algorithm with
$\mu = 0.01$.

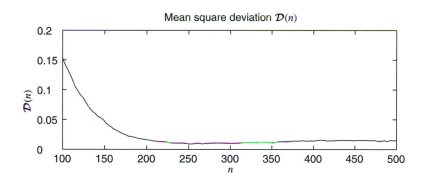

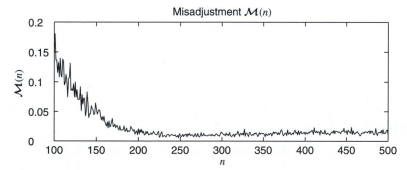

FIGURE 10.46
Learning curves of LMS algorithm for slow adaptation.

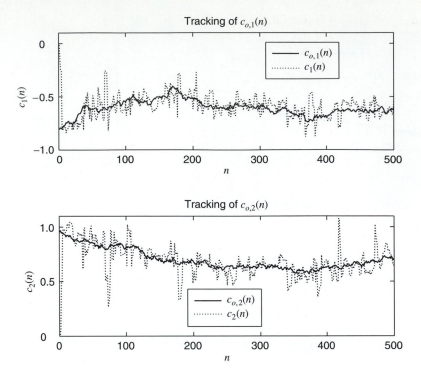

FIGURE 10.47
Fast adaptation of slowly time-varying parameters: LMS algorithm with $\mu = 0.3$.

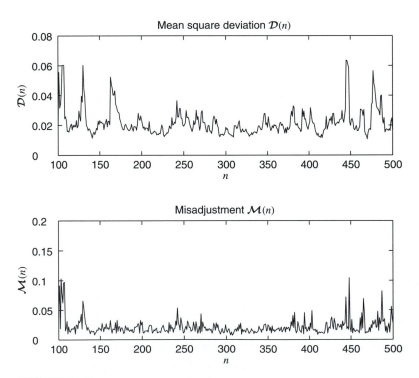

FIGURE 10.48
Learning curves of LMS algorithm for fast adaptation.

and in steady state (assuming $0 < \lambda < 1$)

$$\mathbf{\Phi}_{\text{lag}}(\infty) \simeq \frac{1}{(1-\lambda)^2}\mathbf{R}_\psi \tag{10.8.71}$$

The lag excess MSE is

$$P_{\text{lag}}(\infty) = \text{tr}[\mathbf{R}\mathbf{\Phi}(\infty)] \simeq \frac{1}{(1-\lambda)^2}\text{tr}[\mathbf{R}\mathbf{R}_\psi] \simeq \frac{1}{2(1-\lambda)}\text{tr}[\mathbf{R}\mathbf{R}_\psi] \tag{10.8.72}$$

because $(1-\lambda)^2 = (1+\lambda)(1-\lambda) \simeq 2(1-\lambda)$ for $\lambda \simeq 1$.

The excess MSE due to estimation is $[(1-\lambda)/2]M\sigma_v^2$, hence the total excess MSE is

$$P_{\text{ex}}(\infty) \simeq \frac{1-\lambda}{2}M\sigma_v^2 + \frac{1}{2(1-\lambda)}\sigma_\psi^2\,\text{tr}(\mathbf{R}) \tag{10.8.73}$$

if $\mathbf{R}_\psi = \sigma_\psi^2\mathbf{I}$. Finally, the misadjustment is given by

$$\mathcal{M}(\infty) \simeq \frac{1-\lambda}{2}M + \frac{\sigma_\psi^2\,\text{tr}(\mathbf{R})}{2(1-\lambda)\sigma_v^2} \tag{10.8.74}$$

The first term in (10.8.74) is the estimation misadjustment, which is linearly proportional to $1-\lambda$, while the second term is the lag misadjustment, which is inversely proportional to $1-\lambda$. The optimum value of λ is given by

$$\lambda_{\text{opt}} \simeq 1 - \frac{\sigma_\psi}{\sigma_v}\sqrt{\frac{1}{M}\text{tr}(\mathbf{R})} \tag{10.8.75}$$

and the minimum misadjustment is given by

$$\mathcal{M}_{\text{min}}(\infty) \simeq \frac{\sigma_\psi}{\sigma_v}\sqrt{M\,\text{tr}(\mathbf{R})} \tag{10.8.76}$$

MSD. An analysis similar to the MSD development of the LMS algorithm can be done to obtain

$$D(\infty) \simeq \frac{1-\lambda}{2}\sigma_v^2\,\text{tr}(\mathbf{R}^{-1}) + \frac{\sigma_\psi^2}{2(1-\lambda)} \tag{10.8.77}$$

with

$$\lambda_{\text{opt}} \simeq 1 - \frac{\sigma_\psi}{\sigma_v}\sqrt{\frac{1}{\text{tr}(\mathbf{R}^{-1})}} \tag{10.8.78}$$

and

$$\mathcal{D}_{\text{min}}(\infty) \simeq \frac{\sigma_\psi\sigma_v}{2}\sqrt{\text{tr}(\mathbf{R}^{-1})} \tag{10.8.79}$$

which again highlights the dependence of tracking abilities on λ.

EXAMPLE 10.8.2. To study the tracking performance of the RLS algorithm, we again simulate the slowly time-varying SOE given in Example 10.8.1 whose parameters are repeated here:

$$\mathbf{c}_o(n) \text{ model parameters:} \quad \mathbf{c}_o(0) = \begin{bmatrix} -0.8 \\ 0.95 \end{bmatrix} \quad M = 2 \quad \rho = 0.999$$

$$\boldsymbol{\psi}(n) \sim \text{WGN}(\mathbf{0}, \mathbf{R}_\psi) \quad \mathbf{R}_\psi = (0.01)^2\mathbf{I}$$

Signal $\mathbf{x}(n)$ parameters: $\quad \mathbf{x}(n) \sim \text{WGN}(\mathbf{0}, \mathbf{R}) \quad \mathbf{R} = \mathbf{I}$

Noise $v(n)$ parameters: $\quad v(n) \sim \text{WGN}(0, \sigma_v^2) \quad \sigma_v = 0.1$

For these values, the degree of nonstationarity is $\eta(n) = 0.1414$, which means that the RLS can track the time variations of the SOE.

Three different adaptations (slow, matched, and fast) of the RLS algorithm were designed. Their adaptation results are shown in Figures 10.49 through 10.54. From (10.8.75) and (10.8.77), the optimum misadjustment performance is obtained when

$$\lambda_{\text{opt}} = 0.9 \quad \text{with } \mathcal{M}_{\text{min}}(\infty) = 0.2$$

while from (10.8.78) and (10.8.79), the optimum deviation performance is obtained when

$$\lambda_{\text{opt}} = 0.93 \quad \text{with } \mathcal{D}_{\min}(\infty) = 0.007$$

Hence, the following values for λ were selected for simulation:

$$
\begin{array}{lll}
\text{Slow:} & \lambda = 0.99 \\
\text{Matched:} & \lambda = 0.9 \\
\text{Fast:} & \lambda = 0.5
\end{array}
$$

Figure 10.49 shows the matched adaptation of parameter coefficients while Figure 10.50 shows the resulting $\mathcal{D}(n)$ and $\mathcal{M}(n)$. Clearly, the RLS tracks the varying coefficients nicely with expected small misregistration and deviation errors. Figure 10.51 shows the slow adaptation of parameter coefficients while Figure 10.52 shows the resulting $\mathcal{D}(n)$ and $\mathcal{M}(n)$. In this case, although the RLS algorithm tracks with bounded error variance, the tracking is not very good and the resulting misregistration errors are large. Finally, Figure 10.53 shows the fast adaptation of parameter coefficients while Figure 10.54 shows the resulting $\mathcal{D}(n)$ and $\mathcal{M}(n)$. In this case, although the algorithm is able to keep track of the slowly varying coefficients, the resulting variance is large and hence the estimation errors are large. Once again, the total errors are large compared to those for the matched case.

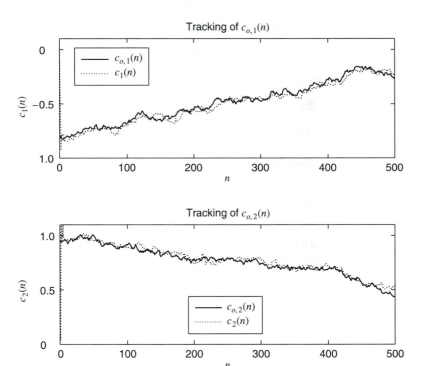

FIGURE 10.49
Matched adaptation of slowly time-varying parameters: RLS algorithm with $\lambda = 0.9$.

10.8.5 Comparison of Tracking Performance

When the optimum filter drifts like a random walk with small increment variance σ_ψ^2, the tracking performance for the LMS algorithm is given by (10.8.54) and (10.8.62) while that for the RLS algorithm is given by (10.8.74) and (10.8.77). Whether the LMS or the RLS algorithm is better depends on matrices $\mathbf{R}$ and $\mathbf{R}_\psi$. A general comparison is difficult to make, but some guidelines have been developed for particular cases. It has been shown that (Haykin 1996)

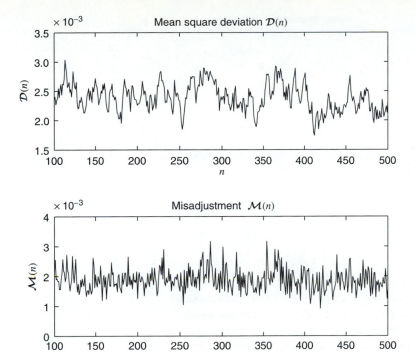

FIGURE 10.50
Learning curves of RLS algorithm for matched adaptation.

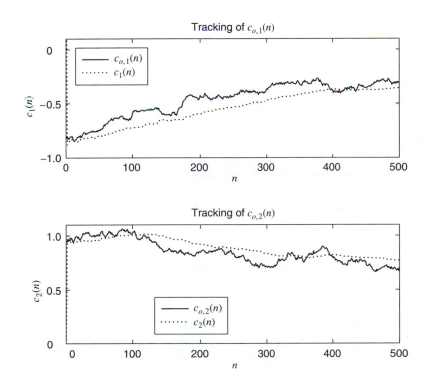

FIGURE 10.51
Slow adaptation of slowly time-varying parameters: RLS algorithm with
$\lambda = 0.99$.

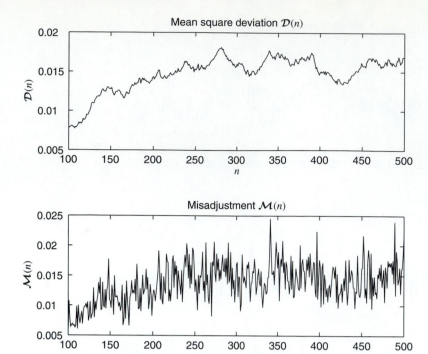

FIGURE 10.52
Learning curves of RLS algorithm for slow adaptation.

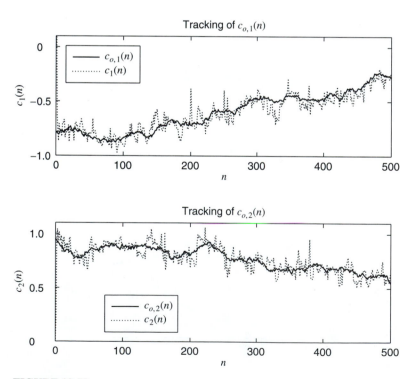

FIGURE 10.53
Fast adaptation of slowly time-varying parameters: RLS algorithm with $\lambda = 0.5$.

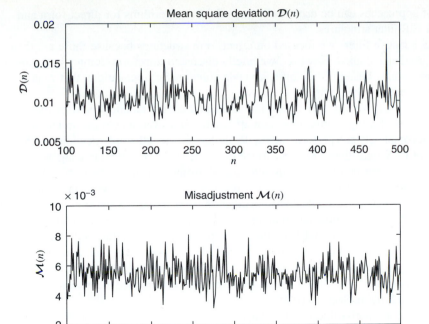

FIGURE 10.54
Learning curves of RLS algorithm for fast adaptation.

- When $\mathbf{R}_\psi = \sigma_\psi^2 \mathbf{I}$, then both the LMS and RLS algorithms produce essentially the same minimum levels of MSD and misadjustment. However, this analysis is true only asymptotically and for slowly varying parameters (small σ_ψ^2).
- When $\mathbf{R}_\psi = \alpha \mathbf{R}$ where α is a constant, then the LMS algorithm produces smaller values of the minimum levels of MSD and misadjustment than the RLS algorithm does.
- When $\mathbf{R}_\psi = \beta \mathbf{R}^{-1}$ where β is a constant, then the RLS algorithm is better than the LMS algorithm in producing the smaller values of the minimum levels of MSD and misadjustment.

In summary, we should state that in practice the comparison of the acquisition and tracking performance of LMS and RLS adaptive filters is a very complicated subject. Although the previous analysis provides some insight only extensive simulations in the context of a specific application can help to choose the appropriate algorithm.

10.9 SUMMARY

In this chapter we discussed the theory of operation, design, performance evaluation, implementation, and applications of adaptive filters. The most significant attribute of an adaptive filter is its ability to incrementally adjust its coefficients so as to improve a predefined criterion of performance over time.

We basically developed and analyzed two families of adaptive filtering algorithms:

- The family of LMS FIR adaptive filters, which are based on a stochastic version of the steepest-descent optimization algorithm.
- The family of RLS FIR adaptive filters, which are based on a stochastic version of the Newton-type optimization algorithms.

Both types of approaches can be used to develop adaptive algorithms for direct-form and lattice-ladder FIR filter structures.

For LMS adaptive filters we focused on direct-form structures because those are the most widely used and studied. However, we briefly discussed transform-domain and sub-band implementations because they offer a viable solution for applications that require adaptive filters with very long impulse responses.

All RLS FIR adaptive filters discussed in this chapter exhibit identical performance if they are implemented using infinite-precision arithmetic. However, they differ in terms of computational complexity and performance under finite-word-length implementations. The various types of RLS algorithms are summarized in Figure 10.40. We stress that algorithms for array processing can be used for FIR filtering (shift-invariant input data vector), but not vice versa. However, such a practice is not recommended because the computational complexity is much higher. The LMS algorithm (Section 10.4), the CRLS algorithm (Section 10.5), and the QR decomposition–based algorithms (Section 10.6) are general and can be used for both array processing and FIR filtering applications. In contrast, the fast RLS algorithms in Section 10.7 can be used only for FIR filtering and prediction applications. The steady-state performance of LMS and RLS algorithms in a stationary environment is discussed in Sections 10.4 and 10.5, whereas their tracking performance in a nonstationary environment is analyzed in Section 10.8.

The treatment of adaptive filters in this chapter has been quite extensive, in both number of topics and depth. However, the following important topics have been omitted:

- IIR adaptive filters (Treichler et al. 1987; Johnson 1984; Shynk 1989; Regalia 1995; Netto et al. 1995; Williamson 1998). Although adaptive IIR filters have the potential to offer the same performance as FIR filters with less computational complexity, they are not widely used in practical applications. The main reasons are related to the nonquadratic nature of their performance error surface (see Section 6.2) and the additional stability problems caused by the presence of poles in their system function.
- Adaptive filters using nonlinear filtering structures and neural networks (Grant and Mulgrew 1995; Haykin 1996; Mathews 1991). The need for such filters arises in applications involving nonlinear input-output relationships, nonlinear detectors (e.g., data equalization), and non-Gaussian or impulsive noise. The optimization required in some of these cases can be performed using genetic optimization algorithms (Tang et al. 1996).
- FIR direct-form and lattice-ladder LS adaptive filters for multichannel signals (Slock 1993; Ling 1993b; Carayannis et al. 1986).

PROBLEMS

10.1 Consider the process $x(n)$ generated using the AR(3) model

$$x(n) = -0.729x(n-3) + w(n)$$

where $w(n) \sim \text{WGN}(0, 1)$. We want to design a linear predictor of $x(n)$ using the SDA algorithm. Let

$$\hat{y}(n) = \hat{x}(n) = c_{0,1}x(n-1) + c_{0,2}x(n-2) + c_{0,3}x(n-3)$$

(a) Determine the 3×3 autocorrelation matrix $\mathbf{R}$ of $x(n)$, and compute its eigenvalues $\{\lambda_i\}_{i=1}^3$.
(b) Determine the 3×1 cross-correlation vector $\mathbf{d}$.
(c) Choose the step size μ so that the resulting response is overdamped. Now implement the SDA

$$\mathbf{c}_k = [c_{k,1} \ c_{k,2} \ c_{k,3}]^T = \mathbf{c}_{k-1} + 2\mu(\mathbf{d} - \mathbf{R}\mathbf{c}_{k-1})$$

and plot the trajectories of $\{c_{k,i}\}_{i=1}^3$ as a function of k.
(d) Repeat part (c) by choosing μ so that the response is underdamped.

10.2 In the SDA algorithm, the index k is an iteration index and not a time index. However, we can treat it as a time index and use the instantaneous filter coefficient vector $\mathbf{c}_k$ to filter data at $n = k$. This will result in an *asymptotically optimum* filter whose coefficents will converge to the optimum one. Consider the process $x(n)$ given in Problem 10.1.

(a) Generate 500 samples of $x(n)$ and implement the asymptotically optimum filter. Plot the signal $\hat{y}(n)$.

(b) Implement the optimum filter $\mathbf{c}_0$ on the same sequence, and plot the resulting $\hat{y}(n)$.

(c) Comment on the above two plots.

10.3 Consider the AR(2) process $x(n)$ given in Example 10.3.1. We want to implement the Newton-type algorithm for faster convergence using

$$\mathbf{c}_k = \mathbf{c}_{k-1} - \mu \mathbf{R}^{-1} \nabla P(\mathbf{c}_{k-1})$$

(a) Using $a_1 = -1.5955$ and $a_2 = 0.95$, implement the above method for $\mu = 0.1$ and $\mathbf{c}_0 = \mathbf{0}$. Plot the locus of $c_{k,1}$ versus $c_{k,2}$.

(b) Repeat part (a), using $a_1 = -0.195$ and $a_2 = 0.95$.

(c) Repeat parts (a) and (b), using the optimum step size for μ that results in the fastest convergence.

10.4 Consider the adaptive linear prediction of an AR(2) process $x(n)$ using the LMS algorithm in which

$$x(n) = 0.95x(n-1) - 0.9x(n-2) + w(n)$$

where $w(n) \sim \text{WGN}(0, \sigma_w^2)$. The adaptive predictor is a second-order one given by $\mathbf{a}(n) = [a_1(n)\, a_2(n)]^T$.

(a) Implement the LMS algorithm given in Table 10.3 as a MATLAB function

$$[\text{a,e}] = \text{lplms}(\text{x,y,mu,M,a0}).$$

which computes filter coefficients in c and the corresponding error in e, given signal x, desired signal y, step size mu, filter order M, and the initial coefficient vector a0.

(b) Generate 500 samples of $x(n)$, and obtain linear predictor coefficients using the above function. Use step size μ so that the algorithm converges in the mean. Plot predictor coefficients as a function of time along with the true coefficients.

(c) Repeat the above simulation 1000 times to obtain the learning curve, which is obtained by averaging the squared error $|e(n)|^2$. Plot this curve and compare its steady-state value with the theoretical MSE.

10.5 Consider the adaptive echo canceler given in Figure 10.25. The FIR filter $c_0(n)$ is given by

$$c_0(n) = (0.9)^n \qquad 0 \le n \le 2$$

In this simulation, ignore the far-end signal $u(n)$. The data signal $x(n)$ is a zero-mean, unit-variance white Gaussian process, and $y(n)$ is its echo.

(a) Generate 1000 samples of $x(n)$ and determine $y(n)$. Use these signals to obtain a fourth-order LMS echo canceler in which the step size μ is chosen to satisfy (10.4.40) and $\mathbf{c}(0) = \mathbf{0}$. Obtain the final echo canceler coefficients and compare them with the true ones.

(b) Repeat the above simulation 500 times, and obtain the learning curve. Plot this curve along with the actual MSE and comment on the plot.

(c) Repeat parts (a) and (b), using a third-order echo canceler.

(d) Repeat parts (a) and (b), using one-half the value of μ used in the first part.

10.6 The normalized LMS (NLMS) algorithm is given in (10.4.67), in which the effective step size is time-varying and is given by $\tilde{\mu}/\|\mathbf{x}(n)\|^2$, where $0 < \tilde{\mu} < 1$.

(a) Modify the function firlms to implement the NLMS algorithm and obtain the function

$$[\text{c,e}] = \text{nfirlms}(\text{x,y,mu,M,c0}).$$

(b) Choose $\tilde{\mu} = 0.1$ and repeat Problem 10.4. Compare your results in terms of convergence speed.

(c) Choose $\tilde{\mu} = 0.1$ and repeat Problem 10.5(a) and (b). Compare your results in terms of convergence speed.

10.7 Another variation of the LMS algorithm is called the *sign-error* LMS algorithm, in which the coefficient update equation is given by

$$\mathbf{c}(n) = \mathbf{c}(n-1) + 2\mu \operatorname{sgn}[e(n)]\mathbf{x}(n)$$

where
$$\operatorname{sgn}[e(n)] = \begin{cases} 1 & \operatorname{Re}[e(n)] > 0 \\ 0 & \operatorname{Re}[e(n)] = 0 \\ -1 & \operatorname{Re}[e(n)] < 0 \end{cases}$$

The advantage of this algorithm is that the multiplication is replaced by a sign change, and if μ is chosen as a negative power of 2, then the multiplication is replaced by a shifting operation that is easy and fast to implement. Furthermore, since $\operatorname{sgn}(x) = x/|x|$, the effective step size $\tilde{\mu}$ is inversely proportional to the magnitude of the error.

(a) Modify the function firlms to implement the NLMS algorithm and obtain the function

$$[c,e] = sefirlms(x,y,mu,M,c0).$$

(b) Repeat Problem 10.4 and compare your results in terms of convergence speed.

(c) Repeat Problem 10.5(a) and (b) and compare your results in terms of convergence speed.

10.8 Consider an AR(1) process $x(n) = ax(n-1) + w(n)$, where $w(n) \sim \text{WGN}(0, \sigma_w^2)$. We wish to design a one-step first-order linear predictor using the LMS algorithm

$$\hat{x}(n) = \hat{a}(n-1)x(n-1)$$
$$e(n) = x(n) - \hat{x}(n)$$
$$\hat{a}(n) = \hat{a}(n-1) + 2\mu e(n)x(n-1)$$

where μ is the adaptation step size.

(a) Determine the autocorrelation $r_x(l)$, the optimum first-order linear predictor, and the corresponding MMSE.

(b) Using the independence assumption, first determine and then solve the difference equation for $E\{\hat{a}(n)\}$.

(c) For $a = \pm 0.95$, $\mu = 0.025$, $\sigma_x^2 = 1$, and $0 \le n < N = 500$, determine the ensemble average of $E\{\hat{a}(n)\}$ using 200 independent runs and compare with the theoretical curve obtained in part (b).

(d) Using the independence assumption, first determine and then solve the difference equation for $P(n) = E\{e^2(n)\}$.

(e) Repeat part (c) for $P(n)$ and comment upon the results.

10.9 Using the a posteriori error $\varepsilon(n) = y(n) - \mathbf{c}^H(n)\mathbf{x}(n)$, derive the coefficient updating formulas for the a posteriori error LMS algorithm. *Note*: Refer to Equations (10.2.20) to (10.2.22).

10.10 Solve the interference cancelation problem described in Example 6.4.1, using the LMS algorithm, and compare its performance to that of the optimum canceler.

10.11 Repeat the convergence analysis of the LMS algorithm for the complex case, using formula (10.4.27) instead of (10.4.28).

10.12 Consider the total transient excess MSE, defined by

$$P_{\text{tr}}^{(\text{total})} = \sum_{n=0}^{\infty} P_{\text{tr}}(n)$$

in Section 10.4.3.

(a) Show that $P_{\text{tr}}^{(\text{total})}$ can be written as $P_{\text{tr}}^{(\text{total})} = \boldsymbol{\lambda}^T (\mathbf{I} - \mathbf{B})^{-1} \Delta\boldsymbol{\theta}(0)$, where $\Delta\boldsymbol{\theta}(0)$ is the initial (i.e., at $n = 0$) deviation of the filter coefficients from their optimum setting.

(b) Starting with the formula in step (a), show that

$$P_{\text{tr}}^{(\text{total})} = \frac{1}{4\mu} \frac{\displaystyle\sum_{i=1}^{M} \frac{\Delta\theta_i(0)}{1 - 2\mu\lambda_i}}{1 - \displaystyle\sum_{i=1}^{M} \frac{\mu\lambda_i}{1 - 2\mu\lambda_i}}$$

(c) Show that if $\mu\lambda_k \ll 1$, then

$$P_{\text{tr}}^{(\text{total})} \simeq \frac{1}{4\mu} \frac{\displaystyle\sum_{i=1}^{M} \Delta\theta_i(0)}{1 - \mu\,\text{tr}(\mathbf{R})} \simeq \frac{1}{4\mu} \sum_{i=1}^{M} \Delta\theta_i(0)$$

which is formula (10.4.62), discussed in Section 10.4.3.

10.13 The frequency sampling structure for the implementation of an FIR filter $H(z) = \sum_{n=0}^{M-1} h(n) \cdot z^{-n}$ is specified by the following relation

$$H(z) = \frac{1 - z^{-M}}{M} \sum_{k=0}^{M-1} \frac{H(e^{j2\pi k/M})}{1 - e^{j2\pi k/M} z^{-1}} \triangleq H_1(z)\, H_2(z)$$

where $H_1(z)$ is a comb filter with M zeros equally spaced on the unit circle and $H_2(z)$ is a filter bank of resonators. Note that $\tilde{H}(k) \triangleq H(e^{j2\pi k/M})$, the DFT of $\{h(n)\}_0^{M-1}$, is the coefficients of the filter. Derive an LMS-type algorithm to update these coefficients, and sketch the resulting adaptive filter structure.

10.14 There are applications in which the use of a non-MSE criterion may be more appropriate. To this end, suppose that we wish to design and study the behavior of an "LMS-like" algorithm that minimizes the cost function $P^{(k)} = E\{e^{2k}(n)\}$, $k = 1, 2, 3, \ldots$, using the model defined in Figure 10.19.

(a) Use the instantaneous gradient vector to derive the coefficient updating formula for this LMS-like algorithm.

(b) Using the assumptions introduced in Section 10.4.2 show that

$$E\{\tilde{\mathbf{c}}(n)\} = [\mathbf{I} - 2\mu k(2k - 1)E\{e_o^{2(k-1)}(n)\}\mathbf{R}]E\{\tilde{\mathbf{c}}(n - 1)\}$$

where $\mathbf{R}$ is the input correlation matrix.

(c) Show that the derived algorithm converges in the mean if

$$0 < 2\mu < \frac{1}{k(2k - 1)E\{e_o^{2(k-1)}(n)\}\lambda_{\max}}$$

where $\lambda_{\max}$ is the largest eigenvalue of $\mathbf{R}$.

(d) Show that for $k = 1$ the results in parts (a) to (c) reduce to those for the standard LMS algorithm.

10.15 Consider the noise cancelation system shown in Figure 10.6. The useful signal is a sinusoid $s(n) = \cos(\omega_0 n + \phi)$, where $\omega_0 = \pi/16$ and the phase ϕ is a random variable uniformly distributed from 0 to 2π. The noise signals are given by $v_1(n) = 0.9\, v_1(n - 1) + w(n)$ and $v_2(n) = -0.75\, v_2(n - 1) + w(n)$, where the sequences $w(n)$ are WGN(0, 1).

(a) Design an optimum filter of order M and choose a reasonable value for M_o by plotting the MMSE as a function of M.

(b) Design an LMS filter with M_o coefficients and choose the step size μ to achieve a 10 percent misadjustment.

(c) Plot the signals $s(n)$, $s(n) + v_1(n)$, $v_2(n)$, the clean signal $e_o(n)$ using the optimum filter, and the clean signal $e_{\text{lms}}(n)$ using the LMS filter, and comment upon the obtained results.

10.16 A modification of the LMS algorithm, known as the *momentum LMS (MLMS)*, is defined by

$$\mathbf{c}(n) = \mathbf{c}(n-1) + 2\mu e^*(n)\mathbf{x}(n) + \alpha[\mathbf{c}(n-1) - \mathbf{c}(n-2)]$$

where $|\alpha| < 1$ (Roy and Shynk 1990).

(a) Rewrite the previous equation to show that the algorithm has the structure of a low-pass ($0 < \alpha < 1$) or a high-pass ($-1 < \alpha < 0$) filter.

(b) Explain intuitively the effect of the momentum term $\alpha[\mathbf{c}(n-1) - \mathbf{c}(n-2)]$ on the filter's convergence behavior.

(c) Repeat the computer equalization experiment in Section 10.4.4, using both the LMS and the MLMS algorithms for the following cases, and compare their performance:
   i. $W = 3.1$, $\mu_{\text{lms}} = \mu_{\text{mlms}} = 0.01$, $\alpha = 0.5$.
   ii. $W = 3.1$, $\mu_{\text{lms}} = 0.04$, $\mu_{\text{mlms}} = 0.01$, $\alpha = 0.5$.
   iii. $W = 3.1$, $\mu_{\text{lms}} = \mu_{\text{mlms}} = 0.04$, $\alpha = 0.2$.
   iv. $W = 4$, $\mu_{\text{lms}} = \mu_{\text{mlms}} = 0.03$, $\alpha = 0.3$.

10.17 In Section 10.4.5 we presented the leaky LMS algorithm [see (10.4.88)]

$$\mathbf{c}(n) = (1 - \alpha\mu)\mathbf{c}(n-1) + \mu e^*(n)\mathbf{x}(n)$$

where $0 < \alpha \ll 1$ is the leakage coefficient.

(a) Show that the coefficient updating equation can be obtained by minimizing

$$P(n) = |e(n)|^2 + \alpha\|\mathbf{c}(n)\|^2$$

(b) Using the independence assumptions, show that

$$E\{\mathbf{c}(n)\} = [\mathbf{I} - \mu(\mathbf{R} + \alpha\mathbf{I})]E\{\mathbf{c}(n-1)\} + \mu\mathbf{d}$$

where $\mathbf{R} = E\{\mathbf{x}(n)\mathbf{x}^H(n)\}$ and $\mathbf{d} = E\{\mathbf{x}(n)y^*(n)\}$.

(c) Show that if $0 < \mu < 2/(\alpha + \lambda_{\max})$, where $\lambda_{\max}$ is the maximum eigenvalue of $\mathbf{R}$, then

$$\lim_{n \to \infty} E\{\mathbf{c}(n)\} = (\mathbf{R} + \alpha\mathbf{I})^{-1}\mathbf{d}$$

that is, in the steady state $E\{\mathbf{c}(\infty)\} \neq \mathbf{c}_o = \mathbf{R}^{-1}\mathbf{d}$.

10.18 There are various communications and speech signal processing applications that require the use of filters with linear phase (Manolakis et al. 1984). For simplicity, assume that m is even.

(a) Derive the normal equations for an optimum FIR filter that satisfies the constraints
   i. $\mathbf{c}_m^{(\text{lp})} = \mathbf{J}\mathbf{c}_m^{(\text{lp})}$ (linear phase)
   ii. $\mathbf{c}_m^{(\text{cgd})} = -\mathbf{J}\mathbf{c}_m^{(\text{cgd})}$ (constant group delay).

(b) Show that the obtained optimum filters can be expressed as $\mathbf{c}_m^{(\text{lp})} = \frac{1}{2}(\mathbf{c}_m + \mathbf{J}\mathbf{c}_m)$ and $\mathbf{c}_m^{(\text{cgd})} = \frac{1}{2}(\mathbf{c}_m - \mathbf{J}\mathbf{c}_m)$, where $\mathbf{c}_m$ is the unconstrained optimum filter.

(c) Using the results in part (b) and the algorithm of Levinson, derive lattice-ladder structure for the constrained optimum filters.

(d) Repeat parts (a), (b), and (c) for the linear predictor with linear phase, which is specified by $\mathbf{a}_m^{(\text{lp})} = \mathbf{J}\mathbf{a}_m^{(\text{lp})}$.

(e) Develop an LMS algorithm for the linear-phase filter $\mathbf{c}_m^{(\text{lp})} = \mathbf{J}\mathbf{c}_m^{(\text{lp})}$ and sketch the resulting structure. Can you draw any conclusions regarding the step size and the misadjustment of this filter compared to those of the unconstrained LMS algorithm?

10.19 In this problem, we develop and analyze by simulation an LMS-type adaptive lattice predictor introduced in Griffiths (1977). We consider the all-zero lattice filter defined in (7.5.7), which is completely specified by the lattice parameters $\{k_m\}_0^{M-1}$. The input signal is assumed wide-sense stationary.

(a) Consider the cost function

$$P_m^{\text{fb}} = E\{|e_m^{\text{f}}(n)|^2 + |e_m^{\text{b}}(n)|^2\}$$

which provides the total prediction error power at the output of the mth stage, and show that

$$\frac{\partial P_m^{\text{fb}}}{\partial k_m} = 2E\{e_m^{\text{f}*}(n)e_{m-1}^{\text{b}}(n-1) + e_{m-1}^{\text{f}*}(n)e_m^{\text{b}}(n)\}$$

(b) Derive the updating formula

$$k_m(n) = k_m(n-1) + 2\mu(n)[e_m^{\text{f}*}(n)e_{m-1}^{\text{b}}(n-1) + e_{m-1}^{\text{f}*}(n)e_m^{\text{b}}(n)]$$

where the normalized step size $\mu(n) = \bar{\mu}/E_{m-1}^{\text{b}}(n)$ is computed in practice by using the formula

$$E_{m-1}^{\text{b}}(n) = \alpha E_{m-1}^{\text{b}}(n-1) + (1-\alpha)[|e_{m-1}^{\text{f}}(n)|^2 + |e_{m-1}^{\text{b}}(n-1)|^2]$$

where $0 < \alpha < 1$. Explain the role and proper choice of α, and determine the proper initialization of the algorithm.

(c) Write a MATLAB function to implement the derived algorithm, and compare its performance with that of the LMS algorithm in the linear prediction problem discussed in Example 10.4.1.

10.20 Consider a signal $x(n)$ consisting of a harmonic process plus white noise, that is,

$$x(n) = A\cos(\omega_1 n + \phi) + w(n)$$

where ϕ is uniformly distributed from 0 to 2π and $w(n) \sim \text{WGN}(0, \sigma_w^2)$.

(a) Determine the output power $\sigma_y^2 = E\{y^2(n)\}$ of the causal and stable filter

$$y(n) = \sum_{k=0}^{\infty} h(k)x(n-k)$$

and show that we can cancel the harmonic process using the ideal notch filter

$$H(e^{j\omega}) = \begin{cases} 1 & \omega = \omega_1 \\ 0 & \text{otherwise} \end{cases}$$

Is the obtained ideal notch filter practically realizable? That is, is the system function rational? Why?

(b) Consider the second-order notch filter (see Problem 2.31)

$$H(z) = \frac{D(z)}{A(z)} = \frac{1 + az^{-1} + z^{-2}}{1 + a\rho z^{-1} + \rho^2 z^{-2}} = \frac{D(z)}{D(z/\rho)}$$

where $-1 < \rho < 1$ determines the steepness of the notch and $a = -2\cos\omega_0$ its frequency. We fix ρ, and we wish to design an adaptive filter by adjusting a.
  i. Show that for $\rho \simeq 1, \sigma_y^2 = A^2|H(e^{j\omega_1})|^2 + \sigma_w^2$, and plot σ_y^2 as a function of the frequency ω_0 for $\omega_1 = \pi/6$.
  ii. Evaluate $d\sigma_y^2(a)/da$ and show that the minimum of $\sigma_y^2(a)$ occurs for $a = -2\cos\omega_1$.

(c) Using a direct-form II structure for the implementation of $H(z)$ and the property $dY(z)/da = [dH(z)/da]X(z)$, show that the following relations

$$s_2(n) = -a(n-1)\rho s_2(n-1) - \rho^2 s_2(n-2) + (1-gr)s_1(n-1)$$

$$g(n) = s_2(n) - \rho s_2(n-2)$$

$$s_1(n) = -a(n-1)\rho s_1(n-1) - \rho^2 s_1(n-2) + x(n)$$

$$y(n) = s_1(n) + a(n-1)s_1(n-1) + s_1(n-2)$$

$$a(n) = a(n-1) - 2\mu y(n)g(n)$$

constitute an adaptive LMS notch filter. Draw its block diagram realization.

(d) Simulate the operation of the obtained adaptive filter for $\rho = 0.9, \omega_1 = \pi/6$, and SNR 5 and 15 dB. Plot $\omega_0(n) = \arccos[-a(n)/2]$ as a function of n, and investigate the tradeoff between convergence rate and misadjustment by experiment with various values of μ.

10.21 Consider the AR(2) process given in Problem 10.4. We will design the adaptive linear predictor using the RLS algorithm. The adaptive predictor is a second-order one given by $\mathbf{c}(n) = [c_1(n)\ c_2(n)]^T$.

(a) Develop a MATLAB function to implement the RLS algorithm given in Table 10.6

```
[c,e] = rls(x,y,lambda,delta,M,c0);
```

which computes filter coefficients in c and the corresponding error in e given signal x, desired signal y, forgetting factor lambda, initialization parameter delta, filter order M, and the initial coefficient vector c0. To update $\mathbf{P}(n)$, compute only the upper or lower triangular part and determine the other part by using Hermitian symmetry.

(b) Generate 500 samples of $x(n)$ and obtain linear predictor coefficients using the above function. Use a very small value for δ (for example, 0.001) and various values of $\lambda = 0.99$, 0.95, 0.9, and 0.8. Plot predictor coefficients as a function of time along with the true coefficients for each λ, and discuss your observations. Also compare your results with those in Problem 10.4.

(c) Repeat each simulation above 1000 times to get corresponding learning curves, which are obtained by averaging respective squared errors $|e(n)|^2$. Plot these curves and compare their steady-state value with the theoretical MSE.

10.22 Consider a system identification problem where we observe the input $x(n)$ and the noisy output $y(n) = y_o(n) + v(n)$, for $0 \le n \le N - 1$. The unknown system is specified by the system function

$$H_o(z) = \frac{0.0675 + 0.1349z^{-1} + 0.0675z^{-2}}{1 - 1.1430z^{-1} + 0.4128z^{-2}}$$

and $x(n) \sim \text{WGN}(0, 1)$, $v(n) \sim \text{WGN}(0, 0.01)$, and $N = 300$.

(a) Model the unknown system using an LS FIR filter, with $M = 15$ coefficients, using the no-windowing method. Compute the total LSE E_{ls} in the interval $n_0 \le n \le N - 1$ for $n_0 = 20$.

(b) Repeat part (a) for $0 \le n \le n_0 - 1$ (do not compute E_{ls}). Use the vector $\mathbf{c}(n_0)$ and the matrix $\mathbf{P}(n_0) = \hat{\mathbf{R}}^{-1}(n_0)$ to initialize the CRLS algorithm. Compute the total errors $E_{\text{apr}} = \sum_{n=n_0}^{N-1} e^2(n)$ and $E_{\text{apost}} = \sum_{n=n_0}^{N-1} \varepsilon^2(n)$ by running the CRLS for $n_0 \le n \le N - 1$.

(c) Order the quantities E_{ls}, E_{apr}, E_{apost} by size and justify the resulting ordering.

10.23 Prove Equation (10.5.25) using the identity $\det(\mathbf{I}_1 + \mathbf{AB}) = \det(\mathbf{I}_2 + \mathbf{BA})$, where identity matrices $\mathbf{I}_1$ and $\mathbf{I}_2$ and matrices $\mathbf{A}$ and $\mathbf{B}$ have compatible dimensions. *Hint:* Put (10.5.7) in the form $\mathbf{I}_1 + \mathbf{AB}$.

10.24 Derive the normal equations that correspond to the minimization of the cost function (10.5.36), and show that for $\delta = 0$ they are reduced to the standard set (10.5.2) of normal equations. For the situation described in Problem 10.22, run the CRLS algorithm for various values of δ and determine the range of values that provides acceptable performance.

10.25 Modify the CRLS algorithm in Table 10.6 so that its coefficients satisfy the linear-phase constraint $\mathbf{c} = \mathbf{Jc}^*$. For simplicity, assume that $M = 2L$; that is, the filter has an even number of coefficients.

10.26 Following the approach used in Section 7.5.1 to develop the structure shown in Figure 7.1, derive a similar structure based on the Cholesky (*not* the LDL^H) decomposition.

10.27 Show that the partitioning (10.7.3) of $\hat{\mathbf{R}}_{m+1}(n)$ to obtain the same partitioning structure as (10.7.2) is possible only if we apply the prewindowing condition $\mathbf{x}_m(-1) = \mathbf{0}$. What is the form of the partitioning if we abandon the prewindowing assumption?

10.28 Derive the normal equations and the LSE formulas given in Table 10.11 for the FLP and the BLP methods.

10.29 Derive the FLP and BLP a priori and a posteriori updating formulas given in Table 10.12.

10.30 Modify Table 10.14 for the FAEST algorithm, to obtain a table for the FTF algorithm, and write a MATLAB function for its implementation. Test the obtained function, using the equalization experiment in Example 10.5.2.

10.31 If we wish to initialize the fast RLS algorithms (fast Kalman, FAEST, and FTF) using an exact method, we need to collect a set of data $\{\mathbf{x}(n), y(n)\}_0^{n_0}$ for any $n_0 > M$.

(a) Identify the quantities needed to start the FAEST algorithm at $n = n_0$. Form the normal equations and use the LDL^H decomposition method to determine these quantities.

(b) Write a MATLAB function `faestexact.m` that implements the FAEST algorithm using the exact initialization procedure described in part (a).

(c) Use the functions `faest.m` and `faestexact.m` to compare the two different initialization approaches for the FAEST algorithm in the context of the equalization experiment in Example 10.5.2. Use $n_0 = 1.5M$ and $n_0 = 3M$. Which value of δ gives results closest to the exact initialization method?

10.32 Using the order-recursive approach introduced in Section 7.3.1, develop an order-recursive algorithm for the solution of the normal equations (10.5.2), and check its validity by using it to initialize the FAEST algorithm, as in Problem 10.31. *Note:* In Section 7.3.1 we could not develop a closed-form algorithm because some recursions required the quantities $\mathbf{b}_m(n-1)$ and $E_m^b(n-1)$. Here we can avoid this problem by using time recursions.

10.33 In this problem we discuss several quantities that can serve to warn of ill behavior in fast RLS algorithms for FIR filters.

(a) Show that the variable

$$\eta_m(n) \triangleq \frac{\alpha_{m+1}(n)}{\alpha_m(n)} = \frac{\lambda E_m^b(n-1)}{E_m^b(n)} = 1 - g_{m+1}^{(m+1)}(n)e_m^{b*}(n)$$

satisfies the condition $0 \le \eta_m(n) \le 1$.

(b) Prove the relations

$$\alpha_m(n) = \lambda^m \frac{\det \hat{\mathbf{R}}_m(n-1)}{\det \hat{\mathbf{R}}_m(n)} \qquad E_m^f(n) = \frac{\det \hat{\mathbf{R}}_{m+1}(n)}{\det \hat{\mathbf{R}}_m(n-1)} \qquad E_m^b(n) = \frac{\det \hat{\mathbf{R}}_{m+1}(n)}{\det \hat{\mathbf{R}}_m(n)}$$

(c) Show that

$$\alpha_m(n) = \lambda^m \frac{E_m^b(n)}{E_m^f(n)}$$

and use it to explain why the quantity $\eta_m^\alpha(n) = E_m^f(n) - \lambda^m E_m^b(n)$ can be used as a warning variable.

(d) Explain how the quantities

$$\eta_{\bar{g}}(n) \triangleq \bar{g}_{M+1}^{(M+1)}(n) - \frac{e^b(n)}{\lambda E^b(n-1)}$$

and

$$\eta_b(n) \triangleq e^b(n) - \lambda E^b(n-1)\bar{g}_{M+1}^{(M+1)}(n)$$

can be used as warning variables.

10.34 When the desired response is $y(j) = \delta(j-k)$, that is, a spike at $j = k$, $0 \le k \le n$, the LS filter $\mathbf{c}_m^{(k)}$ is known as a *spiking filter* or as an LS inverse filter (see Section 8.3).

(a) Determine the normal equations and the LSE $E_m^{(k)}(n)$ for the LS filter $\mathbf{c}_m^{(k)}$.

(b) Show that $\mathbf{c}_m^{(n)} = \mathbf{g}_m(n)$ and $E_m^{(n)}(n) = \alpha_m(n)$ and explain their meanings.

(c) Use the interpretation $\alpha_m(n) = E_m^{(n)}(n)$ to show that $0 \le \alpha_m(n) \le 1$.

(d) Show that $\mathbf{a}_m(n) = \sum_{k=0}^n \mathbf{c}_m^{(k)}(n-1)x(k)$ and explain its meaning.

10.35 Derive Equations (10.7.33) through (10.7.35) for the a posteriori LS lattice-ladder structure, shown in Figure 10.38, starting with the partitionings (10.7.1) and the matrix by inversion by partitioning relations (10.7.7) and (10.7.8).

10.36 Prove relations (10.7.45) and (10.7.46) for the updating of the ladder partial correlation coefficient $\beta_m^c(n)$.

10.37 In Section 7.3.1 we derived order-recursive relations for the FLP, BLP, and FIR filtering MMSEs.

    (*a*) Following the derivation of (7.3.36) and (7.3.37), derive similar order-recursive relations for $E_m^f(n)$ and $E_m^b(n)$.

    (*b*) Show that we can obtain a complete LS lattice-ladder algorithm by replacing, in Table 10.15, the time-recursive updatings of $E_m^f(n)$ and $E_m^b(n)$ with the obtained order-recursive relations.

    (*c*) Write a MATLAB function for this algorithm, and verify it by using the equalization experiment in Example 10.5.2.

10.38 Derive the equations for the a priori RLS lattice-ladder algorithm given in Table 10.16, and write a MATLAB function for its implementation. Test the function by using the equalization experiment in Example 10.5.2.

10.39 Derive the equations for the a priori RLS lattice-ladder algorithm with error feedback (see Table 10.7), and write a MATLAB function for its implementation. Test the function by using the equalization experiment in Example 10.5.2.

10.40 Derive the equations for the a posteriori RLS lattice-ladder algorithm with error feedback (Ling et al. 1986) and write a MATLAB function for its implementation. Test the function by using the equalization experiment in Example 10.5.2.

10.41 The a posteriori and the a priori RLS lattice-ladder algorithms need the conversion factor $\alpha_m(n)$ because the updating of the quantities $E_m^f(n)$, $E_m^b(n)$, $\beta_m(n)$, and $\beta_m^c(n)$ requires both the a priori and a posteriori errors. Derive a double (a priori and a posteriori) lattice-ladder RLS filter that avoids the use of the conversion factor by updating both the a priori and the a posteriori prediction and filtering errors.

10.42 Program the RLS Givens lattice-ladder filter with square roots (see Table 10.18), and study its use in the adaptive equalization experiment of Example 10.5.2.

10.43 Derive the formulas and program the RLS Givens lattice-ladder filter without square roots (see Table 10.18), and study its use in the adaptive equalization experiment of Example 10.5.2.

10.44 In this problem we discuss the derivation of the normalized lattice-ladder RLS algorithm, which uses a smaller number of time and order updating recursions and has better numerical behavior due to the normalization of its variables. *Note:* You may find useful the discussion in Carayannis et al. (1986).

    (*a*) Define the energy and angle normalized variables

$$\bar{e}_m^f(n) = \frac{\varepsilon_m^f(n)}{\sqrt{\alpha_m(n)}\sqrt{E_m^f(n)}} \quad \bar{e}_m^b(n) = \frac{\varepsilon_m^b(n)}{\sqrt{\alpha_m(n)}\sqrt{E_m^b(n)}} \quad \bar{e}_m(n) = \frac{\varepsilon_m(n)}{\sqrt{\alpha_m(n)}\sqrt{E_m(n)}}$$

$$\bar{k}_m(n) = \frac{\beta_m(n)}{\sqrt{E_m^f(n)}\sqrt{E_m^b(n-1)}} \quad \bar{k}_m^c(n) = \frac{\beta_m^c(n)}{\sqrt{E_m(n)}\sqrt{E_m^b(n)}}$$

and show that the normalized errors and the partial correlation coefficients $\bar{k}_m(n)$ and $\bar{k}_m^c(n)$ have magnitude less than 1.

(b) Derive the following normalized lattice-ladder RLS algorithm:

$$E_0^f(-1) = E_0(-1) = \delta > 0$$

For $n = 0, 1, 2, \ldots$

$$E_0^f(n) = \lambda E_0^f(n-1) + |x(n)|^2, \quad E_0(n) = \lambda E_0(n-1) + |y(n)|^2$$

$$\bar{e}_0^f(n) = \bar{e}_0^b(n) = \frac{x(n)}{\sqrt{E_0^f(n)}}, \quad \bar{e}_0(n) = \frac{y(n)}{\sqrt{E_0(n)}}$$

For $m = 0$ to $M - 1$

$$\bar{k}_m(n) = \sqrt{1 - |\bar{e}_m^f(n)|^2}\sqrt{1 - |\bar{e}_m^b(n-1)|^2}\,\bar{k}_m(n-1) + \bar{e}_m^{f*}(n)\bar{e}_m^b(n-1)$$

$$\bar{e}_{m+1}^f(n) = \left(\sqrt{1 - |\bar{e}_m^b(n-1)|^2}\sqrt{1 - |\bar{k}_m(n)|^2}\right)^{-1}[\bar{e}_m^f(n) - \bar{k}_m(n)\,\bar{e}_m^b(n-1)]$$

$$\bar{e}_{m+1}^b(n) = \left(\sqrt{1 - |\bar{e}_m^f(n)|^2}\sqrt{1 - |\bar{k}_m(n)|^2}\right)^{-1}[\bar{e}_m^b(n-1) - \bar{k}_m(n)\,\bar{e}_m^f(n)]$$

$$\bar{k}_m^c(n) = \sqrt{1 - |\bar{e}_m(n)|^2}\sqrt{1 - |\bar{e}_m^b(n)|^2}\,\bar{k}_m^c(n-1) + \bar{e}_m^*(n)\bar{e}_m^b(n)$$

$$\bar{e}_{m+1}(n) = \left(\sqrt{1 - |\bar{e}_m^b(n)|^2}\sqrt{1 - |\bar{k}_m^c(n)|^2}\right)^{-1}[\bar{e}_m(n) - \bar{k}_m^c(n)\bar{e}_m^b(n)]$$

(c) Write a MATLAB function to implement the derived algorithm, and test its validity by using the equalization experiment in Example 10.5.2.

10.45 Prove (10.6.46) by direct manipulation of (10.6.35).

10.46 Derive the formulas for the QR-RLS lattice predictor (see Table 10.18), using the approach introduced in Section 10.6.3 (Yang and Böhme 1992).

10.47 Demonstrate how the systolic array in Figure 10.55, which is an extension of the systolic array structure shown in Figure 10.36, can be used to determine the LS error $e(n)$ and the LS

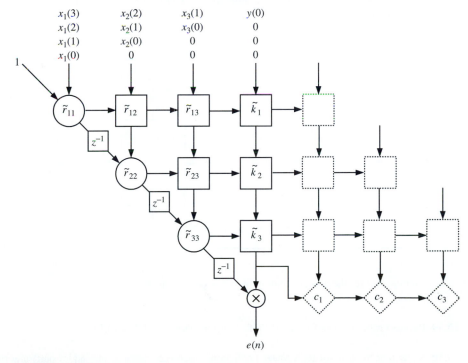

FIGURE 10.55
Systolic array implementation of the extended QR-RLS algorithm.

coefficient vector $\mathbf{c}(n)$. Determine the functions to be assigned to the dotted-line computing elements and the inputs with which they should be supplied.

10.48 The implementation of adaptive filters using multiprocessing involves the following steps: (1) partitioning of the overall computational job into individual tasks, (2) allocation of computational and communications tasks to the processors, and (3) synchronization and control of the processors. Figure 10.56 shows a cascade multiprocessing architecture used for adaptive filtering. To avoid *latency* (i.e., a delay between the filter's input and output that is larger than the sampling interval), each processor should complete its task in time less than the sampling period and use results computed by the preceding processor and the scalar computational unit at the previous sampling interval. This is accomplished by the unit delays inserted between the processors.

(a) Explain why the fast Kalman algorithm, given in Table 10.13, does not satisfy the multi-processing requirements.

(b) Prove the formulas

$$\mathbf{b}(n) = \frac{\mathbf{b}(n-1) - \mathbf{g}_{M+1}^{[M]}(n) e^{b*}(n)}{1 - g_{M+1}^{(M+1)}(n)\, e^{b*}(n)} \tag{k}$$

$$\mathbf{g}(n) = \mathbf{g}_{M+1}^{[M]}(n) - g_{M+1}^{(M+1)}(n)\, \mathbf{b}(n) \tag{l}$$

and show that they can be used to replace formulas (g) and (h) in Table 10.13.

(c) Rearrange the formulas in Table 10.13 as follows: (e), (k), (l), (a), (b), (c), (f), (d). Replace n by $n-1$ in (e), (l), and (k). Show that the resulting algorithm complies with the multiprocessing architecture shown in Figure 10.56.

(d) Draw a block diagram of a single multiprocessing section that can be used in the multiprocessing architecture shown in Figure 10.56. Each processor in Figure 10.56 can be assigned to execute one or more of the designed sections. *Note:* You may find useful the discussions in Lawrence and Tewksbury (1983) and in Manolakis and Patel (1992).

(e) Figure 10.57 shows an alternative implementation of a multiprocessing section that can be used in the architecture of Figure 10.56. Identify the input-output quantities and the various multiplier factors.

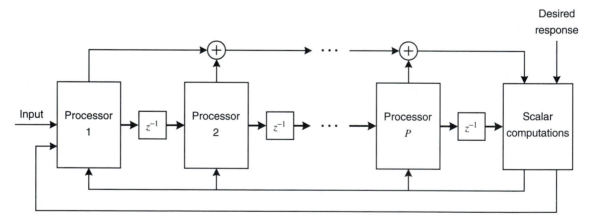

FIGURE 10.56
Cascade multiprocessing architecture for the implementation of FIR adaptive filters.

10.49 Repeat Problem 10.48 for the FAEST algorithm shown in Table 10.14.

10.50 Show that the a priori RLS linear prediction lattice (i.e., without the ladder part) algorithm with error feedback complies with the multiprocessing architecture of Figure 10.56. Explain

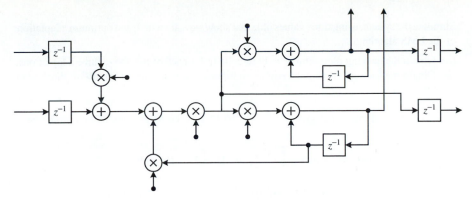

FIGURE 10.57
Section for the multiprocessing implementation of the fast Kalman algorithm.

why the addition of the ladder part violates the multiprocessing architecture. Can we rectify these violations? (See Lawrence and Tewksbury 1983.)

10.51 The fixed-length sliding window RLS algorithm is given in (10.8.4) through (10.8.10).

(a) Derive the above equations of this algorithm (see Manolakis et al. 1987).

(b) Develop a MATLAB function to implement the algorithm

$$[\texttt{c,e}] \; = \; \texttt{slwrls(x,y,L,delta,M,c0)};$$

where L is the fixed length of the window.

(c) Generate 500 samples of the following nonstationary process

$$x(n) = \begin{cases} w(n) + 0.95x(n-1) - 0.9x(n-2) & 0 \le n < 200 \\ w(n) - 0.95x(n-1) - 0.9x(n-2) & 200 \le n < 300 \\ w(n) + 0.95x(n-1) - 0.9x(n-2) & n \ge 300 \end{cases}$$

where $w(n)$ is a zero-mean, unit-variance white noise process. We want to obtain a second-order linear predictor using adaptive algorithms. Use the sliding window RLS algorithm on the data and choose $L = 50$ and 100. Obtain plots of the filter coefficients and mean square error.

(d) Now use the growing memory RLS algorithm by choosing $\lambda = 1$. Compare your results with the sliding-window RLS algorithm.

(e) Finally, use the exponentially growing memory RLS by choosing $\lambda = (L-1)/(L+1)$ that produces the same MSE. Compare your results.

10.52 Consider the definition of the MSD $\mathcal{D}(n)$ in (10.2.29) and that of the trace of a matrix (A.2.16).

(a) Show that $\mathcal{D}(n) = \text{tr}\{\boldsymbol{\Phi}(n)\}$, where $\boldsymbol{\Phi}(n)$ is the correlation matrix of $\tilde{\mathbf{c}}(n)$.

(b) For the evolution of the correlation matrix in (10.8.58), show that

$$\mathcal{D}(\infty) \simeq \mu M \sigma_v^2 + \frac{\text{tr}(\mathbf{R}^{-1} \mathbf{R}_\psi)}{4\mu}$$

10.53 Consider the analysis model given in Figure 10.42. Let the parameters of this model be as follows:

$\mathbf{c}_o(n)$ model parameters: $\quad \mathbf{c}_o(0) = \begin{bmatrix} 0.9 \\ -0.8 \end{bmatrix} \quad M = 2 \quad \rho = 0.95$

$$\boldsymbol{\psi}(n) \sim \text{WGN}(\mathbf{0}, \mathbf{R}_\psi) \quad \mathbf{R}_\psi = (0.01)^2 \mathbf{I}$$

Signal $\mathbf{x}(n)$ parameters: $\quad \mathbf{x}(n) \sim \text{WGN}(\mathbf{0}, \mathbf{R}) \quad \mathbf{R} = \mathbf{I}$

Noise $v(n)$ parameters: $\quad v(n) \sim \text{WGN}(0, \sigma_v^2) \quad \sigma_v = 0.1$

Simulate the system, using three values of μ that show slow, matched, and optimum adaptations of the LMS algorithm.

(a) Obtain the tracking plots similar to Figure 10.43 for each of the above three adaptations.

(b) Obtain the learning curve plots similar to Figure 10.44 for each of the above three adaptations.

10.54 Consider the analysis model given in Figure 10.42. Let the parameters of this model be as follows

$$\mathbf{c}_o(n) \text{ model parameters:} \quad \mathbf{c}_o(0) = \begin{bmatrix} 0.9 \\ -0.8 \end{bmatrix} \quad M = 2 \quad \rho = 0.95$$

$$\boldsymbol{\psi}(n) \sim \text{WGN}(\mathbf{0}, \mathbf{R}_\psi) \quad \mathbf{R}_\psi = (0.01)^2 \mathbf{I}$$

Signal $\mathbf{x}(n)$ parameters: $\quad \mathbf{x}(n) \sim \text{WGN}(\mathbf{0}, \mathbf{R}) \quad \mathbf{R} = \mathbf{I}$

Noise $v(n)$ parameters: $\quad v(n) \sim \text{WGN}(0, \sigma_v^2) \quad \sigma_v = 0.1$

Simulate the system, using three values of μ that show slow, matched, and optimum adaptations of the RLS algorithm.

(a) Obtain the tracking plots similar to Figure 10.49 for each of the above three adaptations.

(b) Obtain the learning curve plots similar to Figure 10.50 for each of the above three adaptations.

(c) Compare your results with those obtained in Problem 10.53.

10.55 Consider the time-varying adaptive equalizer shown in Figure 10.58 in which the time variation of the channel impulse response is given by

$$h(n) = \rho h(n-1) + \sqrt{1-\rho}\, \eta(n)$$

with $\qquad \rho = 0.95 \quad \eta(n) \sim \text{WGN}(0, \sqrt{10}) \quad h(0) = 0.5$

Let the equalizer be a single-tap equalizer and $v(n) \sim \text{WGN}(0, 0.1)$.

(a) Simulate the system for three different adaptations; that is, choose μ for slow, matched, and fast adaptations of the LMS algorithm.

(b) Repeat part (a), using the RLS algorithm.

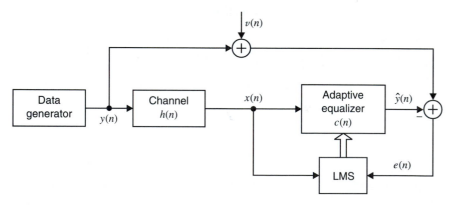

FIGURE 10.58
Adaptive channel equalizer system with time-varying channel in Problem 10.55.

CHAPTER 11

Array Processing

The subject of array processing is concerned with the extraction of information from signals collected using an array of sensors. These signals propagate spatially through a medium, for example, air or water, and the resulting wavefront is sampled by the sensor array. The information of interest in the signal may be either the content of the signal itself (communications) or the location of the source or reflection that produces the signal (radar and sonar). In either case, the sensor array data must be processed to extract this useful information. The methods utilized in most cases are extensions of the statistical and adaptive signal processing techniques discussed in previous chapters, such as spectral estimation and optimum and adaptive filtering, extended to sensor array applications.

Sensor arrays are found in a wide range of applications, including radar, sonar, seismology, biomedicine, communications, astronomy, and imaging. Each of these individual fields contains a wealth of research into the various methods for the processing of array signals. Generally, the type of processing is dictated by the particular application. However, an underlying set of principles and techniques is common to a diverse set of applications. In this chapter, we focus on the fundamentals of array processing with emphasis on optimum and adaptive techniques. To simplify the discussion, we concentrate on linear arrays, where the sensors are located along a line. The extension of this material to other array configurations is fairly straightforward in most cases. The intent of this chapter is to first give the uninitiated reader some exposure to the basic principles of array processing and then apply adaptive processing techniques to the array processing problem. For a more detailed treatment of array processing methods, see Monzingo and Miller (1980), Hudson (1981), Compton (1988), and Johnson and Dudgeon (1993).

The chapter begins in Section 11.1 with a brief background in some array fundamentals, including spatially propagating signals, modulation and demodulation, and the array signal model. In Section 11.2, we introduce the concept of beamforming, that is, the spatial discrimination or filtering of signals collected with a sensor array. We look at conventional, that is, nonadaptive, beamforming and touch upon many of the common considerations for an array that affect its performance, for example, element spacing, resolution, and sidelobe levels. In Section 11.3, we look at the optimum beamformer, which is based on a priori knowledge of the data statistics. Within this framework, we discuss some of the specific aspects of adaptive processing that affect performance in Section 11.4. Then, in Section 11.5, we discuss adaptive array processing methods that estimate the statistics from actual data, first block-adaptive and then sample-by-sample adaptive methods. Section 11.6 discusses other adaptive array processing techniques that were born out of practical considerations for various applications. The determination of the angle of arrival of a spatial signal is the topic of Section 11.7. In Section 11.8, we give a brief description of space-time adaptive processing.

11.1 ARRAY FUNDAMENTALS

The information contained in a spatially propagating signal may be either the location of its source or the content of the signal itself. If we are interested in obtaining this information, we generally must deal with the presence of other, undesired signals. Much as a frequency-selective filter emphasizes signals at a certain frequency, we can choose to focus on signals from a particular direction. Clearly, this task can be accomplished by using a single sensor, provided that it has the ability to spatially discriminate; that is, it passes signals from certain directions while rejecting those from other directions. Such a single-sensor system, shown in Figure 11.1(a), is commonly found in communications and radar applications in which the signals are collected over a continuous spatial extent or *aperture* using a parabolic dish. The signals are reflected to the antenna in such a way that signals from the direction in which the dish is pointed are emphasized. The ability of a sensor to spatially discriminate, known as *directivity*, is governed by the shape and physical characteristics of its geometric structure. However, such a single-sensor system has several drawbacks. Since the sensor relies on mechanical pointing for directivity, it can extract and track signals from only one direction at a time; it cannot look in several directions simultaneously. Also, such a sensor cannot adapt its response, which would require physically changing the aperture, in order to reject potentially strong sources that may interfere with the extraction of the signals of interest.

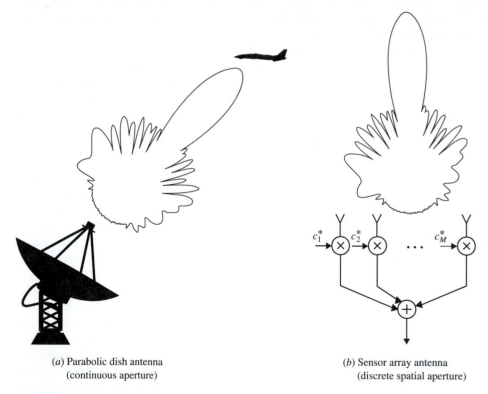

(a) Parabolic dish antenna
(continuous aperture)

(b) Sensor array antenna
(discrete spatial aperture)

FIGURE 11.1
Comparison of a single, directive antenna with multiple sensors that make up an antenna array. In both cases, the response is designed to emphasize signals from a certain direction through spatial filtering, either continuous or discrete.

An array of sensors has the ability to overcome these shortcomings of a single sensor. Figure 11.1 (b) illustrates the use of a sensor array. The sensor array signals are combined in such a way that a particular direction is emphasized. However, the direction in which the array is focused or pointed is almost independent of the orientation of the array. Therefore, the sensors can be combined in distinct, separate ways so as to emphasize different directions, all of which may contain signals of interest. Since the various weighted summations of the sensors simply amount to processing the same data in different ways, these multiple sources can be extracted simultaneously. Also arrays have the ability to adjust the overall rejection level in certain directions to overcome strong interference sources. In this section, we discuss some fundamentals of sensor arrays. First, we give a brief description of spatially propagating signals and the modulation and demodulation operations. Then we develop a signal model, first for an arbitrary array and then by simplifying to the case of a uniform linear array. In addition, we point out the interpretation of a sensor array as a mechanism for the spatial sampling of a spatially propagating signal.

11.1.1 Spatial Signals

In their most general form, spatial signals are signals that propagate through space. These signals originate from a source, travel through a propagation medium, say, air or water, and arrive at an array of sensors that spatially samples the waveform. A processor can then take the data collected by the sensor array and attempt to extract information about the source, based on certain characteristics of the propagating wave. Since space is three-dimensional, a spatial signal at a point specified by the vector $\mathbf{r}$ can be represented either in Cartesian coordinates (x, y, z) or in spherical coordinates $(R, \phi_{az}, \theta_{el})$ as shown in Figure 11.2. Here, $R = \|\mathbf{r}\|$ represents range or the distance from the origin, and ϕ_{az} and θ_{el} are the azimuth and elevation angles, respectively.

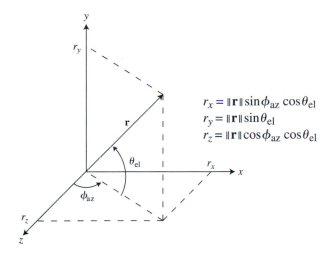

$$r_x = \|\mathbf{r}\|\sin\phi_{az}\cos\theta_{el}$$
$$r_y = \|\mathbf{r}\|\sin\theta_{el}$$
$$r_z = \|\mathbf{r}\|\cos\phi_{az}\cos\theta_{el}$$

FIGURE 11.2
Three-dimensional space describing azimuth, elevation, and range.

The propagation of a spatial signal is governed by the solution to the wave equation. For electromagnetic propagating signals, the wave equation can be deduced from Maxwell's equations (Ishimaru 1990), while for sound waves the solution is governed by the basic laws of acoustics (Kino 1987; Jensen et al. 1994). However, in either case, for a propagating wave emanating from a source located at $\mathbf{r}_0$, one solution is a single-frequency wave given by

$$s(t, \mathbf{r}) = \frac{A}{\|\mathbf{r} - \mathbf{r}_0\|^2} e^{j2\pi F_c\left(t - \frac{\|\mathbf{r}-\mathbf{r}_0\|}{c}\right)} \tag{11.1.1}$$

where A is the complex amplitude, F_c is the carrier frequency of the wave, and c is the speed of propagation of the wave. The speed of propagation is determined by the type of wave (electromagnetic or acoustic) and the propagation medium. For the purposes of this discussion, we ignore the singularity at the source (origin); that is, $s(t, \mathbf{r}_0) = \infty$. This equation suppresses the dependencies on ϕ_{az} and θ_{el} since the wave propagates radially from the source. At any point in space, the wave has the temporal frequency F_c. In (11.1.1) and for the remainder of this chapter, we will assume a lossless, nondispersive propagation medium, that is, a medium that does not attenuate the propagating signal further than predicted by the wave equation, and the propagation speed is uniform so that the wave travels according to (11.1.1). A dispersive medium adds a frequency dependence to the wave propagation (Jensen et al. 1994). Clearly, the signal travels in time where the spatial propagation is determined by the direct coupling between space and time in order to satisfy (11.1.1). We can then define the wavelength of the propagating wave as

$$\lambda = \frac{c}{F_c} \tag{11.1.2}$$

which is the distance traversed by the wave during one temporal period.

Two other simplifying assumptions will be made for the remainder of this chapter. First, the propagating signals are assumed to be produced by a point source; that is, the size of the source is small with respect to the distance between the source and the sensors that measure the signal. Second, the source is assumed to be in the "far field," i.e., at a large distance from the sensor array, so that the spherically propagating wave can be reasonably approximated with a plane wave. This approximation again requires the source to be far removed from the array so that the curvature of the wave across the array is negligible. This concept is illustrated in Figure 11.3. Multiple sources are treated through superposition of the various spatial signals at the sensor array. Although each individual wave radiates from its source, generally the origin ($\mathbf{r} = \mathbf{0}$) is reserved for the position of the sensor array since this is the point in space at which the collection of waves is measured. For more details on spatially propagating signals, see Johnson and Dudgeon (1993).

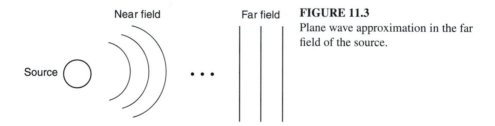

FIGURE 11.3
Plane wave approximation in the far field of the source.

Let us now consider placing a linear array in three-dimensional space in order to sense the propagating waves. The array consists of a series of elements located on a line with uniform spacing. Such an array is known as a *uniform linear array (ULA)*. For convenience, we choose the coordinate system for our three-dimensional space as in Figure 11.2 such that the ULA lies on the x axis. In addition, we have a wave originating from a point $\mathbf{r}$ in this three-dimensional space that is located in the far field of the array such that the propagating signal can be approximated by a plane wave at the ULA. The plane wave impinges on the ULA as illustrated in Figure 11.4. As we will see, the differences in distance between the sensors determine the relative delays in arrival of the plane wave. The point from which the wave originates can be described by its distance from the origin $\|\mathbf{r}\|$ and its azimuth and elevation angles ϕ_{az} and θ_{el}, respectively. If the distance between elements of the ULA is d, then the difference in propagation distance between neighboring elements for a plane wave arriving from an azimuth ϕ_{az} and elevation θ_{el} is

$$d_x = \|\mathbf{r}\| \sin \phi_{az} \cos \theta_{el} \tag{11.1.3}$$

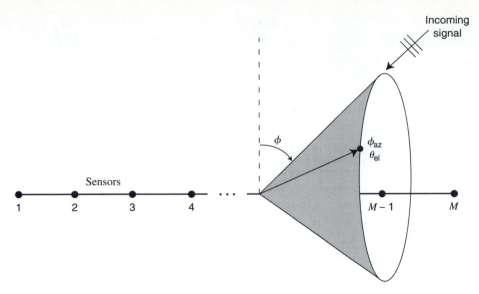

FIGURE 11.4
Cone angle ambiguity surface for a uniform linear array.

These differences in the propagation distance that the plane wave must travel to each of the sensors are a function of a general angle of arrival with respect to the ULA ϕ. If we consider the entire three-dimensional space, we note that equivalent delays are produced by any signal arriving from a cone about the ULA. Therefore, any signal arriving at the ULA on this surface has the same set of relative delays between the elements. This conical ambiguity surface is illustrated in Figure 11.4. For this reason, the angle of incidence to a linear array is commonly referred to as the *cone angle*, ϕ_{cone}. We see that the cone angle is related to the physical angles, azimuth and elevation defined in Figure 11.4, by

$$\sin \phi = \sin \phi_{\text{az}} \cos \theta_{\text{el}} \qquad (11.1.4)$$

where $\phi = 90° - \phi_{\text{cone}}$. In this manner, we can take a given azimuth and elevation pair and determine their corresponding cone angle. For the remainder of this chapter, we use the terms *angle of arrival* and simply *angle* interchangeably.

11.1.2 Modulation-Demodulation

The spatial propagation of signals was described by (11.1.1) using a propagation speed c and a center frequency F_c. For a general class of signals, the signal of interest $s_0(t)$ has a bandwidth that is a small fraction of the center frequency and is modulated up to the center frequency. Since the propagating wave then "carries" certain information to the receiving point in the form of a temporal signal, F_c is commonly referred to as the *carrier frequency*. The process of generating the signal $\tilde{s}_0(t)$ from $s_0(t)$ in order to transmit this information is accomplished by mixing the signal $s_0(t)$ with the carrier waveform $\cos 2\pi F_c t$ in an operation known as *modulation*. The propagating signal is then produced by a high-gain transmitter. The signal travels through space until it arrives at a sensor that measures the signal. Let us denote the received propagating signal as

$$\tilde{s}_0(t) = s_0(t) \cos 2\pi F_c t = \tfrac{1}{2} s_0(t)(e^{j2\pi F_c t} + e^{-j2\pi F_c t}) \qquad (11.1.5)$$

where we say that the signal $s_0(t)$ is carried by the propagating waveform $\cos 2\pi F_c t$. The spectrum of $\tilde{s}_0(t)$ is made up of two components: the spectrum of the signal $s_0(t)$ shifted to F_c and shifted to $-F_c$ and reflected about $-F_c$. This spectrum $\tilde{S}_0(F)$ is shown in Figure 11.5. Here we indicate the signal $s_0(t)$ has a bandwidth B. The baseband signal $s_0(t)$,

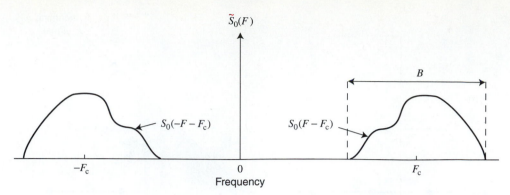

FIGURE 11.5
Spectrum of a bandpass signal.

although originating as a real signal prior to modulation, has a nonsymmetric spectrum due to the asymmetric[†] spectral response of the propagation medium about frequency F_c. The received signal $\tilde{s}_0(t)$, though, is real-valued; that is, its spectrum exhibits even symmetry about $F = 0$. This fact is consistent with actual, physical signals that *are* real-valued as they are received and measured by a sensor.

The reception of spatially propagating signals with a sensor is only the beginning of the process of forming digital samples for both the in-phase and quadrature components of the sensor signal. Upon reception of the signal $\tilde{s}_0(t)$, the signal is mixed back to baseband in an operation known as *demodulation*. Included in the mth sensor signal is thermal noise due to the electronics of the sensor $w_m(t)$

$$\tilde{x}_m(t) = \tilde{s}_0(t) * h_m(t, \phi_s) + \tilde{w}_m(t) \tag{11.1.6}$$

where $h_m(t, \phi_s)$ is the combined temporal and spatial impulse response of the mth sensor. The angle ϕ_s is the direction from which $\tilde{s}_0(t)$ was received. In the case of an omnidirectional sensor with an equal response in all directions, the impulse response no longer is dependent on the angle of the signal. The demodulation process involves multiplying the received signal by $\cos 2\pi F_c t$ and $-\sin 2\pi F_c t$ to form both the in-phase and quadrature channels, respectively. Note the quadrature component is 90° out of phase of the in-phase component. The entire process is illustrated in Figure 11.6 for the mth sensor. This structure is referred to as the *receiver* of the mth channel.

Following demodulation, the signals in each channel are passed through a low-pass filter to remove any high-frequency components. The cutoff frequency of this low-pass filter determines the bandwidth of the receiver. Throughout this chapter, we assume a perfect or ideal low-pass filter, that is, a response of 1 in the passband and 0 in the stopband. In practice, the characteristics of the actual, nonideal low-pass filter can impact the performance of the resulting processor. Following the low-pass filtering operation, the signals in both the in-phase and quadrature channels are critically (Nyquist) sampled at the receiver bandwidth B. Oversampling at greater than the receiver bandwidth is also possible but is not considered here. More details on the signals at the various stages of the receiver, including the sensor impulse response, are covered in the next section on the array signal model. The output of the receiver is a complex-valued, discrete-time signal for the mth sensor with the in-phase and quadrature channels generating the real and imaginary portions of the signal

$$x_m(n) = x_m^{(I)}(n) + j x_m^{(Q)}(n) \tag{11.1.7}$$

For more details on the complex representation of bandpass signals, sampling, and the modulation and demodulation process, see Section 2.1. We should also mention that the sampling process in many systems is implemented using a technique commonly referred

---

[†] The asymmetry can arise from dispersive effects in the transmission medium.

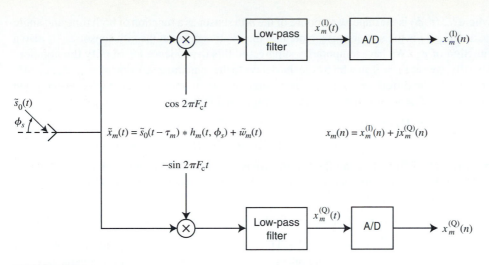

FIGURE 11.6
Block diagram of propagating signal arriving at a sensor with a receiver.

to as *digital in-phase/quadrature* or simply *digital IQ* (Rader 1984), rather than the more classical method outlined in this section. The method is more efficient as it only requires a single analog-to-digital (A/D) converter, though at a higher sampling rate.[†] See Rader (1984) for more details.

11.1.3 Array Signal Model

We begin by developing a model for a single spatial signal in noise received by a ULA. Consider a signal received by the ULA from an angle ϕ_s as in Figure 11.6. Each sensor receives the spatially propagating signal and converts its measured energy to voltage. This voltage signal is then part of the receiver channel from Figure 11.6. In addition, the receiver contains noise due to internal electronics known as *thermal noise*.[‡] Recall from (11.1.6) that $\tilde{x}_m(t)$ is the continuous-time signal in the mth sensor containing both the received carrier-modulated signals and thermal noise. The signal $x_m(t)$ is then obtained by demodulating $\tilde{x}_m(t)$ to baseband and low-pass filtering to the receiver bandwidth, while $x_m(n)$ is its discrete-time counterpart. Since the model is assumed to be linear, the extension to multiple signals, including interference sources, is straightforward.

The discrete-time signals from a ULA may be written as a vector containing the individual sensor signals

$$\mathbf{x}(n) = [x_1(n) \; x_2(n) \; \cdots \; x_M(n)]^T \tag{11.1.8}$$

where M is the total number of sensors. A single observation or measurement of this signal vector is known as an *array snapshot*. We begin by examining a single, carrier-modulated signal $\tilde{s}_0(t) = s_0(t) \cos 2\pi F_c t$ arriving from angle ϕ_s that is received by the mth sensor. We assume that the signal $s_0(t)$ has a deterministic amplitude and random, uniformly distributed phase. The ~ symbol is used to indicate that the signal is a passband or carrier-modulated signal. Here $s_0(t)$ is the baseband signal, and F_c is the carrier frequency. This signal is received by the mth sensor with a delay τ_m

$$\tilde{x}_m(t) = h_m(t, \; \phi_s) * \tilde{s}_0(t - \tau_m) + \tilde{w}_m(t) \tag{11.1.9}$$

[†]This digital IQ technique is very important for adaptive processing as I/Q channel mismatch can limit performance. One A/D converter avoids this source of mismatch.

[‡]Another source of noise may be external background noise. Many times this is assumed to be isotropic so that the overall noise signal is uncorrelated from sensor to sensor.

where $h_m(t, \phi)$ is the impulse response of the mth sensor as a function of both time and angle ϕ, and $\tilde{w}_m(t)$ is the sensor noise. Note that the relative delay at the mth sensor τ_m is also a function of ϕ_s. We have temporarily suppressed this dependence to simplify the notation. Usually, we set $\tau_1 = 0$, in which case the delays to the remaining sensors ($m = 2, 3, \ldots, M$) are simply the differences in propagation time of $\tilde{s}_0(t)$ to these sensors with respect to the first sensor. The sensor signal can also be expressed in the frequency domain as

$$\tilde{X}_m(F) = H_m(F, \phi_s)\tilde{S}_0(F)e^{-j2\pi F\tau_m} + \tilde{W}_m(F)$$
$$= H_m(F, \phi_s)[S_0(F - F_c) + S_0^*(-F - F_c)]e^{-j2\pi F\tau_m} + \tilde{W}_m(F) \quad (11.1.10)$$

by using (11.1.5) and taking the Fourier transform of (11.1.9). Following demodulation and ideal low-pass filtering of the signal from the mth sensor, as shown in Figure 11.6, the spectrum of the signal is

$$X_m(F) = H_m(F + F_c, \phi_s)S_0(F)e^{-j2\pi(F+F_c)\tau_m} + W_m(F) \quad (11.1.11)$$

where $X_m(F) = X_m^{(I)}(F) + jX_m^{(Q)}(F)$. The second term $S_0^*(-F - 2F_c)$ has been removed through the ideal low-pass filtering operation. This ideal low-pass filter has a value of unity across its passband so that $W_m(F) = \tilde{W}_m(F + F_c)$ for $|F| < B/2$.

We now make a critical, simplifying assumption: The bandwidth of $s_0(t)$ is small compared to the carrier frequency; this is known as the *narrowband assumption*. This assumption allows us to approximate the propagation delays of a particular signal between sensor elements with a phase shift. There are numerous variations on this assumption, but in general it holds for cases in which the signal bandwidth is less than some small percentage of the carrier frequency, say, less than 1 percent. The ratio of the signal bandwidth to the carrier frequency is referred to as the *fractional bandwidth*. However, the fractional bandwidth for which the narrowband assumption holds is strongly dependent on the length of the array and the strength of the received signals. Thus, we might want to consider the *time-bandwidth product* (*TBWP*), which is the maximum amount of time for a spatial signal to propagate across the entire array ($\phi_s = \pm 90°$). If TBWP $\ll 1$, then the narrowband assumption is valid. The effects of bandwidth on performance are treated in Section 11.4.2.

In addition to the narrowband assumption, we assume that the response of the sensor is constant across the bandwidth of the receiver, that is, $H_m(F + F_c, \phi_s) = H_m(F_c, \phi_s)$ for $|F| < B/2$. Thus, the spectrum in (11.1.11) simplifies to

$$X_m(F) = H_m(F_c, \phi_s)S_0(F)e^{-j2\pi F_c\tau_m} + W_m(F) \quad (11.1.12)$$

and the discrete-time signal model is obtained by sampling the inverse Fourier transform of (11.1.12)

$$x_m(n) = H_m(F_c, \phi_s)s_0(n)e^{-j2\pi F_c\tau_m} + w_m(n) \quad (11.1.13)$$

The term $w_m(n)$ corresponds to $W_m(F)$, the sensor thermal noise across the bandwidth of the receiver of the mth sensor. Furthermore, we assume that the power spectral density of this noise is flat across this bandwidth; that is, the discrete-time noise samples are uncorrelated. Also, the thermal noise in all the sensors is mutually uncorrelated.[†] If we further assume that each of the sensors in the array has an equal, omnidirectional response at frequency F_c, that is, $H_m(F_c, \phi_s) = H(F_c, \phi_s) = $ constant, for $1 \le m \le M$, then the constant sensor responses can be absorbed into the signal term[‡]

$$s(n) = H(F_c)s_0(n) \quad (11.1.14)$$

[†] In actual systems, thermal noise samples are temporally correlated through the use of antialiasing filters prior to digital sampling. In addition, the thermal noise between sensors may be correlated due to mutual coupling of the sensors.

[‡] In many systems, we can compensate for differences in responses by processing signals from the sensors in an attempt to make their responses as similar as possible. When the data from the sensors are used to perform this compensation, the process is known as *adaptive channel matching*.

For the remainder of the chapter, we use the signal $s(n)$ as defined in (11.1.14). Using (11.1.8) and (11.1.13), we can then write the full-array discrete-time signal model as

$$\mathbf{x}(n) = \sqrt{M}\, \mathbf{v}(\phi_s) s(n) + \mathbf{w}(n) \tag{11.1.15}$$

where
$$\mathbf{v}(\phi) = \frac{1}{\sqrt{M}} [1 \ \ e^{-j2\pi F_c \tau_2(\phi)} \ \ \cdots \ \ e^{-j2\pi F_c \tau_M(\phi)}]^T \tag{11.1.16}$$

is the *array response vector*. We have chosen to measure all delays relative to the first sensor $[\tau_1(\phi) = 0]$ and are now indicating the dependence of these delays on ϕ. We use the normalization of $1/\sqrt{M}$ for mathematical convenience so that the array response vector has unit norm, that is, $\|\mathbf{v}(\phi)\|^2 = \mathbf{v}^H(\phi)\mathbf{v}(\phi) = 1$. The factor is compensated for with the $\sqrt{M}$ term in (11.1.15). The assumption of equal, omnidirectional sensor responses is necessary to simplify the analysis but should always be kept in mind when considering experimentally collected data for which this assumption certainly will not hold exactly. The other critical assumption made is that we have perfect knowledge of the array sensor locations, which also must be called into question for actual sensors and the data collected with them.

Up to this point, we have not made any assumptions about the form of the array, so that the array signal model we have developed holds for arbitrary arrays. Now we wish to focus our attention on the ULA, which is an array that has all its elements on a line with equal spacing between the elements. The ULA is shown in Figure 11.7, and the interelement spacing is denoted by d. Consider the single propagating signal that impinges on the ULA from an angle ϕ. Since all the elements are equally spaced, the spatial signal has a difference in propagation paths between any two successive sensors of $d \sin \phi$ that results in a time delay of

$$\tau(\phi) = \frac{d \sin \phi}{c} \tag{11.1.17}$$

where c is the rate of propagation of the signal. As a result, the delay to the mth element with respect to the first element in the array is

$$\tau_m(\phi) = (m - 1)\frac{d \sin \phi}{c} \tag{11.1.18}$$

and substituting into (11.1.16), we see the array response vector for a ULA is

$$\mathbf{v}(\phi) = \frac{1}{\sqrt{M}} [1 \ \ e^{-j2\pi[(d \sin \phi)/\lambda]} \ \ \cdots \ \ e^{-j2\pi[(d \sin \phi)/\lambda](M-1)}]^T \tag{11.1.19}$$

since $F_c = c/\lambda$.

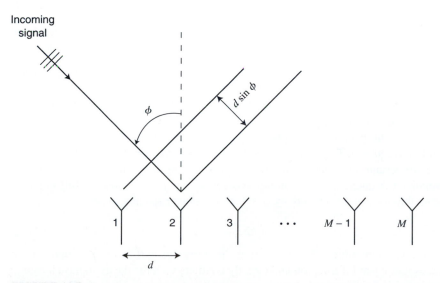

FIGURE 11.7
Plane wave impinging on a uniform linear array.

11.1.4 The Sensor Array: Spatial Sampling

In general, we can think of a sensor array as a mechanism for spatially sampling wavefronts propagating at a certain operating (carrier) frequency. Since in most instances the user either controls or has knowledge of the operating frequency, the sensor array provides a reliable means of interrogating the incoming wavefront for information. Similar to temporal sampling, the sensor array provides discrete (spatially sampled) data that can be used without loss of information, provided certain conditions are met. Namely, the sampling frequency must be high enough so as not to create spatial ambiguities or, in other words, to avoid spatial aliasing. The advantages of discrete-time processing and digital filtering have been well documented (Oppenheim and Schafer 1989; Proakis and Manolakis 1996). In the case of the spatial processing of signals, spatial sampling using an array provides the capability to change the characteristics of a discrete spatial filter, which is not possible for a continuous spatial aperture.

An arbitrary array performs its sampling in multiple dimensions and along a nonuniform grid so that it is difficult to compare to discrete-time sampling. However, a ULA has a direct correspondence to uniform, regular temporal sampling, since it samples uniformly in space on a linear axis. Thus, for a ULA we can talk about a *spatial sampling frequency* U_s defined by

$$U_s = \frac{1}{d} \tag{11.1.20}$$

where the *spatial sampling period* is determined by the interelement spacing d and is measured in cycles per unit of length (meters). Recall from (11.1.19) that the measurements made with a ULA on a narrowband signal correspond to a phase progression across the sensors determined by the angle of the incoming signal. As with temporal signals, the phase progression for uniform sampling is a consequence of the frequency; that is, consecutive samples of the same signal differ only by a phase shift of $e^{j2\pi F}$, where F is the frequency. In the case of a spatially propagating signal, this frequency is given by

$$U = \frac{\sin\phi}{\lambda} \tag{11.1.21}$$

which can be thought of as the *spatial frequency*. The *normalized spatial frequency* is then defined by

$$u \triangleq \frac{U}{U_s} = \frac{d\sin\phi}{\lambda} \tag{11.1.22}$$

Therefore, we can rewrite the array response vector from (11.1.19) in terms of the normalized spatial frequency as

$$\mathbf{v}(\phi) = \mathbf{v}(u) = \frac{1}{\sqrt{M}}[1 \; e^{-j2\pi u} \cdots e^{-j2\pi u(M-1)}]^T \tag{11.1.23}$$

which we note is simply a Vandermonde vector (Strang 1998), that is, a vector whose elements are successive integer powers of the same number, in this case $e^{-j2\pi u}$.

The interelement spacing d is simply the spatial sampling interval, which is the inverse of the sampling frequency. Therefore, similar to Shannon's theorem for discrete-time sampling, there are certain requirements on the spatial sampling frequency to avoid aliasing. Since normalized frequencies are unambiguous for $-\frac{1}{2} \le u < \frac{1}{2}$ and the full range of possible unambiguous angles is $-90° \le \phi \le 90°$, the sensor spacing must be

$$d \le \frac{\lambda}{2} \tag{11.1.24}$$

to prevent spatial ambiguities. Since lowering the array spacing below this upper limit only provides redundant information and directly conflicts with the desire to have as much aperture as possible for a fixed number of sensors, we generally set $d = \lambda/2$. This tradeoff is further explored using beampatterns in the next section.

In many applications, the desired information to be extracted from an array of sensors is the content of a spatially propagating signal from a certain direction. The content may be a message contained in the signal, such as in communications applications, or merely the existence of the signal, as in radar and sonar. To this end, we want to linearly combine the signals from all the sensors in a manner, that is, with a certain weighting, so as to examine signals arriving from a specific angle. This operation, shown in Figure 11.8, is known as *beamforming* because the weighting process emphasizes signals from a particular direction while attenuating those from other directions and can be thought of as casting or forming a beam. In this sense, a beamformer is a spatial filter; and in the case of a ULA, it has a direct analogy to an FIR frequency-selective filter for temporal signals, as discussed in Section 1.5.1. Beamforming is commonly referred to as "electronic" steering since the weights are applied using electronic circuitry following the reception of the signal for the purpose of steering the array in a particular direction.[†] This can be contrasted with mechanical steering, in which the antenna is physically pointed in the direction of interest. For a complete tutorial on beamforming see Van Veen and Buckley (1988, 1998).

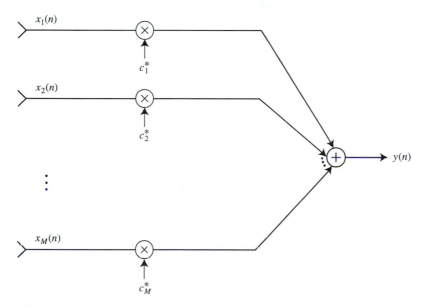

FIGURE 11.8
Beamforming operation.

In its most general form, a beamformer produces its output by forming a weighted combination of signals from the M elements of the sensor array, that is,

$$y(n) = \sum_{m=1}^{M} c_m^* x_m(n) = \mathbf{c}^H \mathbf{x}(n) \tag{11.2.1}$$

where

$$\mathbf{c} = [c_1 \; c_2 \; \cdots \; c_M]^T \tag{11.2.2}$$

is the column vector of beamforming weights. The beamforming operation for an M element array is illustrated in Figure 11.8.

[†] In general, performance does degrade as the angle to which the array is steered approaches $\phi = -90°$ or $\phi = 90°$. Although the array is optimized at broadside ($\phi = 0°$), it certainly can steer over a wide range of angles about broadside for which performance degradation is minimal.

Beam response

A standard tool for analyzing the performance of a beamformer is the response for a given weight vector **c** as a function of angle ϕ, known as the *beam response*. This angular response is computed by applying the beamformer **c** to a set of array response vectors from all possible angles, that is, $-90° \leq \phi < 90°$,

$$C(\phi) = \mathbf{c}^H \mathbf{v}(\phi) \tag{11.2.3}$$

Typically, in evaluating a beamformer, we look at the quantity $|C(\phi)|^2$, which is known as the *beampattern*. Alternatively, the beampattern can be computed as a function of normalized spatial frequency u from (11.1.22). For a ULA with $\lambda/2$ element spacing, the beampattern as a function of u can be efficiently computed using the FFT for $-\frac{1}{2} \leq u < \frac{1}{2}$ at points separated by $1/N_{\text{fft}}$ where $N_{\text{fft}} \geq M$ is the FFT size. Thus, a beampattern can be computed in MATLAB with the command C=fftshift(fft(c,N_fft))/sqrt(M), where the FFT size is selected to display the desired level of detail. To compute the corresponding angles of the beampattern, we can simply convert spatial frequency to angle as

$$\phi = \arcsin \frac{\lambda}{d} u \tag{11.2.4}$$

A sample beampattern for a 16-element uniform array with uniform weighting ($c_m = 1/\sqrt{M}$) is shown in Figure 11.9, which is plotted on a logarithmic scale in decibels. The large mainlobe is centered at $\phi = 0°$, the direction in which the array is steered. Also notice the unusual sidelobe structure created by the nonlinear relationship between angle and spatial frequency in (11.2.4) at angles away from broadside ($\phi = 0°$).

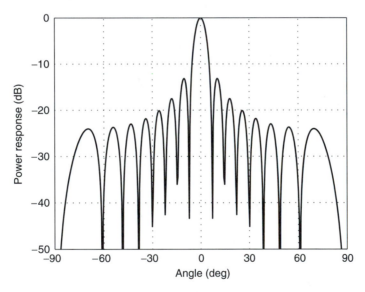

FIGURE 11.9
A sample beampattern of a spatial matched filter for an $M = 16$ element ULA steered to $\phi = 0°$.

Important note. The beampattern is the spatial frequency response of a given beamformer. It should not be confused with the *steered response*, which is the response of the array to a certain set of spatial signals impinging on the array as we steer the array to all possible angles. Since this operation corresponds to measuring the power as a function of spatial frequency or angle, the steered response might be better defined as the *spatial power spectrum*

$$R(\phi) = E\{|\mathbf{c}^H(\phi)\mathbf{x}(n)|^2\} \tag{11.2.5}$$

where the choice of the beamformer $\mathbf{c}(\phi)$ determines the type of spatial spectrum, say, conventional or minimum-variance. Various spectrum estimation techniques were discussed in Chapters 5 and 9, several of which can be extended for the estimation of the spatial spectrum from measurements in practical applications. One interpretation of the estimation of the spectrum was made as a bank of frequency-selective filters at the frequencies at which the spectrum is computed. Similarly, the computation of the spatial spectrum can be thought of as the output of a bank of beamformers steered to the angles at which the spatial spectrum is computed.

Output signal-to-noise ratio

We now look at the signal-to-noise ratio (SNR) of the beamformer output and determine the improvement in SNR with respect to each element, known as the *beamforming gain*. Let us consider the signal model for a ULA from (11.1.15), which consists of a signal of interest arriving from an angle ϕ_s and thermal sensor noise $\mathbf{w}(n)$. The beamformer or spatial filter $\mathbf{c}$ is applied to the array signal $\mathbf{x}(n)$ as

$$y(n) = \mathbf{c}^H \mathbf{x}(n) = \sqrt{M} \mathbf{c}^H \mathbf{v}(\phi_s) s(n) + \bar{w}(n) \qquad (11.2.6)$$

where $\bar{w}(n) = \mathbf{c}^H \mathbf{w}(n)$ is the noise at the beamformer output and is also temporally uncorrelated. The beamformer output power is

$$P_y = E\{|y(n)|^2\} = \mathbf{c}^H \mathbf{R}_x \mathbf{c} \qquad (11.2.7)$$

where
$$\mathbf{R}_x = E\{\mathbf{x}(n)\mathbf{x}^H(n)\} \qquad (11.2.8)$$

is the correlation matrix of the array signal $\mathbf{x}(n)$. Recall from (11.1.15) and (11.1.23) that the signal for the mth element is given by

$$x_m(n) = e^{-j2\pi(m-1)u_s} s(n) + w_m(n) \qquad (11.2.9)$$

where u_s is the normalized spatial frequency of the array signal produced by $s(n)$. The signal $s(n)$ is the signal of interest within a single sensor including the sensor response $H_m(F_c)$ from (11.1.14). Therefore, the signal-to-noise ratio in each element is given by

$$\text{SNR}_{\text{elem}} \triangleq \frac{\sigma_s^2}{\sigma_w^2} = \frac{|e^{-j2\pi(m-1)u_s} s(n)|^2}{E\{|w_m(n)|^2\}} \qquad (11.2.10)$$

where $\sigma_s^2 = E\{|s(n)|^2\}$ and $\sigma_w^2 = E\{|w_m(n)|^2\}$ are the element level signal and noise powers, respectively. Recall that the signal $s(n)$ has a deterministic amplitude and random phase. We assume that all the elements have equal noise power σ_w^2 so that the SNR does not vary from element to element. This SNR_{elem} is commonly referred to as the *element level SNR* or the *SNR per element*.

Now if we consider the signals at the output of the beamformer, the signal and noise powers are given by

$$P_s = E\{|\sqrt{M}[\mathbf{c}^H \mathbf{v}(\phi_s)]s(n)|^2\} = M\sigma_s^2 |\mathbf{c}^H \mathbf{v}(\phi_s)|^2 \qquad (11.2.11)$$

$$P_n = E\{|\mathbf{c}^H \mathbf{w}(n)|^2\} = \mathbf{c}^H \mathbf{R}_n \mathbf{c} = \|\mathbf{c}\|^2 \sigma_w^2 \qquad (11.2.12)$$

because $\mathbf{R}_n = \sigma_w^2 \mathbf{I}$. Therefore, the resulting SNR at the beamformer output, known as the *array SNR*, is

$$\text{SNR}_{\text{array}} = \frac{P_s}{P_n} = \frac{M|\mathbf{c}^H \mathbf{v}(\phi_s)|^2}{\|\mathbf{c}\|^2} \frac{\sigma_s^2}{\sigma_w^2} = \frac{|\mathbf{c}^H \mathbf{v}(\phi_s)|^2}{\|\mathbf{c}\|^2} M \, \text{SNR}_{\text{elem}} \qquad (11.2.13)$$

which is simply the product of the beamforming gain and the element level SNR. Thus, the *beamforming gain* is given by

$$G_{\text{bf}} \triangleq \frac{\text{SNR}_{\text{array}}}{\text{SNR}_{\text{elem}}} = \frac{|\mathbf{c}^H \mathbf{v}(\phi_s)|^2}{|\mathbf{c}|^2} M \qquad (11.2.14)$$

The beamforming gain is strictly a function of the angle of arrival ϕ_s of the desired signal, the beamforming weight vector $\mathbf{c}$, and the number of sensors M.

11.2.1 Spatial Matched Filter

Recall the array signal model of a single signal, arriving from a direction ϕ_s, with sensor thermal noise

$$\mathbf{x}(n) = \sqrt{M}\mathbf{v}(\phi_s)s(n) + \mathbf{w}(n)$$
$$= [s(n)\, e^{-j2\pi u_s}s(n) \, \cdots \, e^{-j2\pi(M-1)u_s}s(n)]^T + \mathbf{w}(n) \tag{11.2.15}$$

where the components of the noise vector $\mathbf{w}(n)$ are uncorrelated and have power σ_w^2, that is, $E\{\mathbf{w}(n)\mathbf{w}^H(n)\} = \sigma_w^2\mathbf{I}$. The individual elements of the array contain the same signal $s(n)$ with different phase shifts corresponding to the differences in propagation times between elements. Ideally, the signals from the M array sensors are added coherently, which requires that each of the relative phases be zero at the point of summation; that is, we add $s(n)$ with a perfect replica of itself. Thus, we need a set of complex weights that results in a perfect phase alignment of all the sensor signals. The beamforming weight vector that phase-aligns a signal from direction ϕ_s at the different array elements is the *steering vector*, which is simply the array response vector in that direction, that is,

$$\mathbf{c}_{\mathrm{mf}}(\phi_s) = \mathbf{v}(\phi_s) \tag{11.2.16}$$

The steering vector beamformer is also known as the *spatial matched filter*[†] since the steering vector is matched to the array response of signals impinging on the array from an angle ϕ_s. As a result, ϕ_s is known as the *look direction*. The use of the spatial matched filter is commonly referred to as *conventional beamforming*.

The output of the spatial matched filter is

$$y(n) = \mathbf{c}_{\mathrm{mf}}^H(\phi_s)\mathbf{x}(n) = \mathbf{v}^H(\phi_s)\mathbf{x}(n)$$

$$= \frac{1}{\sqrt{M}}[1\ e^{j2\pi u_s}\ \cdots\ e^{j2\pi(M-1)u_s}]$$

$$\times \left\{ \begin{bmatrix} s(n) \\ e^{-j2\pi u_s}s(n) \\ \vdots \\ e^{-j2\pi(M-1)u_s}s(n) \end{bmatrix} + \mathbf{w}(n) \right\} \tag{11.2.17}$$

$$= \frac{1}{\sqrt{M}}[s(n) + s(n) + \cdots + s(n)] + \bar{w}(n)$$

$$= \sqrt{M}\,s(n) + \bar{w}(n)$$

where again $\bar{w}(n) = \mathbf{c}_{\mathrm{mf}}^H(\phi_s)\mathbf{w}(n)$ is the beamformer output noise. Examining the array SNR of the spatial matched filter output, we obtain

$$\mathrm{SNR}_{\mathrm{array}} = \frac{P_{\mathrm{s}}}{P_{\mathrm{n}}} = \frac{M\sigma_s^2}{E\{|\mathbf{v}^H(\phi_s)\mathbf{w}(n)|^2\}} \tag{11.2.18}$$

$$= \frac{M\sigma_s^2}{\mathbf{v}^H(\phi_s)\mathbf{R}_{\mathrm{n}}\mathbf{v}(\phi_s)} = M\frac{\sigma_s^2}{\sigma_w^2} = M \cdot \mathrm{SNR}_{\mathrm{elem}}$$

since $P_{\mathrm{s}} = M\sigma_s^2$ and $\mathbf{R}_{\mathrm{n}} = \sigma_w^2\mathbf{I}$. Therefore, the beamforming gain is

$$G_{\mathrm{bf}} = M \tag{11.2.19}$$

that is, equal to the number of sensors. In the case of spatially white noise, the spatial matched filter is optimum in the sense of maximizing the SNR at the output of the beamformer. Thus,

---

[†]The spatial matched filter should not be confused with the optimum matched filters discussed in Section 6.9 that depend on the correlation of the data. However, it is optimum in the case of spatially uncorrelated noise.

the beamforming gain of the spatial matched filter is known as the *array gain* because it is the maximum possible gain of a signal with respect to sensor thermal noise for a given array. Clearly from this perspective, the more elements in the array, the greater the beamforming gain. However, physical reality places limitations on the number of elements that can be used. The spatial matched filter maximizes the SNR because the individual sensor signals are coherently aligned prior to their combination. However, as we will see, other sources of interference that have spatial correlation require other types of adaptive beamformers that maximize the signal-to-interference-plus-noise ratio (SINR).

The beampattern of the spatial matched filter can serve to illustrate several key performance metrics of an array. A sample beampattern of a spatial matched filter was shown in Figure 11.9 for $\phi_s = 0°$. The first and most obvious attribute is the large lobe centered on ϕ_s, known as the mainlobe or *mainbeam*, and the remaining, smaller peaks are known as *sidelobes*. The value of the beampattern at the desired angle $\phi = \phi_s$ is equal to 1 (0 dB) due to the normalization used in the computation of the beampattern. A response of less than 1 in the look direction corresponds to a direct loss in desired signal power at the beamformer output. The sidelobe levels determine the rejection of the beamformer to signals not arriving from the look direction. The second attribute is the *beamwidth*, which is the angular span of the mainbeam. The resolution of the beamformer is determined by this mainlobe width, with smaller beamwidths resulting in better angular resolution. The beamwidth is commonly measured from the half-power (-3-dB) points $\Delta\phi_{3\,\text{dB}}$ or from null to null of the mainlobe $\Delta\phi_{\text{nn}}$. Using the beampattern, we next set out to examine the effects of the number of elements and their spacing on the array performance in the context of the spatial matched filter. However, in the following example, we first illustrate the use of a spatial matched filter to extract a signal from noise.

EXAMPLE 11.2.1. A signal received by a ULA with $M = 20$ elements and $\lambda/2$ spacing contains both a signal of interest at $\phi_s = 20°$ with an array SNR of 20 dB and thermal sensor noise with unit power ($\sigma_w^2 = 1$). The signal of interest is an impulse present only in the 100th sample and is produced by the sequence of MATLAB commands

```
u_s=(d/lambda)* sin(phi_s*pi/180);    s=zeros(M,N);
s(:,100)=(10^(SNR/20))*exp(-j*2*pi*u_s*[(0:(M-1))]/M)/sqrt(M);
```

The uncorrelated noise samples with a Gaussian distribution are generated by

```
w=(randn(M,N)+j*randn(M,N))/sqrt(2);
```

The two signals are added to produce the overall array signal x = s + w. Examining the signal at a single sensor in Figure 11.10(*a*), we see that the signal is not visible at $n = 100$ since the element level SNR is only 7 dB (full-array SNR minus M in decibels). The output power of this sample for a given realization can be more or less than the expected SNR due to the addition of the noise. However, when we apply a spatial matched filter using

```
c_mf=exp(-j*2*pi*u_s*[(0:(M-1))]/M)/sqrt(M);
y=c_mf'*x;
```

we can clearly see the signal of interest since the array SNR is 20 dB. As a rule of thumb, we require the array SNR to be at least 10 to 12 dB to clearly observe the signal.

Element spacing

In Section 11.1.4, we determined that the element spacing must be $d \le \lambda/2$ to prevent spatial aliasing. Here, we relax this restriction and look at various element spacings and the resulting array characteristics, namely, their beampatterns. In Figure 11.11, we show the beampatterns of spatial matched filters with $\phi_s = 0°$ for ULAs with element spacings of $\lambda/4$, $\lambda/2$, λ, and 2λ (equal-sized apertures of 10λ with 40, 20, 10, and 5 elements, respectively).

Element signal

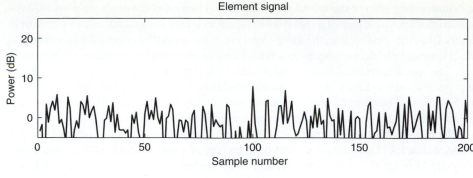

(*a*) Single-sensor signal

Array output signal

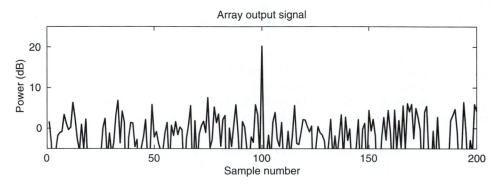

(*b*) Spatial matched filter output signal

FIGURE 11.10
The spatial signals from (a) an individual sensor and (b) the output of a spatial matched filter beamformer.

We note that the beampatterns for $\lambda/4$ and $\lambda/2$ spacing are identical with equal-sized mainlobes and the first sidelobe having a height of -13 dB. The oversampling for the array with an element spacing of $\lambda/4$ provides no additional information and therefore does not improve the beamformer response in terms of resolution. In the case of the undersampled arrays ($d = \lambda$ and 2λ), we see the same structure (beamwidth) around the look direction but also note the additional peaks in the beampattern (0 dB) at $\pm 90°$ for $d = \lambda$ and in even closer for $d = 2\lambda$. These additional lobes in the beampattern are known as *grating lobes*. Grating lobes create spatial ambiguities; that is, signals incident on the array from the angle associated with a grating lobe will look just like signals from the direction of interest. The beamformer has no means of distinguishing signals from these various directions. In certain applications, grating lobes may be acceptable if it is determined that it is either impossible or very improbable to receive returns from these angles; for example, a communications satellite is unlikely to receive signals at angles other than those corresponding to the ground below. The benefit of the larger element spacing is that the resulting array has a larger aperture and thus better resolution, which is our next topic of discussion. The topic of larger apertures with element spacing greater than $\lambda/2$ is commonly referred to as a *thinned array* and is addressed in Problem 11.5.

Array aperture and beamforming resolution

The aperture is the finite area over which a sensor collects spatial energy. In the case of a ULA, the aperture is the distance between the first and last elements. In general, the designer of an array yearns for as much aperture as possible. The greater the aperture, the

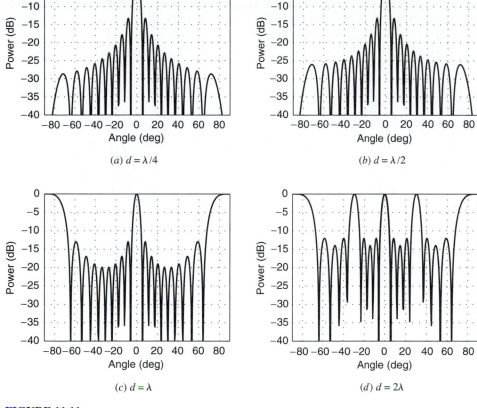

FIGURE 11.11
Beampatterns of a spatial matched filter for different element spacings with an equal-sized aperture $L = 10\lambda$.

finer the resolution of the array, which is its ability to distinguish between closely spaced sources. As we will see in Section 11.7, improved resolution results in better angle estimation capabilities. The angular resolution of a sensor array is measured in beamwidth $\Delta\phi$, which is commonly defined as the angular extent between the nulls of the mainbeam $\Delta\phi_{nn}$ or the half-power points of the mainbeam (-3 dB) $\Delta\phi_{3\,dB}$. As a general rule of thumb, the -3-dB beamwidth for an array with an aperture length of L is quoted in radians as

$$\Delta\phi_{3\,dB} \approx \frac{\lambda}{L} \tag{11.2.20}$$

although the actual -3-dB points of a spatial matched filter yield a resolution of $\Delta\phi_{3\,dB} = 0.89\,\lambda/L$ (the resolution of the conventional matched filter near broadside, $\phi = 0°$). The approximation in (11.2.20) is intended for the full range of prospective beamformers, not just spatial matched filters.[†] Since the resolution is dependent on the operating frequency F_c or equivalently on the wavelength, the aperture is often measured in wavelengths rather than in absolute length in meters. At large operating frequencies, say, $F_c = 10$ GHz or $\lambda = 3$ cm (X band in radar terminology), it is possible to populate a physical aperture of fixed length with a large number of elements, as opposed to lower operating frequencies, say, $F_c = 300$ MHz or $\lambda = 1$ m.

[†]Tapered beamformers, as discussed in Section 11.2.2, may considerably exceed this approximation, particularly for large tapers.

We illustrate the effect of aperture on resolution, using a few representative beampatterns. Figure 11.12 shows beampatterns for $M = 4, 8, 16$, and 32 with interelement spacing fixed at $d = \lambda/2$ (nonaliasing condition). Therefore, the corresponding apertures in wavelengths are $D = 2\lambda, 4\lambda, 8\lambda$, and 16λ. Clearly, increasing the aperture yields better resolution, with a factor-of-2 improvement for each of the successive twofold increases in aperture length. The level of the first sidelobe is always about -13 dB below the mainlobe peak.

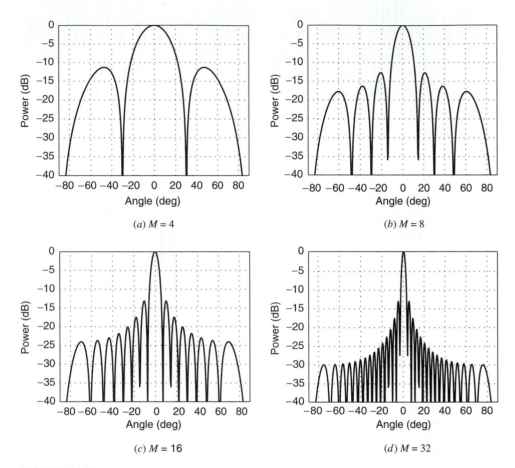

FIGURE 11.12
Beampatterns of a spatial matched filter for different aperture sizes with a common element of spacing of $d = \lambda/2$.

11.2.2 Tapered Beamforming

The spatial matched filter would be perfectly sufficient if the only signal present, aside from the sensor thermal noise, were the signal of interest. However, in many instances we must contend with other, undesired signals that hinder our ability to extract the signal of interest. These signals may also be spatially propagating at the same frequency as the operating frequency of the array. We refer to such signals as *interference*. These signals may be present due to hostile adversaries that are attempting to prevent us from receiving the signal of interest, for example, jammers in radar or communications; or they might be incidental signals that are present in our current operating environment, such as transmissions by other users in a communications system or radar clutter. In Sections 11.3, 11.5, and 11.6, we outline ways in which we can overcome these interferers by using adaptive methods.

However, there are also nonadaptive alternatives that can be employed in certain cases, namely, the use of a taper with the spatial matched filter.

Consider the ULA signal model from (11.2.15), but now including an interference signal $\mathbf{i}(n)$ made up of P interference sources

$$\mathbf{x}(n) = \mathbf{s}(n) + \mathbf{i}(n) + \mathbf{w}(n) = \sqrt{M}\mathbf{v}(\phi_s)s(n) + \sqrt{M}\sum_{p=1}^{P}\mathbf{v}(\phi_p)i_p(n) + \mathbf{w}(n) \quad (11.2.21)$$

where $\mathbf{v}(\phi_p)$ and $i_p(n)$ are the array response vector and actual signal due to the pth interferer, respectively. If we have a ULA with $\lambda/2$ element spacing, the beampattern of the spatial matched filter, as shown in Figure 11.13, may have sidelobes that are high enough to pass these interferers through the beamformer with a high enough gain to prevent us from observing the desired signal. For this array, if an interfering source were present at $\phi = 20°$ with a power of 40 dB, the power of the interference at the output of the spatial matched filter would be 20 dB because the sidelobe level at $\phi = 20°$ is only -20 dB. Therefore, if we were trying to receive a weaker signal from $\phi_s = 0°$, we would be unable to extract it because of sidelobe leakage from this interferer.

The spatial matched filter has weights all with a magnitude equal to $1/\sqrt{M}$. The look direction is determined by a linear phase shift across the weights of the spatial matched filter. However, the sidelobe levels can be further reduced by tapering the magnitudes of the spatial matched filter. To this end, we employ a tapering vector $\mathbf{t}$ that is applied to the spatial matched filter to realize a low sidelobe level beamformer

$$\mathbf{c}_{\text{tbf}}(\phi_s) = \mathbf{t} \odot \mathbf{c}_{\text{mf}}(\phi_s) \quad (11.2.22)$$

where $\odot$ represents the *Hadamard product*, which is the element-by-element multiplication of the two vectors (Strang 1998). We refer to this beamformer as the *tapered beamformer*.

The determination of a taper can be thought of as the design of the desired beamformer where $\mathbf{c}_{\text{mf}}$ simply determines the desired angle. The weight vector of the spatial matched filter from (11.2.16) has unit norm; that is, $\mathbf{c}_{\text{mf}}^{H}\mathbf{c}_{\text{mf}} = 1$. Similarly, the tapered beamformer $\mathbf{c}_{\text{tbf}}$ is normalized so that

$$\mathbf{c}_{\text{tbf}}^{H}(\phi_s)\mathbf{c}_{\text{tbf}}(\phi_s) = 1 \quad (11.2.23)$$

The choices for tapers, or windows, were outlined in Section 5.1 in the context of spectral estimation. Here, we use Dolph-Chebyshev tapers simply for illustration purposes. This taper produces a constant sidelobe level (equiripples in the stopband in spectral estimation), which is often a desirable attribute of a beamformer. The best taper choice is driven by the actual application. The beampatterns of the ULA are used again, but this time the beampatterns of tapered beamformers are also shown in Figure 11.13. The sidelobe levels of the tapers were chosen to be -50 and -70 dB.[†] The same 40-dB interferer would have been reduced to -10 and -30 dB at the beamformer output, respectively.

However, the use of tapers does not come without a cost. The peak of the beampattern is no longer at 0 dB. This loss in gain in the current look direction is commonly referred to as a *tapering loss* and is simply the beampattern evaluated at ϕ_s:

$$L_{\text{taper}} \triangleq |C_{\text{tbf}}(\phi_s)|^2 = |\mathbf{c}_{\text{tbf}}^{H}(\phi_s)\mathbf{v}(\phi_s)|^2 \quad (11.2.24)$$

Since the tapering vector was normalized as in (11.2.23), the tapering loss is in the range $0 \le L_{\text{taper}} \le 1$ with $L_{\text{taper}} = 1$ corresponding to no loss (untapered spatial matched filter). The tapering loss is the loss in SNR of the desired signal at the beamformer output that cannot be recovered. More significantly, notice that the mainlobes of the beampatterns in Figure 11.13 are much broader for the tapered beamformers. The consequence is a loss

---

[†]In practice the tapering sidelobe levels are limited by array element location errors due to uncertainty. This limit is often at -30 dB but may be even higher. For illustration purposes we will ignore these limits in this chapter.

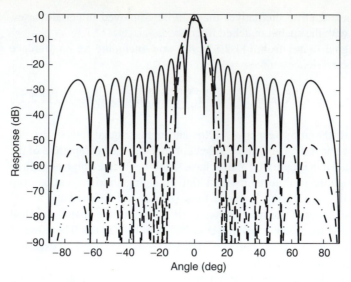

FIGURE 11.13
Beampatterns of beamformers with $M = 20$ with no taper (solid line), -50-dB taper (dashed line), and -70-dB taper (dash-dot line).

in resolution that becomes more pronounced as the tapering is increased to achieve lower sidelobe levels. This phenomenon was also treated within the context of spectral estimation in Section 5.1. However, its interpretation for an array can better be understood by examining plots of the magnitude of the taper vector **t**, shown in Figure 11.14 for the -50- and -70-dB Dolph-Chebyshev tapers. Note that the elements on the ends of the array are given less weighting as the tapering level is increased. The tapered array in effect deemphasizes these end elements while emphasizing the center elements. Therefore, the loss in resolution for a tapered beamformer might be interpreted as a loss in the effective aperture of the array imparted by the tapering vector.

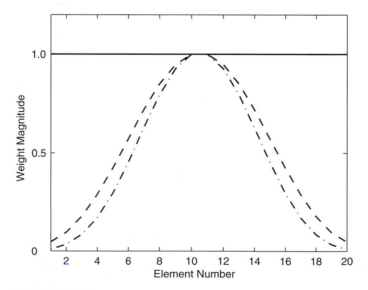

FIGURE 11.14
The magnitude levels of the tapered beamforming weights as a function of element number for $M = 20$ with no taper (solid line), -50-dB taper (dashed line), and -70-dB taper (dash-dot line).

EXAMPLE 11.2.2. We illustrate the use of tapers with the spatial matched filter for the extraction of a radar signal in the presence of a jamming interference source using a ULA with $M = 20$ elements with $\lambda/2$ spacing. The desired radar signal is known as a *target* and is present for only one sample in time. Here the target signal is at time sample (range gate) $n = 100$ and is at $\phi = 0°$ with an array SNR of 20 dB. The jammer transmits a high-power, uncorrelated waveform (white noise). The angle of the jammer is $\phi_i = 20°$, and its strength is 40 dB. The additive, sensor thermal noise has unit power (0 dB). We generate the jammer signal for $N = 200$ samples with the MATLAB commands

```
v_i = exp(-j*pi*[0:M-1]'*sin(phi_i*pi/180))/sqrt(M);
i_x=(10^(40/20))*v_i*(randn(1,N)+j*randn(1,N))/sqrt(2)
```

Similarly, the unit power thermal noise signal is produced by

```
w=(randn(M,N)+j*randn(M,N))/sqrt(2)
```

Two beamformers (steered to $\phi = 0°$) are applied to the resulting array returns: a spatial matched filter and a tapered beamformer with a -50-dB sidelobe level. The resulting beamformer output signals are shown in Figure 11.15. The spatial matched filter is unable to reduce the jammer sufficiently to observe the target signal at $n = 100$. However, the tapered beamformer is able to attenuate the jammer signal below the thermal noise level and the target is easily extracted. The target signal is approximately 18.5 dB with the -1.5 dB loss due to the tapering loss in (11.2.24).

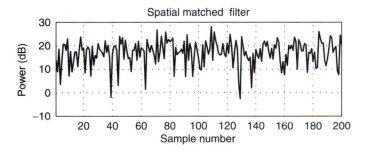

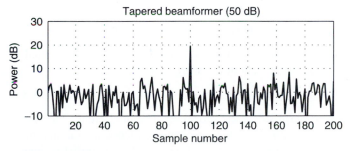

FIGURE 11.15
The output signals of a spatial matched filter and a tapered beamformer (-50-dB).

11.3 OPTIMUM ARRAY PROCESSING

So far, we have only considered beamformers whose weights are determined independently of the data to be processed. If instead we base the actual beamforming weights on the array data themselves, then the result is an *adaptive array* and the operation is known as *adaptive beamforming*. Ideally, the beamforming weights are adapted in such a way as to optimize the spatial response of the resulting beamformer based on a certain criterion. To this end, the criterion is chosen to enhance the desired signal while rejecting other, unwanted signals. This weight vector is similar to the optimum matched filter from Chapter 6. However, the manner in which it is implemented, namely, the methodology of how this equation is successfully applied to the array processing problem, is the topic of this and the next three sections.

This section focuses on optimum array processing methods that make use of the a priori known statistics of the data to derive the beamforming weights. Implicit in the optimization is the a priori knowledge of the true statistics of the array data. In Section 11.5, we discuss techniques for implementing these methods that estimate the unknown statistics from the data. We will use the general term *adaptive* to refer to beamformers that use an estimated correlation matrix computed from array snapshots, while reserving the term *optimum* for beamformers that optimize a certain criterion based on knowledge of the array data statistics. We begin by discussing the array signal model that contains interference in addition to the desired signal and noise. We then proceed to derive the *optimum beamformer*, where the optimality criterion is the maximization of the theoretical signal-to-interference-plus-noise ratio. In addition, we give an alternate implementation of the optimum beamformer: the generalized sidelobe canceler. This structure also gives an intuitive understanding of the optimum beamformer. Various issues associated with the optimum beamformer, namely, the effect of signal mismatch and bandwidth on the performance of an optimum beamformer, are discussed in Section 11.4.

The signal of interest is seldom the only array signal aside from thermal noise present. The array must often contend with other, undesired signals that interfere with our ability to extract this signal of interest, as described in Section 11.2.2. Often the interference is so powerful that even a tapered beamformer is unable to sufficiently suppress it to extract the signals of interest. The determination of the presence of signals of interest is known as *detection*, while the inference of their parameters, for example, the angle of arrival ϕ_s, is referred to as *estimation*. The topic of detection is not explicitly treated here. Rather, we seek to maximize the *visibility* of the desired signal at the array output, that is, the ratio of the signal power to that of the interference plus noise, to facilitate the detection process. There are several textbooks devoted to the subject of detection theory (Scharf 1991; Poor 1994; Kay 1998) to which the interested reader is referred. Parameter estimation methods to determine the angle of the desired signal are the topic of Section 11.7.

Consider an array signal that consists of the desired signal $\mathbf{s}(n)$, an interference signal $\mathbf{i}(n)$, along with sensor thermal noise $\mathbf{w}(n)$, that is,

$$\mathbf{x}(n) = \mathbf{s}(n) + \mathbf{i}(n) + \mathbf{w}(n) = \sqrt{M}\mathbf{v}(\phi_s)s(n) + \mathbf{i}(n) + \mathbf{w}(n) \tag{11.3.1}$$

where $s(n)$ is a signal with deterministic amplitude σ_s and uniformly distributed random phase. The interference-plus-noise component of the array signal is

$$\mathbf{x}_{i+n}(n) = \mathbf{i}(n) + \mathbf{w}(n) \tag{11.3.2}$$

which are both modeled as zero-mean stochastic processes. The interference has spatial correlation according to the angles of the contributing interferers, while the thermal noise is spatially uncorrelated. The interference component of the signal may consist of several sources, as modeled in (11.2.21). The sensor thermal noise is assumed to be uncorrelated with power σ_w^2. The assumption is made that *all of these three components are mutually uncorrelated*. As a result, the array correlation matrix is

$$\mathbf{R}_x = E\{\mathbf{x}(n)\mathbf{x}^H(n)\} = M\sigma_s^2\mathbf{v}(\phi_s)\,\mathbf{v}^H(\phi_s) + \mathbf{R}_i + \mathbf{R}_n \tag{11.3.3}$$

where σ_s^2 is the power of the signal of interest and $\mathbf{R}_i$ and $\mathbf{R}_n$ are the interference and noise correlation matrices, respectively. The interference-plus-noise correlation matrix is the sum of these latter two matrices

$$\mathbf{R}_{i+n} = \mathbf{R}_i + \sigma_w^2\mathbf{I} \tag{11.3.4}$$

where $\mathbf{R}_n = \sigma_w^2\mathbf{I}$ since the sensor thermal noise is spatially uncorrelated.

The ultimate goal of the prospective adaptive beamformer is to combine the sensor signals in such a way that the interference signal is reduced to the level of the thermal noise while the desired signal is preserved. Stated another way, we would like to maximize the ratio of the signal power to that of the interference plus noise, known as the *signal-to-interference-plus-noise ratio* (*SINR*). Maximizing the SINR is the optimal criterion for most detection and estimation problems. Simply stated, *maximizing the SINR seeks to improve the visibility of the desired signal as much as possible in a background of interference.* This criterion should not be confused with maximizing the SNR (spatial matched filter) in the absence of interference.

At the input of the array, that is, in each individual sensor, the SINR is given by

$$\text{SINR}_{\text{elem}} = \frac{\sigma_s^2}{\sigma_i^2 + \sigma_w^2} \tag{11.3.5}$$

where σ_s^2, σ_i^2, and σ_w^2 are the signal, interference, and thermal noise powers in each individual element. The SINR at the beamformer output, following the application of the beamforming weight vector $\mathbf{c}$, is

$$\text{SINR}_{\text{out}} = \frac{|\mathbf{c}^H \mathbf{s}(n)|^2}{E\{|\mathbf{c}^H \mathbf{x}_{i+n}(n)|^2\}} = \frac{M\sigma_s^2 |\mathbf{c}^H \mathbf{v}(\phi_s)|^2}{\mathbf{c}^H \mathbf{R}_{i+n} \mathbf{c}} \tag{11.3.6}$$

We wish to maximize this array output SINR. First, note that the interference-plus-noise correlation matrix can be factored as

$$\mathbf{R}_{i+n} = \mathbf{L}_{i+n} \mathbf{L}_{i+n}^H \tag{11.3.7}$$

where $\mathbf{L}_{i+n}$ is the Cholesky factor of the correlation matrix.[†] See Section 6.3 for details. Thus, defining

$$\tilde{\mathbf{c}} = \mathbf{L}_{i+n}^H \mathbf{c} \qquad \tilde{\mathbf{v}}(\phi_s) = \mathbf{L}_{i+n}^{-1} \mathbf{v}(\phi_s) \tag{11.3.8}$$

we can rewrite (11.3.6) as

$$\text{SINR}_{\text{out}} = \frac{M\sigma_s^2 |\tilde{\mathbf{c}}^H \tilde{\mathbf{v}}(\phi_s)|^2}{\tilde{\mathbf{c}}^H \tilde{\mathbf{c}}} \tag{11.3.9}$$

Using the Schwartz inequality

$$\tilde{\mathbf{c}}^H \tilde{\mathbf{v}}(\phi_s) \le \|\tilde{\mathbf{c}}\| \|\tilde{\mathbf{v}}(\phi_s)\| \tag{11.3.10}$$

and substituting (11.3.10) into (11.3.9), we find that

$$\text{SINR}_{\text{out}} \le M\sigma_s^2 \frac{\|\tilde{\mathbf{c}}\|^2 \|\tilde{\mathbf{v}}(\phi_s)\|^2}{\|\tilde{\mathbf{c}}\|^2} = M\sigma_s^2 \|\tilde{\mathbf{v}}(\phi_s)\|^2 \tag{11.3.11}$$

Thus, the maximum SINR is found by satisfying the upper bound for (11.3.11), which yields

$$\text{SINR}_{\text{out}}^{\max} = M\sigma_s^2 \tilde{\mathbf{v}}^H(\phi_s) \tilde{\mathbf{v}}(\phi_s) = M\sigma_s^2 [\mathbf{v}^H(\phi_s) \mathbf{R}_{i+n}^{-1} \mathbf{v}(\phi_s)] \tag{11.3.12}$$

We also see that the same maximum SINR is obtained if we set $\tilde{\mathbf{c}} = \alpha \tilde{\mathbf{v}}(\phi_s)$ where α is an arbitrary constant. In other words, the SINR is maximized when these two vectors are parallel to each other and α can be chosen to satisfy other requirements. Therefore, using (11.3.8), we can solve for the optimum weight vector (Bryn 1962; Capon et al. 1967; Brennan and Reed 1973)

$$\mathbf{c}_o = \alpha \mathbf{L}_{i+n}^{-H} \tilde{\mathbf{v}}(\phi_s) = \alpha \mathbf{R}_{i+n}^{-1} \mathbf{v}(\phi_s) \tag{11.3.13}$$

where α is an arbitrary constant. Thus, the optimum beamforming weights are proportional to $\mathbf{R}_{i+n}^{-1} \mathbf{v}(\phi_s)$. The proportionality constant α in (11.3.13) can be set in a variety of ways.

[†]Note that any square root factorization $\mathbf{R}_{i+n} = \mathbf{R}_{i+n}^{1/2} \mathbf{R}_{i+n}^{H/2}$ of the correlation matrix can be chosen.

Table 11.1 gives various normalizations for the optimum beamformer. The normalization we adopt throughout this chapter is to constrain the optimum beamformer to have unity gain in the look direction, that is, $\mathbf{c}_o^H \mathbf{v}(\phi_s) = 1$. Therefore,

$$\mathbf{c}_o^H \mathbf{v}(\phi_s) = \alpha [\mathbf{R}_{i+n}^{-1} \mathbf{v}(\phi_s)]^H \mathbf{v}(\phi_s) = 1 \qquad (11.3.14)$$

and the resulting optimum beamformer is given by

$$\mathbf{c}_o = \frac{\mathbf{R}_{i+n}^{-1} \mathbf{v}(\phi_s)}{\mathbf{v}^H(\phi_s)\mathbf{R}_{i+n}^{-1}\mathbf{v}(\phi_s)} \qquad (11.3.15)$$

In general, the normalization of the optimum beamformer is arbitrary and is dictated by the use of the output, for example, measure residual interference power or detection. In any case, the SINR is maximized independently of the normalization. The most commonly used normalizations are listed in Table 11.1.

TABLE 11.1

Optimum weight normalizations for unit gain in look direction, unit gain on noise, and unit gain on interference-plus-noise constraints.

Constraint	Mathematical formulation	Optimum beamformer normalization
MVDR (unit gain in look direction)	$\mathbf{c}_o^H \mathbf{v}(\phi_s) = 1$	$\alpha = [\mathbf{v}^H(\phi_s)\mathbf{R}_{i+n}^{-1}\mathbf{v}(\phi_s)]^{-1}$
Unit noise gain	$\mathbf{c}_o^H \mathbf{c}_o = 1$	$\alpha = [\mathbf{v}^H(\phi_s)\mathbf{R}_{i+n}^{-2}\mathbf{v}(\phi_s)]^{-1/2}$
Unit gain on interference-plus-noise*	$\mathbf{c}_o^H \mathbf{R}_{i+n}\mathbf{c}_o = 1$	$\alpha = [\mathbf{v}^H(\phi_s)\mathbf{R}_{i+n}^{-1}\mathbf{v}(\phi_s)]^{-1/2}$

*This normalization is commonly referred to as the adaptive matched filter normalization (Robey et al. 1992). Its use is primarily for detection purposes. Since the output level of the interference-plus-noise has a set power of unity, a constant detection threshold can be used for all angles.

Alternately, the optimum beamformer can be derived by solving the following constrained optimization problem: Minimize the interference-plus-noise power at the beamformer output

$$P_{i+n} = E\{|\mathbf{c}^H \mathbf{x}_{i+n}(n)|^2\} = \mathbf{c}^H \mathbf{R}_{i+n}\mathbf{c} \qquad (11.3.16)$$

subject to a look-direction distortionless response constraint, that is,

$$\min P_{i+n} \quad \text{subject to} \quad \mathbf{c}^H \mathbf{v}(\phi_s) = 1 \qquad (11.3.17)$$

The solution of this constrained optimization problem is found by using Lagrange multipliers (see Appendix B and Problem 11.7) and results in the same weight vector as (11.3.15). This formulation has led to the commonly used term *minimum-variance distortionless response (MVDR) beamformer*. For a discussion of minimum-variance beamforming, see Van Veen (1992). The optimum beamformer passes signals impinging on the array from angle ϕ_s while rejecting significant energy (interference) from all other angles. This beamformer can be thought of as an optimum spatial matched filter since it provides maximum interference rejection, while matching the response of signals impinging on the array from a direction ϕ_s. The optimal weights balance the rejection of interference with the thermal noise gain so that the output thermal noise does not cause a reduction in the output SINR.

The optimum beamformer maximizes the SINR given by (11.3.12), which is independent of the normalization. Another useful metric is a measure of the performance relative to the interference-free case, that is, $\mathbf{x}(n) = \mathbf{s}(n) + \mathbf{w}(n)$. To gauge the performance of the beamformer independently of the desired signal power, we simply normalize the SINR by the hypothetical array output SNR had there been no interference present, which from (11.2.18) is $\text{SNR}_0 = M\sigma_s^2/\sigma_w^2$. The resulting measure is known as the *SINR loss*, which for the optimum beamformer, by substituting into (11.3.12), is

$$L_{\text{sinr}}(\phi_s) \triangleq \frac{\text{SINR}_{\text{out}}(\phi_s)}{\text{SNR}_0} = \sigma_w^2 \mathbf{v}^H(\phi_s)\mathbf{R}_{i+n}^{-1}\mathbf{v}(\phi_s) \tag{11.3.18}$$

The SINR loss is always between 0 and 1, taking on the maximum value when the performance is equal to the interference-free case. Typically, the SINR loss is computed across all angles for a given interference scenario. In this sense, the SINR loss of the optimum beamformer provides a measure of the residual interference remaining following optimum processing and informs us of our loss in performance due to the presence of interference. We also notice that (11.3.18) is the reciprocal of the minimum-variance power spectrum of the interference plus noise. Minimum-variance power spectrum estimation was discussed in Section 9.5.

> **EXAMPLE 11.3.1.** To demonstrate the optimum beamformer, we consider a scenario in which there are three interference sources and compare it to a conventional beamformer (spatial matched filter). The array is a 20-element ULA with $\lambda/2$ element spacing. These interferers are at the following angles with the corresponding interference-to-noise ratios (INRs) in decibels: $\phi = 20°$ and INR= 35 dB, $\phi = -30°$ and INR= 70 dB, and $\phi = 50°$ and INR= 50 dB. The optimum beamformer is first computed using (11.3.15) for a look direction of $\phi_s = 0°$. The beampattern of this optimum beamformer is computed by using (11.2.3) and is plotted in Figure 11.16(*a*). Notice the nulls at the angles of the interference ($\phi = -30°, 20°, 50°$). These nulls are deep enough that the interference at the beamformer output is below the sensor thermal noise level. The conventional beamformer, however, cannot place nulls on the interferers since it is independent of the data. We also perform optimum beamforming across all angles $-90° < \phi < 90°$ and compute the corresponding SINR loss due to the interference using (11.3.18). The SINR loss is plotted in Figure 11.16(*b*). The notches at the interference angles are simply the negative of the INR of the interferers corresponding to significant losses in performance. However, these performance losses are limited to these angles. The SINR loss at all other angles is almost at its maximum value of 1 (0 dB). The SINR loss of the conventional beamformer is significantly worse at all angles because of the strong interference that makes its way to the beamformer output through its sidelobes.

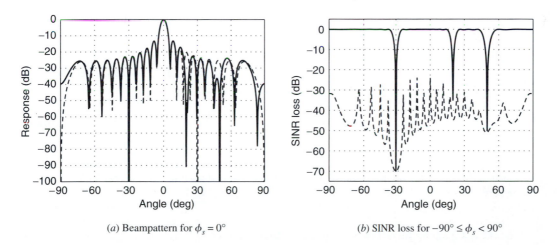

(*a*) Beampattern for $\phi_s = 0°$ (*b*) SINR loss for $-90° \le \phi_s < 90°$

FIGURE 11.16
Beampattern (steered to $\phi = 0°$) and SINR loss plots versus angle. Solid line is the optimum beamformer, and dashed line is the conventional beamformer.

> **EXAMPLE 11.3.2.** We revisit the problem from Example 11.2.2 with a jammer at $\phi_i = 20°$ except the jammer power is now 70 dB. Clearly, the -50-dB tapered beamformer is no longer

capable of sufficiently suppressing this jammer. Rather, we compute the optimum beamformer using (11.3.15), where $\mathbf{R}_{\text{i+n}} = 10^7 \mathbf{v}(\phi_i)\mathbf{v}^H(\phi_i) + \mathbf{I}$. First, we examine the beampattern of the optimum beamformer steered to $\phi = 0°$ in Figure 11.17(a). Notice the null on the jammer at $\phi = 20°$ with a depth of greater than -150 dB. We also plot the SINR loss in Figure 11.17(b) as we scan the look direction from $-90°$ to $90°$. Almost no SINR loss is experienced at angles away from the jammer, while at the jammer angle $\phi = 20°$, the SINR loss corresponds to the jammer power (70 dB). As a similar exercise to that in Example 11.2.2, we can produce a target signal at $\phi = 0°$ and attempt to extract it, using both a spatial matched filter and an optimum beamformer. The output signals are shown in Figure 11.17(c) and (d), respectively. The optimum beamformer is able to successfully extract the signal whereas the ouput of the spatial matched filter is dominated by interference. Notice that we do not suffer the taper loss on the target as we did for the tapered beamformer due to the $\mathbf{c}_o^H \mathbf{v}(\phi_s) = 1$ constraint in the optimum beamformer.

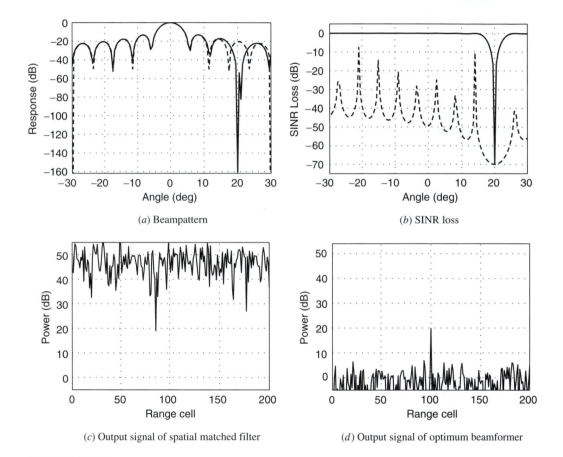

(a) Beampattern

(b) SINR loss

(c) Output signal of spatial matched filter

(d) Output signal of optimum beamformer

FIGURE 11.17
(a) Beampattern and (b) SINR loss of optimum beamformer (solid line) versus spatial matched filter (dashed line), along with (c) the output signals of a spatial matched filter and (d) the optimum beamformer.

11.3.2 Eigenanalysis of the Optimum Beamformer

In many cases, significant insight can be gained by considering the optimum beamformer in terms of the eigenvalues and eigenvectors of the interference-plus-noise correlation matrix

$$\mathbf{R}_{\text{i+n}} = \sum_{m=1}^{M} \lambda_m \mathbf{q}_m \mathbf{q}_m^H \tag{11.3.19}$$

where the eigenvalues have been ordered from largest to smallest, that is, $\lambda_1 \geq \lambda_2 \geq \cdots \geq \lambda_M$. If the rank of the interference is P, then $\lambda_m = \sigma_w^2$ for $m > P$; that is, the remainder of the eigenvalues is equal to the thermal noise power. The eigenvectors are orthonormal ($\mathbf{q}_m^H \mathbf{q}_k = 0$ for $k \neq m$, $\mathbf{q}_m^H \mathbf{q}_m = 1$) and form a basis for the interference-plus-noise subspace that can be split into interference and noise subspaces given by

$$\text{Interference subspace: } \{\mathbf{q}_m \quad 1 \leq m \leq P\} \quad \text{Noise subspace: } \{\mathbf{q}_m \quad P < m \leq M\}$$
$$(11.3.20)$$

The inverse of $\mathbf{R}_{i+n}$ can also be written in terms of the eigenvalues and eigenvectors, λ_m and $\mathbf{q}_m$, of the correlation matrix $\mathbf{R}_{i+n}$, that is,

$$\mathbf{R}_{i+n}^{-1} = \sum_{m=1}^{M} \frac{1}{\lambda_m} \mathbf{q}_m \mathbf{q}_m^H \qquad (11.3.21)$$

We further assume that the rank of the interference is less than the total number of sensors, that is, $P < M$. In this case, the smallest eigenvalues of $\mathbf{R}_{i+n}$ are noise eigenvalues and are equal to the thermal noise power $\lambda_m = \sigma_w^2$ for $m > P$. Substituting (11.3.21) into the optimum beamformer weights in (11.3.15), we have

$$\mathbf{c}_o = \alpha \mathbf{R}_{i+n}^{-1} \mathbf{v}(\phi_s) = \alpha \sum_{m=1}^{M} \frac{1}{\lambda_m} \mathbf{q}_m \mathbf{q}_m^H \mathbf{v}(\phi_s)$$

$$= \alpha \left[\frac{1}{\sigma_w^2} \mathbf{v}(\phi_s) - \frac{1}{\sigma_w^2} \mathbf{v}(\phi_s) + \sum_{m=1}^{M} \frac{\mathbf{q}_m^H \mathbf{v}(\phi_s)}{\lambda_m} \mathbf{q}_m \right] \qquad (11.3.22)$$

$$= \frac{\alpha}{\sigma_w^2} \left\{ \mathbf{v}(\phi_s) - \sum_{m=1}^{M} \frac{\lambda_m - \sigma_w^2}{\lambda_m} [\mathbf{q}_m^H \mathbf{v}(\phi_s)] \mathbf{q}_m \right\}$$

where $\alpha = [\mathbf{v}(\phi_s)^H \mathbf{R}_{i+n}^{-1} \mathbf{v}(\phi_s)]^{-1}$. The resulting beam response is

$$C_o(\phi) = \frac{\alpha}{\sigma_w^2} \left\{ C_q(\phi) - \sum_{m=1}^{M} \frac{\lambda_m - \sigma_w^2}{\lambda_m} [\mathbf{q}_m^H \mathbf{v}(\phi_s)] Q_m(\phi) \right\} \qquad (11.3.23)$$

where
$$C_q(\phi) = \mathbf{v}^H(\phi_s) \mathbf{v}(\phi) = \mathbf{c}_{mf}^H \mathbf{v}(\phi) = C_{mf}(\phi) \qquad (11.3.24)$$

is the response of the spatial matched filter $\mathbf{c}_{mf}(\phi_s) = \mathbf{v}(\phi_s)$ [see (11.2.16)] and is known as the *quiescent response* of the optimum beamformer. However,

$$Q_m(\phi) = \mathbf{q}_m^H \mathbf{v}(\phi) \qquad (11.3.25)$$

is the beam response of the mth eigenvector, known as an *eigenbeam*. Thus, the response of the optimum beamformer consists of weighted eigenbeams subtracted from the quiescent response. The weights for the eigenbeams are determined by the corresponding eigenvalue, the noise power, and the cross-product of the look-direction steering vector and the respective eigenvector. Examining the term $(\lambda_m - \sigma_w^2)/\lambda_m$, we see clearly that for strong interferers $\lambda_m \gg \sigma_w^2$ and $(\lambda_m - \sigma_w^2)/\lambda_m \approx 1$, and the eigenbeam is subtracted from the quiescent response weighted by $\mathbf{q}_m^H \mathbf{v}(\phi)$. This subtraction of properly weighted interference eigenvectors places nulls in the directions of the interference sources. The term $\mathbf{q}_m^H \mathbf{v}(\phi_s)$ in (11.3.23) scales the interference eigenbeam to the quiescent response of the spatial matched filter in the direction of the corresponding interferer. Thus, the null depth for an interferer of the beampattern $|C_o(\phi)|^2$ is determined by the response of the eigenbeam to the quiescent response and the strength of the interferer relative to the noise level. However, for the noise eigenvalues $\lambda_m = \sigma_w^2$ and $(\lambda_m - \sigma_w^2)/\lambda_m = 0$. Therefore, the noise eigenvectors have no effect on the optimum beamformer. Interestingly, for the case of noise only and thus all noise eigenvalues, that is, no interference present, the optimum beamformer reverts to the spatial matched filter $\mathbf{c}_o(\phi_s) = \mathbf{c}_{\mathrm{mf}}(\phi_s) = \mathbf{v}(\phi_s)$, which is the beamformer that maximizes the SNR.

11.3.3 Interference Cancelation Performance

The interference cancelation performance of the optimum beamformer can be determined by examining the beam response at the angles of the interferers. The beam response at these angles indicates the depth of the null that the optimum beamformer places on the interferer. Using the MVDR optimum beamformer from (11.3.15), we see that the response in the direction of an interferer ϕ_p of an optimum beamformer that is steered in direction ϕ_s is

$$C_o(\phi_p) = \mathbf{c}_o^H \mathbf{v}(\phi_p) = \alpha \mathbf{v}^H(\phi_s)\mathbf{R}_{i+n}^{-1}\mathbf{v}(\phi_p) \tag{11.3.26}$$

where ϕ_p is the angle of the pth interferer and $\alpha = [\mathbf{v}^H(\phi_s)\mathbf{R}_{i+n}^{-1}\mathbf{v}(\phi_s)]^{-1}$. Now we note that $\mathbf{R}_{i+n}$ can be split into a component due to the pth interferer and the correlation matrix of the remaining interference-plus-noise $\mathbf{Q}_{i+n}$

$$\mathbf{R}_{i+n} = \mathbf{Q}_{i+n} + M\sigma_p^2\mathbf{v}(\phi_p)\mathbf{v}^H(\phi_p) \tag{11.3.27}$$

where σ_p^2 is the power of the pth interferer in a single element. Using the matrix inversion lemma (Appendix A), we obtain

$$\mathbf{R}_{i+n}^{-1} = \mathbf{Q}_{i+n}^{-1} - M\sigma_p^2 \frac{\mathbf{Q}_{i+n}^{-1}\mathbf{v}(\phi_p)\mathbf{v}^H(\phi_p)\mathbf{Q}_{i+n}^{-1}}{1 + M\sigma_p^2\mathbf{v}^H(\phi_p)\mathbf{Q}_{i+n}^{-1}\mathbf{v}(\phi_p)} \tag{11.3.28}$$

Substituting (11.3.28) into (11.3.26), we find the optimum beamformer response to be (Richmond 1999)

$$\begin{aligned}
C_o(\phi_p) &= \alpha \mathbf{v}^H(\phi_s)\mathbf{R}_{i+n}^{-1}\mathbf{v}(\phi_p) \\
&= \alpha \mathbf{v}^H(\phi_s)\mathbf{Q}_{i+n}^{-1}\mathbf{v}(\phi_p) \\
&\quad - \alpha \mathbf{v}^H(\phi_s)\mathbf{Q}_{i+n}^{-1}\mathbf{v}(\phi_p)\mathbf{v}^H(\phi_p)\mathbf{Q}_{i+n}^{-1}\mathbf{v}(\phi_p) \\
&\quad \times \left[\frac{M\sigma_p^2}{1 + M\sigma_p^2\mathbf{v}^H(\phi_p)\mathbf{Q}_{i+n}^{-1}\mathbf{v}(\phi_p)}\right] \\
&= \underbrace{\frac{\mathbf{v}^H(\phi_s)\mathbf{Q}_{i+n}^{-1}\mathbf{v}(\phi_p)}{\mathbf{v}^H(\phi_s)\mathbf{R}_{i+n}^{-1}\mathbf{v}(\phi_s)}}_{\text{term 1}}\ \underbrace{\frac{1}{1 + M\sigma_p^2\mathbf{v}^H(\phi_p)\mathbf{Q}_{i+n}^{-1}\mathbf{v}(\phi_p)}}_{\text{term 2}}
\end{aligned} \tag{11.3.29}$$

We notice that the optimum beamformer response is made up of the product of two terms. The first term is the response at angle ϕ_p of an optimum beamformer steered in direction ϕ_s formed in the absence of this interferer ($\sigma_p^2 = 0$), that is, the sidelobe level of the optimum beamformer had this interference not been present. However, the power of the interferer is many times significantly greater than this sidelobe level, and the optimum beamformer cancels the interferer by placing a null at the angle of the interferer. The second term produces the null at the angle ϕ_p. By examining this term, it is apparent that the depth of the null is determined by the power of the interferer $M\sigma_p^2$. Clearly, the larger the power of the interferer, the smaller this term becomes and the deeper the null depth of the optimum beamformer is at ϕ_p. The factor $\mathbf{v}^H(\phi_p)\mathbf{Q}_{i+n}^{-1}\mathbf{v}(\phi_p)$ is the amount of energy received from ϕ_p not including the interferer and has as a lower bound equal to the thermal noise power (spatially white). Since the power response of the beamformer is $|C_o(\phi_p)|^2$, the null depth is actually proportional to $M^2\sigma_p^4$, or twice the power of the interferer at the array output, in decibels (Compton 1988).

11.3.4 Tapered Optimum Beamforming

In the derivation of the optimum beamformer, we used the vector $\mathbf{v}(\phi_s)$ that was matched to the array response of a desired signal arriving from an angle ϕ_s. The resulting beamformer weight vector $\mathbf{c}_o$ has unity gain in this direction; that is, $\mathbf{c}_o^H \mathbf{v}(\phi_s) = 1$, owing to the normalization of the weights. However, the sidelobes of the beamformer are still at the same levels as the spatial matched filter (nonadaptive beamformer) from (11.2.16), although with a different structure, as can be seen from a sample beampattern of the optimum beamformer shown in Figure 11.18(a).

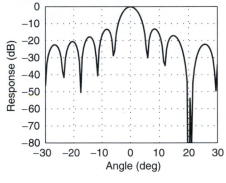

(a) Optimum beamformer (no taper)

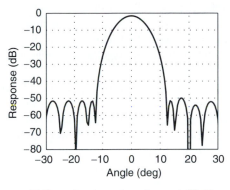

(b) Tapered optimum beamformer (-50-dB taper)

FIGURE 11.18
Beampatterns of an optimum beamformer (a) without and (b) with tapering (-50-dB Dolph-Chebyshev taper) steered to $\phi = 0°$.

The optimum beamformer uses the interference-plus-noise correlation matrix $\mathbf{R}_{i+n}$. Now, although the beamformer weights must be estimated from intervals of the data that contain only interference (no desired signal present), they are presumably applied to segments that contain both interference and a desired signal. What happens when we are searching an angular region for potential desired signals? A desired signal at an angle ϕ_1 may be easily found by using an adaptive beamformer directed to this angle ($\phi_s = \phi_1$), assuming the signal strength after beamforming is significantly larger than the sensor thermal noise. However, we will also be searching other angles for potential desired signals. If we are looking at one of these other angles, say, $\phi_2 \neq \phi_1$, we want to avoid concluding a signal is present when it may actually be due to sidelobe leakage of the signal at ϕ_1. This problem is best illustrated by using the beampattern of an optimum beamformer with an interferer at $\phi = 20°$ in Figure 11.18(a). The optimum beamformer is steered to an angle $\phi_s = 0°$. Let us assume another signal is present at $\phi_1 = -20°$ that was not part of the interference (not accounted for in the interference correlation matrix). The gain of the optimum beamformer at $\phi = -20°$ is approximately -20 dB. If the strength of this signal is significantly greater than 20 dB, the optimum beamformer steered to $\phi_s = 0°$ will pass this sidelobe signal with sufficient strength that we may erroneously conclude a signal is present at $\phi_s = 0°$. This problem is commonly referred to as a *sidelobe target* or *desired signal problem*.

The sidelobe signal problem described above can be cured, at least partially, by reducing the sidelobe levels of the beamformer to levels that sufficiently reject these sidelobe signals. As we described in Section 11.2.2, the application of a taper to a spatial matched filter resulted in a low sidelobe beampattern. The same principle applies to the optimum beamformer. We define a tapered array response vector at an angle ϕ_s as

$$\mathbf{v}_t(\phi_s) = \mathbf{c}_{tbf}(\phi_s) = \mathbf{t} \odot \mathbf{c}_{mf}(\phi_s) \qquad (11.3.30)$$

where $\mathbf{t}$ is the tapering vector and $\odot$ is the Hadamard or element-by-element product. The tapering vector is normalized such that $\mathbf{v}_t^H(\phi_s)\mathbf{v}_t(\phi_s) = 1$ as in (11.2.23). The resulting low sidelobe adaptive beamformer is given by substituting $\mathbf{v}_t(\phi_s)$ for $\mathbf{v}(\phi_s)$ in (11.3.15)

$$\mathbf{c}_{to} = \frac{\mathbf{R}_{i+n}^{-1}\mathbf{v}_t(\phi_s)}{\mathbf{v}_t^H(\phi_s)\mathbf{R}_{i+n}^{-1}\mathbf{v}_t(\phi_s)} \tag{11.3.31}$$

We again use the Dolph-Chebyshev taper for illustration purposes because this choice of taper provides a uniform sidelobe level. Other choices include the window functions discussed in Chapter 5 in the context of spectrum estimation. Consider the optimum beamformer with an interferer at $\phi = 20°$ from Figure 11.18(a) with a potential signal leaking through the sidelobe at $\phi = -20°$. If instead we use a tapered optimum beamformer from (11.3.31) with a -50-dB sidelobe taper, a potential signal at $\phi = -20°$ receives a -50-dB level of attenuation. Figure 11.18(b) shows the beampattern of this tapered optimum beamformer. The sidelobe levels are significantly reduced while the null on the interferer at $\phi = 20°$ has been maintained.

The adaptive beamformer given by (11.3.31) is no longer optimal in any sense [unless it were somehow possible for our desired signal to be spatially matched to $\mathbf{v}_t(\phi_s)$]. However, the resulting adaptive beamformer still provides rejection of unwanted interferers via spatial nulling through the use of $\mathbf{R}_{i+n}^{-1}$ in (11.3.31). In addition, the low sidelobe levels of the beampattern reject signals not contained in the interference that are present at angles other than the angle of look ϕ_s. The penalty to be paid for the robustness provided by these low sidelobes is a small tapering loss in the direction of the look ϕ_s given by

$$L_{\text{taper}} = |\mathbf{c}_{to}^H\mathbf{v}(\phi_s)|^2 = \left|\frac{\mathbf{v}_t^H(\phi_s)\mathbf{R}_{i+n}^{-1}\mathbf{v}(\phi_s)}{\mathbf{v}_t^H(\phi_s)\mathbf{R}_{i+n}^{-1}\mathbf{v}_t(\phi_s)}\right|^2 = \frac{|\mathbf{v}^H(\phi_s)\mathbf{R}_{i+n}^{-1}\mathbf{v}_t(\phi_s)|^2}{\mathbf{v}_t^H(\phi_s)\mathbf{R}_{i+n}^{-2}\mathbf{v}_t(\phi_s)} \tag{11.3.32}$$

and a widening of the mainlobe beamwidth, as can be seen in the beampattern in Figure 11.18. This tapering loss indicates a mismatch between the true signal and the constraint in the optimum beamformer.

11.3.5 The Generalized Sidelobe Canceler

We have shown that the optimum MVDR beamformer maximizes the output SINR and can be formulated as a constrained optimization given by

$$\min \mathbf{c}^H\mathbf{R}_{i+n}\mathbf{c} \quad \text{subject to} \quad \mathbf{c}^H\mathbf{v}(\phi_s) = 1 \tag{11.3.33}$$

which results in the MVDR beamformer weight vector

$$\mathbf{c}_o = \frac{\mathbf{R}_{i+n}^{-1}\mathbf{v}(\phi_s)}{\mathbf{v}^H(\phi_s)\mathbf{R}_{i+n}^{-1}\mathbf{v}(\phi_s)} \tag{11.3.34}$$

This problem formulation can be broken up into constrained and unconstrained components that give rise to both an alternate implementation and a more intuitive interpretation of the optimum beamformer. The resulting structure, known as the *generalized sidelobe canceler* (*GSC*) (Griffiths and Jim 1982), uses a preprocessing stage to transform the optimization from constrained to unconstrained (Applebaum and Chapman 1976; Griffiths and Jim 1982). The GSC structure is illustrated in Figure 11.19.

Consider the array signal $\mathbf{x}(n)$ from (11.3.1) consisting of a signal component $s(n)$ and an interference-plus-noise component $\mathbf{x}_{i+n}(n)$. We are interested in forming the optimum beamformer steered to the angle ϕ_s. Let us start by forming a nonadaptive spatial matched filter in this direction $\mathbf{c}_{\text{mf}} = \mathbf{v}(\phi_s)$. The resulting output is the main channel signal given by

$$y_0(n) = \mathbf{c}_{\text{mf}}^H(\phi_s)\mathbf{x}(n) = \mathbf{v}^H(\phi_s)\mathbf{x}(n) = s_0(n) + i_0(n) + w_0(n) \tag{11.3.35}$$

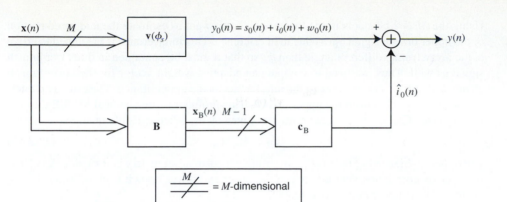

FIGURE 11.19
Generalized sidelobe canceler.

This nonadaptive beamformer makes up the upper branch of the GSC. In addition, let us form a lower branch consisting of $M-1$ channels in which the unconstrained optimization is performed. To prevent signal cancelation according to the unity-gain constraint in (11.3.33), we must ensure that these $M-1$ channels do not contain any signals[†] from ϕ_s. To this end, we form an $(M-1) \times M$ signal blocking matrix $\mathbf{B}$ that is orthogonal to the look-direction constraint $\mathbf{v}(\phi_s)$

$$\mathbf{B}^H \mathbf{v}(\phi_s) = \mathbf{0} \tag{11.3.36}$$

The resulting output of the blocking matrix is the $(M-1) \times 1$ vector signal

$$\mathbf{x}_B(n) = \mathbf{B}^H \mathbf{x}(n) \tag{11.3.37}$$

Thus, several choices for the blocking matrix exist that can perform this projection onto the $(M-1)$-dimensional subspace orthogonal to $\mathbf{v}(\phi_s)$. One choice uses a set of $M-1$ beams that are each chosen to satisfy this constraint. The spatial frequency of the ULA for an angle ϕ_s is

$$u_s = \frac{d}{\lambda} \sin \phi_s \tag{11.3.38}$$

For $\mathbf{v}(\phi_s)$, spatial matched filters at the frequencies

$$u_m = u_s + \frac{m}{M} \tag{11.3.39}$$

for $m = 1, 2, \ldots, M-1$ are mutually orthogonal as well as orthogonal to $\mathbf{v}(\phi_s)$, that is,

$$\mathbf{v}^H(\phi_m)\mathbf{v}(\phi_s) = 0 \tag{11.3.40}$$

where ϕ_m is the angle corresponding to the spatial frequency u_m given by (11.3.39). Thus, we can construct a beamspace signal blocking matrix from these $M-1$ steering vectors

$$\mathbf{B} = [\mathbf{v}(u_1) \, \mathbf{v}(u_2) \, \cdots \, \mathbf{v}(u_{M-1})] \tag{11.3.41}$$

An alternative signal blocking matrix can be implemented, assuming the array is presteered to the angle ϕ_s (Griffiths and Jim 1982). Presteering is accomplished by phase-shifting each element of the array by the corresponding steering vector element to this angle without actually forming a summation. Then any blocking matrix for which the elements of each column sum to zero will satisfy (11.3.36).

Once the nonadaptive preprocessing has been performed for the upper and lower branches of the GSC, an unconstrained optimization can be performed in the lower branch.

[†] Although the optimum beamformer was formulated for a signal-free interference-plus-noise correlation matrix $\mathbf{R}_{i+n}$, it is possible that the presence of the desired signal in the data is unavoidable.

Using the $M - 1$ channels in the lower branch, we want to estimate the undesired portion of the upper branch signal $y_0(n)$ due to interference. This interference is presumed to arrive at the array from a different angle than ϕ_s so that it must be contained in the lower branch signal as well. Thus, we need to compute an adaptive weight vector for the lower branch channels that forms an estimate of the interference in the upper branch. The estimated interference is subtracted from the upper branch. This problem is the classical MMSE filtering problem (see Chapter 6), whose solution is given by the Wiener-Hopf equation

$$\mathbf{c}_B = \mathbf{R}_B^{-1} \mathbf{r}_B \tag{11.3.42}$$

where $\mathbf{R}_B = E\{\mathbf{x}_B(n)\mathbf{x}_B^H(n)\}$ is the lower branch correlation matrix and $\mathbf{r}_B = E\{\mathbf{x}_B y_0^*(n)\}$ is the cross-correlation vector between the upper and lower branch signals. The resulting estimate of the upper branch interference signal is

$$\hat{i}_0(n) = \mathbf{c}_B^H \mathbf{x}_B(n) \tag{11.3.43}$$

and the output of the GSC is

$$y(n) = y_0(n) - \hat{i}_0(n) = y_0(n) - \mathbf{c}_B^H \mathbf{x}_B(n) \tag{11.3.44}$$

As we stated earlier, the GSC is equivalent to the optimum beamformer; that is, it maximizes the SINR at its output for signals arriving at angle ϕ_s. The power of the GSC formulation lies in its interpretation. Whereas for the optimum beamformer, the interference was canceled by forming spatial nulls in the directions of interferers, the GSC can be visualized as estimating the interference component in the upper branch from the lower-branch signals. Of course, the GSC also forms spatial nulls in the directions of the interferers. In terms of an alternate implementation, one must consider that if we are to steer the array to a number of different angles, each direction will require the formation of a new blocking matrix and the computation of a different correlation matrix and cross-correlation vector for the GSC. On the other hand, the optimum beamformer formulation has the same correlation matrix independent of the direction to which it is steered and, therefore, is often preferred for implementation purposes.

11.4 PERFORMANCE CONSIDERATIONS FOR OPTIMUM BEAMFORMERS

In this section we look at some considerations that influence the performance of an optimum beamformer. These considerations are also applicable to the adaptive methods in Section 11.5 that are derived from the optimum beamformer. Since the optimum beamformer serves as an upper bound on the performance of any adaptive method, these considerations can serve as adjustments to this performance bound for the adaptive counterparts to the optimum beamformer.

Two major factors that affect the performance of an optimum beamformer are:

- Mismatch of the actual signal to the assumed signal model used by the optimum beamformer
- Bandwidth of the signal that violates the narrowband assumption.

In the first section, we look at the effects of differences in the actual signal from that assumed for the optimum beamformer, known as *signal mismatch*. In virtually all array processing implementations, some level of mismatch will exist, due to either uncertainty in the exact angle of arrival of the signal of interest or the fact that the locations and characteristics of the individual sensors differ from our assumptions. As we will see, these errors that produce a signal mismatch can have profound implications on performance, particularly when the signal of interest is present in the correlation matrix. Next, we look at the effects of wider bandwidths on the performance of the optimum beamformer. In many applications, certain requirements necessitate the use of larger bandwidths. Their impact and possible means of correction are discussed in this section.

11.4.1 Effect of Signal Mismatch

In our formulation of the optimum beamformer, we assumed that a signal arriving at the array from an angle ϕ_s would produce a response equal to the ideal steering vector for a ULA [see (11.1.19)]. Thus, the optimum beamformer constrained its response to be spatially "matched" to the array response of the signal $\mathbf{v}_s = \mathbf{v}(\phi_s) = \mathbf{v}_0$ where $\phi_0 = \phi_s$

$$\mathbf{c}_o^H \mathbf{v}_0 = \mathbf{c}_o^H \mathbf{v}_s = 1 \tag{11.4.1}$$

that is, to pass it with unity gain. The vector $\mathbf{v}_0$ is the assumed array reponse. However, in reality, the signal may exhibit a different response across the array or may arrive from another angle $\phi_s \neq \phi_0$. The differences in response arise due to distortion of the waveform during propagation, amplitude and phase mismatches between the individual sensors, or errors in the assumed locations of the sensors.[†] These mismatches manifest themselves in a deviation of the array response from that assumed for a ULA in (11.1.19). However, if the angle of arrival of the signal differs from the assumed angle, the result is an array response as in (11.1.19), but for the angle ϕ_s as opposed to the steering angle ϕ_0. In either case, the beamformer is mismatched with the signal of interest and is no longer optimum. In this section, we examine the effect of these mismatches on the performance of the optimum beamformer, for the case of the signal of interest contained in the correlation matrix and absent from it. As we will see, the inclusion of this signal of interest in the correlation matrix has profound implications on the performance of a mismatched optimum beamformer. The analysis that follows was originally reported by Cox (1973).

Consider the case of an array signal consisting of a signal of interest $\mathbf{s}(n)$, interference $\mathbf{i}(n)$, and thermal noise $\mathbf{w}(n)$

$$\mathbf{x}(n) = \mathbf{s}(n) + \mathbf{i}(n) + \mathbf{w}(n) \tag{11.4.2}$$

where the noise is assumed to be uncorrelated, that is, $\mathbf{R}_n = \sigma_w^2 \mathbf{I}$. Now let us assume that the signal of interest is given by

$$\mathbf{s}(n) = \sqrt{M} s(n) \mathbf{u}_s \tag{11.4.3}$$

where $\mathbf{u}_s$, with unit norm ($\mathbf{u}_s^H \mathbf{u}_s = 1$), is the true array response to the signal of interest. For generality, $\mathbf{u}_s$ may be either an ideal or a nonideal array response for a ULA of a signal arriving from angle ϕ_s, but in either case it is mismatched with the assumed response

$$\mathbf{u}_s \neq \mathbf{v}_0 \tag{11.4.4}$$

The correlation matrix of the signal $\mathbf{x}(n)$ is made up of components due to the signal and the interference-plus-noise

$$\mathbf{R}_x = E\{\mathbf{x}(n)\mathbf{x}^H(n)\} = M\sigma_s^2 \mathbf{u}_s \mathbf{u}_s^H + \mathbf{R}_{i+n} \tag{11.4.5}$$

where the signal power is $\sigma_s^2 = |s(n)|^2$. The optimum beamformer with an MVDR constraint for the signal $\mathbf{s}(n)$ in (11.3.15) is

$$\mathbf{c}_o = \frac{\mathbf{R}_{i+n}^{-1} \mathbf{u}_s}{\mathbf{u}_s^H \mathbf{R}_{i+n}^{-1} \mathbf{u}_s} \tag{11.4.6}$$

However, the true array response $\mathbf{u}_s$ is unknown. This optimum beamformer in (11.4.6) yields the maximum output SINR given by

$$\text{SINR}_o = M\sigma_s^2 \mathbf{u}_s^H \mathbf{R}_{i+n}^{-1} \mathbf{u}_s = \text{SNR}_0 \cdot L_{\text{sinr}} \tag{11.4.7}$$

where $\text{SNR}_0 = M\sigma_s^2/\sigma_w^2$ is the matched filter SNR in the absence of interference from (11.2.18) (best performance possible) and $L_{\text{sinr}} = \sigma_w^2 \mathbf{u}_s^H \mathbf{R}_{i+n}^{-1} \mathbf{u}_s$ is the SINR loss from

---

[†] Similar losses also result from using a tapered steering vector. This loss was shown for the tapered optimum beamformer for the case of the signal of interest not present in the correlation matrix. As we will show in this section, the inclusion of the signal of interest in the correlation matrix can cause substantial losses in such a tapered beamformer.

(11.3.18) due to the presence of the interference. Thus, we can evaluate the losses due to signal mismatch and the inclusion of the signal of interest in the correlation matrix with respect to the maximum SINR in (11.4.7).

Loss due to signal mismatch

First, let us consider a mismatched signal $\mathbf{v}_0 \neq \mathbf{u}_s$ without the signal of interest present in the correlation matrix. The mismatch arises due to our lack of knowledge of the true array response to the signal of interest $\mathbf{u}_s$. The computation of the beamformer weights, assuming the array response to the signal to be $\mathbf{v}_0$ with an MVDR normalization, is given by

$$\mathbf{c}_1 = \frac{\mathbf{R}_{i+n}^{-1} \mathbf{v}_0}{\mathbf{v}_0^H \mathbf{R}_{i+n}^{-1} \mathbf{v}_0} \tag{11.4.8}$$

The SINR at the beamformer output for this weight vector is given by

$$\begin{aligned}
\text{SINR}_1 &= \frac{|\mathbf{c}_1^H \mathbf{s}(n)|^2}{\mathbf{c}_1^H \mathbf{R}_{i+n} \mathbf{c}_1} = M\sigma_s^2 \frac{|\mathbf{v}_0^H \mathbf{R}_{i+n}^{-1} \mathbf{u}_s|^2}{\mathbf{v}_0^H \mathbf{R}_{i+n}^{-1} \mathbf{v}_0} \\
&= M\sigma_s^2 \mathbf{u}_s^H \mathbf{R}_{i+n}^{-1} \mathbf{u}_s \frac{|\mathbf{v}_0^H \mathbf{R}_{i+n}^{-1} \mathbf{u}_s|^2}{(\mathbf{v}_0^H \mathbf{R}_{i+n}^{-1} \mathbf{v}_0)(\mathbf{u}_s^H \mathbf{R}_{i+n}^{-1} \mathbf{u}_s)} \\
&= \text{SINR}_o \cdot \cos^2(\mathbf{v}_0, \mathbf{u}_s; \mathbf{R}_{i+n}^{-1})
\end{aligned} \tag{11.4.9}$$

where the term $\cos(\cdot)$ measures the cosine of a generalized angle between two vectors $\mathbf{a}$ and $\mathbf{b}$ weighted by matrix $\mathbf{Z}$ (Cox 1973)

$$\cos^2(\mathbf{a}, \mathbf{b}; \mathbf{Z}) \triangleq \frac{|\mathbf{a}^H \mathbf{Z} \mathbf{b}|^2}{(\mathbf{a}^H \mathbf{Z} \mathbf{a})(\mathbf{b}^H \mathbf{Z} \mathbf{b})} \tag{11.4.10}$$

This term can be shown to have limits of $0 \leq \cos^2(\mathbf{a}, \mathbf{b}; \mathbf{Z}) \leq 1$ through the Schwartz inequality. The SINR from (11.4.9) can be rewritten as

$$\text{SINR}_1 = \text{SNR}_0 \cdot L_{\text{sinr}} \cdot L_{\text{sm}} \tag{11.4.11}$$

where we define the signal mismatch (sm) loss to be

$$L_{\text{sm}} \triangleq \cos^2(\mathbf{v}_0, \mathbf{u}_s; \mathbf{R}_{i+n}^{-1}) \tag{11.4.12}$$

Therefore, the SINR in (11.4.9) is a result of reducing the maximum SNR for a matched filter by the SINR loss due to the interference L_{sinr} as well as the loss due to the mismatch L_{sm}.

To gain some insight into the loss due to mismatch, consider the eigendecomposition of $\mathbf{R}_{i+n}^{-1}$ given by

$$\mathbf{R}_{i+n}^{-1} = \sum_{m=1}^{M} \frac{1}{\lambda_m} \mathbf{q}_m \mathbf{q}_m^H \tag{11.4.13}$$

where λ_m and $\mathbf{q}_m$ are the eigenvalue and eigenvector pairs, respectively. The largest eigenvalues and their corresponding eigenvectors are due to interference, while the small eigenvalues and eigenvectors are due to noise only. Since the eigenvectors form a basis for the M-dimensional vector space, any vector, say, $\mathbf{v}_0$ or $\mathbf{u}_s$, can be written as a linear combination of these eigenvectors. The product of the matrix $\mathbf{R}_{i+n}^{-1}$ with any vector closely aligned with an interference eigenvector will suffer significant degradation. Therefore, the mismatch loss in (11.4.12) should be relatively small for the case of $\mathbf{u}_s$ not closely aligned with interferers. Otherwise, if the signal lies near any of the interference eigenvectors, the beamformer will be more sensitive to signal mismatch.

Intuitively, performance degradation due to a mismatch in the optimum beamformer is relatively insensitive for small mismatches. The beamformer in (11.4.8) attempts to remove any energy that is not contained in its unity-gain constraint for $\mathbf{v}_0$. Since the signal with an array response $\mathbf{u}_s$ is not contained in the correlation matrix, the only losses incurred are due to the degree of mismatch between $\mathbf{u}_s$ and $\mathbf{v}_0$ and the similarity of $\mathbf{u}_s$ to interference components that are nulled through the use of $\mathbf{R}_{i+n}^{-1}$. However, most importantly, the loss due to mismatch is independent of the signal strength σ_s^2.

Loss due to signal in the correlation matrix

To implement the optimum beamformer in (11.4.6) in practice, we must assume that we can estimate $\mathbf{R}_{i+n}$ without the presence of the signal $\mathbf{s}(n)$. However, in many applications the signal is present all the time so that an estimate of a signal-free correlation matrix is not possible. In this case, the optimum beamformer must be constructed with the correlation matrix from (11.4.5) and is given by

$$\mathbf{c}_2 = \frac{\mathbf{R}_x^{-1}\mathbf{v}_0}{\mathbf{v}_0^H \mathbf{R}_x^{-1}\mathbf{v}_0} \tag{11.4.14}$$

Although this beamformer differs from the beamformer $\mathbf{c}_1$ in (11.4.8) that does not include the signal of interest in the correlation matrix, it produces an identical beamforming weight vector in the case when it is perfectly matched to the signal of interest, that is, $\mathbf{v}_0 = \mathbf{u}_s$ (see Problem 11.10). Thus, the beamformer in (11.4.14) also maximizes the SINR in the case of a perfectly matched signal. However, we want to examine the sensitivity of this beamformer to signal mismatches. The SINR of the beamformer from (11.4.14) with the signal present (sp) in the correlation matrix can be shown to be (Cox 1973)

$$\text{SINR}_2 = \frac{|\mathbf{c}_2^H \mathbf{s}(n)|^2}{\mathbf{c}_2^H \mathbf{R}_{i+n} \mathbf{c}_2} = M\sigma_s^2 \frac{|\mathbf{v}_0^H \mathbf{R}_x^{-1} \mathbf{u}_s|^2}{\mathbf{v}_0^H \mathbf{R}_x^{-1} \mathbf{R}_{i+n} \mathbf{R}_x^{-1} \mathbf{v}_0}$$

$$= \frac{\text{SINR}_1}{1 + (2\text{SINR}_o + \text{SINR}_o^2) \cdot \sin^2(\mathbf{v}_0, \mathbf{u}_s; \mathbf{R}_{i+n}^{-1})} \tag{11.4.15}$$

$$= \text{SNR}_0 \cdot L_{\text{sinr}} \cdot L_{\text{sm}} \cdot L_{\text{sp}}$$

where SINR_1 is the SINR of the mismatched beamformer in (11.4.9). The $\sin(\cdot)$ term measures the sine of the generalized angle between $\mathbf{v}_0$ and $\mathbf{u}_s$ and is related to the $\cos(\cdot)$ term from (11.4.10) by

$$\sin^2(\mathbf{v}_0, \mathbf{u}_s; \mathbf{R}_{i+n}^{-1}) = 1 - \cos^2(\mathbf{v}_0, \mathbf{u}_s; \mathbf{R}_{i+n}^{-1}) \tag{11.4.16}$$

Thus, the SINR of a beamformer constructed with the signal of interest present in the correlation matrix suffers an additional loss L_{sp}, beyond the losses associated with the interference L_{sinr} and the mismatch L_{sm} between $\mathbf{u}_s$ and $\mathbf{v}_0$, which is given by

$$L_{\text{sp}} = \frac{1}{1 + (2\text{SINR}_o + \text{SINR}_o^2)\sin^2(\mathbf{v}_0, \mathbf{u}_s; \mathbf{R}_{i+n}^{-1})} \tag{11.4.17}$$

Unlike the mismatch loss from (11.4.12), the loss due to the signal presence in the correlation matrix with signal mismatch is related to the signal strength σ_s^2. In fact, (11.4.17) shows a strong dependence on the signal strength through the terms SINR_o and SINR_o^2 in the denominator. This dependence on signal strength is weighted by the sine term in (11.4.16) that measures the amount of mismatch. Thus for large signals, the losses can be significant. In fact, it can be shown that the losses resulting from strong signals present in the correlation matrix can cause the output SINR to be lower than if the signal had been relatively weak. This phenomenon along with possible means of alleviating the losses is explored in Problem 11.11.

We have shown a high sensitivity to mismatch of strong signals of interest when they are present in the correlation matrix used to compute the beamforming weights in (11.4.14). Since, in practice, a certain level of mismatch is always present, it may sometimes be advisable to use conventional, nonadaptive beamformers when the signal is present at all times and does not allow the estimation of a signal-free correlation matrix $\mathbf{R}_{i+n}$. If the performance of such nonadaptive beamformers is deemed unacceptable, then special measures such as diagonal loading, which is described in Section 11.5.2, must be taken to design a robust beamformer that is less sensitive to mismatch (Cox et al. 1987).

11.4.2 Effect of Bandwidth

So far, we have relied on the narrowband assumption, meaning that the bandwidth B of the received signals is small with respect to the carrier frequency F_c. Previously, we gave a rule of thumb for this assumption, namely, that the *fractional bandwidth*, defined as

$$\bar{B} = \frac{B}{F_c} \tag{11.4.18}$$

is small, say, $\bar{B} \ll 1$ percent. Another measure is the space-time-bandwidth product, which for an array of length L is

$$\text{TBWP} = \frac{LB}{c} \tag{11.4.19}$$

where the time L/c is the maximum amount of time for a plane wave to propagate across the entire array, that is, the maximum propagation delay between the first and last elements ($\phi = \pm 90°$, $\sin \phi = 1$).

However, many real-world applications require increased bandwidths, which cause this assumption to be violated (Buckley 1987; Zatman 1998). The question then is, What is the effect of bandwidth on the performance of an array? Let us begin by examining the narrowband steering vector for a ULA from (11.1.19)

$$\mathbf{v}(\phi) = \frac{1}{\sqrt{M}}[1 \; e^{-j2\pi[(d \sin \phi)/\lambda]} \; \cdots \; e^{-j2\pi[(d \sin \phi)/\lambda](M-1)}]^T \tag{11.4.20}$$

which assumes that λ is constant, that is, the array receives signals only from a frequency F_c. Relaxing this assumption and substituting $\lambda = c/F$ from (11.1.2) gives us a steering vector that makes no assumptions about the bandwidth of the incoming signal

$$\mathbf{v}(\phi, F) = \frac{1}{\sqrt{M}}[1 \; e^{-j2\pi[(d \sin \phi)/c]F} \; \cdots \; e^{-j2\pi[(d \sin \phi)/c](M-1)F}]^T \tag{11.4.21}$$

When we demodulate the received signals by the carrier frequency F_c, we are making an implicit narrowband assumption that allows us to model the time delay between sensor elements as a phase shift. Therefore, a wideband signal arriving from an angle ϕ appears to the narrowband receiver as if it were arriving from an angular region centered at ϕ (provided the spectrum of the incoming signal is centered about F_c), since the approximation of the delay between elements as a single phase shift no longer holds. This phenomenon is known as *dispersion* since the incoming wideband signal appears to disperse in angle across the array.

Let us examine the impact of a wideband interference signal on the performance of an adaptive array. The correlation matrix of a single interference source impinging on the array from an angle ϕ is found by integrating over the bandwidth of the received signal

$$\mathbf{R}_i = \frac{\sigma_p^2}{B} \int_{F_c-B/2}^{F_c+B/2} \mathbf{v}(\phi, F) \mathbf{v}^H(\phi, F) \, dF \tag{11.4.22}$$

where the assumption is made that the spectral response of the signal is flat over the bandwidth, that is, $|R(F)|^2 = 1$ for $F_c - B/2 \leq F \leq F_c + B/2$. Now, focusing on the individual elements of the correlation matrix, namely the (m, n)th element

$$\langle \mathbf{R}_i \rangle_{m,n} = \frac{\sigma_p^2}{B} \int_{F_c - B/2}^{F_c + B/2} e^{j2\pi m[(d \sin \phi)/c]F} e^{-j2\pi n[(d \sin \phi)/c]F} \, dF$$

$$= \frac{\sigma_p^2}{B} \int_{F_c - B/2}^{F_c + B/2} e^{j2\pi (m-n)[(d \sin \phi)/c]F} \, dF \tag{11.4.23}$$

$$= \sigma_p^2 e^{j2\pi (m-n)[(d \sin \phi)/c]F_c} \frac{2 \sin \left[2\pi (m - n) \dfrac{d \sin \phi}{c} \dfrac{B}{2} \right]}{2\pi (m - n) \dfrac{d \sin \phi}{c} B}$$

$$= \sigma_p^2 e^{j2\pi (m-n)[(d \sin \phi)/\lambda]} \operatorname{sinc} \left[(m - n) \frac{d \sin \phi}{c} B \right]$$

where $\operatorname{sinc}(x) = \sin(\pi x)/(\pi x)$. We notice that each element is made up of two terms. The first term is simply the cross-correlation between the mth and nth sensor array elements for a narrowband signal arriving from ϕ

$$\langle \mathbf{R}_i^{(\text{nb})} \rangle_{m,n} = \sigma_p^2 e^{j2\pi (m-n)[(d \sin \phi)/\lambda]} \tag{11.4.24}$$

where the superscript indicates that this is the narrowband correlation matrix. The second term represents the dispersion across the array caused by the bandwidth of the interferer and is given by

$$\langle \mathbf{R}_d \rangle_{m,n} = \operatorname{sinc} \left[(m - n) \frac{d \sin \phi}{c} B \right] \tag{11.4.25}$$

Using (11.4.25), we can construct a matrix that models this dispersion across the entire array, which we refer to as the *dispersion matrix*. Dispersion creates decorrelation of the signal across the array, and this term represents the temporal autocorrelation of the impinging signal. Therefore, we can write the wideband correlation matrix as the Hadamard product of the narrowband correlation and the dispersion matrices

$$\mathbf{R}_i^{(\text{wb})} = \mathbf{R}_i^{(\text{nb})} \odot \mathbf{R}_d \tag{11.4.26}$$

where the Hadamard product is a point-by-point multiplication (Strang 1998).

The dispersion produced by a wideband signal can be compensated or corrected for at its specific angle by using a technique known as *time-delay steering*. Notice that the dispersion term in (11.4.25) is $\langle \mathbf{R}_d \rangle_{m,n} = 1$ for $\phi = 0°$ since the argument of the sinc function is zero. Therefore, for signals arriving from $\phi = 0°$ no dispersion can occur. In other words, for signals arriving from broadside to the array ($\phi = 0°$), the delay between elements is zero, independent of the frequency of the signal or its bandwidth. The dispersion becomes worse as the value of the angle is increased. This suggests a simple remedy to correct for dispersion: refocus the array to angle ϕ. Steering the array in this direction involves time-delaying each element to compensate for its delay between elements explicitly. This time-delay steering can be implemented in analog or digitally and is illustrated in Figure 11.20. The time-delay steered array signal is

$$\mathbf{x}_{\text{td}}(t) = [x_1(t) \, x_2(t - \tau_2) \cdots x_M(t - \tau_M)]^T \tag{11.4.27}$$

where $\tau_m = (d/\lambda)(m - 1) \sin \phi$ with λ being the wavelength of the center frequency F_c. Thus, a signal arriving from this angle will have no delay between elements following the time-delay steering. A convenient means of modeling time-delay steering in the discrete-time signal is through application of the matrix

$$\mathbf{V} = \operatorname{diag}\{\mathbf{v}(\phi)\} \tag{11.4.28}$$

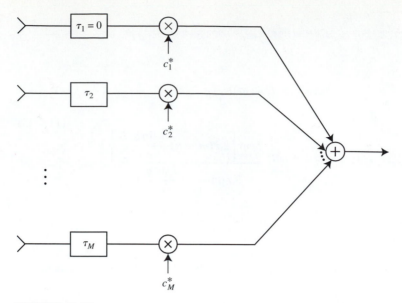

FIGURE 11.20
Time-delay steering prior to beamforming (referenced to the first element,
$\tau_1 = 0$).

to the array signal $\mathbf{x}(n)$ as

$$\mathbf{x}_{\text{td}}(n) = \mathbf{V}^H \mathbf{x}(n) \tag{11.4.29}$$

The resulting interference correlation matrix is

$$\mathbf{R}_i^{(\text{td})} = \mathbf{V}^H \mathbf{R}_i^{(\text{wb})} \mathbf{V} \tag{11.4.30}$$

Time-delay steering will focus signals from angle ϕ but may in fact increase the amount of dispersion from other angles. However, if we are not looking at these other angles, this effect may not be noticed. The underlying phenomenon that is occurring is that an optimum beamformer is forced to use additional adaptive degrees of freedom to cancel the dispersed, wideband interference signals. As long as the optimum beamformer has sufficient degrees of freedom, the effect of dispersion at other angles may not be evident.

EXAMPLE 11.4.1. Consider the radar interference scenario with a single jammer at an angle $\phi = 30°$ with a jammer-to-noise ratio JNR = 50 dB. Again, we have an $M = 10$ element array with $\lambda/2$ spacing. The center frequency of the array is $F_c = 1$ GHz, and the bandwidth is $B = 10$ MHz for a fractional bandwidth of $\bar{B} = 1$ percent. The SINR loss of an optimum beamformer is found by substituting the wideband correlation matrix from (11.4.26) into the SINR loss in (11.3.18)

$$L_{\text{sinr}}(\phi_s) = \mathbf{v}^H(\phi_s)[\mathbf{R}_{i+n}^{(\text{wb})}]^{-1} \mathbf{v}(\phi_s) \tag{11.4.31}$$

where the wideband interference-plus-noise correlation matrix is $\mathbf{R}_{i+n}^{(\text{wb})} = \mathbf{R}_i^{(\text{wb})} + \sigma_w^2 \mathbf{I}$ since the thermal noise is uncorrelated. Scanning across all angles, we can compute the SINR loss, which is shown in Figure 11.21 along with the SINR loss had the signal been narrowband. Notice the increased width of the SINR loss notch centered about $\phi = 30°$, which corresponds to a dropoff in performance in the vicinity of the jammer with respect to the narrowband case. However, at angles farther from the jammer there is no impact on performance; that is, $L_{\text{sinr}}(\phi_s) \approx 0$ dB. Next we look at the performance of an optimum beamformer that incorporates time-delay steering prior to adaptation. In this case, using $\mathbf{R}_i^{(\text{td})} + \sigma_w^2 \mathbf{I}$ from (11.4.30) in place of $\mathbf{R}_{i+n}^{(\text{wb})}$ in the SINR loss equation, we can compute the SINR loss of the optimum beamformer using time-delay steering, which is also plotted in Figure 11.21. The notch around the jammer at $\phi = 30°$ has been restored to the narrowband case for angles immediately surrounding $\phi = 30°$. At the

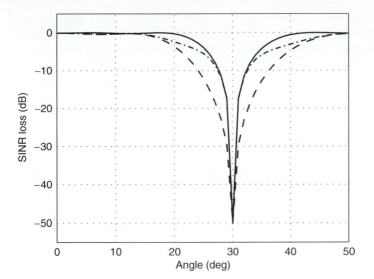

FIGURE 11.21
SINR loss for wideband jammer with JNR $= 50$ dB at an angle of
$\phi = 30°$. The carrier frequency is $F_c = 1$ GHz, and the bandwidth is
$B = 10$ MHz. Solid line is the narrowband signal, dashed line is the
wideband signal, and dash-dot line is the wideband signal with
time-delay steering.

angles a little farther away, the performance is still worse than that for the narrowband case but
still significantly better than that without time-delay steering.

11.5 ADAPTIVE BEAMFORMING

So far, we have only considered the optimum beamformer but have not concerned ourselves with how such a beamformer would be implemented in practice. Optimality was
only achieved because we assumed perfect knowledge of the second-order statistics of the
interference at the array, that is, the interference-plus-noise correlation matrix $\mathbf{R}_{i+n}$. In this
section, we describe the use of adaptive methods that are based on collected data from
which the correlation matrix is estimated. We look at two types of methods: block adaptive
and sample-by-sample adaptive. A block adaptive implementation of the optimum beamformer uses a "block" of data to estimate the adaptive beamforming weight vector and is
known as *sample matrix inversion (SMI)*. The SMI adaptive beamformer is examined in
Section 11.5.1 along with the sidelobe levels and training issues associated with the SMI
adaptive beamformer. Next we introduce the use of diagonal loading within the context
of the block adaptive SMI beamformer in Section 11.5.2. In Section 11.5.3, we discuss
sample-by-sample adaptive methods. These methods, as the block adaptive methods, base
their estimates of the statistics on the data, but update these statistics with each new sample
and are extensions of the adaptive filtering techniques from Chapter 10 to array processing.

11.5.1 Sample Matrix Inversion

In practice, the correlations are unknown and must be estimated from the data. Thus, we turn
to the maximum-likelihood (ML) estimate of the correlation matrix given by the average
of outer products of the array snapshots (Goodman 1963)

$$\hat{\mathbf{R}}_{i+n} = \frac{1}{K} \sum_{k=1}^{K} \mathbf{x}_{i+n}(n_k)\mathbf{x}_{i+n}^H(n_k) \tag{11.5.1}$$

where the indices n_k define the K samples of $\mathbf{x}_{i+n}(n)$ for $1 \le n \le N$ that make up the *training set*. Many applications may dictate that the collected snapshots be split into training data and data to be processed. The ML estimate of the correlation matrix implies that as $K \to \infty$, then $\hat{\mathbf{R}}_{i+n} \to \mathbf{R}_{i+n}$; and it is known as the *sample correlation matrix*. The total number of snapshots K used to compute the sample correlation matrix is referred to as the *sample support*. The larger the sample support, the better the estimate $\hat{\mathbf{R}}_{i+n}$ of the correlation matrix for stationary data. Proceeding by substituting the sample correlation matrix from (11.5.1) into the optimum beamformer weight computation in (11.3.15) results in the adaptive beamformer (Reed et al. 1974)

$$\mathbf{c}_{\text{smi}} = \frac{\hat{\mathbf{R}}_{i+n}^{-1}\mathbf{v}(\phi_s)}{\mathbf{v}^H(\phi_s)\hat{\mathbf{R}}_{i+n}^{-1}\mathbf{v}(\phi_s)} \tag{11.5.2}$$

known as the *sample matrix inversion* adaptive beamformer.[†] As for the optimum beamformer, an SMI adaptive beamformer can be implemented with low sidelobe control through the use of tapers. Simply substitute a tapered steering vector from (11.3.20) for $\mathbf{v}(\phi_s)$ in (11.5.2)

$$\mathbf{c}_{\text{tsmi}} = \frac{\hat{\mathbf{R}}_{i+n}^{-1}\mathbf{v}_t(\phi_s)}{\mathbf{v}_t^H(\phi_s)\hat{\mathbf{R}}_{i+n}^{-1}\mathbf{v}_t(\phi_s)} \tag{11.5.3}$$

Similarly, all the adaptive processing methods that will be discussed in Section 11.6, that is, the linearly constrained beamformer, all the partially adaptive beamformers, and the sidelobe canceler, can be implemented in a similar fashion by substituting the appropriate sample correlation matrix for its theoretical counterpart.

Of course, we cannot expect to substitute an estimate $\hat{\mathbf{R}}_{i+n}$ of the true correlation matrix $\mathbf{R}_{i+n}$ into the adaptive weight equation without experiencing a loss in performance. We begin by computing the output SINR of the SMI adaptive beamformer

$$\text{SINR}_{\text{smi}} = \frac{M\sigma_s^2|\mathbf{c}_{\text{smi}}^H\mathbf{v}(\phi_s)|^2}{E\{|\mathbf{c}_{\text{smi}}^H\mathbf{x}_{i+n}(n)|^2\}} = \frac{M\sigma_s^2|\mathbf{c}_{\text{smi}}^H\mathbf{v}(\phi_s)|^2}{\mathbf{c}_{\text{smi}}^H\mathbf{R}_{i+n}\mathbf{c}_{\text{smi}}}$$
$$= M\sigma_s^2\frac{[\mathbf{v}^H(\phi_s)\hat{\mathbf{R}}_{i+n}^{-1}\mathbf{v}(\phi_s)]^2}{\mathbf{v}^H(\phi_s)\hat{\mathbf{R}}_{i+n}^{-1}\mathbf{R}_{i+n}\hat{\mathbf{R}}_{i+n}^{-1}\mathbf{v}(\phi_s)} \tag{11.5.4}$$

Comparing this to the SINR obtained with the optimum beamformer from (11.3.12), we obtain the loss associated with the SMI adaptive beamformer relative to the optimum beamformer

$$L_{\text{smi}} = \frac{\text{SINR}_{\text{smi}}}{\text{SINR}_o} = \frac{[\mathbf{v}^H(\phi_s)\hat{\mathbf{R}}_{i+n}^{-1}\mathbf{v}(\phi_s)]^2}{[\mathbf{v}^H(\phi_s)\hat{\mathbf{R}}_{i+n}^{-1}\mathbf{R}_{i+n}\hat{\mathbf{R}}_{i+n}^{-1}\mathbf{v}(\phi_s)][\mathbf{v}^H(\phi_s)\mathbf{R}_{i+n}^{-1}\mathbf{v}(\phi_s)]} \tag{11.5.5}$$

This SMI loss is dependent on the array data used to compute $\hat{\mathbf{R}}_{i+n}$, which implies that L_{smi}, like the data, is a random variable. In fact, it can be shown that L_{smi} follows a beta distribution given by (Reed et al. 1974)

$$p_\beta(L_{\text{smi}}) = \frac{K!}{(M-2)!(K+1-M)!}(1-L_{\text{smi}})^{M-2}(L_{\text{smi}})^{K+1-M} \tag{11.5.6}$$

assuming a complex Gaussian distribution for the sensor thermal noise and the interference signals. Here M is the number of sensors in the array, and K is the number of snapshots

---

[†] An adaptive beamformer that is very similar to the SMI adaptive beamformer is known as the *adaptive matched filter (AMF)* (Robey et al. 1992). The difference between the two is actually in the normalization. The AMF requires $\mathbf{c}^H\hat{\mathbf{R}}_{i+n}\mathbf{c} = 1$ rather than $\mathbf{c}^H\mathbf{v}(\phi_s) = 1$ so that the interference-plus-noise has unit power at the beamformer output. As a result, it is straightforward to choose a detection threshold for the output of an AMF beamformer. For this reason, this method is discussed primarily within the context of *adaptive detection*. It is straightforward to show that the relation between the AMF and SMI adaptive weights, as they are defined in (11.5.2), is $\mathbf{c}_{\text{amf}} = [\mathbf{v}^H(\phi_s)\hat{\mathbf{R}}_{i+n}^{-1}\mathbf{v}(\phi_s)]^{1/2}\mathbf{c}_{\text{smi}}$.

used to estimate $\mathbf{R}_{i+n}$. Taking the expectation of this loss yields

$$E\{L_{\text{smi}}\} = \frac{K + 2 - M}{K + 1} \qquad (11.5.7)$$

which can be used to determine the sample support required to limit the losses due to correlation matrix estimation to a level considered acceptable. From (11.5.7), we can deduce the SMI loss will be approximately -3 dB for $K = 2M$ and approximately -1 dB for $K = 5M$.

EXAMPLE 11.5.1. In this example, we study the SMI adaptive beamformer and the loss associated with the number of snapshots used for training. SMI adaptive beamformers are produced with sample supports of $K = 1.5M, 2M$, and $5M$. Consider a ULA with $M = 20$ elements with an interference source at $\phi_i = 20°$ and a power of 50 dB. The thermal noise has unit variance $\sigma_w^2 = 1$. We can generate the interference-plus-noise signal $\mathbf{x}_{i+n}$ as

```
v_i = exp(-j*pi*[0:M-1]'*sin(phi_i*pi/180))/sqrt(M);
x_ipn=(10^(40/20))*v_i*(randn(1,N)+j*randn(1,N))/sqrt(2) + ...
(randn(1,N)+j*randn(1,N))/sqrt(2);
```

The sample correlation matrix is then found from (11.5.2). We compute the SINR at an angle of ϕ by first computing the SMI adaptive weight vector from (11.5.3), and the SINR from (11.5.4) using the actual correlation matrix $\mathbf{R}_{i+n}$, computed by

```
R_ipn = (10^(40/10))*v_i*v_i' + eye(M);
```

and a signal of interest with $M\sigma_s^2 = 1$. We repeat this across all angles $-90° \leq \phi < 90°$ and average over 100 realizations of $\mathbf{x}_{i+n}$. The resulting average SINR for the various sample supports is shown in Figure 11.22 along with the SINR for the optimum beamformer computed from (11.3.12). Note that for a signal of interest with $M\sigma_s^2 = 1$ and unit-variance noise, the SINR of the optimum beamformer is equal to its SINR loss. The jammer null is at $\phi = 20°$, as expected for all the beamformers. However, we notice that the SINR of the SMI adaptive beamformers is less than the optimum beamformer SINR by approximately 4, 3, and 1 dB for the sample supports of $K = 1.5M, 2M$, and $5M$, respectively. These losses are consistent with the SMI loss predicted by (11.5.7).

Sidelobe levels of the SMI adaptive beamformer

In addition to affecting the SINR of the beamformer output, the use of array snapshots to estimate $\mathbf{R}_{i+n}$ has implications for the sidelobe levels of the resulting adaptive beamformer. The following analysis follows directly from Kelly (1989). Consider a signal received from a direction other than the direction of look ϕ_s. The response to such a signal determines the sidelobe level of adaptive beamformer at this angle. For the MVDR optimum beamformer from (11.3.15), the sidelobe level (SLL) at an angle ϕ_u is given by

$$\text{SLL}_o = |C_o(\phi_u)|^2 = \frac{|\mathbf{v}^H(\phi_s)\mathbf{R}_{i+n}^{-1}\mathbf{v}(\phi_u)|^2}{|\mathbf{v}^H(\phi_s)\mathbf{R}_{i+n}^{-1}\mathbf{v}(\phi_s)|^2} \qquad (11.5.8)$$

where ϕ_s is the beamformer steering angle or look direction. Likewise, we can also define the SINR of a signal $\mathbf{s}(n) = \sigma_u\mathbf{v}(\phi_u)$ received from an angle ϕ_u in the sidelobes of the optimum beamformer steered to ϕ_s

$$\begin{aligned}
\text{SINR}_o(\phi_s, \phi_u) &= \frac{|\mathbf{c}_o^H(\phi_s)\mathbf{s}(n)|^2}{E\{|\mathbf{c}_o^H(\phi_s)\mathbf{x}_{i+n}(n)|^2\}} = \frac{\sigma_u^2|\mathbf{v}^H(\phi_s)\mathbf{R}_{i+n}^{-1}\mathbf{v}(\phi_u)|^2}{\mathbf{v}^H(\phi_s)\mathbf{R}_{i+n}^{-1}\mathbf{v}(\phi_s)} \\
&= \frac{\text{SINR}_o(\phi_u, \phi_u)|\mathbf{v}^H(\phi_s)\mathbf{R}_{i+n}^{-1}\mathbf{v}(\phi_u)|^2}{[\mathbf{v}^H(\phi_s)\mathbf{R}_{i+n}^{-1}\mathbf{v}(\phi_s)][\mathbf{v}^H(\phi_u)\mathbf{R}_{i+n}^{-1}\mathbf{v}(\phi_u)]} \\
&= \text{SINR}_o(\phi_u, \phi_u)\cos^2(\mathbf{v}(\phi_s), \mathbf{v}(\phi_u); \mathbf{R}_{i+n}^{-1})
\end{aligned} \qquad (11.5.9)$$

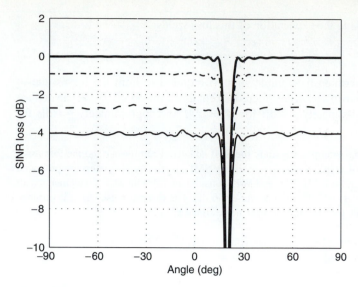

FIGURE 11.22
SINR loss for SMI adaptive beamformer with different numbers of
training snapshots. Thin solid line has 30 snapshots ($K = 1.5M$),
dashed line has 40 snapshots ($K = 2M$), and dash-dot line has 100
snapshots ($K = 5M$). Thick, solid line is SINR loss for the optimum
beamformer.

since $\text{SINR}_o(\phi_u, \phi_u) = \sigma_u^2 \mathbf{v}^H(\phi_u)\mathbf{R}_{i+n}^{-1}\mathbf{v}(\phi_u)$, which is the maximum output SINR possible
for a signal at angle ϕ_u, that is, the SINR if the optimum beamformer had been properly
steered in this direction. The term

$$
\begin{aligned}
\cos(\mathbf{v}(\phi_s), \mathbf{v}(\phi_u); \mathbf{R}_{i+n}^{-1}) &= \frac{\mathbf{v}^H(\phi_s)\mathbf{R}_{i+n}^{-1}\mathbf{v}(\phi_u)}{[\mathbf{v}^H(\phi_s)\mathbf{R}_{i+n}^{-1}\mathbf{v}(\phi_s)]^{1/2}[\mathbf{v}^H(\phi_u)\mathbf{R}_{i+n}^{-1}\mathbf{v}(\phi_u)]^{1/2}} \\
&= \frac{\tilde{\mathbf{v}}^H(\phi_s)\tilde{\mathbf{v}}(\phi_u)}{[\tilde{\mathbf{v}}^H(\phi_s)\tilde{\mathbf{v}}(\phi_s)]^{1/2}[\tilde{\mathbf{v}}^H(\phi_u)\tilde{\mathbf{v}}(\phi_u)]^{1/2}}
\end{aligned} \tag{11.5.10}
$$

where
$$
\tilde{\mathbf{v}}(\phi) = \mathbf{L}_{i+n}^{-1}\mathbf{v}(\phi) \tag{11.5.11}
$$

measures the cosine of a generalized angle between vectors $\mathbf{v}(\phi_s)$ and $\mathbf{v}(\phi_u)$ (Cox 1973).
This last quantity is the cosine of the angle between the whitened vectors $\tilde{\mathbf{v}}(\phi_s)$ and $\tilde{\mathbf{v}}(\phi_u)$
at the respective angles of ϕ_s and ϕ_u. The matrix $\mathbf{L}_{i+n}$ is simply the Cholesky factor of the
correlation matrix, that is, $\mathbf{R}_{i+n} = \mathbf{L}_{i+n}\mathbf{L}_{i+n}^H$. The sidelobe level of the optimum beamformer
from (11.5.8) can also be written in terms of the SINR from (11.5.9)

$$
\text{SLL}_o = |C_o(\phi_u)|^2 = \frac{\text{SINR}_o(\phi_s, \phi_u)}{\text{SINR}_o(\phi_s, \phi_s)} \tag{11.5.12}
$$

From (11.5.9),
$$
\cos^2(\mathbf{v}(\phi_s), \mathbf{v}(\phi_u); \mathbf{R}_{i+n}^{-1}) = \frac{\text{SINR}_o(\phi_s, \phi_u)}{\text{SINR}_o(\phi_u, \phi_u)} \tag{11.5.13}
$$

which is not the same as the sidelobe level in (11.5.12). However, this cosine term is
a measure of the attenuation provided by an optimum beamformer steered in the direction ϕ_s as opposed to the maximum SINR provided by steering to angle ϕ_u. Thus,
$\cos^2(\mathbf{v}(\phi_s), \mathbf{v}(\phi_u); \mathbf{R}_{i+n}^{-1})$ can be thought of as the sidelobe level at an angle ϕ_u of an optimum beamformer steered to ϕ_s in the absence of interference at ϕ_u. As a result, this term
serves as an upper bound on the sidelobe level.

Turning our attention to the SMI adaptive beamformer, we begin by computing the SINR at the beamformer output of a signal received from ϕ_u for a steering angle of ϕ_s

$$
\begin{aligned}
\text{SINR}_{\text{smi}}(\phi_s, \phi_u) &= \frac{|\mathbf{c}_{\text{smi}}^H(\phi_s)\mathbf{s}(n)|^2}{E\{|\mathbf{c}_{\text{smi}}^H(\phi_s)\mathbf{x}_{i+n}(n)|^2\}} \\
&= \frac{\sigma_u^2 |\mathbf{v}^H(\phi_s)\hat{\mathbf{R}}_{i+n}^{-1}\mathbf{v}(\phi_u)|^2}{\mathbf{v}^H(\phi_s)\hat{\mathbf{R}}_{i+n}^{-1}\mathbf{R}_{i+n}\hat{\mathbf{R}}_{i+n}^{-1}\mathbf{v}(\phi_s)} \\
&= \text{SINR}_o(\phi_u, \phi_u) \\
&\quad \times \frac{|\mathbf{v}^H(\phi_s)\hat{\mathbf{R}}_{i+n}^{-1}\mathbf{v}(\phi_u)|^2}{[\mathbf{v}^H(\phi_s)\hat{\mathbf{R}}_{i+n}^{-1}\mathbf{R}_{i+n}\hat{\mathbf{R}}_{i+n}^{-1}\mathbf{v}(\phi_s)][\mathbf{v}^H(\phi_u)\mathbf{R}_{i+n}^{-1}\mathbf{v}(\phi_u)]} \\
&= \text{SINR}_o(\phi_u, \phi_u) L(\phi_s, \phi_u)
\end{aligned}
\tag{11.5.14}
$$

where

$$
\begin{aligned}
L(\phi_s, \phi_u) &= \frac{\text{SINR}_{\text{smi}}(\phi_s, \phi_u)}{\text{SINR}_o(\phi_u, \phi_u)} \\
&= \frac{|\mathbf{v}^H(\phi_s)\hat{\mathbf{R}}_{i+n}^{-1}\mathbf{v}(\phi_u)|^2}{[\mathbf{v}^H(\phi_s)\hat{\mathbf{R}}_{i+n}^{-1}\mathbf{R}_{i+n}\hat{\mathbf{R}}_{i+n}^{-1}\mathbf{v}(\phi_s)][\mathbf{v}^H(\phi_u)\mathbf{R}_{i+n}^{-1}\mathbf{v}(\phi_u)]}
\end{aligned}
\tag{11.5.15}
$$

This term is bounded by $0 < L(\phi_s, \phi_u) < 1$ and can be interpreted as the loss of a signal received from the sidelobe angle ϕ_u processed with an SMI adaptive beamformer steered to ϕ_s relative to the maximum SINR possible for this signal. The term in the denominator of (11.5.15) is the SINR of the optimum, not the SMI adaptive beamformer. It is evident that as the number of array snapshots $K \to \infty$, $\hat{\mathbf{R}}_{i+n} \to \mathbf{R}_{i+n}$, $L(\phi_s, \phi_u) \to \cos^2(\mathbf{v}(\phi_s), \mathbf{v}(\phi_u); \mathbf{R}_{i+n}^{-1})$ from (11.5.15). The sidelobe level, however, of the SMI adaptive beamformer is

$$
\text{SLL}_{\text{smi}} = |C_{\text{smi}}(\phi_u)|^2 = \frac{|\mathbf{v}^H(\phi_s)\hat{\mathbf{R}}_{i+n}^{-1}\mathbf{v}(\phi_u)|^2}{|\mathbf{v}^H(\phi_s)\hat{\mathbf{R}}_{i+n}^{-1}\mathbf{v}(\phi_s)|^2}
\tag{11.5.16}
$$

However, unlike the sidelobe level of the optimum beamformer in (11.5.12) which could be related to the SINR of signals in the sidelobes, such a relation does not hold for the SMI adaptive beamformer because

$$
\mathbf{v}^H(\phi_s)\hat{\mathbf{R}}_{i+n}^{-1}\mathbf{R}_{i+n}\hat{\mathbf{R}}_{i+n}^{-1}\mathbf{v}(\phi_s) \neq \mathbf{v}^H(\phi_s)\hat{\mathbf{R}}_{i+n}^{-1}\mathbf{v}(\phi_s)
\tag{11.5.17}
$$

Asymptotically, this relation holds, but for finite sample support it does not. Nonetheless, we can draw some conclusions about the anticipated sidelobe levels using $L(\phi_s, \phi_u)$. The loss in SINR of the sidelobe signal $L(\phi_s, \phi_u)$ is a random variable with a probability distribution (Boroson 1980)

$$
\begin{aligned}
p(L, \Theta) &= \sum_{j=0}^{J} \binom{J}{j} \cos^2(\mathbf{v}(\phi_s), \mathbf{v}(\phi_u); \mathbf{R}_{i+n}^{-1})^{J-j} \\
&\quad \times \sin^2(\mathbf{v}(\phi_s), \mathbf{v}(\phi_u); \mathbf{R}_{i+n})^j p_\beta(L, J+1, M-1)
\end{aligned}
\tag{11.5.18}
$$

where $\sin^2(\mathbf{v}(\phi_s), \mathbf{v}(\phi_u); \mathbf{R}_{i+n}) = 1 - \cos^2(\mathbf{v}(\phi_s), \mathbf{v}(\phi_u); \mathbf{R}_{i+n}^{-1})$. Recall that $\cos^2(\mathbf{v}(\phi_s), \mathbf{v}(\phi_u); \mathbf{R}_{i+n}^{-1})$ depends on the true correlation matrix. The term J is given by

$$
J = K + 1 - M
\tag{11.5.19}
$$

and $p_\beta(x, l, m)$ is the beta probability distribution given by

$$
p_\beta(x, l, m) = \frac{(l+m-1)!}{(l-1)!(m-1)!} x^{l-1}(1-x)^{m-1}
\tag{11.5.20}
$$

From this probability distribution, we can compute the expected value of the loss of a signal received in the sidelobes of the SMI adaptive beamformer

$$E\{L(\phi_s, \phi_u)\} = \frac{1}{K+1}[1 + (K+1-M)\cos^2(\mathbf{v}(\phi_s), \mathbf{v}(\phi_u); \mathbf{R}_{\text{i+n}}^{-1}) \qquad (11.5.21)$$

For the case of perfect alignment ($\phi_u = \phi_s$), equation (11.5.21) measures the loss in SINR in the look direction since $\cos^2(\cdot) = 1$ and $E\{L(\phi_s, \phi_s)\} = L_{\text{smi}} = (K+2-M)/(K+1)$ from (11.5.21), which is the standard SMI SINR loss. In the opposite extreme, if ϕ_u is the angle of a null in the corresponding optimum beamformer, then $\cos^2(\cdot) = 0$ and

$$E\{L(\phi_s, \phi_u)\} = \frac{1}{K+1} \qquad (11.5.22)$$

The expected value of this loss can be interpreted as a bound on the sidelobe level provided that no interference sources were present at angle ϕ_u. The implication of this equation is a lower bound on the sidelobe level that can be achieved by using an SMI adaptive beamformer. Note that all this analysis also applies for tapered SMI adaptive beamformers when we substitute $\mathbf{v}_t(\phi_s) \rightarrow \mathbf{v}(\phi_s)$ as we did for the weights in (11.5.3). As a rule of thumb, we can use (11.5.22) to determine the sample support required for the desired sidelobe level. For example, if we were to design an adaptive beamformer with -40-dB sidelobe levels, we would require on the order of $K = 10,000$ snapshots.

EXAMPLE 11.5.2. We want to explore the effect of the number of training samples on the sidelobe levels of the SMI adaptive beamformer. To this end, we generate an interference signal at $\phi_i = 10°$ with a power of 70 dB and noise with unit variance ($\sigma_w^2 = 1$) for a ULA with $M = 40$ elements. The interference-plus-noise signal $\mathbf{x}_{\text{i+n}}$ is generated by

```
v_i = exp(-j*pi*[0:M-1]'*sin(phi_i*pi/180))/sqrt(M);
x_ipn=(10^(70/20))*v_i*(randn(1,N)+j*randn(1,N))/sqrt(2) + ...
(randn(1,N)+j*randn(1,N))/sqrt(2);
```

The sample correlation matrix is computed using (11.5.1). Then the SMI adaptive beamformer weights are computed from (11.5.2) with a look direction of $\phi_s = 0°$. We can compute the beampattern of the SMI adaptive beamformer using (11.2.3). The resulting beampatterns averaged over 100 realizations for SMI adaptive beamformers with sample support of $K = 100$ and $K = 1000$ are shown in Figure 11.23 for $-10° < \phi < 90°$ along with the beampattern of an optimum beamformer computed using the weight vector in (11.3.15) and a true correlation

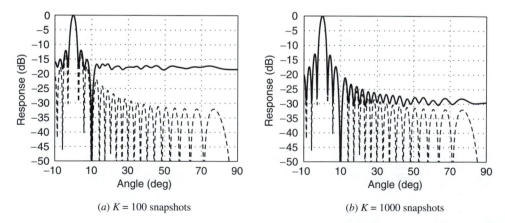

(a) $K = 100$ snapshots

(b) $K = 1000$ snapshots

FIGURE 11.23

Beampatterns of an SMI adaptive beamformer for (a) $K = 100$ snapshots and (b) $K = 1000$ snapshots. The dashed line is the quiescent response (optimum beamformer), and the solid line is the SMI adaptive beamformer.

```
R_ipn = (10^(70/10))*v_i*v_i' + eye(M);
```

Clearly, the sidelobe levels of the SMI adaptive beamformer are limited by the sample support available for training. For the case of $K = 100$, the sidelobe level is approximately -18 dB, whereas for $K = 1000$, the sidelobe level is approximately -30 dB.

Training issues

To implement the SMI adaptive beamformer, we need an estimate of the interference-plus-noise correlation matrix, which of course requires that no desired signal $\mathbf{s}(n)$ be present. The use of $\mathbf{R}_{i+n}$ provided an attractive theoretical basis for the derivation of the optimum beamformer and its subsequent adaptive implementation with the SMI technique. Although it can be shown that the use of a correlation matrix containing the desired signal produces equivalent adaptive weights in the case of perfect steering, this can almost never be accomplished in practice. Usually, we do not have perfect knowledge of the exact array sensor locations and responses. Coupled with the fact that often the angle of the desired signal is not known exactly for cases when we are searching for its actual direction, the presence of the desired signal in the training set results in the cancelation and subsequent loss in performance.

How do we get a signal-free estimate of the correlation matrix from array data in practice? In many applications, such as in certain radar and communications systems, we control when the desired signal is present since it is produced by a transmission that we initiate. In the case of jamming, common to both these applications, we can choose not to transmit for a period of time in order to collect data with which we can estimate $\mathbf{R}_{i+n}$. This type of training is often termed *listen-only*. For other types of interference that are only present at the same time as the desired signal, such as clutter in radar and reverberations in active sonar, the training can be accomplished using a technique known as *split window*. If we use a training set consisting of data samples around the sample of interest (before and after), we can exclude the sample of interest, and possibly some of its neighboring samples, to avoid the inclusion of the desired signal in the training set. This method has significant computational implications because it requires a different correlation matrix and therefore a separate computation of the adaptive beamforming weights for each sample under consideration. This problem can be alleviated somewhat by using matrix update methods, as discussed in Chapter 10; nonetheless, the increase in cost cannot be considered insignificant.

Certain methods have been proposed for the purposes of reducing the computations associated with estimating the correlation matrix. One such method is to assume the correlation matrix is Toeplitz for a ULA. Of course, this assumption is valid if the array consists of elements with equal responses $H_k(F, \phi)$ as a function of both frequency and angle from (11.1.14). However, in practice, this assumption almost never holds. The fact that the spatial signals are measured using different sensors, all with different responses, coupled with the limits on mechanical precision of the sensor placement in the array inevitably will cause these assumptions to be violated. As a result, constraining the correlation matrix to be Toeplitz, which is akin to averaging the correlations down the diagonals of the correlation matrix, will cause performance degradation that can be significant. These methods are well suited for temporal signals that are measured with a common sensor and are sampled at a rate that is very accurately controlled via a single analog-to-digital converter. Unfortunately with arrays, the spatial sampling process is not nearly as precise, and the use of multiple sensors for measurements can produce vastly different signal characteristics.

11.5.2 Diagonal Loading with the SMI Beamformer

Clearly, the ability of an SMI adaptive beamformer to achieve a desired sidelobe level relies on the availability of sufficient sample support K. However, for many practical applications, owing to either the nonstationarity of the interference or operational considerations, a limited number of samples are available to train the SMI adaptive beamformer. How, then, can we achieve this desired low sidelobe behavior? First, recall that the beam response of an optimum beamformer can be written in terms of its eigenvalues and eigenvectors as in (11.3.23). Likewise, for the SMI adaptive beamformer

$$C_{\text{smi}}(\phi) = \frac{\alpha}{\hat{\lambda}_{\min}} \left\{ C_{\text{q}}(\phi) - \sum_{m=1}^{M} \frac{\hat{\lambda}_m - \hat{\lambda}_{\min}}{\hat{\lambda}_m} [\hat{\mathbf{q}}_m^H \mathbf{v}(\phi_s)] \hat{Q}_m(\phi) \right\} \tag{11.5.23}$$

where $\hat{\lambda}_m$ and $\hat{\mathbf{q}}_m$ are the eigenvalues and eigenvectors of $\hat{\mathbf{R}}_{\text{i+n}}$, respectively, and $C_{\text{q}}(\phi)$ and $\hat{Q}_m(\phi)$ are the beampatterns of the quiescent weight vector and the mth eigenvector, known as an eigenbeam, respectively. Therefore, $C_{\text{smi}}(\phi)$ is simply $C_{\text{q}}(\phi)$ minus weighted eigenbeams that place nulls in the directions of interferers. The weights on the eigenbeams are determined by the ratio $(\hat{\lambda}_m - \hat{\lambda}_{\min})/\hat{\lambda}_m$. The noise eigenvectors are chosen to fill the remainder of the interference-plus-noise space that is not occupied by the interference. Ideally, these eigenvectors should have no effect on the beam response because the eigenvalues of the true correlation matrix $\lambda_m = \lambda_{\min} = \sigma_w^2$ for $m > P$. However, this relation does not hold for the sample correlation matrix for which the eigenvalues vary about the noise power σ_w^2 and asymptotically approach this expected value for increasing sample support. Therefore, the eigenbeams affect the beam response in a manner determined by their deviation from the noise floor σ_w^2. Since, as in the case of the sample correlation matrix, these eigenvalues are random variables that vary according to the sample support K, the beam response suffers from the addition of randomly weighted eigenbeams. The result is a higher sidelobe level in the adaptive beampattern.

A means of reducing the variation of the eigenvalues is to add a weighted identity matrix to the sample correlation matrix (Hudson 1981, Carlson 1988)

$$\hat{\mathbf{R}}_\text{l} = \hat{\mathbf{R}}_{\text{i+n}} + \sigma_\text{l}^2 \mathbf{I} \tag{11.5.24}$$

a technique that is known as *diagonal loading*. The result of diagonal loading of the correlation matrix is to add the loading level to all the eigenvalues. This, in turn, produces a bias in these eigenvalues in order to reduce their variation. To obtain the diagonally loaded SMI adaptive beamformer, simply substitute $\hat{\mathbf{R}}_\text{l}$ into (11.5.2)

$$\mathbf{c}_{\text{lsmi}} = \frac{\hat{\mathbf{R}}_\text{l}^{-1} \mathbf{v}(\phi_s)}{\mathbf{v}^H(\phi_s) \hat{\mathbf{R}}_\text{l}^{-1} \mathbf{v}(\phi_s)} \tag{11.5.25}$$

The bias in the eigenvalues produces a slight bias in the adaptive weights that reduces the output SINR. However, this reduction is very modest when compared to the substantial gains in the quality of the adaptive beampattern.

Recommended loading levels are $\sigma_w^2 \leq \sigma_\text{l}^2 < 10\sigma_w^2$. The maximum loading level is dependent on the application, but the minimum should be at least equal to the noise power in order to achieve substantial improvements. The loading causes a reduction in the nulling of weak interferers, that is, interferers with powers that are relatively close to the noise power. The effect on strong interferers is minimal since their eigenvalues only experience a minor increase. One added benefit of diagonal loading is that it provides a robustness to signal mismatch, as described in Section 11.4.1.

> **EXAMPLE 11.5.3.** In this example, we explore the use of diagonal loading of the sample correlation matrix to control the sidelobe levels of the SMI adaptive beamformer using the same set of parameters as in Example 11.5.2. The beampatterns for the SMI adaptive beamformer and the diagonally loaded SMI adaptive beamformer are shown in Figure 11.24 along with the beampattern for the optimum beamformer for $-10° < \phi < 90°$. The diagonal loading level was set to 5 dB above the thermal noise power, that is, $\sigma_\text{l}^2 = 10^{0.5}$, and the sample support was $K = 100$. The sidelobe levels of the diagonally loaded SMI adaptive beamformer

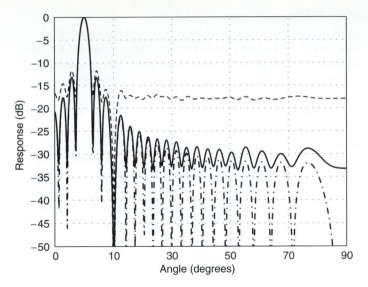

FIGURE 11.24
Beampatterns of an SMI adaptive beamformer for $K = 100$ snapshots without diagonal loading (dashed line), and with $\sigma_l^2 = 5$ dB diagonal loading (solid line). The beampattern of the optimum beamformer is also shown with the dash-dot line.

are very close to those of the optimum beamformer that used a known correlation matrix, while for the SMI adaptive beamformer the sidelobes are at approximately -18 dB. To gain some insight into the higher sidelobe levels of the SMI adaptive beamformer, we compute the eigenvalues of the SMI adaptive beamformer without diagonal loading using the MATLAB command `lambda = eig(Rhat);` where `Rhat` is the sample correlation matrix from (11.5.1). The eigenvalues of the sample and true correlation matrix are shown in Figure 11.25. The largest eigenvalue, corresponding to the 70-dB jammer, is approximately 70 dB but cannot be observed on this plot. We notice that for $K = 100$ training samples, the noise eigenvalues of $\hat{\mathbf{R}}_{i+n}$ are significantly different from those of $\mathbf{R}_{i+n}$, with larger than a 10-dB difference in some cases. As we stated earlier, the effect on the beampattern of the SMI adaptive beamformer is to add a random pattern weighted by this difference in eigenvalues. In the case of diagonal loading, the eigenvalues have as a lower bound the loading level σ_l^2 which, in turn, reduces these errors that are added to the beampatterns. The cost of the diagonal loading is to limit our ability to cancel weak interference with power less than the loading level. However, in the case of strong interference, almost no loss in terms of interference cancelation is experienced by introducing diagonal loading.

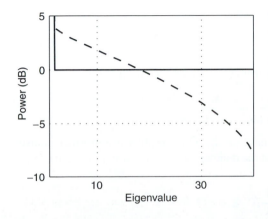

FIGURE 11.25
Noise eigenvalues of the SMI adaptive beamformer without diagonal loading $\sigma_w^2 = 1$ (dashed line) and the optimum beamformer (solid line).

11.5.3 Implementation of the SMI Beamformer

Although the SMI adaptive beamformer is formulated in terms of an estimated correlation matrix, the actual implementation, as with the least-squares methods discussed in Chapter 8, is usually in terms of the data samples directly. In other words, the actual estimate of the correlation matrix is never formed explicitly. Methods that are implemented on the data directly are commonly referred to as *amplitude domain* techniques, whereas if the sample correlation matrix had been formed, the implementation would be said to be performed in the *power domain*. The explicit computation of the sample correlation matrix is undesirable, first and foremost because the squaring of the data requires a large increase in the dynamic range of any processor. Numerical errors in the data are squared as well, and for a large number of training samples this computation may be prohibitively expensive. In this section, we give a brief discussion of the implementation considerations for the SMI adaptive beamformer, where the implementation is strictly in the amplitude domain. The incorporation of diagonal loading in this setting is also discussed since its formulation was given in the power domain.

The SMI beamformer is based on the estimated correlation matrix from (11.5.1). This sample correlation matrix may be written equivalently as

$$\hat{\mathbf{R}}_{i+n} = \frac{1}{K} \sum_{k=1}^{K} \mathbf{x}(n_k)\mathbf{x}^H(n_k) = \frac{1}{K}\mathbf{X}^H\mathbf{X} \tag{11.5.26}$$

where $\mathbf{X}$ is the data matrix formed with the array snapshots that make up the training set for the SMI adaptive weights, presumably containing only interference and noise, that is, no desired signals. This data matrix is

$$\mathbf{X}^H = [\mathbf{x}(n_1)\ \mathbf{x}(n_2)\ \cdots\ \mathbf{x}(n_K)] \tag{11.5.27}$$

$$= \begin{bmatrix} x_1(n_1) & x_1(n_2) & \cdots & x_1(n_K) \\ x_2(n_1) & x_2(n_2) & \cdots & x_2(n_K) \\ \vdots & \vdots & \ddots & \vdots \\ x_M(n_1) & x_M(n_2) & \cdots & x_M(n_K) \end{bmatrix} \tag{11.5.28}$$

where n_k, for $k = 1, 2, \ldots, K$, are the array snapshot indices of the training set. As was shown in Chapter 8, we can perform a QR decomposition on the data matrix to obtain the upper triangular factor

$$\mathbf{X} = \mathbf{Q}\mathcal{R}_x \tag{11.5.29}$$

where $\mathbf{Q}$ is a $K \times M$ orthonormal matrix and $\mathcal{R}_x$ is the $M \times M$ upper triangular factor. If we define the lower triangular factor as

$$\mathbf{L}_x \triangleq \frac{1}{\sqrt{K}}\mathcal{R}_x^H \tag{11.5.30}$$

the sample correlation matrix can then be written as

$$\hat{\mathbf{R}}_{i+n} = \frac{1}{K}\mathbf{X}^H\mathbf{X} = \frac{1}{K}\mathcal{R}_x^H\mathcal{R}_x = \mathbf{L}_x\mathbf{L}_x^H \tag{11.5.31}$$

since $\mathbf{Q}^H\mathbf{Q} = \mathbf{I}$. The SMI adaptive weights from (11.5.2) are then found to be

$$\mathbf{c}_{\text{smi}} = \frac{\hat{\mathbf{R}}_{i+n}^{-1}\mathbf{v}(\phi_s)}{\mathbf{v}^H(\phi_s)\hat{\mathbf{R}}_{i+n}^{-1}\mathbf{v}(\phi_s)} = \frac{\mathbf{L}_x^{-H}\mathbf{L}_x^{-1}\mathbf{v}(\phi_s)}{|\mathbf{L}_x^{-1}\mathbf{v}(\phi_s)|^2} \tag{11.5.32}$$

The implementation of diagonal loading with the SMI adaptive beamformer is also possible in the amplitude domain. Recall that the diagonally loaded correlation matrix from (11.5.24) is given by

$$\hat{\mathbf{R}}_l = \hat{\mathbf{R}}_{i+n} + \sigma_l^2\mathbf{I} = \frac{1}{K}\mathbf{X}^H\mathbf{X} + \sigma_l^2\mathbf{I} \triangleq \frac{1}{K}\mathbf{X}_l^H\mathbf{X}_l \tag{11.5.33}$$

where $\mathbf{X}_1$ is the "diagonally loaded" data matrix. Of course, data matrix $\mathbf{X}$ is not a square matrix, and thus it is not actually diagonally loaded. Instead, we append the data matrix with the square root of the loading matrix as

$$\mathbf{X}_1^H = [\mathbf{X}^H \ \sqrt{K}\sigma_1\mathbf{I}] \tag{11.5.34}$$

The resulting diagonally loaded SMI adaptive weights are found by substituting $\mathbf{X}_1$ for $\mathbf{X}$ in the amplitude-domain implementation of the SMI adaptive beamformer given above. The practical implementation of the SMI adaptive beamformer is performed in the following steps:

1. Compute the QR factorization of data matrix $\mathbf{X} = \mathbf{Q}\mathcal{R}_x$.
2. Find the Cholesky factor by normalizing the upper triagular factor $\mathbf{L}_x = (1/\sqrt{K})\mathcal{R}_x^H$.
3. Solve for $\mathbf{z}_1$ from $\mathbf{L}_x\mathbf{z}_1 = \mathbf{v}(\phi_s)$.
4. Solve for $\mathbf{z}_2$ from $\mathbf{L}_x^H\mathbf{z}_2 = \mathbf{z}_1$.
5. The SMI adaptive weight vector is given by $\mathbf{c}_{\text{smi}} = \mathbf{z}_2/\|\mathbf{z}_1\|^2$.

11.5.4 Sample-by-Sample Adaptive Methods

The SMI adaptive beamformer is a least-squares (LS) block adaptive technique similar to the LS methods discussed in Chapter 8. However, the optimum beamformer can also be implemented by using methods that compute the beamforming weights on a sample-by-sample basis; that is, the weights are updated for each new sample. Such methods are referred to as *sample-by-sample adaptive* and are simply extensions of the adaptive filtering methods from Chapter 10. The manner in which sample adaptive beamformers differ from adaptive filters is that rather than solve an unconstrained LS problem, adaptive beamformers solve a constrained LS problem. The implication of this constraint is that rather than have an estimated cross-correlation in the normal equations $\hat{\mathbf{R}}(n)\mathbf{c} = \hat{\mathbf{d}}(n)$, we have the deterministic steering vector $\mathbf{v}(\phi_s)$. Unlike the cross-correlation vector, the steering vector is known a priori and is *not* estimated from the data. We briefly discuss both techniques based on recursive least-squares (RLS) and steepest-descent methods. Since the derivation of the methods follows that for the adaptive filters in Chapter 10 quite closely, we only give a brief sketch of the algorithms along with some discussion.

An important consideration for sample adaptive methods is whether or not these techniques are appropriate for array processing applications. The problem with these methods is the amount of time required for the adaptive weights to converge. In many applications, the delay associated with the convergence of the adaptive beamformer is not acceptable. For example, a radar system might be attempting to find targets at close ranges. Range corresponds to the time delay associated with the propagation of the radar signal. Therefore, close ranges are the first samples received by the array during a collection period. A sample adaptive method that uses the close ranges to train (converge) could not find the targets at these ranges (samples). In fact, the time needed for convergence may not be insignificant, thus creating a large blind interval that is often unacceptable. However, the sample-by-sample adaptive techniques are appropriate for array processing applications in which the operating environment is nonstationary. Since the sample-by-sample adaptive beamformer alters its weights with each new sample, it can dynamically update its response for such a changing scenario.

Another important distinction between sample and block adaptive methods is the inclusion of the signal of interest in each sample and thus in the correlation matrix. Therefore, for sample adaptive methods, we cannot use a signal-free version of the correlation matrix, that is, the interference-plus-noise correlation matrix $\mathbf{R}_{i+n}$, but rather must use the whole correlation matrix $\mathbf{R}_x$. The inclusion of the signal in the correlation matrix has profound effects on the robustness of the adaptive beamformer in the case of signal mismatch. This effect was discussed in the context of the optimum beamformer in Section 11.4.1.

Recursive least-squares methods

We will not spend a lot of time discussing recursive least-squares (RLS) methods for adaptive beamforming since this topic is treated in Chapter 10. For further details on RLS methods used in array processing, the interested reader is referred to Schreiber (1986), McWhirter and Shepherd (1989), Yang and Böhme (1992). An important difference between the methods discussed here and those in Section 10.6 is in the normal equations that solve a constrained rather than an unconstrained optimization. The output signal of the adaptive beamformer is

$$y(n) = \mathbf{c}^H \mathbf{x}(n) \qquad (11.5.35)$$

However, $y(n)$ is not the desired response. In Section 10.6, we developed techniques based on the normal equations $\hat{\mathbf{R}}\mathbf{c} = \hat{\mathbf{d}}$, where $\hat{\mathbf{d}}$ is the estimated cross-correlation. However, for the adaptive beamformer we use the steering vector $\mathbf{v}(\phi_s)$, which is deterministic, in place of $\hat{\mathbf{d}}$. Algorithms based on RLS methods can be implemented such that the output $y(n)$ is computed directly (direct output extraction) or the adaptive beamformer weights are computed and then applied to determine the output (see Section 10.6). The simplifications for the beamformer case are discussed in Yang and Böhme (1992) and Haykin (1996).

The RLS methods are based on the update equation of the estimate of the correlation matrix

$$\hat{\mathbf{R}}_x(n + 1) = \lambda \hat{\mathbf{R}}_x(n) + \mathbf{x}(n + 1)\mathbf{x}^H(n + 1) \qquad (11.5.36)$$

where $0 < \lambda \leq 1$ is a scalar sometimes referred to as the forgetting factor. From the updated sample correlation matrix, an update for its inverse can be found by using the matrix inversion lemma from Appendix A. The adaptive beamformer weight vector is then found by modifying the solution to the MVDR adaptive weights with the updated inverse sample correlation matrix. In practice, these updatings are implemented by slightly modifying any of the algorithms described in Section 10.6.

Steepest-descent methods

The LMS algorithm from Section 10.4 is based on the method of steepest descent. However, the desired response used to form the LMS adaptive weights is not clear for the adaptive beamforming application. Instead, there is the steering vector $\mathbf{v}(\phi_s)$ that specifies the direction to which the adaptive beamformer is steered, namely, the angle ϕ_s. The resulting constrained optimization produced the optimum MVDR beamformer from Section 11.3. The sample adaptive implementation of this constrained optimization problem based on steepest descent was first proposed by Frost. The resulting algorithm uses a projection operation to separate the constrained optimization into a data-independent component and an adaptive portion that performs an unconstrained optimization (Frost 1972). The original algorithm was formulated using multiple linear constraints, as will be discussed in Section 11.6.1. However, in this section we focus on its implementation with the single unity-gain look-direction constraint for the MVDR beamformer. Note that the separation of the constrained and unconstrained components proposed by Frost provided the motivation for the generalized sidelobe canceler (GSC) structure (Griffiths and Jim 1982) discussed in Section 11.3.5. Below, we simply give the procedure for implementing the Frost algorithm. The interested reader is referred to Frost (1972) for further details.

Formally, the MVDR adaptive beamformer is attempting to solve the following constrained optimization

$$\min \mathbf{c}^H \mathbf{R}_x \mathbf{c} \qquad \text{subject to} \qquad \mathbf{c}^H \mathbf{v}(\phi_s) = 1 \qquad (11.5.37)$$

where the entire correlation matrix $\mathbf{R}_x$ including the desired signal is used in place of the interference-plus-noise correlation matrix $\mathbf{R}_{i+n}$ since we assume the signal of interest is always present. The correlation matrix is unknown and must be estimated from the data. To start the algorithm, we can form an $M \times M$ projection matrix $\mathbf{P}$ that projects onto a subspace orthogonal to the data-independent steering vector $\mathbf{v}(\phi_s)$. This projection matrix is given by (see Chapter 8)

$$\mathbf{P} = \mathbf{I} - \mathbf{v}(\phi_s)\mathbf{v}^H(\phi_s) \tag{11.5.38}$$

We can then define the nonadaptive beamformer weight vector as

$$\mathbf{c}_{na} = \mathbf{v}(\phi_s) \tag{11.5.39}$$

which is simply the spatial matched filter from Section 11.2. The update equation for the sample adaptive beamformer based on Frost's steepest-descent (sd) algorithm is then written as

$$\mathbf{c}_{sd}(n + 1) = \mathbf{c}_{na} + \mathbf{P}[\mathbf{c}_{sd}(n) - \mu y^*(n)\mathbf{x}(n)] \tag{11.5.40}$$

where μ is the step-size parameter and

$$y(n) = \mathbf{c}_{sd}^H\mathbf{x}(n) \tag{11.5.41}$$

is the output of the steepest-descent sample adaptive beamformer.

Since the projection matrix $\mathbf{P}$ maintains orthogonality between the $\mathbf{c}_{na}$ and the adapted portion of (11.5.40), the nonadaptive beamformer weights from (11.5.39) maintain the unity-gain constraint from (11.5.37). In fact, since the adaptation is performed on the component orthogonal to $\mathbf{c}_{na}$ in an unconstrained manner, the Frost algorithm is essentially using an LMS adaptive filter in the GSC architecture from Section 11.3.5. The convergence of the Frost adaptive beamformer, as for the LMS adaptive filter, is controlled by the step-size parameter μ. In order for the adaptive beamformer weights to converge, the step-size parameter must be chosen to be

$$0 < \mu < \frac{1}{\tilde{\lambda}_{max}} \tag{11.5.42}$$

where $\tilde{\lambda}_{max}$ is the maximum eigenvalue of the matrix

$$\tilde{\mathbf{R}} = \mathbf{P}\mathbf{R}_x\mathbf{P} \tag{11.5.43}$$

More details about the algorithm can be found in Frost (1972).

The sample adaptive beamformer based on the Frost algorithm maintains a look direction of ϕ_s through the constraint $\mathbf{c}_{sd}^H\mathbf{v}(\phi_s) = 1$. This constraint is easily seen by interpreting the adaptive weight update equation in (11.5.40) as the steering vector $\mathbf{c}_{na} = \mathbf{v}(\phi_s)$ updated by a component orthogonal to $\mathbf{v}(\phi_s)$. In the case of a signal received from a direction ϕ_s, the adaptive beamformer will immediately track this signal, since it is constrained to observe signals at ϕ_s and is not part of the adaptation. The convergence of this sample adaptive beamformer in terms of interference rejection is very similar to the LMS algorithm. See Chapter 10 for details on the LMS algorithm.

11.6 OTHER ADAPTIVE ARRAY PROCESSING METHODS

In this section, we consider various other adaptive array processing methods. First, we look at the use of multiple constraints in an adaptive array beyond the single constraint of distortionless response for the MVDR optimum beamformer. Second, we consider partially adaptive arrays that are methods that perform deterministic preprocessing prior to adaptation, in order to reduce the adaptive degrees of freedom. These methods are commonly used in practice for both computational reasons as well as limited sample support. Third, we describe the sidelobe canceler that was the first proposed adaptive array processing method. In addition to its historical significance, the sidelobe canceler is still a viable technique for certain array processing applications. Throughout this section, we use the word *adaptive* to indicate that the various methods are based on training data. However, the derivations are all in terms of known statistics. Although none of these methods can really be called optimum, each one satisfies an optimization criterion in the case of known statistics. The implementation of the methods using actual data samples in place of assuming known statistics follows directly from the techniques described in Section 11.5.

11.6.1 Linearly Constrained Minimum-Variance Beamformers

In Section 11.3, we discussed the optimum beamformer that maximizes the signal-to-interference-plus-noise ratio (SINR). This optimum beamformer was also formulated as the solution to a constrained optimization problem, namely,

$$\min \mathbf{c}^H \mathbf{R}_{i+n} \mathbf{c} \quad \text{subject to} \quad \mathbf{c}^H \mathbf{v}(\phi_s) = 1 \tag{11.6.1}$$

where $\mathbf{v}(\phi_s)$ is the array response vector for a signal arriving from an angle ϕ_s. Due to this alternate formulation, the optimum beamformer is commonly referred to as the minimum-variance distortionless response (MVDR) beamformer.

However, some applications may require additional conditions on the beamformer. As with the optimum beamformer, we want to minimize the output power $\mathbf{c}^H \mathbf{R}_{i+n} \mathbf{c}$, but with additional constraints on the response of the beamformer. The imposition of further constraints on the minimum-variance beamformer results in suboptimum performance in terms of SINR. However, if designed properly, the constraints should have little effect on SINR while yielding some desirable attributes. One common use of constraints is for the case when the angle of an interference source ϕ_i is known a priori. In this case, we want to reject all energy received from this angle, that is,

$$\mathbf{c}^H \mathbf{v}(\phi_i) = 0 \tag{11.6.2}$$

The result of the null constraint is an adaptive beamformer that rejects all energy from the angle ϕ_i. Another type of constraint is to require the beamformer to pass signals not only from the angle ϕ_s, but also from another angle ϕ_1. As for the MVDR beamformer, this constraint is formulated as

$$\mathbf{c}^H \mathbf{v}(\phi_1) = 1 \tag{11.6.3}$$

In this manner, multiple angles can be specified to pass signals of interest with unity gain. Such amplitude constraints can also be used to preserve the response of the beamformer in an angular region about ϕ_s (Steele 1983, Takao et al. 1976). These additional constraints help to make the resulting adaptive beamformer more robust to signal mismatches, as discussed in Section 11.4.1, that result from the actual angle of the desired signal ϕ_0 slightly differing from its presumed angle ϕ_s. Therefore, if we choose a pair of angles slightly offset from ϕ_s

$$\phi_1 = \phi_s - \Delta\phi \quad \phi_2 = \phi_s + \Delta\phi \tag{11.6.4}$$

the response of the beamformer steered to ϕ_s broadens. The effect in terms of mainlobe width is similar to tapering the MVDR beamformer when the angle offset $\Delta\phi$ is small. An alternative approach to robust adaptive beamforming is the use of derivative constraints. See Applebaum and Chapman (1976), Er and Cantoni (1983), and Steele (1983) for details.

Once we have determined a set of constraints, for example, the desired responses at a set of angles, we can solve for the constrained adaptive beamformer. The result is known as the *linearly constrained minimum-variance (LCMV)* beamformer (Applebaum and Chapman 1976; Buckley 1987). As we stated earlier, we want to minimize the output energy of the beamformer subject to a set of constraints. This problem is formulated as

$$\min \mathbf{c}^H \mathbf{R}_{i+n} \mathbf{c} \quad \text{subject to} \quad \mathbf{C}^H \mathbf{c} = \boldsymbol{\delta} \tag{11.6.5}$$

where $\mathbf{C}$ is known as the constraint matrix and $\boldsymbol{\delta}$ is the constraint response vector. For example, if we want to pass signals from an angle ϕ_s as well as preserve its response with a pair of amplitude constraints at the angles $\phi_s \pm \Delta\phi$, the constraint matrix and constraint response vectors are given by

$$\mathbf{C} = [\mathbf{v}(\phi_s) \, \mathbf{v}(\phi_s - \Delta\phi) \, \mathbf{v}(\phi_s + \Delta\phi)] \quad \boldsymbol{\delta} = [1 \; 1 \; 1]^T \tag{11.6.6}$$

As for the MVDR beamformer, the solution for the LCMV beamformer is found by using Lagrange multipliers (see Appendix B). The LCMV beamformer weight vector is given by

$$\mathbf{c}_{\text{lcmv}} = \mathbf{R}_{i+n}^{-1} \mathbf{C} (\mathbf{C}^H \mathbf{R}_{i+n}^{-1} \mathbf{C})^{-1} \boldsymbol{\delta} \tag{11.6.7}$$

As for the MVDR beamformer, the LCMV beamformer can also be formulated using a generalized sidelobe canceler architecture (Griffiths and Jim 1982), discussed in Section 11.3.5. In fact, the MVDR beamformer is simply a special case of the LCMV beamformer with $\mathbf{C} = \mathbf{v}(\phi_s)$ and $\boldsymbol{\delta} = 1$.

In this section, we have described the use of *linear* constraints in a minimum-variance beamformer. However, the use of *quadratic constraints* within the context of a minimum-variance beamformer is also possible. The primary motivation for using these quadratic constraints is for robustness purposes against signal mismatch, as discussed in Section 11.4.1. One such quadratic constraint adds a constraint on the norm of the weight vector of the adaptive beamformer in addition to the MVDR constraint (Cox et al. 1987; Maksym 1979)

$$\min \; \mathbf{c}^H \mathbf{R}_{i+n} \mathbf{c} \quad \text{subject to} \quad \mathbf{c}^H \mathbf{v}(\phi_s) = 1 \quad \text{and} \quad \|\mathbf{c}\|^2 \leq \kappa^2 \tag{11.6.8}$$

whose solution is given by

$$\mathbf{c} = \eta (\mathbf{R}_{i+n} + \sigma_\kappa^2 \mathbf{I})^{-1} \mathbf{v}(\phi_s) \tag{11.6.9}$$

where η is a constant and σ_κ^2 is a scaling term on the identity matrix. Thus, mimimizing the norm of the adaptive beamforming weight vector is equivalent to adding a weighted identity matrix to the interference-plus-noise correlation matrix. The solution to this quadratic constraint bears a striking resemblence to diagonal loading as discussed in the context of the SMI adaptive beamformer, in Section 11.5.2. In fact, the use of some level of diagonal loading is generally a recommended practice for implementing an adaptive beamformer to reduce its sensitivity to mismatch, and for the purposes of low sidelobe levels.

11.6.2 Partially Adaptive Arrays

The optimum beamformer maximizes output SINR by placing a null in the direction of any interference sources while maintaining gain in the direction of interest. Recall the optimum beamforming weights from (11.3.13)

$$\mathbf{c}_o = \alpha \mathbf{R}_{i+n}^{-1} \mathbf{v}(\phi_s) \tag{11.6.10}$$

where we choose the MVDR normalization $\alpha = [\mathbf{v}^H(\phi_s) \mathbf{R}_{i+n}^{-1} \mathbf{v}(\phi_s)]^{-1}$. The correlation matrix $\mathbf{R}_{i+n}$ is an $M \times M$ matrix where M is the number of elements in the ULA. The optimum weights adapt to the statistics of the data in an M-dimensional space where M is referred to as the *adaptive degrees of freedom*. However, in many applications, the number of elements in the array exceeds the adaptive degrees of freedom that can be practically implemented. The implementation of such a beamformer requires the estimation of the correlation matrix from collected data. As shown in Section 11.5, the estimation of $\mathbf{R}_{i+n}$ requires a certain number of data samples to maintain a desired level of performance. Many times, the number of data samples is limited, due to either finite regions over which the data are stationary or restrictions on the length of the collection interval. Likewise, the number of adaptive degrees of freedom that can be implemented may be limited for computational reasons. These restrictions motivate the use of methods that reduce the degrees of freedom prior to adaptation. An array implemented using a reduced number of degrees of freedom is referred to as a *partially adaptive array*.

Consider an array signal vector $\mathbf{x}(n)$ consisting of a desired signal, interference, and noise components

$$\mathbf{x}(n) = \mathbf{s}(n) + \mathbf{i}(n) + \mathbf{w}(n) = \sqrt{M}\mathbf{v}(\phi_s)s(n) + \sqrt{M}\sum_{p=1}^{P}\mathbf{v}(\phi_p)i_p(n) + \mathbf{w}(n) \quad (11.6.11)$$

where ϕ_p and $i_p(n)$ are the angle and signal, respectively, of the pth interferer with a total of P interferers. Usually, the number of interferers is limited; yet the number of elements in the array M may be quite large, that is, $M \gg P$. In general, one adaptive degree of freedom is required for each interferer.[†] Therefore, we only require some number of adaptive degrees of freedom $Q > P$, not the full dimensionality provided by the number of elements M. We want to use a large number of elements in order to have an aperture that achieves the desired angular resolution. Therefore, we do not want to limit the number of elements in order to reduce the degrees of freedom; rather, we want to project the array data into a lower-dimensional subspace in which we can perform our optimization (Morgan 1978). The projection is accomplished using a nonadaptive preprocessor and is modeled as a rank-reducing transformation matrix $\mathbf{T}$ with dimensions $M \times Q$ applied to the array signal

$$\tilde{\mathbf{x}}(n) = \mathbf{T}^H \mathbf{x}(n) \quad (11.6.12)$$

where $\tilde{\mathbf{x}}(n)$ is a signal vector of dimension Q. Likewise, the interference-plus-noise signal in the lower-dimensional space is

$$\tilde{\mathbf{x}}_{i+n}(n) = \mathbf{T}^H \mathbf{x}_{i+n}(n) \quad (11.6.13)$$

and has a correlation matrix

$$\tilde{\mathbf{R}}_{i+n} = E\{\tilde{\mathbf{x}}_{i+n}(n)\tilde{\mathbf{x}}_{i+n}^H(n)\} = \mathbf{T}^H \mathbf{R}_{i+n} \mathbf{T} \quad (11.6.14)$$

The partially adaptive beamforming weights are then given by

$$\tilde{\mathbf{c}} = \alpha \tilde{\mathbf{R}}_{i+n}^{-1} \tilde{\mathbf{v}}(\phi_s) \quad (11.6.15)$$

where

$$\tilde{\mathbf{v}}(\phi_s) = \mathbf{T}^H \mathbf{v}(\phi_s) \quad (11.6.16)$$

is the projection of the M-dimensional steering vector $\mathbf{v}(\phi_s)$ onto the same Q-dimensional subspace. The output of the partially adaptive beamformer is then obtained by applying the beamforming weights in (11.6.15) to the reduced-dimension array signal from (11.6.12)

$$y(n) = \tilde{\mathbf{c}}^H \tilde{\mathbf{x}}(n) \quad (11.6.17)$$

The resulting partially adaptive beamformer, shown in Figure 11.26, is no longer optimal in the sense of the full M-dimensional beamformer, but is optimal given the nonadaptive preprocessing transformation onto the Q-dimensional subspace. Thus, the overall performance of the partially adaptive beamformer is governed by how much information was preserved by the nonadaptive preprocessor $\mathbf{T}$. The performance of the partially adaptive beamformer can be assessed relative to the full-dimensional processor by reconstructing the effective $M \times 1$ beamforming weight vector with the transformation matrix and the partially adaptive (pa) weights

$$\mathbf{c}_{\text{pa}} = \mathbf{T}\tilde{\mathbf{c}} \quad (11.6.18)$$

In addition, we must consider the effect of this preprocessing transformation on the noise correlation matrix. For the array signal, we have assumed that the noise has a power of σ_w^2 and is uncorrelated, that is, $\mathbf{R}_n = \sigma_w^2 \mathbf{I}$. Therefore, the noise following the application of the preprocessing transformation has a correlation matrix given by

$$\tilde{\mathbf{R}}_n = \mathbf{T}^H \mathbf{R}_n \mathbf{T} = \sigma_w^2 \mathbf{T}^H \mathbf{T} \quad (11.6.19)$$

[†]The assumption is that the interferers are narrowband and are well separated in angle.

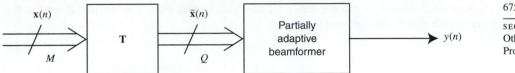

FIGURE 11.26
Partially adaptive array using data transformation.

In the case of an SMI adaptive beamformer, this different structure of the noise correlation matrix has implications for diagonal loading. The diagonal loading of the sample correlation matrix of the full array was performed by adding a weighted diagonal matrix to the sample correlation matrix in (11.5.24). For a partially adaptive array that already has had a preprocessing transform performed, the diagonal loading of a sample correlation matrix becomes

$$\hat{\tilde{\mathbf{R}}}_l = \hat{\tilde{\mathbf{R}}}_{i+n} + \sigma_l^2 \mathbf{T}^H \mathbf{T} \tag{11.6.20}$$

where σ_l^2 is the loading level. Since the thermal noise is not necessarily uncorrelated after the preprocessing transformation, diagonal loading must account for the transformed noise correlation. Otherwise, performance degradation can occur.

So far, we have only stated that the adaptation for a partially adaptive array must take place in a lower-dimensional space using a nonadaptive preprocessor, but we have not given any explicit means of performing this task. Below we discuss two commonly used preprocessing methods used for partially adaptive arrays.

Subarray partially adaptive arrays

Many times, the number of elements in an array can be very large. Thus, one means of reducing the adaptive degrees of freedom is to split the array into a number of smaller arrays, process the smaller arrays in a nonadaptive manner, and perform adaptation on the outputs of these smaller arrays. Let us consider the case in which we are looking for signals from a direction ϕ_s, and the full-dimensional steering vector is $\mathbf{v}_M(\phi_s)$, where we use the subscript M to denote the length of the steering vector. The full array may be divided into Q equal-sized intervals[†] of nonoverlapping subarrays of length

$$\tilde{M} = \frac{M}{Q} \tag{11.6.21}$$

where we have assumed that M is an integer multiple of Q. The rank-reducing transformation for the subarrays then can be written as a sparsely populated matrix made up of length-$\tilde{M}$ steering vectors $\mathbf{v}_{\tilde{M}}(\phi_s)$

$$\mathbf{T} = \begin{bmatrix} \mathbf{v}_{\tilde{M}}(\phi_s) & \mathbf{0} & \cdots & \mathbf{0} \\ \mathbf{0} & \mathbf{v}_{\tilde{M}}(\phi_s) & \cdots & \mathbf{0} \\ \vdots & \vdots & \ddots & \vdots \\ \mathbf{0} & \mathbf{0} & \cdots & \mathbf{v}_{\tilde{M}}(\phi_s) \end{bmatrix} \tag{11.6.22}$$

Each subarray consists of an $\tilde{M}$-dimensional conventional beamformer steered to ϕ_s and can be viewed as a highly directional element as opposed to the omnidirectional elements assumed for the individual sensors of the array.

Beamspace partially adaptive arrays

Another approach to constructing a partially adaptive beamformer is to produce a set of beams using the full array. The ensuing adaptation is performed in a reduced-dimension

---

[†]Subarrays need not necessarily have equal length or be nonoverlapping. This restriction is placed on the formulation only to simplify the discussion.

beamspace, that is, a space spanned by the nonadaptive beams. If we use B beams, the rank-reducing transformation matrix is

$$\mathbf{T} = [\mathbf{v}(\phi_1)\, \mathbf{v}(\phi_2)\, \cdots\, \mathbf{v}(\phi_B)] \qquad (11.6.23)$$

where $\phi_1, \phi_2, \ldots, \phi_B$ are the angles of these beams. These beamformers are typically steered in directions around the angle of interest, ϕ_s. For example, if the angle of interest is $\phi_s = 0°$, beams might be steered to angles $\phi = -5°, -4°, \ldots, 0°, \ldots, 4°, 5°$. The spacing of the beams depends on the full aperture of the array and the angular extent of interest. One can also steer beams in other directions away from the angle of interest, which may contain interference sources that we will want to cancel in the partially adaptive processor.

We have modeled the rank-reducing transformation as a matrix. Usually, the rank of the reduced-dimension space is dictated by the number of digital channels that can be formed due to hardware limitations. Therefore, the rank reduction process is performed prior to sampling using analog beamformers, either across a reduced or full array aperture for the subarray or beamspace partially adaptive array processors, respectively.

11.6.3 Sidelobe Cancelers

The *sidelobe canceler* is actually one of the first implementations of an adaptive array (Howells 1959), and it was originally proposed by Howells and Applebaum. The method uses a main channel along with a single auxiliary, or an array of auxiliary channels, as shown in Figure 11.27. The main channel generally has a high gain in the direction of the desired signal and is produced by either a highly directional sensor, for example, a parabolic dish, or the output of a nonadaptive beamformer, such as a spatial matched filter. The auxiliary channels, however, are low-gain elements often with omnidirectional responses that are used to augment the main channel. The auxiliary channels can be in a ULA configuration. The idea behind the sidelobe canceler is that interference is assumed to be present in both main and auxiliary channels, but the desired signal, though present in the main channel due to its high gain in the direction of the signal, is below the sensor thermal noise in the auxiliary channels. The auxiliary channels are used to form an estimate of the main channel interference that can be used for cancelation purposes. The philosophy behind the sidelobe canceler is shown in Figure 11.28, using representative beampatterns weighted by their directional gains.

Consider a main channel (mc) signal

$$x_{\mathrm{mc}}(n) = g_s s(n) + i_{\mathrm{mc}}(n) + w_{\mathrm{mc}}(n) \qquad (11.6.24)$$

consisting of the desired signal $s(n)$ with a gain of g_s, an interference signal $i_{\mathrm{mc}}(n)$ that may be due to several interferers arriving from various angles, and noise $w_{\mathrm{mc}}(n)$ that is

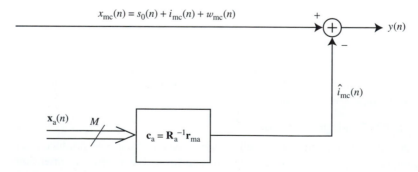

FIGURE 11.27
Sidelobe canceler.

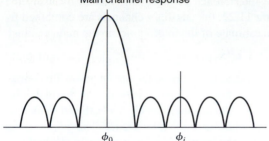

Main channel response

ϕ_0 ϕ_i

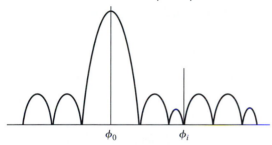

Auxiliary channel net response

ϕ_0 ϕ_i

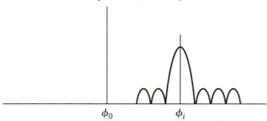

Sidelobe canceler output response

ϕ_0 ϕ_i

FIGURE 11.28

677

SECTION 11.6
Other Adaptive Array
Processing Methods

Illustration of the sidelobe canceler
channel and auxiliary channel
beampatterns.

temporally uncorrelated. All three of these signals are assumed to be mutually uncorrelated. The interference in this main channel is often so strong that it dominates the desired signal even though it has a large gain in the direction of this desired signal. However, the auxiliary channel signals may be written as a signal vector

$$\mathbf{x}_a(n) = s(n)\mathbf{v}(\phi_s) + \sum_{p=1}^{P} i_p(n)\mathbf{v}(\phi_p) + \mathbf{w}(n) \tag{11.6.25}$$

where $\mathbf{v}(\phi)$ is the array response vector at an angle ϕ that was given by (11.1.19) for the case of a ULA. The desired signal impinges on the auxiliary array from the angle ϕ_s, and the sensor thermal noise $\mathbf{w}(n)$ is temporally and spatially uncorrelated. Recall that $s(n)$ is usually considered weak enough that it is well below the sensor noise power. The interference can be made up of several sources. Here we have chosen a model consisting of P interferers with signals $i_p(n)$ and angles of arrival of ϕ_p. Note that the main channel interference $i_{mc}(n)$ is made up of contributions from the same P interferers weighted by the spatial response of the main channel in the directions of the interference sources. These angles of arrival of the interferers, as well as the exact response of the main channel in these directions, are generally unknown and lead to an adaptive solution for the auxiliary channel weight vector.

The sidelobe canceler estimates the interference in the main channel by using the auxiliary channels. As illustrated in Figure 11.27, the auxiliary channels are combined by using a set of adaptive weights to form an estimate of the interference in the main channel.

$$\hat{i}_{mc}(n) = \mathbf{c}_a^H \mathbf{x}_a(n) \tag{11.6.26}$$

where the adaptive weight vector $\mathbf{c}_a$ is chosen so as to minimize the output power. Of course, the implicit assumption has been made that the signal of interest is below the thermal noise level in $\mathbf{x}_a(n)$. Otherwise, if $s(n)$ is strong enough in the auxiliary channels, then the sidelobe canceler will cancel this signal of interest in addition to the interference. The output signal is then obtained by subtracting the estimate of the interference from the main channel

$$y(n) = x_{mc}(n) - \hat{i}_{mc}(n) \tag{11.6.27}$$

The output power is given by

$$P_{out} = \sigma_m^2 - E\{|\mathbf{c}_a^H \mathbf{x}_a(n)|^2\} = \sigma_m^2 - \mathbf{c}_a^H \mathbf{R}_a \mathbf{c}_a \tag{11.6.28}$$

where

$$\mathbf{R}_a = E\{\mathbf{x}_a(n)\mathbf{x}_a^H(n)\} \tag{11.6.29}$$

is the auxiliary array correlation matrix. The solution for these weights is simply the linear MMSE estimator from Section 6.2, given by

$$\mathbf{c}_a = \mathbf{R}_a^{-1} \mathbf{r}_{ma} \tag{11.6.30}$$

where

$$\mathbf{r}_{ma} = E\{\mathbf{x}_a(n)x_{mc}^*(n)\} \tag{11.6.31}$$

is the cross-correlation vector between the auxiliary array and the main channel. The output signal of the sidelobe canceler is

$$y(n) = x_{mc}(n) - \mathbf{c}_a^H \mathbf{x}_a(n) \tag{11.6.32}$$

Hence, the minimum output power is obtained by substituting (11.6.30) into (11.6.28)

$$P_{out}^{(min)} = \sigma_m^2 - \mathbf{r}_{ma}^H \mathbf{R}_a^{-1} \mathbf{r}_{ma} \tag{11.6.33}$$

Of course, all this analysis has considered the case in which the signal of interest is below the thermal noise level in the auxiliary array. Larger signal amplitudes will result in the cancelation of the signal of interest using a sidelobe canceler structure. This topic is treated in Problem 11.15.

11.7 ANGLE ESTIMATION

In this section, we consider the topic of angle estimation, that is, given a spatially propagating signal $\mathbf{s}(n)$, the determination of its angle of arrival at the array. In the formulation of the beamformers in Sections 11.2 through 11.6, the assumption was always made that the beamformer was steered to the angle of the desired signal. However, in practice, the actual angle from which the signal arrives is not precisely known. Instead, an amount of uncertainty exists with respect to the exact angle, even when the signal of interest is within the beam. The beamformer is steered to angle ϕ_0 while the actual signal arrives from ϕ_s. The purpose of an angle estimation algorithm is to attempt to determine this angle ϕ_s. We begin with a discussion of the maximum-likelihood (ML) angle estimator. Next we give a brief sketch of the Cramér-Rao lower bound on angle accuracy, which provides a measure against which the performance of any algorithm can be compared. Then we consider angle estimation algorithms, commonly referred to as *beamsplitting*. In the case of a ULA, a spatially propagating signal is equivalent to a complex exponential in the temporal domain. Hence, we briefly discuss the use of the frequency estimation techniques from Section 9.6 that were based on the model of a complex exponential contained in noise.

In this section, we give a brief discussion of the maximum-likelihood estimator of the angle of a signal arriving at a ULA. Consider a spatially propagating signal of interest

$$\mathbf{s} = \sqrt{M}\sigma_s\mathbf{v}(\phi_s) \tag{11.7.1}$$

where M is the number of sensors in the ULA, σ_s is the complex amplitude of the signal, and ϕ_s is the angle of the signal. The complex signal amplitude has a deterministic magnitude and uniformly distributed random phase. The signal is received by the ULA along with interference $\mathbf{i}$ and spatially uncorrelated thermal noise $\mathbf{w}$, that is,

$$\mathbf{x} = \sqrt{M}\sigma_s\mathbf{v}(\phi_s) + \mathbf{i} + \mathbf{w} = \sqrt{M}\sigma_s\mathbf{v}(\phi_s) + \mathbf{x}_{i+n} \tag{11.7.2}$$

We have dropped the discrete-time index n since we are assuming the signal is present[†] and we are interested in a single snapshot only. The interference-plus-noise correlation matrix of the snapshot $\mathbf{x}$ is given by

$$\mathbf{R}_{i+n} = E\{\mathbf{x}_{i+n}\mathbf{x}_{i+n}^H\} = \mathbf{R}_i + \sigma_w^2\mathbf{I} \tag{11.7.3}$$

Furthermore, we assume that the interference-plus-noise signal $\mathbf{x}_{i+n}$ has a complex Gaussian density function with zero mean. Thus, the probability density function of the snapshot $\mathbf{x}$ is a complex Gaussian function with a mean determined by the signal of interest

$$p(\mathbf{x}; \sigma_s, \phi_s) = \frac{1}{\pi^M \det(\mathbf{R}_{i+n})} \exp\{-[\mathbf{x} - \sqrt{M}\sigma_s\mathbf{v}(\phi_s)]^H \mathbf{R}_{i+n}^{-1}[\mathbf{x} - \sqrt{M}\sigma_s\mathbf{v}(\phi_s)]\} \tag{11.7.4}$$

The peak in this probability density function corresponds to the mean given by the signal of interest $\sqrt{M}\sigma_s\mathbf{v}(\phi_s)$, which is the "most likely" event. The ML angle estimate is the angle $\hat{\phi}_s$ for which this probability density function of the snapshot takes on its maximum value, that is,

$$\hat{\phi}_s = \arg\max_\phi \ p(\mathbf{x}; \sigma_s, \phi_s) \tag{11.7.5}$$

The resulting ML estimator of ϕ_s is then given by (Kay 1993)

$$\hat{\phi}_s = \arg\max_\phi \frac{|\mathbf{v}^H(\phi)\mathbf{R}_{i+n}^{-1}\mathbf{x}|^2}{\mathbf{v}^H(\phi)\mathbf{R}_{i+n}^{-1}\mathbf{v}(\phi)} \tag{11.7.6}$$

Interestingly, this ML estimate can be interpreted as

$$\hat{\phi}_s = \arg\max_\phi = |\mathbf{c}_{amf}^H(\phi)\mathbf{x}|^2 \tag{11.7.7}$$

where $\mathbf{c}_{amf}(\phi)$ is the optimum beamformer given by (11.3.13) with adaptive matched filter (AMF) normalization from Table 11.1

$$\mathbf{c}_{amf}(\phi) = \frac{\mathbf{R}_{i+n}^{-1}\mathbf{v}(\phi)}{\sqrt{\mathbf{v}^H(\phi)\mathbf{R}_{i+n}^{-1}\mathbf{v}(\phi)}} \tag{11.7.8}$$

as shown in Robey et al. (1992). This normalization is in contrast to MVDR normalized optimum beamformer in (11.3.15) that we have considered for the remainder of this chapter. Therefore, the ML angle estimator is the angle to which an AMF normalized optimum beamformer is steered that maximizes the output power for a given snapshot $\mathbf{x}$. In terms of the angle accuracy that might be achieved, the ML estimator can be approximated by forming a dense grid of optimum beamformers in angle with angular spacing at the desired minimum acceptable accuracy (Baranoski and Ward 1997). In many applications, we might

---

[†]In many applications, this assumption may be based on an up-front processing stage that determines the presence of the signal, known as *detection*.

want to achieve a much finer resolution than the beamwidth of the ULA, say, one-tenth of a beamwidth accuracy, known as 10:1 beamsplitting. Thus, this level of angle accuracy would require the computation of roughly $10M$ AMF optimum beamformers, where M is the number of sensors in the ULA. Generally, this requirement is computationally excessive, and we desire an alternative angle estimation algorithm that can achieve performance comparable to the ML estimator. This topic is addressed in Section 11.7.3. However, let us first consider the performance of the ML angle estimator that can be used as a bound for other angle estimation algorithms, which is the topic of the next section.

11.7.2 Cramér-Rao Lower Bound on Angle Accuracy

The *Cramér-Rao bound* (*CRB*) places a lower bound on the performance of an unbiased estimator (Kay 1993). We provide a sketch of the derivation of the CRB for angle accuracy (Ward 1996). This derivation is a simplification of the derivation by Ward (1996) that was done for two-dimensional angle and frequency estimation. Note that the CRB provides the minimum variance of an unbiased estimator. If an estimator can achieve the CRB, then it is the maximum-likelihood estimator. The CRB is found by solving for the diagonal elements of the inverse of the Fisher information matrix. For more details see Kay (1993) and Ward (1996).

Let us start by redefining the beamformer for a ULA from the spatial matched filter in (11.1.19) that has its phase center moved from the first element to the center of the array

$$\mathbf{v}_\Sigma(\phi) = e^{-j2\pi \frac{M-1}{2} \frac{d}{\lambda} \sin\phi} \mathbf{v}(\phi)$$

$$= \frac{1}{\sqrt{M}} \left[e^{-j2\pi \frac{M-1}{2} \frac{d}{\lambda} \sin\phi} \; e^{-j2\pi \frac{M-3}{2} \frac{d}{\lambda} \sin\phi} \; \cdots \; e^{j2\pi \frac{M-1}{2} \frac{d}{\lambda} \sin\phi} \right]^T \tag{11.7.9}$$

which we will refer to as the *sum beamformer*.[†] This choice of a phase center provides the tightest bound on accuracy (Rife and Boorstyn 1974). We can define a second beamformer based on the derivative of $\mathbf{v}_\Sigma(\phi)$ given by

$$\mathbf{v}_\Delta(\phi) = j\boldsymbol{\delta} \odot \mathbf{v}_\Sigma(\phi) \tag{11.7.10}$$

where
$$\boldsymbol{\delta} = \left[-\frac{M-1}{2} \quad -\frac{M-3}{2} \quad \cdots \quad \frac{M-1}{2} \right]^T \tag{11.7.11}$$

which can be thought of as a difference taper. The steering vector $\mathbf{v}_\Delta(\phi)$, however, provides a difference pattern beamformer steered to the angle ϕ, as is commonly used in monopulse radar (Levanon 1988) for angle estimation purposes. For this reason, we refer to it as the *difference beamformer*. In relation to the sum beamformer, we can easily verify that

$$\mathbf{v}_\Delta^H(\phi)\mathbf{v}_\Sigma(\phi) = 0 \tag{11.7.12}$$

that is, the two beamformers are orthogonal to each other. The fact that the two beamformers are orthogonal to each other means that, in terms of the signal $\mathbf{s}$, the two beamformers can make two independent measurements of the signal. These independent measurements allow for the discrimination of the angle.

Using these two steering vectors $\mathbf{v}_\Delta(\phi)$ and $\mathbf{v}_\Sigma(\phi)$, we can form an adaptive sum beamformer

$$\mathbf{c}_\Sigma(\phi) = \mathbf{R}_{i+n}^{-1} \mathbf{v}_\Sigma(\phi) \tag{11.7.13}$$

[†]We use the term *beamformer* for interpretation of the Cramér-Rao bound only. No actual beams are formed since the CRB is only a performance bound and not a processing technique.

and an adaptive difference beamformer

$$\mathbf{c}_\Delta(\phi) = \mathbf{R}_{i+n}^{-1}\mathbf{v}_\Delta(\phi) \tag{11.7.14}$$

which both have not been normalized to satisfy any particular criteria. Proceeding, we can compute the power of the interference-plus-noise output of these two beamformers

$$P_\Sigma = \mathbf{c}_\Sigma^H \mathbf{R}_{i+n}\mathbf{c}_\Sigma \qquad P_\Delta = \mathbf{c}_\Delta^H \mathbf{R}_{i+n}\mathbf{c}_\Delta \tag{11.7.15}$$

Similarly, we can measure the normalized cross-correlation $\rho_{\Sigma\Delta}$ of the interference-plus-noise outputs of these adaptive sum and difference beamformers $\mathbf{R}_{i+n}$

$$\rho_{\Sigma\Delta}^2 = \frac{|\mathbf{c}_\Sigma^H \mathbf{R}_{i+n}\mathbf{c}_\Delta|^2}{P_\Sigma P_\Delta} \tag{11.7.16}$$

Using (11.7.15) and (11.7.16), the CRB on angle estimation for a ULA is given by [†]

$$\sigma_\phi^2 \geq \frac{1}{2\pi^2 \cdot \text{SNR}_0 \cdot P_\Delta(1 - \rho_{\Sigma\Delta}^2)\cos^2\phi} \tag{11.7.17}$$

where SNR_0 is the SNR for a spatial matched filter from (11.2.16) in the absence of interference, that is, noise only, which is given by

$$\text{SNR}_0 = M\frac{\sigma_s^2}{\sigma_w^2} \tag{11.7.18}$$

The CRB on angle accuracy has several interesting interpretations. First and foremost, as the signal power increases in value, SNR_0 increases; as a result, angle accuracy improves. Intuitively, this result makes sense as the stronger the signal of interest, the better the angle estimate should be. Likewise, the term $\cos^2\phi$ simply represents the increase in beamwidth of the ULA as we steer away from broadside ($\phi = 0°$). The interpretation of the other terms P_Δ and $1 - \rho_{\Sigma\Delta}^2$ may be less obvious, but also provides insight. Here P_Δ provides a measure of the received power aligned with the adaptive difference beamformer. On the other hand, $\rho_{\Sigma\Delta}$ is the cross-correlated energy between the adaptive sum and difference beamformers. Ideally, $\rho_{\Sigma\Delta}$ is zero, since $\mathbf{c}_\Sigma$ and $\mathbf{c}_\Delta$ beamformers are derived from $\mathbf{v}_\Sigma$ and $\mathbf{v}_\Delta$, respectively, which are orthogonal to each other. In the case of the two adaptive beamformers, the adaptation will remove this orthogonality, but the beamformers should be different enough that $\rho_{\Sigma\Delta} \ll 1$. Otherwise, angle accuracy will suffer.

11.7.3 Beamsplitting Algorithms

Let us consider the scenario with a single beamformer steered to an angle ϕ_0 with our signal of interest at angle ϕ_s. The beamformer passes all signals within its beamwidth with only slight attenuation of signals that are not directly at the center of the beam steered to ϕ_0. Clearly, this single beamformer cannot discriminate between signals received within its beamwidth. However, we desire a more accurate estimate of the angle of the signal of interest than simply the beamwidth of our beamformer. Thus, any angle estimator must achieve finer accuracy than the beamwidth, and as a result angle estimation algorithms are commonly referred to as *beamsplitting algorithms*.

To construct an angle estimation algorithm, it is necessary to obtain different measurements of the signal of interest in order to determine its angle. These measurements allow an angle estimation algorithm to discriminate between returns that arrive at the array from different angles. To this end, we use a set of beamformers steered in the general direction of

---

[†]This formulation assumes unit-variance thermal noise power. Therefore, if signals have different thermal noise power, the correlation matrix must be normalized by the thermal noise power prior to computing P_Δ and $\rho_{\Sigma\Delta}$.

the signal of interest but with different spatial responses, that is, beampatterns. One means of obtaining different measurements of the signal of interest is to slightly offset the steering direction of two beamformers. For example, we might form two beams at angles

$$\phi_1 = \phi_0 - \Delta\phi \qquad \phi_2 = \phi_0 + \Delta\phi \qquad (11.7.19)$$

where $\Delta\phi$ is a fraction of a beamwidth, for example, half a beamwidth. Let the weight vectors for these two beamformers be $\mathbf{c}_1$ and $\mathbf{c}_2$, respectively. These two beamformers can be either nonadaptive, as in the case of the conventional beamformers discussed in Section 11.2, or one of the various adaptive beamformers from Section 11.3, 11.5, or 11.6. Ideally, a pair of adaptive beamformers is used for applications in which interference is encountered. Since the two beamformers are slightly offset from angle ϕ_0, they may be thought of as "left" and "right" beamformers. Using the beamformer weight vectors, we can then form the ratio

$$\gamma_x = \frac{\mathbf{c}_1^H \mathbf{x}}{\mathbf{c}_2^H \mathbf{x}} \qquad (11.7.20)$$

where recall that $\mathbf{x}$ is the snapshot under consideration that contains the signal of interest $\mathbf{s} = \sqrt{M}\sigma_s \mathbf{v}(\phi_s)$. Similarly, we can also hypothesize this ratio for any angle ϕ to form a discrimination function

$$\gamma(\phi) = \frac{\mathbf{c}_1^H \mathbf{v}(\phi)}{\mathbf{c}_2^H \mathbf{v}(\phi)} \qquad (11.7.21)$$

Comparing the value of the measured ratio in (11.7.20) for the snapshot $\mathbf{x}$ to this angular discrimination function in (11.7.21), we obtain an estimate of the angle of the signal of interest ϕ_s. The key requirement for the discrimination function is that it be monotonic over the angular region in which it is used; that is, there is a one-to-one correspondence of the function in (11.7.21) and every angle in this region. The angular region typically encompasses the beamwidths of the two beamformers. This requirement on the discrimination function $\gamma(\phi)$ means that the two beamformers $\mathbf{c}_1$ and $\mathbf{c}_2$ must have different spatial responses.

We have simply given an example of how an angle estimation algorithm might be constructed. The topic of angle estimation is a very large area, and the choice of algorithm should be determined by the particular application. In the example given, we constructed a beamsplitting algorithm with left and right beams. Similarly, we could have chosen sum and difference beams, as is commonly done in radar in a technique known as monopulse (Sherman 1984). In fact, sum and difference beams can be formed from left and right beams by taking their sum and difference, respectively. In this case, a simple linear transformation exists that provides a mapping between the two beam stategies, and as a result one would anticipate equivalent performance. For further material on angle estimation algorithms, the interested reader is referred to Davis et al. (1974), McGarty (1974), Zoltowski (1992), and Nickel (1993).

11.7.4 Model-Based Methods

In Section 9.6, we discussed frequency estimation techniques based on a model of a complex exponential contained in noise. Certainly all these techniques could also be applied to the angle estimation problem, particularly for a ULA that has a direct correspondence to a discrete-time uniformly sampled signal. In this case, the angle is determined by the spatial frequency of the ULA. These methods are commonly referred to as *superresolution* techniques because they are able to achieve better resolution than traditional, nonadaptive methods. In fact, many of these techniques were originally proposed for array processing applications. However, certain considerations must be taken into account when one is trying to apply these methods for use with a sensor array. First, a certain amount of uncertainty exists with respect to the exact spatial location of all the sensors. All these methods exploit the structure imposed by regular sampling where knowledge of the sampling instance is very precise. In the case of a temporally sampled signal, this assumption is very reasonable; but in the case of an array with these uncertainties, the validity of this assumption must be called into question. In addition, for a sensor array, all the signals are measured by different sensors with slightly different characteristics, as opposed to a temporally sampled signal for which all the samples are measured by the same sensor (analog-to-digital converter). Although these channel mismatches can be corrected for in theory, a perfect correction is never possible. For this reason, caution is in order when using these model-based methods for the purposes of angle estimation with an array.

11.8 SPACE-TIME ADAPTIVE PROCESSING

Space-time adaptive processing (*STAP*) is concerned with the two-dimensional processing of signals in both the spatial and temporal domains. The topic of STAP has received a lot of attention recently as it is a natural extension of array processing (Ward 1994, 1995; Klemm 1999). Although discussions of STAP date back to the early 1970s (Brennan and Reed 1973), the realization of STAP in an actual system was not possible until just recently, due to advances that were necessary in computing technology. We give a brief overview of the principles of STAP and cite some of the considerations for its practical implementation. Although STAP has also been proposed for use in communications systems (Paulraj and Papadias 1997), we primarily discuss it in the context of the airborne radar application for the purposes of clutter cancelation (Brennan and Reed 1973; Ward 1995; Klemm 1999).

A general STAP architecture is shown in Figure 11.29. Consider a ULA of sensors as we have discussed throughout this chapter. We choose a ULA for the sake of simplicity, but note that STAP techniques can be extended for arbitrary array configurations. In addition, the signal from each sensor consists of a set of time samples or delays that make up a

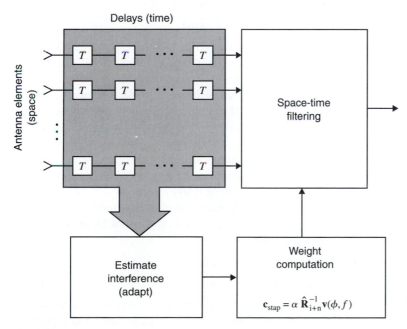

FIGURE 11.29
Space-time adaptive processing.

time window. In radar applications, the time samples represent the returns from a set of transmitted pulses. For an airborne radar that moves with a certain velocity, the reflected signals from moving and nonmoving objects have a single frequency across the pulses. The pulse frequency results in a complex exponential across the pulses. The frequency is produced by the relative velocity of the objects with respect to the array and is known as the *Doppler frequency*. Thus, we wish to construct a space-time model for a signal received from a certain angle ϕ_s at a frequency f_s. We model the spatial component of the signal using the *spatial (sp) steering vector* for a ULA with M sensors from (11.1.23)

$$\mathbf{v}_{\text{sp}}(\phi_s) = \frac{1}{\sqrt{M}}[1 \; e^{-j2\pi[(d/\lambda)\sin\phi]} \; \cdots \; e^{-j2\pi[(d/\lambda)\sin\phi](M-1)}]^T \tag{11.8.1}$$

Likewise, the temporal component of a signal that is a complex exponential can be modeled using a data window frequency vector, which technically is a *temporal frequency steering vector*. This temporal steering vector is given by

$$\mathbf{v}_{\text{time}}(f) = \frac{1}{\sqrt{L}}[1 \; e^{-j2\pi f} \; \cdots \; e^{-j2\pi f(L-1)}]^T \tag{11.8.2}$$

where L is the number of time samples or pulses. Both $\mathbf{v}_{\text{sp}}$ and $\mathbf{v}_{\text{time}}$ have unit norm, that is, $\mathbf{v}_{\text{sp}}^H\mathbf{v}_{\text{sp}} = 1$ and $\mathbf{v}_{\text{time}}^H\mathbf{v}_{\text{time}} = 1$. Using these two one-dimensional steering vectors, we can form the two-dimensional $LM \times 1$ steering vector known as a *space-time steering vector*

$$\mathbf{v}_{\text{st}}(\phi, f) = \mathbf{v}_{\text{time}}(f) \otimes \mathbf{v}_{\text{sp}}(\phi) \tag{11.8.3}$$

where $\otimes$ is the Kronecker product (Golub and Van Loan 1996). This vector, like the two one-dimensional steering vectors, has unit norm. Using this space-time steering vector, we can then model a spatially propagating signal arriving at the ULA from an angle ϕ_s with a frequency f_s as

$$\mathbf{s}(n) = \sqrt{LM}\mathbf{v}_{\text{st}}(\phi_s, f_s)s(n) \tag{11.8.4}$$

Of course, this signal of interest $\mathbf{s}(n)$ is not the only signal since the ULA at the very least will have thermal noise from the sensors. However, let us consider the case where in addition to the signal of interest, the ULA receives other spatially propagating signals that constitute interference $\mathbf{i}(n)$. Thus, the overall space-time signal in the ULA is

$$\mathbf{x}(n) = \mathbf{s}(n) + \mathbf{i}(n) + \mathbf{w}(n) \tag{11.8.5}$$

where $\mathbf{w}(n)$ is the sensor thermal noise space-time signal that is both temporally and spatially uncorrelated, that is, $E\{\mathbf{w}(n)\mathbf{w}^H(n)\} = \sigma_w^2\mathbf{I}$. The interference component is made up of spatially propagating signals that may be temporally uncorrelated or consist of complex exponentials in the time domain, just as the signal of interest. In the case of an airborne radar, jamming interference is temporally uncorrelated while spatially correlated; that is, the jamming signal consists of uncorrelated noise that arrives from a certain angle ϕ. However, clutter returns are produced by reflections of the radar signal from the ground and have both spatial and temporal correlation. Due to the nature of the airborne radar problem, these clutter returns exhibit a certain structure that can be exploited for the purposes of implementing a STAP algorithm (Ward 1995; Klemm 1999).

As we did for the optimum beamformer, we want to find the optimum STAP weight vector. The optimality condition is again the maximization of the output SINR. The space-time interference-plus-noise correlation matrix is

$$\mathbf{R}_{\text{i+n}} = E\{\mathbf{x}_{\text{i+n}}(n)\mathbf{x}_{\text{i+n}}^H(n)\} \tag{11.8.6}$$

where

$$\mathbf{x}_{\text{i+n}}(n) = \mathbf{i}(n) + \mathbf{w}(n) \tag{11.8.7}$$

is the interference-plus-noise component of the signal. The availability of data that do not contain the signal of interest is a training issue for the implementation of the STAP algorithm that we do not consider here. See Borsari and Steinhardt (1995) and Rabideau and Steinhardt (1999).

The optimum STAP weight vector is found in a similar fashion to the optimum beam-former in Section 11.3. Using a unit gain on target constraint, the optimum STAP weight vector is

$$\mathbf{c}_{\text{stap}} = \frac{\mathbf{R}_{i+n}^{-1}\mathbf{v}(\phi_s, f_s)}{\mathbf{v}^H(\phi_s, f_s)\mathbf{R}_{i+n}^{-1}\mathbf{v}(\phi_s, f_s)} \qquad (11.8.8)$$

where the space-time steering vector $\mathbf{v}(\phi_s, f_s)$ specifies the angle and frequency of the presumed signal of interest $\mathbf{s}(n)$. The implementation of STAP requires the estimation of $\mathbf{R}_{i+n}$ from data samples in order to compute the sample correlation matrix $\hat{\mathbf{R}}_{i+n}$. This block adaptive implementation is also known as sample matrix inversion (SMI). SMI was discussed in the context of adaptive beamforming in Section 11.5.

The adaptive degrees of freedom of full STAP as specified in (11.8.8) are LM. For most applications, computational considerations as well as a limited amount of data to train the adaptive weights make the implementation of fully dimensional STAP impractical. Thus, we must consider reduced-dimension versions of STAP (Ward 1994, 1995). To this end, a preprocessing stage precedes the adaptation that reduces the degrees of freedom to an acceptable level. The most commonly considered approaches use a partially adaptive array implementation, as discussed in Section 11.6.2, either beamspace or subarrays, to reduce the spatial degrees of freedom. Temporal degrees of freedom can be reduced by using a frequency-selective temporal or Doppler filter (Ward and Steinhardt 1994). Alternatively, a subset of the total number of pulses can be used where the subsets of pulses are then combined following adaptive processing (Baranoski 1995).

A brief mention should be given to the application of STAP to the communications problem (Paulraj and Papadias 1997). Unlike in the radar application, it is generally not possible to separate the signal of interest from the interference-plus-noise in communications applications. In addition, although as in radar the signals are spatially propagating and thus arriving at the sensor array at a specific angle, for communications the temporal signals are not necessarily complex exponential signals. Instead, many times the signals consist of coded sequences. In this case, STAP must incorporate the codes rather than complex exponentials into its processing for the proper extraction of these signals.

11.9 SUMMARY

In this chapter, we have given a brief overview of array processing, starting with array fundamentals and covering optimum and adaptive beamforming. Throughout the chapter, we focused on the ULA, but it is possible to extend these concepts to other array configurations. The background material included spatially propagating signals, the concepts of modulation and demodulation, and the model for a spatial signal received by a ULA. We drew the analogy between the ULA, in terms of spatial sampling, and the discrete-time sampling of temporal signals. Next, we introduced the topic of conventional beamforming for which we discussed the spatial matched filter, which maximizes the SNR in the absence of interference, and tapered low sidelobe beamformers. Within this context, we looked at the characteristics of an array in terms of resolution and ambiguities known as grating lobes.

The remainder of the chapter dealt with optimum and adaptive beamforming techniques related to the processing introduced earlier in the book for use with discrete-time signals. These methods are concerned with adapting to the characteristics of the data, either assuming knowledge of these characteristics (optimum) or estimating them from the data (adaptive). One might say that the *fundamental equation* in adaptive beamforming is $\mathbf{c} = \mathbf{R}^{-1}\mathbf{v}(\phi_s)$, where $\mathbf{v}(\phi_s)$ determines the direction ϕ_s in which we are steering and $\mathbf{R}^{-1}$, the inverse of the array correlation matrix, performs the adaptation to the array data. Within this context, we looked at various issues, such as sidelobe levels and interference cancelation, and the effects of signal mismatch and bandwidth. The more advanced topics of angle estimation and STAP were also discussed.

Throughout this chapter, we have tried to remain general in our treatment of the array processing principles. Ultimately, specific issues and concerns related to the application will dictate the type of processing that is needed. Areas in which arrays are commonly employed include radar, sonar, and communications. In parts of the chapter, we have used examples based on radar applications since they tend to be the easiest to simplify and describe. Other important issues not discussed that arise in radar include the nonstationarity of the signals as well as training strategies. The sonar application is rich with issues that make the implementation of adaptive arrays a very challenging task. Propagation effects, including signal multipath, lead to complicated models that must be used to estimate the steering vectors. In addition, signals of interest tend to be present at all times, so that the adaptive beamformer must be trained with the signal of interest present. As we described in Section 11.4.1, this situation leads to a heightened sensitivity to signal mismatch. For more details see Baggeroer et al. (1993). Arrays for communications applications have also become a very popular field owing to the rapid growth of the wireless communications industry. The fundamental issue for wireless communications is the number of users that can be simultaneously accommodated. The limitations arise from the interference produced by other users. Arrays can help to increase the capacity in terms of the number of users. For more details, see Litva and Lo (1996).

We have presented some material on the more advanced topics of angle estimation and STAP. Another extension of adaptive beamforming is the subject of adaptive detection. This topic is concerned with the determination of the presence of signals of interest in which the detector is determined adaptively from the data. References on this subject include Kelly (1986), Steinhardt (1992), Robey et al. (1992), Bose and Steinhardt (1995), Scharf and McWhorter (1997), Kreithen and Steinhardt (1996), Conte et al. (1996), and Richmond (1997).

PROBLEMS

11.1 Consider a narrowband spatially propagating signal with a speed of propagation c. The signal impinges on an $M = 2$ element ULA from an angle $\phi = 0°$ with a spacing d between the elements. For illustration purposes, let the temporal content of the signal be a pulse.

(a) Let the time of arrival of the pulse at the first sensor be $t = 0$. At what time does the signal arrive at the second sensor?

(b) Do any other angles ϕ produce the same delay between the two sensors? Why?

(c) Suppose now that we only have a single sensor. Can we determine the angle from which a signal impinges on this sensor?

11.2 We want to investigate the use of a mechanically steered versus an electronically steered array. Consider a spatial matched filter with $M = 20$ elements. Now consider that the array is steered to $\phi = 45°$. In the case of mechanical steering, the pointing direction is always broadside to the array. To compute the beampattern of the mechanically steered array, simply take the beampattern computed at $\phi = 0°$ and shift it by the mechanical steering angle, that is, $\phi' = \phi + \phi_{\text{mech}}$. However, the beampattern of an electronically steered array is simply the beampattern of the spatial matched filter steered to the desired angle. Compare the beampattern of the mechanically steered array to that of the electronically steered array. What do you observe? Repeat this for $\phi = 60°$ steering, both electronic and mechanical.

11.3 In this problem, we want to explore the use of beampatterns and steered or spatial responses of a ULA. Consider a signal $\mathbf{x}(n)$ consisting of two spatially propagating signals from $\phi_1 = -10°$ and $\phi_2 = 30°$, both made of random, complex Gaussian noise. The respective powers of the two signals are 20 and 25 dB. The number of sensors in the ULA is 50, and its thermal noise level is 0 dB. The ULA has interelement spacing of $d = \lambda/2$.

(a) Compute one realization of $\mathbf{x}(n)$ for $N = 1000$ samples, that is, $1 \le n \le N$. Using a spatial matched filter, compute a steered response for this signal from the beamformer outputs, and plot it in decibels versus angle. What do you observe? Compare the result to the expected steered response using the true correlation matrix.

(b) Compute and plot the beampattern for the spatial matched filter steered to $\phi = 30°$. How can you relate the power levels you observed in part (a) at the angles of the two signals to the beampattern?

(c) Change the power level of the signal at $\phi = -10°$ to 60 dB, and compute the steered response. What do you observe? What do you recommend in order to distinguish these two signals? Implement your idea and plot the estimated steered response from the beamformer outputs.

11.4 Suppose that we have an $M = 30$ element ULA with a thermal noise level of 0 dB.

(a) Generate a realization of the ULA signal $\mathbf{x}(n)$ consisting of two random, complex Gaussian signals at $\phi = 0°$ and $\phi = 1.5°$ both with power 20 dB, along with the sensor thermal noise. The interelement spacing is $d = \lambda/2$. Let the number of samples you generate be $N = 1000$. Compute and plot the steered response of $\mathbf{x}(n)$ using a spatial matched filter. What do you observe?

(b) Repeat part (a) for an $M = 60$ element ULA. What do you observe?

(c) Now using the $M = 30$ element ULA again, but with interelement spacing $d = \lambda$, compute the steered response and comment on the result.

(d) Compute the beampatterns for the spatial matched filter steered to $\phi = 0°$ for the three array configurations in parts (a), (b), and (c).

11.5 In this problem, we want to invesigate the use of randomly thinned arrays. Note that the $M = 30$ element ULA with $d = \lambda$ spacing from Problem 11.4 is simply the $M = 60$ element ULA with every other element deleted. Such an array is often referred to as a *thinned array*. Using an $M = 60$ element array, randomly thin the array. (*Hint:* Use a random number generator.) First thin to 75 percent (45 elements) and then to 50 percent (30 elements). Compute and plot the steered response, using a spatial matched filter for the signal in Problem 11.4. Note that the spatial matched filter must take into account the positions of the elements; that is, it is no longer a Vandermonde steering vector. Compute the beampatterns of these two randomly thinned arrays. Repeat this process 3 times. What do you observe?

11.6 The spatial matched filter from (11.2.16) is the beamformer that maximizes the SNR in the absence of interference. For this spatial matched filter, the beamforming or array gain was shown to be $G_{bf} = M$. Suppose now that we have an $M = 20$ element ULA in which the elements have unequal gain. In other words, the spatial matched filter no longer has the same amplitude in each element. Find the spatial matched filter for the case when all even-numbered elements have a unity gain, while all the odd-numbered elements have a gain of $\frac{3}{4}$. What is the beamforming gain for this array?

11.7 The optimum beamformer weights with MVDR normalization are found by solving the following optimization

$$\min P_{i+n} \quad \text{subject to} \quad \mathbf{c}^H \mathbf{v}(\phi_s) = 1$$

Using Lagrange multipliers discussed in Appendix B, show that the MVDR optimum beamformer weight vector is

$$\mathbf{c}_o = \frac{\mathbf{R}_{i+n}^{-1}\mathbf{v}(\phi_s)}{\mathbf{v}^H(\phi_s)\mathbf{R}_{i+n}^{-1}\mathbf{v}(\phi_s)}$$

11.8 In this problem, we want to investigate the different normalizations of the optimum beamformer from Table 11.1. We refer to the three normalizations as MVDR ($\alpha = [\mathbf{v}^H(\phi_s)\mathbf{R}_{i+n}^{-1}\mathbf{v}(\phi_s)]^{-1}$), adaptive matched filter or AMF ($\alpha = [\mathbf{v}^H(\phi_s)\mathbf{R}_{i+n}^{-1}\mathbf{v}(\phi_s)]^{-1/2}$), and unit gain on noise ($\alpha = [\mathbf{v}^H(\phi_s)\mathbf{R}_{i+n}^{-2}\mathbf{v}(\phi_s)]^{-1/2}$). Let the interference-plus-noise signal consist of two interference sources at $\phi = 45°$ and $\phi = 20°$ with powers of 30 and 15 dB, respectively. The noise power is $\sigma_w^2 = 1$. Now compute the steered response of the optimum beamformers with the three different normalizations between $-90° < \phi < 90°$ (using 1° increments), using the true correlation matrix. What is the difference in the outputs of the optimum beamformers with the different normalizations? For what purposes are the different normalizations useful?

11.9 The generalized sidelobe canceler (GSC) was derived as an alternative implementation of the MVDR optimum beamformer. Show that the overall end-to-end weights associated with the GSC are equivalent to the MVDR optimum beamformer weight vector in (11.3.15).

11.10 In the formulation of the optimum beamformer, we used the interference-plus-noise correlation matrix $\mathbf{R}_{i+n}$. However, in many applications it is not possible to have interference-plus-noise-only data, and the signal of interest is always present. Thus, the beamformer must be implemented using the correlation matrix

$$\mathbf{R}_x = \mathbf{R}_{i+n} + \sigma_s^2 \mathbf{v}(\phi_s)\mathbf{v}^H(\phi_s)$$

Show that the use of this correlation matrix will have no effect on the optimum beamformer weight vector for the case of no signal mismatch. *Hint:* Use the matrix inversion lemma (see Appendix A).

11.11 In this problem, we want to look at the effect of signal mismatch on the performance of the optimum beamformer. Of course, the resulting beamformer is no longer really optimum, but instead is optimized to our presumptions about the signal. Consider the case with three interference sources at $\phi = 5°$, $20°$, and $-30°$ with powers of 25, 35, and 50 dB, respectively. Compute the optimum beamformer steered to $\phi = 0°$. Now consider the case where the signal of interest is not at $\phi = 0°$ but rather at $\phi = -1°$. The array consists of an $M = 50$ element ULA with a noise power of $\sigma_w^2 = 1$.

(a) Find the signal mismatch loss when the signal of interest is not in the correlation matrix. Vary the strength of the signal from 0 to 30 dB.

(b) Find the signal mismatch loss when the signal is in the correlation matrix. Vary the strength of the signal from 0 to 30 dB.

11.12 Let us again consider a set of three interference sources at $\phi = 5°$, $20°$, and $-30°$ with powers of 25, 35, and 50 dB, respectively. Now consider the case where the signal of interest is not at $\phi = 0°$ but rather at $\phi = -1°$ and has a power of $M\sigma_s^2 = 30$ dB. The array consists of an $M = 50$ element ULA with a noise power of $\sigma_w^2 = 1$. However, instead of computing the optimum beamformer with the correlation matrix $\mathbf{R}_{i+n}$, use the diagonally loaded interference-plus-noise matrix

$$\mathbf{R}_l = \mathbf{R}_{i+n} + \sigma_l^2 \mathbf{I}$$

where σ_l^2 is the loading level.

(a) Find the signal mismatch loss when the signal of interest is not in the correlation matrix. Vary diagonal loading from $\sigma_l^2 = 0, 5, 10,$ and 20 dB.

(b) Find the signal mismatch loss when the signal is in the correlation matrix. Vary diagonal loading from $\sigma_l^2 = 0, 5, 10,$ and 20 dB.

11.13 The Frost sample-by-sample adaptive beamformer was derived for the MVDR beamformer. Extend the Frost sample-by-sample adaptive beamformer for the case of multiple constraints in an LCMV adaptive beamformer.

11.14 The LCMV beamformer weight vector is given in (11.6.7) and was found by using Lagrange multipliers, which are discussed in Appendix B. Verify this result; that is, using Lagrange multipliers, show that the LCMV beamformer weight vector is given by

$$\mathbf{c}_{\text{lcmv}} = \mathbf{R}_{i+n}^{-1}\mathbf{C}(\mathbf{C}^H \mathbf{R}_{i+n}^{-1}\mathbf{C})^{-1}\boldsymbol{\delta}$$

where $\mathbf{C}$ and $\boldsymbol{\delta}$ are defined as in Section 11.6.1.

11.15 Let us consider the sidelobe canceler, as discussed in Section 11.6.3. We restrict the problem to a single interferer that has an angle ϕ_i with respect to a ULA that makes up the auxiliary channels. The main channel consists of the signal

$$x_{\mathrm{mc}}(n) = g_s s(n) + i_{\mathrm{mc}}(n) + w_{\mathrm{mc}}(n)$$

where $i_{\mathrm{mc}}(n) = g_i i(n)$ is the temporally uncorrelated signal $i(n)$ with unit variance that has a main channel gain of g_i. The main channel thermal noise $w_{\mathrm{mc}}(n)$ is temporally uncorrelated with a power of σ_0^2. The auxiliary channels make up an M-element ULA with thermal noise σ_w^2. The auxiliary channel signal vector is given by

$$\mathbf{x}(n) = s(n)\mathbf{v}(\phi_s) + \sigma_i^2 i(n)\mathbf{v}(\phi_i) + \mathbf{w}(n)$$

where ϕ_s and ϕ_i are the angles of the signal of interest and the interferer with respect to the ULA, respectively.

(a) Form the expressions for the auxiliary channel correlation matrix $\mathbf{R}_a$ and cross-correlation vector $\mathbf{r}_{\mathrm{ma}}$ that include the signal of interest in the auxiliary channels.
(b) Compute the output power of the interference-plus-noise.
(c) Compute the output power of the signal. What conclusions can you draw from your answer?

11.16 The MVDR optimum beamformer is simply a special case of the LCMV beamformer. In this case, the constraint matrix is $\mathbf{C} = \mathbf{v}(\phi_s)$ and the constraint response vector is $\boldsymbol{\delta} = 1$.

(a) Using the LCMV weight vector given in (11.6.7), substitute this constraint and constraint response and verify that the resulting beamformer weight vector is equal to the MVDR optimum beamformer.
(b) Find an expression for the output power of the LCMV beamformer.

11.17 The optimum beamformer could also be formulated as the constrained optimization problem that resulted in the MVDR beamformer. This beamformer can be implemented as a generalized sidelobe canceler (GSC), as shown in Section 11.3.5. Similarly, the LCMV beamformer can be implemented in a GSC architecture. Derive the formulation of a GSC with multiple linear constraints.

11.18 Consider the case of an $M = 20$ element array with $d = \lambda/2$ interelement spacing and thermal noise power $\sigma_w^2 = 1$. An interference source is present at $\phi = 30°$ with a power of 50 dB. Generate one realization of 1000 samples of this interferer. In addition, a signal of interest is present at $\phi_s = 0°$ with a power of $\sigma_s = 100$ (20 dB) in the $n = 100$th sample only.

(a) Using an SMI adaptive beamformer for the full array, compute the output signal. Is the signal of interest visible?
(b) Using a partially adaptive beamformer with $Q = 4$ nonoverlapping subarrays with $\tilde{M} = 5$ elements, compute the output of an SMI adaptive beamformer. What can you say about the signal of interest now?
(c) Repeat part (b) with $Q = 2$ and $\tilde{M} = 10$. What are your observations now?

11.19 Consider the case of an $M = 40$ element array with $d = \lambda/2$ interelement spacing and thermal noise power $\sigma_w^2 = 1$. An interference source is present at $\phi = 20°$ with a power of 50 dB. Generate one realization of 1000 samples of this interferer. In addition, a signal of interest is present at $\phi_s = 0°$ with a power of $\sigma_s = 100$ (20 dB) in the $n = 100$th sample only.

(a) Using an SMI adaptive beamformer for the full array, compute the output signal. Is the signal of interest visible?
(b) Using a beamspace partially adaptive beamformer consisting of 11 beams at the angles $-5° \leq \phi \leq 5°$ at $1°$ increments, compute the output of a partially adaptive SMI beamformer. What can you say about the signal of interest now?
(c) Repeat part (b) with beams only at $\phi = -1°, 0°$, and $1°$. What are your observations now?

11.20 Compute the SINR loss for a partially adaptive beamformer with a general preprocessing transformation $\mathbf{T}$. You need to start with the general definition of SINR loss

$$L_{\mathrm{sinr}} \triangleq \frac{\mathrm{SINR}_{\mathrm{out}}(\phi_s)}{\mathrm{SNR}_0}$$

where $\text{SINR}_{\text{out}}(\phi_s)$ is the output SINR of the partially adaptive beamformer at angle ϕ_s and SNR_0 is the SNR of the spatial matched filter in the absence of interference.

11.21 Consider the case of an interference source at $\phi = 30°$ with a power of 40 dB. The ULA is a 20-element array with $d = \lambda/2$ interelement spacing and has unit-variance thermal noise $(\sigma_w^2 = 1)$.

(a) Compute the SINR loss for the optimum beamformer.

(b) Let us consider the case of the GSC formulation of the optimum beamformer. If we choose to use a beamspace blocking matrix $\mathbf{B}$ in (11.3.41), what are the spatial frequencies of the spatial matched filters in this blocking matrix for an optimum beamformer steered to $\phi = 0°$?

(c) To implement a reduced-rank or partially adaptive beamformer, use only the two spatial matched filters in the beamspace blocking matrix with spatial frequencies closest to the interference source (spatial frequency $u = \frac{1}{2}\sin\phi_i = 0.25$). Compute the SINR loss of this partially adaptive beamformer, and compare it to the SINR loss of the optimum beamformer found in part (a).

CHAPTER 12

Further Topics

The distinguishing feature of this book, up to this point, is the reliance on random process models having finite variance and short memory and specified by their second-order moments. This chapter deviates from this path by focusing on further topics where there is an explicit or implicit need for higher-order moments, long memory, or high variability.

In the first part (Section 12.1), we introduce the area of *higher-order statistics (HOS)* with emphasis on the concepts of cumulants and polyspectra. We define cumulants and polyspectra; we analyze the effect of linear, time-invariant systems upon the HOS of the input process; and we derive the HOS of linear processes. Higher-order moments, unlike second-order moments, are shown to contain phase information and can be used to solve problems in which phase is important.

In the second part (Sections 12.2 through 12.4), we illustrate the importance of HOS for the blind deconvolution of non-minimum-phase systems, and we show how the underlying theory can be used to design unsupervised adaptive filters for symbol-spaced and fractionally spaced equalization of data communication channels.

In the third part (Sections 12.5 and 12.6), we introduce two types of random signal models characterized by long memory: fractional and self-similar, or random, fractal models. We conclude with rational and fractional models with symmetric α-stable (SαS) excitations and self-similar processes with SαS increments. These models have long memory and find many applications in the analysis and modeling of signals with long-range dependence and impulsive or spiky behavior.

12.1 HIGHER-ORDER STATISTICS IN SIGNAL PROCESSING

The statistics of a Gaussian process are completely specified by its second-order moments, that is, correlations and power spectral densities (see Section 3.3). Since non-Gaussian processes do *not* have this property, their higher-order statistics contain additional information that can be used to measure their deviation from normality. In this section we provide some background definitions and properties of higher-order moments, and we discuss their transformation by linear, time-invariant systems. More detailed treatments can be found in Mendel (1991), Nikias and Raghuveer (1987), Nikias and Mendel (1993), and Rosenblatt (1985).

12.1.1 Moments, Cumulants, and Polyspectra

The first four moments of a complex-valued stationary stochastic process are defined by

$$r_x^{(1)} \triangleq E\{x(n)\} = \mu_x \tag{12.1.1}$$

$$r_x^{(2)}(l_1) \triangleq E\{x^*(n)x(n+l_1)\} = r_x(l_1) \tag{12.1.2}$$

$$r_x^{(3)}(l_1, l_2) \triangleq E\{x^*(n)x(n+l_1)x(n+l_2)\} \qquad (12.1.3)$$

$$r_x^{(4)}(l_1, l_2, l_3) \triangleq E\{x^*(n)x^*(n+l_1)x(n+l_2)x(n+l_3)\} \qquad (12.1.4)$$

although other definitions are possible by conjugating different terms. We note that the first two moments are the mean and the autocorrelation sequence, respectively.

In Section 3.2.4 we showed that the cumulant of a linear combination of IID random variables can be determined by a linear combination of their cumulants. In addition, in Section 3.1.2, we noted that the kurtosis of a random variable measures its deviation from Gaussian behavior. For these reasons, we usually prefer to work with cumulants instead of moments. Since higher-order cumulants are invariant to a shift of the mean value, we define them under a zero-mean assumption.

The first four cumulants of a zero-mean stationary process are defined by

$$\kappa_x^{(1)} = E\{x(n)\} = \mu_x = 0 \qquad (12.1.5)$$

$$\kappa_x^{(2)}(l_1) = E\{x^*(n)x(n+l_1)\} = r_x(l_1) \qquad (12.1.6)$$

$$\kappa_x^{(3)}(l_1, l_2) = E\{x^*(n)x(n+l_1)x(n+l_2)\} \qquad (12.1.7)$$

$$\kappa_x^{(4)}(l_1, l_2, l_3) = E\{x^*(n)x^*(n+l_1)x(n+l_2)x(n+l_3)\}$$

$$- \kappa_x^{(2)}(l_2)\kappa_x^{(2)}(l_3 - l_1) - \kappa_x^{(2)}(l_3)\kappa_x^{(2)}(l_2 - l_1) \qquad (12.1.8)$$

(complex-valued case)

$$\kappa_x^{(4)}(l_1, l_2, l_3) = E\{x(n)x(n+l_1)x(n+l_2)x(n+l_3)\} - \kappa_x^{(2)}(l_1)\kappa_x^{(2)}(l_3 - l_2)$$

$$- \kappa_x^{(2)}(l_2)\kappa_x^{(2)}(l_3 - l_1) - \kappa_x^{(2)}(l_3)\kappa_x^{(2)}(l_2 - l_1) \qquad (12.1.9)$$

(real-valued case)

and can be obtained by using the cumulant-generating function discussed in Section 3.1.2 (Mendel 1991). It can be shown that

$$\kappa_x^{(k)}(l_1, l_2, \ldots, l_{k-1}) = m_x^{(k)}(l_1, l_2, \ldots, l_{k-1}) - m_g^{(k)}(l_1, l_2, \ldots, l_{k-1}) \qquad k = 3, 4 \qquad (12.1.10)$$

where $x(n)$ is a non-Gaussian process and $g(n)$ is a Gaussian process with the same mean and autocorrelation sequence. The negative terms in (12.1.8) and (12.1.9) express the fourth-order cumulant of the Gaussian process in terms of second-order ones. In this sense, in addition to higher-order correlations, cumulants measure the distance of a process from Gaussianity. Note that if $x(n)$ is Gaussian, $\kappa_x^{(k)}(l_1, l_2, \ldots, l_{k-1}) = 0$ for all $k \geq 3$ even if Equation (12.1.10) holds only for $k = 3, 4$.

If we assume that $\mu_x = 0$ and set $l_1 = l_2 = l_3 = 0$ in (12.1.6) through (12.1.8), we obtain

$$\kappa_x^{(2)}(0) = E\{|x(n)|^2\} = \sigma_x^2 \qquad (12.1.11)$$

$$\kappa_x^{(3)}(0, 0) = \gamma_x^{(3)} \qquad (12.1.12)$$

$$\kappa_x^{(4)}(0, 0, 0) = E\{|x(n)|^4\} - 2\sigma_x^4 \qquad \text{complex} \qquad (12.1.13)$$

$$= E\{x^4(n)\} - 3\sigma_x^4 \qquad \text{real} \qquad (12.1.14)$$

which provide the variance, unnormalized skewness, and unnormalized kurtosis of the process (see Section 3.1.2).

If the probability distribution of a process is symmetric (e.g., uniform, Gaussian, Laplace), its third-order cumulants are zero. In such cases, we need to consider fourth-order cumulants. Higher-order cumulants ($k > 4$) are seldom used in practice.

If the cumulants are absolutely summable, we can define the kth-order *cumulant spectra, higher-order spectra,* or *polyspectra* as the $(k-1)$-dimensional Fourier transform of the kth-order cumulant. More specifically, the power spectral density (PSD), bispectrum, and trispectrum of a zero-mean stationary process are defined by

$$R_x^{(2)}(e^{j\omega}) \triangleq \sum_{l_1=-\infty}^{\infty} \kappa_x^{(2)}(l_1)e^{-j\omega l_1} = R_x(e^{j\omega}) \tag{12.1.15}$$

$$R_x^{(3)}(e^{j\omega_1}, e^{j\omega_2}) \triangleq \sum_{l_1=-\infty}^{\infty} \sum_{l_2=-\infty}^{\infty} \kappa_x^{(3)}(l_1, l_2)e^{-j(\omega_1 l_1 + \omega_2 l_2)} \quad \text{(bispectrum)} \tag{12.1.16}$$

and $$R_x^{(4)}(e^{j\omega_1}, e^{j\omega_2}, e^{j\omega_3}) \triangleq \sum_{l_1=-\infty}^{\infty} \sum_{l_2=-\infty}^{\infty} \sum_{l_3=-\infty}^{\infty} \kappa_x^{(4)}(l_1, l_2, l_3)e^{-j(\omega_1 l_1 + \omega_2 l_2)}$$

$$\text{(trispectrum)} \tag{12.1.17}$$

where ω_1, ω_2, and ω_3 are the frequency variables. In contrast to the PSD, which is real-valued and nonnegative, both the bispectrum and the trispectrum are complex-valued. Since the higher-order cumulants of a Gaussian process are zero, its bispectrum and trispectrum are zero as well.

Many symmetries exist in the arguments of cumulants and polyspectra of both real and complex stochastic processes (Rosenblatt 1985; Nikias and Mendel 1993). For example, from the obvious symmetry

$$r_x^{(3)}(l_1, l_2) = r_x^{(3)}(l_2, l_1) \tag{12.1.18}$$

we obtain $$R_x^{(3)}(e^{j\omega_1}, e^{j\omega_2}) = R_x^{(3)}(e^{j\omega_2}, e^{j\omega_1}) \tag{12.1.19}$$

which is a basic property of the bispectrum.

For real-valued processes, we have the additional symmetries

$$r_x^{(3)}(l_1, l_2) = r_x^{(3)}(-l_2, l_1 - l_2) = r_x^{(3)}(-l_1, l_2 - l_1)$$
$$= r_x^{(3)}(l_2 - l_1, -l_1) = r_x^{(3)}(l_1 - l_2, -l_2) \tag{12.1.20}$$

$$r_x^{(4)}(l_1, l_2, l_3) = r_x^{(4)}(l_2, l_1, l_3) = r_x^{(4)}(l_1, l_3, l_2)$$
$$= r_x^{(4)}(-l_1, l_2 - l_1, l_3 - l_1) \tag{12.1.21}$$

which can be used to simplify the computation of cumulants. It can be shown that the nonredundant region for $r_x^{(3)}(l_1, l_2)$ is the wedge $\{(l_1, l_2) : 0 \le l_2 \le l_1 \le \infty\}$ and for $r_x^{(4)}(l_1, l_2, l_3)$ is the cone $\{(l_1, l_2, l_3) : 0 \le l_3 \le l_2 \le l_1 \le \infty\}$. The symmetries of cumulants impose symmetry properties upon the polyspectra. Indeed, by using (12.1.20) it can be shown that

$$R_x^{(3)}(e^{j\omega_1}, e^{j\omega_2}) = R_x^{(3)}(e^{j\omega_2}, e^{j\omega_1}) = R_x^{(3)}(e^{j\omega_1}, e^{-j\omega_1 - j\omega_2})$$
$$= R_x^{(3)}(e^{-j\omega_1 - j\omega_2}, e^{j\omega_2}) = R_x^{(3)*}(e^{-j\omega_1}, e^{-j\omega_2}) \tag{12.1.22}$$

which implies that the nonredundant region for the bispectrum is the triangle with vertices at $(0, 0)$, $(2\pi/3, 2\pi/3)$, and $(\pi, 0)$. The trispectrum of a real-valued process has 96 symmetry regions (Pflug et al. 1992).

Finally, we note that if in (12.1.7) we replace $x(n + l_1)$ by $y(n + l_1)$ and $x(n + l_2)$ by $z(n + l_2)$, we can define the cross-cumulant and then take its Fourier transform to find the cross-bispectrum. These quantities are useful for joint signal analysis (Nikias and Mendel 1993).

12.1.2 Higher-Order Moments and LTI Systems

Consider a BIBO stable linear, time-invariant (LTI) system with impulse response $h(n)$ and input-output relation given by

$$y(n) = \sum_{k=-\infty}^{\infty} h(k)x(n - k) \tag{12.1.23}$$

If the input $x(n)$ is stationary with zero mean, the output autocorrelation is

$$r_y(l) = \sum_{k_0} \sum_{k_1} h(k_1)h^*(k_0)r_x(l - k_1 - k_0) \tag{12.1.24}$$

where the range of summations, which is from $-\infty$ to ∞, is dropped for convenience (see Section 3.4). Also we have

$$R_y(e^{j\omega}) = |H(e^{j\omega})|^2 R_x(e^{j\omega}) \tag{12.1.25}$$

which shows that the output PSD is *insensitive* to the phase response of the system.

Using (12.1.23) and (12.1.7), we can show that the input and output third-order cumulants are related by

$$\kappa_y^{(3)}(l_1, l_2) = \sum_{k_0} \sum_{k_1} \sum_{k_2} h^*(k_0)h(k_1)h(k_2)\kappa_x^{(3)}(l_1 - k_1 + k_0, l_2 - k_2 + k_0) \tag{12.1.26}$$

To obtain the fourth-order cumulant of the output, we first determine its fourth-order moment

$$r_y^{(4)}(l_1, l_2, l_3) = \sum_{k_0} \sum_{k_1} \sum_{k_2} \sum_{k_3} h^*(k_0)h^*(k_1)h(k_2)h(k_3)$$

$$\times r_x^{(4)}(l_1 - k_1 + k_0, l_2 - k_2 + k_0, l_3 - k_3 + k_0) \tag{12.1.27}$$

using (12.1.23) and (12.1.4) (see Problem 12.1). Then using (12.1.8), (12.1.9), and (12.1.24), we have

$$\kappa_y^{(4)}(l_1, l_2, l_3) = \sum_{k_0} \sum_{k_1} \sum_{k_2} \sum_{k_3} h^*(k_0)h^*(k_1)h(k_2)h(k_3)$$

$$\times \kappa_x^{(4)}(l_1 - k_1 + k_0, l_2 - k_2 + k_0, l_3 - k_3 + k_0) \tag{12.1.28}$$

which holds for both real- and complex-valued processes. An interesting interpretation of this relationship in terms of convolutions is given in Mendel (1991) and in Therrien (1992).

We now compute the bispectrum of the output signal $y(n)$ in terms of the bispectrum of the input $x(n)$ and the frequency response $H(e^{j\omega})$ of the system. Indeed, taking the two-dimensional Fourier transform of (12.1.26) and interchanging the order of summations, we have

$$R_y^{(3)}(e^{j\omega_1}, e^{j\omega_2}) = \sum_{l_1} \sum_{l_2} \kappa_y^{(3)}(l_1, l_2)e^{-j(\omega_1 l_1 + \omega_2 l_2)}$$

$$= \sum_{l_1} \sum_{l_2} \sum_{k_0} \sum_{k_1} \sum_{k_2} h^*(k_0)h(k_1)h(k_2)$$

$$\times \kappa_x^{(3)}(l_1 - k_1 + k_0, l_2 - k_2 + k_0)e^{-j(\omega_1 l_1 + \omega_2 l_2)}$$

$$= \sum_{k_0} h^*(k_0)e^{j(\omega_1 + \omega_2)k_0} \sum_{k_1} h(k_1)e^{-j\omega_1 k_1} \sum_{k_2} h(k_2)e^{-j\omega_2 k_2}$$

$$\times \sum_{l_1} \sum_{l_2} \kappa_x^{(3)}(l_1 - k_1 + k_0, l_2 - k_2 + k_0)e^{-j\omega_1(l_1 - k_1 + k_0)}e^{-j\omega_2(l_2 - k_2 + k_0)}$$

Rearranging terms and using (12.1.16), we obtain

$$R_y^{(3)}(e^{j\omega_1}, e^{j\omega_2}) = H^*(e^{j\omega_1 + j\omega_2})H(e^{j\omega_1})H(e^{j\omega_2})R_x^{(3)}(e^{j\omega_1}, e^{j\omega_2}) \tag{12.1.29}$$

which shows that the bispectrum, in contrast to the PSD in (12.1.25), is sensitive to the phase of the system.

In a similar, but more complicated way, we can show that the output trispectrum is given by

$$R_y^{(4)}(e^{j\omega_1}, e^{j\omega_2}, e^{j\omega_3}) = H^*(e^{j\omega_1 + j\omega_2 + j\omega_3})H^*(e^{-j\omega_1})H(e^{j\omega_2})$$

$$\times H(e^{j\omega_3})R_x^{(4)}(e^{j\omega_1}, e^{j\omega_2}, e^{j\omega_3}) \tag{12.1.30}$$

which again shows that the trispectrum is sensitive to the phase of the system.

12.1.3 Higher-Order Moments of Linear Signal Models

In Section 4.1 we discussed linear signal models and the innovations representation of stationary processes with given second-order moments. This representation is given by

$$x(n) = \sum_{k=0}^{\infty} h(k)w(n-k) \tag{12.1.31}$$

$$r_x(l) = \sigma_w^2 \sum_{k=0}^{\infty} h(k)h^*(k-l) \tag{12.1.32}$$

$$R_x(e^{j\omega}) = \sigma_w^2 |H(e^{j\omega})| \tag{12.1.33}$$

where $w(n)$ is a white noise process. If $w(n)$ is Gaussian, $x(n)$ is Gaussian and this representation provides a complete description of the process.

If the excitation $w(n)$ is IID and *non-Gaussian* with cumulants

$$\kappa_w^{(k)}(l_1, l_2, \ldots, l_{k-1}) = \begin{cases} \gamma_w^{(k)} & l_1 = l_2 = \cdots = l_{k-1} = 0 \\ 0 & \text{otherwise} \end{cases} \tag{12.1.34}$$

the output of the linear model is also non-Gaussian. The cumulants and the polyspectra of process $x(n)$ are

$$\kappa_x^{(k)}(l_1, l_2, \ldots, l_{k-1}) = \gamma_w^{(k)} \sum_{n=0}^{\infty} h(n)h(n+l_1)\cdots h(n+l_{k-1}) \tag{12.1.35}$$

and

$$R_x^{(k)}(e^{j\omega_1}, e^{j\omega_2}, \ldots, e^{j\omega_{k-1}}) = \gamma_w^{(k)} H(e^{j\omega_1})H(e^{j\omega_2})\cdots H(e^{j\omega_{k-1}})H^*(e^{j\sum_{i=1}^{k-1}\omega_i}) \tag{12.1.36}$$

respectively. The cases for $k = 3, 4$ follow easily from Equations (12.1.26), (12.1.28) to (12.1.30), and (12.1.34). A simpler, direct derivation is discussed in Problem 12.2.

Setting $k = 3$ into (12.1.36), we obtain

$$\angle R_x^{(3)}(e^{j\omega_1}, e^{j\omega_2}) = \angle \gamma_w^{(3)} - \angle H(e^{j\omega_1+j\omega_2}) + \angle H(e^{j\omega_1}) + \angle H(e^{j\omega_2}) \tag{12.1.37}$$

which shows that we can use the bispectrum of the output to determine the phase response of the system if the input is a non-Gaussian IID process. From (12.1.33) we see that this is not possible using the PSD.

EXAMPLE 12.1.1. For $0 < a < 1$ and $0 < b < 1$, consider the MA(2) systems

$$H_{\min}(z) = (1 - az^{-1})(1 - bz^{-1})$$

$$H_{\max}(z) = (1 - az)(1 - bz)$$

$$H_{\min}(z) = (1 - az)(1 - bz^{-1})$$

which obviously are minimum-phase, maximum-phase, and mixed-phase, respectively. All these systems have the same output complex PSD

$$R_x(z) = \sigma_w^2 H_{\min}(z)H_{\min}(z^{-1}) = \sigma_w^2 H_{\max}(z)H_{\max}(z^{-1}) = \sigma_w^2 H_{\min}(z)H_{\min}(z^{-1})$$

and hence the same autocorrelation. As a result, we cannot correctly identify the phase response of an MA(2) model using the PSD (or equivalently the autocorrelation) of the output signal. However, we can correctly identify the phase by using the bispectrum. The output third-order cumulant $\kappa_x^{(3)}(l_1, l_2)$ for the above MA(2) models can be computed by using either the complex bispectrum (the z transform of the third-order cumulant)

$$R_x^{(3)}(z_1, z_2) = \gamma_w^{(3)} H(z_1)H(z_2)H(-z_1^{-1}z_2^{-1})$$

or the formula

$$\kappa_x^{(3)}(l_1, l_2) = \gamma_w^{(3)} \sum_{n=0}^{2} h(n)h(n+l_1)h(n+l_2) \tag{12.1.38}$$

for all values of l_1, l_2 that lead to overlapping terms in the summation. The results are summarized in Table 12.1. The values shown are for the principal region (see Figure 12.1); the remaining ones are computed using the symmetry relations (12.1.20). Using the formula

$$R_x^{(3)}(e^{j\omega_1}, e^{j\omega_2}) = \gamma_w^{(3)} H(e^{j\omega_1})H(e^{j\omega_2})H^*(e^{j(\omega_1+\omega_2)}) \tag{12.1.39}$$

we can numerically compute the bispectrum, using the DFT (see Problem 12.4). The results are plotted in Figure 12.2. We see that the cumulants and bispectra of the three systems are different. Hence, the third-order moments can be used to identify both the magnitude and the phase response of the MA(2) model.

TABLE 12.1

Minimum-, maximum-, and mixed-phase MA(2) systems with the same autocorrelation (or PSD) but with different third-order moments ($0 < a < 1$, $0 < b < 1$) (Nikias and Raghuveer 1987).

Cumulants	Minimum-phase MA(2)	Maximum-phase MA(2)	Mixed-phase MA(2)
$\kappa_x^{(3)}(0,0)$	$1 - (a+b)^3 + a^3b^3$	$1 - (a+b)^3 + a^3b^3$	$(1+ab)^3 - a^3 - b^3$
$\kappa_x^{(3)}(1,1)$	$(a+b)^2 - (a+b)a^2b^2$	$-(a+b) + ab(a+b)^2$	$-a(1+ab)^2 + (1+ab)b^2$
$\kappa_x^{(3)}(2,2)$	a^2b^2	ab	$-ab^2$
$\kappa_x^{(3)}(1,0)$	$-(a+b) + ab(a+b)^2$	$(a+b)^2 - (a+b)a^2b^2$	$a^2(1+ab) - (1+ab)^2b$
$\kappa_x^{(3)}(2,0)$	ab	a^2b^2	$-a^2b$
$\kappa_x^{(3)}(2,1)$	$-(a+b)ab$	$-(a+b)ab$	$ab(1+ab)$
$r_x(0)$	$1 + a^2b^2 + (a+b)^2$	$1 + a^2b^2 + (a+b)^2$	$1 + a^2b^2 + (a+b)^2$
$r_x(1)$	$-(a+b)(1+ab)$	$-(a+b)(1+ab)$	$-(a+b)(1+ab)$
$r_x(2)$	ab	ab	ab

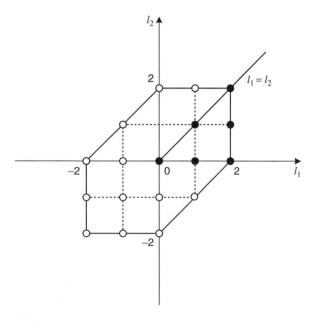

FIGURE 12.1
Region of support for the third-order cumulant of the MA(2) model. The solid circles indicate the primary samples, which can be utilized to determine the remaining samples using symmetry relations.

From the previous example and the general discussion of higher-order moments and their transformation by linear systems, we conclude that HOS can be useful when we deal with non-Gaussian signals or Gaussian signals that have passed through nonlinear

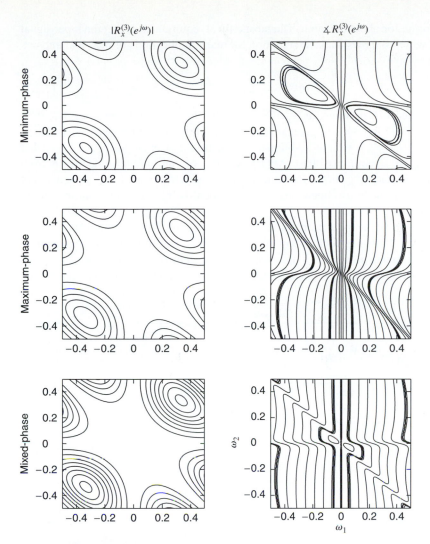

FIGURE 12.2
Bispectrum magnitude and phase for minimum-, maximum-, and mixed-phase
MA(2) models with $a = 0.5$ and $b = 0.9$.

systems. More specifically, the use of HOS is beneficial in the following cases: suppression
of additive Gaussian colored noise, identification of the phase response of a system using
output data only, and characterization of non-Gaussian processes or nonlinear systems.
More details and applications are discussed in Nikias and Mendel (1993). However, note
that the application of HOS-based methods to real-world problems is very difficult because
(1) the computation of reliable estimates of higher-order moments requires a large amount
of data and (2) the assessment and interpretation of the results require a solid statistical
backgound and extensive practical experience.

12.2 BLIND DECONVOLUTION

In Section 6.7, we discussed optimum inverse filtering and deconvolution using the minimum mean square error (MMSE) criterion under the assumption that all required statistical
moments are known. In the case of *blind deconvolution* (see Figure 6.23), the goal is to
retrieve the input of a system $G(z)$ by using only the output signal and possibly some

statistical information about the input. The most critical requirement is that the input signal $w(n)$ be IID, which is a reasonable assumption for many applications of practical interest. In this case, we have

$$R_x(e^{j\omega}) = \sigma_w^2 |G(e^{j\omega})|^2 \tag{12.2.1}$$

which can be used to determine, at least in principle, the magnitude response $|G(e^{j\omega})|$ from the output PSD $R_x(e^{j\omega})$. In general, it is impossible to obtain the phase response of the system from $R_x(e^{j\omega})$ without additional information. For example, if we know that $G(z) = 1/A(z)$ is a minimum-phase AP(P) system, we can uniquely identify it from $r_x(l)$ or $R_x(e^{j\omega})$, using the method of linear prediction. However, if the system is not minimum-phase, the method of linear prediction will identify it as minimum-phase, leading to erroneous results.

The importance of the input probability density function in deconvolution applications is illustrated in the following figures. Figure 12.3 shows a random sequence generated by filtering white Gaussian noise with a minimum-phase system $H(z)$ and the sequences obtained by deconvolution of this sequence with the minimum-phase, maximum-phase, and

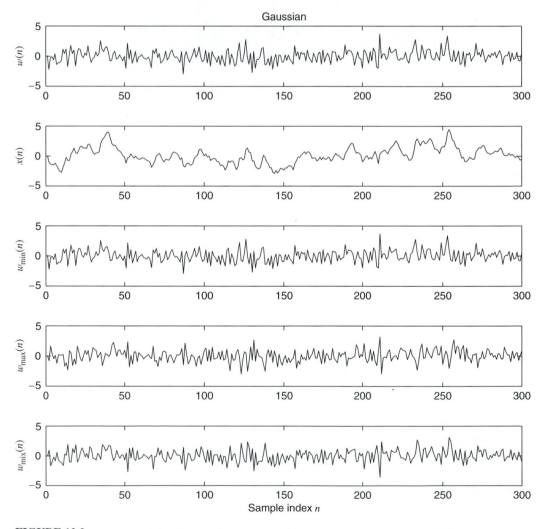

FIGURE 12.3
A minimum-phase Gaussian random sequence and its deconvolution by the corresponding minimum-, maximum-, and mixed-phase inverse systems.

mixed-phase inverse systems corresponding to $H(z)$. The three deconvolved sequences, which look visually similar, are all uncorrelated and statistically indistinguishable, because in the Gaussian case uncorrelatedness implies statistical independence. Figure 12.4 shows the results of the same experiment repeated with the same systems and a white noise sequence with an exponential probability density function. It is now clear that only the minimum-phase inverse system provides the corect answer, although all three deconvolved sequences have the same second-order statistics (Donoho 1981). More details about the generation of these figures and further discussion of their meaning are given in Problem 12.5.

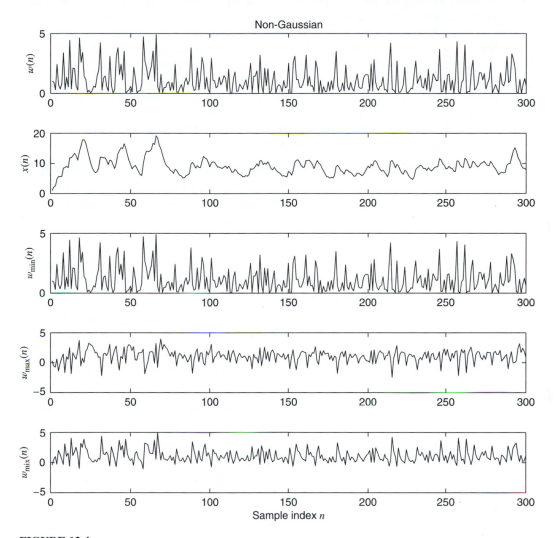

FIGURE 12.4

A minimum-phase non-Gaussian random sequence and its deconvolution by the corresponding minimum-, maximum-, and mixed-phase inverse systems.

We conclude that complete identification of $G(z)$, and therefore correct retrieval of the input signal $w(n)$, requires the identification of the phase response $\angle G(e^{j\omega})$ of the system; failure to do so may lead to erroneous results.

If the input $w(n)$ is IID and non-Gaussian, the bispectrum of the output is [see (12.1.36)]

$$R_x^{(3)}(e^{j\omega_1}, e^{j\omega_2}) = \kappa_w^{(3)} G(e^{j\omega_1})G(e^{j\omega_2})G^*(e^{j\omega_1+j\omega_2}) \tag{12.2.2}$$

and it can be used to determine both the magnitude and phase response of $G(e^{j\omega})$ from the magnitude and argument of the bispectrum. If the bispectrum is identically zero, we can use some nonzero higher-order spectrum. Therefore, HOS can be used in many ways to obtain unbiased estimates of $\angle G(e^{j\omega})$ provided that the polyspectra are not all identically zero, or equivalently the input probability density function (pdf) is non-Gaussian (Matsuoka and Ulrych 1984; Mendel 1991; Nikias and Mendel 1993). We emphasize that, in practice, polyspectra estimators have high variance, and therefore reliable phase estimation requires very long data records. In conclusion, *blind deconvolution is always possible provided a stable inverse $1/G(z)$ exists and $w(n)$ is non-Gaussian.* If $w(n)$ is Gaussian, we cannot correctly identify the phase response of the inverse system using only the second-order moments of the output signal.

As we have already mentioned, MMSE linear prediction solves the blind deconvolution problem for minimum-phase systems with Gaussian inputs using the autocorrelation of the output signal. In essense, the inverse system retrieves the input by restoring its flat PSD, which has been colored by the system $G(z)$. This suggests the following question: *Is it possible to uniquely determine the inverse system $h(n)$ by restoring some property of the input signal (besides spectral flatness) that has been distorted by the system $G(z)$?* To address this question, let us consider the effects of an LTI system upon the probability density function of the input signal. We recall that

- If the input pdf is Gaussian, then the output pdf is Gaussian. In general, if the input pdf is stable, then the output pdf is also stable. This follows from the fact that only stable random variables are invariant under linear transformations (see Section 3.2.4). However, we limit our discussion to Gaussian signals because they have finite variance.
- If the input pdf is non-Gaussian, then the output pdf tends to Gaussian as a result of the central limit theorem (see Section 3.3.7). The "Gaussianization" capability of the system depends on the length and amplitude of its impulse response.[†] This is illustrated in Example 3.2.4, which shows that the sum of uniform random variables becomes "more Gaussian" as their number increases.

We see that filtering of a non-Gaussian IID sequence increases its Gaussianity. The only system that does *not* alter a non-Gaussian input pdf has impulse response with one nonzero sample, that is, $b_0\delta(n - n_0)$. In any other case, the input and output distributions are different, except if the input is Gaussian. A strict proof is provided by the following theorem (Kagan et al. 1973).

> **THEOREM 12.1.** Consider a random variable x defined by the linear combination of IID random variables w_k
>
> $$x = \sum_k c_k w_k \qquad (12.2.3)$$
>
> with coefficients such that $\sum_k |c_k|^2 < \infty$. The random variable x is Gaussian if and only if (*a*) x has finite variance, (*b*) $x \stackrel{d}{=} w_k$ for all k, (*c*) at least two coefficients c_k are not zero.

If we define the overall system (see Figure 12.5)

$$c(n) = g(n) * h(n) \qquad (12.2.4)$$

the signals $y(n)$ and $w(n)$ can have the same non-Gaussian distribution if and only if $c(n)$ has only one nonzero coefficient. Hence, if we know the input pdf, we can determine the inverse system $h(n)$ by restoring the pdf of $y(n)$ to match the pdf of the input $w(n)$. However, it turns out that instead of restoring the pdf, that is, all moments (Benveniste et al. 1980), we only need to restore the moments up to order 4 (Shalvi and Weinstein 1990). This is shown in the following theorem.

[†]In many practical applications (e.g., seismology), the underlying data are non-Gaussian; however, unavoidable filtering operations (e.g., recording instruments) tend to "Gaussianize" their distribution. As a result, many times the non-Gaussianity of the data becomes apparent after proper deconvolution (Donoho 1981).

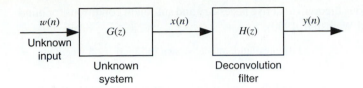

FIGURE 12.5
Basic blind deconvolution model.

THEOREM 12.2. Consider a stable LTI system

$$y(n) = \sum_k c(k)w(n-k) \tag{12.2.5}$$

with an IID input $w(n)$ that has finite moments up to order 4. Then we have

$$E\{|y(n)|^2\} = E\{|w(n)|^2\} \sum_k |c(k)|^2 \tag{12.2.6}$$

$$E\{y^2(n)\} = E\{w^2(n)\} \sum_k c^2(k) \tag{12.2.7}$$

and

$$\kappa_y^{(4)} = \kappa_w^{(4)} \sum_k |c(k)|^4 \tag{12.2.8}$$

where

$$\kappa_y^{(4)} = E\{|y(n)|^4\} - 2E^2\{|y(n)|^2\} - |E\{y^2(n)\}|^2 \tag{12.2.9}$$

is the fourth-order cumulant of $y(n)$ and $\kappa_w^{(4)}$ is the fourth-order cumulant of $w(n)$.

Proof. Relations (12.2.6) and (12.2.7) can be easily shown by using (12.2.5) and the independence assumption. To prove (12.2.8), we start with (12.2.5); then by interchanging the order between expectation and summations, we have

$$E\{|y(n)|^4\} = E\left\{ \left| \sum_k c(k)w(n-k) \right|^2 \right\}$$

$$= \sum_{k_1}\sum_{k_2}\sum_{k_3}\sum_{k_4} c(k_1)c^*(k_2)c(k_3)c^*(k_4) \tag{12.2.10}$$

$$\times \underbrace{E\{w(n-k_1)w^*(n-k_2)w(n-k_3)w^*(n-k_4)\}}_{W}$$

where

$$W = \begin{cases} E\{|w(n)|^4\} & k_1 = k_2 = k_3 = k_4 \\ E^2\{|w(n)|^2\} & k_1 = k_2 \neq k_3 = k_4,\, k_1 = k_4 \neq k_2 = k_3 \\ |E\{w^2(n)\}|^2 & k_1 = k_3 \neq k_2 = k_4 \\ 0 & \text{otherwise} \end{cases} \tag{12.2.11}$$

by invoking the independence assumption. If we substitute (12.2.11) into (12.2.10), we obtain

$$E\{|y(n)|^4\} = E\{|w(n)|^4\} \sum_k |c(k)|^2$$

$$+ 2E^2\{|w(n)|^2\} \left[\left[\sum_k |c(k)|^2 \right]^2 - \sum_k |c(k)|^4 \right] \tag{12.2.12}$$

$$+ |E\{w^2(n)\}|^2 \left[\left| \sum_k c^2(k) \right|^2 - \sum_k |c(k)|^4 \right]$$

Finally, substituting (12.2.6) and (12.2.7) into (12.2.12) and rearranging the various terms, we obtain (12.2.8).

We now use the previous theorem to derive necessary and sufficient conditions for blind deconvolution (Shalvi and Weinstein 1990).

THEOREM 12.3. Consider the blind deconvolution model shown in Figure 12.5 where $c(n) = g(n) * h(n)$. If $E\{|y(n)|^2\} = E\{|w(n)|^2\}$, then

1. $|\kappa_y^{(4)}| \leq |\kappa_w^{(4)}|$, that is, the kurtosis of the output is less than or equal to the kurtosis of the input.
2. $|\kappa_y^{(4)}| = |\kappa_w^{(4)}|$ if and only if $c(n) = e^{j\theta}\delta(n - n_0)$. Hence, if the kurtosis of the output is equal to the kurtosis of the input, the inverse system is given by $H(z) = e^{j\theta}z^{-n_0}/G(z)$.

Proof. The proof can be easily obtained by using the inequality

$$\sum_k |c(k)|^4 \leq \left[\sum_k |c(k)|^2\right]^2 \tag{12.2.13}$$

where equality holds if and only if $c(k)$ has at most one nonzero component. The condition $E\{|y(n)|^2\} = E\{|w(n)|^2\}$ in conjunction with (12.2.6) implies that $\sum_k |c(k)|^2 = 1$. Therefore, $\sum_k |c(k)|^4 \leq 1$ and $|\kappa_y^{(4)}| \leq |\kappa_w^{(4)}|$ due to (12.2.8). Clearly, if $\sum_k |c(k)|^2 = 1$, we can have $\sum_k |c(k)|^4 = 1$ if and only if $c(n) = e^{j\theta}\delta(n - n_0)$.

This theorem shows that a necessary and sufficient condition for the correct recovery of the inverse system $h(n)$, that is, for successful blind deconvolution, is that $E\{|y(n)|^2\} = E\{|w(n)|^2\}$ and $|\kappa_y^{(4)}| = |\kappa_w^{(4)}|$. Therefore, we can determine $h(n)$ by solving the following constrained optimization problem:

$$\max_{h(n)} |\kappa_y^{(4)}| \quad \text{subject to} \quad E\{|y(n)|^2\} = E\{|w(n)|^2\} \tag{12.2.14}$$

It has been shown that for FIR inverse filters of sufficient length, $\kappa_y^{(4)}$ has no spurious local maxima over $E\{|y(n)|^2\} = E\{|w(n)|^2\}$, and therefore gradient search algorithms converge to the correct solution regardless of initialization (Shalvi and Weinstein 1990). We should stress that the IID property of the input $w(n)$ is a key requirement for blind deconvolution methods to work.

By using the normalized cumulants $\tilde{\kappa}_y^{(4)} = \kappa_y^{(4)}/\sigma_y^4$ it has been shown for real signals (Donoho 1981) that

$$\tilde{\kappa}_y^{(4)} = \tilde{\kappa}_w^{(4)} \frac{\sum_k |c(k)|^4}{\left[\sum_k |c(k)|^2\right]^2} \tag{12.2.15}$$

which implies that $|\kappa_y^{(4)}| \leq |\kappa_w^{(4)}|$, a result attributed to Granger (1976). Furthermore, Donoho (1981) showed that if $\tilde{\kappa}_w^{(4)} \neq 0$, then maximization of $|\kappa_y^{(4)}|$ provides a solution to the blind deconvolution problem (Tugnait 1992). An elaborate discussion of cumulant maximization criteria and algorithms for blind deconvolution is given in Cadzow (1996). A review of various approaches for blind system identification and deconvolution is given in Abed-Meraim et al. (1997). In the next section, we apply these results to the design of adaptive filters for blind equalization.

12.3 UNSUPERVISED ADAPTIVE FILTERS—BLIND EQUALIZERS

All the adaptive filters we have discussed so far require the availability of a desired response signal that is used to "supervise" their operation. What we mean by that is that, at each time instant, the adaptive filter compares its output with the desired response and uses this information to improve its performance. In this sense, the desired response serves as a training signal that provides the feedback needed by the filter to improve its performance. However, as we discussed in Sections 1.4.1 and 10.1, there are applications such as blind equalization and blind deconvolution in which the availability of a desired response signal is either impossible or inpractical. In this section we discuss adaptive filters that circumvent this problem; that is, they can operate *without* a desired response signal. These filters are called *unsupervised adaptive filters* to signify the fact that they operate without "supervision," that is, without a desired response signal. Clearly, unsupervised adaptive filters need additional information to make up for the lack of a desired response signal. This information depends on the particular application and has a big influence on the design and performance of the adaptive algorithm. The most widely used unsupervised adaptive filters are application-specific and operate by exploiting (1) the higher-order statistics, (2) the cyclostationary statistics, or (3) some invariant property of the input signal. Most unsupervised adaptive filtering algorithms have been developed in the context of blind equalization, which provides the most important practical application of these filters.

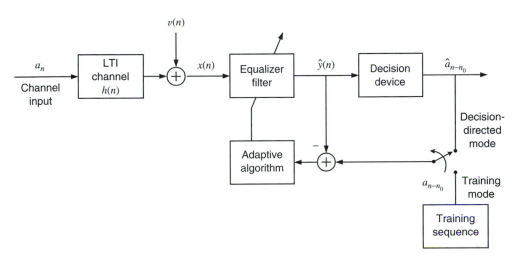

FIGURE 12.6
Conventional channel equalizer with training and decision-directed modes of operation.

12.3.1 Blind Equalization

Figure 12.6 shows the traditional approach to adaptive equalization.[†] When the adaptive equalizer starts its operation, the transmitter sends a known training sequence over the unknown channel. Since the training sequence can be used as a desired response signal, we can adjust the equalizer's coefficients by using the standard LMS or RLS algorithms. The LMS equalization algorithm with a training sequence is

$$\mathbf{c}(n) = \mathbf{c}(n-1) + 2\mu\mathbf{x}(n)e^*(n) \qquad (12.3.1)$$

where
$$e(n) = a_{n-n_0} - \hat{y}(n) = a_{n-n_0} - \mathbf{c}^H(n-1)\mathbf{x}(n) \qquad (12.3.2)$$

is the a priori error. If, at the end of the training period, the MSE $E\{|e(n)|^2\}$ is so small that $\hat{y}(n) \simeq a_{n-n_0}$, then we can replace a_{n-n_0} by the decision $\hat{a}_{n-n_0} \triangleq Q[\hat{y}(n)]$ and switch the equalizer to decision-directed mode. The resulting algorithm is

$$\mathbf{c}(n) = \mathbf{c}(n-1) + 2\mu\mathbf{x}(n)\{Q[\hat{y}(n)] - \hat{y}(n)\}^* \qquad (12.3.3)$$

and its performance depends on how close $\mathbf{c}(n)$ is to the optimum setting $\mathbf{c}_o$ according to the

---

[†]This approach has been discussed in Sections 6.8 and 10.4.4.

MSE or the zero-forcing criterion. If $\mathbf{c}(0)$ is close to $\mathbf{c}_o$, the intersymbol interference (ISI) is significantly reduced (i.e., the eye is open), the decision device makes correct decisions with low probability of error, and the algorithm is likely to converge to $\mathbf{c}_o$. However, if $\mathbf{c}(0)$ is not close to $\mathbf{c}_o$, that is, when the eye is closed (which is when we need an equalizer), then the error surface can be multimodal and the decision-directed equalizer fails to converge or converges to a local minimum (Mazo 1980).

The training session should be repeated each time the channel response changes or after system breakdowns, which results in a reduction of the data throughput. However, there are digital communications applications in which the start-up and retraining of the adaptive equalizer have to be accomplished *without* a training sequence. Adaptive equalizers that operate without the aid of a training signal are known as *blind equalizers,* although the term *unsupervised* would be more appropriate. The need for blind equalization is enormous in digital point-to-multipoint and broadcast networks, such as high-definition and cable television. In all these applications, the transmitter should be able to send its content unaffected by the joining or withdrawal of client receivers or their need for training data (Treichler et al. 1998).

Clearly, blind equalization is a special case of blind deconvolution with input from a finite alphabet. When we deal with blind equalization, we should recall the following facts:

1. The second-order statistics of the output provide information about the magnitude response of an LTI channel. Therefore, mixed-phase channels *cannot* be identified using second-order statistics only.
2. Mixed-phase LTI channels with IID Gaussian inputs cannot be identified from their output because all statistical information is contained in the second-order moments.
3. The inverse of a mixed-phase LTI channel is IIR and unstable. Hence, only an FIR causal approximation can be used for its equalization.
4. Channels with zeros on the unit circle cannot be equalized by using zero-forcing equalizers (Section 6.8).
5. Since $|H(e^{j\omega})|^2 = |H(e^{j\omega})e^{j\theta}|^2$ and for perfect equalization $H(z)C(z) = b_0 z^{-n_0}$, $b_0 \neq 0$, the channel can be identified up to a rotational factor and a constant time shift.
6. The structure of the finite symbol alphabet improves the detection process, which can be thought as an unsupervised pattern classification problem (Fukunaga 1990).

All equalizers (blind or not blind) use the second-order statistics (autocorrelation or power spectrum) of the channel output, to obtain information about the channel's magnitude response. However, blind equalizers need additional information to determine the phase response of the channel and to compensate for the absense of the desired response sequence. Phase information can be obtained from the HOS or the second-order and higher-order cyclostationary moments of the channel output. The cyclostationarity property results from the modulation of the transmitted signal (Gardner 1991).

The above types of information can be exploited, either individually or in combination, to obtain various blind equalization algorithms. The available blind equalization methods can be categorized into two groups:

1. *HOS-based methods.* These can be further divided into two groups:

    a. *Implicit HOS algorithms* implicitly explore HOS by iteratively minimizing a *non-MSE* criterion, which does not require the desired response but reflects the amount of residual ISI in the received signal.

    b. *Explicit HOS algorithms* compute explicitly the block estimates of the power spectrum to determine the magnitude response and block estimates of the trispectrum, to determine the phase response of the channel.

2. *Cyclostationary statistics–based methods,* which exploit the second-order cyclostationary statistics of the received signal.

Since the number of samples required to estimate the mth-order moment, for a given level of bias and variance, increases almost exponentially with order m (Brillinger 1980), both implicit and explicit HOS-based methods have a slow rate of convergence. Indeed, since channel identification requires at least fourth-order moments, HOS-based algorithms require a large number, typically several thousand, of data samples (Ding 1994).

Explicit HOS methods originated in geophysics to solve blind deconvolution problems with non-Gaussian inputs (Wiggins 1978; Donoho 1981; Godfrey and Rocca 1981). A complete discussion of the application of HOS techniques to blind equalization is given in Hatzinakos and Nikias (1991). Because HOS algorithms require a large number of data samples and have high computational complexity, they are not used in practice for blind equalization applications. In contrast to symbol rate blind equalizers that require the use of HOS, the input of fractionally spaced equalizers (which is sampled higher than the symbol rate) contains additional cyclostationarity-based second-order statistics (SOS) that can be exploited to identify the channel. Since SOS requires fewer data samples for estimation, we can exploit cyclostationarity to obtain algorithms that converge faster than HOS-based algorithms. Furthermore, channel identification using cyclic SOS does not preclude inputs with Gaussian or nearly Gaussian statistics. More information about these methods can be found in Gardner (1991), Ding (1994), Tong et al. (1994a, b), and Moulines et al. (1995). We focus on implicit HOS methods because they are easy to implement and are widely used in practice.

12.3.2 Symbol Rate Blind Equalizers

The basic structure of a blind equalization system is shown in Figure 12.7. The key element is a scalar zero-memory nonlinear function $\tilde{\psi}$, which serves to generate a desired response signal $\tilde{\psi}[\hat{y}(n)]$ for the adaptive algorithm.

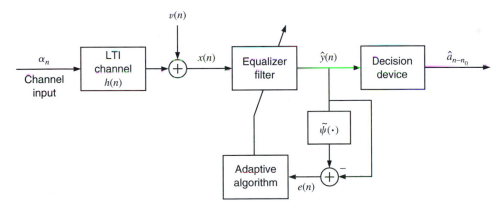

FIGURE 12.7
Basic elements of an adaptive blind equalization system.

We wish to find the function $\tilde{\psi}$ that provides a good estimate of the desired response a_n. To this end, suppose that we have a good initial guess $c(n)$ of the equalizer coefficients. Then we assume that the convolution of the channel and equalizer impulse responses can be decomposed as

$$h(n) * c(n) = \delta(n) + h_{\text{ISI}}(n) \tag{12.3.4}$$

where $h_{\text{ISI}}(n)$ is the component creating the ISI. The output of the equalizer is

$$\hat{y}(n) = c(n) * x(n) = c(n) * [h(n) * a_n + v(n)]$$
$$= a_n + h_{\text{ISI}}(n) * a_n + c(n) * v(n) \triangleq a_n + \tilde{v}(n) \tag{12.3.5}$$

where $h_{\text{ISI}}(n) * a_n$ is the residual ISI and $c(n) * v(n)$ is additive noise. By invoking the central limit theorem, we can show that the convolutional noise $\tilde{v}(n)$ can be modeled as white Gaussian noise (Godfrey and Rocca 1981; Haykin 1996). Since a_n is IID and since a_n and $\tilde{v}(n)$ are statistically independent, the minimum MSE estimate $z(n)$ of a_n based on $\hat{y}(n)$ is

$$z(n) = E\{a_n | \hat{y}(n)\} \triangleq \tilde{\psi}[\hat{y}(n)] \tag{12.3.6}$$

which is a nonlinear function of $\hat{y}(n)$ because a_n has a non-Gaussian distribution. Then the a priori error is

$$e(n) = \tilde{\psi}[\hat{y}(n)] - \hat{y}(n) \tag{12.3.7}$$

where
$$\hat{y}(n) = \sum_{k=-L}^{L} c_k^*(n-1)x(n-k) \triangleq \mathbf{c}^H(n-1)\mathbf{x}(n) \tag{12.3.8}$$

is the output of the equalizer. This leads to the following a priori stochastic gradient algorithm for blind equalization

$$\mathbf{c}(n) = \mathbf{c}(n-1) + \mu \mathbf{x}(n)e^*(n) \tag{12.3.9}$$

where μ is the adaptation step size.

Another approach used to derive (12.3.9) is to start with the cost function

$$P(n) \triangleq E\{\Psi[\hat{y}(n)]\} \tag{12.3.10}$$

where
$$\psi(y) \triangleq \tilde{\psi}(y) - y \tag{12.3.11}$$

is the derivative
$$\psi(y) \triangleq \Psi'(y) = \frac{\partial \Psi(y)}{\partial y} \tag{12.3.12}$$

of a nonlinear function Ψ. The nonlinearity of Ψ creates the dependence of the cost function on the HOS of $\hat{y}(n)$ and a_n. The cost function (12.3.10) should not require the input sequence a_n; it should reflect the amount of current ISI, and its minimum should correspond to the minimum ISI or minimum MSE condition. In contrast to the MSE criterion, which depends on the SOS and is a quadratic (convex) function of the equalizer parameters, the cost function (12.3.10) is nonconvex and may have local minima. If we compute the gradient of $P(n)$ with respect to $\mathbf{c}$ and drop the expectation operation, we obtain the stochastic gradient algorithm (12.3.9).

Equations (12.3.8), (12.3.7), and (12.3.9) provide the general form of LMS-type blind equalization algorithms. Different choices for the nonlinear function $\tilde{\psi}$ result in various algorithms for blind equalization. Because the output $\hat{y}(n)$ is approximately a Bussgang process, these algorithms are sometimes called *Bussgang algorithms* for blind equalization (Haykin 1996). A process is called *Bussgang* (Bussgang 1952; Bellini 1986) if it satisfies the property

$$E\{\hat{y}(n)\hat{y}^*(n-l)\} = E\{\hat{y}(n)\tilde{\psi}[\hat{y}^*(n-l)]\} \tag{12.3.13}$$

that is, its autocorrelation is equal to the cross-correlation between the process and a non-linear transformation of the process.

Sato algorithm. The first blind equalizer was introduced by Sato (1975) for one-dimensional multilevel pulse amplitude modulation (PAM) signals. It uses the error function

$$\psi_1(n) = R_1 \operatorname{sgn}[\hat{y}(n)] - \hat{y}(n) = e(n) \tag{12.3.14}$$

where
$$R_1 \triangleq \frac{E\{|a_n|^2\}}{E\{|a_n|\}} \tag{12.3.15}$$

and sgn(x) is the signum function. Integration of $\psi_1(n)$ gives

$$\Psi_1[\hat{y}(n)] = \tfrac{1}{2}[R_1 - \hat{y}(n)]^2 \tag{12.3.16}$$

whose expectation provides the cost function for the Sato algorithm. The complex version of the algorithm, used for quadrature amplitude modulation (QAM) constellations, uses the error

$$e(n) = R_1 \, \text{csgn}[\hat{y}(n)] - \hat{y}(n) \tag{12.3.17}$$

where
$$\text{csgn}(x) = \text{csgn}(x_r + jx_i) = \text{sgn}(x_r) + j\,\text{sgn}(x_i) \tag{12.3.18}$$

is the complex signum function.

Godard algorithms. The most widely used algorithms, in practical blind equalization applications, were developed by Godard (1980) for QAM signal constellations. Godard replaced the function Ψ_1 with the more general function

$$\Psi_p[\hat{y}(n)] = \frac{1}{2p}[R_p - |\hat{y}(n)|^p]^p \tag{12.3.19}$$

where p is a positive integer and R_p is the positive real constant

$$R_p \triangleq \frac{E\{|a_n|^{2p}\}}{E\{|a_n|^p\}} \tag{12.3.20}$$

which is known as the *dispersion of order p*. The family of Godard stochastic gradient algorithms is described by

$$\mathbf{c}(n) = \mathbf{c}(n-1) + \mu\mathbf{x}(n)e^*(n) \tag{12.3.21}$$

where
$$e(n) = \hat{y}(n)|\hat{y}(n)|^{p-2}[R_p - |\hat{y}(n)|^p] \tag{12.3.22}$$

is the error signal. This is an LMS-type algorithm obtained by computing the gradient of (12.3.19) and dropping the expectation operator.

Other algorithms for blind equalization include (Ding 1998) the extensions of the Sato algorithm in Benveniste et al. (1980), the stop-and-go algorithms (Picchi and Prati 1987), and the Shalvi and Weinstein algorithms (Shalvi and Weinstein 1990).

12.3.3 Constant-Modulus Algorithm

The Godard algorithm for $p = 2$ was independently introduced by Treichler and Agee (1983) with the name *constant-modulus algorithm (CMA)* and used the property restoral approach. The resulting cost function

$$P(n) = E\{[R_2 - |\hat{y}(n)|^2]^2\} \tag{12.3.23}$$

depends on the amount of ISI plus noise at the output of the equalizer. Godard (1980) has shown that the coefficient values that minimize (12.3.23) are close to the values that minimize[†] the MSE $E\{[|a_n|^2 - |\hat{y}(n)|^2]^2\}$. The criterion is independent of the carrier phase because if we replace $\hat{y}(n)$ by $\hat{y}(n)e^{j\phi}$ in (12.3.23), then $P(n)$ remains unchanged. As a result, the adaptation of the CMA can take place independently of and simultaneously with

---

[†]More precisely, we wish to minimize $E\{[|a_{n-n_0}|^2 - |\hat{y}(n)|^2]^2\}$ for a particular choice of the delay n_0. As we have seen in Section 6.8, the value of n_0 has a critical effect on the performance of the equalizer.

operation of the carrier recovery system. The CMA is summarized in Table 12.2. Note that for 128-QAM, $R_2 = 110$. If we choose $R_2 \neq 110$, the CMA converges to a linearly scaled 128-QAM constellation that satisfies (12.3.23). However, choosing an unreasonable value for R_2 may cause problems when we switch to decision-directed mode (Gitlin et al. 1992).

TABLE 12.2

Summary of Godard or constant-modulus algorithm.

Operation	Equation
Equalizer	$\hat{y}(n) = \displaystyle\sum_{k=-L}^{L} c_k^*(n-1)x(n-k)$
Error	$e(n) = \hat{y}(n)[R_2 - |\hat{y}(n)|^2]$
Updating	$\mathbf{c}(n) = \mathbf{c}(n-1) + \mu\mathbf{x}(n)e^*(n)$
Godard constant	$R_2 \triangleq \dfrac{E\{|a(n)|^4\}}{E\{|a(n)|^2\}}$

Because of its practical success and its computational simplicity, the CMA is widely used in blind equalization and blind array signal processing systems.

The CMA in Table 12.2 performs a stochastic gradient minimization of the constant-modulus performance surface (12.3.1). In contrast to the unimodal MSE performance surface of trained equalizers, the constant-modulus performance surface of blind equalizers is multimodal. The multimodality of the error surface and the lack of a desired response signal have profound effects on the convergence properties of the CMA (Johnson et al. 1998). A detailed analysis of the local convergence of the CMA algorithm is provided in Ding et al. (1991).

1. *Initialization.* Since the CMA error surface is nonconvex, the algorithm may converge to undesirable minima, which indicates the importance of the initialization procedure. In practice, almost all blind equalizers are initialized using the *tap-centering* approach: All coefficients are set to zero except for the center (reference) coefficient, which is set larger than a certain constant.
2. *Convergence rate.* The trained LMS algorithm has a bounded convergence rate $(1 - 2\mu\lambda_{\max})^{-1} < \tau < (1 - 2\mu\lambda_{\min})^{-1}$, because the Hessian matrix (which determines the curvature) of the quadratic error surface is constant. Since the error surface of the constant-modulus criterion is multimodal and includes saddle points, the convergence rate of the CMA is slow at the neighborhood of saddle points and comparable to that of the trained LMS in the neighborhood of a local minimum.
3. *Excess MSE.* In the trained LMS algorithm, the excess MSE is determined by the step size, attainable MMSE, number of filter coefficients, and power of the input signal. In addition, the excess MSE of the CMA depends on the kurtosis of the source signal (Fijalkow et al. 1998).

EXAMPLE 12.3.1. To illustrate the key characteristics of the adaptive blind symbol or baud-spaced equalizer (BSE) using the CMA algorithm, we used BERGULATOR, a public-domain interactive MATLAB-5 program that allows experimentation with the constant-modulus criterion and various implementations of the CMA (Schniter 1998). The system function of the channel is $H(z) = 1 + 0.5z^{-1}$; the input is an IID sequence with four equispaced levels (PAM); the SNR = 50 dB; the equalizer has two coefficients c_0 and c_1; and the step size of the CMA is $\mu = 0.005$. Figure 12.8 shows contours of the constant-modulus criterion surface in the equalizer coefficient space, where the location of the MMSE is indicated by the asterisk * and the local

MSE locations by $\times$. Since the constant-modulus surface is multimodal, the equalizer converges at a different minimum depending on the initial starting point. This is illustrated by the two different coefficient trajectories shown in Figure 12.8, which demonstrates the importance of initialization in adaptive algorithms with nonquadratic cost functions. Figure 12.9 shows the learning curves for smoothed versions of the error and the square of the error for the trajectories in Figure 12.8.

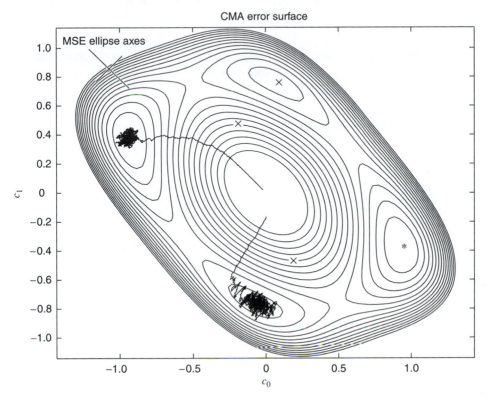

FIGURE 12.8
Contours of the constant-modulus cost function and coefficient trajectories for a blind BSE using the CMA.

12.4 FRACTIONALLY SPACED EQUALIZERS

The input to a *fractionally spaced equalizer (FSE)* (see Figure 12.10) is obtained by sampling the channel output at a rate faster than the symbol or baud rate $\mathcal{R}_B = 1/T_B$, where T_B is the symbol duration. For simplicity and because they are extensively used in practice, we focus on $T_B/2$ spaced FSE. However, all results can be extended to any rational fraction of T_B. One of the most attractive features of an FSE is that under ideal conditions, a finite impulse response (FIR) FSE can perfectly equalize an FIR channel (Johnson et al. 1998). Referring[†] to Figure 6.26 (a), we see that the continuous-time output of the channel is

$$\tilde{x}(t) = \sum_{k=-\infty}^{\infty} a_k \tilde{h}_r(t - kT_B - t_0) + \tilde{v}(t) \tag{12.4.1}$$

where $\tilde{h}_r(t)$ is the continuous-time impulse response and where we have incorporated the channel delay t_0 in $\tilde{h}_r(t)$. The discrete-time model of Figure 6.30 is no longer valid since

---

[†]The material in this section requires familiarity with the notation and concepts developed in Section 6.8.

Smoothed constant-modulus-error history

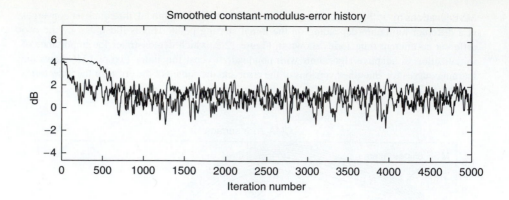

Squared-error history

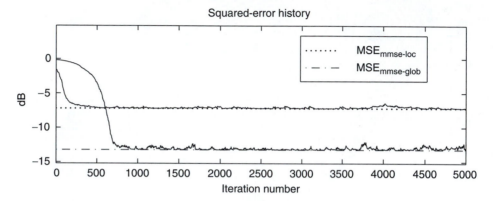

FIGURE 12.9
Learning curves for a blind BSE using the CMA for the two coefficient trajectories in Figure 12.8.

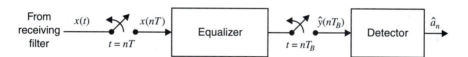

FIGURE 12.10
Block diagram of data communications receiver with a fractionally spaced equalizer.

$T = T_B/2$. However, if we extend the development leading to Figure 6.30 for $t = nT_B/2$, we obtain the discrete-time signal

$$x(n) = \sum_{k=0}^{\infty} a_k h_{\text{r}}(n - 2k) + v(n) \tag{12.4.2}$$

where $h_{\text{r}}(n)$ is the equivalent discrete-time impulse response and $v(n)$ is the equivalent white Gaussian noise (WGN). The output of an FIR $T_B/2$ spaced FSE is

$$y_{\text{f}}(n) = \sum_{k=0}^{2M-1} c_k x(n - k) \tag{12.4.3}$$

where we have chosen the even-order $2M$ for simplicity. If we decimate the output of the equalizer by retaining the *odd*-indexed samples $2n + 1$, we have

$$\hat{y}(n) \triangleq y_f(2n + 1) = \sum_{k=0}^{2M-1} c_k x(2n + 1 - k)$$

$$= \sum_{k=0}^{M-1} c_{2k} x(2n + 1 - 2k) + \sum_{k=0}^{M-1} c_{2k+1} x(2n - 2k)$$

or
$$\hat{y}(n) = \sum_{k=0}^{M-1} c_k^e x^o(n - k) + \sum_{k=0}^{M-1} c_k^o x^e(n - k) \tag{12.4.4}$$

where
$$c_k^e = c_{2k} \qquad c_k^o = c_{2k+1} \qquad x^e(n) = x(2n) \qquad x^o(n) = x(2n + 1) \tag{12.4.5}$$

are known as the *even* (e) and *odd* (o) parts of the equalizer impulse responses and the received sequences, respectively. Equation (12.4.4) expresses the decimated symbol rate output of the equalizer as the sum of two symbol rate convolutions involving the even and odd two-channel subequalizers.

If we define the even and odd symbol rate subchannels

$$h^e(n) = h_r(2n) \qquad \text{and} \qquad h^o(n) = h_r(2n + 1) \tag{12.4.6}$$

we can show that the combined impulse response $\tilde{h}(n)$ from the transmitted symbols a_n to the symbol rate output $\hat{y}(n)$ of the FSE is given by

$$\tilde{h}(n) = c_n^e * h^o(n) + c_n^o * h^e(n) \tag{12.4.7}$$

in the time domain or

$$\tilde{H}(z) = C^e(z)H^o(z) + C^o(z)H^e(z) \tag{12.4.8}$$

in the z domain. The resulting two-channel system model is illustrated in Figure 12.11.

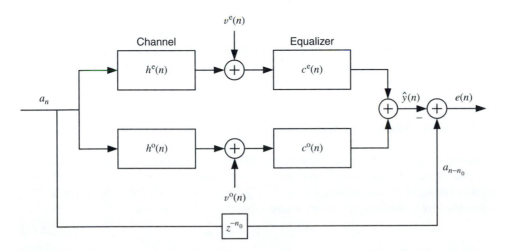

FIGURE 12.11
Two-channel representation of a $T_b/2$ spaced equalizer.

12.4.1 Zero-Forcing Fractionally Spaced Equalizers

If we define the $(M + L - 1) \times M$ even subchannel matrix (we assume a $2L$ FIR channel)

$$\mathbf{H}_e \triangleq \begin{bmatrix} h^e(0) & 0 & \cdots & 0 \\ h^e(1) & h^e(0) & \cdots & \vdots \\ \vdots & h^e(1) & \ddots & 0 \\ h^e(L-1) & \vdots & \ddots & h^e(0) \\ 0 & h^e(L-1) & \ddots & h^e(1) \\ \vdots & \vdots & \ddots & \vdots \\ 0 & \cdots & 0 & h^e(L-1) \end{bmatrix} \qquad (12.4.9)$$

the even subequalizer vector

$$\mathbf{c}_e \triangleq [c_0^e \; c_1^e \; \cdots \; c_{M-1}^e]^T \qquad (12.4.10)$$

and their counterparts $\mathbf{H}_o$ and $\mathbf{c}_o$, we can express the convolution equation (12.4.7) in matrix form as

$$\tilde{\mathbf{h}} = \mathbf{H}\mathbf{c} \qquad (12.4.11)$$

where

$$\mathbf{H} \triangleq [\mathbf{H}_e \; \mathbf{H}_o] \qquad \mathbf{c} \triangleq \begin{bmatrix} \mathbf{c}_e \\ \mathbf{c}_o \end{bmatrix} \qquad (12.4.12)$$

and $\tilde{\mathbf{h}} \triangleq [\tilde{h}(0) \; \tilde{h}(1) \; \cdots \; \tilde{h}(M + L - 1)]^T$ is the symbol-spaced overall system response. In the absence of noise, the system is free of ISI if $\tilde{\mathbf{h}}$ is equal to

$$\boldsymbol{\delta}_{n_0} \triangleq [0 \; \cdots \; 0 \; 1 \; 0 \; \cdots \; 0]^T \qquad (12.4.13)$$

where n_0, $0 \leq n_0 \leq M + L - 1$, indicates the location of the nonzero coefficient. Equivalently, the z domain zero-ISI condition from (12.4.8) is given by

$$z^{-n_0} = \tilde{H}(z) = C^e(z)H^o(z) + C^o(z)H^e(z) \qquad (12.4.14)$$

The zero-forcing FIR equalizer is specified by the system of linear equations $\mathbf{H}\mathbf{c} = \boldsymbol{\delta}_{n_0}$, which has a solution if $\mathbf{H}$ is full row rank. This condition is also known as *strong perfect equalization*, and it holds if the number of columns is equal to or larger than the number of rows, that is, if $2M \geq M + L - 1$ or $M \geq L - 1$. Furthermore, the $T_B/2$ spaced full-rank condition implies that the system functions $H_e(z)$ and $H_o(z)$ have no common roots. These topics are discussed in detail in Johnson et al. (1998).

The main advantage of the zero-forcing FSE over the corresponding synchronous equalizer is that, in the absence of noise, a zero-ISI elimination is possible using a finite-order FSE. In the case of the synchronous equalizer, a similar zero-ISI elimination is possible only when the equalizer is of infinite length.

12.4.2 MMSE Fractionally Spaced Equalizers

When the channel noise $v(n)$ is present, then perfect equalization, even for an FSE, is not possible. Hence, the emphasis shifts to the best possible compromise between ISI and noise amplification (which is present in a zero-forcing equalizer) in a minimum MSE sense. This is obtained by minimizing the mean square value of the data symbol error

$$e(n) \triangleq \hat{y}(n) - a_{n-n_0} \qquad (12.4.15)$$

for a particular choice of delay n_0. To obtain an expression for $\hat{y}(n)$ using the vector $\tilde{\mathbf{h}}$ in (12.4.11), we first define

$$\mathbf{a}_n \triangleq [a_n \; a_{n-1} \; \ldots \; a_{n-(M+L-1)}]^T \qquad (12.4.16)$$

and

$$\mathbf{v}(n) = [v(n-1)\, v(n-3) \,\cdots\, v(n-2L+1) \, v(n) \, v(n-2) \,\cdots\, v(n-2L+2)]^T$$
(12.4.17)

where the samples of the noise sequence are arranged as odd samples followed by the even samples so as to be consistent with the definitions of $\mathbf{H}$ and $\mathbf{c}$. We then substitute (12.4.2) into (12.4.4) and obtain

$$\hat{y}(n) = \mathbf{a}_n^T \mathbf{H}\mathbf{c} + \mathbf{v}^T(n)\mathbf{c}$$
(12.4.18)

Using $\boldsymbol{\delta}_{n_0}$ in (12.4.13), we see the desired symbol a_{n-n_0} is equal to $\mathbf{a}_n^T \boldsymbol{\delta}_{n_0}$. Hence from (12.4.15) and (12.4.18), the symbol error is

$$e(n) = \mathbf{a}_n^T(\mathbf{H}\mathbf{c} - \boldsymbol{\delta}_{n_0}) + \mathbf{v}^T(n)\mathbf{c}$$
(12.4.19)

Assuming that the symbol sequence $\{a_n\}$ is IID with variance σ_a^2 and is uncorrelated with the noise sequence $v(n) \sim \mathrm{WN}(0, \sigma_v^2)$, the mean square value of the error $e(n)$ is given by

$$\mathrm{MSE}(\mathbf{c}, n_0) = E\{|e(n)|^2\} = \sigma_a^2(\mathbf{H}\mathbf{c} - \boldsymbol{\delta}_{n_0})^H(\mathbf{H}\mathbf{c} - \boldsymbol{\delta}_{n_0}) + \sigma_v^2 \mathbf{c}^H \mathbf{c}$$
(12.4.20)

which is a function of two minimizing parameters $\mathbf{c}$ and n_0. Following our development in Section 6.2 on linear MSE estimation, the equalizer coefficient vector that minimizes (12.4.20) is given by

$$\hat{\mathbf{c}} = \left(\mathbf{H}^H\mathbf{H} + \frac{\sigma_v^2}{\sigma_a^2}\mathbf{I}\right)^{-1} \mathbf{H}^H \boldsymbol{\delta}_{n_0}$$
(12.4.21)

which is the classical Wiener filter. Also compare (12.4.21) with the frequency-domain Wiener filter given in (6.8.29). The corresponding minimum MSE with respect to $\hat{\mathbf{c}}$ is given by

$$\min_{\hat{\mathbf{c}}} \mathrm{MSE}\,(\mathbf{c}, n_0) = \mathrm{MSE}(n_0) = \boldsymbol{\delta}_{n_0}^T \left[\mathbf{I} - \mathbf{H}\left(\mathbf{H}^H\mathbf{H} + \frac{\sigma_v^2}{\sigma_a^2}\mathbf{I}\right)^{-1}\mathbf{H}^H\right]\boldsymbol{\delta}_{n_0}$$
(12.4.22)

Finally, the optimum value of n_0 is obtained by determining the index of the minimum diagonal element of the matrix in square brackets in (12.4.22), that is,

$$\hat{n}_0 = \arg\min_{n_0}\left\{\left[\mathbf{I} - \mathbf{H}\left(\mathbf{H}^H\mathbf{H} + \frac{\sigma_v^2}{\sigma_a^2}\mathbf{I}\right)^{-1}\mathbf{H}^H\right]_{n_0,n_0}\right\}$$
(12.4.23)

Once again, similar to the synchronous equalizer, the MMSE fractionally spaced equalizer is more robust to both the channel noise and the large amount of ISI. Additionally, it provides insensitivity to sampling phase and an ability to function as a matched filter in the presence of severe noise. Therefore, in practice, FSEs are preferred to synchronous equalizers.

12.4.3 Blind Fractionally Spaced Equalizers

Fractionally spaced equalizers have just about dominated practical equalization applications because they are insensitive to sampling phase, they can function as matched filters, they can compensate severe band-edge delay distortion, they provide reduced noise enhancement, and they can perfectly equalize an FIR channel under ideal conditions (Gitlin et al. 1992; Johnson et al. 1998).

The CMA for an FSE is given by

$$\hat{y}(n) = \sum_{k=0}^{M-1} c_k^{\mathrm{e}}(n-1)x^{\mathrm{o}}(n-k) + \sum_{k=0}^{M-1} c_k^{\mathrm{o}}(n-1)x^{\mathrm{e}}(n-k) \triangleq \mathbf{c}^T(n-1)\mathbf{x}(n)$$
(12.4.24)

$$e(n) = \hat{y}(n)[R_2 - |\hat{y}(n)|^2] \qquad (12.4.25)$$

$$\mathbf{c}(n) = \mathbf{c}(n-1) + \mu\mathbf{x}(n)e^*(n) \qquad (12.4.26)$$

where $\mathbf{c}(n-1)$ and $\mathbf{x}(n)$ are concatenated even and odd sample vectors. The blind FSE adaptive structure is shown in Figure 12.12. The value of R_2 depends on the input symbol constellation. This algorithm and its convergence are discussed in Johnson et al. (1998).

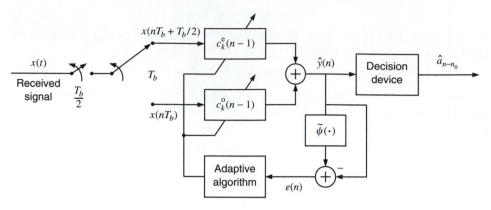

FIGURE 12.12
Basic elements of an FS adaptive blind equalization system.

EXAMPLE 12.4.1. To illustrate the superiority of the blind FSE over the blind BSE, we have used the BERGULATOR to simulate a 16-QAM data transmission system. The channel system function is $H(z) = 0.2 + 0.5z^{-1} + z^{-2} - 0.1z^{-3}$, the SNR = 20dB, and the equalizer has $M = 8$ coefficients. Figure 12.13 shows the constellation of the received signal at the input of the equalizer, where it is clear that the combined effect of ISI and noise makes detection extremely

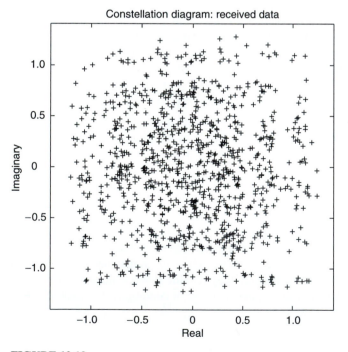

FIGURE 12.13
Constellation of the received signal symbols at the input of the equalizer.

difficult, if not impossible. Figures 12.14 and 12.15 show the symbol constellations at the output of a BSE and an FSE, respectively. We can easily see that the FSE is able to significantly remove ISI. Figure 12.16 shows the learning curves for the blind adaptive FSE using the CMA.

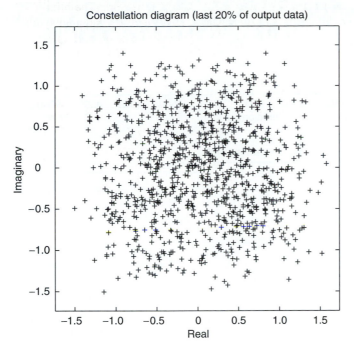

FIGURE 12.14
Constellation of the equalized signal symbols at the output of the
BSE equalizer.

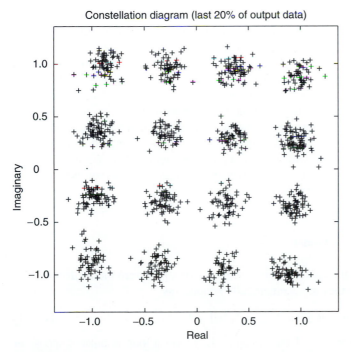

FIGURE 12.15
Constellation of the equalized signal symbols at the output of the FSE.

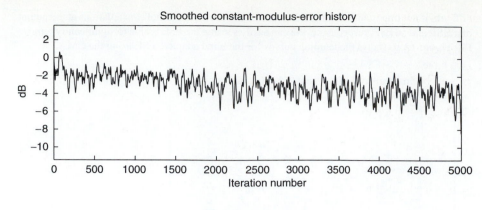

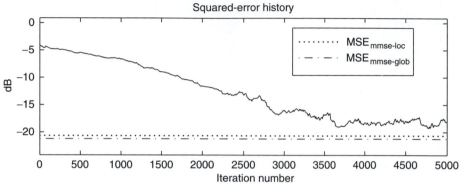

FIGURE 12.16
Learning curves for the blind FSE adaptive equalizer using the CMA.

12.5 FRACTIONAL POLE-ZERO SIGNAL MODELS

In this section we show how to obtain models with hyperbolically decaying autocorrelation, and hence long memory, by introducing fractional poles at zero frequency (fractional pole models) or nonzero frequency (harmonic fractional pole models). Cascading fractional with rational models results in mixed-memory models, known as *fractional pole-zero models*. We explore the properties of both types of models and introduce techniques for their practical implementation. Special emphasis is placed on the generation of discrete fractional pole noise, which is the random process generated by exciting a fractional pole model with white Gaussian noise. We conclude with a brief introduction to pole-zero and fractional pole models with SαS IID inputs, which result in processes with high variability and short or long memory, respectively. Fractional models are widely used in areas such as hydrology, data network traffic analysis, heart rate analysis, and economics.

12.5.1 Fractional Unit-Pole Model

The impulse response and the autocorrelation sequence of a pole-zero model decay exponentially with time, that is, they are geometrically bounded as

$$|h(n)| \leq C_h \, \zeta^{-n} \qquad |\rho(l)| \leq C_\rho \, \zeta^{-l} \tag{12.5.1}$$

where $C_h, C_\rho > 0$ and $0 < \zeta < 1$ (see Chapter 4). To get a long impulse response or a long autocorrelation, at least one of the poles should move very close to the unit circle. However, in many applications we need models whose autocorrelation decays more slowly than ζ^{-l} as $l \to \infty$, that is, models with long memory (see Section 3.2.4). In this section, we introduce a class of models, known as fractional pole models, whose autocorrelation asymptotically exhibits a geometric decay.

We have seen in Chapter 4 that by restricting some "integral" poles to being on the unit circle, we obtain models that are useful in modeling some types of nonstationary behavior. The *fractional pole model* FP(d) was introduced in Granger and Joyeux (1980) and Hosking (1981), and is defined by

$$H_d(z) = \sum_{k=0}^{\infty} h_d(k)z^{-k} \triangleq \frac{1}{(1 - z^{-1})^d} \tag{12.5.2}$$

where d is a nonintegral, that is, a fractional parameter. See Figure 12.17.

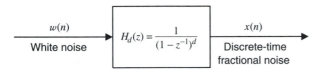

FIGURE 12.17
Block diagram representation of the discrete-time fractional noise model.

The characteristics of the model depend on the value of parameter d. Since d is not an integer, $H_d(z)$ is not a rational function. It is the nonrationality that gives this model its long-memory properties. Although we can approximate a fractional model by a PZ(P, Q) model, the orders P and Q that are needed to obtain a good approximation can be very large. This makes the estimation of pole-zero model parameters very difficult, and in practice it is better to use an FP(d) model.

Impulse response. To obtain the impulse response of the fractional pole model, we expand the system function $H_d(z) = (1 - z^{-1})^{-d}$ in a power series using the binomial series expansion. This gives

$$H_d(z) = \frac{1}{(1 - z^{-1})^d} = 1 + dz^{-1} + \frac{d(d+1)}{2!}z^{-2} + \cdots \tag{12.5.3}$$

The impulse response is given by

$$h_d(n) = \frac{d(d-1)\cdots(d+n-1)}{n!} = \frac{(d+n-1)!}{n!(d-1)!} = \frac{\Gamma(n+d)}{\Gamma(n+1)\Gamma(d)} \tag{12.5.4}$$

for $n \geq 0$ and $h_d(n) = 0$ for $n < 0$. $\Gamma(\cdot)$ is the gamma function defined as

$$\Gamma(\alpha) \triangleq \begin{cases} \displaystyle\int_0^{\infty} t^{\alpha-1}e^{-t}\, dt & \alpha > 0 \\ \infty & \alpha = 0 \\ \alpha^{-1}\,\Gamma(1+\alpha) & \alpha < 0 \end{cases} \tag{12.5.5}$$

with $\Gamma(\alpha + 1) = \alpha\Gamma(\alpha)$ for any α and $\Gamma(n + 1) = n!$ for n an integer. Note that $h_d(n)$ can be easily computed by using the recursion

$$h_d(n) = \frac{d+n-1}{n} h_d(n-1) \qquad n = 1, 2, \ldots \tag{12.5.6}$$

with $h_d(0) = 1$.

The system function of the inverse model is

$$H_I(z) \triangleq \sum_{n=0}^{\infty} h_I(n)z^{-n} = \frac{1}{H_d(z)} = (1 - z^{-1})^d \qquad (12.5.7)$$

Hence

$$h_I(n) = \frac{(-d + n - 1)!}{n!(-d - 1)!} = \frac{\Gamma(n - d)}{\Gamma(n + 1)\Gamma(-d)} = h_{-d}(n) \qquad (12.5.8)$$

As expected, $h_I(n)$ is obtained from $h(n)$ by simply replacing d by $-d$.

Minimum-phase. To understand the behavior of the model, we look at the impulse response as $n \to \infty$. Using Sterling's approximation (Abramowitz and Stegun 1970)

$$\frac{(n + d - 1)!}{n!} \sim n^{d-1} \qquad \text{as } n \to \infty \qquad (12.5.9)$$

we have

$$h(n) \sim \frac{1}{(d - 1)!} n^{d-1} \qquad \text{as } n \to \infty \qquad (12.5.10)$$

As a result of this geometric decay, the sum $\sum_{n=0}^{\infty} |h(n)|$ does not exist for $d > 0$. Therefore, the system is not BIBO stable. However, if $d < \frac{1}{2}$, the sum $\sum_{n=0}^{\infty} h^2(n) < \infty$, and the input $w(n)$ has finite variance, then the output of the system

$$x(n) = \sum_{k=0}^{\infty} h_d(k)w(n - k) \qquad (12.5.11)$$

exists in the mean square sense. In a similar way, the output of the inverse system exists in mean square if $d > -\frac{1}{2}$. In view of this mean square convergence, we say that the fractional pole model is minimum-phase if $-\frac{1}{2} < d < \frac{1}{2}$, even if $h_d(n)$ does not converge absolutely.

Spectrum. The complex power spectrum of the model is $R_x(z) = \sigma_w^2 R_h(z)$, where

$$R_h(z) = H(z)H(z^{-1}) = \frac{1}{(1 - z^{-1})^d(1 - z)^d} \qquad (12.5.12)$$

For $z = e^{j\omega}$ we obtain the power spectrum

$$R_h(e^{j\omega}) = \frac{1}{[2 \sin(\omega/2)]^{2d}} \qquad -\pi < \omega \le \pi \qquad (12.5.13)$$

We see that $R_h(0) = \sum_{l=-\infty}^{\infty} r(l)$ is finite only if $d \le 0$. Also as the frequency $\omega \to 0$, the power spectrum becomes

$$R_h(e^{j\omega}) \sim \frac{1}{\omega^{2d}} \qquad \text{as } \omega \to 0 \qquad (12.5.14)$$

because $\sin\theta \simeq \theta$ as $\theta \to 0$.

Autocorrelation. The autocorrelation $r_x(l) = \sigma_w^2 r_h(l)$ of the model can be found by using the inverse Fourier transform of $R_h(e^{j\omega})$, that is,

$$r_h(l) = \frac{1}{2\pi} \int_{-\pi}^{\pi} R_h(e^{j\omega}) e^{-j\omega l} d\omega = \frac{1}{2\pi} \int_0^{\pi} (\cos \omega l) \left(2 \sin\frac{\omega}{2}\right)^{-2d} d\omega \qquad (12.5.15)$$

Using the identity (Gradshteyn and Ryzhik 1994)

$$\int_0^{\pi} \cos ax \, \sin^{\nu-1} x \, dx = \frac{\pi \cos(a\pi/2)\Gamma(\nu + 1)2^{1-\nu}}{\nu\Gamma[(\nu + a + 1)/2)]\Gamma[(\nu - a + 1)/2)]}$$

we obtain

$$r_h(l) = \frac{(-1)^l \, \Gamma(1 - 2d)}{\Gamma(1 + l - d) \, \Gamma(1 - l - d)} \qquad l = 0, 1, 2, \ldots \qquad (12.5.16)$$

for the autocorrelation and

$$\rho_h(l) = \frac{r_h(l)}{r_h(0)} = \frac{\Gamma(1-d)\Gamma(l+d)}{\Gamma(d)\Gamma(l+1-d)} = \frac{(d+l-1)!}{(d-1)!\,(l-d)!} \qquad (12.5.17)$$

for the normalized autocorrelation. Using Sterling's formula, we obtain the following asymptotic approximation

$$\rho_h(l) \sim C_d l^{2d-1} \qquad \text{as } l \to \infty \qquad (12.5.18)$$

which again verifies the long memory of the model. From (12.5.16) and the definition of power spectrum, we have

$$r_h(0) = \sum_{n=0}^{\infty} h^2(n) = \frac{1}{2\pi} \int_{-\pi}^{\pi} |H(e^{j\omega})|^2 \, d\omega = \frac{\Gamma(1-2d)}{\Gamma^2(1-d)} \qquad (12.5.19)$$

Thus, for $d < \frac{1}{2}$ we have $\int_{-\pi}^{\pi} |H(e^{j\omega})|^2 d\omega < \infty$. Hence, the inverse transform $h(n)$ converges in mean square.

Partial autocorrelation. To determine the partial autocorrelation sequence, we can show, using (12.5.17) and the algorithm of Levinson-Durbin, that the AP(m) model parameters are given by

$$a_k^{(m)} = \binom{m}{k} \frac{(k-d-1)!(m-d-k)!}{(-d-1)!(m-d)!} \qquad (12.5.20)$$

Therefore, since $k_m = -a_m^{(m)}$, we have

$$k_m = \frac{d}{m-d} \qquad m = 1, 2, 3, \dots \qquad (12.5.21)$$

The details of the derivation are the subject of Problem 12.6.

Model memory. From Equations (12.5.10), (12.5.18), and (12.5.14) and from the long-memory definitions in Section 3.4, we conclude that the minimum-phase fractional pole model has long memory. More specifically, we arrive at the following conclusions:

- *Long memory.* For $0 < d < \frac{1}{2}$ the autocorrelation and partial correlation sequences decay monotonically and hyperbolically to zero. Although $\sum_{l=-\infty}^{\infty} |\rho(l)| = \infty$ and $R(e^{j\omega}) \to \infty$ as $\omega \to 0$, the integral (12.5.19) of $R(e^{j\omega})$ is finite. The spectrum is dominated by low-frequency components (low-pass), and the divergence at $\omega = 0$ causes the long-memory behavior. The system acts as a *fractional integrator*.
- *Short memory.* For $-\frac{1}{2} < d < 0$ the autocorrelation and partial autocorrelation sequences decay monotonically and hyperbolically to zero. In this case $\sum_{l=-\infty}^{\infty} |\rho(l)| < \infty$, $R(e^{j0}) = \sum_{l=-\infty}^{\infty} \rho(l) = 0$, and the spectrum is dominated by high-frequency components (high-pass). Sometimes we say that this model exhibits short-memory behavior. The system acts as a *fractional differentiator*.

Figures 12.18 and 12.19 show the impulse response, autocorrelation, partial autocorrelation, and power spectrum of the FP(d) model for various values of d. The short-memory and long-memory behavior of the model, as a function of parameter d, is clearly evident.

Discrete-time fractional pole noise. If we drive an FP(d) model with white Gaussian noise (see Figure 12.20), the resulting process is known as *discrete-time fractional Gaussian noise (DTFGN)*. Since the impulse response of an FP(d) system decays hyperbolically, its system function cannot be accurately approximated by a rational function. Hence, its practical implementation is not straightforward. Short sequences can be generated using the LDL^H or Cholesky decompositions of the process correlation using the (12.5.16) matrix, as explained in Section 3.5. This approach guarantees that the correlation of the generated

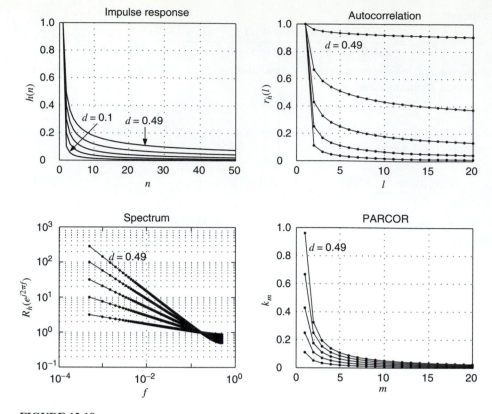

FIGURE 12.18

Impulse response, autocorrelation, partial autocorrelation, and power spectrum of the FP(d)
model for $d = 0.1, 0.2, 0.3, 0.4, 0.49$.

sequence matches the theoretical autocorrelation. Since the correlation matrix is Toeplitz, its
triangular factors can be computed efficiently by using the Schür algorithm (see Section 7.7).
Careful inspection of Figure 12.19 shows that the impulse response of the inverse system
decays extremely rapidly. Therefore, we can obtain a very accurate recursive implementation
of the FP(d) system by following the approach discussed in Example 4.5.1.

A practical algorithm for the generation of DTFGN is derived in Hosking (1984) using
the following result: For any stationary process with zero mean value, the conditional mean
and variance of $x(n)$ given $\{x(j)\}_0^{n-1}$ are given by

$$\mu_x(n) = E\{x(n)|x(n-1), \ldots, x(0)\} = -\sum_{j=1}^{n} a_j^{(n)*} x(n-j) \triangleq -\mathbf{a}_n^H \mathbf{x}_n \qquad (12.5.22)$$

and

$$v_x(n) = \text{Var}\{x(n)|x(n-1), \ldots, x(0)\} = \sigma_x^2 \prod_{j=1}^{n} (1 - |k_j|^2) \qquad (12.5.23)$$

where $\mathbf{a}_n$ is the forward linear predictor (FLP) with lattice parameters k_j and $\sigma_x^2 = E\{|x(n)|^2\}$ (Ramsey 1974). This result implies that we can use the Levinson-Durbin al-
gorithm to recursively determine $\mu_x(n)$ and $v_x(n)$, starting at $n = 0$ and generating
$x(n) \sim \text{WGN}[\mu_x(n), v_x(n)]$ at each step. For the FP(d) model this algorithm is simplified
because $k_m = d/(m - d)$ is known. The algorithm is initialized with $x(0) \sim \text{WGN}(0, \sigma_x^2)$
and continues with repeating the following recursions

$$k_n = \frac{d}{n - d} \qquad (12.5.24)$$

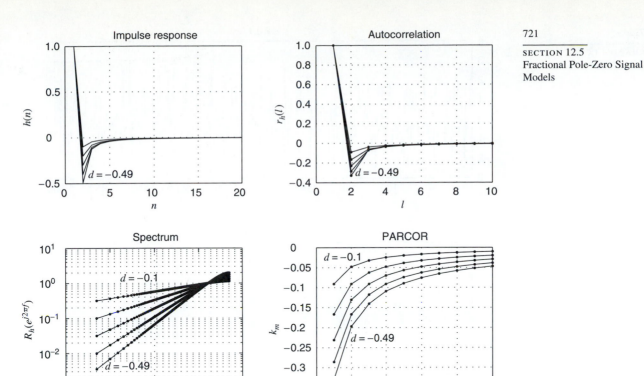

FIGURE 12.19
Impulse response, autocorrelation, partial autocorrelation, and power spectrum of the FP(d) model
for $d = -0.1, -0.2, -0.3, -0.4, -0.49$.

$$\mathbf{a}_{n+1} = \begin{bmatrix} \mathbf{a}_n \\ 0 \end{bmatrix} + \begin{bmatrix} \mathbf{Ja}_n^* \\ 1 \end{bmatrix} k_n \tag{12.5.25}$$

$$\mu_x(n+1) = \mathbf{a}_{n+1}^H \mathbf{x}_{n+1} \tag{12.5.26}$$

$$v_x(n+1) = v_x(n)(1 - |k_n|^2) \tag{12.5.27}$$

$$x(n+1) \sim \text{WGN}[\mu_x(n+1), v_x(n+1)] \tag{12.5.28}$$

for $n = 1, 2, \ldots, N$. The algorithm is implemented by the function x = dtfgn(d,N). Figure 12.20 shows sample realizations of discrete fractional noise, for various values of d, generated by using the above algorithm. A simplified, numerically robust algorithm, using the lattice structure, is introduced in Problem 12.9. The estimation of long memory is discussed in Section 12.6.

12.5.2 Fractional Pole-Zero Models: FPZ(P, d, Q)

Since the behavior of the FP(d) model is controlled by the single parameter d, it is not flexible enough to model the wide variety of short-term (small-lag) autocorrelation structures encountered in practical applications. A more powerful model capable of modeling both short-term and long-term correlation structures can be obtained by cascading a PZ(P, Q) model (to handle short memory) with an FP(d) (to handle long memory). This can be viewed

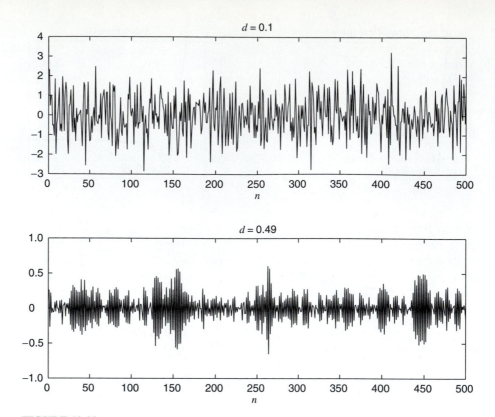

FIGURE 12.20
Sample realizations of discrete-time fractional Gaussian noise for two different values of d.

as filtering discrete-time fractional noise with a pole-zero filter. The resulting model is known as the *fractional pole-zero model* and is denoted by FPZ(P, d, Q). The system function is

$$H_{\text{fpz}}(z) = \frac{1}{(1 - z^{-1})^d} \frac{D(z)}{A(z)} \tag{12.5.29}$$

The FPZ(P, d, Q) is minimum-phase if $-\frac{1}{2} < d < \frac{1}{2}$ and PZ(P, Q) is minimum-phase.

With regard to the long-range behavior of the model, we can show that as $l \to \infty$,

$$\rho(l) \sim C_\rho l^{2d-1} \tag{12.5.30}$$

where $C_\rho \neq 0$, and as $\omega \to 0$

$$R(e^{j\omega}) = \frac{1}{|1 - e^{-j\omega}|^{2d}} \frac{|D(e^{j\omega})|^2}{|A(e^{j\omega})|^2} \sim \frac{|D(0)|^2}{|A(0)|^2} \frac{1}{\omega^{2d}} \tag{12.5.31}$$

Parameter d controls the impulse response and the autocorrelation of the model at large lags and the spectrum at low frequencies. Parameters a_k and d_k control the impulse response and the autocorrelation of the model at small lags, and the spectrum at high frequencies.

Autoregressive fractionally integrated moving-average models. Fractional pole-zero models driven by white noise [*autoregressive fractionally integrated moving-average models (ARFIMA) models*] generate random signals whose samples are significantly dependent even if they are too far apart. In practice (e.g., geophysics, hydrology, economics) there are many time series in which the dependence between samples that are too far away, though small, is still too significant to be ignored. Such signals with *long-term persistence* can be effectively modeled using ARFIMA models, because of their flexibility in dealing with both short-term and long-term correlation structures. An alternative family of random fractal models for modeling long memory behavior is discussed in the next section.

Harmonic fractional pole-zero models. The FP(d) models with $0 < d < \frac{1}{2}$ exhibit long memory, but their spectrum peaks at zero frequency and their autocorrelation does not have any periodicity. We next discuss a class of *harmonic* models with long memory, periodic autocorrelations, and power spectra that resonate at any frequency in the interval $0 \le \omega \le \pi$. Such models are more appropriate for the modeling of data with strong periodicities because they exhibit long memory and pseudoperiodic behavior.

Let $e^{\pm j\theta}$ be a pair of complex conjugate poles on the unit circle and at angles $\pm\theta$ from the real axis. Then we have $(1 - e^{j\theta}z^{-1})(1 - e^{-j\theta}z^{-1}) = 1 - (2\cos\theta)z^{-1} + z^{-2}$. The *harmonic fractional pole model*, denoted by HFP(d, θ), is a causal system defined by

$$H_{\theta,d}(z) = \frac{1}{(1 - 2z^{-1}\cos\theta + z^{-2})^d} = \sum_{n=0}^{\infty} h_{\theta,d}(n)z^{-n} \qquad (12.5.32)$$

where d is a fractional parameter and θ is an angle controlling the location of the peak of the spectrum. For $\theta = 0$, Equation (12.5.32) reduces to a standard FP($2d$) model. The properties of this model are discussed in Problem 12.10. The minimum-phase HFP(d, θ) model can be cascaded with a minimum-phase PZ(P, Q) model to obtain an HFPZ(P, d, Q, θ) model that offers greater flexibility in controlling both the short-term and long-term correlation structure.

12.5.3 Symmetric α-Stable Fractional Pole-Zero Processes

Up to this point we have studied linear signal models driven by a sequence of IID Gaussian or non-Gaussian random variables with *finite* variance. However, many practical time series including isolated sharp spikes or bursts of spikes can be better described by random signal models with *infinite* variance. To ensure that some signal samples take large values with high probability, we need a probability density function with fat or heavy tails. We focus on the family of SαS random variables because of their heavy tails and the fact that they are invariant under linear transformations.

As we have seen in Chapters 4 and 5, the linear process

$$x(n) = \sum_{k=0}^{\infty} h(k)\, w(n - k) \qquad (12.5.33)$$

is strictly stationary if (1) $w(n) \sim \text{IID}(0, \sigma_w^2)$ with $\sigma_w^2 < \infty$ (finite variance) and (2) $\sum_{k=-\infty}^{\infty} |h(k)| < \infty$, that is, the system is BIBO stable. However, to ensure stationarity when the input is SαS with $\sigma_w = \infty$ (power law tails), the sequence $|h(k)|$ should decay exponentially. Since the impulse response of a stable pole-zero system decays exponentially, its response to an SαS IID sequence is strictly stationary and SαS stable.

So far, we have discussed the properties of fractional pole-zero models and their response to white noise with finite variance. The following proposition specifies under what conditions the output of a PZ($0, d, 0$) model with stable excitation is defined.

THEOREM 12.4. Consider the following fractional pole model FP(d)

$$x(n) = \sum_{k=0}^{\infty} h(k)w(n - k) \quad \text{with} \quad h(k) = \frac{(d + k - 1)!}{k!(d - 1)!} \qquad (12.5.34)$$

where $w(n)$ is IID and SαS. A necessary condition for the series (12.5.34) to converge is

$$-\infty < d < 1 - \frac{1}{\alpha} \qquad (12.5.35)$$

When (12.5.35) holds, the series converges in the following sense:

1. $0 < \alpha \le 1$: absolutely almost surely
2. $1 < \alpha \le 2$: absolutely almost surely if $d \le 0$ and absolutely surely if $d > 0$ and $\mu = 0$

Proof. See Samorodnitsky and Taqqu (1994).

We note that because both $h(n)$ and the tails of the input distribution decay as a power law, the stability of the model depends on α. Recall that no dependence on the input signal exists if the input signal has finite variance, because for $E\{w^2(n)\} < \infty$ the stability requirement is $\sum_{k=-\infty}^{\infty} |h(k)|^2 < \infty$.

The output of the inverse model $g(n) = h(n)|_{d \leftarrow -d}$ is defined for $-\infty < -d < 1 - 1/\alpha$ or $-(1 - 1/\alpha) < d < \infty$; hence, the model is minimum-phase if

$$-\left(1 - \frac{1}{\alpha}\right) < d < 1 - \frac{1}{\alpha} \tag{12.5.36}$$

The stability and minimum-phase regions for the FP(d) model with SαS IID excitations are shown in Figure 12.21. Theorem 12.4 applies for the model FPZ(P, d, Q) assuming it is stable, because it behaves asymptotically as the PZ(0, d, 0) model.

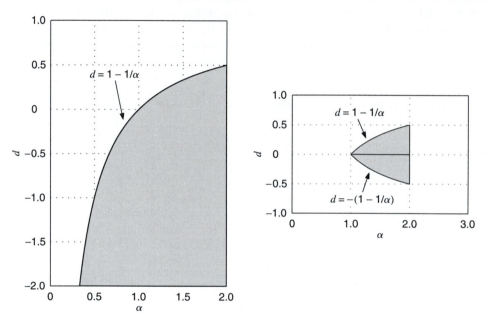

FIGURE 12.21
Stability (left) and minimum-phase (right) regions for a fractional pole model driven by an SαS IID sequence.

Although a linear stable process is strictly stationary, it is not second-order stationary because $E\{|x(n)|^2\} = \infty$. Therefore, the autocorrelation and the PSD of the process $x(n)$ do not exist. However, we can use the normalized autocorrelation of the signal model (12.5.33)

$$\rho(l) = \frac{\displaystyle\sum_{n=-\infty}^{\infty} h(n)h(n-l)}{\displaystyle\sum_{n=-\infty}^{\infty} h^2(n)} \tag{12.5.37}$$

and its Fourier transform to characterize the linear stable process $x(n)$. Clearly, this is a legitimate characterization for processes with finite variance and provides a reasonable characterization for stable linear processes because of the IID nature of the excitation $w(n)$. We can estimate $\rho(l)$ from a set of data $\{x(n)\}_0^{N-1}$ using the consistent estimator (Brockwell

$$\hat{\rho}(l) = \frac{\sum_{n=0}^{N-1+|l|} x(n)x(n-l)}{\sum_{n=0}^{N-1} x^2(n)} \tag{12.5.38}$$

12.6 SELF-SIMILAR RANDOM SIGNAL MODELS

In this section, we introduce the family of statistically self-similar or random fractal models, which are based on self-similar stochastic processes. Any segment of a self-similar process looks similar, in a statistical sense, to a scaled version of a larger segment of the process. Because of their practical importance, we focus on self-similar processes with stationary increments. We show that the stationary-increments requirement leads to processes whose autocorrelation sequences decay hyperbolically, that is, to models with long memory. We mainly focus on the fractional Brownian motion (nonstationary) and the fractional Gaussian noise (stationary) models, as well as their properties, simulation, and applications. However, we provide a brief introduction to self-similar processes with SαS increments, which result in random signal models with long memory and high variability.

12.6.1 Self-Similar Stochastic Processes

Each time a geologist takes a photograph of a geological object, say, a fossil, she or he includes in the picture an object with *known scale* (e.g., a coin or a ruler), because without the scale, it is impossible to determine whether the photograph covers 10 cm or 10 m. For this reason we say that geological phenomena are *scale-invariant*, or that they do not have a *characteristic scale*.

If we can reproduce an object by magnifying some portion of it, we say that the object is *scale-invariant*, or *self-similar*. Thus, self-similarity is invariance with respect to scaling. Such self-similar geometric objects are known as *fractals* (Mandelbrot 1982).

A signal $x(t)$ is self-similar if[†] $x(ct) = c^H x(t)$. It can be easily seen that a signal described by a power law $x(t) = \alpha t^{\beta}$ is self-similar. However, such signals are of limited interest. A more interesting and useful type of signal is that exhibiting a weaker, that is, statistical, version of self-similarity. A random signal is called *(statistically) self-similar* if its statistical properties are scale-invariant, meaning that its statistics do not change under magnification or reduction. Self-similar random signals are also known as *random fractals*.

Statistical self-similarity means that small fluctuations at small scales become larger fluctuations at larger scales. Therefore, as we analyze more and more data, these ever-larger fluctuations increase the value of the measured variance, which in the limit becomes infinite. This increase of variance with the length of the data has been observed in the analysis of various practical time series that exhibit self-similar behavior. Figure 12.22 provides a visual illustration of the self-similar behavior of the variable-rate video traffic time series (Garrett and Willinger 1994).

These ideas can be formalized within the context of the theory of stochastic processes by using the following definition.

[†] The superscript H is an index and not a conjugate transposition operator. For lack of better notation, we will continue to use the accepted notation.

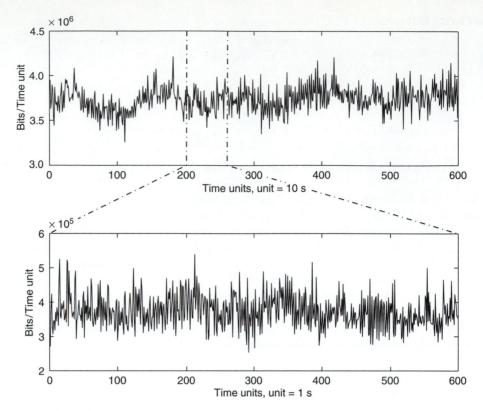

FIGURE 12.22

Pictorial illustration of self-similarity for the variable-bit-rate video traffic time series. The bottom series is obtained from the top series by expanding the segment between the two vertical lines. Although the two series have lengths of 600 and 60s, they are remarkably similar visually and statistically. (*Courtesy of M. Garrett and M. Vetterli.*)

DEFINITION 12.1. A continuous-time stochastic process $x(t)$ is said to be (statistically) *self-similar* with (self-similarity) index[†] H (H-ss) if and only if, for any scaling parameter $c > 0$, the processes $x(ct)$ and $c^H x(t)$ are statistically equivalent, that is, they have the same finite-dimensional distributions. Symbolically

$$x(ct) \overset{d}{=} c^H x(t) \tag{12.6.1}$$

where the symbol $\overset{d}{=}$ denotes equality in distribution and, more specifically, equality of finite-dimensional joint probability distributions.

It should be emphasized that individual realizations of the process are not necessarily deterministically scale-invariant. The above definition of self-similarity has several implications, which can be summarized as follows:

- A change in the time scale is *statistically* equivalent to a change in the amplitude scale. Hence, the statistic of $x(t)$ is invariant under the transformation

$$x(t) \rightarrow c^{-H} x(ct) \tag{12.6.2}$$

- To obtain statistically equivalent processes, the time axis must be scaled *differently* from the amplitude axis. In the language of fractals, we say that the graphs $\{t, x(t)\}$ and $\{t, c^{-H} x(ct)\}$, $0 \le t < \infty$, are *statistically self-affine* because the scaling factor is

---

[†] Also known as the *Hurst exponent*.

different for the time and amplitude axes. An example of such a self-similar process for $H = \frac{1}{2}$ is shown in Figure 12.23. This process, whose distribution at each t is Gaussian, is known as (ordinary) *Brownian motion*. (A detailed discussion of Brownian motion is given in the next section.) The time trace shown in the top plot in Figure 12.23 is generated as a discrete equivalent of $x(t)$, using 16,384 samples over unit time interval. When it is plotted as a continuous curve, we lose sight of its discrete nature and view it as a fractal curve that is indistinguishable from a continuous Brownian—a true fractal curve possessing self-similarity at all levels of magnification. Statistical self-affinity of $x(t)$ is evident as we zoom into it. The zooming area in the top plot is shown as a box, and the scaled curve is shown in the middle plot. Note that we scaled the middle one-fourth of the time axis while the amplitude axis was magnified by 2 since $4^H = 4^{1/2} = 2$. This retained the statistical similarity of the middle curve to the original one. Further scaling of time axis by 4 and the amplitude axis by 2 is shown in the bottom plot of Figure 12.23. Once again the resulting plot is statistically similar to the original one. This Brownian motion displayed at different levels of resolution demonstrates the concept of statistical self-affinity.

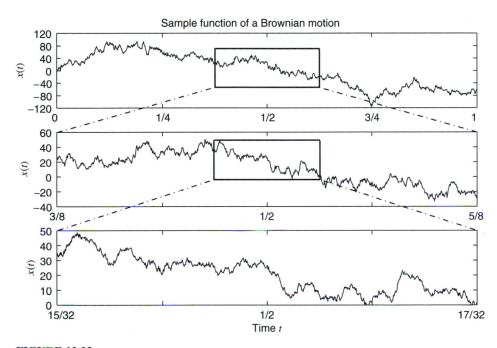

FIGURE 12.23
Statistical self-affine property of the Brownian motion trace.

- If we set $ct = 1$ in (12.6.2), we have

$$x(t) \stackrel{d}{=} t^H x(1) \qquad t > 0 \tag{12.6.3}$$

Therefore, self-similar processes cannot be stationary, except for $H = 0$. This nonstationarity of the Brownian motion trace $x(t)$ of Figure 12.23 is shown in Figure 12.24, which illustrates the spreading of signal values about the mean value of zero as time increases. For display purposes, 10 sample functions of $x(t)$ are shown, all of which begin at $x(0) = 0$. This spreading is in a statistical sense, in that some traces return to zero and some cases return even more than once. To determine this statistical spreading, 100 sample functions were used, and the sample standard deviation $\sigma_x(t)$ at each t was

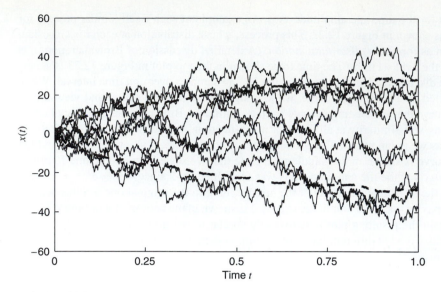

FIGURE 12.24
The diffusion property of the Brownian motion trace.

computed. This $\pm\sigma_x(t)$, shown as dashed lines in Figure 12.24, clearly indicates the diffusion (or nonstationarity) property of Brownian motion. Note that since the standard deviation is proportional to $E\{|x(t+\Delta) - x(t)|\}$, we have

$$\sigma_x(t) \propto t^H = t^{1/2} \tag{12.6.4}$$

for the Brownian motion, and the dashed line in Figure 12.24 confirms it.

- For *strict-sense* self-similar processes, all finite-dimensional distributions are equal. However, for *wide-sense* self-similar processes, only second-order moments are equal. From (12.6.2) these moments are given as

$$\mu_x(t) \triangleq E\{x(t)\} = c^{-H}\mu_x(ct) \tag{12.6.5}$$

$$r_x(t_1, t_2) \triangleq E\{x(t_1)x(t_2)\} = c^{-2H}r_x(ct_1, ct_2) \tag{12.6.6}$$

Clearly, for Gaussian processes the two types of self-similarity are equivalent.

Because of their practical importance, we focus on self-similar stochastic processes that have stationary increments.

DEFINITION 12.2. A real-valued process $x(t)$ has *stationary increments* if

$$x(t+\tau) - x(\tau) \overset{d}{=} x(t) - x(0) \qquad \text{for all } \tau \tag{12.6.7}$$

In practical applications, the nature of processes with stationary increments is analyzed using a quantity known as the *semivariogram*, defined by

$$v_x(\tau) \triangleq \frac{1}{2}E\{[x(t+\tau) - x(t)]^2\} \tag{12.6.8}$$

which, for stationary processes, reduces to

$$v_x(\tau) = 2[r_x(0) - r_x(\tau)] \tag{12.6.9}$$

We next turn our attention to self-similar processes with stationary increments

DEFINITION 12.3. A continuous-time stochastic process is *self-similar with stationary increments* (H-sssi) if and only if

- It is *self-similar* with index H.
- It has *stationary increments*.

As shown in the following theorem, the requirements for self-similarity and stationary increments completely specify the second-order moments of the underlying process $x(t)$.

THEOREM 12.5. The mean value, variance, and autocorrelation of an H-sssi process are given by, respectively

$$\mu_x(t) = 0 \tag{12.6.10}$$

$$\sigma_x^2(t) = t^{2H}\sigma_H^2 \tag{12.6.11}$$

$$r_x(t_1, t_2) = \frac{1}{2}\sigma_H^2(|t_1|^{2H} - |t_1 - t_2|^{2H} + |t_2|^{2H}) \tag{12.6.12}$$

where $\sigma_H^2 = E\{x^2(1)\}$.

Proof. From (12.6.2) we have, for $t = 0$,

$$x(0) \overset{d}{=} c^{-H}x(c0) = c^{-H}x(0) \Rightarrow x(0) = 0 \tag{12.6.13}$$

Also from (12.6.2) and (12.6.3), we conclude that

$$\mu_x(t) = E\{x(t)\} = E\{c^{-H}x(ct)\} = c^{-H}E\{x(ct)\} = t^H E\{x(1)\} \tag{12.6.14}$$

Using the stationary increment property (12.6.7), (12.6.13), and (12.6.14), we obtain

$$E\{x(t+\tau) - x(\tau)\} = E\{x(t) - x(0)\} = E\{x(t)\} = t^H E\{x(1)\} \tag{12.6.15}$$

Using the self-similarity definition, however, we have

$$E\{x(t+\tau) - x(\tau)\} = [(t+\tau)^H - \tau^H]E\{x(1)\} \tag{12.6.16}$$

Comparing (12.6.15) and (12.6.16), we conclude that $E\{x(1)\} = 0$; hence from (12.6.14)

$$\mu_x(t) = 0$$

which proves (12.6.10). Similarly, since $x(t) \overset{d}{=} t^H x(1)$, for $t > 0$, we have

$$\sigma_x^2(t) = E\{x^2(t)\} = t^{2H}E\{x^2(1)\} = t^{2H}\sigma_H^2$$

which proves (12.6.11). Finally, again using stationarity of the increments and (12.6.11), we obtain

$$E\{[x(t_1) - x(t_2)]^2\} = E\{[x(t_1 - t_2) - x(0)]^2\} = \sigma_H^2(t_1 - t_2)^{2H} \tag{12.6.17}$$

or

$$E\{[x(t_1) - x(t_2)]^2\} = E\{x^2(t_1)\} + E\{x^2(t_2)\} - 2E\{x(t_1)x(t_2)\}$$

$$= \sigma_H^2 t_1^{2H} + \sigma_H^2 t_2^{2H} - 2r_x(t_1, t_2) \tag{12.6.18}$$

where $r_x(t_1, t_2)$ is the autocorrelation function of $x(t)$. Combining the last two equations, we obtain

$$r_x(t_1, t_2) = \frac{1}{2}\sigma_H^2[t_1^{2H} - (t_1 - t_2)^{2H} + t_2^{2H}] \tag{12.6.19}$$

which completes the proof of the theorem.

Self-similar processes with stationary increments are well-defined if $H > 0$ and $x(0) = 0$ with probability 1 (Vervaat 1987). For $H = 1$, $x(t) = |t|x(1)$; that is, the realizations are lines through the origin, and the process is of no interest. For $H = 0$, we have $x(t) = 0$, which is a trivial process. For $H < 0$ the process is not mean square continuous, and for $H > 1$ the increments are nonstationary. The permissible range of H is determined by the existence of moments: If $x(t)$ is H-sssi with finite variance, then $0 < H \leq 1$ (Samorodnitsky and Taqqu 1994).

The autocorrelation (12.6.12) shows that H-sssi processes are nonstationary. Despite this nonstationarity, we can define a time-averaged spectrum. Since small scales correspond to large frequencies and large scales to small frequencies, the amplitude of the fluctuations is small at high frequencies and large at low frequencies. In light of the previous discussion, it should not come as a surprise that the power spectrum of self-similar processes follows a power law, that is, is proportional to $1/|F|^\beta$. Indeed, it has been shown (Flandrin 1989) that the time-averaged power spectrum of an H-sssi process is given by

$$R_x(F) = \frac{\sigma_H^2}{|F|^{2H+1}} \qquad (12.6.20)$$

where F is the frequency in cycles per unit of time. As we can easily see, $R_x(cF) = c^{-(2H+1)}R_x(F)$, which shows that the process is wide-sense self-similar.

12.6.2 Fractional Brownian Motion

If we restrict the probability distribution of an H-sssi process to being Gaussian, we obtain a unique process known as the *fractional Brownian motion*, abbreviated as FBM (Mandelbrot and Van Ness 1968). These FBMs have Hurst exponents in the range $0 < H < 1$. The (ordinary) Brownian motion of Figure 12.23 is a special case of FBM when $H = \frac{1}{2}$.

> **DEFINITION 12.4.** A Gaussian H-sssi process, $0 < H \leq 1$, is called *fractional Brownian motion (FBM)* and is denoted by $B_H(t)$.

There are several equivalent definitions of FBM process, which are summarized by the following theorem (Samorodnitsky and Taqqu 1994; Beran 1994).

> **THEOREM 12.6.** If $0 < H \leq 1$ and $\sigma_H^2 = E\{x^2(1)\}$, the following statements are equivalent
>
> 1. $B_H(t)$ is Gaussian and H-sssi.
> 2. $B_H(t)$ is fractional Brownian motion with self-similarity index H.
> 3. $B_H(t)$ is Gaussian and has mean zero for $H < 1$ and autocorrelation function
>
> $$r_{B_H}(t_1, t_2) = \tfrac{1}{2}\sigma_H^2(|t_1|^{2H} - |t_1 - t_2|^{2H} + |t_2|^{2H}) \qquad (12.6.21)$$

EXAMPLE 12.6.1. In Figure 12.25 we show time traces of FBMs for $H = 0.2, 0.5$, and 0.8. Clearly, in these traces there is a qualitative difference between each trace that is very noticeable. For a low value of $H = 0.2$, the trace shows more fractured or crinkled behavior. This behavior occurs for $0 < H < 0.5$, and the corresponding traces have tendencies to turn back upon themselves (negative correlation). The corresponding property is known as *antipersistence*. A

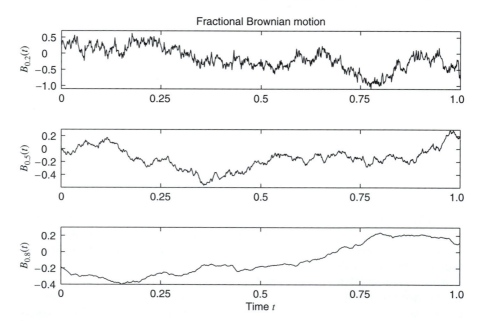

FIGURE 12.25
Time traces of fractional Brownian motion for $H = 0.2$, $H = 0.5$, and $H = 0.8$.

stock market fluctuation is a good example of this process. As H increases, the amount of crinkle reduces. For $H = 0.5$ we have the (ordinary) Brownian motion for which the correlation is zero, and the trace shows no preferred tendency to turn back or persist in the same direction (neutral in persistence). For a high value of $H = 0.8$, the trace is smoother, and in fact for $0.5 < H < 1$, the FBM traces show persistence in the direction in which they are moving (positive correlation). This property is known as *persistence*. Typical coastlines (boundaries between land and water) are good examples of such traces.

From the above example, we note that the fractal behavior of traces diminishes as H increases from 0 to 1. Hence there must be an inverse relationship between H and the fractal dimension D (also known as the Haussdorff dimension). The concept of dimension is closely related to the property of self-similarity or scaling. For the purpose of discussion, let us consider our natural Euclidean dimensions. In one dimension, a line segment possesses a scaling property. If it is subdivided into N identical line segments, then each segment is scaled down by the ratio $r = 1/N$ from the whole, or $Nr = 1$. A square is a two-dimensional plane that possesses the scaling property. If it is subdivided into N equal squares, then each square side is scaled down by a factor of $r = 1/\sqrt{N}$, or $Nr^2 = 1$. Carrying this analysis to the cube in three dimensions, we observe that if a cube is subdivided into N identical cubes, then each subcube side is scaled down by the factor $r = 1/\sqrt[3]{N}$, or $Nr^3 = 1$. Now we can generalize this analysis to an arbitrary noninteger dimension D. If a D-dimensional object is subdivided into N identical copies of itself, then the side of each copy is scaled down by the ratio $r = 1/\sqrt[D]{N}$, or $Nr^D = 1$. Thus we obtain

$$D = \frac{\log N}{\log(1/r)} \tag{12.6.22}$$

The above approach can also be used to determine the fractal dimension D of the FBM traces and to relate it to the Hurst exponent H. One interesting technique for determining the fractal dimension is known as *box counting*. The basic idea is to compute the total number N of enclosing boxes (or rectangles) needed to cover all identical subtraces that have been scaled down by the ratio r from the whole trace and then use formula (12.6.22) to estimate the fractal dimension. Refer to the top plot of Figure 12.23. The enclosing box shows that if the whole trace is divided into 4 identical subtraces, then the box height is scaled down by $(\frac{1}{4})^{1/2} = \frac{1}{2}$. Thus the area of each rectangular box is

$$\left(\frac{1}{4}\right)\left(\frac{1}{2}\right) = \frac{1}{4^{3/2}} = \frac{1}{4^{1+1/2}} \tag{12.6.23}$$

However, we have to relate the scaling of smaller (identical) *square* boxes to the original box since the original is a square box of unit side length (this implicitly assumes that the amplitude axis in Figure 12.23 is unity, which is not unreasonable since we are using fractions). The smaller square boxes of side length $r = \frac{1}{4}$ have area equal to $1/4^2$. Thus the number of square boxes required to cover each subinterval is (note the box counting)

$$\frac{1/4^{3/2}}{1/4^2} = \frac{1}{4^{1/2-1}} \tag{12.6.24}$$

Since there are 4 subintervals in Figure 12.23, the total number of square boxes required to cover the whole trace is

$$N = 4\left(\frac{1}{4^{1/2-1}}\right) = \frac{1}{4^{1/2-2}} \tag{12.6.25}$$

Hence substituting (12.6.25) into (12.6.22), and using $r = \frac{1}{4}$, we obtain

$$D = \frac{\log\left(1/4^{1/2-2}\right)}{\log\left(1/\frac{1}{4}\right)} = \frac{\log 4^{2-1/2}}{\log 4} = 2 - \frac{1}{2} = 1.5 \tag{12.6.26}$$

which is the fractal dimension of (ordinary) Brownian motion. Generalizing to $0 < H < 1$, we can show that (see Problem 12.12)

$$D = 2 - H \tag{12.6.27}$$

Thus the sample paths of fractional Brownian motion are fractal curves with Haussdorff dimension $D = 2 - H$ (Falconer 1990). Referring to Figure 12.25, we see the fractal dimensions of the fractional Brownian motions are $D = 1.8$ for $H = 0.2$ (antipersistent), $D = 1.5$ for $H = 0.5$ (Brownian motion), and $D = 0.8$ for $H = 1.2$ (persistent). Thus the more wiggly the trace, the higher the dimension.

Continuous-time fractional pole systems. In Section 12.5 we used a discrete-time fractional pole to obtain a system with long memory. When this system is driven by a WGN process, the result is a long-memory process called discrete-time FGN. This leads to the following question: Can we use a continuous-time fractional pole to obtain a long-memory system that could be used to generate a long-memory process in general and fractional Brownian motion in particular? The answer is yes, so now we provide an intuitive engineering explanation.

For any $d > 0$, we have the following Laplace transform pair (Abramowitz and Stegun 1970)

$$h_d(t) = \frac{1}{\Gamma(d)} t^{d-1} u(t) \quad \overset{\mathcal{L}}{\Longleftrightarrow} \quad H_d(s) = \frac{1}{s^d} \tag{12.6.28}$$

where $\Gamma(\cdot)$ is the gamma function. Note that for $d = 1$, $h_1(t)$ corresponds to an ideal integrator. However, for fractional d, the function $h_d(t)$ has a hyperbolic decay. The result is a system with long memory called the *fractional integrator*. These topics are the subject of a discipline known as fractional calculus (Oldham and Spanier 1974).

The output of the fractional integrator is provided by the convolution integral

$$x(t) = \frac{1}{\Gamma(d)} \int (t - \tau)^{d-1} u(t - \tau) w(\tau) \, d\tau \tag{12.6.29}$$

which satisfies the scaling property

$$y(t) = \frac{1}{\Gamma(d)} \int (t - \tau)^{d-1} w(c\tau) \, d\tau = \frac{c^{-d}}{\Gamma(d)} \int (ct - \lambda)^{d-1} w(\lambda) \, d\lambda = c^{-d} x(ct) \tag{12.6.30}$$

where $\lambda = c\tau$. Linear systems that satisfy (12.6.30) are said to be *linear, scale-invariant* systems (Wornell 1996). We emphasize that while linear, shift-invariant systems with rational system functions have memory that decays exponentially, linear, scale-invariant systems exhibit self-similarity and long (hyperbolically decaying) memory.

Intuition suggests that the output of scale-invariant systems, driven by white noise, should exhibit statistical self-similarity. Indeed, it can be shown that linear, scale-invariant systems can be used to generate fractional Brownian motion processes (Samorodnitsky and Taqqu 1994). More specifically, the fractional Brownian motion process can be generated by passing white noise through a linear, scale-invariant system

$$B_H(t) = \int_{-\infty}^{\infty} h_t(\tau) w(\tau) \, d\tau \tag{12.6.31}$$

with the following causal impulse response

$$h_t(\tau) = \frac{1}{C(H)} \{[(t - \tau)_+]^{H-1/2} - [(-\tau)_+]^{H-1/2}\} \tag{12.6.32}$$

where

$$C(H) = \left\{ \int_0^{\infty} [(1 + \tau)^{H-1/2} - \tau^{H-1/2}]^2 \, d\tau + \frac{1}{2H} \right\}^{1/2} \tag{12.6.33}$$

and

$$u_+ = \begin{cases} u & \text{if } u \geq 0 \\ 0 & \text{if } u < 0 \end{cases} \tag{12.6.34}$$

We note that the change from the impulse response (12.6.28) to (12.6.32) was introduced by Mandelbrot (1982) to ensure that $B_H(t)$ has the required properties (Wornell 1993; Kasdin 1995). An equivalent *harmonizable* representation of fractional Brownian motion in the frequency domain is also derived in Samorodnitsky and Taqqu (1994).

12.6.3 Fractional Gaussian Noise

The discrete fractional Gaussian noise is a stationary sequence obtained by periodically sampling the fractional Brownian motion process $B_H(t)$ and then computing the first difference. The resulting random sequence is $x(nT) \triangleq B_H(nT) - B_H(nT - T)$, where T is the sampling interval. Since the fractional Brownian motion process is statistically scale-invariant, we set $T = 1$. Therefore, the *discrete fractional Gaussian noise* process is defined by

$$x(n) \triangleq B_H(n) - B_H(n - 1) \tag{12.6.35}$$

and it is simply referred to as *FGN*.

We next determine the second-order moments, that is, the autocorrelation and PSD of the FGN process.

THEOREM 12.7. The autocorrelation sequence of the discrete fractional Gaussian noise is

$$r_x(l) = \tfrac{1}{2}\sigma_H^2 (|l - 1|^{2H} - 2|l|^{2H} + |l + 1|^{2H}) \tag{12.6.36}$$

Since the correlation depends only on the distance l between the samples, the process is wide-sense stationary.

Proof. Using (12.6.21) and (12.6.35), we can easily show that

$$E\{x(n)x(n - l)\} = E\{[B_H(n) - B_H(n - 1)][B_H(n - l) - B_H(n - l - 1)]\}$$

$$= \tfrac{1}{2}\sigma_H^2 [(l - 1)^{2H} - 2l^{2H} + (l + 1)^{2H}]$$

which leads to (12.6.36).

Figure 12.26 shows the autocorrelation sequence for various values of the self-similarity index H. Note that for $H = \tfrac{1}{2}$ we have $r_x(l) = \delta(l)$, which shows that the FGN process is white noise.

THEOREM 12.8. The power spectrum of the FGN process $x(n)$ is given by

$$R_x(e^{j\omega}) = \sum_{l=-\infty}^{\infty} r_x(l)e^{-j\omega l} = \sigma_H^2 C_H |1 - e^{-j\omega}|^2 \sum_{k=-\infty}^{\infty} \frac{1}{|\omega + 2\pi k|^{2H+1}} \tag{12.6.37}$$

where
$$C_H = 2H\Gamma(2H)\sin(\pi H) \tag{12.6.38}$$

is a constant dependent on the self-similarity index.

Proof. A rigorous proof can be found in Samorodnitsky and Taqqu (1994). Here we provide a more heuristic proof. The sequence $x(n)$ is obtained by sampling the FBM process $B_H(t)$ every $T = 1$ time unit, that is, evaluating $s(n) = B_H(nT)$, and then computing the first difference $x(n) = s(n) - s(n - 1)$. From the sampling theorem (see Chapter 2) we have

$$R_s(e^{j\omega}) = \frac{1}{T} \sum_{k=-\infty}^{\infty} R_B\left(\frac{\omega}{T} + \frac{2\pi k}{T}\right)$$

where $\omega = \Omega T$. The frequency response of the first-difference filter is $H(e^{j\omega}) = 1 - e^{-j\omega}$, or

$$|H(e^{j\omega})|^2 = 2(1 - \cos\omega) = 4\sin^2\left(\frac{\omega}{2}\right)$$

Since, from (12.6.20),

$$R_B(F) = C_H \frac{\sigma_H^2}{|F|^{2H+1}}$$

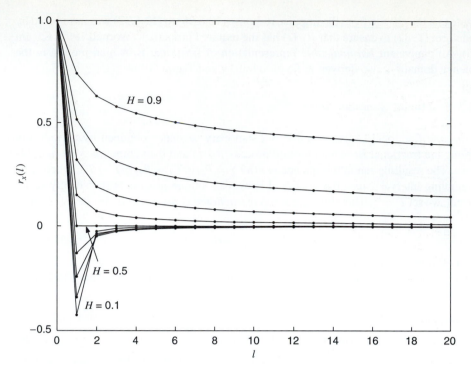

FIGURE 12.26
Autocorrelation sequence of FGN for $H = 0.1$ to $H = 0.9$ at 0.1 increments.

the power spectrum of $x(n)$ is

$$R_x(e^{j\omega}) = |H(e^{j\omega})|^2 R_s(e^{j\omega}) = 2\sigma_H^2 C_H (1 - \cos\omega) \frac{1}{T} \sum_{k=-\infty}^{\infty} \frac{1}{|\omega + 2\pi k|^{2H+1}}$$

which results in (12.6.37) for $T = 1$.

Figure 12.27 shows the PSD of FGN for various values of the self-similarity index H. Note that for $H = \frac{1}{2}$ we have have a flat PSD, which shows that the FGN process is white noise.

Self-similarity. The discrete FGN process is asymptotically (i.e., at large scales) self-similar. Indeed, the autocorrelation

$$r(l) \sim \sigma_H^2 H(2H - 1)|l|^{2H-2} \qquad \text{as } |l| \to \infty, \ H \neq \tfrac{1}{2}$$

decays hyperbolically for large lags, and the PSD

$$R(e^{j\omega}) \sim C_H \frac{\sigma_H^2}{|\omega|^{2H-1}} \qquad \text{as } |\omega| \to 0, \ H \neq \tfrac{1}{2}$$

follows a power law as the frequency becomes very small, that is, as the period becomes very large.

Process memory. The FGN process has long memory for $\frac{1}{2} < H < 1$ because the summation $\sum_{l=-\infty}^{\infty} r(l) = \infty$, or equivalently $R(e^{j\omega}) \to \infty$ as $|\omega| \to 0$. In this case the autocorrelation decays very slowly, the frequency response resembles a low-pass filter, and the resulting realizations look smooth. In contrast, the process exhibits short memory for $0 < H < \frac{1}{2}$, because $\sum_{l=-\infty}^{\infty} |r(l)| < \infty$ and $\sum_{l=-\infty}^{\infty} r(l) = 0$, or equivalently $R(e^{j\omega}) \to 0$ as $|\omega| \to 0$. In addition, for $0 < H < \frac{1}{2}$, the correlation is negative, that is,

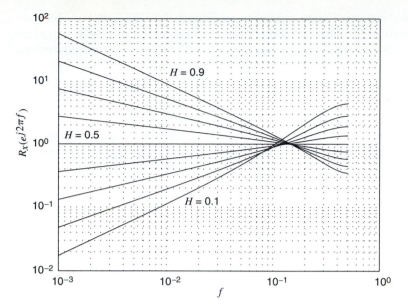

FIGURE 12.27
PSD function of FGN for $H = 0.1$ to $H = 0.9$ at 0.1 increments.

$r(l) < 0$ for $l \neq 0$; and the process exhibits negative dependence, or antipersistence. In this case the autocorrelation decays very rapidly, the frequency response resembles a high-pass filter, and the resulting realizations look rough.

Comparison between FGN and FPN. The discrete-time FGN and FPN processes have been independently introduced and have been developed using different approaches. However, close inspection of their second-order statistics reveals some interesting similarities and differences, which are summarized in Table 12.3. The most interesting feature is that both processes become asymptotically self-similar at large scales.

12.6.4 Simulation of Fractional Brownian Motions and Fractional Gaussian Noises

Although statistical self-similar processes are relatively easy to describe notationally [see (12.6.1)], they are not easy to generate since there is no explicit (or compact) mathematical formula to do so. The FBMs and FGNs are special cases of these processes that have independent increments with underlying distribution that is Gaussian. Although an explicit formula exists for FBM [see (12.6.31)], the additional complication is that we cannot generate a continuous trace (this would require infinitely long memory). We can only hope to generate an approximate, sampled version of the process on a computer. Thus, as explained before, these simulations are not self-similar at all scales. Nevertheless, we provided plots of these processes in Figure 12.25 for various values of the self-similarity index H. This can be done via techniques that either use properties of the processes or employ indirect approaches. In this section, we provide a brief summary of some of these techniques. For more detailed discussion see Samorodnitsky and Taqqu (1994) and Barnsley et al. (1988). We begin with the simulation of ordinary Brownian motion, which is easy to generate.

Cumulative-sum method. This technique is a direct method that is suitable for generating FBM for $H = 0.5$. We note that the increments of this process not only are stationary but also are uncorrelated with one another. These increments then form the WGN process, which can be simulated on a computer. Thus by integrating WGN we can obtain the ordinary Brownian motion. For discrete FBM, this requires taking the cumulative sum of the generated WGN sequence. Therefore, the steps in generating the ordinary Brownian motion are as follows:

1. Subdivide the time axis into a sufficiently fine grid. Let the number of grid points be N.
2. Generate N independent Gaussian random numbers with mean 0 and variance σ^2. In MATLAB this can be done using the `randn(N,1)` function.
3. Obtain a cumulative sum of the random numbers obtained in step 2 above. In MATLAB use the `cumsum` function. The resulting sequence is a discrete approximation of ordinary Brownian motion.

TABLE 12.3

Similarities and differences between discrete fractional Gaussian noise and discrete fractional pole noise.

	Discrete fractional Gaussian noise	Discrete fractional pole noise
Definition	$x(n) \triangleq B_H(n) - B_H(n-1)$	$x(n) \triangleq \sum_{k=0}^{\infty} \dfrac{\Gamma(k+d)}{\Gamma(k+1)\Gamma(d)} w(n-k)$ $w(n) \sim \text{WN}(0, \sigma^2) \quad -\frac{1}{2} < d < \frac{1}{2}$
Autocorrelation	$r(l) = \frac{1}{2}\sigma_H^2 (\lvert l-1\rvert^{2H} - 2\lvert l\rvert^{2H} + \lvert l+1\rvert^{2H})$	$r(l) = \dfrac{\sigma^2(-1)^l \Gamma(1-2d)}{\Gamma(1+l-d)\Gamma(1-l-d)}$
Power spectrum	$R(e^{j\omega}) = \sum_{k=-\infty}^{\infty} \dfrac{C_H \sigma_H^2 (1-\cos\omega)}{\lvert\omega+2\pi k\rvert^{2H+1}}$	$R(e^{j\omega}) = \dfrac{\sigma^2}{[2\sin(\omega/2)]^{2d}}$
Self-similarity (as $\lvert l\rvert \to \infty$)	$r(l) \sim \sigma_H^2 H(2H-1)\lvert l\rvert^{2H-2} \quad H \neq \frac{1}{2}$	$r(l) \sim C_d \lvert l\rvert^{2d-1} \Rightarrow d = H - \frac{1}{2}$
Self-similarity (as $\lvert\omega\rvert \to 0$)	$R(e^{j\omega}) \sim C_H \dfrac{\sigma_H^2}{\lvert\omega\rvert^{2H-1}} \quad H \neq \frac{1}{2}$	$R(e^{j\omega}) \sim \dfrac{\sigma^2}{\lvert\omega\rvert^{2d}} \Rightarrow d = H - \frac{1}{2}$
Long-memory $\sum_{l=-\infty}^{\infty} r(l) = R(0) = \infty$	$\frac{1}{2} < H < 1$	$0 < d < \frac{1}{2}$
Short-memory $\sum_{l=-\infty}^{\infty} r(l) = R(0) = 0$	$0 < H < \frac{1}{2}$ $\sum_{l=-\infty}^{\infty} \lvert r(l)\rvert < \infty \quad r(l) < 0, \text{ for } l \neq 0$	$-\frac{1}{2} < d < 0$
Partial correlation		$k_m = \dfrac{d}{m-d} \quad m = 1, 2, 3, \ldots$

A MATLAB function to implement the above steps is explored in Problem 12.13. Since it is difficult to generate properly correlated random numbers, the cumulative-sum method is not suitable for H other than 0.5.

Spectral synthesis method. This method can be used to generate FBM with an index $0 < H < 1$. The basic principle behind the spectral synthesis approach is that if we can construct its spectral density function $R_x(F)$, then we can obtain the corresponding FBM through inverse transformation. From (12.6.20), we have

$$R_x(F) \propto \frac{1}{\lvert F\rvert^{\beta}} \qquad \beta = 2H + 1 \tag{12.6.39}$$

Also, similar to the spectral density function relation (5.3.2) for discrete-time stochastic processes, we have

$$R_x(F) = \lim_{T \to \infty} \left[\frac{1}{T} |X_c(F)|^2 \right] \qquad (12.6.40)$$

Thus from (12.6.39) and (12.6.40) it is possible to obtain a frequency-domain method for approximating samples of an FBM with $0 < H < 1$. Let $\{x(n)\}$ be the sample functions of an FBM with Hurst parameter H. Then its DTFT magnitude $|X(e^{j\omega})|$ has the form

$$|X(e^{j\omega})| \propto \frac{1}{|\omega|^{\beta/2}} \qquad -\pi < \omega \leq \pi \qquad (12.6.41)$$

Since this is a continuous function, we use the DFT approach to obtain samples in the time domain. If we sample $X(e^{j\omega})$ at N equispaced frequencies $\omega_k = 2\pi k/N, 0 \leq k \leq N-1$, then the DFT magnitude has the form

$$|\tilde{X}(k)| \propto \begin{cases} \dfrac{1}{k^{\beta/2}} & 0 \leq k \leq \dfrac{N}{2} \\ |\tilde{X}(N-k)| & \dfrac{N}{2} < k \leq N-1 \end{cases} \qquad (12.6.42)$$

The phase of $\tilde{X}(k)$ can be chosen to be random, uniformly distributed over $[-\pi, \pi]$ subject to the constraint of odd symmetry. Finally, taking the IDFT of $\tilde{X}(k)$ results in a sequence that approximates samples of the FBM with $H = (\beta - 1)/2$. The steps of this spectral synthesis method can be summarized as follows:

1. Given H, determine $\beta = 2H + 1$.
2. Choose sufficiently large N, and use a suitable proportionality constant to generate $|\tilde{X}(k)|$ according to (12.6.42).
3. Randomize phase $\theta(k)$; that is, generate phase values according to

$$\theta(k) = \begin{cases} \text{uniform random number over } [-\pi, \pi] & 0 \leq k \leq \dfrac{N}{2} \\ -\theta(N-k) & \dfrac{N}{2} < k \leq N-1 \end{cases} \qquad (12.6.43)$$

4. Assemble $\tilde{X}(k) = |\tilde{X}(k)| \exp j\theta(k), 0 \leq k \leq N-1$, and determine the IDFT to obtain $x(n)$.

One major problem with this technique is that the resulting sequence is periodic with period N due to the DFT operation (or sampling in the frequency domain). Therefore, to avoid these boundary problems, a middle third of the sequence is used as a representative FBM trace. The FBM traces shown in Figure 12.25 were generated using the above steps. A MATLAB function to implement the above steps is explored in Problem 12.14.

Note that the corresponding FGN sequence is obtained by taking a first-order difference of the generated FBM sequence, that is,

$$w(n) = x(n) - x(n-1) \qquad 1 \leq n \leq N-1 \qquad (12.6.44)$$

Random midpoint replacement method. This is another direct method to produce FBM and is based on the scaling property of the increments [from (12.6.11)] that

$$\text{var}[\Delta B_H(t)] = |\Delta t|^{2H} \sigma_H^2 \qquad (12.6.45)$$

The approach is to begin generating random sequence values at the endpoints of the interval and then successively decimate the interval and generate a random value at the midpoint of the smaller interval according to (12.6.45). Therefore, this method can be implemented recursively. To generate an FBM over the interval $[0, 1]$ with parameter H, the following steps can be used:

1. Choose $B_H(0) = 0$ and select $B_H(1)$ equal to a Gaussian random number with mean 0 and variance σ_H^2 since

$$\sigma_H^2 = E\{B_H^2(1)\}$$

Clearly, $\text{var}[B_H(1) - B_H(0)] = 1^{2H}\sigma_H^2 = \sigma_H^2$.

2. For the first stage, set $B_H(\frac{1}{2})$ to be the average of $B_H(0)$ and $B_H(1)$ plus some independent Gaussian number offset δ_1 with mean zero and variance σ_1^2, that is,

$$B_H\left(\tfrac{1}{2}\right) = \tfrac{1}{2}[B_H(1) - B_H(0)] + \delta_1 \tag{12.6.46}$$

Thus $B_H(\frac{1}{2}) - B_H(0)$ and $B_H(1) - B_H(\frac{1}{2})$ have mean 0 and variance

$$\text{var}\left[B_H\left(\frac{1}{2}\right) - B_H(0)\right] = \frac{1}{4}\text{var}[B_H(1) - B_H(0)] + \text{var}(\delta_1) \tag{12.6.47}$$

$$\left(\frac{1}{2}\right)^{2H}\sigma_H^2 = \frac{1}{4}\sigma_H^2 + \text{var}(\delta_1)$$

or

$$\text{var}(\delta_1) = \left[\left(\frac{1}{2}\right)^{2H} - \frac{1}{4}\left(\frac{1}{2^0}\right)^{2H}\right]\sigma_H^2 \tag{12.6.48}$$

3. At the second stage, we generate $B_H(\frac{1}{4})$ and $B_H(\frac{3}{4})$, using the above method specialized to $\Delta t = \frac{1}{4}$, that is,

$$B_H\left(\frac{1}{4}\right) = \frac{1}{2}\left[B_H\left(\frac{1}{2}\right) - B_H(0)\right] + \delta_{21}$$

$$B_H\left(\frac{3}{4}\right) = \frac{1}{2}\left[B_H(1) - B_H\left(\frac{1}{2}\right)\right] + \delta_{22}$$

with

$$\text{var}(\delta_{21}) = \text{var}(\delta_{22}) = \left[\left(\frac{1}{2^2}\right)^{2H} - \frac{1}{4}\left(\frac{1}{2^1}\right)^{2H}\right]\sigma_H^2 \tag{12.6.49}$$

4. Continuing in this fashion, at stage r we generate 2^{r-1} midpoints as the average of their respective endpoints plus a Gaussian random number offset $\delta_{r,k}$, $k = 1, 2, \ldots, 2^{r-1}$, with variance

$$\text{var}(\delta_{r,k}) = \left[\left(\frac{1}{2^r}\right)^{2H} - \frac{1}{4}\left(\frac{1}{2^{r-1}}\right)^{2H}\right]\sigma_H^2 = \left(\frac{1}{2^r}\right)^{2H}(1 - 2^{2H-2})\sigma_H^2 \tag{12.6.50}$$

$$= \frac{1}{2^{2H}}\text{var}(\delta_{r-1,k}) \tag{12.6.51}$$

Thus, as expected for an FBM, at time scale $1/2^r$ we add randomness with mean 0 and variance proportional to $(1/2^r)^{2H}$ according to (12.6.50). Also from (12.6.51) we can recursively generate the variance at each stage.

5. Stop the procedure when a sufficient number of trace points are generated.

This method also suffers from a few shortcomings. The most troublesome problem is that once a given midpoint is generated, its value remains unchanged in all later stages. Thus points generated at different stages have different statistical properties in their neighborhood. This produces a visible trace that does not seem to go away even if more stages are added, and the artifact is more pronounced as $H \to 1$. A MATLAB function implementing the above steps in a recursive fashion is explored in Problem 12.15. Once again, the corresponding FGN sequence is obtained by taking a first-order difference of the generated FBM sequence.

The generation of one- and higher-dimensional FBM is a very popular subject in engineering, sciences, and computer graphics. More information and additional references can be found in Mandelbrot (1982), Maeder (1995), Peitgen et al. (1988), and Samorodnitsky and Taqqu (1994) and in the vast literature on fractals.

12.6.5 Estimation of Long Memory

The estimation of the self-similarity index H or the long-memory parameter $d = H - \frac{1}{2}$ is a very difficult task. A summary of the most widely used methods, including an empirical evaluation, is provided in Taqqu et al. (1995). Additional information can be found in Beran (1994), Beran et al. (1995), and Brockwell and Davis (1991). We next present two simple methods that exploit the definition of self-similarity in the time and frequency domains (Pentland 1984; Beran 1994).

For any self-similar process $x(n)$ and any integer $\Delta > 0$, the increments $\Delta x(n) \triangleq x(n + \Delta) - x(n)$ have zero mean and satisfy the relation

$$E\{[\Delta x(n)]^2\} = C\Delta^{2H} \tag{12.6.52}$$

where C is a constant. Taking the natural logarithm of both sides, we have

$$\ln E\{[\Delta x(n)]^2\} = \ln C + 2H \ln \Delta \tag{12.6.53}$$

which can be used to estimate H using linear regression on a log-log plot. The expectation on the left side of (12.6.53) can be estimated by using the mean value of $[\Delta x(n)]^2$.

In practice, to avoid the influence of outliers, we use the quantity $E\{|\Delta x(n)|\}$, which leads to

$$\ln E\{|\Delta x(n)|\} = \ln C + H \ln \Delta \tag{12.6.54}$$

where C is a constant. The expectation in (12.6.52) is estimated by the mean absolute value, and H is determined by linear regression. This approach is illustrated in Figure 12.28, which shows the estimation of the self-similarity index H for two realizations of an FBM process (for details see Problem 12.16). We note that in practice the range of scales extends from 1 to $0.1N$, where N is the length of the used data record.

We have seen that for $|f| \to 0$, the PSD of FBM, FGN, and FPN follows a power law $1/f^\beta$, where $\beta = 2H + 1$ (FBM), $\beta = 2H - 1$ (FGN), and $\beta = 2d = 2H - 1$ (FPN). Therefore, another method for estimating the long-memory parameter H, is to compute an estimate of the PSD (see Chapter 5), and then determine H by linear regression of the logarithm of the PSD on the logarithm of the frequency. In practice, we only use the lowest 10 percent of the PSD frequencies for the linear regression because the power law relationship holds as $|f| \to 0$ (Taqqu et al. 1995). The PSD estimation of power law processes using the multitaper PSD estimation method is discussed in McCoy et al. (1998), which shows that using this method provides better estimates of long memory than the traditionally used periodogram estimator.

In practice, data are *scale-limited*: The sampling interval determines the lowest scale, and the data record length determines the highest scale. Furthermore, the scaling behavior for a certain statistical moment may change from one range of scales to another. When we try to make predictions from an adoption of a scale-invariant model, there are certain discrepancies between theory and practice. In theory, the power increases with wavelength without limit, and the variance increases with profile length without limit. In practice, the power for long wavelengths is not as large as predicted by extrapolating the power law trend observed at short wavelengths (frequency domain), and the variance does not increase without bounds as the profile length increases (spatial domain).

12.6.6 Fractional Lévy Stable Motion

If we assume that the probability density function of the stationary increments is SαS, the resulting self-similar process is known as *fractional Lévy stable motion (FLSM)*. However, unlike the FBM process, the second-order moments of the FLSM process do not exist because SαS distributions have infinite variance. The realizations of FLSM resemble more

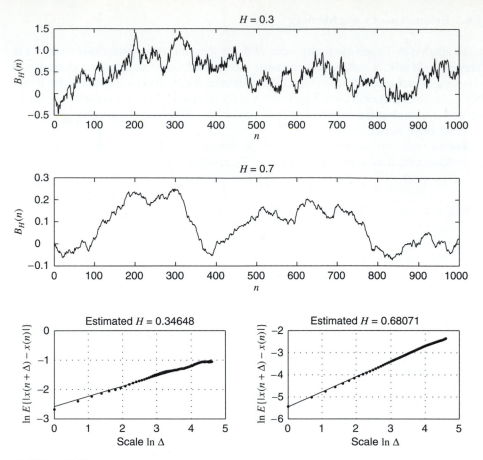

FIGURE 12.28
Sample realizations of an FBM process and log-log plots for estimation of H using linear regression.

spiky versions of FBM realizations because of the heavy tails of the stable distribution. Hence, FLSM processes provide an excellent model for signals with long memory and high variability.

Formally, an FLSM process $L_{H,\alpha}(t)$ is best formulated in terms of its increment process $x_{H,\alpha}(n)$, known as *fractional Lévy stable noise (FLSN)*. The FLSN is defined by the stochastic integral (Samorodnitsky and Taqqu 1994)

$$x_{H,\alpha}(n) = L_{H,\alpha}(n+1) - L_{H,\alpha}(n) = C \int_{-\infty}^{n} [(n+1-s)^{H-1/\alpha} - (n-s)^{H-1/\alpha}] w_\alpha(s)\, ds$$

$$(12.6.55)$$

where C is a constant, α is the characteristic exponent of the SαS distribution, and $w_\alpha(s)$ is white noise from an SαS distribution. Notice that for $\alpha = 2$, Equation (12.6.55) provides an integral description of FGN. From Figure 12.29, which shows several realizations of FLSM for $H = 0.7$ and various values of α, it is evident that the lower the value of α, the more impulsive the process becomes. The techniques described above for generating FBM can be modified to simulate FLSM, by replacing the Gaussian random generator with the SαS one described in Chambers et al. (1976) and Samorodnitsky and Taqqu (1994).

The long-memory parameter H can be estimated by using (12.6.54) and linear regression. The PSD method cannot be used because the second-order moments of the FLSM process do not exist. The estimation of the characteristic exponent α of the SαS increments

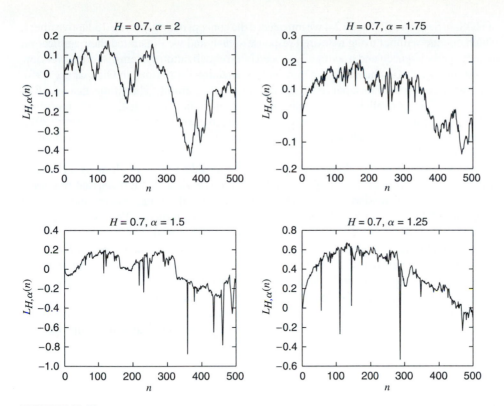

FIGURE 12.29
Sample realizations of the FLSM process for $H = 0.7$ and various values of α. The spikes increase as we go from a Gaussian ($\alpha = 2$) to a Cauchy ($\alpha = 1$) distribution.

is a very difficult task because (1) SαS distributions have infinite variance and (2) the increments are not IID owing to the long-range dependence structure. Further discussion of these topics, which are beyond the scope of this book, is provided in Adler et al. (1998), McCulloch (1986), and Koutrouvelis (1980).

   Some interesting applications of FLSM to the modeling and interpolation of natural signals and images are discussed in Kogon and Manolakis (1994, 1996), Peng et al. (1993), Painter (1996), and Stuck and Kleiner (1974).

12.7 SUMMARY

In this chapter we introduced the basic concepts of three very important areas of statistical and adaptive signal processing that are the subject of extensive research. The goal was to help appreciate the limits of second-order statistical techniques, open a window to the exciting world of modern signal processing, and help the navigation through the ever-increasing literature.

   In Section 12.1 we introduced the basics of higher-order statistics and pointed out the situations in which their use may be beneficial. In general, the advantages of HOS become more evident as the non-Gaussianity and nonlinearity of the underlying models increase. Also HOS is of paramount importance when we deal with non-minimum-phase systems. Concise reviews of several aspects of HOS are given in Swami (1998) and Tugnait (1998), and a comprehensive bibliography is given in Swami et al. (1997).

   Section 12.2 provided a brief introduction to the principles of blind deconvolution and demonstrated that the blind deconvolution of non-minimum-phase systems requires the use of HOS. In Sections 12.3 and 12.4 we introduced the concept of unsupervised adaptive filters, which operate without using a desired response signal; and we illustrated their application to both symbol-spaced and fractionally spaced blind equalization systems. A brief overview of current research in channel estimation and equalization is provided in Giannakis (1998). There are three types of unsupervised adaptive filtering algorithms: algorithms that use HOS either implicitly or explicitly, algorithms that use cyclostationary statistics, and algorithms that use information-theoretic concepts (Bell and Sejnowski 1995; Pham and Garrat 1997). We have focused on the widely used family of Bussgang-type algorithms that make implicit use of HOS.

In the last part of this chapter, we provided an introduction to random signal models with long memory and low or high variability. More specifically, we discussed fractional pole models with Gaussian or SαS IID excitations and self-similar process models with Gaussian (FBM, FGN) or SαS (FLSM, FLSN) stationary increments. The recent discovery that Ethernet traffic data are self-similar and SαS (Willinger et al. 1994) established long-memory models as a very useful tool in communication systems engineering. Finally, we note that the wavelet transform, which decomposes a signal into a superposition of scaled and shifted versions of a single basis function known as the *mother wavelet*, provides a natural tool for the analysis of linear self-similar systems and self-similar random signals. The discrete wavelet transform facilitates, to a useful degree, the whitening of self-similar processes and can be used to synthesize various types of practical self-similar random signals (Mallat 1998; Wornell 1996).

PROBLEMS

12.1 Prove (12.1.27), which relates the output and input fourth-order cumulants of a linear, time-invariant system.

12.2 Derive (12.1.35) and (12.1.36), using the formulas for the cumulant of the sum of IID random variables, developed in Section 3.2.4.

12.3 If $x(n)$ is a stationary Gaussian process, show that $E\{x(n)x^2(n-l)\} = \rho_x^2(l)$ and explain how it can be used to investigate the presence of nonlinearities.

12.4 In this problem we use an MA(2) model to explore some properties of cumulants and bispectra.

 (*a*) Write a MATLAB function k=cuma(b) that computes the cumulant $\kappa_x^{(3)}(l_1, l_2)$ of the MA(2) model $H(z) = b_0 + b_1 z^{-1} + b_2 z^{-2}$ for $-L \le l_1, l_2 \le L$.
 (*b*) Use the functions k=cuma(b), X=fft(x), and X=shiftfft(X) to compute the bispectra of the three MA(2) models in Table 12.1. Plot your results and compare with those in Figure 12.2.
 (*c*) Compute the bispectra of the models using the formula

$$R_x^{(3)}(e^{j\omega_1}, e^{j\omega_2}) = \kappa_w^{(3)} H(e^{j\omega_1}) H(e^{j\omega_2}) H^*(e^{j(\omega_1+\omega_2)})$$

 for $\omega_1 = \omega_2 = 2\pi k/N, 0 \le k \le N-1$. Compare with the results in part *b* and Figure 12.2.
 (*d*) Show that the bispectrum can be computed in MATLAB using the following segment of code:
```
H=freqz(h,1,N,'whole');
Hc=conj(H);
R3x=(H*H').*hankel(Hc,Hc([N,1:N-1]));
R3x=shiftfft(R3x);
```

12.5 Using the minimum-, maximum-, and mixed-phase systems discussed in Example 12.1.1, write a MATLAB program to reproduce the results shown in Figures 12.3 and 12.4. Use $a = 0.4, b = 0.8$, and $N = 300$ samples.

12.6 Use the Levinson-Durbin algorithm, developed in Chapter 7, to derive expressions (12.5.20), direct-form coefficients, and (12.5.21) for the lattice parameters of the fractional pole model.

12.7 Consider the FPZ$(1, d, 0)$ model

$$H_{\text{fpz}}(z) = \frac{1}{(1 - z^{-1})^d} \frac{1}{(1 + az^{-1})}$$

where $-\frac{1}{2} < d < \frac{1}{2}$ and $-1 < a < 1$. Compute and plot the impulse response, autocorrelation, and spectrum for $a = \pm 0.9$ and $d = \pm 0.2, \pm 0.4$. Identify which models have long memory and which have short memory.

12.8 Compute and plot the PSD of the FGN process, using the following approaches, and compare the results.

(a) The definition $R_x(e^{j\omega}) = \sum_{l=-\infty}^{\infty} r_x(l)e^{-j\omega l}$ and formula (12.6.36) for the autocorrelation.

(b) The theoretical formula (12.6.37).

12.9 Use the algorithm of Schür to develop a more efficient implementation of the fractional pole noise generation method described by Equations (12.5.24) to (12.5.28).

12.10 In this problem we study the properties of the harmonic fractional unit-pole model specified by the system function given by (12.5.32). The impulse response is given by (Gray et al. 1989)

$$h_{\theta,d}(n) = \sum_{k=0}^{\lfloor n/2 \rfloor} \frac{(-1)^k \Gamma(d + n - k)(2 \cos \theta)^{n-2k}}{k!(n - 2k)!\Gamma(d)}$$

where $\Gamma(\cdot)$ is the gamma function.

(a) Compute and plot $h_{\theta,d}(n)$ for various values of θ and d.

(b) Demonstrate the validity of the above formula by evaluating $h_{\theta,d}(n)$ from $H_{\theta,d}(z)$ for the same values of θ and d.

(c) Illustrate that the model is minimum-phase if $|\cos \theta| < 1$ and $-\frac{1}{2} < d < \frac{1}{2}$ or $\cos \theta = \pm 1$ and $-\frac{1}{4} < d < \frac{1}{4}$.

(d) Illustrate that the harmonic minimum-phase model, like the FPZ$(0, d, 0)$ one, exhibits long-memory behavior only for positive values of d.

(e) Show that for $0 < d < \frac{1}{4}$ and $\cos \theta = 1$, the autocorrelation equals that of the FPZ$(0, 2d, 0)$ model [multiplied by $(-1)^l$ if $\cos \theta = -1$]. When $|\cos \theta| < 1$ and $0 < d < \frac{1}{2}$, illustrate numerically that the autocorrelation can be approximated by $\rho(l) \sim -l^{2d-1} \sin(\theta l - \pi d)$ as $l \to \infty$.

(f) Compute and plot the spectrum of the model for $\theta = \pi/3$ and various values of d.

(g) Generate and plot realizations of Gaussian HFPZ noise for $\theta = \pi/6$ and $d = -0.3, 0.1$, and 0.4.

12.11 Determine the variogram of the process $x(n)$ obtained by exciting the system

$$H(z) = \frac{1}{(1 - z^{-1})(1 - az^{-1})} \qquad |a| < 1$$

with white noise $w(n) \sim \text{WGN}(0, \sigma_w^2)$.

12.12 Following the steps leading to (12.6.26), show that the fractal (Haussdorff) dimension D is related to the Hurst exponent H by

$$D = 2 - H$$

12.13 Develop a MATLAB function to generate the ordinary Brownian motion trace according to the steps given for the cumulative sum method in Section 12.6.3. The format of the function should be x = obm_cumsum(N).

(a) Generate 16,384 samples of the Brownian motion $x(t)$ over $0 \le t \le 1$.

(b) Investigate the self-affine property of $x(t)$ by reproducing a figure similar to Figure 12.23.

12.14 Develop a MATLAB function to generate the fractional Brownian motion trace according to the steps given for the spectral synthesis method in Section 12.6.3. The format of the function should be x = fbm_spectral(N).

(a) Generate 1024 samples of the FBM $B_H(t)$ over $0 \leq t \leq 1$ for $H = 0.3$. Investigate the self-affine property of $B_{0.3}(t)$.

(b) Generate 1024 samples of the FBM $B_H(t)$ over $0 \leq t \leq 1$ for $H = 0.7$. Investigate the self-affine property of $B_{0.7}(t)$.

12.15 Develop a MATLAB function to generate the fractional Brownian motion trace according to the steps given for the random midpoint replacement method in Section 12.6.3. The format of the function should be x = fbm_replace(N).

(a) Generate 1024 samples of the FBM $B_H(t)$ over $0 \leq t \leq 1$ for $H = 0.5$. Compare visually $B_{0.5}(t)$ with that obtained by using the cumulative-sum method. Comment on your observations.

(b) Generate 1024 samples of the FBM $B_H(t)$ over $0 \leq t \leq 1$ for $H = 0.99$. Investigate the artifact discussed in the chapter for $H \rightarrow 1$.

12.16 Based on Equation (12.6.54), develop a MATLAB function [H,sigmaH] = est_H_mad(x) that computes an estimate of the self-similarity index H and the variance σ_H^2 of an FBM process.

(a) Use function x = fbm_replace(N) to generate $N = 1024$ samples of an FBM process with $H = 0.3$, and use the function [H,sigmaH] = est_H_mad(x) to estimate H and σ_H.

(b) Repeat the previous task for $H = 0.7$.

(c) Perform a Monte Carlo simulation using 100 trials and compute the mean and standard deviation of the estimates for H and σ_H in (a) and (b).

12.17 Repeat Problem 12.16 by developing a function that estimates the self-similarity index H by determining the slope of the first 10 percent values of the periodogram in a log-log plot.

Matrix Inversion Lemma

The matrix inversion lemma is a useful formula that is employed extensively in signal processing. The purpose of this formula is to express the inverse of a matrix in terms of the inverse of one of its additive components, so as to facilitate an efficient computation of the inverse. To motivate this lemma, consider the inverse of the following scalar quantity

$$(a + xy)^{-1} = \frac{1}{a + xy} \qquad a + xy \neq 0, \ a \neq 0$$

in terms of the inverse of a. Since $a + xy \neq 0$ and $a \neq 0$, we also have

$$|xya^{-1}| \neq 1 \quad \text{and} \quad |ya^{-1}x| \neq 1 \tag{A.1}$$

Using the convergence of the geometric series formula

$$1 - xya^{-1} + (xya^{-1})^2 - \cdots = \frac{1}{1 + xya^{-1}} \qquad |xya^{-1}| \neq 1 \tag{A.2}$$

we obtain

$$
\begin{aligned}
\frac{1}{a + xy} &= \frac{a^{-1}}{1 + xya^{-1}} \\
&= a^{-1}[1 - xya^{-1} + (xya^{-1})^2 - \cdots] \\
&= a^{-1} - a^{-1}xya^{-1} + a^{-1}x(ya^{-1}x)ya^{-1} - a^{-1}x(ya^{-1}x)^2 ya^{-1} + \cdots \\
&= a^{-1} - a^{-1}xya^{-1}[1 - ya^{-1}x + (ya^{-1}x)^2 - \cdots] \\
&= a^{-1} - \frac{a^{-1}xya^{-1}}{1 + ya^{-1}x} \qquad |ya^{-1}x| \neq 1
\end{aligned}
\tag{A.3}
$$

which is the desired result. We begin with a special case of the lemma in which a is a matrix and x and y are vectors. This result then can be generalized to the case in which x and y are also matrices.

LEMMA A.1 (SHERMAN-MORRISON'S FORMULA). Let $\mathbf{A}$ be an $N \times N$ invertible matrix and let $\mathbf{x}$ and $\mathbf{y}$ be two $N \times 1$ vectors such that $(\mathbf{A} + \mathbf{xy}^H)$ is invertible. Then we have

$$(\mathbf{A} + \mathbf{xy}^H)^{-1} = \mathbf{A}^{-1} - \frac{\mathbf{A}^{-1}\mathbf{xy}^H \mathbf{A}^{-1}}{1 + \mathbf{y}^H \mathbf{A}^{-1}\mathbf{x}} \tag{A.4}$$

Proof. Consider

$$\mathbf{A} + \mathbf{xy}^H = \mathbf{A}(\mathbf{I} + \mathbf{A}^{-1}\mathbf{xy}^H)$$

Hence
$$(\mathbf{A} + \mathbf{xy}^H)^{-1} = (\mathbf{I} + \mathbf{A}^{-1}\mathbf{xy}^H)^{-1}\mathbf{A}^{-1} \tag{A.5}$$

Using the result that if the matrix $(\mathbf{I} + \mathbf{A})$ is invertible, then $(\mathbf{I} + \mathbf{A})^{-1} = \mathbf{I} - \mathbf{A} + \mathbf{A}^2 - \mathbf{A}^3 + \cdots$, we obtain

$$
\begin{aligned}
(\mathbf{I} + \mathbf{A}^{-1}\mathbf{x}\mathbf{y}^H)^{-1} &= \mathbf{I} - \mathbf{A}^{-1}\mathbf{x}\mathbf{y}^H + (\mathbf{A}^{-1}\mathbf{x}\mathbf{y}^H)^2 - (\mathbf{A}^{-1}\mathbf{x}\mathbf{y}^H)^3 + \cdots \\
&= \mathbf{I} - \mathbf{A}^{-1}\mathbf{x}\mathbf{y}^H + \mathbf{A}^{-1}\mathbf{x}\mathbf{y}^H\mathbf{A}^{-1}\mathbf{x}\mathbf{y}^H - \cdots
\end{aligned}
\tag{A.6}
$$

since from (A.5) $(\mathbf{I} + \mathbf{A}^{-1}\mathbf{x}\mathbf{y}^H)$ is invertible. Substituting (A.6) into (A.5), we obtain

$$
\begin{aligned}
(\mathbf{A} + \mathbf{x}\mathbf{y}^H)^{-1} &= \mathbf{A}^{-1} - \mathbf{A}^{-1}\mathbf{x}\mathbf{y}^H\mathbf{A}^{-1} + \mathbf{A}^{-1}\mathbf{x}\underbrace{(\mathbf{y}^H\mathbf{A}^{-1}\mathbf{x})}_{\text{scalar}}\mathbf{y}^H\mathbf{A}^{-1} - \cdots \\
&= \mathbf{A}^{-1} - \mathbf{A}^{-1}\mathbf{x}\mathbf{y}^H\mathbf{A}^{-1}[1 - \mathbf{y}^H\mathbf{A}^{-1}\mathbf{x} + (\mathbf{y}^H\mathbf{A}^{-1}\mathbf{x})^2 - \cdots] \\
&= \mathbf{A}^{-1} - \frac{\mathbf{A}^{-1}\mathbf{x}\mathbf{y}^H\mathbf{A}^{-1}}{1 + \mathbf{y}^H\mathbf{A}^{-1}\mathbf{x}}
\end{aligned}
$$

since the scalar $\mathbf{y}^H\mathbf{A}^{-1}\mathbf{x} \neq 1$ due to the invertibility of $(\mathbf{I} + \mathbf{A}^{-1}\mathbf{x}\mathbf{y}^H)$ [see also (A.1)]. This completes the proof.

The generalization of (A.4), known as Woodbury's formula, is given by

$$
(\mathbf{A} + \mathbf{B}\mathbf{C}\mathbf{D})^{-1} = \mathbf{A}^{-1} - \mathbf{A}^{-1}\mathbf{B}(\mathbf{C}^{-1} + \mathbf{D}\mathbf{A}^{-1}\mathbf{B})^{-1}\mathbf{D}\mathbf{A}^{-1}
\tag{A.7}
$$

If matrix $\mathbf{A}$ is partitioned as

$$
\mathbf{A} = \begin{bmatrix} \mathbf{A}_{11} & \mathbf{A}_{12} \\ \mathbf{A}_{21} & \mathbf{A}_{22} \end{bmatrix}
\tag{A.8}
$$

then (A.7) can be used in determining inverses of submatrices contained in

$$
\mathbf{A}^{-1} = \begin{bmatrix} (\mathbf{A}_{11} - \mathbf{A}_{12}\mathbf{A}_{22}^{-1}\mathbf{A}_{21})^{-1} & -(\mathbf{A}_{11} - \mathbf{A}_{12}\mathbf{A}_{22}^{-1}\mathbf{A}_{21})^{-1}\mathbf{A}_{12}\mathbf{A}_{22}^{-1} \\ -(\mathbf{A}_{22} - \mathbf{A}_{21}\mathbf{A}_{11}^{-1}\mathbf{A}_{12})^{-1}\mathbf{A}_{21}\mathbf{A}_{11}^{-1} & (\mathbf{A}_{22} - \mathbf{A}_{21}\mathbf{A}_{11}^{-1}\mathbf{A}_{12})^{-1} \end{bmatrix}
\tag{A.9}
$$

where inverses $\mathbf{A}_{11}^{-1}$ and $\mathbf{A}_{22}^{-1}$ are assumed to exist.

APPENDIX B

Gradients and Optimization in Complex Space

In the development of many signal processing algorithms, it is necessary to compute the gradient of a real or complex function with respect to a complex vector **w**. The concepts involved in this gradient operation and the application of the gradient in optimization are described in this section. For more details see Gill et al. (1981), Kay (1993), and Luenberger (1984).

B.1   GRADIENT

We begin with a simplest case. Let $g(\mathbf{x})$ be a real scalar function of real parameter vector $\mathbf{x}$. Then we define the gradient of $g(\mathbf{x})$ with respect to vector $\mathbf{x}$ as a row vector

$$\nabla_{\mathbf{x}}(g) \triangleq \frac{\partial g(\mathbf{x})}{\partial \mathbf{x}} = \left[\frac{\partial g(\mathbf{x})}{\partial x_1} \quad \frac{\partial g(\mathbf{x})}{\partial x_2} \quad \cdots \quad \frac{\partial g(\mathbf{x})}{\partial x_N} \right] \tag{B.1}$$

This definition extends naturally to a vector function $\mathbf{g}(\mathbf{x})$ of parameter vector $\mathbf{x}$ as

$$\nabla_{\mathbf{x}}(\mathbf{g}) \triangleq \frac{\partial \mathbf{g}(\mathbf{x})}{\partial \mathbf{x}} = \begin{bmatrix} \dfrac{\partial g_1(\mathbf{x})}{\partial \mathbf{x}} \\[2mm] \dfrac{\partial g_2(\mathbf{x})}{\partial \mathbf{x}} \\[2mm] \vdots \\[2mm] \dfrac{\partial g_M(\mathbf{x})}{\partial \mathbf{x}} \end{bmatrix} = \begin{bmatrix} \dfrac{\partial g_1(\mathbf{x})}{\partial x_1} & \dfrac{\partial g_1(\mathbf{x})}{\partial x_2} & \cdots & \dfrac{\partial g_1(\mathbf{x})}{\partial x_N} \\[2mm] \dfrac{\partial g_2(\mathbf{x})}{\partial x_1} & \dfrac{\partial g_2(\mathbf{x})}{\partial x_2} & \cdots & \dfrac{\partial g_2(\mathbf{x})}{\partial x_N} \\[2mm] \vdots & \vdots & \ddots & \vdots \\[2mm] \dfrac{\partial g_M(\mathbf{x})}{\partial x_1} & \dfrac{\partial g_M(\mathbf{x})}{\partial x_2} & \cdots & \dfrac{\partial g_M(\mathbf{x})}{\partial x_N} \end{bmatrix} \tag{B.2}$$

Thus $\nabla_{\mathbf{x}}(\mathbf{g})$ is an $M \times N$ matrix. Finally, consider a scalar function $g(\mathbf{A})$ of an $M \times N$ matrix $\mathbf{A}$. We define the gradient of $g(\mathbf{A})$ with respect to $\mathbf{A}$ as a matrix

$$\nabla_{\mathbf{A}}(g) \triangleq \frac{\partial g(\mathbf{A})}{\partial \mathbf{A}} = \begin{bmatrix} \dfrac{\partial g(\mathbf{A})}{\partial a_{11}} & \dfrac{\partial g(\mathbf{A})}{\partial a_{12}} & \cdots & \dfrac{\partial g(\mathbf{A})}{\partial a_{1N}} \\[2mm] \dfrac{\partial g(\mathbf{A})}{\partial a_{21}} & \dfrac{\partial g(\mathbf{A})}{\partial a_{22}} & \cdots & \dfrac{\partial g(\mathbf{A})}{\partial a_{2N}} \\[2mm] \vdots & \vdots & \ddots & \vdots \\[2mm] \dfrac{\partial g(\mathbf{A})}{\partial a_{M1}} & \dfrac{\partial g(\mathbf{A})}{\partial a_{M2}} & \cdots & \dfrac{\partial g(\mathbf{A})}{\partial a_{MN}} \end{bmatrix} \tag{B.3}$$

Using these definitions, we see it is easy to prove the following results:

$$\nabla_{\mathbf{x}}(\mathbf{y}^T \mathbf{A} \mathbf{x}) = \mathbf{y}^T \mathbf{A} \tag{B.4}$$

$$\nabla_{\mathbf{x}}(\mathbf{x}^T \mathbf{A} \mathbf{y}) = \mathbf{y}^T \mathbf{A}^T \tag{B.5}$$

$$\nabla_{\mathbf{x}}(\mathbf{x}^T \mathbf{A} \mathbf{x}) = \mathbf{x}^T (\mathbf{A} + \mathbf{A}^T) \tag{B.6}$$

$$\nabla_{\mathbf{A}}(\mathbf{x}^T \mathbf{A} \mathbf{y}) = \mathbf{x} \mathbf{y}^T \tag{B.7}$$

$$\nabla_{\mathbf{A}}(\mathbf{x}^T \mathbf{A} \mathbf{x}) = \mathbf{x} \mathbf{x}^T \tag{B.8}$$

Now we consider the case of a complex-valued scalar function $g(z, z^*)$ of a complex variable z and its complex conjugate z^*. We assume that the function is analytic with respect to z and z^* independently[†] (in the sense of partial differentiation). An example of such a function is

$$g(z, z^*) = a|z^2| + bz^* + c = azz^* + bz^* + c \tag{B.9}$$

Let $f(x, y)$ be the complex function of the real and imaginary parts x and y of the variable $z = x + jy$, such that $g(z, z^*) = f(x, y)$. Again consider the function in (B.9), then

$$f(x, y) = a(x^2 + y^2) + b(x - jy) + c \tag{B.10}$$

$$= a|z^2| + bz^* + c = g(z, z^*) \tag{B.11}$$

The partial derivative of $g(z, z^*)$ with respect to z (keeping z^* as a constant) is given by

$$\frac{\partial}{\partial z} g(z, z^*) = \frac{1}{2} \left[\frac{\partial}{\partial x} f(x, y) - j \frac{\partial}{\partial y} f(x, y) \right] \tag{B.12}$$

Similarly, the partial derivative of $g(z, z^*)$ with respect to z^* (keeping z as a constant) is given by

$$\frac{\partial}{\partial z^*} g(z, z^*) = \frac{1}{2} \left[\frac{\partial}{\partial x} f(x, y) + j \frac{\partial}{\partial y} f(x, y) \right] \tag{B.13}$$

These results can be easily verified for $g(z, z^*)$ in (B.9):

$$\frac{\partial}{\partial z} a|z^2| + bz^* + c = \frac{\partial}{\partial z}[azz^* + bz^* + c] = az^* = a(x - jy)$$

and
$$\frac{1}{2} \left[\frac{\partial}{\partial x} f(x, y) - j \frac{\partial}{\partial y} f(x, y) \right] = \frac{1}{2} \left\{ \frac{\partial}{\partial x}[a(x^2 + y^2) + b(x - jy) + c] \right.$$

$$\left. - j \frac{\partial}{\partial y}[a(x^2 + y^2) + b(x - jy) + c] \right\}$$

$$= ax + \frac{b}{2} - jy - \frac{b}{2} = a(x - jy) = az^*$$

Let $f(\mathbf{x})$ be a real-valued scalar function of the complex vector $\mathbf{x}$ expressed as

$$f(\mathbf{x}) = g(\mathbf{x}, \mathbf{x}^*) \tag{B.14}$$

where $g(\cdot)$ is a real-valued function of $\mathbf{x}$ and $\mathbf{x}^*$, analytic with respect to $\mathbf{x}$ and $\mathbf{x}^*$ independently (in the sense of partial differentiation). The necessary and sufficient condition to obtain an equilibrium (optimum) point of $f(\mathbf{x})$ is that

$$\nabla_{\mathbf{x}}(g) = \nabla_{\mathbf{x}^*}(g) = \mathbf{0} \tag{B.15}$$

The necessary gradient $\nabla_{\mathbf{x}}(g)$ can be computed by using (B.13). In particular, for any complex vector $\mathbf{y}$, $\mathbf{x}$, and matrix $\mathbf{A}$, we have

$$\nabla_{\mathbf{x}^*}(\mathbf{x}^H \mathbf{y}) = \mathbf{y} \tag{B.16}$$

$$\nabla_{\mathbf{x}^*}(\mathbf{y}^H \mathbf{x}) = \mathbf{0} \tag{B.17}$$

$$\nabla_{\mathbf{x}^*}(\mathbf{x}^H \mathbf{A} \mathbf{y}) = \mathbf{A} \mathbf{y} \tag{B.18}$$

$$\nabla_{\mathbf{x}^*}(\mathbf{x}^H \mathbf{A} \mathbf{x}) = \mathbf{A} \mathbf{x} \tag{B.19}$$

[†] In this approach, the quantities z and z^* are considered to be independent of each other. Clearly they are not, since z is uniquely determined by its conjugate. Nevertheless, this technique works.

$$\nabla_{\mathbf{x}}(\mathbf{x}^H \mathbf{A}\mathbf{x}) = \mathbf{x}^H \mathbf{A} \tag{B.20}$$

$$\nabla_{\mathbf{A}}(\mathbf{x}^H \mathbf{A}\mathbf{y}) = \mathbf{x}^* \mathbf{y}^T \tag{B.21}$$

$$\nabla_{\mathbf{A}}(\mathbf{x}^H \mathbf{A}\mathbf{x}) = \mathbf{x}^* \mathbf{x}^T \tag{B.22}$$

B.2 LAGRANGE MULTIPLIERS

The procedure of using *Lagrange multipliers* is an elegant technique of obtaining optimum values of a function of several variables subject to one or more constraints. Suppose we want to determine the minimum of a function $f(\mathbf{x})$ of N variables $\mathbf{x} = [x_1, \ldots, x_N]$, subject to a constraint relating x_1 through x_N given in the form

$$g(\mathbf{x}) = 0 \tag{B.23}$$

One straightforward approach would be to solve (B.23) for one of the variables, say x_i, in terms of the remaining ones and then eliminate x_i from $f(\mathbf{x})$. The minimization of $f(\mathbf{x})$ can then be carried out in a usual way to determine the minimum point in the N-dimensional space. In practice, this approach is all but impossible to carry out, especially if $f(\mathbf{x})$ is highly nonlinear.

A simpler yet elegant approach is to introduce an additional parameter λ, called a *Lagrange multiplier.*[†] To motivate this technique through a geometric viewpoint, consider a two-dimensional function

$$f(x_1, x_2) = x_1^2 + x_2^2 \tag{B.24}$$

which is a bowl-shaped surface whose minimum is at the origin $x_1 = x_2 = 0$. Thus minimizing $f(\mathbf{x})$ is the same as minimizing the length of vector $\mathbf{x}$. If there is no constraint, the zero vector is the best $\mathbf{x}$. Now let the constraint be a line

$$x_2 = -\tfrac{1}{2}x_1 + \tfrac{5}{2} \tag{B.25}$$

in the (x_1, x_2) plane. Thus

$$g(\mathbf{x}) = x_1 + 2x_2 - 5 = 0 \tag{B.26}$$

This constraint and the bowl-shaped surface are shown in Figure B.1. The constraint plane cuts through the bowl, creating a parabolic edge, as shown in the figure. Since the point $\mathbf{x}$ is restricted to the constraint line (B.26), the minimization function $f(\mathbf{x})$ is constrained to the parabolic edge. Thus the minimization of (B.24) becomes a problem of finding the point on the parabolic curve that is nearest to the origin. This is also the point on the constraint line that is nearest to the origin and is obtained by drawing a perpendicular ray, as shown in Figure B.1. This point is $x_1 = 1$ and $x_2 = 2$. At this point the parabolic edge achieves its minimum.

How is all this related to the Lagrange multiplier? Referring to Figure B.1, we see at any point P on the constraint surface, the gradient of $f(\mathbf{x})$ is given by vector ∇f. To find the minimum point of $f(\mathbf{x})$ within the constraint surface, we have to find the component $\nabla_{\parallel} f$ of ∇f that lies in the surface and to set it equal to zero, that is,

$$\nabla_{\parallel} f = 0 \tag{B.27}$$

Consider the constraint function $g(\mathbf{x})$ and perturb $\mathbf{x}$ to $\mathbf{x} + \delta\mathbf{x}$ within the surface. Then using the Taylor expansion, we can write

$$g(\mathbf{x} + \delta\mathbf{x}) = g(\mathbf{x}) + \delta\mathbf{x}^T \nabla g(\mathbf{x}) = g(\mathbf{x}) \tag{B.28}$$

since $\mathbf{x} + \delta\mathbf{x}$ is chosen to lie within the surface $g(\mathbf{x}) = 0$. This implies that $\nabla g(\mathbf{x}) = 0$, which means that the gradient $\nabla g(\mathbf{x})$ is normal to the constraint surface. As shown in

---

[†]Although we have reserved λ for eigenvalues, we will follow the tradition and use λ also as a Lagrange multiplier.

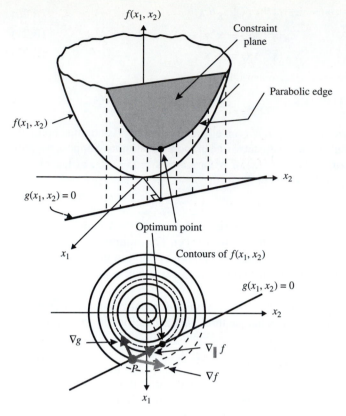

FIGURE B.1
Geometric interpretation of Lagrange multiplier.

Figure B.1, we can now obtain the component $\nabla_{\|} f$ by adding a suitable scaled vector $\nabla g(\mathbf{x})$ to the gradient in the form

$$\nabla_{\|} f = \nabla f + \lambda \nabla g(\mathbf{x}) \tag{B.29}$$

where λ is a Lagrange multiplier. Using linearity of the gradient operator, we introduce the *Lagrangian function*

$$\mathcal{L}(\mathbf{x}, \lambda) \triangleq f(\mathbf{x}) + \lambda g(\mathbf{x}) \tag{B.30}$$

so that the gradient $\nabla \mathcal{L}$ is given by (B.29).

Therefore, to find the minimum of $f(\mathbf{x})$ subject to $g(\mathbf{x}) = 0$, we first define the Lagrangian (B.30) and then find the minimum point of $\mathcal{L}(\mathbf{x}, \lambda)$ by differentiating it with respect to both $\mathbf{x}$ and λ. This results in $N + 1$ equations that can be solved to determine the optimum $\mathbf{x}_o$ and λ_o from which the minimum $f(\mathbf{x}_o)$ can be found. Note that $\partial \mathcal{L}/\partial \lambda = 0$ leads to the constraint $g(\mathbf{x}) = 0$. Thus Lagrange multiplier technique leads to the equations for a constrained minimum, and it does not require us to solve for $g(\mathbf{x}) = 0$.

This technique can be extended to more than one, say K, constraints simply by using one Lagrange multiplier λ_k for each of the constraints $g_k(\mathbf{x}) = 0, k = 1, \dots, K$, and constructing a Lagrangian function of the form

$$\mathcal{L}(\mathbf{x}, \lambda_1, \dots, \lambda_K) = f(\mathbf{x}) + \sum_{k=1}^{K} \lambda_k g_k(\mathbf{x}) \tag{B.31}$$

This Lagrangian is then minimized with respect to $\mathbf{x}$ and $\{\lambda_k\}_1^K$.

EXAMPLE B.1. Consider the problem of fitting the largest (areawise) rectangle inside an ellipse given by

$$\frac{x_1^2}{a^2} + \frac{x_2^2}{b^2} = 1 \tag{B.32}$$

The ellipse and an inscribed rectangle are shown in Figure B.2. Thus the objective function that we want to maximize is

$$f(x_1, x_2) = (2x_1)(2x_2) = 4x_1x_2 \tag{B.33}$$

subject to the constraint

$$g(x_1, x_2) = \frac{x_1^2}{a^2} + \frac{x_2^2}{b^2} - 1 \tag{B.34}$$

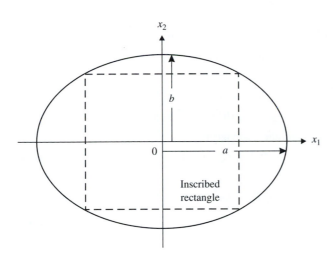

FIGURE B.2
Ellipse and the inscribed rectangle in Example B.1.

Method 1. Solving (B.34) for x_2, we obtain

$$x_2 = \pm \frac{b}{a}\sqrt{a^2 - x_1^2} \tag{B.35}$$

Since the area is positive, choosing the plus sign and substituting in (B.33), we have

$$f(x_1, x_2) = 4\frac{b}{a}x_1\sqrt{a^2 - x_1^2} \tag{B.36}$$

which is a function of x_1 *alone*. Now to obtain the maximum value of $f(x_1, x_2)$, we set

$$\frac{df}{dx_1} = 0 = 4\frac{b}{a}\left(\sqrt{a^2 - x_1^2} - \frac{x_1^2}{\sqrt{a^2 - x_1^2}}\right) \tag{B.37}$$

Thus from (B.37) we get the optimum value of x_1 and subsequently from (B.35) the optimum value for x_2

$$x_{1,o} = \frac{a}{\sqrt{2}} \quad \text{and} \quad x_{2,o} = \frac{b}{\sqrt{2}} \tag{B.38}$$

Method 2. Let us form the Lagrangian

$$\mathcal{L}(x_1, x_2, \lambda) = f(x_1, x_2) + \lambda g(x_1, x_2) = 4x_1x_2 + \lambda\left(\frac{x_1^2}{a^2} + \frac{x_2^2}{b^2} - 1\right) \tag{B.39}$$

Now to find the optimum point, we set

$$\frac{\partial \mathcal{L}}{\partial x_1} = 0 = 4x_2 + \lambda\frac{2x_1}{a^2} \tag{B.40}$$

$$\frac{\partial \mathcal{L}}{\partial x_2} = 0 = 4x_1 + \lambda\frac{2x_2}{b^2} \tag{B.41}$$

$$\frac{\partial \mathcal{L}}{\partial \lambda} = \frac{x_1^2}{a^2} + \frac{x_2^2}{b^2} - 1 \tag{B.42}$$

Solving (B.40) through (B.42), we obtain the optimum values

$$x_{1,o} = \frac{a}{\sqrt{2}} \qquad x_{2,o} = \frac{b}{\sqrt{2}} \qquad \lambda_o = -2ab \tag{B.43}$$

Clearly, the second method is more convenient.

EXAMPLE B.2. Let a real-valued random vector $\mathbf{y}$ be given by

$$\mathbf{y} = \alpha \mathbf{x} + \mathbf{v} \tag{B.44}$$

where $\mathbf{x}$ is a deterministic vector, α is a constant, and $\mathbf{v}$ is a zero-mean random vector with covariance matrix $\mathbf{R_v}$. We want to determine a *best linear unbiased estimator* (*BLUE*) of α, given $\mathbf{y}$. Let

$$\hat{\alpha} = \mathbf{h}^T \mathbf{y} \tag{B.45}$$

Since the estimator must be unbiased, we have

$$\alpha = E\{\hat{\alpha}\} = E\{\mathbf{h}^T \mathbf{y}\} = E\{\mathbf{h}^T(\alpha \mathbf{x} + \mathbf{v})\} = \alpha E\{\mathbf{h}^T \mathbf{x}\} = \alpha \mathbf{h}^T \mathbf{x} \tag{B.46}$$

which implies that $\mathbf{h}^T \mathbf{x} = 1$. Hence the constraint $g(\mathbf{h})$ is

$$g(\mathbf{h}) = \mathbf{h}^T \mathbf{x} - 1 \tag{B.47}$$

Next we want to minimize the variance in the estimation

$$\text{var}(\hat{\alpha}) = \text{var}(\mathbf{h}^T \mathbf{y}) = \text{var}(\mathbf{h}^T \mathbf{v}) = \mathbf{h}^T \mathbf{R_v} \mathbf{h} \tag{B.48}$$

Now to obtain the BLUE of α, consider the Lagrangian

$$\mathcal{L}(\mathbf{h}, \lambda) = \mathbf{h}^T \mathbf{R_v} \mathbf{h} + \lambda(\mathbf{h}^T \mathbf{x} - 1) \tag{B.49}$$

Using (B.5) and (B.6), we obtain

$$\nabla_{\mathbf{h}}(\mathcal{L}) = 2\mathbf{h}_o^T \mathbf{R_v} + \lambda \mathbf{x}^T = \mathbf{0}^T \tag{B.50}$$

or

$$\mathbf{h}_o = \frac{-\lambda}{2} \mathbf{R_v}^{-1} \mathbf{x} \tag{B.51}$$

Substituting (B.51) into (B.47) and solving for λ, we obtain

$$\lambda = -\frac{2}{\mathbf{x}^T \mathbf{R_v} \mathbf{x}}$$

Finally, the optimum estimator becomes

$$\mathbf{h}_o = \frac{\mathbf{R_v}^{-1} \mathbf{x}}{\mathbf{x}^T \mathbf{R_v} \mathbf{x}} \tag{B.52}$$

which can be recognized as a whitening filter and a matched filter.

EXAMPLE B.3. Consider a complex-valued case of the above example. We want to minimize

$$f(\mathbf{h}) = \mathbf{h}^H \mathbf{R_v} \mathbf{h} \tag{B.53}$$

where $\mathbf{R_v}$ is a real-valued symmetric matrix so that $f(\mathbf{h})$ is real, subject to

$$\text{Re}\{\mathbf{h}^H \mathbf{x}\} = b \tag{B.54}$$

Consider $f(\mathbf{h})$ and the constraint function $g(\mathbf{h})$ as

$$f(\mathbf{h}, \mathbf{h}^H) = \mathbf{h}^H \mathbf{R_v} \mathbf{h}$$
$$g(\mathbf{h}, \mathbf{h}^H) = \mathbf{h}^H \mathbf{x} + \mathbf{x}^H \mathbf{h} - 2b \tag{B.55}$$

Thus the Lagrangian is

$$\mathcal{L}(\mathbf{h}, \mathbf{h}^H, \lambda) = \mathbf{h}^H \mathbf{R_v} \mathbf{h} - \lambda(\mathbf{h}^H \mathbf{x} + \mathbf{x}^H \mathbf{h} - 2b) \tag{B.56}$$

Now using (B.20), we get

$$\nabla_{\mathbf{h}}(\mathcal{L}) = \mathbf{h}_o^H \mathbf{R_v} - \lambda \mathbf{x}^H = \mathbf{0}^T \Rightarrow \mathbf{h}_o = \lambda \mathbf{R_v}^{-1} \mathbf{x} \tag{B.57}$$

From the constraint (B.55)

$$\lambda(\mathbf{x}^H \mathbf{R_v}^{-1} \mathbf{x}) = b$$

which gives

$$\mathbf{h}_o = \frac{b \mathbf{R_v}^{-1} \mathbf{x}}{\mathbf{x}^H \mathbf{R_v}^{-1} \mathbf{x}} \tag{B.58}$$

MATLAB Functions

In this appendix, we provide a brief one-line description of MATLAB functions that were referred to in this book. The source of each function is given in parentheses where detailed information can be found. Page numbers for functions explicitly discussed in the text are also given.

TABLE C.1
MATLAB functions.

Function	Description	Page
a2r	Direct parameters to autocorrelation conversion	367
aplatest	Estimation of all-pole lattice parameters (Book toolbox)	460
arls	AR model estimation using the LA criterion without windowing (Book toolbox)	451
armals	ARMA model estimation using the LA criterion without windowing (Book toolbox)	466
arwin	AR model estimation using the LA criterion without windowing (Book toolbox)	451
autoc	Computation of autocovariance sequence (Book toolbox)	210
autocfft	Computation of autocovariance sequence using the FFT (Book toolbox)	210
bartlett	Computation of Bartlett window coefficients (MATLAB)	230
boxcar	Computation of rectangular window coefficients (MATLAB)	206
bt_psd	Blackman-Tukey power spectral density computation (Book toolbox)	227
chebwin	Computation of Chebyshev window coefficients (MATLAB)	206
chol	Computation of Cholesky decomposition (Book toolbox)	278
cohere	Coherence function estimation (MATLAB SP toolbox)	241
conv	Convolution sum computation (MATLAB)	48
corr	Computation of cross-correlation sequence (MATLAB)	
csd	Cross-spectral density computation (MATLAB SP toolbox)	240
cumsum	Cumulative-sum computation (MATLAB)	
df2latcf	Direct-form to lattice-form conversion (Book toolbox)	67
df2ldrf	Direct-form to lattice/ladder-form conversion (Book toolbox)	
dpss	Discrete prolate spheroidal sequence window coefficient computation (MATLAB SP toolbox)	248
dtfgn	Generation of discrete fractional Gaussian noise (Book toolbox)	721
durbin	Implementation of Durbin algorithm (Book toolbox)	358
eig	Computes eigenvalues and eigenvectors of a matrix (MATLAB)	
esprit_ls	Least-squares ESPRIT for frequency estimation (Book toolbox)	493
esprit_tls	Total least-squares ESPRIT for frequency estimation (Book toolbox)	493
ev_method	Eigenvector method for frequency estimation (Book toolbox)	488
faest	FAEST RLS algorithm (Book toolbox)	576
filter	Direct-form-II filter implementation (MATLAB)	50
filtic	Computation of direct-form-II filter initial conditions (MATLAB SP toolbox)	50
firlms	FIR LMS adaptive filtering algorithm (Book toolbox)	526
hamming	Computation of Hamming window coefficients (MATLAB)	206
hanning	Computation of Hann window coefficients (MATLAB)	206

TABLE C.1
MATLAB functions. (Con't)

Function	Description	Page
invschur	Implementation of inverse Schür algorithm (Book toolbox)	375
invtoepl	Computation of $\mathbf{R}^{-1}$ when $\mathbf{R}$ is Toeplitz (Book toolbox)	378
k2r	Lattice parameters to autocorrelation sequence conversion (Book toolbox)	367
kaiser	Computation of Kaiser window coefficients (MATLAB)	208
ladrfilt	Lattice/ladder filter implementation (Book toolbox)	
latcf2df	Lattice to direct-form conversion (Book toolbox)	67
latcfilt	Lattice filter implementation (Book toolbox)	68
ldlt	Computes the LDU decomposition (Book toolbox)	277
ldltchol	Computes LDL^T using chol	278
ldrf2df	Lattice/ladder to direct-form conversion (Book toolbox)	
lduneqs	Solution of normal equations using LDU decomposition (Book toolbox)	277
levins	Implementation of Levinson's algorithm (Book toolbox)	359
lsigest	Computation of LS signal estimators (Book toolbox)	288
lsmatvec	Computation of $\mathbf{R}$ and $\mathbf{d}$ for FIR LS filtering (Book toolbox)	408
lu	LU decomposition (MATLAB)	
mgs	Implementation of modified GL algorithm (Book toolbox)	430
minnorm	Minimum-norm method for frequency estimation (Book toolbox)	488
music	MUSIC frequency estimation (Book toolbox)	485
phd	Pisarenko harmonic decomposition (Book toolbox)	484
pmtm	Power spectrum estimation via Thomson multitaper method (MATLAB SP toolbox)	248
psd	Power spectrum estimation via Welch's method (MATLAB SP toolbox)	213, 232
pzls	Pole-zero coefficient estimation using the LS criterion (Book toolbox)	463
qr	Computation of QR decomposition (MATLAB)	424
rand	Generates pseudorandom numbers that are uniformly distributed over $(0, 1)$ (MATLAB)	83
randn	Generates $\mathcal{N}(0, 1)$ pseudorandom numbers (MATLAB)	83
rls	Implementation of conventional RLS algorithm (Book toolbox)	
rootmusic	Root-MUSIC frequency estimation (Book toolbox)	485
schurlg	Schür algorithm (Book toolbox)	370
stablepdf	Computes pdf plots of stable distributions numerically (Book toolbox)	95
stepdown	Lattice-form to direct-form conversion in Levinson algorithm (Book toolbox)	366
stepup	Direct-form to lattice-form conversion in Levinson algorithm (Book toolbox)	366
svd	Computation of SVD (MATLAB)	436
tfe	Transfer function estimation (MATLAB SP toolbox)	243
toeplitz	Toeplitz matrix from first row and column (MATLAB)	48
triang	Computation of triangular window coefficients (MATLAB)	
udut	Computation of UDU^H decomposition (Book toolbox)	

Useful Results from Matrix Algebra

In this appendix, we review the fundamental concepts of linear algebra in complex-valued space. The aim is to present as many possible concepts as are necessary to understand the book. For a complete treatment, refer to many excellent references in literature including Leon (1998), Strang (1980), and Gill et al. (1981).

D.1 COMPLEX-VALUED VECTOR SPACE

The *unitary complex space* $\mathbb{C}^N$ is defined as the space of all the N-dimensional complex-valued vectors, which are denoted by a boldface letter, or by the N-tuple of its component, for example,

$$\mathbf{x} = [x_1 \ x_2 \ \cdots \ x_N]^T = [x_1^* \ x_2^* \ \cdots \ x_N^*]^H \tag{D.1.1}$$

where we use the following notation for the superscripts: T means transpose, $*$ means conjugate, and H means conjugate (of the) transpose, or adjoint. In the case of real-valued vectors, the real space is denoted by $\mathbb{R}^N$ and is also known as the *Euclidean* space.

Some Definitions

1. The *inner product* between two vectors $\mathbf{x}$ and $\mathbf{y}$ is defined by

$$\langle \mathbf{x}, \mathbf{y} \rangle = \mathbf{x}^H \mathbf{y} = \sum_{i=1}^{N} x_i^* y_i \tag{D.1.2}$$

2. Two vectors $\mathbf{x}$ and $\mathbf{y}$ are *orthogonal* if their inner product is zero, that is,

$$\mathbf{x}^H \mathbf{y} = 0 \tag{D.1.3}$$

The zero vector $\mathbf{0}$ is orthogonal to any vector in the same space.

3. The norm of a vector provides a measure of the "size" of a vector. It is a nonnegative number $\|\mathbf{x}\|$ that satisfies the following properties:

   a. $\|\mathbf{x}\| > 0$ for $\mathbf{x} \neq \mathbf{0}$ and $\|\mathbf{0}\| = 0$.
   b. $\|\alpha \mathbf{x}\| = |\alpha| \|\mathbf{x}\|$ for any complex number α.
   c. $\|\mathbf{x} + \mathbf{y}\| \leq \|\mathbf{x}\| + \|\mathbf{y}\|$ (triangle inequality).

   The *p norm* of $\mathbf{x}$ is defined as

$$\|\mathbf{x}\|_p = \left(\sum_{i=1}^{N} |x_i|^p \right)^{1/p} \tag{D.1.4}$$

which satisfies all three properties given above. For $p = 2$, we obtain the *Euclidean norm* $\|\mathbf{x}\|_2$ which, for simplicity, is denoted by $\|\mathbf{x}\|$. It is defined as

$$\|\mathbf{x}\| = \sqrt{\mathbf{x}^H\mathbf{x}} = \sqrt{\sum_{i=1}^{N} |x_i|^2} \tag{D.1.5}$$

4. An *orthonormalized set* is a set of L vectors $\mathbf{x}_l, l = 1, 2, \ldots, L$, such that

$$\mathbf{x}_l^H\mathbf{x}_k = \begin{cases} 1 & l = k \\ 0 & l \neq k \end{cases} \tag{D.1.6}$$

5. *Cauchy–Schwartz inequality*: Two vectors $\mathbf{x}$ and $\mathbf{y}$ belonging to the same space satisfy

$$|\mathbf{x}^H\mathbf{y}| \leq \|\mathbf{x}\| \cdot \|\mathbf{y}\| \tag{D.1.7}$$

where the equality applies when $\mathbf{x} = a\mathbf{y}$, with a being a (real- or complex-valued) scalar.

6. The *angle θ between two vectors* is defined as

$$\cos\theta = \frac{\mathbf{x}^H\mathbf{y}}{\|\mathbf{x}\| \cdot \|\mathbf{y}\|} \tag{D.1.8}$$

D.2 MATRICES

A rectangular array of $N \times M$ complex numbers ordered in N rows and M columns is called a *matrix* and is denoted by capital boldface letters, for example,

$$\mathbf{A} = [a_{i,k}] \quad 1 \leq i \leq N, 1 \leq k \leq M \tag{D.2.1}$$

Any linear transformation from space $\mathbb{C}^N$ into space $\mathbb{C}^M$ can be represented by a suitable $N \times M$ matrix, if two bases in $\mathbb{C}^N$ and $\mathbb{C}^M$ are already defined. Linear transformations from space $\mathbb{C}^N$ into space $\mathbb{C}^N$ are given by square $N \times N$ non-singular matrices, in which case the transformation can be considered as a change of basis. We consider square matrices for the following development.

D.2.1 Some Definitions

1. A system of *linearly independent* vectors $\mathbf{e}_1, \mathbf{e}_2, \ldots, \mathbf{e}_N$ in a complex space $\mathbb{C}^N$ is called a *basis* for $\mathbb{C}^N$ if it is possible to express any vector $\mathbf{x} \in \mathbb{C}^N$ by means of N coefficients $a_1, a_2, \ldots, a_N$ as

$$\mathbf{x} = a_1\mathbf{e}_1 + a_2\mathbf{e}_2 + \cdots + a_N\mathbf{e}_N = \sum_{i=1}^{N} a_i\mathbf{x}_i \tag{D.2.2}$$

If a vector has the components $x_1, x_2, \ldots, x_N$ in a given basis, then the linearly transformed vector $\mathbf{y}$ has components

$$y_1 = a_{11}x_1 + \cdots + a_{1N}x_N$$
$$\vdots \tag{D.2.3}$$
$$y_N = a_{N1}x_1 + \cdots + a_{NN}x_N$$

in the basis defined by the transformation

$$\mathbf{A} = \begin{bmatrix} a_{11} & \cdots & a_{1N} \\ a_{21} & \cdots & a_{2N} \\ \vdots & \ddots & \vdots \\ a_{N1} & \cdots & a_{NN} \end{bmatrix} \tag{D.2.4}$$

This transformation can be expressed, using the well-known row-by-column product between matrices and vectors, as

$$\mathbf{y} = \mathbf{Ax} \tag{D.2.5}$$

2. The transformation from $\mathbf{y}$ to $\mathbf{x}$ is called an *inverse transformation*, which is again *linear*. It is written as

$$\mathbf{x} = \mathbf{A}^{-1}\mathbf{y} \tag{D.2.6}$$

where $\mathbf{A}^{-1}$ (if it exists) is a matrix and is known as an inverse of $\mathbf{A}$, defined in (D.3.5).
3. The transformation that leaves unchanged the vector basis is said to be the *identity transformation*, and the related matrix is indicated generally by $\mathbf{I}$, which is given by

$$\mathbf{I} = \begin{bmatrix} 1 & 0 & 0 & \cdots & 0 \\ 0 & 1 & 0 & \cdots & 0 \\ 0 & 0 & 1 & \cdots & 0 \\ \vdots & \vdots & \vdots & \ddots & \vdots \\ 0 & 0 & 0 & \cdots & 1 \end{bmatrix} \tag{D.2.7}$$

4. Two linear transformations of C^N into itself can be applied to a vector, obtaining a third transformation, called the *product transformation*

$$\mathbf{y} = \mathbf{Ax} \qquad \mathbf{z} = \mathbf{By} = \mathbf{B}(\mathbf{Ax}) = (\mathbf{BA})\mathbf{x} \qquad \Rightarrow z = \mathbf{Cx} \tag{D.2.8}$$

where the matrix $\mathbf{C}$ is the product of $\mathbf{B}$ and $\mathbf{A}$. In general, the matrix product is not *commutative*, that is, $\mathbf{AB} \neq \mathbf{BA}$.
5. The operation of *transposition* of a matrix inverts the orders of rows and columns; that is, element a_{ij} takes the place of a_{ji} in the new matrix. Similarly, the conjugate transpose of a matrix $\mathbf{A}$ is a matrix in which element a_{ij}^* takes the place of a_{ji}. The operations of conjugation and transposition are commutative, that is,

$$\mathbf{A}^H = (\mathbf{A}^*)^T = (\mathbf{A}^T)^* \tag{D.2.9}$$

6. A *matrix norm* $\|\mathbf{A}\|$ satisfies the following properties:

   a. $\|\mathbf{A}\| > 0$ for $\mathbf{A} \neq \mathbf{0}$ and $\|\mathbf{0}\| = 0$.
   b. $\|\alpha\mathbf{A}\| = |\alpha|\|\mathbf{A}\|$ for any complex number α.
   c. $\|\mathbf{A} + \mathbf{B}\| \leq \|\mathbf{A}\| + \|\mathbf{B}\|$ (triangle inequality).
   d. $\|\mathbf{A}\mathbf{B}\| \leq \|\mathbf{A}\|\|\mathbf{B}\|$, which is needed because the matrix multiplication operation creates new matrices.

An important matrix norm is the *Frobenius norm*, defined as

$$\|\mathbf{A}\|_F \triangleq \sqrt{\sum_{i=1}^{N}\sum_{k=1}^{N} |a_{ik}|^2} \tag{D.2.10}$$

which treats the matrix as a "long vector." Using any vector p norm, we can obtain the matrix norm

$$\|\mathbf{A}\|_p \triangleq \max_{\mathbf{x}\neq\mathbf{0}} \frac{\|\mathbf{Ax}\|_p}{\|\mathbf{x}\|_p} \tag{D.2.11}$$

which measures the *amplification power* of matrix $\mathbf{A}$. The matrix norm for $p = 2$ is known as the *spectral* norm and is of great theoretical significance, and it is simply denoted by $\|\mathbf{A}\|$. When a matrix acts upon a vector $\mathbf{x}$ of length $\|\mathbf{x}\|_p$, it transforms $\mathbf{x}$ into vector $\mathbf{Ax}$ of length $\|\mathbf{Ax}\|_p$. The ratio $\|\mathbf{Ax}\|_p/\|\mathbf{x}\|_p$ provides the magnification factor of the linear transformation $\mathbf{Ax}$. The number $\|\mathbf{A}\|_p$ is the maximum magnification caused by $\mathbf{A}$. Similarly, the minimum magnification due to $\mathbf{A}$ is given by

$$\min |\mathbf{A}|_p \triangleq \min_{\mathbf{x}\neq\mathbf{0}} \frac{\|\mathbf{Ax}\|_p}{\|\mathbf{x}\|_p} \tag{D.2.12}$$

and the ratio $\|\mathbf{A}\|_p / \min |\mathbf{A}|_p$ characterizes the *dynamic range* of the linear transformation performed by matrix $\mathbf{A}$. This interpretation provides a nice geometric picture for the concept of condition number (see Section D.3.2).

7. A matrix $\mathbf{A}$ is called *Hermitian* if

$$\mathbf{A}^H = \mathbf{A} \tag{D.2.13}$$

and a *Hermitian form* $H(\mathbf{x}, \mathbf{x})$ is the second-order real homogeneous polynomial

$$H(\mathbf{x}, \mathbf{x}) = \sum_{i=1}^{N} \sum_{k=1}^{N} h_{ik} x_i^* x_k \qquad h_{ik} = h_{ki}^* \tag{D.2.14}$$

8. A real-valued matrix $\mathbf{A}$ is called *symmetric* if $\mathbf{A}^T = \mathbf{A}$ and a *quadratic* form $Q(\mathbf{x}, \mathbf{x})$ is the second-order real homogeneous polynomial

$$Q(\mathbf{x}, \mathbf{x}) = \sum_{i=1}^{N} \sum_{k=1}^{N} h_{ik} x_i x_k \qquad h_{ik} = h_{ki} \tag{D.2.15}$$

9. Matrix $\mathbf{L}$ is called a *lower triangular* matrix if all elements above the principal diagonal are zero. Similarly, matrix $\mathbf{U}$ is called an *upper diagonal* matrix if all elements below the principal diagonal are zero.

10. The *trace* of a matrix is the sum of the elements of its principal diagonal, that is,

$$\text{tr}(\mathbf{A}) = \sum_{i=1}^{N} a_{ii} \tag{D.2.16}$$

with the property $\qquad \text{tr}(\mathbf{AB}) = \text{tr}(\mathbf{BA}) = \text{tr}(\mathbf{A}^H \mathbf{B}^H) \tag{D.2.17}$

for any square matrices $\mathbf{A}$ and $\mathbf{B}$.

11. A *diagonal* matrix is a square $N \times N$ matrix with $a_{ij} = 0$ for $i \neq j$; that is, all elements off the principal diagonal are zero. It appears as

$$\mathbf{A} = \begin{bmatrix} a_{11} & 0 & \cdots & 0 \\ 0 & a_{22} & \cdots & 0 \\ \vdots & \vdots & \ddots & \vdots \\ 0 & 0 & \cdots & a_{NN} \end{bmatrix} \tag{D.2.18}$$

12. A *Toeplitz* matrix is defined as

$$\mathbf{A} = [a_{i,k}] = [a_{i-k}] \qquad 1 \leq i \leq N, \, 1 \leq k \leq M \tag{D.2.19}$$

A square Toeplitz matrix appears as

$$\mathbf{A} = \begin{bmatrix} a_0 & a_{-1} & a_{-2} & \cdots & a_{1-N} \\ a_1 & a_0 & a_{-1} & \cdots & a_{2-N} \\ a_1 & a_1 & a_0 & \cdots & a_{3-N} \\ \vdots & \vdots & \vdots & \ddots & \vdots \\ a_{N-1} & a_{N-2} & a_{N-3} & \cdots & a_0 \end{bmatrix} \tag{D.2.20}$$

13. A matrix is called *persymmetric* if it is symmetric about the cross-diagonal, that is, $a_{ij} = a_{N-j+1, N-i+1}, 1 \leq i \leq N, 1 \leq j \leq N$.

14. The *exchange* matrix $\mathbf{J}$ is defined by

$$\mathbf{J} \triangleq \begin{bmatrix} 0 & \cdots & 0 & 0 & 1 \\ 0 & \cdots & 0 & 1 & 0 \\ 0 & \cdots & 1 & 0 & 0 \\ \vdots & & \vdots & \vdots & \vdots \\ 1 & \cdots & 0 & 0 & 0 \end{bmatrix}$$

and has the following properties

$$\mathbf{J}^2 = \mathbf{J}$$
$$\mathbf{J}^T = \mathbf{J}$$
$$\mathbf{JA} = \texttt{flipud(A)}$$
$$\mathbf{AJ} = \texttt{fliplr(A)}$$

where the MATLAB functions `flipud(A)` and `fliplr(A)` reverse the order of rows and columns of a matrix $\mathbf{A}$, respectively.

15. A matrix is called *centrosymmetric* if it is both symmetric and persymmetric. It can be easily seen that a centrosymmetric matrix has the property $\mathbf{J}^T \mathbf{AJ} = \mathbf{A}$ when $\mathbf{A}$ is real or $\mathbf{J}^T \mathbf{AJ} = \mathbf{A}^*$ when $\mathbf{A}$ complex.

16. A matrix is called *Hankel* if the elements along the secondary diagonals, that is, the diagonals that are perpendicular to the main diagonal, are equal. If $\mathbf{A}$ is Hankel, then $\mathbf{JA}$ is Toeplitz.

17. The inverse of a triangular, symmetric, Hermitian, persymmetric and centrosymmetric matrix has the same structure. The inverse of a Toeplitz matrix is persymmetric and the inverse of a Hankel matrix is symmetric.

18. A *partition* of an $N \times M$ matrix $\mathbf{A}$ is a notational rearrangement in terms of its submatrices. For example, a 2×2 partitioning of $\mathbf{A}$ is

$$\mathbf{A} = \begin{bmatrix} \mathbf{A}_{11} & \mathbf{A}_{12} \\ \mathbf{A}_{21} & \mathbf{A}_{22} \end{bmatrix} \tag{D.2.21}$$

where each "element" $\mathbf{A}_{ik}$ is a submatrix of $\mathbf{A}$.

D.2.2 Properties of Square Matrices

1. The operations of transposition T, conjugation $*$, or both H are distributive, that is,

$$(\mathbf{A} + \mathbf{B})^T = \mathbf{A}^T + \mathbf{B}^T$$
$$(\mathbf{A} + \mathbf{B})^* = \mathbf{A}^* + \mathbf{B}^*$$
$$(\mathbf{A} + \mathbf{B})^H = \mathbf{A}^H + \mathbf{B}^H \tag{D.2.22}$$

2. For the operators T, H, or -1 (inversion), we have

$$(\mathbf{AB})^T = \mathbf{B}^T \mathbf{A}^T$$
$$(\mathbf{AB})^H = \mathbf{B}^H \mathbf{A}^H$$
$$(\mathbf{AB})^{-1} = \mathbf{B}^{-1} \mathbf{A}^{-1} \tag{D.2.23}$$

3. The operators $*$, T, H, and -1 are commutative, for example,

$$(\mathbf{A}^H)^{-1} = (\mathbf{A}^{-1})^H \tag{D.2.24}$$

Thus we can use the compact notation $\mathbf{A}^{-T}$, or $\mathbf{A}^{-*}$, etc.

4. Given any matrix $\mathbf{A}$, matrix $\mathbf{B} = \mathbf{A}^H \mathbf{A}$ is Hermitian [see (D.2.13)] and if $\mathbf{A}$ is invertible, then for such a $\mathbf{B}$, we have

$$\mathbf{A}^{-H} \mathbf{B} \mathbf{A}^{-1} = \mathbf{A}^{-H} \mathbf{A}^H \mathbf{A} \mathbf{A}^{-1} = \mathbf{I} \tag{D.2.25}$$

5. If $\mathbf{H}$ is the matrix of the coefficients h_{ik}, the Hermitian form (D.2.14) can be written as

$$H(\mathbf{x}, \mathbf{x}) = \mathbf{x}^H \mathbf{Hx} = \langle \mathbf{x}, \mathbf{Hx} \rangle \tag{D.2.26}$$

Similarly, If $\mathbf{H}$ is the real-valued matrix of the coefficients h_{ik}, the quadratic form (D.2.15) can be written as

$$H(\mathbf{x}, \mathbf{x}) = \mathbf{x}^T \mathbf{Hx} = \langle \mathbf{x}, \mathbf{Hx} \rangle \tag{D.2.27}$$

6. A Hermitian matrix $\mathbf{A}$ is called

Positive definite if $\quad\quad \mathbf{x}^H\mathbf{A}\mathbf{x} > 0$

Positive semidefinite if $\quad \mathbf{x}^H\mathbf{A}\mathbf{x} \geq 0 \quad$ (also nonnegative definite)

Negative definite if $\quad\quad \mathbf{x}^H\mathbf{A}\mathbf{x} < 0$ $\quad\quad\quad\quad\quad\quad\quad\quad$ (D.2.28)

Negative semidefinite if $\quad \mathbf{x}^H\mathbf{A}\mathbf{x} \leq 0 \quad$ (also nonpositive definite)

for all $\mathbf{x} \neq \mathbf{0}$.

7. The operation of the trace of a matrix satisfies

$$\mathrm{tr}(\mathbf{A} \pm \mathbf{B}) = \mathrm{tr}(\mathbf{A}) \pm \mathrm{tr}(\mathbf{B}) \tag{D.2.29}$$

$$\mathrm{tr}(k\mathbf{A}) = k\,\mathrm{tr}(\mathbf{A}) \tag{D.2.30}$$

$$\mathrm{tr}(\mathbf{A}\mathbf{B}) = \mathrm{tr}(\mathbf{B}\mathbf{A}) \tag{D.2.31}$$

$$\mathrm{tr}(\mathbf{B}^{-1}\mathbf{A}\mathbf{B}) = \mathrm{tr}(\mathbf{A}) \tag{D.2.32}$$

$$\mathrm{tr}(\mathbf{A}\mathbf{A}^H) = \sum_{i=1}^{N}\sum_{j=1}^{N}|a_{ij}|^2 \tag{D.2.33}$$

D.3 DETERMINANT OF A SQUARE MATRIX

The determinant of a square matrix $\mathbf{A}$ is denoted by

$$\det(\mathbf{A}) \triangleq \begin{vmatrix} a_{11} & a_{12} & \cdots & a_{1N} \\ a_{21} & a_{22} & \cdots & a_{2N} \\ \vdots & \vdots & \ddots & \vdots \\ a_{N1} & a_{N2} & \cdots & a_{NN} \end{vmatrix} \tag{D.3.1}$$

and is equal to the sum of the products of the elements of any row or column and their respective cofactors, that is,

$$\det(\mathbf{A}) = a_{i1}C_{i1} + a_{i2}C_{i2} + \cdots + a_{iN}C_{iN} \tag{D.3.2}$$

or $\quad\quad\quad\quad \det(\mathbf{A}) = a_{1k}C_{1k} + a_{2k}C_{2k} + \cdots + a_{Nk}C_{Nk} \tag{D.3.3}$

where the C_{ik} are called *cofactors*, given by

$$C_{ik} = (-1)^{i+k}\det(\mathbf{A}_{ik}) \tag{D.3.4}$$

where $\mathbf{A}_{ik}$ is an $(N-1)$st-order square matrix obtained by deleting the ith row and kth column. Thus the determinant needs to be computed recursively; that is, the Nth-order determinant is computed from the $(N-1)$st-order determinant, which in turn is computed from the $(N-2)$nd-order, and so on. If $\det(\mathbf{A}) \neq 0$, then the inverse $\mathbf{A}^{-1}$ of $\mathbf{A}$ exists and is unique. The $\mathbf{A}^{-1}$ matrix is given by

$$\mathbf{A}^{-1} = \frac{1}{\det(\mathbf{A})} \begin{vmatrix} C_{11} & C_{21} & \cdots & C_{N1} \\ C_{12} & C_{22} & \cdots & C_{N2} \\ \vdots & \vdots & \ddots & \vdots \\ C_{1N} & C_{2N} & \cdots & C_{NN} \end{vmatrix} \tag{D.3.5}$$

D.3.1 Properties of the Determinant

Below we provide some useful properties of the determinant.

1. If a row (or column) of a matrix is a linear combination of other rows (or columns), then $\det(\mathbf{A}) = 0$. In particular, if (a) a row (or column) is proportional or equal to another row (or column) or (b) a row (or column) is identically zero, then $\det(\mathbf{A}) = 0$.

2. If two rows (or columns) are exchanged with each other, then the determinant changes its sign.
3. For a triangular matrix (upper or lower) $\mathbf{A}$, the determinant is obtained by multiplying all the elements of its principal diagonal, that is,

$$\det(\mathbf{A}) = \prod_{n=1}^{N} a_{nn} \tag{D.3.6}$$

4. The $\det(\mathbf{A})$ is unchanged if $\mathbf{A}$ is replaced by its transpose $\mathbf{A}^T$; that is,

$$\det(\mathbf{A}) = \det(\mathbf{A}^T) \tag{D.3.7}$$

5. Using the above property, we also claim that the determinant of a Hermitian matrix is real, since

$$\det(\mathbf{A}) = \det(\mathbf{A}^H) = \det(\mathbf{A}^T) \Rightarrow \det(\mathbf{A}) = \det(\mathbf{A}^*) = \det(\mathbf{A})^* \tag{D.3.8}$$

6. The determinant of a product of matrices is the product of their determinants; that is,

$$\det(\mathbf{AB}) = \det(\mathbf{A})\det(\mathbf{B}) \tag{D.3.9}$$

7. If matrix $\mathbf{A}$ is nonsingular, that is, its inverse $\mathbf{A}^{-1}$ exists, then

$$\det(\mathbf{A}^{-1}) = [\det(\mathbf{A})]^{-1} = \frac{1}{\det(\mathbf{A})} \tag{D.3.10}$$

8. Given an arbitrary constant c (possibly complex-valued), we have

$$\det(c\mathbf{A}) = c^N \det(\mathbf{A}) \tag{D.3.11}$$

D.3.2 Condition Number

One of the important equations in signal processing is the linear equation $\mathbf{Rc} = \mathbf{d}$, where $\mathbf{R}$ is a matrix of known values, $\mathbf{d}$ is a vector of known quantities, and $\mathbf{c}$ is a vector of unknown coefficients. The investigation of how the solution of $\mathbf{Rc} = \mathbf{d}$ is affected by small changes (perturbations) in the elements of $\mathbf{R}$ and $\mathbf{d}$ leads to an important characteristic number of matrix $\mathbf{R}$, called the *condition number*.

If vector $\mathbf{d}$ is perturbed to $\mathbf{d} + \delta\mathbf{d}$, the exact solution $\mathbf{c}$ is perturbed to $\mathbf{c} + \delta\mathbf{c}$. Therefore,

$$\mathbf{R}(\mathbf{c} + \delta\mathbf{c}) = \mathbf{d} + \delta\mathbf{d}$$

which implies that $\qquad \delta\mathbf{c} = \mathbf{R}^{-1}\delta\mathbf{d} \qquad$ since $\mathbf{Rc} = \mathbf{d}$ $\qquad$ (D.3.12)

or using property 4 of matrix norm

$$\|\delta\mathbf{c}\| \leq \|\mathbf{R}^{-1}\|\,\|\delta\mathbf{d}\| \tag{D.3.13}$$

From the same norm property and $\mathbf{d} = \mathbf{Rc}$, we obtain

$$\|\mathbf{d}\| \leq \|\mathbf{R}\|\,\|\mathbf{c}\| \tag{D.3.14}$$

Multiplying (D.3.13) by (D.3.14) and solving, we obtain

$$\frac{\|\delta\mathbf{c}\|}{\|\mathbf{c}\|} \leq \|\mathbf{R}\|\,\|\mathbf{R}^{-1}\|\,\frac{\|\delta\mathbf{d}\|}{\|\mathbf{d}\|} \tag{D.3.15}$$

Similarly, keeping d constant and perturbing $\mathbf{R}$ to $\mathbf{R} + \delta\mathbf{R}$, we have

$$(\mathbf{R} + \delta\mathbf{R})(\mathbf{c} + \delta\mathbf{c}) = \mathbf{d}$$

from which, after ignoring the second-order product term $\delta\mathbf{R}\,\delta\mathbf{c}$, we obtain

$$\frac{\|\delta\mathbf{c}\|}{\|\mathbf{c}\|} \leq \|\mathbf{R}\|\,\|\mathbf{R}^{-1}\|\,\frac{\|\delta\mathbf{R}\|}{\|\mathbf{R}\|} \tag{D.3.16}$$

A careful inspection of (D.3.15) and (D.3.16) shows that the *relative error* in the exact solution is bounded by the number

$$\text{cond }(\mathbf{R}) \triangleq \|\mathbf{R}\| \, \|\mathbf{R}^{-1}\| \qquad (D.3.17)$$

which is known as the *condition number* of matrix $\mathbf{R}$, multiplied by the relative perturbation in the data ($\mathbf{R}$ or $\mathbf{d}$). When relatively small perturbations in $\mathbf{R}$ cause relatively small (large) perturbations in the solution of $\mathbf{Rc} = \mathbf{d}$, matrix $\mathbf{R}$ is said to be *well* (*ill*) *conditioned*. Clearly, ill-conditioned matrices have large condition numbers, and therefore their large magnification power amplifies small perturbations to the extent that makes the obtained solution totally inaccurate.

Since the norm of the identity matrix $\|\mathbf{I}\| = 1$, we have

$$\|\mathbf{I}\| = \|\mathbf{R}\,\mathbf{R}^{-1}\| \le \|\mathbf{R}\| \, \|\mathbf{R}^{-1}\| = \text{cond }(\mathbf{R})$$

that is,
$$\text{cond }(\mathbf{R}) \ge 1 \qquad (D.3.18)$$

The best possible condition number is 1.

D.4 UNITARY MATRICES

A matrix $\mathbf{A}$ is called a *unitary* matrix if its inverse is equal to its conjugate transpose, that is,

$$\mathbf{A}^{-1} = \mathbf{A}^H \Rightarrow \mathbf{A}^H \mathbf{A} = \mathbf{I} \qquad (D.4.1)$$

For a real-valued matrix, $\mathbf{A}$ is called an *orthogonal* matrix if its inverse is equal to its transpose, that is,

$$\mathbf{A}^{-1} = \mathbf{A}^T \Rightarrow \mathbf{A}^T \mathbf{A} = \mathbf{I} \qquad (D.4.2)$$

If we write the unitary matrix $\mathbf{A}$ as a set of N column vectors, that is,

$$\mathbf{A} = [\mathbf{a}_1 \ \mathbf{a}_2 \ \cdots \ \mathbf{a}_N] \qquad (D.4.3)$$

then we can show that

$$\mathbf{a}_i^H \mathbf{a}_k = \begin{cases} 1 & i = k \\ 0 & i \ne k \end{cases} \triangleq \delta_{ik} \qquad (D.4.4)$$

that is, the column vectors of a unitary matrix are orthonormal.

A transformation is called a *unitary transformation* if the transformation matrix is unitary. Vector inner products, vector norms, and angles between two vectors are *invariant* (i.e., they are preserved) under unitary transformation. Thus given two vectors $\mathbf{x}$ and $\mathbf{y}$ and a unitary matrix $\mathbf{A}$, we have

$$\langle \mathbf{x}, \mathbf{y} \rangle = \langle \mathbf{Ax}, \mathbf{Ay} \rangle \qquad (D.4.5)$$

and
$$\|\mathbf{x}\|^2 = \|\mathbf{Ax}\|^2 \qquad (D.4.6)$$

This implies that the absolute value of the determinant of a unitary matrix is unity, or

$$|\det(\mathbf{A})| = 1 \qquad \mathbf{A} \text{ unitary} \qquad (D.4.7)$$

since from (D.4.1), (D.3.9), and (D.3.8), we have

$$\det(\mathbf{I}) = \det(\mathbf{A}^H \mathbf{A}) = \det(\mathbf{A}^H) \det(\mathbf{A}) = \det(\mathbf{A})^* \det(\mathbf{A}) = |\det(\mathbf{A})|^2 = 1 \qquad (D.4.8)$$

D.4.1 Hermitian Forms after Unitary Transformations

Let $H(\mathbf{y}, \mathbf{y}) = \langle \mathbf{y}, \mathbf{Ry} \rangle = \mathbf{y}^H \mathbf{Ry}$ be an arbitrary Hermitian form for any matrix $\mathbf{R}$. Define a transformation $\mathbf{y} = \mathbf{Ax}$ for any unitary matrix $\mathbf{A}$. Then we can write $H(\mathbf{y}, \mathbf{y})$ as

$$H(\mathbf{y}, \mathbf{y}) = \mathbf{x}^H \mathbf{A}^H \mathbf{RAx} = \mathbf{x}^H \mathbf{Px} \qquad (D.4.9)$$

where
$$\mathbf{P} = \mathbf{A}^H \mathbf{R} \mathbf{A} = \mathbf{A}^{-1} \mathbf{R} \mathbf{A} \tag{D.4.10}$$

Matrix $\mathbf{R}$ can be reduced to a diagonal form by unitary transformation
$$\mathbf{U}^H \mathbf{R} \mathbf{U} = \mathbf{\Lambda} = \operatorname{diag}(\lambda_1, \lambda_2, \ldots, \lambda_N) \tag{D.4.11}$$

Hence the Hermitian form $H(\mathbf{y}, \mathbf{y})$ can be written as
$$H(\mathbf{y}, \mathbf{y}) = \mathbf{y}^H \mathbf{R} \mathbf{y} = \mathbf{y}^H \mathbf{U} \mathbf{\Lambda} \mathbf{U}^H \mathbf{y} = \mathbf{x}^H \mathbf{\Lambda} \mathbf{x} = \langle \mathbf{x}, \mathbf{\Lambda} \mathbf{x} \rangle \tag{D.4.12}$$

where $\mathbf{x} \triangleq \mathbf{A}\mathbf{y} \triangleq \mathbf{U}^H \mathbf{y}$. Therefore, we can write
$$H(\mathbf{y}, \mathbf{y}) = \sum_{i=1}^{N} \sum_{k=1}^{N} r_{ik} y_i^* y_k = \sum_{i=1}^{N} \lambda_i |x_i|^2 \tag{D.4.13}$$

D.4.2 Significant Integral of Quadratic and Hermitian Forms

Consider a quadratic form $Q(\mathbf{x}, \mathbf{x}) = \langle \mathbf{x}, \mathbf{A}\mathbf{x} \rangle$. The indefinite integral of the exponential of $Q(\mathbf{x}, \mathbf{x})$ is given by
$$I_N \triangleq \int_{-\infty}^{\infty} \cdots \int_{-\infty}^{\infty} \exp(-\mathbf{x}^T \mathbf{A} \mathbf{x}) \, d\mathbf{x} \tag{D.4.14}$$

where $d\mathbf{x} = dx_1 \, dx_2 \cdots dx_N$, and it has many applications. Using (D.4.12) and (D.4.13) (specialized to the real case), we obtain
$$\langle \mathbf{x}, \mathbf{A}\mathbf{x} \rangle = \langle \mathbf{y}, \mathbf{\Lambda}\mathbf{y} \rangle = \sum_{i=1}^{N} \lambda_i y_i^2 \tag{D.4.15}$$

where $\lambda_i, i = 1, 2, \ldots, N$, are eigenvalues of $\mathbf{A}$. Thus (D.4.14) becomes
$$I_N = \int_{-\infty}^{\infty} \cdots \int_{-\infty}^{\infty} \exp\left(-\sum_{i=1}^{N} \lambda_i y_i^2\right) d\mathbf{y} = \prod_{i=1}^{N} \int_{-\infty}^{\infty} \exp(-\lambda_i y_i^2) \, dy_i \tag{D.4.16}$$

Now by using the result
$$\int_{-\infty}^{\infty} \exp(-\alpha x^2) \, dx = \sqrt{\frac{\pi}{\alpha}} \tag{D.4.17}$$

Equation (D.4.14) becomes
$$I_N = \prod_{i=1}^{N} \int_{-\infty}^{\infty} \sqrt{\frac{\pi}{\lambda_i}} = \sqrt{\frac{\pi^N}{\lambda_1 \lambda_2 \cdots \lambda_N}} \tag{D.4.18}$$

Finally, using the fact that $\det(\mathbf{A}) = \prod_{i=1}^{N} \lambda_i$, we obtain
$$I_N = \sqrt{\frac{\pi^N}{\det(\mathbf{A})}} \tag{D.4.19}$$

The result in (D.4.19) can be extended to the complex case. Let $H(\mathbf{z}, \mathbf{z})$ be the Hermitian form of a complex-valued vector $\mathbf{z} = \mathbf{x} + j\mathbf{y}$. Then the indefinite integral of the exponential of $H(\mathbf{z}, \mathbf{z})$ is given by
$$J_N \triangleq \int_{-\infty}^{\infty} \cdots \int_{-\infty}^{\infty} \exp(-\mathbf{z}^T \mathbf{A} \mathbf{z}) \, d\mathbf{z} \tag{D.4.20}$$
$$= \frac{\pi^N}{\det(\mathbf{A})} \tag{D.4.21}$$

where $d\mathbf{z} = dx_1 \, dx_2 \cdots dx_N \, dy_1 \, dy_2 \cdots dy_N$. Thus sometimes we get slightly different results for the complex case.

TABLE D.1

Summary of properties of vectors and matrices in real and complex spaces.

Real versus Complex		
$\mathbb{R}^N$: N-dimensional Euclidean space	$\leftrightarrow$	$\mathbb{C}^N$: N-dimensional complex space
Norm: $\|\mathbf{x}\|^2 = x_1^2 + \cdots + x_N^2$	$\leftrightarrow$	Norm: $\|\mathbf{x}\|^2 = |x_1|^2 + \cdots + |x_N|^2$
Transpose: $\mathbf{A}^T = [a_{ji}]$	$\leftrightarrow$	Hermitian: $\mathbf{A}^H = [a_{ji}^*]$
$(\mathbf{AB})^T = \mathbf{B}^T\mathbf{A}^T$	$\leftrightarrow$	$(\mathbf{AB})^H = \mathbf{B}^H\mathbf{A}^H$
Inner product: $\langle \mathbf{x}, \mathbf{y} \rangle = \mathbf{x}^T\mathbf{y}$	$\leftrightarrow$	Inner product: $\langle \mathbf{x}, \mathbf{y} \rangle = \mathbf{x}^H\mathbf{y}$
Orthogonality: $\mathbf{x}^T\mathbf{y} = 0$	$\leftrightarrow$	Orthogonality: $\mathbf{x}^H\mathbf{y} = 0$
Symmetric matrices: $\mathbf{A} = \mathbf{A}^T$	$\leftrightarrow$	Symmetric matrices: $\mathbf{A} = \mathbf{A}^H$
Orthogonal matrices: $\mathbf{Q}^T = \mathbf{Q}^{-1}$	$\leftrightarrow$	Unitary matrices: $\mathbf{U}^T = \mathbf{U}^{-1}$
$\mathbf{A} = \mathbf{Q}\Lambda\mathbf{Q}^{-1} = \mathbf{Q}\Lambda\mathbf{Q}^{-T}$ (real Λ)	$\leftrightarrow$	$\mathbf{A} = \mathbf{U}\Lambda\mathbf{U}^{-1} = \mathbf{U}\Lambda\mathbf{U}^{-H}$ (real Λ)
Norm invariance: $\|\mathbf{Qx}\| = \|\mathbf{x}\|$	$\leftrightarrow$	Norm invariance: $\|\mathbf{Ux}\| = \|\mathbf{x}\|$
$(\mathbf{Qx})^T(\mathbf{Qy}) = \mathbf{x}^T\mathbf{y}$	$\leftrightarrow$	$(\mathbf{Ux})^H(\mathbf{Uy}) = \mathbf{x}^H\mathbf{y}$

Table D.1 summarizes various properties described above as they relate to both complex-valued and real-valued matrices.

D.5 POSITIVE DEFINITE MATRICES

Positive definite matrices play an important role in signal processing in general and least-squares (LS) estimation in particular, and they deserve some attention. A conjugate symmetric $M \times M$ matrix $\mathbf{R}$ is called *positive definite* if and only if the Hermitian form

$$\mathbf{x}^H\mathbf{Rx} = \sum_{i,j}^{M} r_{ij}x_i^* x_j > 0 \tag{D.5.1}$$

for every $\mathbf{x} \neq \mathbf{0}$. For example, the symmetric matrix

$$\mathbf{R} = \begin{bmatrix} 2 & -1 & 0 \\ -1 & 2 & -1 \\ 0 & -1 & 2 \end{bmatrix} \tag{D.5.2}$$

is positive definite because the quadratic form

$$\mathbf{x}^T\mathbf{Rx} = x_1^2 + (x_1 - x_2)^2 + (x_2 - x_3)^2 + x_3^2 > 0 \tag{D.5.3}$$

can be expressed as a sum of squares that is positive unless $x_1 = x_2 = x_3 = 0$.

From this simple example it is obvious that using the definition to find out whether a given matrix is positive definite is very tedious. Fortunately, use of this approach is not necessary because other criteria can be used to make a faster decision (Strang 1980; Horn and Johnson 1985; Nobel and Daniel 1988). We next summarize some positive definiteness tests that are useful in LS estimation.

Positive definiteness criterion

An $M \times M$ matrix $\mathbf{R}$ is positive definite if and only if it satisfies any one of the following criteria:

1. $\mathbf{x}^H\mathbf{Rx} > 0$ for all nonzero vectors $\mathbf{x}$.
2. All eigenvalues of $\mathbf{R}$ are positive.
3. All principal submatrices $\mathbf{R}_m$, $1 \le m \le M$, have positive determinants. The principal

submatrices of $\mathbf{R}$ are determined as follows:

$$\mathbf{R}_1 = [r_{11}] \quad \mathbf{R}_2 = \begin{bmatrix} r_{11} & r_{12} \\ r_{21} & r_{22} \end{bmatrix} \quad \mathbf{R}_3 = \begin{bmatrix} r_{11} & r_{12} & r_{13} \\ r_{21} & r_{22} & r_{23} \\ r_{31} & r_{32} & r_{33} \end{bmatrix} \quad \cdots \quad \mathbf{R}_M = \mathbf{R}$$

(D.5.4)

It is important to stress that this criterion applies also to the lower right submatrices or any chain of submatrices that starts with a diagonal element r_{ii} as the first submatrix and then expands it by adding a new row and column at each step.

4. There exists an $L \times M$, $M > L$, matrix $\mathbf{S}$ with linearly independent columns such that $\mathbf{R} = \mathbf{S}^H \mathbf{S}$. This requirement for the columns of $\mathbf{S}$ to be linearly independent implies that $\mathbf{S}$ has rank M.

5. There exists a nonsingular $M \times M$ matrix $\mathbf{W}$ such that $\mathbf{R} = \mathbf{W}^H \mathbf{W}$. The choices for the matrix $\mathbf{W}$ are a triangular matrix obtained by Cholesky's decomposition (see Section 6.3) or an orthonormal matrix obtained from the eigenvectors of $\mathbf{R}$ (see Section 3.5).

6. There exists a nonsingular $M \times M$ matrix $\mathbf{P}$ such that the matrix $\mathbf{P}^H \mathbf{R} \mathbf{P}$ is positive definite.

Properties of positive definite matrices. A positive definite matrix $\mathbf{R}$ has the following properties:

1. The diagonal elements of $\mathbf{R}$ are positive.
2. $r_{ii} r_{jj} > |r_{ij}|^2 \quad (i \neq j)$
3. The element of $\mathbf{R}$ with the largest absolute value lies on the diagonal.
4. The $\det \mathbf{R} > 0$. Hence $\mathbf{R}$ is nonsingular.
5. The inverse matrix $\mathbf{R}^{-1}$ is positive definite.
6. The matrix obtained by deleting a row and the corresponding column from $\mathbf{R}$ is positive definite.

APPENDIX E

Minimum Phase Test for Polynomials

In this appendix we prove a theorem that provides a test for checking if the zeros of a polynomial are inside the unit circle (minimum phase condition) using the lattice parameters. The required lattice parameters can be obtained from the coefficients of the polynomial using the algorithm (2.5.28) in Section 2.5.

THEOREM E.1. The polynomial

$$A_P(z) = 1 + a_1^{(P)} z^{-1} + \cdots + a_P^{(P)} z^{-P} \tag{E.1}$$

is minimum-phase, that is, has all its zeros inside the unit circle if and only if

$$|k_m| < 1 \qquad 1 \le m \le P \tag{E.2}$$

Proof. We will prove the sufficiency part first, followed by the necessary part. Also we will make use of property (2.4.16)–(2.4.17) of the all-pass systems.

Sufficiency. We will prove by induction that if $|k_m| < 1, 1 \le m \le P$, then $A_P(z)$ is minimum-phase. For $P = 1$ we have

$$A_1(z) = 1 + a_1^{(1)} z^{-1} = 1 + k_1 z^{-1}$$

Clearly if $|k_1| < 1$, then $A_1(z)$ is minimum-phase. Assume now that $A_{m-1}(z)$ is minimum-phase. It can be then expressed as

$$A_{m-1}(z) = \prod_{i=1}^{m-1} (1 - z_i^{(m-1)} z^{-1}) \tag{E.3}$$

where

$$|z_i^{(m-1)}| < 1 \qquad 1 \le i \le m - 1 \tag{E.4}$$

However, from the recursion (2.5.9),

$$A_m(z) = A_{m-1}(z) + k_m z^{-1} B_{m-1}(z) \tag{E.5}$$

Hence

$$A_m(z_i^{(m)}) = A_{m-1}(z_i^{(m)}) + k_m \left(\frac{1}{z_i^{(m)}}\right) B_{m-1}(z_i^{(m)}) = 0 \qquad 1 \le i \le m$$

or

$$k_m = \frac{A_{m-1}(z_i^{(m)})}{(1/z_i^{(m)}) B_{m-1}(z_i^{(m)})} \qquad 1 \le i \le m \tag{E.6}$$

But

$$B_{m-1}(z) = z^{-(m-1)} A_{m-1}(z^{-1}) = z^{-(m-1)} \prod_{i=1}^{m-1} (1 - z_i^{(m-1)} z)$$

$$\tag{E.7}$$

$$= \prod_{i=1}^{m-1} (z^{-1} - z_i^{(m-1)})$$

Since $A^{(m-1)}(z)$ has real coefficients, either its zeros are real or they appear in complex conjugate pairs. Thus, a zero and its complex conjugate can always be grouped in the numerator and denominator of (E.6) as

$$|k_m|^2 = \prod_{i=1}^{m-1} \left| \frac{1 - z_i^{(m-1)}/z_i^{(m)}}{1/z_i^{(m)} - (z_i^{(m-1)})^*} \right|^2 |z_i^{(m)}|^2 \qquad 1 \le i \le m \qquad (E.8)$$

Applying property (2.4.17) to every factor of (E.8), with $a = z_i^{(m-1)}$, gives

$$|k_m| \begin{cases} < 1 & |z_i^{(m)}| < 1 \\ = 1 & |z_i^{(m)}| = 1 \\ > 1 & |z_i^{(m)}| > 1 \end{cases} \qquad (E.9)$$

Thus, if $|z_i^{(m)}| < 1, 1 \le i \le m$, then $|k_M| < 1$.

Necessity. We will prove that if $A_P(z)$ is minimum-phase, then $|k_m| < 1, 1 \le m \le P$. To this end we will show that if $A_m(z)$ is minimum-phase, then $|k_m| < 1$ and $A_{m-1}(z)$ is minimum-phase. From

$$A_m(z) = \prod_{i=1}^{m}(1 - z_i^{(m)}z^{-1})$$

we see by inspection that the coefficient of the highest power z^{-m} is

$$k_m = \alpha_m^{(m)} = \prod_{i=1}^{m}(-z_i^{(m)}) \qquad (E.10)$$

Thus,
$$|k_m| \le \prod_{i=1}^{m}|z_i^{(m)}| < 1 \qquad (E.11)$$

To show that $A^{(m-1)}(z)$ is minimum-phase, we recall that

$$A_{m-1}(z) = \frac{A_m(z) - k_m B_m(z)}{1 - |k_m|^2} \qquad (E.12)$$

If $z_i^{(m-1)}$ is a zero of $A^{(m-1)}(z)$, we have

$$A_{m-1}(z_i^{(m-1)}) = \frac{A_m(z_i^{(m-1)}) - k_m B_m(z_i^{(m-1)})}{1 - |k_m|^2} = 0 \qquad (E.13)$$

If $|k_m| \ne 1$, then (E.13) implies that

$$k_m = \frac{A_m(z_i^{(m-1)})}{B_m(z_i^{(m-1)})} \qquad 1 \le i \le m - 1 \qquad (E.14)$$

Applying again the property (2.4.17) to $|k_m|^2$ in (E.14) shows that since $|k_m| < 1$, then $|z_i^{(m-1)}| < 1$ for $1 \le i \le m - 1$. Hence, $A^{(m-1)}(z)$ is minimum-phase.

Bibliography

Abed-Meraim, K., W. Qiu, and Y. Hua. 1997. Blind system identification. *Proc. IEEE*, 85(8):1310–1322, August.

Abramowitz, M., and I. Stegun, Eds. 1970. *Handbook of Mathematical Functions*. Dover Publications, New York.

Adler, R., R. Feldman, and M. Taqqu, Eds. 1998. *A Practical Guide to Heavy Tails: Statistical Techniques and Applications*. Birkhäuser, Boston.

Akaike, H. 1969. Fitting autoregressive models for prediction. *Annals Inst. Statistical Mathematics*, 21:243–247.

_____. 1970. Statistical predictor identification. *Annals Inst. Statistical Mathematics*, 22:203–217.

_____. 1974. A new look at the statistical model identification. *IEEE Trans. Automatic Control*, 19:716–722.

Alexander, S. T. 1987. Transient weight misadjustment properties for the precision LMS algorithm. *IEEE Trans. Acoustics, Speech and Signal Processing*, ASSP(35):1250–1258.

_____. 1993. A method for recursive least-squares filtering based upon an inverse QR decomposition. *IEEE Trans. Signal Processing*, 41(1):20–30.

Anderson, T. 1971. *The Statistical Analysis of Time Series*. Wiley, New York.

Applebaum, S. P., and D. J. Chapman. 1976. Antenna arrays with mainbeam constraints. *IEEE Trans. Antennas and Propagation*, 24(9):650–662.

Atal, B. S., and M. Schroeder. 1970. Adaptive predictive coding of speech signals. *Bell System Tech. J.*, pp. 1973–1987, October.

_____ and _____. 1978. Linear prediction analysis of speech based on a pole-zero representation. *J. Acoust. Soc. Am.*, 64(5):1310–1318, November.

_____ and _____. 1979. Predictive coding of speech signals and subjective error criteria. *IEEE Trans. Acoustics, Speech and Signal Processing*, 27(3):247–254.

Baggeroer, A. B., W. A. Kuperman, and P. N. Mikhalevsky. 1993. An overview of matched field methods in ocean acoustics. *IEEE J. Ocean Engineering*, 18(4):401–424.

Barabell, A. J. 1983. Improving the resolution performance of eigenstructure-based direction-finding algorithms. *Proc. International Conference on Acoustics, Speech and Signal Processing*, pp. 336–339.

Baranoski, E. J. 1995. Improved pre-Doppler STAP algorithm for adaptive nulling in airborne radars. *Proc. Asilomar Conf. Signals, Systems, and Computers*, pp. 1173–1177.

_____ and J. Ward. 1997. Source localization using adaptive subspace beamformer outputs. *Proc. International Conference on Acoustics, Speech and Signal Processing*, pp. 3773–3776.

Barnsley, M. F., R. L. Devaney, B. B. Mandelbrot, H. O. Peitgen, D. Saupe, and R. F. Voss. 1988. The science of fractal images. In H. Peitgen and D. Saupe, Eds., *Fractals*. Springer-Verlag, New York.

Bartlett, M. S. 1948. Smoothing periodograms for time series with continuous spectra. *Nature*, 161:686–687.

_____. 1950. Periodogram analysis and continuous spectra. *Biometrica*, 31:1–16.

Bell, A. J., and T. J. Sejnowski. 1995. An information maximization approach to blind separation and blind deconvolution. *Neural Computation*, 6:1129–1159.

Bellini, S. 1986. Bussgang techniques for blind equalization. In *GLOBECOM*, pp. 1634–1640.

_____. 1994. Bussgang techniques for blind deconvolution and equalization. In S. Haykin, Ed., *Blind Deconvolution*. Prentice Hall, Englewood Cliffs, NJ.

Bendat, J. S., and A. G. Piersol. 1980. *Engineering Applications of Correlation and Spectral Analysis*. Wiley-Interscience, New York.

_____ and _____. 1986. *Random Data: Analysis and Measurement Procedures*, 2nd ed., Wiley, New York.

Benveniste, A., and M. Goursat. 1984. Blind equalizers. *IEEE Trans. Communications*, 32:871–883.

_____, _____, and G. Ruget. 1980. Robust identification of a non-minimum phase system: Blind adjustment of a linear equalizer in data communications. *IEEE Trans. Automatic Control*, 25:385–399.

_____, M. Metivier, and P. Priouret. 1987. *Adaptive Algorithms and Stochastic Approximations*. Springer-Verlag, New York.

Beran, J. 1994. *Statistics for Long-Memory Processes*. Chapman and Hall, New York.

_____, R. Sherman, M. S. Taqqu, and W. Willinger. 1995. Long-range dependence in variable-bit-rate video traffic. *IEEE Trans. Communications*, 43(2):1566–1579, February.

Bienvenu, G., and L. Kopp. 1983. Optimality of high resolution array processing using the eigensystem approach. *IEEE Trans. Acoustics, Speech and Signal Processing*, 31(10):1235–1248.

Bierman, G. J. 1977a. *Factorization Methods for Discrete Sequential Estimation*. Academic Press, New York.

_____. 1977b. Numerical comparison of Kalman filter algorithmic determination case study. *Automatica*, 13:23–35.

_____ and C. L. Thornton. 1977. Numerical comparison of Kalman filter algorithms: Orbit determination case study. *Automatica*, 13:23–35.

Bingham, J. A. C. 1988. *The Theory and Practice of Modem Design*. Wiley, New York.

Björck, A. 1967. Solving linear least-squares problems by Gram-Schmidt orthogonalization. *BIT*, 7:1–21.

Blackman, R., and J. Tukey. 1958. *The Measurement of Power Spectra*. Dover Publications, New York.

Bode, H., and C. Shannon. 1950. A simplified derivation of linear least squares smoothing and prediction theory. *Proc. IRE*, 38:417–425.

Böhme, J. F., and B. Yang. 1992. Rotation-based RLS algorithms. *IEEE Trans. Signal Processing*, 40(5):1151–1167.

Bolt, B. 1993. *Earthquakes and Geological Discovery*. Freeman and Company, New York.

Borkar, V. S. 1995. *Probability Theory: An Advanced Course*. Springer-Verlag, New York.

Boroson, D. M. 1980. Sample size considerations in adaptive arrays. *IEEE Trans. Aerospace and Electronic Systems*, 16(4):446–451.

Borsari, G. K., and A. O. Steinhardt. 1995. Cost-efficient training strategies for space-time adaptive processing algorithms. *Proc. Asilomar Conf. on Signals, Systems, and Computers*, pp. 650–654.

Bose, S., and A. O. Steinhardt. 1995. A maximal invariant framework for adaptive detection with structured and unstructured covariance matrices. *IEEE Trans. Signal Processing*, 43(9):2164–2175.

Box, G. E. P., and G. M. Jenkins. 1976. *Time Series Analysis: Forecasting and Control*, rev. ed. Holden-Day, San Francisco, CA.

_____, _____, and G. C. Reinsel. 1994. *Time Series Analysis: Forecasting and Control*, 3rd ed. Prentice Hall, Englewood Cliffs, NJ.

Bozic, S. M. 1994. *Digital and Kalman Filtering*, 2nd ed. Halsted Press, New York.

Brennan, L. E., and I. S. Reed. 1973. Theory of adaptive radar. *IEEE Trans. Aerospace and Electronic Systems*, 9(3):237–252.

Brillinger, D. R. 1965. An introduction to polyspectra. *Ann. Math. Statist.*, 36:1351–1374.

_____. 1980. *Time Series: Data Analysis and Theory*. Holden-Day, San Francisco, CA.

Brockwell, P. J., and R. A. Davis. 1991. *Time Series: Theory and Methods*, 2nd ed. Springer-Verlag, New York.

_____ and _____. 1996. *Introduction to Time Series and Forecasting*. Springer-Verlag, New York.

Brown, R. G., and P. Y. C. Hwang. 1997. *Introduction to Random Signals and Applied Kalman Filtering*, 3rd ed. Wiley, New York.

Bryn, F. 1962. Optimum signal processing of three-dimensional arrays operating on Gaussian signals and noise. *J. Acoustical Soc. Am.*, 34:289–297.

Bucklew, J. A., T. Kurtz, and W. A. Sethares. 1993. Weak convergence and local stability properties of fixed step size recursive algorithms. *IEEE Trans. Information Theory*, 39:966–978.

Buckley, K. M. 1987. Spatial/spectral filtering with linearly constrained minimum variance beamformers. *IEEE Trans. Acoustics, Speech and Signal Processing*, 35(3):249–266.

Burg, J. 1975. *Maximum Entropy Spectral Analysis*. Ph.D. thesis, Stanford University, Stanford, CA.

Burg, J. P. 1978. Maximum entropy spectral estimation. In D. G. Childers, Ed., *Modern Spectral Analysis*. IEEE Press, New York. Originally appeared in *Proceedings of the 37th Annual SEG Meeting*, Dec. 31, 1967.

Bussgang, J. 1952. Cross correlation functions of amplitude-distorted Gaussian signals. *Tech. Report 216*, MIT Research Lab. of Electronics, Cambridge, MA.

Butterweck, H. J. 1995. A steady-state analysis of the LMS adaptive algorithm without use of the independence assumption. *Proc. Int. Conf. Acoustics, Speech and Signal Processing*, pp. 1404–1407.

Cadzow, J. A. 1996. Blind deconvolution via cumulant extrema. *IEEE Signal Processing Magazine*, pp. 24–42, May.

_____ and X. Li. 1995. Blind deconvolution. *Digital Signal Processing J.*, 5(1):3–20.

Capon, J. 1969. High-resolution frequency-wavenumber spectrum analysis. *Proc. IEEE*, 57:1408–1418, August.

_____, R. J. Greenfield, and R. J. Kolker. 1967. Multidimensional maximum-likelihood processing of large aperture seismic arrays. *J. Acoustical Soc. Am.*, 55(2):192–211.

Caraiscos, C., and B. Liu. 1984. A roundoff error analysis of the LMS adaptive algorithm. *IEEE Trans. Acoustics, Speech and Signal Processing*, ASSP(32):34–41.

Carayannis, G., N. Kalouptsidis, and D. G. Manolakis. 1982. Fast recursive algorithms for a class of linear equations. *IEEE Trans. Acoustics, Speech and Signal Processing*, 30(2):227–239, April.

_____, E. Koukoutsis, and C. C. Halkias. 1991. Hardware implementation of partitioned parallel algorithms in linear prediction. *Signal Processing*, pp. 257–259, September.

_____, _____, D. G. Manolakis, and C. C. Halkias. 1985. A new look on the parallel implementation of the Schür algorithm for the solution of Toeplitz equations. *Proc. Int. Conf. Acoustics, Speech and Signal Processing*, pp. 1858–1861.

_____, D. Manolakis, and N. Kalouptsidis. 1986. A unified view of parametric processing algorithms for prewindowed signals. *Signal Processing*, 10(4):335–368, June.

_____, D. G. Manolakis, and _____. 1983. A fast sequential algorithm for least-squares filtering and prediction. *IEEE Trans. Acoustics, Speech and Signal Processing*, 31(6):1394–1402.

Carlson, B. D. 1988. Covariance matrix estimation errors and diagonal loading in adaptive arrays. *IEEE Trans. Aerospace and Electronic Systems*, 24(4):397–401.

Carter, C. G., and A. H. Nuttall. 1980a. A brief summary of a generalized framework for spectral estimation. *Signal Processing*, 2(4):387–390, October.

_____ and _____. 1980b. On the weighted overlapped segment-averaging method for power spectral estimation. *Proc. IEEE*, 68(10):1352–1354, October.

Chambers, J. M., C. L. Mallows, and B. W. Stuck. 1976. A method for simulating stable random variables. *J. Am. Stat. Assn.*, 71:340–344, June.

Chan, T. F. 1982. An improved algorithm for computing the SVD. *ACM Trans. Mathematical Software*, 8:72–88.

Childers, D. G. 1978. *Modern Spectral Analysis*. IEEE Press, New York.

Chiuppesi, F., G. Galati, and P. Lombardi. 1980. Optimization of rejection filters. *IEE Proc. Part F: Communications, Radar and Signal Processing*, 127(5):354–360, October.

Cioffi, J. M. 1987. Limited-precision effects in adaptive filtering. *IEEE Trans. Circuits and Systems*, 34(7):821–833.

_____ and T. Kailath. 1984. Fast, recursive-least-squares transversal filters for adaptive filtering. *IEEE Trans. Acoustics, Speech and Signal Processing*, 32(2):304–337.

Claasen, T., and W. Mecklenbrauker. 1985. Adaptive techniques for signal processing in communications. *IEEE Communications Magazine*, 23(11):8–19, November.

Claasen, T. A. C. M., and W. F. G. Mecklenbrauker. 1981. Comparison of the convergence of two algorithms for adaptive FIR digital filters. *IEEE Trans. Circuits and Systems*, 28(6):510–518, June.

Claerbout, J. F., and E. A. Robinson. 1963. The error in least-squares inverse filtering. *Geophysics*, 29(1):118–120, January.

Colwell, R., Ed. 1983. *Manual of Remote Sensing*. American Society for Photogrametry and Remote Sensing, Falls Church, VA.

Compton, R. T. 1988. *Adaptive Antennas*. Prentice Hall, Englewood Cliffs, NJ.

Conte, E., M. Lops, and G. Ricci. 1996. Adaptive matched filter detection in spherically invariant noise. *IEEE Signal Processing Letters*, 3(8):248–250.

Cox, H. 1973. Resolving power and sensitivity to mismatch of optimum array processors. *J. Acoustical Soc. Am.*, 54:771–785.

———, R. M. Zeskind, and M. M. Owen. 1987. Robust adaptive beamforming. *IEEE Trans. Acoustics, Speech and Signal Processing*, 35(10):1365–1377.

Crawford, D., R. Stewart, and E. Toma. 1997. Digital signal processing strategies for active noise control. *Electronics and Communications Engineering J.*, pp. 81–89, April.

Dahlquist, G., and A. Bjorck. 1974. *Numerical Methods*. Prentice Hall, Englewood Cliffs, NJ. Translated by N. Anderson.

Daniell, P. J. 1946. Discussion of "On the theoretical specification and sampling properties of auto-correlated time series." *J. Royal Stat. Soc.*, 8:88–90.

Davenport, W. B., Jr. 1970. *Probability and Random Processes*. McGraw-Hill, New York.

Davis, R. C., L. E. Brennan, and I. S. Reed. 1976. Angle estimation with adaptive arrays in external noise fields. *IEEE Trans. Aerospace and Electronic Systems*, 12(2):179–186.

Delsarte, P., and Y. Genin. 1986. The split Levinson algorithm. *IEEE Trans. Acoustics, Speech and Signal Processing*, 34(3):470–478, June.

Dempster, A. P., N. M. Laird, and D. B. Rubin. 1977. Maximum likelihood from incomplete data via the EM algorithm. *Ann. Royal Stat. Soc.*, pp. 1–38.

Dennis, J. E., and R. B. Schnabel. 1983. *Numerical Methods for Unconstrained Optimization and Nonlinear Equations*. Prentice Hall, Englewood Cliffs, NJ.

Deriche, M., and A. H. Tewfik. 1993. Signal modeling with filtered discrete fractional noise processes. *IEEE Trans. Signal Processing*, 41(9):2839–2849, September.

Ding, Z. 1994. Blind channel identification and equalization using spectral correlation measurements, part 1: Frequency-domain approach. In W. A. Gardner, Ed., *Cyclostationarity in Communications and Signal Processing*, pp. 417–437. IEEE Press, New York.

———. 1998. Adaptive filters for blind equalization. In V. Madisetti and D. Williams, Eds., *The Digital Signal Processing Handbook*. CRC Press, New York.

———, C. R. Johnson, Jr., and R. A. Kennedy. 1994. Global convergence issues with linear blind adaptive equalizers. In S. Haykin, Ed., *Blind Deconvolution*. Prentice Hall, Englewood Cliffs, NJ.

———, R. Kennedy, B. Anderson, and C. Johnson. 1991. Ill-convergence of Godard blind equalizers in data communication systems. *IEEE Trans. Communications*, 39:1313–1327, September.

Diniz, P. S. 1997. *Adaptive Filtering*. Kluwer Academic Publishers, Boston, MA.

Dobrin, M. 1988. *Introduction to Geophysical Prospecting*, 4th ed. McGraw-Hill, New York.

Dongara, J. J., et al. 1979. *LINPACK User's Guide*. SIAM, Philadelphia, PA.

Donoho, D. L. 1981. On minimum entropy deconvolution. In B. F. Findlay, Ed., *Applied Time Series Analysis II*. Academic Press, New York.

Douglas, S., and M. Rupp. 1998. Convergence issues in the LMS adaptive filter. In V. Madisetti and D. Williams, Eds., *The Digital Signal Processing Handbook*. CRC Press, New York.

Duffy, F., V. Iyer, and W. Surwillo. 1989. *Clinical Electroencephalography and Topographic Brain Mapping*. Springer-Verlag, New York.

Durbin, J. 1960. The fitting of time-series models. *Rev. Int. Stat. Inst.*, 28:233–244.

Duttweiler, D. L. 1982. Adaptive filter performance with nonlinearities in the correlation multiplier. *IEEE Trans. Acoustics, Speech and Signal Processing*, 30(4):578–586, August.

Eleftheriou, E., and D. D. Falconer. 1986. Tracking properties and steady state performance of RLS adaptive filter algorithms. *IEEE Trans. Acoustics, Speech and Signal Processing*, 34:1097–1110.

El-Jaroudi, A., and J. Makhoul. 1989. Discrete pole-zero modeling and applications. *Proc. Int. Conf. Acoustics, Speech and Signal Processing*, pp. 2162–2165.

——— and ———. 1991. Discrete all-pole modeling. *IEEE Trans. Signal Processing*, 39(2):411–423, February.

Elliot, S., and P. Nelson. 1993. Active noise control. *IEEE Signal Processing Magazine*, 10(4):12–35, October.

Embree, P. M., and B. Kimble. 1991. *C Language Algorithms for Digital Signal Processing*. Prentice Hall, Englewood Cliffs, NJ.

Er, M. H., and A. Cantoni. 1983. Derivative constraints for broad-band element space antenna array processors. *IEEE Trans. Acoustics, Speech, and Signal Processing*, 31(12):1378–1393.

Falconer, D. D., and L. Ljung. 1978. Application of fast Kalman estimation to adaptive equalization. *IEEE Trans. Communications*, 26(10):1439–1446.

Falconer, K. 1990. *Fractal Geometry: Mathematical Foundations and Applications*. Wiley, New York.

Feder, J. 1988. *Fractals*. Plenum Press, New York.

Feller, W. 1957. *An Introduction to Probability Theory and Its Applications*, 2nd ed., vol. 1. Wiley, New York.

———. 1971. *An Introduction to Probability Theory and Its Applications*, 2nd ed., vol. 2. Wiley, New York.

Feuer, A., and E. Weinstein. 1985. Convergence analysis of LMS filters with uncorrelated Gaussian data. *IEEE Trans. Acoustics, Speech and Signal Processing*, 33(1):222–229, February.

Figueiras-Vidal, A. R., Ed. 1996. *Digital Signal Processing in Telecommunications*. Springer-Verlag, London.

Fijalkow, I., C. Manlove, and R. Johnson. 1998. Adaptive fractionally spaced blind CMA equalization: Excess MSE. *IEEE Trans. Signal Processing*, 46(1):227–231, January.

Flanagan, J. L. 1972. *Speech Analysis, Synthesis and Perception*. Springer-Verlag, New York.

Flandrin, P. 1989. On the spectrum of fractional Brownian motions. *IEEE Trans. Information Theory*, 35(1):197–199, January.

Foschini, G. J. 1985. Equalizing without altering or detecting data. *AT&T Tech. J.*, 64:1885–1911.

Friedlander, B. 1982. Lattice filters for adaptive processing. *Proc. IEEE*, 70(8):829–867, August.

——— and M. Morf. 1982. Least-squares algorithms for adaptive linear-phase filtering. *IEEE Trans. Acoustics, Speech and Signal Processing*, 30(3):381–390, June.

Frost, O. L. 1972. An algorithm for linearly constrained adaptive array processing. *Proc. IEEE*, 60(8):962–935.

Fukunaga, K. 1990. *Statistical Pattern Recognition*, 2nd ed. Academic Press, New York.

Furui, S. 1989. *Digital Speech Processing, Synthesis, and Recognition*. Marcel Dekker, New York.

——— and M. M. Sondhi, Eds. 1992. *Advances in Speech Signal Processing*. Marcel Dekker, New York.

Gardner, W. 1991. Exploitation of spectral redundancy in cyclostationary signals. *IEEE Signal Processing Magazine*, 8(2):14–36, April.

Gardner, W. A. 1984. Learning characteristics of stochastic-gradient-descent algorithm: A general study, analysis, and critique. *Signal Processing*, 6:113–133.

———. 1994. *Cyclostationarity in Communications and Signal Processing*. IEEE Press, New York.

Garrett, M. W., and W. Willinger. 1994. Analysis, modeling and generation of self-similar VBR video traffic. *Proc. ACM Sigcomm '94*, pp. 269–280.

Gelb, A. 1974. *Applied Optimal Estimation*. MIT Press, Cambridge, MA.

Gentleman, W. M., and H. T. Kung. 1981. Matrix triangularization by systolic arrays. *Proc. SPIE*, (Real time signal processing IV), vol. 298, pp. 19–26.

Giannakis, G. 1998. Channel estimation and equalization. *IEEE Signal Processing Magazine*, 15(5): 37–40, September.

Gill, P. E., G. H. Golub, W. Murray, and M. A. Saunders. 1974. Methods of modifying matrix factorizations. *Mathematics of Computation*, 28:505–535.

———, W. Murray, and M. H. Wright. 1981. *Practical Optimization*. Academic Press, London.

Gilloire, A., E. Moulines, D. Slock, and P. Duhamel. 1996. State of the art in acoustic echo cancellation. In A. R. Figueiras-Vidal, Ed., *Digital Signal Processing in Telecommunications*. Springer-Verlag, London.

——— and M. Vetterli. 1992. Adaptive filtering in subbands with critical sampling: Analysis, experiments and application to acoustic echo control. *IEEE Trans. Signal Processing*, 40:1862–1875.

Gitlin, R., J. Hayes, and S. Weinstein. 1992. *Data Communications*. Plenum Press, New York.

Gitlin, R. D., J .E. Mazo, and M. G. Taylor. 1973. On the design of gradient algorithms for digitally implemented adaptive filters. *IEEE Trans. Circuit Theory*, 20(2):125–136, March.

———, H. C. Meadors, and S. B. Weinstein. 1982. The tap leakage algorithm: An algorithm for the stable operation of a digitally implemented fractionally spaced adaptive equalizer. *Bell System Tech. J.*, 61(8):1817–1839, October.

——— and S. B. Weinstein. 1979. On the required tap-weight precision for digitally implemented, adaptive, mean-squared equalizers. *Bell System Tech. J.*, 58(2):301–321, February.

Givens, W. 1958. Computation of plane unitary rotations transforming a general matrix to triangular form. *SIAM J. Applied Math.*, 6:26–50.

Godard, D. N. 1980. Self-recovering equalization and carrier tracking in a two-dimensional data communication system. *IEEE Trans. Communications*, 28(11):1867–1875.

Godfrey, R., and F. Rocca. 1981. Zero memory non-linear deconvolution. *Geophysics Prospect*, 29:189–228.

Golub, G. H. 1965. Numerical methods for solving linear least-squares problems. *Numerical Methods*, 7:206–216.

_____ and C. Reinsch. 1970. Singular value decomposition and least-squares problems. *Numer. Math.*, 14:403–420.

_____ and C. F. Van Loan. 1996. *Matrix Computations*, 3rd ed. The Johns Hopkins University Press, Baltimore, MD.

Goodman, N. R. 1957. On the joint estimation of spectra, cospectrum of a two-dimensional stationary Gaussian process. Scientific paper no. 10, College of Engineering, New York University, New York. ASTIA Doc. AD-134919.

_____. 1963. Statistical analysis based on a certain multivariate complex Gaussian distribution. *Ann. Math. Statist.* 34(1):152–177.

Goodwin, G. C., and R. L. Payne. 1977. *Dynamic System Identification: Experiment Design and Data Analysis*. Academic Press, New York.

Gradshteyn, I. S., and I. M. Ryzhik. 1994. *Tables of Integrals, Series and Products*, 5th ed. Academic Press, London.

Granger, C. W. J., and R. Joyeux. 1980. An introduction to long-memory time series models and fractional differencing. *J. Time Series Analysis*, 1:15–29.

Grant, P., and B. Mulgrew. 1995. Nonlinear adaptive filter: Design and application. *Proc. IFAC (Adaptive Systems in Control and Signal Processing)*, pp. 31–42.

Gray, A. H., and J. D. Markel. 1973. Digital lattice and ladder filter synthesis. *IEEE Trans. Audio Electroacoustics*, 21(6):491–500, December.

Gray, H. L., N. Zhang, and W. A. Woodword. 1989. On generalized fractional processes. *J. Time Series Analysis*, 10:233–257.

Gray, R. M. 1972. On the asymptotic eigenvalue distribution of Toeplitz matrices. *IEEE Trans. Information Theory*, 18:725–730.

_____ and L. D. Davisson. 1986. *Random Processes: A Mathematical Approach for Engineers*. Prentice Hall, Englewood Cliffs, NJ.

Green, W. H. 1993. *Econometric Analysis*, 2nd ed. Macmillan, New York.

Grenander, U., and G. Szegö. 1984. *Topelitz Forms and Their Applications*, 2nd ed. Chelsea Publishing Company, New York.

Griffiths, L. J. 1977. A continuously adaptive filter implemented as a lattice filter. *Proc. Int. Conf. Acoustics, Speech and Signal Processing*, pp. 683–686.

_____ and C. W. Jim. 1982. An alternative approach to linearly constrained adaptive beamforming. *IEEE Trans. Antennas and Propagation*, 30(1):27–34.

Grossman, P. Personal communication.

_____, L. Watkins, F. Wilhelm, D. Manolakis, and B. Lown. 1996. Cardiac vagal control and dynamic responses to psychological stress among patients with coronary artery disease. *Am. J. Cardiology*, 78:1424–1427, December.

Hager, W. W. 1988. *Applied Numerical Linear Algebra*. Prentice Hall, Englewood Cliffs, NJ.

Hamilton, J. D. 1994. *Time Series Analysis*. Princeton University Press, Princeton, NJ.

Harris, F. 1978. On the use of windows for harmonic analysis with a discrete Fourier transform. *Proc. IEEE*, 66(1):51–83, January.

Hassibi, B., A. H. Sayed, and T. Kailath. 1996. LMS is H^∞ optimal. *IEEE Trans. Signal Processing*, 44(2):267–280, February.

Hastings, H. M., and G. Sugihara. 1993. *Fractals: A User's Guide for the Natural Sciences*. Oxford University Press, New York.

Hatzinakos, D., and C. L. Nikias. 1991. Blind equalization using a tricepstrum based algorithm. *IEEE Trans. Communications*, 39:669–682.

_____ and _____. 1994. Blind equalization based on higher-order statistics (HOS). In S. Haykin, Ed., *Blind Deconvolution*. Prentice Hall, Englewood Cliffs, NJ.

Hayes, M. H. 1996. *Statistical Digital Signal Processing and Modeling*. Wiley, New York.

Haykin, S. 1989. *Modern Filters*. Macmillan, New York.

_____, Ed. 1991. *Advances in Spectrum Analysis and Array Processing*, vol. 1. Prentice Hall, Englewood Cliffs, NJ.

_____, Ed. 1994. *Blind Deconvolution*. Prentice Hall, Englewood Cliffs, NJ.

_____. 1996. *Adaptive Filter Theory*, 3rd ed. Prentice Hall, Englewood Cliffs, NJ.

Helstrom, C. W. 1991. *Probability and Stochastic Processes for Engineers*, 2nd ed. Macmillan, New York.

_____. 1995. *Elements of Signal Detection and Estimation*. Prentice Hall, Englewood Cliffs, NJ.

Hewer, G. A., R. D. Martin, and J. Zeh. 1987. Robust preprocessing for Kalman filtering of glint noise. *IEEE Trans. Aerospace and Electronic Systems*, 23(1):120–128, January.

Hinich, M. J. 1982. Testing for Gaussianity and linearity of a stationary time series. *J. Time Series Analysis*, 3(3):169–176.

Horowitz, L. L., and K. D. Senne. 1981. Performance advantage of complex LMS for controlling narrow-band adaptive arrays. *IEEE Trans. Circuits and Systems*, 28(6):562–576.

Hosking, J. R. M. 1981. Fractional differencing. *Biometrica*, 68:165–176.

_____. 1984. Modeling persistence in hydrological time series using fractional differencing. *Water Resources Research*, 20(10):1898–1908.

Householder, J. 1958. Unitary triangularization of a nonsymmetric matrix. *J. ACM*, 5:339–342.

Howells, P. W. Intermediate frequency sidelobe canceler. U.S. Patent 3202990.

Hsiao, J. 1974. On the optimization of MTI clutter rejection. *IEEE Trans. Aerospace and Electronic Systems*, 10(5):622–629, September.

Hsieh, S., K. Liu, and K. Yao. 1993. A unified square-root-free approach for QRD-based recursive least-squares estimation. *IEEE Trans. Signal Processing*, 41(3):1405–1409, March.

Hubing, N. E., and S. T. Alexander. 1991. Statistical analysis of initialization methods for RLS adaptive filters. *IEEE Trans. Signal Processing*, 39(8):1793–1804, August.

Hudson, J. E. 1981. *Adaptive Array Principles*. Peter Peregrinus, New York.

Iltis, R. A. 1991. Interference rejection and channel estimation for spread-spectrum communications. In N. Kalouptsidis and S. Theodoridis, Eds., *Adaptive System Identification and Signal Processing Algorithms*, pp. 466–511. Prentice Hall, Englewood Cliffs, NJ.

Ingle, V. K., and J. G. Proakis. 1996. *Digital Signal Processing Using MATLAB*. PWS Publishing Company, Boston. MA.

Ishimaru, A. 1990. *Electromagnetic Wave Propagation*. Prentice Hall, Englewood Cliffs, NJ.

Itakura, F., and S. Saito. 1971a. Digital filtering techniques for speech analysis and synthesis. *Proc. 7th Int. Congress on Acoustics*, 25-C-1:261–264, Budapest.

_____ and _____. 1971b. A statistical method for estimation of speech spectral density and formant frequencies. *Electr. Commun. Japan*, 53-A(1):36–43.

Jain, A. K. 1989. *Digital Image Processing*. Prentice Hall, Englewood Cliffs, NJ.

Janicki, A., and A. Weron. 1994. *Simulation and Chaotic Behavior of α-Stable Stochastic Processes*. Marcel Dekker, New York.

Jayant, N. S., and P. Noll. 1984. *Digital Coding of Waveforms*. Prentice Hall, Englewood Cliffs, NJ.

Jenkins, G. M., and D. G. Watts. 1968. *Spectral Analysis and Its Applications*. Holden-Day, San Francisco, CA.

Jenkins, W., and D. Marshall. 1998. Transform domain adaptive filtering. In V. Madisetti and D. Williams, Eds., *The Digital Signal Processing Handbook*. CRC Press, New York.

Jensen, F. B., W. A. Kuperman, M. B. Porter, and H. Schmidt. 1994. *Computational Ocean Acoustics*. Springer-Verlag, New York.

Johnson, C., and M. Larimore. 1977. Comments on and additions to "An adaptive recursive LMS filter." *Proc. IEEE*, 65(9):1399–1402, September.

Johnson, C. R., Jr. 1984. Adaptive IIR filtering: Current results and open issues. *IEEE Trans. Information Theory*, 30(2), part 1:237–250.

Johnson, D. H., and S. R. DeGraaf. 1982. Improving the resolution of bearing in passive sonar arrays by eigenvalue analysis. *IEEE Trans. Acoustics, Speech and Signal Processing*, 30(4):638–647, August.

_____ and D. E. Dudgeon. 1993. *Array Signal Processing: Concepts and Techniques*. Prentice Hall, Englewood Cliffs, NJ.

Johnson, R., et al. 1998. Blind equalization using the constant modulus criterion: A review. *Proc. IEEE*, 86(10):1927–1949, October.

Kagan, A., Y. V. Linnik, and C. R. Rao. 1973. *Characterization Problems in Mathematical Statistics*. Wiley-Interscience, New York.

Kailath, T. 1974. A view of three decades of linear filtering theory. *IEEE Trans. Information Theory*, 20:146–181.

_____. 1981. *Lectures on Linear Least-Squares Estimation*. Springer-Verlag, New York.

Kalouptsidis, N., G. Carayannis, and D. Manolakis. 1984. Efficient recursive-in-order least-squares FIR filtering and prediction. *IEEE Trans. Acoustics, Speech and Signal Processing*, 33:1175–1187, October.

_____ and S. Theodoridis, Eds. 1993. *Adaptive System Identification and Signal Processing Algorithms*. Prentice Hall, Englewood Cliffs, NJ.

Kamen, E. W., and B. S. Heck. 1997. *Fundamentals of Signals and Systems Using Matlab*. Prentice Hall International, Upper Saddle River, NJ.

Kasdin, N. J. 1995. Discrete simulation of colored noise and stochastic processes and $1/f^{\alpha}$ power law noise generation. *Proc. IEEE*, 83(5):802–827, May.

Kassam, S. A., and H. V. Poor. 1985. Robust techniques for signal processing: A survey. *Proc. IEEE*, 73(3):433–481, March.

Kaveh, M., and A. Barabell. 1986. The statistical performance of the MUSIC and minimum-norm algorithms in resolving plane waves in noise. *IEEE Trans. Acoustics, Speech and Signal Processing*, 34(2):331–341.

Kay, S. M. 1988. *Modern Spectral Estimation*. Prentice Hall, Englewood Cliffs, NJ.

_____. 1993. *Fundamentals of Statistical Signal Processing: Estimation Theory*. Prentice Hall, Englewood Cliffs, NJ.

_____. 1998. *Fundamentals of Statistical Signal Processing Detection Theory*. Prentice Hall, Upper Saddle River, NJ.

Kelly, E. J. 1986. An adaptive detection algorithm. *IEEE Trans. Aerospace and Electronic Systems*, 22(1):115–127.

_____. 1989. Performance of an adaptive detection algorithm: Rejection of unwanted signals. *IEEE Trans. Aerospace and Electronic systems*, 25(2):122–133.

Kendall, M. G., and A. Stuart. 1983. *Advanced Theory of Statistics*, 4th ed. Macmillan Publishing Company, New York.

Kino, G. S. 1987. *Acoustic Waves*. Prentice Hall, Englewood Cliffs, NJ.

Klema, V. C., and A. J. Laub. 1980. The singular value decomposition: Its computation and some applications. *IEEE Trans. Automatic Control*, 25:164–176.

Klemm, R. 1999. *Space-Time Adaptive Processing*. IEE, London.

Kogon, S. M., and D. G. Manolakis. 1994. Fractal-based modeling and interpolation of non-Gaussian images. *Proc. SPIE: Visual Communications and Image Processing '94*, vol. 2308, part 1:467–477.

_____ and _____. 1996. Signal modeling with self-similar α-stable processes: The fractional Lévy stable motion model. *IEEE Trans. Signal Processing*, 44(4):1006–1010, April.

Kok, A., D. G. Manolakis, and V. K. Ingle. 1993. Symmetric noncausal spatial model for 2-D signals with applications in stochastic texture modeling. *Multidimensional Systems and Signal Processing*, 4:125–147.

Kolmogorov, A. 1939. Sur l'interpolation et extrapolation des suites stationaires. *C. R. Acad. Sci.*, 208:2043–2045.

Kondoz, A. M. 1994. *Digital Speech: Coding for Low Bit Rate Communication Systems*. Wiley, New York.

Koopmans, L. H. 1974. *The Spectral Analysis of Time Series*. Academic Press, New York.

Koukoutsis, K., G. Carayannis, and C. C. Halkias. 1991. Superlattice/superladder computational organization for linear prediction and optimal FIR filtering. *IEEE Trans. Signal Processing*, 39(10):2199–2215, October.

Koutrouvelis, I. A. 1980. Regression-type estimation of the parameters of stable laws. *J. Am. Stat. Assn.*, 75:918–928.

_____. 1981. An iterative procedure for the estimation of the parameters of the stable law. *Communications in Statistics—Computation and Simulation*, 10:17–28.

Kreithen, D. E., and A. O. Steinhardt. 1995. Target detection in post-STAP undernulled clutter. *Proc. Asilomar Conf. Signals, Systems, and Computers*, pp. 1203–1207.

Kullback, S. 1959. *Information Theory and Statistics*. Wiley, New York.

Kumaresan, R., and D. W. Tufts. 1983. Estimating the angles of arrival of multiple plane waves. *IEEE Trans. Aerospace and Electronic Systems*, 19:134–139, January.

Kung, S., and Y. Hu. 1983. A highly concurrent algorithm and pipelined architecture for solving Toeplitz systems. *IEEE Trans. Acoustics, Speech and Signal Processing*, 31(1):66–76, February.

Kuo, S., and D. Morgan. 1996. *Active Noise Control Systems*. Wiley, New York.

Kushner, H. J. 1984. *Approximation and Weak Convergence Methods for Random Processes with Applications to Stochastic System Theory*. MIT Press, Cambridge, MA.

Lacoss, R. T. 1971. Data adaptive spectral analysis methods. *Geophysics*, 36:661–675, August.

Lamperti, J. W. 1996. *Probability: A Survey of Mathematical Theory*, 2nd ed. Wiley, New York.

Lapsley, P., J. Bier, A. Shoham, and E. A. Lee. 1997. *DSP Processor Fundamentals: Architectures and Features*. IEEE Press, New York.

Lawnson, C. L., and R. D. Hanson. 1974. *Solving Least-Squares Problems*. Prentice Hall, Englewood Cliffs, NJ.

Lawrance, A. 1991. Directionality and reversibility in time series. *Int. Stat. Rev.*, 59(1):67–79.

Lawrence, V. B., and S. K. Tewksbury. 1983. Multiprocessor implementation of adaptive digital filters. *IEEE Trans. Communications*, 31(6):826–835, June.

Le Roux, J., and C. Gueguen. 1977. A fixed-point computation of partial correlation coefficients. *IEEE Trans. Acoustics, Speech and Signal Processing*, pp. 257–259, June.

Lee, D. T. L., M. Morf, and B. Friedlander. 1981. Recursive least-squares ladder estimation algorithms. *IEEE Trans. Circuits and Systems*, 28(6):467–481.

Lee, E. A., and D. G. Messerschmitt. 1994. *Digital Communication*, 2nd ed. Kluwer Academic Publishers, Boston, MA.

Leon, S. J. 1995. *Linear Algebra with Applications*, 5th ed. Macmillan Publishing Company, New York.

Levanon, N. 1988. *Radar Principles*. Wiley, New York.

Levinson, N. 1947. The Wiener RMS (root-mean-square) error criterion in filter design and prediction. *J. Math. Physics*, 25:261–278.

Lii, K. S., and M. Rosenblatt. 1982. Deconvolution and estimation of transfer function phase and coefficients for non-Gaussian linear processes. *Ann. Stat.*, 10:1195–1208.

Lin, D. W. 1984. On the digital implementation of the fast Kalman algorithm. *IEEE Trans. Acoustics, Speech and Signal Processing*, 32:998–1005.

Ling, F. 1991. Givens rotation based least-squares lattice and related algorithms. *IEEE Trans. Signal Processing*, 39:1541–1551.

_____. 1993a. Echo cancellation. In N. Kalouptsidis and S. Theodoridis, Eds., *Adaptive System Identification and Signal Processing Algorithms*, pp. 407–465. Prentice Hall. Englewood Cliffs, NJ.

_____. 1993b. Lattice algorithms. In N. Kalouptsidis and S. Theodoridis, Eds., *Adaptive System Identification and Signal Processing Algorithms*, pp. 191–259. Prentice Hall, Englewood Cliffs, NJ.

_____, D. Manolakis, and J. G. Proakis. 1986. Numerically robust least-squares lattice-ladder algorithm with direct updating of the reflection coefficients. *IEEE Trans. Acoustics, Speech and Signal Processing*, 34(4):837–845.

_____ and J. G. Proakis. 1986. A recursive modified Gram-Schmidt algorithm with applications to least-squares and adaptive filtering. *IEEE Trans. Acoustics, Speech and Signal Processing*, 34(4):829–836.

Litva, J., and T. Lo. 1996. *Digital Beamforming for Wireless Communications*. Artech House, Boston, MA.

Liu, K. J. R., S. F. Hsieh, and K. Yao. 1992. Systolic block Householder transformation for RLS algorithm with two-level pipelined implementation. *IEEE Trans. Signal Processing*, 40:946–958.

Ljung, L. 1987. *System Identification Theory for the User*. Prentice Hall, Englewood Cliffs, NJ.

_____ and T. Soderstrom. 1983. *Theory and Practice of Recursive Identification*. MIT Press, Cambridge, MA.

Ljung, S., and L. Ljung. 1985. Error propagation properties of recursive least-squares adaptation algorithms. *Automatica*, 21:157–167.

Lucky, R. W. 1965. Automatic equalization for digital communications. *Bell System Tech. J.*, 44:547–588, April.

_____. 1966. Techniques for adaptive equalization of digital communication systems. *Bell System Tech. J.*, 45:255–286, February.

_____, J. Salz, and E. J. Weldon. 1968. *Principles of Data Communications*. McGraw-Hill, New York.

Luenberger, D. G. 1984. *Linear and Nonlinear Programming*, 2nd ed. Addison-Wesley, Reading, MA.

Lundahl, T., W. J. Ohley, S. M. Kay, and R. Siffert. 1986. Fractional Brownian motion: A maximum likelihood estimator and its application to image texture. *IEEE Trans. Medical Imaging*, 5(3):152–161, September.

Macchi, O. 1995. *Adaptive Processing: The LMS Approach with Applications in Transmission*. Wiley, New York.

———. 1996. The theory of adaptive filtering in a random time-varying environment. In A. R. Figueiras-Vidal, Ed., *Digital Signal Processing in Telecommunications*. Springer-Verlag, London.

Maeder, R. E. 1995. Fractional Brownian motion. *The Mathematica Journal*, 6(1):38–48.

Makhoul, J. 1975a. A class of all-zero lattice digital filters: Properties and applications. *IEEE Trans. Acoustics, Speech and Signal Processing*, 26(4):304–314, August.

———. 1975b. Linear prediction: A tutorial review. *Proc. IEEE*, 63(4):561–580.

———. 1976. New lattice methods for linear prediction. *Proc. Int. Conf. Acoustics, Speech and Signal Processing*, pp. 462–465.

———. 1977. Stable and efficient lattice methods for linear prediction. *IEEE Trans. Acoustics, Speech and Signal Processing*, 25:423–428, October.

———. 1978. A class of all-zero lattice digital filters: Properties and applications. *IEEE Trans. Acoustics, Speech and Signal Processing*, 26:304–314, August.

———. 1981. On the eigenvectors of symmetric Toeplitz matrices. *IEEE Trans. Acoustics, Speech and Signal Processing*, 29:868–872.

———. 1986. Maximum confusion spectral analysis. *IEEE Workshop on Spectral Estimation and Modeling*, pp. 6–9, November.

Maksym, J. N. 1979. A robust formulation of an optimum cross-spectral beamformer for line arrays. *J. Acoust. Soc. Am.*, 65(4):971–975.

Malik, M., and A. J. Camm, Eds. 1995. *Heart Rate Variability*. Futura Publishing Co.

Mallat, S. 1998. *A Wavelet Tour of Signal Processing*. Academic Press, Boston, MA.

Mammone, R. J., Zhang X, and R. P. Ramachandran. 1996. Robust speaker recognition: A feature based approach. *IEEE Signal Processing Magazine*, pp. 58–71, September.

Mandelbrot, B. B. 1982. *The Fractal Geometry of Nature*. W. H. Freeman and Company, New York.

——— and B. J. Van Ness. 1968. Fractional Brownian motion, fractional Gaussian noises, and applications. *SIAM Review*, 10(4):422–438.

Manolakis, D., G. Carayannis, and N. Kalouptsidis. 1983. Fast algorithms for discrete-time Wiener filters with optimum lag. *IEEE Trans. Acoustics, Speech and Signal Processing*, 21:168–179, February.

———, ———, and ———. 1984. Fast design of direct and ladder Wiener filters with linear phase. *IEEE Trans. Circuits and Systems*, 33:1175–1187, October.

———, T. Conley, et al. 1994. Comparison of visual and IR imagery. *1994 Meeting of IRIS Specialty Group on Targets, Background, and Discrimination*, Monterey, CA.

———, F. Ling, and J. G. Proakis. 1987. Efficient time-recursive least-squares algorithms for finite-memory adaptive filtering. *IEEE Trans. Circuits and Systems*, 34(4):400–408.

——— and M. Patel. 1992. Implementation of least squares adaptive filters using multiple processors. *Proc. IEEE Int. Symp. on Circuits and Systems*, pp. 2172–2175.

Markel, J. D., and A. H. Gray, Jr. 1980. *Linear Prediction of Speech*. Springer-Verlag, New York.

Marple, S. L., Jr. 1987. *Digital Spectral Analysis with Applications*. Prentice Hall, Englewood Cliffs, NJ.

Martin, R. D., and D. J. Thomson. 1982. Robust-resistant spectral estimation. *Proc. IEEE*, 70(9):1097–1114, September.

Mathews, V. J. 1991. Adaptive polynomial filters. *IEEE Signal Processing Magazine*, 8(3):10–26, July.

Matsuoka, T., and T. J. Ulrych. 1984. Phase estimation using the bispectrum. *Proc. IEEE*, 72:1403–1411.

Mazo, J. E. 1979. On the independence theory of equalizer convergence. *Bell System Tech. J.*, 58:963–993.

———. 1980. Analysis of decision directed equalizer convergence. *Bell System Tech. J.*, 59(10):1857–1876, December.

McCoy, E. J., A. T. Walden, and D. B. Percival. 1998. Multitaper spectral estimation of power law processes. *IEEE Trans. Signal Processing*, 46(3):655–668, March.

McCulloch, J. H. 1986. Simple consistent estimators of stable distribution parameters. *Communications in Statistics—Computation and Simulation*, 15:1109–1136.

McDonough, R. N., and A. D. Whelen. 1995. *Detection of Signals in Noise*, 2nd ed. Academic Press, San Diego, CA.

McGarty, T. P. 1974. The effect of interfering signals on the performance of angle of arrival estimators. *IEEE Trans. Aerospace and Electronic Systems*, 10(1):70–77.

McWhirter, J. G. 1983. Recursive least-squares minimization using a systolic array. In *Real-Time Signal Processing VI*, vol. 431, pp. 105–112.

_____ and I. K. Proudler. 1993. The QR family. In N. Kalouptsidis and S. Theodoridis, Eds., *Adaptive System Identification and Signal Processing Algorithms*, pp. 260–321. Prentice Hall, Englewood Cliffs, NJ.

_____ and T. J. Shepherd. 1989. Systolic array processor for MVDR beamforming. *IEE Proc. Part F: Radar and Signal Processing*, 136(2):75–80.

Mendel, J. M. 1991. Tutorial on higher-order statistics (spectra) in signal processing and system theory: Theoretical results and some applications. *Proc. IEEE*, 79:278–305.

Messerschmitt, D. G. 1984. Echo cancellation in speech and data transmission. *IEEE J. Selected Areas in Communications*, 2(2):283–297, March.

Michiel, H., and K. Laevens. 1997. Teletraffic engineering in a broadband era. *Proc. IEEE*, 85(12): 2007–2033, December.

Miller, K. S. 1974. *Complex Stochastic Processes: An Introduction to Theory and Application*. Addison-Wesley, Reading, MA.

Miller, T., L. Potter, and J. McCorkle. 1997. RFI suppression for ultra wideband radar. *IEEE Trans. Aerospace and Electronic Systems*, 33(4):1142–1156, October.

Mitra, S. K. 1998. *Digital Signal Processing*. McGraw-Hill, New York.

_____ and J. F. Kaiser, Eds. 1993. *Handbook for Digital Signal Processing*. Wiley, New York.

Montgomery, D. C., and E. A. Peck. 1982. *Introduction to Linear Regression Analysis*. Wiley Series in Probability and Mathematical Statistics, Wiley, New York.

Monzingo, R. A., and T. W. Miller. 1980. *Introduction to Adaptive Arrays*. Wiley, New York.

Morf, M. 1974. Fast Algorithms for Multivariable Systems. Ph.D. dissertation, Stanford University, Stanford, CA.

_____, T. Kailath, A. Vieira, and B. Dickinson. 1977. Efficient solution of covariance equations for linear prediction. *IEEE Trans. Acoustics, Speech and Signal Processing*, ASSP-25:423–433, October.

Morgan, D. R. 1978. Partially adaptive array techniques. *IEEE Trans. Antennas and Propagation*, 26(6):823–833.

_____ and S. G. Kratzer. 1996. On a class of computationally efficient, rapidly converging, generalized NLMS algorithms. *IEEE Signal Processing Letters*, 3(8):245–247, August.

Moulines, E., P. Duhamel, J. F. Cardoso, and S. Mayrargue. 1995. Subspace methods for the blind identification of multichannel FIR filters. *IEEE Trans. Signal Processing*, 43:516–525, February.

Muirhead, R. J. 1982. *Aspects of Multivariate Statistical Theory*. Wiley-Interscience, New York.

Murano, K., et al. 1990. Echo cancellation and applications. *IEEE Communications Magazine*, 28: 49–55.

Nathanson, F. 1991. *Radar Design Principles*, 2nd ed. McGraw-Hill, New York.

Netto, S., P. S. R. Diniz, and P. Agathaklis. 1995. Adaptive IIR filtering algorithms for system identification: A general framework. *IEEE Trans. Education*, 38(1):54–66, February.

Newman, T. G., and P. L. Odell. 1971. *The Generation of Random Variates*. Hafner Publishing Company, New York.

Ng, S., et al. 1996. The genetic search approach. *IEEE Signal Processing Magazine*, 13(6):38–46, November.

Nickel, U. 1993. Monopulse estimation with adaptive arrays. *IEE Proc. Part F: Radar, Sonar, and Navigation*, 140(5):303–308.

Niedermeyer, E., and F. H. Lopes Da Silva, Eds. 1998. *Electroencephalography: Basic Principles, Clinical Applications, and Related Fields*, 4th ed. Lippincott, Williams, and Wilkins, Philadelphia, PA.

Nikias, C. L., and J. M. Mendel. 1993. Signal processing with higher-order spectra. *IEEE Signal Processing Magazine*, 10:10–37.

_____ and A. P. Petropulu. 1993. *Higher-Order Spectra Analysis*. Prentice Hall, Englewood Cliffs, NJ.

———— and M. R. Raghuveer. 1987. Bispectrum estimation: A digital signal processing framework. *Proc. IEEE*, 75(7):869–891.

Noble, B., and J. W. Daniel. 1988. *Applied Linear Algebra*, 3rd ed. Prentice Hall, Englewood Cliffs, NJ.

Nuttall, A. H. 1976. Multivariate linear predictive spectral analysis employing weighted forward and backward averaging: A generalization of Burg's algorithm. *Tech. Rep.*, Naval Underwater Systems Center, New London, CT, October.

———— and G. C. Carter. 1982. Spectral estimation using combined time and lag weighting. *Proc. IEEE*, 70(9):1115–1125, September.

Oldham, K. B., and J. Spanier. 1974. *The Fractional Calculus*. Academic Press, New York.

Oppenheim, A. V., and R. W. Schafer. 1975. *Digital Signal Processing*. Prentice Hall, London.

———— and ————. 1989. *Discrete-Time Signal Processing*, 2nd ed. Prentice Hall, Englewood Cliffs, NJ.

————, A. S. Willsky, and S. H. Nawab. 1997. *Signals and Systems*. Prentice Hall, Upper Saddle River, NJ.

Ottersten, B., M. Viberg, and T. Kailath. 1991. Performance analysis of the total least squares ESPRIT algorithm. *IEEE Trans. Signal Processing*, 39(5):1122–1135.

Ozeki, K., and T. Umeda. 1984. An adaptive filtering algorithm using an orthogonal projection to an affine subspace and its properties. *Electr. Commun. Japan*, 67-A:19–27.

Painter, S. 1996. Evidence of non-Gaussian scaling behavior in heterogeneous sedimentary formations. *Water Resources Research*, 32(5):1183–1195.

Pan, C. T., and R. J. Plemmons. 1989. Least-squares modifications with inverse factorizations: Parallel implications. *Comput. Appl. Math.*, 27:109–127.

Papoulis, A. 1985. Levinson algorithm, Wold's decomposition and spectrum estimation. *SIAM Rev.*, 27(3):405–441, September.

————. 1991. *Probability, Random Variables, and Stochastic Processes*, 3rd ed. McGraw-Hill, New York.

Parzen, E. 1960. *Modern Probability Theory and Its Applications*. Wiley, New York.

————. 1977. Multiple time series modeling: Determining the order of approximating autoregressive schemes. In P. R. Krishnaiah, Ed., *Multivariate Analysis*, vol. 4, pp. 283–295. North-Holland Publishing Company, New York. Originally published by Academic Press, New York, 1969.

Paulraj, A., R. Roy, and T. Kailath. 1986. A subspace rotation approach to signal parameter estimation. *Proc. IEEE*, 74:1044–1045.

Paulraj, A. J., and C. B. Papadias. 1997. Space-time adaptive processing for wireless communications. *IEEE Signal Processing Magazine*, 14(6):49–83.

Peebles, P. Z., Jr. 1987. *Probability, Random Variables, and Random Signal Principles*, 2nd ed. McGraw-Hill, New York.

Peng, C.-K., et al. 1993. Long-range anticorrelations and non-Gaussian behavior of the heartbeat. *Physical Rev. Letters*, 70(9):1343–1346.

Pentland, A. P. 1984. Fractal-based description of natural scenes. *IEEE Trans. Pattern Analysis and Machine Intelligence*, 6:661–674, November.

Percival, D. B., and A. T. Walden. 1993. *Spectral Analysis for Physical Applications*. Cambridge University Press, New York.

Pflug, A. L., G. E. Ioup, and R. L. Field. 1992. Properties of higher-order correlation and spectra for bandlimited, deterministic transients. *J. Acoustical Society of America*, 91(2):975–988.

Pham, D. T., and P. Garrat. 1997. Blind separation of mixture of independent sources through a quasi-maximum likelihood approach. *IEEE Trans. Signal Processing*, 45:1712–1725.

Picchi, G., and G. Prati. 1987. Blind equalization and carrier recovery using a "stop-and-go" decision-directed algorithm. *IEEE Trans. Communications*, 35:877–887, September.

Pindyck, R., and D. Rubinfeld. 1998. *Econometric Models and Economic Forecasts*. McGraw-Hill, New York.

Pisarenko, V. F. 1973. The retrieval of harmonics from a covariance function. *Geophysical J. Royal Astro. Soc.*, 33:347–366.

Poor, H. V. 1994. *An Introduction to Signal Detection and Estimation*. Springer-Verlag, New York.

———— and G. W. Wornell, Eds. 1998. *Wireless Communications*. Prentice Hall, Upper Saddle River, NJ.

Porat, B. 1994. *Digital Processing of Random Signals*. Prentice Hall, Englewood Cliffs, NJ.

Press, W. H., B. P. Flannery, S. A. Teukolsky, and W. T. Vetterling. 1992. *Numerical Recipes in C: The Art of Scientific Computing*. Cambridge University Press, Cambridge.

Priestley, M. B. 1981. *Spectral Analysis and Time Series*. Academic Press, London.

Proakis, J. G. 1995. *Digital Communications*, 3rd ed. McGraw-Hill, New York.

_____ and D. G. Manolakis. 1996. *Digital Signal Processing*, 3rd ed. Prentice Hall, Englewood Cliffs, NJ.

_____ and M. Salehi. 1998. *Contemporary Communication Systems*. PWS Publishing Company, Boston, MA.

Proudler, I. K., J. G. McWhirter, and T. J. Shepherd. 1989. QRD-based lattice filter algorithms. *Proc. SPIE*, vol. 1152, pp. 56–67.

Qureshi, S. 1985. Adaptive equalization. *Proc. IEEE*, 73(9):1349–1387, September.

Rabideau, D. J., and A. O. Steinhardt. 1999. Improved adaptive clutter cancellation through data-adaptive training. *IEEE Trans. Aerospace and Electronic Systems*, 35(3):879–891.

Rabiner, L., and B. Juang. 1993. *Fundamentals of Speech Recognition*. Prentice Hall, Englewood Cliffs, NJ.

Rabiner, L. R., and R. W. Schafer. 1978. *Digital Processing of Speech Signals*. Prentice Hall, Englewood Cliffs, NJ.

Rader, C. M. 1984. A simple method for sampling in-phase and quadrature components. *IEEE Trans. Aerospace and Electronic Systems*, 20(6):821–824.

_____. 1996. VLSI systolic arrays for adaptive nulling. *IEEE Signal Processing Magazine*, 13(4):29–49, July.

_____ and A. O. Steinhardt. 1986. Hyperbolic Householder transformations. *IEEE Trans. Acoustics, Speech and Signal Processing*, ASSP(34):1589–1602.

Ramsey, F. L. 1974. Characterization of the partial autocorrelation function. *Ann. Stat.*, 2:1296–1301.

Rao, B. D., and K. V. S. Hari. 1989. Performance analysis of root-MUSIC. *IEEE Trans. Acoustics, Speech and Signal Processing*, 37(12):1939–1949.

Rao, T. S., and M. M. Gabr. 1984. *An Introduction to Bispectral Analysis and Bilinear Time Series Models*. Lecture Notes in Statistics, 24, Springer-Verlag, New York.

Rechtschaffen, A., and A. Kales. 1968. *A Manual of Standardized Technology, Techniques and Scoring System for Sleep Stages of Human Subjects*. Public Health Service, US. Dept. of Health, Education and Welfare, Bethesda, MD.

Reed, I. S., et al. 1974. Rapid convergence in adaptive arrays. *IEEE Trans. Aerospace and Electronic Systems*, 10(6):853–863.

Regalia, P. A. 1995. *Adaptive IIR Filtering in Signal Processing and Control*. Marcel Dekker, New York.

Reidle, K., and A. Sidorenko. 1995. Minimum bias multiple taper spectral estimation. *IEEE Trans. Signal Processing*, 43(1):188–195, January.

Richmond, C. 1999. Statistics of adaptive nulling and use of the generalized eigen-relation (GER) for modeling inhomogeneities in adaptive processing. *IEEE Trans. Signal Processing*, to appear.

Richmond, C. D. 1997. Statistical performance analysis of the adaptive sidelobe blanker detection algorithm. *Proc. Asilomar Conf. Signals, Systems, and Computers*, pp. 562–565.

Rife, D. C., and R. R. Boorstyn. 1974. Single-tone parameter estimation from discrete-time observations. *IEEE Trans. Information Theory*, 20:591–598.

Risannen, J. 1978. Modeling by shortest data description. *Automatica*, 14:465–471.

Robey, F. C., et al. 1992. A CFAR adaptive matched filter. *IEEE Trans. Aerospace and Electronic Systems*, 28(1):208–216.

Robinson, E. 1984. Statistical pulse compression. *Proc. IEEE*, 72(10):1276–1289, October.

_____ and S. Treitel. 1980. Maximmum entropy and the relationship of the partial autocorrelation to the reflection coefficients of a layered system. *IEEE Trans. Acoustics, Speech and Signal Processing*, 28(2):224–235, April.

Robinson, E. A., and S. Treitel. 1980. *Geophysical Signal Analysis*. Prentice Hall, Englewood Cliffs, NJ.

Rosenblatt, M. 1985. *Stationary Sequences and Random Fields*. Birkhäuser, Stuttgart, Germany.

Ross, S. 1998. *A First Course in Probability*, 5th ed. Prentice Hall, Upper Saddle River, NJ.

Roy, R., and T. Kailath. 1989. ESPRIT—estimation of signal parameters via rotational invariance techniques. *IEEE Trans. Acoustics, Speech and Signal Processing*, 37(7):984–995, July.

_____, A. Paulraj, and T. Kailath. 1986. ESPRIT—a subspace rotation approach to estimation of parameters of cisoids in noise. *IEEE Trans. Acoustics, Speech and Signal Processing*, 34(4):1340–1342.

Roy, S., and J. J. Shynk. 1990. Analysis of the momentum LMS algorithm. *IEEE Trans. Acoustics, Speech and Signal Processing*, 38(12):2088–2098, December.

Rupp, M. 1993. The behavior of LMS and NLMS algorithms in the presence of spherically invariant processes. *IEEE Trans. Signal Processing*, 41(3):1149–1160, March.

———. 1995. Bursting in the LMS algorithm. *IEEE Trans. Signal Processing*, 43(10):2414–2417, October.

Sabins, F. 1987. *Remote Sensing*. W. H. Freeman, New York.

Saltzberg, B. R. 1968. Intersymbol interference error bounds with application to ideal bandlimited signaling. *IEEE Trans. Information Theory*, IT-14:263–268, July.

Samorodnitsky, G., and M.S. Taqqu. 1994. *Stable Non-Gaussian Random Processes*. Chapman and Hall, New York.

Sato, Y. 1975. Two extensional applications of zero-forcing equalization method. *IEEE Trans. Communications*, 23(6):684–687.

Saul, J. P. 1990. Beat-to-beat variations of heart rate reflect modulation of cardiac autonomic outflow. *Int. Union Physiol. Sci.*, 5:32–37.

Sayed, A. H., and T. Kailath. 1994. A state-space approach to adaptive RLS filtering. *IEEE Signal Processing Magazine*, 11:18–60.

——— and ———. 1998. Recursive least-squares adaptive filters. In V. Madisetti and D. Williams, Eds., *The Digital Signal Processing Handbook*. CRC Press, New York.

——— and M. Rupp. 1996. Error-energy bounds for adaptive gradient algorithms. *IEEE Trans. Signal Processing*, 44(8):1982–1989, August.

——— and ———. 1998. Robustness issues in adaptive filtering. In V. Madisetti and D. Williams, Eds., *The Digital Signal Processing Handbook*. CRC Press, New York.

Scales, L. E. 1985. *Introduction to Nonlinear Optimization*. Springer-Verlag, New York.

Scharf, L. L. 1991. *Statistical Signal Processing: Detection, Estimation, and Time Series Analysis*. Addison-Wesley, Reading, MA.

——— and L. T. McWhorter. 1997. Adaptive matched substance detectors and adaptive coherence estimators. *Proc. Asilomar Conf. Signals, Systems, and Computers*, pp. 1114–1117.

Schetzen, M. 1989. *The Volterra and Wiener Theories on Nonlinear Systems*, 2nd ed. Krieger Publishing Company, Malabar, FL.

Schmidt, R. 1986. Multiple emitter location and signal parameter estimation. *IEEE Trans. Antennas and Propagation*, 34:276–290. Originally appeared in *Proc. RADC, Spectral Estimation Workshop*, pp. 243–258, Rome, NY, 1979.

Schniter, P. 1998. The BERGULATOR. *http://backhoe.ee.cornell.edu/BURG/*.

Schreiber, R. J. 1986. Implementation of adaptive array algorithms. *IEEE Trans. Acoustics, Speech, and Signal Processing*, 34(10):1038–1045.

Schür, I. 1917. On power series which are bounded in the interior of the unit circle. *Journal für die Reine und Angewandte Mathematik*, 147:205–232.

Sethares, W. A. 1993. The least mean square family. In N. Kalouptsidis and S. Theodoridis, Eds., *Adaptive System Identification and Signal Processing Algorithms*, pp. 84–122. Prentice Hall, Englewood Cliffs, NJ.

Shalvi, O., and E. Weinstein. 1990. New criteria for blind equalization of non-minimum phase systems (channels). *IEEE Trans. Information Theory*, 36:312–321.

——— and ———. 1994. Universal methods for blind deconvolution. In S. Haykin, Ed., *Blind Deconvolution*. Prentice Hall, Englewood Cliffs, NJ.

Shao, M., and C. L. Nikias. 1993. Signal processing with fractional lower order moments: Stable processes and their applications. *Proc. IEEE*, 81(7):986–1010, July.

Shaughnessy, D. O. 1987. *Speech Communication*. Addison-Wesley, Reading, MA.

Shepherd, T. J., and J. G. McWhirter. 1993. Systolic adaptive beamforming. In S. Haykin and T. J. Shepherd, Eds., *Radar Array Processing*, pp. 153–243. Springer-Verlag, New York.

Sheriff, R. 1994. *Exploration Seismology*, 2nd ed. Cambridge University Press, Cambridge.

Sherman, S. M. 1984. *Monopulse Principles and Techniques*. Artech House, Dedham, MA.

Shiavi, R. 1991. *Introduction to Applied Statistical Signal Analysis*. Irwin, Burr Ridge, IL.

Shynk, J. J. 1989. Adaptive IIR filtering. *IEEE ASSP Magazine*, 6:4–21.

———. 1992. Frequency-domain and multirate adaptive filtering. *IEEE Signal Processing Magazine*, 9(1):14–37.

Siller, C. A., Jr. 1984. Multipath propagation. *IEEE Communications Magazine*, 22(2):6–15.

Skolnik, M. 1980. *Introduction to Radar Systems*, 2nd ed. McGraw-Hill, New York.

_____, Ed. 1990. *Radar Handbook*, 2nd ed. McGraw-Hill, New York.

Slock, D. T. M. 1993. On the convergence behavior of the LMS and the normalized LMS algorithms. *IEEE Trans. Signal Processing*, 45(12):2811–2825, September.

_____ and T. Kailath. 1991. Numerically stable fast transversal filters for recursive least-squares adaptive filtering. *IEEE Trans. Signal Processing*, 39(1):92–114.

_____ and _____. 1993. Fast transversal RLS algorithms. In N. Kalouptsidis and S. Theodoridis, Eds., *Adaptive System Identification and Signal Processing Algorithms*, pp. 123–190. Prentice Hall, Englewood Cliffs, NJ.

Solo, V. 1997. The stability of LMS. *IEEE Trans. Signal Processing*, 45(12):3017–3026, December.

_____ and X. Kong. 1995. *Adaptive Signal Processing Algorithms*. Prentice Hall, Englewood Cliffs, NJ.

Sondhi, M., and D. A. Berkley. 1980. Silencing echoes in the telephone network. *Proc. IEEE*, 68:948–963.

Sorenson, H. 1970. Least-squares estimation: From Gauss to Kalman. *IEEE Spectrum*, 7:63–68.

Stark, H., and J. W. Woods. 1994. *Probability, Random Processes, and Estimation Theory for Engineers*, 2nd ed. Prentice Hall, Englewood Cliffs, NJ.

Steele, A. K. 1983. Comparison of directional and derivative constraints for beamformers subject to multiple linear constraints. *IEE Proc. Parts H*. 130(1):41–45.

Steinhardt, A. O. 1988. Householder transforms in signal processing. *IEEE ASSP Magazine*, 5:4–12.

_____. 1992. Adaptive multisensor detection and estimation. In S. Haykin and A. O. Steinhardt, Eds., *Adaptive Radar Detection and Estimation*, pp. 91–160, Wiley, New York.

Stewart, G. W. 1973. *Introduction to Matrix Computations*. Academic Press, New York.

Stewart, R. W., and R. Chapman. 1990. Fast stable Kalman filter algorithms utilizing the square root. *Proc. Int. Conf. Acoustics, Speech and Signal Processing*, pp. 1815–1818.

Stockham, T., T. Cannon, and R. Ingerbretsen. 1975. Blind deconvolution through digital signal processing. *Proc. IEEE*, 63:678–692.

Stoer, J., and R. Bulirsch. 1980. *Introduction to Numerical Analysis*. Springer-Verlag, New York.

Stoica, P., and R. L. Moses. 1997. *Introduction to Spectral Analysis*. Prentice Hall, Upper Saddle River, NJ.

_____ and A. Nehorai. 1989. MUSIC, maximum likelihood, and Cramer-Rao bound. *IEEE Trans. Acoustics, Speech and Signal Processing*, 37(5):720–741.

Strang, G. 1980. *Linear Algebra and Its Applications*. Academic Press, New York.

_____. 1998. *Introduction to Linear Algebra*. Wellesley-Cambridge Press, Wellesley, MA.

Stuck, B. W. 1978. Minimum error dispersion linear filtering of scalar symmetric stable processes. *IEEE Trans. Automatic Control*, 23(3):507–509, June.

_____ and B. Kleiner. 1974. A statistical analysis of telephone noise. *Bell Systems Tech. J.*, 53:1262–1320.

Swami, A. 1998. Non-Gaussian processes. *IEEE Signal Processing Magazine*, 15(5):40–42, September.

_____, G. Giannakis, and G. Zhou. 1997. Bibliography on higher-order statistics. *Signal Processing*, 60(1):65–126, July.

Takao, K., et al. 1976. An adaptive array under directional constraint. *IEEE Trans. Antennas and Propagation*, 24(9):662–669.

Tang, K., K. Man, S. Kwong, and Q. He. 1996. Genetic algorithms and their applications. *IEEE Signal Processing Magazine*, 13(6):22–37, November.

Taqqu, M. S., V. Teverovsky, and W. Willinger. 1995. Estimators for long-range dependence: An empirical study. *Fractals*, 3(4):795–798.

Theodoridis, S., and N. Kalouptsidis. 1993. Spectral analysis. In N. Kalouptsidis and S. Theodoridis, Eds., *Adaptive System Identification and Signal Processing Algorithms*, pp. 322–387. Prentice Hall, Englewood Cliffs, NJ.

Therrien, C. W. 1992. *Discrete Random Signals and Statistical Signal Processing*. Prentice Hall, Englewood Cliffs, NJ.

Thomson, D. J. 1982. Spectrum estimation and harmonic analysis. *Proc. IEEE*, 72(9):1055–1096.

Tong, L., G. Xu, and T. Kailath. 1994a. Blind channel identification and equalization using spectral correlation measurements, part II: A time-domain approach. In W. A. Gardner, Ed., *Cyclostationarity in Communications and Signal Processing*, pp. 437–454. IEEE Press, New York.

_____, _____, and _____. 1994b. Blind identification and equalization based on second-order statistics: A time-domain approach. *IEEE Trans. Information Theory*, 40(2):340–349, March.

Treichler, J., and B. Agee. 1983. A new approach to multipath correction of constant modulus signals. *IEEE Trans. Acoustics, Speech and Signal Processing*, 31(2):459–472, April.

——, I. Fijalkow, and R. Johnson. 1996. Fractionally spaced equalizers: How long should they really be? *IEEE Signal Processing Magazine*, 13(5):65–81, May.

——, C. R. Johnson, and M. G. Larimore. 1987. *Theory and Design of Adaptive Filters*. Wiley-Interscience, New York.

——, M. Larimore, and J. Harp. 1998. Practical blind demodulators for high-order QAM signals. *Proc. IEEE*, 86(10):1907–1925, October.

Trench, W. F. 1964. An algorithm for the inversion of finite Toeplitz matrices. *SIAM J. Appl. Math.*, 12:515–521.

Tsypkin, Ja. Z., 1971. *Adaptation and Learning in Automatic Systems*, vol. 73 of *Mathematics in Science and Engineering*. Academic Press, New York.

——. 1973. *Foundations of the Theory of Learning Systems*, vol. 101 of *Mathematics in Science and Engineering*. Academic Press, New York.

Tugnait, J. K. 1992. Comments on "New criteria for blind deconvolution of non-minimum phase systems (channels)." *IEEE Trans. Information Theory*, 38(1):210–213, January.

——. 1998. System identification and tests for non-Gaussianity and linearity. *IEEE Signal Processing Magazine*, 15(5):42–43, September.

Ulrych, T. J., and R. W. Clayton. 1976. Time series modeling and maximum entropy. *Phys. Earth Planet. Inter.*, 12:188–200, August.

Vaidyanathan, P., J. Tugan, and A. Kirac. 1997. On the minimum phase property of prediction-error polynomials. *IEEE Signal Processing Letters*, 4(5):126–127, May.

Van Loan, C. 1995. *Introduction to Scientific Computing*. Prentice Hall, Upper Saddle River, NJ.

Van Trees, H. L. 1968. *Detection, Estimation, and Modulation Theory*, vol. 1. McGraw-Hill, New York.

Van Veen, B. 1991. Minimum variance beamforming. In S. Haykin and A. Steinhardt, Eds., *Adaptive Radar Detection and Estimation*, pp. 161–236. Wiley, New York.

—— and K. M. Buckley. 1988. Beamforming: A versatile approach to spatial filtering. *IEEE Acoustic, Speech, and Signal Processing Magazine*, pp. 4–24, April.

—— and ——. 1998. Beamforming techniques for spatial filtering. In V. Madisetti and D. Williams, Eds., *The Digital Signal Processing Handbook*, CRC Press, New York.

Verhaegen, M. H. 1989. Round-off error propagation in four generally-applicable, recursive, least-square estimation schemes. *Automatica*, 25:437–444.

Verhoeckx, N. A. M., H. C. Van Den Elzen, F. A. M. Snijders, and P. J. Van Gerwen. 1979. Digital echo cancellation for baseband data transmission. *IEEE Trans. Acoustics, Speech and Signal Processing*, 27(6), part 2:768–781, December.

Vervaat, W. 1987. Properties of general self-similar processes. *Bull. Int. Statist. Inst.*, 52(4):199–216.

Viswanathan, R., and J. Makhoul. 1975. Quantization properties of transmission parameters in linear predictive systems. *IEEE Trans. Acoustics, Speech and Signal Processing*, 23(3):309–321, June.

Walpole, R. E., R. H. Myers, and S. L. Myers. 1998. *Probability and Statistics for Engineers and Scientists*, 6th ed. Prentice Hall, Upper Saddle River, NJ.

Ward, J. 1994. Space-time adaptive processing for airborne radar. *Tech. Rep. TR-1015*, MIT Lincoln Laboratory, Lexington, MA.

——. 1995. Space-time adaptive processing for airborne radar. *Proc. Int. Con. Acoustics, Speech and Signal Processing*, pp. 2809–2812.

——. 1996. Cramér-Rao bounds for target angle and Doppler estimation with space-time adaptive processing radar. *Proc. Asilomar Conf. Signals, Systems, and Computers*, pp. 1198–1202.

—— and A. O. Steinhardt. 1994. Multiwindow post-Doppler space-time adaptive processing. *Proc. Seventh Workshop on Statistical Signal and Array Processing*, pp. 461–464.

Watkins, D. S. 1991. *Fundamentals of Matrix Computations*. Wiley, New York.

Weinstein, S. B. 1977. Echo cancellation in the telephone network. *IEEE Communications Magazine*, 15(1):9–15, January.

Weiss, G. 1975. Time-reversibility of linear stochastic processes. *J. Appl. Probability*, 12:831–836.

Welch, P. W. 1967. The use of fast Fourier transform for the estimation of power spectra: A method based on time averaging over short, modified periodograms. *IEEE Trans. Audio Electroacoustics*, 15(2):70–76.

Widrow, B., and M. E. Hoff, Jr. 1960. Adaptive switching circuits. *IRE WESCON Conv. Rec.*, part 4:96–104.

_____ and S. D. Sterns. 1985. *Adaptive Signal Processing*. Prentice Hall, Englewood Cliffs, NJ.

_____ and E. Walach. 1994. *Adaptive Inverse Control*. Prentice Hall, Englewood Cliffs, NJ.

_____ et al. 1975. Adaptive noise cancelling: Principles and applications. *Proc. IEEE*, 63:1692–1716.

_____ et al. 1976. Stationary and nonstationary learning characteristics of the LMS adaptive filter. *Proc. IEEE*, 64:1151–1162.

Wiener, N. 1949. *Extrapolation, Interpolation, and Smoothing of Stationary Time Series, with Engineering Applications*. MIT Press, Cambridge, MA.

Wiggins, R. A. 1978. On minimum entropy deconvolution. *Geoexploration*, 16:21–35.

Williamson, G. 1998. Adaptive IIR filters. In V. Madisetti and D. Williams, Eds., *The Digital Signal Processing Handbook*. CRC Press, New York.

Wong, E. 1971. *Stochastic Processes in Information and Dynamical Systems*. McGraw-Hill, New York.

Wornell, G. W. 1993. Wavelet-based representation for the $1/f$ family of fractal processes. *Proc. IEEE*, 81(10):1428–1450, October.

_____. 1996. *Signal Processing with Fractals: A Wavelet-Based Approach*. Prentice Hall, Englewood Cliffs, NJ.

Yang, B., and J. F. Böhme. 1992. Rotation-based RLS algorithms: Unified derivations, numerical properties and parallel implementations. *IEEE Trans. Signal Processing*, 40:1151–1167.

Yokoya, N., K. Yamamoto, and N. Funakubo. 1989. Fractal based analysis and interpolation of 3D natural surface shapes and their application to terrain modeling. *Computer Vision, Graphics and Image Processing*, 46:284–302.

Zadeh, L. A., and J. R. Ragazzini. 1950. An extension of Wiener's theory of prediction. *J. Appl. Phys.*, 21:645–655, July.

Zatman, M. 1998. How narrow is narrowband? *IEE Proc. Part F: Radar, Sonar and Navigation*, 145(2):85–91.

Zoltowski, M. 1992. Beamspace ML bearing estimation for adaptive phased array radar. In S. Haykin and A. Steinhardt, Eds., *Adaptive Radar Detection and Estimation*, pp. 237–332. Wiley, New York.

Index

Serving Dual-Career Couples and Aging Baby Boomers

Now that dual-career couples have become the norm, the market for franchises offering convenience and timesaving devices is booming. Customers are willing to pay for products and services that will save them time or trouble, and franchises are ready to provide them. For instance, Maid Brigade, a residential cleaning franchise with nearly 500 locations across the United States and Canada that has been franchising since 1979, aims its cleaning service at busy professionals who prefer to "spend their time pursuing careers, hobbies and enjoying family and friends" rather than cleaning their homes.[73] Other areas in which franchising is experiencing rapid growth include home delivery of meals, pet day care centers, continuing education and training (especially computer and business training), leisure activities (such as hobbies, health spas, and travel-related activities), products and services aimed at home-based businesses, and health care.

A number of franchises are aiming at one of the nation's largest population segments: aging Baby Boomers. About 40.2 million people, or 12.4 percent of the U.S. population, are 65 or older, and by 2030 that number is expected to double to 72 million. A survey by the American Association of Retired Persons shows that 90 percent of senior citizens want to remain in their homes as they age, which is creating a great business opportunity for franchises such as Home Instead Senior Care, a company with nearly 900 franchises that provides in-home non–health care services to senior citizens. The company expects its domestic sales to grow at 10 percent a year and its international sales to grow at 30 percent a year for the foreseeable future.[74]

Franchising as a Growth Strategy

Entrepreneurs with established and tested business models can use franchising as a growth strategy by becoming franchisors. Franchising enables business owners to use other people's money to grow their businesses with minimal capital investment on the part of the franchisor. Franchisees put up the funds to start their businesses, infuse capital into the franchising operation through franchise fees, and generate ongoing cash flow for the franchisor from ongoing royalty fees and other charges. In short, franchising accelerates a small company's growth. However, selling franchises changes the scope of a business and requires entrepreneurs to take on new and different roles. According to one franchise expert, "Franchising is more than selling services or products. You will be an educator, trainer, psychologist, minister, and perpetual hand-holder to your franchisees. You also will be a fee collector, extracting an initial fee and then collecting royalties for the life of the franchise."[75]

To create a successful franchise operation, a business must meet the criteria outlined in the following sections.

> **10.**
> _____
> Describe the potential of franchising a business as a growth strategy.

Unique Concept

To make a successful franchise operation, a business must have a unique concept that gives it a competitive edge in the marketplace. For instance, an entrepreneur may develop a new twist on fast food, a better way to exercise, or a new process for removing dents from cars.

Replicable

To make a successful franchise operation, an entrepreneur's business model must be replicable. Can potential franchisees reproduce the success of the original unit regardless of location? Is there a business system in place that an entrepreneur can teach to franchisees? In addition, is the owner willing to relinquish some control to these new owners? Franchising requires an entrepreneur to leave the business's daily operations (and its reputation) largely in the hands of franchisees.

ENTREPRENEURIAL PROFILE: Chelsea and Scott Sloan: Uptown Cheapskates While studying for her business degree at the University of Utah, Chelsea Sloan began trading e-mails with her brother Scott about starting a used fashion clothing business. They built a business plan that helped refine their concept and in 2008 launched Uptown Cheapskate, a business that buys and resells trendy, name-brand clothing that customers exchange for cash or credit at the store. Their concept relies on a proprietary inventory management system that took eight months to develop and that includes prices for thousands of items. The business was an immediate hit, and realizing that their concept was easily replicable, the entrepreneurs began selling franchises almost immediately. Today, Uptown Cheapskates has 16 franchises in 12 states with eight more under development.[76]

Expansion Plan

When entrepreneurs make the decision to franchise, they must develop a sound expansion plan. New franchisors must consider issues such as the speed of growth, territorial development, support services, staffing, and fee structure. The entire plan demands a well-conceived strategy for supporting franchisees and a rigorous financial analysis.

Due Diligence

Launching a successful franchise requires undertaking an extensive due diligence process that includes researching legal issues, preparing a franchise disclosure document, registering necessary trademarks, creating a Web site, and writing training manuals for franchisees. New franchisees discover that they have two new roles: selling franchises and servicing franchisees. "Franchising is like starting an entirely new business venture within the existing business structure," says Jim Thomas, a former top manager of the Taco Time International franchise. "The business now becomes a legally responsible support system with an entirely new set of responsibilities."[77]

Legal Guidance

Enlisting professional assistance from a franchise attorney is essential. One of the most important roles of the franchise attorney is to prepare the FDD. Every franchisor must provide to prospective franchisees an FDD that covers the 23 items discussed earlier in this chapter. Recall that 15 states require registration of the FDD before the franchisor can sell franchises. Obtaining legal approval, producing audited financial statements, and marketing the franchise concept is not cheap. Entrepreneurs can expect to invest a minimum of $100,000 to $750,000 to launch a franchise business.[78]

Support for Franchisees

Once a franchise operation is running, the franchisor must have the resources available to train franchisees in the operation of the business system, assist them through the start-up phase, and provide ongoing product support for them.

Conclusion

Franchising has proved its viability in the U.S. economy and has become a key part of the small business sector because it offers many would-be entrepreneurs the opportunity to own and operate a business with a greater chance for success. Despite its impressive growth rate to date, the franchising industry still has a great deal of room left to grow, especially globally. Current trends combined with international opportunities indicate that franchising will continue to play a vital role in the global economy.

Chapter Review

1. Explain the importance of franchising in the U.S. economy.
   - Through franchised businesses, consumers can buy nearly every good or service imaginable—from singing telegrams and computer training to tax services and waste-eating microbes.
   - More than 757,000 franchise outlets operate in the United States, generating more than $802 billion in total economic output. Franchises also employ more than 8.2 million people and contribute more than $460 billion to U.S. GDP.

2. Define the concept of franchising and describe the different types of franchises.
   - Franchising is a method of doing business involving a continuous relationship between a franchisor and a franchisee. The franchisor retains control

of the distribution system, whereas the franchisee assumes all of the normal daily operating functions of the business.

- There are three types of franchising: trade-name franchising, where the franchisee purchases only the right to use a brand name; product distribution franchising, which involves a license to sell specific products under a brand name; and pure franchising, which provides a franchisee with a complete business system.

3. Describe the benefits and limitations of buying a franchise.

- The franchisor has the benefits of expanding his business on limited capital and growing without developing key managers internally. The franchisee also receives many key benefits: management training and counseling, customer appeal of a brand name, standardized quality of goods and services, national advertising programs, financial assistance, proven products and business formats, centralized buying power, territorial protection, and greater chances for success.
- Potential franchisees should be aware of the disadvantages involved in buying a franchise: franchise fees and profit sharing, strict adherence to standardized operations, restrictions on purchasing, limited product lines, potentially ineffective training programs, and less freedom.

4. Describe the legal aspects of franchising, including the protection offered by the FTC's Trade Regulation Rule.

- The FTC's Trade Regulation Rule is designed to help the franchisee evaluate a franchising package. It requires each franchisor to disclose information covering 23 topics at least 10 days before accepting payment from a potential franchisee. This document, the franchise disclosure document (FDD), is a valuable source of information for anyone considering investing in a franchise.

5. Explain the right way to buy a franchise.

- To buy a franchise the right way requires that you evaluate yourself, research your market, consider your franchise options, get a copy of the franchisor's FDD and study it, talk to existing franchisees, ask the franchisor some tough questions, and make your choice.

6. Describe a typical franchise contract and some of its provisions.

- The amount of franchisor–franchisee litigation has risen steadily over the past decade. Three terms are responsible for most franchisor–franchisee disputes: termination of the contract, contract renewal, and transfer and buyback provisions.

7. Explain current trends shaping franchising.

- Trends influencing franchising include international opportunities; the emergence of smaller, nontraditional locations; conversion franchising; multiple-unit franchising; master franchising; cobranding franchising; and products and services targeting busy dual career couples and aging Baby Boomers.

8. Explain the advantages and challenges franchising offers a business as a growth strategy.

- Franchising a business can be an effective method to grow a business using the investments of the franchisees. It does involve a highly litigious and regulated process that demands specialized legal professions and imposes an entirely new set of administrative demands on the business to establish and administer this complex system.

Discussion Questions

6-1. What is franchising?

6-2. Describe the three types of franchising and provide an example of each.

6-3. How does franchising benefit franchisees? Franchisors?

6-4. Discuss the advantages and the disadvantages of franchising for the franchisee.

6-5. Joe Libava, The Franchise King, says, "If you are comfortable following someone else's rules and have a strong desire to be in business for yourself, franchise ownership is an option you should explore. However, if you don't have a very good track record of toeing the line—and instead prefer to make your own rules—becoming a franchise owner may not be the way to get into business for yourself." Do you agree? Explain.

6-6. How beneficial to franchisees is a quality training program? What types of entrepreneurs may benefit most from this training?

6-7. Why might an independent entrepreneur be dissatisfied with a franchising arrangement?

6-8. What are the clues in detecting an unreliable franchisor?

6-9. Should a prospective franchisee investigate before investing in a franchise? If so, how and in what areas?

6-10. What is the function of the FTC's Trade Regulation Rule? What function does the FDD perform?

6-11. Outline the rights the Trade Regulation Rule gives all prospective franchisees.

6-12. What is the source of most franchisor–franchisee litigation? Whom does the standard franchise contract favor?

6-13. Describe the current trends affecting franchising within the United States and internationally.

6-14. One franchisee says, "Franchising is helpful because it gives you somebody (the franchisor) to get you going, nurture you, and shove you along a little. However, the franchisor won't make you successful. That depends on what you bring to the business, how hard you are prepared to work, and how committed you are to finding the right franchise for you." Do you agree? Explain.

6-15. Why might an entrepreneur consider franchising as an attractive growth strategy for his or her business?

6-16. What should an entrepreneur be prepared for when considering franchising his or her business?

CHAPTER 7
Buying an Existing Business

Opportunity is missed by most people because it is dressed in overalls and looks like work.

—Anonymous

The pessimist sees difficulty in every opportunity, the opportunist sees an opportunity in every difficulty.

— L. P. Jacks

The entrepreneurial experience always involves risk. Buying a franchise, as discussed in Chapter 6, is one approach that can help reduce the risk for an entrepreneur. Another way to reduce the risk associated with entrepreneurship is to purchase an existing business rather than start a new venture. Purchasing an existing business can be a good approach, particularly as the entrepreneurs in the Baby Boom Generation are now reaching retirement age and are seeking to sell their businesses. "I tell the kids at school . . . you will soon be talking to a 67-year-old guy who is a little bit tired of what he's doing, who's already talked to his son and his daughter, and the family options don't exist," says Dan Steppe, director the University of Houston's Wolff Center for Entrepreneurship. "The seller will enable the buyer to buy his company fairly inexpensively."[1] In addition, the recent recession created a buyer's market for acquiring small businesses due to the weak economic conditions that continue to this day.

Buying an existing business requires a great deal of analysis and evaluation to ensure that what the entrepreneur is purchasing meets his or her needs and expectations. Exercising patience and taking the necessary time to research a business before buying it are essential to getting a good deal. Research conducted by Pepperdine University's Graziadio School of Business and Management found that 87 percent of the purchases of small to medium-size business took less than one year to complete, with most taking six to eight months.[2] In too many cases, the excitement of being able to implement a "fast entry" into the market causes an entrepreneur to rush into deal and make unnecessary mistakes in judgment.

Before buying any business, an entrepreneur must conduct a thorough analysis of the business and the opportunity that it presents. According to Russell Brown, author of *Strategies for Successfully Buying or Selling a Business*, "You have access to the company's earnings history, which gives you a good idea of what the business will make, and an existing business has a proven track record; most established organizations tend to stay in business and keep making money."[3] If vital information such as audited financial statements and legal clearances are not available, an entrepreneur must be especially diligent before buying a business.

Wise entrepreneurs conduct thorough research before negotiating a purchase price for a business. The following questions provide a good starting point:

- Is this the type of business that you would like to operate?
- Does this business match your experience, knowledge, and talents?
- Will this business offer a lifestyle you find attractive?
- What are negative aspects of owning this type of business?
- Are there any skeletons in the company closet that might come back to haunt you?
- Is this the best market and the best location for this business?
- Are there important demographic, population, or political changes in the community where the business is located that could affect future sales favorably or unfavorably?
- Do you know the critical factors that must exist for this business to be successful?
- Do you have the experience required to operate this type of business? If not, will the current owner be willing to stay on for a time to teach you the "ropes"?
- Does the present building meet all state and federal accessibility guidelines? If not, what will it take to bring the facility up to code?
- If the business is profitable, why does the current owner(s) want to sell? Can you verify the current owner's reason for selling?
- Does the business have a good reputation with its customers and in the community?
- If the business is currently in decline, do you have a plan to return the business to profitability? How confident are you that your turnaround plan will work?
- Have you examined other similar businesses that are currently for sale or that have sold recently to determine what a fair market price for the company is?

The time and energy invested in the evaluation of an existing business will earn significant dividends by allowing an entrepreneur to acquire a business that will continue to be successful or to avoid purchasing a business that is heading for failure.

Buying an Existing Business

Advantages of Buying an Existing Business

The following are some of the most common advantages of purchasing an existing business.

1. _____
Understand the advantages of buying an existing business.

SUCCESSFUL BUSINESSES OFTEN CONTINUE TO BE SUCCESSFUL A business that has been profitable for some time often reflects an owner who has established a solid customer base, developed successful relationships with critical suppliers, and mastered the day-to-day operation aspects of the business. All of these factors are positive and may be keys to continued success. When things have gone well, it is important for a new owner to make changes slowly and retain the relationships with customers, suppliers, and staff that have made the business a success. This advantage often accompanies the second advantage, using the experience of the previous owner.

LEVERAGING THE EXPERIENCE OF THE PREVIOUS OWNER In cases in which the business has a history of success, a new owner may negotiate with the current owner to stay on as a consultant for a time. This allows a smooth transition during which the seller introduces the new owner to customers and suppliers and shows the new owner the secrets of making the company work. The previous owner can also be very helpful in unmasking the unwritten rules of business—whom to trust, expected business behavior, and many other critical intangibles. Hiring the previous owner as a consultant for the first few months can be a valuable investment for both parties. Learning from the previous owner's experience is extremely helpful.

OWNING A BUSINESS GUARANTEES A JOB As long as you work for someone else, you are at the mercy of that employer and that business. Businesses can be sold or closed down, leaving employees in a state of uncertainty or out of work. Owning a business puts the entrepreneur in charge of his or her own destiny.

ENTREPRENEURIAL PROFILE: Linda Jamerson and Ken McDonald: Aluminum Case Company Linda Jamerson and her husband, Ken McDonald, had been trying to purchase a business for three years. Although they wanted to work for themselves, they needed high profits rather quickly to generate the income they needed to meet their personal monthly expenses. That is why they chose to buy an existing business rather pursue a start-up. During their search, two promising deals to buy companies fell through. However, when they found the ideal opportunity, the Aluminum Case Company, they were eager to move ahead. "Our criteria were a manufacturing company with a good reputation and growth potential that had been ignored," said Jamerson. "Aluminum Case Company had a large and varied customer base, a unique product niche, a good reputation and capacity for huge growth. They had no Web presence and had not automated their engineering or equipment, so we felt we could make a quick impact on sales." However, they ran into a roadblock on the purchase when they tried to find traditional bank financing to help fund the purchase. "We were discouraged at how little funding our regular bank—and a few others—were willing to lend," says Jamerson. Then they came across a company that specializes in self-directed IRAs and alternative small business financing. Jamerson and McDonald set up a new retirement plan through which they were able to buy Aluminum Case Co. The company's sales grew 60 percent the first year after the couple purchased the business.[4]

THE TURNKEY BUSINESS Starting a company can be a daunting, time-consuming task, and buying an existing business is one of the fastest pathways to entrepreneurship. When things go well, purchasing an existing business saves the time and energy required to plan and launch a new business. The buyer gets a business that is already generating cash and perhaps profits as well. The day the entrepreneur takes over the ongoing business is the day revenues begin. Tom Gillis, an entrepreneur, and management consultant in Houston, Texas, says, "Acquiring an established company becomes attractive in three situations: when you haven't found 'the idea' that really turns you on and you find it in an existing business; when you have more money than you have time to start a business from scratch; and when you want to grow but lack a compatible product, service, location or particular advantage that is available from an owner who wants out." According to Gillis, the critical question is, "What do I gain by acquiring this business that I would not be able to achieve on my own?"[5]

SUPERIOR LOCATION When the location of the business is critical to its success, purchasing a business that is already in the right location may be the best choice. In fact, the existing business's greatest asset may be its location. A location that provides a significant competitive advantage may be reason enough for an entrepreneur to decide to buy instead of launch. Opening a second-class location and hoping to draw customers often proves fruitless.

EMPLOYEES AND SUPPLIERS ARE IN PLACE Experienced employees who choose to continue to work for the business are a significant resource because they can help the new owner learn the business. In addition, an existing business has an established set of suppliers with a history of business transactions. Vendors can continue to supply the business while the new owner assesses the products and services of other vendors. Thus, the new owner can take the time needed to evaluate alternate suppliers.

INSTALLED EQUIPMENT WITH KNOWN PRODUCTION CAPACITY Acquiring and installing new equipment exerts a tremendous strain and uncertainty on a fledgling company's financial resources. The buyer of an existing business can determine the condition of the plant and equipment, its capacity, its life expectancy, and its value before buying the business. In many cases, the entrepreneur can purchase the existing physical facilities and equipment at prices that are significantly below their replacement costs. In some businesses, purchasing these assets may be the best part of the deal.

INVENTORY IN PLACE The proper mix and amount of inventory is essential to both cost control and sales volume. A business with too little inventory imposes limitations on satisfying customer demand, and too much inventory ties up excessive amounts capital, increases costs, reduces profitability, and increases the likelihood of cash flow problems. Many successful established business owners have learned a proper balance of inventory. Knowing the "right" amount of inventory to keep on hand can be extremely valuable, especially for buyers of businesses that experience seasonal fluctuations, that sell perishable items, or that must meet the needs of high-volume customers.

ESTABLISHED TRADE CREDIT Previous owners also have established trade credit relationships of which the new owner can take advantage. The business's proven track record gives the new owner leverage in negotiating favorable trade credit terms. No supplier wants to lose a good customer.

EASIER ACCESS TO FINANCING Investors and bankers often perceive the risk associated with buying an existing business with a solid history of performance to be lower than that of an unknown start-up. This may make it easier for the new owner to secure financing. A buyer can point to the existing company's track record and to the plans for improving it to convince potential lenders to finance the purchase. Many lenders will finance 50 to 75 percent of the purchase price of a business, depending on a number of factors such as the industry in which it operates, its track record of success, and it profits, cash flow, assets, and collateral.[6] In addition, in many business purchases, buyers use a built-in source of financing: the seller.

HIGH VALUE Some existing businesses are real bargains. If the current owner must sell quickly, he or she may have established a bargain price for the company that is below its actual worth. Any special skills or training required to operate the business limit the number of potential buyers; therefore, the more specialized the business is, the greater the likelihood is that a buyer can find a bargain. If the owner wants a substantial down payment or the entire selling price in cash, there may be few qualified buyers, but those who do qualify may be able to negotiate a good deal.

Source: Bob Thaves/Universal Uclick.

Disadvantages of Buying an Existing Business

Buying an existing business does have disadvantages that are important to consider.

CASH REQUIREMENTS One of the most significant challenges to buying a business is acquiring the necessary funds for the initial purchase price. "[Because] the business concept, customer base, brands, and other fundamental work have already been done, the financial cost of acquiring an existing business is usually greater then starting one from nothing," observes to the Small Business Administration.[7]

THE BUSINESS IS LOSING MONEY A business may be for sale because it is no longer—or never has been—profitable. Owners can use various creative accounting techniques that make a company's financial picture appear to be much more positive than it actually is. The maxim "let the buyer beware" is sound advice in the purchase of a business. Any buyer who is unwilling to conduct a thorough analysis of the business usually ends up paying a much higher price down the road when the business turns out to be struggling.

Although buying a money-losing business is risky, it is not necessarily taboo. If a business analysis indicates that the company is poorly managed, suffering from neglect, or overlooking a prime opportunity, a new owner may be able to turn it around. However, buying a struggling business without a well-defined plan for solving the problems it faces is an invitation to disaster.

ENTREPRENEURIAL PROFILE: Philip Schram: Buffalo Wings and Rings While working for an auto parts maker in Cincinnati, Ohio, Philip Schram learned that a coworker's father was selling an underperforming restaurant franchise, Buffalo Wings and Rings, that he had started in 1988. After analyzing the six-store chain and developing a plan for turning it around, Schram purchased the chicken wings and onion rings franchise. "Since I was a boy, I dreamed of owning a business," he says. Schram worked with franchisees to increase the company's marketing, promotion, and branding efforts; refurbished the chain's stores to give them a fresher, consistent look; and expanded the menu. The changes worked. Within two years, the number of outlets had grown to 43, and sales increased to $20 million from $6 million. Schram continues to expand the chain across the Midwest and recently opened a new franchisee training headquarters in Cincinnati.[8]

Unprofitable businesses often result from at least one of the following problems:

- Excessively high wage and salary expenses due to excess pay or inefficient use of personnel
- Excessively high compensation for the owner
- Excessively high rental or lease rates
- High-priced maintenance costs or service contracts
- Poor location or too many locations for the business to support
- Inefficient equipment
- Intense competition from rivals

2. _____

Understand the disadvantages of buying an existing business.

- Prices that are too low
- Low profit margins

If the business is profitable but does not have adequate cash flow, the following potential problems often are the cause:

- High inventory levels
- Inadequate accounts receivable collection efforts
- Losses due to employee theft, shoplifting, and fraud

Like Philip Schram, a potential buyer usually can trace the causes of a company's lack of profitability by analyzing a company and its financial statements. The question is, Can the new owner take steps to resolve the problems and return the company to profitability and positive cash flow?

PAYING FOR ILL WILL Just as proper business dealings can create goodwill, improper business behavior or unethical practices can create ill will. A business may look great on the surface, but customers, suppliers, creditors, or employees may have negative feelings about their dealings with it. Too many business buyers discover—after the sale—that they have inherited undisclosed credit problems, poor supplier relationships, soon-to-expire leases, lawsuits, building code violations, and other problems created by the previous owner. Vital business relationships may have begun to deteriorate, but their long-term effects may not yet be reflected in the company's financial statements. Ill will can permeate a business for years. The only way to avoid these problems is to investigate a prospective purchase target thoroughly *before* moving forward in the negotiation process.

CURRENT EMPLOYEES ARE UNSUITABLE If a new owner plans to make changes in a business, current employees may not suit the company's needs. Some workers may have a difficult time adapting to the new owner's management style and the new vision for the company. Previous managers may have kept marginal employees because they were close friends or had been with the company for a long time. The new owner, therefore, may have to make some very unpopular termination decisions. For this reason, employees may feel threatened by new ownership. In some cases, employees who may have wanted to buy the business themselves but could not afford it are resentful. They may see the new owner as the person who "stole" their opportunity. Bitter employees are not likely to be productive workers and may have difficulty fitting into the new management structure.

LOCATION HAS BECOME UNSATISFACTORY What was once an ideal location may no longer be because of changing demographic patterns. Recently opened malls and shopping centers, new competitors, or traffic pattern changes can spell disaster, especially for a small retail shop. Prospective buyers must evaluate the current market in the area surrounding the business as well as its potential for future growth and expansion. Researching all zoning, traffic, and land development plans with appropriate jurisdictions, such as the city, county, or state, is important as well.

OBSOLETE OR INEFFICIENT EQUIPMENT AND FACILITIES Potential buyers sometimes neglect to have an expert evaluate a company's building and equipment before they purchase it. They may discover too late that the equipment is obsolete and inefficient, which increases operating expenses to excessively high levels. Modernizing equipment and facilities is seldom inexpensive.

CUSTOMERS MAY BE LOYAL TO PREVIOUS OWNERS Customers can base their purchasing decisions on personal loyalties to the owner of the business. For some businesses, this can have a modest impact on the performance of these businesses after they are sold. For example, customers of restaurants and retail businesses may have gotten to know the previous owners over the years and may be skeptical about how the new owners will operate the business going forward. In other businesses, customer loyalty and trust of previous owners can make the retention of old customers highly uncertain. This is particularly true with service businesses in which the owners are highly active in the day-to-day operation of the company and interact directly with the customers. For example, businesses providing ongoing services such as lawn care, bookkeeping, house cleaning, car repair, and others often experience a high rate of turnover in customers when a business is sold to new owners. It is important to understand the degree of any personal relationship that the previous owner has with customers and their loyalty to the previous owners before making a final decision to buy a business.

THE CHALLENGE OF IMPLEMENTING CHANGE Planning for change is much easier than implementing it. Methods and procedures the previous owner used create precedents that can be difficult or awkward for a new owner to change. For example, if the previous owner granted volume-based discounts to customers, it may be difficult to eliminate that discount without losing some of those customers. The previous owner's policies—even those that are unwise—can influence the changes the new owner can make. Implementing changes to reverse a downward sales trend in a turnaround situation can be just as difficult as eliminating unprofitable procedures. Convincing alienated customers to return can be an expensive and laborious process that may take years.

OBSOLETE INVENTORY Inventory has value only when it is salable. Too many potential owners make the mistake of trusting a company's balance sheet to provide them with the value of its inventory. The inventory value reported on a company's balance sheet is seldom an accurate reflection of its real market value. A company's inventory may reflect the value at the time of purchase, but inventory, especially technologically related inventory, can depreciate quickly. The value reported on the balance sheet reflects the original cost of the inventory, *not* its actual market value. In fact, inventory and other assets reported as having value may be completely worthless because they are outdated and obsolete. It is the buyer's responsibility to discover the *real* value of the assets before negotiating a purchase price for the business.

VALUING ACCOUNTS RECEIVABLE Like inventory, accounts receivable rarely are worth their face value. The prospective buyer should age the accounts receivable to determine their collectability. The older the receivables are, the less likely they are to be collected and, consequently, the lower their actual value. Table 7.1 shows a simple but effective method of evaluating accounts receivable once the buyer ages them.

THE BUSINESS MAY BE OVERPRICED Most business sales involve the purchase of the company's assets rather than its stock. A buyer must be sure which assets are included in the deal and what their real value is. Many people purchase businesses at prices far in excess of their true value. If a buyer accurately values a business's accounts receivable, inventories, and other assets, he or she will be in a better position to negotiate a price that will allow the business to be profitable. Making payments on a business that was overpriced is a millstone around the new owner's neck, making it difficult to keep the business afloat.

Purchasing an existing business can be a time-consuming process that requires a great deal of effort and is often difficult to complete. Repeated studies report that more than half of all business acquisitions fail to meet the buyer's expectations. This statistic alone should provide a warning about the need to conduct a systematic and thorough analysis prior to negotiating any deal.

TABLE 7.1 Valuing Accounts Receivable

A prospective buyer asked the current owner of a business about the value of her accounts receivable.

The owner's business records showed $101,000 in accounts receivable. However, when the prospective buyer aged them and then multiplied the resulting totals by his estimated probabilities of collection, he discovered their *real* value.

Age of Accounts (Days)	Amount	Probability of Collection	Value (Amount × Probability of Collection)
0–30	$40,000	95%	$38,000
31–60	$25,000	88%	$22,000
61–90	$14,000	70%	$9,800
91–120	$10,000	40%	$4,000
121–150	$7,000	25%	$1,750
151+	$5,000	10%	$500
Total	**$101,000**		**$76,050**

Had he blindly accepted the "book value" of these accounts receivable, this prospective buyer would have overpaid by nearly $25,000 for them!

FIGURE 7.1

Process of Buying a Business

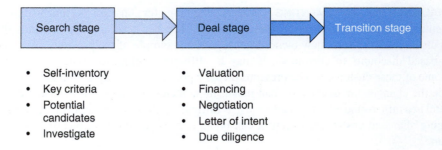

The remainder of this chapter examines the stages that entrepreneurs go through when buying a business: (1) search stage, (2) deal stage, and (3) transition stage (see Figure 7.1).

The Search Stage

3.

Explain the four steps in the search stage for buying the *right* business.

When buying a business, entrepreneurs must search for a business that fits best with their background and personal aspirations. There are four steps to conduct an effective search for the right business to buy:

- Conduct a self-inventory.
- Develop a list of the key criteria that define the "ideal business."
- Seek the help of others in developing a list of potential candidates for acquisitions that meet your criteria.
- Thoroughly investigate the potential acquisition targets that best meet your criteria.

Self-Inventory

The first step in buying a business is conducting a "self-inventory" to determine the ideal business. The following questions produce valuable insights into the best type of business an entrepreneur should consider buying. The answers to these questions provide an important personal guide that help avoid a costly mistake:

- What business activities do you enjoy most? What activities do you enjoy the least?
- Which industries interest you most? Which interest you the least?
- What kind of business do you want to buy?
- What kinds of businesses do you want to *avoid*?
- In what geographic area do you want to live and work?
- What do you expect to get out of the business in terms of income, wealth, and the ability to meet your nonfinancial goals in life?
- What are your long-term goals for work and retirement? What types of businesses will help you meet these goals?
- How much time do you want to set aside for your interests outside of work—family, hobbies, volunteer work, community organizations, and so forth—and how well will different kinds of businesses allow you to continue with these activities?
- How much can you put into the business—in both time and money?
- What business skills and experience do you have? Which ones do you lack?
- How easily can you transfer your existing skills and experience to other types of businesses? In what kinds of businesses would that transfer be easiest?
- How much risk are you willing to take?
- What size company do you want to buy?
- How much can you afford to spend on a company?

Answering those and other questions *beforehand* will allow you to develop a list of criteria that a company must meet before it becomes a purchase candidate.

Develop a List of Criteria

Based on the answers to the self-inventory questions, the next step is to develop a list of criteria that a potential business acquisition must meet. Investigating every business that you find for sale is a waste of time. The goal is to identify the characteristics of the "ideal business" for you so that you can focus on the most viable candidates as you wade through a multitude of business opportunities. These criteria provide specific parameters against which you can evaluate potential acquisition candidates.

Potential Candidates

Once you know the criteria and parameters for the ideal candidate, you can begin your search. One technique is to start at the macro level and work down. Drawing on the resources of the Internet and the library, government publications, and industry trade associations and reports, buyers can discover which industries are growing fastest and offer the greatest potential for future growth. For entrepreneurs with a well-defined idea of what they are looking for, another effective approach is to begin searching in an industry in which they have experience or knowledge.

Typical sources for identifying potential acquisition candidates include the following:

- The Internet, where several sites such as *Bizbuysell.com*, Bizquest, and others, include listings of companies for sale
- Bankers
- Accountants
- Attorneys
- Investment bankers
- Trade associations
- Industry contacts such as suppliers, distributors, customers, and others
- Contacting owners of businesses you would like to buy (even if they're not advertised "for sale")
- Newspaper and trade journal listings of businesses for sale (e.g., the "Business Opportunities" section of the *Wall Street Journal*)
- "Networking" through social and business contact with friends and relatives

Business brokers, professionals who help buy and sell business for others, can be a valuable resource. Business brokers typically charge a commission based on the value of the sale, ranging from 6 to 10 percent.[9] Much like the role a real estate agent plays when purchasing a home, a good business broker handles the many details and helps facilitate the transaction. The broker is, however, working for the *seller*, not the buyer. Always ask brokers about their success rates. "My standard answer is that I aim for 80 percent or better," says business broker Barbara Taylor. "Beware of a broker who tries to justify a low close ratio by quoting dismal industry statistics." Low success rates may mean that the broker is not effective. However, it may also mean that they go for quantity of deals over quality. Either way, steer clear of brokers with low closing rates.[10]

Buyers should consider all businesses that meet their criteria—even those that may not be listed for sale. Just because a business does not have a "for sale" sign in the window does not mean it is not for sale. In fact, the hidden market of companies that might be for sale but are not advertised as such is one of the richest sources of top-quality businesses. Getting the word out that a buyer has an interest in buying a particular type of business often leads to the discovery of many rich business opportunities.

ENTREPRENEURIAL PROFILE: Sean Bandawat: Jacob Bromwell, Inc. Sean Bandawat started two Internet businesses while he was a student at the University of Southern California Marshall School of Business. However, when he graduated, he decided to buy an existing company. The company he chose to purchase was Jacob Bromwell, which was established in 1819. Jacob Bromwell is the 34th-oldest continuously operated company in the United States. "I was attracted to the idea of having an established customer base, established revenue figures, and established assets," says Bandawat. Jacob Bromwell produces handcrafted, historically correct American products, including the Original Popcorn Popper, Frontier Frying Pan, and Classic Tin Cup. "For the last decade, the company suffered major losses in customers and sales," says Sean Bandawat. "It took six months of

Sean Bandawat.
Source: Jacob Bromwell Inc.

due diligence to truly understand why. After months of analyzing the business from every angle, we actually discovered that the company's decline had nothing to do with a decline in demand for these products." The new owner has expanded distribution by securing agreements with the QVC network, catalog companies such as the Vermont Country Store, and specialty retailers nationwide. "With the economy being so uncertain, consumers now more than ever want to buy American-made products that remind them of the past and bring them back to simpler times. That's the real story behind the products we're selling."[11]

Investigation

The final step in the search process for buying a business is to investigate the businesses that meet the key criteria and are of interest to the entrepreneur. Most people selling a business require the prospective buyer to sign a confidentiality agreement, also known as a nondisclosure agreement (NDA). An NDA is a legally binding contract that defines the information that is covered under the agreement and the purpose of supplying this information. An NDA is a specific promise not to disclose any of the information to other parties and not to use it in any other way (e.g., starting a competing company). The NDA also includes how long the buyer and seller must keep the information confidential and what the parties must do with the information provided if the deal does not go forward. There are three parts to the initial investigation of the business deal once both parties sign the NDA: researching the customer base, analyzing competitors, and determining the seller's motivation for selling.

RESEARCH THE CUSTOMER BASE An entrepreneur should analyze both existing and potential customers before purchasing an existing business. Discovering why customers buy from the business and developing a profile of the existing customer base allows the buyer to identify a company's strengths and weaknesses. The entrepreneur should answer the following questions:

- Does the business have a well-defined customer base? Is it growing or shrinking?
- Who are its customers in terms of race, age, gender, and income level?
- What do customers expect the business to do for them?
- What needs are they satisfying when they buy from the company?
- How often do customers buy?
- Do they buy in seasonal patterns?
- How loyal are present customers?
- Why do some potential customers *not* buy from the business?
- How easily can the company attract new customers? Will the new customers be significantly different from existing customers?
- Is the customer base from a large geographic area, or do they all live near the business?

Analyzing the answers to those questions helps the potential owner develop a marketing plan. Ideally, the entrepreneur will keep the business attractive to existing customers and change features of its marketing plan to attract new customers.

COMPETITOR ANALYSIS A potential buyer must identify the company's direct competitors that are the businesses in the immediate area that sell the same or similar products or services. The potential profitability and survival of the business may depend on the strategies of these competitors. In addition to analyzing direct competitors, buyers should evaluate the trend in the level of competition. Answering the following questions provides valuable insight:

- How many similar businesses have entered the market in the last five years?
- How many similar businesses have closed in the past five years?
- What caused them to fail?
- Has the market already reached the saturation point? Being a late comer in a saturated market is plagued with challenges.

When evaluating the competitive environment, the prospective buyer should answer additional questions:

- What are the characteristics that have led to the success of the company's direct competitors?

- How do competitors' sales volumes compare with those of the business the entrepreneur is considering?
- What unique services do competitors offer?
- How well organized and coordinated are the marketing efforts of competitors?
- How strong are competitors' reputations?
- What are their strengths and weaknesses?
- How can you gain market share in this competitive environment? Can you offer the product or service better, faster, or cheaper?

The intent of the competitor analysis is to determine the company's current competitive situation and the competitive landscape facing the firm in the future. In addition, gathering information about how the company stacks up against its competitors through articles written about the business and interviews with industry insiders and experts provides meaningful insight.

MOTIVATION OF THE SELLER Why does the owner want to sell? Every prospective business owner should investigate the *real* reason the business owner wants to sell. In addition to a planned retirement, the most common reasons businesses are for sale usually fall into three categories:[12]

- The seller is not making enough money in the business.
- The seller has a personal reason for selling, such as health, boredom, or burnout.
- The seller is aware of pending changes in the business or the business environment that will adversely affect its future.

These changes may include a major competitor entering the market, a degraded location, lease problems, cash flow issues, supplier shifts, or a declining customer base. In other cases, owners decide to cash in their business investments and diversify into other types of assets. Every prospective buyer should investigate *thoroughly* the reason a seller gives for selling a business. Remember: Let the buyer beware!

ENTREPRENEURIAL PROFILE: Doug Bolton: The Floor Show When Doug Bolton moved from Wisconsin to Lake Tahoe, California, in 1983, he obtained his independent contractor's license and opened The Floor Show, a flooring and remodeling retail store. Because Lake Tahoe is made up primarily of vacation homes, the properties have high turnover rates. New owners want to make their vacation homes fit their tastes, so there is a steady demand for carpets, flooring, kitchen cabinets, countertops, and window treatments. Bolton and his wife wanted to work fewer hours and spend more time outdoors, so they put the business up for sale for $318,000. The building is not included in the sales price, but it is available for purchase separately from Bolton's landlord. Sales at The Floor Show had declined by over 25 percent in the two years prior to putting the business up for sale, from more than $1 million to less than $750,000. The Boltons managed to cut expenses, so their profit margins remained at about 15 percent.[13]

Businesses do not last forever, and most owners know when the time has come to sell. Some owners do not feel obliged to disclose to potential buyers the whole story of their motivation for selling. In every business sale, the buyer bears the responsibility of determining whether the business is a good value. Visiting local business owners may reveal general patterns about the area and its overall vitality. The local chamber of commerce also may have useful information. Suppliers and competitors may be able to shed light on why a business is for sale. Combining this information with an analysis of the company's financial records, a potential buyer should be able to develop a clear picture of the business and its real value.

REVIEW FINANCES Because both parties have signed an NDA, the buyer should seek detailed financial information including at least three years of past financial statements. The buyer should also tour all facilities and do an initial inspection of equipment and inventory. The prospective buyer can conduct a more detailed analysis of the finances and facilities if the acquisition moves ahead from the search stage to the deal stage.

Lessons from the Street-Smart Entrepreneur

Don't Get Burned When You Buy a Business

Rather than experience the expense, sweat, and toil of starting a new business, many entrepreneurs buy an existing business from someone who has gone through the process of starting a venture and proving its worth. Buying a business, however, is rife with potential pitfalls, and an unprepared entrepreneur can easily get burned. The Street-Smart Entrepreneur offers the following tips for buying an existing business:

- **Recognize that you are not just buying a company; you are buying a livelihood and a lifestyle.** Buyers of small businesses are not just buying assets and inventories and leases; they are also choosing a lifestyle. When you buy a business you are buying two things—a job running the business and an asset that should provide you with a return. The business should provide a reasonable salary for the new owner. It also should provide a return on the investment in the business over time that comes from the profits the business earns after paying a salary to the owner. The business may be the single most significant determinant of your future lifestyle; therefore, choose it carefully.

- **Explore seller financing.** Recent turbulence in the financial industry has made banks hesitant to make loans to business buyers. Fortunately, buyers have a built-in source of capital available: the seller. In a typical deal, the buyer makes a down payment to the seller that ranges from 20 to 70 percent of the purchase price. The seller takes a note for the balance, which the buyer repays over 3 to 10 years. When Alex Shlepakov, founder of Network One, a business based in Elk Grove Village, Illinois, that provides network support services to small businesses, decided to sell the company, he offered to finance a portion of the selling price. Shlepakov accepted a large down payment from the buyers and agreed to finance the balance over two and a half years. Shlepakov says that providing financing is a tangible way for sellers to demonstrate confidence in the business that they are selling.

- **Use professional advisers.** When it comes to conducting due diligence on a potential target company, smart entrepreneurs turn to professionals—accountants, attorneys, business brokers, and others—for valuable insight and advice. Choose professionals who specialize in buying and selling businesses. When buying a business the devil is in the details. Experienced advisors will protect the buyer from financially disastrous terms and unnecessary risks when structuring the purchase agreement.

- **Link the final price to customer retention.** In many small businesses, particularly service businesses, a significant part of what a buyer is receiving is the existing client or customer base. In sales of these businesses, the agreed-on price often depends on the company's ability to retain a certain percentage of customers. If the company's customer base declines after the buyer takes over, the agreement calls for a reduction in the selling price.

- **Get the seller to stick around.** Buyers usually benefit from having the previous owner stay on during the transition period following the sale. The complexity of the business and the new owner's familiarity with the industry determine whether the time frame is a few weeks or a few years. The previous owner may have years of experience and knowledge of the industry and the local community that prove to be highly valuable to the buyer. Negotiating a deal for the seller to stay on for a time takes a great deal of pressure off of an inexperienced buyer.

- **Do your homework on valuation techniques.** As you have learned in this chapter, valuing a business is partly an art and partly a science. Smart entrepreneurs educate themselves in advance about the various methods practitioners in the industry use to value businesses. Remember that a common technique for estimating the value of a business is to apply a multiple to its earnings before interest and taxes (EBIT). However, the multiples used vary significantly across industries.

In many ways, buying a business is easier than starting a business from scratch, but buying a business poses a unique set of challenges and potential pitfalls. Following these tips from the Street-Smart Entrepreneur lowers the probability that you will get burned when you buy a business.

Source: Based on Arden Dale and Simona Covel, "Sellers Offer a Financial Hand to Their Buyers," *Wall Street Journal*, November 13, 2008, p. B6; Joseph Anthony, "Seven Tips for Buying a Business," Microsoft Small Business Center, *www .microsoft.com/smallbusiness/resources/startups/business_opportunities/7_tips_ for_buying_a_business.mspx#bio1.*

The Deal Stage

4.

Describe the five steps of the deal stage for buying a business the *right* way.

Once the entrepreneur has identified the specific business to be purchased, the process moves into the deal stage. The deal stage includes a valuation of the business, formalizing the financing of the purchase, negotiating the details of the purchase, signing a letter of intent, and conducting due diligence. This section examines each of these steps.

Methods for Determining the Value of a Business

Business valuation is partly an art and partly a science. It is important for the prospective buyer to understand that valuation is the point in the process of buying a business at which many deals fail. A survey of business brokers found that a gap between the price that the seller wanted versus the actual valuation was the largest cause of deals being terminated before closing.[14] The sheer number of variables that influence the value of a privately owned business makes establishing a price difficult. These factors include the nature of the business itself, its position in the market or industry, the outlook for the market or industry, the company's financial status and stability, its earning capacity, intangible assets (such as patents, trademarks, and copyrights), the value of similar companies that are publicly owned, and many others. The median selling price of a private company is $420,000 according to a database compiled by Business Valuation Resources, a company that tracks private company transactions (see Figure 7.2).[15]

Assessing the value of the company's tangible assets usually is straightforward, but assigning a price to the intangible assets, such as goodwill, almost always creates controversy. The seller expects the value of the goodwill to reflect the hard work and long hours invested in building the business. Valuing goodwill often is an emotional issue for sellers because their businesses are tied closely to their egos. The buyer, however, is willing to pay only for those intangible assets that produce extra income. How can a buyer and a seller arrive at a fair price? There are few hard-and-fast rules in establishing the value of a business, but the following guidelines can help:

- There is no single best method for determining a business's worth because each business sale is unique. A practical approach is to estimate a company's value using several techniques, review those values, and then determine the range in which most of the values converge.

- The deal must be financially feasible for both parties to be viable. The seller must be satisfied with the price received for the business, and the buyer cannot pay an excessively high price that requires heavy borrowing that strains cash flow from the outset.

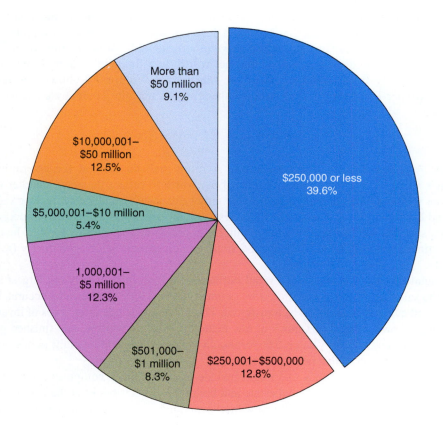

FIGURE 7.2

Selling Prices of Private Companies

Source: Based on Business Valuation Resources, January 16, 2013; *www.bvmarketdata.com/defaulttextonly.asp?f=Database%20Chart.*

- The buyer should have access to all business records.
- Valuations should be based on facts, not feelings or fiction.
- The two parties should deal with one another openly, honestly, and in good faith.

The main reason that buyers purchase existing businesses is to get their future earning potential. The second most common reason is to obtain an established asset base; it is much easier to buy assets than to build them. Although some valuation methods take these goals into consideration, many business sellers and buyers simplify the process by relying on rules of thumb to estimate the value of a business. For instance, one rule for valuing sporting goods stores is 30 percent of its annual sales plus its inventory.[16] Other rules use multiples of a company's net earnings to value the business. Although the multipliers vary by industry, most small companies sell for 2 to 12 times their EBIT, with an average selling price of between 6 and 7 times EBIT.[17] For instance, a study by Business Valuation Resources of 2,168 business sales over a recent three-year period shows that the median selling price of a restaurant is 2.45 times EBIT, the median price of a car wash is 6.27 times EBIT, and the median price of business consulting service is 11.56 times EBIT.[18] Factors that increase the value of the multiplier include proprietary products and patents; a strong, diversified customer base; above-average growth rate; a strong, balanced management team; and dominant market share. Factors that decrease the value of the multiplier include generic, "me-too" products; dependence on a single customer or a small group of customers for a significant portion of sales; reliance on the skills of a single manager (e.g., the founder); declining market share; and dependence on a single product for generating sales.[19]

This section describes three basic techniques— the balance sheet method, the earnings approach, the market approach,—and several variations on them for determining the value of a hypothetical business, Kuyper Electronics.

BALANCE SHEET TECHNIQUE The balance sheet method computes the book value of a company's **net worth**, or **owner's equity** (net worth = assets − liabilities) and uses this figure as the value. A common criticism of this technique is that it oversimplifies the valuation process. The problem with this technique is that it fails to recognize reality: Most small businesses have market values that exceed their reported book values.

The first step is to determine which assets are included in the sale. In most cases, the owner has some personal assets that he or she does not want to sell. Professional business brokers can help the buyer and the seller arrive at a reasonable value for the collection of assets included in the deal. Remember that net worth on a financial statement will likely differ significantly from actual net worth in the market. Figure 7.3 shows the balance sheet for Kuyper Electronics. This balance sheet shows that the company's net worth is as follows:

$$\$278,990 - \$114,325 = \$151,766$$

Variation: Adjusted Balance Sheet Technique A more realistic method for determining a company's value is to adjust the book value of net worth to reflect the actual market value. The values reported on a company's books may either overstate or understate the true value of assets and liabilities. Typical assets in a business sale include notes and accounts receivable, inventory, supplies, and fixtures. If a buyer purchases notes and accounts receivable, he or she should estimate the likelihood of their collection and adjust their value accordingly (refer to Table 7.1).

In manufacturing, wholesale, and retail businesses, inventory is usually the largest single asset in the sale. Taking a physical inventory count is the best way to determine accurately the condition and quantity of goods to be transferred. The sale may include three types of inventory, each having its own method of valuation: raw materials, work in process, and finished goods. Before accepting any inventory value, a buyer should evaluate the condition of the goods to avoid being stuck with inventory that he or she cannot sell.

Fixed assets transferred in a sale might include land, buildings, equipment, and fixtures. Business owners frequently carry real estate and buildings on their books at their original purchase prices, which typically are well below their actual market value. Equipment and fixtures,

ASSETS

Current Assets

Cash	$11,655	
Accounts receivable	15,876	
Inventory	56,523	
Supplies	8,574	
Prepaid insurance	5,587	
Total current assets		$98,215

Fixed Assets

Land		$24,000	
Buildings	$141,000		
less accumulated depreciation	51,500	89,500	
Office equipment	$12,760		
less accumulated depreciation	7,159	5,601	
Factory equipment	$59,085		
less accumulated depreciation	27,850	31,235	
Trucks and autos	$28,730		
less accumulated depreciation	11,190	17,540	
Total fixed assets			$167,876
Total Assets			**$266,091**

LIABILITIES

Current Liabilities

Accounts payable	$19,497	
Mortgage payable	5,215	
Salaries Payable	3,671	
Note Payable	10,000	
Total current liabilities		$38,383

Long-Term Liabilities

Mortgage payable	$54,542	
Note payable	21,400	
Total long-term liabilities		$75,942
Total Liabilities		**$114,325**

OWNER'S EQUITY

Owner's Equity (Net Worth)	$151,766
Total Liabilities + Owner's Equity	**$266,091**

FIGURE 7.3

Balance Sheet for Kuyper Electronics, June 30, 2015

depending on their condition and usefulness, may increase or decrease the value of the business. Appraisals of these assets on insurance policies are helpful guidelines for establishing market value. In addition, business brokers can be useful in determining the current value of fixed assets. Some brokers use an estimate of what it would cost to replace a company's physical assets (less a reasonable allowance for depreciation) to determine their value. As indicated by the adjusted balance sheet in Figure 7.4, the adjusted net worth for Kuyper Electronics is $279,738 − $114,325 = $165,413, which indicates that some of the entries on its books did not accurately reflect market value.

Business valuations based on any balance sheet methods suffer one major drawback: They do not consider the future earnings potential of the business. These techniques value assets at

FIGURE 7.4

Adjusted Balance Sheet for Kuyper Electronics, June 30, 2015

ASSETS

Current Assets

Cash		$11,655
Accounts receivable		10,051
Inventory		39,261
Supplies		7,492
Prepaid insurance		5,587
Total current assets		**$74,046**

Fixed Assets

Land			$52,000
Buildings	$177,000		
less accumulated depreciation	51,500	115,500	
Office equipment	$11,645		
less accumulated depreciation	7,159	4,486	
Factory equipment	$50,196		
less accumulated depreciation	27,850	22,346	
Trucks and autos	$22,550		
less accumulated depreciation	11,190	11,360	
Total fixed assets			**$205,692**
Total Assets			**$279,738**

LIABILITIES

Current Liabilities

Accounts payable		$19,497
Mortgage payable		5,215
Salaries Payable		3,671
Not Payable		10,000
Total current liabilities		**$38,383**

Long-Term Liabilities

Mortgage payable		$54,542
Note payable		21,400
Total long-term liabilities		**$75,942**
Total Liabilities		**$114,325**

OWNER'S EQUITY

Owner's Equity (Net Worth)		$165,413
Total Liabilities + Owner's Equity		**$279,738**

current prices and do not consider them as tools for creating future profits. An additional omission is that balance sheet methods do not attach value to intangible assets of the business such as goodwill.

EARNINGS APPROACH The earnings approach is an approach to valuation that is favored by finance professionals and experienced entrepreneurs because it considers the future income potential of the business. That is what an entrepreneur really is buying with an existing business—its ability to generate returns on their investment into the future.

Variation 1: Adjusted Earnings Method The easiest approach to calculating the value of a business based on its earnings is known as the adjusted earnings method of valuation.

Although it does not have the sophistication of the other methods, it is quite commonly used and is accurate enough for many small businesses. This simple method starts with earnings before interest, taxes, depreciation, and amortization (EBITDA) because this is a good measure of the true cash earnings of a business. The buyer deducts interest and taxes because they likely will change with a new owner. Depreciation and amortization will also change because the new owner's accountants will reset the values and time lines that generate these noncash expenses. The buyer normally adjusts the EBITDA reported by the seller to help determine the true value of the business going forward. Because entrepreneurs rarely pay themselves market-based salaries, the buyer replaces the salary of the previous owner with a market-based salary for a typical CEO of a similar-size company. Next, the buyer adds back all personal expenses (such as an auto lease or country club membership) that the current owner runs through the business. Finally, the buyer deducts the costs required to bring inventories up to necessary levels and updating equipment. Suppose that a prospective buyer discovers the following values for Kuyper Electronics:

Reported EBITDA	$200,000
Add: Current owner's salary	50,000
Subtract: Market salary for CEO	(75,000)
Add: Personal expenses	10,000
Subtract: Inventory needed	(25,000)
Subtract: Equipment updates	(40,000)
Adjusted EBITDA	$120,000

Note that the adjusted EBITDA is significantly lower than the EBITDA reported on Kuyper Electronics' income statement. Had the buyer simply used the reported EBITDA, the price paid for buying the business would have been significantly higher!

The buyer then uses the adjusted EBITDA to project future potential earnings using an *earnings multiple*. Recall that the typical earnings multiple is from two to twelve times EBITDA with an average multiple of between 6 and 7. Five factors determine the actual multiple used in a business sale. First, the recent trend of earnings in the business influence the multiple. A business with flat or declining earnings over the three years leading up to the sale will have a much lower multiple than one that has shown a consistently strong growth in recent earnings. Second, the buyer must take into account the growth trends of the overall industry. Third, the buyer must give consideration to market conditions in the industry by looking at multiples from comparable companies that have sold recently. Fourth, the buyer must consider strategic factors such as unique patents or proprietary process into determining the earnings multiple. For example, an engineering firm in Minnesota recently sold for a multiple of *25 times EBITDA* because the company had a proprietary patented technology that several larger firms wanted to control. Finally, the structure of the deal itself will have an impact. Buyers who pay with cash typically pay a lower multiple than buyers who rely on owner financing or a stock purchase. There is no hard-and-fast rule for establishing an earnings multiple. Most businesses sell for a multiple of six to seven times their earnings. Buyers make adjustments in the multiple, higher *or lower*, based on the factors discussed above. Suppose that taking into account the factors described earlier, the buyer who is considering Kuyper Electronics uses a multiple of 4. The estimated value of the business is as follows:

$$\$120,000 \times 4 = \$480,000$$

Although the adjusted earnings method tends to be an approach to valuation that is much more art than science, it is a commonly used method of valuation for small businesses.

Variation 2: Excess Earnings Method This method combines both the value of a company's existing assets (less its liabilities) and an estimate of its future earnings potential to determine the selling price for the business. One advantage of the **excess earnings method** is that it offers an estimate of goodwill. **Goodwill** is the difference between an established, successful business and one that has yet to prove itself. Goodwill is based on the company's reputation and its ability to attract customers. This intangible asset often creates problems in a business sale. A common method of valuing a business is to compute its tangible net worth and then to add an often arbitrary adjustment for goodwill. However, a buyer should not accept blindly the seller's arbitrary adjustment for goodwill because it is likely to be inflated.

The excess earnings method provides a reasonable approach for determining the value of goodwill. It measures goodwill by the amount of profit the business earns above that of the average firm in the same industry. It also assumes that the owner is entitled to a reasonable return on the company's adjusted tangible net worth.

Step 1. *Compute adjusted tangible net worth.* The buyer computes the company's adjusted tangible net worth. Total tangible assets (adjusted for market value) minus total liabilities yields adjusted tangible net worth. In the Kuyper Electronics example shown in Figure 7.4, the adjusted tangible net worth is $279,738 − $114,325 = $165,413.

Step 2. *Calculate the opportunity costs of investing in the business.* **Opportunity costs** represent the cost of forgoing a choice; what income does the potential buyer give up by purchasing the business? If the buyer chooses to purchase the assets of a business, he or she cannot invest his or her money elsewhere. Therefore, the opportunity cost of the purchase is the amount that the buyer could have earned by investing the same amount *in a similar risk investment*.

Three components determine the rate of return used to value a business: (1) the basic, risk-free return; (2) an inflation premium; and (3) the risk allowance for investing in the particular business. The basic, risk-free return and the inflation premium are reflected in investments such as U.S. Treasury bonds. To determine the appropriate rate of return for investing in a business, the buyer must add to this base rate a factor reflecting the risk of purchasing the company. The greater the risk involved, the higher the rate of return. An average-risk business typically indicates a 20 to 25 percent rate of return. For Kuyper Electronics, the opportunity cost of the investment is $165,413 × 25% = $41,353.

The second part of the buyer's opportunity cost is the salary that he or she could have earned working for someone else. For the Kuyper Electronics example, if the buyer purchases the business, he or she must forgo a salary of, say, $75,000 that he or she could have earned working elsewhere. Adding these amounts yields a total opportunity cost of 41,353 + 75,000 = $116,353.

Step 3. *Project net earnings.* The buyer must estimate the company's net earnings for the upcoming year *before* subtracting the owner's salary. Averages can be misleading; therefore, the buyer must be sure to investigate the *trend* of net earnings. Have the earnings risen steadily over the past five years, dropped significantly, remained relatively constant, or fluctuated wildly? Past income statements provide useful guidelines for estimating earnings, but, as you have seen, the buyer often must adjust the earnings. In the Kuyper Electronics example, the buyer and an accountant project the buyer's adjusted net earnings to be $195,000 (earnings of $120,000 plus the buyer's anticipated salary of $75,000).

Step 4. *Compute extra earning power.* A company's **extra earning power** is the difference between forecasted earnings (step 3) and total opportunity costs of investing (step 2). Many small businesses that are for sale do not have extra earning power (i.e., excess earnings), and they show marginal or no profits. The extra earning power of Kuyper Electronics is: $195,000 − $116,353 = $78,647.

Step 5. *Estimate the value of intangibles.* The buyer can use the business's extra earning power to estimate the value of its intangible assets. Multiplying the extra earning power by a years-of-profit figure yields an estimate of the intangible assets' value. The years-of-profit figure for a normal-risk business typically ranges from three to four. A high-risk business may have a years-of-profit figure of one, whereas a well-established firm might use a figure of seven.

Rating the company on a scale of 1 (low) to 7 (high) on the following factors allows an entrepreneur to calculate a reasonable years-of-profit figure to use to estimate the value of the intangibles:[20]

Factor	Score						
	1	2	3	4	5	6	7
1. Risk	More risky				Less risky		
2. Degree of competition	Intense competition				Few competitors		
3. Industry attractiveness	Fading				Attractive		
4. Barriers to entry	Low				High		
5. Growth potential	Low				High		
6. Owner's reason for selling	Poor performance				Retiring		
7. Age of business	Young				10+ years old		
8. Current owner's tenure	Short				10+ years		
9. Profitability	Below average				Above average		
10. Location	Problematic				Desirable		
11. Customer base	Limited and shrinking				Diverse and growing		
12. Image and reputation	Poor				Stellar		

To calculate the years-of-profit figure, the entrepreneur adds the score for each factor and divides by the number of factors (12). For Luxor Electronics, the scores are as follows:

Risk	3
Degree of competition	2
Industry attractiveness	4
Barriers to entry	2
Growth potential	4
Owner's reason for selling	6
Age of business	6
Owner's tenure	6
Profitability	4
Location	4
Customer base	3
Image and reputation	5
Total	**49**

Thus, for Kuyper Electronics, the years-of-profit figure is: $49 \div 12 = 4.1$, and the value of intangibles is $\$78,647 \times 4.1 = \$321,141$.

Step 6. ***Determine the value of the business.*** To determine the value of the business, the buyer simply adds together the adjusted tangible net worth (step 1) and the value of the intangibles (step 5). Using this method, the value of Kuyper Electronics is $\$165,413 + \$321,141 = \$486,554$.

Both the buyer and the seller should consider the tax implications of transferring goodwill. Because the *buyer* can amortize both the cost of goodwill and a covenant over 15 years, the tax treatment of either would be the same. However, the *seller* would prefer to have the amount of the purchase price in excess of the value of the assets allocated to goodwill, which is a capital asset. The gain on the capital asset is at lower capital gains rates. If that same amount were allocated to a restrictive covenant (which is negotiated with the seller personally, not the business), the seller must treat it as ordinary income, which is taxed at regular rates that currently are higher than the capital gains rates.

Variation 3: Capitalized Earnings Approach Another earnings approach capitalizes expected net earnings to determine the value of a business. The buyer should prepare his own pro forma income statement and should ask the seller to prepare one also. Many appraisers use a five-year weighted average of past sales (with the greatest weights assigned to the most recent years) to estimate sales for the upcoming year.

Once again, the buyer must evaluate the risk of purchasing the business to determine the appropriate rate of return on the investment. The greater the perceived risk, the higher the return the buyer will require. Risk determination is always somewhat subjective, but it is a necessary consideration for proper evaluation.

The **capitalized earnings approach** divides estimated net earnings (after subtracting the owner's reasonable salary) by the rate of return that reflects the risk level. For Kuyper Electronics, the capitalized value (assuming a reasonable salary of $75,000) is as follows:

$$\frac{\text{Net earnings (after deducting owner's salary)}}{\text{Rate of return}} = \frac{\$195,000 - \$75,000}{25\%} = \$480,000$$

Companies with lower risk factors offer greater certainty and, therefore, are more valuable. For example, a lower rate of return of 10 percent yields a value of $1,200,000 for Kuyper Electronics, whereas a 50 percent rate of return produces a value of $240,000. Most normal-risk businesses use a rate-of-return factor ranging from 20 to 25 percent. The lowest risk factor that most buyers would accept for any business ranges from 15 to 18 percent.

Variation 4: Discounted Future Earnings Approach This variation of the earnings approach assumes that a dollar earned in the future is worth less than that same dollar today. Using the **discounted future earnings approach**, the buyer estimates the company's net income for several years into the future and then discounts these future earnings back to their present value. The resulting present value is an estimate of the company's worth. The present value represents the cost of the buyers' giving up the opportunity to earn a reasonable rate of return by receiving income in the future instead of today.

To visualize the importance of present value and the time value of money, consider two $1 million sweepstakes winners. Rob wins $1 million in a sweepstakes, and he receives it in $50,000 installments over 20 years. If Rob invests every installment at 8 percent interest, he will have accumulated $2,288,098 at the end of 20 years. Lisa wins $1 million in another sweepstakes, but she collects her winnings in one lump sum. If Lisa invests her $1 million today at 8 percent, she will have accumulated $4,660,957 at the end of 20 years. The dramatic difference in their wealth—Lisa is now worth nearly $2,373,000 more—is the result of the time value of money.

The discounted future earnings approach has five steps:

Step 1. ***Project earnings for five years into the future.*** One way is to assume that earnings will grow by a constant amount over the next five years. Perhaps a better method is to develop three forecasts—pessimistic, most likely, and optimistic—for each year and find a weighted average using the following formula:

$$\text{Forecasted earnings for year } i = \frac{\text{Pessimistic earnings for year } i + 4 \times \text{Most Likely earnings for year } i + \text{Optimistic earnings years } i}{6}$$

The most likely forecast is weighted four times greater than the pessimistic and optimistic forecasts; therefore, the denominator is the sum of the weights ($1 + 4 + 1 = 6$). For Kuyper Electronics, the buyer's earnings forecasts are as follows:

Year	Pessimistic	Most Likely	Optimistic	Weighted Average
2016	$100,000	$120,000	$135,000	**$119,167**
2017	108,000	128,000	145,000	**127,500**
2018	115,000	140,000	160,000	**139,167**
2019	119,000	150,000	170,000	**148,167**
2020	125,000	162,000	182,000	**159,167**

The buyer must remember that the further into the future he or she forecasts, the less reliable the estimates will be.

Step 2. ***Discount these future earnings using the appropriate present value factor.*** The appropriate present value factor can be found by looking in published present value tables or by solving the equation

$$1/(1 + k)^t$$

where k = rate of return

and t = time (year 1, 2, 3 . . . n).

The rate that the buyer selects should reflect the rate he or she could earn on a similar risk investment. Because Kuyper Electronics is a normal-risk business, the buyer chooses 25 percent.

Year	Income Forecast (Weighted Average)	Present Value Factor (at 25 Percent)	Net Present Value
2016	$119,167	.8000	$95,333
2017	127,500	.6400	81,600
2018	139,167	.5120	71,253
2019	148,167	.4096	60,689
2020	159,167	.3277	52,156
Total			**$361,031**

Step 3. ***Estimate the income stream beyond five years.*** One technique suggests multiplying the fifth-year income by 1/(rate of return). For Kuyper Electronics, the estimate is as follows:

Income beyond year 5 = $159,167 × (1/25%) = $636,667

Step 4. ***Discount the income estimate beyond five years using the present value factor for the sixth year.*** For Kuyper Electronics,

Present value of income beyond year 5: $636,667 × 0.2621 = $166,898

Step 5. ***Compute the total value of the business.***

Total value: $361,031 + $166,898 = $527,929

The primary advantage of this technique is that it values a business solely on the basis of its future earnings potential, but its reliability depends on making accurate forecasts of future earnings and on choosing a realistic present value factor. The discounted future earnings approach is especially well suited for valuing service businesses, whose asset bases are often small, and for companies experiencing high growth rates.

MARKET APPROACH The **market (or price/earnings) approach** uses the price/earnings ratios of similar businesses to establish the value of a company. The buyer must use businesses whose stocks are publicly traded in order to get a meaningful comparison. A company's price/earnings ratio (P/E ratio) is the price of one share of its common stock in the market divided by its earnings per share (after deducting preferred stock dividends). To get a representative P/E ratio, the buyer should average the P/Es of as many similar businesses as possible.

The buyer multiplies the average price/earnings ratio by the private company's estimated earnings (*after* deducting the owner's salary) to compute a company's value. For example, suppose that the buyer found four companies comparable to Kuyper Electronics but whose stock is publicly traded. Their price/earnings ratios are as follows:

Company 1	3.9
Company 2	3.8
Company 3	4.3
Company 4	4.1
Average	4.025

This average P/E ratio produces a value of $483,000:

$$\text{Value average P/E ratio} \times \text{Estimated net earnings}$$
$$4.025 \times \$120,000 = \$483,000$$

The most significant advantage of the market approach is its simplicity. However, the market approach method does have several disadvantages, including the following:

1. ***Necessary comparisons between publicly traded and privately owned companies.*** The stock of privately owned companies is illiquid, and therefore the P/E ratio used is often subjective and lower than that of publicly held companies.

2. ***Unrepresentative earnings estimates.*** The private company's net earnings may not realistically reflect its true earnings potential. To minimize taxes, owners usually attempt to keep profits low and rely on benefits to make up the difference.

3. ***Finding similar companies for comparison.*** Often, it is extremely difficult for a buyer to find comparable publicly held companies when estimating the appropriate P/E ratio.

4. ***Applying the after-tax earnings of a private company to determine its value.*** If a prospective buyer is using an after-tax P/E ratio from a public company, he or she also must use the after-tax earnings from the private company.

Despite its drawbacks, the market approach is useful as a general guideline to establishing a company's value.

THE BEST METHOD? Which of these methods is best for determining the value of a small business? Simply stated, there is no single best method. These techniques will yield a range of values, and buyers should look for values that might cluster together and then use their best judgment to determine their offering price. The final price will be based on both the valuation used and the negotiating skills of both parties. Like all assets, a business is ultimately worth what the highest bidder is willing to pay with terms and conditions that are most acceptable to the seller.

In the Entrepreneurial Spotlight

Bond Coffee

Mel Bond's career included entrepreneurship, mergers and acquisitions, and wealth management in Nashville, Tennessee. His daughter Kathleen followed her father's footsteps, graduating with a degree in entrepreneurship from Belmont University in 2011.

A year after Kathleen graduated from college, Mel found an advertisement for a coffee shop for sale in Nashville.

"Hey, wouldn't it be fun to buy it? Kathleen could manage it," said Mel. "We could sell baked goods and play James Bond movies on the TV while we made lattes for customers."

But the $100,000 price seemed too high.

Several months later, Mel noticed the business listing again. The price had dropped to $60,000. "This time, the family got excited," recalls Kathleen. "This was something we might actually be able to do!"

They contacted the broker in charge of selling the coffee shop to set up a meeting. The owner explained his reasons for selling. The business was not profitable. Although sales had increased monthly for more than a year, the owner and his partner had been losing about $4,000 per month for some time. The shop had been open for about four years. They had done minimal advertising, mostly in print ads that did little to increase business. Over time, the shop had expanded the menu to include

Mel and Kathleen Bond, owners of Bond Coffee.
Source: **Kathleen Bond.**

sandwiches and salads, which did increase sales but still did not result in reaching the break-even point. The owner shared financial summaries for the past four years.

The owners had not updated the decor of the coffee shop since they opened it, and the furniture was beginning to look worn. There were some problems with management and customer service. It was clear that if the Bonds bought the shop, they would have to make some physical improvements to it. In addition, they would have to change the culture to a more customer-focused approach.

The location was in a growing part of Nashville near downtown known as the Gulch, where there were 900 residential apartments and condominiums within walking distance of the coffee shop. An apartment complex with 295 units was under construction, and another one was due to be completed within a year. A local developer also was building a hotel. The Gulch was home to several bars and restaurants, an organic grocery store, and other retail stores. On the weekends, Nashville tourists frequented the Gulch.

After reviewing the financial statements provided by the current owner, the Bond family estimated that they would have to increase sales by 30 percent to reach the shop's breakeven point. The price of $60,000 included all equipment and furniture in the business, from the espresso machine to the tables and chairs. The estimated value of the existing equipment was $8,000. The rest of the price was goodwill, based on the business's location, established brand, and consumer base. The Bonds estimated that it would require about another $60,000 to make the necessary improvements in the coffee shop.

"Purchasing a coffee shop, as opposed to opening a new location, gave us some unique advantages," says Kathleen. "First, the purchase saved us a significant amount in start-up costs. To start a coffee shop from scratch, we would have needed at least $100,000 in renovation and equipment alone. By purchasing the shop, we were able to obtain all of the equipment needed and a location for $60,000 plus the cost of renovations. Another advantage was the lease. Many businesses rent commercial property, signing leases that range from three- to five-year commitments. With the purchase of the business, we were able to assume a lease—one with only one year left before renewal. If we found that we could not break even, this gave us the ability to exit after only one year. Purchasing a business also saved us from many of the headaches of learning the best strategies for a particular market. The current owner had already tried many things with advertising, hiring, equipment, and recipes, scheduling employees and, of course, the menu. These were mistakes that the previous owners had made and that we would be able to avoid."

The Bonds made an offer to the owner, which he accepted. Due diligence, which took two weeks, included examining all of the financial data, reviewing payroll and employee turnover records, learning more about existing business practices, and carefully reviewing tax returns to verify the shop's earnings. The owner agreed to spend two weeks training Kathleen Bond, who would manage the business for the family when the store reopened after they made the necessary renovations.

1. What was the motivation behind the owner's decision to sell the coffee shop? Why do you think it took a drop in the sale price to entice the Bond family to make an offer? Explain.

2. How common is it to buy a small business without seeking owner financing? How are these deals typically structured?

3. Do you expect that the Bonds will be able to turnaround the coffee shop and make it profitable? What steps do you recommend they take to ensure a good return on their investment?

Negotiating the Deal

Once an entrepreneur has established a reasonable value for the business, the next step in making a successful purchase is negotiating a suitable deal. Most buyers do not realize that the price they pay for a company often is not as crucial to its continued success as the terms of the purchase. In other words, *the structure of the deal—the terms and conditions of payment—is more important than the actual price the seller agrees to pay.*

Wise business buyers attempt to negotiate the best price they can, but they pay more attention to negotiating favorable terms: how much cash they pay out and when, how much of the price the seller is willing to finance and for how long, the interest rate at which the deal is financed, and others. The buyer's primary concern is to ensure that the deal does not endanger the company's financial future and that it preserves the company's cash flow.

On the surface, the negotiation process may appear to be strictly adversarial. Although each party may be trying to accomplish objectives that are at odds with those of the opposing party, the negotiation process does not have to be conflict oriented. The process goes more smoothly and faster if the two parties work to establish a cooperative relationship based on honesty and trust from the outset. A successful deal requires both parties to examine and articulate their respective positions while trying to understand the other party's position. Recognizing that neither of them will benefit without a deal, both parties must work to achieve their objectives while making certain concessions to keep the negotiations alive.

To avoid a stalled deal, both buyer and seller should go into the negotiation with a list of objectives ranked in order of priority. Prioritizing desired outcomes increases the likelihood that both parties will get most of what they want from the bargain. Knowing which terms are most important (and which ones are least important) enables the parties to make concessions without regret and avoid getting bogged down in unnecessary details. If, for instance, the seller insists on a term that the buyer cannot agree to, the seller can explain why he cannot concede on that term and then offer to give up something in exchange.

THE "ART OF THE DEAL" Both buyers and sellers must recognize that no one benefits without an agreement. Both parties must work to achieve their goals while making concessions to keep the negotiations alive.

Figure 7.5 is an illustration of two individuals prepared to negotiate for the purchase and sale of a business. The buyer and seller both have high and low bargaining points in this example:

- The buyer would like to purchase the business for $900,000 but would not pay more than $1,300,000.
- The seller would like to get $1,500,000 for the business but would not take less than $1,000,000.
- If the seller insists on getting $1,500,000, they will not sell the business to this buyer.
- Likewise, if the buyer stands firm on an offer of $900,000, there will be no deal.

The bargaining process may eventually lead both parties into the *bargaining zone*, the area within which the two parties can reach an agreement. It extends from above the lowest price the seller is willing to take to below the maximum price the buyer is willing to pay. The dynamics of this negotiation process and the needs of each party ultimately determine whether the buyer and seller can reach an agreement and for what price.

The following negotiating tips can help parties reach a mutually satisfying deal:

- *Establish the proper mind-set.* Successful negotiations are built on a foundation of trust. The first step in any negotiation should be to establish a climate of trust and communication. Too often, buyers and sellers rush into putting their chips on the bargaining table without establishing a rapport with one another.

- *Know what you want to have when you walk away from the table.* What will it take to reach your business objectives? What would the perfect deal be? Although it may not be possible to achieve it, defining the perfect deal may help to identify which issues are most important to you.

- *Develop a negotiating strategy.* One of the biggest mistakes business buyers can make is entering negotiations with only a vague notion of the strategies they will employ. To be successful, it is necessary to know how to respond to a variety of situations that are likely to arise. Once you know where you want to finish, decide where you will start and remember to leave some room to give. Try not to be the first one to mention price. Let the other party do that and negotiate from there. Every strategy has an upside and a downside, and effective negotiators know what they are.

FIGURE 7.5

**Identifying the
Bargaining Zone**

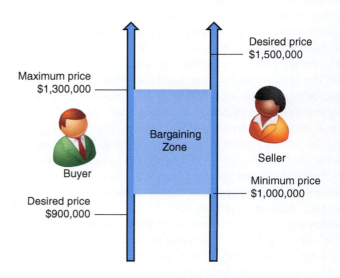

- *Recognize the other party's needs.* For a bargain to occur, both parties must believe that they have met at least some of their goals. Asking open-ended questions can provide insight to the other side's position and enable you to understand why it is important.

- *Be an empathetic listener.* To truly understand what the other party's position is, you must listen attentively.

- *Focus on the issue, not on the person.* If the negotiation reaches an impasse, a natural tendency is to attack the other party. Instead, focus on developing a workable solution to accomplish your goals.

- *Avoid seeing the other side as "the enemy."* This attitude reduces the negotiation to an "I win, you lose" mentality that only hinders the process.

- *Educate; don't intimidate.* Rather than trying to bully the other party into accepting your point of view, explain the reasoning and the logic behind your proposal.

- *Be creative.* When negotiations stall or come to an impasse, negotiators must seek creative alternatives that benefit both parties or, at a minimum, get the negotiations started again.

- *Keep emotions in check.* A short temper and an important negotiation make ill-suited partners. The surest way to destroy trust and to sabotage a negotiation is to lose one's temper and lash out at the other party. Anger leads to poor decisions.

- *Be patient.* Sound negotiations often take a great deal of time, especially when one is buying a business from the entrepreneur who founded it. The seller's ego is woven throughout the negotiation process, and wise negotiators recognize this. Persistence and patience are the keys to success in any negotiation involving the sale of the business.

- *Don't become a victim.* Well-prepared negotiators are not afraid to walk away from deals that are not right for them.

- *Remember that "no deal" is an option.* What would happen if the negotiations failed to produce a deal? In most negotiations, walking away from the table is an option. In some cases, it may be the best option.

The Buyer's Goals The buyer seeks to realize the following goals:

- Get the business at the lowest price possible.
- Negotiate favorable payment terms, preferably over time.
- Get assurances that he is buying the business he thinks it is.
- Avoid enabling the seller to open a competing business.
- Minimize the amount of cash paid up front.

The Seller's Goals Entrepreneurs who are most effective at acquiring a business know how important it is to understand the complex emotions that influence the seller's behavior and decisions. For some sellers, this business has been their life and has defined their identity. They may have been the original founder and the person who created the business. For many entrepreneurs, their business defines who they are. They nurtured the business through its infancy and matured it, and now it is time to "let go." Sellers may be asking themselves, What will I do now? Where will I go each morning? Who will I be without "my business"? The negotiation process may bring these questions to light because it requires sellers to, in effect, put a price tag on their life's work. For these reasons, the potential buyer must negotiate in a manner that displays sensitivity and respect.

In general, the seller of the business is looking to accomplish the following goals:

- Get the highest price possible for the company.
- Sever all responsibility for the company's liabilities.
- Avoid unreasonable contract terms that might limit future opportunities.
- Maximize the cash from the deal.
- Minimize the tax burden from the sale.
- Make sure the buyer will make all future payments.

THE STRUCTURE OF THE DEAL To make a negotiation work, the two sides must structure the deal in a way that is acceptable to both parties. Following are typical ways that parties structure business sales:

Straight Business Sale A straight business sale may be best for a seller who wants to step down and turn over the reins of the company to someone else. A study of small business sales in 60 categories found that 94 percent were asset sales. In an asset sale, the seller keeps all liabilities—those that are on the books and any that might emerge in the future due to litigation. That is why buyers favor asset sales. The remaining 6 percent involved the sale of stock. About 22 percent were for cash, and 75 percent included a down payment with a note carried by the seller. The remaining 3 percent relied on a note from the seller with no down payment. When the deal included a down payment, it averaged 33 percent of the purchase price. Only 40 percent of the business sales studied included covenants not to compete.[21]

Although selling a business outright is often the safest exit path for an entrepreneur, it is usually the most expensive. Sellers who want cash and take the money up front may face a significant tax burden. They must pay a capital gains tax on the sale price less their investments in the company. Nor is a straight sale an attractive exit strategy for those who want to stay on with the company or for those who want to surrender control of the company gradually rather than all at once.

Ideally, a buyer has already begun to explore the options available for financing the purchase. (Recall that many entrepreneurs include bankers on their teams of advisers.) If traditional lenders shy away from financing the purchase of an existing business, buyers often find themselves searching for alternative sources of funds. Fortunately, most business buyers discover an important source of financing built into the deal: the seller. Typically, a deal is structured so that the buyer makes a down payment to the seller, who then finances a note for the balance. The buyer makes regular principal and interest payments over time—perhaps with a larger balloon payment at the end—until the note is paid off. A common arrangement involves a down payment with the seller financing the remaining 20 to 70 percent of the purchase price over time, usually 3 to 10 years.

Sellers must be willing to finance a portion of the purchase price, particularly when credit is tight.

ENTREPRENEURIAL PROFILE: Rick Hunt: Risk Removal During a recent recession, Rick Hunt knew that he and his partners would have to accept a note for at least part of the purchase price of their Fort Collins, Colorado–based environmental services company, Risk Removal. With annual sales of $3 million and a good reputation in a lucrative niche market (the company removes asbestos and lead paint from buildings), Risk Removal had attracted attention from several buyers, but none had been able to close a financing deal. Hunt hired a business broker, and within months, he and his partners had accepted an offer from a buyer who had experience in the business and could pay 25 percent of the purchase price in cash. A bank financed 50 percent of the price, and Hunt and his partners accepted a five-year promissory note at 7 percent interest for the remaining 25 percent.[22]

The terms of the deal are vital to both buyer and seller. They cannot be so burdensome that they threaten the company's continued existence; that is, the buyer must be able to make the payments to the seller out of the company's cash flow.

Use a Two-Step Sale For owners wanting the security of a sales contract now but not wanting to step down from the company's helm for several years, a two-step sale may be ideal. The buyer purchases the business in two phases, getting 20 to 70 percent today and agreeing to buy the remainder within a specific time period. Until the final transaction takes place, the original owner retains at least partial control of the company.

Letter of Intent

Once the buyer and seller have negotiated the deal, they put the details of the structure of the sale into a *letter of intent*. The letter of intent is a firm commitment by both sides that they are ready to move toward closing the sale of the business. The letter clearly sets the price and any factors that may affect the actual price paid at closing. It also outlines the terms of the sale, including whether it is an asset sale or a stock sale and the nature of owner financing if it is part of the deal.

Any other specific deal points that are important to either side also should be included in the letter of intent. Finally, the letter of intent gives specific dates when the deal must close. Although both parties may be committed to closing the sale of the business, only *25 percent* of deals make it from the letter of intent stage to the final closing. The old saying "the devil is in the details" holds true when buying a business. Many deals fall apart during the process of due diligence and during the creation of the closing documents for the sale of the business.

The Due Diligence Process

Due diligence involves studying, reviewing, and verifying all of the relevant information concerning the acquisition. The goal of the due diligence process is to discover exactly what the buyer is purchasing and avoid any unpleasant surprises *after* the deal is closed. Exploring a company's character and condition through the Better Business Bureau, credit-reporting agencies, its bank, its vendors and suppliers, your accountant, your attorney, and through other resources is a vital part of making certain the entrepreneur is going to get a good deal on a business with the capacity to succeed. It is important invest in the due diligence process—you may choose to pay now or pay later.[23]

ENTREPRENEURIAL PROFILE: Emma Cerulli and Sebastien Barthe: Em and Seb Emma Cerulli and Sebastien Barthe purchased the assets of a failing catering company in Montreal, Quebec. They believed that it had a strong client base and a great deal of potential. Their client base grew to include area hospitals, law firms, and university departments. Barthe, who had been a head chef prior to buying the business, is responsible for the business operations. Cerulli brought her experience from culinary school at the Pacific Institute of Culinary Arts in Vancouver as well as what she had learned working in several well-known Montreal restaurants. However, after they purchased the business, they discovered that their overhead was much higher than they had estimated in their budgets based on company history because the financial information that they had received from the previous owner was inaccurate. Careful due diligence would have given them a more accurate picture of the real financial situation of the business. In addition, their labor costs were too high. "Although I believe labor cost is the main issue," says Cerulli, "I'm afraid that if we lay people off, we won't have enough staff to meet demand." Despite with their best efforts, Cerulli and Barthe ultimately decided to close the business.[24]

A thorough analysis of an intended acquisition usually requires an entrepreneur to assemble a team of advisers. Buying a business involves many complex legal, financial, tax, and business issues, and good advice can be a valuable tool. Many entrepreneurs involve an accountant, an attorney, an insurance agent, a banker, and sometimes a business broker to serve as consultants during the due diligence process.

The due diligence process involves investigating three critical areas of the business and the potential deal beyond those already evaluated earlier in the search and deal processes:

1. Confirming valuation: What is the real value of the business?
2. Legal issues: What legal aspects of the business are known or hidden risks?
3. Financial state: Is the business financially sound?

CONFIRMING VALUATION What is the true nature of the firm's assets? A prospective buyer should evaluate the business's assets to determine their true value. The buyer bases the valuation used to negotiate the deal on the financial statements provided by the seller. The buyer and the advising team must verify the actual value of the business through careful inspection of the business and its assets. Questions to ask about assets include the following:

- Are the assets really useful, or are they obsolete?
- Will the assets require replacement soon?
- Do the assets operate efficiently?

The potential buyer should check the condition of all equipment and the building. Even if the building is not part of the sale price, its condition will determine the ability of the business to operate effectively into the future. It may be necessary to hire a professional to evaluate the major components of the building—its structure and its plumbing, electrical, and heating and cooling systems. Renovations are often expensive and time consuming and can have a negative impact on a buyer's budget.

What is the status of the firm's existing inventory? Can the buyer sell all of it at full price, or is some of it damaged or outdated? How much of it would the buyer have to sell at a loss? Is it consistent with the image the new owner wants to project? A potential buyer may need an independent appraisal to determine the value of the firm's inventory and other assets because the current owner may have priced them far above their actual value. These items typically constitute the largest portion of a business's value, and a potential buyer should not accept the seller's asking price blindly. Remember that *book value is not the same as market value.* Value is determined in the market, not on a balance sheet. A buyer often can purchase equipment and fixtures at substantially lower prices than book value.

Other important factors that the potential buyer should investigate include the statues of accounts receivable, the lease, business records, intangible assets, and the business location:

1. *Accounts receivable.* If the sale includes accounts receivable, the buyer must check their quality before purchasing them. How creditworthy are the accounts? What portion of them is past due? By aging the accounts receivable, the buyer can judge their quality and determine their *real* value. (Refer to Table 7.1.)

2. *Lease arrangements.* If the seller does not own the building, is the lease included in the sale? When does it expire? What restrictions does it have on renovation or expansion? What is the status of the relationship with the property owner? The buyer should determine beforehand any restrictions the landlord has placed on the lease. Does the lease agreement allow the seller to assign the lease to a buyer? The buyer must negotiate all necessary changes with the landlord and get them in writing prior to buying the business.

3. *Business records.* Accurate business records can be a valuable source of information and can tell the story of the company's pattern of success—or lack of it! Unfortunately, many business owners are sloppy record keepers. Consequently, a potential buyer and his or her team may have to reconstruct critical records. It is important to verify as much information about the business as possible. For instance, does the owner have current customer direct mail and e-mail lists? These can be valuable marketing tools for a new business owner.

4. *Intangible assets.* As we have seen, determining the value of intangible assets is much more difficult than computing the value of the tangible assets, yet intangible assets can be one of the most valuable parts of a business acquisition. Does the sale include intangible assets such as trademarks, patents, copyrights, or goodwill? Edward Karstetter, director of valuation services at USBX says, "The value placed on intangible assets such as people, knowledge, relationships, and intellectual property is now a greater proportion of the total value of most businesses than is the value of tangible assets such as machinery and equipment."[25]

5. *Location and appearance.* The location and appearance of the building are important to most businesses because they send clear messages to potential customers. Every buyer should consider the location's suitability for today and for the near future. Potential buyers also should check local zoning laws to ensure that the changes they want to make are permissible. In some areas, zoning laws are very difficult to change and can restrict a business's growth.

LEGAL ISSUES   What legal aspects of the business present risk? Business buyers face myriad legal pitfalls. The most significant legal issues involve liens, contract assignments, covenants not to compete, and ongoing legal liabilities.

Liens   The key legal issue in the sale of any asset is typically the proper transfer of good title from seller to buyer. However, because most business sales involve a collection of assorted assets, the transfer of a good title is complex. Some business assets may have **liens** (creditors' claims) against them, and unless those liens are satisfied before the sale, the buyer must assume them and become financially responsible for them. One way to reduce this potential problem is to include a clause in the sales contract that states that any liability not shown on the balance sheet at the time of sale remains the responsibility of the seller. A prospective buyer should have an attorney thoroughly investigate all of the assets for sale and their lien status before buying any business.

Contract Assignments A buyer must investigate the rights and the obligations he or she would assume under existing contracts with suppliers, customers, employees, lessors, and others. To continue the smooth operation of the business, the buyer must assume the rights of the seller under existing contracts. For example, the current owner may have four years left on a 10-year lease that he or she will assign to the buyer (if the lease allows assignment). A seller can assign most contractual rights unless the contract specifically prohibits the assignment or the contract is personal in nature. For instance, loan contracts sometimes prohibit assignments with **due-on-sale clauses**. These clauses require the buyer to pay the full amount of the remaining loan balance or to finance the balance at prevailing interest rates. Thus, the buyer cannot assume the seller's loan at a lower interest rate. In addition, a seller usually cannot assign his or her credit arrangements with suppliers to the buyer because they are based on the seller's business reputation and are personal in nature. If contracts such as these are crucial to the business operation and cannot be assigned, the buyer must negotiate new contracts.

The prospective buyer also should evaluate the terms of other contracts the seller has, including the following:

- Patent, trademark, or copyright registrations
- Exclusive agent or distributor contracts
- Insurance contracts
- Financing and loan arrangements
- Union contracts

Covenants Not to Compete One of the most important and most often overlooked legal considerations for a prospective buyer is negotiating a **covenant not to compete** (or a **restrictive covenant**) with the seller. Under a restrictive covenant, the seller agrees not to open a competing business within a specific time period and geographic area of the existing one. (The buyer must negotiate the covenant directly with the owner, not the corporation; if the corporation signs the agreement, the owner may not be bound by it.) However, the covenant must be a part of a business sale and must be reasonable in scope in order to be enforceable. Without this protection, a buyer may find his new business eroding beneath his or her feet.

Ongoing Legal Liabilities Finally, a potential buyer must look for any potential legal liabilities the purchase might expose. These typically arise from three sources:

1. Physical premises
2. Product liability claims
3. Labor relations

Physical Premises The buyer must first examine the physical premises for safety. Is the employees' health at risk because of asbestos or some other hazardous material? If the business is a manufacturing operation, does it meet Occupational Safety and Health Administration and other regulatory agency requirements?

Product Liability Claims The buyer must consider whether the product contains defects that could result in **product liability lawsuits**, which claim that a company is liable for damages and injuries caused by the products or services it sells. Existing lawsuits might be an omen of more to follow. In addition, the buyer must explore products that the company has discontinued because he or she might be liable for them if they prove to be defective. The final bargain between the parties should require the seller to guarantee that the company is not involved in any product liability lawsuits.

Labor Relations The relationship between management and employees is a key to a successful transition of ownership. Does a union represent employees in a collective bargaining agreement? The time to discover sour management–labor relations is before the purchase, not after.

The existence of liabilities such as these does not necessarily eliminate a business from consideration. Insurance coverage can shift risk from the potential buyer, but the buyer should check to see whether the insurance covers lawsuits resulting from actions taken before the purchase. The buyer can also insist on a hold back of a percentage of the purchase price for a period of time to protect against unknown liabilities and other discrepancies. Despite conducting a thorough search, a buyer may purchase a business only to discover later the presence of hidden liabilities, such as unpaid back taxes or delinquent bills, unpaid pension fund contributions, undisclosed lawsuits, or others. Including a clause in the purchase agreement that imposes the responsibility for such hidden liabilities on the seller can protect a buyer from unpleasant surprises after the sale.

FINANCIAL STATE Is the business financially sound? Any investment in a company should produce a reasonable salary for the owner and a healthy return on the money invested. Otherwise, it makes no sense to purchase the business. Therefore, every serious buyer must analyze the records of the business to determine its true financial health. Accounting systems and methods can vary tremendously from one company to another, and buyers usually benefit from enlisting the assistance of an accountant. Some business sellers know all of the tricks to make profits appear to be higher than they actually are. For example, a seller might lower costs by gradually eliminating equipment maintenance or might boost sales by selling to marginal customers who will never pay for their purchases. Techniques such as these can artificially inflate a company's earnings, but a well-prepared buyer will be able to see through them. For a buyer, the most dependable financial records are audited statements, those prepared by a certified public accountant in accordance with generally accepted accounting principles. Unfortunately, most small businesses that are for sale do not have audited financial statements.

A buyer also must remember that he or she is purchasing the future earning potential of an existing business. To evaluate a company's earning potential, a buyer should review past sales, operating expenses, and profits as well as the assets used to generate those profits. The buyer must compare current balance sheets and income statements with previous ones and then develop pro forma statements for the next two or three years. Sales tax records, income tax returns, and financial statements are valuable sources of information.

Earnings trends are another area to analyze. Are profits consistent over time, or have they been erratic? If there are fluctuations, what caused them? Is this earnings pattern typical in the industry, or is it a result of unique circumstances or poor management? If poor management has caused these fluctuations, can a new manager make a difference? Some of the financial records that a potential buyer should examine include the income statement, balance sheet, tax returns, owner's compensation, and cash flow.

Income Statements and Balance Sheets for at Least Three Years It is important to review data from several years because creative accounting techniques can distort financial data in any single year. Even though buyers are purchasing the future earnings of a business, they must remember that many businesses intentionally show low profits to minimize the owners' tax bills. Low profits should prompt a buyer to investigate their causes. Specific entries should be verified using company records for all major categories of expenses and revenues. If the financial statements are not audited, the buyer should take care to validate the financial statements. It is best to seek the assistance of a reputable accountant with experience in acquisitions.

Income Tax Returns for at Least Three Years Comparing basic financial statements with tax returns can reveal discrepancies of which the buyer should be aware. Some small business owners "skim" from their businesses; that is, they take money from sales without reporting it as income. Owners who skim will claim their businesses are more profitable than their tax returns show. However, buyers should not pay for "phantom profits."

Cash Flow Most buyers understand the importance of evaluating a company's earnings history, but few recognize the need to analyze its cash flow. They assume that if earnings are adequate, there will be sufficient cash to pay all of the bills and to fund an adequate salary for them. *That is not necessarily the case.* Before closing any deal, a buyer should review the information with an accountant and convert the target company's financial statements into a cash flow forecast. This forecast must take into account not only existing debts and obligations but also any modifications or additional debts the buyer plans to make in the business. It should reflect the repayment of financing the buyer arranges to purchase the company. The critical questions are the following: Can the company generate sufficient cash to be self-supporting? How much cash will it generate for the buyer?

A potential buyer must look for suspicious deviations from the average (in either direction) for sales, expenses, profits, assets, and liabilities. Are sales increasing or decreasing? Does the equipment's value on the balance sheet reflect its real value? Are advertising expenses unusually high? How is depreciation reflected in the financial statements?

Finally, a potential buyer should always be wary of purchasing a business if the present owner refuses to disclose the company's financial records. In that case, the buyer's best course of action is to walk away from the deal. Without access to information about the business, a buyer cannot conduct a proper analysis.

Entrepreneurship in Action

What's the Deal?

Dividend.com

Paul Rubillo launched *Dividend.com* in 2008 after a 15-year career on Wall Street. The Web site ranks 1,600 dividend-paying stocks on their relative strength, overall yield attractiveness, dividend reliability, and earnings growth. The site also includes research reports and news updates written by Rubillo, who uses a freemium model, in which users can access some of the material for free, but much of the more valuable information is available for a $99 annual subscription.

Dividend.com has more than 85,000 unique visitors each month. It is ranked in the top five among Google searches that use the words "dividend" or "dividend stock." The site has 1,300 premium subscribers, generating approximately $160,000 in annual subscriptions, and more than 8,000 readers who have signed up for free electronic newsletters that Rubillo publishes weekly.

Rubillo wants to sell *Dividend.com* to an individual or a company with experience selling stocks and bonds that would like access to his database of wealthy subscribers and sees the benefit of the domain name and the traffic it generates each month. A business valuation adviser has told him that his asking price of $1.975 million is in line with comparable high-value Internet domains linked to finance. For example, *Invest.com* sold for $1 million in 2007.

Rubillo wants to sell *Domain.com* to launch a baseball news site called *Hotstove.com*. "Goodbye, Wall Street; hello, sports," says Rubillo.

Corbin-Pacific

Mike Corbin started building handmade motorcycle seats in 1965. He transformed his passion into a business, Corbin-Pacific, which sells more than $14 million worth of custom seats, saddlebags, fenders, and other motorcycle gear. About 75 percent of sales are direct to consumers, and the remaining products are sold through motorcycle dealerships.

Revenues had declined over the years prior to Corbin's decision to sell the company he had founded. In fact, sales had declined almost $2 million from their peak of more than $16 million in 2006. In addition, cash flow from the business had declined over that same time period from more than $2.7 million in 2006 to a little more than $700,000 in 2008.

Corbin-Pacific has 115 employees who work out of his 82,000-square-foot facility in Hollister, California. Corbin is selling the land on which the company is located separately for $7.2 million, and its showroom in Daytona Beach, Florida, is for sale for $3.2 million

The purchase price of $11.5 million is based on the value of the 1,500 seat molds and 82 patents and copyrights that are included in the sale. The molds are valued at $9 million and equipment at $1.5 million. The remainder of the price includes the value of inventory. Corbin has many loyal customers including celebrities such as Arnold Schwarzenegger and Jay Leno.

Corbin, 65, is ready to cash out of the business he has managed for more than 40 years. He hopes to find a buyer with the same enthusiasm he has for motorcycles. "You can't be successful in the motorcycle business unless you love these machines and the people that ride them," said Corbin.

1. Assume the role of a prospective buyer for these two businesses. How would you conduct the due diligence necessary to determine whether they would be good investments?

2. Do you notice any red flags or potential sticking points in either of these deals? Explain.

3. Which techniques for estimating the value of a business described in this chapter would be most useful to a prospective buyer of these businesses? Are the owners' asking prices reasonable?

Sources: Based on Darren Dahl, "Business for Sale: A Financial Website," *Inc.*, April 1, 2010, *www.inc.com/magazine/20100401/business-for-sale-a-financial-website .html*; Darren Dahl, "Business for Sale: A Motorcycle-Seat Manufacturer," *Inc.*, May 1, 2009, *www.inc.com/magazine/20090501/business-for-sale-a-motorcycle-seat- manufacturer.html?nav=next*.

The Transition Stage

5.

Understand the transition
stage for buying a business
the *right* way.

Once the deal stage is completed, the transition stage begins with the actual closing of the purchase. Closing the sale of a business is a complex legal process. Many deals fall apart at the closing table due to unforeseen surprises or last minute legal maneuvering by either the buyer or seller. Closing documents include the following:

- Asset purchase agreement—the formal agreement of the deal
- Bill of sale—transfers ownership
- Asset list—all assets that are included in the sale including tangible assets and intellectual property
- Buyer's disclosure statement
- Allocation of purchase price—a formal document that must be filed with the Internal Revenue Service at the end of the tax year that allocates the price among the various assets.
- Non-compete agreement
- Consulting/Training agreement
- Transfer of subsidiaries associated with business
- Transfer of utilities
- Transfer of Web sites, social media addresses, and phone numbers
- Documentation of new entity that will own the business and documentation of new bank account for that business
- Transfer of merchant accounts
- Notice to creditors
- Lease assignments
- Financing documents, security agreement, promissory note, and UCC Financing Statement if seller is financing all or part of the sale
- Sales tax and payroll tax clearance
- Escrow instructions
- Closing adjustments/proration
- Transfer of any third party contracts
- Corporate resolution authorizing sale of the corporate assets

Once the parties finalize the closing, the challenge of facilitating a smooth transition is next. No matter how well planned the sale is, there are always surprises. For instance, the new owner may have ideas for changing the business—perhaps radically—that cause a great deal of stress and anxiety among employees and with the previous owner. Charged with such emotion and uncertainty, the transition phase may be difficult and frustrating—and sometimes painful. To avoid a bumpy transition, a business buyer should do the following:

- Concentrate on communicating with employees. Business sales are fraught with uncertainty and anxiety, and employees need reassurance. Take the time to explain your plans for the company.
- Be honest with employees. Avoid telling them only what they want to hear.
- Listen to employees. They have intimate knowledge of the business and its strengths and weaknesses and usually can offer valuable suggestions. Keep your door and your ears open and come in as somebody who is going to be good for the entire organization.
- Devote time to selling the vision for the company to its key stakeholders, including major customers, suppliers, bankers, and others.
- Consider asking the seller to serve as a consultant until the transition is complete. The previous owner can be a valuable resource.

Chapter Review

1. Understand the advantages and disadvantages of buying an existing business.
   - The *advantages* of buying an existing business include the following: A successful business may continue to be successful, the business may already have the best location, employees and suppliers are already established, equipment is installed and its productive capacity known, inventory is in place and trade credit established, the owner hits the ground running, the buyer can use the expertise of the previous owner, and the business may be a bargain.
   - The *disadvantages* of buying an existing business include the following: An existing business may be for sale because it is deteriorating, the previous owner may have created ill will, employees inherited with the business may not be suitable, its location may have become unsuitable, equipment and facilities may be obsolete, change and innovation are hard to implement, inventory may be outdated, accounts receivable may be worth less than face value, and the business may be overpriced.

2. Describe the four steps in the search stage for buying the *right* business.
   - Conduct a self-inventory to determine what businesses fit with your experience, knowledge, background, and lifestyle preferences.
   - Develop a list of the key criteria that define the "ideal business." These criteria will provide specific parameters against which the entrepreneur can evaluate potential acquisition candidates.
   - Seek the help of others in developing a list of potential candidates for acquisitions that meet your criteria. While you can start with Internet research, it is important to talk to experts from various professional disciplines, including bankers, attorneys, and accountants and with industry insiders and your professional network to get a good list of candidates. You may also want to contact specific companies directly to see if they have an interest in talking about a sale. Business brokers, professionals who help with selling private companies, can also be a source of acquisition candidates.
   - Thoroughly investigate the potential acquisition targets that best meet your criteria. Carefully research the business to learn more about its reputation and uncover any details about its past operations that may be of concern. Assess the motivation of the seller to determine the specific reasons, both personal and business, that are leading them to want to sell the company. Also, review the financial condition of the business in more detail.

3. Describe the five steps of the deal stage for buying a business the *right* way.
   - Valuation—Placing a value on a business is partly an art and partly a science. There is no single best method for determining the value of a business. The following techniques (with several variations) are useful: the earnings approach (excess earnings method, capitalized earnings approach, and discounted future earnings approach), the market approach, and the balance sheet technique (adjusted balance sheet technique).
   - Financing—determine how much, if any, of the purchase price will be financed by the previous owner and where the remaining financing will come from.
   - Negotiation:
     - Selling a business takes time, patience, and preparation to locate a suitable buyer, strike a deal, and make the transition. Sellers must always structure the deal with tax consequences in mind.
     - The first rule of negotiating is never confuse price with value. The party who is the better negotiator usually comes out on top. Before beginning negotiations, a buyer should identify the factors that are affecting the negotiations and then develop a negotiating strategy. The best deals are the result of a cooperative relationship based on trust.
   - Letter of intent—a firm commitment by both sides that they are ready to move toward closing the sale of the business.
   - Due diligence—studying, reviewing, and verifying all of the relevant information concerning the acquisition. The process of due diligence includes the following:
     - Confirming Valuation: What is the real value of the business?
     - Legal Issues: What legal aspects of the business are known or hidden risks?
     - Financial State: Is the business financially sound?

4. Describe the transition stage for buying a business the *right* way.
   - The closing process involves a significant set of legal documents that must all be carefully negotiated.
   - The transition process to the new owner involves clear, honest communication with employees; listening to employees to hear concerns and learn about the business; selling the vision to all key stakeholders; and seeking advice from the previous owner when necessary.

Discussion Questions

7-1. What advantages can an entrepreneur who buys a business gain over one who starts a business from scratch?

7-2. How would you go about determining the value of the assets of a business if you were unfamiliar with them?

7-3. Why do so many entrepreneurs run into trouble when they buy an existing business? Outline the steps involved in the *right* way to buy a business.

7-4. When evaluating an existing business that is for sale, what areas should an entrepreneur consider? Briefly summarize the key elements of each area.

7-5. How should a buyer evaluate a business's goodwill?

7-6. What is a restrictive covenant? Is it fair to ask the seller of a travel agency located in a small town to sign a restrictive covenant for one year covering a 20-square-mile area? Explain.

7-7. How much negative information can you expect the seller to give you about the business? How can a prospective buyer find out such information?

7-8. Why is it so difficult for buyers and sellers to agree on a price for a business?

7-9. Which method of valuing a business is best? Why?

7-10. Explain the buyer's position in a typical negotiation for a business. Explain the seller's position. What tips would you offer a buyer about to begin negotiating the purchase of a business?

7-11. What benefits might you realize from using a business broker? What are the disadvantages?

CHAPTER 8

New Business Planning Process: Feasibility Analysis, Business Modeling, and Crafting a Winning Business Plan

Learning Objectives

On completion of this chapter, you will be able to:

1. Present the steps involved in conducting a feasibility analysis.

2. Identify the components of the business model canvas and explain how to use them to develop a viable business model.

3. Explain the benefits of building an effective business plan.

4. Describe the elements of a solid business plan.

5. Explain the three tests every business plan must pass.

6. Explain the "five Cs of credit" and why they are important to potential lenders and investors.

7. Understand the keys to making an effective business plan presentation.

Businesses are like dogs at a fancy dog show. Each one may look different, but they all have the same essential moving parts.

—Patrick van der Pijl

A wise man will make more opportunities than he finds.

—Francis Bacon

For many entrepreneurs, the easiest part of launching a business is coming up with an idea for a new business concept or approach. Business success, however, requires much more than just a great new idea. Once entrepreneurs develop an idea for a business, the next step is to subject it to a feasibility analysis to determine whether they can transform the idea into a viable business. A **feasibility analysis**, which is the first of three steps in planning for a new business, is the process of determining whether an entrepreneur's idea is a viable foundation for creating a successful business. Its purpose is to determine whether a business idea is worth pursuing. If the idea passes the feasibility analysis, the entrepreneur moves on to the next step of the new business planning process. If the idea fails to pass muster, the entrepreneur drops it and moves on to the next opportunity. He or she has not wasted valuable time, money, energy, and other resources launching a business that is destined to fail because it is based on a flawed concept. Although it is impossible for a feasibility study to guarantee an idea's success, conducting a study reduces the likelihood that entrepreneurs will waste their time pursuing fruitless business ventures.

Conducting a feasibility study is *not* the same as developing a business model or building a full-blown business plan; all three play important but separate roles in the start-up process. A feasibility study answers the question "*Should* we proceed with this business idea?" Its role is to serve as a filter, screening out ideas that lack the potential for building a successful business, *before* an entrepreneur commits the necessary resources to develop and test a business model or to build a business plan. A feasibility study is primarily an *investigative* tool. It is designed to give an entrepreneur a picture of the market, sales, and profit potential of a particular business idea. Will a ski resort located here attract enough customers to be successful? Will customers in this community support a sandwich shop with a retro rock-and-roll theme? Can we build the product at a reasonable cost and sell it at a price customers are willing and able to pay? Does this entrepreneurial team have the ability to implement the idea successfully?

The business model answers the question "*How* would we proceed with this business idea?" Developing a business model, which is the second step in planning a new business, helps an entrepreneur to fully understand all that will be required to launch and build the business. Business modeling is another step that determines the potential for success for the new venture. It is a visual process that examines how all the moving parts of the business must fit together in a unified whole to build a successful venture. It is the step in the planning process in which entrepreneurs test their concepts and use what they learned from real customers to refine their business models before committing the resources to grow the business to its full potential.

A business plan is a tool that builds on the foundation of the feasibility study and business model but provides a more comprehensive and detailed analysis than the first two steps in the new business planning process. Together with a well-developed business model, it functions primarily as a planning tool, describing in greater detail how to transform an idea into a successful business. Its primary goals are to expand on the business model to guide entrepreneurs as they launch and operate their businesses and to help them acquire the financing needed to launch.

Feasibility studies are particularly useful when entrepreneurs have generated multiple ideas for business concepts and must winnow their options down to the "best choice." They enable entrepreneurs to explore quickly the practicality of each of several potential paths for transforming an idea into a successful business venture. Sometimes the result of a feasibility study is the realization that an idea simply won't produce a viable business—no matter how it is organized. In other cases, the process of developing a business model and business plan shows an entrepreneur that the business idea is a sound one but that it must be organized in a different fashion to be profitable.

If an idea proves feasible, the entrepreneur's next step is to leverage the findings of the feasibility analysis to develop a business model and then build a business plan. The primary goals of the business plan are to guide entrepreneurs as they launch and operate their businesses and to help them acquire the necessary financing to launch. A well-constructed business plan may be the best possible insurance against failure. Research suggests that, whatever their size, companies that engage in business planning outperform those that do not. A business plan offers the following:

- A systematic, realistic evaluation of a venture's chances for success in the market
- A way to determine the principal risks facing the venture
- A "game plan" for managing the business successfully
- A tool for comparing actual results against targeted performance
- An important tool for attracting capital in the challenging hunt for money

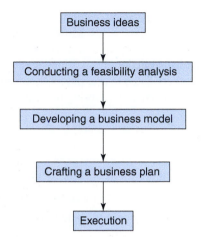

FIGURE 8.1

New Business Planning Process

The feasibility study, the business model, and the business plan all play important but separate roles in the start-up process (see Figure 8.1). This chapter describes how to build and use these vital business documents, and it will help entrepreneurs create business plans that will guide them on their entrepreneurial journey and help them attract the capital they need to launch and grow their businesses.

Conducting a Feasibility Analysis

A feasibility analysis consists of four interrelated components: an industry and market feasibility analysis, a product or service feasibility analysis, a financial feasibility analysis, and an entrepreneur feasibility analysis (see Figure 8.2). "A feasibility analysis is a chance to open your eyes, ask yourself some very tough questions, then check to see whether your idea, as originally conceived, needs to be modified, refocused, or changed dramatically. (Or perhaps even scrapped altogether)," says Rhonda Abrams, nationally syndicated columnist, author, and successful entrepreneur.[1]

1.

Present the steps involved in conducting a feasibility analysis.

Industry and Market Feasibility Analysis

When evaluating the feasibility of a business idea, an analysis of the industry and targeted market segments serves as the starting point for the remaining three components of a feasibility analysis. The focus in this phase is twofold: (1) to determine how attractive an industry is overall as a "home" for a new business and (2) to identify possible niches a small business can occupy profitably.

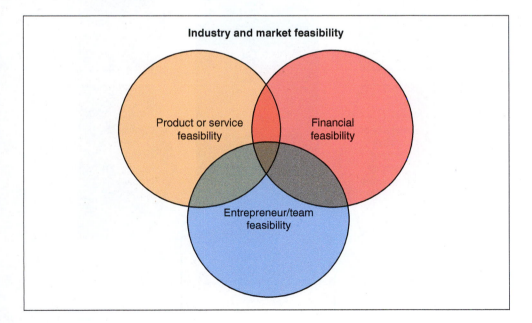

FIGURE 8.2

Elements of a Feasibility Analysis

The first step in assessing industry attractiveness is to paint a picture of the industry with broad strokes, assessing it from a "macro" level. Most opportunities for new businesses within an industry are due to changes taking place in that industry. For example, although the Internet disrupted business for the established major corporations in the music industry, the fundamental changes it created in how customers listen to and purchase music have created many opportunities for entrepreneurs within this industry. Answering the following questions will help establish this perspective:

- How large is the industry?
- How fast is it growing?
- Is the industry as a whole profitable?
- Is the industry characterized by high profit margins or razor-thin margins?
- How essential are its products or services to customers?
- What trends are shaping the industry's future?
- What threats does the industry face?
- What opportunities does the industry face?
- How crowded is the industry?
- How intense is the level of competition in the industry?
- Is the industry young, mature, or somewhere in between?

Addressing these questions helps entrepreneurs determine whether the potential exists for sufficient demand for their products and services.

A useful tool for analyzing an industry's attractiveness is Porter's "five forces" model developed by Michael E. Porter of the Harvard Business School (see Figure 8.3). Five forces interact with one another to determine the setting in which companies compete and hence the attractiveness of the industry: (1) the rivalry among competing firms, (2) the bargaining power of suppliers, (3) the bargaining power of buyers, (4) the threat of new entrants, and (5) the threat of substitute products or services.

FIGURE 8.3

Porter's Five Forces Model

Source: Based on Michael E. Porter, "How Competitive Forces Shape Strategy," *Harvard Business Review* 57, no. 2 (March–April 1979), pp. 137–145. Reprinted by permission of Harvard Business School Publishing.

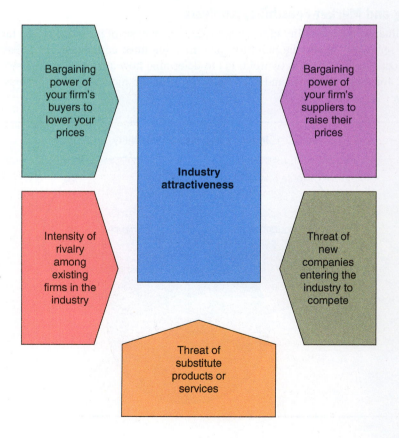

RIVALRY AMONG COMPANIES COMPETING IN THE INDUSTRY The strongest of the five forces in most industries is the rivalry that exists among the businesses competing in a particular market. Much like the horses running in the Kentucky Derby, businesses in a market are jockeying for position in an attempt to gain a competitive advantage. When a company creates an innovation or develops a unique strategy that transforms the market, competing companies must adapt or run the risk of being forced out of business. This force makes markets a dynamic and highly competitive place. Generally, an industry is more attractive under the following conditions:

- The number of competitors is large or, at the other extreme, quite small (fewer than five).
- Competitors are not similar in size or capability.
- The industry is growing at a fast pace.
- The opportunity to sell a differentiated product or service is present.

BARGAINING POWER OF SUPPLIERS TO THE INDUSTRY The greater the leverage that suppliers of key raw materials or components have, the less attractive is the industry. For instance, because they supply the chips that serve as the "brains" of PCs and because those chips make up a sizable portion of the cost of a computer, chip makers such as Intel and Advanced Micro Devices exert a great deal of power over computer manufacturers such as Dell, Toshiba, and Acer. Generally, an industry is more attractive under the following conditions:

- Many suppliers sell a commodity product to the companies in it.
- Substitute products are available for the items suppliers provide.
- Companies in the industry find it easy to switch from one supplier to another or to substitute products (i.e., "switching costs" are low).
- When the items suppliers provide the industry account for a relatively small portion of the cost of the industry's finished products.

BARGAINING POWER OF BUYERS Just as suppliers to an industry can be a source of pressure, buyers also have the potential to can exert significant power over businesses, making it less attractive. When the number of customers is small and the cost of switching to competitors' products is low, buyers' influence on companies is high. Famous for offering its customers low prices, Walmart, the largest company in the world, is also well known for applying relentless pressure to its 21,000 suppliers for price concessions, which it almost always manages to get.[2] Generally, an industry is more attractive under the following conditions:

- Industry customers' "switching costs" to competitors' products or to substitutes are relatively high.
- The number of buyers in the industry is large.
- Customers demand products that are differentiated rather than purchase commodity products that they can obtain from any supplier (and subsequently can pit one company against another to drive down price).
- Customers find it difficult to gather information on suppliers' costs, prices, and product features—something that is becoming much easier for customers in many industries to do by using the Internet.
- When the items companies sell to the industry account for a relatively small portion of the cost of their customers' finished products.

THREAT OF NEW ENTRANTS TO THE INDUSTRY The larger the pool of potential new entrants to an industry, the greater is the threat to existing companies in it. This is particularly true in industries where the barriers to entry, such as capital requirements, specialized knowledge, access to distribution channels, and others, are low. Generally, an industry is more attractive to new entrants under the following conditions:

- The advantages of economies of scale are absent. Economies of scale exist when companies in an industry achieve low average costs by producing huge volumes of items (e.g., computer chips).
- Capital requirements to enter the industry are low.
- Cost advantages are not related to company size.

- Buyers are not extremely brand loyal, making it easier for new entrants to the industry to draw customers away from existing businesses.

- Governments, through their regulatory and international trade policies, do not restrict new companies from entering the industry.

THREAT OF SUBSTITUTE PRODUCTS OR SERVICES Substitute products or services can turn an entire industry on its head. For instance, many makers of glass bottles have closed their doors in recent years as their customers—from soft-drink bottlers to ketchup makers—have switched to plastic containers, which are lighter, less expensive to ship, and less subject to breakage. Printed newspapers have seen their readership rates decline as new generations of potential readers turn to their iPads and smart phones for sources of news that are constantly updated. Generally, an industry is more attractive under the following conditions:

- Quality substitute products are not readily available.

- The prices of substitute products are not significantly lower than those of the industry's products.

- Buyers' cost of switching to substitute products is high.

After surveying the power that these five forces exert on an industry, entrepreneurs can evaluate the potential for their companies to generate reasonable sales and profits in a particular industry. In other words, they can answer the question "Is this industry a good home for my business?" Table 8.1 provides a matrix that allows entrepreneurs to assign quantitative scores to the five forces influencing industry attractiveness. Note that the *lower* the score for an industry, the *more* attractive it is.

The next step in assessing an industry is to identify potentially attractive niches that exist in the industry. Many small businesses prosper by sticking to niches in a market that are too small to attract the attention of large competitors. The key question that entrepreneurs address is "Can we identify a niche that is large enough to produce a profit, or can we position our company uniquely in the market to differentiate it from the competition in a meaningful way?" Entrepreneurs who have designed successful focus or differentiation strategies for their companies can exploit these niches to their advantage.

TABLE 8.1 Five Forces Matrix

Assign a value to rate the importance of each of the five forces to the industry on a 1 (not important) to 5 (very important) scale. Then assign a value to reflect the extent to which each force poses a threat to the industry. Multiply the importance rating in column 2 by the threat rating in column 3 to produce a weighted score. Add the weighted scores in column 3 to get a total weighted score. This score measures the industry's attractiveness. The matrix is a useful tool for comparing the attractiveness of different industries.

Force	Importance (1 = Not Important, 5 = Very Important)	Threat to Industry (1 = Low, 3 = Medium, 5 = High)	Weighted Score Col 2 × Col 3
Rivalry among companies competing in the industry	5	2	10
Bargaining power of suppliers in the industry	2	2	4
Bargaining power of buyers	2	4	8
Threat of new entrants to the industry	3	4	12
Threat of substitute products or services	4	1	4
		Total	**38**

Minimum Score = 5 (Very attractive)
Maximum Score = 125 (Very unattractive)

Questions entrepreneurs should address in this portion of the feasibility analysis include the following:

- Which niche(s) in the market will we occupy?
- How large is this market segment, and how fast is it growing?
- What is the basis for differentiating our product or service from competitors?
- Do we have a superior business model that will be difficult for competitors to reproduce?

Companies can shield themselves from some of the negative impact of these five forces by finding a niche and occupying it.

Aloompa co-founders Kurt Nelson, Tyler Seymour, and Drew Burchfield.

Source: Belmont University.

ENTREPRENEURIAL PROFILE: Kurt Nelson, Tyler Seymour, and Drew Burchfield: Aloompa Kurt Nelson, Tyler Seymour, and Drew Burchfield have found success in a niche in the rapidly growing and highly competitive smart phone app development business. Their company, Aloompa, creates a variety of apps for the music industry. Their first big success came from a music festival called Bonnaroo, which attracts 60,000 rock music fans to its rural Tennessee venue. Their success with this first festival led to 25 additional music festival customers within their first two years of business. While other app developers are trying to create apps that reach a wide audience of smart phone users, the owners of Aloompa are finding success within a very specific niche within the music industry. Although Aloompa has not become a brand name to smart phone users, the company is well known and has a strong position within the market niche of music festival organizers.[3]

Generally, a niche strategy is a good way to enter the market for a new business. It usually takes fewer resources for the start-up due to lower marketing costs and the ability to start on a smaller scale. Success rates tend to be higher for niche businesses because they have less direct competition. Without much competition, niche businesses can charge higher prices, which allows for quicker positive cash flow during the start-up phase and better margins once they become profitable.

However entrepreneurs should be aware of the following risks that are associated with a niche strategy:

- *Entering a niche requires adaptability in your initial plan.* While developing their business models, entrepreneurs can misjudge what customers in a niche market actually need. Unless they are willing and able to adjust their business model to react to the realities of the market niche, the business will fail if it does not offer the customers what they really want.

- *Niches change.* Even if an entrepreneur evaluates the market correctly in the beginning, niche markets (like any market) change over time. Success in a niche requires that entrepreneurs adapt as the market changes. Too many entrepreneurs get stuck doing the same thing or offering the same product while their customers' needs and wants evolve. Even though it is a niche, the market is not isolated and is subject to the same forces and trends that can impact any market.

- *Niches can go away.* Although many niches can last for years, no market is forever. Niches can dry up—sometimes quite suddenly. Again, adaptation can offer some hope, but if the decline is too rapid, niche businesses can fail.

- *Niches can grow.* Although significant growth in a market may not sound bad, it can attract more competitors. If a market grows large enough, it can attract some very large competitors. At some point, an entrepreneur's cozy market niche can become quite crowded. Eventually, it may no longer be a true market niche, which requires that the entrepreneur adapt his or her business strategies to meet this more competitive market. Growing competition forces prices downward, while at the same time pushes the cost of business upward due to increased marketing costs, greater expectations from customers, and higher labor costs due to more competition for qualified staff.

Although finding a market niche is the most common and a relatively safe entry strategy for new entrepreneurial businesses, it still requires continued attention to competitive forces in the market.

Product or Service Feasibility Analysis: Is There a Market?

Once entrepreneurs discover that sufficient market potential for their product or service idea actually exists, they sometimes rush in with exuberant enthusiasm ready to launch a business without actually considering whether they can actually produce the product or provide the service at a reasonable cost. A product or service feasibility analysis determines the degree to which a product or service idea appeals to potential customers and identifies the resources necessary to produce the product or provide the service. This portion of the feasibility analysis addresses the question "Are customers willing to purchase our goods and services?" Entrepreneurs need feedback from potential customers to answer this question. Conducting **primary research** involves collecting data firsthand and analyzing it; **secondary research** involves gathering data that has already been compiled and is available, often at a very reasonable cost or sometimes even free. In both types of research, gathering both quantitative and qualitative information is important to drawing accurate conclusions about a product's or service's market potential. Primary research tools include customer surveys, focus groups, building prototypes, conducting in-home trials, and "windshield" research. One goal of primary and secondary research to find support to validate the business concept. However, an overly optimistic entrepreneur may overlook important information about the true feasibility of the business if he or she searches only for information that affirms starting the business. It also is important to search for information that does *not* support the concept. Depending on the nature and severity of any negative evidence, the entrepreneur can either (1) adapt the concept if it is possible to do so or (2) abandon the idea entirely if necessary.

CUSTOMER SURVEYS AND QUESTIONNAIRES Keep them short. Word your questions carefully so that you do not bias the results and use a simple ranking system (e.g., a 1-to-5 scale, with 1 representing "definitely would not buy" and 5 representing "definitely would buy"). Test your survey for problems on a small number of people before putting it to use. Web surveys are inexpensive, easy to conduct, and provide feedback fast. Do not survey only people you know or those who are convenient to reach. Rather, survey people who represent the target market of the business.

FOCUS GROUPS A focus group involves enlisting a small number of potential customers (usually 8 to 12) to provide feedback on specific issues about a product or service (or the business idea itself). Listen carefully for what focus group members like and don't like about your product or service as they tell you what is on their minds. The founders of one small snack food company that produced apple chips conducted several focus groups to gauge customers' acceptance of the product and to guide many key business decisions, ranging from the product's name to its packaging. Consider creating virtual focus groups on the Web; one small bicycle retailer conducts 10 online focus groups each year at virtually no cost and gains valuable marketing information from them. Feedback from online customers is fast, convenient, and in real time.

PROTOTYPES An effective way to gauge the viability of a product is to build a prototype of it. A prototype is an original, functional model of a new product that entrepreneurs can put into the hands of potential customers so that they can see it, test it, and use it. Prototypes usually point out potential problems in a product's design, giving inventors the opportunity to fix them even before they put the prototype into customers' hands. The feedback customers give entrepreneurs based on prototypes often leads to design improvements and new features, some of which the entrepreneurs might never have discovered on their own. Makers of computer software frequently put prototypes of new products into customers' hands as they develop new products or improve existing ones. Known as *beta tests*, these trials result in an iterative design process in which software designers collect feedback from users and then incorporate their ideas into the product for the next round of tests.

IN-HOME TRIALS One technique that reveals some of the most insightful information into how customers actually use a product or service is also the most challenging to coordinate: in-home trials. An **in-home trial** involves sending researchers into customers' homes to observe them as they use the company's product or service. However, in-home trials can be expensive to conduct and therefore may not be affordable for the budgets of most entrepreneurs.

"WINDSHIELD" RESEARCH A good source of information is to observe customers interacting with existing businesses within an industry. For example, before one potential investor was willing to commit funding for a new coffee shop, he required that the entrepreneur get traffic counts at local competitors' outlets. He observed heavy demand and often long lines, which helped provide support for the need for a new coffee shop.

Secondary research should be used to support, not replace, primary research. Secondary research, which is usually less expensive to collect than primary data, includes the sources discussed in the following sections.

TRADE ASSOCIATIONS AND BUSINESS DIRECTORIES To locate a trade association, use *Business Information Sources* (University of California Press) or the *Encyclopedia of Associations* (Gale Research). To find suppliers, use the *Thomas Register of American Manufacturers* (*www .thomasnet.com/* or *Standard and Poor's Net Advantage* (*www.netadvantage.standardandpoors .com/NetAd/demo/pubLibrary.htm*). The *American Wholesalers and Distributors Directory* includes information on more than 28,000 wholesalers and distributors.

INDUSTRY DATABASES Several online business databases are available through university libraries, such as BizMiner, Encyclopedia of American Industries, Encyclopedia of Emerging Industries, Encyclopedia of Global Industries, Encyclopedia of Products & Industries—Manufacturing, IBISWorld, Manufacturing & Distribution USA: Industry Analyses, Statistics and Leading Companies, and Market Share Reporter. These databases offer a rich variety of information on specific industries, including statistical analyses, geographic reports, trend analyses, and profiles.

DEMOGRAPHIC DATA To learn more about the demographic characteristics of customers in a particular area, entrepreneurs should tap the treasure trove of data available from the U.S. Census Bureau (*www.census.gov*). There, entrepreneurs can learn about a population's age, income levels, housing status, education attainment, travel time to work, languages spoken, and many other characteristics.

FORECASTS Many government agencies, including the Department of Commerce, offer forecasts on topics ranging from interest rates to the number of housing starts. A government librarian can help you find what you need.

MARKET RESEARCH Someone may already have compiled the market research you need. Web sites that provide access to market research reports include Market Research (*www.marketresearch .com*), Companies and Markets (*www.companiesandmarkets.com/Home/ContactUs*), and Transparency Market Research (*www.transparencymarketresearch.com*). Many entrepreneurs find the market research they need by using an Internet search engine to conduct focused searches.

ARTICLES Magazine and journal articles pertinent to your business are a great source of information. In addition to online resources, entrepreneurs can use the *Reader's Guide to Periodical Literature*, the *Business Periodicals Index* (similar to the *Reader's Guide* but focuses on business periodicals), and *Ulrich's Guide to International Periodicals* to locate the ones you need.

LOCAL DATA Your state department of commerce and your local chamber of commerce will very likely have useful data on the local market of interest to you. Call to find out what is available.

THE INTERNET Entrepreneurs can benefit from the vast amount of market research information available on the Internet. This is an efficient resource with up-to-date information, and much of it is free. Entrepreneurs must use caution, however, to ensure the credibility of online sources.

Financial Feasibility Analysis: Is There Enough Margin?

The third component of a feasibility analysis involves assessing the financial feasibility of a proposed business venture. At this stage of the process, a broad financial analysis that examines the basic economic feasibility is sufficient. This component of the feasibility analysis answers the question "Can this business generate adequate profits?" If the business concept passes the overall feasibility analysis, an entrepreneur should conduct a more thorough financial analysis when developing the business model and creating a full-blown business plan. The major elements to be included in a financial feasibility analysis include the initial capital requirement, estimated earnings, time out of cash, and the resulting return on investment.

CAPITAL REQUIREMENTS Just as a boy scout needs fuel to start a fire, an entrepreneur needs capital to start a business. Some businesses require large amounts of capital, but others do not. Typically, service businesses require less capital to launch than do manufacturing or retail businesses. Start-up companies often need capital to purchase equipment, buildings, technology, and other tangible assets as well as to hire and train employees, promote their products and services, and establish a presence in the market. A good feasibility analysis provides an estimate of the amount of start-up capital an entrepreneur needs to get the business up and running.

Larry Leith, founder
of Tokyo Joe's.
Source: Tokyo Joe's.

ENTREPRENEURIAL PROFILE: Larry Leith: Tokyo Joe's When preparing to launch his first restaurant, Larry Leith, founder of Tokyo Joe's, estimated that he would need $330,000. About half of that money would go to build out the restaurant in the space he rented—costs associated with construction, architecture, and engineering materials and services. The other half he would use to cover the remainder of the start-up costs, which would include a simple marketing plan, furnishings, supplies, and enough cash to cover operating expenses until the restaurant reached its break-even point, which occurred in the second month of operation.[4]

It is important to keep in mind that the typical start-up in the United States is able to successfully launch with an average of only $10,000,[5] and one in five start-ups get launched with no funding at all.[6] How do entrepreneurs get started with so little funding? Most entrepreneurs employ a variety of techniques called bootstrapping. **Bootstrapping** is the process of finding creative ways exploit opportunities to launch and grow businesses with the limited resources available for most start-up ventures. It includes a variety of strategies and techniques that cover all of the functions of running a business, including marketing, staffing, inventory and production management, cash flow management, and the administrative processes needed to keep a business operating.[7]

You will learn more about finding sources of business funding, both debt and equity, in Chapters 16 and 17.

ESTIMATED EARNINGS In addition to producing an estimate of the start-up company's capital requirements, an entrepreneur also should forecast the earning potential of the proposed business. Industry trade associations and publications such as the *RMA Annual Statement Studies* offer guidelines on preparing sales and earnings estimates for specific types of businesses. From these, entrepreneurs can estimate the financial results they and their investors can expect to see from the business venture if the start-up is executed according to plan.

TIME OUT OF CASH A common cause of business failure is running out of cash before the business is able to reach breakeven and support itself through the cash flow from operations. According to a study by U.S. Bank, four out of five small business failures can be attributed to "starting out with too little money."[8] During the planning stage, an entrepreneur should estimate the total cash required to sustain the business until the business achieves break-even cash flow. This estimate should be based on a less than optimistic scenario because there are almost always unexpected costs and delays in the start-up and growth phases of a new business. For an operating business, to calculate the time until the business runs out of cash, simply take the negative cash flow from the business each month (its "burn rate") and divide it by the amount of available cash in the business. This gives the number of months the business can survive at its current rate of negative cash flow. Ideally, the business will be able to grow quickly enough to avoid reaching a cash crisis.

RETURN ON INVESTMENT The final aspect of the financial feasibility analysis combines the estimated earnings and the capital requirements to determine the rate of return the venture is expected to produce. One simple measure is the rate of return on the capital invested, which is calculated by dividing the estimated earnings the business yields by the amount of capital invested in the business. This aspect of financial feasibility is generally of most concern to investors. Although financial estimates at the feasibility analysis stage typically are rough, they are an important part of the entrepreneur's ultimate "go" or "no-go" decision about the business ventures. A venture must produce an attractive rate of return relative to the level of risk it requires. This risk–return tradeoff means that the higher the level of risk a prospective business involves,

Developing and Testing a Business Model

In their groundbreaking study of how successful entrepreneurs develop business models, Osterwalder and Pigneur found that most entrepreneurs use a visual process, such as whiteboarding, when developing their business models.[9] They do not start writing text; rather, they create a visual map of the key components required to make their businesses successful. A business model adds more detail to the feasibility analysis by graphically depicting the "moving parts" of the business and ensuring that they are all working together. In building a business model, an entrepreneur addresses a series of key questions that define how to create a successful business. Who is the target customer? What value does the business offer the customer? What do customers expect from the business? How do I get information to them, and how do they want to get the product? What are the key activities required to make all of the pieces come together, and what will they cost? What resources must I have to make this happen (including money)? Which key partners must I attract to be successful? In their study, Osterwalder and Pigneur found a pattern in the way that entrepreneurs used a visual representation of their business models to answer these questions. They used these findings to develop a business model canvas that is comprised of nine segments that help guide entrepreneurs through the process of developing a business model (see Figure 8.5).

2.

Identify the components of the business model canvas and explain how to use them to develop a viable business model.

Value Proposition

A compelling value proposition is at the heart of every successful business. The value proposition is the collection of products and/or services that the business offers to meet the needs of the customers. It is all the things that will set the business apart from its competitors, such as pricing, quality, service, features, product availability, and other features. Most value propositions for new businesses come from fundamental trends within the economy, such as demographics, technology, or cultural patterns. Trends lead to changes within industries. For example, the widespread use of the Internet forever changed industries such as music and newspapers. Inflation in the economy led to both soaring health care costs and rising food costs, shaking up both these industries. Entrepreneurs uncover trends such as these in the industry and market feasibility analysis. Recall from Chapter 1 that a fundamental role of entrepreneurs is finding solutions for problems and needs that customers have as a result of the changes that follow disruptive trends.

It is best to identify and focus on one or two benefits that will make the new business stand out to customers and motivate them to purchase its products or services. The best way to develop the key benefits that are at the heart of a strong value proposition is to listen to customers. Although an entrepreneur may think he or she knows what the customer wants, most of the time they will not have it quite right. In most cases, entrepreneurs must adjust their products or services to fit with what the customer actually wants or needs.

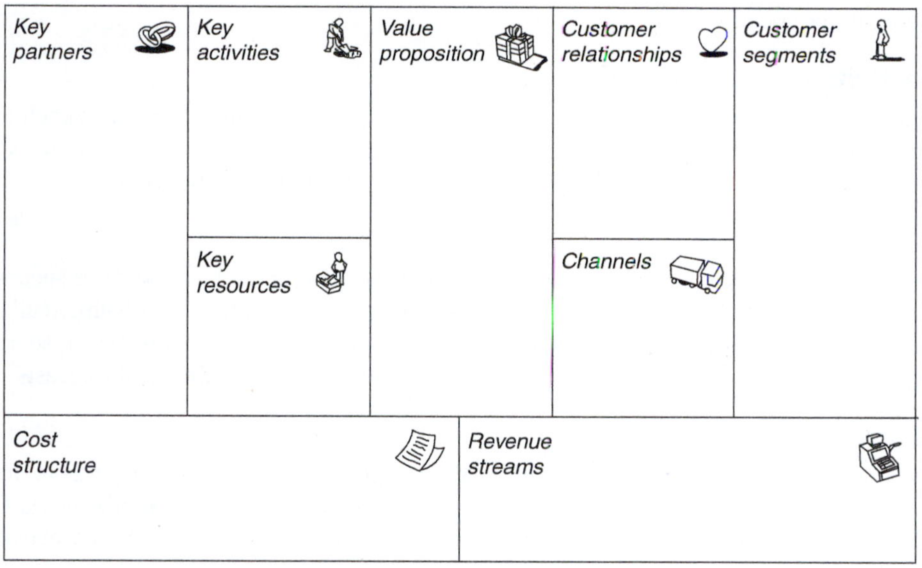

FIGURE 8.5

Business Model Canvas

Source: A. Osterwalder and Y. Pigneur, *Business Model Generation* (Hoboken, NJ: Wiley, 2010).

Customer Segments

Defining the customer segments in the business model requires an entrepreneur to define the company's target market. Narrowing the target market enables a small company to focus its limited resources on serving the needs of a specific group of customers rather than attempting to satisfy the desires of the mass market. Creating a successful business depends on an entrepreneur's ability to attract real customers who are willing and able to spend real money to buy its products or services. Perhaps the worst marketing error an entrepreneur can commit is failing to define his or her target market and trying to make his or her business "everything to everybody." Small companies usually are much more successful focusing on a specific market niche or niches where they can excel at meeting customers' special needs or wants. At whose needs are the new business's products and/or services aimed? It may be a market niche, a mass market, or a segmented market based on age, gender, geography, or socioeconomic grouping.

Customer Relationships

Not every business provides the same type and same level of customer service. This is what defines the customer relationship component in the business model. For example, there are many effective business models that provide meals to consumers, ranging from a vending machine or a fast-food restaurant to a fast-casual restaurant or a fine-dining establishment. Each of these business models takes a different approach to providing customer service. The vending business offers quick, convenient, and impersonal service. The fine-dining restaurant works closely and personally with customers to ensure that they get exactly what they want. Each approach is effective and appropriate for its particular target market. When developing this segment of the business model, an entrepreneur must answer several questions: How do customers want to interact with the business? Do they want intensive personal service, or would they rather have limited engagement or even an automated interaction? There is no one best approach to customer relationships for all businesses, but there is always one best approach for a particular business model.

Channels

In the business model canvas, *channels* refer to both *communication channels* and *distribution channels*. Communication channels define how the customers seek out information about this type of product. Where do potential customers go when they want to get information about products and services? It could be Web sites, their social network, blogs, advertisements, experts, and so forth. Again, there is no one best way to communicate for all businesses, but there will be one or more ways that are most effective with the specific target market for a given business model.

The distribution channel defines the most effective way to get products to the customers for this type of business. It may be best to use in-home sales. Customers may want to order online from the comfort of their living rooms, or they may want to see the product, touch it, and feel it in an exciting new retail location. The entrepreneur must determine where the customer wants to make the purchase and then determine the most effective way to get it to the customer at that location.

Key Activities

What are the key activities that an entrepreneur must do to ensure a successful launch and to sustain the growth of the business? In the business model, the goal is to build a basic checklist of what needs to be done. The business plan expands on this list in much greater detail.

Key Resources

What are the human, capital, and intellectual resources needed for the business to be successful? Again, this will serve as an initial checklist to ensure that an entrepreneur has identified all of the key resources required. The business plan provides an opportunity to explain these resources in much greater detail and to develop the necessary cost estimates for the financial forecasts.

Key Partners

This segment of the business model includes key suppliers, key outsourcing partners, and all other external businesses or entities that are critical to making the business model work. Entrepreneurs cannot expect to become successful all by themselves. They must build a network of relationships when launching and growing their businesses.

Revenue Streams

How will the value proposition generate revenue? Will it be a one-time sale, ongoing fees, advertising, or some other sources of cash into the business? Entrepreneurs must answer these questions using the entrepreneur's definitions of the value proposition, customer segments, customer relationship, and channel components of the business model (the right side of the business model canvas). The revenue streams information serves as the framework for the more detailed revenue forecasts contained in the business plan.

Cost Structure

What are the fixed and variable costs that are necessary to make the business model work? The key activities, key resources, and key partners components of the plan (the left side of the business model canvas) help to identify the basic types of costs and provide an estimate of their scope. Just like the revenue streams, the cost structure of the business model becomes the framework to develop more detailed cost estimates that become part of the financial forecasts of the business plan.

Developing a business model is a four-phase process.[10] The first phase is creating an initial business model canvas. It is best to do this on a whiteboard or on the wall using sticky notes. As the entrepreneur goes through the next three steps, the business model will change. At this point, the business model is only a series of hypotheses to be tested.

The second phase in designing the business model is to test the problem that the entrepreneur thinks the business solves for customers. The best way to do this is to use primary research data. By engaging potential customers early in developing of a new business and listening to what they have to say, entrepreneurs have a much better chance of creating a successful business model that solves a real customer need.

The third phase is testing the business model *in the market*. One technique to test the solution offered by the business model involves **business prototyping**, in which entrepreneurs test their business models on a small scale before committing serious resources to launch a business that might not work. If the test supports the hypothesis and its accompanying assumptions, the entrepreneur moves forward and develops a business plan. If the prototype flops, the entrepreneur scraps the business idea with only minimal losses and turns to the next idea.

The Internet is a valuable business prototyping tool because it gives entrepreneurs easy and inexpensive access to real live potential customers. Entrepreneurs can test their ideas by selling their products on established sites, such as eBay, or by setting up their own Web sites to gauge customers' response. Testing early, often rough, versions of a product or service is part of a **lean start-up**, a process of rapidly developing simple prototypes to test key assumptions by engaging real customers.[11] To launch a business using the lean start-up process, entrepreneurs begin with what is called a **minimum viable product**, the simplest version of a product or service on which to build a sustainable business.

ENTREPRENEURIAL PROFILE: Drew Houston and Arash Ferdowsi: Dropbox Dropbox, founded in 2007 by Drew Houston and Arash Ferdowsi, is a free service that lets users easily share photos, documents, and videos among any devices using their software. When developing the idea for their business, Houston and Ferdowsi knew that creating the software for Dropbox would take years and that developing a simple working prototype was not practical. To test their idea, they created a three-minute video that demonstrated how their new product would work once it was fully developed. The video drove hundreds of thousands of viewers to their Web site and created a long waiting list of people willing to use the product in its beta version. The strong interest convinced the founders and the investment community that Dropbox was a product for which there was a strong market. Dropbox now has more than 200 million users around the globe and has raised $257 million in funding.[12]

The fourth phase of designing a business model is making changes and adjustments in the business, called **pivots** (refer to Chapter 1), based on what the entrepreneur learns from engaging customers about the business's proposed products and services. Some pivots are subtle adjustments to the business model, while others are fundamental changes to key parts of the model, including in the value proposition, markets served, or ideal revenue streams.

Building a sound business model in a start-up venture improves the chances that the business will survive the launch, gain acceptance in the market, and grow. However, there is one more

Entrepreneurship in Action

The Evolution of CoolPeopleCare's Business Model[13]

Sam Davidson and Stephen Moseley first met while working at a nonprofit in Nashville, Tennessee. The original idea for CoolPeopleCare came to Davidson after he had taken a trip to Washington, D.C., for a Save Darfur Rally in April 2006. While at the rally, he remembers seeing a man holding a cardboard sign with "Cool People Care" written on it. He remembers being struck by how powerful this phrase was. It had a really persuasive ring to it. Returning to Nashville, Davidson couldn't stop thinking about the phrase "Cool People Care." Initially, he thought it would be appealing to build a Web site and sell "Cool People Care" T-shirts and to ask people what they care about through the site. He approached Moseley to inquire about building a Web site. A 15-minute conversation turned into a three-hour planning session in which Davidson and Moseley mapped out what the first version of the Cool People Care Web site.

Davidson had observed from his experience working in nonprofits that "there are two main excuses people give for not volunteering or making a difference: one—there is not enough time; and, two—I do not know how." He believed that the CoolPeopleCare Web site could eliminate both excuses. Through the CoolPeopleCare Web site, Davidson wanted to enlighten and connect his audience. "CoolPeopleCare essentially pairs people who want to make a difference with organizations and opportunities to do so."[14] He believed that the Web site would be a good way to connect young people with causes that they could help and support.

In May 2007, Davidson decided it was time to leave his full-time job and dedicate 100 percent of his time to building and growing CoolPeopleCare. The business was set up as a social for-profit venture. When asked why he did this, Davidson says, "We didn't want to compete for funds with those they wanted to help. [In addition, we] wanted to prove that a for-profit entity can have an explicit purpose of helping the community. In other words, we wanted to prove that corporations making a difference didn't have to be (and shouldn't be) an afterthought."[15]

The Web site initially had two main features. The first was *5 Minutes of Caring*, which highlights a single action that one can take in less than five minutes to make the world a better place and to make a difference. The second feature is customized content for more than 40 cities, which includes a community calendar and volunteer and job opportunities specific to each city and a nonprofit directory for that city.

After Davidson committed to working on CoolPeopleCare full-time, he realized that the business model would have to change for the business to be viable. Over time, Davidson has added several new revenue streams to the business. The first was having Davidson offer speaking engagements. He soon added training sessions to help other nonprofits create revenue streams that alleviate the need for traditional fund-raising. Mosley also rolled his nonprofit consulting practice into CoolPeopleCare. Next, Mosley and Davidson wrote a book, *New Day Revolution: How to Save the World in 24 Hours*, and began to sell it through the Web site. They have sold thousands of copies of their book.

Mosley and Davidson also realized that they could offer more value to the nonprofit community. They added "Partner Pages," where nonprofits could highlight their organizations, events and volunteer listings, contact information, link to their own Web sites, donation buttons, national e-mail announcements, and job listings. Nonprofits pay $1 a day to be listed on this page. Dozens of nonprofits soon took advantage of this new feature. The entrepreneurs also added "Cool Pages" to allow corporate sponsors to highlight their support of social causes.

The most significant change to the business model was adding merchandise to the Web site, although at the time they did not realize its eventual impact. Besides their book, the CoolPeopleCare store sells merchandise with the CoolPeopleCare logo including T-shirts, coffee mugs, and reusable shopping bags. They also sell fair-trade coffee. After Nashville was hit by a devastating flood, CoolPeopleCare began selling "We Are Nashville" T-shirts to help raise money for flood relief. Mosley and Davidson had hoped to sell 1,000 shirts, but they sold that many the very first day. Eventually, CoolPeopleCare made more than $100,000 in profits from the T-shirt, which they donated to the Nashville Community Foundation for flood relief. They also sold "We Are Nashville" prints and bumper stickers that raised more money for this cause. In addition to the good that this new product was able to do for Nashville, it helped put CoolPeopleCare on the map on a national and even global scale, as orders came in from around the world for "We Are Nashville" products. This visibility helped take CoolPeopleCare to another level and reach a much broader audience.

1. Why did the business model of CoolPeopleCare change over time? Explain why each of the changes in the business model was successful.

2. Develop a business model canvas for the current operations of CoolPeopleCare.

3. What specific recommendations would you make to change the business model of CoolPeopleCare going forward? What is your rationale for each of these recommendations?

---

critical step in the business planning process. Crafting a business plan that is based on what the entrepreneur has learned through the feasibility process and developing the business model is the third step in the new business planning process. It adds necessary detail and provides the final test of the business concept.

The Benefits of Creating a Business Plan

When based on the foundation of a fully developed and tested business model, a well-conceived and factually based business plan increases a new venture's probability of success. For decades, research has proved that companies that engage in business planning outperform those that do not. A study by the Small Business Administration reports that entrepreneurs who write business plans early on are two and a half times more likely to actually start their businesses than those who do not.[16] Unfortunately, many entrepreneurs never take the time to engage in the new business planning process, of which the business plan is an important element. The implications of the lack of planning are all too evident in the high failure rates that small companies experience.

A **business plan** is a written summary of an entrepreneur's proposed business venture, its operational and financial details, its marketing opportunities and strategy, and its managers' skills and abilities. There is no substitute for a well-prepared business plan, and there are no shortcuts to creating one. The plan serves as an entrepreneur's road map on the journey toward building a successful business. A business plan describes the direction the company is taking, what its goals are, where it wants to be, and how it intends to get there. The plan is written proof that an entrepreneur has performed the necessary research, has studied the business opportunity adequately, and is prepared to capitalize on it with a sound business model. Crafting a business plan is an entrepreneur's best insurance against launching a business destined to fail or mismanaging a potentially successful company.

A business plan serves two essential functions. First, it helps to guide the company's growth and development by charting its future course and devising a strategy for following it. The plan provides a battery of tools—a mission statement, goals, objectives, budgets, financial forecasts, target markets, and strategies—to help entrepreneurs lead the company successfully. A solid business plan provides managers and employees with a sense of direction when everyone is involved in creating and updating it. As more team members become committed to making the plan work, it takes on special meaning. It gives everyone targets to shoot for, and it provides a yardstick for measuring actual performance against those targets, especially in the crucial and chaotic start-up phase of the business. Creating a plan also forces entrepreneurs to, once again, subject their ideas to the test of reality. The greatest waste of a completed business plan is to let the plan go unused. When properly done, a plan becomes an integral and natural part of a company. In other words, successful entrepreneurs actually *use* their business plans to help them build strong companies.

ENTREPRENEURIAL PROFILE: Rhonda Abrams: The Planning Shop Rhonda Abrams, founder of The Planning Shop, a small publisher of books and tools for entrepreneurs, says that the business plan that she and her employees craft every year has played an important role in her company's success. "Developing a business plan is a key to long-term business survival and success," she says. Abrams credits her company's strategic plan for helping her team identify a new market opportunity that later allowed the business to survive the bankruptcy of its former distributor.[17]

The second function of the business plan is to attract lenders and investors. A business plan must prove to potential lenders and investors that a venture will be able to repay loans and produce an attractive rate of return. They want proof that an entrepreneur has evaluated the risk involved in the new venture realistically and has a strategy for addressing it. Unfortunately, many small business owners approach potential lenders and investors without having prepared to sell their business concepts. Given the increased challenges in funding small businesses, being thoroughly prepared to pitch a business plan to potential lenders and investors has become even more important. "We've had 3,000 entrepreneurs come to us needing money," says Tim Williamson, cofounder of The Idea Village, an incubator for entrepreneurs. "And 99 percent of them were not even ready to get the money." A collection of figures scribbled on a notepad to support a loan application or investment request is not enough. Applying for loans or attempting to attract investors without a solid business plan rarely attracts needed capital. The best way to secure the necessary capital is to prepare a sound business plan. The quality of an entrepreneur's business plan weighs heavily in the final decision to lend or

3.

Explain the benefits of building an effective business plan.

invest funds. It is also potential lenders' and investors' first impression of the company and its managers. Therefore, the finished product should be highly polished and professional in both form and content.

Three Tests That Every Business Plan Must Pass

4.
Describe the elements of a solid business plan.

Preparing a sound business plan clearly requires time and effort, but the benefits greatly exceed the costs. Building a plan forces a potential entrepreneur to look at his or her business idea in the harsh light of reality by refining the concepts developed in the business model and defining them in more detail. To get external financing, an entrepreneur's plan must pass three tests with potential lenders and investors: (1) the reality test, (2) the competitive test, and (3) the value test. Entrepreneurs develop the information used for these three tests in the feasibility analysis discussed earlier in this chapter. The first two tests, discussed in the following sections, have both an external and an internal component.

Reality Test

The external component of the reality test revolves around proving that a market for the product or service really does exist. It focuses on industry attractiveness, market niches, potential customers, market size, degree of competition, and similar factors. Entrepreneurs who pass this part of the reality test prove in the marketing portion of their business plan that there is strong demand for their business idea. Evidence they gather during the testing of the business model should be an integral part of the marketing plan to bolster the proof for the idea using real customers.

The internal component of the reality test focuses on the product or service itself. Can the company *really* build it for the cost estimates in the business plan? Is it truly different from what competitors are already selling? Does it offer customers something of value?

Competitive Test

The external part of the competitive test evaluates the company's position relative to its key competitors. How do the company's strengths and weaknesses match up with those of the competition? Do competitors' actions threaten the new company's success and survival?

The internal competitive test focuses on management's ability to create a company that will gain an edge over existing rivals. To pass this part of the competitive test, a plan must prove the quality, skill, and experience of the venture's management team. What other resources does the company have that can give it a competitive edge in the market?

Value Test

To convince lenders and investors to put their money into the venture, a business plan must prove to them that it offers a high probability of repayment or an attractive rate of return. Entrepreneurs usually see their businesses as good investments because they consider the intangibles of owning a business—gaining control over their own destinies, freedom to do what they enjoy, and others. Lenders and investors, however, look at a venture in colder terms: dollar-for-dollar returns. A plan must convince lenders that the business will repay the money they lend to the business, and it must convince investors that they will earn an attractive return on their money.

Even after completing a feasibility analysis and building a business model, entrepreneurs sometimes do not come to the realization that "this business just won't work" until they build a business plan. Have they wasted valuable time? Not at all! The time to find out that a business idea will not succeed is in the planning stages *before* committing significant money, time, and effort to the venture. It is much less expensive to make mistakes on paper than in reality. In other cases, a business plan reveals important problems to overcome before launching a company. Exposing these flaws and then addressing them enhances a venture's chances of success. Business plans help nascent entrepreneurs nail down important aspects of their concepts and prevent costly mistakes.

The real value in preparing a plan is not as much in the plan itself as it is in the *process* the entrepreneur goes through to create the plan—from the feasibility analysis, through the development and testing of the business model, and finally with the crafting of the written business plan. Although the finished product is extremely useful, the process of building the plan requires entrepreneurs to explore all areas of the business and subject their ideas to an objective, critical evaluation from many different angles. What entrepreneurs learn about their industry, target customers, financial requirements, competition, and other factors is essential to making their ventures successful. Building a business plan is one controllable factor that can reduce the risk and uncertainty of launching a company.

 In the Entrepreneurial Spotlight

The Battle of the Plans

The Burton D. Morgan Center for Entrepreneurship at Purdue University runs four entrepreneurship competitions each year. Two of these competitions are for entrepreneurs to compete for prize money to support their life science or nanotechnology ventures. One of their competitions is an elevator pitch competition. The Burton D. Morgan Business Plan Competition is the oldest of these events. The 2012 competition marked the 25th anniversary of this competition. Even though it is limited to Purdue University students, its rich history and significant prize money has made it a major competitive event.

The event has two divisions. The Gold division is the open division with teams comprised undergraduate students, graduate students, or a combination of both. Staff and faculty of Purdue University may be included on the teams. The winning team of the Gold division receives a $30,000 first prize. The Black division is only for teams wholly made up of currently enrolled undergraduate students. The winning team in the Black division receives a $20,000 first-place prize. The goal of the competition is to provide an opportunity for Purdue students to learn about entrepreneurship by developing and presenting a business idea to a panel of judges. Participants in the competition have an opportunity to define their ideas in commercial terms and compete for substantial cash prizes that they can use to commercialize their business ideas.

The competition has three stages of judging. The first phase is based on an executive summary of the business plan. After organizers narrow that group of submissions, teams submit business plans. Competition organizers evaluate the plans and select five finalists in each division to compete in the final competition, which includes a presentation to a panel of judges. The judges evaluate the students' executive summaries, business plans, and presentations using the criteria of commercial viability, technical viability, financial viability, and strength of the management team.

In a recent competition, one of the finalists in the Black division was Azzip Pizzeria, a fast-casual restaurant concept aimed at the fastest-expanding sector of the industry. The team was led by Brad Niemeier, a senior studying hospitality and tourism management and a member of the Boilermaker football team. Niemeier came up with his idea from his experience making pizzas for his teammates. Niemeier's teammates loved his pizzas and ate all that he made for them.

Niemeier made the most of both his hospitality classes and his entrepreneurship classes. His entrepreneurship classes gave him the confidence and skills to open a restaurant at such a young age. His hospitality classes taught him the skills he needed to manage his business day-to-day.

Niemeier's team faced stiff competition in the finals. The other four finalists were Dentural (all-natural adhesive for dentures), Kyk Energy (flavor-neutral energy powder that can turn any drink into an energy drink), PlayitSafe (head impact monitoring system), and Gamers' Esc (a video gaming center).

Past finalists in this competition have had great success as entrepreneurs. In a survey of past finalists in the Burton D. Morgan Business Plan Competition, 67 percent of the respondents indicated that they currently view themselves as entrepreneurs, 35 percent indicated that they started companies as a result of participating in the business plan competition, and 67 percent of the companies launched following the competition, which began in 1987, remain in existence.

In the end, Azzip Pizza won $20,000 as the top finisher in the Black division. Although he played in only two games during his football career at Purdue, Niemeier is proving to be a winner in game of business.

1. What benefits do entrepreneurs who compete in business plan competitions such as the one at Purdue University gain?

2. Work with a team of your classmates to brainstorm ideas for establishing a business plan competition on your campus. How would you locate judges? What criteria would you use to judge the plans? What prizes would you offer the winners, and how would you raise the money to give those prizes? Who would you allow to compete in your competition?

3. Using the ideas you generated in question 2, create a two-page proposal for establishing a business plan competition at your school.

Sources: Stacy Clardie, "Dream Realized," *Rivals.com*, May 15, 2012, *http://purdue.rivals.com/content.asp?CID=1365657&PT=4&PR=2*, *www.purdue.edu/newsroom/students/2012/120222CosierBizPlan.html*, and *www.insideindianabusiness.com/newsitem.asp?ID=52293*.

The Elements of a Business Plan

5.
_____

Explain the three tests every business plan must pass.

Wise entrepreneurs recognize that every business plan is unique and must be tailored to the specific needs of their business. They avoid the off-the-shelf, "cookie-cutter" approach that produces a look-alike business plan. The elements of a business plan may be standard, but the way entrepreneurs tell their stories should be unique and reflect the strengths of their business models, the experience of their team, their personalities and how they will shape the culture of the business, and their enthusiasm for the new venture. In fact, the best business plans usually are those that tell a compelling story in addition to the facts. For those making a first attempt at writing a business plan, seeking the advice of individuals with experience in this process often proves helpful. Accountants, business professors, attorneys, advisers working with local chapters of the Service Corps of Retired Executives (SCORE), and consultants with Small Business Development Centers (SBDCs) are excellent sources of advice when creating and refining a plan. (For a list of SBDCs, see the Small Business Administration's Web SBDC Web page at _www.sba.gov/content/small-business-development-centers-sbdcs_ and for a list of SCORE chapters, see their Web site at _www.score.org._) Remember, however, that you should be the one to author your business plan, not someone else.

Initially, the prospect of writing a business plan may appear to be overwhelming. Many entrepreneurs would rather launch their companies and "see what happens" than invest the necessary time and energy defining and researching their target markets, defining their strategies, and mapping out their finances. After all, building a plan is hard work—it requires time, effort, and thought. However, in reality, the entrepreneur should do _both_. By getting started and seeing what happens, the entrepreneur is able to test and improve the basic business model. The plan is essential as the entrepreneur gets ready to build the business and scale its growth. The business plan is hard work that pays many dividends, and not all of these are immediately apparent. Entrepreneurs who invest their time and energy building plans are better prepared to face the hostile environment in which their companies will compete than those who do not.

Entrepreneurs can use business planning software available from several companies to create their plans. Some of the most popular programs include Business Plan Pro (Palo Alto Software*), PlanMaker (Power Solutions for Business), and Plan Write (Business Resources Software). Business Plan Pro, for example, covers every aspect of a business plan from the executive summary to the cash flow forecasts. These packages help entrepreneurs organize the material they have researched and gathered, and they provide helpful tips on plan writing with templates for creating financial statements. Business planning software may help to produce professional-looking business plans with a potential drawback: The plans they produce may look as if they came from the same mold. That can be a turnoff for professional investors who review hundreds of business plans each year.

In the past, conventional wisdom was that business plans should be 20 to 40 pages in length, depending on the complexity of the business. More recently, experts have begun to recommend that plans should be shorter, typically suggesting that they be limited to about 10 pages. There is mixed opinion on how complex the financial forecasts should be. If the forecasts are based on evidence that is substantiated by testing the business model, more detail will strengthen the entrepreneur's case. If the numbers appear to be unsubstantiated or even fabricated, more detail can actually hurt the presentation. In many ways, having to write shorter business plans can make writing them even more of a challenge. A shorter business plan does not mean that an

Source: Scott Adams/Universal Uclick.

_____
* Business planning software from Palo Alto Software is available at a nominal cost with this textbook.

entrepreneur should omit any of the elements of the plan. Instead, the entrepreneur must work hard to communicate all of the key aspects of the plan as succinctly as possible. Although entrepreneurs find it difficult to communicate all of the important elements of their stories within the shorter-page-length recommendations, they run the risk of never getting used or read if their plans get too long! This section explains the most common elements of a business plan. However, entrepreneurs must recognize that, like every business venture, every business plan is unique. An entrepreneur should use the following elements as the starting point for building a plan and should modify them as needed to better tell the story of his or her new venture.

Title Page and Table of Contents

A business plan should contain a title page with the company's name, logo, and address as well as the names and contact information of the company founders. Many entrepreneurs also include the copy number of the plan and the date on which it was issued on the title page. Business plan readers appreciate a table of contents that includes page numbers so that they can locate the particular sections of the plan in which they are most interested.

The Executive Summary

To summarize the presentation to each potential financial institution or investors, the entrepreneur should write an executive summary. It should be concise—a maximum of one page—and should summarize all of the relevant points of the proposed deal. After reading the executive summary, anyone should be able to understand the entire business concept, what differentiates the company from the competition, and the amount of financing it requires. The executive summary is a synopsis of the entire plan, capturing its essence in a capsulized form. It should explain the basic business model and the problem the business will solve for customers, briefly describing the owners and key employees, target market(s), financial highlights (e.g., sales and earnings projections, the loan or investment requested, how the funds will be used, and how and when any loans will be repaid or investments cashed out), and the company's competitive advantage. Much like Abraham Lincoln's Gettysburg Address, which at 256 words lasted just two minutes and is hailed as one of the greatest speeches in history, a good executive summary provides a meaningful framework for potential lenders and investors of the essence of a company.

The executive summary is a written version of what is known as "the elevator pitch." Imagine yourself on an elevator with a potential lender or investor. Only the two of you are on the elevator, and you have that person's undivided attention for the duration of the ride, but the building is not very tall! To convince the investor that your business idea is a great investment, you must condense your message down to its essential elements—key points that you can communicate in a matter of no more than two minutes. In the Babcock Elevator Competition at Wake Forest University, students actually ride an elevator 27 floors with a judge, where they have the opportunity to make their elevator pitches in just two minutes. "The competition was designed to simulate reality," says Stan Mandel, creator of the event and director of the Angell Center for Entrepreneurship. Winners receive the chance to make 20-minute presentations of their business plans to a panel of venture capitalists, who judge the competition using criteria that range from the attractiveness of the business idea and the value proposition it offers to the quality of the plan's marketing and financial elements.[18]

Like a good movie trailer, an executive summary is designed to capture readers' attention and draw them into the plan. If it misses, the chances of the remainder of the plan being read are minimal. The difference between an executive summary and a movie trailer, however, is that the executive summary should give away the ending! A coherent, well-developed summary of the full plan establishes a favorable first impression of the business and the entrepreneur behind it and can go a long way toward obtaining financing. A good executive summary should allow the reader to understand the business concept and how it will make money as well as answering the ultimate question from investors or lenders: "What's in it for me?" Although the executive summary is the first part of the business plan, it should be the last section written to ensure that it truly captures all of the important points as they appear in the full plan.

Mission and Vision Statement

As you learned in Chapter 4, a mission statement expresses an entrepreneur's vision for what his or her company is and what it is to become. It is the broadest expression of a company's purpose and defines the direction in which it will move. It anchors a company in reality and serves as the thesis statement for the entire business plan by answering the question "What business are we in?" Every good plan captures an entrepreneur's passion and vision for the business, and the mission statement is the ideal place to express them. Entrepreneurs should avoid using too much business jargon and business clichés. It should clearly state what the business sells, its target market, and the basic nature of the business (i.e., manufacturing, consulting, service, outsourcing, and so forth).

Company History

The owner of an existing small business should prepare a brief history of the operation, highlighting the significant financial and operational events in the company's life. This section should describe when and why the company was formed, how it has evolved over time, and what the owner envisions for the future. It should highlight the successful accomplishment of past objectives and should convey the company's image in the marketplace.

 In the Entrepreneurial Spotlight

A Business Plan: Don't Launch Without It

A recent study by the Small Business Administration reports that entrepreneurs who create business plans in the early stages of the start-up process are more likely to actually launch companies and complete typical start-up activities such as acquiring patents, attracting capital, and assembling start-up teams more quickly than entrepreneurs who do not. "Early formal planners are doers," write the study's authors. "Challenging prospective entrepreneurs to accomplish a formal business plan early in the venture creation process enables them to engage in additional start-up behaviors that further the process of venture creation."

In 1991, days before his 30th birthday, Bob Bernstein quit his job at the *Tennessee Journal* even though did not really know what he wanted to do next. However, he had been thinking about opening a coffeehouse. Nashville did not really have a true coffeehouse. The closest thing to a coffeehouse was located in a strip mall bookstore.

Bob was not prepared for the shock of the transition from employee to entrepreneur. There were mornings when it was almost unbearable to get out of bed and face the day. He had been living on his savings but soon found the $10,000 he had set aside to fund whatever he decided was his next step in life was all that was remaining.

Bob hoped that with the $10,000 and his great idea, he would be off and running to start up his coffeehouse. Realistically, he knew he would need more funding, so he decided to approach some local banks. They were not interested in financing his start-up. Not only was he proposing a food service business, which bankers are not generally eager to lend to, but Bob had no experience. He mentioned his idea to a friend and former colleague from his journalism days, Chuck Kane. Chuck thought he knew someone, Brad Green, who might be interested in investing in Bob's idea.

Bob wrote an "offbeat" business plan to present to Brad Green, which mixed traditional writing with cartoons. Green liked the plan and introduced Bob to another person who would eventually become one of his investors. Bob found his next investors from his Sunday football hangout. In a casual conversation with two men from Bob's hometown of Chicago, he shared his vision for a coffeehouse. Both men were interested in the idea. Bob also began to find people who were willing to invest through his volunteer work in various community projects, friends he made at a sports bar where he went to watch Chicago Bears football games, and his personal network of friends in Nashville. As he talked about his idea with these people, he found that many were quite interested in the coffee business. Although they loved the idea, they were not interested in running the business. They all agreed that Nashville was in need of a "funky little coffeehouse."

Each of Bob's various investors contributed up to $9,000. Although individual investments were relatively small, Bob was humbled by the trust his investors placed in him and his new venture. Through various investors and his own $10,000, Bob raised the $80,000 he needed to open the first store.

After 20 years in business, Bernstein now has four coffee shops in various parts of the city. Although he has had many proposals to franchise his concept, he always insists that it is "too weird to franchise." The offbeat nature of the original business plan—cartoons and all—is apparent in the atmosphere and the culture of all of his stores. The work he put into his original business plan not only helped raise the money he needed to launch the business but also became the document that helped guide the development of the business and its culture.

1. Some entrepreneurs claim that creating a business plan is not necessary for launching a successful business venture. Do you agree? Explain.

2. What benefits do entrepreneurs who create business plans before launching their companies reap?

3. Suppose that a friend who has never taken a course in entrepreneurship tells you about a business that he or she is planning to launch. When you ask about a business plan, the response is, "Business plan? I don't have time to write a business plan! I know this business will succeed." Write a one-page response to your friend's comment.

Source: Cornwall, J. "Bongo Bob," United States Association for Small Business and Entrepreneurship, Proceedings, 2006, *www.bongojava.com.*

Description of Firm's Product or Service

An entrepreneur should describe the company's overall product line, giving an overview of how customers use its goods or services. Drawings, diagrams, and illustrations may be required if the product is highly technical. It is best to write product and service descriptions so that laypeople can understand them. A statement of a product's position in the product life cycle might also be helpful. An entrepreneur should include a summary of any patents, trademarks, or copyrights that protect the product or service from infringement by competitors.

One danger entrepreneurs must avoid in this part of the plan is the tendency to dwell on the features of their products or services. This problem is the result of the "fall-in-love-with-your-product" syndrome, which often afflicts inventors. Customers, lenders, and investors care less about how much work, genius, and creativity went into a product or service than about what it will do for them. The emphasis of this section should be on defining the benefits customers get by purchasing the company's products or services rather than on just a "nuts-and-bolts" description of the features of those products or services. A *feature* is a descriptive fact about a product or service (e.g., "an ergonomically designed, more comfortable handle"). A *benefit* is what the customer gains from the product or service feature (e.g., "fewer problems with carpal tunnel syndrome and increased productivity"). Benefits are at the core of the value proposition of the business model. Advertising legend Leo Burnett once said, "Don't tell the people how good you make the goods; tell them how good your goods make them."[19] This part of the plan must describe how a business will transform tangible product or service features into important but often intangible customer benefits—for example, lower energy bills, faster access to the Internet, less time paying monthly bills, greater flexibility in building floating structures, shorter time required to learn a foreign language, or others. Remember: *Customers buy benefits, not product or service features*. Table 8.2 offers an easy exercise designed to help entrepreneurs translate their products' or services' features into meaningful customer benefits.

ENTREPRENEURIAL PROFILE: Ami Kassar: Multifunding Ami Kassar came up with the idea for his Philadelphia-based business, Multifunding, from his experience in the small business credit industry. He had seen the difficulty small businesses have finding financing, yet he knew that there was financing available for many of these businesses if they could just get connected to the right source. Although the value of the service was apparent from the beginning, it took Kassar several pivots of his business model before he offered small business customers what they wanted in

---

TABLE 8.2 Transforming Features into Meaningful Benefits

For many entrepreneurs, there's a big gap between what a business is selling and what its customers are buying. The following exercise is designed to eliminate that gap.

First, develop a list of the features your company's product or service offers. List as many as you can think of, which may be 25 or more. Consider features that relate to price, performance, convenience, location, customer service, delivery, reputation, reliability, quality, features, and other aspects.

The next step is to group features that have similar themes together by circling them in the same color ink. Then translate those groups of features into specific benefits to your customers by addressing the question "What's in it for me?" from the customer's perspective. (Note: It usually is a good idea to ask actual customers why they buy from you. They usually give reasons that you never thought of.) As many as six or eight product or service (or even company) features may translate into a single customer benefit, such as saving money or time or making life safer. Don't ignore intangible benefits, such as increased status; they can be more important than tangible benefits.

Finally, combine all of the benefits you identify into a single benefit statement. Use this statement as a key point in your business plan and to guide your company's marketing strategy.

Product or Service *Features*	Product or Service *Benefits*

Benefit Statement:

Source: Based on Kim T. Gordon, "Position for Profits," *Business Start-Ups*, February 1998, pp. 18–20.

the way they wanted it. "There is a fundamental difference between a vision—and a business model," says Kassar. "The core model of how we make our money has not changed from day one—we are constantly testing it and looking for ways to improve it, evolve it, and grow it. Every few months, an entrepreneur should take a cold shower and take the time to look in the rearview mirror and come up with some new things to test." After getting a profile of a small business and an assessment of its financing needs, Multifunding puts together a report that gives the business owner various funding options. Multifunding gets paid only when the financing is completed, taking a small percentage of the approved amount of financing. Multifunding helped seven small businesses find funding in its first year and 48 in its second year. As part of the process of testing and refining his business model, Kassar began looking into new sources of revenue streams during the company's third year.[20]

Business and Industry Profile

If an entrepreneur intends to use the plan to raise funding, he or she should include a section that acquaints lenders and investors with the industry in which a company competes. Industry data, such as key trends or emerging developments within the industry, market size and its growth or decline, and the relative economic and competitive strength of the major firms in the industry, set the stage for a better understanding of the viability of a new business. Strategic issues, such as ease of market entry and exit, the ability to achieve economies of scale or scope, and the existence of cyclical or seasonal economic trends, further help readers evaluate the new venture. This part of the plan also should describe significant industry trends and key success factors as well as an overall outlook for its future. Information about the evolution of the industry helps the reader comprehend its competitive dynamics. A useful resource of industry and economic information is the *Summary of Commentary on Current Economic Conditions*, more commonly known as the *Beige Book*. Published eight times a year by the Federal Reserve, the *Beige Book* provides detailed statistics and trends in key business sectors and in the overall economy. It offers valuable information on topics ranging from tourism and housing starts to consumer spending and wage rates. Entrepreneurs can find this wealth of information at their fingertips on the Web at *www .federalreserve.gov/monetarypolicy/beigebook/*. This section should cover all of the relevant information the entrepreneur uncovered during the market and industry feasibility analysis.

Goals and Objectives

This section contains a statement of the business goals and the more specific objectives that the entrepreneur seeks to achieve. Together, goals and objectives define the targets that the company strives for. **Goals** are broad, long-range statements of what a company plans to achieve in the future that guide its overall direction. **Objectives** are short-term, specific performance targets that are measurable, attainable, and controllable. Every objective should include a technique for measuring progress toward its accomplishment. Recall from Chapter 4 that to be meaningful, an objective must have a time frame for achievement. Every objective should be tied to a business goal, which, in turn, should be tied to the company's overall mission. In other words, accomplishing each objective should move a business closer to achieving its goals, which, in turn, should move it closer to its mission.

Business Strategy

An even more important part of the business plan is the owner's view of the strategy needed to meet—and beat—the competition. In the previous section, an entrepreneur defined where he or she wants to take the business by establishing goals and objectives. This section addresses the question of how to get there—business strategy. Here, an entrepreneur explains how he or she plans to gain a competitive edge in the market and how his or her value proposition sets the business apart from the competition. A key component of this section is defining what makes the company unique in the eyes of its customers. One of the quickest routes to business failure is trying to sell "me-too" products or services that offer customers nothing newer, better, bigger, faster, or different. The foundation for this part of the business plan comes from the material presented in Chapter 4.

This section of the business plan should outline the methods the company will use to meet the key success factors in the industry. If, for example, making sales to repeat customers is critical to success, an entrepreneur must devise a plan of action for achieving a customer retention rate that exceeds that of existing companies in the market.

Competitor Analysis

An entrepreneur should describe the new venture's competition and how his or her chosen business strategy will allow him or her to effectively compete with key competitors. Failing to assess competitors realistically makes entrepreneurs appear to be poorly prepared, naive, or dishonest, especially to potential lenders and investors. The plan should include an analysis of each significant competitor and how well they are meeting the important criteria that target customers are currently using to make their decisions among the various companies. Entrepreneurs who believe they have no competitors are only fooling themselves and are raising a red flag to potential lenders and investors. Gathering information on competitors' market shares, products, and strategies is usually not difficult. Trade associations, customers, industry journals, marketing representatives, and sales literature are valuable sources of data. This section of the plan should focus on demonstrating that the entrepreneur's company has an advantage over its competitors and address these questions:

- Who are the company's key competitors?
- What are their strengths and weaknesses?
- What are their strategies?
- What images do they have in the marketplace?
- How successful are they?
- What distinguishes the entrepreneur's product or service from others already on the market, and how will these differences produce a competitive edge?

Firsthand competitor research is particularly valuable.

Marketing Strategy

One of the most important tasks a business plan must fulfill is proving that a viable market exists for a company's goods or services. The business modeling process identified and described a company's target customers and their characteristics and habits. Defining the target audience and its potential is one of the most important—and most challenging—parts of the business planning process.

Proving that a profitable market exists involves two steps: showing customer interest and documenting market claims, both of which were part of the business modeling process.

SHOWING CUSTOMER INTEREST An important element of any business plan is showing how a company's product or service provides a customer benefit or solves a customer problem. Entrepreneurs must be able to prove that their target customers actually need or want their goods or services and are willing to pay for them. This is why including customer feedback to validate the business model is so important. Venture capitalist Kathryn Gould, who has reviewed thousands of business plans, says that she looks for plans that focus on "target customers with a compelling reason to buy. The product must be a 'must-have.' "[21]

Proving that a viable market exists for a product or service is relatively straightforward for a company already in business but can be quite difficult for an entrepreneur with only an idea. In this case, the key is to get primary customer data. The feasibility analyses and the process of validating the value proposition during the development of the business model provide real data from real customers. Two of the most reliable techniques include building a prototype of a product so that customers can see how it works and producing a small number of products that customers can actually use. The entrepreneur could sell the product to several customers, perhaps at a discount on the condition that they provide evaluations of it. Doing so proves that there are potential customers for the product and allows customers to experience the product in operation. Getting a product into customers' hands is also an excellent way to get valuable feedback that can lead to significant design improvements and increased sales down the road. Integrating this type of primary data into a business plan demonstrates that the business has a good chance of success.

ENTREPRENEURIAL PROFILE: Charley Moore: RocketLawyer Charley Moore originally launched RocketLawyer, a Web site that targets entrepreneurs with a variety of legal documents ranging from basic contracts and corporate documents to noncompete agreements and Web site design contracts, using a fee-per-downloaded-document business model. After several months in business, the site's analytics showed Moore that even though the site was attracting large numbers of visitors, its abandonment rate was stifling the company's revenue. "We were charging [our customers] a couple of hundred dollars less than other sites or the cost of hiring a lawyer," says Moore, "but we were asking them to pay for every document they wanted to download. It was still a lot of money for most folks." Based on feedback from customers, Moore changed the business model that he had developed in his business plan. "We decided to go with a 'first-one-free' model to accommodate customers who wanted to try our service," he explains. RocketLawyer also added a $40 monthly membership that gives customers unlimited access to legal documents and to online document storage. The revamped business model produced results immediately. Sales increased from $1 million to $5 million in just one year, and the number of RocketLawyer customers climbed from 150,000 to 900,000.[22]

DOCUMENTING MARKET CLAIMS Too many business plans rely on vague generalizations, such as "This market is so huge that if we get just 1 percent of it, we will break even in eight months." Statements such as these usually reflect nothing more than an entrepreneur's unbridled optimism and in most cases are quite unrealistic! In *The Art of the Start*, entrepreneur and venture capitalist Guy Kawasaki calls this the Chinese Soda Lie: "If just 1 percent of the people in China drink our soda, we will be more successful than any company in the history of mankind."[23] The problems with this reasoning are (1) few markets, especially the niche markets that small businesses often pursue, are as large as that, and (2) capturing 1 percent of a big market is extremely difficult to do, especially for a small company. Capturing a large share of a small, well-defined niche market is much more realistic for a small company than is winning a small share of a huge market.

Entrepreneurs must support claims of market size and growth rates with *facts*, and that requires market research. Results of market surveys, customer questionnaires, and demographic studies developed in the feasibility analyses and business modeling steps in the business planning process lend credibility to an entrepreneur's frequently optimistic sales projections contained within the formal business plan. (Refer to the market research techniques and resources in Chapter 9.) Quantitative market data are important because they form the basis for all of the company's financial projections in the business plan. Fortunately, entrepreneurs who follow the business planning process will already have much of this data from their feasibility analyses and from building and testing their business models.

One of the main purposes of the marketing section of the plan is to lay the foundation for the financial forecasts that follow. Sales, profit, and cash forecasts must be founded on more than wishful thinking. An effective market analysis should address the items discussed in the following sections in detail based on the framework developed in the business model.

Target Market   Who are the company's target customers? How many of them are in the company's trading area? What are their characteristics (age, gender, educational level, income, and others)? What do they buy? Why do they buy? When do they buy? What expectations do they have about the product or service? Will the business focus on a niche? How does the company seek to position itself in the market(s) it will pursue? Knowing my customers' needs, wants, and habits, what should be the basis for differentiating my business in their minds?

Advertising and Promotion   Only after entrepreneurs understand their companies' target markets can they design a promotion and advertising campaign to reach those customers most effectively and efficiently. Which media are most effective in reaching the target market? How will they be used? How much will the promotional campaign cost? How will the promotional campaign position the company's products or services? How can the company benefit from publicity? How large is the company's promotional budget?

Market Size and Trends   Assessing the size of the market is a critical step. How large is the potential market? Is it growing or shrinking? Why? Are customers' needs changing? Are sales seasonal? Is demand tied to another product or service?

ENTREPRENEURIAL PROFILE: Ludwick Marishane: DryBath Clean water is an extremely scarce resource in many parts of the globe. Water shortages, which are common throughout much of the developing world, making bathing difficult for people who live in regions with water shortages. Ludwick Marishane of Cape Town, South Africa, a 21-year-old student, was named the 2011 Global Student Entrepreneur of the Year at the Global Student Entrepreneur Awards for his product DryBath, the only non–water-based bath substitute lotion for the whole body. "I invented it to benefit people from the poorest communities in the world and also for people in the developed world," said Marishane. "For people with limited water supplies, DryBath provides an affordable tool to achieve lifesaving personal hygiene."[24]

Location For many businesses, choosing the right location is a key success factor. For retailers, wholesalers, and service companies, the best location usually is one that is most convenient to their target customers. Using census data and other market research, entrepreneurs can determine the sites with the greatest concentrations of their customers and locate there. Which sites put the company in the path of its target customers? Maps that show customer concentrations (available from census maps and other sources), traffic counts, the number of customers using a particular train station and when, and other similar types of information provide evidence that a solid and sizable customer base exists. Do zoning regulations restrict the use of a site? For manufacturers, the location issue often centers on finding a site near their key raw materials or near their primary customers. Using demographic reports and market research to screen potential sites takes the guesswork out of choosing the "right" location for a business. We will discuss the location decision in more detail in Chapter 18.

Pricing What does the product or service cost to produce or deliver? Before opening a restaurant, for example, an entrepreneur should know *exactly* what it will cost to produce each item on the menu. Failing to know the total cost (including the cost of the food as well as labor, rent, advertising, and other indirect costs) of putting a plate in front of a customer is a recipe for failure. As we will discover in Chapter 11, cost is just one part of the pricing equation. Another significant factor to consider is the image a company is trying to create in the market. "Price really is more of a marketing tool than it is a vehicle for cost recovery," says Peter Meyer, author of *Creating and Dominating New Markets*. "People will pay more for a high value product or solution, so be sure to research your [product's or service's] total value."[25] Pricing helps communicate and reinforce key elements of the value proposition, such as quality and value.

Other pricing issues that a plan should address include the following: What is the company's overall pricing strategy? Will the planned price support the company's strategy and desired image? Given the company's cost structure, will the price produce a profit? How does the planned price compare to those of similar products or services? Are customers willing to pay it? What price tiers exist in the market? How sensitive are customers to price changes? Will the business sell to customers on credit? Will it accept credit cards? Will the company offer discounts? All of these questions help develop the revenue forecasts in the business plan. Remember that revenues are calculated with a simple formula: price times quantity. Therefore, understanding the proper pricing strategy is half the battle of developing accurate revenue forecasts.

Distribution Using the distribution channels defined in the business model, this portion of the plan should describe the specific channels of distribution that the business will use (the Internet, direct mail, in-house sales force, sales agents, retailers, or others) to distribute its products and services. Will distribution be extensive, selective, or exclusive? What is the average sale? How large will the sales staff be? How will the company compensate its sales force? What are the incentives for salespeople? How many sales calls does it take to close a sale? What can the company do to make it as easy as possible for customers to buy?

DESCRIPTION OF THE MANAGEMENT TEAM The most important factor in the success of a business venture is its management, and financial officers and investors weight heavily the ability and experience of a company's managers in financing decisions. A plan should include the résumés of managers, key directors, and any person with at least 20 percent ownership in the company. This is the section of the plan in which entrepreneurs have the chance to sell the qualifications and the experience of their management team. Remember that *lenders and investors prefer experienced managers.* Ideally, they look for managers with at least two years of operating experience in the industry they are targeting. In a technology business, investors are looking for partners that have both management and technology expertise.

A résumé should summarize each individual's education, work history (emphasizing managerial responsibilities and duties), and relevant business experience. Lenders and investors look for the experience, talent, and integrity of the people who will breathe life into the plan. This portion of the plan should show that the company has the right people organized in the right fashion for success. One experienced private investor advises entrepreneurs to remember the following:

- Ideas and products don't succeed; people do. Show the strength of your management team. A top-notch management team with a variety of proven skills is crucial. "We're investing in people, not companies," says J. Skyler Fernandes, a New York venture capitalist at Centripetal Capital Partners. "We're not investing in ideas—people always think if you have a great idea you can always raise venture capital. In the end we are investing in people who can execute."[26]

- Show the strengths of key employees and how you will retain them. Most small companies cannot pay salaries that match those at large businesses, but stock options and other incentives can improve employee retention.

- Enhance the strength of the management team with a capable, qualified board of advisers. A board of directors or advisers consisting of industry experts lends credibility and can complement the skills of the management team.

PLAN OF OPERATION To complete the description of the business, an entrepreneur should construct an organization chart identifying the business's key positions and the people who occupy them. Assembling a management team with the right stuff is difficult, but keeping it together until the company is established can be even harder. Thus, entrepreneurs should describe briefly the steps taken to encourage important officers to remain with the company. Employment contracts, shares of ownership, and perks are commonly used to keep and motivate key employees. A plan of operation should also begin to detail how the business operates, including space requirements, inventory management if applicable, staffing plans, and accounting processes and policies.

Finally, a description of the form of ownership (e.g., sole proprietorship, partnership, C corporation, S corporation, or limited liability company) and of any leases, contracts, and other relevant agreements pertaining to the operation is helpful.

PRO FORMA (PROJECTED) FINANCIAL STATEMENTS One of the most important sections of the business plan is an outline of the proposed company's financial statements—the "dollars and cents" of the proposed venture. In fact, one survey found that 74 percent of bankers say that financial documentation is the most important aspect of a business plan for entrepreneurs who are seeking loans.[27] For an existing business, lenders and investors use past financial statements to judge the health of the company and its ability to repay loans or generate adequate returns; therefore, an owner should supply copies of the company's financial statements from the past three years. Ideally, these statements should be audited by a certified public accountant because most financial institutions prefer that extra reliability, although a financial review of the statements by an accountant sometimes may be acceptable.

Whether assembling a plan for an existing business or for a start-up, an entrepreneur should carefully prepare projected (pro forma) financial statements for the operation for the next year using past operating data, published statistics, and research to derive forecasts of the income statement, balance sheet, cash forecast (always!), and a schedule of planned capital expenditures. (You will learn about creating projected financial statements in Chapter 14.) Although including only most likely forecasts in the business plan is acceptable, entrepreneurs also should develop forecasts for pessimistic and optimistic conditions that reflect the uncertainty of the future in case potential lenders and investors ask for them.

It is essential for financial forecasts be realistic. Entrepreneurs must avoid the tendency to "fudge the numbers" just to make their businesses look good. Experienced lenders and investors can detect unrealistic forecasts easily. In fact, some venture capitalists automatically discount an entrepreneur's financial projections by as much as 50 percent. One experienced angel investor says that when looking at the financial forecasts compiled by an entrepreneur, he always "doubles the start-up costs and triples the time it will take to launch."

After completing these forecasts, an entrepreneur should perform a break-even analysis for the business. The break-even point is critical for an entrepreneurial venture because it signals the point at which the business is able to sustain itself through cash generated by operations and should not need any additional start-up capital. It is also the point when the entrepreneur is able to get paid by the business!

It is also important to include a statement of the *assumptions* on which these financial projections are based. Potential lenders and investors want to know how an entrepreneur derived forecasts for sales, cost of goods sold, operating expenses, accounts receivable, collections, accounts payable, inventory, taxes, and other items. Spelling out realistic assumptions gives a plan more credibility and reduces the tendency to include overly optimistic estimates of sales growth and profit margins. Greg Martin, a partner in the venture capital company Redpoint Ventures, says, "I have problems with start-ups making unrealistic assumptions—how much money they need or how quickly they can ramp up revenue. Those can really kill a deal for me."[28]

In addition to providing valuable information to potential lenders and investors, projected financial statements help entrepreneurs run their businesses more effectively and more efficiently after start-up. They establish important targets for financial performance and make it easier for an entrepreneur to maintain control over routine expenses and capital expenditures. Entrepreneurs can use their projections to construct financial dashboards to track the progress of the business and assess how well the actual outcomes match the key assumptions made in the business plan.

THE LOAN OR INVESTMENT PROPOSAL The loan or investment proposal section of the business plan should state the purpose of the financing, the amount requested, and the plans for repayment or, in the case of investors, an attractive exit strategy. When describing the purpose of the loan or investment, an entrepreneur must specify the planned use of the funds. Entrepreneurs should state the precise amount requested and include relevant supporting data, such as vendor estimates of costs. The proposal should include all sources of funding for the business from all intended sources, including money the entrepreneur is investing in the business. Most bankers and investors want to see evidence that the entrepreneur is willing to "put skin in the game," putting some of his or her own money at risk in the venture.

Another important element of the loan or investment proposal is the repayment schedule or exit strategy. A lender's main consideration when granting a loan is the reassurance that the applicant will repay, whereas an investor's major concern is earning a satisfactory rate of return. Financial projections must reflect a company's ability to repay loans and produce adequate returns. Without this proof, a request for funding stands little chance of being approved. Entrepreneurs must provide tangible evidence that shows their ability to repay loans or to generate attractive returns. Developing an exit strategy, such as the option to cash out through an acquisition or a public offering, is important. Including examples of other firms in the same industry that have already exited proves to investors that a viable path for them to exit the business and realize a return on their investments exists.

Finally, an entrepreneur should include a realistic timetable for implementing the proposed plan. This should include a schedule showing the estimated start-up date for the project and noting all significant milestones along the way.

A business plan must present an honest assessment of the risks facing the new venture. Evaluating risk in a business plan requires an entrepreneur to walk a fine line, however. Dwelling too much on everything that can go wrong discourages potential lenders and investors from financing the venture. Ignoring the project's risks makes those who evaluate the plan see the entrepreneur as either naive, dishonest, or unprepared. The best strategy is to identify the most significant risks the venture faces and then to describe the plans the entrepreneur has developed to avoid them altogether or to overcome the negative outcome if the event does occur. Figure 8.6 explains how two simple diagrams communicate effectively to investors both the risks and the rewards of a business venture.

Visualizing a Venture's Risks and Rewards

There is a difference between a *working* business plan—the one the entrepreneur is using to guide her business—and the *presentation* business plan—the one he or she is using to attract capital.

FIGURE 8.6

Visualizing a Venture's Risks and Rewards

(a) This diagram shows the amount of money an entrepreneur needs to launch the business, the time required to reach the point of positive cash flow, and the anticipated amount of the payoff.

(b) The second diagram shows investors the range of possible returns and the probability of achieving them. This diagram portrays what investors intuitively understand: Most companies either fail big or achieve solid success.

Source: Based on William A. Sahlman, "How to Write a Great Business Plan," *Harvard Business Review,* July/August 1997, pp. 98–108.

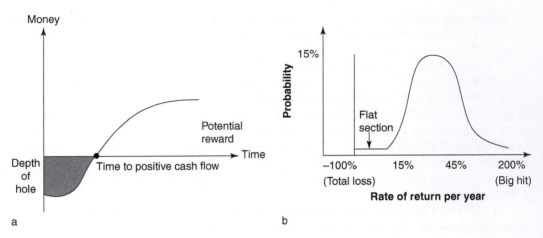

Although coffee rings and penciled-in changes in a working plan don't matter (in fact, they're a *good* sign that the entrepreneur is actually using the plan), they have no place in a plan going to someone outside the company. A plan is usually the tool that an entrepreneur uses to make a first impression on potential lenders and investors. To make sure that impression is a favorable one, an entrepreneur should follow these tips:

- Realize that first impressions are crucial. Make sure the plan has an attractive (but not an expensive) cover.

- Make sure the plan is free of spelling and grammatical errors and "typos." It is a professional document and should look like one.

- Make it visually appealing. Use color charts, figures, and diagrams to illustrate key points. Don't get carried away, however, and end up with a "comic book" plan.

- Include a table of contents with page numbers to allow readers to navigate the plan easily. Reviewers should be able to look through a plan and quickly locate the sections they want to see.

- Make it interesting. Boring plans seldom get read; a good plan tells an interesting story.

- A plan must prove that the business will make money. In one survey of lenders, investors, and financial advisers, 81 percent said that, first and foremost, a plan should prove that a venture will earn a profit.[29] Start-ups do not necessarily have to be profitable immediately, but sooner or later (preferably sooner), they must make money.

- Use computer spreadsheets to generate a set of realistic financial forecasts. They allow entrepreneurs to perform valuable "what if" (sensitivity) analysis in just seconds.

- *Always* include cash flow projections. Entrepreneurs sometimes focus excessively on their proposed venture's profit forecasts and ignore cash flow projections. Although profitability is important, lenders and investors are much more interested in cash flow because they know that's where the money to pay them back or to cash them out comes from.

- The ideal plan is "crisp," long enough to say what it should but not so long that it is a chore to read.

- Tell the truth. Absolute honesty is always critical when preparing a business plan.

Business plans are forecasts about the future that an entrepreneur plans to create, something that one expert compares to "taking a picture of the unknown," which is a challenging feat! As uncertain and difficult to predict as the future may be, an entrepreneur who launches a business without a plan arguing that "trying to forecast the future is pointless" is misguided. In the *Harvard Business Review,* William Sahlman says that "the best business plans . . . are like movies of the future. They show the people, the opportunity, and the context from multiple angles. They offer a plausible, coherent story of what lies ahead. They unfold the possibilities of action and reaction."[30] That's the kind of "movie" an entrepreneur should strive to create in a plan.

What Lenders and Investors Look for in a Business Plan

To increase their chances of success when using their business plans to attract capital, entrepreneurs must be aware of the criteria lenders and investors use to evaluate the creditworthiness of entrepreneurs seeking financing. Lenders and investors refer to these criteria as the five Cs of credit: capital, capacity, collateral, character, and conditions.

6.

Explain the "five Cs of credit" and why they are important to potential lenders and investors.

Capital

A small business must have a stable capital base before any lender will grant a loan. Otherwise the lender would be making, in effect, a capital investment in the business. Most lenders refuse to make loans that are capital investments because the potential for return on the investment is limited strictly to the interest on the loan, and the potential loss would probably exceed the reward. In fact, the most common reasons that banks give for rejecting small business loan applications are undercapitalization or too much debt. Investors also want to make sure that entrepreneurs have invested enough of their own money into the business to survive the tenuous start-up period. Lenders and investors see capital formation as a risk-sharing strategy with entrepreneurs.

Capacity

A synonym for *capacity* is *cash flow*. Lenders and investors must be convinced of a company's ability to meet its regular financial obligations and to repay bank loan, and that takes cash. More small businesses fail from lack of cash than from lack of profit. It is possible for a company to be earning a profit and still run out of cash. Lenders expect a business to pass the test of liquidity; they study closely a small company's cash flow position to decide whether it has the capacity required to succeed. Most lenders have become extremely cautious when evaluating cash flow since the financial crisis of 2008.

Collateral

Collateral includes any assets an entrepreneur pledges to a lender as security for repayment of the loan. If an entrepreneur defaults on the loan, the bank has the right to sell the collateral and use the proceeds to satisfy the loan. Typically, lenders make very few unsecured loans (those not backed by collateral) to business start-ups. Bankers view an entrepreneur's willingness to pledge collateral (personal or business assets) as an indication of dedication to making the venture a success. Bankers will always look first at the entrepreneur's personal assets because that is the simplest way for them to get repaid on a loan if the business fails. Business assets are their last resort because selling a company's inventory, equipment, and buildings to repay loans is not an easy or an effective means of repayment for the bank.

Character

Before putting money into a small business, lenders and investors must be satisfied with the owner's character. An evaluation of character frequently is based on intangible factors such as honesty, competence, polish, determination, knowledge, experience, and ability. Although the qualities judged are abstract, this evaluation plays a critical role in a lender's or investor's decision. Banks have also begun to use the comments that potential borrowers post on social networking sites such as Facebook and Twitter to assess character. Posts and links that are not professional hurt an entrepreneur's character and reduce the likelihood of an entrepreneur securing a loan.[31]

Lenders and investors know that most small businesses fail because of poor management, and they try to avoid extending loans to high-risk entrepreneurs. Preparing a solid business plan and a polished presentation can go far in convincing potential lenders and investors of an entrepreneur's ability to manage a company successfully.

Conditions

The conditions surrounding a loan request also affect the owner's chance of receiving funds. Banks consider factors relating to the business operation, such as potential growth in the market, competition, location, form of ownership, and loan purpose. Again, the best way to provide this relevant information is in a business plan. Another important condition influencing the banker's decision is the shape of the overall economy, including interest rates, the inflation rate, and the demand for money. Although these factors are beyond an entrepreneur's control, they still are an important component in a lender's decision. Overall conditions have not been as favorable for bank loans since the financial crisis in 2008, which has made getting business loans more difficult for all small businesses.

Jerell Harris, President of QuickMed Collect, pitching his company at a business plan competition.
Source: Belmont University.

7. _____

Understand the keys to making an effective business plan presentation.

The higher a small business scores on these five Cs, the greater its chance will be of receiving a loan or an investment.

The Pitch: Making the Business Plan Presentation

Lenders and investors are impressed by entrepreneurs who are informed and prepared when requesting a loan or investment. When entrepreneurs try to secure funding from lenders or investors, the written business plan most often precedes the opportunity to meet face-to-face. The written plan must first pass muster before an entrepreneur gets the opportunity to present the plan in person. Usually, the time for presenting a business opportunity is short, often no more than just a few minutes. (When presenting a plan to a venture capital forum, the allotted time is usually less than 20 minutes and rarely more than 30.) When the opportunity arises, an entrepreneur must be well prepared. It is important to rehearse, rehearse, and then rehearse some more. It is a mistake to begin by leading the audience into a long-winded explanation about the technology on which the product or service is based. Within minutes most of the audience will be lost, and so is any chance the entrepreneur has of obtaining the necessary financing for the new venture. A business plan presentation should cover five basic areas:

- Your company and its products and services. The presentation should answer in simple terms the first question that every potential lender and investor has: What does your company do?

- The problem to be solved, preferably told in a personal way through a compelling story. Is it eliminating the time, expense, and anxiety of waiting for the results of medical tests with a device that instantly reads blood samples or making hearing aids more effective at filtering out background noise while enhancing the dominant sound for the user?

- A description (again in simple terms) of your company's solution to the problem. Ideally, the solution your company has developed is unique and serves as the foundation of your company's competitive edge in the marketplace.

- Your company's business model. This part of the presentation explains how your company makes money and includes figures such as revenue per sale, expected gross profit and net profit margins, and other relevant statistics. This is your opportunity to show lenders and investors how your company will produce an attractive payback or payoff.

- Your company's competitive edge. Your presentation should identify clearly the factors that set your company apart from the competition.

ENTREPRENEURIAL PROFILE: NuMat Technologies NuMat Technologies recently won two of the most prestigious business plan competitions in the United States. NuMat develops new materials to enhance and make more practical the use of alternative fuels. Their team was made of two chemists (one of whom is an associate professor) and two JD/MBAs from Northwestern University. The team won the grand prizes at both the University of Texas Global Venture Labs Investment Competition and Rice University's business plan competition. "The NuMat pitch which was sensational," say Tim Berry, president and founder of Palo Alto Software. "The key was explaining the science just enough to be credible, focusing on the business, and keeping it clear and flowing from point to point."[32]

No matter how good a written business plan is, entrepreneurs who stumble through the presentation will lose the deal. Entrepreneurs who are successful raising the capital their companies need to grow have solid business plans and make convincing presentations of them. Some helpful tips for making a business plan presentation to potential lenders and investors include the following:

- Prepare. Good presenters invest in preparing their presentations and knowing the points they want to get across to their audiences.

- Practice your delivery and then practice some more.

- Demonstrate enthusiasm about the business but don't be overemotional. Be genuine and be yourself.

- Focus on communicating the dynamic opportunity your idea offers and how you plan to capitalize on it. Fight the temptation to launch immediately into a lengthy discourse about the details of your product or service or how much work it took to develop it. Otherwise, you'll never have the chance to describe the details to lenders and investors.

- Hook investors quickly with an up-front explanation of the new venture, its opportunities, and the anticipated benefits to them. For some businesses, a story of its impact can be a good hook to start a presentation.

- Use visual aids. They make it easier for people to follow your presentation. Don't make the mistake of making the visuals the "star" of the presentation, however. Visual aids should punctuate your spoken message and focus the audience's attention on the plan's key points.

- Follow Guy Kawasaki's 10/20/30 rule for PowerPoint presentations. Use 10 slides that you can cover in 20 minutes and use fonts no smaller than 30 points to ensure that you do not try to put too many words on each slide.[33]

- Explain how your company's products or services solve some problem and emphasize the factors that make your company unique.

- Offer proof. Integrate relevant facts into your presentation to prove your plan's claims, customers' satisfaction with your company's products or services, and its profit potential.

- Hit the highlights. Specific questions will bring out the details later. Don't get caught up in too much detail in early meetings with lenders and investors.

- Keep the presentation "crisp" just like your business plan. Otherwise, says one experienced investor, "Information that might have caused an investor to bite gets lost in the endless drone."[34]

- Avoid the use of technical terms that will likely be above most of the audience. Do at least one rehearsal before someone who has no special technical training. Tell him or her to stop you anytime he or she does not understand what you are talking about. When this occurs (and it likely will), rewrite that portion of your presentation.

- Remember that every potential lender and investor you talk to is thinking "What's in it for me?" Be sure to answer that question in your presentation.

- Close by reinforcing the potential of the opportunity. Be sure you have sold the benefits the investors will realize when the business succeeds.

- Be prepared for questions. In many cases, there is seldom time for a long question-and-answer session, but interested investors may want to get you aside to discuss the details of the plan.

- Anticipate the questions the audience is most likely to ask and prepare for them in advance.

- Be sensitive to the issues that are most important to lenders and investors by reading the pattern of their questions. Focus your answers accordingly. For instance, some investors may be interested in the quality of the management team, whereas others are more interested in marketing strategies. Be prepared to offer details on either.

- Follow up with every investor to whom you make a presentation. Don't sit back and wait; be proactive. They have what you need—investment capital. Demonstrate that you have confidence in your plan and have the initiative necessary to run a business successfully.

Ross Hill, founder of PhotoBoothCo, pitching his company at a business plan competition.
Source: Belmont University.

Conclusion

Although there is no guarantee of success when launching a business, the best way to protect against failure is to follow the business planning process that includes feasibility analysis, building and testing a business model, and crafting a strong business plan. A well-developed business model ensures that all of the working parts of a business will be in place and will mesh smoothly. The business model serves as the foundation for a good plan, which serves as a strategic compass that keeps a business on course as it travels into an uncertain future. In addition, a solid plan is essential to raising the capital needed to start a business; lenders and investors demand it. "There may be no easier way for an entrepreneur to sabotage his or her request for capital than by failing to produce a comprehensive, well-researched, and, above all, credible business plan," says one small business expert.[35] Of course, following the three-phase business planning process is just one step along the path to launching a business. Creating a successful business requires entrepreneurs to put the plan into action. The remaining chapters in this book focus on putting your business model and plan to work.

Suggested Business Plan Elements

Although every company's business plan will be unique, reflecting its individual circumstances, certain elements are universal. The following outline summarizes these components.

I. Executive Summary (not to exceed one page)
   A. Company name, address, and phone number
   B. Name(s), addresses, and phone number(s) of all key people
   C. Brief description of the business, its products and services, the customer problems they solve, and the company's competitive advantage
   D. Brief overview of the market for your products and services
   E. Brief overview of the strategies that will make your company successful
   F. Brief description of the managerial and technical experience of key people
   G. Brief statement of the financial request and how the money will be used
   H. Charts or tables showing highlights of financial forecasts

II. Vision and Mission Statement
   A. Entrepreneur's vision for the company
   B. "What business are we in?"
   C. Values and principles on which the business stands
   D. What makes the business unique? What is the source of its competitive advantage?

III. Company History (for existing businesses only)
   A. Company founding
   B. Financial and operational highlights
   C. Significant achievements

IV. Company Products and Services
   A. Description
      1. Product or service features
      2. Customer benefits
      3. Warranties and guarantees
      4. Unique selling proposition (USP)
   B. Patent or trademark protection
   C. Description of production process (if applicable)
      1. Raw materials
      2. Costs
      3. Key suppliers
      4. Lead times
   D. Future product or service offerings

   V. Industry Profile and Overview

      A. Industry analysis

         1. Industry background and overview

         2. Significant trends

         3. Growth rate

         4. Barriers to entry and exit

         5. Key success factors in the industry

      B. Outlook for the future

      C. Stage of growth (start-up, growth, maturity)

  VI. Competitor Analysis

      A. Existing competitors

         1. Who are they? Create a competitive profile matrix

         2. Strengths

         3. Weaknesses

      B. Potential competitors: Companies that might enter the market

         1. Who are they?

         2. Impact on your business if they enter

 VII. Business Strategy

      A. Desired image and position in market

      B. Company goals and objectives

         1. Operational

         2. Financial

         3. Other

      C. SWOT analysis

         1. Strengths

         2. Weaknesses

         3. Opportunities

         4. Threats

      D. Competitive strategy

         1. Cost leadership

         2. Differentiation

         3. Focus

VIII. Marketing Strategy

      A. Target market

         1. Problem to be solved or benefit to be offered

         2. Demographic profile

         3. Other significant customer characteristics

      B. Customers' motivation to buy

      C. Market size and trends

         1. How large is the market?

         2. Is it growing or shrinking? How fast?

      D. Personal selling efforts

         1. Sales force size, recruitment, and training

         2. Sales force compensation

         3. Number of calls per sale

         4. Amount of average sale

    E. Advertising and promotion

       1. Media used—reader, viewer, listener profiles

       2. Media costs

       3. Frequency of usage

       4. Plans for generating publicity

    F. Pricing

       1. Cost structure

          a. Fixed

          b. Variable

       2. Desired image in market

       3. Comparison against competitors' prices

       4. Discounts

       5. Gross profit margin

    G. Distribution strategy (if applicable)

       1. Channels of distribution used

       2. Sales techniques and incentives for intermediaries

    H. Test market results

       1. Surveys

       2. Customer feedback on prototypes

       3. Focus groups

IX. Location and Layout

    A. Location

       1. Demographic analysis of location versus target customer profile

       2. Traffic count

       3. Lease/rental rates

       4. Labor needs and supply

       5. Wage rates

    B. Layout

       1. Size requirements

       2. Americans with Disabilities Act compliance

       3. Ergonomic issues

       4. Layout plan (suitable for an appendix)

X. Description of management team

    A. Key managers and employees

       1. Their backgrounds

       2. Experience, skills, and know-how they bring to the company

    B. Résumés of key managers and employees (suitable for an appendix)

    C. Future additions to management team

    D. Board of directors or advisers

XI. Plan of Operation

    A. Form of ownership chosen and reasoning

    B. Company structure (organization chart)

    C. Decision-making authority

    D. Compensation and benefits packages

    E. Staffing plans

XII. Financial Forecasts (suitable for an appendix)

  A. Key assumptions

  B. Financial statements (year 1 by month, years 2 and 3 by quarter)

    1. Income statement

    2. Balance sheet

    3. Cash flow statement

  C. Break-even analysis

  D. Ratio analysis with comparison to industry standards (most applicable to existing businesses)

XIII. Loan or Investment Proposal

  A. Amount requested

  B. Purpose and uses of funds

  C. Repayment or "cash out" schedule (exit strategy)

  D. Timetable for implementing plan and launching the business

XIV. Appendices (supporting documentation, including market research, financial statements, organization charts, résumés, and other items)

Chapter Review

1. Present the steps involved in conducting a feasibility analysis.
- A feasibility analysis consists of three interrelated components: an industry and market feasibility analysis, a product or service feasibility analysis, and a financial feasibility analysis. The goal of the feasibility analysis is to determine whether an entrepreneur's idea is a viable foundation for creating a successful business.

2. Identify the components of the business model canvas and explain how to use them to develop a viable business model.
- The business model canvas helps entrepreneurs define their business models by defining the relationships among nine factors: value proposition, customer segments, customer relationships, channels, key activities, key resources, key partners, revenue streams, and cost structure.

3. Describe the benefits of building an effective business plan.
- A business plan serves two essential functions. First, and more important, it guides the company's operations by charting its future course and devising a strategy for following it. The second function of the business plan is to attract lenders and investors. Applying for loans or attempting to attract investors without a solid business plan rarely attracts needed capital. Rather, the best way to secure the necessary capital is to prepare a sound business plan.

4. Describe the elements of a solid business plan.
- Although a business plan should be unique and tailor-made to suit the particular needs of a small company, it should cover these basic elements: an executive summary, a mission statement, a company history, a business and industry profile, a description of the company's business strategy, a profile of its products or services, a statement explaining its marketing strategy, a competitor analysis, owners' and officers' résumés, a plan of operation, financial data, and the loan or investment proposal.

5. Explain the three tests every business plan should pass.
- Reality test. The external component of the reality test revolves around proving that a market for the product or service really does exist. The internal component of the reality test focuses on the product or service itself.
- Competitive test. The external part of the competitive test evaluates the company's relative position to its key competitors. The internal competitive test focuses on the management team's ability to create a company that will gain an edge over existing rivals.
- Value test. To convince lenders and investors to put their money into the venture, a business plan must prove to them that it offers a high probability of repayment or an attractive rate of return.

6. Explain the "five Cs of credit" and why they are important to potential lenders and investors.
   - Small business owners needs to be aware of the criteria bankers use in evaluating the creditworthiness of loan applicants—the five Cs of credit: capital, capacity, collateral, character, and conditions.
   - Capital: Lenders expect small businesses to have an equity base of investment by the owner(s) that will help support the venture during times of financial strain.
   - Capacity: A synonym for *capacity* is *cash flow*. The bank must be convinced of the firm's ability to meet its regular financial obligations and to repay the bank loan, and that takes cash.
   - Collateral: Collateral includes any assets the owner pledges to the bank as security for repayment of the loan.
   - Character: Before approving a loan to a small business, the banker must be satisfied with the owner's character.

   - Conditions: The conditions—interest rates, the health of the nation's economy, industry growth rates, and so on—surrounding a loan request also affect the owner's chance of receiving funds.

7. Understand the keys to making an effective business plan presentation.
   - Lenders and investors are favorably impressed by entrepreneurs who are informed and prepared when requesting a loan or investment.
   - Tips include the following: Demonstrate enthusiasm about the venture but don't be overemotional; "hook" investors quickly with an up-front explanation of the new venture, its opportunities, and the anticipated benefits to them; use visual aids; hit the highlights of your venture; don't get caught up in too much detail in early meetings with lenders and investors; avoid the use of technological terms that will likely be above most of the audience; rehearse your presentation before giving it; close by reinforcing the nature of the opportunity; and be prepared for questions.

Discussion Questions

8-1. What is involved in a feasibility analysis, and what value does it provide?

8-2. Explain the nine components of the business model canvas.

8-3. Why should an entrepreneur develop a business plan?

8-4. Why do entrepreneurs who are not seeking external financing need to prepare business plans?

8-5. Describe the major components of a business plan.

8-6. How can an entrepreneur seeking funds to launch a business convince potential lenders and investors that a market for the product or service really does exist?

8-7. What are the five Cs of credit? How do lenders and investors use them when evaluating a request for financing?

8-8. How would you prepare to make a formal presentation of your business plan to a venture capital forum?

CHAPTER 9

Building a Bootstrap Marketing Plan

Learning Objectives

On completion of this chapter, you will be able to:

1. Describe the components of a bootstrap marketing plan and explain the benefits of preparing one.

2. Explain how small businesses can pinpoint their target markets.

3. Explain how to determine customer needs through market research.

4. Outline the steps in the market research process.

5. Describe the bootstrap marketing strategies on which a small business can build a competitive edge in the marketplace.

6. Discuss the "four Ps" of marketing—product, place, price, and promotion—and their role in building a successful bootstrap marketing strategy.

I am a bootstrapper. I have initiative and insight and guts, but not much money. I will succeed because my efforts and my focus will defeat bigger and better-funded competitors.

—Seth Godin

Bootstrapping is not about not spending money—it's about only spending money where it matters.

—Len Kendall

To be effective, both the business model and the business plan must be based on both a financial plan and a marketing plan. Like the financial plan, an effective marketing plan includes forecasts and analysis but from a different perspective. Rather than focus on cash flow, net income, and owner's equity, the marketing plan concentrates on a company's target customers, their needs, their buying power, and their buying behavior. Both plans should tell the same story. It is just that the financial plan tells the story in numbers, while the marketing plan tells it in words.

This chapter is devoted to creating an effective marketing plan. Before producing computer-generated spreadsheets with financial projections, entrepreneurs must determine what to sell, to whom and how often, on what terms and at what price, and how to get the product or service to the customer. In short, a marketing plan identifies a company's target customers and describes how it will attract and keep them. The process does not have to be complex. Figure 9.1 explains how to build a seven-sentence marketing strategy.

Creating a Bootstrap Marketing Plan

1.
_____

Describe the components of a bootstrap marketing plan and explain the benefits of preparing one.

Marketing is the process of creating and delivering desired goods and services to customers and involves all of the activities associated with winning and retaining loyal customers. The secret to successful marketing is to understand the company's target customers' needs, demands, and wants; to offer them the products and services that will satisfy those needs, demands, and wants; and to provide those customers with quality, service, convenience, and value so that they will keep coming back rather than choosing a competitor. The marketing function cuts across the entire organization, affecting every aspect of its operation—from finance and production to hiring and purchasing.

Marketing strategies are not just for megacorporations competing in international markets; small companies require effective marketing strategies as much as their largest rivals do. A recent study of small businesses by research firm Hurwitz and Associates reports a positive correlation between small companies that are experiencing increases in revenue and their expenditures on marketing.[1] Because their entire marketing budgets may be nothing more than rounding errors on larger competitors' marketing budgets, however, small companies must develop creative approaches and invest their marketing dollars wisely to reach their target customers. By developing **bootstrap marketing strategies**—unconventional, low-cost, creative techniques—small companies can wring as much or more "bang" from their marketing bucks as their larger rivals. Bootstrap marketing also is sometimes known as **guerrilla marketing**.

ENTREPRENEURIAL PROFILE: Steve Lichtman: Fitness Together The importance of bootstrap marketing hit Steve Lichtman, owner of four franchised Fitness Together centers near Boston, Massachusetts, after he spent $7,000 to send a direct-mail ad to 20,000 households in his trading area. The results were unimpressive; the campaign generated just enough new business to pay for itself. Lichtman realized that "we had to be a lot more strategic and creative" in the

FIGURE 9.1

A Seven-Sentence Marketing Strategy

Source: Based on Alan Lautenslager, "Write a Creative Marketing Plan in Seven Sentences," _Entrepreneur_, April 24, 2006, _www.entrepreneur.com/marketing/marketingideas/guerrillamarketingcolumnistallautenslager/article159486.html._

Building a successful marketing plan does not have to be a complex process. One marketing expert says that entrepreneurs can create the foundation of a marketing plan with just seven sentences:

1. What is the goal of your marketing efforts?
2. What target market are you trying to reach with your message?
3. What do you offer that is unique?
4. What is the specific need that the market? Specifically how are you addressing that need?
5. How does your business stand out from the competitors? What are you known for?
6. What are the strategies and specific steps you will use to implement your marketing plan?
7. What budget will you need to sell and promote your business?

Answering these seven questions will give you an outline of your company's marketing plan. Implementing a successful marketing plan boils down to two essentials:

1. Having a thorough understanding of your target market, including what customers want and expect from your company and its products and services.
2. Identifying the obstacles that stand in your way of satisfying customers (competitors, barriers to entry, outside influences, budgets, knowledge, and others) and eliminating them.

Pretty strange term, market share, considering the whole object is not to.

company's marketing efforts. He worked with his management team to define the company's value proposition, "One client, one trainer, one goal," which involves offering customized, one-on-one personal training sessions in small, suite-like settings, before looking for the most effective ways to communicate that message to customers. As part of his bootstrap marketing strategy, Lichtman formed a partnership with a local child care provider so that parents know that their children are cared for while they are exercising. He also began building a network of contacts with local health care providers and has conducted exercise programs for rehab patients. To raise his company's visibility in the local community, Lichtman launched a "Lunch and Learn" program aimed at local businesses in which he talks with employees about health, nutrition, and exercise. Fitness Together also sends electronic newsletters with articles of interest to customers and potential customers and has a Facebook page. The company's Web site also includes testimonials (some in the form of short videos) from satisfied customers. Lichtman's most comprehensive bootstrap marketing tactic was to organize a cooperative advertising program with other Fitness Together franchisees in New England. The franchisees pool their resources to purchase advertising in traditional media such as cable television ads and billboards at discounted prices.[2]

A marketing plan focuses a company's attention on the customer and recognizes that satisfying the customer is the foundation of every business. Indeed, the customer is the central player in the cast of every business venture. According to marketing expert Ted Levitt, the primary purpose of a business is not to earn a profit; instead, it is "to create and keep a customer. The rest, given reasonable good sense, will take care of itself."[3] Every area of the business must practice putting the customer first in planning and actions. A **bootstrap marketing plan** should accomplish four objectives:

1. It should pinpoint the target markets the small company will serve.
2. It should determine customer needs, wants, and characteristics through market research, that is, find the "pain points" in the market.
3. It should analyze a company's competitive advantages and build a strong value proposition and an effective, cost-efficient marketing strategy around them.
4. It should help create a marketing mix that meets customer needs and wants.

This chapter focuses on building a customer orientation into these four objectives of the small company's marketing plan.

Bootstrap Marketing in the New Music Industry

The music industry has undergone tremendous change over the last decade. Digital distribution of music has sent the traditional music corporation reeling. In the past, large companies in the music industry controlled everything from the creation of music through its production and its distribution. Consumers no longer have to buy an album produced and distributed by a major record company. Now music fans can buy just the songs they want by downloading them directly to their phones, tablets, and computers, often directly from their favorite artists. People's musical tastes have changed, as the long tail of marketing has created many artists will fewer but intensely loyal consumers. While the major record labels have fallen on tough times, a whole new generation of music industry entrepreneurs has emerged, including entrepreneurs who are artists, managers, software developers, and writers.

Erin O. Anderson is founder and artist manager of Olivia Management. She and the artists she works with are a part of the

Tin Cup Gypsy.
Source: Tin Cup Gypsy Entertainment Co.

Ben Cooper.
Source: Ben Cooper.

entrepreneurial revolution that is changing the structure of the music industry. Anderson works with indie (independent) musicians, which means these artists aren't signed to record deals, and they don't have a big music label backing them and pushing their music into the market. The typical indie band tours the country in a van and sells merchandise in the back of a club, hoping to make enough to afford gas to the next venue and trying to scrape enough dollars together to make an album. These artists have learned to be bootstrap entrepreneurs with their music.

"As a band, it can be tough to get in front of the right audience," says Tyler Oban, CEO and drummer of the band Tin Cup Gypsy. "However, we were in a unique situation. The four of us in the band all had side gig experience. This opened doors for our band by making trades with the artists who called us to play with them. The deal was simple—let our band open for you, and we will give you a discounted rate as musicians." Although it was a lot of extra work to prepare both their own music and the music of the established bands they were playing for, this strategy allowed them to quickly get in front of big audiences and gain fans. "It also helped us to define our target market," says Oban. "Playing in different venues and in different musical genres quickly allowed us to focus on the fans we wanted to as our business gets bigger."

Producing music has also gone through significant changes. "I've had to get creative doing record promotion without a big budget," says Anderson. "Luckily, in the music world, there are a lot of technology start-ups. I do the research and find out what free and cheap tools exist to assist the artists in marketing their music."

Many other entrepreneurs have discovered that indie musicians need help and have developed an abundance of applications, programs, sites, and services to help artists reach their fans and market their music at a very low price or even sometimes for free. For example, Jamplify encourages fans of a band to share unique links to their music and earn special prizes for driving traffic to their Web sites.

"I love the idea that my bands' fans could help grow their fan base, so I volunteered to be a part of Jamplify's beta testing and got to use their service for free," says Anderson. "As a result, I have been able to drive thousands of new listeners to a music video my band had just premiered."

Another new tool for artists to bootstrap the growth of their customer bases is Noisetrade, an online music discovery site. Noisetrade enables artists to give their music away for free while collecting their fans' e-mail addresses and ZIP codes. "We have amassed great, active e-mail list of fans who enjoy the music and enjoy getting updates from my artists," says Anderson. "And they are people who are likely to buy any new albums my artists put out! My other favorite free music service is Next Big Sound, which sends a weekly dashboard that shows popularity, activity, and other metrics for musicians across social media, sales, and events. This free weekly report is sent to my in-box and contains an overview of what is happening on the Internet in regards to the artists I work with," adds Anderson. "It saves me time and money and enables me to make better decisions and have quick reliable feedback on the success of our marketing."

Songwriters have also had to become bootstrapping entrepreneurs in the highly competitive music industry. "After three years of writing songs professionally, I noticed an opportunity to increase my chances of getting songs cut by artists," said Ben Cooper, songwriter and producer. Most songs that get picked up by artists are professionally produced. However, technology has developed to the point that a home studio recording can reach almost the same quality as any professional studio found in New York, Los Angeles, or Nashville. "I decided to buy a textbook on home recording techniques, and I invested a few thousand dollars in a good setup," says Cooper. "After I spend time with another songwriter composing the song, the other songwriter goes home while I work a few extra hours polishing our demo recording. Even though I'm spending the additional time, I'm saving hundreds of dollars per song by not paying other musicians to record. By the end of each year, I'm able to have up to 10 times as many fully recorded songs as the next songwriter."

1. Even though the music industry has gone through tremendous change, there are still opportunities in this industry. Explain.

2. Assume that you or your friends have a band. How would you develop a plan to enter and succeed in today's music industry?

Sources: Mark Mulligan, "Why the Music Industry Must Change Its Strategy to Reach Digital Natives," Mashable, February 04, 2011, *http://mashable.com/2011/02/04/music-industry-digital-natives*; "A Change of Tune," *The Economist*, July 5, 2007, *www.economist.com/node/9443082*; Jeff Price, "The End of the New Music Industry Transformation; How Technology Destroyed the Traditional Music Industry," *ArtistCore*, September 14, 2012, *http://artistcore.blogspot.com/2012/09/the-end-of-new-music-industry.html.*

Market Diversity: Pinpointing the Target Market

One of the first steps in building a marketing plan is identifying a small company's **target market**, the group of customers at whom the company aims its products and services. The more a business learns from market research about its local markets, its customers, and their buying habits and preferences, the more precisely it can focus its marketing efforts on the group(s) of prospective and existing customers who are most likely to buy its products or services.

2. _____

Explain how small businesses can pinpoint their target markets.

ENTREPRENEURIAL PROFILE: Jonathan and James Murrell: Candygalaxy.com Jonathan and James Murrell have launched several online stores, but **Candygalaxy.com**, an online candy store specializing in bulk and nostalgic candy used for events and parties, is their most successful online store to date. They sell specialized, hard-to-find candy to professional event planners, moms planning parties and corporate clients. For example, if a customer wants to plan an event around a certain color scheme, Candygalaxy.com can offer them an assortment of bulk candy in the specific colors being used for their event. Enter "pink" into their search engine, and two pages of pink-colored candies are displayed, ranging from pink M&Ms and pink lollypops to pink Gummy Bears and pink candied almonds. Candygalaxy.com has many repeat customers who know they can find exactly the candy they are looking for through this online retailer.[4]

Unfortunately, most marketing experts contend that the greatest marketing mistake that small businesses make is failing to define clearly the target market to be served. Too many entrepreneurs identify a new product they would like to offer and then try to find a market for their product. Instead, it is always best to start with a target market that has a specific "pain point"—that is, a need or a want that no business is fulfilling. "It is amazing how many people assume they know what customers want without actually asking customers," says Hunter Phillips, CEO of PRSM Healthcare in Nashville, Tennessee. "Present it as you are trying to solve a problem for them. Remember, this is about their needs rather than your idea."

Failing to pinpoint their target markets is especially ironic because small firms are ideally suited to reaching market segments that their larger rivals overlook or consider too small to be profitable. Why, then, do so many small businesses fail to pinpoint their target markets? Because identifying, defining, and researching a target market requires market research and a marketing plan, both of which involve hard work! The result is that these companies follow a sales-driven rather than a customer-driven marketing strategy. To be customer driven, an effective marketing strategy must be based on a clear, well-defined understanding of a company's target customers.

A "one-size-fits-all" approach to marketing no longer works because the mass market is rapidly disappearing. Much of this change is a result of how people get information about products and services. Technology is replacing the power of **mass marketing**—where companies find products and services that appeal to an entire market—with the power of online word of mouth and social media, which creates many much smaller niche markets. In his book *The Long Tail*, Chris Anderson offers evidence that suggests that this has made a fundamental shift in how businesses think about the customers to whom they market and how they provide them with goods and services. "Now, with online distribution and retail, we are entering a world of abundance," says Anderson. "The differences are profound."[5] Figure 9.2 displays the **long tail of marketing**,

FIGURE 9.2

**The Long Tail
of Marketing**

Source: Based on Chris Anderson,
The Long Tail (New York: Hyperion,
2008).

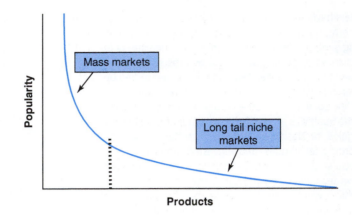

where a few products that target the mass market are replaced by many products that are targeted at specific niche markets. There are three driving forces when marketing to niche markets in the long tail:

1. *Tools of production.* Technology has put the tools of production into the hands of everyone. For example, new tools for recording music can turn a bedroom into a recording studio with sound quality that rivals that of traditional studies with millions of dollars of equipment. Software empowers small businesses to reach and sell to a global market with Web sites and blogs that give them the means for promotion and distribution. The result has been an explosion in niche products.

2. *Internet aggregators.* Studies show that most consumers now begin their shopping by going to online aggregators to compare prices and features when making purchases. This behavior opens many niche markets to even the smallest businesses.

3. *Filtering software connects supply and demand.* Search engines and aggregators now allow consumers to search for a product, see the ads for the various businesses selling that product, and purchase that product with a "click." Companies such as Amazon are now reaching out to specific customer segments with e-mail offers for products and services, many offered by small niche-focused businesses. These advances in technology allow small businesses to not only reach niche markets but also do business with them wherever they are through the Internet.

The most successful businesses have well-defined portraits of the customers they are seeking to attract. From market research, they know their customers' income levels, ages, lifestyles, buying patterns, education levels, likes and dislikes, and even their psychological profiles. At successful companies, the target customer permeates the entire business—from the merchandise it purchases and the ads it uses to the layout and decor of the store. They have an advantage over their rivals because the images they have created for their companies appeal to their target customers, and that's why they prosper. Without a clear picture of its target market, a small company will try to reach almost everyone and usually will end up appealing to almost no one.

Determining Customer Needs and Wants Through Market Research

3.
Explain how to determine customer needs through market research.

In addition to the changes in technology that have created the long tail of marketing, the changing nature of the U.S. population is a potent force altering the landscape of business. Shifting patterns in age, income, education, race, and other population characteristics (which are the subject of demographics) have a major impact on companies, their customers, and the way they do business with those customers. Entrepreneurs who recognize demographic, social, and cultural trends as they emerge have the opportunity to differentiate their companies from the competition in meaningful ways. Those who fail to spot important trends and adjust their strategies accordingly run the risk of their companies becoming competitively obsolete as their target customers pass them by. How can entrepreneurs spot these significant trends?

• Read a variety of current publications, including those that might ordinarily be outside of your areas of interest. Reading industry-related publications also allows you to identify significant trends that affect your business directly.

- Watch the top 10 television shows (at least periodically). They are great indicators of customers' attitudes and values and give valuable insight into the products and services they are interested in buying. AMC's hit series *Mad Men*, a show about the advertising industry set in the early 1960s, proved to be a major influence on men's and women's fashions, home decor, eyeglasses, and other aspects of life.

- See the top 10 movies. They also influence consumer behavior. After Tommy Lee Jones and Will Smith donned Rayban sunglasses in the original *Men in Black*, sales of the company's sunglasses skyrocketed.

- Look to the past. Like clothing styles, some trends are recycled from the past. Organic food? That's what your grandparents ate! Gabriela Hernandez drew the inspiration for the line of retro makeup that her company, Bésame Cosmetics, sells from the classic colors and styles used by celebrity icons of the 1940s, such as Grace Kelly and Audrey Hepburn.[6]

- Talk to at least 100 customers each year to find out what they are buying and why. Ask them about other features they would like or other problems you could help them solve. "When in doubt, go to your customers," says veteran entrepreneur and author Norm Brodsky. "They will tell you what they want and lead you to solutions you'd never come up with on your own. Just about every successful new initiative I've taken in business has come from listening to customers."[7]

- Sensitize yourself to trend tracking. Be on the lookout for emerging trends every day by stepping outside of your normal routine and noticing what's happening around you.

- Monitor social networking sites for evidence of emerging trends. The discussions of people who gather online can reveal significant shifts in attitudes and interests.

For entrepreneurs, the key to success is to align their businesses with as many demographic, social, and cultural trends as possible. This is how successful businesses find their niche target markets. Staying on trend means staying in synchronization with the market as it shifts and changes over time. The more trends a business converges with, the more likely it is to be successful. Conversely, a business moving away from significant trends in society is in danger of losing its customer base.

By performing some basic market research, entrepreneurs can detect key demographic, social, and cultural trends and zero in on the needs, wants, preferences, and desires of its target customers. Indeed, every business can benefit from a better understanding of its market, customers, and competitors. **Market research** is the vehicle for gathering the information that serves as the foundation for the marketing plan. It involves systematically collecting, analyzing, and interpreting data pertaining to the small company's market, customers, and competitors. Businesses face the challenge of reaching the highly fragmented markets that have emerged today, and market research can help them. Market research allows entrepreneurs to answer questions such as the following: Who are my customers and potential customers? To which age-group(s) do they belong? What is their income level? Where do they live? Do they rent or own their own homes? What features are they looking for in the products or services I sell? How often do they buy these products or services? What models, styles, colors, or flavors do they prefer? What radio stations do they listen to? Which Web sites do they visit? What factors are most important to their buying decisions? How do the strengths of my product or service serve their needs and wants? What hours do they prefer to shop? How do they perceive my business? Which advertising media are most likely to reach them? How do customers perceive my business versus competitors? This information is an integral part of developing an effective marketing plan.

A small company must avoid mistakes when marketing its goods and services because there is little margin for error when funds are scarce and budgets are tight. Small businesses simply cannot afford to miss their target markets, and market research can help them zero in on the bull's-eye. Entrepreneurs should conduct market research when developing their business models and constructing their business plans, *before* launching their companies.

ENTREPRENEURIAL PROFILE: Ross Hill: PictureBoothCo Ross Hill, founder and CEO of PictureBoothCo, engaged customers to help develop his idea of a portable photo booth. His idea was to develop a photo booth that was compact enough to ship to and from events anywhere in the country and easy enough to use that it could be set up and operated by anyone by just following a few simple instructions. "When I was in the process of creating the first photo booth prototype,

I was constantly bouncing ideas off of future customers," says Hill. "Their feedback allowed me to avoid wasting my time on unnecessary aspects and focus all of my time on the parts of the business that were important." What prospective customers told him was that they were more concerned about the backgrounds and image quality than they were with the look and design of the booth. This allowed him to develop a simple booth design and put most of his energy into the "user experience" aspects of the booth. "You can't be scared to talk to future or past customers about their experience with your product," said Hill. "If you treated them fairly, they will want to help you succeed."

One of the worst—and most common—mistakes entrepreneurs make is *assuming* that a market exists for their products or services. The time to find out whether customers are likely to buy a product or a service is *before* investing thousands of dollars to launch it! Market research can tell entrepreneurs whether a sufficient customer base exists and how likely those customers are to buy their products and services. In addition to collecting and analyzing demographic data about their target customers, entrepreneurs can learn a great deal by actually observing, mingling with, and interviewing customers as they shop.

Market research for a small business can be informal; it does *not* have to be time consuming, complex, or expensive to be valuable. Many entrepreneurs are discovering the speed, the convenience, and the low cost of conducting market research over the Internet. Online surveys, customer opinion polls, and other research projects are easy to conduct, cost virtually nothing, produce quick responses, and help companies connect with their customers. Insight Express, an online market research firm, estimates that online surveys cost just 20 percent of what it costs to conduct a mail survey and only 10 percent of what it costs for a telephone survey.[8] With Web-based surveys, businesses can get real-time feedback from customers, often using surveys they have designed themselves. Web sites such as Survey Monkey and Zoomerang allow entrepreneurs to conduct low-cost (in some cases free) online surveys of existing or prospective customers. One comparison study of online and mail surveys reports that response rates are higher for online surveys than for mail surveys.[9]

ENTREPRENEURIAL PROFILE: Tim Weber: GoodMusicAllDay LLC When Tim Weber, founder and CEO of GoodMusicAllDay LLC, first attempted to sell merchandise through his music blog site, he spent almost $1,000 for one shirt design. Unfortunately, he only sold about $200 worth of the product. "About 10 months later, I was ready to try again but not willing to put up any capital," said Weber. Weber reached out to three start-up apparel companies and proposed that they front all costs for his next T-shirt. He would make sure that they would recoup their expenses first, and after that they would split the profits 50/50. "We created a poll on our Web site to see which kinds of clothing our audience would actually buy and were soon up and running with virtually no financial risk," said Weber.

How to Conduct Market Research

4.

Outline the steps in the market research process.

The marketing approach that companies of all sizes strive to achieve is **individualized (or one-to-one) marketing**, a system of gathering data on individual customers and then developing a marketing plan designed specifically to appeal to their needs, tastes, and preferences. Its goal is not only to attract customers but also to keep them and to increase their purchases. In a society in which people feel so isolated and transactions are so impersonal, one-to-one marketing gives a business a competitive advantage. Companies following this approach know their customers, understand how to give them the value they want, and, perhaps most important, know how to make them feel special and important. The goal is to treat each customer as an individual. The Ritz Carlton hotel group uses a centralized computer network called Mystique that tracks its guests' preferences and spending habits with the company and provides information to all of the hotels in the chain so that they can offer guests individual attention. When a guest checks in, the desk clerk will know, for instance, that he or she prefers a queen-size bed with foam pillows, a stock of Le Bleu bottled water in the minibar, and a glass of orange juice and a copy of the *Wall Street Journal* with breakfast. Offering these "extras" without requiring the customer to ask for them makes hotel guests feel as though the hotel is catering specifically to their unique needs and preferences.[10]

Individualized marketing requires business owners to gather and assimilate detailed information about their customers. Fortunately, owners of even the smallest companies now have access to affordable technology that creates and manages computerized databases, allowing them to develop close, one-to-one relationships with their customers. Much like gold nuggets waiting to be discovered, significant amounts of valuable information about customers and their buying habits is hidden *inside* many small businesses, tucked away in computerized databases. For most business owners, collecting useful information about their customers and potential new products and markets is simply a matter of sorting and organizing data that are already floating around somewhere in their companies. "Most companies are data rich and information poor," claims one marketing expert.[11] The key is to mine these data and turn them into useful information that allows the company to "court" its customers with special products, services, ads, and offers that appeal most to them.

Entrepreneurs have at their disposal two basic types of market research: conducting *primary research* (data you collect and analyze yourself) and gathering *secondary research* (data that has already been compiled and is available), often at a very reasonable cost or even free. Although secondary data on customers are necessary, they are not sufficient. Entrepreneurs must get firsthand data from real customers! Primary research techniques include the following:

- *Customer surveys and questionnaires.* Keep them short. Word your questions carefully so that you do not bias the results and use a simple ranking system (e.g., a 1-to-5 scale, with 1 representing "unacceptable" and 5 representing "excellent"). Test your survey for problems on a small number of people before putting it to use. Web surveys are inexpensive and easy to conduct and provide feedback fast. Femail Creations, a mail-order company that sells clothing, accessories, and gifts to women, uses Web surveys to gather basic demographic data about its customers and to solicit new product ideas as well. Customer responses have led to profitable new product lines for the small company.[12]

- *Social media.* Small companies have discovered that social media such as Facebook, Twitter, Yelp, and others are easy, inexpensive, and effective tools for gathering feedback from their customers. The owners of Liberty Market, a popular restaurant and food store in Gilbert, Arizona, routinely send tweets to their regular customers after they visit to find out about their dining experience. Not only do the entrepreneurs receive useful feedback and suggestions for improvement, but they also stay close to their core customers.[13]

- *Focus groups.* Enlist a small number of customers to give you feedback on specific issues in your business—quality, convenience, hours of operation, service, and so on. Listen carefully for new marketing opportunities as customers or potential customers tell you what is on their minds. Once again, consider using the Web; one small bicycle company conducts 10 online focus groups each year at virtually no cost and gains valuable marketing information from them.

- *Daily transactions.* Sift as much data as possible from existing company records and daily transactions—customer warranty cards, personal checks, frequent-buyer clubs, credit applications, and others.

- *Other ideas.* Set up a suggestion system (for customers and employees) and use it. Establish a customer advisory panel to determine how well your company is meeting needs. Talk with suppliers about trends they have spotted in the industry. Contact customers who have not bought anything in a long time and find out why. Contact people who are no longer customers and find out why. Teach employees to be good listeners and then ask them what they hear.

As you learned in Chapter 8, entrepreneurs have access to many useful sources of secondary research, which is usually less expensive to collect than primary data. Using secondary data, entrepreneurs can learn about their target customers' demographic profiles, purchasing power, media preferences, and many other characteristics and gain insight into their buying habits and tendencies.

Thanks to advances in computer hardware and software, data mining, once available only to large companies with vast computer power, is now possible for even very small businesses. **Data mining** is a process in which computer software that uses statistical analysis, database technology, and artificial intelligence finds hidden patterns, trends, and connections in "big data" so that business owners can make better marketing decisions and predictions about their customers' behavior. Finding relationships among the many components of a data set, identifying clusters of customers with similar buying habits, and predicting customers' buying patterns, data mining gives entrepreneurs incredible marketing power. Popular data mining software packages include Clementine, DataScope Pro, GoldMine, MineSet, and many others.

FIGURE 9.3

How to Become an Effective One-to-One Marketer

Source: Based on Susan Greco, "The Road to One-to-One Marketing," *Inc.*, October 1995, pp. 56–66.

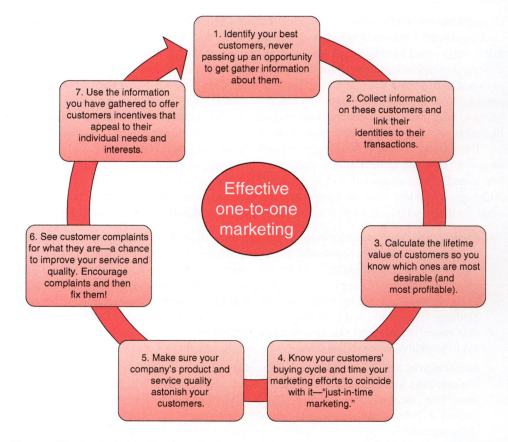

For an effective individualized marketing campaign to be successful, business owners must collect and mine three types of information:

1. *Geographic.* Where are my customers located? Do they tend to be concentrated in one geographic region?

2. *Demographic.* What are the characteristics of my customers (age, education levels, income, gender, marital status, and many other features)?

3. *Psychographic.* What drives my customers' buying behavior? Are they receptive to new products or are they among the last to accept them? What values are most important to them?

Figure 9.3 explains how to become an effective individualized marketer.

In the Entrepreneurial Spotlight

Bootstrap Marketing Using Databases

Although social media has grown in its importance for attracting customers, a recent study shows that e-mail marketing is still the most common tool for customer acquisition (used by 86.3 percent of both B2B and B2C companies) followed by direct mail (used by 69.0 percent of businesses). Although they may seem a bit old-fashioned, e-mail and direct mail are still the workhorses of promotional channels. Businesses can purchase lists of contacts through a variety of services that can offer highly targeted lists that are specific to the type of customer, size, and geographic location. However, buying access to these services is not cheap for a small business. Many of them charge as much as several hundred dollars a month. Clever bootstrapping entrepreneurs

find ways to build a database of potential customers without subscribing to costly database services.

Jerell Harris, president of QuickMed Collect, built his own database for his company that collects medical waste from private physician practices, funeral homes, surgical centers, and other professionals who accumulate regulated waste. Rather than purchase data through a service, which Harris could not afford, he turned instead to the local library. "I found a Web site called Referenceusa.com, which is accessible, without charge, at the library. The Web site offers access to a verifiable database that has 14 million existing businesses and 4 million new businesses. In addition, one can search for businesses using NAICS codes,

Julie Zaloba, founder of Julie Zaloba Consulting, LLC.

SIC codes, city, state, and so on. Moreover, the information can be customized before it is exported." The only limitation to this service is that the user can export only 50 contacts per download. However, the download process can be repeated multiple times. "The data extracted from Referenceusa.com has allowed me develop a list of contacts that resulted in many of them being converted into clients without incurring expenses," adds Harris.

Julie Zaloba runs a small business consulting practice and also uses free databases available through the library. She develops leads not only from these databases but also uses them as tools to work with her clients. "As a business research consultant, I need to access various industry resources that I can reference in my reports to clients. Instead of paying hundreds of dollars in subscription fees to access these online databases, I go to the university library or public library. At the library, I have free access to a wide variety of online databases in which I can research relevant information for my projects. Since I do not have to incur the cost of utilizing these resources, I can offer my consulting services at a more affordable rate."

Jake Jorgovan is the cofounder of an event design company operating in the entertainment industry that he and his partner launched while they were in college. "When we were looking to do market research and grow our sales pipeline, I was faced with the problem that I had little experience in the industry and did not know who all of the players were," said Jorgovan. "We couldn't afford a sales rep, so the only choice was to learn the industry myself." Jorgovan subscribed to variety of industry trade magazines. He carefully read through every article to identify all the names of the key players involved in each production. He then would research these people further using Google. "I started seeing trends in the industry, who the big players were and who worked with whom," explained Jorgovan. "My document ended up growing to more than 60 pages long! It covered profiles of nearly every company and key individual in the event production industry." Jorgovan found that developing his database yielded two important outcomes. First, it gave him a contact list for the company's marketing efforts. It also helped him to be more credible when he made contacts from this list because when he talked to someone else in the industry, he knew whom they were talking about. "That created the perception that I had been in the industry for years," said Jorgovan.

In addition to actively seeking lists of contacts from free or low-cost sources, entrepreneurs can add to their database by making it easy for interested people to give them their contact information. This can be something as simple as a guest visitor book on the counter of a store or as high tech as embedding e-mail sign-up through a quick-response code. Because e-mail is fast becoming the preferred method of direct contact for many people, building an e-mail database can be a highly effective tool for attracting new customers and for retaining existing customers. Experts recommend that any direct contact with customers be as personal as possible. People tend to ignore messages that appear to be part of a large e-mail blast. Many effective marketers are using their databases to send customized messages to customers. Because social media are growing in their marketing applications, small companies can develop strategies to tie traditional e-mail and direct mail promotions into their social media marketing strategies, allowing the customers they contact to sign up for their companies' Facebook and Twitter accounts. Although direct-mail and e-mail marketing may seem a bit "old school," effective bootstrap marketers know that a list of potential customers can still be a gold mine for growing a business.

1. Are direct-mail and e-mail marketing still effective marketing tools? How can a small business owner with a limited budget develop a database when he or she is on a limited budget?

2. How can a small business owner use a database of potential and existing customers to help grow his or her business? In what ways can entrepreneurs tie e-mail marketing into their social media marketing strategies?

3. Develop a specific plan to develop a database of prospective customers for a small business you have already started or a business you hope to start in the near future. How would you go about gathering data? What type of data would you need? How would you target your data collection to ensure it is most effective for your business?

Sources: Based on Hollis Thomases, "Email Marketing Isn't Dead (Yet)," *Inc.*, August 28, 2012, *www.inc.com/hollis-thomases/email-marketing-strategies.html*; Brian Halligan, "Segments of One: Smarter Targeting for Your Customers," *Inc.*, January 26, 2012, *www.inc.com/brian-halligan/audience-segmenting-targeting-your-customers.html.*

Plotting a Bootstrap Marketing Strategy: Building a Competitive Edge

Small companies simply do not have the resources to devote to marketing that their larger rivals do. Marketing a small company effectively requires entrepreneurs to rely on bootstrap marketing strategies, those that use creative, low-cost marketing techniques that hit home with their target customers. Independent bookstores have discovered that large chains and online competitors use their buying power to get volume discounts and undercut the independents' prices. Individual shop

5. _____

Describe the bootstrap marketing strategies on which a small business can build a competitive edge in the marketplace.

owners are finding new ways, such as special ordering, adult reading groups, educational classes, children's story hours, author events, newsletters, autograph parties, and targeting unique niches, to retain loyal customers. These entrepreneurs are finding that the best way to gain a competitive advantage is to create value by giving customers what they really want and cannot get elsewhere.

Successful businesses often use the special advantages they have to gain a competitive edge over their larger rivals. Their close contact with the customer, personal attention, focus on service, and organizational and managerial flexibility provide a solid foundation from which to build a towering competitive edge in the market. Small companies can exploit their size to become more effective than their larger rivals at **relationship marketing** or **customer relationship management** (CRM)—developing, maintaining, and managing long-term relationships with customers so that they will want to keep coming back to make repeat purchases. CRM puts the customer at the center of a company's thinking, planning, and action and shifts the focus from a product or service to customers and their needs and wants. CRM requires business owners to take the following steps:

- Collect meaningful information about existing customers and compile it in a database.

- Mine the database to identify the company's best and most profitable customers, their needs, and their buying habits. In most companies, a small percentage of customers account for the majority of sales and profits. These are the customers on whom a business should focus its attention and efforts.

- Focus on developing lasting relationships with these customers. This often requires entrepreneurs to "fire" some customers that require more attention, time, and expense than they generate in revenue for the business. Failing to do so reduces a company's return on its CRM effort.

- Attract more customers who fit the profile of the company's best customers.

Business owners are discovering that even though they may provide their customers with satisfactory service and value, many of their customers do not remain loyal, choosing instead to buy from other companies. Businesses that provide poor customer service are in grave danger. Hepworth, a consulting firm that specializes in customer retention, measures its clients' **revenue at risk**, which calculates the sales revenue a company stands to lose by measuring the percentage of customers who would leave because of poor service. According to Verde Group, another company that has conducted extensive research on customer satisfaction, the typical North American company is at risk of losing 21 percent of its customers because of high levels of dissatisfaction. What is more alarming is that more than 40 percent of customers who experience a customer service problem never tell the company about it; 70 percent of customers defect because of a problem that, if the company knew about it, would have been easy to fix.[14] Today, earning customers' loyalty requires businesses to take customer focus and service to unprecedented levels, and that requires building long-term relationships with customers.

To be successful bootstrap marketers, entrepreneurs must be as innovative in creating their marketing strategies as they are in developing new product and service ideas. The following bootstrap marketing principles help business owners build a competitive edge:

Find a Niche and Fill It

As we saw in Chapter 4, many successful small companies choose their niches carefully and defend them fiercely rather than compete head-to-head with larger rivals. A niche strategy allows a small company to maximize the advantages of its smallness and to compete effectively even in industries dominated by giants. It also allows them to find their place on the long tail of marketing. Focusing on niches that are too small to be attractive to large companies or in which entrepreneurs have unique expertise are common recipes for success among thriving small companies.

ENTREPRENEURIAL PROFILE: Matt Fiedler and Tyler Barstow: Vinyl Me, Please Matt Fiedler and Tyler Barstow, cofounders of Chicago-based Vinyl Me, Please, are seeking to tap into the renewed interest in vinyl records. According to Nielsen SoundScan, the demand for vinyl records has grown for the past five years. There were 4.6 million newly pressed vinyl records sold in 2012, which is a 17.7 percent increase from the previous year. Music enthusiasts love vinyl records for the warmth and depth of their sound. The niche that Vinyl Me, Please fills is bringing new and interesting music to a new generation of vinyl record enthusiasts. Each month, the subscribers to Vinyl Me,

Please receive a brand new, hand-wrapped vinyl album from an undiscovered artist. In addition to the monthly vinyl record, the company assigns subscribers a personal music consultant who gets to know their musical tastes and preferences. Every month, the consultant creates a personalized play list for each subscriber. Vinyl Me, Please brings together in one service what today's young music enthusiasts want. The company's customers love the sound of vinyl, like to interact on social media with friends about new music to try, and enjoy the surprise factor they get from services such as Pandora.[15]

"Small business is uniquely positioned for niche marketing," says marketing expert Phil Kotler. "If a small business sits down and follows the principles of targeting, segmenting, and differentiating, it doesn't have to collapse to larger companies."[16]

Retain Existing Customers

Keeping an existing customer is cheaper and easier than attracting a new customer. Loyal, long-term customers are the bedrock of every business. High customer retention rates translate into superior financial performance and more efficient marketing budgets. Studies by the Boston Consulting Group show that companies with high customer retention rates produce above-average profits and superior growth in market share.[17] Increasing a company's retention rate by just 2 percent has the same impact as cutting expenses by 10 percent![18]

Because about 20 to 30 percent of a typical companies' customers account for about 70 to 80 percent of its sales, it makes more sense to focus resources on keeping the best (and most profitable) customers than to spend them trying to chase "fair-weather" customers who will defect to any better deal that comes along. Suppose that a company increases its customer base by 20 percent each year, but it retains only 85 percent of its existing customers. Its effective growth rate is just 5 percent per year [20% − (100% − 85%) = 5%]. If this same company can raise its customer retention rate to 95 percent, its net growth rate *triples* to 15 percent [20% − (100% − 95%) = 15%].[19]

Shrewd entrepreneurs recognize that the greatest opportunity for new business often comes from existing customers.

ENTREPRENEURIAL PROFILE: Sharon McRill: The Betty Brigade Sharon McRill, the owner of The Betty Brigade, believes in the power of personal relationships with existing customers. Her Ann Arbor, Michigan–based personal assistance and concierge firm keeps track of important dates in its clients' lives using a low-cost online service called Benchmarkemail.com. McRill can send out automatic e-mails using templates from the site to wish her clients a happy birthday, congratulate them on an anniversary, or thank them for being loyal customers. "This keeps our company in the customers' minds in a positive way," she says.[20]

Although winning new customers keeps a company growing, keeping existing ones is essential to success. Research shows that repeat customers spend 67 percent more than new customers. In addition, attracting a new customer actually costs the typical business *seven to nine times* as much as keeping an existing one.[21] Table 9.1 shows the high cost of lost customers and the steps entrepreneurs can take to improve their customer retention rates.

The formula for marketing success is simple—retain existing customers, enhance relationships with them, and attract new customers like them. However, it takes hard work and constant attention to put this simple formula into practice. Although business owners devote much attention to finding new customers, keeping existing customers coming back is the most cost-effective part of marketing efforts. Entrepreneurs are better off asking, "How can we improve customer value and service to encourage our existing customers to do more business with us?" rather than "How can we increase our market share by 10 percent?" One way that small companies can entice current customers to keep coming back is with a loyalty program (e.g., a car wash offering a punch card that gives customers one free wash after they purchase nine washes). Perhaps the most effective way for a business to build customer loyalty is to sell quality products and to offer outstanding customer service. A study by Forrester Research shows that a good customer experience correlates to customers' willingness to make repeat purchases from a company, reluctance to switch to a competing company, and likelihood of recommending the company to friends and colleagues.[22]

Concentration on Innovation

Because the opportunities for most entrepreneurial ventures came from change and disruption in the market, innovation is the key to continued success. Change has become a part of the global

TABLE 9.1 The High (Annual) Cost of Lost Customers

If you lose ...	Spending $5 Weekly	Spending $10 Weekly	Spending $50 Weekly	Spending $100 Weekly	Spending $200 Weekly	Spending $300 Weekly
1 customer a day	$94,900	$189,800	$949,000	$1,898,000	$3,796,000	$5,694,000
2 customers a day	189,800	379,600	1,898,000	3,796,000	7,592,000	11,388,000
5 customers a day	474,500	949,000	4,745,000	9,490,000	18,980,000	28,470,000
10 customers a day	949,000	1,898,000	9,490,000	18,980,000	37,960,000	56,940,000
20 customers a day	1,898,000	3,796,000	18,980,000	37,960,000	75,920,000	113,880,000
50 customers a day	4,745,000	9,490,000	47,450,000	94,900,000	189,800,000	284,700,000
100 customers a day	9,490,000	18,980,000	94,900,000	189,800,000	379,600,000	569,400,000

When entrepreneurs understand the actual cost of losing a customer, they see their existing customers in a different way. What steps can business owners take to improve their customer retention rates?

1. **Contact your company's best customers regularly.** Identify the customers that account for 75 to 80 percent of your company's sales and call them (better yet, visit them) at least once every quarter.

2. **Keep your company's name in front of your customers.** You can accomplish this by consistently advertising, sending useful newsletters or e-mails, sponsoring workshops, seminars, or special events, or visiting customers to learn how your company can serve them better.

3. **Reward existing customers, especially longtime customers, with special deals exclusively for them.** It might be a special sale or a bonus discount.

4. **Surprise existing customers by giving them something extra.** In Louisiana, locals call it a lagniappe ("lan-yap"), a small gift that a merchant gives to a customer. Send loyal customers a special gift or include an extra "bonus" in their next order. It does not have to be expensive to be effective. For instance, when customers make a sizable purchase at Wilson Creek Outfitters, a fly-fishing shop in Morganton, North Carolina, the owner includes a dozen flies in the order for free. The cost of the lagniappe is minimal, but the goodwill and loyalty it garners are significant.

5. **Keep track of your customers and their needs.** Take the time to build a database of your customers, their contact information, and other relevant information about them and their needs.

6. **Don't take your company's customers for granted.** Your competitors are trying to lure them away; don't give them a reason to go! Avoid the tendency to become so inwardly focused that your company forgets about the importance of its customers.

Sources: Based on Customer Service Institute, 1010 Wayne Avenue, Silver Spring, MD 20910; Rhonda Abrams, "Strategies: Make Customer Retention Priority," *USA Today,* May 29, 2009, *www.usatoday.com/money/smallbusiness/columnist/abrams/2009-05-29-customer-retention_N.htm.*

economy. For small companies to remain competitive, they must constantly innovate. Because of their organizational and managerial flexibility, small businesses often can detect and act on new opportunities faster than large companies. As you learned in Chapter 3, innovation is one of the greatest strengths of entrepreneurs, and it shows up in the new products, unique techniques, and unusual approaches they introduce. A study by the Product Development and Management Association of the top-performing companies across more than 400 industries revealed that new products accounted for 49 percent of profits, more than twice as much as their less innovative competitors.[23] Because product life cycles are growing shorter, innovation, even incremental innovation that makes small improvements in existing products, is essential to long-term business success.

Innovation is an important source of competitive advantage for small companies in any industry, not just those in high-tech sectors.

Drew Hanlen, founder of Pure Sweat Basketball.

Source: Belmont University.

ENTREPRENEURIAL PROFILE: Drew Hanlen: Pure Sweat Basketball Drew Hanlen, National Basketball Association (NBA) skills coach and consultant and founder of Pure Sweat Basketball, built his business to bring together his passion for entrepreneurship and his love of basketball. Hanlen runs high school camps and clinics all across the country. His main focus, however, is helping NBA players improve their games and helping top collegiate players get ready for the NBA. Hanlen has developed a variety of programs and products centered on basketball training. Recently, Hanlen decided to expand into a line of clothing. "I partnered with an up-and-coming clothing company, Elevate, to launch my own clothing line. I sent them apparel that I liked from existing clothing companies, then they researched and picked fabrics, did all of the product research and developing, customized the fabrics and fit to my likings, and let me collaborate with their designer to finalize the products that they'll bring to the market. I have my own athletic apparel line, which cost me no money in product research or development. The only thing it cost me was a bit of time working with their developers, designers and marketing team."

Large corporations achieve innovation by devoting large budgets to research and development. How do small businesses manage to maintain their leadership role in innovating new products and services? They use their size to their advantage, maintaining their speed and flexibility much like a martial arts expert does against a larger opponent. And like Drew Hanlen, they bootstrap whenever possible. Their closeness to their customers enables them to read subtle shifts in the market and to anticipate trends as they unfold. Their ability to concentrate their efforts and attention in one area also gives small businesses an edge in innovation. "Small companies have an advantage: a dedicated management team focused solely on a new product or market," says one venture capitalist.[24]

To be an effective innovator, an entrepreneur should do the following:

- *Make innovation a strategic priority in the company by devoting management time and energy to it.* Smart entrepreneurs know that their business models are works in progress. Consistently testing the assumptions of the business model with current information from the market enhances competitiveness in a dynamic market.

- *Set goals and objectives for innovation.* Establishing targets and rewarding employees for achieving them can produce amazing results. Innovation is not just related to products. Encourage innovation in every aspect of how the company does business.

- *Encourage new product and service ideas among employees.* Workers have many incredible ideas, but they will lead to new products or services only if someone takes the time to listen to them.

- *Listen to customers.* A recent survey found that customers are eager to use social media to provide ideas on product innovation. Leading companies use social media to stay on top of the changes and improvements that customers want from their products.[25]

- *Always be on the lookout for new product and service ideas.* They can come to you (or to anyone inside or outside the company) at any time.

- *Keep a steady stream of new products and services flowing.* Even before sales of her safety-handle children's toothbrush took off, Millie Thomas, founder of RGT Enterprises, had developed other children's products using the same triangular-shaped handle, including a crayon holder, paintbrushes, and fingernail brushes.[26]

Table 9.2 describes a screening device for testing the viability of new product ideas.

TABLE 9.2 Testing the Viability of a New Product Idea

Testing the viability of new product ideas in their early stages of development can help entrepreneurs avoid expensive product failures later—after they have already invested significant amounts of cash in developing and launching them. The Chester Marketing Group, Inc., of Washington Crossing, Pennsylvania, has developed the following test to determine the viability of a new product idea at each stage in the product development process. To calculate a new product idea's score, entrepreneurs simply multiply the score for each criterion by its weight and then add up the resulting weighted scores. For a product to advance to the next stage in the development process, its score should be at least 16.

Criterion	Score			Weight	Weighted Score
Extent of Target Market Need	Below Average 1	Average 2	Above Average 3	2	
Potential Profitability	Below Average 1	Average 2	Above Average 3	2	
Likely Emergence of Competition	Below Average 1	Average 2	Above Average 3	1	
Service Life Cycle	Below Average 1	Average 2	Above Average 3	1	
Compatibility with Company Strengths	Below Average 1	Average 2	Above Average 3	2	
				Total Weighted Score	

Source: Roberta Maynard, "Test Your Product Idea," *Nation's Business,* October 1997, p. 23.

The Marketing Mix

5.

Discuss the "four Ps" of marketing—product, place, price, and promotion—and their role in building a successful bootstrap marketing strategy.

Implementing a marketing strategy requires entrepreneurs to determine how they will use the "four Ps" of marketing—product, place, price, and promotion. Think of these four elements of marketing like ingredients in soup. All of them must work together to give a soup a delicious flavor. Entrepreneurs must integrate these elements into a coherent strategy to maximize the impact of their product or service on the consumer. Using the highest-priced components and promoting the highest quality in the market while pricing the product far below all competitors results in an inconsistent message in a marketing mix. All four Ps must reinforce the same image of the product or service the company presents to potential customers.

Product

The product is the foundation of the marketing mix. What makes up the product is much more than the physical item that a company sells. For example, a coffee shop sells much more than a cup of coffee. It sells a place to hang out with friends, meet with customers, or read a good book. Every element of the customer's experience makes up the product in a marketing mix. Products can have form and shape, or they can be services with no physical form.

Products travel through various stages of development. The **product life cycle** (see Figure 9.4) measures these stages of growth, and these measurements enable the company's management to make decisions about whether to continue selling the product, when to introduce new follow-up products, and when to introduce changes to an existing product. The length of a product's life cycle depends on many variables, including the type of product. Fashion clothing may have a short product life cycle, lasting only four to six weeks, but a video game console's life cycle typically lasts about four years.[27] Products that are more stable, such as appliances, may take years to complete a life cycle. Product life cycles are growing shorter, however. For example, the life cycle for golf equipment has shrunk over the last decade from three or four years to less than one year today.[28]

In the *introductory stage*, marketers present their product to potential consumers. Initial high levels of acceptance are rare. Generally, new products must break into existing markets and compete with established products. Traditionally, companies use advertising and promotion to help the new product gain recognition among potential customers, who must get information about the product and the needs it can satisfy. Small businesses must intensify their marketing efforts at this level of the life cycle to overcome customer resistance and inertia—customer buying habits are hard to change. Entrepreneurs must find ways to bootstrap their marketing efforts during the entry stage of a new product because they do not have the marketing budgets of their large competitors. "Start by selling yourself and leverage your network to rise to the top," recommends Andy Tabar, founder of Cleveland-based Bizooki. Personal selling, public relations efforts to get media stories, and creative low-cost forms of promotion can overcome the normal high cost of introducing a new product.

FIGURE 9.4

The Product Life Cycle

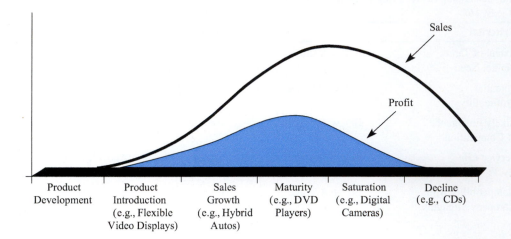

| Product Development | Product Introduction (e.g., Flexible Video Displays) | Sales Growth (e.g., Hybrid Autos) | Maturity (e.g., DVD Players) | Saturation (e.g., Digital Cameras) | Decline (e.g., CDs) |

After the introductory stage, the product enters the *growth and acceptance stage*. In this stage, customers begin to purchase the product in large enough numbers for sales to rise and profits to increase. Products that reach this stage, however, do not necessarily become successful. If in the introductory or the growth stage the product fails to meet consumer needs, it does not generate adequate sales volume and eventually disappears from the marketplace. For successful products, sales and profit margins continue to rise through the growth stage.

In the *maturity and competition stage*, sales volume continues to rise, but profit margins peak and then begin to fall as competitors enter the market. Normally, this causes a reduction in the product's selling price to meet competitor's prices and to hold its share of the market. Sales peak in the *market saturation stage* of the product life cycle and give the marketer fair warning that it is time to introduce a new product innovation.

The final stage of the product life cycle is the *product decline stage*. Sales continue to drop, and profit margins fall drastically. However, when a product reaches this stage of the cycle, it does not mean that it is doomed to failure. Products that have remained popular are always being revised. No company can maintain its sales position without product innovation and change. Although Radio Flyer, a company started in 1917 by immigrant cabinetmaker Antonio Pasin (and managed by his grandson, Robert), still sells the classic all-metal, rubber-wheeled red wagon that made the company famous, it has introduced several models for twenty-first-century children. The Ultimate Family Wagon includes five-way flip-and-fold seats, storage compartments, cup holders (two for kids and two for adults), a tray table, and a sun canopy. Radio Flyer's latest prototype, the Cloud 9, includes enough updates to appear on *Pimp My Ride*. The high-tech wagon that features reclining upholstered seats with five-point safety harnesses; cup holders; foldout storage compartments; foot brakes; a digital handle that tracks temperature, time, distance, and speed; a slot for an MP3 player; and stereo speakers. These innovations are designed to draw new generations of children to the classic toy.[29]

Understanding the product life cycle helps a business owner plan the introduction of new products to the company's product line. Too often, companies wait too late into the life cycle of one product to introduce another. The result is that they are totally unprepared when a competitor produces a "better mousetrap" and their sales decline. The ideal time to develop new products is early on in the life cycle of the current product (see Figure 9.5). Waiting until the current product is in the saturation or decline stages is like living on borrowed time.

There are a number of ways that a small business can enhance its product offerings.

FOCUS ON THE CUSTOMER The relationship a company forges with its customers is a crucial part of how customers perceive its products. Companies that focus on the customer experience generate 60 percent higher profits than their competitors.[30] A study by market research firm Genesys reports that businesses in the United States lose $83 billion in sales each year in abandoned sales or defections to competitors because of poor customer service.[31] The *Retail Customer*

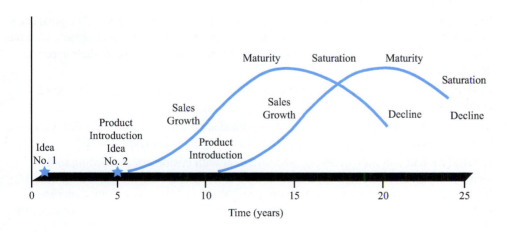

FIGURE 9.5

Time Between Introductions of Products

Source: Based on Constant Contact 2009 Small Business Attitudes and Outlook Survey, June 18, 2009.

Dissatisfaction Study, conducted by the Jay H. Baker Retailing Initiative at the University of Pennsylvania and consulting firm Verde Group, reports the following:

- Seventy-one percent of customers in the United States have defected to a competing business because of a poor customer service experience with a company.[32]

- For every complaint that a company receives from a customer, 17 other complaints exist, but the company never hears about them.

- Thirty-one percent of dissatisfied customers tell family members, friends, and colleagues about their negative experience with a company.

- Six percent of those people tell their "horror stories" to six or more people.

- Negative word-of-mouth has exponential power. For every 100 customers who have a negative experience with a business, the company stands to lose 32 to 36 current customers or potential customers.[33]

Because most of a company's sales come from existing customers, no business can afford to alienate them.

The most successful small businesses have developed a customer orientation and have instilled an attitude of customer satisfaction throughout the company. Companies with world-class customer attitudes set themselves apart by paying attention to the little things.

ENTREPRENEURIAL PROFILE: Donna Flanagin: Flanagin's Bulk Mail When customers with children enter Flanagin's Bulk Mail, a company in Valparaiso, Indiana, that specializes in mailing lists and bulk mail services, owner Donna Flanagin offers the children crayons and a coloring sheet so that they can create pictures that she displays on the front door. Flanagin sends children their drawings with a card on their birthdays. "It costs virtually nothing," observes Bruce Jones, director of the Disney Institute, "yet it reminds the parents and grandparents about her business and helps make a connection with her customers."[34]

How do these companies focus so intently on their customers? They follow basic principles:

- When you create a dissatisfied customer, fix the problem fast. One study found that, given the chance to complain, 95 percent of customers will buy again if a business handles their complaints promptly and effectively.[35] The worst way to handle a complaint is to ignore it, to pass it off to a subordinate, or to let a lot of time slip by before dealing with it.

- *Encourage* customer complaints. You can't fix something if you don't know it's broken, but once you know, be sure to fix it!

- Ask employees for feedback on improving customer service. A study by Technical Assistance Research Programs, a customer service research firm, found that frontline service workers can predict nearly 90 percent of the cases that produce customer complaints.[36] Put that expertise to work the by involving frontline employees in process improvement efforts. A successful chain of medical clinics made clinic receptionists key members of their marketing teams.

- Entrepreneurs and their managers should occasionally wait on customers. It's a great dose of reality. Dell CEO Michael Dell and his team of top managers meet periodically with the company's major customers to get a better understanding of how to serve their needs more effectively.[37]

- Develop a service theme as part of your culture to communicate your attitude toward customers.

- Reward employees "caught" providing exceptional service to customers. At Pitney Bowes, employees recognized by name in customer surveys receive a gift certificate.[38]

- Carefully select and train everyone who will deal with customers. According to the Genesys study of customer satisfaction, the most important factor in providing a satisfying customer service experience is competent sales and customer service representatives.[39] Smart entrepreneurs view employee training for what it is: an investment rather than an expense.

DEDICATION TO SERVICE AND CUSTOMER SATISFACTION Small companies that lack the financial resources of their larger rivals have discovered that offering exceptional customer service is one of the most effective ways to differentiate their products and to attract and maintain a growing customer base. "It doesn't take money to [provide] good customer service," says the head of one retail company. "It takes a commitment."[40] Unfortunately, the level of service in most companies is poor. A study by Accenture shows that 67 percent of customers globally had switched at least one service provider within the last year because of poor customer service. The study also reports that only one in four customers feels "very loyal" to current providers.[41] This creates opportunities to small companies ready to earn loyalty through exceptional service.

Successful businesses recognize that superior customer service is only an intermediate step toward the goal of customer satisfaction. These companies seek to go beyond customer satisfaction, striving for *customer astonishment*! They concentrate on providing customers with quality, convenience, and service as their customers define those terms. Certainly, the least expensive—and the most effective—way to achieve customer satisfaction is through friendly, personal service. Numerous surveys of customers in a wide diversity of industries, from manufacturing and services to banking and high technology, conclude that the most important element of service is "the personal touch." Indeed, a study conducted by market research firm NFO WorldGroup found that friendly service, not the food, is the primary reason customers return to a restaurant![42] Whatever the nature of the business, calling customers by name; making attentive, friendly contact; and truly caring about customers' needs and wants are more essential than any other factor, even convenience, quality, and speed!

How can a company achieve stellar customer service and satisfaction?

Hire the Right Employees The key ingredient in the superior service equation is *people*. There is no substitute for friendly, courteous sales and service representatives. A customer service attitude requires hiring employees who believe in and embrace customer service. When it comes to the impact of customer retention on a company's profitability, a responsive, customer-centric employee is worth many times the value of an employee who provides average (or, worse yet, below-average) customer service. "You hire people for their inherent skill," says Gary Danko, owner of a restaurant that recently won a prestigious customer service award. "You can teach them the mechanics."[43]

Train Employees to Deliver Superior Service According to customers, the single most important factor in providing good customer service is having knowledgeable employees who are well informed, polite, and friendly.[44] Successful businesses train every employee who deals directly with customers; they don't leave the art of customer service to chance.

Listen to Customers The best companies constantly listen to their customers and respond to what they hear! Social media have created a more direct channel to receive customer opinions, and successful businesses pay attention to this feedback. This allows them to keep up with customers' changing needs and expectations. The only way to find out what customers really want and value is to ask them. In addition to social media, businesses still rely on a number of more traditional techniques, including surveys, focus groups, telephone interviews, comment cards, suggestion boxes, toll-free hot lines, and regular one-on-one conversations with customers (perhaps still the best technique). Marie Moody, founder of Stella & Chewy's, a company that makes an all-natural line of premium frozen pet food, changed the freezing process to reduce the formation of ice crystals on the product and the company's packaging in response to customer feedback. After making the changes, Stella & Chewy's sales skyrocketed from $500,000 to more than $5 million in just two years.[45]

Define Superior Service Based on what customers say, managers and employees must decide exactly what "superior service" means in the company. Such a statement should (1) be a strong statement of intent, (2) differentiate the company from others, and (3) have value to customers. Deluxe Corporation, a printer of personal checks, defines superior service quite simply: "Forty-eight hour turnaround; zero defects."

Set Standards and Measure Performance To be able to deliver on its promise of superior service, a business must establish specific standards and measure overall performance against them. Satisfied customers should exhibit at least one of three behaviors: loyalty (increased customer retention rate), increased purchases (climbing sales and sales per customer), and resistance to rivals' attempts to lure them away with lower prices (market share and price tolerance).[46] Companies must track performance on these and other service standards and reward employees accordingly.

Examine Your Company's Service Cycle What steps must a customer go through to get your product or service? Business owners often are surprised at the complexity that has seeped into their customer service systems as they have evolved over time. One of the most effective techniques is to map each step a customer must go through to purchase a product—down to each click of the mouse—using sticky notes on a wall and identifying any steps that can be eliminated or improved. The goal is to look for steps and procedures that are unnecessary, redundant, or unreasonable and then to eliminate them.

Empower Employees to Offer Superior Service One of the most important variables in employees delivering superior service is whether they perceive that they have permission to do so. The goal is to push decision making down the organization to the employees who have contact with customers. This includes giving them the freedom to circumvent company policy if it means improving customer satisfaction. At Ritz-Carlton Hotels, every employee is authorized to spend up to $2,000 to resolve a customer's complaint.[47] The Apple Store has benchmarked its customer service expectations to the standards set by Ritz-Carlton. When a customer walks into an Apple Store, the associate who greats that customer "owns" that relationship and is expected to do everything he or she can to make the customer experience right.[48] To be empowered, employees need knowledge and information, adequate resources, and managerial support.

Use Technology to Provide Improved Service Technological innovations provide new tools to help enhance customer service. Integrated systems that employees can access through smart phones and tablets allow employees to provide real-time solutions to customers' problems and to answer customers' questions. Chat features on Web sites turn static Web pages into live, interactive customer service tools. Even small companies can use these features by using smart phones to answer customer questions entered into the chat boxes on Web sites in real time from anywhere.

Ensure Top Management's Support The drive toward superior customer service will fall far short of its target unless the entrepreneur and the management team support it fully. Success requires more than just a verbal commitment; it calls for everyone's involvement and dedication. Periodically, the entrepreneur and all managers should spend time in customer service positions to maintain contact with customers, frontline employees, and the challenges of providing good service.

Give Customers an Unexpected (and Pleasant) Surprise Companies can make a lasting, favorable impression on their customers by providing them with an unexpected surprise periodically. The surprise does not have to be expensive to be effective.

ENTREPRENEURIAL PROFILE: Josh Gilreath: Aperture Construction, LLC Josh Gilreath, president and CEO of Aperture Construction, LLC, has discovered that an old tool still has power. Gilreath frequently sends thank-you notes to customers and key partners of his general contracting firm. "A simple box of thank-you cards can be purchased at an office supply store for less than $20 and are a great way to show potential clients/subcontractors/suppliers that you care about what they have to say or that you value their help and opinions," says Gilreath. "In today's digital world, a handwritten thank-you stands so far out from the rest of the crowd that people can't help but notice you. In addition to being noticed and remembered, a handwritten card speaks volumes for the customer service you can offer a potential client, so in a sense you hit two birds with one stone!"

DEVOTION TO QUALITY In this intensely competitive global business environment, quality goods and services are a prerequisite for success—and even survival. According to one marketing axiom, the worst of all marketing catastrophes is to have great advertising and a poor-quality product. Customers have come to expect and demand quality goods and services, and those businesses that provide them consistently have a distinct competitive advantage.

Today, quality is more than just a slogan posted on the company bulletin board; world-class companies treat quality as a strategic objective—an integral part of the company culture. This philosophy is called **total quality management** (TQM)—quality not just in the product or service itself but also in *every* aspect of the business and its relationship with the customer and in continuous improvement in the quality delivered to customers. Companies achieve continuous improvement by using statistical techniques to discover problems, determine their causes, and solve them; then they must incorporate what they have learned into improving the process. The ultimate goals of TQM are to *avoid* quality problems, reduce cycle time (the time between a customer's order and delivery of the finished product), reduce costs, and continuously improve the process. TQM's focus on continuous improvement is built on the "define, measure, analyze, improve, and control (DMAIC) process illustrated in Figure 9.6.

Companies on the cutting edge of the quality movement are developing new ways to measure quality. Manufacturers were the first to apply TQM techniques, but retail, wholesale, and service organizations have seen the benefits of becoming champions of quality. They are tracking customer complaints, contacting "lost" customers, and finding new ways to track the cost of quality and their return on quality (ROQ). ROQ recognizes that although any improvement in quality may improve a company's competitive ability, only those improvements that produce a reasonable rate of return are worthwhile. In essence, ROQ requires managers to ensure that the quality improvements they implement will more than pay for themselves. Using basic quality principles, Allen Edmonds invested more than $1.5 million to redesign its Port Washington, Wisconsin, factory using principles of lean manufacturing (which focus on maximizing value for customers and minimizing waste), a move that not only improved quality but also increased worker productivity.[50]

Companies that are successful in building a reputation for top-quality products and services follow certain guidelines to "get it right the first time":

- Build quality into the process; don't rely on inspection to obtain quality.
- Emphasize simplicity in the design of products and processes; it reduces the opportunity for errors to sneak in.
- Foster teamwork and dismantle the barriers that divide disparate departments.

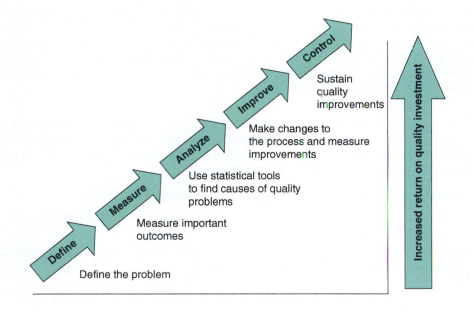

FIGURE 9.6

The Quality DMAIC Process

Source: Adapted from Walter H. Ettinger, MD, "Six Sigma," *Trustee*, September 2001, p. 14.

- Establish long-term ties with select suppliers; don't award contracts on low price alone.

- Provide managers and employees the training needed to participate fully in the quality improvement program.

- Empower workers at all levels of the organization; give them the authority and the responsibility for making decisions that determine quality.

- Get managers' commitment to the quality philosophy. Otherwise, the program is doomed. Describing his role in his company's TQM philosophy, one CEO says, "People look to see whether you just talk about it or actually do it."[51]

- Rethink the processes the company uses now to get its products or services to customers. Employees at Analog Devices redesigned its production process and significantly lowered the defect rate on its silicon chips, saving $1.2 million a year.[52]

- Reward employees for quality work. Ideally, employees' compensation is linked clearly and directly to key measures of quality and customer satisfaction.

- Develop a company-wide strategy for constant improvement of product and service quality.

ATTENTION TO CONVENIENCE If you ask customers what they want from the businesses they deal with, one of the most common responses is "convenience," which has become an important tool for differentiating a product from those offered by the competition. In this busy, fast-paced world of dual-career couples and lengthy commutes to and from work (the average commute time is now 25.4 minutes), consumers have more disposable income but less time in which to enjoy it. Anything a business can do to enhance convenience for its customers will give it an edge. Several studies have found that customers rank easy access to goods and services at the top of their purchase criteria. Unfortunately, many businesses fail to understand that convenience has become such an important characteristic of how the customer views their products; as a result, they fail to attract and retain customers. Some companies make it a chore to do business with them, almost as if their owners have created an obstacle course for customers to negotiate. In an effort to defend themselves against a few unscrupulous customers, these businesses have created elaborate procedures for exchanges, refunds, writing checks, and other basic transactions that frustrate legitimate customers.

Successful companies go out of their way to make it easy for customers to do business with them. Many restaurants have created online tools and social networking applications that give customers the convenience and the speed of placing their meal orders online or from their mobile devices. Other restaurateurs enhance customer convenience by placing their restaurants on wheels and driving them to their customers.

ENTREPRENEURIAL PROFILE: Hayden Coleman: Moovers and Shakers Hayden Coleman started his food truck, Moovers and Shakers, when he was a sophomore in college using funds from a business plan competition and from a Kickstarter campaign. Moovers and Shakers is an old-fashioned soda parlor on wheels that serves milk shakes and floats. Coleman uses Twitter to let his regular customers know where the food truck is heading next and Twitter and Facebook to promote his ever-changing flavor of the week. In its second season, Coleman began to set up his mobile soda shop at weddings and other private events in addition to the traditional public events where food trucks typically congregate.[53]

Many small companies have had success by finding simple ways to make it easier for customers to do business with them. How can entrepreneurs boost the convenience levels of their businesses? By conducting a "convenience audit" from the customer's point of view to get an idea of its ETDBW ("Easy To Do Business With") index:

- Is your business located near your customers? Does it provide easy access?

- Are your business hours suitable to your customers? Should you be open evenings and weekends to serve them better?

- Would customers appreciate pickup and delivery service? John Pugh, owner of Prosperity Drug Company, a small pharmacy in tiny Prosperity, South Carolina, has built a loyal following, especially among older customers, by offering a free prescription delivery service using a classic Volkswagen Beetle affectionately known as "The Drug Bug."

- Does your business provide a sufficient number of checkout stations so that shoppers do not have to stand in long lines to make their purchases? Does your company make it easy for customers to make purchases on credit or with credit and debit cards?

- Are you using technology to enhance customer convenience? Jim Olekszy, owner of 10 North, a restaurant in Oakland, California, uses iPads and iPods to take and process orders utilizing a touch-based ordering system. "Everybody thinks it's cool," Olekszy says. The system is much more cost-effective than the usual computer systems used in other restaurants and speeds up customer service. An attachment for the iPads allows severs to scan credit cards tableside, speeding up the checkout process. Guests sign for their purchases on the screen and have their receipts either printed or e-mailed.[54]

- Are your employees trained to handle business transactions quickly, efficiently, and politely? Waiting while rude, poorly trained employees fumble through routine transactions destroys customer goodwill.

- Do your employees use common courtesy when dealing with customers?

- Does your company offer "extras" that make customers' lives easier? With a phone call to one small gift store, customers in need of a special gift simply tell how much they want to spend, and the owner takes care of the rest—selecting the gift, wrapping it, and shipping it. All customers have to do is pay the invoice when it arrives in the mail.

- Can you adapt existing products to make them more convenient for customers? When J.M. Smucker Company began test-marketing a premade frozen peanut butter and jelly sandwich with no crust, CEO Tim Smucker was amazed at the results. The sandwiches, called Uncrustables, generated $20 million in sales, and Smucker added them to its product line.[55]

- Does your company handle telephone calls quickly and efficiently? Long waits "on hold," transfers from one office to another, and too many rings before answering signal customers that they are not important.

EMPHASIS ON SPEED We live in a world of instantaneous expectations. Technology that produces immediate results at the click of a mouse and that allows for real-time communication has altered our sense of time and space. Speed reigns. Customers now expect companies to serve them at the speed of light! In such a world, speed has become an important competitive weapon. With their smaller, nimbler, more flexible organizations, small companies have a natural advantage in speed over large companies, which are burdened with lumbering bureaucracies and self-absorbed cultures. World-class companies recognize that reducing the time it takes to provide a service or to develop, design, manufacture, and distribute a product reduces costs, increases quality, and increases market share. Service companies also know that they must build speed into their business systems if they are to satisfy their impatient, time-sensitive customers.

This philosophy of speed is called **time compression management** (TCM), and it involves three aspects: (1) speeding new products to market, (2) shortening customer response time in manufacturing and delivery or providing a service, and (3) reducing the administrative time required to fill an order. In the hypercompetitive market that businesses now face, TCM has been shown to be a critical strategic tool.[56] Studies show plenty of room for improvement; most businesses waste 85 to 99 percent of the time it takes to produce products or services without ever realizing it![57] Although speeding up the manufacturing process is a common goal, companies using TCM have learned that manufacturing takes only 5 to 10 percent of the total time between taking an order and getting the product into the customer's hands. The rest is consumed by clerical and administrative tasks. The primary opportunity for TCM lies in its application to the administrative process.

Companies that rely on TCM to help them turn speed into a competitive edge should do the following:

- "Reengineer" the entire process rather than attempt to do the same things in the same way, only faster.

- Study every phase of the business process—whether it involves manufacturing, shipping, administration, or some other function—looking for small improvements that speed up the entire process.

- Create cross-functional teams of workers and give them the power to attack and solve problems. In world-class companies, product teams include engineers, manufacturers, sales people, quality experts—even customers.

- Share information and ideas across the company. Easy access to meaningful information can reduce a company's customer response time.

- Set aggressive goals for time reduction and stick to the schedule. Some companies using TCM have been able to reduce cycle time from several weeks to just a few hours!

- Instill speed in the culture. At Domino's Pizza, kitchen workers watch videos of the fastest pizza makers in the country.

- Use technology to find shortcuts wherever possible. Rather than build costly, time-consuming prototypes, many time-sensitive businesses use computer-aided design and computer-assisted manufacturing to speed up product design and testing.

Promotion

Promotion is how a business communicates with customers using advertising, public relations, and personal selling. For an entrepreneur, bootstrapping is the key to cost effective promotion. Advertising communicates to potential customers through mass media the benefits of a good or service. Public relations involves getting media, including social media outlets such as blogs, to write stories about a business. Personal selling involves the art of persuasive sales on a one-to-one basis.

The goals of a small company's promotional efforts are to create a brand image, to persuade customers to buy, and to develop brand loyalty. Promotion can take many forms and is put before the public through a variety of media. Entrepreneurs often must find ways to use low-cost bootstrap tactics to create promotions that get their companies noticed by both local and national media. What follows are other effective components of a bootstrap marketing strategy. Chapter 10 is devoted to more details on how to create an effective advertising and promotion campaign for a small company.

USE THE POWER OF PUBLICITY **Publicity** is any commercial news covered by the media that boosts sales but for which a small company does not pay. Publicity has power; because it is from an unbiased source, a news feature about a company or a product that appears in a newspaper or magazine has more impact on people's buying decisions than an advertisement does. Exposure in any medium raises a company's visibility and boosts sales, and, best of all, publicity is free! It does require some creativity and effort, however. Entrepreneurs generate publicity in a multitude of ways, ranging from sponsoring a seminar and supporting a charity to writing an article for publication on and creating an offbeat contest.

ENTREPRENEURIAL PROFILE: Joe Keeley: College Nannies and Tutors Joe Keeley founded College Nannies and Tutors in 2000 when he was a student in business school. Keeley funded the start-up by the winnings he received from entering business plan competitions. During his initial start-up, he paid nothing for advertising. Instead, Keeley used the novelty of a former high school hockey player running a nanny business while still in college to get publicity through stories in newspapers and magazines and on radio and television. Today, College Nannies and Tutors has expanded and has franchisees in 27 states.

Don't Just Sell—Entertain Numerous surveys have shown that consumers are bored with shopping and that they are less inclined to spend their scarce leisure time shopping than ever before. Winning customers today requires more than low prices and wide merchandise selection; increasingly, businesses are adopting strategies based on **entertailing**, the notion of drawing customers into a store by creating a kaleidoscope of sights, sounds, smells, and activities, all designed to entertain—and, of course, sell (think Disney). The primary goal of entertailing is to catch customers' attention and engage them in some kind of entertaining experience so that they shop longer and buy more goods or services. Entertailing involves "making [shopping] more fun, more educational, more interactive," says one retail consultant.[58] Research supports the benefits of entertailing's hands-on, interactive, educational approach to selling; one study reports that, when making a purchase, 34 percent of consumers are driven more by emotional factors, such as fun and excitement, than by logical factors, such as price and convenience.[59]

Connect with Customers on an Emotional Level Some of the most powerful marketers are those companies that have a clear sense of who they are, what they stand for, and why they exist. Defining their vision for their companies in a meaningful way is one of the most challenging tasks facing entrepreneurs. As we learned in Chapter 4, that vision stems from the beliefs and values of the entrepreneur and is reflected in a company's culture, ethics, and business strategy. Although it is intangible, this vision is a crucial ingredient in a successful bootstrap marketing campaign. Once this vision is firmly planted, bootstrap marketers can use it to connect with their customers. Harley-Davidson, the maker of classic motorcycles with that trademark throaty rumble, has established an emotional connection with its customers that many other businesses only dream of. Clint Harris of Buckeye Lake, Ohio, was such a devoted fan of the motorcycle maker that when he died, he had a replica of his motorcycle (complete with his and his wife's biker nicknames, Heavy and Ruffy, engraved on it) made to serve as a headstone.[60]

Companies that establish a deeper relationship with their customers rather than one based merely on making a sale have the capacity to be exceptional bootstrap marketers. These businesses win because customers receive an emotional boost every time they buy the company's product or service. Companies connect with their customers emotionally by supporting causes that are important to their customer base, sponsoring events that are of interest to their customers, taking exceptional care of their customers, and making it fun and enjoyable to do business with them.

ENTREPRENEURIAL PROFILE: Clint Smith: Emma The e-mail marketing company Emma grew at a healthy pace in its start-up years by developing loyal customers who loved the way Emma provided personalized service. Cofounder Clint Smith intentionally built a culture that fosters strong ties among Emma, its employees, and its customers. As Emma grew, however, customers began to experience problems with the service. The volume of the several hundred small business customers was overwhelming the software's ability. However, a more functional and more scalable new version was still in development. Emma's leadership decided to switch over to the unfinished new version of the software rather than continue to battle the many shortcomings of the original software. When they switched over, the new system didn't work at all for almost a week. Once the system finally went online, employees discovered that some functions of the system were not operational and spent six months patching it so that it would run reliably. Amazingly, because of the company's strong relationship with its customers, Emma lost only one client, despite the rocky changeover in software systems.[61]

Build a Consistent Branding Strategy Establishing an emotional bond with customers is the first step to building a successful brand. Branding involves creating a distinct identity for a business and requires a well-coordinated effort at every touch point that a company has with its customers. A brand represents a company's "personality," and entrepreneurs should spell out their companies' brand strategy in the business plan. In an age when companies find standing out from the crowd of competitors increasingly difficult, branding strategies have taken on much greater importance. The foundation of a successful brand is providing a quality product or superior customer service that meets or, preferably, *exceeds* customers' expectations. One way to do this is by defining exactly how your company's product or service solves a problem your customers face (preferably in a unique fashion) and communicating it to your customers. Green Mountain Coffee Roasters in Waterbury, Vermont, has built a strong brand not only by creating a better cup of coffee but also by doing its part to create a better world by embracing environmental sustainability and organic growing methods and supporting local communities through charitable donations. Although entrepreneurs don't have the resources to invest in building a brand that Apple and Google do (Apple's brand alone is estimated to be worth more than *$87 billion*[62]), they can take steps to add value to their companies' images through branding. One important step is to develop consistent logos, letterheads, graphics, packaging, and decor that serve as visual ambassadors for the company, communicating its desired image, values, and personality at a glance. Caribou Coffee, the nation's second-largest chain of coffee stores, recently introduced a brand makeover built around the theme "Life is short. Stay awake for it."[63]

ENTREPRENEURIAL PROFILE: Tim Weber: GoodMusicAllDay LLC Branding is another part of the marketing process that can benefit from bootstrapping. The original logo for the music blog site GoodMusicAllDay LLC was one that Tim Weber, founder and CEO, designed himself using Microsoft Word. "It was terrible," says Weber. "It was a giant lime green 'G' next to a light purple 'M' and the words '*GoodMusicAllDay.com*' with a red line under it from Word thinking it was a misspelling." Six months after founding his blog site, Weber had 40,000 views a day but knew that he needed a more professional logo. However, the site still was not generating revenues, so Weber had to find a way to bootstrap his new logo once again. "I opened a contest to our 40,000 visitors to design our new logo in return for a free shirt," said Weber. "Within two weeks I received hundreds of submissions. I let the fans vote on our Facebook page and narrow it down to five choices." From the final five, Weber selected the winning design. "This bootstrapping technique accomplished three things for the cost of one shirt," said Weber. "We got a professional logo, we increased our social media traffic, and I formed a relationship with a top-notch designer who submitted the winning logo."

EMBRACE SOCIAL MEDIA Social networking sites such as Facebook, LinkedIn, and Twitter allow companies and their customers to engage in ways that were not possible before, and smart entrepreneurs are using those sites to their advantage. Americans between the ages of 18 and 64 who use social media spend 3.2 hours *per day* on social media sites.[64] One recent study reports that companies that are highly engaged in social media significantly outperform in both revenues and profits those companies that are less engaged.[65] Yet another study finds that only 29 percent of small businesses are actively using social media but that another 23 percent want to learn how to use it.[66] Clearly, small businesses that are already successfully using social media have a clear edge!

Lessons from the Street-Smart Entrepreneur

Marketing to Millennials

Ross Chandler Hill, CEO, PictureBoothCo, has built social media into his business model. Hill rents out his proprietary picture booths to people running events. The booth's simple design means that it can be broken down and stored in a case the size of a small suitcase, and its simple operating system allows almost anyone to set it up and operate it. Therefore, Hill is able to ship his booths to people running corporate events and conferences anywhere in the country. Once they are done, they simply pack it back up and ship it back. The booth's operating system posts all of the pictures to the users' social media sites so that people can immediately share their pictures with colleagues and friends.

"When we are part of an event, the users of the photo booth create marketing content for us to use," says Hill. The pictures taken with PictureBoothCo's booths are tagged and sent out through Facebook. "One of our first customers, RedBull, found us through a past event album that we had tagged ourselves in on Facebook," says Hill. "It is amazing to think about where we would be had that not happened." Tagging their company in all of the pictures taken in the booth was a strategic decision from the outset. Social Media is not simply another sales channel for PhotoBoothCo; it is a customer referral outlet. When someone posts something about a company, they are giving their digital stamp of approval. With the Millennial Generation, recommendations sell—advertising does not.

Social media have become the main channel of communication for most Millennials. A recent study by the Pew Research Center found that 75 percent of Millennials have a social networking profile. Almost half of Millennials who use social networking sites check them at least once a day, and 88 percent of Millennials use text messaging to communicate with friends and family, sending an average of 20 text messages a day. They embrace each new development in technology that comes along.

Because Millennials use social media platforms extensively, businesses cannot reach them using the same marketing channels that they use to reach earlier generations of customers. "These unique factors make them very savvy consumers, who pay great attention to the value of what they buy and require a different way to interact with brands," says Robert Polet, president and CEO of Gucci Group. "At Gucci Group, we recognize their transformative power in the way they engage with luxury brands. We are embracing different ways of creating dialogue through social media. Some of our brands have launched Facebook and Twitter pages and iPhone applications."

Kraft Food Group recently discovered the power of effectively marketing to Millennials when they first introduced their new zero-calorie beverage called MiO. "A lot of Kraft products traditionally target the stereotypical soccer mom," says Barry Calpino, vice president of breakthrough innovation at Kraft Foods Group. "For Mio, we decided on a Millennial target. We did things differently. For example, we launched digitally before we were even shipping product and ran a campaign with a character from The Second City (a popular comedy club based in Chicago). The campaign was viral and had high viral content before Mio was on shelf. We were on Facebook offering free samples. We also focused our resources. Mio was the biggest launch in the history of Kraft."

The Street-Smart Entrepreneur offers the following principles for effectively marketing to Millennials:

- Understand that social media is where Millennials go to get information that they trust. Millennials do not trust or respond to traditional advertising. They trust their network of friends and listen to what their friends have to say about brands through social media.

- Millennials want to interact with their brands. They want to "communicate" with their brands just like they do with their friends on social media. They want to be able to rate products, make comments about products without censorship, and talk with their friends via social media about products. Smart entrepreneurs create forums for Millennials to interact with their brands and launch conversations by posting questions and ideas on various social media platforms.

- Millennials trust their "influencers." Popular musicians, sports stars, or other celebrities talking about your product in social media influence Millennials' opinions faster than data travel over the Internet!

- Build trust through frequent and consistent messages sent via social media. Building a brand means building a story. Actively manage your social media reputation. If people complain, address their concerns honestly and sincerely.

- Be creative, funny, and quirky. Millennials do not trust brands that take themselves too seriously.

- Use multiple social media channels, including Facebook, Twitter, YouTube, LinkedIn, and others. Social media channels are complementary rather than competitive.

- Use the latest innovations in technology to make your content more interesting.

- Keep in mind that Millennials have a strong social conscience. The more you are able to engage in the causes they care about and show your company's engagement via social media, the more likely they are to become loyal customers. But remember—you have to be sincere!

Sources: Based on Suzy Menkes, "Marketing to the Millennials," *New York Times*, March 2, 2010, *www.nytimes.com/2010/03/03/fashion/03iht-rmil .html?pagewanted=all&_r=0*; Michal Clements, "Kraft's Breakthrough Innovation with Mio: Marketing to Millennials," *Chicago Now*, February 6, 2013, *www .chicagonow.com/marketing-strategist/2013/02/krafts-breakthrough-innovation- with-mio-marketing-to-millennials*; "Marketing to Millennials: You'd Better Learn to Keep Up," *Allbusiness*, April 16, 2012, *smallbusiness.yahoo.com/advisor/marketing- millennials-youd-better-learn-140000612.html*; Michael Fleishner, "5 Tips for Marketing to Millennials," *www.businessknowhow.com/marketing/millennials.htm*; Marc Koenig, "3 Tips for Marketing to Millennials," *Firespring*, February 19, 2013, *blog.firespring.com/2013/02/19/3-tips-for-marketing-to-millennials.*

ENTREPRENEURIAL PROFILE: Cassie Schreiner: CSN Photography and Design Cassie Schreiner, owner of CSN Photography and Design, shoots weddings all over the central part of the United States. "One of the main ways I have bootstrapped is through free marketing tools," said Schreiner. "Facebook has been a huge part of my marketing strategy." Schreiner uses Facebook to promote word-of-mouth marketing. Her clients are excited about the images and designs she creates for their weddings. Placing them on Facebook is an easy way to have them share her work with friends and family. Some of her customers' albums or posts go viral through tagging and commenting by fellow Facebook users. "I have been able to spread the views to over 3,000 people in less than two days," said Schreiner, "which is huge for a small business like mine."

Cassie Schreiner.
Source: Cassie Schreiner.

Facebook and Twitter Although sites such as Facebook and Twitter are better known for their social applications, they also can be powerful—and inexpensive—marketing tools for small companies. More than 44 percent of companies using these social networking sites report landing at least one customer from their efforts (see Figure 9.7). Because implementing an effective social marketing strategy demands a significant investment of time (50 percent of business owners who use social

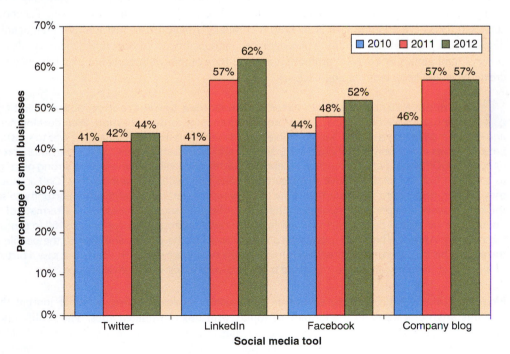

FIGURE 9.7

Percentage of Companies that Have Acquired at Least One Customer from Social Networking Sites or a Blog

Source: Based on *The State of Inbound Marketing 2011*, HubSpot, p. 10 and *The State of Inbound Marketing 2012*, Hubspot, p. 23.

media report that doing so demands more time than they had expected), some entrepreneurs outsource their social marketing efforts to companies that specialize in social media.[67]

Twitter users send more than 1.2 billion tweets per week, and the average user spends 21 minutes per month on Twitter.[68] About one third of small businesses use Twitter to market their products to more than 200 million active monthly Twitter users.[69] According to research by HubSpot, businesses that use Twitter generate twice as many sales leads as companies that do not.[70] Small companies also use Twitter to listen to feedback from their customers or to update them about special events, sales, and promotions.

ENTREPRENEURIAL PROFILE: Dotti Townsend and Matt Townsend: Roots Coffeehouse Dotti Townsend and her husband Matt, owners of Roots Coffeehouse in Fort Worth, Texas, used social media to help build traffic in their first store. They use Twitter, Facebook, FourSquare, and Instagram to engage their customers and send them free mail-newsletters using programs such as MailChimp. "We've found that asking our customers questions via these networks draws the most attention and visibility," says Dotti. "A question like 'How do you brew your coffee at home?' will get customers engaging both with us and with other customers as well as popping up on a lot of feeds." They also feature promotions where customers get a reward if they post pictures of their Roots drink on Instagram or Twitter. "Using these outlets (without paying for their advertising programs) has enabled us to capture our digital audience and widen our customer base without costing us a penny in the marketing budget," said Dotti.

The following tips help entrepreneurs use Twitter successfully as a bootstrap marketing tool:

- Engage in conversations. Twitter is a two-way communication tool, not an outlet for sending one-way marketing messages such as advertisements and press releases.
- Give people a reason to follow you. Reveal the "inside story" of your company, ask customers for feedback, and offer special deals to followers.
- Link Twitter to your company's Web site. Refer followers to your company's Web site, blog, or a video about your company and its products or services.[71]

Facebook now has more than 1 billion active monthly users and is available in more than 70 languages.[72] Two-thirds of Internet users (72 percent of women and 62 percent of men) have Facebook accounts.[73] Because its volume of online traffic is immense, Facebook offers several business-oriented features, including a survey tool that allows business users to conduct market research and an option that lets businesses create Facebook pages ("public profiles") and connect with potential customers, post photographs, and promote events. In addition to reaching potential customers, establishing a business presence on social networking sites increases a company's visibility because search engines are able to locate social network pages. One key to using Facebook successfully as a marketing tool is to keep a company's page fresh, just like the merchandise displays in a physical store. Adding photographs, announcements of upcoming events, polls and surveys, and games or contests promoting a cause the company supports are excellent ways to create buzz and keep fans coming back.

The Team at Golden Spiral Creative.
Source: Golden Spiral Creative.

ENTREPRENEURIAL PROFILE: Peter Smith: Golden Spiral Creative Golden Spiral Creative is a marketing company that provides branding, content creation, Web development, and online marketing services. "Our main objective as we began to market ourselves was to substantiate ourselves as not only a legitimate company but also an industry leader," says Peter Smith, president and COO. "Social networks such as Facebook, Twitter, and LinkedIn have been a strong place for us to start." Golden Spiral Creative first created profiles filled with relevant content. On Facebook, they created graphics to populate their time line, giving for potential clients an opportunity to see the kind of work they perform. "One push that was particularly successful was in the aftermath of an open house we held at our office," said Smith. "We hired a photographer for under $100 to capture the evening. After the party, we posted an album with all of the photos and tagged the people at the party. This brought us two direct leads from people who were not at the party but saw a picture of a friend at the party and were impressed by the crowd and our office environment."

Blogging Blogs (Web logs) started out as a frequently updated online personal journal that contains a writer's ideas on a multitude of topics with links to related sites. The proliferation of blogs has been stupendous; everyone from teenagers to giant corporations has created blogs. There are about 42 million blogs that publish about 500,000 posts a day. A recent survey of small business owners found that 55 percent have blogs.[74] However, about 65 percent of businesses with blogs have not made a post on their blog in the past year.[75] The most successful small business blogs are not just remakes of a company's Web site with thinly veiled marketing messages but instead those that tell interesting stories from the perspective of an industry (and company) insider. The key to successful business blogging is to create a blog that provides useful industry information but that also is entertaining. Web sites such as WordPress, Blogger, or LiveJournal make it easy for entrepreneurs to start blogging.

Business blogging can be an effective part of a bootstrap marketing strategy, enabling entrepreneurs to reach large numbers of potential customers economically. Blogs help establish a business owner as an expert in the field, attract the attention of potential customers, and boost a company's visibility and its sales. Companies post their blogs, promote them on their Web sites and on other blogs, and then watch as the viral nature of the Web takes over with visitors posting comments and e-mailing their friends about the blog. In fact, many small companies allow customers to contribute to their blogs, offering the potential for one of the most valuable marketing tools: unsolicited endorsements from satisfied users. Blogging's informal dialogue is an ideal match for small companies whose cultures and style tend to be informal.

Blogs can serve many business purposes, including keeping customers updated on new products, enhancing customer service, and promoting the company. Increasingly, they are becoming mainstream features on business Web sites. If monitored regularly, blogs also can give entrepreneurs keen insight into customers' viewpoints and preferences in ways that few other techniques can. One business writer says that blogs are "like never-ending focus groups."[76] Creating a blog is not risk free, however. Companies must be prepared to deal with negative feedback from some visitors.

Online Videos Video hosting sites such as YouTube give creative entrepreneurs the opportunity to promote their businesses at no cost by creating videos that feature their companies' products and services in action. Unlike television ads, uploading a video to YouTube costs nothing, and in some cases the videos reach millions of potential customers. Online videos do not have to be of professional quality to be effective; in fact, some of the most successful videos boast a distinctive "amateur" look. Eighty-five percent of Internet users watch videos online each month, and the average online viewer watches 17.4 hours of content video per month. More importantly for business, viewing of video ads has increased dramatically. More than 50 percent of Americans view 9.9 billion video ads each month.[77]

ENTREPRENEURIAL PROFILE: George Livingston and Ian Raffalovich: Sweet Meat Jerky George Livingston and Ian Raffalovich, cofounders of Sweet Meat Jerky, use videos to connect with their customers. When they were launching their fundraising campaign on Kickstarter, they filmed the steps required for two bootstrapping entrepreneurs to prepare a large order for shipment from their apartment using a time-lapse video. "When I posted it to social media, people loved it," said Livingston. "Everyone knew that shipping 400 packages of jerky was a lot of work—but when you show them the process of packing, sealing, labeling, addressing, and shipping packages over the course of an entire day, they see exactly how much work it was. People like feeling special, so creating unique video content that you share with only certain people (our Kickstarter backers in this case) can be a powerful tool." The entrepreneurs also post their videos on the company's Facebook page.

The goal of an online video is to increase awareness of a company, drive traffic to its Web site or store, and increase sales. To market their companies successfully on YouTube, entrepreneurs should do the following:

- *Develop a well-defined channel.* A "channel" on YouTube is a home page for an account. *RevZilla.com* is a motorcycle apparel, parts, and accessories online retailer whose YouTube channel has more than 2,000 videos designed to answer common customer questions about sizes and fit for many of its products by showing their employees trying on various products and discussing how to pick out the right styles and sizes.[78]

- *Use the right key words.* YouTube works just like any search engine. People find content by searching key words. Tie your video to the most popular topics that people are searching on YouTube.

- *Think "edutainment."* Some of the most successful online videos combine both educational content and entertainment. Many businesses that sell cooking equipment and accessories have YouTube channels that give cooking lessons using their products. People who watch YouTube don't just want to see commercials—they are seeking information.

- *Be funny.* A common denominator among many successful online videos is humor. For businesses, the key is to link the humor in the video to the company's product or service and its customer benefits.

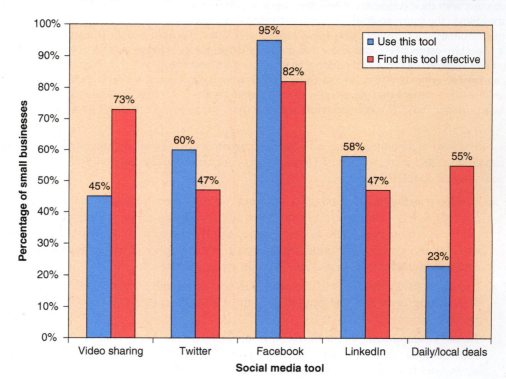

ENTREPRENEURIAL PROFILE: Dr. Robert Wagstaff: Orabrush Orabrush is a Provo, Utah, company that makes a tongue cleaner. Developed by Dr. Robert Wagstaff, the device, which is designed to eliminate bad breath, was selling only 10 tongue cleaners a month at $3 each. Wagstaff decided to focus his marketing efforts on a YouTube video. The result was a video titled "How to Tell When Your Breath Stinks." A friend he hired to act in the video wears a white lab coat and protective goggles and educates viewers about bad breath and how the Orabrush tongue cleaner could cure it. Wagstaff posted the video on YouTube and purchased video ads using the key word "bad breath." The video was also posted on Orabrush's Facebook page and sent out via Twitter. Within the next six weeks, more than 900,000 people viewed the video, and 20 percent of them clicked on the link to the Orabrush Web site. As a result, the company sold about 10,000 tongue cleaners. Over the next two years, Orabrush sold more than 1 million products, and its video was viewed more than 15 million times. The company now posts weekly videos on its YouTube channel. The total budget for the campaign was $1,260.[79]

- *Post videos on multiple social media sites.* Customers are more likely to see a video when a company makes it available on multiple social media sites. TubeMogul is a free site that pushes video content out to multiple social media sites. However, YouTube is still by far the most watched online source for video.
- *Use other social media tools to promote new videos.* Active users of online video for promotion post many new videos each year. Some try to post a new video as often as every week. In fact, more video content is posted on YouTube each month than the three major television networks produced in the last 60 years! With so much content, finding ways to point customers to your videos using tools such as social media is essential.

Figure 9.8 shows how owners of small companies view the use of these social media tools. Small business owners say that Facebook and online videos are their most effective social media marketing tools

Price

Almost everyone agrees that the price of the product or service is a key factor in the decision to buy. Price affects both sales volume and profits, and without the right price, both sales and profits will suffer. But pricing is also a powerful part of the marketing mix. As we will see in Chapter 11,

FIGURE 9.8

Percentage of Small Business Owners Who Say They . . .

Source: Based on *Small Business Attitudes and Outlook Survey*, Spring 2011, Constant Contact, p. 1.

the right price for a product or service depends on three factors: (1) a small company's cost structure, (2) an assessment of what the market will bear, and (3) the desired image the company wants to create in its customers' minds.

For many small businesses, nonprice competition, focusing on factors other than price, is a more effective strategy than trying to beat larger competitors in a price war. Nonprice competition, such as free trial offers, free delivery, lengthy warranties, and money-back guarantees, intends to play down the product's price and stress its durability, quality, reputation, or special features.

Place

Place (or method of distribution) has grown in importance as customers expect greater service and more convenience from businesses. Because of this trend, mail-order houses, home shopping channels, home shopping parties, and the Internet's offering the ultimate in convenience—shop at home—have experienced booming sales in recent years. In addition, many traditionally stationary businesses have added wheels, becoming mobile animal clinics, computer repair shops, and dentist offices. Any activity involving movement of goods to the point of consumer purchase provides place utility. Place utility is directly affected by the marketing channels of distribution, that is, the path that goods or services and their titles take in moving from producer to consumer.

ENTREPRENEURIAL PROFILE: Louis and Peter Tomassetti: ShaveMate Louis and Peter Tomassetti invented a razor, ShaveMate, that contains the shaving cream in the handle. Their first customer was the military, which bought the product directly from the brothers, but they wanted to expand sales to the retail market. However, convincing retailers to give shelf space to a brand other than Gillette and Schick was difficult. They needed to build brand awareness before they could ever hope to sell their product through traditional retail channels, so they decided on a marketing plan that included both a low-cost specific public relations effort and an advertising blitz on cable television. They were able to gain local and national media exposure and were eventually invited to try selling their product through infomercials. The media attention and growing brand awareness opened the door for the Tomassettis to offer ShaveMate through national online retailers including Walgreens and Target. Because of the success of sales through the online retail distribution channel, Walgreens decided to sell ShaveMate in its retail stores.[80]

Channels typically involve a number of intermediaries who perform specialized functions that add valuable utility to the goods or service. Specifically, these intermediaries provide time utility (making the product available when customers want to buy it) and place utility (making the product available where customers want to buy it).

For consumer goods, there are four common channels of distribution:

1. *Manufacturer to consumer.* In some markets, producers sell their goods or services directly to consumers. Services, by nature, follow this channel of distribution. Dental care and haircuts, for example, go directly from creator to consumer.

2. *Manufacturer to retailer to consumer.* Another common channel involves a retailer as an intermediary. Many clothing items, books, shoes, and other consumer products are distributed in this manner.

3. *Manufacturer to wholesaler to retailer to consumer.* This is the most common channel of distribution for consumer goods. Prepackaged food products, hardware, toys, and other items are commonly distributed through this channel.

4. *Manufacturer to wholesaler to wholesaler to retailer to consumer.* A few consumer goods (e.g., agricultural goods and electrical components) follow this pattern of distribution.

Two channels of distribution are common for industrial goods:

1. *Manufacturer to industrial user.* The majority of industrial goods are distributed directly from manufacturers to users. In some cases, the goods or services are designed to meet the user's specifications.

2. *Manufacturer to wholesaler to industrial user.* Most expense items (paper clips, paper, rubber bands, and cleaning fluids) that firms commonly use are distributed through wholesalers. For most small manufacturers, distributing goods through established wholesalers and agents is often the most effective route. With their limited resources, entrepreneurs sometimes have to rely on nontraditional distribution channels and use their creativity to get their products into customers' hands.

Chapter Review

1. **Describe the components of a bootstrap marketing plan and explain the benefits of preparing one.**
   - A major part of the entrepreneur's business plan is the marketing plan, which focuses on a company's target customers and how best to satisfy their needs and wants. A solid marketing plan should pinpoint the specific target markets the company will serve, determine customer needs and wants through market research, analyze the firm's competitive advantages and build a marketing strategy around them, and create a marketing mix that meets customer needs and wants.

2. **Explain how small businesses can pinpoint their target markets.**
   - Sound market research helps the owner pinpoint his or her target market. The most successful businesses have well-defined portraits of the customers they are seeking to attract.

3. **Explain how to determine customer needs through market research and the steps in the market research process.**
   - Market research is the vehicle for gathering the information that serves as the foundation of the marketing plan. Good research does not have to be complex and expensive to be useful. The steps in conducting market research include the following:
     - Defining the problem: "What do you want to know?"
     - Collecting the data from either primary or secondary sources
     - Analyzing and interpreting the data
     - Drawing conclusions and acting on them

4. **Describe the bootstrap marketing strategies on which a small business can build competitive edge in the marketplace.**
   - When plotting a marketing strategy, owners must strive to achieve a competitive advantage—some way to make their companies different from and better than the competition. Successful small businesses rely on niche picking, retaining their existing customers, and concentrating on innovation to implement a bootstrapped marketing strategy.

5. **Explain the "four Ps" of marketing—product, place, price, and promotion—and their role in building a successful bootstrap marketing strategy.**
   - The marketing mix consists of the "four Ps":
     Product. Entrepreneurs should understand their product's place in the product life cycle. Positioning the product includes focusing on the customers' needs, dedication to service and customer satisfaction, devotion to quality, dedication to service, and emphasizing speed.
     Promotion. Bootstrapped promotion included the use of publicity, an understanding the entertaining must be part of selling, the power of branding, and embracing social media.
     Price. Price is an important factor in customers' purchase decisions, but many small businesses find that non-price competition can be profitable. But, pricing must fit with the other three "Ps".
     Place. The focus here is on choosing the appropriate channel of distribution and using it most efficiently.

Discussion Questions

9-1. What is a marketing plan? What lies at its center?

9-2. What objectives should a marketing plan accomplish?

9-3. How can market research benefit a small business owner? List some possible sources of market information.

9-4. Does market research have to be expensive and sophisticated to be valuable? Explain.

9-5. Why is it important for small business owners to define their target markets as part of their marketing strategies?

9-6. What is a competitive edge? How might a small company gain a competitive edge?

9-7. Describe how a small business owner could use the following sources of a competitive advantage: niche picking, entertailing, building a consistent branding strategy, emphasizing their uniqueness, connecting with their customers, focusing on customers' needs, emphasizing quality, paying attention to convenience, concentrating on innovation, dedicating themselves to service, and emphasizing speed.

9-8. Explain the concept of the marketing mix. What are the four Ps? How can an entrepreneur use bootstrapping in the marketing mix?

9-9. List and explain the stages in the product life cycle. How can a small firm extend its product's life?

9-10. With a 70 percent customer retention rate (average for most U.S. firms according to the American Management Association), every $1 million of business in 2014 will grow to more than $4 million by 2024. If you retain 80 percent of your customers, the $1 million will grow to a little over $6 million. If you can keep 90 percent of your customers, that $1 million will grow to more than $9.5 million. What can the typical small business do to increase its customer retention rate?

CHAPTER 10

Creative Use of Advertising and Promotion

Learning Objectives

On completion of this chapter, you will be able to:

1. Define your company's unique selling proposition.
2. Explain the differences among promotion, publicity, personal selling, and advertising.
3. Describe the advantages and disadvantages of the various advertising media.
4. Identify four basic methods for preparing an advertising budget.
5. Explain practical methods for stretching an entrepreneur's advertising budget.

Advertising is what you do when you can't go see somebody.

—Fairfax Cone

In our factory, we make lipstick.
In our advertising, we sell hope.

—Peter Nivio Zarlenga

Advertising is not just a business expense; it is an investment in a company's future. Without a steady advertising and promotional campaign, a small business's customer base will soon dry up. Advertising can be an effective means of increasing sales by telling customers about a business and its goods or services, by improving the image of the firm and its products, and by persuading customers to purchase its goods or services. A megabudget is not a prerequisite for building an effective advertising campaign. With a dose of creativity and ingenuity, a small company can make its voice heard above the clamor of its larger competitors—and stay within a limited budget! A company's promotional strategy, which is comprised of publicity, personal selling, and advertising, must deliver the same clear, consistent, and compelling message about the business and its products or services. Customers respond best to a positive message that is delivered consistently by each component of the strategy. One goal of a company's promotional strategy is to build **brand equity**, which is measured by customer loyalty and customers' willingness to pay a premium for its products and services.

Developing an effective advertising program has become more challenging for business owners. Because of media overflow, overwhelming ad clutter, increasingly fragmented audiences, the popularity of social media, more advertising options, and more skeptical consumers, companies have had to become more innovative and creative in their advertising campaigns. Rather than merely turning up the advertising volume on their campaigns, companies are learning to change their frequencies, trying out new approaches in different advertising media.

A company's promotional efforts must differentiate its products and services from those of competitors. Some of the most effective advertisers have enhanced their brand loyalty by emphasizing in their promotional strategies the unique customer benefits that their products or services provide. For example, Nordstrom department stores are defined by friendly customer service, Volvo is known for automotive safety, and FedEx is recognized for guaranteed overnight delivery. One of the first steps is to carefully and thoughtfully define the message that a company's promotional campaign will emphasize by defining its *unique selling proposition*.

Define Your Company's Unique Selling Proposition

1.

Define your company's unique selling proposition.

Entrepreneurs should build their advertising messages on a **unique selling proposition** (USP), a key customer benefit or a product or service that sets it apart from its competition. To be effective, a USP must actually *be* unique—something the competition does not (or cannot) provide—and compelling enough to encourage customers to buy. One technique for testing uniqueness is to replace your company's name and logo in one of your advertisements with those of your top competitor. Does the ad still make sense? If so, the ad is not based on your company's USP! Unfortunately, many business owners never define their companies' USPs, and the result is an uninspiring "me-too" message that cries out "buy from us" without offering customers any compelling reasons to do so.

A successful USP answers the critical question every customer asks: "What's in it for me?" A successful USP should express in no more than 10 words what a business can do for its customers. Can your product or service save your customers time or money, make their lives easier or more convenient, improve their self-esteem, or make them feel better? If so, you have the foundation for building a USP. The most effective ads are *not* just about a company's products and services; instead, they focus on the company's customers and how its products and services can improve *their* lives.

The most effective USPs are simple, concrete, believable, emotional, and easy to communicate to prospective customers. The best way to identify a meaningful USP is to describe the primary benefits a product or service offers customers and then to list other secondary benefits it provides. Most businesses will have no more than three primary benefits. Smart entrepreneurs look beyond the physical characteristics of their products or services, recognizing that sometimes the most powerful foundation for a USP is the *intangible or psychological* benefit a product or service offers customers, such as safety, security, acceptance, status, prestige, and others. The key is to identify a gap that customers typically experience and explain how your company's product or service can fill it. Before creating advertisements, entrepreneurs should develop a brief list of the facts that support the company's USP, such as an unconditional guarantee; 24-hour service; a fully trained, experienced staff; industry awards won; and others. The final step is to consolidate the gap-filling benefits the company offers into a single statement: the USP.

The USP becomes the heart of a company's advertising message because it has the ability to cut through all of the advertising clutter. For instance, the owner of a quaint New England bed-and-breakfast came up with a four-word USP that captures the essence of the escape her business offers guests from their busy lives: "Delicious beds, delicious breakfasts." Sheila Paterson, cofounder of Marco International, a marketing consulting firm, says her company's USP is "Creative solutions for impossible marketing problems."[1] Dave Munson, founder of Saddleback Leather, a company that manufactures premium bags made from exclusive leather from the best tannery in the Western Hemisphere and featuring the finest details, stitching, and hardware, has a witty (but true) USP: "They'll fight over it when you're dead" (Munson is so confident in the superior quality of his company's bags, which carry a 100-year guarantee against all defects in workmanship and materials, that he includes links to his competitors' Web sites on the Saddleback Web site!).[2]

By focusing a company's advertising message on these top benefits and the facts that support them, entrepreneurs can communicate their USPs to their target audiences in meaningful, attention-getting ways. Building a firm's marketing message around a USP spells out for customers the benefits they can expect if they buy the company's product or service and why they should do business with a company rather than with its competition. However, a company must be able to *deliver* on its USP; otherwise, the advertising effort is futile!

Creating a Promotional Strategy

The terms *advertising* and *promotion* are often confused. **Promotion** is any form of persuasive communication designed to inform consumers about a product or service and to influence them to purchase these goods or services. It includes publicity, personal selling, and advertising.

Publicity

Publicity is any commercial news covered by the media that boosts sales but for which the small business does not pay. "[Publicity] is telling your story to the people you want to reach—namely the news media, potential customers, and community leaders," says the head of a public relations firm. "It is not haphazard. . . . It requires regular and steady attention."[3] Because it originates from an external source, some entrepreneurs worry about losing control of the message and the medium. However, smart entrepreneurs realize that what they lose in control they gain in credibility. Publicity has power; because it is from an unbiased source, a news feature about a company or a product appearing in a newspaper or magazine has more impact on people's buying decisions than an advertisement does. Exposure in any medium raises a company's visibility and boosts sales, and, best of all, publicity is low cost!

ENTREPRENEURIAL PROFILE: Kim Nelson: Daisy Cakes Kim Nelson, owner of Daisy Cakes, a small bakery based in tiny Pauline, South Carolina, learned to bake delicious cakes from her grandmothers and was slowly building a base of loyal customers, almost all of them local. After Nelson appeared on *Shark Tank,* a television show in which entrepreneurs in search of capital pitch their business ideas to potential angel investors, and on the food segment of NBC's *Today Show,* sales skyrocketed, causing the company's Web site to crash. (Barbara Corcoran, one of the *Shark Tank* angels, invested $50,000 in Daisy Cakes in exchange for 25 percent of the company.) Before the publicity that Nelson's television appearances generated, Daisy Cakes sold 500 cakes per year; today, Daisy Cakes sells 10,000 cakes a week, most of them through the company's Web site, *www .ilovedaisycakes.com.*[4]

A sound marketing plan includes strategies for both publicity and advertising; each complements the other. Arguing whether one is more important than the other "is like debating what's more important in football—offense or defense," says entrepreneur Freddy Nager.[5] The following tactics can help entrepreneurs stimulate publicity for their companies:

Write an article that will interest your customers or potential customers. One investment adviser writes a monthly column for the local newspaper on timely topics such as "Retirement Planning," "Minimizing Your Tax Bill," and "How to Pay for College." Not only do the articles help build her credibility as an expert, but they also have attracted new customers to her business.

2. _____

Explain the differences among promotion, publicity, personal selling, and advertising.

Sponsor an event designed to attract attention. Even local events, for which sponsorships can be quite inexpensive, garner press coverage for sponsors.

Involve celebrities "on the cheap." Few small businesses can afford to hire celebrities as spokespersons for their companies. Some companies have discovered other ways to get celebrities to promote their products, however. For instance, when Karen Neuburger, owner of Karen Neuburger's Sleepwear, learned that Oprah Winfrey is a "pajama connoisseur," she sent the talk show host a pair of her pajamas. The move paid off; Neuburger appeared on Oprah's popular television show on three separate occasions, and each time, sales for her sleepwear increased.[6]

Contact local television and radio stations and offer to be interviewed. Many local news or talk shows are looking for guests to talk about topics of interest to their audiences (especially in January and February). Even local shows can reach new customers.

Publish a newsletter. With inexpensive software, any entrepreneur can publish a professional-looking newsletter. Freelancers can offer design and editing advice. The key is to make newsletters useful and interesting to existing or potential customers.

ENTREPRENEURIAL PROFILE: David Sanford: Crowne Point Historic Inn and Spa
David Sanford, owner of the Crowne Point Historic Inn and Spa in Provincetown, Massachusetts, used e-mail newsletters to increase sales during the traditionally slow off-season in New England. Sanford had tried using "traditional" advertising techniques but found the costs to be high and the results they produced to be minimal. He began sending regular e-mail newsletters with the goal of increasing the inn's customer base and attendance at the special events the inn hosts. Analytics allow Sanford to track recipients' responses to the newsletters and to determine which features produce the greatest results. "I've increased my revenue while decreasing my marketing costs," he says. "I can stay in front of customers 12 months a year with very little expense."[7]

Contact local business and civic organizations and offer to speak to them. A powerful, informative presentation can win new business. (Be sure your public speaking skills are up to par first! If not, consider joining Toastmasters.)

Offer or sponsor a seminar. Teaching people about a subject you know a great deal about builds confidence and goodwill among potential customers. The owner of a landscaping service and nursery offers a short course in landscape architecture and always sees sales climb afterward.

Write news releases e-mail them to media outlets. The key to having a news release picked up and printed is finding a unique angle on your business or industry that would interest an editor. Keep it short, simple, and interesting. E-mail press releases should be shorter than printed ones—typically four or five paragraphs rather than one or two pages—and they should include a link to the company's Web site. For reasonable fees, entrepreneurs can get news releases in front of thousands of journalists around the world using services such as PR Newswire and Business Newswire.

Volunteer to serve on community and industry boards and committees. You can make your town a better place to live and work and raise your company's visibility at the same time.

Sponsor a community project or support a nonprofit organization or charity. Not only will you be giving something back to the community, but you will also gain recognition, goodwill, and, perhaps, customers for your business.

ENTREPRENEURIAL PROFILE: Rocky and Courtney Shanower: Park Street Pizza Every December, Rocky and Courtney Shanower, owners of Park Street Pizza in Sugarcreek, Ohio, partner with their friend Joel McKinnon, owner of a local insurance agency, in a "Pizza with a Purpose" event that supports the local United Way. The Shanowers donate 15 percent of their sales on the day of the event to the charity. Not only do the entrepreneurs take pride in giving back to the community that supports their businesses, but they also receive the equivalent of thousands of dollars worth of advertising from the publicity generated by the Pizza with a Purpose event.[8]

Support a cause. By engaging in **cause marketing**, entrepreneurs can support and promote a nonprofit cause or charity and raise the visibility of their companies in the community at the same time. The key is choosing a cause that is important to your customers. One marketing expert offers the following formula for selecting the right cause: Mission statement + personal passion + customer demographics = ideal cause.[9] Charity Navigator (*www.charitynavigator.org*) is an excellent resource for finding the right charity for your business to support.

Promote your company's publicity. When your company receives good publicity, promote it by posting the article or video on your company's Web site, posting links to the article or video on Facebook, tweeting about the coverage on Twitter, and using excerpts from it in your company's marketing materials. When Blue Ridge Log Cabins, a Spartanburg, South Carolina–based company that manufactures log homes, was featured on *Extreme Makeover Home Edition* and HGTV, owner Chip Smith placed links to the videos in a prominent place on the company's Web site and referred to the coverage in all of the company's marketing materials, including outdoor ads directing customers to the company's showroom.

Personal Selling

Advertising often marks the beginning of a sale, but personal selling usually is required to close the sale. **Personal selling** is the personal contact between salespeople and potential customers that comes from sales efforts. Effective personal selling can give a small company a definite advantage over its larger competitors by creating a feeling of personal attention. Personal selling deals with the salesperson's ability to match customer needs to the company's goods and services. Top salespeople are characterized as follows:

- They are enthusiastic and are alert to opportunities. Star sales representatives demonstrate deep concentration, high energy, and drive.

- They are experts in the products or services they sell. They understand how their product lines or services can help their customers.

- They concentrate on their best accounts. They focus on customers with the greatest sales potential first. They understand the importance of the 80/20 rule: Approximately 80 percent of their sales comes from about 20 percent of their customers.

- They plan thoroughly. On every sales call, the best representatives act with a purpose to close the sale.

- They use a direct approach. They get right to the point with customers.

- They approach the sales call from their customers' perspectives. They have empathy for their customers and know their customers' businesses and their needs. Rather than sell the features of a product or service, they emphasize the benefits those features offer their customers.

- They offer proof of the benefits their product or service provides. The best salespeople provide tangible evidence such as statistics, facts, and testimonies from other customers about how their product or service will benefit the customer.

- They are good listeners. They ask questions and listen. By listening, sales representatives are able to identify customers' "hot buttons," key issues that drive their purchase decisions. "Questions are the key to selling," says one experienced salesperson. "Nobody ever listened themselves out of a sale!"[10]

- They use past success stories. They encourage customers to express their problems and then present solutions using examples of past successes.

- They leave sales material with clients. The material gives the customer the opportunity to study company and product literature in more detail.

- They see themselves as problem solvers, not just vendors. Their perspective is "How can I be a valuable resource for my customers?" In fact, smart salespeople take the time to ask their existing customers, "Is there anything I am not doing that I could be doing to serve you better?" A study by Cahners Research found that sales representatives who understand the business needs and pressures that their customers face are 69 percent more likely to close a sale.[11]

- They measure their success not only by sales volume but also by customer satisfaction.

FIGURE 10.1

**How Sales
Representatives Spend
Their Time**

Source: 2011 Sales Optimization
Report, CSO Insights, 2011.

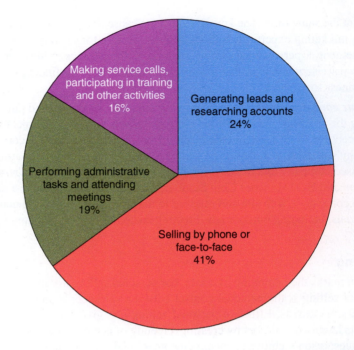

One extensive study of salespeople found that just 20 percent of all salespeople have the ability to sell and are selling the "right" product or service. That 20 percent of sales reps makes 80 percent of all sales. The study also concluded that 55 percent of sales representatives have "absolutely no ability to sell;" the remaining 25 percent have sales ability but are selling the wrong product or service.[12]

A study by Cahners Research found that it takes an average of 5.12 sales calls to close a deal.[13] Common causes of sales rejections include the representative's failure to determine customers' needs, talking too much, and neglecting to ask for the order. Given the high cost of making a sales call (an average of $412), those missed opportunities are quite costly. Figure 10.1 shows how sales representatives spend their time. (Note that they spend just 41 percent of their time engaged in active selling.)

Small business owners can improve their sales representatives' "closing averages" by following some basic guidelines:

Hire the right people. A successful sales effort starts well before a sales representative calls on a potential customer. The first step is hiring capable salespeople who demonstrate empathy for customers and are motivated, persistent, and focused.

Train sales representatives. Too often, business owners send sales representatives out into the field with little or no training and then wonder why they cannot produce. Training starts with teaching salespeople every aspect of the products or services they will be selling before moving on to teach them how to build relationships with customers. Training must also include the two most important selling skills of all: listening to the customer and closing the sale. Many business owners find that role-playing exercises are an effective sales training technique.

Develop a selling system. To be successful, sales representatives must develop an effective selling system. To build a winning selling system, entrepreneurs can take the following steps:

1. *Prepare.* The best sales representatives know that what they do *before* they make a sales call significantly influences their success. In fact, the top complaint about sales representatives among buyers is a salesperson who is unprepared. Unfortunately, according to a study by Knowledge Anywhere, nearly 63 percent of sales representatives spend less than 20 minutes preparing for a sales call.[14] Smart salespeople take the time to research their customers (most often using the Internet) and to learn about the companies where their customers work.

2. *Approach.* Establish rapport with the prospect. Customers seldom buy from salespeople they dislike or distrust.

3. *Interview.* Get the prospect to do most of the talking; the goal is to identify his or her needs, preferences, and problems. The key is to *listen* and then to ask follow-up questions that help determine exactly what the customer needs or wants before proposing a solution. Norm Brodsky, founder of six companies, including a highly successful records storage business, says, "When I call on a prospect for the first time, I don't even talk about our company. I spend the whole visit just trying to learn all I can about the people I'm dealing with. I look to build rapport and understand how the customer likes to do business."[15]

4. *Demonstrate, explain, and show.* Make clear the features and benefits of your product or service and point out how they meet the prospect's needs or solve his or her problems.

5. *Validate.* Prove the claims about your product or service. If possible, offer the prospect names and numbers of other satisfied customers (with their permission, of course). Testimonials really work.

6. *Negotiate.* Listen for objections from the prospect. Objections can be the salesperson's best friend; they tell him or her what must be "fixed" before the prospect will commit to an order. The key is to determine the *real* objection and confront it.

7. *Close.* Ask for a decision. Good sales representatives know when the prospect flashes the green light on a sale. They stop talking and ask for the order.

Be empathetic. The best salespeople look at the sale from the customer's viewpoint, not their own! Doing so encourages the sales representative to stress *value* to the customer.

Set multiple objectives. Before making a sales call, salespeople should set three objectives:

1. *The primary objective* is the most reasonable outcome expected from the meeting. It may be to get an order or to learn more about a prospect's needs.

2. *The minimum objective* is the very least the salesperson will leave with. It may be to set another meeting or to identify the prospect's primary objections.

3. *The visionary objective* is the most optimistic outcome of the meeting. This objective forces the salesperson to be open-minded and to shoot for the top.

Monitor sales efforts and results. Selling is just like any other business activity and must be controlled. At a minimum, entrepreneurs should know the following numbers for their companies:

1. Actual sales versus projected sales

2. Sales generated per call made

3. Average cost of a sales call

4. Total sales costs

5. Sales by product, salesperson, territory, customer, and so on

6. Profit contribution by product, salesperson, territory, customer, and so on

 Entrepreneurship in Action

Should an Ambulance Service Hire Its First Sales Representative?

Mike Woronka had worked at Action Ambulance Service, a small business in Wilmington, Massachusetts, that provides ambulance and transportation services for hospitals, nursing homes, rehabilitation centers, and special education programs, for 10 years before becoming the company's CEO in 2006. Since its founding in 1977 by brothers David and Stanley Portman, Action Ambulance has grown to employ 260 employees and to generate $17 million in annual sales. However, the company faces intense competition from seven rivals in the greater Boston area, and rising prices and wages have driven up its cost of doing business—all while its customers are demanding faster response times. Woronka wants to differentiate Action Ambulance from its competition and believes that a more aggressive selling approach would expand the company's territory and increase sales and profits.

(continued)

Entrepreneurship in Action (continued)

Currently, Action Ambulance follows the standard model in the industry, using its account managers, who in addition to serving current customers, spend some of their time trying to develop new accounts. At conferences, the account managers meet with representatives of health care providers who are considering switching ambulance companies and with their current customers to land more of their transportation business. Currently, the account managers have no goals for landing new customers or increasing sales and are paid straight salaries with no commissions.

Woronka has considered retraining the company's existing account managers to use more aggressive direct-sales techniques. Even though doing so is the least expensive option, he is concerned that the managers may not have the skills or the motivation to become part-time salespeople. Adding to their job duties also runs the risk that they will spend less time providing quality service to existing customers.

Woronka wonders whether hiring a full-time sales representative to generate new sales would be worth the additional cost, which he estimates would include $100,000 in salary plus a commission and 30 percent for benefits. "Many people say that you can't sell ambulance services," he says, "and that we are crazy to want to do things differently." If Woronka hires a sales representative, that person would report directly to him, which creates another potential problem because he has no sales experience.

Another option is to hire both a sales representative and a sales manager, whom he estimates would be paid $150,000

in salary. That would add to the company's cost structure, but the sales representative and the two current account managers would report directly to the sales manager, giving Woronka more time to focus on his duties as CEO.

If Woronka decides to hire a sales representative, he wonders how he should design the job. How many sales calls should the rep make each week? How much should he expect the sales rep to generate in new sales each year? What type of base salary and commission structure should he establish? From a broader perspective, how would potential customers perceive a sales representative selling ambulance services in an industry that traditionally has not used sales reps?

1. Should Woronka retrain the company's existing account managers or hire a sales representative? Should he hire a sales manager? Explain the advantages and the disadvantages of each option.

2. If Woronka hires a sales representative and/or a sales manager, how should he design the job? What system should he put in place to make sure that the sales representative produces results?

Sources: Based on Pamela Ryckman, "When the Soft Sell Needs a Hard Look," *New York Times*, February 15, 2012, *www.nytimes.com/2012/02/16/business/smallbusiness/trying-to-ramp-up-sales-in-an-industry-that-prefers-the-soft-sell.html?ref=casestudies&_r=0*; Pamela Ryckman, "A Reluctant CEO Hires His First Sales Rep," *New York Times*, February 22, 2012, *http://boss.blogs.nytimes.com/2012/02/22/a-reluctant-c-e-o-hires-his-first-sales-rep*; "About Action," Action Ambulance Services, 2012, *www.actionambulance.com/index-2.html#whos.*

Advertising

Advertising is any sales presentation that is nonpersonal in nature and is paid for by an identified sponsor. A company's target audience and the nature of its message determine the advertising media it will use. However, the process does not end with creating and broadcasting an ad. Entrepreneurs also must evaluate an ad campaign's effectiveness. Did it accomplish the objectives it was designed to accomplish? Immediate-response ads can be evaluated in a number of ways. For instance, a business owner can include coupons that customers redeem to get price reductions on products and services. Some firms use "hidden offers," statements hidden somewhere in an ad that offer customers special deals if they mention an ad or bring in a coupon from an ad. For example, Scott Fiore, owner of the Herbal Remedy, an all-natural pharmacy in Littleton, Colorado, uses a "bring this ad in for 10 percent off" message in his print ads so that he can track each ad's success rate and adjust his advertising expenditures accordingly.

Business owners can also gauge an ad's effectiveness by measuring the volume of store traffic generated. Effective advertising should increase store traffic, which boosts sales of advertised and nonadvertised items. Of course, if an advertisement promotes a particular bargain item, the owner can judge its effectiveness by comparing sales of the items to preadvertising sales levels. Remember: The ultimate test of an ad is whether it increases sales!

Ad tests allow entrepreneurs to determine the most effective methods of reaching their target customers. An owner can design two different ads (or use two different media or broadcast times) that are coded for identification and see which one produces more responses. For example, a business owner can use a split run of two different e-mail ads, each with a different subject line. Then he or she can measure the response level to each ad to determine which one generates the greatest response. Table 10.1 offers 12 tips for creating an effective advertising campaign.

TABLE 10.1 Twelve Tips for Effective Advertising

1. *Plan more than one advertisement at a time.* An advertising campaign is likely to be more effective if you build it from a comprehensive plan for a specific time period. A piecemeal approach produces ads that lack continuity and a unified theme. One goal of advertising is to build a consistent image for your business in the minds of your customers, and that requires a comprehensive plan.

2. *Set long-run advertising objectives and measure performance against them.* One cause of inadequate planning is the failure to establish specific objectives for the advertising program. If an entrepreneur never defines what is expected from advertising, the program is likely to lack a sense of direction. Measuring an ad campaign's performance against objectives allows entrepreneurs to improve the performance of future campaigns by learning what works—and what does not.

3. *Use advertisements, themes, and media that appeal to your target customers.* Although personal judgment influences every business decision, business owners cannot afford to let bias interfere with advertising decisions. For example, you should not advertise on a particular radio station simply because you like it. What matters is whether your company's target customers listen to the station. One of the worst advertising mistakes entrepreneurs make is to create ads that attempt to make their businesses "everything to everyone." Focus instead on your company's USP and on what your company can do for its target customers.

4. *View advertising expenditures as investments, not as expenses.* From an accounting perspective, advertising is a business expense, but the money you spend on ads produces sales and profits over time that might not be possible without advertising. An effective advertising program generates more in sales than it costs. You must ask, "Can I afford *not* to advertise?"

5. *Use advertising that is different from your competitors' advertising.* Some entrepreneurs tend to "follow the advertising crowd" because they fear being different from their competitors. "Me-too" advertising frequently is ineffective because it fails to create a unique image for a business. Don't be afraid to be bold and try something different! Although entrepreneurs must adjust their advertising expenditures to coincide with the seasonal nature of their businesses, investing consistently in advertising produces better results than sporadic fits of spending.

6. *Choose the media vehicle that is best for your business even if it's not "number one."* It is not uncommon for several media within the same geographic region to claim to be "number one." Different media offer certain advantages and disadvantages. Entrepreneurs must understand who their target customers are and select the media that are best for reaching their target audiences most effectively and efficiently.

7. *Consider using someone else as the spokesperson on your television and radio commercials.* Although being your own spokesperson may lend a personal touch to your ads, the ads may be seen as nonprofessional or "homemade" and may detract from your company's image rather than improve it.

8. *Focus ads on your company's USP.* Some entrepreneurs think that to get the most for their advertising dollars, they must pack their ads full of facts and information. However, overcrowded ads only confuse customers and usually produce poor results. Simple, well-designed ads that focus on your USP and have a single call to action are much more effective.

9. *Devise ways of measuring your ads' effectiveness that don't depend on just two or three customers' responses.* Measuring the effectiveness of advertising is an elusive art at best, but the opinions of a small unrepresentative sample of customers, whose opinions may be biased, is not a reliable gauge of an ad's effectiveness. Asking new customers how they learned about your business often provides insight into the effectiveness of your advertising campaign. Online advertising media not only are low-cost but also offer the benefit of providing easily measurable analytics that tell entrepreneurs how successful their ads are. With sufficient planning, entrepreneurs can use focus groups to test the effectiveness of ads *before* they run them.

10. *Don't simply drop an ad because nothing happens immediately.* Some ads are designed to produce immediate results, but many ads require more time because of the lag effect they experience. One of advertising's rules is this: It's not the size; it's the frequency. The head of one advertising agency claims, "The biggest waste of money is stop-and-start advertising." With advertising, patience is essential, and entrepreneurs must give an advertising campaign a reasonable time to produce results. One recent study concluded that sales increases are most noticeable four to six months after an advertising campaign begins. One advertising expert claims that successful advertisers "are not capricious ad-by-ad makers; they're consistent ad campaigners."

11. *Emphasize the benefits that the product or service provides to the customer.* Too often, ads emphasize only the features of the products or services a company offers without mentioning the benefits they provide customers. Customers really don't care about a product's or service's "bells and whistles"; they are much more interested in the *benefits* those features can give them! Their primary concern is "What's in it for me?" Your ads should focus on your target customers and how your company's products or services fill their needs rather than on your company.

12. *Evaluate the cost of different advertising media.* Remember the difference between the absolute and relative cost of an ad. The medium that has a low absolute cost may actually offer a high relative cost if it does not reach your intended target audience. Evaluate the cost of different media by looking at the cost per thousand customers reached. Remember: No medium is a bargain if it fails to connect you with your intended customers.

Sources: Based on "Top 10 Tips for an Effective Advertising Campaign," *All Business,* December 2012, *www.allbusiness.com/10-tips-effective-advertising/16566950-1.html#axzz2F2BbyMNm*; Sue Clayton, "Advertising," *Business Start-Ups,* December 1995, pp. 6–7; *Marketing for Small Business* (Athens: University of Georgia Small Business Development Center, 1992), p. 69; "Advertising Leads to Sales," *Small Business Reports,* April 1988, p. 14; Shelly Meinhardt, "Put It in Print," *Entrepreneur,* January 1989, p. 54; Danny R. Arnold and Robert H. Solomon, "Ten 'Don'ts' in Bank Advertising," *Burroughs Clearing House* 16, no. 12 (September 1980), pp. 20–24, 43–43; Howard Dana Shaw, "Success with Ads," *In Business,* November/December 1991, pp. 48–49; Jan Alexander and Aimee L. Stern, "Avoid the Deadly Sins in Advertising," *Your Company,* August/September 1997, p. 22.

FIGURE 10.2

Characteristics of a Successful Ad

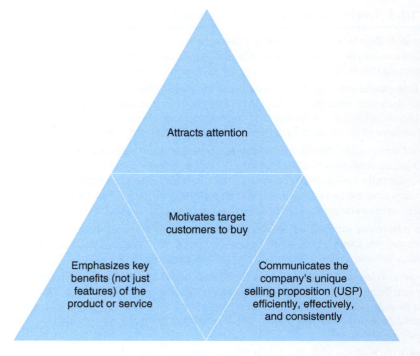

The remainder of this chapter will focus on selecting advertising media, developing an advertising plan, and creating an advertising budget. Figure 10.2 illustrates the characteristics of a successful ad.

Selecting Advertising Media

3.

Describe the advantages and disadvantages of the various advertising media.

Entrepreneurs quickly discover a wide array of advertising media options, including newspapers, magazines, radio, television, direct mail, and the Internet, as well as many specialty options. One of the most important decisions an entrepreneur must make is which media to use to disseminate the company's message. The medium used to transmit the message influences the customer's perception—and reception—of it. The right message broadcast in the wrong medium will miss its mark. Before selecting the vehicle for the message, entrepreneurs should consider several important questions:

- *How large is my company's trading area?* How big is the geographical region from which the firm will draw its customers? The size of this area influences the choice of advertising media.

- *Who are my target customers, and what are their characteristics?* Until they know who their target customers are, business owners cannot select the proper advertising media to reach them.

- *What budget limitations do I face?* Every business owner must direct his or her company's advertising program within the restrictions of its operating budget. The goal is to generate the greatest impact while staying within the company's budget limitations.

- *Which media do my competitors use?* Business owners should know which advertising media that their competitors use; however, they should *not* automatically assume that those media are best. Sometimes an approach that differs from the traditional one produces superior results. Figure 10.3 illustrates the results of a recent study that shows the media that adults use to get information about local small businesses.

- *How important is repetition and continuity of my advertising message?* Generally, an ad becomes effective only after it is repeated several times, and many ads must be continued for some time before they produce results. Some experts suggest that an ad must appear at least six times in most mass media before it becomes effective.

- *How does each medium compare with others in its audience, its reach, and its frequency?* **Audience** measures the number of paid subscribers a particular medium attracts

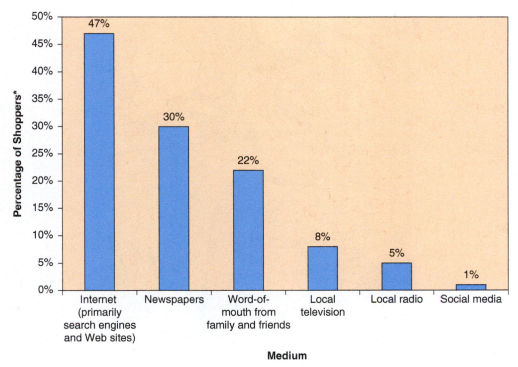

FIGURE 10.3

How Shoppers Gather Information About Local Small Businesses

Source: Based on Lee Rainie, Kristen Purcell, Amy Mitchell, and Tom Rosenstiel, "Where People Get Information About Restaurants and Other Local Businesses," Pew Internet and American Life Project, December 14, 2011, *http://pewinternet .org/Reports/2011/Local-business-info .aspx.*

and is called *circulation* in most print media, such as newspapers and magazines. **Reach** is the total number of people exposed to an ad at least once in a period of time, usually four weeks. **Frequency** is the average number of times a person is exposed to an ad in that same time period.

- *What does the advertising medium cost?* There are two types of advertising costs the that entrepreneurs must consider: absolute cost and relative cost. **Absolute cost** is the actual dollar outlay a business owner must make to place an ad in a particular medium for a specific time period. An even more important measure is an ad's **relative cost**, the ad's cost per potential customer reached. Relative cost is most often expressed as **cost per thousand** (CPM), the cost of the ad per 1,000 customers reached. Suppose an entrepreneur decides to advertise his product in one of two newspapers in town. The *Sentinel* has a circulation of 21,000 and charges $1,200 for a quarter-page ad. The *Independent* has a circulation of 18,000 and charges $1,300 for the same space. Reader profiles of the two papers suggest that 25 percent of *Sentinel* readers and 37 percent of *Independent* readers are potential customers. Using this information, the manager computes the following relative costs:

	Sentinel	Independent
Circulation	21,000	18,000
Percentage of readers who are potential customers	× 25%	× 37%
Potential customers reached	5,250	6,660
Absolute cost of ad	$1,200	$1,300
Relative cost of ad (CPM)	$1,200/5,250 = .22857 or $228.57 per thousand potential customers reached	$1,300/6,660 = .19520 or $195.20 per thousand potential customers reached

Although the *Sentinel* has a larger circulation and a lower absolute cost for running the ad, the *Independent* offers this entrepreneur a better advertising deal because of its lower cost per thousand potential customers (CPM) reached. It is important to note that this technique does not give a reliable comparison across media; it is a meaningful comparison only *within* a single

Before we begin tonight's dream,
a word from our sponsor...

medium. Differences among the format, presentation, and coverage of ads in different media are so vast that cross-comparisons are not meaningful.

Media Options

The world of advertising is undergoing significant changes. The lines that once separated the various advertising media are now blurring. Features that once were unique to a specific medium now operate across multiple media. Video, once the distinctive signature of television, now appears on companies' Web sites, in e-mail ads, on YouTube, on smart phones, and through other devices. Traditional methods of advertising are not as effective as they once were because of increased advertising clutter, the growth in the time that customers spend online, and intense competition for buyers' attention. Small businesses are steadily shifting their advertising expenditures away from traditional media, such as newspapers, television, direct mail, radio, magazines, and directories, and toward digital media, such as e-mail campaigns, search engines, social media, online and mobile device ads, and others (see Figure 10.4). Entrepreneurs are looking to supplement or even replace traditional methods of advertising with inexpensive online tools and innovative, sometimes offbeat techniques that capture shoppers' attention.

Choosing advertising media is no easy task because each has particular advantages, disadvantages, and costs. Figure 10.5 shows recent trends in advertising expenditures by medium. Let's examine the features of various advertising media.

WORD-OF-MOUTH ADVERTISING Perhaps the most effective and certainly the least expensive form of advertising is **word-of-mouth advertising** in which satisfied customers recommend a business to friends, family members, and acquaintances. Unsolicited testimonials are powerful; because they are impartial, they score high on importance and credibility among potential customers. According to the Word of Mouth Marketing Association, 54 percent of U.S. consumers say that the primary factor that drives their purchasing decisions is word-of-mouth. In addition, 59 percent of consumers believe that offline (face-to-face or voice-to-voice) word-of-mouth is highly credible.[16] Business owners recognize the power of word-of-mouth advertising; they rate word-of-mouth advertising as the most beneficial form of advertising to their businesses (79 percent), ahead of traditional advertising (44 percent) and social media (35 percent).[17]

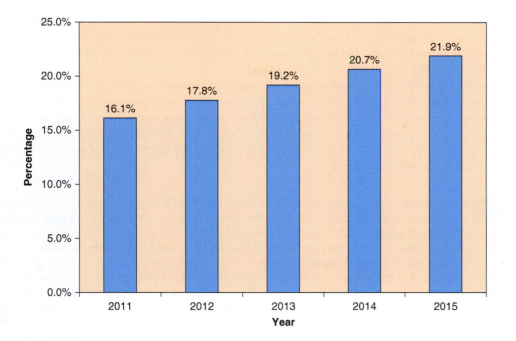

FIGURE 10.4

Global Online Advertising Expenditures as a Percentage of Total Advertising Expenditures

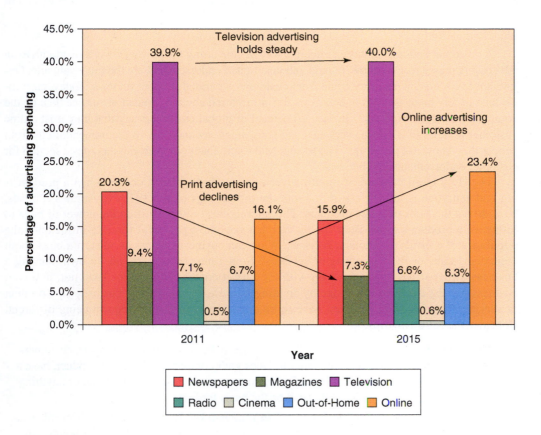

FIGURE 10.5

Advertising Expenditures by Medium Percentage of Advertising Expenditures 2011 and 2015

Source: Based on "Global Online Ad Spending Forecast to Exceed Print in 2015," Marketing Charts, December 3, 2012, *www.marketingcharts.com/wp/ television/global-online-ad-spend-forecast-to-exceed-print-in-2015-25105.*

Word-of-mouth advertising takes place naturally (the average American mentions specific brand names 60 times per week in online and offline conversations), but entrepreneurs can take steps to encourage customers to promote their businesses to others.[18] The best ways for a company to generate positive word-of-mouth advertising are to provide superior quality, offer outstanding service, and reward customer loyalty—all of which give customers a reason to recommend the company to others. A stellar experience leads to loyal customers who become walking advertisements for the company. In an age of social networking, the quality of the experience customers have with a business has more impact than any form of advertising.

Word-of-mouth advertising can make or break a business because *dissatisfied* customers also speak out against businesses that treat them poorly. To ensure that the word-of-mouth advertising a company generates is positive, business owners must actually do what they want their customers to say they do.

Good word-of-mouth advertising is based on five characteristics: credible (a message that is honest and authentic), respectful (a company's behavior is transparent and trustworthy), social (a company that listens to, participates in, and encourages conversations with and among customers both online and offline), measurable (a company's ability to define, monitor, and evaluate success), and repeatable (a company's ability to generate consistent word-of-mouth advertising by becoming a "talkable" brand).[19]

ENTREPRENEURIAL PROFILE: Tom Jenkins: Big Juan's Tacos y Burros To get the word out about Big Juan's Tacos y Burros, a fast-service Mexican restaurant (now with two locations in Tucson, Arizona), to the students at the nearby University of Arizona who make up his primary target audience, Tom Jenkins hired a corps of student "brand ambassadors." The six students, most of whom are business or marketing majors at the University of Arizona, hand out free food samples, plan and coordinate special on-campus events, and chat up Big Juan's on social media. In one of their most successful campaigns, the students set up a photo backdrop featuring a big mustache, provided students with a variety of themed props, and invited them to take photos. "We posted the photos on Facebook, and the person whose photo received the most likes received a $25 gift certificate to Big Juan's," says one student. The word-of-mouth campaign has increased sales and profits at Big Juan's, "and the costs are minimal," says Jenkins. "This is a very effective way to reach a very specific target."[20]

A customer endorsement is an effective way of converting the power of word-of-mouth to an advertising message. Of course, unpaid and unsolicited endorsements are the most valuable. Online, these endorsements often come from customer-generated product reviews. Today, customers tend to rely more on customer reviews for information about a product or service than on the company's own descriptions.[21] In fact, 70 percent of global customers say that they trust online customer reviews of products, services, and companies, second only to recommendations from people they know (92 percent).[22] The lesson: Make sure that your Web site includes a section for customer endorsements and reviews.

The Holy Grail of word-of-mouth advertising is "buzz." Buzz occurs when a product is hot and everyone is talking about it. From the mood rings of the 1970s to Apple's iPhones, buzz drives the sales of many products. The Internet has only magnified the power of buzz to influence a product's sales. Buzz on the Web has become a powerful force in influencing the popularity of a firm's products or services. What can business owners do to start a buzz about their companies or their products or services? Sometimes buzz starts on its own, leaving a business owner struggling to keep up with the fury it creates. More often than not, however, business owners can give it a nudge by creating interest, mystique, and curiosity in a product or service. Creating buzz does not have to be expensive, but it does require being different. Consider the following tips:

- *Make your business buzz-worthy.* If your company has nothing to set it apart, customers have no incentive to create buzz about it. Does your company sell a novel product, have a unique marketing approach, offer stellar customer service, use a wacky logo, or anything else that can set it apart? If so, that can be the basis for buzz.

- *Promote your company to "influencers" in your market.* Influencers are high-profile customers who are on the front edge of every trend. They are the first to wear the hottest athletic shoe, master the coolest video game, or make the hippest restaurant their new hangout; they also are willing to tell their friends. Promoting your company's products and services to influencers increases the likelihood that your company will be the subject of buzz.

- *Make it easy for satisfied customers to spread the word about your company.* Ask customers periodically to tell a friend about your business and their positive experience with it. Put a "Tell-a-friend" link on every page of your company's Web site. Reward customers who refer other customers to your business by offering them something special in return.

- *Use e-mail and social media to encourage viral marketing and amplify your company's word-of-mouth advertising.* One of the easiest ways to accomplish this is through e-mail because it is so easy for people to pass along to their friends. Another technique is to publicize news about your company on a blog and include links to your company's Web site. Entrepreneurs also can use social media sites such as Facebook, Twitter, Pinterest, and others to engage customers and encourage buzz.

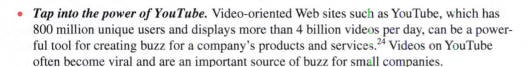
ENTREPRENEURIAL PROFILE: Luna and Larry Kaplowitz: Coconut Bliss In 2004, copreneurs Luna and Larry Kaplowitz launched Coconut Bliss, their organic frozen dessert company in Eugene, Oregon, with a hand-cranked ice cream churn they bought for $1.50 at a Goodwill store and the idea of making premium ice cream from coconut milk. As their company has grown, the Kaplowitzes have been able to build an impressive word-of-mouth marketing program with strategic use of social media, primarily Facebook and Twitter, where they have amassed a following of 12,000 and 8,300 fans, respectively. They also raise their company's profile by posting photos on Tumblr and videos on their own YouTube channel, which enhances their ranking in search engine results. Coconut Bliss features its customers enjoying their organic frozen desserts in customer-submitted photos and videos on its Web site, Facebook page, Twitter feed, and YouTube channel. They also keep their customers engaged—and talking about the company—by sponsoring interesting contests, sharing information about organic foods, and constantly talking with them on social media.[23]

Larry and Luna Kaplowitz, cofounders of Coconut Bliss.
Source: Bliss Unlimited LLC.

- *Tap into the power of YouTube.* Video-oriented Web sites such as YouTube, which has 800 million unique users and displays more than 4 billion videos per day, can be a powerful tool for creating buzz for a company's products and services.[24] Videos on YouTube often become viral and are an important source of buzz for small companies.

INTERNET ADVERTISING Just as the Internet has become a common tool for conducting business, it also has become a popular medium for advertisers. Internet advertising is growing rapidly because advertisers recognize that that is where their target customers are spending more of their time and because advertisers can track the effectiveness of their advertising campaigns. By 2016, U.S. companies are expected to spend $62 billion on online advertising.[25]

The Internet provides opportunity for small businesses to reach customers with inexpensive video ads. The Web's multimedia capabilities make it an ideal medium for companies to demonstrate their products and services with full motion, color, and sound and to involve customers in discussions about them. Businesses that traditionally used direct mail can bring the two-dimensional photos and product descriptions in their print catalogs to life in video, avoid the expense of mailings, and attract new customers that they might otherwise miss.

Online advertisements take five basic forms: banner ads, display ads, contextual ads, pay-per-click ads, and e-mail ads. **Banner ads** are small rectangular ads that reside on Web sites, much like roadside billboards, touting a company's product or service. When visitors to a site click on the banner ad, they go straight to the advertiser's home page. One measure of a banner ad's effectiveness is the number of impressions it produces. An **impression** occurs every time an ad appears on a Web page, whether or not the user clicks on the ad to explore it. Another common way of judging the effectiveness of banner ads is the **click-through rate**, which is calculated by dividing the number of times customers actually click on the banner ad by the number of impressions for that ad. For instance, if an ad is displayed 1,000 times and 12 customers actually click on the ad and go to the advertiser's Web site, the ad's click-through rate is 1.2 percent (12 ÷ 1,000). Banner ads suffer from a very low click-through rate—just 0.1 percent compared to an average click-through rate of 2.5 percent for search engines.[26] The cost of a banner ad to an advertiser depends on the number of prospects who actually click on it.

The primary disadvantages of banner ads is that Web users dislike them and can easily ignore them. These ads have become such a part of the landscape of the Web that users tend to ignore them. Web designers search for the best page placement for banner ads and add unique features that will catch users' attention and encourage them to click through. Another form of Web advertising that is more difficult to ignore are **display ads**, which include both pop-up, interstitial ads, and contextual ads. A **pop-up ad** is a separate window in which an ad pops up spontaneously, blocking the site behind it. It is designed to grab consumers' attention for the few nanoseconds it takes them to close the window. One danger is the negative view that Internet users have of pop-up ads, which they perceive as an annoying intrusion. A slight variation on this ad is the "pop-under" ad that immediately goes behind the active screen but stays open until the browser window is closed. An **interstitial ad** is an ad page that appears for a short time before a user-requested page appears. These ads are also called transition ads, splash pages, and flash pages.

Contextual ads appear on users' screens when they download information such as news, sports, or entertainment from another site, and the ad is correlated to the user's interest or online behavior. For instance, a Web user downloading sports information might receive an ad for athletic shoes or T-shirts with the information, and one conducting a search for vitamins might receive an ad for green tea or herbal remedies. To catch the attention of Web users, many advertisers, particularly those companies who aim their products at young customers, are using video ads rather than traditional display ads. Retargeted ads are display ads that pop up after a user visits a company's Web site or types a key word into a search engine.

Pay-per-click ads require companies to bid on top-ranking search engine listings using key words that they expect Internet users to type into a search engine when they are interested in purchasing a particular product or service. The higher a company's bid is for a key word, the more prominent is the location of its ad on the results that the search engine returns. Companies pay for an ad only when a prospect actually clicks on it. Entrepreneurs who advertise on the Internet should consider making pay-per-click ads a part of their advertising strategies. According to a study by the Pew Internet and American Life Project, 73 percent of all Americans use search engines.[27] Google, Bing, and Yahoo!, the three leading search engines, account for about 95 percent of all searchers, and advertisers who want to increase the odds of reaching their target audiences should advertise on all three.[28]

To calculate the maximum amount a company can afford to bid on a key word, use the following formula:

$$\text{Maximum key word bid} = \text{Conversion rate} \times \text{Profit per sale}$$

Conversion rates vary from one online industry to another, but the average conversion rate is about 3.2 percent. That means that for every 1,000 visitors to a company's site, 32 of them actually make a purchase. For a company that has a conversion rate of 3.2 percent and an average profit per order of $12, the maximum amount the owner should bid on a key word is the following:

$$\text{Maximum key word bid} = 3.2\% \times \$12 = 38.4 \text{ cents}$$

Used properly, pay-per-click ads can drive customers to a company's Web site even before the search engines discover it and include it in their natural or organic listings. Pay-per-click ads also allow advertisers to test the effectiveness of different ads by running several variations at once (e.g., one version might include a discounted price, and another might include a free accessory). These ads are efficient because advertisers pay for an ad only when a customer actually clicks on it. **Click fraud**, which occurs when a person or a computer program generates ad clicks even though they have no interest in the advertiser's product or service, is a danger to entrepreneurs who use pay-per-click ads. The click fraud rate on search engines ranges from 13 and 18 percent, with an average of 14.7 percent.[29]

E-MAIL ADVERTISING E-mail is the most common application on the Internet, and e-mail advertising capitalizes on that popularity. The Radicati Group estimates that Internet users have more than 3.3 billion e-mail accounts, and only 14 percent of them originate in North America![30] **E-mail advertising**, in which companies broadcast their advertising messages by e-mail, is growing rapidly because it is so effective and so inexpensive. Retailers send an average of almost 15 e-mail promotions to their subscribers each month (177 per year). The busiest month for e-mail promotions, of course, is December (with 22 campaigns), and February is the slowest (with 12 campaigns). The most popular days for sending promotional e-mails are Friday and Monday, and Saturday is the least popular day.[31]

E-mail advertising takes two forms: permission e-mail and spam. As its name suggests, **permission e-mail** involves sending e-mail ads to customers with their permission; **spam** is unsolicited commercial e-mail. The Radicati Group also estimates that despite modern anti-spam technology, 15 percent of all delivered e-mail messages are spam.[32] Because most e-mail users see spam as a nuisance, they often view companies that use it in a negative fashion. Smart entrepreneurs do *not* rely on spam in their marketing strategies; they know that the quality of their companies' e-mail list is more important than its quantity. Permission e-mail messages typically produce very high response rates and attractive returns on investment because recipients are more engaged. According to a global study by Silverpop, an e-mail marketing services company, the average open rate for e-mail ads is 20.1 percent, and the average click-through rate is 5.2 percent.[33] **Triggered e-mails**, those that companies send in response to a potential customer's actions, such as registering on a company's Web site or abandoning a shopping cart, have even higher open and click-through rates at 48 and 10 percent, respectively.[34]

Forrester Research reports that 38 percent of e-mail users open their e-mail on a mobile device such as a smart phone; by 2017, 78 percent of users will access their e-mail via a mobile device. In addition, 56 percent of smart phone users in the United States who have made a purchase from their phones did so in response to an e-mail message.[35] As the number of smart phone users continues to grow, entrepreneurs must recognize that customers' in-boxes are moving targets and adapt their e-mail marketing strategies to accommodate smaller screen sizes. Keeping message size below 50 kilobytes so that they load quickly and optimizing them with simple designs and attention-getting call-to-action buttons so that mobile users can navigate them easily increase click-through rates. Scanning printed quick response (QR) codes allows shoppers to use their smart phones or tablets to go directly to a Web site linked to the QR code without having to type in a URL.

Building an e-mail list simply requires attention to the basics of marketing. The goal is to encourage potential buyers to share their e-mail addresses. Monument Lane, a restaurant in New York City's West Village, has been successful gathering customers' e-mail addresses by including a card with customers' bills.[36] Offering a reward, such as a white paper report, a one-time discount, a special offer, or an entry in drawing for a prize, increases the likelihood that customers will share their e-mail addresses. Once a small company obtains potential customers' e-mail addresses, the next step is to send messages that are useful and interesting to them. The message must be geared to their interests, highlight the product's USP, and include a link to a landing page that allows them to purchase the product easily. Entrepreneurs also must track the success and the returns that their e-mail marketing campaigns deliver with appropriate e-mail analytics.

Many companies have success with e-mail ad campaigns that produce immediate results and are very inexpensive to conduct.

ENTREPRENEURIAL PROFILE: Eric Shamban and Gilbert Johnson: The Chocolate Bar Eric Shamban and Gilbert Johnson launched The Chocolate Bar (TCB) in 2001 in Houston, Texas, as a haven for chocolate lovers. The original store, located near Rice University, offers a variety of freshly made chocolate-themed treats, including candy, cookies, cakes, pies, ice cream, and a unique chocolate pizza. Customers also can select a specific chocolate blend and choose from a menu of nuts, fruits, and other ingredients to make their own chocolate bars. The entrepreneurs had been collecting customers' e-mail addresses as part of a weekly free dessert contest but had not engaged in any e-mail marketing until they opened their second Houston location and hired Lynne Singerman, TCB's Ambassador of Chocolate. "My goal is to create awareness of our new location and introduce the happy spirit of The Chocolate Bar to all chocolate lovers," says Singerman. She worked with e-mail marketing company Constant Contact to more than triple the size of TCB's e-mail list and to create clever e-mail marketing campaigns to make customers aware of the company's new location, special products, gift ideas, and calendar of events, including live music nights. TCB's first e-mail newsletter was a hit with customers, not only generating an open rate of nearly 50 percent but also increasing in-store traffic. Singerman enjoys the flexibility and speed of e-mail ad campaigns. "Within moments of an e-mail going out, we are getting responses from people," she says. Because of the informational reports that Singerman receives after each ad goes out, TCB constantly fine-tunes its e-mail campaign and has an impressive e-mail ad open rate of 42.7 percent.[37]

Lessons from the Street-Smart Entrepreneur

E-Mail Ads That Produce Results

Launched in 2007 as a small jewelry company selling beads on e-Bay, Ana Silver Company grew quickly, ultimately specializing in selling jewelry made from the finest sterling silver and semi-precious stones. Today, Ana Silver is the top seller of jewelry on e-Bay, stocking more than 100,000 pieces of unique silver jewelry, including bracelets, earrings, necklaces, pendants, and rings. Within two years, sales had grown to $1 million, but co-owner George Liang believed that the company could accelerate its sales growth dramatically by connecting more closely with its customers. "We didn't really communicate with our customers after their purchases," he says. "E-bay sent an automatic newsletter to our customers, but we didn't have any ongoing outreach."

Liang wanted a low-cost, effective way to drive customers to Ana Silver's e-Bay store and to inform them about new products and special sales. He saw e-mail marketing as the ideal solution, one that would allow the company to create efficient, personalized ad campaigns targeting its base of 50,000 customers. Not wanting to overwhelm customers with too many messages, Ana Silver settled on sending one e-mail ad to them each week, but customers can (and many do) opt in to receive more frequent messages from the company. "Every time we send out an e-mail, 60 to 80 customers make a purchase," says Liang. "We may sell several hundred items from one e-mail." One recent e-mail ad went out at 6 a.m., and by that afternoon, Ana Silver had received 50 orders and sold more than 200 pieces of jewelry. Just one year after launching its e-mail advertising campaign, Ana Silver's annual sales reached $10 million.

Businesses spend $1.7 billion a year on e-mail marketing campaigns, and experts expect that number to grow to $2.5 billion by 2016. What makes e-mail marketing appealing, especially to entrepreneurs such as George Liang, is that it is inexpensive, it is measurable, and it works. Studies by the Direct Marketing Association suggest that every $1 that companies spend on e-mail marketing produces a return of $44. In addition, according to ExactTarget's Channel Preference Survey, 77 percent of Internet users of all ages say that their preferred method of receiving permission-based promotional messages is e-mail. (The second most preferred method is direct mail at just 9 percent.) Entrepreneurs who want to reproduce the success that Ana Silver has created using an e-mail advertising campaign should consider the following tips from the Street-Smart Entrepreneur:

Make a concerted effort to collect customers' (and potential customers') e-mail addresses. Every contact that anyone in your company has with a customer presents an opportunity to collect another e-mail address. Seize them! Ensure that everyone in the company understands the importance of building an accurate and reliable e-mail list because that's where e-mail marketing success starts.

Recognize the power of simplicity. Ads that generate the highest response rates focus on a single call to action, which allows the advertiser to keep the message simple and focused. Don't try to accomplish too much in an e-mail advertisement. The best place to put your call to action is in the e-mail's second (short) paragraph; the first paragraph should explain the benefits that your company's product or service can provide the recipient.

Make sure the e-mail's subject line emphasizes your company's USP. Without the right subject line, e-mail recipients may never open the e-mail. The most effective subject line messages are short; however, one study of more than 1.2 billion e-mail marketing messages by MailerMailer concludes that the highest click-through rates occur when the subject line included fewer than 40 characters, including spaces. The best subject lines suggest the product's or service's USP (refer to Chapter 9). For example, "Fresh lunch FAST—just $5.99!" in the subject line of an e-mail expresses several product features to potential customers. Other subject line words that produce responses include "new," "sale," "save," "you," "instant," "newsletter," "news," and "free." Including links in the e-mail also increases a message's click-through rate; the more links you include, the more likely the recipient is to click through on at least one of them.

Make sure your e-mails' look and feel are consistent with your company's overall image. Every component of a company's advertising campaign should have a look and feel that is consistent with its brand, even though the ads may appear in many diverse media. The design of a company's e-mail ads should rely on the same colors, themes, slogans, and look as its ads in other media.

Send e-mails when customers are most likely to make their purchases. Proper timing of e-mail ads can improve customer response rates dramatically. Messages sent on Sunday produce the highest click-through rates. January, July, August, and December are the months in which customers are *least* likely to pay attention to e-mail ads (although there are exceptions). E-mail messages sent between 4 P.M. and 8 A.M. yield the highest click-through rates, and those sent in the middle of the day produce the lowest response rates. Companies must time their e-mail ads to correspond to their customers' demand for their products and services.

Write copy that produces the results you seek. Start by concentrating on one idea. Before writing any copy, develop a mental picture of your target customer. Give him or her a name and try to envision how your company's product or service can benefit him or her. This will help you keep your ad copy focused on the USP you put in the subject line of the e-mail. When appropriate, consider including an endorsement from an existing customer (perhaps with a photo) to add credibility to your claims. Be sure to provide clearly visible links to your company's Web site at several places in the e-mail ad, including at the top and the bottom of the page. Always include a prominent call to action: How do you want the customer to respond?

Incorporate social networking links and video into e-mails. Adding "Like us on Facebook" or "Follow us on Twitter" links to e-mails encourages customers to engage your company in other media and to promote it to their friends. The E-Mail Marketing Trends Survey reports that 53.8 percent of small business owners say that including video in e-mails increases their company's click-through and open rates. Buttons that allow customers to download videos, forward them to friends, or post them on their Facebook pages encourage the viral nature of the Web to expand their advertising efforts many times over.

Use value-added items to increase your campaign's response rate. In the typical in-box, there are dozens (if not more) of messages competing for the person's attention. One way to boost your campaign's response rate is to offer recipients something of value—for example, a coupon, a newsletter, or a white paper. Betsy Harper, CEO of Sales and Marketing Search, an executive search firm that specializes in sales and marketing positions, says that publishing a monthly e-mail newsletter that includes hiring trends and tips has improved her company's visibility, reputation, and sales. Recently, says Harper, "within two minutes (literally!) of sending our e-mail newsletter, I got an e-mail from a fellow in New York. He asked me to call him right away about filling a senior sales position. We talked, signed an agreement to work together the very next day, and started the search. We finished the search in record time and received a $22,000 fee."

Always comply with the CAN-SPAM Act. The CAN-SPAM Act, a law that sets the rules for commercial e-mail, establishes requirements for commercial messages, gives recipients the right to opt out of e-mails, and spells out tough penalties for violators.

Never stop testing. Successful e-mail advertisers never stop experimenting with variations on their ads, testing to see which products, pitches, subject lines, timing, and other elements produce the highest response rates.

Sources: Based on "Case Study: Ana Silver Company," Campaigner, 2012, *www .campaigner.com/solutions/success-stories/ana-silver.aspx*; Bob Krummert, "E-Mail Rules Social Media, Even for Fans," Restaurant Hospitality, April 17, 2012, *http://restaurant-hospitality.com/social-media/email-rules-social-media-even-fans*; Darrell Zahorsky, "The 6 Laws of Small Business Advertising Success," About, 2012, *sbinformation.about.com/cs/advertising/a/aa022303a.htm*; "The Case for Short and Sweet," *Marketing Profs*, August 22, 2012, *www.marketingprofs .com/short-articles/2627/the-case-for-short-and-sweet*; David Schwartz, "E-Mail Marketing Earns a Staggering $44 Return on Investment per Dollar Spent. Beats Wall Street Yields," SOS Marketing, May 8, 2011, *www.sosemarketing .com/2011/05/08/e-mail-blasts-earn-a-staggering-44-return-on-investment-per-dollar-spent*; "Our Twelfth E-Mail Metrics Report," MailerMailer, July 2012, *www.mailermailer.com/resources/metrics/index.rwp*; Lisa Barone, "E-Mail Marketing Success Is About Relevance," *Small Business Trends*, June 18, 2009, *http:// smallbiztrends.com/2009/06/email-marketing-success.html*; Peter Prestipino and Mike Phillips, "E-Mail Marketing's Future . . . Right Now," *Website*, November 2009, pp. 26–29; Michelle Keegan, "Real Life Small Business Newsletter Tips," Constant Contact, *www.constantcontact.com/learning-center/hints-tips/volume10-issue2.jsp*; Gail Goodman, "Writing Compelling Promotional Copy," Constant Contact, *www.constantcontact.com/learning-center/hints-tips/ht-2006-07.jsp*; Ivan Levison, "Five Common E-Mail Mistakes and How to Avoid Them," *Levison Letter* 17, no. 2 (April 2002), *www.levison.com/email-advertising.htm*; "You've Got the Power," *MarketingProfs* 1, no. 23 (May 22, 2008); 2010 E-Mail Marketing Trends Survey, GetResponse, 2010, p. 5.

SOCIAL MEDIA ADVERTISING Although search engine advertising (those ads that appear next to search results on search engines) account for the greatest percentage of companies' online advertising expenditures (more on those techniques in Chapter 13), growing numbers of entrepreneurs are investing in social media advertising, placing pay-per-click ads on popular social media sites such as Facebook, Twitter, and others (see Figure 10.6).

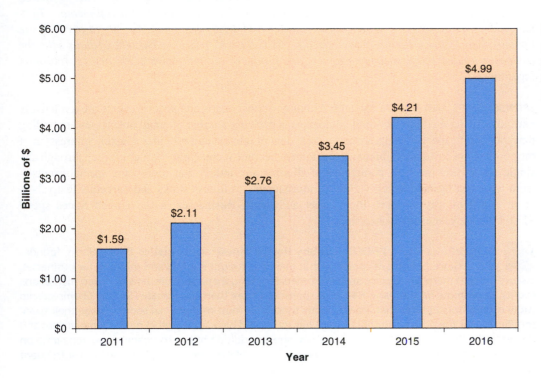

FIGURE 10.6

Social Media Advertising Expenditures in the United States (in Billions of $)

Facebook Ads Entrepreneurs can purchase ads on Facebook and tap into the potential customer base of more than 1 billion Facebook users worldwide. Once you decide what to advertise (e.g., a specific page on the company's Web site, a Facebook post, or a registration page for an event), you create your ad by writing a brief headline and a short description and uploading a thumbnail photograph. Then you choose your target audience (ads for narrowly defined audiences are more likely to be displayed) by defining a geographic region (country, state, city, or ZIP code), age, gender, and broad categories (e.g., parents or people interested in country music or cooking) or specific interests, such as the products or services that you sell. For instance, a wedding planner in Fort Worth, Texas, could target her ads to women within the city whose relationship status is "engaged." Facebook also gives you the ability to bid on pay-per-click ads (you pay for the ad only when a user actually clicks on it) and to limit the total amount you spend on the campaign (either a daily budget or the maximum amount for the entire campaign). Finally, Facebook gives advertisers useful feedback on their ads' performances and allows them to generate useful reports that make comparing different advertisements simple.

ENTREPRENEURIAL PROFILE: Brian Spaly and Andy Dunn: Bonobos Bonobos, an online retailer of tailored, fashionable men's clothing founded by Brian Spaly and Andy Dunn, has had great success with Facebook ads in attracting new customers and selling specific styles of pants. With carefully crafted and targeted ads, Bonobos increased its click-through rates, sold out of what had been hard-to-move merchandise, and, at the peak of the campaign, was drawing 10 percent of the visitors to its Web site from its Facebook ads.[38]

Twitter Ads On Twitter, small businesses can advertise their businesses through promoted accounts and promoted tweets. By signing up for a promoted account, Twitter identifies people who have interests that are similar to a company's current followers and promotes the company in these prospects' "Who to follow" sections, allowing the business to gain new followers who can spread the word about the company. With promoted tweets, Twitter constantly monitors a company's best, most engaging tweets and promotes them to people both on the Web and on mobile devices, which allows business to reach customers on the go. Like Facebook, you pay only when someone follows your account or engages in your promoted tweet. Twitter allows entrepreneurs to focus their ads within a specific geographic location and to limit the amount they spend per campaign.

Bonobos, the online clothing retailer, also has had great success with Twitter ads. "On short notice, we can launch campaigns on Twitter with minimal effort and low costs compared to traditional media channels," says David Fudge, the company's head of social media. "Twitter gives us quick access to an engaged audience and drives customers directly to our Web site for purchases." When Bonobos advertised a "Twixclusive" (only on Twitter) 24-hour sale, the campaign landed 100 first-time customers and produced an impressive 1,200 percent return on investment.[39]

SPONSORSHIPS AND SPECIAL EVENTS Although sponsorships and special events are a relatively new advertising medium for small companies, a growing number of small businesses are finding that sponsoring special events attracts a great deal of interest and provides a lasting impression of the company in customers' minds. As customers become increasingly harder to reach through any single advertising medium, companies of all sizes are finding that sponsoring special events— from wine tastings and beach volleyball tournaments to fitness walks and barbeque festivals— is an excellent way to reach their target audiences. Businesses in the United States spend $13.6 billion a year event sponsorships.[40]

ENTREPRENEURIAL PROFILE: Bobby Harris: BlueGrace Logistics Bobby Harris, founder of BlueGrace Logistics, a company that manages freight hauling and shipping for businesses, discovered that many of his company's customers (shipping managers) are fans of mixed martial arts events. To promote his company, Harris has sponsored more than 20 Ultimate Fighting Championship (UCF) competitors at rates that are far below those that most professional sports charge their sponsors. Several of the athletes that BlueGrace has sponsored have appeared in national magazines and on network television sporting the company's logo. In addition to the company's logo appearing on competitors' clothing, the company benefits from UCF athletes mentioning its name in their frequent social media posts.[41]

A wide range of sponsorships exists. Sponsoring a hole at a charitable golf outing may cost as little as $100, but landing the name of your business or product on the hood of a car driven by a NASCAR racer can cost $8 million or more. Sponsorships and participation in special events can be very cost-effective if an entrepreneur selects events whose attendees are potential customers. For instance, a company whose target audience is women between the ages of 35 and 54 with household incomes that exceed $150,000 might consider sponsoring equestrian events because they draw that demographic. Local festivals and events gain a sponsor a great deal of positive publicity for a modest sponsorship fee. Support for charity functions enhances the sponsor's community image, increases sales, and often attracts new customers.

Small companies do not have to rely on other organizations' events to generate advertising opportunities; they can create their own special events. The owner of Quadrille Quilting in North Haven, Connecticut, partnered with the owners of two other quilting shops to create Shop Hops, an event in which customers buy "passports" to all three stores that entitle them to refreshments and special prizes. The first Shop Hop, which took place on a Super Bowl weekend, generated an entire month's sales in just one day for Quadrille Quilting.[42]

Creativity and uniqueness are essential ingredients in any special event promotion, and entrepreneurs excel at those. The following tips help entrepreneurs get the greatest promotional impact from event sponsorships:

- Do not count on sponsorships for your entire advertising campaign. Sponsorships are most effective when they are part of a coordinated advertising effort. Most sponsors spend no more than 10 percent on their advertising budgets or sponsorships.

- Look for an event that is appropriate for your company and its products and services. The owner of a small music store in an upscale mountain resort sponsors a local jazz festival every summer during the busy tourist season and generates lots of business among both residents and tourists. Ideally, an event's audience should match the sponsoring company's target audience. Otherwise, the sponsorship will be a waste of money.

- Research the event and the organization hosting it *before* agreeing to become a sponsor. How well attended is the event? What is the demographic profile of the event's visitors? Is it well organized?

- Try to become a dominant (or, ideally, the only) sponsor of the event. A small company can be easily lost in a crowd of much larger companies sponsoring the same event. If sole sponsorship is too expensive, make sure that your company is the only one from its industry sponsoring the event.

- Clarify the costs and level of participation required for sponsorship up front.

- Get involved. Do not simply write a check for the sponsorship fee and then walk away. Find an event that is meaningful to you, your company, and its employees and take an active role in it. Your sponsorship dollars will produce a higher return if you do.

TELEVISION In advertising dollars spent, television ranks first in popularity of all media. Although the cost of national television ads precludes their use by most small businesses, local spots on cable television can be an extremely effective means of broadcasting a small company's message. A 30-second commercial on network television may cost more than $500,000 (a 30-second spot during the Super Bowl sells for $3.5 million, up from $600,000 in 1987), but a 30-second spot on local cable television, which is in 60.4 percent of U.S. homes, may go for as little as $10 in small markets.[43]

ENTREPRENEURIAL PROFILE: Bill Geiger: Oregon Mint Company Bill Geiger, founder of Oregon Mint Company, a company that makes Mint Snuff, a chewable mint designed to help people stop using chewing tobacco, sells his products through select retail stores nationwide and through the company's Web site. Geiger knows that his target customers are the wives, girlfriends, and mothers of chewing tobacco users and for years had used cable television ads that promoted the benefits of Mint Snuff. Geiger used an application called SpotMixer to create a television commercial from his computer. "It was incredibly easy," he says. "I built my ad at 1:30 in the morning in just 25 minutes." He also used Google TV Ads to decide which cable networks across the nation to use and which programs to target while placing caps on costs. Geiger continues to test the effectiveness of his ads on different networks and television programs to determine which ones generate the greatest response. Since Geiger began running his television ads with the help of Google TV Ads, sales of Mint Snuff have increased fivefold. "Google TV Ads and SpotMixer have allowed me to reach my audience affordably," he says.[44]

Television advertising offers a number of distinct *advantages*:

Broad coverage. Television ads provide extensive coverage of a sizable region, and they reach a significant portion of the population. Television reaches 88.3 percent of adults every day, exceeding the reach of all other major advertising media.[45] In fact, the average household spends 8 hours and 21 minutes each day tuned in to television.[46] The typical adult sees 26 commercial breaks a day, for a total of 73 minutes of advertisements.[47]

Ability to focus on a target audience. Because many cable channels focus their broadcasting in topical areas—from home and garden or food to science or cartoons—cable television offers advertisers the ability to reach specific target markets much as radio ads do. Because an inverse relationship exists between time spent in television viewing and education level, television ads overall are more likely to reach people with lower educational levels.

Visual advantage. The primary benefit of television is its capacity to present the advertiser's product or service visually. With television ads, entrepreneurs are not limited to mere descriptions of a product or service; instead, they can demonstrate their uses and show firsthand their advantages. For instance, a small retail store selling a hydraulic log splitter can design a television commercial to show how easily the machine works. The ability to use sight, sound, and motion makes television ads a powerful selling tool.

Flexibility. Television ads can be modified quickly to meet the rapidly changing conditions in the marketplace. Advertising on television is a close substitute for personal selling. Like a sales representative's call, television commercials can use "hard-sell" techniques, attempt to convince through logic, appeal to viewers' emotions, persuade through subtle influence, or use any number of other strategies. In addition, advertisers can choose the length of the spot, its time slot, and even the program during which to broadcast the ad.

Design assistance. Few entrepreneurs have the skills to prepare an effective television commercial. Although professional production firms might easily charge $50,000 to produce a commercial, the television station from which an entrepreneur purchases airtime often will to help design and produce an ad very inexpensively.

Television advertising also has several *disadvantages*:

Brief exposure. Most television ads are on the screen for only a short time (30-second spots are most common) and require substantial repetition to achieve the desired effect. One of the realities is that television viewers often avoid or ignore the commercial messages.

Clutter. By the age of 65, the average person has seen more than 2 million television commercials, and more ads are on the way![48] With so many television ads beaming across the airwaves, a small company's advertising message easily could become lost in the shuffle.

Zapping and zipping. **Zappers**, television viewers who flash from one channel to another during commercials, and **zippers**, those who use digital video recording devices such as TiVo to fast-forward through commercials, pose a real threat to television advertisers. Zapping and zipping can cut deeply into an ad's reach, preventing television advertisers from reaching the audiences they hope to reach. Zipping poses a significant threat because 43 percent of U.S. households now own a DVR device (compared to just 12.3 percent in 2007).[49]

Fragmented audience. As the number of channels that cable and satellite providers offer proliferates, the question of where to advertise becomes more difficult to answer. The typical television viewer now has about 145 channels from which to choose. The dramatic increase in the number of channels available has fragmented the audience that an ad run on a single channel will reach.

Costs. Television commercials can be expensive to create. A professionally done 30-second ad can cost several thousand dollars to develop, even before an entrepreneur purchases

TABLE 10.2 Guidelines for Creative Television Ads

- *Keep it simple.* Avoid confusing the viewer by sticking to a simple concept.
- *Have one basic idea.* The message should focus on a single, important benefit to the customer. Why should people buy from your business?
- *Make your point clear.* The customer benefit should be obvious and easy to understand.
- *Make it unique.* To be effective, a television ad must reach out and grab the viewer's attention. Take advantage of television's visual experience.
- *Get viewers' attention.* Unless viewers watch the ad, its effect is lost.
- *Involve the viewer.* To be most effective, an ad should portray a situation to which the viewer can relate. Common, everyday experiences are easiest for people to identify with.
- *Use emotion.* The most effective ads evoke an emotion from the viewer—a laugh, a tear, or a pleasant memory. A recent study by Adobe reports that 73 percent of consumers say that an advertisement should tell a unique story rather than merely attempt to sell.
- *Consider production values.* Television offers vivid sights, colors, motions, and sounds. Use them to your advantage!
- *Prove the benefit.* Television allows an advertiser to prove a product's or service's customer benefit by actually demonstrating it.
- *Identify your company well and often.* Make sure your store's name, location, Web site, and product line stand out. The ad should portray your company's image.

Source: Based on Television Bureau of Advertising, *How to Make a Creative Television Commercial*, New York, NY; Adobe, *Click Here: The State of Online Advertising* (San Jose, CA: October 2012), p. 14.

airtime. The cost per thousand households reached for network ads is $14.50, and for cable ads it is $6.[50] Advertising agencies and professional design firms offer design assistance—sometimes at hefty prices—leading many small business owners hire less expensive free-lance ad designers or turn to the stations on which they buy airtime for help with their ads. Table 10.2 offers some suggestions for developing creative television commercials.

USING TELEVISION CREATIVELY Although television ads are not affordable for every small business, many entrepreneurs have found creative ways to use the power of television advertising without spending a fortune. Two popular methods include creating infomercials and using home shopping networks. **Infomercials** (also called **direct-response television**) come in two lengths: short-form, two- to three-minute pitches and long-form, 30-minute television full-length commercials packed with information, testimonials, and a sales pitch asking for an immediate response. The length of these ads allows entrepreneurs to demonstrate and explain their products in detail and to show customers the benefits of using them, a particularly important consideration for a new or complex product. Producing and airing a half-hour infomercial can be expensive, often costing $300,000 to $1 million, depending on its production quality, format, content, celebrity involvement, and broadcast schedule. Short-form infomercials cost about $15,000 to $20,000 to produce. Because most infomercials ask for an immediate response from viewers, entrepreneurs can gauge their success at landing customers, sometimes within minutes of airing them and almost always within one week. Products such as the Shamwow, Snuggie, PedEgg, and Total Gym have reached millions of units in sales with the help of infomercials.

ENTREPRENEURIAL PROFILE: Tom Campanaro: Total Gym Inc. Tom Campanaro, a former bodybuilder, launched Total Gym Inc. in 1974 to sell the pulley-based home workout system that he had developed. Campanaro created a 30-minute infomercial starring Total Gym user and celebrity Chuck Norris and Christie Brinkley. The infomercial, which at 15 years and counting holds the record for the longest-running infomercial in history, has proved to be highly successful, leading to sales of more than $1 billion of the Total Gym, which retails for between $600 and $900.[51]

Only one in 10 products that rely on infomercials succeeds.[52] To become an infomercial star, a product should meet the following criteria:

- Be unique and of good quality.
- Solve a common problem.

- Be easy to use and easy to demonstrate.
- Appeal to a mass audience.
- Have an "aha! factor" that makes customers think "What a great idea!"

Shopping networks such as QVC, HSN, and ShopNBC, which reach about 100 million homes in the United States, offer entrepreneurs another route to television. Although time on these networks is free, getting a product accepted is difficult; networks accept only about 3 to 4 percent of the items they review. Shopping networks look for items that have "demonstration appeal" and offer viewers attractive benefits. Entrepreneurs must exercise caution, however, because sales are on consignment (the network pays only for the items it actually sells), and returns may pose problems because most networks offer 100-percent-satisfaction guarantees. In addition, the networks pay for the items that sell 30 to 60 days after the sale, which sometimes causes cash flow problems for small businesses. The upside is that entrepreneurs have the potential sell thousands of products in just a matter of minutes.

RADIO   With 11,000 on-air stations, 7,000 streaming stations, and 1,800 digital stations, radio permits advertisers to reach specific audiences over large geographic areas. By choosing the appropriate station, program, and time for an ad, a small company can reach virtually any target audience.

Radio advertising offers several *advantages*:

Extensive reach. Radio's nearly universal presence gives advertisements in this medium a major advantage. Nearly every home and car in the United States is equipped with a radio, which means that radio ads receive a tremendous amount of exposure in the target market. Although the myriad entertainment options available has reduced the time that customers spend with any particular advertising media, the average adult spends nearly 15 hours each week listening to the radio. According to the Radio Advertising Bureau, radio reaches 92.5 percent of adults each week.[53]

Audience delivery. Radio commercial breaks retain on average 92 percent of the lead-in listening audience, which means fewer commercial zappers than television ads experience.[54]

Market segmentation. Radio advertising is flexible and efficient because advertisers can choose stations aimed at a specific market within a broad geographic region. Radio stations design their formats to appeal to specific types of audiences. (Ever notice how the stations you listen to are not the same ones your parents listen to?) AM stations, which once ruled the airways, now specialize mainly in "talk formats," such as call-in, news, religion, sports, and automotive shows. On the FM dial, country, urban contemporary, classical, classic rock, rhythm and blues, Hispanic, and "oldies" stations have listener profiles that give entrepreneurs the ability to pinpoint practically any advertising target.

Flexibility and timeliness. Radio commercials have short closing times and can be changed quickly. Small firms that sell seasonal merchandise or advertise special sales or events can change their ads on short notice to match changing market conditions.

Friendliness. Radio ads are more "active" than ads in printed media because they use the spoken word to influence customers. Vocal subtleties used in radio ads are impossible to convey through printed media. Spoken ads can suggest emotions and urgency, and they lend a personalized tone to the message.

Radio advertisements also have some *disadvantages*:

Poor listening. Radio's intrusiveness into the public life almost guarantees that customers will hear ads, but they may not listen to them. Listeners are often engaged in other activities while the radio is on and may ignore the message.

Need for repetition. Listeners usually do not respond to radio ads after a single exposure to them. Radio ads must be broadcast repeatedly to be effective. Consistency in radio ads is the key to success.

Limited message. Radio ads are limited to one minute or less, which requires that business owners keep their messages simple, covering only one or two points. In addition, radio spots do not allow advertisers to demonstrate their products or services. Although listeners can hear the engine purr, they can't see the car; spoken messages can only describe the product or service.

TABLE 10.3 Guidelines for Effective Radio Copy

- *Mention the business often.* This is the single most important and inflexible rule in radio advertising. Also make sure listeners know how to find your business. If the address is complicated, use landmarks.
- *Stress the benefit to the listener.* Don't say "Dixon's has new fall fashions." Say "Dixon's fall fashions make you look fabulous."
- *Use attention-getters.* One key to a successful radio ad is grabbing listeners' attention from the start and holding it. Radio gives the options of music, sound effects, and unusual voices. Crack the barrier with sound.
- *Zero in on your audience.* Know to whom you're selling. Radio's selectivity attracts the right audience. It's up to you to communicate in the right language.
- *Keep the copy simple and to the point.* Don't try to impress listeners with vocabulary. "To be or not to be" may be the best-known phrase in the language—and the longest word has just three letters.
- *Sell early and often.* Don't back into the selling message. At most, you've got 60 seconds. Make the most of them. Don't be subtle.
- *Write for the ear.* Forget the rules of grammar; write conversationally.
- *Prepare your copy.* Underline words you want to emphasize so that the announcer knows how the ad should read.
- *Triple space.* Type clean, legible copy. Make sure the announcer rehearses the ad.
- *Use positive action words.* Use words such as *now* and *today*, particularly when you're writing copy for a sale. Radio has qualities of urgency and immediacy. Take advantage of them by including a time limit or the date the sale ends.
- *Put the listener in the picture.* Radio's theater of the mind means you don't have to talk about a new car. With sounds and music, you can put the listener behind the wheel.
- *Focus the spot on getting a response.* Make it clear what you want the listener to do. Don't try to get a mail response. Use phone numbers or a Web site address only and repeat the number at least three times. End the spot with the phone number or the Web address.
- *Don't stay with a loser.* Direct-response ads produce results right way—or not at all. Don't stick with a radio spot that is not generating sales. Change it.

Sources: Kim T. Gordon, "Turn It Up," *Entrepreneur*, January 2004, pp. 80–81; Radio Advertising Bureau, *New York, NY, Radio Basics.*

BUYING RADIO TIME A small business owner can zero in on a specific advertising target by using the appropriate radio station. Stations following various formats—from rap to rhapsodies—appeal to specific audiences. Radio advertising time usually sells in 15-second, 30-second, and 60-second increments. Many radio stations now offer five-second spots called "adlets" and even super-short one- or two-second "blinks" that are designed to increase the awareness of a brand among listeners. Fixed spots are guaranteed to be broadcast at the times specified in the owner's contract with the station. Preemptible spots are cheaper than fixed spots, but the advertiser risks being preempted by an advertiser willing to pay the fixed rate for a time slot. Floating spots are the least expensive, but the advertiser has no control over broadcast times. Many stations offer package plans, using flexible combinations of fixed, preemptible, and floating spots. Table 10.3 offers a guide to producing effective radio copy.

Radio rates vary depending on the time of day they are broadcast, and, like television, there are prime-time slots knows as drive-time spots. Although exact hours may differ from station to station, the following classifications are common (listed in descending order of cost):

Class AA: Morning drive time—6 A.M. to 10 A.M.

Class A: Evening drive time—4 P.M. to 7 P.M.

Class B: Home worker time—10 A.M. to 4 P.M.

Class C: Evening time—7 P.M. to midnight

Class D: Nighttime—midnight to 6 A.M.

Some stations may have different rates for weekend time slots.

ENTREPRENEURIAL PROFILE: Robin and Chris Sorensen: Firehouse Subs Robin and Chris Sorensen, former firefighters and cofounders of Firehouse Subs, a chain of sandwich shops with nearly 600 locations, watched as a recession caused sales to decline. "In our entire history, we had never had a period when our entire system was running negative sales," says CEO Don Fox. Research showed that customers were still buying from lower-priced restaurants such as Firehouse Subs and indicated that the company's declining sales problem stemmed from a lack of brand awareness. Executives launched a brand-building radio campaign that emphasized the company's USP and included the Sorensens as spokesmen. "Our way beats their way," the ads said. "If you don't agree, it's free." The ads "turned things around on a dime," says Fox, pointing out that sales increased by more than 10 percent. "We immediately went to positive sales for the entire system."[55]

NEWSPAPERS For decades, local print newspapers were the medium that most small businesses relied on to get their advertising messages out to customers. Online media have reduced significantly the money that businesses of all sizes spend on newspaper ads. Both the circulation and the number of newspapers in the United States have declined, as has the share of total advertising dollars the medium attracts. The circulation of nation's 1,382 daily newspapers has declined from 62.2 million in 1980 to 44.4 million today, and many papers have added online versions of their content in an attempt to offset the decline.[56] Newspapers provide several *advantages* to small business advertisers:

Selected geographical coverage. Newspapers are geared to a specific geographic region, and they reach potential customers across all demographic classes. Local newspapers, in particular, provide extensive coverage of a company's immediate trading area.

Flexibility. A business can change its newspaper advertisements on very short notice. Entrepreneurs can select the size of the ad, its location in the paper, and the days on which it runs. For instance, auto repair shops often advertise their tune-up specials in the sports section on weekends, and party shops display their ads in the entertainment section as the weekend approaches.

Timeliness. Newspapers almost always have very short closing times, the publication deadline prior to which the advertising copy must be submitted. Many newspapers allow advertisers to submit their copy as late as 24 hours before the ad runs.

Communication potential. Newspaper ads can convey a great deal of information by employing attractive graphics and copy. Properly designed, they can be very effective in attracting attention and persuading readers to buy.

Low costs. Newspapers normally offer advertising space at low absolute cost and, because of their blanket coverage of a geographic area, at low relative cost as well.

Prompt responses. Newspaper ads typically produce relatively quick customer response. A newspaper ad is likely to generate sales the very next day, and advertisers who use coupons can track the response to an ad. This advantage makes newspapers an ideal medium for promoting special events, such as sales, grand openings, or the arrival of a new product.

Attractive target audience. Adults who read newspapers are an attractive target market for many small businesses. Newspaper readers tend to be well-educated, older people with high household incomes.

Newspaper advertisements also have *disadvantages*:

Wasted readership. Because newspapers reach a wide variety of people, at least a portion of an ad's coverage will be wasted on readers who are not potential customers. This non-selective coverage makes it more difficult for newspapers to reach specific target markets than ads in other media.

Reproduction limitations. The quality of reproduction in newspapers is limited, especially when it is compared with that of magazines and direct mail. Recent technological advances, however, are improving the quality of reproduction in newspaper ads.

Lack of prominence. One frequently cited drawback of newspapers is that they carry so many ads that a small company's message might be lost in the crowd. The typical newspaper is 63 percent advertising.[57] This disadvantage can be overcome by increasing the size of the ad or by adding color to it. Color can increase the reading of ads by as much as 80 percent over black-and-white ads. Studies show that two-color ads do "pull" better than black-and-white ones but only by a small margin. The *real* increase in ad recall and response comes from using full four-color ads. Bold headlines, illustrations, and photographs also increase an ad's prominence. Proper ad placement in the newspaper can increase an ad's effectiveness. The best locations are on a right-hand page, near the right margin, above the half-page mark, or next to editorial articles. The most read sections in a newspaper are the front page, the local news section, and the sports section.[58]

Less effective in reaching young adults. Newspaper circulation as a percentage of U.S. households has declined from 98 percent in 1970 to 40 percent today as readers have migrated to the Internet.[59] Print newspaper ads are less effective in reaching young adults; just 24 percent of 18- to 34-year-old adults read a daily newspaper.[60] Young people, in particular, read news online rather than from a printed newspaper.

Short ad life. The typical newspaper is soon discarded, and as a result, an ad's life is extremely short. Business owners can increase the effectiveness of their ads by giving them greater continuity. Spot ads can produce results, but maintaining a steady flow of business requires some degree of continuity in advertising.

BUYING NEWSPAPER ADS Print newspapers typically sell ad space by lines and columns or inches and columns. For instance, a 4-column × 100-line ad occupies four columns and 100 lines of space (14 lines = 1 column inch). For this ad, the small business owner would pay the rate for 400 lines. If the newspaper's line rate is $3.50, this ad would cost $1,400 (400 lines × $3.50 per line). Most papers offer discounts for bulk, long-term and frequency contracts, and full-page ads. Advertising rates vary from one paper to another, depending on factors such as circulation and focus. Entrepreneurs should investigate the circulation statements, advertising rates, and reader profiles of a newspaper to see how well it matches the company's target audience before selecting one as an advertising medium.

MAGAZINES Another advertising medium available to the small business owner is magazines. Today, customers have nearly 7,200 magazine titles from which to choose, many of them in electronic form. With a total circulation of more than 312 million adults in the United States, magazines have a wide reach, and their readers tend to be more educated and have higher incomes than consumers of other advertising media such as television.[61]

Magazines offer several *advantages* for advertisers:

Long life spans. Magazines have a long reading life because readers tend to keep them longer than other printed media. Few people read an entire magazine at one sitting. Instead, most pick it up, read it at intervals, and come back to it later. The result is that each magazine ad has a good chance of being seen several times.

Multiple channels of engagement. Growing numbers of magazine ads include action codes, such as QR codes, that take users to a company's Web site, Facebook page, or online video, giving advertisers the ability to engage customers across multiple channels. Some online magazines include apps that allow readers to click on an ad to purchase the advertised product. In addition, 69 percent of magazine readers say that they have posted a magazine article on Facebook, and 75 percent have followed a magazine on Twitter and Pinterest.[62]

Reader engagement. The average magazine reader spends 41 minutes with a magazine.[63] Print magazines have a high "pass-along" rate, being handed down front reader to reader.

Target marketing. Within the last 25 years, magazines have become increasingly focused. Advertisers can select magazines aimed at customers with specific interests—from wooden boats and black-and-white photography to container gardening and bodybuilding. By selecting the appropriate special-interest periodical, small business owners can reach those customers with a high degree of interest in their goods or service. Once business owners define their target markets, they can select magazines whose readers most closely match their customer profiles. For instance, *House and Garden* magazine reaches a very different audience than *Rolling Stone.*

Ad quality. Magazine ads tend to be of high quality. Consumers rank magazine ads ahead of ads in all other media on capturing readers' attention and in trustworthiness.[64] Advertisers can choose the location of their ads in a magazine. The most effective locations for magazine ads are the back cover, the inside front cover, and the inside back cover. Multiple page spreads also increase ad recall among readers.[65]

Magazines also have several *disadvantages*:

Costs. Magazine advertising rates vary according to their circulation rates; the higher the circulation, the higher the rate. Thus, local magazines, whose rates are often comparable to newspaper rates, may be the best bargains for small businesses.

Long closing times. Another disadvantage of magazines is the relatively long closing times they require. For a weekly print magazine, the closing date for an ad may be several weeks before the actual publication date, making it difficult for advertisers to respond quickly to changing market conditions.

Lack of prominence. Another disadvantage of magazine ads arises from their popularity as an advertising vehicle. The effectiveness of a single ad may be reduced because of a lack of prominence; 46.2 percent of the typical magazine content is devoted to advertising.[66] Proper ad positioning, therefore, is critical to an ad's success. Research shows that readers "tune out" right-hand pages and look mainly at left-hand pages.

Declining circulation rates. Circulation rates for most magazines have declined over the last decade, prompting many of them to launch digital versions of their content. According to the Pew Research Center, only 17 percent of adults say they read a print magazine yesterday, down from 23 percent in 2002.[67]

SPECIALTY ADVERTISING As advertisers have shifted their focus to "narrowcasting" their messages to target audiences and away from "broadcasting," specialty advertising has grown in popularity. Businesses spend more than $20 billion annually on promotional gift items such as pens, shirts, caps, memory sticks, umbrellas, and calendars that are imprinted with a company's name, address, telephone number, Web address, logo, and slogan.[68] The most popular promotional items are apparel, writing instruments, bags, and calendars.[69] Specialty items are best used as reminder ads to supplement other forms of advertising and help to create goodwill among existing and potential customers.

Specialty advertising offers several *advantages*:

Reaching select audiences. Advertisers have the ability to reach specific audiences with well-planned specialty items. Lindner, one of Europe's leading makers of coffins, made a memorable splash with its customers by giving away calendars featuring photographs of models sporting nothing but body paint posing with the company's coffins! The calendar was such a hit that the family-owned company was inundated with requests and has made the calendar an annual tradition.[70]

Personalized nature. By carefully choosing a specialty item, business owners can "personalize" their advertisements. When choosing advertising specialties, business owners should use items that are unusual, related to the nature of the business, and meaningful to customers. In addition, the item should be something that customers will keep and use (preferably in the environment in which they make decisions about using the company's products and services) and reinforce the company's USP.

Versatility. The rich versatility of specialty advertising is limited only by the business owner's imagination. Advertisers print their logos on everything from pens and golf balls to key chains and caps.

They work. Specialty advertising cuts through ad clutter because people enjoy receiving promotional products. One study found that 76.2 percent of people who had received a promotional product within the last two years were able to recall the specific product, the advertiser, and the message.[71]

There are *disadvantages* to specialty advertising:

Potential for waste. Unless entrepreneurs choose the appropriate specialty item for their businesses and their target audiences, they will be wasting time and money.

Cost. Some specialty items can be quite expensive. In addition, some owners have a tendency to give advertising materials to anyone—even to those people who are not potential customers. Proper distribution of giveaway items is an important aspect of enhancing the effectiveness of and controlling the cost of specialty advertising.

POINT-OF-PURCHASE ADS In-store advertising has become popular as a way of reaching the customer at a crucial moment—the point of purchase. Smart entrepreneurs "plan the in-store experience to win over shoppers where it matters most—the point of purchase," says Richard Winter, president of Point of Purchase Advertising International. Research suggests that shoppers make 76 percent of all buying decisions at the point of sale and that end-cap (those displays at the end of aisles) and freestanding displays of items are most effective.[72] Self-service stores are especially well suited for point-of-purchase ads because they remind people of the products as they walk the aisles. These in-store ads are not just simple signs or glossy photographs of the product in use. Some businesses use in-store music interspersed with household hints and, of course, ads. Another technique involves shelves that contain tiny devices that sense when a customer passes by and triggers a prerecorded sales message. Some self-service stores use floor graphics, point-of-purchase ads that transform their floors into advertising space.

OUTDOOR ADVERTISING Outdoor (or out-of-home) advertising is one of the oldest forms of advertising in existence. Archaeological evidence shows that merchants in ancient Egypt chiseled advertising messages on stone tablets and placed them along major thoroughfares. Outdoor advertising remains popular today; advertisers spend $6.4 billion on this medium annually.[73] The United States is a highly mobile society, and outdoor advertising takes advantage of this mobility. Outdoor advertising is popular among small companies, especially retailers, because well-placed ads serve as reminders to shoppers that the small business is nearby and ready to serve their needs. In addition, the brevity required for an effective outdoor ad is ideal for directing viewers to a company's Web site. Very few small businesses rely solely on outdoor advertising; instead, they supplement other advertising media with outdoor ads such as billboards and transit ads. With a creative out-of-home ad campaign, a small company can make a big impact with only a small budget.

ENTREPRENEURIAL PROFILE: Wade and Betty Lindsey: Wade's Southern Cooking Wade's Southern Cooking, a restaurant that Wade and Betty Lindsey founded in 1947 in Spartanburg, South Carolina, and that specializes in fresh meats and vegetables, created an outdoor ad in 1995 that depicted a green bean, the pun "Bean me up, Scottie," and the restaurant logo and address. The ad was an instant hit, and the company has since created more than 70 billboards that use puns and comical themes based on vegetables and other items on its menu. (One ad featured a stalk of broccoli reclining on a couch wearing a Snuggie and the phrase "Comfort food.") The outdoor campaign has played a significant role in the second-generation family business's growth and success.[74]

One of Wade's Southern Cooking's outdoor advertisements.
Source: Wade's Restaurant Inc.

Outdoor advertising offers certain *advantages* to a small business:

High exposure. Outdoor advertising offers high-frequency exposure, especially among people who commute to work. The average one-way commute to work in the United States is more than 25 minutes.[75] Most people tend to follow the same routes in their daily traveling, and billboards are there waiting for them when they pass by.

Broad reach. The typical outdoor ad reaches an adult 29 to 31 times each month. The nature of outdoor ads makes them effective devices for reaching a large number of potential customers within a specific area. Not only has the number of cars on the road increased, but the number of daily vehicle trips people take has also climbed. In addition, the people outdoor ads reach tend to be younger, wealthier, and better educated than the average person.

Attention-getting. The introduction of new technology such as 3-D, fiber optics, and other creative special effects to outdoor advertising has transformed billboards from flat, passive signs to innovative, attention-grabbing promotions that passersby cannot help but notice.

Flexibility. Advertisers can buy outdoor advertising units separately or in a number of packages. Through its variety of graphics, design, unique features, and choice of location, outdoor advertising enables the small advertiser to match his or her message to the particular audience.

Cost efficiency. Outdoor advertising offers one of the lowest costs per thousand customers reached of all advertising media. The cost per thousand customers (CPM) for out-of-home ads is $2.26, compared to $4.54 for radio, $5.50 for newspaper ads, $6.98 for magazine ads, and $6.00 to $14.50 for television commercials.[76]

Out-of-home ads also have several *disadvantages*:

Brief exposure. Because billboards are immobile, the reader is exposed to the advertiser's message for only a short time—typically only one or two seconds. As a result, the message must be short and to the point.

Limited ad recall. Because customers often are zooming past outdoor ads at high speed, they are exposed to an advertising message very briefly, which limits their ability to retain the message.

Legal restrictions. In many cities, outdoor ads are subject to strict regulations; some cities place limitations on the number, size, content, and type of billboards allowed along road-sides. Some cities have banned digital billboards altogether in the name of traffic safety and beautification.

Lack of prominence. Some heavily traveled routes are so cluttered with out-of-home ads that the effectiveness of any single ad is reduced.

USING OUTDOOR ADS Consumers are spending more time in their cars than ever before (18.5 hours per week), and outdoor advertising is an effective way to reach them.[77] Technology has changed the face of outdoor advertising dramatically in recent years. Computerized printing techniques that render truer, crisper, and brighter colors, billboard extensions, and three-dimensional effects have improved significantly the quality of standard billboards (known as posters or bulletins in the industry). New vinyl surfaces display print-quality images and are extremely durable. Digital billboards, giant computer screens that rotate messages every 6 to 10 seconds, allow companies to create vibrant, high-resolution, eye-catching ads that capture viewers' attention at reasonable cost.

Because the outdoor ad is stationary and the viewer is in motion, a small business owner must pay special attention to its design. An outdoor ad should do the following:

- Allow viewers to identify the product and the company clearly and quickly.
- Use a simple background. The background should not compete with the message.
- Rely on large illustrations that jump out at the viewer.
- Include clear, legible fonts. All lowercase or a combination of uppercase and lowercase letters works best. Very bold or very thin typefaces become illegible at a distance. Select simple fonts that are easy to read from a distance.
- Use black-and-white designs. Research shows that black-and-white outdoor ads are more effective than color ads. If color is important to the message, pick color combinations that contrast both hue and brightness, such as black on yellow.
- Emphasize simplicity. Short copy and short words are best. Don't try to cram too much onto a billboard. Because of their brief window of exposure, ads with just three to five words are most effective, and ads containing more than 10 words are ineffective.
- Use illumination so that passersby can read them at night. By using illuminated billboards, advertisers can increase the reach of outdoor ads by 16 percent.[78]
- Be located on the right-hand side of the highway.

Two of the latest trends in outdoor advertising are Internet-connected digital boards and billboards that send messages to customers' cell phones. With digital billboards, ad content is virtually unlimited; advertisers can include eye-catching graphics and streaming media in their ads. (Giant digital billboards, called spectaculars, are common in New York's Times Square, where large auto-free plazas cater to pedestrians, whose travel speeds are slower than cars.) Digital billboards ads, which cost as little as $600 to more than $15,000 per month, offer advertisers great flexibility. For instance, a restaurant could change the messages it displays to advertise its breakfast offerings in the morning, lunch specials at midday, and dinner menu in the afternoon and evening.

The latest outdoor ads include a computer chip that interacts with a Web browser that is common to many cell phones, which enables advertisers to send messages to the cell phones of passersby. For instance, a movie theater's smart billboard could send show times to customers' cell phones for the feature films it is running.

Many entrepreneurs use company vehicles as rolling outdoor ads, posting business signs on them or transforming the entire car into a rolling advertisement with vinyl wrap. These rolling ads should be consistent with a company's advertising strategy, tasteful, and include ways for people who see them to contact the company, including its Web site and telephone number. Outdoor gear retailer L.L.Bean generated buzz with its "Bootmobile," a vehicle created to celebrate the company's one-hundredth anniversary featuring a larger-than-life (13 feet

L.L.Bean's Bootmobile, a rolling advertisement.
Source: L.L.Bean Bootmobile.

tall and 20 feet long) replica of L.L.Bean's first and most iconic product, the Maine hunting boot, which L.L.Bean began selling in 1912. The Bootmobile travels to various stores to host special events, which its drivers promote through various social media such as Twitter (@Bootmobile).[79]

TRANSIT ADVERTISING A variation of outdoor advertising is transit advertising, which includes advertising signs on the inside and outside of the public transportation vehicles such as trains, buses, and subways throughout the country's urban areas. The medium is poised for growth as more cities look to public transit systems to relieve traffic congestion.

Transit ads offer a number of *advantages*:

Wide coverage. Transit advertising offers advertisers mass exposure to a variety of customers. The message literally goes to where the people are.

Repeat exposure. Transit ads provide lengthy and repeated exposure to a message, particularly for inside cards, the ads that appear inside the vehicle.

Low cost. Even small business owners with limited budgets can afford transit advertising.

Flexibility. Transit ads come in a wide range of sizes, numbers, and duration. With transit ads, an owner can select an individual market or any combination of markets across the country.

Transit ads also have several *disadvantages*:

Generality. Although entrepreneurs can choose the specific transit routes on which to advertise, they cannot target a particular segment of the market through transit advertising as effectively as they can with other media. The effectiveness of transit ads depends on the routes that public vehicles travel and on the people they reach, which, unfortunately, the advertiser cannot control. Overall, transit riders tend to be young, affluent, and culturally diverse.

Limited appeal. Unlike many media, transit ads are not beamed into the potential customer's residence or business. The result is that customers cannot keep them for future reference.

Brief message. Transit ads do not permit advertisers to present a detailed description or a demonstration of the product or service for sale. Although inside ads have a relatively long exposure (the average ride lasts 22.5 minutes), outside ads must be brief and to the point.

DIRECT MAIL Direct mail has long been a popular method of direct marketing and includes tools such as letters, postcards, catalogs, discount coupons, brochures, and other items that businesses mail to customers' or potential customers' homes or businesses. The earliest known catalogs were printed by fifteenth-century printers. Although online sales long ago surpassed direct-mail catalog sales, companies spend $21 billion annually on direct-mail ads, selling virtually every kind of product imaginable, from Christmas trees and lobsters to furniture and clothing (the most popular mail-order purchase).[80] Direct mail offers some distinct *advantages* to entrepreneurs:

Flexibility. Direct mail gives advertisers the capacity to tailor a message to target customers using variable data printing. Rather than send a blanket mail blast to 100,000 addresses, advertisers can target 5,000 high-potential customers with a mailing. An advertiser's presentation to customers can be as simple or as elaborate as necessary. One custom tailor shop achieved a great deal of success with fliers it mailed to customers and included a swatch of material from the fabrics for the upcoming season's line of suits. With direct mail, the tone of the message can be personal, creating a positive psychological effect. In addition, advertisers control the timing of their campaigns, sending ads when they are most appropriate.

Reader attention. With direct mail, an advertiser's message does not have to compete with other ads for the reader's attention. Most people enjoy getting mail, and studies show that 80 percent of people read some or all of their advertising mail.[81] Unlike many e-mail messages, direct mail gets a recipient's undivided attention at least for a moment. If the message is on the mark and sent to the right audience, direct-mail ads can be a powerful advertising tool. Catalogs are particularly effective at capturing recipients' attention. The average household receives three catalogs per week, and recipients typically look at two of them. In addition, 92 percent of people who receive catalogs say that they have purchased something from at least one of them.[82]

Rapid feedback. Direct-mail advertisements produce quick results. In most cases, an ad will generate sales within three or four days after customers receive it. Business owners should know whether a mailing has produced results within a relatively short time period.

Measurable results and testable strategies. Because they control their mailing lists, direct marketers can readily measure the results their ads produce. Also, direct mail allows advertisers to test different ad layouts, designs, and strategies (often within the same "run") to see which one pulls the greatest response. The best direct marketers are always fine-tuning their ads to make them more effective. Table 10.4 offers guidelines for creating direct mail ads that really work.

Effectiveness. The right message targeted at the right mailing list can make direct mail one of the most efficient forms of advertising. Direct mail to the right people produces results.

Direct-mail ads also suffer from several *disadvantages*:

Inaccurate mailing lists. The key to the success of the entire mailing is the accuracy of the customer list. Using direct-mail ads with a poor mailing list is a guaranteed waste of money. Experienced direct mail marketers cite the 60-30-10 rule, which says that 60 percent of a campaign's success depends on the quality of the list, 30 percent on the offer, and 10 percent on the creativity of the ad.[83] Make sure the mailing list you use is accurate and up to date.

Clutter. The typical household receives on average 624 pieces of direct mail annually.[84] With that volume of direct mail, it can be difficult for an advertisement to get customers' attention.

High relative costs. Relative to the size of the audience reached, the cost of designing, producing, and mailing an advertisement via direct mail is high. Rising paper and postage costs pose real threats to companies that use direct mail. However, if a mailing is well planned and properly executed, it can produce a high percentage of responses, making direct mail one of the least expensive advertising methods in terms of return on investment.

High throwaway rate. Often called junk mail, direct-mail ads become "junk" when an advertiser selects the wrong audience or broadcasts the wrong message. According to the Direct Mail Association, the average response rate for a direct-mail campaign is 4.4 percent.[85] By supplementing traditional direct-mail pieces with toll-free (800) numbers and carefully timed follow-up phone calls, companies have been able to increase their response rates.

HOW TO USE DIRECT MAIL The key to a direct mailing's success is the right mailing list. Even the best direct-mail ad will fail if sent to the "wrong" customers. Owners can develop lists themselves, using customer accounts, telephone books, city and trade directories, and other sources, including companies selling complementary but not competing products, professional organizations' membership lists, business or professional magazines' subscription lists, and

TABLE 10.4 Guidelines for Creating Direct-Mail Ads That Really Work

In many industries, the average direct-mail campaign is one that produces a response rate of at least 4.4 percent, which means that 95.6 percent of the customers who received the ad did *not* respond to it! What steps can entrepreneurs take to improve the results of their direct mail campaigns?

Realize that repetition is one key to success.

Experts estimate that customers must receive at least three direct-mail pieces per month from a business before they really notice the ad.

Provide meaningful incentives.

Direct mail succeeds by getting prospects to respond to a written offer. To do that, a direct-mail ad must offer potential customers something of value—a free sample, a special price, a bonus gift, or anything that a company's target customers value. Twenty percent of prospects who do not open the direct-mail ads they receive say that they have no reason to open them. Make sure your offer gives them a reason!

Write copy that will get results.

Try the following proven techniques:

- Write catchy headlines and openers. Promise readers your most important benefit in the headline or first paragraph.
- Use short "action" words and paragraphs and get to the point quickly.
- Make the copy look easy to read with lots of "white space."
- Use eye-catching words such as *free*, *you*, *save*, *guarantee*, *new*, *profit*, *benefit*, *improve*, and others.
- Consider using computerized "handwriting" somewhere on the page or envelope; it attracts attention.
- Forget grammatical rules; write as if you were speaking to the reader.
- Repeat the offer three or more times in various ways.
- Back up claims and statements with proof and endorsements whenever possible.
- Always include a clear call to action—ask for the order or a specific response. "People are more likely to respond when you specifically tell them what to do," says marketing expert Dean Rieck.
- Ask attention-getting questions such as "Would you like to lower your home's energy costs?" in the copy.
- Use high-quality copy paper and envelopes (those with windows are best) because they stand a better chance of being opened and read. Brown envelopes that resemble government correspondence work well.
- Envelopes that resemble bills almost always get opened.
- Address the envelope to an individual, not "Occupant."
- Avoid mailing labels, which shout "direct-mail ad piece." The best campaigns print addresses directly on the envelopes.
- Use stamps if possible. They get more letters opened than metered postage.
- Use a postscript (P.S.)—always; they are the most often read part of a printed page. Make sure the P.S. contains a "hook" that will encourage the recipient to read on. This is the perfect place to restate offer's USP.
- Include a separate order form that passes the following "easy" test:
  - *Easy to find.* Consider using brightly colored paper or a unique shape.
  - *Easy to understand.* Make sure the offer is easy for readers to understand. Marketing expert Paul Goldberg says, "Confuse 'em and you lose 'em."
  - *Easy to complete.* Keep the order form simple and unconfusing.
  - *Easy to pay.* Direct-mail ads should give customers the option to pay by whatever means is most convenient.
  - *Easy to return.* Including a postage-paid return envelope (or at a minimum a return envelope) will increase the response rate.

Build and maintain a quality mailing list over time.

The right mailing list is the key to a successful direct-mail campaign. You may have to rent lists to get started, but once you are in business, use every opportunity to capture information about your customers. Constantly focus on improving the quality of your mailing list.

Integrate direct-mail campaigns with other advertising media.

Use direct mail to direct customers to your company's Web site, Facebook page, or blog (perhaps by including a QR code in the mailing).

Test your campaigns and track their results.

Successful direct-mail marketers constantly tweak their ads and test the results. "Testing is everything," says the founder of a company that used direct-mail ads as part of a marketing strategy that led his company to $10 million in annual sales. Monitoring the response rate from each mailing is essential for knowing which ads and which lists actually produce results.

Sources: Based on Dean Rieck, "How to Use Direct Mail to Drive Targeted Web Site Traffic," MarketingProfs, August 3, 2010, *www.marketingprofs.com/articles/2010/3796/how-to-use-direct-mail-to-drive-targeted-website-traffic*; Chrisanne Sternal, "23 Great Direct Mail Tips for Entrepreneurs," Understanding *Marketing*, March 23, 2010, *http://understandingmarketing.com/2010/03/23/23-great-direct-mail-tips-for-entrepreneurs*; *What's in the Mailbox? The Impact of One-to-One Marketing on Consumer Response*, Winterberry Group, January 2007, p. 7; "Direct Mail Tips for Manufacturers' Letters," Koch Group, *www.kochgroup.com/directmail.html*; Kim T. Gordon, "Copy Right," *Business Start-Ups*, June 1998, pp. 18–19; Paul Hughes, "Profits Due," *Entrepreneur*, February 1994, pp. 74–78; "Why They Open Direct Mail, " *Communications Briefings*, December 1993, p. 5; Ted Lammers, "The Elements of Perfect Pitch," *Inc.*, March 1992, pp. 53–55; "Special Delivery," *Small Business Reports*, February 1993, p. 6; Gloria Green and James W. Peltier, "How to Develop a Direct Mail Program," *Small Business Forum*, Winter 1993/1994, pp. 30–45; Susan Headden, "The Junk Mail Deluge," *U.S. News & World Report*, December 8, 1997, pp. 40–48; Joanna L. Krotz, "Direct-Mail Tips for Sophisticated Marketers," Microsoft Small Business Center, *www.microsoft.com/smallbusiness/resources/marketing/customer_service_acquisition/direct_mail_tips_for_sophisticated_marketers.mspx*.

mailing list brokers who sell lists for practically any need. Advertisers can locate list brokers through *The Direct Marketing List Source* from the Standard Rate and Data Service found in most public libraries. The key to success with a direct mail campaign is to get your ad noticed, and the right mailing list is the ideal starting point.

ENTREPRENEURIAL PROFILE: Clare Meehan: AlphaGraphics in the Cultural District Clare Meehan, CEO of AlphaGraphics in the Cultural District, a print and marketing communications company in Pittsburgh, Pennsylvania, and her team created an unusual direct-mail ad campaign that involved mailing a coconut ("Did you know you could mail a coconut?" said the hangtag), material on the power of direct mail, and a link to a personalized Web site to 200 carefully selected high-level marketing executives in local businesses. The goal was to direct the prospects to the Web site, invite them to a marketing workshop hosted by AlphaGraphics, and build a list of potential clients. The direct-mail campaign was a huge success, generating a 300 percent return on investment, drawing 41 prospects to the marketing workshop, and producing a 20 percent increase in sales for the company.[86]

TRADE SHOWS Trade shows provide manufacturers and distributors with a unique opportunity to advertise to a preselected audience of potential customers who are inclined to buy. Thousands of trade shows take place each year, and carefully evaluating and selecting the right shows can produce profitable results for a business owner. Companies spend $24.5 billion on trade shows annually, but trade show success does *not* depend on how much an exhibitor spends; instead, success is a function of planning, preparation, and follow-up.[87]

Trade shows offer the following *advantages*:

A natural market. Trade shows bring together buyers and sellers in a setting in which exhibitors can explain and demonstrate their products. Converting a prospective customer at a trade show into an actual sale costs 38 percent less than converting a prospect on a sales call.[88]

Preselected audience. Trade exhibits attract potential customers with a genuine interest in the goods or services on display. There is a high probability that trade show attendees will make a purchase. According to one survey, 83 percent of trade show attendees have the authority to make purchasing decisions, and 91 percent of attendees say they get the most useful buying information from trade shows.[89]

New customer market. Trade shows offer exhibitors a prime opportunity to reach new customers and to contact people who are not accessible to sales representatives.

Cost advantage. As the cost of making a field sales call continues to escalate, companies are realizing that trade shows are an economical method of generating leads and making sales presentations.

There are, however, certain *disadvantages* associated with trade shows:

Increasing costs. The cost of exhibiting at trade shows is rising. Registration fees, travel and setup costs, sales salaries, and other expenditures may be a barrier to some small firms.

Wasted effort. A poorly planned exhibit ultimately costs the small business more than its benefits are worth. Too many firms enter exhibits in trade shows without proper preparation, and they end up wasting their time, energy, and money on unproductive activities.

To avoid these disadvantages, entrepreneurs should do the following:

- Verify that the audience is a good fit for your company's products or services. Research trade shows to find the ones that will put you in front of the best prospects for your company.

- Establish objectives for every trade show. Do you want to generate 100 new sales leads, make new product presentations to 500 potential customers, or generate $5,000 in sales?

- Communicate with key potential customers *before* the show; send them invitations or invite them to stop by your booth for a special gift.

- Plan your display with your target audience in mind and make it memorable. Be sure your exhibit shows your company and its products or services in the best light. Do everything to maximize the visibility of your exhibit and keep the display neat.

- Staff your booth with knowledgeable salespeople. Attendees appreciate meeting face-to-face with knowledgeable and friendly staff. Every staff member should have an elevator pitch and questions with which to engage prospects in conversation.

- Do something to attract a crowd to your booth. Demonstrate your product or service so that customers can see it in action, sponsor a drawing for a prize, or set up an interactive display. Drawing a crowd creates "buzz" for your company among attendees.

- Learn to distinguish between serious customers and "tire kickers."

- Make it easy for potential customers to get information about your company and its products and services. Distribute literature that clearly communicates the benefits of your products or services.

- Project a professional image at all times. Salespeople who man the booth should engage prospects in conversation and should ask qualifying questions.

- Capture prospective customers' contact information, including e-mail addresses, and follow up with prompt "welcome" e-mails.

- Follow up promptly on every sales lead. The most common mistake trade show participants make is failing to follow up on the sales leads the show generated. If you are not going to follow up leads, why bother to attend the show in the first place?

Few small businesses rely on a single advertising medium to communicate their advertising messages to potential customers, choosing instead to employ **cross-channel advertising strategies** in which they communicate with potential customers using a variety of media. For instance, Gary Lindsey, head of marketing at The Parent Company, a consumer product business that targets young parents, says that although the company is primarily Web based, it relies heavily on direct-mail catalogs and e-mail marketing to drive sales. "There's real return on investment from those catalogs," says Lindsey. "People love shopping on the Internet, but there is something powerful when you combine print and Internet."[90]

How to Prepare an Advertising Budget

One of the most challenging decisions confronting a small business owner is how much to spend on advertising. The amount entrepreneurs want to spend and the amount they can afford to spend usually differ significantly. There are four methods of creating an advertising budget: *what is affordable*, *matching competitors*, *percentage of sales*, and *objective and task*.

4.
Identify four basic methods for preparing an advertising budget.

Under the what-is-affordable method, business owners see advertising as a luxury. They view it completely as an expense rather than as an investment that generates sales and profits in the future. As the name implies, entrepreneurs who use this method spend on advertising whatever their companies can afford. Too often, business owners determine their advertising budgets after they have funded all of the other budget items. The result is an advertising budget that is inadequate for getting the job done. This method also fails to relate the marketing communications budget to the marketing communications objective.

Another approach is to match the advertising budget of the company's competitors, either in a flat dollar amount or as a percentage of sales. This method assumes that a company's advertising needs and strategies are the same as those of its competitors, which is rarely the case. Although competitors' actions can be helpful in establishing a floor for marketing communications expenditures, relying on this technique can lead to blind imitation instead of a budget suited to a small company's circumstances.

The most commonly used method of establishing an advertising budget is the simple percentage-of-sales approach. This method relates advertising expenditures to actual sales results. Tying advertising expenditure to sales rather than to profits creates greater consistency in advertising because most companies' sales tend to fluctuate less than profits. One rule of thumb for establishing an advertising budget is spending 10 percent of projected sales the first year of business, 7 percent the second year, and at least 5 percent in each successive year. Relying totally on broad rules like these can be dangerous, however. They may not be representative of a small company's advertising needs.

The objective-and-task method is the most difficult and least used technique for establishing an advertising budget. It also is the method most often recommended by advertising experts.

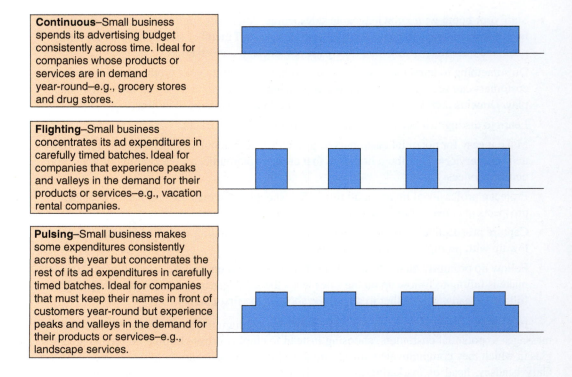

FIGURE 10.7

Advertising Scheduling Strategies

Continuous–Small business spends its advertising budget consistently across time. Ideal for companies whose products or services are in demand year-round–e.g., grocery stores and drug stores.

Flighting–Small business concentrates its ad expenditures in carefully timed batches. Ideal for companies that experience peaks and valleys in the demand for their products or services–e.g., vacation rental companies.

Pulsing–Small business makes some expenditures consistently across the year but concentrates the rest of its ad expenditures in carefully timed batches. Ideal for companies that must keep their names in front of customers year-round but experience peaks and valleys in the demand for their products or services–e.g., landscape services.

With this method, an entrepreneur links advertising expenditures to specific business objectives. The objective-and-task method builds up an advertising budget from the bottom up by analyzing what it will cost to accomplish an entrepreneur's business objectives. For example, suppose that an entrepreneur wants to boost sales of a particular product 10 percent by attracting local college students. He may determine that a nearby rock radio station would be the best advertising medium to use. Then he must decide on the number and frequency of the ads and estimate their costs. Entrepreneurs follow this process for each advertising objective.

Once they establish their advertising objectives and budgets, many entrepreneurs find it useful to use a calendar to plan the timing of their advertising campaigns and expenditures across the year. An advertising calendar helps entrepreneurs ensure that they create a continuous presence in customers' minds throughout the year by creating a steady flow of interesting and useful information flowing their way. Figure 10.7 illustrates three common advertising scheduling strategies.

How to Advertise Big on a Small Budget

5. _____

Explain practical methods for stretching an entrepreneur's advertising budget.

The typical small business does not have the luxury of an unlimited advertising budget. Most cannot afford to hire a professional ad agency. This does not mean, however, that a small company should assume a second-class advertising posture. Most advertising experts say that, unless a small company spends more than $10,000 to $15,000 a year on advertising, it probably doesn't need an ad agency. For most, hiring freelance copywriters and artists on a per-project basis is a much better bargain. With a little creativity and a dose of ingenuity, small business owners can stretch their advertising dollars and make the most of what they spend. Three useful techniques are cooperative advertising, shared advertising, and stealth advertising.

Cooperative Advertising

In **cooperative advertising**, a manufacturing company shares the cost of advertising with a retailer if the retailer features its products in those ads. Both the manufacturer and the retailer get more advertising per dollar by sharing expenses. Cooperative advertising not only helps small businesses stretch their advertising budgets but also offers another source of savings: the free advertising packages that many manufacturers supply to retailers. These packages usually include photographs and illustrations of the product as well as professionally prepared ads to use in different media.

Shared Advertising

In **shared advertising**, a group of similar businesses forms a syndicate to produce generic ads that allow the individual businesses to dub in local information. The technique is especially useful for small businesses that sell relatively standardized products or services such as legal assistance, autos, and furniture. Because the small firms in the syndicate pool their funds, the result usually is higher-quality ads and significantly lower production costs.

Stealth Advertising

In Chapter 9, you learned about bootstrap marketing principles, offbeat, low-cost techniques for marketing a small company's goods and services. In advertising, these techniques are called **stealth advertising**, which includes innovative ads that do not necessarily look like traditional ads and often are located in unexpected places. Ads now appear on electrical outlets in airport terminals, on eggs (gently printed directly onto the shells with lasers), on clothes hangers from laundries, in urinals in public restrooms (using a device called Wizmark that plays sounds and pictures when a guest arrives), and in other unusual places.[91] One consumer products company achieved success with a campaign that involved painting manhole covers in New York City to look like steaming hot cups of coffee. Bamboo Lingerie attracted a great deal of attention for its brand by stenciling on New York City sidewalks the message "From here it looks like you could use some new underwear" and its name and logo.[92] Some cities, including New York, Chicago, Los Angeles, San Diego, San Francisco, and others, have outlawed stealth advertising, calling it logo graffiti.[93]

Other Ways to Save

Other cost-saving suggestions for advertising expenditures include the following:

- *Repeat ads that have been successful.* In addition to reducing the cost of ad preparation, repetition may create a consistent image in a small firm's advertising program.
- *Use identical ads in different media.* If a billboard has been an effective advertising tool, an owner should consider converting it to a newspaper or magazine ad or a direct-mail flier.
- *Hire independent copywriters, graphic designers, photographers, and other media specialists.* Many small businesses that cannot afford a full-time advertising staff buy their advertising services à la carte. They work directly with independent specialists and usually receive high-quality work that compares favorably with that of advertising agencies without paying a fee for overhead.
- *Concentrate advertising during times when customers are most likely to buy.* Some small business owners make the mistake of spreading an already small advertising budget evenly—and thinly—over a 12-month period. A better strategy is to match advertising expenditures to customers' buying habits.

Chapter Review

1. Define your company's unique selling proposition (USP).
   - Branding a company's products or services depends on communicating the correct USP.
   - Answers the customer's ultimate question: What's in it for me?
2. Explain the differences among promotion, publicity, personal selling, and advertising.
   - Promotion is any form of persuasive communication designed to inform consumers about a product or service and to influence them to purchase these goods or services. It includes publicity, personal selling, and advertising.

- Publicity is any commercial news covered by the media that boosts sales but for which the small business does not pay.
- Personal selling is the personal contact between salespeople and potential customers that comes from sales efforts.
- Advertising is any sales presentation that is nonpersonal in nature and is paid for by an identified sponsor. A company's target audience and the nature of its message determine the advertising media it will use.

3. Describe the advantages and disadvantages of various advertising media.
   - The medium used to transmit an advertising message influences customers' perception—and reception—of it.

- Media options include word-of-mouth, the Internet, e-mail, social media, newspapers, radio, television, magazines, direct mail, the Internet, outdoor advertising, transit advertising, directories, trade shows, special events and promotions, and point-of-purchase ads.
4. Discuss the four basic methods for preparing an advertising budget.
   - Establishing an advertising budget presents a real challenge to the small business owner.

- There are four basic methods: what is affordable, matching competitors, percentage of sales, and objective and task.
5. Explain practical methods for stretching a business owner's advertising budget.
   - Despite their limited advertising budgets, small businesses do not have to take a second-class approach to advertising. Three techniques that can stretch a small company's advertising dollars are cooperative advertising, shared advertising, and stealth advertising.

Discussion Questions

10-1. What are the three elements of promotion? How do they support one another?

10-2. What factors should an entrepreneur consider when selecting advertising media?

10-3. What is a unique selling proposition (USP)? What role should it play in a company's advertising strategy?

10-4. One writer describes the unique selling proposition of a company from which she received an e-mail: "Combining the strategy, business processes, implementation, and technical support skills of a CRM systems integrator with the data management, analytic, and marketing skills of a database marketing service provider to deliver and operate a close-looped marketing and sales environment." How do you rate the effectiveness of this USP? Explain. What are the characteristics of an effective USP?

10-5. Summarize the advantages and disadvantages of the following advertising media:
- Word-of-mouth
- Social media
- Sponsorships
- Television
- Radio
- Newspapers
- Internet advertising
- Magazines
- Specialty advertising
- Direct mail
- Out-of-home advertising
- Transit advertising
- Directories
- Trade shows

10-6. Assume you are a small business owner who has an advertising budget of $1,500 to invest in a campaign promoting a big July Fourth "Blowout" sale. Where would you be most likely to invest your advertising budget if you were trying to reach customers in the 25- to 45-year-old age range with higher-than-average disposable income who are likely to be involved in boating activities in a local resort town? Explain. How would you generate free publicity to extend your advertising budget?

10-7. What are fixed spots, preemptible spots, and floating spots in radio advertising?

10-8. Describe the characteristics of an effective outdoor advertisement.

10-9. Briefly outline the steps in creating an advertising plan. What principles should the small business owner follow when creating an effective advertisement?

10-10. Describe the common methods of establishing an advertising budget. Which method is most often used? Which technique is most often recommended? Why?

10-11. What techniques can small business owners use to stretch their advertising budgets?

10-12. Use a search engine to locate the most recent "E-Mail Marketing Trends Survey." Using the information in it, work with a team of your classmates to select a local small business with which you are familiar to design an effective e-mail advertising campaign. What are the advantages and the disadvantages of using e-mail as an advertising medium?

CHAPTER 11

Pricing and Credit Strategies

Price is what you pay.
Value is what you get.

—Warren Buffett

The moment you make a mistake in pricing, you're eating into your reputation or your profits.

—Katharine Paine

Pricing: A Creative Blend of Art and Science

1.

Explain why pricing is both an art and a science.

One of the most challenging yet most important decisions entrepreneurs must make involves pricing their products and services. Studies by consulting firm Accenture show that increasing prices by just 1 percent can produce an 11 percent increase in a company's profit, a result that is much greater than that produced by a comparable 1 percent decrease in costs.[1] "There's nothing you can do as quickly to improve profitability—and nothing you can do as quickly to destroy profitability—as change your pricing," says one consultant who specializes in pricing.[2] Prices that are too high can drive customers away and hurt a small company's sales. Pricing products and services too low, a common tendency among first-time entrepreneurs, robs a business's ability to earn a profit, leaves customers with the impression that its goods and services are of inferior quality, and threatens its long-term success. Improper pricing has created serious problems for many business owners who mistakenly believed that their prices were high enough to generate a reasonable profit when, in reality, they were not.

ENTREPRENEURIAL PROFILE: Jill Caren: *OutdoorPlayToys.com* When Jill Caren started *OutdoorPlayToys.com*, an online retailer of toys designed to allow children to have a fun day of play outside, she set her prices based on each toy's cost and an estimate of shipping charges. However, shipping charges, particularly on heavy items such as pedal cars and playhouses, were higher than she had anticipated, and her company lost money on sales of those items. Caren conducted an analysis of her company's total costs of operation (including shipping costs), raised her prices accordingly, and turned her company around.[3]

Another complicating factor is a holdover from the Great Recession: Customers have become more price sensitive. "The recession changed people's perceptions and expectations about value significantly," says Kurt Kane, chief marketing officer for Pizza Hut, a company that has reformulated its pricing strategy to emphasize value to its customer base. "A generation of consumers has been born under a new value mindset, and I don't see that changing anytime soon."[4] A recent survey of shoppers' behavior by Parago reports that 70 percent of shoppers say that they are more price sensitive than they were just 12 months before. Reflecting this sensitivity, 95 percent of shoppers look for sales, deals, rebates, or lowest advertised prices at least some of the time before they buy.[5] Shoppers are using technology such as smart phones and tablet PCs with price comparison apps to shop for the best deals—often right in the middle of a store's aisles. In a recent survey by Perception Research Services International, 76 percent of smart phone owners use their phones while shopping, 53 percent of these mobile shoppers use their phones to compare product prices, and 48 percent use them to look for discount coupons or sales.[6] The result is a new age of price transparency that entrepreneurs have never before experienced and an environment that makes setting the right prices all the more important—and difficult (see Figure 11.1).

Price is the monetary value of a good or service; it is a measure of what a customer must give up to obtain a good or service. For shoppers, price is a reflection of value. Customers often look to a product's or service's price for clues about its value. Consider the following examples, which illustrate the sometimes puzzling connection between price and perceived value:

- After a stint in the advertising industry, Brent Black launched The Panama Hat Company of the Pacific, a small shop in Kailua, Hawaii, from which he sells some of the world's finest Panama hats. Made by skilled artisans in Ecuador, the hats sport weaves of exceptionally thin straw that range from 19 to 60 rows per inch. Black's basic line of hats require 150 hours of work by a talented team of craftsmen, and his ultrapremium hats can take up to 1,000 hours to make. Black purchases the hats directly from Ecuadorian weavers and then blocks them by hand into dozens of styles using another labor-intensive process in his shop. Prices for Black's hats range from $650 for a basic panama hat with 19 rows per inch to $25,000 for one made by premier hatmaker Simón Espinal.[7]

- Shoppers know that they can find a smart phone for as little as $99 (sometimes for free) if they sign a contract with a wireless carrier. Yet Swiss watchmaker Ulysse Nardin, founded in 1846, cannot keep up with demand for its Chairman smart phone, which carries the official title of the most expensive smart phone in the world. The company's base model sells for $14,000, and its premier model, which features 3,000 hand-cut diamonds on the faceplate, goes for a lofty $130,000! When Ulysse Nardin introduced its ultraluxurious phone, it expected to receive at most 400 orders; instead, customers placed more than 8,000

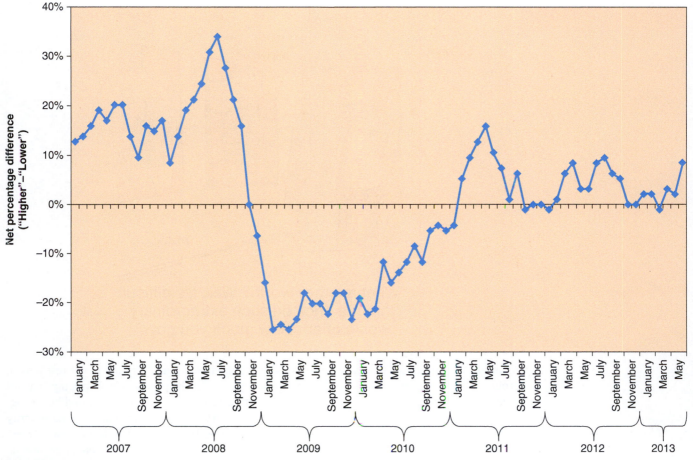

FIGURE 11.1

The Reality of Pricing Decisions—Net Difference in the Percentage of Small Business Owners Who Set Prices *Higher* and Those Who Set Prices *Lower* Than in the Previous 3 Months

Source: Based on William Dunkelberg and Holly Wade, "NFIB Small Business Economic Trends," National Federation of Independent Businesses, July 2013, p. 8.

orders. In addition to its touch screen, the feature-packed Chairman boasts a rose gold case, a thumbprint reader (for unlocking the phone), a physical numeric keypad, and a winding stem on the side (a throwback to the company's watchmaking heritage) that supplies it with backup power.[8]

- 3 × 1, a shop in Manhattan's SoHo District started by fashion entrepreneur Scott Morrison, sells bespoke, custom-made, and limited-edition jeans to upscale clients. 3 × 1's limited-edition jeans, which are already made and ready to wear, range from $295 to $495 per pair. 3 × 1's custom process allows customers to select the denim fabric and the hardware that go into their jeans that sell for between $535 and $750 per pair. Customers who choose the bespoke option get to work with Morrison and his pattern maker to create a pair of jeans made to their measurements and specifications. The shop offers more than 135 different denim fabrics, and the most expensive denim that Morrison uses costs more than $100 per meter. Making a typical pair of jeans requires at least 3 meters of fabric. Morrison, his pattern makers, and tailors produce 22 to 30 pairs of jeans per day on site in full view of customers using old-fashioned single-needle sewing machines and years of cutting and sewing experience and skill. Every pair of jeans is numbered. Sales have exceeded Morrison's expectations, and he is considering opening a second store in Japan.

As you can see, setting higher prices sometimes can *increase* the appeal of a product or service ("If you charge more, you must be worth it").

Entrepreneurs must develop a keen sensitivity to both the psychological and economic impact of their pricing decisions. A product's or service's price must exceed the cost of providing

"What are they worth? Well, it's difficult to put a price on such extraordinary works of art. How much you got?"

it, and it must be compatible with customers' perceptions of value. "Pricing is not just a math problem," says one business writer. "It's a psychology test."[9] The psychology of pricing is an art much more than it is a science. It focuses on creating value in the customer's mind but recognizes that value is what the customer perceives it to be. At the outset, the goal is not necessarily to determine *the* ideal price but rather an ideal price *range*. This **price range** is the area between the **price floor** that is established by a company's total cost to produce the product or provide the service and the **price ceiling**, which is the most the target customers are willing to pay (see Figure 11.2). The final price within this range depends on the image the company wants to create in the minds of its customers.

The price floor depends on a company's cost structure, which can vary considerably from one business to another, even though they may be in the same industry. Although their cost structures may be different from those of their competitors, many entrepreneurs play follow the leader with their prices, simply charging what their competitors do on similar or identical products or services. Although this strategy simplifies the pricing decision, it can be very dangerous. Determining the price floor for a product or service requires entrepreneurs to have access to timely, accurate information about the cost of producing or selling a product or providing a service.

The price ceiling depends on entrepreneurs' ability to understand their customers' characteristics and buying behavior, the benefits that the product or service offers customers, and the prices of competing products. The best way to learn about customers' buying behavior is to conduct ongoing market research and to spend time with customers, listening to the feedback they

FIGURE 11.2

What Determines Price?

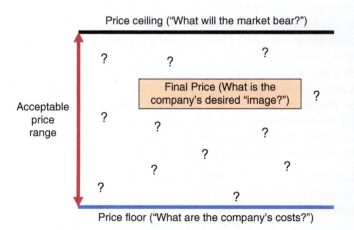

offer. Small companies with effective pricing strategies tend to have a clear picture of their target customers and how their companies' products and services fit into their customers' perception of value. A company that begins losing valued customers who complain that its prices are too high has bumped into the price ceiling, and the owner should consider cutting prices.

An entrepreneur's goal is to position the company's prices within this acceptable price range that falls between the price floor and the price ceiling. The final price that entrepreneurs set depends on the desired image they want to create for their products or services: discount (bargain), middle of the road (value), or prestige (luxury). A prestige pricing strategy is not necessarily better or more effective than a no-frills, value pricing strategy. What matters most is that the company's pricing strategy matches the image the owner wants to create for it and the prices its target customers are willing and able to pay. Surveying potential customers, either in person or online, about the price they expect to pay for a particular product or service can be helpful. Serial entrepreneur Bryan Janeczko, cofounder of Nu-Kitchen, an online food retailer, surveyed customers using Survey Monkey about the price they would expect to pay for a day's worth of healthy meals. Responses ranged from $20 to $40 per day, leading Janeczko to set a price of $30.[10]

Entrepreneurs often find themselves squeezed by rising operating and raw material costs but are hesitant to raise prices because they fear losing customers. Businesses faced with rising operating and raw material costs should consider the following strategies:

- *Communicate with customers.* Let your customers know what's happening. Danny O'Neill, owner of The Roasterie, a wholesale coffee business that sells to upscale restaurants, coffeehouses, and supermarkets, operates in a market in which the cost of raw material and supplies can fluctuate wildly due to forces beyond his control. When coffee prices nearly doubled in just three months, O'Neill was able to pass along the rising costs of his company's raw material to customers without losing a single one. He sent his customers a six-page letter and copies of newspaper articles about the increases in coffee prices. The approach gave the Roasterie credibility and showed customers that the necessary price increases were beyond his control.[11]

- *Rather than raise the price of the good or service, include a surcharge.* Price increases tend to be permanent, but if higher costs are the result of a particular event (e.g., a hurricane that disrupted the nation's ability to process oil and resulted in rapidly rising fuel costs), a company can include a temporary surcharge. If the pressure on its costs subsides, the company can eliminate the surcharge. Before gasoline prices climbed rapidly, Ricky Eisen, who owns Between the Bread, a catering business in New York City, and uses two trucks and a van to make deliveries, paid $40 to $60 for gas for each vehicle; now she pays $75 to fill up each one. "It's enough to put a dent in our profit," she says. Eisen recently added a 5 percent fuel surcharge to offset the increased cost of doing business.[12]

- *Rather than raise prices, consider eliminating customer discounts, coupons, and "freebies."* Over time, discounts, coupons, and other freebies add up and can cut into a company's profits, sometimes significantly. Eliminating them is an invisible way of raising prices that can add significantly to a small company's profit margin.

- *Offer products in smaller sizes or quantities.* As food costs soared, many restaurants introduced "small plates," reduced portion items that enabled them to keep their prices in check. In the quick-service sector, mini-burgers billed as "fun food" and offered in bundles became a popular item on many menus.

- *Focus on improving efficiency everywhere in the company.* Although raw materials costs may be beyond a business owner's control, other costs within the company are not. One way to cope with the effects of a rapid increase in costs is to find ways to cut costs and to improve efficiency in other areas. These improvements may not totally offset higher raw materials costs, but they can dampen their impact. Rather than raise prices, the owners of Jen-Mor Florists, a family-run flower shop in Dover, Delaware, decided to cut the number of deliveries to the edge of their territory to just one per day to reduce the company's delivery expenses.[13]

- *Consider absorbing raw material cost increases to keep customers with long-term importance to the company.* Saving a large account might be more important than keeping pace with rising costs. Companies that absorb the rising cost of raw materials often find ways to cut costs in other areas and to operate more efficiently.

- *Emphasize the value your company provides to customers.* Unless a company reminds them, customers can forget the benefits and value its products offer. "If you provide great value to your customers, a little price increase isn't going to scare them away," says Elizabeth Gordon, a small business consultant.

- *Raise prices incrementally and consistently rather than rely on large periodic increases.* Companies that do so are less likely to experience resistance due to customers' sticker shock. In the restaurant business, about 33 cents of each $1 of sales goes to cover food and beverage costs. Recent increases in food costs have had a direct (and negative) impact on a restaurant's profits, which average between 3 and 5 percent of sales. When the cost of chicken wings increased by 70 percent in just one year, Buffalo Wild Wings, a chain of more than 850 restaurants that specializes in wings, gradually raised the price of its wings from 45 to 50 cents and then to 60 cents. "We have not seen any resistance to that price increase at all," says CEO Sally Smith.[14]

- *Shift to less expensive raw materials if possible.* When seafood and beef prices increased, many restaurants added more chicken dishes to their menus. When gold prices tripled within a four-year period, jeweler John Christian, based in Austin, Texas, began creating more designs in silver and gold, all-silver, and even steel to keep costs and prices under control. The company also launched a separate line of products called Carved Creations priced well below the average $750 price for the John Christian line. Within two years, Carved Creations accounted for 30 percent of the company's sales.[15]

- *Modify the product or service to lower its cost.* At quick-service restaurants, value menu items priced at $1 account for about 10 percent of sales. However, rapidly rising food and energy costs have squeezed or eliminated franchisees' profits on these items, forcing chains to modify the items by eliminating a slice of cheese (which saves six cents) or shaving two ounces of beef from the patty to maintain the $1 price.[16] Companies using this strategy must exercise caution, taking care not to reduce the quality of their products and services so much that they damage their reputations.

- *Anticipate rising materials costs and try to lock in prices early.* It pays to keep tabs on raw materials prices and be able to predict cycles of inflation. Entrepreneurs who can anticipate rising prices may be able to make purchases early or lock in long-term contracts before prices take off.

Three Powerful Pricing Forces: Image, Competition, and Value

Price Conveys Image

2.

Discuss the relationships among pricing, image, competition, and value.

A company's pricing policy can be a powerful tool for establishing a brand and for creating a desired image among its target customers. Whether they are seeking an image of exclusivity or one that reflects bargain basement deals, companies use price to enhance the image of their brands. Some companies emphasize low prices, but others establish high prices to convey an image of quality, exclusivity, and prestige, all of which appeal to a particular market segment.

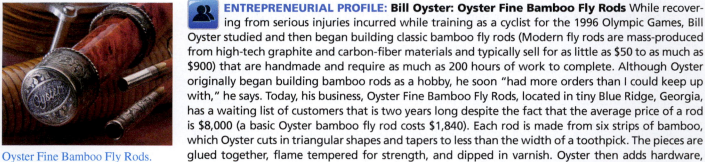

Oyster Fine Bamboo Fly Rods.
Source: Oyster Fine Bamboo Flyrods.

ENTREPRENEURIAL PROFILE: Bill Oyster: Oyster Fine Bamboo Fly Rods While recovering from serious injuries incurred while training as a cyclist for the 1996 Olympic Games, Bill Oyster studied and then began building classic bamboo fly rods (Modern fly rods are mass-produced from high-tech graphite and carbon-fiber materials and typically sell for as little as $50 to as much as $900) that are handmade and require as much as 200 hours of work to complete. Although Oyster originally began building bamboo rods as a hobby, he soon "had more orders than I could keep up with," he says. Today, his business, Oyster Fine Bamboo Fly Rods, located in tiny Blue Ridge, Georgia, has a waiting list of customers that is two years long despite the fact that the average price of a rod is $8,000 (a basic Oyster bamboo fly rod costs $1,840). Each rod is made from six strips of bamboo, which Oyster cuts in triangular shapes and tapers to less than the width of a toothpick. The pieces are glued together, flame tempered for strength, and dipped in varnish. Oyster then adds hardware, which he engraves himself to customers' specifications. Customers from around the world, including former presidents, have ordered fly rods from Oyster for prices of up to $17,000.[17]

Value for products such as Oyster bamboo rods does not reside solely in their superior technical performance or handmade beauty but rather in their scarcity and uniqueness and the resulting image ("wow" factor) they create for the buyer. Although entrepreneurs must recognize the extremely limited market for ultraluxury items such as these, the ego-satisfying ownership of limited-edition fly rods, watches, shoes, handbags, cars, jewelry, and other items is a potent psychological force that supports a premium price strategy.

Too often, small companies underprice their products and services, believing that low prices are the only way they can gain an advantage in the marketplace. According to management consulting firm McKinsey and Company, 80 to 90 percent of the pricing mistakes that companies make involve setting prices that are too low.[18] Companies that fall into this trap fail to recognize the extra value, convenience, service, and attention that they give their customers—things that many customers are willing to pay extra for. These entrepreneurs forget that price is just *one* element of the marketing mix and for many customers is not the most important factor. A global study by Accenture reports that 67 percent of customers are not willing to compromise on quality just to get a lower price, and 54 percent of customers are not willing to compromise on the level of customer service they receive in exchange for a lower price.[19]

The secret to setting prices properly is understanding a company's target market: the customer groups at which the it aims its goods or services. Target market, business image, and price are interwoven.

ENTREPRENEURIAL PROFILE: Mark Kronenberg: Math 1-2-3 Mark Kronenberg, founder of Math 1-2-3, an in-home math tutoring and test preparation company in 21 cities across the United States, targets upscale customers who are interested more in results than in price. Some clients have balked at the rates (up to $200 per hour) that Math 1-2-3 charges for its highly qualified, carefully selected tutors, but Kronenberg says that attracting and retaining high-end clients has been more profitable for his company. "It's a misconception that if you raise prices too much, you'll have no business," he says. "There are many customers who shop based on quality, not lowest price. I think it's best to avoid a race to the bottom. It's an easy race to win, but you won't have a lot of profit to show for it."[20]

Competition and Prices

An important part of setting appropriate prices is tracking competitors' prices regularly; however, the prices that competitors are charging is just *one* variable in the pricing mix (and often not the most important one at that). When setting prices, entrepreneurs should take into account their competitors' prices, but they should *not* automatically match or beat them. Businesses that offer customers extra quality, value, service, or convenience can charge higher prices as long as customers recognize the "extras" they are getting. In other words, companies that implement a successful differentiation strategy (refer to Chapter 3) can charge higher prices for their products and services.

ENTREPRENEURIAL PROFILE: Naomi Poe: Better Batter Gluten Free Flour Naomi Poe, founder of Better Batter Gluten Free Flour (BBGFF) in Altoona, Pennsylvania, initially tried to match competitors' prices, but because her flour and baking mixes contain no gluten, they cost more to produce than traditional products that contain gluten. As a result, BBGFF was losing money. Poe knew that she had to raise prices but feared that her company's sales would plummet—until she realized that her products offered customers extra value that she could promote. Not only were they gluten free, a significant benefit for the 2.4 million Americans who suffer from celiac disease, but they also tasted good! "In blind taste tests on people, not just those who are gluten intolerant, we heard consistently that our products were superior," she says. "We also offer an unconditional guarantee as well as education and counseling services." Over the course of the next three years, Poe raised prices on BBGFF's products by one-third and actually saw sales increase. Her company is now profitable.[21]

Two factors are vital to studying the effects of competition on a small firm's pricing policies: the location of the competitors and the nature of the competing goods. In most cases, unless a company can differentiate the quality and the quantity of extras it provides, it must match the prices charged by nearby competitors for identical items. For example, if a self-service station charges a nickel more for a gallon of gasoline than the self-service station across the street charges, customers will simply go across the street to buy. Without the advantage of a unique business image—quality of goods sold, value of service provided, convenient location, favorable credit terms, and others—a small company must match local competitors' prices or else lose sales. Before matching any competitor's price change, however, entrepreneurs should consider the rivals' strategies. The competition may be establishing its prices using a unique set of criteria and a totally different strategy.

The nature of competitors' goods also influences a small company's pricing policies. Entrepreneurs must recognize those products that are direct substitutes for those they sell and strive to keep prices in line with them. For example, the local sandwich shop should consider the hamburger restaurant, the taco shop, and the roast beef shop as competitors because they all serve fast foods. Although none of them offers the identical menu of the sandwich shop, they are all competing for the same quick-meal dollar.

Whenever possible, entrepreneurs should avoid head-to-head price competition with other firms that can more easily offer lower prices because of their lower cost structures. Most locally owned drugstores cannot compete with the prices of large national drug chains. However, many local drugstores operate successfully by using nonprice competition by offering personal service, free delivery, credit sales, and other extras that the chains have eliminated. Nonprice competition can be an effective strategy for a small business in the face of larger, more powerful enterprises because there are many dangers in experimenting with prices. For instance, price shifts cause fluctuations in sales volume that the small firm may not be able to tolerate. In addition, frequent price changes may muddle a company's image and damage customer relations.

One of the deadliest games a small business can get into with competitors is a price war. Price wars can eradicate profit margins, force companies out of business, and scar an entire industry for years. The tablet computer market has been characterized by price wars. Apple, which relies on a premium pricing strategy for it leading edge products, quickly became the dominant player in the market when in introduced its iPad in 2010 at a starting price of $499. In 2011, as the tablet market became more crowded, online giant Amazon decided to grab market share by offering a lower-priced tablet and introduced its Kindle Fire at just $199. Within a year, Amazon had reduced the price of an entry-level Kindle to $159 and introduced updated versions of a high-definition Kindle Fire at prices that ranged from $199 to $599, prices that were $230 to $349 less than a comparable iPad. "Amazon did what it had to do to compete with Apple, Google, and other tablet makers, putting some pressure on them, particularly on price," says an industry analyst. Other companies, including Google and Microsoft entered the fray, introducing their own tablets at prices similar to Amazon's. Although Apple's iPad remains the dominant tablet with 55 percent market share, Amazon managed to capture 21 percent of the market with its low-price strategy but had to rely on sales of apps and other services, such as its Prime membership for profits. "We want to make money when people use our devices, not when they buy our devices," says Amazon CEO Jeff Bezos.[22]

Price wars usually begin when one competitor believes that it can achieve a higher volume through lower price or that it can exert enough pressure on competitors' profits to drive them out of business. In most cases, entrepreneurs overestimate the power of price cuts to increase sales sufficiently to improve their profitability. In a price war, a company may cut its prices so severely that it is impossible to achieve the volume necessary to offset the lower profit margins. If a company that has a 25 percent gross profit margin cuts prices by 10 percent, it would have to *triple* its sales volume just to break even. Even when price cuts work, their effects are often temporary. Customers lured by the lowest price usually have almost no loyalty to a business. The lesson: The best way to survive a price war is to stay out of it by emphasizing the unique features, benefits, advantages, and value your company offers its customers!

Focus on Value

Ultimately, the "right" price for a product or service depends on one factor: the value that it provides for customers. There are two aspects of value, however. Entrepreneurs may recognize the *objective* value of their products and services, which is the price customers would be willing to pay if they understood perfectly the benefits that a product or service delivers for them. Unfortunately, few if any customers can see a product's or a service's true objective value; instead, they see only its *perceived* value, which determines the price they are willing to pay for it. "Customers see value in more than product and price," says Jim Barnes, a business owner, consultant, and author. "They spend where they believe they get the best total value, and that does not mean always opting for the lowest price." Small companies that find creative ways to add value to their products and services—for instance, by making it easy for customers to buy from them (Barnes calls these "I'll-look-after-that-for-you" moments), impressing them with stellar service, and providing unexpected extras (such as an electronics retailer consolidating a customer's various remote controls into one when installing a new television)—do not have to resort to price cuts as often as companies that fail to do these things. By offering extra value, these companies encourage their customers to look beyond mere price to determine value. "We must offer the customer something that allows us to earn the prices we charge," concludes Barnes. "We must go above and beyond, creating new and different forms of value that will compensate for the pressure on customers to obtain more for less."[23]

Businesses that underprice their products and services or run constant sales and discount price promotions may be short-circuiting the value proposition they are trying to communicate to their customers. Customers may respond to price cuts, but companies that rely on them to boost sales risk undermining the perceived value of their products and services. In addition, once customers grow accustomed to buying products and services during special promotions, the habit can be difficult to break. They simply wait for the next sale. Many retailers now face this problem as customers accustomed to buying items on sale postpone buying them until the next special sale arrives. The result has been fluctuating sales and a diminished value of those stores' brands.

In some economic conditions, companies have little choice but to offer lower-priced products. Techniques that companies can use to increase customers' perception of value and essentially lower their prices with less risk of diminishing their brands include offering coupons and rebates that are not as closely connected to the product as direct price cuts. **Limited-time-only (LTO) discounts** used sparingly also increase short-term sales without causing long-term damage to a brand. Another strategy that some companies have used successfully is to launch a **fighter brand**, a less expensive, no-frills version of a company's flagship product that is designed to confront lower-priced competitors head-on, satisfy the appetites of value-conscious customers, and preserve the image of the company's premium product. Rather than lower the price of its Pentium computer chip, Intel introduced the lower-cost Celeron chip to stave off rival AMD's line of value-priced chips.

The good news is that companies can influence through marketing and other efforts customers' perception of value. "The price you get for a product is a function of what it's truly worth—and how good a job you do communicating that value to the end user," says one entrepreneur.[24] Indeed, setting a product's or a service's price is another way a company can communicate value to its customers. For most shoppers, three reference points define a fair price: the price that they have paid for the product or service in the past, the prices that competitors charge for the same or similar product or service, and the costs that a company incurs to provide the product or service. The price that customers have paid in the past for an item serves as a baseline reference point, but people often forget that inflation causes a company's costs to rise from year to year. Therefore, it is important for business owners to remind customers periodically that they must raise prices to offset the increased cost of doing business. "Over time, costs always go up," says Norm Brodsky, owner of a successful document storage company. "I'd rather raise prices a little every year or with every new contract than be forced to demand a big increase down the road."[25]

As we have seen already, companies often find it necessary to match competitors' prices on the same or similar items unless they can establish a distinctive image in customers' minds. One of the most successful strategies for companies facing direct competition is to differentiate their products or services by adding value for customers and then charging for it. For instance, a company might offer faster delivery, a longer product warranty, extra service, or something else that adds value to an item for its customers and allows the business to charge a higher price.

Perhaps the least understood of the three reference points is a company's cost structure. Customers often underestimate the costs businesses incur to provide products and services, whether it is a simple cotton T-shirt on a shelf in a beachfront shop or a lifesaving drug that may have cost hundreds of millions of dollars and many years to develop. They forget that business owners must make or buy the products they sell, market them, pay their employees, and cover a host of other operating expenses, ranging from health care to legal fees.

Pricing Strategies and Tactics

There is no limit to the number of variations in pricing strategies and tactics. The wide variety of options is exactly what allows entrepreneurs to be so creative with their pricing. This section examines some of the most commonly used tactics under a variety of conditions. Pricing always plays a critical role in a firm's overall strategy; pricing policies must be compatible with a company's total marketing plan.

3.

Describe effective pricing strategies for both new and existing products and services.

New Product Pricing: Penetration, Skimming, or Sliding

Most entrepreneurs approach setting the price of a new product with a great deal of apprehension because they have no precedent on which to base their decisions. As a result, they often look at the prices their competitors charge and set prices that are slightly lower. If a new product's price is too high, it is in danger of failing because of low sales volume. However, if its price is too low, the product's sales revenue might not cover costs. *Establishing a price that is too low is far more dangerous than setting one that is too high.* Not only does the company forgo revenues and profits, but it also limits the product's perceived value in the eyes of its target customers. "When you build a business around the lowest price, you soon find that there's always someone else who can offer an even lower one," says serial entrepreneur Norm Brodsky. "As a result, you are under constant pressure to keep reducing your price. At best, you end up with an unsustainable commodity business that's no fun to run and an obstacle to achieving the goal of becoming economically self-sufficient."[26] To avoid the trap of setting their prices too low, entrepreneurs should consider the total value that they provide their customers, including intangibles such as additional service, convenience, speed, and others, and set a price that reflects that value.

ENTREPRENEURIAL PROFILE: Andrew Schmertz: Hopscotch Air In 2009, Andrew Schmertz, an avid aviator, started Hopscotch Air, an air-taxi service that uses small planes to take passengers to about 10,000 general aviation airports across the United States. When setting prices, Schmertz took into account the cost of fuel, airplane maintenance, pilots' salaries, and other operating costs, but only later did he realize that those prices neglected to reflect the value and convenience that Hopscotch Air provided customers. "Customers book when they want and show up when they want to go," he says. "That's time and convenience that our prices needed to reflect." Schmertz raised Hopscotch Air's prices, and customers never flinched.[27]

When pricing any new product, an entrepreneur must satisfy three objectives:

1. *Get the product accepted.* No matter how unique a product is, its price must be acceptable to a company's potential customers. The price a company can charge depends, in part, on the type of product it introduces:
   - **Revolutionary products** are so new that they transform an industry. Companies that introduce these innovative products usually have the ability to charge prices that are close to the price ceiling although they may have to educate customers about the product's benefits.
   - **Evolutionary products** involve making enhancements and improvements to products that are already on the market. Companies that introduce these products do not have the ability to charge premium prices unless they can use the enhancements they have made to differentiate their products from those of competitors. Establishing a price that is too low for an evolutionary product can lead to a price war.
   - **Me-too products** are products that companies introduce just to keep up with competitors. Because they offer customers nothing new or unique, me-too products offer companies the least amount of pricing flexibility. Achieving success with these products means focusing on cost control and targeting the right market segments.

2. *Maintain market share as competition grows.* If a new product is successful, competitors will enter the market, and a small company must work to expand or at least maintain its market share. Continuously reappraising a product's price in conjunction with special advertising and promotion techniques helps the company maintain market share.

3. *Earn a profit.* A small company must establish a price for the new product that is higher than its cost. Entrepreneurs should not introduce a new product at a price below cost because it is much easier to lower the price than to increase it once the product is on the market. Pricing their products too low is a common and often fatal mistake for new businesses; entrepreneurs are tempted to underprice their products and services when they enter a new market to ensure its acceptance.

Entrepreneurs have three basic strategies to choose from in establishing a new product's price: penetration, skimming, and life cycle pricing.

PENETRATION If a small company introduces a new product into a highly competitive market in which a large number of competitors are competing for acceptance, the product must penetrate the market to be successful. To gain quick acceptance and build market share quickly, some entrepreneurs use a penetration pricing strategy, introducing the product at a low price. Setting the price just above total unit cost allows the business to develop a wedge in the market and quickly achieve a high volume of sales. The resulting low profit margins may discourage other competitors from entering the market with similar products.

A penetration pricing strategy is ideal when introducing relatively low-priced goods into a market in which no elite segment and little opportunity for differentiation exist. This strategy works best when customers' switching costs (the cost of switching to a lower priced competitor's product) is high (e.g., video game consoles). Penetration pricing also works when a company's competitors are locked into high cost structures that result from the channels of distribution they use, labor agreements, or other factors. For instance, since its inception, Southwest Airlines has relied on its lower cost structure to compete with older, "legacy" carriers by emphasizing low prices. For a penetration pricing strategy to be successful, a company should have a cost advantage over its rivals; otherwise, it risks starting a no-win price war.

Entrepreneurs must recognize that a penetration pricing may take time to be effective; until a company achieves customer acceptance for the product, profits are likely to be small. When a young college student launched a carpet cleaning business to help pay for his education, he decided to be the low-cost provider in his area. Although he landed plenty of work for his part-time business, he found that his company generated very little profit after deducting the expenses of doing business. Realizing that his customers would be willing to pay more for quality work, he raised his prices and began earning a reasonable profit.[28]

A danger of a penetration pricing strategy is that it attracts customers who know no brand loyalty. Companies that garner customers by offering low introductory prices must wonder what will become of their customer bases if they increase their prices or if a competitor undercuts their prices. If a penetration pricing strategy succeeds and the product achieves mass-market penetration, sales volume increases, economies of scale result in lower unit cost, and the company earns attractive profits. The objectives of the penetration strategy are to achieve quick acceptance among customers and to generate high sales volume as soon as possible.

SKIMMING Companies often use a skimming pricing strategy to introduce a new product or service into a market with little or no competition or to establish a product or service as unique and superior to those of its competitors. Sometimes a company uses this tactic when introducing a product into a competitive market that contains an elite group that is willing and able to pay a premium price or when introducing a revolutionary product. A company sets a higher-than-normal price in an effort to quickly recover its initial developmental and promotional costs of the product. The idea is to set a price well above the product's total unit cost and to promote the product heavily to appeal to the segment of the market that is not sensitive to price. This pricing tactic often reinforces the unique, prestigious image of a company and projects a high-quality image of the product. If a product's price proves to be too low under a penetration strategy, raising the price can be very difficult. If a company using a skimming strategy sets a price that is too high to generate sufficient volume, it can easily lower the price. Successful skimming strategies require a company to differentiate its products or services from those of competitor to justify the above-average price.

ENTREPRENEURIAL PROFILE: Billy Lowe Billy Lowe, who owns a hair salon in Hollywood, California, has developed an "A-list" clientele of celebrities and has a reputation for treating his not-so-famous clients like celebrities as well. Lowe reinforces his salon's high-quality image ("Every day is a good hair day!") with a premium pricing strategy for his styling services that include a cut and style starting at $125 and full highlights starting at $175. "Clients are glad to pay your prices when there is perceived value and excellence," he says. "The spirit in which you offer your products and services also can help determine their attitudes toward your premium prices."[29]

LIFE CYCLE PRICING A variation of the skimming pricing strategy is called life cycle pricing. Using this technique, a small company introduces a product at a high price. Then technological advances, the learning curve effect, and economies of scale enable the company to lower its costs and reduce the product's price faster than its competitors can. By beating other businesses in a price decline, the company discourages competitors and, over time, becomes a high-volume producer. Blu-Ray players are a prime example of a product introduced at a high price that quickly cascaded downward as companies forged important technological advances and took

TABLE 11.1 Tips for Avoiding Pricing Mistakes

Tip 1. Be careful with cost-plus pricing. When companies base their prices on costs rather than on customers' perception of value, the result is almost always prices that are either too low or too high.

Tip 2. Recognize that "me-too" pricing gives a company no pricing power. A much better strategy is to differentiate your company's products or services by creating additional value for customers or by targeting market niches.

Tip 3. Realize that you cannot achieve the same profit margin across every product line your company sells. The profit margin for paper clips is likely to be quite different from the profit margin for printers.

Tip 4. Recognize that your customer base is made up of different customer segments and that some of them are more sensitive to price than others. Even if a company sells a single product or service, its value proposition differs among its different customer segments. That means that by adding extra value to its offerings aimed at customers who are willing to pay for it, a company can charge higher prices.

Tip 5. Do not put off raising prices out of fear of a customer backlash. If your costs of providing a product or service go up and you never raise prices, your profit margins shrink until you can no longer stay in business. Perpetually absorbing cost increases by holding prices the same is the pricing equivalent of sticking your head in the sand. The outcome is certain: business failure.

Tip 6. Do not compensate sales representatives solely on sales volume. Doing so encourages them to sell at any price, particularly low prices that destroy the company's profitability. Create profit-based incentives for your sales force.

Tip 7. Avoid launching a price war. As you learned in this chapter, no one "wins" a price war, and they can devastate an industry's profits for years.

Tip 8. Realize that although discounts have their place in a company's pricing strategy, they can be as addictive as drugs. "Companies that get hooked on discounts do little more than drive down their value proposition, sometimes past the point of no return," says one pricing consultant. If you decide to use discounts, use them sparingly, briefly, and creatively.

Tip 9. Recognize that some customers are more valuable to your business than others. Customers who always demand the lowest prices and the highest level of service often are a company's least profitable customers. Do not waste a disproportionate amount of time and energy catering to them; instead, identify your company's most profitable customers, focus on serving them well, and attract more customers like them.

Tip 10. Remember that price is just one variable in the sales equation. Costs, customers' perception of value, and image are important factors as well. Use them!

Sources: Based on "Eradicate Pricing Errors," *Sales & Marketing Management*, August 5, 2009, p. 1; Steve McKee, "How to Discount (If You Insist)," *Business Week*, August 14, 2009, *www.businessweek.com/smallbiz/content/aug2009/sb20090814_425078.htm.*

advantage of economies of scale. When Blu-Ray players were first introduced in 2006, they sold for $800; three years later, they were selling for just $220. Today, shoppers can purchase Blu-Ray players that have more features for less than $100. Life cycle pricing assumes that competition will emerge over time. Even if no competition arises, companies almost always lower the product's price to attract a larger segment of the market. In a life cycle pricing strategy, the initial high price contributes to the rapid return of start-up or development costs and generates a pool of funds to finance expansion and technological advances.

Table 11.1 offers useful tips for avoiding common pricing mistakes.

Pricing Techniques for Established Products and Services

Entrepreneurs have a variety of pricing techniques or tactics available to them to apply to established products and services. Entrepreneurs must examine each of these techniques or tactics to determine their effectiveness under different circumstances and situations.

ODD PRICING Although studies of consumer reactions to prices are mixed and generally inconclusive, many entrepreneurs use the technique known as **odd pricing**. They set prices that end in odd numbers (frequently 5, 7, or 9) because they believe that an item selling for $12.69 appears to be much cheaper than an item selling for $13.00. Psychological techniques such as odd pricing are designed to appeal to certain customer interests, but research on their effectiveness is mixed. Some studies show no benefits from using odd pricing, but others have concluded that the technique can produce significant increases in sales. Omitting the "$" symbol from prices may help, too. Researchers at Cornell University have discovered that restaurants that list menu prices without the "$" symbol ("12") achieved higher sales on average than those whose menu prices were written in script ("twelve dollars") or included the "$" symbol ("$12").[30]

PRICE LINING Price lining or tiered pricing is a technique that greatly simplifies the pricing decision. Under this system, an entrepreneur sells a product in several different price tiers or price lines. Each category of merchandise contains items that are similar in appearance, quality, cost, performance, or other features. Many lined products appear in sets of three—good, better, and best—at prices designed to satisfy different market segments' needs and incomes. Apple uses price lining for many of its products, including the iPad Mini, which it introduced at prices of $329 (16 GB), $429 (32 GB), and $529 (64 GB). Price lining can boost a store's sales because it makes goods available to a wide range of shoppers, simplifies the purchase decision for customers, and allows them to keep their purchases within their budgets.

 ## In the Entrepreneurial Spotlight

How to Compete with Cheap Knockoffs of Your Successful Product

Sarah and Jenifer Kaplan, cofounders of Footzyrollupz.
Source: Rollashoe LLC.

Sisters Jenifer and Sarah Kaplan admit that the designer high heels and stilettos that they wear often end up hurting their feet and causing them to quietly slip off their shoes whenever they can. They, like many women, were willing to put up with some degree of discomfort to be able to sport the latest "must-have" shoe. Little did they know that their fashion persistence would lead them to become entrepreneurs. After graduating from college, the sisters realized that many women suffered from the same uncomfortable shoe problem and launched Footzyrollupz, a company based in Miami Beach that sells comfortable flat shoes that women can roll up and discreetly slip into a small purse, clutch, or handbag, car glove box, or desk drawer.

As growing numbers of women discovered the simplicity and usefulness of Footzyrollupz, the company's sales increased quickly, which attracted competitors. Large retail stores began selling cheap knockoffs of their rollable yet comfortable and stylish shoes at much lower prices. "We had to differentiate ourselves from the $10 version at Target," says Sarah, "so we went with tiered pricing." The sisters decided to introduce a lower-priced Everyday Collection that sold for $20 per pair and a higher-end line

called Lux that sold for $30 per pair. The strategy was successful, and the impact was immediate. "We had a 100 percent increase in revenue," says Sarah. "We actually have had the most interest in our higher-priced shoes," says Sarah. Buoyed by the success of their initial pricing strategy, the Kaplans now offer Footrollupz shoes at three general price levels: basic models that start at just $22, mid-range shoes that are priced around $36, and a luxury collection that sell at prices from $55 to $69.

By refusing to engage in a price war with competitors and using tiered pricing, the Kaplans were able to generate additional sales without portraying their company as a bargain-basement discount seller. When their company's sales began to decline due to a weak economy and competition from cheap knockoffs, the Kaplans knew that they had to reevaluate the pricing strategy of Footrollupz but refused to be drawn into a price war. What are the signs that mean it is time to consider changing your company's pricing strategy?

1. ***Unit sales growth slows or declines.*** When sales volume stalls or declines, the market may be saturated, the economy may be struggling, competitors could be stealing away your customers, or your prices are out of line with customers' perceived value of your products and services.

2. ***Discounts fail to increase sales.*** The reason that companies offer price discounts is to increase sales, ideally by a greater percentage than the discount. If a discount fails to produce results, continuing to offer it is a recipe for disaster. "Price cutting usually is not the best strategy for a small business—especially a business that serves a target market that cares more about value and service than paying the lowest possible price," explains one business writer. "Not only does discounting generally fail to help you acquire new customers, but it may also result in your making less money from the customers you already have."

3. ***Competitors introduce new products or services.*** Innovations by competitors can change—sometimes

(continued)

In the Entrepreneurial Spotlight *(continued)*

dramatically—the price–value equation in the market. "If the competition has leapfrogged you on value, you may not be able to maintain your current pricing strategy," explains one pricing expert.

4. ***Low-cost competitors enter the market.*** When a market is experiencing high growth, it often attracts new entrants. If those new competitors have lower cost structures and utilize penetration pricing strategies, their entry can muddle the entire industry's pricing structure. When faced with this situation, some companies engage in a price war, but others take different approaches, such as introducing fighter brands or moving into less-price-sensitive niche markets.

5. ***Gross profit margin declines.*** As you will learn in Chapter 14, that a company's gross profit margin = (Sales − Cost of good sold) ÷ Sales. The only ways to repair a gross profit margin that is too low is to either reduce the company's cost of goods sold or increase its prices.

1. Explain the dangers of discounting as a pricing strategy for increasing sales.

2. Use the Internet to research price wars. What conditions usually prompt price wars? What impact do price wars have on an industry and the companies in it? What outcomes are typical in a price war?

3. Many small companies compete successfully without focusing on providing the lowest prices, even in industries in which customers view product or service prices as important purchasing criteria. What tactics do these companies use to compete successfully without relying on the lowest prices?

Sources: Based on Eilene Zimmerman, "Real-Life Lessons in the Delicate Art of Setting Prices," *New York Times*, April 20, 2011, *www.nytimes.com/2011/04/21/business/smallbusiness/21sbiz.html?pagewanted=all&_r=0*; Ryan McCarthy, "Pricing: How Low Can You Really Go?," *Inc.*, March 2009, pp. 91–92; Vincent Ryan, "The Price Is Wrong," *CFO*, December 2009, p. 52; Rosalind Resnick, "Hold the Line on Price," *Washington Post*, March 9, 2009, *www.washingtonpost.com/wp-dyn/content/article/2009/03/11/AR2009031103668.html*; "About Us," ePromos, *www.epromos.com/AboutePromos/AboutUs.jsp.*

FREEMIUM PRICING Companies that use **freemium pricing** provide a basic product or service to customers for free but charge a premium for expanded or upgraded versions of the product or service. Products or services that customers must use or experience to appreciate their value, such as software, are ideal candidates for a freemium pricing strategy. The goals of a freemium pricing strategy are to gain rapid and extensive adoption of a product or service and to give potential customers a chance to discover the value that it offers, particularly in its upgraded versions. The key to a successful freemium strategy is to expose users to enough free product or service features while reserving the most valuable benefits for customers who are willing to pay for the expanded versions. Typically, only about 2 to 4 percent of customers who use the free version of a product actually purchase its upgraded versions; therefore, for a freemium strategy to be successful, the potential market must be sizable.[31] Suppose, for example, that a company's revenue target is $10 million and that its average annual revenue per *paying* customer is $120. The company would need 83,333 paying customers ($10 million ÷ $120 per customer) to reach its target. Assuming a conversion rate of 3 percent, the company would have to attract 2,777,778 free users (83,333 customers ÷ 3 percent) to generate $10 million in annual sales.

One of the greatest dangers of a freemium pricing strategy is underestimating the cost of providing service and support for free users.

ENTREPRENEURIAL PROFILE: Drew Houston and Arash Ferdowsi: Dropbox "The big lesson [with] a freemium business model is that your marketing cost is the free users," says Drew Houston, who with Arash Ferdowsi cofounded Dropbox, a company that provides cloud storage and universal access to and sharing of files. Houston and Ferdowsi decided to use a freemium pricing strategy because "Dropbox was offering a product that people didn't know they needed until they tried it," explains Houston. Dropbox allows users to store up to 2 gigabytes of data for free (with incentives of up to 18 gigabytes of space for referring other users). Customers who need more storage space can purchase plans that start at just $9.99 per month for 100 gigabytes of space and go up to $6,420 per year for larger corporate users.[32]

DYNAMIC PRICING For many businesses, the pricing decision has become more challenging because the Internet gives customers access to incredible amounts of information about the prices of items ranging from cars to computers. Increasingly, customers are using the Internet to find the lowest prices available. To maintain their profitability, companies have responded with

dynamic (or customized) pricing, in which they set different prices on the same products and services for different customers using the information they have collected about their customers. Rather than sell their products at fixed prices, companies using dynamic pricing rely on fluid prices that may change based on supply and demand, the prices that competitors are charging, and which customer is buying or when a customer makes a purchase. For instance, a first-time customer making a purchase at an online store may pay a higher price for an item than a regular customer who shops there frequently pays for that same item.

Online retailers, especially those that operate stores through eBay or Amazon, are heavy users of dynamic pricing, often updating the prices of the products they sell minute by minute. Their goal is to establish the lowest price on an item, sometimes by just a penny, so that they can achieve the top spot in search engine results for shoppers who are searching for the lowest prices. The software they use allows retailers, even small ones, to define how often to update their prices, which competitors' prices to monitor, how much to beat competitors prices by, and how low the total price can go.

Dynamic pricing is not a new concept. The standard practice in ancient bazaars involved merchants and customers haggling until they came to a mutually agreeable price, which meant that different customers paid different prices for the same goods. Although the modern version of dynamic pricing often involves sophisticated market research and powerful computer software that can change millions of products in just minutes, the goal is the same: to charge the right customer the right price at the right time. The products for which price changes occur most frequently are consumer electronics, clothing, shoes, jewelry, and household staples. Research shows that prices actually go down about as often as they go up.

ENTREPRENEURIAL PROFILE: Al Falack: Cookie's Cookie's, an online children's clothing store that Al Falack operates through Amazon, uses dynamic pricing software from Mercent Corporation to update its prices every 15 minutes in an attempt to make it into Amazon's recommended "buy box," which shoppers who purchase select 95 percent of the time. Falack says that even with the software, monitoring prices requires a commitment, but the payoff is dramatically higher sales. Falack sets the software's parameters to beat the prices of certain competitors (only those with at least two stars out of Amazon's five-star rating) by a specific percentage and to establish a minimum price floor. The almost constant adjustments mean that Falack often sells clothing at lower prices in his Amazon store than he does in his physical store in Brooklyn, New York. "The long-term implication is that a price is no longer a price," says Mercent's CEO.[33]

LEADER PRICING Leader pricing is a technique in which a retailer marks down the customary price (i.e., the price consumers are accustomed to paying) of a popular item in an attempt to attract more customers. The company earns a much smaller profit on each unit because the markup is lower, but purchases of other merchandise by customers seeking the leader item often boost sales and profits. In other words, the incidental purchases that consumers make when shopping for the leader item boosts sales revenue enough to offset a lower profit margin on the leader. Grocery stores often use leader pricing. For instance, during the holiday season, stores often use turkeys as a price leader, knowing that they will earn higher margins on the other items that shoppers purchase with their turkeys.

GEOGRAPHIC PRICING Small businesses whose pricing decisions are greatly affected by the costs of shipping merchandise to customers across a wide range of geographic regions frequently employ one of the **geographic pricing** techniques. For these companies, freight expenses constitute a substantial portion of the cost of doing business and often cut deeply into already narrow profit margins. One type of geographic pricing is **zone pricing**, in which a company sells its merchandise at different prices to customers located in different territories. For example, a manufacturer might sell at one price to customers east of the Mississippi and at another to those west of the Mississippi. A small business must be able to show a legitimate basis (e.g., difference in selling or transportation costs) for the price discrimination or risk violating Section 2 of the Clayton Act.

Another variation of geographic pricing is the **uniform delivered pricing**, a technique in which a company charges all of its customers the same price regardless of their location, even though the cost of selling or transporting merchandise varies. The company calculates freight charges for each region in which it sells and combines them into a uniform fee. The result is that local customers subsidize the firm's charge for shipping merchandise to distant customers.

A final variation of geographic pricing is **F.O.B. factory**, in which a small company sells its merchandise to customers on the condition that they pay all shipping costs. Using this technique, a company can set a uniform price for its product and let each customer cover the freight cost.

DISCOUNTS Many small businesses use **discounts**, or **markdowns**, reductions from normal list prices, to move stale, outdated, damaged, or slow-moving merchandise. A seasonal discount is a price reduction designed to encourage shoppers to purchase merchandise before an upcoming season. For instance, many retail clothiers offer special sales on winter coats in late summer. Some companies grant discounts to special groups of customers, such as senior citizens or college students, to establish a faithful clientele and to generate repeat business. One study suggests that for items other than luxury goods, placing discount signs close to merchandise displays and promoting dollar discounts rather than percentage discounts increases the probability of making a sale.[34]

As tempting as discounts are to businesses when sales are slow, they also carry risks. Because price is an important signal of quality and image to customers, businesses that turn to discounts too often create the impression that they may be lowering their quality standards, thereby diluting the value of their brand and image in the marketplace. "For the sake of a short-term increase in sales, you can wreck the long-term value of your brand," says Rafi Mohammed, author of *The Art of Pricing*.[35] Many restaurants, from quick-service chains to fast-casual outlets, relied heavily on price discounts in an attempt to attract customers during a recent recession. For many, however, the traffic that the discounts offered failed to offset the price cuts, resulting in the same lower total revenues they were trying to avoid. In addition, when the economy improved, many restaurants faced difficulty weaning customers from their discount price expectations. "They've trained customers to eat $5 foot-long sandwiches," says one industry analyst about a popular chain's discounts.[36] As the economy improved, restaurants reduced their use of discount coupons and introduced new, higher-priced menu items.[37] One less visible way to offer discounts is to enroll customers in a loyalty program that entitles them to **earned discounts**, discounts that customers receive when their purchases reach a minimum threshold.

Limited-time offers (LTOs) are discounts that retailers run for a limited amount of time ("Regular price: $150. Sale price $120 *for three days only.*") with the goal of creating a sense of urgency and excitement among customers. Although limited-time offers are a common pricing tool for many retailers, quick-service restaurants are perhaps the most frequent users of limited-time offers. To create a successful limited-time offer, retailers should emphasize the end date of the offer and include a distinct call to action in their advertising, promote the offer on social media as well as in traditional advertising channels, and end the offer on the advertised date. Toppers Pizza, a pizza chain with 41 locations in the Midwest and South, has had great success running periodic, aggressively priced limited-time offers (such as a one-topping large pizza for just $5) by promoting them among its 50,000 Facebook fans. "We do double the volume of transactions during the LTO compared to a normal week," says Scott Iversen, the company's vice president of marketing. "The volume we do in ancillary sales of drinks and sides makes up for the lower [profit] margin on the LTO."[38]

Recent research suggests that using a **steadily decreasing discount (SDD)**, a limited-duration discount that declines over time, is superior to a standard (hi-lo) discount, a common tactic in which a company offers frequent discounts off of its standard prices. When one company used a hi-lo discount of 20 percent for three days before returning to the items to full price, sales increased by 75 percent. For the same items, a steadily decreasing discount of 30 percent the first day, 20 percent the second day, and 10 percent the third day (which yielded the same average discount of 20 percent) produced an increase in sales of 200 percent. The researchers conclude that the steadily decreasing discount is more effective because it creates a sense of urgency, especially among wary or indecisive customers.[39]

Multiple unit pricing is a promotional technique that offers customers discounts if they purchase in quantity. Many products, especially those with a relatively low unit value, are sold using multiple pricing. For example, instead of selling an item for 50 cents, a small company might offer five for $2.

Entrepreneurship in Action

Social Coupons: Beneficial—or Just Plain Bad—for Business?

Lisa Bridge bought a successful yoga studio in New York City and relocated it to a new, less expensive location; completing the renovation and the move took longer than expected, however, and sales fell precipitously. In an attempt to generate sales quickly, Bridge began offering discount coupons on Groupon, LivingSocial, and other daily deal sites. Some of the coupons offered discounts of more than 50 percent off the regular price of a yoga session. Although Bridge was concerned about offering such large discounts, she needed customers; her goal was to use the coupons to entice first-time customers to join her studio. Unfortunately, she was offering so many deals through so many sites that shoppers could become regular customers without having to pay full price.

The discount coupons worked. New customers, almost all of whom were paying significantly reduced rates, began signing up, but Bridge could not afford to hire new instructors to accommodate the demand for classes that the discount coupons had created. Existing customers were crowded out, and confusion reigned at the front desk: How much is your discount? Did the Groupon deal include a free yoga mat, or was that the LivingSocial deal? "Rather than focus on members, management catered to the 'deals' people," says one former employee. "But people who bought deals rarely stayed on as customers. If they did, they bought an offer from a different [daily deal] site and never paid full price." Several months later, the yoga studio closed.

When sales are slow, business owners are tempted to reduce prices to get customers in the front door and move merchandise, and daily deal sites such as Groupon, Living Social, Amazon Local, Google Offers, and others that offer shoppers coupons that yield significant discounts at local businesses are ready to promote the deal (think social media meets coupons). To promote a small company's special, Groupon, which claims to have 33 million active users worldwide, collects 50 percent of the discounted price (which usually at least 50 percent of the normal price) for each coupon it sells, which means that business owners get only 25 percent of a product's or service's full price. In addition, Groupon does not pay merchants their portion of the coupons' sales immediately; businesses receive three equal payments spread across 60 days. Mark Grohman, owner of Meridian Restaurant in Winston-Salem, North Carolina, says that he does not plan to offer daily discounts through Groupon again because the slow payments put too much pressure on his restaurant's cash flow. "The payment timing is so erratic you can't count on any of that money to pay your bills," he says. "With small margins in restaurants, you need that cash in the bank as fast as possible." To protect their cash flow and profits, business owners must be extremely careful concerning the deals and discounts they offer.

Daily deal sites have the power to drive significant volumes of traffic to a business, but making a profit on the increased volume can be a challenge. A study by the Jesse H. Jones Graduate School of Business at Rice University reports that 27 percent of businesses lose money on Groupon discount deals. The study also reports that although 80 percent of coupon users are first-time customers, only 20 percent of them become repeat customers.

Given the nature of daily discount sites such as Groupon, the customers that companies attract with the discount coupons tend to be bargain shoppers and deal seekers who are looking for low prices. "In their current form, social coupons are not ideally suited to ensure customer acquisition and yield business profits," conclude two professors at the Massachusetts Institute of Technology who studied the impact of social coupons on small businesses.

What steps can business owners take to ensure that they reap the benefits of promotions on daily deal sites and avoid the pitfalls?

- *Use daily deal to build relationships with new customers.* Use the deals as a way to find new customers, then stay connected with them through social media channels and other marketing avenues. The goal is to convert them into long-term, loyal customers.

- *Emphasize customer service.* The best way to create repeat customers is to provide them with excellent service and make sure that their first experience with your company is a positive one.

- *Be ready for a surge in business.* Coupon offers through daily deal sites can create significant increases in the volume of business for companies in a short period of time. Make sure that you have the staff, cash, and raw materials ready to handle it. "If people come in the door and get frustrated by long lines, it's a total backfire," says Brendan Shapiro, owner of outdoor store Potomac River Running and a regular Groupon user.

- *Limit your promotion.* To avoid nasty surprises, smart daily deal site users place time and quantity limits and blackout dates on their offers.

- *Run specials on overstocked merchandise or underutilized services.* A daily deal special can be a useful tool for selling slow-moving merchandise.

- *Use the new customer contact opportunity to cross-sell.* For example, a yoga studio might cross-sell yoga clothing and accessories to new customers who purchase discount coupons.

- *Know your costs and the revenue that each coupon will generate.* The typical Groupon sale generates only 25 percent of the normal revenue that a company earns, which makes earning a profit on a deal challenging. Nearly three-fourths of companies make money on their Groupon offers; your goal is to be one of them.

- *Limit offers to new customers only.* Providing social coupons only to new customers avoids the problem of cannibalizing a company's existing revenue.

- *View your daily deal offer as a marketing expense.* Heather Speizman, owner of Bottles and Brushes, an art studio in Mount Pleasant, South Carolina, offered a $15 Groupon for an art class normally priced at $35 to

(continued)

Entrepreneurship in Action *(continued)*

generate buzz about a new location she was opening in nearby Summerville. The promotion worked, generating lots of buzz for her new store. "It was a great experiment in social media at its best," she says.

- **Track the results.** The only way to determine whether a Groupon deal was successful is to monitor the results it produces. How many coupons did customers purchase? How many did they redeem? What is the average sale for customers who used the coupons? What proportion of them bought additional products or services? How many of them become repeat customers?

1. Use the Internet to find examples of companies that have experienced great success with and great failure with daily

deal sites such as Groupon and LivingSocial. What lessons can you draw from them?

2. Work with a group of your classmates to brainstorm local businesses that could benefit from an offer on a daily deal site. What deal do you suggest the business offer?

Sources: Based on Dan Slater, "Are Daily Deals Done?," *Fast Company*, April 2012, pp. 43–44; "Beware the Cost of Social Coupons," *Build*, Fall 2012, Section D.13; Sarah E. Needleman and Shayndi Raice, "Groupon Holds Cash Tight," *Wall Street Journal*, November 10, 2011, pp. B1, B4; Tim Gray, "How to Find Success Using Groupon," *The ROI Factor*, August 3, 2012, *www.bluefountainmedia.com/blog/how-to-find-success-using-groupon*; Tim Donnelly, "How Groupon Can Boost Your Company's Exposure," *Inc.*, January 24, 2011, *www.inc.com/guides/201101/how-groupon-works-for-small-businesses.html*; Duff McDonald, "The High Value of Discount Prices," *Wall Street Journal*, June 23–24, 2012, p. C9; Gwen Moran, "Sealing the Deal-Seekers," *Entrepreneur*, June 2012, p. 77.

BUNDLING Many small businesses have discovered the marketing benefits of **bundling**, grouping together several products or services (or both) into a package that offers customers extra value at a special price. Rather than cut into their already thin profit margins with price discounts during a recent recession, some restaurants used a bundling strategy, offering customers value-priced groupings of items. Pizza Hut has had great success with its "$10 Dinner Box," a limited-time offer that includes one medium one-topping pizza, five breadsticks with marinara dipping sauce, and 10 cinnamon sticks with icing. The items in the dinner box would cost $16 if customers purchased them separately. "Our brand has come to recognize that there is a much higher standard for value than there was just a few years ago," says Kurt Kane, the company's chief marketing officer. "[Customers] are looking to spend their money wisely."[40]

OPTIONAL-PRODUCT PRICING **Optional-product pricing** involves selling the base product for one price but selling the options or accessories at a much higher percentage markup. Automobiles are often sold at a base price with a multitude of options available at separate prices. In many cases, automakers bundle together the most popular options in specially priced packages. Apple relies on optional-product pricing, offering an extensive selection of accessories for its smart phones, tablets, and computers. For instance, smart phone accessories include cases and armbands, speakers, car chargers and adapters, headsets, docking stations, credit card readers, and others and together can cost as much or more as the phone itself.

CAPTIVE-PRODUCT PRICING **Captive-product pricing** is the granddaddy of all pricing tactics in which the basic product is useless without the appropriate accessories. King Gillette, the founder of the company that manufactures Gillette razors, taught the business world that the *real* money is not in the razor (the product) itself but in the blades (the accessory). Today, we see the same pricing strategy used by Nintendo and other electronic game manufacturers that have a very small profit margin on the product but substantially higher margins on the game cartridges. When Nintendo launched its popular Wii game station, the company's strategy was to sell a simpler game station with games that players could enjoy without having to invest dozens of hours to learn them. This strategy enabled Nintendo to introduce its game station at a price of just $249, well below the $500 price tag on Sony's PlayStation 3 and Microsoft's $400 Xbox 360. Nintendo's real moneymaker, however, is the games that it sells to Wii owners that are priced at $50 each (still below the $60 price tag on most PlayStation games). Nintendo's pricing strategy worked, and sales of Wii stations and games outstripped those of Sony's and Microsoft's products.[41]

BY-PRODUCT PRICING **By-product pricing** is a technique in which the revenues from the sale of by-products allow a firm to be more competitive in its pricing of the main product. For years, owners of sawmills considered bark chips to be a nuisance. Today, they package them and sell them as ground cover to home owners, gardeners, and landscapers. Zoos across the globe offer one of the most creative examples of by-product pricing, packaging once-worthless exotic animal droppings and marketing it as fertilizer under the clever name "Zoo Doo."

SUGGESTED RETAIL PRICES Many manufacturers print suggested retail prices on their products or include them on invoices or in wholesale catalogs. Small business owners frequently follow these suggested retail prices because doing so eliminates the need to make a pricing decision. Nonetheless, following prices established by a distant manufacturer may create problems for a small business. For example, a clothing retailer may try to create a high-quality, exclusive image through a prestige pricing policy, but manufacturers may suggest discount outlet prices that are incompatible with the small firm's image. Another danger of accepting the manufacturer's suggested price is that it does not take into consideration a small company's cost structure or competitive situation. A controversial U.S. Supreme Court decision in 2007 overturned a nearly 100-year-old ruling and allows manufacturers to set and enforce minimum prices that retailers can charge for the manufacturer's products as long as doing so does not reduce competition. In its decision, the Court upheld the right of Leegin Creative Leather Products to refuse to sell its Brighton brand belts to Kays Kloset, a small company in Lewisville, Texas, that was selling the belts discount prices that were 20 percent below the suggested retail price. However, more than 30 states are considering passing new antitrust laws that explicitly ban all minimum price agreements in an attempt to preempt the court's decision.[42]

FOLLOW-THE-LEADER PRICING Some small businesses make no effort to be price leaders in their immediate geographic areas and simply follow the prices that their competitors establish. Maintaining a follow-the-leader pricing policy may not be healthy for a small business because it robs the company of the opportunity to create a distinctive image in its customers' eyes.

A small company's pricing strategy must be compatible with its marketing objectives, its marketing mix, and its cost structure. In addition, the pricing strategy must be consistent with the competitive realities of the marketplace and the shifting forces of supply and demand. The forces that shape the pricing decision can change rapidly, and therefore a company's pricing strategy is never completely fixed. Pricing decisions must take into account a company's cost, the special value the product or service creates for buyers, and the pricing tactics of competitors.

The underlying forces that dictate how a business prices its goods or services vary greatly among industries. The next three sections investigate pricing techniques used in retailing, manufacturing, and service firms.

 Lessons from the Street-Smart Entrepreneur

Enhancing Your Company's Pricing Power

As the economy slowed in a recent recession, businesses of all sizes and across myriad industries began offering price discounts to encourage customers to buy their goods and services. Fast-casual restaurants such as Applebee's and Ruby Tuesday offered a selection of dinner specials priced at two for $20. Quick-service restaurants expanded their value menus and introduced bundled items at rock-bottom prices. Subway's $5 foot-long sandwiches were a hit, appealing to a broad base of customers.

Then many of these companies realized that they had created a challenging problem: How do we raise prices when the economy improves? Once customers become accustomed to value deals, how does a business change their perceptions of the relationship between price and value? The following pricing power matrix can help:

The horizontal axis of the matrix measures the extent to which customers view a product or service as a necessity or a discretionary purchase. The vertical axis describes a product's or service's level of uniqueness, which ranges from an undifferentiated commodity to a completely unique item. Where on the matrix does your company's product or service fall?

The best quadrant for a company to operate in is the upper left corner, a unique necessity, a situation in which customers

have a high need for a product or service that is unique and highly differentiated from competing products and services. An individual's brand of shampoo or razor blades that fit his or her razor offer good examples. These products have the greatest degree of pricing power, even during economic downturns.

The lower-right quadrant is discretionary commodities. With a discretionary commodity, a company's products or services are very similar to those of competitors, and customers do not have to have them or can postpone their purchases of them at least for a while. The airline industry finds itself in this unenviable position because many customers can choose alternative methods of travel, postpone their trips, or choose a less expensive flight on a competing airline.

The remaining two quadrants offer in-between positions of pricing power. In the lower left corner are products and services that, although necessary, offer little opportunity for differentiation. Lightbulbs and lumber provide good examples, and companies that produce them often end up matching competitors' prices because customers see them as the same.

The upper right corner contains products and services that are quite unique but are highly discretionary; customers simply do not have to have them, which means that companies that operate in this sector do not have maximum pricing power. Luxury watches

(continued)

Lessons from the Street-Smart Entrepreneur *(continued)*

Pricing power matrix.

(Matrix diagram with axes: Unique (top), Commodity (bottom), Necessary (left), Discretionary (right). Items plotted: Brand name shampoo, Ultrapremium watches, Luxury cars, Razor blades, iPhone, Lumber, PCs, Light bulbs, Toaster ovens, Air travel.)

that sell for as much as $650,000 and cars such as Rolls Royce are good examples. During a recent economic downturn, for instance, Rolls Royce introduced the $245,000 Ghost, a smaller, more "affordable" car (at least compared to the rest of its line, including the $380,000 Phantom) that is, in the Rolls Royce tradition, hand built and offers amenities such as night-vision cameras, inch-thick lamb's wool carpet, and a cashmere head liner. Located in the same quadrant, Apple recently cut the price of its entry-level iPhone in half and introduced a next-generation model to maintain its sales in the face of competing smart phones and more frugal customers.

Companies have the ability to move from one quadrant to a more desirable one by executing the proper strategy. The following questions help entrepreneurs consider their strategic options to increase their pricing power:

1. Can you offer a product or service that your customers consider a necessity? Doing so enhances a small company's pricing power.

2. Can you offer an "affordable luxury"? Even in austere economic conditions, customers are willing to splurge on small

luxuries such as gourmet chocolates, premium ice cream, and luxury muffins. These affordable luxuries give companies a great deal of pricing power.

3. What steps can you take to differentiate your company's products or services from those of your competitors? The greater the degree of differentiation of a company's products and services, the less price sensitive customers tend to be and the more pricing power a company has.

4. Can you offer customers something that will save your customers money? If so, your company has the ability to increase its pricing power.

5. Can you reduce the role of price in customers' buying decisions by, for example, offering superior customer service? Most small companies have the ability to deemphasize the role that price plays in customers' buying decisions.

Sources: Based on Geoff Colvin, "Yes, You Can Raise Prices," *Fortune*, March 2, 2009, p. 20; Sara Wilson, "When to Lower Your Price Point," *Entrepreneur*, April 2009, pp. 28–29; Hannah Elliott, "Stealth Wealth," *Forbes*, January 18, 2010, p. 62; Dan Neil, "Rolls Royce Builds a Real Car," *Wall Street Journal*, April 10–11, 2010, p. W6; Yukari Iwatani Kane, "To Sustain iPhone, Apple Halves Price," *Wall Street Journal*, June 9, 2009, p. B1.

Pricing Techniques for Retailers

4.
_____

Explain the pricing techniques used by retailers.

Because retail customers have become more price conscious and the Internet has made prices more transparent, many retailers have changed their pricing strategies to emphasize value. This value–price relationship allows for a wide variety of highly creative pricing and marketing practices. Delivering high levels of perceived value in products and services is one key to retail customer loyalty. To justify paying a higher price than those charged by competitors, customers must perceive a company's products or services as giving them greater value.

Markup

The basic premise of a successful business operation is selling a good or service for more than it costs to produce or purchase it. The difference between the cost of a product or service and its selling price is called **markup** (or markon). Markup can be expressed in dollars or as a percentage of either cost or selling price:

$$\text{Dollar markup} = \text{Retail price} - \text{Cost of the merchandise}$$

$$\text{Percentage (of retail price) markup} = \frac{\text{Dollar markup}}{\text{Retail price}}$$

$$\text{Percentage (of cost) markup} = \frac{\text{Dollar markup}}{\text{Cost of unit}}$$

For example, if a man's shirt costs $14 and the manager plans to sell it for $30, the markup is as follows:

$$\text{Dollar markup} = \$30 - \$14 = \$16$$

$$\text{Percentage (of retail price) markup} = \frac{\$16}{\$30}$$

$$= 53.3\%$$

$$\text{Percentage (of cost) markup} = \frac{\$16}{\$14}$$

$$= 114.3\%$$

The cost of merchandise used in computing markup includes not only the wholesale price of the merchandise but also all other costs (such as shipping and transportation charges) that the retailer incurs minus any discounts (such as quantity or cash discounts) that the wholesaler or manufacturer offers. Markups vary across industries and product lines. Tobacco products typically carry a markup (of cost) of 18 percent, jeans have a markup of 115 percent (although the markup on some designer jeans is 350 percent or more), and popcorn in a movie theater has a markup of 1,275 percent.[43] Table 11.2 shows markup calculations for Apple's iPhone 5 and Nokia's Lumia 900 smart phone.

Once entrepreneurs have a financial plan in place, including sales estimates and anticipated expenses, they can compute their companies' **average markup**, the average markup required on all merchandise to cover the cost of the items, all incidental expenses, and a reasonable profit:

$$\text{Initial markup} = \frac{\text{Operating expenses} + \text{Reductions} + \text{Profits}}{\text{Net sales} + \text{Reductions}}$$

Operating expenses are the cost of doing business, such as rent, utilities, and depreciation; reductions include markdowns, special sales, employee discounts, and the cost of stockouts. For

TABLE 11.2 Markup Calculations for Apple's 16 GB iPhone 5 and Nokia's Lumia 900 Smartphone

Component	iPhone 5	Lumia 900
Memory	$20.85	$27.00
Display and touch screen	$44.00	$58.00
Processor	$17.50	$17.00
Cameras	$18.00	$18.00
Wireless hardware	$39.00	$38.00
User interface and sensors	$6.50	$14.00
Power management device	$8.50	$9.00
Battery	$4.50	$4.50
Mechanical/electromechanical components	$33.00	$18.00
Box contents	$7.00	$5.50
Assembly cost	$8.00	$8.00
Total cost	$206.85	$217.00
Price without a contract*	$649.00	$449.99
$ Markup = Price − cost	$442.15	$232.99
Percentage (of cost) markup =	213.8%	107.4%
*Price with a contract (Carriers typically subsidize the full price of phones for their customers.)	$199.00	$99.99

Sources: HIS iSuppli Research, September 2012; Arik Hesseldahl, "iPhone 5 Costs Stay in Check," *Wall Street Journal*, September 24, 2012, p. B4; Adrian Kingsley-Hughes, "iPhone 5 16GB Cost an Estimated $207 to Build," *ZDNet*, September 19, 2012, *www.zdnet.com/iphone-5-16gb-costs-an-estimated-207-to-build-7000004476*.

example, if a small retailer forecasts sales of $980,000, operating expenses of $540,000, and $24,000 in reductions and expects a profit of $58,000, the initial markup percentage is as follows:

$$\text{Initial markup percentage} = \frac{\$540,000 + \$24,000 + \$58,000}{\$980,000 + \$24,000}$$

$$= 62\%$$

Any item in the store that carries a markup of at least 62 percent covers costs and meets the owner's profit objective. Any item that carries a markup less than 62 percent reduces the company's net profit.

Once an entrepreneur determines the average markup percentage, he or she can compute the appropriate retail price to achieve that markup using the following formula:

$$\text{Retail price} = \frac{\text{Dollar cost}}{(1 - \text{Average markup percentage})}$$

For instance, applying the markup of 62 percent to an item that cost the retailer $17.00 gives the following retail price:

$$\text{Retail price} = \frac{\$17.00}{(1 - 62\%)} = \$44.74$$

The owner establishes a retail price of $44.74 for the item using a 62 percent markup.

Finally, a retailer must verify that the computed retail price is consistent with the company's overall image. Is the final price congruent with the company's strategy? Is it within an acceptable price range? How does it compare with the prices charged by competitors? Perhaps most important, are the customers willing and able to pay this price?

Pricing Techniques for Manufacturers

5.

Explain the pricing techniques used by manufacturers.

For manufacturers, the pricing decision requires the support of accurate, timely accounting records. The most commonly used pricing technique for manufacturers is **cost-plus pricing**. Using this method, manufacturers establish a price composed of direct materials, direct labor, factory overhead, selling and administrative costs, plus the desired profit margin. Figure 11.3 illustrates the components of cost-plus pricing.

The main advantage of the cost-plus pricing method is its simplicity. Given the proper cost accounting data, computing a product's final selling price is relatively easy. In addition, because this technique adds a profit onto the top of the firm's costs, a manufacturer is likely to achieve the desired profit margin. This strategy does not encourage manufacturers to use their resources efficiently, however. Because manufacturers' cost structures vary so greatly, cost-plus pricing also fails to consider the competition adequately. Finally, cost-plus pricing fails to recognize the important links among price, value, and image. "The price that consumers are willing to pay has little to do with manufacturing costs," says pricing expert Rafi Mohammed. He says that a better pricing strategy is to capture "the value of a product, not simply mark up its costs."[44] Despite its

FIGURE 11.3

Components of Cost-Plus Pricing

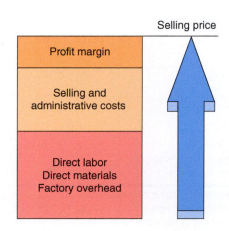

drawbacks, the cost-plus method of establishing prices remains prominent in many industries, such as construction and printing.

Direct Costing and Pricing

One requisite for a successful pricing policy in manufacturing is a reliable cost accounting system that can generate timely reports to determine the costs of processing raw materials into finished goods. The traditional method of product costing is called **absorption costing** because all manufacturing and overhead costs are absorbed into the finished product's total cost. Absorption costing includes direct materials and direct labor, plus a portion of fixed and variable factory overhead costs, in each unit manufactured. Full-absorption financial statements are used in published annual reports and in tax reports and are very helpful in performing financial analysis. However, full-absorption statements are of little help to a manufacturer when determining prices or the impact of price changes.

A more useful technique for managerial decision making is **variable (or direct) costing**, in which the cost of the products manufactured includes only those costs that vary directly with the quantity produced. In other words, variable costing encompasses direct materials, direct labor, and factory overhead costs that vary with the level of the company's output of finished goods. Factory overhead costs that are fixed (such as rent, depreciation, and insurance) are *not* included in the costs of finished items. Instead, they are considered to be expenses of the period.

A manufacturer's goal when establishing prices is to discover the cost combination of selling price and sales volume that exceeds the variable costs of producing a product and contributes enough to cover fixed costs and earn a profit. Full-absorption costing clouds the true relationships among price, volume, and costs by including fixed expenses when calculating unit cost. Direct costing, however, yields a constant unit cost of the product no matter what the volume of production is. The result is a clearer picture of the price–volume–costs relationship.

The starting point for establishing product prices is the direct-cost income statement. As Table 11.3 indicates, the direct-cost statement yields the same net income as does the full-absorption income statement. The only difference between the two statements is the format. The full-absorption statement allocates costs such as advertising, rent, and utilities according to the activity that caused them, but the direct-cost income statement separates expenses into fixed and variable costs. Fixed expenses remain constant regardless of the production level, but variable expenses fluctuate according to production volume.

When variable costs are subtracted from total revenues, the result is the manufacturer's **contribution margin**, the amount remaining that contributes to covering fixed expenses and earning a profit. Expressing this contribution margin as a percentage of total revenue yields the company's contribution percentage. Computing the contribution percentage is a critical step in establishing prices through the direct-costing method. This manufacturer's contribution percentage is 36.5 percent, which is calculated as follows:

$$\text{Contribution percentage} = 1 - \frac{\text{Variable expenses}}{\text{Revenues}}$$

$$= 1 - \frac{\$502,000}{\$790,000} = 36.5\%$$

Computing a Break-Even Selling Price

A manufacturer's contribution percentage tells what portion of total revenue remains after covering variable costs to contribute toward meeting fixed expenses and earning a profit. This manufacturer's contribution percentage is 36.5 percent, which means that variable costs absorb 63.5 percent of total revenues. In other words, variable costs represent 63.5 percent $(1.00 - 0.365 = 0.635)$ of the product's selling price. If this manufacturer plans to produce 50,000 units, its variable costs include the following:

Material	**$5.01/unit**
Direct labor	$3.80/unit
Variable factory overhead and selling expenses	$1.23/unit
Total variable cost	$10.04/unit

TABLE 11.3 Full-Absorption Versus Direct-Cost Income Statement

Full-Absorption Income Statement

Sales revenue		$790,000
Cost of goods sold		
Materials	250,500	
Direct labor	190,200	
Factory overhead	120,200	560,900
Gross profit		$229,100
Operating expenses		
General and administrative	66,100	
Selling	112,000	
Other	11,000	
Total operating expenses		189,100
Net income (before taxes)		$40,000

Direct-Cost Income Statement

Sales revenue (100%)		$790,000
Variable costs		
Materials	250,500	
Direct labor	190,200	
Variable factory overhead	13,200	
Variable selling expenses	48,100	
Total variable costs (63.5%)		502,000
Contribution margin (36.5%)		288,000
Fixed costs		
Fixed factory overhead	107,000	
Fixed selling expenses	63,900	
General and administrative	66,100	
Other fixed expenses	11,000	
Total fixed expenses (31.4%)		248,000
Net income (before taxes) (5.1%)		$40,000

The minimum price at which the manufacturer would sell the item is $10.04. Any price below that does not cover variable costs. To compute the break-even selling price for his product, find the selling price using the following equation:

$$\text{Profit} = \frac{\left(\begin{array}{c}\text{Selling} \\ \text{price}\end{array} \times \begin{array}{c}\text{Quantity} \\ \text{produced}\end{array}\right) + \left(\begin{array}{c}\text{Variable cost} \\ \text{per unit}\end{array} \times \begin{array}{c}\text{Quantity} \\ \text{produced}\end{array}\right) + \text{Total fixed cost}}{\text{Quantity produced}}$$

which becomes:

$$\text{Break-even selling price} = \frac{\text{Profit} + \left(\begin{array}{c}\text{Variable cost} \\ \text{per unit}\end{array} \times \begin{array}{c}\text{Quantity} \\ \text{produced}\end{array}\right) + \text{Total fixed cost}}{\text{Quantity produced}}$$

To break even, the manufacturer assumes $0 profit. Given that its plans are to produce 50,000 units of the product and that fixed costs will be $248,000, the break-even selling price is as follows:

$$\text{Break-even selling price} = \frac{\$0 + (\$10.04/\text{unit} \times 50,000 \text{ units}) + \$248,000}{50,000 \text{ units}}$$

$$= \frac{\$750,000}{50,000 \text{ units}}$$

$$= \$15.00 \text{ per unit}$$

Thus, $4.96 ($15.00/unit − $10.04/unit) of the $15.00 break-even price goes toward meeting fixed production costs. But suppose the manufacturer wants to earn a $75,000 profit. Then the required selling price is calculated as follows:

$$\text{Selling price} = \frac{\$75,000 + (10.04/\text{unit} \times 50,000 \text{ units}) + \$248,000}{50,000 \text{ units}}$$

$$= \frac{\$825,000}{50,000 \text{ units}}$$

$$= \$16.50/\text{unit}$$

Now the manufacturer must determine whether customers will purchase 50,000 units at $16.50. If the manufacturer thinks they won't, managers must decide to either produce a different, more profitable product or lower the selling price by lowering either its cost or its profit target. Any price above $15.00 will generate some profit although less than that desired. In the short run, the manufacturer could sell the product for less than $15.00 if competitive factors dictate but *not* below $10.04 because a price below $10.04 would not cover the variable costs of production.

Because the manufacturer's capacity in the short run is fixed, pricing decisions should be aimed at using resources most efficiently. The fixed cost of operating the plant cannot be avoided, and the variable costs can be eliminated only if the firm ceases to offer the product. Therefore, the selling price must be at least equal to the variable costs (per unit) of making the product. Any price above that amount contributes to covering fixed costs and providing a reasonable profit.

Of course, over the long run, the manufacturer cannot sell below total costs and continue to survive. A product's selling price must cover total product costs—both fixed and variable—and generate a reasonable profit.

Pricing Techniques for Service Businesses

Service businesses must establish their prices on the basis of the materials used to provide the service, the labor employed, an allowance for overhead, and a profit. As in a manufacturing operation, a service firm must have a reliable and accurate accounting system to track the total costs of providing the service. Most service firms base their prices on an hourly rate, usually the actual number of hours required to perform the service. Some companies, however, base their fees on a standard number of hours, determined by the average number of hours needed to perform the service. For most businesses, labor and materials constitute the largest portion of the cost of providing a service. To establish a reasonable and profitable price for service, entrepreneurs must know the cost of materials, direct labor, and overhead for each unit of service they provide. Using these basic cost data and a desired profit margin, an owner of a small service firm can determine the appropriate price for the service.

6.

Explain the pricing techniques used by service firms.

Consider a simple example for pricing a common service—computer repair. Ned's Computer Repair Shop uses the direct-costing method to prepare an income statement for exercising managerial control (see Table 11.4). Ned estimates that he and his employees spend about 9,250 hours in the actual production of computer repair service. The total cost per productive hour for Ned's TV Repair Shop is as follows:

$$\text{Total cost per hour} = \frac{\$104,000 + 68,000}{9,250 \text{ hours}} = \$18.59/\text{hour}$$

Now Ned must add in an amount for his desired profit. He expects a net operating profit margin of 18 percent of sales. To compute the final price, he uses the following equation:

Price per hour = Total cost per productive hour ÷ (1 − net profit target as % of sales)

$$= \$18.59 \div (1 - .18)$$

$$= \$22.68/\text{hour}$$

TABLE 11.4 Direct-Cost Income Statement, Ned's Computer Repair Shop

Sales revenue		$199,000
Variable expenses		
Labor	52,000	
Materials	40,500	
Variable factory overhead	11,500	
Total variable expenses		104,000
Fixed expenses		
Rent	2,500	
Salaries	38,500	
Fixed overhead	27,000	
Total fixed expenses		68,000
Net income		$27,000

A price of $22.68 per hour will cover Ned's costs and generate the desired profit. Smart service shop owners compute the cost per production hour at regular intervals throughout the year because they know that rising costs can eat into their profit margins very quickly. Rapidly rising labor costs and materials prices dictate that service company owners compute the price per hour even more frequently. As in the case of the retailer and the manufacturer, Ned must evaluate the pricing policies of competitors and decide whether his price is consistent with the firm's image.

Of course, the price of $22.68 per hour assumes that all jobs require the same amount of materials. If this is not a valid assumption (and it probably is not), Ned must recalculate the price per hour without including the cost of materials:

$$\text{Cost per productive hour} = \frac{\$172,000 - \$40,500}{9,250 \text{ hours}}$$

$$= \$14.22/\text{hour}$$

Adding in the desired 18 percent net operating profit on sales yields the following:

$$\text{Price per hour} = \$14.22/\text{hour} \times \frac{1.00}{(1 - 0.18)}$$

$$= \$14.22/\text{hour} \times 1.219$$

$$= \$17.34/\text{hour}$$

Under these conditions, Ned would charge $17.34 per hour plus the actual cost of materials used and a markup on the cost of materials. For instance, a repair job that takes four hours to complete would have the following price:

Cost of service (4 hours × $17.34/hour)	$69.36
Cost of materials	$41.00
Markup on materials (60%)	$24.60
Total price	$134.96

Because services are intangible, their pricing offers more flexibility than do tangible products. One danger that entrepreneurs face is pricing their services too low because prospective customers' perceptions of a service are influenced heavily by its price. In other words, establishing a low price for a service actually may harm a service company's sales!

ENTREPRENEURIAL PROFILE: Reid Carr: Red Door Interactive To avoid this problem, Reid Carr, founder of Red Door Interactive, a San Diego–based company that specializes in Web services, prices each project that his company takes on by estimating the number of hours it will take to complete and multiplying that number by an hourly rate and then including some "wiggle room" for unforeseen cost overruns. If the flow of work slows, Carr allows his employees to work on pro bono projects to raise the visibility of his company and to show the quality of work his employees create. "Pro bono work is free advertising," explains Carr.[45]

The Impact of Credit on Pricing

In today's business environment, linking a company's pricing strategy with its credit strategy has become essential. Consumers crave convenience when they shop, and one of the most common conveniences they demand is the ability to purchase goods and services on credit. Small businesses have three options for selling to customers on credit: credit cards, installment credit, and trade credit.

7. _____

Describe the impact of credit on pricing.

Entrepreneurship in Action

To Accept Credit Cards or Not: That Is the Question

In 2008, when Miki Mishihata opened his unique bike shop, Hello Bicycle, in the Beacon Hill neighborhood in Seattle, Washington, he faced a difficult decision: whether to accept credit card payments. He wanted to provide potential customers with as many convenient ways as possible to pay for the bikes and accessories that they purchased from his small shop, but he worried about the "exorbitant" fees he would pay on each credit card transaction, the flat monthly charges, and the cost of the equipment he would have to lease or purchase. Ultimately, he decided to forgo the added cost of accepting credit card payments, but he often wondered just how many sales Hello Bicycle missed because he did not accept credit cards.

Mishihata's dilemma is one that is all too familiar to millions of small business owners across the United States. In fact, a recent survey by Intuit Inc. reports that 55 percent of the country's 27 million small businesses do not accept credit card payments. The survey also estimates that by not accepting credit card payments, the average small business misses out on $7,000 in sales annually. That means that these small businesses give up the potential to generate a total of $189 billion more in sales annually!

Three years after starting his bicycle shop, Mishihata signed up with Square, a company that provides a digital credit card system that charges merchants a 2.75 percent swipe fee (or a flat rate of $275 per month) and supplies them with a free card reader. Once he began accepting credit card payments, Mishihata began to notice a difference in his business almost immediately. In just one year, the number of transactions at Hello Bicycle doubled, and sales revenue nearly tripled. In addition, Mishihata is pleased with the Square's environmentally friendly paperless system and his ability to track the shop's transactions online from almost anywhere. "Small businesses shouldn't be asking how much new credit card processing technology costs but how it's going to make them more money," says Merrell Sheehan, a top manager at a merchant services firm. Omar

Green, Intuit's director of strategic mobile initiatives, was surprised when he discovered that a savvy group of Girl Scouts was using his company's GoPayment mobile payment services and smart phones to process customers' credit card purchases of Girl Scout cookies. "It became obvious to me that there was a sea change in the world of payments when the Girl Scouts started using GoPayment," he says. "It was crazy how much these little girls were bringing in."

The benefits of accepting credit card payments are significant. Studies show that credit card users spend more (up to 50 percent more) than people who make purchases only with cash. Small business owners who accept credit card payments say that they not only generate more sales but also benefit from faster cash receipts and reduced losses to bad debts resulting from customers who fail to pay their invoices. Customers also appreciate the convenience of mobile payments and are adopting the technology rapidly. Globally, more than 212 million people make mobile payment purchases totaling nearly $172 billion annually; by 2016, Gartner estimates that 416 million people will purchase $617 billion in goods in services with mobile payments. In the United States, mobile commerce sales generate $5.3 billion in sales but are expected to reach $31 billion by 2016.

Many companies now offer simple mobile payment services that accept credit and debit cards through smart phones or tablet computers. Rather than apply for a merchant account to be able to accept credit card payments through a bank or independent sales organization (which sometimes takes months), these companies allow merchants to begin processing credit card transactions almost instantly, and the money from the transactions is deposited into the small company's account the day after the transaction takes place. Pay Anywhere, a service provided by payment processing company North American Bancard, is growing fast, says company founder Marc Gardner. The company is signing up thousands of small businesses

(continued)

Entrepreneurship in Action *(continued)*

each month, charging them a transaction fee of 2.69 percent and providing them with free card readers and software that analyzes transactions.

Lance Boyd, founder of Bucking Bull pro.
Source: Ursula O'Hara.

For years, Lance Boyd, founder of Bucking Bull Pro, a small business that sells products to bull riders both online and at bull-riding events, accepted only checks and cash but realized that he was missing out on many sales. He recently began accepting mobile credit card payments through AppNinjas' Swipe service, which charges him $24.95 per month plus 24 cents and 1.74 percent per swipe. Not only have sales increased, but Boyd also experiences fewer write-offs due to uncollectible debts. "I use it 10 to 20 times a day during an event," he says.

1. Why do 55 percent of small businesses not accept credit card payments? What are the disadvantages of forcing customers to pay with cash or checks?

2. Have you ever decided not to make a purchase because a small business did not accept credit or debit card payments? Describe the event.

3. What benefits can small companies reap by accepting credit card payments?

Sources: Based on Jennifer Wang, "Power Pay," *Entrepreneur,* January 2012, pp. 25–32; Steven Henn, "What's in Your Wallet? Wait, You Don't Need One," *National Public Radio,* August 16, 2012, *www.npr.org/blogs/alltechconsidered/2012/08/16/158928044/whats-in-your-wallet-wait-you-dont-need-one*; Kathy Ames Carr, "Small Businesses Weigh Fee or Free When Accepting Credit Cards," *Crain's Cleveland Business,* September 24, 2012, *www.crainscleveland.com/article/20120924/SUB1/309249991*; "GoPayment Survey Estimates $100 Billion in Missed Sales for Small Businesses That Deny Plastic," Intuit Corporation, May 22, 2012, *http://about.intuit.com/about_intuit/press_room/press_release/articles/2012/GetBusinessGrowing.html*.

Credit Cards

Since 1992, the share of purchases made with cash or checks has declined from 85 percent to just 39 percent; shoppers today prefer using credit and debit cards to make purchases.[46] Consumers around the globe hold more than 2 billion credit cards and use them to purchase nearly $6.1 trillion in goods and services annually—more than $11.5 million in purchases per minute.[47] The message is clear: Customers expect to make purchases with credit cards, and small companies that fail to accept credit cards run the risk of losing sales to competitors who do. Research shows that customers who use credit cards make purchases that are 112 percent higher than if they had used cash.[48] Accepting credit cards broadens a small company's customer base and closes sales that it would lose if customers had to pay in cash.

Before a business can accept credit cards, it must obtain authorization and merchant status from either a bank or an independent sales organization. Companies that accept credit cards incur additional expenses for offering this convenience, however. Businesses must pay to use the system—typically 1.5 to 3 percent of each transaction, which they in turn must factor into the prices of their products or services. They also pay a transaction fee of 5 to 50 cents per charge (the average fee is 10 cents per transaction) and must purchase or lease equipment to process transactions. Credit card processing fees average about 2 percent of each transaction and operate on a multistep process.[49] On a $100 Visa or MasterCard purchase at a typical business, the bank that issued the customer's credit card collects $1.80, an amount that consists of a 1.70 percent processing fee called the **interchange fee**, the fee that banks collect from retailers whenever customers use a credit or debit card to make a purchase, and a 10-cent flat transaction fee. The retailer's bank, called the processing bank, receives a processing fee of 0.4 percent of the purchase amount (40 cents in this example), leaving the retailer with $97.80. The prices entrepreneurs charge must reflect the higher costs associated with credit card transactions.

Credit and debit card processing fees, known as "swipe fees," costs businesses $50 billion per year, and small businesses typically pay higher credit card processing fees than their larger counterparts.[50] These fees, especially on small purchases, can eradicate any profit that a small company might have earned. To minimize the fees associated with credit card transactions, some entrepreneurs offer incentives to customers to pay with cash. For Luan Schooler, owner of Forster & Dobbs, a specialty food store in Portland, Oregon, annual interchange fees total $10,000, an amount she considers part of her company's overhead cost. "People will pay for a $1.50 soda with a credit card," she says. "In those circumstances, we are losing money—or making so little that it wasn't worth selling it in the first place."[51]

Credit card readers that attach to tablets and smart phones allow entrepreneurs to accept debit and credit card payments on the go. Square, a start-up company cofounded by Twitter cofounder Jack Dorsey, offers merchants a credit card reader that plugs into a smart phone, giving businesses the convenience of a mobile cash register. Square provides its reader to merchants at no cost and charges a flat 2.75 percent transaction fee (which includes banks' interchange fees). Mobile payments made with smart phones or tablets are common in Europe and Asia and are growing rapidly in the United States. Start-up companies such as Square as well as established companies such as PayPal, Google, and Microsoft also offer "digital wallet" apps that allow shoppers to pay for purchases (either debit or credit) with their phones or tablets rather than by swiping a credit card.[52]

E-COMMERCE AND CREDIT CARDS When it comes to online business transactions, the most common method of payment is the credit card. Internet vendors are constantly challenged by the need to provide secure methods of transacting business in a safe environment. As you will learn in Chapter 13, many shoppers remain suspicious of online transactions for reasons of security and privacy. Therefore, online merchants must ensure their customers' privacy and the security of their credit card transactions by using computer encryption software.

Online merchants also face another obstacle: credit card fraud. Because they lack the face-to-face contact with their customers, online merchants face special challenges to avoid credit card fraud. According to a study by the CyberSource, online merchants lose 1.0 percent of their annual revenue, about $3.4 billion, to fraud each year.[53] Because small and midsize companies are less likely than large businesses to use high-tech online fraud detection tools, they are more likely to be victims of e-commerce fraud. The following steps can help online merchants reduce the probability that they will become victims of credit card fraud:

- Use an address verification system to compare every customer's billing information on the order form with the billing information in the bank or credit card company's records.
- Require customers to provide the CVV2 number from the back of the credit card. Although crooks can get access to this number, it can help screen out some fraudulent orders.
- Check customers' Internet Protocol (IP) addresses. If an order contains a billing address in California, but the IP address from which the order is placed is in China, chances are that the order is fraudulent.
- Monitor activity on your Web site with the help of a Web analytics software package. Many packages are available, and analyzing log files can help online entrepreneurs pinpoint the sources of fraud.
- Verify large orders. Large orders are a cause for celebration but only if they are legitimate. Check the authenticity of large orders, especially if the order is from a first-time customer.
- Post notices on your Web site that your company uses antifraud technology to screen orders. These notices make legitimate customers feel more confident about placing their orders and crooks trying to commit fraud tentative about running their scams.
- Contact the credit card company or the bank that issued the card. If you suspect that an order may be fraudulent, contact the company *before* processing it. Taking this step could save a small company thousands of dollars in losses.[54]

DEBIT CARDS Consumers around the world carry more than 4.5 billion debit cards that act as electronic checks, automatically deducting the purchase amount immediately from a customer's checking account. Globally, shoppers conduct nearly 98 billion debit card transactions, totaling almost $5.1 trillion each year.[55] As customers' use of debit cards continues to grow, more small businesses are equipping their stores to handle debit card transactions. The equipment is easy to install and to set up, and the cost to the company is negligible. The payoff can be big, however, in the form of increased sales, improved cash flow, and decreased losses from bad checks. In addition, interchange fees on debit cards are lower than those on credit cards.

INSTALLMENT CREDIT Small companies that sell big-ticket consumer durables—major appliances, cars, and boats—frequently rely on installment credit. Because very few customers can purchase these items in a single lump-sum payment, small businesses finance them over time. The time horizon may range from just a few months up to 25 or more years. Most companies require the customer to make an initial down payment for the merchandise and then finance the balance for the life of the loan. The customer repays the loan principal plus interest on the loan. One advantage of installment loans for a small business is that the owner retains a security interest as collateral on the loan. If the customer defaults on the loan, the owner still holds the title to the merchandise. Because installment credit absorbs a small company's cash, many entrepreneurs rely on financial institutions such as banks and credit unions to provide the installment credit. When a business has the financial strength to "carry its own paper," the interest income from the installment loan contract often yields more than the initial profit on the sale of the product. For some businesses, such as auto dealerships and furniture stores, financing is an important source of revenue and profit.

TRADE CREDIT Many small companies, especially those that sell to other businesses, offer their customers trade credit; that is, they create customer charge accounts. The typical small business invoices its credit customers monthly. To speed collections, some offer cash discounts if customers pay their balances early; others impose penalties on late payers. Before deciding to use credit as a competitive weapon, a small business owner must make sure that the company's cash position is strong enough to support that additional pressure. Trade credit can be a double-edged sword. Small businesses must be willing to grant credit to purchasers to get and keep their business, but they must manage credit accounts carefully to make sure that their customers pay in full and on time.

LAYAWAY Although technically not a form of credit, layaway plans, like trade credit, enable customers to purchase goods over time. In the typical layaway plan, a customer selects an item, pays a deposit on it, and makes regular payments on the item until it is paid in full. Unlike trade credit, the retailer keeps the item until the customer has finished paying. Most stores establish minimum payments and maximum payoff dates, and some charge a service fee. Created during the Great Depression as a way to help shoppers purchase goods, layaway has become popular once again, especially around the holiday season, as stubborn unemployment and slow economic growth have posed challenges for both merchants and shoppers.

Chapter Review

1. Explain why pricing is both an art and a science.
   - Pricing requires a knowledge of accounting to determine the firm's cost, strategy to understand the behavior of competitors, and psychology to understand the behaviors of customers.

2. Discuss the relationships among pricing, image, and competition.
   - Company pricing policies offer potential customers important information about the firm's overall image. Accordingly, when developing a marketing approach

to pricing, business owners must establish prices that are compatible with what their customers expect and are willing to pay. Too often, small business owners *underprice* their goods and services, believing that low prices are the only way they can achieve a competitive advantage. They fail to identify the extra value, convenience, service, and quality they give their customers—all things many customers are willing to pay for.

- An important part of setting appropriate prices is tracking competitors' prices regularly; however, what the competition is charging is just one variable in the pricing mix. When setting prices, business owners should take into account their competitors' prices, but they should not automatically match or beat them. Businesses that offer customers extra quality, value, service, or convenience can charge higher prices as long as customers recognize the "extras" they are getting. Two factors are vital to studying the effects of competition on the small firm's pricing policies: the location of the competitors and the nature of the competing goods.

3. Discuss effective pricing strategies for both new and existing products and services.
- Pricing a new product is often difficult for the small business manager, but it should accomplish three objectives: getting the product accepted, maintaining market share as the competition grows, and earning a profit.
- There are three major pricing strategies generally used to introduce new products into the market: penetration, skimming, and sliding down the demand curve.
- Pricing techniques for existing products and services include odd pricing, price lining, leader pricing,

geographic pricing, opportunistic pricing, discounts, multiple pricing, bundling, and suggested retail pricing.

4. Explain the pricing techniques used by retailers.
- Pricing for the retailer means pricing to move merchandise. Markup is the difference between the cost of a product or service and its selling price.
- Some retailers use retail price, but others put a standard markup on all their merchandise; more frequently, they use a flexible markup.

5. Explain the pricing techniques used by manufacturers.
- A manufacturer's pricing decision depends on the support of accurate cost accounting records. The most common technique is cost-plus pricing, in which the manufacturer charges a price that covers the cost of producing a product plus a reasonable profit. Every manufacturer should calculate a product's break-even price, the price that produces neither a profit nor a loss.

6. Explain the pricing techniques used by service firms.
- Service firms often suffer from the effects of vague, unfounded pricing procedures and frequently charge the going rate without any idea of their costs. A service firm must set a price based on the cost of materials used, labor involved, overhead, and a profit. The proper price reflects the total cost of providing a unit of service.

7. Describe the impact of credit on pricing.
- Offering customer credit enhances a small company's reputation and increases the probability, speed, and magnitude of customers' purchases. Small firms offer three types of customer credit: credit cards, installment credit, and trade credit (charge accounts).

Discussion Questions

11-1. Stuart Frankel, a Subway franchisee, came up with the idea for Subway's $5 foot-long sandwich to combat slow weekend sales at his restaurants. It was such a hit that Subway introduced the idea to all of its 33,000 outlets, and in one year it generated $3.8 billion in sales. One marketing consultant asks, "Is the $5 foot-long just a flash in the pan, or is it a function of consumer price points and price elasticity that affect all markets?" What do you think?

11-2. What does the price of a good or service represent to the customer? Why is a customer orientation to pricing important?

11-3. How does pricing affect a small firm's image?

11-4. What competitive factors must the small firm consider when establishing prices?

11-5. Describe the strategies a small business could use in setting the price of a new product. What objectives should the strategy seek to achieve?

11-6. Define the following pricing techniques: odd pricing, price lining, leader pricing, geographic pricing, and discounts.

11-7. Why do many small businesses use the manufacturer's suggested retail price? What are the disadvantages of this technique?

11-8. What is markup? How is it used to determine prices?

11-9. What is follow-the-leader pricing? Why is it risky?

11-10. What is cost-plus pricing? Why do so many manufacturers use it? What are the disadvantages of using it?

11-11. Explain the difference between full-absorption costing and direct costing. How does absorption costing help a manufacturer determine a reasonable price?

11-12. Explain the techniques for a small service firm setting an hourly price.

11-13. What is the relevant price range for a product or service?

11-14. What advantages and disadvantages does offering trade credit provide to a small business?

11-15. What are the most commonly used methods to purchase online using credit?

11-16. What advantages does accepting credit cards provide a small business? What costs are involved?

CHAPTER 12
Global Marketing Strategies

Learning Objectives

On completion of this chapter, you will be able to:

1. Explain why "going global" has become an integral part of many small companies' strategies.

2. Describe the nine principal strategies small businesses can use to go global.

3. Explain how to build a successful export program.

4. Discuss the major barriers to international trade and their impact on the global economy.

5. Describe the trade agreements that have the greatest influence on foreign trade.

American small businesses that don't go global today will be playing catch-up tomorrow.

—Dan Brutto

There's been a fundamental shift. Years ago, the competition for business was across the street, across the town, and sometimes across the state. Now it's across the world.

—Michael Masserman

Until recently, the world of international business was much like the world of astronomy before Copernicus, who revolutionized the study of the planets and the stars with his theory of planetary motion. In the sixteenth century, his Copernican system replaced the Ptolemaic system, which held that the earth was the center of the universe with the sun and all the other planets revolving around it. The Copernican system, however, placed the sun at the center of the solar system with all of the planets, including the earth, revolving around it. Astronomy would never be the same.

In the same sense, business owners across the globe were guilty of having Ptolemaic tunnel vision when it came to viewing international business opportunities. Like their pre-Copernican counterparts, owners saw an economy that revolved around the nations that served as their home bases. Market opportunities stopped at their homeland's borders. Global trade was only for giant corporations that had the money and the management to tap foreign markets and enough resources to survive if the venture flopped. That scenario no longer holds true in the twenty-first century. Managers must focus on expanding their companies internationally "because that's the only way their businesses can survive," says Mona Pearl, author of *Grow Globally: Opportunities for Your Middle-Market Company Around the World*. "They're playing in a global field no matter what they think. Instead of feeling panic and fear, they should look at the global market as an opportunity."[1]

Twenty-five years ago, if a company was considered to be multinational, everyone knew that it was a giant corporation; today, that is no longer the case. The global marketplace is as much the territory of small, upstart companies as it is that of giant multinational corporations. Powerful, affordable technology, the Internet, increased access to information on conducting global business, and the growing interdependence of the world's economies have made it easier for companies of all sizes, many of which had never before considered going global, to engage in international trade. Only 24 percent of U.S. small businesses are engaged in global commerce, but those that have made the transition often reap significant benefits.[2]

ENTREPRENEURIAL PROFILE: Alexis Maybank and Alexandra Wilkis Wilson: Gilt Groupe In 2007, Alexis Maybank and Alexandra Wilkis Wilson, who had become best friends at Harvard Business School, started Gilt Groupe, a Web site that sponsors flash sales, brief sales that sometimes last only a few hours, of top designer labels at prices that are up to 60 percent off of regular retail prices. Their business model caught on, and sales grew quickly. Within 16 months, Gilt Group, which is based in New York City, went global when Maybank and Wilson launched a Gilt Groupe Web site for Japan, where they faced "an untapped market, no competitors, and access to the biggest luxury spenders in the world," says Wilson. "The biggest challenge in Japan was learning that what worked in the United States wouldn't work there for cultural reasons." To support its e-commerce platform, the company soon opened a technology headquarters in Dublin, Ireland, a popular destination for many Internet-based companies because of its sizable population of talented

Source: CartoonStock.

"I don't call expanding to New Jersey thinking globally."

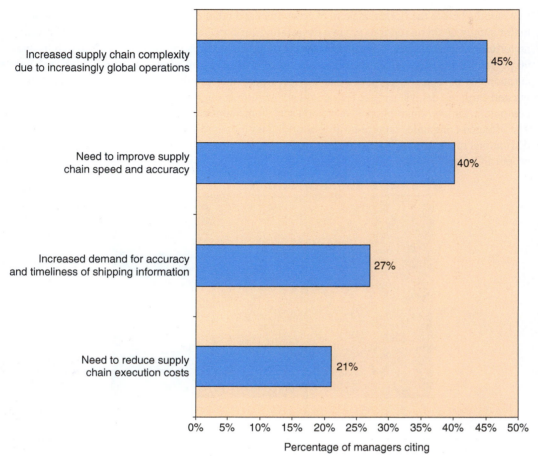

FIGURE 19.1

Top Supply Chain Pressures

Source: Bob Heaney, "Supply Chain Visibility in Consumer Markets," Aberdeen Group, March 2013, p. 2.

Depending on the type of business involved, the purchasing function can consume anywhere from 25 to 85 percent of each dollar of sales. By shaving just 2 percent off of its cost of goods sold, a typical small company can increase its net income by more than 25 percent! One classic study of various industries by IBM determined that to match the bottom-line impact of a $1 savings in purchasing, a company must increase its sales revenue by an average of 19.7 percent.[6] To realize these savings, however, entrepreneurs must create a purchasing plan, establish well-defined measures of product or service quality, and select vendors and suppliers using a set of relevant, objective criteria.

When a company's supply chain breaks down, the result can be devastating in both immediate and future costs, such as recalling dangerous or defective products and lost sales from customers who turn to substitute products. Some of those lost customers never return, and, worse yet, the company's name and reputation are tarnished forever. For instance, Yum! Brands, owner of the KFC, Taco Bell, and Pizza Hut brands, saw sales at its restaurants in China decline 25 percent in the months after news reports revealed excessive levels of antibiotics in samples of chicken that the company purchased from some of its Chinese suppliers. Yum! Brands, with more than 39,000 restaurants in 125 countries, quickly realigned its supply chain in China, eliminating more than 1,000 small poultry producers (where almost all of the problems originated) and reinforced its quality control standards. Despite the company's quick response to the supply problems, overcoming the negative publicity from the incident took many months.[7]

A company's supply chain is an integral part of its operations, and managers must identify the primary risks that their supply chains face (see Figure 19.2) and develop contingency plans for managing them. A survey by the World Economic Forum reports that 93 percent of managers around the globe say that managing supply chain risks has become a higher priority in their companies over the last five years.[8] Both man-made and natural disasters pose a threat to companies' supply chains. Massive floods in Thailand that left 900 factories in seven industrial parks under 15 feet of water disrupted the supply chains of companies around the world in a variety of industries. The personal computer industry was especially hard hit because factories located near Bangkok produce one-third of the world's hard drives and up to 80 percent of other computer components. Because of the interconnected nature of supply chains, computer makers around the world faced shortages of hard drives and other parts for more than a year after the floods. Just a

few months before the floods, makers of computers and other electronic devices had experienced major supply chain disruptions when a devastating earthquake and resulting tsunamis shut down factories in Japan, which accounts for 60 percent of the world's production of silicon chips and 90 percent of BT resin, a substance used to make printed circuit boards, both of which are vital components in almost every electronic device. Companies around the world faced manufacturing and shipping delays because they could not acquire enough silicon chips to produce personal computers, tablets, smart phones, television sets, home appliances, and other electronic devices.[9] The affected companies incurred millions of dollars in lost revenue. Minimizing problems from disruptions in a company's supply chains as a result of disasters and unexpected events requires proper SCM and a sound purchasing plan.

Massive floods in Thailand disrupted the supply chains of companies around the globe.

Source: Pornchai Kittiwongsakul/ Getty Images.

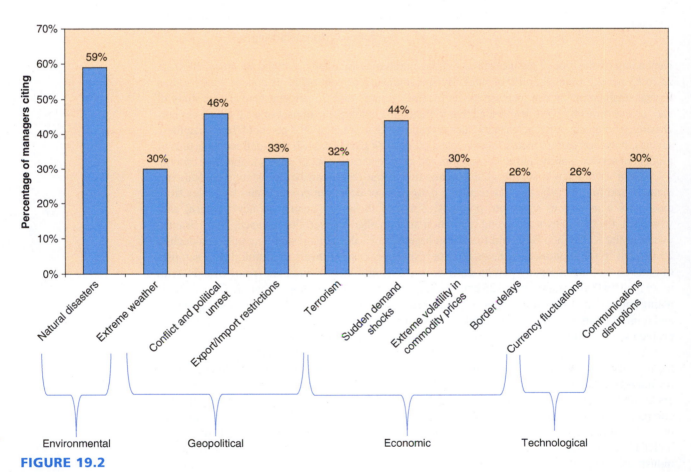

FIGURE 19.2

Greatest Threats to Supply Chains

Source: New Models for Addressing Supply Chain and Transport Risk, World Economic Forum, 2012, p. 8.

Lessons from the Street-Smart Entrepreneur

How to Manage Supply Chain Risk

Many small businesses pay prices for goods and services that are too high, and they find that their options are limited by the lack of competent suppliers. The broader the base of potential suppliers a business has, the more supply flexibility and security it has and the greater is its ability to get the best prices on its purchases. Use the following tips from the Street-Smart Entrepreneur to build meaningful, long-term relationships with the best suppliers.

1. ***Establish your company's critical criteria for selecting a vendor.*** What characteristics would the ideal vendor have? You must know up front what you are looking for in a vendor.

2. ***Research thoroughly prospective vendors before you purchase anything from them.*** The Internet is a useful tool for conducting a preliminary screening analysis.

3. ***Interview prospective vendors with the same level of intensity that you interview prospective employees.*** Both relationships influence how successful a company is at achieving quality objectives. "It's not easy to put aside the time," says James Walker, president of Octagon Research Solutions, a 60-person software company, "but for us, it's all about whether we can have a good relationship with [our vendors.]" Use the criteria you established in step 1 to establish a list of questions to ask potential vendors.

4. ***Be assertive.*** Ask tough questions and be a knowledgeable buyer. Don't allow suppliers to do their typical "sales pitches" before you have time to ask your questions.

5. ***Check potential vendors' credit ratings.*** Judging a potential vendor's financial strength and stability is important. Dealing with a vendor that is undergoing financial woes creates unnecessary complications for business owners.

6. ***Get referrals.*** Ask all potential vendors to supply a list of referrals of businesses that they have served over the past five years or more. Then make the necessary contacts.

7. ***Visit potential vendors' businesses.*** The best way to judge a vendor's ability to meet your company's needs is to see the operation firsthand. Nancy Connolly, president of Lasertone Corporation, a maker of copier and laser toner, insists on "a personal meeting between me and the president of the company," she says, before establishing a relationship with a vendor. The goal is to judge the level of the potential vendor's commitment to meeting Lasertone's needs.

8. ***Evaluate potential suppliers' plans for dealing with risks and interruptions in their own supply chains.*** Remember that if your vendor's supply of materials is interrupted, your supply also will be interrupted.

9. ***Don't fixate on price.*** Look for value in what they sell. If your only concern is lowest price, vendors will push their lowest-priced (and often lowest-quality) product lines.

10. ***Ask "What if?"*** The real test of a strong vendor–customer relationship occurs when problems arise. Smart entrepreneurs ask vendors how they will handle particular types of problems when they arise.

11. ***Attend trade shows.*** Work the room. A visit to a trade show is not a vacation; it's business. Find out whether the next booth has a valuable new vendor who has the potential to increase your company's profits!

12. ***Don't forget about local vendors.*** Because of their proximity, local vendors can sometimes provide the fastest service. Solving problems often is easier because local vendors can make on-site service calls.

13. ***Test a vendor before committing completely.*** Susan Gilbert, owner of Café in the Park, a restaurant in San Diego, says, "Inventory is cash flow. When I'm dealing with a vendor, I need to know how quickly they can deliver, how quickly I can turn over the inventory, and keep it all tight." Rather than place an order for 100 dozen Danish pastries with a new vendor, for instance, Gilbert starts with an order of several dozen pastries to judge the vendor's performance.

14. ***Find an appropriate balance between sole sourcing, which reduces economies of scale, and multisourcing, which reduces the risk of interrupted supplies.*** Recall that a common strategy uses Pareto's Law, the 80-20 Rule.

15. ***Work with the vendors.*** Tell your suppliers what you like and don't like about their products and service. In many cases, they can resolve your concerns. Most vendors want to build long-term relationships with their customers. Give them a chance to do so.

16. ***Do unto others . . .*** Treat your vendors well. Be selective but pay on time and treat them with respect.

1. Why do so many firms focus solely on selecting vendors that offer the lowest prices? What are the dangers of doing so?

2. Develop a list of 10 questions you would ask a potential vendor on a product you select.

Sources: Based on Nancy Germond, "Supply Chain Risk Management a Must as Global Sourcing Intensifies," *All Business,* June 11, 2010, *www.allbusiness.com/ company-activities-management/management-risk-management/14609006-1.html*; Kelly L. Frey, "Selecting a Vendor: RFPs and Responses to RFPs," Baker, Donelson, Bearman, Caldwell, and Berkowitz, *www.bakerdonelson.com/Documents/Selecting %20a%20Vendor.pdf*; Allison Stein Wellner, "Finding the Right Vendor," *Inc.,* July 2003, pp. 88–95; Jan Norman, "How to Find Suppliers," *Business Start-Ups,* October 1998, pp. 44–47; *Physical Risks to the Supply Chain: A View from Finance,* CFO Research Services and FM Global, February 2009, pp. 6–7.

Creating a Purchasing Plan

A **purchasing plan** involves planning for the acquisition of needed materials, supplies, services, and equipment of the right quality, in the proper quantities, for reasonable prices, at the appropriate time, and from the right vendor. A major objective of creating a purchasing plan is to establish a reliable, efficient supply chain through which a business acquires enough (but not too much) inventory to generate smooth, uninterrupted production or sales and ensures that finished goods are delivered on time to customers. "A more responsive supply chain is required to improve customer service levels," says Stephanie Ranada, manager of supply and logistics at Tropical Retail Company. "It is the foundation of fulfilling the mandate of 'right product at the right place at the right time.'"[10] Companies are purchasing goods and supplies across the globe, making their supply chains longer and more challenging to manage. Coordinating the many pieces of the global puzzle requires a comprehensive purchasing plan. The plan must identify a company's quality requirements, its cost targets, and the criteria for determining the best supplier, considering such factors as reliability, service, delivery, and cooperation.

McCormick and Company purchases its spices from suppliers in 24 countries.

Source: Kesu/Shutterstock.

 **ENTREPRENEURIAL PROFILE: McCormick and Company** McCormick and Company, a business based in Sparks, Maryland, that sells spices ranging from allspice to turmeric, manages a global supply chain that stretches back more than 2,000 years to when the Nabataeans established a hub-and-spoke network of suppliers on the Arabian Peninsula. McCormick's supply chain literally spans the globe (see Table 19.1), reaching into 50 locations in 24 countries to purchase from hundreds of suppliers the raw materials it requires for its product line. Since 1889, company buyers have traveled to Uganda and Madagascar for vanilla, to China and Nigeria for ginger, to Yugoslavia and Albania for sage, and to India, Turkey, Pakistan, and Syria for cumin seed. McCormick makes significant investments to find suppliers that can deliver quality materials in a timely manner and engages in extensive testing and security practices to ensure the quality and the safety of the raw materials it purchases. Recently, McCormick was put to the test when in the span of a matter of months, its supply chain was disrupted by a revolution in Egypt, devastating floods in Thailand, and computer system and shipping problems with a key vendor in Peru.[11]

A purchasing plan is closely linked to the other functional areas of managing a small business: production, marketing, sales, engineering, accounting, finance, and others. A purchasing plan should recognize this interaction and help integrate the purchasing function into the total organization. A small company's purchasing plan should focus on the five key elements of purchasing: quality, quantity, price, timing, and vendor selection (see Figure 19.3).

FIGURE 19.3

Components of a Purchasing Plan

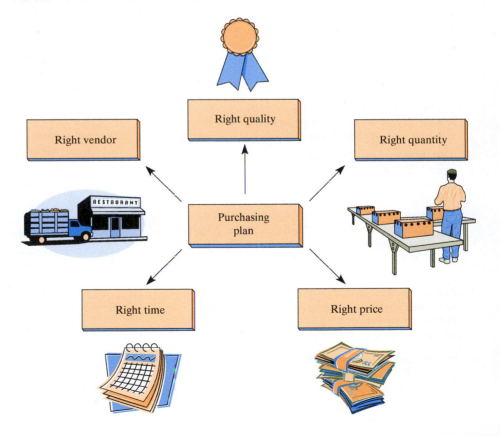

TABLE 19.1 McCormack and Company's Global Supply Chain

Albania	Australia	Brazil	Canada	China	Comoros	Croatia	Egypt	France	Greece	Guatemala/Honduras	Hungary	India	Indonesia
Oregano	Poppy seed	Arrowroot	Caraway	Celery seed	Vanilla beans	Sage	Anise	Basil	Oregano	Allspice	Paprika	Cardamom	Cinnamon
Rosemary		Black pepper	Coriander	Chives	Cloves		Basil	Chervil			Poppy seed	Celery seed	Cloves
Sage		Cloves	Mustard	Cinnamon/Cassia			Caraway	Fennel seed				Cumin seed	Mace
Savory		White pepper		Coriander			Coriander	Rosemary				Dill seed	Nutmeg
				Cumin seed			Dill seed	Savory				Fennel seed	Vanilla beans (Bali; Java)
				Fennel seed			Fennel seed	Tarragon				Fenugreek	Black pepper (Lampong)
				Garlic			Marjoram	Thyme				Nutmeg	White pepper (Muntok)
				Ginger			Mint flakes					Red pepper	
				Onion								Vanilla beans	
				Oregano									
				Parsley									
				Red pepper									
				Shallots									
				Star Anise									
				Szechwan pepper									
				Tumeric									
				White pepper									

Iran	Ivory & Cameroon	Jamaica	Madagascar	Malabar (S. India)	Malaysia	Mexico	Morocco	Netherlands	Nigeria	Pakistan	Peru	Poland	Romania
Cumin	Kola Nuts	Allspice	Clove	Black pepper	Cloves (Pedang)	Allspice	Coriander	Caraway seed	Ginger	Cumin seed	Paprika	Poppy seed	Coriander
		Kola nuts	Vanilla beans	Turmeric	Black pepper (Sarawak)	Ancho chile	Fenugreek	Chervil		Dill seed	Turmeric	Thyme	
				Cumin	White pepper (Sarawak)	Oregano	Oregano	Poppy seed		Red pepper			
				Ginger		Vanilla beans	Parsley						
				Celery seed			Rosemary						
							Thyme						

Seychelles	Spain	Sri Lanka	Syria	Tahiti	Thailand	Tonga	Turkey	Uganda	United States	Vietnam	West Indies (Granada)
Cinnamon	Anise	Cinnamon	Cumin seed	Vanilla beans	Lemongrass	Vanilla beans	Anise	Vanilla beans	Basil	Black pepper	Mace
	Paprika				Turmeric		Bay leaves		Cilantro	White pepper	Nutmeg
	Rosemary						Cumin seed		Dill weed	Cinnamon	
	Saffron						Fennel seed		Garlic		
	Thyme						Oregano		Marjoram		
							Poppy seed		Mint		
							Sage		Parsley		
									Paprika		
									Red pepper		
									Tarragon		

Quality

2.

Explain the principles of total quality management, including Lean, 5S, and Six Sigma, and their impact on quality.

Not long ago, businesses saw quality products and services as luxuries for customers who could afford them. Many companies mistakenly believed that producing or purchasing high-quality products and services was too costly. Business owners understand that quality goods and services are absolutely *essential* to remaining competitive. The benefits that companies earn by creating quality products, services, and processes come not only in the form of fewer defects but also in lower costs, higher productivity, and higher customer retention rates. W. Edwards Deming, one of the founding fathers of the modern quality movement, always claimed that higher quality resulted in lower costs. Internally, companies with a quality focus report significant improvements in work-related factors such as increased employee morale, lower employee turnover, and enhanced quality of work life. Benefits such as these can result in earning a significant competitive advantage over rivals of *any* size.

Total quality companies believe in and manage with the attitude of continuous improvement, a concept the Japanese call *kaizen*. The kaizen philosophy holds that small improvements made continuously over time accumulate into a radically reshaped and improved process. When defective items do occur, managers and employees who are engaged in continuous improvement do not simply rework or repair them. Instead, they see defectives as an opportunity to improve the entire process. Their goal is to identify the root cause of the defects and to change the entire process so that the same problem does not occur again. Kaizen also encourages managers to focus on improving the entire system, not just its individual components. World-class organizations in the twenty-first century have made continuous improvement a fundamental element of their competitive strategies.

Quality has an impact on both costs and revenues. Improved quality leads to less scrap and rework time, lower warranty costs, and increased worker productivity. On the revenue side of the equation, quality improves the firm's reputation, attracts customers, and often gives a firm the opportunity to charge higher prices. The bottom-line impact of quality is increased profitability.

TOOLS FOR ENSURING QUALITY: LEAN PRINCIPLES, 5S PRINCIPLES, SIX SIGMA, AND TOTAL QUALITY MANAGEMENT No matter which industry they are in, small companies have at their disposal four basic tools to improve quality in their operations: lean principles, 5S principles, six sigma, and total quality management.

Lean Principles Originally applied to manufacturing, Lean principles also have produced significant quality improvements in both the retail and service sectors. The fundamental idea behind the Lean philosophy is to eliminate waste in a company's activities, whether they are in manufacturing, distribution, inventory control, customer service, human resources, or other areas, and to make a company lean (and efficient) in its operation. Lean is built on five principles:

- *Value.* Companies must create products and services that add value from the customer's perspective.
- *Value stream.* Businesses must identify the essential steps that allow them to create an efficient production or service work flow.
- *Flow.* Companies must eliminate every step in the value stream that adds no value or creates delays, bottlenecks, and wasted effort. The goal is to create a smoothly flowing, efficient process that produces value for customers.
- *Pull.* Companies produce only when customer demand pulls products and services through the system. Attempting to push them through the system results in inefficiency in the form of excess inventory, costs, and waste. Customer demand, the "pull" in the system, drives the entire supply chain.

Source: Bob Thaves, Tom Thaves/ Cartoonistgroup.com

- *Perfection.* Companies should strive for perfection by eliminating waste and inefficiency everywhere it arises and by providing exactly the products and services customers want.

The principles of Lean are aimed at eliminating seven wastes. Figure 19.4a shows the seven wastes in a manufacturing environment, and Figure 19.4b shows the seven wastes in a service operation. Some experts have added an eighth waste to both lists: the underutilization of people. Some companies treat their workers as if they know very little. World-class companies see their employees as a valuable source of ideas, creativity, and solutions, and they create systems that capture their employees' knowledge and apply it to the process of continuous improvement.

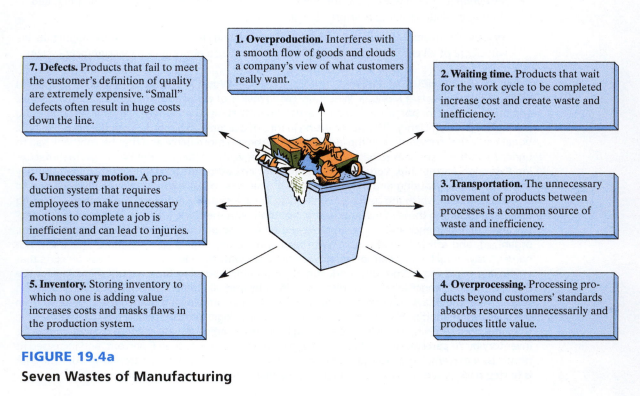

7. Defects. Products that fail to meet the customer's definition of quality are extremely expensive. "Small" defects often result in huge costs down the line.

6. Unnecessary motion. A production system that requires employees to make unnecessary motions to complete a job is inefficient and can lead to injuries.

5. Inventory. Storing inventory to which no one is adding value increases costs and masks flaws in the production system.

1. Overproduction. Interferes with a smooth flow of goods and clouds a company's view of what customers really want.

2. Waiting time. Products that wait for the work cycle to be completed increase cost and create waste and inefficiency.

3. Transportation. The unnecessary movement of products between processes is a common source of waste and inefficiency.

4. Overprocessing. Processing products beyond customers' standards absorbs resources unnecessarily and produces little value.

FIGURE 19.4a
Seven Wastes of Manufacturing

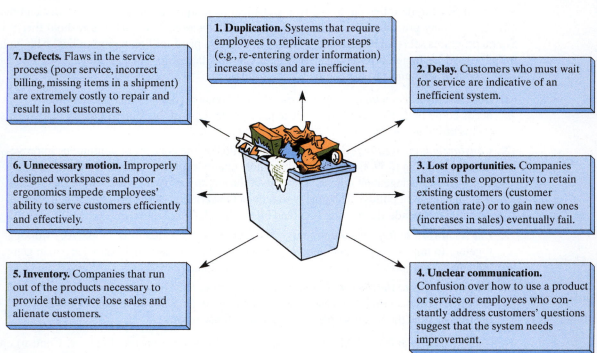

7. Defects. Flaws in the service process (poor service, incorrect billing, missing items in a shipment) are extremely costly to repair and result in lost customers.

6. Unnecessary motion. Improperly designed workspaces and poor ergonomics impede employees' ability to serve customers efficiently and effectively.

5. Inventory. Companies that run out of the products necessary to provide the service lose sales and alienate customers.

1. Duplication. Systems that require employees to replicate prior steps (e.g., re-entering order information) increase costs and are inefficient.

2. Delay. Customers who must wait for service are indicative of an inefficient system.

3. Lost opportunities. Companies that miss the opportunity to retain existing customers (customer retention rate) or to gain new ones (increases in sales) eventually fail.

4. Unclear communication. Confusion over how to use a product or service or employees who constantly address customers' questions suggest that the system needs improvement.

FIGURE 19.4b
Seven Wastes of Service

5S Principles Often used in conjunction with lean principles, the 5Ss are simple but effective ways to improve quality by organizing employees' work environment:

- *Sort.* Companies sort items in the workplace into two categories—necessary and unnecessary—and eliminate everything that is unnecessary.
- *Straighten.* Companies arrange the tools, equipment, and materials employees need to do their jobs in the most efficient manner that minimizes wasted motion and effort.
- *Shine.* Once organized, employees keep their work spaces neat and clean.
- *Systematize.* Businesses strive for continuous improvement in sorting, straightening, and shining by standardizing best-practice processes.
- *Sustain.* Companies sustain their drive toward continuous improvement by encouraging the involvement of all employees, self-discipline, and creating a performance measurement system.

ENTREPRENEURIAL PROFILE: Bill Vogel: Vogel Wood Products Bill Vogel, CEO of Vogel Wood Products, a Monona, Wisconsin–based maker of wood and laminate cabinetry, fixtures, and office systems for home and commercial use, knew that his 15-employee company was operating well below peak efficiency. "We were putting in tons of overtime to get orders out the door," he says. Vogel knew that competition in the industry was becoming more fierce and that his company had to undergo dramatic changes to gain a competitive edge. With the help of the Wisconsin Manufacturing Extension Partnership, Vogel and his employees embarked on a quality improvement journey using lean manufacturing and 5S principles. The goal was to get every employee involved and to create a culture in which the customer comes first. Their first step was to map the company's existing processes, including those in manufacturing, assembly, and the front office, so that they could identify areas of inefficiency and waste. Employees then focused on using 5S principles to create a clean, organized, and efficient workplace. "With lean and 5S, the goal is to work more effectively, not harder," says Vogel. One team of employees came up with the idea of placing tools on carts that employees could move from one work station to another, reducing time wasted searching for the right tool. The changes allowed employees to identify and eliminate bottlenecks in the production process that Vogel says "unlocked the flow" of work. The company has reaped many benefits from the changes. In addition to improved employee morale, Vogel Wood Products has cut its cycle time (the time from receiving an order to delivering it to the customer) in half, reduced its work-in-process inventory by 70 percent, reduced overtime and labor costs by more than 6 percent, and increased its profits by 4 percent. "The principles of lean are simple," says sales manager Denita Ward. "The trick is to step back to see where basic changes can be made that result in real improvements."[12]

Six Sigma Six Sigma relies on data-driven statistical techniques to improve the quality and the efficiency of any process and to increase customer satisfaction. The quality threshold that Six Sigma programs set is high: an average of only 3.4 defects per 1 million opportunities! Although initially used by large corporations, Six Sigma can be adapted to work in small businesses as well. The four key tenets of Six Sigma are the following:

1. *Delight customers with quality and speed.* Six Sigma recognizes that the customer's needs come first. The goal is to produce products that are of the highest quality in a process that is efficient and fast.
2. *Constantly improve the process.* Six Sigma builds on the concept of continuous improvement. According to W. Edwards Deming, most quality problems are the result of the *process* (which management creates) rather than the *employees* (who work within the process that management builds). The goal is to reduce the variation of the process, which is measured by the standard deviation (denoted by the Greek letter sigma, σ).
3. *Use teamwork to improve the process.* Six Sigma counts on teams of employees working together to improve a process. People working together to share their knowledge can generate better solutions to quality problems than individuals can.
4. *Make changes to the process based on facts, not guesses.* To improve a process, employees must have quantifiable measures of results (e.g., quality of output) and of the process itself (e.g., how the process operated to produce those results).[13]

Table 19.2 explains the DMAIC process (Define, Measure, Analyze, Improve, Control) on which the Six Sigma approach is built. For small companies, the goal of this process is to understand their core business processes better so that managers and employees can work together to make significant improvements to them over time.

TABLE 19.2 The Six Sigma DMAIC Approach

Principle	Process Improvement Technique
Define	Identify the problem. Define the requirements. Set the goal for improvement.
Measure	Validate the process problem by mapping the process and gathering data about it. Refine the problem statement and the goal. Measure current performance by examining the relevant process inputs, steps, and output to establish a baseline.
Analyze	Develop a list of potential root causes. Identify the vital few. Use data analysis tools to validate the cause and effect connections between root causes and the quality problem.
Improve	Develop potential solutions to remove root causes by making changes to the process. Test potential solutions and develop a plan for implementing those that are successful. Measure the results of the improved process.
Control	Establish standard measures for the new process. Establish standard procedures for the new process. Review performance periodically and make adjustments as needed.

Source: Adapted from Andrew Spanyi and Marvin Wurtzel, "Six Sigma for the Rest of Us," *Quality Digest*, July 2003, *www.qualitydigest.com/july03/ articles/01_article.shtml.*

ENTREPRENEURIAL PROFILE: Phil Parduhn: Pelco Products Pelco Products, a family-owned company founded in 1985 and based in Edmond, Oklahoma, has achieved success in a unique niche—manufacturing traffic signal hardware—by producing high-quality products and providing superior customer service. For Pelco, which sells its products in all 50 states as well as in international markets, speedy delivery is essential, especially when a city has experienced a natural disaster such as a hurricane or a tornado and must have replacement traffic lights, parts, and hardware quickly. Phil Parduhn, president of Pelco, implemented Lean principles in the company years ago and recently introduced Six Sigma methodology to continue the company's emphasis on continuous improvement. An analysis showed Parduhn that one of the areas in which the company needed to improve was its on-time delivery record. A team of employees studied the problem, identified the primary causes of "stuck" orders (those that were held up in the manufacturing process), and implemented a series of Lean- and Six Sigma–based techniques to resolve them. In less than four months, Pelco saw the volume of "stuck" orders decrease by 88 percent and the number of late deliveries to customers decline by 73 percent. Increased productivity that resulted from the streamlined process totaled more than $15,500, but the higher level of customer satisfaction that the company experienced is priceless.[14]

Total Quality Management Under the total quality management (TQM) philosophy, companies define a quality product as one that conforms to predetermined standards that satisfy customers' demands. That means getting *everything* from delivery and invoicing to installation and follow-up right the first time. Although these companies know that they may never reach their targets of perfect quality, they never stop striving for perfection, recognizing that even a 99.9 percent level of quality is not good enough (see Table 19.3). The businesses that have effectively implemented these programs understand that the process involves a total commitment from strategy to practice and from the top of the organization to the bottom.

Rather than trying to inspect quality into products and services after they are completed, TQM instills the philosophy of doing the job right the first time. Although the concept is simple, implementing such a process is a challenge that requires a very different kind of thinking and very different culture than most organizations are comfortable with. Because the changes TQM requires are so significant, patience is a must for companies adopting the philosophy. Consistent quality improvements rarely occur over night, yet too many small business managers think, "We'll implement TQM today and tomorrow our quality will soar." TQM is *not* a "quick-fix," short-term program that can magically push a company to world-class-quality status overnight. Because it requires such fundamental, often drastic changes in the way a company does business, TQM takes time both to implement and to produce results. Although some small businesses that use TQM begin to see some improvements within just a matter of weeks, the *real* benefits take longer to realize. It takes at least three or four years before TQM principles gain acceptance among employees and as much as eight years are necessary to fully implement TQM in a company.

TABLE 19.3 Why 99.9 Percent Quality Isn't Good Enough

Most companies willingly accept a certain percentage of errors and defects. For many businesses, an acceptable range of defects is from 1 to 5 percent. In some companies, defects are regarded as a normal part of daily operations. However, quality consultants say that even 99.9 percent quality isn't good enough.

- To improve quality, many companies are limiting the number of suppliers they use for their raw materials and components, sometimes to just one or two.
- Partnering with suppliers so closely means that those suppliers must strive for 100 percent quality and performance.

What would be the result if some things were done right only 99.9 percent of the time? Consider the implications in the United States:

- 102,740 cell phone calls dropped per hour
- 2,283 credit card transactions charged to the wrong accounts every hour
- 3 unsafe landings at Atlanta's Hartsfield-Jackson Airport per day
- 18,265 pieces of mail in the postal system lost per hour
- 8.555 incorrectly filled drug prescriptions per day
- 15,800 packages misdelivered or lost by UPS each day
- 922 incorrect surgical procedures performed each week
- 4,326 debit card transactions deducted from the wrong accounts every hour
- 11 babies delivered to the wrong parents each day
- 251,142 text messages delivered to the wrong people every hour

If you are in the unlucky one-tenth of 1 percent, the error affects you 100 percent. In addition, unless a company strives for 100 percent product or service quality, there is little chance that it will ever achieve 99.9 percent quality.

To implement TQM successfully, a small business owner must rely on these fundamental principles:

- *Employ benchmarking to achieve quality outcomes.* **Benchmarking** is the process of identifying world-class processes or procedures that other companies (often in other industries) currently are using and building higher-quality standards around these for your business. This search for best practices is ongoing. As part of their quality initiative, employees at Scotsman Ice Systems, a small company in Fairfax, South Carolina, that makes high-end commercial and residential ice machines and refrigerators, benchmark the production processes of other manufacturing companies. To make sure that employee teams do not miss anything, teams have developed a benchmarking booklet that contains a list of key questions to take with them on benchmarking trips.[15]

- *Shift from a management-driven culture to a participative, team-based one.* Two basic tenets of TQM are employee involvement and teamwork. Business owners must be willing to push decision-making authority down the organization to where the real experts are. Teams of employees working together to identify and solve problems can be a powerful force in an organization of any size. Experience with TQM has taught entrepreneurs that the combined knowledge and experience of workers is much greater than that of only one person. Tapping into the problem-solving capabilities of the team produces profitable results.

- *Modify the reward system to encourage teamwork and innovation.* Because the team, not the individual, is the building block of TQM, companies often have to modify their compensation systems to reflect team performance. Traditional compensation methods pit one employee against another, undermining any sense of cooperation. Often they are based on seniority rather than on how much an employee contributes to the company. Compensation systems under TQM usually rely on incentives, linking pay to performance. However, rather than tying pay to individual performance, these systems focus on team-based incentives. Each person's pay depends on whether the entire team (or, sometimes, the entire company) meets a clearly defined, measurable set of performance objectives.

- *Train workers constantly to give them the tools they need to produce quality and to upgrade the company's knowledge base.* One of the most important factors in making long-term, constant improvements in a company's processes is teaching workers the philosophy and the tools of TQM. Admonishing employees to "produce quality" or offering them rewards for high quality is futile unless a company gives them the tools and know-how to achieve that end. Managers must be dedicated to making their companies "learning organizations" that encourage people to upgrade their skills and give them the opportunities and

incentives to do so. The most successful companies spend anywhere from 1 to 5 percent of their employees' time on training, most of it invested in workers, not managers. To give employees a sense of how the quality of their job fits into the big picture, many TQM companies engage in **cross training**, teaching workers to do other jobs in the company.

- *Train employees to measure quality with the tools of statistical process control (SPC).* The only way to ensure gains in quality is to measure results objectively and to trace the company's progress toward its quality objectives. That requires teaching employees how to use statistical process control techniques such as fishbone charts, Pareto charts, control charts, and measures of process capability. Without knowledgeable workers using these quantitative tools, TQM cannot produce the intended results.

- *Use Pareto's Law to focus TQM efforts.* One of the toughest questions managers face in companies embarking on TQM for the first time is "Where do we start?" The best way to answer that fundamental question is to use Pareto's Law (also called the 80/20 Rule), which states that 80 percent of a company's quality problems arise from just 20 percent of all causes. By identifying this small percentage of causes and focusing quality improvement efforts on them, a company gets maximum return for its efforts. This simple yet powerful rule forces workers to concentrate resources on the most significant problems first, where payoffs are likely to be biggest, and helps build momentum for successful TQM effort.

- *Share information with everyone in the organization.* Asking employees to make decisions and to assume responsibility for creating quality necessitates that the owner share information with them. Employees cannot make sound decisions consistent with the company's initiative if managers are unwilling to give them the information they need to make those decisions.

- *Focus quality improvements on astonishing the customer.* The heart of TQM is customer satisfaction—better yet, customer astonishment. Unfortunately, some companies focus their quality improvement efforts on areas that never benefit the customer. Quality improvements with no customer focus (either internal or external customers) are wasted.

- *Don't rely on inspection to produce quality products and services.* The traditional approach to achieving quality was to create a product or service and then to rely on an army of inspectors to "weed out" all of the defects. Not only is such a system a terrible waste of resources (consider the cost of scrap, rework, and no-value-added inspections), but it gives managers no opportunity for continuous improvement. The only way to improve a process is to discover the cause of poor quality, fix it (the sooner, the better), and learn from it so that workers can *avoid* the problem in the future. Using the statistical tools of the TQM approach allows a company to learn from its mistakes with a consistent approach to constantly improving quality.

- *Avoid using TQM to place blame on those who make mistakes.* In many firms, the only reason managers seek out mistakes is to find someone to blame for them. The result is a culture based on fear and the unwillingness of workers to take chances to innovate. The goal of TQM is to improve the processes in which people work, *not* to lay blame on workers. Searching out "the guilty party" doesn't solve the problem. The TQM philosophy sees each problem that arises as an opportunity for improving the company's system.

- *Strive for continuous improvement in processes as well as in products and services.* There is no finish line in the race for quality. A company's goal must be to improve the quality of its processes, products, and services constantly, no matter how high it currently stands!

Many of these principles are evident in quality guru W. Edwards Deming's 14 points, a capsulized version of how to build a successful TQM approach (see Table 19.4).

Implementing a TQM program successfully begins at the top. If the owner or chief executive of a company doesn't actively and visibly support the initiative, the employees who must make it happen will never accept it. TQM requires change: change in the way a company defines quality, in the way it sees its customers, in the way it treats employees, and in the way it sees itself. Successful implementation involves modification in an organization's culture as much as in its work processes.

Quantity: The Economic Order Quantity

The typical small business has its largest investment in inventory. However, an investment in inventory is not profitable because dollars spent return nothing until the inventory is sold. In a sense, a small company's inventory is its largest non–interest-bearing investment. Entrepreneurs must focus on controlling this investment and on maintaining proper inventory levels. A primary objective of

3.
Conduct economic order quantity analysis to determine the proper level of inventory.

TABLE 19.4 Deming's 14 Points

TQM cannot succeed as a piecemeal program or without true commitment to its philosophy. W. Edwards Deming, the man most visibly connected to TQM, drove home these concepts with his 14 points, the essential elements for integrating TQM successfully into a company. Deming's message was straightforward. Companies must transform themselves into customer-oriented, quality-focused organizations in which teams of employees have the training, the resources, and the freedom to pursue quality on a daily basis. The goal is to track the performance of a process, whether manufacturing a clock or serving a bank customer, and to develop ways to minimize variation in the system, eliminate defects, and spur innovation. The 14 points are as follows:

1. *Constantly strive to improve products and services.* This requires total dedication to improving quality, productivity, and service—*continuously.*

2. *Adopt a total quality philosophy.* There are no shortcuts to quality improvement; it requires a completely new way of thinking and managing.

3. *Correct defects as they happen.* Rather than relying on mass inspection of end products, companies should correct defects as they happen. Real quality comes from improving the process, not from inspecting finished products and services. At that point, it's too late. Statistical process control charts can help workers detect when a process is producing poor-quality goods or services. Then they can stop it, make corrections, and get the process back on target.

4. *Don't award business on price alone.* Rather than choosing the lowest-cost vendor, businesses should work toward establishing close relationships with the vendors who offer the highest quality.

5. *Constantly improve the system of production and service.* Managers must focus the entire company on customer satisfaction, measure results, and make adjustments as necessary.

6. *Institute training.* Workers cannot improve quality and lower costs without proper training to erase old ways of doing things.

7. *Institute leadership.* The supervisor's job is not to boss workers around; it is to lead. The nature of the work is more like coaching than controlling.

8. *Drive out fear.* People often are afraid to point out problems because they fear the repercussions. Managers must encourage and reward employee suggestions.

9. *Break down barriers among staff areas.* Departments within organizations often erect needless barriers to protect their own turf. Total quality requires a spirit of teamwork and cooperation across the entire organization.

10. *Eliminate superficial slogans and goals.* These only offend employees because they imply that workers could do a better job if only they would try.

11. *Eliminate standard quotas.* They emphasize quantity over quality. Not everyone can move at the same rate and still produce quality.

12. *Remove barriers to pride of workmanship.* Most workers want to do quality work. Eliminating "demotivators" frees them to achieve quality results.

13. *Institute vigorous education and retraining.* Managers must teach employees the new methods of continuous improvement, including statistical process control techniques.

14. *Take demonstrated management action to achieve the transformation.* Although success requires involvement of all levels of the organization, the impetus for change must come from the top.

These 14 interrelated elements contribute to a chain reaction effect. As a company improves its quality, costs decline, productivity increases, the company gains additional market share because of its ability to provide high-quality products at competitive prices, and the company and its employees prosper.

Source: Based on the W. Edwards Deming Institute, *www.deming.org/deminghtml/wedi.html.*

this part of the purchasing plan is to generate an adequate turnover of merchandise by purchasing proper quantities. Tying up capital in extra inventory limits a company's working capital and exerts unnecessary pressure on its cash flow. In addition, a business runs the risk of being stuck with obsolete or spoiled merchandise, an extremely serious problem for many small businesses. Excess inventory also takes up valuable storage or selling space that could be used for items with higher turnover rates and more profit potential. On the other hand, maintaining too little inventory can be extremely costly. An owner may be forced to reorder merchandise too frequently, escalating total inventory costs. In addition, inventory stockouts occur when customer demand exceeds a company's supply of merchandise, causing customer ill will. Persistent stockouts are inconvenient for customers, and many customers eventually abandon the store to shop elsewhere. Because many companies now employ lean and just-in-time inventory practices, interruptions in their supply chains can cause them to shut down temporarily and incur huge costs. "The very flexibility that provides the supply chain with its cost advantages has also caused its inherent vulnerability," says Paul Carter, head of global risk consulting for financial services company Allianz.[16] After an explosion at Evonik Industries AG, an auto parts supplier in Marl, Germany, incapacitated one of the few factories in the world that makes nylon-12, a special resin used to manufacture automotive fuel and brake lines, carmakers around the world faced shortages and plant shutdowns. The Evonik Industries factory produced 25 percent of the global supply of nylon-12, and its closure forced automakers to scramble to find substitute resins and alternative sources of supply.[17] A recent survey by Deloitte Consulting reports

that 53 percent of executives say that interruptions in their companies' supply chains such as the one caused by the explosion have become more expensive over the last three years.[18]

Carrying either too much or too little inventory are expensive mistakes that lead to serious problems in other areas of the business, particularly cash management. The goal is to maintain enough inventory to meet customer orders and to satisfy production needs but not so much that storage costs and inventory investments are excessive. The analytical techniques used to determine **economic order quantities** (EOQs) help business owners determine the amount of inventory to purchase in an order or to produce in a production run to minimize total inventory costs. To compute the EOQ, an entrepreneur must first determine the three principal elements of total inventory costs: the cost of the units, the holding (or carrying) cost, and the setup (or ordering) cost.

Cost of Units

The cost of the units is simply the number of units demanded for a particular time period multiplied by the cost per unit. Suppose that a small manufacturer of lawn mowers forecasts demand for the upcoming year to be 100,000 mowers. He must order enough wheels at $1.55 each to supply the production department. He computes the following:

$$\text{Total annual cost of units} = D \times C$$

where

$$D = \text{Annual demand (in units)}$$
$$C = \text{Cost of a single unit (\$)}$$

In this example,

$$D = 100{,}000 \text{ mowers} \times 4 \text{ wheels per mower} = 400{,}000 \text{ wheels}$$
$$C = \$4.15/\text{wheel}$$

$$\text{Total annual cost of units} = D \times C$$
$$= 400{,}000 \text{ wheels} \times \$4.15$$
$$= \$1{,}660{,}000$$

Holding (Carrying) Costs

An excessive inventory investment ties up a large amount of a company's cash unproductively in the form of holding costs. The typical costs of holding inventory include the costs of storage, insurance, taxes, interest, depreciation, damage or spoilage, obsolescence, and pilferage. The expense involved in physically storing the items in inventory is usually substantial, especially if the inventories are large. An entrepreneur may have to rent or build additional warehousing facilities, pushing the cost of storing the inventory even higher. The company also may incur expenses in transferring items into and out of inventory. The cost of storage also includes the expense of operating the facility (e.g., heating, lighting, and refrigeration) as well as the depreciation, taxes, and interest on the building. Most small business owners purchase insurance on their inventories to shift the risk of fire, theft, flood, and other disasters to an insurer. The premiums paid for this coverage also are included in the cost of holding inventory. In general, the larger a company's average inventory, the greater is its storage cost. For most companies, holding costs for an item range from 15 to 35 percent of its actual cost.

Depreciation costs represent the reduced value of inventory over time. Some businesses are strongly influenced by the depreciation of inventory. For example, an auto dealer's inventory is subject to depreciation because he or she must sell models left over from the previous year at reduced prices.

Spoilage, obsolescence, and pilferage also add to the costs of holding inventory. Some small firms, especially those that deal in trendy merchandise, assume an extremely high risk of obsolescence. For example, a fashion merchandiser with a large inventory of the latest styles may be left with worthless merchandise when styles change. In addition, unless the entrepreneur establishes sound inventory control procedures, the business will suffer losses from employee theft and shoplifting.

Let us return to the lawn mower manufacturer example to illustrate the cost of holding inventory:

$$\text{Total annual holding (carrying) costs} = \frac{Q}{2} \times H$$

where

$$Q = \text{Quantity of inventory ordered}$$
$$H = \text{Holding cost per unit per year}$$

TABLE 19.5 Holding (Carrying) Costs

If Q Is	$Q/2$, Average Inventory, Is	$Q/2 \times H$, Holding Cost, Is
500	250	$312.50
1,000	500	625.00
2,000	1,000	1,250.00
3,000	1,500	1,875.00
4,000	2,000	2,500.00
5,000	2,500	3,125.00
6,000	3,000	3,750.00
7,000	3,500	4,375.00
8,000	4,000	5,000.00
9,000	4,500	5,625.00
10,000	5,000	6,250.00

The greater the quantity ordered, the greater is the inventory carrying costs. This relationship is shown in Table 19.5, assuming that the cost of carrying a single lawn mower wheel for one year is $1.25.

Setup (Ordering) Costs

The various expenses incurred in actually ordering materials and inventory or in setting up the production line to manufacture them determine the level of setup or ordering costs of a product. The costs of obtaining materials and inventory typically include preparing purchase orders; analyzing and choosing vendors; processing, handling, and expending orders; receiving and inspecting items; and performing all of the required accounting and clerical functions. Even if the small company produces its own supply of goods, it encounters most of these same expenses. Ordering costs are usually relatively fixed, regardless of the quantity ordered.

Setup or ordering costs are found by multiplying the number of orders made in a year (or the number of production runs in a year) by the cost of placing a single order (or the cost of setting up a single production run). In the lawn mower manufacturing example, the annual requirement is 400,000 wheels per year, the cost to place an order is $9.00, and the ordering costs are as follows:

$$\text{Total annual setup (ordering) costs} = \frac{D}{Q} \times S$$

where

$$D = \text{Annual demand}$$
$$Q = \text{Quantity of inventory ordered}$$
$$S = \text{Setup (ordering) costs for a single run (or order)}$$

The greater the quantity ordered, the smaller the number of orders placed. This relationship is shown in Table 19.6, assuming an ordering cost of $9.00 per order.

Solving for EOQ

If carrying costs were the only expense involved in obtaining inventory, the business owner would purchase the smallest number of units possible in each order to minimize the cost of holding the

TABLE 19.6 Setup Costs

If Q Is	D/Q, Number of Orders per Year, Is	$D/Q \times S$, Setup (Ordering) Cost, Is
500	800	$7,200.00
1,000	400	3,600.00
5,000	80	720.00
10,000	40	360.00

inventory. For example, if the lawn mower manufacturer purchased just four wheels per order, carrying cost would be minimized:

$$\text{Carrying cost} = \frac{Q}{2} \times H$$

$$= \frac{4}{2} \times \$1.25$$

$$= \$2.50$$

However, ordering cost would be outrageous:

$$\text{Ordering cost} = \frac{D}{Q} \times S$$

$$= \frac{400,000}{4} \times \$9$$

$$= \$900,000$$

Obviously this is not the small manufacturer's ideal inventory solution.

Similarly, if ordering costs were the only expense involved in procuring inventory, the business owner would purchase the largest number of units possible in order to minimize the ordering cost. In our example, if the lawn mower manufacturer purchased 400,000 wheels per order, ordering cost would be minimized:

$$\text{Ordering cost} = \frac{D}{Q} \times S$$

$$= \frac{400,000}{400,000} \times \$9$$

$$= \$9$$

However, carrying cost would be tremendously high:

$$\text{Carrying cost} = \frac{Q}{2} \times H$$

$$= \frac{400,000}{2} \times \$1.25$$

$$= \$250,000$$

A quick inspection shows that neither of those solutions minimizes the total cost of the manufacturer's inventory. Total cost is composed of the cost of the unit, carrying cost, and ordering costs:

$$\text{Total cost} = (D \times C) + \left(\frac{Q}{2} \times H\right) + \left(\frac{D}{Q} \times S\right)$$

These costs are illustrated in Figure 19.5. Notice that as the quantity ordered increases, the ordering costs decrease and the carrying costs increase.

The EOQ formula simply balances the ordering cost and the carrying cost of the small business owner's inventory so that total costs are minimized. Table 19.6 summarizes the total costs for various values of Q for our lawn mower manufacturer.

As Table 19.7 and Figure 19.3 illustrate, the EOQ formula locates the minimum point on the total cost curve, which occurs where the cost of carrying inventory ($Q/2 \times H$) equals the cost of ordering inventory ($D/Q \times S$). As we have seen, if a small business places the smallest number of orders possible each year, its ordering cost is minimized, but its carrying cost is maximized. Conversely, if the company orders the smallest number of units possible in each order, its carrying cost is minimized, but its ordering cost is maximized. Total inventory cost is minimized when carrying cost and ordering costs are balanced.

FIGURE 19.5

Economic Order Quantity

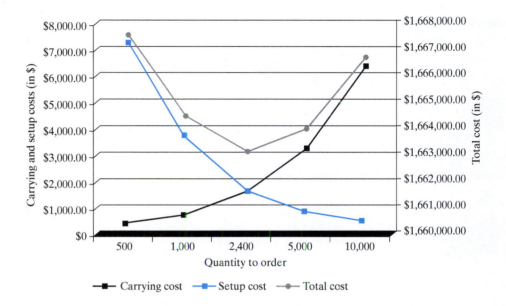

TABLE 19.7 EOQ and Total Cost

If Q Is	$D \times C$, Cost of Units, Is	$Q/2 \times H$, Carrying Cost, Is	$D/Q \times S$, Ordering Cost, Is	TC, Total Cost, Is
500	$1,660,000	$312.50	$7,200.00	$1,667,512.50
1,000	$1,660,000	$625.00	$3,600.00	$1,664,225.00
2,400	**$1,660,000**	**$1,500.00**	**$1,500.00**	**$1,663,000.00**
5,000	$1,660,000	$3,125.00	$720.00	$1,663,845.00
10,000	$1,660,000	$6,250.00	$360.00	$1,666,610.00

Let us return to our lawn mower manufacturer and compute its EOQ using the following formula:

$$EOQ = \sqrt{\frac{2 \times D \times S}{H}}$$

$$= \sqrt{\frac{2 \times 400,000 \times \$9.00}{\$1.25}}$$

$$= 2,400 \text{ wheels}$$

To minimize total inventory cost, the lawnmower manufacturer should order 2,400 wheels at a time. Furthermore,

$$\text{Number of orders per year} = \frac{D}{Q}$$

$$= 400,000/2,400$$

$$= 166.67 \text{ orders}$$

This manufacturer will place approximately 167 orders this year at a minimum total cost of $1,663,000, computed as follows:

$$\text{Total cost} = (D \times C) + \left(\frac{Q}{2} \times H\right) + \left(\frac{D}{Q} \times S\right)$$

$$= (400,000 \times \$4.15) + (2,400/2 \times \$1.25) + (400,000/2,400 \times \$9.00)$$

$$= \$1,660,000 + \$1,500 + \$1,500$$

$$= \$1,663,000$$

EOQ with Usage

The preceding EOQ model assumes that orders are filled instantaneously; that is, fresh inventory arrives all at once. Because that assumption does not hold true for many small manufacturers, it is necessary to consider a variation of the basic EOQ model that allows inventory to be added over a period of time rather than instantaneously. In addition, a manufacturer is likely to be taking items from inventory for use in the assembly process over the same time period. For example, the lawn mower manufacturer may be producing blades to replenish his or her supply, but at the same time, assembly workers are reducing the supply of blades to make finished mowers. The key feature of this version of the EOQ model is that inventories are used while inventories are being added.

Using the lawn mower manufacturer as an example, we can compute the EOQ for the blades. To make the calculation, we need two additional pieces of information: the usage rate for the blades, U, and the factory's capacity to manufacture the blades, P. Suppose that the maximum number of lawn mower blades the company can manufacture is 480 per day. We know from the previous illustration that annual demand for mowers is 100,000 units (therefore, 100,000 blades). If the factory operates 5 days per week for 50 weeks (250 days), its usage rate is as follows:

$$U = \frac{100,000 \text{ units per year}}{250 \text{ days}} = 400 \text{ units per day}$$

It costs \$325 to set up the blade manufacturing line and \$8.71 to store one blade for one year. The cost of producing a blade is \$24.85. To compute EOQ with usage, we use the following formula:

$$\text{EOQ} = \sqrt{\frac{2 \times D \times S}{H \times \left(1 - \dfrac{U}{P}\right)}}$$

For the lawn mower manufacturer,

$$D = 100,000 \text{ blades}$$
$$S = \$325 \text{ per production run}$$
$$H = \$8.71 \text{ per blade per year}$$
$$U = 400 \text{ blades per day}$$
$$P = 480 \text{ blades per day}$$

$$\text{EOQ} = \sqrt{\frac{2 \times 100,000 \times \$325}{8.71 \times \left(1 - \dfrac{400}{480}\right)}}$$

$$= 6,691.50 \text{ blades} = 6,692 \text{ blades}$$

Therefore, to minimize total inventory cost, the lawn mower manufacturer should produce 6,692 blades per production run. In addition,

$$\text{Number of production runs per year} = \frac{D}{Q}$$

$$= \frac{100,000 \text{ blades}}{6,692 \text{ blades per run}}$$

$$= 14.9 \approx 15 \text{ runs}$$

The manufacturer will make 15 production runs during the year at a total cost of the following:

$$\text{Total cost} = (D \times C) + \left(\left(1 - \frac{U}{P}\right) \times \frac{Q}{2} \times H\right) + \left(\frac{D}{Q} \times S\right)$$

$$= (100,000 \times \$24.85) + \left(\left(1 - \frac{400}{480}\right) \times \frac{6,692}{2} \times \$8.71\right) + \left(\frac{100,000}{6,692} \times \$325\right)$$

$$= \$2,485,000 + \$4,857 + \$4,857$$

$$= \$2,494,714$$

Business owners must remember that the EOQ analysis is based on estimations of cost and demand. The final result is only as accurate as the input used. Consequently, this analytical tool serves only as a guideline for decision making. The final answer may not be the ideal solution because of intervening factors, such as opportunity costs or seasonal fluctuations. Knowledgeable entrepreneurs use EOQ analysis as a starting point in making a decision and then use managerial judgment and experience to produce a final decision.

Price

For the typical small business owner, price is always a substantial factor when purchasing inventory and supplies. In many cases, an entrepreneur can negotiate price with potential suppliers on large orders of frequently purchased items. In other instances, perhaps when small quantities of items are purchased infrequently, the small business owner must pay list price. The typical entrepreneur shops around to find the supplier that offers the best price. However, this does not mean that a business owner should always purchase inventory and supplies at the lowest price available. The best purchase price is the lowest price at which the owner can obtain goods and services *of acceptable quality*. As quality guru W. Edwards Deming said, "Price has no meaning without a measure of the quality being purchased." Companies that are lured by low prices on key products or components from suppliers in foreign countries sometimes discover that currency exchange rates, shipping costs, customs fees, and the additional costs and challenges of coordinating long-distance shipments more than offset the goods' lower prices.

Recall that one of Deming's 14 points is *"Don't award business on price alone."* Without proof of quality, an item with the lowest initial price actually may produce the *highest* total cost. Deming condemned the practice of constantly switching suppliers in search of the lowest initial price because it increases the variability of a process and lowers its quality. Instead, he recommended that businesses establish long-term relationships built on mutual trust and cooperation with a single supplier. Some companies use long-term contracts that lock in the price of key raw materials for an extended time period to minimize the risks that inflation imposes on them. Texas Roadhouse Inc., a chain of steak restaurants that Kent Taylor started in 1993 in Clarksville, Indiana, creates annual contracts with its suppliers for 80 percent of the beef that it purchases for its nearly 400 restaurants. The company buys the remaining 20 percent of its meats, those beef products for which the company forecasts price decreases, on the open market.[19]

When evaluating a supplier's price, small business owners must consider not only the actual price of goods and services but also the selling terms accompanying them. In some cases, the selling terms can be more important than the price itself. Some vendors' selling terms include some type of purchase discount. Vendors typically offer three types of discounts: trade discounts, quantity discounts, and cash discounts.

Trade Discounts

4.

Differentiate among the three types of purchase discounts that vendors offer.

Trade discounts are established on a graduated scale and depend on a small company's position in the channel of distribution. In other words, trade discounts recognize the fact that manufacturers, wholesalers, and retailers perform a variety of vital functions at various stages in the channel of distribution and compensate them for providing these needed activities. Figure 19.6 illustrates a typical trade discount structure.

Quantity Discounts

Quantity discounts are designed to encourage businesses to order large quantities of merchandise and supplies. Vendors are able to offer lower prices on bulk purchases because the cost per unit is lower than for handling small orders. Quantity discounts normally exist in two forms: noncumulative and cumulative. Noncumulative quantity discounts are granted only if a business purchases a certain volume of merchandise in a single order. For example, a wholesaler may offer small retailers a 3 percent discount only if they purchase 10 gross of Halloween masks in a single order. Table 19.8 shows a typical noncumulative quantity discount structure.

Cumulative quantity discounts are offered if a company's purchases from a particular vendor exceed a specified quantity or dollar value over a predetermined time period. The time frame varies, but a year is most common. For example, a manufacturer of appliances may offer a small business a 3 percent discount on subsequent orders if its purchases exceed $50,000 per year.

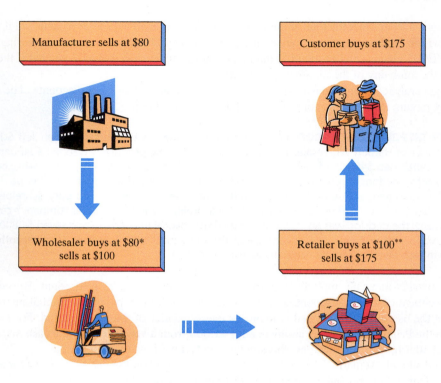

FIGURE 19.6
Trade Discount Structure

*Wholesale discount = 54% of suggested retail price.
** Retail discount = 43% of suggested retail price.

TABLE 19.8 Noncumulative Quantity Discount Structure

Order Size	Price
1–1,000 units	List price
1,001–5,000 units	List price—2%
5,001–10,000 units	List price—4%
10,001 units and more	List price—6%

Some small business owners who normally buy in small quantities and are unable to qualify for quantity discounts can earn quantity discounts by joining group purchasing organizations (GPOs), purchasing pools, or buying cooperatives. According to the National Cooperative Business Association, more than 29,000 cooperatives operate in the United States.[20] GPOs and cooperatives leverage the combined purchasing power of their members to negotiate the lowest prices and best discounts for their members.

ENTREPRENEURIAL PROFILE: EATFLEET EATFLEET is a group purchasing organization based in St. Paul, Minnesota, that bands together hundreds of independent restaurants across the United States and generates significant savings for members by negotiating quantity discounts for them. Member restaurateurs earn discounts on a variety of purchases, including food, beverages, linen services, office supplies, payroll services, secret shopper services, and others. "My food costs since joining EATFLEET have been reduced nearly $30,000 annually," says one satisfied restaurant owner.[21]

GPOs not only save their members money but also save them time by eliminating the need to constantly track and compare vendors' prices to secure the best deals.

Cash Discounts

Cash discounts are offered to customers as an incentive to pay for merchandise promptly. Many vendors grant cash discounts to avoid being used as an interest-free bank by customers who purchase merchandise and then fail to pay by the invoice due date. To encourage prompt payment of invoices, many vendors allow customers to deduct a percentage of the purchase amount if they pay within a specified time. Cash discount terms "2/10, net 30" are common in many industries.

This notation means that the total amount of the invoice is due 30 days after its date, but if the bill is paid within 10 days, the buyer may deduct 2 percent from the total. A discount offering "2/10, EOM" (EOM means "end of month") indicates that the buyer may deduct 2 percent if the bill is paid by the tenth day of the month after purchase.

In general, it is sound business practice to take advantage of cash discounts. The money saved by paying invoices promptly is freed up for use elsewhere.

ENTREPRENEURIAL PROFILE: Jeff Schreiber: Hansen Wholesale When Jeff Schreiber, owner of Hansen Wholesale, a small distributor of home products, attended a January trade show, he purchased $40,000 of ceiling fans from a manufacturer. The contract gave Schreiber until July to pay for the fans, but the manufacturer also included a cash discount: If Schreiber paid before May 1, he could earn a 3 percent discount on the purchase. By paying in February, Schreiber could save another 1.5 percent of the purchase price. For Schreiber, who manages his company's cash flow meticulously, the decision was an easy one; he paid the invoice in February and saved $1,800. "Your money works better if you take advantage of the discounts," says Schreiber, who recently saved $15,000 in cash discounts for his company in just one year.[22]

Businesses incur an implicit (opportunity) cost of forgoing a cash discount. By failing to take advantage of a cash discount, a business owner is, in effect, paying an annual interest rate to retain the use of the discounted amount for the remainder of the credit period. For example, suppose the Print Shop receives an invoice for $1,000 from a vendor offering a cash discount of 2/10, net 30. Figure 19.7 illustrates this situation and shows how to compute the cost of forgoing the cash discount. Actually, it costs the Print Shop $20 to retain the use of its $980 for an extra 20 days. Translating this into an annual interest rate gives the following result:

$$I = P \times R \times T$$

where

$$I = \text{Interest (\$)}$$
$$P = \text{Principle (\$)}$$
$$R = \text{Rate of interest (\%)}$$
$$T = \text{Time (number of days/360)}$$

To compute R, the annual interest rate,

$$R = \frac{I}{P \times T}$$

In our example,

$$R = \frac{\$20}{\left(\$980 \times \dfrac{20}{360}\right)}$$
$$= 36.7\%$$

The cost to the Print Shop of forgoing the cash discount is 36.7 percent per year! If there is $980 available on day 10 of the trade credit period, the entrepreneur should pay the invoice unless

FIGURE 19.7

A Cash Discount

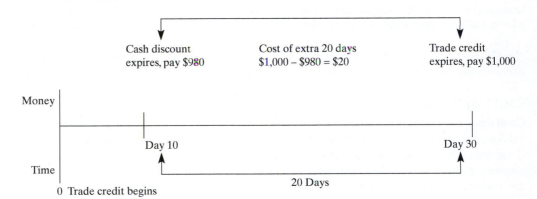

Cash discount expires, pay $980

Cost of extra 20 days $1,000 − $980 = $20

Trade credit expires, pay $1,000

Money

Day 10

Day 30

Time

0 Trade credit begins

20 Days

TABLE 19.9 Cost of Forgoing Cash Discounts

Cash Discount Terms	Cost of Forgoing Cash Discounts (Annually)
2/10, net 30	36.7%
2/30, net 60	34.50%
2/10, net 60	13.7%
3/10, net 30	55.7%
3/10, net 60	22.3%

he or she is able to return greater than 36.7 percent on that money. If the entrepreneur does not have $980 on day 10 but can borrow it at less than 36.7 percent, he or she should do so to take advantage of the cash discount. Table 19.9 summarizes the cost of forgoing cash discounts offering various discount terms.

Although business owners should take advantage of cash discounts, they should not stretch their accounts payable to suppliers beyond the payment terms specified on the invoices. Letting payments become past due can destroy the trusting relationship a small company has built with its vendors.

Timing—When to Order

Timing the purchase of merchandise and supplies is also a critical element of a purchasing plan. Entrepreneurs must schedule delivery dates so that their companies do not lose customers because they run out of inventory ("stockouts"). In addition, they must concentrate on maintaining proper control over the firm's inventory investment without tying up an excessive amount of working capital. There is a tradeoff between the cost of running out of stock and the cost of carrying additional inventory.

When planning delivery schedules for inventory and supplies, owners must consider the **lead time** for an order, the time gap between placing an order with a vendor and actually receiving the goods. In general, entrepreneurs cannot expect instantaneous delivery of merchandise. As a result, they must plan reorder points for inventory items with lead times in mind. To determine when to order merchandise for inventory, entrepreneurs must calculate the reorder point for key inventory items. Developing a reorder point model involves determining the lead time for an order, the usage rate for the item, the minimum level of stock allowable, and the EOQ. The lead time may be as little as a few hours or as long as several weeks to process purchase requisitions and orders, contact the supplier, receive the goods, and add them to the company's inventory stock. Entrepreneurs who purchase from local vendors encounter shorter lead times than those who rely on distant suppliers. More companies are relying on third-party logistics providers (3PLs), companies that specialize in value-added supply chain services, to manage their supply chain activities,

5.

Calculate a company's reorder point.

Entrepreneurs must coordinate both the quantity and the timing of their inventory shipments to ensure that they have enough inventory to meet customer demand but not so much that they tie up cash unnecessarily.

Source: © Jerry Bernard/Alamy.

including inventory planning, distribution, storage, and others. A recent survey by Capgemini Consulting shows that 56 percent of companies report seeing benefits from using 3PLs, including lower logistics and inventory costs and higher on-time order fill rates and order accuracy rates.[23]

Business owners can determine the usage rate for a particular product from past inventory and accounting records. Business owners must estimate the speed at which the supply of merchandise will be depleted over a given time. The anticipated usage rate for a product determines how long the supply will last. For example, if an entrepreneur projects that she will use 900 units in the next six months, the usage rate is five units per day (900 units/180 days). The simplest reorder point model assumes that the firm experiences a linear usage rate; that is, depletion of the firm's stock continues at a constant rate over time.

Business owners also must determine the minimum level of stock allowable. If a firm runs out of a particular item (i.e., incurs stockouts), customers will lose faith in the business and begin to shop elsewhere. To avoid stockouts, many firms establish a minimum level of inventory greater than zero. In other words, they build a cushion, called **safety (or buffer) stock**, into their inventories in case demand runs ahead of the anticipated usage rate. If that occurs, the owners can dip into the safety stock to fill customer orders until a shipment of goods arrives. Wayne Rosenbaum, director of operations for Papaya King, a three-unit chain of stores that sell hot dogs and tropical drinks made from papayas, typically keeps a two-week supply of papaya products on hand in case problems arise in its supply chain. "When you are the 'Papaya King,' you'd better have papaya," says Rosenbaum. The company's safety stock proved to be valuable when a salmonella outbreak associated with papayas from Mexico created an interruption in its supply of fresh papayas. Papaya King turned to other suppliers, adapted its recipes, and used its safety stock to meet customer demand for its tropical drinks.[24]

To compute the reorder point for an item, the owner must combine this inventory information with the product's EOQ. The following example illustrates the reorder point technique:

$$L = \text{Lead time for an order} = 5 \text{ days}$$
$$U = \text{Usage rate} = 18 \text{ units per day}$$
$$S = \text{Safety stock (minimum level)} = 75 \text{ units}$$
$$\text{EOQ} = 540 \text{ units}$$

The formula for computing the reorder point is:

$$\text{Reorder point} = (L \times U) + S$$

In this example,

$$\text{Reorder point} = (5 \text{ days} \times 18 \text{ units/day}) + 75 \text{ units}$$
$$= 165 \text{ units}$$

This business owner should order 540 more units when inventory drops to 165 units. Figure 19.8 illustrates the reorder point situation for this small business.

FIGURE 19.8

Reorder Point Model

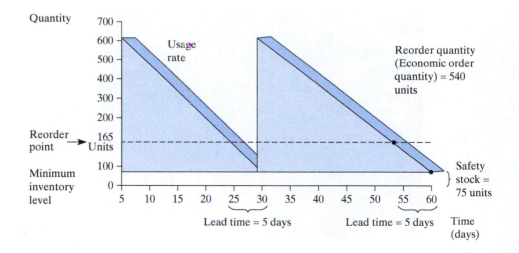

FIGURE 19.9
Demand During Leading

Below average   Average demand   Above average
demand   demand
$\overline{D}_L$

The simple reorder technique makes assumptions that may not be valid in particular situations. First, the model assumes that the firm's usage rate is constant when in fact for most small businesses demand varies daily. Second, the model assumes that lead time for an order is constant when in fact few vendors deliver precisely within lead time estimates. Third, in this model, the owner never taps safety stock; however, late deliveries or accelerated demand often force owners to dip into their inventory reserves. More advanced models relax some of these assumptions, but the simple model can be a useful inventory guideline for making inventory decisions in a small company.

Another popular reorder point model assumes that the demand for a product during its lead time is normally distributed (see Figure 19.9). The area under the normal curve at any given point represents the probability that a particular demand level will occur. Figure 19.10 illustrates the application of this normal distribution to the reorder point model *without* safety stock. The model recognizes that three different demand patterns can occur during a product's lead time. Demand pattern 1 is an example of below-average demand during lead time, demand pattern 2 is an example of average demand during lead time, and demand pattern 3 is an example of an above-average demand during lead time.

If the reorder point for this item is normal for the product during lead time, 50 percent of the time, demand will be below average (note that 50 percent of the area under the normal curve lies below average). Similarly, 50 percent of the time, demand during lead time will exceed the average, and the firm will experience stockouts (note that 50 percent of the area under the normal curve lies above average).

To reduce the probability of inventory shortage, a business owner can increase the reorder point above $\overline{D}_L$ (average demand during the lead time). But how much should the owner increase the reorder point? Rather than attempt to define the actual costs of carrying extra inventory versus the costs of stockouts (remember the tradeoff described earlier), this model allows the small business owner to determine the appropriate reorder point by setting a desired customer service level.

FIGURE 19.10
Reorder Point without Safety Stock

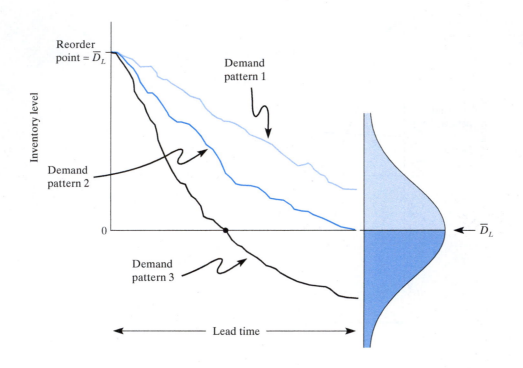

TABLE 19.10 Service Level Factors and Z Scores

Target Customer Service Level	Service Level Factor (Z Score*)
99%	2.33
97.5%	1.96
95%	1.645
90%	1.275
80%	0.845
75%	0.675

Any basic statistics book provides a table of areas under the normal curve, which gives the appropriate Z score for any service level factor.

For example, the owner may want to satisfy 95 percent of customer demand for a product during lead time. This service level determines the amount of increase in the reorder point. In effect, these additional items serve as a safety stock:

$$\text{Safety stock} = SLF \times SD_L$$

where

SLF = Service level factor (the appropriate Z score)

SD_L = Standard deviation of demand during lead time (a common measure of the variability in the demand during lead time)

Table 19.10 shows the appropriate service level factor (Z score) for some of the most popular target customer service levels.

Figure 19.11 shows the shift to a normally distributed reorder point model with safety stock. In this case the manager has set a 95 percent customer service level; that is, the manager wants to meet 95 percent of the demand during lead time. The normal curve in the model without safety stock (from Figure 19.10) is shifted *up* so that 95 percent of the area under the curve lies above

FIGURE 19.11

Reorder Point with Safety Stock

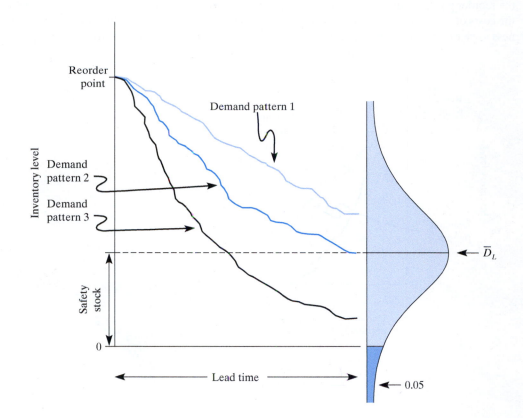

the zero inventory level. The result is a reorder point that is higher than the original reorder point by the amount of the safety stock:

$$\text{Reorder point} = \overline{D}_L + (SLF \times SD_L)$$

where

$\overline{D}_L$ = Average demand during lead time (original reorder point)

SLF = Service level factor (the appropriate Z score)

SD_L = Standard deviation of demand during lead time

To illustrate, suppose that the average demand for a product during its lead time (one week) is 325 units with a standard deviation of 110 units. If the desired service level is 95 percent, the service level factor (from Table 19.9) would be 1.645. The reorder point is as follows:

$$R = 325 + (1.645 \times 110) = 325 + 181 = 506 \text{ units}$$

Figure 19.12 illustrates the shift from a system without safety stock to one with safety stock for this example. With a reorder point of 325 units ($\overline{D}_L$), this small business owner will experience inventory shortages during the lead time 50 percent of the time. With a reorder point of 506 units (i.e., a safety stock of 181 units), the business owner will experience inventory stockouts during the lead time only 5 percent of the time.

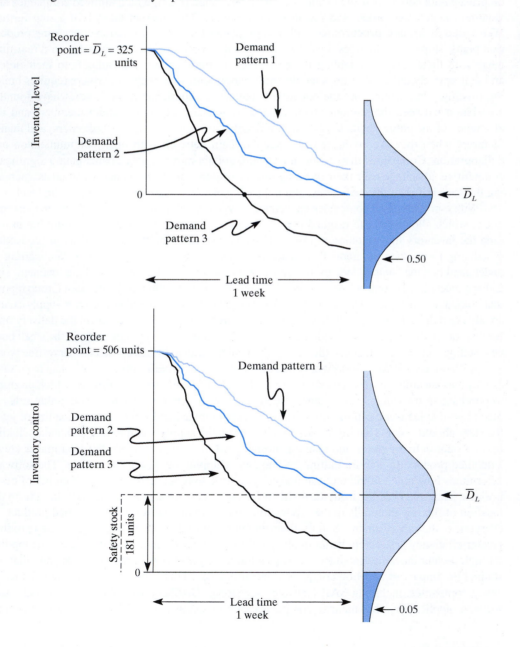

FIGURE 19.12

Shift from a No-Safety Stock System to a Safety Stock System

Managing the Supply Chain:
Vendor Analysis and Selection

6.
<hr/>

Develop a vendor rating scale.

Businesses have discovered that managing their supply chains for maximum effectiveness and efficiency not only can increase their profitability but also can provide them with an important competitive advantage in the marketplace. Proper SCM enables companies to reduce their inventories, get products to market much faster, increase quality, and improve customer satisfaction. "We're seeing a level of sophistication in supply chain management that didn't exist five years ago," says Dave Donnan, a consultant who specializes in SCM. "The separation of those [companies] that will succeed and those that will fail will be based on attention to detail."[25] SCM requires businesses to forge long-term partnerships with reliable suppliers rather than to see vendors merely as "someone trying to sell me something." Doing so can produce an impressive payoff; experts note that implementing a successful SCM system yields an average savings of 15 percent by controlling unnecessary spending, negotiating lower prices, maintaining lower inventory levels, and reducing waste and inefficiency in the purchasing process, all of which save companies vast amounts of money.[26]

Companies are learning that, to make SCM work, they must share information with their suppliers and make their entire supply chains transparent to everyone involved in them. In the early days, that meant linking suppliers and companies as if they were parts of a single business on private data networks using electronic data interchange (EDI), which allowed companies and vendors to exchange orders and invoices electronically. The Internet takes EDI a step further. Web-based SCM, or **e-procurement**, allows companies to share information concerning production plans, shipment schedules, inventory levels, sales forecasts, and actual sales on a real-time basis with their vendors, enabling the companies to make instant adjustments to their orders and delivery schedules. Sharing sensitive information with supply chain partners requires building trusting relationships, but the potential benefits are significant. Many studies have found a correlation between the amount of information shared among supply chain members and the efficiency of the entire chain. Costs are significantly lower for retailers, wholesalers, and manufacturers when they are connected in a supply chain network with open communication and collaboration. Companies that establish efficient, collaborative supply chains gain a significant competitive advantage over their rivals by minimizing inventory investments and costs, increasing the speed of customer deliveries, and improving customer service and satisfaction levels.

With e-procurement, companies are connected via the Web to their customers and to their suppliers, which allows them to respond rapidly to changing buyer preferences by modifying in real time the inventory they purchase. In turn, suppliers can make the fast adjustments in production scheduling to produce the items that customers actually are buying. Companies "are starting to understand that the 'supply' side of the supply chain isn't worth a hill of beans if the 'demand' side is disconnected," says one industry expert.[27] A study by the consulting firm Aberdeen Group reports that companies that use e-procurement to connect seamlessly the members of their supply chains are able to reduce the prices that they pay for their purchases by 7 percent and cut the delivery time for their orders by 67 percent.[28] With these systems, valuable information flows from the small business selling suits to customers up the supply chain all the way to the sheep shearer harvesting wool!

A Web-based SCM process works like this: Software at a retail store captures data from sales as they happen and looks for underlying trends for use in calculating quantities of which products to purchase in the future. For instance, SCM software at a retail store may notice that sales of black low-rise jeans are selling more briskly than anticipated and forecasts the quantity of jeans the store should order. That information then goes up the supply chain to the jean maker. Taking into account delivery times and manufacturing speed, the software helps the jean maker create a detailed production plan for cutting, sewing, and shipping the garments on time. The software determines how much fabric the jean maker must have to produce the required number of black low-rise jeans and orders it from the textile producer. The software can track everything from the location of the raw materials in the production process to the quality of the finished product for everyone in the supply chain. "Supply chain analytics boost the bottom line because they produce greater efficiency, less scrap, better quality, and lower production costs and improve the top line through greater customer satisfaction," says a top manager at IDC, a research company that has studied the impact of SCM on companies' performance. "This is basic business made better."[29] Many companies, including SAP, Netsuite, Salesforce, BizSlate, and others, offer cloud-based software applications that manage companies' supply chains; some of the software, including

BizSlate's, is aimed specifically at small and midsize companies with annual sales between $1 million and $200 million.[30]

To function smoothly, a small company's supply chain should follow the "triple As": agile, adaptable, and aligned.[31] An *agile* supply chain is one that is fast, flexible, and responsive to changes in demand. Agile supply chains are able to deal with the inevitable disruptions and fluctuations by creating strong partnerships with suppliers, adequate but not excessive levels of safety stock, contingency plans for catastrophic events, and an information system that provides everyone in the chain with timely information. Companies that sell products such as fashion merchandise or video games that have unpredictable demand and high costs associated with stockouts and that require large end-of-season markdowns to move leftover items require an agile supply chain. Companies that sell basic products such as groceries and cosmetics that have predictable demand, carry low profit margins, and require small end-of-season markdowns require supply chains that are lean and efficient. An *adaptable* supply chain is one that changes as a company's needs change and is able to accommodate a small company's growth. Adaptable supply chains are predictive, able to anticipate changes in companies' buying and selling process and helping them to adapt to the changes in real time. A supply chain is properly *aligned* when all of companies in it work together as a team to improve the chain's performance for the benefit of the entire group. In the past, some companies were hesitant to share information with the businesses in their supply chains. Success today requires that companies not only share information seamlessly but also synchronize their efforts to maximize efficiency throughout the entire chain. Agile, adaptable, aligned supply chains reduce costs and improve performance by increasing the speed at which companies get products into customers' hands, reducing excess inventory and decreasing the use of price markdowns that erode companies' profit margins.

In the Entrepreneurial Spotlight

Integrating Local Suppliers into a Complex Supply Chain

One of the hottest trends in the restaurant industry is local sourcing, purchasing fresh ingredients from local growers and producers. A recent survey by the National Restaurant Association shows that 70 percent of people say that they are more likely to visit a restaurant that offers locally produced menu items, and more than 60 percent of people say that locally sourced menus are a key consideration when they are choosing a restaurant. Local sourcing is a "macro trend that will maintain its momentum," predicts a top official at the National Restaurant Association. A study conducted by Rabobank America, a bank that specializes in loans to farmers, ranchers, and other types of agribusiness, says that the local sourcing movement is "a permanent mainstream trend" that is generating opportunities for regional growers to take market share from larger, national suppliers. "National growers [must] adapt their business models to accommodate the desire for fresh, local produce," the study concludes.

Finding reliable, consistent local suppliers poses a challenge for restaurateurs, however. Restaurants in the United States spend more than $225 billion annually on food and beverage purchases, most of which arrive through a complex supply chain that includes value-added intermediaries at every stage. Many restaurants and restaurant chains have revamped their supply chains to accommodate customers' preferences for locally grown ingredients, but the process usually is not easy. In addition to controlling food and beverage costs, a restaurant's supply chain must meet high standards of safety and traceability. In addition, managing a complex restaurant supply chain efficiently requires an integrated system capable of managing the details necessary

to ensure quality and cost control and of providing the restaurateur with a view of "the big picture." Small restaurants often get by with a piecemeal system of spreadsheets, paper invoices, and telephone orders, but as food supply chains become longer and more complex, those systems can create inefficiencies that are difficult to detect. "It's like a miles-long pipeline that's leaking from many tiny holes that can be serviced only on foot," explains one supply chain expert.

Sysco, one of the nation's largest food suppliers, recently redesigned its system to enable the company to meet restaurateurs' demand for local food. Doing so required three years; many meetings with farmers and ranchers to learn about their growing seasons, costs, and business models; and integrating their smaller-volume supplies into Sysco's distribution system, which traditionally was designed to handle large volumes of standard food items in bulk. The process began when a chef at Michigan State University asked Sysco to help him source more locally grown produce for the campus dining operation. The chef noted that Mike Passorelli grew basil in a greenhouse located just two hours west of the Michigan State campus; however, he also knew that buying from local suppliers would create a logistical nightmare that would be far too complex to organize. The chef approached Sysco about incorporating local produce, such as the basil that Passorelli grew in his greenhouse, into the company's supply chain. Two of Sysco's managers, Rich Dachman and Denis Jennisch began working on a system that would allow Sysco to incorporate locally—grown produce into their supply chain. The following diagram shows how they did it.

(continued)

In the Entrepreneurial Spotlight (continued)

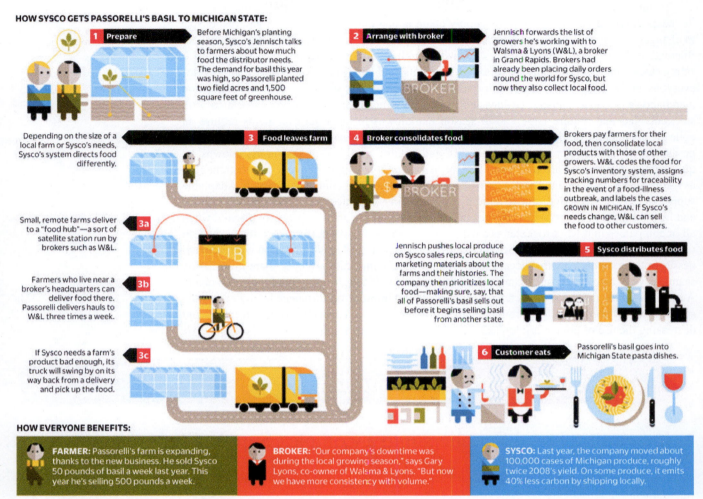

HOW SYSCO GETS PASSORELLI'S BASIL TO MICHIGAN STATE:

1 Prepare — Before Michigan's planting season, Sysco's Jennisch talks to farmers about how much food the distributor needs. The demand for basil this year was high, so Passorelli planted two field acres and 1,500 square feet of greenhouse.

2 Arrange with broker — Jennisch forwards the list of growers he's working with to Walsma & Lyons (W&L), a broker in Grand Rapids. Brokers had already been placing daily orders around the world for Sysco, but now they also collect local food.

Depending on the size of a local farm or Sysco's needs, Sysco's system directs food differently.

3 Food leaves farm

4 Broker consolidates food — Brokers pay farmers for their food, then consolidate local products with those of other growers. W&L codes the food for Sysco's inventory system, assigns tracking numbers for traceability in the event of a food-illness outbreak, and labels the cases GROWN IN MICHIGAN. If Sysco's needs change, W&L can sell the food to other customers.

3a Small, remote farms deliver to a "food hub"—a sort of satellite station run by brokers such as W&L.

3b Farmers who live near a broker's headquarters can deliver food there. Passorelli delivers hauls to W&L three times a week.

3c If Sysco needs a farm's product bad enough, its truck will swing by on its way back from a delivery and pick up the food.

Jennisch pushes local produce on Sysco sales reps, circulating marketing materials about the farms and their histories. The company then prioritizes local food—making sure, say, that all of Passorelli's basil sells out before it begins selling basil from another state.

5 Sysco distributes food

6 Customer eats — Passorelli's basil goes into Michigan State pasta dishes.

HOW EVERYONE BENEFITS:

FARMER: Passorelli's farm is expanding, thanks to the new business. He sold Sysco 50 pounds of basil a week last year. This year he's selling 500 pounds a week.

BROKER: "Our company's downtime was during the local growing season," says Gary Lyons, co-owner of Walsma & Lyons. "But now we have more consistency with volume."

SYSCO: Last year, the company moved about 100,000 cases of Michigan produce, roughly twice 2008's yield. On some produce, it emits 40% less carbon by shipping locally.

How Sysco Get Passorelli's Basil to Michigan State.
Source: Lou Lou & Tummie.

1. What advantages can a restaurant gain by offering locally sourced products on its menu?

2. What makes purchasing food and beverage supplies from local growers and suppliers so difficult for restaurateurs?

3. What steps can restaurateurs take to increase the selection of locally grown items on their menus?

Sources: Based on "Local Sourcing, Nutrition to Top 2013 Restaurant Trends," *Food Business News*, December 11, 2012, www.foodbusinessnews.net/articles/news_home/Food-Service-Retail/2012/12/Local_sourcing_nutrition_to_to.aspx?ID={6B48F136-3E21-4016-88E1-7B502BB981A1}&cck=1; Barney Wolf, "10 Trends for 2013," *QSR Magazine*, January 2013, www.qsrmagazine.com/reports/10-trends-2013; Janet Erickson, "Top 5 Supply Chain Management Issues for Restaurants," *Nation's Restaurant News*, September 17, 2012, http://nrn.com/latest-headlines/top-5-supply-chain-management-issues-restaurants; Kristen Hinman, "Sysco's New System," *Fast Company*, November 2011, p. 50; "Restaurant Supply Chain Management," *Nation's Restaurant News*, p. 1.

If a company's supply chain is to conform to the triple As, it must be built on the foundation of reliable vendors that can supply it with quality merchandise, equipment, supplies, and services at reasonable prices in a timely manner. Finding the right vendors sometimes can be a challenge, but the Internet is a useful tool for tracking down potential suppliers. Online supplier networks, such as Ariba Discovery (with more than 730,000 global suppliers listed) and *Thomas.net* (with more than 610,000 global suppliers listed), provide a marketplace where entrepreneurs can tap into a network of global suppliers. Supplier networks provide the businesses who use them with many benefits, including higher on-time delivery rates (77 percent for companies that use networks vs. 69 percent for those that do not), lower "cost of poor quality" expenses (7 percent for companies that use networks vs. 10 percent for those that do not), and the ability to screen potential suppliers for financial

stability and reputation.[32] Web sites such as Alibaba allow entrepreneurs to gain access to potential suppliers around the globe for almost any item—from thumb drives to thumbtacks.

ENTREPRENEURIAL PROFILE: Jonathan Shriftman and Jake Medwell: Solé Bicycles
Jonathan Shriftman and Jake Medwell cofounded Solé Bicycles, an online company that markets single-speed, fixed-gear bicycles (referred to as "fixies"), while they were students at the University of Southern California. Early on, they realized that one key to success was finding quality suppliers capable of providing the bicycles and accessories they had designed. The least expensive fixies sold for between $800 and $1,000, but the young entrepreneurs wanted to offer sturdy, durable bikes at a lower price point. Shriftman and Medwell turned to Alibaba, which gave "us access to countless suppliers that we would not have been able to reach on our own," says Medwell. They began working with a reliable bicycle manufacturer in China that could provide quality bicycles at affordable prices and with which the company has forged a long-term relationship. "We're very happy with our price point and the products we offer," says Shriftman. The entrepreneurs recently traveled to China to visit their supplier. "It was great to meet the people who helped us launch our business," says Medwell. "We knew that they could produce great bikes, but seeing the quality workmanship in person gave us a sense of the value they place on their product."[33]

As Shriftman's and Medwell's experience proves, selecting the right vendors or suppliers for a business has an impact well beyond simply obtaining goods and services at the lowest cost. Although searching for the best price always is an important factor, successful small business owners must always consider other factors in vendor selection, such as reliability, reputation, quality, support services, speed, and proximity.

Vendor Certification

To add objectivity to the vendor selection process, many firms are establishing vendor certification programs, agreements to give one supplier the majority of their business if that supplier meets rigorous quality and performance standards. Today, businesses of all sizes and types are forging long-term partnerships with vendors that meet their certification standards. When creating a vendor certification program, entrepreneurs should remember the three Cs: commitment, communication, and control. *Commitment* to consistently meeting the company's quality standards must be paramount. No company can afford to do business with vendors that cannot meet its quality targets. Second, a company must establish two-way *communication* with vendors. Communication implies trust, and trust creates working relationships that are long term and mutually beneficial. Treating suppliers like partners can reveal ways to boost quality and lower costs for both parties. Finally, a company must make sure that its vendors and suppliers have in place the *controls* that enable them to produce quality results and to achieve continuous improvements in their processes. In today's competitive marketplace, entrepreneurs expect all vendors to demonstrate that they operate processes built on continuous improvement.

Creating a vendor certification program requires entrepreneurs to develop a vendor rating scale that allows them to evaluate the advantages and disadvantages of each potential vendor. The scale allows entrepreneurs to score potential vendors using a measurement of the purchasing criteria that are most important to their companies' success. The first step to developing a scale is to determine the criteria that are most important to selecting a vendor (e.g., price, quality, and prompt delivery). The next step is to assign weights to each criterion to reflect its relative importance. The third step involves developing a grading scale for comparing vendors on the criteria. Developing a usable scale requires that the owner maintain proper records of past vendor performances. Finally, the owner must compute a weighted total score for each vendor and select the vendor that scores the highest on the set of criteria. Consider the following example. Bravo Bass Boats, Inc., is faced with choosing from among several suppliers of a critical raw material. The company's owner has decided to employ a vendor rating scale to select the best vendor using the following procedure.

Step 1. *Determine important criteria.* The owner of Bravo has selected the following criteria:

Quality

Price

Prompt delivery

Service

Assistance

Step 2. *Assign weights to each criterion to reflect its relative importance.*

Criterion	Weight
Quality	35
Price	30
Prompt delivery	20
Service	10
Assistance	5
Total	100

Step 3. *Develop a grading scale for each criterion.*

Criterion	Grading Scale
Quality	$\dfrac{\text{Number of acceptable lots from Vendor X}}{\text{Total number of lots from Vendor X}}$
Price	$\dfrac{\text{Lowest quoted price of all vendors}}{\text{Price offered by Vendor X}}$
Prompt Delivery	$\dfrac{\text{Number of on-time deliveries from Vendor X}}{\text{Total number of deliveries from Vendor X}}$
Service	A subjective evaluation of the variety of service offered by each vendor
Assistance	A subjective evaluation of the advice and assistance provided by each vendor

Step 4. *Compute a weighted score for each vendor.*

Criterion	Weight	Grade	Weighted Score (weight × grade)
Vendor 1			
Quality	35	9/10	31.5
Price	30	12.50/13.00	28.8
Prompt delivery	20	10/10	20.0
Service	10	8/10	8.0
Assistance	5	5/5	5.0
Total weighted score			93.3
Vendor 2			
Quality	35	8/10	28.0
Price	30	12.50/13.50	27.8
Prompt delivery	20	8/10	16.0
Service	10	8/10	8.0
Assistance	5	4/5	4.0
Total weighted score			83.8
Vendor 3			
Quality	35	7/10	24.5
Price	30	12.50/12.50	30.0
Prompt delivery	20	6/10	12.0
Service	10	7/10	7.0
Assistance	5	1/5	1.0
Total weighted score			74.5

Using this analysis of the three suppliers, Bravo should purchase the majority of this raw material from Vendor 1.

The Final Decision

Once business owners identify potential vendors and suppliers, they must decide which one (or ones) to do business with. Entrepreneurs should consider the following factors before making the final decision about the right supplier.

NUMBER OF SUPPLIERS One important question entrepreneurs face is "Should I buy from a single supplier or from several different sources?" Concentrating purchases at a single supplier (or sole sourcing) results in special attention from the supplier, especially if orders are substantial. Second, a business may be able to negotiate quantity discounts if its orders are large enough. Finally, a small company can cultivate a closer, more cooperative relationship with the supplier. Suppliers are more willing to work with companies that prove to be loyal customers. The result of this type of partnership can be better-quality goods and services. Stratsys, a company that makes plastic prototypes for the aerospace, automotive, and medical industries, purchases some of its most important raw materials from a single source. Company managers admit that doing so involves risk, but they believe that their company produces better-quality products by eliminating the variability that multiple sources of supply would introduce into their production process.[34]

However, using a single vendor also has disadvantages. A company can experience shortages of critical materials if its only supplier suffers a catastrophe, such as bankruptcy, a fire, a strike, or a natural disaster. To offset the risks of sole sourcing, many companies rely on the 80/20 rule. They purchase 80 percent of their supplies from their premier supplier and the remaining 20 percent from several "backup" vendors. If a catastrophe shuts down the company's principal supplier, the business can shift its orders to its "minor" suppliers with whom it has established relationships. Although this strategy may require a compromise on getting the lowest prices, it removes the risk of sole sourcing and lets a company's primary suppliers know that they have competition.

RELIABILITY Business owners must evaluate a potential vendor's ability to deliver adequate quantities of quality merchandise on time. One common complaint that small businesses have against their suppliers is late delivery. Late deliveries and the resulting shortages they often cause result in lost sales and customer ill will. Large customers often take precedence over small ones when it comes to service.

PROXIMITY A supplier's physical proximity is an important factor when choosing a vendor. The cost of transporting merchandise can increase significantly the total cost of merchandise to a buyer. Foreign manufacturers require longer delivery times, and because of the distance that shipments must travel, a hiccup anywhere in the distribution channel often results in late deliveries. In addition, entrepreneurs can solve quality problems more easily with nearby suppliers than with distant vendors. Some companies that once outsourced production of products or components to factories in foreign countries because they offered lower costs than domestic factories are bringing their orders back to companies in the United States, a trend called **reshoring**. Rapidly rising labor costs in countries such as China and India, the high cost of oil that makes shipping goods around the world more expensive, and the complexity of dealing with suppliers located 7,000 miles away have made domestic suppliers much more attractive to U.S. businesses.

ENTREPRENEURIAL PROFILE: Sonja Zozula and Jerry Anderson: LightSaver Technologies When Sonja Zozula and Jerry Anderson started LightSaver Technologies in 2009, they outsourced production of the company's emergency lights for home owners to a factory in China. Two years later, Zozula and Anderson decided to shift production back to a company in Carlsbad, California, located just 30 miles from their headquarters in San Clemente, California. "It's 30 percent cheaper to manufacture in China," says Anderson. "But [you must] factor in shipping and all the other [problems] you have to endure." Neither Zozula nor Anderson has ever been to China, which made communicating with their suppliers there more difficult, and shipments of the company's emergency lights often were stuck in customs for some reason, sometimes for several weeks. Sometimes, product quality was an issue, and Anderson would spend hours on the phone with managers at the Chinese factory trying to explain necessary changes to their products. "[Now] if we have an issue in manufacturing, we can walk down to the plant floor," he says. "We can't do that in China." Anderson estimates that the total cost of producing LightSaver Technologies' products in the United States is 2 to 5 percent cheaper than producing them in China.[35]

production of palm oil, and forced labor in mining operations in the Democratic Republic of the Congo.[39]

PRICE Small firms usually must pay list price for items that they purchase infrequently or in small quantities. However, this is not the case for goods that they purchase regularly in large quantities. Entrepreneurs should always attempt to negotiate the best prices and terms of sale with their vendors, especially on the products that they purchase in volume.

Legal Issues Affecting Purchasing

7.

Describe the legal implications of the purchasing function.

When a small business purchases goods from a supplier, ownership passes from seller to buyer, but when do title to and the risk associated with the goods pass from one party to the other? The answer is important because any number of things can happen to the merchandise after a customer orders it but before a company delivers it. When entrepreneurs order merchandise and supplies from their vendors, they should know when the ownership of the merchandise—and the risk associated with it—shifts from supplier to buyer.

Title

Before the Uniform Commercial Code (UCC) was enacted, the concept of title—the right to ownership of goods—determined where responsibility for merchandise fell. Today, however, three other concepts, identification, risk of loss, and insurable interest, play a more important role when disputes arise over lost, damaged, or destroyed shipments of goods.

IDENTIFICATION Identification is important because it gives the buyer an insurable interest in the goods. Before title can pass to the buyer, the goods must already be in existence and must be identifiable from all other similar goods. Specific goods already in existence are identified at the time the sales contract is made. For example, if Graphtech, Inc., orders a laser printer, the goods are identified at the time the contract is made. Fungible goods, those that cannot be separated from a larger mass (e.g., wheat in a silo), are identified when they are marked, shipped, or otherwise designated as the goods in the contract. For example, an order of oil may not be identified until it is loaded into a transfer truck for shipment.

Risk of Loss

Risk of loss determines which party incurs the financial risk if the goods are damaged, destroyed, or lost while in transit. Risk of loss does *not* always pass with title. Three particular rules govern the passage of title and the transfer of risk or loss:

Rule 1: Agreement. A supplier and a small business owner can agree (preferably in writing) to shift the risk of loss at any time during the transaction. In other words, any explicit agreement between buyer and seller determines when risk of loss passes to the buyer. For example, if an entrepreneur whose business is located in Freeport, Maine, orders goods from a vendor in St. Louis, Missouri, the contract may specify that risk of loss transfers from seller to buyer as soon as the truck carrying the goods crosses the Missouri border.

Rule 2: F.O.B. Seller. Under a sales contract designated F.O.B. ("free on board") seller, title and risk of loss pass to the buyer as soon as the seller delivers the goods into the care of a carrier or shipper. In addition, an **F.O.B. seller contract** (also known as a **shipment contract**) requires that the buyer pay all shipping and transportation costs. For example, a North Carolina manufacturer sells 100,000 capacitors to a buyer in Ohio with terms "F.O.B. North Carolina." Under this contract, the Ohio firm (buyer) pays all shipping costs, and title and risk of loss pass from the manufacturer as soon as the carrier takes possession of the shipment. If the goods are lost or damaged in transit, the *buyer* suffers the loss. Of course, the buyer can purchase insurance (see insurable interest below) and has legal recourse against the carrier if the carrier is at fault. If a contract is silent on shipping terms, the courts assume that the contract is a shipment contract (F.O.B. seller), and the buyer bears the risk of loss while the goods are in transit.

Rule 3: F.O.B. Buyer. A sales contract designated F.O.B. buyer requires that the seller deliver the goods to the buyer's place of business (or to a place that the buyer designates,

such as a warehouse). Risk of loss transfers to the buyer when the carrier delivers the goods to the buyer's business or to a destination that the buyer designates. In addition, an **F.O.B. buyer contract** (also called a **destination contract**) requires the seller to pay all shipping and transportation costs. In the example above, if the contract were "F.O.B. Ohio," the North Carolina manufacturer would pay the cost of shipping the order, and title and risk of loss would pass to the Ohio company only when the shipment is delivered to its place of business. In this case, the seller would bear any losses due to goods that are lost or damaged in transit.

Breaching a contract influences when risk of loss transfers from seller to buyer. If a seller ships goods that do not comply with the terms of the contract (e.g., they are defective, the wrong color, or the incorrect size or create some other problem), risk of loss does not pass to the buyer until the seller "cures" the problem with the goods or the buyer accepts the goods despite their problems. If the buyer breaches the contract, risk of loss shifts immediately from the seller to the buyer.

Insurable Interest

Insurable interest ensures the right of either party to the sales contract to obtain insurance to protect against lost, damaged, or destroyed merchandise as long as that party has "sufficient interest" in the goods. In general, once goods are identified, the buyer has an insurable interest in them. The seller has a sufficient interest as long as the seller retains title to the goods. However, under certain circumstances, both the buyer and the seller have insurable interests even after title has passed to the buyer.

 Entrepreneurship in Action

Who Bears the Loss for a Shipment of Missing Watches?

On March 4, Pedro Pestana, a resident of Chetumal, Mexico, entered into a contract in which he agreed to purchase from the Karinol Corporation, a company based in Miami, Florida, 64 watches for $6,006. An employee of Karinol Corporation wrote the contract in Spanish, and at the bottom of the agreement was a notation that said, "Please send the merchandise in cardboard boxes duly strapped with metal bands via air parcel post to Chetumal. Documents to Banco de Commercio De Quintano Roo S.A." The contract contained no provisions for allocating the risk of loss on the goods sold while they were in transit; it also did not include any specific shipping terms (such as F.O.B., F.A.S., C.I.F., or others). Pestana paid Karinol a 25 percent deposit on the watches prior to their shipment.

On April 11, a Karinol employee took the watches, which were packaged in two cardboard cartons, to the freight forwarding company that Karinol typically used to make international shipments, American International Freight Forwarders. Karinol also purchased insurance on the watches from the Fidelity & Casualty Company of New York. An employee of American International Freight Forwarders put metal straps on the two cartons of watches and delivered the packages to TACA International Airlines for shipment to Belize City, Belize, where Bernard Smith, an agent for Pestana, was to pick them up and take them to Pestana in Chetumal.

On April 15, the packages arrived in Belize City. TACA International Airlines placed them in storage in a customs warehouse

and notified Smith that the packages were available for pickup. On May 2, Smith picked up the packages, which were still bound by the metal straps specified in the original contract, but when he opened them for customs officials to inspect, the watches were missing. Pestana contacted Karinol Corporation about the missing watches, but a manager at Karinol told Pestana that Karinol had no liability for the lost watches and that Pestana must bear the loss and requested payment of the remaining $4,504 balance due. Pestana filed a lawsuit against Karinol Corporation, claiming that the watches were lost or stolen while the packages were in Karinol's care. He also cited the notation at the bottom of the contract that required Karinol to ship the watches "to Chetumal," which Karinol failed to do. Therefore, Pestana claimed, Karinol Corporation should bear the loss of the watches.

1. What mistakes did Karinol Corporation and Pedro Pestana make when they created the contract for the sale of the watches?

2. In the absence of an agreement on risk of loss and specific shipping terms in the parties' contract, what kind of contract will the court rule this is?

3. Who will prevail in this case? What could the losing party have done to protect against the loss or theft of the watches?

Source: Based on *Pestana v. Karinol Corporation*, 367 So.2d 1096 (1979), District Court of Appeal of Florida, Third District.

Receiving Merchandise

Once the merchandise is received, the buyer must verify its identity and condition. When the goods are delivered, entrepreneurs should check the number of cartons unloaded against the carrier's delivery receipt to make sure that the shipment is complete. It is also a good idea to examine the boxes for damage; if shipping cartons are damaged, the carrier should note the damage on the delivery receipt. Entrepreneurs should open all cartons immediately after delivery, inspect the merchandise for quality and condition, and check it against the invoices for discrepancies. If merchandise is damaged or faulty, the buyer should contact the supplier immediately and follow up with a written report. Entrepreneurs should never destroy or dispose of damaged or flawed merchandise unless the supplier specifically authorizes it. Proper control techniques in receiving merchandise prevent the small business owner from paying for suppliers' and shippers' mistakes.

Selling on Consignment

Entrepreneurs who lack the necessary capital to invest or are unwilling to assume the risk of investing in inventory may be able to sell goods on consignment. Selling on **consignment** means that the entrepreneur does not actually purchase the merchandise from the supplier (called the consignor); instead, the entrepreneur pays the consignor only for the merchandise actually sold. In exchange for providing the supplier with a marketplace to sell his or her goods, the entrepreneur typically receives a percentage of the revenue on each item sold. The entrepreneur (called the **consignee**) may return any unsold merchandise to the supplier (the **consignor**) without obligation. Under a consignment agreement, title and risk of loss do not pass to the consignee unless the contract specifies these terms. In other words, the supplier bears the financial costs of lost, damaged, or stolen merchandise. Entrepreneurs who sell merchandise on consignment realize the following advantages:

- The entrepreneur does not have to invest money in these inventory items, but the merchandise is on hand and available for sale.
- The entrepreneur does not pay the consignor until the item is sold.
- Because the consignment relationship is founded on the law of agency, the consignee never takes title to the merchandise and does not bear the risk of loss for the goods.
- The supplier normally plans and sets up displays for the merchandise and is responsible for maintaining them.

Before selling items on consignment, the entrepreneur and the supplier should create a written contract that should include the following items:

- A list of items to be sold and their quantities
- Prices to be charged
- Location of the merchandise in store
- Duration of contract
- Commission charged by the consignee
- Policy on defective items and rejects
- Schedule for payments to consignor
- Delivery terms and merchandise storage requirements
- Responsibility for items lost to pilferage and shoplifting
- Provision for terminating consignment contract

If managed properly, selling goods on consignment can be beneficial to both the consignor and the consignee.

Chapter Review

1. Understand the components of a purchasing plan.
   - The purchasing function is vital to every small business's success because it influences a company's ability to sell quality goods and services at reasonable prices. Purchasing is the acquisition of needed materials, supplies, services, and equipment of the right quality, in the proper quantities, for reasonable prices, at the appropriate time, and from the right suppliers.

2. Explain the principles of total quality management, including Lean, 5S, and Six Sigma and their impact on quality.
   - Under the total quality management (TQM) philosophy, companies define a quality product as one that conforms to predetermined standards that satisfy customers' demands. The goal is to get delivery and invoicing to installation and follow-up—right the first time.
   - To implement TQM successfully, a small business owner must rely on 10 fundamental principles: shift from a management-driven culture to a participative, team-based one; modify the reward system to encourage teamwork and innovation; train workers constantly to give them the tools they need to produce quality and to upgrade the company's knowledge base; train employees to measure quality with the tools of statistical process control; use Pareto's Law to focus TQM efforts; share information with everyone in the organization; focus quality improvements on astonishing the customer; don't rely on inspection to produce quality products and services; avoid using TQM to place blame on those who make mistakes; and strive for continuous improvement in processes as well as in products and services.
   - Like TQM, Six Sigma relies on quantitative tools to improve quality. At its core is the DMAIC process: **D**efine the quality problem, **M**easure current process performance to establish a baseline, **A**nalyze the process and identify potential root causes of the problem, **I**mprove the system by implementing solutions to remove the root causes, and **C**ontrol the new process by establishing standards of measurement and procedure.

3. Conduct economic order quantity analysis to determine the proper level of inventory.
   - A major goal of the small business is to generate adequate inventory turnover by purchasing proper quantities of merchandise. A useful device for computing the proper quantity is economic order quantity (EOQ) analysis, which yields the ideal order quantity, the amount that minimizes total inventory costs. Total inventory costs consist of the cost of the units, holding (carrying) costs, and ordering (setup) costs. EOQ analysis balances the costs of ordering and of carrying merchandise to yield minimum total inventory cost.

4. Differentiate among the three types of purchase discounts vendors offer.
   - Trade discounts are established on a graduated scale and depend on a small firm's position in the channel of distribution.
   - Quantity discounts are designed to encourage businesses to order large quantities of merchandise and supplies.
   - Cash discounts are offered to customers as an incentive to pay for merchandise promptly.

5. Calculate a company's reorder point.
   - There is a time gap between the placing of an order and actual receipt of the goods. The reorder point model tells the owner when to place an order to replenish the company's inventory.

6. Develop a vendor rating scale.
   - Creating a vendor analysis model involves four steps: determine the important criteria (i.e., price, quality, prompt delivery, service, and so on), assign a weight to each criterion to reflect its relative importance, develop a grading scale for each criterion, and compute a weighted score for each vendor.

7. Describe the legal implications of the purchasing function.
   - Important legal issues involving purchasing goods involve title, or ownership of the goods; identification of the goods, risk of loss, and when it shifts from seller to buyer; and insurable interests in the goods. Both the buyer and the seller can have an insurable interest in the same goods at the same time.

Discussion Questions

19-1. What is SCM? Why is it important to entrepreneurs?

19-2. Explain the elements of a purchasing plan.

19-3. Explain Lean and 5S principles. What is Six Sigma? What is TQM? How can these tools help entrepreneurs achieve the quality goods and services they require?

19-4. One top manager claims that to implement TQM successfully, "You have to change your company culture as much as your processes." Do you agree? Explain.

19-5. Visit the Web site of the National Institute of Standards and Technology (*www.nist.gov/baldrige/*), the organization that grants the Malcolm Baldrige National Quality Award, the highest quality award in the United States. Research one of the companies that received the Baldrige Award and prepare a one-page summary of its quality initiative and the results that it produced.

19-6. List and briefly describe the three components of total inventory costs.

19-7. What is the EOQ? How does it minimize total inventory costs?

19-8. Should a small business owner always purchase the products with the lowest prices? Why or why not?

19-9. Briefly outline the three types of purchase discounts. Under what circumstances is each the best choice?

19-10. What is lead time? Outline the procedure for determining a product's reorder point.

19-11. Explain how an entrepreneur launching a company could locate suppliers and vendors.

19-12. What factors are commonly used to evaluate suppliers?

19-13. Explain the procedure for developing a vendor rating scale.

19-14. Explain briefly the three concepts that have replaced the concept of title. When do title and risk of loss shift under an F.O.B. seller contract? An F.O.B. buyer contract?

19-15. What should a small business owner do when merchandise is received?

19-16. Explain how a small business would sell goods on consignment. What should be included in a consignment contract?

Managing Inventory

If a product isn't selling, I want to get it out of there because it's taking up space that can be devoted to another part of my line that moves. Besides, having a product languish on the shelves doesn't do much for our image.

—Norman Melnick, chairman of Pentech International

Honesty pays, but it doesn't seem to pay enough to suit some people.

—F. M. Hubbard

Supply chain management and inventory control are closely linked. The previous chapter focused on managing a company's supply chain—purchasing the correct materials, of the proper quality, in the correct quantity, at the best price, from the best vendors. This chapter will continue that process by discussing various inventory control methods, how to move "slow" inventory items, and how to protect inventory from theft. An entrepreneur's goal is to maximize the value of a company's inventory while reducing both the cost and the risks of owning inventory. The issue is significant; the largest expenditure for many small businesses is in inventory (see Figure 20.1).

Although many business owners understand the dangers of carrying excess inventory, they face a constant battle to avoid the problem. A study by the Aberdeen Group reports that 70 percent of businesses rate themselves "average" or "below average" on inventory management.[1] In addition, 91 percent of them report that they are making changes to their businesses, such as improving sales forecasting, enhancing replenishment strategies, and increasing the level of cooperation with vendors to improve their ability to manage inventory more effectively.[2] For years, businesses maintained high levels of inventory so that their manufacturing or sales processes ran smoothly. Managers now realize that inventory simply masks other problems that a company may have, such as poor quality, sloppy supply chain management, improper pricing, inadequate marketing, inefficient layout, low productivity, and others. Reducing the amount of inventory a company carries exposes these otherwise hidden problems; only then can managers and employees solve them.

As you learned in earlier chapters, excess inventory ties up a company's cash unnecessarily, putting it at greater risk of failure. "Inventory is money sitting around in another form," explains one business writer. "You can't use the money tied up in inventory to grow your business. You can't spend that money on marketing or hiring a new salesperson, and you can't use it to pay down your line of credit."[3] Businesses that carry excess inventory also incur high inventory carrying costs. American businesses incur $418 billion in carrying costs each year (20.3 percent of the total value

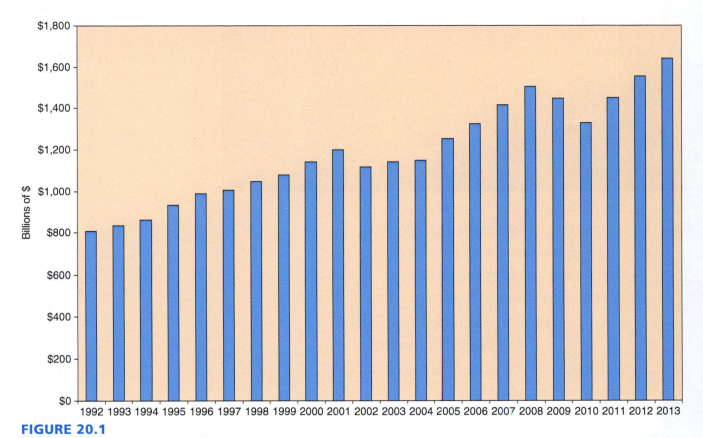

FIGURE 20.1

Total U.S. Business Inventories, 1992–2013 (in Billions of $)

Source: Based on "Manufacturing and Trade Inventories and Sales," U.S. Census Bureau, *www.census.gov/mtis.*

of the inventory they hold), the majority of it in the form of taxes, depreciation, insurance, and obsolescence.[4] Holding excess inventory has serious consequences for a company, including renting or purchasing additional warehouse space, higher labor costs, increased borrowing, greater risk of being stuck with obsolescent merchandise, and tying up a company's valuable cash unnecessarily. Companies with lean inventory levels lower their costs of operation, and those savings go straight to the bottom line. The potential payoff for managing inventory efficiently is huge; companies that switch to lean inventory systems can increase their profitability by 20 to 50 percent.

ENTREPRENEURIAL PROFILE: Li Ning: Li Ning Company Li Ning Company, a business based in Beijing and started by its namesake, former Chinese Olympic gymnast Li Ning, incurred significant losses after the company built up significant stocks of inventory in anticipation of strong demand for the sportswear and athletic shoes it sells under several brand names after China hosted the summer Olympic Games. Unfortunately, because of intense competition from industry giants Nike and Adidas and from companies such as Gap and H&M that have entered the sportswear market, its stores were overflowing with inventory, and Li Ning was forced to offer sizable discounts to move the excess merchandise. Even at the discounted prices, the company struggled to move the inventory off the shelves of its more than 6,400 stores, which prevented it from being able to introduce new styles and products (a key to success in the sportswear industry) to customers, further harming its sales. As part of its revitalization effort, the company refocused on its core Li Ning brand and reformulated its inventory management process to increase the number of times it turns its inventory and to replenish inventory in its stores faster. Li Ning also signed marketing agreements with the Chinese Basketball Association and with National Basketball Association star Dwayne Wade to create a Wade-endorsed "Dynasty" sportswear line.[5]

The information age has made techniques such as just-in-time inventory systems available to even the smallest of businesses. Internet-based networks that connect a company seamlessly with its suppliers have dramatically reduced the time for needed parts or material to arrive and the need to hold inventory. At the other end of the pipeline, a company's customers expect to have what they need when they need it. In today's competitive market, few customers will wait beyond a reasonable time for items they want. Managing inventory properly requires business owners to master an intricate balancing act, keeping enough inventory on hand to meet customers' expectations but maintaining inventory levels low enough to avoid excessive costs.

Managing inventory effectively requires an entrepreneur to implement the following seven interrelated steps:

1. *Develop an accurate sales forecast.* The proper inventory level for each item is directly related to customers' demand for that item. A business cannot sell merchandise that it does not have, and, conversely, an entrepreneur does not want to stock inventory that customers will not buy.

ENTREPRENEURIAL PROFILE: Dan Richardson: Northway Sports Dan Richardson, owner of Northway Sports, a retailer in East Bethel, Minnesota, that sells Polaris all-terrain vehicles (ATVs), snowmobiles, and motorcycles, says that he has lost several sales of Polaris's most popular ATV because the manufacturer had not been able to deliver the units he ordered five months before. When a recession hit, Polaris, like most manufacturers, slashed production to keep its inventory lean and to avoid putting pressure on its dealers to purchase products that they could not sell. Polaris's forecasts showed sluggish sales. Several months later, however, surging demand for the company's products caught managers by surprise, and Polaris was unable to fill its dealers' orders.[6]

Economic uncertainty and volatility make forecasting sales difficult, but top-performing companies make accurate forecasting a priority. A study by the Aberdeen Group reports that 75 percent of "best-in-class" companies create accurate demand forecasts versus only 26 percent of "laggard" companies. The study shows that these top-performing companies demonstrate an average forecast accuracy level of 82.3 percent three months into the future.[7] Best-in-class companies rely on **multiechelon inventory management**, a technique that views a company's supply chain holistically and balances its inventory levels against desired customer service levels, lead times, suppliers' production capacity, forecasted and actual customer demand, seasonality, and other factors across all product lines and at every level of the supply chain. Table 20.1 explains the differences between the traditional approach to inventory management and the multiechelon approach.

TABLE 20.1 Multiechelon Versus Traditional Inventory Policy Setting

	Traditional Approaches	Multiechelon Approaches
Target setting	Independently or sequentially set inventory target for each location or item (set target at one location or level first, then the next, and so on).	Simultaneously optimize inventory targets across all stock keeping units (SKUs), echelons, and locations to meet global service level objective (e.g., customer fill rate).
Variability	Use a normal distribution of demand to describe variability.	Allow for different measures of forecast/demand variability, such as forecast error; also takes into account supply variability and replenishment frequency impact on overall variability, often modeling actual distribution probabilities.
Inventory interdependencies	Assign inventory levels without regard to upstream or downstream inventory.	When setting inventory targets, take into account postponement opportunities and upstream/downstream inventory risk pooling.
Mix optimization	Single echelon inventory calculations assign each individual SKU location a target based on the factors in the calculation versus rules of thumb (e.g., 21 days of supply) being applied across whole production groups.	Multiechelon inventory mix optimization sets individual SKU-location targets versus broad rules of thumb, accounting for more factors in the calculation. It does this simultaneously for the different forms of inventory across time.
Goal-based optimization	Set inventory policies to meet service target defined for individual level. Typically only for finished goods and unit based.	Set inventory policies to meet global service-level objective with minimum total supply chain inventory investment.
Calculations	Calculations take into account expected demand, forecast error, and lead times for one point in the supply chain.	Optimization takes into account expected demand, forecast error, finished good service level, service times, inventory policies, cycle stock impacts, time-phased yield factors, production or distribution capacity constraints; production and handling lead times and lead time variability by location; transportation lead times and deviations by lane; handling/manufacturing/transportation cost factors; and so on.
Time sensitivity	Static inventory targets are changed when there is a shift in a product's life cycle stage.	Time-varying inventory targets account for seasonality and other time variations in demand, supply, or capacity; timing of inventory levels are coordinated across time between raw materials, and work in process through finished goods.
Demand propagation	Traditional approaches have no formal mechanism for coordinating the demand upstream for proper target setting.	Demand and demand/forecast uncertainty are propagated upstream so that inventory can be positioned properly throughout the supply chain.
Multilevel cost accumulation	Traditional approaches often do not include cost as a part of their calculation and cannot optimize based on multilevel costs.	By accounting for value added at each step along the supply chain, the multiechelon optimization can position inventory before expensive costs are added into an item while still satisfying service-level goals.
Service time coordination	No coordination of service times across the supply chain.	Coordination of service times in multi-level bills of materials and distribution routing so that excess inventory is not waiting for other long lead time components unnecessarily.

Source: Based on Aberdeen Group, December 2010, *www.aberdeen.com.*

2. ***Develop a plan to make inventory available when and where customers want it.*** Inventory will not sell if customers have a difficult time finding it. If a company is constantly running out of items customers expect to find, its customer base will dwindle over time as shoppers look elsewhere for those items. An important component of superior customer service is making sure that adequate quantities of items are available when customers want them. Two ways of measuring this aspect of customer service include calculating the percentage of customer orders that a company ships on time and the percentage of the dollar volume of orders that it ships on time. Tracking these numbers over time gives business owners sound feedback on how well they are managing their inventory levels from the customer's perspective. Polaris, the maker of all-terrain vehicles, snowmobiles, and motorcycles, once used a system that required dealers to place large orders of inventory twice a year. However, when sales slowed, the "stack 'em high, watch 'em fly" system caused problems with dealers and at Polaris. Today Polaris has switched to a system that allows

dealers to order smaller quantities of inventory more frequently, which allows them to maintain leaner inventories and match supply with customer demand more effectively.[8]

3. ***Build relationships with your most critical suppliers to ensure that you can get the merchandise you need when you need it.*** Business owners must keep suppliers and vendors aware of how their merchandise is selling and communicate their needs to them. Vendors and suppliers can be an entrepreneur's greatest allies in managing inventory. Increasingly, the word that describes the relationship between world-class companies and their suppliers is *partnership*.

4. ***Set realistic inventory turnover objectives.*** Keeping in touch with their customers' likes and dislikes and monitoring their inventory enable owners to estimate the most likely buying patterns for different types of merchandise. As you learned in Chapter 7, one of the factors having the greatest impact on a company's sales, cash flow, and ultimate success is its inventory turnover ratio.

5. ***Compute the actual cost of carrying inventory.*** Many business owners do not realize how expensive carrying inventory actually is. Recall that the average carrying cost for all businesses is 20.3 percent of the value of their inventory. Without an accurate cost of carrying inventory, it is impossible to determine an optimal inventory level. Carrying costs include items such as interest on borrowed money, insurance expenses associated with the inventory, inventory-related personnel expenses, obsolescence, and others. When new product introductions make existing products obsolete, companies must hold inventory to an absolute minimum. For instance, in the computer industry, the onrush of new technology causes the value of a personal computer held in inventory to decline 1 percent each week! This gives computer makers big incentives to keep their inventories as lean as possible.

ENTREPRENEURIAL PROFILE: George Falzon: G. Falzon & Company George Falzon, owner of G. Falzon & Company, a small jewelry store in Holliston, Massachusetts, watched the cost of holding inventory, particularly engagement rings and wedding bands, increase as the prices of precious metals soared. A weak economy demanded that Falzon avoid overinvesting in inventory, which is a common cause of cash crises, but potential customers expect to see a wide selection of items when shopping. Falzon developed a clever solution: He began stocking replicas of jewelry pieces made from plated silver and cubic zirconium rather than platinum, silver, gold, and diamonds. Because most engagement and wedding pieces are special orders, customers do not mind looking at the replicas. Falzon says that stocking the replicas saves him about $75,000 in inventory, allowing him to display four times as many styles of engagement rings and wedding bands.[9]

6. ***Use the most timely and accurate information system the business can afford to provide the facts and figures necessary to make critical inventory decisions.*** Modern inventory control systems and point-of-sale terminals that are linked to a company's inventory records enable business owners to know exactly which items are selling and which are not. The owner of a chain of baby products stores uses a computer network to link all of his stores to the computer at central headquarters. Every night, after the stores close, the point-of-sale terminals in each store download the day's sales to the central computer, which compiles an extensive sales and inventory report. When he walks into his office every morning, the owner reviews the report and knows exactly which items are moving fast, which are moving slowly, and which are not selling at all. He credits the system with the company's above-average inventory turnover ratio and much of its success.

7. ***Teach employees how inventory control systems work so that they can contribute to managing the firm's inventory on a daily basis.*** All too often, the employees on the floor have no idea of how the various information systems and inventory control techniques operate or interact with one another. Consequently, the people closest to the inventory contribute little to controlling it. Well-trained employees armed with information can be one of an entrepreneur's greatest weapons in the battle to control inventory. Fast changes in product demand require inventory control systems that are capable of responding quickly, allowing entrepreneurs and their employees to make adjustments to inventory levels in real time.

The goal is to find and maintain the proper balance between the cost of holding inventory and the requirement to have merchandise on hand when customers demand it. Either extreme can be costly. If entrepreneurs focus solely on minimizing cost, they will undoubtedly incur

stockouts, lost sales, and customer ill will because they cannot satisfy their customers' needs. For instance, researchers studying inventory control systems at Bulgari, a jewelry manufacturer headquartered in Rome, Italy, discovered that stockouts of just one popular item had lowered the company's profits by 5 percent of sales.[10] At the other extreme, entrepreneurs who attempt to hold enough inventory to meet every peak customer demand find that high inventory costs diminish their chances of remaining profitable. "There's a fine line between not having racks of clearance [merchandise] from leftover inventory and cutting to the bone," says one retail expert.[11] Walking this inventory tightrope is never easy, but the following inventory control systems can help business owners strike a reasonable balance between the two extremes.

Inventory Control Systems

Regardless of the type of inventory control system business owners choose, they must recognize the importance of **Pareto's Law** (or the **80/20 rule**), which holds that about 80 percent of a company's sales revenue is generated by 20 percent of the items in its inventory. Sometimes a company's best-selling items are its highest-priced items, but more often they are low-priced items that sell in high volume. Because most sales are generated by a small percentage of items, entrepreneurs should focus the majority of their inventory control efforts on this 20 percent. Observing this simple principle ensures that entrepreneurs will spend time controlling only the most productive—and, therefore, most valuable—inventory items. With this technique in mind, we now examine three basic types of inventory control systems: perpetual, visual, and partial.

Perpetual Inventory Systems

Perpetual inventory systems are designed to maintain a running count of the items in inventory. Although a number of different perpetual inventory systems exist, they all have a common element: They all keep a continuous tally of each item added to or sold from the company's stock of merchandise. A manual system that uses a perpetual inventory sheet that includes fundamental product information, such as the item's name, stock number, description, economic order quantity (EOQ), and reorder point, can be time consuming to maintain and produces inaccurate results if employees add items to or take items out of stock without recording them. Fortunately, many companies offer perpetual inventory management software that connect with small companies' accounting and financial packages, such as QuickBooks and Sage 50, to create an integrated control system that is more accurate and more efficient to operate. Many of these software packages include automatic reorder point triggers that generate purchase orders for the appropriate quantities of goods to replenish the inventory.

ENTREPRENEURIAL PROFILE: Justin Bosshardt: The Mending Shed Started in 1971 in Orem, Utah, as a mom-and-pop repair shop, The Mending Shed has expanded into a repair service for almost any item and a brick-and-mortar and online wholesaler and retailer of replacement parts. "Our inventory includes parts for washers, dryers, dishwashers, hair dryers, curling irons, shavers, power wheels, power tools, lawn equipment, and the list goes on," says Justin Bosshardt, The Mending Shed's director of IT. The company uses QuickBooks, one of the most popular small business accounting software packages, but The Mending Shed had outgrown its inventory management capability. "We were constantly fighting shortages on inventory, which made customers upset when we had to put their orders on back order," says Bosshardt. "We also had inventory on the shelves that was not listed on our Web site because there was no way to keep track of it." After several months of screening, Bosshardt purchased Fishbowl Inventory, the most popular inventory management software for QuickBooks users. Fishbowl Inventory integrated seamlessly with QuickBooks, and its inventory management, order fulfillment, automatic reorder points, and automatic purchase order generation features enabled the company to reduce its average inventory level while increasing its order fulfillment and customer service levels. "Fishbowl Inventory has improved every aspect of our company in some way," says Bosshardt. The time required to prepare purchase orders decreased from six hours a day to only one hour a day. In addition, the time that workers spent picking orders from warehouse shelves decreased from two hours per 100 orders to just 15 minutes per 100 orders. "We invested a significant amount of money into Fishbowl Inventory, and it paid for itself in the first year," says Bosshardt.[12]

Many small retailers now use computerized **point-of-sale (POS) systems** that perform all of the functions of a traditional cash register and maintain and up-to-the minute inventory count. Although POS systems are not new (major retailers have been using them for more than 30 years), their affordable prices are. Not so long ago, most systems required large investments in hardware and software. Today, small business owners can set up cloud-based POS systems, such as Shopkeep, Kronos, or Revel Systems, on personal computers and tablets for as little as $1,000. Setting up POS systems on tablets allows entrepreneurs to create mobile cash registers that are capable of taking customers' orders anywhere in the store. To prevent customers from having to wait in checkout lines (a common customer complaint), some stores dispatch employees with POS-equipped tablets to process customers' purchases and e-mail them receipts.

Combining a POS system with Universal Product Code (bar code) labels and high-speed scanners or radio frequency identification tags gives a small business a state-of-the-art checkout system that feeds vital information into its inventory control system. These systems rely on an inventory database; as items are rung up on the register, product information is recorded and inventory balances are adjusted. Using the system, business owners can tell how quickly each item is selling and how many items are in stock at any time. In addition, inventory records are accurate and are always current. Modern POS systems also allow business owners to generate an array of inventory reports instantly. Entrepreneurs can slice and dice data in a multitude of ways, allowing them to determine which items are selling the fastest and which are moving the slowest. Timely reports such as these give entrepreneurs the ability to make sound decisions about scheduling advertising, running special promotions, offering discounts, and arranging store displays. Computerized POS systems also make it possible for entrepreneurs to use perpetual inventory systems for a large number of inventory items, a task that, if performed manually, would be virtually impossible.

ENTREPRENEURIAL PROFILE: David Steingard and Hugh Jackman: Laughing Man Coffee & Tea David Steingard, who cofounded Laughing Man Coffee & Tea with actor Hugh Jackman in 2011, was shopping for a POS system for their unique coffee and tea bar in New York City's Tribeca neighborhood but was shocked at the $15,000 to $30,000 price tag that traditional POS systems carried. Then Steingard discovered ShopKeep, a cloud-based POS system that runs on iPads and iPhones created by entrepreneur Jason Richelson after he could not find a suitable POS system for his small wine and food store. "We needed something that was affordable but had enough robust inventory features," says Steingard. ShopKeep provides a "back office" that allows entrepreneurs to enter inventory items, descriptions, and prices and through which they can generate customized and detailed inventory and sales reports. "It's a very easy interface," says Steingard. Because the system is cloud-based, entrepreneurs do not have to purchase servers. "You pay a monthly fee and then pay for your transactions through your merchant account," he says. At peak hours, Laughing Man employees take their iPads, greet customers, and take their orders both inside and outside the shop. "The iPad POS system helps move the line, but more important, it sends a message to our customers that we are continually working to make their coffee experience better," says Steingard.[13]

Laughing Man Marketplace.
Source: Aaron Showalter/Newscom.

VISUAL INVENTORY CONTROL SYSTEMS The most common method of controlling inventory in a small business is the **visual control system**, in which managers simply conduct periodic visual inspections to determine the quantity of various items they should order. This system suits businesses that stock a large number of low-value items with low dollar volume. Unfortunately, this method is also the least effective for ensuring accuracy and reliability. Oversights of key items often lead to stockouts and resulting lost sales. The biggest disadvantage of the visual control system is its inability to detect and to foresee shortages of inventory items.

In general, a visual inventory control system works best in firms in which daily sales are relatively consistent, the entrepreneur is closely involved with the inventory, the variety of merchandise is small, and items can be obtained quickly from vendors. For example, small firms dealing in perishable goods use visual control systems very successfully and rarely, if ever, rely on analytical inventory control tools. For these firms, shortages are less likely to occur under a visual system; when they do occur, they are not likely to create major problems. Entrepreneurs who rely on visual systems must be alert to shifts in customer buying patterns that alter required inventory levels.

In the Entrepreneurial Spotlight

An Ideal Inventory Solution

Tim Balsimo, founder of Quality Pet Products, a wholesale distributor of pet foods to veterinary clinics and retail pet stores in six states, is a 20-year veteran of the pet food industry. Before starting Quality Pet Products, Balsimo owned a successful chain of pet stores that he sold in 2001. Based in Woodbury, Minnesota, Quality Pet Products carries more than 500 different pet foods in inventory to serve more than 400 customers in Minnesota, North Dakota, South Dakota, Wisconsin, Iowa, and Nebraska. "We purchase everything from our manufacturing partners in truckloads," says Balsimo, who offers customers a 24-hour delivery guarantee on all orders placed before 10 A.M. "We have customers from Rapid City to Green Bay," he says, meaning that the company must manage its inventory carefully to be able to meet its 24-hour delivery guarantee 100 percent of the time. As a former retail pet store owner, Balsimo understands how important fast delivery is to his customers and that the guarantee is one important way of differentiating his company from the competition.

As the company's customer base grew and stretched into a broader geographic area, Balsimo began looking for ways to improve his company's efficiency, manage its inventory investment more effectively, and improve the level of customer service it provides. He began shopping for comprehensive inventory tracking software that would integrate seamlessly with QuickBooks, the popular accounting software package that Quality Pet Products uses. He also wanted to create a Web-based order-entry system for his customers that would funnel orders directly into the company's inventory control system. Finally, he wanted Quality Pet Products to earn certification from the American Institute of Baking (AIB), which requires a company to track every edible item that comes through its warehouse by lot number or bar code in case of a product recall. The software would have to keep track of every item in stock, pinpoint its location in the warehouse to maximize workers' efficiency when "picking" customers' orders, create bills of lading, track shipment weights and delivery dates, and perform other essential inventory control functions. Balsimo found software that would perform these individual functions but not as a comprehensive package. "What was missing was the ability to handle serial and lot numbers," he says, both of which were crucial to earning

AIB certification. "[AIB certification] is a safety standard," explains Balsimo. "We must have a process to track every single item that is edible in and out of the warehouse by lot number or bar code."

Balsimo found Fishbowl Inventory and conducted a test to see whether the software could meet the company's needs and measure up to his expectations. After just one month of testing, Balsimo decided to completely implement Fishbowl Inventory. "It integrated with QuickBooks so well that we were able to move our [product] information quickly to Fishbowl," he says. "It took only a week or two to fine-tune everything. We were able to write a custom ordering report that takes into account our previous sales history, quantities on hand, and 'what's on order' to develop our own ordering process based on our lead times."

According to Balsimo, Quality Pet Products's new inventory control system has produced impressive results at a very reasonable price. "We've been able to make things run as efficiently as possible from order taking right on through to our picking and shipping process," he says. "The time it saves in daily operations is the equivalent of two office people per day—16 hours per day. It saves time, allows us to do more with less staff, integrates into one piece of software, and prevents us from having to set up more workstations or servers." The system also enabled Quality Pet Products to establish a convenient, integrated, Web-based customer order-entry system and to achieve its goal of becoming AIB compliant. Balsimo also is pleased that the company now boasts a 0.001 percent inventory variance, compared to the 1 to 2 percent variance that the typical pet food distributor has. "For us, the investment certainly was worth it," he says.

1. What benefits has Quality Pet Products reaped from improving its ability to manage its inventory more accurately?

2. Go to your favorite search engine and conduct a search of the most common problems that small businesses face when managing their inventory. What steps can entrepreneurs take to avoid these problems?

Sources: Based on Christina DesMarais, "Choosing the Best Inventory Tracking Software," *Inc.*, May 24, 2011, *www.inc.com/articles/201105/best-inventory-tracking-software.html*; Fishbowl Case Studies: Quality Pet Products, Fishbowl, *http://marketplace.intuit.com/CaseStudyDL.asp?CaseStudyID=0x0000000000000012*.

Partial Inventory Control Systems

For small business owners with limited time and money, the most viable option for inventory management is a partial inventory control system. These systems rely on the validity of the Pareto's Law, the 80/20 rule. For example, if a small business carries 5,000 different items in stock, roughly 1,000 of them account for about 80 percent of the company's sales volume, another 1,500 items account for 15 percent of its sales, and the remaining 2,500 items account for only 5 percent of its sales. Experienced business owners focus their inventory control efforts on the 1,000 items that make up 80 percent of their companies' sales. Unfortunately, many owners seek to maintain tight control over the remaining 4,000 items, a frustrating and wasteful practice. Smart entrepreneurs design their inventory control systems with Pareto's Law in mind. One of the most popular partial inventory control systems is the ABC system.

THE ABC METHOD OF INVENTORY CONTROL Partial inventory systems such as the ABC method minimize the expense involved in analyzing, processing, and maintaining records, a substantial

cost of any inventory control system. The ABC method focuses control efforts on that small percentage of items that accounts for the majority of a company's sales. The typical **ABC system** divides a firm's inventory into three major categories:

A *items* account for a high dollar usage volume.

B *items* account for a moderate dollar usage volume.

C *items* account for low dollar usage volume.

The **dollar usage volume** of an item measures the relative importance of that item in a company's inventory. Note that value is *not* necessarily synonymous with high unit cost. In some instances, a high-cost item that generates only a small dollar volume can be classified as an A item. More frequently, however, A items are those that are low to moderate in cost and high volume by nature.

The initial step in establishing an ABC classification system is to compute the annual dollar usage volume for each product (or product category). **Annual dollar usage volume** is simply the cost per unit of an item multiplied by the annual quantity sold. For instance, the owner of a music supply house may find that she sold 190 pairs of a popular brand of speakers during the previous year. If the speakers cost $75 per unit, their annual dollar usage volume would be as follows:

$$190 \times \$75 = \$14,250$$

The next step is to arrange the products in descending order on the basis of their computed annual dollar usage volume. Once so arranged, they can be divided into appropriate classes by applying the following rule:

A *items*: roughly the top 15 percent of the items listed

B *items*: roughly the next 35 percent

C *items*: roughly the remaining 50 percent

For example, Florentina's small retail shop is interested in establishing an ABC inventory control system to lower losses from stockouts, theft, or other hazards. Florentina has computed the annual dollar usage volume for the store's merchandise inventory, as shown in Table 20.2. (For simplicity, we show only 12 inventory items.)

TABLE 20.2 Calculating Annual Dollar Usage Volume and an ABC Inventory Analysis for Florentina's

Item	Annual Dollar Usage Volume	% of Annual Dollar Usage
Paragon	$374,100	42.00
Excelsior	294,805	33.10
Avery	68,580	7.70
Bardeen	54,330	6.10
Berkeley	27,610	3.10
Tara	24,940	2.80
Cattell	11,578	1.30
Faraday	9,797	1.10
Humboldt	8,016	0.90
Mandel	7,125	0.08
Sabot	5,344	0.06
Wister	4,453	0.05
Total	$890,678	100.00

Classification	Items	Annual Dollar Usage	% of Total
A	Paragon, Excelsior	$668,905	75.1
B	Avery, Bardeen, Berkeley, Tara	175,460	19.7
C	Cattell, Faraday Humboldt		
	Mandel, Sabot, Wister	46,313	5.2
Total		$890,678	100.00

FIGURE 20.2

ABC Inventory Control

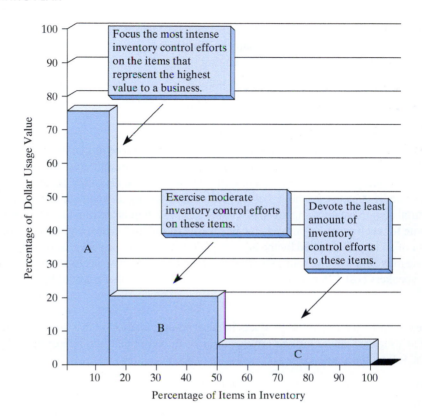

The ABC inventory control method divides the firm's inventory items into three classes depending on the items' value. Figure 20.2 graphically portrays the segmentation of the items listed in Table 20.2.

The purpose of classifying items according to their annual dollar usage volume is to establish the proper degree of control over each item held in inventory. Clearly, exercising the same level of control over C items and A items is wasteful and inefficient.

ENTREPRENEURIAL PROFILE: Marc Isaacson: Village Green Apothecary Marc Isaacson, CEO of Village Green Apothecary, a full-service pharmacy that Mickey Weinstein and Irv Rosenberg started in Bethesda, Maryland, in 1965, was frustrated because the business was consistently running out of its best-selling products. "We lost those particular sales, and in some cases, we lost the customers," says Isaacson. He estimates that the stockouts cost Village Green Apothecary at least 2 to 3 percent of its sales, caused employees to become frustrated because they could not fill customers' orders, and damaged the pharmacy's image as the "go-to" pharmacy among its customers. To solve the problem, Isaacson worked with his staff of 50 employees to conduct an ABC inventory analysis and identify Village Green's best-selling items, which turned out to be nearly 1,000 of the items it stocked. He drew attention to each of these "A" items by placing a stick-on blue dot on the shelf next to each one and established a safety stock for each one that is equal to three weeks of sales. Before conducting the ABC analysis, about 3 percent of Village Green's inventory was out of stock at any given time; after making the inventory control changes, that number dropped to just 0.5 percent.[14]

Entrepreneurs should control items in the A classification under a perpetual inventory system with as much detail as necessary. Analytical tools and frequent counts may be required to ensure accuracy, but the extra cost of tight control for these valuable items is justified. Entrepreneurs should not retain a large supply of reserve or safety stock because doing so ties up excessive amounts of money in inventory, but they must monitor the stock closely to avoid stockouts and the lost sales that result.

Control of B items should rely more on periodic control systems and basic analytical tools, such as EOQ and reorder point analysis (recall the discussion in Chapter 17). Entrepreneurs can maintain moderate levels of safety stock for these items to guard against shortages and can afford monthly or even bimonthly merchandise inspections. Because B items are not as valuable to the business as A items, they require less rigorous control systems.

C items typically constitute a minor proportion of the small firm's inventory value and, as a result, require the least effort and expense to control. These items are usually large in number and small in total value. The most practical way to control them is to use uncomplicated records and procedures. Large levels of safety stock for these items are acceptable because the cost of carrying them is usually minimal. Substantial order sizes often enable the business to take advantage of quantity discounts without having to place frequent orders. The cost involved in using detailed record keeping and inventory control procedures greatly outweighs the advantages gleaned from strict control of C items.

One practical technique for maintaining control over C items simply is the **two-bin system**, which keeps two separate bins full of material. The first bin is used to fill customer orders, and the second bin is filled with enough safety stock to meet customer demand during the lead time. When the first bin is empty, the owner places an order with the vendor large enough to refill both bins. During the lead time for the order, the manager uses the safety stock in the second bin to fill customer demand.

When storage space or the type of item makes a two-bin system impractical, an entrepreneur can use a **tag system**. Based on the same principles as the two-bin system, which is suitable for many manufacturers, the tag system applies to most retail, wholesale, and service firms. Instead of placing enough inventory to meet customer demand during lead time into a separate bin, the owner marks this inventory level with a brightly colored tag. When the supply is drawn down to the tagged level, the owner reorders the merchandise. Figure 20.3 illustrates the two-bin and tag systems of controlling C items.

In summary, business owners minimize total inventory costs when they spend time and effort controlling items that represent the greatest inventory value. Some inventory items require strict, detailed control techniques; other items simply do not justify the additional cost of tight controls. Because of its practicality, the ABC inventory system is commonly used in industry. In addition, the technique is easily computerized, speeding up the analysis and lowering its cost. Table 20.3 summarizes the use of the ABC control system.

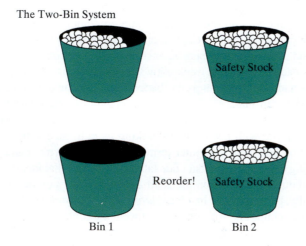

FIGURE 20.3

The Two-Bin and Tag Systems of Inventory Control

TABLE 20.3 ABC Inventory Control Features

Feature	A Items	B Items	C Items
Level of control	Monitor closely and maintain tight control.	Maintain moderate control.	Maintain loose control.
Reorder point	Based on forecasted requirements.	Based on EOQ calculations and past experience.	When level gets low, reorder.
Record keeping	Keep detailed records of receipts and disbursements.	Use periodic inspections and control procedures.	No records required.
Safety stock	Keep low levels of safety stock.	Keep moderate levels of safety stock.	Keep high levels of safety stock.
Inspection frequency	Monitor schedule changes frequently.	Check on changes in requirements periodically.	Make few checks on requirements.

Physical Inventory Count

Regardless of the type of inventory control system used, every small business owner must conduct a periodic physical inventory count. Even when a company uses a perpetual inventory system, the owner still must count the actual number of items on hand because errors will occur. A physical inventory count allows owners to reconcile the actual amount of inventory in stock with the amount reported through the inventory control system. These counts give managers a fresh start when determining the actual number of items on hand and enable them to evaluate the effectiveness and the accuracy of their inventory control systems.

The manual method of taking inventory involves two employees; one calls out the relevant information for each inventory item, and the other records the count on a tally sheet. There are two basic methods of conducting a physical inventory count. One alternative is to take inventory at regular intervals. Many businesses take inventory at the end of the year. In an attempt to minimize counting, many managers run special year-end inventory reduction sales. This **periodic count** generates the most accurate measurement of inventory. The other method of taking inventory, called **cycle counting**, involves counting a number of items on a continuous cycle. Instead of waiting until year-end to tally the entire inventory of items, an entrepreneur counts a few types of items each day or each week and checks the numbers against the inventory control system. Performing a series of "mini-counts" each day or each week allows for continuous correction of mistakes in inventory control systems and detects inventory problems faster than an annual count does.

Once again, technology can make the job of taking inventory much easier for small business owners. Cloud-based systems enable business owners to track their inventories and to place orders with vendors quickly and with few errors by linking them to their vendors electronically. These systems often rely on handheld computer terminals equipped with scanning devices. An employee runs the scanning device across a bar code label on the shelf that identifies the inventory item; then the employee counts the items on the shelf and enters that number using the number pad on the terminal. Finally, by linking the handheld terminal to a computer, the employee downloads the physical inventory count into the company's inventory control software in seconds.

In the past, suppliers simply manufactured a product, shipped it, and then sent the customer an invoice. To place an order, employees or managers periodically estimated how many units of a particular item they needed and when they needed them. Today, however, with cloud- or Web-based supply chain management systems, a vendor is tied directly into a company's POS system, monitoring it constantly; when the company's supply of a particular item drops to a preset level, the vendor automatically sends a shipment to replenish its stock to an established level. Information that once traveled by mail (or was never shared at all), such as inventory balances, unit sales, purchase orders, and invoices, now travel instantly between businesses and their suppliers. The result is a much more efficient system of purchasing, distribution, and inventory control.

Radio Frequency Identification Tags

Inventory control systems that use bar codes to track the movement of inventory through the supply chain have been around for years. Increasingly, businesses are replacing their bar code

systems with more flexible systems based on **radio frequency identification** (RFID) tags that are attached to individual items or to shipments and transmit data to a company's inventory management system. Each tag, which is about the size of a grain of sand, contains a tiny microchip that stores a unique electronic product code and a tiny antenna. Because the tags use short-range radio frequencies, they can transmit information under almost any condition, avoiding the line-of-sight restrictions that bar code systems experience. Once activated, the tags perform like talking bar codes and enable business owners to identify, count, and track the inventory items to which they are attached, providing them with highly accurate, real-time information constantly. When a shipment arrives at a warehouse or retail store, the RFID tags signal an inventory system reader, an object about the size of a coin that records the identity, quantity, and characteristics of each item now in stock. The reader relays the information to a central inventory control system so that business owners can have access to all of this information online. Some stores have installed "smart" shelves equipped with readers that detect the identity and quantity of the items placed on them. When a customer makes a purchase, the smart shelf sends a message to the inventory control system, telling it to reduce the number on hand by the number of items the customer buys. In essence, RFID technology allows business owners to locate and track an item at any point in the supply chain—from the raw material stage to the store shelf.

Retailers use RFID technology to make inventory counts a breeze. Employees simply walk the aisles of the store holding a special reader that scans the RFID tags of the items on the shelves. A study conducted by the RFID Research Center at the University of Arkansas comparing the effectiveness of bar code and RFID inventory counting systems in a retail setting shows that the RFID system improved the accuracy of inventory records by 27 percent over the bar code system. Not only was the RFID system more accurate, but the time required to complete the inventory count using the RFID system was significantly less as well, counting 10,000 items in 53 hours with the bar code reader compared to just two hours with the RFID system, a 96 percent reduction.[15] At some stores, fully integrated RFID systems allow cashiers to ring up customers' purchases by scanning the contents of an entire shopping cart in just seconds, minimizing the time that customers have to stand in checkout lines. The cost of RFID tags is declining and their reliability improving, meaning that more businesses will adopt the technology to improve the degree of control they have over their inventory.

ENTREPRENEURIAL PROFILE: Billabong The Billabong retail store in Iguatemi Alphaville mall in Barueri, Brazil, uses RFID technology to track inventory accurately, provide customers with detailed product information, and enhance customers' shopping experience. When a shipment of inventory arrives, employees simply wave a handheld reader near the box of products, which records the identity and the quantity of each item in the shipment and uploads the information to the store's computerized inventory system. The store's RFID system also allows customers to learn about product details by picking an item from a shelf and holding the tag near an RFID receiver that reads the information on the tag and uses a Wi-Fi connection to display more detailed information about the product. When customers take garments into the store's Smart Dressing Rooms, the system automatically reads the information on their tags and displays on a touch screen images of and information about coordinating products and their availability in various sizes and colors in the store. Customers can touch a prompt on the screen, sending a text message to a staff member's smart phone requesting any or all of the recommended items. When customers are ready to check out, an RFID receiver under the counter reads the information on each item's tag, deducts the items from the store's computerized inventory records, and totals their prices automatically. Finally, when customers leave the store, RFID readers capture the information from the tags on their purchases and check them against the store's inventory records. If someone attempts to leave the store with an item for which he or she has not paid, the system triggers an audible alarm and turns on a video camera to record the incident.[16]

As Billabong's experience suggests, the impact of RFID technology, which actually dates back to World War II, on inventory control is enormous. "This is an innovative technology similar to the Internet," says Mark Roberti, editor of *RFID Journal*. "You can now make any object smart."[17] International consulting firm McKinsey and Company estimates that once in use, RFID technology has the ability to increase companies' revenues by as much as 6 percent by improving the availability of items that customers want to buy and reducing the time and energy that staff spend looking for merchandise.[18]

Lessons from the Street-Smart Entrepreneur

Best Practices in Inventory Management

Many entrepreneurs have discovered the dangers of excess inventory. Not only does it tie up a company's valuable cash unnecessarily, but it also hides a host of other operating problems that a company has and must address. When it comes to managing inventory, small business owners often face four problems:

1. They have too much of some products.
2. They have too little of other products.
3. They don't know what they have in stock.
4. They know what they have in stock but cannot find it.

Addressing these four problems requires business owners to create a system of inventory management based on best practices. "Effective inventory management allows a distributor to meet or to exceed customers' expectations of product availability by maintaining the amount of each item that will also maximize their company's net profit," says one expert. The following tips from the Street-Smart Entrepreneur about inventory management best practices help accomplish that goal:

Recognize the difference between your company's "stock" and its "stuff." "Stock" is made up of the inventory that customers want and expect a company to have available. "Stuff" is everything else that is in the warehouse or stockroom and typically includes slow-moving items. The goal is to manage the stock in such a way that the company can meet customers' demand for items and make a profit and to get rid of everything else—the "stuff."

Set up an inventory management process that recognizes the value of your company's stock. Remember that Pareto's Law, the 80/20 rule, applies to many situations, particularly inventory control. About 20 percent of the items in a typical company's inventory account for about 80 percent of its sales. The idea is to set up a system that exercises the greatest degree of control over the most valuable 20 percent of the company's items. Using the principles of ABC inventory analysis ensures that a company is applying the proper level of control to each item in its inventory.

Focus on sales forecasting accuracy. One of the most important determinants of how well a company manages its inventory is how accurate its sales forecasts are. Using sales data and past experience allows entrepreneurs to build accurate forecasting models over time. Without an accurate sales forecast, a business suffers lost sales because of too little inventory available for some items and cash shortages because of too much inventory in stock of others. Experienced entrepreneurs also know that their sales follow seasonal patterns, and so should their inventory levels. Tracking forecast error and setting goals for improving it are key ingredients in managing inventory more effectively.

Work with vendors and suppliers to keep the inventory of essential items as lean as possible. Even companies that utilize just-in-time techniques find it necessary to carry inventory; however, they keep their levels of stock to a minimum. Look for suppliers that can meet your company's quality requirements and provide rapid deliveries on short notice. Sharing information with the members of your company's supply chain and connecting with them electronically are excellent ways to shorten the lead time on the items you order.

Use computerized inventory control systems to monitor your company's stock. Computerized inventory control systems that are linked to POS terminals allow entrepreneurs to know which items they have in stock at any time. The reports that these systems generate also help them know which items are selling best and which items are not selling at all. This information leads to improved inventory decisions in the future and allows entrepreneurs to adjust their buying decisions on the fly. Antolini Luigi & Company, a producer of granite, marble, and other stone located in Verona, Italy, uses an RFID-based system to track and control its inventory of uncut stone blocks and the 900,000 finished stone slabs that it sells in a typical year. With the RFID system, an employee walks up and down each aisle of the company's large stockyard carrying a handheld reader that collects the unique identification number from each block or each slab's tag to get an accurate inventory count. Before Antolini Luigi implemented the RFID system, taking inventory required two employees who spent two days manually identifying, counting, and logging each item. The new system allows one employee to do the same job in less than three hours. When the company sells a finished slab, employees use the handheld readers to collect the information from the RFID tag on the slab, which the system uses to update inventory records automatically. The owners of the family business, started in the 1950s by its namesake, estimate that the RFID system has reduced its labor costs associated with inventory control by 40 percent, increased the accuracy of its inventory records, and reduced its average investment in inventory.

Track the inventory metrics that are most important to your company's success. A business can improve only what it measures. Common metrics for inventory control include the inventory turnover ratio (for the business as a whole and for individual products or product lines), gross profit margin, age of inventory, measures of overstock, customer order fill rate, number of back orders, and many others. Modern inventory control systems allow businesses of all sizes to track these metrics for almost every product line. Know which metrics are most important to your business, establish goals for them, and track your company's progress toward achieving them.

Organize your warehouse or stockroom to make it easy to find the items you need. Organizing a warehouse or stockroom based on the knowledge of which items are in highest demand and which ones are seldom needed allows businesses to minimize the cost of filling orders. Placing the fastest-moving items in the most accessible location, preferably nearest the packing and shipping area, minimizes the time that employees spend walking around the warehouse or stockroom. Once again, computerized inventory control systems can help by printing the warehouse "address" of the items that employees must "pick" to fill orders.

Get rid of the "stuff." Eliminating unwanted or unnecessary inventory, especially SLOBs (slow-moving, obsolete

items), frees up valuable cash and simplifies the inventory management process. Possible ideas include the following:

- Reducing the price of the items to get rid of them.
- Offer incentives to sales people to sell slow-moving "stuff."
- Offer the items for sale at an online auction site.

Sources: Based on Mary Catherine O'Conner, "Italian Stone Company Carves Out Savings with RFID," *RFID Journal*, August 13, 2010, *www.rfidjournal.com/articles/view?7808*; Jon Schreibfeder, *The First Steps to Achieving Effective Inventory Control*, Microsoft Business Solutions, *http://download.microsoft.com/download/b/f/3/bf334d7f-ad07-458e-a716-fdf46a0cf63c/eimwp1_invcontrol.pdf*; "Inventory Best Practices," *The Bottom Line*, Manufacturing Extension Partnership, June 2003, pp. 1–2; Matt Gruhn, "16 Steps to Inventory Success," *Boating Industry White Paper*, June 2009, pp. 5–6; "Ten Common Inventory Mistakes and How to Avoid Them," Quality Digest, January 10, 2012, *www.qualitydigest.com/inside/quality-insider-article/ten-common-inventory-mistakes-and-how-avoid-them.html*.

Just-in-Time Inventory Control Techniques

Just-in-Time Techniques

Many U.S. businesses have turned to a popular inventory control technique called **just in time** (JIT) to reduce costly inventories and turn around their financial fortunes. Until recently, these companies had accepted the following long-standing principles of manufacturing: Long production runs of standard items are ideal, machines should be up and running as much as possible, machines must produce a large number of items to justify long setup times and high costs, similar processes should be consolidated into single departments, tasks should be highly specialized and simplified, and inventories (raw materials, work in process, and finished goods) should be large enough to avoid emergencies, such as supply interruptions, strikes, and breakdowns.

The JIT philosophy, however, views excess inventory as a blanket that masks problems and as a source of unnecessary costs that inhibit a company's competitive position. Tim Cook, CEO of Apple Inc., says that inventory is "fundamentally evil" because in the technology industry, inventory loses one to two percent of its value per week. "You want to manage [inventory] like you're in the dairy business," he says. "If it gets past its freshness date, you have a problem." By managing its inventory and supply chain so closely, Apple Inc. is able to keep just 5.3 days' worth of inventory on hand at any time, almost half that of its closest competitor, Dell, which keeps 10.2 days' worth of inventory on hand.[19] In contrast, Research in Motion, the company that makes the Blackberry smart phone and the PlayBook tablet, saw its inventory swell from $618 million to more than $1 billion in just one year—even after it had already written off $485 million worth of PlayBooks from its balance sheet. In an attempt to move its stockpile of inventory, the struggling company cut the PlayBook's price from $299 to $199, but its inventory levels and those of the Blackberry continued to increase.[20]

Under a JIT system, materials and inventory arrive at the appropriate location just in time instead of becoming part of a costly inventory stockpile. JIT is a philosophy that seeks to improve a company's efficiency. One key measure of efficiency is the level of inventory on hand; the lower the level of inventory, the more efficient is the production system. The heart of the JIT philosophy is eliminating waste in whatever form it may take—time wasted moving work in process from one part of a factory to another, money wasted when employees must scrap or rework an item because of poor quality, cash tied up unnecessarily in excess inventory because of a poorly designed process, and many others.

Companies using JIT successfully embrace a broader philosophy of continuous improvement ("kaizen"), which was discussed in the previous chapter. These companies encourage employees to find ways to improve processes by simplifying them, making them more efficient, and redesigning them to make them more flexible. A cornerstone of the JIT philosophy is making waste in a company visible. The idea is that hidden waste is easy to ignore; visible waste gives everyone an incentive to eliminate it. Managers at a small company that manufactures fabrics for use in the papermaking industry set off an area in the middle of the production floor and put all of

2.

Describe how just-in-time (JIT) and JIT II inventory control techniques work.

the wasted fabrics there on display. The not-so-subtle message was "help us find ways to reduce this waste." Within a matter of months, with the help of suggestions from both individuals and teams of employees, the pile of waste shrank dramatically.

In the past, only large companies could reap the benefits of computerized JIT and inventory control software, but now a proliferation of inexpensive programs gives small companies that ability. The most effective businesses know that what is required is not simply the technology but also the critical strategic alliances with suppliers who are themselves technologically sophisticated enough to interact on a real-time basis to deliver what is needed when it is needed. The ultimate goal is to drive excess inventory to as close to zero as possible.

Motorcycle maker Harley Davidson implemented a JIT system that reduced its investment in inventory and increased the number of times it turns its inventory each year.

Source: MCT/Getty Images.

ENTREPRENEURIAL PROFILE: Harley-Davidson For years, iconic motorcycle manufacturer Harley-Davidson relied on maintaining high levels of inventory to ensure that its assembly lines never shut down for lack of parts. Not only did the company's sizable investment in excess inventory tie up valuable cash, but it also covered up serious structural problems in the manufacturing process and in the overall business. "Parts were made in large batches for long production runs, stored until needed, then loaded onto the 3.5-mile conveyor that rattled endlessly around the plant," says Peter Reid, author of a book on Harley-Davidson's transformation. "Sometimes we couldn't even find the parts we needed, or if we found them they were rusted or damaged, or there had been an engineering change since the parts were made and they didn't even fit," admits Tom Gelb, the company's manager of operations. Then CEO Vaughn Beals converted Harley-Davidson's manufacturing plant in York, Pennsylvania, to a JIT inventory system. "With JIT, as each problem is exposed, you are forced to identify its cause, fix it, and move on to the next problem that is revealed," explains Reid. The results were impressive: Harley-Davidson's average inventory decreased by 75 percent, the space required to store inventory declined by 25 percent, the number of inventory turns increased from 5 to 20 per year, and the percentage of completed motorcycles coming off of the production line improved from 76 to 99 percent.[21]

Just as with Harley-Davidson, when JIT is successfully implemented, companies experience six positive results:

1. Lower investment in inventory
2. Reduced inventory carrying and handling costs
3. Reduced cost from obsolescence of inventory
4. Lower investment in space for inventories and production
5. Reduced total manufacturing costs from the better coordination needed between departments to operate at lower inventory levels
6. Higher inventory turnover ratios

Despite the many benefits that JIT systems offer, they do carry risks. Because the inventory that a company has in stock is so small, any disruption in its supply chain, even for inexpensive, common items, can cause the entire operation to come to a halt. After an earthquake and resulting tsunamis devastated the infrastructure in northern Japan and many factories that produce silicon used to make silicon chips, production at companies in many industries, ranging from computers and personal electronics to autos and appliances, was paralyzed for weeks. Because of the global and interconnected nature of modern supply chains, a disruption at any link in the chain, even though it occurs thousands of miles away, is likely to have a serious impact on a company's ability to provide its products and services.

For JIT systems to be most productive, entrepreneurs must consider the human component of the equation as well. Two elements are essential:

1. *Mutual trust and teamwork.* Managers and employees view each other as equals, have a commitment to the organization and its long-term effectiveness, and are willing to work as a team to find and solve problems.
2. *Empowerment.* Effective organizations provide their employees with the authority to take action to solve problems. The objective is to have the problems dealt with at the lowest level and as quickly as possible.

JIT is most effective in repetitive manufacturing operations where companies traditionally have relied on holding significant levels of inventory, where production requirements can be

forecast accurately, and where suppliers and customers work together as partners throughout the supply chain. Experience shows that companies with the following characteristics have the greatest success with JIT:

- Reliable deliveries of all parts and supplies
- Short distance between a company and its vendors
- Consistently high quality of vendors' products
- Stable and predictable product demand that allows for accurate production schedules

JUST-IN-TIME II TECHNIQUES In the past, some companies that adopted JIT techniques discovered an unwanted side effect: increased hostility resulting from the increased pressure they put on their suppliers to meet tight and often challenging schedules. To resolve that conflict, many businesses have turned to an extension of JIT, just-in-time II (JIT II), which focuses on creating a close, harmonious relationship with a company's suppliers so that both parties benefit from increased efficiency. Lance Dixon, who created the JIT II concept when he was a manager at Bose Corporation, a manufacturer of audio equipment, sought to create a working environment that empowered the supplier within the customer's organization. To work successfully, JIT II requires suppliers and their customers to share what was once closely guarded information in an environment of trust and cooperation. Under JIT II, customers and suppliers work hand in hand, acting more like partners than mere buyers and sellers.

In many businesses practicing JIT II, suppliers' employees work on-site at the customer's plant, factory, or warehouse almost as if they were employees of the customer. These on-site workers are responsible for monitoring, controlling, and ordering inventory from their own companies. While at Bose, Dixon decided to try JIT II because it offered the potential to reduce sharply the company's inventories of materials and components, to cut purchasing costs, and to generate cost-cutting design and production tips from suppliers who understood Bose's process. This new alliance between suppliers and their customers formed a new supply chain that lowered costs at every one of its links. To protect against leakage of confidential information, Dixon had all of the employees from Bose's suppliers who would work in its plant sign confidentiality agreements. Dixon also put a ceiling on the amount that each supplier's employee could order without previous authorization from Bose.

Manufacturers are not the only companies benefiting from JIT II. In a retail environment, the concept is more commonly known as **efficient consumer response (ECR)**, but the principles are the same. Rather than build inventories of merchandise that might sit for months before selling (or, worse, never sell at all), retailers that use efficient consumer response replenish their inventories constantly on an as-needed basis. Because vendors are linked electronically to the retailer's POS system, they can monitor the company's inventory and keep it stocked with the right merchandise mix in the right quantities. Both parties reduce the inventories they must carry and experience significant reductions in paperwork and ordering costs. JIT II works best when two companies transact a significant amount of business that involves many different parts or products. Still, maintaining trust is the biggest barrier the companies must overcome.

Turning Slow-Moving Inventory into Cash

Managing inventory effectively requires an entrepreneur to monitor the company's inventory turnover ratio and to compare it to those of other companies of similar size in the same industry. As you recall from Chapter 14, the inventory turnover ratio is calculated by dividing a company's cost of goods sold by its average inventory. This ratio expresses the number of times per year the business turns over its inventory. In most cases, the higher the inventory turnover ratio, the better a small company's financial position will be. A below-average inventory turnover ratio indicates that a company's inventory is stale and obsolete or that its inventory investment is too large.

Because of the variability in demand and the cyclical nature of the market, auto dealers often struggle to maintain adequate number of inventory turns and to keep the number of cars on their lots from ballooning, which drives up their operating costs. The longer a car sits on a dealer's lot, the greater is the cost of borrowing to pay for it (recall the discussion of floor planning in Chapter 17). Auto companies consider 50 to 60 days' worth of cars to be an adequate inventory (54 days is the

3.

Describe methods for reducing losses from slow-moving inventory.

average). When gasoline prices increase, however, dealers see large gas-guzzling SUVs languish on their lots, with inventories increasing to more than 100 days' worth, while inventories of fuel-efficient and hybrid models fall to record lows. After one spike in gas prices, dealers held large cars in inventory for an average of 84 days, but small cars stayed on their lots for an average of only 46 days. Inventories of certain popular models, particularly fuel-efficient hybrids, such as Toyota's Highlander SUV and Ford's Fusion, stood at just five days' worth.[22]

Slow-moving items carry a good chance of loss resulting from spoilage or obsolescence. Companies that sell trendy fashion merchandise or highly seasonal items often experience losses as a result of being stuck with unsold inventory for long periods of time. Some small business owners are reluctant to sell these slow-moving items by cutting prices, but it is much more profitable to dispose of this merchandise quickly at a lower profit margin than it is to hold it indefinitely in inventory.

ENTREPRENEURIAL PROFILE: Mickey Gee: Pants Store Sales during the all-important holiday season at the Pants Store, a small chain of clothing stores near Birmingham, Alabama, were slower than owner Mickey Gee expected, leaving him with 20 percent more merchandise than normal. Gee, whose father, Taylor, started the business in 1950, used markdowns, some as much as 80 percent, to turn the slow-moving merchandise into cash. He also invested in an inventory control system that helps him identify which brands and which items are selling best (and which are not) in his four locations.[23]

A business owner who postpones marking down stale merchandise, fearing that it will reduce profits and hoping that the goods will sell eventually at the regular price, is making a mistake. The longer the merchandise sits, the dimmer are the prospects of ever selling it, much less selling it at a profit. Pricing these items below regular price or even below cost is difficult, but it is much better than having valuable working capital tied up in unproductive assets.

The technique that Mickey Gee used, the markdown, is the most common technique for liquidating slow-moving merchandise. Not only is the markdown effective in eliminating slow-moving goods, but it also is a successful promotional tool. Advertising special prices on such merchandise helps a small business garner a larger clientele and contributes to establishing a favorable business image. Using special sales to promote slow-moving items helps create a functional program for turning over inventory more quickly. To get rid of a large supply of out-of-style neckties, one small business offered a "one-cent sale" to customers purchasing neckwear at the regular price. One retailer of electronic and sound equipment chooses an unusual holiday—Presidents' Day—to sponsor an all-out blitz, including special sales, prices, and promotions, to reduce its inventory. Other techniques that help eliminate slow-moving merchandise include the following:

- Creating middle-of-the-aisle display islands that attract customer attention
- Offering one-day-only sales
- Giving quantity discounts for volume purchases
- Creating bargain tables with a variety merchandise for customers to explore
- Using eye-catching lights and tickets marking sale merchandise
- Setting up an online store on eBay
- Using an inventory liquidation company to get rid of excess merchandise

As inventory control techniques become increasingly sophisticated and accurate, slow-moving inventory will never be "lost" in the supply chain. Aggressive methods of selling slower-moving inventory allow business owners to convert inventory into cash and to produce an acceptable inventory turnover ratio. The inventory management tools described in this chapter also play an important role in avoiding slow-moving merchandise. They highlight those items that are slow moving, enabling business owners to avoid the mistake of ordering them again. In effect, the information on the items that *aren't* selling influences entrepreneurs' decisions about the merchandise they order in the future as much as information on those items that *are* selling well. The ability to avoid slow-moving items in the first place means that business owners can invest their working capital more effectively and produce faster inventory turnover ratios, lower costs, and higher profits.

Protecting Inventory from Theft

Small companies are a big target for crime. Businesses worldwide lose 5 percent of their total sales, nearly $9.6 billion *per day*, to criminals, although the actual loss may be even greater because so many business crimes go unreported.[24] Whatever the actual loss is, its effect is staggering. If a company operates at a 5 percent net profit margin, it must generate an additional $20 in sales to offset every $1 lost to theft. Because small businesses often lack the sophistication to identify early on the illegal actions of employees or professional thieves and the controls to prevent theft and fraud, they are particularly vulnerable to theft and fraud. A study by the Association of Certified Fraud Examiners reports that small companies are the most frequent victims of theft and fraud and that the median loss for small companies is $147,000.[25] When a company has a small asset base, a loss from theft and fraud can be a crippling blow, threatening its very existence.

Many entrepreneurs believe that the primary sources of theft originate outside the business. In reality, most firms are victimized by their own employees.

4.

Discuss employee theft and shoplifting and how to prevent them.

Employee Theft

Ironically, the greatest criminal threat to small businesses comes from the *inside*. Employee theft accounts for the greatest proportion of the criminal losses that businesses suffer and costs companies $42.4 million per day.[26] One-third of all business bankruptcies are the result of employee theft.[27] Because employees have access to the inner workings of a business, they can inflict more damage than shoplifters. One study reports that dishonest employees steal on average 5.9 times more per incident than shoplifters ($665.77 vs. $113.30).[28] The types of employee theft schemes that affect small businesses are different from those that large companies encounter. Figure 20.4 shows the most common employee theft schemes in small businesses.

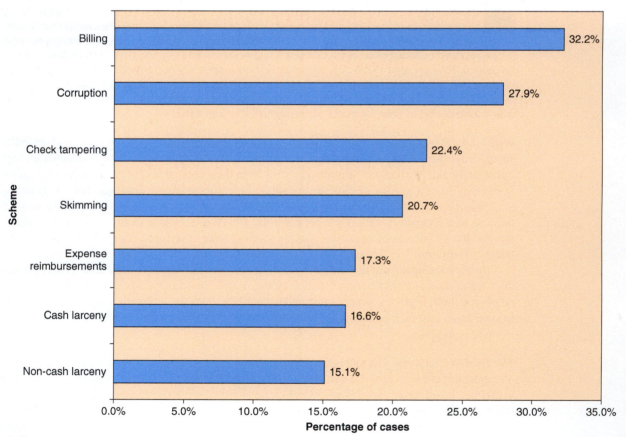

FIGURE 20.4

Employee Theft Schemes in Small Businesses

Source: Based on *2012 Report to the Nations on Occupational Fraud and Abuse*, Association of Certified Fraud Examiners, 2013, p. 27.

Unfortunately, employee theft is more prevalent than ever. Tim Dimoff, president of Mogadore, Ohio–based SACS Consulting & Investigative Services Inc., gives one reason for the increased prevalence. "I call the attitude employees take in the workplace 'entitlement,'" he says. "They justify in their minds that they are entitled to take things because they work so hard." Dimoff adds that some businesses all but encourage employee theft. How? By failing to file criminal charges against employees caught stealing. Only 65 percent of businesses that have been victimized by employee theft report the incident to law enforcement officials. The main reason business owners do not report thefts: They do not want the negative publicity that results from prosecuting employee thieves.[29] Others worry about the cost to the company to prosecute, how the time away from management will affect the organization, and the impact that the incident will have employee morale. Often it is easier just to ask the guilty employee to leave.[30]

The median length of time it takes employers to catch an employee who is stealing is 18 months, and managers usually discover the theft when another employee tips them off. (Managers detect just 14 percent of employee theft schemes in small companies.[31]) How can thefts go undetected for so long? Most thefts occur when employees take advantage of the opportunities to steal that small business owners unwittingly give them. Typically, small business owners are so busy building their companies that they rarely even consider the possibility of employee theft—until disaster strikes.

In addition, many small companies do not have adequate financial, audit, and security procedures in place. Fewer than 33 percent of small companies use internal audit teams as a loss prevention tool, compared to nearly 85 percent of large companies. Even though tips from employees are the most common way of discovering employee theft and fraud, only 20 percent of small companies have installed anti-fraud hotlines for employees to report suspicious activities.[32] Add to this mix of lax control procedures the high degree of trust that most small business owners place in their employees, and you have a perfect recipe for employee theft.

ENTREPRENEURIAL PROFILE: Ennis-Flint Ennis-Flint, a company based in Dallas, Texas, that makes road marking and pavement surface treatments, recently discovered that a longtime employee had embezzled more than $700,000. For more than four years, the employee regularly submitted fictitious invoices for Outlook Environmental & Safety Solutions, a sham business that he owned, for work that Outlook Environmental had supposedly performed. In reality, Outlook Environmental had done no work for Ennis-Flint and had no employees. "This defendant took advantage of and violated his employer's trust, using his position to embezzle money on a routine basis over a long period of time," says U.S. Attorney Sally Quillian Bates.[33]

WHAT CAUSES EMPLOYEE THEFT? Security experts estimate that 30 percent of workers pilfer small items from their employers and that 60 percent of employees will steal if given enough opportunity and motivation.[34] Employees steal from their companies for any number of reasons. Some may have a grudge against the company; others may have a drug, alcohol, or gambling addiction to support. Still others succumb to the temptation of an easy opportunity to steal because of a company's lack of proper controls. The Association of Certified Fraud Examiners reports that 87 percent of employees caught stealing have never been charged or convicted of a prior theft offense.[35]

Employees steal from the company for four reasons: need, greed, temptation, and opportunity. A business owner can control only temptation and opportunity. To minimize their losses to employee theft, business owners must understand how both the temptation and the opportunity to steal creep into their companies. The following are conditions that lead to major security gaps in small companies.

The Trusted Employee The fact is that *any* employee can be a thief, although most are not. About 48 percent of the workers who steal from their companies have been employed for less than five years, but longtime employees who steal cause more damage (a median loss of $100,000 vs. $229,000).[36] Many entrepreneurs see their longtime employees almost as partners, a view that, although not undesirable, can result in security breaches. "More times than not, it's the most trusted person in the office who's perpetrating the fraud," says James Ratley, president of the Association of Certified Fraud Examiners.[37] Many owners refuse to believe that their most trusted employees present the greatest security threat, but these workers have the greatest accessibility to keys, cash registers, records, and even safe combinations. A strong correlation exists between

an employee's level of authority and the losses that a company incurs from his or her theft or fraud. In the United States, the median loss to an employee's theft or fraud (43 percent of cases) is $50,000; in theft or fraud by a manager (34.3 percent of cases), the median loss is $150,000; and in an executive's theft of fraud (18.5 percent of cases), the median loss is $373,000.[38] Because of their seniority, these employees hold key positions, are familiar with a company's operations, and know where gaps in control and security procedures lie.

ENTREPRENEURIAL PROFILE: Michael and John Koss: Koss Corporation Michael and John Koss, second-generation owners of Koss Corporation, a Milwaukee, Wisconsin–based maker of stereo headphones, discovered that the company's vice president of finance had used her position to steal more than $31 million from the company over six years. She covered her theft for so long by inducing employees in the finance department to make fraudulent accounting entries that made her theft transactions appear to be legitimate. The employee, who had held her position at Koss for 18 years, used the money to finance purchases of cars, clothing, jewelry, home renovations, household furnishings, trips, and other personal expenditures.[39]

Business owners also should be wary of "workaholic" employees. Is this worker really dedicated to the company, or is he or she working so hard to cover up theft? Employee thieves are unwilling to take extended breaks from their jobs for fear of being detected. As long as a dishonest employee remains on the job, he or she can cover up theft. As a security precaution, business owners should require every employee to take vacations long enough so that someone else has to take over their responsibilities (at least five consecutive business days). Most schemes are relatively simple and require day-to-day maintenance to keep them going. Business failure records are filled with stories of firms in which the "ideal" employee turned out to be a thief. "In 90 percent of the cases in which people steal from their companies, the employer would probably have described this person, right up to the time the crime was discovered as a trusted employee," says one expert.[40]

Disgruntled Employees Business owners also must monitor the performance of disgruntled employees. Employees are more likely to steal if they believe that their company treats them unfairly, and the probability of their stealing goes even higher if they believe that they themselves have been treated unfairly. Employees dissatisfied with their pay or their promotions may retaliate against an employer by stealing. Dishonest employees make up the difference between what they are paid and what they believe they are worth by stealing. Many believe pilfering is a well-deserved "perk."

Organizational Atmosphere Many entrepreneurs unintentionally create an atmosphere that encourages employee dishonesty. Failing to establish formal controls and procedures invites

theft. Nothing encourages dishonest employees to steal more than knowing they are unlikely to be caught. Four factors encourage employee theft:

1. The need or desire to steal (e.g., to support a habit or to cope with a sudden financial crisis)
2. A rationalization for the act (e.g., "They owe me this.")
3. The opportunity to steal (e.g., access to merchandise or complete control of financial functions)
4. The perception that there is a low probability of being caught (e.g., "Nobody will ever know.")

Entrepreneurs must recognize that they set the example for security and honesty in the business. Employees place more emphasis on what owners *do* than on what they *say*. Entrepreneurs who install a complete inventory control system and then ignore it are telling employees that security is unimportant. No one should remove merchandise, materials, or supplies from inventory without recording them properly. There should be no exceptions to the rules, even for owners and their relatives. Managers should develop clear control procedures and establish penalties for violations. The single biggest deterrent (to employee theft) is a strong, top-down policy that is well communicated to all employees that theft will not be tolerated and that anyone caught stealing will be prosecuted—*no exceptions.*

Entrepreneurs must constantly emphasize the importance of security. Business owners must use every available opportunity to reduce employees' temptation to steal. One business owner relies on payroll inserts to emphasize to employees how theft reduces the funds available for growth, expansion, and higher wages. Another useful tool is a written code of ethics that spells out penalties for violations that every worker signs. Workers must understand that security is a team effort. Security rules and procedures must be reasonable, and owners must treat workers equitably. Unreasonable rules are no more effective—and may even be more harmful—than poorly enforced procedures. A work environment that fosters honesty at every turn serves as an effective deterrent to employee theft.

Physical Breakdowns Another major factor contributing to employee theft is weak physical security. Entrepreneurs who pay little attention to the distribution of keys, safe combinations, and other entry devices are inviting theft. In addition, those who fail to lock doors and windows or to install reliable alarm systems literally are leaving their businesses open to thieves both inside and outside the organization.

Open windows and unattended doors give dishonest employees a prime opportunity to slip stolen merchandise out of the plant or store. One security expert worked with a small manufacturing operation that was experiencing high levels of employee theft during the night shift. His investigation revealed that employees could exit the building through 14 different doors with little or no supervision. The company closed most of the exits, installed security cameras at those that remained open, and assigned managers to supervise the night shift. After implementing these simple changes, employee theft plummeted to nearly zero.[41]

Many businesses find that their profits go out with the trash, literally. When collecting trash, a dishonest employee may stash valuable merchandise in with the refuse and dump it in the receptacle. After the store closes, the thief returns to collect the loot. One drugstore owner lost more than $7,000 in merchandise in just six months through trash thefts.

Improper Cash Control May small business owners encourage employee theft by failing to implement proper cash control procedures. Without a system of logical, practical audit controls on cash, a small business will likely suffer internal theft. Dishonest employees quickly discover that there is a low probability of detection and steal cash with impunity.

Cashiers clearly have the greatest accessibility to the firm's cash and, consequently, experience the greatest temptation to steal, most often by "skimming" money from cash sales. Suppose that a customer makes a purchase with the exact amount of cash and leaves quickly. The cashier simply fails to ring up the purchase and pockets the cash without anyone's knowledge. Some small business owners create a cash security problem by allowing too many employees to operate cash registers and handle customer payments. If a cash shortage develops, the owner is unable to trace responsibility.

A daily inspection of cash register transactions can point out potential employee theft problems. When a cashier rings up an excessive number of voided transactions or no-sale transactions, the owner should investigate. A no-sale transaction could mean that the register

was opened to give a customer change or to steal cash. A large number of incorrect register transactions also are a sign of foul play. Employees may be camouflaging thefts by voiding transactions, underringing sales amounts, or pretending to scan items at checkout without actually ringing them up (a problem known as "sweethearting").

PREVENTING EMPLOYEE THEFT Many incidents of employee theft go undetected, and of those employees who are caught stealing, only a small percentage is prosecuted. Because nearly half of businesses that are victims of employee theft never recover any of their losses, entrepreneurs must focus on *preventing* employee theft.[42] Although business owners cannot eliminate the possibility of employee theft and fraud, they can reduce its likelihood by maintaining accurate inventory records and implementing basic loss prevention strategies.

Screen Employees Carefully Statistics show that, on average, one out of every 36 employees is caught committing employee theft.[43] Perhaps a business owner's greatest weapon against crime is a thorough preemployment screening process. The best time to weed out prospective criminals is before hiring them! One security company conducted an analysis of more than 19,000 applicants for retail jobs and rated 19.3 percent of them as "high-risk" candidates for employee theft.[44] Although state and federal regulations prohibit employers from invading job applicants' privacy and from using discriminatory devices in the selection process, employers have a legitimate right to determine job candidates' integrity and qualifications. A comprehensive selection process and reliable screening devices greatly reduce the chances that an entrepreneur will hire a thief. Smart entrepreneurs verify the information applicants provide on their résumés because they know that some of them will either exaggerate or misrepresent their qualifications. A thorough background check with references and previous employers also is essential. (One question that sheds light on a former employer's feelings toward a former employee is "Would you hire this person again?")

Some security experts recommend the use of integrity tests, paper-and-pencil tests that offer valuable insight into job applicants' level of honesty. Business owners can buy integrity tests for $20 or less that are already validated (to avoid charges of discrimination) and that they can score on their own. Because drug addictions drive many employees to steal, employers also should administer drug tests consistently to all job applicants. The most reliable drug tests cost the company from $35 to $50 each, a small price to pay given the potential losses that can result from hiring an employee with a drug habit. In addition, business owners should conduct criminal background checks on every candidate they are considering hiring.

Create an Environment of Honesty Creating an environment of honesty and integrity starts at the top of an organization. This requires business owners to set an impeccable example for everyone else in the company. In addition to creating a standard of ethical behavior, business owners should strive to establish high morale among workers. A positive work environment in which employees see themselves as an important part of the team is an effective deterrent to employee theft. Establishing a written code of ethics and having employees sign "honesty clauses" offers tangible evidence of a company's commitment to honesty and integrity.

Establish a System of Internal Controls The basis for maintaining internal security on the job is establishing a set of reasonable internal controls designed to prevent employee theft. An effective system of checks and balances goes a long way toward deterring internal crime; weak or inconsistently enforced controls are an open invitation for theft. The most basic rule is to separate among several employees related duties that might cause a security breach if assigned to a single worker. For instance, owners should avoid letting the employee who issues checks reconcile the company's bank statement. Similarly, the person who orders merchandise and supplies should not be the one who also approves those invoices for payment. Spreading these tasks among a number of employees makes organizing a theft more difficult. The owner of a small retail art shop learned this lesson the hard way. After conducting an inventory audit, he discovered that more than $25,000 worth of art supplies was missing. The owner finally traced the theft to the company bookkeeper, who was creating fictional invoices and then issuing checks to herself for the same amount.

Business owners should insist that all company records be kept up to date. Sloppy record keeping makes theft difficult to detect. All internal documents—shipping, ordering, invoicing, and collecting—should be numbered. Missing numbers should arouse suspicion. One subtle way

FIGURE 20.5

Behaviors Exhibited by Perpetrators of Employee Theft

Source: Based on *2010 Report to the Nations on Occupational Fraud and Abuse*, Association of Certified Fraud Examiners, Austin, Texas, 2010, p. 70.

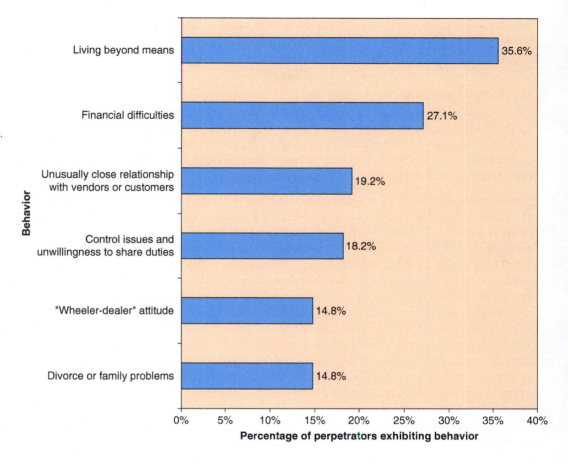

to test employees' honesty is to commit deliberate errors occasionally to see whether employees detect them. If you send an extra case of merchandise to the loading dock for shipment, does the supervisor catch it, or does it disappear?

Use Technology to Control Theft A variety of technology tools help business owners minimize losses to employee theft and fraud at very reasonable prices. Simple video camera systems, such as the ones used on the Food Network's show *Restaurant Stakeout*, are responsible for nabbing many employee thieves, especially cameras that are focused on checkout stations and cash registers. Kevin Donahue, owner of a Planet Beach franchise in McLean, Virginia, uses a security system that gives him access to his store's alarm system and security cameras from almost anywhere in the world over the Internet. "It gives me the ability to travel and manage my staff remotely," says Donahue, who paid $100 to install the system and pays a monthly fee of $39.[45]

Watch for Signs of Employee Theft Research shows that employees who are stealing tend to exhibit certain behavior patterns (see Figure 20.5). Watch for them.

Set Up a Hotline One of the most effective tools for preventing employee theft is to encourage employees to report suspicious activity. Perhaps the easiest way to encourage reporting it is to establish a hotline that allows employees provide tips anonymously.

Embrace a Zero-Tolerance Policy Business owners should demonstrate zero tolerance for theft. They must adhere strictly to company policy when dealing with employees who violate the company's trust. When business owners catch an employee thief, the best course of action is to fire the perpetrator and to prosecute. Too often, owners take the attitude "Resign, return the money, and we'll forget it." Letting thieves off, however, only encourages them to move on to other businesses where they will steal again. Prosecuting a former employee for theft is never easy, but it does send a clear signal about how the company views employee crime.

Notice in Figure 20.6 that although the primary cause of inventory shrinkage is employee theft, shoplifting also is a common problem.

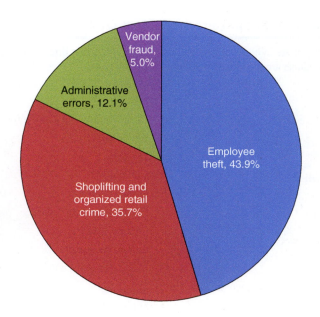

FIGURE 20.6

Causes of Inventory Shrinkage

Source: Based on 2012 National Retail Security Survey, National Retail Federation.

 Entrepreneurship in Action

Misplaced Trust

Barbara Wein Allen, founder of Multi-Point Communications, a provider of Web- and audio-conference services based in Birmingham, Alabama, trusted her employees so much that at one time she made loans to them so that they could purchase homes, cars, and other "big-ticket" items. Then Wein Allen discovered that an employee who had held a senior finance position in Multi-Point Communications for more than seven years had embezzled $250,000 and decided to stop making loans to employees. "I trusted her implicitly," says Wein Allen.

The embezzlement was difficult to detect because the former employee had access to all of the company's financial records, knew the weaknesses in its financial control systems, and stole the money in small amounts over the seven years she had worked for Multi-Point Communications. In addition, the company's financial performance was solid and improving, giving Wein Allen no hint that something was amiss. "Business was just that much better, and we didn't realize it," she says.

The theft went on for years and came to light only when the former employee went on vacation and was not able to cover the trail of her embezzlement. Wein Allen hired a forensic accountant who confirmed that the former employee had recorded several suspicious expenses that were part of her embezzlement scheme and found evidence that the theft started shortly after the company hired her.

Small businesses are the most common targets of employee theft and fraud because they lack the control procedures that most large companies have implemented. In addition, many entrepreneurs put their companies at risk with lax hiring practices, failing to exercise sufficient caution in screening dishonest employees. Karen Howard, owner of Dunham's Bay Boat Company, a boat retail and repair business in Lake George, New York, received an envelope from an anonymous source that contained a copy of an article that showed that one of her 15 employees had been arrested for workplace theft. Howard already had suspicions that the employee was stealing cash, but after receiving the article, she contacted state police. An undercover officer watched as one of Howard's friends posing as a customer made a $2,200 cash payment to the employee, who promptly stole the money. The employee, who had been with Dunham's Bay Boat Company only 10 months, was sentenced to jail time because of her theft. "She had stuffed the cash in her underwear," says Howard, who admits that she neglected to perform a thorough background check on the employee, a step that most likely would have saved her company the $30,000 that the employee stole in her brief tenure. "She was so friendly," says Howard. "If I didn't know what she had done, I would've believed her."

1. What factors led to the thefts at Multi-Point Communications and Dunham's Bay Boat Company?

2. Do you agree with the view that small businesses are more vulnerable to employee theft because they lack the systems to detect the signs of theft? Explain.

3. List at least five steps business owners can take to prevent employee theft.

Source: Based on Sarah E. Needleman, "Business Owners Get Burned by Sticky Fingers," *Wall Street Journal*, March 11, 2010, *http://online.wsj.com/article/SB1 0001424052748703862704575099661793568310.html.*

TABLE 20.4 **Items That Shoplifters Steal Most Often**

Shoplifters target items that are in high demand and that they can resell at prices that are close to full retail. The following products are the most commonly shoplifted items:

Grocery Items
- Cigarettes
- Energy drinks
- High-end liquor
- Infant formula

Over-the-Counter Medicine
- Allergy medicine
- Diabetic testing strips
- Pain relievers
- Weight loss pills

Health and Beauty Items
- Electric toothbrushes and replacement heads
- Lotions and creams
- Pregnancy tests

Clothing
- Jeans
- Designer clothing and denim
- Handbags

Electronics
- Cell phones
- Digital cameras
- Digital recorders
- GPS devices
- Laptop computers
- LCD televisions and monitors

Home Items
- High-end vacuum cleaners
- Kitchen Aid mixers

Source: Based on *2012 Organized Retail Crime Survey*, National Retail Federation, 2012, p. 12.

Shoplifting

The most frequent business crime is shoplifting. In fact, research shows that 1 out of 10 adults in the United States has shoplifted but that only 1 in 49 is arrested for the crime.[46] Retail businesses in the United States lose an estimated $37.1 billion annually to shoplifters each year, and the toll on small businesses is especially heavy because they usually have the weakest lines of defense against shoplifters.[47] Shoplifting exacts a price on shoppers as well. The National Retail Federation estimates that the average U.S. household pays nearly $500 annually to cover merchants' shoplifting losses.[48] Table 20.4 shows the items that shoplifters steal most often.

TYPES OF SHOPLIFTERS Anyone who takes merchandise from a store without paying for it, no matter what the justification, is a shoplifter. Shoplifters look exactly like other customers. They can be young children in search of a new toy or elderly people who are short of money. *Anyone* can be a shoplifter, given the opportunity, the ability, and the desire to steal. Police have apprehended people from all walks of life—including wealthy socialites and famous celebrities— for shoplifting. One former senior adviser to the president of the United States who was earning $161,000 per year was caught shoplifting several items, including a $525 sound system, a mop that cost $12, massage gloves priced at $5, and two bras valued at $10.[49]

Several factors lead to increased shoplifting losses for businesses, including economic recessions, increased organized retail crime activity, stores that reduce operating costs by scheduling minimal numbers of staff and supervisors, and stores that use open display sales strategies that allow customers to interact with goods. Fortunately for small business owners, most shoplifters are amateurs who steal because the opportunity presents itself. Many steal on impulse, and the theft is the first criminal act. Many of those caught have the money to pay for their "five-finger discounts." Observant business owners supported by trained store personnel can spot potential shoplifters and deter many shoplifting incidents; however, they must understand the shoplifter's profile. Experts identify five types of shoplifters.

Juveniles Juveniles account for approximately one-half of all shoplifters. Many juveniles steal as a result of peer pressure. Most have little fear of prosecution, assuming that they can hide behind their youth. When owners detect juvenile shoplifters, they must not let sympathy stand in the way of good judgment. Many hard-core criminals began their careers as shoplifters, and business owners who fail to prosecute the youthful offender do nothing to discourage a life of crime. Entrepreneurs should prosecute juvenile offenders through proper legal procedures just as they would any adult shoplifter.

Juveniles now use social media to coordinate their crimes, plotting "flash robs" on Twitter or Facebook or via text messages. The National Retail Federation estimates that 10 percent of retailers

Shoplifters cost U.S. businesses $37.1 billion annually, and small companies bear a disproportionate share of shoplifting losses.
Source: © Meritzo/Alamy.

have been victims of flash robs and that police or retail security personnel are successful in apprehending the perpetrators in only half of the incidents.[50] In Wicker Park, a suburb of Chicago, a group of about 20 teen boys descended on Mildblend Supply Company, a retail store selling designer clothing, and stole more than $3,000 worth of jeans despite the owner's attempts to stop them.[51]

Impulse Shoplifters Impulse shoplifters steal on the spur of the moment when they succumb to temptation. These shoplifters do not plan their thefts, but when a prime opportunity to shoplift arises, they take advantage of it. For example, a salesperson may be showing a customer several pieces of jewelry. If the salesperson is called away, the customer might pocket an expensive ring and leave the store before the employee returns.

The most effective method of fighting impulse shoplifting is prevention. To minimize losses, the owner should remove the opportunity to steal by implementing proper security procedures and devices.

Shoplifters Supporting Other Criminal Behaviors Shoplifters motivated to steal to support a drug or alcohol habit often are easy to detect because their behavior is usually unstable and erratic. One recently apprehended shoplifter was supporting a $100-a-day heroin habit by stealing small items from local retailers and then returning the merchandise for refunds. (The stores almost never asked for sales receipts.) Small business owners should exercise great caution in handling these shoplifters because they can easily become violent. Criminals deranged by drugs or alcohol might be armed and could endanger the lives of customers and employees if they are detained. It is best to let the police apprehend these shoplifters.

Kleptomaniacs Kleptomaniacs have a compulsive need to steal even though they have little, if any, need for the items they shoplift. In many cases, these shoplifters could afford to purchase the merchandise they steal. Kleptomaniacs account for less than 5 percent of shoplifters, but their disease costs business owners a great deal. They need professional counseling, and the owner helps them only by seeing that they are apprehended.

Professionals A study by the National Retail Federation reports that 96 percent of businesses have been victims of organized retail crime, thefts by professional shoplifters, within the last year.[52] Professional shoplifters are individuals, groups, or gangs who steal merchandise in significant quantities as part of a criminal enterprise with the intent to resell it for financial gain rather than keep it for personal use. When it comes to shoplifting, "most people think about little Johnny stealing a pack of bubble gum," says one loss prevention expert. "This is anything but that; these are professional criminals."[53] Known as organized retail crime rings, these criminal operations rely on "boosters," people who are paid to steal merchandise from stores, and "fences," those who specialize in converting stolen goods into cash or drugs. "This is a completely different type of perpetrator, doing this strictly for profit, and often working in very organized groups of two, three, or more individuals," says one experienced police officer.[54] Although professional shoplifters account for 32 percent of shoplifting incidents, the dollar impact of their thefts is

disproportionately large. Police in Indiana and the head of a retail chain's organized crime division recently arrested the people behind a professional shoplifting ring that stole cosmetics, baby formula, over-the-counter drugs, medical supplies, and other items that, when resold, netted the thieves $17 million per year.[55] In Austin, Texas, police broke up an eight-person organized retail crime ring that specialized in stealing cosmetics and household goods, including razor blades, batteries, shampoo, nail polish, lotion, hair clips, Tide detergent (known as "liquid gold" among fences), and other items. The ringleader, Maria Patricia Villegas, had seven boosters shoplifting the merchandise, which she resold at local flea markets or smuggled into Mexico. When police raided her home, they found more than $35,000 worth of stolen goods. Another organized retail crime ring in Austin specialized in shoplifting meat from grocery stores, which the fence then resold to three local restaurants.[56] Many organized retail crime operations are so sophisticated that the thieves create schedules of the stores they have hit to avoid stealing from any one store too often and raising suspicion. Table 20.5 provides some interesting facts about shoplifting.

TABLE 20.5 Shoplifting Facts

- Approximately $37 billion worth of goods are stolen from retailers each year. That's more than $4.2 million worth of merchandise per hour.

- There are approximately 27 million shoplifters (or 1 in 11 people) in the United States. More than 10 million people have been caught shoplifting in the last five years.

- Shoplifting affects more than the offender. It overburdens the police and the courts, adds to a store's security expenses, increases the cost of goods for legitimate shoppers, reduces sales tax dollars that go to communities, and hurts children and families.

- There is no such thing as a "typical" shoplifter. *Anyone* can be a shoplifter. Men and women shoplift about equally as often.

- Approximately 25 percent of shoplifters are kids. Fifty-five percent of adult shoplifters say they started shoplifting in their teens.

- Many shoplifters buy and steal merchandise in the same visit. The average value of items that shoplifters steal per incident is $129.

- Shoplifting is often an impulse crime: 73 percent of adult and 72 percent of juvenile shoplifters don't plan to steal in advance.

- Eighty-six percent of kids say they know other kids who shoplift, and 66 percent say they hang out with those kids.

- Shoplifters say they are caught an average of only once in every 48 times they steal. They are turned over to the police just 50 percent of the time.

- Approximately 3 percent of shoplifters are "professionals" who steal solely for resale or profit as a business. These include drug addicts who steal to feed their habit, hardened professionals who steal as a lifestyle, and international shoplifting gangs that steal for profit as a business. Eight-eight percent of retailers say that shoplifting losses from professional shoplifters has increased over the last three years.

- The majority of shoplifters are nonprofessionals who steal not out of financial need or greed but as a response to social and personal pressures in their lives.

- The excitement generated from "getting away with it" produces a chemical reaction that results in what shoplifters describe as an incredible "rush" or "high" feeling. Many shoplifters say that this high is their true reward rather than the merchandise itself.

- Drug addicts who have become addicted to shoplifting describe shoplifting as equally addicting as drugs.

- Even after getting caught, 57 percent of adults and 33 percent of juveniles say it is hard for them to stop shoplifting.

- Most nonprofessional shoplifters don't commit other types of crimes. They'll never steal an ashtray from your house and will return to you a $20 bill that you drop. Their criminal activity is restricted to shoplifting, and therefore any rehabilitation program should be "offense specific" for this crime.

- The typical shoplifter steals an average of 1.6 times per week.

Sources: Based on National Association for Shoplifting Prevention (NASP), *www.shopliftingprevention.org*; *2012 Organized Retail Crime Survey*, National Retail Federation, Washington, DC, 2012, p. 5; Richard C. Hollinger and Amanda Adams, *2012 National Retail Security Survey*, University of Florida, pp. 9–10.

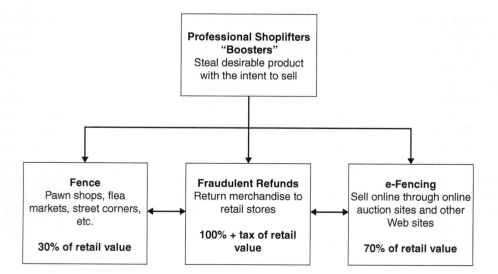

FIGURE 20.7

How Professional Shoplifters Work

Source: 2012 Organized Crime Survey, National Retail Federation, Washington, DC, p. 6.

Because professional shoplifters' business is theft, they are very difficult to detect and deter. Professional shoplifters tend to focus on expensive merchandise that they can sell quickly to their fences, such as stereo equipment, appliances, guns, or jewelry. Usually the fences don't keep the stolen goods long, often selling them on online auction sites ("e-fencing"), in pawn shops, or at flea markets at a fraction of their value. Therefore, apprehending and prosecuting professional shoplifters is quite difficult. Police have apprehended professional shoplifters with detailed maps of a city's shopping districts, showing target stores and the best times to make a "hit." Figure 20.7 shows how professional shoplifters operate.

DETECTING SHOPLIFTERS Although shoplifters can be difficult to detect, business owners who know what to look for can spot them in action. Entrepreneurs must always be on the lookout for shoplifters, but they should be especially vigilant on Saturdays and around holidays, when shoplifters can hide their thefts more easily in the frenzy of a busy shopping day.

Shoplifters can work alone or in groups. In general, impulse shoplifters prefer solitary thefts, whereas juveniles and professionals operate in groups. A common tactic for group shoplifters is for one member of the gang to create some type of distraction while other members steal the merchandise. Business owners should be wary of loud, disruptive groups that enter their stores. Professional shoplifters work in teams with each person filling a particular role, such as lookout, driver, decoy, picker, and packer. They use hand signals and smart phones to coordinate their thefts and often work from "fence lists," shopping lists of desirable merchandise provided to them by their fences.

Some shoplifters avoid working with fences and engage in receipt fraud scams. They steal merchandise from a store, create counterfeit receipts for the stolen goods, and then return them to the store for a full refund. Receipt fraud nets shoplifters the full retail price of the stolen goods rather than the 10 to 30 percent of retail price that a fence typically pays them.

Solitary shoplifters are usually quite nervous. They avoid crowds and shy away from store personnel, preferring privacy to ply their trade. To make sure they avoid detection, they constantly scan the store for customers and employees. These shoplifters spend more time nervously looking around the store than examining merchandise. In addition, they shop when the store is most likely to be understaffed, during early morning, lunch, or late evening hours. Shoplifters frequently linger in the same area for an extended time without purchasing anything. Customers who refuse the help of sales personnel or bring in large bags and packages (especially empty ones) also arouse suspicion.

Shoplifters have their own arsenal of tools to assist them in plying their trade. They often shop with booster boxes, shopping bags, umbrellas, bulky jackets, baby strollers, or containers disguised as gifts. These props often have hidden compartments that can be tripped easily, allowing the shoplifter to fill them with merchandise quickly. "Booster (or magic) bags," foil-lined bags or purses that block the signals from electronic surveillance tags attached to goods, are another commonly used shoplifting tool.

Some shoplifters use specially designed coats with hidden pockets and compartments that can hold even large items. Small business owners should be suspicious of customers wearing out-of-season clothing (e.g., heavy coats in warm weather or rain gear on clear days) that could conceal stolen goods. Hooked belts also are used to enable the shoplifter to suspend items from hangers without being detected.

Another common tactic is "ticket switching," in which the shoplifter exchanges price tickets on items and pays a very low price for an expensive item. An inexperienced or unobservant cashier may charge $9.95 for a $30.00 item that the shoplifter re-marked while no one was looking. A more elaborate scheme is one in which shoplifters create counterfeit bar codes that they paste over existing bar codes on packages so that when the item is scanned, it rings up at a much lower price. After three years, police finally nabbed a shoplifter who used this technique to steal more than $600,000 worth of toy Legos from dozens of stores in five western states. His phony bar codes caused $100 Lego sets to ring up for just $19 at checkout counters. He then resold the Lego sets at a markup on a Web site for toy collectors.[57]

Entrepreneurship in Action

A Tale of Two Shoplifters

Anyone can be a shoplifter. The following confessions from two shoplifters offer valuable insight into what makes shoplifters tick and what merchants can do to deter them.

Lilit: Never Caught

I was a teenage shoplifter. As freshmen in high school, my best friends—we'll call them Emma and Katie—and I spent almost every weekend at the local mall. Katie's and Emma's parents gave them money to go shopping, and I was jealous. They ended every Saturday with a stack of new tank tops, jeans, and dangly earrings while I was lucky to score a new T-shirt with what was left of my babysitting money. I started stealing because I wanted to balance things out. When you're 14, every problem in life can be solved with a new sweater or a tube of lip gloss, and I fancied myself as a teenage Robin Hood, stealing from the rich (stores) to give to the poor (me).

I was good at it.

I was small and unassuming, hanging out with two friends who always bought stuff while I waited outside the dressing room. I knew which stores had cameras and which ones didn't have security tags on their clothes. I limited myself to stealing one item per store, dancing on the invented line of morality: Either the earrings or the skirt is fine, I told myself, but not both. Shoplifting was my way of balancing it out, of making up for the perceived deficiencies in my life. At my school, rich kids were popular, and poor kids weren't. Stealing clothes I couldn't afford was my way of trying to make up the inequity. If I could steal clothes, I could work my way up to stealing happiness.

I never got caught. At some point, I stopped. I realized that new clothes couldn't make a new girl. I still have one dress from my shoplifting days, a black mini-dress with spaghetti straps. I've worn it three times. When I stole it, I had an idea of the sort of woman I was going to grow up to be—somebody who was no longer self-conscious about money and status but somebody who was sexy and confident.

Although I have enough money to pay for my clothes now, there's still a part of me that, when I go into stores, looks around for cameras and checks to see whether the clerks are paying attention to peoples' bags when they leave. Like any addiction, my desire to shoplift bubbles up now and again, wending through my veins and trying to make me reach for things. But I think about that black dress in my closet, and I keep walking.

Samantha: Busted!

I wish I could say that being arrested for shoplifting was a moment of madness—one of those irrational actions so far removed from my law-abiding behavior that it can be dismissed as an out-of-character blip in an otherwise upstanding, middle-class life. In reality, I had it coming to me. I'd been stealing stuff for months—for no other reason than I wanted things and didn't have the money to pay for them. I stole audaciously, unforgivably, and because I knew I could get away with it. Nobody suspects somebody like me of being a thief, which is probably why I was able to stroll out of those shops without paying for so long. I'm a middle-class mother of three in my late thirties, living in a beautiful four-bedroom Georgian house.

My first shoplifting experience was an innocent mistake. I walked out of a shop holding a hair band for my daughter that I'd grabbed at the last minute and forgot to pay for. I should have walked back inside and admitted what had happened. But in that moment of deliberation, I felt no remorse. Far worse, it gave me a buzz—an actual high of getting something for nothing. I wasn't a proper criminal; I was just taking a bit extra to redress the ridiculous markup of the stores. That was the beginning of my spiral into shoplifting, but when you've gotten away with something once, it becomes all too easy to do again. My behavior is inexcusable. It was greediness, pure and simple.

A pattern soon emerged. I never stole anything without first paying for something else. It was a case of buy one, get

one free—but not, I suspect, in the way the store campaign intended. At Christmas, I acquired all of my children's stocking stuffers without paying for them. Most people would feel tainted on Christmas morning, watching their children's eyes widen with wonder as they unwrap their gifts. I didn't. I just thought, "Thank God I got away with it."

On the day I was caught, my thirty-seventh birthday, I'd gone into town to buy something special for supper. I had time on my hands and went into a large department store. I wandered around, selecting items that caught my fancy (and had no security tags) and slinging them over my arm. The amount totaled just over £100. For the first time, I hadn't paid for anything to get my crucial "carrier bag." I walked out of the store blatantly holding clothes and not even attempting to hide what I was taking. Seconds later, a security guard and three policemen appeared out of nowhere. They had been following me and waiting for me to leave the store. In my panic, I tried to sound indignant: "I was just getting my purse from the car." But they'd heard it all before. They slapped handcuffs on my wrists and pushed me into the back of a police van and banged the doors shut. At the police station, they led me down an endless, smelly corridor to a cell. All of a sudden, I was alone behind a heavy, locked door. That's when the enormity of what I had done hit me. I thought

of my children, waiting at home with the chocolate cake they had baked for me, ready to sing "Happy Birthday," and I felt worse than a criminal. It was 10 hours before I was released. An attorney told me that the store had closed-circuit television evidence and that the chances of a jury believing I was innocent were highly unlikely. His advice was to admit guilt, apologize, and be thankful for a warning. This is what I did.

The arrest was a long-overdue slap in the face that forced me to take an uncomfortable audit of everything I could lose. I'll be forever grateful for those 10 hours in a lonely cell because they reminded me just how valuable freedom is.

1. What do the profiles of these two shoplifters say about assumptions that many business owners make about the "typical shoplifter"?

2. List and explain at least five steps businesses can take to minimize losses to shoplifting.

Sources: Adapted from Lilit Marcus, "Confessions of a Former Teenage Shoplifter," *Crushable,* June 8, 2011, *www.crushable.com/2011/06/08/other-stuff/confessions-of-a-former-teenage-shoplifter;* Samantha Booker, "Confessions of a Middle-Class Shoplifter: Stealing for Kicks, Even Down to Gifts for Her Children. How One Woman's Obsession Ended in a Police Cell," *Daily Mail,* March 1, 2011, *www.dailymail.co.uk/femail/article-1361614/Confessions-middle-class-shoplifter-Stealing-kicks-womans-obsession-ended-police-cell.html.*

One variation of traditional shoplifting techniques is the "grab-and-run," in which a shoplifter grabs an armload of merchandise located near an exit and then dashes out the door into a waiting getaway car. The element of surprise gives these thieves an advantage, and they are often gone before anyone in the store realizes what has happened. Other shoplifters are more brazen, simply walking out of the store with merchandise as if they had paid for it. One shoplifter says that he often stole kayaks from sporting goods stores by merely carrying them out in full view of store personnel, none of whom ever questioned him.[58]

DETERRING SHOPLIFTERS The problem of shoplifting is worsening. Every year, business losses due to customer theft increase, and many companies are declaring war on shoplifting. Funds allocated for fighting shoplifting losses are best spent on *prevention.* By focusing on preventing shoplifting rather than on prosecuting violators after the fact, business owners take a strong stand in protecting their firms' merchandise. Of course, no prevention plan is perfect. When violations occur, owners must prosecute; otherwise, the business becomes known as an easy target. Retailers say that when a store gets a reputation for being tough on shoplifters, thefts drop off.

Knowing what to look for dramatically improves a business owner's odds in combating shoplifting:

- *Watch the eyes.* Amateurs spend excessive time looking at the merchandise they're about to steal. Their eyes, however, are usually checking to see who (if anyone) is watching them.

- *Watch the hands.* Experienced shoplifters, like good magicians, rely on sleight of hand.

- *Watch the body.* Amateurs' body movements reflect their nervousness; they appear to be unnatural.

- *Watch the clothing.* Loose, bulky clothing is the uniform of the typical shoplifter.

- *Watch for devices.* Anything a customer carries is a potential concealing device.

- *Watch for loiterers.* Many amateurs must work up the nerve to steal.

- *Watch for switches.* Working in pairs, shoplifters will split duties; one will lift the merchandise, and, after a switch, the other will take it out of the store.

Store owners can take other steps to discourage shoplifting.

Provide Good Customer Service Shoplifters need privacy to ply their trade and prefer to avoid sales personnel. Employees should walk the sales floor and interact with customers. When employees approach potential shoplifters, the shoplifters know they are being watched. Even when all salespeople are busy, an alert employee should approach the customer and say, "I'll be with you in a moment." Honest customers appreciate the clerk's politeness, and shoplifters are put off by the implied surveillance.

Train Employees to Spot Shoplifters One of the best ways to prevent shoplifting is to train store personnel to be aware of shoplifters' habits and to be alert for possible theft. In fact, most security experts agree that alert employees are the best defense against shoplifters. Employees should watch for suspicious people, especially those carrying the props of concealment. Employees in clothing stores must keep a tally of the items being taken into and out of dressing rooms. Some clothing retailers prevent unauthorized use of dressing rooms by locking them; customers who want to try on garments must check with a store employee first.

An alert cashier can be a tremendous boon to a store owner attempting to minimize shoplifting losses. A cashier who knows the store's general pricing policy and is familiar with the prices of many specific items is the best insurance against the ticket-switching shoplifter. A good cashier also should inspect all containers being sold; tool boxes, purses, briefcases, and other items can conceal stolen merchandise.

Employees should be trained to watch for group shoplifting tactics. A group of shoppers that enters the store and then disperses in all directions may be attempting to distract employees so that some gang members can steal merchandise. Sales personnel should watch closely the customer who lingers in one area for an extended time, especially one who examines a lot of merchandise but never purchases anything.

The sales staff should watch for customers who consistently shop during the hours when most employees are on breaks. Managers can help eliminate this cause of shoplifting by ensuring that their stores are well staffed at all times. Coordinating work schedules to ensure adequate coverage is a simple but effective method of discouraging shoplifting.

The cost of training employees to be alert to shoplifting "gimmicks" can be recouped many times over by preventing losses from retail theft. The local police department or chamber of commerce may be able to conduct training seminars for small business owners and their employees, or security consulting firms might sponsor a training course on shoplifting techniques and protective methods. Refresher courses every few months can help keep employees sharp in spotting shoplifters.

Pay Attention to Store Layout A well-planned store layout also can be an effective obstacle in preventing shoplifting losses. Proper lighting throughout the store makes it easier for employees to monitor shoppers, whereas dimly lit areas give dishonest customers a prime opportunity to steal without detection. In addition, keeping displays of items low gives store personnel a clear line of sight of the entire store.

Business owners should keep small, expensive items, such as jewelry, silver, and personal electronics, inside display counters with a sales clerk nearby. Retailers should keep valuable items out of customers' reach and should not display them near exits, where shoplifters can pick them up and quickly step outside. All merchandise displays should be neat and organized so that employees can easily notice missing items. Other protective measures include prominently posting anti-shoplifting signs describing the penalties involved and keeping unattended doors locked (within fire regulations). Exits that cannot be locked because of fire regulations should be equipped with noise alarms to detect any attempts at unauthorized exit.

Use Technology to Deter and Detect Theft Another option business owners have in the attempt to reduce shoplifting losses is to use loss prevention technology. A complete monitoring system can be expensive, but failure to implement one is usually more expensive. Closed-circuit television cameras help entrepreneurs combat both shoplifting and employee theft. Not every small business can afford to install a closed-circuit camera system, but one clever entrepreneur got the benefit of such a system without the high cost. He installed one "live" camera and several "dummy" cameras. The cameras worked because potential shoplifters thought that they were all live.

An electronic surveillance system is one of the most effective weapons in retailers' battle against shoplifting.
Source: Pavel L/Shutterstock.

An owner can deter ticket-switching shoplifters by using tamper-proof price tickets: perforated gummed labels that tear away if a customer tries to remove them or price tags attached to merchandise by hard-to-break plastic strips. Some owners use multiple price tags concealed on items to deter ticket switchers. One of the most effective weapons for combating shoplifting is the electronic article surveillance system, small tags that are equipped with electronic sensors that set off sound and light alarms if customers take them past a store exit. These tags are attached to the merchandise and can be removed only by employees with special shears. Owners using these electronic tags must make sure that all cashiers are consistent in removing them from items purchased legitimately; otherwise, they may be liable for false arrest or, at the very least, may cause customers embarrassment.

APPREHENDING SHOPLIFTERS Despite all of the weapons that business owners use to curtail shoplifting, the sad reality is that most of the time shoplifters are successful at plying their trade. Shoplifters say that they are caught an average of only once in every 48 times they steal and that they are turned over to the police just 50 percent of the time. Of those shoplifters who do get caught, less than half are prosecuted. The chance that any shoplifter will actually go before

Source: CartoonStock.

© Mike Baldwin / Cornered

CACTUS WORLD

"Stopped a shoplifter yesterday.
Caught him red-handed."

a judge is about 1 in 100.[59] Building a strong case against a shoplifter is essential; therefore, small business owners must determine beforehand the procedures to follow once they detect a shoplifter. The store owner has to be certain that the shoplifter has taken or concealed the merchandise and has left the store with it. Although state laws vary, owners must do the following to make the charges stick:

1. *See* the person take or conceal the merchandise.
2. *Identify* the merchandise as the belonging to the store.
3. *Testify* that it was taken with the intent to steal.
4. *Prove* that the merchandise was not paid for.

Most security experts agree that an owner should never apprehend the shoplifter if he or she has lost sight of the suspect even for an instant. In that time, the person may have dumped the merchandise.

Another primary consideration in apprehending shoplifters is the safety of store employees. Retailers report that 15 percent of shoplifting apprehensions lead to some level of violence.[60] In general, employees should never directly accuse a customer of shoplifting and should never try to apprehend the suspect. The wisest course of action when a shoplifter is detected is to alert the police or store security personnel and let them apprehend the suspect. Apprehension *outside* the store is safest. This tactic strengthens the owner's case and eliminates unpleasant in-store scenes that upset other customers or that might be dangerous. Of course, if the stolen merchandise is very valuable or if the criminal is likely to escape once outside, the owner may have no choice but to apprehend the shoplifter in the store.

Once business owners detect and apprehend a shoplifter, they must decide whether to prosecute. Many small business owners fail to prosecute because they fear legal entanglements or negative publicity. However, failure to prosecute encourages shoplifters to try again and gives the business the image of being an easy target. Of course, each case is an individual matter. For example, the owner may choose not to prosecute elderly or senile shoplifters or those who are mentally incompetent. In most cases, however, prosecuting the shoplifter is the best option, especially for juveniles and first-time offenders. The business owner who prosecutes shoplifters consistently soon develops a reputation for toughness that most shoplifters hesitate to test.

Conclusion

Inventory control is one of those less-than-glamorous activities that business owners must perform if their businesses are to succeed. Although it doesn't offer the flash of marketing or the visibility of customer service, inventory control is no less important. In fact, business owners who invest the time and the resources to exercise the proper degree of control over their inventory soon discover that the payoff is huge!

Chapter Review

1. Explain the various inventory control systems and the advantages and disadvantages of each.
   - Inventory represents the largest investment for the typical small business. Unless properly managed, the cost of inventory will strain the firm's budget and cut into its profitability. The goal of inventory control is to balance the cost of holding and maintaining inventory with meeting customer demand.
   - Regardless of the inventory control system selected, business owners must recognize the relevance of the 80/20 rule, which states that roughly 80 percent of the value of the firm's inventory is in about 20 percent of

   the items in stock. Because only a small percentage of items account for the majority of the value of the firm's inventory, managers should focus control on those items.
   - Three basic types of inventory control systems are available to the small business owner: perpetual, visual, and partial. Perpetual inventory control systems are designed to maintain a running count of the items in inventory. Although they can be expensive and cumbersome to operate by hand, affordable computerized POS terminals that deduct items sold from inventory on hand make perpetual systems feasible for small

companies. The visual inventory system is the most common method of controlling merchandise in a small business. This system works best when shortages are not likely to cause major problems. Partial inventory control systems are most effective for small businesses with limited time and money. These systems operate on the basis of the 80/20 rule.

- The ABC system is a partial system that divides a firm's inventory into three categories depending on each item's dollar usage volume (cost per unit multiplied by quantity used per time period). The purpose of classifying items according to their value is to establish the proper degree of control over them. A items are most closely controlled by perpetual inventory control systems, B items use basic analytical tools, and C items are controlled by very simple techniques, such as the two-bin system, the level control method, or the tag system.

2. Describe how JIT and JIT II inventory control techniques work.
   - The JIT system of inventory control sees excess inventory as a blanket that masks production problems and adds unnecessary costs to the production operation. Under a JIT philosophy, the level of inventory maintained is the measure of efficiency. Materials and parts should not build up as costly inventory. They should flow through the production process without stopping, arriving at the appropriate location just in time.
   - JIT II techniques focus on creating a close, harmonious relationship with a company's suppliers so that both parties benefit from increased efficiency. To work successfully, JIT II requires suppliers and their customers to share what was once closely guarded information in an environment of trust and cooperation. Under JIT II, customers and suppliers work hand in hand, acting more like partners than mere buyers and sellers.

3. Describe methods for reducing losses from slow-moving inventory.
   - Managing inventory requires monitoring the company's inventory turnover ratio; slow-moving items result in losses from spoilage or obsolescence.

- Slow-moving items can be liquidated by markdowns, eye-catching displays, or quantity discounts.

4. Discuss employee theft and shoplifting and how to prevent them.
   - Employee theft accounts for the majority of business losses due to theft. Most small business owners are so busy managing their companies' daily affairs that they fail to develop reliable security systems. Thus, they provide their employees with prime opportunities to steal.
   - The organizational atmosphere may encourage employee theft. The owner sets the organizational tone for security. A complete set of security controls, procedures, and penalties should be developed and enforced. Physical breakdowns in security invite employee theft. Open doors and windows, poor key control, and improper cash controls are major contributors to the problem of employee theft. Employers can build security into their businesses by screening and selecting employees carefully. Orientation programs also help the employee to get started in the right direction. Internal controls, such as division of responsibility, spot checks, and audit procedures, are useful in preventing employee theft.
   - Shoplifting is the most common business crime. Fortunately, most shoplifters are amateurs. Juveniles often steal to impress their friends, but prosecution can halt their criminal ways early on. Impulse shoplifters steal because the opportunity suddenly arises. Simple prevention is the best defense against these shoplifters. Alcoholics, vagrants, and drug addicts steal to supply some need and are usually easiest to detect. Kleptomaniacs have a compelling need to steal. Professionals are in the business of theft and can be very difficult to detect and quite dangerous.
   - Three strategies are most useful in deterring shoplifters. First, employees should be trained to look for signs of shoplifting. Second, store layout should be designed with theft deterrence in mind. Finally, anti-theft devices should be installed in the store.

Discussion Questions

20-1. Describe some of the incidental costs of carrying and maintaining inventory for the small business owner.

20-2. What is a perpetual inventory system? How does it operate? What are the advantages and disadvantages of using such a system?

20-3. What advantages and disadvantages does a visual inventory control system have over other methods?

20-4. For what type of business product line is a visual control system most effective?

20-5. What is the 80/20 rule, and why is it important in controlling inventory?

20-6. Outline the ABC inventory control procedure. What is the purpose of classifying inventory items using this procedure?

20-7. Briefly describe the types of control techniques that should be used for A, B, and C items.

20-8. What is the basis for the JIT philosophy? Under what condition does a JIT system work best?

20-9. What is JIT II? What is its underlying philosophy? What risks does it present to businesses?

20-10. Outline the two methods of taking a physical inventory count. Why is it necessary for every small business manager to take inventory?

20-11. Why are slow-moving items dangerous to the small business? What can be done to liquidate them from inventory?

20-12. Why are small companies more susceptible to business crime than large companies?

20-13. Why is employee theft a problem for many small businesses? Briefly describe the reasons for employee theft.

20-14. Construct a profile of the employee most likely to steal goods or money from an employer. What four elements must be present for employee theft to occur?

20-15. Briefly outline a program that could help the typical small business owner minimize losses due to employee theft.

20-16. List and briefly describe the major types of shoplifters.

20-17. Outline the characteristics of a typical shoplifter that should arouse a small business manager's suspicions. What tools and tactics is a shoplifter likely to use?

20-18. Describe the major elements of a program designed to deter shoplifters.

20-19. How can proper planning of store layout reduce shoplifting losses?

20-20. What must an owner do to have a good case against a shoplifter? How should a suspected shoplifter be apprehended?

CHAPTER 21

Staffing and Leading a Growing Company

Learning Objectives

On completion of this chapter, you will be able to:

1. Explain the challenges involved in the entrepreneur's role as leader and what it takes to become a successful leader.

2. Describe the importance of hiring the right employees and how to avoid making hiring mistakes.

3. Explain how to build the kind of company culture and structure to support the entrepreneur's mission and goals and to motivate employees to achieve them.

4. Understand the potential barriers to effective communication and describe how to overcome them.

5. Discuss the ways in which entrepreneurs can motivate their employees to achieve higher levels of performance.

A leader is one who knows the way, goes the way, and shows the way.

—John C. Maxwell

The way management treats their associates is exactly how the associates will then treat the customers.

—Sam M. Walton

The Entrepreneur's Role as Leader

1.

Explain the challenges involved in the entrepreneur's role as leader and what it takes to become a successful leader.

To be successful, an entrepreneur must assume a wide range of roles, tasks, and responsibilities, but none is more important than the role of leader. Some entrepreneurs are uncomfortable assuming this role, but they must learn to be effective leaders if their companies are to grow and reach their potential. **Leadership** is the process of influencing and inspiring others to work to achieve a common goal and then giving them the power and the freedom to achieve it. Without leadership ability, entrepreneurs—and their companies—never rise above mediocrity. Entrepreneurs can learn to be effective leaders, but the task requires dedication, discipline, and hard work. In the past, business owners often relied on an autocratic management style, one built on command and control. Today's workforce is more knowledgeable, has more options, and is more skilled and, as a result, expects a different, more sophisticated style of leadership. Companies that fail to provide that leadership are at risk of losing their best employees. Leaders of small companies must gather information and make decisions with lightning-fast speed, and they must give workers the resources and the freedom to solve problems and exploit opportunities as they arise. Effective leaders empower and embolden employees to act in the best interest of the business.

Until recently, experts compared a leader's job to that of a symphony orchestra conductor. Like the symphony leader, an entrepreneur made sure that everyone in the company played from the same score, coordinated individual efforts to produce a harmonious sound, and directed the orchestra members as they played. The conductor (entrepreneur) retained virtually all of the power and made all of the decisions about how the orchestra would play the music without any input from the musicians themselves. Today's successful entrepreneur, however, is more like the leader of a jazz band, which is known for its improvisation, innovation, creativity, and freewheeling style. "The success of a small [jazz band] rests on the ability to be agile and flexible, skills that are equally central to today's business world," says Michael Gold, founder of Jazz Impact, a company that teaches management skills through jazz.[1] Business leaders, like the leaders of jazz bands, should exhibit the following characteristics:

Innovative. Entrepreneurial leaders must step out of their own comfort zones to embrace new ideas; they avoid the comfort of complacency. They understand the very changes that created the opportunity for their new venture will force them to continue to adapt and innovate.

Passionate. One of the greatest strengths of entrepreneurs is their passion for their businesses. Members of their team feed off that passion and draw inspiration from it.

Willing to take risks. "[Taking] risk is not an option in jazz or for any company that wants to be solvent ten years from now," says Gold.[2] Entrepreneurs understand that there is risk associated with both pursuing new things but also in not pursuing something new that actually has great promise (opportunity cost).

Adaptable. Although leaders must stand on a bedrock of resolute values, like jazz band leaders, they must adapt their leadership styles to fit the situation and the people involved.

Management and leadership are not the same, yet both are essential to a company's success. Leadership without management is unbridled; management without leadership is uninspired. Leadership gets a small business going; management keeps it going. In other words, leaders are the architects of small businesses; managers are the builders.

Some entrepreneurs are good managers yet poor leaders; others are powerful leaders but weak managers. Although both are skills that can be developed, the best bet for entrepreneurs who are strong leaders is to hire people with solid management skills to help them to execute the vision they have for their companies. Stephen Covey, author of *Principle-Centered Leadership*, explains the difference between management and leadership in this way:

Leadership deals with people; management deals with things. You manage things; you lead people. Leadership deals with vision; management deals with logistics toward that vision. Leadership deals with doing the right things; management focuses on doing things right. Leadership deals with examining the paradigms on which you are operating; management operates within those paradigms. Leadership comes first, then management, but both are necessary.[3]

Leadership and management are intertwined; one without the other means that a small business is going nowhere. Leadership is especially important for companies in the growth phase,

when entrepreneurs are hiring employees (often for the first time) and must keep the company and everyone in it focused on its mission as growth tests every seam in the organizational structure. At this stage, selling everyone in the company on the mission, goals, and objectives for which the leader is aiming is crucial to a business's survival and success. "People don't want to be managed," says one CEO. "They want to be led."[4] Effective leaders exhibit certain behaviors. They do the following:

- *Create a set of values and beliefs for employees and passionately pursue them.* Employees look to their leaders for guidance in making decisions. True leaders focus attention on the principles, values, and beliefs on which they founded their companies.

- *Establish a culture of ethics.* One of the most important tasks facing leaders is to intentionally mold a highly ethical culture for their companies that reflects their values. They also must demonstrate the character and the courage necessary to stick to the ethical standards that they create—especially in the face of difficulty. Leaders' words ring hollow if they fail to "practice what they preach." Workers detect few contradictions faster than the hypocrisy of leaders who sell employees on one set of values and principles and then act according to a different set. Real leaders know that they set the ethical tone in the organization. Even small lapses in a leader's ethical standards can have a significant impact on a company's ethical climate. Workers know that they can trust leaders whose actions support their words. Similarly, they quickly learn not to trust leaders whose day-to-day dealings belie the principles they preach.

- *Define and then constantly reinforce the vision they have for the company.* Effective leaders have a clear vision of where they want their companies to go, and they concentrate on communicating that vision to those around them. Unfortunately, this is one area in which employees say their leaders could do a better job. Clarity of purpose is essential to a successful organization because people want to be a part of something that is bigger than they are; however, the purpose must be more than merely achieving continuous quarterly profits. As an entrepreneurial venture grows, communicating the vision of the business becomes the most important part of an entrepreneur's job as leader.

- *Respect and support employees.* To gain the respect of their employees, leaders must first respect those who work for them.

- *Create a climate of trust in the organization.* Leaders who demonstrate integrity soon win the trust of their employees, an essential ingredient in the success of any organization. Honest, open communication and a consistent pattern of leaders doing what they say they will do serve to build trust in a business. Research suggests that building trust among employees is another one of the most important tasks of leaders, wherever they may work. Building trust demands that leaders rely on three "Cs": competence (the leader is able to get the job done), consistency (the leader's actions are reliable, whatever the situation), and caring (the leader demonstrates compassion for those he or she leads).[5] Employees at small businesses are more likely to trust their leaders than employees at large companies.[6]

- *Focus employees' efforts on challenging goals and keep them driving toward those goals.* Effective leaders have a clear vision of where they want their companies to go, and they are able to communicate their vision to those around them. Leaders must repeatedly reinforce the goals they set for their companies.

- *Create an environment in which people have the motivation, the training, the resources, and the freedom to achieve the goals they have set.* Leaders know that *their* success is determined by the success of their followers. Effective leaders know that workers cannot do their jobs well unless they have the tools they need. They provide workers with not only the physical resources they need to excel but also the necessary intangible resources, such as training, coaching, and mentoring.

- *Communicate with employees.* Leaders recognize that helping workers see the company's overarching goal is just one part of effective communication. Encouraging employee feedback and then listening is just as vital. In other words, they know that communication is a two-way street.

- *Value the diversity of workers.* Smart business leaders recognize the value of their workers' varied skills, abilities, backgrounds, and interests. When channeled in the right direction, diversity can be a powerful weapon in achieving innovation and maintaining a competitive edge.

- *Celebrate workers' successes.* Effective leaders recognize that workers want to be winners and do everything they can to encourage top performance among their people. The rewards they give are not always financial; in many cases, a reward may be as simple as a handwritten congratulatory note. A true leader strives to make employees look good rather than expecting employees to make him or her look good.

- *Understand that leadership is multidimensional.* Smart leaders know that there is no single "best" style of leadership. The dimensions of leadership change depending on the people participating, the conditions and circumstances of the situation, and the desired outcome.

- *Value new ideas from employees.* Successful leaders know that because employees work every day on the front lines of the business, they see ways to improve quality, customer service, and business systems.

- *Understand that success really is a team effort.* Small companies typically depend more on their founding entrepreneurs than on anyone else. After all, someone has to take responsibility for the toughest decisions. However, effective leaders understand that their roles are only a small piece of the entire company puzzle. When they recognize this, they also build credibility with their workers. *Undercover Boss*, a television series on CBS, features CEOs of companies ranging from Waste Management and Choice Hotels to White Castle and Hooters who are disguised as new employees as they take on jobs on the front lines of their companies, where the "real work" is performed. In one episode, Larry McDonnell, CEO of Waste Management, cleans toilets and rides a garbage truck route. Along the way, he encounters hardworking people who do everything they can to help "the new guy" succeed and learns about the struggles workers face every day. In addition to seeing firsthand just how difficult many jobs can be, all of the CEOs receive a superb refresher course in how important every worker's role is in the success of a company and how the policies that they and other senior managers create often make workers' jobs harder.

- *Encourage creativity and risk taking throughout the organization.* Effective leaders recognize that in a rapidly changing competitive environment, everyone in the organization must make decisions with incomplete information and must be willing to take risks to succeed. Rather than punish workers who take risks and fail, effective leaders are willing to accept failure as a natural part of innovation and creativity. They know that innovative behavior is the key to future success. They do everything they can to encourage creativity among workers.

- *Become a catalyst for change.* With market and competitive conditions changing so rapidly, entrepreneurs must reinvent their companies constantly. Although leaders must cling to the values and principles that form the bedrock of their companies, they must be willing to change, sometimes radically, the policies, procedures, and processes within their businesses. If a company is headed in the wrong direction, the leader's job is to recognize that and get the company moving in the right direction. "No leader knows enough about the future to make the optimal decision every time, but it's better to set a clear course today and tackle problems that arise tomorrow," says Andy Grove, former CEO of Intel, the computer chip maker.[7]

- *Develop leadership talent.* Effective leaders look beyond themselves to spot tomorrow's leaders and take the time to help them grow into their leadership potential.

- *Maintain a sense of humor.* One of the most important tools a leader can have is a sense of humor. Without it, work can become dull and unexciting for everyone.

- *Keep an eye on the horizon.* Effective leaders are never satisfied with what they and their employees accomplished yesterday. They know that yesterday's successes are not enough to sustain their companies indefinitely. They see the importance of building and maintaining sufficient momentum to carry their companies to the next level.

Leading an organization, whatever its size, is one of the biggest challenges any entrepreneur faces, yet for an entrepreneur, leadership success is a key determinant of a company's success. Research suggests that there is no single "best" style of leadership; the style a leader uses depends, in part, on the situation at hand. Some situations are best suited for a participative leadership style, but in others, an authoritarian style may be best. Research by Daniel Goleman and others suggests that today's workers tend to respond more to adaptive, humble leaders who are results oriented and

who take the time to cultivate other leaders in the organization.[8] The practice is known as **servant leadership**, a phrase coined by Robert Greenleaf in 1970. Servant leaders are servants *first* and leaders second, putting their employees and their employees' needs ahead of their own. They are more concerned about empowering others in the organization than they are at enhancing their own power bases. "Servant leadership is about getting people to a higher level by leading people at a higher level," says author Ken Blanchard.[9] Researchers have discovered that three dimensions of leader behavior best define effective servant leaders: service, humility, and vision.[10]

Entrepreneurs cannot bestow the mantle of "leader" on themselves. Managers may inherit their subordinates, but leaders have to *earn* their followers. An entrepreneur's employees determine whether he or she is worthy of leadership. *Without followers, there are no leaders*. Astute leaders know that their success depends on their employees' success. After all, the employees actually do the work, implement the strategies, and produce the results. Successful leaders establish an environment in which their followers can achieve success. Joe Tortorice Jr., founder and CEO of Jason's Deli, a privately owned company with 225 restaurants in 28 states, says, "People are looking for three things out of their leaders: direction, trust, and hope."[11]

To be effective, entrepreneurial leaders must perform four vital tasks:

- Hire the right employees and constantly improve their skills.
- Build an organizational culture and structure that allows both workers and the company to reach their potential.
- Communicate the vision and the values of the company effectively and create an environment of trust among workers.
- Motivate workers to higher levels of performance.

Hiring the Right Employees: The Company's Future Depends on It

The decision to hire a new employee is an important one for every business but is *especially* important for small businesses because the impact of a single hire on a small company is significant. "As an entrepreneur, every single hire is critical," says Stephen Fairley, CEO of the Rainmaker Institute, a business coaching firm.[12] Every new employee a business owner hires determines the heights to which the company can climb—or the depths to which it will plunge. "Bad hires" can poison a small company's culture for years.

Hiring mistakes also are expensive. Although some employee turnover is healthy, high employee turnover rates cost companies billions of dollars annually. According to the American Management Association, the cost of employee turnover falls within a range between 25 percent of annual salary and benefits for entry-level employees up to as much as 200 percent for executive, managerial, and sales positions.[13] Consider a small business with 20 employees that has an employee turnover rate of 21 percent, the national average. Assuming that the employees earn wages that are consistent with the national average (about $42,980) and receive benefits that total 25 percent of their wages, the *minimum* turnover cost (above employees' salaries and benefits) to the company is $56,411, but it could cost as much as $451,290!

Unfortunately, hiring mistakes in business are all too common. The culprit in most cases is the company's selection and hiring process. A recent survey by SurePayroll reports that 69 percent of business owners say that they have made *at least* one bad hire in the last year.[14] One study reported in the *Harvard Business Review* concludes that bad hiring decisions cause 80 percent of employee turnover.[15] Some small business owners invest more time and effort into deciding which copy machine to lease than which employee to hire for a key position.

The most common causes of a company's poor hiring decisions include the following:

- Managers who rely on candidates' descriptions of themselves rather than requiring candidates to demonstrate their abilities.
- Managers who fail to follow a consistent, evidence-based selection process. Forty-seven percent of managers admit that they make hiring decisions in minutes or less, and 44 percent of managers say that they rely on their intuition to make hiring decisions.
- Managers who fail to provide candidates with sufficient information about what the jobs for which they are hiring entail.[16]

2.

Describe the importance of hiring the right employees and how to avoid making hiring mistakes.

Source: Scott Adams/Universal Uclick.

As crucial as finding good employees is to a small company's future, it is no easy task because entrepreneurs face a labor shortage, particularly among knowledge-based workers. The severity of this shortage will worsen as Baby Boomers retire in increasing numbers and the growth rate of the U.S. labor force slows.

The following guidelines can help small business managers avoid costly hiring mistakes.

How to Hire Winners

Even though the importance of hiring decisions is magnified in small companies, small businesses are most likely to make hiring mistakes because they lack the human resources experts and the disciplined hiring procedures that large companies have. In many small businesses, the hiring process is informal, and the results often are unpredictable. The following guidelines can help entrepreneurs to hire winners and avoid making costly hiring mistakes as they build their team of employees.

COMMIT TO HIRING THE BEST TALENT Smart entrepreneurs follow the old adage, "A players hire A players; B players hire C players." They are not threatened by hiring people who may be smarter and more talented than they are. In fact, they recognize that doing so is the best way to build a quality team.

ELEVATE RECRUITING TO A STRATEGIC POSITION IN THE COMPANY Assembling a quality workforce begins with a sound recruiting effort. By investing time and money at this crucial phase of the staffing process, entrepreneurs can generate spectacular savings down the road by hiring the best talent. The recruiting process is the starting point for building quality into a company. Recruiting is so important that many entrepreneurs choose to become actively involved in the process themselves. Visionary entrepreneurs *never* stop recruiting because top-quality talent is hard to find and is extremely valuable. Tom Bonney, founder of CMF Associates, a fast-growing financial consulting firm in Philadelphia, knows that finding superior talent is essential to the success of his service business. "I never stop recruiting," he says. "Even if I don't have a need, I am always looking."[17]

Attracting a pool of qualified job candidates requires not only constant attention but also creativity, especially among smaller companies that often find it difficult to match the more generous offers that large companies make. With a sound recruiting strategy and a willingness to look in new places, however, smaller companies *can* hire and retain high caliber employees. The following techniques will help.

Look Inside the Company First One of the best sources for top prospects is right inside the company itself. According to a report from the Saratoga Institute, the average cost of finding and hiring someone from outside the company is 1.7 times more than an internal hire. Additionally, research shows that between 40 and 60 percent of external hires are successful, compared to 75 percent of internal hires.[18] A promotion-from-within policy serves as an incentive for existing workers to upgrade their skills and to produce results. In addition, an entrepreneur already knows the employee's work habits and the employee already understands the company's culture. "Internal hires tend to do better than outsiders, so if you promote from within you're likely to reduce your risk," says Margaret Heffernan, successful entrepreneur and author. "You want to encourage the talent you already have, so work hard to discover what you have before you go looking for more."[19]

Encourage Employee Referrals To cope with the shortage of available talent, many companies offer their employees (and others) bonuses for referring candidates who come to work and prove to be valuable employees. Employees serve as reliable screens because they do not want to jeopardize their reputations with their employer. Employee referrals from social networks such as LinkedIn, Facebook, and Twitter allow employers to tap into their employees' network of contacts. A recent study by Jobvite, an online employee-recruiting platform, found that employee referrals now account for 40 percent of all hires. Employee hires have a higher conversion rate to successful hiring, get on the job more quickly, have higher job satisfaction once on the job, and are less expensive to recruit.[20] Sodexo, a food service and facilities management company that works with many universities across the country, reports that referred employees are 10 times more likely to be hired than other applicants. "We're focusing on what will be most efficient," says Arie Ball, vice president for talent acquisition at Sodexo. "And it's just easier to connect on social networks than it used to be."[21]

Use Multiple Channels to Recruit Talent Although newspaper ads have historically been the most common source of recruiting, employers are now using a variety of channels to recruit employees (Figure 21.1). In addition to referrals, posting positions with online job boards, utilizing job listings on career-oriented Web sites, and social media have become common source of finding employees. In fact, when it comes to recruiting, 93 percent of employers use Linked-in, 66 percent use Facebook, and 54 percent use Twitter; 70 percent of employers say that they have hired an employee whom they found through these sites.[22]

Recruit on Campus For many employers, college and university campuses remain an excellent source of workers, especially for entry-level positions. After screening résumés, a recruiter can interview a dozen or more high-potential students in just one day. Companies must be sure that the recruiters they send to campuses are professional, polished, and prepared to represent the company well because 42 percent of students say that their impression of a recruiter is the primary determinant of their perception of a company.[23]

Forge Relationships with Schools to Gain Access to Interns Some employers have found that forging long-term relationships with schools and other institutions can provide a valuable source of workers. As colleges and universities actively seek internship opportunities for their students, small businesses can identify potential employees by hosting one or more students for a semester or for the summer. The company has an opportunity to observe the student's work habits and, if positive, sell the student on a permanent position on his or her graduation.

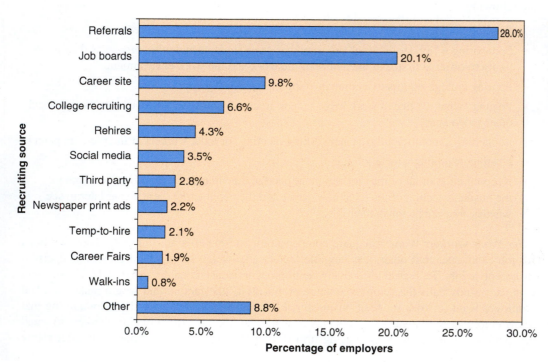

FIGURE 21.1

Sources of Recruiting Employees

Source: "2012 Sources of Hire: Channels That Influence," CareerXroads, *www.careerxroads .com/news/slideshows.asp.*

ENTREPRENEURIAL PROFILE: Kelsey Meyer: Digital Talent Agents Kelsey Meyer, co-founder of the public relations firm Digital Talent Agents, likes to test interns by giving them loose directions and then setting them free to complete their projects. "We want them to figure out the best way to do things on their own," says Meyer. "It's interesting to watch what they come up with when we aren't breathing over their shoulders." Meyer will occasionally ask interns for ideas for improving the business. "We make it into a contest," she says. "If we are having a problem with a client, for example, I will task my interns with coming up with at least one solution by 2:00 P.M. The person whose idea we choose to use will receive a gift certificate for a smoothie at a nearby restaurant." This type of exercise has proved to be a valuable way to assess interns for full-time positions. "When you are looking to hire someone full-time, you want to know how they will add value to your company," Meyer says. "This exercise teaches them how to think strategically; the more strategically an employee thinks, the less they will think about having to leave at 5:00 P.M. They will instead be thinking about how they can contribute to making your company better."[24]

Recruit "Retired" Workers By 2016, nearly 35 percent of workers in the United States will 55 or older, and in 2019, the youngest members of the Baby Boom generation will turn 55.[25] According to the American Association of Retired Persons, three-fourths of these Baby Boomers plan to continue working after reaching retirement age. Some do this to maintain their lifestyles, but many are choosing to do this because they want to continue working.[26] To avoid labor shortages, small businesses must be ready to hire them, perhaps as part-time employees. With a lifetime of work experience, time on their hands, and a strong work ethic, retired workers can be the ideal solution to many entrepreneurs' labor problems. One survey by WorldatWork, an international association of human resource professionals, reports that just 49.4 percent of employers proactively pursue older workers in their recruiting efforts.[27] Older employees can be a valuable asset to small firms.

ENTREPRENEURIAL PROFILE: Kevin Dent: Dentco Kevin Dent, CEO of Dentco, has turned to older workers when hiring new employees for his exterior services management firm. A growing part of Dentco's workforce is the part-time quality service inspectors whose ages range from the mid-fifties to the mid-eighties. These inspectors work three to 10 days a month traveling from one job site to another, walking the grounds, taking photos, filing computer reports, and meeting clients. "It's worked miraculously well," says Dent. "There's a whole different work ethic with senior citizens. They know how to handle people at the sites, make great ambassadors, and have empathy and discipline. They meet their commitments."[28]

Consider Using Offbeat Recruiting Techniques To attract the workers they need to support their growing businesses, some entrepreneurs have resorted to creative recruiting techniques such as the following:

- Sending young recruiters to mingle with college students on spring break.
- Sponsoring a "job shadowing" program that gives students and other prospects the opportunity to observe firsthand the nature of the work and the work environment.
- Inviting college seniors to a company tailgating party at a sports event.
- Posting "what it's like to work here" videos created by current employees on YouTube and other video sites.
- Launching a monthly industry networking meeting for local workers at Internet companies.
- Hosting or joining a local job fair.
- Keeping a file of all of the workers mentioned in the "People on the Move" column in the business section of the local newspaper and then contacting them a year later to see whether they are happy in their jobs.[29]

Offer What Workers Want Adequate compensation and benefits are important considerations for job candidates, but other, less tangible factors also weigh heavily in a prospect's decision to accept a job. To recruit effectively, entrepreneurs must consider what a McKinsey and Company study calls the "employee value proposition," the factors that would make the ideal employee want to work for their businesses. Flexible work schedules and telecommuting that allow employees to balance the demands of work and life can attract quality workers to small companies. In fact, a study by staffing firm Robert Half International reports that after salary

and benefits, flexible work schedules and telecommuting are the most important incentives in attracting employees.[30]

CREATE PRACTICAL JOB DESCRIPTIONS AND JOB SPECIFICATIONS Business owners must recognize that what they do *before* they ever start interviewing candidates for a position determines to a great extent how successful they will be at hiring winners. The first step is to perform a **job analysis**, the process by which a firm determines the duties and nature of the jobs to be filled and the skills and experience required of the people who are to fill them. Without a proper job analysis, a hiring decision is, at best, a coin toss. The first step in conducting a job analysis is to develop a **job description**, a written statement of the duties, responsibilities, reporting relationships, working conditions, and methods and techniques as well as materials and equipment used in a job. A results-oriented job description explains what a job entails and the duties the person filling it is expected to perform. A detailed job description includes a job title, job summary, duties to be performed, nature of supervision, job's relationship to others in the company, working conditions, and definitions of job-specific terms.

Preparing job descriptions may be one of the most important parts of the hiring process because it creates a blueprint for the job. Without this blueprint, managers tend to hire the person with experience whom they like the best. Useful sources of information for writing job descriptions include the manager's knowledge of the job, the workers currently holding the job, and the *Dictionary of Occupational Titles*, available at most libraries and online. The *Dictionary of Occupational Titles,* published by the Department of Labor, lists more than 20,000 job titles and descriptions and serves as a useful tool for getting an entrepreneur started when writing job descriptions.

The second objective of a job analysis is to create a **job specification**, a written statement of the qualifications and characteristics needed for a job stated in terms such as education, skills, and experience. A job specification shows a small business manager what kind of person to recruit and establishes the standards that an applicant must meet to do the job well. In essence, it is a written "success profile" of the ideal employee. Does the person have to be a good listener, empathetic, well organized, decisive, or a "self-starter"? Should he or she have experience in Java programming? For example, an entrepreneur about to hire a new employee who will be telecommuting from home would look for someone with excellent communication skills, problem-solving ability, a strong work ethic, and the ability to use technology comfortably. One of the best ways to develop this success profile is to study the top performers currently working for the company and to identify the characteristics that make them successful. Table 21.1 provides an example that links the tasks for a sales representative's job (drawn from the job description) to the traits or characteristics that an entrepreneur identified as necessary to succeed in that job. These traits become the foundation for writing the job specification.

SCREEN RÉSUMÉS Reviewing candidates' résumés is the starting point for screening prospective employees. A survey from the Corporate Executive Board reports that the average number of

TABLE 21.1 Linking Tasks from the Job Description to the Traits Needed to Perform a Sales Representative's Job

Job Task	Trait or Characteristic
Generate new leads and close new sales.	Outgoing, strong communication skills, persuasive, friendly
Make 15 "cold calls" per week.	A self-starter, determined, optimistic, independent, confident
Analyze customer needs and recommend proper equipment.	Good listener, intuitive, patient, empathetic
Counsel customers about options and features required.	Organized, polished speaker, "other" oriented
Prepare and explain financing methods; negotiate finance contracts.	Honest, mathematically oriented, comfortable with numbers, understands basics of finance, computer literate
Retain existing customers.	Relationship builder, customer focused

TABLE 21.2 Résumé Bloopers

All of the following résumé bloopers are real. Would you hire someone who committed these blunders?

- A job candidate listed God as one of her references (but, strangely enough, did not list a telephone number).
- A woman listed "alligator watching" as one of her hobbies.
- A man included "Master of Time and the Universe" as part of his work experience.
- One candidate's résumé was 24 pages long, and she had been in the workforce only five years.
- A candidate wrote that he was "looking for a full-time position with minimal time commitment."
- Under "Accomplishments," one man listed that he "finished eighth in his class of 10."
- An applicant claimed, "You will want me to be Head Honcho in no time."
- A job seeker claimed that he spoke "English and Spinach."
- One résumé listed the applicant's skills as follows: "Strong Work Ethic, Attention to Detail, Team Player, Self Motivated, Attention to Detail."
- A woman who sent her résumé and cover letter failed to delete comments from someone who proofed the materials, which included such comments as "I don't think you want to say this about yourself here."
- Under "Qualifications," one woman explained that her "twin sister has an accounting degree."
- A candidate listed as skills, "written communication = 3 years; verbal communication = 5 years."
- Under education, one young applicant said, "I have a bachelorette degree in computers."
- One job applicant claimed that he had 28 dog years of experience in sales (four human years, we assume).
- A candidate included a video with his résumé. The purpose of the video: to hypnotize the manager and persuade him to hire the man.
- One applicant used colored paper and glued glitter around the edges.
- One job candidate claimed that he possessed "demonstrated ability at multi-tasting."
- Under "Education," one applicant claimed to have "repeated courses repeatedly."

Sources: Based on "Hiring Managers Share Most Memorable Resume Mistakes in New CareerBuilder Survey," *CareerBuilder*, September 15, 2010, *www.careerbuilder.com/share/aboutus/pressreleasesdetail.aspx?id=pr58 6&sd=9/15/2010&ed=9/15/2010*; "150 Funniest Resume Mistakes, Bloopers, and Blunders Ever," Job Mob, *jobmob.co.il/blog/funniest-resume-mistakes*.

applicants for each job opening is 118.[31] In addition, 38 percent of managers say that they spend less than one minute reviewing a résumé, and 18 percent spend less than 30 seconds![32] Table 21.2 describes some unusual items lifted from actual résumés.

CHECK REFERENCES Entrepreneurs should take the time to check *every* applicant's references. Although many entrepreneurs see checking references as a formality and pay little attention to it, others realize the need to protect themselves (and their customers) from hiring unscrupulous workers. A reference check is necessary because more than half of job seekers lie in their résumés, often by inflating their job titles.[33] A recent study by OfficeTeam reports that employers eliminate 21 percent of job applicants after they conduct reference checks.[34] Rather than rely only on the references that candidates list on their résumés, wise employers call an applicant's previous employers and talk to their immediate supervisors to get a clear picture of the applicant's job performance, character, and work habits.

ENTREPRENEURIAL PROFILE: Andy Levine: Development Counselors International Andy Levine, president of Development Counselors International, now requires 12 references for the final stage of the interview process. "It can be pretty amusing when you ask for 12 references. Some candidates have an email to us within an hour; some we never hear from again" says Levine. "When I call references, I start by trying to get them comfortable. I make it clear that what they say will not travel back to the person. Then I often ask, 'If you had to pick three words to describe this person, what are the first that come to mind?' It's very interesting, the picture that emerges after you've done eight or nine of these interviews."[35]

PLAN AN EFFECTIVE INTERVIEW Once an entrepreneur knows what to look for in a job candidate, he or she can develop a plan for conducting an informative job interview. Too often, business owners go into an interview unprepared, and as a result, they fail to get the information they need to judge the candidate's qualifications, qualities, and suitability for the job. A common symptom of failing to prepare for an interview is that the interviewer rather than the candidate does most of the talking. Effective interviewers spend about 25 percent of the interview talking and about 75 percent listening. This can be hard for entrepreneurs who are excited about their businesses and who are used to giving a "pitch" to everyone who will listen! Despite their popularity, interviews are less reliable in predicting job performance than samples of a candidate's work, job-knowledge tests, and peer ratings of past job performance.[36] The following tips improve the quality of the interview process.

Involve Others in the Interview Process Solo interviews are prone to errors. A better process is to involve other employees, particularly employees with whom the prospect would be working, in the interview process either individually or as part of a panel.

Develop a Series of Core Questions and Ask Them of Every Candidate Entrepreneurs benefit by relying on a set of relevant questions they ask in every interview. Doing so gives the screening process consistency, but entrepreneurs should customize each interview using impromptu questions based on an individual's responses. These questions should relate to the most important job requirements (tasks, demands, and organizational cultural) and employee requirements (knowledge, skills, and personality/values).

Ask Open-Ended Questions Open-ended questions demanding more than a yes-or-no response are most effective because they encourage candidates to talk about their work experience in a way that will disclose the presence or the absence of the traits and characteristics entrepreneurs are seeking. Questions that begin with phrases such as "Tell me about a time . . ." encourage candidates to tell stories that allow the interviewer to assess their fit with the organization's culture. Peter Bregman, CEO of Bregman Partners, a company that helps businesses implement change, says that one of the most revealing questions that an interviewer can ask candidates is "What do you do in your spare time?" To emphasize the importance of a candidate's hobbies, Bregman points to Captain C. B. "Sully" Sullenberger, the airline pilot who safely landed a disabled jet with 155 passengers on the Hudson River using skills that he learned from his hobby: flying gliders.[37]

Present Hypothetical Situations Building the interview around job-specific hypothetical situations gives the owner a preview of the candidate's actual work habits and attitudes. Rather than telling interviewers about what candidates might do, these scenarios give them insight into what candidates actually do (or have done) in job-related situations.

Probe for Specific Examples in the Candidate's Past Work Experience That Demonstrate the Necessary Traits and Characteristics A common mistake interviewers make is failing to get candidates to provide the detail they need to make an informed decision.

Inquire About Recent Successes and Failures Smart interviewers look for candidates who describe them both with equal enthusiasm because they know that peak performers put as much into their failures as they do their successes and usually learn something valuable from their failures. Ask the candidates to provide examples of their successes and failures.

Create an Informal Setting Select a "noninterview" location that allows several employees to observe the candidate in an informal setting. Taking candidates on a plant tour or setting up a coffee break gives everyone a chance to judge a candidate's interpersonal skills and personality outside the formal interview process. These informal settings can be revealing. At Zappos, the online shoe store, recruiters often interview shuttle service drivers and administrative assistants to discover how job candidates treated them. "I want to know about that interaction," says recruiter Andrew Kovacs.[38]

Table 21.3 shows an example of some interview questions the manager might use to uncover the traits and characteristics he or she seeks in a top-performing sales representative (refer to Table 21.1).

TABLE 21.3 Interview Questions for Candidates for a Sales Representative Position

Trait or Characteristic	Question
Outgoing, persuasive, friendly, a self-starter, determined, optimistic, independent, confident	How do you persuade reluctant prospects to buy? Can you give an example?
Good listener, patient, empathetic, organized, polished speaker, other oriented	What would you say to a fellow salesperson that was getting more than her share of rejections and was having difficulty getting appointments?
Honest, customer oriented, relationship builder	How do you feel when someone questions the truth of what you say?
	Can you give an example of how you handled this situation?

Other questions:

- If you owned a company, why would you hire yourself?
- If you were head of your department, what would you do differently? Why?
- How do you acknowledge the contributions of others in your department?

 Entrepreneurship in Action

Facebook Interview Process

Facebook has developed a rigorous application process for prospective employees that ensures that the company gets top talent and finds people who fit the Facebook culture. For those interested in technical jobs, Facebook offers an online timed coding challenge, open to all, where the best performers automatically win a phone interview.

The phone interview is the first major step in the hiring process. It covers the candidate's résumé and questions about previous work experiences, especially about the leadership roles they have held in the past. Facebook recruiters pay close attention to candidates who have been successful in previous positions and show a passion for the work they did on that job. The length of the résumé is not important at Facebook. What recruiters look for is demonstrated excellence and accomplishments. For some positions, applicants may go through second, third, and even fourth telephone interviews. Some of these telephone screenings can be quite lengthy and may involve collaborative online problem-solving exercises with the person conducting the interview.

Those who pass the telephone screening are flown to the company's Silicon Valley headquarters for a series of on-site interviews with the hiring manager and several team members who are part of the work group. These interviews are designed to determine candidates' skills and their fit with the Facebook's unique culture. Team members ask many questions about the candidate's experience working in teams and make note of how they interact with the team members doing the interview. If a candidate is applying for a technical position, he or she faces more skill-based challenges and a take-home test. The team assesses not only the candidate's skills and abilities but also the approach they take to problem solving. Creative solutions are particularly important.

Candidates have the opportunity to offer suggestions on how they might make the Facebook product or user experience better. Interviewers want to see specific solutions and metrics to assess how the improvements are working.

Some of the typical questions asked in Facebook interviews are not that typical. Questions such as, "If you were an animal what kind would you be and why?," "What is the difference between Facebook ads and Google Ads?," "Should Facebook be available in China?," and "What do you see as Facebook's biggest challenge in the next five years?" are common. The goal of these questions is to see how each candidate responds when faced with questions that they could not prepare for in advance.

Facebook employees who are involved in the interview process then make a collective decision on the candidate's fit for the position.

Interview teams expect applicants for nontechnical positions, such as business operations, sales, marketing, or analytics, to have done a great deal of research into Facebook before they arrive for the interview. "If you are going to work for Facebook tomorrow, what project do you want to work on?" was one question posed to a recent applicant for a market research position.

Facebook managers recognize that once they have made the decision to hire, the candidate also must make a decision about his or her fit with the company's culture. "After the interview, I wasn't sure if I would be happy working at Facebook," says one software engineer candidate, "so they let me come back and speak with my would-be manager and director, as well as some coworkers, so I could make a good decision."

"We're primarily looking for builders," says Thomas Arnold, head of recruitment at Facebook, which prides itself on its entrepreneurial spirit. The company has maintained a flat organizational structure that is best for those who seek to be empowered in their jobs and are highly self-motivated.

Strong applicants also "just get the social space," says Arnold, and interviewers confirm this by making sure that the

candidate is an active Facebook user. If an applicant has not used his or her Facebook account for several weeks, recruiters take this is a sign that he or she will not be a good fit.

1. What are the principles behind Facebook's interview process? Do you think it is too complex? Explain.

2. What does the interview process tell you about Facebook's culture? Explain.

3. Can Facebook's interview process be applied in companies that are not technology based? How might it be applied to a company that makes golf clubs?

4. Why is hiring new employees so important for a small business? Why is it so difficult? How does the process at Facebook address some of the challenges that companies face when hiring?

Sources: Based on Claire Gordon, "Getting a Job at Facebook: Inside the 'Meritocratic' Hiring Process," *AOL Jobs,* October 5, 2012, *http://jobs.aol.com/articles/2012/10/05/want-to-get-a-job-at-facebook-weve-demystified-the-hiring-proc*; Jacquelyn Smith, "Ten Things to Know Before Interviewing at Facebook," *Forbes,* February 2, 2012, *www.forbes.com/sites/jacquelynsmith/2012/02/02/ten-things-to-know-before-interviewing-at-facebook*; Ben Parr, "How to: Land a Job at Facebook," *Mashable,* April 3, 2011, *mashable.com/2011/04/03/facebook-jobs-2.*

Conducting the Interview

An effective interview contains three phases: breaking the ice, asking questions, and selling the candidate on the company.

BREAKING THE ICE In the opening phase of the interview, the entrepreneur's primary goal is to create a relaxed environment. Icebreakers—questions about a hobby or special interest—get the candidate to relax. These icebreaker questions also allow the interviewer an opportunity to gain valuable insight into the person. These questions generate little or no pressure, allowing the interviewee to feel free to expound on something he or she knows a great deal about.

ASKING QUESTIONS During the second phase of the interview, employers ask the questions from their question bank to determine the applicant's suitability for the job. Employers' primary job at this point is to *listen.* They also take notes during the interview to help them ask follow-up questions based on a candidate's comments and to evaluate a candidate after the interview is over. Experienced interviewers also pay close attention to a candidate's nonverbal clues, or body language, during the interview. They know that candidates may be able to say exactly what they want with their words but that the candidate's body language does not lie!

Some of the most valuable interview questions attempt to gain insight into the candidate's ability to reason, be logical, and be creative. In a **puzzle interview**, the goal is to determine how job candidates think by asking them offbeat, unexpected questions, such as "How would you weigh an airplane without scales?," "Why are man-hole covers round?" or "How would you determine the height of a building using only a barometer?" At Zappos, interviewers ask candidates which superhero they would like to be and why.[39] Usually, the logic and creativity that a candidate uses to derive an answer is much more important than the answer itself.

ENTREPRENEURIAL PROFILE: Bob Bernstein: Bongo Java Coffee When Bob Bernstein founded Bongo Java Coffee, he had a clear vision for the kind of culture he wanted to create in his coffee shops, one that reflects his strongly held political convictions and his rather offbeat sense of humor. Bernstein decided that to achieve a common culture across his stores, he should implement a selection process that would ensure that he hired employees who would embrace and sustain the culture. The employee application became one of the vehicles to help identify people who would fit into Bongo Java's culture. Although the application begins like most others, gathering basic information about the person's experience, it soon moves into some rather nontraditional questions. Because community engagement and activism are important to Bernstein, he asks, "What volunteer activities are you currently or have you recently been involved in?" Other questions include "If you could have a one hour conversation with anybody living or dead, who would it be and what would you talk about?" and "If you were to write your autobiography what would the title be? Explain." Although these are not typical questions one might ask a potential employee, Bernstein believes that they have helped him find people who fit in his company's culture. As a result, Bongo Java's employee turnover is well below that of a typical coffee shop.[40]

Another interview format is the **situational interview**, in which the interviewer gives candidates a typical job-related situation (sometimes in the form of a role-playing exercise) and presents a series of open-ended questions to assess how the candidates might respond. One entrepreneur had a candidate deal with an "angry customer" who was played by a fellow interviewer.

Studies show that situational interviews have a 54 percent accuracy rate in predicting future job performance, much higher than the 7 percent accuracy rate of the traditional interview.[41]

The **peer-to-peer interview** may provide a closer look at how prospective employees will get along with other staff. Applicants meet one-on-one with potential peers to ask questions about the job and the company. The employees then share their assessments with the manager. This interviewing technique is becoming more common in companies, especially those in which work is team based. "In a small organization, you're going to spend a lot of time together," says Michael Harris, an expert in peer-to-peer interviews. "It becomes even more important for the entrepreneur to share some of the [hiring] responsibility with the other employees."[42] Because employees are involved in the hiring process, they feel empowered and "buy into" the hiring process, which can be good for morale and productivity. Peer interviews also allow applicants to gain insight into an organization's culture.

ENTREPRENEURIAL PROFILE: Clint Smith and Sara McManigal: Emma When applying to Emma, an e-mail marketing business in Nashville, Tennessee, prospective employees go through an intense screening process, and peer-to-peer interviews are a critical aspect of it. "It gets the team invested in the new hire before they ever start," says Sara McManigal, director of talent at Emma. "If staffers are able to participate in the interview and help make the decision as to whom we hire, they're more committed to the success of that hire once they start. The early commitment aids in keeping staffers around longer, and it allows the staffers to build a great team culture." During the early part of the interview process, all team members take part in interviewing prospective new members for their teams, but they expand the use of peer-to-peer interviewing as a last step in the interviewing process. "At the very end of the process," explains Clint Smith, cofounder and CEO of Emma, "We tend to have something called 'all hands.' The candidate has already been vetted by the team and the team leader. Everybody's feeling pretty good, so we get a hodgepodge of folks from around the company and look for that final cultural fit." Peer-to-peer interviewing plays a fundamental role in screening prospective employees at Emma throughout the hiring process because these interviews carry as much weight as the actual hiring manager's recommendation.[43]

Managers should conduct training sessions with employees who participate in the interviews to make certain that they know which questions are illegal and keep their questions job related.[44] The Equal Employment Opportunity Commission does not outlaw specific interview questions; rather, it recognizes that some questions can result in employment discrimination. If a candidate files charges of discrimination against a company, the burden of proof shifts to the employer to prove that all preemployment questions were job related and nondiscriminatory. Table 21.4 offers a quiz to help entrepreneurs understand the types of questions are most likely to result in charges of discrimination.

The goal of the interview process is to find someone who is qualified to do the job well. By steering clear of questions about subjects that are peripheral to the job itself, employers are less likely to ask questions that will land them in court. Wise entrepreneurs ask their attorneys to review their bank of questions before using them in an interview. Table 21.5 describes a simple test for determining whether an interview question might be considered discriminatory.

SELLING THE CANDIDATE ON THE COMPANY "A" players want to play for "A" teams. In the final phase of the interview, when employers have an attractive candidate, they should sell the benefits of working for the company. This phase begins by allowing the candidate to ask questions about the company, the job, or other issues. Experienced interviewers note the nature of these questions and the insights they give into the candidate's personality. This part of the interview offers employers a prime opportunity to explain to the candidate why the company is an attractive place to work. The best candidates will have other offers, and it is up to the entrepreneur to make sure they leave the interview wanting to work for the company. Finally, before closing the interview, employers should thank the candidates and tell them what happens next, such as "We will be contacting you about our decision within two weeks."

BACKGROUND CHECKS Background checks are essential. In addition to turning up convictions for criminal activity, a background check can show whether a job candidate has been convicted of stealing from a previous employer. A check of a candidate's driving records will show convictions for DUI and other traffic violations. This information can save an entrepreneur thousands of dollars by avoiding a bad hire at a cost of $50 or less for a basic criminal records check. BeenVerified

TABLE 21.4 Is It Legal?

Legal	Illegal	Interview Question
❏	❏	1. Are you currently using illegal drugs?
❏	❏	2. Have you ever been arrested?
❏	❏	3. Do you have any children or do you plan to have children?
❏	❏	4. When and where were you born?
❏	❏	5. Is there any limit on your ability to work overtime or travel?
❏	❏	6. How tall are you? How much do you weigh?
❏	❏	7. Do you drink alcohol?
❏	❏	8. How much alcohol do you drink each week?
❏	❏	9. Would your religious beliefs interfere with your ability to do the job?
❏	❏	10. What contraceptive practices do you use?
❏	❏	11. Are you HIV positive?
❏	❏	12. Have you ever filed a lawsuit or worker's compensation claim against a former employer?
❏	❏	13. Do you have physical/mental disabilities that would interfere with doing your job?
❏	❏	14. Are you a U.S. citizen?
❏	❏	15. What is your Facebook password?

Answers: 1. Legal. 2. Illegal. Employers cannot ask about an applicant's arrest record, but they can ask whether a candidate has ever been convicted of a crime. 3. Illegal. Employers cannot ask questions that could lead to discrimination against a particular group (e.g., women, physically challenged, and so on). 4. Illegal. The Civil Rights Act of 1964 bans discrimination on the basis of race, color, sex, religion, or national origin. 5. Legal. 6. Illegal. Unless a person's physical characteristics are necessary for job performance (e.g., lifting 100-pound sacks of mulch), employers cannot ask candidates such questions. 7. Legal. 8. Illegal. Notice the fine line between question 7 and question 8; this is what makes interviewing challenging. 9. Illegal. This question would violate the Civil Rights Act of 1964. 10. Illegal. What relevance would this have to an employee's job performance? 11. Illegal. Under the Americans with Disabilities Act, which prohibits discrimination against people with disabilities, people who are HIV positive or who have AIDS are considered "disabled." 12. Illegal. Workers who file such suits are protected from retribution by a variety of federal and state laws. 13. Illegal. This question also would violate the Americans with Disabilities Act. 14. Illegal. This question violates the Civil Rights Act of 1964. 15. Currently legal—but creepy—and creates the possibility that employers would have access to information about which they cannot legally ask, such as religion, marital status, and others, which creates a potential legal liability.

TABLE 21.5 A Guide for Interview Questions

Small business owners can use the "OUCH" test as a guide for determining whether an interview question might be considered discriminatory:

- Does the question *Omit* references to race, religion, color, sex, or national origin?
- Does the question *Unfairly* screen out a particular class of people?
- Can you *Consistently* apply the question to every applicant?
- Does the question *Have* job relatedness and business necessity?

offers an app for smart phones that gives one free basic background check a month, with each additional check costing $9.95. "A lot of times, employers argue that a background check is too expensive," says Zuni Corkerton, president of Hilliard, Ohio-based RefCheck Information Services Inc. "But the litigation that comes as a result of not having done their due diligence and having been negligent in their hiring process can be far greater."[45] Brad Carlson, vice president of sales and marketing for Minneapolis-based Orange Tree Employment Screening says, "All it takes is one theft, and you can pay for a background screening."[46] Some employers also conduct credit checks on potential employees if the position involves handling large amounts of cash or if they will have access to confidential financial information within the company.

A background check is a basic step in avoiding charges of negligent hiring, the failure to investigate the background of a prospective employee who proves to be dangerous to customers or other employees. For example, a Nebraska delivery driver for a pizza chain attacked a woman after delivering a pizza to her home. The employee had a previous sexual-assault conviction that a simple background check would have detected, preventing a tragedy, damage to the company's reputation, and the resulting litigation. A court ordered the pizza chain to pay the victim $175,000.[47]

Checking potential employees' social networking pages such as Facebook and Twitter also can provide a revealing look at their character. A study by CareerBuilder reports that 37 percent of employers investigate job candidates' Facebook pages and that one-third have discovered something there that caused them to reject a candidate.[48] Figure 21.2 displays the most common reasons that employers do not hire an employee based on what they discovered when conducting background checks on applicants' social media pages.

EMPLOYMENT TESTS   Although various state and federal laws have made using employment tests as screening devices more difficult in recent years, many companies find them quite useful. To avoid charges of discrimination, entrepreneurs must be able to prove that the employment tests they use are both valid and reliable. A **valid test** is one that measures what it is intended to measure, for example, aptitude for selling, creativity, or integrity. A **reliable test** is one that measures consistently over time. Employers must also be sure that the tests they use measure aptitudes and factors that are job related. Many testing organizations offer ready-made tests that have been proved to be both valid and reliable, and entrepreneurs can use these tests safely. In today's environment, if a test has not been validated and proved to be reliable or is not job related, it is best not to use it.

ENTREPRENEURIAL PROFILE: Richard Linder: PCA Skin Inc. PCA Skin Inc., a company based in Scottsdale, Arizona, that develops clinical skin care products, recently began using personality-based employment tests, and managers say that the tests, which take only 10 minutes to administer, have helped the company improve its hiring process. "We have had 65 candidates take the test," says CEO Richard Linder. The company has hired 17 new employees, and "so far, every hire we have made using the test has resulted in a successful placement."[49]

FIGURE 21.2

Why Employers Do Not Hire Based on Content of Candidates' Social Media Pages

Source: Based on *www.careerbuilder .com/share/aboutus/pressreleasesdetail .aspx?id=pr691&sd=4%2F18%2F20 12&ed=4%2F18%2F2099.*

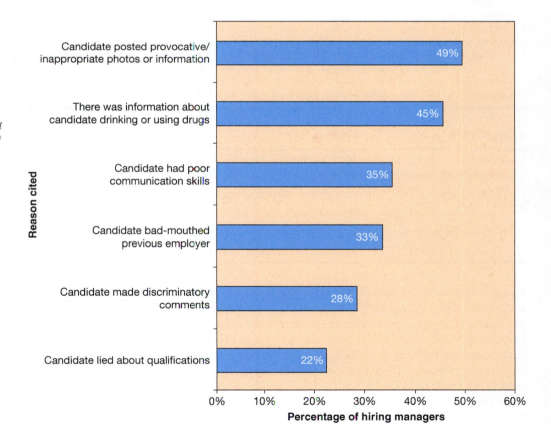

According to the U.S. Department of Health and Human Services, 65.7 percent of illegal drug users are employed, and those who are employed are most likely to work for small companies, which rely less on drug tests than large businesses.[50] In an attempt to avoid hiring illegal drug users, 84 percent of employers use preemployment drug tests, and 39 percent administer postemployment random drug tests.[51] Although administering drug tests adds expense to the hiring process, the cost is far less than that of the potential problems that an employee with a drug habit causes. Employers who use drug tests should establish a policy and follow it consistently.

Experienced entrepreneurs do not rely on any one element in the employee selection process. They look at the total picture painted by each part of a candidate's portfolio. They know that the hiring process provides them with one of the most valuable raw materials their companies count on for success—capable, hardworking people. They also recognize that hiring an employee is not a single event but the beginning of a long-term relationship.

Entrepreneurship in Action

Building an Intentional Culture at Zappos

Tony Hsieh first joined Zappos, an online retailer of shoes, as an investor and adviser. He had recently sold the company he founded, LinkExchange, to Microsoft for $265 million. The main reason he sold his company? It was no longer a fun place to work! Hsieh's involvement with Zappos grew, and he soon became its CEO. His top priority was from the start—and still is today—getting the culture of Zappos right.

The core of the Zappos culture is based on one word: happiness. Hsieh focuses on making his employees and his customers feel really good. Everything at Zappos focuses on sustaining a culture of happiness.

Building and sustaining the Zappos culture begins with the hiring process. A human resource manager conducts the initial interview, which is an assessment of the cultural fit of the candidate. If the person fails the cultural fit interview, he or she is not invited to meet with the hiring manager.

Every person who passes the cultural fit interview then goes through a process in which the prospective employee meets with the supervisor and team members several times. Interviewers use behaviorally based questions that assess a candidate's potential ability to fit within the culture and to exhibit the necessary skills. Every prospect is also invited to at least one company event, often a party of some sort, to allow those who are not part of the formal interviewing process to meet the candidate. For some positions, the hiring process can last for months.

Once an employee is hired, the training process is also structured to support and sustain the Zappos culture. Every new hire—from a call center employee to a senior executive—goes through the same four-week employee orientation and training program. A team trains each new employee on each of the core values that are at the heart of the Zappos culture:

- Deliver "wow" through service
- Embrace and drive change
- Create fun and a little weirdness
- Be adventurous, creative, and open-minded
- Pursue growth and learning
- Build open and honest relationships with communication
- Build a positive team and family spirit
- Do more with less
- Be passionate and determined
- Be humble

"At the end of that first week of training we make an offer to the entire class that we'll pay you for the time you've already spent training plus a bonus of $2,000 to quit and leave the company right now," says Hsieh.

Zappos rarely ever has to write a check. The company often goes more than a year before a candidate takes the buyout offer and leaves. Zappos uses the "quitting bonus" to ensure that the employees it hires are engaged and committed to the company and are a good fit with its unique culture.

As CEO of Zappos, Hsieh focuses on seven key levers to ensure that once people are integrated into the company, they help play a role in sustaining the Zappos culture:

1. ***Build a company, not a product.*** "I've never been interested in shoes," says Hsieh. "My passion has always been customer service, company culture, and community."

2. ***Motivate through inspiration.*** Zappos uses its mission and culture to inspire performance. "If you can inspire employees through a higher purpose beyond profits, that you're doing something that can help change the world," says Hsieh, "you can accomplish so much more."

3. ***View your company as a greenhouse.*** Hsieh says that his role is to build a greenhouse that allows his employees to flourish and grow.

4. ***Encourage "collisions."*** At Zappos, offices are set up to encourage interactions. Hsieh's office is right in the middle of all of the other cubicles.

5. ***Make company values flexible.*** Zappos has its headquarters in Las Vegas, an office in China, and a warehouse in Kentucky. The core values have to be flexible enough to be effective regardless of an employee's job function or geography.

(continued)

Entrepreneurship in Action *(continued)*

6. ***Encourage employees to continuously learn.*** Zappos offers employees classes, some taught by Hsieh himself, on inspirational business books and other topics that challenge every employee to grow.

7. ***Offer a clear career path.*** Help employees see that they can advance in the organization and understand what it takes to do so. "Set expectations on both sides," Hsieh says.

Zappos has created a book on its culture and core values. Each year, all employees are encouraged to submit essays that reflect on what it means to work at Zappos. The book is now almost 500 pages long. Hsieh uses the book as a way to preserve the culture, to get employees thinking about the meaning of their work at Zappos, and as way to demonstrate to the outside world the essence of Zappo's culture.

1. Why is it so important to be intentional about creating and sustaining a culture within a business venture? What does Zappos do to sustain its culture? What are some other ways that entrepreneurs can instill and sustain a culture within their businesses?

2. List the core values that you would use to build a culture in your business. Why are each of these important to you? Where do these values come from? What steps would you take to sustain the culture in your business as it grows?

Sources: Based on Tony Hsieh, "How Zappos Infuses Culture Using Core Values," *Harvard Business Review*, May 24, 2010, *http://blogs.hbr.org/cs/2010/05/how_zappos_infuses_culture_using_core_values.html*; Sam Narisi, "You Hired Them—Now Pay Them to Quit?," *HR Recruiting Alert*, June 16, 2008, *www.hrrecruitingalert.com/new-onboarding-twist-bribing-hires-to-quit*; Gwen Moran, "Zappos' Secrets to Building an Empowering Company Culture," *Entrepreneur*, March 6, 2013, *www.entrepreneur.com/article/226003*; Allison Fass, "Grow Your Company as Big as *Zappos.com*: 7 Tips from Tony Hsieh," *Inc.*, November 15, 2012, *www.inc.com/allison-fass/tony-hsieh-zappos-growth-strategies.html*; Max Chafkin, "The Zappos Way of Managing," *Inc.*, May 1, 2009, *www.inc.com/magazine/20090501/the-zappos-way-of-managing.html*; Susan M. Heathfield, "20 Ways Zappos Reinforces Its Company Culture," *http://humanresources.about.com/od/organizationalculture/a/how-zappos-reinforces-its-company-culture.htm*; Steven Rosenbaum, "The Happiness Culture: Zappos Isn't a Company—It's a Mission," *Fast Company*, June 4, 2010, *www.fastcompany.com/1657030/happiness-culture-zappos-isnt-company-its-mission*.

Building the Right Culture and Organizational Structure

3.

Explain how to build the kind of company culture and structure to support the entrepreneur's mission and goals and to motivate employees to achieve them.

Company culture is the distinctive, unwritten code of conduct that governs the behavior, attitudes, relationships, and style of an organization. It is the essence of "the way we do things around here." In many entrepreneurial companies, culture plays an important role in gaining a competitive edge. Many surveys examining why entrepreneurs started their businesses have found that creating a particular culture is among the top reasons cited for launching their businesses. A company's culture has a powerful impact on the way people work together in a business, how they do their jobs, and how they treat their customers. Company culture manifests itself in many ways—from how workers dress and act to the language they use. At some companies, the unspoken dress code requires workers to wear suits and ties, but at many high-technology companies, employees routinely show up in jeans, T-shirts, and flip-flops. In many companies, the culture creates its own language. At Disney theme parks, workers are not "employees"; they are "cast members." They do not merely go to work; their jobs are "parts in a performance." Customers are referred to as "guests." When a cast member treats someone to lunch, it's "on the mouse." Anything negative—such as a cigarette butt on a walkway—is "a bad Mickey," and anything positive is "a good Mickey."

An important ingredient in a company's culture is the performance objectives an entrepreneur sets and against which employees are measured. If entrepreneurs want integrity, respect, honesty, customer service, and other important values to be the foundation on which a positive culture can flourish, they must establish measures of success that reflect those core values. *Effective executives know that building a positive organizational culture has a direct, positive impact on the financial outcomes of an organization.* The intangible factors that make up an organization's culture have an influence, either positive or negative, on the tangible outcomes of profitability, cash flow, return on equity, employee productivity, innovation, and cost control. An entrepreneur's job is to create a culture that creates a positive influence on the company's tangible outcomes. Companies that focus on creating a positive corporate culture have better financial performance than those that do not, according to the San Francisco–based Great Places to Work Institute.[52]

Sustaining a company's culture begins with the hiring process. Beyond the normal requirements of competitive pay and working conditions, the hiring process must focus on finding employees who share the values of the organization. "Companies are realizing that culture is as important as strategy and that they can't just look at the short term anymore," says Barbara Bilodeau, a manager at Boston-based Bain & Co.[53] Nurturing the right culture in a company enhances a company's competitive position by improving its ability to attract and retain quality workers and by creating an environment in which workers can grow and develop. As a new

generation of employees enters the workforce, companies are discovering that more relaxed, open cultures have an edge in attracting the best workers. These companies embrace nontraditional, fun cultures that incorporate concepts such as casual dress, team-based assignments, telecommuting, flexible work schedules, free meals, company outings, and many other unique options. Modern organizational culture relies on the following principles to create a productive, fun workplace.

1. ***Respect for the quality of work and a balance between work life and home life.*** Modern companies must recognize that their employees have lives away from work. These businesses offer flexible work schedules, part-time work, job sharing, telecommuting, sabbaticals, and conveniences such as on-site day care or concierge services that handle employees' errands. Work–life balance issues are becoming more important to employees, and companies that address them have an edge when it comes to recruiting and retaining a quality workforce. "Employers realize that by offering work-life programs, they are getting a lot in return in terms of productivity and commitment to the organization," says one consultant.[54]

ENTREPRENEURIAL PROFILE: Jim Goodnight: SAS Dr. Jim Goodnight, cofounder and CEO of SAS software company in Cary, North Carolina, believes that by treating its employees well, SAS benefits as a company. "Creativity is especially important to SAS because software is a product of the mind. As such, 95 percent of my assets drive out the gate every evening. It's my job to maintain a work environment that keeps those people coming back every morning. The creativity they bring to SAS is a competitive advantage for us." SAS offers its employees heavily subsidized Montessori child care, unlimited sick time, and a free health care center staffed by physicians, nutritionists, physical therapists, and psychologists. SAS employees and their families also have free access to a sports complex that includes tennis and basketball courts, a weight room, and a heated pool. Employees also benefit from free "work–life" counseling that helps them manage the stresses of everyday life.[55]

Jim Goodnight, cofounder of SAS.
Source: SAS.

A sense of purpose. These companies rely on a strong sense of purpose to connect employees to the company's mission. At motorcycle legend Harley-Davidson, employees are so in tune with the company's mission that some of them have tattoos of the company's name.

Diversity. The U.S. workforce is becoming more diverse; by 2039, the majority of the workforce will consist of minorities. Companies with appealing cultures embrace cultural diversity in their workforces, actively seeking out workers with different backgrounds. They recognize that a workforce with a rich mix of cultural diversity gives their companies more talent, skills, and abilities from which to draw. Because the entire world is now a potential market for many small companies, having a workforce that looks, acts, and thinks like their customers, with all of their ethnic, racial, religious, and behavioral variety, is a strength.

Integrity. Employees want to work for a company that stands for honesty and integrity. They do not want to have to check their personal value systems at the door when they report to work. Indeed, many workers take pride in the fact they work for a company that is ethical and socially responsible. Chapter 2 discussed the issues of ethics, integrity, and social responsibility in detail.

Participative management. Modern managers recognize that employees expect a participative management style to be part of a company's culture. Today's workforce does not respond well to the autocratic management styles of the past. To maximize productivity and encourage commitment to accomplishing the company's mission, entrepreneurs must trust and empower employees at all levels of the organization to make decisions and to take the actions necessary for doing their jobs well.

Learning environment. Progressive companies encourage and support lifelong learning among their employees. They are willing to invest in their employees, improving their skills and helping them to reach their full potential. That attitude is a strong magnet for the best and brightest workers who know that, to stay at the top of their fields, they must always be learning.

ENTREPRENEURIAL PROFILE: Ara Krafian: Symmes Maini & McKee Associates Ara Krafian, president and CEO of Symmes Maini & McKee Associates (SMMA), believes in the importance of employee learning. "It's just the culture we have—it's a learning culture," says Krafian. SMMA, an architectural firm located in Cambridge, Massachusetts, brings in academics and other experts to lecture and train employees on the latest best practices. SMMA pays for employees to take courses at local universities, such as Harvard and the Massachusetts Institute of Technology.

The company also has developed its own educational programs, covering current topics including sustainable design. In addition, SMMA supports a strong mentoring program that matches senior leaders with junior employees as part of its learning culture. Krafian believes that his commitment to employee learning is one of the key reasons his company has continued to be successful in the construction sector even during recessions.[56]

A sense of fun. Children laugh an average of 400 times a day; however, by the time a person reaches age 35, he or she laughs on average just 15 times a day.[57] At many successful small companies, the lines between work and play are blurred, and laughter is common. The founders of these businesses see no reason for work and fun to be mutually exclusive. In fact, they believe that a workplace that creates a sense of fun makes it easier to recruit quality workers and encourages them to be more productive and more customer oriented. "Healthy and sustainable organizations focus on the fundamentals: quality, service, fiscal responsibility, leadership—but they didn't forget to add fun to that formula," says Leslie Yerkes, a consultant and author.[58] At Insomniac Games, in addition to testing out the newest video games, employees get breaks for massage therapy, a quick happy hour, and silly celebrations. The company also offers company-wide movie days when the company pays for everyone to go to the movie theater and even buys them all popcorn.[59]

Engagement. Employees who are fully engaged in their work take pride in making valuable contributions to the organization's success and derive personal satisfaction from doing so. Although engaged employees are a key ingredient in superior business performance, just 30 percent of employees in the United States are fully engaged in their work, and 18 percent of them actually are disengaged. Experts estimate that actively disengaged employees cost companies between $450 and $550 billion per year.[60] What can managers do to improve employee engagement?

- Constantly communicate the purpose and vision of the organization and why it matters.
- Challenge employees to learn and advance in their careers and give them the resources and the incentives to do so.
- Create a culture that encourages and rewards engagement.

Companies that build their cultures on these principles have an edge when it comes to attracting, retaining, and motivating workers. In other words, creating the right culture helps a small company compete more effectively.

No screening process is perfect, which is why small companies must make sure that every new hire is an appropriate fit with their culture. "Most [employee] turnover is from a lack of cultural match," says Julie Godshall Brown, president of Godshall and Godshall Personnel Consultants.[61] The partners of a Minneapolis-based engineering firm take every new prospect bowling once the candidate makes it through the first interview. The partners have tried to build a culture in which none of the employees takes himself or herself too seriously because they did not want a firm full of "uptight engineers." If prospective employees can relax and have fun at a bowling alley, there is a good chance that they will fit into the company's culture.

Managing Growth and a Changing Culture

As companies grow, they often experience dramatic changes in their cultures. Procedures become more formal, operations grow more widespread, jobs take on more structure, communication becomes more difficult, and the company's personality often begins to change. As more workers come on board, employees find it more difficult to know everyone in the company and to understand how their jobs connect with others. This transition presents a new set of demands for the entrepreneur. Unless entrepreneurs work hard to maintain their companies' unique culture, they may wake up one day to find that they have sacrificed that culture—and the competitive edge that went with it—in the name of growth. Entrepreneurs must be aware of the challenges rapid growth brings with it; otherwise, they may find their companies crumbling around them as they reach warp speed. An entrepreneur's challenge is to walk a fine line between retaining the small company traits that are the seeds of the company's success and incorporating the elements of infrastructure that are essential to supporting and sustaining the company's growth.

What a Great Place to Work!

Ruby Receptionists

Ruby Receptionists is at its core an answering service, which is a business model that has been around for many years. What makes Ruby Receptionists noteworthy is that even though it operates as a simple business in a mature industry, it has been ranked among the best places to work in the country!

To help foster its culture, Ruby Receptionists focuses on benefits and perks that make it a fun place to work. Ruby employees receive funding for art classes, book clubs, and yoga sessions and are encouraged to be involved in activities outside of work. Several employees play in bands or act in local theater. Each month, one employee makes a presentation during an all-employee meeting on any topic of his or her choice.

"We have long been creating a remarkable and rewarding workplace culture," says Jill Nelson, founder and CEO. "We all come to work each day with a real passion and purpose for the work we do—and, more importantly, for the people we work with. In an era focused on technology, we're keeping real human connections alive and well."

Ruby Receptionists holds quarterly games and competitions to reward employees for creating meaningful connections within their company. The company sponsors several legendary employee parties each year. Employees work together on various community service activities that foster team building and camaraderie.

Ruby gives employees access to company-sponsored gift cards and a prepaid Amazon.com account so that they can spontaneously send clients small, thoughtful gifts. "Smile Files" are used to store compliments and handwritten thank-you cards from customers. Ruby's "Happiness Journal" program asked employees to write daily about their gratitude over a six-week period. Ruby recently launched its "Five at Five" sabbatical program, which provides five weeks of paid sabbatical after five years of employment at Ruby. (See the "Street-Smart Entrepreneur" feature later in this chapter on company sabbaticals.)

Ruby has three full-time positions dedicated to fostering the company culture—director of culture, office champion, and the "Rubyinator." In addition, Ruby retains the services of a psychologist to help keep staff positive, motivated, and happy.

"We live our core values with our employees the same way we ask our employees to live them with our clients," says Nelson. "This enables us to deliver superior customer service while fostering an empowering, positive, and enjoyable workplace for our employees."

InQuicker

InQuicker provides an online waiting room service for emergency rooms and urgent care centers. Its software enables patients to check in and wait at home based on projected treatment times. InQuicker also offers online appointment scheduling services for medical clinics.

Michael Brody-Waite, cofounder and CEO, wants to make InQuicker a cool, fun place to work. The first thing visitors see when they walk into the InQuicker office is a ping-pong table. The workspace is open, with no cubicles, to facilitate collaboration and teamwork.

InQuicker has a "culture club" that is responsible for leading activities that facilitate building a strong culture. Team members of InQuicker have weekly lunch meetings at a local restaurant. Once a

Michael Brody-Waite, cofounder of InQuicker.
Source: InQuicker, LLC.

month, one of the employees hosts the rest of the staff for a cookout at his or her home. Employees dress up and eat sweets for each person's birthday and for various holidays, and each year, the team spends a week together away from the office on a company retreat.

Brody-Waite offers employees flextime and flexspace and takes great care is to ensure that employees keep family time and work time separate. When employees are with family and friends, they are expected *not* to answer e-mails and phone calls from work.

To ensure that the company retains its culture, Brody-Waite's founders have rejected the opportunity to grow their business more quickly and to take on outside investments.

"I need someone to tell me why we need venture capital money," says Brody-Waite. "When you take on an investor, the number one priority is selling the company so that they can get a return on investment. For us, it's about being ambitious and (saying) 'no' to things so that you have the time and space to say 'yes' to the right things. Our approach is to stay extremely simple."

1. Many entrepreneurs believe that corporate culture can be a strong competitive advantage. Do you agree? Explain.

2. Why do company cultures like the ones at Ruby Receptionists and InQuicker appeal to employees, particularly members of Generation X and Generation Y? Would you want to work for Ruby Receptionists or InQuicker? Explain.

Sources: Based on *www.greatplacetowork.com/2012-best-workplaces/ruby-receptionists*; "The 25 Best Small Companies to Work For," *CNN Money*, October 25, 2012, *http://money.cnn.com/gallery/news/companies/2012/10/25/best-small-companies .fortune/index.html*; "Great Places to Work and Fortune Name Ruby Receptionists the #1 Best Place to Work in the U.S.," October 26, 2012, *www.prweb.com/ releases/2012/10/prweb10058588.htm*; Monica Metz, "Ruby Receptionists Named #3 Best Company to Work for in Oregon," March 6, 2012, *http://blog.oregonlive.com/ business-watch/2012/03/ruby_receptionists_named_3_bes.html*; "InQuicker, LLC," February 22, 2013, *http://eonashville.files.wordpress.com/2013/04/eo-first-2.jpg*; E. J. Boyer, "What I Learned About InQuicker in 10 Minutes," *Nashville Business Journal*, February 15, 2013, *www.bizjournals.com/nashville/blog/2013/02/what-i-learned-about-inquicker-in-10.html*; Annie Johnson, "Case Study: For InQuicker, Success Means Taking It Slow," *Nashville Business Journal*, July 13, 2012, *www.bizjournals .com/nashville/print-edition/2012/07/13/for-inquicker-nashville.html?page=all*; Zina Moukheiber, "Health Tech Startup InQuicker Shuns Venture Money and Hype," *Forbes*, August 20, 2012, *www.forbes.com/sites/zinamoukheiber/2012/08/20/ health-tech-start-up-inquicker-shuns-venture-money-and-hype.*

Team-Based Management

As a company grows, its success may lie in the founder's willingness to shift from a top-down, single-leader structure to one that is team based. Unlike the early days of a company when the founder handled much of the work alone, he or she must accept that the magnitude and complexity of work requires delegating authority and empowering employees to make decisions. Leaders who build successful teams understand that each team member has a role to play and that every role plays a part in a bigger picture. Companies now rely more on team-based job designs as competition and complexity increase and business problems cross departmental and geographic boundaries. Even though converting from a traditional management style to a team approach requires a major change in management style, it is often easier to implement with a small number of workers.

A **self-directed work team** is a group of workers from different functional areas of a company who work together as a unit. The team operates largely without supervision, making decisions and performing tasks that once belonged only to managers. Some teams may be temporary, attacking and solving a specific problem, but many are permanent components of an organization's structure. As their name implies, these teams manage themselves, performing functions such as setting work schedules, ordering raw materials, evaluating and purchasing equipment, developing budgets, hiring and firing team members, and solving customers' problems. Teams function best in environments in which the work is interdependent and people must interact to accomplish their goals. The goal is to get people working together to serve customers better. Johnsonville Sausage, a privately owned company founded in 1945 and based in Sheboygan Falls, Wisconsin, uses a team-based structure rather than traditional departments. The company hires "members" rather than employees, and supervisors have the title of "coaches," a not-so-subtle reminder of the role that the company expects them to fill.[62]

Managers in companies using teams are just as involved as before, but the nature of their work changes dramatically. Before teams, managers were bosses who made most of the decisions affecting their subordinates alone. They often hoarded information and power for themselves. In a team environment, managers take on the role of coaches. They empower those around them to make decisions affecting their work and share information with their workers. As facilitators, their job is to support and to serve the teams' functioning in the organization and to make sure that the teams produce results.

Companies have strong, competitive reasons for using team-based management. Businesses that use teams effectively report significant gains in quality, reductions in cycle time, lower costs, increased customer satisfaction, and improved employee motivation and morale. A team-based approach is not appropriate for every organization, however. Although teams have saved some companies from extinction, for others the team approach has failed. A team-based management system is *not* easy to start. Switching from a traditional organizational structure to a team-based one is filled with potential pitfalls. A common criticism of teams is groupthink, a concept identified by Yale psychology professor Irving Janis in his classic book *Victims of Groupthink*. Janis observed that **groupthink** sometimes leads groups to build a false sense of confidence that produces unsound decisions that team members would not have made individually.[63] Years later, Jerry Harvey described another danger of group decisions that he called the **Abilene paradox**, a situation in which a group makes a decision that is precisely the *opposite* of what its individual members want to do. What makes the difference? What causes teams to fail? The following errors are common in team-oriented environments:

- Assigning a team an inappropriate task, one in which the team members may lack the necessary skills to be successful (lack of training and support).
- Creating work teams but failing to provide the team with meaningful performance targets.
- Failing to deal with known underperformers and assuming that being part of a group will solve the problem—it doesn't.
- Failing to compensate the members of the team equitably.

To ensure the success of the teams approach, entrepreneurs must do the following:

- *Make sure that teams are appropriate for the company and the nature of the work.*
  A good starting point is to create a "map" of the company's work flow that shows how workers build a product or deliver a service. Is the work interdependent, complex, and interactive? If so, teamwork is likely to improve the company's performance.

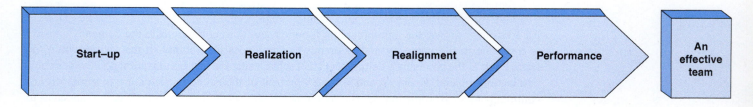

Start–up	Realization	Realignment	Performance	An effective team

Description

- High expectations
- Unclear goals and roles
- Anxiety and reliance on leader
- Avoidance of tasks

- Recognition of time and effort required
- Roadblocks
- Frustration
- Conflict

- Resetting of goals and roles
- Development of trust and cooperation
- Progress
- Structure

- Involvement, openness, and teamwork
- Commitment both to process and to task achievement

Leadership focus

- Help team focus on task
- Provide goals and structure
- Supervise and define accountability

- Emphasize task and process
- Clarify expectations and roles
- Encourage open discussion and address concerns
- Ensure proper skills and resources

- Focus on process
- Promote participation and team decision making
- Encourage peer support
- Provide feedback

- Focus on monitoring and feedback
- Let team take responsibility for solving problems and making decisions

FIGURE 21.3

The Stages of Team Development

- *Form teams around the natural work flow and give them specific tasks to accomplish.* Teams can be effective only if managers challenge them to accomplish specific, measurable objectives. They need targets to shoot for.

- *Provide adequate support and training for team members and leaders.* Team success requires a new set of skills. Workers must learn how to communicate, resolve conflict, support one another, and solve problems as a team. Smart managers see that team members get the training they need.

- *Involve team members in determining how their performances will be measured, what will be measured, and when it will be measured.* Doing so gives team members a sense of ownership and pride about the tasks they are accomplishing.

- *Make at least part of team members' pay dependent on team performance.* Companies that have used teams successfully still pay members individually, but they make successful teamwork a major part of an individual's performance review.

Figure 21.3 illustrates the four stages teams go through on their way to performing effectively and reaching set goals.

Communicating Effectively

Effective communication is the lifeblood of a successful company. It reinforces the organization's vision, connects employees to the business, fosters process improvement, facilitates change, and drives business results by changing employee behavior. An important and highly visible part of entrepreneurs' role is to communicate the values, beliefs, and principles for which their businesses stand. Entrepreneurs also must help employees understand the importance of their roles and how they fit into the "big picture" of the company's success. Studies confirm that effective communication makes a difference in a company's performance. Management consulting firm Towers Watson reports that the return on investment over the last five years in companies with the most effective communications is 47 percent higher than the return for those with the least effective communications. The study also found a strong correlation between a company's communication effectiveness and its employee engagement level and retention rate.[64]

4.

Understand the potential barriers to effective communication and describe how to overcome them.

Improving Communication

As a business grows, the entrepreneur's foremost job is to communicate the company's vision to everyone in the company and to empower employees to accomplish the vision within the framework of the company's culture. Much of what leaders in any organization do involves communication; indeed, leaders spend about 80 percent of their time participating in some form of communication. To some managers, communicating means only one thing: sending messages to others. Although talking to people both inside and outside the organization is an important part of an entrepreneur's job, so is encouraging communication throughout the company at all levels and across all functional areas. "Communicators are evolving from crafting the content [of the message] to facilitating the discussion," says Sharon McIntosh, director of global communications at Pepsico.[65] As more small businesses embrace the virtual office as a means of bootstrapping overhead costs, communication can become an even greater challenge. However, with the wide array of communication tools, including Skype, FaceTime, texting, e-mail, meeting software platforms, and conference calling, it is possible to build an effective system for communication even when workers do not share a common office space.

ENTREPRENEURIAL PROFILE: Andy Miller: CardStar Andy Miller, CEO of CardStar, has embraced technology as a means to keep workers in his smart phone application company connected. CardStar employees are located in three different cities, and most work from home. All of the company's employees are equipped with laptops, smart phones, and USB modems so that they can stay connected with each other anywhere and at any time. Texting and e-mail are the basic tools for quick communication, just as they can be in any organization. Video meetings using Skype substitute for face-to-face meetings. "When I was in corporate America," says Miller, "I realized that it was highly inefficient. There were so many meetings you didn't have to be at but were required anyway. You end up working harder and doing less."[66]

COMMUNICATING EFFECTIVELY One of the most frustrating experiences for entrepreneurs occurs when they ask an employee to do something and nothing happens. Although entrepreneurs are quick to perceive the failure to respond as the employee's lack of motivation or weak work ethic, often the culprit is improper communication. The primary reasons employees usually don't do what they are expected to do have little to do with their motivation and desire to work. Instead, workers often fail to do what they need to do because of the following:

- They don't know what to do.
- They don't know how to do it.
- They don't have the authority to do it.
- They get no feedback on how well or how poorly they are doing it.
- They are either ignored or punished for doing it right.
- They realize that no one ever notices even if they *are* doing it right.

The common thread running through all of these causes is poor communication between the entrepreneur and employee. What barriers to effective communication must entrepreneurs overcome?

- *Managers and employees don't always feel free to say what they really mean.* CEOs and top managers in companies of any size seldom hear the truth about problems and negative results from employees. This less-than-honest feedback results from the hesitancy of subordinates to tell "the boss" bad news. Over time, this tendency paralyzes the upward communication in a company.

- *Ambiguity blocks real communication.* The same words can have different meanings to different people, especially in modern companies, where the workforce is likely to be highly diverse. For instance, an entrepreneur may tell an employee to "take care of this customer's problem as soon as you can." The owner may have meant "solve this problem by the end of the day," but the employee may think that fixing the problem by the end of the week will meet the owner's request.

- *Information overload causes the message to get lost.* With information from text messages, telephone, e-mail, Skype, social media, face-to-face communication, and other sources, employees in modern organizations are literally bombarded with messages. With

such a large volume of information washing over workers, it is easy for some messages to get lost.

- *Selective listening interferes with the communication process.* Sometimes people hear only what they want to hear, selectively tuning in and out on a speaker's message. The result is distorted communication.

- *Defense mechanisms block a message.* When people are confronted with information that upsets them or conflicts with their perceptions, they immediately put up defenses. Defense mechanisms range from verbally attacking the source of the message to twisting perceptions of reality to maintain self-esteem.

- *Conflicting verbal and nonverbal messages confuse listeners.* Nonverbal communication includes a speaker's mannerisms, gestures, posture, facial expressions, and other forms of body language. When a speaker sends conflicting verbal and nonverbal messages, research shows that listeners will believe the nonverbal message almost every time.

How can entrepreneurs overcome these barriers to become better communicators? The following tips will help:

- *Clarify your message before you attempt to communicate it.* Identify exactly what you want the receiver to think and do as a result of the message and focus on getting that point across clearly and concisely.

- *Use face-to-face communication whenever possible.* Although not always practical, face-to-face communication reduces the likelihood of misunderstandings because it allows for immediate feedback and nonverbal clues. Even in a virtual organization, face-to-face meetings can be held using Skype, FaceTime, or Web-based conferencing platforms, such as *Gotomeeting.com.*

- *Be empathetic.* Put yourself in the place of those who will receive your message and develop it accordingly.

- *Match your message to your audience.* An entrepreneur would be very unlikely to use the same words, techniques, and style to communicate his or her company's financial position to a group of industry analysts as he or she would to a group of workers on the factory floor.

- *Be organized.* Effective communicators organize their messages so that their audiences can understand them easily.

- *Encourage feedback.* Good leaders actively seek honest feedback from as many employees as possible. At computer chip maker Intel, managers routinely hold "skip-level meetings," in which managers meet with employees who are two levels down the organization. "It's a powerful tool for getting information," says Patricia Murray, the company's director of human resources.[67]

- *Get out of the office and talk to employees.* Some of the most meaningful conversations managers have take place when they leave their offices to "wander" through the workplace. Management author Tom Peters calls it "MBWA, management by wandering around."

- *Tell the truth.* The fastest way to destroy your credibility as a leader is to lie.

- *Don't be afraid to tell employees about the business, its performance, and the forces that affect it.* Too often, entrepreneurs assume that employees don't care about such details. Employees *are* interested in the business that employs them and want to understand where it is headed and how it plans to get there.

Listening

When one thinks about communication, listening typically does not come to mind, yet listening is an essential part of the communication process. Entrepreneurs must listen to what employees on the front line are learning about customers' needs and demands. "The key to success and growth is getting employees to tell you what's really going on," says Vineet Nayar, CEO of HCL Technologies and author of *Employees First, Customers Second.*[68] The employees who serve customers are the *real* experts in the company's day-to-day activities. They are in closer contact with potential problems and opportunities at the operating level than anyone else in the company, particularly managers. Entrepreneurs and their managers who take the time to listen to employees who are in direct contact with customers can prevent many potential problems. In addition, by

encouraging employees to develop creative solutions to problems and innovative ideas for capitalizing on opportunities and then listening to and acting on them, entrepreneurs can make their companies more successful.

ENTREPRENEURIAL PROFILE: Scott Moorehead: The Cellular Connection Scott Moorehead, CEO of The Cellular Connection, has developed a creative way to ensure that he is able to listen to the employees who work in his more than 800 stores that sell Verizon products. "I was sitting in front of my computer, trying to come up with something I could tell all these smart people in my company that would help them do their jobs better," says Moorehead, "and I realized that what I really should be doing is asking them what I should do." He sent out an e-mail to all of his regional managers asking them one simple question: If today you became the CEO, what would you do to make the company better? "Overall I loved the feedback," says Moorehead, "but, it was also depressing because a number of people said, 'I would do whatever I could to bring back the family atmosphere we used to have in the company.' Those responses made me feel like such a fraud. Every day I was talking about how our business is a family and about really knowing our employees . . . and that's not how employees in the field felt." By listening to his employees, Moorehead made the decision to stop focusing on growing his company and instead to spend more time making the business run better and ensuring that the culture is what he had intended it to be within The Cellular Connection.[69]

Improvements such as these depend on entrepreneurs' ability to listen. To improve listening skills, entrepreneurs can use the PDCH formula: identify the speaker's *purpose*, recognize the *details* that support that purpose, see the *conclusions* they can draw from what the speaker is saying, and identify the *hidden* meanings communicated by body language and voice inflections.

The Informal Communication Network: The "Grapevine"

Despite all of the modern communication tools available, the grapevine, the informal lines of communication that exist in every company, remains an important link in a company's communication network. The grapevine carries vital information—and sometimes rumors—through every part of the organization with incredible speed. The grapevine kicks into overdrive when the information in a company's formal communication network is scarce. It is not unusual for employees to hear about important changes in an organization by the grapevine well before official communication channels transmit the news. Research shows that up to 70 percent of all organizational communication comes by way of the grapevine, yet many managers are not aware of the efficiency with which this informal communication channel operates.[70] Text, social media, and e-mail increase the speed at which the grapevine transmits informal communications, all under the radar of management.

Knowing that employees are connected through the grapevine allows entrepreneurs to send out ideas to obtain reactions without making a formal announcement. When management is in the loop, the grapevine can be an excellent source of informal feedback. Smart small business owners recognize the grapevine's existence and use it as a communication tool to both send and receive meaningful information.

The Challenge of Motivating Workers

5.

Discuss the ways in which entrepreneurs can motivate their employees to achieve higher levels of performance.

Motivation is the degree of effort an employee exerts to accomplish a task; it shows up as excitement about work. Motivating workers to higher levels of performance is one of the most difficult and challenging tasks facing a small business manager. Few things are more frustrating to an entrepreneur than an employee with a tremendous amount of talent who lacks the desire to use it. This section discusses four aspects of motivation: empowerment, job design, rewards and compensation, and feedback.

Empowerment

One motivating principle is empowerment. **Empowerment** involves giving workers at every level of the organization the authority, the freedom, and the responsibility to control their own work, to make decisions, and to take action to meet the company's objectives. Research indicates that employees experience increased initiative and motivation when they are empowered. Empowerment affects their self-confidence and the level of tenacity they display when faced with

setbacks. Empowered employees take responsibility for making decisions and following them through to completion. "The credit for Virgin's enduring and varied success is often attributed to me," says Richard Branson, successful entrepreneur and founder of Virgin Group. "But it's actually due to the people who piloted those businesses. My decision to give them autonomy and encourage them to take risks has allowed us to grow while keeping costs down."[71] Empowerment complements the team-based management style discussed earlier.

Empowerment builds on what real business leaders already know: that the people in their organizations bring with them to work an amazing array of talents, skills, knowledge, and abilities. Workers are willing—even anxious—to put these to use; unfortunately, in too many businesses, suffocating management styles and poorly designed jobs quash workers' enthusiasm and motivation. Enlightened entrepreneurs recognize their workers' abilities, develop them, and then give workers the freedom and the power to use them. Entrepreneurs who share information, responsibility, authority, and power soon discover that their success (and their companies' success) is magnified many times over.

When implemented properly, empowerment can produce impressive results not only for the business but also for newly empowered employees. For the business, benefits typically include significant productivity gains, quality improvement, more satisfied customers, improved morale, and increased employee motivation. For workers, empowerment offers the chance to do a greater variety of work that is interesting and challenging. Empowerment challenges workers to make the most of their creativity, imagination, knowledge, and skills.

Not every worker *wants* to be empowered, however. Some will resist, wanting only to "put in their eight hours and go home." Companies that move to an empowerment philosophy will lose about 5 percent of their workforce because they simply are unwilling or are unable to make the change. Another 75 percent of the typical workforce will accept empowerment and thrive under it, and the remaining 20 percent will pounce on it eagerly because they want to contribute their talents and their ideas.

Empowerment works best when entrepreneurs do the following:

- *Are confident enough to give workers all the authority and responsibility they can handle.* Initially, this may involve giving workers the power to tackle relatively simple assignments. As their confidence and ability grow, most workers are eager to take on additional responsibility.

- *Play the role of coach and facilitator.* Smart owners empower their workers and then get out of the way so that they can do their jobs.

- *Recognize that empowered employees will make mistakes.* The worst thing an owner can do when empowered employees make mistakes is to hunt them down and punish them. That teaches everyone in the company to avoid taking risks and to always play it safe— something that no innovative small business can afford.

- *Hire people who can blossom in an empowered environment.* Empowerment is not for everyone. Owners quickly learn that as costly as hiring mistakes are, such errors are even more costly in an empowered environment. Ideal candidates are high-energy self-starters who enjoy the opportunity to grow and enhance their skills.

- *Train workers continuously to upgrade their skills.* Empowerment demands more of workers than traditional work methods. Managers are asking workers to solve problems and make decisions they have never made before. To handle these problems well, workers need training, especially in effective problem-solving techniques, communication, teamwork, and technical skills.

- *Trust workers to do their jobs.* Once workers are trained to do their jobs, owners must learn to trust them to assume responsibility for their jobs. After all, they are the real experts; they face the problems and challenges every day.

- *Listen to workers when they have ideas, solutions, or suggestions.* Because they are the experts on the job, employees often come up with incredibly insightful, innovative ideas for improving them—*if* entrepreneurs give them the chance. Surveying employees, for example, can become a critical part of companies' efforts to bolster employees' commitment to their jobs, a concept called **employee engagement.** Engaged workers are more willing to help bosses and coworkers solve problems, take initiative, promote the company outside

of work, and offer ideas for improving the company. Failing to acknowledge or act on employees' ideas sends them a clear message: Your ideas really don't count.

- *Recognize workers' contributions.* One of the most important tasks an entrepreneur can perform is to recognize positive employee performance. In *The Carrot Principle*, authors Adrian Gostick and Chester Elton say that recognition must be frequent, specific and timely, and, of course, deserved.[72] Some businesses reward workers with monetary awards, others rely on recognition and praise, and still others use a combination of money and praise. Whatever system an owner chooses, the key to keeping a steady flow of ideas, improvements, suggestions, and solutions is to recognize the people who supply them.

ENTREPRENEURIAL PROFILE: Don Zietlow: Kwik Trip Don Zietlow, owner and CEO of the $2.6 billion Kwik Trip convenience store chain, is intentional about building a culture based on sharing and recognizing the results that employees produce. Zietlow returns nearly half of the company's profits to employees, gives them equity ownership in the company's real estate, and provides generous wages and health care benefits. However, a company survey revealed that employees wanted more recognition for their efforts, so Kwik Trip implemented a program that formally recognizes workers at four months, one year, three years, and five years of service. "Recognizing people fosters respect and a sense of ownership," says Zietlow. "I get a lot of thank-you notes from new employees who are surprised by recognition they've witnessed or received. It makes them want to stay." Since the inception of the employee recognition program, Kwik Trip has reduced its turnover rate by 15 percent and achieved one of the lowest turnover figures in the industry.[73]

- *Share information with workers.* For empowerment to succeed, entrepreneurs must make sure workers get adequate information, the raw material for good decision making. Some companies have gone beyond sharing information to embrace **open-book management**, in which employees have access to *all* of a company's records, including its financial statements. The goal of open-book management is to enable employees to understand why they need to raise productivity, improve quality, cut costs, and improve customer service. Under open-book management, employees do the following:
  - Review and learn to understand the company's financial statements and other critical numbers in measuring its performance.
  - Learn that a significant part of their jobs is making sure that those critical numbers move in the right direction.
  - Have a direct stake in the company's success through profit sharing, equity-like compensation such as employee stock ownership plans, or performance-based bonuses.

Job Design

A recent survey by the Conference Board shows that only 47 percent of employees are satisfied with their jobs, a significant decrease from 61 percent in 1987. About the same percentage find their work interesting.[74] Managers have learned that the job itself and the way it is designed can make it more interesting and can be a source of satisfaction and motivation for workers. During the industrial age, work was organized on the principle of **job simplification**, which involves breaking the work down into its simplest form and standardizing each task. Assembly-line operations are based on job simplification. The scope of workers' jobs is extremely narrow, resulting in impersonal, monotonous, and boring work that creates little challenge or motivation for workers. The result is apathetic, unmotivated workers who care little about quality, customers, or costs.

To break this destructive cycle, some companies have redesigned workers' jobs. The following strategies are common: job enlargement, job rotation, job enrichment, flextime, job sharing, and flexplace.

Job enlargement (or **horizontal job loading**) adds more tasks to a job to broaden its scope. For instance, rather than an employee simply mounting four screws in computers coming down an assembly line, a worker might assemble, install, and test the entire motherboard (perhaps as part of a team). The idea is to make the job more varied and to allow employees to perform a more complete unit of work.

Job rotation involves cross training employees so that they can move from one job in the company to others, giving them a greater number and variety of tasks to perform. As employees learn other jobs within an organization, both their skills and their understanding of the company's

purpose and processes rise. Cross-trained workers are more valuable because they give a company the flexibility to shift workers from low-demand jobs to those where they are most needed. As an incentive for workers to learn to perform other jobs within an operation, some companies offer skill-based pay, a system under which the more skills workers acquire, the more they earn.

Job enrichment (or **vertical job loading**) involves building motivators into a job by increasing the planning, decision making, organizing, and controlling functions—traditionally managerial tasks—that workers perform. The idea is to make every employee a manager or at least a manager of his or her own job.

To enrich employees' jobs, a business owner must build five core characteristics into them:

- *Skill variety* is the degree to which a job requires a variety of different skills, talents, and activities from the worker. Does the job require the worker to perform a variety of tasks that demand a variety of skills and abilities, or does it force him or her to perform the same task repeatedly?

- *Task identity* is the degree to which a job allows the worker to complete a whole or identifiable piece of work. Does the employee build an entire piece of furniture (perhaps as part of a team), or does he or she merely attach four screws?

- *Task significance* is the degree to which a job substantially influences the lives or work of others—employees or final customers. Does the employee get to deal with customers, either internal or external? One effective way to establish task significance is to put employees in touch with customers so that they can see how customers use the product or service they make.

- *Autonomy* is the degree to which a job gives a worker the freedom, independence, and discretion in planning and performing tasks. Does the employee make decisions affecting his or her work, or must he or she rely on someone else (the owner, a manager, or a supervisor) to "call the shots?"

- *Feedback* is the degree to which a job gives the worker direct, timely information about the quality of his performance. Does the job give employees feedback about the quality of their work, or does the product (and all information about it) simply disappear after it leaves the worker's station?

A study conducted by researchers at the University of New Hampshire and the Bureau of Labor Statistics concludes that employees of companies that use job enrichment principles are more satisfied than those who work in jobs designed using principles of simplification.[75]

Flextime is an arrangement under which employees work a normal number of hours but have flexibility about when they start and stop work. Several recent studies suggest that employees show improved mental and physical health when they work in a company that offers flextime. Most flextime arrangements require employees to build their work schedules around a set of "core hours," such as 10 A.M. to 2 P.M., but give them the freedom to set their schedules outside of those core hours. For instance, one worker might choose to come in at 7 A.M. and leave at 3 P.M. to attend her son's soccer game, and another may work from 11 A.M. to 7 P.M. Flextime not only raises worker morale but also makes it easier for companies to attract high-quality young workers who want rewarding careers without sacrificing their lifestyles. In addition, companies using flextime schedules experience higher levels of employee engagement and lower levels of tardiness, turnover, and absenteeism.

ENTREPRENEURIAL PROFILE: Sara Sutton Fell: FlexJobs Sara Sutton Fell, CEO and founder of FlexJobs, a job-listings Web site based in San Francisco, allows employees to work out of their homes from cities across the United States and Europe. FlexJobs employees work the same kind of flextime schedule as the clients they serve: FlexJobs employees set their own schedules. However, they still have an "up-front understanding of the range of hours they should be working each week," says Fell. "They track their own hours, account for the work they do, and clock in and out as needed."[76]

Flextime is becoming an increasingly popular job design strategy. A recent survey by the Families and Work Institute reports that 77 percent of the nation's workers have flexible schedules, up from 66 percent in 2005.[77] The number of companies using flextime is likely to continue to grow as companies find recruiting capable, qualified full-time workers more difficult and as

Sara Sutton Fell, founder of FlexJobs.
Source: Jamie Kripke.

technology makes working from a dedicated office space less important. Research shows that when considering job offers, candidates, particularly members of Generation Y, weigh heavily the flexibility of the work schedule that companies offer.

Job sharing is a work arrangement in which two or more people share a single full-time job. For instance, two college students might share the same 40-hour-a-week job, one working mornings and the other working afternoons. Salary and benefits are prorated between the workers sharing a job. Because job sharing is a simple solution to the growing challenge of life–work balance, it is becoming more popular. Companies already using it are finding it easier to recruit and retain qualified workers. "Employers get the combined strengths of two people, but they only have to pay for one," says one hotel sales manager, herself a job sharer.[78]

Flexplace is a work arrangement in which employees work at a place other than a traditional office, such as a satellite branch closer to their homes or, in many cases, at home. Flexplace is an easy job design strategy for companies to use because of **telecommuting**. Using modern communication technology such as iPads, smart phones, texting, e-mail, and laptop computers, employees have more flexibility in choosing where they work. Today, it is simple for workers to connect electronically to their workplaces (and to all of the people and the information there) from practically anywhere on the planet.

ENTREPRENEURIAL PROFILE: Ben Kirshner: Elite SEM Ben Kirshner, CEO of Elite SEM, a search-engine marketing firm headquartered in New York, has been allowing his staff to work remotely whenever they choose for six years. "It gives my employees the feeling of being entrepreneurs, which will make them more loyal and dedicated," Kirshner says. Kirshner believes that small businesses are the best equipped and have the most to gain by offering flexplace work arrangements. Small businesses usually are more flexible than large companies, and small businesses are always looking for ways to bootstrap overhead costs. Flexplace work reduces the number of offices required in the company's headquarters because not all the employees will be there at any one time. Elite SEM has been rated one of the best companies to work for, with many employees citing the flexibility given to them for work schedules and locations.[79]

Before implementing telecommuting, entrepreneurs must address the following important issues:

- Does the nature of the work fit telecommuting? Obviously, some jobs are better suited for telecommuting than others.
- Have you selected the right employees for telecommuting? Telecommuting is not suitable for every job or for every worker. Experienced managers say that employees who handle it best are experienced workers who know their jobs well, are self-disciplined, and are good communicators.
- Can you monitor compliance with federal wage and hour laws for telecommuters? Generally, employers must keep the same employment records for telecommuters that they do for traditional office workers.
- Have you provided the necessary computer, communications, and ergonomically designed office equipment for employees to work offsite? Trying to "make do" with substandard equipment creates problems and frustration and undermines any telecommuting effort from the outset.
- Are you adequately insured? Employers should be sure that the telecommuting equipment that employees use in their homes is covered under their insurance policies.
- Can you keep in touch? Telecommuting works well as long as long-distance employees stay in touch with headquarters.
- Have you created an equitable telecommuting policy that defines under what conditions telecommuting is acceptable? One danger of telecommuting is that it can create resentment among employees who remain office bound.

Rewards and Compensation

The rewards an employee receives from the job itself are intrinsic, but managers use a wide variety of extrinsic rewards to motivate workers at their disposal. The key to using rewards to

motivate involves tailoring the reward system to the needs and characteristics of the workers. Effective reward systems tap into the values and issues that are important to people. Smart entrepreneurs take the time to learn what makes their employees "tick" and then build their reward system around those motivational factors. For instance, a technician making $30,000 a year may consider a chance to earn a $5,000 bonus to be a powerful motivator; an executive earning $200,000 a year may not.

Research by Globoforce, a Boston-based company that specializes in rewards and incentives, shows that small, frequent awards are more effective than periodic cash bonuses, which is good news for small companies that cannot always afford financial rewards. The study suggests that 80 to 90 percent of a company's employees should get some type of reward every year and that every week a company should be giving rewards to 5 percent of its employees (a concept known as continuous reinforcement). "Small awards all the time are a way to constantly touch people," he says. Jennifer Lepird, who works in the human resources department at software developer Intuit, recently spent several weeks and many long days integrating into Intuit's salary structure the employees at a company that Intuit had purchased. Her manager sent her a congratulatory e-mail thanking her for her quality work and a gift certificate worth $200. Lepird was thrilled. "The fact that somebody took the time to recognize the effort made the long hours just melt away," she says.[80]

One of the most popular rewards is money. Cash is an effective motivator—up to a point; its effects tend to be short-term. Many companies have moved to **pay-for-performance compensation systems**, in which employees' pay depends on how well they perform their jobs. In other words, extra productivity equals extra pay. By linking employees' compensation directly to the company's financial performance, an entrepreneur increases the likelihood that workers will achieve performance targets that are in their best interest and in the company's best interest. A common application of the pay-for-performance concept is a **profit-sharing system** in which a company shares a portion of its profit with the employees who work to produce it.

ENTREPRENEURIAL PROFILE: Curt Richardson: OtterBox OtterBox, a Fort Collins, Colorado, company that designs and manufactures rugged cases to protect smart phones and other mobile devices, has a profit-sharing program that pays out bonuses during monthly company meetings to every employee, including part-time employees and interns. "We have a hybrid program that's similar to profit sharing," says Curt Richardson, company founder and CEO. "This discretionary cash bonus is available on a monthly basis and is based largely on the financial performance of the company. The elements that contribute to higher or lower bonuses are highlighted each month. This represents a better way to engage employees. They understand the business and the personal impact they have within the organization." Employees pay attention to how their actions impact expenses and how these expenses in turn impact profitability. It has become a part of the culture at OtterBox.[81]

Curt Richardson, founder of OtterBox.
Source: Otter Products LLC.

Pay-for-performance systems work only when employees see a clear correlation between their performance and their pay. This offers an advantage for small companies when the employees can see clearly the impact that their performance has on the company's profitability and ultimate success compared to their counterparts at large corporations. To be successful, however, pay-for-performance systems should meet the following criteria:

- *Performance based.* Employees' incentive pay must be clearly and closely linked to their performances.
- *Relevant.* Entrepreneurs must set up the system so that employees see the connection between what they do every day on the job—selling to customers, producing a product, or anything else—and the rewards they receive under the system.
- *Simple.* The system must be simple enough so that employees understand and trust it. Complex systems that employees have difficulty understanding will not produce the desired results.
- *Equitable.* Employees must consider the system fair.
- *Inclusive.* The system should be inclusive. Entrepreneurs are finding creative ways to reward all employees no matter what their jobs might be.

- *Timely.* The company should make timely payouts to employees. A single annual payout is ineffective—employees have long since forgotten what they did to earn the incentive pay. The closer a reward payment is to the action that prompted it, the more effective it will be.

Money is not the only motivator entrepreneurs have at their disposal. Nonfinancial incentives can be more important sources of employee motivation. With a little creativity, small businesses can provide meaningful rewards that motivate employees without breaking the bank. Often the most meaningful motivating factors are the simplest—and least costly—ones, such as praise, recognition, respect, feedback, job security, and promotions. When an employee has done an exceptional job, an entrepreneur should be the first to recognize that accomplishment and to say "thank you." Praise is a simple and powerful motivational tool. "Praise is the most powerful driver of performance known to mankind," says Bob Nelson, a workplace consultant.[82] People enjoy getting praise and recognition; it is just human nature. As Mark Twain once said, "I can live for two months on a good compliment."

ENTREPRENEURIAL PROFILE: Jay Love: Slingshot SEO Slingshot SEO, an Indianapolis, Indiana, Web marketing consulting firm, established a program for employees to recognize the efforts of their colleagues. "Each month, I solicit open nominations for Slingshot SEO's Outstanding Team Member of the Month," says Jay Love, the company's CEO. "Each employee with at least 60 seconds to spare can e-mail me with their recommendations. Although just two are publicly honored at each monthly meeting, many others are encouraged by this program: I always forward the e-mails of the remarkable kudos to all the nominees along with a few comments of my own."[83]

One sure way to kill high performance is failing to recognize the performance and the employees responsible for it. Failing to praise good work eventually conveys the message that the owner either doesn't care about exceptional performance or cannot distinguish between good work and poor work. In either case, through inaction, the manager destroys employees' motivation to excel.

Rewards do *not* have to be expensive to be effective, but they should be creative and should have a direct link to employee performance. Consider how the following rewards for exceptional performance both recognize the employee's contribution and build a positive organizational culture:

- Frima Studio, a video game development company, recognizes creative ideas from employees with "Frima Points," which contributors can trade in for payment for babysitters, home repair services, and other things that enhance work–family balance—a core value for Frima Studio.[84]
- Beryl, a call-center company based in Texas, throws a pizza party when the company meets its monthly goals.[85]
- The AAA Fair Credit Foundation in Salt Lake City, Utah, involves all of its employees in rewarding excellent performance. Employees recognize the extra efforts and special accomplishments of their coworkers by recommending them for "Dollar Days." When an employee earns eight Dollar Days, he or she cashes them in for a day off.[86]

Whatever system of rewards they use, entrepreneurs will be most successful if they match rewards to employees' interests and tastes. For instance, the ideal reward for one employee might be tickets to a sports event; to another, it might be tickets to a theatrical performance. The better entrepreneurs know their employees' interests and tastes, the more effective they will be at matching rewards with performance.

As Generation Y enters the workforce, entrepreneurs will rely more on nonmonetary rewards—praise, recognition, letters of commendation, and others—to create a work environment that fits what this generation values. Under this system, employees enjoy what they do and find their work challenging, exciting, and rewarding. The benefit to the company is that these employees are more likely to act like owners of the business themselves. The goal of nonmonetary rewards is to let employees know that every person is important and that the company notices, appreciates, and recognizes excellent performance.

Sabbaticals

A benefit that both large companies and small businesses have begun to offer their employees is one that used to be found only in academia—paid sabbaticals. Sabbaticals offer an extended break from work that is beyond the traditional family or sick leave. Sabbaticals are considered a time to get away to recharge, to study, or to do something different than the traditional duties of a job. One study found that 16 percent of large corporations offer paid sabbaticals to employees. The study also found that 11 percent of small companies and 19 percent of medium enterprises offer sabbaticals, although most of these businesses offer them only as unpaid leave.

Sabbaticals can help in the recruiting process. Sabbatical leave programs are a good tool to prevent employee burnout, reduce turnover, and improve health. They also lead to greater creativity and productivity when employees return and enable employees to expand their skills and knowledge and to cross train employees who fill in during the leave.

Red Frog Events, a company that offers unusual sporting and entertainment events such as obstacle courses, citywide scavenger hunts, and bar crawls, offers its 80 employees paid sabbaticals. Every five years, Red Frog employees and a guest of their choice get a fully paid one-month trip to the destination of their choice. However, the trip cannot be to a location in North America or Australia, where the company does most of its business.

"This isn't a cocktail-umbrella-on-the-beach sort of trip," says Red Frog founder and CEO Joe Reynolds. "It's a push-yourself-outside-of-your-comfort-zone, culture-drenched, that-just-changed-my-life trip. Those trips bring home game-changing ideas. Those are sabbaticals."

Reynolds believes that the benefits from sabbaticals for the company and its employees are well worth the costs. He finds that employees return recharged and ready to work, in a much more creative frame of mind, and with a stronger worldly perspective. He also thinks that because he expects high performance and long hours when they are at work, time to connect with a loved one is important. His employees show great appreciation for the opportunity to travel with a loved one to a new and exciting location.

There are several important considerations before a small business implements a sabbatical program for its employees:

- **Can we afford it?** Even an unpaid sabbatical has a cost associated with it because a company may have to hire temporary help or ask other employees to cover the slack when an employee goes on leave.
- **Should the sabbatical be paid or unpaid?** If the sabbatical is unpaid, the employee actually may be eligible for unemployment insurance in some states. In addition, will the company extend benefits to employees while they are on an unpaid sabbatical? If not, they may be eligible for COBRA health insurance.
- **Do our jobs fit with a sabbatical program?** Some companies may not have a work flow or jobs that are easily covered if employees take a sabbatical leave, especially in a smaller firm where one person may be the sole expert in a key area of responsibility. "A person on sabbatical can be hard to replace, so you'll have to find workarounds that will prevent an employee from working too hard in the months leading up to the sabbatical and their coworkers from doing much more work to compensate for the loss," says Lance Haun, a human resource specialist and author.

- **Who will be eligible for sabbatical leaves?** At Ruby Receptionists, all employees are eligible after five years of work to apply for a five-week paid sabbatical. REI offers employees paid sabbatical leaves after 15 years of service. They get four weeks off in addition to their earned vacation time for that year. At every five-year anniversary thereafter, employees earn another sabbatical with one additional week added to the sabbatical each time.
- **What happens if the employee does not return?** Does the employee have to come back to work for a specified time? If not, what are the consequences? It can be difficult to enforce any monetary penalty on a worker who quits at the end of a sabbatical.
- **What are the expectations of the employee while on sabbatical?** Some companies view a sabbatical as purely time to rest. Others require that the employee write a white paper discussing what they did on sabbatical and how it benefited them personally and professionally.

Although a small business can offer sabbaticals on a case-by-case basis, it is advisable to implement a formal policy if sabbaticals are going to become a regular occurrence in the company. The policy should address the following questions:

- How often is an employee eligible?
- Is there a limit as to how many sabbaticals can be granted at any given time?
- How long will the sabbatical leave be?
- What are the expectations while on sabbatical and when the employee returns?
- What guarantees are there, if any, regarding position and pay on return?
- What are the consequences for failing to return to work?
- Will sabbatical time count toward seniority and pay increases?
- Are employees going to be paid during the sabbatical?
- Will standard employee benefits be extended during the sabbatical?

1. Do you think that sabbaticals are a good benefit for small businesses to offer employees? Explain.
2. If you were to offer a sabbatical program in your company, what would you set as the rules for this program? Explain.
3. Joe Reynolds believes that every employee deserves a sabbatical. Do you agree? Explain.

Sources: Based on *www.lifemeetswork.com/knowledge-center/work-redesign/ways-to-flex/corporate-sabbatical-how-to*; Alexandra Levit, "Should Companies Offer Sabbaticals?," *CNN Money*, January 3, 2011, *http://management.fortune.cnn.com/2011/01/03/should-companies-offer-sabbaticals*; Elizabeth Wilson, "Voluntary Benefits 101," *Entrepreneur*, March 31, 2009, *www.entrepreneur.com/article/201000*; "Pay and Benefits," REI, *www.rei.com/jobs/pay-benefits.html*; Joe Reynolds, "All Employees Deserve Paid Sabbaticals," *Inc.*, February 15, 2012, *www.inc.com/joe-reynolds/let-your-employees-unplug-and-take-sabbaticals.html*.

Performance Feedback

Entrepreneurs not only must motivate employees to excel in their jobs but also focus employees' efforts on the right business targets. Providing feedback on progress toward those targets can be a powerful motivating force in a company. To strengthen the link between the entrepreneur's vision for the company and its operations, he or she must build a series of specific performance measures that serve as periodic monitoring points. For each critical element of the organization's performance—quality, financial performance, market position, productivity, and employee development—he or she should develop specific measures that connect daily operational responsibilities with the company's overall strategic direction. These measures establish the benchmarks for measuring employees' performance and the company's progress. The adage "what gets measured and monitored gets done" is true for most organizations. An entrepreneur defines for everyone in the company what is most important by connecting the company's long-term strategy to its daily operations and measuring performance.

Providing feedback implies that entrepreneurs have established meaningful targets that serve as standards of performance for them, their employees, and the company as a whole. One characteristic that successful people have in common is that they set goals and objectives—usually challenging ones—for themselves. Entrepreneurs are no different. Successful entrepreneurs usually set targets for performance that make them stretch to achieve, and then they encourage their employees to do the same. The result is that they keep their companies constantly moving forward.

For feedback to serve as a motivating force in a business, entrepreneurs should follow the procedure illustrated in Figure 21.4, the feedback loop.

DECIDING WHAT TO MEASURE The first step in the feedback loop is deciding what to measure. Every business has a set of numbers that are critical to its success, and these "critical numbers" are what entrepreneurs should focus on. Obvious critical numbers include sales, profits, profit margins, cash flow, and other standard financial measures. However, supporting these measurements is an additional set of critical numbers that are unique to a company's operations. In most cases, these are the numbers that actually drive profits, cash flow, and other financial measures—they are the company's *real* critical numbers (refer to Chapter 14).

DECIDING HOW TO MEASURE Once an entrepreneur identifies his or her company's critical numbers, he or she must decide how to measure them. In some cases, identifying the critical numbers defines the measurements that owners must make, and measuring them simply becomes a matter of collecting and analyzing data. In other cases, the method of measurement is not as obvious—or as tangible. For instance, in some businesses, social responsibility is a key factor,

FIGURE 21.4
The Feedback Loop

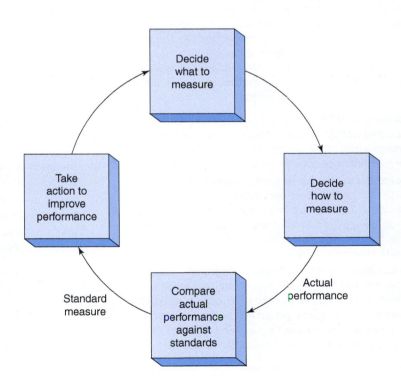

but how should managers measure their company's performance on such an intangible concept? One of the best ways to develop methods for measuring such factors is to use brainstorming sessions involving employees, customers, and even outsiders. For example, one company used this technique to develop a "fun index," which used the results of an employee survey to measure how much fun employees are having at work. The index is an indication of how satisfied they are with their work, the company, and their managers.

COMPARING ACTUAL PERFORMANCE AGAINST STANDARDS In this stage of the feedback loop, the goal is to look for variances *in either direction* from company performance standards. In other words, opportunities to improve performance arise when there is a gap between "what should be" and "what is." The most serious deviations usually are those in which actual performance falls far below the standard. Entrepreneurs must focus their efforts on figuring out why actual performance is substandard. The goal is *not* to hunt down the guilty party (or parties) for punishment but to discover the cause of the poor performance and fix it. Managers should not ignore deviations in the other direction, however. If actual performance consistently exceeds the company's standards, it may be an indication that the standards are set too low.

TAKING ACTION TO IMPROVE PERFORMANCE When managers or employees detect a performance gap, their next challenge is to decide on a course of action that will eliminate it. Typically, several suitable alternatives to solving a performance problem exist; the key is finding an acceptable solution that solves the problem quickly, efficiently, and effectively.

Performance Appraisal

One of the most common methods of providing feedback on employee performance is through **performance appraisal**, the process of evaluating an employee's actual performance against desired performance standards. Most performance appraisal programs strive to accomplish three goals:

1. To give employees feedback about how they are performing, which can be an important source of motivation
2. To provide entrepreneurs and employees the opportunity to create a plan for developing employee skills and abilities and for improving their performance
3. To establish a basis for determining promotions and salary increases

The primary purpose of performance appraisals is to encourage and to help employees improve their performance. Unfortunately, they can turn into uncomfortable confrontations that do nothing more than upset the employees, aggravate the entrepreneur, and destroy trust and morale. This may occur because the entrepreneur does not understand how to conduct an effective performance appraisal. Although U.S. businesses have been conducting performance appraisals for at least 75 years, most companies, their managers, and their employees are dissatisfied with the process. A survey by *Salary.com* shows that 60 percent of workers say that performance appraisals do not produce any useful feedback and fail to help them set meaningful objectives.[87] Common complaints include unclear standards and objectives, managers who lack information about employees' performances, managers who are unprepared or who lack honesty and sincerity, and managers who use general, ambiguous terms to describe employees' performances.

One complaint is that the performance appraisal happens only periodically: in most cases, just once a year. Employees do not have the opportunity to receive any ongoing feedback on a regular basis. All too often, managers save up all of the negative feedback to give employees and then dump it on them in the annual performance review. Doing so destroys employees' motivation and does *nothing* to improve their performance. What good does it do to tell an employee that six months earlier, he or she botched an assignment and caused the company to lose a customer? Performance reviews that occur once or twice a year in an attempt to improve employees' performance are similar to working out once or twice a year in an attempt to get into top physical condition!

The lack of ongoing feedback is similar to asking employees to bowl in the dark. They can hear some pins falling, but they have no idea how many are down or which ones are left standing for the next frame. How motivated would you be to keep bowling? How do you know your score as you bowl? Managers should address problems when they occur rather than wait until the performance appraisal session. Continuous feedback, both positive and negative, is a much more effective way to improve employees' performance and to increase their motivation than once-a-year feedback in a performance appraisal session.

Performance appraisals require planning and preparation on the entrepreneur's part. The following guidelines can help an entrepreneur create a performance appraisal system that actually works:

- *Link the employee performance criteria to the job description discussed earlier in this chapter.* To evaluate an employee's performance effectively, a manager must fully understanding of the responsibilities of the employee's position.

- *Establish meaningful, job-related, measurable, and results-oriented performance criteria.* The criteria should describe behaviors and actions, not traits and characteristics. What kind of behavior constitutes a solid performance in the job? Criteria that are quantifiable, such as customer satisfaction scores, the percentage of on-time shipments, and other specific measurements, rather than subjective criteria, such as leadership potential, initiative, and problem-solving ability, form the foundation of a meaningful performance evaluation.

- *Prepare for the appraisal session by outlining the key points you want to cover with employees.* Important points to include are employees' strengths and weaknesses and developing a plan for improving their performance.

- *Invite employees to provide an evaluation of their own job performance based on the performance criteria.* In one small company, workers rate themselves on a 1-to-5 scale in categories of job-related behavior and skills as part of the performance appraisal system. Then they meet with their supervisor to compare their evaluations with those of their supervisor and discuss them.

- *Be specific.* One of the most common complaints employees have about the appraisal process is that managers' comments are too general to be of any value. Offer the employees specific examples of their desirable or undesirable behavior.

- *Keep a record of employees' critical incidents—both positive and negative.* The most productive evaluations are those based on managers' direct observation of their employees' on-the-job performance. These records also can be vital in case legal problems arise.

- *Discuss employees' strengths and weaknesses.* An appraisal session is not the time to "unload" about everything that employees have done wrong over the past year. Use it as an opportunity to design a plan for improvement and to recognize employees' strengths, efforts, and achievements.

- *Incorporate employees' goals into the appraisal.* Ideally, the standard against which to measure employees' performance is the goal that they have played a role in setting. Workers are more likely to be motivated to achieve—and buy into—goals that they have helped establish.

- *Keep the evaluation constructive.* Avoid the tendency to belittle employees. Do not dwell on past failures. Instead, point out specific things they should do better and help them develop meaningful goals for the future and a strategy for getting there.

- *Praise good work.* Avoid focusing only on what employees do wrong. Take the time to express your appreciation for hard work and solid accomplishments.

- *Focus on behaviors, actions, and results.* Problems arise when managers move away from tangible results and actions and begin to critique employees' abilities and attitudes. Such criticism creates a negative tone for the appraisal session and undercuts its primary purpose.

- *Avoid surprises.* If entrepreneurs are doing their jobs well, performance appraisals should contain no surprises for employees or the owner. The ideal time to correct improper behavior or slumping performance is when it happens, not months later. Managers should provide employees with continuous feedback on their performance and use the appraisal session to keep employees on the right track.

- *Plan for the future.* Smart entrepreneurs use appraisal sessions as gateways to workers' future success. They spend only about 20 percent of the time discussing past performance; they use the remaining 80 percent of the time developing goals, objectives, and a plan for the future.

ENTREPRENEURIAL PROFILE: Brian Roth: Trufast Brian Roth, CEO of Trufast, a small maker of fasteners located in Bryan, Ohio, realized that his company's performance appraisal process was ineffective. "The only thing the review did was cover the previous two weeks of performance," he says. "Basically, it was worthless." Roth revamped the entire process to focus on more

Selling to Outsiders

As you learned in Chapter 7, selling a business to an outsider is no simple task. Done properly, it takes time, patience, and preparation to locate a suitable buyer, strike a deal, and make the transition. Advance preparation, maintaining accurate financial records, and timing are the keys to a successful sale. Too often, however, business owners, like some famous athletes, stay with the game too long until they and their businesses are well past their prime. They postpone selling until the last minute when they reach retirement age or when they face a business crisis. Such a "fire-sale" approach rarely yields the maximum value for a business.

A straight sale may be best for entrepreneurs who want to step down immediately and turn the reins of the company over to someone else. However, selling a business outright is not an attractive exit strategy for those who want to stay with the company or for those who want to surrender control of the company gradually rather than all at once.

ENTREPRENEURIAL PROFILE: Ralph Ostrove and Paul Stuart Recently, members of the Ostrove family, descendants of Ralph Ostrove, who in 1938 founded Paul Stuart, the retailer of classic suits, shirts, ties, and shoes preferred by Wall Street executives, decided to sell the family business. Clifford Grodd, Ralph Ostrove's son-in-law, joined the business in 1952 and served as its president until his death in 2010 at the age of 86, when Michael Ostrove took over. The family voted to sell 100 percent of Paul Stuart to Mitsui & Company, the company's longtime distributor in Japan. Mitsui plans to broaden the product line with more casual clothing, expand the number of retail outlets in the United States and in Asia, and use the company's "man on the fence" logo more strategically.[38]

The financial terms of a sale also influence the selling price of the business and the number of potential bidders. Does the owner want "clean, cash only, 100 percent at closing" offers, or is he or she willing to finance a portion of the sale? A 100 percent, cash-only requirement dramatically reduces the number of potential buyers. On the other hand, the owner can exit the business "free and clear" and does not incur the risk that the buyer may fail to operate the business profitably and be unable to complete the financial transition.

Selling to Insiders

When entrepreneurs have no family members to whom they can transfer ownership or who want to assume the responsibilities of running a company, selling the business to employees is often the preferred option. In most situations, the options available to owners are (1) sale for cash plus a note, (2) a leveraged buyout, and (3) an employee stock ownership plan.

A SALE FOR CASH PLUS A NOTE Whether entrepreneurs sell their businesses to insiders, outsiders, or family members, they often finance a portion of the sales price. The buyer pays the seller a lump-sum amount of cash up front, and the seller holds a promissory note for the remaining portion of the selling price, which the buyer pays off in installments. Because of its many creative financial options, this method of selling a business is popular with buyers. They can buy promising businesses without having to come up with the total purchase price all at one time. Sellers also appreciate the security and the tax implications of accepting payment over time. They receive a portion of the sale up front and have the assurance of receiving a steady stream of income in the future. In addition, they can stretch their tax liabilities from the capital gains on the sale over time rather than having to pay them in a single year. In many cases, sellers' risks are lower because they may even retain a seat on the board of directors to ensure that the new owners are keeping the business on track.

LEVERAGED BUYOUTS In a **leveraged buyout** (LBO), managers and/or employees borrow money from a financial institution and pay the owner the total agreed-on price at closing; then they use the cash generated from the company's operations to pay off the debt. The drawback of this technique is that it creates a highly leveraged business. Because of the high levels of debt they take on, the new management team has very little room for error. Too many management mistakes or a slowing economy has led many highly leveraged businesses into bankruptcy.

If properly structured, LBOs can be an attractive to both buyers and sellers. Because they get their money up front, sellers do not incur the risk of loss if the buyers cannot keep the business operating successfully. The managers and employees who buy the company have a strong incentive to make sure the business succeeds because they own a piece of the action and some of their capital is at risk in the business. The result can be a highly motivated workforce that works hard and makes sure that the company operates efficiently.

> **ENTREPRENEURIAL PROFILE: Jack Stack: Springfield Remanufacturing Corporation**
> In one of the most successful LBOs in history, Jack Stack and a team of 12 other managers purchased an ailing subsidiary of International Harvester in an attempt to save their jobs and those of the 120 employees they managed. The new company, Springfield Remanufacturing Corporation (SRC), which specializes in engine remanufacturing for automotive, trucking, agricultural, and construction industries, began with an astronomically high debt to equity ratio of 89:1, but the team of motivated managers and employees turned the company around. Today, SRC has more than 1,200 employees and 26 divisions that range from automotive engines to home furnishings.[39]

EMPLOYEE STOCK OWNERSHIP PLANS Unlike LBOs, **employee stock ownership plans** (ESOPs) allow employees and/or managers (i.e., the future owners) to purchase the business gradually, freeing up enough cash to finance the venture's growth. With an ESOP, employees contribute a portion of their earnings over time toward purchasing shares of the company's stock from the founder until they own the company outright. In a leveraged ESOP, the ESOP itself borrows the money to buy the owner's stock up front. Then, using employees' contributions, the ESOP repays the loan over time. One advantage of a leveraged ESOP is that the principal and the interest that the ESOP borrows to buy the business are tax deductible, and this can save thousands or even millions of dollars in taxes. Transferring ownership to employees through an ESOP is a long-term exit strategy that benefits everyone involved. The owner sells the business to the people he or she can trust the most—his or her managers and employees. The managers and employees buy a business they already know how to run successfully. In addition, because they own the company, the managers and employees have a huge incentive to see that it operates effectively and efficiently. One study of employee stock ownership plans in privately held companies found that the ESOPs increased sales, employment, and sales per employee by 2.4 percent a year.[40] Figure 22.3 shows the trend in the number of ESOPs and the number of employee owners.

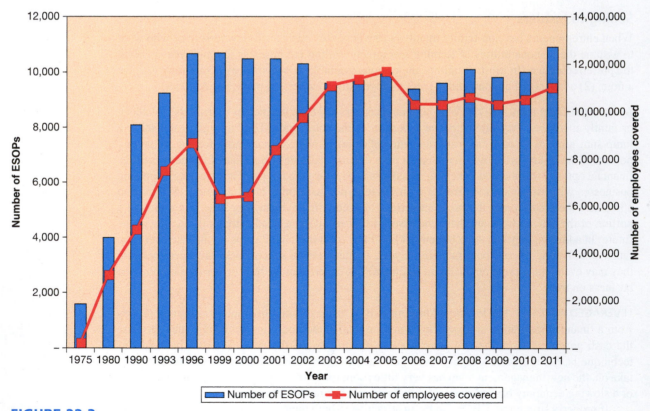

FIGURE 22.3

Employee Stock Ownership Plans

Source: Based on National Center for Employee Ownership, 2013.

ENTREPRENEURIAL PROFILE: Tom and Adele Connors: Adworkshop Tom and Adele Connors started Adworkshop, a marketing agency located in Lake Placid, New York, in 1977. In 2009, the Connorses established an ESOP and turned over 33 percent of the ownership in the business to their 30 employees with the goal of transferring 100 percent of ownership to employees through the ESOP over the next several years. Adele says that since the employees became owners in the business, they are more engaged and understand "that the harder they work, the more impact they have on the business's success." Kelly Frady, an account supervisor at Adworkshop says that the ESOP has fostered a stronger team spirit among the employees. "It's a driving factor in making the company succeed in the long term," she says.[41]

The third exit strategy available to company founders is transferring ownership to the next generation of family members with the help of a comprehensive management succession plan.

Lessons from the Street-Smart Entrepreneur

How to Set Up an ESOP

Dansko.
Source: Dansko.

In 1990, Mandy Cabot and her husband Peter Kjellerup discovered incredibly comfortable barn shoes while traveling in Europe and started a company, Dansko, from their farm in West Grove, Pennsylvania, to sell them in the United States. In the early days, Cabot sold Dansko shoes to individuals from the back of her Volvo station wagon before landing retail shoe stores as customers. By 1996, Dansko was growing so fast that Cabot and Kjellerup moved the company from their farm to an office and warehouse. Two years later, the company had outgrown its space once again, and the copreneurs broke ground on a larger 26,000-square-foot office and warehouse. By 2002, Dansko had sold 1 million pairs of shoes, and its sales continued to grow at an impressive 46 percent annual rate. The couple began to consider the future of their company and decided against selling to a competitor or to a private equity firm that might institute layoffs in a cost-cutting initiative. They decided that the best way to ensure Dansko's continued success was to transfer ownership to the people who had played a vital role in building it: their employees. In 2005, they created an ESOP and began transferring ownership of the company, which is an S corporation, to

their employees, who at the time numbered more than 100. In 2012, Cabot and Kjellerup, both of whom were approaching retirement age, transferred 100 percent of the company to their 180 employees through the ESOP, giving them total control of its future. "This is our baby," says Cabot, "but at some point we have to cease being parents and become grandparents."

Dawn Huston started working at Dansko when she was just 20 years old, sorting shoes for delivery. Today, she is a warehouse processor and likes the idea of owning the business for which she works, although she admits that she was nervous about it at first. Huston, who now refers to Dansko as "our company," says, "They [Cabot and Kjellerup] consider us family, and it feels like a family."

Like Cabot and Kjellerup, many Baby Boomer entrepreneurs are transferring ownership of their businesses to their employees through ESOPs. "A company's success isn't driven [solely] by the brilliance of the CEO," says Michael Keeling, president of the ESOP Association, "but also by its employees, and more owners feel that their employees deserve something more for that." What steps should an entrepreneur who is interested in setting up an ESOP take?

Step 1. Conduct a feasibility analysis to determine whether an ESOP is right for you and your company

A company should be profitable and should have at least 20 to 30 employees to make an ESOP work and have a minimum valuation of between $5 million and $10 million. Only companies organized as C corporations or S corporations can create ESOPs. Creating the necessary plan documents and filing them with the proper government agencies costs about $10,000. A business valuation, which can range from $5,000 to $10,000 for a small company, is a necessity. Fixed costs of administering the ESOP and complying with the Employee Retirement Income Security Act run about $2,000 plus $20 to $30 per employee participant per year. A final consideration is whether the company is able to generate sufficient revenue consistently over time to be able to repay the loan in a leveraged ESOP.

(continued)

Lessons from the Street-Smart Entrepreneur *(continued)*

Step 2. Hire an attorney who specializes in ESOPs to help you develop a plan for creating and implementing an ESOP

ESOPs can take many different forms, and an expert can help you determine the advantages and disadvantages of each one so that you can determine the one that is best for you and your company.

Step 3. Find the money to fund the ESOP

About 75 percent of ESOPs are leveraged, which means that they borrow the money to purchase the owner's stock from the ESOP trust. Banks and other financial institutions usually find loans to ESOPs quite attractive.

Step 4. Establish a process to operate the ESOP

Companies most often create an ESOP committee of managers and employees to provide guidance to the ESOP trust for managing the ESOP. The team also is responsible for communicating the details of the ESOP and the benefits of investing in it to company employees.

In 1991, Kim Jordan cofounded New Belgium Brewing Company in Fort Collins, Colorado, with her husband, Jeff Lebesch, who was inspired by the full-bodied local beers he encountered while bicycling through Europe. Early on, Jordan and Lebesch created an ESOP that allows employees to begin purchasing

shares of ownership in the company on the first anniversary of their employment date. (They also receive a cruiser bicycle as a gift.) Over the years, the employees purchased 41 percent of the company through the ESOP, but on December 29, 2012, Jordan and her family sold the remaining shares to the employees. New Belgium Brewing is now 100 percent employee owned. Jordan says that the goal is not only to reward employees but also to foster creativity by enhancing a culture in which employees think and act like entrepreneurs. "There are few times in life where you get to make choices that will have multigenerational impact, and this is one of those times," says Jordan. "We have an opportunity to write the next chapter of this incredible story and we're really excited about that. We have always had a high-involvement ownership culture, and this allows us to take that to the next logical level. It will provide an elegant succession framework that keeps the executive team intact ensuring our vision stays true going forward."

Sources: Based on Toddi Gutner, "What You Need to Know About ESOPs," *Fox Business Small Business Center*, December 12, 2012, *http://smallbusiness.foxbusiness.com/biz-on-main/2012/12/12/what-need-to-know-about-esops*; Angus Loten, "Founders Cash Out, but Do Workers Gain?," *Wall Street Journal*, April 18, 2013, p. B4; Karen E. Klein, "ESOPs on the Rise Among Small Businesses," *Bloomberg Business Week*, March 26, 2010, *www.businessweek.com/smallbiz/content/mar2010/sb20100325_591132.htm*; "ESOP Statistics," ESOP Association, *www.esopassociation.org/media/media_statistics.asp*; "How Small Is Too Small for an ESOP?," National Center for Employee Ownership, *www.nceo.org/library/howsmall.html*; "Steps to Setting Up an ESOP," National Center for Employee Ownership, *www.nceo.org/library/steps.html*.

Management Succession

3.

Discuss the stages of management succession.

By 2050, practically every family-owned business in the United States will lose its primary owner(s) to retirement or death. Experts estimate that by 2040, $10.4 trillion in wealth will be transferred from one generation to the next, much of it funneled through family businesses.[42] Sixty percent of small business owners are members of the Baby Boomer generation, born before 1964, and many of the entrepreneurs who have not yet transferred ownership to the next generation now are at or past retirement age and are ready to pass the torch of leadership. A Baby Boomer business owner retires, on average, every 57 seconds, a trend that will continue for the next 17 years.[43]

For a smooth transition from one generation to the next, family businesses need a succession plan. Although 95 percent of small business owners acknowledge the need for a succession plan, only one in eight actually has a written plan in place for leadership continuity.[44] Without a succession plan, family businesses face an increased risk of faltering or failing in the next generation. Those businesses with the greatest probability of surviving are the ones whose owners prepare a succession plan well before it is time to transfer control to the next generation. Succession planning also allows business owners to minimize the impact of taxes on their businesses, their estates, and their successors' wealth as well and to avoid saddling the next generation of ownership with burdensome debt.

Why, then, do so many entrepreneurs postpone succession planning until it is too late? One expert explains,

> Many business owners simply refuse to quit. Some would like to quit but don't trust anyone to do the job well after them. Some are tangled up by unrealistic expectations from family and business partners. Others wonder how they'll be treated once they retire and what they will do with themselves. Many simply don't see how they can attract a buyer, arrange a buyout, or realize other acceptable transactions.[45]

"My business is my passion," says Dave Hale, who cofounded ScaleTronix, a designer and manufacturer of medical scales, in 1975. "The idea of ending what I do is just a terrible thought. Exiting seems like dying to me. What am I, if not my business?"[46] Like Hale, many business founders hesitate to let go of their businesses because their personal identities are so wrapped up in their companies. Over time, an entrepreneur's identity becomes so intertwined in the business that, in his or her mind, there is no distinction between the two. Many entrepreneurs share Hale's feelings; 52 percent of business owners say that they have no plans to retire and intend to work indefinitely.[47]

Another barrier to succession planning is that, in planning the future of the business, owners must accept the painful reality of their own mortality. In addition, turning over the reins of a business they have sacrificed for, fretted over, and dedicated themselves to for so many years is extremely difficult to do—even if the successor is a son or daughter! Paul Snodgrass, son of the founder of Pella Products, a maker of apparel for work and outdoor activities who accepted leadership of the company from his father, explains, "Dad loves you and wants you to take over the business, but he also put heart and soul into that business, and he's not going to let anybody screw it up—not even you."[48] Finally, many family business founders believe that controlling the business also gives them a degree of control over family members and family behavior.

Planning for management succession protects not only the founder's, successor's, and company's financial resources, but it also preserves what matters most in a successful business: its heritage, tradition, and legacy. "Real succession planning involves developing a strategy for transferring the trust, respect, and goodwill built by one generation to the next," explains Andy Bluestone, who took over as president of the financial services company his father founded.[49] Management succession planning requires, first, an attitude of trusting others. It recognizes that other family members have a stake in the future of the business and want to participate in planning its future. Planning demonstrates an attitude that shows that decisions made with open discussion are more constructive than those without family input. Second, management succession is an evolutionary process and must reconcile an entrepreneur's inevitable anguish with the successors' desire for autonomy. Owners' emotional ties to their businesses usually are stronger than their financial ties. On the other side of the equation are the successors, who yearn to have the autonomy to run the business their way. These inherent conflicts can—and often do—result in skirmishes.

Succession planning reduces the tension and stress created by these conflicts by gradually "changing the guard." A well-developed succession plan is like the smooth, graceful exchange of a baton between runners in a relay race. The new runner still has maximum energy; the concluding runner has already spent his or her energy by running at maximum speed. The athletes never come to a stop to exchange the baton; instead, the handoff takes place on the move. The race is a skillful blend of the talents of all team members—an exchange of leadership so smooth and powerful that the business never falters but accelerates, fueled by a new source of energy at each leg of the race. At L.L. Bean, Leon Gorman, who led the company for 40 years before handing leadership over to current CEO Chris McCormick, recognized the importance of a seamless transition of leadership. He set up a three-member family governance committee made up of fourth-generation family members and encouraged family members to stay active in the business. Today, the fifth generation of family members is involved in orientation sessions in different areas of the company, and at the executive level, a structured leadership development process is in place. L.L. Bean has "moved the process down through the organization with leaders at all levels expected to contribute to succession planning," says McCormick.[50]

Management succession involves a lengthy series of interconnected stages that begin very early in the life of the owner's children and extends to the point of final ownership transition (see Figure 22.4). If management succession is to be effective, it is necessary for the process to begin early in the successor's life (stage I). For instance, the owner of a catering business recalls putting his son to work in the family owned company at age seven. On weekends, the boy would arrive at dawn to baste turkeys and was paid in his favorite medium of exchange—doughnuts![51] In most cases, family business owners involve their children in their businesses while they are still in junior high or high school. "I grew up in the business," recalls Anthony Halas, now CEO of Seafolly, the swimwear company his father cofounded. "I spent my school holidays working in the warehouse, packing boxes. I always felt that [the business] was part of the family."[52] Just as Anthony Halas experienced, the tasks in this phase may be routine, but the child is learning many important lessons, including the basics of how the business operates and an appreciation of the role the business plays in the life of the family. They learn firsthand about the values and responsibilities of running the company.

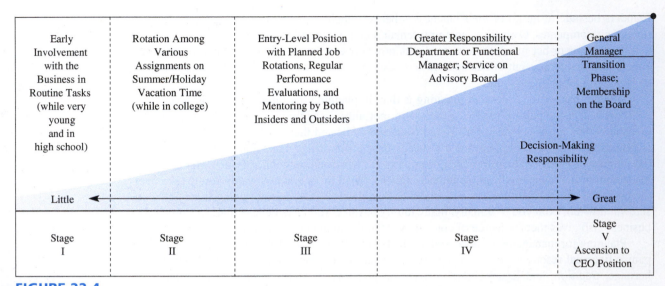

FIGURE 22.4

Stages in Management Succession

While in college, the successor moves to stage II of the continuum. During this stage, the individual rotates among a variety of job functions to both broaden his or her base of understanding of the business and permit the parents to evaluate his or her skills. On graduation from college, the successor enters stage III. At this point, the successor becomes a full-time worker and ideally has already begun to earn the respect of coworkers through his or her behavior in the first two stages of the process. In some cases, the successor may work for a time outside of the family business to gain experience and to establish a reputation for competency that goes beyond "being the boss's kid." Stage III focuses on the successor's continuous development, often through a program designed to groom the successor using both family and nonfamily managers as mentors.

ENTREPRENEURIAL PROFILE: Danny Joyner: C. Dan Joyner Real Estate From the time that Danny Joyner was a young boy "helping" his father, Dan, in the real estate business that the elder Joyner had started with a $2,000 loan, he knew that he wanted to join his father's company. "I've always known that this is what I would do," says Danny, who joined the company full-time after graduating from college. Danny managed the successful company's commercial real estate division, and Dan's two daughters and son-in-law, David Crigler, also worked in the family business. Because Dan had the foresight to put a succession plan in place and to help his children grow into leadership positions in the company, the company made a smooth transition into the second generation of ownership when Dan died in 2012. Danny took over as president, and Crigler assumed the role of chief operating officer. Carmen Feemster, representing the third generation of the Joyner family, recently began working for the company.[53]

As the successor develops his or her skills and abilities, he or she moves to stage IV, in which real decision-making authority grows rapidly. Stage IV of the succession continuum is the period when the founder makes a final assessment of the successor's competence and ability to take full and complete control over the firm. The skills the successor needs include the following:

- *Financial abilities.* Understanding the financial aspects of a business, what its financial position is, and the managerial implications of that position are crucial to success.
- *Technical knowledge.* Every business has its own body of knowledge, ranging from how the distribution system works to the trends shaping the industry, that an executive must master.
- *Negotiating ability.* Much of business, whether buying supplies and inventory or selling to customers, boils down to negotiating, and a business owner must be adept at it.
- *Leadership qualities.* Leaders must be bold enough to stake out the company's future and then give employees the resources, the power, and the freedom to pursue it.
- *Communication skills.* Business leaders must communicate the vision they have for their businesses; good listening skills also are essential for success as a top manager.

- *Juggling skills.* Business owners must be able to handle multiple projects effectively. Like a juggler, they must maintain control over several important assignments simultaneously.

- *Integrity.* To be an effective leader of a family business, a successor must demonstrate honesty and integrity in business dealings.

- *Commitment to the business.* It helps if a successor has a genuine passion for the business. Leaders who have enthusiasm for what they do create a spark of excitement throughout the entire organization.[54]

The final stage in the management succession process involves the ultimate transition of leadership. It is during this stage that the founder's role as mentor is most crucial. "One characteristic of stable family businesses is consistent leadership," says one family business expert.[55]

 ENTREPRENEURIAL PROFILE: Laura Michaud: Beltone Electronics Corporation Laura Michaud, who in 1980 became the third-generation owner of Beltone Electronics Corporation, a very successful maker of hearing aids that was founded in 1940, says that the training and mentoring that her father provided was key to her success to managing the family business. Her father insisted on an extensive training program that involved Michaud in all aspects of the company, including having her actually build a hearing aid. "You need to work in all areas of operations," says Michaud, who ran the company for 17 years before selling it to a larger company.[56]

In stage IV, the successor may become the organization's CEO, while the former CEO retains the title of chairman of the board. In other cases, the best solution is for the founder to step out of the business entirely and give the successor the chance to establish his or her own identity within the company. "Any leader's final legacy is building the next generation," says one business consultant.[57]

In the Entrepreneurial Spotlight

Can Your Family Business Survive for Centuries?

In 1535, Henry VIII was King of England and was still married to Anne Boleyn; the first English translation of the Bible was printed in Antwerp, Belgium; and R. J. Balson began selling meat from a stall in a market on South Street in Bridport, Dorset, England. Nearly 500 years and 25 generations later, R.J. Balson and Sons is still thriving in Dorset and holds the title of the oldest family business in Great Britain. In 1880, the butcher shop moved to its current location, not far from the original South Street stall, having passed from one generation of the family to the next. Current owners Richard Balson and Rudi and Jane Boulay took over the family business after Richard and Jane's father, Donald, died in 2011 at age 88. R.J. Balson and Sons has survived for centuries by sticking to the business that the family knows best—selling a variety of meats and sausages ("bangers" in local parlance) made out of exotic meats such as elk, wild boar, duck, and ostrich, rusk (ground bread crumbs), and fresh herbs. To thrive as it has, the company also has changed over time, adapting to the dynamic environment in which it operates. Once a purely local business, R.J. Balson and Sons now sells its exotic meats to customers around the world through its Web site (*www.balsonbutchers.com*). "My son has come into the business to run the online side," says Richard. "He's got a son, too, so hopefully it'll keep going. The family joke is that we've just never made enough money to be able to retire."

In 2007, after a career as a professional soccer player, coach, and referee, Mike Balson, Richard's brother, returned to the family business and opened a production facility in the United States

with his son, Oliver. "They have survived many more transitions than most and have naturally had to evolve to take into account the significant economic developments that they have been part of to still be in business today," says Paul Andrews, founder of Family Business United.

What allows a family business such as R.J. Balson and Sons to succeed for 25 generations when so many fail to make the transition from the first generation to the second? The following tips increase the chances of a smooth transition from one generation to the next:

Develop a passion for passing the torch of leadership. Management succession does not happen by accident; it requires commitment from each generation of leadership. "Family business leaders feel a certain duty to act as stewards of the family business for the next generation," says Paul Andrews.

Cultivate a passion for the family business in each successive generation. Leaders of successful family businesses usually involve their children in the business early on and allow them to have fun while they work and develop an appreciation for how special a family business can be. "The love of the job, which has been passed down from generation to generation, is one of the main reasons we have been successful," says Richard Balson of R.J. Balson and Sons.

(continued)

In the Entrepreneurial Spotlight *(continued)*

Develop a succession plan. As you have learned in this chapter, passing a business to the next generation requires a sound succession plan. Many family business owners work with consultants, who provide important objectivity, to create succession plans. Building and implementing a succession plan takes time, and family business leaders must resist the natural temptation to postpone this vital task until it is too late.

Select the right successor. Family business owners must look beyond seniority to determine which family member—or whether *any* family member—of the next generation is most qualified to lead the business. Leaders of some enduring family businesses, such as L.L. Bean and Zambelli Fireworks, have been bold enough to bring in managers from outside the family to run the business. One family business consultant suggests that family business owners create a list of all possible successors and rank them on a 1-to-10 scale in areas such as education, work experience, ability to learn, desire to advance, interpersonal skills, decision-making ability, knowledge of the industry and the business, character, and others.

Groom the successor over time. Once a family business leader identifies the right successor, his or her duty is to prepare the successor for the leader's job. Once again, starting early is the key because grooming a successor usually takes years. Ensuring that the successor gains experience in all of the functional areas of the business and understands the "big picture" that determines the company's ability to succeed are vital tasks.

Allow the successor to put his or her "stamp" on the business. When a successor takes the helm of the family business, the former leader must avoid undermining his or her leadership by unofficially remaining in power. Some business leaders have difficulty letting go of the reins that they have held for so long, but doing so only weakens the new leader's authority. Outgoing leaders must recognize that the new generation of leadership has its own ideas for improving the business. If the outgoing leaders have properly groomed their successors, they can be confident that the changes that the new leadership makes are consistent with the vision and mission of the company.

Establish a mechanism for dealing with conflict. Conflict is a natural part of any family and any business, but families must manage conflict if a business is to succeed. A family council that is comprised of family members from different generations and trusted business advisers from outside the family is one of the best ways to handle conflicts. Regular meetings in which everyone feels free to speak out on important issues facing the company not only help resolve conflict but also help business leaders make good decisions by hearing all sides of the issues. "The family agreed that a lack of transparent communication would be the only thing that could pull us apart," says Jack Mitchell, second-generation CEO of Mitchells Family of Stores. "We have scheduled weekly family meetings in which we discuss in a 'safe haven' any issues that are on the working family members' minds."

1. Use a search engine to learn more about family businesses that have thrived across many generations (e.g., Zambelli Fireworks, Omaha Steaks, Martin Guitars, George R. Ruhl & Son, Antoine's Restaurant, Verdin Bells and Clocks, and Hicks Nurseries). What factors have allowed these companies to be successful across many generations of owners?

2. Select one of the long-lasting family businesses that you discovered and write a one-page summary of the business, what it does, how it handles the issue of management succession, and what has enabled the business to succeed for so long.

Sources: Based on Leah Golob, "Ten Reasons Why Family Businesses Fail," *Globe and Mail*, July 20, 2012, *www.theglobeandmail.com/report-on-business/small-business/sb-tools/top-tens/ten-reasons-why-family-businesses-fail/article4219703*; Jack Mitchell, "Family Business: How to Pass the Baton," *CNN Money*, July 9, 2012, *http://management.fortune.cnn.com/2012/07/09/family-business-succession*; Jean Scheid, "10 Business Success Stories: Keeping It in the Family," *Bright Hub*, October 27, 2011, *www.brighthub.com/office/entrepreneurs/articles/126179.aspx*; Lois Lang, "5 Tips for Family Business Succession Planning," *Landscape Management*, April 4, 2013, *http://landscapemanagement.net/2013/04/04/march-web-extra-who-gets-dads-office*; Fraser McAlpine, "Introducing Britain's Oldest Family-Run Business," *BBC America*, November 4, 2011, *www.bbcamerica.com/anglophenia/2011/11/introducing-britains-oldest-family-run-business*; Luke Salkeld, "Butcher's That's Been Around Since Henry VIII Celebrates Being Oldest Family Business in Britain," *Daily Mail*, November 3, 2011, *www.dailymail.co.uk/news/article-2057236/Butchers-RJ-Balson-Sons-crowned-oldest-family-business-Britain.html*; Shane Schutte, "A Glimpse Behind the History of Britain's Oldest Family Business, Dating Back to 476 Years Ago," *Real Business*, December 3, 2012, *http://realbusiness.co.uk/article/16716-the-oldest-family-businesses-in-britain*; "History of RJ Balson and Son," RJ Balson and Son, *www.balsonbutchers.com/pages/about*.

Developing a Management Succession Plan

4.

Explain how to develop an effective management succession plan.

Families that are most committed to ensuring that their businesses survive from one generation to the next exhibit four characteristics: (1) They believe that owning the business helps achieve their families' missions, (2) they are proud of the values their businesses are built on and exemplify, (3) they believe that the business is contributing to society and makes it a better place to live, and (4) they rely on management succession plans to ensure the continuity of their companies.[58] Unfortunately, 55 percent of small business owners have not established a formal succession plan for their companies (including 43 percent of owners who are 67 or more years old).[59] Developing a plan takes time and dedication, yet the benefits are well worth the cost. A sound succession plan enables a company to maintain its momentum and sense of purpose and direction.

It is important to start the planning process early, well before the founder's retirement. Succession planning is not the kind of activity an entrepreneur can do in a hurry, and the sooner an entrepreneur starts, the easier it will be. Sandy and Sheena Noce, owners of Pro 6 Cycle, a company in Etobicoke, Canada, that provides motorbike racing experiences for cycling fans, already have begun making plans to include their daughter, Claire, who recently turned one, in the family business—if she chooses that route when she is old enough. "She has influenced all of our plans," says Sandy.[60] Unlike the Noces, too many entrepreneurs put off succession planning until it's too late. "Very few privately owned business make it through several generations, and one reason is the failure of the senior generation to do any planning at all until it is too late in the game," says one expert.[61] Creating a succession plan involves the following steps.

Step 1. ***Select the successor.*** Families often grow faster than a family business. A family in which each member of each generation has three children with a business that survives to the fourth generation would have 52 members (including spouses), with 27 potential successors in the fourth generation. The numbers alone dictate that a founder commits to selecting a successor long before he or she is ready to step down. The average tenure of the founder of a family business is 25 years, compared to just seven years for the CEO of a publicly held company.[62] Yet there comes a time for even the most dedicated founder to step down and hand the reins of the company to the next generation. Entrepreneurs should never assume that their children want to take control of the business, however. Above all, they should not be afraid to ask the question "Do you really want to take over the family business?" Too often, children in this situation tell Mom and Dad what they want to hear out of loyalty, pressure, or guilt. It is critical to remember at this juncture in the life of a business that children do not necessarily inherit their parents' entrepreneurial skills and desires. By leveling with their children about the business and their options regarding a family succession, owners can know which heirs, if any, are willing to assume leadership of the business.

One of the worst mistakes entrepreneurs can make is to postpone naming a successor until just before they are ready to step down. "I remember saying to our family business advisor, 'If my parents do not provide me and my siblings with a clear picture of my future in the family business soon, including whether I will ever be an owner, I am out of here,'" says the daughter of a family business owner. Fortunately, her parents created a succession plan and selected her as their successor.[63] The problem is especially acute when more than one family member works for the company and is interested in assuming leadership of it. Sometimes founders avoid naming successors because they don't want to hurt the family members who are not chosen to succeed them. However, both the business and the family will be better off if, after observing family members as they work in the business, the founder picks a successor based on skill and ability.

ENTREPRENEURIAL PROFILE: David de Rothschild: Rothschild Group David de Rothschild, the sixth-generation CEO of the Rothschild Group, one of the world's largest financial advisory groups founded in the early 1800s, has been grooming his son, Alexandre, for years to become the next leader of the family business. After graduating from college, Alexandre worked in other financial services companies around the world, learning various aspects of the financial and investment industries before joining the family business. None of David's three daughters had an interest in the family business, leaving Alexandre as the natural choice to manage the company. Alexandre remains humble about his selection. "The last thing I want to do is impose myself as leader just because I carry the [family] name." Alexandre, who has served in a variety of roles as Rothschild Group since joining the business, will assume the leadership role within five years.[64]

When naming a successor, merit is a better standard to use than gender or birth order. The key is to establish standards of performance, knowledge, education, and ability and then to identify the person who best meets those standards. As part of his company's succession plan, Joe De La Torre selected his daughter Gina to take over Jaunita's Foods rather than his two sons because her financial skills and her ability to solve problems were what the company needed most.[65] Gina La Torre is part of a growing trend among family businesses; 34 percent of family business founders expect the next CEO to be a woman, quite a change from just a generation ago.[66]

ENTREPRENEURIAL PROFILE: Harry and Kirsten Vold: Harry Vold Rodeo Company After 60 years, Harry Vold, founder of Harry Vold Rodeo Company, a business based in Pueblo West, Colorado, that supplies rodeos with bucking broncos and bulls, turned over the reins of the

Source: CartoonStock.

"I want you to meet my son Edward.
One day, this company will belong to him
and I think it's time for all of you to get to know
each other."

family business to the youngest of his six children, Kirsten, when she was just 25 years old. Although Harry expected one of his children to take over the business, he was a bit surprised that Kirsten was the one who showed the greatest potential and stepped forward. Kirsten, too, was somewhat surprised. "I was going to be a lawyer, drive a sports car, and live in L.A," she says with a laugh. After graduating from the University of Southern Colorado, Kirsten took a job in sports marketing but soon felt the tug of the family business and returned home to take an active role in managing it. Kirsten worked hard to prove herself in an industry that men tend to dominate. "She's earned their respect," says her mother proudly. Transferring control from one generation to the next taxed the skills, patience, and ability of both Harry and Kirsten. "It was a struggle, a power struggle, along the way, but it built character for both of us," she says. Harry is pleased with the way things worked out. "I was a bit surprised to begin with, but I'm not now," he says. "I am totally confident that she makes the right decisions. As far as I'm concerned, she's all the cowgirl I need."[67]

Even in China, traditionally a male-dominated society, entrepreneurs are turning over their businesses to their daughters more frequently. Zong Qinhou, China's wealthiest person with a net worth of $20.1 billion, recently named his daughter Kelly as his successor at Hangzhou Wahaha Group, the company that he started and built into China's largest beverage company. Kelly, a graduate of California's Pepperdine University with a degree in international business, joined her father's company in 2005 and became president in 2012.[68]

Step 2. *Create a survival kit for the successor.*   Once an entrepreneur identifies a successor, he or she should prepare a survival kit and then brief the future leader on its contents, which should include all of the company's critical documents (wills, trusts, insurance policies, financial statements, bank accounts, key contracts, corporate bylaws, and so forth). The founder should be sure that the successor reads and understands all of the relevant documents in the kit. Other important steps the owner should take to prepare the successor to take over leadership of the business include the following:

- Create a strategic analysis for the future. Working with the successor, entrepreneurs should identify the primary opportunities and the challenges facing the company and the requirements for meeting them.

- Share with the successor their vision of the business's future direction, describing key factors that have led to its success and those that will bring future success.

- Be open and listen to the successor's views and concerns.

- Teach and learn at the same time.

- Identify the industry's key success factors and link them to the company's performance and profitability.

- Explain the company's overall strategy and how it creates a competitive advantage.

- Discuss the values and philosophy of the business and how they have inspired and influenced past actions.

- Discuss the people in the business and their strengths and weaknesses.

- Describe the philosophy underlying the company's compensation policy.

- Make a list of the company's most important customers and its key suppliers or vendors and review the history of all dealings with the parties on both lists.

- Discuss how to treat the company's key stakeholders to ensure the company's continued success.

- Develop a job analysis by taking an inventory of the activities involved in leading the company. This analysis can show successors those activities on which they should be spending most of their time.

- Document as much process knowledge—"how we do things"—as possible. After many years in their jobs, business owners are not even aware of their vast reservoirs of knowledge. For them, making decisions is a natural part of their business lives. They do it effortlessly because they have so much knowledge and experience. It is easy to forget that a successor will not have the benefit of those years of experience unless the founder communicates it.

- Include an ethical will, a document that explains to the next generation of leaders the ethical principles on which the company operates. An ethical will gives company founders the chance to bequeath to their heirs not only a business but also the wisdom and ethical lessons learned over a lifetime.

ENTREPRENEURIAL PROFILE: Paul Weber: Weber's Hamburgers In 1963, Paul Weber started Weber's Hamburgers, a small restaurant that has become a landmark, in Ontario, Canada, known for its fresh, tasty burgers. Targeting vacationers and city dwellers looking for an escape, Weber selected a location near the small town of Orillia, about two hours north of Toronto. On a typical Saturday during the peak summer season, Weber's serves about 800 hamburgers per hour, a pace that tests the restaurant's systems and employees. Managing the family business is second-generation owner Paul Weber Jr., who grew up working in the restaurant and took over the family business in 1989. Paul Jr. credits much of the company's success to the systems that his father established, refined, and documented over more than a quarter of a century. In fact, the Weber family continues to document for future generations every part of their business, including the order processing procedure and the techniques they use to entertain guests when lines stretch across the parking lot.[69]

Step 3. *Groom the successor.* The process by which business founders transfer their knowledge to the next generation is gradual and often occurs informally as they spend time with their successors. Grooming the successor is the founder's greatest teaching and development responsibility and takes time, usually 5 to 10 years. To implement the succession plan, the founder must be as follows:

- Patient, realizing that the transfer of power is gradual and evolutionary and that the successor should earn responsibility and authority one step at a time until the final transfer of power takes place

- Willing to accept that the successor will make mistakes

- Skillful at using the successor's mistakes as a teaching tool

- Able to build the successor's confidence as his or her abilities grow

- An effective communicator and an especially tolerant listener

- Capable of establishing reasonable expectations for the successor's performance

- Able to articulate the keys to the successor's performance

Teaching is in reality the art of assisting discovery and requires letting go rather than controlling. When problems arise in the business, the founder should consider delegating some of them to the successor in training. The founder also must resist the tendency to wade in and fix the problem unless it is beyond the scope of the successor's ability. Most great teachers and leaders are remembered more for the success of their students than for their own success.

Step 4. *Promote an environment of trust and respect.* Another priceless gift a founder can leave a successor is an environment of trust and respect. Trust and respect on the part of the founder and others fuel the successor's desire to learn and excel and build the successor's confidence in making decisions. Empowering the successor by gradually delegating responsibilities creates an environment in which all parties can objectively view the growth and development of the successor. Customers, creditors, suppliers, and staff members can gradually develop confidence in the successor. The final transfer of power is not a dramatic, wrenching change but a smooth, coordinated passage.

A problem for some founders at this phase is the meddling retiree syndrome, in which they continue to show up at the office after they have officially stepped down and get involved in business issues that no longer concern them. This tendency merely undermines the authority of the successor and confuses employees as to who really is in charge. Helen Dragas, who succeeded her father at the Dragas Company, a residential construction business, praises her father for handing the reins of the company over to her and then trusting her to handle them. "He gave me the authority and then he stepped back," she says of the successful transfer of leadership.[70]

Step 5. *Cope with the financial realities of estate and gift taxes.* The final step in developing a workable management succession plan structuring is the transition to minimize the impact of estate and gift taxes on family members and the business. Entrepreneurs who fail to consider the impact of these taxes (which have been as high as 55 percent) may force their heirs to sell a successful business just to pay the estate's tax (commonly known as the "death tax") bill. Despite facing potentially large tax bills, only 41 percent of business owners have created a comprehensive estate plan![71]

ENTREPRENEURIAL PROFILE: Clayton Leverett: Stillwater Farm Just a few hours after Clayton Leverett's son, Whit, was born, the young father began to wonder whether Whit would be able to retain the family's 150-year-old Stillwater Ranch, a cattle ranch in Llano, Texas, now in its fifth generation of ownership. Estate taxes have cut deeply into the Leverett family's landholdings and cattle operations twice before. The family was forced to sell thousands of acres of valuable land to pay estate taxes when Clayton's grandmother died in 2006. When Clayton's father died later that same year, the family faced the estate tax a second time and once again had to sell acreage and lay off employees. Clayton also had to take on a second job to pay the estate tax. "We will only be able to sell only so much [land] before the ranch becomes unprofitable and we are forced to sell the entire operation," explains a frustrated Clayton.[72]

Congress eliminated the estate tax in 2010 but reinstated it in 2011 (see Table 22.2). A study by Antony Davies of Duquesne University reports that for every 4.5 percent increase in the estate tax (the average increase in the tax since 1993), 6,000 small businesses are liquidated or absorbed into large companies. Conversely, Davies's research suggests that repealing the tax would create 100,000 new businesses that would employ 2 million workers with a total payroll of $80 billion.[73]

Entrepreneurs who fail to engage in proper estate planning subject their family members to a painful tax bite when they inherit the family business. Entrepreneurs should be actively engaged in estate planning no later than age 45; those who start businesses early in their lives or whose businesses grow rapidly may need to begin as early as age 30. A variety of options exist that may prove to be helpful in reducing the estate tax liability. Each operates in a different fashion, but their objective remains the same: to remove a portion of business owners' assets from their estates so that when they die, those assets will not be subject to estate taxes. Many of these estate planning tools need time to work their magic, so the key is to put them in place early on in the life of the business.

Buy-Sell Agreement

One of the most popular estate planning techniques is the buy-sell agreement. A **buy-sell agreement** is a contract that co-owners often rely on to ensure the continuity of a business. In a typical arrangement, the co-owners create a contract stating that each agrees to buy the others out in case

TABLE 22.2 Changes in the Estate and Gift Taxes

After years of complaints from family business owners, Congress finally overhauled the often punishing structures of estate and gift taxes. The gift tax applies to gifts that a donor makes during his or her lifetime; the estate tax applies to assets that a person leaves at his or her death. The federal estate tax is actually interwoven with the gift tax but for a time, the impact of the two taxes differed. Congress repealed the estate tax in 2010, but the tax reappeared in 2011. When George Steinbrenner, owner of the New York Yankees, died in 2010, the year Congress repealed the estate tax, his heirs paid no estate taxes. Had Steinbrenner died in 2009, however, his heirs would have faced a stiff tax bill of $500 million! The following table shows the trends in the exemptions and the maximum tax rates for the estate and gift taxes:

Year	Estate Tax Exemption	Gift Tax Exemption	Maximum Tax Rate
2001	$675,000	$675,000	55%
2002	$1 million	$1 million	50%
2003	$1 million	$1 million	49%
2004	$1.5 million	$1 million	48%
2005	$1.5 million	$1 million	47%
2006	$2 million	$1 million	46%
2007	$2 million	$1 million	45%
2008	$2 million	$1 million	45%
2009	$3.5 million	$1 million	45%
2010	Tax repealed	$1 million	35% (gifts only)
2011	$1 million	$1 million	55%
2012	$5.12 million	$5.12 million	35%
2013	$5.25 million	$5.25 million	40%

No matter how the federal laws governing estate taxes may change over the next few years, entrepreneurs whose businesses have been successful must not neglect estate planning. Even though the federal estate tax burden has eased somewhat (at least for a while), many states have *increased* their estate tax rates.

Sources: Based on Brad Hamilton and Jeane MacIntosh, "Death'$ Perfect Timing," *New York Post*, July 14, 2010, *www.nypost.com/p/news/local/death_perfect_timing_NusLyGlMu8cn8kyepprVJP?CMP=OTC-rss&FEEDNAME*; Tom Herman, "Estate Taxes Will Turn Sharply Lower on Jan. 1," *Wall Street Journal*, November 20, 2003, p. D2; Jeanne Lee, "Death and Estate Taxes," *FSB*, April 2004, p. 96.

of the death or disability of one. That way, the heirs of the deceased or disabled owner can "cash out" of the business while leaving control of the business in the hands of the remaining owners. The buy-sell agreement specifies a formula for determining the value of the business at the time the agreement is to be executed. One problem with buy-sell agreements is that the remaining co-owners may not have the cash available to buy out the disabled or deceased owner. To resolve this issue, many businesses buy life and disability insurance for each of the owners in amounts large enough to cover the purchase price of their respective shares of the business. Without the support of adequate insurance policies, a buy-sell agreement offers virtually no protection.

Lifetime Gifting

The owners of a successful business may transfer money to their children (or other recipients) from their estate throughout the parents' lives. Current federal tax regulations allow individuals to make gifts of $14,000 per year, per parent, per recipient, that are exempt from federal gift taxes. Another benefit: gift recipients do not have to pay taxes on the gift. For instance, husband-and-wife business owners could give $1.68 million worth of stock to their three children and their spouses over a period of 10 years without incurring any estate or gift taxes at all.

Setting Up a Trust

A **trust** is a contract between a grantor (the founder) and a trustee (generally a bank officer or an attorney) in which the grantor gives to the trustee legal title to assets (e.g., stock in the company), which the trustee agrees to hold for the beneficiaries (the founder's children). The beneficiaries can receive income from the trust, the property in the trust, or both at some specified time. More entrepreneurs could be using trusts to minimize the impact of estate taxes on their heirs; only 52 percent of business owners have transferred assets into a trust established for future generations.[74]

Trusts can take a wide variety of forms, but two broad categories of trusts are available: revocable trusts and irrevocable trusts. A **revocable trust** is one that the grantor can change or revoke during his or her lifetime. Under present tax laws, however, the only trust that provides a tax benefit is an **irrevocable trust**, in which the grantor cannot require the trustee to return the assets held in trust. The value of the grantor's estate is lowered because the assets in an irrevocable trust are excluded from the value of that estate. However, an irrevocable trust places severe restrictions on the grantor's control of the property placed in the trust. Business owners use several types of irrevocable trusts to lower their estate tax liabilities.

BYPASS TRUST The most basic type of trust is the bypass trust (or A-B trust), which allows a business owner to put assets into a trust and name his or her spouse as the beneficiary on the owner's death. The spouse receives the income from the trust throughout his or her life, but the principal in the trust bypasses the surviving spouse's estate and goes to the couple's heirs free of estate taxes on the spouse's death. A bypass trust is particularly useful for couples who plan their estates together. By leaving assets to one another in bypass trusts, they can make sure that their assets are taxed only once between them. However, entrepreneurs should work with experienced attorneys to create bypass trusts because the IRS requires that they contain certain precise language to be valid. Because the American Taxpayer Relief Act increased the estate tax exemption to $5.25 million ($10.5 million per married couple), fewer entrepreneurial couples may need the tax relief that bypass trusts offer.

IRREVOCABLE LIFE INSURANCE TRUST This type of trust allows a business owner to keep the proceeds of a life insurance policy out of his or her estate and away from estate taxes, freeing up that money to pay the taxes on the remainder of the estate. To get the tax benefit, business owners must be sure that the business or the trust (rather than themselves) owns the insurance policy. The primary disadvantage of an irrevocable life insurance trust is that if the owner dies within three years of establishing it, the insurance proceeds *do* become part of the estate and *are* subject to estate taxes. Because the trust is irrevocable, it cannot be amended or rescinded once it is established. Like most trusts, irrevocable life insurance trusts must meet stringent requirements to be valid, and entrepreneurs should use experienced attorneys to create them.

IRREVOCABLE ASSET TRUST An irrevocable asset trust is similar to a life insurance trust except that it is designed to pass the assets in the parents' estate on to their children. The children do not have control of the assets while the parents are still living, but they do receive the income from those assets. On the parents' death, the assets in the trust go to the children without being subjected to the estate tax.

GRANTOR RETAINED ANNUITY TRUST A grantor retained annuity trust (GRAT) is a special type of irrevocable trust and has become a popular tool for entrepreneurs to transfer ownership of a business to their heirs while maintaining control over it and minimizing estate taxes. Under a GRAT, an owner can put property in an irrevocable trust for a maximum of ten years. While the trust is in effect, the grantor (owner) retains the voting power and receives the interest income from the property in the trust. At the end of the trust (not to exceed 10 years), the property passes to the beneficiaries (heirs). The beneficiaries are required to pay the gift tax on the value of the assets placed in the GRAT but no estate tax on them. However, the IRS taxes GRAT gifts only according to their discounted present value because the heirs did not receive use of the property while it was in trust. The primary disadvantage of using a GRAT in estate planning is that if the grantor dies during the life of the GRAT, its assets pass back into the grantor's estate. These assets then become subject to the full estate tax.

Establishing a trust requires meeting many specific legal requirements and is not something business owners should do on their own. It is much better to hire experienced attorneys, accountants, and financial advisers to assist in creating them. Although the cost of establishing a trust can be high, the tax savings they generate are well worth the expense.

ESTATE FREEZE An **estate freeze** attempts to minimize estate taxes by having family members create two classes of stock for the business: (1) preferred voting stock for the parents and (2) nonvoting common stock for the children. The value of the preferred stock is frozen, whereas the common stock reflects the anticipated increased market value of the business. Any appreciation in the value of the business after the transfer is not subject to estate taxes. However, the parents must pay gift tax on the value of the common stock given to the children. The value of the common stock is the total value of the business less the value of the voting preferred stock retained by the parents. The parents also must accept taxable dividends at the market rate on the preferred stock they own.

FAMILY LIMITED PARTNERSHIP Creating a **family limited partnership** (FLP) allows business-owning parents to transfer their company to their children (thus lowering their estate taxes) while still retaining control over it for themselves. To create an FLP, the parents (or parent) set up a partnership among themselves and their children. The parents retain the general partnership interest, which can be as low as 1 percent, and the children become the limited partners. As general partners, the parents control both the limited partnership and the family business. In other words, nothing in the way the company operates has to change. Over time, the parents can transfer company stock into the limited partnership, ultimately passing ownership of the company to their children.

One of the principal tax benefits of an FLP is that it allows discounts on the value of the shares of company stock the parents transfer into the limited partnership. Because a family business is closely held, shares of ownership in it, especially minority shares, are not as marketable as those of a publicly held company. As a result, company shares transferred into the limited partnership are discounted at 20 to 50 percent of their full market value, producing a large tax savings for everyone involved. The average discount is 40 percent, but that amount varies based on the industry and the individual company involved. A business owner should consider an FLP as part of a succession plan "when there has been a buildup of substantial value in the business and the older generation has a substantial amount of liquidity," says one expert.[75]

Because of their ability to reduce estate and gift taxes, FLPs have become one of the most popular estate planning tools in recent years. However, a Tax Court ruling against a Texas entrepreneur who, two months before he died, established an FLP that contained both business and personal assets cast a pall over the use of FLPs as estate planning tools. Another case, however, calmed estate planners' fears and reestablished the use of FLPs as legitimate estate planning tools as long as entrepreneurs create them properly. The following tips will help entrepreneurs establish an FLP that will withstand legal challenges:

- Establish a legitimate business reason other than avoiding estate taxes—such as transferring a business over time to the next generation of family members—for creating the FLP and document it on paper.

- Make sure all members of the FLP make contributions and take distributions according to a predetermined schedule. "Don't allow partners to use partnership funds to pay for personal expenses and do not time partnership distributions with personal needs for cash," says one attorney.[76]

- Do not allow members to put their personal assets (such as a house, automobiles, or personal property) into the FLP. Commingling personal and business assets in an FLP raises a red flag to the IRS.

- Maintain proper records for establishing and operating the FLP and conduct all required meetings.

- Expect an audit of the FLP. The IRS tends to scrutinize FLPs; be prepared for a thorough audit.[77]

Developing a succession plan and preparing a successor requires a wide variety of knowledge and skills, some of which the business founder will not have. That's why it is important to bring into the process experts when necessary. Entrepreneurs often call on their attorneys, accountants, insurance agents, and financial planners to help them build a succession plan that works best for their particular situations. Because the issues involved can be highly complex and charged with emotion, bringing in trusted advisers to help improves the quality of the process and provides an objective perspective.

Risk Management Strategies

Insurance is an important part of creating a management succession plan because it can help business owners minimize the taxes on the estates they pass on to their heirs and can provide much-needed cash to pay the taxes the estate does incur. However, insurance plays an important role in many other aspects of a successful business—from covering employee injuries to protecting against

5. _____

Understand the four risk management strategies.

natural disasters that might shut a business down temporarily. When most small business owners think of risks such as these, they automatically think of insurance. However, insurance companies are the first to point out that insurance does not solve all risk problems. A more comprehensive strategy is risk management, which takes a proactive approach to dealing with the risks that businesses face daily. One recent survey shows that just 37 percent of small companies have an emergency preparedness plan for dealing with either a natural or man-made disaster.[78] In addition, only one-third of companies say that they are well prepared to cope with a threat or disaster.[79] "Small companies often spend more time planning the company picnic than planning for an event that could put them out of business," says one insurance expert.[80] Dealing with risk successfully requires a combination of four risk management strategies: avoiding, reducing, anticipating, and transferring risk.

Avoiding risk requires a business to take actions to shun risky situations. For instance, conducting credit checks of customers helps decrease losses from bad debts. Wise entrepreneurs know that they can avoid some risks simply by taking proactive management actions. Workplace safety improves when business owners implement programs designed to make all employees aware of the hazards of their jobs and how to avoid being hurt. Business owners who have active risk identification and prevention programs can reduce their potential insurance costs as well as create a safer, more attractive work environment for their employees. Because avoiding risk altogether usually is not practical, however, a strategy of reducing risk becomes necessary.

A risk reducing strategy requires a company to take steps to lower the level of risk associated with a situation.

ENTREPRENEURIAL PROFILE: George Pauli: Great Embroidery LLC George Pauli, owner of Great Embroidery LLC in Mesa, Arizona, recently installed backup generators for the two machines that he uses to stitch logos on 12 garments at a time. Pauli's business experiences power interruptions about six times a year, and before he installed the generators, each incident cost him at least $120 in lost merchandise because the machines' needles could not resume exactly where they had left off when the power shut off.[81]

Risk reduction strategies do not eliminate risk, but they lessen its impact. Even with avoidance and reduction strategies, the risk is still present; thus, losses can occur.

Risk anticipation strategies promote self-insurance. Knowing that some element of risk still exists, a business owner puts aside money each month to cover potential losses. Sometimes a self-insurance fund may not be large enough to cover the losses from a particular situation. When this happens, a business stands to lose despite the best efforts to anticipate risk, especially in the first few years before the fund is fully funded and able to cover large claims. Most businesses, therefore, include in their risk strategies some form of insurance to transfer risk. The Patient Protection and Affordable Care Act, the sweeping health care reform legislation passed in 2010, makes self-insurance a more attractive option for small businesses that provide health care to their employees. When a company is self-insured, it assumes the financial risk of paying for its employees' medical claims. If several workers suffer catastrophic injuries or illnesses at once, a company could face a cash crisis. If employees' claims are low in a given year, however, the company saves money over what it would have paid an insurance company in premiums. Recognizing that a self-insurance strategy alone could be risky, many self-insured businesses purchase "stop-loss" policies, which take over payment if any individual employee's health care costs exceed a specific cap that can be as low as $10,000. Self-insured companies usually hire a benefits administration company to handle all of the insurance paperwork; many providers now offer administration services for companies with as few as 10 employees. In addition, the Patient Protection and Affordable Care Act allows self-insured companies to switch to full insurance if several of their employees develop costly illnesses. Under the law, companies can reenter the fully insured market with no penalty because insurance companies must accept every employer or individual who applies for coverage.

Self-insurance for health care provides small companies with many benefits. They can customize their health care plans to meet the needs of their workers, maintain control over the health care reserves that they put aside, avoid prepaying insurance companies for coverage, circumvent state health insurance premium taxes, and save money.[82]

ENTREPRENEURIAL PROFILE: NorthBay Adventure NorthBay Adventure, a company based in North East, Maryland, that provides outdoor educational experiences for children, recently switched to a self-insurance health care plan in an attempt to reduce the rapidly escalating costs of the premiums its former insurance company was charging. The company's 60 employees, most of whom are young and healthy, hardly noticed the change, yet executive director George Comfort says that self-insurance saves the company 45 percent per year over its former fully insured coverage.[83]

Risk transfer strategies depend on using insurance. Purchasing insurance is a risk transfer strategy because an individual or a business shifts some of the costs of a particular risk to an insurance company that is set up to spread out the financial burdens of risk. During a specific time period, the insured business pays money (a premium) to an insurance carrier (either a private company or a government agency). In return, the carrier promises to pay the insured a certain amount of money in the event of a loss. Small companies across the United States are feeling the pinch of rapidly escalating insurance costs and are devising creative ways to control them.

Captive insurance, which is a hybrid of self-insurance and risk transfer strategies, is a technique that large businesses have used for years but is gaining popularity among small businesses. To implement a captive insurance strategy, a small company sets up its own insurance company or bands together with other small businesses to create their own insurance company and contribute enough capital to cover a defined level of risk. The group outsources the daily management of the insurance company to a business that specializes in that area and then purchases reinsurance to cover losses above the amount that they have contributed. Over time, if the group experiences no large losses, the excess capital paid into the insurance company goes back to the businesses as dividends. Experts estimate that creating a captive insurance company costs between $30,000 and $50,000 (including a feasibility study) and $35,000 to $100,000 per year to maintain. To make captive insurance viable, a company should have annual sales of at least $15 million.[84]

ENTREPRENEURIAL PROFILE: Ken Sturm: Iridium Jazz Club Ken Sturm, owner of several restaurants and clubs in New York City, including the Iridium Jazz Club, that employ about 450 people, recently established a captive insurance company after an incident in which a crane collapse forced one of his restaurants in Times Square to close for three days, costing him tens of thousands of dollars in lost sales. Sturm had business interruption insurance with a traditional insurance company, but the policy had a waiting period of 96 hours before it paid to cover any losses. Sturm spent $25,000 to set up his captive insurance company and pays Artex Risk Solutions, a captive insurance administration company, about $36,000 a year to maintain it.[85]

The Basics of Insurance

Insurance is the transfer of risk from one entity (an individual, a group, or a business) to an insurance company. Without insurance, many of the products and services that businesses provide would be impossible because the risk of overwhelming financial loss would be too great given the litigious society in which we live, yet many small business owners fail to buy adequate insurance coverage protect their companies from the most basic risks such as property damage, fire, theft, and liability. A survey of conducted by Travelers Insurance reports that 57 percent of business owners say that they are not confident at all or are only somewhat confident that their businesses are protected against insurable risks that could result in significant financial losses or cause them to go out of business altogether.[86]

To be insurable, a situation or hazard must meet the following requirements:

1. It must be possible to calculate the actual loss being insured. For example, it would probably not be possible to insure an entire city against fire because too many variables are involved. It is possible, however, to insure a specific building.

2. It must be possible to select the risk being insured. No business owner can insure against every potential hazard, but insurance companies offer a wide variety of policies. Famous insurer Lloyd's of London has sold more than 400,000 alien abduction policies ($150 a year for $150 million of coverage) and has actually paid one claim! Another offers werewolf insurance, but the policy pays only if the insured turns into a werewolf.[87] Lloyd's of London insured winemaker Ilja Gort's nose for $8 million. Gort, who is prohibited under the policy from boxing, skiing, riding a motorcycle, or working as a knife thrower's assistant, owns the Chateau la Tulipe de la Garde winery in France. Gort purchased the insurance because he has "a chateau of people hanging on my nose," he says. Lloyd's also once wrote a policy to insure professional football player Troy Polamalu's hair, which plays a prominent role in a shampoo commercial for Procter and Gamble. Singer Mariah Carey's legs are insured for $1 billion.[88]

6.

Discuss the basics of insurance for small businesses.

Lloyds of London wrote an insurance policy on Troy Polamalu's hair.

Source: © Everett Collection Inc/ Alamy.

3. There must be enough potential policyholders to assume the risk. A tightrope walker who specializes in walking between tall downtown buildings would have difficulty purchasing insurance because there are not enough people engaging in that activity to spread the risk sufficiently.

Perhaps the biggest barrier facing entrepreneurs is the difficulty of understanding the nature of the risks that they and their businesses face. The risk management pyramid (see Figure 22.5) helps entrepreneurs decide how they should allocate their risk management dollars. Begin by identifying the primary risks your company faces, such as a fire in a manufacturing plant, a lawsuit from a customer injured by your company's product, a computer system meltdown, an earthquake, and so on. Then rate each event on three factors:

1. *Severity.* How much would the event affect your company's ability to operate?

2. *Probability.* How likely is the event to occur?

3. *Cost.* How much would it cost your company if the event occurred?

Rate the event on each of these three factors using a simple scale: A (high) to D (low). For instance, a small technology company might rate a fire in its offices as BDA. On the other hand, that same company might rank a computer system crash as ABA. Using the risk management pyramid, a business owner sees that the event rated ABA is higher on the risk scale than the event rated BDA. Therefore, this company would focus more of its risk management dollars on preventing a computer system crash than on protecting against an office fire.

Types of Insurance

No longer is the cost of insurance an inconsequential part of doing business. Now the ability to get adequate coverage and to pay the premiums is a significant factor in starting and running a

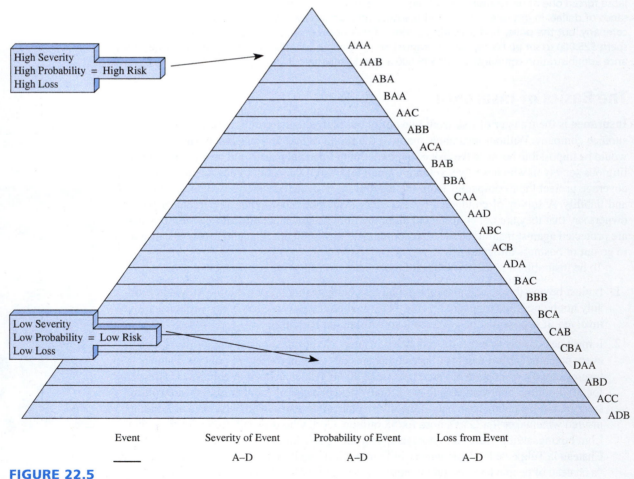

FIGURE 22.5

The Risk Management Pyramid

small business. Sometimes just *finding* coverage for their businesses is a challenge for entrepreneurs, especially those whose companies cater provide extreme entertainment.

ENTREPRENEURIAL PROFILE: Jane Reifert: Incredible Adventures For Incredible Adventures, a small Sarasota, Florida–based company founded in 1993 that offers customers the opportunity to fly in a Russian MiG jet fighter; experience weightlessness; go to the edge of space; swim with sharks; make a high-altitude, low-opening sky dive; and experience many other exciting adventures, purchasing insurance is a challenge. Prices for the company's expeditions range from $375 (to swim with sharks) to $95,000 (for a suborbital flight, many of which take place outside the United States. "You can do more high adrenaline activities outside the country because of the legal climate," says president Jane Reifert, who claims that she spends as much time with lawyers and insurance agents as she does with customers. "You can't buy a ride in a fighter jet in the U.S. It's illegal. You can insure people swimming in shark-infested waters, but you can't insure people inside shark cages," she says incredulously.[89]

A wide range of business, individual, and group insurance is available to small business owners, and deciding which ones are necessary can be difficult. Some types of insurance are essential to providing a secure future for the company; others may provide additional employee benefits. The four major categories of insurance are property and casualty insurance, life and disability insurance, health insurance and workers' compensation coverage, and liability insurance. Each category is divided into many specific types, each of which has many variations offered by insurance companies. Business owners should begin by purchasing a basic **business owner's policy** (BOP), which typically bundles together basic property and casualty insurance, liability insurance, and perhaps basic crime, business interruption, and vehicle insurance. BOPs alone usually are not sufficient to meet most small business owners' insurance needs, however. Entrepreneurs should start with BOPs and then customize their insurance coverage to suit their companies' special needs by purchasing additional types of coverage.

PROPERTY AND CASUALTY INSURANCE Property and casualty insurance covers a company's tangible assets, such as buildings, equipment, tools, inventory, machinery, signs, and others that might be damaged, destroyed, or stolen. Business owners should be sure that their policies cover the replacement cost of their property, not just its value at the time of the loss, even if this coverage costs extra.

ENTREPRENEURIAL PROFILE: John Carney: Express Oil Change and Service Center After John Carney's Express Oil Change and Service Center was destroyed by a tornado that swept through Tuscaloosa, Alabama, his insurance policy paid to rebuild the building, but because building codes had changed, Carney had to absorb the $300,000 in additional costs the new codes required to relocate parking lot entrances and exits, add parking spaces, and use specific building materials. "There was nothing but rubble left," says Carney. "We had been in business 25 years, and it took a tornado 45 seconds to erase that." Carney was able to reopen his business within four months, and two months after that, sales were back to 90 percent of pre-tornado volume.[90]

Specific types of property and casualty insurance include property, surety, marine and inland marine, crime, liability, business interruption, and motor vehicle insurance.

Property insurance protects a company's assets against loss from damage, theft, or destruction. It applies to automobiles, boats, homes, office buildings, stores, factories, and other items of property. Some property insurance policies are broadly written to include all of an individual's property up to some maximum amount of loss, whereas other policies are written to cover only one building or one specific piece of property, such as a company car. Many natural disasters, such as floods and earthquakes, are not covered under standard property insurance; business owners must buy separate insurance policies for those particular events.

Within the last decade, business owners across the United States have suffered billions of dollars in losses from natural disasters ranging from tornadoes and hurricanes to floods and ice storms. Many of the businesses that lacked proper insurance coverage were forced to close for good, and others are still struggling to recover. After a tornado ripped away the awning, blew out the windows, and destroyed a chandelier in Maria's Pizzeria and Seafood, in Monson, Massachusetts, owner Maria Makopoulos's insurance company paid to repair the $15,000 in damage. Not far away, Daniel O'Conner & Sons Wire Rope, a company that makes cables for ski lifts, took a direct hit from the twister and lost three buildings and many specialized tools.

Damages to the 30-year-old company totaled more than $100,000. Rebuilding took six months, and O'Conner says that he has "gotten back about 80 percent of the value of the damage" from his insurance company and still hopes to receive the remaining 20 percent.[91]

Risk management experts advise entrepreneurs to create an emergency preparation plan as part of their risk reduction strategies. Unfortunately, 48 percent of small businesses have no disaster preparedness plan in place.[92] The Federal Emergency Management Administration offers a helpful outline for preparing an emergency preparedness plan on its Web site (*www.ready.gov/business*). Charlie Williamson, vice president of the Ralph Brennan Restaurant Group, says that the company's disaster preparation plan grew from two pages to a 68-page booklet that covers not only operating plans but also technology and communications plans after a hurricane devastated the historic Commander's Palace restaurant in New Orleans, knocking it out of operation for more than a year.[93]

BOPs do not include **flood insurance**; businesses must purchase flood insurance separately from the National Flood Insurance Program, which the federal government runs. Even relatively small amounts of water can cause significant damage; the average flood damage claim is $85,000.[94] Some business owners learn after the fact that traditional property insurance does not cover damage caused by floods. Others say that flood insurance premiums are too expensive for them to afford.

Although Jackie Summers, founder of Jack from Brooklyn, had no flood insurance, he was able to rebuild his business after floods from Hurricane Sandy all but destroyed it.

Source: Dom Gervast.

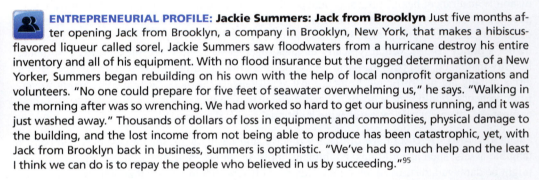

ENTREPRENEURIAL PROFILE: Jackie Summers: Jack from Brooklyn Just five months after opening Jack from Brooklyn, a company in Brooklyn, New York, that makes a hibiscus-flavored liqueur called sorel, Jackie Summers saw floodwaters from a hurricane destroy his entire inventory and all of his equipment. With no flood insurance but the rugged determination of a New Yorker, Summers began rebuilding on his own with the help of local nonprofit organizations and volunteers. "No one could prepare for five feet of seawater overwhelming us," he says. "Walking in the morning after was so wrenching. We had worked so hard to get our business running, and it was just washed away." Thousands of dollars of loss in equipment and commodities, physical damage to the building, and the lost income from not being able to produce has been catastrophic, yet, with Jack from Brooklyn back in business, Summers is optimistic. "We've had so much help and the least I think we can do is to repay the people who believed in us by succeeding."[95]

A company's BOP may insure the buildings and contents of a factory for loss from fire or natural disaster, but the owner may also buy insurance, called extra expense coverage, to cover expenses that occur while the destroyed factory is being rebuilt. **Extra expense coverage** pays for the costs of temporarily relocating workers and machinery so that a business can continue to operate while it rebuilds or repairs its factory. A similar type of insurance, called **business interruption insurance**, covers business owners' lost income and ongoing expenses in case their companies cannot operate for an extended period of time. Interruptions can be devastating to a small company; about 40 percent of small businesses that are forced to close after a major disaster never reopen.[96] To collect on a business interruption policy, a small business must incur some kind of physical damage to its property that causes it to suspend or reduce its operations. After power failures caused by Hurricane Sandy shut down many businesses in New Jersey and New York, their owners were shocked to learn that their business interruption policies would not pay for the business they lost because their businesses incurred no physical damage.

ENTREPRENEURIAL PROFILE: George Forgeis: Café Noir, Bar Tabac, and Cercle Rouge George Forgeis, owner of three French bistros in New York City, lost at least $160,000 in sales because of the power outage that lasted for several days. He filed a claim under his business interruption policy, but his insurer denied his claim because the power outage was caused by a flood at a Con Edison substation and his policy included a flood damage exclusion. "We'll be absorbing the loss," says Forgeis.[97]

Business owners should know exactly what their business interruption policies cover—and what they do not.

After the terrorist attacks on September 11, 2001, Congress passed the Terrorism Risk Insurance Act to make **terrorism insurance** available to businesses. To be covered against losses due to a terrorist act (as defined by the federal government), businesses must purchase a special terrorism insurance rider. If they fail to purchase the terrorism insurance coverage, losses from a terrorist act are not covered under their regular business policies. A study by the Congressional

Research Service concludes that only 65 percent of businesses in the United States have purchased terrorism insurance; most of those that have not are small businesses.[98] When two brothers set off bombs near the finish line at the Boston Marathon, hundreds of businesses shut down for days as police investigated one of the largest crime scenes in the history of the United States and engaged in a five-day manhunt. Ball and Buck, a retail clothing store located two blocks away from the finish line, suffered no physical damage but was closed for several days after the bombing. Owner Mark Bollman had the foresight to purchase terrorism insurance, and his loss of business is covered. "It's not something that you would ever want to need," he says, "but being located downtown in a major city, we wanted to be safe rather than sorry."[99]

Machinery and equipment insurance is a common addition for many businesses and covers a wide range of problems with equipment such as production machinery; electrical systems; heating, ventilating, and air-conditioning systems; and others. For instance, a restaurant that loses thousands of dollars' worth of food when a freezer breaks down would be covered for its loss under machinery and equipment insurance.

Auto insurance policies offer liability coverage that protects against losses resulting from injuries, damage, or theft involving the use of company vehicles. Not every BOP includes liability coverage for automobiles; business owners often must purchase a separate policy for auto insurance. The automobiles a business owns must be covered by a commercial policy, not a personal one.

A business may also purchase **surety insurance**, which protects against losses to customers that occur when a company fails to complete a contract on time or completes it incorrectly. Surety protection guarantees customers that they will get either the products or services they purchased or the money to cover losses from contractual failures.

Businesses also buy insurance to protect themselves from losses that occur when either finished goods or raw materials are lost or destroyed while being shipped. **Marine insurance** is designed to cover the risk associated with goods in transit. The name of this insurance goes back to the days when a ship's cargo was insured against high risks associated with ocean navigation.

Crime insurance does not deter crime, but it can reimburse the small business owner for losses from the "three Ds": dishonesty, disappearance, and destruction. Business owners should ask their insurance brokers or agents exactly what their crime insurance policies cover; after-the-fact insurance coverage surprises are seldom pleasant. Premiums for crime policies vary depending on the type of business, store location, number of employees, quality of the business's security system, and the business's history of losses. Coverage may include fidelity bonds, which are designed to reimburse business owners for losses from embezzlement and employee theft. Forgery bonds reimburse owners for losses sustained from the forgery of business checks.

LIFE AND DISABILITY INSURANCE Unlike most forms of insurance, life insurance does not pertain to avoiding risk because death is a certainty for everyone. Rather, **life insurance** protects families and businesses against loss of income, security, or personal services that results from an individual's untimely death. Life insurance policies are usually issued with a face amount payable to a beneficiary on the death of the insured. Life insurance for business protection, although not as common as life insurance for family protection, is becoming more popular. As you learned in the section on management succession, life insurance policies are an important part of many estate planning tools. In addition, many businesses insure the lives of key executives to offset the costs of having to make a hurried and often unplanned replacement of important managers.

When it comes to assets that are expensive to replace, few are more costly than the key people in a business, including the owner. What would it take to replace a company's top sales representative? Its production supervisor? Although money alone cannot solve the problem, it does allow a business to find and train key employees' replacements and to cover the income lost because of their untimely deaths or disabilities. That is the idea behind **key-person insurance**, which provides valuable working capital to keep a business on track while it reorganizes and searches for the right person to replace the loss of someone in a key position in the company. Although 71 percent of small businesses report being "very dependent" on a few key employees, only 22 percent of them have key person insurance policies.[100]

Pensions and annuities are special forms of life insurance policies that combine insurance with a form of saving. With an annuity or pension plan, the insured person's premiums go partly to provide standard insurance coverage and partly to a fund that is invested by the insurance company. The interest from the invested portion of the policy is then used to pay an income to the policyholder when he or she reaches a certain age. If the policyholder dies before reaching that age, either the policy converts to income for the spouse or family of the insured or the insurance proceeds (plus interest) go to the beneficiary as they would in ordinary life insurance.

Disability insurance, like life insurance, protects an individual in the event of unexpected and often very expensive disabilities. Because a sudden disability limits a person's ability to earn a living, the insurance proceeds are designed to help make up the difference between what that person could have expected to earn if the accident had not occurred. Sometimes called income insurance, these policies usually guarantee a stated percentage of an individual's income—usually around 60 percent—while he or she is recovering and is unable to run a business. Short-term disability policies cover the 90-day gap between the time a person is injured and when workers' compensation payments begin. Long-term disability policies pay for lost income after 90 days or longer. In addition to the portion of income a policy will replace, another important factor to consider when purchasing disability insurance is the waiting period, the time gap between when the disability occurs and the disability payments begin. Although many business owners understand the importance of maintaining adequate life insurance coverage, fewer see the relevance of maintaining proper coverage for disabilities. The likelihood that a person will become disabled is 2.2 to 3.5 times greater than the risk of death, depending on age; 26 percent of today's 20-year-olds will become disabled before reaching age 67.[101]

Business owners can supplement traditional disability policies with **business overhead expense insurance**. Designed primarily for companies with fewer than 15 employees, a business overhead expense policy replaces 100 percent of a small company's monthly overhead expenses, such as rent, utilities, insurance, taxes, and others, if the owner is incapacitated. Payments typically begin 30 days after the owner is incapacitated and continue for up to two years.

In the Entrepreneurial Spotlight

The Aftermath of a Storm

Superstorm Sandy, which bashed the shores of New Jersey, New York, and Connecticut with high winds and floodwaters, was the deadliest hurricane to hit the Northeast in 40 years and with damage estimated at $50 billion was the second costliest storm to hit the United States. A survey by Hartford Small Business Pulse reports that 74 percent of the affected businesses had to close temporarily, and the average time required to reopen was seven days. Many businesses suffered power outages, damage to their property, destruction of their inventory, disrupted supply chains, and lost sales. Most of the affected businesses were able to reopen, but some are gone forever. Many entrepreneurs who were able to reopen their businesses had to overcome significant obstacles and learned about the importance of proper insurance coverage. Unfortunately, many of the businesses affected by the storm had no flood insurance, and much of the damage they incurred was the result of flood waters driven by the intense winds. Some business owners, such as Noni Signoretti and her cousin Liz Hanna, co-owners of Brown's Hardware, a family-owned hardware store in Rockaway Park, New York, started in 1968, tried to purchase flood insurance but were denied coverage. The flooding "wiped out our whole selling floor," says Signoretti, noting that the storm destroyed the building's electrical and heating system as well. "We lost all of our merchandise." However, the determined business owners managed to restock and reopen their store within five days of the storm so that people could purchase the tools and supplies they needed to repair and rebuild their homes and businesses. To help small businesses in the area recover, the U.S. Small Business Administration has approved more than 2,700 disaster loans totaling more than $279 million.

In the first month after Sandy hit, Debbie Lane, owner of Testa Wines of the World, a family-owned wine importing company in Oyster Bay, New York, lost nearly $200,000 in sales to her wholesale and retail customers located across New York City. The storm's timing—October 31—could not have been worse; more than 40 percent of Testa Wines's sales occur in November and December. One shipment of 16,000 bottles of Italian table wines that normally sell for between $8 and $13 was sitting in a shipping container in New York City. Floodwaters from the storm breached the container, covered the bottles, and washed off their labels. Lane's insurance company agreed to pay $10,000 to relabel and repackage the wine; however, the wine was good only for use in cooking, meaning that Testa Wines could sell it for just $4 per case. The result: a loss of $92,400.

Lane's biggest concern was her six-story warehouse in South Kearney, New Jersey, where she stores the company's constantly revolving stock of 40,000 cases of wine worth about $2 million. It was nearly three weeks after the storm before authorities allowed Lane and her son to go to the six-story warehouse that was built in the 1930s. At the height of the storm, a 10-foot surge of water from the nearby Passaic River flooded the first two floors. Fortunately, Lane had the foresight to store almost all of the wine on the warehouse's fourth floor, where it was undamaged. Only 1 percent of the wine in the warehouse—400 cases of Chiantis, Pinot Grigios, Chardonnays, and Merlots stored on the lower floors—was lost.

Once Lane and her 30 employees were able to return to the warehouse, they began to focus on filling customers' orders to salvage what was left of the busy holiday sales season. She knew that her regular customers were counting on having the wines they had ordered in stock; otherwise, their sales would suffer. With the power out and no prospect of it coming back on for weeks, employees worked overtime to set up portable generators to power lights and lined the warehouse's stairwells with plywood so that they could slide cases of wine by hand down to ground level for shipping. Filling orders after such a devastating blow was a small miracle even though the lack of equipment slows workers' progress. Lane says that filling customers' orders took a day longer than it did under normal conditions. The trucking company that the family business had used for decades was also hit by the storm, and all but three of its trucks were ruined by floodwaters, a situation that further complicated deliveries. The company was able to lease smaller trucks to make deliveries, but space limitations necessitated more trips and more time loading and unloading.

Lane, who started working at Testa Wines at age 14, purchased the family business from her father, who started the company in 1963, when he retired in 2000 and began distributing wines produced by small family vineyards around the world. Despite the trials that Hurricane Sandy imposed on Testa Wines, Lane remains determined and optimistic about rebuilding her business.

1. What types of insurance should Testa Wines of the World have had in place? Explain.

2. Why do many small businesses lack adequate insurance coverage?

3. Work with a team of your classmates to select a local small business and interview its owner about the company's insurance coverage. Do you notice any significant gaps in the company's coverage? What recommendations can you make for filling the gaps?

Sources: Based on "Small Business Owners Hit Hard by Sandy Outages, The Hartford Finds," *The Hartford*, March 19, 2013, *http://newsroom.thehartford.com/News-Releases/Small-Business-Owners-Hit-Hard-By-Sandy-Outages-The-Hartford-Finds-5bc.aspx*; "Brian Patrick Eha, "Six Months After Hurricane Sandy, Many Businesses Are Still Struggling to Recover," *Entrepreneur*, April 29, 2013, *www.entrepreneur.com/article/226520*; Clare Trapasso, "Many Struggling Rockaway Merchants Miss Out on Lucrative Holiday Season," *New York Daily News*, November 28, 2012, *www.nydailynews.com/new-york/queens/rockaway-merchants-holiday-shopping-article-1.1207121*; Angus Loten, "Three Weeks After Sandy, Setbacks for Small Firms," *Wall Street Journal*, November 15, 2012, pp. B1, B4; Angus Loten, "Makeshift Ramps, Sympathy Orders: Small Businesses Cope After Sandy," *Wall Street Journal*, November 23, 2012, pp. B1, B7; Angus Loten, "A Month After the Floods, Owners Still Feel the Hurt," *Wall Street Journal*, November 29, 2012, pp. B1, B6; Angus Loten, "Six Months Later, They're Still Struggling," *Wall Street Journal*, May 1, 2013, *http://online.wsj.com/article/SB10001424127887324582004578457052012071148.html*; "Family Wine Business Uncovers Loss from Sandy," *Wall Street Journal*, November 22, 2012, *http://live.wsj.com/video/family-wine-business-uncovers-loss-from-sandy/17C4B747-B070-4112-BBE6-6379F3992033.html#!17C4B747-B070-4112-BBE6-6379F3992033*.

HEALTH INSURANCE AND WORKERS' COMPENSATION During World War II, a shortage of workers coupled with wage and price controls that prevented companies from raising wages led businesses to offer health care insurance as a way to attract workers. Today, 56.8 percent of private sector employees get their health insurance from their employers, but the passage of the Patient Protection and Affordable Care Act is creating sweeping changes in employees' health care coverage and emphasizing consumer-driven health care options.[102] Rising health care costs are a constant problem for businesses of all sizes and their employees. According to the National Federation of Independent Businesses, small business owners' greatest concern for the last 25 years has been the skyrocketing cost of health insurance.[103] Currently, health care spending in the United States accounts for 17.9 percent of GDP, an amount that will increase to 19.6 percent by 2017.[104] The average small company spends on average $5,615 per year on health care insurance premiums for an employee. Because of the high cost of providing health care coverage for employees, only 61 percent of small businesses offer health insurance to their employees compared to 99 percent of large companies (see Figure 22.6).[105] As health care costs continue to climb and the average age of the workforce has increased, small companies are having more difficulty providing coverage for their employees (see Figure 22.7). The primary reason cited by small business owners who do not offer health care coverage is high cost.[106] Small businesses actually pay 18 percent more than large companies for the same health insurance because of higher broker fees and costs of administering health care plans for a smaller number of employees.[107]

Health insurance has become an extremely important benefit to most workers (see Table 22.3). Small companies that offer thorough health care coverage often find that it gives them an edge in attracting and retaining a quality workforce. In fact, 84 percent of employees say that health care is the most important benefit that employers offer.[108] "Small business owners are in a Catch-22 situation," says one expert. "Even though it is increasingly difficult to offer health care benefits, a company that offers little or no health care benefits is putting up a red flag to potential talent. Businesses find it harder to attract talented individuals without offering benefits." One small business owner laments, "I can't afford it, but I can't afford *not* to have it."[109]

ENTREPRENEURIAL PROFILE: Mark Hodesh: Downtown Home and Garden Mark Hodesh, owner of Downtown Home and Garden, a retail shop in Ann Arbor, Michigan, says that the premiums he pays for health benefits for his 12 employees have risen more than 300 percent

FIGURE 22.6

Percentage of Companies Offering Health Benefits by Company Size

Source: Based on Employer Health Benefits: 2012 Summary of Findings, Kaiser Family Foundation an Health Research Educational Trust, p. 5.

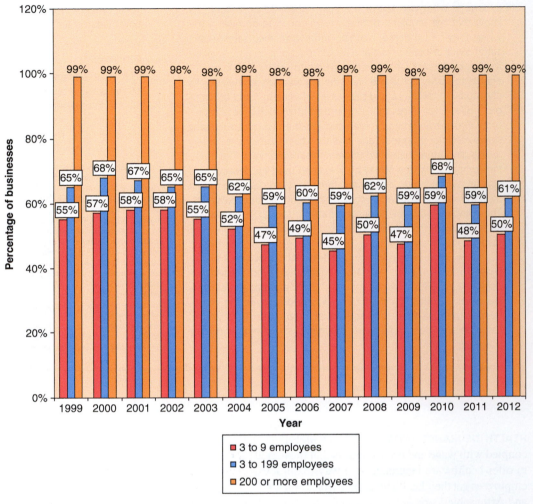

FIGURE 22.7

Average Annual Health Insurance Premiums for Family Coverage

Source: Based on Employer Health Benefits: 2012 Summary of Findings, Kaiser Family Foundation and Health Research and Education Trust, p. 1.

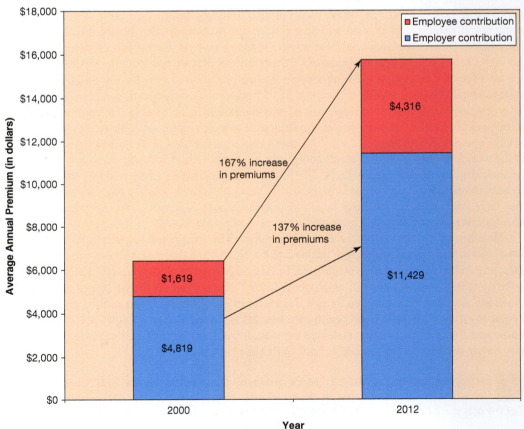

TABLE 22.3 Most Important Factors in Retaining Employees by Age-Group

Employees Less Than 40 Years Old	Employees Between 40 and 49 Years Old	Employees 50 Years Old and Older
1. Job security	1. **Health care benefits**	1. Job security
2. Base pay	2. Base pay	2. **Health care benefits**
3. **Health care benefits**	3. Job security	3. Base pay
4. Vacation time	4. Vacation time	4. Retirement benefits
5. Reputation of company as a great place to work	5. Challenging work	5. Length of commute

Source: Based on Joshua Bjerke, "Survey Reveals Most Important Factors in Attracting and Retaining Employees," *Recruiter*, April 27, 2012, *www.recruiter.com/i/factors-in-attracting-and-retaining-employees*.

since 1999. Yet he is committed to retaining the employee benefit (he pays 75 percent of the premiums for his employees) because his longtime employees are key to his company's success. "We're a 100-year-old downtown business," he says. "We're able to compete with the big box stores thanks to a knowledgeable and stable workforce. You don't keep better people if they're afraid they might lose their houses after a bad weekend at the emergency room."[110]

Although it will take business owners years to figure out how to comply with the Affordable Care Act, the law requires companies that have at least 50 full-time-equivalent workers who work at least 30 hours a week to provide "minimum essential" and "affordable" health care coverage. Employers are required to pay at least 60 percent of the total cost of a plan's benefits. Businesses that fall under the act's provisions can choose not to offer their employees insurance and pay a $2,000 penalty for each uncovered worker beyond 30 employees. Those employees and employees of companies that have fewer than 50 full-time-equivalent workers can purchase their own coverage from health insurance companies that sell individual or small-group policies on Small Business Health Options Programs exchanges that either the states or the federal government operate. Only 14 states and the District of Columbia have created their own exchanges; the federal government operates the exchanges in the remaining 36 states. The exchanges are open only to individuals buying their own coverage (some with the help of federal subsidies) and employees of companies with 100 or fewer employees (50 or fewer in a few states with their own exchanges). The exchanges are designed to simplify health care premium payments for small businesses, which pay a lump sum to the exchange, which, in turn, distributes the money to each insurance company that covers the businesses' employees. Businesses that provide health care benefits that fail to meet the government's definition of "affordable" must pay a $3,000 penalty for each affected worker. Abbey Duke, founder of Sugarsnap, a food retailer and caterer in Burlington, Vermont, worries about complying with the act. "What I can afford is not very good insurance," she says. "My approach is to pay them as much as I can so that they, in turn, can purchase health care coverage through the exchange."[111]

Many small business owners say that cannot afford the cost that the Affordable Care Act imposes on them and are choosing to keep their workforces under the 50 full-time equivalent worker total. Many are reducing their workforces, switching full-time workers to part-time positions, and paying independent contractors to perform certain jobs rather than hiring employees.

ENTREPRENEURIAL PROFILE: Elizabeth Turley: Meesh and Mia Corporation Elizabeth Turley, founder of Meesh and Mia Corporation, an apparel company based in Sandpoint, Idaho, that makes women's clothing with licensed college colors and logos, is on the cusp of the total of 50 full-time-equivalent workers. Turley's fast-growing company needs employees, "but we have to find another way to get there," she says, referring to the cost her company would incur under the Affordable Care Act. Her strategy is to pay independent contractors rather than hire employees to perform certain jobs even though the strategy is not ideal. Businesses do not pay taxes on wages or provide benefits to independent contractors; "you have less control over the hours they work and how much involvement they have in the business," says Turley. "Employees take more pride and ownership."[112]

Although the Affordable Care Act continues to change the mechanics of health care in the United States, employers face four basic health care options:

Traditional Indemnity Plans Under these plans, employees choose their own health care providers, and the insurance company either pays the provider directly or reimburses employees for the covered amounts. Only 2 percent of small companies use traditional indemnity plans.[113]

Managed Care Plans As part of employers' attempts to put a lid on escalating health care costs, these plans have become increasingly popular. Three variations, the health maintenance organization (HMO), the preferred provider organization (PPO), and the point of service (POS), are most common. An HMO is a prepaid health care arrangement under which employees must use health care providers who are employed by or are under contract with the HMO their company uses. Although they lower health care costs, employees have less freedom in selecting physicians under an HMO. Under a PPO, an insurance company negotiates discounts for health care with certain physicians and hospitals. If employees choose a health care provider from the approved list, they pay only a small fee for each office visit (often just $10 to $25). The insurance company pays the remainder. Employees may select a provider outside the PPO, but they pay more for the service. A POS is a hybrid of an HMO and a PPO that gives employees the freedom to select their health care providers (as with a PPO) and lowers costs (as with a HMO). As long as employees choose a primary care physician within the approved network, the POS will pay for that care and for care by specialists, even those outside the network, as long as the primary care physician makes the referral. PPOs are the most common managed care plans; 56 percent of covered employees are enrolled in PPOs, compared to 16 percent of covered employees in HMOs and 9 percent of covered employees in POSs.[114] Before the Affordable Care Act, Joel Whitehurst, owner of a funeral home started by his grandfather in 1921, covered his 13 employees under an HMO. When the act went into effect, however, Whitehurst could no longer afford the higher premiums that the HMO charged and reluctantly switched everyone to a PPO, which, he says, offers coverage that is not as good as the HMO provided.[115]

Health Savings Accounts and High-Deductible Health Plan with Savings Option Created as part of a major Medicare overhaul in 2003, health savings accounts (HSAs) and high-deductible health plans with a savings option (HDHP/SOs) are similar to independent retirement accounts except employees' contributions are used for medical expenses rather than for retirement. An HSA is a special savings account coupled with a high-deductible (typically $1,250 to $6,250 for an individual) insurance policy that covers major medical expenses. Employees or employers contribute pretax dollars (up to a defined ceiling) from their paychecks into the fund and use them as they need to. Because an employee's deductible amount (the amount the employee actually pays out of pocket) is so high, monthly premiums for insurance coverage are lower than with other types of plans. Withdrawals from an HSA or HDHP/SO are not taxed as long as the money is used for approved medical expenses. Unused funds can accumulate indefinitely and earn tax-free interest. HSAs and HDHP/SOs offer employees incentives to contain their health care costs, but the employer must choose both an insurance carrier to provide coverage and a custodial firm to manage employees' accounts. Although critics contend that consumer-driven plans push a greater portion of health care expenses onto employees, these plans have grown in popularity among small businesses because of their potential to rein in escalating costs. The average annual premium for an HDHP/SO for a small company is $4,928, which is 13.2 percent lower than the cost of traditional managed care plans.[116] Although self-employed individuals find HDHP/SOs attractive, employers are adding them to their menu of health care options for employees. The percentage of employees who are covered by HDHP/SOs has increased from 8 percent in 2009 to 19 percent today.[117]

Self-Insurance As you learned earlier in this chapter, some business owners choose to insure themselves for health coverage rather than to incur the costs of fully insured plans offered by outsiders. The benefits of self-insurance include greater control over the plan's design and the coverage it offers, fewer paperwork and reporting requirements, and, in most cases, lower costs. Experts estimate that some employers can reap 30 to 40 percent cost savings by switching to self-insurance from traditional health insurance.[118] The primary disadvantage, of course, is the possibility of having to pay large amounts to cover treatments for several employees' major illnesses at the same time, which can strain a small company's cash flow. Many self-insured businesses limit their exposure to such losses by purchasing **stop-loss insurance**, under which the business owner pays for health care expenses up to a predetermined point; beyond that, point the stop-loss policy takes over the expenses. Although just 15 percent of covered workers in small companies are in self-insured plans, the fastest growth in self-insured plans is among small businesses.[119]

Another type of health-related coverage is **workers' compensation**, which is designed to cover employees who are injured on the job or who become sick as a result of a work environment. Worker's compensation is a mandatory insurance program; companies that fail to offer workers' compensation must pay out of pocket for workers' claims and face penalties from the state. Before passage of workers' compensation legislation in 1911, an employee who was injured on the job had to bring a lawsuit to prove the employer was liable for the worker's injury. Because of the red tape and expenses involved in these lawsuits, many employees never received compensation for job-related accidents and injuries. Although the details of coverage vary from state to state, workers' compensation laws require employers to purchase insurance that provides benefits and medical and rehabilitation costs for employees injured on the job. The amount of compensation an injured employee receives depends on a fixed schedule of payment benefits based on three factors: the wages or salary that the employee was earning at the time of the accident or injury, the seriousness of the injury, and the extent of the disability to the employee.

Only two states, New Jersey and Texas, do not require companies to purchase workers' compensation coverage once they reach a certain size (usually three or more employees). Usually, the state sets the rates businesses pay for workers' compensation coverage, and business owners purchase their coverage from private insurance companies. Rapidly escalating workers' compensation rates, driven in large part by rising medical expenses, have become a major concern for small businesses across the nation. Rates vary by industry, business size, and the number of claims a company's workers make. For instance, workers' compensation premiums are higher for a timber-cutting business than for a retail gift store. Figure 22.8 shows the 10 most dangerous occupations in the United States.

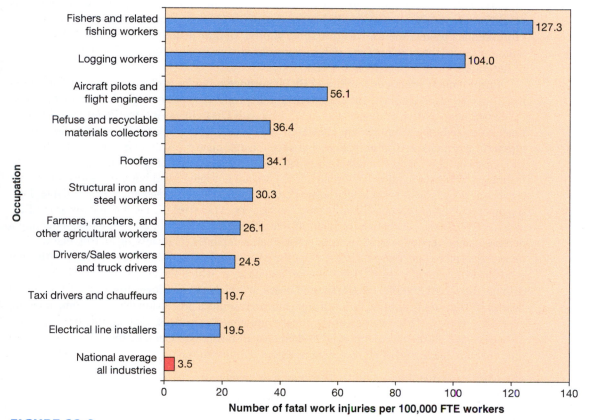

FIGURE 22.8

Most Dangerous Jobs in the US: Number of Fatal Work Injuries per 100,000 FTE Workers

Source: Based on Bureau of Labor Statistics, U.S. Department of Labor, Current Population Survey, Census of Fatal Occupational Injuries and U.S. Census Bureau, 2013.

LIABILITY INSURANCE One of the most common types of insurance coverage is **liability insurance**, which protects a business against losses resulting from accidents or injuries people suffer on the company's property and from its products or services and from damage the company causes to others' property. Most BOPs include basic liability coverage; however, the limits on the typical policy are not high enough to cover the potential losses many small business owners face. For example, one "slip-and-fall" case involving a customer who is injured by slipping and falling on a wet floor could easily exceed the standard limits on a basic BOP. Claims from customers injured by a company's product or service also are covered by its liability policy. Although most product liability lawsuits are settled out of court, the median award for those that go to court is more than $2 million.[120]

Even though courts often dismiss frivolous lawsuits, some small companies have been victims of frivolous lawsuits because they are seen as easy targets. Frivolous lawsuits can cost a small company thousands of dollars to defend, however. Jin and Soo Chung, owners of Custom Cleaners in Washington, D.C., were hit with a $54 million lawsuit by a customer after the dry cleaner lost a pair of the customer's pants! The trial court ruled for Custom Cleaners, but plaintiff Roy Pearson filed an appeal, extending the legal nightmare for the small business owners, who incurred $83,000 in legal fees to defend themselves.[121] With jury awards in product liability cases often reaching into the millions of dollars, entrepreneurs who fail to purchase sufficient liability coverage may end up losing their businesses. Most insurance experts recommend purchasing an additional commercial general liability policy that provides coverage of at least $2 million to $3 million for the typical small business.

Another important type of liability insurance for many small businesses is **professional liability insurance** or **errors and omissions coverage**. This insurance protects against damage a business causes to customers or clients as a result of an error an employee makes or an employees' failure to take proper care and precautions. For instance, a land surveyor may miscalculate the location of a customer's property line. If the landowner relies on that property line to build a structure on what he thinks is his land and it turns out to be on his neighbor's land, the surveyor is liable for damages. Doctors, dentists, attorneys, and other professionals protect themselves through a similar kind of insurance, malpractice insurance, which protects them against the risk of lawsuits arising from errors in professional practice or judgment.

Employment practices liability (EPL) insurance provides protection against claims arising from charges of employment discrimination, improper discipline, wrongful termination, sexual harassment, and violations of the Americans with Disabilities Act, the Family and Medical Leave Act, and other employment legislation (see Figure 22.9). The most common charges are wrongful termination, discrimination, and sexual harassment.[122] Although more than half of all EPL cases are filed against small companies, fewer than 2 percent of small businesses have EPL insurance.[123] Most violations of employment laws are not intentional but are the result of either carelessness or lack of knowledge; however, the company that violates them is still liable. Losing an employment practices liability case can be very expensive; the median jury award in EPL cases is $325,000.[124] Although the plaintiff wins in just 51 percent of all EPL cases, mounting a defense can be very expensive, especially for small companies.[125] Resolving an EPL claim takes, on average, more than one year. Cases that the parties settle out of court (the average settlement is $100,000) cost on average $15,000 to defend, cases in which the employer wins when courts dismiss them cost $60,000 to defend, and cases in which the employer loses cost $125,000 to defend.[126] Because they often lack full-time human resources professionals, small companies are especially vulnerable to charges of improper employment practices, making this type of insurance coverage all the more important to them. Small businesses can take the following steps to minimize the likelihood of employment practices liability suits:

- Create and distribute an up-to-date employee handbook.
- Develop a code of ethics and review it with every employee.
- Create accurate job descriptions for every job in the business.
- Conduct training sessions for employees, particularly managers and supervisors, on how to handle employment practices complaints from workers.

As more companies engage in e-commerce, the extensive databases of customer information that they manage have grown rapidly. Unfortunately, hackers are always developing new cyberattacks

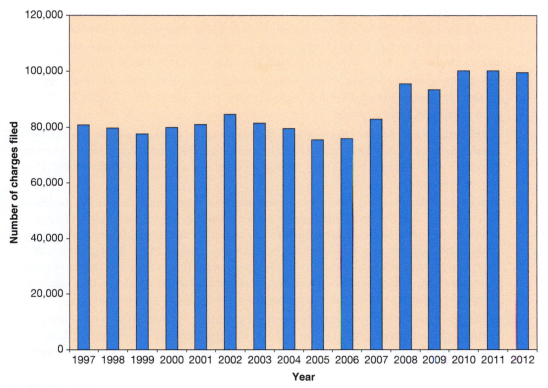

FIGURE 22.9

Number of Employment Charges Filed with EEOC

Source: Based on EEOC, *http://www.eeoc.gov/eeoc/statistics/enforcement/all.cfm.*

designed to illegally access that data and use it for illicit purposes such as identity theft and fraud, which affects 12.6 million people and cost $20.9 billion annually.[127] Because they often lack the sophisticated data security infrastructure and tools that large businesses put in place, cyberattacks on small businesses are growing much faster; small businesses are now the victims of nearly one-third of all cyberattacks.[128] **Cyber liability insurance** is designed to protect businesses from the crippling costs that accompany data breaches and cyberattacks. Cyber insurance policies cover the costs associated with lost trade secrets and intellectual property ("first-party claims"), damages a business must pay if customers file claims over lost or compromised personal information ("third-party claims"), lost revenue a company incurs if it is shut down by a denial-of-service attack, and the cost of a forensic investigation to determine the source of the attack. Coverage for these costs "can be a life-or-death issue for smaller companies," says Ethan Miller, an insurance attorney.[129] According to the Ponemon Institute, the cost of a data breach to a business is a whopping $194 per lost or stolen customer record.[130]

Every business's insurance needs are somewhat unique, which means that entrepreneurs must customize the insurance coverage they need to protect their businesses. Entrepreneurs also must keep their insurance coverage updated as their companies grow; when companies expand, so do their insurance needs. The Insurance Industry Institute (*www.iii.org*) offers a helpful book, *Insuring Your Business: Small Business Owners' Guide to Insurance*, for a small fee.

Controlling Insurance Costs

Small business owners face constantly rising insurance premiums. Entrepreneurs can take steps to lower insurance costs, however. To control the cost of insurance, owners should take the following steps:

1. *Pursue a loss-control program by making risk reduction a natural part of all employees' daily routine.* As discussed earlier in this chapter, risk reduction minimizes claims and eventually lowers premiums. Establishing a loss-control program means taking steps such as installing modern fire alarms, sprinkler systems, safety programs, and sophisticated security systems.

2. *Increase their policies' deductibles.* If a business can afford to handle minor losses, the owner can save money by raising the deductible to a level that protects the business against catastrophic events but, in effect, self-insures against minor losses. Business owners must determine the amount of financial exposure they can reasonably accept.

3. *Work with qualified professional insurance brokers or agents.* Business owners should do their homework before choosing insurance brokers or agents. This includes checking their reputation, credentials, and background by asking them to supply references.

4. *Work actively with brokers to make sure they understand business owners' particular needs.* Brokers need to know about entrepreneurs' businesses and objectives for insurance coverage. They can help only if they know their clients' needs and the degree of risk they are willing to take.

5. *Work with brokers to find competitive companies that want small companies' insurance business and have the resources to cover losses when they arise.* The price of the premium should never be an entrepreneur's sole criterion for selecting insurance. The rating of the insurance company should always be a primary consideration. What good is it to have paid low premiums if, after a loss, a business owner finds that the insurance company is unable to pay? Many small business owners learned costly lessons when their insurance companies, unable to meet their obligations, filed for bankruptcy protection.

6. *Utilize the resources of your insurance company.* Many insurers provide risk management inspections designed to help business owners assess the level of risk in their companies either for free or for a minimal fee. Smart entrepreneurs view their insurance companies as partners in their risk management efforts.

7. *Conduct a periodic insurance audit.* Reviewing your company's coverage annually can ensure that insurance coverage is adequate and can lead to big cost savings as well.

8. *Compile discrimination, harassment, hiring, and other employment polices into an employee handbook and train employees to use them.* Companies that take an active approach to avoiding illegal employment practices have less exposure to lawsuits and therefore, may be able to negotiate lower premiums.

To control the cost of health insurance, small business owners should consider the following steps:

1. *Increase the dollar amount of employee contributions and the amount of the employee deductibles.* Neither option is desirable, but rising medical costs have resulted in individuals becoming more responsible for their own health insurance and self-insuring to cover high deductibles.

2. *Switch to PPOs, POSs, or HMOs.* Higher premium costs have encouraged some small business owners to reevaluate PPOs, POS plans, and HMOs as alternatives to traditional health insurance policies. Although some employees resent limitations on their choice of providers, PPOs, POS, and HMOs, have become the primary vehicles for companies to provide health care coverage to their employees.

3. *Consider joining an insurance pool.* Small businesses can lower their insurance premiums by banding together to purchase coverage. In many states, chambers of commerce, trade associations, and other groups form insurance pools that small businesses can join, spreading risk over a larger number of employees.

4. *Keep employees informed.* By giving employees information about the costs and the benefits of various treatment alternatives and medications, employers empower their workers to make informed decisions that can lower health care costs.

5. *Conduct a yearly utilization review.* A review may reveal that your employees' use of their policies is statistically lower, which may provide you leverage to negotiate lower premiums or to switch to an insurer that wants a business with your track record and offers lower premiums.

6. *Make sure your company's health plan fits the needs of your employees.* One of best ways to keep health care costs in check is to offer only those benefits that employees actually need. Getting employee input is essential to the process.

7. *Create a wellness program for all employees.* We have all heard the old adage that an ounce of prevention is worth a pound of cure, but when it comes to the high cost of medical expenses, this is especially true! Companies that have created wellness programs report cost savings of up to $3 for every $1 they invest.[131] Employees involved in wellness programs not only incur lower health care expenses but also tend to be more productive. "Companies are reforming their own health care costs by recognizing that healthy workers cost less, are more productive, and are better for the company as a whole," says the president of a nonprofit health and productivity company.[132] Providing a wellness program does not mean building an expensive gym, however. Instead, it may be as simple as a providing routine checkups from a nurse, incentives for quitting smoking, weight-loss counseling, or after-work athletic games that involve as many employees as possible. One recent study reports that 71 percent of companies offer their employees financial incentives to participate in wellness programs.[133]

8. *Conduct a safety audit.* Reviewing the workplace with a safety professional to look for ways to improve its safety has the potential for saving some businesses thousands of dollars a year in medical expenses and workers' compensation claims. The National Safety Council offers helpful information on creating a safe work environment.

9. *Create a safety manual and use it.* Incorporating the suggestions for improving safety into a policy manual and then using it reduces the number of on-the-job accidents. Training employees, even experienced ones, in proper safety procedures is also effective.

10. *Create a safety team.* Assigning the responsibility for workplace safety to workers themselves can produce amazing results. When one small manufacturer turned its safety team over to employees, the plant's lost time due to accidents plummeted to zero for three years straight! The number of accidents is well below what it was when managers ran the safety team, and managers say that's because employees now "own" safety in the plant.

The key to controlling insurance costs is aggressive prevention. Entrepreneurs who actively manage the risks to which their companies are exposed find that they can provide the insurance coverage their businesses need at a reasonable cost. Finding the right insurance coverage to protect their businesses is no easy matter for business owners. The key to dealing with those differences is to identify the risks that represent the greatest threat to a company and then to develop a plan for minimizing their risk of occurrence and insuring against them if they do.

Chapter Review

1. Explain the factors necessary for a strong family business.
   - More than 90 percent of all companies in the United States are family owned. Family businesses generate 64 percent of the U.S. GDP, account for 63 percent of employment, and pay 65 percent of all wages. Several factors are important to maintaining a strong family business, including shared values, shared power, tradition, a willingness to learn, behaving like families, and strong family ties.

2. Understand the exit strategy options available to an entrepreneur.
   - Family business owners wanting to step down from their companies can sell to outsiders, sell to insiders, or transfer ownership to the next generation of family members. Common tools for selling to insiders (employees or managers) include sale for cash plus a note, LBOs, and ESOPs.
   - Transferring ownership to the next generation of family members requires a business owner to develop a sound management succession plan.

3. Discuss the stages of management succession.
   - Unfortunately, 70 percent of first-generation businesses fail to survive into the second generation, and, of those that do, only 13 percent make it to the third generation. One of the primary reasons for this lack of continuity is poor succession planning. Planning for management succession not only protects the founder's, successor's, and company's financial resources but also preserves what matters most in a successful business: its heritage and tradition. Management succession planning can ensure a smooth transition only if the founder begins the process early on.

4. Explain how to develop an effective management succession plan.
   - A succession plan is a crucial element in transferring a company to the next generation. Preparing a succession plan involves five steps: (1) select the successor, (2) create a survival kit for the successor, (3) groom the successor, (4) promote an environment of trust and respect, and (5) cope with the financial realities of estate taxes.

- Entrepreneurs can rely on several tools in their estate planning, including buy-sell agreements, lifetime gifting, trusts, estate freezes, and FLPs.

5. Understand the four risk management strategies.
   - Four risk strategies are available to the small business: avoiding, reducing, anticipating, and transferring risk.

6. Discuss the basics of insurance for small businesses.
   - Insurance is a risk transfer strategy. Not every potential loss can be insured. Insurability requires that it be possible to estimate the amount of actual loss being insured against and identify the specific risk and that there be enough policyholders to spread out the risk.
   - The four major types of insurance small businesses need are property and casualty insurance, life and disability insurance, health insurance and workers' compensation coverage, and liability insurance.
   - Property and casualty insurance covers a company's tangible assets, such as buildings, equipment, inventory, machinery, signs, and others that have been damaged, destroyed, or stolen. Specific types of property and casualty insurance include extra expense coverage, business interruption insurance, surety insurance, marine insurance, crime insurance, fidelity insurance, and forgery insurance.
   - Life and disability insurance also comes in various forms. Life insurance protects a family and a business against the loss of income and security in the event of the owner's death. Disability insurance, like life insur-

ance, protects an individual in the event of unexpected and often very expensive disabilities.
   - Health insurance is designed to provide adequate health care for business owners and their employees. The Patient Protection and Affordable Care Act is creating sweeping changes in health care for small companies. Workers' compensation is designed to cover employees who are injured on the job or who become sick as a result of a work environment.
   - Liability insurance protects a business against losses resulting from accidents or injuries people suffer on the company's property and from its products or services and from damage the company causes to others' property. Typical liability coverage includes professional liability insurance or "errors and omissions" coverage, which protects against damage a business causes to customers or clients as a result of an error an employee makes or an employee's failure to take proper care and precautions. Doctors, dentists, attorneys, and other professionals protect themselves through a similar kind of insurance, malpractice insurance, which protects them against the risk of lawsuits arising from errors in professional practice or judgment. Employment practices liability insurance provides protection against claims arising from charges of employment discrimination, sexual harassment, and violations of the Americans with Disabilities Act, the Family and Medical Leave Act, and other employment legislation.

Discussion Questions

22-1. What factors must be present for a strong family business?

22-2. Discuss the stages of management succession in a family business.

22-3. What steps are involved in building a successful management succession plan?

22-4. What exit strategies are available to entrepreneurs wanting to step down from their businesses?

22-5. What strategies can business owners employ to reduce estate and gift taxes?

22-6. Can insurance eliminate risk? Why or why not?

22-7. Outline the four basic risk management strategies and give an example of each.

22-8. What problems occur most frequently with a risk anticipating strategy?

22-9. What is insurance? How can insurance companies bear such a large risk burden and still be profitable?

22-10. Describe the requirements for insurability.

22-11. Briefly describe the various types of insurance coverage available to small business owners.

22-12. What kinds of insurance coverage would you recommend for the following businesses?
- A manufacturer of steel beams
- A retail gift shop
- A small accounting firm
- A limited liability partnership involving three dentists

22-13. What can business owners do to keep their insurance costs under control?

CHAPTER 23
The Legal Environment: Business Law and Government Regulation

Learning Objectives

On completion of this chapter, you will be able to:

1. Explain the basic elements required to create a valid, enforceable contract.

2. Outline the major components of the Uniform Commercial Code governing sales contracts.

3. Discuss the protection of intellectual property rights using patents, trademarks, and copyrights.

4. Explain the basics of the law of agency.

5. Explain the basics of bankruptcy law.

6. Explain some of the government regulations affecting small businesses, including those governing trade practices, consumer protection, consumer credit, and the environment.

If you have ten thousand regulations, you destroy all respect for the law.

—Winston Churchill

Regulations grow at the same rate as weeds.

—Norman Ralph Augustine

The legal environment in which small businesses operate is becoming more complex, and entrepreneurs must understand the basics of business law if they are to avoid legal entanglements. Situations that present potential legal problems arise every day in most small businesses, although the majority of small business owners never recognize them. Routine transactions with customers, suppliers, employees, government agencies, and others have the potential to develop into costly legal problems. For example, a manufacturer of lawn mowers might face a lawsuit if a customer injures himself while using the product, or a customer who slips on a wet floor while shopping could sue the retailer for negligence. A small manufacturer who reneges on a contract for a needed raw material when he finds a better price elsewhere may be open to a breach-of-contract suit. Even when they win a lawsuit, small businesses often lose because the costs of defending themselves can run quickly into thousands of dollars, depleting their already scarce resources.

Amy Brooks, founder of Bubbles by Brooks.
Source: Bubbles By Brooks.

ENTREPRENEURIAL PROFILE: Amy Brooks: Bubbles by Brooks   Amy Brooks, of Rochester, Minnesota, is a cancer survivor. Based on her experience going through treatment, she developed handcrafted soaps designed to reduce skin irritation to cancer patients and began selling her soaps under the name Bubbles by Brooks. However, Brooks Brothers, a Connecticut-based clothing company that also sells a line of fragrances and cleansing products, filed a trademark objection against Brooks. In the suit, Brooks Brothers demanded that Bubbles by Brooks withdraw its trademark application with the U.S. Patent and Trademark Office to trademark Bubbles by Brooks or face "potentially costly litigation." In its cease and desist letter to Brooks demanding that she withdraw her trademark application, attorneys for Brooks Brothers said, "Although 'Brooks' may be your surname, it does not give you the right to infringe on the Brooks Brothers trademark or otherwise compete with Brooks Brothers." Brooks's attorneys advised her that fighting Brooks Brothers's in court would cost $200,000. "I've grown every year for the last 10 years by word of mouth," says Brooks. "I could change my company's name, but then there's the domino effect. I have custom printed boxes, Web site domains. What does that affect? Absolutely everything."[1]

Even if small companies have the resources to endure a legal battle, lawsuits are bothersome distractions that prevent entrepreneurs from focusing their energy on running their businesses. In addition, one big judgment against a small company in a legal case could force it out of business. Judgments, the financial penalties that a company must pay if it loses a lawsuit, take three forms: compensatory, consequential, and punitive damages. As the name implies, **compensatory damages** are the monetary damages that are designed to place the plaintiff in the same position he or she would have been in had a contract been performed. In other words, compensatory damages require the defendant to pay the actual amount of loss the plaintiff incurred because of the defendant's actions. Suppose that a small manufacturer creates a contract to deliver 1,000 plastic barrels for $80 per unit by a particular date. If it fails to do so and the customer must purchase the barrels from another supplier for $88 per unit and pay an additional $500 for rush delivery, the customer's compensatory damages are $8,500 (1,000 barrels × $8 price difference plus $500 rush delivery charges). **Consequential damages** are awarded to offset the losses suffered by the plaintiff that go beyond simple compensatory damages because of lasting effects of the damage. If the customer in the previous example lost $15,000 in sales because it did not receive the barrels on time, it could request consequential damages in that amount. For a party to recover consequential damages, the breaching party must have known the consequences of the breach. Courts typically award **punitive damages** in cases in which the defendant engages in intentionally wrongful behavior or behavior that is so negligent or reckless that it is considered intentional. As the name suggests, punitive damages are intended to punish the wrongdoer. The due process clause of the Fourteenth Amendment to the U.S. Constitution prohibits grossly excessive punitive awards, and many states impose limits on punitive damages in court cases.

ENTREPRENEURIAL PROFILE: Dena Lockwood: Professional Neurological Services   When Professional Neurological Services (a Chicago-area company that sells medical tests to doctors) hired Dena Lockwood, she mentioned to them that she had children. The interviewer asked whether her parental responsibilities would get in the way of working 70 hours a week. Lockwood said that it would not be a problem. Professional Neurological Services hired Lockwood at a salary of $25,000 plus 10 percent sales commission. After being hired, Lockwood learned that the company paid other female sales employees who did not have children a base of $45,000 with the same 10 percent commission. After learning of this disparity, Lockwood renegotiated her compensation. Although she negotiated the $45,000 base, the company cut her commission to 5 percent. Her supervisors told her that if she could sell $300,000, her commission would be raised to 10 percent and she would receive five vacation days a year. Lockwood reached her sales goal of $300,000. However, her employer did not pay her a commission and told her that she would then have to reach a higher level of sales to earn any commission in the future. When Lockwood tried to take a day off to stay home with one of her children, her manager informed her that if she did not resign, she would be fired, saying that it "just wasn't working out." Lockwood filed a complaint with Chicago's Human Rights Commission, which found that she was the victim of blatant discrimination against employees with children. The Commission awarded Lockwood $213,000, which included $100,000 in punitive damages, plus another $87,000 for her legal fees.[2]

Small business owners should know the basics of the laws that govern business practices to minimize the chances that their decisions and actions lead to costly lawsuits. This chapter is designed not to make you an expert in business law or the regulations that govern businesses but to make you aware of the fundamental legal issues of which every business owner should know. Entrepreneurs should consult their attorneys for advice on legal questions involving specific situations.

The Law of Contracts

Contract law governs the rights and obligations among the parties to an agreement (contract). It is a body of laws that affects virtually every business relationship. A **contract** is simply a legally binding agreement. It is a promise or a set of promises for the breach of which the law gives a remedy, or the performance of which the law enforces. A contract arises from an agreement, and it creates an obligation among the parties involved. Although almost everyone has the capacity to enter into a contractual agreement (freedom of contract), not every contract is valid and enforceable. A **valid contract** has four elements:

1. **Agreement.** An agreement is composed of a valid offer from one party that is accepted by the other.
2. **Consideration.** Consideration is something of legal (not necessarily economic) value that the parties exchange as part of their bargain.
3. **Contractual capacity.** The parties must be adults capable of understanding the consequences of their agreement.
4. **Legality.** The parties' contract must be for a legal purpose.

In addition, to be enforceable, a contract must meet two supplemental requirements: *genuineness of assent* and *form*. *Genuineness of assent* is a test to make sure that the parties' agreement is genuine and not subject to problems such as fraud, misrepresentation, or mistakes. *Form* involves the writing requirement for certain types of contracts. Although not every contract must be in writing to be enforceable, the law does require some contracts to be evidenced by a writing.

Agreement

Agreement requires a "meeting of the minds" and is established by an offer and an acceptance. One party must make an offer to another who must accept that offer. Agreement is governed by the **objective theory of contracts**, which states that a party's intention to create a contract is measured by outward facts—words, conduct, and circumstances—rather than by subjective, personal intentions. When settling contract disputes, courts interpret the objective facts surrounding the contract from the perspective of an imaginary reasonable person. Agreement requires that one of the parties to a contract make an offer and the other an acceptance.

ENTREPRENEURIAL PROFILE: Republic Bank v. West Penn Allegheny Health System
Republic Bank had taken ownership of a CT scanner, a CT workstation, an ultrasound machine, and an ultrasound table when the buyer defaulted on a lease. Republic hired Tetra Financial Services to find potential buyers for the equipment. Mark Loosli, a Tetra employee acting on behalf of Republic, offered in an e-mail to sell the CT scanner to West Penn Allegheny Health System for $750,000. He also offered to sell the company the ultrasound equipment for an additional $30,000. Michele Hutchison, West Penn's negotiator, sent Loosli an e-mail that said, "We are interested in the 64 slice scanner, CT work station, ultrasound and ultrasound table. Our offer is as follows: Scanner—$600,000 CT Workstation—$50,000 Ultrasound and ultrasound table—$26,500. If there is a good time for us to

1. _____

Explain the basic elements required to create a valid, enforceable contract.

talk live, let me know." Loosli e-mailed Hutchison, stating that he had conveyed her offer to Republic's president, Boyd Lindquist, and that he hoped to have "something concrete in the next day or so." Loosli later e-mailed Hutchison to let her know the deal had been approved. A few weeks later, the deal for the CT scanner fell apart. West Penn claimed it was not contractually bound to purchase any of the items because West Penn had not yet signed a purchase order or sales agreement. Republic maintained that there was offer and acceptance via e-mail. Republic auctioned the equipment for $350,303.76, which was a difference of $299,694.24 from the agreed-on price between West Penn and Republic. Republic then sued West Penn for the difference. The Tenth Circuit Court of Appeals found that given the e-mail communications between West Penn and Tetra/Republic, Republic's response was a valid acceptance and that West Penn understood it as such. Therefore, West Penn was liable for breach of contract.[3]

OFFER An **offer** is a promise or commitment to do or refrain from doing some specified thing in the future. For an offer to stand, there must be an intention to be bound by it. The terms of the offer must be defined and reasonably certain, and the offeror (the party making the offer) must communicate the offer to the offeree (the party to whom the offer is made). The offeror must genuinely intend to make an offer, and the offer's terms must be definite, not vague. The following terms must either be expressed or be capable of being implied in an offer: the parties involved, the identity of the subject matter (which goods or services), and the quantity. Other terms of the offer should specify price, delivery terms, payment terms, timing, and shipping terms. Although these elements are not required, the more terms a party specifies, the more likely it is that an offer exists.

Courts often supply missing terms in a contract when there is a reliable basis for doing so. For instance, the court usually supplies a time term that is reasonable for the circumstances. It supplies a price term (a reasonable price at the time of delivery) if a readily ascertainable market price exists; otherwise, a missing price term defeats the contract. On rare occasions, courts supply a quantity term, but a missing quantity term usually defeats a contract. For example, a small retailer who mails an advertising circular to a large number of customers is not making an offer because one major term—quantity—is missing. Most ads are not offers but are invitations for an offer. Similarly, price lists and catalogs sent to potential customers are not offers.

In general, an offeror can revoke an offer at any time prior to acceptance, but two exceptions to this rule exist: an option contract and a merchant's firm offer. In an **option contract**, the parties create a separate contract to keep an offer open for a particular time period. Option contracts are common in real estate transactions. For instance, the owner of a fast-food franchise created an option contract with the owner of a piece of land that the franchisee was considering purchasing. The landowner made an offer to sell the property to the franchisee, who wanted time to study the demographics, traffic count, and other data at the potential location but did not want to lose a promising piece of real estate by having the owner sell it to someone else. The franchisee and the landowner created an option contract; the franchisee paid the landowner $5,000 for a six-month option on the land, meaning that the landowner could not revoke his offer to sell the property during the six-month option.

The other exception to the revocation-before-acceptance rule is a **merchant's firm offer**. If a merchant seller (a merchant is defined later in this chapter in the section on the Uniform Commercial Code) makes a promise or assurance to hold an offer open in a signed writing, the offer is irrevocable for the stated time period or, if no time is stated, for a reasonable time period. Neither time period can exceed 90 days, however.

An offeror must communicate the offer to the other party because one cannot agree to a contract unless he or she knows it exists. The offeror may communicate an offer verbally, in writing, or by action.

Offers do not last forever. Several actions by either the offeror or the offeree can cause an offer to terminate. In addition, the law itself can cause an offer to cease to exist. As you have learned, an offeror can revoke an offer as long as he or she does so before the offeree accepts it. The offeree can cause an offer to terminate by rejecting the offer (e.g., saying "no" to it) or by making a counteroffer. For instance, suppose that an entrepreneur offers to purchase a piece of land for $175,000. The landowner responds, "Your price is too low, but I'll sell it to you for $190,000." When the landowner made the counteroffer, the entrepreneur's original offer terminated. An offer terminates by operation of the law if the time specified in the offer has elapsed ("This offer is good until noon on October 7"), if the subject matter of the offer is destroyed before the offeree accepts, or if either the offeror or the offeree dies or becomes incapacitated before the offeree accepts the offer.

ACCEPTANCE Only the person to whom the offer is made (the offeree) can accept an offer and create a contract. The offeree must accept voluntarily, agreeing to the terms exactly as the offeror presents them. When an offeree suggests alternative terms or conditions to those in the original offer, he or she is implicitly rejecting the original offer and making a counteroffer. Common law requires that the offeree's acceptance exactly match the original offer. This is called the **mirror image rule**, which says that an offeree's acceptance must be the mirror image of the offeror's offer.

Generally, silence by the offeree cannot constitute acceptance, even if the offer contains statements to the contrary. For instance, when an offeror claims, "If you do not respond to this offer by Friday at noon, I conclude your silence to be your acceptance," no acceptance exists even if the offeree does remain silent. The law requires an offeree to act affirmatively to accept an offer in most cases.

An offeree must accept an offer by the means of communication authorized by and within the time limits specified by the offeror. Generally, offers accepted by alternative media or after specified deadlines are ineffective. If the offeror specifies no means of communication, the offeree must use the same medium used to extend the offer (or a faster method). According to the **mailbox rule**, if an offeree accepts by mail, the acceptance is effective when the offeree drops the letter in the mailbox, even if it never reaches the offeror. In addition, all offers must be properly dispatched; that is, they must be properly addressed, noted, and stamped. Most courts have extended the mailbox rule to electronic communications, which means acceptance occurs instantaneously at the time the offeree accepts via e-mail or other Web-based communication.

CONSIDERATION Contracts are based on promises, and because it is often difficult to distinguish between promises that are serious and those that are not, courts require that consideration be present in virtually every contract. **Consideration** is something of *legal* value (*not* necessarily economic value) that the parties to a contract bargain for and exchange as the "price" for the promises given. Consideration can be money, but parties most often swap promises for promises. For example, when a buyer promises to buy an item and a seller promises to sell it, the parties have exchanged valuable consideration. The buyer's promise to buy and the seller's promise to sell constitute the consideration for their contract. To comprise valuable consideration, a promise must impose a liability or create a duty.

For a contract to be binding, the two parties involved must exchange valuable consideration. The absence of consideration makes a promise unenforceable. A promise to perform something that one is already legally obligated to do is not valuable consideration. Because consideration is something that a promisor requires in exchange for his promise, past consideration is not valid. In addition, under the common law, new promises require new consideration. For instance, if two businesspeople have an existing contract for performance of a service, any modifications to that contract must be supported by new consideration. In many states, promises made in exchange for "love and affection" are not enforceable because the contract lacks valuable consideration.

One important exception to the requirement for valuable consideration is **promissory estoppel**. Under this rule, a promise that induces another party to act can be enforceable without consideration if the promisee substantially and justifiably relies on the promise. Thus, promissory estoppel is a substitute for consideration.

ENTREPRENEURIAL PROFILE: Joseph Hoffman v. Red Owl Stores Joseph Hoffman owned a bakery in Wautoma, Wisconsin, but wanted to open a Red Owl grocery store. He approached Edward Lukowitz, a division manager for Red Owl, and told him that he had $18,000 to invest in a Red Owl franchise. Lukowitz assured Hoffman that $18,000 was sufficient to set him up in business as a Red Owl franchisee. Lukowitz suggested that Hoffman needed experience running a grocery store before he became a Red Owl franchisee, and Hoffman purchased a small grocery store in Wautoma. After several months, Red Owl confirmed that Hoffman was operating the store at a profit. Lukowitz then told Hoffman that he would have to sell the grocery store to purchase a Red Owl franchise, and Hoffman sold the store to one of his employees. In a meeting, Lukowitz assured Hoffman that "everything is ready to go. Get your money together and we are set." Shortly after this meeting, Lukowitz told Hoffman that he would have to sell his bakery business and building, and that this was the only "hitch" that remained. Hoffman sold the bakery and the building and moved to Chilton, Wisconsin, where Red Owl had found a potential site for a store. During this time, however, Red Owl Stores raised the price of the franchise from $18,000 to $24,100, and later to $26,100.

Hoffman ended negotiations with Red Owl and filed a lawsuit, claiming that although Hoffman had not given any consideration, he had justifiably relied on Red Owl's promises to his detriment. The court applied the doctrine of promissory estoppel and ruled in favor of Hoffman.[4]

In most cases, courts do not evaluate the adequacy of consideration given for a promise. In other words, there is no legal requirement that the consideration the parties exchange be of approximately equal value. Even if the value of the consideration one party gives is small compared to the value of the bargain to the other party, the bargain stands. Why? The law recognizes that people have the freedom to contract and that they are just as free to enter into "bad" bargains as they are to enter into "good" ones. Only in extreme cases (e.g., cases affected by mistakes, misrepresentation, fraud, duress, and undue influence) will the court examine the value of the consideration provided in a trade.

Contractual Capacity

The third element of a valid contract requires that the parties involved in it must have contractual capacity for it to be enforceable. Not every person who attempts to enter into a contract has the capacity to do so. Under the common law, minors, intoxicated people, and insane people lack or have limited contractual capacity. As a result, contracts these people attempt to enter are *voidable*—that is, the party can annul or disaffirm the contract at his option.

MINORS Minors constitute the largest group of individuals without contractual capacity. In most states, anyone under the age 18 is a minor. With a few exceptions, any contract made by a minor is voidable at the minor's option. In addition, a minor can avoid a contract during minority and for "a reasonable time" afterward. The adult involved in the contract cannot avoid it simply because he or she is dealing with a minor.

In most states, if a minor receives the benefit of a completed contract and then disaffirms that contract, he or she must fulfill the **duty of restoration** by returning the benefit. In other words, the minor must return any consideration he or she has received under the contract to the adult and is entitled to receive any consideration he or she gave the adult under the contract. The minor must return the benefit of the contract no matter what its condition is. For instance, suppose that Brighton, a 16-year-old minor, purchases a mountain bike for $415 from Cycle Time, a small bicycle shop. After riding the bike for a little more than a year, Brighton decides to disaffirm the contract. Under the law, all he must do is return the mountain bike to Cycle Time, whatever condition it is in (pristine, used, wrecked, or rubble), and he is entitled to get all of his money back. In most states, he does not have to pay Cycle Time for the use of the bike or the damage done to it. A few states impose an additional duty on minors. The **duty of restitution** requires that minors who disaffirm contracts return any consideration they received to the adult and must pay a "reasonable value" for the depreciation of or damage to the item (which is usually less than the actual value of the depreciation of or damage to the item). Adults enter into contracts with minors at their own risk.

Parents are usually not liable for any contracts made by their children, although a cosigner is bound equally with a minor. Entrepreneurs can protect themselves when dealing with minors by requiring an adult to cosign. If the minor disaffirms the contract, the adult cosigner remains bound by it.

INTOXICATED PEOPLE A contract entered into by an intoxicated person can be either voidable or valid, depending on the person's condition when entering into the contract. If a person's reason and judgment are impaired so that he or she does not realize that he or she is making a contract, the contract is voidable (even if the intoxication was voluntary) and the intoxicated person must return the benefit. However, if the intoxicated person understands that he is forming a contract, although it may be foolish, the contract is valid and enforceable.

PEOPLE WITH MENTAL INCAPACITIES A contract entered into by a person with a mental incapacity can be void, voidable, or valid, depending on the mental state of the person. Those people who have been judged to be so mentally incompetent that a guardian is appointed for them cannot enter into a valid contract. If such a person does make a contract, it is *void* (i.e., it does not exist). A person who has not been legally declared insane and appointed a guardian (e.g., someone suffering from Alzheimer's disease) is bound by a contract if he or she was lucid enough at the time of the contract to comprehend its consequences. On the other hand, if at the time of entering the contract that same person was so mentally incompetent that he or she could not realize what was happening or could not understand the terms, the contract is voidable. Just as with minors, he or she must return any benefit received under the contract.

Legality

The final element required for a valid contract is legality. The purpose of the parties' contract must be legal. Because society imposes certain standards of conduct on its members, contracts that are illegal (criminal or tortuous) or against public policy are void. Examples of these situations include contracts in which the stated interest rate exceeds the rate allowed by a state's usury laws, interstate gambling that is conducted in states where the type of gambling is illegal (e.g., casino games via the Internet), business transactions that violate a state's blue laws (creating certain types of contracts on Sunday), activities that require a practitioner to have a license to perform (e.g., attorneys, real estate brokers, contractors, and others), and freestanding contracts that restrain competition and trade.

If a contract contains both legal and illegal elements, courts will enforce the legal parts as long as they can separate the legal portion from the illegal portion. However, in some contracts, certain clauses are so unconscionable that the courts will not enforce them. Usually, the courts do not concern themselves with the fairness of a contract between parties because individuals are supposed to be intelligent. However, in the case of unconscionable contracts, the terms are so harsh and oppressive to one party that the courts often rule the clause to be void. These clauses, called **exculpatory clauses**, frequently attempt to free one party of all responsibility and liability for an injury or damage that might occur. For instance, suppose that Miguel Ferras signs an exculpatory clause when he leaves his new BMW with an attendant at a parking garage. The clause states that the garage is "not responsible for theft, loss, or damage to cars or articles left in cars due to fire, theft, or other causes." The attendant leaves Miguel's car unattended with the keys in the ignition, and a thief steals the car. A court would declare the exculpatory clause void because the garage owes a duty to its customers to exercise reasonable care to protect their property, a duty it breached because of gross negligence.

GENUINENESS OF ASSENT AND THE FORM OF CONTRACTS A contract that contains the four elements just discussed—agreement, consideration, capacity, and legality—is *valid*, but a valid contract may be unenforceable because of two possible defenses against it: genuineness of assent and form. **Genuineness of assent** serves as a check on the parties' agreement, verifying that it is genuine and not subject to mistakes, misrepresentation, fraud, duress, or undue influence. The existence of a contract can be affected by mistakes that one or both parties to the contract make. Different types of mistakes exist, but only mistakes of *fact* permit a party to avoid a contract. Suppose that a small contractor submits a bid on the construction of a bridge but the bidder mistakenly omits the cost of some materials. The client accepts the contractor's bid because it is $32,000 below all others. If the client knew or should have known of the contractor's mistake, the contractor can avoid the contract; otherwise, he must build the bridge at the bid price.

Fraud also voids a contract because no genuineness of assent exists. **Fraud** is the intentional misrepresentation of a material fact, justifiably relied on, that results in injury to the innocent party. The misrepresentation with the intent to deceive can result from words, silence, or conduct. Suppose a small retailer purchases a new security system from a dealer who promises it will provide 20 years of reliable service and lower the cost of operation by 40 percent. The dealer then knowingly installs a used, unreliable system. In this case, the dealer has committed fraud, and the retailer can either rescind the contract with his original position restored or enforce it and seek damages for injuries.

Duress, forcing an individual into a contract by fear or threat, eliminates genuineness of assent. The innocent party can choose to carry out the contract or to avoid it. For example, if a supplier forces the owner of a small video arcade to enter a contract to lease his machines by threat of personal injury, the supplier is guilty of duress. Blackmail and extortion used to induce another party to enter a contract also constitute duress.

Generally, the law does not require contracts to follow a prescribed form; a contract is valid whether it is written or oral. Most contracts do *not* have to be in writing to be enforceable, but for convenience and protection, a small business owner should insist that every contract be in writing. If a contract is oral, the party attempting to enforce it must first prove its existence and then establish its actual terms. Although each state has its own rules, the Statute of Frauds, generally requires the following contracts to be in writing:

- Contracts for the sale of land
- Contracts involving lesser interests in land (e.g., rights-of-way or leases that last more than one year)
- Contracts that cannot by their terms be performed within one year
- Collateral contracts, such as promises to answer for the debt or duty of another
- Promises by the administrator or executor of an estate to pay a debt of the estate personally
- Contracts for the sale of goods (as opposed to services) priced above $500

Breach of Contract

Both parties fully performing the terms of their agreement discharges the majority of contracts. Occasionally, however, one party fails to perform as agreed. This failure is called *breach of contract*, and the injured party has certain remedies available. A breach of contract can be either a minor breach in which substantial but not complete performance occurs or a material breach of contract associated with nonperformance or inferior performance. In cases where there exists a minor breach of contract, the party "in breach" may agree to complete the specific terms of the contract or compensate the other party for the unperformed component of the contract. If these two remedies are not accepted, the next step is legal action to recover the cost to repair the defect.

In contrast, a *material breach* occurs when a party renders inferior performance that impairs or destroys the essence of the contract. The nonbreaching party may either rescind the contract and recover restitution or affirm the contract and recover damages. Of course, the injured party must make a reasonable effort to minimize the damages incurred by the breach.

ENTREPRENEURIAL PROFILE: Gaia Healthcare Systems v. National Renal Alliance Gaia Healthcare Systems filed a claim against National Renal Alliance seeking more than $178,000 in unpaid software license fees and other damages. Gaia claimed that National Renal stopped making payments on its system but continued to use its software and make "untrue and disparaging statements" about Gaia. National Renal denied Gaia's allegations and issued a counterclaim against the vendor, alleging breach of contract, fraud, and misrepresentation. National Renal argued that Gaia's software did not perform as promised, causing billing delays that interfered with its ability to collect payments and forced it to draw on its line of credit, pay its employees overtime, and "hire temporary employees to correct errors generated by the software." An arbitrator with the American Arbitration Association sided with National Renal, denying Gaia's claim in its entirety and awarding $305,000 to National Renal.[5]

In some cases, monetary damages are inadequate to compensate an injured party for a defendant's breach of contract. The only remedy that would compensate the nonbreaching party might be specific performance of the act promised in the contract. **Specific performance** is usually the remedy for breached contracts dealing with unique item (antiques, land, and animals). For example, if an antique auto dealer enters a contract to purchase a rare Corvette and the seller breaches the contract, the dealer may sue for specific performance. That is, she may ask the court to order the breaching party to sell the antique car. Courts rarely invoke the remedy of specific performance. Generally, contracts for performance of personal services are not subject to specific performance.

The Uniform Commercial Code

2.

Outline the major components of the Uniform Commercial Code governing sales contracts.

For many years, sales contracts relating to the exchange of goods were governed by a loosely defined system of rules and customs called the *Lex Mercatoria* (Merchant Law). Many of these principles were assimilated into the U.S. common law through court opinions, but they varied widely from state to state and made interstate commerce difficult and confusing for businesses. In 1952, the commission on Uniform State Laws created the **Uniform Commercial Code** (or the UCC or the Code) to replace the hodgepodge collection of confusing, often conflicting state laws that governed basic commercial transactions with a document designed to provide uniformity and consistency. The UCC replaced numerous statutes governing trade when each of the states, the District of Columbia, and the Virgin Islands adopted it. (Louisiana has adopted only Articles 1, 3, 4, and 5.)

The Code does not alter the basic tenets of business law established by the common law; instead, it unites and modernizes them into a single body of law. In some cases, however, the Code changes some of the specific rules under the common law. The Code consists of 10 articles, but we will discuss the general principles relating to one of its most common sections, Article 2, which governs the sale of goods. The UCC creates a "caste system" of merchants and nonmerchants and requires merchants to have a higher degree of knowledge and understanding of the Code.

Sales and Sales Contracts

Every sales contract is subject to the basic principles of law that govern all contracts—agreement, consideration, capacity, and legality. However, when a contract involves the sale of goods, the UCC imposes rules that may vary slightly or substantially from basic contract law. Article 2 governs *only* contracts for the *sale of goods*, but it pertains to *every* sale of goods, whether the good involved is a 99-cent pen or a billion-dollar battleship. To be considered a good, an item must be personal property that is tangible and movable (e.g., not real estate or services), and a "sale" is "the "passing of title from the seller to the buyer for a price" (UCC Sec. 2-106[1]). The UCC does *not* cover the sale of services, although certain "mixed transactions," such as the sale by a garage of car parts (goods) and repairs (a service), fall under the Code's jurisdiction if the goods are the dominant element of the contract.

In addition to the rules it applies to the sale of goods in general, the Code imposes special standards of conduct in certain instances when merchants sell goods to one another. Usually, a person is considered a professional **merchant** if he "deals in goods of the kind" involved in the contract and has special knowledge of the business or of the goods, employs a merchant agent to conduct a transaction for him, or holds him- or herself out to be a merchant.

Although the UCC requires that the same elements outlined in common law be present in forming a sales contract, it relaxes many of the specific restrictions. For example, the UCC states that a contract exists even if the parties omit one or more terms (price, delivery date, place of delivery, or quantity), as long as they intended to make a contract and there is a reasonably certain method for the court to supply the missing terms. Suppose a manufacturer orders a shipment of needed raw materials from her usual supplier without asking the price. When the order arrives, the price is substantially higher than she expected, and she attempts to disaffirm the contract. The Code verifies the existence of a contract and assigns to the shipment a price that is reasonable at the time of delivery.

Common law requires that acceptance of an offer to be exactly the same as the offer; an acceptance that adds some slight modification is no acceptance at all, and no contract exists. Any modification constitutes a counteroffer. However, the UCC states that as long as an offeree's response (words, writing, or actions) indicates a sincere willingness to accept the offer, it is a legitimate acceptance even if the offeree adds terms. This section of the UCC is known as "the battle of the forms." In dealings between nonmerchant buyers and sellers, these added terms become "proposals for addition." In other words, a contract is formed on the offeror's original terms. Between merchants, however, these additional proposals *automatically* become part of the contract unless they materially alter the original contract, the offer expressly states that no terms other than those in the offer will be accepted, or the offeror objects to the particular terms. In other words, the contract is formed on the offeree's modified terms. For example, suppose that an appliance wholesaler offers to sell a retailer a shipment of appliances for $5,000 plus freight. The retailer responds, "I accept," but includes an additional term by stating, "Delivery within three days." A contract exists, and the addition will become part of the contract unless the wholesaler objects within a reasonable time.

When the offeree includes a term in the acceptance that *contradicts* a term in the offeror's original offer, the UCC says that the two terms cancel out each other. What, then, are the terms of the resulting contract? The UCC turns to its gap-filling rules, which establish reasonable terms for prices, delivery dates, warranties, payment times, and other topics, to supply the disputed term.

ENTREPRENEURIAL PROFILE: Superior Boiler Works v. R.J. Sanders Company The R.J. Sanders Company won a contract to install the heating system at a federal prison and negotiated a contract with Superior Boiler Works to purchase three large commercial boilers for the project. On March 27, Superior sent an offer to Sanders in which it offered to sell three boilers for $156,000 with an estimated delivery time of four weeks. After several discussions, Sanders sent a purchase order to Superior on July 20 for three boilers, agreeing to pay $145,827 and stating a delivery date of four weeks (August 20). Superior responded by sending Sanders a sales order in which it agreed to the price but stated a shipping date of October 1. Superior shipped the boilers on October 1, just as it had promised, and they arrived at Sanders on October 5. This delivery date forced Sanders to rent temporary boilers at a cost of $45,315, and Sanders sent Superior a check for $100,000 with a note explaining that the deduction was to offset the cost of the rented boilers. Superior sued Sanders for the $45,000 difference, claiming that the October 1 shipping date was reasonable. The Supreme Court of Rhode Island ruled that the parties' conflicting delivery terms canceled out each other. The court then applied the UCC's gap-filling rules (boilers are goods), which state that the time for delivery of the goods must be within "a reasonable time." The court ruled in favor of Superior, stating that the October 1 shipping date was within a reasonable time.[6]

The UCC significantly changes the common law requirement that a contract modification requires new consideration. ("New promises require new consideration."). Under the Code, modifications to contract terms are binding *without* new consideration if they are made in good faith. ("New promises do *not* require new consideration.") For example, suppose that a small building contractor forms a contract to purchase a supply of lumber for $12,000. After the agreement but before the lumber is delivered, a hurricane forces the price of the lumber to double, and the supplier notifies the contractor that it must raise the price of the lumber shipment to $24,000. The contractor reluctantly agrees to the additional cost but later refuses to pay. According to the UCC, the contractor must pay the higher price because the contract modification requires no new consideration.

The Code also has its own Statute of Frauds provision relating to the form of contracts for the sale of goods. If the price of the goods is $500 or more, the contract must be in writing to be enforceable. Of course, the parties can agree orally and then follow up with a written memorandum. The Code does not require both parties to sign the written agreement, but it must be signed by the party against whom enforcement is sought (which is impossible to tell before a dispute arises, so it is a good idea for *both* parties to sign the agreement at the outset).

The UCC includes a special provision involving the writing requirement in contracts between merchants. If merchants form a verbal contract for the sale of goods priced at more than $500 and one of them sends a written confirmation of the deal to the other, the merchant receiving the confirmation must object to it *in writing* within 10 days. Otherwise, the contract is enforceable against *both* merchants even though the merchant receiving the confirmation has not actually signed anything.

Once the parties create a sales contract, they are bound to perform according to its terms. Both the buyer and the seller have certain duties and obligations under the contract. Generally, the Code assigns the obligations of "good faith" (defined as "honesty in fact in the conduct or transaction concerned") and "commercial reasonableness" (commercial standards of fair dealing) to both parties.

The seller must make delivery of the items involved in the contract, but "delivery" is not necessarily physical delivery. The seller simply must make the goods available to the buyer. The contract normally outlines the specific details of the delivery, but occasionally the parties omit this provision. In this instance, the place of delivery will be the seller's place of business if one exists; otherwise, it is the seller's residence. If both parties know the usual location of the identified goods, that location is the place of delivery (e.g., a warehouse). In addition, the seller must make the goods available to the buyer at a reasonable time and in a reasonable manner. All goods covered by the contract must be tendered in one delivery unless the parties' agreement states otherwise.

A buyer must accept the delivery of conforming goods from the seller. Of course, the buyer has the right to inspect the goods in a reasonable manner and at any reasonable time or place to ensure that they are conforming goods before making payment. However, cash-on-delivery terms prohibit the right to advance inspection unless the contract specifies otherwise. The UCC also says that if goods or tender of delivery fail, in any respect, to conform to the contract, the buyer is not required to accept them.

A buyer can indicate his acceptance of the goods in several ways. Usually the buyer indicates acceptance by an express statement that the goods are suitable. This expression can be by words or by conduct. Suppose that a small electrical contractor orders a truck to use in her business. When she receives it, she equips it to suit her trade, including a company decal on each door. Later the contractor attempts to reject the truck and return it. By customizing the truck, the buyer has indicated her acceptance of the truck. In addition, the Code assumes acceptance if the buyer has a reasonable opportunity to inspect the goods and fails to reject them within a reasonable time.

A buyer has the duty to pay for the goods on the terms stated in the contract when they are received. A seller cannot require payment before the buyer receives the goods. Unless otherwise stated in the contract, payment must be in cash.

Breach of Sales Contracts

As we have seen, when a party to a sales contract fails to perform according to its terms, that party is said to have breached the contract. The law provides the innocent (nonbreaching) party numerous remedies, including damage awards and the right to retain possession of the goods. The object of these remedies is to place the innocent party in the same position as if the contract had been carried out. The parties to the contract may specify their own damages in case of breach. These provisions, called **liquidated damages**, must be reasonable and cannot be in the nature of a penalty. For example, suppose that Alana Mitchell contracts with a local carpenter to build a booth from which she plans to sell crafts. The parties agree that if the carpenter does not complete the booth by September 1, Mitchell will receive $500. If the liquidated damages had been $50,000, they would be unenforceable because such a large amount of money is clearly a penalty.

An unpaid seller has certain remedies available under the terms of the Code. Under a seller's lien, every seller has the right to maintain possession of the goods until the buyer pays for them. In addition, if the buyer uses a fraudulent payment to obtain the goods, the seller has the right to recover them. If the seller discovers that the buyer is insolvent, the seller can withhold delivery of the goods until the buyer pays in cash. If a seller ships goods to an insolvent buyer, the seller can require their return within 10 days after receipt. In some cases, the buyer breaches a contract while the goods are still unfinished in the production process. When this occurs, the seller must use "reasonable commercial judgment" to decide whether to sell them for scrap or complete them and resell them elsewhere. In either case, the buyer is liable for any loss the seller incurs. Of course, the seller has the right to withhold performance when the buyer breaches the sales contract.

When the seller breaches a contract, the buyer also has specific remedies available. For instance, if the goods do not conform to the contract's terms, the buyer has the right to reject them. If the seller fails to deliver the goods, the buyer can sue for the difference between the contract price and the market price at the time that the buyer discovers the breach. When the buyer accepts goods and then discovers that they are defective or nonconforming, he or she must notify the seller of the breach. In this instance, damages amount to the difference between the value of the goods delivered and their value if they had been delivered as promised. If a buyer pays for goods that the seller retains, the buyer can take possession of the goods if the seller becomes insolvent within 10 days after receiving the first payment. If the seller unlawfully withholds the goods from the buyer, the buyer can recover them. Under certain circumstances, a buyer can obtain specific performance of a sales contract; that is, the court orders the seller to perform according to the contract's terms. As mentioned earlier, specific performance is a remedy only when the goods involved are unique or unavailable on the market. Finally, if the seller breaches the contract, the buyer has the right to rescind the contract; if the buyer has paid any part of the purchase price, the seller must refund it.

Whenever a party breaches a sales contract, the innocent party must bring suit within a specified period of time. The Code sets the statute of limitations at four years. In other words, any action for a breach of a sales contract must begin within four years after the breach occurred.

Sales Warranties and Product Liability

The U.S. economy once emphasized the philosophy of *caveat emptor*, "let the buyer beware," but today the marketplace enforces a policy of *caveat venditor*, "let the seller beware." **Tort law** deals with cases in which one party commits a wrong against another party and causes injury or damage to the person and/or his or her property. Tort law covers a wide range of topics, including defamation of character, false imprisonment (e.g., wrongly detaining a suspected shoplifter), fraud, wrongful interference with a contractual relationship, and others. Tort liability represents a significant risk for small companies. A study by the U.S. Chamber Institute for Legal Reform reports that small companies bear 81 percent of total tort liability costs for a total of $105 billion per year but take in only 22 percent of total commercial revenue. Small businesses pay $35.6 billion of their total cost of tort liability out of pocket (not paid by insurance)! Tort liability costs businesses with less than $10 million in annual sales $14.59 for every $1,000 of revenue; in other words, a small company with $5 million in revenue incurs, on average, $72,950 in tort related costs each year.[7] Entrepreneurs must be aware of two general categories related to torts that involve the quality and reliability of the products they sell: sales warranties and product liability.

SALES WARRANTIES Simply stated, a **sales warranty** is a promise or a statement of fact by the seller that a product will meet certain standards. Because a breach of warranty is a breach of promise, the buyer has the right to recover damages from the seller. Several different types of warranties can arise in a sale. A seller creates an **express warranty** by making statements about the condition, quality, and performance of the good that the buyer substantially relies on. Sellers create express warranties by words or actions. For example, a manufacturer selling a shipment of cloth to a customer with the promise that "it will not shrink" is creating an express warranty. Similarly, the jeweler who displays a watch in a glass of water for promotional purposes creates an express warranty that "this watch is waterproof" even though no such promise is ever spoken. Generally, an express warranty arises if the seller indicates that the goods conform to any promises of fact the seller makes, to any description of them (e.g., printed on the package or statements of fact made by salespersons), or to any display model or sample (e.g., a floor model used as a demonstrator).

Whenever someone sells goods, the UCC automatically implies certain types of warranties unless the seller specifically excludes them. These **implied warranties** take several forms. Sellers, simply by offering goods for sale, imply a **warranty of title**, which promises that their title to the goods is valid (i.e., no liens or claims exist) and that transfer of title is legitimate. A seller can disclaim a warranty of title only by using very specific language in a sales contract.

An implied **warranty of merchantability** applies to every merchant seller, and the only way to disclaim it is by mentioning the term "warranty of merchantability" in a conspicuous manner. An implied warranty of merchantability assures the buyer that the product will be of average quality—not the best and not the worst. In other words, merchantable goods are "fit for the ordinary purposes for which such goods are used."[8] For example, a commercial refrigeration unit that a food store purchases should keep food cold.

ENTREPRENEURIAL PROFILE: Michael Williams v. O'Charley's Michael Williams ordered a chicken dinner at an O'Charley's restaurant located in Concord, North Carolina. Williams had felt fine before eating the meal and ate nothing afterward. The next morning, Williams became severely ill. He went to the hospital, where he spent seven days recovering. Williams filed a lawsuit against O'Charley's in which he described that the chicken he had eaten, which was stuck to the plate and dry, had a foul aftertaste. Although the evidence of the actual chicken was long gone, the jury relied on the expert physician witness, who testified with a reasonable degree of certainty that Williams's illness was caused by the chicken dinner at O'Charley's. The jury found in favor of Williams and awarded him $140,000, finding that O'Charley's was liable for breach of implied warranty of merchantability. Although O'Charley's appealed the verdict, arguing that there was insufficient evidence to establish a defect in the chicken, the North Carolina Court of Appeals found numerous precedents in previous foodborne illness cases to warrant finding in favor of Williams.[9]

An implied **warranty of fitness for a particular purpose** arises when a seller knows the particular reason for which a buyer is purchasing a product and knows that the buyer is depending on the seller's judgment to select the proper item. For example, suppose a contractor asks the owner of a paint store for a paint that adheres to metal roofs. The store owner sells the contractor paint that he says will do the job, but two months later, the paint is peeling off. The owner has violated the warranty of fitness for a particular purpose.

The Code also states that the only way a merchant can disclaim an implied warranty is to include the words "sold as is" or "with all faults," stating that the buyer purchases the product as it is, without any guarantees. The following statement is usually sufficient to disclaim most warranties, both express and implied: "Seller hereby disclaims all warranties, express and implied, including all warranties of merchantability and all warranties of fitness for a particular purpose." To protect a business, the statement must be printed in bold letters and placed in a conspicuous place on the product or its package.

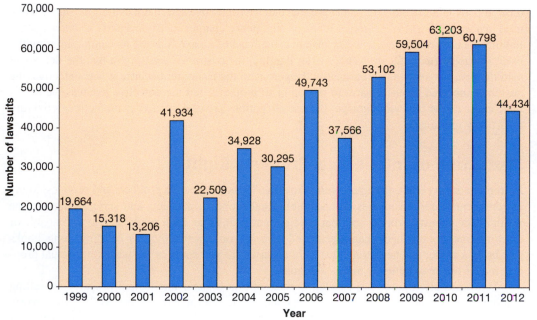

FIGURE 23.1

Number of Product Liability Lawsuits

Source: Based on Judicial Business of the United States Courts, *www .uscourts.gov/Statistics/JudicialBusiness/ JudicialBusiness.aspx?doc=/uscourts/ Statistics/JudicialBusiness/2009/ tables/S10Sep09.pdf; www.uscourts.gov/ uscourts/Statistics/JudicialBusiness/2012/ appendices/C02ASep12.pdf.*

PRODUCT LIABILITY   At one time, only the parties directly involved in the execution of a contract were bound by the law of sales warranties. Today, the UCC and the states have expanded the scope of warranties to include any person (including bystanders) incurring personal or property damages caused by a faulty product. In addition, most states allow an injured party to sue *any* seller in the chain of distribution for breach of warranty (a concept known as joint and several liability). Product liability is built on the principle that a person who introduces a product into the stream of commerce owes a duty of care not only to the person who first purchases the product but also to anyone else who might foreseeably come into contact with it. A company that may be responsible for only a small percentage of a person's injury may end up bearing the majority of the damage award in the case. If a small company is hit with a product liability lawsuit, the results can be devastating. Figure 23.1 shows the number of product liability lawsuits filed in recent years.

Many customers who ultimately file suit under product liability laws base their claims on **negligence,** when a manufacturer or distributor fails to do something that a "reasonable" person would do. Typically, negligence claims arise from one or more of the following charges:

Negligent design.   In suits based on negligent design, a buyer claims that an injury occurred because the manufacturer designed the product improperly. To avoid liability charges, a company does not have to design products that are 100 percent safe, but it must design products that are free of "unreasonable" risks.

Negligent manufacturing.   In cases claiming negligent manufacturing, a buyer claims that a company's failure to follow proper manufacturing, assembly, or inspection procedures allowed a defective product to get into the customer's hands and cause injury. A company must exercise "due care" (including design, assembly, and inspection) to make its products safe when they are used for their intended purpose.

Failure to warn.   Although manufacturers do not have to warn customers about obvious dangers of using their products, they must warn them about the dangers of normal use and of foreseeable misuse of the product. (Have you ever read the warning label on a stepladder?) Many businesses hire attorneys to write the warning labels they attach to their products and include in their instructions.[10]

Another common basis for product liability claims against businesses is **strict liability,** which states that a manufacturer is liable for its actions no matter what its intentions or the extent of its negligence. Unlike negligence, a claim of strict liability does not require the injured party to prove that the company's actions were unreasonable. The injured person must prove only that the company manufactured or sold a product that was defective and that it caused the injury when used in a way that was foreseeable. Whereas negligence charges focus on a party's *conduct*, strict liability focuses on the *product*. For instance, the head of an axe flies off its handle, injuring the user. To sue the manufacturer under strict liability, the customer must prove that the defendant sold the axe, the axe was unreasonably dangerous to the customer because it was defective, the customer incurred physical harm to person or to property, and the defective axe was the proximate cause of the injury or damage. If these allegations are true, the axe manufacturer's liability is virtually unlimited.

Protection of Intellectual Property Rights

3.

Discuss the protection of intellectual property rights using patents, trademarks, and copyrights.

Entrepreneurs excel at coming up with innovative ideas for creative products and services. Many entrepreneurs build businesses around intellectual property, products and services that are the result of the creative process and that have commercial value. New methods that are capable of teaching foreign languages at an accelerated pace, hit songs with which we can sing along, books that bring a smile, and new drugs that fight diseases are just some of the ways intellectual property makes our lives better or more enjoyable.

Unfortunately, thieves are escalating their efforts to steal intellectual property by selling counterfeit merchandise. The U.S. Customs and Border Patrol seizes about $1.25 billion worth of counterfeit goods each year, but sales of counterfeit goods continue to cost U.S. businesses between $500 and $600 billion per year.[11] The problem extends far beyond pirated software, fake shoes and handbags, and knockoffs of expensive watches or the latest styles of designer clothing. Authorities have discovered pirates selling counterfeit helicopter, airplane, and auto parts as well as prescription medications (including blood pressure medication and birth control pills) and many other products. A tidal wave of counterfeit products, originating mostly in China (which accounts for 72 percent of the counterfeit items that authorities confiscate), Hong Kong, India, and Taiwan, is flooding the world.[12] The International Chamber of Commerce estimates that by 2015 there will be $1.7 trillion in counterfeit goods sold around the globe, which is more than 2 percent of the world's total economic output.[13] Table 23.1 shows the top 10 counterfeit products that U.S. Customs agents seize based on their value.

Entrepreneurs can protect their intellectual property from unauthorized use with the help of three important tools: patents, trademarks, and copyrights.

ENTREPRENEURIAL PROFILE: Catherine Simms: Whiner and Diner Catherine Simms is the owner of Whiner and Diner, which makes pet beds and other products out of used wine crates. She has had a significant problem with copycat products and knockoffs showing up on the Web site Etsy, which allows people to buy and sell handmade and vintage products. "It's terrible,

TABLE 23.1 Top 10 Products Seized by U.S. Customs Agents

In one recent year, U.S. Customs agents conducted more than 22,000 seizures of counterfeit goods coming into the United States. Unfortunately, these seizures represent only a portion of the total traffic in pirated goods. Which items do pirates target for counterfeiting?

Rank	Item	Percentage of Seized Value
1.	Handbags and wallets	40%
2.	Watches and jewelry	15%
3.	Apparel and accessories	11%
4.	Consumer electronics and parts	8%
5.	Footwear	8%
6.	Pharmaceuticals and personal care items	8%
7.	Optical media	3%
8.	Computers and accessories	3%
9.	Labels and tags	2%
10.	Toys	1%

Source: Based on *Intellectual Property Rights: Fiscal Year 2012 Seizure Statistics*, U.S. Customs and Border Protection, Office of International Trade, 2013, p. 18.

horrible," says Sims. "They copy the product, they re-write all of our descriptions and titles. One store just copied and pasted the whole Web site. I work on the Web site myself. It took me years to learn about [search engine] optimization. But if I optimize my text and then they use that text, they may show up [on search engine results] before we do!" Most of the violators deny that they have done anything wrong. "They send a very polite letter saying: We just happen to manufacture the same thing and we're not using your company name, so it's okay," says Simms. When a cease-and-desist letter does not get the copycat to stop selling the knockoff, Simms goes to the offending Web site's hosting company with a request to remove the questionable content, which will often comply with the Simms's requests. "When it's obvious, the hosting companies will just shut them down," she says.[14]

Patents

A **patent** is a grant from the U.S. Patent and Trademark Office (PTO) to the inventor of a product, giving the exclusive right to make, use, or sell the invention in the United States for 20 years from the date of filing the patent application. The purpose of giving an inventor a 20-year monopoly over a product is to stimulate creativity and innovation. After 20 years, the patent expires and cannot be renewed. Most patents are granted for new product inventions, but **design patents**, issued for 3½, 7, or 14 years beyond the date the patent is issued, are given to inventors who make new, original, and ornamental changes in the design of existing products that enhance their sales. Inventors who develop a new plant can obtain a **plant patent** (issued for seven years), provided that they can reproduce the plant asexually (e.g., by grafting or crossbreeding rather than planting seeds). To be patented, a device must be new (but not necessarily better; see Figure 23.2), not obvious to a person of ordinary skill or knowledge in the related field, and useful. An inventor cannot patent a device if it has been publicized in print anywhere in the world or if it has been used or offered for sale in the United States prior to the date of the patent application. A U.S. patent is granted only to the true inventor, not to a person who discovers another's invention. No one can copy or sell a patented invention without getting a license from its creator. A patent does not give one the right to make, use, or sell an invention but does give one the right to exclude others from making, using, or selling it unless they purchase the rights to it.

In recent years, the PTO has awarded companies, primarily Web-based businesses, patents on their business methods. Rather than giving them the exclusive rights to a product or an invention, a business method patent protects the way a company conducts business. For instance, *Amazon .com* earned a patent on its "1-Click" Web-based checkout process, precluding other e-tailers from using it. *Priceline.com* has a patent on its business model of "buyer-driven commerce," in which customers name the prices they are willing to pay for airline tickets, hotel rooms, and other items.

Although inventors have no guarantee of getting a patent, they can enhance their chances considerably by following the basic steps suggested by the PTO. Before beginning the lengthy and involved procedure, inventors should obtain professional assistance from a patent practitioner—a patent attorney or a patent agent—who is registered with the PTO. Only attorneys and agents who are officially registered may represent an inventor seeking a patent. Approximately 98 percent of all inventors rely on these patent experts to steer them through the convoluted process.[15] Experienced patent attorneys say that the cost to obtain a patent ranges from $5,000 for a simple invention to $25,000 or more for a highly complex invention and possibly up to $100,000 for an international patent.

THE PATENT PROCESS Since George Washington signed the first patent law in 1790, the PTO has issued patents on everything imaginable (and some unimaginable items, too), including mousetraps, animals (genetically engineered mice), games, and various fishing devices. To date, the PTO has issued more than 8 million patents, and it receives more than 400,000 new applications each year. The first patent was issued to Samuel Hopkins on July 31, 1790, for an improved method for making potash, an ingredient in fertilizer and other products.[16] Patent number 8 million went to Second Sight Medical Products, Inc. for a visual prosthesis (which provides electrical stimulation to the retina of those who have gone blind to simulate visual perceptions of patterns of light). Figure 23.3 shows the trend in the number of patent applications and number of patents actually granted in recent years.

To receive a patent, an inventor must follow these steps:

Establish the invention's novelty. An invention is not patentable if it is known or has been used in the United States or has been described in a printed publication in the United States or a foreign country.

US007484328B1

(12) **United States Patent**

Daugherty

(10) **Patent No.:** US 7,484,328 B1

(45) **Date of Patent:** Feb. 3, 2009

(54) **FINGER MOUNTED INSECT DISSUASION DEVICE AND METHOD OF USE**

(76) Inventor: **John Richard Daugherty**, 1647 N. Woodhollow Way, Flagstaff, AZ (US) 86004

(*) Notice: Subject to any disclaimer, the term of this patent is extended or adjusted under 35 U.S.C. 154(b) by 122 days.

(21) Appl. No.: **11/080,023**

(22) Filed: **Mar. 15, 2005**

Related U.S. Application Data

(63) Continuation-in-part of application No. 10/839,590, filed on May 5, 2004, now abandoned.

(51) **Int. Cl.**
A01M 3/02 (2006.01)
A01M 3/00 (2006.01)

(52) **U.S. Cl.** **43/137**; 43/134

(58) **Field of Classification Search** 43/137, 43/134

See application file for complete search history.

(56) **References Cited**

U.S. PATENT DOCUMENTS

36,652	A	10/1862	Jacobs	
97,161	A *	11/1869	Buttles	43/137
160,606	A *	3/1875	Marsh	43/134
542,464	A *	7/1895	Chase	43/137
599,404	A *	2/1898	Robertson	43/134
609,160	A *	8/1898	McWithey	43/134
648,336	A *	4/1900	Bellamy	273/317.2
974,887	A	11/1910	Huddle	
1,099,342	A *	6/1914	Copenhaver	43/137
1,206,976	A	12/1916	Barth	
1,354,775	A *	10/1920	Moore	43/137
1,412,312	A *	4/1922	Little	43/137
1,479,046	A *	1/1924	Herbert	43/137
1,500,442	A *	7/1924	Cooper	43/137
1,639,559	A *	8/1927	Gatch	43/137
1,650,548	A *	11/1927	Sullivan	43/137
1,656,969	A *	1/1928	Adolph	43/137
1,660,011	A *	2/1928	Linding	43/137
1,662,264	A *	3/1928	Henderson	43/137
1,763,205	A *	6/1930	Winbigler	273/317.2
1,820,360	A *	8/1931	Meggitt	43/137
1,861,688	A *	6/1932	Crawford	43/137
1,942,252	A *	1/1934	Martin	43/137
1,967,384	A *	7/1934	Urbanek	43/137

(Continued)

FOREIGN PATENT DOCUMENTS

DE 29712704 U1 * 11/1997

(Continued)

Primary Examiner—Darren W Ark
(74) *Attorney, Agent, or Firm*—John R. Daugherty

(57) **ABSTRACT**

An insect dissuasion method that incorporates a miniature fly swatter adapted to be fixed onto an end of a human finger. An insect can be discouraged by simply flexing, slowly encroaching upon and then "flicking" the finger with the attached device to strike the insect. Devices of the present invention are designed to be removably attached to a finger by a ring-like structure. The ring-like structure is tailored to slip onto and engage a finger in various positions and remain attached to the finger when the finger is flicked. Joined to the ring-like structure is an extension shaft that terminates in an insect engagement head. The length and/or cross-sectional profile of the extension shaft can be altered as well as the shape of the head portion or ring-like structure.

3 Claims, 13 Drawing Sheets

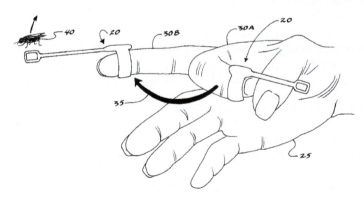

FIGURE 23.2

Design Patent # 7,484,328, Finger-Mounted Insect Dissuasion Device

Source: U.S. Patent and Trademark Office.

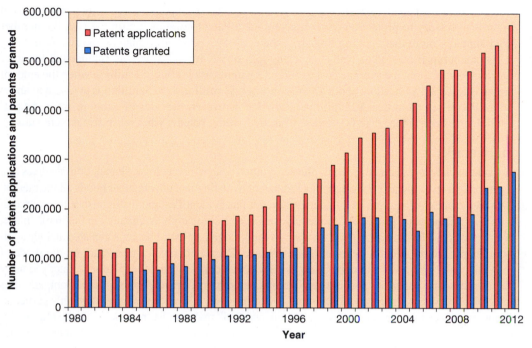

FIGURE 23.3

Patent Applications and Patents Granted, 1980–2012

Source: Based on U.S. Patent and Trademark Office, *www.uspto.gov.*

File a provisional patent. To protect their patent claims, inventors should be able to verify the date on which they first conceived the idea for their inventions. Inventors can document a device by keeping dated records (including drawings) of their progress on the invention and by having knowledgeable friends witness these records. Inventors also can file a provisional patent with the PTO—a process that includes submitting a written description of the invention and any supporting drawings along with the provisional patent cover sheet and filing fee to the PTO. A provisional patent gives protection for 12 months.

Search existing patents. To verify that the invention truly is new, nonobvious, and useful, inventors must conduct a search of existing patents on similar products. The purpose of the search is to determine whether the inventor has a chance of getting a patent. Most inventors hire professionals trained in conducting patent searches to perform the research. Inventors themselves can conduct an online search of all patents granted by the PTO since 1976 from the office's Web site. An online search of these patents does not include sketches; however, subscribers to Delphion's Research Intellectual Property Network can access patents, including sketches, as far back as 1971 at *www.delphion.com.*

Study search results. Once the patent search is finished, inventors must study the results of the search to determine their chances of getting a patent. To be patentable, a device must be sufficiently different from what has already been patented and must not be obvious to a person having ordinary skill in the area of technology related to the invention.

Submit the patent application. An inventor must file an application describing the invention with the PTO. This description, called the patent's claims, should be broad enough that others cannot easily engineer around the patent, rendering it useless. However, they cannot be so narrow as to infringe on patents that other inventors already hold. The typical patent application runs 20 to 40 pages, although some, especially those for biotech or high-tech products, are tens of thousands of pages long.

Prosecute the patent application. Before the PTO will issue a patent, one of its examiners studies the application to determine whether the invention warrants a patent. If the PTO rejects the application, the inventor can amend the application and resubmit it to the PTO. The average time for a patent to be issued is 29 months.

Defending a patent against "copycat producers" can be expensive and time consuming but often is necessary to protect an entrepreneur's idea. Patent lawsuits are on the rise, the number filed annually more than quadrupling since the early 1990s.[17] Unfortunately, the cost of defending a patent has increased as well. According to a survey by the American Intellectual Property Owners Association, the average cost to defend a patent case is about $1 million when the amount at risk is less than $1 million, about $3 million when $1 million to $25 million is at risk, and about $6 million when more than $25 million is at risk.[18] However, the odds of winning are in the patent holder's favor; more than 60 percent of those holding patents win their infringement suits.[19]

Trademarks

A **trademark** is any distinctive word, phrase, symbol, design, name, logo, slogan, or trade dress that a company uses to identify the origin of a product or to distinguish it from other goods on the market. (A **servicemark** is the same as a trademark except that it identifies and distinguishes the source of a service rather than a product.) A trademark serves as a company's "signature" in the marketplace. A trademark can be more than just a company's logo, slogan, or brand name; it can also include symbols, shapes, colors, smells, or sounds. For instance, Coca-Cola holds a trademark on the shape of its bottle, and NBC owns a trademark on its three-toned chime. Components of a product's identity such as these are part of its **trade dress**, the unique combination of elements that a company uses to create a product's image and to promote it. For instance, a Mexican restaurant chain's particular decor, color schemes, design, and overall "look and feel" make up its trade dress. To be eligible for trademark protection, trade dress must be inherently unique and distinctive to a company, and another company's use of that trade dress must be likely to confuse customers.

ENTREPRENEURIAL PROFILE: Thermal-Wise Insulation v. Questar Gas Co. Rob Linden owns a small residential and commercial insulation company in Columbia Heights, Minnesota, called Thermal-Wise Insulation, that provides insulation services to commercial and residential customers. Questar Gas Co., a Utah-based natural gas utility, is suing Linden's small business, claiming that the name "Thermal-Wise" infringes on the name of a rebate program it offers its customers called "ThermWise." "I didn't think what they are claiming is even possible," says Linden. "I'm an insulation company. They sell gas." Questar Gas Co. claims that Linden's continued use of the Thermal-Wise name will "damage [its] reputation and impair and dilute [its] goodwill." The financial toll on Linden's business is mounting. He has already spent $15,000 in legal fees fighting Questar, and his business has dropped by 70 percent, as people are hesitant to do business with a contractor in the middle of litigation. He has had to abandon his business expansion plans and has indefinitely delayed increasing his workforce. "I've invested a lot into this business as it is," says Linden. "And for me to start over with a new name would be devastating. A change of name is often a sign of a failed business." Linden offered to give up the Thermal-Wise name if Questar ever came into the Minnesota; Questar rejected that offer. "Thousands search the Internet for ThermWise to get rebates, and if they turned up something with a negative connotation that would reflect on our brand," says Chad Jones, a spokesman for Questar. "The world is all connected on the Internet now."[20]

There are more than1.8 million trademarks registered in the United States. Federal law permits a manufacturer to register a trademark, which prevents other companies from employing a similar mark to identify their goods. Businesses aggressively protect their trademarks. Each year, there are between 9,000 and 12,000 lawsuits filed regarding contested trademarks. Before 1989, a business could not reserve a trademark in advance of use. Today, the first party that either uses a trademark in commerce or files an application with the PTO has the ultimate right to register that trademark. The PTO takes an average of 10 months to process a trademark application.[21] Unlike patents and copyrights, which are issued for limited amounts of time, trademarks last indefinitely as long as the holder continues to use it. However, a trademark cannot keep competitors from producing the same product and selling it under a different name. It merely prevents others from using the same or confusingly similar trademark for the same or similar products.

Many business owners are confused by the use of the symbols ™ and ®. Anyone who claims the right to a particular trademark (or servicemark) can use the ™ (or SM) symbols without having to register the mark with the PTO. The claim to that trademark or servicemark may or may not be valid, however. Only those businesses that have registered their marks with the PTO can use the ® symbol. Entrepreneurs do not have to register trademarks or servicemarks to establish their rights to those marks; however, registering a mark with the PTO does give entrepreneurs greater power to protect their marks. Filing an application to register a trademark or servicemark is relatively easy, but it does require a search of existing names. Entrepreneurs can use the Trademark Electronic Search System (TESS) at the U.S. Patent and Trademark Office's Web site (*www .uspto.gov*) to determine whether a business or product name is already trademarked.

An entrepreneur may lose the exclusive right to a trademark if it loses its unique character and becomes a generic name or if the company abandons its trademark by failing to market the brand adequately. Aspirin, escalator, thermos, brassiere, super glue, yo-yo, and cellophane were once enforceable trademarks that have become common words in the English language. These generic terms can no longer be licensed as a company's trademark.

Entrepreneurship in Action

Protecting Your Brand

Wendi Levi and Kim Etheredge, cofounders of Mixed Chicks.
Source: Mixed Chicks LLC.

Wendi Levy and Kim Etheredge, both of whom are biracial, came up with the idea to develop a hair product for women of mixed race. Both women had struggled with their curly and hard-to-manage hair for most of their lives by using a combination of hair products. "We would have to use 10 different products instead of one," says Etheredge. In 2003, Levy and Etheredge decided to take their problem to a chemist in hopes that they could find out which ingredients from all of the various products were the ones that helped get their hair under control. With the chemist's help and lots of experimentation, they developed their product, which they branded as Mixed Chicks.

"The name Mixed Chicks was chosen because we both grew up in the seventies, and we were always forced to choose between our ethnicities rather than embracing both of them," says Etheredge. "We weren't allowed to check two boxes; we had to pick one on any application. We wanted our own box— one that is inclusive. Our products are meant for all curly-haired women, whether black, white, Asian, Latin, Mediterranean, or any glorious combination."

Mixed Chicks first sold its products online in 2004. Based on the success of its online sales, the company was soon able to place its products in salons and beauty-supply stores nationwide. Mixed Chicks received a significant boost in sales when Halle Berry endorsed its product.

In 2009, Etheredge and Levy were at a trade show. A representative from Sally Beauty Supply approached Etheredge and Levy about selling Mixed Chicks in its stores. Sally Beauty Supply is a $3.5 billion publicly traded company with more than 3,000 stores located in cities throughout North America and Europe. Although the entrepreneurs initially were excited about the prospects of selling Mixed Chicks products to a large national chain, they decided not to pursue the opportunity when they learned about Sally Beauty Supply's strict return and liberal discount policies. Neither would be good for their small company. "We wanted to make sure we had control of our merchandise and inventory," says Etheredge.

By 2011, Mixed Chicks was doing quite well, even without a contract with Sally Beauty Supply. The company's sales had grown to $5 million, and it had introduced a variety of new products, including some for infants and toddlers. Soon, however, Levy and Etheredge began to hear from clients and customers that Sally Beauty Supply had rolled out its own product line for multiracial women called Mixed Silk. The bottles were the same shape, the packaging used the same colors and fonts, and the ads used similar language to those developed for Mixed Chicks. In fact, the Sally Beauty Supply's products looked so much like Mixed Chicks that some customers assumed that Mixed Chicks introduced a new low-cost product to segment the market. Mixed Chicks sold for $14 to $20 a bottle, while Mixed Silk sold for only $8. Etheredge decided to do her own competitive research and visited a Sally Beauty store. When she asked about the Mixed Silk product being offered by Sally Beauty, a store clerk told her that Mixed Silk was a "generic" version of Mixed Chicks.

Etheredge and Levy knew that they had an uphill battle but decided to fight to protect the brand that they had worked so hard to develop. Fighting a large company like Sally Beauty Supply is the type of battle that often results in failure for many small businesses. However, Etherege and Levy decided to forge ahead, and Mixed Chicks filed a lawsuit charging Sally Beauty Supply with trademark infringement, trade dress infringement, and unfair competition. In its lawsuit, Mixed Chicks alleged that Sally Beauty Supply marketed, sold, and advertised imitations of its products. "They messed with the wrong broads," says Etheredge.

After two years, the case finally went to trial. In 2012, after deliberating for only six hours, the jury ruled in favor of Mixed Chicks, awarding them $839,535 in actual damages. In addition,

(continued)

Entrepreneurship in Action *(continued)*

the jury awarded Mixed Chicks $7.27 million in punitive damages in the trademark and trade dress dispute.

"Others will enter the market, but we're not afraid of competition—as long as it's fair competition," says Etheredge. "We knew we were in the right, and we're happy we can now move forward."

1. Why do you think that the jury found in favor of Mixed Chicks? Explain.

2. What are steps that a small business can take to protect its brand? Explain.

3. Even though Mixed Chicks won the lawsuit, what were the risks that they took when they went after a large company for trademark and trade dress infringements? Explain.

Sources: Based on "Wendi Levy Kaaya and Kim Etheredge Co-Owners: Mixed Chicks," *Our Ventura Blvd*, September/October 2012, *www.ourventurablvd.com/ September-October-2012/Wendi-Levy-Kaaya-and-Kim-Etheredge-Co-owners*; Jennifer Alseaver, "Case Study: To Sue or Not to Sue," *Inc.*, January 24, 2012, *www.inc.com/magazine/201202/case-study-the-rival-mixed-chicks-sally-beauty .html*; Lisa Shuchman, "Mixed Chicks Gets $8.5M Jury Award for Infringing Mixed-Race Hair Products," *Corporate Counsel*, December 5, 2012, *www.law .com/corporatecounsel/PubArticleCC.jsp?id=1202580228393&Mixed_Chicks_ Gets_85M_Jury_Award_for_Infringing_MixedRace_Hair_Products&slreturn= 20130405104009*.

Copyrights

A **copyright** is an exclusive right that protects the creators of original works of authorship, such as literary, dramatic, musical, and artistic works (e.g., art, sculptures, literature, software, music, videos, video games, choreography, motion pictures, recordings, and others). The internationally recognized symbol © denotes a copyrighted work. A copyright protects only the form in which an idea is expressed, not the idea itself. A copyright on a creative work comes into existence the moment its creator puts that work into a tangible form. Just as with a trademark, obtaining basic copyright protection does not require registering the creative work with the U.S. Copyright Office; doing so, however, gives creators greater protection over their work. When author J. K. Rowling wrote the manuscripts for the immensely popular *Harry Potter* series, she automatically had a copyright on her creation. To secure her works against infringement, however, Rowling registered the copyright with the U.S. Copyright Office. Copyright applications must be filed with the Copyright Office in the Library of Congress for a fee of $35 per application. A valid copyright on a work lasts for the life of the creator plus 70 years after his or her death. (A copyright lasts 75 to 100 years if the copyright holder is a business.) When a copyright expires, the work becomes public property and can be used by anyone free of charge.

Because they are easy to duplicate, copyright infringers most often pirate computer software, CDs, and DVDs. Copyright piracy costs tens of billions of dollars each year in lost sales. The global software industry alone loses $63.4 billion annually to counterfeiters; experts estimate that as much as 41 percent of the software installed on PCs is pirated.[22] The U.S. music industry loses $12.5 billion annually to thieves, and the U.S. motion picture industry loses $20.5 billion to piracy.[23]

Protecting Intellectual Property

Acquiring the protection of patents, trademarks, and copyrights is useless unless an entrepreneur takes action to protect those rights in the marketplace. Unfortunately, some businesspeople do not respect others' rights of ownership to products, processes, names, and works and infringe on those rights with impunity. In other cases, the infringing behavior simply is the result of a lack of knowledge about other's rights of ownership. After acquiring the proper legal protection through patents, copyrights, or trademarks, entrepreneurs must monitor the market (and the Internet in particular) for unauthorized copycat users. If an entrepreneur has a valid patent, trademark, or copyright, stopping an infringer usually requires nothing more than a stern "cease-and-desist" letter from an attorney. Offenders usually want to avoid expensive legal battles and agree to stop their illegal behavior. If that tactic fails, the entrepreneur may have no choice but to bring an infringement lawsuit, many of which end up being settled out of court.

The primary weapon an entrepreneur has to protect patents, trademarks, and copyrights is the legal system. The major problem with relying on the legal system to enforce ownership rights is the cost of infringement lawsuits, which can quickly exceed the budget of most small business. Legal battles usually are expensive. Before bringing a lawsuit, an entrepreneur must consider the following issues:

- Can the opponent afford to pay if you win?

- Do you expect to get enough from the suit to cover the costs of hiring an attorney and preparing a case?

- Can you afford the loss of time, money, and privacy from the ensuing lawsuit?

The Law of Agency

An **agent** is one who stands in the place of and represents another in business dealings. Although an agent has the power to act for the principal, he or she remains subject to the principal's control. Many small business managers do not realize that their employees are agents while performing job-related tasks. Employers are liable only for those acts that employees perform within the scope of employment. For example, if an employee loses control of a flower shop's delivery van while making a delivery and crashes into several parked cars, the owner of the flower shop (the principal) and the employee (the agent) are liable for any damages caused by the crash. Even if the accident occurred while the employee was on a small detour of his own (e.g., to stop by his house), the owner is still liable for damages as long as the employee is working "within the scope of his employment." Normally, an employee is considered to be within the scope of his employment if he is motivated in part by the principal's action and if the place and time for performing the act is not significantly different from what is authorized.

4.

Explain the basics of the law of agency.

Any person, even those lacking contractual capacity, can serve as an agent, but a principal must have the legal capacity to create contracts. Both the principal and the agent are bound by the requirements of a fiduciary relationship, one characterized by trust and good faith. In addition, each party has specific duties to the other. An agent's duties include the following:

- *Loyalty.* Every agent must be faithful to the principal in all business dealings.
- *Performance.* An agent must perform his or her duties according to the principal's instructions.
- *Notification.* The agent must notify the principal of all facts and information concerning the subject matter of the agency.
- *Duty of care.* An agent must act with reasonable care when performing duties for the principal.
- *Accounting.* An agent is responsible for accounting for all profits and property received or distributed on the principal's behalf.

A principal's duties include the following:

- *Compensation.* Unless a free agency is created, the principal must pay the agent for his or her services.
- *Reimbursement.* The principal must reimburse the agent for all payments made for the principal or any expenses incurred in the administration of the agency.
- *Cooperation.* Every principal has the duty to indemnify the agent for any authorized payments or any loss or damages incurred as a result of the agency, unless the liability is the result of the agent's mistake.
- *Safe working conditions.* The law requires a principal to provide a safe working environment for all agents. Workers' compensation laws cover an employer's liability for injuries that agents receive on the job.

As agents, employees can bind a company to agreements, even if the owner did not intend for them to do so. An employee can create a binding obligation, for instance, if the business owner presents him or her as authorized to perform such transactions. For example, the owner of a flower shop who routinely permits a clerk to place orders with supplier has given that employee *apparent authority* for purchasing. Similarly, employees have *implied authority* to create agreements when performing the normal duties of their jobs. For example, the chief financial officer of a company has the authority to create binding agreements when dealing with the company's bank.

One issue related to agency that many confronts many businesses is whether their workers are employees who are directly under their control or independent contractors who are hired temporarily by contract to perform a job. Because employers do not have to incur payroll taxes or provide health care or other benefits to independent contractors, paying an independent contractor is less expensive than hiring an employee to do the same job. In addition, an employer is liable for negligent acts by an employee but is not liable for the negligent acts of an independent contractor. However, over the past decades, the IRS has significantly narrowed the definition of who can actually be considered a true independent contractor. The determination is now made by three main tests:

1. *Behavioral control.* Does the company control or have the right to control what the worker does and how the worker does his or her job?

2. *Financial control.* Are the business aspects of the worker's job controlled by the company (these include things such as how worker is paid, whether expenses are reimbursed, who provides tools/supplies, and so on)?

3. *Type of relationship between the parties.* Are there written contracts or employee type benefits (i.e., pension plan, insurance, vacation pay, and so on)? Will the relationship continue, and is the work performed a key aspect of the business?

Some businesses have experienced disputes with the IRS over the status of workers they claim are independent contractors and the IRS considers employees. In general, the more control that an employer exercises over a worker, the more likely it is that he or she is an employee. If, however, the employer controls only the final result of the work, the worker is most likely an independent contractor. Given that the cost in back taxes and penalties can be quite severe if an employer designates someone incorrectly as an independent contractor, it is always advisable to consult with a CPA or attorney who has expertise in employment law on such matters. The IRS provides guidelines for determining the difference between employees and independent contractors at its Web site at *www.irs.gov.*

Bankruptcy

5.
_____

Explain the basics of bankruptcy law.

Bankruptcy occurs when a business is unable to pay its debts as they come due. Although filing for bankruptcy traditionally has had a social stigma attached to it, today it has become an accepted business strategy for troubled companies (see Figure 23.4). Companies such as American Airlines, Houghton Mifflin Harcourt (book and textbook publishing company), Reader's Digest (magazine company), Atari (video game company), Eastman Kodak (film and camera manufacturer), Hawker Beechcraft (manufacturer of business and other aircraft), Borders (book and music stores), and many others have filed for bankruptcy recently. Some of these companies have vanished, but others have emerged from bankruptcy and continue to operate. The most recent peak in the number of business bankruptcies occurred in 2009 and 2010, due to the fallout from the recession and financial crisis that began in 2008.

FIGURE 23.4

Number of Business Bankruptcies

Source: Based on *www.uscourts.gov/ Statistics/JudicialBusiness/2012/us- bankruptcy-courts.aspx.*

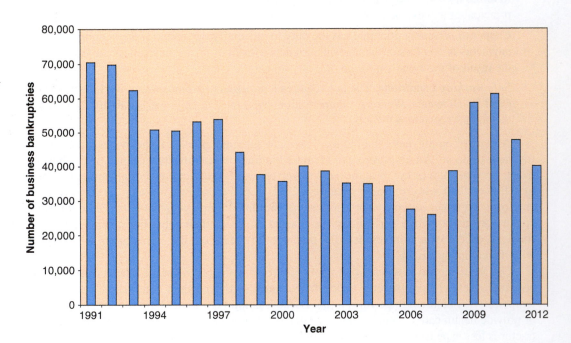

Forms of Bankruptcy

Many people who file for bankruptcy are small business owners seeking protection from creditors under one of the eight chapters created by the Bankruptcy Reform Act of 1978, which was amended in 2005. The Bankruptcy Reform Act of 2005 requires debtors to pay as many of their debts as possible rather than having them discharged by bankruptcy. Under the act, three chapters (7, 11, and 13) govern the majority of bankruptcies related to small businesses. Usually, small business owners in danger of failing choose from two types of bankruptcies: **liquidation** (Chapter 7, in which an owner files for bankruptcy and the business ceases to exist) and **reorganization** (Chapter 11, in which after filing for bankruptcy, the owner formulates a reorganization plan under which the business continues to operate).

CHAPTER 7: LIQUIDATIONS The most common type of bankruptcy is filed under Chapter 7 (called straight bankruptcy), which accounts for more than 70 percent of all filings. A company that completes Chapter 7 bankruptcy is liquidated and ceases to exist. The business simply declares all of its debts and turns over all of its assets to a trustee who is elected by the creditors or appointed by a court. The trustee sells the assets and distributes all proceeds first to secured creditors and then to unsecured creditors (which include stockholders). Depending on the outcome of the asset sale, creditors can receive anywhere between 0 and 100 percent of their claims against the bankrupt company. Once the bankruptcy proceeding is complete, any remaining debts are discharged, and the company disappears.

> **ENTREPRENEURIAL PROFILE: Maclaren USA (American Baby Products)** Maclaren was long considered an elite brand of baby strollers. Parents in some of the most affluent neighborhoods could be seen walking their babies in Maclaren strollers that cost hundreds of dollars. The U.K.-based company sold its strollers in the United States through its distributor, Maclaren USA. Maclaren pioneered the lightweight, collapsible baby stroller (also called umbrella strollers). Then in 2009 Maclaren announced a recall for 1 million of its strollers. There were reports that more than 20 children had fingers amputated by the hinges on Maclaren strollers, and more than 100 more had serious injuries to their fingers. In 2011, Maclaren USA changed its name to American Baby Products and then filed for Chapter 7 bankruptcy. Parents who had filed claims against Maclaren USA were uncertain whether they would ever receive any compensation related to their children's injuries. However, in 2012, the U.K.-based Maclaren announced that it would pay all product liability claims that were filed against Maclaren USA regardless of the outcomes of its bankruptcy case.[24]

A business begins the straight bankruptcy proceedings by filing either a voluntary or an involuntary petition. A voluntary case starts when the debtor files a petition with a bankruptcy court, stating the names and addresses of all creditors, the debtor's financial position, and all property the debtor owns. On the other hand, creditors start an involuntary petition by filing with the bankruptcy court. If there are 12 or more creditors, at least three of them whose unsecured claims total $14,425 or more must file the involuntary petition. If a debtor has fewer than 12 creditors, only one of them having a claim of $14,425 or more is required to file. As soon as a petition (voluntary or involuntary) is filed in a bankruptcy court, all creditors' claims against the debtor are suspended. Called an **automatic stay**, this provision prevents creditors from collecting any of the debts the debtor owed them before filing the petition. In other words, no creditor can begin or continue to pursue debt collection once the petition is filed.

Not every asset an individual bankrupt debtor owns is subject to court attachment; certain assets are exempt, although each state establishes its own exemptions. Most states make an allowance for equity in a home, interest in an automobile, interest in a large number of personal items, and other personal assets. Federal law allows a $22,975 exemption for a home, an $12,250 exemption for household items and clothing, a $3,675 exemption for equity in a car, and several other exemptions.

The law does not allow a debtor to transfer the ownership of property to others to avoid its seizure in a bankruptcy. If a debtor transfers property within one year of the filing of a bankruptcy petition, the trustee can ignore the transfer and claim the assets. In addition, a court will overturn any transfer of property made for the express purpose of avoiding repayment of debts (called **fraudulent conveyance**). The new law also enables a judge to dismiss a Chapter 7 bankruptcy petition if it is a "substantial abuse" of the bankruptcy code.

CHAPTER 11: REORGANIZATION For a small business weakened by a faltering economy, excessive debt load, or management mistakes, Chapter 11 provides a second chance for success. The philosophy behind this form of bankruptcy is that ailing companies can prosper again if given a fresh start with less debt. A Chapter 11 bankruptcy filing protects a company's assets from creditors' legal actions while it formulates a plan for reorganization and repaying or settling its debts or for selling the business. In most cases, a small business and its creditors negotiate a settlement in which the company repays a percentage of its debts with the remainder of them dismissed. The business continues to operate under the court's direction, but creditors cannot foreclose on it, nor can they collect any prebankruptcy debts the company owes.

The average duration of bankruptcies is declining as companies realize the benefits of exiting bankruptcy as quickly as possible. Unlike a typical bankruptcy, which may take two or more years to complete, Section 363 bankruptcy allows a bankrupt company to emerge from bankruptcy in as little as 30 to 60 days. Because of Section 363, automaker Chrysler completed its Chapter 11 bankruptcy in just 42 days. Another exemption allows a fast-track version of Chapter 11 bankruptcy for small businesses with liabilities that do not exceed $2 million that streamlines the process and is less expensive.

A Chapter 11 bankruptcy filing can be either voluntary or involuntary. Once the petition is filed, an automatic stay goes into effect, and the debtor has 120 days to file a reorganization plan with the court. Usually, the court does not replace management with an appointed trustee; instead, the bankrupt party, called the debtor in possession, serves as trustee. If the debtor fails to file a plan within the 120-day limit, any party involved in the bankruptcy, including creditors, may propose a plan. The plan must identify the various classes of creditors and their claims, outline how each class will be treated, and establish a method to implement the plan. It also must spell out the debts that the company cannot pay, those that it can pay, and the methods the debtor will use to pay them.

Once the plan is filed, the court must decide whether to approve it. A court will approve a plan if a majority of each of the three classes of creditors—secured, priority, and unsecured— votes in favor of it. The court will confirm a plan if it has a reasonable chance of success, is submitted in good faith, and is "in the best interest of the creditors." If the court rejects the plan, the creditors must submit a new one for court approval.

Filing under Chapter 11 offers a weakened small business a number of advantages, the greatest of which is a chance to survive (although most of the companies that file under Chapter 11 ultimately are liquidated). In addition, employees keep their jobs, and customers get an uninterrupted supply of goods and services. However, there are costs involved in bankruptcy proceedings. Customers, suppliers, creditors, and often employees lose confidence in a company's ability to succeed. Creditors frequently incur substantial losses in Chapter 11 bankruptcies, receiving payments of just pennies for every dollar they are owed.

ENTREPRENEURIAL PROFILE: Frank McCourt: The Los Angeles Dodgers Frank McCourt, a successful commercial real estate developer, purchased the Los Angeles Dodgers from News Corp. in 2004 for $430 million. However, after years of financial entanglement with his personal finances, overspending on payroll that he could not support with cash flow, and finally a bitter divorce, McCourt was left with no choice but to file for Chapter 11 bankruptcy for his company that owned the Dodgers. The team had amassed more than $500 million in debt and was no longer able to fund operations from its cash flow. In a document filed with the bankruptcy court, the team said that the Chapter 11 filing would allow it to meet payroll, sign players, pay vendors, and continue playing baseball. In 2012, McCourt agreed to sell the Dodgers to a group of investors that included former Los Angeles Laker Magic Johnson for a record price of $2 billion, the highest ever paid for a professional sports team.[25]

CHAPTER 13: INDIVIDUAL'S REPAYMENT PLANS Chapter 13 bankruptcy is the consumer version of Chapter 11 proceedings. Individual debtors (not businesses) with a regular income who owe unsecured debts of less than $360,475 or secured debts of less than $1,081,400 may file for bankruptcy under Chapter 13. Many debtors who have the choice of filing under Chapter 11 or 13 find that Chapter 13 is less complicated and less expensive. Chapter 13 proceedings must begin voluntarily. Once the debtor files a petition, creditors cannot start or continue legal action to collect payment. Under Chapter 13, only the debtor can file a repayment plan, whose terms cannot exceed five years. If the court approves the plan, the debtor may pay off the obligations—either in full or partially—on an installment basis. The plan is designed with the debtor's future income in mind, and when the debtor completes the payments under the plan, all debts are discharged.

In the Entrepreneurial Spotlight

A Second Chance at Success

Curt Jones, founder of Dippin Dots.
Source: Dippin' Dots.

Curt Jones grew up in a small town where churning ice cream was a summer tradition. After college, Jones went to work in a lab that was developing a new kind of animal feed through a process of flash freezing using liquid nitrogen. One day while making some homemade ice cream as he had when he was young, Jones decided to try speed up the process by flash freezing it using liquid nitrogen. Not only did the process result in a unique small pellets of ice cream, but also the flash freezing process created a product with a very smooth texture.

Within six months, Jones quit his job to begin selling his "pelletized" ice cream in a small store in Lexington, Kentucky, that became the first Dippin' Dots. Because the product must be stored at −40 degrees Fahrenheit, it could not be distributed through grocery stores and other retail outlets that did not have freezers that kept product at such a low temperature, so Jones decided to grow the business through retail outlets and by selling it through kiosks in stadiums and theme parks. Dippin' Dots touted its unique frozen treat as the "ice cream of the future." Jones established the company's headquarters in Paducah, Kentucky.

Although the company experienced significant growth, the recession that began in 2008 took its toll on sales. Revenues dropped from $33.9 million in 2009 to $26.7 million in 2010. In 2011, Regions Bank foreclosed on the $11.1 million in loans it had outstanding with Dippin' Dots. The company had no choice but to declare Chapter 11 bankruptcy to keep its cash available to support its operations and attempt to reorganize its debt. During the bankruptcy process, Jones stepped down as CEO of the company he founded.

Then in 2012, an investment group out of Oklahoma City bought Dippin' Dots LLC out of bankruptcy for $12.7 million. Scott Fischer and his father Mark Fischer, who had made their fortune in the energy business, led the investment group. Fisher recalled that he enjoyed Dippin' Dots as a child visiting Six Flags and Sea World and looked forward to reviving the company with so many memories for him.

The new owners reinstated company founder Curt Jones as CEO and Scott Fischer became president of the company. The new owners also committed to keeping the company headquarters in Paducah and invested in a new 125,000-square-foot factory built in Kentucky.

After emerging from bankruptcy, Dippin' Dots began to implement several new strategies aimed at growing the brand. Dippin' Dots introduced a new line of nonfat yogurt dots called YoDots which have only 70 calories. The company plans not only to sell the product through its traditional outlets but also to begin to market it to schools. The product meets the health standards that the U.S. Department of Agriculture has established for schools, so Dippin' Dots plans to market the new product to school districts across the country.

Dippin' Dots also began to experiment with a treat for pets called Doggie Dots. The product development team is using banana and peanut butter flavors (without the sugar) as dog treats. Although they taste bland to humans, dogs seem to love them.

When the Fischers bought Dippin' Dots, they identified international growth as a key strategy for the company. The company has signed licensing agreements with sellers in Russia and Greece. They also have a new license agreement with a company from Japan that will purchase products from the existing Dippin' Dots facility in South Korea.

1. What advantages does Chapter 11 bankruptcy offer troubled companies? What disadvantages do companies that file for Chapter 11 bankruptcy incur?

2. Use the Internet to research other companies that have declared Chapter 11 bankruptcy. Select one company that is of interest to you and prepare a one-page report that describes the sequence of events that led to the bankruptcy, the company's plan for reorganizing, and the outcome for the company.

Sources: Based on Derek Thompson, "Dippin' Dots: What's the Future of the 'Ice Cream of the Future'?," *The Atlantic,* August 24, 2011, *www.theatlantic.com/ business/archive/2011/08/dippin-dots-whats-the-future-of-the-ice-cream-of-the-future/244040*; "Dippin' Dots Maker Declares Bankruptcy; 'Ice Cream of the Future' Files for Chapter 11 Reorganization," *New York Daily News,* November 7, 2011, *www.nydailynews.com/life-style/eats/dippin-dots-maker-declares-bankruptcy-ice-cream-future-files-chapter-11-reorganization-article-1.973683*; Kim Peterson, "Chilly Business Pushes Dippin' Dots into Bankruptcy," *MSN Money,* November 4, 2011, *http://money.msn.com/top-stocks/post.aspx?post=357090ea-6c66-4259-945d-3c1250f68258*; Don Mecoy, "Dippin' Dots Deal Is Done," *Oklahoman,* May 18, 2012, *http://newsok.com/dippin-dots-deal-done/article/3676556#ixzz1vFQI2tWI*; G. Chambers Williams III, "Dippin' Dots Hopes for a Comeback After Bankruptcy Filing," *USA Today,* April 29, 2012, *http://usatoday30.usatoday.com/money/companies/story/2012-04-29/dippin-dots-bankruptcy-comeback/54613920/1*; Katy Stech, "After Bankruptcy, Dippin' Dots Aims for a Healthier Future," *Wall Street Journal,* March 26, 2013, *http://blogs.wsj.com/bankruptcy/2013/03/26/after-bankruptcy-dippin-dots-aims-for-healthier-future.*

Government Regulation

6.

Explain some of the government regulations affecting small businesses, including those governing trade practices, consumer protection, consumer credit, and the environment.

Although most entrepreneurs recognize the need for some government regulation of business, most believe that the process is overwhelming and out of control. Government regulation of business is far from new; in fact, Congress created the first regulatory agency, the Interstate Commerce Commission in 1887. The Great Depression of the 1930s triggered a great deal of regulation of business. From the 1930s on, laws regulating business practices and the creation of government agencies to enforce the regulations have expanded continuously. Not to be outdone by the federal regulators, most states have created their own regulatory agencies to create and enforce a separate set of rules and regulations. Small business owners often feel overwhelmed by the paperwork required to respond to all the governmental agencies trying to regulate and protect them. For instance, an entrepreneur who wants to start an auto repair shop must contend with 38 sets of regulations from 18 federal, state, and local agencies.[26]

The major complaint that small business owners have concerning government regulation concerns the cost of compliance. The Small Business Administration's Office of Advocacy estimates that complying with government regulation cost businesses $1.75 trillion per year.[27] Because many of the costs of complying with regulations are fixed, the impact of the regulatory burden is greater on small businesses than on big businesses. Large companies can spread the cost of compliance over a larger number of employees and consequently have a lower regulatory per employee cost. The Small Business Administration study shows that the cost of compliance per employee for small companies with 1 to 20 workers is $10,585, which is 36 percent higher than the $7,755 cost per employee at companies with more than 500 workers.[28] Figure 23.5 shows the cost of complying with federal regulations by company size.

In a competitive market, small companies cannot simply pass these additional costs forward to their customers, and consequently they experience a squeeze on their profit margins. The

FIGURE 23.5

Federal Regulatory Compliance Cost by Company Size

Source: Based on Nicole V. Crain and Mark Crain, "The Impact of Regulatory Costs on Small Firms," Small Business Administration Office of Advocacy, September 2010, p. 7.

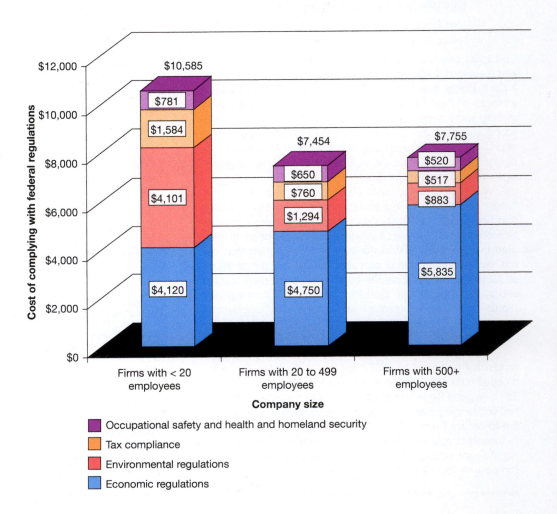

Small Business Regulatory Enforcement and Fairness Act (SBREFA) offers business owners some hope. SBREFA amended the Regulatory Flexibility Act of 1980, which Congress passed in response to small business owner's frustration at an ever-increasing burden of federal regulation. SBREFA's purpose is to require government agencies to consider the impact of their regulations on small companies and gives business owners more input into the regulatory process.

Most business owners agree that some government regulation is necessary. There must be laws governing working safety, environmental protection, package labeling, consumer credit, and other relevant issues because some dishonest, unscrupulous managers abuse the opportunity to serve the public's interest. It is not the regulations that protect workers and consumers and achieve social objectives to which business owners object but rather those that produce only marginal benefits relative to their costs. Owners of small companies, especially, seek relief from wasteful and meaningless government regulations, charging that the cost of compliance exceeds the benefits gained.

ENTREPRENEURIAL PROFILE: Richard and Eileen Bergmann v. City of Lake Elmo
Richard and Eileen Bergmann had been operating their family farm in Lake Elmo, Minnesota, for almost 40 years. The Bergmann's have a stand where they sell pumpkins to local residents during the Halloween season. To keep their stand stocked, the Bergmanns must add to their inventory with produce grown outside the city limits of Lake Elmo. One of their sources is a pumpkin farm they operate just a few miles away in Wisconsin. However, Lake Elmo had established a law that banned the Bergmanns and other farmers in the city from bringing in and selling farm goods grown outside the city limits. Under this local ordinance, violators could face 90 days in jail and a $1,000 fine, so the Bergmanns and other local farmers filed a lawsuit in U.S. District Court for the District of Minnesota challenging Lake Elmo's trade ban as a violation of fundamental constitutional rights. In response to the lawsuit and a preliminary victory before a magistrate judge, the city changed its laws to allow sales of outside products from farms if the farm obtained a permit. The Bergmanns then obtained that permit, allowing them to freely trade across local and state lines and continue to operate their pumpkin stand on their farm.[29]

Richard Bergmann.
Source: Institute for Justice.

Trade Practices

SHERMAN ANTITRUST ACT Contemporary society places great value on free competition in the marketplace, and antitrust laws reflect this. The notion of laissez-faire—that the government should not interfere with the operation of the economy—that once dominated U.S. markets no longer prevails. One of the earliest trade laws was the Sherman Antitrust Act, which was passed in 1890 to promote competition in the U.S. economy. This act is the foundation on which antitrust policy in the United States is built and was aimed at breaking up the most powerful monopolies of the late nineteenth century. The Sherman Antitrust Act contains two primary provisions affecting growth and trade among businesses.

Section I forbids "every contract, combination in the form of trust or otherwise, or conspiracy, in restraint of trade or commerce among the several states, or with foreign nations." This section outlaws any agreement among sellers that might create an unreasonable restraint on free trade in the marketplace. For example, a group of small and medium-size regional supermarkets formed a cooperative association to purchase products to resell under private labels only in restricted geographic regions. The U.S. Supreme Court ruled that their action was an attempt to restrict competition by allocating territories and had "no purpose except stifling of competition."[30]

Section II of the Sherman Antitrust Act makes it illegal for any person to "monopolize or attempt to monopolize any part of the trade or commerce among the several states, or with foreign nations." The primary focus of Section II is on preventing the undesirable effects of monopoly power in the marketplace.

CLAYTON ACT Congress passed the Clayton Act in 1914 to strengthen federal antitrust laws by spelling out specific monopolistic activities. The major provisions of the Clayton Act forbid the following activities:

1. *Price discrimination.* A firm cannot charge different customers different prices for the same product, unless the price discrimination is based on an actual cost savings, is made to meet a lower price from competitors, or is justified by a difference in grade, quality, or quantity sold.

2. *Exclusive dealing and tying contracts.* A seller cannot require a buyer to purchases only her product to the exclusion of other competitive sellers' products (an exclusive dealing agreement). In addition, the act forbids sellers to sell a product on the condition that the buyer agrees to purchase another product the seller offers (a tying agreement). For example, a computer manufacturer could not sell a computer to a business and, as a condition of the sale, require the firm to purchase software as well.

3. *Purchasing stock in competing corporations.* A business cannot purchase the stock or assets of another business when the effect may be to substantially lessen competition. This does not mean that a corporation cannot hold stock in a competing company; the rule is designed to prevent horizontal mergers that would reduce competition. The Federal Trade Commission (FTC) and the Antitrust Division of the Justice Department enforce this section, evaluating the market shares of the companies involved and the potential effects of a horizontal merger before ruling on its legality.

4. *Interlocking directorates.* The act forbids interlocking directorates—a person serving on the board of directors of two or more competing companies.

FEDERAL TRADE COMMISSION ACT To supplement the Clayton Act, Congress passed the Federal Trade Commission Act in 1914, which created its namesake agency and gave it a broad range of powers. Section 5 gives the FTC the power to prevent "unfair methods of competition in commerce and unfair or deceptive acts or practices in commerce." To be considered deceptive, a company's activity must involve a material misrepresentation that is likely to mislead a consumer who is acting in a reasonable manner.

Recent amendments have expanded the FTC's powers. The FTC's primary targets are those businesses that engage in unfair trade practices, often brought to the surface by consumer complaints. In addition, the agency has issued a number of trade regulation rules defining acceptable and unacceptable trade practices in various industries. Its major weapon is a "cease-and-desist order," commanding the violator to stop its unfair trade practices.

The FTC Act and the Lanham Trademark Act of 1988 (plus state laws) govern the illegal practice of deceptive advertising. In general, the FTC can review any advertisement that might mislead people into buying a product or service they would not buy if they knew the truth. For instance, if a small business advertised a "huge year-end inventory reduction sale" but kept its prices the same as its regular prices, it is violating the law.

ROBINSON-PATMAN ACT Although the Clayton Act addressed price discrimination and the FTC forbade the practice, Congress found the need to strengthen the law because many businesses circumvented the original rules. In 1936, Congress passed the Robinson-Patman Act, which further restricted price discrimination in the marketplace. The act forbids any seller "to discriminate in price between different purchases of commodities of like grade and quality" unless there are differences in the cost of manufacture, sale, or delivery of the goods. Even if a price-discriminating firm escaped guilt under the Clayton Act, it violated the Robinson-Patman Act. Traditionally, the FTC has had the primary responsibility of enforcing the Robinson-Patman Act.

OTHER LEGISLATION The Celler-Kefauver Act of 1950 gave the FTC the power to review certain proposals for mergers so that it could prevent too much concentration of power in any particular industry. Congress created the Miller-Tydings Act in 1937 to introduce an exception to the Sherman Antitrust Act. This act made it legal for manufacturers to use fair trade agreements that prohibit sellers of the manufacturer's product from selling it below a predetermined fair trade price. This form of price fixing was outlawed when Congress repealed the Miller-Tydings Act in 1976. Manufacturers can no longer mandate minimum or maximum prices on their products to sellers.

Consumer Protection

Since the early 1960s, legislators have created many laws aimed at protecting consumers from unscrupulous sellers, unreasonable credit terms, and mislabeled or unsafe products. Early laws focused on ensuring that food and drugs sold in the marketplace were safe and of proper quality. The first law, the Pure Food and Drug Act, passed in 1906, regulated the labeling of various food and drug products. Later amendments empowered government agencies to establish safe levels of food additives and to outlaw carcinogenic (cancer-causing) additives. In 1938, Congress passed the Food, Drug, and Cosmetics Act, which created the Food and Drug Administration, the agency responsible for establishing standards of safe over-the-counter drugs; inspecting food and drug manufacturing operations; performing research on food, additives, and drugs; regulating drug labeling; and other, related tasks.

Congress has also created a number of laws to establish standards pertaining to product labeling for consumer protection. Since 1976, manufacturers have been required to print accurate information about the quantity and content of their products in a conspicuous place on the package. Generally, labels must identify the raw materials used in the product, the manufacturer, the distributor (and its place of business), the net quantity of the contents, and the quantity of each serving if the package states the number of servings. The law also requires labels to be truthful. For example, a candy bar labeled "new, bigger size" must actually be bigger. These requirements, created by the Fair Packaging and Labeling Act of 1976, were designed to improve customers' ability to comparison shop. A 1970 amendment to the Fair Packaging and Labeling Act, the Poison Prevention Packaging Act, required manufacturers to install childproof caps on all products that are toxic.

With the passage of the Consumer Products Safety Act in 1972, Congress created the Consumer Product Safety Commission (CPSC) to control potentially dangerous products sold to consumers, and it has broad powers over manufacturers and sellers of consumer products. For instance, the CPSC can set safety requirements for consumer products, and it has the power to ban the production of any product it considers hazardous to consumers. It can also order vendors to remove unsafe products from their shelves. In addition to enforcing the Consumer Product Safety Act, the CPSC is charged with enforcing the Refrigerator Safety Act, the Federal Hazardous Substance Act, the Child Protection and Toy Safety Act, the Poison Prevention Package Act, and the Flammable Fabrics Act.

The Magnuson-Moss Warranty Act, passed in 1975, regulates written warranties that companies offer on the consumer goods they sell. The act does not require companies to offer warranties; it regulates only the warranties that companies choose to offer. It also requires businesses to state warranties in easy-to-understand language and defines the conditions warranties must meet before they can be designated as "full warranties."

The Telemarketing and Consumer Fraud and Abuse Protection Act of 1994 put in place the following restrictions on telemarketers:

- Calling a person's residence at any time other than 8:00 A.M. to 8:00 P.M.
- Claiming an affiliation with a government agency where such an affiliation does not exist
- Claiming an ability to improve a customer's credit record or obtain a loan for a person regardless of that person's credit history
- Not telling the receiver of the call that it is a sales call
- Claiming an ability to recover goods or money lost by a consumer

Lessons from the Street-Smart Entrepreneur

Are Your Ads Setting You Up for Trouble?

Lynda and Stewart Resnick have launched several successful brands, including Fiji Water, Teleflora, and the Franklin Mint, but it was their pomegranate juice, sold as Pom Wonderful, which got them in trouble with the FTC. The company advertised Pom Wonderful as a product that could help reduce the risk of heart disease, prostate cancer, and impotence. The FTC charged Pom Wonderful and the company's owners with making false and unsubstantiated claims about the power of its product. The owners of Pom Wonderful appealed the ruling, but the commission unanimously upheld an administrative law judge's decision that the marketers of POM Wonderful deceptively advertised the products and did not have adequate clinical support for its health claims. The FTC barred the POM Wonderful marketers from making any future claims unless the claims are supported by two randomized, well-controlled, human clinical trials.

Entrepreneurs sometimes run afoul of the laws concerning advertising because they do not know how to comply with legal requirements. The FTC is the federal agency that regulates advertising and deals with problems created by deceptive ads. Under federal and state laws, an advertisement is unlawful if it misleads or deceives a reasonable customer, even if the business owner responsible for it had no intention to deceive. Any ad containing a false statement is in violation of the law although the entrepreneur may not know that the statement is false. The FTC judges an ad by the overall impression it creates and not by the technical truthfulness of its individual parts.

What can entrepreneurs do to avoid charges of deceptive advertising? The following tips from the Street-Smart Entrepreneur will help:

- ***Make sure that your ads are accurate.*** Avoid creating ads that promise more than a product or service can deliver. Take the time to verify the accuracy of every claim or statement in your ads. If a motor oil protects an engine from damage, don't claim that it will repair damage that already exists in an engine—unless you can prove that it actually does.

- ***Understand the difference between sales 'puffery' and false advertising.*** The distinction is not always clear. Sales puffery involves claims that are so general or so exaggerated that they would not confuse customers. (How many times have you seen a small restaurant advertising that it has "the best hot dog in town"?) The more specific and fact based the claims in an ad are, the more likely they are to pose problems for a company if they are false or if the company has no factual basis for making them. When Pizza Hut ("Best Pizza Under One Roof") filed a false advertising claim against Papa John's Pizza over a Papa John's ad that claimed "Better Ingredients, Better Pizza," a federal court of appeals ruled that Papa John's claim was puffery and that the company could continue to use it *if* it stopped making specific fact-based claims in the same ad that its

tomato paste and dough were superior. For instance, the ad for Papa John's claimed that its sauce, which was made from "vine-ripened tomatoes," was superior to Pizza Hut's "remanufactured tomato sauce." Because Papa John's had no facts to prove this claim, the court ruled that this was false advertising.

- ***Get permission to use quotations, pictures, and endorsements.*** Never use material in an ad from an outside source unless you get written permission to do so. Some business owners find themselves in trouble when they use copyrighted material (such as images, videos, photographs, and other content) in online ads or in posts that promote their companies that they find online using a search engine without first securing permission. One business owner got into trouble when he inserted a photograph of a famous athlete without his permission into an ad for his company's service.

- ***Be careful when you compare competitor's products or services to your own.*** False statements that harm the reputation of a competitor's business, products, or services may result not only in charges of false advertising but also in claims of trade libel. Make sure that any claims in your ads comparing your products to competitors' are fair and accurate. You can use a competitor's trademark in your advertising (e.g., for purposes of comparison) as long as it does not cause confusion among customers concerning the origin of the product or its affiliation with the competitor.

- ***Stock sufficient quantities of advertised items.*** Businesses that advertise items for sale must be sure to have enough units on hand to meet anticipated demand. If you suspect that demand may outstrip your supply, state in the ad that quantities are limited.

- ***Avoid "bait-and-switch" advertising.*** This illegal technique involves advertising an item for sale at an attractive price when a business has no real intention of selling that product at that price. Companies using this technique often claim to have sold out of the advertised special. Their goal is to lure customers in with the low price and then switch them over to a similar product at a higher price.

- ***Use the word "free" carefully and accurately.*** Every advertiser knows that one of the most power words in advertising is "free." However, anything you advertise as being free must actually be free. For instance, suppose a business advertises a free paintbrush to anyone who buys a gallon of a particular type of paint for $19.95. If the company's regular price for the paint is less than $19.95, the ad is deceptive because the paintbrush is not really free.

- ***Be careful of what your ad does not say.*** Omitting information in an ad that leaves customers with a false impression about a product or service and its performance is also a violation of the law.

- **Describe sale prices and "savings" carefully.** Business owners sometimes get into trouble with false advertising when they advertise items at prices that offer huge "savings" over their "regular" prices. One jeweler violated the law by advertising a bracelet for $299, a savings of $200 from the item's regular $499 price. In reality, the jeweler had never sold the item at its $499 "regular" price; the item's normal price was the $299 that he advertised as the "sale" price.

Sources: Adapted from *Guides Against Bait Advertising*, Federal Trade Commission (Washington, DC), *www.ftc.gov/bcp/guides/baitads-gd.htm*; *Advertising FAQ's: A Guide for Small Business*, Federal Trade Commission, *http://business.ftc.gov/documents/bus 35-advertising-faqs-guide-small-business*; James Astrachan, "False Advertising Primer," Astrachan, Gunst, Thomas, PLC, 2006, *www.aboutfalseadvertising.com/index1_files/ False%20Advertising%20Primer.pdf*, p. 14; "Seven Rules for Legal Advertising," *Inc.* (n.d.), *http://www.inc.com/articles/2000/05/20153.html*; "Consumer Protection Laws," *Inc.*, May 12, 2000, *www.inc.com/search/19691.html*; Edward Wyatt, "Regulators Call Health Claims in Pom Juice Ads Deceptive," *New York Times*, September 28, 2010, p. B1; Stuart Pfeifer, "Billionaires Behind Pom Wonderful Push Back Against FTC Ruling," *Los Angeles Times*, June 20, 2012, *http://articles.latimes.com/2012/jun/20/ business/la-fi-pom-response-20120621*; "FTC Commissioners Uphold Trial Judge Decision that POM Wonderful, LLC; Stewart and Lynda Resnick; Others Deceptively Advertised Pomegranate Products by Making Unsupported Health Claims," Federal Trade Commission, January 16, 2013, *http://ftc.gov/opa/2013/01/pom.shtm*.

Consumer Credit

Another area subject to intense government regulation is consumer credit. This section of the law has grown in importance as credit has become a major part of many consumer purchases. The primary law regulating consumer credit is the Truth-in-Lending Act of 1969. This law requires sellers who extend credit and lenders to fully disclose the terms and conditions of credit arrangements. The FTC is responsible for enforcing the Truth-in-Lending Act. The law outlines specific requirements that any firm that offers, arranges, or extends credit to customers must meet. The two most important terms of the credit arrangement that lenders must disclose are the finance charge and the annual percentage rate. The finance charge represents the total cost—direct and indirect—of the credit, and the annual percentage rate is the relative cost of credit stated in annual percentage terms.

The Truth-in-Lending Act applies to any consumer loan for less than $25,000 (or loans of any amount secured by mortgages on real estate) that includes more than four installments. Merchants extending credit to customers must state clearly the following information, using specific terminology:

- The price of the product
- The down payment and any trade-in allowance made
- The unpaid balance owed after the down payment
- The total dollar amount of the finance charge
- Any prepaid finance charges or required deposit balances, such as points, service charges, or lenders' fees
- Any other charges not include in the finance charge
- The total amount to be financed
- The unpaid balance
- The deferred payment price, including the total cash price and finance and incidental charges
- The date on which the finance charge begins to accrue
- The annual percentage rate of the finance charge
- The number, amount, and due dates of payments
- The penalties imposed in case of delinquent payments
- A description of any security interest the creditor holds
- A description of any penalties imposed for early repayment of principal

Another provision of the Truth-in-Lending Act limits a credit card holder's liability in case the holder's card is lost or stolen. As long as the holder notifies the company of the missing card, he or she is liable for only $50 of any amount that an unauthorized user might charge on the card (or zero if the holder notifies the company before any unauthorized use of the card).

In 1974, Congress passed the Fair Credit Billing Act, an amendment to the Truth-in-Lending Act. Under this law, a credit card holder may withhold payment on a faulty product, provided that he or she has made a good-faith effort to settle the dispute first. A credit card holder can also withhold payment to the issuing company if he or she believes that his or her bill is in error. The cardholder must notify the issuer within 60 days but is not required to pay the bill until the dispute is settled. The creditor cannot collect any finance charge during this period unless there was no error.

Another credit law designed to protect consumers is the Equal Credit Opportunity Act of 1974, which prohibits discrimination in granting credit based on race, religion, national origin, color, gender, marital status, or whether the individual receives public welfare payment.

In 1970, Congress created the Fair Credit Reporting Act to protect consumers against the circulation of inaccurate or obsolete information pertaining to credit applications. Under this act, the consumer can request the nature of any credit investigation, the type of information assembled, and the identity of those persons receiving the report. The law requires that any obsolete or misleading information contained in the file be updated, deleted, or corrected.

Congress enacted the Fair Debt Collection Practices Act in 1977 to protect consumers from abusive debt collection practices. The law does not apply to business owners collecting their own debts but only to debt collectors working for other businesses. The act prevents debt collectors from doing the following:

- Contacting the debtor at his or her workplace if the employer objects
- Using intimidation, harassment, or abusive language to pester the debtor
- Calling on the debtor at inconvenient times (before 8 A.M. or after 9 P.M.)
- Contacting third parties (except parents, spouses, and financial advisers) about the debt
- Contacting the consumer after receiving notice of refusal to pay the debt (except to inform the debtor of the involvement of a collection agency)
- Making false threats against the debtor

The Consumer Leasing Act of 1976 amended the Truth-In-Lending Act for the purpose of providing meaningful disclosure to consumers who lease goods. The Uniform Consumer Leases Act (UCLA) amended the Consumer Leasing Act. To be covered by the UCLA, a lease must run for more than four months, or the dollar value of the lease obligation must exceed $150,000.

In 2003, Congress passed the Fair and Accurate Transactions Act (the FACT Act) to address the fastest-growing crime in the United States: identity theft. On average, someone in the United States becomes a victim of identity theft every three seconds. Experts estimate that 12.6 million people in the United States are victims of identity theft each year, most often in the form of credit card fraud. The total amount of fraud is $21 billion per year.[31] The FACT Act allows victims of identity theft to file theft reports with credit reporting agencies and requires those agencies to include "fraud alerts" in their credit reports.

Environmental Law

In 1970, Congress created the Environmental Protection Agency (EPA) and gave it the authority to create laws that would protect the environment from pollution and contamination. Although the EPA administers a number of federal environmental statutes, three in particular stand out: the Clean Air Act, the Clean Water Act, and the Resource Conservation and Recovery Act.

THE CLEAN AIR ACT To reduce the problems associated with global warming, acid rain, and airborne pollution, Congress passed the Clean Air Act in 1970 (and several amendments since then). The act targets everything from coal-burning power plants to automobiles. The Clean Air Act assigned the EPA the task of developing national air quality standards for carbon monoxide, hydrocarbons, sulfur oxide, ozone, lead, and other harmful substances. The agency works with state and local governments to enforce compliance with these standards.

THE CLEAN WATER ACT The Clean Water Act, passed in 1972, set out to make all navigable waters in the United States suitable for fishing and swimming by 1983 and to eliminate the discharge of pollutants into those waters by 1985. Although the EPA has made progress in cleaning up many bodies of water, it has yet to achieve these goals. The Clean Water Act requires each state to establish water quality standards and to develop plans to reach them. The act also prohibits the draining, dredging, or filling wetlands without a permit. The Clean Water Act also addresses the issues of providing safe drinking water and cleaning up oil spills in navigable waters.

THE RESOURCE CONSERVATION AND RECOVERY ACT Congress passed the Resource Conservation and Recovery Act in 1976 to deal with solid waste disposal. The act, which was amended in 1984, sets guidelines by which solid waste landfills must operate, and it establishes rules governing the disposal of hazardous wastes. The act's goal is to prevent solid waste from contaminating the environment. What about those waste disposal sites that are already contaminating the environment? In 1980, Congress passed the Comprehensive Environmental Response, Compensation, and Liability Act to deal with those sites. The act created the Superfund, a special federal fund set up to finance and to regulate the cleanup of solid-waste disposal sites that are polluting the environment.

The Pollution Prevention Act of 1990 set forth a public policy statement that offered rewards to firms that reduced the creation of pollution. The federal government provides matching funds to states for programs that promote the use of "source reduction techniques" dealing with pollution problems. This is a milestone piece of legislation because it replaces the regulatory "stick" approach resented by business with a "carrot" approach that rewards businesses for positive actions that reduce pollution.

Entrepreneurship in Action

Small Businesses and Eminent Domain

Susette Kelo.

Source: Institute for Justice.

Susette Kelo had dreamed of owning a home that looked out over the water in her hometown of New London, Connecticut. In 1997, she purchased and restored a little house in a quiet residential neighborhood called Fort Turnbull, where the Thames River meets the Long Island Sound. She enjoyed the great view of the water from its windows. Then, the day before Thanksgiving in 2000, she received a notice of condemnation

of her property from the New London Development Corporation (NLDC), which was seizing her property under the city's eminent domain rights.

Eminent domain is the right of the government to seize property for public use and public good. Historically, the definition of "public use" had been limited by court decisions to three categories:

1. Roads, hospitals, or military bases
2. Railroads, public utilities, or stadiums
3. Blighted areas or areas that are injurious to public safety

In New London, the NLDC was using eminent domain to secure a large tract of property for the Pfizer Corporation, which had been developing plans to build a research park. The NLDC and the city wanted to help encourage Pfizer to develop the project within the city limits, because officials believed that commercial development would follow and would create economic growth in the city.

Susette Kelo and her neighbors, all of whom lived on the land being condemned for this project, decided to fight the condemnation in court. Kelo's attorneys built their case from what is known as the "taking clause" of the Fifth Amendment to the U.S. Constitution, which states that "nor shall private property be taken for public use, without just compensation." The argument was that the condemnation of the property to allow Pfizer to build a research park was not "public use."

The case (*Kelo v. New London*) eventually made its way to the U.S. Supreme Court. In a 5-to-4 decision, the Court ruled in favor the NLDC. In effect, this decision created a fourth category of "public use" in which local or state government could use eminent domain as a tool for economic development.

In her dissenting opinion, Justice Sandra Day O'Connor stated, "Today the Court abandons this long-held, basic

(continued)

Entrepreneurship in Action *(continued)*

limitation on government power. Under the banner of economic development, all private property is now vulnerable to being taken and transferred to another private owner, so long as it might be upgraded—i.e., given to an owner who will use it in a way that the legislature deems more beneficial to the public—in the process."

What followed was a significant public and political backlash against the Court's decision. Several states and cities passed legislation that limited the local use of eminent domain, which was permitted under the Court's decision. However, many other local governmental entities used the expanded power of eminent domain to attempt to spur new development projects. Some examples follow:

- Local officials identified as "blighted" well-established and safe St. Louis, Missouri, residential neighborhood that was home to more than 100 families so that they could turn the property over to developers who offered the city the promise of a higher tax base.

- The city of Belmar, New Jersey labeled the Belmar Mall "blighted." The owners of the mall had been fighting to keep their property from being condemned for a redevelopment project. The Belmar Mall contains 20 small businesses, along with three major chain tenants.

- The owners of Country International Records, a small music studio located in Nashville, Tennessee, lodged a legal and public relations battle to fight the city's move to condemn their property after they refused to sell to a Houston developer planning to build a new hotel complex on the site.

- Central Radio is a small business that is among more than 170 residential, institutional, and business buildings condemned by the Norfolk Redevelopment and Housing Agency after its owners refused to voluntarily sell their properties. The properties are in an area near Old Dominion University that the city wanted to redevelop with a private–public development partnership.

Proponents of the use of eminent domain cite the need for cities to use economic strategies that can help spur growth that will benefit the larger community. Many proponents argue that this is particularly true during times of economic decline.

1. Develop an argument for the Supreme Court that supports the NLDC's case.

2. Develop an argument for the Supreme Court that supports Susette Kelo's case.

3. Which of the two arguments to you find more compelling? In other words, do you agree with the Supreme Court's decision in this case? Explain.

Sources: Based on "Kelo et al. v. City of New London et al.," *FindLaw,* June 23, 2005, *http://caselaw.lp.findlaw.com/scripts/getcase.pl?court=us&vol=000&in vol=04-108*; "Kelo v. New London: Lawsuit Challenging Eminent Domain Abuse in New London, Connecticut," Institute for Justice, (n.d.), *www.ij.org/kelo-v-new-london*; "Eminent Domain Being Abused?," *CBS News,* February 11, 2009, *www.cbsnews.com/8301-18560_162-575343.html*; "BMIA LLC v. Borough of Belmar—Eminent Domain Strikes Another Business," National Federation of Independent Business, (n.d.), *www.nfib.com/press-media/press-media-item?cmsid=31363*; *www.ij.org/nashville-tenn-eminent-domain-release7-21-2008*; George Will, "Latest Eminent Domain Outrage," *Newsmax.com,* January 25, 2013, *www.newsmax.com/GeorgeWill/eminent-domain-Central-Radio/2013/01/24/id/472801*; David M. Lewis, "Eminent Domain: Still a Useful Tool Despite Its Recent Thrashing," *Planetizen.com,* September 5, 2006, *www.planetizen.com/node/21109.*

The Affordable Care Act

The Patient Protection and Affordable Care Act of 2010 (also known as "Obamacare") requires that all Americans obtain health insurance and that all insurance companies provide coverage regardless of preexisting conditions and begins to go into effect in January 2014. Small businesses with the equivalent of 50 or more full-time employees face fines if they fail to provide full-time workers with affordable health insurance that meets certain standards. Small businesses with fewer than 50 employees, while not required to provide health insurance, will be required to fully inform employees about the law. If these small firms offer health care, their plans must comply with the mandates of the Affordable Care Act. The financial impact of this law on small businesses has created significant debate. Businesses that have not provided health insurance in the past will be mandated to do so or face penalties that can exceed $100,000 a year. Although entrepreneurs may be tempted to create several small companies, keeping each under 50 employees, the IRS intends to evaluate privately owned companies and count employment among all companies with common ownership as part of the 50-employee threshold. For example, an entrepreneur who owns three small businesses, each with 20 employees, will be treated as if it were a single company with 60 employees and thus will be subject to the requirements of the Affordable Care Act.

Chapter Review

1. Explain the basic elements required to create a valid, enforceable contract.
   - A valid contract must contain these elements: agreement (offer and acceptance), consideration, capacity, and legality. A contract can be valid and yet unenforceable because it fails to meet two other conditions: genuineness of assent and proper form.
   - Both parties performing their promised actions fulfill most contracts; occasionally, however, one party fails to perform as agreed, thereby breaching the contract. Usually, the nonbreaching party is allowed to sue for monetary damages that would place him or her in the same position he or she would have been in had the contract been performed. In cases where money is an insufficient remedy, the injured party may sue for specific performance of the contract's terms.

2. Outline the major components of the UCC governing sales contracts.
   - The UCC was an attempt to create a unified body of law governing routine business transactions. Of the 10 articles in the UCC, Article 2 on the sale of goods affects many business transactions.
   - Contracts for the sale of goods must contain the same four elements of a valid contract, but the UCC relaxes many of the specific restrictions that the common law imposes on contracts. Under the UCC, once the parties create a contract, they must perform their duties in good faith.
   - The UCC also covers sales warranties. A seller creates an express warranty when he or she makes a statement about the performance of a product or indicates by example certain characteristics of the product. Sellers automatically create other warranties—warranties of title, implied warranties of merchantability, and, in certain cases, implied warranties of fitness for a particular purpose—when they sell a product.

3. Discuss the protection of intellectual property rights using patents, trademarks, and copyrights.
   - A patent is a grant from the federal government that gives an inventor exclusive rights to an invention for 20 years. To submit a patent, an inventor must establish novelty, document the device, search existing patents, study the search results, submit a patent application to the PTO, and prosecute the application.
   - A trademark is any distinctive word, symbol, or trade dress that a company uses to identify its product or to distinguish it from other goods. It serves as the company's "signature" in the marketplace.
   - A copyright protects original works of authorship. It covers only the form in which an idea is expressed and not the idea itself and lasts for 50 years beyond the creator's death.

4. Explain the basic workings of the law of agency.
   In an agency relationship, one party (the agent) agrees to represent another (the principal). The agent has the power to act for the principal but remains subject to the principal's control. While performing job-related tasks, employees play an agent's role.
   - An agent has the following duties to a principal: loyalty, performance, notification, duty of care, and accounting. The principal has certain duties to the agent: compensation, reimbursement, cooperation, indemnification, and safe working conditions.

5. Explain the basics of bankruptcy law.
   - Entrepreneurs whose businesses fail often have no other choice but to declare bankruptcy under one of three provisions: Chapter 7 liquidations, where the business sells its assets, pays what debts it can, and disappears; Chapter 11, reorganizations, where the business asks that its debts be forgiven or restructured and then reemerges; and Chapter 13, straight bankruptcy, which is for individuals only.

6. Explain some of the government regulations affecting small businesses, including those governing trade practices, consumer protection, consumer credit, and the environment.
   - Businesses operate under a multitude of government regulations governing many areas, including trade practices, where laws forbid restraint of trade, price discrimination, exclusive dealing and tying contracts, purchasing controlling interests in competitors, and interlocking directorates.
   - Other areas subject to government regulations include consumer protection (the Food, Drug, and Cosmetics Act and the Consumer Product Safety Act) and consumer credit (the Consumer Credit Protection Act, the Fair Debt Collection Practices Act, and the Fair Credit Reporting Act), and the environment (the Clean Air Act, the Clean Water Act, the Resource Conservation and Recovery Act), and health (the Affordable Care Act).

Discussion Questions

23-1. What is a contract? List and describe the four elements required for a valid contract. Must a contract be in writing to be valid?

23-2. What constitutes an agreement?

23-3. What groups of people lack contractual capacity? How do the courts view contracts that minors create? Intoxicated people? Insane people?

23-4. What circumstances eliminate genuineness of assent in the parties' agreement?

23-5. What is breach of contract? What remedies are available to a party injured by a breach?

23-6. What is the UCC? To which kinds of contracts does the UCC apply? How does it alter the requirements for a sales contract?

23-7. Under the UCC, what remedies does a seller have when a buyer breaches a sales contract? What remedies does a buyer have when a seller breaches a contract?

23-8. What is a sales warranty? Explain the different kinds of warranties that sellers offer.

23-9. Explain the different kinds of implied warranties the UCC imposes on sellers of goods. Can sellers disclaim these implied warranties? If so, how?

23-10. What is product liability? Explain the charges that most often form the basis for product liability claims. What must a customer prove under these charges?

23-11. What is intellectual property? What tools do entrepreneurs have to protect their intellectual property?

23-12. Explain the differences among patents, trademarks, and copyrights. What does each protect? How long does each last?

23-13. What must an inventor prove to receive a patent?

23-14. Briefly explain the patent application process.

23-15. What is an agent? What duties does an agent have to a principal? What duties does a principal have to an agent?

23-16. Explain the differences among the three major forms of bankruptcy: Chapter 7, Chapter 11, and Chapter 13.

23-17. Explain the statement "For each benefit gained by regulation, there is a cost."

23-18. What are the benefits and costs of the Affordable Care Act? Explain from both the perspective of the small business and that of an employee.

The picturebooth co.

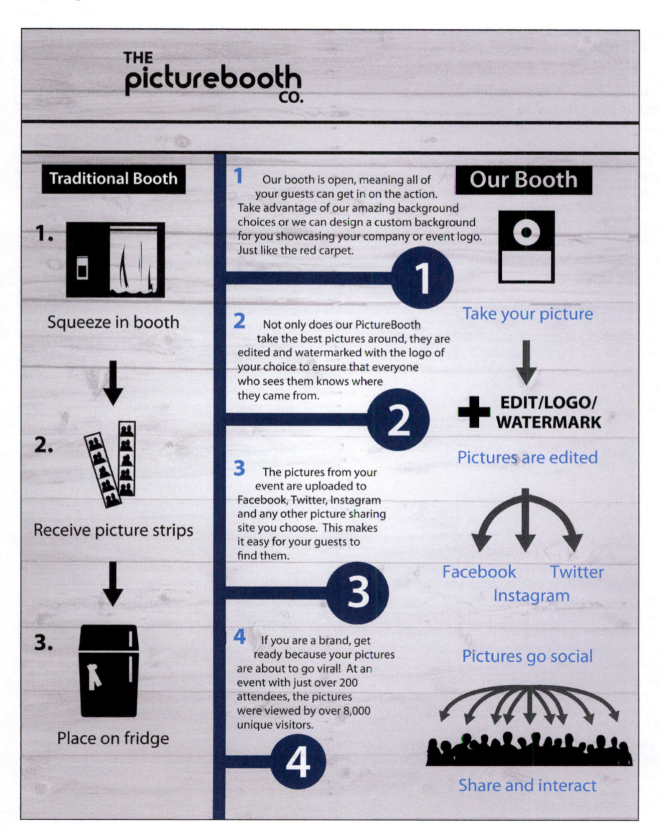

THE **picturebooth** CO.

Traditional Booth

1.

Squeeze in booth

2.

Receive picture strips

3.

Place on fridge

1 Our booth is open, meaning all of your guests can get in on the action. Take advantage of our amazing background choices or we can design a custom background for you showcasing your company or event logo. Just like the red carpet.

2 Not only does our PictureBooth take the best pictures around, they are edited and watermarked with the logo of your choice to ensure that everyone who sees them knows where they came from.

3 The pictures from your event are uploaded to Facebook, Twitter, Instagram and any other picture sharing site you choose. This makes it easy for your guests to find them.

4 If you are a brand, get ready because your pictures are about to go viral! At an event with just over 200 attendees, the pictures were viewed by over 8,000 unique visitors.

Our Booth

Take your picture

+ EDIT/LOGO/ WATERMARK

Pictures are edited

Facebook   Twitter
Instagram

Pictures go social

Share and interact

Executive Summary

PictureBooth is a photo booth start-up based in Nashville, Tennessee. I have been in the event industry for about five years as a DJ, specializing in weddings. In September, as my standard photo booth rental numbers started to climb, I realized that it was time to start my own photo booth business. By asking some of my potential clients simple questions, I found that there is a need in the market for a more affordable photo booth rental. To provide them what they wanted, I decided to design my own.

Our Mission Is to Provide High-Quality Photo Booths at an Affordable Price

PictureBooth is not your typical photo booth business that operates in one city or a few cities within close distances of each other. PictureBooth is designed to be the most portable booth on the market and can operate anywhere. PictureBooth is designed so that an individual of any size or stature can receive the booth, read its simple assembly instructions, and have the booth set up within 5 to 10 minutes. PictureBooth rentals start at $400, compared to the standard $800 to $1,400 rental within the industry. We chose the $500 price point because the brides to whom we talked told us it was a "no-brainer" price point for almost any wedding reception.

Our company has three different streams of revenue that we can tap. The first is special event rentals, such as weddings, proms, birthday parties, office parties, and so on. Our booth will be a lower priced option in this space; however, because of the technology we use, it will produce the highest quality content. The second stream of revenue is contracted branding deals. We have a mission to create organic and interactive social media content for businesses through our booths. The third stream of revenue is permanent booth installations in businesses. There are many different types of businesses that are looking for a way to create a social media presence through organic content.

WEDDINGS The wedding industry in the United States is a $54 billion industry, with $23 billion being spent on event rentals and about $6 billion spent on photography services. Although PictureBooth is not limited to the wedding industry, these numbers provide concrete evidence of PictureBooth's target market, the special event industry. Our target market is weddings and special events that typically occur on weekends, but we also will have booths that are available throughout the week. We will reach our customers primarily through the Internet. Our initial market is in our home base of Nashville but will expand nationwide by the end of 2013.

We will market PictureBooth to the wedding industry through several online wedding vendor distribution sites, such as WeddingWire, which receives 2.8 million unique visitors per month, and Perfect Wedding guide, which receives thousands of unique local visitors per month. We will also employ a part-time SEO specialist located in Nashville to implement our strategy. We will differentiate ourselves from our competitors at the outset by creating a professional Web site and professional quality promotional videos. In addition to these online marketing strategies, we will use other standard means of reaching bridal customers, such as bridal shows and cold calling.

BRANDING The second major market that PictureBooth will appeal to is corporations that want to promote their brands. Our booths are designed so that companies can create custom design booths inspired by their brands. Our booths take advantage of the digital space by creating interactive social media content for brands at an event. In the appendix to this business plan is an example of a proposal [not included] we send to prospective clients explaining why our booth is better for them in the short term and the long term.

We are currently building a relationship with one of the most recognized brands in the world, RedBull. We attracted RedBull as a customer because of the quality of the photos we provide as well as our booth rental rates. RedBull offered us nearly 40 events in the Nashville/Middle Tennessee area in exchange for a price concession. We set the price at $275 per event. In return, we are now in the position to work with RedBull on a national scale and provide booths for hundreds of their events.

PERMANENT INSTALLATIONS The third market is permanent installations. PictureBooth will lease or sell custom-designed booths to clients who are looking for a way to generate content through their permanent locations. For example, for pediatric dentistry, it is difficult to continuously create relevant social media content. By installing a photo booth, it is now the patients and their parents creating interactive social media content and not the office. We are no longer in the age of information; we are in the age of recommendation. As parents tag their children in Facebook pictures, not only does it expose the dentist's office to potential new customers, but it also gives a digital "stamp of approval" from parents, telling their friends, "My kids can have fun here, and so can yours."

What Makes Us Stand Out?

What makes our booth unique for our customers? For weddings, our booth is unique in that it provides brides with an alternative option to the standard expensive photo booth rental and gives them exceptionally high quality pictures that they can print and/or share with friends and family digitally. For companies, it provides a way to create organic interactive social media content with their brands. When an individual takes a picture in the booth, that picture is edited and watermarked with the company's logo. Instead of allowing individuals to post pictures directly to their Facebook pages, they must instead go to the specified page (the brand's Facebook page) and tag themselves in the photo. When individuals tag themselves in photos, it shows up on their timelines/news feeds, allowing their friends to view the photos and thereby increase brand awareness because of the watermarked logos. Another interactive quality is that individuals must seek out the brand's Facebook page. If companies properly maintain their pages, they have the opportunity to engage individuals as regular visitors to their pages.

Ross Hill will serve as president and CEO of PictureBooth throughout the first three years. My main purpose right now is to manage the growth of our company as well as the sales and distribution of our product. I have hired someone who serves in a multifunctional, part-time role, including photo editing, transporting the booths to and from upcoming events, and other jobs.

I have invested countless hours and around $10,000 into PictureBooth, but to continue growing the business, I need more money. I have estimated the start-up cost of this business at around $40,000, but it could easily be more. I plan to get a loan for this start-up capital. This money will go toward building three to five booths as well as covering some of our other start-up expenses. We estimate that we will reach our break-even point in May. At that point, we plan to use internally generated cash to grow at a constant rate. It may be necessary for us to find investors if the company's internal cash engine cannot keep up with its growth and capital needs. My salary will be supplemented by income earned through other businesses during the first year, but I will give myself a $20,000 bonus at the end of year one if we hit our benchmark goals. Our original financial estimates show that in year one, we will hit the $150,000 mark in sales and grow to $300,000 by year three.

MISSION STATEMENT PictureBooth is a high-quality, portable photo booth designed to create interactive social media content for events, artists, and brands.

ADDITIONAL CONCEPT DESCRIPTION PictureBooth's business model is simple; we call it "Ship, Use, Ship." PictureBooth is a portable photo booth, about the same size as a carry-on suitcase that is designed to be shipped to our clients or representatives, used at an event, and shipped back to a central warehouse.

CORE VALUES In every venture that I enter into, I strive to bring four values along with me: hard work, serving others, providing opportunity, and fairness. These four values will help in shaping the culture of my organization.

KEY OBJECTIVES My primary objectives with PictureBooth are designing a career that allows me to spend an ample amount of time with my family, making enough money to provide a great life for myself and my family, giving other people an opportunity to succeed in life and learn lessons that I learned at an early age, and creating a product that brings immense value to its clients.

Value Proposition

PictureBooth was invented out of necessity. I have been in the event industry for almost five years, and one concept that has always held true in the event production realm is product setup efficiency. The individual executing the setup and teardown of the event wants to spend the least possible amount of time doing the job. Photo booths are notorious for having long setup times because of their size and their lack of innovative design.

PictureBooth solves many problems with standard photo booths, but it solves primarily two major problems in corporate branding as well as social media marketing.

Value Proposition #1

PictureBooth was designed with the user in mind. PictureBooth is about the same size as a carry-on suitcase and is very light. A person of any size or stature can use the booth with ease. It takes between 5 and 10 minutes to set up the booth, and teardown is just as easy. Users do not have to drive vans or trucks to transport the booth; they can now fit multiple PictureBooths inside their cars.

Another value that PictureBooth offers is its affordability, coupled with high-quality shareable content. Many of our clients want to use our booth to increase brand awareness. An event with just 200 attendees created more than 8,000 unique Facebook interactions, meaning that 8,000 different people interacted with the branded pictures.

Proof of Concept

Photo booths at events have been a growing trend since 2008. A simple Google search including the name of a city and photo booth will show that there is a lot of competition. This is both good and bad, but it proves that there is a market. There are millions of events each year that could benefit from the use of a photo booth. Some different types of events where photo booths could be deployed include weddings, proms, university events, fraternity and sorority socials, corporate branding initiatives, and artist tours.

Why Is This a Bad Thing?

Ideally, you would enter a market that has little competition. We will be entering a market that is proven but has a lot of competition.

Why Is This a Good Thing?

Our competition cannot compete with us easily. PictureBooth differentiates itself by offering a product that is affordable, but it is also portable and does not require an operator. For some of our high-cost competitors to compete with us, they will have to radically change their business models. For other similar business models to compete with us, they will have to enter the market and lose the first-mover advantage.

Competitive Analysis

We differentiate ourselves from our competitors by giving customers four things that most photo booth companies do not focus on. One is portability and ease of use. Most photo booths weigh a couple of hundred pounds, take an hour to set up, and require an operator throughout the event. PictureBooth can be handled by one person of any size or stature, set up in about 5 to 10 minutes at an event, and taken down in the same amount of time. Our booth weighs only 26 pounds and is about the same size as a carry-on suitcase.

Another differentiator is affordability. The average photo booth rental is between $800 and $1,400. Our photo booth rentals start at $500. We set our price point at $500 to gain customers that wouldn't normally choose to have a photo booth at their event, but because of the low price point, this makes it an easier choice to have at an event.

Furthermore, PictureBooth sets itself apart from the competition because of its high-quality shareable content creation. Many photo booths in the industry take low-quality pictures, and they lack the ability to share their content.

The photo booth industry is very competitive, which is both positive and negative. From the negative perspective, there is a lot of competition; therefore, it may be more difficult to break into certain parts of industry because of already established relationships. On the positive side, it will be very difficult for already existing photo booth companies to compete with us. Our operating costs are substantially lower than those of our competitors, and our booths are much more technologically advanced.

Market Entry Strategy

Target Market

Our target market is the special events industry, including weddings, proms, corporate events, and artist tours. Our customer is someone who produces events or has his or her own event. We have gained the greatest traction in corporate branding and plan to continue to nurture these relationships. We have booked more than 50 dates with only one corporate brand and have the potential to book hundreds more.

Promotion

PictureBooth will be promoted through several streams, one of which will be strategic networking. Thus far, we have gained the most traction through personal contacts and meetings. We will always employ this principle throughout the growth of our business. Another form of promotion we will be using is creating extremely high quality digital content, including videos and product animations for our pitches. We will take full advantage of our first-mover strategy and plan to gain market share in the beginning before competitors have a chance to catch up. A unique aspect of our business is that the booth itself creates quality content for us to use in promotion. When other companies find out that some of the largest and most recognized brands in the world use our booths, they will be attracted to our PictureBooths.

Place

It is our goal to get PictureBooth in front of as many people as possible. We will do this by becoming a part of as many events as possible. We currently have an opportunity for our product to be utilized in more than 3,000 events this year. These events will create buzz for our company, and we believe that being a part of this many events in our first year will give us instant credibility. Another place where customers will discover us is online. Our Web site is designed for ease of use, just like our product. Our site will include e-commerce, giving customers the ability to book Picture-Booths for their events instantly.

Operating Plan

Currently, we are acquiring the majority of our customers through personal connections. These connections have helped us gain immense traction, and we have hundreds of events on the table. Strategic networking will always be part of our customer acquisition strategy. Another way we will acquire customers is through a unique distribution strategy using new and existing vendors in different cities as well as direct to consumer.

For delivery of our product, we will be partnering with J3BLLC, an online retailer based in Nashville. Because of their sales volume, they receive some of the best shipping rates possible. J3BLLC will handle all order fulfillment functions for PictureBooth. We will have a part-time employee receiving and prepping booths at their warehouse, but J3BLLC will be responsible for shipping.

Our e-commerce sales will be cash payments, collected immediately. Our account receivables orders will be under a 2/10, net 30 arrangement. This arrangement will be used strictly with our large corporate clients that book a substantial number of events. We will devote energy into creating an engaging online store experience for our customers, and we expect that by doing this, we can generate a significant cash sales that will fund other business activities.

We are currently in the product development phase of our business, but we have hard deadlines that we must meet. The cost of each booth is roughly $2,500. The estimated lifetime of each booth is around 100 events, meaning that each booth has the potential to generate $50,000 in revenue. Once we are manufacturing these booths on a larger scale, our cost per booth will decrease substantially.

In the beginning, we will lower our initial equipment investment costs by using a just-in-time build philosophy. We create contracts for events well in advance, giving us the time we need to build more booths if needed. We plan to employ this strategy during the growth phase of our company.

We are in the process of hiring a bookkeeper, an accountant, and a lawyer. As we begin to book more events and are spending and receiving money on a daily basis, we need to hire someone who is in charge of that aspect of the business.

During the first year, I will assume the position of CEO, CMO, and COO. We plan to hire one or two people in April to get our systems in place for our May launch. I will not be paying myself until the end of the first year, when we have hit our benchmark goals. My advisers and investors include my dad, Vaughn Hill; my girlfriend's dad, Dallen Wendt; my brother-in-law, Morgan Dillow; and an outside investment firm, Byte Ventures. I keep in constant contact with them about the majority of my business decisions to bounce ideas off of and allow them to give me their opinions on my business activities.

We will grow our company on a lean philosophy. I will not hire unnecessary employees. I want to employ as many people as we need but no more than that.

In the beginning, PictureBooth will have very humble space needs. We need a small warehouse space or storage unit to hold product until it is time to ship, and we also need an office space to call home and a place to meet with clients.

Growth

Currently, PictureBooth has an opportunity to be a part of about 1,000 events in its first year of operation. These events have come from just three different clients. The potential for PictureBooth is huge, and our goal is to have our booths utilized at 4,000 to 5,000 events in our second year. We will be using our first-mover advantage to gain maximum market share early on.

Financial Plan

Sources of Financing

PictureBooth is receiving a total investment of about $100,000.

PictureBooth generates revenue through online sales and through large contracts. Our photo booth rental price starts at $500 and increases to $1,200, depending on the customization and options that customers choose.

Worst-Case Scenario

In our worst-case scenario, the RedBull and Krispy Kreme events disappear, as does the hope for more events through those connections. We built five fully functioning booths, and we know how to make more and the time we need to accomplish that. We begin launching our product distribution strategy (licensing strategy) through other vendors in other states who are looking to add a photo booth rental feature to their rental inventory or to their "additional services." After contacting 1,000 DJs and event rental companies, we have found 50 of them willing to carry and promote our product. We will collect a licensing fee from them of $1,000. In return for their licensing fees, they get the ability to sell our product and a custom subdomain on our Web site built for their companies. In this worst-case scenario, our per event rental cost is $500 for the base model; for the first 10 events, we collect $200, and for every rental beyond that, we collect only $150 per event. In this scenario, we book 100 of our own events between the months of May and December, but we also do 220 events through our "distributors." We stay on a constant growth pattern for the next two years with our distributors, but we start to book some big corporate contracts along the way.

Most-Likely Scenario

Once we have our product built and ready to distribute, I will begin to use my network to find opportunities. One of my strengths is connecting with people. RedBull and Krispy Kreme are *ready* to move forward; we need to meet the demand. I will also begin using my connections in the music industry to get in front of people who can use PictureBooth. At the same time, starting in June, we will begin to reach out to our potential distributors, show them how the business model works, send them product demonstration videos, and possibly even ship them booths. Our goal is to build a network of "regulars." At the end of year one, we have successfully booked 560 events. A majority of those events are from large contracts that we have established. In year two, as our wedding distribution business begins to pick up and we continue to grow our base of large clients, we do 880 events. In year three, we have 1,180 events.

Best-Case Scenario

All cylinders are firing perfectly; we are ramping up growth immediately out of the gate because we have our product ready. Our first year is comparable to our most-likely scenario because we believe that the majority of our growth will happen in year two, when our distribution model is in full swing. We want to be in 40 cities around the United States but could potentially be in many more. We finish year one with 630 events, year two with 1,300, and year three with 1,900. This may seem like a lot, but some of the big companies do close to 3,000 events a year through their franchisees.

Investment Proposal

Based on the current market, I am placing a $250,000 valuation on PictureBooth. This number is based on current intent and interest, which accounts for almost 1,000 events; however, because we have signed no long-term contracts, I cut the revenue generated from those potential events in half for my evaluation. PictureBooth is a start-up venture with no previous track record, meaning that there is substantial risk in investing. No one can predict the future, but I will do everything in my power to make this happen and for everyone to at the very least break even and hopefully generate an attractive return.

Year 1 Income Statement Accrual Basis

	Month 1	Month 2	Month 3	Month 4	Month 5	Month 6	Month 7	Month 8	Month 9	Month 10	Month 11	Month 12	Total
REVENUES													
Cash sales	–	–	20,000	20,000	25,000	40,000	35,000	35,000	25,000	30,000	30,000	20,000	280,000
Charge sales	–	–	–	–	–	–	–	–	–	–	–	–	–
TOTAL SALES	–	–	20,000	20,000	25,000	40,000	35,000	35,000	25,000	30,000	30,000	20,000	280,000
DIRECT EXPENSES													
Direct Costs	–	–	–	–	–	–	–	–	–	–	–	–	–
Salaries	–	1,000	1,000	1,000	1,000	1,000	1,000	1,000	1,000	1,000	1,000	20,000	29,000
Benefits	–	187	187	187	187	187	187	187	187	187	187	3,730	5,409
Rent	–	–	–	–	–	–	–	–	–	–	–	–	–
Utilities	–	–	–	–	–	–	–	–	–	–	–	–	–
Telephone	–	–	120	120	120	120	120	120	120	120	120	120	1,200
Transportation	–	–	–	–	–	–	–	–	–	–	–	–	–
Insurance	100	100	100	100	100	100	100	100	100	100	100	100	1,200
Bad debt expense	–	–	–	–	–	–	–	–	–	–	–	–	–
TOTAL DIRECT EXPENSES	100	100	1,407	1,407	1,407	1,407	1,407	1,407	1,407	1,407	1,407	23,950	36,809
OPERATING MARGIN	(100)	(100)	18,594	18,594	23,594	38,594	33,594	33,594	23,594	28,594	28,594	(3,950)	243,192
General & Admin. Expenses													
Salaries	1,200	1,200	4,800	4,800	4,800	4,800	4,800	4,800	4,800	4,800	4,800	4,800	50,400
Benefits	224	224	895	895	895	895	895	895	895	895	895	895	9,400
Rent	500	500	1,500	1,500	1,500	1,500	1,500	1,500	1,500	1,500	1,500	1,500	16,000
Utilities	–	–	–	–	–	–	–	–	–	–	–	–	–
Telephone	–	–	60	60	60	60	60	60	60	60	60	60	600
Transportation	–	–	100	100	100	100	100	100	100	100	100	100	1,000
Insurance	1,000	–	–	–	–	–	–	–	–	–	–	–	1,000
Legal & Accounting	–	–	200	200	200	200	200	200	200	200	200	200	2,000
Software and Website	15,000	–	–	–	–	–	–	–	–	–	–	–	15,000
Marketing	1,100	750	150	1,200	1,200	1,200	1,200	2,600	1,700	1,200	1,200	1,200	14,700
Office supplies	30	30	30	30	30	30	30	30	30	30	30	30	360
Equipment leases	–	–	–	–	–	–	–	–	–	–	–	–	–
Depreciation-Building	–	–	–	–	–	–	–	–	–	–	–	–	–
Depreciation-Equipment	278	278	278	556	556	556	1,111	1,111	1,667	2,222	2,778	2,778	14,167
TOTAL G&A	19,332	2,982	8,013	9,341	9,341	9,341	9,896	11,296	10,952	11,007	11,563	11,563	124,626
EBIT	(19,432)	(3,082)	10,581	9,253	14,253	29,253	23,697	22,297	12,642	17,586	17,031	(15,513)	118,565
Interest Expense	–	–	–	–	–	–	–	–	–	–	–	–	–
EARNINGS BEFORE TAXES	(19,432)	(3,082)	10,581	9,253	14,253	29,253	23,697	22,297	12,642	17,586	17,031	(15,513)	118,565

BALANCE SHEET—Year 1	Balance	Month 1	Month 2	Month 3	Month 4	Month 5	Month 6	Month 7	Month 8	Month 9	Month 10	Month 11	Month 12
Cash	—	15,846	13,042	23,901	23,709	38,517	68,326	73,134	96,542	90,851	90,659	90,467	77,732
Accounts Receivable	—	—	—	—	—	—	—	—	—	—	—	—	—
Total Current Assets	—	15,846	13,042	23,901	23,709	38,517	68,326	73,134	96,542	90,851	90,659	90,467	77,732
Land	—	—	—	—	—	—	—	—	—	—	—	—	—
Building	—	—	—	—	—	—	—	—	—	—	—	—	—
Equipment	—	10,000	10,000	10,000	20,000	20,000	20,000	40,000	40,000	60,000	80,000	100,000	100,000
-LESS Accum. Depreciation	—	(278)	(556)	(833)	(1,389)	(1,944)	(2,500)	(3,611)	(4,722)	(6,389)	(8,611)	(11,389)	(14,167)
Net Fixed Assets	—	9,722	9,444	9,167	18,611	18,056	17,500	36,389	35,278	53,611	71,389	88,611	85,833
TOTAL ASSETS	—	25,568	22,487	33,067	42,320	56,573	85,826	109,523	131,820	144,462	162,048	179,078	163,565
LIABILITIES													
Accounts payable	—	—	—	—	—	—	—	—	—	—	—	—	—
Short-term loan inc. interest	—	—	—	—	—	—	—	—	—	—	—	—	—
Interest on long-term	—	—	—	—	—	—	—	—	—	—	—	—	—
TOTAL CURRENT	—	—	—	—	—	—	—	—	—	—	—	—	—
Long-term loans	—	—	—	—	—	—	—	—	—	—	—	—	—
Total liabilities	—	—	—	—	—	—	—	—	—	—	—	—	—
OWNERS' EQUITY													
Investment by owner	—	45,000	45,000	45,000	45,000	45,000	45,000	45,000	45,000	45,000	45,000	45,000	45,000
Retained earnings (loss)	—	(19,432)	(22,513)	(11,933)	(2,680)	11,573	40,826	64,523	86,820	99,462	117,048	134,078	118,565
Net equity	—	25,568	22,487	33,067	42,320	56,573	85,826	109,523	131,820	144,462	162,048	179,078	163,565
TOTAL LIAB AND OWNERS	—	25,568	22,487	33,067	42,320	56,573	85,826	109,523	131,820	144,462	162,048	179,078	163,565

CASH FLOW—Year 1	Month 1	Month 2	Month 3	Month 4	Month 5	Month 6	Month 7	Month 8	Month 9	Month 10	Month 11	Month 12	Total
Cash flow from operations													
Receipts													
Cash sales	–	–	20,000	20,000	25,000	40,000	35,000	35,000	25,000	30,000	30,000	20,000	280,000
Accounts Receivable collections	–	–	–	–	–	–	–	–	–	–	–	–	–
Total receipts	–	–	20,000	20,000	25,000	40,000	35,000	35,000	25,000	30,000	30,000	20,000	280,000
Disbursements													
Direct expenses except bad debt	100	100	1,407	1,407	1,407	1,407	1,407	1,407	1,407	1,407	1,407	23,950	36,809
G&A except depreciation	19,054	2,704	7,735	8,785	8,785	8,785	8,785	10,185	9,285	8,785	8,785	8,785	110,460
Interest on long-term	–	–	–	–	–	–	–	–	–	–	–	–	–
Total disbursements	19,154	2,804	9,142	10,192	10,192	10,192	10,192	11,592	10,692	10,192	10,192	32,735	147,268
Net cash flow from operations	(19,154)	(2,804)	10,858	9,808	14,808	29,808	24,808	23,408	14,308	19,808	19,808	(12,735)	132,732
Cash flow from investing activities													
Purchase of Land	–	–	–	–	–	–	–	–	–	–	–	–	–
Purchase of Building	–	–	–	–	–	–	–	–	–	–	–	–	–
Purchase of Equipment	(10,000)	–	–	(10,000)	–	–	(20,000)	(20,000)	(20,000)	(20,000)	(20,000)	–	(100,000)
Net cash flow from investing activities	(10,000)	–	–	(10,000)	–	–	(20,000)	(20,000)	(20,000)	(20,000)	(20,000)	–	(100,000)
Cash flow from financing activities													
Investment by owners	45,000	–	–	–	–	–	–	–	–	–	–	–	45,000
Long-term loan additions (payments)	–	–	–	–	–	–	–	–	–	–	–	–	–
Net cash flow from long-term financing activities	45,000	–	–	–	–	–	–	–	–	–	–	–	45,000
Net cash increase (decrease)	15,846	(2,804)	10,858	(192)	14,808	29,808	4,808	23,408	(5,692)	(192)	(192)	(12,735)	77,732
Short-term Loan increase (decrease)	–	–	–	–	–	–	–	–	–	–	–	–	–
Beginning cash	–	15,846	13,042	23,901	23,709	38,517	68,326	73,134	96,542	90,851	90,659	90,467	–
Ending cash	15,846	13,042	23,901	23,709	38,517	68,326	73,134	96,542	90,851	90,659	90,467	77,732	77,732

Year 2 Income Statement Accrual Basis

	Month 13	Month 14	Month 15	Month 16	Month 17	Month 18	Month 19	Month 20	Month 21	Month 22	Month 23	Month 24	Total
REVENUES													
Cash sales	30,000	30,000	35,000	40,000	45,000	50,000	50,000	35,000	35,000	500,000	30,000	25,000	905,000
Charge sales	–	–	–	–	–	–	–	–	–	–	–	–	–
TOTAL SALES	30,000	30,000	35,000	40,000	45,000	50,000	50,000	35,000	35,000	500,000	30,000	25,000	905,000
DIRECT EXPENSES													
Direct Costs	–	–	–	–	–	–	–	–	–	–	–	–	–
Salaries	5,000	5,000	5,000	5,000	5,000	5,000	5,000	5,000	5,000	5,000	5,000	5,000	60,000
Benefits	933	933	933	933	933	933	933	933	933	933	933	933	11,190
Rent	800	800	800	800	800	800	800	800	800	800	800	800	9,600
Utilities	150	150	150	150	150	150	150	150	150	150	150	150	1,800
Telephone	120	120	120	120	120	120	120	120	120	120	120	120	1,440
Transportation	100	100	100	100	100	100	100	100	100	100	100	100	1,200
Insurance	200	200	200	200	200	200	200	200	200	200	200	200	2,400
Bad debt expense	–	–	–	–	–	–	–	–	–	–	–	–	–
TOTAL DIRECT EXPENSES	7,303	7,303	7,303	7,303	7,303	7,303	7,303	7,303	7,303	7,303	7,303	7,303	87,630
OPERATING MARGIN	22,698	22,698	27,698	32,698	37,698	42,698	42,698	27,698	27,698	492,698	22,698	17,698	817,370
General & Admin. Expenses													
Salaries	9,800	9,800	9,800	9,800	9,800	9,800	9,800	9,800	9,800	9,800	9,800	9,800	117,600
Benefits	1,828	1,828	1,828	1,828	1,828	1,828	1,828	1,828	1,828	1,828	1,828	1,828	21,932
Rent	–	–	–	–	–	–	–	–	–	–	–	–	–
Utilities	–	–	–	–	–	–	–	–	–	–	–	–	–
Telephone	120	120	120	120	120	120	120	120	120	120	120	120	1,440
Transportation	50	50	50	50	50	50	50	50	50	50	50	50	600
Insurance	1,000	–	–	–	–	–	–	–	–	–	–	–	1,000
Legal & Accounting	200	200	200	200	200	200	200	200	200	200	200	200	2,400
Marketing	2,600	1,600	1,500	1,500	1,500	1,500	1,500	1,500	1,500	1,500	1,500	1,500	19,200
Office supplies	50	50	50	50	50	50	50	50	50	50	50	50	600
Equipment leases	300	300	300	300	300	300	300	300	300	300	300	300	3,600
Depreciation- Building	–	–	–	–	–	–	–	–	–	–	–	–	–

(Continued)

	Month 13	Month 14	Month 15	Month 16	Month 17	Month 18	Month 19	Month 20	Month 21	Month 22	Month 23	Month 24	Total
Depreciation-Equipment	3,333	3,333	3,889	3,889	4,444	4,444	4,444	4,444	5,000	5,000	5,556	5,556	53,333
TOTAL G&A	19,281	17,281	17,737	17,737	18,292	18,292	18,292	18,292	18,848	18,848	19,403	19,403	221,706
EBIT	3,416	5,416	9,961	14,961	19,405	24,405	24,405	9,405	8,850	473,850	3,294	(1,706)	595,664
Interest Expense	–	–	–	–	–	–	–	–	–	–	–	–	–
EARNINGS BEFORE TAXES	3,416	5,416	9,961	14,961	19,405	24,405	24,405	9,405	8,850	473,850	3,294	(1,706)	595,664

BALANCE SHEET—Year 2

	Month 12	Month 13	Month 14	Month 15	Month 16	Month 17	Month 18	Month 19	Month 20	Month 21	Month 22	Month 23	Month 24
Cash	77,732	64,482	73,232	67,081	85,931	89,781	118,631	147,481	161,330	155,180	634,030	622,880	626,730
Accounts Receivable	–	–	–	–	–	–	–	–	–	–	–	–	–
Total Current Assets	77,732	64,482	73,232	67,081	85,931	89,781	118,631	147,481	161,330	155,180	634,030	622,880	626,730
Land	–	–	–	–	–	–	–	–	–	–	–	–	–
Building	–	–	–	–	–	–	–	–	–	–	–	–	–
Equipment	100,000	120,000	120,000	140,000	140,000	160,000	160,000	160,000	160,000	180,000	180,000	200,000	200,000
-LESS Accum. Depreciation	(14,167)	(17,500)	(20,833)	(24,722)	(28,611)	(33,056)	(37,500)	(41,944)	(46,389)	(51,389)	(56,389)	(61,944)	(67,500)
Net Fixed Assets	85,833	102,500	99,167	115,278	111,389	126,944	122,500	118,056	113,611	128,611	123,611	138,056	132,500
TOTAL ASSETS	163,565	166,982	172,398	182,359	197,320	216,725	241,131	265,536	274,941	283,791	757,641	760,935	759,230
LIABILITIES													
Accounts payable	–	–	–	–	–	–	–	–	–	–	–	–	–
Short-term loan inc. interest	–	–	–	–	–	–	–	–	–	–	–	–	–
Interest on long-term	–	–	–	–	–	–	–	–	–	–	–	–	–
TOTAL CURRENT	–	–	–	–	–	–	–	–	–	–	–	–	–
Long-term loans	–	–	–	–	–	–	–	–	–	–	–	–	–
Total liabilities	–	–	–	–	–	–	–	–	–	–	–	–	–
OWNERS' EQUITY													
Investment by owner	45,000	45,000	45,000	45,000	45,000	45,000	45,000	45,000	45,000	45,000	45,000	45,000	45,000
Retained earnings (loss)	118,565	121,982	127,398	137,359	152,320	171,725	196,131	220,536	229,941	238,791	712,641	715,935	714,230
Net equity	163,565	166,982	172,398	182,359	197,320	216,725	241,131	265,536	274,941	283,791	757,641	760,935	759,230
TOTAL LIAB AND OWNERS	163,565	166,982	172,398	182,359	197,320	216,725	241,131	265,536	274,941	283,791	757,641	760,935	759,230

CASH FLOW—Year 2	Month 13	Month 14	Month 15	Month 16	Month 17	Month 18	Month 19	Month 20	Month 21	Month 22	Month 23	Month 24	Total
Cash flow from operations													
Receipts													
Cash sales	30,000	30,000	35,000	40,000	45,000	50,000	50,000	35,000	35,000	500,000	30,000	25,000	905,000
Accounts Receivable collections	—	—	—	—	—	—	—	—	—	—	—	—	—
Total receipts	30,000	30,000	35,000	40,000	45,000	50,000	50,000	35,000	35,000	500,000	30,000	25,000	905,000
Disbursements													
Direct expenses except bad debt	7,303	7,303	7,303	7,303	7,303	7,303	7,303	7,303	7,303	7,303	7,303	7,303	87,630
G&A except depreciation	15,948	13,948	13,848	13,848	13,848	13,848	13,848	13,848	13,848	13,848	13,848	13,848	168,372
Interest on long-term	—	—	—	—	—	—	—	—	—	—	—	—	—
Total disbursements	23,250	21,250	21,150	21,150	21,150	21,150	21,150	21,150	21,150	21,150	21,150	21,150	256,002
Net cash flow from operations	6,750	8,750	13,850	18,850	23,850	28,850	28,850	13,850	13,850	478,850	8,850	3,850	648,998
Cash flow from investing activities													
Purchase of Land	—	—	—	—	—	—	—	—	—	—	—	—	—
Purchase of Building	—	—	—	—	—	—	—	—	—	—	—	—	—
Purchase of Equipment	(20,000)	—	(20,000)	—	(20,000)	—	—	—	(20,000)	—	(20,000)	—	(100,000)
Net cash flow from investing activities	(20,000)	—	(20,000)	—	(20,000)	—	—	—	(20,000)	—	(20,000)	—	(100,000)
Cash flow from financing activities													
Investment by owners	—	—	—	—	—	—	—	—	—	—	—	—	—
Long-term loan additions (payments)	—	—	—	—	—	—	—	—	—	—	—	—	—
Net cash flow from long-term financing activities	—	—	—	—	—	—	—	—	—	—	—	—	—
Net cash increase (decrease)	(13,250)	8,750	(6,150)	18,850	3,850	28,850	28,850	13,850	(6,150)	478,850	(11,150)	3,850	548,998
Short-term loan increase (decrease)	—	—	—	—	—	—	—	—	—	—	—	—	—
Beginning cash	77,732	64,482	73,232	67,081	85,931	89,781	118,631	147,481	161,330	155,180	634,030	622,880	77,732
Ending cash	64,482	73,232	67,081	85,931	89,781	118,631	147,481	161,330	155,180	634,030	622,880	626,730	626,730

Year 3 Income Statement Accrual Basis

	Month 25	Month 26	Month 27	Month 28	Month 29	Month 30	Month 31	Month 32	Month 33	Month 34	Month 35	Month 36	Total
REVENUES													
Cash sales	45,000	45,000	45,000	45,000	62,500	62,500	45,000	45,000	50,000	50,000	45,000	50,000	590,000
Charge sales	–	–	–	–	–	–	–	–	–	–	–	–	–
TOTAL SALES	45,000	45,000	45,000	45,000	62,500	62,500	45,000	45,000	50,000	50,000	45,000	50,000	590,000
DIRECT EXPENSES													
Direct Costs	–	–	–	–	–	–	–	–	–	–	–	–	–
Salaries	6,000	6,000	6,000	6,000	6,000	6,000	6,000	6,000	6,000	6,000	6,000	6,000	72,000
Benefits	1,119	1,119	1,119	1,119	1,119	1,119	1,119	1,119	1,119	1,119	1,119	1,119	13,428
Rent	1,500	1,500	1,500	1,500	1,500	1,500	1,500	1,500	1,500	1,500	1,500	1,500	18,000
Utilities	200	200	200	200	200	200	200	200	200	200	200	200	2,400
Telephone	120	120	120	120	120	120	120	120	120	120	120	120	1,440
Transportation	100	100	100	100	100	100	100	100	100	100	100	100	1,200
Insurance	200	200	200	200	200	200	200	200	200	200	200	200	2,400
Bad debt expense	–	–	–	–	–	–	–	–	–	–	–	–	–
TOTAL DIRECT EXPENSES	9,239	9,239	9,239	9,239	9,239	9,239	9,239	9,239	9,239	9,239	9,239	9,239	110,868
OPERATING MARGIN	35,761	35,761	35,761	35,761	53,261	53,261	35,761	35,761	40,761	40,761	35,761	40,761	479,132
General & Admin. Expenses													
Salaries	13,200	13,200	13,200	13,200	13,200	13,200	13,200	13,200	13,200	13,200	13,200	13,200	158,400
Benefits	2,462	2,462	2,462	2,462	2,462	2,462	2,462	2,462	2,462	2,462	2,462	2,462	29,542
Rent	–	–	–	–	–	–	–	–	–	–	–	–	–
Utilities	–	–	–	–	–	–	–	–	–	–	–	–	–
Telephone	120	120	120	120	120	120	120	120	120	120	120	120	1,440
Transportation	–	–	–	–	–	–	–	–	–	–	–	–	–
Insurance	1,000	–	–	–	–	–	–	–	–	–	–	–	1,000
Legal & Accounting	250	250	250	250	250	250	250	250	250	250	250	250	3,000
Marketing	2,600	1,600	1,600	1,600	1,600	1,600	1,600	2,600	2,000	1,600	1,600	1,600	21,600
Office supplies	100	50	50	50	50	50	50	50	50	50	50	50	650
Equipment leases	–	–	–	–	–	–	–	–	–	–	–	–	–
Depreciation-Building	–	–	–	–	–	–	–	–	–	–	–	–	–
Depreciation-Equipment	6,172	6,172	6,789	6,789	7,097	7,097	7,097	7,097	7,097	7,097	7,097	7,097	82,700
TOTAL G&A	25,904	23,854	24,471	24,471	24,779	24,779	24,779	25,779	25,179	24,779	24,779	24,779	298,332
EBIT	9,857	11,907	11,290	11,290	28,482	28,482	10,982	9,982	15,582	15,982	10,982	15,982	180,800
Interest Expense	–	–	–	–	–	–	–	–	–	–	–	–	–
EARNINGS BEFORE TAXES	9,857	11,907	11,290	11,290	28,482	28,482	10,982	9,982	15,582	15,982	10,982	15,982	180,800

BALANCE SHEET – Year 3

	Month 24	Month 25	Month 26	Month 27	Month 28	Month 29	Month 30	Month 31	Month 32	Month 33	Month 34	Month 35	Month 36
Cash	626,730	620,559	638,638	634,517	652,596	677,076	712,655	730,734	747,813	770,492	793,572	811,651	834,730
Accounts Receivable	–	–	–	–	–	–	–	–	–	–	–	–	–
Total Current Assets	626,730	620,559	638,638	634,517	652,596	677,076	712,655	730,734	747,813	770,492	793,572	811,651	834,730
Land	–	–	–	–	–	–	–	–	–	–	–	–	–
Building	–	–	–	–	–	–	–	–	–	–	–	–	–
Equipment	200,000	222,200	222,200	244,400	244,400	255,500	255,500	255,500	255,500	255,500	255,500	255,500	255,500
-LESS Accum. Depreciation	(67,500)	(73,672)	(79,844)	(86,633)	(93,422)	(100,519)	(107,617)	(114,714)	(121,811)	(128,908)	(136,006)	(143,103)	(150,200)
Net Fixed Assets	132,500	148,528	142,356	157,767	150,978	154,981	147,883	140,786	133,689	126,592	119,494	112,397	105,300
TOTAL ASSETS	759,230	769,086	780,993	792,284	803,574	832,056	860,538	871,520	881,502	897,084	913,066	924,048	940,030
LIABILITIES													
Accounts payable	–	–	–	–	–	–	–	–	–	–	–	–	–
Short-term loan inc. interest	–	–	–	–	–	–	–	–	–	–	–	–	–
Interest on long-term	–	–	–	–	–	–	–	–	–	–	–	–	–
TOTAL CURRENT	–	–	–	–	–	–	–	–	–	–	–	–	–
Long-term loans	–	–	–	–	–	–	–	–	–	–	–	–	–
Total liabilities	–	–	–	–	–	–	–	–	–	–	–	–	–
OWNERS' EQUITY													
Investment by owner	45,000	45,000	45,000	45,000	45,000	45,000	45,000	45,000	45,000	45,000	45,000	45,000	45,000
Retained earnings (loss)	714,230	724,086	735,993	747,284	758,574	787,056	815,538	826,520	836,502	852,084	868,066	879,048	895,030
Net equity	759,230	769,086	780,993	792,284	803,574	832,056	860,538	871,520	881,502	897,084	913,066	924,048	940,030
TOTAL LIAB AND OWNERS	759,230	769,086	780,993	792,284	803,574	832,056	860,538	871,520	881,502	897,084	913,066	924,048	940,030

CASH FLOW–Year 3	Month 25	Month 26	Month 27	Month 28	Month 29	Month 30	Month 31	Month 32	Month 33	Month 34	Month 35	Month 36	Total
Cash flow from operations													
Receipts													
Cash sales	45,000	45,000	45,000	45,000	62,500	62,500	45,000	45,000	50,000	50,000	45,000	50,000	590,000
Accounts Receivable collections	—	—	—	—	—	—	—	—	—	—	—	—	—
Total receipts	45,000	45,000	45,000	45,000	62,500	62,500	45,000	45,000	50,000	50,000	45,000	50,000	590,000
Disbursements													
Direct expenses except bad debt	9,239	9,239	9,239	9,239	9,239	9,239	9,239	9,239	9,239	9,239	9,239	9,239	110,868
G&A except depreciation	19,732	17,682	17,682	17,682	17,682	17,682	17,682	18,682	18,082	17,682	17,682	17,682	215,632
Interest on long-term	—	—	—	—	—	—	—	—	—	—	—	—	—
Total disbursements	28,971	26,921	26,921	26,921	26,921	26,921	26,921	27,921	27,321	26,921	26,921	26,921	326,500
Net cash flow from operations	16,029	18,079	18,079	18,079	35,579	35,579	18,079	17,079	22,679	23,079	18,079	23,079	263,500
Cash flow from investing activities													
Purchase of Land	—	—	—	—	—	—	—	—	—	—	—	—	
Purchase of Building	—	—	—	—	—	—	—	—	—	—	—	—	
Purchase of Equipment	(22,200)	—	(22,200)	—	(11,100)	—	—	—	—	—	—	—	(55,500)
Net cash flow from investing activities	(22,200)	—	(22,200)	—	(11,100)	—	—	—	—	—	—	—	(55,500)
Cash flow from financing activities													
Investment by owners	—	—	—	—	—	—	—	—	—	—	—	—	
Long-term loan additions (payments)	—	—	—	—	—	—	—	—	—	—	—	—	
Net cash flow from long-term financing activities	—	—	—	—	—	—	—	—	—	—	—	—	—
Net cash increase (decrease)	(6,171)	18,079	(4,121)	18,079	24,479	35,579	18,079	17,079	22,679	23,079	18,079	23,079	208,000
Short-term loan increase (decrease)	—	—	—	—	—	—	—	—	—	—	—	—	—
Beginning cash	626,730	620,559	638,638	634,517	652,596	677,076	712,655	730,734	747,813	770,492	793,572	811,651	626,730
Ending cash	620,559	638,638	634,517	652,596	677,076	712,655	730,734	747,813	770,492	793,572	811,651	834,730	834,730

Case 1

Big Bottom Market

How Can a Small Restaurant and Specialty Food Store Cope with Highly Seasonal Sales?

After Michael Volpatt, co-owner with Kate Larkin of a successful public relations firm, moved to Guerneville, California, a small town in Sonoma County in the heart of the Russian River valley's wine country, he decided to realize his longtime dream of owning a specialty food and wine store. After meeting Crista Luedtke, the owner of two Guerneville businesses, the Boon Hotel and Spa and Boon Eat and Drink, Volpatt convinced Larkin to join him and Luedtke in opening the Big Bottom Market, a restaurant and specialty food and wine store that also sells local crafts. The three entrepreneurs invested $100,000 of their own money to convert a 1,500-square-foot storefront on Guerneville's Main Street into a restaurant and retail store. They decided to name the restaurant Big Bottom Market after the once-booming logging town's original name, which was inspired by its location in the alluvial flats of the Russian River. "The unique, quirky, and inspiring name fits not only the historical nature of days gone by but also the unique aspects and incredible environment of our restaurant-market," says Volpatt. Originally settled in 1860, the well-to-do town is a now popular destination in the spring, summer, and fall for tourists and visitors who are drawn by the availability of activities on the river, hiking among the giant redwoods in nearby Armstrong Woods, or visiting the valley's many boutique wineries.

Volpatt and Larkin, who are based in New York City, would keep their day jobs but would handle the marketing and finance functions, respectively, for Big Bottom Market. They would count on Luedtke, who has experience in both the food and hospitality industries to manage the business's day-to-day operations.

The renovated storefront features hardwood floors, walls made of barn wood, funky metal chairs at its nine tables, bar stools at the counter, and a "communal table" for large parties or for people who want to mix and mingle. "Think gourmet deli meets farmer's market meets a modern-day general store—now add in 'lumberjack chic' style—that's us," explains Volpatt. Currently, Big Bottom Market serves only breakfast and lunch, but executive chef Tricia Brown, who came to Big Bottom Market from New York City's famous Gramercy Tavern, is pushing the owners to begin offering dinner. Brown's menu is varied and, except for several staple dishes, changes frequently. Some of her most popular dishes include chilled cucumber soup, wild salmon Niçoise salad, green chile cheddar turkey meatloaf, chipotle sweet potatoes, and baguette sandwiches. The Big Bottom Market's signature item is its homemade Big Bottom Biscuits, which come in a multitude of flavors, including regular, cheddar and thyme, ham and cheese, and sea biscuit (house-smoked salmon, capers, and red onions). The biscuit recipe came from Luedtke's mother, who is one of Big Bottom Market's 20 employees and who oversees their baking.

Big Bottom Market opened in July, and in its first year of operation, sales during the busy summer months were strong, averaging between $20,000 and $24,000 per week. In September, sales began tapering off, and by November, sales were down 80 percent. "We thought we'd lose about 30 to 40 percent of our business [in the off-season]," says Volpatt, "but not 80 percent. This is my first time at the rodeo, and we were freaking out." After analyzing their financial statements, their accountant had grim news. "At the rate you're going, you're going to have to close your doors," he said.

Volpatt, Larkin, and Luedtke began to trim Big Bottom Market's expenses immediately. They closed on Mondays and Tuesdays, the slowest days, and trimmed their staff. They gathered their remaining employees and conducted a brainstorming session designed to generate ideas to keep the business afloat during the slow off-season until sales picked up again in the spring. Some of the ideas they came up with included introducing a Big Bottom Market food truck to increase sales and build the company's brand name in the area, adding a catering service, opening for dinner, focusing more on marketing to Guerneville residents, and emphasizing the Big Bottom Market's signature biscuits and selling them through a gourmet wholesaler, such as Bi-Rite in San Francisco.

Questions

C1-1. What steps do you recommend that the owners of Big Bottom Market take to manage cash flow in their seasonal business?

C1-2. Identify the advantages and the disadvantages of each of the ideas that the owners and their employees came up with to help Big Bottom Market survive the slow off-season. Based on your analysis, do you recommend that they pursue any of these options? Explain your reasoning.

C1-3. Identify at least two other options that the owners should consider to get through the slow off-season and the advantages and disadvantages of each one.

C1-4. Develop a two-page marketing strategy for Big Bottom Market. Which social media tools should the owners use? How, specifically, should they put them to work?

Sources: Based on John Grossman, "A Seasonal Business Aims to Survive the Off-Season," *New York Times*, July 11, 2012, *www.nytimes.com/2012/07/12/business/smallbusiness/a-seasonal-business-aims-to-survive-the-off-season.html?pagewanted=all&_r=0;* Carey Sweet, "Guerneville's Big Bottom Market Rolls Out Dinner Service," *Inside Scoop SF*, June 8, 2012, *http://insidescoopsf.sfgate.com/blog/2012/06/08/guerneville%E2%80%99s-big-bottom-market-rolls-out-dinner-service;* "About," Big Bottom Market, *www.bigbottommarket.com/about.html;* John Grossman, "A Tourist-Dependent Business Decides to Reboot," *New York Times*, July 18, 2012, *http://boss.blogs.nytimes.com/2012/07/18/a-tourist-dependant-business-decides-to-reboot/?pagewanted=print.*

MyBizHomepage

Can an Online Company that Provides Easy-to-Use Financial Metrics for Small Companies Recover from a Devastating Cyberattack?

A few years after Peter Justen sold the financial services company that he and several cofounders had started to TD Ameritrade, the serial entrepreneur was ready to launch another company. A self-confessed "numbers guy," Justen recognized that most small business owners struggled to understand their companies' financial statements and the valuable information that the statements could provide them. Justen's research also showed that many business owners used the accounting software QuickBooks to manage the financial aspects of their companies. That led Justen to create MyBizHomepage, a free Web-based financial service that includes MyBizDashboard, a free online financial dashboard aimed at small businesses that extracts the necessary data from their QuickBooks records and presents it in an easy-to-understand dashboard that features simple charts and graphs. "My idea was to simplify things for business owners to give them an easy way to see the problems and opportunities in the numbers of their businesses," he says.

Justen hired a team of programmers and spent the next two years developing the prototype for MyBizHomepage, which uses a series of algorithms to analyze the data extracted from small companies' QuickBooks records to create performance indicators and comparisons against industry standards. The dashboard was designed to reveal meaningful information that entrepreneurs could use to make better business decisions and that might otherwise stay buried in a company's financial records. MyBizHomepage automatically pulls the required values from a company's QuickBooks records and generates easy-to-understand reports on its vital financial components, including accounts payable, accounts receivable, cash available, sales, cost of goods sold, payroll, and working capital. The dashboard also includes alerts that communicate time-sensitive key business indicators to business owners using e-mail and text messaging. "The idea was that by checking the numbers every day, a business owner could see where he was headed," says Justen.

When MyBizHomepage went live in 2008, the site, which was free to small business owners, attracted a great deal of media attention, driving significant traffic to the company's Web site. Justen's business model called for generating a profit by selling advertising to businesses that wanted to reach small business owners using MyBizHomepage. As traffic grew, Justen turned to several private investors, including Joe Silbaugh, a former real estate developer, and Bryan Elicker, an entrepreneur who had recently sold his coffin manufacturing business, to raise the capital he needed to expand the company. The investors also served on the company's board of directors.

Within a few months, Justen and his board received an offer from a large company to purchase the business for nearly $100 million, but they declined the offer. "We hadn't yet tapped the potential of the product, especially among the global audience," says Justen. At that point, the company had only 6,000 customers, and Justen and the board believed that they could increase that number significantly.

Shortly after declining the offer to purchase the company, Justen learned that his chief technology officer (CTO), with whom he had worked for several years, was working with two other managers at MyBizHomepage to launch a similar, competing company. Furious, Justen fired the men and had his attorney send them a "cease-and-desist" letter about starting the competing company. Almost immediately, MyBizHomepage's Web site began to crash regularly, causing problems for its small business customers and credibility problems for the company. Someone also hacked into the e-mail accounts of Justen and his board members and sent false e-mails to everyone in their address books accusing them of unethical and improper business practices. The messages implied that MyBizHomepage was defrauding investors. "It hurts your reputation when someone Googles your name and finds that," says Silbaugh, who had invested more than $1 million in the company.

Justen contacted authorities about the cyberattacks and told them that he suspected that his former CTO was behind them. Only then did he discover that the former CTO was not the person he claimed to be. In fact, he had no official identity at all. He had no driver's license and no credit cards in his name and had filed no tax returns—all of which made tracking him down virtually impossible. Justen and his IT staff ultimately determined that the former CTO had built multiple hidden "backdoor" entrances into MyBizHomepage's software that he could exploit undetected whenever he wanted. Justen knew that the only way to make the site and the software safe again was to shut down the company and rebuild the software from scratch, which would require a capital investment, but he was hesitant to go back to his original investors for more money. If he declared bankruptcy, he and his investors would lose all of the money they had put into the business. He also worried about how much information about the incident to make public because his small business customers had trusted his company with very sensitive information about their companies. He was considering simply shutting down MyBizHomepage and walking away.

Questions

C2-1. Is the way that Peter Justen spotted this business opportunity typical of the way that entrepreneurs come up with creative ideas for the businesses they start?

C2-2. What steps can online companies such as MyBizHomepage take to minimize the effects of cyberattacks?

C2-3. Why did Justen and his board decline the offer from the larger business to purchase MyBizHomepage? Do you think that they made the right decision at the time? Explain.

C2-4. If Justen decides to rebuild his company's software and start a new company, what sources of financing do you recommend that he use? What steps should he take to attract either debt or equity capital? Which sources of financing do you recommend that he avoid? Explain.

C2-5. MyBizHomepage's selection process was obviously flawed. What steps should entrepreneurs take to avoid hiring dishonest employees who have the potential to damage or destroy their companies?

Sources: Based on Darren Dahl, "Struggling to Recover from a Cyberattack," *New York Times*, August 22, 2012, *www.nytimes.com/2012/08/23/business/small-business/struggling-to-recover-from-a-cyberattack.html?pagewanted=all&_r=0*; "MyBizHomepage," Crunchbase, 2013, *www.crunchbase.com/company/mybizhomepage*; "Five Plus," 2013, *www.fiveplus.co*.

Jacquii LLC

Should a Young Entrepreneur Accept a Potential Investor's Terms that Require Her to Give Up Control of Her Business?

Jacqui Rosshandler grew up in Australia but was drawn to New York City, where she worked as legal counsel for an interior design company. She put in long hours at her job, but her goal was to one day own a business of her own, just like her father did back in Australia. "If I am going to work this hard, I want to do it for myself," she recalls thinking. One New Year's Day, with her mouth feeling less than fresh, Rosshandler recalled Odor-Go, a breath mint sold in Australia that really worked. She had never seen a similar product in the United States and decided to start a company to produce and market one. She realized that the best way to eliminate bad breath was to treat the source of the problem, the stomach, rather than its symptoms, which appear in the mouth, as most breath mints do. Rosshandler decided to launch a company, Jacquii LLC, and began working with a contract manufacturer to develop a unique breath-freshening product. "Parsley has been used for generations to freshen breath," she says, "but freshening the mouth only, especially after consuming pungent foods, doesn't get rid of the smell that comes from the stomach. We found that a combination of concentrated peppermint and parsley oils, when dissolved in the stomach, provides this fresh feeling from within. Your breath actually smells good from deep inside, not just superficially from the mouth."

The result of several months of work was a two-step breath-freshening product that Rosshandler named Eatwhatever to give her product a trendy, fun image. Customers swallow a gel cap filled with an all-natural concentration of peppermint and parsley oils and then pop one of the package's small white mints into their mouths for instantly fresh breath. Rosshandler came up with a clever tagline, "2 Steps to Kissable Breath," aimed squarely at her target audience—young people—and hired a package designer to create a clever package. She began marketing her new breath freshener herself, walking boldly into the flagship C.O. Bigelow apothecary store in Manhattan and asking, "Who does the buying here?" She actually met with a buyer and left the store with her first sale. "I had no idea what I was doing," she recalls with a laugh. A month later, a friend who worked in public relations convinced DailyCandy, a popular Web site that focuses on fashion, food, and fun, to mention Eatwhatever, generating $20,000 in orders on her Web site in just 12 hours. With a distributor's help, Rosshandler was able to get Eatwhatever in retail stores such as Zitomer, Ricky's, and Joe Coffee in New York City; Collette in Paris; Terry White Chemists in Sydney; and online at Amazon, Victoria Health, an Shopmasc. Sales volume for the company's first three years of operation was small, never exceeding $40,000.

Rosshandler had used her own money to create her product and bring it to market, but getting widespread distribution and generating significant sales would require a lot more money than she could invest in her small business. The promising business was about to run out of cash, and Rosshandler was considering shutting it down and getting another job. Then, through her network of contacts, Rosshandler met Arthur Shorin, who had recently sold his business, the Topps Company, which is famous for selling bubble gum packaged with collectible baseball cards. Shorin had extensive knowledge and experience in a similar industry and had an impressive network of contacts. Shorin was impressed with Rosshandler and Eatwhatever and offered to invest a minimum of $250,000 (more if necessary) to propel the company's growth. There was a catch, however, and it was a big one. In return for his investment, Shorin would own 75 percent of Jacquii LLC leaving Rosshandler with minority ownership of just 25 percent. He also offered terms that would allow her to regain 15 percent of the company, bringing her total ownership to 40 percent, if Eatwhatever met certain financial and performance benchmarks. The offer also included a job for Rosshandler at Artuitive, Shorin's business incubator for start-up companies.

Rosshandler talked to several friends about the deal, and they advised her to reject Shorin's offer, citing what one friend called "draconian terms"; even if the company met the performance benchmarks, she would still own just 40 percent of what was once "her company." Another pointed out that by giving up 75 percent of her company for an investment of $250,000, she was saying that her company was worth just $333,333 ($250,000 ÷ 75%). Rosshandler listened to her friends' advice but kept thinking, "Isn't owning 25 percent of something better than owning 100 percent of nothing?"

Questions

C3-1. What other potential sources of financing for Jacquii LLC do you recommend Rosshandler explore? Explain.

C3-2. What are the advantages and the disadvantages of using equity capital and debt capital to finance a small business's growth?

C3-3. What steps could Rosshandler have taken to avoid her company's cash flow problem?

C3-4. Should Jacqui Rosshandler accept the investment offer from Arthur Shorin? Explain.

Sources: Based on John Grossman, "Help for a Start-Up, but at a High Price," *New York Times,* January 2, 2013, *www.nytimes.com/2013/01/03/business/smallbusiness/a-start-ups-dilemma-a-lack-of-capital-or-lack-of-control.html?_r=0;* John Grossman, "Why the Founder of a Start-Up Chose to Give Up Control," *New York Times,* January 9, 2013, *http://boss.blogs.nytimes.com/2013/01/09/why-the-founder-of-a-start-up-chose-to-give-up-control;* Caroline Dowd-Higgins, Jacqui Rosshandler, "Eatwhatever," Caroline Dowd-Higgins, March 14, 2010, *http://carolinedowdhiggins.com/2011/05/jacqui-rosshandler-eat-whatever-2;* "Who We Are," Jacquii LLC, *www.eatwhatever.com/who/who-we-are.*

Red Iguana

Should a Family-Owned Restaurant Open a Second Location Nearby to Accommodate the Crowds that Appeared after It Was Featured on the Food Network?

In 1965, Rámon and María Cardenas, immigrants from Mexico, left their jobs in a restaurant in San Francisco and moved to Salt Lake City, Utah, where they purchased a restaurant called Casa Grande. They served authentic Mexican dishes that they learned to make in their native state of Chihuahua, unlike the "Americanized" version of Mexican food that most restaurants serve. "People were not used to their kind of Mexican food," says Lucy Cardenas, the couple's daughter, who now runs the family business with her husband, Bill Coker. Initially, sales were slow, but over time, the restaurant built a loyal customer following, and in 1970 Rámon and Maria moved the restaurant to a downtown location.

By 1985, Casa Grande was struggling, prompting the Cardenases to shutter it and open a new restaurant named Red Iguana in Salt Lake City's working-class west side. With its moderate prices and enormous—and enormously varied—menu, Red Iguana thrived, generating $300,000 in sales in its first year. Red Iguana's sales have increased every year, reaching $1.9 million in 2003. María died in 2002, and an exhausted Rámon was ready to close the restaurant despite its success. Lucy and Bill were eager to take over the family business, but Rámon's traditional views made him hesitant to turn the restaurant over to a woman, even if she was his daughter. "My father would have given my brother the business [simply] because he was a man," recalls Lucy, "but we *bought* it from my Dad. He wanted to sell the business to me, not to me and my husband, because it's part of the family. It was at times painful."

After they purchased Red Iguana for $560,000, Lucy and Bill embarked on a modernization initiative, introduced a computerized restaurant management system, upgraded the electrical system, and purchased the parking lot next door. To improve food consistency across shifts, Lucy had her father cook every dish on the menu in front of her and had the chefs write down every recipe, something that Rámon had never done. By 2008, annual revenue at Red Iguana had reached $3.8 million. The colorful restaurant, with its green and red walls, mismatched furniture, and plastic floral-print tablecloths, had a distinct family vibe and drew a wide variety of customers, ranging from skiers straight off the slopes and families to businesspeople and hipsters.

In 2008, Guy Fiero featured Red Iguana on his Food Network television show, *Diners, Drive-Ins, and Dives*, and sales accelerated. Crowds pushed the restaurant, with room to seat just 100 people, beyond its capacity. Even in the winter off-season during the middle of the week, customers often wait for more than an hour to get a table. On winter weekends, the wait stretches to two hours. During the busy summer months, wait times are even longer. On a typical day, about 700 customers eat lunch or dinner at Red Iguana. Longtime local customers began telling Coker that they had stopped coming to the restaurant because of the long wait times and large number of out-of-town diners. "My experience has taught me that that kind of popularity can flip and become a negative," says Bill.

Lucy and Bill soon learned about another complication for Red Iguana: The city would soon be starting construction on a light rail line, and the bridge on the street on which the restaurant was located that connected the west side of Salt Lake City with the downtown district would be closed for about four months. Because many of their customers, especially businesspeople at lunch, came from downtown and used that bridge to get to the Red Iguana, they estimated that they would lose between 10 and 20 percent of their sales during the construction.

To Lucy and Bill, the solution to their problem was to open a second Red Iguana location. "Anecdotal evidence was that we had gotten too busy, too crowded, and too successful and that [customers] would indeed fill a new restaurant," says Bill. But where should the second Red Iguana be located? Conventional wisdom said that the couple should choose a site far enough away from the original Red Iguana so that the second location would not cannibalize sales at the original restaurant. That would be relatively easy to do because officials in several nearby towns had approached Lucy and Bill within the last several years about opening locations within their city limits. Several mall owners also had been trying to convince them to open locations in their shopping malls. However, the copreneurs lived only a few blocks from Red Iguana and saw that as a significant advantage. "I love being able to get to my business so fast," says Lucy. Opening a second location farther away meant giving up some control over its operations.

While investigating potential sites for a second location, Lucy and Bill heard about an old warehouse located just two blocks away from the original Red Iguana that was for sale for $259,000. They saw a great deal of potential in the old barrel-roofed building, which had a large concrete pad that they could use for patio dining in the warmer months. Preliminary plans indicated that a restaurant in the renovated warehouse could seat 119 diners, slightly more than the original restaurant. However, to purchase the building and renovate it, they would have to convince the Salt Lake City Office of Economic Development and the loan officers at Zions Bank that opening a second location just two blocks from their original restaurant would work. In the initial meeting, the lenders were skeptical.

Questions

C4-1. What management succession issues do you detect in this case? What steps could Rámon Cardenas have taken to avoid them?

C4-2. Lucy and Bill eventually were able to purchase Red Iguana from her father for $560,000. Describe at least three methods that the family could use to establish the value of the business. What factors make placing a value on a business difficult?

C4-3. Should Lucy and Bill open a second location of Red Iguana? If so, what factors should they consider when selecting a location?

C4-4. Where should they locate the second restaurant? Explain.

C4-5. What steps should Lucy and Coker take when they make their loan request to officers at Zions Bank?

Sources: Based on Ian Mount, "Build a Second Restaurant in the First's Shadow?," *New York Times*, August 24, 2011, *www.nytimes.com/2011/08/25/business/smallbusiness/red-iguana-facing-disruptions-ponders-opening-a-2nd-restaurant.html?_r=0*; Ian Mount, "Why Red Iguana Built Red Iguana 2 Right Next Door," *New York Times*, August 30, 2011, *http://boss.blogs.nytimes.com/2011/08/30/why-red-iguana-built-red-iguana-2-right-next-door*; Ian Mount, "You're the Boss; Second Restaurant: Following Up," *New York Times*, August 30, 2012, *http://boss.blogs.nytimes.com/2012/08/28/a-restaurant-makes-the-most-of-being-forced-to-close*; Larry Olmstead, "Mexican Food in Utah? Red Iguana Is the Real Deal," *USA Today*, March 28, 2013, *www.usatoday.com/story/travel/columnist/greatamericanbites/2013/03/28/great-american-bites-mexican-food-utah-red-iguana-real-deal/2027295*.

Baked in the Sun

Should a Small, Wholesale Baker Offer Employees Health Care Coverage at a Significant Cost or Pay a Penalty and Let Employees Buy Coverage on a Government Exchange?

The Affordable Care Act requires businesses that employ 50 or more full-time-equivalent employees to provide adequate health care insurance to all employees who work at least 30 hours per week. Companies that choose not to provide health care coverage must pay a penalty of $2,000 for each worker, excluding the first 30 workers. If businesses provide their employees with health care coverage but the insurance fails to meet the law's minimum requirements, they face a penalty of $3,000 for each worker who gets a federal subsidy through state insurance exchanges. Owners of businesses that have between 50 and 200 full-time-equivalent employees find themselves facing numerous dilemmas: Should they provide health insurance even though they face significant increases in the cost of coverage? Should they forgo offering their employees health insurance, pay the penalty, and allow workers to buy health care coverage on state exchanges? If they fail to provide insurance coverage, will they lose valuable employees? Should they use part-time workers and independent contractors to shrink their workforces enough that their companies fall below the law's 50-employee floor?

Businesses that employ between 50 and 200 employees fall under Affordable Care Act, but their owners say that they lack the purchasing power that large companies have to negotiate the best rates with insurance companies. These entrepreneurs say that they face increasing insurance premiums and declining profits. Rachel Shein and Steve Pilarski, a husband-and-wife team that own Baked in the Sun, a wholesale baker and distributor of fresh pastries located in San Marcos, California, employ 95 workers and must decide what to do about the Affordable Care Act. They estimate that providing health care insurance to their employees could cost an additional $108,000 per year. "Our [annual] revenues are about $8 million, but the food business is a low-margin industry," says Shein. Cutting $108,000 out of our profits, which are just over $200,000, is a big deal."

Shein and Pilarski purchased Baked in the Sun in 1997 and over time expanded their product line to include more than 200 items, including muffins, croissants, cinnamon rolls, bear claws, Danish, scones, bagels, cookies, brownies, crumb cakes, cupcakes, and breads. Their core coffee shop business contracted during the Great Recession, but the copreneurs attracted new customers, such as hotels and hospitals, to keep their business going. Although the bakery now turns out nearly 20,000 items per day in an 18,500-square-foot modern bakery,

Shein and Pilarski are still working to rebuild Baked in the Sun's profitability. In light of the Affordable Care Act, they are considering three options.

Option 1: Offer Every Employee Health Insurance

Because the company already provides managers with insurance, Shein and Pilarski estimate the cost of covering the bakery's additional workers to be $108,000, plus the cost of administering the plan or paying a third-party administrator to manage it. Shein and Pilarski estimate that providing coverage would cost $200 per employee per month, half of which the company would pay and half of which each employee would pay. They suspect that some employees would not sign up for the insurance because they already have coverage through a spouse. "We have offered health insurance to our employees in the past," says Shein. However, many workers chose not to participate in the coverage because they would have had to contribute some of their earnings to cover the premiums. "They are mostly young and healthy," says Shein. "They don't have a lot of extra money, and they would rather have more in their paychecks than health insurance." Covering the additional costs of providing health insurance would require Shein and Pilarski to raise prices by 2 to 4 percent, but they are concerned about the impact of higher prices on sales. Currently, their prices are similar to those of their competitors. Would raising their prices cause a significant decrease in sales?

Option 2: Do Not Provide Health Insurance to Employees and Pay the Penalty

Under this option, Baked in the Sun would pay a $2,000 penalty for every employee (excluding the first 30 workers) to the federal government as part of an "employer-shared responsibility payment." Employees would purchase health care coverage elsewhere, probably on one of the government-operated exchanges. For Baked in the Sun, the penalty would be $130,000, but the company would not incur the additional $10,000 cost of administering an insurance plan. Not providing health insurance goes against the copreneurs' principles, however. Both of them believe that all workers should have health insurance and have wrestled with the problem of providing it for years.

Option 3: Outsource Certain Jobs

By outsourcing work or paying independent contractors to perform certain jobs, Shein and Pilarski could shrink their company enough to fall under the 50-employee threshold and be exempt from the requirements of the Affordable Care Act. "We can outsource the cleaning and make the drivers independent contractors," says Shein. "We can cut the least profitable delivery routes, eliminate the least profitable accounts, or reduce the variety of items we create."

"Our employees will have access to health insurance," says Shein, "whether we provide it or we pay the penalty and they purchase it using a subsidy on the government exchange."

Questions

C5-1. Evaluate the costs and benefits, including both financial and nonfinancial, of each of the three options that Shein and Pilarski have identified.

C5-2. Which option do you recommend for Baked in the Sun? Explain.

C5-3. Are there other options that Shein and Pilarski should consider? Explain.

Sources: Based on Emily Maltby and Sarah E. Needleman, "Sizing Up Health Costs," *Wall Street Journal*, May 30, 2013, pp. B1, B4; Julie Weed, "Questions Abound in Learning to Adjust to Health Care Overhaul," *New York Times*, March 30, 2013, *www.nytimes.com/2013/03/21/business/smallbusiness/a-bakery-with-95-employees-confronts-the-new-health-care-law.html?pagewanted=all*; Julie Weed, "Bakery Owner Talks About Coping with Health Insurance Changes," *New York Times*, March 26, 2013, *http://boss.blogs.nytimes.com/2013/03/26/bakery-owner-talks-about-coping-with-health-insurance-changes*; Nate C. Hindman, "Bakery Owners: Obamacare Will Cut Our Profits in Half," *Huffington Post*, March 22, 2013, *www.huffingtonpost.com/2013/03/22/bakery-obamacare_n_2926322.html*.

Case 6

Bluffton Pharmacy—Part 1

What Can Two New Pharmacy Owners Learn about Their Business from Its Financial Statements?

It has been a little more than two years since Angela Crawford and Martin Rodriguez purchased the Bluffton Pharmacy from Frank White, the previous owner and founder, who had started the pharmacy in 1969. The two had spent many long hours in the store and had learned many valuable lessons as business owners that they had not had the opportunity to learn as employees of large chain pharmacies where they had previously worked.

Crawford and Rodriguez just received an e-mail from their accountant that contained the balance sheet and the income statement for Bluffton Pharmacy for the fiscal year that had just ended. The two financial statements appear below.

Bluffton Pharmacy

Balance Sheet, December 31, 20XX

Assets

CURRENT ASSETS	
Cash	$74,473
Accounts receivable	$112,730
Inventory	$224,870
Supplies	$21,577
Other assets	$10,202
Total current assets	$443,851
FIXED ASSETS	
Autos, net	$33,156
Equipment, net	$35,706
Furniture and fixtures, net	$16,323
Total fixed assets	$85,185
Total assets	$529,036

Liabilities

CURRENT LIABILITIES	
Accounts payable	$29,585
Notes payable	$70,902
Line of credit payable	$32,136
Total current liabilities	$132,623
LONG-TERM LIABILITIES	
Note payable	$170,880
Loan	$93,346
Total long-term liabilities	$264,226

Owner's Equity

Crawford and Rodriguez, capital	$132,187
Total liabilities and owner's equity	$529,036

Bluffton Pharmacy

Income Statement December 31, 20XX

Prescription sales revenue		$2,228,767
All other sales revenue		$167,757
Total sales		$2,396,524
COST OF GOODS SOLD		
Beginning inventory, 1/1/xx	$169,578	
+ Purchases	$1,938,097	
Goods available for sale	$2,107,675	
− Ending inventory, 12/31/xx	$224,870	
Cost of goods sold		$1,882,805
Gross profit		$513,719
OPERATING EXPENSES		
Utilities	$10,305	
Rent	$35,948	
Advertising	$9,586	
Insurance	$9,586	
Depreciation	$5,033	
Salaries and benefits	$321,134	
Computer and e-commerce	$11,983	
Repairs and maintenance	$28,758	
Travel	$4,793	
Professional fees	$3,595	
Supplies	$5,991	
Total operating expenses		$446,712
OTHER EXPENSES		
Interest expense	$24,879	
Miscellaneous expense	$374	
Total other expenses		$25,253
Total expenses		$471,965
Net income		$41,754

To see how their pharmacy's financial position has changed since their first full year of operation, they want to calculate 12 financial ratios. They also want to compare Bluffton Pharmacy's ratios to those of the typical small pharmacy in the industry. The table below shows the value of each of the 12 ratios from last year and the industry median for small pharmacies.

Ratio Comparison

Ratio	Bluffton Pharmacy Current Year	Bluffton Pharmacy Last Year	Pharmacy Industry Median*
LIQUIDITY RATIOS			
Current ratio		3.41	4.71
Quick ratio		1.72	2.42
LEVERAGE RATIOS			
Debt ratio		0.70	0.62
Debt to net worth ratio		2.23	2.1
Times interest earned ratio		3.04	3.9
OPERATING RATIOS			
Average inventory turnover ratio		10.90	11.7 times/year
Average collection period ratio		14.0	15.0 days
Average payable period ratio		5.0	14.0 days
Net sales to total assets ratio		4.75	4.68
PROFITABILITY RATIOS			
Net profit on sales ratio		1.94%	2.9%
Net profit to assets ratio		9.20%	8.2%
Net profit to equity ratio		29.21%	48.0%

* From Risk Management Association Annual Statement Studies and National Community Pharmacists Association.

"Let's see how our ratios compare to last year's numbers," said Angela.

"I hope that we're headed in the right direction," said Martin.

"There's only one way to find out," said Angela with a slight hint of tension in her voice.

Questions

C6-1. Calculate the 12 ratios for the Bluffton Pharmacy for this year.

C6-2. How do the ratios you calculated for this year compare to those for the pharmacy last year? What factors are most likely to account for those changes?

C6-3. How do the ratios you calculated for this year compare to those of the typical company in the industry? Do you spot any areas that could cause the company problems in the future? Explain.

C6-4. Develop a set of specific recommendations for improving the financial performance of Bluffton Pharmacy using the analysis you conducted in questions 1 to 3.

Case 7

Bluffton Pharmacy—Part 2

How Should the Owners of a Small Pharmacy Create a Cash Flow Forecast for Their Business?

It has been a little more than two years since Angela Crawford and Martin Rodriguez purchased the Bluffton Pharmacy from Frank White, the previous owner and founder, who had started the pharmacy in 1969. Although Crawford and Rodriguez have prepared budgets for Bluffton Pharmacy and have analyzed their financial statements using ratio analysis, they have not created a cash flow forecast. During a recent meeting, their banker explained the importance of a reliable cash flow forecast, telling them that banks traditionally are "cash flow lenders." Bankers appreciate strong balance sheets and income statements, but

they are most interested in a company's cash flow because they know that positive cash flow is required to repay a loan.

Crawford and Rodriguez expect sales to increase 4.5 percent next year to $2,504,368. Credit sales account for 79 percent of total sales, and the company's collection pattern for credit sales is 11 percent in the same month in which the sale is generated, 63.5 percent in the first month after the sale is generated, and 22 percent in the second month after the sale is generated. The pharmacy's cost of goods sold is 77.4 percent, and vendors grant "net 30" credit terms, which means that the pharmacy pays for the goods it purchases every month in the following month. Crawford and Rodriguez have been working with their accountant to develop the following estimates for their sales and expenses for the upcoming year:

	Jan	Feb	Mar	Apr	May	Jun	Jul	Aug	Sep	Oct	Nov	Dec
Sales	$230,402	$237,915	$215,376	$177,810	$175,306	$172,801	$162,784	$167,793	$197,845	$247,932	$255,445	$262,959
Other cash receipts	105	55	60	75	85	55	65	60	65	85	95	110
Rent	3,083	3,083	3,083	3,083	3,083	3,083	3,083	3,083	3,083	3,083	3,083	3,083
Utilities	1,049	1,083	980	809	798	787	741	764	901	1,129	1,163	1,197
Advertising	1,150	1,188	1,075	888	875	863	813	838	988	1,238	1,275	1,313
Insurance	–	–	2,700	–	–	2,700	–	–	2,700	–	–	2,700
Salaries, wages, and benefits	27,404	27,515	27,182	26,627	26,590	26,553	26,405	26,479	26,923	27,663	27,774	27,885
Computer system and E-commerce	1,042	1,042	1,042	1,042	1,042	1,042	1,042	1,042	1,042	1,042	1,042	1,042
Repairs and maintenance	2,000	2,000	2,000	2,000	2,000	2,000	2,000	2,000	2,000	2,000	2,000	2,000
Travel	–	–	150	–	5,000	–	–	–	200	–	–	–
Professional fees	–	–	–	–	–	–	–	–	–	–	–	3,900
Supplies	644	665	602	497	490	483	455	469	553	693	714	735
Loan payments	2,073	2,073	2,073	2,073	2,073	2,073	2,073	2,073	2,073	2,073	2,073	2,073
Other	50	50	50	40	40	40	40	45	50	50	50	50

Actual sales for the last two months, November and December, were $272,357 and $315,458. The company's cash balance as of January 1 is $74,473. The interest rate on Bluffton Pharmacy's current line of credit is 8.25 percent, and whatever the pharmacy borrows must be repaid the following month (with interest), even if it must borrow again in that month. The entrepreneurs have established a minimum cash balance of $15,000.

Questions

C7-1. Develop a monthly cash budget for Bluffton Pharmacy for the upcoming year.

C7-2. What recommendations can you offer Angela Crawford and Martin Rodriguez to improve their pharmacy's cash flow?

C7-3. If you were Bluffton Pharmacy's banker, would you be comfortable extending a line of credit to the pharmacy? Explain.

Case 8

United By Blue

Can an Eco-Friendly Apparel Company Afford to Stay Mission Focused in the Face of High Costs?

Brian Linton grew up in Singapore and Japan and, even as a child, was fascinated by water, particularly the ocean. "I was fortunate to travel the world and go to a lot of different beaches," he says. "I loved all things aquatic and had about 30 fish tanks in my room. These experiences put me in beautiful locations as well as places that were so littered with trash that you couldn't see the sand. Seeing trash and polluted oceans was especially discouraging for me." In 2006, while still a student at Temple University, Linton started Sand Shack, a business that sells beach-themed jewelry to stores along the East Coast (and that he still owns). He donated a portion of the company's sales to nonprofit organizations dedicated to ocean conservation but wanted to "do something that was more concrete and tangible than giving money to nonprofits," he says.

In 2010, Linton launched United By Blue, an eco-friendly apparel company, with the idea that this business would be different. The name Linton gave his company is inspired by his belief that water unites all of humankind. "The water of the world is what we all need to live," he says. "It's what unites everything, and it's often the most mistreated part of the world." In fact, people dump an estimated 14 billion tons of trash into the earth's oceans each year, and Linton wanted to do something about it. Rather than merely give away money that may or may not accomplish worthy environmental goals, United By Blue, which is based in Philadelphia, would associate a tangible environmental action for every sale that it makes. Linton pledged to sponsor cleanups that would remove one pound of trash (which primarily is plastic) from the world's oceans and waterways for every product that United By Blue sells. "Our cleanups are the bedrock of our company and allow us to engage with thousands of volunteers and inspire participation in the 'blue' movement," he says. So far, the company has hosted nearly 100 beach or waterway cleanups and has removed more than 140,000 pounds of trash.

Two years after starting United By Blue, Linton was analyzing the company's financial statements and noticed something alarming: Its wholesale gross profit margin had shrunk from 60 percent to just 15 percent, a pattern that was unsustainable. T-shirts are the company's best-selling products, and customers who believe in the company's mission are buying them through retailers such as Urban Outfitters, Whole Foods, and other small, independent shops for a retail price of $29.50. United By Blue sells the shirts to retailers for $14.50, but some of the company's larger customers receive discounts and pay less. The environmentally friendly T-shirts, which are produced in India, cost more to make and package than standard T-shirts because Linton used softer, more expensive slub cotton rather than the traditional jersey cotton that most companies use to make T-shirts. He also insisted on avoiding the use of plastic in any of the company's packaging because of its negative impact on the environment, choosing instead to use biodegradable packaging made from banana fiber and paper hangtags made from recycled elephant dung that is infused with bluebell flower seeds so that customers can actually plant the tags and grow flowers. The banana fiber wrapping alone costs 50 cents each—50 times the cost of a plastic bag. In addition, each cleanup that United By Blue sponsored costs between $2,000 and $5,000, a significant expense for a small company with annual sales of less than $1 million.

Linton turned to his spreadsheet and calculated that he would have to raise its wholesale price to $16.50, which in turn would mean that retailers applying the standard markup would charge customers a final price of $34 for United By Blue T-shirts. He realized that raising the wholesale price to $16.50 probably would cause two of the company's largest customers, Urban Outfitters and Whole Foods, which together account for 25 percent of United By Blue's sales, to drop the shirts. "Large retailers want discounts to maintain [gross profit] margins around 60 percent," he says. Losing major retailers also would make selling to small, independent shops more difficult as well. "[Having big-name retailers as customers] helped us build legitimacy," says Linton. If he raised prices, the company's small retail shops might drop the line as well. If he did not raise prices, he estimated that United By Blue would run out of cash in less than six months. Another option is to reduce the shirts' cost by using less expensive plastic bags and hangtags and traditional jersey cotton, which also has a negative impact on the environment. United By Blue has 50 cleanup events planned for the upcoming year, and eliminating them would save between $100,000 and $250,000. However, all of these changes would go against the company's mission and the reason that Linton started United By Blue in the first place. "Our cleanups are the bedrock of our company," he says. "My grand vision is to use the power of business to leave a positive impact on this world."

Questions

C8-1. What risks does United By Blue face if Linton raises the prices of its T-shirts? What risks does the company face if it fails to raise its T-shirt prices?

C8-2. Should Linton abandon United By Blue's mission as an active, eco-friendly apparel company to lower the cost of the company's T-shirts? What are risks of taking this approach?

C8-3. Should United By Blue raise the wholesale price of its T-shirts from $14.50 to $16.50? Explain.

C8-4. If Linton decides to raise the price of United By Blue's T-shirts, what steps should he take to communicate the price increases to the company's retailers and its final customers?

Sources: Based on Issie Lapowsky, "Case Study: United By Blue's Eco-Friendly Values Sent Costs Soaring. Was It Time to Put Money Before Mission?," *Inc.*, April 2012, pp. 99–101; Mike Sullivan, "The Grand Vision Is to Use the Power of Business to Leave a Positive Impact on This World," *M.O.*, May 26, 2011, *www.mo.com/brian-linton-united-by-blue*; "Entrepreneur Profile: Brian Linton, Founder of United By Blue," *Elite Daily*, September 27, 2012, *http://elitedaily.com/money/entrepreneurship/entrepreneur-profile-brian-linton-founder-united-blue*; Peter Key, "Temple Alums Do Good: United By Blue," *Philadelphia Business Journal*, March 8, 2013, *www.bizjournals.com/philadelphia/print-edition/2013/03/08/temple-alums-do-good.html?page=all*; "Our Story," United By Blue, *www.unitedbyblue.com/our-story*.

Socedo

Should an Entrepreneur Close or Sell One of His Businesses to Focus on the Other or Split His Time between the Two Companies?

At age 16, Aseem Badshah showed his entrepreneurial prowess by starting a social networking Web site that allowed students to collaborate on their schoolwork outside the classroom. Several years later, when he graduated from the Foster School of Business at the University of Washington, Badshah collaborated with Kevin Yu, whom he met at a University of Washington entrepreneurship event, to launch Uptown Treehouse, a business based in Los Angeles that creates social media campaigns on Facebook, Twitter, LinkedIn, StumbleUpon, and Outbrain for media and product companies, including several large companies such as Microsoft and Guess. Within three years, Uptown Treehouse had grown to seven employees and was generating a profit of $300,000 on annual sales of $1.3 million.

While Badshah and Yu, who serves as Uptown Treehouse's chief technology officer, were building the company, they created cloud-based software that searches social media and other Internet sources to generate potential sales leads for companies. "Our algorithms find relevancy in the unstructured nature of social media content," explains Badshah. The software scours various social media outlets for key words that can identify potential customers; companies then start conversations with those potential customers through social media. The product identifies "the right individuals to build relationships with and helps make a connection with them via social media," says Badshah. Uptown Treehouse's clients were so interested in the software that Badshah and Yu launched another company, Socedo, to market the product. "Inspiration hit us when we executed a social media campaign that relied on building personal relationships with potential customers through a sales pipeline rather than just posting content to social networks," says Badshah. "The results of the campaign far exceeded our expectations, and we knew we were on to something. We quickly realized that the technology to identify and engage with potential customers on social media did not exist yet, and Socedo was born." They selected the name for their company by combining the words "social" and "succeed" to create a short name that they considered easily "brandable." Beta tests indicate that 20 percent of the sales leads that Socedo generates for clients are potential customers.

Badshah is torn between his two companies. He could invest his time solely in building Uptown Treehouse. Doing so involves less risk because the company is established, profitable, and growing. Although Uptown Treehouse does not offer as much growth potential as Socedo does, it would require a smaller capital investment over time and would not need as many employees to operate efficiently.

Badshah could sell or close Uptown Treehouse and devote his time and energy to building Socedo. This option is riskier because Socedo is a high-tech start-up in the early stage; the company also would require Badshah to invest (and raise) much more capital and hire more employees than Uptown Treehouse. However, Socedo offers the potential of much higher returns than Uptown Treehouse because the company sells a software product that is aimed at businesses and that could achieve large volume quickly rather than a customized service the revenue stream of which is limited by the number of clients it can serve.

Badshah could hire a professional manager to run Uptown Treehouse and use the profits it generates to finance Socedo's growth. There is danger in this option, however, that involves tying the two companies together financially. Losing just one client at Uptown Treehouse could have a serious impact on the financing available for Socedo during a time when the start-up is most vulnerable. Badshah also wonders whether the cash that Uptown Treehouse generates would be sufficient to fuel what he expects will be very rapid growth at Socedo.

The final option that Badshah is considering is raising money from equity investors to finance Socedo's growth so that the company can grow quickly. "We have bootstrapped until now and are currently looking for angel financing," he says. "Angel funding will help us accelerate and take advantage of the huge opportunity in front of us." Raising capital would allow Badshah to reinvest the earnings from Uptown Treehouse back into that company to keep it healthy and growing.

Badshah realizes that he is at an entrepreneurial crossroads. He wonders whether handing the leadership of Uptown Treehouse over to a manager will be good for the company's future. Will he regret putting a successful company at risk to take a chance at making a much riskier venture a success? Will he have the time, energy, and capital to build both companies at the same time?

Questions

C9-1. Which option should Aseem Badshah pursue? Why?

C9-2. If he chooses to pursue equity investors to finance Socedo's growth, what steps should he take to find investors and convince them to put equity capital into the company?

C9-3. What advice can you offer Badshah about his decision?

Sources: Based Julie Weed, "When Your First Company Is Working, but Another Is Beckoning," *New York Times*, May 29, 2013, *www.nytimes .com/2013/05/30/business/smallbusiness/when-your-first-company-is-working-but-you-may-have-a-better-idea.html?pagewanted=all&_r=0*; Julie Weed, "Which Start-Up Should This Entrepreneur Pursue?," *New York Times*, May 29, 2013, *http://mobile.nytimes.com/blogs/boss/2013/05/29/can-one-owner-build-two-start-ups-at-once*; John Cook, "Start-Up Spotlight: Socedo Dives into Social Media to Help You Find Sales Leads," *Geek Wire*, May 9, 2013, *www .geekwire.com/2013/socedo*; "About" Uptown Treehouse, *http://uptowntree-house.com/about.*

Case 10

EasyLunchboxes

How Should an Entrepreneur Use Social Media to Market Her Home-Based Business that Sells Lunch Boxes?

Kelly Lester is a talented singer, actress, wife, mother to three daughters, and entrepreneur. Lester started her first company, a business that sold decorative light switch covers, in 1996 when the Web was in its infancy. Even then, Lester saw the power and the marketing potential that the Web provided businesses, particularly small businesses that lack the massive marketing budgets that their larger rivals have. In addition to selling switch plates through retail shops and museum gift stores, she built a Web site, *www.switchplates .com*, and began selling online. In those days, competition in e-commerce was slim; when users typed the phrase "switch plate" into a search engine, Lester's Web site was listed first, a benefit that generated significant online sales for her company.

Lester sold the business, and in 2009, with growing daughters, she was inspired to start a second company. "I'm sad when I see what a lot of kids bring to school for lunch," she says. "I'm even sadder when I see what school districts offer our kids for 'school lunch.'" All three of her daughters were in school, and Lester always packed healthy lunches for them but found herself spending too much time preparing and packing lunches. "I'm very concerned about my family's health and nutrition, but as a busy mom, I like to spend as little time as possible in the kitchen," says Lester. "I'm all about fresh, healthy, and fast."

When Lester was in school herself, she often packed lunches for her brothers, but that involved tossing a few cheese sandwiches, potato chips, and Oreo cookies into brown paper bags. Lester wanted something better for her children and to pack the same lunch for all three girls "so I didn't have to think so hard," she says. She soon found that washing, filling, and packing nine separate containers (three for each girl) with food every weekday was driving her "absolutely insane." She began searching store shelves for lunchboxes that made organizing easy but found nothing. Web searches proved no more fruitful. "That's why I created the EasyLunchbox System," she says.

She began calling plastics manufacturers in the United States, none of whom showed much interest in her idea. Finally, one manufacturer told her that his company could produce a single-lid bento-style (with compartments) plastic lunchbox, but he would first have to create a mold, which would cost $75,000. Lester formed a company, EasyLunchboxes, and found an international broker, who began contacting plastic manufacturers in China. There she found a company that would produce the lunchbox she had designed for far less than any of the domestic companies could. "I'm a thrifty shopper," she explains.

Transactions with the foreign supplier did not always go smoothly, however. Lester had to send one of the first shipments back three times because of quality issues and failure to meet

the Food and Drug Administration's standards. Today, Lester has an independent company test samples of every shipment to make sure they meet all safety standards. The polypropylene plastic used to manufacture the EasyLunchboxes are BPA free, and the cooler bags that keep items in the lunchboxes cold are tested for lead. Because of the time required to ship products from China and unpredictable interruptions in the supply chain, Lester has learned to keep a large inventory of EasyLunchboxes in a warehouse in the United States. On one occasion, the company that manufactures the lunchboxes was closed for three months. "They just shut the power grid off," says Lester. No one knew when the power would come back on so that the factory could get finish her order. "It was very stressful," she says.

Lester decided to use the Web as the primary marketing tool for her company. When Lester launched the Web site for EasyLunchboxes, she quickly discovered that the world of e-commerce had changed dramatically. Her company's Web site did not appear near the top of any of the major search engines. "If computers had crickets," she says, "we had crickets." If you're not on page one of search engines, you don't exist. Lester, who bills herself as "Mom and CEO" on the site, is determined to generate "buzz" for her company and believes that social media is one of the best tools for accomplishing that. Her primary target customer is a busy mom with children who are in school. She also wants to build a recognizable brand for EasyLunchboxes. The question in her mind is "How do I go about doing that?"

Questions

C10-1. Do you believe that Kelly Lester's business is a candidate for reshoring? Explain.

C10-2. Develop a social media marketing plan for EasyLunchboxes. On which social media should Lester focus her efforts? Why? What specific tactics should she employ in each one of the social media that you recommend?

C10-3. What steps should she take to transform EasyLunchboxes into a recognizable brand name?

C10-4. What bootstrap marketing techniques should Lester use to promote her company?

Sources: Based on Sarah Maraniss Vander Schaaf, "How Do You Become #1 on Amazon? Ask the Mother of Reinvention," *Lunch Box Mom*, July 10, 2011, *http://lunchboxmom.blogspot.com/2011/07/how-do-you-become-1-on-ama-zon-ask.html*; Phil Mershon, "9 Small Business Social Media Success Stories," *Social Media Examiner*, January 18, 2012, *www.socialmediaexaminer.com/9-small-business-social-media-success-stories*; Juan Felix, "Discover 'Social Secrets' While Having Lunch with Kelly Lester," *Mirror Yourself*, September 8, 2011, *www.mirroryourself.nl/social-media-secrets-of-lunch-with-kelly-lester*; Ellyn Davidson, "How Social Media Helped a Little Lunch Box Hit the Big Time," Brogan Partners, September 5, 2012, *www.brogan.com/blog/how-social-media-helped-little-lunch-box-hit-big-time*; Connie Bensen, "Social Success Story: EasyLunchboxes Lessons in Branding, Content Creation, and Community," *Dell Social Business Connection*, December 9, 2012, *http://en.community.dell.com/dell-groups/sbc/b/weblog/archive/2012/12/09/social-success-story-easylunchboxes-lessons-in-branding-content-creation-amp-community.aspx*.

Endnotes

Chapter 1

1. Robert W. Farlie, Kauffman Index of Entrepreneurial Activity, 1996–2011, March 2012, Kauffman Foundation, p. 2.
2. Akbar Sadeghi, "The Births and Deaths of Business Establishments in the United States," *Monthly Labor Review,* December 2008, p. 3.
3. Valerie J. Calderon, "U.S. Students' Entrepreneurial Energy Waiting to Be Tapped," *Gallup*, October 13, 2011, www.gallup.com/poll/150077/Students-Entrepreneurial-Energy-Waiting-Tapped.aspx.
4. Alyson Shontell, "Tons of College Students Are Starting Businesses—and a Lot Are Doing It Because They Can't Find Jobs," *Business Insider*, February 17, 2011, www.businessinsider.com/youth-entrepreneurship-council-2011-2?op=1.
5. Donna J. Kelley, Slavica Singer, and Mike Herrington, *Global Entrepreneurship Monitor: 2011 Global Report*, Babson College, Universidad del Desarollo, and Universiti Tun Abdul Razak, pp. 10–11.
6. Katie Morell, "Fourth of July Special: Small Businesses of the Founding Fathers," *Open Forum*, July 4, 2012, www.openforum.com/articles/fourth-of-july-special-small-businesses-of-the-founding-fathers.
7. "The Woman Who Took on Zimbabwe's Security Men and Won," *BBC News*, July 6, 2012, www.bbc.co.uk/news/world-africa-17896466.
8. Adrian Woolridge, "Global Heroes," *The Economist*, March 14, 2009, p. 8.
9. Jeffry A. Timmons, "An Obsession with Opportunity," *Nation's Business*, March 1985, p. 68.
10. Michelle Juergen, "Entrepreneur: A Powerful Team Gains a Cultural Foothold," *Entrepreneur*, June 2012, p. 144.
11. "Age Groups and Sex: 2010 Census Summary File 2," U.S. Census Bureau, http://factfinder2.census.gov/faces/tableservices/jsf/pages/productview.xhtml?pid=DEC_10_SF2_QTP1&prodType=table; Robert W. Farlie, *Kauffman Index of Entrepreneurial Activity*, 1996–2011, March 2012, Kauffman Foundation, p. 4.
12. Eric Schurenberg, "What's an Entrepreneur? The Best Answer Ever," *Inc.*, January 9, 2012, www.inc.com/eric-schurenberg/the-best-definition-of-entepreneurship.html.
13. Dave Mosher, "Inside Brooklyn's DIY Spacesuit Startup," *Wired*, July 9, 2012, www.wired.com/wiredscience/2012/07/new-york-spacesuits; Evan Rodgers, "Final Frontier Design's Third-Generation Spacesuit Reaches Kickstarter Goal," *The Verge*, July 18, 2012, www.theverge.com/2012/7/18/3167671/final-frontier-design-third-generation-spacesuit-kickstarter; Brad Hamilton, "Runway to Liftoff," *New York Post*, April 3, 2011, www.nypost.com/p/news/local/runway_to_liftoff_tfvuyf8x12vJi0M3g2yqhO?CMP=OTC-rss&FEEDNAME=.
14. David McClellan, *The Achieving Society* (Princeton, NJ: Van Nostrand, 1961), p. 16; Nancy Michaels, "Entrepreneurship: An Alternative Career Choice," *U.S. News & World Report*, March 24, 2003, p. 45; Susan Ward, "So You Want to Start a Small Business? Part 2: The Personality of the Entrepreneur," *Small Business Canada*, http://sbinfocanada.about.com/library/weekly/aa082900b.htm.
15. Ned Smith, "Gutsy Entrepreneur's Rum Hasn't Lost Its Punch," *Business News Daily*, July 9, 2012, www.businessnewsdaily.com/2808-entrepreneur-rum-brand.html.
16. Charles Gerena, "Nature vs. Nurture," *Region Focus*, Fall 2005, p. 19.
17. Max Chafkin, Donna Fenn, April Joyner, Ryan McCarthy, Kate Pastorek, and Nitasha Tiku, "Cool, Determined, and Under 30," *Inc.*, October 2008, p. 101.
18. Teri Evans, "Boston Beer Co.'s Jim Koch on Self Reliance," *Entrepreneur*, November 30, 2011, www.entrepreneur.com/article/220792.
19. Jeffrey Shuman and David Rottenberg, "Famous Failures," *Small Business Start-Ups*, February 1999, pp. 32–33; Francis Huffman, "A Dairy Tale," *Entrepreneur*, February 1999, p. 182; "Gail Borden," Famous Texans, www.famoustexans.com/GailBorden.htm.
20. Sabin Russell, "Being Your Own Boss in America," *Venture*, May 1984, p. 40.
21. Jeff Stibel, "Entrepreneurs Don't Need Work-Life Balance," *Harvard Business Review*, April 2, 2012, http://blogs.hbr.org/cs/2012/04/work-life_balance_is_overrated.html.
22. Jennifer Wang, "Business as Art," *Entrepreneur*, March 2012, p. 51.
23. Leigh Buchanan, "40,882 Percent Three-Year Growth Is the New Black," *Inc.*, September 2011, pp. 81–84.
24. Stephanie N. Mehta, "Young Entrepreneurs Are Starting Business After Business," *Wall Street Journal*, March 19, 1997, p. B2.
25. Roger Rickleffs and Udayan Gupta, "Traumas of a New Entrepreneur," *Wall Street Journal*, May 10, 1989, p. B1.
26. Teri Evans, "Boston Beer Co.'s Jim Koch on Self Reliance," Entrepreneur, November 30, 2011, www.entrepreneur.com/article/220792.
27. Kris Frieswick, "I Once Was Lost," *Inc.*, June 2012, pp. 92–98; Stephen Kurczy, "Catherine Rohr Helps Ex-Cons Return to Society by Learning to Start Businesses," *Christian Science Monitor*, April 23, 2012, www.csmonitor.com/layout/set/print/content/view/print/491922.
28. John Case, "The Origins of Entrepreneurship," *Inc.*, June 1989, p. 52.
29. Adrian Woolridge, "Global Heroes," *The Economist*, March 14, 2009, p. 4.
30. Jennifer Wang, "Business as Art," *Entrepreneur*, March 2012, p. 51.
31. Gwen Moran, "Independent Living," *Entrepreneur*, March 2012, pp. 70–71; Brian Dolan, "Aging in Place Startup Independa Tops Up $2.35 Round," *MobiHealthNews*, April 25, 2012, http://mobihealthnews.com/17142/aging-in-place-startup-independa-tops-up-2-35m-round.
32. Jessica Shambora, "Eileen Fisher's Timeless Vision," *Fortune*, September 26, 2011, pp. 49–52.
33. Tina Traster, "Romantic Disappointment Sparks a Business," *Crain's New York Business*, July 6, 2012, www.crainsnewyork.com/article/20120706/SMALLBIZ/120709950.
34. Jia Lynn Yang, "Gang Green," *Fortune*, April 27, 2009, pp. 12–13.
35. Nick Carbone, "A Super-Sized Sandbox Just for Adults," *Time*, September 4, 2011, http://newsfeed.time.com/2011/09/04/a-super-sized-sandbox-just-for-adults; "The Dig This Story," Dig This, www.digthisvegas.com/the-dig-this-story.
36. Kara Ohngren, "Flight Plan," *Entrepreneur*, January 2012, p. 66.
37. Jodi Helmer, "Success in a Flash," *Entrepreneur*, June 2012, p. 86.
38. Catherine Kaputa, "5 Contrarian Lessons from Successful Entrepreneurs," *Fast Company*, July 6, 2012, www.fastcompany.com/1842005/five-contrarian-lessons-from-successful-entrepreneurs.
39. Dinah Eng, "The Creative Spark Behind the Viking Oven," *Fortune*, January 17, 2011, p. 46.
40. Darren Dahl, "Top 10 Reasons to Run Your Own Business," *Inc.*, January 21, 2011, www.inc.com/guides/201101/top-10-reasons-to-run-your-own-business.html.
41. "Entrepreneurship: A Question of Gender," *The Exchange*, Summer 2012, p. 6.
42. Carolyn Horowitz, "The View from the Top," *Entrepreneur*, January 2012, pp. 44–48.
43. Roger P. Levin, "You've Got to Love It or Leave It," *Success*, December 2000/January 2001, p. 22.

44. "The Knowledge," *FSB*, May 2009, p. 90.

45. "The Forbes 400: The Richest People in America," *Forbes*, September 21, 2011, www.forbes.com/forbes-400/#p_1_s_arank_All%20industries_All%20states_All%20categories_.

46. "Most Middle Class Millionaires Are Entrepreneurs," *Small Business Labs*, May 13, 2008, http://genylabs.typepad.com/small_biz_labs/2008/05/most-middle-cla.html; Thomas Kostigen, "The 'Middle Class Millionaire,'" *MarketWatch*, March 5, 2008, www.marketwatch.com/news/story/rise-middle-class-millionaire-reshaping-us/story.aspx?guid=%7B6CF2AF9B-7A4C-487E-8AD1-8B49A6A87104%7D.

47. "40 Under 40," *Fortune*, November 7, 2011, pp. 129–138; "The World's Billionaires: Kevin Plank," *Forbes*, April 24, 2012, www.forbes.com/profile/kevin-plank; Kevin Plank and Mark Hyman, "How I Did It," *Inc.*, December 2003, pp. 102–104.

48. *What Americans Think About Business: 2011 American Affairs Pulse Survey*, Public Affairs Council and Princeton Survey Research Associates International, p. xxii.

49. "Distrust, Discontent, Anger, and Partisan Rancor: The People and Their Government," Pew Research Center, April 18, 2010, http://people-press.org/2010/04/18/section-3-government-challenges-views-of-institutions.

50. Dennis Jacobe, "Nearly Half of Small Business Owners May Never Retire," *Gallup Economy*, October 1, 2010, www.gallup.com/poll/143351/nearly-half-small-business-owners-may-retire.aspx.

51. Margaret Littman, "The Wave Riders," *Entrepreneur*, August 2011, pp. 16–17; "Paddleboard," *Florida Travel + Life*, March/April 2012, pp. 34–36.

52. Gayle Sato-Stodder, "Never Say Die," *Entrepreneur*, December 1990, p. 95.

53. "Small Business Owner Salary," PayScale, July 21, 2012, www.payscale.com/research/US/Job=Small_Business_Owner/Salary; "Usual Weekly Earnings of Wage and Salary Workers: Second Quarter 2012," Bureau of Labor Statistics, July 18, 2012, p. 1.

54. "Percentage of U.S. Small Business Owners Reporting Positive Business Conditions Nearly Doubles Since 2010," Business Wire, June 6, 2012, www.businesswire.com/news/home/20120606005835/en/Percentage-U.S.-Small-Business-Owners-Reporting-Positive.

55. Anne Fisher, "Is Your Business Ruining Your Marriage?" *FSB*, March 2003, pp. 63–71.

56. "FAQs," U.S. Small Business Administration, 2010, http://web.sba.gov/faqs/faqIndexAll.cfm?areaid=24.

57. Meg Hirshberg, "Picking Up the Pieces," *Inc.*, July/August 2011, p. 43.

58. "The Employment Situation," Bureau of Labor Statistics, June 2012, p. 2.

59. Carrie Ghose, "Small Business Owners Skipping Summer Vacation, Survey Finds," *Columbus Business First*, July 17, 2012, www.bizjournals.com/columbus/news/2012/07/17/small-business-owners-skipping-summer.html.

60. Greg Selkoe and Liz Welch, "The Way I Work," *Inc.*, May 2012, p. 124.

61. Anne Fisher, "Make Sleep Work for You," *FSB*, September 2008, pp. 85–90.

62. Jennifer Wang, "Not Ready to Retire?" *Entrepreneur*, March 2009, pp. 73–79.

63. Geoff Williams, "Guiding Light," *Entrepreneur B.Y.O.B.*, August 2003, p. 84.

64. Mark Henricks, "Parent Trap?" *Entrepreneur*, September 2005, pp. 17–18.

65. Pattie Simone, "You Can Do It!" *Entrepreneur*, July 2006, pp. 94–101.

66. "Citibank Survey Finds Small Business Owners Taking Action to Grow in 2011," Citigroup, May 4, 2011, www.citigroup.com/citi/press/2011/110504a.htm.

67. *Global Entrepreneurship Week 2011 Impact Report*, Unleashing Ideas, p. 6.

68. Judith Cone, "Teaching Entrepreneurship in Colleges and Universities: How (and Why) a New Academic Field Is Being Built," Ewing Marion Kauffman Foundation, www.kauffman.org/entrepreneurship/teaching-entrepreneurship-in-colleges.aspx.

69. "In Slow Economy, Service Sector Showed Growth in August," *New York Times*, September 6, 2011, www.nytimes.com/2011/09/07/business/economy/service-sector-grew-in-august.html; "Gross Domestic Product: First Quarter 2012," Bureau of Economic Analysis, June 28, 2012, p. 8.

70. "AT&T Small Business Technology Poll: Survey Highlights," AT&T, February 15, 2012, www.att.com/gen/press-room?pid=22312.

71. Angus Loten, "Dropbox Seeks Big Solutions," *Wall Street Journal*, March 15, 2012, p. B6; "About Dropbox," Dropbox, https://www.dropbox.com/about; Victorian Barret, "Dropbox: The Inside Story of Tech's Hottest Start-up," *Forbes*, October 18, 2011, www.forbes.com/sites/victoriabarret/2011/10/18/dropbox-the-inside-story-of-techs-hottest-startup; Sarah Lacy, "Dropbox Raising Massive Round at a $5 Billion-Plus Valuation," *TechCrunch*, July 12, 2011, http://techcrunch.com/2011/07/12/dropbox-raising-massive-round-at-a-5b-plus-valuation.

72. Thad Rueter, "E-Retail Spending to Increase by 62% by 2016," *Internet Retailer*, February 27, 2012, www.internetretailer.com/2012/02/27/e-retail-spending-increase-45-2016.

73. "Apparel Drives U.S. Retail E-Commerce Growth," *eMarketer*, April 5, 2012, www.emarketer.com/Articles/Print.aspx?R=1008956.

74. "84% of U.S. Small Businesses Using Mobile Marketing See Increase in New Business Activity," *Web.com*, May 3, 2012, http://ir.web.com/releasedetail.cfm?ReleaseID=670104.

75. Tamara Thiessen, "French Shoe e-Retailer Spartoo.com Laces Up Rapid Growth," *Internet Retailer*, May 8, 2012, http://www.internetretailer.com/2012/05/08/french-shoe-e-retailer-spartoocom-laces-rapid-growth.

76. Vangie Beal, "E-Commerce Profile: PillowDecor.com," *E-Commerce*, March 4, 2009, www.ecommerce-guide.com/news/trends/article.php/3808511.

77. "U.S. Trade Officials to Urge Small Businesses to Export," *Fox Business*, January 27, 2011, www.foxbusiness.com/markets/2011/01/27/trade-officials-urge-small-businesses-export; Paul Davidson, "Small Businesses Look Across Borders to Add Markets," *USA Today*, April 12, 2011, www.usatoday.com/money/economy/2011-04-06-small-businesses-go-international.htm.

78. John Larsen, "Trade Promotion Coordinating Committee," U.S. Department of Commerce, International Trade Administration, 2011, p. 10.

79. *Small and Medium-Sized Enterprises: Characteristics and Performance*, U.S. International Trade Commission, Publication 4189, November 2010, pp. xi–xiii.

80. Robert W. Farlie, *Kauffman Index of Entrepreneurial Activity, 1996–2011*, March 2012, Kauffman Foundation, p. 12.

81. Daniel McGinn, "Inheriting the Start-Up Gene," *Inc.*, May 2012, p. 26.

82. Jessica Bruder, "A 13-Year-Old Enlists MBA Students to Build Her Start-Up," *New York Times*, May 1, 2012, http://boss.blogs.nytimes.com/2012/05/01/a-13-year-old-enlists-m-b-a-students-to-build-her-start-up.

83. *The American Express OPEN State of Women-Owned Businesses Report: A Summary of Important Trends, 1997–2011*, American Express OPEN, 2012, p. 2.

84. Whitney Pastorak, "How to Woo an Eco-Conscious Crowd," *Fast Company*, June 2012, pp. 96–97; Darren Dahl, "Close-Up: Jessica Alba," *Inc.*, April 2012, p. 28.

85. *The American Express OPEN State of Women-Owned Businesses Report: A Summary of Important Trends, 1997–2011*, American Express OPEN, 2012, p. 2.

86. "Census Bureau Reports Hispanic-Owned Businesses Increased at More Than Double the National Rate," U.S. Census Bureau, September 21, 2010, www.census.gov/newsroom/releases/archives/business_ownership/cb10-145.html.

87. Robert W. Farlie, *Kauffman Index of Entrepreneurial Activity, 1996–2012*, April 2013, Kauffman Foundation, p. 9.

88. *Minority Biz Fast Facts*, U.S. Department of Commerce, Minority Business Development Agency, www.mbda.gov/node/562; *2007 SRP Arizona Business Study: Focus on Minority Owned Businesses*, SRP and Arizona State University Hispanic Research Center, 2007, p. 1.

89. Cindy Kent, "Williams' Star Power Drives Design Firm," *Greenville News*, March 17, 2012, pp. 1D, 8D; Michael D'Estries, "Venus Williams Launches New Jamba Juice Store," *Mother Nature Network*, July 23, 2012, www.mnn.com/health/fitness-well-being/blogs/venus-williams-launches-new-jamba-juice-store; "Venus Williams Brings Jamba Juice to DC's Dupont Circle," *QSR*, July 9, 2012, www.qsrmagazine.com/news/venus-williams-brings-jamba-juice-dc-s-dupont-circle.

90. "New Kauffman Video Features America's Great Job Creators: Immigrant Entrepreneurs," Kauffman Foundation, July 11, 2012, www.kauffman.org/newsroom/new-kauffman-videos-feature-americas-great-job-creator-immigrant-entrepreneurs.aspx; Janie Reyes,

"Immigrant Entrepreneurs Ask Government to Get in Step with Cutting Edge Start-Ups," *Small Business Advocate*, May–June 2012, p. 4.

91. David Dyssegaard Kallick, "Report Breaks New Ground on Immigrant Businesses," Fiscal Policy Institute, June 14, 2012, p.1.

92. Betty Joyce Nash, "Immigrant Entrepreneurs: Talent, Technology, and Jobs," *Region Focus*, Fall 2008, pp. 20–22.

93. Miriam Jordan, "Migrants Keep Small Business Faith," *Wall Street Journal*, June 14, 2012, p. B4.

94. Mitchell York, "Mompreneur Hits 7 Figures as Only Employee and Part-Time," About.com *Entrepreneurs*, 2012, http://entrepreneurs.about.com/od/casestudies/a/Mompreneur-Hit-7-Figures-As-Only-Employee-And-Part-Time.htm; Mitchell York, "Entrepreneur Finds a Niche in Juvenile Products Market," About.com *Entrepreneurs*, 2012, http://entrepreneurs.about.com/od/casestudies/a/Entrepreneur-Finds-A-Niche-In-Juvenile-Products-Market.htm.

95. Sarah E. Needleman, "A Toe in the Water," *Wall Street Journal*, February 23, 2009, p. R7.

96. Scott Brown, "10 Encouraging Facts About Home Based Businesses," *Christian Home Industry*, February 2011, http://christianhomeindustry.com/2011/02/10-encouraging-facts-about-home-based-businesses.

97. Scott Brown, "10 Encouraging Facts About Home Based Businesses," *Christian Home Industry*, February 2011, http://christianhomeindustry.com/2011/02/10-encouraging-facts-about-home-based-businesses.

98. Scott Brown, "10 Encouraging Facts About Home Based Businesses," *Christian Home Industry*, February 2011, http://christianhomeindustry.com/2011/02/10-encouraging-facts-about-home-based-businesses.

99. Laurence Baxter, "Home-Based Business Facts," *eZine Articles*, June 12, 2009, http://ezinearticles.com/?Home-Based-Business-Facts-and-Statistics&id=2469758.

100. Amanda Kwan, "Furniture Finds," *Greenville News*, January 21, 2012, pp. 1D, 4D.

101. *Annual Family Business Survey General Results and Conclusions*, Family Enterprise USA, March 2011, p. 1; Karen E. Klein, "Fathers and Daughters: Passing on the Family Business," *Bloomberg Businessweek*, December 27, 2011, www.businessweek.com/small-business/fathers-and-daughters-passing-on-the-family-business-12272011.html; Veronica Dagher, "Who Will Run the Family Business?," *Wall Street Journal*, March 12, 2012, p. R6.

102. Chris Arnold, "Wiffle Ball: Born and Still Made in the USA," *National Public Radio*, September 5, 2011, www.npr.org/2011/09/05/140145711/wiffle-ball-born-and-still-made-in-the-usa.

103. Erick Calonius, "Blood and Money," *Newsweek*, Special Issue, p. 82.

104. Maureen Farrell, "Why Minority Entrepreneurs Matter in America," *Forbes*, April 11, 2011, www.forbes.com/forbes/2011/0411/entrepreneurs-minority-report-rene-diaz-immigration-southern-spice.html.

105. "Facts and Perspectives on Family Business in the U.S.," The Family Firm Institute, www.ffi.org/looking/fbfacts_us.pdf.

106. "Facts About Family Business," S. Dale High Center for Family Business at Elizabethtown College," www.centerforfamilybusiness.org/about-us/family-business-facts.

107. "Love and the Bottom Line: Couples in Business Find High Rewards and Risks," *Nando Times News*, http://archive.nandotimes.com/newsroom/nt/0212bizcpl.html.

108. Colleen Taylor, "PetBoarding Marketplace DogVacay Fetches $15 Million Series B Led by Foundation Capital," TechCrunch, October 30 2013, http://techcrunch.com/2013/10/10/pet-boarding-marketplace-dogvacay-fetches-15-million-series-b-led-by-foundation-capital/. "Dog Vacay: Husband and Wife Entrepreneurs Make Dog Boarding More Awesome," *YFS Magazine*, April 25, 2012, http://yfsentrepreneur.com/2012/04/25/dog-vacay-husband-and-wife-entrepreneurs-make-dog-boarding-more-awesome; Matthew Wong, "U.S. Venture Capital Financings," *Wall Street Journal*, July 2, 2012, http://topics.wsj.com/article/SB20001424052702304708604577503033132130466.html; Tomio Geren, "Dog Sitter Site Dog Vacay Expands Nationwide as Kennel Alternative," *Forbes*, July 2, 2012, www.forbes.com/sites/tomiogeron/2012/07/02/dog-sitter-site-dogvacay-expands-nationwide-as-kennel-alternative; Leena Rao, "Dog bnb: Dog Vacay Wants to Help You Find a Boarding Alternative for Your 4-Legged Friends,"

TechCrunch, March 1, 2012, http://techcrunch.com/2012/03/01/dogbnb-dogvacay-wants-to-help-you-find-a-boarding-alternative-for-your-four-legged-friends.

109. Janet Civitelli, "Lessons from Reluctant Entrepreneurs: Success Story #4," *Career Thought Leaders*, October 10, 2010, www.careerthoughtleaders.com/blog/lessons-from-reluctant-entrepreneurs-success-story-4.

110. "Franchise News," Liberty Tax Service, November 9, 2010, www.libertytaxfranchise.com/franchise-news.html?a=800156869.

111. "Workplace Redefined," Robert Half Management Resources, July 14, 2010, http://rhmr.mediaroom.com/workplace_redefined.

112. Ian Austen, "66-Year-Old Sets Up Business Selling Shaving Products," *Croydon Advertiser*, February 4, 2011, p. 46; Mike Didymus, 'Retired Croydon Entrepreneur's Soaps Scrub Well," *Croydon Guardian*, June 17, 2010, www.croydonguardian.co.uk/news/localnews/8220015.Retired_entrepreneur_s_soaps_scrub_up_well.

113. "Data on Small Business and the Economy," U.S. Small Business Administration Office of Advocacy, 2012, Table A.1.

114. "Frequently Asked Questions," U.S. Small Business Administration Office of Advocacy, January 2011, p. 1; "Small Business Profile," U.S. Small Business Administration Office of Advocacy, January 2012, p. 1; *The Small Business Economy: A Report to the President*, Washington, DC, 2010, p. 114.

115. *NFIB Small Business Policy Guide*, (Washington, DC: NFIB Education Foundation, 2000), p. 30; "Help Wanted," *Inc. Special Report: The State of Small Business 1997*, pp. 35–41; "The Job Factory," *Inc. Special Report: The State of Small Business 2001*, pp. 40–43; "The Gazelle Theory," *Inc. Special Report: The State of Small Business 2001*, pp. 28–29.

116. Zoltan Acs, William Parsons, and Spencer Tracy, "High Impact Firms" Gazelles Revisited," Small Business Administration, Office of Advocacy, June 2008, pp. 1–2.

117. Jules Lichtenstein, "Small Businesses Provide Flexible Training to Job Market Entrants," *Small Business Advocate*, April 2009, p. 4; Preston McLaurin, "Small Businesses Are Winners," *S.C. Business Journal*, May 2000, p. 10.

118. "Frequently Asked Questions," U.S. Small Business Administration Office of Advocacy, January 2011, p. 1; Kathryn Kobe, *Small Business GDP: Update 2002–2010*, Small Business Research Summary, Small Business Administration, Office of Advocacy, January 2012, p. 1.

119. Frequently Asked Questions," U.S. Small Business Administration Office of Advocacy, January 2011, p. 1.

120. Michael Warshaw, "Great Comebacks," *Success*, July/August 1995, p. 43.

121. Megan Korn, "From Failure to Success: What Steve Jobs, J. K. Rowling, and Jack Bogle All Have in Common," *The Daily Ticker*, May 10, 2012, http://finance.yahoo.com/blogs/daily-ticker/failure-success-steve-jobs-j-k-rowling-jack-134815058.html.

122. Megan Korn, "From Failure to Success: What Steve Jobs, J. K. Rowling, and Jack Bogle All Have in Common," *The Daily Ticker*, May 10, 2012, http://finance.yahoo.com/blogs/daily-ticker/failure-success-steve-jobs-j-k-rowling-jack-134815058.html; "Famous Rejections #7: Harry Potter Was, Yup, Rejected Too," One Hundred Famous Rejections, July 21, 2010, www.onehundredrejections.com/2010/07/harry-potter-was-yup-rejected-too.html.

123. J. K. Rowling, "The Fringe Benefits of Failure, and the Importance of Imagination," *Harvard Magazine*, June 5, 2008, http://harvardmagazine.com/2008/06/the-fringe-benefits-failure-the-importance-imagination.

124. "The Disrupters," *Fast Company*, December 2011/January 2012, pp. 108–110.

125. Clint Willis, "Try, Try Again," *Forbes ASAP*, June 2, 1997, p. 63.

126. Chloe Gotsis, "Newton Native Launches Clothing Line for 'Real Women,'" *Wicked Local*, July 13, 2012, www.wickedlocal.com/newton/news/x1806564287/Newton-native-launches-new-clothing-line-for-real-women#axzz228sSzNwo; "About Dobbin," Dobbin Clothing, www.dobbinclothing.com/about-dobbin.htm.

127. Stephanie Barlow, "Hang On!" *Entrepreneur*, September 1992, p. 156.

128. Jared Sandberg, "Counting Pizza Slices, Cutting Water Cups—You Call This a Budget?," *Wall Street Journal*, January 21, 2004, p. B1.

129. Rhonda Abrams, "Building Blocks of Business: Great Faith, Great Doubt, Great Effort," *Business*, March 4, 2001, p. 2.

Chapter 2

1. Dick Youngblood, "A Firm That Means What It Says About Ethical Conduct," *Star Tribune* (St. Paul, MN), December 28, 1992.
2. Mark Lee, "Facebook Reaches Deal for China Site With Baidu," *Bloomberg*, April 11, 2011, www.bloomberg.com/news/2011-04-11/facebook-reaches-deal-for-china-site-with-baidu-sohu-com-says.html; Michael Kan, "Google Blocked in China by Censors, Unclear How Long It Will Last," *InfoWorld*, November 9, 2012, www.infoworld.com/d/the-industry-standard/google-blocked-in-china-censors-unclear-how-long-it-will-last-206792; Bruce Einhorn, "Facebook, Twitter Growth in China Has Lots of Caveats," *Bloomberg Businessweek*, September 28, 2012, www.businessweek.com/articles/2012-09-28/facebook-twitter-growth-in-china-has-lots-of-caveats; Meg Roggensack, "Facebook Faces China Censorship Dilemma," *Human Rights First*, April 21, 2011, www.humanrightsfirst.org/2011/04/21/facebook-faces-china-censorship-dilemma.
3. Vernon R. Loucks Jr., "A CEO Looks at Ethics," *Business Horizons*, March/April 1987, p. 2.
4. Susan Caminiti, "The Payoff from a Good Reputation," *Fortune*, February 10, 1992, p. 74.
5. Kevin Gray, "Once Prominent Miami Businessman Arrested on Fraud Charges," *Chicago Tribune*, December 7, 2012, http://articles.chicagotribune.com/2012-12-07/news/sns-rt-us-usa-florida-fraudbre8b700e-20121207_1_innovida-holdings-fraud-scheme-million-investment-fraud.
6. Richard C. Morais, "A River Runs Through It," *Forbes*, March 16, 2009, p. 90.
7. Bridget Hilton, "Five Keys to Success for Social Entreprenuers," *Incentivize*, May 3, 2012, http://incentivize.us/2012/05/03/5-keys-to-success-for-social-entrepreneurs.
8. *2010 Edelman Trust Barometer*, Edelman, 2010, p. 6.
9. *Edelman Trust Barometer 2013: Annual Global Survey*, Edelman, 2013, p. 16.
10. Brent Freeman, "5 Great Companies That Make Money & Do Good," *Inc.*, August 16, 2012, www.inc.com/brent-freeman/social-entrepreneurs-5-great-companies-that-make-money-and-do-good_1.html; Kelli Korducki, "Whirlwind Year for Oakville Shoe Company Oliberté," *CityNews Toronto*, November 24, 2012, www.citytv.com/toronto/citynews/life/money/article/237758—whirlwind-year-for-oakville-shoe-company-oliberte.
11. Kate O'Sullivan, "Virtue Rewarded," *CFO*, October 2006, p. 51; Jennifer Reingold, "Walking the Walk," *Fast Company*, November 2005, pp. 81–85; Joseph Pereira, "Doing Good and Doing Well at Timberland," *Wall Street Journal*, September 9, 2003, pp. B1–B10.
12. Richard McGill Murphy, "Why Doing Good Is Good for Business," *Fortune*, February 8, 2010, pp. 91–95.
13. Joseph Pereira, "Doing Good and Doing Well at Timberland," *Wall Street Journal*, September 9, 2003, pp. B1–B10.
14. Robbie Whelan, "Insect Chic: In Colorado, Beetles Create Decor Trend," *Wall Street Journal*, November 30, 2012, http://online.wsj.com/article/SB10001424127887324469304578145753792576048.html; http://azurefurniture.com/azure_furniture_-_green_modern_furniture/the_azure_furniture_company,_inc.html.
15. *Integrity Survey 2008–09*, KPMG LLC, 2009, p. iii.
16. *The Importance of Ethical Culture: Increasing Trust and Driving Down Risks* (Arlington, VA: Ethics Resource Center, 2010), p. 5.
17. Pablo Ruiz-Palomino, Ricardo Martínez-Cañas, and Joan Fontrodona, "Ethical Culture and Employee Outcomes: The Mediating Role of Person-Organization Fit," *Journal of Business Ethics*, August 2012, http://link.springer.com/article/10.1007/s10551-012-1453-9/fulltext.html.
18. Joshua Kurlantzick, "Liar, Liar," *Entrepreneur*, October 2003, pp. 68–71.
19. Gene Laczniak, "Business Ethics: A Manager's Primer," *Business*, January–March 1983, pp. 23–29.
20. Patricia Wallington, "Honestly?!" *CIO*, March 15, 2003, p. 42.
21. John Rutledge, "The Portrait on My Wall," *Forbes*, December 30, 1996, p. 78.
22. Patricia Wallington, "Total Leadership—Ethical Behaviour Is Essential," *CIO*, March 15, 2003, www.cio.com/article/31779/Total_Leadership_Ethical_Behaviour_Is_Essential.
23. Michael Josephson, "Teaching Ethical Decision Making and Its Principled Reasoning," *Ethics: Easier Said Than Done*, Winter 1988, p. 28.
24. Mark Henricks, "Well, Honestly!," *Entrepreneur*, December 2006, p. 103.
25. Robert S. Kaplan, "On Ethics, You Set the Tone," *Inc.*, February 21, 2012, www.inc.com/robert-kaplan/on-ethics-you-set-the-tone.html.
26. Arthur C. Brooks, *Social Entrepreneurship* (Upper Saddle River, NJ: Prentice Hall, 2009).
27. Niels Bosma and Jonathan Levie, *Global Entrepreneurship Monitor 2009 Global Report*, www.gemconsortium.org, p. 49.
28. "What Is a Social Entrepreneur?," Ashoka, www.ashoka.org/social_entrepreneur.
29. Lucas Kavner, "Kohl Crecelius Crochets for Communities," *Huffington Post*, May 16, 2011, www.huffingtonpost.com/2011/05/16/huffpost-greatest-person-_n_862542.html; www.krochetkids.org/who-we-are/our-story.
30. SurePayroll Insights Survey: Economy Woes Squeeze Small Business Charity, SurePayroll, May 28, 2008, www.surepayroll.com/spsite/press/releases/2008/release052808.asp; "The Big Impact of Small Business," SurePayroll report, June 19, 2012, http://blog.surepayroll.com/the-big-impact-of-small-business.
31. http://myemma.com/meet-us/giving-back.
32. Survey conducted by Penn Schoen Berland, in conjunction with Burson-Marsteller and Landor, from interviews conducted February 10–12, 2010, www.slideshare.net/BMGlobalNews/csr-branding-survey-2010-final.
33. Edward Iwata, "Businesses Grow More Socially Conscious," *USA Today*, February 14, 2007, www.usatoday.com/money/companies/2007-02-14-high-purpose-usat_x.htm.
34. Brent Freeman, "5 Great Companies That Make Money & Do Good," *Inc.*, August 16, 2012, www.inc.com/brent-freeman/social-entrepreneurs-5-great-companies-that-make-money-and-do-good_1.html; Katie Lobosco, "Doing Good and Making Money," *Forbes*, October 15, 2012, www.forbes.com/sites/nyuentrepreneurchallenge/2012/10/15/doing-good-and-making-money; www.indosole.com/content/theprocess.
35. Chris Pentilla, "Shades of Green," *Entrepreneur*, August 2007, pp. 19–20; "Nature Meet Nurture," Pangea Organics, www.pangeaorganics.com/home.html.
36. Jeffrey Cornwall and Michael Naughton, *Bringing Your Business to Life* (Ventura, CA: Regal Publishing, 2008), pp. 120–122; www.leecompany.com.
37. Jennifer J. Salopek, "The 2020 Workplace," *Workforce Management*, June 2010, pp. 36–40.
38. Juan Rodriguez, "U.S. Workforce Will Be Smaller, More Diverse in 2050," *Diversity Jobs*, August 14, 2008, http://diversityjobs.com/news/us-workforce-will-be-smaller-more-diverse-in-2050.
39. Keith H. Hammonds, "Difference Is Power," *Fast Company*, July 2000, p. 58.
40. Jessica Stillman, "Diversity: It's About Improving Your Business, Not Checking Boxes," *Inc.*, June 11, 2012, www.inc.com/jessica-stillman/diversity-its-about-improving-your-business-not-checking-boxes.html.
41. Best Practices of Private Sector Employers, Equal Employment Opportunity Commission, 2003, www.eeoc.gov/abouteeoc/task_reports/prac2.html.
42. Martha Lagace, "Racial Diversity Pays Off," Harvard Business School: *Working Knowledge*, June 21, 2004, http://hbsworkingknowledge.hbs.edu/item.jhtml?id=4207&t=organizations.
43. Mitra Toossi, "Labor Force Projections to 2020: A More Slowly Growing Workforce," *Monthly Labor Review*, January 2012, pp. 43–64.
44. Rebekah K. Hirsch and Royer F. Cook, "Workplace Substance Abuse Prevention: What the Evidence Tells Us," United Nations Office on Drugs and Crime, 2012, pp. 3–4.

45. Gina Ruiz, "Expanded EAPs Lend a Hand to Employer's Bottom Line," *Workforce Management*, January 16, 2006, pp. 46–47.

46. "EAP Facts," Northeast Georgia Employee Assistance Program, 2013, http://negeap.com/facts.htm.

47. *The President's National Drug Control Strategy*, February 2007, www .whitehousedrugpolicy.gov/publications/policy/ndcs07, pp. 14–15; "Nationwide Survey Shows Most Illicit Drug Users and Heavy Alcohol Users Are in the Workplace and May Pose Special Problems," Substance Abuse and Mental Health Services Administration, U.S. Department of Health and Human Services, July 17, 2007, http://oas .samhsa.gov/work2k7/press.htm.

48. "Statistics Overview," Centers for Disease Control and Prevention, 2013, http://www.cdc.gov/hiv/statistics/basics/index.html.

49. Belinda Goldsmith, "Indians Most Likely to Report Sexual Harassment at Work—Poll," *Reuters*, August 12, 2010, http://in.reuters.com/article/ idINIndia-50803120100812.

50. "Sexual Harassment Charges," Equal Employment Opportunity Commission, 2010, www.eeoc.gov/eeoc/statistics/enforcement/sexual_ harassment.cfm.

51. www.eeoc.gov/eeoc/statistics/enforcement/sexual_harassment.cfm.

52. National Federation of Independent Business, "Sexual Harassment 101 for Small Business," www.nfib.com/business-resources/business-resources-item?cmsid=49311

53. *Sexual Harassment Manual for Managers and Supervisors* (Chicago: Commerce Clearing House, 1992), pp. 25–26.

54. Lori A. Carter, "$2 Million Harassment Verdict Against Petaluma Card Room," *Press Democrat*, August 5, 2010, www.petaluma360.com/ article/20100805/COMMUNITY/100809800?p=3&tc=pg&tc=ar; Lori A. Carter, "Judge Reduces Jury Award in Petaluma Card Room Sexual Harassment Case," *Press Democrat*, October 8, 2010, www .pressdemocrat.com/article/20101008/ARTICLES/101009447.

55. *Burlington Industries v. Ellerth* (97-569) 123 F.3d 490; "Employer Liability for Harassment," Equal Employment Opportunity Commission, www.eeoc.gov/types/harassment.html.

56. Nicole P. Cantey, "High Court Rules Same Sex Harassment Is Against the Law," *South Carolina Business Journal*, August 1998, p. 3; Jack Corcoran, "Of Nice and Men," *Success*, June 1998, pp. 64–67.

57. Sarah Radicati and Justin Levenstein, "E-Mail Statistics Report, 2013–2017," Radicati Group, April 2013, p. 3.

58. Nancy Flynn, "Survey Reveals 52% of Bosses Fire Workers for E-mail and Internet Misuse," *The Examiner*, July 27, 2009, http://www .examiner.com/article/survey-reveals-52-of-bosses-fire-workers-for-e-mail-and-internet-misuse.

59. "13th Annual Wacky Warning Labels Contest Winners Selected on National Television," Wacky Warning Labels, July 12, 2010, www .wackywarninglabelstv.com/press-releases/2010/7/12/13th-annual-wacky-warning-labels-contest-winners-selected-on.html; "M-Law's Tenth Annual Wacky Warning Label Contest," Wacky Warning Labels, January 4, 2007, www.wackywarnings.com; Things People Said: Warning Labels," Rinkworks.com, www.rinkworks.com/said/ warnings.shtml.

60. "FTC Settlement Requires Oreck Corporation to Stop Making False and Unproven Claims That Its Ultraviolet Vacuum and Air Cleaner Can Prevent Illness," Federal Trade Commission, April 7, 2011, www.ftc.gov/opa/2011/04/oreck.shtm.

61. "Examples of Corporate Fraud Investigations 2010: Corporate CEO, Vice President and Accountant Sentenced for Defrauding Maryland Company of Over $1.4 Million," Internal Revenue Service, www.irs .gov/compliance/enforcement/article/0,,id=213768,00.html.

62. Ruben Hernandez-Murillo and Christopher J. Martinek, "Corporate Social Responsibility Can Be Profitable," *Regional Economist*, April 2009, pp. 4–5.

63. Corporate Social Responsibility Branding Survey, Penn, Schoen, Berland and Burson-Marseller, 2010, www.psbresearch.com/files/ CSR%20Branding%20Survey%202010%20EXTERNAL%20FINAL .pdf, p. 1.

64. Elise Reinemann, "Chilly Dogs," *FSB*, September 2008, pp. 18–19.

Chapter 3

1. "How to Come Up with a Great Idea," *Wall Street Journal*, April 29, 2013, p. R1.

2. Jeremy Quittner, "A One-Acre Farm in a 320-Square-Foot Box," *Inc.*, March 2013, pp. 36–37.

3. *2013 Global R&D Funding Forecast*, Battelle and *R&D Magazine*, December 2012, p. 3.

4. "Frequently Asked Questions," U.S. Small Business Administration, Office of Advocacy, September 2012, p. 3.

5. Jessie Romero, "What We Don't Know About Innovation," *Region Focus*, First Quarter 2012, p. 13.

6. "Innovation Quotes," Think Exist, http:/thinkexist.com/quotations/ innovation.

7. Intuit Future of Small Business Report: Defining Small Business Innovation, Intuit, March 2009, pp. 4–5.

8. *GE Global Innovation Barometer: Global Research Findings and Insights*, General Electric, January 2013, pp. 10, 17.

9. Doreen Lorenzo, "Time to Redefine 'Innovation,'" *Fortune*, March 12, 2012, http://tech.fortune.cnn.com/2012/03/12/time-to-redefine-innovation.

10. Warren Bennis, "Cultivating Creative Collaboration," *Industry Week*, August 18, 1997, p. 86.

11. Amy Clarke Burns, "New Product Helps Golfers Get Into the Swing of Things," *Greenville News*, March 16, 2013, pp. 1B, 5B.

12. Roger von Oech, *A Whack on the Side of the Head* (New York: Warner Books, 1990), p. 108.

13. Matt Villano, "Pure Energy," *Entrepreneur*, November 2012, p. 51.

14. "How to Come Up with a Great Idea," *Wall Street Journal*, April 29, 2013, p. R1.

15. Michael Maiello, "They Almost Changed the World," *Forbes*, December 23, 2002, p. 217.

16. John Bussey, "Expensive Pipeline: Developing New Drugs," *Wall Street Journal*, February 26, 2013, p. B12.

17. Kevin McGourty, "What Percentage of Sales Should New Product Launches from the Last Three Years Represent?," *INPD Center*, September 10, 2012, http://inpdcenter.com/blog/percentage-sales-product-launches-3-years-represent.

18. Charlie Farrell, "A Penny for Your Thoughts," *Business & Economic Review*, October–December 2006, p. 25.

19. David H. Freedman, "Freeing Your Inner Think Tank," *Inc.*, May 2005, pp. 65–66.

20. Robert Fulghum, "Time to Sacrifice the Queen," *Reader's Digest*, August 1993, pp. 136–138.

21. Jay Walker, "The Power of Imagination," *Wall Street Journal*, February 26, 2013, p. B11.

22. Jonah Lehrer, "How to Be Creative," *Wall Street Journal*, March 10–11, 2012, p. C1.

23. Carla Goodman, "Sparking Your Imagination," *Entrepreneur*, September 1997, p. 32.

24. Holly Finn, "The Slow Road to Invention," *Wall Street Journal*, November 2, 2012, http://online.wsj.com/article/SB1000142405297020 3880704578089233956502680.html.

25. "A Passion for Great Conversation," Table Topics, 2013, www .tabletopics.com.

26. Chuck Salter, "Failure Doesn't Suck," *Fast Company*, May 2007, p. 44; James Dyson, "Cleaning Up in His Industry," *Fortune*, January 22, 2007, p. 33.

27. Kasey Wehrun, "Twice Blessed," *Inc.*, April 2011, p. 120; Stephen Miller, "Helicopter Designer and Guitar Hero," *Wall Street Journal*, February 2, 2011, p. A6.

28. Betty Edwards, *Drawing on the Right Side of the Brain* (Los Angeles: J. P. Tarcher, 1979), p. 32.

29. Roger von Oech, *A Whack on the Side of the Head* (New York: Warner Books, 1990), pp. 21–167; "Obstacles to Creativity," Creativity Web, www.ozemail.com.au/~caveman/Creative/Basics/obstacles.htm.

30. Erin Zagursky, "Professor Discusses America's Creativity Crisis in *Newsweek*," William and Mary News and Events, July 14, 2010, www .wm.edu/news/stories/2010/professor-discusses-americas-creativity-crisis-in-newsweek-123.php; Po Bronson and Ashley Merryman, "The Creativity Crisis," *Newsweek*, July 10, 2010, www.newsweek.com/2010/07/10/the-creativity-crisis.html.

31. Peter Gray, "As Children's Freedom Has Declined, So Has Their Creativity," *Psychology Today*, September 17, 2012, www .psychologytoday.com/blog/freedom-learn/201209/children-s-freedom-has-declined-so-has-their-creativity.

32. "Innovation: Companies on the Cutting Edge," *Inc.*, December 2010/January 2011, pp. 52–53; "Lyric Semiconductor Named in Technology Review's 2011 TR50 List of the World's Most Innovative Companies," Lyric Semiconductor, February 22, 2011, www.lyricsemiconductor.com/news.htm.

33. Rachel Z. Amdt, "The United States of Innovation: Georgia Chopsticks," *Fast Company*, May 2012, p. 92.

34. Karen Axelton, "Imagine That," *Entrepreneur*, April 1998, p. 96; "Thomas Edison Biography," http://edison-ford-estate.com/ed_bio.htm.

35. Michelle Juergen, "Sleep Like a Champ," *Entrepreneur*, March 2012, p. 74.

36. Jonah Lehrer, "How to Be Creative," *Wall Street Journal*, March 10–11, 2012, pp. C1–C2.

37. Laurie Tarkan, "Work Hard, Play Harder," *Fox News*, September 15, 2012, www.foxnews.com/health/2012/09/13/work-hard-play-harder-fun-at-work-boosts-creativity-productivity.

38. Tim McKeough, "Toy Factory," *Fast Company*, December 2011/January 2012, p. 84; Mary Timmins, "Rules of the Game," *Illinois Alumni Magazine*, November 20, 2012, www.uiaa.org/illinois/news/blog/index .asp?id=491.

39. Naveen Jain, "Rethinking the Concept of 'Outliers': Why Non-Experts Are Better at Disruptive Innovation," *Forbes*, July 12, 2012, www .forbes.com/sites/singularity/2012/07/12/rethinking-the-concept-of-outliers-why-non-experts-are-better-at-disruptive-innovation.

40. Jason Daley, "If It's Broke, Decorate It," *Entrepreneur*, July 2009, p. 17; "About Casttoo," Casttoo, www.casttoo.com/Casttoo.com/About_Casttoo.html.

41. Joseph Schumpeter, "The Creative Response in Economic History," *Journal of Economic History*, November 1947, pp. 149–159.

42. J. J. McCorvey, "Innovation: Companies on the Cutting Edge," *Inc.*, February 2012, pp. 42–43.

43. Laurie McCabe, "Seven Daily Inspirations from Dell's Women Entrepreneur Network Event," *Laurie McCabe's Blog*, June 10, 2011, http://lauriemccabe.wordpress.com/2011/06/10/seven-daily-inspirations-from-dell%E2%80%99s-women-entrepreneur-network-event; Daniel McGinn, "How I Did It: Arianna Huffington," *Inc.*, February 1, 2010, www.inc.com/magazine/20100201/how-i-did-it-arianna-huffington .html; Jay Yarow, "Huffington Post Traffic Zooms Past the New York Times," *Business Insider*, June 9, 2011, www.businessinsider .com/chart-of-the-day-huffpo-nyt-unique-visitors-2011-6?utm_source=twbutton&utm_medium=social&utm_term=&utm_content=&utm_campaign=sai.

44. *Bits & Pieces*, January 1994, p. 6.

45. Hal Gregersen, "The Entrepreneur's DNA," *Wall Street Journal*, February 26, 2013, p. B13.

46. "Harnessing Your Team's Creativity," *BNET*, June 7, 2007, www.bnet .com/2403-13059_23-52990.html.

47. Jens Martin Skribsted and Rasmus Bech Hanson, "Do Consultants Kill Innovation?," *Fast Company Design*, January 2012, www.fastcodesign .com/1665764/do-innovation-consultants-kill-innovation.

48. Frank T. Rothaermel and Andrew M. Hess, "Innovation Strategies Combined," *Sloan Management Review* 51, no. 3 (Spring 2010), pp. 13–15.

49. Rob Goffee and Gareth Jones, "The Odd Clever People Every Organization Needs," *Forbes*, August 13, 2009, www.forbes.com/2009/08/13/clever-employees-talent-leadership-managing-recruiting .html.

50. Anya Kamenetz, "The Power of the Prize," *Fast Company*, May 2008, pp. 43–45.

51. John Bessant, Kathrin Möslein, and Bettina Von Stamm, "In Search of Innovation," *Wall Street Journal*, June 22, 2009, p. R4.

52. Carol Tice, "Fueling Change," *Entrepreneur*, November 2007, p. 47.

53. Beth Kowitt, "Dunkin' Brands' Kitchen Crew," *Fortune*, May 24, 2010, pp. 72–74.

54. Nadine Heintz, "Employee Creativity Unleashed," *Inc.*, June 2009, pp. 101–102.

55. Rachel Emma Silverman, "The Science of Serendipity in the Workplace," *Wall Street Journal*, May 1, 2013, p. B6; Zach Brand and David Gorsline, "Happy Accidents: The Joy of Serendipity Days," *Inside NPR*, October 14, 2011, www.npr.org/blogs/inside/2011/10/14/141312774/happy-accidents-the-joy-of-serendipity-days.

56. Matthew Carmichael, "Best Places to Work No. 1: Radio Flyer," *Crain's Chicago Business*, March 29, 2010, www.radioflyer.com/skin/frontend/blank/radioflyer/docs/media/crains_best_il.pdf; Amelia Forczak, "Rolling Down the Path Toward Success," *HR Solutions eNews*, www.hrsolutionsinc.com/enews_1010/RadioFlyer_1010.html.

57. Cheryl Strauss Einhorn, "Dance Troupe Markets Creativity to Cube-Dwellers," *CNN Money*, April 25, 2013, http://money.cnn .com/2013/04/23/technology/innovation/trey-mcintyre-project-hewlett-packard/index.html.

58. Carol Tice, "Fueling Change," *Entrepreneur*, November 2007, p. 47.

59. Nicholas Bakalar, "Self-Adjustable Eyeglass Lenses," *New York Times*, September 26, 2011, www.nytimes.com/2011/09/27/health/27glasses .html; Mary Jordan, "From a Visionary English Physicist, Self-Adjusting Lenses for the Poor," *Washington Post*, January 10, 2009, http://articles .washingtonpost.com/2009-01-10/world/36786562_1_glasses-plastic-lenses-syringes; Robin McKie, "British Inventor's Spectacles Revolution for Africa," *The Guardian*, May 21, 2011, www.guardian.co.uk/global-development/2011/may/22/joshua-silver-glasses-self-adjusting.

60. Rachel Emma Silverman, "The Science of Serendipity in the Workplace," *Wall Street Journal*, May 1, 2013, p. B6.

61. Laurie Tarkan, "Work Hard, Play Harder," *Fox News*, September 15, 2012, www.foxnews.com/health/2012/09/13/work-hard-play-harder-fun-at-work-boosts-creativity-productivity; Julia Benton, "Tour Zygna's Virtual Reality-Themed Office," *California Home + Designs*, 2013, www.californiahomedesign.com/house-tours/tour-zyngas-virtual-reality-themed-office/slide/5468.

62. Erik Sofge, "Clean-Burning Motorcycles," *Fast Company*, February 2013, p. 34; "U-M Students Launch Small Engine Efficiency Start-Up," *Concentrate*, March 14, 2012, www.concentratemedia.com/innovationnews/picosprayannarbor0186.aspx.

63. Robert Epstein, "How to Get a Great Idea," *Reader's Digest*, December 1992, p. 102.

64. James E. Burroughs, Darren W. Dahl, C. Page Moreau, Amitava Chattopadhyay, and Gerald R. Gorn, "A One-Two Punch to Foster Creativity," *Strategy+Business*, November 4, 2011, www.strategy-business.com/article/re00166?gko=4b582.

65. Georgia Flight, "How They Did It: Seven Intrapraneur Success Stories," *BNET*, April 18, 2007, www.bnet.com/2403-13070_23-196890.html.

66. Sarah E. Needleman and Angus Loten, "3 Ways Small Firms Can Drive Innovation," *Wall Street Journal*, August 30, 2011, http://blogs.wsj.com/in-charge/2011/08/30/how-small-firms-can-drive-innovation.

67. Sarah E. Needleman and Angus Loten, "3 Ways Small Firms Can Drive Innovation," *Wall Street Journal*, August 30, 2011, http://blogs.wsj.com/in-charge/2011/08/30/how-small-firms-can-drive-innovation; "What Is SeatGeek?," SeatGeek, 2013, http://seatgeek.com/about.

68. Patrick Seitz, "Using a Double Dose of Product Creativity," *Investor's Business Daily*, June 12, 2009, www.investors.com/NewsAndAnalysis/Article/479404/200906121817/Using-A-Double-Dose-Of-Product-Creativity.aspx.

69. Nichole L. Torres, "Industrial Revolution," *Entrepreneur*, November 2007, pp. 142–143.

70. Lydia Dishman, "Watermelons, Chuck Taylors, and How Caribou Encourages a Culture of Innovation," *Fast Company*, July 24, 2012,

www.fastcompany.com/1843448/watermelons-chuck-taylors-and-how-caribou-coffee-encourages-culture-innovation.

71. Jason Fried, "It's Time for You to Go," *Inc.*, July/August 2012, p. 35; Jason Fried, "Starting Over," *Inc.*, February 2012, p. 40.

72. Mike Figliuolo, "How to Innovate Like Captain Jack Sparrow," *Thought Leaders*, April 25, 2011, www.thoughtleadersllc.com/2011/04/how-to-innovate-like-captain-jack-sparrow.

73. Jay Walker, "The Power of Imagination," *Wall Street Journal*, February 26, 2013, p. B11.

74. Stephanie Kang, "Fashion Secret: Why Big Designers Haunt Vintage Shops," *Wall Street Journal*, April 2, 2007, pp. A1, A10.

75. "The 100 Most Creative People in Business," *Fast Company*, June 2011, p. 114.

76. "The Keys to Making History," *Fast Company*, December 2012/January 2013, p. 30.

77. McKenna Grant, "Spark Your Creativity!," *USA Weekend*, June 29–July 1, 2012, p. 4.

78. "How to Come Up with a Great Idea," *Wall Street Journal*, April 29, 2013, p. R1.

79. Rudolph Bell, "Local Inventor Puts Finger to the Wind," *Greenville News*, November 13, 2011, pp. 1E, 3E.

80. Mattew Smith, "The Pressure's Off," *Stand Firm*, April 2012, p. 29.

81. "Louis-Jacque Mande Daguerre," The Robinson Library, www.robinsonlibrary.com/technology/photography/biography/daguerre.htm; Mary Bellis, "Daguerreotype," www.robinsonlibrary.com/technology/photography/biography/daguerre.htm.

82. Robert J. Toth, "Help Wanted? Eureka!," *Wall Street Journal*, April 29, 2013, p. R5.

83. Navi Radjou, Jaideep Prabhu, and Simone Ahuja, "Millenials Are the McGyvers of Business," *Harvard Business Review*, March 13, 2012, http://blogs.hbr.org/cs/2012/03/millennials_are_the_macgyvers.html; Peter Frost, "SwipeSense Gets a Hand with $50,000 in Seed Capital, Guidance from HealthBox, a Healthcare Technology Accelerator Program," *Chicago Tribune*, April 4, 2012, http://articles.chicagotribune.com/2012-04-04/business/ct-biz-0405-health-pitch-20120404_1_hospital-acquired-infections-angel-investors-wellness-program.

84. Don Debelak, "Ideas Unlimited," *Business Start-Ups*, May 1999, pp. 57–58.

85. Mark Evans, "Truck Washer Expands into Home Ice Rinks," *Globe and Mail*, December 26, 2010, www.theglobeandmail.com/report-on-business/small-business/start/mark-evans/truck-washer-expands-into-home-ice-rinks/article1850030.

86. Katherine Duncan, "Sweet Dreams," *Entrepreneur*, August 2012, p. 43.

87. "How to Come Up with a Great Idea," *Wall Street Journal*, April 29, 2013, p. R2; Dinah Eng, "How Maxine Clark Built Buiild-a-Bear," *CNN Money*, March 19, 2012, http://money.cnn.com/2012/03/16/smallbusiness/build-bear-maxine-clark.fortune/index.htm.

88. Julie Sloane, "Inside the Mind of a (Rich) Inventor," *FSB*, November 2007, pp. 90–102.

89. Gwen Moran, "Lessons Learned," *Entrepreneur*, November 2012, p. 50; Judy Berman, "50 Up-and Coming Culture-Makers to Watch in 2013," *Flavorwire*, December 19, 2012, http://flavorwire.com/356385/50-up-and-coming-new-york-culture-makers-to-watch-in-2013/22.

90. Nadine Heintz, "Employee Creativity Unleashed," *Inc.*, June 2009, pp. 101–102.

91. Rosa Alphonso, "Small Business Optimism Is on an Upswing, According to the OPEN from American Express Small Business Monitor," American Express, May 24, 2007, http://home3.americanexpress.com/corp/pc/2007/monitor.asp.

92. Jonah Lehrer, "Mom Was Right: Go Outside," *Wall Street Journal*, May 26–27, 2012, p. C18.

93. Geoff Williams, "Innovative Model," *Entrepreneur*, September 2002, p. 66.

94. Diana Lodderhose and Marc Graser, "J. K. Rowling Unveils 'Pottermore,'" *Variety*, June 23, 2011, www.variety.com/article/VR1118039008?refcatid=1009.

95. Roy Rowan, "Those Hunches Are More Than Blind Faith," *Fortune*, April 23, 1979, p. 112.

96. Michael Waldholz, "A Hallucination Inspires a Vision for AIDS Drug," *Wall Street Journal*, September 29, 1993, pp. B1, B5.

97. "Fly Goes Hi-Tech with New S. A. Sharkskin Lines and Sage Targets Bassers," *Tackle Tour*, January 20, 2008, www.tackletour.com/reviewise08flycoverage.html.

98. Alison Overholt, "From Idea to Innovation," *Fast Company*, February 2013, pp. 49–50.

99. Siri Schubert, "Folate Is Gr-r-reat!," *Business 2.0*, November 2004, p. 72.

100. Josh Dean, "Saul's House of Cool Ideas," *Inc.*, February 2010, p. 71.

101. Shivani Vora, "Wallpaper That's Temporary," *Inc.*, November 2010, p. 124.

102. Nicole Marie Richardson, "The Answer Is Blowing in the (Very Gentle) Wind," *Inc.*, October 2009, pp. 38–39.

103. Nick D'Alto, "Think Big," *Business Start-Ups*, January 2000, pp. 61–65.

104. Sue Shellenbarger, "Tactics to Spark Creativity," *Wall Street Journal*, April 3, 2013, pp. B1–B2.

105. Brian Nadel, "The Art of Innovation," Advertising Insert, *Fortune*, December 13, 2004, pp. S1–S22.

106. Sue Shellenbarger, "Tactics to Spark Creativity," *Wall Street Journal*, April 3, 2013, pp. B1–B2.

107. Jonah Lehrer, "Bother Me, I'm Thinking," *Wall Street Journal*, February 19, 2011, http://online.wsj.com/article/SB1000142405274870358480457614419213214506.html.

108. Sue Shellenbarger, "Tactics to Spark Creativity," *Wall Street Journal*, April 3, 2013, pp. B1–B2.

109. Sarah Kessler, "Walk, Talk, Ride, Recharge," *Inc.*, May 2010, p. 26; "About Us," nPower PEG, www.npowerpeg.com/index.php/our-story.

110. "What Are Dreams?," *Nova*, PBS, June 29, 2011, www.pbs.org/wgbh/nova/body/what-are-dreams.html.

111. Thea Singer, "Your Brain on Innovation," *Inc.*, September 2002, pp. 86–88.

112. "The 100 Most Creative People in Business," *Fast Company*, June 2011, p. 88.

113. Ben Baldwin, "Ben Baldwin: Stop and Smell the Flowers," *Wall Street Journal Blogs*, May 1, 2013, http://blogs.wsj.com/accelerators/2013/05/01/ben-baldwin-stop-and-smell-the-flowers; "How to Come Up with a Great Idea," *Wall Street Journal*, April 29, 2013, p. R1.

114. Paul Bagne, "When to Follow a Hunch," *Reader's Digest*, May 1994, p. 77.

115. Susan Hansen, "The Action Hero," *Inc.*, September 2002, pp. 82–84.

116. Thea Singer and Lea Buchanan, "Who? What? Where? Why? When? How?," *Inc.*, August 2002, p. 66.

117. Michael Waldholz, "A Hallucination Inspires a Vision for AIDS Drug," *Wall Street Journal*, September 29, 1993, pp. B1, B5.

118. Theunis Bates, "Supertiny Power Plants," *Fast Company*, June 2010, p. 38; "Technology," Innowattech, www.innowattech.co.il/technology.aspx; Diane Pham, "Pavegen: Energy-Generating Pavement Hits the Streets," *Inhabitat*, October 28, 2009, http://inhabitat.com/energy-generating-pavement.

119. Robert Epstein, "How to Get a Great Idea," *Reader's Digest*, December 1992, p. 104.

120. Michael Waldholz, "A Hallucination Inspires a Vision for AIDS Drug," *Wall Street Journal*, September 29, 1993, pp. B1, B5.

Chapter 4

1. Joan Magretta, "The Most Common Strategy Mistakes," *Harvard Business School Working Knowledge*, December 21, 2011, http://hbswk.hbs.edu/item/6737.html.

2. Robert Safian, "Generation Flux," *Fast Company*, February 2012, pp. 59–71, 97.

3. Robert Safian, "Generation Flux," *Fast Company*, February 2012, pp. 59–71, 97; Chris Foresman, "Samsung, Apple Continue Smartphone Market Share Tug-of-War," *Ars Technica*, May 1, 2012, http://arstechnica.com/gadgets/2012/05/samsung-apple-continue-smartphone-marketshare-tug-of-war.

4. "iTunes Store Sets New Record with 25 Billion Songs Sold," Apple, February 6, 2013, http://www.apple.com/pr/library/2013/02/06iTunes-Store-Sets-New-Record-with-25-Billion-Songs-Sold.html; Evan Niu, "Five Things You May Have Missed in Apple's Earnings Last Week," *Daily Finance*, April 30, 2012, www.dailyfinance.com/2012/04/30/5-things-you-may-have-missed-in-apples-earnings-la; Donald Melanson, "Apple: 16 Billion iTunes Songs Downloaded, 300 Million iPods Sold," *Engadget*, October 4, 2011, www.engadget.com/2011/10/04/apple-16-billion-itunes-songs-downloaded-300-million-ipods-sol; L. Gordon Crovitz, "How the Music Industry Can Get Digital Satisfaction," *Wall Street Journal*, January 12, 2009, p. A11; Fred Vogelstein, "Mastering the Art of Disruption," *Fortune*, February 6, 2006, pp. 23–24.

5. Alvin Toffler, "Shocking Truths About the Future," *Journal of Business Strategy*, July/August 1996, p. 6; Russ Juskalian, "Knowledge Drives Future, Creates Wealth, Authors Say," *USA Today*, May 15, 2006, p. 5B.

6. Norm Brodsky, "Be Prepared," *Inc.*, January 2006, pp. 53–54.

7. Thomas A. Stewart, "You Think Your Company's So Smart? Prove It," *Fortune*, April 30, 2001, p. 188.

8. Andrew J. Sherman, "Business Rx: Building Intellectual Capital," *Washington Post*, December 25, 2011, www.washingtonpost.com/business/capitalbusiness/business-rx-best-practices-and-common-mistakes-in-optimizing-your-intellectual-capital/2011/12/19/gIQAt6IUHP_story.html.

9. Zeynep Ton, "Retailers Should Invest More in Employees," *Harvard Business Review*, December 20, 2011, http://blogs.hbr.org/cs/2011/12/retailers_should_invest_more_i.html; Christopher Matthews, "Future of Retail: Companies That Profit by Investing in Employees," *Time Business*, June 8, 2012, http://business.time.com/2012/06/18/future-of-retail-companies-that-profit-by-investing-in-employees; Beth Kowitt, "Inside the Secret World of Trader Joe's," *Fortune*, August 23, 2010, http://money.cnn.com/2010/08/20/news/companies/inside_trader_joes_full_version.fortune/index.htm; Elliot Zwiebach, "2012 Power 50: #23 Dan Bane," *Supermarket News*, July 17, 2012, http://supermarketnews.com/trader-joe039s-market/2012-power-50-no-23-dan-bane.

10. Geoffrey Colvin, "Managing in Chaos," *Fortune*, October 2, 2006, pp. 76–82.

11. Margaret Littman, "Pay Dirt," *Entrepreneur*, December 2011, p. 17.

12. Jennifer Miller, "Novel Ideas," *Fast Company*, September 2011, p. 32; "From Our Store," Square Books, www.squarebooks.com.

13. Jim McCraigh, "Value Propositions," *Direct Marketer's Digest*, 2011, www.mccraigh.com/Archives/value_proposition.html.

14. Rachael Z. Arndt, "Picture Perfect," *Fast Company*, February 2012, p. 56.

15. "John Naisbitt Quotes," ThinkExist, http://thinkexist.com/quotation/strategic-planning-is-worthless-unless-there-is/363708.html.

16. Ray Smilor, *Daring Visionaries: How Entrepreneurs Build Companies, Inspire Allegiance, and Create Wealth* (Avon, MA: Adams Media Corporation, 2001), pp. 12–13.

17. Thomas A. Stewart, "Why Values Statements Don't Work," *Fortune*, June 10, 1996, p. 137.

18. "Core Values," Turner Construction Company, www.turnerconstruction.com/corporate/content.asp?d=5736; "Turner News: Special Centennial Issue," Turner Constrction Company, March 2002, pp. 1, 19.

19. Danny Meyer, "The Saltshaker Theory," *Inc.*, October 2006, p. 70.

20. "About Us," Five Guys Burgers and Fries, www.fiveguys.com/about-us.aspx; Elissa Elan, "Fast-Grown Five Guys Chain Beefs Up, Eyes Meaty Growth," Nation's Restaurant News, November 18, 2007, http://nrn.com/article/fast-grown-five-guys-chain-beefs-eyes-meaty-growth; *Five Guys Corporate Identify and Brand Manual*, Five Guys Burgers and Fries, 2009, p. 6.

21. Diane Hamilton, "Top 10 Company Mission Statements in 2012," Dr. Diane Hamilton, http://drdianehamilton.wordpress.com/2011/01/13/top-10-company-mission-statements-in-2011.

22. "Our Starbucks Mission Statement," Starbucks Corporation, www.starbucks.com/about-us/company-information/mission-statement.

23. "Our Mission," Zahner's Clothiers, http://zahnersclothiers.com/home.htm.

24. "Inc. 500 Mission Statements," MissionStatements.com, www.missionstatements.com/inc_500_mission_statements.html.

25. Jessica Shambora, "David vs. Goliath," *Fortune*, February 28, 2011, p. 26; "SimpliSafe," SimpliSafe, http://simplisafe.com.

26. Sarah E. Needleman, "Dial-a-Mattress Retailer Blames Troubles on Stores, Executive Team," *Wall Street Journal*, July 14, 2009, http://online.wsj.com/article/SB124752875953235607.html.

27. Erik Rhey, "Bottles to Bridges," *Fortune*, April 30, 2012, p. 35; *The Fix We're in For: The State of Our Nation's Bridges* (Washington, DC: Transportation for America, 2011), p. 5.

28. Chuck Raasch, "Drought Takes Toll on Business," *Greenville News*, August 5, 2012, p. 6E.

29. *Theatrical Market Statistics 2012*, Motion Picture Association of America, 2013, p. 9.

30. "Business Strategy: What Is a Luxury Movie Theater?" *Quora*, May 12, 2012, www.quora.com/Business-Strategy/What-is-a-luxury-movie-cinema; Lauren A. E. Shuker, "Double Feature: Dinner and a Movie," *Wall Street Journal*, January 5, 2011, pp. D1–D2; Peggy Edersheim Kalb, "A Movie Theater as Comfy as Our Sofa," *Wall Street Journal*, April 24, 2008, p. D2; Andy Serwer, Corey Hajim, and Susan M. Kaufman, "Movie Theaters: Extreme Makeover," *Fortune*, May 23, 2006, http://money.cnn.com/2006/05/19/magazines/fortune/theater_futureof_fortune/index.htm; Paul Donsky, "New Theaters to Offer One-Stop Dinner and a Movie," *Access Atlanta*, April 25, 2008, www.accessatlanta.com/movies/content/movies/stories/2008/04/25/movie_0425.html?cxntlid=homepage_tab_newstab.

31. Nicole L. Torres, "Roast of the Town," *Entrepreneur B.Y.O.B.*, March 2003, p. 118; "Company History," Mayorga Coffee Roasters, www.mayorgacoffee.com/about_01_2.html.

32. Andria Cheng, "Daffy's, a Discounter, Falls Victim to Its Rivals," *Wall Street Journal*, July 17, 2012, p. B3; Adrianne Pasquarelli, "Off the Rack: Discounter Daffy's Is Done," *Craine's New York Business*, July 16, 2012, www.crainsnewyork.com/article/20120716/RETAIL_APPAREL/120719920.

33. Carolyn Z. Lawrence, "Know Your Competition," *Business Startups*, April 1997, p. 51.

34. Brady MacDonald, "Pat's vs. Geno's: Cheesesteak War Still Simmers in Phladelphia," *Los Angeles Times*, July 25, 2011, http://articles.latimes.com/2011/jul/25/news/la-trb-philadelphia-cheesesteak-pats-genos-07201125; "Our History," Geno's Steaks, www.genosteaks.com; "History," Pat's King of Steaks, www.patskingofsteaks.com/history.htm.

35. Julia Boorstin, "Cruising for a Bruising?," *Fortune*, June 9, 2003, pp. 143–150; Martha Brannigan, "Cruise Lines Look to the Land to Get Boomers on Board," *Wall Street Journal*, December 6, 1999, p. B4.

36. Shari Caudron, "I Spy, You Spy," *Industry Week*, October 3, 1994, p. 36.

37. Stephen D. Solomon, "Spies Like You," *FSB*, June 2001, pp. 76–82.

38. Dan Brekke, "What You Don't Know Can Hurt You," *Smart Business*, March 2001, pp. 64–76.

39. Stephanie Faris, "Perch: The One-Stop Social Media Tracking App for Small Businesses Is Coming Soon to Your Neighborhood," *Small Biz Technology*, April 12, 2012, www.smallbiztechnology.com/archive/2012/04/perch-the-one-stop-social-media-tracking-app-for-small-businesses-is-coming-soon-to-your-neighborhood.html.

40. David Witt, "Only 14% of Employees Understand Their Company's Strategy and Direction," *Blanchard LeaderChat*, May 21, 2012, http://leaderchat.org/2012/05/21/only-14-of-employees-understand-their-companys-strategy-and-direction.

41. Verne Harnish, "Five Ways to Get Your Stretegy Right," *Fortune*, April 11, 2011, p. 42.

42. Mark Henricks, "In the BHAG," *Entrepreneur*, August 1999, pp. 65–67.

43. Jim Collins and Jerry Poras, *Built to Last: Successful Habits of Visionary Companies* (New York: HarperBusiness, 1994), p. 232.

44. Debbie Stocker, "Four Approaches to Creating BHAGs—What's Yours?," *Matt Stocker Ltd.*, March 16, 2011, www.mattstocker.com/blog/four-approaches-to-creating-a-bhag-whats-yours.

45. Joseph C. Picken and Gregory Dess, "The Seven Traps of Strategic Planning," *Inc.*, November 1996, p. 99.

46. Nancy Averett, "Behind the Scenes: Companies at the Heart of Everyday Life," *Inc.*, July/August 2011, pp. 24–25; "Who We Are," Advanced Flexible Materials, www.afminc.com/company.asp.

47. Ian Mount, "Myth Busters," *FSB*, November 2008, p. 30.

48. Kambiz Foroohar, "Step Ahead—and Avoid Fads," *Forbes*, November 4, 1996, pp. 172–176.

49. Gwen Moran, "Vision Quest," *Entrepreneur*, January 2012, pp. 56–57; Andrew Maclean, "Warby Parker Disrupts the Eyewear Industry," *Inc.*, November 24, 2011, www.inc.com/video/2011/success-stories-warby-parker.html; Colin Magee, "Taking on the Big Guys: Multichannel Strategies for Midsize Brands," *Salt Digital*, June 10, 2012, http://web.saltdigitalinc.com/blog/bid/165013/Taking-on-the-big-guys-multichannel-strategies-for-midsize-brands; Alexander Taub, "Warby Parker: The Human Referral Effect," *Forbes*, August 2, 2012, www.forbes.com/sites/alextaub/2012/08/02/warby-parker-the-human-referral-effect.

50. Elaine Pofeldt, "David vs. Goliath," *Fortune*, March 19, 2012, p. 54; Kristina Knight, "3 Questions for Mad Mimi's Gary Levitt," *BizReport*, August 18, 2011, www.bizreport.com/2011/08/3-questions-with-mad-mimis-gary-levitt.html; Tony Case, "Q&A: Gary Levitt, Founder and CEO, Mad Mimi," *Direct Marketing News*, November 8, 2011, www.dmnews.com/qa-gary-levitt-founder-and-ceo-mad-mimi/article/216301.

51. Debra Phillips, "Leaders of the Pack," *Entrepreneur*, September 1996, p. 127.

52. Darren Dahl, "Fido Turns Foodie," *Inc.*, February 2012, p. 28; Kimberly Weisul, "Start-Ups 2010: A Dog Lover Builds a $5 Million Business," *Inc.*, October 1, 2010, www.inc.com/magazine/20101001/start-ups-2010-a-dog-lover-builds-a-5-million-business.html; Jeanette Hurt, "From Fashionista to Pet Food Mogul," *MSNBC*, June 25, 2010, www.msnbc.msn.com/id/37694300/ns/business-small_business/t/fashionista-pet-food-mogul/#.UCVxKRo_kis; Sharon L. Peters, "Are Americans Crazy for Treating Our Pets Like Kids?," *USA Today*, December 20, 2011, http://yourlife.usatoday.com/parenting-family/pets/story/2011-12-18/Are-Americans-crazy-for-treating-our-pets-like-kids/52054058/1; "Industry Statistics and Trends," American Pet Products Association, www.americanpetproducts.org/press_industrytrends.asp.

53. Phillips, "Leaders of the Pack," p. 127.

54. Brian Dumaine, "Save a Tree: Log Underwater," *Fortune*, July 25, 2011, p. 18; Stephen Messenger, "New Logging Equipment Harvests Underwater Forests," *Tree Hugger*, July 21, 2011, www.treehugger.com/clean-technology/new-logging-equipment-harvests-underwater-forests.html.

55. Erika Napoletano, "Narrowing the Focus," *Entrepreneur*, April 2012, p. 17; "Our Story," Lehman's, www.lehmans.com/store/util/about?Args=.

56. Dyan Castle, "All I Do Is Make Pipes," J. M. Boswell and Son Handmade Pipes, www.boswellpipes.com/allidoismakepipes.html; Paige Reddinger, "J. M. Boswell and Son Pipes," *Forbes Life*, July 2011, p. 44.

57. Geoff Williams, "Rico Elmore: When Regular Glasses Just Won't Fit," *AOL Small Business*, July 29, 2011, http://smallbusiness.aol.com/2011/07/29/rico-elmore-fatheadz-when-regular-glasses-just-wont-fit; Rico Elmore, "I Knew I Couldn't Be the Only Person in the World with This Issue," *M.O.*, 2012, www.mo.com/Rico-Elmore-Fatheadz-Eyewear.

58. Kasey Wehrum, "Behind the Scenes," *Inc.*, November 2011, pp. 18–19; "About Us," Peerless Handcuff Company, www.peerless.net/about-us.html.

59. Joel Kurtzman, "Is Your Company Off Course? Now You Can Find Out Why," *Fortune*, February 17, 1997, p. 128.

60. Robert S. Kaplan and David P. Norton, "The Balanced Scorecard—Measures That Drive Performance," *Harvard Business Review*, January–February 1992, pp. 71–79.

61. John Jantsch, "Effective Planning Is About What to Leave Out," *Duct Tape Marketing*, December 19, 2011, www.ducttapemarketing.com/blog/2011/12/19/effective-planning-is-about-what-to-leave-out.

Chapter 5

1. "Starting a Business," Doing Business, World Bank, June 2013, http://www.doingbusiness.org/data/exploretopics/starting-a-business.

2. Ian Mount, "Business Licenses to Go," *FSB*, March 2007, p. 14; "FAQs," Business Licenses, www.businesslicenses.com/faqs.php#q2.

3. Joel Holland, "From College Students to Million-Dollar Partners," *Entrepreneur*, January 18, 2010, www.entrepreneur.com/article/204718.

4. Matthew Bandyk, "Five Things Entrepreneurs Should Know About Business Partners," *U.S. News and World Report*, May 6, 2008, www.usnews.com/articles/business/small-business-entrepreneurs/2008/05/06/5-things-entrepreneurs-should-know-about-business-partners.html.

5. Norm Brodsky, "Sam and Me," *Inc.*, June 2006, p. 65.

6. Stephanie Clifford, "Until Death, or Some Other Sticky Problem, Do Us Part," *Inc.*, November 2006, pp. 104–110.

7. Samantha Cortez, "How to Come Back from a Failed Partnership," *Business Insider*, September 26, 2012, www.businessinsider.com/how-to-come-back-from-a-failed-partnership-2012-9.

8. Henry R. Chesseman, *Business Law*, 5th ed. (Upper Saddle River, NJ: Pearson Prentice Hall, 2004), p. 675.

9. Richard Reinis, "Be Wary of Personal Guarantees," *Bloomberg BusinessWeek*, June 15, 2011, www.businessweek.com/smallbiz/tips/archives/2011/06/be_wary_of_personal_guarantees.html.

10. Jill Andresky Fraser, "Perfect Form," *Inc.*, www.inc.com/magazine/19971201/1368.html, p. 3.

11. Ryan Roberts, "Why the Corporation Is King for Getting Venture Capital," *Startup Lawyer*, July 17, 2008, http://startuplawyer.com/venture-capital/why-the-corporation-is-king-for-getting-venture-capital.

12. Jessica Bruder, "A Harvard Professor Analyzes Why Start-Ups Fail," *Inc.*, May, 25, 2012, http://boss.blogs.nytimes.com/2012/05/25/a-harvard-professor-analyzes-why-start-ups-fail/?_r=0.

13. Jane Gravenstine Brown, "For Profit Social Ventures," www.caseatduke.org/articles/0205/research.htm.

14. "Failed Bank Information: ShoreBank, Chicago, Illinois," Federal Deposit Insurance Corporation, www.fdic.gov/bank/individual/failed/shorebank.html.

Chapter 6

1. "Jiffy Lube Welcomes Youngest Franchisee," *PR Newswire*, June 5, 2012, www.prnewswire.com/news-releases/jiffy-lube-welcomes-youngest-franchisee-157264215.html; "Exclusive Interview with Miles Blauvelt, Jiffy Lube's Youngest Franchisee to Date," *FranchiseChatter*, July 22, 2012, www.franchisechatter.com/2012/07/22/exclusive-interview-with-miles-blauvelt-jiffy-lubes-youngest-franchisee-to-date.

2. *2012 Franchise Business Economic Outlook*, International Franchise Association, January 2012, p. 1; *Economic Impact of Franchised Businesses*, Volume 3, International Franchise Association, p. 4.

3. Philip Zeidman, "Mapping the Franchise World from the IFA Convention," *Franchise Times*, April 2011, www.franchisetimes.com/content/story.php?article=02125.

4. Gregory Matusky, "The Franchise Hall of Fame," *Inc.*, April 1994, pp. 86–89.

5. Personal contact with Nicholas A. Bibby, The Bibby Group, www.bibbbygroup.com, May 2, 2007.

6. *Economic Impact of Franchised Businesses*, Volume 3, International Franchise Association, p. 7.

7. David Kaplan, "They're Giving a Franchise a Go," *Houston Chronicle*, June 11, 2012, www.chron.com/business/article/They-re-giving-a-franchise-a-go-3621231.php.

8. David Kaplan, "They're Giving a Franchise a Go," *Houston Chronicle*, June 11, 2012, www.chron.com/business/article/They-re-giving-a-franchise-a-go-3621231.php.

9. Nancy Cook, "Left Behind by Lehman," *The Daily Beast*, September 13, 2009, www.thedailybeast.com/newsweek/2009/09/13/left-behind-by-lehman.html.

10. Patricia Orsini, "Everything I Need to Know I Learned at McDonald's: CEO," *CNBC*, July 8, 2012, www.cnbc.com/id/48097904/Everything_I_Need_to_Know_I_Learned_at_McDonald_s_CEO.

11. Mark Henricks, "Masters of Franchising," *Inc.*, June 2012, advertising insert.

12. Janean Huber, "Franchise Forecast," *Entrepreneur*, January 1993, p. 73.

13. "Hamburger University," McDonald's, www.aboutmcdonalds.com/mcd/corporate_careers/training_and_development/hamburger_university.html.

14. Maya Norris, "Franchisees Benefit from Learning 2.0 at Zaxby's," *Chain Leader*, January 15, 2009, www.chainleader.com/article/CA6629891.html.

15. Maya Norris, "Franchisees Benefit from Learning 2.0 at Zaxby's," *Chain Leader*, January 15, 2009, www.chainleader.com/article/CA6629891.html.

16. Patricia Calhoun, "Cool Jobs: From the Colorado Legislature to Red Mango," Denver Westword Blogs, July 6, 2012, http://blogs.westword.com/cafesociety/2012/07/red_mango.php.

17. Janean Huber, "Franchise Forecast," *Entrepreneur*, January 1993, p. 73.

18. Mark Henricks, "Million Dollar Concepts," *Inc.*, July/August 2012, pp. 106–110.

19. *The Profile of Franchising 2006* (Washington, DC: International Franchise Association, 2007), p. 67.

20. *The Profile of Franchising 2006* (Washington, DC: International Franchise Association, 2007), p. 67.

21. *Franchise Business Economic Outlook: March 2012*, International Franchise Association and IHS Global Insight, March 22, 2011, p. 13.

22. *Small Business Lending Matrix and Analysis*, FRANdata, Volume V, March 2013, p. 18.

23. *The Profile of Franchising 2006* (Washington, DC: International Franchise Association, 2007), p. 70.

24. "Huddle House Franchising," Huddle House, September 2012, www.huddlehousefranchising.com; "About Us," Huddle House, September 2012, www.huddlehouse.com/about-us.aspx.

25. Julie Bennett, "Operators Are Finding Financing Less Daunting," *Wall Street Journal*, June 14, 2012, p. B6.

26. Darrell Johnson and John Reynolds, "A Study of Franchise Loan Performance in the SBA Guaranty Programs," *Franchising World*, September 2007, pp. 53–56; Richard Gibson, "How to Finance a Franchise," *Wall Street Journal*, March 17, 2008, p. R8.

27. "Success Stories: Funding Great Clips—Start-Up Loan," BoeFly, www.boefly.com/boefly_success_stories.cfm?cat=retail&cn=greatClips#.

28. Stephanie Barlow, "Sub-Stantial Success," *Entrepreneur*, January 1993, p. 126.

29. Julie Bennet and Cheryl Babcock, *Franchise Times Guide to Selecting, Buying, and Owning a Franchise* (Minneapolis: Franchise Times and Sterling Publishing, February 2008), p. 255.

30. Richard Gibson, "Why Franchisees Fail," *Wall Street Journal*, April 30, 2007, p. R9.

31. "2010 Franchise SBA Loan Default and Charge-Off Data," Coleman Publishing, www.colemanpublishing.com/public/387.cfm; "SBA Franchise Failure Rates by Brand," Blue Mau Mau, June 15, 2012, www.bluemaumau.org/sba_franchisee_failure_rates_brand_2012.

32. Julie Bennett, "Operators Are Finding Financing Less Daunting," *Wall Street Journal*, June 14, 2012, p. B6.

33. Steven C. Michael and James G. Combs, "Entrepreneurial Failure: The Case of Franchisees," *Journal of Small Business Management* 46, no. 1 (January 2008), pp. 75–90.

34. Jason Daley, "Back from the Brink," *Entrepreneur*, May 2012, pp. 94–99.

35. *The Profile of Franchising 2006* (Washington, DC: International Franchise Association, 2007), p. 62.

36. *The Profile of Franchising* (Washington, DC: FRANDATA Corp and the IFA Educational Foundation, 2000), p. 123.

37. *The Profile of Franchising 2006* (Washington, DC: International Franchise Association, 2007), p. 66.

38. *The Profile of Franchising 2006* (Washington, DC: International Franchise Association, 2007), p. 68.

39. Colleen Curry, "Donut Cops Sniff Out Frauds in Fast Food Restaurants," *ABC News*, June 6, 2012, http://abcnews.go.com/Business/donut-cops-sniff-frauds-fast-food-restaurants/story?id=16476480.

40. Jim Coen, "Quiznos Franchisees Walloped by Recession," *Let's Talk Franchising*, October 21, 2008, www.franchiseperfection.com/blog/?p=327; Julie Creswell, "When Disillusion Sets In," *New York Times*, February 24, 2007, www.nytimes.com/2007/02/24/business/24quiznos.html.

41. Richard Gibson, "Burger King Franchisees Can't Have It Their Way," *Wall Street Journal*, January 21, 2010, http://online.wsj.com/article/SB10001424052748704320104575014941842011972.html; Elaine Walker, "BK Franchisees Lose Pricing War," *Miami Herald*, November 23, 2010, www.miamiherald.com/2010/11/23/1939144/bk-franchisees-lose-pricing-war.html; Elaine Walker, "Burger King, Franchisees Clash over $1 Burger Deal," *Miami Herald*, February 3, 2010, www.miamiherald.com/2010/02/03/1459724/burger-king-franchisees-clash.html; Elaine Walker, "BK to Switch Out $1 Double Cheeseburger," *Nation's Restaurant News*, February 17, 2010, www.nrn.com/article/bk-switch-out-1-double-cheeseburger.

42. Richard Gibson, "The Inside Scoop," *Wall Street Journal*, June 16, 2008, http://wsj.com/article/SB121321718319265569.html?mod=Financing_1.

43. Jason Daley, "Downsizing," *Entrepreneur*, October 2011, p. 160.

44. "McDonald's Egg McMuffin Turns 40," *Yahoo! Finance*, August 29, 2012, http://finance.yahoo.com/news/mcdonald-egg-mcmuffin-turns-40-150000656.html; Janet Adamy, "For McDonald's, It's a Wrap," *Wall Street Journal*, January 30, 2007, pp. B1–B2; "The Birth of the Egg McMuffin," McDonald's, March 15, 2009, www.aboutmcdonalds.com/mcd/students/amazing_stories/the_birth_of_the_egg_mcmuffin.html.

45. Christopher Mims, "Only Place in America More Than 100 Miles from a McDonald's," *Smart Planet*, November 9, 2011, www.smartplanet.com/blog/cities/only-place-in-america-more-than-100-miles-from-a-mcdonalds-video/1185.

46. Jason Daley, "A Perfect Fit?," *Entrepreneur*, August 2012, pp. 91–92.

47. Elaine Pofeldt, "Success Franchisee Satisfaction Survey," *Success*, April 1999, p. 59.

48. Julie Bennett, "Deciphering the FDD," *Entrepreneur*, January 2012, pp. 106–111.

49. Julie Bennett, "Deciphering the FDD," *Entrepreneur*, January 2012, pp. 106–111.

50. Lillia Callum-Penso, "Taking the Subway from Iran," *Greenville News*, April 29, 2012, pp. 1E, 3E.

51. Mark Henricks, "Masters of Franchising," *Inc.*, June 2012, advertising insert.

52. Carol Tice, "How to Research a Franchise," *Entrepreneur*, January 2009, p. 117.

53. Amy Covington, "Microtel Inns & Suites: The Rocco Valluzo Story," *Franchise Prospector*, April 2007, www.franchiseprospector.com/success/microtel-rocco-valluzo.php.

54. Sarah E. Needleman, "Tough Times for Franchising," *Wall Street Journal*, February 9, 2012, http://online.wsj.com/article/SB10001424052970204136404577211391192172770.html.

55. "New Research Shows Changes in Franchise Ownership Demographics Among Women, Minorities," International Franchise Association, December 9, 2011, www.franchise.org/Franchise-News-Detail.aspx?id=55497.

56. Jan Fletcher, "Leveraging Franchise Diversity," *QSR Magazine*, May 2011, www.qsrmagazine.com/franchising/leveraging-franchisee-diversity; "Our Commitment to Diversity," Domino's Pizza, www.dominosbiz.com/Biz-Public-EN/Site+Content/Secondary/About+Dominos/Diversity.

57. Christian Conte, "Former SunTrust Executive Finds Happiness in Franchising," *Jacksonville Business Journal*, June 19, 2009, http://jacksonville.bizjournals.com/jacksonville/stories/2009/06/22/story3.html.

58. Jason Daley, "Capital Ideas," *Entrepreneur*, October 2011, pp. 135–139.

59. Carol Tice, "Running the Numbers," *Entrepreneur*, July 2009, pp. 87–95.

60. Julie Bennett, "Multi-Unit Operators Spur Growth of Sector," *Wall Street Journal*, February 13, 2012, p. B6.

61. Tracy Stapp, "The Making of a Multi-Unit Franchise Maven," *Entrepreneur*, July 16, 2012, www.entrepreneur.com/article/223654.

62. Tracy Stapp, "The Making of a Multi-Unit Franchise Maven," *Entrepreneur*, July 16, 2012, www.entrepreneur.com/article/223654.

63. Philip Zeidman, "Mapping the Franchise World from the IFA Convention," *Franchise Times*, April 2011, www.franchisetimes.com/content/story.php?article=02125.

64. David Novak, "Letter from Our CEO," *Yum! Brands Annual Report 2011*, www.yum.com/annualreport; "Yum! Brands Responsibility," www.yum.com/responsibility.

65. Jason Daley, "And the Number One Franchise Is . . . Hampton Inns," *Entrepreneur*, December 14, 2011, www.entrepreneur.com/article/222421.

66. Alan J. Liddle, "10 Non-Traditional Subway Restaurants," *Nation's Restaurant News*, July 26, 2011, www.nrn.com/article/10-non-traditional-subway-restaurants; Geoff Williams, "Subway Opens First Gravity-Defying Restaurant at the Freedom Tower," Daily Finance, January 4, 2010, www.dailyfinance.com/2010/01/04/subway-opens-first-restaurant-at-the-freedom-tower-restaurant.

67. Stephanie Simon, "Franchises on a Smaller Scale," *Wall Street Journal*, January 12, 2012, p. B5.

68. "Dunkin' Donuts Opens 10 College Campus Locations," *QSR Magazine*, September 12, 2012, www.qsrmagazine.com/news/dunkin-donuts-opens-10-college-campus-locations.

69. Sarah E. Needleman and Angus Loten, "Fast-Food Franchisees Bulking Up," *Wall Street Journal*, April 12, 2012, http://online.wsj.com/article/SB10001424052702304587704577333443052487330.html.

70. "Pizza Patron Signs Agreement for Eight San Antonio Units," *QSR Magazine*, August 21, 2012, www.qsrmagazine.com/news/pizza-patron-signs-agreement-eight-san-antonio-units.

71. "Papa Murphy's Signs First Franchise Deal Outside N.A.," *QSR Magazine*, September 11, 2012, www.qsrmagazine.com/news/papa-murphys-signs-1st-franchise-deal-outside-na.

72. Jason Daley, "Playing Well Together," *Entrepreneur*, April 2012, pp. 87–91; Ron Ruggless, "Focus Brands Grows Tri-Branded Locations," *Nation's Restaurant News*, April 5, 2011, http://nrn.com/article/focus-brands-grows-tri-branded-locations.

73. "Do You Need Maid Brigade?," Maid Brigade, www.maidbrigade.com/whoneedsmaidbrigade.php.

74. Laura Kulikowski, "5 Successful Senior Care Franchisees," *The Street*, July 29, 2011, www.thestreet.com/story/11203654/2/5-successful-senior-care-franchises.html; Raymund Flandez, "A Look at High-Performing Franchises," *Wall Street Journal*, February 12, 2008, p. B5; Lindsay Holloway, James Park, Nichole L. Torres, and Sara Wilson, "This Just In . . ." *Entrepreneur*, January 2008, pp. 100–110.

75. Erwin J. Keup and Peter Keup, "Can Your Business Be Franchised?," *Entrepreneur*, September 2012, www.entrepreneur.com/article/224323?cam=Dev&ctp=Carousel&cdt=13&cdn=224323.

76. Jason Daley, "New Life for Old Castoffs," *Entrepreneur*, August 2012, p. 110.

77. Personal contact with Jim Thomas, May 16, 2007.

78. Amy Joyce, "The Art of the Successful Franchise," *Washington Post*, Monday, April 30, 2007, p. D1.

Chapter 7

1. Elaine Appleton Grant, "How to Buy a Small Business Without Getting Taken," *US News and World Report*, February 26, 2008, http://money.usnews.com/money/business-economy/small-business/articles/2008/02/26/how-to-buy-a-small-business-without-getting-taken.

2. John Paglia, Pepperdine Private Capital Markets Project—Capital Markets Report 2011–2012, http://bschool.pepperdine.edu/appliedresearch/research/pcmsurvey/content/PPCMP_Capital_Markets2012FIN.pdf.

3. Julie Bawden Davis, "Buying an Existing Business? You'd Better Shop Around," *Entrepreneur*, August 1999, www.entrepreneur.com/magazine/entrepreneur/1999/august/18132.html.

4. Catherine Clifford, "Top Sources of Small-Business Financing in 2012," *Entrepreneur*, January 3, 2012, www.entrepreneur.com/article/222540; Barbara Taylor, "Using Your 401(k) to Buy a Small Business," *New York Times*, June 13, 2012, http://boss.blogs.nytimes.com/2012/06/13/using-your-401k-to-buy-a-small-business.

5. Lin Grensing-Pophal, "Decide Whether You'll Buy an Existing Business, a Business Opportunity, a Franchise, or Go It Alone", *Business Start-Ups*, December 2000.

6. Lil Sawyer, "Buying a Business: The Safer Alternative," About.com, http://entrepreneurs.about.com/od/buyingabusiness/a/buyingabusiness.htm.

7. "Buying a Business," Small Business Administration, www.sba.gov/smallbusinessplanner/start/buyabusiness/SERV_SBP_S_BUYB.html.

8. Mina Kimes, "Sales Take Wing," *FSB*, December 2008/January 2009, p. 27; David Farkas, "Buffalo Wings and Rings Is Taking Flight," *Chain Leader*, September 1, 2009, www.chainleader.com/article/CA6686882.html.

9. Russell L. Brown, *Strategies for Successfully Buying or Selling a Business*, 2nd ed. (Niantic, CT: The Business Book Press, 2002), pp. 1–2.

10. Barbara Taylor, "Four Questions to Ask a Business Broker," *New York Times*, October 20, 2011, http://boss.blogs.nytimes.com/2011/10/20/four-questions-to-ask-before-hiring-a-business-broker.

11. http://usamade.files.wordpress.com/2011/03/usc_graduates.pdf; Mike Sullivan, "You Could Have The Greatest Vision and the Greatest Business Plan in the World, but It's Just an Idea Until You Execute It," *mo.com*, April 6, 2012, www.mo.com/sean-bandawat-jacob-bromwell; Cindy Vanegas, "5 Questions to Ask Before Buying a Business," *FOXBusiness*, May 21, 2012, http://smallbusiness.foxbusiness.com/finance-accounting/2012/05/21/5-questions-to-ask-before-buying-business.

12. Robert F. Klueger, *Buying and Selling a Business: A Step by Step Guide* (New York: John Wiley & Sons, 2004), p. 12.

13. Darren Dahl, "Business for Sale: A Flooring and Remodeling Store," *Inc.*, August 24, 2010, www.inc.com/magazine/20100901/business-for-sale-a-flooring-and-remodeling-store.html.

14. John Paglia, Pepperdine Private Capital Markets Project—Capital Markets Report 2011–2012, http://bschool.pepperdine.edu/appliedresearch/research/pcmsurvey/content/PPCMP_Capital_Markets2012FIN.pdf.

15. "BV Market Data Database Summary," Business Valuation Resources, June 21, 2013, http://www.bvmarketdata.com/defaulttextonly.asp?f=bvmdtable.

16. "Valuation Rule of Thumb," BizStats, www.bizstats.com/reports/valuation-rule-thumb.asp.

17. Darren Dahl, "The Most Valuable Companies in America," *Inc.*, April 2008, pp. 97–105.

18. Ryan McCarthy, "A Buyer's Market," *Inc.*, June 2009, pp. 82–92.

19. James Laabs, "What Is Your Company Worth?," *The Business Sale Center*, www.businesssalecenter.com/new_page_3.htm.

20. *Business Planning Tools: Buying and Selling a Small Business*, MasterCard Worldwide, www.mastercard.com/us/business/en/smallbiz/businessplanning/businessplanning.html, p. 11.

21. Ryan McCarthy, "A Buyer's Market," *Inc.*, June 2009, p. 85.

22. Jeanne Lee, "Exit Strategies," *FSB*, April 2009, p. 49.

23. Mark Blayney, *Buying a Business and Making It Work* (London: How To Books, 2007), p. 420.

24. Donna Nebenzahl, "The Duo: Em & Seb," Urban Expressions, November 26, 2012, www.urbanexpressions.ca/print/story/the-duo-em-seb; Norm Brodsky, "Buying a Business? Expect the Unexpected, *Inc.*, May 1, 2012, www.inc.com/magazine/201205/norm-brodsky/norm-on-buying-a-business.html.

25. Edward Karstetter, "How Intangible Assets Affect Business Value," *Entrepreneur*, May 6, 2002, http://entrepreneur.com/article/0,4621,299514,00.html.

Chapter 8

1. Rhonda Abrams, "Before Jumping into a New Venture, Do a Feasibility Study," *USA Today*, April 15, 2010, www.usatoday.com/money/smallbusiness/columnist/abrams/2010-04-15-new-venture_N.htm.

2. Charles Fishman, "The Wal-Mart You Don't Know," *Fast Company*, December 2003, www.fastcompany.com/magazine/77/walmart.html.

3. www.aloompa.com.

4. Jason Del Rey, "How to Start a Restaurant," *Inc.*, July 1, 2009, www.inc.com/magazine/20090701/how-to-start-a-restaurant.html.

5. Wells Fargo, "How Much Money Does It Take to Start a Small Business?," news release, August 15. www.wellsfargo.com/press/20060815_Money?year=2006.

6. U.S. Census Bureau, "Survey of Business Owners," www.census.gov/econ/sbo/methodology.html.

7. J. Cornwall, *Bootstrapping* (Englewood Cliffs, NJ: Pearson/Prentice Hall, 2009), p. 2.

8. Brad Sugars, "The 6 Biggest Mistakes in Raising Startup Capital," *Entrepreneur*, September 20, 2007, www.entrepreneur.com/article/184350.

9. A. Osterwalder and Y. Pigneur, *Business Model Generation* (Hoboken, NJ: Wiley, 2010).

10. Steve Blank and Bob Dorf, *The Startup Owner's Manual* (Pescadero, CA: K&S Ranch, 2012).

11. Eric Ries, *The Lean Startup: How Today's Entrepreneurs Use Continuous Innovation to Create Radically Successful Businesses* (New York: Crown Business, 2011).

12. Eric Ries, "How DropBox Started as a Minimal Viable Product," *Techcrunch.com*, October 19, 2011, http://techcrunch.com/2011/10/19/dropbox-minimal-viable-product.

13. J. Cornwall, J. Gonzalez, and S. Brown, "CoolPeopleCare: A Social Venture," United States Association for Small Business and Entrepreneurship, *Proceedings*, 2009; "CoolPeopleCare, Inc. Business Plan," 2008; "Cool People Care Does Way More Than Sell T-Shirts", November 19, 2010, www.wkrn.com/Global/story.asp?S=13539331; *Ray Chung's Online Journal*, February 13, 2008, http://raychung22.com/blog/2008/02/11/young-people-spotlight-sam-davidson.

14. *Ray Chung's Online Journal*, February 13, 2008, http://raychung22.com/blog/2008/02/11/young-people-spotlight-sam-davidson.

15. *Ray Chung's Online Journal*, February 13, 2008, http://raychung22.com/blog/2008/02/11/young-people-spotlight-sam-davidson.

16. Mark Henricks, "Do You Really Need a Business Plan?," *Entrepreneur*, December 2008, pp. 93–95; Kelly Spors, "Advance Planning Pays Off for Start-Ups," *Wall Street Journal*, February 9, 2009, http://blogs.wsj.com/independentstreet/2009/02/09/advance-planning-pays-off-for-start-ups.

17. Rhonda Abrams, "Get Going on a New Business Plan," *Greenville News*, January 25, 2009, pp. 1E, 3E.

18. "The Business Plan Competition," Wake Forest Schools of Business, http://business.wfu.edu/default.aspx?id=279; "Stan Mandel, "The Elevator Pitch (EP): Engage People, Move to Action . . . in Two Minutes," *Office of Entrepreneurship and Liberal Arts Newsletter*, January 2006, http://entrepreneurship.wfu.edu/newsletter/the-elevator-pitch.html.

19. "Advice from the Great Ones," *Communication Briefings*, January 1992, p. 5.

20. Mike Hanlon, "Bespoke Apartments Created in 24 Hours," *GizMag*, September 15, 2006, www.gizmag.com/go/6324; Adam McCulloch, "Prefab with a View," *Business 2.0*, May 2005, p. 70; "The Company," First Penthouse, www.firstpenthouse.com/new/company.

21. Edward Clendaniel, "The Professor and the Practitioner," *Forbes ASAP*, May 28, 2001, p. 57.

22. Karen E. Klein, "To Beat the Recession, Reinvent Your Business," *BusinessWeek*, October 23, 2009, www.businessweek.com/smallbiz/content/oct2009/sb20091023_229168.htm.

23. Guy Kawasaki, *The Art of the Start* (New York: Penguin, 2004), p. 130.

24. Carol Roth, "Interview with Award-Winning Social Student Entrepreneur Ludwick Marishane," *Huffington Post Small Business Amercia*, January 30, 2012, www.huffingtonpost.com/carol-roth/social-student-entrepreneur_b_1222586.html?ref=fb&src=sp&comm_ref=false.

25. Jeff Wuorio, "Get an 'A' in Researching a Business Idea," *Microsoft bCentral*, www.bcentral.com/articles/wuorio/140.asp.

26. Lauren Keyson, "Inside the Minds of NY Venture Capitalists: People, Not Ideas," *NYConvergence*, January 13, 2012, http://nyconvergence.com/2012/01/inside-the-mind-of-the-venture-capitalist-it%E2%80%99s-people-not-ideas.html.

27. "Raising Money," *Entrepreneur*, July 2005, p. 58.

28. Michael V. Copeland, "How to Make Your Business Plan the Perfect Pitch," *Business 2.0*, September 2005, p. 88.

29. Karen Axelton, "Good Plan, Stan," *Business Start-Ups*, March 2000, p. 17.

30. William Sahlman, "How to Write a Great Business Plan," p. 105.

31. Adrianne Jeffries, "As Banks Start Nosing Around Facebook and Twitter, the Wrong Friends Might Just Sink Your Credit," *BetaBeat*, December 12, 2011, http://betabeat.com/2011/12/as-banks-start-nosing-around-facebook-and-twitter-the-wrong-friends-might-just-sink-your-credit.

32. Tim Berry, "True Story: A Great Presentation Wins Big," *Business Insider*, May 8, 2012, www.businessinsider.com/true-story-a-great-presentation-wins-big-2012-5.

33. Guy Kawasaki, "The 10/20/30 Rule of Powerpoint," *How to Change the World*, December 20, 2005, http://blog.guykawasaki.com/2005/12/the_102030_rule.html#axzz1vnw0foX3.

34. David R. Evanson, "Capital Pitches That Succeed," *Nation's Business*, May 1997, p. 41.

35. Jill Andresky Fraser, "Who Can Help Out with a Business Plan?," *Inc.*, June 1999, p. 115.

Chapter 9

1. Sanjeev Aggarwal and Laurie McCabe, *Small Business Marketing Health Check* (Newton, MA: Hurwitz and Associates, 2009), p. 5.

2. Sean McFadden, "Healthy Strategy: Lifestyle Fitness Corp. Owner Gets Strategic with Marketing," *Boston Business Journal*, March 31, 2009, www.bizjournals.com/boston/stories/2009/03/30/smallb1.html.

3. Howard Fana Shaw, "Customer Care Checklist," *In Business*, September/October 1987, p. 28.

4. Jeff Cornwall, "Sweet Success," *Tennessean*, February 4, 2013, p. 3D.

5. Chris Anderson, *The Long Tail* (New York: Hyperion, 2008), p. 18.

6. Sara Wilson, "Capitalize on Nostalgia," *Entrepreneur*, March 2009, p. 94.

7. Norm Brodsky, "My First Year," *Inc.*, April 2009, p. 34.

8. Kim Komando, "Three Reasons to Use Online Customer Surveys," Microsoft Small Business Center, www.microsoft.com/smallbusiness/resources/marketing/market_research/3_reasons_to_use_online_customer_surveys.mspx.

9. Damon Brown, "Online Customer Surveys for Small Business," *Inc. Technology*, December 2006, http://technology.inc.com/internet/articles/200612/onlinesurveys.html.

10. Avery Johnson, "Hotels Take 'Know Your Customer' to New Level," *Wall Street Journal*, February 7, 2006, pp. D1, D3.

11. Shari Caudron, "Right on Target," *Industry Week*, September 2, 1996, p. 45.

12. Angela Garber Wolf, "Million-Dollar Questionnaire," *Small Business Computing*, January 2002, pp. 47–48.

13. Samuel Axon, "How Small Businesses Are Using Social Media for Real Results," *Mashable*, March 22, 2010, http://mashable.com/2010/03/22/small-business-social-media-results.

14. "Converting Customer Dissatisfaction into Loyalty and Profits at Cornerstone of New Boulder Business," Verde Group, October 3, 2005, http://verdegroup.ca/default.asp?action=article&ID=36.

15. Jeff Cornwall, "What Was Old Is New Again," *Tennessean*, April 1, 2013, p. D3.

16. Roberta Maynard, "Rich Niches," *Nation's Business*, November 1993, p. 41.

17. Patricia Sellers, "Companies That Serve You Best," *Fortune*, May 31, 1993, p. 75.

18. Robert B. Tucker, "Earn Your Customers' Loyalty," Economics Press Techniques, Strategies, and Inspiration for the Sales Professional, www.epic.co/SALES/selltips.htm#earn_loyalty.

19. William A. Sherden, "The Tools of Retention," *Small Business Reports*, November 1994, pp. 43–47.

20. Daniel Bortz, "Three Ways to Keep Your Customers Coming Back," *Entrepreneur*, December 29, 2011, www.entrepreneur.com/article/222520.

21. Richard Stone, "Retaining Customers Requires Constant Contact," *Small Business Computing*, January 11, 2005, www.smallbusinesscomputing.com/biztools/print.pho/3457221.

22. Bruce D. Temkin, William Chu, and Steven Geller, "Customer Experience Correlates to Loyalty," *Forrester Research*, February 17, 2009, www.forrester.com/rb/Research/customer_experience_correlates_to_loyalty/q/id/53794/t/2; Bruce D. Temkin, "Customer Experience Correlates to Loyalty," *Customer Experience Matters*, February 18, 2009, http://experiencematters.wordpress.com/2009/02/18/customer-experience-correlates-to-loyalty.

23. Mike Doherty, "The Best vs. the Rest," *Innovation.net*, November 1, 2004, http://venture2.typepad.com/innovationnet/2004/11/the_best_vs_the.html.

24. Alan Deutschman, "America's Fastest Risers," *Fortune*, October 7, 1991, p. 58.

25. *The New Realities of "Dating" in the Digital Age*, Accenture 2011 Global Consumer Research Study, www.accenture.com/us-en/Pages/insight-acn-global-consumer-research-study.aspx.

26. Roberta Maynard, "The Heat Is On," *Nation's Business*, October 1997, pp. 14–23.

27. Jeffrey M. O'Brien, "Wii Will Rock You," *Fortune*, June 4, 2007, http://money.cnn.com/magazines/fortune/fortune_archive/2007/06/11/100083454/index.htm.

28. Brett Nelson, "Fore!-Boding," *Forbes*, April 14, 2003, p. 60.

29. Jeremy Bradley, "Classic Radio Flyer Wagon Updated for 2.0 World," *CNN*, December 24, 2008, www.cnn.com/2008/TECH/12/24/radio.flyer.wagon/index.html; Lucas Mearian, "Opinion: Radio Flyer Is Pimpin' Its Wagons," *Computer World*, October 27, 2008, www.computerworld.com/s/article/9118086/Opinion_Radio_Flyer_is_pimpin_its_wagons.

30. Emmit C. Murphy and Mark A. Murphy, *Leading on the Edge of Chaos* (New York: Prentice Hall, 2002).

31. Jack Loechner, "Poor Customer Service Costs Companies $83 Billion Annually," *Media Post*, February 18, 2010, www.mediapost.com/publications/?fa=Articles.showArticle&art_aid=122502.

32. Jack Loechner, "Poor Customer Service Costs Companies $83 Billion Annually," *Media Post*, February 18, 2010, www.mediapost.com/publications/?fa=Articles.showArticle&art_aid=122502.

33. "Beware of Dissatisfied Customers: They Like to Blab," *Knowledge @ Wharton*, March 8, 2006, http://knowledge.wharton.upenn.edu/article.cfm?articleid=1422.

34. Carmine Gallo, "How Disney Works to Win Repeat Customers," *BusinessWeek*, December 1, 2009, www.businessweek.com/smallbiz/content/nov2009/sb20091130_866423.htm.

35. "Encourage Customers to Complain," *Small Business Reports*, June 1990, p. 7.

36. Dave Zielinski, "Improving Service Doesn't Require a Big Investment," *Small Business Reports*, February 1991, p. 20.

37. Brian Caufield, "How to Win Customer Loyalty," *Business 2.0*, March 2004, pp. 77–78.

38. Paul Hagen, "Nine Ways to Reward Employees to Reinforce Customer-Centric Behaviors," Forrester, May 11, 2012, http://blogs.forrester.com/paul_hagen/12-05-11-9_ways_to_reward_employees_to_reinforce_customer_centric_behaviors.

39. Jack Loechner, "Poor Customer Service Costs Companies $83 Billion Annually," *Media Post*, February 18, 2010, www.mediapost.com/publications/?fa=Articles.showArticle&art_aid=122502.

40. Susan Greco, "Fanatics," *Inc.*, April 2001, p. 38.

41. *The New Realities of "Dating" in the Digital Age*, Accenture 2011 Global Consumer Research Study, www.accenture.com/us-en/Pages/insight-acn-global-consumer-research-study.aspx.

42. Richard Gibson, "Can I Get a Smile with My Burger and Fries?," *Wall Street Journal*, September 23, 2003, p. D6.

43. Janet Adamy, "A Menu of Options," *Wall Street Journal*, October 30, 2006, p. B6.

44. *High Performance in the Age of Customer Centricity*, Accenture 2008 Global Customer Satisfaction Report, Accenture, p. 16.

45. Marie Moody, "How to Profit from Complaints," *FSB*, December 2009, http://money.cnn.com/2009/12/23/smallbusiness/profiting_from_customer_complaints.fsb/index.htm.

46. Thomas A. Stewart, "After All You've Done for Your Customers, Why Are They Still NOT HAPPY?," *Fortune*, December 11, 1995, pp. 178–182; Gile Gerretsen, "Special Tools Are Used by Super Markets," *Upstate Business*, June 14, 1998, p. 4.

47. Robert Reiss, "How Ritz-Carlton Stays at the Top," *Forbes*, October 30, 2009, www.forbes.com/2009/10/30/simon-cooper-ritz-leadership-ceonetwork-hotels.html.

48. Carmine Gallo, "How the Ritz-Carlton Inspired the Apple Store," *Forbes*, April 10, 2012, www.forbes.com/sites/carminegallo/2012/04/10/how-the-ritz-carlton-inspired-the-apple-store-video.

49. Patricia Neale, "John Ratzenberger's 'Made in America' Season Premiere Features Allen Edmonds Shoe Corporation, company press release, November 21, 2005, www.allenedmonds.com/wcsstore/AllenEdmonds/about/Made%20in%20America%20PR%2011.21.05.pdf.

50. Patricia Neale, "John Ratzenberger's 'Made in America' Season Premiere Features Allen Edmonds Shoe Corporation, company press release, November 21, 2005, www.allenedmonds.com/wcsstore/AllenEdmonds/about/Made%20in%20America%20PR%2011.21.05.pdf.

51. Rahul Jacobs, "TQM: More Than a Dying Fad," *Fortune*, October 18, 1993, p. 67.

52. Rahul Jacobs, "TQM: More Than a Dying Fad," *Fortune*, October 18, 1993, p. 67.

53. "Things We Like: Moovers and Shakers, A Soda Parlor on Wheels," *Nashvillest*, April 11, 2011, http://nashvillest.com/2011/04/11/things-we-like-moovers-shakers-a-soda-parlor-on-wheels.

54. Paul Kampe,"Novi Restaurant Ditches Tradition, Using iPads in Customer Service," *Oakland Press*, October 9, 2011, http://www.theoaklandpress.com/articles/2011/10/09/business/doc4e92458a2ead5089945284.txt.

55. Emily Nelson, "Marketers Push Individual Portions and Families Bite," *Wall Street Journal*, July 23, 2002, pp. A1, A6.

56. Gonçalo Pacheco-de-Almeida, "Erosion, Time Compression, and Self-Displacement of Leaders in Hypercompetitive Environments," *Strategic Management Journal* 31, no. 13 (2010), pp. 1498–1526.

57. Mark Henricks, "Time Is Money," *Entrepreneur*, February 1993, p. 44.

58. Dale D. Buss, "Entertailing," *Nation's Business*, December 1997, p. 18.

59. Dale D. Buss, "Entertailing," *Nation's Business*, December 1997, pp. 12–18.

60. Beth Kowitt, "True Obsessions," *Fortune*, May 11, 2009, pp. 84–91.

61. John Wark and Jeffrey Cornwall, "Emma," *Proceedings* (Nashville, TN: United States Association for Small Business and Entrepreneurship, 2010).

62. "Best Global Brands 2013," Brandirectory, July 2013, http://brandirectory.com/.

63. Mark Brandau, "Caribou's Brand Makeover," *Nation's Restaurant News*, March 2, 2010, www.nrn.com/breakingNews.aspx?id=379916.

64. "Social Networking Eats Up 3+ Hours Per Day for the Average American User," *Marketing Charts*, January 9, 2013, www.marketingcharts.com/wp/interactive/social-networking-eats-up-3-hours-per-day-for-the-average-american-user-26049.

65. "Social Media Engagement Directly Linked to Financial Success," *Marketing Charts*, July 21, 2009, www.marketingcharts.com/interactive/social-media-engagement-directly-linked-to-financial-success-9858.

66. The 2012 Hiscox DNA of an Entrepreneur Study, October 2012, p. 9.

67. Matt McGee, "More Small Businesses Using Social Media," *Small Business Search Marketing*, February 24, 2010, www.smallbusinesssem.com/more-small-businesses-using-social-media/2847.

68. Craig Smith, "By the Numbers: 16 Amazing Twitter Stats," *Digital Marketing Ramblings*, May 25, 2013, http://expandedramblings .com/index.php/march-2013-by-the-numbers-a-few-amazing-twitter-stats/; Zoe Fox, "This Is How Much Time You Spend on Facebook, Twitter, Tumblr," *Mashable*, November 28, 2012, http://mashable .com/2012/11/28/social-media-time/.

69. Leyl Master Black, "Getting Started with Twitter Ads for Small Business," *Mashable*, September 22, 2012, http://mashable .com/2012/09/22/twitter-ads-small-business; Craig Smith, "How Many People Use the Top Social Media, Apps & Services?," *Digital Marketing Ramblings*, March 23, 2013, http://expandedramblings.com/ index.php/resource-how-many-people-use-the-top-social-media.

70. "More Than 100 Twitter Followers Boost B2C Leads," *Marketing Charts*, April 19, 2010, www.marketingcharts.com/direct/more-than-100-twitter-followers-boost-b2c-leads-12625.

71. Robert Gourley, "Twitter 101: Seven Tips for Effective Marketing," *Marketing Profs*, June 2010, www.marketingprofs.com/ articles/2010/3704/twitter-101-seven-tips-for-effective-marketing.

72. Craig Smith, "By the Numbers: 32 Amazing Facebook Stats," *Digital Marketing Ramblings*, June 23, 2013, http://expandedramblings.com/ index.php/by-the-numbers-17-amazing-facebook-stats.

73. Maeve Duggan and Joanna Brenner, "The Demographics of Social Media Users—2012," *Pew Internet*, February 14, 2013, http://pewinternet.org/Reports/2013/Social-media-users/Social-Networking-Site-Users/Demo-portrait.aspx.

74. "How Much Do Small Businesses Spend on Social Media?" Vertical Response Marketing, October 31, 2012, www.verticalresponse.com/ blog/how-much-time-and-money-do-small-businesses-spend-on-social-media-infographic.

75. John Rampton, State of the Blogging World, Blogging.org, July 13, 2012, http://blogging.org/blog/blogging-stats-2012-infographic/.

76. John Nardini, "Create a Blog to Boost Your Business," *Entrepreneur*, September 27, 2005, www.entrepreneur.com/article/0,4621,323598,00 .html.

77. "ComScore Releases February 2013 U.S. Online Video Rankings," March 14, 2013, www.comscore.com/Insights/Press_Releases/2013/3/ comScore_Releases_February_2013_U.S._Online_Video_Rankings.

78. Kim Lachance Shandrow, "10 Questions to Ask When Creating Your Company's YouTube Channel," *Entrepreneur*, March 25, 2013, www .entrepreneur.com/article/226148#.

79. J. J. McCorvey, "Translating Viral YouTube Videos into Sales," *Inc.*, November 2011, www.inc.com/magazine/201111/translating-viral-youtube-videos-into-sales.html.

80. Mickey Meece, "Bringing an Innovative Razor to the Masses," *New York Times*, April 29, 2010, p. B5.

Chapter 10

1. Lin Grensing-Pophal, "Who Are You?," *Business Start-Ups*, September 1997, pp. 38–44.

2. Corbett Barr, "10 Examples of Killer USPs on the Web," *Think Traffic*, August 24, 2010, http://thinktraffic.net/10-examples-of-killer-unique-selling-propositions-on-the-web; "Warranty," Saddleback Leather, www.saddlebackleather.com/Warranty.

3. Meg Whittemore, "PR on a Shoestring," *Nation's Business*, January 1991, p. 31.

4. Amanda Phipps, "Pauline-Based Daisy Cakes Sees Owner to Appear on 'Today' Show, Other Media Outlets," *Go Upstate*, July 10, 2012, www.goupstate.com/article/20120709/ARTICLES/120709681; "New Product, Daisy Cakes, Keeps It Hot in the Oven for Sales After 'Shark Tank,'" New Product Consulting, March 6, 2012, http:// newproductconsulting.me/2012/03/06/new-product-daisy-cakes-keeps-it-hot-in-the-oven-for-sales-after-shark-tank.

5. Freddy J. Nager, "Public Relations vs. Advertising: No Contest," *Cool Rules Pronto*, September 4, 2008, http://coolrulespronto.wordpress .com/2008/09/04/public-relations-vs-advertising.

6. Debra Phillips, "Fast Track," *Entrepreneur*, April 1999, p. 42.

7. "Historic Inn and Spa Uses E-Mail Marketing to Enhance Customer Loyalty and Drive Revenue," Constant Contact, www.constantcontact .com/email-marketing/customer-examples/crowne-pointe.jsp.

8. Stacey Politi, "Three Successful Small Business Cause Marketing Campaigns," *American Express Open Forum*, January 17, 2012, https:// www.openforum.com/articles/3-successful-small-business-cause-marketing-campaigns.

9. Peggy Linial, "Small Business and Cause Related Marketing: Getting Started," Cause Marketing Forum, www.causemarketingforum.com/ framemain.asp?ID=189.

10. Barry Farber, "Sales Shape-Up," *Entrepreneur*, August 2006, p. 72.

11. Barbara K. Mednick, "Behavior Counts in Sales," *Minneapolis St. Paul Star Tribune*, May 28, 2004, www.startribune.com/working.

12. "Most Salespeople Can't Sell," *Small Business Reports*, September 1990, p. 10.

13. Eric Anderson and Bob Trinkle, *Outsourcing the Sales Function: The Real Cost of Field Sales* (Mason, OH: Thomson Publishing, 2005), p. 8.

14. "Salespeople Don't Prep Enough for Calls," *Sales & Marketing Management's Performance Newsletter*, December 12, 2005, p. 2.

15. Norm Brodsky, "Keep Your Customers," *Inc.*, September 2006, pp. 57–58.

16. "The Word and the World of Customers: Word of Mouth Marketing Offline and Online," Word of Mouth Marketing Association, 2011.

17. "Small Business Tepid on Social Media, Prefer WOM and Advertising," *MarketingProfs*, December 7, 2012, www.marketingprofs.com/ charts/2012/9651/small-businesses-tepid-on-social-media-prefer-wom-advertising.

18. "The Word and the World of Customers: Word of Mouth Marketing Offline and Online," Word of Mouth Marketing Association, 2011.

19. "WOMMA Defines WOM," *YouTube*, November 15, 2011, www .youtube.com/watch?v=Fz22PfPxoXI.

20. Linda Formichelli, "Extra Credit," *QSR*, November 2012, www .qsrmagazine.com/exclusives/extra-credit.

21. Dave Kissel, "Shaping the 90 Percent," *Media Myths and Realities*, Ketchum Perspectives, 2009, http://ketchumperspectives.com/ archives/2009_i1/90percent.php.

22. *Global Trust in Advertising and Brand Messages*, Nielsen, April 2012, p. 3.

23. Connie Bensen, "Social Media Wins: How Coconut Bliss Boosts Fan Engagement," *Dell Social Business Connection*, September 23, 2012, http://en.community.dell.com/dell-groups/sbc/b/weblog/ archive/2012/09/23/social-media-wins-how-coconut-bliss-boosts-fan-engagement.aspx; Phil Mershon, "9 Small Business Social Media Success Stories," *Social Media Examiner*, January 18, 2012, www .socialmediaexaminer.com/9-small-business-social-media-success-stories.

24. Megan O'Neill, "Video Infographic Reveals the Most Impressive YouTube Statistics of 2012," *Social Times*, September 10, 2012, http:// socialtimes.com/video-youtube-statistics-2012_b104480.

25. "Global Online Advertising Spending Statistics," *Go-Gulf*, May 2, 2012, www.go-gulf.com/blog/online-ad-spending.

26. "In-Stream Video AD CTR Far Higher in Rich Media Than Banners, Rich Media in H1," *e-Marketer*, September 14, 2012, www .marketingcharts.com/wp/direct/in-stream-video-ad-ctr-far-higher-than-banners-rich-media-in-h1-23396; "Travel SEM Is at a Crossroads," *e-Marketer*, December 5, 2012, www.emarketer.com/Article .aspx?R=1009522.

27. Kristin Purcell, Joanna Brenner, and Lee Rainie, *Search Engine Use 2012*, Pew Internet and American Life Project, March 9, 2012, p. 3.

28. Frederic Lardinois, "Comscore: Google's Search Engine Market Share Increased in September, Yahoo Down Another .6 Percentage Points," *Tech Crunch*, October 11, 2012, http://techcrunch.com/2012/10/11/ comscore-googles-search-engine-market-share-increased-in-september-yahoo-down-another-0-6-percentage-points.

29. *Click Fraud Report, Q1 2010*, Click Forensics, April 2010, p. 3.

30. Sara Radicati, *E-Mail Statistics Report, 2012–2016* Palo Alto, CA: Radicati Group, April 2012), p. 2.

31. "Promotional Retail E-Mail Volume Up 16% in '11," *Marketing Charts*, January 17, 2012, www.marketingcharts.com/wp/direct/promo-retail-email-volume-grew-16-y-o-y-20759.

32. Sara Radicati, *E-Mail Statistics Report, 2012–2016* (Palo Alto, CA: Radicati Group, April 2012), p. 3.

33. *2012 Silverpop E-Mail Marketing Metrics Benchmark Study*, Silverpop, pp. 3, 6.

34. "E-Mail Open, Click Rates Recover in Q3; Triggered E-Mails Continue Strong Showing," *Marketing Charts*, December 10, 2012, www.marketingcharts.com/wp/direct/email-open-click-rates-recover-in-q3-triggered-emails-continue-strong-showing-25307.

35. Jordie van Rijn, "Mobile E-Mail Usage Statistics," *EMailMonday*, 2012, www.emailmonday.com/mobile-email-usage-statistics.

36. Laura Key, "A Spotlight on Three Restaurants in NYC," Emma, July 17, 2012, http://myemma.com/blog/category/customer-stories.

37. "Whipping Up a Delicious E-Mail," Constant Contact, www.constantcontact.com/email-marketing/customer-examples/chocolate-bar.jsp; Maya Fastoff, "Chocolate Heaven in Rice Village," *My Fox Houston*, January 27, 2012, www.myfoxhouston.com/story/18179594/chocolate-heaven-in-rice-village.

38. Jason Keath, "105 Facebook Advertising Case Studies," *Social Fresh*, June 19, 2012, http://socialfresh.com/facebook-advertising-examples.

39. Case Studies: Bonobos," Twitter for Business, 2012, https://business.twitter.com/en/optimize/case-studies/bonobos.

40. "Global Sponsorship Spending by Region from 2009 to 2011," *Statista*, 2012, www.statista.com/statistics/196898/global-sponsorship-spending-by-region-since-2009.

41. Issie Lapowsky, "'Can't Afford the NFL?,'" *Inc.*, June 2012, pp. 108–109.

42. Kim T, Gordon, "Tips for Event Sponsorship," *Entrepreneur*, March 2006, www.entrepreneur.com/magazine/entrepreneur/2006/march/83672.html.

43. *TV Basics*, Television Bureau of Advertising, June 2012, p. 10.

44. "Oregon Mint Company," Google AdWords, www.google.com/ads/tv/success.html.

45. *TV Basics*, Television Bureau of Advertising, June 2012, p. 23.

46. "Consumers Allocate High Percentage of Total Daily Media Hours to Television," Television Bureau of Advertising, www.tvb.org/nav/build_frameset.aspx.

47. "TV Viewers Watch Commercials," *Marketing Charts*, May 14, 2010, www.marketingcharts.com/television/tv-viewers-watch-commericals-12877.

48. "Television and Health," The Sourcebook for Teaching Science, www.csun.edu/science/health/docs/tv&health.html.

49. *TV Basics*, Television Bureau of Advertising, June 2012, p. 5.

50. Jessica E. Vascellaro and Sam Schechner, "TV Lures Ads but Viewers Drop Out," *Wall Street Journal*, September 11, 2011, pp. B1–B2.

51. Travers Korch, "Top 5 Successful Infomercial Products," *Bankrate*, 2012, www.bankrate.com/finance/personal-finance/infomercial-products.aspx#slide=3.

52. Shelly Banjo, "As Seen (Often) on TV," *Wall Street Journal*, May 11, 2009, http://online.wsj.com/article/SB10001424052970204475004574126820963295620.html.

53. *Why Radio Fact Sheet*, Radio Advertising Bureau, 2012, www.rab.com/public/marketingGuide/DataSheet.cfm?id=1.

54. *Radio Marketing Guide*, Radio Advertising Bureau, 2009, www.rab.com/public/marketingGuide/rabRmg.html.

55. "Radio Case Study: Firehouse Subs," Radio Advertising Bureau, 2012, www.rab.com/public/adchannel/casestudies.cfm.

56. "Newspaper Circulation Volume," Newspaper Association of America, September 2012, www.naa.org/Trends-and-Numbers/Circulation/Newspaper-Circulation-Volume.aspx.

57. "Newspapers Ignoring More 'Traditional Wisdom,'" *Inland Press*, September 10, 2012, www.inlandpress.org/articles/2012/12/10/research/current_research/doc504e238d7d0cf714832132.txt.

58. "A Guide to Competitive Media: Newspaper," Radio Advertising Bureau, 2012, www.rab.com/whyRadio/mfdetails.cfm?id=8.

59. Ian Tenant, "Sixty Years of Declining Circulation Suggests Newspapers Will Perish, Report Says," Knight Center for Journalism in the Americas, University of Texas at Austin, May 30, 2011, http://knightcenter.utexas.edu/blog/sixty-years-declining-circulation-suggests-newspapers-will-perish-says-report.

60. Rick Edmonds, Emily Guskin, Tom Resenstiel, and Amy Mitchell, *The State of the News Media: An Annual Report on American Journalism*, Pew Research Center Project for Excellence in Journalism, 2012, http://stateofthemedia.org/2012/newspapers-building-digital-revenues-proves-painfully-slow/newspapers-by-the-numbers.

61. *The Magazine Factbook* (New York: Magazine Publishers of America, 2012), pp. 85, 89.

62. *The Magazine Factbook* (New York: Magazine Publishers of America, 2012), p. 64.

63. *The Magazine Factbook* (New York: Magazine Publishers of America, 2012), p. 12.

64. *The Magazine Factbook* (New York: Magazine Publishers of America, 2012), p. 12.

65. *A Guide to Competitive Media: Magazines*, Radio Advertising Bureau, 2012, www.rab.com/whyRadio/mfdetails.cfm?id=5.

66. *The Magazine Handbook* (New York: Magazine Publishers of America, 2009), p. 13.

67. "In Changing News Landscape, Even Television Is Vulnerable," Pew Research Center for the People and the Press, September 27, 2012, www.people-press.org/2012/09/27/in-changing-news-landscape-even-television-is-vulnerable.

68. M. P. Mueller, "The Surprising Power of Promotional Products," *New York Times*, July 18, 2011, http://boss.blogs.nytimes.com/2011/07/18/the-surprising-power-of-promotional-products.

69. "2011 Estimate of Promotional Products Distributors Sales," Promotional Products Association International, 2012, p. 1.

70. Jan Cienski, "Sex and Caskets Make for a Lively Market," *Financial Times*, January 5, 2012, www.ft.com/cms/s/0/ef2dbf38-3799-11e1-897b-00144feabdc0.html#axzz2FtlTtR7d.

71. "The Effectiveness of Promotional Products as an Advertising Medium," Promotional Products Association International, 2009, p. 2.

72. "Shoppers Are Making More Purchasing Decisions In-Store Than Ever Before," 2012 Shopper Engagement Study, Point of Purchase Advertising International, www.popai.com/engage/?p=52.

73. "Facts & Figures," Outdoor Advertising Association of America, www.oaaa.org/marketingresources/factsandfigures.aspx.

74. "See Our Billboards," Wade's Southern Cooking, 2012, http://eatatwades.com/seeo.php.

75. Joanna Lin, "California Commute Times Rank 10th Longest in U.S.," *Huffington Post*, November 1, 2012, www.huffingtonpost.com/2012/11/01/california-commute-times-rank_n_2060062.html.

76. "U.S. Advertising CPM, by Media," *eMarketer*, February 2009, www.emarketer.com/Results.aspx?N=785.

77. *Arbitron National In-Car Study*, New York, 2009, p. 5.

78. *Arbitron National In-Car Study*, New York, December 2003, www.oaaa.org/pdf/Incarstudy_summary.pdf, p. 2.

79. "Foot Traffic," *Wall Street Journal*, January 19, 2012, p. B6.

80. "Direct Mail Success Tips from the U.S. Postal Service," United States Postal Service, August 20, 2012, about.usps.com/news/national-releases/2012/pr12_096.htm.

81. "Frequently Asked Questions About Advertising Mail," *Mail Moves America*, 2012, p. 1.

82. *Catalogs: The Consumers' Point of View* (Washington, DC: American Catalog Mailers Association, 2012), p. 2.

83. "Direct Mail Tips for Manufacturers' Letters," Koch Group, www.kochgroup.com/directmail.html.

84. "Direct Mail Success Tips from the U.S. Postal Service," United States Postal Service, August 20, 2012, http://about.usps.com/news/national-releases/2012/pr12_096.htm.

85. Allison Schiff, "DMA: Direct Mail Response Rates Beat Digital," *Direct Marketing News*, June 14, 2012, www.dmnews.com/dma-direct-mail-response-rates-beat-digital/article/245780.

86. Mindi Charsky, "Going Coconuts," *Deliver*, June 19, 2012, https://delivermagazine.com/2012/06/going-coconuts.

87. Charlotte Woolard, "CIER Study Finds Little Change in Exhibitor Spending," *B to B*, September 17, 2012, www.btobonline.com/article/20120917/EVENT08/309179996/ceir-study-finds-little-chang...

88. Katie Marsack, "10 Impressive Statistics to Back Your Trade Show Marketing Program," *Graphicolor Exhibits*, August 14, 2012, http://info.graphicolor.com/blog/bid/153404/10-impressive-statistics-to-back-your-trade-show-marketing-program.

89. Katie Marsack, "10 Impressive Statistics to Back Your Trade Show Marketing Program," *Graphicolor Exhibits*, August 14, 2012, http://info.graphicolor.com/blog/bid/153404/10-impressive-statistics-to-back-your-trade-show-marketing-program.

90. "'Dead Tree Medium' No Longer: For Many Marketers, Print Outperforms Digital," *Knowledge@Wharton*, March 19, 2008, http:// knowledge.wharton.upenn.edu/article.cfm?articleid=1919&CFID=1919 3179&CFTOKEN=81243060&jsessionid=a830557b30537714f7f0107e 5da2e19272f2.

91. Max Chafkin, "Ads and Atmospherics: Outdoor Campaigns Are Suddenly Hip," *Inc.*, February 2007, pp. 39–41; Jennifer Pollock," Can Your Banner Ad Do This?," *Fast Company*, July/August 2006, p. 51; Sara Wilson, "Hawking on Eggshells," *Entrepreneur*, February 2007, p. 75.

92. Steven Heller, "Going Overground," *Metropolis*, March 14, 2007, www.metropolismag.com/cda/story.php?artid=2557.

93. Jim Carlton, "Logo Graffitti Gets Scrubbed," *Wall Street Journal*, December 4, 2010, http://online.wsj.com/article/SB1000142405274870 437700457565079309580228.2html.

Chapter 11

1. Christopher T. Heun, "Dynamic Pricing Boosts Bottom Line," *Information Week*, October 29, 2001, www.informationweek.com/story/showArticle.jhtml?articleID=6507202.

2. Vincent Ryan, "Price Fixing," *CFO*, December 2009, p. 50.

3. Emily Maltby, "How Much?," *Wall Street Journal Classroom Edition*, September 2011, http://wsjclassroomedition.com/cre/articles/11sep_pricingECON.htm.

4. Ron Ruggless, "Pizza Hut Targets Value with $10 Dinner Box," *Nation's Restaurant News*, February 13, 2012, http://nrn.com/archive/pizza-hut-targets-value-10-dinner-box.

5. *The 2012 Parago Shopping Behavior Insights Study* (Lewisville, TX: Parago, 2012), pp. 5–6.

6. Bill Siwicki, "For Most Smartphone Owners, the Device Is a Personal Shopping Assistant," *Internet Retailer*, December 31, 2012, www.internetretailer.com/2012/12/31/most-smartphone-owners-device-personal-shopping.

7. Ann Abel, "Brent Black Panama Hats," *Forbes Life*, May 3, 2010, pp. 40–44.

8. "Get a Charge," *Forbes Life*, December 2011, p. 28; Jason Gilbert, "Ulysse Nardin Chairman, 'World's Most Expensive Mobile Phone,' Is Totally Decadent," *Huffington Post*, February 13, 2012, www.huffingtonpost.com/2012/02/09/ulysse-nardin-chairman_n_1266794.html.

9. Howard Scott, "The Tricky Art of Raising Prices," *Nation's Business*, February 1999, p. 32.

10. "Bryan Janeczko, "How Should I Price My Product?," SCORE: Small Business Success Blog, September 5, 2012, http://blog.score.org/2012/bryan-janeczko/how-should-i-price-my-product.

11. Rick Bruns, "Tips for Coping with Rising Costs of Key Commodities," *Fast Company*, December 1997, pp. 27–30.

12. Joyce M. Rosenberg, "Small Businesses Cope with Rising Gas Prices," *Huffington Post*, February 23, 2012, www.huffingtonpost.com/2012/02/24/small-businesses-gas-prices_n_1298992.html?view=print&comm_ref=false.

13. "Gas Prices Could Affect the Price of Pizzas, Flowers," *Greenville News*, April 29, 2006, p. 3A.

14. Mark Brandau, "Commodity Inflation Curbs 3Q Income for Buffalo Wild Wings," *Nation's Restaurant News*, October 24, 2012, http://nrn.com/latest-headlines/commodity-inflation-curbs-3q-income-buffalo-wild-wings; Mark Brandau, "Buffalo Wild Wings 2Q Results Positive Despite Inflation," *Nation's Restaurant News*, July 25, 2012, http://nrn.com/latest-headlines/buffalo-wild-wings-2q-results-positive-despite-inflation.

15. Malika Zouhali-Worrall, "Ring in the Profits," *FSB*, December 2009/January 2010, p. 15.

16. Candice Choi, "The Secrets of the Value Menu," *Yahoo! Finance*, September 14, 2012, http://finance.yahoo.com/news/smart-spending-secrets-value-menu-180303014.html; Sam Oches, "The Value Equation," *QSR Magazine*, February 2010, www.qsrmagazine.com/competition/value-equation?microsite=9342;microsite=9342.

17. "A Guy Thing," *Forbes Life*, December 2011, p. 24; Erica Jackson Curran, "Bill Oyster's Bamboo Fly Rods," *Breathe*, September 14, 2012, www.readbreathe.com/outdoors/bill-oysters-artisan-fly-fish-rods; "Oyster Bamboo Fly Rods," *CNNMoney*, September 5, 2008, http://money.cnn.com/galleries/2008/fsb/0809/gallery.great_outdoors.fsb/2.html.

18. Michael V. Marn, Eric V. Roegner, and Craig C. Zawada, "Pricing New Products," *The Mckinsey Quarterly*, Number 3, 2003, www.mckinseyquarterly.com/article_abstract.aspx?ar=1329&l2=16&l3=19&srid=190&gp=0.

19. "Accenture Study Finds Fewer Consumers Switch Service Providers Despite Decreasing Satisfaction, Reversing a Trend," Accenture, February 16, 2011, http://newsroom.accenture.com/article_display.cfm?article_id=5150.

20. Eilene Zimmerman, "Real-Life Lessons in the Delicate Art of Setting Prices," *New York Times*, April 20, 2011, www.nytimes.com/2011/04/21/business/smallbusiness/21sbiz.html?pagewanted=all&_r=0.

21. Eilene Zimmerman, "Real-Life Lessons in the Delicate Art of Setting Prices," *New York Times*, April 20, 2011, www.nytimes.com/2011/04/21/business/smallbusiness/21sbiz.html?pagewanted=all&_r=0.

22. "Greg Besinger, "Amazon Stirs Up Price War," *Wall Street Journal*, September 7, 2012, p. B3; Stab Schroeder, "iPad Is Still the King of Tablets, but Kindle Fire and Others Are Catching Up," *Mashable*, March 14, 2012, http://mashable.com/2012/03/14/ipad-kindle-fire-market-share.

23. Jim Barnes, "Stop Cutting Prices! Retain Customers with Four Types of Experiential Value," Customer Think, April 24, 2009, www.customerthink.com/article/retain_customers_with_four_types_experiential_value.

24. Alison Stein Wellner, "Is It Time to Raise Prices?," *Inc.*, June 2005, p. 80.

25. Norm Brodsky, "Dealing with Cost Hikes," *Inc.*, August 2005, p. 49.

26. Norm Brodsky, "Street Smarts: The Commodity Pricing Trap," *Inc.*, April 2012, p. 37.

27. Emily Maltby, "How Much?," *Wall Street Journal Classroom Edition*, September 2011, http://wsjclassroomedition.com/cre/articles/11sep_pricingECON.htm.

28. Gladys Edmunds, "Price Is Right? Not If It's Too Low," *USA Today*, June 8, 2005, www.usatoday.com/money/smallbusiness/columnist/edmunds/2005-06-08-price_x.htm.

29. "Pricing Strategy for a Start-Up Business," *Start-Up Nation*, www.startupnation.com/business-articles/1483/1/pricing-strategy-startup.asp; "Services," Billy Lowe, www.billylowe.com/Services.html.

30. "Restaurant Price Presentation Influences Check Averages, New Cornell Research Shows," *Hospitality.net*, May 6, 2009, www.hospitalitynet.org/news/4041281.html.

31. Uzi Shmilovici, "The Complete Guide to Freemium Business Models," *Tech Crunch*, September 4, 2011, http://techcrunch.com/2011/09/04/complete-guide-freemium; Tom Tunguz, "Your Start-Up's Pricing Strategy," Tomas Tunguz, May 25, 2012, http://tomasztunguz.com/2012/05/25/your-startups-pricing-strategy.

32. Liz Gannes, "Case Studies in Freemium: Pandora, Dropbox, Evernote, Automattic, and Chimpnote," *Gigaom*, March 26, 2010, http://gigaom.com/2010/03/26/case-studies-in-freemium-pandora-dropbox-evernote-automattic-and-mailchimp; "Pricing," Dropbox, https://www.dropbox.com/pricing.

33. Julia Angwin and Dana Mattioli, "Don't Like This Price? Wait a Minute," *Wall Street Journal*, September 5, 2012, pp. A1–A2; Dana Mattioli, "Holiday Price War Rages in Real Time," *Wall Street Journal*, November 24–25, 2012, pp. A1–A2.

34. "New Study on Retail Discounting: What Works for Some Products Might Be a Bust for Others," Kelley School of Business, Indiana University, November 18, 2009, http://info.kelley.iu.edu/news/page/normal/12630.html.

35. Josh Hyatt, "And in this Corner, the Price-Fighter," *CFO*, December 2008, p. 29.

36. Lisa Baertlein, "Subway's $5 Deal a Hard Habit to Break, *Reuters*, September 17, 2009, www.reuters.com/article/idUSTRE58G6NG20090917.

37. "NPD: Restaurants Likely to Wean Customers Off Discounts/Deals," *Fast Casual*, May 20, 2010, www.fastcasual.com/article.php?id=18348.

38. John Morell, "The Art of the LTO," *QSR Magazine*, August 2012, www.qsrmagazine.com/promotions/art-lto.

39. "Don't Let This One Get Away," *Get to the Point: Customer Insight*, January 20, 2010, pp. 1–2.

40. Ron Ruggless, "Pizza Hut Targets Value with $10 Dinner Box," *Nation's Restaurant News*, February 13, 2012, http://nrn.com/archive/pizza-hut-targets-value-10-dinner-box.

41. Hiroshi Suzuki, "Nintendo's Japan Wii Sales Double Those of Sony PlayStation 3," *Bloomberg News*, April 2, 2007, www.bloomberg.com/apps/news?pid=conewsstory&refer=conews&tkr=NTDOY:US&sid=a_A5anLn.xE8; Brian Bremmer, "Will Nintendo's Wii Strategy Score?," *BusinessWeek*, September 20, 2006, www.businessweek.com/globalbiz/content/sep2006/gb20060920_163780.htm; Kathleen Sanders and Casamassina, "U.S. Wii Price, Launch Date Revealed," *IGN Entertainment*, September 13, 2006, http://wii.ign.com/articles/732/732669p1.html; "Playing a Different Game," *Economist*, October 26, 2006, www.economist.com/business/displaystory.cfm?story_id=8080787.

42. Joan Biskupic, "States Try to Counter Supreme Court's Minimum Price Ruling," *USA Today*, December 22, 2010, http://usatoday30.usatoday.com/news/washington/judicial/2010-12-22-robertscourt22_CV_N.htm; Joseph Pereira, "State Law Targets 'Minimum Pricing,'" *Wall Street Journal*, April 28, 2009, p. D1; Lee Aronson, "Why Some Brands Never Go on Sale," *Best of Times News*, February 2009, www.thebestoftimesnews.com/articleprint.php?article=469.

43. Renee Morad, "Overpriced Products: 20 Most Insane Markups," *Huffington Post*, September 28, 2012, www.huffingtonpost.com/2012/09/28/overpriced-products-insane-markups_n_1922573.html.

44. Rafi Mohammed, "Apple's iPad Can Capitalize on Fragmented Marketplace," *The Wrap*, February 8, 2010, www.thewrap.com/blog-entry/apples-ipad-can-capitalize-fragmenting-marketplace-14006.

45. David Worrell, "Time Well Spent," *Entrepreneur*, June 2006, p. 63.

46. Miguel Helft, "The Death of Cash," *Fortune*, July 23, 2012, pp. 118–128.

47. *The Nilson Report* (Carpinteria, CA: The Nilson Report, April 17, 2012), Issue 992, p. 1.

48. "Credit Counseling Statistics," Consumer Credit Counseling Service, http://creditcounselingbiz.com/credit_counseling_statistics.htm.

49. J. Craig Shearman, "NRF Says Swipe Fee Settlement Still Unacceptable to Retailers," National Retail Federation, October 19, 2012, www.nrf.com/modules.php?name=News&op=viewlive&sp_id=1439.

50. J. Craig Shearman, "NRF Says Swipe Fee Settlement Still Unacceptable to Retailers," National Retail Federation, October 19, 2012, www.nrf.com/modules.php?name=News&op=viewlive&sp_id=1439.

51. Emily Maltby, "The 'Swipe Fee' Conundrum," *Wall Street Journal*, July 19, 2012, p. B1.

52. Miguel Helft, "The Death of Cash," *Fortune*, July 23, 2012, pp. 118–128.

53. *2012 Online Fraud Report* (San Francisco: CyberSource Corporation, 2012), p. 1.

54. Michael Bloch, "Preventing Credit Card Chargebacks—Anti-Fraud Strategies," Taming the Beast, www.tamingthebeast.net/articles2/card-fraud-strategies.htm.

55. *The Nilson Report* (Carpinteria, CA: The Nilson Report, April 17, 2012), Issue 992, p. 1.

Chapter 12

1. Marta Bright, Bobbie Hartman, Monica Mehta, Christopher Null, and Kate Pavao, "Business Beyond Borders," *Profit*, November 2011, p. 7.

2. "UPS 'Perceptions of Global Trade' Survey," UPS International, September 2011, p. 1.

3. "Gilt Groupe Entrepreneurs Tell Their Story: By Invitation Only," *Endeavor Global*, April 16, 2012, www.endeavor.org/blog/giltgroupe; Dan O'Shea, "Think Global, Act Local," *Entrepreneur*, July 2012, p. 47; Neal Ungerleider, "Why Guilt Group Went Irish," *Fast Company*, April 9, 2012, www.fastcompany.com/1830160/why-gilt-groupe-went-irish; Susan S. Nichols, "Exclusive: Guilt Group—5 Million Members and Growing," *Apparel*, April 12, 2012, http://apparel.edgl.com/case-studies/Exclusive--Guilt-Groupe--Five-Million-Members-and-Growing79623; Ben Popper, "Gilt Groupe Expands Internationally to 90 Countries," *Beta Beat*, November 8, 2011, http://betabeat.com/2011/11/gilt-groupe-expands-internationally-to-90-countries.

4. "Trade Growth to Slow in 2012 After Strong Deceleration in 2011," World Trade Organization, www.wto.org/english/res_e/statis_e/its2009_e/its09_toc_e.htm.

5. "World Economic Outlook Update: Gradual Upturn in Economic Growth in 2013," International Monetary Fund, January 23, 2013, www.imf.org/external/pubs/ft/weo/2013/update/01.

6. "Has China ALREADY Passed the U.S. as the World's Largest Economy?," *Washington's Blog*, April 5, 2012, www.washingtonsblog.com/2012/04/has-china-already-passed-the-u-s-as-the-worlds-largest-economy.html.

7. *UPS Snapshot for Small Businesses: Doing Business in China*, UPS International, 2011, p. 2.

8. "The World's Shifting Center of Gravity," *The Economist*, June 28, 2012, www.economist.com/blogs/graphicdetail/2012/06/daily-chart-19.

9. Jack Stack with Bo Burlingham, "My Awakening," *Inc.*, April 2007, pp. 93–97.

10. *UPS Snapshot for Small Businesses: Doing Business in Vietnam*, UPS International, 2011, p. 3.

11. Roger F. Noriega and José R. Cardenás, "An Action Plan for U.S. Policy in the Americas," American Enterprise Institute, December 5, 2012, www.aei.org/outlook/foreign-and-defense-policy/regional/latin-america/an-action-plan-for-us-policy-in-the-americas; "Exporting Is Good for Your Bottom Line," International Trade Administration, 2012, www.trade.gov/cs/factsheet.asp.

12. Ted Miller, "Can America Compete in the Global Economy?," *Kiplinger's Personal Finance Magazine*, November 1991, p. 8.

13. Bernard Wysocki Jr., "Going Global in the New World," *Wall Street Journal*, September 21, 1990, p. R3.

14. Matt Glynn, "Setting Sites on Small Business Exports," *Buffalo News*, April 21, 2010, www.buffalonews.com/2010/04/20/1025651/setting-sights-on-small-business.html.

15. "UPS 'Perceptions of Global Trade' Survey," UPS International, September 2011, p. 1.

16. *Small and Medium Size Businesses Export Insights and Opportunities: Executive Summary*, CompTIA, January 2010, pp. 3, 4.

17. "Globesmanship," *Across The Board*, January/February 1990, p. 26.

18. Michael Barrier, "Why Small Looms Large in the Global Economy," *Nation's Business*, February 1994, p. 9; Vivian Pospisil, "Global Paradox: Small Is Powerful,"*Industry Week*, July 18, 1994, p. 29.

19. Michael Barrier, "A Global Reach for Small Firms," *Nation's Business*, April 1994, p. 66.

20. Jeremy Main, "How to Go Global—and Why," *Fortune*, August 28, 1989, p. 70.

21. Mike D. Smith, "For Small Businesses Getting into Global Trade, Know Your Rights and Protections, Experts Say," *Caller.com*, August 19, 2012, www.caller.com/news/2012/aug/18/for-small-business-getting-into-global-trade-and/?partner=RSS.

22. "Internet Usage Statistics: The Internet Big Picture," Internet World Stats 2012, www.internetworldstats.com/stats.htm; "Internet Users by World Language: Top 10 Languages," Internet World Stats 2012, www.internetworldstats.com/stats7.htm.

23. "eBay Study Shows Businesses Selling Online Exporting at Record Levels," *World Trade WT100*, November 1, 2012, www.worldtradewt100.com/articles/88966-ebay-study-shows-businesses-selling-online-exporting-at-record-levels?WT.rss_f=Economic+Development+&WT.rss_a=eBay+Study+Shows+Businesses+Selling+Online+Exporting+at+Record+Levels&WT.rss_ev=a.

24. Carmine Gallo, "Why Today's Grads Will Become Tomorrow's Greatest Innovators," *Forbes*, May 31, 2012, www.forbes.com/sites/carminegallo/2012/05/31/why-todays-grads-will-become-tomorrows-greatest-innovators; Cindy Vanegas, "Tips for Expanding Your Small Business Internationally," *Fox Business*, February 14, 2012, http://smallbusiness.foxbusiness.com/starting-a-business/2012/02/14/tips-for-expanding-your-business-internationally.

25. "Who We Are," eBay, www.ebayinc.com/who; Stefany Moore, "Sales on eBay Jump 10%," *Internet Retailer*, July 18, 2012, www.internetretailer.com/2012/07/18/sales-ebay-jump-10.

26. Darren Dahl, "Meet the eBay Millionaires," *AOL Small Business*, March 23, 2011, http://smallbusiness.aol.com/2011/03/23/meet-the-ebay-millionaires; "About Us," AfricaDirect, www.africadirect.com/about-us.

27. "Foodservice," Dorian Drake International, 2013, http://doriandrake.com/food_frameset.htm; "History," Michigan Maple Block Company, 2013, http://www.butcherblock.com/about-the-wood-welded-companies.

28. Ian Mount, "Tips for Increasing Sales in International Markets," *New York Times*, April 21, 2010, www.nytimes.com/2010/04/22/business/smallbusiness/22sbiz.html.

29. Jennifer LeClaire, "How to Take Your Small Business Global," *E-Commerce Times*, June 20, 2006, www.ecommercetimes.com/story/50910.html?wlc=1276800833.

30. Joseph E. Pattison, "Global Joint Ventures," *Overseas Business*, Winter 1990, p. 25.

31. Polly Larson, "Opening Doors to Emerging Markets," International Franchise Association, www.ifa.org/intl/News/Prjf6.asp.

32. "Wolverine Worldwide Announces Joint Venture in India," *Reuters*, April 23, 2012, www.reuters.com/article/2012/04/23/idUS106200+23-Apr-2012+PRN20120423.

33. "ObjectVideo Enters into Global Patent Licensing Agreement with Panasonic System Networks," CNBC, January 9, 2013, www.cnbc.com/id/100365953/ObjectVideo_Enters_into_Global_Patent_License_Agreement_with_Panasonic_System_Networks; "Company Overview," ObjectVideo, 2013, www.objectvideo.com/company/overview.html.

34. "Franchising Facts," Export.gov, 2013, http://export.gov/industry/franchising/index.asp; Ryan Underwood, "Bridging the Gulf When Franchisees Are Far, Far Away," *Inc.*, February 2011, pp. 98–100.

35. William Edwards, "International Expansion: Do Opportunities Outweigh Challenges?," *Franchising World*, February 2008, www.franchise.org/Franchise-News-Detail.aspx?id=37992.

36. "Six Tips for Taking Your Franchise Global," *Entrepreneur*, September 30, 2011, www.entrepreneur.com/article/220386.

37. Trefis Team, "Dunkin' Hits 10,000 Stores on China Push, Stock Going to $29," *Forbes*, January 24, 2012, www.forbes.com/sites/greatspeculations/2012/01/04/dunkin-donuts-hits-10000-stores-on-china-push-stock-going-to-29; Walter Hamilton, "Dunkin' Donuts Adds Pork to Its Menu in China," *Los Angeles Times*, March 6, 2012, http://articles.latimes.com/2012/mar/06/business/la-fi-mo-dunkin-donuts-pork-20120306; Keith Richburg, "Doughnut Wars Give Shanghai a Sugar Jolt," *Washington Post*, July 18, 2011, www.washingtonpost.com/world/asia-pacific/donut-wars-give-shanghai-a-sugar-jolt/2011/07/14/gIQA7DTrLI_story.html.

38. "Fast Foods Gone Global," *Travel Channel*, www.travelchannel.com/video/fast-foods-gone-global; "The Hotlist: 10 Unusual Items on McDonald's Menus Around the World," *Daily Mail*, January 23, 2009, www.dailymail.co.uk/femail/food/article-1126655/The-hotlist-10-unusual-items-McDonalds-menus-world.html; Geoffrey A. Fowler and Ramin Setoodeh, "Outsiders Get Smarter About China's Tastes," *Wall Street Journal*, August 4, 2004, pp. B1, B2.

39. Mark Brandau, "McD to Franchise in Russia for First Time," *Nation's Restaurant News*, April 10, 2012, http://nrn.com/archive/mcd-franchise-russia-first-time; Natalia Ishchenko, "McDonald's in Franchising Deal with Russia's Rosinter," *Reuters*, April 10, 2012, www.reuters.com/article/2012/04/10/us-mcdonalds-russia-franchisee-idUSBRE8390OL20120410; "The Global 30," *QSR Magazine*, January 22, 2013, www.qsrmagazine.com/content/global-30.

40. Robin Van Tan, "Mission: International," *QSR Magazine*, April 7, 2011, www.qsrmagazine.com/exclusives/mission-international.

41. "Krispy Kreme (U.S.) Signs New Development Agreement for Russia," *Restaurant Magazine*, April 23, 2012, www.restaurantmagazine.com/krispy-kreme-signs-new-development-agreement-for-russia.

42. "Tasti D-Lite Prepares for Expansion in Middle East," *QSR Magazine*, January 6, 2012, www.qsrmagazine.com/news/tasti-d-lite-prepares-expansion-middle-east.

43. *Small and Medium Size Business Export Insights and Opportunities*, CompTIA, January 2010, p. 3.

44. Jake Colvin, "Mom and Pop Go Global," *Huffington Post*, October 25, 2012, www.huffingtonpost.com/jake-colvin/mom-and-pop-go-global_b_2012014.html.

45. "Charles Gerena, "The Decision to Export," *Region Focus*, First Quarter 2011, p. 37.

46. "Exporting Is Good for Your Bottom Line," International Trade Administration, 2013, www.trade.gov/cs/factsheet.asp; *Internationalisation of European SMEs*, Entrepreneurship Unit, European Commission: Enterprise and Industry, 2010, p. 8.

47. Paul C. Hsu, "Profiting from A Global Mind-Set," *Nation's Business*, June 1994, p. 6.

48. "Emily Libman, "Small Business Rekluse Motor Sports Gives Racers a Competitive Edge," September 14, 2012, UPS International, http://blog.ups.com/2012/09/14/small-business-rekluse-motor-sports-gives-racers-a-competitive-edge; Bill Roberts, "Boise Company Takes Victory Lap," *Idaho Statesman*, April 26, 2012, www.idahostatesman.com/2012/04/26/2092100/boise-company-takes-victory-lap.html; "2012 National Exporter of the Year," U.S. Small Business Administration, 2012, www.sba.gov/about-offices-content/2/3115/success-stories/149611.

49. Ian Mount, "Tips for Increasing Sales in International Markets," *New York Times*, April 21, 2010, www.nytimes.com/2010/04/22/business/smallbusiness/22sbiz.html.

50. Frances Huffman, "Hello, World!," *Entrepreneur*, August 1990, p. 108.

51. UPS 'Perceptions of Global Trade' Survey," UPS International, September 2011, p. 1.

52. Ryan Bradley, "Where Grills Are Born," *Fortune*, August 13, 2012, pp. 13–14.

53. Eric Decker, "The Art of the Trade Mission," *Small Business Times*, June 8, 2007, www.biztimes.com/news/2007/6/8/the-art-of-the-trade-mission.

54. Whit Richardson, "Maine Group Buoyed by Trade Mission to China," *Bangor Daily News*, September 17, 2012, http://bangordailynews.com/2012/09/17/business/maine-group-buoyed-by-trade-mission-to-china; "Destination China: Governor LePage Leads Trade Delegation Overseas," *The Maine Wire*, September 2012, www.themainewire.com/2012/09/destination-china-governor-lepage-leads-trade-delegation-overseas.

55. "Pennsylvania Company Growing with 16 Years of Ex-Im Support," Export-Import Bank, 2013, www.exim.gov/about/whatwedo/successstories/Aquatech-International.cfm.

56. Charlotte Mulhern, "Fast Forward," *Entrepreneur*, October 1997, p. 34.

57. *Export-Import Bank of the United States, 2012 Annual Report*, Export-Import Bank, p. 13.

58. *Growing Beyond: A Place for Integrity*, 12th Global Fraud Survey, Ernst and Young, 2012, p. 2.

59. Daniel Kaufmann and Shang-Jin Wei, "Does 'Grease Money' Speed Up the Wheels of Commerce?," World Bank, www.worldbank.org/wbi/governance/pdf/grease.pdf.

60. Stephen Cotterill, "'Hello, World,' J. Crew Says Via the Web," *Internet Retailer*, June 27, 2012, www.internetretailer.com/2012/06/27/hello-world-jcrew-says-web; Dana Mattioli, "J. Crew Suits Up for Overseas," *Wall Street Journal*, March 22, 2012, p. B8.

61. "U.S. Trade in International Goods and Services," November 2012, U.S. Census Bureau, January 11, 2013, p. 1.

62. Edu Lopez, "Global Outsourcing Industry Seen Growing 9% to $464 B in 2011," *MB.com.ph*, September 10, 2011, http://mb.com.ph/node/333788/global-out#.URECR_KN-So.

63. "Small Business Success Video Channel: Entrepreneur Success Stories," Making It TV, www.makingittv.com/Entrepreneur-Success-Story-Small-Business3.htm; Caleb Melby, "Moscow Beats New York, London in List of Billionaire Cities," *Forbes*, March 16, 2012, www.forbes.com/sites/calebmelby/2012/03/16/moscow-beats-new-york-london-in-list-of-billionaire-cities.

64. Mark Henricks, "The New China?," *Entrepreneur*, February 2006, pp. 17–18.

65. Michelle Wu, "How I . . . Hopscotched the Globe to Build My Business," *Wall Street Journal*, January 15, 2010, http://online.wsj.com/article/SB10001424052748704281204575002943447096122.html.

66. "International Trade Statistics 2012," World Trade Organization, p. 23.

67. *Harmonized Tariff Schedule of the United States 2012* (Washington, DC: U.S. International Trade Commission, 2012), p. 1760.

68. *The Economic Effects of Significant U.S. Import Restraints, Seventh Edition* (Washington, DC: U.S. International Trade Commission, February 2011), p. ix.

69. "Tariff Rate, Applied, Weighted Mean, All Products (%)," World Bank, 2013, http://data.worldbank.org/indicator/TM.TAX.MRCH.WM.AR.ZS.

70. "Sugar and Sweeteners Yearbook Tables," U.S. Department of Agriculture, July 1, 2013, www.ers.usda.gov/data-products/sugar-and-sweeteners-yearbook-tables.aspx#25442; Bob Meyer, "Sugar Import Quota Increase Being Considered," *Brownfield Ag News for America*, April 14, 2010, http://brownfieldagnews.com/2010/04/14/sugar-import-quota-increase-being-considered; Alexandra Wexler, "Sugar Users Want U.S. to Ease Import Curbs," *Wall Street Journal*, April 1, 2012, http://online.wsj.com/article/SB10001424052702303816504577314034112332296.html.

71. Ryan Naka Shima, "Hollywood in China? Country's New Foreign Film Quotas Make the Industry Optimistic," *Huffington Post*, April 16, 2012, www.huffingtonpost.com/2012/04/17/hollywood-in-china-countr_n_1431395.html; Chi-Chi Zhang, "China: Theaters Must Meet Domestic Movie Quota," *ABC News*, January 27, 2010, http://abcnews.go.com/Entertainment/wireStory?id=9672337.

72. "U.S. Trade in Goods with Vietnam: 2012," U.S. Census Bureau, www.census.gov/foreign-trade/balance/c5520.html.

73. "Steel Wire Garment Hangers from Taiwan: Antidumping Duty Order," *Federal Register*, December 10, 2012, www.federalregister.gov/articles/2012/12/10/2012-29765/steel-wire-garment-hangers-from-taiwan-antidumping-duty-order.

74. Kremena Krumova and Yue Pang, "Fake Products Swallow Economy on a Global Scale," *The Epoch Times*, September 11, 2012, www.theepochtimes.com/n2/world/fake-products-swallow-economy-on-global-scale-290673.html.

75. *Intellectual Property Rights: Fiscal Year 2012 Seizure Statistics*, U.S. Customs and Border Protection Office of International Trade, pp. 6, 10.

76. Jason Hanna, "Five Years After Iraq Abduction, Family Tries Making Own Closure," *CNN*, April 11, 2010, www.cnn.com/2010/LIVING/04/11/missing.iraq.ake.altaie/index.html.

77. Mark Johanson, "These Countries Get the Most Vacation Days (Hint: America Isn't on the List)," *International Business Times*, August 28, 2012, www.ibtimes.com/these-countries-get-most-vacation-days-hint-america-isnt-list-758021.

78. John L. Graham and N. Mark Lam, "The Chinese Negotiation," *Harvard Business Review*, October 2003, p. 87.

79. Stephen J. Simurda, "Trade Secrets," *Entrepreneur*, May 1994, p. 120.

80. Edward T. Hall, "The Silent Language of Overseas Business," *Harvard Business Review*, May–June 1960, pp. 5–14.

81. Anton Piech, "Lost in the Translation," *Inc.*, June 2003, p. 50.

82. Maura Judkis, "'Black and Tan' Shoes Force Nike Apology," *Washington Post*, March 15, 2012, www.washingtonpost.com/blogs/arts-post/post/black-and-tan-shoes-force-nike-apology/2012/03/15/gIQAlYXGES_blog.html.

83. "Going Global: Resources for Entrepreneurs and Small Businesses," Small Business & Entrepreneurship Council, November 2011, www.sbecouncil.org/resources/going-global.

84. "Trade in Goods with CAFTA-DR: 2012," U.S. Census Bureau, www.census.gov/foreign-trade/balance/c0017.html.

85. "John Murphy, "Accord Is a 'Win-Win' for Workers, Farmers, and Companies," *ICOSA*, July 6, 2011, www.icosa.co/2011/07/cafta-dr-a-resounding-success.

86. John S. McClenahen, "Sound Thinking,"*Industry Week*, May 3, 1993, p. 28.

87. Jeremy Main, "How to Go Global—and Why," *Fortune*, August 28, 1989, p. 70.

88. "Punita Kumar Sinha, Arvind Subramanian, and Richard Titherington, "BRIC's Share of Global GDP Will Go Up from 18% to 26% over the Next Decade: Arvind Subramanian," *The Economic Times*, September 6, 2012, http://articles.economictimes.indiatimes.com/2012-09-06/news/33650208_1_bric-countries-brics-share-punita-kumar-sinha.

89. *The EU in the World in 2013: A Statistical Portrait*, Eurostat European Commission, 2013, pp. 9, 17.

90. Orit Gadiesh and Jean-Marie Pean, "Think Globally, Market Locally," *Wall Street Journal*, September 9, 2003, p. B2.

91. Jennifer LeClaire, "How to Take Your Small Business Global," *E-Commerce Times*, June 20, 2006, www.ecommercetimes.com/story/50910.html?wlc=1276800833.

Chapter 13

1. "E-Commerce Sales Topped $1 Trillion for First Time in 2012," eMarketer, February 5, 2013, www.emarketer.com/Article/Ecommerce-Sales-Topped-1-Trillion-First-Time-2012/1009649.

2. Samuel Wagreich, "'Showrooming' on the Rise," *Inc.*, January 16, 2013, www.inc.com/samuel-wagreich/showrooming-a-renewed-threat.html.

3. Brad Tuttle, "It's Official: We're Comfortable Ordering Pizza Online," *Time*, June 13, 2012, http://business.time.com/2012/06/13/its-official-were-comfortable-ordering-pizza-online; "Domino's Hits $1 Billion in Digital Sales in One Year," *Pizza Marketplace*, June 11, 2012, www.pizzamarketplace.com/article/195681/Domino-s-hits-1-billion-in-digital-sales-in-one-year; Karl Flinders, "Web Takes a Bigger Slice of Domino's," February 16, 2010, *Computer Weekly*, www.computerweekly.com/Articles/2010/02/16/240320/Web-takes-a-bigger-slice-of-Domino39s-Pizza.htm; "Papa John's Surpasses $1 Billion in Online Pizza Sales," Fresh News, May 8, 2008, www.freshnews.in/papa-johns-surpasses-1-billion-in-online-pizza-sales-26218; "Online Ordering Leader Papa John's First to Surpass $2 Billion in Online Sales," *QSR Magazine*, May 3, 2010, www.qsrmagazine.com/articles/wire/story/20100503006043en.

4. Tiffany Hsu, "Online Pizza Sales Break Records as Mobile Ordering Takes Off," *Los Angeles Times*, December 8, 2011, http://latimesblogs.latimes.com/money_co/2011/12/online-pizza-sales-break-records-as-mobile-ordering-takes-off.html.

5. "E-Commerce Sales Topped $1 Trillion for First Time in 2012," eMarketer, February 5, 2013, www.emarketer.com/Article/Ecommerce-Sales-Topped-1-Trillion-First-Time-2012/1009649.

6. Lauren Indvik, "U.S. Online Retail Sales to Reach $327 Billion by 2016," *Mashable*, February 27, 2012, http://mashable.com/2012/02/27/ecommerce-327-billion-2016-study.

7. "Top Selling Internet Items," *Statistic Brain*, July 24, 2012, www.statisticbrain.com/top-selling-internet-items.

8. Jennifer Wang, "On Her Toes," *Entrepreneur*, December 2011, p. 53.

9. "Living in Exponential Times," Get to the Point: Marketing Inspiration, *Marketing Profs*, May 11, 2009, pp. 1–2.

10. "E-Commerce/Online Sales Statistics," *Statistic Brain*, August 24, 2012, www.statisticbrain.com/total-online-sales.

11. Allison Enright, "M-Commerce Promises to Extend the Web's Influence over Retail," *Internet Retailer*, June 13, 2012, www.internetretailer .com/2012/06/13/m-commerce-promises-extend-webs-influence-over-retail.

12. E. B. Boyd, "Google Goes After Your Business," *Fast Company*, January 9, 2012, www.fastcompany.com/1805995/google-goes-after-your-local-small-business.

13. Lauren Simonds, "Web Sites: They're Not Just for E-Commerce," *Small Business Computing*, February 15, 2007, www.smallbusinesscomputing .com/news/article.php/3660186.

14. *The Foresee 2012 E-Retail Satisfaction Index (U.S. Holiday Edition)*, Foresee, December 27, 2012, www.foreseeresults.com/news-events/ press-releases/us-holiday-eretail-2012-foresee.shtml.

15. "Customer Retention: Keep Customers by Growing Relationships Online," pb Smart for Small Business, Pitney Bowes, www .pbsmartessentials.com/customer-satisfaction/customer-retention-keep-customers-by-growing-relationships-online.

16. "Small Business Market Survey Shows Small Businesses Underutilize E-Commerce," *Small Business Trends*, May 1, 2011, http://smallbiztrends .com/2011/05/small-business-market-survey-e-commerce.html.

17. "Internet 2012 in Numbers," *Pingdom*, January 16, 2013, http://royal .pingdom.com/2013/01/16/internet-2012-in-numbers.

18. Robert McGarvey, "Connect the Dots," *Entrepreneur*, March 2000, pp. 78–85.

19. Carol Tice, "Small Businesses Don't Have Time for Social Media—and Don't Track Results," *Forbes*, October 31, 2012, www.forbes.com/sites/ caroltice/2012/10/31/small-businesses-dont-have-time-for-social-media-and-dont-track-results/2.

20. Phil Mershon, "9 Facebook Marketing Success Stories You Should Model," *Social Media Examiner*, September 13, 2011, www .socialmediaexaminer.com/9-facebook-marketing-success-stories-you-should-model; Lynette Young, "Can Wall Paint Be Social?," *Purple Stripe*, November 30, 2011, www.purplestripe.com/tag/idea-paint.

21. Carol Tice, "Small Businesses Don't Have Time for Social Media—and Don't Track Results," *Forbes*, October 31, 2012, www.forbes.com/sites/ caroltice/2012/10/31/small-businesses-dont-have-time-for-social-media-and-dont-track-results/2; Zac Johnson, "State of the Blogging World in 2012," *Blog World*, July 25, 2012, www.blogworld.com/2012/07/25/ state-of-the-blogging-world-in-2012.

22. Lou Dubois, "7 Blogging Mistakes That Small Businesses Make," *Inc.*, March 24, 2011, www.inc.com/guides/201103/7-blogging-mistakes-that-small-businesses-make.html; "Company Overview," Wegmans, www.wegmans.com/webapp/wcs/stores/servlet/CategoryDisplay? categoryId=281152&storeId=10052&catalogId=10002&langId=-1.

23. "Many Americans Won't Forgive a Faulty Web Site," 1&1 Internet, July 11, 2012, http://press.1and1.com/xml/article?article_id=1131.

24. *Online Shopping Customer Experience Study*, comScore and UPS, May 2012, pp. 5, 7.

25. "Fits.me Launches World's First Virtual Fitting Room," Fits.me, May 18, 2010, http://fits.me/news/fitsme-launches-world%E2%80%99s-first-virtual-fitting-room-0; "Online Retail Virtual Fitting Room Solution," Slideshare, August 8, 2012, www.slideshare.net/fitsme/ online-retail-virtual-fitting-room-solution-fitsme-august-2012.

26. *The Seven Habits of Highly Effective Web Sites*, Sitecore, 2012, p. 2.

27. "9 in 10 Americans Concerned About Online Privacy," *Marketing Charts*, February 17, 2012, www.marketingcharts.com/wp/direct/9-in-10-americans-concerned-about-online-privacy-21156.

28. Dave Deasy, "A Booming Market for Mobile Commerce Where Trust Is Essential," TRUSTe, January 28, 2013, www.truste.com/ blog/2013/01/28/a-booming-market-for-mobile-commerce-where-trust-is-essential.

29. "Survival of the Fastest," *Inc. Technology*, no. 4, 1999, p. 57.

30. Darren Dahl, "Among Online Entrepreneurs, Subscriptions Are All the Rage," *New York Times*, March 7, 2012, www.nytimes.com/2012/03/08/ business/smallbusiness/selling-online-products-by-subscription-is-all-the-rage.html?_r=0; Stu Woo, "Pets.com 2.0," *Wall Street Journal*, March 6, 2012, http://online.wsj.com/article/SB1000142405297020477 8604577240261600840678.html; Arlene Weintraub, "PetFlow Paws Its Way to the Top of Facebook, Sniffs Out Growth Path," *Xconomy*, July 5, 2012, www.xconomy.com/new-york/2012/07/05/petflow-paws-its-way-to-the-top-of-facebook-sniffs-out-growth-path.

31. Stu Woo, "Pets.com 2.0," *Wall Street Journal*, March 6, 2012, http:// online.wsj.com/article/SB10001424052970204778604577240261600840 678.html.

32. Stu Woo, "Pets.com 2.0," *Wall Street Journal*, March 6, 2012, http:// online.wsj.com/article/SB10001424052970204778604577240261600840 678.html.

33. Steve Bennett and Stacey Miller, "The E-Commerce Plunge," *Small Business Computing*, February 2000, p. 50.

34. Alanna Finck, "The Anatomy of Convenience: The 2012 Service Design Report," *Continuum*, January 7, 2013, http://continuuminnovation.com/ the-anatomy-of-convenience-continuums-2012-service-design-report; Paule Demery, "IRCE 2012 Report: Consumers Explain How They Shop Online," *Internet Retailer*, June 7, 2012, www.internetretailer .com/2012/06/07/irce-2012-report-consumers-explain-how-they-shop-online.

35. Khalid Saleh, "Shopping Cart Abandonment Rate Statistics (Infographic), *InvespBlog*, May 16, 2012, www.invesp.com/blog/cro/ shopping-cart-abandonment-rate-statistics-infographic.html.

36. *Conversion Rate Optimization Report 2012*, Econsultancy and Redeye, October 2012, p. 6.

37. Julie Rains, "3 Killer E-Commerce Web Site Features," *American Express Open Forum*, www.openforum.com/articles/3-killer-e-commerce-website-features.

38. Khalid Saleh, "Shopping Cart Abandonment Rate Statistics (Infographic)," *InvespBlog*, May 16, 2012, www.invesp.com/blog/cro/ shopping-cart-abandonment-rate-statistics-infographic.html.

39. "Unexpected Delivery Costs the Top Reason Consumers Abandon Online Purchases," *Marketing Charts*, January 31, 2013, www .marketingcharts.com/wp/topics/e-commerce/unexpected-delivery-costs-the-top-reason-consumers-abandon-online-purchases-26625.

40. *Consumer Views of Live Online Help 2012: A Global Perspective*, March 2012, Oracle, p. 4.

41. "E-Mail Service Provider Listrak Conducts 'Shop and Abandon' Cart Study Using Internet Retailer 500 List," August 13, 2009, Listrak, www.listrak.com/News/Listrak-Conducts-Abandon-Cart-Study.

42. *The Remarketing Report: Benchmark Data and Analysis on Connecting Web Behavior to E-Mail Marketing*, Experian CheetahMail, January 2010, p. 2.

43. Christian Gulliksen, "Optimize E-Mail for Abandoned Carts," *Marketing Profs*, January 3, 2013, www.marketingprofs.com/ pics/2013/9784/optimize-email-for-abandoned-carts-slide-show.

44. "Case Studies: Rockler Automated Cart Abandonment Program Drives 2% of Sales," Blue Hornet Digital River, www.bluehornet.com/case-studies/full/rockler; "E-Mail Service Provider Listrak Conducts 'Shop and Abandon' Cart Study Using Internet Retailer 500 List," August 13, 2009, Listrak, www.listrak.com/News/Listrak-Conducts-Abandon-Cart-Study.

45. "The Biggest Mistakes in Web Design," *Website Magazine*, September 6, 2012, www.websitemagazine.com/content/blogs/posts/ archive/2012/09/06/the-biggest-mistakes-in-web-design.aspx.

46. "Akamai Reveals 2 Seconds as the New Threshold of Acceptability for e-Commerce Web Page Response Times," Akamai, September 14, 2009, www.akamai.com/html/about/press/releases/2009/press_091409. html; Kristina Knight, "Study: Consumers Abandon Slow-Loading Web Sites," *BizReport*, April 27, 2010, www.bizreport.com/2010/04/ study_consumers_abandon_slow_loading_websites.html#; "TagMan Solves Slow Page Load/Audience Loss Problems with Introduction of New TagMan ServerTags," TagMan, May 10, 2010, http://blog.tagman .com/2010/05/tagman-solves-slow-page-loadaudience-loss-problems-with-introduction-of-new-tagman-servertags.

47. "How Loading Time Affects Your Bottom Line," *Kissmetrics*, November 2011, http://blog.kissmetrics.com/loading-time/?wide=1.

48. "Web Site Performance Management," *Internet Retailer*, July 1, 2010, www.internetretailer.com/2010/07/01/web-site-performance-management.

49. Fred Vogelstein, "A Cold Bath for Dot-Com Fever," *U.S. News & World Report*, September 13, 1999, p. 37.

50. Darren Dahl, "Among Online Entrepreneurs, Subscriptions Are All the Rage," *New York Times*, March 7, 2012, www.nytimes.com/2012/03/08/business/smallbusiness/selling-online-products-by-subscription-is-all-the-rage.html?_r=0; Stu Woo, "Pets.com 2.0," *Wall Street Journal*, March 6, 2012, http://online.wsj.com/article/SB10001424052970204778604577240261600840678.html.

51. Erin Lynch, "Survey Says Consumers Want a Crosschannel Shopping Experience," *Multichannel Merchant*, May 8, 2012, http://multichannelmerchant.com/news/hybris-crosschannel-shopping-0508e11.

52. Greg Sterling, "Survey: 91 Percent Have Gone into Stores Because of Online Promotion," *Marketing Land*, December 14, 2012, http://marketingland.com/survey-91-percent-have-gone-into-stores-because-of-online-promotion-28796.

53. Bronwyn Fryer and Lee Smith, ".com or Bust," *Forbes Small Business*, December 1999/January 2000, p. 41.

54. Martha McCarthy, "Milwaukee Teen to Sell Over 1,000 Ugly Christmas Sweaters," *PR Leap*, December 7, 2012, www.prleap.com/pr/194474; Dinesh Ramde, "Ugly, Gaudy Christmas Sweaters All the Rage," *Greenville News*, December 24, 2011, p. 3D.

55. Shelly Banjo, "Firms Take Online Reviews to Hear," *Wall Street Journal*, July 30, 2012, p. B7.

56. Ralph F. Wilson, "The Five Mutable Laws of Web Marketing," *Web Marketing Today*, April 1, 1999, www.wilsonweb/com/wmta/basic-principles.htm, pp. 1–7.

57. "E-Mail Marketing Wags the Tail of Washington DC Area Pet Business," Constant Contact, www.constantcontact.com/email-marketing/customer-examples/furget-me-not.jsp.

58. "E-Mail Open and Click Rates: Benchmarks and Trends," *Marketing Profs*, July 30, 2012, www.marketingprofs.com/charts/2012/8560/email-open-and-click-rates-benchmarks-trends.

59. Hanna Andrzejewska, "Optimizing Your Campaigns: Best Day to Send E-Mails," GetResponse, November 2012, http://blog.getresponse.com/optimizing-your-campaigns-best-days-to-send-emails.html.

60. "Best Time to Send E-Mail," GetResponse, October 9, 2012, http://blog.getresponse.com/best-time-to-send-email-infographic.html.

61. "Spam Statistics," Trustwave, March 3, 2013, www.trustwave.com/support/labs/spam_statistics.asp.

62. Daniel Burstein, "Automatic for the People: The Pros and Cons of Triggered E-Mails," *Marketing Sherpa Blog*, September 6, 2012, http://sherpablog.marketingsherpa.com/email-marketing/pros-cons-triggered-emails; Sara Rand, "Three Best Practices for E-Mail, Derived from the Top Retailers," National Retail Federation, September 14, 2011, http://blog.shop.org/2011/09/14/three-best-practices-for-email-of-the-top-retailers.

63. Peter Prestipino, "E-Commerce Emergency," *Website Magazine*, August 2012, pp. 24–27.

64. Julie Rains, "3 Killer E-Commerce Web Site Features," *American Express Open Forum*, December 15, 2011, www.openforum.com/articles/3-killer-e-commerce-website-features.

65. "Web Site Design and Credibility," *CMS Intelligence*, January 26, 2013, www.cmsintelligence.com/site/blog/2013/01/26/website-design-credibility.

66. David Port, "How to Make Your Web Site Really Sell," *Entrepreneur*, September 2009, p. 84.

67. "Top Ten Languages Used in the Internet," *Internet World Stats*, June 30, 2012 www.internetworldstats.com/stats.htm.

68. Ryan Underwood, "Clicks from Around the World," *Inc.*, December 2010/January 2011, pp. 146–147.

69. "Mobile Takeover," *Website Magazine*, November 2011, p. 13.

70. "As U.S. Smartphone Penetration Grows, So Does Apple's Market Share," Marketing Charts, March 7, 2013, www.marketingcharts.com/wp/topics/signs-of-whats-to-come/as-us-smartphone-penetration-grows-so-does-apples-market-share-27600; Lee Rainie, "25% of American Adults Own Tablet Computers," Pew Internet and American Life Project, October 4, 2012, p. 2.

71. Allison Howen, "The Mobile Impact: Stats, States, and Data," *Website Magazine*, November 21, 2012, www.websitemagazine.com/content/blogs/posts/archive/2012/11/21/the-mobile-impact.aspx.

72. Allison Howen, "SMBs Not Ready for Mobile," *Website Magazine*, July 2, 2012, www.websitemagazine.com/content/blogs/posts/archive/2012/07/02/study-shows-smb-not-ready-for-mobile.aspx.

73. "84% of U.S. Small Businesses Using Mobile Marketing See Increase in New Business Activity," *PR Newswire*, May 3, 2012, www.prnewswire.com/news-releases/84-of-us-small-businesses-using-mobile-marketing-see-increase-in-new-business-activity-149993195.html.

74. "What Users Want Most from Mobile Sites," *Think With Google*, September 2012, www.thinkwithgoogle.com/insights/library/studies/what-users-want-most-from-mobile-sites-today.

75. "How Loading Time Affects Your Bottom Line," Kissmetrics, 2013, http://blog.kissmetrics.com/loading-time; Lakshmi Harikumar, "Nine Tips for Building a Solid Mobile Web Site," *Marketing Profs*, October 15, 2012, www.marketingprofs.com/articles/2012/9123/nine-tips-for-building-a-solid-mobile-website.

76. Bill Siwicki, "A Mobile Commerce Site Overachieves for an Underwear Retailer," *Internet Retailer*, February 27, 2013, www.internetretailer.com/2013/02/27/mobile-commerce-site-over-achieves-underwear-retailer.

77. Kristin Purcell, "The State of Online Video," Pew Internet and American Life Project, June 3, 2010, www.pewinternet.org/Reports/2010/State-of-Online-Video.aspx; Joanna Brenner, "Pew Internet: Social Networking," Pew Internet and American Life Project, February 14, 2013, http://pewinternet.org/Commentary/2012/March/Pew-Internet-Social-Networking-full-detail.aspx; "comScore Releases August 2012 Online Video Rankings," comScore, September 19, 2012, www.comscore.com/Insights/Press_Releases/2012/9/comScore_Releases_August_2012_U.S._Online_Video_Rankings.

78. "SMBs Struggle to Adopt, Integrate Social Media," *eMarketer*, September 10, 2012, www.emarketer.com/%28S%28gjrgkh45dmewndatycqrprnz%29%29/Article/SMBs-Struggle-Adopt-Integrate-Social-Media/1009332.

79. Ari Herzog, "90 Percent of Small Businesses Use Social Media," *Social Media Today*, September 18, 2012, http://socialmediatoday.com/ariherzog/820046/90-percent-small-business-use-social-media.

80. Jenn Garbee, "Cookbook of the Week: Humphrey Slocombe Ice Cream Book Is Not Your Mother's (Damn) Ice Cream Book," *LA Weekly Blogs*, July 5, 2012, http://blogs.laweekly.com/squidink/2012/07/humphry_slocombe_ice_cream.php; Renata Sternfeld-Allon, "Small Business Twitter Success Stories," *Fee Fighters*, August 26, 2010, https://feefighters.com/blog/small-business-twitter-success-stories.

81. Dan Briody, "Puppy Power," *Inc.*, November 2007, pp. 55–56.

82. Lee Rainie, Kristen Purcell, Amy Mitchell, and Tom Rosensteil, "Where People Get Information About Restaurants and Other Local Businesses," Pew Internet and American Life Project, December 14, 2011, http://pewinternet.org/Reports/2011/Local-business-info/Overview.aspx.

83. "Less Than 20% of SMB Web Sites Link to Social Presence, According to SMB DigitalScape," *PR Newswire*, April 16, 2012, www.prnewswire.com/news-releases/less-than-20-of-smb-websites-link-to-social-presence-according-to-smb-digitalscape-147592525.html.

84. "Agency News: 2012 Digital Influence Index Shows Internet as Leading Influence in Consumer Purchasing Choices," Fleishman-Hillard, January 31, 2012, http://fleishmanhillard.com/2012/01/31/2012-digital-influence-index-shows-internet-as-leading-influence-in-consumer-purchasing-choices.

85. "Search Still Drives More E-Commerce Traffic, Higher Order Values Than E-Mail, Social," *Marketing Charts*, February 12, 2013, www.marketingcharts.com/wp/interactive/search-still-drives-more-e-commerce-traffic-higher-order-values-than-email-social-26905.

86. Sarah E. Needleman and Emily Maltby, "As Google Tweaks Searches, Some Get Lost in the Web," *Wall Street Journal*, May 16, 2012, pp. B1, B6.

87. Jason Prescott, "How to Optimize Your Site for Google in 2010," *iMedia Connection*, March 23, 2010, www.imediaconnection.com/content/26275.asp.

88. Marta Kagan, "100 Awesome Marketing Charts and Graphs," HubSpot, May 20, 2011, http://blog.hubspot.com/blog/tabid/6307/bid/14416/100-Awesome-Marketing-Stats-Charts-Graphs-Data.aspx.

89. Miranda Miller, "53% of Organic Search Clicks Go to First Link," *Search Engine Watch*, October 10, 2012, http://searchenginewatch.com/article/2215868/53-of-Organic-Search-Clicks-Go-to-First-Link-Study.

90. James A. Martin, "Search Engine Optimization: SEO Tips for Small Business," *Small Business Computing*, September 29, 2009, www.smallbusinesscomputing.com/buyersguide/article.php/3841381/Search-Engine-Optimization-SEO-Tips-for-Small-Business.htm.

91. Nate Elliot, "The Easiest Way to a First-Page Ranking on Google," *Forrester Blogs*, January 8, 2009, http://blogs.forrester.com/interactive_marketing/2009/01/the-easiest-way.html.

92. Matt McGee, "Google Hits 67% Market Share (Again), Bing Hits Another All-Time High," *Search Engine Land*, February 13, 2013, searchengineland.com/comscore-january-2013-search-rankings-148478.

93. "Google's Going Cost per Click," Radio Advertising Bureau, 2013, www.rab.com/public/rst/article.cfm?article=1&id=2611; Justin Martin, "Get Right with Google," *FSB*, September 2006, pp. 70–78.

94. "Google Adwords Success Story: Happy Hounds," *Ajax Union*, January 5, 2012, www.ajaxunion.com/2012/01/google-adwords-success-story-happy-hounds.

95. Clem Chambers, "Is Click Fraud a Ticking Time Bomb Under Google?," *Forbes*, June 18, 2012, www.forbes.com/sites/investor/2012/06/18/is-click-fraud-a-ticking-time-bomb-under-google/2.

96. "Santa's Helpers," *The Economist*, May 15, 2004, pp. 5–8.

97. "Consumers Say Free (Not Same-Day) Delivery Option Would Spur Increased Online Shopping," *Marketing Charts*, March 7, 2013, www.marketingcharts.com/wp/topics/e-commerce/consumers-say-free-not-same-day-delivery-option-would-spur-increased-online-shopping-27595.

98. "How to Sell an Old Product in a New Way," *Fast Company*, June 2012, p. 153.

99. Anick Jesdanun, "Businesses Dominate Bids for Internet Suffixes," *Greenville News*, June 17, 2012, p. 3E; Daniel P. Smith, "King of Your Domain," *QSR*, December 2011, www.qsrmagazine.com/exclusives/king-your-domain.

100. "Facebook 'Likes' More Important Than Ever in Purchase Decision-Making," *IT Analysis*, October 4, 2011, www.it-analysis.com/channels/online/news_release.php?rel=27417.

101. Gwen Moran, "Target Markets," *Entrepreneur*, July 2012, p. 44.

102. "About Us—American Pearl, www.americanpearl.com/aboutus.html.

103. *Revolutionizing Web Site Design: The New Rules of Usability* (Traverse City, MI: OneupWeb, 2010), p. 11.

104. Melissa Campanelli, "The Right Stuff," *Entrepreneur*, April 2007, p. 54.

105. Carol Stavraka, "There's No Stopping E-Business. Are You Ready?," *Forbes*, December 13, 1999, Special Advertising Section.

106. *2012 National Cyber Security Alliance/Symantec National Small Business Study*, NCSA/Symantec, October 2012, p. 11.

107. Julie Rains, "3 Killer E-Commerce Web Site Features," *American Express Open Forum*, December 15, 2011, www.openforum.com/articles/3-killer-e-commerce-website-features.

108. Paul Demery, "Online Shoppers Prefer to Pay with Credit Cards," *Internet Retailer*, April 26, 2012, www.internetretailer.com/2012/04/26/online-shoppers-prefer-pay-credit-cards.

109. Alex Konrad, "Tablets Storm the Corner Office," *Fortune*, October 13, 2011, p. 29.

110. Fiona Swerdlow, "2009 State of Retailing Online: Marketing Report," May 5, 2009, Shop.org, http://blog.shop.org/2009/05/05/2009-state-of-retailing-online-marketing-report.

111. *TRUSTe 2013 U.S. Consumer Privacy Confidence Report: What Consumers Think, Business Impact, and Recommended Actions*, TRUSTe, p. 6.

112. *2012 National Cyber Security Alliance/Symantec National Small Business Study*, NCSA/Symantec, October 2012, p. 5.

113. Shashi Bellamkonda, "Kikscore Online Trust Survey Finds Information Sharing Leads to Trust," *Tech Cocktail*, January 11, 2012, http://tech.co/kikscore-online-trust-survey-2012-01.

114. *2012 Norton Cybercrime Report*, Norton by Symantec, 2013, p. 3.

115. *Trustwave 2013 Global Security Report*, Trustwave, March 2013, pp. 9–10.

116. "Small Business Online Security Infographic," National Cyber Security Alliance, www.staysafeonline.org/stay-safe-online/resources/small-business-online-security-infographic.

117. *2012 National Cyber Security Alliance/Symantec National Small Business Study*, NCSA/Symantec, October 2012, pp. 4, 14.

118. Small Business Online Security Infographic," National Cyber Security Alliance, www.staysafeonline.org/stay-safe-online/resources/small-business-online-security-infographic.

119. Larry Greenemeier, "Largest Data Breach Ever," *Information Week*, April 2, 2007, p. 21.

120. John Patrick Pullen, "Smooth Criminals," *Entrepreneur*, February 2013, pp. 50–53.

121. "Data Breaches Often Not Detected for Months, Report Finds," *Globalscape*, February 14, 2013, www.globalscape.com/blog/2013/2/14/data-breaches-often-not-detected-for-months-report-finds.

122. *2012 National Cyber Security Alliance/Symantec National Small Business Study*, NCSA/Symantec, October 2012, p. 5.

123. *Online Fraud Report: Thirteenth Annual Edition* (Mountain View, CA: Cybersource Corporation, 2012), p. 1.

124. Emily Maltby, "Shady Shoppers Beware," *Wall Street Journal*, November 24, 2009, http://online.wsj.com/article/SB10001424052748704533904574548210301039526.html#.

Chapter 14

1. Benjamin B. Gaunsel, "Toward a Framework of Financial Planning in New Venture Creation," paper presented at the annual meeting of the United States Association for Small Business and Entrepreneurship, January 2005, Palm Springs, California, www.sbaer.uca.edu/research/usasbe/2005/pdffiles/papers/25.pdf.

2. Eileen Davis, "Dodging the Bullet," *Venture*, December 1988, p. 78.

3. C. J. Prince, "Number Rustling," *Entrepreneur*, March 2003, pp. 43–44.

4. Darren Dahl, "Basics of Accounting Are Vital to Survival for Entrepreneurs," *New York Times*, August 4, 2011, p. B9.

5. Norm Brodsky, "Our Irrational Fear of Numbers," *Inc.*, January/February 2009, p. 33.

6. Mike Hogan, "Pocket Books," *Entrepreneur*, March 2006, pp. 42–43.

7. Norm Brodsky, "The Magic Number," *Inc.*, September 2003, pp. 43–46.

8. Diana Ransom, "Fixing a Nearly Fatal Flaw," *Entrepreneur*, March 7, 2011, www.entrepreneur.com/article/219265.

9. Geoff Colvin, "How to Manage Your Business in a Recession," *Fortune*, January 19, 2009, p. 91.

10. "When the Going Gets Tough, the Tough Don't Skimp on the Ad Budgets," *Knowedge@Wharton*, November 26, 2008, http://knowledge.wharton.upenn.edu/article.cfm?articleid=2101.

11. Jared Sandberg, "Counting Pizza Slices, Cutting Water Cups—You Call This a Budget?" *Wall Street Journal*, January 21, 2004, p. B1.

12. Diedrich Von Soosten, "The Roots of Financial Destruction," *Industry Week*, April 5, 1993, pp. 33–34.

13. Lori Ioannou, "He's Preaching the Power of Thrift," *Fortune*, October 30, 2000, p. 208[P].

14. Gwen Moran, "How One Business Bounced Back from Bankruptcy," *Entrepreneur*, January 4, 2013, www.entrepreneur.com/blog/225402.

15. Peter Lattman and Lauren A. E. Schuker, "MGM Recasts Leadership in Bid to Dig Out of Debt," *Wall Street Journal*, August 19, 2009, p. B1, B2;

Brett Pulley and Kelly Riddell, "MGM Said to Seek $3.7 Billion Debt Restructuring, Forbearance," *Bloomberg*, September 25, 2009, www .bloomberg.com/apps/news?pid=20601087&sid=aPV.ZqDG6FRY.

16. "Analyzing Creditworthiness," *Inc.*, November 1991, p. 196.

17. http://purebelly.com/sample-page; Joe Robinson, "Managing the Stress of Starting Your Own Business," *Entrepreneur*, September 7, 2011, www.entrepreneur.com/article/220235#.

18. Jill Andresky Fraser, "Giving Credit to Debt," *Inc.*, November 2000, p. 125.

19. Cliff Banks, "New Ways to Move Used Cars," *Ward's Dealer Business*, November 1, 2005, http://wardsdealer.com/ar/auto_new_ways_move; Alex Taylor III, "Survival on Dealer's Row," *Fortune*, March 31, 2008, p. 24.

20. Jeff Cornwall, "Franklin Brothers Build Successful Business While in College," *Tennessean*, February 3. 2013, p. B3.

21. Darren Dahl, "Basics of Accounting Are Vital to Survival for Entrepreneurs," *New York Times*, August 4, 2011, p. B9.

22. "The Cash Conversion Cycle," *Forbes*, March 10, 2012, http:// www.forbes.com/sites/ycharts/2012/03/10/the-cash-conversion-cycle/. www.investopedia.com/articles/06/cashconversioncycle.asp.

23. William C. Dunkelberg and Holly Wade, *NFIB Small Business Economic Trends*, NFIB Research Foundation, January 2013, p. 1.

24. Darren Dahl, "The Truth About Profits," *Inc.*, October 2009, pp. 91–96.

25. www.jellio.com; Adriana Gardella, "Solutions for Jellio," *New York Times*, June 7, 2012, B7; Adriana Gardella, "Meeting the Demand for Quirky Objects," *New York Times*, May 31, 2012, p. B4.

26. Adriana Gardella, "A Start-Up's Financial Reckoning," *New York Times*, March 29, 2011, http://boss.blogs.nytimes.com/2011/03/29/a-start-ups-financial-reckoning.

27. Joshua Hyatt, "Planes, Trains, and . . . Buses," *FSB*, February, 2004, p. 20.

28. Neil Parmar, "Crafty Ways Restaurants Cut Costs," *SmartMoney*, October 2, 2009, www.smartmoney.com/spending/rip-offs/crafty-ways-restaurants-cut-costs.

29. Dirk Smillie, "What Recession?," *Forbes*, August 11, 2008, p. 64.

30. "STR Reports U.S. Hotel Performance for Week Ending 5 December 2009," *Hospitality.net*, December 14, 2009, www.hospitalitynet.org/ news/154000320/4044710.html; Terrie Lloyd, "Hotel Occupancies Fall Below Breakeven, eBiz News from Japan," *Japan Inc.*, August 3, 2009, www.japaninc.com/terries_take_528.

31. www.gomez.com/us-retail-web-mobile-site-performance-index/.

32. Norm Brodsky, "The Magic Number," *Inc.*, September 2003, pp. 43–46.

33. William F. Doescher, "Taking Stock," *Entrepreneur*, November 1994, p. 64.

34. William F. Doescher, "Taking Stock," *Entrepreneur*, November 1994, p. 64.

35. Darren Dahl, "Basics of Accounting Are Vital to Survival for Entrepreneurs," *New York Times*, August 4, 2011, p. B9.

36. "LimoLiner," *Executive Travel*, June 25, 2009, www .executivetravelmagazine.com/page/LimoLiner; Joshua Hyatt, "Planes, Trains, and . . . Buses," *FSB*, February, 2004, p. 20.

Chapter 15

1. Philip Campbell, "Cash Flow Projections Made Easy," *Inc.*, www.inc .com/resources/finance/articles/20041001/cashprojection.html.

2. Jerry Useem, "The Icon That Almost Wasn't," *Inc: The State of Small Business 1998*, p. 142; "History," Ford Motor Company, http://www .ford.com/en/heritage/history/default.htm.

3. "Are You Ready for the Major Leagues?," *Inc.*, February 2001, p. 106.

4. Daniel Kehrer, "Big Ideas for Your Small Business," *Changing Times*, November 1989, p. 58.

5. Rick Desloge, "Cerf Brothers Grandchildren Can't Hold On," *St. Louis Business Journal*, November 13, 2009, http://stlouis.bizjournals .com/stlouis/stories/2009/11/16/story2.html; "Cerf Bros. Files for Bankruptcy, Puts Itself Up for Sale," *8264.net*, November 12, 2009, http://www.8264.net/html/Sports_News/International_Sports_ Industry/200911/12-6877.html.

6. Daniel Lyons, "Wool Gatherer," *Forbes*, April 16, 2001, p. 310.

7. Bramble Berry, 2013, www.brambleberry.com/About-Us-W9C146 .aspx; Lisa Girard, "Three Common Accounting Mistakes," *Entrepreneur*, April 27, 2011, www.entrepreneur.com/article/219523#.

8. Jason Leopold, "Enron but Not Forgotten," *Entrepreneur*, January 2003, p. 63.

9. Douglas Bartholomew, "4 Common Financial Mistakes . . . and How to Avoid Them," *Your Company*, Fall 1991, p. 9.

10. Carolyn Crummey, "10 Tips To Help You Make It Big in Small Business," *Small Biz Technology*, February 6, 2013, www .smallbiztechnology.com/archive/2013/02/10-tips-to-help-you-make-it-big-in-small-business.html.

11. David Evans, "Cash Is King: 5 Simple Rules for Creating a Cash Flow Plan," *Inc.*, April 30, 2012, www.inc.com/david-evans/5-rules-for-making-cash-a-figurehead-king.html.

12. Douglas Bartholomew, "4 Common Financial Mistakes . . . and How to Avoid Them," *Your Company*, Fall 1991, p. 9.

13. Jill Andresky Fraser, "Monitoring Daily Cash Trends," *Inc.*, October 1992, p. 49.

14. Kelly K. Spors and Simona Covel, "Slow Payments Squeeze Small-Business Owners," *Wall Street Journal*, October 31, 2008, http://online .wsj.com/article/SB122540968964686203.html.

15. "2009 Fact Sheet: Intuit Billing Manager 'Get Paid Survey' Results, Intuit, November 13, 2008, p. 1.

16. Mark Henricks, "Losing Stream," *Entrepreneur*, September 2003, pp. 77–78.

17. C. J. Prince, "Give 'Em Credit," *Entrepreneur*, April 2004, pp. 59–60.

18. Michael Selz, "Big Customers' Late Bills Choke Small Suppliers," *Wall Street Journal*, June 22, 1994, p. B1.

19. Kelly K. Spors and Simona Covel, "Slow Payments Squeeze Small-Business Owners," *Wall Street Journal*, October 31, 2008, http://online .wsj.com/article/SB122540968964686203.html.

20. Richard G. P. McMahon and Scott Holmes, "Small Business Financial Practices in North America: A Literature Review," *Journal of Small Business Management*, April 1991, p. 21.

21. Raymund Flandez, "Three Best Ways to Make Sure Customers Pay," *Wall Street Journal*, November 13, 2009, http://online.wsj.com/article/ SB10001424052748703683804574533482721614374.html?mod=WSJ_ FinancingAndInvesting_LEFTTopHeadlines.

22. William M. Bulkeley, "The Battle Against Slow Payers," *Wall Street Journal*, February 13, 2011, http://online.wsj.com/article/SB100014240 5274870398930457550423888347154.html.

23. Howard Muson, "Collecting Overdue Accounts," *Your Company*, Spring 1993, p. 4.

24. 2009 Fact Sheet: Intuit Billing Manager 'Get Paid Survey' Results, Intuit, November 13, 2008, p. 1.

25. 2009 Fact Sheet: Intuit Billing Manager 'Get Paid Survey' Results, Intuit, November 13, 2008, p. 1.

26. Kimberly Stansell, "Tend to the Business of Collecting Your Money," *Inc.*, March 2, 2000, http://www.inc.com/articles/2000/03/17568.html; Frances Huffman, "Calling to Collect," *Entrepreneur*, September 1993, p. 50.

27. *The Impact of Third Party Debt Collection on the National and State Economies*, Ernst & Young, February 2012, p. 2.

28. "Time Shrinks Value of Debts," *Collection*, Winter 1992, p. 1.

29. National Federation of Independent Businesses, "Small Business, Credit Access, and a Lingering Recession," January 2012.

30. Raymund Flandez, "Three Best Ways to Make Sure Customers Pay," *Wall Street Journal*, November 13, 2009, http://online.wsj.com/article/ SB10001424052748703683804574533482721614374.html?mod=WSJ_ FinancingAndInvesting_LEFTTopHeadlines.

31. Elaine Pofeldt, "Collect Calls," *Success*, March 1998, pp. 22–24.

32. "Make Them Pay!," *Inc.*, August 2003, p. 50.
33. John Gorham, "Revenge of the Lightweight," *Forbes*, March 6, 2000, p. 54.
34. C. J. Prince, "Vulture Capital," *Entrepreneur*, February 2003, pp. 47–48.
35. "Challenges with Customer Payments Growing Concern for Young Firms, Kauffman Study Shows," Kauffman Foundation, May 9 2012, www .kauffman.org/newsroom/challenges-with-customer-payments-growing-concern-for-young-firms-kauffman-foundation-study-shows.aspx.
36. Angus Loten, "Small Firms' Big Customers Are Slow to Pay," *Wall Street Journal*, June 6, 2012, http://online.wsj.com/article/SB100014240 5270230329660457745056143449668.html.
37. "Apple Inc (Consolidated Issue Listed on NASDAQ Global Select," *Bloomberg/Businessweek,* http://investing.businessweek.com/research/ stocks/financials/financials.asp?ticker=AAPL.
38. Serena Ng and Cari Tuna, "Big Firms Are Quick to Collect, Slow to Pay," *Wall Street Journal*, August 31, 2009, http://online.wsj.com/ article/SB125167116756270697.html.
39. Jill Andresky Fraser, "How to Get Paid," *Inc.*, March 1992, p. 105.
40. "Hot Tip: Pay Bills Early, Earn Money," *Inc.*, November 2000, www .inc.com/articles/2000/11/21082.html.
41. William G. Shepherd Jr., "Internal Financial Strategies," *Venture*, September 1985, p. 68.
42. Jay Goltz, "Surviving the Recession," *FSB*, February 2009, p. 27.
43. Roberta Maynard, "Can You Benefit from Barter?," *Nation's Business*, July 1994, p. 6.
44. "33 Ways to Increase Your Cash Flow and Manage Cash Balances," *The Business Owner*, February 1988, p. 8.
45. Jeffrey Cornwall. *Bootstrapping* (Englewood Cliffs, NJ: Pearson/ Prentice Hall, 2009).
46. "12 Creative Money-Saving Tactics for Small-Business Owners," *Entrepreneur*, December 21, 2012, www.entrepreneur.com/ slideshow/224914#10.
47. *U.S. Equipment Finance Market Study 2012-2013 Fact Sheet,* Equipment Leasing and Finance Foundation, Washington, DC, 2012.
48. Bob Violino, "What's the Deal?," *CFO-IT*, Spring 2004, pp. 15–18.
49. "12 Creative Money-Saving Tactics for Small-Business Owners," *Entrepreneur*, December 21, 2012, www.entrepreneur.com/ slideshow/224914#3.
50. Bruce G. Posner, "Skipped Loan Payments," *Inc.*, September 1992, p. 40.
51. "12 Creative Money-Saving Tactics for Small-Business Owners," *Entrepreneur*, December 21, 2012, www.entrepreneur.com/ slideshow/224914#6.
52. Mikal E. Belicove, "How Fine Art America Built Its Business by Bootstrapping," *Entrepreneur*, January 20, 2012, www.entrepreneur .com/blog/222670.
53. Roger Thompson, "Business Copes with the Recession," *Nation's Business*, January 1991, p. 20.
54. Roger Thompson, "Business Copes with the Recession," *Nation's Business*, January 1991, p. 20.
55. Eric Spitznagel, "Rise of the Barter Economy," *Bloomberg Business Week*, April 26, 2012, www.businessweek.com/articles/2012-04-26/ rise-of-the-barter-economy.
56. Justin Martin, "Fair Trade," *FSB*, June 2009, pp. 76–79; "Entrepreneurs Say Managing Through Recession Makes Them Sharper Business Owners," *American Express OPEN Small Business Monitor*, April 15, 2009, http://ir.americanexpress.com/Mobile/file. aspx?IID=102700&FID=7648975.
57. Richard J. Maturi, "Collection Dues and Don'ts," *Entrepreneur*, January 1992, p. 328.
58. Barbara Taylor, "How I Learned to Love Bartering," *New York Times*, June 21, 2011, http://boss.blogs.nytimes.com/2011/06/21/how-i-learned-to-love-bartering.
59. "12 Creative Money-Saving Tactics for Small-Business Owners," *Entrepreneur*, December 21, 2012, www.entrepreneur.com/ slideshow/224914#8.
60. *The 2010 Federal Reserve Payments Study*, Federal Reserve System, 2010, pp. 13–24.
61. Diana Ransom, "Innovative Ways to Reel in Cash," *Wall Street Journal*, June 26, 2009, http://online.wsj.com/article/SB124603061762161137.html.
62. Jill Andresky Fraser, "Better Cash Management," *Inc.*, May 1993, p. 42.
63. C. J. Prince, "Money to Burn?," *Entrepreneur*, July 2004, pp. 51–52.
64. *2012 Report to the Nation on Occupational Fraud and Abuse*, Association of Certified Fraud Examiners, Austin, Texas, 2012, pp. 4–5, 26–27.
65. Robert A. Mamis, "Money In, Money Out," *Inc.*, March 1993, p. 103.

Chapter 16

1. Jennifer Wang, "Dearly Beloved, Please Send Cash," *Entrepreneur*, December 2009, p. 76.
2. Elizabeth Holmes, "Show Me the Money—Maybe," *Wall Street Journal*, June 25, 2007, p. R6.
3. Mark Brandau, "Protein Bar," *Nations Restaurant News*, June 27, 2011, www.theproteinbar.com/ftpprotein/ProteinBar_6.27.11_ NationsRestaurantNews.pdf; Thomas Heath, "Capital Buzz: A New Source of Protein for Washington," *Washington Post*, May 13, 2012, www .washingtonpost.com/business/capitalbusiness/capital-buzz-this-new-business-is-built-on-a-diet-of-protein/2012/05/11/gIQAfnPvMU_story.html.
4. Mark Henricks, "The Money Market," *Entrepreneur*, July 2006, p. 72.
5. Sven Wehrwein, "Wipeout," *Twin Cities Business*, September 2010, http://tcbmag.com/Industries/Manufacturing/Wipeout; "Excelsior-Henderson Down to $1 Million," *Minneapolis St. Paul Business Journal*, November 24, 1999, www.bizjournals.com/twincities/stories/1999/11/22/ daily6.html; "Excelsior-Henderson: Motorcycle Dream Fades in Bankruptcy," *Minneapolis StarTribune*, December 23, 1999, www .minneapolisfed.org/publications_papers/pub_display.cfm?id=2390.
6. "Frequently Asked Questions About Small Business Finance," *SBA Office of Advocacy*, September 2011, www.sba.gov/sites/default/files/ files/Finance%20FAQ%208-25-11%20FINAL%20for%20web.pdf.
7. Darren Dahl, "The Cost of Starting Up a Retail Shop," *Inc.*, August 8, 2011, www.inc.com/articles/201108/business-start-up-costs-retail-store .html.
8. Jeffrey R. Cornwall, David O. Vang, and Jean M. Hartman, *Entrepreneurial Financial Management* (Armonk, NY: M. E. Sharpe, 2009), pp. 165–167.
9. Rosalind Resnick, "Hanky Panky: Building a Business on Empowering Women," *Entrepreneur*, September 21, 2011, www.entrepreneur.com/ article/220270; Daria Meoli, "Measure Twice, Cut Once—Interview with Lida Orzeck, CEO of Lingerie Company Hanky Panky," *New York Enterprise Report*, January 22, 2010, http://www.nyreport.com/ cover_stories/articles/72819/measure_twice_cut_once_interview_with_ lida_orzeck_ceo_of_lingerie_compa.
10. Leigh Buchanan, "How to Start an Aerospace Company," *Inc.*, July 1, 2009, www.inc.com/magazine/20090701/how-to-start-an-aerospace-company.html.
11. Leigh Buchanan, "How to Start an Adventure Travel Company," *Inc.*, July 1, 2009, www.inc.com/magazine/20090701/how-to-start-an-adventure-travel-company.html.
12. Leigh Buchanan, "How to Start a T-Shirt Company," *Inc.*, July 1, 2009, www.inc.com/magazine/20090701/how-to-start-a-t-shirt-company.html.
13. Max Chafkin, "Case Study #1: The Reluctant Entrepreneur," *Inc.*, July 1, 2007, www.inc.com/magazine/20070701/features-start-up-reluctant-entrepreneur.html; Nanda Home, Inc., www.nandahome.com/ story/index.php; Abigail Pesta, "I Am the Boss of Me," *Marie Claire*, September 12, 2008, www.marieclaire.com/career-money/jobs/female-business-owners-entrepreneurs-4.
14. Frequently Asked Questions About Small Business Finance," *SBA Office of Advocacy*, September 2011, www.sba.gov/sites/default/files/ files/Finance%20FAQ%208-25-11%20FINAL%20for%20web.pdf;
15. "Frequently Asked Questions About Small Business Finance," *SBA Office of Advocacy*, September 2011, www.sba.gov/sites/default/files/ files/Finance%20FAQ%208-25-11%20FINAL%20for%20web.pdf.

16. Paul Kvinta, "Frogskins, Shekels, Bucks, Moolah, Cash, Simoleans, Dough, Dinero: Everybody Wants It. Your Business Needs It. Here's How to Get It," *Smart Business*, August 2000, pp. 74–89.

17. Dylan Love, "Meet the Man Behind Pebble: The Smart Watch You Didn't Even Know You Wanted," *Business Insider*, May 7, 2012, www.businessinsider.com/eric-migicovsky-pebble-2012-5; John McDermott, "Pebble 'Smartwatch' Funding Soars on Kickstarter," *Inc.*, April 20, 2012, www.inc.com/john-mcdermott/pebble-smartwatch-funding-sets-kickstarter-record.html; Kickstarter, www.kickstarter.com/projects/597507018/pebble-e-paper-watch-for-iphone-and-android.

18. Daniel Winterfeldt and Carole Kilgore, "United States: US Securities Law Update: Jumpstart Our Business Startups Act 2012," *Mondaq.com*, May 8, 2012, www.mondaq.com/unitedstates/article.asp?articleid=176256&login=true&nogo=1.

19. Catherine Clifford, "3 Rules for Successful Crowdfunding," *Entrepreneur*, May 23, 2012, www.entrepreneur.com/blog/223608.

20. Christina DesMarais, "Accelerator vs. Incubator: What's the Difference?," *Inc.*, February 7, 2012, www.inc.com/christina-desmarais/difference-between-startup-accelerator-and-incubator.html.

21. "How Incubators Speed the Start-Up Process," *Inc.*, July 1, 2010, www.inc.com/magazine/20100701/more-startup-incubators.html.

22. Ali Kousari, "New Solutions to the Funding Dilemma of Technology Startups," *Technology Innovation Management Review*, June 2011, http://timreview.ca/article/449.

23. Chris Silva, "Startup DAIO Raises $400,000 in Funding," *Nashville Business Journal*, May 21, 2012, www.bizjournals.com/nashville/news/2012/05/21/startup-daio-raises-400000-in-funding.html; Brad McCarty, "The First 6 Startups from Nashville's Jumpstart Foundry Accelerator," *Next Web*, August 25, 2011, http://thenextweb.com/insider/2011/08/25/the-first-6-startups-from-nashvilles-jumpstart-foundry.

24. Missy Frederick, "Posh Brood Website Caters to Travel with Family in Mind," *Washington Business Journal*, March 30, 2012, www.bizjournals.com/washington/print-edition/2012/03/30/posh-brood-website-caters-to-travel.html; J. J. McCorvey, "Elevator Pitch: Can Poshbrood Earn $500,000?," *Inc.*, February 28, 2012, www.inc.com/magazine/201203/jj-mccorvey/elevator-pitch-can-poshbrood-earn-500000.html.

25. Jeffrey Sohl, "The Angel Investor Market in 2011: The Recovery Continues," Center for Venture Research, April 3, 2012.

26. Robert Wiltbank and Warren Boeker, "Returns to Angel Investors in Groups," Angel Capital Education Foundation, November 2007, http://www.kauffman.org/uploadedFiles/angel_groups_111207.pdf.

27. Susan L. Preston, *Angel Financing for Entrepreneurs: Early Stage Funding for Long-Term Success* (New York: John Wiley & Sons, 2007).

28. "Raising Funds," *Inc.*, November 2008, pp. 69–70.

29. Jeanne Lee, "Building Wealth," *FSB*, June 2006, p. 43.

30. Marco R. della Cava, "Boomers Roll with the Punches," *USA Today*, January 18, 2007, www.usatoday.com/life/lifestyle/2007-01-17-boomer-boxing_x.htm; "Elevator Pitch: Prime Time Boxing," *Inc.*, February 1, 2010, www.inc.com/magazine/20100201/elevator-pitch-prime-time-boxing.html.

31. Robert E. Wiltbank and Warren Boeker, "Angel Performance Project," Angel Capital Education Foundation, November 2007, http://www.kauffman.org/uploadedFiles/angel_groups_111207.pdf; Lori Wright, "Angel Investor Market Holds Steady in 2009 but Changes Seen in Types of Deals UNH Center for Venture Research Finds," University of New Hampshire, March 31, 2010, www.unh.edu/news/docs/2009angelanalysis.pdf.

32. Dee Power and Brian E. Hill, "Angel Power - An Angel Survey," *Angel Investor News*, http://angelinvestornews.com/ART_angelpower.htm.

33. Lori Wright, "Angel Investor Market Holds Steady in 2009 but Changes Seen in Types of Deals UNH Center for Venture Research Finds," University of New Hampshire, March 31, 2010, www.unh.edu/news/docs/2009angelanalysis.pdf.

34. Mark Henricks, "The Money Market," *Entrepreneur*, July 2006, p. 72.

35. Tomio Geron, "'Super Angels' Rise to Fore," *Wall Street Journal*, May 6, 2010, http://online.wsj.com/article/SB10001424052748704423504575212792672226992.html?KEYWORDS=Super+Angels+Rise+to+Fore&mg=com-wsj; Bambi Francisco Roizen, "How to Get Funded by Super Angel Mike Maples," *Vator News*, May 13, 2010, http://vator.tv/news/2010-05-13-how-to-get-funded-by-super-angel-mike-maples.

36. "FAQ: The Value of Angel Investors and Angel Groups," Angel Capital Association, 2009, p. 1.

37. "ACA Member Landscape: 2009 Different Than 2008 (and Not)," Angel Capital Association, April 16, 2009, pp. 2, 3, and 7.

38. Annie Chang, "Golden Seeds' Managing Director Stephanie Hanbury-Brown on Women Entrepreneurs," *Huffington Post*, August 10, 2011, www.huffingtonpost.com/angie-chang/golden-seeds-women-entrepreneurs_b_923256.html; Alexis Leondis, "Wall Street Women of Golden Seeds Give Cash to Female CEOs," *Bloomberg*, April 15, 2011, http://www.bloomberg.com/news/2011-04-15/wall-street-women-of-golden-seeds-use-angel-cash-to-cultivate-female-ceos.html.

39. Bruce J. Blechman, "Step Right Up," *Entrepreneur*, June 1993, pp. 20–25.

40. Robert Wiltbank and Warren Boeker, "Returns to Angel Investors in Groups," Angel Capital Education Foundation, November 2007, http://www.kauffman.org/uploadedFiles/angel_groups_111207.pdf.

41. "Corporate Venture Capital Activity on Three-Year Upward Trend," report issued by PriceWaterhouseCooper, February 21, 2012, www.pwc.com/us/en/press-releases/2012/q4-2011-corp-vc-press-release.jhtml.

42. Quentin Hardy, "Cisco Announces Its $850 Million Spin-In," *New York Times*, April 19, 2012, http://bits.blogs.nytimes.com/2012/04/19/cisco-announces-its-850-million-spin-in; Timothy Hay, "What Cisco's 'Tribes' Look for In Start-Ups," *Wall Street Journal*, October 15, 2009, http://blogs.wsj.com/venturecapital/2009/10/15/what-ciscos-tribes-look-for-in-start-ups.

43. "Investments by Region," *PriceWaterhouseCoopers MoneyTree Survey*, Q1 2013, https://www.pwcmoneytree.com/MTPublic/ns/nav.jsp?page=region.

44. David Worrell, "School Ties," *Entrepreneur*, November 2006, pp. 88–90.

45. P. D. Reynolds, S. M. Camp, W. D. Bygrave, A. Autio, and M. Hay, *The Global Entrepreneurship Monitor 2003 Report* (Babson Park, MA, and London: Babson College and London Business School, 2003).

46. Gwen Moran, "The Voices of Venture Capital," *Entrepreneur*, June 21, 2010, www.entrepreneur.com/article/207186.

47. William D. Bygrave with Mark Quill, *Global Entrepreneurship Monitor: 2006 Financing Report*, Global Entrepreneurship Research Association, 2006, p. 23.

48. Dee Power and Brian E. Hill, "Venture Capital Survey," *The Capital Connection*, 2008, www.capital-connection.com/survey-close.html.

49. Sharon Kahn, "The Venture Game," *FSB*, May 2009, pp. 60–95.

50. William D. Bygrave with Mark Quill, *Global Entrepreneurship Monitor: 2006 Financing Report*, Global Entrepreneurship Research Association, 2006, p. 14; Cara Cannella, "Where Seed Money Really Comes From," *Inc.*, August 2003, p. 26.

51. "Investments by Stage of Development," *PriceWaterhouseCoopers MoneyTree Survey*, Q1 2013, https://www.pwcmoneytree.com/MTPublic/ns/nav.jsp?page=stage.; National Venture Capital Association, www.nvca.org.

52. Jason Del Ray, "The Ticker," *Inc.*, September 2008, p. 30.

53. Dan Bigman, "On the Hunt," *Forbes*, August 3, 2009, pp. 56–59; Michael J. Roberts and Noam T. Wasserman, "The Founding CEO's Dilemma: Stay or Go?," *Working Knowledge*, August 15, 2005, http://hbswk.hbs.edu/item/4948.html.

54. Dalia Fahmy, "Financing, with Strings Attached," *New York Times*, January 29, 2009, www.nytimes.com/2009/01/29/business/smallbusiness/29sbiz.html.

55. Gwen Moran, "The Voices of Venture Capital," *Entrepreneur*, June 21, 2010, www.entrepreneur.com/article/207186.

56. *Results from the 2010 Global Venture Capital Survey*, National Venture Capital Association and Deloitte, July 13, 2010, p. 14.

57. "N.H. Tech Company Started in an Attic on April Fools Day," *New England Sun Journal*, November 7, 2007, http://www.sunjournal.com/node/628192.

58. Mark Henricks, "The Money Market," *Entrepreneur*, July 2006, p. 72.

59. Gwen Moran, "Toward Safer, Greener Roads," *Entrepreneur*, May 2010, p. 65; "GreenRoad Company Overview," GreenRoad Technologies, www.greenroad.com/company_overview.html.

60. Dave Pell, "What's Old Is New Again," *FSB*, July/August 2000, p. 122.

61. "Saying Goodbye: New Exit Strategies for Today's Venture Capitalists," *Knowledge@Wharton*, March 3, 2010, http://knowledge.wharton.upenn.edu/article.cfm?articleid=2440.

62. David A. Fox, "PayMaxx Agrees to Merge with Larger Miami Company," *Nashville Post*, June 24, 2005, http://nashvillepost.com/news/2005/6/24/paymaxx_agrees_to_combine_with_larger_miami_company; Lynn Stephens and Robert C. Schwartz, "The Chilling Effects of Sarbanes-Oxley: Myth or Reality?," *CPA Journal*, June 2006, www.nysscpa.org/cpajournal/2006/606/infocus/p14.htm.

63. Ritter, "Initial Public Offerings: Updated Statistics," University of Florida, June 30, 2013, p. 16.

64. Jay Ritter, "Initial Public Offerings: Updated Statistics," University of Florida, June 30, 2013, p. 20.

65. Jay Ritter, "Initial Public Offerings: Median Age of IPOs Through 2012," University of Florida, January 4, 2013, p. 3; *Yearbook 2013*, Venture Capital Association, http://www.nvca.org/index.php?option=com_content&view=article&id=257&Itemid=103.

66. Stephanie Clifford, "How Fast Can This Thing Go, Anyway?," *Inc.*, March 1, 2008, www.inc.com/magazine/20080301/how-fast-can-this-thing-go-anyway.html; Trefis Team, "Zipcar: A Maturing Business Model That Holds Promise," *Forbes*, August 8, 2011, www.forbes.com/sites/greatspeculations/2011/08/09/zipcar-a-maturing-business-model-that-holds-promise; Kyle Alspach, "Zipcar IPO Raises $173M, Beating Expectations," *Boston Business Journal*, April 14, 2011, www.bizjournals.com/boston/news/2011/04/14/zipcar-ipo-raises-173m-beating.html.

Chapter 17

1. Victoria Williams, "Small Business Lending in the United States 2010–2011," U.S. Small Business Administration, Office of Advocacy, p. 4.

2. William C. Dunkelberg and Holly Wade, "NFIB Small Business Economic Trends," National Federation of Independent Business, March 2013.

3. "Small Business Borrowers Poll," New York Federal Reserve, www.newyorkfed.org/smallbusiness/2012/#dayinthelife.

4. Catherine Rampell, "Small Businesses Still Struggle, and That's Impeding a Recovery," *New York Times*, February 14, 2013, p. B1.

5. Cynthia E. Griffin, "Something Borrowed," *Entrepreneur*, February 1997, p. 26.

6. *The Small Business Credit Crunch and the Impact of the TARP*, Congressional Oversight Panel, May 2010, p. 13.

7. *2010 Small Business Economy: A Report to the President*, U.S. Small Business Administration, January 2011, p. 74.

8. Bonnie Kavoussi, "Small Businesses Struggling to Get Loans: Federal Reserve Study," *Huffington Post*, August 15, 2012; "Small Business Borrowers Poll," New York Federal Reserve, www.newyorkfed.org/smallbusiness/2012/#dayinthelife.

9. Conor Dougherty and Pu-Wing Tam, "Start-Ups Chase Cash as Funds Trickle Back," *Wall Street Journal*, April 1, 2010, p. B1.

10. *Small Business Lending in the United States 2012*, U.S. Small Business Administration, Office of Advocacy, July 2013, p. 6.

11. Biz2Credit Small Business Lending Index for May 2013 Reports a 70 Percent Increase in Big Bank Loan Approvals Over Past 12 Months," Biz2Credit, June 12, 2013, http://www.biz2credit.com/small-business-lending-index/may-2013.html.

12. Daniel M. Clark, "Banks and Bankability," *Venture*, September 1989, p. 29.

13. Rosalind Resnick, "Loan Woes," *Entrepreneur*, April 2007, p. 96.

14. William J. Dennis Jr., *Small Business Credit in a Deep Recession*, National Federation of Independent Businesses, February 2010, p. 1.

15. Janean Chun, "More Small Businesses Are Pulling Their Accounts Out of Big Banks," *Huffington Post*, May 22, 2012, www.huffingtonpost.com/2011/11/04/small-businesses-switching-from-big-banks-to-local-banks_n_1070547.html; Tyrone Beason, "Seattle's Dave Meinert is a Nightlife Entrepreneur and a Political Player," *Seattle Times*, August 6, 2010, www.seattletimes.com/html/pacificnw/2012302278_pacificpmeinert18.html.

16. "Thrifty Names President of Thrifty Car Sales; Announces Strategic Alliances with Bank of America, APCO, Manheim, and Others," Dollar Thrifty Automotive Group, February 8, 2005, www.dtag.com/phoenix.zhtml?c=71946&p=irol-newsArticle&ID=27723&highlight=.

17. Juan Hovey, "Want Easy Money? Look for Lenders Who Say Yes," *FSB*, November 2000, pp. 41–44.

18. Jeremy Quittner, "How to Get a Loan? Let Us Count the Ways," *Inc.*, February 4, 2013, www.inc.com/jeremy-quittner/alternatives-to-banking.html.

19. Carol Tice, "Can a Purchase Order Loan Keep Your Business Growing?," *Entrepreneur*, June 17, 2010, www.entrepreneur.com/article/207058#.

20. Rebel Cole, *Bank Credit, Trade Credit, or No Credit: Evidence from the Surveys of Small Business Finances*, U.S. Small Business Administration, Office of Advocacy, June 2010, p. 21.

21. Emily Maltby, "Vendors Can Help Financing," *Wall Street Journal*, February 18, 2010, p. B5; Ina Steiner, "eFashion Solutions Powers Brands on eBay, *Fashion Vault*, Amazon.com," Auction Bytes, July 8, 2010, www.auctionbytes.com/cab/abn/y10/m07/i08/s01.

22. *Small Business and Micro Business Lending in the United States 2005*, p. 6.

23. David Worrell, "The Other Colors of Money," *Entrepreneur*, July 2004, p. 67.

24. David Worrell, "The Other Colors of Money," *Entrepreneur*, July 2004, p. 67.

25. "P.C. Richard & Son." GE Capital, www.gecapital.com/en/our-customers/pc-richard.html.

26. John R. Walter, "Not Your Father's Credit Union," *Economic Quarterly* (Federal Reserve Bank of Richmond) 92, no. 4 (Fall 2006), pp. 353–377.

27. Donna Fuscaldo, "Out-of-the-Box Loan Options to Start a Business," *Fox Business,* April 23, 2013, http://smallbusiness.foxbusiness.com/finance-accounting/2013/04/23/out-box-loan-options-to-start-business/.

28. Michelle Samaad, "Credit Unions Issue Record $12 Billion in Business Loans in 2011: NAFCU," *Credit Union Times,* May 24, 2012, http://www.cutimes.com/2012/05/24/credit-unions-issue-record-12-billion-in-business.

29. Biz2Credit Small Business Lending Index for May 2013 Reports a 70 Percent Increase in Big Bank Loan Approvals Over Past 12 Months," Biz2Credit, June 12, 2013, http://www.biz2credit.com/small-business-lending-index/may-2013.html.

30. Adam Belz, "Credit Unions Growing Commercial Lending Business," *USA Today*, July 10, 2011, http://usatoday30.usatoday.com/money/industries/banking/2011-07-11-credit-unions-small-business_n.htm.

31. "NGK Overview," NGK Spark Plugs, www.ngksparkplugs.com/About_nGK/index.asp?mode=nml; *West Virginia Economic Development Agency Annual Report 2008*, West Virginia Economic Development Agency, p. 9; "Success Stories," West Virginia Department of Commerce, www.wvcommerce.org/business/successstories/default.aspx.

32. "Small Business Investment Company Annual Report FY2012," U.S. Small Business Administration Investment Division, www.sba.gov/sites/default/files/files/SBIC%20Program%20FY%202012%20Annual%20Report.pdf.

33. "History and Current Highlights," National Association of Small Business Investment Companies, www.nasbic.org/?page=SBIC_Program_History.

34. *The Small Business Investment Company Annual Report FY 2012*, U.S. Small Business Administration Investment Division, Washington, DC, p. 14.

35. "A Profile in Success: JSI Store Fixtures, Inc.," *The Small Business Investment Company Annual Report FY 2012*, U.S. Small Business Administration Investment Division, Washington, DC, p. 16.

36. Eric Boudinet, "Eric Boudinet Working the Earth: Morgan Man Hopes Farm Becomes Shining Example of Sustainability," *Knoxville News Sentinel*, April 18, 2012, www.knoxnews.com/news/2012/apr/18/working-the-earth-morgan-man-hopes-farm-becomes.

37. "Covenant Doors Takes Back Market Share from Foreign Competitors," Rocky Mountain Trade Adjustment Assistance Center, www.rmtaac.org/content/covenant-doors-takes-back-market-share-foreign-competitors.

38. "Section 108 Case Studies," U.S. Department of Housing and Urban Development, http://portal.hud.gov/hudportal/HUD?src=/program_offices/comm_planning/communitydevelopment/programs/108/casestudies.

39. Charles Wessner, "An Assessment of the SBIR Program," National Research Council, http://books.nap.edu/openbook.php?record_id=11989&page=12, pp. 91–107.

40. "Pathfinder Therapeutics," SBIR/STTR, www.sbir.gov/sbirsearch/detail/263815; "Pathfinder," National Institute of Health, http://grants.nih.gov/grants/funding/sbir_showcase/Pathfinder.pdf.

41. Emily Maltby, "A Credit Crunch That Lingers," *Wall Street Journal*, June 21, 2010, http://online.wsj.com/article/NA_WSJ_PUB:SB10001424052748704852004575257970246239874.html.

42. Diana Ransom, "Status Report: Small Business Lending," *Wall Street Journal*, September 23, 2009, http://online.wsj.com/article/NA_WSJ_PUB:SB125372460362634601.html.

43. *Summary of Performance and Financial Information 2012 Fiscal Year*, U.S. Small Business Administration, www.sba.gov/sites/default/files/files/FY%202012%20Summary%20of%20Performance%20and%20Financial%20Information.pdf.pdf.

44. Julie Monahan, "Quick Fix," *Entrepreneur*, April 2004, p. 27.

45. "Veteran-owed Business Receives Patriot Express Loan, Contracting Opportunities," U.S. Small Business Administration, www.sba.gov/about-offices-content/2/3159/success-stories/3987.

46. "Small Business Owner Receives 7(a) Loan for Start-up, Growth," U.S. Small Business Administration, www.sba.gov/content/small-business-owner-receives-7a-loan-moves-new-building.

47. Ronni Newton, "May Cookie Co. Is This Entrepreneur's Sweet Dream," *West Hartford Patch*, February 14, 2012, http://westhartford.patch.com/groups/business-news/p/may-cookie-co-is-this-entrepreneur-s-sweet-dream; Small Business Person of the Year 2013, U.S. Small Business Administration, www.sba.gov/about-offices-content/2/3105/success-stories.

48. "Service-disabled Veteran Small Business Owner Receives 7(a) Loan for Start-up Capital," U.S. Small Business Administration, www.sba.gov/about-offices-content/2/3159/success-stories/3959.

49. "On the Cutting Edge," CDC Small Business Finance, http://cdcloans.com/wp-content/uploads/2011/10/successstory_laser-tronics.pdf.

50. "Small Business Administration Microloan Program," Congressional Research Service, https://www.fas.org/sgp/crs/misc/R41057.pdf.

51. "Mitesh and Chetna's Story," Opportunity Fund, www.opportunityfund.org/index/success-stories-home/mitesh-and-chetna.

52. Gwendolyn Bounds, "Risky Businesses May Find Loans Even Scarcer," *Wall Street Journal*, April 13, 2004, p. B8.

53. Mae Anderson, "Sam's Club Will Offer Small Business Loans," *MSNBC*, July 6, 2010, www.msnbc.msn.com/id/38103657; Stephanie Clifford, "Retailers Devise Novel Ways to Revive Sales," *New York Times*, July 4, 2010, www.nytimes.com/2010/07/05/business/05loan.html.

54. "Missouri Company Increases Exports 70 Percent Due to Working Capital Funds," BusinessUSA, http://business.usa.gov/success_story/missouri-company-increases-exports-70-percent-due-working-capital-funds.

55. Adolfo Flores, "SBA Approves $2 Billion in Superstorm Sandy Disaster Loans," *Los Angeles Times*, April 5, 2013, http://articles.latimes.com/2013/apr/05/business/la-fi-mo-sba-hurricane-sandy-20130405.

56. Ziona Austrian and Zhongcai Zhang, "An Inventory and Assessment of Pollution Control and Prevention Financing Programs," Great Lakes Environmental Finance Center, Levin College of Urban Affairs, Cleveland State University, http://urban.csuohio.edu/research/pubs/abstracts/inventor.htm.

57. Sean P. Melvin, "Hidden Treasure," *Entrepreneur*, February 2002, pp. 56–58.

58. "Entrepreneurial Survival Stories," *Entrepreneur*, February 6, 2008, www.entrepreneur.com/slideshow/173762#6.

59. Ian Mount, "When Banks Won't Lend, There Are Alternatives, Though Often Expensive," *New York Times*, August 1, 2012, www.nytimes.com/2012/08/02/business/smallbusiness/for-small-businesses-bank-loan-alternatives.html.

60. Ian Mount, "When Banks Won't Lend, There Are Alternatives, Though Often Expensive," *New York Times*, August 1, 2012, www.nytimes.com/2012/08/02/business/smallbusiness/for-small-businesses-bank-loan-alternatives.html.

61. Robert Scott, *The Use of Credit Card Debt by New Firms*, The Kauffman Firm Survey, August 2009, pp. 2, 4.

Chapter 18

1. Betty Joyce Nash, "West Virginia Glass Houses," *Region Focus*, Winter 2009, pp. 43–46.

2. Mark Henricks, "Hot Spots," *Entrepreneur*, October 2005, pp. 68–74.

3. "E-Commerce/Online Sales Statistics," *Statistic Brain*, August 24, 2012, www.statisticbrain.com/total-online-sales.

4. Michael Hartnett, "Successful Site Selection," *Stores*, July 2012, www.stores.org/STORES%20Magazine%20July%202012/successful-site-selection; "Smoothie King Blends Up Healthy Lifestyles and Healthy Businesses," *QSR Magazine*, http://www.qsrmagazine.com/content/smoothie-king.

5. William Atkinson, "Middle of Somewhere," *Site Selection*, September 2012, pp. 134–137.

6. Paula Scommegna, "U.S. Megalopolises 50 Years Later," Population Reference Bureau, www.prb.org/Articles/2011/us-megalopolises-50-years.aspx.

7. "2011 Inc. 5000: Company Profile: Paylocity," *Inc.*, www.inc.com/inc5000/profile/paylocity; "Paylocity Talks About ZoomProspector and Fast Growth," *YouTube*, September 25, 2008, www.youtube.com/watch?v=QrbI3XmhoMc.

8. Daniel Brooks, "New IKEA Factory Hums," *Region Focus*, Fall 2009, p. 2.

9. John W. McCurry, "Moving Cheese," *Site Selection*, September 2012, pp. 197–198.

10. "Employer Costs for Employee Compensation for the Regions," September 2012, Bureau of Labor Statistics, www.bls.gov/ro5/ececmid.htm.

11. Ron Starner, "What Makes a Business Climate Good?," *Site Selection*, January 2013, pp. 14–19.

12. Ron Starner, "Back to School," *Site Selection*, November 2012, pp. 142–149.

13. Jacquelyn Lynn, "Location Is Key," *Entrepreneur*, November 2008, p. 26.

14. Peter Cohan, "6 Great Reasons to Relocate Your Start-Up," *Forbes*, November 28, 2012, www.forbes.com/sites/petercohan/2012/11/28/6-great-reasons-to-relocate-your-start-up; Hiawatha Bray, "Video Game Industry a Bright Spot in Mass.," *Boston Globe*, September 17, 2012, www.bostonglobe.com/business/2012/09/16/massachusetts-video-game-industry-growing-score/5FaQlI4Mdb5Te6Mo6ipwrJ/story.html; "Video Game Companies in Massachusetts," Boston.com, 2013, www.boston.com/jobs/galleries/video_game_companies.

15. Ron Starner, "What Makes a Business Climate Good?," *Site Selection*, January 2013, pp. 14–19.

16. Ralph Bartholdt, "Couple: High Fees Preventing Dream of Sandpoint Restaurant," *Bonner County Daily Bee*, August 5, 2010, http://bonnercountydailybee.com/articles/2010/08/05/news/doc4c5a57423fa76606867788.txt.

17. State Individual Income Taxes 2013, Taxadmin.org, http://taxadmin.org/fta/rate/ind_inc.pdf.

18. William La Jeunesse, "California Residents, Businesses Consider Bailing on Golden State over Taxes," *Fox News*, January 23, 2013, www.foxnews.com/politics/2013/01/23/california-residents-businesses-consider-bailing-on-golden-state-over-taxes; Adam Nagourney, "Two-Tax Rise Tests Wealthy in California," *New York Times*, February 6, 2013, www.nytimes.com/2013/02/07/us/millionaires-consider-leaving-california-over-taxes.html?pagewanted=1&_r=0.

19. *The Impact of Broadband Speed and Price on Small Business*, Columbia Telecommunications Corporation, U.S. Small Business Administration Office of Advocacy, November 2010, p. 2.

20. David Farkas, "Site Selection: Where Good Food Means the Most," *Chain Leader*, September 1, 2009, pp. 14–17.

21. Khalil AlHajal, "Detroit Has Half the Median Income, Three Times the Poverty Rate of Nation, New Census Number Show," *Michigan Live*, September 21, 2012, www.mlive.com/news/detroit/index.ssf/2012/09/detroit_has_half_the_median_in.html; Alex P. Kellogg, "Black Flight Hits Detroit," *Wall Street Journal*, June 5–6, 2010, pp. A1, A12; Kate Davidson, "Detroit Has Tons of Vacant Land. But 40 Square Miles?," *Michigan Radio*, April 18, 2012, www.michiganradio.org/post/detroit-has-tons-vacant-land-forty-square-miles.

22. Adam Jones-Kelley, "Utah: Alter Your Perception," *Site Selection*, March 2013, pp. 27–30; Kurt Badenhausen, "The Best for Business," *Forbes*, July 16, 2012, pp. 100–103; Tom Van Riper, "Best Cities for Raising a Family," *Forbes*, April 12, 2012, www.forbes.com/pictures/eddf45gihi/no-3-provo-utah; "10 Fastest Growing U.S. Cities," *CNNMoney*, 2012, http://money.cnn.com/galleries/2012/real_estate/1204/gallery.US-Cities/index.html.

23. "Megaregions," America2050, www.america2050.org/content/megaregions.html#more.

24. *Anytime Fitness Media Guide*, Anytime Fitness, 2013, p. 3.

25. "Clusters and Cluster Development," Institute for Strategy and Competitiveness, Harvard Business School, www.isc.hbs.edu/econ-clusters.htm.

26. "Cities Growing Apart," *Region Focus*, Second/Third Quarter 2012, p. 43.

27. Ron Starner, "Speed Racer," *Site Selection*, March 2013, pp. 106–113.

28. David Van Den Berg, "Cinemas Prosper During Recession," *Region Focus*, Winter 2009, p. 2.

29. "Small Business Owners Say Simpler Tax Codes, Licensing Requirements Would Help Them Succeed and Create Jobs, Kauffman Paper Shows," Kauffman Foundation, October 24, 2012, www.kauffman.org/newsroom/small-business-owners-say-simpler-tax-codes-licensing-requirements-would-help-them-succeed-and-create-jobs-kauffman-paper-shows.aspx.

30. "SF's Own Planning Department Satirizes Convoluted Regulations Deterring Small Businesses," *California City News*, February 2012, www.californiacitynews.org/2012/02/sf%E2%80%99s-own-planning-department-satirizes-convoluted-regulations-deterring-small-businesses.htm.

31. Julie V. Iovine, "Zoning Laws Grow Up," *Wall Street Journal*, January 19, 2012, p. D6.

32. Ben Szobody, "Obscure Panel Makes Key Bar Decisions," *Greenville News*, August 12, 2012, pp. 1A, 3A.

33. Catherine Saint Louis, "Sleepwalk to Work," *New York Times*, August 24, 2012, www.nytimes.com/2012/08/26/realestate/running-a-home-business-in-new-york.html?pagewanted=2&_r=3&adxnnl=1&adxnnlx=1363536734-K3rvukdIgzKBJb6aZs326A.

34. Ron Starner, "What Makes a Business Climate Good?," *Site Selection*, January 2013, p. 16.

35. Mark Arend, "Water Works," *Site Selection*, September 2012, pp. 182–185; Molly Ryan, "Niagara Bottling to Open Bottling Plant in Missouri City," *Houston Business Journal*, August 6, 2012, www.bizjournals.com/houston/news/2012/08/06/niagara-bottling-to-open-bottling.html.

36. Rudolph Bell, "Return to Downtown," *Greenville News*, May 13, 2012, pp. 1A, 6A.

37. "Worker Relocation Worries," *Inside Training Newsletter*, November 29, 2007, p. 1.

38. Mark Brandau, "Meatheads Burgers and Fries Heads to the City," *Nation's Restaurant News*, August 22, 2012, http://nrn.com/latest-headlines/meatheads-burgers-fries-heads-city.

39. Kris Hudson and Dana Mattioli, "Fifth Avenue's Eye-Popping Rents," *Wall Street Journal*, November 21, 2012, pp. C1, C10.

40. Andrew Moore, "The Clock on Wade Hampton to Close After Six Decades," *Greer-TaylorsPatch*, March 14, 2013, http://taylors.patch.com/articles/the-clock-on-wade-hampton-to-close-after-six-decades; Aaron Barker and Casey Vaughn, "After Nearly 60 Years, Greenville Landmark to Close," *Fox Carolina*, March 15, 2013, www.foxcarolina.com/story/21635630/after-nearly-60-years-greenville-landmark-to-close.

41. Ron Ruggless, "BK Debuts Whopper Bar Concept, Eyes On-Site Arena in Plan to Beef Up Growth," *Nation's Restaurant News*, http://nrn.com/landingPage.aspx?menu_id=1424&coll_id=676&id=364528; "BK to Debut Whopper Bar Next Year," *Nation's Restaurant News*, October 7, 2008, http://nrn.com/breakingNews.aspx?id=359160.

42. "Freshëns Yogurt Debuts Fresh Kiosk, Items, and Branding," *QSR Magazine*, August 1, 2012, www.qsrmagazine.com/news/fresh-ns-debuts-fresh-kiosk-items-and-branding.

43. Shannon Perez, "6 Tips to Finding the Perfect Location for Your Practice," Massage Therapy, www.massagetherapy.com/articles/index.php/article_id/1377.

44. Matt Rosenberg, "About Reilly's Law of Retail Gravitation," About.com, http://geography.about.com/cs/citiesurbangeo/a/aa041403a.htm; G. I. Thrall and J. C. del Valle, "The Calculation of Retail Market Areas: The Reilly Model," *GeoInfoSystems* 7, no. 4 (1997), pp. 46–49.

45. Kris Hudson, "For Malls, Occupancy Firms Up," *Wall Street Journal*, January 9, 2012, http://online.wsj.com/article/SB10001424052970203436904577148813815182788.html.

46. Michelle Davey, "Downtown Retail," *Columbus CEO*, July 18, 2012, www.columbusceo.com/industry_news/commercial_real_estate/article_d98b2f1e-d116-11e1-92eb-001a4bcf6878.html.

47. David A. Kaplan, "Shake Shack's New Adventure," *Fortune*, November 7, 2011, p. 45; Keith Loria, "Breakout Brands: Shake Shack," *Nation's Restaurant News*, January 28, 2013, http://nrn.com/nrn-50/breakout-brands-shake-shack.

48. Paul Lukas, "Our Malls, Ourselves," *Fortune*, October 18, 2004, pp. 243–256.

49. "Shopping Center Facts and Stats," International Council of Shopping Centers, www.icsc.org/research/stats.php.

50. "What's the Direction of Retail Development—Malls or Lifestyle Centers?," SS&C, 2013, www.ssctech.com/eBriefings/eBriefingArticle/tabid/597/Default.aspx?V=8&A=578.

51. Tony Tagliavia, "Ground Broken on New 'Lifestyle' Mall," *Wood Television*, October 6, 2009, www.woodtv.com/dpp/news/local/kent_county/knapp_crossing_groundbreaking.

52. "Randyl Drummer, "Can This Mall Be Saved? Elements Needed for a Turnaround Include Lower Debt, Deep Pockets," *CoStar*, October 10, 2012, www.costar.com/News/Article/Can-This-Mall-Be-Saved-Elements-Needed-for-a-Turnaround-Include-Lower-Debt-Deep-Pockets/142143.

53. Emily Badger, "The Shopping Mall Turns 60 (and Prepares to Retire)," *The Atlantic Cities*, July 13, 2012, www.theatlanticcities.com/arts-and-lifestyle/2012/07/shopping-mall-turns-60-and-prepares-retire/2568.

54. "Industry Fun Facts," International Council of Shopping Centers, www.icsc.org/srch/about/impactofshoppingcenters/Did_You_Know.pdf; "About Mall of America," Mall of America, www.mallofamerica.com/about_moa_faqs.aspx.

55. Jacquelyn Gutc, "Summer Brings Life to Restaurant Row," *Worcester Business Journal*, June 25, 2012, www.wbjournal.com/article/20120625/PRINTEDITION/306229987/summer-brings-life-to-%E2%80%98restaurant-row%E2%80%99.

56. Anne Moore, "Share the Wealth," *Entrepreneur*, May 12, 2012, pp. 70–72.

57. Mark Brandau, "Restaurant Chains to Drive Growth Through Nontraditional Locations," *Nation's Restaurant News*, January 2, 2013, http://nrn.com/latest-headlines/restaurant-chains-drive-growth-through-nontraditional-locations.

58. "Subway Non-Traditional Development Hits 8,000 Store Milestone," *PR Newswire*, July 13, 2011, www.prnewswire.com/news-releases/subway-non-traditional-development-hits-8000-store-milestone-125503313.html.

59. "Frequently Asked Questions," Small Business Administration, Office of Advocacy, September 2012, p. 2; Ying Lowrey, *Start-Up Business Characteristics and Dynamics: A Data Analysis of the Kauffman Firm Survey*, Small Business Administration Office of Advocacy, August 2009, p. 20.

60. Ron Starner, "The Sweet Spot of Global Trade," *Site Selection*, November 2012, pp. 106–109.

61. Ron Starner, "Room to Grow," *Site Selection*, November 2012, pp. 96–99.

62. *72nd Annual Report of the Foreign Trade Zones Board to the Congress of the United States*, Foreign Trade Zone Board, December 2011, pp. 9, 12.

63. Elizabeth Whiteman, Marc Chittum, and Michael Masserman, "Foreign Trade Zones Record Increase in Shipments," International Trade Administration, http://trade.gov/press/publications/newsletters/ita_0110/shorttakes_0110.asp.

64. "Business Incubation FAQ," National Business Incubation Association, www.nbia.org/resource_center/bus_inc_facts/index.php.

65. James Haggerty, "10 Years Later, Incubator Shares Success Stories," *Scranton Times Tribune*, December 23, 2012, http://thetimes-tribune.com/news/business/10-years-later-incubator-shares-success-stories-1.1420163.

66. John B. Vinturella, "Business Incubators," Entrepreneurship University, www.jbv.com/lessons-CES/incubators.php.

67. "New Incubator Focuses on Female Founders," *Inc.*, March 2012, p. 28.

68. What Is Business Incubation?," National Business Incubation Association, March 31, 2006, www.nbia.org/resource_center/what_is/index.php.

69. *2012 State of the Business Incubation Industry*, National Business Incubation Association, www.nbia.org/resource_library/faq/index.php.

70. *The Gensler Design + Performance Index: The U.S. Workplace Study*, Gensler Inc., October 22, 2008, p. 12.

71. "Workplace Design = Job Performance?," *Inside Training*, October 29, 2008, p. 1.

72. *The Integrated Workplace* (Washington, DC: Office of Governmentwide Policy, Office of Real Property, 2008), pp. 8–9.

73. Christopher Hann, "Space, the Final Frontier," *Entrepreneur*, October 2012, p. 28.

74. Andrew Shafer, "Data Bank: Crunching the Numbers," *Inc.*, May 2012, p. 24.

75. Melanie Turner, "Open Office Layout Can Boost Creativity, Cut Costs," *Sacramento Business Journal*, May 4, 2012, www.bizjournals.com/sacramento/print-edition/2012/05/04/open-office-layout-boost-creativity.html?page=all.

76. Laura Tiffany, "The Rules of . . . Retailing," *Business Start-Ups*, December 1999, p. 106; Paul Keegan, "The Architect of Happy Customers," *Business 2.0*, August 2002, pp. 85–87.

77. Sam Oches, "2012 QSR Drive-Thru Study," *QSR Magazine*, October 2012, www.qsrmagazine.com/reports/2012-qsr-drive-thru-study.

78. Erin Dostal, "Huddle House Redesign Spurs Sales," *Nation's Restaurant News*, March 13, 2013, http://nrn.com/latest-headlines/huddle-house-redesign-spurs-sales.

79. Laura Tiffany, "The Rules of . . . Retailing," *Business Start-Ups*, December 1999, p. 106.

80. Matthew W. Brault, "Americans with Disabilities 2010," U.S. Census Bureau, July 2012, p. 4.

81. "Educational Kit," President's Committee on Employment of People with Disabilities, www50.pcepd.gov/pcepd/archives/pubs/ek99/wholedoc.htm#decisions.

82. Laura Tiffany, "The Rules of . . . Retailing," *Business Start-Ups*, December 1999, p. 106; Paul Keegan, "The Architect of Happy Customers," *Business 2.0*, August 2002, pp. 85–87.

83. Brian Amble, "Poor Workplace Design Damages Productivity," *Management-Issues*, May 23, 2006, www.management-issues.com/2006/8/24/research/poor-workplace-design-damages-productivity.asp.

84. *Nonfatal Occupational Injuries and Illnesses Requiring Days Away from Work, 2011*, Bureau of Labor Statistics, November 8, 2012, p. 6.

85. Tiffany Hsu, "Fast Food: Drive-Thru Patronage Up 2 Percent in 2011," *Colorado Springs Business Gazette*, May 30, 2012, www.gazette.com/articles/food-139469-patronage-percent.html.

86. Sarah Pruitt, "Ancient 'Fast-Food' Window Discovered," *History*, November 4, 2011, www.history.com/news/ancient-fast-food-window-discovered.

87. Lisa Jennings, "Seattle's Best Coffee to Add Drive-Thru-Only Units," *Nation's Restaurant News*, March 18, 2013, http://nrn.com/latest-headlines/seattles-best-coffee-add-drive-thru-only-units.

88. Amanda MacArthur, "Retail Atmospherics: Can You Really Influence Customer Buying Habits?," *Swipely*, September 19, 2011, http://blog.swipelyworks.com/restaurant-store-atmospherics/retail-atmospherics-can-you-really-influence-customer-buying-habits.

89. Eric Markowitz, "How Cinnamon Smells Will Save Holiday Sales," *Inc.*, November 3, 2011, www.inc.com/articles/201111/how-cinnamon-smells-will-save-holiday-sales.html.

90. Ned Smith, "Consumers Will Pay More for Products They Can Touch," *Business News Daily*, September 13, 2010, www.businessnewsdaily.com/203-consumers-will-pay-more-for-products-they-can-touch.html.

91. "Paco Underhill: Shopping Scientist," *CBC News*, November 7, 2000, www.cbc.ca/consumers/market/files/home/shopping/index.html.

92. Amanda MacArthur, "Retail Atmospherics: Can You Really Influence Customer Buying Habits?," *Swipely*, September 19, 2011, http://blog.swipelyworks.com/restaurant-store-atmospherics/retail-atmospherics-can-you-really-influence-customer-buying-habits.

93. Martin Lindstrom, "How Whole Foods 'Primes' You to Shop," *Fast Company*, September15, 2011, www.fastcompany.com/1779611/priming-whole-foods-derren-brown.

94. Alex Athanassoulas, "The Smell of Success: How Scent Can Boost Your Sales and Brand Value," Strixis Group, May 12, 2012, www.athanassoulas.com/?p=355; *Dollars and Sense: The Impact of Multi-Sensory Marketing*, 4Imprint, Oshkosh, Wisconsin,2009, p. 3.

95. Eric Sorensen, "WSU Researchers Tie Simple Scent to Increased Retail Sales," *WSU News*, November 26, 2012, http://news.wsu.edu/Pages/Publications.asp?Action=Detail&PublicationID=34039&PageID.

96. Andrew M. Goldsmith, "When Scents Make Cents," *Dental Economics*, 2013, www.dentaleconomics.com/articles/print/volume-97/issue-2/features/when-scents-make-cents.html.

97. Suzanne Hoppough, "What's That Smell?," *Forbes*, October 2, 2006, p. 76.

98. Alex Athanassoulas, "The Smell of Success: How Scent Can Boost Your Sales and Brand Value," Strixis Group, May 12, 2012, www.athanassoulas.com/?p=355.

99. James Vlahos, "Scent and Sensibility," *New York Times*, September 9, 2007, http://query.nytimes.com/gst/fullpage.html?res=9D07EFDC1E3AF93AA3575AC0A9619C8B63&sec=&spon=&pagewanted=2.

100. "How Can In-Store Music Increase Sales?," *Tune One*, February 19, 2013, www.tune-one.com/tune-one-news/1186.

101. Diane Toroian, "Eat to the Beat: Restaurants Cook Up the Right Recipe for Music Playlists," *St. Louis Post-Dispatch*, January 10, 2013, www.stltoday.com/entertainment/dining/restaurants/eat-to-the-beat-restaurants-cook-up-the-right-recipe/article_95037d45-7577-5836-bf8e-d4736ad8e99a.html.

102. *Dollars and Sense: The Impact of Multi-Sensory Marketing*, 4Imprint, Oshkosh, Wisconsin, 2009, p. 3.

103. Diane Toroian, "Eat to the Beat: Restaurants Cook Up the Right Recipe for Music Playlists," *St. Louis Post-Dispatch*, January 10, 2013, www.stltoday.com/entertainment/dining/restaurants/eat-to-the-beat-restaurants-cook-up-the-right-recipe/article_95037d45-7577-5836-bf8e-d4736ad8e99a.html; Michele Kayal, "The Sounds of Great Food: Restaurant Music Playlists," *Greenville News*, June 3, 2012, p. 3E.

104. Colleen Bazdarich, "In the Buying Mood? It's the Muzak," *Business 2.0*, March 2002, p. 100.

105. Nadine Heintz, "Play Bach, Boost Sales," *Inc.*, January 2004, p. 23.

106. Theunis Bates, "Volume Control, Time.com, August 2, 2007, www.time.com/time/printout/0,8816,1649304,00.html.

107. "The Five Guys Mistake," *QSR Magazine*, August 2010, www.qsrmagazine.com/articles/second_location/144/fiveguys-1.phtml.

108. "Buildings Overview," Center for Climate and Energy Solutions, 2013, www.c2es.org/technology/overview/buildings.

109. "LED Light Bulbs: Comparison Charts," Eartheasy, 2013, http://eartheasy.com/live_led_bulbs_comparison.html.

110. Ted Reguly, "San Diego Restaurant Hodad's Sees the Light," *Restaurant Hospitality*, August 7, 2012, http://restaurant-hospitality.com/operations/san-diego-restaurant-hodads-sees-light.

111. Jennifer Alsever, "Showing Products in a Better Light," *Business 2.0*, September 2005, p. 62.

112. Josh Gould, "Sustainable Workplace Design Creates Innovation Opportunities," *Buildings*, July 2009, www.buildings.com/tabid/3413/ArticleID/8617/Default.aspx.

113. Josh Gould, "Sustainable Workplace Design Creates Innovation Opportunities," *Buildings*, July 2009, www.buildings.com/tabid/3413/ArticleID/8617/Default.aspx.

114. Julie Gerstein, "11 Surprising LEED-Certified Restaurants," The Daily Green, 2013, www.thedailygreen.com/environmental-news/latest/leed-certified-restaurants; "Gurnee Mills—LEED Highlights," Chipotle, 2013, pp. 1–2; "Chipotle's Vision for Sustainability," U.S. Green Building Council, 2010, pp. 1–2.

115. Paul Keegan, "The Architect of Happy Customers," *Business 2.0*, August 2002, pp. 85–87.

116. Russell Boniface, "I Spy a Shopper," *AIArchitect*, June 23, 2006, www.aia.org/aiarchitect/thisweek06/0623/0623paco.cfm.

117. "Survey: 43 Percent of Shoppers Rely on Lists," *Retail Shopping Experience*, August 19, 2010, www.retailcustomerexperience.com/article/139799/Survey-43-percent-of-customers-rely-on-shopping-lists.

118. Annette Elton, "I'll Take That Too: Increasing Impulse Buys," *Gift Shop*, Spring 2008, www.giftshopmag.com/2008/spring/unique_giftware/increasing_impulse_buys.

119. "Survey: 43 Percent of Shoppers Rely on Lists," *Retail Shopping Experience*, August 19, 2010, www.retailcustomerexperience.com/article/139799/Survey-43-percent-of-customers-rely-on-shopping-lists.

120. Jeffrey A. Trachtenberg, "How a Children's Book Got a Christmas Break," *Wall Street Journal*, December 5, 2005, pp. B1, B5.

121. "Reference Question of the Week," *Swiss Army Librarian*, March 10, 2012, www.swissarmylibrarian.net/2012/03/10/reference-question-of-the-week-3412.

122. Paul Keegan, "The Architect of Happy Customers," *Business 2.0*, August 2002, pp. 85–87; Kenneth Labich, "This Man Is Watching You," *Fortune*, July 19, 1999, pp. 131–134.

123. Elizabeth Holmes and Day A. Smith, "Why Are Fitting Rooms So Awful?," *Wall Street Journal*, April 6, 2011, pp. D1–D2.

124. Kris Hudson and Ann Zimmerman, "Big Boxes Aim to Speed Up Shopping," *Wall Street Journal*, June 27, 2007, pp. B1, B8; Tom Ryan, "Checkout Time Limit Around Four Minutes," *Retail Wire*, July 8, 2008, http://www.retailwire.com/discussion/13077/checkout-time-limit-around-four-minutes.

125. Rachel Tobin Ramos, "Old Navy Store Go Under the Knife," *Atlanta Journal Constitution*, March 17, 2010, www.ajc.com/business/old-navy-stores-go-378109.html.

126. Russell Boniface, "I Spy a Shopper," *AIArchitect*, June 23, 2006, www.aia.org/aiarchitect/thisweek06/0623/0623paco.cfm.

127. John W. McCurry, "Economic Juggernaut," *Site Selection*, July 2012, www.siteselection.com/issues/2012/may/south-carolina.cfm; "BMW Manufacturing: Racing Ahead," *Industry Today* 13 no. 3 (2013), http://industrytoday.com/article_view.asp?ArticleID=2677.

Chapter 19

1. Gina Paglucia Morrison and Anca van Assendelft, *Charting a New Course: The Retail Merchandising-Supply Network*, IBM, November 1, 2006, p. 1.

2. Ian Mount and Brian Caulfield, "The Missing Link," *eCompany*, May 2001, p. 84.

3. *Global Supply Chain Survey 2013: Next Generation Supply Chains, Efficient, Fast, and Tailored*, PriceWaterhouseCoopers, 2013, pp. 4, 8.

4. Arie Y. Lewin, *Global Sourcing of Business Services: Key Findings and Trends from ORN Research*, 2012 Outsourcing World Summit, February 20–22, 2012, p. 6.

5. *Risk Ready: New Approaches to Environmental and Social Change*, PriceWaterhouseCoopers, November 2012, p. 2.

6. *Low-Cost Country Sourcing Can Benefit a Company's Bottom Line*, IBM, 2006, p. 6.

7. Mark Brandau, "Yum China Overhauls Supply Chain," *Nation's Restaurant News*, February 25, 2013, http://nrn.com/latest-headlines/yum-china-overhauls-supply-chain-marketing; "Yum Profits Hit as China Renews Consumer Protection Push," *Bloomberg*, February 5, 2013, www.bloomberg.com/news/2013-02-05/yum-profits-hit-as-china-strengthens-consumer-protection-push.html.

8. *New Models for Addressing Supply Chain and Transport Management Risk*, World Economic Forum, 2012, p. 7.

9. Rudolph Bell, "Volcano Disrupts BMW Supply Chain to S.C," *The State*, April 20, 2010, www.thestate.com/2010/04/20/1251405/volcano-disrupts-bmw-supply-chain.html; "Iceland Volcano: Nissan and BMW Suspend Some Production," *BBC News*, April 20, 2010, http://news.bbc.co.uk/2/hi/8631676.stm.

10. Bob Heaney, "Supply Chain Visibility in Consumer Markets," Aberdeen Group, March 2013, p. 2.

11. Robert J. Bowman, "A Bit Too Much Spice in McCormick's Supply Chain," *Supply Chain Brain*, October 8, 2012, www.supplychainbrain.com/content/featured-content/single-article/article/a-bit-too-much-spice-in-mccormicks-supply-chain; Randy Myers, "The Spice Trade," *CFO*, June 2007, p. 74; Bradley Z. Hull, "Frankincense, Myrrh, and Spices," *Journal of Macromarketing* 28, no. 3 (September 2008), p. 275.

12. "Vogel Wood Slices Through Lead Times with Lean Transformation," Wisconsin Manufacturing Extension Partnership, www.wmep.org/SuccessStories/vogel.aspx.

13. Joelle Dick, Caroline Kvitka, Aaron Lazenby, and Rich Schwerin, "Four Keys to Lean Six Sigma," *Profit*, November 2004, p. 9.

14. Sheila R. Poling, "Case Study: Lean Six Sigma Delivers for Pelco Products, Inc.," *Quality Magazine*, May 4, 2012, www.qualitymag.com/articles/88534-case-study--lean-six-sigma-delivers-for-pelco-products--inc-.

15. Jusko, "Scotsman Ice Systems: IW Best Plants Profile 2006," *Industry Week*, October 1, 2006, www.industryweek.com/ReadArticle.aspx?ArticleID=12682.

16. *Managing Disruptions*, Allianz Global Corporate & Specialty, November 2012, p. 4.

17. Jeff Bennett and Jan Hromadko, "'Nylon-12' Haunts Car Makers," *Wall Street Journal*, April 18, 2012, p. B1.

18. "Supply Chain Interruptions Getting More Expensive: Survey," *Canadian Underwriter*, February 14, 2013, www.canadianunderwriter.ca/news/supply-chain-interruptions-getting-more-expensive-survey/1002073177.

19. Julie Jargon, "Eateries' New Way to Shop," *Wall Street Journal*, April 1, 2011, p. B5.

20. "Co-Op Research/Economic Impact," National Cooperative Business Association, www.ncba.coop/ncba/about-co-ops/research-economic-impact.

21. "About Us," EATFLEET, 2013, www.eat-fleet.com/aboutus.htm.

22. Crystal Detamore-Rodman, "Cash In, Cash Out," *Entrepreneur*, June 2003, www.entrepreneur.com/magazine/entrepreneur/2003/june/61916.html.

23. *2013 Third Party Logistics Study: The State of Logistics Outsourcing*, Capgemini Consulting, 2013, p. 9.

24. Alan J. Liddle, "New York's Papaya King on Papaya Supply Difficulties," *Nation's Restaurant News*, August 29, 2011, http://nrn.com/latest-headlines/new-yorks-papaya-king-papaya-supply-difficulties.

25. Julie Jargon, "Eateries' New Way to Shop," *Wall Street Journal*, April 11, 2011, p. B5.

26. Connie Winkler, "Where Does the Money Go?," *CFO-IT*, Spring 2005, pp. 45–49.

27. Bob Evans, "Supply Chains Hit Home (Sweet Home)," *Information Week*, February 7, 2007, p. 68.

28. Brian Nadel, "Show Me the Money," Special Advertising Feature in *Fortune*, April 3, 2006, pp. S1–S5.

29. Ian Mount and Brian Caulfield, "The Internet-Based Supply Chain," *ecompany*, May 2001, p. 85.

30. Amy Dusto, "New Supply Chain Software Is Designed for Small Business," *Internet Retailer*, August 17, 2012, www.internetretailer.com/mobile/2012/08/17/new-supply-chain-software-designed-small-business.

31. Hau Lee, "The Three A's of Supply Chain Excellence," *Electronics Supply and Manufacturing*, October 1, 2004, www.my-esm.com/showArticle.jhtml?articleID=47903369; Tracy Mayor, "The Supple Supply Chain," *CIO*, June 13, 2007, www.cio.com/article/119301.

32. David Rosenbaum, "Network Power," *CFO*, November 2011, pp. 33–34.

33. "'Alibaba.com Allowed Us Access to Countless Suppliers That We Would Not Be Able to Reach on Our Own," Alibaba.com, 2013, http://news.alibaba.com/article/detail/america-member/100580796-1-%25E2%2580%259Calibaba.com-allowed-us-access-countless.html; Eric Markowitz, "Bringing Fixed-Gear Bikes to the Masses," *Inc.*, 2011, www.inc.com/coolest-college-start-ups-2011/sole-bicycles.html.

34. Ian MacMillan, "A Few Good Suppliers," *CFO*, October 2004, p. 26.

35. David Rocks and Nick Leiber, "Small U.S. Manufacturers Give Up on 'Made in China,'" *Bloomberg Businessweek*, June 21, 2012, www.omaha.com/article/20130409/NEWS/704099899/1707.

36. Cecelie Rohwedder, "Zara Grows as Retail Rivals Struggle," *Wall Street Journal*, March 26, 2009, pp. B1, B5.

37. Michael Fitzgerald, "Turning Vendors into Partners," *Inc.*, August 2005, pp. 94–100.

38. Michael Fitzgerald, "Turning Vendors into Partners," *Inc.*, August 2005, pp. 94–100.

39. Donald Reed and Cope Willis, "Sustaining the Supply Chain," *Resilience: Winning with Risk*, PriceWaterhouseCoopers, Volume 1, Number 1, June 2012, p. 40; Jessica E. Vascellaro and Owen Fletcher, "Apple Navigates China Maze," *Wall Street Journal*, January 14–15, 2012, p. B1.

Chapter 20

1. Sahir Anand and Chris Cunane, *Inventory Optimization: Retail Strategies for Eliminating Stockouts and Overstocks*, Aberdeen Group, May 2009, p. 4.

2. Nari Viswanathan, *Inventory Management: 3 Keys to Freeing Working Capital*, Aberdeen Group, May 2009, p. 4.

3. Rhonda Abrams, "When You Reduce Your Stuff, Profits Increase," *Greenville News*, May 29, 2011, p. 1E.

4. Patrick Burnson, "Slow and Steady: 23rd Annual State of Logistics Report, *Logistics Management*, July 2012, pp. 26–27.

5. Laurie Burkett, "Li Ning Sees Loss," *Wall Street Journal*, December 18, 2012, p. B9; Kathy Chu, "Li Ning Scaling Back After 2012 Loss," *Wall Street Journal*, March 27, 2013, http://online.wsj.com/article/SB1000142412788732478950457838333215 8202140.html.

6. Joann S. Lublin, "Polaris, Maker of Sport Vehicles, Races to Catch Up with Business," *Wall Street Journal*, May 24, 2010, pp. B1–B2.

7. "Inventory Optimization Technology Strategies for the Chief Supply Chain Officer," Aberdeen Group, December 2010, pp. 2–3.

8. Joann S. Lublin, "Polaris, Maker of Sport Vehicles, Races to Catch Up with Business," *Wall Street Journal*, May 24, 2010, pp. B1–B2.

9. Simona Covel, "Looking for Cost Cuts in Lots of New Places," *Wall Street Journal*, October 16, 2008, p. B5.

10. "With Billions of Bytes of Customer Data, How Can Retailers Be 'Starved for Information?,'" *Knowledge@Wharton*, August 2000, http://pf.inc.com/articles/2000/08/20043.html.

11. Pallavi Gogoi, "Retailers Pull Back, at a Cost," *Bloomberg Business Week*, August 7, 2008, www.businessweek.com/bwdaily/dnflash/content/aug2008/db2008087_200562.htm?chan=top+news_top+news+index_news+++analysis.

12. "Case Study: Mendingshed.com," *Fishbowl*, https://www.fishbowlinventory.com/case-studies/mending-shed.

13. Stephanie Miles, "Case Study: The Benefits of Running a POS System in the Cloud," *Street Fight*, May 24, 2012, http://streetfightmag.com/2012/05/24/case-study-the-benefits-of-running-a-pos-system-in-the-cloud; Suzi Harkola, "Put Retail POS in the Palm of Your Hand," *Point of Sale News*, February 18, 2013, http://pointofsale.com/201302181302/Point-of-Sale-News/iPad-Minis-Put-Retail-POS-in-the-Palm-of-Your-Hand.html.

14. Lisa Gerard, "Five Steps to Painless Inventory Management," *Entrepreneur*, November 3, 2011, www.entrepreneur.com/article/220631.

15. Mary Catherine O'Conner, "Bloomingdale's Test Item-Level RFID," *RFID Journal*, August 26, 2009, www.rfidjournal.com/articles/view?5160.

16. Claire Swedburg, "RFID Brings Intelligence to Billabong Store in Brazil," *RFID Journal*, August 22, 2011, www.rfidjournal.com/articles/view?8722.

17. Mark Henricks, "Tell and Show," *Entrepreneur*, April 2004, pp. 77–78.

18. Alex Niemeyer, Minsok H. Pak, and Sanjay Ramaswamy, "Smart Tags for Your Supply Chain," *McKinsey Quarterly*, Number 4, 2003, www.mckinseyquarterly.com/article_page.aspx?ar=1347&L2=1&L3=26.

19. Evan Niu, "Apple Lesson of the Day: Inventory Is Evil," *The Motley Fool*, March 23, 2012, www.fool.com/investing/general/2012/03/23/apple-lesson-of-the-day-inventory-is-evil.aspx; "What Is Apple's Manufacturing Secret?," *QuickBooks Manufacturing Blog*, April 17, 2012, http://quickbooksmanufacturing.wordpress.com/2012/04/17/apple-inventory-management-secret.

20. Sam Oliver, "RIM's Unsold Inventory of Blackberry's, PlayBooks Swells to $1B Value," *Apple Insider*, May 29, 2012, http://appleinsider.com/articles/12/05/29/rims_unsold_inventory_of_blackberrys_playbooks_swells_to_1b_value.

21. Peter C. Reid, *Well Made in America* (New York: McGraw-Hill Ryerson, 1991), pp. 148–151.

22. "October 2012 TrueTrends: Shortest and Longest Days in Vehicle Inventory," TrueTrends, October 17, 2012, http://blog.truecar.com/2012/10/17/october-2012-truetrends-shortest-and-longest-days-in-vehicle-inventory.

23. Dana Mattioli, "Little Shops Make Big Plays for the Holidays," *Wall Street Journal*, October 27, 2009, p. B5.

24. *2012 ACFE Report to the Nation on Occupational Fraud and Abuse* (Austin, TX: Association of Certified Fraud Examiners, 2013), pp. 4, 8.

25. *2012 ACFE Report to the Nation on Occupational Fraud and Abuse* (Austin, TX: Association of Certified Fraud Examiners, 2013), pp. 26–27.

26. "National Retail Security Survey Reveals U.S. Retail Industry Lost More Than $35.28 Billion to Theft in 2011," *Yahoo! Finance*, November 27, 2012, http://finance.yahoo.com/news/national-retail-security-survey-reveals-140000867.html.

27. "Employee Theft Statistics," *Statistic Brain*, September 18, 2012, www.statisticbrain.com/employee-theft-statistics.

28. "Theft Surveys," Jack L. Hayes International, 2012, http://hayesinternational.com/news/annual-retail-theft-survey.

29. *2012 ACFE Report to the Nation on Occupational Fraud and Abuse* (Austin, TX: Association of Certified Fraud Examiners, 2013), p. 62.

30. *2012 ACFE Report to the Nation on Occupational Fraud and Abuse* (Austin, TX: Association of Certified Fraud Examiners, 2013), pp. 61–62.

31. *2012 ACFE Report to the Nation on Occupational Fraud and Abuse* (Austin, TX: Association of Certified Fraud Examiners, 2013), pp. 13, 14.

32. *2012 ACFE Report to the Nation on Occupational Fraud and Abuse* (Austin, TX: Association of Certified Fraud Examiners, 2013), p. 34.

33. "Former Ennis Paint Employee Charged with Embezzling More Than $700,000," Federal Bureau of Investigation, Atlanta Division, April 24, 2013, www.fbi.gov/atlanta/press-releases/2013/former-ennis-paint-employee-charged-with-embezzling-more-than-700-000.

34. Nancy Germond, "Your Company Isn't Immune to Employee Theft," *All Business*, February 2, 2008, www.allbusiness.com/crime-law-enforcement-corrections/crime-prevention/6623333-1.htmlv.

35. *2012 ACFE Report to the Nation on Occupational Fraud and Abuse* (Austin, TX: Association of Certified Fraud Examiners, 2013), p. 4.

36. *2012 ACFE Report to the Nation on Occupational Fraud and Abuse* (Austin, TX: Association of Certified Fraud Examiners, 2013), p. 49.

37. Sarah E. Needleman, "Business Owners Get Burned by Sticky Fingers," *Wall Street Journal*, March 11, 2010, http://online.wsj.com/article/SB10001424052748703862704575099661793568310.html.

38. *2012 ACFE Report to the Nation on Occupational Fraud and Abuse* (Austin, TX: Association of Certified Fraud Examiners, 2013), p. 41.

39. "Koss Corp. Hires Chief Financial Officer," *Business Journal of Milwaukee*, January 18, 2010, http://milwaukee.bizjournals.com/milwaukee/stories/2010/01/18/daily2.html; Rich Kirchen, "Former Koss Exec Sachdeva Indicted on Six Counts," *Business Journal of Milwaukee*, January 20, 2010, http://milwaukee.bizjournals.com/milwaukee/stories/2010/01/18/daily39.html.

40. Robert T. Gray, "Clamping Down on Worker Crime," *Nation's Business*, April 1997, p. 44.

41. Scott Wescott, "Are Your Staffers Stealing?," *Inc.*, October 2006, pp. 33–35.

42. *2012 ACFE Report to the Nation on Occupational Fraud and Abuse* (Austin, TX: Association of Certified Fraud Examiners, 2013), p. 61.

43. Mark Doyle, "24th Annual Retail Theft Survey: Shoplifter and Dishonest Employee Apprehensions and Recovery Dollars," Jack L. Hayes International, June 15, 2012, http://hayesinternational.com/2012/06/24th-annual-retail-theft-survey-shoplifters-dishonest-employee-apprehensions-and-recovery-dollars.

44. "Annual Retail Theft Survey," Jack L. Hayes International, 2012, http://hayesinternational.com/news/annual-retail-theft-survey.

45. Raymund Flandez, "Stop That Thief," *Wall Street Journal*, June 12, 2008, http://online.wsj.com/article/NA_WSJ_PUB:SB121322091260765769.html.

46. "Shoplifting Statistics," National Association for Shoplifting Prevention, 2013, www.shopliftingprevention.org/whatnaspoffers/nrc/publiceducstats.htm.

47. "Annual Retail Theft Survey," Jack L. Hayes International, 2012, http://hayesinternational.com/news/annual-retail-theft-survey.

48. Bill Laitner, "Retail Fraud Is More Than Just Shoplifting—Now It's a Felony," *Detroit Free Press*, April 17, 2013, www.freep.com/article/20130330/NEWS06/303300100/Retail-fraud-is-more-than-just-shoplifting-and-now-it-s-a-felony.

49. Jennifer Weigel, "Author Gives Insight into Shoplifting," *Greenville News*, July 31, 2011, p. 2D.

50. Ann Zimmerman and Miguel Bustillo, "'Flash Robs' Vex Retailers," *Wall Street Journal*, October 21, 2011, pp. B1–B2.

51. Laura Cox, "Flash Rob as 20 Teens Are Caught on Camera Stealing $3,000 in Designer Jeans in Flash Rob Trend," *Daily Mail*, July 30, 2012, www.dailymail.co.uk/news/article-2181041/Flash-ROB-20-teens-caught-camera-stealing-3-000-designer-jeans-in.html.

52. *2012 Organized Retail Crime Survey* (Washington, DC: National Retail Federation, 2012), p. 7.

53. Andrea Chang, "Retailers Battle Sophisticated Shoplifting Teams, 'CSI' Style," *Los Angeles Times*, July 2, 2011, http://articles.latimes.com/2011/jul/02/business/la-fi-0702-retail-crime-20110702.

54. Bill Laitner, "Retail Fraud Is More Than Just Shoplifting—Now It's a Felony," *Detroit Free Press*, April 17, 2013, www.freep.com/article/20130330/NEWS06/303300100/Retail-fraud-is-more-than-just-shoplifting-and-now-it-s-a-felony.

55. Dan Lieberman and Lauren Effron, "Busting Underground Shoplifting Rings: Inside Organized Retail Crime Raids," *ABC News*, February 26, 2013, http://abcnews.go.com/US/busting-underground-shoplifting-crime-rings-inside-organized-retail/story?id=18600080.

56. "Huge Shoplifting Operation in Austin Caught Stealing Paper Towels, Tide Detergent," *Huffington Post*, April 3, 2012, www.huffingtonpost.com/2012/04/03/austin-shoplifting-ring-texas-tide-detergent_n_1400096.html.

57. Ann Zimmerman, "As Shoplifters Use High-Tech Scams, Retail Losses Rise," *Wall Street Journal*, October 25, 2006, pp. A1, A12.

58. Jennifer Weigel, "Author Gives Insight into Shoplifting," *Greenville News*, July 31, 2011, p. 2D.

59. "Shoplifting Statistics," National Association for Shoplifting Prevention, www.shopliftingprevention.org/WhatNASPOffers/NRC/PublicEducStats.htm; Jennifer Weigel, "Author Gives Insight into Shoplifting," *Greenville News*, July 31, 2011, p. 2D.

60. *2012 Organized Retail Crime Survey* (Washington, DC: National Retail Federation, 2012), p. 13.

Chapter 21

1. Michael Gold, "Jazzin' CEO," *Manage Smarter*, January 9, 2008, p. 1.

2. Michael Gold, "Jazzin' CEO," *Manage Smarter*, January 9, 2008, p. 1.

3. Francis Huffman, "Taking the Lead," *Entrepreneur*, November 1993, p. 101.

4. Sam Allman, "Leadership vs. Management," *Successful Meetings*, October 2009, p. 12.

5. Matthew E. May, "The 3 C's of Trust," *Open Forum*, September 24, 2010, www.openforum.com/idea-hub/topics/the-world/article/the-3-cs-of-trust-matthew-e-may.

6. "Trust in Managers in Short Supply," *Right Management*, May 27, 2010, www.right.com/news-and-events/press-releases/2010-press-releases/item8361.aspx.

7. Jeffrey Pfeffer, "Executive-in-Chief," *Business 2.0*, March 1, 2005, http://money.cnn.com/magazines/business2/business2_archive/2005/03/01/8253107/index.htm.

8. Dave Zielinski, "New Ways to Look at Leadership," *Presentations*, June 2005, pp. 26–33.

9. Ken Blanchard, *Leading at a Higher Level* (Upper Saddle River, NJ: FT Press, 2010), p. 279.

10. Jeff R. Hale and Dail Fields. "A Cross-Cultural Measure of Servant Leadership Behaviors," in *Online Instruments, Data Collection, and Electronic Measurements: Organizational Advancements*, Mihai C. Bocarnea, Rodney A. Reynolds, and Jason D. Baker (Editors), (Hershey, PA: IGI Global, 2012), p. 152.

11. Sam Oches, "It's the Customers, Stupid. Or Is It?," *QSR Magazine*, September 16, 2010, www.qsrmagazine.com/exclusives/it-s-customers-stupid-or-it.

12. Chris Penttila, "Hire Away," *Entrepreneur*, June 2007, p. 20.

13. Leigh Branham, *Keeping the People Who Keep You in Business: 24 Ways to Hang On to Your Most Valuable Talent* (New York: Amacom Books, 2001).

14. "Nearly Seven in Ten Businesses Affected by a Bad Hire in the Past Year, According to CareerBuilder Survey," *CareerBuilder*, December 13, 2012, www.careerbuilder.com/share/aboutus/pressreleasesdetail.aspx?sd=12/13/2012&id=pr730&ed=12/31/2012.

15. David Meyer, "Nine Recruiting and Selection Tips to Ensure Successful Hiring," About.com, http://humanresources.about.com/od/selectemployees/a/staff_selection_2.htm.

16. "Hiring Decisions Miss the Mark 50% of the Time," Corporate Executive Board, October 24, 2008, http://ir.executiveboard.com/phoenix.zhtml?c=113226&p=irol-newsArticle&ID=1205091&highlight=; "2 Out of 3 Managers Still Fear a Hiring Decision They'll Regret," DDI, March 16, 2009, www.ddiworld.com/about/pr_releases_en.asp?id=211.

17. Chris Pentilla, "Talent Scout," *Entrepreneur*, July 2008, p. 19.

18. Dan Schawbel, "The Power Within: Why Internal Recruiting and Hiring Are on the Rise," *Time*, August 15, 2012, http://business.time.com/2012/08/15/the-power-within-why-internal-recruiting-hiring-are-on-the-rise.

19. Margaret Heffernan, "9 Secrets of Highly Successful Hiring," *Inc.*, September 17, 2012, www.inc.com/margaret-heffernan/hiring-recruiting-secrets-of-success.html.

20. "Why Employee Referrals are the Best Source of Hire," *Undercover Recruiter*, http://theundercoverrecruiter.com/infographic-employee-referrals-hire; www.jobvite.com.

21. Nelson D. Schwartz, "In Hiring, a Friend in Need Is a Prospect, Indeed," *New York Times*, January 28, 2013, p. A1.

22. Megan O'Neill, "92% of U.S. Companies Now Using Social Media for Recruitment," *SocialTimes*, September 6, 2012, http://socialtimes.com/social-media-recruitment-infographic_b104335.

23. Jennifer J. Salopek, "Recruiters Look to Be Big Man on Campus," *Workforce Management*, September 2010, p. 12.

24. Jessica Stillman, "Turn Your Interns into Killer Employees," *Inc.*, May 23, 2012, www.inc.com/jessica-stillman/how-to-convert-interns-into-killer-junior-employees.html.

25. Mitra Toossi, "Labor Force Projections to 2016: More Workers in Their Golden Years," *Monthly Labor Review*, November 2007, p. 33.

26. Tamara Lytle, "Are You Planning to Work Into Your Retirement Years?," *AARP Blog*, May 24, 2013, http://blog.aarp.org/2013/05/24/working-past-retirement-age-retiring-later-in-life-building-nest-egg/.

27. *The Real Talent Debate: Will Aging Boomers Deplete the Workforce?* (Scottsdale, AZ: WorldatWork, 2007), p. 12.

28. Kay McFadden, "7 Tips for Hiring Older Workers," *Inc.*, April 29, 2011, www.inc.com/guides/201104/7-tips-for-hiring-older-workers.html.

29. "Innovating Human Resources," *BrainReactions*, January 16, 2007, www.brainreactions.com/whitepapers/brainreactions_hr_innovation_paper.pdf, pp. 11–14; Christopher Caggiano, "Recruiting Secrets," *Inc.*, October 1998, pp. 30–42.

30. Amy Barrett, "Making Telecommuting Work," *BusinessWeek*, October 17, 2008, www.businessweek.com/stories/2008-10-16/making-telecommuting-work.

31. "Applicants are Fewer and Many are Lacking," *Wall Street Journal*, February 13, 2012, http://online.wsj.com/article/SB1000142405297020 46426045772153720105 43642.html.

32. "Survey: Up to 25 Applications Reviewed per Job," *Workforce Management*, September 16, 2010, www.workforce.com/articles/survey-up-to-25-applications-reviewed-per-job.

33. "Smart Questions for You Hiring Manager," *Inc.*, February 2007, p. 47.

34. "Survey: 21 Percent of Job Seekers Dropped After Reference Checks," *Workforce Management*, June 23, 2010, www.workforce.com/articles/sustainability-no-buzzword-in-running-a-21st-century-company.

35. Andy Levine, "Dig We Must," *Inc. Magazine*, August 2006, p. 97.

36. Dan Heath and Chip Heath, "Hold the Interview," *Fast Company*, June 2009, pp. 51–52.

37. Peter Bregman, "The Interview Question You Should Always Ask," Harvard Business Publishing, January 27, 2009, http://blogs.harvardbusiness.org/cs/2009/01/the_interview_question_you_sho.html.

38. Ed Frauenheim, "Serious Hiring Keeps Zappos in a Fun Mood," *Workforce Management*, September 14, 2009, p. 20.

39. Ed Frauenheim, "Serious Hiring Keeps Zappos in a Fun Mood," *Workforce Management*, September 14, 2009, p. 20.

40. Jeffrey Cornwall, "Bongo Bob," United States Association for Small Business and Entrepreneurship, *Proceedings*, 2006.

41. Chris Pentilla, "Testing the Waters," *Entrepreneur*, January 2004, pp. 72–73.

42. Chris Penttila, "Peering In," *Entrepreneur*, January 2005, pp. 70–71.

43. "Emma Co-Founder Discusses Company's Hiring Process," May 1, 2012, www.daveramsey.com/article/emma-co-founder-discusses-companys-hiring-process/lifeandmoney_business.

44. Chris Penttila, "Peering In," *Entrepreneur*, January 2005, p. 71.

45. Richard Slawsky, "Reducing Risk: The Search for Reputable Employees," QSRWeb.com, August 9, 2007, www.qsrweb.com/article/104930/Reducing-risk-The-search-for-reputable-employees.

46. Richard Slawsky, "Reducing Risk: The search for reputable employees," *QSRWeb*, August 9, 2007, www.qsrweb.com/article/104930/Reducing-risk-The-search-for-reputable-employees.

47. Richard Slawsky, "Reducing Risk: The search for reputable employees," QSRWeb.com, August 9, 2007, http://www.qsrweb.com/article/104930/Reducing-risk-The-search-for-reputable-employees.

48. "Thirty-seven percent of Companies Use Social Networks to Research Potential Job Candidates, According to New Career Builder Survey, CareerBuilder.com, April 18, 2012, www.careerbuilder.com/share/aboutus/pressreleasesdetail.aspx?id=pr691&sd=4%2F18%2F2012&ed=4%2F18%2F2099.

49. Emily Maltby, "To Find Best Hires, Firms Become Creative," *Wall Street Journal*, November 17, 2009, http://online.wsj.com/article/SB100 01424052748704538404574539971535489470.html.

50. "Results from the 2011 National Survey on Drug Use and Health: Summary of National Findings," Substance Abuse and Mental Health Services Administration, www.samhsa.gov/data/nsduh/2k11results/nsduhresults2011.htm.

51. McGuire Woods, "But I Have a Prescription! . . . Drug Testing in the Age of Medical Marijuana," Society of Human Resource Managers, July 1, 2010, www.shrm.org/LegalIssues/StateandLocalResources/Pages/ButIHaveaPrescription.aspx.

52. Jessica Marquez, "Kindness Pays . . . Or Does It?," *Workforce*, June 25, 2007, pp. 40–49.

53. Jessica Marquez, "Kindness Pays . . . Or Does It?," *Workforce*, June 25, 2007, pp. 40–49.

54. Julia Chang, "Balancing act," *Sales & Marketing Management*, February 2004, p. 16.

55. "Jim Goodnight, Chief Executive Officer," SAS, www.sas.com/company/about/bios/jgoodnight.html; Mark C. Crowley, "How SAS Became the World's Best Place to Work," *FastCompany*, January 22, 2013, www.fastcompany.com/3004953/how-sas-became-worlds-best-place-work.

56. April Joyner, "Learning Company," *Inc.*, May 1, 2011, www.inc.com/winning-workplaces/articles/201105/the-learning-company.html.

57. Jody Urquhart, "Creating a Fun Workplace . . . 13 Ways to Have Fun at Work," I Do Inspire, www.idoinspire.com/?q=node/15.

58. Nichole L. Torres, "Let the Good Times Roll," *Entrepreneur*, November 2004, p. 57.

59. Vivian Giang, "25 Small Companies with Better Perks Than Google," *Business Insider*, November 7, 2012, www.businessinsider.com/the-25-best-small-companies-to-work-for-right-now-2012-10?op=1.

60. State *of the American Workplace: Employee Engagement Insights for U.S. Business Leaders*, Gallup, Washington, DC: 2013, pp. 9, 13.

61. Noelle Coyle, "Fish or Cut Bait?," *Black Box*, Quarter 1, 2010, p. 60.

62. Ralph Stayer, "How I Learned to Let My Workers Lead," Western Kentucky University, http://people.wku.edu/rich.patterson/CFS-452/Readings/stayer.htm.

63. David Freedman, "The Idiocy of Crowds," *Inc.*, September 2006, pp. 61–62.

64. *Capitalizing on Effective Communication: How Courage, Innovation, and Discipline Drive Business Results in Challenging Times, Communication ROI Study Report 2009/2010*, Towers Watson, November 2009, pp. 2–3; How to Communicate with Employees, *Inc. Guidebook*, Vol. 2, No. 2, p. 1.

65. "Employee Engagement," *Motiv8*, May 26, 2010, www.motiv8comm.com/IdeasandTrends/index_may26.html.

66. Diana Ransom, "6 Ways to Manage a Virtual Workforce," *Entrepreneur*, April 19, 2010, www.entrepreneur.com/article/206214v.

67. Mina Kimes, "How Can I Get Candid Feedback from My Employees?," *Fortune*, April 13, 2009, p. 24.

68. Kathy Gurchiek, "U.S. Rank-and-File Workers Feel Undervalued by Managers," Society for Human Resource Managers, July 16, 2010, www.shrm.org/Publications/HRNews/Pages/WorkersFeelUndervalued.aspx.

69. Jeff Haden, "Best Way to Make Employees Better at Their Jobs," *Inc.*, July 23, 2012, www.inc.com/jeff-haden/motivating-employees-with-simple-question.html.

70. "I Heard It Through the Grapevine," American Management Association, November 21, 2005, www.amanet.org/training/articles/I-Heard-It-Through-the-Grapevine.aspx.

71. Richard Branson, "How Can I Empower Employees?," *Entrepreneur*, January 17, 2011, www.entrepreneur.com/article/217880.

72. Daniel Akst, "The Rewards of Recognizing a Job Well Done," *Wall Street Journal*, January 31, 2007, p. D9.

73. "Kwik Trip Fuels Results with Recognition," www.octanner.ca/insight/client-stories/kwik-trip-fuels-results-with-recognition.

74. "Job Satisfaction: 2012 Edition," The Conference Board, June 27, 2012, www.conference-board.org/publications/publicationdetail.cfm?publicationid=2258.

75. Robert D. Mohr and Cindy Zoghi, *Is Job Enrichment Really Enriching?* (Washington, DC: U.S. Department of Labor, U.S. Bureau of Labor Statistics, Office of Productivity and Technology, January 2006), pp. 13–15.

76. "Keys to a Good Flex-Time Policy," National Federation of Independent Business, www.nfib.com/business-resources/business-resources-item?cmsid=56050.

77. Stephen Miller, "2012 National Study of Employers Reveals Increased Workplace Flexibility," April 30, 2012, www.shrm.org/hrdisciplines/benefits/articles/pages/2012nse.aspx.

78. Carol Kleiman, "Job Sharing Working Its Way into Mainstream," *Greenville News*, August 6, 2000. p. 3G.

79. "Elite SEM," Glass Door, www.glassdoor.co.uk/Reviews/Elite-SEM-Reviews-E412035.htm; Stephen Miller, "2012 National Study of Employers Reveals Increased Workplace Flexibility," April 30, 2012, www.shrm.org/hrdisciplines/benefits/articles/pages/2012nse.aspx.

80. Telis Demos, "Motivate Without Spending Millions," *Fortune*, April 12, 2010, pp. 37–38.

81. Curt Richardson, "Influence Your Company Culture (for the Better)," *Inc.*, October 9, 2012, www.inc.com/curt-richardson/how-to-influence-your-company-culture-for-the-better.html; "Best Medium Businesses to Work for 2011," *Entrepreneur*, www.entrepreneur.com/gptw/medium/index.html.

82. Kevin Gray, "Can't Pay Your Employees What You'd Like? Praise Them Instead," *BNet*, January 19, 2010, www.bnet.com/article/cant-pay-your-employees-what-youd-like-praise-them-instead/385221.

83. Jay Love, "5 Ways to Reward Your All-Star Employees," *Inc.*, February 9, 2012, www.inc.com/jay-love/5-ways-to-boost-employee-recognition.html.

84. Drew Gannon, "How to Reward Great Ideas," *Inc.*, July 19, 2011, www.inc.com/guides/201107/how-to-reward-employees-great-ideas.html.

85. Jacqueline M. Hames, "10 Tips for Boosting Employee Morale," *Inc.*, January 7, 2011, www.inc.com/guides/2011/01/10-tips-for-boosting-employee-morale.html.

86. Ellen Galinsky, Shanny L. Peer, and Sheila Eby, *When Work Works: 2009 Guide to Bold New Ideas for Making Work Work* (New York: Families and Work Institute, 2009), pp. 20–21.

87. Scott Westcott, Putting an End to End-of-Year Reviews," *Inc.*, December 2007, pp. 58–59.

88. Scott Westcott, Putting an End to End-of-Year Reviews," *Inc.*, December 2007, pp. 58–59.

Chapter 22

1. "Facts About Family Business," S. Dale High Center for Family Business at Elizabethtown College," www.centerforfamilybusiness.org/facts.asp; MassMutual Family Business Network, www.massmutual.com/fbn/index.htm; "Family Business Facts," Family Business Institute, www.ffi.org/looking/fbfacts_us.pdf; *Annual Family Business Survey*, Family Enterprise USA, March 2011, p. 1.

2. "Facts About Family Business," Center for Family Business, Mihaylo College of Business and Economics, California State University Fullerton, http://business.fullerton.edu/centers/cfb/facts.htm.

3. "Global Data Points: United States," Family Firm Institute, 2013, www.ffi.org/?page=GlobalDataPoints.

4. "Wal-Mart Stores Inc.," Reuters, May 6, 2013, www.reuters.com/finance/stocks/overview?symbol=WMT.N; "List of Countries by GDP 2012, Exploredia, May 2013, http://exploredia.com/list-of-countries-by-gdp-2012.

5. Tony Taylor, "Small Businesses Show Relative Strength," *GSA Business*, September 4, 2006, p. 22; Jim Lee, "Family Firm Performance: Further Evidence," *Family Business Review* 19, no. 2 (June 2006), pp. 103–114.

6. "Research Reveals: Family Firms Perform Better," *Family Business Advisor* 12, no. 3 (March 2003), p. 1.

7. "Family Business Facts," University of Vermont School of Business Administration, www.uvm.edu/business/vfbi/?Page=facts.html.

8. Caroline Bayley, "Staedtler and Faber-Castel's Productive Pencil Rivalry," *BBC News*, April 13, 2011, www.bbc.co.uk/news/business-13019777.

9. "Equipped to Success: L.L. Bean," *Family Firm: A Resilient Model for the 21st Century, PWC Family Business Survey*, PWC, October 2012, p. 9.

10. Adriana Gardella, "Family Businesses Learn to Adapt to Keep Thriving," *New York Times*, April 4, 2012, www.nytimes.com/2012/04/05/business/smallbusiness/how-they-beat-the-odds-to-keep-family-businesses-healthy.html?pagewanted=all; Stacy Perman, "Learning from the Great Depression," *Bloomberg Business Week*, October 17, 2008, www.businessweek.com/stories/2008-10-17/learning-from-the-great-depressionbusinessweek-business-news-stock-market-and-financial-advice.

11. Stacy Perman, "Learning from the Great Depression," *Bloomberg Business Week*, October 17, 2008, www.businessweek.com/stories/2008-10-17/learning-from-the-great-depressionbusinessweek-business-news-stock-market-and-financial-advice.

12. *Annual Family Business Survey*, Family Enterprise USA, March 2011, p. 6.

13. Charles Paikert, "Stepping Up at Family Firms," *New York Times*, November 8, 2012, www.nytimes.com/2012/11/09/giving-more-family-firms-make-philanthropy-their-business.html.

14. Ken Otterbourg, "The Rift—A Family Dynasty Fights over the Future of Luray Caverns," *Washington Post*, March 14, 2013, www.washingtonpost.com/blogs/liveblog/wp/2013/03/14/magazine-the-rift-a-family-dynasty-fights-over-the-future-of-luray-caverns; "Judge Tosses Caverns' Family Lawsuit from Federal Court," *Northern Virginia Daily*, February 13, 2013, www.nvdaily.com/news/2013/02/judge-tosses-caverns-family-lawsuit-from-federal-court.php.

15. "Facts and Perspectives on Family Business Around the World: United States," Family Business Institute, www.ffi.org/looking/fbfacts_us.pdf.

16. "Facts and Perspectives on Family Business Around the World: United States," Family Business Institute, www.ffi.org/looking/fbfacts_us.pdf.

17. "Facts and Perspectives on Family Business Around the World: United States," Family Business Institute, www.ffi.org/looking/fbfacts_us.pdf.

18. Alix Stewart, "Holding on to Family Ties," *CFO*, January 31, 2012, www3.cfo.com/article/2012/1/credit-capital_crane-currency-kittredge-family-business-lindsay-goldberg-devisscher; Mary Furash, "The State of One Small Family Business: Crane & Company," *Inc.*, May 15, 1996, www.inc.com/magazine/19960515/2093.html; Eliza Browning, "Joining the Family Business: 6 Tips," *Inc.*, March 26, 2012, www.inc.com/eliza-browning/joining-the-family-business-6-tips.html.

19. *Family Firm: A Resilient Model for the 21st Century*, PriceWaterhouseCoopers Family Business Survey, 2012, p. 8.

20. *Kin in the Game*, PriceWaterhouseCoopers Family Business Survey, 2010/11, p. 22.

21. "Facts and Perspectives on Family Business Around the World: United States," Family Business Institute, www.ffi.org/looking/fbfacts_us.pdf.

22. Sharon Nelton, "Ten Keys to Success in Family Business," *Nation's Business*, April 1991, pp. 44–45.

23. *Family Firm: A Resilient Model for the 21st Century*, PriceWaterhouseCoopers Family Business Survey, 2012, p. 5.

24. *Family Firm: A Resilient Model for the 21st Century*, PriceWaterhouseCoopers Family Business Survey, 2012, p. 9.

25. Adriana Gardella, "Family Businesses Learn to Adapt to Keep Thriving," *New York Times*, April 4, 2012, www.nytimes.com/2012/04/05/business/smallbusiness/how-they-beat-the-odds-to-keep-family-businesses-healthy.html?pagewanted=all.

26. Facts about Family Business," Center for Family Business, Mihaylo College of Business and Economics, California State University Fullerton, business.fullerton.edu/centers/cfb/facts.htm.

27. *Family Firm: A Resilient Model for the 21st Century*, PriceWaterhouseCoopers Family Business Survey, 2012, p. 6.

28. Jack Mitchell, "Family Business: How to Pass the Baton," *CNN Money*, July 9, 2012, http://management.fortune.cnn.com/2012/07/09/family-business-succession.

29. "Family Members Fight over Control of Texas Pete Hot Sauce Empire," *Greenville News*, May 17, 1997, p. 11D.

30. Nicholas Stein, "The Age of the Scion," *Fortune*, April 2, 2001, p. 124.

31. *Family Firm: A Resilient Model for the 21st Century*, PriceWaterhouseCoopers Family Business Survey, 2012, p. 15; "Interview: Seafolly CEO Anthony Halas," *Lingerie Insight*, August 6, 2012, www.lingerieinsight.com/article-2545-interview-seafolly-ceo-anthony-halas; Gillian Tan, "Seafolly CEO Anthony Halas on the Technicalities of Selling Bikinis," *The Australian*, May 22, 21012,

www.theaustralian.com.au/business/wall-street-journal/seafolly-ceo-anthony-halas-on-the-technicalities-of-selling-bikinis/story-fnay3vxj-1226363530957.

32. *On the Minds of Family Enterprise Owners*, Family Enterprise USA, 2013 Survey of Family Firms, p. 4.

33. "Fine Tuning Your Legacy," Hubler for Business Families, www.hublerfamilybusiness.com/OwnershipPlanning/FineTuningYourLegacy.aspx.

34. Bob Dotson, "How a 390-Year-Old Family Business Avoids Layoffs," *Today*, March 8, 2013, www.today.com/news/how-390-year-old-family-business-avoids-layoffs-1C8711297.

35. Adriana Gardella, "Family Businesses Learn to Adapt to Keep Thriving," *New York Times*, April 4, 2012, www.nytimes.com/2012/04/05/business/smallbusiness/how-they-beat-the-odds-to-keep-family-businesses-healthy.html?pagewanted=all.

36. Thomas L. Kalaris, "Family Business: In Safe Hands?" *Barclays Wealth Insights*, 2009, p. 10.

37. Nicholas Stein, "The Age of the Scion," *Fortune*, April 2, 2001, p. 124.

38. Chester Dawson, "Paul Stuart Broadens Its Horizons," *Wall Street Journal*, April 18, 2013, p. D3.

39. Bo Burlingham, "Why a CEO Needs to Have a Plan B," *Inc.*, May 2009, www.inc.com/magazine/20090501/why-a-ceo-needs-to-have-a-plan-b.html.

40. "Employee Ownership and Corporate Performance," National Center for Employee Ownership, www.nceo.org/library/esop_perf.html.

41. Angus Loten, "Founders Cash Out, but Do Workers Gain?," *Wall Street Journal*, April 18, 2013, p. B4.

42. "Family Business Statistics," American Management Service, www.amserv.com/index.cfm/page/Family-Business-Statistics/pid/10715.

43. Karl Moore, "Family Business—Your Most Important Issue—Successfully Passing It On," *Forbes*, December 7, 2012, www.forbes.com/sites/karlmoore/2012/12/07/family-business-your-most-important-issue-successfully-passing-it-on.

44. Karl Moore, "Family Business—Your Most Important Issue—Successfully Passing It On," *Forbes*, December 7, 2012, www.forbes.com/sites/karlmoore/2012/12/07/family-business-your-most-important-issue-successfully-passing-it-on.

45. Karl Moore, "Family Business—Your Most Important Issue—Successfully Passing It On," *Forbes*, December 7, 2012, www.forbes.com/sites/karlmoore/2012/12/07/family-business-your-most-important-issue-successfully-passing-it-on.

46. Bo Burlingham, "What Am I, If Not My Business?," *Inc.*, November 2010, p. 92.

47. "U.S. Trust Survey Finds Majority of Business Owners Focused on Keeping People Employed and Creating Jobs: However, Less Than Half Are Planning for Their Future," *Business Wire*, November 9, 2012, www.businesswire.com/news/home/20121109005635/en/U.S.-Trust-Survey-Finds-Majority-Business-Owners.

48. TCPN Quotations Center, www.cyber-nation.com/victory/quotations/subject/quotes_subjects_f_to_h.html#f.

49. Andy Bluestone, "Succession Planning Isn't Just About Money," *Nation's Business*, November 1996, p. 6.

50. *Family Firm: A Resilient Model for the 21st Century*, PriceWaterhouseCoopers Family Business Survey, 2012, p. 9.

51. Shelly Branch, "Mom Always Liked You Best," *Your Company*, April/May 1998, pp. 26–38.

52. "Interview: Seafolly CEO Anthony Halas," *Lingerie Insight*, August 6, 2012, www.lingerieinsight.com/article-2545-interview-seafolly-ceo-anthony-halas.

53. Lillia Callum-Penso, "Carrying On," *Greenville News*, September 9, 2012, pp. 1E–2E; Tony Taylor, "Small Businesses Show Relative Strength," *GSA Business*, September 4, 2006, pp. 22, 24.

54. Patricia Schiff Estess, "Heir Raising," *Entrepreneur*, May 1996, pp. 80–82.

55. *On the Minds of Family Enterprise Owners*, Family Enterprise USA, 2013 Survey of Family Firms, p. 4.

56. Carol Tice, "Lost in Transition," *Entrepreneur*, November 2006, pp. 101–103.

57. Jacquelyn Lynn, "What Price Successor?," *Entrepreneur*, November 1999, p. 146.

58. Craig E. Aronoff and John L. Ward, "Why Continue Your Family Business," *Nation's Business*, March 1998, pp. 72–74.

59. "U.S. Trust Survey Finds Majority of Business Owners Focused on Keeping People Employed and Creating Jobs: However, Less than Half Are Planning for Their Future," *Business Wire*, November 9, 2012, www.businesswire.com/news/home/20121109005635/en/U.S.-Trust-Survey-Finds-Majority-Business-Owners.

60. Paul Dalby, "Family Business Succession Planning Starts with a Family Conference," *The Star*, March 31, 2013, www.thestar.com/business/personal_finance/retirement/2013/03/31/family_business_succession_planning_starts_with_a_family_conference.html.

61. Jeremy Quittner, "Creating a Legacy," *BusinessWeek*, June 25, 2007, www.businessweek.com/magazine/content/07_26/b4040443.htm?chan=search.

62. Sydney L. Barton, "How Strategic Management Can Help Family Firms Succeed," *Family Business*, Autumn 2007, www.familybusinessmagazine.com/index.php?/articles/single/how_strategic_management_can_help_family_firms_succeed2.

63. *Strengthen Your Family Business*, KPMG Enterprise, July 25, 2011, p. 1.

64. Jacqueline Simmons and Anne-Sylvaine Chassany, "Rothschild Anoints Alexandre Heir as Family Cements Reign," *Bloomberg*, July 19, 2012, www.bloomberg.com/news/2012-07-19/rothschild-anoints-alexandre-heir-as-family-cements-reign.html.

65. Annetta Miller, "You Can't Take It with You," *Your Company*, April 1999, pp. 28–34.

66. "Family Business Facts," Family Business Institute, www.ffi.org/looking/fbfacts_us.pdf.

67. Frank Silverstein, "Daddy's Cowgirl Takes Over the Rodeo Business," *MSNBC*, August 30, 2009, www.msnbc.msn.com/id/32573940/ns/business-small_business; Courtney Elam, "Kirsten Vold," Harry Vold Rodeo Company, www.harryvoldrodeo.com/staff.html.

68. Chen Man-nung, "More Family Businesses in China Anointing Female Successors," *WantChinaTimes*, February 16, 2013, www.wantchinatimes.com/news-subclass-cnt.aspx?id=20130216000076&cid=1502; "Zong Tops China's Billionaires as Communist-to-Capitalist," *Bloomberg*, October 31, 2012, www.bloomberg.com/news/2012-10-31/zong-tops-china-billionaires-as-communist-to-capitalist.html.

69. John Warrillow, "Leave the Business to the Kids? Maybe Not," *Wall Street Journal*, June 10, 2010, http://online.wsj.com/article/SB10001424052748704575304575296523166009344.html.

70. Sharon Nelton, "Why Women Are Chosen to Lead," *Nation's Business*, April 1999, p. 51.

71. *U.S. Trust Insights on Wealth and Worth*, U.S. Trust, 2012, p. 19.

72. Dick Patten, "The Death Tax Is Killing Family Businesses," *Daily Caller*, October 8, 2010, http://dailycaller.com/2010/10/08/the-death-tax-is-killing-family-businesses; Amanda Hill, "Estate Taxes Could Mean 'Death' of Family Farms and Ranches," *Texas Agriculture News*, September 27, 2010, www.nodeathtax.org/news/estate-tax-could-mean-death-of-family-farms-and-ranches.

73. Dick Patten, "The Death Tax Is Killing Family Businesses," *Daily Caller*, October 8, 2010, http://dailycaller.com/2010/10/08/the-death-tax-is-killing-family-businesses.

74. *U.S. Trust Insights on Wealth and Worth*, U.S. Trust, 2012, p. 20.

75. Joan Szabo, "Spreading the Wealth," *Entrepreneur*, July 1997, pp. 62–64.

76. Gay Jervey, "Family Ties," *FSB*, March 2006, p. 60.

77. "Your Family Limited Partnership: Hoping It Worked . . .?" *Calamos*, March 2013, www.calamos.com/~/media/documents/wm/calamos-wealth-management-2013-03-04-your-family-partnership.ashx; Gay Jervey, "Family Ties," *FSB*, March 2006, p. 60; Tom Herman, "Court Ruling Bolsters Estate Planning Tool," *Wall Street Journal*, May 27, 2004, p. D1.

78. Angus Loten, "For Small Firms, Disaster Prep Starts at Home," *Wall Street Journal*, August 29, 2011, http://blogs.wsj.com/in-charge/2011/08/29/for-small-firms-disaster-prep-starts-at-home.

79. Rita Pyrillis, "Report: Preparedness Lacking for Many Firms as September 11 Anniversary Nears," *Workforce Management*, September 2, 2011, www.workforce.com/article/20110902/NEWS01/309029999/report-preparedness-lacking-for-many-firms-as-sept-11-anniversary-nears#.

80. Daniel Tynan, "In Case of Emergency," *Entrepreneur*, April 2003, p. 60.

81. Sarah E. Needleman, "Lights Out Means Lost Sales," *Wall Street Journal*, July 22, 2010, p. B6.

82. Rita Pyrillis, "Self-Thought: Companies Ponder Bringing Insurance In-House," *Workforce Management*, April 2012, pp. 3–4.

83. Jay Hancock, "Some Small Businesses Choose to Self-Insure," *USA Today*, March 14, 2013, www.usatoday.com/story/money/business/2013/03/14/some-small-businesses-choose-to-self-insure/1988481.

84. Alec Foege, "The DIY Approach to Small Business Insurance," *Crain's New York Business*, January 20, 2013, www.crainsnewyork.com/article/20130120/SMALLBIZ/301209994.

85. Paul Sullivan, "An Insurer of One's Own? It's Possible, with Caveats," *New York Times*, July 13, 2012, www.nytimes.com/2012/07/14/your-money/a-captive-insurance-company-offers-financial-benefits-if-not-abused-wealth-matters.html?pagewanted=all; Alec Foege, "The DIY Approach to Small Business Insurance," *Crain's New York Business*, January 20, 2013, www.crainsnewyork.com/article/20130120/SMALLBIZ/301209994.

86. "Traveler's Questionnaire Reveals Nearly Half of Small Business Owners Aren't Prepared for Hurricane Season," Traveler's Insurance, June 1, 2012, http://investor.travelers.com/phoenix.zhtml?c=177842&p=irol-newsArticle&ID=1701641&highlight=.

87. "The Most Unusual Insurance Policies—We Couldn't Make This Up," Southern States Insurance, 2013, http://southernstatesinsurance.com/the-most-unusual-insurance-policies-we-couldnt-make-this-up; Kimberly Lankford, "Weird Insurance," *Kiplinger's Personal Finance Magazine*, October 1998, pp. 113–116.

88. Natahn Erb, "Unusual Insurance Policies," Geico, 2012, www.geico.com/information/publications/newsletter/2012/unusual-insurance-policies; "Top 10 Most Unusual Insurance Policies," *TopTenz*, September 13, 2012, www.toptenz.net/top-10-most-unusual-insurance-policies.php.

89. "Businesses That Cater to Adrenaline Junkies," *CNBC*, August 16, 2012, http://finance.yahoo.com/news/businesses-cater-adrenaline-junkies-135015162.html; John Fried, "Having Fun Yet?," *Inc.*, March 2006, pp. 75–77; "Incredible Adventures: Our Story," Incredible Adventures, www.incredible-adventures.com/about_us.html; Esther Dyson, "I Live and Die by Waivers," Esther Dyson's Flight School, May 11, 2007, www.edventure.com/flightschool/blog/?p=4.

90. John Carney, "How I Did It," *Inc.*, January 2013, p. 118.

91. Jason Staeck, "Disaster Recovery a Lesson in Planning," *Telegram*, January 27, 2013, www.telegram.com/article/20130127/NEWS/101279984.

92. Traveler's Questionnaire Reveals Nearly Half of Small Business Owners Aren't Prepared for Hurricane Season," Traveler's Insurance, June 1, 2012, http://investor.travelers.com/phoenix.zhtml?c=177842&p=irol-newsArticle&ID=1701641&highlight=.

93. Joyce M. Rosenberg, "Preparing a Disaster Plan Gets Serious," *Los Angeles Times*, August 16, 2007, www.latimes.com/business/la-fi-smalldisaster16aug16,1,4272739.story?coll=la-headlines-business&ctrack=1&cset=true; "Famed Restaurant Reopens Today," *Greenville News*, October 1, 2006, p. 4B.

94. Protecting Your Business," Federal Emergency Management Administration, March 1, 2013, www.fema.gov/protecting-your-businesses.

95. Stacy Cowley, "Sandy's Small Business Victims: We Don't Want Loans," *CNN Money*, November 2, 2012, http://money.cnn.com/2012/11/02/smallbusiness/sandy-loans/index.html; "Getting Back to Business," New York City Economic Development Corporation, 2013, www.nycedc.com/gettingbacktobusiness/jackfrombrooklyn.html.

96. "Protecting Your Business," Federal Emergency Management Administration, March 1, 2013, www.fema.gov/protecting-your-businesses.

97. Lisa Flickenscher, "Insurers Stick Restaurants with Sandy Tab," *Crain's New York Business*, April 5, 2013, www.crainsnewyork.com/article/20130405/HOSPITALITY_TOURISM/130409920.

98. Baird Webel, *Terrorism Risk Insurance: Issue Analysis and Overview of Current Program*, Congressional Research Service, April 26, 2013, p. 9.

99. Susanna Kim, "Insurance Payout May Depend on Whether Boston Bombing Was 'Terrorist Act,'" *ABC News*, April 26, 2013, http://abcnews.go.com/Business/boston-firms-wait-terrorism-certification-insurance-payout/story?id=19043385#.UY_2ykqWyuM.

100. Bill Harris, "Irreplaceable You," *Inc.*, February 2013, p. 35.

101. "Disability Insurance Statistics," Disability Insurance Resource Center, 2013, http://di-resource-center.com/statistics.php; "Social Security Basic Facts," Social Security Administration, February 7, 2013, www.ssa.gov/pressoffice/basicfact.htm.

102. Elizabeth Mendes, "Fewer Americans Getting Health Insurance from Employer," *Gallup*, February 22, 2013, www.gallup.com/poll/160676/fewer-americans-getting-health-insurance-employer.aspx.

103. Holly Wade, *Small Business Problems and Priorities*, National Federation of Independent Businesses, August 2012, p. 12.

104. Alex Wayne, "Healthcare Spending to Reach 20% of U.S. Economy by 2021," *Bloomberg*, June 13, 2012, www.bloomberg.com/news/2012-06-13/health-care-spending-to-reach-20-of-u-s-economy-by-2021.html; *National Healthcare Expenditure Projections 2011–2021*, Centers for Medicare and Medicaid Services, www.cms.gov/Research-Statistics-Data-and-Systems/Statistics-Trends-and-Reports/NationalHealthExpendData/NationalHealthAccountsProjected.html.

105. *Employer Health Benefits 2010 Summary of Findings*, Kaiser Family Foundation and Health Research and Educational Trust, p. 5.

106. *Employer Health Benefits 2012 Survey*, Kaiser Family Foundation and Health Research and Educational Trust, p. 45.

107. Daniel Wityk, "Small Business Health Insurance," National Center for Policy Analysis, February 11, 2009, p. 1.

108. *Principal Financial Well Being Index*, Principal Financial Group, First Quarter 2013, p. 24.

109. "SurePayroll Insights Survey: Fewer Small Business Owners Can Afford to Offer Healthcare, but Split on Solution," SurePayroll, 2012, www.spinsurance.com/press_release/surepayroll-insights-survey-fewer-small-business-owners-can-afford-offer-healthcare-sp.

110. Reed Richardson, "The Small Business Case for Offering Health Benefits," Bank of America Small Business Community, February 9, 2012, https://smallbusinessonlinecommunity.bankofamerica.com/community/growing-your-business/employeebenefitsandretirementplanning/blog/2012/02/09/the-small-business-case-for-offering-health-benefits.

111. Joyce M. Rosenberg, "Businesses May Get Sticker Shock on Health Care," Associated Press, May 1, 2013, http://bigstory.ap.org/article/businesses-may-get-sticker-shock-health-care.

112. Emily Maltby, "A Health Scare for Small Businesses," *Wall Street Journal*, January 17, 2013, pp. B1, B4.

113. *Employer Health Benefits 2012 Survey*, Kaiser Family Foundation and Health Research and Educational Trust, p. 63.

114. *Employer Health Benefits 2012 Survey*, Kaiser Family Foundation and Health Research and Educational Trust, p. 4.

115. Yesenia Amaro, "Small Businesses: Health Costs Strangling Us," *Modesto Bee*, May 9, 2011, www.modbee.com/2011/05/08/1680083/small-businesses-health-costs.html.

116. *Employer Health Benefits 2012 Survey*, Kaiser Family Foundation and Health Research and Educational Trust, p. 2.

117. "Census Shows 8 Million People Covered by HSA/High-Deductible Health Plans," America's Health Insurance Plans, Center for Policy Research, January 2009, p. 4; *Behind the Numbers: Medical Cost Trends for 2010*, PriceWaterhouseCoopers, Health Research Institute, 2010, p. 7.

118. Cynthia Bunting, "The Alternative to Buying Small Business Health Insurance," *Business News Daily*, May 6, 2011, www.businessnewsdaily.com/935-small-business-health-insurance.html.

119. *Employer Health Benefits 2012 Survey*, Kaiser Family Foundation and Health Research and Educational Trust, p. 161; Cynthia Bunting, "The Alternative to Buying Small Business Health Insurance," *Business News Daily*, May 6, 2011, www.businessnewsdaily.com/935-small-business-health-insurance.html.

120. "Litigiousness," Insurance Information Institute, 2013, www.iii.org/facts_statistics/litigiousness.html.

121. Marc Fisher, "Judge Who Seeks Millions for Lost Pants Has His (Emotional) Day in Court," *Washington Post*, June 13, 2007,

www.washingtonpost.com/wp-dyn/content/article/2007/06/12/
AR2007061201667.html; "Dry Cleaner Raises Enough Cash to Pay
Legal Fees, *WUSA9.com*, August 13, 2007, www.wusa9.com/news/
news_article.aspx?storyid=61690.

122. Mark Centoni, "Employment Practices Claims Are More Likely Than
Property or General Liability Claims," Adreini and Company, February 15,
2012, www.andreini.com/blog/2012/02/15/employment-practices-
claims-are-more-likely-than-property-or-general-liability-claims.

123. "Why Small Businesses in California Need Employment Practices
Liability Insurance," ERM Insurance Brokers, March 20, 2013, www
.erminsurance.com/blog/why-small-businesses-in-california-need-
employment-practices-liability-insurance-epli.

124. "Litigiousness," Insurance Information Institute, 2013, www.iii.org/
facts_statistics/litigiousness.html.

125. "Employment Practice Liability Jury Award Trends Hit New High,"
HR Works, January 17, 2013, www.hrthatworksblog.com/2013/01/17/
employment-practice-liability-jury-award-trends-hit-new-high; "Why
Small Businesses in California Need Employment Practices Liability
Insurance," ERM Insurance Brokers, March 20, 2013, www
.erminsurance.com/blog/why-small-businesses-in-california-need-
employment-practices-liability-insurance-epli.

126. Employment Practice Liability Jury Award Trends Hit New High,"
HR Works, January 17, 2013, www.hrthatworksblog.com/2013/01/17/
employment-practice-liability-jury-award-trends-hit-new-high.

127. "More Than 12 Million Identity Fraud Victims in 2012 According
to Latest Javelin Strategy and Research Report," Javelin Strategy
and Research, February 20, 2013, https://www.javelinstrategy.com/
news/1387/92/More-Than-12-Million-Identity-Fraud-Victims-
in-2012-According-to-Latest-Javelin-Strategy-Research-Report/
d,pressRoomDetail.

128. Taylor Provost, "Should You Consider Cyber Liability Insurance?"
CFO, April 24, 2013, www3.cfo.com/article/2013/4/data-security_
cyber-attacks-cybersecurity-liability-insurance-smb-growth-companies-
risk-hogan-lovells.

129. Taylor Provost, "Should You Consider Buying Cyber-Liability
Insurance?" *CFO*, April 24, 2013, www3.cfo.com/article/2013/4/data-
security_cyber-attacks-cybersecurity-liability-insurance-smb-growth-
companies-risk-hogan-lovells.

130. *2011 Cost of Data Breach Study: United States*, Symantec and Ponemon
Research Institute, March 2012, p. 5.

131. "A Closer Look: Wellness ROI," International Foundation of Employee
Benefit Plans, September 2012, www.ifebp.org/bookstore/wellnessroi
.htm.

132. Susan Caminiti, "Keeping America Fit," *Fortune*, May 3, 2010, p. S1.

133. Patricia F. Weisberg, "Wellness Programs: Legal Requirements
and Risks," Workforce Management, March 2010, www.workforce
.com/section/legal/feature/wellness-programs-legal-requirements-
risks.

Chapter 23

1. David Phelps, "For Small Businesses, Trademark Battles Are Life
or Death," *Star Tribune*, June 30, 2012, www.startribune.com/
business/160880395.html?refer=y.

2. Emily Friedman, "Dena Lockwood Was Fired When She Called in Sick
to Care for Her Daughter," *ABC News*, January 29, 2010, http://abcnews.
go.com/Business/SmallBiz/single-mother-wins-court-losing-job-care-
sick/story?id=9689779#.UYEeUJX3CRt; Courtney Rubin, "Single
Mother Wins $200,000 in Job Bias Case," *Inc.*, January 25, 2010, www
.inc.com/news/articles/2010/01/chicago-parents-rights-case.html.

3. Robin Hagan Cain, "Court Finds Offer and Acceptance in Email,"
April 27, 2012, http://blogs.findlaw.com/tenth_circuit/2012/04/court-
finds-offer-and-acceptance-in-email.html.

4. "McIntosh v. Murphy," Leagle.com, *May 11, 1970,* http://www.leagle
.com/decision?q=1970646469P2d177_1646.xml/McINTOSH%20v
.%20MURPHY; "McIntosh v. Murphy—Case Brief," Lawnix.com,
(n.d.), www.lawnix.com/cases/mcintosh-murphy.html; "Promissory
Estoppel," *Contract Law Cases,* November 1, 2008, http://contract-law-
cases.blogspot.com/2008/11/promissory-estoppel.html.

5. Erin Lawley, "Dialysis Firm Wins Breach of Contract Suit," *Nashville
Post,* September 20, 2010, http://nashvillepost.com/news/2010/9/20/
dialysis_firm_wins_breach_of_contract_case.

6. *Superior Boiler Works v. R.J. Sanders, Inc.*, 1998 R.I. Lexis 153,
Supreme Court of Rhode Island, 1998.

7. *Tort Liability Costs for Small Businesses*, U.S. Chamber Institute for
Legal Reform, July 2010, pp. 2, 10.

8. UCC Section 2-314[1-C].

9. Gordon Gibb, "Allegedly Bad Chicken Worth $140,000 to Foodborne
Illness Plaintiff," LawyersandSettlements.com, July 21, 2012, www
.lawyersandsettlements.com/articles/food_poisoning/foodborne-illness-
food-poisoning-16-17916.html#.UYJXwpX3A6U; "Food Poisoning and
North Carolina's Implied Warranty of Merchantability," Maginnis Law,
July 19, 2012, www.maginnislaw.com/2012/07/food-poisoning-and-
north-carolinas-implied-warranty-of-merchantability.

10. "Product Liability Basics," *Inc.*, February 2000, www.inc.com/
articles/2000/02/17249.html.

11. "U.S. Seizes Counterfeit Goods Worth $1.26 Billion in 2012, *World
Intellectual Property Review*, January 22, 2013, www.worldipreview
.com/news/us-seizes-counterfeit-goods-worth-1-26-billion-in-2012;
"U.S. Firms Paying High Price for Global IP Theft," A.E. Feldman,

August 4, 2009, http://blog.aefeldman.com/2009/08/04/us-firms-paying-
high-price-for-global-ip-theft.

12. Doug Palmer, "U.S. Seizes Web Sites in Fake Goods Crackdown,"
Reuters, November 29, 2010, www.reuters.com/article/
idUSTRE6AS4PW20101129; *Intellectual Property: Observations on
Efforts to Quantify the Economic Effects of Counterfeit and Pirated
Goods*, U.S. Government Accountability Office, April 2010, p. 8.

13. Steve Hargreaves, "Counterfeit Goods Becoming More Dangerous,"
CNN Money, September 27, 2012, http://money.cnn.com/2012/09/27/
news/economy/counterfeit-goods/index.html.

14. "How 4 Small Businesses Protected Their Intellectual Property from
Chinese Manufacturers and Etsy Copycats," NFIB, www.nfib.com/
business-resources/business-resources-item?cmsid=62517.

15. Anne Field, "How to Knock Out Knock Offs," *BusinessWeek*, March 14,
2005, www.businessweek.com/stories/2005-03-13/how-to-knock-out-
knock-offs.

16. "U.S. Patent History," The Great Idea Finder, www.ideafinder.com/
history/inventions/uspatent.htm; "U.S. Patent and Trademark Office
Issues 8 Millionth Patent," U.S. Department of Commerce, August 16,
2011, www.commerce.gov/blog/2011/08/16/us-patent-and-trademark-
office-issues-its-8-millionth-patent.

17. "2012 Patent Litigation Study," Pricewaterhousecoopers, LLC, (n.d.),
www.pwc.com/en_US/us/forensic-services/publications/assets/2012-
patent-litigation-study.pdf.

18. Jeremy P. Oczek, "Rethinking Defense in 'Patent Troll' Cases,"
Corporate Counsel, March 27, 2013, www.law.com/corporatecounsel/
PubArticleCC.jsp?id=1202593233566&Rethinking_Defense_in_
Patent_Troll_Cases&slreturn=20130402102214.

19. Kris Frieswick, "License to Steal?," *CFO*, September 2001, pp. 89–91;
Megan Barnett, "Patents Pending," *U.S. News & World Report*, June 10,
2002, pp. 33–34; Tomima Edmark, "On Guard," *Entrepreneur*, August
1997, pp. 92–94; Tomima Edmark, "On Guard," *Entrepreneur*, February
1997, pp. 109–111.

20. David Phelps, "For Small Businesses, Trademark Battles Are Life
or Death," *Star Tribune*, June 30, 2012, www.startribune.com/
business/160880395.html?refer=y; "Small Businesses Seek to
Address Trademark Bullying," Minnesota House of Representatives,
February 25, 2013, www.house.leg.state.mn.us/hinfo/sessiondaily
.asp?storyid=3439.

21. *Summary of Financial and Performance Highlights*, U.S. Patent and Trademark Office, 2012, www.uspto.gov/about/stratplan/ar/USPTOFY2012PAR.pdf.

22. "Cost of Software Copyright Infringement Tops $63 Billion," *Global Journal*, May 15, 2012, http://theglobaljournal.net/news/technology/cost-of-software-copyright-infringement-tops-$63-billion.html; "The True Cost of Piracy," *Venture Beat*, 2012, http://venturebeat.files.wordpress.com/2012/03/cost-of-piracy-800.png.

23. Michael Gazdecki, "Study Shows That Consumers Understand the Importance of Protecting Creators," Copyright Alliance, July 2013, https://copyrightalliance.org/2013/07/study_shows_american_consumer_attitudes_are_firmly_side_ip_holders#.UhphkH_pySo.

24. Stephanie Gleason, "Maclaren to Cover Bankrupt Distributor's Product-Liability Suits," *Wall Street Journal*, June 28, 2012, http://blogs.wsj.com/bankruptcy/2012/06/28/maclaren-to-cover-bankrupt-distributor's-product-liability-suits Alice Hines, "Maclaren Stroller Maker Pledges Settlement with Children Whose Fingers Were Amputated," *Huffington Post*, July 3, 2012, www.huffingtonpost.com/2012/07/02/maclaren-stroller-settlement_n_1644108.html; Peter Lattman and Andrew Martin, "For Stroller Maker, Bankruptcy Is Latest Setback," *New York Times*, March 2, 2012, http://dealbook.nytimes.com/2012/03/02/maclaren-a-stroller-maker-in-bankruptcy.

25. Jonathan Stempel and Steve Gorman, "Los Angeles Dodgers Baseball Team Files Bankruptcy," *Reuters*, June 27, 2011, www.reuters.com/article/2011/06/27/us-baseball-dodgers-bankruptcy-idUSTRE75Q2M120110627; Jason Fell, "What Small Business Owners Can Learn from the Dodger's Financial Woes," *Entrepreneur*, June 28, 2011, www.entrepreneur.com/blog/219920; "Dodgers Sold to Magic Johnson Group," *ESPN.com*, March 28, 2012, http://espn.go.com/los-angeles/mlb/story/_/id/7745566/los-angeles-dodgers-selling-team-magic-johnson-group-2b.

26. Andrea James, "Small Business Owners Vent Their Regulation Frustrations," *Seattle Post-Intelligencer*, August 20, 2007, www.sba.gov/advo/research/rs264.pdf.

27. Nicole V. Crain and Mark Crain, *The Impact of Regulatory Costs on Small Firms*, Small Business Administration Office of Advocacy, September 2010, p. iv.

28. Nicole V. Crain and Mark Crain, *The Impact of Regulatory Costs on Small Firms*, Small Business Administration Office of Advocacy, September 2010, p. 6.

29. "Richard Bergmann et al. v. City of Lake Elmo--Freeing Small Farms Through Fair Trade," Institute for Justice, (n.d.), www.ij.org/richard-bergmann-et-al-v-city-of-lake-elmo.

30. *United States v. Topco Associates Inc.*, 405 U.S. 596, (1972).

31. "Identity Theft/Fraud Statistics, Statistic Brain, June 18, 2013, www.statisticbrain.com/identity-theft-fraud-statistics/; "Research and Statistics," Identity Theft Assistance Center, 2013, www.identitytheftassistance.org/pageview.php?cateid=47

Index